2018 International Power Electronics Conference (IPEC-Niigata 2018 –ECCE Asia-)

Niigata, Japan
20-24 May 2018

Pages 3549-4238

IEEE Catalog Number: CFP1854I-POD
ISBN: 978-1-5386-4190-3

Copyright © 2018, IEEJ Industry Applications Society
All Rights Reserved

*** This is a print representation of what appears in the IEEE Digital
Library. Some format issues inherent in the e-media version may also
appear in this print version.*

IEEE Catalog Number: CFP1854I-POD
ISBN (Print-On-Demand):   978-1-5386-4190-3
ISBN (Online):   978-4-88686-405-5

Additional Copies of This Publication Are Available From:

Curran Associates, Inc
57 Morehouse Lane
Red Hook, NY 12571 USA
Phone: (845) 758-0400
Fax: (845) 758-2633
E-mail: curran@proceedings.com
Web: www.proceedings.com

TABLE OF CONTENTS

THREE-PHASE INDUCTIVE POWER TRANSFER SYSTEM WITH 12 COILS FOR RADIATION NOISE REDUCTION 69
Keisuke Kusaka ; Jun-Ichi Itoh

SECONDARY-SIDE-ONLY CONTROL FOR SMOOTH VOLTAGE STABILIZATION IN WIRELESS POWER TRANSFER SYSTEMS WITH CONSTANT POWER LOAD 77
Giorgio Lovison ; Takehiro Imura ; Hiroshi Fujimoto ; Yoichi Hori

CONSTANT CURRENT CHARGING AND THE MAXIMUM SYSTEM EFFICIENCY TRACKING FOR WIRELESS CHARGING SYSTEMS EMPLOYING DUAL-SIDE CONTROL 84
Zhenjie Li ; Xiaoliang Huang ; Kai Song ; Jinhai Jiang ; Chunbo Zhu ; Zhijiang Du

ELECTRIC FIELD COUPLING TYPE HIGH POWER WIRELESS POWER TRANSFER WITH LEAKAGE ELECTRIC FIELD STRUCURE 88
Mitsuru Masuda

TRANSFER POWER ANALYSIS OF CAPACITIVELY ISOLATED OUTLET AND PLUG (CAPISOP) USING SERIES RESONANCE 94
Hirohito Funato ; Koki Amano ; Takuya Hatsumi ; Junnosuke Haruna

WIDE VOLTAGE GAIN RANGE LLC DC/DC TOPOLOGIES: STATE-OF-THE-ART 100
Qi Cao ; Zhiqing Li ; Haoyu Wang

DUAL HALF-BRIDGE LLC RESONANT CONVERTER WITH HYBRID-SECONDARY-RECTIFIER (HSR) FOR WIDE-OUPUT-VOLTAGE APPLICATIONS 108
Jae-Il Baek ; Chong-Eun Kim ; Keon-Woo Kim ; Min-Su Lee ; Gun-Woo Moon

A STUDY ON THE ANALYSIS AND CONTROL OF NO-LOAD CHARACTERISTICS OF LLC RESONANT CONVERTER FOR PLASMA PROCESS 114
Min-Jun Kwon ; Woo-Cheol Lee

MECHANISM OF CURRENT IMBALANCE IN LLC RESONANT CONVERTER WITH CENTER TAPPED TRANSFORMER 118
Mitsuru Sato ; Shingo Nagaoka ; Takeshi Uematsu ; Toshiyuki Zaitsu

PERFORMANCE STUDY OF HIGH-POWER HALF-BRIDGE INTERLEAVED LLC CONVERTER 123
Hung-I Hsieh ; Hui-Lung Chiu ; Guan-Chyun Hsieh

MULTI-CHIP SIC MOSFET POWER MODULES FOR STANDARD MANUFACTURING, MOUNTING AND COOLING 130
Alberto Castellazzi ; Asad Fayyaz ; Emre Gurpinar ; Abdallah Hussein ; Jianfeng Li ; Bassem Mouawad

AN ALTERNATIVE METHOD TO ACCURATELY DETERMINE THE THERMAL RESISTANCE OF SIC MOSFET STRUCTURES WITH DISCRETE DIODES 137
Andras Vass-Varnai ; Young Joon Cho ; Gabor Farkas ; Marta Rencz

HEAT-RESISTANT PACKAGING TECHNOLOGY FOR WIDE BANDGAP POWER DEVICES AND THERMAL RELIABILITY TESTING 142
K. Suganuma ; H. Zhang ; S. Nagao ; C. Chen ; T. Sugahara ; A. Shimoyama ; A. Suetake

VERIFICATION OF IDENTIFICATION ACCURACY OF LOSS CALCULATED BY INVERSE THERMAL ANALYSIS 148
Yuki Ikari ; Kazushige Nakao

PACKAGING ARCHITECTURES FOR SILICON CARBIDE POWER ELECTRONIC MODULES 153
H. Alan Mantooth ; Simon S. Ang

DEVELOPMENT OF A HOMO-POLAR BEARINGLESS MOTOR WITH CONCENTRATED WINDING FOR HIGH SPEED APPLICATIONS 157
Dai Suzuki ; Takaaki Oiwa

HIGH-SPEED SLOTLESS PERMANENT MAGNET MACHINES: MODELLING AND DESIGN FRAMEWORKS 161
S. Jumayev ; K.O. Boynov ; E.A. Lomonova ; J. Pyrhonen

DEVELOPMENT AND PERFORMANCE OF HIGH-SPEED SPM SYNCHRONOUS MACHINE 169
Kota Kawanishi ; Keisuke Matsuo ; Takayuki Mizuno ; Koji Yamada ; Takashi Okitsu ; Kouki Matsuse

1.2KW 100,000RPM HIGH SPEED MOTOR FOR AIRCRAFT 177
Takehiro Jikumaru ; Gen Kuwata

COMPARATIVE EVALUATION OF Y-INVERTER AGAINST THREE-PHASE TWO-STAGE BUCK-BOOST DC-AC CONVERTER SYSTEMS 181
Michael Antivachis ; Dominik Bortis ; David Menzi ; Johann W. Kolar

DC-POWERED OFFICE BUILDINGS AND DATA CENTRES : THE FIRST 380 VDC MICRO GRID IN A COMMERCIAL BUILDING IN GERMANY 190
Tilo Pueschel

RECENT TREND IN POWER ELECTRONICS FOR ICT SYSTEMS 196
Hiroshi Nakao ; Yu Yonezawa ; Yoshiyasu Nakashima

GREEN BASE STATION USING ROBUST SOLAR SYSTEM AND HIGH PERFORMANCE LITHIUM ION BATTERY FOR NEXT GENERATION WIRELESS NETWORK (5G) AND AGAINST MEGA DISASTER 201
M. Nakamura ; K. Takeno

OPTIMIZATION OF MAINTENANCE BY FAILURE PREDICTION CONSIDERING INSTANTANEOUS AND CUMULATIVE EFFECTS OF EXTERNAL ENVIRONMENTS 207
Kaisei Kanetani ; Masahiro Yamazaki ; Tadatoshi Babasaki ; Hideaki Kim ; Tatsushi Matsubayashi

HYBRID CONVERTERS WITH REDUCED INDUCTOR LOSS FOR INTEGRATABLE POWER CONVERSION 213
Gab-Su Seo ; Hanh-Phuc Le

ENERGY SAVING SYSTEM TREND FOR HARBOR CRANE WITH LITHIUM ION BATTERY 219
Hidemasa Yoshihara

INVERTER DRIVE OF DYNAMOMETERS FORAUTOMOTIVE EVALUATION SYSTEM 227
Shizunori Hamada ; Toshimichi Takahashi ; Nobutaka Kezuka ; Masaju Kouketsu ; Shingo Ishigaki

EXPERIMENTAL INVESTIGATION OF PROTOTYPE ALL-SIC CONVERTER FOR ULTRA-HIGH-SPEED ELEVATOR 233
Kazuhisa Mori ; Kaoru Katoh ; Yohei Matsumoto ; Tatsushi Yabuuchi ; Naoto Ohnuma

HIGH-VOLTAGE, LARGE-CAPACITY CONVERTER TECHNOLOGIES AND THEIR APPLICATIONS 238
Daisuke Yoshizawa ; Paul Bixel ; Masahiko Tsukakoshi

HIGHER RADIAL SUSPENSION FORCE OF MAGNETIC BEARING ON CENTRIFUGAL COMPRESSOR FOR HVAC 244
Yuji Nakazawa ; Yusuke Irino ; Atsushi Sakawaki ; Kazunobu Ohyama

NOVEL SWITCHING CONTROL METHOD FOR FULL-BRIDGE DC-DC CONVERTERS FOR IMPROVING LIGHT-LOAD EFFICIENCY USING REVERSE RECOVERY CURRENT 250
Fumihiro Sato ; Takae Shimada ; Takayuki Ouchi

A 800V/14V SOFT-SWITCHED CONVERTER WITH LOW-VOLTAGE RATING OF SWITCH FOR XEV APPLICATIONS 256
Byeongwoo Kim ; Kangsan Kim ; Sewan Choi

HIGH SPEED CONTROL METHOD FOR SUPERPOSING HIGH-FREQUENCY-HIGH-SINUSOIDAL-CURRENT WITH DC CURRENT TO ANALYZE BATTERY AC IMPEDANCE 261
Jin Xu ; Toshihiko Kishimoto ; Noboru Shimosato

EV BMS WITH TIME-SHARED ISOLATED CONVERTERS FOR ACTIVE BALANCING AND AUXILIARY BUS REGULATION 267
Z. Gong ; B.A.C. Van De Ven ; Y. Lu ; Y. Luo ; K. Gupta ; C. Da Silva ; H.J. Bergveld ; O. Trescases

A DRIVING CIRCUIT WITH PARTIAL POWER REGULATION FOR RGB LED LAMPS 275
You-Chun Huang ; Yu-Jen Chen ; Yong-Jyun Li ; Chin-Sien Moo

FPGA-BASED DYNAMIC DUTY CYCLE AND FREQUENCY CONTROLLER FOR A CLASS-E2DC-DC CONVERTER 282
Sanghyeon Park ; Juan Rivas-Davila

DESIGN METHODOLOGY OF 3 KW INDUCTION HEATING SYSTEM FOR BOTH LOW RESISTANCE AND HIGH RESISTANCE CONTAINERS IN A SINGLE BURNER 289
Si-Hoon Jeong ; Hwa-Pyeong Park ; Jee-Hoon Jung

MULTI-RESONANT INVERTER REALIZING DOWNSIZING AND LOSS REDUCTION FOR ALL-METALLIC IH COOKTOP 296
Takayuki Hirokawa ; Makoto Imai ; Atsushi Fujita

TEMPERATURE ESTIMATION OF ALUMINUM ELECTROLYTIC CAPACITOR UNDER ACTUAL CIRCUIT OPERATION 302
Kazuki Urata ; Toshihisa Shimizu

DESIGN AND EVALUATION OF CURRENT DISTRIBUTION IN POWER MODULE 309
Takaaki Ibuchi ; Eisuke Masuda ; Tsuyoshi Funaki

DEVELOPMENT OF IMPEDANCE-SOURCE INVERTER USING SIC-MOSFET 313
Ryuji IIjima ; Thilak Senanayake ; Takanori Isobe ; Hiroshi Tadano

CONTROL METHODOLOGY FOR REALIZATION OF 100KW HEECS CHOPPER WITH 99.5% EFFICIENCY 318
Yukinori Tsuruta ; Atsuo Kawamura

IRON LOSS REDUCTION IN THE CORES OF INDUCTION HEATING COILS FOR SMALL-FOREIGN-METAL PARTICLE DETECTOR WITH A 400-KHZ SIC-MOSFETS HIGH-FREQUENCY INVERTER 324

Takuya Shijo ; Yuki Uchino ; Yujiro Noda ; Hiroaki Yamada ; Toshihiko Tanaka

FREQUENCY TRACKING BURST-MODE PDM-CONTROLLED CLASS-D ZERO VOLTAGE SOFT-SWITCHING RESONANT CONVERTER FOR INDUCTIVE POWER TRANSFER APPLICATIONS 329

Yoichiro Tabata ; Tomokazu Mishima ; Tatsuya Kido

REDUCED-ORDER DYNAMICAL MODELS OF TUNED WIRELESS POWER TRANSFER SYSTEMS 337

Hongchang Li ; Jingyang Fang ; Yi Tang

DYNAMIC MODELLING AND CLOSED LOOP CONTROL OF TRANSMITTER PARALLEL AND RECEIVER SERIES COMPENSATED IPT TOPOLOGY FOR EV APPLICATIONS 342

Suvendu Samanta ; Akshay Kumar Rathore

DEVELOPMENT OF INDUCTIVE POWER TRANSFER SYSTEM FOR EXCAVATOR UNDER LARGE LOAD FLUCTUATION : CONSIDERATION OF RELATIONSHIP BETWEEN LOAD VOLTAGE AND RESONANCE PARAMETER 348

Jun-Ichi Itoh ; Kent Inoue ; Keisuke Kusaka

WIRELESS POWER TRANSFER SYSTEM USING THREE-PHASE TO SINGLE-PHASE MATRIX CONVERTER 356

Yuji Hayashi ; Hiromasa Motoyama ; Takaharu Takeshita

DESIGN OF A REDUCED-ORDER OBSERVER FOR SENSORLESS CONTROL OF DUAL-ACTIVE-BRIDGE CONVERTER 363

Nguyen Duy Dinh ; Goro Fujita

IMPROVED LOAD TRANSIENT RESPONSE OF A DUAL-ACTIVE-BRIDGE CONVERTER 370

Sheng-Zhi Zhou ; Chuan Sun ; Song Hu ; Guo Chen ; Xiaodong Li

MODULATION AND ACTIVE MIDPOINT CONTROL OF A THREE-LEVEL THREE-PHASE DUAL-ACTIVE BRIDGE DC-DC CONVERTER UNDER NON-SYMMETRICAL LOAD 375

Philipp Joebges ; Anton Gorodnichev ; Rik W. De Doncker

A NOVEL SWITCHING ALGORITHM TO IMPROVE EFFICIENCY AT LIGHT LOAD CONDITIONS FOR THREE-PHASE DAB CONVERTER IN LVDC APPLICATION 383

Hyun-Jun Choi ; Si-Hoon Jung ; Jee-Hoon Jung

DESIGN OF A HIGH-FREQUENCY DUAL-ACTIVE BRIDGE CONVERTER WITH GAN DEVICES FOR AN OUTPUT POWER OF 3.7 KW 388

Philipp Schülting ; Christian Winter ; Rik W. De Doncker

EXPLORATION OF THE DESIGN AND PERFORMANCE SPACE OF A HIGH FREQUENCY 166 KW/10 KV SIC SOLID-STATE AIR-CORE TRANSFORMER 396

Piotr Czyz ; Thomas Guillod ; Florian Krismer ; Johann W. Kolar

NOVEL CALCULATION METHOD OF IRON LOSS OF GAPPED INDUCTORS USING LOSS MAP 404

Yoshihiro Miwa ; Toshihisa Shimizu

VERIFICATION OF THE REDUCTION OF THE COPPER LOSS BY THE THIN COIL STRUCTURE FOR INDUCTION COOKERS 410

Morimasa Hataya ; Koki Kamaeguchi ; Eiji Hiraki ; Kazuhiro Umetani ; Takayuki Hirokawa ; Makoto Imai ; Hideki Sadakata

CONDITION MONITORING OF ELECTROLYTIC CAPACITOR BASED ON ESR ESTIMATION AND THERMAL IMPEDANCE MODEL USING IMPROVED POWER LOSS COMPUTATION 416

Sundararajan Prasanth ; Mohamed Halick ; Mohamed Sathik ; Firman Sasongko ; Tan Chuan Seng ; Peng Yaxin ; Rejeki Simanjorang

TEST SETUP FOR CHARACTERISATION OF BIASED MAGNETIC HYSTERESIS LOOPS IN POWER ELECTRONIC APPLICATIONS 422

Min Luo ; Drazen Dujic ; Jost Allmeling

A FAST OPEN-CIRCUIT FAULT DIAGNOSIS SCHEME FOR MODULAR MULTILEVEL CONVERTERS WITH MODEL PREDICTIVE CONTROL 428

Dehong Zhou ; Shunfeng Yang ; Yi Tang

AN ONLINE OPEN-CIRCUIT FAULT DIAGNOSIS AND FAULT TOLERANT SCHEME FOR THREE-PHASE AC-DC CONVERTERS WITH MODEL PREDICTIVE CONTROL 434

Dehong Zhou ; Yi Tang

THE LIFETIME ASSESSMENT OF A MICRO-INVERTER FOR PV APPLICATIONS 439

Tohihiro Shimao ; Koji Kato ; Youichi Ito ; Akio Iwabuchi ; Yongheng Yang ; Frede Blaabjerg

ONLINE HEALTH MONITORING OF MULTIPLE MOSFETS IN A GRID-TIED PV INVERTER USING SPREAD SPECTRUM TIME DOMAIN REFLECTOMETRY (SSTDR) 446
Sourov Roy ; Faisal Khan

AN IMPROVED EQUIVALENT MODEL FOR A LONG PV STRING UNDER PARTIAL SHADING CONDITIONS 453
Xiaoyang Wang ; Huiqing Wen ; Xingshuo Li

OPTIMIZED FLUX-WEAKENING CONTROL OF INDUCTION MOTOR FOR TORQUE ENHANCEMENT IN VOLTAGE EXTENSION REGION 459
Zhen Dong ; Yong Yu ; Bo Wang ; Qinghua Dong ; Dianguo Xu

IMPROVED PERFORMANCE OF CFTC-BASED DIRECT TORQUE CONTROL OF INDUCTION MACHINES BY INCREASING TORQUE LOOP BANDWIDTH 466
Ibrahim Mohd Alsofyani ; June-Hee Lee ; Byung-Moon Han ; Kyo-Beum Lee

μ-ANALYSIS EVALUATION OF A NOVEL COMBINED CURRENT-AND-SPEED CONTROL FOR INDUCTION MOTORS VIA ILQ DESIGN METHOD 471
Shuto Omori ; Hiroshi Takami ; Masashi Nakamura

LOSS MINIMIZATION CONTROL OF SENSORLESS SCALAR-CONTROLLED INDUCTION MOTOR DRIVES CONSIDERING IRON LOSS 478
Nguyen Anh Tan ; Dong-Choon Lee

TUNING OF INDUCTION MOTOR DRIVE WITH TORQUE SENSOR 483
Hajime Kubo ; Yugo Tadano

QUASI-TWO-LEVEL CONVERTER FOR OVERVOLTAGE MITIGATION IN MEDIUM VOLTAGE DRIVES 488
F. Bertoldi ; M. Pathmanathan ; R. S. Kanchan ; K. Spiliotis ; J. Driesen

A MEDIUM-VOLTAGE THREE-PHASE AC-DC CONVERTER CONSISTING OF CASCADED THREE-LEVEL BOOST-TYPE RECTIFIERS AND AN OPEN-END WINDING TRANSFORMER 495
Ryoji Tsuruta ; Hiromitsu Suzuki ; Ritaka Nakamura

A FAULT TOLERANT CONTROL STRATEGY FOR THE DELTA-CONNECTED CASCADED CONVERTER 503
Ping-Heng Wu ; Po-Tai Cheng

COOLING PERFORMANCE IMPROVEMENT OF HEAT SINK BY OSCILLATING HEAT PIPE ADDITION AND DESIGN FOR ENVIRONMENT OF OSCILLATING HEAT PIPE REFRIGERANT 511
Kuan-Chung Tey ; Kenichiro Suzuki

COMPACT LARGE CAPACITY GAS TURBINE STATIC STARTER 517
Hironori Kawaguchi ; Shigeyuki Nakabayashi ; Akinobu Ando ; Hiroshi Ogino ; Yasuaki Matsumoto ; Ikuto Udagawa ; Takahiro Ohta

VOLTAGE REFERENCE MODIFICATION SCHEME FOR RESONANCE SUPPRESSION IN LCL-FILTERED INVERTERS WITH DISCONTINUOUS PWM METHOD 521
Hyeon-Sik Kim ; Seung-Ki Sul

PARAMETRIC ROBUSTNESS ANALYSIS FOR PARALLEL FEEDFORWARD COMPENSATION BASED ACTIVE DAMPING OF LCL GRID CONNECTED INVERTER 528
Muhammad Talib Faiz ; Muhammad Mansoor Khan ; Xu Jianming ; Muhammad Ali ; Houjun Tang

OPEN-LOOP-BASED ISLAND-MODE VOLTAGE CONTROL METHOD FOR SINGLE-PHASE GRID-TIED INVERTER WITH MINIMIZED LC FILTER 534
Satoshi Nagai ; Jun-Ichi Itoh

EXPERIMENTAL VALIDATION OF ADAPTIVE CURRENT INJECTING METHOD FOR GRID-SYNCHRONIZATION IMPROVEMENT OF GRID-TIED REGS DURING SHORT-CIRCUIT FAULT 542
Shaokang Ma ; Hua Geng ; Geng Yang ; Bo Liu

ADAPTIVE CONTROL OF GRID-VOLTAGE FEEDFORWARD FOR GRID-CONNECTED INVERTERS BASED ON REAL-TIME IDENTIFICATION OF GRID IMPEDANCE 547
Roni Luhtala ; Tuomas Messo ; Tomi Roinila

MODEL BASED TUNING OF PROPORTIONAL RESONANT CONTROLLERS FOR VOLTAGE SOURCE INVERTERS 555
Stefan Almér ; Thomas Besselmann ; Mario Schweizer

AN SOC-BASED PLATFORM FOR INTEGRATED MULTI-AXIS MOTION CONTROL AND MOTOR DRIVE 560
Yongping Sun ; Ming Yang ; Yangyang Chen ; Wangpin He ; Dianguo Xu

VARIABLE SWITCHING FREQUENCY STRATEGY FOR ENHANCED SETTLING PERFORMANCE OF POSITION CONTROL WITHIN INVERTER LOSS LIMIT 565
Choongin Lee ; Jung-Ik Ha

TWO-WHEEL CANE FOR WALKING ASSISTANCE..571
Phi Van Lam ; Yasutaka Fujimoto

FALL PREVENTION AND VIBRATION SUPPRESSION OF WHEELCHAIR USING RIDER MOTION STATE..575
Isseki Takahashi ; Toshiyuki Murakami

STABILIZATION METHOD FOR RESIDENTIAL DC SYSTEM BASED ON PASSIVITY CRITERION..583
Hiroaki Kakigano

A NOVEL CONTROL APPROACH TO MULTI-TERMINAL POWER FLOW CONTROLLER FOR NEXT-GENERATION DC POWER NETWORK..588
Kenji Natori ; Yuta Nakao ; Yukihiko Sato

DC MICROGRID FOR TELECOMMUNICATIONS SERVICE AND RELATED APPLICATION..593
Keiichi Hirose

MVDC DISTRIBUTION GRIDS FOR ELECTRIC VEHICLE FAST-CHARGING INFRASTRUCTURE..598
Marco Stieneker ; Benedict J. Mortimer ; Arne Hinz ; Adolf Müller-Hellmann ; Rik W. De Doncker

REVIEW OF RESONANT GATE DRIVER IN POWER CONVERSION..607
Bainan Sun ; Zhe Zhang ; Michael A.E. Andersen

A LOW PROFILE HIGH FREQUENCY LED DRIVING SYSTEM BASED ON AIRCORE PLANAR INDUCTOR..614
Yueshi Guan ; Xihong Hu ; Shu Zhang ; Yijie Wang ; Dianguo Xu ; Wei Wang

ANALYSIS AND COMPENSATION OF DEAD-TIME EFFECT IN SIC-DEVICE-BASED HIGH-SWITCHING-FREQUENCY INVERTERS..619
Qingzeng Yan ; Xibo Yuan ; Xiaojie Wu ; Yiwen Geng

CONTROL AND PERFORMANCE OF NEW ASYMMETRICAL OPERATION FOR SWITCHED-CAPACITOR-BASED RESONANT CONVERTERS..626
Hadi Setiadi ; Hideaki Fujita

HIGH-FREQUENCY RESONANT CONVERTER WITH SYNCHRONOUS RECTIFICATION FOR HIGH CONVERSION RATIO AND VARIABLE LOAD OPERATION..632
Lei Gu ; Kawin Surakitbovorn ; Juan Rivas-Davila

SMART PV INVERTERS FOR SMART GRID APPLICATIONS..639
Cheng-Jhen Yang ; Terng-Wei Tsai ; Yi-Chan Li ; Cheng-Yu Tang ; Yaow-Ming Chen ; Yung-Ruei Chang

HIGH-VOLTAGE BI-DIRECTIONAL HALF-BRIDGE THREE-LEVEL SERIES RESONANT CONVERTER WITH FREQUENCY MODULATION CONTROL..645
Lee Sih-Yi ; Jhang Jynu-Jhe ; Lin Jing-Yuan ; Hsieh Yao-Ching ; Chiu Haung-Jen

A CONTROL STRATEGY FOR FLYING-START OF SHAFT SENSORLESS PERMANENT MAGNET SYNCHRONOUS MACHINE DRIVE..651
Zih-Cing You ; Sheng-Ming Yang

CONTACTLESS EV POWER TRACK SYSTEM WITH SEGMENT-EXCITED INDUCTIVELY COUPLED STRUCTURE..657
Jia-You Lee ; Yu-Chi Wang ; Chih-Yi Liao

DRIVING TEST EVALUATION OF SENSORLESS VEHICLE DETECTION METHOD FOR IN-MOTION WIRELESS POWER TRANSFER..663
Katsuhiro Hata ; Kensuke Hanajiri ; Takehiro Imura ; Hiroshi Fujimoto ; Yoichi Hori ; Motoki Sato ; Daisuke Gunji

A SYSTEM DESIGN METHOD OF HIGH-FREQUENCY CLASS-D INVERTER FOR WIDEBAND CURRENT CONTROL..669
Hiroki Kurumatani ; Seiichiro Katsura

ANALYSIS OF INTERIOR PERMANENT MAGNET TWO DEGREES OF FREEDOM MOTOR BASED ON CROSS-COUPLED STRUCTURE..675
Yoshiyuki Hatta ; Tomoyuki Shimono

STUDY COMPARISON BETWEEN FIREFLY ALGORITHM AND PARTICLE SWARM OPTIMIZATION FOR SLAM PROBLEMS..681
Mounia Janah ; Yasutaka Fujimoto

BANDWIDTH LIMITATIONS IN FORCE CONTROL OF A SERIES ELASTIC ACTUATOR WITH BACKLASH AND QUANTIZATION..688
Hanul Jung ; Chan Lee ; Sehoon Oh

ROTOR SHAPE OPTIMIZATION OF INTERIOR PERMANENT MAGNET SYNCHRONOUS MOTORS WITH CONCENTRATED WINDINGS BY CONSIDERING END-LEAKAGE FLUX..693
Katsumi Yamazaki ; Hiroki Narushima

LOSS ANALYSIS OF PERMANENT-MAGNET SYNCHRONOUS MACHINES CONSIDERING IN-PLANE EDDY CURRENT IN ELECTRICAL STEEL SHEETS...... 699

Hideki Ohguchi ; Satoshi Imamori ; Katsumi Yamazaki ; Haiyan Yui ; Masao Shuto

STUDY ON INFLUENCE OF DIFFERENCE IN STRUCTURE OF CONCENTRATED WINDING IPMSMS OBTAINED BY AUTOMATIC DESIGN 704

A. Ura ; M. Sanada ; S. Morimoto ; Y. Inoue

CARRIER HARMONIC LOSS REDUCTION TECHNIQUE ON DUAL THREE-PHASE PERMANENT-MAGNET SYNCHRONOUS MOTORS WITH PHASE-SHIFT PWM...... 711

Yoshihiro Miyama ; Haruyuki Kometani ; Kan Akatsu

FLUX INTENSIFYING PM-MOTOR WITH VARIABLE LEAKAGE MAGNETIC FLUX TECHNIQUE...... 718

Masahiro Aoyama ; Toshihiko Noguchi

CONTINUOUS OPERATION CONTROL OF PMSM IN THE CASE OF DC POWER SUPPLY LOSS...... 726

Jongwon Heo ; Keiichiro Kondo

MODEL PREDICTIVE CONTROL FOR MULTIPHASE MOTOR DRIVES – A TECHNOLOGY STATUS REVIEW 732

A. Tenconi ; S. Rubino ; R. Bojoi

INFLUENCE OF FAST SWITCHING SEMICONDUCTORS ON THE WINDING INSULATION SYSTEM OF ELECTRICAL MACHINES 740

Kay Hameyer ; Andreas Ruf ; Florian Pauli

CENTRALIZED CONTROL OF MODULAR MULTI RECTIFIER FOR MOTOR DRIVE APPLICATIONS UNDER UNBALANCED GRID...... 746

Yipeng Song ; Pooya Davari ; Frede Blaabjerg

VECTOR CONTROL OF MAGNETICALLY MODULATED MOTOR FOR POWER SPLITTING OF HEV APPLICATION 753

Toshihiko Noguchi ; Sawanth Krishna Machavolu ; Masahiro Aoyama ; Yuto Motohashi

IMPEDANCE-BASED STABILITY EVALUATION OF VIRTUAL SYNCHRONOUS MACHINE IMPLEMENTATIONS IN CONVERTER CONTROLLERS 759

Eneko Unamuno ; Atle Rygg ; Mohammad Amin ; Marta Molinas ; Jon Andoni Barrena

STABLE POWER SUPPLY METHOD FOR HOUSEHOLD APPLIANCES VIA VIRTUAL SYNCHRONOUS GENERATOR IN SINGLE-PHASE THREE-WIRE MICROGRID 767

Yuko Hirase ; Hidehiko Nakagawa ; Eiji Yoshimura ; Shogo Katsura ; Kensho Abe ; Osamu Noro ; Kazushige Sugimoto ; Kenichi Sakimoto

A NOVEL OSCILLATION DAMPING METHOD OF VIRTUAL SYNCHRONOUS GENERATOR CONTROL WITHOUT PLL USING POLE PLACEMENT 775

Jia Liu ; Yushi Miura ; Toshifumi Ise

OPERATION OF A MODULAR MULTILEVEL CONVERTER CONTROLLED AS A VIRTUAL SYNCHRONOUS MACHINE...... 782

Salvatore D'arco ; Giuseppe Guidi ; Jon Are Suul

ASSESSMENT OF VIRTUAL SYNCHRONOUS MACHINE BASED CONTROL IN GRID-TIED POWER CONVERTERS...... 790

Chi Li ; Igor Cvetkovic ; Rolando Burgos ; Dushan Boroyevich

RESEARCH ON THE BLOCKCHAIN-BASED INTEGRATED DEMAND RESPONSE RESOURCES TRANSACTION SCHEME 795

Shengnan Zhao ; Yang Li ; Beibei Wang ; Huiling Su

INDIRECT CURRENT CONTROL FOR SEAMLESS TRANSFER OF UTILITY INTERACTIVE INVERTER 803

Kyungbae Lim ; Injong Song ; Jaeho Choi

STUDY OF AC POWER INTERCHANGE AND DC POWER INTERCHANGE FOR MICRO GRID SYSTEMS 809

Kazuto Yukita ; Daiki Owaki ; Shunsuke Horie ; Toshiro Matsumura ; Yasuyuki Goto

STABILITY ENHANCEMENT STRATEGY FOR ISLANDING MICROGRID WITH MULTI-TYPE INVERTERS BASED ON HYBRID IMPEDANCE MODELLING 815

Meiqin Mao ; Yong Ding ; Yatao Shen ; Liuchen Chang

DC POWERED DATA CENTER WITH 200 KW PV PANELS 822

Keiichi Hirose

INFLUENCES OF DETERIORATION IN CAPACITOR AND INDUCTOR ON CURRENT SENSORLESS STATIC MODEL DC-DC CONVERTER...... 826

Fujio Kurokawa ; Masashi Taguchi ; Jizhe Wang ; Hidenori Maruta ; Nobumasa Matsui

CAPACITIVE DIVIDER BASED PASSIVE START-UP METHODS FOR FLYING CAPACITOR STEP-DOWN DC-DC CONVERTER TOPOLOGIES...831
Michael Halamicek ; Tom Moiannou ; Nenad Vukadinovic ; Aleksandar Prodic

HIGH VOLTAGE GAIN INTERLEAVED ACTIVE-CLAMP FORWARD (IACF) CONVERTER HAVING REDUCED PRIMARY CONDUCTION LOSS...838
Yeonho Jeong ; Mu-Hyun Park ; Gun-Woo Kim ; Byoung-Hee Lee ; Gun-Woo Moon

CONTROL OF SWITCHING-CAPACITOR BASED BUCK-BOOST CONVERTER...................845
M. Veerachary ; Vasudha Khubchandani

IMPROVEMENT OF UPLOAD TRANSIENT RESPONSES FOR ULTRA HIGH STEP-DOWN CONVERTER...851
Y.T. Yan ; K.I. Hwu

POWER ELECTRONICS AND CONTROL TECHNOLOGIES FOR HOUSEHOLD WASHER.........856
Toru Niki

DEVELOPMENT OF ROOM AIR CONDITIONER WITH TWIN-PROPELLER FANS................860
Takamasa Uemura ; Tomoya Fukui ; Kenichi Sakoda

ELECTROLYTIC CAPACITOR-LESS SINGLE-PHASE TO THREE-PHASE INVERTER WITH HARMONICS SUPPRESSION CONTROL FOR AIR CONDITIONER...866
Nobuo Hayashi ; Takuro Ogawa ; Tomoisa Taniguchi ; Morimitsu Sekimoto

LATEST DEVELOPMENT OF SIC POWER MODULE-BASED SINGLE-STAGE AC-AC RESONANT CONVERTER FOR HIGH-FREQUENCY INDUCTION HEATING APPLICATIONS...........872
Tomokazu Mishima

AN OPTIMIZED CONTROL STRATEGY TO IMPROVE THE CURRENT ZERO-CROSSING DISTORTION IN BIDIRECTIONAL AC/DC CONVERTER BASED ON V2G CONCEPT...................878
Lei Jing ; Xiaoqing Wang ; Bodong Li ; Maohang Qiu ; Bo Liu ; Min Chen

PER-PHASE CONTROL STRATEGY OF THE THREE-PHASE FOUR-WIRE INVERTER...........883
Yi-Chun Li , Terng-Wei Tsui , Cheng-Jhen Yung , Yaow-Ming Chen ; Yung-Ruei Chang

OPPORTUNITIES FOR PERFORMANCE IMPROVEMENT OF SINGLE-PHASE POWER CONVERTERS THROUGH ENHANCED AUTOMATIC-POWER-DECOUPLING CONTROL...........889
Huawei Yuan ; Sinan Li ; Wenlong Qi ; Siew-Chong Tan ; S. Y. Ron Hui

ZERO VOLTAGE SWITCHING SCHEME FOR FLYBACK CONVERTER TO ENSURE COMPATIBILITY WITH ACTIVE POWER DECOUPLING CAPABILITY.................................896
Hiroki Watanabe ; Jun-Ichi Itoh

MODEL PREDICTIVE FAULT TOLERANT CONTROL OF BIDIRECTIONAL AC/DC CONVERTER WITH VOLTAGE BALANCE OF SPLIT CAPACITOR...904
Nan Jin ; Chongyan Zhao ; Leilei Guo

PWM STRATEGY FOR PARALLEL OPERATION OF THREE PHASE CONVERTERS TIED TO GRID..911
Hyun-Sam Jung ; Seung-Ki Sul

PRACTICAL ISSUES AND IMPLEMENTATION CIRCUITS OF THE DIGITAL-ANALOG HYBRID FULL FEED-FORWARD METHOD WITH UNIPOLAR AND BIPOLAR MODULATIONS...917
Xin Zhang ; Henry S. H. Chung ; Zhixun Ma

AN AC-DC POWER CONVERTER FOR ELECTROLYTIC CAPACITOR-LESS LED DRIVER WITH HIGH LUMINOUS EFFICACY...922
Kwon-Sik Park ; Byuong-Jun Seo ; Kyoung-Suk Kang ; Eui-Cheol Nho

AN IMPROVED CASCADED DUAL-BUCK INVERTER...927
Usman Ali Khan ; Honnyong Cha ; Ashraf Ali Khan ; Heung-Geun Kim ; Wilson Eberle ; Liwei Wang

A SINGLE-SWITCH INTEGRATED-STAGE LED DRIVER BASED ON CUK AND CLASS-E CONVERTER..934
Shu Zhang ; Yijie Wang ; Xiaosheng Liu ; Yan Zhou ; Dianguo Xu

A FAULT-TOLERANT PARALLEL INVERTER APPLIED TO MICRO-GRID.........................939
Xiangyue Shi ; Jinjie Peng ; Zhifeng Qiu ; Wei Xiong

STABILITY ANALYSIS OF GRID-CONNECTED CONVERTERS WITH ADD-ON VOLTAGE SUPPORT FUNCTIONALITY USING REPETITIVE CONTROL...946
Y. Zhang ; M. G. L. Roes ; M. A. M. Hendrix ; J. L. Duarte

ADAPTIVE SERIES STABILIZER MODULE FOR THE GRID CONNECTED INVERTER UNDER VARIABLE GRID CONDITIONS...953
Xin Zhang

AN IMPROVED DROOP CONTROL BASED SMOOTH TRANSFER CONTROL STRATEGY.........957
Xin Meng ; Jinjun Liu ; Zeng Liu ; Ronghui An

FREQUENCY RESPONSE ANALYSIS OF LOAD EFFECT ON DYNAMICS OF GRID-FORMING INVERTER 963

Matias Berg ; Tuomas Messo ; Teuvo Suntio

A NEW CONTROL METHOD FOR TRIPLE-ACTIVE BRIDGE CONVERTER WITH FEED FORWARD CONTROL 971

Takanobu Ohno ; Nobukazu Hoshi

ANALYSIS OF PFM OPERATION MODEL FOR CAPACITOR CHARGER RESONANT TOPOLOGY WITH ENERGY DOSAGE 977

Pengyu Jia ; Yiqin Yuan ; Shengwen Fan ; Zhenyu Shan

AN ACTIVE-CLAMPED CURRENT-FED HALF-BRIDGE DC-DC CONVERTER WITH THREE SWITCHES 982

Truong-Duy Duong ; Minh-Khai Nguyen ; Young-Cheol Lim ; Joon-Ho Choi

A HIGH GAIN QUASI SINGLE STAGE LLC RESONANT DC/DC CONVERTER WITH COUPLED INDUCTOR AND PARTIAL ACTIVE CLAMP 987

Chongcan Huo ; Xiaogao Xie ; Shuai Jiang ; Hanjing Dong

SUPPRESSION OF RIPPLE CURRENT IN HIGH STEP-UP DC-DC CONVERTER UTILIZING COCKCROFT-WALTON CIRCUIT WITH INDUCTOR 992

Takumi Yasuda ; Masataka Minami ; Shin-Ichi Motegi ; Masakazu Michihira

AN OPTIMAL DESIGN METHOD CONSIDERING TRANSFORMER PARASITIC CAPACITANCE OF LLC RESONANT CONVERTERS 998

Naizeng Wang ; Xu Yang ; Mofan Tian ; Haiyang Jia ; Guangzhao Xu ; Zhenwei Li

COMPARISON OF HARMONIC LINEARIZATION AND HARMONIC STATE SPACE METHODS FOR IMPEDANCE MODELING OF MODULAR MULTILEVEL CONVERTER 1004

Jing Lyu ; Xin Zhang ; Jingjing Huang ; Jianwen Zhang ; Xu Cai

AN IMPROVED PHASE-SHIFTED PWM FOR A FIVE-LEVEL HYBRID-CLAMPED CONVERTER 1010

Kui Wang ; Nianzhou Liu ; Zedong Zheng ; Yongdong Li

INTEGRATED CONTROL METHODS FOR ASYMMETRICAL CASCADED H-BRIDGE RECTIFIER 1015

Wenjing Dai ; Jie Chen ; Xin Chen ; Chunying Gong

TRANSIENT VOLTAGE STRESS MODELING FOR SUBMODULES OF MODULAR MULTILEVEL CONVERTERS UNDER GRID VOLTAGE SAGS 1021

Zhijian Yin ; Yongheng Yang ; Huai Wang

SVPWM STRATEGY BASED ON MULTILEVEL 3LNPC-CR 1027

Xiaoqiong He ; Pengcheng Han ; Xiaolan Lin ; Yi Wang ; Xu Peng

THE MULTIPLE DEGREE OF FREEDOM BASED NEUTRAL POINT POTENTIAL CONTROL OF THREE LEVEL NEUTRAL POINT CLAMPED CONVERTERS 1032

Bo Guan ; Shinji Doki

A MODIFIED PHASE-SHIFTED PWM TECHNIQUE FOR THE GRID-CONNECTED HYBRID CASCADED CONVERTER 1038

Yu-Chen Su ; Po-Tai Cheng

NOVEL T-TYPE DUAL-BUCK INVERTER WITH MINIMUM NUMBER OF INDUCTORS 1046

Tien-The Nguyen ; Honnyong Cha ; Bang Le-Huy Nguyen ; Heung-Geun Kim

CONTROL OF DIRECT AC/AC MODULAR MULTILEVEL CONVERTER IN RAILWAY POWER SUPPLY SYSTEM 1051

Shuguang Song ; Jinjun Liu ; Shaodi Ouyang ; Xingxing Chen ; Baojin Liu

WIRELESS POWER TRANSFER: CRITICAL REVIEW OF RELATED STANDARDS 1062

Mohamad Abou Houran ; Xu Yang ; Wenjie Chen ; Mehdi Samizadeh

COMPARATIVE STUDY OF SINGLE-PHASE FUNDAMENTAL COMPONENT FREQUENCY ESTIMATION SCHEMES UNDER TIME-VARYING HARMONIC DISTORTION OPERATION 1067

E. B. Kapisch ; J. L. Duarte ; C. A. Duque

A COMPREHENSIVE DEAD-TIME COMPENSATION METHOD FOR A THREE-PHASE DUAL-ACTIVE BRIDGE CONVERTER WITH HYBRID MODULATION SCHEMES 1073

Jingxin Hu ; Zhiqing Yang ; Rik W. De Doncker

EVALUATION OF A HIGH-FREQUENCY REACTOR WITH A NEW WIRE GUIDE FOR A TOROIDAL CORE 1080

Hideki Ayano ; Akira Fujimura ; Yoshihiro Matsui

CORE LOSS EVALUATION IN POWDER CORES: A COMPARATIVE COMPARISON BETWEEN ELECTRICAL AND CALORIMETRIC METHODS 1087

Yuki Ishikura ; Jun Imaoka ; Mostafa Noah ; Masayoshi Yamamoto

MODELING, MAGNETIC DESIGN, AND SIMULATION METHODS CONSIDERING DC SUPERIMPOSITION CHARACTERISTIC OF POWDER CORES USED IN POWER CONVERTERS 1095

Jun Imaoka ; Kenkichiro Okamoto ; Masahito Shoyama ; Yuki Ishikura ; Mostafa Noah ; Masayoshi Yamamoto

MODELLING AND DESIGN OF A MEDIUM FREQUENCY TRANSFORMER FOR HIGH POWER DC-DC CONVERTERS 1103

Miloš Stojadinovic ; Jürgen Biela

EVALUATION OF INDUCTOR LOSSES ON Z-SOURCE INVERTER CONSIDERING AC AND DC COMPONENTS 1111

Ryuji IIjima ; Naoki Kamoshida ; Rene Alexander Barrera Cardenas ; Takanori Isobe ; Hiroshi Tadano

AN INTEGRATING STRUCTURE OF OUTPUT FILTER FOR GRID CONNECTED INVERTER BASED ON FMLF TECHNIQUE 1118

Jie Ma ; Yenan Chen ; Pingping Chen ; Wenxing Zhong ; Dehong Xu

NEW SCREENING METHOD FOR IMPROVING TRANSIENT CURRENT SHARING OF PARALLELED SIC MOSFETS 1125

Junji Ke ; Zhibin Zhao ; Peng Sun ; Huazhen Huang ; James Abuogo ; Xiang Cui

PSPICE MODELING AND APPLICATION FOR SIC POWER MOSFET TO EVALUATE THE POWER LOSS IN FULL-BRIDGE CONVERTER 1131

Juan Wei ; Fei Lin ; Zhongping Yang ; Xianjin Huang ; Chanjuan Xiao ; Hao Zhang ; Wencai Liang

ALL-SIC MODULE PACKAGING TECHNOLOGY 1137

Kento Shirata ; Norihiro Nashida ; Hideyo Nakamura ; Yoshitaka Nishimura

A NEW SMALLEST 1200V INTELLIGENT POWER MODULE FOR THREE PHASE MOTOR DRIVES 1141

Minsub Lee ; Miran Baek ; Junbae Lee ; Daewoong Chung

DESIGN AND ENHANCEMENT OF ESD RELIABILITY IN CIRCULAR UHV 300-V NLDMOS POWER COMPONENTS 1145

Shen-Li Chen ; Yi-Hao Chao ; Chih-Ying Yen ; Jen-Hao Lo ; Chun-Ting Kuo ; Yu-Lin Lin ; Yi-Hao Chiu ; Pei-Lin Wu ; Yu-Lin Jhou

A TECHNOLOGY ANALYSIS OF VOLTAGE SHARING IN SERIES CONNECTED POWER DEVICES 1149

Z Davletzhanova ; O Alatise ; R Bonyadi ; J Ortiz-Gonzalez ; T Dai ; M Jennings ; L Ran ; P Mawby

FAILURE MECHANISM ANALYSIS AND PHYSICS-OF-FAILURE LIFETIME PREDICTION METHOD FOR PRESS-PACK THYRISTOR OF CONVERTER VALVE 1157

Ning Liang ; Zhigang Zhang ; Yating Gou ; Cuicui Liu ; Zebin Yang ; Jiangnan Chen ; Fang Zhuo ; Feng Wang

SURGE VOLTAGE ABSORPTION BY A SILICON CARBIDE AVALANCHE-DIODE WITH P-N STRUCTURE 1162

K. Koseki ; Y. Tanaka

CALCULATION OF THYRISTOR RELIABILITY PARAMETER OF UHVDC CONVERTER VALVE IN HEMP ENVIRONMENT 1167

Zhigang Zhang ; Yating Gou ; Cuicui Liu ; Zebin Yang ; Xiaotong Du ; Jiangnan Chen ; Fang Zhuo ; Feng Wang ; Yuanliang Lan ; Caiwang Sheng

GENERALIZED STACKELBERG GAME-THEORETIC APPROACH FOR JOINTED ENERGY AND RESERVE COORDINATION OF ELECTRIC VEHICLES 1172

Tianyang Zhao ; Xuewei Pan ; Lei Li ; Fei Zhao ; Can Wang

IMPEDANCE INFLUENCE ANALYSIS OF PHASE-LOCKED LOOPS ON THREE-PHASE GRID-CONNECTED INVERTERS 1177

Yuncheng Wang ; Xin Chen ; Yang Zhang ; Jie Chen ; Chunying Gong

PULSE-INJECTION-BASED SENSORLESS CONTROL METHOD WITH IMPROVED DYNAMIC CURRENT RESPONSE FOR PMSM 1183

Hechao Wang ; Kaiyuan Lu ; Dong Wang ; Frede Blaabjerg

INFLUENCE OF PARAMETER VARIATIONS ON OPERATING CHARACTERISTICS OF MTPF CONTROL FOR DTC-BASED PMSM DRIVE SYSTEM 1189

Keisuke Fujii ; Yukinori Inoue ; Shigeo Morimoto ; Masayuki Sanada

A QUIET POSITION SENSORLESS CONTROL FOR AN IPMSM BASED ON EXTENDED EMF AND VOLTAGE INJECTION SYNCHRONIZED WITH PWM CARRIER 1196

Yuki Ishii ; Hiroki Yamashita ; Hisao Kubota

STUDY OF TORQUE RIPPLE REDUCTION AND TORQUE BOOST BY MODIFIED TRAPEZOIDAL MODULATION 1202

Satoshi Joryo ; Kazuto Tatsumi ; Toshimitsu Morizane ; Katsunori Taniguchi ; Noriyuki Kimura ; Hideki Omori

FAULT DIAGNOSIS METHOD OF CURRENT SENSOR FOR PERMANENT MAGNET SYNCHRONOUS MOTOR DRIVES 1206

Guoqiang Zhang ; Guoxin Wang ; Gaolin Wang ; Junya Huo ; Lianghong Zhu ; Dianguo Xu

SENSORLESS SPEED CONTROL OF DIESEL-GENERATOR SYSTEMS BASED ON MULTIPLE SOGI-FLLS...1212

Ngoc Dat Dao ; Dong-Choon Lee ; Dae-Sik Lim

ROBUSTNESS OF SIMPLIFIED SPEED-SENSORLESS VECTOR CONTROL FOR INDUCTION MOTOR...1217

Naoki Akao ; Mineo Tsuji ; Shin-Ichi Hamasaki

MAXIMUM TORQUE CONTROL REFERENCE FRAME BASED ON A TORQUE MAP FOR IPMSMS WITH LARGE INDUCTANCE VARIATION...1223

Kazuki Ohta ; Takumi Ohnuma ; Shinji Doki

PMSM MODEL DISCRETIZATION IN CONSIDERATION OF PARK TRANSFORMATION FOR CURRENT CONTROL SYSTEM...1228

Masamichi Inoue ; Shinji Doki

PSEUDO-RANDOM HIGH-FREQUENCY SINUSOIDAL VOLTAGE INJECTION BASED SENSORLESS CONTROL FOR IPMSM DRIVES...1234

Guoqiang Zhang ; Huiying Wang ; Gaolin Wang ; Junya Huo ; Lianghong Zhu ; Dianguo Xu

AT-NPC 3-LEVEL INVERTER-FED INDUCTION MOTOR VECTOR CONTROL WITH NEUTRAL POINT VOLTAGE CONTROL...1240

K. Sudo ; M. Tsuji ; S. Hamasaki ; T. Fukuoka ; H. Ichinose

INVESTIGATION OF VARIOUS POSITION ESTIMATION ACCURACY ISSUES IN PULSE-INJECTION-BASED SENSORLESS DRIVES...1246

Hechao Wang ; Kaiyuan Lu ; Dong Wang ; Frede Blaabjerg

POSITION SENSORLESS CONTROL OF SWITCHED RELUCTANCE MOTOR USING ESTIMATED PWM PHASE VOLTAGE...1253

Y. Nakazawa ; K. Ohyama ; H. Fujii ; H. Uehara ; Y. Hyakutake

EXPERIMENTAL CONFIRMATION OF THRUST AND ATTRACTIVE FORCE CONTROL OF LINEAR INDUCTION MOTOR BY TWO DIFFERENT FREQUENCY COMPONENTS......................1259

Kenta Sannomiya ; Toshimitsu Morizane ; Noriyuki Kimura ; Hideki Omori

GA BASED OPTIMIZED TRAJECTORIES OF ROTATING SPEED AND D-Q AXIS CURRENTS FOR AN IPMSM...1264

Shuta Kumagai ; Kaoru Inoue ; Toshiji Kato

2-DEGREE-OF-FREEDOM DEADBEAT CONTROL WITH DISTURBANCE COMPENSATION FOR PMSM DRIVE SYSTEM USING FPGA...1270

Arata Takahashi ; Shotaro Takakura ; Tomoki Yokoyama

EXTENDED EMF-BASED SIMPLE IPMSM SENSORLESS VECTOR CONTROL USING COMPENSATED CURRENT CONTROLLER...1276

Takatoshi Inoue ; Yasumasa Hamabe ; Mineo Tsuji ; Shin-Ichi Hamasaki

FULL-BAND OUTPUT IMPEDANCE MODEL OF VIRTUAL SYNCHRONOUS GENERATOR IN DQ FRAMEWORK...1282

Li Wenbing ; Wang Jianhua ; Song Jingyu ; Luo Fangfang ; Gao Shang ; Wu Zaijun

AN MTPA CONTROL METHOD OF A PMSM AND A SYNRM BASED ON A DTC IN THE STATOR FLUX LINKAGE SYNCHRONOUS FRAME...1289

Gimpei Itoh ; Yukinori Inoue ; Shigeo Morimoto ; Masayuki Sanada

EEMFS EXCITED BY SIGNAL INJECTION FOR POSITION SENSORLESS CONTROL OF PMSMS AND THEIR PERFORMANCE COMPARISON BY USING IMAGINARY ELECTROMOTIVE FORCE...1295

Takumi Nimura ; Shota Kondo ; Shinji Doki ; Mutuwo Tomita

HARMONIC CURRENT CANCELLATION METHOD FOR PMSM DRIVE SYSTEM USING RESONANT CONTROLLERS...1301

Dongsheng Li ; Yoshitaka Iwaji ; Yasuo Notohara ; Ken Kishita

ESTIMATION ERROR ANALYSIS OF STATOR FLUX OBSERVER FOR DTC-BASED PMSM DRIVES...1308

Atsushi Shinohara ; Kichiro Yamamoto

APPLICATION OF FICTITIOUS REFERENCE ITERATIVE TUNING TO CONTROLLER DESIGN FOR VARIOUS MACHINES...1315

Hidehiro Ikeda ; Kazuya Goto ; Feili Zhang ; Kazuya Kayashima ; Tsuyoshi Hanamoto

HIGH EFFICIENCY CONTROL FOR PERMANENT MAGNET MOTOR DRIVE SYSTEM WITH FUEL CELLS CONNECTED IN SERIES WITH ELECTRIC DOUBLE-LAYER CAPACITORS......................1322

Kichiro Yamamoto ; Fumiya Ohdera ; Atsushi Shinohara

COMPARATIVE STUDY OF SPEED RIPPLE REDUCTION BY VARIOUS CONTROL METHODS IN PMSM DRIVE SYSTEMS WITH PULSATING LOAD...1329

Yuma Komaru ; Yukinori Inoue ; Shigeo Morimoto ; Masayuki Sanada

ESTIMATION OF THE PARAMETERS OF THE SERVO DRIVE SYSTEM USING PARTICLE SWARM OPTIMIZATION ALGORITHM 1336
Helin Zhu ; Jae Hyuk Choi ; Sang Uk Park ; Jusuk Lee ; Hyong Gun Lee ; Hyung Soo Mok

A PROGRAMMABLE BATTERY TEST SYSTEM WITH ENERGY RECYCLING FEATURE BASED ON SINUSOIDAL LOADING TECHNIQUE 1341
Chang-Hua Lin ; Guan-Jung Chen ; Hwa-Dong Liu ; Kun-Feng Chen

DEVELOPMENT OF LARGE-CAPACITY CONVERTER FOR BATTERY ENERGY STORAGE SYSTEMS 1346
Hiroyoshi Komatsu ; Tatsuji Katayama ; Noriko Kawakami

ANALYSIS AND COMPARISON OF DC/DC TOPOLOGIES IN PARTIAL POWER PROCESSING CONFIGURATION FOR ENERGY STORAGE SYSTEMS 1351
Maria C. Mira ; Zhe Zhang ; A. E. Michael Andersen

TWO-STAGE PROTECTION FOR MULTI-CHANNEL POWER ELECTRONIC CONVERTERS FED LARGE ASYNCHRONOUS HYDRO-GENERATING UNIT 1358
R. R. Semwal ; Anto Joseph

CURRENT SHARING CONTROL FOR SERIES-PARALLEL CHANGEOVER USING BATTERY AND ELECTRIC DOUBLE-LAYER CAPACITOR BANK 1364
Taisei Nishino ; Keisaku Isozaki ; Naoki Kogai ; Kyungmin Sung

CONTROL METHOD OF ENERGY STORAGE SYSTEM TO IMPROVE OUTPUT POWER OF PCS 1370
Mikiya Ishibashi ; Hitoshi Haga ; Kenji Arimatsu ; Koji Kato

A CONTROL STRATEGY OF MMC BATTERY ENERGY STORAGE SYSTEM BASED ON ARM CURRENT CONTROL 1376
Liu Danqing ; Wang Guangzhu ; Ou Zhujian ; Liu Jiaxing

EQUIVALENT RESISTANCE CONTROL FOR MAXIMUM POWER TRANSFER METHOD OF PIEZOELECTRIC ELEMENT IN VIBRATION POWER GENERATION 1381
Kenya Takamura ; Hiroaki Yamada ; Toshihiko Tanaka ; Tomoharu Yada ; Hajime Fujiwara

DC BUS VOLTAGE STABILIZATION FOR CASCADED POWER CONVERTER BY INTEGRATING AN EXTRA PORT INTO LOAD SIDE PSFB 1386
Jiang You ; Weiyan Fan ; Mengyan Liao

COMMON MODE CURRENT REDUCTION OF THREE-PHASE CASCADED MULTILEVEL TRANSFORMERLESS INVERTER FOR PV SYSTEM 1391
Wenjie Wang ; Ke Chen ; Lijun Hang ; Anping Tong ; Yiliang Gan

CURRENT SHARING/VOLTAGE SHARING CONTROL STRATEGY FOR CASCADED DC/DC CONVERTER IN PHOTOVOLTAIC DC COLLECTION SYSTEM 1397
Bo Chen ; Yi Wang ; Yanjun Tian ; Shilei Wei

PCC VOLTAGE COMPENSATION OF PV INVERTER WITH ACTIVE POWER DECOUPLING CIRCUIT 1403
Duck-Hwan Hwang ; Jung-Yong Lee ; Younghoon Cho

A NOVEL PARTIAL SHADING DETECTION ALGORITHM UTILIZING POWER LEVEL MONITORING OF PHOTOVOLTAIC PANELS 1409
Thusitha Randima Wellawatta ; Sung-Jin Choi

BOOST INTEGRATED THREE-PHASE SOLAR INVERTER USING CURRENT UNFOLDING AND ACTIVE DAMPING METHODS 1414
N. Ha Pham ; Tomoyuki Mannen ; Keiji Wada

LINEAR ACTIVE DISTURBANCE REJECTION CONTROL FOR ISOLATED THREE-PORT CONVERTER 1421
Jiang You ; Mengyan Liao ; Weiyan Fan

STABILITY CONSTRAINED GAIN OPTIMIZATION OF DROOP CONTROLLED CONVERTERS IN DC NANOGRIDS 1426
Soumya Bandyopadhyay ; Laura Ramirez-Elizondo ; Pavol Bauer

SIC BASED SSPC FOR HIGH VOLTAGE SPACE APPLICATIONS 1435
D. Marroquí ; A. Garrigós ; José M. Blanes ; R. Gutiérrez

AN IMPROVED VOLTAGE-TYPE GRID-CONNECTED CONTROL STRATEGY FOR COMPENSATING UNBALANCED VOLTAGE 1442
Liu Hongpeng ; Zhou Jiajie ; Wang Wei

DUAL TWO-STAGE ISOLATED BIDIRECTIONAL DC-DC CONVERTER FOR DC GRID STORAGE 1447
Gabriel Tibola ; Jorge L. Duarte

MODULAR MULTILEVEL CONVERTER WITH CAPACITOR VOLTAGE SELF-BALANCING USING REDUCED NUMBER OF VOLTAGE SENSORS 1455
Taiyuan Yin ; Yue Wang ; Xiaolei Wang ; Shiyuan Yin ; Shumin Sun ; Guanglei Li

PLUG AND OUTLET IN HOUSEHOLD DC LOW VOLTAGE MICRO-GRID POWER DISTRIBUTION........1460
Worapong Pairindra ; Surin Khomfoi

PERFORMANCE PROGRAMMING TECHNIQUE FOR MULTI-STAGE DC POWER DISTRIBUTION SYSTEMS........1465
Syam Kumar Pidaparthy ; Hansang Kim ; Yeonjung Kim ; Byungcho Choi

COORDINATION CONTROL FOR PARALLELED INVERTERS BASED ON VSG FOR PV/BATTERY MICROGRID........1472
Meiqin Mao ; Cheng Qian ; Liuchen Chang ; Yan Du

ADAPTIVE VOLTAGE CONTROL SCHEME FOR DAB BASED MODULAR CASCADED SST IN PV APPLICATION........1478
Tao Liu ; Yang Xuan ; Xu Yang ; Peng Xu ; Yang Li ; Lang Huang ; Xiang Hao

SIX-STEP MMC-BASED HIGH POWER DC-DC CONVERTER........1484
Stefan Milovanovic ; Dražen Dujic

COMBINED DC POWER FLOW CONTROLLER FOR DC GRID........1491
Yongning Chi ; Xizhou Du ; Siqi Liu ; Xu Cai

AN APPROACH FOR THE EMULATION OF DC GRID ADMITTANCES: IMPLEMENTATION ON A BUCK CONVERTER........1498
Enrique Rodriguez-Diaz ; Fracisco D. Freijedo ; Drazen Dujic ; Juan C. Vasquez ; Josep M. Guerrero

A COMPOUND CONTROLLER FOR POWER FLOW AND SHORT-CIRCUIT FAULT IN DC GRID........1504
Han Ye ; Wu Chen ; Pengpeng Pan ; Xiaokun He

DESIGN PROCEDURE AND CONTROL OF A HYBRID CIRCUIT BREAKER WITH ADAPTABLE PULSE CURRENT INJECTION........1509
Andreas Jehle ; Jürgen Biela

A PRAGMATIC SOH AND SOC CO-ESTIMATOR FOR LITHIUM-ION BATTERIES IN SMART GRID APPLICATIONS........1517
Kaiyuan Li ; King Jet Tseng ; Feng Wei ; Boon-Hee Soong

MODELING AND STABILITY ANALYSIS OF PARALLEL DROOP-CONTROLLED AND CURRENT-CONTROLLED INVERTERS........1524
Shike Wang ; Zeng Liu ; Jinjun Liu ; Ronghui An

DIRECT WIRELESS BATTERY CHARGING SYSTEM........1530
Woo-Seok Lee ; Jin-Hak Kim ; Shin-Young Cho ; Il-Oun Lee

AN IMPROVED PWM SCHEME TO ACHIEVE ZERO-VOLTAGE SWITCHING FOR ALL DEVICES IN THREE-PHASE ISOLATED MATRIX RECTIFIER........1537
Xuerui Lin ; Yunwei Ryan Li ; Jahangir Afsharian ; Dewei David Xu

FIXED-FREQUENCY HF GATE DRIVER BY A PUSH-PULL SELF-EXCITATION LC OSCILLATOR HAVING A CAPACITANCE TRANSISTOR........1543
Naoyuki Ishibashi ; Takuya Mizushima ; Masahiko Hirokawa ; Akihiko Katsuki

A FLEXIBLE REDUCED CAPACITOR VOLTAGES STRATEGY FOR VARIABLE-SPEED DRIVES WITH MODULAR MULTILEVEL CONVERTER........1549
Fangzhou Zhao ; Guochun Xiao ; Daoshu Yang ; Zhiqian Wu ; Xin Meng

A LEAKAGE FLUX CANCELLATION TECHNIQUE FOR SERIES-PARALLEL COMBINED RESONANT CIRCUITS WITH ASYMMETRIC ROTARY TRANSFORMERS USED FOR ULTRASONIC SPINDLE DRIVE........1554
Jun Imaoka ; Masahito Shoyama

A NOVEL STRUCTURAL HEALTH MONITORING SYSTEM WITH WIRELESS POWER AND BI-DIRECTIONAL DATA TRANSFER........1562
Yujin Jangs ; Keon-Woo Kim ; Moo-Hyun Park ; Nayoung Lee ; Gun-Woo Moon

CONTROL STRATEGY FOR STARTER GENERATOR IN UAV WITH MICRO JET ENGINE........1567
Jun-Ichi Itoh ; Kazuki Kawamura ; Hiroyuki Koshikizawa ; Kazuyuki Abe

STUDY ON THE INFLUENCE OF VOLTAGE VARIATIONS FOR NON-INTRUSIVE LOAD IDENTIFICATIONS........1575
Yu-Hsiu Lin ; Shun-Kang Hung ; Men-Shen Tsai

BASIC EXPERIMENT OF A MAGLEV SYSTEM FOR A FLEXIBLE STEEL PLATE WITH CURVATURE: FUNDAMENTAL CONSIDERATION ON LEVITATION STABILITY UNDER DISTURBANCE........1580
Makoto Tada ; Kazuki Ogawa ; Takayoshi Narita ; Hideaki Kato ; Hiroyuki Moriyama

PERFORMANCE OF HYBRID MAGNETIC LEVITATION CONTROL SYSTEM FOR THIN STEEL PLATE BY EMS AND PMS: EXPERIMENTAL EVALUATION OF APPLYING OPTIMAL GAP AND ARRANGEMENT OF PMS...................1586

Yasuaki Ito ; Yoshiho Oda ; Kengo Okuno ; Toshiki Suzuki ; Masahiro Kida ; Takayoshi Narita ; Hideaki Kato ; Hiroyuki Moriyama

A PRACTICAL LITHIUM-ION BATTERY MODEL BASED ON THE BUTLER-VOLMER EQUATION...................1592

Kaiyuan Li ; King Jet Tseng ; Feng Wei ; Boon-Hee Soong

BONDING TECHNOLOGY USING COLD-ROLLED AG SHEET IN DIE-ATTACHMENT APPLICATIONS...................1598

Seungjun Noh ; Chanyang Choe ; Chuantong Chen ; Hao Zhang ; Katsuaki Suganuma

HIGH-FREQUENCY SELF-DRIVEN SYNCHRONOUS RECTIFIER CONTROLLER FOR WPT SYSTEMS...................1602

Akihiro Konishi ; Kazuhiro Umetani ; Eiji Hiraki

AUTOMATIC RESONANCE FREQUENCY TUNING METHOD FOR REPEATER IN RESONANT INDUCTIVE COUPLING WIRELESS POWER TRANSFER SYSTEMS...................1610

Masataka Ishihara ; Kazuhiro Umetani ; Eiji Hiraki

INDUCTIVE POWER TRANSFER FOR T5 FLUORESCENT LAMP LIGHTING SYSTEM...................1617

Chung-Chuan Hou ; Tang-Jung Chen ; Ching-Chen Chen ; Chen-Wei Chang ; Po-Wei Wang

AN IMPLEMENT 1.5 MHZ OF INDUCTION HEATING FOR ALUMINUM BASED ON VACUUM TUBE OSCILLATOR CIRCUIT...................1622

A. Bilsalam ; P. Chanmontree ; S. Supanyapong ; V. Chunkag

SINGLE-INDUCTOR MULTIPLE-OUTPUTS DIMMABLE LED DRIVER WITH BUCK CONVERTER...................1626

Ta-Wei Huang ; Wei-Jing Tseng ; Jun-Xian Huang

A SOFT-SWITCHED THREE-LEVEL T-TYPE INVERTER WITH AUXILIARY COMMUTATED POLES...................1634

Apollo Charalambous ; Xibo Yuan

CARRIER-BASED REALIZATION OF ARBITRARY SPACE-VECTOR PWM METHODS FOR THREE-LEVEL INVERTERS...................1642

Somboon Sangwongwanich ; Supakorn Paiboon

MULTI-LEVEL TOPOLOGY BASED LINEAR AMPLIFIER FAMILY FOR REALIZATION OF NOISE-LESS INVERTERS...................1649

Hidemine Obara ; Tatsuki Ohno ; Atsuo Kawamura

A NEW ZERO-VOLTAGE SWITCHING THREE-LEVEL CONVERTER WITH REDUCED RECTIFIER VOLTAGE STRESS...................1655

Keon-Woo Kim ; Cheon-Yong Lim ; Dong-Kwan Kim ; Yu-Jin Jang ; Gun-Woo Moon

MODEL PREDICTIVE CONTROL OF A THREE-LEVEL NPC RECTIFIER WITH A SLIDING MANIFOLD TERM...................1661

Xiaonan Gao ; Wei Tian ; Xicai Liu ; Zhenbin Zhang ; Ralph Kennel

H∞ CONTROL-BASED VIBRATION SUPPRESSION IN ROBOT ARM WITH STRAIN WAVE GEARING...................1666

Tran Vu Trung ; Makoto Iwasaki

FINE FORCE SENSORLESS FORCE CONTROL BASED ON FRICTION-FREE DISTURBANCE OBSERVER...................1673

Ohishi Kiyoshi ; Naoki Kamiya ; Toshimasa Miyazaki ; Yuki Yokokura

KINEMATICS AND TRACKING CONTROL OF A FOUR AXIS ANTENNA FOR SATCOM ON THE MOVE...................1680

Oguz Kaan Hancioglu ; Mustafa Celik ; Ugur Tumerdem

POSITION SENSORLESS POSITION CONTROL FOR DUAL SOLENOID ACTUATOR...................1687

Sakahisa Nagai ; Atsuo Kawamura

CAE TECHNOLOGY APPLICATION TREND FOR LARGE-CAPACITY POWER ELECTRONICS DEVELOPMENT...................1692

Teruo Yoshino ; Kuniaki Nagasaka ; Shigeaki Nakabayashi ; Ikuto Udagawa ; Isamu Tominaga ; Junya Konno

XILINX SYSTEM GENERATOR BASED MODELLING OF FINITE STATE MPC...................1698

Vijay Kumar Singh ; Ravi Nath Tripathi ; Tsuyoshi Hanamoto

POWER HARDWARE-IN-THE-LOOP SETUP FOR STABILITY STUDIES OF GRID-CONNECTED POWER CONVERTERS...................1704

Tommi Reinikka ; Henrik Alenius ; Tomi Roinila ; Tuomas Messo

PASSIVITY-BASED LCL FILTER DESIGN OF GRID-CONNECTED VSCS WITH CONVERTER SIDE CURRENT FEEDBACK...................1711

Shih-Feng Chou ; Xiongfei Wang ; Frede Blaabjerg

ADAPTIVE CONTROL OF DC POWER DISTRIBUTION SYSTEMS: APPLYING PSEUDO-RANDOM SEQUENCES AND FOURIER TECHNIQUES...... 1719
  Tomi Roinila ; Hessamaldin Abdollahi ; Silvia Arrua ; Enrico Santi

AN IMPROVED FINITE-SET MODEL PREDICTIVE TORQUE CONTROL FOR INTERIOR PERMANENT MAGNET SYNCHRONOUS MOTOR DRIVES...... 1724
  Xinan Zhang ; Gilbert Foo ; Tung Ngo

PREDICTIVE TORQUE CONTROL FOR FIVE PHASE INDUCTION MOTOR DRIVE WITH COMMON MODE VOLTAGE REDUCTION 1730
  Apekshit Bhowate ; Mohan Aware ; Sohit Sharma ; Yogesh Tatte

INDIRECT MATRIX CONVERTER FOR PERMANENT-MAGNET-SYNCHRONOUS-MOTOR DRIVES BY IMPROVED TORQUE PREDICTIVE CONTROL...... 1736
  Yun Jang ; Yeongsu Bak ; Kyo-Beum Lee

PREDICTIVE DC-LINK CURRENT CONTROL BASED ON IPMSM DISCRETE STATE EQUATION FOR INVERTER WITHOUT INDUCTOR OR ELECTROLYTIC CAPACITOR 1741
  Yousuke Akama ; Kodai Abe ; Kiyoshi Ohishi ; Yuki Yokokura ; Koji Kobayashi ; Tatsuki Kashihara

NEW SEARCH ALGORITHM OF MODEL PREDICTIVE CONTROL TO REDUCING CALCULATION AMOUNT FOR IMPROVING STEADY CURRENT CONTROL PERFORMANCE...... 1747
  Masahiro Shimaoka ; Shinji Doki

DISTRIBUTED POWER SHARING STRATEGY FOR ISLANDED MICROGRIDS WITHOUT FREQUENCY AND VOLTAGE DEVIATIONS...... 1752
  Tuan V. Hoang ; Hong-Hee Lee

LIFETIME-ORIENTED DROOP CONTROL STRATEGY FOR AC ISLANDED MICROGRIDS...... 1758
  Yanbo Wang ; Dong Liu ; Fujin Deng ; Dao Zhou ; Zhe Chen

EXPERIMENT ON HIERARCHICAL CONTROL BASED POWER QUALITY ENHANCEMENT FOR STANDALONE MICROGRID...... 1764
  Darith Leng ; Sompob Polmai ; Kittichot Soontorntaweesub

A DISTRIBUTED PREDICTIVE CONTROL STRATEGY BASED ON STATE ESTIMATOR FOR ISLANDED MICROGRID 1771
  Mi Dong ; Li Li ; Xiaoyu Tian

MAXIMUM POWER POINT TRACKING METHOD FOR PV MODULE UNDER WIDE RANGE VARYING IRRADIANCE LEVELS...... 1777
  Hwa-Dong Liu ; Chang-Hua Lin

DUAL MPPT CONTROL AND FIELD TESTING FOR SWITCHED CAPACITOR-BASED CELL-LEVEL POWER BALANCING UTILIZING DIFFUSION CAPACITANCE OF PHOTOVOLTAIC CELLS...... 1782
  Masatoshi Uno ; Yota Saito ; Masaya Yamamoto ; Shinichi Urabe

SERIES RESONANT DC-DC CONVERTER WITH DUAL-MODE RECTIFIER FOR PV MICROINVERTERS...... 1788
  Yanfeng Shen ; Huai Wang ; Zhan Shen ; Yongheng Yang ; Frede Blaabjerg

VOLTAGE-REFERENCE ACTIVE POWER DECOUPLING BASED ON BOOST CONVERTER FOR SINGLE-PHASE BRIDGE INVERTER...... 1793
  Shuang Xu ; Meiqin Mao ; Riming Shao ; Liuchen Chang

A SINGLE-PHASE COMMON GROUND BOOST INVERTER FOR PHOTOVOLTAIC APPLICATIONS 1799
  Tan-Tai Tran ; Minh-Khai Nguyen ; Young-Cheol Lim ; Joon-Ho Choi

STUDY FOR FURTHER INTRODUCTION OF THE ELECTRONIC FREQUENCY CONVERTERS TO THE TOKAIDO SHINKANSEN 1803
  Toshimasa Shimizu ; Ken Kunomura ; Masahiko Kai ; Hiroki Miyajima ; Teruhisa Matsui

COUNTERMEASURE FOR PARTIAL TURN-OFF OF THYRISTOR CHANGEOVER SWITCH INTRODUCED TO TOHOKU SHINKANSEN SHIN-YONO SECTIONING POST 1810
  Yuki Mizumoto ; Nobuhito Kurosawa

HARDWARE–IN–THE–LOOP REAL–TIME SIMULATION EXPERIMENT PLATFORM FOR TRACTION POWER SUPPLY SYSTEM BASED ON DSPACE-XSIM 1816
  Runze Zhang ; Fei Lin ; Zhongping Yang ; Hu Cao ; Yuping Liu

EVALUATING THE NON-SINUSOIDAL AND NON-SYMMETRIC REGIMES FROM A RAILWAY SUPPLYING SUBSTATION 1822
  Ileana-Diana Nicolae ; Petre-Marian Nicolae ; Radu-Florin Marinescu

A FUNDAMENTAL TRAIN RUNNING EXPERIMENT FOR A BASIC PERFORMANCE VERIFICATION OF A TRAIN POWER DEMAND CONTROL SYSTEM BY DECENTRALIZED CONTROL ALGORITHM 1828
  Yusuke Oki ; Tomoyuki Ogawa ; Yoko Takeuchi ; Tatsuhito Saito ; Jun'ichiro Kawaguchi

VERIFICATION OF SIC BASED MODULAR MULTILEVEL CASCADE CONVERTER (MMCC) FOR HVDC TRANSMISSION SYSTEMS 1834

Y. Ishii ; T. Jimichi

CONTROL OF A 6.6-KV TRANSFORMERLESS STATCOM BASED ON THE MMCC-SDBC USING SIC MOSFETS 1840

Laxman Maharjan ; Toshihisa Tajyuta ; Hiroshi Shinohara ; Akio Suzuki ; Akio Toba

ISOLATED THREE–PHASE AC/DC CONVERTER USING A SOFT–SWITCHING TECHNIQUE FOR BATTERY CHARGER 1847

Yuto Matsui ; Kazuma Suzuki ; Takaharu Takeshita ; Wataru Kitagawa

IMPLEMENTATION OF A MINIATURIZED SIC INVERTER 1854

Hideaki Fujita ; Cristian Andres Garces Guajardo

DESIGN CONSIDERATION OF FLYING CAPACITOR MULTILEVEL INVERTERS USING SIC MOSFETS 1860

Yukihiko Sato ; Kenji Natori

A CONTROL METHOD OF OVERVOLTAGE SUPPRESSION ACROSS THE DC CAPACITOR IN A GRID-CONNECTION CONVERTER USING LEG SHORT-CIRCUIT OF POWER MOSFETS DURING THE INITIAL CHARGE 1866

Tomoyuki Mannen ; Keiji Wada

THE ESSENTIAL RELATIONSHIP BETWEEN DEADBEAT PREDICTIVE CONTROL AND CONTINUOUS-CONTROL-SET MODEL PREDICTIVE CONTROL FOR PWM CONVERTERS 1872

Bi Liu ; Tao Chen ; Wensheng Song

DEADBEAT CONTROL FOR MULTI-LEVEL INVERTER USING 1MHZ MULTISAMPLING METHOD FOR UTILITY INTERACTIVE SYSTEM 1877

Ryosuke Kikuchi ; Ryunosuke Araumi ; Tomoki Yokoyama

1MHZ MULTISAMPLING DEADBEAT CONTROL WITH DISTURBANCE COMPENSATION METHOD FOR THREE PHASE PWM INVERTER 1883

Hiroaki Ueta ; Tomoki Yokoyama

MODULAR MULTILEVEL CONVERTER REPLACED ONE MODULE WITH HIGH VOLTAGE IGBT 1890

Kazunobu Oi ; Kenta Takasho ; Yugo Tadano

INCREASED EFFICIENCY AND REDUCED REALIZATION EFFORT OF DSBC AND DSCC MODULAR MULTILEVEL CONVERTERS (MMCS) 1896

A. Hillers ; J. Biela

COMMON-MODE VOLTAGE INJECTION TECHNIQUES FOR QUASI TWO-LEVEL PWM-OPERATED MODULAR MULTILEVEL CONVERTERS 1904

Jakub Kucka ; Axel Mertens

CURRENT TRACKING AND CELL-VOLTAGE LIMITATIONS OF MODULAR MULTILEVEL CONVERTERS WITH DIRECT DIGITAL CONTROL 1912

T.-F. Wu ; T.-C. Chou ; K.-E. Lin ; T.-Y. Li

SWITCHING LOSS ANALYSIS OF SIC-MOSFET BASED ON STRAY INDUCTANCE SCALING 1919

Keiji Wada ; Masato Ando

MODELING AND OPTIMIZATION OF DISPLACEMENT WINDINGS FOR TRANSFORMERS IN DUAL ACTIVE BRIDGE CONVERTERS 1925

Zhan Shen ; Yanfeng Shen ; Zian Qin ; Huai Wang

OPTIMIZED SELECTION AND UTILIZATION OF DC-LINK CAPACITOR IN A SINGLE-PHASE PV GRID INVERTER SYSTEM 1931

Caspar Collins ; Li Ran

AN EVALUATION CIRCUIT FOR DC-LINK CAPACITORS USED IN A HIGH-POWER THREE-PHASE INVERTER WITH CONDITION MONITORING 1938

Kazunori Hasegawa ; Ichiro Omura ; Shin-Ichi Nishizawa

RECENT MARKET AND TECHNICAL TRENDS IN COPPER ROTORS FOR HIGH-EFFICIENCY INDUCTION MOTORS 1943

Daniel Liang ; Victor Zhou

OVERVIEW OF THE LATEST RESEARCH AND DEVELOPMENT FOR COPPER DIE-CAST SQUIRREL-CAGE ROTORS 1949

Shu Yamamoto

A NOVEL HEAT-RESISTANT INSULATION-PROCESSING AGENT APPLICABLE TO COPPER DIE-CAST SQUIRREL-CAGE ROTORS 1955

Junichi Uchida ; Yuki Sueuchi ; Naosumi Kamiyama

INSULATION-PROCESSING OF COPPER DIE-CAST SQUIRREL-CAGE ROTOR ON MOTOR EFFICIENCY IN HIGH-SPEED OPERATION OVER 10,000 R/MIN 1960

Hideaki Hirahara ; Akira Tanaka ; Shu Yamamoto

HIGH-PRECISION ROTOR POSITION ESTIMATION FOR HIGH-SPEED SPMSM DRIVE BASED ON STATE OBSERVER AND HARMONIC ELIMINATION 1966
Peng Yang ; Xi Xiao ; Meng Zhang ; Shkodyrev Vyacheslav

HARMONIC LOSS REDUCTION IN HIGH SPEED MOTOR DRIVE SYSTEMS BY FLYING CAPACITOR MULTILEVEL INVERTER 1972
Anudari Tumurbaatar ; Sae Mochidate ; Koji Yamaguchi ; Tomohiro Matsuda ; Yukihiko Sato

CURRENT SOURCE TYPE PMSG WIND TURBINE SYSTEM WITH THREE-PHASE THREE-SWITCH BUCK-TYPE RECTIFIER FOR MACHINE-SIDE CONVERTER 1977
Beomseok Chae ; Tahyun Kang ; Yongsug Suh

A STUDY OF 10MW LOAD COMMUTATED INVERTER FOR GAS-TURBINE START-UP 1985
An Hyunsung ; Cha Hanju

PROTOTYPING OF 500 KVA MEDIUM FREQUENCY TRANSFORMER FOR OFFSHORE DIRECT-CURRENT COLLECTION GRID 1991
Tomoyuki Hatakeyama ; Naoyuki Kurita ; Mamoru Kimura

PSCAD/EMTDC AND RTDS SIMULATION ANALYSIS OF MULTIVENDOR MULTI-TERMINAL HVDC SYSTEM CONNECTED TO OFFSHORE WINDFARMS 1997
Hiroshi Suwa ; Takuro Arai ; Takahiro Ishiguro ; Tohru Yoshihara ; Mamoru Kimura ; Tsuneshisa Wachi ; Takahiro Horikoshi ; Tatsuhito Nakajima

INTEROPERABILITY OF MODULAR MULTILEVEL CONVERTERS AND 2-LEVEL VOLTAGE SOURCE CONVERTERS IN A LABORATORY-SCALE MULTI-TERMINAL DC GRID 2003
Salvatore D'arco ; Atsede G. Endegnanew ; Giuseppe Guidi ; Jon Are Suul

PRINCIPLE EXPERIMENT OF CURRENT COMMUTATED HYBRID DCCB FOR HVDC TRANSMISSION SYSTEMS 2011
Ryuta Hasegawa ; Kazuhisa Kanaya ; Yushi Koyama ; Toshiaki Matsumoto ; Takahiro Ishiguro

A THREE-INPUT CENTRAL CAPACITOR DC/DC CONVERTER 2016
Jiaxin Liu ; Feng Gao

SERIES/PARALLEL SWITCHING CIRCUITS USING POWER MOSFETS FOR PHOTOVOLTAIC MODULES 2022
Masamichi Tanemo ; Koki Matsudate ; Shinichi Nomura

MODULARIZED EQUALIZATION ARCHITECTURE BASED ON SWITCHED CAPACITOR CONVERTER TO VIRTUALLY UNIFY MISMATCHED PHOTOVOLTAIC PANEL CHARACTERISTICS 2030
Masatoshi Uno ; Masaya Yamamoto

BUCK-BOOST TYPE MPPT CIRCUIT SUITABLE FOR PHOTOVOLTAIC GENERATION OF VEHICLE INSTALLATION 2036
Fumihisa Kano ; Yuji Kasai ; Hideki Kimura ; Kouhei Sagawa ; Junnosuke Haruna ; Hirohito Funato

VERIFICATION TEST OF ENERGY-EFFICIENT OPERATIONS AND SCHEDULING UTILIZING AUTOMATIC TRAIN OPERATION SYSTEM 2042
Shoichiro Watanabe ; Yasuhiro Sato ; Takafumi Koseki ; Eisuke Isobe ; Jun Kawashita

THE DIRECT BENEFIT OF SIC POWER SEMICONDUCTOR DEVICES FOR RAILWAY VEHICLE TRACTION INVERTERS 2047
Shingo Makishima ; Kazuki Fujimoto ; Keiichiro Kondo

THE LOSS CHARACTERISTICS OF PSFB ZVS DC-DC CONVERTER APPLIED TO THE AUXILIARY POWER SYSTEM 2051
Xianjin Huang ; Juan Zhao ; Fei Lin

SURVEY ON ELECTROMAGNETIC INTERFERENCE ANALYSIS FOR TRACTION CONVERTERS IN RAILWAY VEHICLES 2058
Zhichang Yang ; Hong Li ; Chao Feng ; Yanfeng Jiang ; Fei Lin ; Zhongping Yang

DEVELOPMENT OF TRACTION MOTOR FOR NEW ZERO - EMISSION VEHICLE 2066
Akinobu Iwai ; Satoshi Honjo ; Hirofumi Suzumori ; Toshio Okazawa

EMC DESIGN AND DEVELOPMENT METHODOLOGY FOR TRACTION POWER INVERTERS OF ELECTRIC VEHICLES 2073
Isao Hoda ; Jia Li ; Hiroki Funato

SIMULATION-DRIVEN DESIGN OPTIMIZATION OF A MULTILAYER EMC INPUT FILTER 2078
Fatou Diouf ; Nadim Sakr ; Anna Gheonjian

EV TRACTION INVERTER EMPLOYING DOUBLE-SIDED DIRECT-COOLING TECHNOLOGY WITH SIC POWER DEVICE 2082
Takashi Hirao ; Masami Onishi ; Yusuke Yasuda ; Akihiro Namba ; Kinya Nakatsu

AN OVERVIEW OF STABILITY IMPROVEMENT METHODS FOR WIDE-OPERATION-RANGE FLYBACK CONVERTER WITH VARIABLE FREQUENCY PEAK-CURRENT-MODE CONTROL 2086

Ching-Hsiang Cheng ; Ching-Jan Chen ; Shinn-Shyong Wang

DESIGN AND IMPLEMENTATION OF A HIGH POWER DENSITY ACTIVE-CLAMPED FLYBACK CONVERTER 2092

Yu-Chen Liu ; Bing-Siang Huang ; Cheng-Hung Lin ; Katherine A. Kim ; Huang-Jen Chiu

OPTIMIZED VARIABLE ON-TIME CONTROL FOR LED LIGHTING DRIVER 2097

Jizhe Wang ; Haruhi Eto ; Fujio Kurokawa

DESIGN OF MULTIMODE BATTERY CHARGER WITH DYNAMIC VOLTAGE TRACKING CONTROL 2102

Pang-Jung Liu ; Lin-Hao Chien ; Song-Kai Lee ; Ang-Tung Chen

DUAL-SLOT POWER-PICKUP STRUCTURE FOR CONTACTLESS STRIP INDUCTIVE POWER TRACK SYSTEM 2107

Jia-You Lee ; I-Lin Chen ; Chien-Tzu Ko

DISCONTINUOUS SVM TECHNIQUE FOR THREE-LEG VSI FED BALANCED/UNBALANCED TWO-PHASE LOADS 2113

Supanut Charoensuksirikul ; Yuttana Kumsuwan

REDUCTION OF POWER LOSSES BASED ON GENERALIZED TWO-LEVEL PWM ALGORITHM FOR A NINE-SWITCH VSI 2121

Neerakorn Jarutus ; Yuttana Kumsuwan

SIC-BASED THREE-PHASE QUASI-Z-SOURCE INVERTER VERSUS THE TWO-STAGE TOPOLOGY - A COMPARISON 2129

Kornel Wolski ; Mariusz Zdanowski ; Jacek Rabkowski

DC-SIDE CIRCUIT IMPLEMENTATION OF A THREE-PHASE INVERTER FOR BALANCING PHASE-LEG CAPACITOR CURRENTS 2137

Takashi Hirao ; Keiji Wada ; Toshihisa Shimizu

A THREE-PHASE HYBRID SWITCHED-BOOST INVERTER 2145

Minh-Khai Nguyen ; Tan-Tai Tran ; Hoan-Tien Luong ; Kyoung-Won Lee ; Youn-Ok Choi ; Geum-Bae Cho

THE EFFECT OF BUILT-IN CR SNUBBER CAPACITOR INTO THE POWER MODULE 2149

Ryotaro Hata ; Shigeki Nishiyama

EVALUATION OF NOVEL HYBRID PROTECTION BASED ON PYROSWITCH AND FUSE TECHNOLOGIES 2153

Tomokazu Sakuraba ; Rémy Ouaida ; Song Chen ; Thibaut Chailloux

OPTIMAL DESIGN OF A MAGNETICALLY COUPLED FILTER FOR HIGH EFFICIENCY, LOW COST AND LOW VOLUME DC-DC BATTERY STORAGE CONVERTER 2158

Timothé Delaforge ; Robert Pasterczyk ; Mickaël Robert ; Hervé Chazal ; Jean-Luc Schanen ; Sébastien Mariethoz

HIGH POWER/CURRENT INDUCTOR LOSS MEASUREMENT WITH SHUNT RESISTOR CURRENT-SENSING METHOD 2165

Pin Yu Huang ; Toshihisa Shimizu

SENSITIVITY ANALYSIS OF MEDIUM FREQUENCY TRANSFORMER DESIGN 2170

Marko Mogorovic ; Drazen Dujic

STANDARD MODELS FOR POWER ELECTRONIC SYSTEM SIMULATION 2176

Koichi Shigematsu ; Hiroki Ishikawa ; Taku Noda ; Kentarou Fukushima ; Yoichi Sekiba ; Yusuke Kouno ; Takashi Abe ; Takayuki Sekisue ; Shinji Katoh

MODELING AND MODEL PARAMETER EXTRACTION OF WIDE BANDGAP POWER SEMICONDUCTOR DEVICE, PACKAGE, AND CIRCUIT FOR SIMULATING FAST SWITCHING BEHAVIOR 2181

Tsuyoshi Funaki

STABILITY ANALYSIS METHODS OF A GRID-CONNECTED INVERTER IN TIME AND FREQUENCY DOMAINS 2186

Toshiji Kato ; Kaoru Inoue ; Taiki Sakiyama

FINITE ELEMENT METHODS FOR MULTI-OBJECTIVE OPTIMIZATION OF A HIGH STEP-UP INTERLEAVED BOOST CONVERTER 2193

Wilmar Martinez ; Camilo Cortes ; Ahmad Bilal ; Jorma Kyyra

HIGH FIDELITY REAL-TIME SIMULATION OF MULTI-LEVEL CONVERTERS 2199

Jost Allmeling ; Niklaus Felderer ; Min Luo

AN ENHANCED HIGH FREQUENCY PULSATING VOLTAGE INJECTION METHOD BASED ON IMMUNE ALGORITHM FOR SENSORLESS IPMSM DRIVES 2204

Yanping Zhang ; Zhonggang Yin ; Chao Du ; Youyun Wang ; Xiangdong Sun

POSITION ESTIMATION ACCURACY IMPROVEMENT FOR MAGNETIC SALIENCY BASED SENSORLESS CONTROL INCLUDING CROSS-COUPLING FACTOR 2210

Keita Shimamoto ; Shinya Morimoto

SENSORLESS DRIVE IN THE LOW SPEED REGION AND AUTO-TUNING METHOD FOR PERMANENT MAGNET SYNCHRONOUS MOTORS 2216

Naofumi Nomura ; Shinichi Higuchi

HIGH STABILITY V/F CONTROL OF PMSM USING STATE FEEDBACK CONTROL BASED ON N-T COORDINATE SYSTEM 2224

Yosuke Matsuki ; Shinji Doki

STABILIZATION METHOD USING EQUIVALENT RESISTANCE GAIN BASED ON V/F CONTROL FOR IPMSM WITH LONG ELECTRICAL TIME CONSTANT 2229

Jun-Ichi Itoh ; Takato Toi ; Koroku Nishizawa

SINGLE-PHASE SOLID-STATE TRANSFORMER USING MULTI-CELL WITH AUTOMATIC CAPACITOR VOLTAGE BALANCE CAPABILITY 2237

Jun-Ichi Itoh ; Kazuki Aoyagi ; Keisuke Kusaka ; Masakazu Adachi

A DEVELOPED DUAL MMC ISOLATED DC SOLID STATE TRANSFORMER AND ITS MODULATION STRATEGY 2245

Yan Li ; Chao Liu ; Xu Cai

DC FAULT RIDE-THROUGH OF A THREE-PHASE DUAL-ACTIVE BRIDGE CONVERTER FOR DC GRIDS 2250

Jingxin Hu ; Shenghui Cui ; Rik W. De Doncker

A COMPOUND 10KV DVR SYSTEM BASED ON SOLID STATE TRANSFORMER STRUCTURE 2262

Yaqian Zhang ; Jianzhong Zhang ; Xing Hu ; Zakiud Din

A DUAL-ENERGY-SOURCE UNINTERRUPTIBLE POWER SUPPLY (UPS) 2270

Hao Wang ; Dehong Xu ; Binci Xu ; Haijin Li ; Ye Zhu

INFLUENCE OF WIND POWER FORECASTS ON EQUITABLE DISTRIBUTION METHOD OF WIND POWER CURTAILMENT 2278

Daisuke IIoka ; Hiroumi Saitoh

COMPARISON OF OPTIMIZED DEMAND OF EGS FOR MINIMIZING FUEL CONSUMPTION AND EGS MODEL WITH POWER GRID FREQUENCY USING A HPSPITAL LOAD WITH PV 2283

Yuji Mizuno ; Teppei Baba ; Fujio Kurokawa ; Nobumasa Matsui

COORDINATED DFIG WIND TURBINES AND SOLAR PV GENERATORS FOR INTER-AREA OSCILLATION DAMPING 2287

Tossaporn Surinkaew ; Issarachai Ngamroo

ENERGY MANAGEMENT USING A QUICK CHARGER WITH STORAGE BATTERIES FOR ELECTRIC VEHICLES 2292

Taku Ishibashi ; Toyonari Shimakage ; Norikazu Takeuchi ; Takaaki Kikuchi ; Midori Nonogaki

A METHOD FOR JUNCTION TEMPERATURE ESTIMATION UTILIZING TURN-ON SATURATION CURRENT FOR SIC MOSFET 2296

Hui-Chen Yang ; Rejeki Simanjorang ; Kye Yak See

FIELD BUS FOR DATA EXCHANGE AND CONTROL OF MODULAR POWER ELECTRONIC SYSTEMS WITH HIGH SYNCHRONISATION ACCURACY 2301

Stefan Rietmann ; Simon Fuchs ; André Hillers ; Jürgen Biela

ANALYTICAL INVESTIGATION ON ASYMMETRIC LCC COMPENSATION CIRCUIT FOR TRADE-OFF BETWEEN HIGH EFFICIENCY AND POWER 2309

Kodai Takeda ; Takafumi Koseki

PROBABILISTIC PCA-SUPPORT VECTOR MACHINE BASED FAULT DIAGNOSIS OF SINGLE PHASE 5-LEVEL CASCADED H-BRIDGE MLI 2317

Nagendra Vara Prasad Kuraku ; Yigang He ; Murad Ali

A STUDY ON EDGE SUPPORTED ELECTROMAGNETIC LEVITATION SYSTEM: FUNDAMENTAL CONSIDERATION ON LEVITATION PERFORMANCE OF THIN STEEL PLATE 2324

Yoshiho Oda ; Yasuaki Ito ; Kengo Okuno ; Masahiro Kida ; Toshiki Suzuki ; Takayoshi Narita ; Hideaki Kato ; Hiroyuki Moriyama

APPLICATION OF FACTS DEVICES FOR A DYNAMIC POWER SYSTEM WITHIN THE USA 2329

Jan Paramalingam ; Fuminori Nakamura ; Akihiro Matsuda ; Daisuke Yamanaka ; Taichiro Tsuchiya

CAPACITOR VOLTAGE BALANCING IN SEMI-FULL-BRIDGE SUBMODULE WITH DIFFERENTIAL-MODE CHOKE : (INVITEDPAPER) 2335

Kalle Ilves ; Yuhei Okazaki ; Nan Chen ; Muhammad Nawaz ; Antonios Antonopoulos

RESEARCH ON KEY TECHNOLOGY AND EQUIPMENT FOR ZHANGBEI 500KV DC GRID 2343

Hui Pang ; Xiaoguang Wei

WHAT LED TO SUCCESS IN ACADEMIC RESEARCH ON THE FAMILY OF MODULAR MULTILEVEL CASCADE CONVERTERS? 2352
Hirofumi Akagi

OPERATING PRINCIPLE OF CURRENT RESONANT CONVERTER USING AIR CORE TRANSFORMER FOR ISOLATED POWER SUPPLY ON CHIP 2360
Seiya Abe ; Hikaru Kaishakuji ; Satoshi Matsumoto

ANALYSIS FOR HIGH-FREQUENCY LLC RESONANT CONVERTER WITH PLANAR TRANSFORMER AT LIGHT-LOAD CONDITION 2365
Keon-Woo Kim ; Jae-Il Baek ; Yeonho Jeong ; Ki-Mok Kim ; Gun-Woo Moon

A NOVEL FULL DIGITAL CONTROL H-BRIDGE DC-DC CONVERTER FOR POWER SUPPLY ON CHIP APPLICATIONS 2370
Shigeki Nakano ; Toshiomi Oka ; Seiya Abe ; Satoshi Matsumoto

A HIGH-EFFICIENCY POWER SUPPLY FROM MAGNETIC ENERGY HARVESTERS 2376
Cheon-Yong Lim ; Yeonho Jeong ; Keon-Woo Kim ; Feel-Soon Kang ; Gun-Woo Moon

OPPORTUNITIES FOR LEVERAGING LOW-VOLTAGE GAN DEVICES IN MODULAR MULTI-LEVEL CONVERTERS FOR ELECTRIC-VEHICLE CHARGING APPLICATIONS 2380
Mojtaba Ashourloo ; Mohammad Shawkat Zaman ; Miad Nasr ; Olivier Trescases

A NEW CONTROL STRATEGY FOR MODULAR MULTILEVEL CONVERTER OPERATING IN QUASI TWO-LEVEL PWM MODE 2386
Chao Wang ; Kui Wang ; Zedong Zheng ; Yongdong Li

A CURRENT-SOURCE TYPE MMC WITH DELTA-CONNECTED ARMS FOR SMES 2393
Yushi Miura ; Toshifumi Ise

NEW MODULE WITH ISOLATED HALF BRIDGE OR ISOLATED FULL BRIDGE FOR MODULAR MEDIUM VOLTAGE CONVERTER 2400
Yunpeng Si ; Yifu Liu ; Qin Lei

DEVELOPMENT OF A 700-V-CLASS REVERSE-BLOCKING IGBT FOR ADVANCED T-TYPE NEUTRAL POINT-CLAMPED POWER CONVERSION SYSTEM 2404
Hiroki Wakimoto ; Haruo Nakazawa ; David H. Lu ; Takashi Matsumoto ; Yoichi Nabetani

CERAMIC EMBEDDING AS PACKAGING SOLUTION FOR FUTURE POWER ELECTRONIC APPLICATIONS 2410
Hoang Linh Bach ; Tobias Maximilian Endres ; Daniel Dirksen ; Sigrid Zischler ; Christoph Friedrich Bayer ; Andreas Schletz ; Martin März

MICROELECTROMECHANICAL SYSTEM (MEMS) RESONATOR: A NEW ELEMENT IN POWER CONVERTER CIRCUITS FEATURING REDUCED EMI 2416
A N M Wasekul Azad ; Sourov Roy ; Abu Saleh Imtiaz ; Faisal Khan

A LUMPED THERMAL MODEL INCLUDING THERMAL COUPLING EFFECTS AND BOUNDARY CONDITIONS FOR CAPACITOR BANKS 2421
Qiusheng Wang

HYSTERESIS MODELING OF MAGNETIC DEVICES BASED ON RELUCTANCE NETWORK ANALYSIS 2426
Yoshiki Hane ; Kenji Nakamura

OPTIMAL SIZING AND PLACEMENT OF SOLAR POWERED CHARGING STATION UNDER EV LOADS PENETRATION USING ARTIFICIAL BEE COLONY TECHNIQUE 2430
Yuttana Kongjeen ; Kulsomsup Yenchamchalit ; Krischonme Bhumkittipich

A COMPARISON OF AVERAGE MODEL, SAMPLED-DATA MODEL AND MULTI-FREQUENCY MODEL BASED ON DC/DC CONVERTERS 2435
Xiangpeng Cheng ; Jinjun Liu ; Zeng Liu ; Yiming Tu ; Danhong Xue

SMALL-SIGNAL DISCRETE-TIME MODELING AND DIGITAL CONTROL OF THE BI-DIRECTIONAL DC/DC CONVERTERS 2441
Jia Yaoqin ; Xu Yingchun ; Hou Yijie

ENERGY MANAGEMENT OF HYDROGEN-STORAGE PHOTOVOLTAIC GENERATION SYSTEM WITH A FUNCTION OF SUPPRESSING SHORT-PERIOD COMPONENTS 2449
Yuuki Machida ; Akihisa Goto ; Akiko Takahashi ; Shigeyuki Funabiki

A DYNAMIC BATTERY CHARGING APPROACH FOR ENERGY TRADING IN THE SMART GRID 2456
Avinash Sharma ; Akshay Kumar Rathore ; Rajesh Kumar

A FORCED COMMUTATION METHOD OF THE SOLID-STATE TRANSFER SWITCH IN THE UNINTERRUPTED POWER SUPPLY APPLICATIONS 2462
Meng-Jiang Tsai ; Jiuyang Zhou ; Po-Tai Cheng

ONLINE INTERNAL IMPEDANCE MEASUREMENTS OF LI-ION BATTERY USING PRBS BROADBAND EXCITATION AND FOURIER TECHNIQUES: METHODS AND INJECTION DESIGN......2470

Jussi Sihvo ; Tuomas Messo ; Tomi Roinila ; Roni Luhtala

A DC CURRENT FLOW CONTROLLER FOR MESHED HVDC GRIDS......2476

Viktor Hofmann ; Mark-M. Bakran

AN ISOLATED SOFT-SWITCHING HYBRID-SOURCE DC-DC CONVERTER FOR DC OFFSHORE WIND FARMS......2484

Shenghui Cui ; Jingxin Hu ; Marco Stieneker ; Rik W. De Doncker

A TRANSFORMERLESS MULTI-CELL SOLID-STATE FAULT CURRENT LIMITER FOR MEDIUM VOLTAGE POWER SYSTEM......2490

Pantarote Techama ; Sompob Polmai ; Chanin Bunlaksananusorn

A NOVEL DC POWER FLOW CONTROLLER FOR HVDC GRIDS WITH DIFFERENT VOLTAGE LEVELS......2496

Ya'nan Wu ; Han Ye ; Wu Chen ; Xiaokun He

DESIGN AND CONTROL OF SINGLE-PHASE GRID-CONNECTED PHOTOVOLTAIC MICROINVERTER WITH REACTIVE POWER SUPPORT CAPABILITY......2500

Geon-Hong Min ; Kyung-Hwan Lee ; Jung-Ik Ha ; Myong Hwan Kim

OPTIMAL SIZE AND MULTI-OBJECTIVE CONTROL OF BATTERY ENERGY STORAGES IN DISTRIBUTION SYSTEM WITH HIGH PENETRATION OF DISTRIBUTED PV GENERATORS......2505

Meiqin Mao ; Lei Zhou ; Yangyang Wang ; Liuchen Chang

MISSION PROFILE-ORIENTED CONTROL FOR RELIABILITY AND LIFETIME OF PHOTOVOLTAIC INVERTERS......2512

Ariya Sangwongwanich ; Yongheng Yang ; Dezso Sera ; Frede Blaabjerg

DISCONTINUOUS CURRENT MODE CONTROL FOR MINIMIZATION OF THREE-PHASE GRID-TIED INVERTER IN PHOTOVOLTAIC SYSTEM......2519

Hoai Nam Le ; Jun-Ichi Itoh

A THEORETICAL ANALYSIS ON STATIC CHARACTERISTICS OF VOLTAGE BASED CONTROL METHOD AND CURRENT BASED CONTROL METHOD FOR THE WAYSIDE ENERGY STORAGE SYSTEM IN DC-ELECTRIFIED RAILWAY......2527

Hiroyasu Kobayashi ; Keiichiro Kondo ; Diego Iannuzzi

IMPROVEMENT OF A DC ELECTRICAL RAILWAY SIMULATOR USING ARTIFICIAL INTELLIGENCE......2534

Alvaro J. Lopez-Lopez ; Ramon R. Pecharroman ; Antonio Fernandez-Cardador ; Asuncion P. Cucala

FEEDING-LOSS REDUCTION BY HIGHER-VOLTAGE DC RAILWAY FEEDING SYSTEM WITH DC-TO-DC CONVERTER......2540

Hidenori Shigeeda ; Hiroaki Morimoto ; Kazuhiko Ito ; Toshiyuki Fujii ; Naoki Morishima

MODELING AND SIMULATION OF NOVEL RAILWAY POWER SUPPLY SYSTEM BASED ON POWER CONVERSION TECHNOLOGY......2547

Minwu Chen ; Ruofei Liu ; Shaofeng Xie ; Xiaofang Zhang ; Yimin Zhou

COMPARATIVE STUDY ON FRONT-END PARAMETER IDENTIFICATION METHODS FOR WIRELESS POWER TRANSFER WITHOUT WIRELESS COMMUNICATION SYSTEMS......2552

Sinan Li ; S. Y. Ron Hui

A NEW TYPE OF WIRELESS V2X SYSTEM WITH A DUAL-ACTIVE BIDIRECTIONAL SINGLE-ENDED CONVERTER AND OPTIMIZED SIC-MOSFET......2558

Hideki Omori ; Aoto Yamamoto ; Naoki Mukaiyama ; Masahito Tsuno ; Kenji Fukuda ; Hisato Michikoshi ; Noriyuki Kimura ; Toshimitsu Morizane

METAL OBJECT DETECTION SYSTEM WITH PARALLEL-MISTUNED RESONANT CIRCUITS AND NULLIFYING INDUCED VOLTAGE FOR WIRELESS EV CHARGERS......2564

Seog Y. Jeong ; Van X. Thai ; Jun H. Park ; Chun T. Rim

WIRELESS EV CHARGING SYSTEM WITHOUT AIR-GAP AND MISALIGNMENT......2569

Wenxing Zhong ; Dehong Xu

FIXED SLOPE CARRIER PWM FOR INDIRECT MATRIX CONVERTER......2576

Tzung-Lin Lee ; Chun-Yao Hung ; Yen-Wen Chen ; Wen-Mei Huang

CARRIER-BASED OVERMODULATION STRATEGY FOR MATRIX CONVERTERS......2581

Paiboon Kiatsookkanatorn ; Somboon Sangwongwanich

THREE-PHASE TO HIGH-FREQUENCY SINGLE-PHASE MATRIX CONVERTER : A FREQUENCY CONTROL SUITABLE FOR SOFT SWITCHING......2589

Wataru Kodaka ; Satoshi Ogasawara ; Koji Orikawa ; Masatsugu Takemoto ; Takashi Hyodo ; Hiroyuki Tokusaki

TWO-STEP COMMUTATION FOR ISOLATED DC-AC CONVERTER WITH MATRIX CONVERTER......2596

Shunsuke Takuma ; Jun-Ichi Itoh

A DC-LINK CAPACITOR VOLTAGE OSCILLATION REDUCTION METHOD FOR A MODULAR MULTILEVEL CASCADE CONVERTER WITH SINGLE DELTA BRIDGE CELLS (MMCC-SDBC) 2604

Takaaki Tanaka ; Huai Wang ; Frede Blaabjerg

OPTIMIZED DECOUPLING CONTROL OF FLYING CAPACITOR IN ANPC FIVE-LEVEL INVERTER 2611

Fusheng Wang ; Deyou Zheng ; Jianing Wang ; Fei Li ; Fang Liu ; Shuying Yang ; Zhen Xie

CASCADED DUAL-BUCK AC-AC CONVERTER USING COUPLED INDUCTORS 2619

Sanghun Kim ; Duekjin Jang ; Heung-Geun Kim ; Honnyong Cha

INSTANTANEOUS POWER LOSS CALCULATION FOR MMC BASED ON VIRTUAL ARM MATHEMATICAL MODEL 2625

Yin Shiyuan ; Wang Yue ; Yin Taiyuan ; Nie Cheng ; Duan Guozhao ; Wang Zhang

COMPARISON OF CURRENT CONTROL STRATEGIES IN MODULAR MULTILEVEL CONVERTER 2630

Jianzhao Wei ; Anirudh Budnar Acharya ; Lars Norum ; Pavol Bauer

MODEL PREDICTIVE CONTROL OF A MODULAR MULTILEVEL CONVERTER WITH AN IMPROVED CAPACITOR BALANCING METHOD 2638

Shichong Zhang ; Baodong Bai ; Dezhi Chen

HIGH STEP-UP DC-DC CONVERTER BASED ON MULTI-CELL COUPLED INDUCTOR DIODE-CAPACITOR NETWORK 2646

Xinying Li ; Yan Zhang ; Jinjun Liu ; Pengxiang Zeng

NOVEL ACTIVE CLAMPING STEP-DOWN DC-DC CONVERTER WITH LOWER VOLTAGE STRESS 2653

Chi-Hsuan Hsu ; Jun-Min Jian ; Jiann-Fuh Chen ; Hsuan Liao

DESIGN AND EVALUATION OF A MAGNETICALLY-LOOSELY-COUPLED INDUCTOR FOR A FOUR-PHASE INTERLEAVED BOOST CHOPPER 2660

Hiroki Kowatari ; Toshinori Kitamura ; Nobukazu Hoshi

A SYNCHRONOUS-REFERENCE-FRAME I-V DROOP CONTROL METHOD FOR PARALLEL-CONNECTED INVERTERS 2668

Mingshen Li ; Yonghao Gui ; Zheming Jin ; Yajuan Guan ; Josep M. Guerrero

TRANSIENT STABILITY IMPACT OF THE PHASE-LOCKED LOOP ON GRID-CONNECTED VOLTAGE SOURCE CONVERTERS 2673

Heng Wu ; Xiongfei Wang

COMPREHENSIVE ANALYSIS OF VIRTUAL IMPEDANCE-BASED ACTIVE DAMPING FOR LCL RESONANCE IN GRID-CONNECTED INVERTERS 2681

Teng Liu ; Zeng Liu ; Jinjun Liu ; Yiming Tu ; Zipeng Liu

A COMPARATIVE STUDY OF THE TRADITIONAL FS-MPC AND THE PROPOSED CSF-PCC FOR THE THREE-PHASE GRID-CONNECTED INVERTERS 2688

Zhixun Ma ; Xin Zhang ; Jingjing Huang

CONSTANT SWITCHING-FREQUENCY PREDICTIVE- CURRENT-CONTROL METHOD WITH A DICHOTOMY SOLUTION FOR THE GRID-TIED INVERTERS 2692

Zhixun Ma ; Xin Zhang ; Jingjing Huang ; Zhao Bin ; Lyu Jing

OBSERVER-BASED ACTIVE DAMPING FOR GRID-CONNECTED CONVERTERS WITH LCL FILTER 2697

Y. Zhang ; M. G. L. Roes ; M. A. M. Hendrix ; J. L. Duarte

CONDUCTION LOSS ANALYSIS AND OPTIMIZATION DESIGN OF FULL BRIDGE LLC RESONANT CONVERTER 2703

Yugang Yang ; Lifei Zhang ; Tianshu Ma

FULL-BRIDGE T-TYPE ISOLATED DC/DC CONVERTER WITH WIDE INPUT VOLTAGE RANGE 2708

Dong Liu ; Yanbo Wang ; Fujin Deng ; Zhe Chen

RESEARCH ON HIGH EFFICIENCY LLC DC-DC CONVERTER BASED ON SIC MOSFET 2714

Pengcheng Han ; Xiaoqiong He ; Haijun Ren ; Zhiqing Zhao ; Xu Peng

AN IMPROVED DUAL PHASE SHIFT CONTROL STRATEGY FOR DUAL ACTIVE BRIDGE DC-DC CONVERTER WITH SOFT SWITCHING 2718

Miao Hong ; Gao Xuanjie ; Zeng Chengbi ; Duan Shujiang

DEVELOPMENT OF AN SIC HIGH-FREQUENCY PWM INVERTER USING A THICK MULTILAYER PCB TO MINIMIZE STRAY INDUCTANCE 2725

Kohsuke Ishikawa ; Satoshi Ogasawara ; Masatsugu Takemoto ; Koji Orikawa

FAST SWITCHING PLANAR POWER MODULE WITH SIC MOSFETS AND ULTRA-LOW PARASITIC INDUCTANCE 2732

Arash Edvin Risseh ; Hans-Peter Nee ; Konstantin Kostov

EXPERIMENTAL EVALUATION OF INVERTER SYSTEM CONSISTING OF 4-PARALLEL GAN DEVICES UNIT 2738

Yoshiya Ohnuma ; Satoshi Miyawaki ; Fumiya Hattori ; Masayoshi Yamamoto

IMPACT OF THE THERMAL-INTERFACE-MATERIAL THICKNESS ON IGBT MODULE RELIABILITY IN THE MODULAR MULTILEVEL CONVERTER 2743

Yi Zhang ; Huai Wang ; Zhongxu Wang ; Yongheng Yang ; Frede Blaabjerg

NANOSCALE INVESTIGATION OF THE POWER MOSFET BY THE AFM/KFM/SCFM 2750

Mizuki Nakajima ; Yuuki Uchida ; Nobuo Satoh ; Hidekazu Yamamoto

SIMULATION ANALYSIS OF OPTIMUM GATE DRIVING CONDITIONS OF IGBTS 2756

Satoshi Sugahara ; Masaki Kawakami ; Kousuke Kamakura

IMPROVEMENT OF THE I2T CAPABILITY FOR XEV ACTIVE SHORT CIRCUIT PROTECTION BY COMBINATION OF RC-IGBT AND LEADFRAME TECHNOLOGIES 2764

Keiichi Higuchi ; Hayato Nakano ; Akihiro Osawa ; Akio Kitamura ; Shunji Takenoiri ; Daisuke Inoue ; Souichi Yoshida ; Hiromichi Gohara

INVESTIGATION OF SWITCHING BEHAVIOR OF AN IGBT UNDER SOFT TURN-OFF IN APPLICATION FOR DUAL-ACTIVE BRIDGE CONVERTERS 2768

Eri Ogawa ; Yuichi Onozawa ; Rik W. De Doncker

600 V HIGH VOLTAGE GATE DRIVER IC (HVIC) WITH 1.0 MHZ HIGH FREQUENCY OPERATION FOR LLC CURRENT RESONANT POWER SUPPLY 2774

Masaharu Yamaji ; Masashi Akahane ; Takahide Tanaka ; Akihiro Jonishi ; Hidetomo Ohashi ; Masahiro Sasaki ; Hitoshi Sumida

AN INTEGRATED VOLTAGE AND CURRENT BALANCING STRATEGY OF SERIES-PARALLEL CONNECTED IGBTS 2780

Xiaotong Du ; Fang Zhuo ; Haotian Sun ; Hao Yi ; Yanlin Zhu

THERMAL DESIGN AND ANALYSIS OF A CABLE CHARGER USED FOR PORTABLE ELECTRONICS 2785

Mofan Tian ; Xu Yang ; Naizeng Wang ; Yang Chen ; Laili Wang

PARASITIC INDUCTANCE DESIGN CONSIDERATIONS TO SUPPRESS GATE VOLTAGE OSCILLATION OF FAST SWITCHING POWER SEMICONDUCTOR DEVICES 2789

Yusuke Sugihara ; Kimihiro Nanamori ; Masayoshi Yamamoto ; Yasuki Kanazawa

THE EXAMINATION OF INCREASING OPERATION SPEED OF CONSEQUENT POLE TYPE AXIAL GAP MOTOR FOR HIGHER OUTPUT POWER DENSITY 2796

Toru Ogawa ; Tomohira Takahashi ; Masatsugu Takemoto ; Satoshi Ogasawara ; Hideaki Arita ; Akihiro Daikoku

BASIC STUDY OF PMASYNRM WITH BONDED MAGNETS FOR TRACTION APPLICATIONS 2802

Marika Kobayashi ; Shigeo Morimoto ; Masayuki Sanada ; Yukinori Inoue

STUDY ON ROTOR STRUCTURE SUITABLE FOR IMPROVING POWER DENSITY AND EFFICIENCY IN IPMSMS FOR AUTOMOTIVE APPLICATIONS 2808

R. Imoto ; M. Sanada ; S. Morimoto ; Y. Inoue

EXAMINATION OF THE DEMAGNETIZATION SUPPRESSION EFFECT OF PLACING FLUX BARRIERS IN AN IPMSM USING RARE-EARTH BONDED MAGNETS 2814

Takashi Umeda ; Masayuki Sanada ; Shigeo Morimoto ; Yukinori Inoue

A NOVEL POLE-CHANGING METHOD WITH A MULTIPLE THREE-PHASE INVERTER 2820

Yuki Hidaka ; Taiga Komatsu ; Hideaki Arita

STARTING CHARACTERISTICS OF AN ULTRA-LIGHTWEIGHT MOTOR USING MAGNETIC RESONANCE COUPLING 2826

Kenta Takishima ; Kazuto Sakai

DESIGN AND BASIC CHARACTERISTICS ANALYSIS OF TOROIDAL WINDING AXIAL GAP INDUCTION MOTOR 2832

Ryosuke Sakai ; Yukihiro Yoshida ; Katsubumi Tajima

MAGNET ARRANGEMENT SUITABLE FOR LARGE AIR GAP LENGTH IN LINEAR PM VERNIER MOTOR 2836

Tatsuya Ninomiya ; Abdulaziz Gasim ; Shoji Shimomura

MICRO ELECTROMAGNETIC VIBRATION ENERGY HARVESTER WITH MECHANICAL SPRING AND IRON FRAME FOR LOW FREQUENCY OPERATION 2842

Yecheng Shen ; Kaiyuan Lu ; Yongming Xia

MEASUREMENT OF TWO-LEVEL INVERTER INDUCED CURRENT SLOPES AT HIGH SWITCHING FREQUENCIES FOR CONTROL AND IDENTIFICATION ALGORITHMS OF ELECTRICAL MACHINES 2848

Simon Decker ; Andreas Liske ; Daniel Schweiker ; Johannes Kolb ; Michael Braun

A NEW TOPOLOGY OF SWITCHED-CAPACITOR MULTILEVEL INVERTER FOR SINGLE-PHASE GRID-CONNECTED WITH ELIMINATING LEAKAGE CURRENT 2854

Mehdi Samizadeh ; Xu Yang ; Bagher Karami ; Wenjie Chen ; Mohamad Abou Houran ; Adib Abrishamifar ; Abdolreza Rahmati

AN INTERLEAVED BUCK-CASCADED BUCK-BOOST INVERTER FOR PV GRID-CONNECTION APPLICATIONS 2860

Chien-Hsuan Chang ; Chun-An Cheng ; Hung-Liang Cheng

A NOVEL PV ARRAY CONNECTION STRATEGY WITH PV-BUCK MODULE TO IMPROVE SYSTEM EFFICIENCY 2866

Chi Shao ; Wenjie Wang ; Lijun Hang ; Anping Tong ; Shitao Wang

A COMMON-MODE VOLTAGE REDUCTION FOR TWO-STAGE THREE-PHASE TRANSFORMERLESS PV INVERTERS 2871

Adisak Promyoo ; Surapong Suwankawin

A GRID-CONNECTED PV-ENERGY STORAGE SYSTEM WITH SYNCHRONOUS GENERATOR CHARACTERISTICS 2877

Huadian Xu ; Jianhui Su ; Ning Liu ; Yong Shi ; Yan Du

A TRANSFORMERLESS BIDIRECTIONAL DC-DC CONVERTER BASED ON POWER UNITS WITH UNIPOLAR AND BIPOLAR STRUCTURE FOR MVDC INTERCONNECTION 2882

Lejia Sun ; Fang Zhuo ; Feng Wang ; Hao Yi ; Baohui Ma

NEW MODULATION CONTROL OF CONVERTER SYSTEM APPLIED FOR OFFSHORE WIND FARMS 2887

Naoki Kawabata ; Noriyuki Kimura ; Toshimitsu Morizane ; Hideki Omori

SPHERE DECODING BASED LONG-HORIZON PREDICTIVE CONTROL OF THREE-LEVEL NPC BACK-TO-BACK PMSG WIND TURBINE SYSTEMS 2895

Ferdinand Grimm ; Zhenbin Zhang ; Ralph Kennel

BASED ON PCHD AND IIPSO SLIDING MODE CONTROL OF D-PMSG WIND POWER SYSTEM 2901

Lijun Hou ; Xuemei Zheng ; Chao Wang ; Yangman Li ; Haoyu Li

ESTABLISHMENT AND DYNAMIC CONTROL OF WIND INDUCTION GENERATOR 2907

M. Z. Lu ; V. K. Ganisetti ; C. M. Liaw

MIDDLE FREQUENCY SOLID STATE TRANSFORMER FOR HVDC TRANSMISSION FROM OFFSHORE WINDFARM 2914

Noriyuki Kimura ; Toshimitsu Morizane ; Isao Iyoda ; Kazushige Nakao ; Tomoki Yokoyama

SIMULATION OF WIND POWER GENERATION SYSTEM USING SWITCHED RELUCTANCE GENERATOR AND CAPACITOR-LESS AC-AC CONVERTER 2921

Guyuan Ji ; Kazuhiro Ohyama

VARIABLE FREQUENCY CONTROL AND FILTER DESIGN FOR OPTIMUM ENERGY EXTRACTION FROM A SIC WIND INVERTER 2932

Abdallah Hussein ; Alberto Castellazzi

EXPERIMENTAL VERIFICATIONS OF UPFC USING DEADBEAT CONTROL WITH 3-PHASE UNBALANCED COMPENSATION 2938

Shin-Ichi Hamasaki ; Hiroto Fukuda ; Syohei Tokumaru ; Mineo Tsuji

A CONTROL METHOD FOR TWO TYPES OF THREE-PHASE TRANSFORMERLESS UNIFIED POWER QUALITY CONDITIONER 2944

Fujian Li ; Guochun Xiao ; Fangzhou Zhao ; Shuai Zhang ; Baojin Liu

DESIGN OF CUSTOMER-END CONVERTER SYSTEMS FOR LOW VOLTAGE DC DISTRIBUTION FROM A LIFE CYCLE COST PERSPECTIVE 2948

A. Mattsson ; P. Nuutinen ; T. Kaipia ; P. Peltoniemi ; J. Karppanen ; V. Tikka ; A. Lana ; P. Pinomaa ; P. Silventoinen ; J. Partanen

A CONTROL METHOD OF DC CAPACITOR VOLTAGE IN MMC FOR HVDC SYSTEM USING NEGATIVE SEQUENCE CURRENT 2956

Hanis Afiqah Binti Jaffar ; Ahmad Arif Bin Abd Rahman ; Hiroaki Kakigano

A COORDINATE AND DISTRIBUTED CONTROL SCHEME FOR MULTILEVEL AND MULTI-STAGE MEDIUM VOLTAGE SOLID STATE TRANSFORMER 2963

Jintong Nie ; Liqiang Yuan ; Qing Gu ; Jianning Sun ; Zhengming Zhao

AN IMPROVED HARMONIC POWER SHARING SCHEME OF PARALLELED INVERTER SYSTEM 2969

Liu Hongpeng ; Liu Xiaoxi ; Zhang Wei ; Wang Wei

THE GRID IMPEDANCE ADAPTATION DUAL MODE CONTROL STRATEGY IN WEAK GRID 2973

Ming Li ; Xing Zhang ; Ying Yang ; Pengpeng Cao

TRANSMISSION POWER ANALYSIS AND CONTROL OF THE DC TRANSFORMER IN HYBRID AC/DC MICROGRID 2980
Jingjin Huang ; Xin Zhang ; Tengfei Zhang

A NOVEL FLEXIBLE INTERCONNECTION SCHEME FOR MICROGRID TO OPTIMIZE THE CAPACITY OF ENERGY STORAGE SYSTEM (ESS) 2986
Zhou Jianqiao ; Zhang Jianwen ; Cai Xu ; Li Zhuyong ; Wang Jiacheng ; Zang Jiajie

VSC CONTROL AND PARAMETERS DESIGN BASED ON VIRTUAL SYNCHRONOUS GENERATOR 2992
Fang Liu ; Meng Wang ; Zhen Xie ; Fusheng Wang ; Jinxin Deng ; Xing Zhang

MULTI-TARGET VIRTUAL RESISTANCE CONTROL STRATEGY IN A 400 HZ LOW VOLTAGE MICROGRID 2997
Yuze Li ; Xuejun Pei ; Zhi Chen ; Hanyu Wang ; Yong Kang

AN ADAPTIVE POWER COMPENSATION STRATEGY FOR THE VOLTAGE STABILIZATION OF LCL-VSC BASED MICROGRIDS 3002
Sheng Xu ; Wu Cao ; Dongchen Fan ; Jianfeng Zhao ; Shunyu Wang

RESONANCE DETECTION STRATEGY FOR MULTIPLE GRID-CONNECTED INVERTERS-BASED SYSTEM USING CASCADED SECOND-ORDER GENERALIZED INTEGRATOR 3010
Wu Cao ; Dongchen Fan ; Kangli Liu ; Jianfeng Zhao ; Liheng Ruan ; Xiaojun Wu

HARMONIC STABILITY ASSESSMENT BASED ON GLOBAL ADMITTANCE FOR MULTI-PARALLELED GRID-CONNECTED VSIS USING MODIFIED NYQUIST CRITERION 3015
Wu Cao ; Dongchen Fan ; Kangli Liu ; Jianfeng Zhao ; Liheng Ruan ; Xiaojun Wu

THE AC TRACTION POWER SUPPLY SYSTEM FOR URBAN RAIL TRANSIT BASED ON NEGATIVE SEQUENCE CURRENT COMPENSATOR 3020
Tianshu Zhao ; Xu Peng

GRID CONNECTED POWER GENERATION CONTROL METHOD FOR Z-SOURCE INTEGRATED BIDIRECTIONAL CHARGING SYSTEM 3025
Xu Jia ; Guoming Chuai ; Haonan Niu ; Qianfan Zhang

AN ISOLATED PFC CONVERTER WITH HARMONIC MODULATION TECHNIQUE FOR EV CHARGERS 3030
Byung-Kwon Lee ; Jun-Young Lee ; Dong-Hun Kang

HIGHLY DYNAMIC SWITCHING FREQUENCY-BASED CALCULATION OF POWER QUANTITIES, FUNDAMENTAL WAVEFORMS, AND RMS VALUES OF INVERTER-FED ELECTRICAL MACHINES 3034
Alexander Stock ; Johannes Teigelkötter ; Johannes Büdel

DESIGN AND ANALYSIS OF HIGH VOLTAGE POWER SUPPLY FOR INDUSTRIAL ELECTROSTATIC PRECIPITATORS 3040
Shengwen Fan ; Yiqin Yuan ; Pengyu Jia ; Zhigang Chen ; Haisi Li

LOAD SHARING OPERATION IN N+1 UPS SYSTEM BY USING HARMONIC SHARING CONTROL METHOD 3046
Prashant Patel ; Sagar Naina ; Utsav Patel ; Premal Patwa

RESEARCH ON CAPACITY OPTIMIZATION OF PV-WIND-DIESEL-BATTERY HYBRID GENERATION SYSTEM 3052
Cailing Zhu ; Furong Liu ; Sheng Hu ; Shu Liu

A NUMERICAL ANALYSIS AND IMPROVEMENT OF OUTPUT CHARACTERISTICS IN DIFFERENT PASSIVE RECTIFIERS BASED ON VIBRATION GENERATORS 3058
Tomoki Sakabe ; Masataka Minami ; Shin-Ichi Motegi ; Masakazu Michihira

CIRCUIT MODELING APPROACH FOR ANALYZING TRIBOELECTRIC NANOGENERATORS FOR ENERGY HARVESTING 3063
Bo-Kyung Yoon ; Jeong Min Baik ; Katherine A. Kim

GENERAL POWER ELECTRIC CONVERTER MODEL 3069
Jingwen Xie

A MODULAR CONVERTER- AND SIGNAL-PROCESSING-PLATFORM FOR ACADEMIC RESEARCH IN THE FIELD OF POWER ELECTRONICS 3074
Rüdiger Schwendemann ; Simon Decker ; Marc Hiller ; Michael Braun

CONTROL IC FOR BOOST-FLYBACK CONVERTER FOR ENERGY HARVESTING APPLICATIONS 3081
Jhih-Sian Li ; Kai-Hui Chen ; Jui-Hung Lai ; Jun-Xian Huang

NEW CONCEPT OF THE DC-DC CONVERTER CIRCUIT APPLIED FOR THE SMALL CAPACITY UNINTERRUPTIBLE POWER SUPPLY 3086
Dang Minh Huynh ; Yoichi Ito ; Shinji Aso ; Koji Kato ; Kenji Teraoka

COMPARATIVE STUDY ON THE PERFORMANCE OF DUAL-PHASE TAPPED-INDUCTOR BOOST CONVERTER AND INTERLEAVED BOOST PARALLEL-INPUT SERIES-OUTPUT CONVERTER IN 40 TO 400V APPLICATIONS 3092
Niño Christopher Ramos ; Tsuyoshi Funaki

A NEW STANDBY STRUCTURE INTEGRATED WITH BOOST PFC CONVERTER FOR SERVER POWER SUPPLY 3100
Jae-Il Baek ; Jae-Kuk Kim ; Jae-Bum Lee ; Moo-Hyun Park ; Gun-Woo Moon

NONISOLATED TWO-CHANNEL LED DRIVER WITH SIMPLE SNUBBER 3107
Jong-Woo Kim ; Jung-Kyu Han ; Jih-Sheng Lai

DESIGN AND IMPLEMENTATION OF SINGLE-PHASE ASYMMETRIC MULTILEVEL STATCOM 3112
Hao Chen ; Yang Han ; Ping Yang ; Congling Wang ; Josep M. Guerrero

SUBMODULE VOLTAGE BALANCING AND LOSS EQUALISATION IN ALTERNATE ARM CONVERTERS BASED ON VIRTUAL VOLTAGES 3117
Georgios Konstantinou ; Harith R. Wickramasinghe ; Salvador Ceballos ; Josep Pou

BALANCED CONDUCTION LOSS DISTRIBUTION AMONG SMS IN MODULAR MULTILEVEL CONVERTERS 3123
Zhongxu Wang ; Huai Wang ; Yi Zhang ; Frede Blaabjerg

SIMPLIFICATION OF MODEL PREDICTIVE CONTROL FOR MODULAR MULTILEVEL CONVERTER THROUGH DIRECT VOLTAGE LEVEL SELECTION 3129
Xingxing Chen ; Jinjun Liu ; Shaodi Ouyang ; Shuguang Song ; Rui Luo

FAMILY OF INTEGRATED MULTI-INPUT MULTI-OUTPUT DC-DC POWER CONVERTERS 3134
Bang Le-Huy Nguyen ; Honnyong Cha ; Tien-The Nguyen ; Heung-Geun Kim

LOW-COMPLEXITY STATE-SPACE BASED SYSTEM IDENTIFICATION AND CONTROLLER AUTO-TUNING METHOD FOR MULTI-PHASE DC-DC CONVERTERS 3140
Marc Kanzian ; Harald Gietler ; Christoph Unterrieder ; Matteo Agostinelli ; Michael Lunglmayr ; Mario Huemer

A PHASE-SHIFT DOUBLE FULL-BRIDGE (PSDB) CONVERTER WITH THREE SHARED LEADING-LEGS 3145
Junjie Zhu ; Qinsong Qian ; Shengli Lu ; Weifeng Sun ; Le Zhang

DUAL ACTIVE BRIDGE SYNCHRONOUS RECTIFIED STEP-DOWN CONVERTER 3151
Chien-Chun Huang ; Chang-Lin Tsai ; Tsung-Lin Tsai ; Yao-Ching Hsieh ; Huang-Jen Chiu ; Jing-Yuan Lin

ACCURATE IMPEDANCE MODEL OF GRID-CONNECTED INVERTER FOR SMALL-SIGNAL STABILITY ASSESSMENT IN HIGH-IMPEDANCE GRIDS 3156
Tuomas Messo ; Roni Luhtala ; Aapo Aapro ; Tomi Roinila

MODELING OF UNBALANCED THREE-PHASE GRID-CONNECTED CONVERTERS WITH DECOUPLED TRANSFER FUNCTIONS 3164
Wei Liu ; Xiongfei Wang ; Frede Blaabjerg

PREDICTING VOLTAGE CHARACTERISTIC OF CHARGING MODEL FOR LI-ION BATTERY WITH ANN FOR REAL TIME DIAGNOSIS 3170
Minella Bezha ; Naoto Nagaoka

IMPEDANCE MODELING AND STABILITY ANALYSIS OF THE CASCADED THREE-PHASE SYMMETRIC SYSTEMS USING COMPLEX TRANSFER FUNCTIONS 3176
Teng Liu ; Zeng Liu ; Jinjun Liu ; Yiming Tu ; Zipeng Liu

ACOUSTIC NOISE REDUCTION OF 12/8 POLES SRM WITHOUT EFFICIENCY DROP USING SIMPLE CURRENT WAVEFORMS 3182
Kyohei Kiyota ; Kenji Amei ; Takahisa Ohji ; Jun Jisaki ; Masanobu Nakai

STUDY OF SWITCHED RELUCTANCE MOTOR DIRECTLY DRIVEN BY COMMERCIAL THREE-PHASE POWER SUPPLY 3186
Masaki Takahashi ; Kohei Aiso ; Kan Akatsu

DOUBLE STATOR AXIAL-FLUX SWITCHED RELUCTANCE MOTOR FOR ELECTRIC CITY COMMUTERS 3192
Hiroki Goto

TORQUE RIPPLE REDUCTION USING ASYMMETRIC FLUX BARRIERS IN SYNCHRONOUS RELUCTANCE MOTOR 3197
Yuuto Yamamoto ; Shigeo Morimoto ; Masayuki Sanada ; Yukinori Inoue

ON-BOARD SINGLE-PHASE ELECTRIC VEHICLE CHARGER WITH ACTIVE FRONT END 3203
Theodore Soong ; Peter W. Lehn

A BIDIRECTIONAL BUFFERED CHARGING UNIT FOR EV'S (BBCU) 3209
Gabriel Fernandez

RECONFIGURABLE CONVERTER WITH MULTIPLE-VOLTAGE MULTIPLE-POWER FOR E-MOBILITY CHARGING 3215
Mohamed S A Dahidah ; He Liu ; Vassilios G. Agelidis

DEVELOPMENT OF A SERIES HYBRID ELECTRIC VEHICLE LABORATORY TEST BENCH WITH HARDWARE-IN-THE-LOOP CAPABILITIES...... 3223
Poria Fajri ; Nima Lotfi ; Mehdi Ferdowsi

NEW THREE-PHASE STATIC TRANSFER SWITCH USING AC SSCB...... 3229
Seung-Min Song ; Jin-Young Kim ; In-Dong Kim

HARMONICS COMPENSATION IN HIGH FREQUENCY RANGE OF ACTIVE POWER FILTER WITH SIC-MOSFET INVERTER IN DIGITAL CONTROL SYSTEM...... 3237
Shin-Ichi Hamasaki ; Kengo Nakahara ; Mineo Tuji

CONTROL OF BUCK-BOOST DIRECT MATRIX CONVERTER WITH LOW VOLTAGE RIDE-THROUGH CAPABILITY...... 3243
Nico Remus ; Martin Leubner ; Wilfried Hofmann

AN IMPROVED PLL BASED SEAMLESS TRANSFER CONTROL STRATEGY...... 3251
Xin Meng ; Jinjun Liu ; Zeng Liu ; Ronghui An

EFFICIENT URBAN RAILWAY DESIGN INTEGRATING TRAIN SCHEDULING, ONBOARD ENERGY STORAGE, AND TRACTION POWER MANAGEMENT...... 3257
Warayut Kampeerawar ; Takafumi Koseki ; Fulin Zhou

OPTIMAL CONTROL METHOD OF AN ENERGY STORAGE SYSTEM FOR ENERGY SAVING...... 3265
Yoko Takeuchi ; Tomoyuki Ogawa ; Keisuke Sato ; Hiroaki Morimoto ; Tatsuhito Saito

START-UP AND TRANSIENT OPERATION OF A BIDIRECTIONAL CHOPPER WITH AN AUXILIARY CONVERTER...... 3273
Hamzeh J. Ahmad ; Haruna Ohnishi ; Makoto Hagiwara

EXPERIMENTAL RESULTS OF QUASI-OPTIMAL CHARGING CURRENT PATTERNS TO REDUCE THE INTERNAL HEAT GENERATION OF THE LITHIUM-ION BATTERY...... 3280
Yoshiaki Taguchi ; Gaku Yoshikawa

DEVELOPMENT OF TEST METHODS AND EVALUATION RESULTS FOR 500KV HVDC CONVERTER...... 3286
Keisuke Hattori ; Asuka Ohtake ; Takayoshi Kamejima ; Haruhisa Wada

DISSIPATION LOOP FOR SHOOT-THROUGH FAULTS IN HVDC CONVERTER CELLS...... 3292
Keijo Jacobs ; Staffan Norrga ; Hans-Peter Nee

A SUPPRESSION METHOD OF HARMONIC INSTABILITY IN LINE-COMMUTATED CONVERTERS APPLYING ACTIVE HARMONIC FILTERS...... 3299
Kenichiro Sano ; Toshiaki Kikuma ; Tatsuhito Nakajima ; Junya Kanno

EXPERIMENT OF SEMICONDUCTOR BREAKER USING SERIES-CONNECTED IEGTS FOR HYBRID DCCB...... 3304
Kazuyasu Takimoto ; Hiroshi Takenaka ; Toshiaki Matsumoto ; Takahiro Ishiguro

STUDY OF EMI CAUSED BY BUCK CONVERTER ON CONTROLLER AREA NETWORK...... 3309
Ryo Shirai ; Toshihisa Shimizu

A STUDY ON REDUCTION TECHNIQUES OF A WIDEBAND COMMON-MODE VOLTAGE PRODUCED BY A PWM INVERTER...... 3315
Shotaro Takahashi ; Satoshi Ogasawara ; Masatsugu Takemoto ; Koji Orikawa ; Michio Tamate

A MODIFIED DISCONTINUOUS PWM FOR COMMON-MODE VOLTAGE ELIMINATION IN 3-LEVEL 4-LEG PWM CONVERTER SYSTEM...... 3323
Seon-Ik Hwang ; Jun-Hyung Jung ; In-Ho Cho ; Jang-Mok Kim ; Yung-Deug Son

EMI ANALYSIS OF FULL-SIC INTEGRATED POWER MODULE...... 3329
Xiliang Chen ; Wenjie Chen ; Yu Ren ; Liang Qiao ; Yilin Sha ; Xu Yang

EXPERIMENTAL VERIFICATION OF COUPLING EFFECT AND POWER TRANSFER CAPABILITY OF DYNAMIC WIRELESS POWER TRANSFER...... 3332
Chan Anyapo ; Nithiphat Teerakawanich ; Chowarit Mitsantisuk ; Kiyoshi Ohishi

NEIGHBORING EFFECTS ON THE DEACTIVATED INVERTER IN A SEGMENTED DYNAMIC WIRELESS EV CHARGING SYSTEM...... 3338
Qingwei Zhu ; Yanjie Guo ; Lifang Wang ; Shufan Li ; Chenglin Liao

MULTIPLE EXCITING VOLTAGE CONTROL FOR MAXIMIZATION OF MULTI-HOP WIRELESS POWER TRANSFER EFFICIENCY...... 3344
Masato Sasaki ; Masayoshi Yamamoto

GENERAL ANALYTICAL MODEL FOR INDUCTIVE POWER TRANSFER SYSTEM WITH EMF CANCELING COILS...... 3349
Keita Furukawa ; Keisuke Kusaka ; Jun-Ichi Itoh

STABILITY INFLUENCE OF FILTER COMPONENTS PARASITIC RESISTANCE ON LCL-FILTERED GRID CONVERTERS...... 3357
Hiroaki Matsumori ; Toshihisa Shimizu ; Frede Blaabjerg ; Xiongfei Wang ; Dongsheng Yang

REAL-TIME ESTIMATION CONTROL OF INDUCTANCE PARAMETERS USING DUST CORE MATERIALS FOR PWM INVERTER..3363

Kazu Imai ; Takuma Yoshino ; Ohasi Shunsuke ; Tomoki Yokoyama

CONTROL DESIGN OF OUTPUT-STAGE FILTERLESS SINUSOIDAL-WAVE INVERTER......................3369

Shinichi Hiroshige ; Kenji Yamanaka ; Masahide Hojo

SERIES REACTIVE POWER COMPENSATOR WITH REDUCED CAPACITANCE FOR HYBRID TRANSFORMER..3375

Yuki Takahashi ; Takanori Isobe ; Hiroshi Tadano

AN INSIGHT INTO THE VOLTAGE RISING BEHAVIOR DURING TURN-OFF PROCESS OF SERIES CONNECTED SIC MOSFETS ON CIRCUIT LEVEL..3383

Panrui Wang ; Feng Gao ; Yang Jing ; Yufeng Chen ; Lei Zhang

PARALLELING SIX 320A 1200V ALL-SIC HALF-BRIDGE MODULES FOR A LARGE CAPACITY POWER STACK..3390

David Hongfei Lu ; Hiromu Takubo ; Sho Takano ; Yuhei Suzuki

3.3KV ALL-SIC MODULE FOR ELECTRIC DISTRIBUTION EQUIPMENT..3396

Ryohei Takayanagi ; Katsumi Taniguchi ; Satoshi Kaneko ; Naoyuki Kanai ; Keishirou Kumada ; Motohito Hori ; Yoshinari Ikeda ; Kouji Maruyama ; Itsuo Kawamura

PRESENT STATUS OF SIC BASED POWER CONVERTERS AND GATE DRIVERS – A REVIEW..............3401

Abhijit Choudhury

METHOD OF APPLYING FORCE DISTRIBUTION FUNCTION FOR LINEAR SWITCHED RELUCTANCE MOTOR DRIVEN BY CURRENT SOURCE INVERTER..3406

Tadashi Hirayama ; Shuma Kawabata

A NOVEL DRIVE CIRCUIT FOR SWITCHED RELUCTANCE MOTORS WITH BIPOLAR CURRENT DRIVE..3412

Hiroki Ishikawa ; Yuma Uesugi ; Seiya Sakurai

TORQUE RIPPLE MINIMIZATION CONTROL OF SRM BASED ON NOVEL MOTOR MODEL CONSIDERING MUTUAL COUPLING EFFECT..3418

Sungyong Shin ; Naruse Hikaru ; Takashi Kosaka ; Nobuyuki Matsui

COMPARISON OF HIGH FREQUENCY VOLTAGE INJECTION METHODS FOR SHAFT SENSORLESS CONTROL OF WOUND-FIELD FLUX SWITCHING MACHINE..3426

Hong-Quan Nguyen ; Sheng-Ming Yang

DESIGN AND EXPERIMENTAL VERIFICATION OF A DAB MEDIUM FREQUENCY TRANSFORMER FOR A 6.6KV/200V SOLID STATE TRANSFORMER..3431

Rene Barrera-Cardenas ; Takanori Isobe ; Terazono Katsushi ; Tadano Hiroshi

RESEARCH ON THE UNBALANCED COMPENSATION RANGE OF DELTA-CONNECTED CASCADED H-BRIDGE MULTILEVEL SVG..3439

Rui Luo ; Yingjie He ; Yiming Tu ; Xingxing Chen ; Jinjun Liu

STATIC SYNCHRONOUS COMPENSATOR TO STABILIZE GRID VOLTAGE FOR WIND AND PHOTOVOLTAIC POWER PLANT..3450

Ryota Okuyama ; Naoki Morishima ; Yusuke Ashizaki ; Yohei Itaya

LARGE EQUALIZATION CURRENT CONTROL STRATEGY FOR SERIES CONNECTED BATTERY PACKS BASED ON BUCK-BOOST CONVERTER..3455

Xinbo Liu ; Zhuo Gao ; Xuehao Huang ; Yaohan Zou

A MULTI-PORT BIDIRECTIONAL POWER CONVERSION SYSTEM FOR REVERSIBLE SOLID OXIDE FUEL CELL APPLICATIONS..3460

Xiang Lin ; Kai Sun ; Jin Lin ; Zhe Zhang ; Wei Kong

SELF-PREHEATING METHOD FOR LI-ION BATTERY USING BATTERY IMPEDANCE ESTIMATOR..3466

Dong-Kwan Kim ; Young-Dal Lee ; Sang-Hyun Ha ; Yu-Jin Jang ; Gun-Woo Moon

ACTIVE ANTI-ISLANDING TECHNIQUE WITH REDUCED NON-DETECTION ZONE FOR CENTRALIZED INVERTERS..3471

Prashant Jain ; Vivek Agarwal ; Bishnu Prasad Muni ; Eswar Rao ; Deepak Gehlot ; S. Gautam Kumar

DEVELOPMENT OF SIC APPLIED TRACTION SYSTEM FOR SHINKANSEN HIGH-SPEED TRAIN..3478

Kenji Sato ; Hirokazu Kato ; Takafumi Fukushima

DEVELOPMENT OF A HIGH POWER DENSITY AUXILIARY CONVERTER BASED ON 1700V 225A SIC MOSFET FOR TRAMS..3484

Liu Hao ; Fei Lin ; Zhongping Yang ; Hu Cao ; Meng Xia

EXPERIMENTAL TESTS RESULTS OF DAMPING CONTROL WITH OVER VOLTAGE RESISTOR FOR REGENERATIVE BRAKE CONTROL OF RAILWAY VEHICLE..3490

Natsuki Kawagoe ; Febry Pandu Wijaya ; Hiroyasu Kobayashi ; Keiichiro Kondo ; Tetsuya Iwasaki ; Akihiko Tsumura ; Takumi Nagashima ; Yoshinori Yamashita ; Ryota Gondo

COILS LAYOUT OPTIMIZATION OF DYNAMIC WIRELESS POWER TRANSFER SYSTEM TO REALIZE OUTPUT VOLTAGE STABLE......3495
Yi Wang ; Fei Lin ; Zhongping Yang ; Panpan Cai ; Zhiyuan Liu

QUICK CHARGER FOR A BATTERY USING MODULAR MATRIX CONVERTER (MMXC)......3501
Kazuma Suzuki ; Takaharu Takeshita

VARIABLE OUTPUT VOLTAGE CONTROL OF AN ISOLATED BI-DIRECTIONAL AC/DC CONVERTER WITH A SOFT-SWITCHING TECHNIQUE......3507
Takumi Hamaguchi ; Kazuma Suzuki ; Wataru Kitagawa ; Takaharu Takeshita

A NEW MODULATION METHOD APPLYING OPTIMAL DUTY CYCLE AND PHASE SHIFT FOR BIDIRECTIONAL ISOLATED THREE-PHASE AC/DC CONVERTER BASED ON MATRIX CONVERTER......3514
Koji Shigeuchi ; Jin Xu ; Noboru Shimosato ; Yukihiko Sato

DECOUPLING CONTROL METHOD FOR ELIMINATING DC BIAS FLUX OF HIGH FREQUENCY TRANSFORMER IN A BIDIRECTIONAL ISOLATED AC/DC CONVERTER......3522
Kensuke Sakuma ; Koji Shigeuchi ; Jin Xu ; Noboru Shimosato ; Yukihiko Sato

INTERLEAVED VOLTAGE-DOUBLER BOOST CONVERTER FOR POWER FACTOR CORRECTION......3528
Bo-Jia Huang

ZVS INTERLEAVED TOTEM-POLE BRIDGELESS PFC CONVERTER WITH PHASE-SHIFTING CONTROL......3533
Moo-Hyun Park ; Jae-Il Baek ; Jung-Kyu Han ; Cheon-Yong Lim ; Gun-Woo Moon

A ZERO-VOLTAGE-SWITCHING TOTEM-POLE BRIDGELESS BOOST POWER FACTOR CORRECTION RECTIFIER HAVING MINIMIZED CONDUCTION LOSSES......3538
Young-Dal Lee ; Chong-Eun Kim ; Jae-Il Baek ; Dong-Kwan Kim ; Gun-Woo Moon

POWER-FACTOR-CORRECTION WITH POWER DECOUPLING FOR AC-TO-DC CONVERTER......3544
Wan-Jung Chen ; Tsung-Hsi Wu ; Yao-Ching Hsieh ; Chin-Sien Moo ; Po-Hsiang Wen

DESIGN AND ANALYSIS OF THE DISTRIBUTED CONTROLLER FOR THE MODULAR MULTILEVEL CASCADED CONVERTER......3549
Ping-Heng Wu ; Yu-Chen Su ; Po-Tai Cheng

ASYMMETRIC MIXED MODULAR MULTILEVEL CONVERTER TOPOLOGY IN HYBRID BIPOLAR HVDC TRANSMISSION SYSTEMS......3557
Joon-Hee Lee ; Jae-Jung Jung ; Seung-Ki Sul

HIGH POWER MEDIUM VOLTAGE 10 KV SIC MOSFET BASED BIDIRECTIONAL ISOLATED MODULAR DC–DC CONVERTER......3564
Sayan Acharya ; Ritwik Chattopadhyay ; Anup Anurag ; Satish Rengarajan ; Yos Prabowo ; Subhashish Bhattacharya

MULTI-LEVEL POWER CONVERTER USING SERIES-CONNECTED SOLID-STATE TRANSFORMERS......3572
Yuichi Mabuchi ; Yuki Kawaguchi ; Kimihisa Furukawa ; Mitsuhiro Kadota ; Mizuki Nakahara ; Akihiko Kanoda

CAPACITOR VOLTAGE CONTROL OF MMC-STATCOM DURING UNBALANCED AC SYSTEM FAULT......3578
Kaho Nada ; Takeshi Kikuchi ; Tsuguhiro Takuno ; Toshiyuki Fujii ; Ryosuke Uda ; Takashi Sugiyama

SIC BASED POWER SEMICONDUCTOR IN APPLICATIONS - ASPECTS AND PROSPECTS......3584
Peter Friedrichs

ELECTROMAGNETIC MODELING APPROACHES TOWARDS VIRTUAL PROTOTYPING OF WBG POWER ELECTRONICS......3588
Ivana Kovacevic-Badstübner ; Daniele Romano ; Giulio Antonini ; Jonas Ekman ; Ulrike Grossner

SILICON BASED DEVICES FOR DEMANDING HIGH POWER APPLICATIONS......3596
A. Kopta ; J. Vobecky ; M. Rahimo ; T. Wikström ; U. Vemulapati ; C. Papadopoulos ; C. Corvasce ; M. Andenna ; F. Dugal ; F. Fischer ; S. Hartmann

RECENT PROGRESS IN HIGH TO ULTRA-HIGH-VOLTAGE SIC POWER DEVICES: DEVELOPMENT AND APPLICATION......3603
Y. Yonezawa

DYNAMIC DRIFT EFFECTS IN GAN POWER TRANSISTORS: CORRELATION TO DEVICE TECHNOLOGY AND MISSION PROFILE......3607
Joachim Würfl ; Eldad Bahat-Treidel ; Oliver Hilt ; Maria Troppenz ; Mihaela Wolf ; Jan Böcker ; Carsten Kuring ; Sibylle Dieckerhoff

COMPENSATION METHOD OF RADIAL UNBALANCE FORCE AT FAILURE OF A MOTOR SECTION IN A D-Q AXIS CURRENT CONTROL BEARINGLESS MOTOR......3613
Masahide Ooshima

A BEARINGLESS SYNCHRONOUS RELUCTANCE SLICE MOTOR WITH ROTOR FLUX BARRIERS 3619
Thomas Holenstein ; Thomas Nussbaumer ; Johann W. Kolar

PARAMETER IDENTIFICATIONS OF CURRENT-FORCE FACTOR AND TORQUE CONSTANT IN SINGLE-DRIVE BEARINGLESS MOTORS 3627
Hiroya Sugimoto ; Akira Chiba

DAMPENING OF AXIAL VIBRATIONS IN A BEARINGLESS FLUX-SWITCHING SLICE MOTOR BY FIELD CURRENT REGULATION 3632
Bianca Klammer ; Karlo Radman ; Wolfgang Gruber

ANALYSIS AND DESIGN OF A BEARINGLESS AXIAL-FORCE/TORQUE MOTOR WITH FLEX-PCB WINDINGS 3640
Nobuyuki Kurita ; Walter Bauer ; Gerald Jungmayr ; Wolfgang Gruber ; Wolfgang Amrhein

A PLOTTER-BASED AUTOMATIC MEASUREMENT AND STATISTICAL CHARACTERIZATION OF MULTIPLE DISCRETE POWER DEVICES 3644
Michihiro Shintani ; Benjamin Dauphin ; Kazuki Oishi ; Masayuki Hiromoto ; Takashi Sato

A NOVEL HIGH-SPEED SIC MOSFET DRIVER WITH A LOW SWITCH-VOLTAGE STRESS 3650
Xiuqin Wei ; Yuchong Sun ; Hiroo Sekiya

ENHANCEMENT OF DRIVING CAPABILITY OF GATE DRIVER USING GAN HEMTS FOR HIGH-SPEED HARD SWITCHING OF SIC POWER MOSFETS 3654
Takafumi Okuda ; Takashi Hikihara

DESIGN AND EXPERIMENTAL VERIFICATION OF ROBOT ARM OPERATION FOR POWER PACKET DISPATCHING SYSTEM 3658
Tomoki Yokoyama ; Ryunosuke Araumi ; Kazunori Asada ; Takashi Ando

A RESOURCE SHARING MODEL IN A POWER PACKET DISTRIBUTION NETWORK 3665
H. Ando ; R. Takahashi ; S. Azuma ; M. Hasegawa ; T. Yokoyama ; T. Hikihara

DECOUPLED DSOGI-PLL FOR IMPROVED THREE PHASE GRID SYNCHRONISATION 3670
A. A. Nazib ; D. G. Holmes ; B. P. Mcgrath

A DEVIATION ELIMINATION CONTROL BASED ON AUTONOMOUS CURRENT-SHARING CONTROLLER FOR THE PARALLEL-CONNECTED INVERTERS IN AC MICROGRIDS 3678
Yajuan Guan ; Wei Feng ; Baoze Wei ; Wenzhao Liu ; Mingshen Li ; C. Juan Vasquez ; M. Josep Guerrero

SISO TRANSFER FUNCTIONS FOR STABILITY ANALYSIS OF GRID-CONNECTED VOLTAGE-SOURCE CONVERTERS 3684
Hongyang Zhang ; Lennart Harnefors ; Xiongfei Wang ; Jean-Philippe Hasler ; Hans-Peter Nee

A COMMUNICATION-INDEPENDENT REACTIVE POWER SHARING SCHEME WITH ADAPTIVE VIRTUAL IMPEDANCE FOR PARALLEL CONNECTED INVERTERS 3692
Ronghui An ; Zeng Liu ; Jinjun Liu ; Shike Wang

DESIGN AND INTEGRATION OF THE BI-DIRECTIONAL ELECTRIC VEHICLE CHARGER INTO THE MICROGRID AS EMERGENCY POWER SUPPLY 3698
Yang Song ; Pengcheng Li ; Yuanliang Zhao ; Shuai Lu

STABILITY IMPACT OF PV INVERTER GENERATION ON MEDIUM VOLTAGE DISTRIBUTION SYSTEMS 3705
Ye Tang ; Rolando Burgos ; Chi Li ; Dushan Boroyevich

1MW POWER CONDITIONING SYSTEM WITH MULTIPLE DC INPUTS FOR PVS AND BATTERIES 3711
Yasuaki Furusho ; Yasuyuki Noto ; Kansuke Fujii

A ROBUST AND FLEXIBLE DC-LINKED 3-PHASE ENERGY MANAGEMENT SYSTEM WITH ADAPTIVE DROOP CONTROL STRATEGY 3717
Yue Ma ; Yuki Ishikura ; Hitoshi Tsuji ; Kazuaki Mino

MAXIMUM POWER POINT TRACKING CONTROL FOR SMALL HYDROELECTRIC GENERATION 3723
Kazuya Azegami ; Masashi Takiguchi ; Junya Yano ; Hirohiko Tsutsumi ; Toshitake Masuko

DESIGN AND EXPERIMENTAL VERIFICATION OF A THREE-PHASE DUAL-ACTIVE BRIDGE CONVERTER FOR OFFSHORE WIND TURBINES 3729
Takushi Jimichi ; Murat Kaymak ; Rik W. De Doncker

OPTIMIZED BIDIRECTIONAL PFC RECTIFIERS & INVERTERS - SI VS. SIC VS. GAN IN 2L AND 3L TOPOLOGIES - 3734
Jonas Wyss ; Jürgen Biela

A STANDARD BLOCK OF "SERIES CONNECTED SIC MOSFET" FOR MEDIUM/HIGH VOLTAGE CONVERTER 3742
Qin Lei ; Chunhui Liu ; Yunpeng Si ; Yifu Liu

DESIGN AND TESTING OF 1 KV H-BRIDGE POWER ELECTRONICS BUILDING BLOCK BASED ON 1.7 KV SIC MOSFET MODULE 3749
Jun Wang ; Rolando Burgos ; Dushan Boroyevich ; Zeng Liu

A FLYBACK CONVERTER WITH SIC POWER MOSFET OPERATING AT 10 MHZ: REDUCING LEAKAGE INDUCTANCE FOR IMPROVEMENT OF SWITCHING BEHAVIORS 3757
Kazuki Hashimoto ; Takafumi Okuda ; Takashi Hikihara

A STUDY ON LOAD FLUCTUATION OF ISOLATED DC-DC CONVERTER WITH CLASS PHI-2 INVERTER USING GAN-HFET 3762
Yuta Yanagisawa ; Yushi Miura ; Hiroyuki Handa ; Tetsuzo Ueda ; Toshifumi Ise

SINGLE-INDUCTOR MULTIPLE-OUTPUT CURRENT-SOURCE CONVERTER WITH IMPROVED CROSS REGULATION AND SIMPLE CONTROL STRATEGY 3768
Zheng Dong ; Xiaolu Lucia Li ; Chi K. Tse

LIMIT OPERATING FREQUENCY OF PEAK CURRENT-MODE CONTROL DC-DC CONVERTER CONSIDERING TURN-OFF DELAY TIME 3773
Ryo Ute ; Kazuya Fujiwara ; Jun Imaoka ; Masahito Shoyama

A NOVEL SINGLE SWITCH HIGH FREQUENCY DC/DC CONVERTER AND ITS MATHEMATIC MODEL 3780
Yueshi Guan ; Xihong Hu ; Shu Zhang ; Yijie Wang ; Dianguo Xu ; Wei Wang

ANALYSIS OF CLOSED LOOP OPERATION OF AN ISOLATED BIDIRECTIONAL DAB DC-DC CONVERTER WITH LC COUPLING 3785
Bruno Yukio Enomoto ; Kelly C. M. Carvalho ; Lourenço Matakas Junior ; Wilson Komatsu

ISOLATED AC/DC CONVERTER USING SIMPLE PWM STRATEGY 3791
Naoki Hirose ; Yuto Matsui ; Takaharu Takeshita

ANALYSIS OF ONE PHASE LOSS OPERATION OF THREE-PHASE ISOLATED BUCK MATRIX-TYPE RECTIFIER WITH EIGHT-SEGMENT PWM SCHEME 3797
Jahangir Afsharian ; Dewei David Xu ; Bin Wu ; Bing Gong ; Zhihua Yang ; Jun-Ichi Itoh

NOVEL ISOLATED BIDIRECTIONAL INTEGRATED DUAL THREE-PHASE ACTIVE BRIDGE (D3AB) PFC RECTIFIER 3805
F. Krismer ; E. Hatipoglu ; J. W. Kolar

LOAD VOLTAGE REGULATION METHOD FOR AN ISOLATED AC-DC CONVERTER WITH POWER DECOUPLING OPERATION 3813
Shohei Komeda ; Hideaki Fujita

OPTIMAL DESIGN OF A LOW COST 20KW 99.1% EFFICIENCY ACTIVE ZCS ISOLATED DC-DC CONVERTER 3820
Timothé Delaforge ; Sébastien Mariéthoz

SOFT-SWITCHING ANALYSIS AND PFM CONTROL METHOD OF BIDIRECTIONAL DC/DC CONVERTER TOPOLOGY 3825
Yijie Wang ; Haoyu Wang ; Hongyu Song ; Dianguo Xu

A FULLY SOFT-SWITCHED PWM DC-DC CONVERTER USING AN ACTIVE-SNUBBER-CELL 3833
Hai N. Tran ; Adhistira M. Naradhipa ; Sunju Kim ; Ali Tausif

FLYING CAPACITOR RESONANT POLE INVERTER WITH DIRECT INDUCTOR CURRENT FEEDBACK 3840
Sjef J. Settels ; Jorge L. Duarte ; Jeroen Van Duivenbode

DESIGN OF A GAN-BASED WIRELESS POWER TRANSFER SYSTEM AT 13.56 MHZ TO REPLACE CONVENTIONAL WIRED CONNECTION IN A VEHICLE 3848
Kawin Surakitbovorn ; Juan Rivas-Davila

EFFICIENCY MAXIMIZATION OF INDUCTIVE POWER TRANSFER SYSTEM BY IMPEDANCE AND SWITCHING FREQUENCY CONTROL IN SECONDARY-SIDE CONVERTER 3855
Ryosuke Ota ; Dannisworo S. Nugroho ; Nobukazu Hoshi

ANALYSIS OF OPTIMAL OPERATION FREQUENCY RANGE FOR BATTERY CHARGING IN WPT SYSTEM 3863
Yongbin Jiang ; Min Wu ; Junwen Liu ; Yue Wang ; Laili Wang ; Hailong Zhang

INITIAL CURRENT INJECTION METHOD OF A DIRECT THREE-PHASE TO SINGLE-PHASE AC/AC CONVERTER FOR INDUCTIVE CHARGER 3870
Ferdi Perdana Kusumah ; Jorma Kyyrä

MISSION PROFILE EMULATOR FOR PERMANENT MAGNET SYNCHRONOUS MACHINE BASED ON THREE-PHASE POWER ELECTRONIC CONVERTER 3877
Yubo Song ; Ran Cheng ; Ke Ma

A VARIABLE DC BUS VOLTAGE BASED POWER HARDWARE-IN-THE-LOOP EMULATION OF ELECTRIC MOTORS WITH WIDE VARIATION IN INTERFACE FILTER INDUCTANCE 3884
Tsai-Fu Wu ; Mitradatta Misra ; Ying-Yi Jhang ; Chang-Jun Yang ; Yin-Chi Xu

COPPER LOSS MINIMIZATION CONTROL AT ZERO OUTPUT VOLTAGE FOR ELECTROLYTIC CAPACITOR-LESS INVERTER 3890
Kodai Abe ; Haruya Kada ; Kiyoshi Ohishi ; Hitoshi Haga ; Yuki Yokokura

ARMATURE TEMPERATURE ESTIMATION INSENSITIVE TO ROTOR FLUX VARIATION FOR SPMSM 3896
Toshiki Sano ; Kiyoshi Ohishi ; Yuki Yokokura ; Hiroki Iwata ; Yuji Ide ; Daigo Kuraishi ; Akihiko Takahashi

VIRTUAL SYNCHRONOUS GENERATOR CONTROL WITH RELIABLE FAULT RIDE-THROUGH CAPABILITY BY ADOPTING MODEL PREDICTIVE CONTROL 3902
Jonggrist Jongudomkarn ; Jia Liu ; Toshifumi Ise

RESHAPING QUADRATURE-AXIS IMPEDANCE OF THREE-PHASE GRID-CONNECTED CONVERTERS FOR LOW-FREQUENCY STABILITY IMPROVEMENT 3910
Yi Tang ; Jingyang Fang ; Xiaoqiang Li ; Hongchang Li

COMPARISON BETWEEN TRADITIONAL DROOP AND A NEW AUTONOMOUS CONTROL SCHEME FOR PARALLEL INVERTERS 3916
Mohammad Bani Shamseh ; Teruo Yoshino ; Atsuo Kawamura

A NOVEL MICROGRID POWER SHARING SCHEME ENHANCED BY A NON-INTRUSIVE FEEDER IMPEDANCE ESTIMATION METHOD 3924
Baojin Liu ; Zeng Liu ; Jinjun Liu ; Ronghui An ; Shuguang Song

DEVELOPMENT OF A 3.2MW PHOTOVOLTAIC INVERTER FOR LARGE-SCALE PV POWER PLANTS 3929
Naoya Shibata ; Tsuguhiro Tanaka ; Masahiro Kinoshita

IMPEDANCE-BASED STABILITY ANALYSIS OF LARGE-SCALE PV STATION UNDER WEAK GRID CONDITION CONSIDERING SOLAR RADIATION FLUCTUATION 3934
Yiming Tu ; Jinjun Liu ; Teng Liu ; Xiangpeng Cheng

EXPERIMENTAL VERIFICATION OF GRID-CONNECTION OF A PV CONVERTER USING A SYMMETRICALLY CONNECTED BOOST CONVERTER FOR A HIGH-LEG DELTA TRANSFORMER 3940
Daiki Yamaguchi ; Hideaki Fujita

A NOVEL SINGLE- STAGE HIGH-FREQUENCY BOOST INVERTER CASCADED BY RECTIFIER-INVERTER SYSTEM FOR PV GRID-TIE APPLICATIONS 3945
Hamdy Radwan ; Mahmoud A. Sayed ; Takaharu Takeshita ; Adel A. Elbaset ; G. Shabib

NINE SWITCHES MATRIX CONVERTER USING BI-DIRECTIONAL GAN DEVICE 3952
Takashi Hirota ; Kentaro Inomata ; Daisuke Yoshimi ; Masato Higuchi

A MODEL PREDICTIVE DUAL CURRENT CONTROL METHOD FOR INDIRECT MATRIX CONVERTER FED INDUCTION MOTOR DRIVES 3958
Mei Yang ; Chen Lisha ; Liang Wang ; Yunwei Li

FAULT TOLERANT PREDICTIVE CONTROL OF THREE-LEVEL NEUTRAL-POINT-CLAMPED BACK-TO-BACK POWER CONVERTERS 3965
Zhenbin Zhang ; Xicai Liu ; Kejun Cai ; Feng Gao ; Ralph Kennel

TWO-STAGE OPTIMIZATION BASED PREDICTIVE TORQUE CONTROL WITH REDUCED COMPLEXITY FOR A THREE-LEVEL INVERTER DRIVEN INDUCTION MOTOR 3971
Ilham Osman ; Dan Xiao ; Faz Rahman

DESIGN CHALLENGES OF SIC DEVICES FOR LOW- AND MEDIUM-VOLTAGE DC-DC CONVERTERS 3979
Georges Engelmann ; Alexander Sewergin ; Markus Neubert ; Rik W. De Doncker

DESIGN AND TESTING OF 6 KV H-BRIDGE POWER ELECTRONICS BUILDING BLOCK BASED ON 10 KV SIC MOSFET MODULE 3985
Jun Wang ; Slavko Mocevic ; Jiewen Hu ; Yue Xu ; Christina Dimarino ; Igor Cvetkovic ; Rolando Burgos ; Dushan Boroyevich

HIGH POWER MEDIUM VOLTAGE CONVERTERS ENABLED BY HIGH VOLTAGE SIC POWER DEVICES 3993
Sanket Parashar ; Ashish Kumar ; Subhashish Bhattacharya

SOFT-SWITCHING – THE KEY TO HIGH POWER WBG CONVERTERS 4001
Deepak Divan ; Zheng An ; Prasad Kandula

SIC: TECHNOLOGY ENABLER FOR MV DC/DC GALVANICALLY INSULATED MODULAR CONVERTERS 4009
S. Alvarez ; M. Bellini ; U. Vemulapati ; F. Canales ; M. Rahimo

A BEARINGLESS SLICE MOTOR WITH A SOLID IRON ROTOR FOR DISPOSABLE CENTRIFUGAL BLOOD PUMP 4016
Tadahiko Shinshi ; Ryo Yamamoto ; Yoshiki Nagira ; Junichi Asama

REDUCED HARDWARE PARALLEL DRIVE FOR NO VOLTAGE BEARINGLESS MOTORS 4020
Eric L. Severson

DUAL FIELD-ORIENTED CONTROL OF BEARINGLESS MOTORS WITH COMBINED WINDING SYSTEM 4028
Wolfgang Gruber ; Siegfried Silber

OPEN-CIRCUIT FAULT TOLERANT STUDY OF BEARINGLESS MULTI-SECTOR PERMANENT MAGNET MACHINES 4034
G. Valente ; L. Papini ; A. Formentini ; C. Gerada ; P. Zanchetta

BALANCE CONTROL OF SPLIT CAPACITOR POTENTIAL FOR MAGNETICALLY LEVITATED MOTOR SYSTEM USING ZERO-PHASE CURRENT 4042
Takaaki Oiwa

ASYMMETRICAL HALF-BRIDGE CONVERTER WITH ZERO DC-OFFSET CURRENT IN TRANSFORMER USING NEW RECTIFIER STRUCTURE 4049
Jung-Kyu Han ; Jong-Woo Kim ; Seung-Hyun Choi ; Jih-Sheng Lai ; Gun-Woo Moon

CIRCULATING CURRENT-LESS PHASE-SHIFTED FULL-BRIDGE CONVERTER WITH NEW RECTIFIER STRUCTURE 4054
Jung-Kyu Han ; Gun-Woo Moon

A BI-DIRECTIONAL CURRENT DETECTION USING CURRENT TRANSFORMERS FOR BI-DIRECTIONAL DC-DC CONVERTER 4059
Seiji Iyasu ; Yuji Hahashi ; Yuuichi Handa ; Kimikazu Nakamura ; Keiji Wada

A 10 MHZ GANFET BASED ISOLATED HIGH STEP-DOWN DC-DC CONVERTER 4066
Prasanth Thummala ; Dorai Babu Yelaverthi ; Regan Zane ; Ziwei Ouyang ; Michael A. E. Andersen

ANALYSIS AND DESIGN OF A PARALLEL RESONANT CONVERTER FOR CONSTANT CURRENT INPUT TO CONSTANT VOLTAGE OUTPUT DC-DC CONVERTER OVER WIDE LOAD RANGE 4074
Tarak Saha ; Hongjie Wang ; Baljit Riar ; Regan Zane

NOVEL SINUSOIDAL INPUT CURRENT SINGLE-TO-THREE-PHASE Z-SOURCE BUCK+BOOST AC/AC CONVERTER 4080
M. Haider ; D. Bortis ; J. W. Kolar ; Y. Ono

SIMPLE PWM STRATEGY OF A MATRIX CONVERTER FOR MINIMIZING OUTPUT VOLTAGE HARMONICS 4088
Takuya Oshima ; Takaharu Takeshita

NOVEL THREE-LEVEL BACK-TO-BACK CONVERTERS: STRUCTURE, MODULATION METHOD, AND EXPERIMENT 4096
S. Sangwongwanich ; K. Niyomsatian ; S. Samermurn ; S. Nuchnoi ; S. Suwankawin

MODEL PREDICTIVE CONTROL USING SUBDIVIDED VOLTAGE VECTORS FOR CURRENT RIPPLE REDUCTION IN AN INDIRECT MATRIX CONVERTER 4104
Keon Young Kim ; Yeongsu Bak ; Jin-Hyuk Park ; Kyo-Beum Lee

DC-LINK RIPPLE CURRENT REDUCTION IN BACK-TO-BACK CONVERTERS WITH DPWM 4109
Anatolii Tcai ; Kyo-Beum Lee

AN ANALYSIS OF CLASS DE VOLTAGE-SOURCE PARALLEL RESONANT INVERTER 4114
Takeshi Kondo ; Tsuyoshi Inaba ; Yoshikazu Sakai ; Hirotaka Koizumi

AN IMPROVEMENT ON EXTENDED IMPEDANCE METHOD TOWARDS EFFICIENT STEADY-STATE ANALYSIS OF HIGH-FREQUENCY CLASS-E RESONANT INVERTERS 4122
Junrui Liang

OUTPUT POWER CAPABILITY COMPARISONS OF CLASS-E POWER AMPLIFIERS WITH HARMONIC RESONANCE 4127
Hiroo Sekiya ; Xiuqin Wei ; Yuchong Sun

A CLASS Φ2 RESONANT BUCK CONVERTER WITH RIPPLE INJECTION BURST CONTROL METHOD 4133
Min Lin ; Masahiko Hirokawa

PRACTICAL DESIGN TECHNIQUE FOR HIGH POWER DENSITY LLC RESONANT CONVERTER 4139
Shingo Nagaoka ; Hiroyuki Onishi ; Koji Takatori ; Toshiyuki Zaitsu ; Takeshi Uematsu

OPERATIONAL STUDY AND PROTECTION OF A SERIES RESONANT CONVERTER WITH DC CURRENT INPUT APPLIED IN DC CURRENT DISTRIBUTION SYSTEMS 4145
Hongjie Wang ; Tarak Saha ; Baljit Riar ; Regan Zane

A STUDY ON IMPROVEMENT OF POWER UTILIZATION RATE OF ENERGY SYSTEMS WITH PVS AND BATTERIES 4151
Hiroaki Endo ; Masakatsu Kurisaka ; Tsutomu Ueno ; Yusuke Yoshioka ; Kaoru Inoue ; Toshiji Kato

A NOVEL DC DISTRIBUTION NETWORK WITH MULTI-LEVEL BUS VOLTAGES AND ITS ENERGY MANAGEMENT SYSTEM DESIGN 4157
Jingjin Huang ; Xin Zhang ; Zhixun Ma ; Jianfang Xiao

A NOVEL DC-SIDE-PORT IMPEDANCE MODELING OF MODULAR MULTILEVEL CONVERTERS BASED ON HARMONIC STATE SPACE METHOD 4162
Jing Lyu ; Xin Zhang ; Zhixun Ma ; Xu Cai

AN IMPROVED MASTER-SLAVE CONTROL FOR THREE-PORT CONVERTER BASED DISTRIBUTED DC GRID-CONNECTED PV SYSTEM 4168
Siyue Jiang ; Kai Sun ; Hongfei Wu ; Haixu Shi ; Xiaofeng Dong ; Syed Muhammad Raza Kazmi

SENSORLESS POSITION ESTIMATION, PARAMETER IDENTIFICATION AND CONTROL INTEGRATION FOR PERMANENT MAGNET SYNCHRONOUS MACHINES USING CURRENT DERIVATIVE MEASUREMENTS 4174
M.X. Bui

DYNAMIC PERFORMANCE IMPROVEMENT OF BIDIRECTIONAL SWITCHED-CAPACITOR DC/DC CONVERTER BY RIGHT-HALF-PLANE ZERO ELIMINATION 4181
Ding Kaicheng ; Zhang Yan ; Liu Jinjun ; Zeng Pengxiang ; Zhang Jinshui

A MATRIX BASED ISOLATED BIDIRECTIONAL AC-DC CONVERTER WITH LCL TYPE INPUT FILTER FOR ENERGY STORAGE APPLICATION 4186
Prathamesh Pravin Deshpande ; Amit Kumar Singh ; Sanjib Kumar Panda

ON A STUDY OF VOLTAGE DIVIDING CLASS Φ AMPLIFIER 4193
Katsutoshi Hirayama ; Tadashi Suetsugu ; Yudai Furukawa ; Fujio Kurokawa

A DPWM BASED CONTROL STRATEGY TO INTEGRATE PHOTOVOLTAIC SYSTEM AND BATTERY STORAGE USING GRID CONNECTED THREE-LEVEL T-TYPE INVERTER 4198
Mohammad M. Hashempour ; Yue-Ting Tsai ; T. L. Lee

IMPEDANCE MEASUREMENT OF MEGAWATT-LEVEL RENEWABLE ENERGY INVERTERS USING GRID-FORMING AND GRID-PARALLEL CONVERTERS 4205
Matias Berg ; Tuomas Messo ; Tomi Roinila ; Henrik Alenius

IMPROVED VIRTUAL INDUCTANCE BASED CONTROL STRATEGY OF DFIG UNDER WEAK GRID CONDITION 4213
Ran Fang ; Wenjia Chen ; Xueguang Zhang ; Dianguo Xu

CONTROL OF VSC-HVDC FOR WIND FARM INTEGRATION WITH REAL-TIME FREQUENCY MIRRORING AND SELF-SYNCHRONIZING CAPABILITY 4220
Renxin Yang ; Chen Zhang ; Xu Cai ; Gang Shi ; Jing Lyu

A STUDY ON STEADY-STATE CHARACTERISTICS OF SERIES-CONNECTED WIND FARM USING AN EXPERIMENTAL SET OF LABORATORY SIZE 4227
Fujio Tatsuta ; Shoji Nishikata

A NOVEL ISLANDING DETECTION METHOD WITH TWO-PHASE MAGNIFICATION INSPECTION 4233
Jian-Tang Liao ; Shun-Hao Yeh ; Hong-Tzer Yang

Author Index

Design and Analysis of the Distributed Controller for the Modular Multilevel Cascaded Converter

Ping-heng Wu[1*], *Student Member, IEEE*, Yu-chen Su[1*], *Student Member, IEEE*, and Po-tai Cheng[1], *Fellow, IEEE*

[1]Center for Advanced Power Technologies, Department of Electrical Engineering
National Tsing Hua University, Hsinchu, Taiwan
*E-mail: super497415008@gmail.com

Abstract—The modular multilevel cascaded converter (MMCC) has in recent years drawn interest for applying in the utility-scale photovoltaic (PV) and the battery energy storage (BES) systems. However, the large numbers of the cascaded modules lead to the difficulty of the implementation with a centralized control architecture. The paper [1] has proposed a distributed control strategy which significantly reduces the complexity and increases the flexibility to expand the system. This paper continues the research which analyzes the operation range of each module and its fault tolerant capability. The analysis provides a guideline to design the MMCC controller based on the distributed control approach. Simulation and experimental results are carried out on a star-connected cascaded converter to validate the analysis.

I. INTRODUCTION

The modular multilevel cascaded converters (MMCCs) have become an attractive topology for medium-voltage (MV) high-power applications due to their simple and modular structure. In recent years, they have drawn considerable interests in the utility-scale photovoltaic (PV) and battery energy storage (BES) systems [1]–[8]. For example, Fig. 1 shows a star-connected cascaded converter, also known as the MMCC with single-star bridge cells (MMCC-SSBC) [9], as an interface converter in the PV system.

Since all of the dc capacitor voltages are floating, the dc voltages balancing control is an important issue for MMCC. A hierarchical control proposed in [10] provides an effective solution of dc voltages balancing control of every module. Several papers have achieved voltages balancing under the imbalanced power generations of each module based on this control algorithm [4]–[8], and some of them even maintain stable operation during the grid faults [7]. However, these methods are all based on the centralized control architecture. This means the measured signals, including grid voltages, currents, and every module dc bus voltages are required to be sensed to compute the entire control algorithms in the main controller. This requires a sophisticated communication with high voltage insulation, which is costly and complicated. When the cascaded number increases, numerous analog-to-digital (AD) channels and the burden of the control computation restrict the expansion capability of the implementation.

Several papers discussed the master-slave dc voltages balancing control strategy to increase the system flexibility [11]–[13]. Every dc voltage is measured by the master controller for the overall current control, then sends the voltage commands to the slave controller for the individual voltage regulation. In [14], [15], the controller of one of the bridge cells receives global capacitor voltages to achieve the overall power control and others are operated as an individual voltage regulator. The proposed techniques in [3], [16] are a duality of the master-slave structure of parallel connected converters. This approach is suitable for single-phase MMCC. As the system comes into the utility-grid application, the converter is more likely implemented in three-phase where the voltages balancing control is much more critical.

The paper [1] has proposed a distributed control strategy for the three-phase cascaded converter. In this approach, only a few modules are in centralized control architecture, which are defined as the total-current-regulator (TCR) modules, and others operate as an individual voltage regulator, which is defined as the autonomous-voltage-regulator (AVR) module. The AVR modules in this control architecture can operate without communicating with the information of other cells. Significantly improve the system's expansion capability. The tested results show the dc voltages are maintained balanced even under the sudden change of dc loads based on the distributed control architecture. Their expanded discussion should be how to decide the numbers of TCR and AVR module in this distributed control system. Therefore, the detailed reference voltage calculations are investigated in this paper. The objective is to analyze the operation limitation of each module. The simulation and experimental results are given to validate the analysis and control method.

II. SYSTEM CONFIGURATION AND DISTRIBUTED CONTROL BLOCK DIAGRAM

When cascaded converter applied in the PV systems, each PV panel is linked to an isolated dc/dc converter so that each module can produce the maximum power by its MPPT function [17]. The behavior of the dc/dc converter and the PV panel can be considered as that of a variable current source feeding a dc power P_{mn} ($m \in \{a, b, c\}, n \in \{1...N\}$) to the dc link capacitor [5]. Fig. 1 shows a MMCC-SSBC applied in a PV system. The conventional centralized control architecture is shown in Fig. 1(a), where the grid voltages v_{mO}

The 2018 International Power Electronics Conference

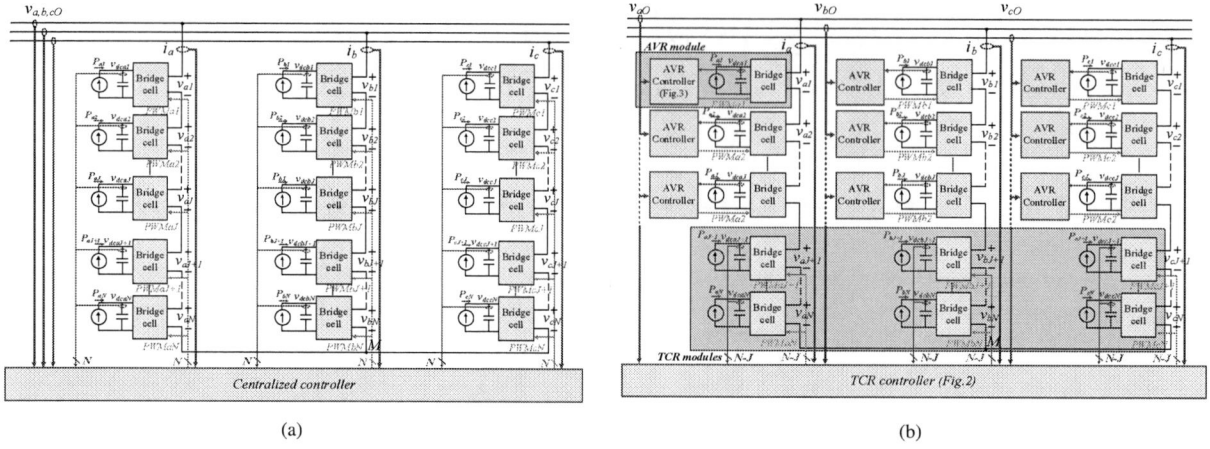

Fig. 1. The system configuration diagram. (a)The conventional centralized control architecture. (b)The proposed distributed control architecture in [1].

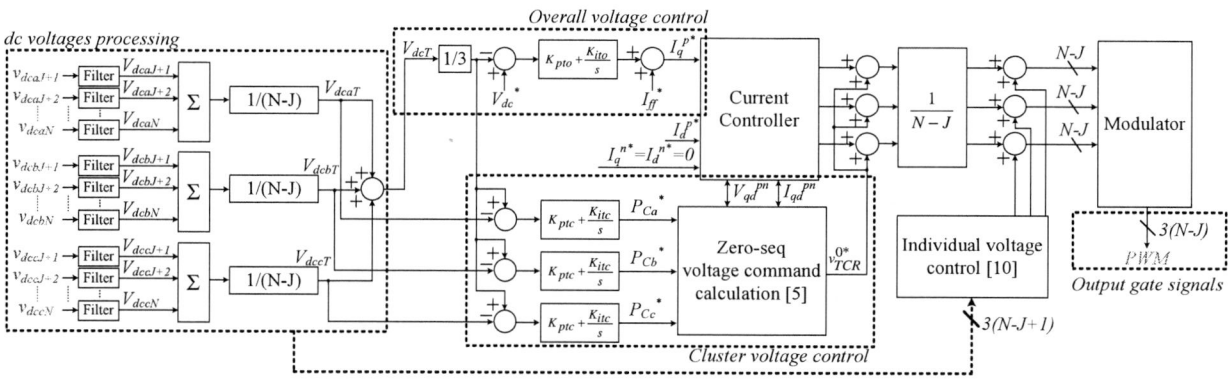

Fig. 2. The controller of TCR modules.

($m \in \{a, b, c\}$), currents i_m ($m \in \{a, b, c\}$) and all of the dc capacitor voltages v_{dcmn} ($m \in \{a, b, c\}, n \in \{1...N\}$) are measured into the central controller. In this architecture, the signal communication becomes complexity as the increasing of the cascaded number.

The distributed control architecture proposed in [1] is shown in Fig. 1(b). To decentralized toe conventional centralized control structure, specific functional objectives are assigned to the particular modules. This distributed control architecture categorizes the bridge cells in the layer 1 to J as the autonomous-voltage-regulator (AVR) modules, which the local capacitor voltage is controlled by its controller, and the bridge cells in the layer $J + 1$ to N as the total-current-regulator (TCR) modules, which regulate the output currents.

Fig. 2 shows the control block diagram of the TCR modules. The controller measures the grid voltages and currents, and the commands are calculated in positive ($V_q^p, V_d^p, I_q^p, I_d^p$) and negative sequence ($V_q^n, V_d^n, I_q^n, I_d^n$) forms. In this paper, the "q-axis" current represents the active power current, and "d-axis" current represents the reactive power current. The

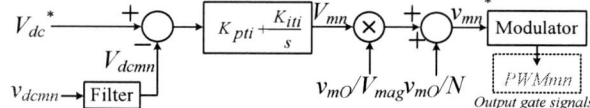

Fig. 3. The controller of the AVR module.

TCR controller manages the output currents of the entire system, and meanwhile, maintains the balancing of the cluster and individual voltages of the TCR modules [1]. Therefore, the reference voltage of other modules can follow the grid current to regulate their dc capacitor voltage. Fig. 3 shows the AVR control block diagram. The controller measures the grid voltage of the local phase as its phase angle information and the feed-forward of the voltage command. Assuming that the phase angle of grid voltages is the same as the grid currents in the steady-state operation, the voltage command is in-phase with the converter current to produce the maximum active power to regulate its dc voltage.

3550

In [1], the distributed controller has been tested under the imbalanced dc power generations and ac grid fault. To better understand the operation limitation of the distributed control strategy. The theoretical analysis of the reference voltages of AVR and TCR modules is presented in this paper. This analysis provides a guideline to design the module number of the AVR and TCR modules.

III. REFERENCE VOLTAGE CALCULATION

Assuming that the system is lossless and the filter inductor is neglected. Besides, the grid voltages and currents are balanced in the following analysis, which the definitions are shown as:

$$v_{aO} = V\cos(\omega t), v_{bO} = V\cos\left(\omega t - \frac{2\pi}{3}\right), v_{cO} = V\cos\left(\omega t + \frac{2\pi}{3}\right) \\ i_a = I\cos(\omega t), i_b = I\cos\left(\omega t - \frac{2\pi}{3}\right), i_c = I\cos\left(\omega t + \frac{2\pi}{3}\right). \quad (1)$$

The output voltage of each H-bridge module in Fig. 1(b) is expressed as v_{mn} ($m \in \{a,b,c\}, n \in \{1...N\}$). Note that the v_{mn} is composed of many harmonic components, but only its foundational component is considered in the following analysis. In other words, the analysis focuses on the reference voltage of each module. The cluster reference voltage (v_{mM}) can be expressed in positive (v_m^p), negative (v_m^n), and zero (v_{OM}) sequence voltages [7]:

$$v_{mM} = \sum_{n=1}^{N} v_{mn} = v_{mO} + v_{OM} = v_m^p + v_m^n + v_{OM}. \quad (2)$$

A. Reference voltages of overall system

In Fig. 1(b), each module is fed by a PV power P_{mn} which means that the equal power should be transferred to ac side. The average active power of one module can be calculated as:

$$P_{mn} = \frac{\omega}{2\pi}\int_0^{\frac{2\pi}{\omega}} v_{mn} \cdot i_m dt, \ m\in\{a,b,c\}, n\in\{1...N\}. \quad (3)$$

The phase power P_m is defined as the summation of the PV power in one cluster ($P_m = P_{m1} + ... + P_{mN}$). Then the overall power P_T can be calculated as the total of the three-phase powers ($P_T = P_a + P_b + P_c$). In MMCC-SSBC, the zero sequence voltage injection influences the distribution of the cluster powers [7]. The cluster powers P_{Cm} ($m \in \{a,b,c\}$) are defined as the difference between phase powers and one-third overall power ($P_{Cm} = P_m - \frac{1}{3}P_T$), and the zero sequence voltage v_{oM} can be calculated based on these power equations.

B. Reference voltages of AVR modules

The AVR controller in Fig. 3 shows the voltage reference v_{a1} can be derived as:

$$v_{a1} = \frac{v_a}{N} + V_{a1}\frac{v_a}{V_{mag}} = \left(\frac{V}{N} + V_{a1}\right)\cos\omega t = \left(3V\frac{P_{a1}}{P_T}\right)\cos\omega t. \quad (4)$$

In (4), the reference voltage of the AVR module depends on the ratio of its PV power and overall power. In order to operate in the linear modulation, the magnitude of the reference voltage cannot exceed its dc capacitor voltage V_{dc}:

$$3V\frac{P_{a1}}{P_T} \leq V_{dc} \rightarrow \frac{P_{a1}}{P_T} \leq \frac{V_{dc}}{3V}. \quad (5)$$

Then, the phase reference voltages ($v_{mAVR} = v_{a1} + ... + v_{aJ}, m \in \{a,b,c\}$) of the AVR modules can be calculated:

$$v_{mAVR} = v_{mAVR}^p + v_{mAVR}^n + v_{AVR}^0 \quad (6)$$

where v_{mAVR}^p and v_{mAVR}^n represent the positive and negative sequence components of the reference voltages respectively. The v_{AVR}^0 is the zero sequence voltage.

C. Reference voltages of TCR modules

The output reference voltage of the entire system is shown in (2) in the steady-state operation. The reference voltages of the TCR modules is to calculate the difference between the overall reference voltages in (2) and the AVR reference voltages in (6). Therefore, the reference voltages of the TCR modules are displayed as follows:

$$v_{mTCR} = v_{mM} - v_{mAVR} = v_{mTCR}^p + v_{mTCR}^n + v_{TCR}^0 \quad (7)$$

where v_{mTCR}^p and v_{mTCR}^n represent the positive and negative sequence components of the reference voltages respectively. The v_{TCR}^0 is the zero sequence voltage.

D. Detail information of reference voltages

Note that the reference voltages in positive sequence ($v_m^p, v_{mAVR}^p, v_{mTCR}^p$) can be expressed in q-, and d-axis values ($v_q^p, v_{qAVR}^p, v_{qTCR}^p$). The reference voltages in negative sequence ($v_m^n, v_{mAVR}^n, v_{mTCR}^n$) can be expressed in q-, and d-axis values ($v_q^n, v_{qAVR}^n, v_{qTCR}^n$). The reference voltages in zero sequence ($v_m^n, v_{mAVR}^n, v_{mTCR}^n$) can also be expressed in coefficients of cos, and sin angles The TABLE I shows the detail information of the reference voltages.

From TABLE I, it is worth mention that the AVR reference voltages of AVR only relate to powers in AVR modules. On the other hand, since the TCR reference voltages produce the minus AVR negative sequence voltage to maintain balanced output currents, the TCR reference voltages are influenced by the powers on AVR modules.

IV. DEGREE OF DISTRIBUTION ANALYSIS

One of the design issues in the distributed control architecture is to decide how many AVR and TCR modules in the converter. To theoretically analyze this problem, the degree of distribution is induced in this paper. In Fig. 1(b), the modules in layer 1 to J are the AVR modules. This paper defines a degree of distribution D to theoretically analyze the numbers of TCR and AVR modules. The larger D means the system is more decentralized. The definition is shown as:

$$D = \frac{J}{N} \times 100\%. \quad (8)$$

The 2018 International Power Electronics Conference

TABLE I
DETAIL INFORMATION OF REFERENCE VOLTAGES

Components	Entire system	AVR modules	TCR modules
$v_q^p, v_{qAVR}^p, v_{qTCR}^p$	$\frac{V}{P_T}(P_a+P_b+P_c)$	$\frac{V}{P_T}(P_{aAVR}+P_{bAVR}+P_{cAVR})$	$\frac{V}{P_T}(P_{aTCR}+P_{bTCR}+P_{cTCR})$
$v_d^p, v_{dAVR}^p, v_{dTCR}^p$	0	0	0
$v_q^n, v_{qAVR}^n, v_{qTCR}^n$	0	$\frac{V}{P_T}\left(P_{aAVR}-\frac{1}{2}P_{bAVR}-\frac{1}{2}P_{cAVR}\right)$	$\frac{V}{P_T}\left(-P_{aAVR}+\frac{1}{2}P_{bAVR}+\frac{1}{2}P_{cAVR}\right)$
$v_d^n, v_{dAVR}^n, v_{dTCR}^n$	0	$\frac{V}{P_T}\left(\frac{\sqrt{3}}{2}P_{bAVR}-\frac{\sqrt{3}}{2}P_{cAVR}\right)$	$\frac{V}{P_T}\left(-\frac{\sqrt{3}}{2}P_{bAVR}+\frac{\sqrt{3}}{2}P_{cAVR}\right)$
$V^0\cos\gamma, V_{AVR}^0\cos\gamma_{AVR},$ $V_{TCR}^0\cos\gamma_{TCR}$	$\frac{V}{P_T}(2P_a-P_b-P_c)$	$\frac{V}{P_T}\left(P_{aAVR}-\frac{1}{2}P_{bAVR}-\frac{1}{2}P_{cAVR}\right)$	$\frac{V}{P_T}(P_{aAVR}-\frac{1}{2}P_{bAVR}-\frac{1}{2}P_{cAVR}$ $+2P_{aTCR}-P_{bTCR}-P_{cTCR})$
$V^0\sin\gamma, V_{AVR}^0\sin\gamma_{AVR},$ $V_{TCR}^0\sin\gamma_{TCR}$	$\frac{V}{P_T}\left(-\sqrt{3}P_b+\sqrt{3}P_c\right)$	$\frac{V}{P_T}\left(-\frac{\sqrt{3}}{2}P_{bAVR}+\frac{\sqrt{3}}{2}P_{cAVR}\right)$	$\frac{V}{P_T}(-\frac{\sqrt{3}}{2}P_{bAVR}+\frac{\sqrt{3}}{2}P_{cAVR}$ $-\sqrt{3}P_{bTCR}+\sqrt{3}P_{cTCR})$

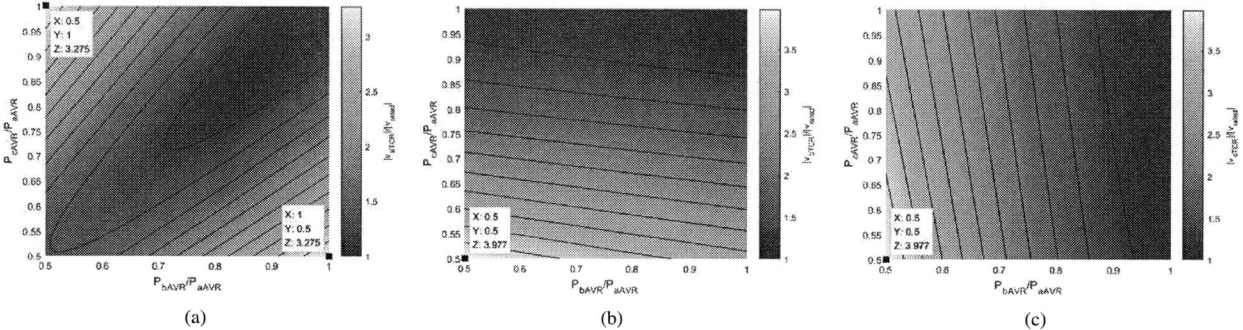

Fig. 4. Normalized peak reference voltage when degree of distribution is 90% (D=90%). (a)a-phase. (b)b-phase. (c)c-phase.

The operation limitation of the AVR module has already calculated in (5). The limitation only relates to the ratio between the local power in the AVR module and the overall power. In other words, the unbalanced TCR powers do not affect the operation limitation of the AVR modules. On the other hand, the reference voltage of the TCR module in (7) is influenced by both the unbalanced dc powers on AVR and TCR modules. Therefore, the TCR modules require more modulation utilization. The following analysis focuses on the increase of TCR reference voltage when the unbalanced dc powers on AVR modules.

The reference voltage in (6) and (7) related to the phase powers of AVR ($P_{mAVR} = P_{m1} + ... + P_{mJ}$) and TCR ($P_{mTCR} = P_{mJ+1} + ... + P_{mN}$) modules. To investigate the degree of distribution D in different unbalanced conditions, the analysis is based on relative power ratios with respect to the highest performing cluster (assuming a-phase is the highest performing cluster). The worst condition occurs in the powers in b- and c-phase reduce to half of their rated power. The maximum normalized peak cluster reference voltage is calculated under different cases as follows.

A. Case I: imbalanced powers in AVR modules

Assuming that the dc powers on the TCR modules are balanced, Fig. 4 shows the normalized peak cluster reference voltage of the TCR module when the degree of distribution is 90% ($D = 90\%$). For example, in a-phase reference voltage, it can be observed that the worst condition occurs in one

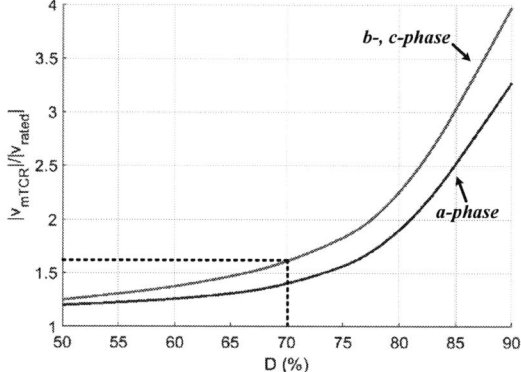

Fig. 5. $V_{TCR,max}$ in Case I.

of the cluster powers reduce to 50% rated power. The worst condition in b- and c-phase is two cluster powers reduce to 50% rated power. Therefore, the maximum normalized peak cluster reference voltage of the TCR module is calculated as:

$$V_{TCR,max} = \frac{|v_{mTCR,max}|}{|v_{TCR,rated}|},$$
$$|v_{mTCR,max}| = Max\left(|v_{mTCR}|_{\frac{P_{bAVR}}{P_{aAVR}}=0.5\sim1;\frac{P_{cAVR}}{P_{aAVR}}=0.5\sim1}\right)$$
(9)

3552

The 2018 International Power Electronics Conference

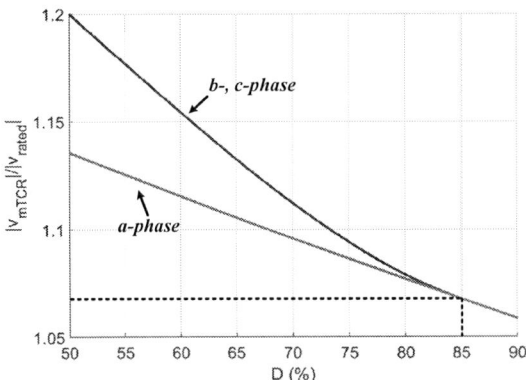

Fig. 6. $V_{TCR,max}$ in Case II.

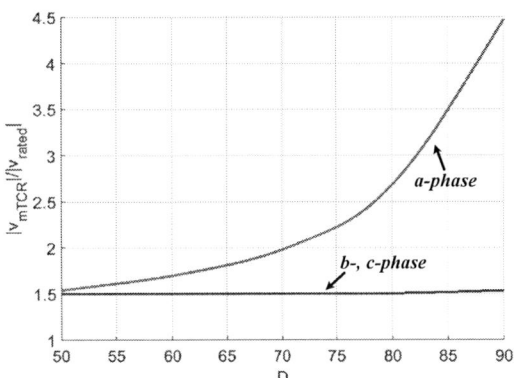

Fig. 7. $V_{TCR,max}$ in Case III.

where $m \in \{a, b, c\}$.

Fig. 5 shows the curve graph of the data arrangements of $V_{TCR,max}$ under the D from 0.5 to 1. Since the $V_{TCR,max}$ means the increase ratio of the TCR reference voltage, the maximum modulation of the TCR module is $\frac{1}{V_{TCR,max}}$. If the modulation of the TCR modules is designed as 0.8, the maximum degree of distribution is 50% which means half of the power modules can operate as AVR modules. On the contrary, if the system is designed as 70%, the modulation of the TCR modules should be designed as 0.6.

B. Case II: imbalanced powers in TCR modules

In case II, assuming that the dc powers on the AVR modules are balanced, the imbalanced powers only occur in TCR modules. The maximum normalized peak cluster reference voltage of the TCR module is calculated as:

$$V_{TCR,max} = \frac{|v_{mTCR,max}|}{|v_{TCR,rated}|},$$
$$|v_{mTCR,max}| = Max\left(|v_{mTCR}|_{\frac{P_{bTCR}}{P_{aTCR}}=0.5\sim1;\frac{P_{cTCR}}{P_{aTCR}}=0.5\sim1}\right) \tag{10}$$

where $m \in \{a, b, c\}$.

Fig. 6 shows the curve graph of the data arrangements of $V_{TCR,max}$ under the D from 0.5 to 1 in this case. As shown in the figure, the increase ratio of the normalized peak reference voltage is decreased when the degree of distribution is getting higher. In other words, the imbalanced TCR powers do not significantly influence the reference voltage in the distributed control architecture.

C. Case III: imbalanced powers in AVR and TCR modules

The Case III considers both imbalanced powers on AVR and TCR modules. Assuming that the dc powers on the TCR modules are balanced, the maximum normalized peak cluster reference voltage of the TCR module is calculated as:

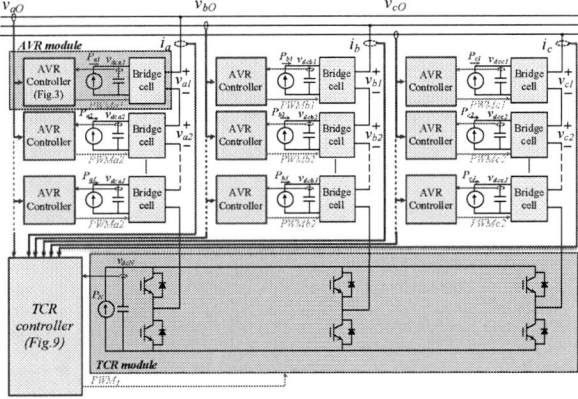

Fig. 8. Hybrid cascaded converter applied in distributed control architecture.

$$V_{TCR,max} = \frac{|v_{mTCR,max}|}{|v_{TCR,rated}|},$$
$$|v_{mTCR,max}| = Max\left(|v_{mTCR}|_{\frac{P_{bTCR}}{P_{aTCR}}=0.5\sim1;\frac{P_{cTCR}}{P_{aTCR}}=0.5\sim1}\right) \tag{11}$$

where $m \in \{a, b, c\}$.

Fig. 7 shows the curve graph of the data arrangements of $V_{TCR,max}$ under the D from 0.5 to 1. As shown in the figure,

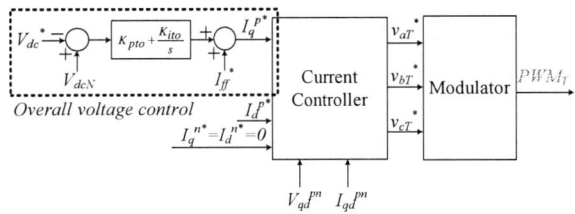

Fig. 9. TCR module controller in the hybrid cascaded converter.

the reference voltage in the a-phase increase significantly when degree of distribution increases.

In Fig. 5 to Fig. 7, the reference voltages on TCR modules under different unbalanced cases show that the is hard to implement high degree of distribution of the distributed control architecture when both AVR and TCR powers unbalanced. If the powers on TCR modules is balanced, as shown in Fig. 5, the better degree of distribution could be obtained. Therefore, this distributed control architecture is considered to be implemented in the hybrid cascaded converter shown in Fig. 8. Note that the TCR controller is shown in Fig. 9, where only the dc voltage control is required since there is only one dc capacitor voltage.

V. TEST RESULTS

In order to verify the distributed control strategy and its reference voltage under unbalanced conditions and different degree of distributions, a 7-level ($N = 3$), 11-level ($N = 5$) MMCC-SSBC, and hybrid cascaded converter hardware test-benches are built in the laboratory, where the grid voltage is $220V$, $60Hz$. The filter inductor is $6.8\ mH$, and the switching frequency is $12\ kHz$. The system configurations are shown in Fig. 1(b) and Fig. 8. In the 7-level converter, the modules of the third layer are selected as the TCR modules. On the other hand, the fourth and fifth layer modules are assigned as the TCR modules in the 11-level converter. Besides, the different combinations of resistors are installed in each bridge cell to simulate the unbalanced PV or storage powers in the system. Each dc load consumes $128\ W$ in the normal operation.

A. Sudden change of dc loads

To illustrate the effectiveness of the dc voltages balancing of the distributed controller, Fig. 10(a) shows the 50% sudden change of dc load in module a1 in a 7-level MMCC-SSBC ($K = 67\%$). It can be observed that the dc voltage at cell a1 has an obviously transient when the power in cell a1 is reduced to 50% of normal operation. Since the unbalanced power among each phase, the TCR module produces the zero sequence voltage to maintain the balanced of cluster voltages. Note that the v_{oM}^{*} in the figure is the zero sequence voltage command in the TCR module. In additions, the output currents are still balanced even under the unbalanced operation which is shown in Fig. 13(b).

To compare the operation under different degree of distribution environment, Fig. 11(a) and Fig. 11(b) show the tested results when sudden change of dc load in module a1 in an 11-level MMCC-SSBC ($D = 60\%$). The dc bus voltages and grid currents are maintained balanced even the imbalanced dc loads. Note that the zero sequence voltage verify the analysis in TABLE I.

Fig. 12(a) and Fig. 12(b) show the experimental results of the hybrid cascaded converter in Fig. 8 in the distributed control architecture. In this cascaded topology, the TCR module consumes $384W$, and without the unbalanced problem since only one dc capacitor voltage in the module.

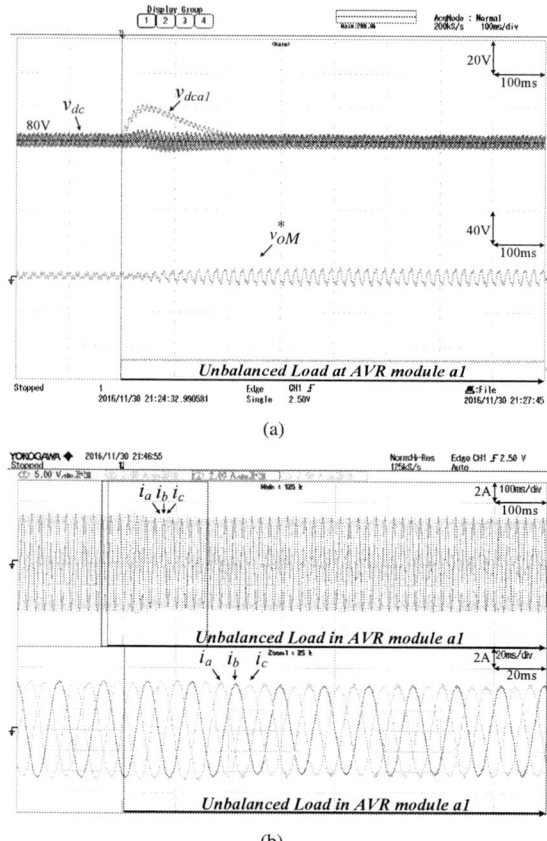

Fig. 10. Sudden change of dc loads in MMCC-SSBC (7-level). (a)V_{dcmn} when sudden change of dc loads of module a1. (b)i_m when sudden change of dc loads of module a1.

B. Fault tolerant operation

Fig. 13(a) shows the dc voltages under the bridge cell a1 is failed and bypassed (7-level MMCC-SSBC), which is equal to the condition of no load at module a1. In the distributed control system, the controller does not require the adjustment. Based on the analysis in Section III, the reference voltage of TCR modules and other AVR modules do not produce the over-modulation problem. Therefore, the dc voltages of other modules can maintain balance in this condition.

VI. CONCLUSION

This paper extends the research of the distributed control technique in [1]. The distributed control architecture increase the extension capability of the cascaded converter. However, their operation limitation do not have detail discussion in the literature. This paper provides the theoretical reference voltage analysis on the AVR and TCR modules. The reference voltage under different unbalanced conditions are arranged in the paper. Moreover, this paper defines the degree of distribution to analyze the increase ratio of the reference voltage. Simulation and experimental results are given to verify the analysis.

The 2018 International Power Electronics Conference

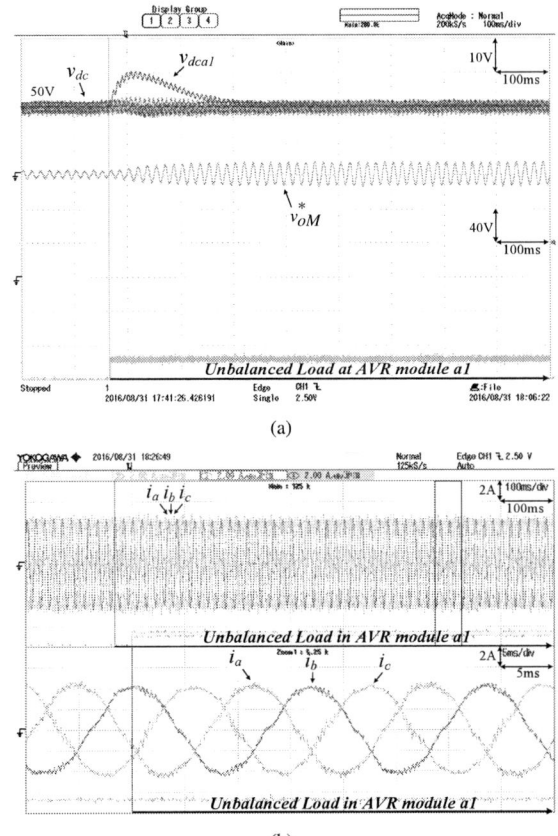

(a)

(b)

Fig. 11. Sudden change of dc loads in MMCC-SSBC (11-level). (a)V_{dcmn} when sudden change of dc loads of module $a1$. (b)i_m when sudden change of dc loads of module $a1$.

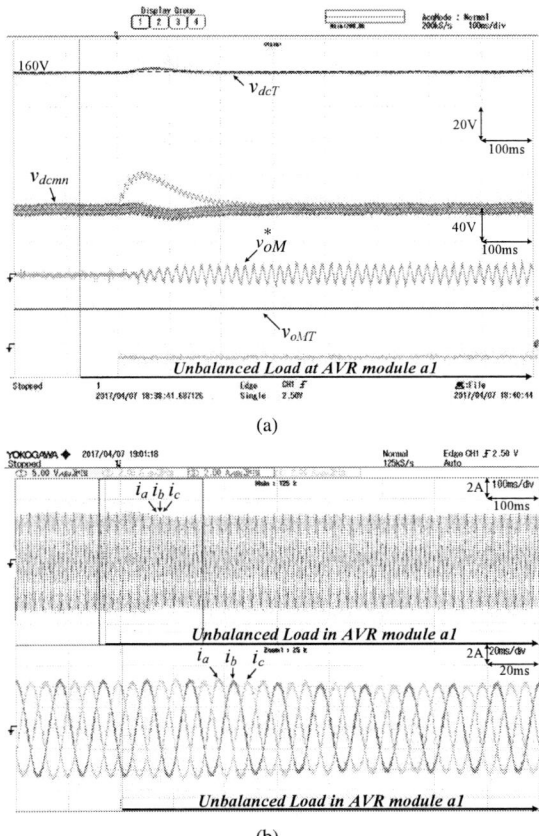

(a)

(b)

Fig. 12. Sudden change of dc loads in hybrid cascaded converter. (a)V_{dcmn} when sudden change of dc loads of module $a1$. (b)i_m when sudden change of dc loads of module $a1$.

REFERENCES

[1] P. Wu, Y. Su, and P. Cheng, "A distributed control technique for the multilevel cascaded converter," in *Energy Conversion Congress and Exposition (ECCE), 2017 IEEE*, Oct 2017.

[2] L. Liu, H. Li, Y. Xue, and W. Liu, "Reactive power compensation and optimization strategy for grid-interactive cascaded photovoltaic systems," *IEEE Transactions on Power Electronics*, vol. 30, no. 1, pp. 188–202, Jan 2015.

[3] H. Jafarian, I. Mazhari, B. Parkhideh, S. Trivedi, D. Somayajula, R. Cox, and S. Bhowmik, "Design and implementation of distributed control architecture of an ac-stacked pv inverter," in *2015 IEEE Energy Conversion Congress and Exposition (ECCE)*, Sept 2015, pp. 1130–1135.

[4] Y. Yu, G. Konstantinou, B. Hredzak, and V. G. Agelidis, "Power balance of cascaded h-bridge multilevel converters for large-scale photovoltaic integration," *IEEE Transactions on Power Electronics*, vol. 31, no. 1, pp. 292–303, Jan 2016.

[5] P. Sochor and H. Akagi, "Theoretical comparison in energy-balancing capability between star- and delta-configured modular multilevel cascade inverters for utility-scale photovoltaic systems," *IEEE Transactions on Power Electronics*, vol. 31, no. 3, pp. 1980–1992, March 2016.

[6] B. Xiao, L. Hang, J. Mei, C. Riley, L. M. Tolbert, and B. Ozpineci, "Modular cascaded h-bridge multilevel pv inverter with distributed mppt for grid-connected applications," *IEEE Transactions on Industry Applications*, vol. 51, no. 2, pp. 1722–1731, March 2015.

[7] H. C. Chen, S. Y. Tsai, P. H. Wu, W. L. Huang, and P. T. Cheng, "Managed dc voltage utilization technique for the renewable energy source based on the star-connected cascaded h-bridges converter," in *2015 IEEE Energy Conversion Congress and Exposition (ECCE)*, Sept 2015, pp. 3726–3733.

[8] L. Maharjan, T. Yamagishi, H. Akagi, and J. Asakura, "Fault-tolerant operation of a battery-energy-storage system based on a multilevel cascade pwm converter with star configuration," *IEEE Transactions on Power Electronics*, vol. 25, no. 9, pp. 2386–2396, Sept 2010.

[9] H. Akagi, "Classification, terminology, and application of the modular multilevel cascade converter (mmcc)," *Power Electronics, IEEE Transactions on*, vol. 26, no. 11, pp. 3119–3130, Nov 2011.

[10] H. Akagi, S. Inoue, and T. Yoshii, "Control and performance of a transformerless cascade pwm statcom with star configuration," *IEEE Transactions on Industry Applications*, vol. 43, no. 4, pp. 1041–1049, July 2007.

[11] S. Huang, L. Mathe, and R. Teodorescu, "A new method to implement resampled uniform pwm suitable for distributed control of modular multilevel converters," in *IECON 2013 - 39th Annual Conference of the IEEE Industrial Electronics Society*, Nov 2013, pp. 228–233.

[12] S. Yang, Y. Tang, M. Zagrodnik, G. Amit, and P. Wang, "A novel distributed control strategy for modular multilevel converters," in *2017 IEEE Applied Power Electronics Conference and Exposition (APEC)*, March 2017, pp. 3234–3240.

[13] L. Mathe, P. D. Burlacu, and R. Teodorescu, "Control of a modular multilevel converter with reduced internal data exchange," *IEEE Transactions on Industrial Informatics*, vol. 13, no. 1, pp. 248–257, Feb 2017.

[14] D. M. Scholten, N. Ertugrul, and W. L. Soong, "Analysis and control of decentralized pv cascaded multilevel modular integrated converters," in *2016 IEEE Energy Conversion Congress and Exposition (ECCE)*, Sept

3555

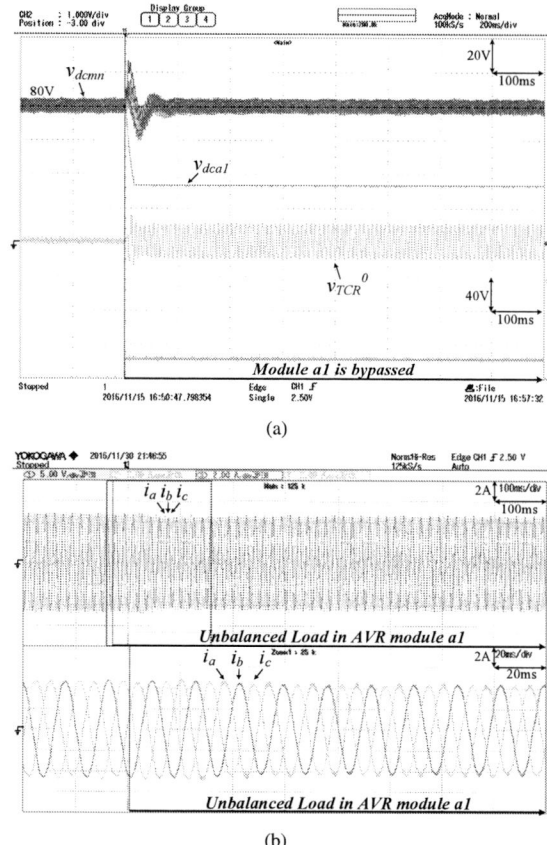

(a)

(b)

Fig. 13. Sudden change of dc loads in MMCC-SSBC (7-level). (a)V_{dcmn} when sudden change of dc loads of module $a1$. (b)i_m when sudden change of dc loads of module $a1$.

2016, pp. 1–9.

[15] S. Huang, R. Teodorescu, and L. Mathe, "Analysis of communication based distributed control of mmc for hvdc," in *2013 15th European Conference on Power Electronics and Applications (EPE)*, Sept 2013, pp. 1–10.

[16] H. Jafarian, B. Parkhideh, J. Enslin, R. Cox, and S. Bhowmik, "On reactive power injection control of distributed grid-tied ac-stacked pv inverter architecture," in *2016 IEEE Energy Conversion Congress and Exposition (ECCE)*, Sept 2016, pp. 1–6.

[17] S. Rivera, S. Kouro, B. Wu, J. I. Leon, J. Rodrguez, and L. G. Franquelo, "Cascaded h-bridge multilevel converter multistring topology for large scale photovoltaic systems," in *2011 IEEE International Symposium on Industrial Electronics*, June 2011, pp. 1837–1844.

[18] C.-T. Lee, B.-S. Wang, S.-W. Chen, S.-F. Chou, J.-L. Huang, P.-T. Cheng, H. Akagi, and P. Barbosa, "Average power balancing control of a statcom based on the cascaded h-bridge pwm converter with star configuration," *Ind. Appl., IEEE Trans. on*, vol. 50, no. 6, pp. 3893–3901, Nov 2014.

[19] H.-C. Chen, P.-H. Wu, C.-T. Lee, and P.-T. Cheng, "Zero-sequence voltage injection for dc capacitor voltage balancing control of the star-connected cascaded h-bridge pwm converter under unbalanced grid," *Ind. Appl., IEEE Trans. on*, vol. PP, no. 99, pp. 1–1, 2015.

[20] "Grid code high and extra high voltage," E.ON Netz GmbH, Bayreuth, April 2006. [Online]. Available: http://www.eon-netz.com

[21] A. Marinopoulos, F. Papandrea, M. Reza, S. Norrga, F. Spertino, and R. Napoli, "Grid integration aspects of large solar pv installations: Lvrt capability and reactive power/voltage support requirements," in *PowerTech, 2011 IEEE Trondheim*, June 2011, pp. 1–8.

The 2018 International Power Electronics Conference

Asymmetric Mixed Modular Multilevel Converter Topology in Hybrid Bipolar HVDC Transmission Systems

Joon-Hee Lee[1]*, Jae-Jung Jung[2]and Seung-Ki Sul[1]
1 Department of Electrical and Computer Engineering, Seoul National University, Seoul, South Korea
2 Manufacturing Technology Center, Samsung Electronics Company, Ltd., Hwaseong, South Korea
*E-mail: lightling123@snu.ac.kr

Abstract— **In this paper, a hybrid bipolar HVDC system consisted of Line Commutated Converter (LCC) and asymmetric mixed MMC is introduced. Among the various MMC topologies for hybrid bipolar HVDC, the asymmetric mixed MMC has advantages in terms of reduced number of switching devices and lower losses. Also, the asymmetric mixed MMC can regulate its DC voltage flexibly from null to the rated value so that it can provide DC fault ride-through capability. Especially, at the post-fault, DC bus voltage of the asymmetric MMC can increases in slow ramp-up manner, which is important to prevent over-voltage and travelling wave generation along the transmission line. Finally, the full-scale simulation result and the down-scaled experimental results demonstrate the validity of the asymmetric mixed MMC based hybrid bipolar HVDC system.**

Keywords— *Modular Multilevel Converter, Hybrid HVDC, Asymmetric MMC, MMC topology*

I. INTRODUCTION

For several decades, HVDC systems based on line commutated converter (LCC) have been developed and commercialized. These days, most of HVDC system in operation employs LCCs because of higher reliability, overload capability, and efficiency. However, it has several shortcomings such as large system size with reactive power and harmonic filter, requirement of strong AC grid and lack of black starting capability [1]. To solve these disadvantages of LCC, a voltage source converter (VSC) based HVDC system has been developed recently. Among the various VSC technologies, the modular multilevel converter (MMC) is the most promising technology over two- or three-level VSC topologies [1-2]. Two submodules types are mostly used in the MMC: one is a half-bridge submodule (HBSM) and the other is a full-bridge submodule (FBSM). The HBSM is used to reduce the switching devices, so that it has cost-effectiveness and less losses. Meanwhile, the output voltage of HBSM is confined to zero and DC link voltage of each module. Contrast to that, the output voltage of the FBSM contains also negative DC link voltage and it can synthesize AC side voltage even with the zero DC bus voltage.

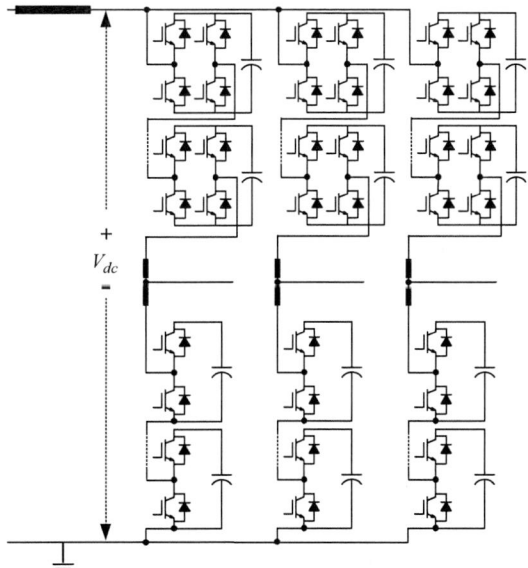

Fig. 1. Circuit diagram of the asymmetric mixed modular multilevel converter (ASYM-MMC)

Among the various applications of the HVDC system, the multi-terminal system connecting a strong AC grid to several loads in distributed locations, such as inter-island HVDC link or offshore oil platform, becomes popular. In this application, the compactness and black starting capability of the sending end converter may not be important concerns. Therefore, the LCC can be the best option for the sending end converter, because its technical maturity and higher operating efficiency. Meanwhile, on the receiving end, the distributed loads require the compact structure and black starting capability. In such an application, a hybrid HVDC structure with LCC-HVDC converter at the strong AC grid side and several medium power MMC-HVDC converters in distributed load sides can be a promising candidate. Therefore, various researches have been conducted to accommodate LCC and MMC simultaneously, which is called a hybrid HVDC system.

3557

As the demand of high power transmission capacity of HVDC system increases, newly constructed HVDC system tries to increase its DC bus voltage. In order to meet this requirement, the bipolar HVDC systems with overhead transmission line are preferred to other system configurations. The bipolar HVDC system consists of two monopole asymmetric converters connected in series and each converter pole can operate independently. Under normal operation, positive and negative poles carry equal current and no current flows through the ground (metallic) return path. Even in the case of the outage of one converter pole, at most a half of the rated power can be transmitted through other healthy pole and the return path.

Among the many concerns in HVDC system with overhead line, the ride-through capability against DC short circuit fault is one of the most important features that MMC should has. Considering that, several topologies have been devised to ensure safety of the MMC in DC fault condition. The HBSM-MMC with high power diode [3], FBSM-MMC, and HBSM-FBSM mixed MMCs [4-5]. Recently, a promising asymmetric mixed MMC (ASYM-MMC) was proposed in [5] due to the cost saving, less losses, DC fault ride-through capability, and fast DC bus voltage regulation. Also, ASYM-MMC was emerged as a hybrid MMC topology which can overcome the submodule unbalance between HBSMs and FBSMs of symmetric mixed MMC (SYM-MMC) [5]. Contrary to SYM-MMC, where equal number of HBSMs and FBSMs are used in an arm of MMC, one arm of ASYM-MMC consists of fully HBSMs and the other arm consists of fully FBSMs, as shown in Fig. 1. The characteristics and advantages of ASYM-MMC have been described in [5] compared to SYM-MMC. Among the several advantages, flexible DC voltage controllability in ASYM-MMC permits a slow ramp-up along the flash arc's dielectric recovery characteristic during the post-fault.

Therefore, this paper deals with the operational principle and validity of ASYM-MMC based hybrid bipolar HVDC system. In the section II, the basic structure of the hybrid HVDC system is introduced. The operating principle and DC fault ride through strategy of the system are discussed in section III. In order to verify the functionality of the proposed system and operating principle, a full scale simulation study is done. Finally, the experimental test with down-scaled version is conducted to support the validity of the proposed system.

II. HYBRID BIPOLAR HVDC TRANSMISSION SYSTEMS

Fig. 2 shows the structure of a point-to-point hybrid bipolar HVDC system. The converters at sending end consist of conventional LCC systems and those at receiving end are MMC systems. In this paper, it is assumed that the LCC regulates the DC current as constant and the MMC controls the DC bus voltage for DC transmission power regulation. The MMC system at receiving end consists of two ASYM-MMCs, i.e. the positive MMC and negative MMC which require no reactive power from AC grid and even can be operated as

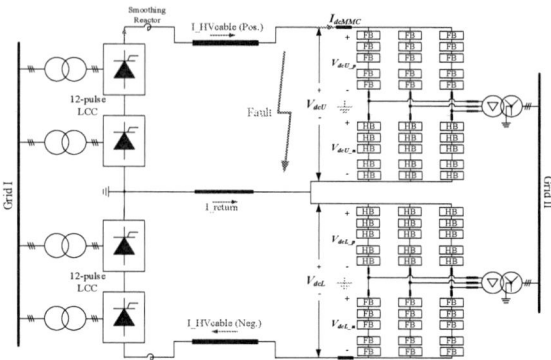

Fig. 2. Circuit diagram of the hybrid biopolr HVDC system based on ASYM-MMC

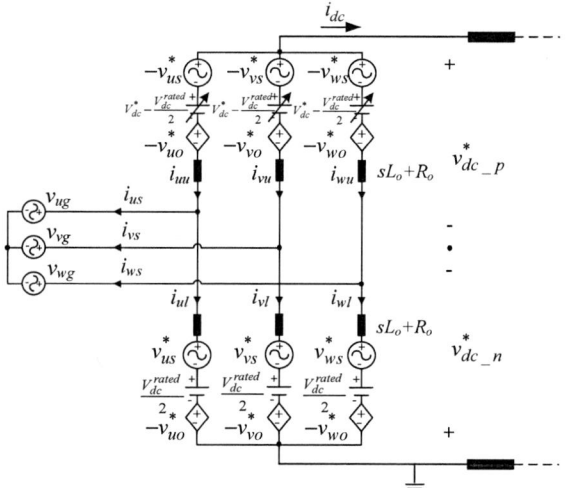

Fig. 3. Modeling of the ASYM-MMC.

a STATCOM. Therefore, the hybrid HVDC system combines both the well-developed technology and low cost of LCC and the desired regulating capability of MMC. The arms in the bipolar ASYM-MMC can be divided into two types: the FBSM-arms and the HBSM-arms. The HBSM-arms are connected to the ground pole, and the FBSM-arms are connected to the positive pole or negative pole as shown in Fig. 2 [6].

III. ASYMMETRIC MIXED MMC IN BIPOLAR HVDC TRANSMISSION SYSTEMS

A. Normal Operating Principle of Asymmetric Mixed MMC

One ASYM-MMC in the bipolar system can be considered as an asymmetric monopole system in Fig. 3. The arm voltage references of ASYM-MMC are given by (1) and (2) where V_{dc}^{rated} stands for the rated DC bus voltage.

$$v_{xP}^* = (V_{dc}^* - \frac{V_{dc}^{rated}}{2}) + v_{xs}^* - v_{xo}^*. \tag{1}$$

$$v_{xN}^* = \frac{V_{dc}^{rated}}{2} - v_{xs}^* - v_{xo}^*. \tag{2}$$

The controllable range of a lower arm (HBSM) output voltage is from 0 to NV_{cap} and that of an upper arm (FBSM) output voltage is from $-NV_{cap}$ to NV_{cap}. The definition of leg current i_{xo} is the average value of the currents in upper arm and lower arm as (3). A circulating current of x-phase, $i_{xo,cir}$, is defined as (4), which is the difference between the leg current i_{xo} and DC bus current that equally flows into each phase.

$$i_{xo} = (i_{xu} + i_{xl}) / 2 . \tag{3}$$

$$i_{xo,cir} = i_{xo} - i_{dc} / 3 . \tag{4}$$

$$i_{xu} = i_{xo} + \frac{1}{2} i_{xs} = \frac{1}{3} i_{dc} + \frac{1}{2} i_{xs} + i_{xo,cir} . \tag{5}$$

$$i_{xl} = i_{xo} - \frac{1}{2} i_{xs} = \frac{1}{3} i_{dc} - \frac{1}{2} i_{xs} + i_{xo,cir} . \tag{6}$$

The upper and lower arm currents can be deduced by (5) and (6), respectively, from i_{xo} and $i_{xo,cir}$ defined by (3) and (4) and the output phase current, i_{xs}.

B. DC Fault Ride-Through (FRT) of Bipolar HVDC System based on Asymmetric Mixed MMC

In HVDC transmission system with the overhead line, the temporary DC faults are frequently caused by lightning strikes or broken branches of trees nearby, which can normally be cleared quickly due to arc flash. However, for a permanent DC fault or multi-terminal HVDC system using cables, the extra operating scheme should be required in the post-fault process [7]. The DC FRT strategy includes isolation of DC fault and continued operation to support the connected AC grid during the fault period [7]. By using the negative voltage output of the FBSMs, ASYM-MMC can operate in fault condition independent to the DC voltage. In other words, ASYM-MMC can be stably operated at a reduced DC voltage, even zero DC voltage. While the FRT process is conducted in the faulty pole, a half of the rated power can still be transmitted through the healthy pole and the ground (metallic) return.

The DC FRT and fault clearance process in ASYM-MMC based hybrid HVDC are illustrated as a flowchart in Fig. 4. As shown in the flowchart, the FRT sequence of the hybrid HVDC is divided into the three stages [7]. During all following stages, LCC operates in constant current control mode. The superior overload capability of LCC and the large smoothing reactor on DC transmission line would enable LCC to suppress and withstand temporary overcurrent.

1) Stage 1: In normal operation, a DC line current flowing into MMC is measured for current regulation. In the meantime, when a DC line current exceeds the predefined threshold value, MMC detects the fault occurrence and starts FRT process. The DC voltage of ASYM-MMC is immediately controlled to be around zero, so DC over-current flowing into MMC is prevented. In order to extinguish and de-ionize the electric arc, the ASYM-MMC is also capable of synthesizing negative DC voltage by using the voltage margin from pre-defined

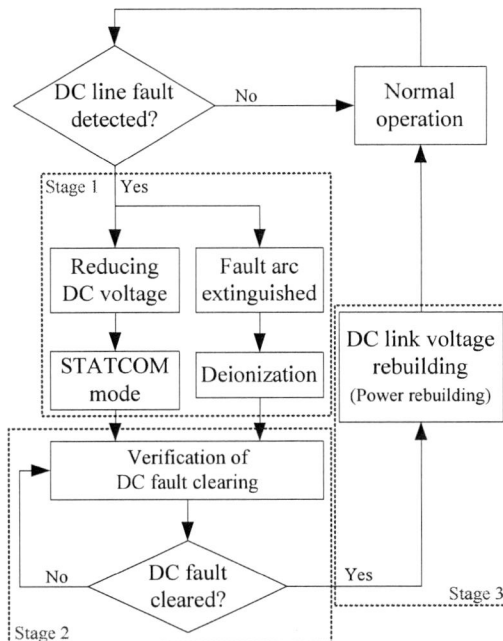

Fig. 4. DC FRT and fault clearance process in ASYM-MMC based HVDC system.

modulation index or redundant cells. By absorbing the inductive energy in electric arc and transferring it the AC grid, the extinguishing process can be accelerated. At the same time the voltages of submodule capacitors can be regulated as its rated value by the energy balancing controller of the MMC. Also, the reference of reactive AC current can be remained or set to a new value to support the AC grid after the system enters the STATCOM mode.

2) Stage 2: After adequate de-ionization time, the fault clearance should be checked by increasing the DC bus voltage with small value before entering the system rebuilding process. If the fault is not cleared, the huge fault current flows into MMC again even with a small DC voltage. In that case, MMC detects the fault again and keeps FRT operation. Then system waits for the next attempt to rebuild the DC voltage after a certain period of time. Otherwise, the successful establishment of DC bus voltage indicates the DC fault clearance of the faulty pole.

3) Stage 3: After fault clearance, the system rebuilding process at post-fault can starts under an initial condition of zero DC bus voltage. A ramp signal can then be set for the DC voltage reference and it can be built up smoothly for ambient fault recovery. As aforementioned in Section II, the DC bus voltage of ASYM-MMC can be controlled smoothly up to the rated value by control of the output voltage of FBSM arms. Consequently, DC power transmission would be reestablished as DC bus voltage increases, because LCC regulates DC current as constant during the entire FRT process.

IV. Full Scaled Simulation Results

The 400MW full scale simulation study is carried out using PSCAD/EMTDC software to validate the ASYM-MMC based hybrid HVDC system, which has the same configuration in Fig. 2. The system parameters are listed in Table I. Some of those, such as AC grid voltage and transmission line impedance, are based on the HVDC system between Korean peninsula and Jeju island.

The scenario of the simulation is as follows. The LCC regulate the DC current as constant and the MMC controls the DC bus voltage. At the same time, MMC monitors its DC current at all times and it detects the DC fault occurrence if DC current exceeds the threshold value. The threshold is set as 1000A (125% of the rated value) and -200A (-25% of the rated value) in the simulation, because the power flow of hybrid HVDC system is unidirectional. If the fault is detected, the FRT process starts. The time duration of the de-ionization is set to 500ms.

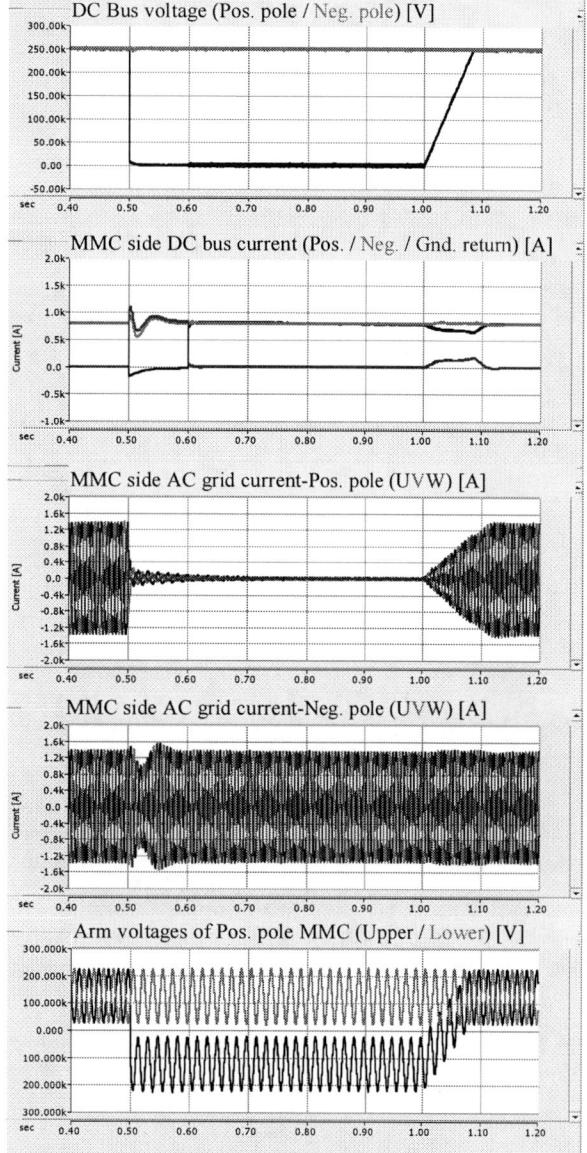

Fig. 5. Simulation results of DC fault ride through in the bipolar hybrid HVDC system at short circuit fault between positive pole and ground - MMC side

TABLE 1
THE PARAMETERS OF THE SIMULATED SYSTEM

Quantity	Values
Rated DC bus voltage (1 pole)	250 kV
Rated DC bus current	800 A
Number of submodules per one arm	130
Rated voltage of module capacitor	2.2 kV
Capacitance of module capacitor	4.5 mF
Sampling frequency	10 kHz
Inductance of arm inductor	15 mH
Resistance of arm inductor	1 mΩ
AC grid voltage (Primary side of Tr.)	154 kV
AC grid voltage (Secondary side of Tr.)	123 kV
Inductance of smoothing reactor (LCC side)	500 mH
Transmission line inductance (Pos. , Neg. pole)	16 mH
Transmission line inductance (Ground return)	11 mH

Fig. 5 shows the simulation results of DC FRT process in MMC side. As shown in the 1st and 2nd trace, before the fault occurs, the DC voltage and current are controlled as its rated value, 250kV and 800A, respectively. It is noted that the current flowing in the ground return path is almost zero, because the positive pole and negative pole carry the same amount of current. Then, the DC pole to ground fault occurs on upper pole of MMC side at the time t=0.5s. The duration of the fault is set to 100ms. Because the DC current exceeds the threshold value, -200A, MMC starts the FRT process and synthesizes the DC voltage as null immediately in order to extinguish the fault current. Meanwhile, the rated current are flowing through the ground return path and the negative DC pole, so that a half of the rated power can be transmitted by the healthy pole. After the fault is cleared at t=0.6s, the DC current in the positive pole gets back to the rated value because LCC regulates the current as constant.

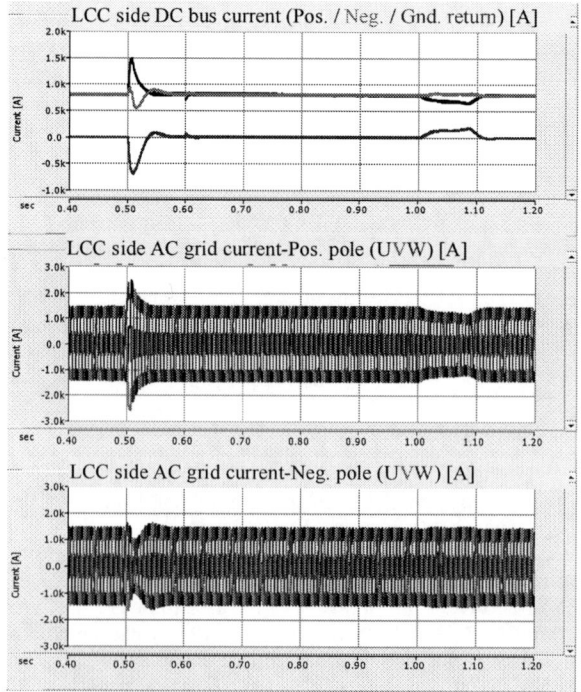

Fig. 6. Simulation results of DC fault ridge through in the bipolar hybrid HVDC system at short circuit fault between positive pole and ground - LCC side

After 0.5s of time delay for the consideration of de-ionization, upper pole MMC enters the stage 3 (power rebuilding) and it ramps up the DC voltage with slew rate to prevent the travelling wave generation along the transmission line. The 3rd and 4th trace show the AC grid current of upper pole and lower pole MMC, respectively. During the FRT process, DC bus voltage and transmitted power of faulty pole are kept as zero, so the AC grid current is regulated as null. However, in the 4th trace, AC current of the negative pole is regulated to the rated value without change, so it is confirmed that a half of the rated power can be transmitted even in the fault situation. The 5th trace shows the U-phase arm voltage of the upper pole MMC, which describes the operation of the ASYM-MMC. Since FBSMs are only at the upper arm, output voltage of the upper arm goes to the negative value, while that of lower arm with HBSM is kept the same as before.

Fig. 6 shows the simulation results in LCC side. Even if the fault occurs at t=0.5s, LCC can operate in constant current regulation mode because of its overload capability as depicted in the 1st trace. Therefore, both of AC grid current of upper pole and lower pole are kept as constant in the 2nd and 3rd trace.

V. EXPERIMENTAL RESULTS

In order to confirm the effectiveness of the ASYM-MMC based bipolar hybrid HVDC system, test with 6kW down-scaled system is conducted. The experimental setup is shown in Fig. 8. The LCCs at the sending end are emulated by two full-bridge circuits operating in constant current control mode, while two ASYM-MMCs are utilized as the receiving end converter. In each MMC, one arm is composed of 6 FBSMs and the other arm is of 6 HBSMs. The other parameters are listed in Table II.

The first scenario of the experiment is as same as that of the simulation in Section IV. At the time indicated in Zoom-in I of Fig. 9 (a), DC pole-to-ground fault occurs between the positive pole and the ground. Because the DC current exceeds the threshold current which is set to 13A in this test, MMC starts FRT process as depicted in Fig. 9 (b). In Zoom-in I of Fig. 9 (a), MMC synthesizes its DC voltage as null to extinguish the fault current. The de-ionization time delay is set to 500ms, so that DC bus voltage ramps-up as shown in Zoom-in II. Since the DC fault is cleared within the de-ionization time, system rebuilding process is successful and the DC voltage is recovered to the rated value. Fig. 9 (b) shows the DC current in positive, negative pole and the ground path. Because LCC regulates the DC current for all time, the current in ground path only flows during the fault and becomes zero right after the fault is cleared. Also, because the negative pole is healthy, the DC bus voltage and current remain for all times independent of FRT process in the positive pole. At the same time, MMC can operates in STATCOM mode even in the FRT process to enhance the transient stability of the connected AC grid. As

TABLE II
PARAMETERS OF THE EXPERIMENTAL SETUP

Quantity	Values
Rated DC bus voltage (1 pole)	300 V
Rated DC bus current	10 A
Rated voltage of module capacitor	50 V
Capacitance of module capacitor	5.4 mF
Inductance of arm inductor	4 mH
Resistance of arm inductor	5 mΩ
AC grid voltage (line to line RMS)	110 V
Sampling frequency	10 kHz
Transmission line inductance	20 mH
Transmission line resistance	0.5 Ω

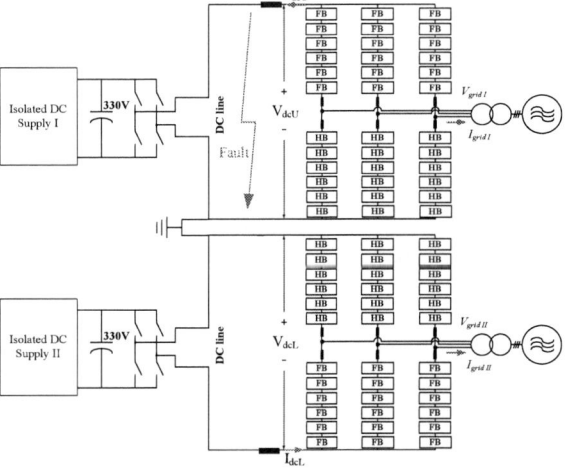

(a) Circuit diagram of experimental setup

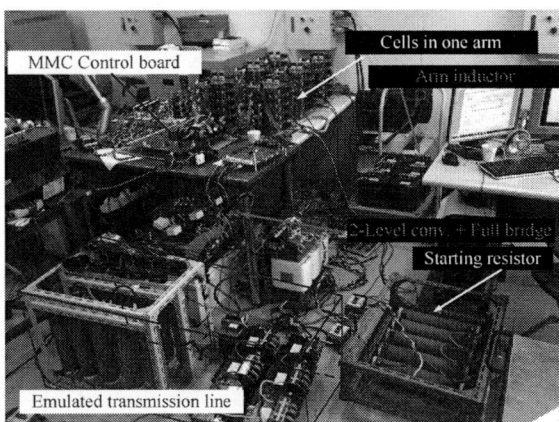

(b) Down scaled prototype hybrid HVDC system
Fig. 8. Experimental set-up

shown in the Fig. 10, MMC injects the reactive current (D-axis current) to the connected AC grid for voltage support during the FRT process while the active current (Q-axis current) is regulated as null. Then, the active current increases in slope after FRT is over, because the power transmission is resumed along with the DC bus voltage rising.

The 2018 International Power Electronics Conference

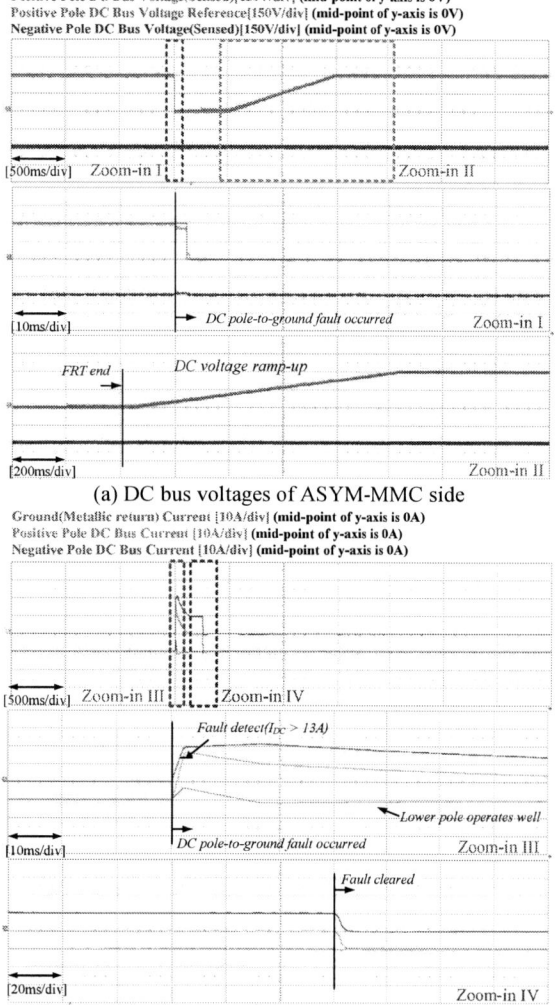

(a) DC bus voltages of ASYM-MMC side

(b) DC bus currents of ASYM-MMC side

Fig. 9. Experimental result of DC short fault ride through process in DC side of ASYM-MMCs

Fig. 10. Experimental result of DC FRT process in the upper pole ASYM-MMC

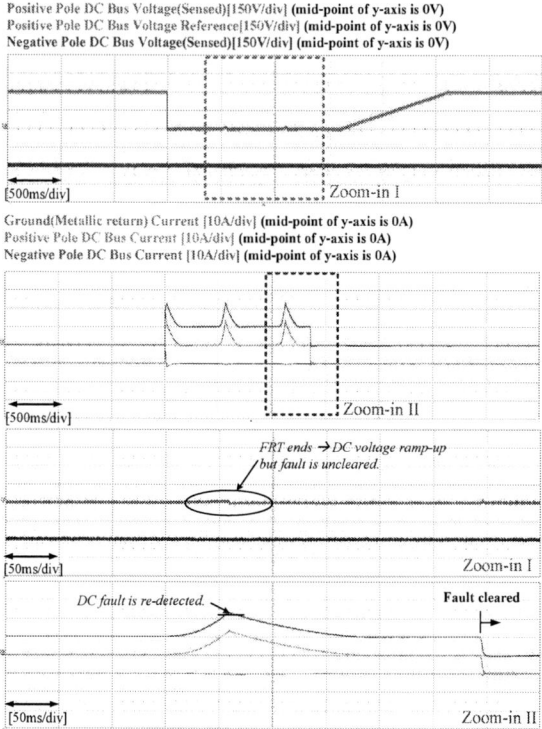

Fig. 11. Experimental result of DC FRT while DC fault continues for a while

Fig. 11 shows another scenario that DC fault continues for longer period (more than 1s). After 500ms of de-ionization time, MMC attempts to rebuild the system by increasing DC bus voltage as shown in the Zoom-in I. However, in the 2nd trace, it fails to rebuild DC voltage and huge current flows into MMC again because of everlasting fault. Since this current exceeds the threshold value again, MMC re-enters the FRT process. After two more attempts to build up the DC voltage, the system finally identify that the fault is cleared and resume the power transmission successfully.

VI. CONCLUSION

In this paper, an asymmetric mixed MMC based hybrid bipolar HVDC system has been introduced. Compared to other topologies, the asymmetric mixed MMC has advantages such as reduced cost, loss and capability of fault ride-through. The asymmetric mixed MMC can regulate the DC voltage from zero to the rated value smoothly, so it can continue the operation during the DC fault and support the connected AC grid. At post-fault, system recovery is smoothly done by the DC voltage regulation of the asymmetric mixed MMC, so there is no over voltage and travelling wave generation along the DC transmission line. The validity of the hybrid bipolar HVDC system has been confirmed by both of the 400MW full scaled simulation results and 6kW down scaled experimental results.

REFERENCES

[1] A. Lesnicar, R. Marquardt, "An innovative modular multilevel converter topology suitable for a wide power range," in Power Tech Conference Proceedings, 2003 IEEE Bologna , vol.3, no., pp.6 pp. Vol.3-, 23-26 June 2003.

[2] B. Gemmell, J. Dorn, D. Retzmann, and D. Soerangr, "Prospects of multilevel VSC technologies for power transmission," in IEEE PES Transmission and Distribution Conference and Exposition, pp. 1-16, 2008.

[3] G. Tang and Z. Xu, "A LCC and MMC hybrid HVDC topology with DC line fault clearance capability," International Journal of Electrical Power & Energy Systems, vol. 62, pp. 419-428, 2014.

[4] S. Inoue, and S. Katosh, "Modular multilevel converter with DC fault protection," European patent application, Jun. 12, 2013.

[5] Jae-Jung Jung, Shenghui Cui, Joon-Hee Lee and Seung-Ki Sul, "A New Topology of Multilevel VSC Converter for Hybrid HVDC Transmission System", in IEEE Trans. on Power Elec., vol. 32, no. 6, pp. 4199-4209, June 2017.

[6] A. Nami, J. Liang, F. Dijkhuizen and P. Lundberg, "Analysis of modular multilevel converters with DC short circuit fault blocking capability in bipolar HVDC transmission systems," *Power Electronics and Applications (EPE'15 ECCE-Europe), 2015 17th European Conference on*, Geneva, 2015, pp. 1-10.

[7] J. Hu, R. Zeng, Z. He, "DC fault ride-through of MMCs for HVDC systems: a review," *The Journal of Engineering*, 2016, open access

High Power Medium Voltage 10 kV SiC MOSFET Based Bidirectional Isolated Modular DC–DC Converter

Sayan Acharya, Ritwik Chattopadhyay, Anup Anurag, Satish Rengarajan, Yos Prabowo and Subhashish Bhattacharya

FREEDM Systems Center, Department of Electrical and Computer Engineering
North Carolina State University, Raleigh, NC 27606
Email: sachary@ncsu.edu, rchatto@ncsu.edu, aanurag2@ncsu.edu, srengar@ncsu.edu, yprabow@ncsu.edu and sbhatta4@ncsu.edu

Abstract—Recent advancement in the packaging technology for the SiC MOSFETs with blocking voltage of 10 kV or higher have opened up opportunities to consider these devices for medium voltage and high power applications. This paper focuses on the design of a modular medium voltage, high power DC–DC converter enabled by 10 kV SiC MOSFETs which aims at increasing the efficiency, power density and inter-operability. The proposed DC–DC converter is suitable for applications like DC distribution for the data centres, sub-sea power transmission, offshore wind farms and photovolatic energy transmission - distribution - coordination, electric ship DC power transmission and distribution solid state transformer.

Index Terms—10 kV SiC MOSFETs, DC–DC converter, Dual Active Bridge, High Frequency Transformer, Medium voltage, Modular converter, Silicon Carbide, XHV-6 module, Wide Band Gap (WBG) devices.

I. INTRODUCTION

NEW modular structured power electronic converters have been introduced particularly for large scale systems with ratings of multi-MVA and above [1] to reduce the overall size of the power conversion systems, increase efficiency and maximize interoperability while satisfying all necessary standards and requirements. Reduced size, efficient and flexible modular power converter topologies providing galvanic isolation are of particular interest for Medium Voltage DC (MVDC) applications [2]. At this voltage and current levels medium voltage high frequency transformers are essential which would eliminate the bulky low frequency transformers. Recent advances in wide bandgap (WBG) based switching devices [3] have enabled new classes of medium voltage (MV) to high voltage (HV) and medium frequency (MF) power electronics converters that can impact MVDC transmission. Modular structured Voltage Source Converters (VSC) have now enabled to form a Multi-Terminal DC (MTDC) transmission system enabling power transmission at medium voltage DC which greatly increases the power transmission efficiency [4]. The Multi-Terminal DC (MTDC) grid has lower capital costs and lower losses than an equivalent AC transmission system. This paper demonstrates design techniques for the modular converters including the topology, device selection, medium voltage medium frequency magnetics design etc. In this paper,

a modular 10 kV SiC MOSFET enabled 80 kV to 11 kV, 1 MVA bidirectional isolated DC–DC converter is presented. Fig. 1, Fig. 2, Fig. 3 demonstrate target applications for such modular high power, medium voltage DC–DC converter.

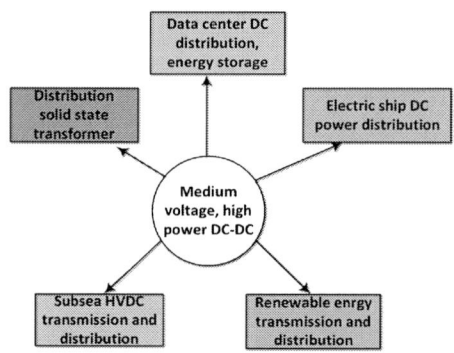

Fig. 1: Target applications for the proposed high power medium voltage DC–DC converter.

However, taking high power, medium voltage and medium frequency effects into account, there are several challenges to be addressed. These challenges are basically related to:

- Control circuit noise immunity (to withstand 20 kV to 100 kV per microsecond voltage transients);
- Utilization of the medium frequency transformer parasitic to enable zero voltage switching;
- Selective harmonic elimination modulator to set the first significant harmonic frequency at more than twice that of the switching frequency to further reduce the size of the medium frequency transformer;
- Surge suppression that is fast enough to protect the proposed converter components
- The losses as a result of eddy current in the magnetic core, excess losses in the windings due to enhanced skin and proximity effects and parasitic elements, i.e., leakage inductance and winding capacitances, causing excess switching losses in the power semiconductors, which are usually the dominant power losses at higher frequencies.

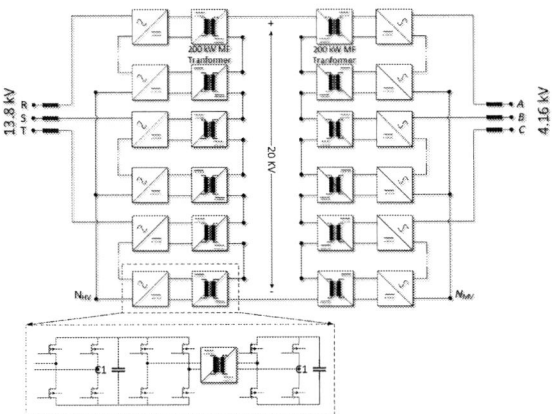

Fig. 2: Back to back converter system for MVDC transmission.

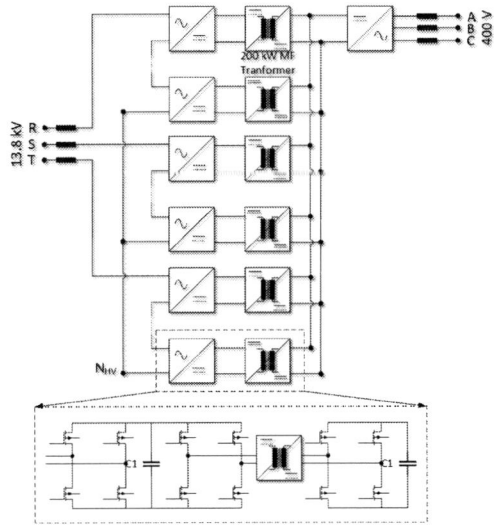

Fig. 3: Medium voltage UPS system

- Management of these extra losses together with the reduced size of the transformer lead to higher loss densities requiring a proper thermal management scheme in order to dissipate these power losses from a smaller component. This would be even more challenging when, unlike for a line-frequency power transformers, an oil cooled design is not a preference and the transformer needs to fulfill MV isolation requirements.

This paper is organized as follows. Section II lays out the system structure and design of the basic building block. Section III summarizes the practical design considerations and key challenges to mitigate common mode current in medium voltage fast switching application. Section IV provides detail insight into medium frequency high power magnetic design. The series connection of SiC-MOSFET and the full load loss analysis for the proposed DC–DC converter is presented in Section V. Based on the loss estimation the converter efficiency is presented in section VI and finally the conclusions are drawn.

II. ISOLATED BIDIRECTIONAL DC/DC CONVERTER

For a proposed 1 MVA, 80 kV - 11 kV bidirectional isolated DC–DC converter power conversion at medium voltage and medium switching frequency is desirable in order to achieve the power transfer with relatively simpler topology and increased power density. To meet these requirements several approaches such as series connection of power semiconductor devices, multilevel or modular topology needs to be considered due to limitation of the power semiconductor devices in terms of their blocking voltage and power handling capability. Taking into consideration the obtainable higher voltage blocking capability and faster switching times, medium voltage SiC MOSFETs in 6.5 kV to 15 kV voltage class have gained increased interest in medium voltage - medium frequency power electronics applications [5], [6]. With recent advancement in the power module packaging technology, SiC MOSFET power modules that are specially tailored for high power and medium voltage applications are becoming available as engineering samples [7]. These technological advancement and advantages offered by medium voltage SiC MOSFET technology make them an ideal choice as a power semiconductor devices over traditional Si based devices for the proposed DC–DC converter.

For the application at hand a modular approach as presented in Fig. 4 is adopted, where four DC–DC basic building blocks enabled by MV SiC MOSFETs are connected in input-series-output-parallel (ISOP) configuration in order to meet the voltage (80 kV – 11 kV) and power ratings (1 MVA) for the DC–DC converter [8].

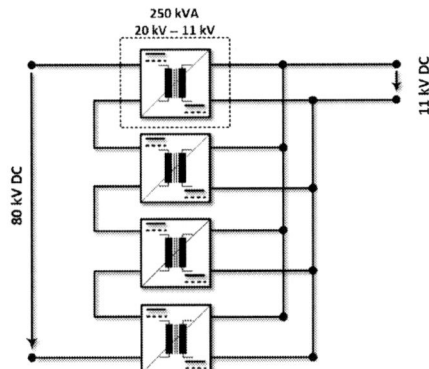

Fig. 4: Schematic of the modular DC–DC converter rated for 1 MVA, 80 kV DC - 11 kV DC

In ISOP configuration, each of these DC–DC basic building block is designed for rated power of 250 kVA with input and ouput voltage ratings of 20 kV and 11 kV respectively. This DC–DC basic building block utilizes a Dual Active bridge (DAB) converter topology as shown in Fig. 5 [9], [10].

The DAB is one of the attractive choices for the high power DC–DC converter due to its attractive features such as: bidirectional power transfer capability with galvanic isolation, enabling lower switching losses as a result of soft switching operation and lower device count. For a DAB converter generation–3 10 kV, 350 mΩ SiC MOSFET is chosen as the switching device for the primary and secondary side [3]. Considering, the blocking voltage requirement of the 20 kV and 11 kV, series connection of three and two 10 kV SiC MOSFETs is utilized on the primary and secondary stage of the DAB converter [11]. With such arrangement a reasonable voltage utilization ratio of 0.67 and 0.55 on the primary and secondary side respectively can be achieved for the 10 kV SiC MOSFETs. The transformer turns ratio (n:1) for the DAB converter is chosen as the ratio of the primary to secondary side DC bus voltage in order to ensure the zero voltage switching (ZVS) for an extended range of loading conditions [12]. For a phase shift control with switching frequency of 10 kHz and considering phase shift angle of 45 $^\circ$ at rated power, electrical specifications for the DAB converter and medium frequency transformer are presented in Table I. Simulation results for the primary and secondary bridge output voltage together with primary transformer winding current for individual basic building blocks are presented in Fig. 5.

As shown in Fig. 6, parallel connection of four 10 kV SiC MOSFETs per switch position is deduced to satisfy the peak current ratings of approximately 17 A and 32 A on the primary and secondary side of the DAB converter respectively. Proposed arrangement for the 10 kV SiC MOSFETs is feasible with the XHV-6 power module. One such XHV-6 half bridge SiC MOSFET power module in EconoDUAL footprint offers possibility of integrating upto eighteen generation-3 10 kV SiC MOSFET dies per switch position with rated current of upto 240 A [7].

TABLE I: Key specification for the proposed 1 MVA medium voltage modular DC-DC converter.

Parameter	Symbol	Value
System specification		
Power rating	S	1 MVA
Voltage rating	V	80 kV – 11 kV
Number of basic building blocks	n_{cell}	4
Specification of a basic building block		
Power rating	S_{cell}	250 kVA
Primary side DC bus voltage	V_{DC-pri}	20 kV
Primary winding peak current	i_{pk-pri}	$\approx$ 16.6 A
Primary winding rms current	$i_{rms-pri}$	$\approx$ 15.2 A
Secondary side DC bus voltage	V_{DC-sec}	11 kV
Secondary winding peak current	i_{pk-sec}	$\approx$ 30 A
Secondary winding rms current	$i_{rms-sec}$	$\approx$ 27.6 A
Switching frequency	f_{sw}	10 kHz
Medium frequency transformer turns ratio	$n:1$	1.81
Leakage inductance	L_{lkg}	81.2 μH

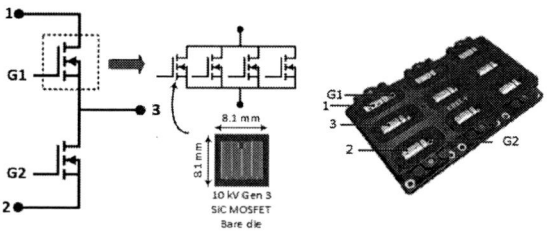

Fig. 6: Proposed die level arrangement for the 10 kV half bridge module (left) and XHV-6 10 kV half bridge SiC MOSFET power module from Wolfspeed (right) [7].

III. MITIGATION OF HIGH dv/dt INDUCED CM CURRENTS

The fast turn on/off switching transition of the medium voltage SiC MOSFETs results in the considerably high dv/dt. The high dv/dt switching transition appears across the isolation barriers, resulting in high frequency common mode(CM) and ground currents of significant magnitudes [13]. These high frequency CM currents needs to be attenuated to achieve reliable operation to comply with electromagnetic compatibility standards [14].

One of the paths for these CM currents is through the isolation capacitance of the high side gate driver power supply [15],[16],[17]. In order to maintain the control signal fidelity a gate driver power supply with low isolation capacitance is required. Together with low isolation capacitance and high common mode noise rejection, medium voltage isolation requirement needs to be fulfilled [18]. In the ISOP modular DC–DC converter the isolation requirement for the gate driver is set by the voltage stress seen by the top most module. This also applies for determining the isolation coordination for the MF transformer of the DAB converter. In addition to the CM currents the fast switching transitions in MV SiC MOSFETs can result in the false turn on of the MOSFET due to the Miller effect resulting in the direct shoot through in a phase leg [19]. Implementing an active Miller clamp in the gate driver circuit is one of the possible approaches to overcome Miller effect induced false turn on. A photo of the designed gate driver with

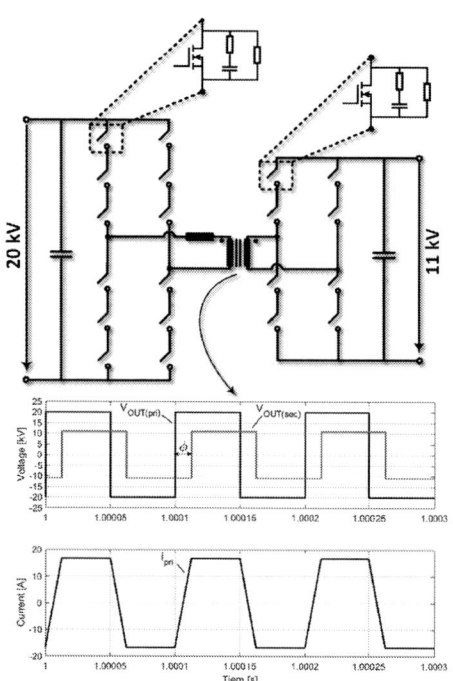

Fig. 5: Schematic of the DC–DC converter basic building block rated for 250 kVA, 20 kV DC - 11 kV DC

very low isolation capacitance ($\approx$ 1.5 pF) and active miller clamp functionality is presented in the Fig. 7 below. The power supply for the gate driver utilizes the full bridge circuit. The secondary side ouput voltage of the gate driver is regulated to +20 and -5 volts respectively, which are the driving voltage for the 10 kV SiC MOSFET. A planar winding based transformer is used to minimize the isolation capacitance.

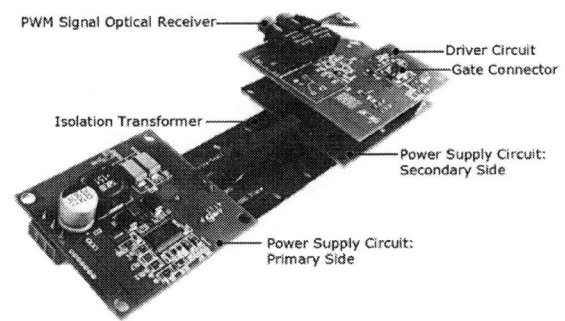

Fig. 7: Low isolation capacitance gate driver for 10 kV half bridge SiC MOSFET power module

The experimental results for the common mode (CM) currents measured at the primary side of the high side gate driver power supply in a double pulse test setup is presented in the Fig. 8. The experiments are performed at DC bus voltage of 4 kV utilizing 10 kV SiC MOSFET in a half bridge configuration with high side MOSFET as a device under test. The peak CM current of approximately 300 mA under dv/dt of 19 kV/μs.

Another path for the CM current/ground currents is provided by the power module parasitic capacitance. High dv/dt appearing across the output terminal of the half bridge module results in the displacement currents due to the capacitive coupling between the top copper layer of the DBC and baseplate [20]. This ground currents have the detrimental impact on the converter in terms of EMI as well as reliability, which needs to be taken care by the proper heatsink grounding and CM mitigation techniques [14].

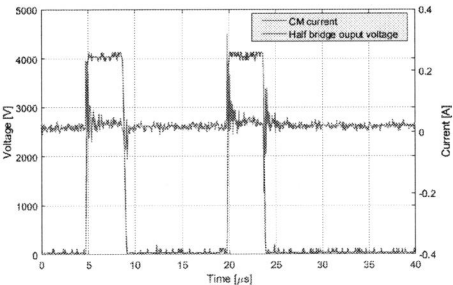

Fig. 8: Common mode current measurement for the designed gate driver in the double pulse test bench.

The CM current value is also influenced by the medium frequency transformer primary and secondary winding coupling capacitance. Hence, the coupling capacitance minimization has to be taken into consideration while designing the transformer [21].

IV. MEDIUM FREQUENCY TRANSFORMER DESIGN

The DC–DC converter block provides isolation between sending and receiving end which is crucial for different applications. Therefore, high power Medium voltage medium frequency transformer is a necessity for these converters. The power flow capability depends on the transformer leakage inductance and switching frequency. The transformer design requires optimized volume and losses with required leakage inductance for rated power flow. Use of fast-switching SiC devices also demand low inter-winding parasitic capacitance for reduced common mode current. A core loss-volume optimized design of high power medium frequency transformer is presented in this section.

A. DAB Transformer equivalent circuit

For these medium voltage applications, the transformer size is limited by the insulation requirements as the switching frequency is increased. As mentioned, the primary winding of the transformer will experience 20 kV. Since three of these power DC/DC block are connected in series to match the 60 kV input voltage level, 300 kV insulation class has to be met for each of the DC/DC converters and the transformers [22]. For this design MICA is chosen as the insulating material which has break down strength of 118 kV/mm. Hence, 5 mm gap between the core and the winding is enough to meet the insulation standard of 300 kV. Fig. 9 presents the equivalent

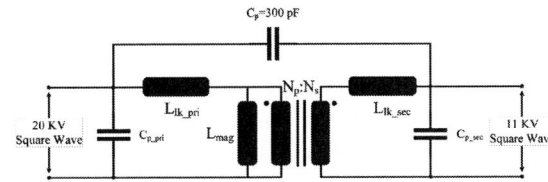

Fig. 9: Electrical equivalent circuit schematic of medium frequency transformer.

electrical circuit of the transformer of DAB. The VA rating is 250 kVA. The power flow equation of DAB is

$$P = \frac{N V_{pri} V_{sec}}{2\pi L} \phi \left(1 - \frac{\phi}{\pi}\right) \qquad (1)$$

where N is the turns ratio, V_{pri} and V_{sec} are the primary and secondary DC bus voltages respectively, ϕ is the phase shift between primary and the secondary square waves in radian and L is the leakage inductance referred to primary side. Considering the rated operating phase shift of $\frac{\pi}{4}$ and rated power rating of 250 kW, L can be determined as 15 mH at the operating Switching frequency of 10 kHz. The core material is chosen to be ferrite which can handle very high frequency operation with low loss.

The 2018 International Power Electronics Conference

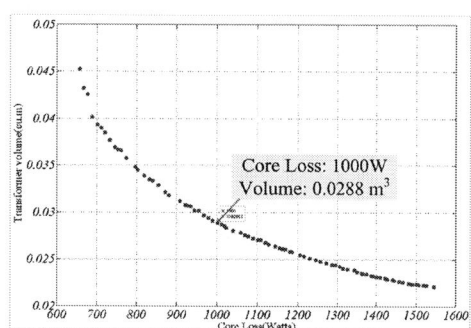

Fig. 10: Pareto optimal front design plot for 10 kHz designs.

B. Optimized Transformer Design for DAB

The leakage inductance for DAB is the element which helps to transfer power between two AC voltage sources [23]. Very high leakage inductance results in limiting the power transfer and very low leakage inductance creates issues in converter dynamic performance and affects ZVS operating range during turn-on. To minimize the system volume the required amount of inductance is designed as the transformer leakage inductance. Based on the design requirements, a pareto optimization based algorithm is followed to minimize the core loss and the transformer volume. The optimization process is carried out with ferrite based DAB transformer design for 250 kW active power transfer rating 10 KHz operating frequency. The input and output DC bus is rated at 20 kV and 11 kV respectively. In Fig. 10 the pareto optimal front core loss-volume optimized design plot is shown for 10 kHz designs. The optimized plots provide several locally optimized designs over the range. The pareto front provides several designs which have very low losses but high volume and several other designs which have higher losses but lower volume. Hence from practical considerations, designs which do not have very high losses and the volume is also not too high are considered feasible solutions. The specific design for 10kHz is selected near the bend portion of the pareto fronts, having equal weighted losses. The transformer design is selected from the design specs of the pareto front shown in Fig. 10.

C. Winding Arrangements

Another output of the optimization is winding configuration. Based the design number of turns are chosen for both the windings. For the 20 kV side the number of turns are designed as 108 arranged in four layers. Whereas, for the 11 kV side the number of turns are 60 arranged in 4 layers. Since the current carried by the windings are of 10 KHz, litz wires are chosen. The winding configurations are depicted in Fig. 11. As can be pointed out that for the 11 kV side, the wire cross section are more due to higher current and this side needs 114 parallel AWG 30 wires to carry 27 A of rms current. For the 20 KV side, 64 parallel AWG 30 wires are needed to meet the current requirement (15 Arms). With this arrangements primary winding achieves 890 mΩ of winding resistance and secondary winding has 290 mΩ resistance at 10 kHz. To meet the insulation class 5 mm of gap is kept between any point of winding and core. MICA is placed as the dielectric medium. Based on the core dimension the parasitic coupling capacitance of the transformer primary and secondary winding (Fig. 9) is estimated to be 300 pF. It is critical to minimize this capacitance for this application since the coupling capacitance along with high $\frac{dV}{dt}$ of SiC-MOSFET will determine the CM current.

D. Finite Element Analysis (FEA) of the Designed Transformer

The optimization algorithm provides the dimension of the required ferrite core as shown in Fig. 12. In this section detailed Finite Element Analysis (FEA) is done to validate the design. Fig. 13 presents the magnetic flux distribution at no load. It can be pointed out that the magnetizing flux density is around 0.28 T which links both the winding. Also, there are not much leakage flux flowing outside the ferrite core. In contrast with Fig. 13, when FEA is performed at full load the flux distribution changes quite significantly because of the leakage flux. This is presented in Fig. 14. It is depicted that at full load there are lot of leakage flux that changes the flux distribution. But even at full load the maximum flux density

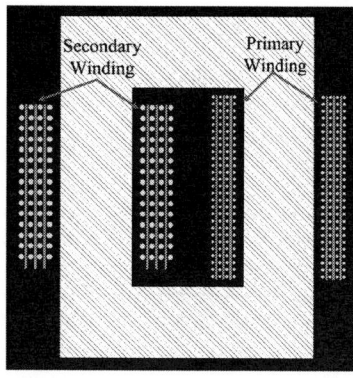

Fig. 11: Primary and secondary winding arrangements for a DAB transformer.

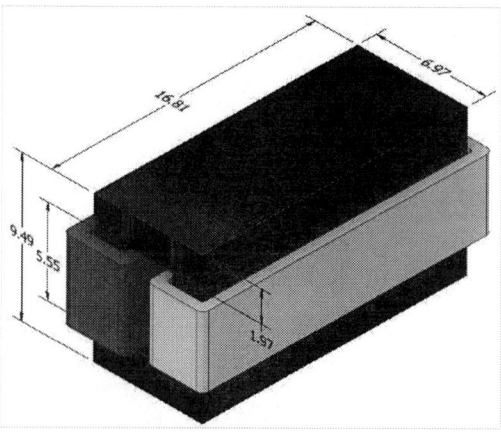

Fig. 12: Core dimension (in inch) for the optimized transformer design.

3568

is maintained bellow 0.08 T which indicates that there is no magnetic saturation happening in the core at full load. Fig. 15 shows the leakage magnetic field intensity is concentrated near the window area and the outer part of each of the windings.

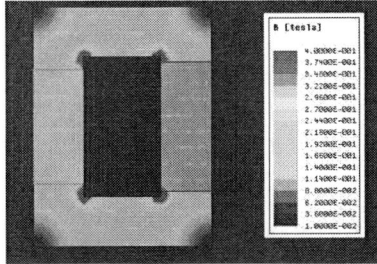

Fig. 13: MF transformer flux density distribution in at No load.

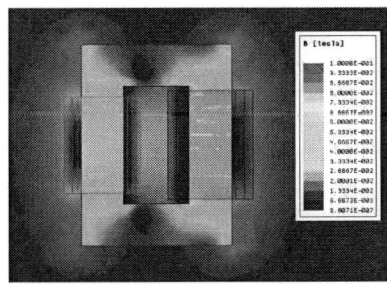

Fig. 14: MF transformer flux density distribution at full load.

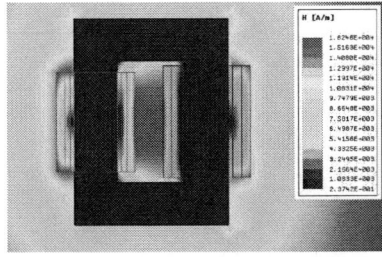

Fig. 15: MF transformer Leakage magnetic field intensity distribution at full load.

E. Transformer parasitic elements

For any high frequency transformer, it is critical to minimize the parasitic components specially the coupling capacitances which determine the common mode performance of the system. Care is taken to minimize these components to enhance the common mode performances of the system.

V. VOLTAGE BALANCING SCHEME FOR SERIES CONNECTION OF MV SiC MOSFETS

Some of the major challenges with the series connection of the power semiconductor devices are the unequal voltage sharing amongst the series connected devices in a phase leg during voltage transition and blocking state. This voltage imbalance results due to the variability in a device off-state drain-source leakage current ($I_{DS(lkg)}$) as well as difference in their dynamic switching characteristics [24]. The difference in switching characteristics may arise due to the slight mismatch in the device characteristics as well as imperfect synchronization of the driving signals and mismatch in thermal/electrical characteristics of the packages and drift in other hardware components. One of the most common method to approach this problem is to utilize an active or passive voltage balancing scheme for the dynamic as well as static voltage balancing for the series connected power devices [25], [26]. The method chosen for the application at hand utilizes voltage sharing resistor and RC snubber in parallel with the power semiconductor device as presented in Fig. 16.

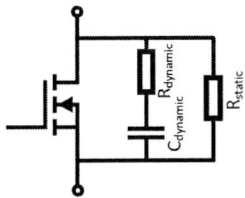

Fig. 16: Voltage balancing scheme for the series connected MV SiC MOSFETs

The value for resistance R_{static} for the voltage balancing circuit is calculated from the drain-source leakage current measurements of the 10 kV SiC MOSFETs. For the dynamic voltage balancing, $C_{dynamic}$ is chosen approximately five to ten times the device output capacitance and the value of resistance $R_{dynamic}$ is determined such that the $R_{dynamic} \cdot C_{dynamic}$ time constant is small enough in order to discharge the snubber capacitor $C_{dynamic}$ within the turn on transition [27].

The snubber circuit helps in controlling the dV/dt at the output terminal of the composite half bridge. The snubber capacitance $C_{snubber}$, reduces the rate of rise of voltage across the device during the turn off transition since the value of $C_{snubber}$ is significantly higher than that of the device output capacitance. During the turn off switching transition occurs at nearly zero voltage thus reducing the switching losses. In hard switching application, the turn on switching losses are increased since stored energy in the capacitor $C_{dynamic}$ discharges through the resistor $R_{dynamic}$ and MOSFET channel. However, since the DAB converter operates in the ZVS during turn on switching losses are not amplified by the snubber circuit. The optimum value for the dynamic voltage balancing circuit can be obtained such that total switching losses of the device, snubber losses and difference in dynamic voltage sharing amongst the series connected device is minimized. Key parameters of the voltage balancing circuit for the composite 10 kV half bridge SiC MOSFET power module are presented in the Table II.

A. Experimental results for the series connection of the 10 kV SiC MOSFETs

The aforementioned voltage balancing is verified utilizing a composite half bridge module with two series connected

TABLE II: Parameters for the dynamic and voltage balancing circuit for 10 kV SiC MOSFETs.

Parameter	Symbol	Value
Static Voltage balancing resistor	R_{static}	1 MΩ
Dynamic voltage balancing resistor	$R_{dynamic}$	4.7 Ω
Dynamic voltage balancing capacitor	$C_{dynamic}$	1 nF

SiC MOSFET in a clamped inductive test for 12 kV DC bus voltage as shown in Fig. 17.

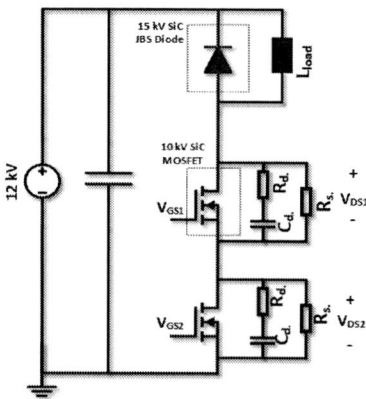

Fig. 17: Schematic of the experimental clamped inductive test circuit for evaluating voltage balancing performance of two series connected 10 kV SiC MOSFETs.

Fig.18 shows, the experimental results of the turn off switching transition for two series connected 10 kV SiC MOSFETs in clamped inductive test without voltage balancing circuit. The static drain-source voltage imbalance between the two series connected SiC MOSFETs is 800 V for DC bus voltage of 6 kV and drain current of 10 A. Whereas, the dv/dt across the two sereis connected device is higher than 50 kV/μs.

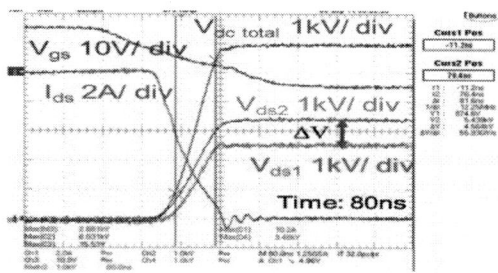

Fig. 18: Experimental results for the series connection of two 10 kV/10 A SiC MOSFETs without passive voltage balancing circuit for DC bus voltage of 6 kV. [28]

Fig.19 presents, the experimental results of the turn off switching transition for two series connected 10 kV SiC MOSFETs in clamped inductive test with voltage balancing circuit. The same voltage balancing circuit as presented in Fig. 16 is utilized with R_{static} = 20 MΩ - 50 MΩ, $R_{dynamic}$ = 15 Ω, $C_{dynamic}$ = 2.2 nF. Experimental results show a good dynamic as well as static balancing for the DC bus voltage of 12 kV and drain current of 10 A. The dv/dt in this case is less than 5 kV/μs.

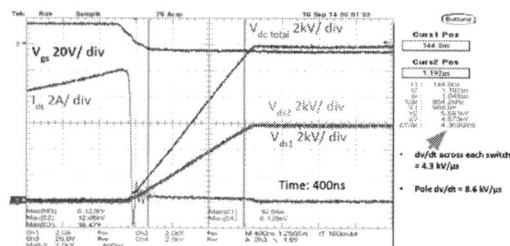

Fig. 19: Experimental results for the series connection of two 10 kV/10 A SiC MOSFETs with passive voltage balancing circuit for DC bus voltage of 12 kV. [28]

VI. EFFICIENCY CALCULATION FOR A BASIC BUILDING BLOCK

In this section the full load efficiency calculation and loss data for the 20 kV -11 kV, 250 kVA DC–DC basic building block is summarized briefly.

The power semiconductor switching losses for the DAB converter are almost negligible due to ZVS operation for an extended loading range. The calorimatric switching loss measurement for the 10 kV SiC MOSFETs show that the soft switching losses for the 10 kV SiC MOSFETs are significantly lower than compared to the hard switching losses [29]. Considering this fact into account for the full load efficiency evaluation power semiconductor switching losses in the DAB converter are neglected.

The full load conduction losses for the power semiconductor devices, transformer copper and core losses are calculated analytically for the rated load conditions. The results are summarized in Fig. 20.

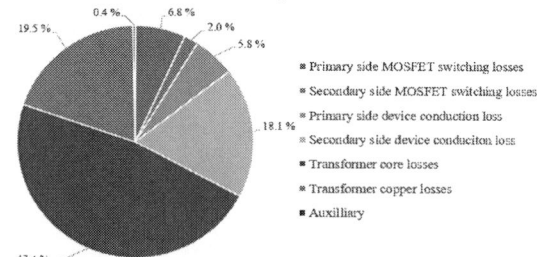

Fig. 20: Full load loss distribution for the 250 kVA DC–DC basic building block.

The conduction losses for the DAB primary side and secondary side are calculated to be 121.56 W and 382.64 W considering the primary and secondary side rms current of 15.2 A and 27.6 A respectively. Transformer design is optimized for the core losses of 1000 W. Based on the primary and secondary side winding resistance summarized in Section IV, the copper losses are calculated to be 200 W and 211 W on the DAB primary and secondary side respectively. Considering the DAB power semiconductor device condition and switching, transformer core and copper losses into account the efficiency for the 50 kVA DC–DC building block is calculated to be approximately 99.1 %.

VII. CONCLUSION

In this paper a modular 1 MVA 80 kV/11 kV modular DC–DC converter is proposed. A detail design and practical challenges for the basic building block utilizing the latest generation 10 kV SiC MOSFETs is discussed. Key challenges for mitigating the common mode currents in medium voltage fast switching applications are briefly summarized. Paper provides insight into the high power medium frequency magnetic design with pareto optimization. The transformer design has been evaluated with finite element analysis. Analytic loss analysis of the proposed DC–DC building block is presented with statistical loss distribution.

REFERENCES

[1] A. Lesnicar and R. Marquardt, "An innovative modular multilevel converter topology suitable for a wide power range," in *2003 IEEE Bologna Power Tech Conference Proceedings,*, vol. 3, June 2003, pp. 6 pp. Vol.3–.

[2] S. Acharya, A. Azidehak, K. Vechalapu, M. Kashani, G. Chavan, S. Bhattacharya, and N. Yousefpoor, "Operation of hybrid multi-terminal dc system under normal and dc fault operating conditions," in *2015 IEEE Energy Conversion Congress and Exposition (ECCE)*, Sept 2015, pp. 5386–5393.

[3] J. B. Casady, V. Pala, D. J. Lichtenwalner, E. V. Brunt, B. Hull, G. Y. Wang, J. Richmond, S. T. Allen, D. Grider, and J. W. Palmour, "New generation 10kv sic power mosfet and diodes for industrial applications," in *Proceedings of PCIM Europe 2015; International Exhibition and Conference for Power Electronics, Intelligent Motion, Renewable Energy and Energy Management*, May 2015, pp. 1–8.

[4] N. Flourentzou, V. G. Agelidis, and G. D. Demetriades, "Vsc-based hvdc power transmission systems: An overview," *IEEE Transactions on Power Electronics*, vol. 24, no. 3, pp. 592–602, March 2009.

[5] S. Sabri, E. V. Brunt, A. Barkley, B. Hull, M. O'Loughlin, A. Burk, S. Allen, and J. Palmour, "New generation 6.5 kv sic power mosfet," in *2017 IEEE 5th Workshop on Wide Bandgap Power Devices and Applications (WiPDA)*, Oct 2017, pp. 246–250.

[6] A. Q. Huang, Q. Zhu, L. Wang, and L. Zhang, "15 kv sic mosfet: An enabling technology for medium voltage solid state transformers," *CPSS Transactions on Power Electronics and Applications*, vol. 2, no. 2, pp. 118–130, 2017.

[7] B. Passmore, Z. Cole, B. McGee, M. Wells, J. Stabach, J. Bradshaw, R. Shaw, D. Martin, T. McNutt, E. VanBrunt, B. Hull, and D. Grider, "The next generation of high voltage (10 kv) silicon carbide power modules," in *2016 IEEE 4th Workshop on Wide Bandgap Power Devices and Applications (WiPDA)*, Nov 2016, pp. 1–4.

[8] H. Fan and H. Li, "A high-frequency medium-voltage dc-dc converter for future electric energy delivery and management systems," in *8th International Conference on Power Electronics - ECCE Asia*, May 2011, pp. 1031–1038.

[9] J. Walter and R. W. D. Doncker, "High-power galvanically isolated dc/dc converter topology for future automobiles," in *Power Electronics Specialist Conference, 2003. PESC '03. 2003 IEEE 34th Annual*, vol. 1, June 2003, pp. 27–32 vol.1.

[10] M. N. Kheraluwala, R. W. Gascoigne, D. M. Divan, and E. D. Baumann, "Performance characterization of a high-power dual active bridge dc-to-dc converter," *IEEE Transactions on Industry Applications*, vol. 28, no. 6, pp. 1294–1301, Nov 1992.

[11] K. Vechalapu, A. K. Kadavelugu, and S. Bhattacharya, "High voltage dual active bridge with series connected high voltage silicon carbide (sic) devices," in *2014 IEEE Energy Conversion Congress and Exposition (ECCE)*, Sept 2014, pp. 2057–2064.

[12] V. M. Iyer, S. Gulur, and S. Bhattacharya, "Optimal design methodology for dual active bridge converter under wide voltage variation," in *2017 IEEE Transportation Electrification Conference and Expo (ITEC)*, June 2017, pp. 413–420.

[13] A. Tripathi, S. Madhusoodhanan, K. Mainali, A. Kadavelugu, D. Patel, S. Bhattacharya, and K. Hatua, "Grid connected cm noise considerations of a three-phase multi-stage sst," in *2015 9th International Conference on Power Electronics and ECCE Asia (ICPE-ECCE Asia)*, June 2015, pp. 793–800.

[14] N. Christensen, A. B. Jrgensen, D. Dalal, S. D. Sonderskov, S. Bczkowski, C. Uhrenfeldt, and S. Munk-Nielsen, "Common mode current mitigation for medium voltage half bridge sic modules," in *2017 19th European Conference on Power Electronics and Applications (EPE'17 ECCE Europe)*, Sept 2017, pp. P.1–P.8.

[15] D. N. Dalal, N. Christensen, A. B. Jrgensen, S. D. Snderskov, S. Bczkowski, C. Uhrenfeldt, and S. Munk-Nielsen, "Gate driver with high common mode rejection and self turn-on mitigation for a 10 kv sic mosfet enabled mv converter," in *2017 19th European Conference on Power Electronics and Applications (EPE'17 ECCE Europe)*, Sept 2017, pp. P.1–P.10.

[16] A. Kadavelugu and S. Bhattacharya, "Design considerations and development of gate driver for 15 kv sic igbt," in *2014 IEEE Applied Power Electronics Conference and Exposition - APEC 2014*, March 2014, pp. 1494–1501.

[17] J. Gottschlich, M. Schfer, M. Neubert, and R. W. D. Doncker, "A galvanically isolated gate driver with low coupling capacitance for medium voltage sic mosfets," in *2016 18th European Conference on Power Electronics and Applications (EPE'16 ECCE Europe)*, Sept 2016, pp. 1–8.

[18] K. Mainali, S. Madhusoodhanan, A. Tripathi, K. Vechalapu, A. De, and S. Bhattacharya, "Design and evaluation of isolated gate driver power supply for medium voltage converter applications," in *2016 IEEE Applied Power Electronics Conference and Exposition (APEC)*, March 2016, pp. 1632–1639.

[19] S. Yin, K. J. Tseng, C. F. Tong, R. Simanjorang, C. J. Gajanayake, A. Nawawi, Y. Liu, Y. Liu, K. Y. See, A. Sakanova, K. Men, and A. K. Gupta, "Gate driver optimization to mitigate shoot-through in high-speed switching sic half bridge module," in *2015 IEEE 11th International Conference on Power Electronics and Drive Systems*, June 2015, pp. 484–491.

[20] A. B. Jrgensen, N. Christensen, D. N. Dalal, S. D. Snderskov, S. Bczkowski, C. Uhrenfeldt, and S. Munk-Nielsen, "Reduction of parasitic capacitance in 10 kv sic mosfet power modules using 3d fem," in *2017 19th European Conference on Power Electronics and Applications (EPE'17 ECCE Europe)*, Sept 2017, pp. P.1–P.8.

[21] A. Tripathi, S. Madhusoodhanan, K. Mainali, A. Kadavelugu, D. Patel, S. Bhattacharya, and K. Hatua, "Grid connected cm noise considerations of a three-phase multi-stage sst," in *2015 9th International Conference on Power Electronics and ECCE Asia (ICPE-ECCE Asia)*, June 2015, pp. 793–800.

[22] T. Guillod, J. E. Huber, G. Ortiz, A. De, C. M. Franck, and J. W. Kolar, "Characterization of the voltage and electric field stresses in multi-cell solid-state transformers," in *2014 IEEE Energy Conversion Congress and Exposition (ECCE)*, Sept 2014, pp. 4726–4734.

[23] R. Chattopadhyay, M. A. Juds, G. Gohil, S. Gulur, P. R. Ohodnicki, and S. Bhattacharya, "Optimized design for three port transformer considering leakage inductance and parasitic capacitance," in *2017 IEEE Energy Conversion Congress and Exposition (ECCE)*, Oct 2017, pp. 3247–3254.

[24] K. Vechalapu, S. Bhattacharya, and E. Aleoiza, "Performance evaluation of series connected 1700v sic mosfet devices," in *2015 IEEE 3rd Workshop on Wide Bandgap Power Devices and Applications (WiPDA)*, Nov 2015, pp. 184–191.

[25] S. Hong, V. Chitta, and D. A. Torrey, "Series connection of igbt's with active voltage balancing," *IEEE Transactions on Industry Applications*, vol. 35, no. 4, pp. 917–923, Jul 1999.

[26] T. Lu, Z. Zhao, S. Ji, H. Yu, and L. Yuan, "Parameter design of voltage balancing circuit for series connected hv-igbts," in *Proceedings of The 7th International Power Electronics and Motion Control Conference*, vol. 2, June 2012, pp. 1502–1507.

[27] K. Vechalapu, A. Negi, and S. Bhattacharya, "Comparative performance evaluation of series connected 15 kv sic igbt devices and 15 kv sic mosfet devices for mv power conversion systems," in *2016 IEEE Energy Conversion Congress and Exposition (ECCE)*, Sept 2016, pp. 1–8.

[28] ———, "Comparative performance evaluation of series connected 15 kv sic igbt devices and 15 kv sic mosfet devices for mv power conversion systems," in *2016 IEEE Energy Conversion Congress and Exposition (ECCE)*, Sept 2016, pp. 1–8.

[29] D. Rothmund, D. Bortis, and J. W. Kolar, "Accurate transient calorimetric measurement of soft-switching losses of 10-kv sic mosfets and diodes," *IEEE Transactions on Power Electronics*, vol. 33, no. 6, pp. 5240–5250, June 2018.

Multi-Level Power Converter Using Series-Connected Solid-State Transformers

Yuichi Mabuchi, Yuki Kawaguchi, Kimihisa Furukawa, Mitsuhiro Kadota, Mizuki Nakahara and Akihiko Kanoda
Power Electronics System Research Department
Research and Development Group, Hitachi, Ltd.
1-1, Omika-cho, 7-chome, Hitachi-shi, Ibarakiken, 319-1292 Japan
Email: yuichi.mabuchi.wz@hitachi.com

Abstract—**Solid state transformers (SSTs) and multi-level power conversion systems have been researched actively. These techniques are able to control voltage and current easier than conventional systems and have a potential to reduce the volume and weight of the power conversion systems. We develop a multi-level converter with the series-connected SST units for solar power conditioning system (PCS). In each SST unit, an LLC converter driven by SiC-MOSFETs are included, and we propose a control method of the LLC converter to keep the efficiency for solar PCS in which the input voltage form photovoltaic panel changes. The proposed method is evaluated with a prototype SST unit, and its efficiency shows 98% or more even if the case of the input voltage from the panel is relatively high.**

I. Introduction

Along with the progress of the practical use of the wide band gap devices such as SiC and GaN, the development of power electronics systems using these low loss devices has been advanced. Solid state transformer (SST) is one of power electronics systems using such semiconductor devices[1][2]. SST is an electric equipment which change the voltage levels between input and output using insulated DC/DC converter. In the insulated DC/DC converter, a transformer is driven by switching devices with higher frequency compared to 50 / 60 Hz which is commercial frequency. Therefore, the power conversion system using the SST can downsize the volume of the system itself as compared with a system including conventional commercial transformer. Furthermore, SST can change the input / output voltage ratio arbitrarily by controlling, and SST is expected to be applied to the next generation power systems such as smart grid[3][4]. In SST, the efficiency of the system can be improved by using a SiC device whose switching and conductive losses are smaller than that of Si for an insulated DC/DC converter.

Recently, in high voltage and large capacity power electronic systems, multi-level power converters such as modular multi-level converter (MMC) have been actively researched. The MMC constitutes a power conversion system by connecting units composed of half bridge or full bridge circuits in series for each phase. The MMC operates to control the output of each unit adequately and make a multi-level voltage. As a result, a sinusoidal wave with few harmonics can be realized. In MMCs, each unit shears a part of the output voltage, therefore, it is possible to use a lower withstand voltage device in case of MMCs as compared with a usual power converter

whose output is 2 or 3 levels. Because the production volume of the power semiconductor devices with lower withstand voltage are more than those with higher withstand voltage, their price tends to be proportional to their withstand voltage. For this reason, there is a potential that the cost of multi-level power converter can be lowered as compared with a power converters using high withstand voltage devices.

In these situations, we research a multi-level power converter with multiple SST units connected in series. A high frequency transformer is implemented in each SST unit, and the output of the multi-level power converter is insulated from the input. Therefore, it is possible to realize a compact and lightweight system as compared with a power conversion system using a conventional commercial transformer for insulation and conversion of voltage level. We apply the multi-level power converter with multiple SST units to a power conditioning system (PCS) for solar power generation. In this paper, the developed system, unit and its control method to improve the efficiency is shown. In the SST unit, an LLC resonance type DC/DC converter is used and its control method for efficient operation in solar PCS is explained. Then, the effect of the proposed control method is verified by measurement, and the result and conclusion are shown.

II. Multi-Level Power Converter with SST Units for Solar PCS

In this section, the system configuration of the multi-level power converter for solar PCS is shown. And the circuit and structure of the SST unit used in the multi-level power converter are also explained.

A. Configuration of multi-level power converter system

In the conventional system, DC power generated by the photovoltaic (PV) panels is inputted into the PCS. The PCS controls the DC voltage level so as to maximize the generated power in the panels by using the method like maximum power point tracking (MPPT). The PCS converts the power from DC to AC and three phase AC power whose voltage is several hundred volts is outputted. Then, the three phase AC power is boosted up to several kV by a commercial sub-transformer driven and connected to commercial main transformer. These transformers are by 50 / 60 Hz. In case of the multi-level converter, inputted DC power is directly boosted up in the

The 2018 International Power Electronics Conference

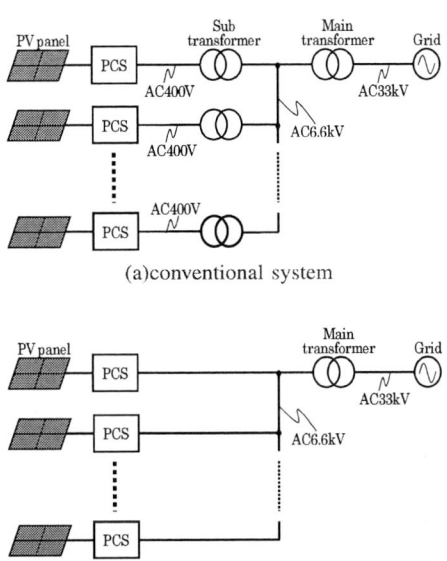

(a)conventional system

(b)multi-level power converter

Fig. 1. Comparison of the solar generation systems using between a conventional one which uses a commercial transformer and a multi-level power converter. (a) corresponds to a conventional system. In this case, there is a commercial transformer to boost up the voltage outputted by the PCS. (b) is the case of the multi-stage power converter, and the voltage is directly boosted up in this converter.

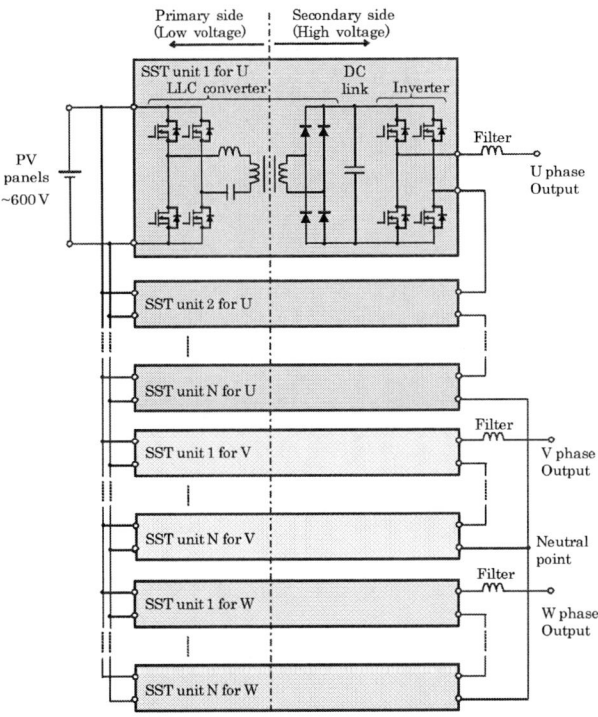

Fig. 2. A system configuration of multi-level power converter for solar PCS. The low voltage sides of the units are input form PV panels and connected in parallel. The high voltage sides of the units are output to grid and connected in series for each phase. All SST units have the same circuit configuration.

multi-level converter, and the three phase AC power whose voltage is several kV is outputted from the converter to the commercial main transformer. In this case, the commercial sub-transformers are not need. Therefore, the space and cost can be saved.

A system configuration of the multi-level power converter for solar PCS using the SST units is shown in Fig.2. The voltage range of the inputted DC power from the PV panels is assumed to be up to 600 V. The output of the system is AC three phases and its phase-to-phase voltage is also assumed to be 6.6 kVrms. In this system, the input side of the SST unit from the PV panels is connected in parallel and the output side of the SST unit to the AC grid is connected in series. The SST unit is composed of a low voltage side circuit (PV side), a high voltage side circuit (grid side), and a DC/DC converter. In the conventional photovoltaic power generation system, in order to connect to the grid of AC 6.6 kV, the commercial transformer driven at 50 / 60 Hz was used to boost up the output voltage from the PCS. And, with this commercial transformer, PCS is insulated from the grid. On the other hand, by using a multi-level power converter with the SST units, it is possible to insulate the PCS from the grid and boost up the voltage level to the grid without a commercial transformer. The transformers in SST units are driven at the frequency of several kHz. The volume of the transformer tends to be proportional to the drive frequency. The drive frequency of the transformers in SST units are several hundred times faster than that of the commercial transformer, therefore, the

system volume of the multi-level power converter can be much smaller than the conventional system using the commercial transformer. Furthermore, since it is a multi-level output, the harmonic filter can be miniaturized.

B. Circuit and insulation structure of SST unit

In the SST unit, an LLC resonance type converter is used as an insulated DC/DC converter. The LLC converter can reduce the switching current by using the characteristics of the series current resonance of the inductance (L) and the capacitance (C), and soft switching becomes possible, therefore high-frequency switching can be performed with high efficiency[5]. In case of the circuit configuration shown in Fig.2, in the SST unit, the maximum voltage difference between the low voltage (LV) and high voltage (HV) sides is obtained by adding the DC voltage in LV side to the peak phase voltage of the HV side, which corresponds to about 6,000V. Therefore, it is necessary to ensure an insulation withstand voltage of several kV between circuits on the LV side and the HV side. In order to miniaturize the system, a unit structure considering both insulation and cooling becomes an issue. Fig.3 shows a prototype of the SST unit. A schematic view of the sectional structure of this SST unit as seen from the X axis direction is shown in Fig.4. This SST unit cooled by forced air. The air is flowed from the left side of unit in Fig.4, because the semiconductor device should be cooled

3573

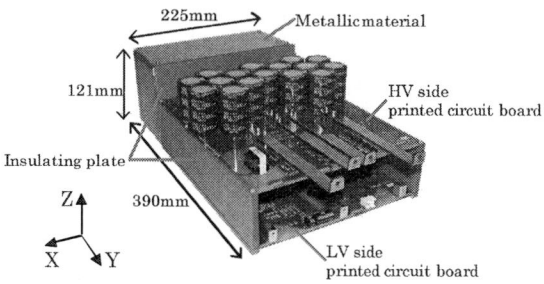

Fig. 3. An image of the developed SST unit.

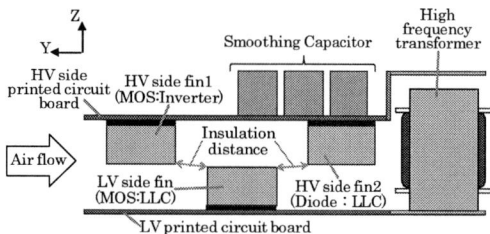

Fig. 4. A schematic view of the sectional structure of the SST unit.

first by fresh air from outside. The printed circuit boards on the LV and the HV side are arranged above and below each other, and each printed circuit board is held by an insulating plate. Furthermore, by arranging the cooling fins for the power devices on the LV side facing those on HV side, it is possible to cool the power devices on both LV and HV side by the same air duct while keeping the insulation distance. With this structure, miniaturization and efficient cooling of the SST unit are realized. TABLE I shows the specifications of the system and SST unit. The rated power of the system is 300 kW and 8 units are connected per phase (There are 24 units in all). The rated power of the SST unit is 12.5 kW. In the LLC converter of the SST unit, SiC-MOSFET and SiC-SBD are used for the switching and rectifier, respectively.

III. HIGH EFFICIENCY CONTROL METHOD OF LLC CONVERTER

In this section, control method realizing high efficiency of the multi-level power converter for solar power generation is explained from the viewpoint of an LLC converter operation.

TABLE I. SPECIFICATIONS OF THE DEVELOPED SST UNIT

Item	Value
Rated power of the system	300 kW
Nummber of the units per phase	8
Nummber of the units (total)	24
Rated power of the unit	12.5 kW
DC input voltage	320 ~ 600 V
MPPT range	400 ~ 540 V
Insulation voltage (between LV and HV side)	AC 6.6 kV

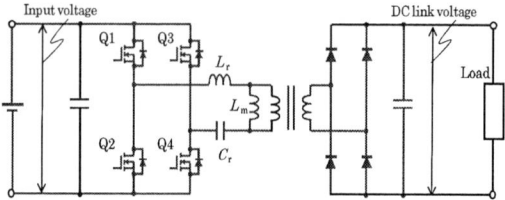

Fig. 5. The circuit of the LLC converter implemented in the SST units. The high-frequency transformer is driven by the switching devices $Q_1 \sim Q_4$. And resonance frequency F_r of the circuit is determined by the resonance inductance L_r of the high-frequency transformer and the resonance capacitor C_r.

An LLC converter is implemented in the SST unit and it controls the DC link voltage. The main circuit of the LLC converter is shown in Fig.5. Generally, the LLC converter controls the output voltage so that it becomes the target level by means of changing the driving frequency F_{sw} of the high-frequency transformer. The resonance frequencies F_r of the LLC converter is determined by the inductances L_r, L_m of the high-frequency transformer and the resonance capacitance C_r that are implemented in the converter. L_r and L_m are correspond to the resonance and excitation inductances, respectively. F_r usualy varies according to the load capacitance between $F_r = \sqrt{L_r \cdot C_r}$ and $F_r = \sqrt{(L_r + L_m) \cdot C_r}$. And the operation mode of the converter varies according to the input / output voltage ratio. There are three operation modes. The current waveforms and switching statuses of the LLC converter according to each mode are shown in Fig.6. In these figures, V_{gs} is the voltage between the gate and source of the switching device, and I corresponds to the currents of switching device of Q1 and high-frequency transformer. When the driving frequency of the high-frequency transformer F_{sw} is lower than the resonance F_r, the output voltage (DC link voltage) is boosted from the voltage determined by the turn ratio of the high-frequency transformer (boost mode). In this mode, the periods while the switching devices Q1~Q4 are being ON states are longer than the resonance period of $1/F_r$. Therefore, the current flowing through the switching device converges to a certain value. This current value is relatively small compared to the peak of the current, and the switching losses of the devices can be suppressed to low level. When F_{sw} is equal to F_r, the DC link voltage takes the value determined by the turn ratio of the high-frequency transformer (resonance mode). In this case, the current waveform of the high-frequency transformer becomes close to sinusoidal one. The current value when the switching device turns off is as small as that of the boost mode, and the switching loss is also suppressed. On the other hand, if F_{sw} is higher than F_r, the DC link voltage is stepped down from the inputted voltage (buck mode). In this case, the switching period is shorter than the resonance period of $1/F_r$, and the switching device turns off before the current becomes enough small. Therefore, in the buck mode, the switching loss increases compared to the boost and resonance mode, and thus the efficiency of the system is decreased.

The 2018 International Power Electronics Conference

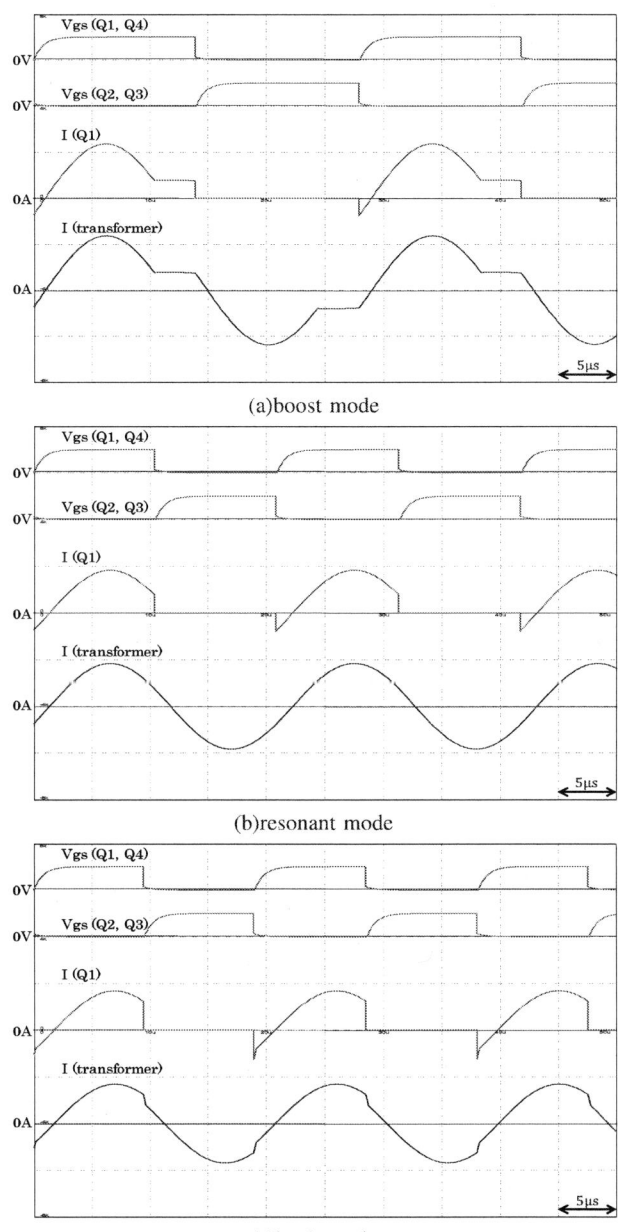

(a)boost mode

(b)resonant mode

(c)buck mode

Fig. 6. Operation modes of the LLC converter according to the input / output voltage ratio. (a) is boost mode. The driving frequency of the high-frequency F_{sw} is lower than resonance frequency F_r, and the output voltage is boosted up from the turn ratio of the transformer. (b) is resonance mode. F_{sw} is driven at F_r, and the output voltage is equal to the turn ratio. (c) is buck mode. F_{sw} is higher than F_r, and the output voltage is decreased from the turn ratio.

We propose a DC link voltage cooperative control method to enhance the efficiency compared to the conventional control method. In the conventional control method, when the inputted voltage becomes relatively high, the LLC converter operates on buck mode, and the efficiency is decreased because the switching loss increases. In the proposed method, the input

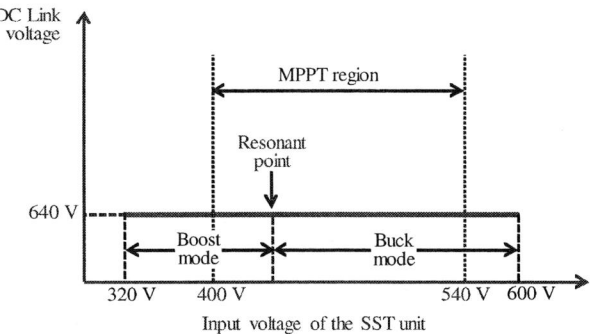

Fig. 7. The relationship between the input voltage and the DC link voltage in the conventional control. DC link voltage is controlled so as to be a constant value regardless of the input voltage change.

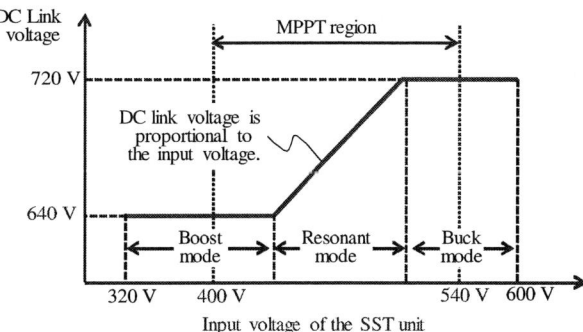

Fig. 8. The relationship between the input voltage and the DC link voltage in the proposed control. DC link voltage varies in proportion to the input voltage in some voltage range, and resonance mode operation in wider voltage range becomes possible compared to the conventional method.

/ output voltage ratio of the DC/DC converter is maintained under some input voltage range, which means that the DC link voltage is changed in proportion to the input voltage under some voltage range. By doing so, the LLC converter can operate on resonance mode for wider input voltage range than that of the conventional method, and it is possible to enhance the efficiency. The relationship between the input voltage and the DC link voltage in the case of conventional and proposed are shown in Fig.7 and 8, respectively. In the conventional case shown in Fig.7, since the DC link voltage takes constant value with respect to the change of the input voltage, in the MPPT range, the input voltage range where the LLC converter can operates with high efficiency on the boost and resonance mode occupies less than half of the MPPT range. On the other hand, in the case of the proposal shown in Fig.8, since the DC link voltage changes in proportion to the input voltage when the input voltage is within a certain range, even if the input voltage changes, the LLC converter can continues to operate on the resonance mode. Therefore, it is possible to maintain high efficiency operation over a wider voltage range as compared with the conventional control method.

3575

The 2018 International Power Electronics Conference

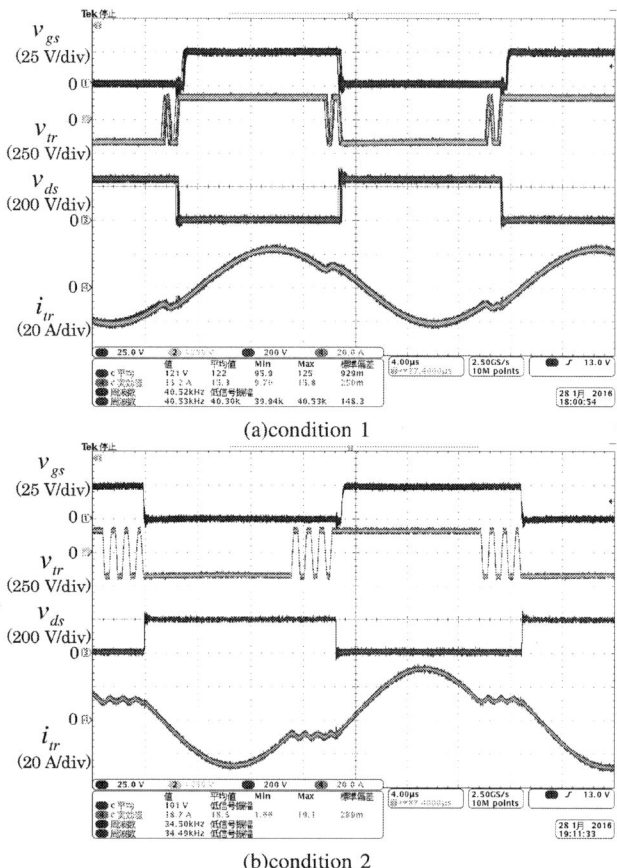

(a)condition 1

(b)condition 2

Fig. 9. Measured voltage and current waveforms. (a) and (b) correspond to resonance and boost mode operation, respectively.

IV. EVALUATION ON PROTOTYPE SST UNIT

We evaluate the proposed control method of the LLC converter with the prototype unit. In this section, evaluation of efficiency according to load characteristics and operation under rated voltage condition are shown.

A. Experimental conditions

In order to confirm the effect of the proposed control method, we experimented with the prototype unit under two input / output voltage conditions of the LLC converter. These two experimental conditions are shown in TABLE II. In these two cases, to examine the effect to the efficiency, the input voltages are set to the half of the rated voltage of the unit. Condition 1 is a case where the input voltage is relatively high and the LLC converter operates on the resonance mode.

TABLE II.   MEASUREMENT CONDITIONS

Condition No.	Input voltage	DC link voltage
1	250 V	360 V
2	200 V	320 V

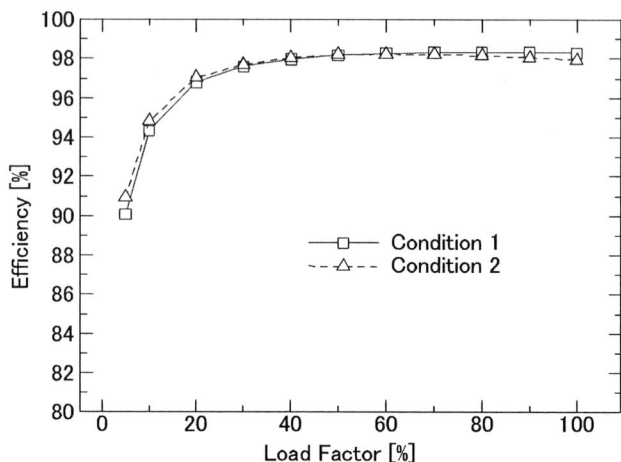

Fig. 10.   Measured efficiency of the develped LLC converter when the proposed control method is applied.

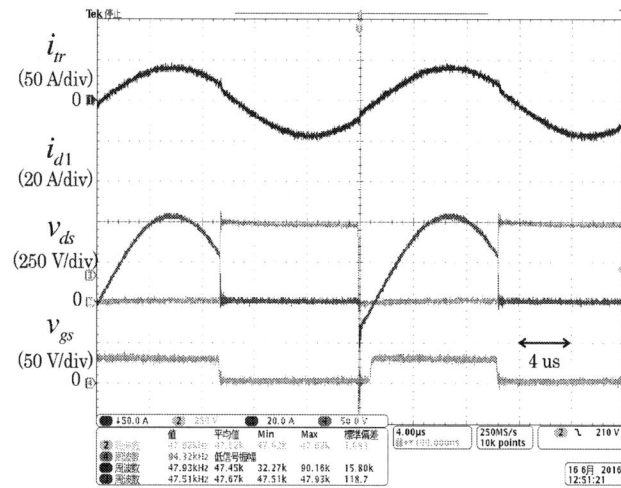

Fig. 11.   Measured voltage and current waveforms at the rated voltage condition.

In this case, the DC link voltage is increased in proportion to the input voltage. Condition 2 is a case where the input voltage is relatively low and the LLC converter operates on the boost mode. Under these two conditions, efficiency was measured by varying the load factor from 5% to 100%. In the evaluation under the rated voltage condition of the LLC converter, the efficiency was also measured by changing the load ratio from 5% (625 W) to 100% (12.5 kW). In this case, the input and DC link voltages are 475 V and 640 V, respectively, and the LLC converter operates on resonance mode.

B. Experimental results

Fig.9 shows the waveforms of each voltage and current when the load factor is 100% under the conditions 1 and 2. In these graphs, v_{gs}, v_{tr}, v_{ds} and i_{tr} are voltages between

The 2018 International Power Electronics Conference

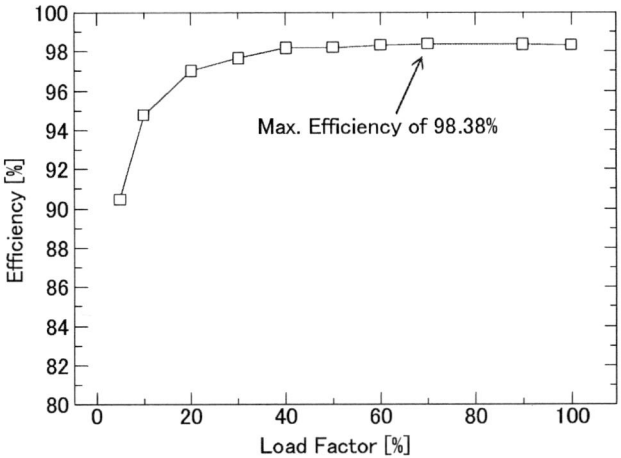

Fig. 12. Measured efficiency of the developed LLC converter under rated voltage condition.

gate and source, terminals of the transformer in LV side, drain and source, current of the transformer in LV side, respectively. From the waveform of the condition 1, the current i_{tr} of the high frequency transformer becomes a sinusoidal wave, and even though the input voltage is high, the LLC converter is operating near the resonance point. The measured efficiencies of the conditions 1 and 2 are shown in Fig.7. Under both conditions, the efficiency at the rated load was over 98%, confirming highly efficient operation. Even at light load with a load factor of 10%, efficiency of 94% or more was obtained under both conditions. The voltage and current waveforms at the rated voltage are shown in Fig.11. From this waveform, even at the rated voltage, the LLC converter operates as assumed. The measured efficiencies according to the load factor is shown in Fig.12. The maximum efficiency at the rated voltage was 98.38% (at 70% load), efficiency at rated 12.5 kW output was 98.34%, and high efficiency operation with efficiency of 98% or more was confirmed even at rated voltage. Moreover, the efficiency was 90% or more even at light load of 5%.

V. CONCLUSION

In this paper, the concept of multilevel power conversion system for solar power generation using SST unit is shown. And we propose link voltage cooperative control method to improve efficiency of LLC converter of SST unit. Experimental results with the prototype unit confirmed that the proposed control enables the LLC converter to operate with high efficiency in a wide input voltage range. Furthermore, even under rated voltage conditions, we confirmed highly efficient operation with maximum efficiency of 98.38% and light load efficiency of 90% or more.

REFERENCES

[1] G. Ortiz, M. Georg, J. E. Huber, and J. W. Kolar, "Design and Experimental Testing of a Resonant DC-DC Converter for Solid-State-Transformers," *IEEE Transaction on Power Electronics*, vol. 32, no. 10, pp. 7534–7542, Oct. 2017.

[2] X. She, R. Burgos, G. Wang, and A. Q. Huang, "Review of Solid State Transformer in the Distribution system: From Components to Field Application," *International Symposium on IEEE Energy Conversion Congr. And Expo. (ECCE)*, pp. 4077–4084, Sept. 2012.

[3] A. J. Watson, P. W. Wheeler, and J. C. Clare, "Field programmable gate array based control of Dual Active Bridge DC/DC Converter for the UNIFLEX-PM project," *14th European Conference on Power Electronics and Applications (EPE)*, pp. 1–9, Sept. 2011.

[4] S. Bifaretti, P. Zanchetta, A. Watson, L. Tarisciotti, and J. C. Clare, "Advanced Power Electronic Conversion and Control System for Universal and Flexible Power Management," *IEEE Transactions on Smart Grid*, vol. 2, no. 2, pp. 231–243, 2011.

[5] J. F. Lazar and R. Martinelli, "Steady-State Analysis of the LLC Series Resonant Converter," *International Symposium on IEEE Applied Power Electronics Conference (APEC)*, pp. 728–735, 2012.

Capacitor Voltage Control of MMC-STATCOM during Unbalanced AC System Fault

Kaho Nada[1], Takeshi Kikuchi[2], Tsuguhiro Takuno[1], Toshiyuki Fujii[1], Ryosuke Uda[2] and Takashi Sugiyama[3]
1 Advanced Technology R&D Center, Mitsubishi Electric Corporation, Amagasaki, Japan
2 Power Systems Engineering Project Group, Mitsubishi Electric Corporation, Kobe, Japan
3 Power Electronics Department, Toshiba Mitsubishi-Electric Industrial Systems Corporation, Kobe, Japan

*E-mail: Nada.Kaho@da.MitsubishiElectric.co.jp

Abstract— This paper describes a capacitor voltage control method of a modular multi-level cascade (MMC) converter configuring delta connection for a STATCOM. Each converter cell has its capacitor in DC-side, therefore this circuit configuration requires controlling the capacitor voltage level individually. Especially, controlling the voltage of each phase is important because the grid voltage can be a sudden unbalance condition in a transient event such as a system fault. The proposed method of the phase balance of cell capacitor voltages adjusts both the negative-sequence output current and the circulating current during unbalanced AC power system faults. The 5 kVA experimental system shows that the proposed method is effective during such a system fault.

Keywords— *Multi-level inverters, STATCOM, Multi-level cascaded converters, AC system faults.*

I. INTRODUCTION

Static synchronous compensators (STATCOMs) are playing an important role such as stabilization of power system or voltage regulation of an electrical grid. Since the 80MVA GTO STATCOM has been introduced to the Inuyama switching station of the Kansai Electric Power Co., Inc. in 1991[1], installations of STATCOM have been increased as a means of power quality improvement. Even several hundreds MVA of high capacity STATCOMs are realized [2]-[4].

There are several main-circuit topologies have been introduced for STATCOMs, for instance multiple converter connection with multi-stage transformer or reactors, two or three-level converter with series connection of switching devices. These topologies require bulky transformers or ac power filters. Recently, modular multi-level cascade converters become attractive solution to high capacity STATCOM and HVDC since it has features of flexibility of capacity design because of multi-cell configuration and applicability of ac filter-less design by sinusoidal output voltages [5]~ [8].

The modular multi-level cascade converter consists of series connected modular cells for three phase legs configuring star or delta connection. Each cell contains DC capacitor independently without any power sources. Therefore, capacitor voltage must be controlled inherently with the topology. Due to such a three-phase configuration using single-phase converters, it is difficult to maintain capacitor voltage under unbalanced voltage conditions in a power system. A voltage imbalance occurs during an AC system fault which may cause sudden change of active power and deterioration of capacitor voltage level in each phase. The capacitor voltage control should keep the level especially during such a fault.

Previous works proposed control method of modular multi-level converter during AC system faults such as phase balance control by circulating current [7]. However this method requires high current ratings. This paper describes novel capacitor voltage control which adjusts output negative-sequence current component of the converter in addition to circulating current during unbalanced AC system faults. Performance of the proposed method is verified by experiments using a 5 kVA system and simulation results.

II. SYSTEM CONFIGURATION

A. Circuit Configuration

Figure 1 shows a circuit configuration of the delta-connected MMC-STATCOM. Each cell is a single-phase H-bridge voltage source converter using IGBTs. A single phase leg consists of n cells and a reactor in series. The three-phase output terminals of delta-connected legs are connected to the power system via tie-transformer.

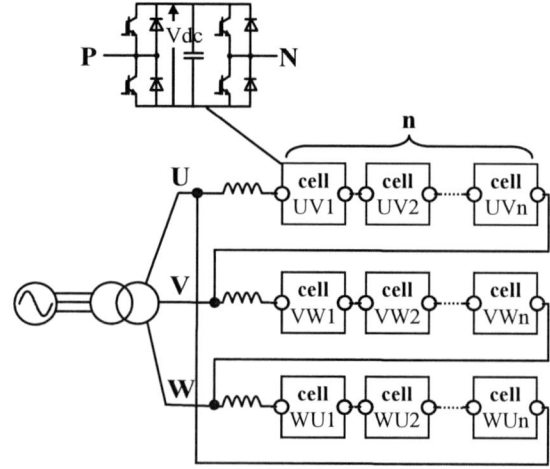

Fig. 1. Circuit configuration of the MMC-STATCOM.

The MMC-STATCOM can produce low harmonic sinusoidal output voltages. Output voltage states of each H-bridge cell are zero and two polarities of DC capacitor voltage (0, ±Vdc). With n cells connected in series the each phase leg can output (2n+1)-level voltages, which can give sinusoidal waveform to its output voltage.

Another feature of delta-connected MMC-STATCOM allows power transferring among phases by circulating current flow in order to have energy balancing. A grid imbalance causes unbalance of capacitor voltage level among phases. Since the circulating current can be controlled independently, it can be used for capacitor voltage phase balance control. Thus the delta-connected configuration is suitable for STATCOM.

B. Control Configuration

A basic control scheme is shown in Fig. 2. Considering AC grid voltage stabilization which is typical role of the STATCOM, AC voltage control (AVR) is introduced in the scheme. In addition to the AVR, the control scheme of MMC-STATCOM includes output and circulating current controls and three kinds of DC capacitor voltage controls. The three DC capacitor voltage controls are average voltage control, phase balance control and individual voltage control. Further details of each control are described below.

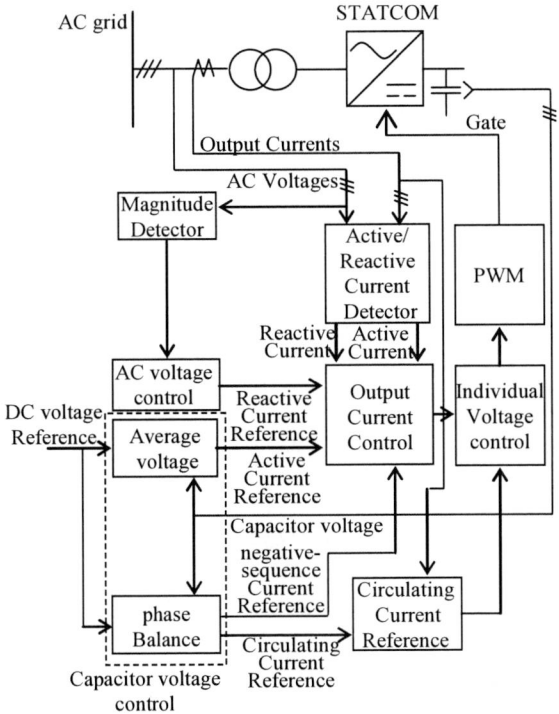

Fig. 2. Block diagram of proposed control method for MMC-STATCOM.

The average voltage control maintains the average voltage of all cells' capacitor. The average voltage is obviously controlled by adjusting the active component of the output current because it depends on total active

power flow of the STATCOM. The phase balance control keeps voltage balance among phases. The voltage balance is controlled by adjusting the circulating current and the output negative-sequence current in proposed control scheme as shown in Fig. 2. The output active and reactive currents are controlled using rotational synchronous reference frame to the AC grid voltage as usual manner for voltage-source converters. The reactive current is adjusted by the AVR. The output current is controlled by adjusting AC components of converter voltages. The circulating current can be controlled by adjusting the zero-sequence or common-mode of converter voltages since the circulating current is the common component in three phase currents flowing in delta-connection with reactors in phase legs.

III. CONTROL METHOD FOR UNBALANCED AC SYSTEM FAULTS

Since cascaded STATCOM has a three-phase configuration using single-phase converters, it is difficult to maintain capacitor voltage under unbalanced voltage conditions in a power system. A voltage unbalance fault may cause deterioration of capacitor voltage level in each phase as mentioned above. Because a STATCOM is required continuous operation under such a condition, the STATCOM control should keep the level of capacitor voltage especially during unbalance faults. Previous works proposed control method of modular multi-level converter during AC system faults, for instance, phase balance control by circulating current. However, the method using circulating current requires excessive current flow in the MMC-STATCOM under severe unbalance fault conditions. In proposed method, in addition to circulating current, negative-sequence current of MMC-STATCOM output is used in order to reduce amplitude of the phase leg currents.

It is easy to understand by using vector diagram that effectiveness of negative-sequence output current to phase balance control. First the case of no phase balance control will be explained. Figure 3 shows vector diagrams of AC grid voltages and STATCOM output currents. Figure 3(a) shows vector diagram during balanced voltage condition. It can be seen that the phase difference of each phase current and each voltage is 90 degrees. Since active power of each phase is zero with 90-degree phase difference between current and voltage, capacitor voltage does not change. The vector diagram shown in Fig. 3(b) shows a case of unbalanced voltage condition due to an asymmetrical AC power system fault, such as a single-phase to ground fault (1LG). In the phase UV or WU, the phase difference between current and voltage is not 90 degrees, in case of the same positive-sequence output current as before the fault. Thus active power is not zero in each phase of UV or WU in which the capacitor of phase UV will be discharged and that of phase WU will be charged.

Next the method using circulating current will be explained. Figure 3(c) shows a vector diagram when the

3579

MMC-STATCOM outputs positive-sequence current and controls circulating current at the 1LG. The positive-sequence reactive current is the same as the cases of Fig. 3(a) and (b). In order to keep 90-degree phase difference between current and voltage, the circulating current should be controlled as shown in Fig. 3(c). The current vectors of phases can be derived by adding the positive-sequence current vectors and the circulating current (zero-sequence component) vector. Then each phase current is given by (1).

$$I_{uv} = I_{wu} = \sqrt{3}I_{pos} \, ,$$

$$I_{vw} = 0 \qquad\qquad (1)$$

Where I_{pos} is a positive-sequence current. Hence phase currents of phase UV and WU are 1.7 times of positive-sequence component, rated current must be designed large enough to operating during unbalanced AC power system fault.

Finally the method using both circulating current and negative-sequence output current will be explained. Figure 3(d) shows a vector diagram when the STATCOM outputs positive-sequence current and controls the negative-sequence output current and the circulating current to have 90-degree phase difference between current and voltage. The control method using negative-sequence current allows much smaller phase current than the method using only the circulating current. Therefore the rated current can be designed much smaller as well.

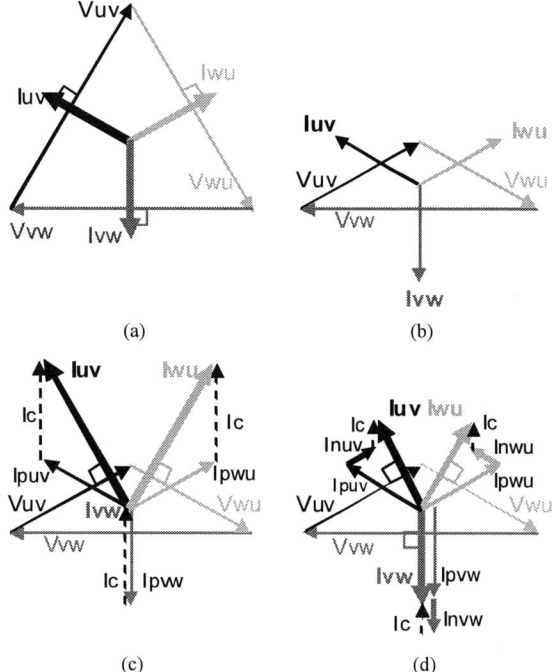

(a)

(b)

(c)

(d)

Fig. 3. Vector diagrams of AC grid voltages and output currents. (a) balanced condition; (b) positive-sequence current output during 1LG; (c) positive-sequence and circulating current output; (d) positive-, negative-sequence and circulating current output; Vuv, Vvw, Vwu: line-to-line voltages; Iuv, Ivw, Iwu: phase leg currents; Ipuv, Ipvw, Ipwu: positive sequence of output currents; Inuv, Invw, Inwu: negative sequence of output currents; Ic: circulating current.

IV. EXPERIMENT

A. Configuration

Performance of the proposed method is verified by the 5 kVA MMC-STATCOM with eight cells in each phase leg. Configuration of the experimental system is shown in Fig. 4. As an AC power system, voltage controllable three phase 200V-AC power supply is used by which unbalanced voltage can be applied to the MMC-STATCOM. The output terminals of delta-connected legs are connected to the AC power supply via reactors which are modeling impedances of a transformer. Three phase voltages and currents of the power supply, currents of the legs and capacitor voltages are all measured and sent to the digital signal processor (DSP) system. The gate signals of switching devices are generated by the DSP system and sent to the gate drivers.

Figure 5 shows a photograph of the rack for the phase legs and the controller. Top of the rack contains the DSP system as a controller, and the second to fifth rows store eight three-phase cells using H-bridge MOSFET converters. The reactors of three legs are placed in the bottom of the rack. The rated values of this system are listed in Table I.

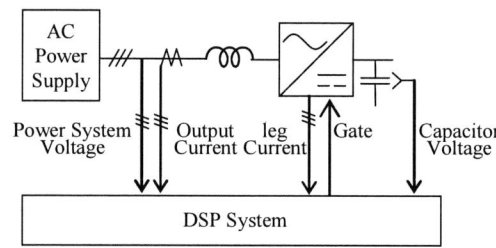

Fig. 4. Configuration of the experimental system.

TABLE I
RATED OF EXPERIMENTAL SYSTEM

Rated capacity	5 kVA
Rated frequency	60Hz
Rated AC voltage	200V
Rated leg current	8.33A
Number of cells	8×3

B. Experimental Results

In order to make confirmation of performance of proposed control method, a single-phase voltage-sag is demonstrated. Figure 6 shows the MMC-STATCOM response of 100% voltage sag in 10 cycles at phase U as a simulation of 1LG fault at the connected point. When 1LG fault occurred, the MMC-STATCOM started to output rated reactive power according to the AVR output to support grid voltage. The phase averages of capacitor voltages are almost balanced and stable during the fault. In transient within two cycles after the fault occur or clearing the fault, voltage imbalances are observed. Although the maximum and the minimum averaged voltages are not excessive.

3580

The 2018 International Power Electronics Conference

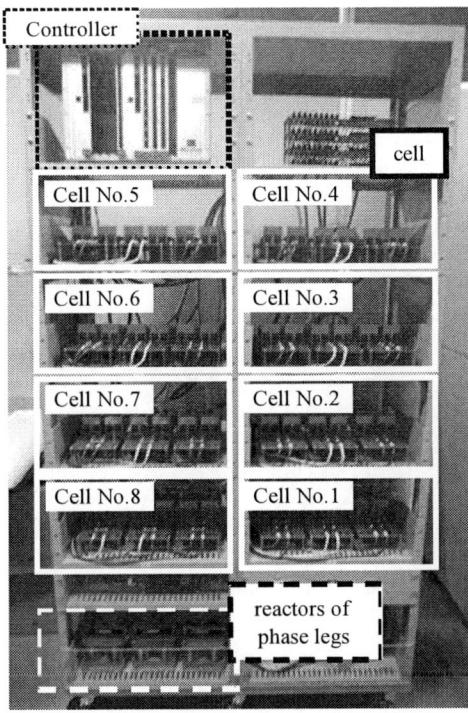

Fig. 5. A photograph of the 5 kVA experimental system of MMC-STATCOM.

The maximum current of the leg current during the fault is about 1.6 (p.u.). The steady state leg current during fault is about 1.3 (p.u.) which is less than theoretical value of 1.7 (p.u.) with the method using circulating current only as mentioned in the section III. The current includes positive- and negative- sequence components of the output current and the circulating current flowing in delta-connected legs. The positive-sequence current is about -1.0 (p.u.) as of the DC component of the capacitive reactive current (the fifth graph in Fig. 6). The negative-sequence current is about 0.4 (p.u.) which can be derived by measuring peak value of the AC components having two times of fundamental frequency in the active and the reactive current. And the circulating current is about 0.25 (p.u.) as in the sixth graph of Fig. 6.

The bottom three graphs in Fig. 6 show the capacitor voltages of all cells. Each graph shows the voltages of eight cells of each leg. It is also verified that the voltage balance among the cells is achieved by individual voltage control during the fault. The AC components of the voltages are mainly second harmonics which are naturally observed in MMC-STACOMs.

As a consequence, Fig. 6 shows that the capacitor voltages of all cells are controlled under the unbalance condition by three voltage controls; average voltage control, phase balance control and individual control. Thus, it is confirmed that the proposed control method is performing well during the voltage unbalance fault.

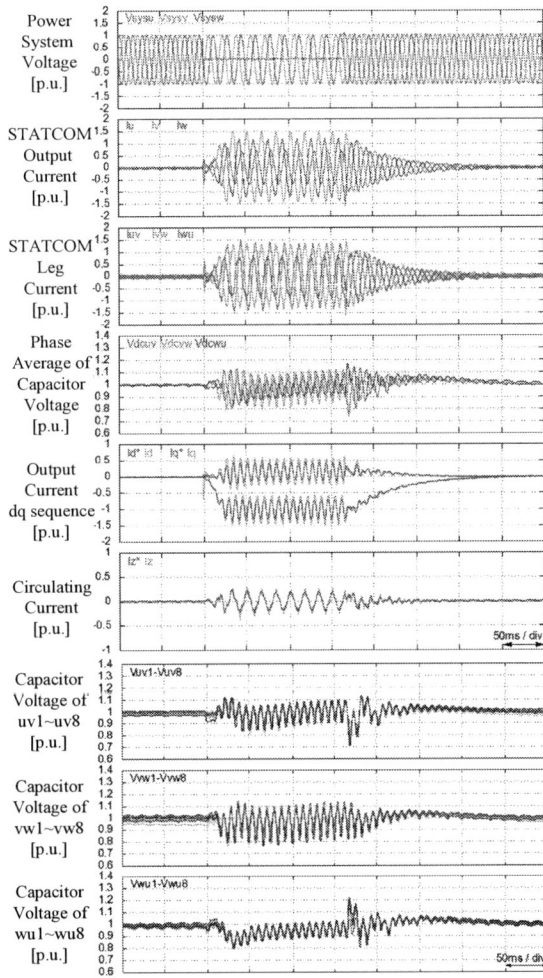

Fig. 6. Waveforms of 5 kVA MMC-STATCOM for a single-phase voltage-sag (1LG fault)

V. SIMULATION

In this section, the effectiveness of the proposed control method is also confirmed by simulation results with comparison of the conventional control method using only a circulating current to control imbalance of the capacitor voltages among phases.

A. Arm Current and Capacitor Voltage

Figure 7 shows the simulated waveforms of MMC-STATCOM response during a single-phase to ground fault. The STATCOM outputs capacitive reactive current following to its reference signal produced by the AVR during fault. The waveforms of the conventional method show the current peak of the circulating current of 1.0 (p.u.), and that of the leg current of 1.7 (p.u.) as shown in the vector diagram in Fig. 3. On the other hand, the waveforms of the proposed method show the current peak of the circulating current of 0.25 (p.u.), and that of the leg current of 1.3 (p.u.), which proves the effectiveness of the leg current reduction in the proposed method.

3581

The 2018 International Power Electronics Conference

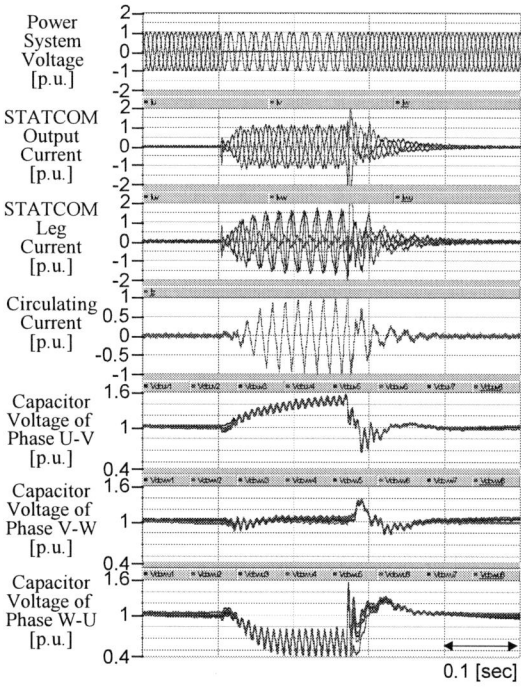

(a) Conventional Method

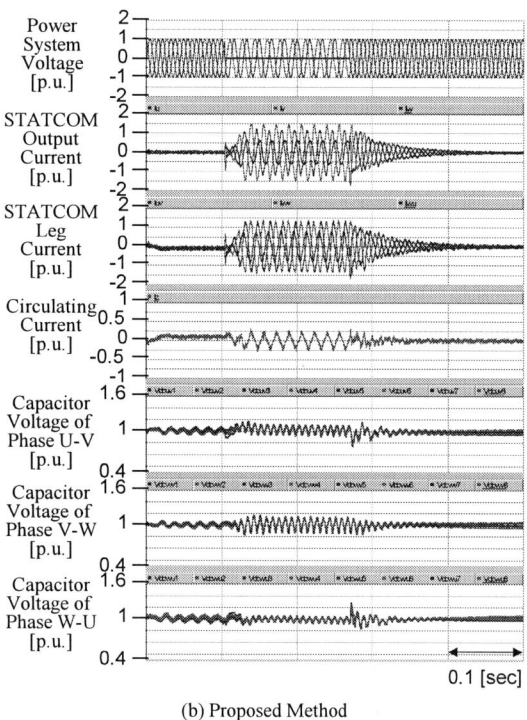

(b) Proposed Method

Fig. 7 Simulation Results.

The comparison in the waveforms of the capacitance voltages of each legs show the noticeable balancing of the proposed method. Since the current difference between before and during the fault is much smaller in the

proposed method, it is easier to control the current by using the current feedback control scheme as shown in Fig. 2 with smaller leg voltages. The result of the conventional method in Fig. 7 (a) is not the optimum parameter design, however, 1.0 (p.u.) of circulating current requires a higher voltage margin to the leg specification as well as the current ratings.

In summary, the simulated waveforms of Fig. 7 clearly show the better stability using the proposed method. Considering the ratings of phase legs and performance of control of capacitor voltages, proposed method is very attractive to the MMC-STATCOM for improvement of voltage stability or even transient stability of power systems [3].

B. Effects to Power Systems

The conventional control method has no influence to the voltage compensation operation of the STATCOM because it utilizes the circulating current that circulates solely inside of the STATCOM. On the other hand, the proposed control method utilizes the negative sequence currents in addition to the circulating currents, it may be a discussion item that is there any possibility of a negative effect to the voltage compensation of the STATCOM. Thus, the above possibility is investigated using the simulation results.

Table II shows the simulated results of the ratios of the RMS values of the compensated voltages to that of the rated voltages in case of power system short circuit ratio is 10. As shown in table 2, the proposed control method has better compensation ratio by a few percent, which indicates the proposed method contributes to maintain the system voltage better than the conventional method during the AC fault. In other words, the proposed method does not bring the negative effects to the connected power system in terms of voltage control.

TABLE II
COMPENSATED VOLTAGE LEVEL OF POWER SYSTEM

Circulating current only (conventional method)	+3%
Circulating current and Negative sequence current (proposed method)	+5%

VI. CONCLUSIONS

This paper describes the capacitor voltage control method of the MMC-STATCOM configuring delta connection. The proposed method of the phase balance of cell capacitor voltages adjusts both the negative-sequence output current and the circulating current during unbalanced AC power system fault. With the control method, leg current can be reduced reasonable level compared to the conventional method using circulating current only. Further the performance of the voltage control is quite stable during the fault, therefore the proposed method is attractive to the MMC-STATCOM for power systems. The 5 kVA experimental results and comparative simulation results show that the proposed method is effective during unbalanced AC power system fault.

REFERENCES

[1] S. Mori, et. al., "Development of A Large Static Var Generator using Self-Commutated Inverters for Improving Power System Stability", IEEE Trans. on Power Systems, Vol. 8, No. 1, Feb., 1993.

[2] D. J. Hanson, C. Horwill, J. Loughran, D. R. Monkhouse, "The Application of a Relocatable STATCOM-based SVC on the UK National Grid Sytem", IEEE PES Transmission and Distribution Conference and Exhibition, pp1202-1207, Vol. 2, 2002.

[3] T. Imanishi, Y. Nagatomo, S. Iwasaki, K. Masaki, T. Fujii and J. Ieda, "130MVA-STATCOM for Transient Stability Improvement", Conference Record of IPEC-Hiroshima, 2014.

[4] T. Fujii, K. Temma, N. Morishima, T. Akedani, T. Shimonosono and H. Harada, "450MVA GCT-STATCOM for Stability improvement and Over-Voltage Suppression", International Power Electronics Conference, pp. 1766-1772, 2010.

[5] M. Pereira, D. Retzmann, J. Littoes, M. Wiesinger, G. Wong, "SVC PLUS: An MMC STATCOM for Network and Grid Access Applications", *IEEE Trondheim PowerTech*, pp. 1-5, 2011.

[6] J.D.Ainsworth, M. Davies, P. J. Fits, K.E. Qwen, D.R. Trainer, "Static VAr compensator (STATCOM) based on single-phase chain circuit converters," *IEEE Proc.-Gener. Transm. Distrib*, vol. 145, no. 4, pp. 381-386, 1998.

[7] M. Hagiwara, R. Maeda, H. Akagi, "Negative-Sequence Reactive-Power Control by a PWM STATCOM Based on a Modular Multilevel Cascade Converter (MMCC-SDBC)" IEEE, pp. 3728-3735, 2011.

[8] H. Akagi, S. Inoue and T. Yoshii, "Control and Performance of a Transformerless Cascade PWM STATCOM with Star Configuration", IEEE Trans. on Industrial Applications, Vol. 43, No. 4, pp1041-1049, 2007.

SiC based power semiconductor in applications - aspects and prospects

Peter Friedrichs

Infineon Technologies AG, Schottkystrasse 10, D-91058 Erlangen, Germany
E-mail: peter.friedrichs@infineon.com

Abstract: **SiC based power devices become more and more accepted in a wide range of applications. While diodes are in the field since more than 15 years in the recent years transistors are just entering the field. The contribution will focus on two aspects – on the one hand side the design of SiC components in order to match application needs and allow easy to use plus quick design in. On the other hand it will be shown on related examples how the today still more expensive technology can be beneficially utilized in existing applications in an economically and technically attractive manner.**

I. INTRODUCTION

Outstanding material properties of SiC expose this material besides GaN as an ideal semiconductor for power devices enabling new horizons in power density and efficiency:

• A critical breakdown field strength in the area of >2MV/cm allows the design of fast unipolar devices at several kV blocking capability with still very attractive on resistance

• The bandgap of 3.2eV leads to negligible intrinsic carrier density even at device temperatures of several hundred degrees centigrade.

The world market of SiC power devices today is already in the range of several 100 Million $ and driven by diode technologies which are on the market since 2001. Switch technologies suffered so far from weak performance (Bipolar Junction Transistor - BJT), questionable robustness (planar Melat Oxide Semiconductor Field Effect Transistor -MOSFETs) or inconvenient device operating modes (Junction Field Effect Transistor - JFETs).

For MOSFET based devices significant progress was achieved in the recent years. Trench based modern device concepts allow an excellent reliability in combination with very attractive performance as explained in the next chapters.

Practical use of SiC devices is still dominated by diodes and thus, the first section will briefly describe the major device technologies and its application impact.

In the 2nd part of the contribution the focus will be on SiC MOSFET designs and applications

II. SIC DIODE TECHNOLOGY AND USE CASES

When SiC diodes have been commercially released in 2001 (600V and 1200V class) they have been initially constructed as pure Schottky diodes. Besides the fact, that they allowed for the 1st time in these voltage classes virtually zero reverse recovery loss commutation (see figure 1) those devices had deficiencies with respect to overload capability [1]. For their initial key application (PFC), this ended up with an unusual over-sizing of the diodes in order to cope with related failure modes.

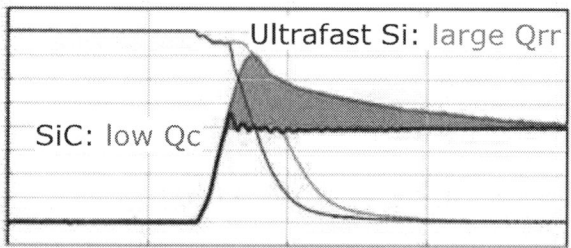

Fig 1 : Advantage of SiC diodes over Si pin diodes – reverse recovery free turn off

This item was solved only few years later by the roll out of so called MPS (merged-PIN-Schottky) rectifiers [1]. These devices are pure majority carrier devices in normal operation, but allow minority carrier based resistivity modulation in surge current operation (> 5000A/cm² for 10ms pulse time for 650V devices, see figure 2). This allowed a significant shrink of the device size and respective added reliability in the systems.

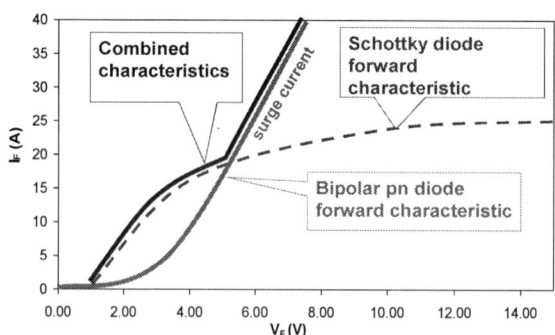

Fig 2 : Surge current behavior of MPS diodes compared to standard SBD's

Meanwhile nearly all available SiC diodes are based on the MPS or JBS (Junction-Barrier-Schottky) concept. They are considered as the best concept for voltage classes up-to 1.7kV. At higher voltages the SBD concept becomes more and more unattractive due to high differential on-resistance and thus, the need for large areas. PN based diodes are suffering from a higher knee voltage (2.7V for pn vs ~1V for MPS). Thus, at least for today it turned out that above 2kV the concept of integrated diodes (body diode) is more favorable than designing separate devices.

Several measures like thin wafer technology, smart assembly techniques and optimized cell layouts made SiC diodes an affordable choice for all compact and fast switching applications. The cost ratio towards fast Si diodes is meanwhile clearly below 3. Impact of diodes is traditionally focused on PFC application, however, over the recent years solar power conversion systems, mainly the booster stages, benefit significantly from the added value coming along with SiC diodes. Currently in more or less all new 1200V components based designs the booster diode is selected from the scope of SiC SBD's. But also in other areas significant added value can be derived. One example was demonstrated in a European funding project called SPEED. The teams showed that by implementing 1700V SiC diodes as freewheeling diodes for IGBTs instead of the standard silicon based pin diode the filter size of the wind power converter system could be drastically reduced and thus, one third of the system volume could be eliminated (see figure 3).

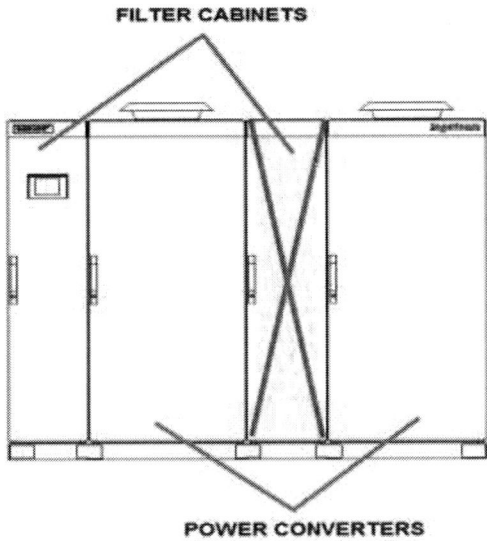

Fig 3 : Potential to reduce the system dimensions by a simple replacement of Si freewheeling diodes with SiC based components

The final paper will give a few more typical examples for the beneficial implementation of modern silicon carbide Schottky barrier diodes, e.g. in power supply applications or in the already mentioned solar power conversion systems.

III. SIC POWER SWITCHES

Si as well as SiC has a native thermal oxide which is an ideal prerequisite to design devices based on a high quality oxide interface like MOSFETs. In the case of SiC, however, there are some obstacles when realizing a powerful and robust SiC MOSFET. Defects are located directly at and close to the interface and cause significant lower field effect channel mobility due to a much higher density of interface states compared to Si, mainly at the planar silicon surface which is commonly used for device processing. Furthermore, as SiC devices allow roughly 10 times higher electric fields than their Si counterparts, the electric field in the gate oxide has to be limited in order to maintain a required reliability of the device.

The upper mentioned difficulties limit the options to design a powerful and reliable component in the planar case and thus, trench based structures seem to be superior with respect to performance (see figure 4).

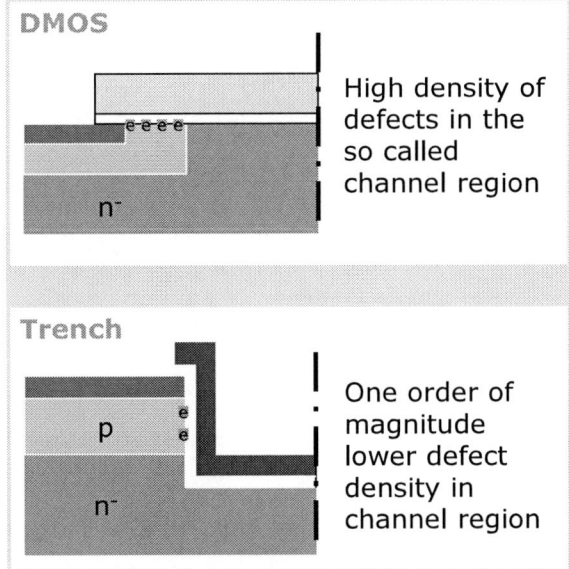

Fig 4 : Major difference in the interface quality between planar (DMOS) and trench based SiC MOSFETs.

The recently introduced CoolSiC™ MOSFET is a trench SiC MOSFET based on a novel asymmetric concept [2].

In this trench concept, only one side of the groove serves as MOS channel which is well aligned to the preferred crystal plane by a dedicated process. This favorite crystal plane is seen as essential for getting a minimum of interface states. At the same time, all these measures minimize the amount of threshold voltage shift with temperature and time and thus, have a positive impact on the device reliability. Since now the electric field required to open up the channel and make it conductive enough is much lower than in case of the planar concepts. The on-state ruggedness of the oxide can be designed according to levels known from modern silicon power devices. The oxide thickness is designed to

TABLE I.   KEY PARAMETERS OF THE COOLSIC™ MOSFET

Parameter	Value	Unit	Condition
$R_{DS(on),\ typ}$	45 (75)	mΩ	Single die, T_j = 25°C (175°C) I_D = 20A, V_{GS} = 15V
V_{DSS}	> 1200	V	-55°C < T_j < 175°C
V_{GS}	-5 / +15	V	recommended range
V_{GS}	-10 / +20	V	maximum rating
V_{GSth}	4.5 (3.8)	V	T_j = 25°C (175°C) I_D = 10mA, V_{GS} = V_{DS}
V_{SD}	3.3 (3.1)	V	T_j = 25°C (175°C) I_D = 20A, V_{GS} = 0V

keep the electric field low enough for the commonly used on-state gate source voltage of V_{GS} = + 15V.

Deep p-wells are implemented in order to protect the gate oxide at the bottom and at the corners of the trench against high electric fields. These p-type regions are intended to function also as emitters for the body diode. Therefore, an external freewheeling diode is no longer needed. In fact, with the option to open the channel in freewheeling mode (synchronous rectification) even lower losses compared to an SBD can be achieved due to the lack of a knee voltage in the I-V curve. This cell construction inherently owns a favorable small ratio of the Miller charge Q_{GD} related to the gate source charge Q_{GS}. This enables well-controlled switching with very low dynamic losses [3]. In particular this ratio is essential to limit unwanted losses caused by a parasitic turn-on in topologies using half bridges.

The cell design further enables the implementation of a certain short circuit capability. The JFET region lowers the saturation current of the device by adjusting the distance between the p-type regions. A distinct and for the application very important feature of the presented MOSFET design is the controllability of the voltage slopes during switching. The CoolSiC™ MOSFET can be easily controlled via a simple gate resistor during turn on and turn off as shown in figure 5.

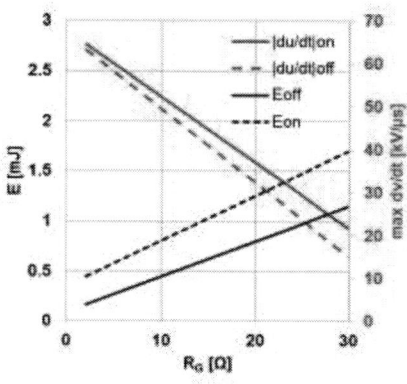

Fig 5 : Controllability of the switching speed (dv/dt) via

the gate resistor R_G for the CoolSiC™ MOSFET

SiC MOSFETs are prone to enable very fast transients in the range above 50kV/µs. However, in a number of use cases it is desirable to adjust the speed, either because driven loads cannot sustain high dv/dt (motor drives) or EMI problems have to be addressed.

III.   Application examples for SiC transistors

Beside concerns about the reliability of the new components on of the major obstacles to implement SiC in a broader manner in applications is the comparably high cost vs. existing silicon based IGBT's e.g. However, meanwhile significant progress can be reported regarding the ruggedness by using trench based switches [4]. Also the cost topic can be addressed as soon as system cost aspects are considered (including saving potential enabled by SiC for example by shrinking magnetic components). Figure 6 compares the situation in a schematic way.

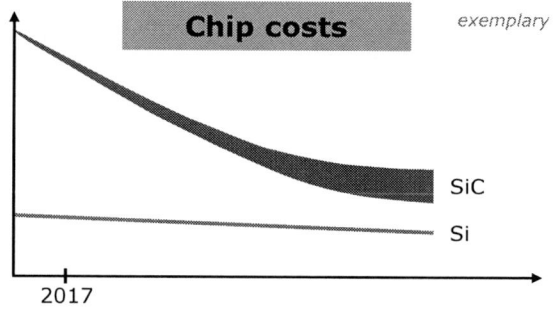

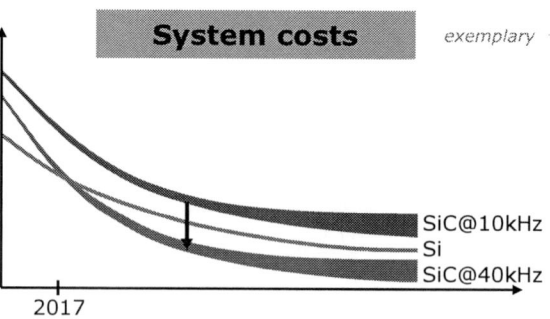

Fig. 6: Cost comparison between silicon and SiC based products – upper picture just assuming pore chip cots, lower picture shows the situation at system level for different switching frequencies

It is visible that the gap at system level is much smaller, and mainly if higher frequencies in the range above 20kHz are assumed already today the semiconductor cost adder can be easily compensated, leading to an actual saving for the supplier of power systems in its bill of material. Thus, SiC transistors are at the hockey stick in adoption in applications benefiting from such operating conditions. Prominent examples are solar string inverters or switch mode power supplies. In other applications the role of reduced losses and the connected benefit of a reduced cost of ownership due to energy saving over the life time of the product can be leveraged. One example is the use of SiC based switches in motor drives. Sahan et al. [5] showed that even under switching conditions as known from IGBT's (low frequency and reduced dv/dt) a significant reduction in power losses compared to silicon based systems can be achieved. Figure 7 summarizes the findings.

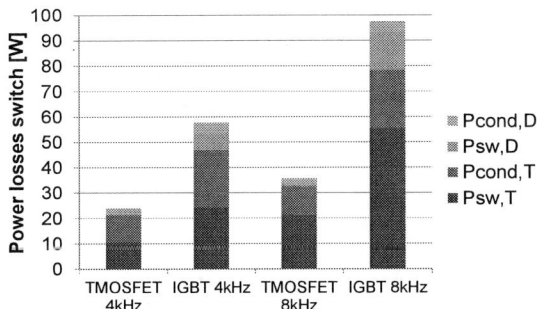

Fig. 7: Comparison of power semi losses between IGBT's and SiC MOSFETs in a motor drive application

Evidently the loss reduction arises from both, lower conduction and lower switching losses. The latter finding seem snot to be evident on a first glance since slopes and frequency are the same, however, still reduced turn on overshoot due to lack of reverse recovery and the lack of tail currents lead to reduced switching losses even under IGBT like switching conditions (see figure 8).

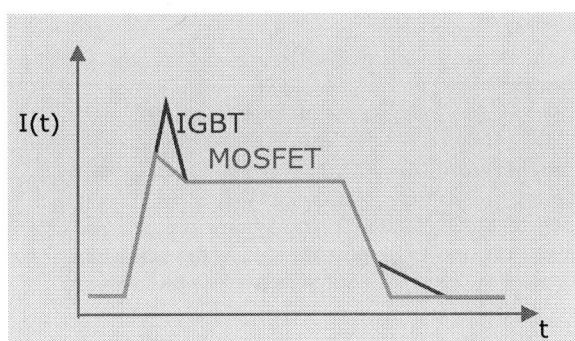

Fig. 8: Reduced current overshoot at turn on and lack of tail currents in turn off lead to switching loss advantages for SiC MOSFETs compared to IGBT'S even at the same dv/dt and frequency

In the final contribution more details presented and further applications will be analyzed with respect to the applicability of SiC based switches either on short term or on long term scale. Aspects like driving the components with suitable circuits or IC's as well as the relevance of smart board designs and the suitable packaging technology will be explained as well. Finally, an outlook will be given regarding SiC transistors at very high blocking voltages.

IV. SUMMARY

In summary it was shown that silicon carbide based components are ready to enter various application seven despite the higher cost position since sufficient saving potential exists at system level. However, an application specific assessment is mandatory and the corresponding device technologies need to be designed in order to meet various application needs besides a pure performance. SiC diodes are already well established components in power supplies and solar power conversion systems, as a next steps with significant impact the introduction of SiC transistors is seen. For their success it is mandatory to merge performance, ruggedness and ease of use. One example fulfilling those criteria is the new CoolSiC™ MOSFET from Infineon. Several examples show that such a component can be cost performance wise effectively implemented in applications, mainly if higher frequencies are considered or if cost of ownership in the final application are major driving forces.

ACKNOWLEDGEMENT

The author would like to thank the whole Infineon team working on the silicon carbide technology for contributing to the presented results. Parts of the work were funded by the German Government under various contracts and the European Union as well.

REFERENCES

[1] R. Rupp, M. Treu, S. Voss, F. Dahlquist, T. Reimann, "2nd Generation SiC Schottky diodes: A new benchmark in SiC device ruggedness", Proc. ISPSD 2006.

[2] D. Peters, R. Siemieniec, T. Aichinger, T. Basler, Romain Esteve, W. Bergner, D. Kueck, "Performance and Ruggedness of 1200V SiC Trench MOSFETs" Proc. ISPSD 2017 to be published

[3] D. Heer, D. Domes and D. Peters, "Switching performance of a 1200 V SiC-Trench-MOSFET in a Low-Power Module", Proc. PCIM Nuremberg 2016

[4] M. Beier-Möbius, J.Lutz, „Breakdown of gate oxide of SiC-MOSFETs and Si-IGBTs under high temperature and high gate voltage", PCIM Europe 2017, 16-18May, Nuremberg. - Berlin Offenbach : VDE Verlag GmbH, 2017, S. 365-372

[5] B.Sahan, A.Brodt, D.Heer, U.Schwarzer, M.Slawinski, T.Villbusch, K.Vogel, "Enhancing Power Density and Efficiency of Variable Speed Drives with 1200 V SiC T-MOSFETs" PCIM Europe 2017, 16-18May, Nuremberg. - Berlin Offenbach : VDE Verlag GmbH, 2017, S. 196-203

Electromagnetic Modeling Approaches Towards Virtual Prototyping of WBG Power Electronics

Ivana Kovačević-Badstübner[1], Daniele Romano[2], Giulio Antonini[2], Jonas Ekman[3], Ulrike Grossner[1]

1 Advanced Power Semiconductor Laboratory, ETH Zurich, Switzerland
2 University of L'Aquila, L'Aquila, Italy
3 Luleå University of Technology, Sweden
*E-mail: kovacevic@aps.ee.ethz.ch

Abstract—**High frequency power electronics utilizing wide-band gap semiconductor devices imposes more stringent requirements for highly accurate extraction of parasitics of power electronics systems in a wide frequency range. This paper presents the state-of-the-art modeling approaches used to predict the electromagnetic behavior of power electronic systems and components in terms of accuracy and computational cost. The potential of the Partial Element Equivalent Circuit (PEEC) technique for virtual prototyping of power electronic systems is assessed. The main advantage of this numerical technique is its capability for direct coupling between the circuit and electromagnetic domains provided by the PEEC meshing of three-dimensional geometries in partial elements. The aim of this paper is to provide a more comprehensive understanding of PEEC-based modeling for power electronics packaging.**

Keywords—*parasitic extraction, PEEC, multiphysics modelling, electromagnetic modelling*

I. INTRODUCTION

Emerging wide-band-gap (WBG) semiconductor technologies, such as silicon carbide (SiC) and gallium nitride (GaN) [1], have been pushing the limits of silicon (Si)-based power electronics (PE) towards higher power densities and higher energy conversion efficiency [2]. As a result, the design of power electronic converters becomes an even more challenging task by increased coupling between electrical, thermal, electromagnetic, and mechanical aspects. Hence, a system has to be carefully optimized in order to fully benefit from the new technology. Accurate virtual prototyping can potentially enhance the design of power modules and converters from the trial-and-error approach to higher levels that allows PE engineers to gain a deep insight into the actual system behavior prior to the final hardware implementation stage. This can reduce both development time and cost. With the increase of the computational power of todays personal computers, the extensive multi-physics modeling of PE systems becomes feasible, and hence, it has been integrated in the design process of power electronic components and systems. Accordingly, multiphysics modeling as an engine for virtual prototyping has gained in importance in the last decades. This led to the development and improvement of specialized numerical techniques and methods being employed for modeling of different properties of power electronic systems, including temperature, electrical characteristics, electromagnetic interference, and reliability [3]–[7]. A main challenge in

TABLE I. A SELECTION OF COMMERCIAL TOOLS FREQUENTLY USED FOR PARASITIC EXTRACTION IN POWER ELECTRONICS.

Tool	Vendor	Numerical technique(s)	References
Q3D Extractor	ANSYS	BEM, FEM	[10]–[14]
MWS	CST	FIT	[15], [16]
Momentum	Keysight	MoM	[17], [18]

developing a virtual prototyping platform for PE is to achieve the required modeling accuracy as well as an acceptable computational time when coupling different modeling domains.

Together with thermal and mechanical aspects, the electromagnetic (EM) behavior of a power module or system represents an important characteristic and determines its compliance with Electromagnetic Compatibility (EMC) standards. Therefore, it should be properly addressed within the modeling approach by including non-ideal electrical properties such as stray inductances, resistances, mutual inductive and capacitive couplings, and self-parasitics of components. The coupling between electric and electromagnetic domains as a constitutive part of virtual prototyping in PE is, accordingly, closely related to the procedures for the extraction of parasitics in electrical circuits. Novel packaging technologies for advanced power modules housing WBG devices also impose more stringent requirements for highly accurate extraction of parasitic inductances, resistances, and capacitances. Particularly, a modeling error in the estimation of the loop inductance of more than 10 % has a much higher impact for fast switching WBG power electronics applications [8], [9] than it is the case of Si-based power electronics. Accordingly, this paper identifies the modeling challenges for PE applications, and presents the requirements for more accurate EM modeling of advanced power converters utilizing WBG power devices.

II. PARASITICS EXTRACTION FOR WBG POWER ELECTRONICS: LITERATURE SURVEY

The simple equivalent models of power converters can correctly describe their electrical functionality but lack in accuracy in the high frequency (HF) range, where detailed compact models of power switching devices have to be taken into account and the parasitic effects become more pronounced. Moreover, the prediction of HF phenomena that originate from the non-ideal electrical behavior is quite difficult, and so, there has been an ever-increasing need for EM tools which can be used for the purpose of the parasitic extraction and multi-physics modelling. Predicting EM behavior of power modules, EMI filters, and power converters has been a topic of a significant number of publications, e.g. [8], [17], [19], [20]. A selection of commercial simulation tools frequently referenced in literature are summarized in Table I, together with the corresponding numerical methods employed: the Finite Integration Technique (FIT), the PEEC method, the Boundary Element Method (BEM), the Finite Element Method (FEM), and the Method of Moments (MoM). The right selection of a numerical solver depends mainly on the application and, hence, only relative (dis)advantages of the specific method can be discussed. Therefore, the commercially available simulation tools commonly implement several numerical solvers in order to cover a wider range of applications. The EM modeling in PE rather belongs to the frequency domain by its nature, so that the accuracy of the presented methods and tools from Table I was demonstrated using the test structures such as EMI filters, Printed Circuit Boards (PCBs), and power module layouts characterized in the frequency domain. On the other hand, the analysis of power electronic systems, e.g. an estimation of current/voltage ringing during the switching transients, is performed in the time domain. Here, the parasitic inductances, resistances, and capacitances are of interest rather than the transfer functions, impedance curves, and/or signal spectra as in the frequency domain. A small inductance in the range of 1 nH can significantly contribute to e.g. the over-voltage across power WBG semiconductor devices characterized by fast current slopes ($\mathrm{d}i/\mathrm{d}t$). Therefore, WBG power electronics pushes the requirements for highly accurate estimation of parasitics beyond its current stage. Moreover, when including the frequency dependent parasitic elements in the switching transients simulations, coupling of the time-domain (circuit) simulations and the electromagnetic numerical simulations has to be performed. This coupled, i.e. multi-physics, modeling implies a need for a frequency-to-time transformation. Two approaches of including EM problems in the time domain can be distinguished in practice. The first is to export the reduced order S-parameter model description or equivalent circuit of 3D geometries to a circuit simulator so that the HF behavior of the 3D structures can be simulated together with power devices and passive components in the electrical domain (i.e. circuit domain). For example, ANSYS Q3D/Simplorer and Momentum/Saber support this multi-physics approach, where ANSYS Simplorer and Synopsys Saber are two commercial circuit simulators. The second approach is to build a unique modeling environment, enabling the simulation of the HF behavior of 3D structures together with the power switches and other circuit elements. As described in e.g. [3], [21], the PEEC method is a promising numerical technique for building such a modeling environment.

While the development of an accurate modeling environment requires skills from different fields such as numerical mathematics and computational electromagnetics, a PE engineer as a user is not required to possess these skills. Rather, the understanding of the principles of a simulation tool, to some extent, can help in efficiently running simulations and gaining accurate results. The tools summarized in Table I are powerful tools. However, the design of "good" power semiconductor packages and PCB layouts is often performed as a trial-and-error process that requires great practical experience. In power electronics, there is a need to predict the parasitics in a wide frequency, from the kHz to the GHz range, considering the switching frequency and the fast rise time of switching transient signals. The development of a 3D electromagnetic modelling tool, which simultaneously can consider low- and high-frequency effects with the same accuracy, is a very challenging task. As a result, the software tools, e.g. from CST and ANSYS, are commonly specialized for either high frequency or low-frequency simulations. In the following, the commercial tools from Table I are discussed regarding their functionality for parasitic extraction and multi-physics simulations.

1) CST Microwave Studio: CST MWS enables the extraction of a S-parameter matrix representing the HF behavior of a 3D electromagnetic model. The S-parameter matrix is defined by a set of ports modeling the points in the 3D EM model for the external connection of lumped circuit elements. In this way, CST MWS enables the coupling between 3D field and circuit simulations. The built-in circuit simulator allows the import of Spice-based device models. CST MWS was shown to be a very useful tool for estimating the conducted EMI of power converters defined by the actual PCB layout. In [22], a loop inductance within an IGBT power module was estimated by a CST MWS simulation in the frequency range from 1 MHz to 1 GHz and compared with the corresponding impedance measurements. The comparison shows a decreased accuracy of the simulation results in the lower frequency range from 1 MHz to 10 MHz. The work of [22] does not fully demonstrate the capabilities of CST MWS for modeling of power semiconductor packages, since the geometry was rather simplified and the mutual couplings were not taken into account. Accordingly, CST MWS is considered a powerful tool for multi-physics modeling of power converter circuits and PCB layouts in the HF range, above $\approx 10\,\mathrm{MHz}$. However, it is not the best modeling tool for the extraction of parasitics in the full frequency range. Furthermore, a more complete CST MWS modeling of the current loops inside of power modules including all inductive and capacitive couplings is missing in the literature.

2) Keysight Momentum: Momentum is a 3D planar full-wave and quasi-static EM solver used for RF passives, high-frequency interconnects, and parasitic modeling. Similar to CST MWS, Momentum allows determining the frequency-dependent parasitic elements of interconnections by means of a S-parameter matrix, which has to be transformed to an impedance (Z)-matrix in order to couple the EM simulation to a time-domain circuit simulation. In [17], a PCB layout of a buck converter was modelled using a Momentum-Saber and a InCa3D-Saber coupled simulation. The authors presented the comparison between the measured input impedances of three selected terminals and the corresponding Momentum and InCa3D simulation results, showing a limited modelling accuracy in the frequency range from 20 kHz to 30 MHz. They observed that the transformation from the S- to the Z-matrix introduces an additional error. Accordingly, the direct calculation of the parasitics in a form of an equivalent circuit, i.e. a direct calculation of resistances R, inductances L, and capacitances C, is preferable in terms of accuracy. To the best knowledge of the authors of this paper, a 3D EM modeling of the current paths inside of power semiconductor packages determined by the geometries with the thickness much higher than the thickness of PCBs using Momentum has not been demonstrated in the literature.

3) ANSYS Q3D Extractor: The ANSYS Q3D Extractor has been frequently used by PE engineers for the extraction of parasitics in various PE applications [10]–[14]. As a quasi-static 3D EM solver, ANSYS Q3D Extractor calculates R, L, and C matrices, which can be used directly to generate an equivalent circuit model allowing a circuit-electromagnetic coupled modeling. The Q3D model of a 3D structure containing the information on its EM behavior can be imported in the ANSYS Simplorer either as a S-parameter model or as an equivalent circuit for multi-domain simulations. The calculation of partial inductances, resistances and capacitances in Q3D Extractor is based on the assumption of equipotential excitation ports, similar to the modeling approach presented in [23]. The modeling limitations due to this assumption and the resulting accuracy of Q3D Extractor are summarized in [24] for the examples of a TO-247 package and a planar half-bridge all-SiC power module. Similar to Q3D, the modeling approach described in [6] uses Π- and L-equivalent circuits for modeling electrical parasitics, and hence, does not model the actual distributed HF capacitive behavior of conductors. This implies that the accuracy of these modeling tools dedicated for multiphysics modeling has to be further verified for WBG-based power electronics.

With respect to exploring new numerical techniques for accurate 3D electromagnetic modeling, capable of overcoming the existing limitations, the PEEC method represents a promising numerical technique as shown in [3], [19]–[21], [25]. The PEEC-based software tool InCa3D used in e.g. [17], [20], [26]–[28] and the PEEC-solver from [3] cannot calculate distributed capacitive effects of 3D geometries. A comprehensive demonstration of the accuracy of these tools for the parasitic extraction of power semiconductor packages is missing as well. In summary, a PEEC-based tool for multi-physics modeling, which can take into account all design aspects (resistive, inductive, capacitive and thermal couplings) [25], [29] is still under research.

The aim of this paper is to demonstrate the advantage of the PEEC method for the EM modelling of PE geometries, such as PCBs and power semiconductor packages in terms of accuracy required for WBG-based power electronics. Furthermore, the existing challenges of highly accurate PEEC-based modeling in a wide frequency range are addressed in Section III-A and possible solutions how

to improve the PEEC method in terms of computational cost are summarized in Section III-C.

III. PEEC METHOD FOR POWER ELECTRONICS

The PEEC method is derived from the integral formulation of Maxwell's field equations similar to MoM [30]; however, in contrast to MoM, the PEEC method provides a circuit interpretation of the electric field integral equation (EFIE) and the continuity equation [31] in terms of partial elements, namely resistances, partial inductances and coefficients of potential. The resulting equivalent circuit can be then analyzed in both time and frequency domain in a circuit environment such as SPICE-like circuit solvers [32]. The PEEC method was originally developed for the extraction of parasitic inductances and resistances inside integrated circuits. Over the years, different PEEC formulations have been developed: quasi-static and full-wave formulations [33], including non-linear and dispersive magnetic materials [34]–[36], ideal, lossy and dispersive dielectrics [37], [38], magneto-dielectric materials [39], anisotropic dielectrics [40], silicon [41].

Compared to differential equation-based EM numerical techniques, which require a boundary region to be defined around the conductors being modelled, the PEEC method and MoM as integral equation-based methods require the discretization of the volumes and surfaces containing the sources of the electric field only, namely current and charge densities. Accordingly, while differential EM numerical techniques used in computational electromagnetics produce sparse matrices, a PEEC mesh leads to a dense system matrix containing the information about all inductive and capacitive couplings. This implies a significant computational cost for highly accurate modeling of HF effects in conductors, such as the skin and proximity effects, as it will be explained in the following sections.

A. PEEC Modelling Challenges

The first PEEC solvers, e.g. FastHenry [42], were based on the assumption that the current direction is known in advance, as it can be the case for RF circuits and the interconnections in small IC packages. However, this modeling assumption can produce a significant error for geometries that are more complex such as power module packages and/or multi-layer PCBs used in power electronic circuits. In the frequency range of interest for modern PE applications, the current has to be represented as a 3D vector in order to accurately capture the skin and proximity effects, which significantly increases the number of unknowns, and thus, the computational cost. Therefore, most efforts are directed nowadays towards acceleration of the PEEC solvers [43]–[46] in order to allow the analysis of more complex circuits in a wide-frequency range, which will be discussed in more detail in Section III-C. Moreover, PE applications, where the electric field effects – i.e. capacitive couplings – cannot be neglected, require a modeling tool, which can simultaneously take into account Ohmic losses (R matrix), as

3590

well as magnetic (L matrix) and electric (C matrix) field effects. The computational power of the PEEC method also relies on the usage of analytical formulas for fast calculation of the L and P matrices. Namely, the standard PEEC modeling is based on the volume discretization of conductors into orthogonal/non-orthogonal PEEC volume cells. Using an uniform mesh for capturing the skin-effect easily may lead to a huge number of unknowns. This issue can be mitigated by resorting to a non-uniform mesh [33], which was shown also in [47]; however, the adoption of a non-uniform mesh may easily generate PEEC volume cells with high aspect ratios (thin and long cells) for which the analytical formulas can be affected by significant numerical errors.

The PEEC method has been used in PE especially to capture the inductive effects [48], [49]. The resulting PEEC models are typically very large. Thus, demanding storage and computing resources are easily required depending on the problem of interest [50]. To enable multi-physics PEEC modelling, there is a strong need to compress the models and make them integrable in circuit environments [3], [51]. Different options for speeding up the PEEC modelling are further described in Section III-C.

B. Advantage of the PEEC Method

The advantage of the PEEC method will be shown on the example of two test PCBs and a T0-247-3 package housing a single $80\,\Omega$ $1.2\,kV$ SiC power MOSFET. The PCB test structures are used to show the modelling capabilities of a specifically developed (R, L, P) PEEC method for capturing inductive and capacitive effects with a higher accuracy. The T0-247-3 package example demonstrates the (R, L, P) PEEC modelling of a distributed commutation loop inductance inside of a power semiconductor package overcoming the modelling limitations of the ANSYS Q3D Exractor.

In this paper, the so-called (R, L, P) quasi-static PEEC formulation is used, simultaneously including Joule losses, and inductive and capacitive effects in the frequency range from $1\,kHz$ to $200\,MHz$, which corresponds to the signal rise time of $1.75\,ns$. The rise/fall switching times of commercial SiC power MOSFETs is in the range from $5\,ns$ to $10\,ns$, while for GaN HEMTs is in the range from $\approx 1.5\,ns$ to $5\,ns$.

1) PEEC Modelling of a TO-247-3 package:
The PEEC model of a TO-247-3 package is shown in Fig. 2. The 3D geometry is extracted from a disassembled package and a non-uniform PEEC mesh is applied generating 17221 unknowns, divided in 14896 unknown cell currents and 2325 unknown node potentials. The drain-source (DS) impedance, Z_{DS}-Θ_{DS}, was measured with an applied gate-source voltage of $20\,V$ using a Keysight Impedance Analyzer E4990, with the frequency measurement limit of $120\,MHz$. The drain-source inductance L_{DS} was extracted from Z_{DS}-Θ_{DS} measured curves. The verification of the PEEC modeling and the corresponding Q3D modelling of the drain-source current path inside of the TO-247-3 package is shown on Fig. 2.

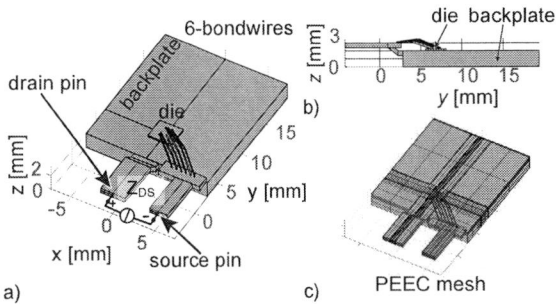

Fig. 1. The developed PEEC model of a TO-247-3 package for the estimation of drain-source loop inductance: a) 3D view, b) YZ view, and c) a non-uniform PEEC mesh.

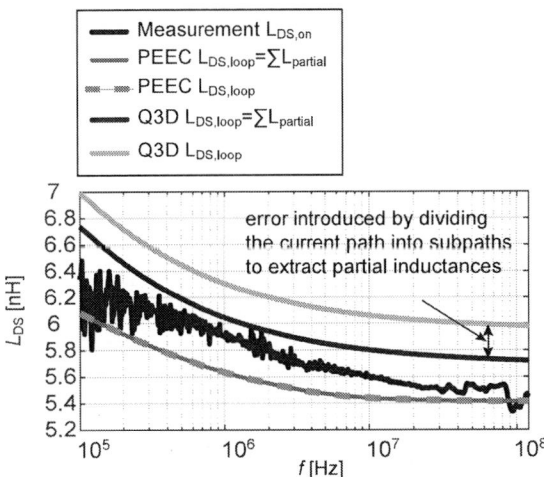

Fig. 2. Comparison between the measurement of the drain-source (DS) inductance, Z_{DS}, the PEEC-based modelling, and the Q3D modeling with and without cutting the current loop for extracting the partial inductances.

As the stray inductances in the drain and source terminals have different impact on the switching properties of power converters, potentially hampering the utilization of WBG power semiconductor devices, it is highly useful to have the information about these inductances and minimize them in an optimal design. As described in [24], to extract the partial inductances seen from the drain and source terminals in Q3D, special cuts of the geometry have to be made at the position of the semiconductors, which modifies the current loop path. On the other hand, using the PEEC method, the partial inductances can be extracted using the internal nodes, i.e. without changing the geometry of the current path. The partial inductances calculated by the PEEC method and Q3D at 50 MHz are summarized in Table II. The influence of the cuts in Q3D on modelling L_{DS} is illustrated on Fig. 2. The Q3D simulation without modifying the current path is supposed to provide more accurate estimation of the drain-source inductance than can be found here. Based on these findings, the PEEC method is expected to overcome the limits of the Q3D Extractor.

TABLE II. PEEC AND Q3D MODELING OF L_{DS} INSIDE OF THE TO-247-3 PACKAGE AT 50 MHz.

L [nH]	L_D	L_S	M_{DS}	$L_{tot} = \sum L_{partial}$	L_{tot}
Q3D	3.53	5.39	-1.58	5.76	5.98
PEEC	4.06	5.37	-2.0	5.42	5.42
Experiment	-	-	-	-	5.54

2) PEEC Modeling of PCBs: A test loop was fabricated as a top layer of two PCBs, without (PCB1) and with (PCB2) the bottom copper layer, as shown in Fig. 3. The previuosly used Keysight Impedance Analyzer E4990, based on the auto-balancing bridge measurement method, cannot be used to accurately measure the loop impedance beyond $\approx 10\,$MHz, i.e. high Q measurements. Accordingly, the measurements were performed using a Keysight Impedance Analyzer E4991 with the operational frequency range above 1 MHz based on radio frequency (RF) I-V measurements [52].

To capture the high frequency behavior of the loop, a non-uniform mesh was applied, resulting in 15657 unknown cell currents and 3764 unknown node potentials for the PCB1 and 25231 unknown cell currents and 5136 unknown node potentials for the PCB2. The Passive Reduced-order Interconnect Macromodeling Algorithm (PRIMA) [53] Model Order Reduction (MOR) [54] approach was applied, generating a reduced order 72×72 PEEC system, which can be exported as a netlist into a circuit simulator in order to simulate the time response of the loop for different transient waveforms. Moreover, the MOR technique allows the calculation of the frequency response in many frequency points in a fast and accurate way. The MOR technique enables to capture all resonances, which would be computationally expensive and practically infeasible if the original PEEC system matrix with an order above 10000 was used. The comparison between the measurements and the modelling results is shown in Fig. 4 and 5. A good matching between the PEEC simulations and measurements was achieved up to 200 MHz.

The Q3D simulation results of the PCB2 show a better agreement with the measurements and the PEEC simulation than it is the case for the PCB0. As it could be seen from Fig. 4, the HF inductance of the PCB0 does not reach a constant AC value even at 200 MHz. The computational core of Q3D requires a frequency sweep from a very low ($f < f_{DC}$, the skin effect is negligible) up to a very high frequency ($f > f_{AC}$, the skin effect fully developed) in order to accurately estimate the partial inductances and resistances at the specific mid-frequency ($f_{DC} < f < f_{AC}$ the skin effect partially developed) [55]. Moreover, Q3D extractor cannot predict the self-capacitance of the PCB layout, which becomes influential above 10 MHz and ≈ 70 MHz for PCB0 and PCB2, respectively. Accordingly, a quasi-static PEEC solver can predict the PCB loop impedance with a higher accuracy than Q3D in a wide frequency range.

The PEEC simulations were performed with and without considering the dielectric layer of the PCBs (FR4,

a glass-reinforced epoxy laminate). According to Fig. 4 and 5, the results of the PEEC simulations without modeling the dielectric layer show a higher mismatch to the measurements, which is even more significant for the PCB2. Accordingly, the PCB dielectric must be properly modelled to accurately predict the parasitics of PCB layouts. The PEEC method has the capability of modeling dielectric materials.

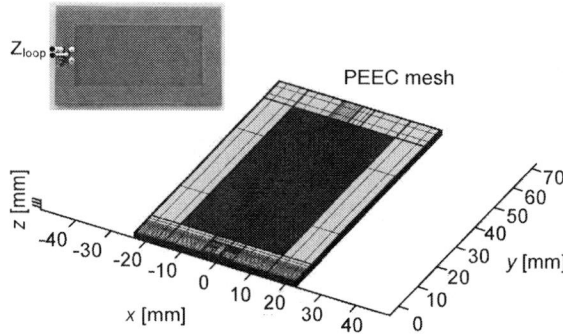

Fig. 3. The test PCB layout: a) a photo of the actual PCB layout illustrating the measurement ports and b) the PEEC meshed model.

C. Future of the PEEC method

As previously mentioned, the weakest point of the PEEC modelling approach is the generation of large dense matrices, implying a high computational cost of the PEEC solvers and the requirement for higher memory storage. Therefore, speeding up of the PEEC-based modelling and compressing the PEEC system matrices is a necessary task in order to enable an efficient PEEC-based multiphysics modelling in the future.

Several options are available for enabling a PEEC solver to be efficiently used for coupling electromagnetic and circuit modelling domains. A first possibility is to resort to high performance computing platforms. A PEEC parallel implementation was developed in the past [56]. A second option resides in the use of acceleration to speed-up the analysis of PEEC circuits. This task can be accomplished by means of fast multipole techniques (FMM) [43], [57] or by adopting algebraic methods to take advantage of the low-rank behavior of electric and magnetic field interactions [44], [46]. The FMM techniques are based on the expansion of the Green's function in terms of spherical or plane waves which allows to significantly speed-up the matrix-vector products required by iterative solvers. The main advantage of Algebraic methods relies on the fact that they are independent on the Green's function adopted for the PEEC implementation (standard free space Green's function [33], layered media Green's function [58]).

A further class of methods to accelerate the solution of large problems is represented by MOR techniques [54]. The PRIMA algorithm [53] is well consolidated and has been used to reduce PEEC models (see also Section

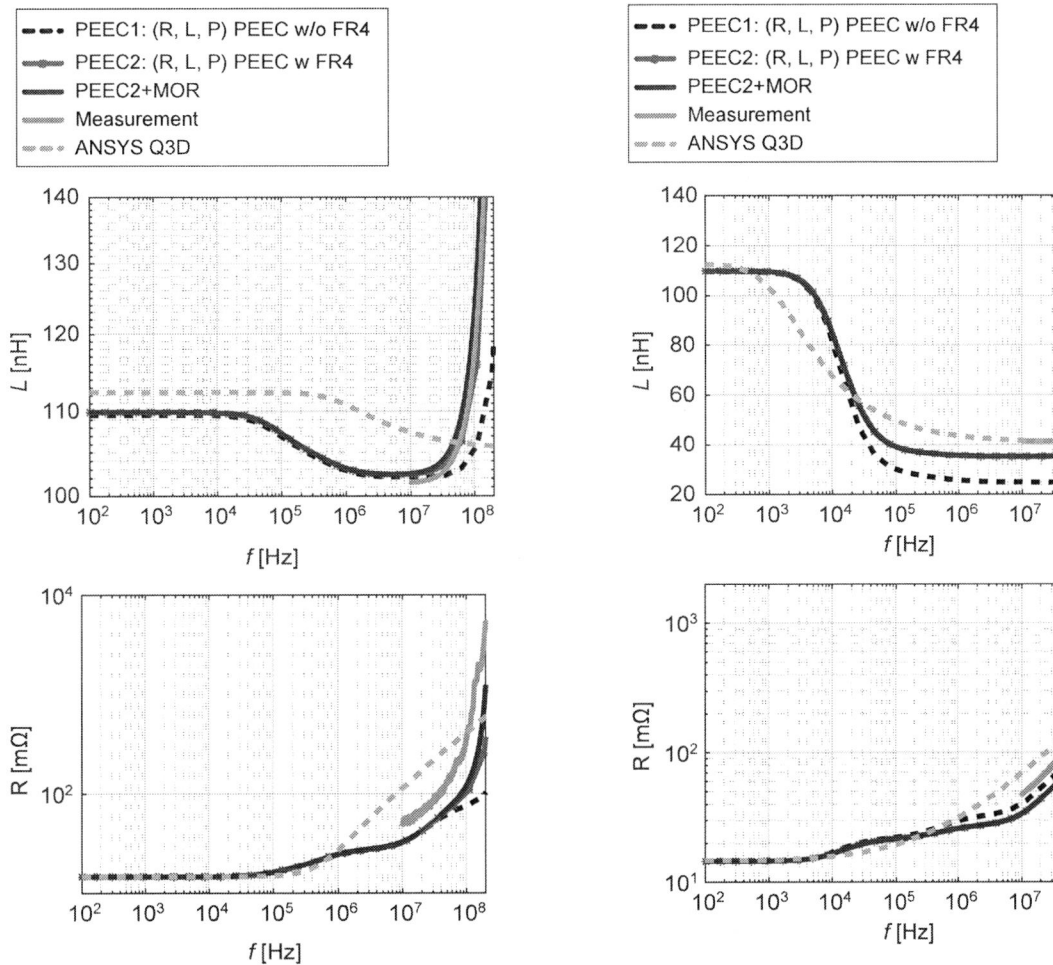

Fig. 4. Verification of the HF electromagnetic modelling of PCB1: inductance (upper panel) and resistance (lower panel).

Fig. 5. Verification of the HF electromagnetic modelling of PCB2: inductance (upper panel) and resistance (lower panel).

III-B2). Nevertheless the generation requires several solutions of the high-order PEEC system, which can be computationally expensive. To mitigate this limitation, a partitioned model order reduction has been proposed in [45] by partitioning the solution and by resorting to algebraic methods to perform matrix manipulations. This allows to optimize the reduction process minimizing the size of the resulting model but, at the same time, guaranteeing a very good accuracy. A different approach for adaptive multipoint MOR scheme for large-scale inductive PEEC circuits was proposed in [59]. In addition, parameterized MOR techniques for PEEC models were proposed in [60]–[63]. The parametrized MOR techniques can be useful for PEEC-based optimizations, which require repeated runs of PEEC models during the optimization process. When electrically large problems are investigated, it is recommended to adopt delayed PEEC models and specialized MOR techniques are needed [64], [65].

The need to speed-up the computation of partial

elements, partial inductances and coefficients of potential for large PEEC models can also be afforded by exploiting the redundancy in their evaluation. Indeed, when dealing with the PEEC mesh of 3D geometries, including Manhattan type meshes, i.e. orthogonal blocks parallel or perpendicular, but also non-orthogonal ones characterized by hexahedra with comparable sizes, a huge number of configurations are repeated under only translation and rotation transformations. Through an efficient identification of these geometrical signatures, the calculation of volume and surface integrals required to fill the partial element matrices can be significantly accelerated as reported in [66].

IV. CONCLUSIONS

Multi-physics modeling represents a technology, which can help to analyze and verify all benefits coming along with the implementation of power WBG semiconductor devices in advanced power electronics systems before hardware prototyping. To properly address the fast switching properties of the WBG devices, coupled electromagnetic and circuit modelling is increasingly important. Additionally, there is a requirement for highly accurate prediction of parasitics inside the power semiconductor packages and PCB layouts. Small parasitic inductances and capacitances have a more severe impact in WBG-based power converters as they operate at higher switching frequencies and produce the current and voltage waveforms with much faster slopes than the Si-based power electronic systems. This paper gives a brief overview of different modelling tools commonly used in power electronics to predict HF behavior of power modules and PCB layouts. Analyzing the limitations of the existing modeling approaches mainly for accurate extraction of parasitics within WBG power modules and PCB layouts, the PEEC method is introduced as a promising numerical technique for 3D electromagnetic-circuit coupled modeling for future power electronics. This paper demonstrates the advantage of the PEEC method for the EM modelling of PE geometries, such as PCBs and power semiconductor packages in terms of accuracy required by WBG power electronics. Furthermore, the existing challenges of highly accurate PEEC-based modeling in a wide frequency range and the potential for improving the PEEC method in future in terms of computational cost are comprehensively described.

REFERENCES

[1] N. Kaminski and O. Hilt, "SiC and GaN devices-wide bandgap is not all the same," *IET Circuits, Devices & Systems*, vol. 8, no. 3, pp. 227–236, 2014.

[2] J. Schuderer et al., "Packaging SiC power semiconductors - challenges, technologies and strategies," in *Proc. of IEEE Workshop on Wide Bandgap Power Devices and Applications (WIPDA)*, 2014.

[3] P. L. Evans et al., "Design tools for rapid multidomain virtual prototyping of power electronic systems," *IEEE Tran. on Power Electronics*, vol. 31, no. 3, pp. 2443–2455, 2016.

[4] N. Hingora et al., "Power-CAD: A novel methodology for design, analysis and optimization of power electronic module layouts," in *Proc. of IEEE Energy Conversion Congress and Exposition (ECCE)*, 2010.

[5] U. Drofenik et al., "Design tools for power electronics: Trends and innovations," *Ingenieurs de l'automobile*, no. 791, pp. 55–62, 2007.

[6] Z. Gong, "Thermal and electrical parasitic modeling for multichip power module layout synthesis," Ph.D. dissertation, University of Arkansas, Fayetteville, 2012.

[7] W. Wang et al., "Mesh-based lumped parameter model with MOR for thermal analysis of virtual prototyping for power electronics systems with comparison to FDM," in *Proc. of 3rd IEEE Int. Future Energy Electronics Conf. and ECCE Asia (IFEEC-ECCE Asia)*, 2017, pp. 1037–1042.

[8] M. Kegeleers et al., "Parasitic inductance analysis of a fast switching 100kW full SiC inverter," in *Proc. of Int. Exhib. and Conf. for Power Electronics, Intelligent Motion, Renewable Energy and Energy Management (PCIM)*, 2017.

[9] D. Reusch and J. Strydom, "Understanding the effect of PCB layout on circuit performance in a high frequency gallium nitride based point of load converter," in *Proc. of 28th IEEE Applied Power Electronics Conference and Exposition (APEC)*, 2013.

[10] Z. Liu et al., "Package parasitic inductance extraction and simulation model development for the high-voltage cascode GaN HEMT," *IEEE Tran. on Power Electronics*, vol. 29, no. 4, 2014.

[11] A. J. Morgan et al., "Decomposition and electro-physical model creation of the CREE 1200V, 50A 3-ph SiC module." in *Proc. of Int. Symp. on 3D Power Electronics Integration and Manufacturing (3D-PEIM)*, 2016.

[12] S. Kicin et al., "Full SiC half-bridge module for high frequency and high temperature operation," in *Proc. of 65th IEEE Electronic Components and Technology Conference (ECTC)*, 2015.

[13] D.-P. Sadik et al., "Analysis of parasitic elements of sic power modules with special emphasis on reliability issues," *IEEE J. of Emerging and Selected Topics in Power Electronics*, vol. 4, no. 3, 2016.

[14] F. Yang et al., "Parasitic inductance extraction and verification for 3D planar bond all module," in *Proc. of Int. Symp. on 3D Power Electronics Integration and Manufacturing (3D-PEIM)*, 2016.

[15] A. Asmanis et al., "High-frequency modelling of EMI filters considering parasitic mutual couplings," in *Proc. of ESA Workshop on Aerospace EMC*, 2016.

[16] G. Chiappori et al., "EM emissions induced by a DC/DC power converter: First experiments on a laboratory system." in *Proc. of Int. Symp. on Electromagnetic Compatibility (EMC Europe),*, 2014, pp. 69–73.

[17] A.-S. Podlejski et al., "Layout modelling to predict compliance with EMC standards of power electronic converters," in *Proc. of IEEE Int. Symp. on Electromagnetic Compatibility*, 2015, pp. 779–784.

[18] S. Ahyoune et al., "Extending the frequency range of quasi-static electromagnetic solvers," in *Proc. of 14th Int. Conf. on Synthesis, Modeling, Analysis and Simulation Methods and Applications to Circuit Design (SMACD)*, 2017, pp. 1–4.

[19] I. F. Kovacevic et al., "3-D electromagnetic modeling of EMI input filters." *IEEE Tran. on Industrial Electronics*, vol. 61, no. 1, pp. 231 – 242, 2014.

[20] V. Ardon et al., "EMC modeling of an industrial variable speed drive with an adapted PEEC method," *IEEE Tran. on Magnetics*, vol. 46, no. 8, pp. 2892–2898, 2010.

[21] K. Li et al., "Developing power semiconductor device model for virtual prototyping of power electronics systems," in *Proc. of IEEE Vehicle Power and Propulsion Conference (VPPC)*, 2016, pp. 1–6.

[22] C. M. Patton. "Inductance modeling and extraction in EMC applications," Master's thesis, Missouri University of Science and Technology, 2009.

[23] Z. Zhu et al., "Algorithms in FastImp: a fast and wide-band impedance extraction program for complicated 3-D geometries," *IEEE Tran. on Computer-Aided Design of Integrated Circuits and Systems*, vol. 24, no. 7, pp. 981–998, 2005.

[24] I. Kovacevic-Badstuebner et al., "Parasitic extraction procedures for SiC power modules," in *Proc. of 10th Int. Conf. on Integrated Power Electronics (CIPS)*, 2018.

[25] K. Li et al., "Using multi time-scale electro-thermal simulation approach to evaluate SiC-MOSFET power converter in virtual prototyping design tool," in *Proc. of 18th Workshop on Control and Modeling for Power Electronics (COMPEL)*, 2017, pp. 1–8.

[26] C. Martin et al., "Analysis of electromagnetic coupling and current distribution inside a power module," *IEEE Tran. on Industry Applications*, vol. 43, no. 4, pp. 893 – 901, 2007.

[27] G. Regnat et al., "Silicon carbide power chip on chip module based on embedded die technology with paralleled dies," in *Proc. of IEEE Energy Conversion Congress and Exposition (ECCE)*, 2015.

[28] R. Robutel et al., "Design and implementation of integrated common mode capacitors for SiC-JFET inverters," *IEEE Tran. on Power Electronics*, vol. 29, no. 7, pp. 3625 – 3636, 2014.

[29] L. Lombardi et al., "Electrothermal formulation of the partial element equivalent circuit method." *Int. Journal of Numerical Modelling: Electronic Networks, Devices and Fields*, 2017.

[30] R. F. Harrington, *Field Computation by Moment Methods*. Malabar: Krieger, 1982.

[31] C. A. Balanis, *Advanced Engineering Electromagnetics.* Wiley, 1989.

[32] L. W. Nagel, "SPICE: A computer program to simulate semiconductor circuits," EECS Department, University of California, Berkeley, Electr. Res. Lab. Report ERL M520, May 1975.

[33] A. E. Ruehli et al., *Circuit Oriented Electromagnetic Modeling Using the PEEC Techniques.* John Wiley & Sons, Inc., Hoboken, New Jersey., 2017.

[34] D. Romano and G. Antonini, "Quasi-static partial element equivalent circuit models of linear magnetic materials," *IEEE Tran. on Magnetics*, vol. PP, no. 99, pp. 1–1, 2014.

[35] ——, "Augmented quasi-static partial element equivalent circuit models for transient analysis of lossy and dispersive magnetic materials," *IEEE Tran. on Magnetics*, vol. 52, no. 5, pp. 1–11, May 2016.

[36] D. Romano et al., "Time-domain partial element equivalent circuit solver including non-linear magnetic materials," *IEEE Tran. on Magnetics*, vol. 52, no. 9, pp. 1–11, Sept 2016.

[37] A. E. Ruehli and H. Heeb, "Circuit models for three-dimensional geometries including dielectrics," *IEEE Tran. on Microwave Theory and Technique*, vol. 40, no. 7, pp. 1507–1516, Jul. 1992.

[38] G. Antonini, A. E. Ruehli and C. Yang, "PEEC modeling of dispersive and lossy dielectrics," *IEEE Tran. on Advanced Packaging*, vol. 31, no. 4, pp. 768–782, Sep. 2008.

[39] D. Romano and G. Antonini, "Quasi-static partial element equivalent circuit models of magneto-dielectric materials," *IET Microwaves, Antennas Propagation*, vol. 11, no. 6, pp. 915–922, 2017.

[40] A. Hartman et al., "Partial element equivalent circuit models of three-dimensional geometries including anisotropic dielectrics," *IEEE Tran. on Electromagnetic Compatibility*, vol. 60, no. 3, pp. 696–704, June 2018.

[41] L. Lombardi et al., "Partial element equivalent circuit method modeling of silicon interconnects," *IEEE Transactions on Microwave Theory and Techniques*, vol. 65, no. 12, pp. 4794–4801, Dec 2017.

[42] M. Kamon et al., "FASTHENRY: A multipole-accelerated 3-D inductance extraction program," *IEEE Tran. on Microwave Theory and Techniques*, vol. 42, no. 9, pp. 1750–1758, 1994.

[43] G. Antonini, "Fast multipole method for time domain PEEC analysis," *IEEE Tran. on Mobile Computing*, vol. 2, no. 4, pp. 275–287, 2003.

[44] G. Antonini and D. Romano, "Efficient frequency-domain analysis of PEEC circuits through multiscale compressed decomposition," *IEEE Tran. on Electromagnetic Compatibility*, vol. 56, no. 2, pp. 454–465, 2014.

[45] D. Romano and G. Antonini, "Partitioned model order reduction of partial element equivalent circuit models," *IEEE Tran. on Components, Packaging and Manufacturing Technology*, vol. 4, no. 9, pp. 1503 – 1514, 2014.

[46] G. Antonini and D. Romano, "Adaptive-cross-approximation-based acceleration of transient analysis of quasistatic partial element equivalent circuits," *IET Microwaves, Antennas & Propagation*, vol. 9, no. 7, pp. 700–709, 2015.

[47] E. Masuda et al., "A study on wiring pattern design for intelligent SiC power module with PEEC method," in *Proc. of Asia-Pacific International Symposium on Electromagnetic Compatibility (APEMC)*, 2017, pp. 213–215.

[48] G. Antonini, "Full wave analysis of power electronic systems through a PEEC state variables method," in *Proc. of the IEEE Int. Symp. on Industrial Electronics*, 2002, pp. 1386–1391.

[49] J. L. Schanen et al., "Impact of the physical layout of high-current rectifiers on current division and magnetic field using PEEC method," *IEEE Tran. on Industry Applications*, vol. 46, no. 2, pp. 892–900, 2010.

[50] M. L. Zitzmann and R. Weigel, "Fast and efficient methods for circuit-based automotive EMC simulation," *Time Domain Methods in Electrodynamics*, pp. 189–210, 2008.

[51] C. E. Tonry and C. Bailey, "Model order reduction in inductors for rapid virtual prototyping in power electronics," in *Proc. of 22nd Int. Workshop on Thermal Investigations of ICs and Systems (THERMINIC)*, 2016, pp. 64–68.

[52] "Keysight Technologies, Impedance Measurement Handbook: A guide to measurement technology and techniques." [Online]. Available: http://literature.cdn.keysight.com/litweb/pdf/5950-3000.pdf

[53] A. Odabasioglu, M. Celik, and L. T. Pileggi, "PRIMA: passive reduced-order interconnect macromodeling algorithm," *IEEE Tran. on Computer-Aided Design of Integrated Circuits and Systems*, vol. 17, no. 8, pp. 645–654, Aug. 1998.

[54] Sheldon X.-D. Tan and Lei He, *Advanced Model Order Reduction Techniques in VLSI Design*, C. U. Press, Ed., Cambridge, 2007.

[55] Ansys. (2018). [Online]. Available: https://www.ansys.com/products/electronics/ansys-q3d-extractor

[56] J. Ekman and P. Anttu, "Parallel implementations of the PEEC method," in *Proc. of IEEE Int. Symp. on Electromagnetic Compatibility*, 2007.

[57] G. Antonini, A. E. Ruehli, "Fast multipole and multi-function PEEC methods," *IEEE Tran. on Mobile Computing*, vol. 2, no. 4, pp. 288–298, 2003.

[58] S. V. Kochetov et al., "PEEC formulation based on dyadic green's functions for layered media in the time and frequency domains," *IEEE Transactions on Electromagnetic Compatibility*, vol. 50, no. 4, pp. 953–965, Nov 2008.

[59] T.-S. Nguyen et al., "Adaptive multipoint model order reduction scheme for large-scale inductive PEEC circuits," *IEEE Tran. on Electromagnetic Compatibility*, vol. 59, no. 4, pp. 1143–1151, 2017.

[60] F. Ferranti et al., "Guaranteed passive parameterized model order reduction of the partial element equivalent circuit (PEEC) method," *IEEE Tran. on Electromagnetic Compatibility*, vol. 52, no. 4, pp. 974–984, 2010.

[61] ——, "Passivity-preserving parametric macromodeling for highly dynamic tabulated data based on Lur'e equations," *IEEE Tran. on Microwave Theory and Techniques*, vol. 58, no. 12, pp. 3688–3696, 2010.

[62] ——, "Physics-based passivity-preserving parameterized model order reduction for PEEC circuit analysis," *IEEE Tran. on Components, Packaging and Manufacturing Technology*, vol. 1, no. 3, pp. 399–409, 2011.

[63] ——, "Multipoint full-wave model order reduction for delayed PEEC models with large delays," *IEEE Tran. on Electromagnetic Compatibility*, vol. 53, no. 4, pp. 959–967, 2011.

[64] ——, "Interpolation-based parameterized model order reduction of delayed systems," *IEEE Tran. on Microwave Theory and Techniques*, vol. 60, no. 3, pp. 431–440, 2012.

[65] E. Samuel et al., "Passivity-preserving parameterized model order reduction using singular values and matrix interpolation," *IEEE Tran. on Components, Packaging and Manufacturing Technology*, vol. 3, no. 6, pp. 1028–1037, 2013.

[66] D. Romano et al., "Acceleration of the partial element equivalent circuit method with uniform tessellation - Part I: Identification of geometrical signatures; Part II: Frequency domain solver with interpolation and reuse of partial elements," *Int. J. of Numerical Modelling: Electronic Networks, Devices and Fields*, to appear, 2017.

The 2018 International Power Electronics Conference

Silicon based devices for demanding high power applications

A. Kopta, J. Vobecky, M. Rahimo, T. Wikström, U. Vemulapati, C. Papadopoulos, C. Corvasce, M. Andenna F. Dugal, F. Fischer, S. Hartmann

ABB Switzerland Ltd, Semiconductors, Fabrikstrasse 3, CH-5600 Lenzburg, Switzerland

Phone: +41 58 586 70 43, e-mail: umamaheswara.vemulapati@ch.abb.com

Abstract

This paper gives an overview of the recent progress of high voltage silicon based power devices for high power grid applications, traction- and industrial drives. The first part of the paper covers the latest developments of thyristor based technologies whereas the second and third part focuses on IGBT-type devices and the corresponding packaging technologies.

Introduction

In this paper, we will focus on applications and devices in the highest power range, ranging from about 10 GW in grid applications like HVDC, down to several MW in converters for traction- and large industrial drives. In contrast to the low power part of the spectrum, where SiC and GaN devices are starting competing with Si based devices, the high voltage / high power domain remains a Si stronghold. Along with the still very high cost of high voltage SiC devices, the Si counterparts have undergone some strong improvement steps over the last years. In combination, this has pushed the brake-even point in system cost where SiC starts overtaking Si further into the future. As an example, in IGBT modules, the power density has nearly doubled over the last 15-20 years. Although SiC based devices have undergone some impressive development steps as well, Si still offers the best total cost/power trade-off in most applications above 1200 V. In this paper we will focus on devices with rated voltages from 3.3 kV and upwards. From today's perspective, but depending on future innovation on system side, Si-based devices will remain dominant in this voltage range for the foreseeable future. We will show the latest development steps and give an outlook what is still to be expected from Si-based power devices.

In the first section, we will look into thyristor-based structures, starting with the Phase Control Thyristor (PCT), the workhorse for HVDC systems with the highest transmission power. We will further look into the progress of the latest generation of Integrated Gated Commutated Thyristors (IGCT), as of today the turn-off device concept with the lowest losses, already today enabling device with rated voltages of up to 10kV. In the second part of the paper, we will present the lat-

est results of IGBT-based devices. We will show the final characteristics of the newly developed 3.3 kV enhanced trench IGBT platform. We will further present our 4.5 kV RC-IGBT device referred to as the Bi-Mode Insulated Gate Transistor (BIGT). The current generation of this device is based on a planar IGBT cell and has been specially developed for grid application, but combined with the enhanced trench cell, it will become the next generation device for all high voltage applications.

General trends

The general device development trends, which are valid for both thyristor- and IGBT-based devices, can be summarized in the following four points:

- Loss reduction by optimizing the device structure and design.
- Increasing the power density for a given foot-print/silicon area by integration.
- Increasing the possible output power of the converter by improving the thermal design.
- Where applicable, increasing the single device footprint to increase the total current of an individual device.

Point number one is the classical battlefield of sophisticated designs. Point number two aims for a better total silicon utilization. The most prominent representative of this trend in the last few years is the integration of the free-wheeling diode into the switch structure creating integrated reverse conducting devices like the BIGT. Point number three mostly aims for increasing the junction temperature, and on packaging side reducing thermal resistance as well as enabling the higher junction temperature by improving the load cycling capability. Point number four is an option when a certain converter output current only can be reached by increasing the available silicon area. Instead of paralleling devices, the system cost can be kept low by using larger area devices.

As current densities and operating temperatures continue increasing, aspects like reliability, turn-off ruggedness and controllable and soft switching behavior to prevent EMI problems all need to be further improved as an integral part of the device development efforts.

Thyristor based devices

PCTs

In order to satisfy the growing demands on HVDC systems, ABB Semiconductors has already developed three generations of phase controlled thyristors (PCTs) with blocking voltages 6.7 kV to 8.5 kV, rating currents 1.5 kA to 6.25 kA in hockey puck packages with pole pieces 100 mm to 138 mm [1]-[4]. Fig. 1 illustrates the design changes of the six inch, 8.5 kV PCT from the generation 1 to the generation 3. The design changes include reduction of the Si thickness by choosing the right resistivity, optimal layout of the amplifying gate and cathode shorts design, and reduction of the p-base as well as p-anode junction depths in the active area to further reduce the Si thickness. The latest generation i.e. Gen. 3 has a thinner Si and has lower overall losses compared to other generations of the same blocking capability (8.5 kV).

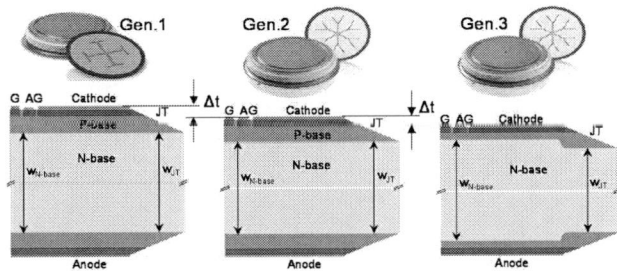

Fig. 1. The evolution of the six inch, 8.5 kV PCT for HVDC in hockey puck package shown together with 150 mm wafer. Left: Gen. 1; Middle: Gen. 2; Right: Gen. 3.

Latest achievements in PCTs with the highest ratings has been driven by the first 12 GW HVDC system with DC voltage of ±1100 kV, DC current 5.5 kA and transmission distance of 3400 km across China. Fig. 2 illustrates the recent development speedup of the highest voltage class of 8.5 kV at 150 mm wafer in the package with 138 mm pole piece [1].

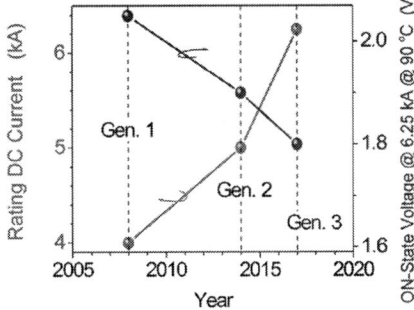

Fig. 2. Development trend of key parameters of six inch, 8.5 kV PCT (current rating and V_T @ 6.25 kA, 90 °C).

The device provides symmetric forward and reverse blocking up to 8.5 kV and ON-state voltage drop below 2 V at 6 kA and T = 90 °C. The fact that no other semiconductor switch provides so low ON-state losses explains the importance of PCTs for HVDC in spite of the missing turn-off capability. As explained before, although the Gen. 3 is processed from thinner Si (shown in Fig. 1), the leakage current is similar to Gen. 2 as shown in Fig. 3. Gen. 3 has a better technology curve (Q_{rr} vs. V_T) compared to Gen. 2 as shown in Fig. 4.

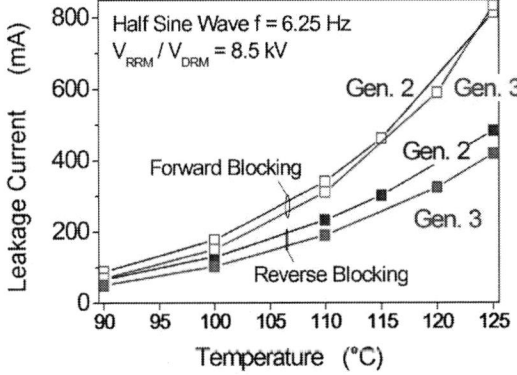

Fig. 3. Leakage current vs. Junction temperature of 8.5 kV Gen. 2 and Gen. 3 PCTs.

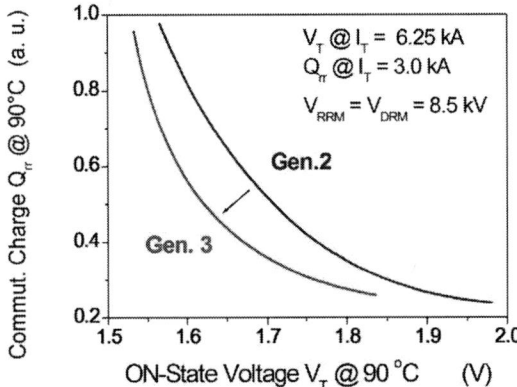

Fig. 4. Commutation charge vs. On-state voltage drop of 8.5 kV Gen. 2 and Gen. 3 PCTs at 90 °C.

Further development trends are not only about the lower ON-state losses. It is also important to further reduce the leakage current, commutation recovery time tq, dV/dt capability and surge current ITSM. This is because on top of the HVDC, these devices are used for pumped-storage hydropower and industry.

IGCTs

IGCT application is increasingly honed towards high performance and power, where the benefits can outweigh its disadvantages:

- There are few IGCT manufacturers [5], [6]. These work with customer- and application-specific device optimizations. As a result, changing suppliers is difficult which contradicts basic supply-chain strategy.
- Using the IGCT optimally is paired with a certain proficiency in electrical and mechanical engineering due to, respectively, the inability of the IGCT to have the controllable turn-on behavior (di/dt limiting inductor is needed to limit the di/dt during turn-on thereby protect the anti-parallel diode from the high power failures), and challenges of assembling press-pack devices.

The aim for the IGCT in future applications is to accentuate its performance benefit over the IGBT: the thyristor-like on-state. It facilitates making devices in silicon with 10kV blocking capability, with good performance [7]-[9]. For lower voltages (2.5kV to 6.5kV), where the IGCT is already well established, the aim is to improve the inverter output as well as the cost and applicability of the device. Similar to RC-IGBT, a Reverse Conducting IGCT (RC-IGCT) concept provides compactness by monolithically integrating a freewheeling diode with the GCT on the same wafer which allows easy way of designing the converter stacks [10], [11].

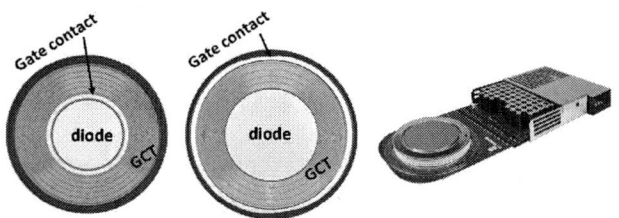

Fig. 5. RC-IGCT. Left: Conventional 91 mm wafer (gate contact is at the separation region between diode and GCT parts); Middle: Newly developed 94 mm wafer (gate contact is at the edge of the wafer); Right: RC-IGCT wafer in a hermetic package and with its integrated gate unit.

The conventional 91 mm, RC-IGCT wafer is shown in Fig. 5 along with the newly developed 94 mm, RC-IGCT [12]. The new RC-IGCT design offers 21% more device active area compared to conventional design and this gained area is mainly used to increase the diode part of the device as shown in Fig. 5. The main improvements of the new RC-IGCT design over the conventional RC-IGCT is as follows:

- Increased device active area
- Reduced thermal resistance
- Reduced gate circuit impedance
- Increased maximum controllable turn-off current

Today, the new RC-IGCTs are available in two voltage ratings 4.5 kV / 3 kA and 6.5 kV / 2.15 kA, which are optimized for low frequency applications employing Modular Multilevel Converter (MMC) topology. Fig. 6 illustrates the turn-off waveforms of the 4.5 kV, 94 mm RC-IGCT in both GCT- and diode-modes beyond the Safe Operation Area (SOA) specifications at 135 °C.

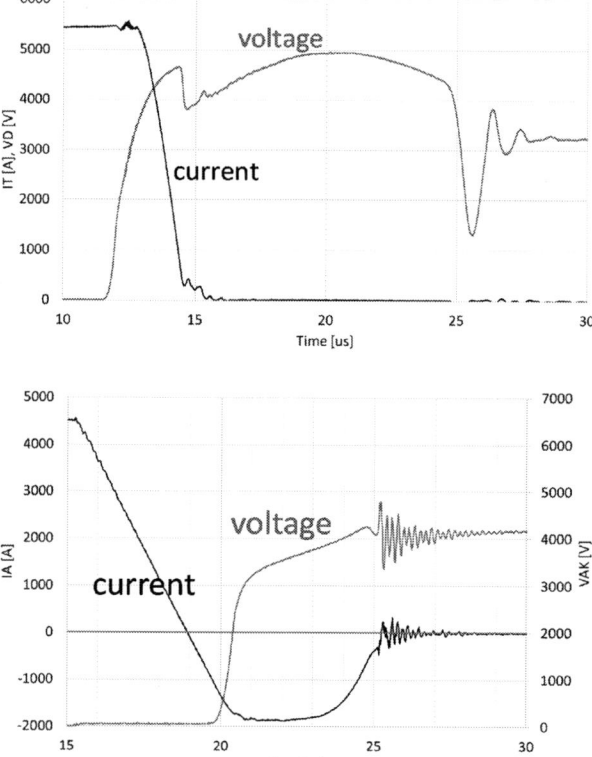

Fig. 6. 94 mm 4.5 kV RC-IGCT turn-off waveforms beyond SOA specifications at 135°C. Top: GCT turn-off at 5.5 kA and 3.2 kV; Bottom: Diode turn-off at 4.5 kA and 3.9 kV.

In RC-IGCTs, the main challenge is to ensure the diode robustness under low forward current and high dc-link voltage, especially at low temperatures. Under the above conditions, there is a tendency for the diode to snap-off at the end of the reverse recovery period leading to over voltage spikes that can cause failure of the device. This is mainly due to the insufficient thickness of the device. The Si thickness is mainly governed by the GCT part in order to reduce the conduction as well as the switching losses thereby increasing the efficiency and power handling capability of the device while maintaining reasonable reverse recovery behavior.

To overcome this limitation, an advanced reverse conducting device, fully integrated device concept referred to as the Bi-mode Gate Commutated Thyristor (BGCT) has been developed. In the new design, the IGCT and diode parts are integrated in an interdigitated way forming a single structure as shown in Fig. 7, which enables the same silicon volume to be utilized better in both GCT- and diode-modes.

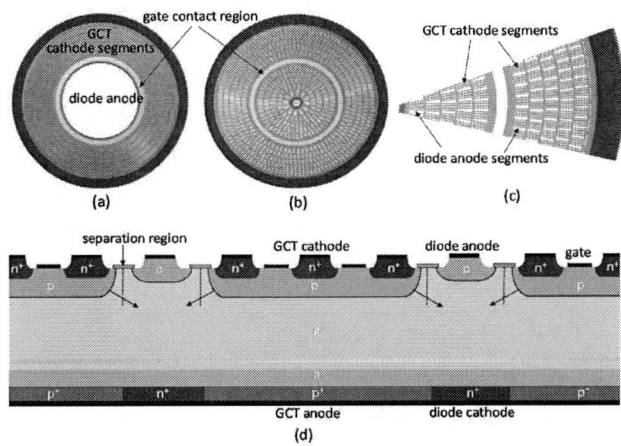

Fig. 7. (a) and (b) are top-view of 91mm, 4.5 kV RC-IGCT and 91mm, 4.5 kV BGCT, (c) zoomed part of the BGCT; the wide white regions are the diode anode segments and the ones between the diode anode segments are the GCT cathode segments, (d) cross section of a BGCT with shallow diode anode.

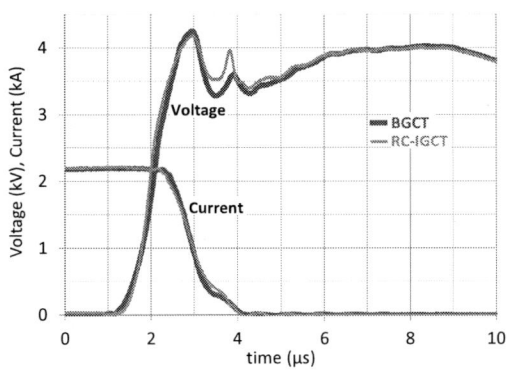

Fig. 8. 91 mm, 4.5 kV devices: Measured turn-off curves of BGCT and RC-IGCT at 2.2 kA, 2.8 kV, 115°C in GCT-mode.

This interdigitated integration results in an improved diode as well as GCT area, better thermal distribution, softer turn-off behavior (as shown in Fig. 8 and Fig. 9) and lower leakage

current compared to conventional RC-IGCTs. The BGCT concept has been demonstrated first experimentally with 38 mm, 4.5kV prototypes [13] and extended to 91 mm wafer size [14]. The BGCT probably represents the maximum possible increase in performance-per-area for silicon-based RC-IGCTs. The performance is higher than IGBTs, which warrants the IGCT's raison-d'être until the next fundamental change in device- or inverter-technology.

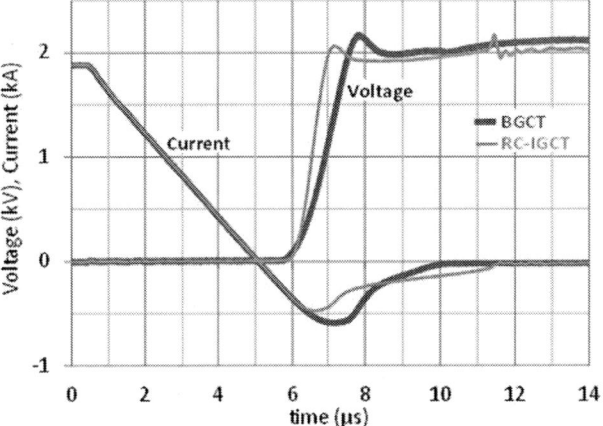

Fig. 9. 91 mm, 4.5 kV devices: Measured reverse recovery waveforms of BGCT and RC-IGCT at 2.0 kA, 1.9 kV, 115°C in diode-mode.

IGBT based devices

IGBT cell improvements

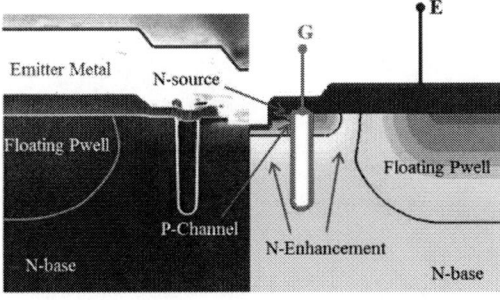

Fig. 10. Cross-section of the 3.3 kV enhanced trench IGBT cell.

The IGBT cell design still remains the main parameter to optimize in order to reduce the overall losses. In this paper we present the newest development results from our next generation 3.3kV enhanced trench IGBT (Fig. 10), which combines low losses with a highly controllable turn-on behavior [15]-[17]. Fig. 11 shows the improvement in the technology curve compared to the previous planar chip generation. In spite of the big improvement, we believe that the trench cell can be further optimized as there are still many design trade-offs that influence the losses in a negative way.

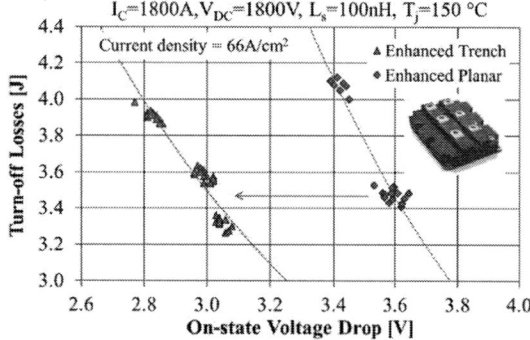

Fig. 11. $E_{off}/V_{CE,on}$ trade-off curve comparison between the 3.3kV enhanced trench and the enhanced planar IGBT. HiPak module results.

Integration of the diode into the IGBT structure

A further improvement step beyond cell optimization can be reached by integrating the free-wheeling diode into the IGBT structure, creating a reverse conducting device (RC-IGBT). In this way, the overall silicon utilization can be improved, increasing the total power density. Fig. 12 shows the cross-section of our recently qualified 4.5kV device, referred to as the Bi-mode Insulated Gate Transistor (BIGT) [18]. The electrical characteristics will be shown in the press pack section below.

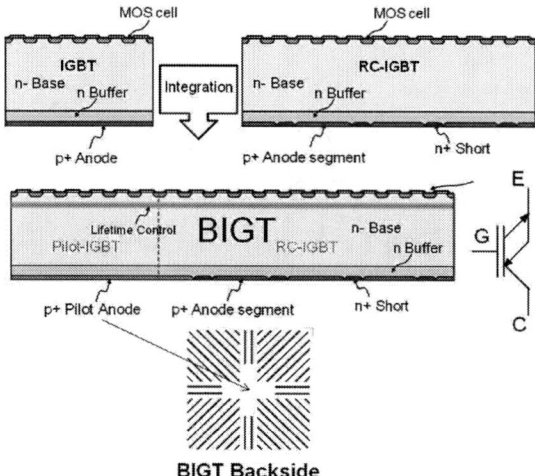

Fig. 12. BIGT (Bimode Insulated Gate Transistor), an advanced Reverse Conducting IGBT. Cross section and it's collector (backside) design.

As explained before, both enhanced trench IGBT cell concept and BIGT concept provide additional increases of output power for a given module. A further improvement step in current rating of the module can be achieved with the enhanced trench BIGT [19], which combines the advantage of the low loss enhanced trench IGBT cell concept and BIGT chip integration concept. The enhanced trench BIGT concept has been demonstrated with 3300 V chip and shows 20% improvement in the current rating compared to enhanced planer BIGT as shown in Fig. 13.

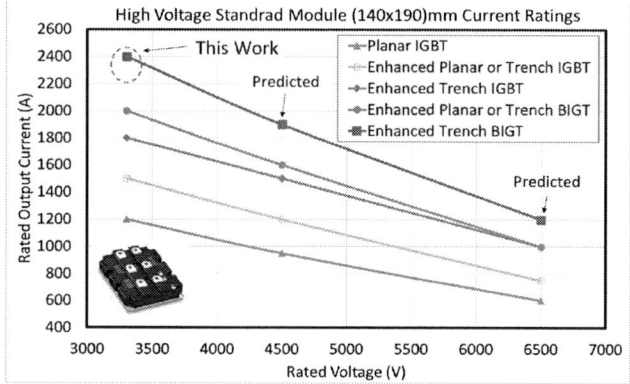

Fig. 13. High voltage standard module (140 mm x 190 mm) current ratings for 3300 V, 4500 V and 6500 V with different generations of IGBT technologies [19].

IGBT packaging

Insulated power modules for traction applications

The 140 x 190mm module (HiPak) shown in Fig. 14a has developed into a de facto standard for traction applications. In spite of its wide usage, this module type has some disadvantages. Especially the fact that it has a single switch configuration leads to relatively high stray inductances in the commutating circuit. To overcome this problem, we have proposed and developed a new half-bridge module, the LinPak, shown in Fig. 14b [20]. The new module not only massively reduces the stray inductance, but also allows for easy paralleling thus enabling modular converter designs that can be used to cover a wide range of output powers using the same basic layout. The LinPak module has been designed to accommodate as much as silicon chip area as possible. As shown in Fig. 15, this higher packing density corresponds to an improvement of approximately one chip generation. The step from a planar to an enhanced trench IGBT corresponds to a power density increase of 17%. In comparison, just shifting from the HiPak- to the LinPak modules using the same chips technology brings still an improvement of 14% (normalized to the same footprint).

3600

The 2018 International Power Electronics Conference

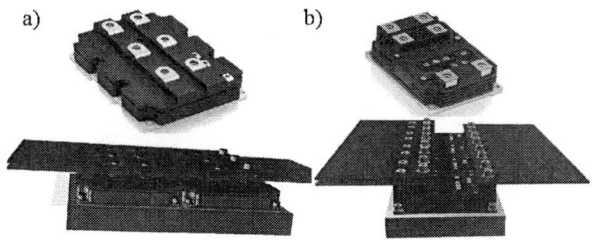

Fig. 14. a) 3.3kV/1500A HiPak Module (single switch). b) The newly developed 3.3kV/450A LinPak module (phase-leg configuration). The lower part of the picture shows the respective converter phase-legs with the buss-bars and the main converter terminals (red=DC+, blue=DC-, green=AC).

Module	IGBT Technology	Size [mm]		Swithes	Rating [A]	Current dens. [A/cm2]	
LinPak	Enh. Trench	100	140	2	525	7.5	17%
LinPak	SPT+	100	140	2	450	6.4	14%
HiPak	SPT+	140	190	1	1500	5.6	

Fig. 15. 3.3kV LinPak and HiPak module ratings and the correspond-ing footprint current density evolution.

Press pack modules for grid applications

Fig. 16 shows our new 4.5 kV StakPak module. It is the first commercially available high voltage module equipped with BIGT-chips. With its current rating of 3 kA, it is at the same time the most powerful IGBT-type device developed up to date having both IGBT and diode mode of operations. The module consists of six submodules each one having eight BIGT chips contacted with individual springs to allow for an easy clamping [21].

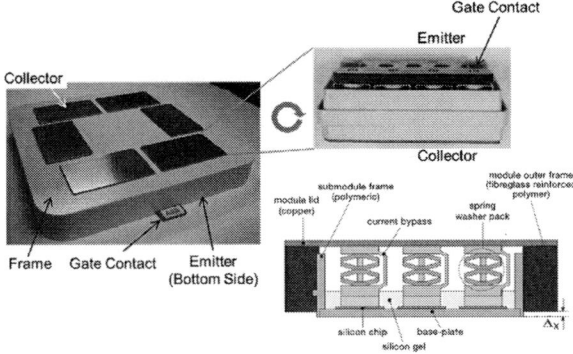

Fig. 16. 4.5 kV / 3 kA StakPak module with BIGT chips

Both the module and the BIGT chips have been optimized for HVDC converters using the MMC concept. One big ad-

vantage of this topology is the low switching frequency. This, together with the asymmetric loading in diode/IGBT mode makes the BIGT the ideal device for this type of applications. Fig. 17 and Fig. 18 show the turn-off waveforms of this module in IGBT- and diode-modes of operation, respectively.

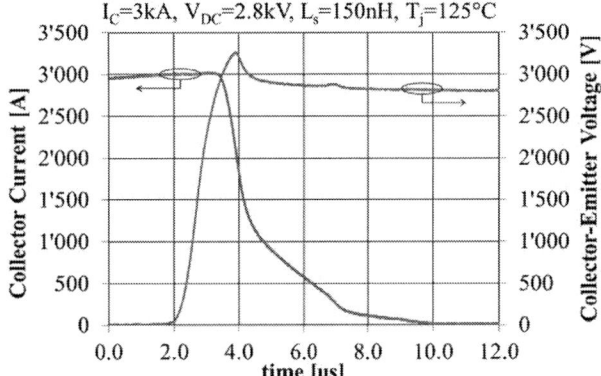

Fig. 17. 4.5 kV BIGT StakPak turn-off under nominal conditions (3 kA / 2.8 kV) in IGBT-mode of operation

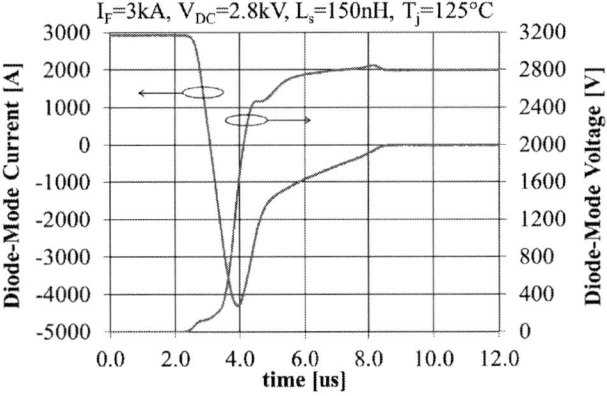

Fig. 18. 4.5 kV BIGT StakPak turn-off under nominal conditions (3 kA / 2.8 kV) in diode-mode of operation

Conclusion

SiC MOSFETs have over the past ten years been developed into mature products and are widely used in different applications where the higher chips costs are overcompensated by savings in the system. In high voltage / high power converters used in grid applications, traction- and large industrial drives these potential savings are as of today too small to motivate the use of SiC. Over the last 15-20 years, the power density of Si-devices has increased by a factor of two with potential for further improvements.

3601

References

[1] J. Vobecky, K. Stiegler, M. Bellini, and U. Meier, "New generation Large Area Thyristor for UHVDC Transmission", Proc. PCIM 2017, Nürnberg, p.761-764.

[2] J. Vobecky, H.-J. Schulze, P. Streit, F.-J. Niedernostheide, V. Botan, J. Przybilla, U. Kellner-Werdehausen, and M. Bellini, "Silicon Thyristors for Ultrahigh Power (GW) Applications", IEEE Trans. El. Dev., 64, 2017, p.760-768.

[3] J. Vobecky, T. Stiasny, V. Botan, and U. Meier, "New Thyristor Platform for UHVDC (>1 MV) Transmission", Proc. PCIM 2014, Nürnberg, p.54-59.

[4] V. Botan, J. Waldmeyer, M. Kunow, and K. Akurati, "Six Inch Thyristors for UHVDC Transmission", Proc. PCIM 2010, Nürnberg p.476-479.

[5] H. Gruening, B. Oedegard, A. Weber, E. Carroll, and S. Eicher, "High power hard-driven GTO module for 4.5 kV/3kA snubberless operation", Proc. PCIM 1996, Nürnberg, Germany, pp.169-183.

[6] K. Satoh, M. Yamamoto, T. Nakagawa, and A. Kawakami, "A new high power device GCT (gate commutated turn-off) thyristor", Proc. EPE, 1997, pp. 2070-2075.

[7] T. Wikström, T. Stiasny, M. Rahimo, D. Cottet and P. Streit, "The Corrugated P-Base IGCT - a New Benchmark for Large Area SOA Scaling, Proc. ISPSD, Jeju, Korea, 2007, pp. 29-32.

[8] I. Nistor, T. Wikström, M. Scheinert, M. Rahimo, and S. Klaka, "10kV HPT IGCT rated at 3200A, a new milestone in high power semiconductors", Proc. PCIM 2010, Stuttgart, Germany, pp. 467-471.

[9] U. Vemulapati, M. Rahimo, M. Arnold, T. Wikström, J. Vobecky, B. Backlund and T. Stiasny, "Recent Advancements in IGCT Technologies for High Power Electronics Applications", Proc. EPE 2015, Geneva.

[10] S. Linder, S. Klaka, M. Frecker and H. Zeller, "A new range of reverse conducting gate-commutated thyristors for high voltage, medium power applications", Proc. EPE 1997, Trondheim.

[11] Y. Yamaguchi et al, "A 6 kV/5 kA reverse conducting GCT", IEEE IAS, Chicago 2001, pp. 1497-1503.

[12] T. Wikström, and M. Alexandrova, "A technology platform for reverse conducting Integrated Gate Commutated Thyristors with 94 mm device diameter", Proc. PCIM 2017, Nürnberg.

[13] U. Vemulapati et al, "An experimental demonstration of a 4.5 kV "Bi-mode Gate Commutated Thyristor" (BGCT)", Proc. ISPSD, Hong Kong, 2015, pp. 109-112

[14] T. Stiasny et al, "Experimental results of a Large Area (91mm) 4.5 kV "Bi-mode Gate Commutated Thyristor" (BGCT)", Proc. PCIM 2016, Nürnberg.

[15] M. Andenna et al, "The Next Generation High Voltage IGBT Modules utilizing Enhanced-Trench ET-IGBTs and Field Charge Extraction FCE-Diodes", EPE 2014

[16] Y. Toyota et al, "Novel 3.3-kV Advanced Trench HiGT with Low Loss and Low dv/dt Noise", ISPSD 2013

[17] C. Corvasce et al, "3300 V HiPak2 modules with Enhanced Trench (TSPT+) IGBTs and Field Charge Extraction Diodes rated up to 1800A", Proc. PCIM 2016, Nürnberg.

[18] M. Rahimo et al, "The Bi-mode Insulated Gate Transistor (BIGT) A Potential Technology for Higher Power Applications", ISPSD 2009

[19] M. Rahimo et al, "Demonstration of an Enhanced Trench Bi-mode Insulated Gate Transistor ET-BIGT", ISPSD 2016, Prague.

[20] S. Hartmann, F. Fischer, A. Baschnagel, H. Beyer, R. Schnell, C. Treier, "The LinPak high power density design and its switching behaviour at 1.7 kV and 3.3 kV", Proc PCIM Europe 2016. Nürnberg

[21] F. Dugal A. Baschnagel, M. Rahimo, and A. Kopta, "The Next Generation 4500 V / 3000 A BIGT Stakpak Modules" PCIM 2017, in Proc. PCIM 2017, Nuremberg.

Recent Progress in High to Ultra-High-Voltage SiC Power Devices: Development and Application

Y. Yonezawa[1]

1 Advanced Power Electronics Research Center,
National Institute of Advanced Industrial Science and Technology, AIST, Tsukuba, Japan
*E-mail: yoshiyuki-yonezawa@aist.go.jp

Abstract- **The current status of silicon carbide (SiC) device development in various voltage ranges is reviewed. Especially for next-generation high to ultra-high-voltage devices, developments in SiC super-junction Metal Oxide Semiconductor Field Effect Transistors (MOSFET, here denoted as SJ-MOS) and SiC Insulated Gate Bipolar Transistors (IGBTs) are introduced. We expect that these next generation devices are going to trigger a paradigm shift in power electronics components, enabling very low conduction and switching losses.**

Keywords— IGBT, power device, power electoronics, silicon carbide.

I. INTRODUCTION

In response to rising energy demand, optimization of the energy value chain is desired for the efficient use of energy and to ensure the stability of energy supply. In particular, countermeasures against global warming and an early realization of a low-carbon emission society are expected with the large-scale introduction of renewable energy. At the same time, to improve the resilience of the electric power system for a safer and more secure society, there is a need to develop energy management technologies to create smart linkages inside and outside the smart grid. Recently, intercontinental super grids have been proposed using high-voltage direct current (HVDC) transmission systems, such as the Asian Super Grid. Energy management systems for local power production and consumption have also been developed. Under these circumstances, the role of power electronics has become increasingly important in the energy value chain with the fusion of energy and information.

II. COMPARISON OF SiC UNIPOLAR AND Si BIPOLAR DEVICES

Power electronics and power devices are the two sides of the ongoing evolution in energy distribution and consumption. The recent power electronics evolution, in particular, has been supported by improvement of the tradeoff between the reduction of conduction loss and switching loss in silicon (Si) IGBTs. However, since the

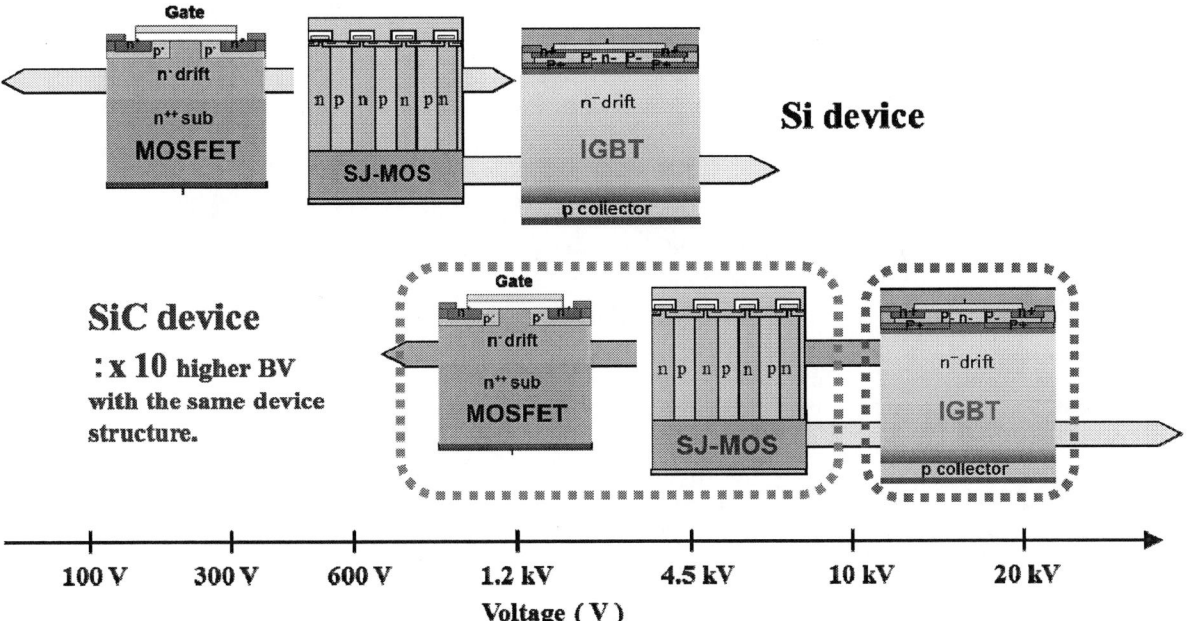

Fig. 1: Comparison of advanced Si and SiC devices.

performance improvement of Si-IGBTs has reached the physical limit, expectations for wide bandgap semiconductor devices are raised. SiC has a band gap three times larger than that of Si and a three times higher thermal conductivity. The breakdown electric field of SiC is 10 times higher than that of Si, allowing SiC devices to achieve 10 times higher breakdown voltages (BVs) than Si devices with the same structure, as shown in Fig. 1, along with a high junction temperature.

By replacing bipolar Si-IGBTs with unipolar 600 V–3.3 kV SiC-MOSFETs, a higher operating frequency will be possible, leading to lower switching losses, as well as significant size and cost reductions in power electronics components [1-3].

Another advantage of the replacement of the Si IGBT with a SiC MOSFET is the possibility to omit the freewheeling diode, i.e., Schottky barrier diode (SBD), by using the body diode of the MOSFET in the respective circuit. Omitting the SBD will contribute to the size and cost reduction of inverters, such as those for in-wheel motors. However, if we use a body diode, countermeasures against forward degradation are required due to bipolar operation. When the surge current enters, basal plane dislocations (BPD) in the substrate could expand, resulting in single shockley stacking faults expansion that cause forward voltage degradation. To prevent this phenomenon, we proposed a recombination enhancement layer that suppressed the number of holes reaching the substrate and, thereby, prevented the expansion of BPDs [4].

For high-voltage unipolar devices, 6.5–13 kV SiC-MOSFETs have been reported [5-6]. A 6.5 kV-class MOSFET is expected to be applied to the traction systems of high-speed trains and a 13 kV MOSFET is aimed at high-voltage power supplies. An SBD-embedded 6.5 kV SiC-MOSFET has also been developed [7]. Using the built-in SBD, the turn-on of the body diode can be suppressed.

III. NEXT GENERATION HIGH TO ULTRA-HIGH-VOLTAGE SiC DEVICES

With regard to next-generation MOSFETs, we are developing a 6.5 kV-class super junction structure that can reduce the drift layer resistance to less than half compared to a regular SiC-MOSFET. The target-specific on-resistance of the SiC SJ-MOS is shown in Figure 2.

If we apply the IGBT structure to SiC, over 10 kV MOS-controlled switching devices can be realized, which cannot be reached by any Si device to date. The specific-on-resistance target of SiC-IGBT is also shown in Fig. 2.

As for the material requirements of 13-33 kV IGBTs and PiN diodes, the withstand voltage layers need to have doping concentrations in the range of $(1–4) \times 10^{14}$ cm^{-3} and a thickness in the range of 100–300 μm [8]. With regard to the carrier lifetime, and in order to obtain sufficient conductivity modulation, a Shockley–Read–Hall carrier lifetime of approximately 30 μs is required because during conductivity modulation, lifetimes are

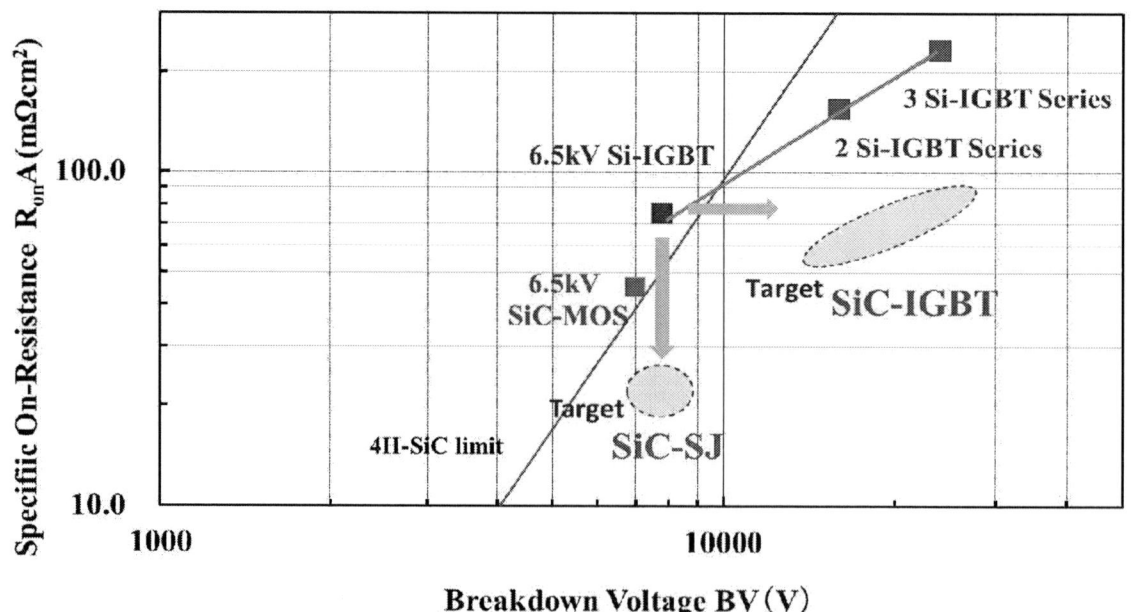

Fig.2: Specific on-resistance of SiC SJ-MOS and SiC-IGBT compared with Si-IGBT and normal SiC-MOS.

suppressed by direct and Auger recombination. For the bipolar device design, we have specifically developed a new calibration and new settings of TCAD parameters, since important simulation parameters such as carrier lifetime are not sufficiently well included in the respective libraries.

To date, BVs of 4.5–27 kV and 6.5–27 kV have been reported for PiN diodes and SiC-IGBTs, respectively, including also switching tests [9-15]. A demonstration of the turn-off switching waveforms of the 5.3 mm × 5.3mm 16kV n-channel SiC-IGBT devices is shown in Figure 3, in dependence of the DC bias voltage, ranging from 2,500 V to 10,000V. Bidirectional switches with two MOS gates on both sides have also been proposed [16].

Concerning conduction losses, the differential on-resistance of PiN diodes is close to the limit value of SiC. For SiC-IGBTs, the value of the differential on-resistance is still far from the theoretical limit. Moreover, improving the tradeoff between conduction losses and switching losses is a challenge in IGBTs. We also have to work on lifetime enhancement and device design optimizations, such as injection enhancement and injection control of the collector side, which is similar to the history of Si, in order to obtain a sufficient level of device performance.

In actual wafer preparation and processing, many problems have occurred such as suppressing the generation of interface dislocation between p^{++} layers and n$^-$ withstand voltage layers, and substrate cracking owing to the large lattice constant mismatch between those. Optimizations for both wafer preparation and device processing are needed.

IV. BIPOLAR SiC POWER DEVICES FOR GRID APPLICATIONS

Ultra-high-voltage SiC devices offer the possibility of drastically changing the design concept of existing power electronics components, not only for high-voltage inverters but also transmission and distribution systems. These include direct linkage of the mega solar systems to distribution systems without transformers, a solid state transformer (SST) and intelligent power switch for smart grids [17], a miniaturization of static var compensators (SVC), downsizing of HVDC valves for offshore wind power stations by reduction of the stages, and highly reliable high-speed circuit breakers. In addition, the new devices can contribute to the innovation of next-generation power systems, such as power system tidal current controls, system voltage boosting, and DC power supplies for realizing the loop system.

Utilizing SiC devices in power systems promotes not only the downsizing of power electronics components but also significant reductions in total cost, including operating cost, compared to the Si-based systems because of the low losses. However, the high initial cost of SiC devices is a major obstacle for the expansion of the SiC market. With improvements in quality to improve yields,

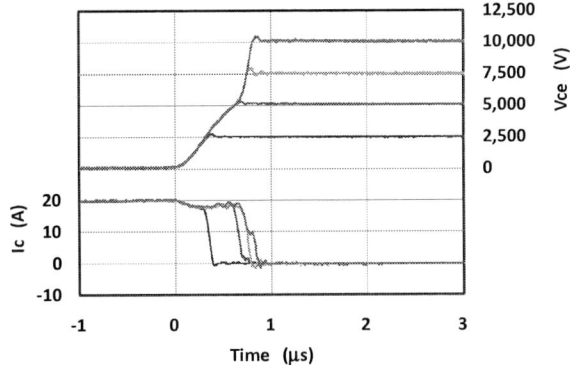

Fig. 3: The turn-off switching waveforms of the 5.3 mm × 5.3mm n-channel SiC-IGBT dependence on DC bas voltage from 2,500 V to 10,000V.

significant cost reductions for SiC substrates and the epitaxial layer growth processes are expected. To take full advantage of the high current density obtained with ultra-high-voltage bipolar devices, heat removal technology and insulation technology must be improved in ultra-high-voltage modules.

V. CONCLUSIONS

Recent progress in SiC device developments in various voltage ranges are introduced. We hope that the next generation SiC SJ-MOS and IGBT device technology will contribute towards the realization of a super smart society, providing the necessary products and services in the required amounts to people who need them when they need them, and fostering efficient energy usage, including energy storage, rather than increasing energy consumption.

ACKNOWLEDGMENT

Part of this work was supported by Council for Science, Technology and Innovation (CSTI), Cross-ministerial Strategic Innovation Promotion Program (SIP), "Next-generation power electronics" (funding agency: NEDO).

REFERENCES

[1] T. Kimoto and J.A. Cooper, Fundamentals of Silicon Carbide Technology (John Wiley & Sons, Singapore, 2014).

[2] Y. Kobayashi, et al., "3.3 kV-class 4H-SiC UMOSFET by Double-trench with Tilt Angle Ion Implantation", *Mater. Sci. Forum*, vol. 858, pp. 974–977, 2016.

[3] D. Peters, et al., "Performance and Ruggedness of 1200V Si – Trench – MOSFET" *Proceedings of ISPSD*, pp. 239–242, 2017.

[4] T. Tawara, et al., "Understanding and reduction of degradation phenomena in SiC power devices", *J. Appl. Phys.* vol. 120, pp. 115101–1, 2016.

[5] M. K. Das, et al., "10 kV SiC Power MOSFETs and JBS Diodes: Enabling Revolutionary Module and Power Conversion Technologies", *Mater. Sci. Forum*, vol. 717-720, pp. 1225-1228, 2012.

[6] H. Kitai, et al., "Low on-Resistance and Fast Switching of 13-kV SiC MOSFETs with Optimized Junction Field-Effect Transistor Region", *Proceedings of ISPSD*, pp. 343–346, 2017

[7] K. Kawahara, et al., "6.5 kV Schottky-Barrier-Diode-Embedded SiC-MOSFET for Compact Full-Unipolar Module", *Proceedings of ISPSD*, pp. 41–44, 2017.

[8] N. Kaji, et al, *IEEE Trans. Electron Devices*, "Ultrahigh-Voltage SiC p-i-n Diodes With Improved Forward Characteristics", vol. 62, no. 2, 374–381, 2015

[9] X. Wang and J. A. Cooper, "High-Voltage n-Channel IGBTs on Free-Standing 4H-SiC Epi layers", *IEEE Trans. Electron Devices*, vol. 57, no. 2, pp. 511–515, 2010.

[10] S. Ryu, et al., *Mater. Sci. Forum*. 717–720, 1135, 2012.

[11] Y. Yonezawa, et al., "Low V_f and highly reliable 16 kV ultrahigh voltage SiC flip-type n-channel implantation and epitaxial IGBT", *Tech. Digest of 2013 Int. Electron Device Meeting*, pp. 6.6.1–6.6.4, 2013.

[12] T. Mizushima, et al., "Dynamic characteristics of large current capacity module using 16-kV ultrahigh voltage SiC flip-type n-channel IE-IGBT", *Proceedings of ISPSD*, p. 277–280, 2014.

[13] Y. Yonezawa, et al., "Device Performance and Switching Characteristics of 16 kV Ultrahigh-Voltage SiC Flip-Type n-channel IE-IGBTs", *Mater. Sci. Forum*, vol. 821-823, pp. 842–846, 2015.

[14] E. van Brunt et al., "27 kV, 20 A 4H-SiC IGBTs", *Mater. Sci. Forum*, vol. 821-823, pp. 847–850, 2015.

[15] N. Watanabe, et al., "6.5 kV n-Channel 4H-SiC IGBT with Low Switching Loss Achieved by Extremely Thin Drift Layer", *Mater. Sci. Forum*, vol. 858, pp. 939-944, 2016.

[16] S. Chowdhury, et al., "Experimental Demonstration of High-Voltage 4H-SiC Bi-Directional IGBTs", *IEEE Electron Device Letters*, vol. 37, no. 8, pp. 1033-1036, 2016.

[17] K. Mainali, et al., "A Transformerless Intelligent PowerSubstation", *IEEE Power Electron. Mag*, vol. 2, no. 3, pp. 31–43, 2015.

Dynamic drift effects in GaN power transistors: Correlation to device technology and mission profile

Joachim Würfl[1*], Eldad Bahat-Treidel[1], Oliver Hilt[1], Maria Troppenz[1+], Mihaela Wolf[1],
Jan Böcker[2], Carsten Kuring[2], Sibylle Dieckerhoff[2]

[1] Ferdinand-Braun-Institut, Leibniz Institut für Höchstfrequenztechnik,
Gustav-Kirchhoff-Strasse 4, 12489 Berlin, Germany
[2] Technical University of Berlin, Einsteinufer 19, 10587 Berlin, Germany
[+] Now with Humboldt Universität zu Berlin, 12489 Berlin, Germany
* E-mail: Joachim.wuerfl@fbh-berlin.de

Abstract— GaN devices for high voltage power switching are facilitating smaller, more light-weighted and more efficient converter systems. In order to provide an optimum design of such systems it is necessary to understand dynamic GaN device performance in dependence on targeted mission profile and technological parameters. The paper shortly introduces to GaN device technology and provides a widely accepted physical interpretation of mechanisms that may adversely influence device switching properties. Then important scenarios of GaN power transistor switching are presented and correlated to biasing conditions relevant for in-system device operation. In detail, dynamic switching properties depending on off- and on-state time and voltage, substrate biasing conditions and temperature are analysed and correlated to different device technologies and manufacturers. The abovementioned parameters are influencing dynamic device properties in quite a complex manner and can often be considered as a characteristic finger print of a specific technological implementation or a specific device or epitaxial manufacturer.

Keywords— *drift effects, dynamic on-state resistance, gallium nitride, power switching*

I. INTRODUCTION

GaN devices for microwave and power switching are enabling new and innovative system applications. The specific material properties of AlGaN/GaN or similar heterojunctions facilitate very compact and extremely fast devices. On system level these properties translate into advantages such as low weight and low volume for a given power handling. Most of the GaN devices implemented so far are relying on lateral architectures (see Fig. 1a). The device current flows in an infinitesimal small sheet layer located at the interface between two adjacent semiconducting materials with different degrees of spontaneous and piezoelectric polarization. Thus a 2-dimensional electron gas (2DEG) forms whose properties depend on the pairing of the adjacent semiconductor materials, their respective mechanical strain and on charged trap states in the vicinity. Traps located close to the 2DEG influence electron population and are thus affecting maximum device current and on-state resistance

(see Fig 1b). In general, trapping and de-trapping depends on specific device biasing conditions. Strictly spoken this means that at a given bias condition the trap population is practically frozen and cannot respond to fast device switching transients. Therefore, any transition to a new bias point may be delayed as traps need to be charged or discharged according to their inherent time constants. For example, if negative charges are trapped in the vicinity of the channel, the device cannot fully turn on immediately after switching from an off-state to an on-state bias point. As an example of this scenario Fig. 1 (b) illustrates a typical situation where negatively charged traps appear in the vicinity of the transistor channel. Due to the demand for overall charge neutrality the existence of these traps leads to a reduction of the number of channel electrons. Thus after turning on the device, the drain current will be compromised initially and will reach its final value only after the traps have been discharged. This phenomenon is known as "dynamic on-state resistance increase" for power switching devices or "gate or drain lagging" in GaN microwave devices. It turns out that the properties of the dynamic on-state resistance strongly depend on history of device biasing, on the time elapsed since device switching, on local electric field in the active device, on temperature and on polarity and magnitude of substrate biasing. Furthermore the dynamics of switching e.g. the switching speed is also decisive.

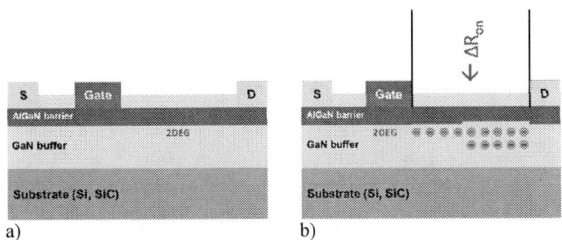

Fig. 1. Basic representation of a GaN field effect transistor showing ideal channel conductivity (a) and reduced channel conductivity (b) because of trapping in the vicinity of the channel [1, 2].

II. ANALYSES OF DYNAMIC DRIFT EFFECTS

It is important to test GaN power switching devices in almost all biasing conditions that might appear during practical in-system applications. It has turned out that GaN devices maybe particularly prone to dynamic drift effects. This means that the device characteristics might change after the devices have experienced a certain electric or thermal load [1,2,3]. Depending on specific technology the devices usually show more or less pronounced memory effects and therefore show switching performance depending on biasing history. For circuit and system designers it is therefore very important to be aware of these properties and to take them into account.

A. Dynamic on-state resistance characterization

As the dynamic on-state resistance depends on device biasing history it is important to set up a characterization tool that takes of this issue in a quantitative manner. This means that device biasing conditions before the switching event have to be set precisely in terms of off-state time and voltage. While turning on the device it is important to provide a precise time dependent measurement of the on-state current.

Fig. 2 shows one of the possibilities of measuring dynamic on-state resistance using a resistive load switching setup. A high voltage capacitor is first charged, and then switched to the off-state biased test transistor (DUT) for a well-defined amount of time. While turning on the DUT the capacitor partly discharges through a load resistor and the DUT itself. The value of on-state resistance is then determined by the source current and the voltage drop across the DUT.

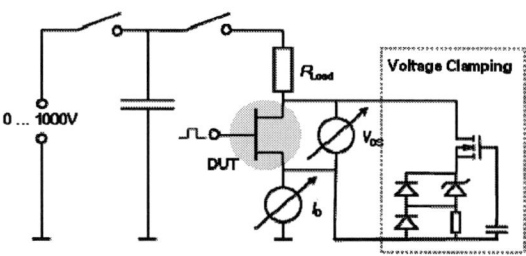

Fig. 2. Measurement set-up used for evaluating dynamic on-state resistance. The voltage clamping circuit has been realized according to [4].

B. Unveiling interdependencies

The measurement set-up according to Fig. 2 has been used to characterize changes of dynamic on-state resistance in dependence on various parameters such as off-state bias voltage, time at off-state before turning on the DUT and temperature.

1) Drain bias voltage

It is well known that the magnitude of dynamic on-state resistance depends on off-state bias voltage [2,3]. In order to analyse this in detail and to find out a possible dependence on the properties of the epitaxial layers wa-

fers from different epitaxial vendors have been compared to each other. The wafers have been processed in the same batch using FBH's normally-off p-GaN gate device technology [5].

Fig. 3 shows the evolution of dynamic on-state resistance in dependence on drain voltage. In the actual case the on-state resistance is normalized to quasi-static properties obtained after switching from a bias of 20 V. Wafers with nominally the same epitaxial structures from different epi vendors (1, 2 and 3) are compared to each other. Throughout measurements the off-state time had been set to 300 ms, the on-state resistance was measured 10 µs after switching. Although the device process itself had been the same for all wafers, wafers form different epi vendors showed quite some individual performance. All devices showed a certain dynamic deviation from the static on-state value however there is obviously not a general rule that the dynamic on-state resistance always monotonically increases with drain bias voltage. For example, devices fabricated on a wafer from Epi vendor 2 depict a clear maximum between 400 V and 450 V drain bias, and show even better properties at higher drain voltage; devices on wafers from Epi vendor 1 degraded at around 500 V whereas devices on wafers from Epi vendor 3 showed a rather constant but small dynamic on-state resistance increase nearly independent from drain voltage.

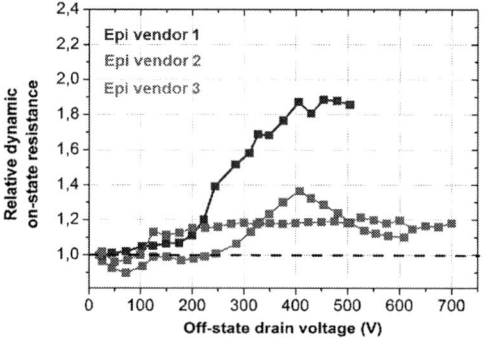

Fig. 3. Relative dynamic on-state resistance normalized to the conditions of 20 V switching in dependence on off-state drain-source voltage before turning on the devices. (off-state time: 300 ms, on-state 10 µs, 25 °C). Wafers from different epi vendors are compared to each other.

The difference between the individuals is mainly due to a different epitaxial implementation of the GaN buffer. This obviously creates individual trap states at different energy levels and different geometrical positions in the device. These traps can even be considered as a typical "fingerprints" of a given epi vendor. The behaviour of Epi vendor 2 can be understood as interplay between increased electron trapping at off-state conditions and increased device leakage to the substrate as the drain voltage is ramped up. This results in a leakage induced de-trapping of previously occupied trap states. The theoretical background for this mechanism can be found in [5].

The 2018 International Power Electronics Conference

2) Long-term off-state biasing

The initial value of the on-state resistance after turning on the device may significantly depend on the time the device has remained at off state. In order to study this effect more in detail, GaN Schottky gate devices depicting a noticeable increase of dynamic on-state resistance have been selected. According to Fig. 4 these devices were biased at 200 V off-state for a defined time and turned on afterwards. The turn-on resistance has then been monitored 200 ms after the switching event and plotted versus off-state time in Fig. 4 a). After each switching event the devices were fully de-trapped by operating them at on-state condition at a gate voltage of +1 V and a drain voltage of +1 V for 1000 s. The devices are then recovering

to the static value of R_{on} (Fig. 4 b), which makes sure that each subsequent off-state trapping test starts from the same conditions. Repeating this measurement sequence leads to the time dependence of R_{on} shown in Fig. 4 (a). Even short trapping times at off-state rapidly increase R_{on} followed by a gradual increase up to ~100 s off-state time. Longer off-state times lead to a further drastic increase of R_{on_dyn}. It should be noted that this effect can be minimized significantly with an appropriate device (epitaxial) technology. As the basic mechanism behind is inherently linked to the material quality of epitaxy and the maturity of processing this feature should be characterized also in devices from commercial vendors in order to cope with long term off-state mission profiles.

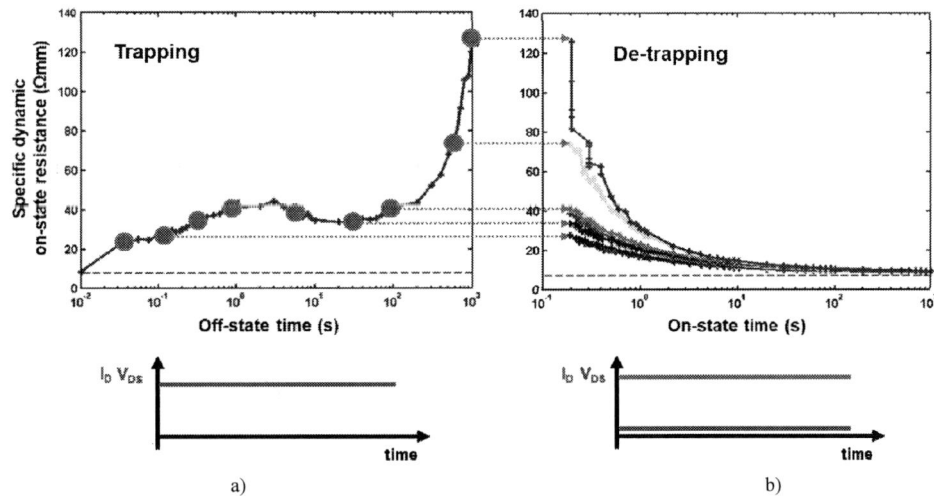

Figure 4: a) Dependency of dynamic on-state resistance (measured 10 µs after turn-on) on duration of off-state bias at 200 V prior to switching event. b) Time dependency of de-trapping at $V_{ds} = 1$ V and $V_{gs} = +1$ V after off-state biasing according to the data points selected in a). Trapping and de-trapping experiments have been performed at 80°C base-plate temperature

3) Off-state timing and temperature

In Fig. 3 the wafer of epi vendor 3 shows the lowest variability of dynamic on-state resistance with drain bias. However, if the temperature dependence of the dynamic on-state resistance is characterized a well pronounced dynamic on-state resistance increase is observed under the following conditions: The off-state drain bias voltage level has to be between 100 and 200 V, the temperature has to be elevated and the off-state drain voltage has to be applied for quite long time before switching (2500 ms). Fig. 5 visualizes this effect.

As evident from Fig 5 the absolute value of on-state resistance increases with temperature. This is due to electron scattering in the channel. However, if the time at off-state bias increases, R_{on} reaches the abovementioned maximum value between 100 V and 200 V. Therefore, at a specific drain voltage (means at a certain internal electrical field in the device) and temperature negative trap charges are crated in channel vicinity. In consequence these charges are compromising (reducing) channel electron population. Localized negative charges can be creat-

ed for example by de-trapping of acceptor states or by electron trapping. At a high level of off-state drain voltage this effect diminishes. This may be due a field dependent trap population statistics or to leakage current effects. In case of devices fabricated on wafers from Epi vendor 3 this effect is not very pronounced but still visible and has to be taken into account when evaluating overall device performance.

Similar investigations have been performed on devices based on wafers from Epi vendor 2 at a drain bias voltage of 200 V. According to Fig. 3 the on-state resistance increase at this drain bias level can be practically neglected for the measurement conditions selected there. However if temperature and timing of off-state bias is varied systematically, a completely different scenario appears. The 3D-plot shown in Fig. 6 visualizes the observed behaviour. First of all, the on-state resistance increases with temperature. As discussed previously this is a very general effect. However at elevated temperatures and in combination with a certain off-state timing (around 10 ms) a distinct maximum of dynamic on-state resistance increase appears. Compared to the properties observed on devices based on wafers from Epi vendor 3 the actual dynamic R_{on} increase is very pronounced; R_{on} rises by a factor of 16. In actual power electronic circuits this increase could give rise to premature burnout if this parameter combination is accidentally hit during in-system device operation. According to [6] the basic physical mechanism causing this effect is due to the interaction of trapping and de-trapping phenomena with different activation energies and effective trap capture cross sections.

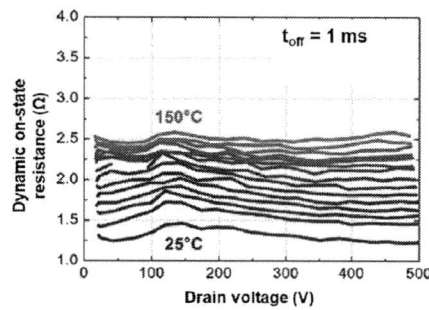

a)

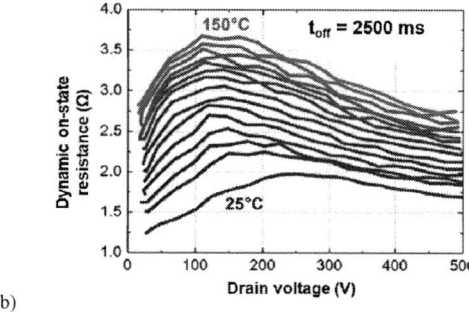

b)

Fig. 5. Absolute dynamic on-state resistance of devices fabricated on wafers from Epi vendor 3 in dependence on off-state drain bias for a duration of 1 ms (a) and 2500 ms (b) with different device temperatures ranging from 25°C to 150°C as parameters.

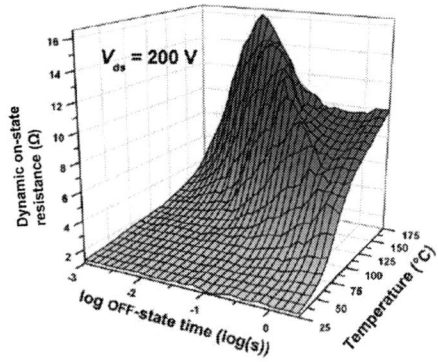

Fig. 6. 3D-Plot of absolute dynamic on-state resistance in dependence on off-state time (log scale) and temperature. The measurements have been taken at a drain bias of 200 V on a wafer with epitaxy from Epi vendor 2.

For even longer time constants exceeding 2500 ms and elevated temperature a second maximum tends to pop up

(right top corner of Fig. 6). This indicates the existence of another trap state that will be noticeable only at higher temperatures and after even longer time intervals at off-state.

4) Influence of substrate bias

In a GaN-on-Si device the active region of the transistor is separated from the conductive substrate by a few micron (typically 4-6 μm) only. Therefore the bias of the substrate with respect to the active transistor elements on top of the wafer may influence device performance. Depending on the interconnection scheme of the devices, for example in half or full bridge this needs to be considered by circuit designers. A simple case backgating may take place if the substrate is biased at a high negative voltage. If the full voltage applied to the vertical epitaxial layer stack (see also Fig. 1) the channel may be partly or totally depleted resulting in a higher turn-on resistance. If the semiconductor region where the voltage drop is taking place contains trap states which can be activated by the electric field, charges can be released which may give rise to an additional dynamic effect.

The sensitivity against backside biasing effects significantly depends on epitaxial wafer properties. Fig. 7 compares the properties of wafers from two different vendors which have been processed together in the same device process. For this test "gateless" transistor tests structures have been used as these devices are suitable to directly characterize the influence of backside bias on the 2DEG region itself without being disturbed by the high field region of a gated structure.

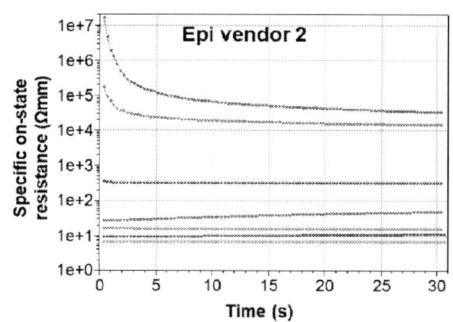

a)

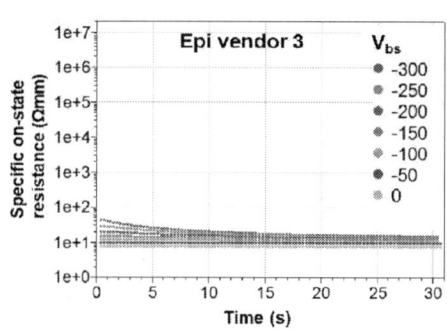

b)

Fig. 7. On-state resistance of a "gateless" transistor test structure after switching the substrate to a given negative bias voltage V_{BS}. Figures a) and b) are comparing epitaxial layers from different commercial GaN-on-Si suppliers, see also Fig. 3.

Both devices show some transient properties lasting for about 10 minutes after turning on the negative substrate bias. Afterwards practically steady-state conditions are obtained. Here the behaviour significantly differs from epi vendor to epi vendor ranging from an on-state resistance increase by almost 4 orders of magnitude for epi vendor 2 to a moderate increase by a factor of 3 for epi vendor 3. In any case, the test clearly shows that device substrate biasing cannot be neglected when setting up power switching circuits. The individual properties of the devices have to be taken into account.

Similarly, the dynamic properties of GaN power switching devices significantly depend on polarity of substrate bias. Fig. 8 compares quasi vertical GaN power switching devices according to the inset given in the same figure. In one case the source is electrically grounded to the substrate and in the other case where the drain is connected to the substrate. Both types of devices where dynamically characterized according to the conditions given in Fig. 3. It is clearly visible that for higher drain biasing the drain-down device shows significantly inferior performance as compared to the source down device. This behaviour may be related to deep acceptor like traps in the epitaxial buffer structure of the devices. Due to the back-side biasing ionized acceptors are created during off-state biasing in a similar manner as depicted in Fig. 1 b) however with a different spatial distribution. In fact, as the whole substrate is at drain potential, also acceptor traps will be activated leading in a much larger volume of the device leading to the observed well pronounce dynamic increase of resistance.

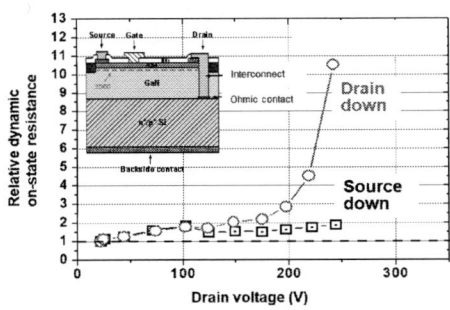

Fig. 8: Example of dependency of dynamic on-state resistance on substrate biasing: The on-state resistance is normalized to 20 V switching conditions.

5) Continuous switching

In real power-conversion systems the devices are continuously switching on and off at a given switching frequency. In terms of device physics this means that the devices are repeatedly driven into a continuous sequence of trapping and de-trapping. Depending on the respective time constants for trapping and de-trapping it is expected that an effective dynamic R_{on} builds up. In particular this should depend on drain bias, on-state current and switching frequency. Fig. 9 depicts this effect. Here, a normally-off GaN device having a static on-state resistance of 75 mΩ and a dynamic on-state resistance increase at 400 V drain bias of only 70% (as measured after 100 µs

off-state and 5 µs after a single switching event) has been operated in a continuous switching mode between 25 kHz and 500 kHz. The dynamic on-state resistance increases with switching frequency and seems to saturate at a rather high value at 500 kHz (increased by a factor 4 compared to its static value). Obviously, the temperature rises with the switching frequency as well due to increased switching losses. Therefore both of the effects, temperature and trapping and their correlation that was shown in Fig. 6 lead to the observed increase of the resistance. A method to separate both effects during a continuous operation is presented in [6].

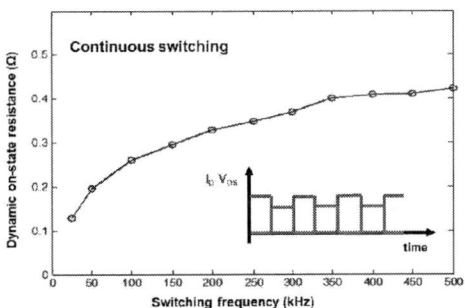

Fig. 9. Dependence of on-state resistance on switching frequency, off-state drain bias 80 V, on-state drain current 4 A drain bias, duty cycle 0.5, measurements taken after 250 ms continuous operation . Device under test: p-GaN gated normally-off GaN HFET with 75 mΩ static on-state resistance.

III. COMPARING COMMERCIAL DEVICES

A systematic comparison of the current switching properties of GaN devices from different vendors has been performed by double pulse testing according to Fig. 10. In contrast to the tests described in Fig. 2 where the transistor is switched to a resistive load at the drain, the device is now switched to an inductive load in a half-bridge arrangement. Thus the current through the device is determined by the bias voltage across the inductor (practically V_{dc} if the DUT is fully turned on) and the on-state time. The test itself proceeds as follows: After a defined off-state bias time the DUT is turned on at the gate for a certain amount of time until the targeted current is reached. Then the transistor is switched to off-state for a short time (1- 2 µs) and turned on again. During the short off-state time between the turn-on pulses the inductor current commutates through the top transistor of the half-bridge. This arrangement allows testing hard switching conditions of the DUT.

Fig. 11 compares the test results obtained from devices of different vendors. As the devices do not have the same specification in terms of voltage and current rating a comparable situation has to be defined. Thus the devices were switched to a drain current that gives the same static drain voltage drop of about 800 mV. The switching voltage has been selected according to the specified safe maximum voltage of each device. This selection should provide fairly comparable internal electric fields in the devices.

The 2018 International Power Electronics Conference

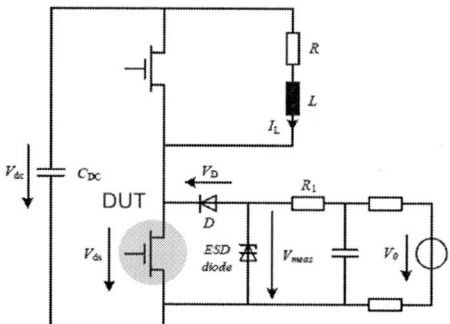

Fig. 10. Half-bridge test circuit with a clamped drain-source voltage measurement. This test mimics hard switching of an inductive load.

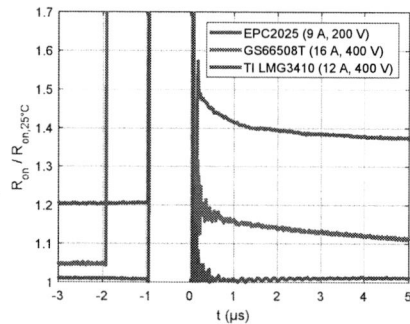

Fig. 11. Comparison of dynamic on-state resistance increase of devices from the different vendors EPC, GaNsystems (GS) and Texas Instruments (TI). All devices, with the TI device as an exception, had a normally-off gate drive characteristic.

Already during ramp-up time of the inductor current through the respective DUT a certain increase of the on-state resistance can be observed. Since all device tests are using the same inductive load, the ramp up time differs depending on the DC-link voltage and the chosen current. Therefore the EPC device has the longest ramp up time of 33 µs, compared to 22 µs for the TI device and 29 µs for the GaN Systems device. Fig. 11 only shows a small fraction of the ramp-up time, the time scale is normalized to the time of second turn-on. The observed on-state resistance increase during current ramping could have a thermal reason; however trapping is not excluded as well.

After turning off the device and switching it on again (at t=0 µs) a very pronounced increase of the dynamic on-state resistance is observed for the EPC and the GS device while the TI device behaves nearly ideal. Although the results would indicate that normally-on devices show a better switching performance it is not possible to really derive the root cause of this behaviour from the actual switching experiments. Charge trapping depends on many other parameters and may be due to specific epitaxial or passivation related properties of the devices from different vendors.

III. CONCLUSIONS

The dynamic properties observed in GaN power switching transistors strongly depend on technology, biasing conditions and temperature. For power electronic circuit designers it is of utmost importance to be aware of these interdependencies. At certain device operation conditions internal trapping constellations might be present which could even endanger safe device operation. The investigations have shown that devices fabricated on epitaxial layers from different vendor depict a very different device performance even if the wafers had been processed together in a stable device process. Furthermore devices fabricated by different vendors show quite different dynamic properties and thus would also behave in a different manner if implemented in power switching systems.

REFERENCES

[1] J. Würfl: „Drift Effects in GaN High-Voltage Power Transistors" in "Power GaN Devices: Materials, Applications and Reliability", Editors: Matteo Meneghini, Gaudenzio Meneghesso, Enrico Zanoni, ISBN 978-3-319-43197-0, pp.295-317, 2017

[2] J. Würfl et al., "Techniques towards GaN power transistors with improved high voltage dynamic switching properties," IEEE International Electron Devices Meeting, Washington, DC, pp. 6.1.1-6.1.4., 2013

[3] M. J. Uren, S. Karboyan, I. Chatterjee, A. Pooth, P. Moens, A. Banerjee, M. Kuball, ""Leaky Dielectric" Model for the Suppression of Dynamic R_{ON} in Carbon-Doped AlGaN/GaN HEMTs," IEEE Transactions on Electron Devices, vol. 64, no. 7, pp. 2826-2834, 2017.

[4] B. Lu, et al., "Extraction of Dynamic On-resistance in GaN Transistors under Soft- and Hard-switching Conditions" Compound Semiconductor Integrated Circuit Symposium (CSICS), 6062461, 2011.

[5] O. Hilt, E. Bahat-Treidel, A. Knauer, F. Brunner, R. Zhytnytska, and J. Würfl, "High-voltage normally OFF GaN power transistors on SiC and Si substrates", MRS Bull., vol. 40, no. 05, pp. 418-424, 2015.

[6] J. Böcker, C. Kuring, M. Tannhäuser and S. Dieckerhoff, "Ron increase in GaN HEMTs - Temperature or trapping effects," 2017 IEEE Energy Conversion Congress and Exposition (ECCE), Cincinnati, OH, 2017, pp. 1975-1981.

Compensation Method of Radial Unbalance Force at Failure of a Motor Section in a d-q Axis Current Control Bearingless Motor

Masahide Ooshima

Department of Electrical and Electronic Engineering, Tokyo University of Science, Suwa, Chino, JAPAN
E-mail: moshima@rs.suwa.tus.ac.jp

Abstract—This paper focuses on a stabilized magnetic suspension control method when a motor section is unfortunately failed during its operation in the d-q axis current control bearingless motor (d-q BELM). The d-q BELM consists of three sections, in which is divided by every 120° part in the stator circumferential direction. In each section, the 3-phase winding current is independently controlled by an inverter. Hence, it is possible that the rotor shaft will be supported by the other two sections without mechanical contact even though a section is failed. The stabilized control method at no load has been proposed by the authors, and the validation of the proposed method has been verified by the simulation and experimental test results. The aim of this paper is to support the rotor shaft even under loaded condition when a section will be failed. In order to solve this problem, the suspension force compensator is newly proposed and the effectiveness is found by FEA (Finite Element Analysis) using a simulation software.

Keywords— magnetic levitation, bearingless motor, permanent magnet synchronous motor, field-strengthening and -weakening controls

I. INTRODUCTION

In the bearingless motors (BELMs), the rotor shaft is supported without mechanical contact in the bearing journal [1]. It is maintenance free and hence the longevity of the motor is increased. The functions of motor and magnetic rotor levitation are successfully integrated so that the rotor shaft can be made shorter and at the same time it requires less number of inverters, controller and electric wires as compared to a motor with magnetic bearings. Thus, the overall size and cost of the bearingless machine is considerably reduced as compared to the machine with magnetic bearings. Furthermore, there is no fear to decrease the critical speed of the BELM by the rotational axis bend. The BELM has been commercially available in the liquid pumps [2].

The d-q axis current control bearingless motor (d-q BELM), which has been proposed by the authors, is one of the BELMs with the combination of the motor and suspension windings [3]-[7]. The advantageous feature of the d-q BELM is as follows. 1) The motor structure is just the same as the conventional brushless dc motors. The stator winding is short-pitched, hence, the winding

arrangement becomes simple. 2)The motor volume is less without reducing the capability of the motor. Hence, the torque and force densities are increased as compared to the conventional BELMs. 3)The control method is similar to that of the interior permanent magnet synchronous motor (IPMSM), i.e., in the d-q BELM, the rotational torque is controlled by the q-axis current and the suspension force is controlled by the d-axis current (the field-weakening or field-strengthening controls). Thus, the control method is also quite simple and the general-use 3-phase inverter can be employed to control the torque and suspension force. Additionally, the controllable region of the torque and suspension force is expanded [6][7].

A prototype machine was built and it was confirmed by the experimental tests that the rotor shaft was stably suspended without mechanical contact on the bearing journals during its operation at no load and under loaded condition.

The bearingless motors can continuously operate for a long time by the advantages mentioned above. However, there is fear that the inverters will be unfortunately failed or the stator winding will be shortened or opened by some reasons during the long time operation. Hence, the fault diagnosis and fault tolerance techniques are needed also in the bearingless drives. However, the literatures with respect to them have not yet published in the area of bearingless drives.

The d-q BELM consists of three sections, in which is divided by every 120° part in the stator circumferential direction. In each section, the 3-phase winding current is independently controlled by an inverter. Hence, it is possible that the rotor shaft will be supported by the other two sections without mechanical contact even though a section is failed by accidents such as a stator winding short or open, a failure in the inverter. If possible, the motor can be able to be continuously operated supporting the rotor shaft without mechanical contact until the failed section will be repaired. The stabilized control method at no load has been proposed by the authors, and the validation of the proposed method has been verified by the simulation and experimental test results [5].

As the next step, in this paper, it is focused how the rotor shaft is supported by non-contact when the accidents will occur while operating the d-q BELM under loaded condition. The unbalance force is applied on the rotor surface in the tangential direction by the q-axis flux, which is generated by two sections under normal operation. In order to support the rotor shaft by non-contact, the unbalance force should be successfully eliminated. The compensation method of the suspension force considering the unbalance force is proposed, and the effectiveness is obviously described by the simulation results.

II. Structure and Principle of D-Q Axis Current Control Bearingless Motor

Fig. 1 shows the cross section of the proposed d-q BELM. The stator core is classified into three sections as the section-α, the section-β and the section-γ. In the section-α, the there-phase three-wire windings $N_{u\alpha}$, $N_{v\alpha}$ and $N_{w\alpha}$ are wound; in the section-β, $N_{u\beta}$, $N_{v\beta}$ and $N_{w\beta}$ are wound; in the section-γ, $N_{u\gamma}$, $N_{v\gamma}$ and $N_{w\gamma}$ are wound. All these windings are short pitch and simple. Each section is controlled by separate general-use 3-phase inverters. Totally, three 3-phase inverters are needed to drive the proposed BELM. The number of inverters is much than that of the conventional BELM. However, the capacity per an inverter may be decreased to 1/3 times that of the conventional motor. Hence the total volume and cost of the inverters are almost compatible with those of the conventional motors.

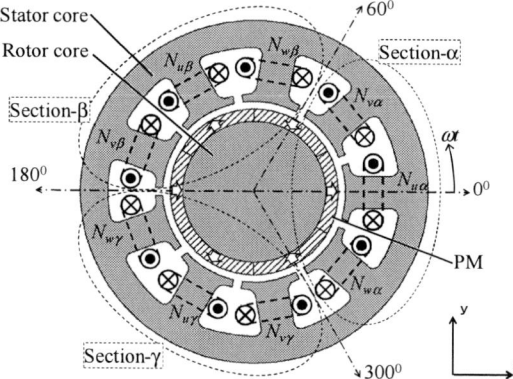

Fig. 1. Cross-section of d-q BELM

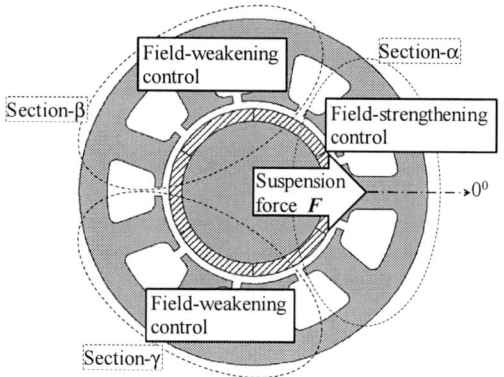

Fig. 2. Principle of the suspension force generation in d-q BELM.

Fig. 2 shows the principle of the suspension force generation in the proposed d-q BELM. The suspension force is generated by unbalanced flux density in air-gap with controlled d-axis currents $i_{d\alpha}$, $i_{d\beta}$ and $i_{d\gamma}$ in each section. For example, in section-α the field-strengthening control is done and then the air-gap flux density is increased; in the section-β and section-γ the field-weakening control is done and then the air-gap flux density is decreased. By the net vector sum in three sections, thus the suspension force is generated in the x-positive direction. By these controlled d-axis currents, the suspension force can be successfully generated in the arbitrary radial direction.

III. Stabilized Control Method at Failure of a Motor Section

Let us suppose that the section-γ is unfortunately failed. Fig. 3 shows the suspension force in the α-β axis coordinate and the coordinate transformation into the x- and y-axis coordinate. Normally, the rotor shaft is stably suspended by the net vector sum of three electromagnetic forces generated in the section-α, the section-β and the section-γ, respectively. However, suspension force is generated by the net vector sum of two electromagnetic forces generated in the section-α and the section-β if the γ-section is failed. In Fig. 3, using the suspension force components F_α and F_β along the α- and β-axis, the x- and y-axis components F_x and F_y of the suspension force F are shown as, respectively,

$$F_x = F_\alpha + F_\beta \cos 120° = F_\alpha - \frac{1}{2}F_\beta \tag{1}$$

$$F_y = F_\beta \sin 120° = \frac{\sqrt{3}}{2}F_\beta \tag{2}$$

Solving for the F_α and F_β in (1) and (2),

$$F_\alpha = F_x + \frac{\sqrt{3}}{3}F_y \tag{3}$$

$$F_\beta = \frac{2\sqrt{3}}{3}F_y \tag{4}$$

In the practical magnetic suspension controller, the required suspension forces F_x and F_y along the x- and y-axes are determined based on the instantaneous rotor radial displacements x and y, respectively. And then, the F_x and F_y are transformed into the suspension forces F_α and F_β along the α- and β-axes using the equations (3)

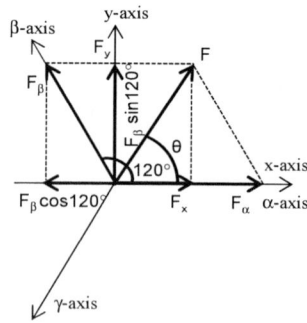

Fig. 3. Coordinate transformation.

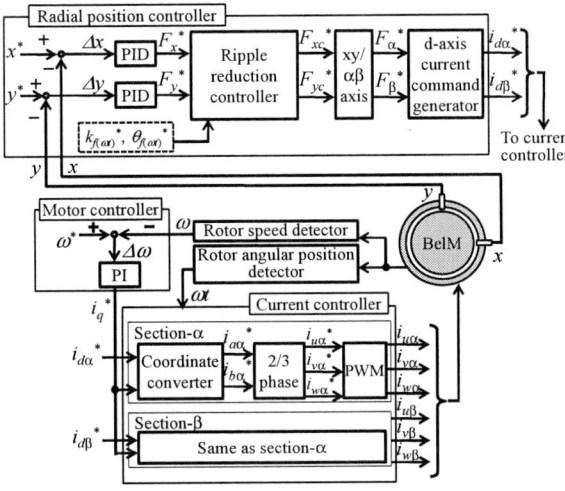

Fig. 4. Configuration of stable suspension control system at failure.

and (4). As a result, the rotor shaft can be successfully suspended without mechanical contact by the resultant suspension force of F_α and F_β, which are generated in the section-α and β under the normal operation. In the next section, the practical control system configuration at failure will be indicated.

Fig. 4 shows the control system configuration when the section-γ is failed and the other two sections, α and β are normally operated. The rotor radial displacements x and y are detected by the eddy current type gap sensors. The difference between the detected displacement and its reference is amplified by the Proportional-Integral-Derivative (PID) controller in the x- and y-axes, respectively, and then the suspension force commands F_x^* and F_y^* along the x- and y-axes are determined by the output of PID controllers, respectively. In the d-q BELM, the magnitude and direction of suspension force are varied in accordance with the rotor angular position based on its principle. In [1], the ripple reduction control method of the suspension force has been proposed to stably levitate the rotor shaft and explained in detail. According to its method, the magnitude compensation coefficient $k_{f(\omega t)}$ and direction compensation coefficient $\theta_{f(\omega t)}$ are defined and successfully identified. If the section-γ is failed, these coefficients need to be changed compared with those at normal operation by three sections, however, the purpose and principle of ripple reduction control is not changed. The coordinate transformation from x-y to α-β is done, i.e., the F_x^* and F_y^* are transformed into the force commands F_α^* and F_β^* along the α- and β-axes using the equations (3) and (4). Furthermore, the d-axis current commands $i_{d\alpha}^*$ and $i_{d\beta}^*$ in the section-α and β are determined to be proportional to the suspension force commands F_α^* and F_β^*, respectively.

In the current controller, the current is independently controlled in each section. In the section-α, for example, the current commands i_q^* and $i_{d\alpha}^*$ are transformed into 2-phase the current commands $i_{a\alpha}^*$ and $i_{b\alpha}^*$ in the stationary

coordinate. Then $i_{a\alpha}^*$ and $i_{b\alpha}^*$ are transformed into $i_{u\alpha}^*$, $i_{v\alpha}^*$ and $i_{w\alpha}^*$. The winding currents $i_{u\alpha}$, $i_{v\alpha}$ and $i_{w\alpha}$ are regulated to follow the current commands in the Pulse Width Modulation (PWM) block. The current controller of the section-β is same as that of the section-α. The current controller block is the same as those at the normal operation by the three sections except for the lack of the section-γ block.

IV. Unbalance Force at Failure of a Motor Section under Loaded Condition

The effectiveness and validation of the proposed stabilized control method when a motor section is unfortunately failed at no load have been verified by the simulation results by FEM and the experimental test results using a prototype machine [5]. Note that the unbalance force by the q-axis flux, which is oriented in radial direction, is generated when a motor section is failed under loaded condition. Fig. 5 shows the unbalance force in the radial direction by the q-axis flux when the γ-section is failed. The torque is generated by the q-axis fluxes in α- and β-sections, which are normally operated. The forces are applied on the rotor surface and oriented in the tangential direction in each section. The applied forces on the rotor are unbalanced as they are not applied in γ-section due to the failure; as a result, the radial force is applied on the rotor and it causes the lack of the rotor stability. In Fig. 5(a), $F_{T\alpha}$ and $F_{T\beta}$ are the sums of

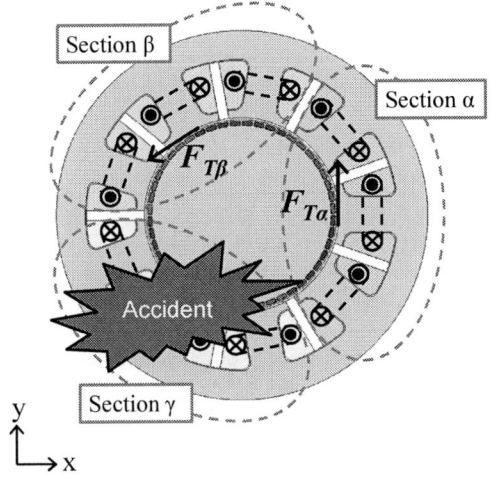

(a) Tangential forces on rotor surface when the section γ breaks down.

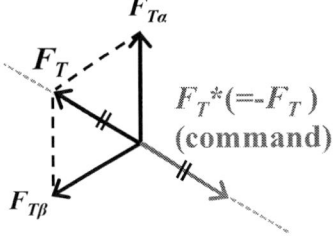

(b) Resultant force of tangential forces.

Fig. 5. Unbalanced force by q-axis flux when the section γ breaks down.

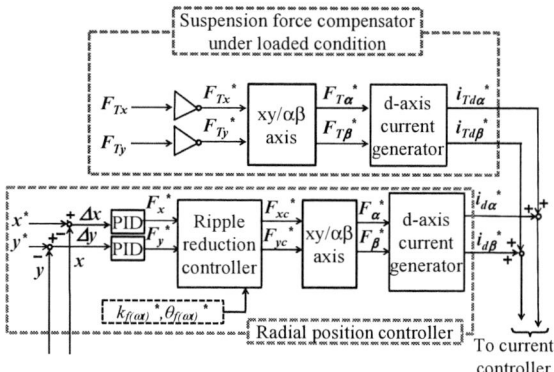

Fig. 6. Radial position controller with suspension force compensator at failure of a motor section under loaded condition.

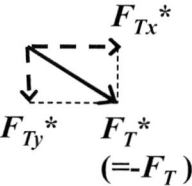

Fig. 7. Updated magnetic force commands to eliminate the influence of unbalance force.

tangential forces applied on the rotor surface in α- and β- sections, respectively. They are picked up in Fig. 5(b). The resultant vector F_T is generated by the net vector sum of $F_{T\alpha}$ and $F_{T\beta}$ and causes the unstable rotor levitation. Therefore, it is possible that the rotor shaft will successfully be levitated without mechanical contact if the magnetic force is newly commanded so as to eliminate the unbalance force F_T.

V. COMPENSATION METHOD OF RADIAL UNBALANCE FORCE AT FAILURE OF A MOTOR SECTION UNDER LOADED CONDITION

A. Radial Position Control with Suspension Force Compensator

In order to eliminate the unbalance force by the q-axis flux and realize the rotor suspension without mechanical contact, as a first step, the radial position controller with suspension force compensator is presented in this section.

Fig. 6 shows the proposed radial position controller with suspension force compensator under loaded condition. The x- and y-axis components, F_{Tx} and F_{Ty}, of the resultant vector F_T are detected or computed, for example, by FEM using a simulation software. In order to eliminate the influence of the unbalance force, as shown in Fig. 7, the magnetic forces F_{Tx}^* and F_{Ty}^* are commanded in opposite direction of F_{Tx} and F_{Ty}, respectively. And then, the magnetic force commands F_{Tx}^* and F_{Ty}^* in x- and y-axis coordinate are transformed into the commands $F_{T\alpha}^*$ and $F_{T\beta}^*$ in α- and β-axis coordinate using the coordinate transformation shown in Fig. 3. The d-axis current commands $i_{Td\alpha}^*$ and $i_{Td\beta}^*$ are determined so as to be proportional to the force commands $F_{T\alpha}^*$ and $F_{T\beta}^*$, respectively. The $i_{Td\alpha}^*$ and $i_{Td\beta}^*$

are added to the outputs of radial position controller, $i_{d\alpha}^*$ and $i_{d\beta}^*$. Consequently, the d-axis current commands are updated and are input to the current controller.

B. Analyses of Radial Unbalance Force by Finite Element Method

In order to confirm the effectiveness of the proposed suspension force compensator under loaded condition and validate it, the suspension forces are demonstrated using an FE model. Fig. 8 and table I show the specification of the FE model. It is a 9-slot, 6-pole IPMSM type, the unbalance forces, the magnitude and direction of the suspension force are computed using FE analysis software, JMAG-Designer (Ver. 14.0, 2D, JSOL Corporation). The rotational speed is set at 4,500 r/min. The q-axis current is set to 2.25A, consequently it is under 77% loaded condition.

Fig. 9 (a) and (b) show the computed x- and y- axis components of the unbalance force, F_{Tx} and F_{Ty}. Fig. 9 (a) shows the F_{Tx} and F_{Ty} when no suspension force compensator is applied, on the other hand, it is additionally applied in Fig. 9 (b). The solid line shows the x-axis component F_{Tx} and the dotted line shows the y-axis component F_{Ty}. In Fig. 9 (a), the F_{Tx} and F_{Ty} are intensely vibrated and the peak to peak values of them are reached to 52 N and 32 N, respectively. They are corresponding to the disturbance applied in the rotor shaft and it is difficult to levitate it without mechanical contact. The dc bias is included only in the y-component F_{Ty}. It is caused by asymmetry of arrangement in three sections. On the other hand, by equipping the proposed compensator, the vibration of F_{Tx} and F_{Ty} is remarkably reduced. The peak to peak value is about 25 N in both F_{Tx} and F_{Ty}. This means that the suspension force compensator is effectively functioned.

In the next simulation, the magnitude and direction of suspension force are derived when the suspension force is

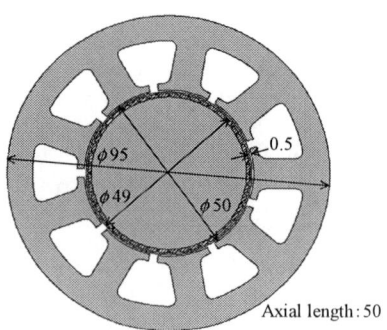

Fig. 8. Dimension of FE model.

TABLE I
SPECIFICATION OF FE MODEL

Rotor core, Stator core	Silicon steel
PM (thickness)	Nd-Fe-B (1 mm)
Air-gap length	0.5 mm
Stator winding	114 turns/tooth
Coil diameter	0.6 mm
Current rating (3-phase, rms)	1.7 A

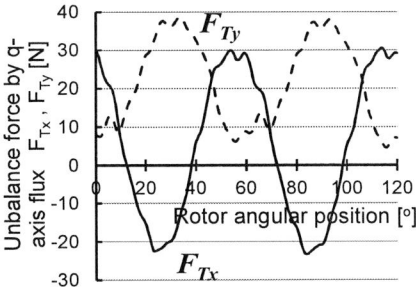

(a) Without suspension force compensator.

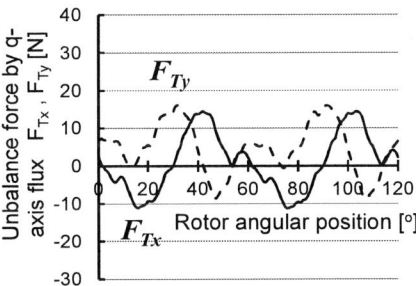

(b) With suspension force compensator.

Fig. 9. Computed unbalance force by q-axis flux at failure of a motor section under loaded condition.

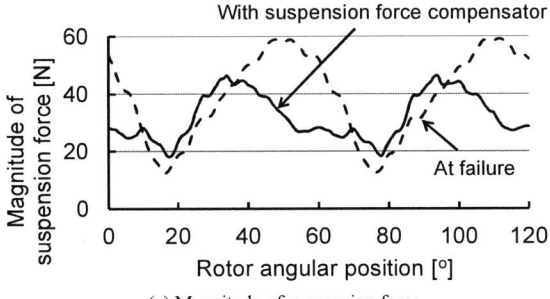

(a) Magnitude of suspension force.

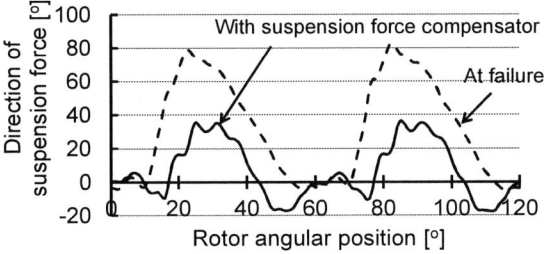

(b) Direction of suspension force.

Fig. 10. Simulation results of suspension force when the proposed compensator is applied.

actually generated and the effectiveness of suspension force compensator is found. Figs. 10 (a) and (b) show the magnitude and direction of suspension force, respectively, when the suspension force of 30 N is commanded in x-positive direction (direction of 0 degree as shown in Fig. 1). The simulation condition of the rotational speed and q-axis current is the same as that in the former FE analyses shown in Fig. 9. Fig. 10 (a) shows the

magnitude of suspension force. The dotted line shows the magnitude of suspension force when the proposed compensator is not applied although the section-γ is failed. The ripple is quite large and it is reached around 120 %. The rotor levitation is considerably unstable. The solid line shows the magnitude when the compensator is applied. The ripple is reduced to 88 %. Fig. 10 (b) shows the direction of suspension force. The theoretical direction is 0 deg as the suspension force is commanded in the x-positive direction. Therefore the value in the vertical axis means the direction error. When the proposed compensator is not applied, the direction error is reached to 82 deg in the maximum. It is quite difficult to support the rotor shaft without mechanical contact. On the other hand, it is seen that the direction error is remarkably reduced by applying the suspension force compensator, and it is about 36 deg even in the maximum. From the simulation results in Fig. 10, the effectiveness of the proposed suspension force compensator is obviously confirmed and also the validation is verified. However, the ripple of suspension force and its direction error should be smaller to guarantee the non-contact suspension of the rotor shaft. In the next section, the ripple reduction controller is applied also in the suspension force compensator.

C. Compensation with Ripple Reduction Controller

Fig. 11 shows the updated suspension force compensator, in which the ripple reduction controller is additionally applied [3]. Similar to the suspension force compensator in Fig. 6, the x- and y-components F_{Tx} and F_{Ty} of the unbalance force are obtained and the magnetic forces F_{Tx}^{*} and F_{Ty}^{*} are commanded in the opposite direction of F_{Tx} and F_{Ty}, respectively. The F_{Tx}^{*} and F_{Ty}^{*} are input in the ripple reduction controller and compensated so as to more stably levitate the rotor shaft. The purpose and principle of the ripple reduction control is the same as that in the radial position controller when three sections are normally operated or a section is failed at no load as shown in Fig. 4. However, it is necessary that the compensation coefficients $k_{f(\omega t)}$ and $\theta_{f(\omega t)}$ are calculated or detected, and then input in the suspension force compensator with ripple reduction controller as

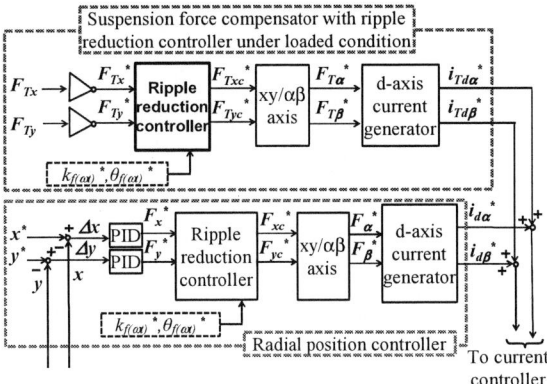

Fig. 11. Suspension force compensator with ripple reduction controller in radial position controller.

The 2018 International Power Electronics Conference

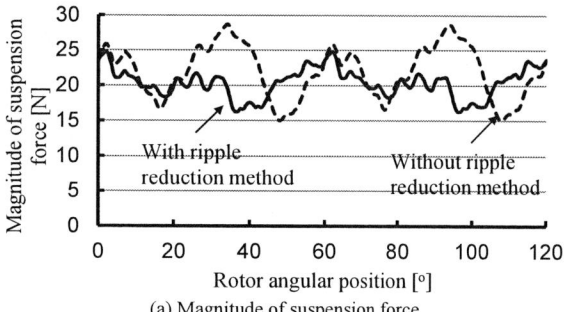

(a) Magnitude of suspension force.

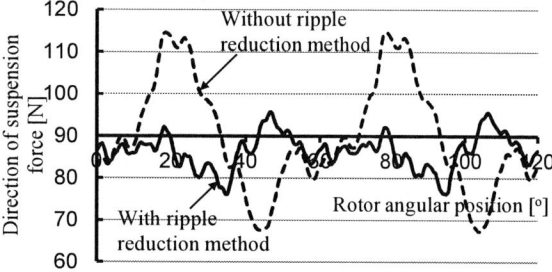

(b) Direction of suspension force.

Fig. 12. Simulation results of suspension force when ripple reduction controller is added in the suspension force compensator.

described in Fig. 11. The compensated F_{Txc}^* and F_{Tyc}^* are transformed into the $i_{Td\alpha}^*$ and $i_{Td\beta}^*$ and superimposed on the d-axis current commands $i_{d\alpha}^*$ and $i_{d\beta}^*$ through the same process as Fig. 6.

In order to verify the effectiveness of the ripple reduction controller updated in the suspension force compensator, the magnitude and direction of suspension force are simulated by FE analysis using the same model as those in Fig. 8 and Table I. The analysis condition is a little bit different from the former one mentioned in Chapter V, Section B. In this analysis, the rotational speed is set at 4,500 r/min and the suspension force of 20 N is commanded in the y-positive direction. The q-axis current is set to 1.6 A, consequently it is under 54 % loaded condition.

Figs. 12 (a) and (b) show the computed magnitude and direction of the suspension force, respectively. It is seen that in Fig. 12 (a) that the average of magnitude is about 20 N regardless of the application of ripple reduction control and it almost agrees with the command. However, the ripple is reduced from 62 % to 42 % by applying the ripple reduction control. The effectiveness of the ripple reduction control is distinctly found. In Fig. 12 (b), the theoretical direction of suspension force is 90 deg as it is commanded in the y-positive direction as described in Fig. 1. Under the condition without the ripple reduction controller, the direction is changed in the wide range and greatly vibrated. The direction error is reached to 22.4 deg in the maximum. However, it is significantly reduced to 13.8 deg by applying the ripple reduction controller. Even under the updated condition, however, the direction error of about 5 deg is regularly generated. The author will be continuously exploring the reason why this error

is caused although it would not influence the stability of rotor shaft.

From the analyzed results in Figs. 12 (a) and (b), the effectiveness of the ripple reduction controller updated in the suspension force compensator under the loaded condition is distinctly found and the validation is verified.

VI. CONCLUSION

In the d-q BELM, the author has presented the radial position controller when a motor section is unfortunately failed. The validation has been verified only at no load condition. However, when a motor section is failed under loaded condition, it is seriously concerned that the rotor levitation will fall into unstable by the unbalance force caused by the q-axis flux in the other two sections under normal operation. In order to solve this problem, in this paper, the suspension force compensator under loaded condition is proposed, in which the unbalance force is successfully eliminated. By the proposed compensator, the suspension force is successfully compensated so that the rotor shaft is supported without mechanical contact even under loaded condition until the failed section will be repaired. The effectiveness of compensation is more remark by additionally equipping the ripple reduction controller. The validation of the updated suspension force compensator with ripple reduction controller is verified by the simulation results of FE analysis.

ACKNOWLEDGMENT

The author would like to thank Mr. Yabana, Mr. Kobayashi and Mr. Kumazaki who were former graduate students at the Tokyo University of Science, and engaged on the FE Analysis of suspension force at failure and the rotor levitation tests using a prototype machine.

REFERENCES

[1] A. Chiba, T. Fukao, O. Ichikawa, M. Ooshima, M. Takemoto and David G. Dorrell, "Magnetic Bearings and Bearingless Drives", Newnes Publishers, ISBN 0-7506-5727-8, March 2005.

[2] Reto Schob and Natale Barletta, "Principle and Application of a Bearingless Slice Motor", *Proceedings of the Fifth International Symposium on Magnetic Bearings*, .pp.313-318, Kanazawa, Japan, August, 1996.

[3] Syunsuke Kobayashi, Masahide Ooshima and M. Nasir Uddin, "A Radial Position Control Method of Bearingless Motor Based on d-q Axis Current Control", *IEEE Transactions on Industry Applications*, vol.49, No.4, pp.1827-1835, (2013).

[4] Masahide Ooshima, Toshiki Karasawa and M. Nasir Uddin, "Stabilized Control Strategy Under Loaded Conditions in a Bearingless Motor Based on d-q Axis Current Control", *IEEE Industry Application Society 2013 Annual Meeting*, 2013-IACC-315, CDROM, @Orlando, 2013.10

[5] Masahide Ooshima and Ayumu Kobayashi, "Stabilized Suspension Control Strategy at failure of a Motor Section in a d-q Axis Current Control", *IEEE Industry Application Society 2015 Annual Meeting*, 2015-IACC-0392, USB, @Dallas, 2015.10

[6] Masahide Ooshima and Yuto Gomi, "Characteristics of Torque and Suspension Force in a d-q Axis Current Control Bearingless Motor", *Proceedings of 14th International Symposium on Magnetic Bearings*, pp.163-167, @Linz, Austria, 2014.8

[7] Masahide Ooshima and Yuto Gomi, "Evaluation of Motor Losses and Efficiency in a d-q Axis Current Control Bearingless Motor", *Proceedings of 15th International Symposium on Magnetic Bearings*, T2C4, 10165, @Kitakyusyu, Japan, 2016.8.

A Bearingless Synchronous Reluctance Slice Motor with Rotor Flux Barriers

Thomas Holenstein[1]*, Thomas Nussbaumer[2], Johann W. Kolar[1]

1 Power Electronic Systems Laboratory. ETH Zurich, Zurich, Switzerland

2 Levitronix GmbH, Zurich, Switzerland

*E-mail: holenstein@lem.ee.ethz.ch

Abstract—This paper presents a bearingless synchronous reluctance slice motor, which contains no permanent magnets. The rotor with four iron poles and flux barriers is levitated and rotated through a stator winding system with six coils wired as two three-phase systems. After applying a constant rotor oriented magnetization current, the system can be controlled just like a bearingless permanent magnet synchronous slice motor, including the passive stabilization of axial and tilting movements. In a first step, the motor geometry is being optimized and the performance characteristics of the designed motor are examined. The motor is then compared to two other designs, which contain permanent magnets either in the rotor or the stator. The comparison includes torque generation, radial force generation, passive axial and tilting stiffnesses and wide air gap suitability. The introduced topology outperforms the others for ultra high process or ambient temperatures and rotor disposable applications with a short exchange interval.

Keywords—bearingless slice motors, synchronous reluctance motor, topology comparison, wide air gap machines

I. INTRODUCTION

Bearingless motors feature magnetically levitated rotors, and a magnetically integrated bearing function [1]. The same iron circuit is used for torque and radial force generation, with either a separated or a combined winding system. If the stator and rotor lengths are chosen to be much smaller than the rotor diameter, as for the so-called slice motor, only two radial degrees of freedom remain to be actively stabilized apart from the rotation [2].

A significant advantage of bearingless slice motors is that the rotor can be completely separated and isolated from the stator in a simple manner. Contactless rotation in its own containment is possible in the widest range of environmental conditions, which makes these motors perfect for ultra-pure, low shear fluid handling, harsh environmental conditions such as aggressive chemicals, abrasive media or extreme ambient temperatures. To take full advantage, a thick, pressure, heat, and chemistry resistant process chamber wall is needed between the stator and the rotor, requiring a wide air gap in the range of several millimeters. A schematical drawing of such an arrangement is given in Fig. 1.

Many conventional motor topologies can also be configured as bearingless motors [3]. For this reason, bearingless motors have undergone similar evolution since their first demonstration as mechanically supported electrical machines, just with a delay of several years due to the

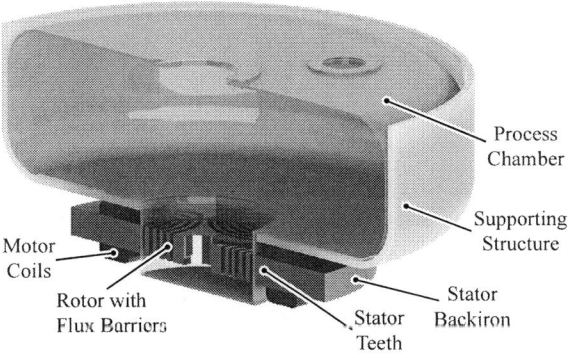

Fig. 1. Bearingless synchronous reluctance slice motor with flux barriers in a process environment.

added complexity. Bearingless induction and reluctance motors were initially demonstrated around 1990 [4], and superseded by rotor permanent magnet (PM) topologies [5], as soon as strong permanent magnet materials became widely available. More recent research has also demonstrated stator-PM topologies [6].

Modern simulation and inverter technologies have lead to a reconsideration of synchronous reluctance motors (SynRM) [7]. This development was also driven by the rare-earth price rally in 2011 and has lead to commercially available magnet free motors, e.g. from ABB and Siemens. These motors feature rotor flux barriers and achieve competitive efficiencies (IE4, super-premium efficiency level class) [8].

Bearingless SynRM with flux barriers were first introduced in [9]. Linear torque and force generation with much smaller fluctuations over the rotor angle was observed, when compared to reluctance topologies with salient rotor poles. In addition, almost no coupling between force and torque generation exists, which allows to obtain stable bearing operation without a decoupling control algorithm such as needed for salient-pole topologies.

Therefore, it is expected, that such a machine is relatively easy to control, with control algorithms being identical to those of a rotor-PM machine. The only modification needed is to set a constant magnetization current $i_{\mathrm{mag}} = i_{\mathrm{drv,d}}$, which is zero for rotor-PM machines. A variety of recent works deal with the bearingless operation of SynRM with flux barriers [10]–[14].

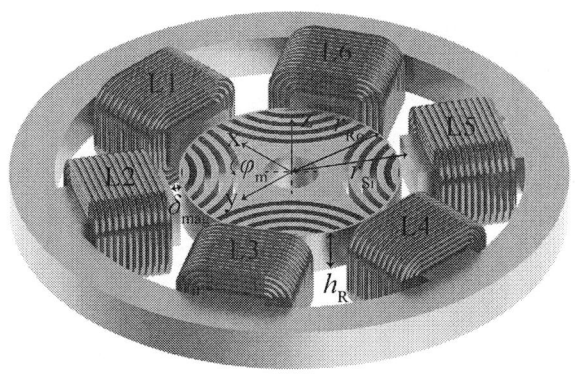

Fig. 2. Introduced bearingless six-slot, four-pole SynRM slice motor with rotor flux barriers and six concentrated motor windings for combined torque and radial force generation.

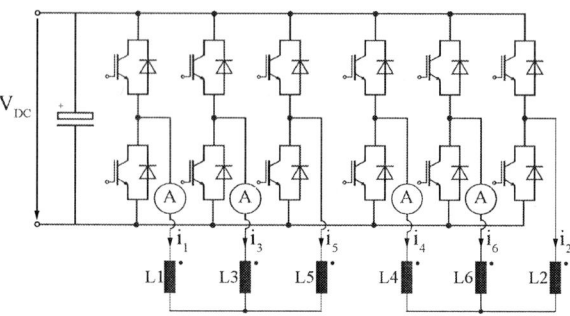

Fig. 3. Connection scheme of the six motor windings to the employed six-phase inverter.

The main focus of this paper lies on bearingless slice motor topologies without PMs in the rotor featuring a wide air gap. The omission of PMs in the rotor is advantageous for very high ambient temperatures, high rotational speeds, and low manufacturing costs of the rotor. A bearingless flux-switching permanent magnet (FSPM) slice motor was presented in an earlier publication [15]. Despite featuring high torque, several disadvantages regarding the bearing operation were described, namely small radial startup distance as well as strong coupling and angle dependency of the force generation, which required additional control algorithms.

These disadvantages can be mitigated with the SynRM slice motor topology with flux barriers presented in this paper, at the cost of slightly lower torque. The introduced topology is explained in detail in the first part of the paper. A special focus is set on aspects specific to slice motors, namely passive axial and tilting stabilization. A thorough comparison to two other topologies with either permanent magnets in the rotor or the stator is presented in the second part of the paper. Advantageous applications for each of the three topologies are pointed out. Finally, topics of further research are indicated.

II. SYNCHRONOUS RELUCTANCE SLICE MOTOR

A. Motor Design

The introduced bearingless synchronous reluctance slice motor topology with rotor flux barriers is shown in Fig. 2. A four-pole reluctance rotor with four flux barriers per pole is used. All flux barriers are circular and concentric. Six stator teeth, each with a concentrated motor winding for combined torque and radial force generation are used. The stator teeth are connected by a circular back-iron.

In order to accommodate a pressure, heat, and chemistry resistant process chamber wall in the air gap in a later stage, an air gap thickness δ_{mag} to rotor outer radius r_{Ro} ratio G, as defined in (1), of 0.1 is used. This is in line with existing rotor-PM bearingless motor topologies [16]. Note that the terminology "magnetic gap" [17] can

be used interchangeably for "air gap", since all materials inside the gap, namely the fluid and process chamber wall, have a relative permeability μ_r very close to that of air.

$$G = \frac{\delta_{\mathrm{mag}}}{r_{\mathrm{Ro}}} = \frac{r_{\mathrm{Si}}}{r_{\mathrm{Ro}}} - 1 \qquad (1)$$

Furthermore, the ratio H of rotor height h_{R} to rotor diameter d_{Ro}, as defined in (2), is set to 0.2 to assure that the axial and tilting movement of the rotor are passively stabilized by the magnetic bias field, which is generated by the constant magnetization current i_{mag}.

$$H = \frac{h_{\mathrm{R}}}{d_{\mathrm{Ro}}} = \frac{h_{\mathrm{R}}}{2r_{\mathrm{Ro}}} \qquad (2)$$

B. Winding Layout and Current Generation

The six motor windings are connected as two three-phase systems with a floating star point each and are powered by a six-phase inverter, as shown in Fig. 3. This arrangement is commonly used (see e.g. [18]), leaves four degrees of freedom to be controlled, and requires four current sensors in the inverter to control all currents, since $i_1 + i_3 + i_5 = 0$ and $i_4 + i_6 + i_2 = 0$ holds.

The four degrees of freedom are used to control the radial position in x and y direction, the rotational speed ω_{m} and the magnetization current i_{mag}. A superimposed control algorithm is used to generate setpoint values for the virtual bearing and drive currents $i_{\mathrm{bng,x}}$, $i_{\mathrm{bng,y}}$, and $i_{\mathrm{drv,q}}$, which are directly proportional to the radial forces F_{x}, F_{y} and the motor torque T_{m} for a given magnetization current $i_{\mathrm{mag}} = i_{\mathrm{drv,d}}$.

Equations (3) and (4) show how these virtual bearing and drive currents are transformed and added to generate the six combined motor winding currents i_1 to i_6. The rotor angle φ_{m}, rotor pole-pair number $p_{\mathrm{drv}} = 2$, and bearing pole-pair number p_{bng} are used for this transformation and Fig. 2 shows the corresponding coordinate system. For bearingless reluctance motors the relation $p_{\mathrm{bng}} = p_{\mathrm{drv}} \pm 1$ has to hold.

A pole-pair number $p_{\mathrm{bng}} = p_{\mathrm{drv}} - 1 = 1$ is used, since $p_{\mathrm{bng}} = 3$ would exhibit single-phase characteristics with

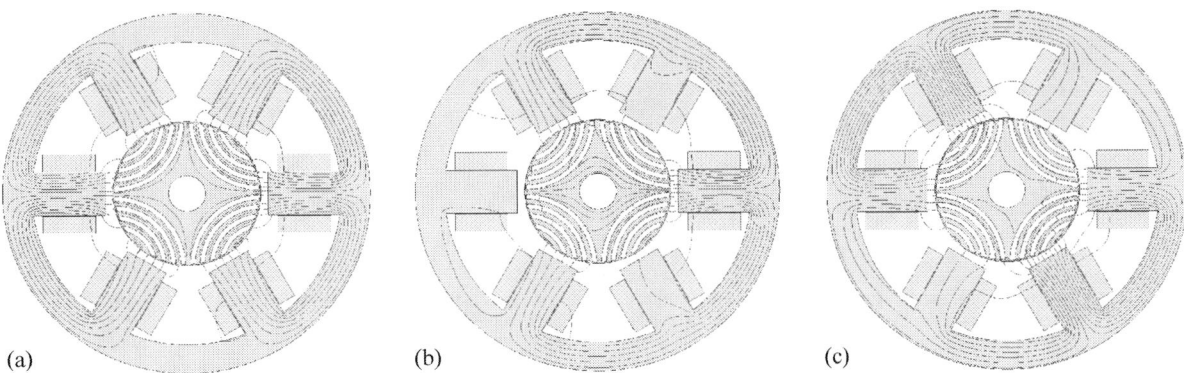

Fig. 4. Bearingless SynRM with field lines shown for: (a) magnetization current ($i_{\mathrm{mag}} = 2000\ \mathrm{AT}$), (b) radial force generation ($i_{\mathrm{mag}} = i_{\mathrm{bng,x}} = 2000\ \mathrm{AT}$), (c) torque generation ($i_{\mathrm{mag}} = i_{\mathrm{drv,q}} = 2000\ \mathrm{AT}$).

six stator teeth, i.e. it would not be possible to generate radial forces at all rotor angles.

It has to be noted that the winding layout and connection scheme described in this subsection, as well as the motor winding current generation formula

$$
\begin{bmatrix} i_1 \\ i_2 \\ i_3 \\ i_4 \\ i_5 \\ i_6 \end{bmatrix} = K(p_{\mathrm{drv}}) \cdot \begin{bmatrix} i_{\mathrm{drv,d}} \\ i_{\mathrm{drv,q}} \end{bmatrix} + K(p_{\mathrm{bng}}) \cdot \begin{bmatrix} i_{\mathrm{bng,x}} \\ i_{\mathrm{bng,y}} \end{bmatrix}, \quad (3)
$$

where

$$
K(p) =
$$
$$
\begin{bmatrix}
\cos(p_{\mathrm{drv}}\varphi) & \cos(p_{\mathrm{drv}}\varphi + \tfrac{\pi}{2}) \\
\cos(p_{\mathrm{drv}}\varphi - p\tfrac{\pi}{3}) & \cos(p_{\mathrm{drv}}\varphi - p\tfrac{\pi}{3} + \tfrac{\pi}{2}) \\
\cos(p_{\mathrm{drv}}\varphi - p\tfrac{2\pi}{3}) & \cos(p_{\mathrm{drv}}\varphi - p\tfrac{2\pi}{3} + \tfrac{\pi}{2}) \\
\cos(p_{\mathrm{drv}}\varphi - p\tfrac{3\pi}{3}) & \cos(p_{\mathrm{drv}}\varphi - p\tfrac{3\pi}{3} + \tfrac{\pi}{2}) \\
\cos(p_{\mathrm{drv}}\varphi - p\tfrac{4\pi}{3}) & \cos(p_{\mathrm{drv}}\varphi - p\tfrac{4\pi}{3} + \tfrac{\pi}{2}) \\
\cos(p_{\mathrm{drv}}\varphi - p\tfrac{5\pi}{3}) & \cos(p_{\mathrm{drv}}\varphi - p\tfrac{5\pi}{3} + \tfrac{\pi}{2})
\end{bmatrix} \quad (4)
$$

holds for all numbers of rotor pole pairs, as well as for PM-rotor topologies with and without stator teeth (with $i_{\mathrm{mag}} = i_{\mathrm{drv,d}} = 0$), as long as a stator with six combined motor windings is used. For FSPM only a very small modification to the matrix $K(p)$ is necessary, as is shown in Section III. Pole pair number and topology configurations can easily be adjusted in software.

In Fig. 4 the field lines in the bearingless SynRM are shown for the three scenarios of magnetization current only (a), radial force generation (b) and torque generation (c). It can be seen that the flux density inside the stator teeth is almost perfectly proportional to the applied current. The field lines do not have to cross the flux barriers for a pure magnetization current. Crossing of the flux barriers results in a reluctance torque.

C. Passive Stabilization, Radial Force, and Torque

A simplified rectangular magnetic circuit with constant cross sectional area A_{fe}, iron length l_{fe}, two air gaps with length l_δ, and a coil with n windings wound around the iron carrying a current i is considered. If it is further assumed that there is no stray flux and that the field lines cross the air gap with the same cross section as the iron circuit ($A_{\mathrm{fe}} = A_\delta$), the following relationship between magnetomotive force ni (MMF in ampere turns AT) and the B- and H-fields is obtained:

$$
\oint H \cdot ds = l_{\mathrm{fe}} H_{\mathrm{fe}} + 2 l_\delta H_\delta = l_{\mathrm{fe}} \frac{B}{\mu_0 \mu_r} + 2 l_\delta \frac{B}{\mu_0} = ni. \quad (5)
$$

Solving (5) for B and assuming infinite permeability of the iron, it can be seen that the B-field is proportional to the coil current divided by the air gap length:

$$
B = \mu_0 \frac{ni}{\frac{l_{\mathrm{fe}}}{\mu_r} + 2 l_\delta} \approx \mu_0 \frac{ni}{2 l_\delta}. \quad (6)
$$

The force acting on the two air gaps is proportional to B^2 (7), and therefore, according to (6) also to i^2 (for more details, refer to e.g. [19], Chapter 3)

$$
f = \frac{B^2 A_{\mathrm{fe}}}{\mu_0} \approx \frac{\mu_0 A_{\mathrm{fe}} n^2 i^2}{2 l_\delta^{\,2}}. \quad (7)
$$

If this simple model is applied to the considered SynRM topology, it can be seen that a magnetization current i_{mag} leads to attracting forces between the stator teeth and the rotor, which increase quadratically with i_{mag}. The sum of all of these forces is zero for a centered rotor due to (3) and (4). For an axial or tilting deflection passive restoring forces are obtained which are pulling the rotor back towards the axial center of the stator. These restoring forces increase linearly with the deflection and stiffness factors k_z, k_α, and k_β can be defined for a given value of i_{mag}.

For a radial deflection, a negative, unstable force pulling the rotor away from the stator center is obtained

The 2018 International Power Electronics Conference

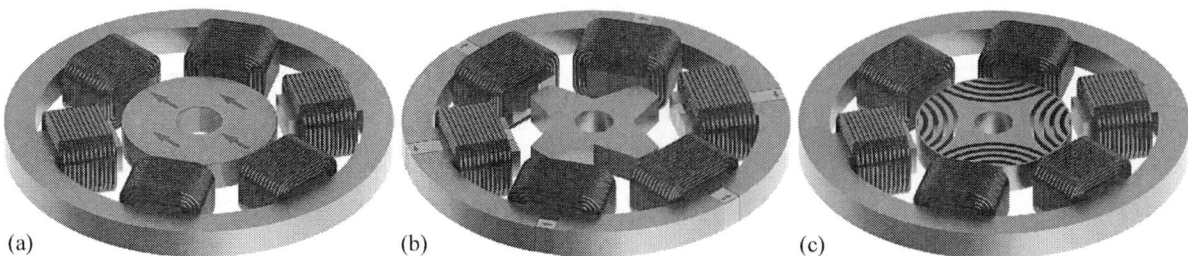

Fig. 5. Compared topologies: (a) bearingless six-slot, two-pole rotor permanent magnet motor (PMSM), (b) bearingless six-slot, four-pole pair flux switching permanent magnet (FSPM) motor, (c) bearingless six-slot, four-pole synchronous reluctance slice motor (SynRM).

and the stiffness factor k_r can be defined accordingly for a given value of i_{mag}.

Considering the motor torque it has to be noted however, that torque increases linearly with $i_{drv,q}$ and also $i_{drv,d}$ as can be seen from the reluctance motor torque equation

$$T = \frac{3}{2}p_{drv}(L_d - L_q)i_{drv,d} \cdot i_{drv,q}. \tag{8}$$

Radial forces increase linearly with i_{bng} and i_{mag} as well, which is best illustrated by the forces of two opposing stator teeth being added to form the resulting radial force, e.g. $F_{x,1} = F_{coil1} + F_{coil2}$ at $\varphi_m = 0\,deg$ yielding

$$F_{x,1} \propto (i_{mag} + i_{bng,x})^2 - (i_{mag} - i_{bng,x})^2$$
$$= 4i_{mag}i_{bng,x}. \tag{9}$$

For slice motors with a wide air gap as in the considered case, the simple model assumption of having no stray flux and straight field lines within the air gap does not hold true any more. Considerable stray flux paths between the stator teeth as well as below and above the motor exist. This leads to a higher than expected B-field magnitude in the iron below the coils and a lower than expected B-field magnitude within the air gap and the rotor. In other words, saturation occurs earlier than expected and forces are lower than expected. For this reason, 3D FEM simulations are carried out to obtain the absolute values of the expected forces and the torque. Nevertheless, the proportionality relations of the simple model hold true.

D. Magnetization Current Considerations

Due to the quadratic relation between the total current and forces as well as torque, respectively, the SynRM topology performs better compared to PM topologies for a high MMF. For this reason, a very high MMF of several thousand AT was chosen for the simulations, for which the stator is already partly saturated. Torque simulations were performed at a drive current angle of $45\,deg$, since the maximum torque per total current is achieved for $i_{mag} = i_{drv,d} = i_{drv,q}$. Therefore, the magnetization current was set to $i_{mag} = MMF/\sqrt{2}$, which causes the ohmic idle losses to be half of the full load losses. It has

to be noted that generating such high MMF constantly will most likely require advanced water cooling for the stator coils. In this paper the notation $\hat{A}T$ was used for a $MMF/\sqrt{2}$.

Having a completely firmware-adjustable magnetization current for a bearingless slice motor provides several new possibilities unknown to PM topologies, namely:

- Adjustable axial steady-state position

- Improved damping of axial oscillations compared to PM motors (e.g. [20])

- Avoidance of resonances during run up, through dynamic stiffness adjustment

- Dynamic prioritization between high dynamics and low iron losses for high speed operation.

These items provide a variety of research topics once a bearingless SynRM prototype is available, and are expected to open new opportunities and applications.

III. PERMANENT MAGNET TOPOLOGIES FOR COMPARISON

To put the performance of the introduced SynRM topology into perspective, it is compared to two other topologies with PMs either in the rotor or the stator. All three topologies are shown in Fig. 5. Identical rotor diameters as well as ratios for G and H according to (1) and (2) respectively, are used for all three topologies. All topologies have six stator teeth, each with a concentrated motor winding for combined torque and force generation.

The rotor permanent magnet synchronous motor (PMSM) topology features a diametrically-magnetized two-pole rotor with identical current generation to the SynRM topology as described in (3) and (4), using $p_{pmsm,drv} = 1$, $p_{pmsm,bng} = 2$, the same coordinate system as in Fig. 2, and setting $i_{pmsm,mag} = i_{pmsm,drv,d} = 0$.

The stator PM topology is an FSPM motor with four rotor teeth. Each stator tooth contains a tangentially-magnetized PM with alternating magnetization direction. For the current generation the coordinate system from Fig. 2 and (3) with $p_{fspm,drv} = 4$ and $p_{fspm,bng} = 5$ are used. However, the matrix K needs to be slightly modified as shown in (10), since alternating current directions due

3622

to the alternating PM bias flux are required, and a rotor tooth in front of an energized stator tooth experiences a tangential instead of a radial force. Due to the fact that mostly tangential forces are generated, this topology is very effective at generating torque, but less effective for generating bearing forces (see e.g. [15]), since both radial and tangential forces are used to generate bearing forces.

$$
K_{\mathrm{fspm}}(p) =
$$
$$
\begin{bmatrix}
\cos(p_{\mathrm{drv}}\varphi + \tfrac{\pi}{2}) & \cos(p_{\mathrm{drv}}\varphi + \pi) \\
\cos(p_{\mathrm{drv}}\varphi - p\tfrac{\pi}{3} - \tfrac{\pi}{2}) & \cos(p_{\mathrm{drv}}\varphi - p\tfrac{\pi}{3}) \\
\cos(p_{\mathrm{drv}}\varphi - p\tfrac{2\pi}{3} + \tfrac{\pi}{2}) & \cos(p_{\mathrm{drv}}\varphi - p\tfrac{2\pi}{3} + \pi) \\
\cos(p_{\mathrm{drv}}\varphi - p\tfrac{3\pi}{3} - \tfrac{\pi}{2}) & \cos(p_{\mathrm{drv}}\varphi - p\tfrac{3\pi}{3}) \\
\cos(p_{\mathrm{drv}}\varphi - p\tfrac{4\pi}{3} + \tfrac{\pi}{2}) & \cos(p_{\mathrm{drv}}\varphi - p\tfrac{4\pi}{3} + \pi) \\
\cos(p_{\mathrm{drv}}\varphi - p\tfrac{5\pi}{3} - \tfrac{\pi}{2}) & \cos(p_{\mathrm{drv}}\varphi - p\tfrac{5\pi}{3})
\end{bmatrix}
\tag{10}
$$

It has to be noted that the usage of stator PMs, both for a homopolar and multipolar stator-PM bias flux, doubles p_{drv} for the same rotor geometry compared to a SynRM. Each rotor tooth yields identical characteristics in front of the same stator tooth for a given PM stator bias flux. For a SynRM, however, a rotating bias flux is applied, which effectively assigns a positive or negative value to each rotor tooth.

IV. Performance Comparison

The following performance comparison is carried out with respect to the target application, which requires achieving high rotor torque densities and high passive axial and tilting stiffnesses with wide air gap bearingless slice motors at relatively low rotational speeds (see e.g. [15]).

As such motors are usually thermally limited, torque and active radial forces are compared at the same motor losses, which consist mostly of ohmic winding losses while iron losses can be neglected due to the low rotational speeds. This is achieved for an identical MMF in AT as the winding space is equal. However, to provide the passive stabilization, the SynRM will generate half of these ohmic full-load losses, while the PMSM and FSPM topologies will both generate no ohmic losses.

For the 3D FEM simulations the material properties of neodym-iron-boron magnets in grade N45 and magnetic steel M330-35A were used.

A. Passive Axial and Tilting Stability

Passive stabilization of the axial position z and for the two tilting degrees of freedom, α and β, is achieved with the rotor or stator PMs for the PMSM and the FSPM topology, respectively, while for the introduced SynRM passive stabilization is achieved through the current i_{mag}.

The axial restoring force versus the axial deflection of the rotor is shown in Fig. 6. While the axial stiffness

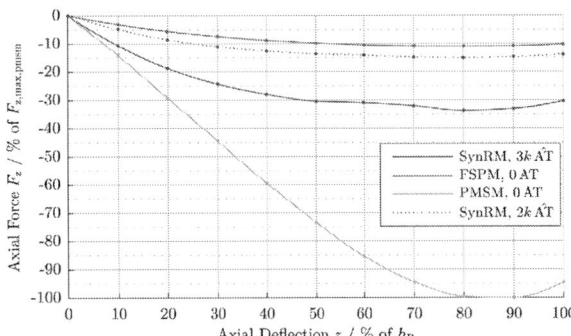

Fig. 6.   Passive axial restoring force F_{z} vs. axial deflection z.

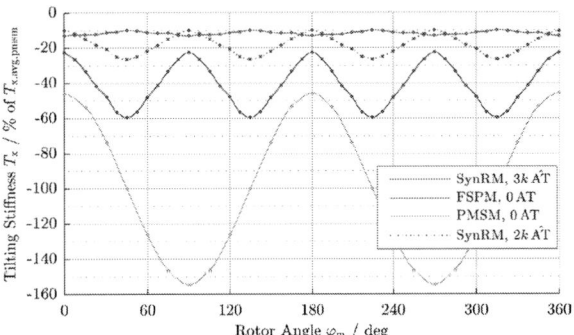

Fig. 7.   Passive tilting stiffness T_{x} vs. rotor angle φ_{m}.

as described by (11) is linear for the PMSM up to approximately 50% of deflection, it decreases immediately for the two other topologies. Table I summarizes the maximum axial stiffness $k_{\mathrm{z,max}}$, maximum axial load $F_{\mathrm{z,max}}$, and the axial equilibrium deflection z_{equ} (11) for all three topologies. The axial equilibrium position results from gravity ($g = 9.81\,\tfrac{\mathrm{m}}{\mathrm{s}^2}$) acting on the rotor of the horizontally-oriented motor, where the relative rotor densities ($\rho_{\mathrm{v,pmsm}} = 91.0\%$, $\rho_{\mathrm{v,fspm}} = 53.6\%$, $\rho_{\mathrm{v,synrm}} = 59.8\%$), material densities ($\rho_{\mathrm{m,ndfeb}} = 7.5\,\tfrac{\mathrm{g}}{\mathrm{cm}^3}$, $\rho_{\mathrm{m,m330}} = 7.65\,\tfrac{\mathrm{g}}{\mathrm{cm}^3}$), and axial restoring forces from Fig. 6 have been used.

$$
k_{\mathrm{z}} = \frac{\mathrm{d}F_{\mathrm{z}}}{\mathrm{d}z} \quad \text{and} \quad z_{\mathrm{equ}} : F_{\mathrm{z,equ}} = -F_{\mathrm{g}} \tag{11}
$$

It should be noted that the following simulations, for which the results are shown in Fig. 7 - 12, do not include the performance degradation due to the axial equilibrium deflection, since the influence would be small and dependent on the mounting orientation and additional rotor load.

TABLE I.   Axial and Tilting Performance

	PMSM	FSPM	SynRM $2k$ ÂT	SynRM $3k$ ÂT
$k_{\mathrm{z,max}}$	100%	20.7%	32.2%	70.8%
$F_{\mathrm{z,max}}$	100%	11.0%	15.1%	33.8%
z_{equ}	7.5%	22.0%	15.2%	6.3%
$k_{\alpha,\beta,\mathrm{avg}}$	100%	11.5%	18.3%	41.1%

3623

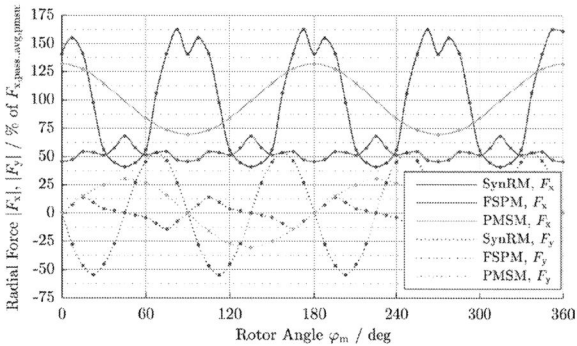

Fig. 8. Passive radial forces F_x and F_y for a deflection in x direction of 40% of the air gap δ_{mag}. SynRM with $i_{mag} = 2k$ AT.

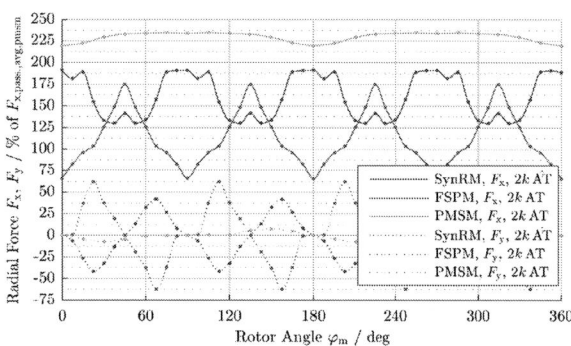

Fig. 9. Active radial forces F_x and F_y generated by current $i_{bng,x}$.

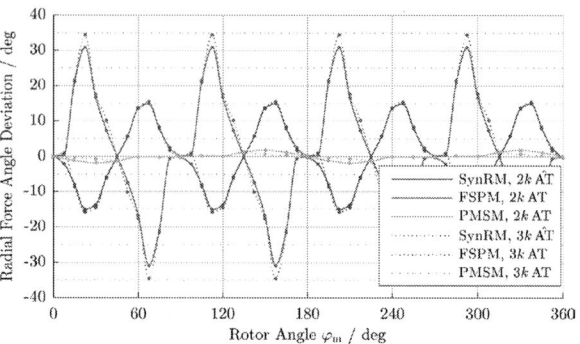

Fig. 10. Radial force angle deviation from x for a current $i_{bng,x}$.

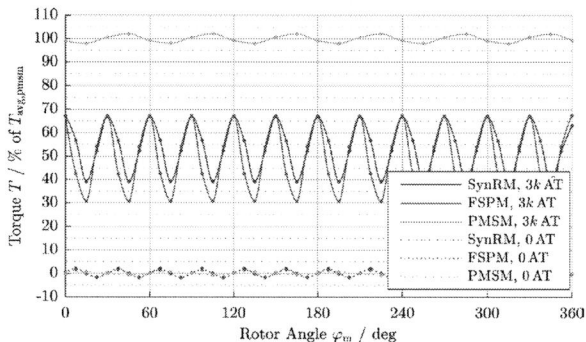

Fig. 11. Torque generation including cogging torque vs. rotor angle.

The passive tilting stiffness versus the rotor angle is shown in Fig. 7 and summarized in Table I. It can be seen how the difference between the weak α-axis and the strong β-axis is very pronounced for small pole pair numbers (i.e. the PMSM), while it is almost negligible for the FSPM topology. A big difference between $k_\alpha = dT_\alpha/d\alpha$ and $k_\beta = dT_\beta/d\beta$ can be problematic for low rotational speeds, since the tilting eigenfrequencies are excited just from having an initial tilting deflection, e.g. through disturbance forces.

B. Radial Forces and Torque

Destabilizing passive radial forces for a radially deflected rotor are shown in Fig. 8. Active radial forces for a centered rotor are shown in Fig. 9. In order to safely achieve radial startup when the motor is switched on, the active forces need to be larger than the passive destabilizing forces at any rotor angle φ_m, as can be seen in (12). This startup condition is satisfied for all three topologies for $x = 0.4 \cdot \delta_{mag}$ and $i_{bng} = 2k$ AT with varying margins.

$$k_x(\varphi_m) \cdot x < k_i(\varphi_m) \cdot i_{bng,x} \quad \forall \varphi_m \quad (12)$$

The angle deviation of the radial force is shown in Fig. 10. The deviation is almost zero for the PMSM and quite large for the FSPM at ± 30 deg, as expected. For the SynRM it is significantly larger than expected at ± 15 deg.

In [9], a stability range of ± 5 deg is given for the bearing control, which is considered to be rather conservative, but for ± 15 deg a decoupling controller is most likely required.

Additional investigations have shown that the stator slot number of six is the root cause for this angle deviation. For a stator with twelve slots (or 24 as in [9]), the same rotor shows constant force and torque generation over φ_m and almost no angle deviation similar to the PMSM topology. However, a stator with twelve slots and a four-pole rotor exhibit a poor winding factor for concentrated coils and would also require separate force and torque windings, since combined torque and force generation would not be possible with a power converter consisting of six half bridges.

Torque generation, including the cogging torque, for all three topologies is shown in Fig. 11. The torque generation for different air gap ratios G is shown in Fig. 12, which reveals that the performance of the PMSM scales with a different exponent with regard to the air gap ratio. Therefore, the PMSM topology is much better suited for even larger air gap ratios G compared to the FSPM and SynRM topologies.

Table II summarizes the radial force and torque performance for the three topologies. From the performance increase between an MMF of $2k$ AT and $3k$ AT a good indication about the average saturation level can be obtained. Without saturation an increase of factor 1.5,

The 2018 International Power Electronics Conference

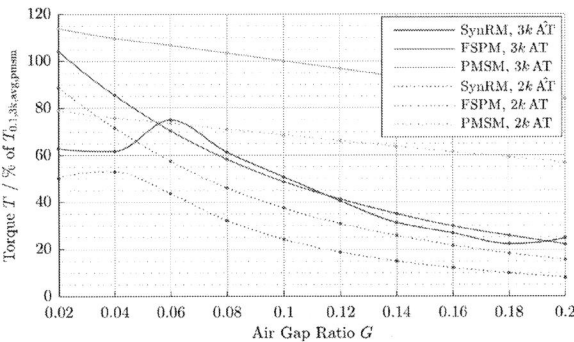

Fig. 12. Torque generation vs. air gap ratio.

TABLE II.   RELATIVE RADIAL FORCE AND TORQUE PERFORMANCE

	PMSM	FSPM	SynRM
F_x 2k AT	0.720	0.353	0.502
F_x 3k AT	1.000	0.397	0.845
F_x increase	1.39	1.12	1.30^2
T 2k AT	0.686	0.374	0.242
T 3k AT	1.000	0.487	0.538
T increase	1.46	1.30	1.49^2

respectively 1.5^2 would be expected, which in terms of the torque is almost reached for the SynRM and the PMSM, but not for the FSPM. Therefore, conclusions regarding the maximum torque and force capability can be drawn from constants obtained for small MMF.

The saturation level for radial force generation is generally higher, since the bearing current $i_{bng,x}$ and magnetization current i_{mag} (or PM bias flux, respectively) coincide at certain rotor angles φ_m, resulting in an MMF which is higher by a factor of $\sqrt{2}$.

V. CONCLUSION

A bearingless synchronous reluctance slice motor (SynRM) with six stator slots and rotor flux barriers was introduced in this paper. The topology was explained in detail and the basic feasibility of the concept was shown through 3D FEM simulations. The evaluated performance of the PM-free topology was compared to two other topologies with PMs in the stator (FSPM) and the rotor (PMSM), respectively.

Strengths and weaknesses of each topology are summarized in Table III. The application-specific suitability is shown in Table IV. It is clear that PMSMs will remain the best option for general applications. For some special applications, the presented SynRM might, however, be an attractive option. These applications include ultra-high process or ambient temperatures and rotor disposable applications with a short exchange interval.

In a next step the authors are planning to build a prototype of such a bearingless SynRM with rotor flux barriers also to explore the new possibilities of dynamic stiffness adjustments, e.g. for axial position adjustments,

TABLE III.   PERFORMANCE COMPARISON: (A) SYNRM, (B) FSPM, (C) PMSM

	(A)	(B)	(C)
Torque Generation Capability	-	+	+ +
Passive Axial and Tilting Stiffness	-	- -	+
Force and Torque Linearity	-	- -	+ +
Expected Power Efficiency	- -	-	+
Wide Air Gap Suitability	-	-	+ +
Manufacturing Cost Rotor	+ +	+ +	- -
Manufacturing Cost Stator	+	- -	+ +
Rare Earth Independence Rotor	+ +	++	- -
Rare Earth Independence Stator	+ +	- -	+ +

TABLE IV.   APPLICATION SUITABILITY: (A) SYNRM, (B) FSPM, (C) PMSM

	(A)	(B)	(C)
General, Allround	- -	- -	+ +
High Speed Rotation	+	-	+ +
High Process Temperatures	+ +	+	-
High Ambient Temperatures	+ +	-	-
Long Usage Rotor Disposable	-	+	+
Short Usage Rotor Disposable	+ +	+	- -

resonance avoidance during run up, or prioritization between performance and iron losses.

REFERENCES

[1] A. Chiba, T. Fukao, O. Ichikawa, M. Oshima, M. Takemoto, and D. G. Dorrell, *Magnetic bearings and bearingless drives.* Elsevier, 2005.

[2] S. Silber, W. Amrhein, P. Bösch, R. Schöb, and N. Barletta, "Design aspects of bearingless slice motors," *IEEE/ASME Transactions on Mechatronics*, vol. 10, no. 6, pp. 611–617, Dec. 2005.

[3] B. Liu, "Survey of bearingless motor technologies and applications," in *Proc. IEEE Int. Conf. Mechatronics and Automation (ICMA)*, Aug. 2015, pp. 1983–1988.

[4] A. Chiba, T. Deido, T. Fukao, and M. A. Rahman, "An analysis of bearingless AC motors," *IEEE Transactions on Energy Conversion*, vol. 9, no. 1, pp. 61–68, Mar. 1994.

[5] X. Sun, L. Chen, and Z. Yang, "Overview of bearingless permanent-magnet synchronous motors," *IEEE Transactions on Industrial Electronics*, vol. 60, no. 12, pp. 5528–5538, Dec. 2013.

[6] W. Gruber, W. Briewasser, M. Rothböck, and R. T. Schöb, "Bearingless slice motor concepts without permanent magnets in the rotor," in *Proc. IEEE Int. Conf. Industrial Technology (ICIT)*, Feb. 2013, pp. 259–265.

[7] G. Pellegrino, T. M. Jahns, N. Bianchi, W. L. Soong, and F. Cupertino, *The rediscovery of synchronous reluctance and ferrite permanent magnet motors: tutorial course notes.* Springer, 2016.

[8] J. Estima and A. Cardoso, "Super premium synchronous reluctance motor evaluation," in *Proc. International Conference on Energy Efficiency in Motor Driven Systems (EEMODS)*, 2013, pp. 213–222.

[9] M. Takemoto, K. Yoshida, N. Itasaka, Y. Tanaka, A. Chiba, and T. Fukao, "Synchronous reluctance type bearingless motors with multi-flux barriers," in *Proc. Power Conversion Conf. - Nagoya*, Apr. 2007, pp. 1559–1564.

[10] V. Mukherjee, J. Pippuri, S. E. Saarakkala, A. Belahcen, M. Hinkkanen, and K. Tammi, "Finite element analysis for bearingless operation of a multi flux barrier synchronous reluctance motor," in *Proc. 18th Int. Conf. Electrical Machines and Systems (ICEMS)*, Oct. 2015, pp. 688–691.

3625

[11] S. E. Saarakkala, M. Sokolov, V. Mukherjee, J. Pippuri, K. Tammi, A. Belahcen, M. Hinkkanen *et al.*, "Flux-linkage model including cross-saturation for a bearingless synchronous reluctance motor," *in Proc. of the ISMB15, Kitakyushu, Japan*, 2016.

[12] H. Ding, H. Zhu, and Y. Hua, "Optimization design of bearingless synchronous reluctance motor," *IEEE Transactions on Applied Superconductivity*, vol. 28, no. 3, Apr. 2018.

[13] X. Diao, H. Zhu, Y. Qin, and Y. Hua, "Torque ripple minimization for bearingless synchronous reluctance motor," *IEEE Transactions on Applied Superconductivity*, vol. 28, no. 3, Apr. 2018.

[14] A. Belahcen, V. Mukhrejee, F. Martin, and P. Rasilo, "Computation of hysteresis torque and losses in a bearingless synchronous reluctance machine," *IEEE Transactions on Magnetics*, vol. 54, no. 3, Mar. 2018.

[15] T. Holenstein, J. Greiner, D. Steinert, and J. W. Kolar, "A high torque, wide air gap bearingless motor with permanent magnet free rotor," in *Proc. IEEE Int. Electric Machines and Drives Conf. (IEMDC)*, May 2017.

[16] J. Asama, T. Tatara, T. Oiwa, and A. Chiba, "Suspension performance of a two-dof actively positioned consequent-pole bearingless motor with a wide magnetic gap," in *Proc. IEEE Int. Electric Machines Drives Conf. (IEMDC)*, May 2015, pp. 786–791.

[17] H. Sugimoto, Y. Uemura, A. Chiba, and M. A. Rahman, "Design of homopolar consequent-pole bearingless motor with wide magnetic gap," *IEEE Transactions on Magnetics*, vol. 49, no. 5, pp. 2315–2318, May 2013.

[18] D. Steinert, T. Nussbaumer, and J. W. Kolar, "Slotless bearingless disk drive for high-speed and high-purity applications," *IEEE Transactions on Industrial Electronics*, vol. 61, no. 11, pp. 5974–5986, Nov. 2014.

[19] H. Bleuler, M. Cole, P. Keogh, R. Larsonneur, E. Maslen, Y. Okada, G. Schweitzer, A. Traxler *et al.*, *Magnetic bearings: theory, design, and application to rotating machinery.* Springer Science & Business Media, 2009.

[20] H. Sugimoto, M. Miyoshi, and A. Chiba, "Axial vibration suppression by field flux regulation in two-axis actively positioned permanent magnet bearingless motors with axial position estimation," *IEEE Transactions on Industry Applications*, vol. 54, no. 2, pp. 1264–1272, Mar. 2018.

Parameter Identifications of Current-Force Factor and Torque Constant in Single-Drive Bearingless Motors

Hiroya Sugimoto[1*], and Akira Chiba[1]

1 Department of Electrical & Electronic Engineering, Tokyo Institute of Technology, Tokyo, Japan
*E-mail: sugimoto@belm.ee.titech.ac.jp

Abstract— One-axis actively positioned bearingless motors have been studied for industry applications such as pumps and cooling fans. In particular, single-drive bearingless motors have a great advantage of simple regulation system. Only one three-phase inverter is required for generating torque and magnetic suspension force by q- and d-axis currents, respectively. The voltage equation is unique because characteristics of both rotating and linear machines are magnetically integrated. The torque constant is derived from the back EMF, then, the current-force factor can be identified. In this paper, a novel parameter identification method of a current-force factor is proposed without direct measurements of the active axial force and the d-axis current.

Keywords—bearingless motor, one degree-of-freedom, single-drive, current-force factor

I. INTRODUCTION

Elimination of mechanical bearings has been required to reduce the mechanical loss, wear, lubricant, maintenance and pollution. To solve these issues bearingless motors have been studied for about twenty-five years. The bearingless motor has a magnetic bearing function which is magnetically integrated in a single motor unit. A lot of bearingless motors have been proposed for various applications, including centrifugal pumps for semiconductor manufacturing and artificial hearts, stirring devices, rotating stages, flywheels, compressors, and cooling fans [1]-[10].

In bearingless motor topologies with reduced number of active positioning axes, one-axis actively positioned bearingless motors and magnetic bearing motors have advantage of low cost and down-sizing [9], [11]-[22]. In particular, single-drive bearingless motors have been proposed to simplify the drive system [9], [19]-[22]. Generally, both a three-phase inverter and a single-phase inverter are required to regulate the motor drive and the magnetic suspension. On the other hand, the single-drive bearingless motor requires only one three-phase inverter because the motor torque and the active axial force can be generated by q- and d-axis currents, respectively. The other four degree of freedoms of the rotor shaft are passively stabilized: radial restoring force and tilting restoring torque are generated in passive magnetic bearings.

In previous studies of the single-drive bearingless motor, several topologies have been proposed. In [19], a four-pole surface permanent magnet (SPM) structure with two axial gaps was proposed with a rotor shaft between two stator cores. Basic principle for generating the active axial force and motor torque by d- and q-axis currents was theoretically discussed in the permeance method. The numerical calculation was verified by the three-dimensional finite-element-method (3D-FEM) analysis and the test results. In [21], a two-pole SPM structure with a cylindrical radial gap was proposed by the authors. The V-shaped winding was unique structure to generate the Lorentz force torque and active axial force simultaneously. However, the rated torque and active axial force in aforementioned structures were considerable low.

To increase the torque and force an eight-pole SPM structure with three axial layers has been proposed by the authors [22]. A liner actuator is integrated in a rotating machine; therefore, the torque and active axial force can be improved. In addition, a unique voltage equation has been proposed. In general permanent magnet machines, the back EMF is generated on q-axis due to the rotor rotational speed. In case of the single-drive bearingless motor, another induced voltage is caused by the rotor axial displacement on d-axis. There are several unique machine parameters in the voltage equation.

In this paper, a novel parameter identification method is proposed. The torque constant is derived from the back EMF, then, the current-force factor can be identified without direct measurements of the active axial force and the d-axis current. It is found that the proposed identification method is effective in the experiment.

II. PROPOSED STRUCTURE AND PRINCIPLE

A. Proposed concept of single-drive bearingless motor

Figure 1 shows a cut view of the fabricated prototype machine of the proposed single-drive bearingless fan motor by the authors in [9]. The cooling fan is one of possible applications for one-axis actively positioned bearingless motors [9], [18]. This machine has a single-drive bearingless motor and two repulsive passive magnetic bearings at both ends of the rotor shaft. At the top of the rotor shaft, a fan blade is installed. At the bottom, a displacement sensor is installed to detect the

The 2018 International Power Electronics Conference

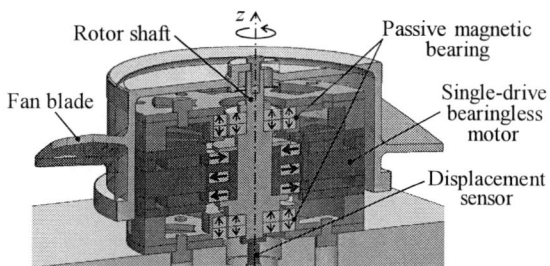

Fig. 1. Proposed single-drive bearingless motor for cooling fan applications.

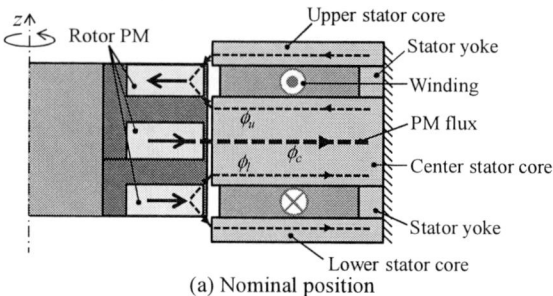

(a) Nominal position

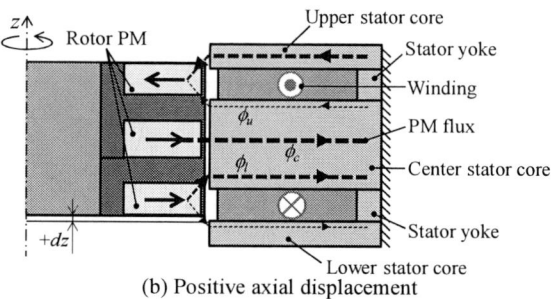

(b) Positive axial displacement

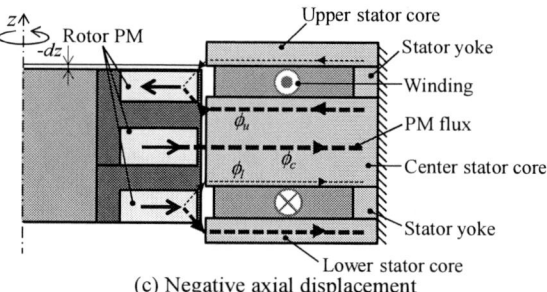

(c) Negative axial displacement

Fig. 2. Flux linkage variation at the rotor axial displacement.

rotor axial displacement. One set of three-phase winding is installed in stator teeth as a concentrated winding structure. The numbers of rotor poles and stator slots are twelve and eight, respectively.

B. Flux linkage variation at axial dispalcement

Figure 2 shows enlarged right half views of the single-drive bearingless motor part. In the three-layer rotor permanent magnets, the motor torque is generated between the center permanent magnet and the center stator core. The upper and lower permanent magnets contribute to generate the active axial force. The upper and lower stator cores are also required to generate the active axial force. The magnetized directions of the center and lower permanent magnets are the same. In contrast, the magnetized direction in the upper permanent magnet is opposite to the other layers to generate the active axial force. The blue part in the rotor is not magnetic material but plastic holder to fix the permanent magnets.

Fig. 2(a) shows that the rotor is nominal position. Let us define three fluxes ϕ_u, ϕ_c, and ϕ_l which are coming from the upper, center, and lower rotor permanent magnets, respectively. When the rotor is nominal position, the fluxes ϕ_u and ϕ_l are the same.

Fig. 2(b) shows that the rotor is displaced in positive axial direction. In this case, the flux ϕ_u is decreased because the upper permanent magnet is apart from the center stator core. On the other hand, the flux ϕ_l is increased because the lower permanent magnet is close to the center stator core. The directions of fluxes ϕ_c and ϕ_l are the same. As a result, the flux linkage is increased when the rotor is displaced in positive axial direction.

Fig. 2(c) shows that the rotor is displaced in negative axial direction. In contrast with Fig. 2(b), the flux ϕ_l is decreased, and the flux ϕ_u is increased. The direction of the flux ϕ_u is opposite to that of the flux ϕ_c. Consequently, the flux linkage is decreased when the rotor is displaced in negative axial direction.

The variation of the flux linkage with respect to the rotor axial displacement is expressed in the voltage equation as a back EMF on the d-axis voltage. In next section, the unique voltage equation and the machine parameter identification method are discussed.

III. DERIVATIONS OF TORQUE CONSTANT AND CURRENT-FORCE FACTOR FROM VOLTAGE EQUATION

The voltage equation of the single-drive bearingless motor is based on that of permanent magnet motors as shown in [22]. In this paper, the proposed structure in Fig. 1 is the surface permanent magnet synchronous motor (SPMSM), and therefore, let us assume that d- and q-axis inductances, which are L_d and L_q, are equal to the self-inductance L.

Equation (1) shows the voltage equation. Let us define v_d and v_q as the d- and q-axis voltages, respectively, R as the winding resistance per phase, ω_e as the electrical angular speed, and P as the time-derivative operator of d/dt. Let us also define $\varPsi_z$ as rate of change of the flux linkage with respect to the axial displacement, $\varPsi_\omega$ as the flux linkage with respect to the electrical angular speed, and k_ω as the coefficient of rate of change of q-axis induced voltage caused by the rotor axial displacement.

3628

$$\begin{bmatrix} v_d \\ v_q \end{bmatrix} = \begin{bmatrix} R + P\dfrac{3}{2}L & -\omega_e\dfrac{3}{2}L \\ \omega_e\dfrac{3}{2}L & R + P\dfrac{3}{2}L \end{bmatrix} \begin{bmatrix} i_d \\ i_q \end{bmatrix} + \begin{bmatrix} P\varPsi_z \\ \omega_e\varPsi_\omega\left(1 + k_\omega z\right) \end{bmatrix} \quad (1)$$

The voltage equation is almost the same with that of the SPMSM, however, the back EMF terms on d- and q-axis are different from typical ones. In case of the single-drive bearingless motor, the d-axis induced voltage of $P\varPsi_z$ is included although it is generally zero in the typical SPMSM. In addition, the second term of $k_\omega z$ is included in the q-axis induced voltage. These terms are generated by the rotor axial displacement, which rarely occurs in conventional motors with mechanical bearings. Therefore, these back EMF terms is unique in the voltage equation of the single-drive bearingless motor.

The theoretical torque equation is derived from the voltage equation. The torque is proportional to the number of pole pairs p, the flux linkage coefficient per electrical angular speed, and the q-axis current as follows:

$$T = p\varPsi_\omega\left(1 + k_\omega z\right)i_q = K_t i_q \quad (2)$$

where K_t is a torque constant. The torque is increased with an increase of the rotor axial position z because the flux linkage is increased as shown in Fig. 2.

Similarly, the theoretical active axial force equation is derived from the back EMF term on d-axis. The back EMF term $P\varPsi_z$ is replaced with the rate of change of flux linkage $\varPsi_z'$ with respect to the axial displacement as follows:

$$\frac{d\varPsi_z}{dt} = \frac{dz}{dt}\frac{d\varPsi_z}{dz} = \frac{dz}{dt}\varPsi_z' \quad (3)$$

The flux linkage derivative $\varPsi_z'$ with respect to the axial displacement z is equal to the current-force factor; the unit of Wb/m can be transformed into N/A. Therefore, the active axial force F_z is obtained by

$$F_z = \varPsi_z' i_d \quad (4)$$

In this paper, let us call the coefficient $\varPsi_z'$ the current-force factor. In addition, the coefficient $\varPsi_z'$ means rate of change of the flux linkage when the rotor is displaced in axial direction. Therefore, $\varPsi_z'$ is equal to a product of coefficients $\varPsi_\omega$ and k_ω. As a result, the coefficient k_ω is expressed with $\varPsi_z'$ and $\varPsi_\omega$ as follows:

$$k_\omega = \frac{\varPsi_z'}{\varPsi_\omega} \quad (5)$$

By substituting (5) into (2), the torque equation is given as

$$T = p\varPsi_\omega i_q + p\varPsi_z' i_q z \quad (6)$$

The current-force factor is included in the torque equation, and therefore, it can be identified in the torque measurement. This characteristic is convenient to identify the current-force factor and the torque constant,

simultaneously.

In addition, we propose much more convenient method to identify the current-force factor and the torque constant from the back EMF. Equation (7) shows the q-axis voltage at open circuit; d- and q-axis currents are equal to 0 A.

$$v_q = \omega_e\varPsi_\omega + \omega_e\varPsi_\omega k_\omega z = \omega_e\varPsi_\omega + \omega_e\varPsi_z' z \quad (7)$$

In (7), the coefficients $\varPsi_z'$ and $\varPsi_\omega$ are included, as a result, the current-force factor and the torque constant can be identified from the measurement of back EMF without measurements of the current, torque, and active axial force.

IV. Experiments

Figure 3 shows the fabricated stator and the rotor shaft. In Fig. 3(a), stator cores are composed of laminated silicon steels. The three-phase winding is installed in the center stator core. The number of turns is 90 per tooth. In Fig. 3(b), the rare-earth material is used in the three-layer permanent magnets and passive magnetic bearings. The rotor outer diameter is 27 mm. The blue part is just plastic material to hold the segment permanent magnets.

Figure 4 shows the active axial force with respect to d-axis current. This is conventional method to identify the current-force factor. The active axial force is proportional to d-axis current in desaturation region, and then, the rate of change of the force is just the current-force factor. In Fig. 4, measured and calculated current-force factors are

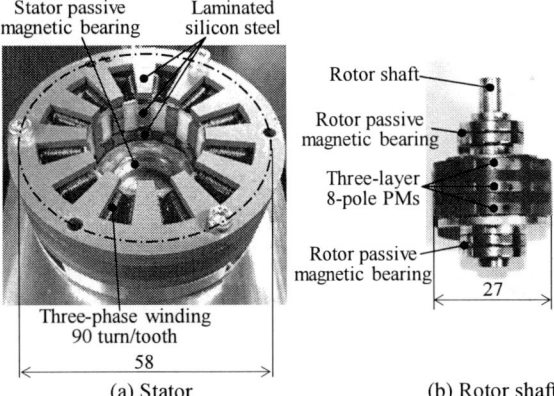

Fig. 3. Fabricated stator and rotor shaft.

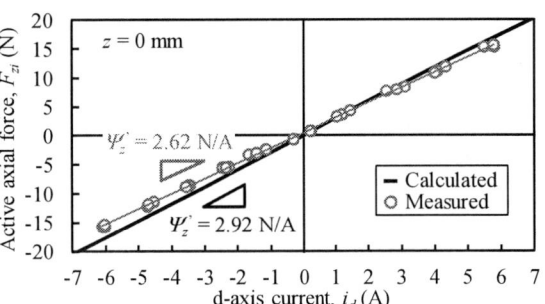

Fig. 4. Calculated and measured active axial force with respect to d-axis current at $z = 0$ mm.

2.62 N/A and 2.92 N/A, respectively.

Figure 5 shows measured average torque with respect to the rotor axial displacement at the rated current of $i_q = 1.73$ A. In this measurement, the rotor shaft is fixed by mechanical ball bearings, and the rotor axial position is adjusted by inserting thin sim tapes between the bearing and base. The static torque is measured by a torque meter. The three-phase currents are provided from a conventional voltage source three-phase inverter. The current-force factor and the torque constant are calculated by comparing coefficients in approximated line with (6). As a result, identified values of Ψ_z' and K_t are 2.58 N/A and 23.33 mNm/A, respectively.

Figure 6 shows measured back EMF on q-axis with respect to the rotor axial displacement when the winding circuit is open. The three phase voltages are measured by a power analyzer. By the aforementioned manner, the current-force factor and the torque constant are calculated with (7). These values are 2.72 N/A and 24.02 mNm/A, respectively.

Table I shows identified machine parameters by each measurement method. The error is less than 5% between two identification methods from the static torque and the back EMF measurements. Therefore, it is experimentally verified that the proposed machine parameter identification is effective.

V. Conclusions

This paper presents novel machine parameter identification methods for single-drive bearingless motors. Both the current-force factor and the torque constant are successfully identified from the back EMF without measuring the active axial force and d-axis current. The theoretical equation is also shown. In the experiment, it is found that the proposed identification method is effective and valuable in single-drive bearingless motors.

Acknowledgment

We would like to thank Mr. Itsuki Shimura and Mr. Hoang Duy Chinh for their great contributions. This work was supported by JSPS KAKENHI Grant Number 17H04916.

References

[1] M. Neff, N. Barletta, and R. Schob, "Bearingless Centrifugal Pump for Highly Pure Chemicals," *in Proc. 8th International Symosium on Magnetic Bearings (ISMB8)*, pp. 283-287, 2002.

[2] M. Oohshima, and C. Takeuchi, "Magnetic Suspension Performance of a Bearingless Brushless DC Motor for Small Liquid Pumps," *IEEE Transactions on Industry Applications*, vol. 47, no. 1, pp. 72-78, 2011.

[3] H. Hoshi, T. Shinshi, and S. Takatani, "Third-generation Blood Pumps With Mechanical Noncontact Magnetic Bearings," *Artificial Organs*, vol. 30, no. 5, pp. 324-338, 2006.

[4] T. Matsuzaki, M. Takemoto, S. Ogasawara, S. Ota, K. Oi, and D. Matsuhashi, "Operation Characteristics of an IPM-type Bearingless Motor with 2-pole Motor Windings and 4-pole Suspension Windings," *IEEE Transactions on Industry Applications*, 2017 (IEEE Early Access Articles).

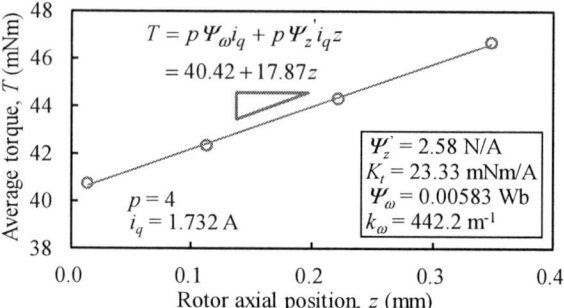

Fig. 5. Measured static average torque with respect to the rotor axial position.

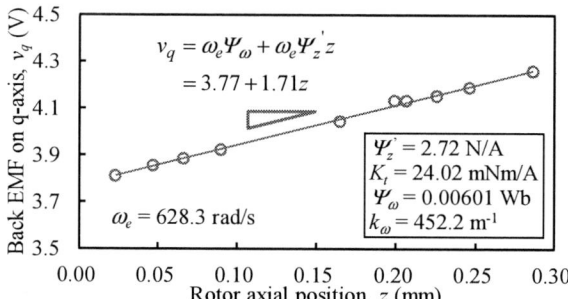

Fig. 6. Measured back EMF on q-axis with respect to the rotor axial position.

TABLE I
IDENTIFIED CURRENT-FORCE FACTOR AND TORQUE CONSTANT IN THREE MEASUREMENT METHODS

	Unit	From axial force	From static torque	From back EMF
Current-force factor, Ψ_z'	N/A =Wb/m	2.62	2.58	2.72
Torque constant, K_t	mNm/A	-	23.33	24.02
Flux linkage, Ψ_ω	Wb	-	0.00583	0.00601
Coefficient, k_ω	m^{-1}	-	442.2	452.2

[5] T. Reichert, T. Nussbaumer, and J. W. Kolar, "Investigation of Exterior Rotor Bearingless Motor Topologies for High-Quality Mixing Applications," *IEEE Transactions on Industry Applications*, vol. 48, no. 6, pp. 2206–2216, 2012.

[6] T. Nussbaumer, P.Karutz, F. Zurcher, and J. W. Kolar, "Magnetically levitated slice motors—An overview," *IEEE Transactions on Industry Applications*, vol. 47, no. 2, pp. 754–766, 2011.

[7] E. Severson, R. Nilssen, T. Undeland, and N. Mohan, "Suspension Force Model for Bearingless AC Homopolar Machines Designed for Flywheel Energy Storage," *in Proc. 2013 7th IEEE GCC Conference and Exhibition*, pp. 274-278, 2013.

[8] H. Mitterhofer, B. Mrak, and W. Gruber, "Comparison of High-Speed Bearingless Drive Topologies With Combined Windings," *IEEE Transactions on Industry Applications*, vol. 51, no. 3, pp. 2126-2122, 2015.

[9] H. Sugimoto, I. Shimura, and A. Chiba, "Principle and Test Results of Energy-Saving Effect of a Single-Drive

Bearingless Motor in Cooling Fan Applications," *IEEJ Journal of Industry Applications*, vol.6, no.6, pp. 456-462, 2017.

[10] A. Chiba, T. Fukao, O. Ichikawa, M. Oshima, M. Takemoto and D.G. Dorrell,"Magnetic Bearings and Bearingless Drives," Newnes Elsevier ISBN 07506 5727 8, 2005.

[11] Y. Okada, N. Yamashiro, K. Ohmori, T. Masuzawa, T. Yamane, Y. Konishi, and S. Ueno, "Mixed Flow Artificial Heart Pump with Axial Self-Bearing Motor," *IEEE/ASME Transactions on Mechatronics*, vol. 10, no. 6, pp. 658–665, 2005.

[12] J. Kuroki, T. Shinshi, L. Li, and A. Shimokohbe, "Miniaturization of a one-axis-controlled magnetic bearing," *Precision Engineering*, vol. 29, no. 2, pp. 208–218, 2005.

[13] A. Yumoto, T. Shinshi, X. Zhang, H. Tachikawa, and A. Shimokohbe, "A One-DOF Controlled Magnetic Bearing for Compact Centrifugal Blood Pumps," *Motion and Vibration Control*, pp. 357–366, 2009.

[14] S. Yang, and M. Huang, "Design and Implementation of a Magnetically Levitated Single-Axis Controlled Axial Blood Pump," *IEEE Transactions on Industrial Electronics*, vol. 56, no. 6, pp. 2213-2219, 2009.

[15] I. D. Silva, J. R. Cardoso, and O. Horikawa, "Design Considerations for Achieving High Radial Stiffness in an Attraction-Type Magnetic Bearing With Control in a Single Direction," *IEEE Transactions on Magnetics*, vol. 47, no. 10, pp. 4112–4115, 2011.

[16] Q. D. Nguen, and S. Ueno, "Modeling and control of salient-pole permanent magnet axial-gap self-bearing motor," *IEEE/ASME Transactions on Mechatronics*, vol. 16, no. 3, pp. 518–526, 2011.

[17] T. Ohji, Y. Katsuda, K. Amei, and M. Sakui, "Structure of One-Axis Controlled Repulsive Type Magnetic Bearing System With Surface Permanent Magnets Installed and Its Levitation and Rotation Tests," *IEEE Transactions on Magnetics*, vol. 47, no. 12, pp. 4734-4739, 2011.

[18] W. Bauer, and W. Amrhein, "Electrical Design Considerations for a Bearingless Axial-Force/Torque Motor," *IEEE Transactions on Industry Applications*, vol. 50, no. 4, pp. 2512–2522, 2014.

[19] J. Asama, Y. Hamasaki, T. Oiwa, and A. Chiba, "Proposal and Analysis of a Single-Drive Bearingless Motor," *IEEE Transactions on Industrial Electronics*, vol. 60, no. 1, pp. 129-138, 2013.

[20] J. Asama, D. Watanabe, T. Oiwa, and A. Chiba, "Development of a One-Axis Actively Regulated Bearingless Motor with a Repulsive Type Passive Magnetic Bearing," *in Proc. International Power Electronics Conference (IPEC-Hiroshima2014-ECCE-ASIA)*, pp. 988-993, 2014.

[21] H. Sugimoto, S. Tanaka, A. Chiba, and J. Asama, "Principle of a Novel Single-Drive Bearingless Motor with Cylindrical Radial Gap," *IEEE Transactions on Industry Applications*, vol. 51, no. 5, pp. 3696-3709, 2015.

[22] H. Sugimoto, I. Shimura, and A. Chiba, "A Novel Stator Structure for Active Axial Force Improvement in a One-Axis Actively Positioned Single-Drive Bearingless Motor," *IEEE Transactions on Industry Applications*, vol. 53, no. 5, pp. 4414-4421, 2017.

The 2018 International Power Electronics Conference

Dampening of Axial Vibrations in a bearingless Flux-Switching Slice Motor by Field Current Regulation

Bianca Klammer[1]*, Karlo Radman[1] and Wolfgang Gruber[2]

1 Linz Center of Mechatronics GmbH, Linz, Austria

2 Department for Electrical Drives and Power Electronics, Johannes Kepler University, Linz, Austria

* E-Mail: bianca.klammer@gmail.com

Abstract- **Disturbance or resonance effects can cause axial vibrations in bearingless slice motors. Typically, these vibrations are only very poorly dampened, as the passive restoring force due the permanent magnets is proportional to the axial deflection (rather than to the axial speed). In this paper, a strategy to actively reduce axial vibrations is presented. In addition to reluctance forces for passive stabilisation, axial damping force is generated by modulating the permanent magnetic air-gap field. To model the rotor's behaviour the equation of motion is set up and analysed. It is shown, that applying field current, which is proportional to the axial velocity, leads to an improved dampening behaviour. Several static and dynamic measurements are conducted to proove the function of the proposed vibration dampening strategy with a bearingless flux-switching slice motor. Finally, also the case without gravitational force is considered, which leads to a different operational behaviour.**

Keywords- bearingless slice motor, field current regulation, vibration dampening

I. INTRODUCTION

A. Bearingless Motors

A bearingless motor is a device which is capable of both creating bearing forces and rotational torque in a common unit [1]. The omission of mechanical bearings has the advantage of contact-free operation. As a result no abrasive wear decreases life expectancy of the motor and no lubrication is needed. Therefore, bearingless motors perform well in harsh environments such as low temperature or vacuum. Industrial applications of bearingless motors are drives for food and pharmacy processes, medical pumps and high speed drives [1].

B. Bearingless Slice Motors

In 1995, when bearingless motors were already used in industrial applications, the bearingless slice motor principle was introduced [2]. The rotors of these devices are disk shaped, which means that their length is small compared to their diameter. Using bearingless slice motors leads to reduced system complexity because the axial position and tilting are passively stabilised by reluctance forces [4]. This stabilisation mechanism is shown in Fig. 1. Only radial directions and the drive torque remain to be actively controlled by the stator coil currents.

C. Bearingless Flux-Switching Slice Motors

The permanent magnets which are necessary for generating the reluctance forces are usually located in the rotor. For implementation of bearingless motors in high temperature or high-speed applications and in areas, where the rotor has to be changed frequently, a permanent magnet-free rotor can be favourable [3]. However, the permanent magnets of the flux-switching motor, depicted in Fig. 2, are located in the midst of the stator teeth. There are two separated winding systems:

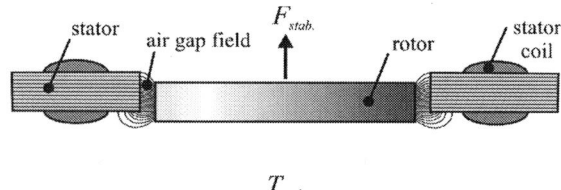

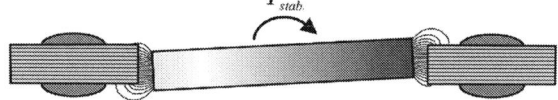

Fig. 1. Principle of passive stabilisation for axial deflection (top) and tilting effects (bottom)

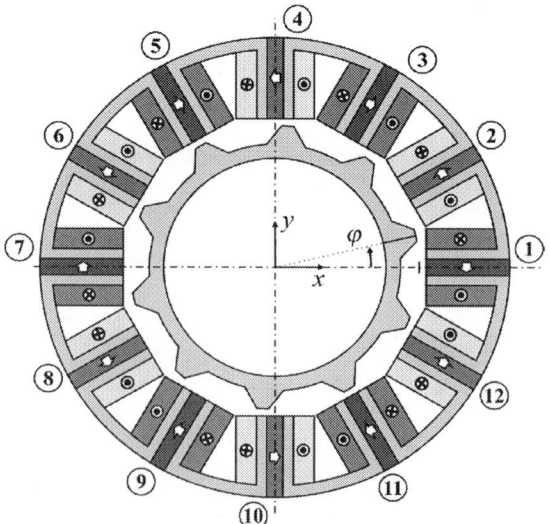

Fig. 2. Cross section of a flux-switching motor with twelve stator and ten rotor teeth. The arrows indicate the direction of the magnetization.

3632

one is used to create bearing forces (indicated by yellow coloured coils in Fig. 2) and another one generates rotational torque (represented by orange coloured coils in Fig. 2) [5], [6].

II. PRINCIPLE OF FIELD-FLUX REGULATION TO GENERATE AXIAL DAMPING FORCE

Disturbance or resonance effects may cause axial vibrations. These are typically poorly dampened, as the passive restoring force of the permanent magnets is only proportional to the axial deflection. An additional axial damping force can be generated by modulating the motor field-flux [4]. Field current components generate no bearing forces and rotational torque but strengthen or weaken the air-gap bias field-flux of the permanent magnets. This is well known from field oriented control.

For an axially decentred rotor there are two cases to consider: axial deflections in and against the direction of the gravitational force. If the rotor moves in the direction of the gravitational force, positive field current should be provided to intensify the field-flux and therefore strengthen the axial stiffness. In the second case, if the rotor moves against the direction of the gravitational force, the axial stiffness should be decreased by energising the coils with negative field current.

III. FINITE ELEMENT SIMULATION RESULTS

3D finite element simulations are performed using Ansys Maxwell. The aim of these simulations is to investigate the influence of field-flux regulation on the axial stiffness of the rotor. Furthermore, the dependence of the stiffness value on the field current is analysed and parameters which are needed for the mathematical model are identified.

The relation between axial stiffness and field current is shown in Figure 3. In this graph the magnetomotive force for field-flux regulation is varied. The passive stiffness due to Maxwell forces features a square dependency on the flux density. Therefore, the axial stiffness also features a quadratic dependency on the flux density in the

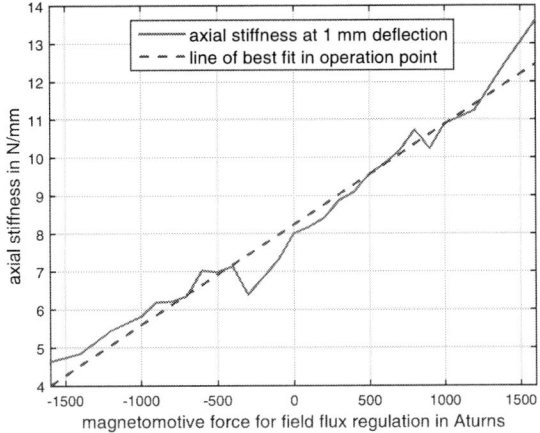

Fig. 3. Axial stiffness over field-flux regulation current with line of best fit

air gap.

Because of the small range of field-flux regulation which is used, the relation between the axial stiffness and the flux density can be considered linear

$$k_z = k_{z,0} + k_{z,i} \cdot i_{s,d} . \tag{1}$$

without significant mistake.

The gradient of the best-fit line correlates to the parameter $k_{z,i}$, which describes the increase of the axial stiffness with increasing field current. The zero crossing of the best-fit line gives $k_{z,0}$, the axial stiffness due to the air gap flux density caused by the permanent magnets. These two values are computed as $k_{z,0} = 8.234 \frac{N}{mm}$ and $k_{z,i} = 0.423 \frac{N}{mm \cdot A}$.

IV. THEORETICAL INVESTIGATIONS

To model the rotor's axial movement the equation of motion for the rotor is set up with

$$m_r \cdot \ddot{z} = -F_d - F_r + F_g \tag{2}$$

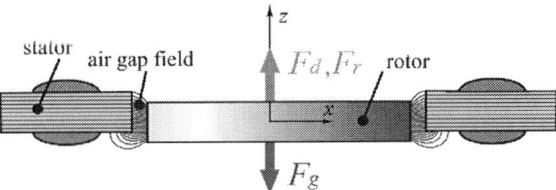

Fig. 4. Conceptual illustration of the forces acting on the rotor.

according to Fig. 4.

The dampening force F_d is modelled proportional to the rotor's axial speed. The restoring force F_r shows a linear dependency on the axial position as long as the rotor remains in a region of sufficiently small deflections. This condition is generally fulfilled in real operation scenarios. F_g denotes the gravitational force acting on the rotor. Therefore (2) is developed into

$$m_r \cdot \ddot{z} + d \cdot \dot{z} + k_z \cdot z = m_r \cdot g \tag{3}$$

Here g denotes the gravitational constant and d is the dampening coefficient of the rotor movement. The dampening coefficient can be determined experimentally by applying a force impulse to the rotor and measuring the axial displacement. The contour of the measured curve can be approximated by a decreasing exponential function, where the time constant of the exponential function corresponds to the dampening coefficient [4]. Table I summarises the parameters of the used bearingless flux-switching slice motor for the measurements.

TABLE I
PARAMETERS OF THE USED SLICE MOTOR

Parameter	Symbol	Value	Unit
axial damping coefficient	d	1.4	kg/s
axial stiffness ($i_d=0$)	$k_{z,0}$	8.234	N/mm
axial stiffness gradient	$k_{z,i}$	0.423	N/mmA
Rotor mass	m_r	0.8	kg

3633

Without any field weakening or strengthening the equation of motion for the rotor is represented by a linear ordinary differential equation of second order given in (3) with constant $k_z = k_{z,0}$ which can be solved analytically

$$z(t) = e^{-\xi \cdot t} \cdot (A \cdot \sin(\omega \cdot t) + B \cdot \cos(\omega \cdot t), \qquad (4)$$

where A and B are constants depending of initial conditions and with

$$\xi = \frac{d}{2 \cdot m_r} \quad \text{and} \quad \omega = \sqrt{\frac{d^2}{4 \cdot m_r^2} - \frac{k_{z,0}}{m_r}} \qquad (5)$$

In a next step, the motion equation with linear field current dependency of the stiffness coefficient according to equation (1) is examined and gives

$$\ddot{z} + \frac{d}{m_r} \cdot \dot{z} + \frac{k_{z,0} + k_{z,i} \cdot i_{s,d}}{m_r} \cdot z = g \qquad (6)$$

To get a direct relationship between the axial speed and the variation of the stiffness, field current can be applied according to

$$i_{s,d} = -k_d \cdot \dot{z} . \qquad (7)$$

The derivative gain k_d is a parameter which can be used for an adjustment of the control performance. A constraint for the field current is that the maximum current levels must not be exceeded. Applying this equation to the equation of motion a nonlinear ordinary differential equation

$$\ddot{z} + \frac{d}{m_r} \cdot \dot{z} + \frac{k_d \cdot k_{z,i}}{m_r} \cdot \dot{z} \cdot z + \frac{k_{z,0}}{m_r} \cdot z = g \qquad (8)$$

results. Because of the nonlinear term with $\dot{z} \cdot z$, it is not possible to solve equation (8) analytically.

For the following investigations concerning the dampening behaviour of the above nonlinear equation of motion a case differentiation between axially centred and decentred rotor rest position is made.

A. Decentred rotor rest position

If the motor's z-axis (axial direction) is located vertically, the rest position of the rotor is not at z = 0. Due to gravity forces the rotor sags. This can also happen due to other external forces, like axial process forces. Therefore, the D-controlled field-flux regulation current as defined in equation (7) is sufficient for stabilising the decentered rotor.

To examine the dampening behaviour of the equation of motion the phase portrait is plotted. In the phase portrait in Fig. 5 the system trajectories with and without D-controlled field current are visualised. On the horizontal axis, the axial deflection z is plotted, whereas the vertical axis depicts the axial speed $\dot{z}$. All graphs in this work show the axial position shifted to a rest position of zero.

From the phase portrait it can be concluded that the D-controlled field current proves beneficial. With the same starting position and speed for both trajectories, the green trajectory converges much faster towards the origin. Moreover, all purple vectors in the phase plot, which refer to the derivative with D-controlled field-flux

regulation current, point more directly towards the origin than the blue vectors of the equation without field current.

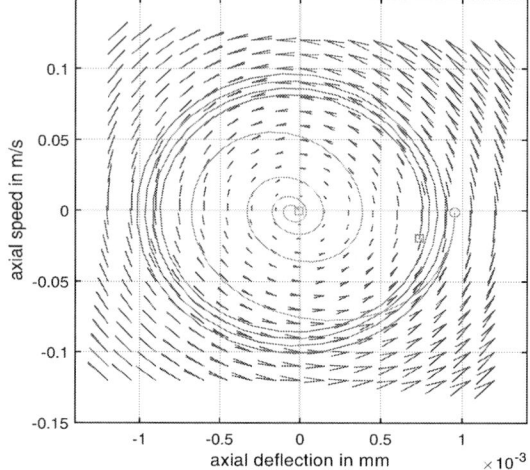

Fig. 5. Phase portrait with velocity proportional field current (green trajectory, purple vectors) and without (red trajectory, blue vectors).

In the lower graph of Fig. 6 the field-flux regulation current is plotted. It is directly proportional to the axial speed, and limited to a maximum current of 5A by the choice of k_d. After fractions of a second no more field-flux regulation current is needed to control the axial position as the oscillation has deceased. Therefore, it can be concluded that the proposed D-controller performs rather fast and efficient. Thus, simulation of the axial movement shows that the damping of the vibration is strongly increased with the D-controlled field-flux regulation current.

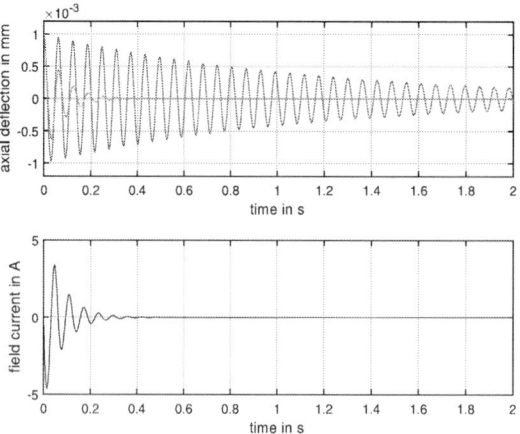

Fig. 6. Axial movement with (green) and without (red) D-controlled field-flux regulation current.

B. Centred rotor rest position

For the centred rotor rest position, e.g. if the motor's z-axis lies horizontally, the simple velocity proportional field current does not show the desired effect, as positive deflections lead to a decreasing damping, which is unfavourable.

For the same value of k_d as used for decentred rotor rest position nearly no benefit is visible for centred rotor rest position. The shape of the system trajectory with field current, which can be seen in Figure 7, is only slightly different from the passive trajectory. For positive deflections the purple vectors point more directly to the origin than the blue ones. But for negative deflections it is just the opposite. Hence, both system trajectories (with and without D-controller) end up at nearly the same point.

For higher derivative gain the behaviour of the nonlinear differential equation can even become instable. Moreover, it is obvious that this kind of field-flux regulation provides hardly any advantages.

A first approach for a better control of the field current in case of centred rotor rest position is to inject a current depending from the sign of the deflection with the control law

$$i_{s,d} = -k_d \cdot \dot{z} \cdot \mathrm{sign}(z) . \tag{9}$$

With this strategy any decrease of the dampening coefficient (for negative deflections) is avoided. This results in the following equation of motion

$$\ddot{z} + \frac{d}{m_r} \cdot \dot{z} + \frac{k_d \cdot k_{z,i}}{m_r} \cdot \dot{z} \cdot |z| + \frac{k_{z,0}}{m_r} \cdot z = 0 \tag{10}$$

From the phase portrait in Fig. 8 it can be seen that the green system trajectory, which is the one with controlled field current, converges faster to the origin than the red one. The increase of the damping is not as significant as in the case with decentred rotor rest position, but still beneficial.

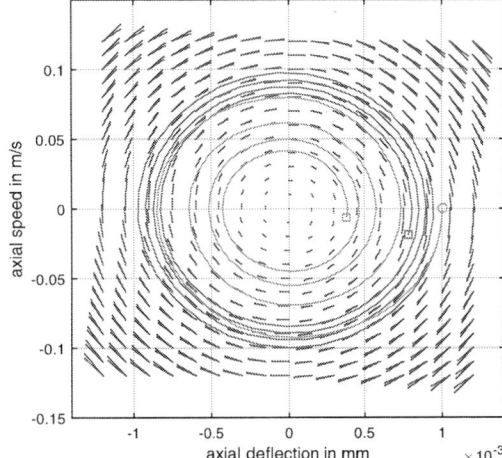

Fig. 8. Phase portrait with field current according to (9) (green trajectory, purple vectors) and without (red trajectory, blue vectors).

In Fig. 9 the time curve of the axial deflection and the field current is plotted. Here, the dominant drawback of the control rule as defined in equation (9) becomes visible. The field current changes stepwise because the

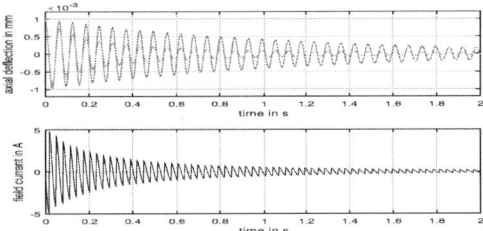

Fig. 9. Axial movement with (green) and without (red) D-controlled field-flux regulation current.

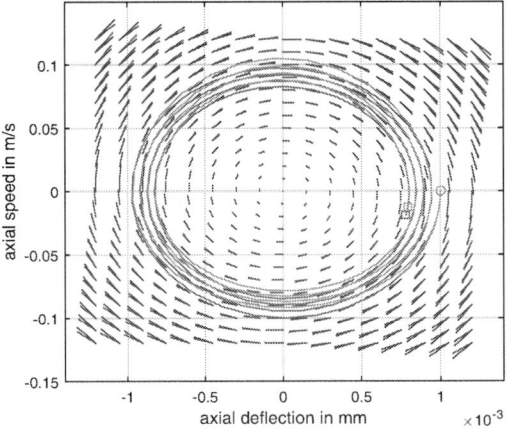

Fig. 7. Phase portrait with velocity proportional field current (green trajectory, purple vectors) and without (red trajectory, blue vectors).

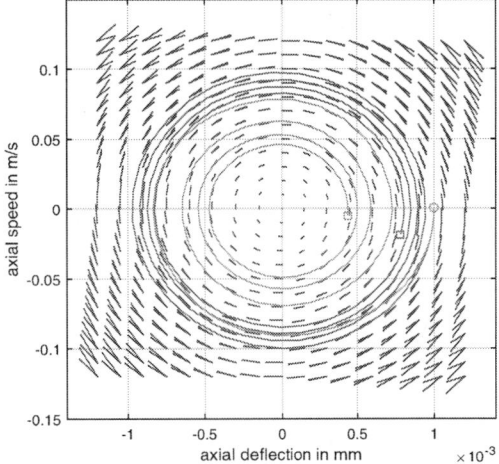

Fig. 10. Phase portrait with field current according to (11) (green trajectory, purple vectors) and without (red trajectory, blue vectors).

TABLE II.
DAMPING CONSTANTS FOR DIFFERENT SCENARIOS

rotor rest position	field current	k_d	damping constant δ in $\frac{1}{s}$	amplitude after 300 ms
decentred	-	-	1.41	0.75 mm
decentred	$i_{s,d}=-k_d \cdot \dot{z}$	60	16.64	0.01 mm
decentred	$i_{s,d}=-k_d \cdot \dot{z} \cdot z$	55000	16.53	0.01 mm
centred	$i_{s,d}=-k_d \cdot \dot{z} \cdot z$	110000	3.52 to 1.73	0.35 mm

maximum of the axial speed occurs when the axial deflection crosses the zero position. For such current slopes extremely high voltages would be necessary. Hence the feasibility of injecting this field current is hardly ever given. The corresponding phase portrait is shown in Fig. 10.

To avoid the steep gradients in the current curve another approach for the regulation of the field current is made according to

$$i_{s,d}=-k_d \cdot \dot{z} \cdot z . \qquad (11)$$

This leads to a quadratic term in the equation of motion with

$$\ddot{z}+\frac{d}{m_r}\cdot\dot{z}+\frac{k_d \cdot k_{z,i}}{m_r}\cdot\dot{z}\cdot z^2+\frac{k_{z,0}}{m_r}\cdot z=0 \qquad (12)$$

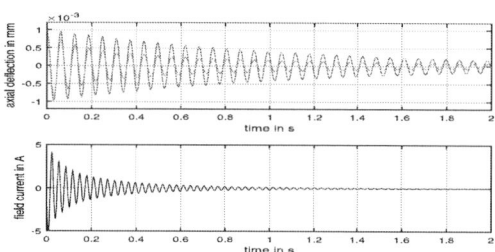

Fig. 11. Axial movement with (green) and without (red) PD-controlled field-flux regulation current.

This regulation method shows very similar convergence behaviour in the phase portrait compared to the method with the sign-function. However, an improvement is visible in the field-current time curve, due to the fact that there are no steps necessary any more. The current slope in Figure 11 shows sinusoidal characteristic with doubled frequency compared to the axial oscillation.

C. Analysis of the dampening behaviour

To achieve a better comparability of the different control strategies the dampening behaviour will be represented by only one constant. For this purpose, an approximated damping constant is calculated for the different curves of the axial oscillation. An exponential falling curve

$$z(t)=z_0 \cdot e^{-\delta \cdot t} \qquad (13)$$

with the time constant δ is fitted to the envelope of the vibrations using a least squares method.

For decentred rotor rest position without field current the damping constant is $\delta_{decentered,uncontr.} = 1.41 \; \frac{1}{s}$ and therefore rather small.

With the D-controlled field current the damping constant can be increased by a factor of ten with $\delta_{decentered,lin.} = 16.6\frac{1}{s}$. The improvement is remarkable. After 300ms the axial oscillation is nearly abated, whereas without field current the deflection is only dampened to 75% of the initial amplitude at the same time.

Similar investigations are conducted for the centred rotor rest position. Due to the nonlinearities in the equation of motion the dampening factor shows a strong dependency on the axial deflection and cannot be expressed by a single damping constant. Therefore, another approach to describe the dampening characteristics of this equation is taken by fitting an exponential function with time-dependent damping constant. Due to the coupling of time and deflection the dependence of the stiffness on the axial deflection can be modelled by this time-dependent damping constant. It turned out by trial that a good estimation of the envelope curve is achieved by adding the fifth root of the time to the damping constant according to

$$z(t)= e^{-\delta_0 \cdot t - \delta_r \cdot \sqrt[5]{t}}. \qquad (14)$$

$\delta_0 = 1.41 \frac{1}{s}$ describes the damping constant without field current. Using a least squares method, the best value for δ_r was evaluated with $\delta_r = 0.49 \frac{1}{\sqrt[5]{s}}$.

Table II shows a summary of the calculated damping constants. The starting deflection is always 1 mm and k_d is applied such that the amplitude of the field current does not exceed 10A.

V. MEASUREMENTS

Figure 12 shows the control scheme for a bearingless flux-switching motor with the additional controller for axial vibration damping. The axial position is measured by digital Eddy current sensors [7]. Static and dynamic measurements for the centred and decentred rotor rest position are performed to examine the functionality and performance of the axial vibration dampening strategies.

A. Measurements for rotor in decentred rest position

First, the rotor in decentred rest position is examined, which means that the motor's z-axis is aligned vertically. In this case, a velocity proportional field current according to equation (7) is used, which is called "standard control" thereafter.

Static measurements

For the static measurements at standstill an external excitation is applied to the rotor. The resulting vibrations and the field-flux regulation current are measured.

The 2018 International Power Electronics Conference

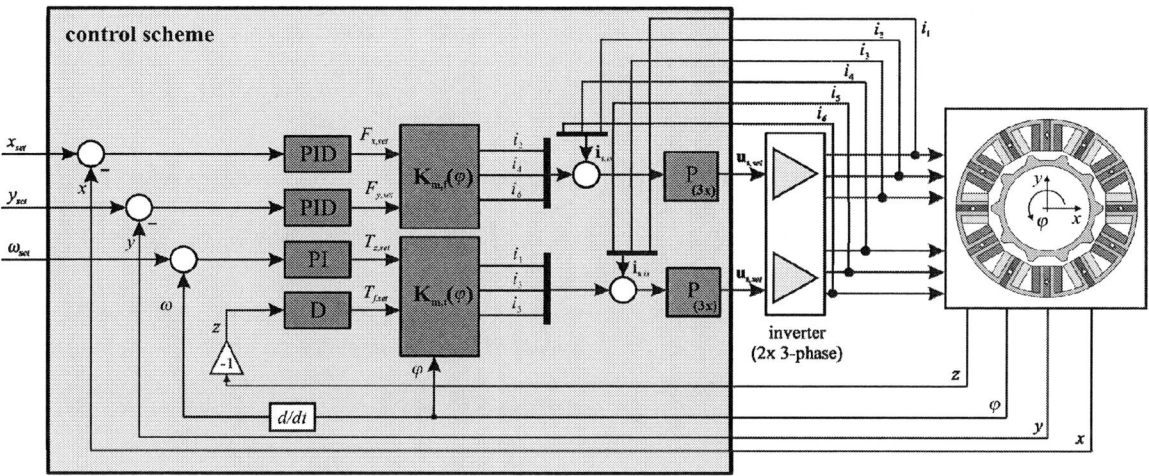

Fig. 12. Control scheme for a bearingless flux-switching motor with axial vibration damping

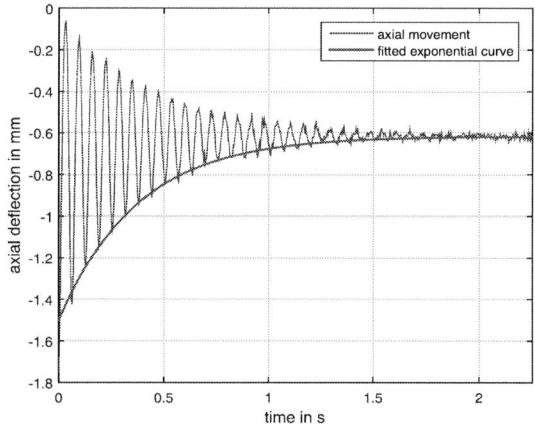

Fig. 13. Axial oscillation with fitted exponential curve for rotor with decentred idle position

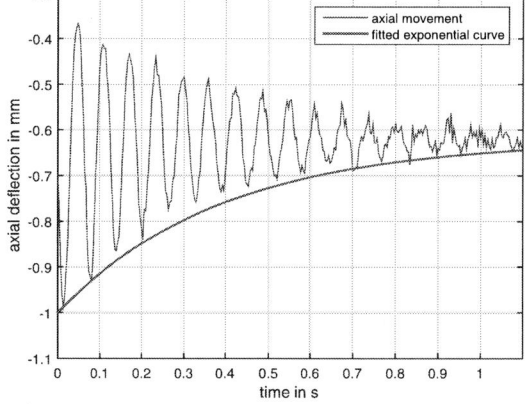

Fig. 14. Axial oscillation with fitted exponential curve for rotor with decentred idle position and standard field current control with $k_d = 0.015$

Figure 13 shows the axial deflection in the first two seconds after applying an external axial excitation to the rotor. No field current is used and the vibrations are therefore dampened weakly by system losses alone. The decay constant of the envelope of this curve was determined by fitting an exponential function through the peaks. We obtain $\delta = 2.41 \frac{1}{s}$. This value differs from the idle decay constant used for the mathematical model, which was $1.4 \frac{1}{s}$. For the evaluation of this previous value, the motor was installed on different feet which featured other dampening behaviour. The same feet could not be used now since the sensor board for the axial position measurement was added and higher feet became necessary. Therefore, a different but nevertheless quite low decay constant was determined.

Figure 14 shows the axial oscillation with field-flux current and standard control with a k_d of 0.015 in the first second. Although the factor k_d is small, the dampening behaviour is improved significantly. The decay constant varies from $\delta_{begin} = 13.7 \frac{1}{s}$ at the beginning to $\delta_{1s} = 3.53 \frac{1}{s}$ after one second. The amplitude of the applied field current, shown in Figure 15, remains below 2 A. The field-current characteristic is shifted by a phase of 90° relative to the axial deflection, due to the proportionality between field current and axial speed.

In order to further improve the disturbance suppressions, k_d is increased to 0.06. In Figure 16 the effect of this field-flux control is visualised. Axial disturbances are dampened within 200 ms. The nonlinear term dominates the motion equation (8) and it is no longer suitable to fit an exponential function. The field-current of this case in Figure 17 shows that about 15A are

3637

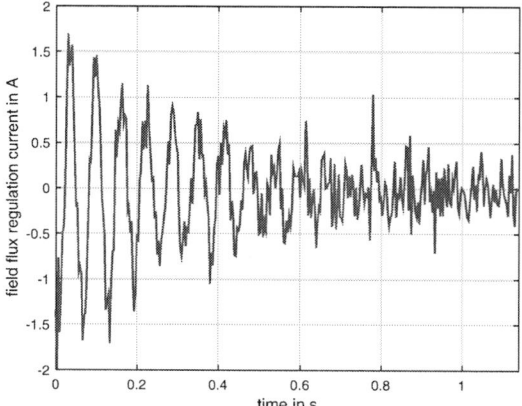

Fig. 15. Field current characteristic for standard control with $k_d = 0.015$

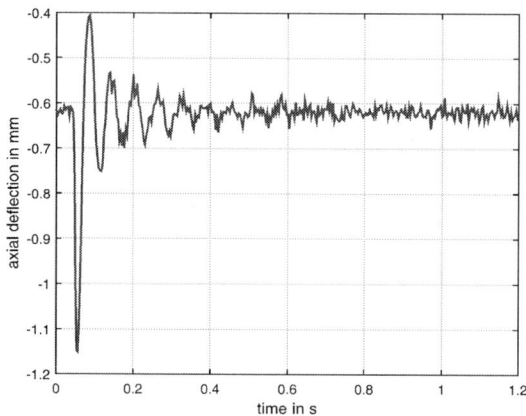

Fig. 16. Axial oscillation with fitted exponential curve for rotor with decentred idle position and standard field current control with $k_d = 0.06$

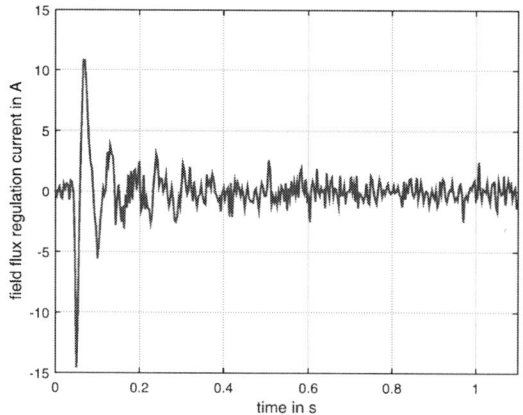

Fig. 17. Field current characteristic for standard control with $k_d = 0.06$

needed. As the high energisation of the coils only lasts for a very short time this dampening strategy has still acceptable losses.

Dynamic measurements

For the dynamic measurements the rotor is driven at a speed of 176 rounds per minute or 2.93 Hz. At this speed the rotor starts to vibrate heavily in axial direction due to resonance effects. In Figure 18 in the first two seconds without any field current, the axial vibrations' amplitude is about 1 mm. After 2.9 s a field current with standard control and $k_d = 0.06$ is applied for dampening. In the beginning, high field currents up to 10 A are needed, but within 200 ms vibrations abate and the field current is significantly reduced. The rotor is also lifted back to its standstill rest position.

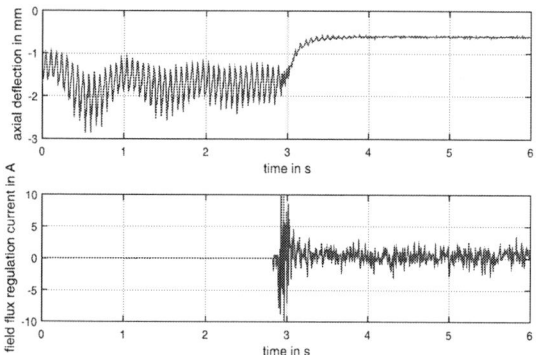

Fig. 18. Axial movement for a rotor with decentred idle position in resonance frequency; standard field current control with $k_d = 0.06$ applied after 2.9 s

B. Measurements for rotor in the centred idle position

After the measurements for the rotor with decentred idle position, which show that the proposed dampening strategy is working effective, the rotor with centred idle position is examined. Instead of a D-controlled field current, a PD-controlled field-current control according to equation (11) is implemented in this case.

Static measurements

Without field-flux current, the vibrations are dampened weakly. The decay constant of the envelope of the axial deflection is determined by fitting an exponential function through the peaks and results in $\delta = 1.63 \, \frac{1}{s}$. In the centred idle position case, the decay constant is smaller than for the decentred rotor, because no damping feet are in use, as the rotor and stator are in upright position.

Figure 19 shows the axial deflection with field-flux control current and a k_d of 0.6 in the first two seconds. k_p is set to one for all measurements. The decay constant varies from $\delta_{begin} = 5.22 \, \frac{1}{s}$ at the beginning to $\delta_{1s} = 1.93 \, \frac{1}{s}$ after one second. The amplitude of the field current, shown in Figure 20, remains below 5 A. The field-flux current features mainly the double frequency of the position oscillation, which correlates to the numerical analysis. It is slightly asymmetric because a band-pass

The 2018 International Power Electronics Conference

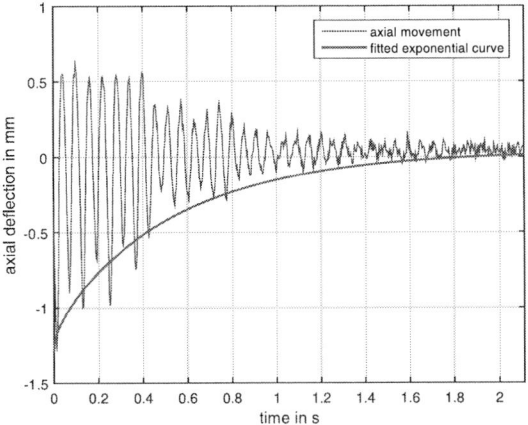

Fig. 19. Axial movement with fitted exponential curve for the rotor with centred idle position and PD-controlled field current with $k_d = 0.6$

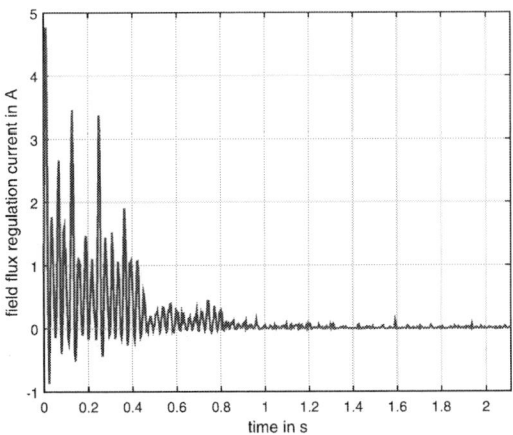

Fig. 20. Field current characteristic for the rotor with centred idle position and PD-controlled field current with $k_d = 0.6$

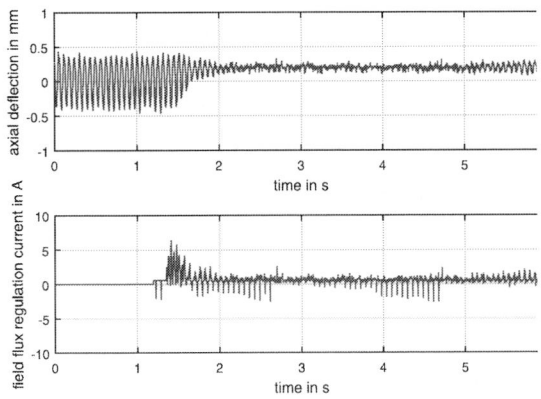

Fig. 21. Axial vibration of a rotor with centred idle position in resonance frequency; PD-controlled field current with $k_d = 5$ and $k_p = 1$ applied after 1.3 s

filter was used to suppress high-frequency disturbances, noise and the constant level of axial deflection which would lead to a DC-level in the field current.

Dynamic measurements

For the dynamic measurements the rotor is driven at a speed of 184 rotations per minute or 3.07 Hz. At this speed the rotor begins to vibrate axially. In Figure 21 it can be seen that in the first 1.5 seconds without any regulation, the axial vibration amplitude is about 0.8 mm. Then a P.D-controlled field current with $k_d = 5$ is applied to dampen the vibrations. The impact of the generated field-flux current can be seen instantaneous, as the axial deflection decreases and finally stabilises by an amplitude of only 0.2 mm. The currents which are needed for this amplitude reduction are below 6 A in the beginning and much lower in the new steady state.

VI. CONCLUSIONS

3D finite-element simulations clearly show a significant dependence of the axial stiffness values on the field current. To increase the axial damping, control strategies for the field current using this axial stiffness dependency are developed. For decentred rotor, a field current proportional to the axial speed is used successfully. In case of centred rotor the field current is chosen proportional to the product of axial speed and axial deflection, yielding a faster decay.

Measurements at zero rotor speed show that the decay of the axial oscillations is highly increased. At axial resonance speed, the D-controlled field current reduces the axial vibration amplitude significantly in case of decentred rotor. For centred rotor vibrations can be dampened using a PD-controlled field current. Due to the fact that the absolute value of the deflection is smaller than for decentred rotor, the improvement in damping is less effective.

ACKNOWLEDGMENT

This work has been supported by the COMET-K2 "Center for Symbiotic Mechatronics" of the Linz Center of Mechatronics (LCM) funded by the Austrian federal government and the federal state of Upper Austria.

REFERENCES

[1] A. Chiba, T. Fukao, O. Ichikawa, M. Oshima, M. Takemoto, D. G. Dorrell, "Magnetic Bearings and Bearingless Drives", *Elsevier, 2005*

[2] R. Schöb, N. Barletta, "Principle and application of a bearingless slice motor", *Proc. 5th Int. Symp. on Magnetic Bearings (ISMB)*, pp. 333-338, 1996

[3] W. Gruber, K. Radman, R. T. Schöb, "Design of a bearingless flux-switching motor", *Education and Research Conference (EDERC), 2014;*

[4] M. Miyoshi, H. Sugimoto, A. Chiba, "Axial vibration suppression by field-flux regulation in two-axis actively positioned permanent magnet bearingless motors with axial position estimation", *Electrical Machines (ICEM), 2016;*

[5] K. Radman, N. Bulic, W. Gruber, "Performance evaluation of a bearingless flux switching slice motor", *Energy Conversion Congress and Exposition (ECCE), 2014; DOI: 10.1109/ECCE.2014.6953919*

[6] K. Radman, W. Amrhein, R.T. Schöb, W. Gruber, W. Bauer, "Considerations regarding bearingless flux switching slice motors", *Proceedings of 1st Brazilian Workshop on Magnetic Bearings*, 10-2013

[7] J. Passenbrunner, S. Silber, W. Amrhein, "Investigation of a digital eddy current sensor", *IEEE International Electric Machines & Drives Conference (IEMDC)*, 2015

Analysis and Design of a Bearingless Axial-Force/Torque Motor with Flex-PCB Windings

Nobuyuki Kurita[1*], Walter Bauer[2], Gerald Jungmayr[3], Wolfgang Gruber[2] and Wolfgang Amrhein[2]

1 Division of electronics and informatics, Gunma University, Kiryu, Japan
2 Institute of Electrical Drives and Power Electronics, Johannes Kepler University, Linz, Austria
3 Linz Center of Mechatronics GmbH, Linz, Austria
*E-mail: nkurita@gunma-u.ac.jp

Abstract— **The axial force/torque motor (AFTM) can generate axial bearing strength and drive torque simultaneously with a conventional 2-phase winding system. Since the end windings provide the active axial suspension forces, there is no necessity of additional windings to control the axial position. Passive magnetic ring bearings stabilize the remaining four degrees of freedom. It is crucial to achieving a high fabrication precision and filling factor of the winding to reach a high efficient AFTM. Therefore, we utilize the technique of directly printing the winding on a flexible printed circuit board (flex-PCB). The focus of this paper lies in the underlying design of the AFTM and the configuration of the flex-PCB windings. Analytic models that describe the influence of the motor dimensions on its force and torque generation capability are presented and validated by finite element method (FEM) analysis. Based on these results, a detailed experimental setup was designed to achieve performance validation.**

Keywords— *Axial-force/torque motor, bearingless motor, flex-PCB winding, magnetic analysis.*

I. INTRODUCTION

The brushless motor drive technology has one major drawback in common, the wear and tear of its mechanical bearings. A promising concept for minimization of friction losses is the utilization of magnetic bearings. However, the electrical and mechanical complexity of magnetically levitated rotors is usually high because either passive or active control strategies must stabilize a rigid rotor the behavior of six degrees of freedom (DOF). Any new desired bearingless drive concept should also achieve full stabilization of all the DOF. The first prototype of a magnetic levitated compact rotor, using PM ring supports, is treated in [1]. The proposed design makes already use of PM ring bearings to support radial forces and for stabilization of the tilting. Also, an additional active magnetic bearing is necessary to compensate the resulting axial unstable dynamics caused by the ring bearings. The drive torque generating part is carried out as axial flux motor.

II. AXIAL FORCE/TORQUE MOTOR

The claim for minimum, constructive complexity and optimization regarding material usage leads to the concept

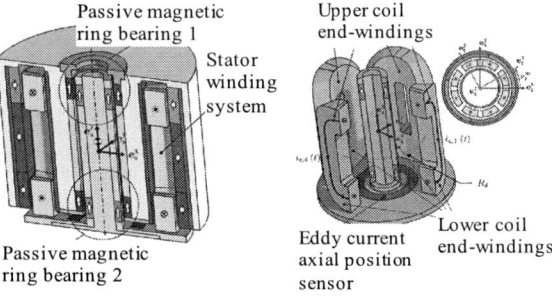

(a) Cross-section of the motor (b) Configuration of the winding

Fig. 1. The bearingless alternating field AFTM

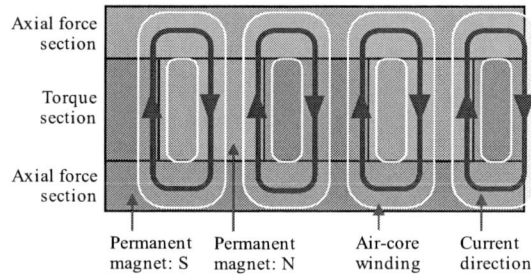

Fig. 2. Arrangement of the permanent magnets and air-core windings

of the Axial-Force/Torque Motor [2-6]. In an approach to integrate the axial force and drive torque generation into a single unit with one common winding system, authors found a new bearingless motor concept. The first prototype, featuring this new technology, was designed as an air-core coil motor. Fig. 1(a) shows the cross-section of the motor and Fig. 1(b) shows the configuration of the air-core cos. The inner back and the outer back iron are both assigned to the rotor, and they rotate together. This arrangement minimizes the eddy current losses and hysteresis losses. Radial translation and tilt motion are stabilized by two sets of passive magnetic ring bearings. Fig. 2 shows the arrangement of the permanent magnets and air-core coils. The green and red part indicate the permanent magnets; the bright yellow part shows the windings, and the brown arrows indicate the direction of the current. The end windings of the air-core stator coils can be used to create suspension forces in the axial direction. This permanent

The 2018 International Power Electronics Conference

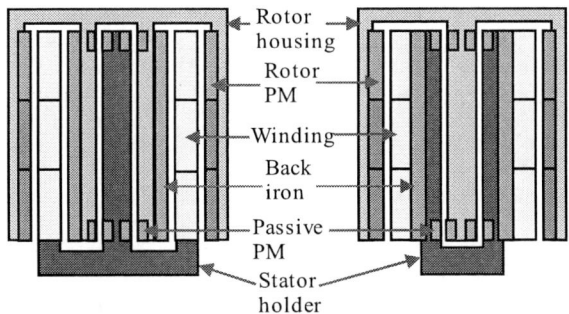

Fig. 3. Comparison between conventional configuration (left) and novel configuration (right)

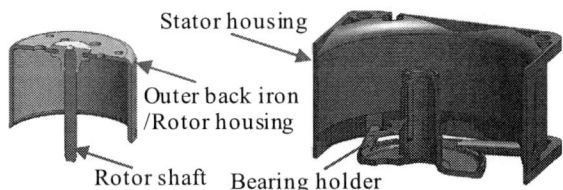

Fig. 4. Cross-section of the comically available small fan

TABLE I DESIGN PARAMETERS OF THE PASSIVE PM

Stator PM outer diameter	ϕ 8 mm (Fixed)
Rotor PM inner diameter	ϕ 3 mm (Fixed)
Distance between upper and lower PM	15 mm (Fixed)
Air gap between rotor and stator PM	0.5 ~ 0.7 mm
Height of the stator and rotor PM	0.5 ~ 1.5 mm
Remanence flux density of PM	1.0 ~ 1.4 T

magnet layout prevents the cancellation of the effects produced by the opposing end-windings of a single air-core coil when using a conventional alternating pole motor magnet design. It is also remarkable that a winding arrangement is required to decouple the bearing force and drive torque generation in a way, that both quantities can be controlled without any destabilizing side effects. In AFTM as mentioned above, inner back iron rotates with permanent magnets to avoid eddy current losses and hysteresis losses; however, the resulting increase in air gap and rise of the rotor mass makes it difficult to miniaturize the device further. Therefore, we propose the novel configuration of the AFTM in this paper. Fig. 3 shows the comparison between conventional and novel arrangement. The windings and the inner back iron are fixed to the stator holder in the left figure. Although, the eddy current loss will probably increase; however, by eliminating the air gap in the middle, further miniaturization of the apparatus becomes possible. The operating principle is the same in both configurations.

III. DESIGN OF PM SIZE OF THE ATFM

Detailed experimental setup design is conducted by using the numerical calculation MagOpt which is developed by Linz Center of Mechatronics, GmbH., and the FEM magnetic analysis ANSYS Maxwell (Cybernet Systems Co., Ltd). In this paper, we implement AFTM into the commercially available small fan. Fig. 4 shows the cross-section of the small fan. The stator housing size is

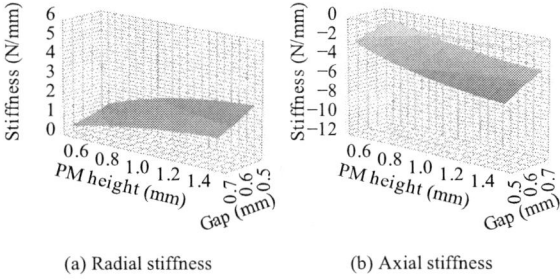

(a) Radial stiffness   (b) Axial stiffness

Fig. 5. Stiffness versus variation of PM height and width with remanence of the PM of 1.0 T

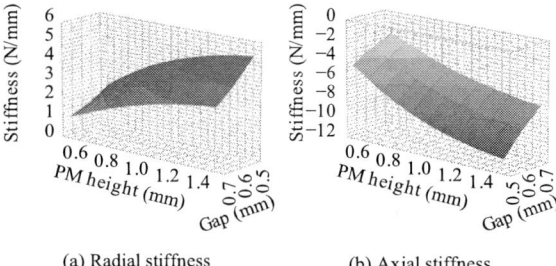

(a) Radial stiffness   (b) Axial stiffness

Fig. 6. Stiffness versus variation of PM height and width with remanence of the PM of 1.4 T

W 60 × L 60 mm × H 32 mm. The outer diameter of the rotor housing is 30 mm. Two sets of upper and lower ball bearings support the rotor shaft, and the ball bearings are fixed to the inside of the bearing holder. PMs for the field magnet are on the inner side of the rotor housing, and the winding is around the bearing holder. The fan blades are around the rotor housing. Although, ball bearings, motor permanent magnets, windings and fan blades are omitted in the figure. Instead of the existing ball bearing, passive magnetic bearing support radial directional translation motion and tilt motion and the AFTM support axial directional translation motion. Also, instead of the brushless DC motor, the AFTM generate rotating torque.

A. Analysis and design of the passive magnetic bearing

According to the commercially available fan, size of the AFTM is limited. The design parameters of the radial passive permanent magnet ring are listed in Table I. The outer diameter of the stator PM is ϕ 8 mm, since they are installed on the inner side of the bearing holder. Similarly, the inner diameter of the rotor PM is ϕ 3 mm, since they are mounted on the outer side of the rotor shaft. It is better to keep the distance between the upper permanent magnet and the lower permanent magnet as far as possible to obtain the maximum restoring torque for the inclination stabilization; thus, the gap between upper and lower PM is 15 mm. Among the above limitations, the parameters for determining the size of the passive PM are the gap distance between the rotor PM and the stator PM (0.5 ~ 0.7 mm), the height of the PM (0.5 ~ 1.5 mm), and the remanence flux density of the PM (1.0, 1.4 T). The thickness of the PM varies depending on the width of the air gap (1 ~ 0.9 mm). The positive stiffness of the radial displacement: k_r

3641

TABLE II DESIGN PARAMETERS OF THE ROTOR AND SO ON

Outer diameter of rotor PM	28.3 mm (Fixed)
Inner diameter of stator back iron	10.4 mm (Fixed)
Height of the rotor PM	20.0 mm (Fixed)
Outer back iron width	0.5 mm (Fixed)
Gap between PM and winding	0.7 mm (Fixed)
Gap between the adjacent PMs	0.2 mm (Fixed)
Thickness of rotor PM	1.0 ~ 4.0 mm
Width of winding	5.8 ~ 3.3 mm
Number of the PM	16 ~ 28

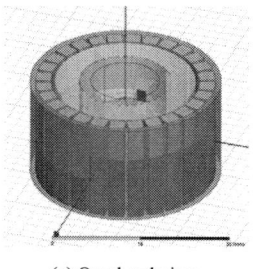 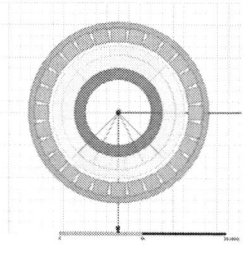

(a) Overhead view   (b) Top view

Fig. 7. Analytical model

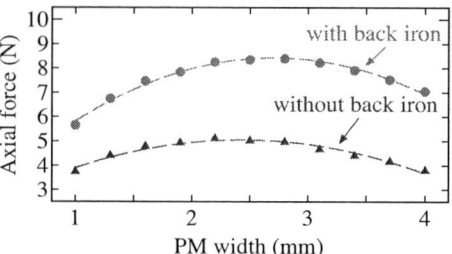

Fig. 8 FEM magnetic analysis result of axial force

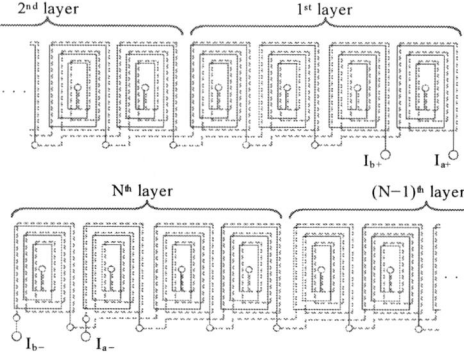

Fig. 9. Conceptual diagram of a winding to be printed on a flex-PCB

N/mm and the negative stiffness of the axial displacement: k_z N/mm are calculated by using MagOpt LCM. Fig. 5 and Fig. 6 shows calculation results of the stiffness with remanence of the PM of 1.0 T and 1.4 T, respectively. As expected, the stronger the residual magnetic flux density, the bigger the height of the permanent magnet and the narrower the air gap, the higher the stiffness. Also, as expected, the relationship of the positive radial stiffness: k_r and the negative axial stiffness: k_z follows Eq. (1).

$$2k_r + k_z = 0 \qquad (1)$$

Maximum stiffness is obtained when the air gap is 0.5 mm, PM height is 1.5 mm, and remanence is 1.4 T; radial and axial stiffness are k_r: 5.25 N/mm, k_z: −10.50 N/mm, respectively. It is estimated that the total mass of the rotor and fan housing with AFTM is about 50 g. And, the thickness of the PM becomes 1 mm or less. Thin PM less than 1mm is difficult to manufacture. Therefore, we choose the air gap of 0.5 mm, the PM height of 1 mm, and the PM thickness of 1 mm. As a result, radial stiffness k_r: 4.37 N/mm, axial negative stiffness k_z: −8.73 N/mm.

B. Analysis and design of the AFTM

The size of the rotor PM is also limited by the commercially available small fan. The design parameters of the rotor PM, winding, and inner back iron are listed in Table II. Fig. 7 shows analytical model built on ANSYS Maxwell. The outer diameter of the rotor PM, Inner diameter of the stator back iron, and height of the rotor PM is fixed since they are installed on the inner side of the rotor housing. The rotor housing is made of steel; therefore, we can eliminate outer back iron; however, we need some holder to support rotor PM to set them the proper position. But, they should very thin to reduce the total mass of the rotor. Thus, we decide the thickness of the outer back iron is 0.5 mm by considering manufacturability. In Fig. 7, the thickness of the outer ring is the sum of the rotor housing

and the rotor back iron. The gap between the adjacent PMs is 0.2 mm by considering manufacturing error. By increasing the magnetization direction thickness of the PM, the width of the PM decreases and the thickness of the winding declines at the same time, and vice versa. Fig. 8 shows the FEM analytical result of the axial force versus width of the PM with and without inner back iron. By changing the thickness of the PM in the direction of magnetization, the axial force changes, and it has a local maximum value. Also, compared to the case without back yoke, the axial force increased by 85% at maximum by providing the inner back yoke. By this magnetic field analysis, we decide the thickness in the direction of magnetization of the PM is 2.5 mm, the width is 3.5 mm, and the thickness of the winding is 3.05 mm. The axial direction force with these values was 8.39 N. While; the negative axial stiffness by passive PM is k_z: −8.73 N/mm; thus, even when the rotor is significantly displaced in the axial direction, the axial force can control it to achieve stable levitation.

C. Design of the flex-PCB winding

According to the analytical results, the maximum thickness of the winding is 3.05 mm, i.e., the inner diameter of the winding, that much to the outer diameter of the inner back iron, is ϕ10.4 mm and the maximum outer diameter of the winding is ϕ16.5 mm. The space available for winding is minimal. Thus, it is essential to achieve a high fabrication precision and filling factor of the winding to reach a high efficient AFTM. Therefore, we utilize the technique of directly printing the winding on a flex-PCB. Fig. 9 shows is a conceptual diagram of a winding to be printed on a flex-PCB. The AFTM is the 2-

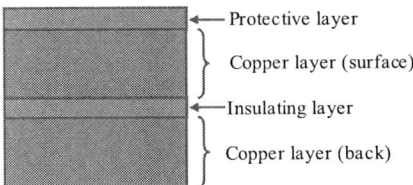

Fig. 10 FEM magnetic analysis result of axial force

phase 4-pole motor. Therefore, it has the four windings, the windings which face each other are connected in series. In the figure, the solid blue line is the front print pattern, and the orange dashed line is the print pattern on the back. Also, the blue circle is an input terminal, and the purple ring is a "through-via" which connects the printed pattern of the front side and the back side through the insulating layer. The input terminal I_{a+} connect to the pattern on the surface from the outside to the inside, passes through the center through via, and to the pattern on the back side. Then, it connects from the inside to the outside; then it is connected to the next profile by the connection part of the lower portion. Four sets of winding patterns constitute one layer of winding. Finally, it is connected to the terminal I_{a-} of the Nth layer. The winding is formed by rolling it into a cylinder shape. Fig. 10 shows the configuration of the flex-PCB. One sheet consists of a protective layer on the surface, a surface copper layer, an intermediate insulating layer, and a copper layer on the back. In this paper, the thickness of the protective and insulating layer is chosen so that the thickness of one sheet is about 0.3 mm. The width of the outer layer increases with the thickness of the layer, so the careful design is required. Since the thickness of the winding space is 3.05 mm, ten set of the winding sheet can fit into the area. As a result, the number of turns of one winding is 110 and filling factor of about 66 % is obtained.

IV. CONCLUSIONS

This paper proposed the novel structure of the bearingless AFTM and explained operating principle of the motor. The size of the PM for passive magnetic bearing was determined by using numerical calculation of MagOpt LCM, and size of the motor PM, windings, and back iron was determined by using FEM magnetic analysis of ANSYS Maxwell. By conducting detailed calculations on PM size, winding size, and back iron size, we were able to design AFTM that can be built into commercially available small fans. Moreover, we designed winding by using the technique of directly printing the winding on a flex-PCB to achieve a high fabrication precision and filling factor of the winding.

The designed AFTM is under manufacture currently. Static and dynamic performance of the fabricated setup will be reported shortly.

ACKNOWLEDGMENT

This work has been supported by the COMET-K2 "Center for Symbiotic Mechatronics" of the Linz Center of Mechatronics (LCM) funded by the Austrian federal government and the federal state of Upper Austria, and JSPS KAKENHI Grant-in-Aid for Scientific Research (C) Grant Number 17K06223.

REFERENCES

[1] G. Jungmayr "Der magnetisch gelagerte Lüfter", PhD Thesis, Johannes Kepler Universität Linz, 2008.

[2] W. Bauer, W. Amrhein, "Design and Sizing Relations for a Novel Bearingless Motor Concept", Proc. of ICEMS 2011, Beijing, China (2011)

[3] W. Bauer, W. Amrhein, "Electrical Design and Winding Selection for a Bearingless Axial-Force/Torque Motor", Proc. of SPEEDAM 2012, Sorrento, Italy (2012)

[4] W. Bauer, W. Amrhein, "Performance Analysis of a Bearingless Axial-Force/Torque Motor", Proc. of GMM/ETG Symposium, Innovative Small Drives and Micro-Motor Systems, Nuremberg, Germany (2013)

[5] W. Bauer, W. Amrhein, "Electrical Design Considerations for a Bearingless Axial-Force/Torque Motor", IEEE T IAS, Vol. 50, No. 4, pp. 2512-2522 (2014)

[6] W. Amrhein, W. Gruber, W. Bauer, and M. Reisinger, "Magnetic Levitation Systems for Cost-Sensitive Applications -Some Design Aspects-", IEEE T IAS, Vol. 52, No. 5, pp. 3739-3752 (2016)

A Plotter-Based Automatic Measurement and Statistical Characterization of Multiple Discrete Power Devices

Michihiro Shintani[1*], Benjamin Dauphin[2], Kazuki Oishi[2], Masayuki Hiromoto[2], and Takashi Sato[2]

[1] Graduate School of Information Science, Nara Institute of Science and Technology, Ikoma, Japan
[2] Graduate School of Informatics, Kyoto University, Kyoto, Japan
*E-mail: shintani@is.naist.jp

Abstract—We propose an automatic measurement system for efficiently characterizing multiple packaged power devices. The proposed system sequentially measures arranged power devices using a robot arm on an XY plotter. The proposed measurement system facilitates the characterization of multiple power devices, eliminating possible human errors. Through experiments using 144 SiC power MOSFETs, we demonstrate that our measurement system achieves comparable accuracy and precision to the conventional measurement results using a commercial curve tracer. We also present statistical characterization of the measurement data of 30 MOSFETs using surface potential based SPICE model.

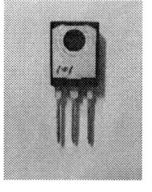

Fig. 1. Appearance of a packaged discrete power device.

Fig. 2. An example curve tracer [7].

I. INTRODUCTION

Power semiconductor devices are the key elements of power converters, which play important roles as electrical switches. The wide band gap semiconductors, such as silicon carbide (SiC) and gallium nitride (GaN), have promising physical properties superior to the conventional silicon (Si) semiconductors [1], [2]. Due to their superiority, SiC and GaN power devices theoretically possess the following advantages: low on-resistance, high ambient temperature operation, and fast switching behavior. Currently, both kinds of semiconductors are commercially available and have attracted increasing attention as new power switching devices.

Designers of power converter circuits utilize a device model in circuit simulations throughout their circuit design [3], [4], [5]. The devices must be characterized accurately, including their variation, before applying them to the power converters. Also, for the device identification, the measurement of discrete devices is a key step [6]. In general, discrete devices are delivered to the designers as packaged in resin mold, as shown in Fig. 1. The packaged semiconductor devices are characterized by using curve tracers such as a commercial one [7] shown in Fig. 2. The manual measurement of a large number of devices using such curve tracers is a very time-consuming process due to the necessity of swapping devices under test (DUT) many times, because insertion and removal of the DUT into and from the socket require human intervention.

With this motivation, in this paper, we propose an automatic measurement system for a large number of packaged devices. The proposed system utilizes a robot arm mounted on an XY plotter for automatic measurement. Before the measurement, DUTs are arranged on a panel under the robot arm. The robot arm has a set of sensor electrodes connected to a commercial curve tracer, and works as an extended probe of the curve tracer. The robot arm is programmed to repeatedly execute the measurement process for each DUT one by one, which includes the operations of (1) moving over the DUT, (2) touching the DUT's terminal electrodes with the sensor electrodes, (3) measuring the DUT characteristics by the curve tracer, and (4) releasing the electrodes and moving on to the next DUT. Because the curve tracer is used for the measurement, the same characteristics as the commercial curve tracer should be obtained. The movement of the robot arm and timing of the measurement by the curve tracer are controlled by a computer.

The measurement data obtained by the proposed measurement system allows us to conduct the statistical analysis, to measure and model process variations. These statistical analysis has become a common practice in silicon-based integrated circuit simulation [8], [9]. We expect the statistical analysis will become increasingly important in characterizing power devices. In this paper, using the measurement data, we also demonstrate that the estimation of device performance at process corners can be estimated accurately through the variation-aware simulation.

The remainder of this paper is organized as follows. Section II describes the design concept of the proposed measurement system for packaged power devices. Then, Section III provides the detailed implementation of our system from the viewpoints of hardware and software. Section IV presents the accuracy and precision of our

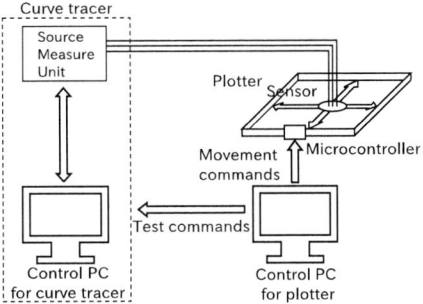

Fig. 3. Proposed automatic measurement system.

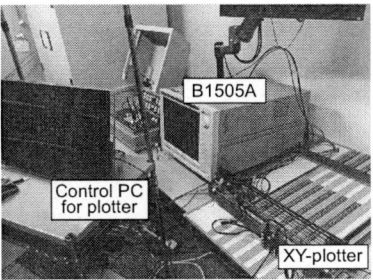

Fig. 4. Implementation of the proposed automatic measurement system.

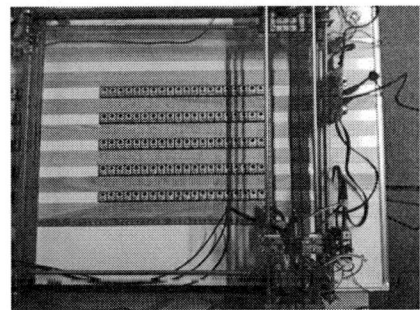

Fig. 5. XY plotter and DUT layout.

measurement system through experiments using a commercial SiC power MOSFETs. The statistical analysis of the device characteristics using the measurement data and a SPICE simulation model [3] are also presented in this section. Finally, we give our conclusions in Section V.

II. DESIGN OF AUTOMATIC MEASUREMENT SYSTEM FOR DISCRETE POWER DEVICES

Conventionally, the measurement of a discrete power device using a curve tracer is conducted by using a transistor socket mounted on the tracer. The types of characterizations, associated measurement ranges, steps of bias voltages, etc., are set manually with the assigned measurement label. For the measurement of multiple devices, repetition of insertion and removal of each device to and from the socket is necessary, which is entirely done manually. This measurement process requires a large effort. This process may also include human error in various stages, such as during the device handling, measurement setup, and labeling.

To overcome these problems, in our proposed measurement system, we utilize an XY plotter with a robot arm having a measurement sensor (probes), which is controlled to move around the DUTs as shown in Fig. 3. The system consists of the XY plotter, a commercial curve tracer, and two computers each of which controls the plotter and the tracer. The measurement sensor is connected to the curve tracer. DUTs to be measured are arranged in a grid manner on the panel under the XY plotter so that they can be measured successively by the plotter.

Our measurement system can measure whatever the device characteristics that the curve tracer can measure, including current, capacitance, and diode characteristics, because the sensor at the tip of the plotter arm is basically an extension of the socket in the commercial curve tracer.

III. IMPLEMENTATION

The implementation of the proposed automatic measurement system is shown in Fig. 4. In this section, hardware and software components of the proposed system are presented.

A. Hardware

In our implementation, power device analyzer B1505A (Keysight Technologies, Inc.) [7] and XY plotter (Makeblock Co. Ltd.) [10] are used as the curve tracer and the XY plotter, respectively.

The XY plotter, which is ordinarily used to print drawings of buildings or machines, has an arm to hold a pen. In order to move the arm in an XY-plane, two stepper motors control the arm in a horizontal and vertical directions, and a servomotor moves a pen up and down. The plotter utilized in this system has an Arduino microcomputer board [11] that executes the control program to send commands to locate and to move the pen up and down.

The arm of the plotter in our system holds a sensor connected to the socket of B1505A instead of a pen. DUTs are arranged in the accessible area of the arm, as illustrated in Fig. 5. The sensor is controlled to be sequentially contacted to the electrodes of a DUT for the measurement. After the contact of the sensor, the computer sends a measurement command to B1505A.

In order to precisely contact the sensor to three electrodes at a time for all DUTs, the calibration of the initial arm position is very important. For this calibration, the micro-switches attached to the plotter are used. When the arm moves to the end of the movable region, the micro-switch is pressed. Detection of the switch to be pressed enables us to locate the arm exactly at the initial position at the beginning of every measurement.

The electrical connections are shown in Fig. 6. In the current characteristic measurement, Kelvin connection is

3645

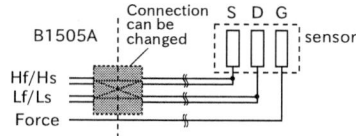

(a) Current characteristic measurement

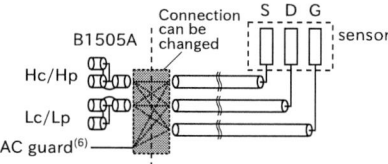

(b) Capacitance characteristic measurement

Fig. 6. Electrical connections of the system.

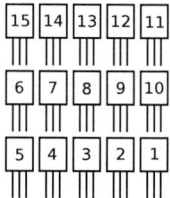

Fig. 7. Layout of the discrete power devices arranged in a zigzag manner.

used to compensate voltage drop associated with the wires between the sensor and the B1505A. However, the contact resistance between the sensor and the electrode cannot be removed by the Kelvin connection. Hence, the result of current measurement may contain some error unless we calibrate the contact resistance.

In the capacitance measurement, coaxial cables are used for the connection between the sensor and B1505A in order to reduce the coupling of the signal line with surrounding electric fields to attain the highest possible precision. In addition, open and short calibrations are conducted to compensate the parasitic capacitance and resistance introduced by the cables. However, short calibration for all DUTs is hard to automate, as it requires the use of a single calibration device. It is impossible with our proposed method to conduct the short calibration at each position of the device layout, because doing so would require to replace the DUT with the calibration device during the automatic measurements. Hence, only a single short calibration at the beginning of the measurement is conducted, and the position of a device in the layout might slightly affect the test results.

B. Software

To control the automatic measurement system, two programs have been developed. The DUTs are measured in a "zigzag" fashion, i.e. rows are measured alternatively from left to right and from right to left as illustrated in Fig 7, to minimize the total moving distance of the arm. Basically, the time required for the movement has to be minimized to maximize time for the actual current or capacitance measurement.

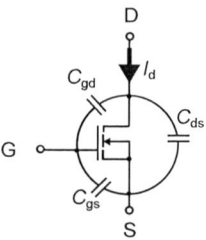

Fig. 8. Equivalent circuit of SiC power MOSFET.

The first program runs on a micro-controller and manages the plotter. This program waits for inputs, corresponding to the different possible movements, on the USB port of the micro-controller that is connected to the control PC. The second program runs on the control PC and has two functions. It sends the direction of the sensor movement to the micro-controller, and controls the commercial curve tracer in order to run the tests and label data correspondingly. Having a single program running those two functions enables to synchronize the movements of the sensor.

IV. Evaluation of the Automatic Measurement System

We conducted the measurements using a commercial SiC power MOSFET to evaluate the proposed measurement system. In our system, up to 144 SiC power MOSFETs can be arranged, hence our measurement system can sequentially measure 144 MOSFETs by a single measurement setup. When the measurement is repeated 10 times for each device with a measurement setup lasting 5 seconds, it takes about 2 hours and 35 minutes to finish the measurements, out of which 2 hours are used for the actual measurements and the remainder is the waiting time between measurements, and the time required to locate the arm at each location of the device.

As shown in Fig. 8, SiC power MOSFET has two major characteristics to measure: drain current I_d and capacitance characteristics. Further, the capacitance characteristics consist of three capacitors: capacitance between gate and source terminals C_{gs}, capacitance between drain and source terminals C_{ds}, and capacitance between gate and drain terminals C_{gd}. In this section, we present the accuracy and precision of the proposed measurement system with measuring the above characteristics compared to the conventional method that only uses a commercial curve tracer [7].

In addition, we demonstrate a statistical analysis using the measurement data obtained by the proposed system on the basis of a surface-potential-based SiC power MOSFET model [3]. By the statistical analysis, we can estimate the range of circuit performance under process variation, thus enable to design robust and reliable power converters.

3646

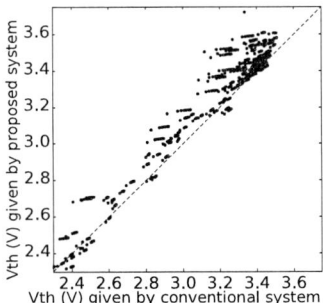

Fig. 9. Comparison of V_{th} values obtained with the socket and with the automatic sensor.

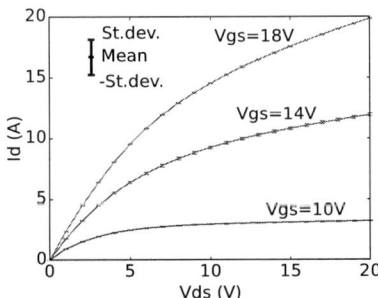

Fig. 10. Series of 20 I_{d}-V_{ds} measurements done on the same device.

A. Current Characteristics

Figure 9 shows the difference in the computed threshold voltage, V_{th}, obtained by the proposed and conventional measurements. For each device, 10 repetitive measurements are conducted. When using our automatic sensor without calibration, slightly higher V_{th} values are obtained. This is due to the increase in the length of the wires used, increasing the overall parasitic resistance. The worst case amounts to a relative error of 11%.

Figure 10 shows I_{d}-V_{ds} characteristics, where V_{ds} is drain-source voltage, with 20 times measurement for one DUT using the proposed system. The error bars represent the standard deviation of the measured I_{d}. In between the 20 measurements, the sensor is detached and attached from the DUT while staying in the same location. From the results, the response of a DUT changes when measured multiple times in a short period of time. Part of this may be because I_{d}-V_{ds} characteristics are influenced by a voltage stress [12]. The maximum standard deviation obtained for a DUT is 0.261 A, and thus, as seen from the figure, the proposed system is sufficiently accurate to measure current characteristics of power devices. Since the stressed devices are known to recover after the removal of stress, a possible solution to remedy the problem is to measure each device one time, then going back to the first device to start the measurement again, in order to give sufficient time for devices to recover.

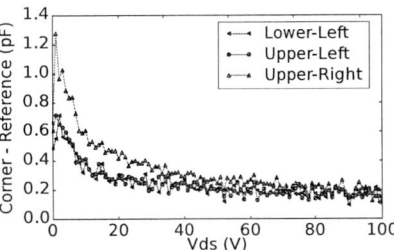

Fig. 11. C_{ds} difference between layout corners of the automatic measurement system.

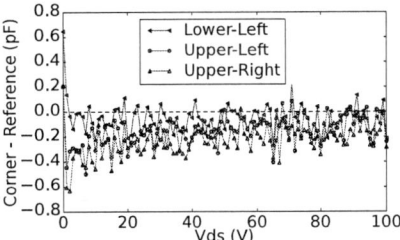

Fig. 12. C_{gd} difference between layout corners of the automatic measurement system.

B. Capacitance Characteristics

As we described in Sec. III-A, the short calibrations for all DUTs were not done in our measurement system. Figures 11 and 12 show the impact of the position of a device on the measured values. The same device was measured at the four corners of the layout, and they are compared being the lower-right corner as the reference measurement. For C_{ds} and C_{gd} the effect induced by the corner is low, around 0.5 pF, which is less than 1% of C_{ds} value, and can be considered as negligibly small in our experiment.

Figures 13 and 14 show the difference in the measured C_{ds} and C_{gd} between the conventional measurements directly using the socket and the automatic sensor. The reversed trend can be seen between C_{ds} and C_{gd}. This inversion is considered to be caused due to the configuration of cables. Between C_{ds} and C_{gs} the connection of gate and source are inverted, hence the effect induced by the cables became opposite in sign. In both cases, the absolute difference is highest around 4 V, and then gradually converges. The maximal relative errors are 3.7% for C_{ds} and 93.6% for C_{gd}. This very high relative error for C_{gd} is explained by the very low absolute values it can take (down to 5 pF at high voltages), so an absolute error of -4 pF becomes important.

The another observation is that, contrary to I_{d}-V_{ds} characteristics, capacitance measurements are not affected significantly by the stress, i.e., capacitance measurement are easier to repeat accurately, or one measurement is sufficient for a device. Figures 15, 16, and 17 show a series of 20 measurements done successively on the same device. Again, the sensor is detached and attached from the DUT between the 20 measurements. Only the C_{gd} measurement varies more at higher voltages. This is also

The 2018 International Power Electronics Conference

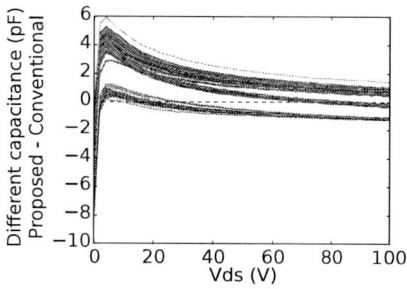

Fig. 13. C_{ds} difference between our automatic sensor and the socket, depending on V_{ds} (70 devices, 10 measurements).

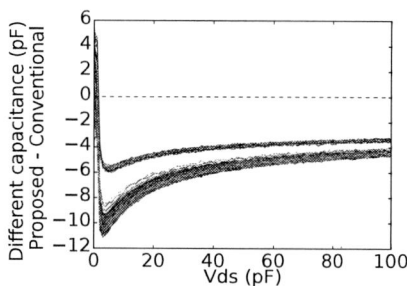

Fig. 14. C_{gd} difference between our automatic sensor and the socket, depending on V_{ds} (70 devices, 10 measurements).

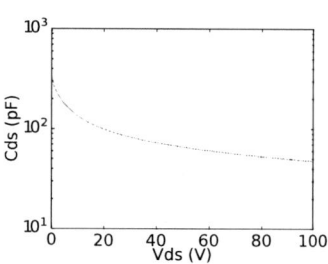

Fig. 15. Series of 20 C_{ds} measurements done on the same device.

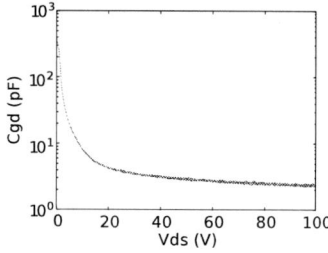

Fig. 16. Series of 20 C_{gd} measurements done on the same device.

due to the very low values that C_{gd} can take. The absolute values of C_{ds} and C_{gs} measurements are not as that low as compared with C_{gd}, and thus they are very precise. The respective maximum standard deviations are: 1.024 pF for C_{gd}, 0.503 pF for C_{ds} and 0.469 pF for C_{gs}.

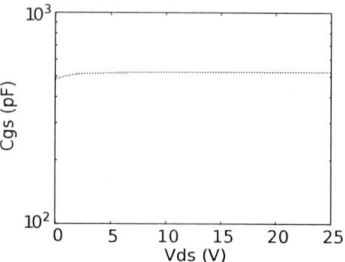

Fig. 17. Series of 20 C_{gs} measurements done on the same device.

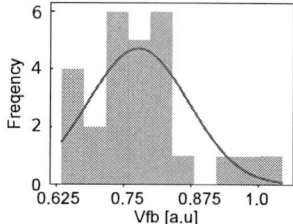

Fig. 18. Distribution of fitted parameter, V_{fb}.

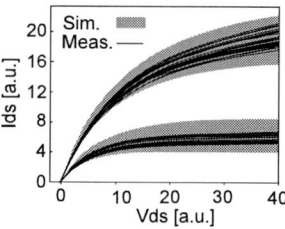

Fig. 19. Measured and simulated I_{ds}-V_{ds} curves.

C. Statistical Analysis

On the basis of the measured I_d-V_{ds} data of a 30 SiC MOSFETs in the same lot, the statistical analysis is conducted using a surface potential based SPICE model [3]. The model parameters are extracted by simulated annealing [13] for each device. The distribution of the estimated parameter, flatband voltage V_{fb}, is shown in Fig. 18. For the other parameters, the distributions can be obtained in the same manner. Figure 19 shows the comparison between measured and simulated I_{ds}-V_{ds} curves. The simulated curve is calculated by the Monte Carlo simulation using the SPICE model. Assuming V_{fb} follows a normal distribution, 1,000 samples are drawn from the distribution, and I_{ds}-V_{ds} curves are calculated for each sample. The measured curves of 30 MOSFETs are plotted and the min-max regions of the simulated 1,000 curves are shown as the filled area. The trend of the measured curves is reproduced in the simulation. Through the proposed automatic measurement system, statistical characterization of the device parameters and the statistical simulation using physics based SPICE model become possible. The circuit simulation with the process variation allows us to estimate the range of circuit performance under variability.

3648

V. CONCLUSION

A novel automatic measurement system for packaged power devices was proposed. In our system, an XY plotter is utilized to measure multiple packaged devices without manually setting devices on a curve tracer. In the experiment conducted using a commercial SiC power MOSFET, the accuracy and precision of the proposed system were compared to the conventional commercial curve tracer. An accuracy of 96% for capacitance (except C_{gd}) and of 89% in the worst case for current characteristics have been achieved with our system. Very good repeatability has been observed with the measurements using our system, achieving a maximal standard deviation of 0.26 A for I_d-V_{ds}, and of 0.5 pF for capacitance measurements. In addition, statistical analysis of the measured I_d-V_{ds} data using a surface potential based SPICE model is demonstrated. Our system enables us to conduct variation-aware circuit simulations for estimating accurate circuit performances.

ACKNOWLEDGEMENT

This work was partially supported by JST Super Cluster Program, NEDO Cross-ministerial Strategic Innovation Promotion Program, and JSPS KAKENHI Grant No. 17H01713.

REFERENCES

[1] B. J. Baliga, *Fundamentals of Power Semiconductor Devices*. Springer, 2008.

[2] T. Kimoto and J. A. Cooper, *Fundamentals of Silicon Carbide Technology*. Wiley, 2014.

[3] Y. Nakamura, M. Shintani, K. Oishi, T. Sato, and T. Hikihara, "A simulation model for SiC power MOSFET based on surface potential," in *Proceedings of International Conference on Simulation of Semiconductor Processes and Devices*, 2016, pp. 121–124.

[4] M. Shintani, K. Oishi, R. Zhou, M. Hiromoto, and T. Sato, "A circuit simulation model for V-groove SiC power MOSFET," in *Proceedings of IEEE Workshop on Wide Bandgap Power Devices and Applications*. 2016, pp. 286–290.

[5] M. Shintani, Y. Nakamura, M. Hiromoto, T. Hikihara, and T. Sato, "Measurement and modeling on gate-drain capacitance of silicon carbide vertical MOSFET," *Japanese Journal of Applied Physics*, vol. 56, no. 45, p. 04CR07, 2017.

[6] M. Shintani, K. Oishi, R. Zhou, M. Hiromoto, and T. Sato, "Device identification from mixture of measurable characteristics," in *Proceedings of IEEE Applied Power Electronics Conference and Exposition*, 2017, pp. 1001–1006.

[7] *B1505A Power Device Analyzer/Curve Tracer*, Keysight Technologies, Inc., 2015.

[8] B. Cheng, D. Dideban, N. Moezi, C. Millar, G. Roy, X. Wang, S. Roy, and A. Asenov, "Statistical-variability compact-modeling strategies for BSIM4 and PSP," *IEEE Design & Test of Computers*, vol. 27, no. 2, pp. 26–35, 2010.

[9] C. C. McAndrew, I. Stevanovic, X. Li, and G. Gildenblat, "Extensions to backward propagation of variance for statistical modeling," *IEEE Design & Test of Computers*, vol. 27, no. 2, pp. 36–43, 2010.

[10] *XY Plotter Robot Kit (with Electronic Version)*. Makeblock Co., Ltd., http://www.makeblock.com/xy-plotter-robot-kit/.

[11] *Arduino Project Official Site*, http://www.arduino.cc.

[12] A. Ibrahim and Z. Khatir, "Power cycling ageing tests at $200\,^\circ$C of SiC assemblies for high temperature electronics," in *Proceedings of European Conference on Power Electronics and Applications*, 2013, pp. 1–10.

[13] S. Kirkpatrick, C. D. Gelatt, Jr., and M. P. Vecchi, "Optimization by simulated annealing," *Science*, vol. 220, no. 4598, pp. 671–680, 1983.

The 2018 International Power Electronics Conference

A Novel High-Speed SiC MOSFET Driver with a Low Switch-Voltage Stress

Xiuqin Wei[1], Yuchong Sun[2], and Hiroo Sekiya[2]

[1]Department of Electrical and Electronic Engineering, Chiba Institute of Technology, Chiba, 275–0016 Japan
[2]Graduate School of Advanced Integration Science, Chiba University, Chiba, 263–8522 Japan
Email: xiuqin.wei@p.chibakoudai.jp

Abstract—**A novel high-speed SiC MOSFET driver with a low switch-voltage stress is presented in this paper. In the proposed driver, a harmonic component is injected to the conventional class-E inverter. Consequently, the class-E driver proposed in this paper not only maintains the strong points of the conventional class-E inverter, such as simple topology and high-frequency high-efficiency operation, but also has its own advantage of low switch-voltage stress. The proposed class-E driver is designed in this paper. Additionally, the PSpice-simulation and laboratory experiment are carried out. It can be seen from the PSpice-simulation and experimental results that all the switch-voltage waveforms satisfy the zero-voltage switching (ZVS) and zero-derivative switching (ZDS) conditions. Therefore, the proposed class-E driver is available in the high-frequency and high-efficiency applications. Moreover, a quite lower switch-voltage stress is obtained in the proposed class-E driver. The simulated and experimental results agreed with the theoretical one well. These results mentioned before demonstrated the proposed driver's validity.**

Index Terms—**Low switch-voltage stress, class-E inverter, SiC MOSFET, harmonic component, ZVS/ZDS**

I. Introduction

Silicon carbide (SiC) MOSFETs have rapidly emerged as the next-generation power semiconductor device because they have many favorable capabilities, such as low on resistance, high breakdown voltage, great high temperature, and so on. However, they have large input capacitance and resistance generally, which is a fatal defect. In order to solve the defect, a driver, which enables rapid charge and discharge of the input capacitance, is required. Recently, several SiC MOSFET drivers have been proposed [1]- [3]. Among them, a driver based on class-E inverter has attracted general concern from both scholars and researchers [3]. The class-E inverter is a circuit topology applied the ZVS/ZDS technologies, which can guarantee a high power-conversion efficiency even at high frequencies [3]- [13]. Meanwhile, there is a problem that the switch-voltage stress of the class-E inverter is high, which results in high cost, high breakdown rate, and so on. In other words, the switch-voltage stress of the SiC MOSFET driver based on the class-E inverter is also high. Therefore, it is essential to put forward a new and effective scheme to tackle the problem of the high switch-voltage stress.

This paper presents a novel high-speed SiC MOSFET driver with a low switch-voltage stress. In the proposed driver, a harmonic resonant filter produced a harmonic current is con-

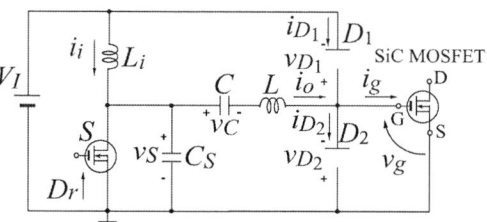

Fig. 1. Circuit topology of the SiC MOSFET driver based on the conventional class-E inverter [3].

nected to the conventional class-E inverter's switch in parallel. Thus, the new class-E driver proposed in this paper can not only obtain high-frequency high-efficiency operation with simple circuit topology, but also yield a low switch-voltage stress. A design example of the new class-E driver is given. Additionally, the PSpice simulation and laboratory experiment are carried out. It can be seen from the simulated and experimental results that all the switch-voltage waveforms achieve the ZVS/ZDS operations. Therefore, it is possible to obtain high frequency and high efficiency via using the proposed class-E driver. Moreover, the switch-voltage stress is significantly lower than one of the conventional class-E driver. The simulated and experimental results agreed with the theoretical ones well, which showed the proposed driver's validity.

II. Conventional Class-E SiC MOSFET Driver

A. Illustrations of Circuit and Operation

The circuit topology of the SiC MOSFET driver based on the conventional class-E inverter [3] is shown in Fig. 1. The conventional class-E SiC MOSFET driver is comprised of a dc-supply source V_I, an input inductor L_i, a MOSFET S, a shunt capacitor C_S, two diodes D_1 and D_2, a L-C resonant filter, and a SiC MOSFET. The MOSFET and diodes work as switching devices. Figure 2 shows the example waveforms of the conventional SiC MOSFET class-E driver. The voltage D_r is the driving signal to drive the MOSFET. In the interval that the switch is off, the sum of currents from the input inductor and the resonant filter is passed through the shunt capacitor C_S and produces the switch voltage v_S. Additionally,

3650

The 2018 International Power Electronics Conference

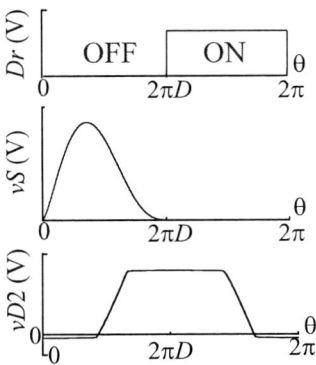

Fig. 2. Example waveforms of the SiC MOSFET driver based on the conventional class-E inverter [3].

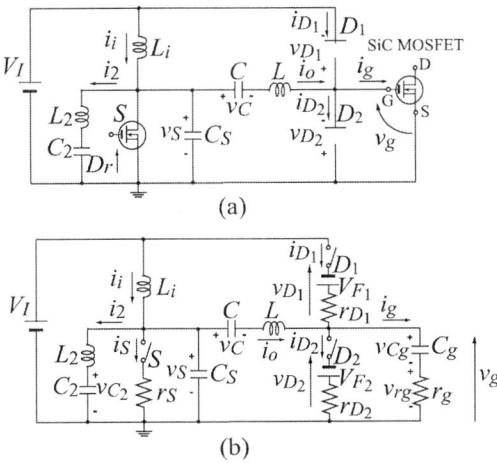

Fig. 3. Circuit topology. (a) Proposed class-E SiC MOSFET drive. (b) Equivalent circuit.

the waveform of the switch voltage meets the ZVS/ZDS at $\theta = 2\pi D$, that is

$$v_S(\theta = 2\pi D) = 0, \tag{1}$$

$$\left.\frac{dv_S}{d\theta}\right|_{\theta=2\pi D} = 0, \tag{2}$$

where D is the MOSFET's on duty ratio. At the instant of $\theta = 2\pi D$, the switch turns on. Therefore, there is no overlap in the switch-voltage-and-current waveforms that the high frequency and high efficiency requirements can be met [3]- [13]. Both two diodes turn on and off when their voltages are lower than their forward voltage and the sum of the voltages across the two diode is always equal to the dc-supply voltage, namely, $V_I = v_{D1} + v_{D2}$. Additionally, the voltage v_{D2} is the target voltage and regarded as the driving voltage of the SiC MOSFET as shown in Fig. 1.

B. Problem Statement

The conventional class-E SiC MOSFET driver can realize the high-frequency and high-frequency operations simultaneously due to the application of the ZVS/ZDS technologies. Therefore, it has being expected to drive the SiC MOS-FET. The conventional class-E driver, however, has a high switch-voltage stress that it is limited for more practical applications. For instance, the switch-voltage stress is almost 4 times as high as its dc-supply voltage for the case of $D = 0.5$. As a result, the switching device is breakdown easily, even collapsing the whole system. To avoid the trouble mentioned above, one way is to use a specific switching device with a high rating voltage. The higher rating-voltage switching device is, the higher cost of the whole system is. Additionally, when the rating voltage of a switching device is large, its on resistance is large. Namely, there is a linear relationship between the rating voltage and on resistance of a switching device. Therefore, the application of the switching device with a high rating voltage will lead to high conduction power losses.

Consequently, it is essential and important to put forward a new and effective scheme to tackle the problem of the high switch-voltage stress.

III. PROPOSED CLASS-E SiC MOSFET DRIVER

In this section, an original class-E SiC MOSFET driver is presented and its design procedure is introduced. In the original class-E driver, a specific resonant filter which generates the 2nd harmonic current is connected to the switch of the conventional class-E driver in parallel. As a result, the proposed class-E driver can meet the high frequency and high efficiency requirements. Moreover, it is possible to yield a low switch-voltage stress with a simple circuit topology.

A. Assumptions

The following assumptions are given for designing the original class-E driver presented in this paper:

1) The MOSFET works as a switch without threshold voltage and parasitic capacitance. But its off resistance is infinite and on resistance is not zero.

2) The output resonant filter L-C and 2nd resonant filter L_2-C_2 are ideal filters for the fundamental frequency and the frequency, which is twice as the fundamental frequency.

3) The loaded quality factors Q_1 and Q_2 of the output resonant filter and harmonic resonant one are high enough to generated pure sinusoidal currents through them.

4) All the inductors and capacitors do not include the equivalent series resistances.

5) The gate-source input port of the SiC MOSFET includes a parasitic resistance r_g and capacitance C_g.

6) The switch voltage satisfies the class-E ZVS/ZDS conditions given in (1) and (2).

7) The diodes also work as switches without threshold voltage and parasitic capacitance. But its off resistance is infinite and on resistance is not zero. Additionally, the diodes include their forward voltages.

According to the above, the equivalent circuit for this design procedure can be obtained as shown in Fig. 3 (b).

3651

B. Parameters

The parameters of the original class-E driver are defined as follows.

$$A_{o1} = f_{o1}/f = 1/(2\pi f \sqrt{LC}), \tag{3}$$

$$A_{o2} = f_{o2}/2f = 1/(4\pi f \sqrt{L_2 C_2}), \tag{4}$$

$$B = C/C_S, \tag{5}$$

$$H = L/L_i, \tag{6}$$

$$Q_1 = \omega L/r_g = 2\pi f L/r_g, \tag{7}$$

$$Q_2 = 2\omega L_2/r_g = 4\pi f L_2/r_g, \tag{8}$$

$$J = C/C_g. \tag{9}$$

C. Circuit Equations

The proposed driver operates in the interval of $0 \leq \theta < 2\pi$. Therefore, we have

$$
\begin{cases}
\dfrac{di_i}{d\theta} = \dfrac{H}{Q_1 r_g}(V_I - v_S) \\[2mm]
\dfrac{dv_S}{d\theta} = A_{o1}^2 B Q_1 r_g \left(i_i - \dfrac{v_S}{r_S} - i_o - i_2 \right) \\[2mm]
\dfrac{dv_C}{d\theta} = A_{o1}^2 Q_1 r_g i_o \\[2mm]
\dfrac{di_2}{d\theta} = \dfrac{2}{Q_2 r_g}(v_S - v_{C2}) \\[2mm]
\dfrac{dv_{C2}}{d\theta} = 2A_{o2}^2 Q_2 r_g i_2 \\[2mm]
\dfrac{dv_{Cg}}{d\theta} = A_{o1}^2 J Q_1 r_g \left[i_o + \dfrac{V_I}{r_{D1}} + V_F \left(\dfrac{1}{r_{D1}} - \dfrac{1}{r_{D2}} \right) \right. \\[2mm]
\qquad \left. - v_{Cg}\left(\dfrac{1}{r_{D1}} + \dfrac{1}{r_{D2}} \right) \right] \Big/ \left[1 + r_g \left(\dfrac{1}{r_{D1}} + \dfrac{1}{r_{D2}} \right) \right] \\[2mm]
\dfrac{di_o}{d\theta} = \dfrac{1}{Q_1 r_g} \left[v_S - v_C - \left(v_{Cg} + \dfrac{1}{A_{o1}^2 J Q_1} \dfrac{dv_{Cg}}{d\theta} \right) \right],
\end{cases}
\tag{10}
$$

where V_F is the forward voltage of the diodes. Additionally, r_S, r_{D1}, and r_{D2} are the on resistances of MOSFET and diodes, respectively, which are defined as

$$
r_S = \begin{cases} r_{on} & \text{for } 0 < \theta \leq \pi \\ \infty & \text{for } \pi < \theta \leq 2\pi, \end{cases} \tag{11}
$$

$$
r_{D1} = \begin{cases} r_D & \text{for } v_{D1} < V_F \\ \infty & \text{for } v_{D1} \geq V_F, \end{cases} \tag{12}
$$

and

$$
r_{D2} = \begin{cases} r_D & \text{for } v_{D2} < V_F \\ \infty & \text{for } v_{D2} \geq V_F. \end{cases} \tag{13}
$$

When we define $\mathbf{y}(\theta)[y_1, y_2, \cdots, y_7]^T = [i_i, v_S, v_C, i_2, v_{C2}, v_{Cg}, i_o]^T \in \mathbf{R}^7$. (10) can be rewritten as

$$\frac{d\mathbf{y}}{d\theta} = f(\theta, \mathbf{y}, \lambda) \tag{14}$$

where $\lambda = [A_{o1}, A_{o2}, B, H, Q_1, Q_2, J, r_S, r_{D1}, r_{D2}, r_g, V_F]^T \in \mathbf{R}^{12}$.

TABLE I
ELEMENT VALUES OF THE NEW CLASS-E DRIVER.

	Theoretical	Experimental	Difference
L_i	5.20 μH	5.20 μH	0.0 %
C_S	539 pF	538 pF	-0.186 %
C	110 pF	110 pF	0.0 %
L	5.20 μH	5.20 μH	0.0 %
C_2	99.5 pF	100 pF	0.503 %
L_2	1.30 μH	1.30 μH	0.0 %
r_g	22.85 Ω	22.87 Ω	0.09 %
C_g	1.11 nF	1.11 nF	0.0 %
V_{Smax}	30.2 V	30.5 V	0.993 %

D. Condition Equations

Generally, the class-E driver should operate in the steady state, achieving the switching conditions. Therefore, we derived the explicit expressions of the boundary conditions of the proposed class-E driver in the steady state as follows,

$$\mathbf{y}(2\pi) - \mathbf{y}(0) = \mathbf{o} \in \mathbf{R}^7. \tag{15}$$

The ZVS/ZDS conditions shown in (1) and (2) are mandatory to achieve.

IV. DESIGN EXAMPLE

To testify the effectiveness of the proposal, a class-E driver with a low switch-voltage stress was designed. Firstly, the design specifications of the proposed class-E driver were given as follows: $V_I = 15$ V, $D = 0.3$, $f = 7$ MHz, $H = 1.0$, $Q_1 = 10$, $Q_2 = 5$, and $A_{o2} = 1.0$. This driver will be used to drive SCT2450KE SiC MOSFET. Therefore, $C_g = 1.11$ nF and $r_g = 22.85$ Ω were obtained by measuring the SCT2450KE SiC MOSFET. Additionally, SUD06N10 Si MOSFET by Vishay Siliconix and SS2040FL Schottky barrier diode by Panjit International were used as the switching devices of the driver that we have $r_{on} = 0.225$ Ω, $r_D = 0.4$ Ω, and $V_F = 0.4$ V from their datasheets. According to the design specifications mentioned the above, and the condition equations shown in (1), (2), and (15), the element values were derived as shown in Table I. Here, V_{Smax} is the switch-voltage stress, which reduced from almost 4 times to twice as high as the dc-supply voltage. These values were obtained by applying the algorithm in [13]. Table I also shows the experimental results. According to Table I, it can say that there is a good agreement between the theoretical and experimental results. the theoretical, simulated, and experimental waveforms were given in Fig. 4. In Fig. 4, all the switch-voltage waveforms met the ZVS/ZDS conditions which can guarantee high-efficiency and high-frequency operations. Additionally, the effectiveness and validity of the proposed driver were demonstrated by the results shown in Table I and Fig. 4.

V. CONCLUSION

In this paper, a new class-E driver with a low switch-voltage stress has been presented. In the new class-E driver, a specific resonant filter generated the 2nd harmonic current is connected to the conventional class-E driver's switch in parallel.

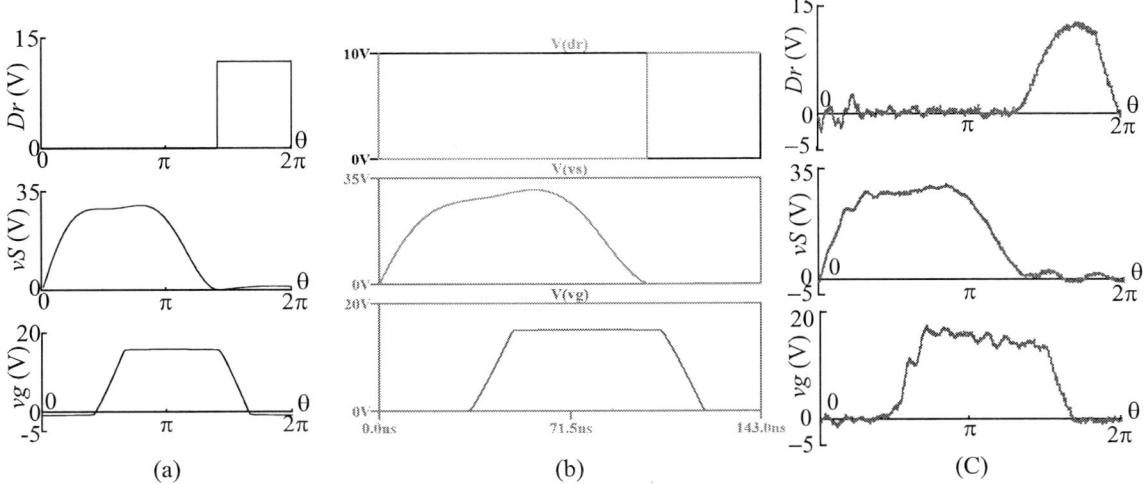

Fig. 4. Waveforms of the new class-E driver. (a) From the theory. (b) From the simulation. (c) From the experiment.

Consequently, the proposed class-E driver can obtain a low switch-voltage stress and high-frequency high-efficiency operations with a simple circuit topology. The new class-E driver was design to prove its effectiveness. The PSpice simulation and laboratory experiment were also carried out. All the switch-voltage waveforms satisfy the ZVS/ZDS. Additionally, the switch-voltage stress is reduced to twice as high as its dc-supply voltage. There was a good agreement among the theoretical, simulated, and experiments waveforms. The results mentioned above show that the proposal is effective.

REFERENCES

[1] A. Orellana and B. Piepenbreier, "Fast gate drive for SiC-JFET using a conventional driver for MOSFETs and additional protections," *in Proc. IEEE IECON*, Busan, Korea, Nov. 2004, pp. 938–943.

[2] K. Nagaoka, K. Chikamatsu, A. Yamaguchi, K. Nakahara, and T. Hikihara, "High-speed gate drive circuit for SiC MOSFET by GaN HEMT," *IEICE Electronics Express*, vol. 12, no. 11, pp. 1–8, June 2015.

[3] Y. Sun, R. Sugano, X. Wei, T. Hikihara, and H. Sekiya, "High-speed driver for SiC MOSFET based on class-E inverter," *in Proc. IEEE ISCAS*, Baltimore, MD, USA , May 2017, 4 pages.

[4] N. O. Sokal and A. D. Sokal, "Class E—A new class of high-efficiency tuned single-ended switching power amplifiers," *IEEE J. Solid-State Circuits*, vol. SC-10, no. 3, pp. 168–176, Jun. 1975.

[5] F. H. Raab, "Effects of circuit variations on the class-E tuned power amplifier," *IEEE J. Solid-State Circuits*, vol. SC-13, no. 2, pp. 239–247, Apr. 1978.

[6] M. K. Kazimierczuk, "Class E tuned power amplifier with shunt inductor," *IEEE J. Solid-State Circuits*, vol. SC-16, no. 1, pp. 2–7, Feb. 1981.

[7] D. K. Choi and S. I. Long, "The effect of transistor feedback capacitance in class-E power amplifiers," *IEEE Trans. Circuits Syst. I*, vol. 50, no. 12, pp. 1556–1559, Dec. 2003.

[8] X. Wei, H. Sekiya, S. Kuroiwa, T. Suetsugu, and M. K. Kazimierczuk, "Effect of MOSFET gate-to-drain parasitic capacitance on class-E power amplifier," *in Proc. IEEE ISCAS*, Paris, France, June 2010, pp. 3200–3203.

[9] X. Wei, H. Sekiya, S. Kuroiwa, T. Suetsugu, and M. K. Kazimierczuk, "Design of class-E amplifier with MOSFET linear gate-to-drain and nonlinear drain-to-source capacitances," *IEEE Trans. Circuits Syst. I*, vol. 58, no.10, pp. 2556–2565, Oct. 2011.

[10] M. J. Chudobiak, "The use of parasitic nonlinear capacitors in class-E amplifiers," *IEEE Trans. Circuits Syst. I*, vol. 41, no. 12, pp. 941–944, Dec. 1994.

[11] T. Suetsugu and M. K. Kazimierczuk, "Analysis and design of class E amplifier with shunt capacitance composed of nonlinear and linear shunt capacitances," *IEEE Trans. Circuits Syst. I*, vol. 51, no.7, pp. 1261–1268, Jul. 2004.

[12] M. K. Kazimierczuk, *RF Power Amplifiers*. New York, NY: John Wiley & Sons, 2008.

[13] H. Sekiya, T. Ezawa, and Y. Tanji, "Design procedure for class-E switching circuits allowing implicit circuit equations," *IEEE Trans. Circuits Syst. I, Reg. Papers*, vol. 55, no. 11, pp. 3688–3696, Nov. 2008.

The 2018 International Power Electronics Conference

Enhancement of Driving Capability of Gate Driver Using GaN HEMTs for High-Speed Hard Switching of SiC Power MOSFETs

Takafumi Okuda[1*], Takashi Hikihara[1]

1 Graduate School of Engineering, Kyoto University, Nishikyo, Kyoto, Japan

*E-mail: t-okuda@dove.kuee.kyoto-u.ac.jp

Abstract—**A high-speed gate driver based on GaN-HEMTs push-pull configuration has been proposed to drive a SiC power MOSFET at high switching frequency. In this study, we investigate the influences of parasitic inductances of the GaN HEMTs on the output voltage of the GaN-based gate driver. The parasitic inductances at the gate, source, and drain terminals of the GaN HEMTs are analyzed with SPICE simulation. It is found that the parasitic inductances at the gate and source terminals of the GaN HEMT have almost no influences on the output waveform of the gate driver, while the parasitic inductances at the drain terminal of the GaN HEMT produce large voltage oscillations of the output voltage.**

Keywords—*GaN-based gate driver, parasitic inductance, numerical analysis, high-frequency switching*

I. INTRODUCTION

A power packet dispatching system is expected to be one of the advanced power distribution systems for controlling electric power and providing energy on demand [1], [2]. The power packet delivers electric power according to the attached information tag through power routers, and power flow can be managed flexibly using the power packet dispatching system. In order to generate a power packet with less power dissipation, low on-resistance must be mentioned with high switching speed characteristics.

Silicon carbide (SiC) is an attractive semiconductor material for power devices owing to its superior material properties such as wide bandgap and high critical electric field [3]–[5]. SiC metal-oxide-semiconductor field-effect transistors (MOSFET) have been intensively developed in the last two decades [6]–[8], and 1-kV-class SiC MOS-FETs are commercially available with low on-resistance. Since SiC MOSFETs are unipolar devices, their switching characteristics are much higher than those of Si IGBTs. In addition, SiC MOSFETs have an enough avalanche capability [9], [10], which is suitable for valve transistors in power electronics circuits. However, conventional gate drivers are optimized for driving Si IGBTs and the maximum switching frequency is limited at approximately 1 MHz [11]. Therefore, it is important to develop a high-speed and high-frequency gate driver for driving SiC MOSFETs at the higher switching frequency applications.

Gallium nitride (GaN) is also a promising semiconductor material owing to high critical electric field.

Although GaN MOSFETs are still immature because of the lack of high-quality bulk crystals for homoepitaxial GaN growth, GaN-based high electron transistors (HEMTs) can be fabricated on Si substrates [12]–[14]. Two-dimensional electron gas (2DEG) is available in GaN HEMTs and fast switching characteristics are obtained, which can enable high-frequency operation with low power consumption. Using n-channel GaN-HEMTs push-pull configuration, a high-frequency gate driver was proposed [11], and 10-MHz hard switching of SiC MOSFET was achieved with the GaN-based gate driver. However, very large surge voltage was observed in the output waveform of the gate driver. In order to enhance the driving capability of the gate driver, it is necessary to investigate the origin of the voltage surge and suppress the oscillations.

In this study, we investigate the influences of parasitic inductances in the GaN-based gate driver using SPICE simulation. As a result, it is found that parasitic inductances at the drain side of the GaN HEMT cause the voltage surge in the GaN-based gate driver.

II. SIMULATION PROCEDURE

The detailed configuration of the GaN-based gate driver is described elsewhere [11]. Two driving signals for the high and low sides are electrically separated from the controller with digital isolators (Silicon Labs., Si8610) and the output signals from the digital isolators are enhanced with complementary Si MOSFETs (ROHM, US6M1). The final output stage is GaN-HEMTs push-pull configuration, which is driven by the complementary Si MOSFETs. In this study, we focus on the influences of parasitic inductances at the GaN-HEMTs push-pull configuration in the gate driver.

The schematic used for SPICE simulation is shown in Fig. 1. GaN-HEMTs push-pull configuration is driven through the gate resistance of $5\,\Omega$. The device model of GaN HEMT (EPC, EPC2014C) is distributed from EPC. The driving signals are shown in Fig. 2. The driving voltage is set at 4 V, the fall and rise times are at 10 ns, and the switching frequency is at 10 MHz in the SPICE simulation. Three kinds of parasitic inductances are assumed for both the high-side and low-side GaN HEMTs, totally six parasitic inductances being investigated (L_{dH}, L_{gH}, L_{sH}, L_{dL}, L_{gL}, and L_{sL}). The driving target is a capacitor of 500 pF through a resistance of

3654

$10\,\Omega$ to simplify the analysis of switching waveforms. SPICE simulation is performed using SIMetrix software.

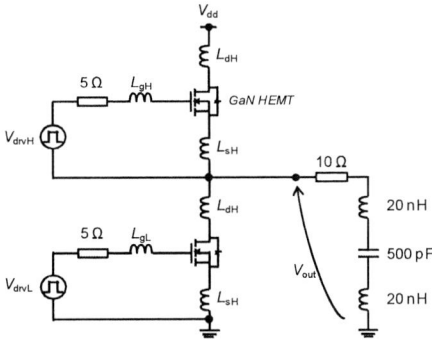

Fig. 1. Schematic used for SPICE simulation. GaN-HEMTs push-pull configurations are driven through resistance of $5\,\Omega$. Driving target is capacitive load of 500 pF with parasitic inductances of 20 nH.

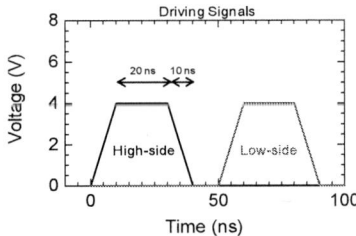

Fig. 2. Driving signals for GaN HEMTs in SPICE simulations. Switching frequency is set at 10 MHz.

III. RESULTS AND DISCUSSION

A. Gate parasitic inductance

Figure 3 shows the simulation results of the output voltage of the gate driver. The gate parasitic inductance of the high-side GaN HEMT (L_{gH}) is varied from 1 to 10 nH, while the other inductances are fixed at 0.1 nH. It is found that the output voltage of the gate driver does not depend on the gate parasitic inductance. The influences of the gate parasitic inductance at the low-side GaN HEMT (L_{gL}) also exhibited almost the same results (not shown). The GaN HEMT (EPC2014C) has very low input capacitance, the value of which is approximately 220 pF. The gate current for driving the GaN HEMT can be reduced and the simulated peak current is approximately 200 mA, so that gate parasitic inductance has little effect on the output voltage in the gate driver.

B. Source parasitic inductance

Figure 4(a) shows the simulation results of the output voltage in the gate driver. The source parasitic inductance of the high-side GaN HEMT (L_{sH}) is varied from 1 to 10 nH, while the other inductances are fixed at 0.1 nH. The turn-on characteristics became slightly slow with increasing the source parasitic inductance and the oscillations were observed at the on-state. The influences

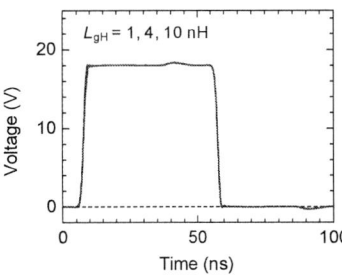

Fig. 3. Simulation results of output voltage in gate driver. Gate parasitic inductance of high-side GaN HEMT (L_{gH}) is varied from 1 to 10 nH. The other inductances are fixed at 0.1 nH.

of the source parasitic inductance of the low-side GaN HEMT (L_{sL}) are shown in Fig. 4(b). The turn-off characteristics became slow with increasing the source parasitic inductance. In these cases, the influences of the source parasitic inductances were small, but too large source parasitic inductance at the source of the GaN HEMT will induce false turn-on and turn-off. In addition, a surge voltage is generated by the source parasitic inductance. In the SPICE simulation at the parasitic inductance of 10 nH, the generated surge voltage is $-4.5\,V$ at the gate-source terminal. The gate oxide of SiC MOSFET will be deteriorated by large reverse bias, so that the reverse surge voltage caused by the source parasitic inductance must be reduced below a rated voltage of the gate oxide.

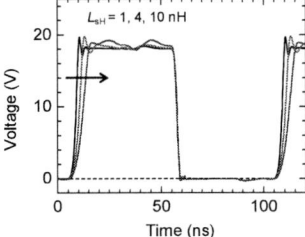

(a) Source parasitic inductance at high side

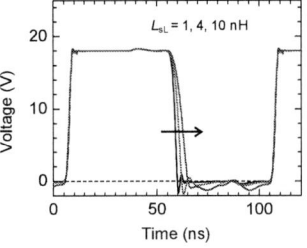

(b) Source parasitic inductance at low side

Fig. 4. Simulation results of output voltage in gate driver. Gate parasitic inductances of (a) high-side GaN HEMT (L_{sH}) and (b) low-side GaN HEMT (L_{sL}) are varied from 1 to 10 nH. The other inductances are fixed at 0.1 nH.

3655

The 2018 International Power Electronics Conference

C. Drain parasitic inductance

Figure 5 shows the simulation results of the output voltage of the gate driver. The drain parasitic inductance of the high-side GaN HEMT (L_{dH}) is varied from 1 to 8 nH, while the other inductances are fixed at 0.1 nH. It is found that the drain parasitic inductance significantly induces the voltage oscillations of the output voltage. At the parasitic inductance of 8 nH, the voltage oscillation frequency is calculated to be 155 MHz. Assuming that the voltage oscillations are caused by LC resonance between the parasitic inductance of 8 nH and the output capacitance of the GaN HEMT of 140 pF at the drain-source voltage of 20 V, the resonant frequency is estimated to be 150 MHz, which is consistent with the simulated value of 155 MHz. Thus, the voltage oscillations may be caused by the drain parasitic inductance with the output capacitance of the GaN HEMT.

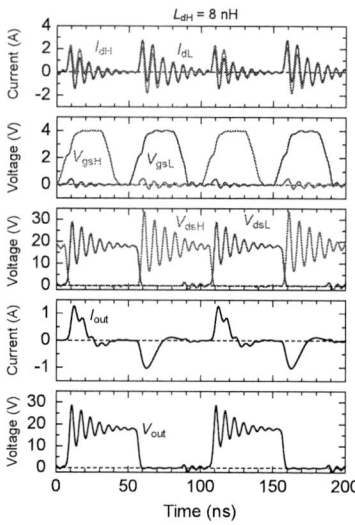

Fig. 6.   Detailed switching waveforms simulated in gate driver at drain parasitic inductance of high-side GaN HEMT of 8 nH. The other inductances are fixed at 0.1 nH.

parasitic inductance induces the voltage oscillations of the gate driver. Hence, the drain parasitic inductance of the GaN HEMTs should be reduced to enhance the driving capability of the gate driver.

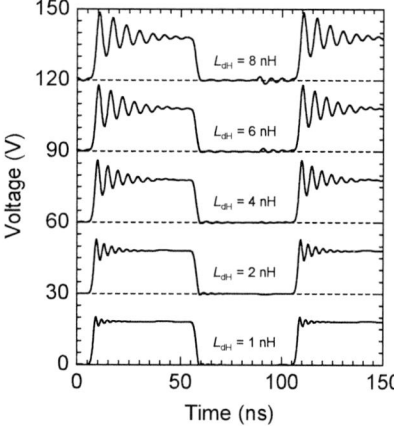

Fig. 5.   Simulation results of output voltage in gate driver. Drain parasitic inductance of high-side GaN HEMT (L_{dH}) is varied from 1 to 8 nH. The other inductances are fixed at 0.1 nH.

The detailed switching waveforms in the SPICE simulations are shown in Fig. 6. The drain parasitic inductance of the high-side GaN HEMT (L_{dL}) is set at 8 nH. V_{gsH} and V_{gsL} denote the gate-source voltage of the GaN HEMTs at the high and low side, I_{dH} and I_{dL} the drain current of the GaN HEMTs at the high and low side, V_{dsH} and V_{dsL} the drain-source voltage of the GaN HEMTs at the high and low side, and I_{out} the output current of the gate driver. It is found that the drain-source voltage of the GaN HEMT has large oscillations, which is caused by the output current with parasitic inductance of 8 nH.

Figure 7 shows the simulation results of the output voltage in the gate driver at the drain parasitic inductance of the low-side GaN HEMT (L_{dL}) varied from 1 to 8 nH. The other inductances are fixed at 0.1 nH. The voltage oscillations were observed, but the oscillations occurred at the off-state of the gate driver. The detailed switching waveforms in the SPICE simulations are shown in Fig. 8. The large oscillations of the drain-source voltage of the GaN HEMT are observed and the resonance with the

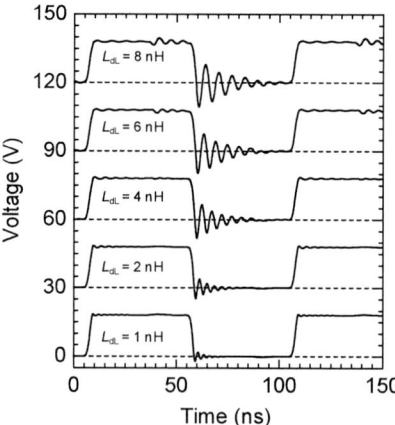

Fig. 7.   Simulation results of output voltage in gate driver. Drain parasitic inductance of low-side GaN HEMT (L_{dL}) is varied from 1 to 8 nH. The other inductances are fixed at 0.1 nH.

IV. CONCLUSIONS

The influences of parasitic inductances are simulated with SPICE simulation in the GaN-based gate driver. The parasitic inductances at the gate and source terminals of the GaN HEMT have almost no influences on the output waveform of the gate driver, while the parasitic inductances at the drain terminal of the GaN HEMT significantly produce the voltage oscillations of the output

3656

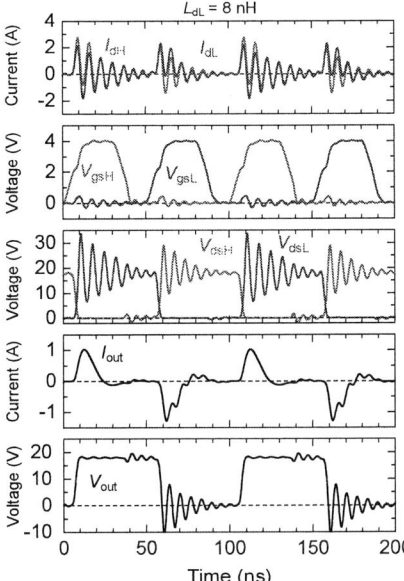

Fig. 8. Detailed switching waveforms simulated in gate driver at drain parasitic inductance of low-side GaN HEMT of 8 nH. The other inductances are fixed at 0.1 nH.

voltage. In particular, the high-side GaN HEMT is connected to the power line for the push-pull stage, which tends to be long and have a large parasitic inductance. In order to improve the driving capability, it is important to design the circuit pattern of the gate driver. Using an improved high-speed gate driver, high frequency switching of power transistors is expected and power packets can be flexibly generated owing to superior device performance of SiC power devices.

ACKNOWLEDGEMENT

This research is partially supported by Kyoto Super Cluster Program (JST) and Cross-ministerial Strategic Innovation Promotion Program (SIP, "Next-generation power electronics" (NEDO)).

REFERENCES

[1] R. Takahashi, K. Tashiro, and T. Hikihara, "Router for Power Packet Distribution Network: Design and Experimental Verification", *IEEE Trans. Smart Grid*, vol. 6, no. 2, pp. 618-626, Mar. 2015.

[2] R. Takahashi, S. Azuma, M. Hasegawa, H. Ando, and T. Hikihara, "Power Processing for Advanced Power Distribution and Control", *IEICE Trans. Communications*, vol. E100.B, no. 6, pp. 941-947, 2017.

[3] J. A. Cooper, Jr, M. R. Melloch, R. Singh, A. Agarwal, J. W. Palmour, "Status and Prospects for SiC Power MOSFETs", *IEEE Trans. Electron Devices*, vol. 49, no. 4, pp. 658-664, Apr. 2002.

[4] J. Millán, P. Godignon, X. Perpiñà, A. Pérez-Tomás, and J. Rebollo, "A Survey of Wide Bandgap Power Semiconductor Devices", *IEEE Trans. Power Electron.*, vol. 29, no. 5, pp. 2155-2163, May 2014.

[5] T. Kimoto, "Material Science and Device Physics in SiC Technology for High-Voltage Power Devices", *Jpn. J. Appl. Phys.*, vol. 54, p. 40103, 2015.

[6] J. A. Carr, D. Hotz, J. C. Balda, H. A. Mantooth, A. Ong, and A. Agarwal, "Assessing the Impact of SiC MOSFETs on Converter Interfaces for Distributed Energy Resources", *IEEE Trans. Power Electron.*, vol. 24, no. 1, pp. 260-270, Jan. 2009.

[7] Q. C. Zhang, R. Callanan, M. K. Das, S.-H. Ryu, A. K. Agarwal, and J. W. Palmour, "SiC Power Devices for Microgrids", *IEEE Trans. Power Electron.*, vol. 25, no. 12, pp. 2889-2896, Dec. 2010.

[8] R. A. Wood and T. E. Salem, "Evaluation of a 1200-V, 800-A All-SiC Dual Module", *IEEE Trans. Power Electron.*, vol. 26, no. 9, pp. 2504-2511, Sep. 2011.

[9] X. Huang, G. Wang, Y. Li, A. Q. Huang, and B. J. Baliga, "Short-Circuit Capability of 1200V SiC MOSFET and JFET for Fault Protection", *Twenty-Eighth Annual IEEE Applied Power Electronics Conference and Exposition*, CA, USA, pp. 197-200, 2013.

[10] B. Hull, S. Allen, Q. Zhang, D. Gajewski, V. Pala, J. Richmond, S. Ryu, M. O'Loughlinm, E. Van Brunt, L. Cheng, A. Burk, J. Casady, D. Grider, and J. Palmour, "Reliability and Stability of SiC Power Mosfets and Next-Generation SiC MOSFETs", *IEEE Workshop on Wide Bandgap Power Devices and Applications*, Knoxville, TN, pp. 139-142, 2014.

[11] K. Nagaoka, K. Chikamatsu, A. Yamaguchi, K. Nakahara, and T. Hikihara, "High-Speed Gate Drive Circuit for SiC MOSFET by GaN HEMT", *IEICE Electron. Express*, vol. 12, p. 20150285, 2015.

[12] S. Nakamura, "GaN Growth Using GaN Buffer Layer", *Jpn. J. Appl. Phys.*, vol. 30, pp. L1705-L1707, 1991.

[13] M. Ishida, T. Ueda, T. Tanaka, and D. Ueda, "GaN on Si Technologies for Power Switching Devices", *IEEE Trans. Electron Devices*, vol. 60, no. 10, pp. 3053-3059, Oct. 2013.

[14] T. Nomura, M. Masuda, N. Ikeda, and S. Yoshida, "Switching characteristics of GaNHFETs in a half bridge package for high temperature applications", *IEEE Trans. Power Electron.*, vol. 23, no. 2, pp. 692-697, Mar. 2008.

[15] R. Motiva, R. Ghosh, U. Mhaskar, D. Klikic, M. X. Wang, and A. Dentella, "Investigations of 600-V GaN HEMT and GaN Diode for Power Converter Applications", *IEEE Trans. Power Electron.*, vol. 29, no. 5, pp. 2441-2452, May. 2014.

The 2018 International Power Electronics Conference

Design and Experimental Verification of Robot Arm Operation for Power Packet Dispatching System

Tomoki Yokoyama[1]*, Ryunosuke Araumi[1]*, Kazunori Asada[1]*, Takashi Ando[1]*

1 Tokyo Denki University, 5 Asahicho senju, Adachiku, Tokyo, Japan

*E-mail: yoko@fr.denda.ac.jp

Abstract—A power packet dispatching system for the robot arm operation is discussed. Recent wide-band-gap power semiconductor technology realizes high speed and low switching loss devices such as SiC and GaN. The power packet dispatching system is the power distribution system in which DC voltage is delivered with the information tag in the same power line. To include the information into the packet, switching speed of the switching device is desired over MHz range. SiC or GaN device is attracting solution to realize the power packet system. Six arm robot operation with the power packet dispatching system is designed and evaluated. Design concept is described to interface the power and the information in simultaneously. Realization of the proposed method was evaluated.

Keywords—FPGA, Power packet, Robot Arm, Routing

I. INTRODUCTION

The power packet dispatching system is one of the advanced power distribution system for the information technology society. To realize highly integrated power distribution system, high reliable and high efficient power distribution system is desired. Information technology is the key issue not only in the network technology but also in the power electronics fields.

The concept of open-electrics-energy-netwok (OEEN) was proposed in [1], in which power flow is controlled by multiple electric routers, power is treated as the packet same as the network information. Data transfer latency is very fast compared with the switching speed of the power modules for the power converter at the time. It was hard to realize to implement the power packet dispatching system due to the limitation of the switching device capability. But recently, wide-band-gap power semiconductor devices began to release to the consumer market like SiC devices or GaN devices [2]. These devices have the potential to switch high power in MHz order range with low switching losses. The power packet dispatching system can be realized with these new semiconductor devices.

Hikihara proposed this concept for the DC power distribution system in [3]-[7], and the experimental verifications were carried out for various applications in [8], [9]. In the latest equipment, the power packet dispatching equipment was built using GaN devices, the fundamental operation of the power distribution system was confirmed in [10]. DC voltage source is delivered with the information tag via a mixer device and a router device. The power and the information can be delivered simultaneously in the one common power line and the different DC power level can be delivered to the different loads. Also the synchronization between the mixer and the router without

the clock wiring was achieved in [9].

The robot system is one of the attracting applications for the power packet dispatching system. To operate the robot system, not only the power to activate the actuator, but also the control information is required. In the conventional robot system, many wiring is required and it prevent the high density integration of the system. If the power dispatching system is adopted to the robot system, the power line and the control information line can be combined to the same wiring, the downsizing of the robot system can be realized.

In this paper, the power packet dispatching system was applied to the robot arm operation. Six arm robot was applied for the verification, the power packet dispatching system was combined with the power line of the robot arm and also the control informations for the robot arm operations is delivered to the servo modules of each axis using the same power line.

Each arm has the servo motor unit with PMSM and the inverter, the DC power and the current reference is delivered from the DC power source and the master controller. To combine with the power packet dispatching system, the current reference is included in the header of the power packet, the power for the servo motor and the current reference for the servo motor can be delivered in one common power line.

Design concept of the proposed robot arm system with the power packet dispatching system was described. Interfaces between the robot arm controller and the mixer and/or the router of the power packet dispatching system is designed. Simulations for the arm operation was carried out. Experiments were also carried out for the basic operation of the robot arm.

II. POWER PACKET DISPATCHING SYSTEM

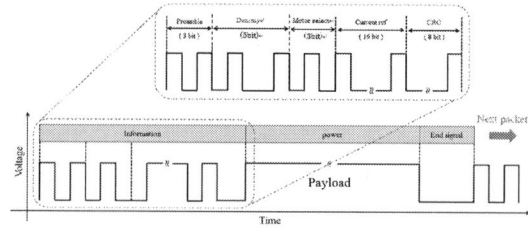

Fig. 1. Configuration of power packet.

The power packet dispatching system is applied to the power supply between the power source and the loads. As shown in Fig. 1, the power packet is a unit of the pulsed power information and the power itself as the payload. In

3658

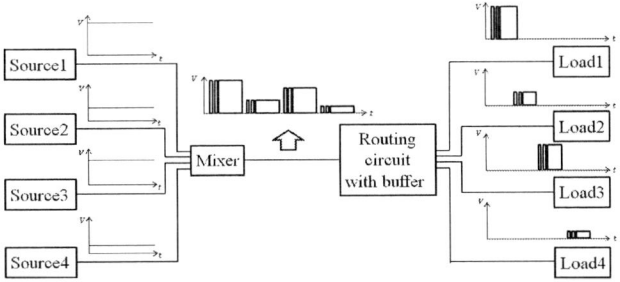

Fig. 2. Configuration of dispatching circuit.

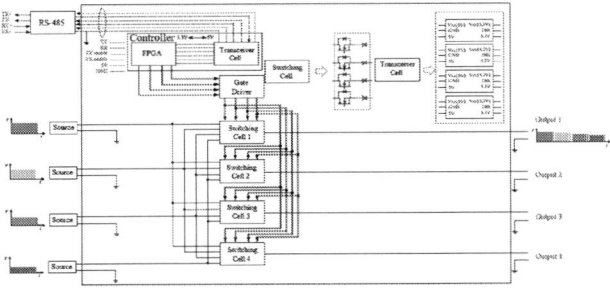

Fig. 3. Configuration of mixer circuit.

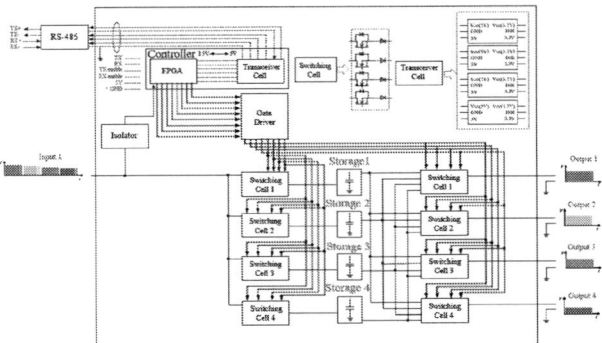

Fig. 4. Configuration of router circuit.

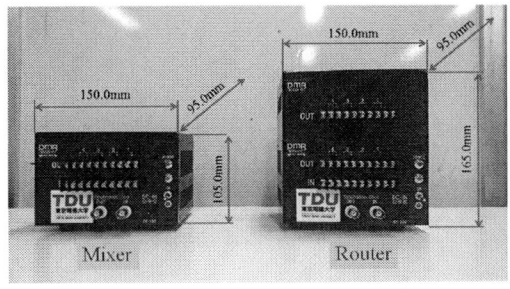

Fig. 5. Appearance of the mixer and the router.

this research, one packet consists of a header, a payload, and a footer. The header is assumed to be the start of packet

information, and it consists of information required for the router and the load. The footer is assumed to be the end of the packet information. The pulsed power information is treated as the information header of the packet. The power packet dispatching system consists of the mixer and the router. As shown in Fig. 2, the mixer generates the power packets from the supplied voltage and transmits for the router. The transmitted packets is distributed to the load by the router. The configuration of the mixer circuit is shown in Fig. 3. The control signal generated by the controller is connected to the gate drive circuit and the switching device is driven according to the packet information. The header, payload and footer of the power packet are generated by this switching device. FPGA (XC3S250E) is used for the controller and GaN (EPC 2014) is used as a switching device. In addition, a clock signal is transmitted from the mixer to the router. The power packet which is created by the mixer is transferred to the router. The configuration of the router circuit is shown in Fig. 4. The router consists of the switching devices, the gate driver, the controller, and the isolator. The isolator consists of a zener diode and a photocoupler when an input voltage is higher than the threshold value of the zener diode, the controller recognizes the input as "1", in other cases, the input is treated as "0". When the router reads the header information, the switch selects the storage, and the switch of the selected output regenerates the information tag based on the header information. In addition, the mixer and the router have a communication module to transmit and receive information to and from other controllers. Fig.5 shows the appearance of the mixer and the router.

III. CONSTRUCTION OF ROBOT ARM BY POWER PACKET DISPATCHING SYSTEM

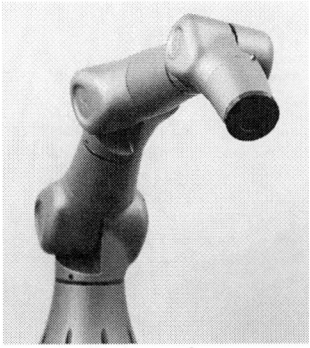

Fig. 6. Six axis robot arm.

In this study, the six axis robot arm was operated using the power packet dispatching system [11]. Fig. 6 indicates the six axis robot arm. Fig. 7 shows the control block diagram for one link module. In the robot arm system, the control parameters of the joint orbit such as position, speed, torque and current reference are transmitted from the user interface of the main PC to the master controller. The master controller transmits the current reference to the slave controller of each axis. The slave controller transmits the current reference to the motor driver, the servo module of PMSM operate the each axis. The power

The 2018 International Power Electronics Conference

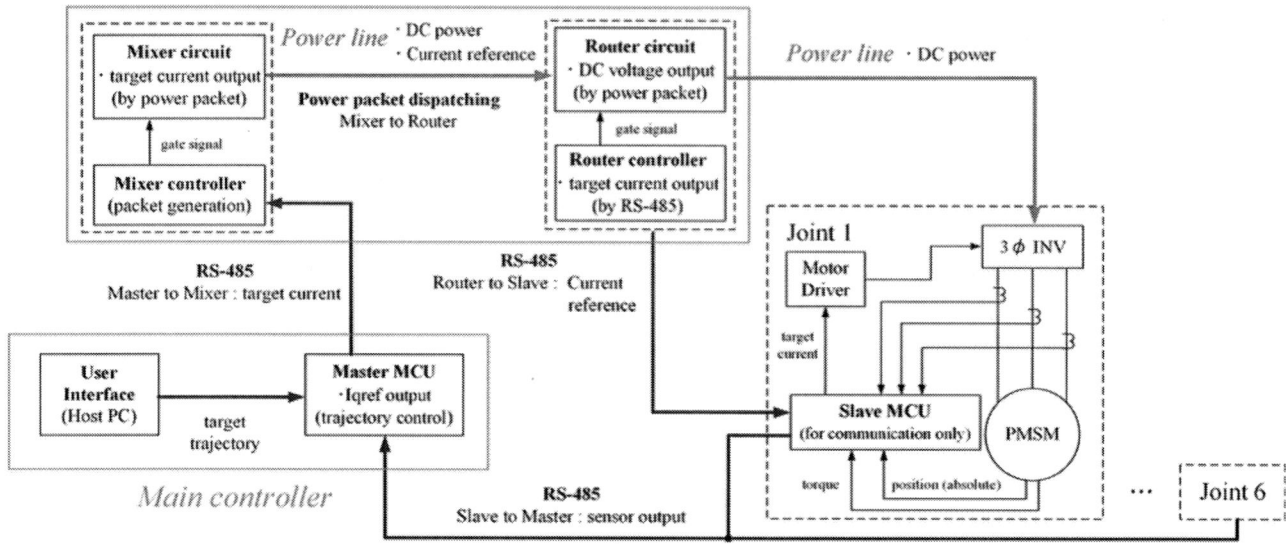

Fig. 7.   Control block diagram for one link.

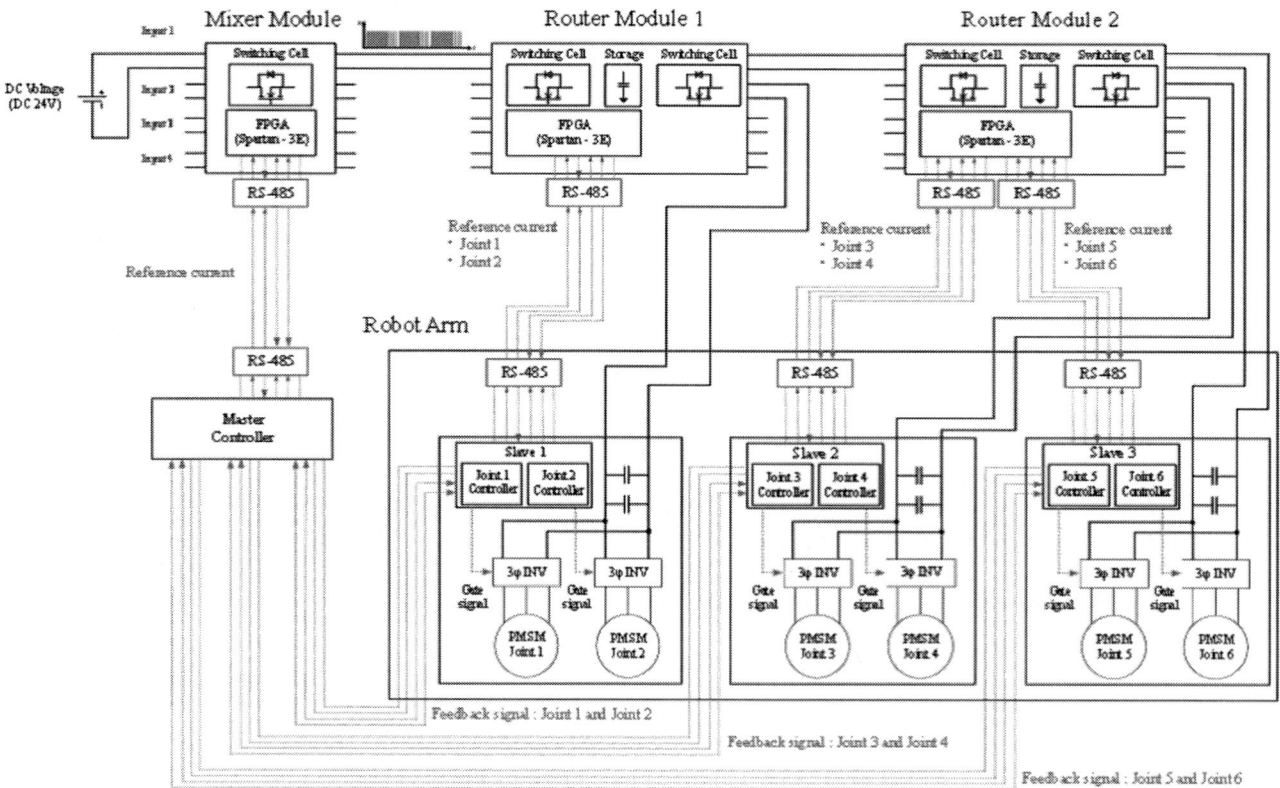

Fig. 8.   Construction of robot arm driven by power packet dispatching system

3660

packet dispatching system transmits the current reference to the power line as an information tag. The construction of robot arm by power packet dispatching system is shown in Fig. 8, and the system specification is shown in TABLE I.

TABLE I. SPECIFICATION OF ROBOT ARM COTROL.

Power Packet	Specification
Header	start signal : 3[bit]
	Dummy bit : 5[bit]
	motor selection : 3[bit]
	Current reference : 16[bit]
	error detection and correction : 8[bit]
payload	127[bit]
Footer	end signal : 4[bit]
Control cycle	1[ms]
Communication standard	RS-485

In this system, the power supply unit is consists of the mixer and two routers, and the control unit is consists of the master controller (AP-RX71M-0A) and three slave controllers (AP-RX71M-0A). The master controller generates the current reference, which is transmitted to the slave controller by the power packet dispatching system and controls the motor for each axis. In this system, to control the robot arm, the motor selection, the current reference and the error detection information is required. Therefore, the header is consists of the control information and the start information. In addition, information is transmitted between the master controller and the mixer, the router and the slave controller is connected by asynchronous serial communication (8-N-1). The 3 bits are used as the motor selection information, and the 16 bits are used as the current reference, and the 8 bits are used as the error detection signal (CRC-8-ATM), and the remaining 5 bits are used as the dummy bit.

The communication speed between the master controller to each slave unit is settled to 1[Mbps]. To operate the robot arm in stable, the current reference of each servo unit should be updated in every 1[ms]. The current reference of the six axis requires 192bits. So the communication standard were implemented by differential transmission (RS - 485) which is capable of high speed communication.

IV. SIMULATION

Simulations were carried out for the robot arm system [12]-[15]. Power flow of each six axis are confirmed. Simulation condition is indicated in TABLE II, and the motor parameter shown in TABLE III.

TABLE II. SIMULATION CONDITION

Parameter	Value
Carrier frequency	50kHz
Voltage inverter	24V
Speed reference	±100rad/s
dead time	0.85μs

Fig. 9 shows the link current when the motor of each axis is operated by the same speed reference. The regenerative current

TABLE III. IPMSM PARAMETER

Parameter	Link1,2	Link3	Link4,5	Link6
Armature resistance(R)	0.157Ω	0.184Ω	0.608Ω	4.83Ω
d axis inductance(L_d)	0.267mH	0.24mH	0.463mH	2.24mH
q axis inductance(L_q)	0.267mH	0.24mH	0.463mH	2.24mH
D axis magnetic flux(ϕ)	0.0107wd	0.0057wd	0.0050wd	0.0115wd
Moment of interia($10^{-6}J$)	21.0 kgm^2	4.9kgm^2	18.1kgm^2	17.2kgm^2
Pole pairs(P)	10	10	8	8

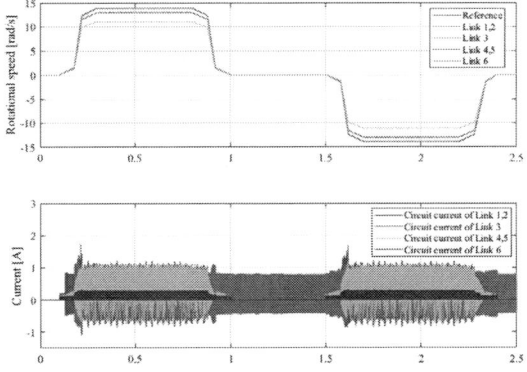

Fig. 9. Link current and rotational speed.

was generated when the speed command value was changed for the motor drive operations at the links 1, 2, 3, 4 and 5.

V. EXPERIMENT

In order to apply the power packet dispatching system to the robot arm system, the regenerative current of each servo module was measured. As shown in Fig. 10, the cycle of the current control for the robot arm is 0.02 [ms], the DC current changes its polarity very quickly, and the regenerative current flows due to the motor operation. The prevention circuit for the regenerative current is required in the power packet dispatching system.

Fig. 11 shows the circuit configuration to prevent the regenerative current. Even though regenerative current is generated by the motor, the router is protected by placing a Zener diode at the output of the router.

Communication between the Master Controller and Mixer is performed, and it is confirmed that the power packet can be transmitted with the communication data. The configuration of the system is shown in Fig. 12, the communication data to be transmitted from the master controller to the mixer is shown in TABLE IV.

As shown in Fig. 13, it can be confirmed that the packets for the six axes are generated within 1 ms of the control cycle of the robot arm. In addition, the enlarged view of each header is shown in Fig. 14. As shown in Fig. 14, it can be confirmed that all packets are generated based on the communication data.

The configuration of the power packet dispatching system is shown in Fig. 15. The router has a configuration of multi-hop system, the power for the axis 1 and 2 are outputted from the router 1, the power for the axis 3, 4, 5, and 6 axis are outputted form the router 2 via the router 1.

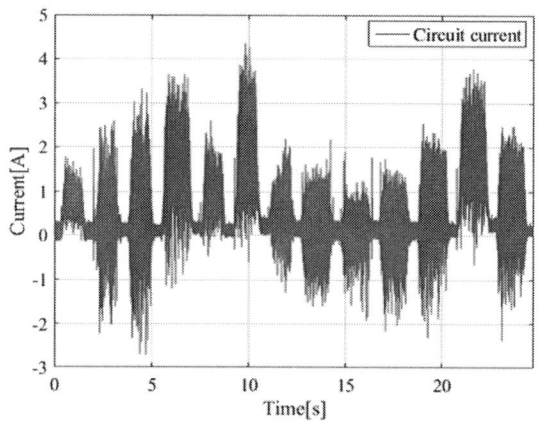

Fig. 10.   DC current waveforms for six axis operation(Link1~6).

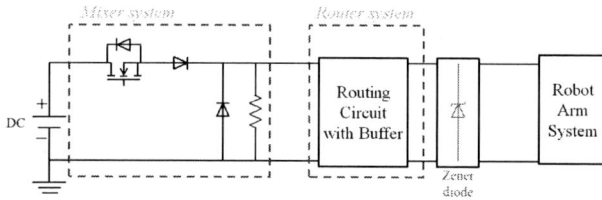

Fig. 11.   Circuit configuration to prevent the regenerative current.

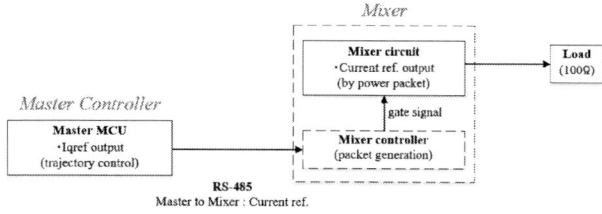

Fig. 12.   Configuration of power packet generation.

TABLE IV.   COMMUNICATION CONDITIONS

Link Number	Dummy and Motor	Current ref.1	Current ref.2	CRC
Link1	X'00'	X'06'	X'66'	X'99'
Link2	X'01'	X'0C'	X'CD'	X'4B'
Link3	X'02'	X'13'	X'33'	X'69'
Link4	X'03'	X'19'	X'99'	X'AE'
Link5	X'04'	X'20'	X'00'	X'4B'
Link6	X'05'	X'26'	X'66'	X'B9'

Fig.16 shows the power packet waveform of the input and the output of the router 1, the upper waveform is the mixer output, the second waveform is the output 1 of the router 1, and the lower waveform is the output 2 of the router 1. The power packets for the axis 1 and 2 were transferred to the load 1 and the other power packets were transferred to the router

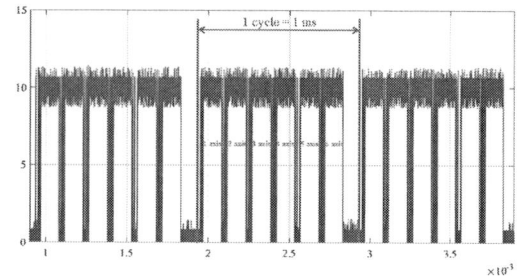

Fig. 13.   Voltage waveform of Mixer output.

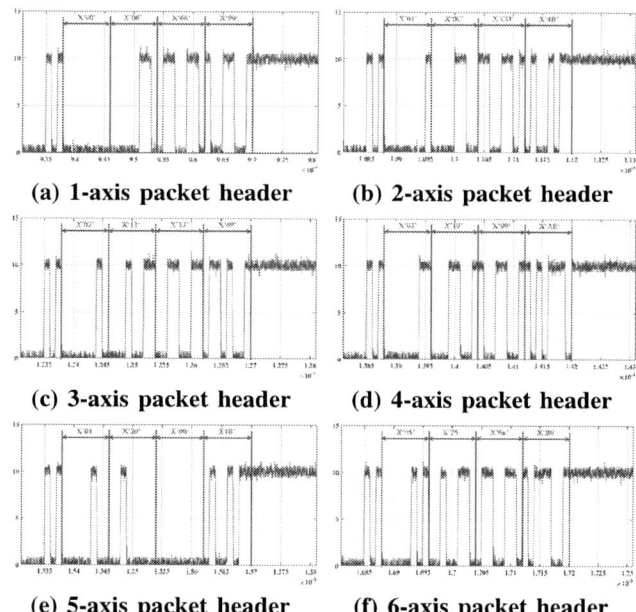

(a) 1-axis packet header   **(b) 2-axis packet header**

(c) 3-axis packet header   **(d) 4-axis packet header**

(e) 5-axis packet header   **(f) 6-axis packet header**

Fig. 14.   Enlargement waveforms of packet header for each axis.

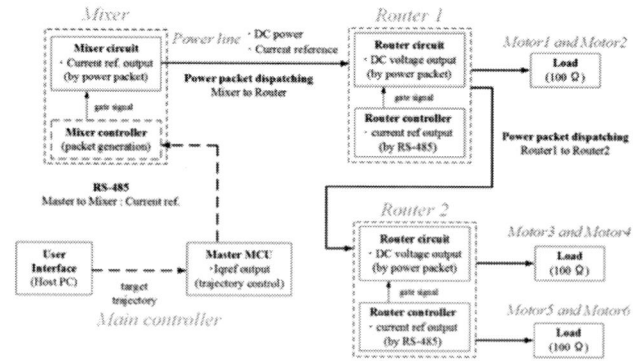

Fig. 15.   Configuration of power packet dispatching system for robot arm.

2. Fig.17 shows the power packet waveform of the input and the output of the router 2, the upper waveform is the mixer output, the second waveforms are the output 2 of the router 1 and the output 1 of the router 2, and the lower waveforms are

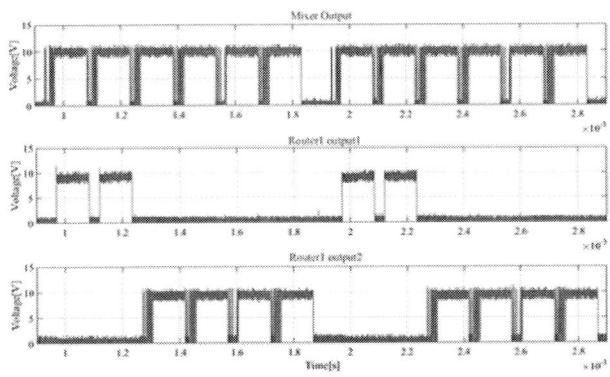

Fig. 16. Output waveform of packet dispatching in router 1.

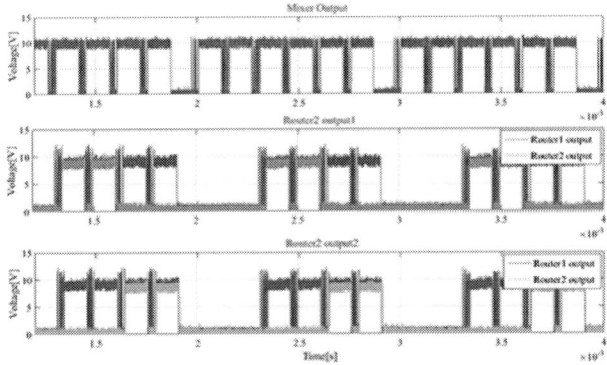

Fig. 17. Output waveform of packet dispatching in router 2.

the output 2 of the router 1 and the output 2 of the router 2. The basic operation of the robot arm system with the power packet dispatching system was confirmed.

Verification was carried out for one axis by power packet dispatching the system. The configuration of the system is shown in Fig. 18. In order to confirm the motor operation by the power packet dispatching system, a current reference is transmitted from the master controller to the slave controller. As shown in Fig. 19, the motor rotated by the power packet dispatching system. The multi axis operations will be reported in near future.

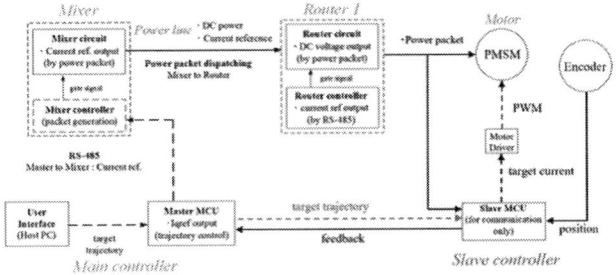

Fig. 18. System configuration for motor drive by power packet dispatching system.

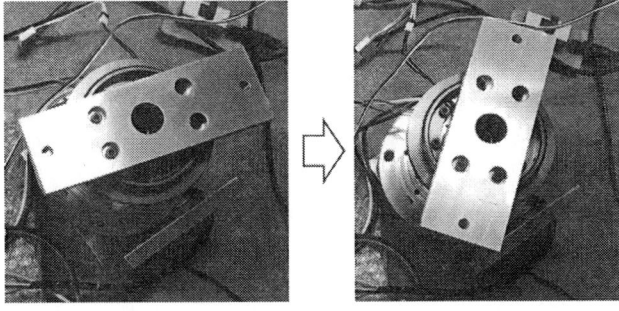

Fig. 19. Motor operation by power packet dispatching system.

VI. CONCLUSION

The six axis robot arm with the power packet dispatching system was proposed. In order to apply the power packet dispatching system to the robot arm system, the design concept was described. The interface specifications were evaluated between the mixer, the router and the servo unit of the robot arm. The system configuration was described in detail. The power packet header information can be generated based on communication data. The power packet dispatching system applied to the robot arm system was realized, and the basic operation was confirmed.

ACKNOWLEDGEMENT

This research was partially supported by the Cross-ministerial Strategic Innovation Promotion Program from New Energy and Industrial Technology Development Organization, Japan.

REFERENCES

[1] Hiroumi Saitoh, Satoshi Miyamori, Toru Shimada, and Junichi Toyoda, "A Study on Autonomous Decentralized Control Mechanism of Power Flow in Open Electric Energy Network", *IEEJ Transactions on Power and Energy*, vol. 117, no. 1, pp. 10-18, 1997.

[2] Noriyuki Iwamuro, Akira Bandoh, Koji Yano, Tetsuya Miyazawa, Hiroomi Eguchi, Yoshinao Miura, Yoji Shikauchi, Nariaki Ikeda, Yasuhiro Uemoto, and Atsushi Hiraiwa, "Development Status of New Material Power Devices", *IEEJ Transactions on Electronics, Information and Systems*, Vol.4, No.6 pp.703-713, 2015.

[3] Takashi Hikihara, "Power Processing by Packetization and Routing", *IEICE Technical Report*, CCS2015-57, pp. 63-67, 2015.

[4] Ryo Takahashi, Shun-ichi Azuma, Keiji Tashiro, and Takashi Hikihara, "Design and experimental verification of power packet generation system for power packet dispatching system", in Proc. Amer. Control Conf., Washington, DC, USA, Jun. 2013, pp.4368-4373.

[5] Ryo Takahashi, Kenji Tashiro, and Takashi Hikihara, "Router for Power Packet Distribution Network: Design and Experimental Verification", *IEEE Transactions On Smart Grid*, Vol.6, No.2, pp.618-626, 2015.

[6] Keiji Tashiro, Ryo Takahashi, and Takashi Hikihara, "Feasibility of power packet dispatching at in-home DC distribution network," in Proc. 3rd IEEE Int. Conf. Smart Grid Commun., Tainan, Taiwan, Nov. 2012, pp.401-405.

[7] Shinya Nawata, Naoaki Fujii, Yanzi Zhou, Ryo Takahashi, and Takashi Hikihara, "A Theoretical Examination of an Unexpected Transfer of Power Packets by Synchronization Failure", *Proc. 2014 International Symposium on Nonlinear Theory and Its Applications*, pp. 60-63, 2014.

[8] Shiu Mochiyama, Ryo Takahashi, Takuya Kajiyama, and Takashi Hiki-hara, "An experiment on trajectory control of manipulator fed by power packets via common power-line ", *IEICE Technical Report*, Vol.116, no.133, pp. 37-42, 2016.

[9] Ryunosuke Araumi, Kazunori Asada, and Tomoki Yokoyama, "An experiment on lighting control of LEDs by power packet dispatching system", *IFEEC 2017 - ECCE Asia*, pp.1120-1125, 2017.

[10] Ryunosuke Araumi, Kazunori Asada, and Tomoki Yokoyama, "Proposal of Instantaneous Clock Synchronization Control for Power Packet Feed System using FPGA", *IEE - Japan Industry Applications Society Conference*, no. 1, pp. 321-324, 2017.

[11] Tokyo Robotics. http://robotics.tokyo/ja/products/torobo_arm/ [9 Mar. 2018]

[12] Yuki Yokokura, and Kiyoshi Ohishi, "Fine Sensorless Force Control using Diode-Clamped Linear Amplifiers", *IEEJ Journal of Industry Applications*, Vol.3, no.3, pp.277-285, 2014.

[13] Thao Tran Phuong, Kiyoshi Ohishi, and Yuki Yokokura, "Motion-Copying System Using FPGA-based Friction-Free Disturbance Observer", *IEEJ Journal of Industry Applications*, Vol.3, No.3, pp.248-259, 2014.

[14] Naoki Oda, and Noriaki Fujinaga, "Vision-based Posture Estimation and Null Space Control for Redundant Manipulator", *IEEJ Journal of Industry Applications*, Vol.2, No.1, pp.48-54, 2013.

[15] Tuyoshi Hanamoto, Hasan Zidan, Ryuichi Oguro, Yoshiaki Tanaka and Teruo Tsuji, "Sensor-less Speed Control of Cylindrical Type PMSM Using Estimated Flux Linkages", *IEEJ Journal of Industry Applications*, Vol.120, No.7, pp.877-883, 2000.

A resource sharing model in a power packet distribution network

H. Ando*, R. Takahashi[†], S. Azuma[‡], M. Hasegawa[§], T. Yokoyama[¶] and T.Hikihara[‖]

* University of Tsukuba, Tsukuba, 305–8573, JAPAN, Email: ando@sk.tsukuba.ac.jp
[†] Aichi University of Technology, Gamagori, 443–0047, JAPAN
[‡]Nagoya University, Nagoya, 464–8603, JAPAN
[§]Tokyo University of Science, Tokyo, 125–8585, JAPAN
[¶]Tokyo Denki University, Tokyo, 120–8551, JAPAN
[‖]Kyoto University, Kyoto, 615–8510, JAPAN

Abstract—**Distributing power packets in a network is an efficient technique for supplying power to distributed loads. In the case that the amount of power is limited in a closed network system, the consensus algorithm can be useful for sharing the limited power resource in the system. For example, we consider a power network consisting of one power source and N($\gg$1) loads. We assume that each load has a buffer that can store the power for driving the load. It is important to efficiently share the limited power from the single source. One possible way for sharing power is as follows. First, all loads are divided into several clusters. Next, the power source sends power packets to each cluster by responding to the feedback information on how much power is required by the loads in the cluster. If the power demand of one load exceeds the power supply from the source, the deficient power is accommodated by the consensus algorithm among the buffers within the cluster. We discuss by numerical simulations how to cluster the loads in terms of the balance between the cost for distributing power from the source and the performance of loads.**

I. INTRODUCTION

Distributed power sources and their autonomous energy management in a networked system are expected to be robust and flexible in multiple layers of power networks ranging from isolated system such as automobiles and robots to large scale power grids. One of promising energy management techniques for those distributed power sources is the power packet dispatching system in which a discretized power is transmitted in a power network with the information of the destination node [1]. The power packet dispatching system has routers and mixers and transmit the discretized powers to distributed loads in order to satisfy the power demand from loads. Moreover, the power packets are transmitted in the scheme of time-division multiplexing with regards to several levels of voltages, which means that a single power wire can transmit several levels of voltage for the power packets. Therefore, the number of wires can be reduced in the whole system.

The function of the router is to receive the power packet and forward it to the target load depending on the tags attached to the packet. Also, the router is able to store the power in its capacitors from received packets. These capacitors can work as distributed buffers of powers in a network. Each router can

be considered as a local power source for the nieghboring load. Recently, the power packets dispatching system has been implemented by electric circuits and discussed in view of its possible applications in a distributed power supply system [2]. In addition, a simple mathematical model for exchanging power packets in a power distributed network system has been modeled by using the consensus dynamics [3]. Along the research line of implementing the power packets in distributed power supply system, bio-inspired approaches can be useful [6], since biological systems reduces the cost of sending power from the main power source by exploiting distributed power storages.

In this study, we consider a power resource sharing model in distributed power supply systems with a single power source. We assume that the power resource sharing is done by using consensus dynamics among buffers of the routers in the scheme of the power packet. Specifically, we focus on the systems in which the number of loads or power buffers is huge. This kind of power system is practical, e.g., the electric vehicle has a lot of loads with a single power source. In those systems, the frequency of transmitting the power packet to each buffer tend to be low. Due to the low frequency, power supply is intermittent and the fluctuation of voltage, i.e. ripple voltage of the buffer, is unavoidable. Therefore, power sharing by the consensus dynamics that supplies power to many buffers at the same time can overcome the voltage fluctuation. Note that the consensus algorithm requires that the number of wires increases. Based on these facts, we investigate the effect of sharing power by the consensus dynamics on the power packet dispatching system in terms of efficiency and cost of power supply. We evaluate the performance of the consensus dynamics by the degree of the ripple voltage.

II. MODEL

As mentioned in the previous section, the power packet dispatching system has a mixer for generating power packets with several levels of voltage and a router for receiving and forwarding the power packets to a destination load in a network. See Figure 1. In this study, we consider a simple graph for modeling a power packet system. In order to model electric circuits and simulate them, we represent the mixers and

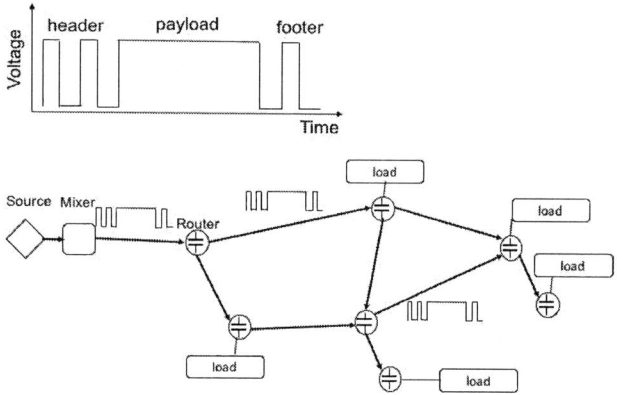

Fig. 1.   System schematic of a power packet dispatching network system.

the routers as nodes in a graph. Then, the links between those nodes corresponds to wires or power lines in the circuits. As a topology of the network model, we assume a tree structure of a directed graph, which implies that there is no loop in the network. We further assume that the mixers and the routers have power buffers and their amount of electric charge, i.e. power is represented by the internal variable of the nodes. In addition, loads can be attached to the nodes and use the power in the nodes.

In one of the simplest cases, we consider a star graph as the network model, in which the center node corresponds to the power source or the mixer and other nodes corresponds to buffers of the routers. Then, we consider additional edges between two router nodes except the source node. The connection between two router nodes of the star implies that electric charges can be transmitted between the two corresponding buffers in the routers. The dynamics of the charge transmission is described in the next subsection. It is important how to design those connection among router nodes for efficient power sharing. That is to say, when one router has less power, other routers gather power efficiently to the less power node regardless of supplying power packets from the mixer.

As for the rule of 1) power supply from the mixer, 2) power consumption by the loads and 3) leaks of power due to the discretized power transmission, we assume the following conditions.

- Each buffer (capacitor) in the router loses its charge due to the switching for the power packets. Namely, the voltage of the capacitor ripples.

- Power consumption by loads loses the charges in the buffer.

- Power packet is transmitted at a single time step to a single buffer whose amount of charges is the smallest in all the buffers connected to the power source.

In addition, we assume a cluster consisting of some buffers in which power is shared by the consensus dynamics. The power

supply to the cluster is to only a single buffer by the power packet from the mixer.

A. Power transmission between router nodes

In order to represent the power dynamics in the network model, we consider the consensus dynamics between one node and its neighboring nodes except power source node [4]. We consider the routers i is connected to the router j by the resistor (resistance $R_{j,i}$) and the router i has a capacitor (capacitance C_i). The dynamics of the voltage of the router i, denoted by V_i is as follows.

$$\dot{V}_i = \sum_{j \in N_i} \frac{1}{C_i R_{j,i}} (V_j - V_i), \tag{1}$$

where N_i is the set of neighboring nodes to the node i. If the capacitance C_i of all nodes are the same as C. The dynamics of charge at node i, denoted by q_i can be written as:

$$\dot{q}_i = \sum_{j \in N_i} \frac{1}{C R_{j,i}} (q_j - q_i). \tag{2}$$

If the $C R_{j,i}$ takes the value 1, the dynamics of (2) is equivalent to the consensus dynamics [5]. According to the eq. (2), the dynamics of the node i depends only on its neighboring nodes, which means that the network dynamics is locally and autonomously determined.

As for modeling the consensus dynamics by the power packet, we additionally introduce switching topology to the consensus network system governed by (2). The power packet is generated by the switch in electric circuits and only one packet can be sent simultaneously in a single power line. Therefore, we consider switching connection in the network model where one node can only connect to at most one node at one time step.

Moreover, we introduce a node connection rule by taking into account that each node demands some fixed amount of charge. The connection rule is as follows.

1) Each node has its own target amount of charge.
2) If the node i has less amount of charge than its target amount, set $\Delta_i < 0$.
3) If the node i has more amount of charge than its target amount, set $\Delta_i > 0$.
4) One node with $\Delta < 0$ can connect with other node with $\Delta > 0$, vice versa.
5) The number of connection of one node is at most one.

Once connection is established between nodes, those nodes transmit electric charge by the consensus dynamics.

B. Power transmission from the source node

Regarding the dynamics of the power packet transmitted from the power source, it is modeled as follows:

$$\dot{q}_i = -a q_i, \tag{3}$$
$$\dot{q}_j = a q_i. \tag{4}$$

This model represents that the source node i transmits the power packet to the node j with an amount of charge determined by that of the node i. As is the case for the switching consensus dynamics, the power packet from the source is transmitted to a single node at one time step.

C. Network topology

As mentioned above, if the topology of the network model is the star without consensus between nodes, the power is supplied only to one node at one time step by a single power packet. This limitation of power packet transmission may yield an intermittency for the interval of timing of receiving the power packet when the number of node increases. In order to overcome the limitation, we consider the network topology as follows.

For the initial condition, the network topology is star and the power is supplied from the source node by the power packet to the router node. We introduce the replacement rate R with which the router nodes are disconnected to the source node (called the terminal node) and re-connected to the router node to which the power is supplied from the source node (called the relay node). It is possible for the terminal node to be re-connected to any number of the relay nodes. In this study, we consider two extreme cases. One is that re-connection from the terminal node is to one relay node (case (i)). The other is that the re-connection is to all relay nodes (case (ii)). Note that the total number of wires is constant in the case (i) regardless of the value of R. We compare the performance of the consensus dynamics by numerical simulation in the next section.

III. NUMERICAL SIMULATION

In the numerical simulation, we determine the number of nodes $N = 50$ and the amount of charge for the consensus dynamics in one time step is half of that for the power packet from the source. The time length for simulation is 200 time steps for N nodes after 100 time steps for transient. We run the simulations for 10 trials with regards to the value of the parameter R to calculate the distribution of the amount of charge at every time step at each node as well as that of CV, i.e. the standard deviation divided by the mean of time series of the amount of charge at each node. We interpret that the amount of CV corresponds to the ripple voltage. The target amount of charge for the consensus dynamics is 0.3. The leaks at the nodes corresponding to power consumption and the ripple effect are regarded as random number $L_t = 0.001 + 0.009\xi$, where ξ is random variable following the uniform distribution in $[0, 1]$. We vary the value of the replacement rate R and investigate the system in the cases (i) and (ii).

A. Case (i): Single connection to relay node

Figure 2 shows the probability distribution of the amount of charge at every time step at each node. For $R = 0$, namely the power is supplied only from the source node by the power packet, the distribution is widely spread. On the other hand, the distribution become sharper when R increases.

Specifically, when $R = 0.8$, there is a peak in the distribution at around 0.3 that is the value for the target amount of charge for the consensus dynamics, which implies that the consensus dynamics strongly influences on the charge dynamics in the network.

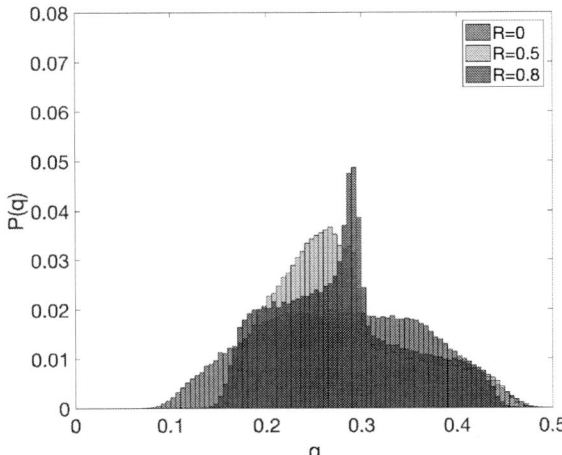

Fig. 2. Probability distribution of charge amount at every time step. $R = 0, 0.5, 0.8$.

Figure 3 shows the probability distribution of the CV for time series of the amount of charge at each node. As shown in Fig. 2, the CV is distributed at relatively large values for $R = 0$ due to the intermittency of power supply. These large CV values can be considered as the large ripple voltage. On the other hand, the CV is smaller for $R = 0.5$ than the case of no consensus, which implies that the ripple voltage is also getting smaller by the consensus dynamics. By detailed observation of the CV distribution, there are a few very small CVs at around 0.1. These small CV values are due to the relay nodes connected to more terminal nodes than the other relay nodes. This is because terminal nodes reduces the amount of charge in the relay node, and then the relay node is more frequent to be supplied by the power packet due to the power supply rule, namely, the power packet is transmitted to the relay node with the minimum amount of charge. With further increase of R, the small CV values are getting smaller, since the number of relay nodes decreases.

B. Case (ii): Full connection to relay nodes

Figure 4 shows the probability distribution of the amount of charge at every time step at each node. As is the case (i), the distribution is sharper when R increases. However, for $R = 0.3$, we observe the peak at around 0.3, which moves leftward for $R = 0.5$. This can be understood as follows. For $R = 3$, the number of connections used for the consensus dynamics is smaller than that in the case of $R = 0.5$. In fact, the number of all connections in the case (ii) is described as $N(1 - R)(1 + NR)$, which is the quadratic function of R. The number of consensus connections is 525 (94% out of all connections) for

The 2018 International Power Electronics Conference

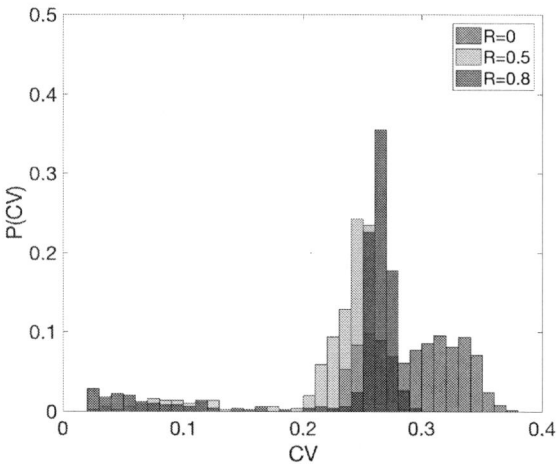

Fig. 3. Probability distribution of CV for time series of charge amount. $R = 0, 0.5, 0.8$.

$R = 0.3$, while that for $R = 0.5$ is 625 connections (96% out of all connections). This comparison implies that the consensus for $R = 0.3$ is easier to be achieved than for $R = 0.5$. Then, for $R = 0.8$, the consensus dynamics is more dominant than for other R, since the number of consensus connections is 400 (96% out of all connections).

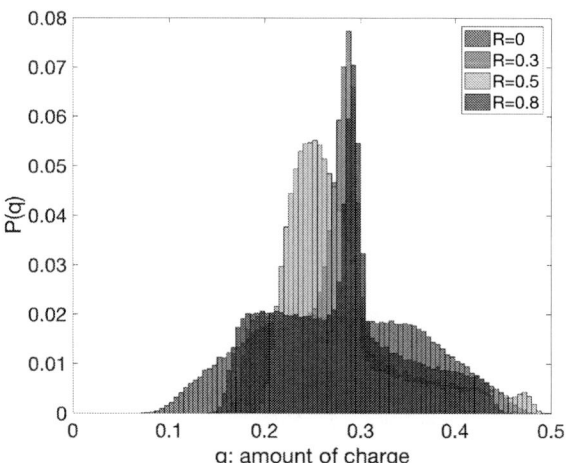

Fig. 4. Probability distribution of charge amount at every time step. $R = 0, 0.3, 0.5, 0.8$.

Finally, Figure 5 shows the probability distribution of the CV for time series of the amount of charge at each node. For $R = 0.3$ and $R = 0.8$, the distribution is split in two large and small CV values. For $R = 0.3$, the number of terminal nodes is not so large that power supply by the consensus dynamics are sufficient from other relay nodes. Therefore, there is small CV values in the distribution. And, the larger CV values in the distribution are due to the relay nodes supplied by the power

packets. On the other hand, $R = 0.8$, the number of relay nodes decreases so that the frequency of the power packets from the source is sufficiently high. Therefore, the small CV values in the distribution are due to the relay nodes, while the larger CV values are due to the terminal nodes composing a larger consensus networks. Regarding $R \sim 0.5$, there is the transition of nodes contributing to the small CV values from the terminal nodes for $R < 0.5$ to the relay nodes for $R > 0.5$. In terms of the ripple voltage, for any values of R larger than 0, the ripple voltage become smaller.

With regards to the assessment of the number of connections, $R = 0$ and $R = 1 - 1/N$ provide similar results. This might be interesting, since in the former case, the power supply is performed by only the power packet. On the other hand, in the latter case, power packet is transmitted to a single node and power supply to the other nodes is conducted by the consensus dynamics with only the single relay node. These results as well as the number of connections depending on R should be considered in more detail as a future work.

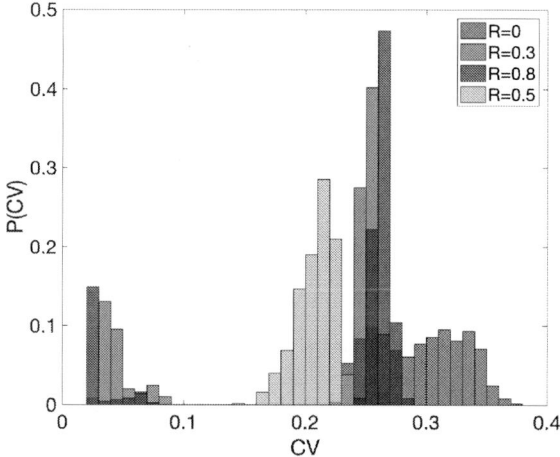

Fig. 5. Probability distribution of CV for time series of charge amount. $R = 0, 0.3, 0.5, 0.8$.

IV. DISCUSSION

In this study, we investigated the power resource sharing model by dividing the power network into some clusters in which power is shared by the consensus dynamics. In addition, the power is supplied to one node in the cluster. We considered a simple graph model and confirmed numerically the performance of the consensus dynamics in terms of reduction of the ripple voltage. In this study, we simply assumed that the leaks and power consumption at loads depends on random number. However, we should introduce several kinds of loads as a practical issue. For example, time scales of power consumption at loads are different from each other. Accordingly, it is not necessary that the capacitors attached loads are homogeneous. Furthermore, we should consider a specific load which reduces large amount of electric charge from buffers very quickly. It

3668

is expected that the power sharing by the consensus dynamics in a network can flexibly respond to such a sudden request of charge.

In order to compare the power resource sharing model in the real world system, we consider the power system of the electric vehicle that has a huge loads in its power system. Also, the voltage rating of those loads are at several levels. The power packet dispatching system with resource sharing by the consensus dynamics could be applicable to that power system. This should be considered as a future work.

ACKNOWLEDGMENTS

The authors would like to thank CSTI, SIP, Next-generation power electronics (NEDO).

REFERENCES

[1] T. Takuno, M. Koyama and T. Hikihara. *Proc. 2010 First IEEE Int. Conf. on Smart Grid Commun.*, pp. 427–430, 2010.

[2] R. Takahashi, et al. *IEICE transanctions*, E100B(6) 948–954, 2017.

[3] H. Ando, S.-I. Azuma, R. Takahashi *IFAC-PapersOnLine*, 49(22):351–354, 2016.

[4] R. Olfati-Saber, R. M. Murray *Proc. of the IEEE*, 95(1): 215–233, 2007.

[5] M. Mesbahi, M. Egerstedt *Graph Theoretic Methods In Multiagent Networks*, Princeton University Press, 2010.

[6] H. Ando, M. Hasegawa, T. Hikihara *Proc. of NOLTA symposium*, 4pages, 2017.

The 2018 International Power Electronics Conference

Decoupled DSOGI-PLL for Improved Three Phase Grid Synchronisation

A. A. Nazib   D. G. Holmes   B. P. McGrath

School of Engineering, RMIT University, Melbourne, Australia

afif.nazib@rmit.edu.au   grahame.holmes@rmit.edu.au   brendan.mcgrath@rmit.edu.au

Abstract — **The DSOGI-PLL structure is widely used for grid synchronization in three phase systems because of its capability to reject voltage harmonics and rapidly calculate the positive sequence fundamental voltage under unbalanced grid conditions. However, its conventional design requires the estimated supply frequency from the PLL to be fed back to the SOGI structure to make it frequency adaptive. This creates an interdependent loop that reduces stability margins and limits the PLL bandwidth. To reduce this interaction, this paper presents an alternative approach that improves the stability margin of the overall system and so allows the PLL bandwidth to be increased without degrading the SOGI harmonic rejection capability. Detailed simulation and experimental results are presented to validate the proposed approach.**

Keywords— *Grid synchronization, second order generalised integrator (SOGI), phase-locked loop (PLL), grid connected converter.*

I. INTRODUCTION

In recent years, higher penetration levels of distributed power generation into the distribution utility networks are challenging the power quality and stability of the power grid [1]-[2]. Consequently, there is an increasing demand for grid-connected converters to contribute to grid-support functions regardless of grid operating conditions, and thus help meet grid codes [3][4]. Almost always, this involves careful synchronizing of the current injected by a grid-connected converter to its Point of Common Coupling (PCC) grid voltage, which requires rapid and accurate estimation of the PCC voltage phase angle and frequency, particularly as these quantities vary [1][3][5]. Hence it is crucial that the grid synchronization technique used is robust against abnormal grid conditions such as distorted/unbalanced voltages, voltage sag/swells, and voltage frequency/phase fluctuations [5][6].

Fig. 1 shows the general structure of a three phase grid-connect inverter, from which it can be seen that precise knowledge of the PCC voltage frequency and phase angle is essential to create a current reference for the current controller that is properly synchronized to the grid. Commonly, phase locked loops (PLL) are used to achieve this task [3][5][7]. For three phase systems, the PLL is usually constructed in the synchronously rotating reference frame (SRF-PLL) since this considerably

simplifies the loop filter structure [7][8]. Under normal grid operating conditions, the performance of this approach is very good [8]. However, if the grid voltage becomes unbalanced or distorted due to nonlinear loads or grid faults, the SRF-PLL can no longer provide accurate frequency and phase angle estimation of the grid voltage at a high bandwidth. Consequently, a number of more advanced grid synchronization techniques have been proposed to address this issue [6],[9]-[16].

A PLL based on a second order generalized integrator (SOGI) is one such approach that is often used for both single- and three-phase synchronization systems [14]-[17]. For three-phase synchronization systems, the dual SOGI (DSOGI) structure is common, since it not only attenuates low-order voltage harmonics but also allows ready estimation of symmetrical components by passing its output through a positive/negative sequence calculator (PSC) prior to feeding into the SRF-PLL [14]. However, the estimated frequency from the SRF-PLL needs to be fed back into the SOGI to make it frequency adaptive, and thus provide accurate voltage magnitude and phase estimation as the grid frequency varies [13]. This feedback path limits the dynamic performance of the SRF-PLL, increases the complexity of tuning the PLL gains and reduces its stability margin [15][17].

One approach that has been proposed for single phase systems to address this issue is to make its SOGI operate at a fixed frequency, and then to add a feedforward term to the output of the PLL estimated phase angle [18] instead of adapting the SOGI frequency as the input

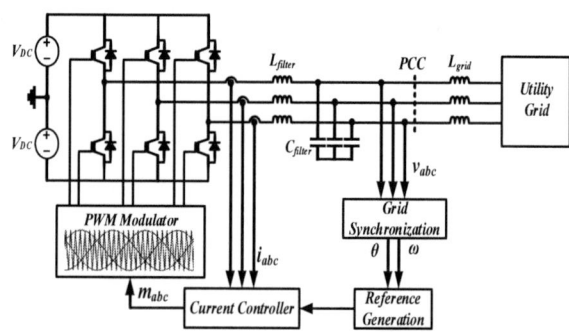

Fig. 1: Three phase grid-connected inverter.

frequency varies. However, the response of this system is significantly limited by the quadrature resonator that is inbuilt within the single phase SOGI-PLL structure, which constrains its harmonic rejection capability and limits the achievable PLL bandwidth [19].

This paper now explores the benefits of using a similar approach for a three phase system, identifying that the normalized positive sequence calculation after the DSOGI quadrature output results in filtered quadrature signals that have their phase angles as the internal DSOGI error signals. Using this knowledge, feedforward compensation for frequency variations can then be implemented, which allows the DSOGI to operate with a fixed center frequency and thus become decoupled from the PLL dynamic response while still accurately synchronising. This enhances the stability of the DSOGI-PLL and allows the bandwidth of the PLL to be significantly increased. A modified decoupling network using the principles of [12] is then used to mitigate the effects of unbalanced voltages. Detailed simulation and experimental results comparing the performance of this new structure against that of a conventional DSOGI-PLL, show the enhanced dynamic performance achieved by this approach for transient grid voltage frequency and phase variations.

II. PRINCIPLES OF A CONVENTIONAL DSOGI-PLL

Fig. 2 shows the structure of a conventional DSOGI-PLL that is commonly used for synchronising to an unbalanced and distorted three phase grid [14]. It consists of three main sections, viz:

- a DSOGI made up of two SOGI-QSGs to eliminate low-order harmonics in the measured voltages;

- a Positive Sequence Calculator (PSC) that extracts the positive sequence voltages;

- a synchronous frame PLL that estimates the phase angle and frequency of the grid voltages at the PCC. This estimated frequency is then fed back into the DSOGI to make it frequency adaptive.

The theoretical principles of these sections are as follows:

A. DSOGI Structure.

The two SOGI-QSG blocks operate on the grid PCC stationary frame α and β voltages, which are created from the measured abc voltages using a conventional Clarke transform according to

$$v_{\alpha\beta} = \left[T_{\alpha\beta} \right] v_{abc} \tag{1}$$

where

$$\left[T_{\alpha\beta} \right] = \frac{2}{3} \begin{bmatrix} 1 & -\dfrac{1}{2} & -\dfrac{1}{2} \\ 0 & \dfrac{\sqrt{3}}{2} & \dfrac{\sqrt{3}}{2} \end{bmatrix} \tag{2}$$

Each SOGI-QSG is formed by a second order generalized integrator with unity feedback, as shown in the top left

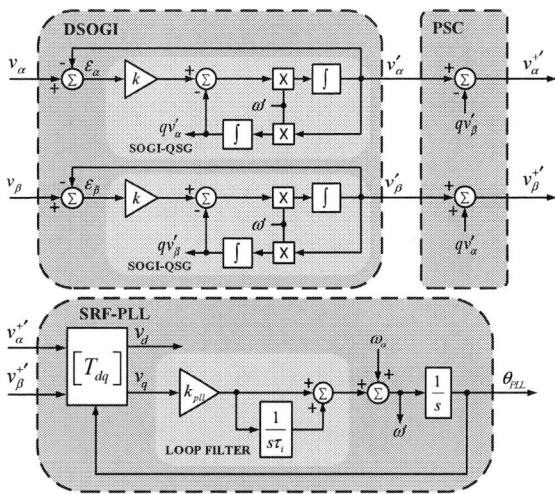

Fig. 2: Conventional DSOGI-PLL structure.

block of Fig. 2. The direct and quadrature transfer functions of these SOGI-OSGs that relate the input to the outputs are given by

$$D_x(s) = \frac{v_x'}{v_x} = \frac{k\omega' s}{s^2 + k\omega' s + \omega'^2} \tag{3}$$

$$Q_x(s) = \frac{qv_x'}{v_x} = \frac{k\omega'^2}{s^2 + k\omega' s + \omega'^2} \tag{4}$$

where $x \in \{\alpha, \beta\}$, q is a 90° lagging phase shift operator, k is the damping factor and ω' is the center frequency.

From (3) and (4), it can be identified that each SOGI-QSG structure has a direct channel band-pass and a quadrature channel low-pass filtering characteristic. This allows it to significantly attenuate voltage harmonics, with the harmonic rejection and dynamic performance depending on the damping factor k [16]. Furthermore, it is important to note that irrespective of the magnitude of k or the value of the center frequency ω', the output voltages v_x' and qv_x' of each SOGI-QSG will always be in quadrature.

B. Positive Sequence Calculation.

For unbalanced systems, it is important to avoid any double frequency phase oscillation in the synchronous frame voltages, caused by an unbalanced set of measured input three phase voltages. This can be achieved by extracting the positive sequence stationary frame voltages from the DSOGI outputs using a PSC.

The sequence extraction proceeds by creating a positive sequence set of fundamental $\alpha\beta$ frame voltages from a positive sequence set of fundamental abc frame voltages using the Clarke transform (2). This positive sequence set of abc voltages can be derived from an unbalanced set of abc voltages using the well known positive sequence transform $[T_+]$, which in turn can be re-created from the unbalanced DSOGI outputs using a

3671

reverse Clarke transform. Mathematically, this can be expressed as

$$v_{\alpha\beta}^+ = \left[T_{\alpha\beta}\right]v_{abc}^+ = \left[T_{\alpha\beta}\right]\left[T_+\right]v_{abc}$$
$$= \left[T_{\alpha\beta}\right]\left[T_+\right]\left[T_{\alpha\beta}\right]^T v'_{\alpha\beta} \qquad (5)$$

where

$$\left[T_+\right] = \frac{2}{3}\begin{bmatrix} 1 & a^2 & a \\ a & 1 & a^2 \\ a^2 & a & 1 \end{bmatrix}, \ a = e^{-j\frac{2\pi}{3}} \qquad (6)$$

After a little manipulation, (5) reduces to

$$\begin{bmatrix} v_\alpha^+ \\ v_\beta^+ \end{bmatrix} = \frac{1}{2}\begin{bmatrix} 1 & -q \\ q & 1 \end{bmatrix}\begin{bmatrix} v'_\alpha \\ v'_\beta \end{bmatrix} \qquad (7)$$

which can be readily implemented using the transfer block structure shown in the top right block of Fig. 2.

Integrating the DSOGI and PSC function blocks by substituting (3) and (4) into (7), gives the input-to-output transfer functions of the combined structure as:

$$\begin{bmatrix} v_\alpha^{+\prime}(s) \\ v_\beta^{+\prime}(s) \end{bmatrix} = \frac{1}{2}\begin{bmatrix} D(s) & -Q(s) \\ Q(s) & D(s) \end{bmatrix}\begin{bmatrix} v_\alpha(s) \\ v_\beta(s) \end{bmatrix}$$
$$= \frac{1}{2}\frac{k\omega'}{s^2 + k\omega's + \omega'^2}\begin{bmatrix} s & -\omega' \\ \omega' & s \end{bmatrix}\begin{bmatrix} v_\alpha(s) \\ v_\beta(s) \end{bmatrix} \qquad (8)$$

Eqn. (8) shows how the combination of the DSOGI and the PSC form a low-pass filter that both filters out harmonic voltages and extracts the positive sequence only, from a distorted and unbalanced input voltage set. Consequently the positive sequence stationary frame quadrature output signals $v_\alpha^{+\prime}$ and $v_\beta^{+\prime}$ will have equal amplitudes, which will eliminate any unbalanced double frequency voltage oscillations prior to feeding the voltages into the SRF-PLL for phase angle and frequency estimation.

C. SRF-PLL

As shown in the lower block of Fig. 2, the positive sequence components of the DSOGI/PSC combination are now fed into a synchronous rotating frame PLL to estimate the grid voltage phase angle and fundamental frequency. The first step is to convert the positive sequence voltages into the PLL synchronous frame using

$$\begin{bmatrix} v_d \\ v_q \end{bmatrix} = \begin{bmatrix} \cos(\theta_{PLL}) & \sin(\theta_{PLL}) \\ -\sin(\theta_{PLL}) & \cos(\theta_{PLL}) \end{bmatrix}\begin{bmatrix} v_\alpha \\ v_\beta \end{bmatrix} \qquad (9)$$

where θ_{PLL} is the estimated phase angle produced by the PLL. The PLL then regulates the q-axis voltage v_q to zero using a simple PI loop filter of the form

$$LF(s) = k_{pll}\left(1 + \frac{1}{s\tau_i}\right) \qquad (10)$$

where k_{pll} is the PLL proportional gain and τ_i is the PLL integral time constant. The loop filter output is added to a nominal initial frequency ω_o, to create an estimated frequency ω', which is then integrated to calculate the estimated grid voltage phase angle. This estimated phase angle feeds back into the PLL Park transform (9) to close the PLL control loop.

For grid synchronization, the DSOGI structure must adapt to any variations in the estimated frequency derived from the PLL. Conventionally, this is done by feeding the estimated PLL frequency ω' back as the center frequency for the DSOGI structure, thus adapting its filter characteristic to track this frequency. However, as a consequence there are two interacting feedback loops within the overall DSOGI-PLL structure, which constrain its dynamic stability limits and limit the PLL bandwidth as a consequence.

III. DECOUPLED DSOGI REALISATION

For the decoupled DSOGI realization presented in this paper, each SOGI-QSG block is permanently tuned to a fixed frequency ω_o, and a feedforward error correction is added to the DSOGI outputs to compensate for any input frequency deviations away from this value. This eliminates the adaptive frequency feedback loop from the PLL back to the DSOGI, which allows the PLL bandwidth to then be significantly increased.

The feedforward error term is determined as follows:

A. DSOGI Error Reconstruction

Fig. 3 shows the structure of the decoupled DSOGI, where the outputs of these filters feed into the SRF-PLL in the same way as for the conventional system shown in Fig 2. Note however that with this arrangement, the positive sequence calculation cannot be added after the DSOGI. Hence for unbalanced input voltages this extraction calculation needs to be placed before the DSOGI, as will be discussed in the next section.

For the system in Fig. 3, when the input frequency ω' deviates away from ω_o, the DSOGI will create an error ε_x between its input and output voltages, because it is no longer exactly tuned to the PCC grid voltage frequency.

From [9], this error can be expressed as

$$\varepsilon_x(s) = \frac{s^2 + \omega_o^2}{s^2 + k\omega_o s + \omega_o^2} * v_x(s), \ x \in \{\alpha, \beta\} \qquad (11)$$

When s is replaced by $j\omega'$, (11) becomes

$$\varepsilon_x(\omega') \approx \frac{\left(\omega'^2 - \omega_o^2\right)}{k\omega_o\omega'} * jv_x$$
$$= \frac{(\omega' - \omega_o)(\omega' + \omega_o)}{k\omega_o\omega'} * jv_x \qquad (12)$$

where the operator j represents a leading 90^0 phase shift

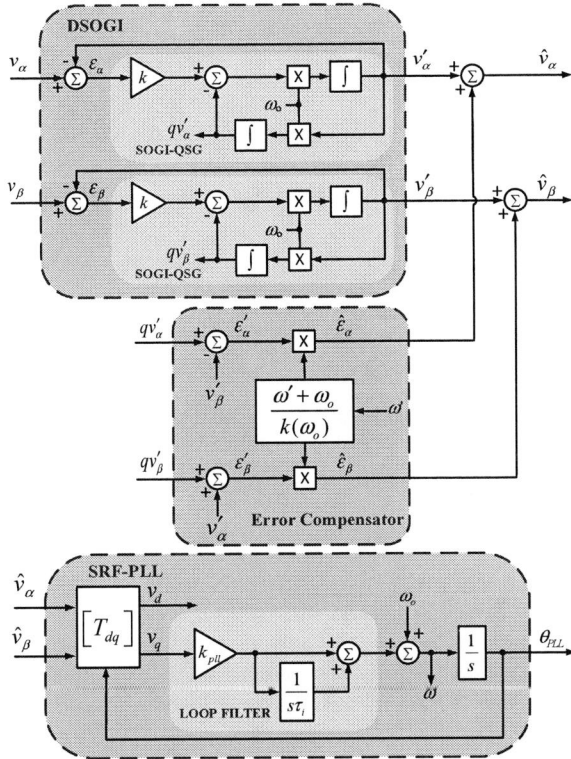

Fig. 3: Decoupled DSOGI structure.

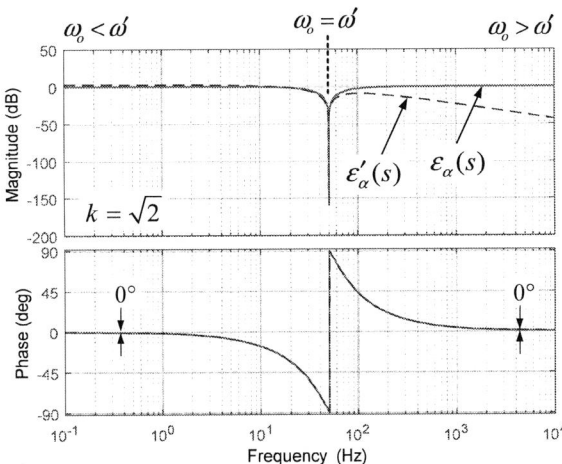

Fig. 4: Frequency response of $\varepsilon_\alpha(s)$ and $\varepsilon'_\alpha(s)$.

between the time domain approximate error voltage and the stationary frame input voltage.

For the error compensators shown in Fig. 3, the error summations are given by

$$\begin{bmatrix} \varepsilon'_\alpha(s) \\ \varepsilon'_\beta(s) \end{bmatrix} = \begin{bmatrix} qv'_\alpha(s) - v'_\beta(s) \\ qv'_\beta(s) + v'_\alpha(s) \end{bmatrix} \tag{13}$$

Substituting from (3) and (4), (13) becomes

$$\begin{bmatrix} \varepsilon'_\alpha(s) \\ \varepsilon'_\beta(s) \end{bmatrix} = \begin{bmatrix} Q(s) & -D(s) \\ D(s) & Q(s) \end{bmatrix} \begin{bmatrix} v_\alpha(s) \\ v_\beta(s) \end{bmatrix}$$

$$= \frac{k\omega_o}{s^2 + k\omega_o s + \omega_o^2} \begin{bmatrix} \omega_o & -s \\ s & \omega_o \end{bmatrix} \begin{bmatrix} v_\alpha(s) \\ v_\beta(s) \end{bmatrix} \tag{14}$$

Substituting $j\omega'$ for s again, gives, with careful phase shift considerations, an error voltage of

$$\varepsilon'_x(\omega') \approx \frac{k\omega_o \omega' - k\omega_o^2}{k\omega_o \omega'} jv'_x \approx \frac{\omega' - \omega_o}{\omega'} jv_x \tag{15}$$

since from (3), v_x and v'_x have essentially the same magnitude.

Now, considering only the α phase and comparing the frequency responses of the $\varepsilon_\alpha(s)$ error functions from (12) and (15), as shown in Fig. 4, it can be seen that the frequency response for $\varepsilon_\alpha(s)$ and $\varepsilon'_\alpha(s)$ have an identical phase response over the entire frequency range. This means that the feedforward error correction term for the

DSOGI needs only to be rescaled to account for magnitude differences between the actual input voltage error and the error determined from the DSOGI variables. This is achieved by multiplying ε'_x by a gain d such that

$$|\varepsilon_x| = d * |\varepsilon'_x|$$

$$\Rightarrow d = \frac{|\varepsilon_x|}{|\varepsilon'_x|} = \frac{\omega'(\omega' - \omega_o)(\omega' + \omega_o)}{k\omega_o \omega'(\omega' - \omega_o)} = \frac{(\omega' + \omega_o)}{k\omega_o} \tag{16}$$

as shown in the Error Compensator section in Fig 3.

B. Compensation for Unbalanced Grid Voltages

For a standard DSOGI-PLL, unbalanced input grid voltages can be compensated by extracting the positive sequence of the voltages from the outputs of the DSOGI, as discussed in Section II. However, for the decoupled DSOGI, this approach is no longer viable since the new strategy requires the magnitude of α and β frame voltages to be equal within the DSOGI. Hence the positive sequence voltages need to be extracted before they are fed into the DSOGI structure. This is done by adopting the basic sequence extraction principles from [12] as shown in Fig. 5, viz:

Assuming an unbalanced three phase system, the grid voltage can be expressed as the summation of the fundamental frequency positive sequence and negative sequence components in the $\alpha\beta$ stationary frame (extracted from the measured abc components using a Clarke transform (2) as before), viz:

$$v_{\alpha\beta} = [T_{\alpha\beta}]v_{abc} = v^+_{\alpha\beta} + v^-_{\alpha\beta}$$

$$\Rightarrow v_{\alpha\beta} = v^+ \begin{bmatrix} \cos(\omega t + \theta^+) \\ \sin(\omega t + \theta^+) \end{bmatrix} + v^- \begin{bmatrix} \cos(\omega t + \theta^-) \\ -\sin(\omega t + \theta^-) \end{bmatrix} \tag{17}$$

where v^+ and v^- are the peak magnitudes of the positive and negative sequence abc frame components, respectively.

The 2018 International Power Electronics Conference

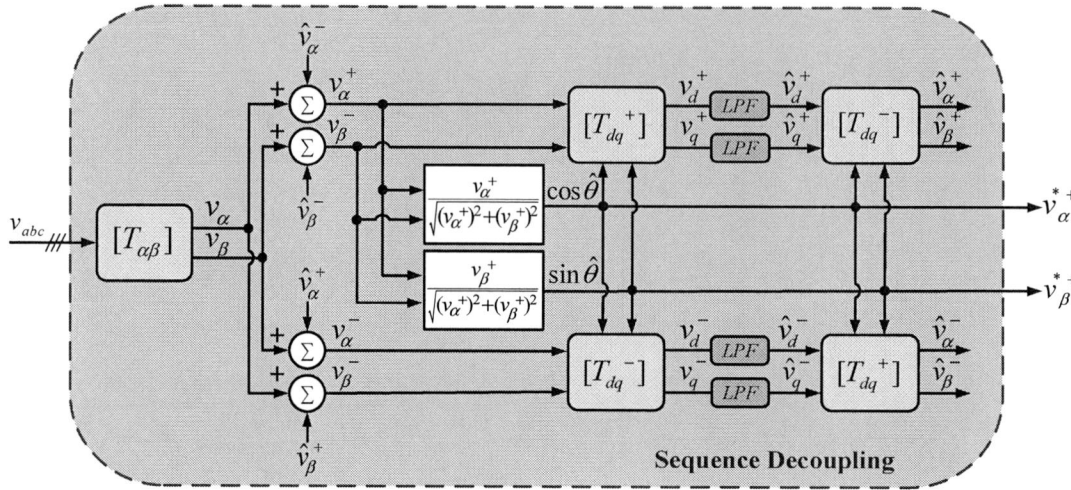

Fig. 5: Sequence Extraction using Decoupling Network.

Transforming the voltage vectors from (17) into the dq rotating frame via a Parks transform (9) gives

$$v_{dq} = [T_{dq}] v_{\alpha\beta}$$
$$= v^+ \begin{bmatrix} \cos(\theta^+) \\ \sin(\theta^+) \end{bmatrix} + v^- \begin{bmatrix} \cos 2\omega t & -\sin 2\omega t \\ -\sin 2\omega t & \cos 2\omega t \end{bmatrix} \begin{bmatrix} \cos(\theta^-) \\ \sin(\theta^-) \end{bmatrix} \quad (18)$$

As shown in (18), the unbalanced voltage vectors create an oscillation-free (dc) component of magnitude v^+ and a double frequency oscillating component of voltage magnitude v^- in the dq rotating frame. The oscillating term can then be eliminated [12] using

$$v_{\alpha\beta}{}^+ = v_{\alpha\beta} - [T_{dq}]^T [LPF(s)][T_{dq}] v_{\alpha\beta}{}^-$$
$$v_{\alpha\beta}{}^- = v_{\alpha\beta} - [T_{dq}]^T [LPF(s)][T_{dq}] v_{\alpha\beta}{}^+ \quad (19)$$

where $LPF(s)$ is the low pass filter needed to estimate the $\alpha\beta$ stationary frame signals of the positive and negative sequence, and ω_c is its cut-off frequency, viz

$$LPF(s) = \frac{\omega_c}{s + \omega_c} \quad (20)$$

The transform angle used for the Park's transform $[T_{dq}]$ and its inverse Park's transform, $[T_{dq}]^T$ in (18) and (19) is $\hat{\theta}$, which is derived from the positive sequence components produced by the sequence extraction process as shown in Fig. 5.

This approach achieves rapid extraction of the positive sequence component, while avoiding any stability issues that might be caused by using the estimated phase angle from the PLL feedback loop. Note also that any harmonics that may be present within the sequence extraction process can also be ignored since the DSOGI attenuates these harmonics as the signal passes through it. Hence they have no influence on the sequence extraction process.

IV. SIMULATION RESULTS

The dynamic performance of the decoupled DSOGI-PLL has been investigated and compared against that of a standard DSOGI-PLL using a detailed PSIM simulation. Three operating conditions were considered:

- a balanced set of ac measured voltages
- balanced ac measured voltages with a significant harmonic content (4% 5th harmonic, 3% 7th harmonic and 1% 11th harmonic)
- significantly unbalanced ac measured voltages.

For all conditions, the PLL for both the standard and the decoupled DSOGI was tuned to have a high bandwidth with parameters of $k_p = 219.91$ and $\tau_i = 1ms$. For each operating condition, the transient performance of the two approaches was compared for a +5Hz frequency jump at 0.2sec, and for a +20⁰ phase jump at the same point in time.

Fig. 6 shows the comparative performance for the two approaches under balanced measured voltage conditions, where the improved transient response of the decoupled DSOGI can be clearly seen (i.e. a reduced transient error magnitude and a significantly reduced settling time for both the frequency jump and the phase jump).

Fig. 7 shows a similar comparative performance for balanced and distorted measured voltages, with a similar improved transient response achieved by the decoupled DSOGI. It is significant to also note that the magnitude of the PLL output frequency continues to retain some ripple from the input harmonic components. However, this ripple is unimportant for the final phase angle output from the PLL, since it is essentially eliminated by the low pass filter characteristic of this part of the PLL process.

Fig. 8 shows the comparative performance for unbalanced measured voltage conditions, where the decoupled DSOGI structure again achieves a significant reduction in transient settling time, with a similar phase transient response as for the conventional strategy.

3674

The 2018 International Power Electronics Conference

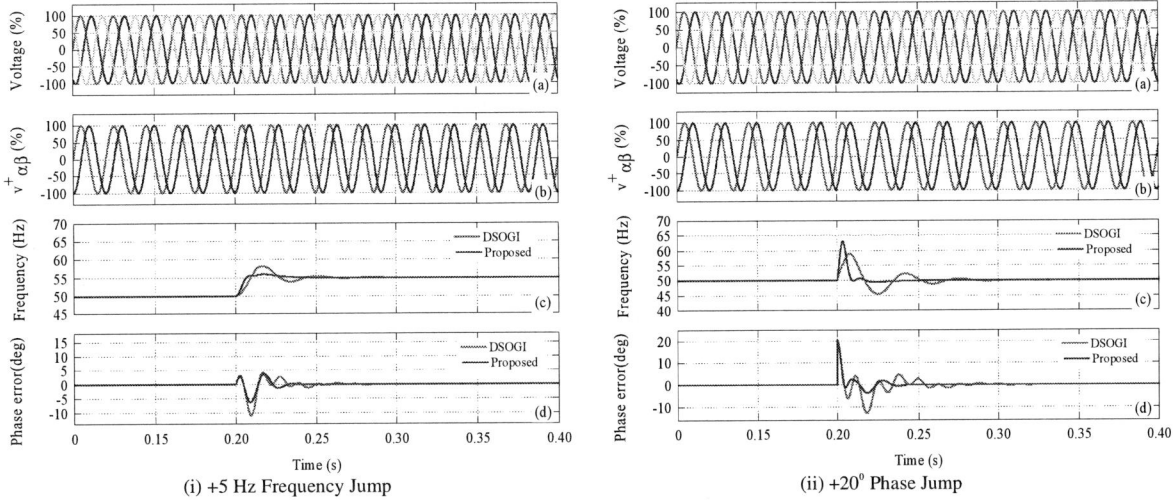

(i) +5 Hz Frequency Jump   (ii) +20⁰ Phase Jump

Fig. 6: Balanced sinusoidal input voltages – **Simulation** performance of conventional DSOGI and decoupled DSOGI strategy.
(a) *abc* phase voltages; (b) *αβ* voltages; (c) PLL output frequency; (d) PLL Output Phase Error.

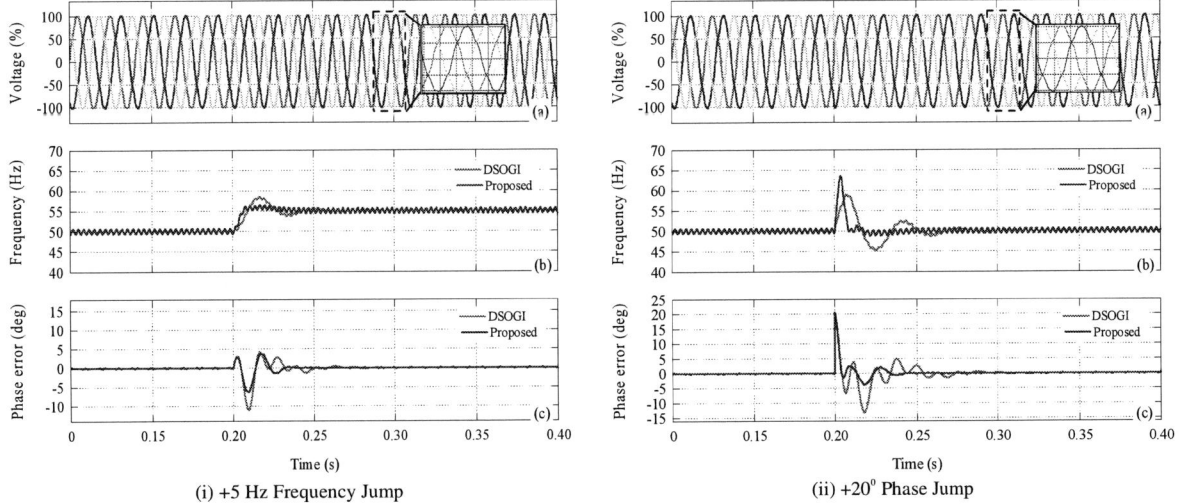

(i) +5 Hz Frequency Jump   (ii) +20⁰ Phase Jump

Fig.7: Harmonic distorted sinusoidal input voltages – **Simulation** performance of conventional DSOGI and decoupled DSOGI strategy.
(a) *abc* phase voltages; (b) PLL output frequency; (c) PLL Output Phase Error.

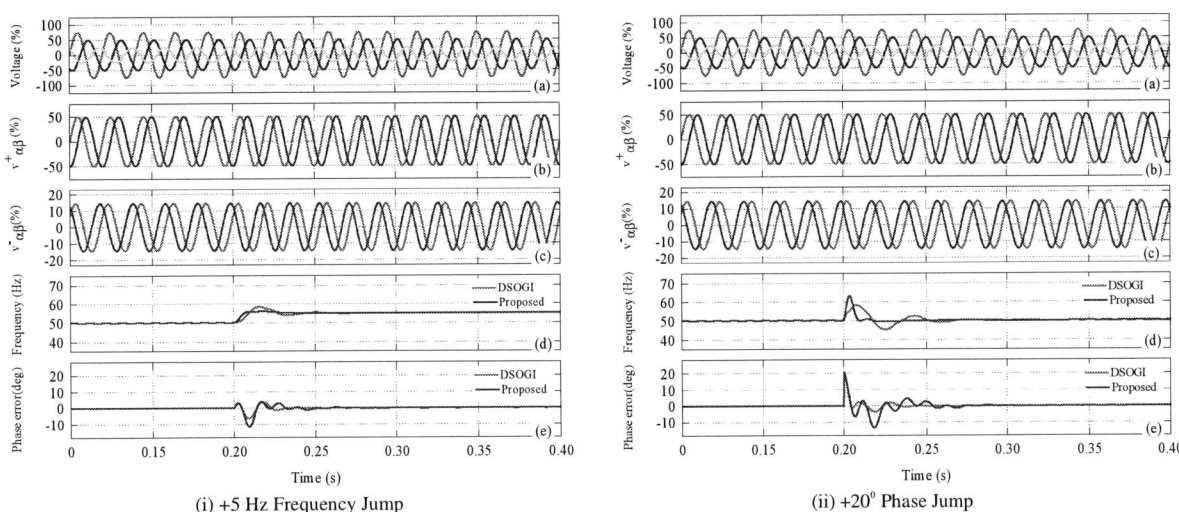

(i) +5 Hz Frequency Jump   (ii) +20⁰ Phase Jump

Fig. 8: Unbalanced sinusoidal input voltages – **Simulation** performance of conventional DSOGI and decoupled DSOGI strategy.
(a) *abc* phase voltages; (b) *αβ* positive sequence voltages; (c) *αβ* negative sequence voltages; (d) PLL output frequency; (e) PLL Output Phase Error.

3675

The 2018 International Power Electronics Conference

V. EXPERIMENTAL RESULTS

The improved performance of the decoupled DSOGI-PLL was confirmed experimentally using a standard laboratory inverter, with the measured voltages generated from a California Instruments MX-series grid emulator. The emulator was programmed to produce nominal 300V peak ac voltages, with harmonics of the same magnitude added to match the simulation results shown in Fig 7. Both the DSOGI-PLL and the decoupled DSOGI-PLL were programmed within the inverter controller using a TI TMS320F2810 digital signal processor, sampling the analogue voltages at 10 kHz.

Fig. 9 shows the experimental results obtained for a balanced measured ac voltage set, where essentially the same result for the PLL frequency output as the simulation results shown in Fig. 6 can be seen. Trace 3 in Fig. 9 shows the PLL output phase angle, within which the phase angle transient error created in response to the

transient frequency jump is too small to see.

Fig. 10 shows experimental results for balanced ac measured voltages with added harmonics, where again the significantly improved performance of the decoupled DSOGI in terms of magnitude overshoot and transient settling time can be clearly seen. From this result it can also be seen how, while the PLL output frequency includes a small level of harmonic voltage ripple that feeds through from the input voltages, this ripple is eliminated by the low pass filter characteristic of the PLL and does not appear in the PLL phase angle output.

Both these experimental results close match their simulation counterparts – Fig. 6(i) compared with Fig. 9, and Fig. 7(i) compared with Fig. 10 – and confirm the improved transient response that can be achieved using the decoupled DSOGI approach compared to a more conventional coupled DSOGI approach.

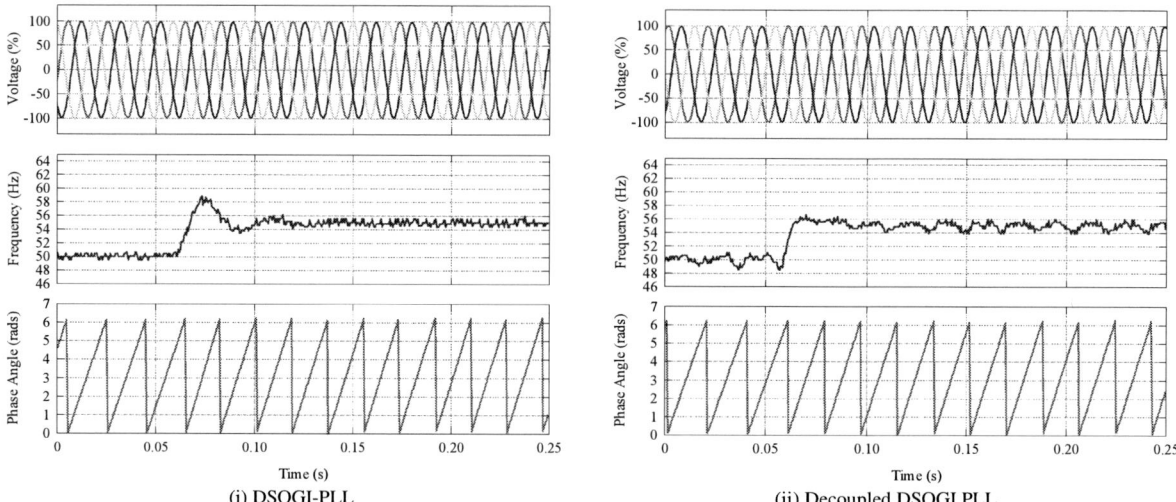

Fig. 9: Balanced sinusoidal input voltages – **Experimental** performance of conventional DSOGI and decoupled DSOGI strategy.
Trace 1 - *abc* phase voltages; Trace 2 - PLL output frequency; Trace 3 - PLL Output Phase.

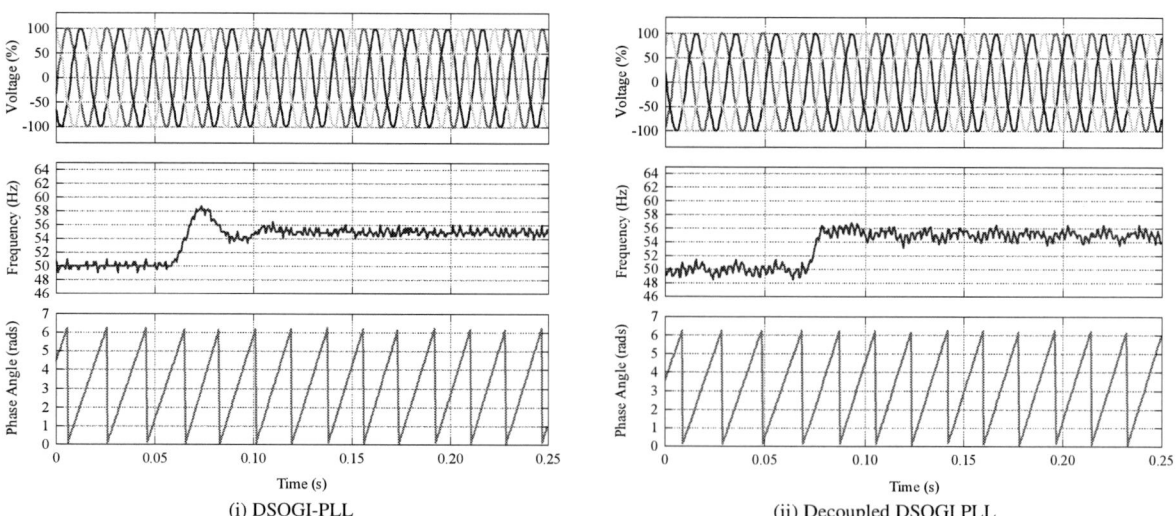

Fig. 10: Harmonic distorted sinusoidal input voltages – **Experimental** performance of conventional DSOGI and decoupled DSOGI strategy.
Trace 1 - *abc* phase voltages; Trace 2 - PLL output frequency; Trace 3 - PLL Output Phase.

VI. CONCLUSION

This paper has presented an improved approach for a three phase grid synchronization strategy that decouples the input DSOGI harmonic filtering subsystem from the main PLL block. This allows the PLL gain to be significantly increased compared to a conventional DSOGI-PLL arrangement, which allows it to track grid voltage frequency changes and phase jumps more effectively. The result is a significantly improved stability margin and a substantially reduced settling time. The theoretical analysis and performance expectations have been confirmed by detailed simulation studies and matching experimental results.

REFERENCES

[1] J. Rocabert, A. Luna, F. Blaabjerg and P. Rodríguez, "Control of power converters in AC microgrids," *IEEE Trans. Power Electron.*, vol. 27, no. 11, pp. 4734–4749, 2012.

[2] F. Blaabjerg, C. Zhe and S. B. Kjaer, "Power electronics as efficient interface in dispersed power generation systems," *IEEE Trans. Power Electron.*, vol. 19, no. 5, pp. 1184–1194, 2004.

[3] F. Blaabjerg, R. Teodorescu, M. Liserre, and A. V. Timbus, "Overview of control and grid synchronization for distributed power generation systems," *IEEE Trans. Ind. Electron.*, vol. 53, no. 5, pp. 1398–1409, 2006.

[4] M. Reyes, P. Rodriguez, S. Vazquez, A. Luna, R. Teodorescu and J. M. Carrasco, "Enhanced decoupled double synchronous reference frame current controller for unbalanced grid-voltage conditions," *IEEE Trans. Power Electron.*, vol. 27, no. 9, pp. 3934–3943, 2012.

[5] A. Timbus, M. Liserre, R. Teodorescu and F. Blaabjerg, "Synchronization methods for three phase distributed power generation systems – An overview and evaluation," in Proc. IEEE Power Electronics Specialists Conference (PESC), pp. 2474–2481, 2005.

[6] P. Rodriguez, A. Luna, M. Ciobotaru, R. Teodorescu and F. Blaabjerg, "Advanced grid synchronization system for power converters under unbalanced and distorted operating conditions," in Proc. IEEE Industrial Electronics Conference (IECON), pp. 5173–5178, 2006.

[7] C. Se-Kyo, "A phase tracking system for three phase utility interface inverters," *IEEE Trans. Power Electron.*, vol. 15, no. 3, pp. 431–438, 2000.

[8] F. D. Freijedo, J. Doval-Gandoy, O. Lopez and E. Acha, "Tuning of phase-locked loops for power converters under distorted utility conditions," *IEEE Trans. Ind. Applicat.*, vol. 45, no. 6, pp. 2039–2047, 2009.

[9] P. Rodríguez, A. Luna, R. S. Muñoz-Aguilar, I. Etxeberria-Otadui, R. Teodorescu and F. Blaabjerg, "A stationary reference frame grid synchronization system for three-phase grid-connected power converters under adverse grid conditions," *IEEE Trans. Power Electron.*, vol. 27, no. 1, pp. 99–112, 2012.

[10] P. Rodriguez, J. Pou, J. Bergas, J. I. Candela, R. P. Burgos and D. Boroyevich, "Decoupled double synchronous reference frame PLL for power converters control," *IEEE Trans. Power Electron.*, vol. 22, no. 3, pp. 584–592, 2007.

[11] F. D. Freijedo, A. G. Yepes, L. Ó, A. Vidal and J. Doval-Gandoy, "Three-phase PLLs with fast postfault retracking and steady-state rejection of voltage unbalance and harmonics by means of lead compensation," *IEEE Trans. Power Electron.*, vol. 26, no. 1, pp. 85–97, 2011.

[12] L. Hadjidemetriou, E. Kyriakides and F. Blaabjerg, "A robust synchronization to enhance the power quality of renewable energy systems," *IEEE Trans. Ind. Electron.*, vol. 62, no. 8, pp. 4858–4868, 2015.

[13] A. Luna, J. Rocabert, J. I. Candela, J. R. Hermoso, R. Teodorescu, F. Blaabjerg and P. Rodriguez, "Grid voltage synchronization for distributed generation systems under grid fault conditions," *IEEE Trans. Ind. Applicat.*, vol. 51, no. 4, pp. 3414–3425, 2015.

[14] P. Rodríguez, R. Teodorescu, I. Candela, A. V. Timbus, M. Liserre and F. Blaabjerg, "New positive-sequence voltage detector for grid synchronization of power converters under faulty grid conditions," in Proc. IEEE Power Electronics Specialists Conference (PESC), pp. 1–7, 2006.

[15] M. S. Reza, M. Ciobotaru and V. G. Agelidis, "Accurate estimation of single-phase grid voltage parameters under distorted conditions," *IEEE Trans. Power Delivery*, vol. 29, no. 3, pp. 1138–1146, 2014.

[16] M. Ciobotaru, R. Teodorescu and F. Blaabjerg, "A new single-phase PLL structure based on second order generalized integrator," in Proc. IEEE Power Electronics Specialists Conference (PESC), pp. 1–6, 2006.

[17] S. Golestan, M. Monfared, F. D. Freijedo and J. M. Guerrero, "Dynamics assessment of advanced single-phase PLL structures," *IEEE Trans. Ind. Electron.*, vol. 60, no. 6, pp. 2167–2177, 2013.

[18] F. Xiao, L. Dong, L. Li and X. Liao, "A frequency-fixed SOGI-based PLL for single-phase grid-connected converters," *IEEE Trans. Power Electron.*, vol. 32, no 3, pp. 1713–1719, 2017.

[19] S. Golestan, S. Y. Mousazadeh, J. M. Guerrero and J. C. Vasquez, "A critical examination of frequency-fixed second-order generalized integrator-based phase-locked loops," *IEEE Trans. Power Electron.*, vol. 32, no. 9, pp. 6666–6672, 2017.

A Deviation Elimination Control Based on Autonomous Current-Sharing Controller for the Parallel-Connected Inverters in AC Microgrids

Yajuan Guan[1], Wei Feng[2], Baoze Wei[1], Wenzhao Liu[1], Mingshen Li[1], Juan C. Vasquez[1], Josep M. Guerrero[1]

Research Programme on Microgrids, Department of Energy Technology, Aalborg University, Denmark

State Key Lab of Power Systems, Department of Electrical Engineering, Tsinghua University, Beijing, China

[1] {ygu, bao, wzl, msh, juq, joz}@et.aau.dk

[2] fwqqrse@163.com

Abstract— Conventional autonomous current-sharing controller (ACSC) will induce inevitable frequency and voltage magnitude deviations due to the output currents and virtual impedances. Thereby secondary control has been proposed to restore the frequency and voltage magnitude deviations by using communication links. In this paper, an autonomous deviation eliminative control method is proposed for the parallel-connected inverters in AC islanded microgrids in order to simplify the control scheme and to improve power quality and control reliability. Two high-pass filters are included in synchronous-reference-frame virtual impedance loop and are activated in steady-state operation to attenuate frequency and voltage magnitude deviations, meanwhile maintaining the autonomous current-sharing performance among the parallel inverters. Simulation results based on two parallel-connected voltage controlled inverters from Matlab/Simulink verify the effectiveness of the proposed approach in different scenarios.

Keywords—ACSC; islanded microgrids; restoration control; deviation;

I. INTRODUCTION

At present, microgrids (MGs) are considered promising electric power systems with decentralized power architectures because of its capability to integrate various kinds of renewable energy sources and power electronics interfaced with distributed generation units (DGs) [1]-[3]. These concepts are growing due not only to environmental aspects but also to social, economic, and political interests [4].

Modeling, analysis, and control design are main concerns for an MG, therefore various control strategies have been proposed and used as the decentralized control of paralleled voltage controlled inverters (VCIs). Power droop control method has been widely used in the last decade as the decentralized control of parallel inverters in a couple of applications such as distributed power systems (DGs), parallel redundant uninterruptible power supplies (UPS) to avoid critical communication links among DGs, MGs, and so forth [5]-[10]. This well-

known control technique aims to proportionally share active and reactive powers which adjusting frequency and output voltage amplitudes of each inverter locally in order to emulate the behavior of a synchronous generator [6]. Several control methodologies with different implementations for conventional droop controller have been also proposed [11]-[14]. A Q – V dot droop control method is mostly used to improve reactive power sharing [15]. An autonomous current sharing controller (ACSC) [16] for the parallel VCIs is proposed to obtain a fast transient response and large stability margin compared with the conventional power droop control.

However, these control strategies produce frequency and voltage magnitude deviations inside the MG because of virtual inertias and output virtual impedances. Therefore, secondary control [17] and some other control strategies have been proposed and applied to MGs to restore voltage magnitude and frequency deviations. An advance robust droop controller and transient droop characteristic were adopted in [18], meanwhile, the frequency and voltage magnitude deviations were mitigated. A consensus algorithm-based distributed frequency control is investigated for voltage and frequency restorations in [19], by allowing these controllers to communicate with neighboring units. However, the secondary control relays on low/high-bandwidth communication links that may affect control reliability. Besides, the aforementioned voltage magnitude and frequency restoration methods were all based on power droop control which has a relatively slow transient response caused by low-pass filters (LPF) and small stability margin.

In this paper, an improved ACSC is proposed for autonomously eliminating voltage magnitude and frequency deviations. With the proposed method, high bandwidth communication links and high-level regulations can be instead by simply using two high-pass filters (HPF) in synchronous-reference-frame (SRF) virtual resistance (VR) loop. In this case, steady-state frequency and magnitude drops can be autonomously

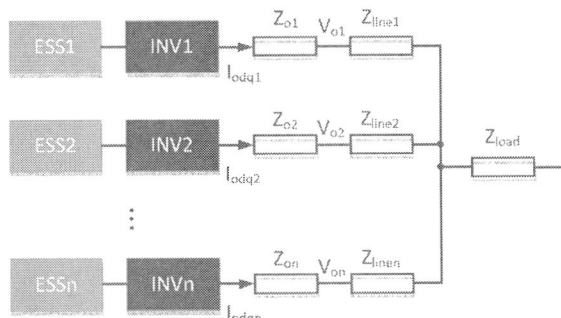

Fig. 1. Electrical scheme of an islanded MG.

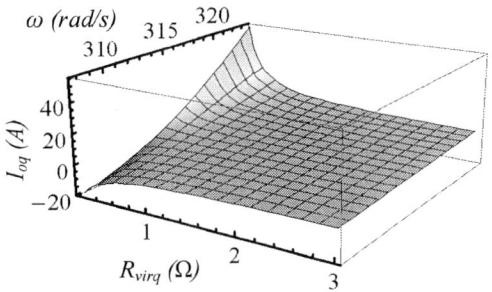

Fig. 2. The relationship of R_{virq}, I_{oq} and ω with different line impedances and conventional ACSC.

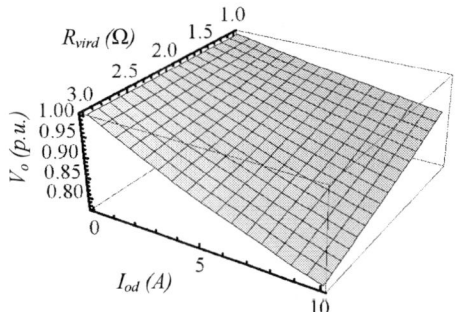

Fig. 3. The relationship of R_{vird}, I_{od} and V_o with different line impedances and conventional ACSC.

attenuated, meanwhile maintaining the current-sharing control performance among the paralleled inverters. Simulation results based on two parallel-connected voltage controlled inverters from Matlab/Simulink verify the effectiveness of the proposed approach in different scenarios.

The paper is organized as follows. Section II analyzes the frequency and voltage magnitude deviations with conventional ACSC. Section III introduces the proposed control structure and control principle. Simulation results are shown in Section IV in order to evaluate the feasibility of the proposed approach and to compare its control performance with the conventional ACSC. Section V concludes the paper.

II. FREQUENCY AND VOLTAGE DEVIATIONS WITH CONVENTIONAL ACSC

A typical islanded MG scheme is shown in Figs. 1, in which each VCI is powered by an energy storage system (ESS). Z_{o1} and Z_{on} are the equivalent output impedances of ACSC controlled VCIs. VCIs are parallel connected at PCC point through line impedance $Z_{line1,...,n}$. A local sensitive load is represented as Z_{load}.

With the conventional ACSC [16], a phase-locked loop (PLL) is used to synchronize the inverter with common AC bus to feed reactive loads. However, the quadrature current (I_{oq}) flowing through the VR will produce a quadrature voltage drop and further lead to an increase in PLL frequency. In this sense, the PLL will compel the inverter to be stabilized at a frequency point with zero phase delay (ZPD) obtained from the transfer function of the system shown as follows:

$$arc\tan\left[\frac{V_o(s)}{V_{ref}(s)}\right]_{s=j\omega} = 0° \tag{1}$$

where $V_o(s)$ and $V_{ref}(s)$ are the output voltage and the reference voltage of ACSC based VCI respectively. The inherent droop relationship of I_{oq} and system frequency (ω) with different quadrature VRs (R_{virq}) is shown in Fig. 2.

In case of supplying active loads, the direct current output I_{od} and d-axis VR will result in an output voltage magnitude drop.

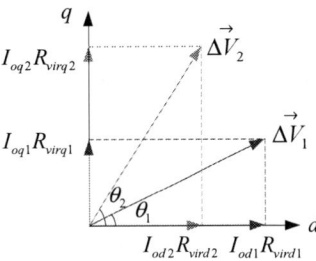

Fig. 4. Vector diagram of the concept of the conventional ACSC.

The voltage drop (ΔV) can be expressed and divided into two parts as follows:

$$\Delta V = I_{od}R_{vird} + |V_{ref}|\left[1 - \left|\frac{V_o(s)}{V_{ref}(s)}\right|_{s=j\omega_o'}\right] \tag{2}$$

where $|V_{ref}|$ is a constant voltage amplitude reference, the ω_o' is the angular frequency of frequency-stable operation point.

In practical, as the system frequency will be stable at ω_o' and the deviations among the parallel inverters will be equal to each other, therefore, the second part of (2) will be equal with each VCIs. Therefore, the first part of (2) becomes the dominant voltage magnitude drop. The droop relationship of I_{od} and bus voltage magnitude V_o in p.u. with different R_{vird} values is shown in Fig. 3. The relation of voltage drops vector and a current output

The 2018 International Power Electronics Conference

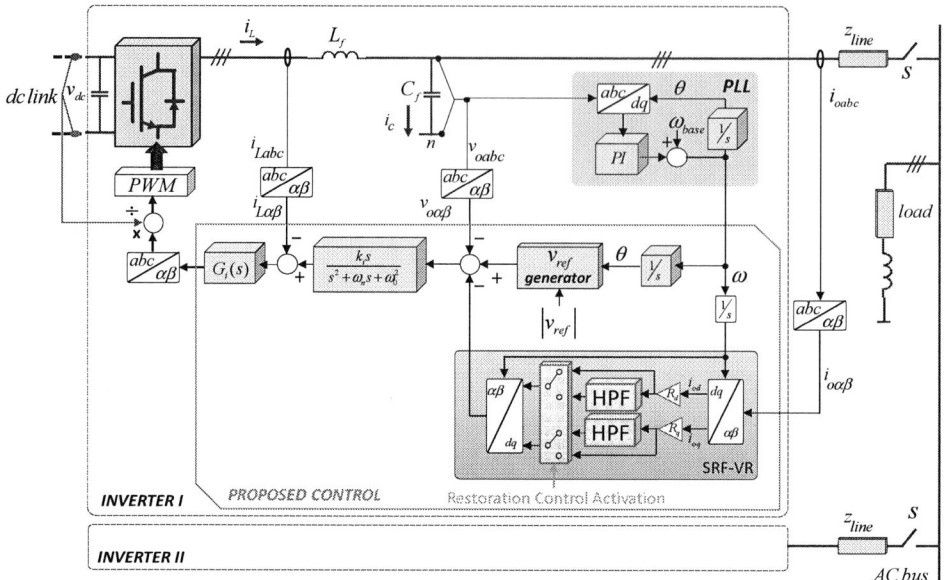

Fig. 5. Improved ACSC for mitigating voltage magnitude and frequency deviations.

vector of two VCIs is presented in Fig. 4. It can be summarized as follows:

$$I_{od1}R_{vird1} = I_{od2}R_{vird2} = ... = I_{odN}R_{virdN} \qquad (3a)$$

$$I_{oq1}R_{virq1} = I_{oq2}R_{virq2} = ... = I_{oqN}R_{virqN} \qquad (3b)$$

As mentioned, the frequency will be raised and voltage magnitude will be dropped as the reactive and active power outputs of a VCI increasing. These deviations will be increased when larger R_{virdq} values are adopted to improve the ACSC power-sharing performance; therefore, deviation issue affects the application of ACSC in high power capacity converter-based MGs.

III. PROPOSED DEVIATION ELIMINATIVE CONTROL STRATEGY

Based on the above analysis, voltage magnitude and frequency deviations will be induced to the common AC bus because of d- and q- axis currents and VRs. Especially in case of high-power converters, large current outputs will result in large frequency and voltage magnitude deviations, thereby deteriorating power quality of ACSC-based isolated MGs.

In order to eliminate the bus frequency and voltage magnitude deviations during current output changes, meanwhile maintaining active and reactive current sharing performance among the paralleled VCIs, a first-order HPF-based restoration control method is proposed and is included in the SRF-VR loop, as shown in Fig. 5. The restoration control will be activated after 0.5s of steady-state operation. The adopted HPF is represented as follows:

$$G_{HPF}(\text{s}) = \frac{s}{s + \omega_c} \qquad (4)$$

where ω_c is the bandwidth of an HPF. The magnitude-frequency characteristics of an HPF with different

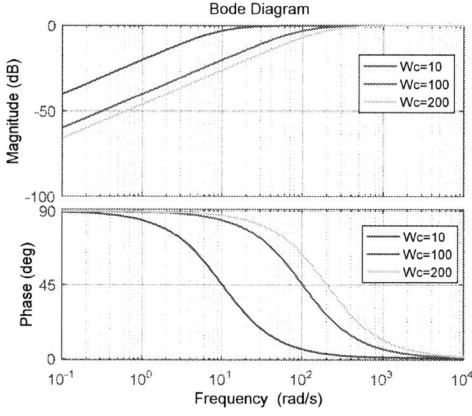

Fig. 6. Bode diagram of the adopted HPF with different bandwidth values.

bandwidth values are shown in Fig. 6

The transfer function of a VCI with the improved ACSC can be derived as follows:

$$T_{plant}(s) = \frac{V_o(s)}{V_{ref}(s)}$$

$$= \frac{G(s)\left[1 + G_{Lequ}(s)Z_{line}(s)\right]}{1 + G_{Lequ}(s)Z_{line}(s) + G_{vir}(s)G(s) + G_{Lequ}(s)Z_o(s)} \qquad (5)$$

being

$$G_{vir}(s) = (\frac{s}{s + \omega_c})(\frac{R_{vird}}{R_{equ}} + \frac{R_{virq}}{sL_{equ}})$$

where $G(s)$ presents the tracking performance of the output voltage following the voltage reference, $Z_o(s)$ is the equivalent output impedance, $Z_{line}(s)$ is the line impedance, $G_{Lequ}(s)$ is the equivalent load admittance, R_{Lequ} and L_{Lequ} are load equivalent resistor and inductance, R_{vird} and R_{virq} are the d-axis and q-axis

3680

The 2018 International Power Electronics Conference

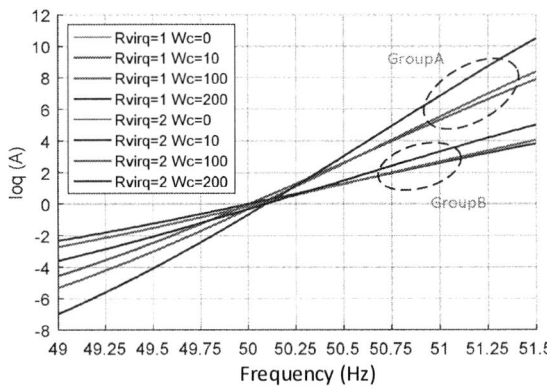

Fig. 7. Relationship of I_{oq} and ω when R_{virq}=1 Ω/2 Ω and with different HPFs.

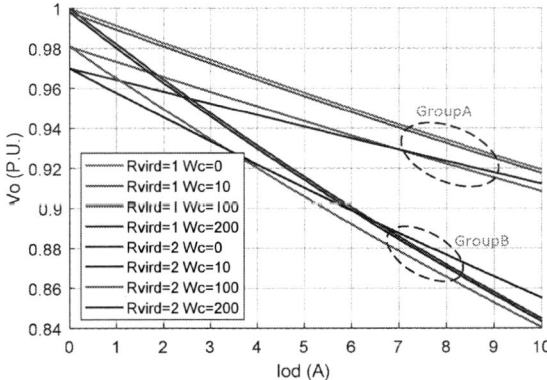

Fig. 8. Relationship of I_{od} and V_o when R_{vird}=1 Ω/2 Ω and with different HPFs.

VRs. Note that, the virtual admittance $G_{vir}(s)$ is updated by including the HPF.

Since the high-frequency components of the product of quadrature current output and q-axis VR will pass the HPF and then be sent back to the voltage control loop, the reactive current sharing performance can be guaranteed during load changes, as shown in Fig. 7. It can be seen that although the adopted HPF changes the slopes of I_{oq}-ω droop curves, the I_{oq} sharing ratio between VCIs is maintained with small bandwidth HPFs (for example ω_c=10 ∼ 40 rad/s).

On the other hand, as the DC components of the product of quadrature current output and q-axis VR is attenuated in steady-state thanks to the HPF, the frequency is automatically stabilized at 50 Hz without any deviations.

In the same way, another HPF is used to compensate voltage magnitude deviations in steady-state operation. As sown in Fig. 8, Group A and Group B represent I_{od}-V droop characteristic of the improved ACSC with different HPFs. R_{vird1} is preset to 1 and R_{vird2} is preset to 2 respectively. It is can be noticed that the original I_{od} sharing ratio can be also maintained when the bandwidth of the adopted HPF is around 10 to 40. Voltage magnitude drop with the proposed method can be derived

TABLE I
ELECTRICAL AND CONTROL PARAMETERS IN SIMULATION

	Parameters		Values
		Description	
Electrical	V_{dc}	DC voltage	650 V
	V_{MG}	MG voltage	311 V
	f	MG frequency	50 Hz
	L_f	Filter inductance	1.8 mH
	C_f	Filter capacitance	25 μF
	L_{line_1}	Resistive line	0.8 Ω
	L_{line_2}	Resistive line	1.1 Ω
Inner Loops	k_{pi}	Current proportional term	0.07
	K_{ii}	Current integral term	0
	K_{pv}	Voltage proportional term	0.04
	K_{iv}	Voltage integral term	94
Improved ACSC loop	K_{p_PLL}	PLL proportional term	1.4
	K_{i_PLL}	PLL integral coefficient	1000
	R_{vird1}	d-axis virtual resistance (DG1)	2 Ω
	R_{virq1}	q-axis virtual resistance (DG1)	1 Ω
	R_{vird2}	d-axis virtual resistance (DG2)	1 Ω
	R_{virq2}	q-axis virtual resistance (DG2)	2 Ω
	ω_c	HPF bandwidth	30

based on (2) and (4):

$$\Delta V = \frac{\omega_c}{s+\omega_c}(I_{od}R_{vird}) + \left|V_{ref}\right|\left[1-\left|T_{plant}(j\omega_0')\right|\right] \quad (6)$$

It can be seen that the first part of (6) will be attenuated to nearly zero in steady-state thanks to the adopted HPF. The second part is caused by the magnitude attenuation that results from the small but nonlinear frequency deviation of inner voltage controller. Although the second part of (6) cannot be directly attenuated by the adopted HPF, it is reduced subsequently when the steady-state frequency (ω_0') is restored to 50 Hz. As a result, the final steady-state voltage magnitude deviation can be effectively mitigated.

IV. SIMULATION RESULTS

In order to validate the proposed bus frequency and voltage magnitude deviation restorative control method, a two-VCIs-based simulation model is established in Matlab/Simulink according to the islanded MG scheme shown in Fig. 1. The electrical and control parameters of the simulation model are listed in Tab I in detail. Two scenarios are designed and are tested to verify the validity of the proposed approach.

A. Turn on/off the proposed restoration control

At the beginning, two conventional ACSC controlled VCIs are parallel connected feeding the local sensitive load (10000 W+2000 Var). Direct and quadrature current outputs sharing ratio between VCI1 and VCI2 are preset to 1:2 and 2:1 respectively. As observed in Figs. 9 and 10, AC bus voltage is dropped to 296 V and frequency is raised to 50.9 Hz resulting from the d- and q-axis current

3681

The 2018 International Power Electronics Conference

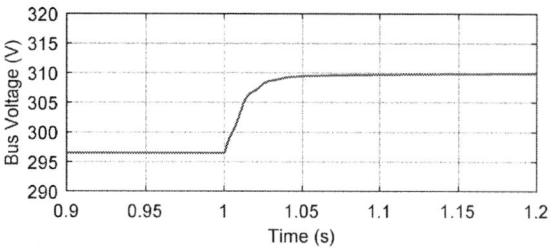

Fig. 9. Dynamic bus voltage response by activating the proposed control.

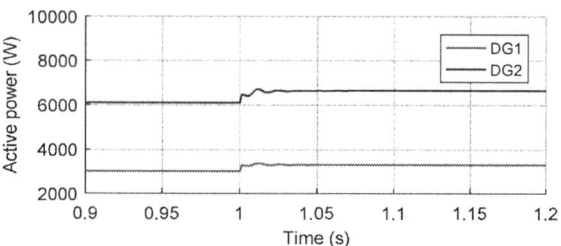

Fig. 11 Dynamic responses of VCIs active power output by activating the proposed control.

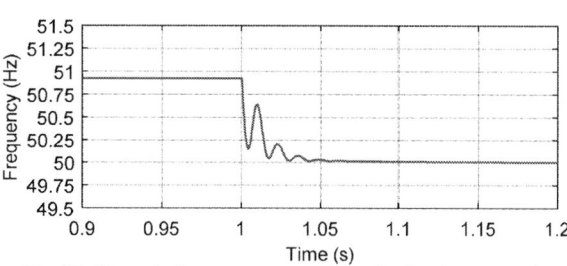

Fig. 10. Dynamic frequency response by activating the proposed control.

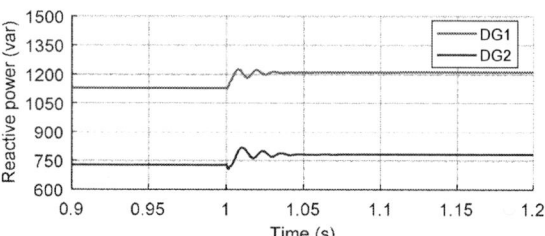

Fig. 12 Dynamic responses of VCIs reactive power output by activating the proposed control.

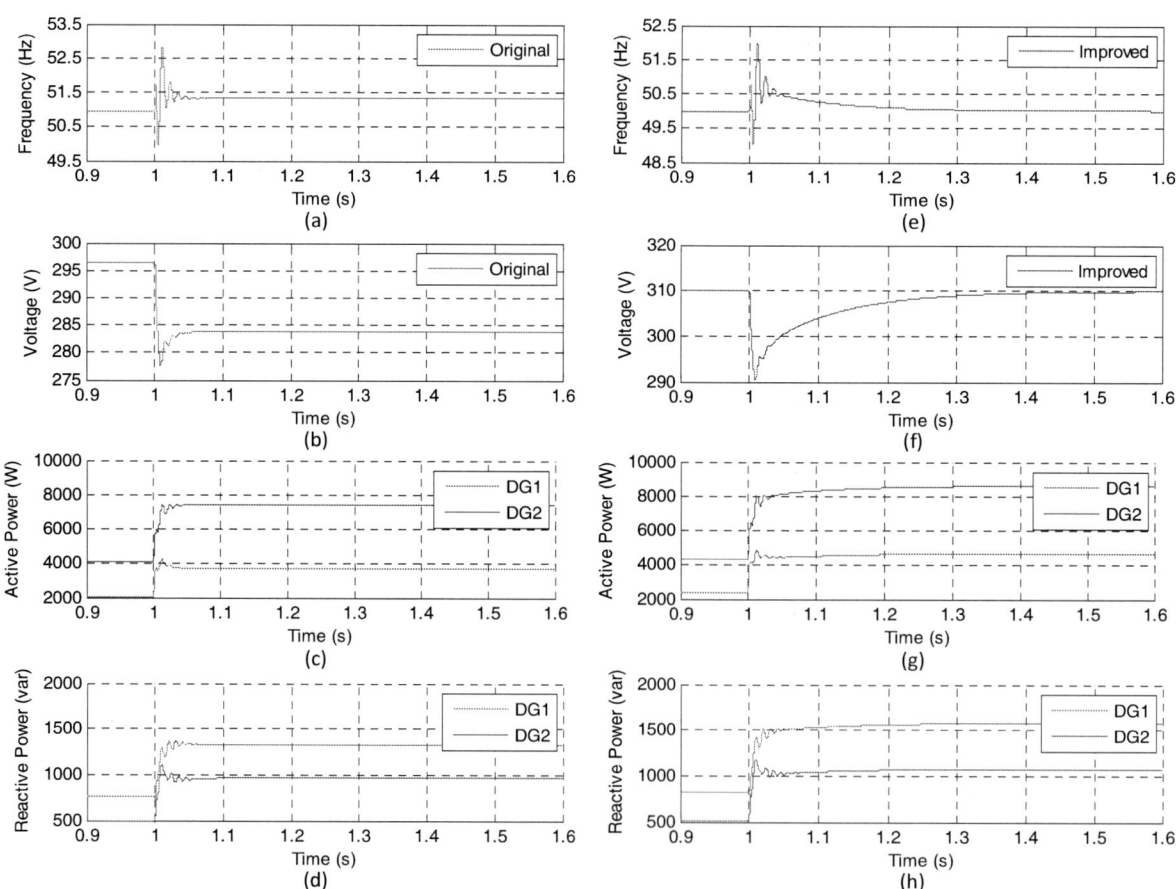

Fig. 13 Simulation results comparison with the conventional ACSC (a)-(d) and the improved ACSC (e)-(h).

3682

outputs and virtual resistances. In this sense, power quality of the isolated MG cannot be maintained by using the conventional ACSC [20]. At 1 s, the proposed voltage magnitude and frequency deviation restorative controls are activated. As shown in Figs. 11 and 12, system frequency and AC bus voltage are restored to normal values respectively. It can be seen that the restoration dynamic response is finished within 0.05 s, meanwhile, active and reactive power sharing ratios between VCI1 and VCI2 are effectively maintained with the improved ACSC.

B. Frequency and voltage magnitude restoration

As observed in Fig. 13, the two paralleled VCIs supply 6 kW and 1.25 kVar to local loads at the beginning, the ratios of R_{vird} and R_{virq} are 2:1 and 1:2 respectively. Figs. 13(a) and 13(b) show the frequency and voltage transient responses during load step up (6 kW+1.25 kVar) changes at 1 s in AC common bus with conventional ACSC. Figs. 13(c) and 13(d) show the active and reactive power outputs with conventional ACSC. After 0.1 s, system frequency and bus voltage are stabilized at 51.4 Hz and 284 V respectively. By contrast, with the proposed method, the frequency and voltage magnitude can automatically restore to 50 Hz and 310 V after transient responses, as shown in Figs. 13(e) and 13(f). Moreover, the active and reactive current sharing ratios can be maintained and the power supply capability can be promoted, as shown in Figs. 13(g) and 13(h).

V. CONCLUSIONS

A simple and improved ACSC is proposed in this paper to autonomously eliminate frequency and voltage magnitude deviations without low/high bandwidth communication links or high-level regulations in AC islanded MGs. Two HPFs are adopted in SRF VR loop to attenuate steady-state frequency and voltage magnitude deviations, meanwhile maintaining the autonomous current-sharing performance. The proposed deviation eliminative method can facilitate the practical implementation of ACSC in high-power converters in real MGs. Simulation results validate the effeteness of the proposed approach.

REFERENCES

[1] Lasseter, R.H., "MicroGrids," Power Engineering Society Winter Meeting, 2002. IEEE, vol.1, no., pp. 305-308, 2002

[2] R. Lasseter, A. Akhil, C. Marnay, J. Stevens, et al, "The certs micr ogrid concept white paper on integration of distributed energy res ources," Technical Report, U.S. Department of Energy, 2002

[3] Li Yunwei, Nejabatkhah F, "Overview of control, integration and energy management of microgrids, Journal of Modern Power Syst ems and Clean Energy, vol. 2, no. 3, pp. 212-222. 2014.

[4] Josep M. Guerrero, Juan C. Vásquez, José Matas, etc. "Control Strategy for Flexible Microgrid Based on Parallel Line-Interactive UPS Systems," Industrial Electronics, IEEE Transactions on, vol.56, no.3, pp.726-736, March 2009

[5] Guerrero, J.M.; Chandorkar, M.; Lee, T.; Loh, P.C.cAdvanced Control Architectures for Intelligent Microgrids—Part I: Decentralized and Hierarchical Control," Industrial Electronics, IEEE Transactions on, vol.60, no.4, pp.1254,-1262, April 2013

[6] Piagi, P., Lasseter, R.H., "Autonomous control of microgrids," Power Engineering Society General Meeting, 2006. IEEE, vol., no., pp., 2006

[7] Yunwei Li, Vilathgamuwa, D.M., Poh Chiang Loh, "Design, Analysis, and Real-Time Testing of a Controller for Multibus Microgrid System," IEEE Transactions on Power Electronics, vol. 19, no. 5, pp. 1195–1204, Sept. 2004.

[8] R. H. Lasseter and P. Paigi, "Microgrid: A conceptual solution," in Proc. IEEE PESC, Aachen, Germany, 2004, pp. 4285-4290.

[9] F. Katiraei and M. R. Iravani, "Power management strategies for a microgrid with multiple distributed generation units," IEEE Trans. Power Syst., vol. 21, no. 4, pp. 1821–1831, Nov. 2006.

[10] Chandorkar, M.C., Divan, D.M., Adapa, R., "Control of parallel connected inverters in standalone AC supply systems," Industry Applications, IEEE Transactions on, vol.29, no.1, pp. 136-143, Jan/Feb 1993

[11] Vandoorn, T.L., De Kooning, J.D.M., Meersman, B., Guerrero, J.M., Vandevelde, L., "Voltage-Based Control of a Smart Transformer in a Microgrid," Industrial Electronics, IEEE Transactions on, vol.60, no. 4, pp. 1291-1305, Apr. 2013.

[12] E. Rokrok and M. E. H. Golshan, "Adaptive voltage droop method for voltage source converters in an islanded multibus microgrid," IET Gen., Trans., Dist., vol. 4, no. 5, pp. 562–578, 2010.

[13] G. Diaz, C. Gonzalez-Moran, J. Gomez-Aleixandre, and A. Diez, "Scheduling of droop coefficients for frequency and voltage regulation in isolated microgrids," IEEE Trans. Power Syst., vol. 25, pp. 489–496, Feb. 2010

[14] Majumder, R., Chaudhuri, B., Ghosh, A., Majumder, R., Ledwich, G., Zare, F., "Improvement of Stability and Load Sharing in an Autonomous Microgrid Using Supplementary Droop Control Loop," Power Systems, IEEE Transactions on, vol. 25, no. 2, 796-808, May. 2010.

[15] Lee, C-.T., Chu, C-.C., Cheng, P-.T., "A New Droop Control Method for the Autonomous Operation of Distributed Energy Resource Interface Converters," Power Electronics, IEEE Transactions on, vol. 28, no. 4, 1980-1993, April. 2013.

[16] Guan, Y., Guerrero, J.M., Zhao, X., Vasquez, J.C., Guo, X. "A New Way of Controlling Parallel-Connected Inverters by Using Synchronous-Reference-Frame Virtual Impedance Loop—Part I: Control Principle," Power Electronics, IEEE Transactions on, vol. 31, no. 6, pp: 4576 - 4593, June. 2016.

[17] Qobad Shafiee, Guerrero, J.M., Vasquez, J.C. "Distributed Secondary Control for Islanded Microgrids—A Novel Approach," Power Electronics, IEEE Transactions on, vol. 29, no. 2, pp: 1018 - 1031, Feb. 2014.

[18] Yu Zeng, Qing-Chang Zhong, "A droop controller achieving proportional power sharing without output voltage amplitude or frequency deviation," 2014 Int. Conf. Energy Conversion Congress and Exposition (ECCE), Pittsburgh, USA, Sept, 2014.

[19] Fanghong Guo, Changyun Wen, Jianfeng Mao, "Distributed Secondary Voltage and Frequency Restoration Control of Droop-Controlled Inverter-Based Microgrids," Industrial Electronics, IEEE Transactions on, vol.62, no.7, pp. 4355-4364, Dec, 2015.

[20] IEEE Std 1159-2009, "IEEE Recommended Practice for Monitoring Electric Power Quality," 26 June, 2009.

SISO Transfer Functions for Stability Analysis of Grid-Connected Voltage-Source Converters

Hongyang Zhang[1*], Lennart Harnefors[2], Xiongfei Wang[3], Jean-Philippe Hasler[1], and Hans-Peter Nee[4]

1 Power Grids Division, ABB, Västerås, Sweden
2 Corporate Research, ABB, Västerås, Sweden
3 Department of Energy Technology, Aalborg University, Aalborg, Denmark
4 Department of Electric Power and Energy Systems, KTH Royal Institute of Technology, Stockholm, Sweden
*E-mail: hongyang.zhang@se.abb.com

Abstract—Converter–grid interaction is of great interest in a weak-grid condition. In this paper, an alternative multi-input multi-output closed-loop system is developed for the stability analysis of grid-connected voltage-source converters. In contrast to the conventional dq-impedance model and the eigenvalue analysis, this model eventually yields a single-input single-output transfer function. This enables to apply a single Nyquist curve for analyzing the overall system stability. The model is validated against time-domain simulations and it shows excellent accuracy for predicting the system stability.

Keywords—*multi-input multi-output (MIMO), Nyquist stability criterion, single-input single-output (SISO), voltage-source converter (VSC), weak grid.*

I. Introduction

The number of grid-connected voltage-source converters (VSCs) has been increased rapidly. Various stability problems have been reported for the operation of VSCs, and many of them occur in a weak-grid condition [1], [2].

Numerous modeling methods have been proposed to study the stability problems in the grid-connected VSC systems. Among them, reference [3] suggests a frequency-domain input-admittance design approach with a dq-impedance model. The method aims to have the converter behave as a passive system, i.e., its conductance should be nonnegative. To fulfill this criterion, the converter controller is thus designed accordingly, at least in certain frequency ranges. In [4], the author concludes that the overall system stability can be satisfied provided that the ratio between the grid impedance and the converter output impedance fulfills the Nyquist stability criterion.

The system stability can also be analyzed by the eigenvalues from the state-space equations. In addition, to investigate the stability impact of individual controller or converters, the participation factor analysis can be used. Such state-space model analysis methods were firstly used for microgrid applications [5]. In addition, the method has also been utilized for analyzing control interactions between rail vehicles and supply converters in a traction network [6].

References [3], [7], and [8] utilize the input-admittance approach to study the destabilizing effect by the increase of phase-locked loop (PLL) bandwidth in the weak-grid condition. A small-signal model utilizing the

eigenvalue analysis in [9] shows the interaction between the PLL and dc-bus-voltage controller (DVC) in the weak-grid condition. In [10], the ac-bus-voltage control (AVC) impact is discussed and the interactions between the DVC and AVC are presented with numerical results.

Although numerous methods enhance the understanding of the stability problems, the multi-input multi-output (MIMO) modeling is unavoidable. The input-admittance-based dq-impedance model requires the generalized Nyquist criterion. The stability margins are not easy or even impossible to obtained. Furthermore, neither the input-admittance approach, nor the eigenvalue analysis can enable the standard Nyquist criterion on a single-input single-output (SISO) system, which facilitates identification of the stability and also stability margins.

The contribution of this paper is to present closed-loop SISO transfer functions for studying the stability of grid-connected VSCs. The closed loops are derived by integrating the current control (CC), PLL, grid impedance, the AVC, and the DVC loops. The derived model eventually yields a SISO open-loop transfer function. Therefore, the overall system stability can be studied easily with a single Nyquist curve by the standard Nyquist stability criterion. Simulations are presented which verify that the analysis of the SISO transfer function accurately predicts system stability and stability margins.

This paper is organized as follows. In Section II, a small-signal modeling method is utilized to linearize the control loops of the VSC. Section III presents the derivation of the MIMO closed-loop system and also the proposed SISO transfer functions. In Section IV, the model is verified against the time-domain simulation.

II. MIMO Closed-Loop Modeling

A. Fundamentals

As demonstrated in Fig. 1, the control system contains the PLL, CC, DVC, and AVC. The converter is modeled as an adjustable voltage source, i.e., the pulsewidth modulator is disregarded. The converter topology is consequently transparent provided that the output voltage of the converter is not affected by its internal dynamics [11], [12]. The control system operates in the synchronous reference (dq) frame. To linearize the control system [3], current/voltage limiters and other nonlinear functions

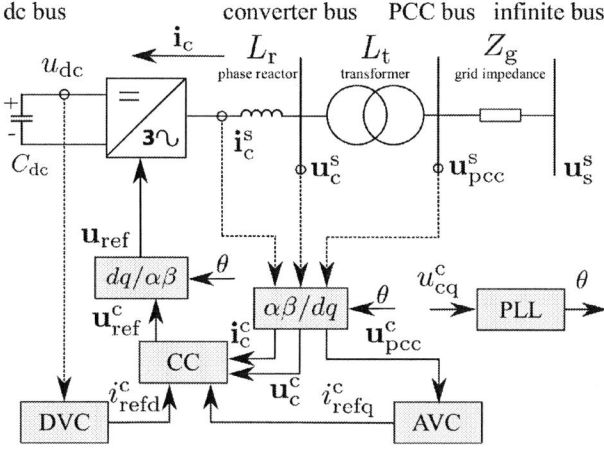

Fig. 1. Schematic of the grid-connected VSC.

are omitted. The grid is considered as a balanced three-phase system which contains only the positive-sequence components. The converter is assumed to be lossless.

1) Stationary Reference Frame and Grid dq Frame:
The stationary reference ($\alpha\beta$) frame is denoted with the superscript 's'. For instance, the converter-bus voltage space vector is given as

$$\mathbf{u}_c^s = \mathbf{u}_c e^{j\omega_1 t}, \quad \mathbf{u}_c = u_{cd} + j u_{cq} \quad (1)$$

where $\mathbf{u}_c$ referred to the grid dq frame, which is not influenced by the PLL, and ω_1 is the angular grid frequency.

2) Converter dq Frame: Space vectors referred to the converter dq frame are impacted by the PLL and are denoted by the superscript 'c'.

3) $\alpha\beta$ to dq Transformation: Three such transformations are used in the control system: for the point-of-common coupling (PCC)-bus voltage $\mathbf{u}_{pcc}^s$, the converter-bus voltage $\mathbf{u}_c^s$, and the converter input current $\mathbf{i}_c^s$.

The angle used in the transformations is the output of the PLL. The angle equals a steady-state angle θ_1 (which is the integral of ω_1) plus an angle deviation $\Delta\theta$

$$\mathbf{u}_c^c = \mathbf{u}_c^s e^{-j(\theta_1+\Delta\theta)}, \quad \mathbf{u}_c^s = (\mathbf{u}_{c0} + \Delta\mathbf{u}_c)e^{j\theta_1}. \quad (2)$$

$\mathbf{u}_c$ in the grid dq frame is in a similar way as the angle, expressed by its steady-state value $\mathbf{u}_{c0}$ plus its deviation value $\Delta\mathbf{u}_c$. Equation (2) can be linearized so that the relation between the deviation variables is expressed as

$$\Delta\mathbf{u}_c^c = -j\mathbf{u}_{c0}\Delta\theta + \Delta\mathbf{u}_c. \quad (3)$$

Similarly, for the PCC voltage and the converter current we obtain the following relations

$$\Delta\mathbf{u}_{pcc}^c = -j\mathbf{u}_{pcc0}\Delta\theta + \Delta\mathbf{u}_{pcc}, \quad \Delta\mathbf{i}_c^c = -j\mathbf{i}_{c0}\Delta\theta + \Delta\mathbf{i}_c. \quad (4)$$

4) dq to $\alpha\beta$ Transformation: The voltage reference space vector in the converter dq frame is transformed to the $\alpha\beta$ frame by the angular output from the PLL. The space vector can be expressed as

$$\mathbf{u}_{ref}^s = (\mathbf{u}_{ref0}^c + \Delta\mathbf{u}_{ref}^c)e^{j(\theta_1+\Delta\theta)}. \quad (5)$$

Linearizing (5), the relation between the deviation variables in the two dq frames becomes

$$\Delta\mathbf{u}_{ref} = \left(\Delta u_{refd}^c - u_{refq0}^c\Delta\theta\right) + j\left(\Delta u_{refq}^c + u_{refd0}^c\Delta\theta\right). \quad (6)$$

B. PLL

A conventional synchronous-reference-frame PLL [13] is considered. The PLL contains a proportional-integral (PI) controller, the input of the PI is the imaginary part of $\mathbf{u}_c^c$. Therefore, the PLL input signal is the imaginary part of

$$\Delta\mathbf{u}_c^c = u_{cq0}\Delta\theta + \Delta u_{cd} + j(\Delta u_{cq} - u_{cd0}\Delta\theta). \quad (7)$$

The output of the PI controller is the angular frequency deviation $\Delta\omega$ and afterwards the signal is integrated into $\Delta\theta$. The PLL open-loop transfer function yields

$$G_{pllol}(s) = F_{pll}(s)\frac{1}{s} \quad (8)$$

where $F_{pll}(s) = k_{ppll} + \frac{1}{sT_{ipll}}$.

The PLL closed-loop transfer function is obtained as:

$$\Delta\theta = \frac{G_{pllol}(s)}{1 + u_{cd0}G_{pllol}(s)}\Delta u_{cq}. \quad (9)$$

C. CC

The CC contains a PI controller and a converter-bus voltage feedforward via a low-pass filter [3]. The following relation is obtained for the deviation variables:

$$\Delta\mathbf{u}_{ref}^c = H_{vff}(s)\Delta\mathbf{u}_c^c - F_{cc}(s)(\Delta\mathbf{i}_{ref}^c - \Delta\mathbf{i}_c^c) \quad (10)$$

where $F_{cc}(s) = k_{pcc} + \frac{1}{sT_{icc}}$ and $H_{vff}(s) = \frac{1}{sT_{vff}+1}$.

D. DVC

The objective is to control the converter dc-link voltage. The DVC output gives the d-axis current reference. The DVC uses a PI controller cascaded with a low-pass filter. The low-pass filter is to suppress the noise in the dc-bus voltage. The PI controller regulates the error of the square value of the dc-bus voltage [3], which is proportional to the total stored energy. The d-axis current references is given as

$$i_{refd}^c = F_{udc}(s)[u_{dcref}^2 - H_{udclp}(s)u_{dc}^2] \quad (11)$$

where $F_{udc}(s) = k_{pudc} + \frac{1}{sT_{iudc}}$ and $H_{udclp}(s) = \frac{1}{sT_{udclp}+1}$.

Equation (11) is linearized around the operating point u_{dc0}, which yields

$$\Delta i_{refd}^c = 2F_{udc}(s)[u_{dcref0}\Delta u_{dcref} - H_{udclp}(s)u_{dc0}\Delta u_{dc}]. \quad (12)$$

The stored energy and dc-bus voltage are related as

$$E = E_0 + \Delta E = \frac{1}{2}C_{dc}(u_{dc0} + \Delta u_{dc})^2. \quad (13)$$

Linearizing (13) yields

$$\Delta E = C_{dc}u_{dc0}\Delta u_{dc}. \quad (14)$$

The energy exchange for a three-phase system yields

$$E = \frac{\mathrm{Re}\{\mathbf{u}_{c0}^c \mathbf{i}_{c0}^{c*} + \mathbf{u}_{c0}^c \Delta \mathbf{i}_c^{c*} + \mathbf{i}_{c0}^{c*}\Delta \mathbf{u}_c^c + \Delta \mathbf{u}_c^c \Delta \mathbf{i}_c^{c*}\}}{sK} \quad (15)$$

where $K = 2/3$ is set as the peak-value space-vector scaling. Linearizing (15) yields

$$\Delta E = \frac{u_{cd0}^c \Delta i_{cd}^c + u_{cq0}^c \Delta i_{cq}^c + \Delta u_{cd}^c i_{cd0}^c + \Delta u_{cq}^c i_{cq0}^c}{sK}. \quad (16)$$

Substituting (14) into (12) further results in

$$\Delta i_{refd}^c = 2F_{udc}(s)\left[u_{dcref0}\Delta u_{dcref} - \frac{H_{udclp}(s)\Delta E}{C_{dc}}\right]. \quad (17)$$

E. AVC

It is assumed that the AVC regulates the d-axis component of the PCC voltage. The output of this controller produces the q-axis current reference. A PI controller is used and a low-pass filter (with the time constant T_{upcclp}) is applied on the PCC voltage to remove harmonics. The q-axis current reference deviation can be described as

$$\Delta i_{refq}^c = F_{uac}(s)\left[\Delta u_{acref} - H_{upcclp}(s)\Delta u_{pccd}^c\right] \quad (18)$$

where $F_{uac}(s) = k_{puac} + \frac{1}{sT_{iuac}}$ and $H_{upcclp}(s) = \frac{1}{sT_{upcclp}+1}$.

III. MIMO SYSTEM AND SISO TRANSFER FUNCTIONS

In this section, MIMO closed-loop transfer functions and expressions for the d- and q-axes current derivations are firstly derived with CC, PLL and the grid impedance. Unlike the input-admittance-based approach in [3], the grid impedance is necessary to be included in the PLL transfer functions. Furthermore, the current reference deviations of the DVC (17) and the AVC (18) are integrated with the derived transfer functions and expressions, to construct the complete MIMO model. The detailed derivations of the transfer functions and expressions are given in Section VI, since they are fairly complex.

Fig. 2 (a) shows the block diagrams with input and output signals of the individual transfer functions $[G_{d1}(s), G_{d2}(s), G_{q1}(s), \text{ and } G_{q2}(s)]$ for the MIMO system without the DVC or the AVC effect. The input signals are the current reference deviation Δi_{refd}^c or Δi_{refq}^c and the cross-coupling current deviation from the other axis Δi_{cq}^c or Δi_{cd}^c. The output signal is the corresponding converter current deviation Δi_{cd}^c or Δi_{cq}^c. The overall closed-loop system is a two-input (Δi_{refd}^c and Δi_{refq}^c), two-output (Δi_{cd}^c and Δi_{cq}^c) system.

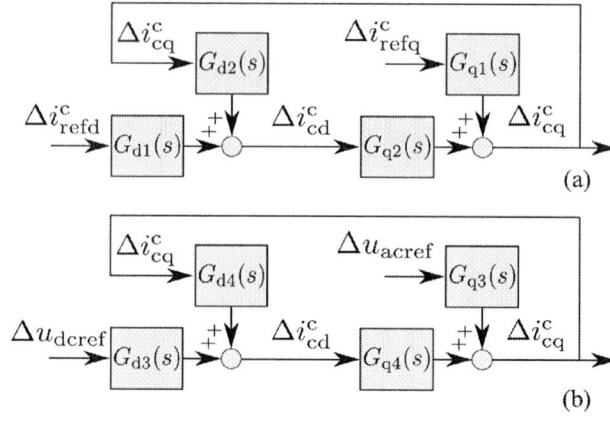

Fig. 2. Block diagrams of the MIMO closed-loop system. (a) d- and q-axis current deviation with CC, PLL and grid impedance. (b) d- and q-axis current deviation with CC, PLL, grid impedance, DVC and AVC.

For the system shown in Fig. 2 (a), the dq-axes current derivations are obtained respectively as

$$\Delta i_{cd}^c = G_{d1}(s)\Delta i_{refd}^c + G_{d2}(s)\Delta i_{cq}^c \quad (19)$$

and

$$\Delta i_{cq}^c = G_{q1}(s)\Delta i_{refq}^c + G_{q2}(s)\Delta i_{cd}^c. \quad (20)$$

In (19) and (20), the two closed loops interact with each other, via Δi_{cd}^c and Δi_{cq}^c.

Similarly, Fig. 2 (b) illustrates an MIMO system including the DVC and the AVC, the input signals of the individual transfer functions $[G_{d3}(s), G_{d4}(s), G_{q3}(s), \text{ and } G_{q4}(s)]$ are the dc-bus-voltage reference deviation Δu_{dcref} or the ac-bus-voltage reference deviation Δu_{acref} and the cross-coupling current deviation from the other axis Δi_{cq}^c or Δi_{cd}^c. The output signal is the corresponding converter current deviation Δi_{cd}^c or Δi_{cq}^c. The overall closed-loop system is a two-input (Δu_{dcref} and Δu_{acref}), two-output (Δi_{cd}^c and Δi_{cq}^c) system.

The above MIMO expression can be derived by combining (17) and (19), as well as combining (18) and (20). The closed-loop dq-axis current deviations with the DVC and the AVC impacts are derived respectively. Thus

$$\Delta i_{cd}^c = G_{d3}(s)\Delta u_{dcref} + G_{d4}(s)\Delta i_{cq}^c \quad (21)$$

and

$$\Delta i_{cq}^c = G_{q3}(s)\Delta u_{acref} + G_{q4}(s)\Delta i_{cd}^c. \quad (22)$$

The two closed loops interact with each other, via Δi_{cd}^c and Δi_{cq}^c.

Furthermore, merging (21) and (22), the closed-loop dq-axes current deviations are obtained as

$$\Delta i_{cd}^c = G_{d5}(s)\Delta u_{dcref} + G_{d6}(s)\Delta u_{acref}, \quad (23)$$

3686

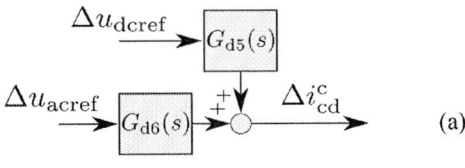

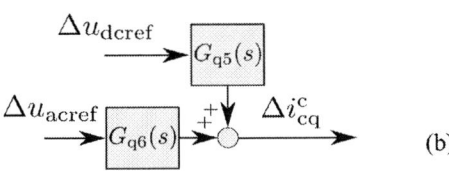

Fig. 3. Block diagrams of the MIMO closed-loop system. (a) d-axis current deviation with CC, PLL, grid impedance, DVC and AVC. (b) q-axis current deviation with CC, PLL, grid impedance, DVC and AVC.

where

$$G_{d5}(s) = \frac{G_{d3}(s)}{1 - G_{d4}(s)G_{q4}(s)}$$

$$G_{d6}(s) = \frac{G_{d4}(s)G_{q3}(s)}{1 - G_{d4}(s)G_{q4}(s)}$$

and

$$\Delta i_{cq}^c = G_{q5}(s)\Delta u_{dcref} + G_{q6}(s)\Delta u_{acref}, \quad (24)$$

where

$$G_{q5}(s) = \frac{G_{d3}(s)G_{q4}(s)}{1 - G_{d4}(s)G_{q4}(s)}$$

$$G_{q6}(s) = \frac{G_{q3}(s)}{1 - G_{d4}(s)G_{q4}(s)}.$$

In (23) and (24), the two closed loops now share the same inputs, i.e., Δu_{dcref} and Δu_{acref}, See Fig. 3(a) and (b). Unlike in the previous equations (19) to (22), the two closed loops in (23) and (24) do not have signals to interact with each other.

Obviously, the overall system stability can thus be analyzed by four individual closed-loop transfer functions $G_{d5}(s)$, $G_{d6}(s)$, $G_{q5}(s)$, and $G_{q6}(s)$. They are seen as SISO transfer functions. Note that all four SISO closed-loop transfer functions have the same denominator $1 - G_{d4}(s)G_{q4}(s)$. Consequently, it is possible to apply a single Nyquist curve to analyze the stability problem for the entire system, including both d and q axes.

The common open-loop transfer function yields

$$G_{ol}(s) = -G_{d4}(s)G_{q4}(s), \quad (25)$$

which is a SISO open-loop transfer function.

IV. VERIFICATION

A. Reference-System Data

A 100-MVA static synchronous compensator (STATCOM) is used as a test system to verify the theory of Section III. The device is connected to a 33-kV (line-to-line rms) converter bus. The control uses peak-value space-vector scaling with the voltage and current base values $U_{base} = 26.94$ kV and $I_{base} = 2.47$ kA respectively. The impedance base value $Z_{base} = 10.9\ \Omega$. The dc-bus capacitance is $C_{dc} = 1.0$ mF and the phase-reactor and transformer inductances are $L_r = L_t = 0.1$ per unit (p.u.). The grid impedance is inductive with $L_g = 1$ p.u. for a weak grid. The nominal angular frequency is $\omega_1 = 100\pi$ rad/s. The dc-bus voltage reference $u_{dcref} = 1.2$ p.u.

B. Case Study

Two cases are considered for the same STATCOM. Both are for the weak-grid condition. In Section IV-B1, it is shown that the controller can be tuned with adequate phase and gain margins. Section IV-B2 demonstrates that the theory accurately predicts instability. Two design cases that respectively cause destabilization of the DVC closed loop and the AVC closed loop are considered.

1) Stable System: Using the method developed in Section III, the VSC controllers are tuned via trial and error so that adequate stability margins are obtained, yielding the following parameters: $k_{ppll} = 30.0$ rad/s, $T_{ipll} = 1$ s, $k_{pcc} = 1.0$ p.u., $T_{icc} = 100$ ms, $k_{pudc} = 1.0$ p.u., $T_{iudc} = 100$ ms, $k_{puac} = 0.1$ p.u., and $T_{iuac} = 10.0$ ms. The following choices of the filter time constant are found to be appropriate: $T_{vff} = 1.0$ ms, $T_{upcclp} = 5.0$ ms, and $T_{udclp} = 10.0$ ms. Unless otherwise mentioned, the above parameters are used in the remainder of the Section IV-B.

The operating point of STATCOM is chosen at its maximum loading condition, in order to represent the worst-case scenario [3]. When the grid is weak, it is reasonably assume that the STATCOM runs at its maximum capacitive reactive current $i_{cq0}^c = 0.1$ p.u. In Fig. 4, a Nyquist plot for the open-loop transfer functions $G_{ol}(s)$ is shown. The Nyquist curve does not encircle $-1 + j0$, so the corresponding closed-loop systems are stable, and $G_{ol}(s)$ has adequate stability margins.

2) Unstable System: Fig. 5 shows the Nyquist plots of $G_{ol}(s)$ with $k_{pudc} = 1.6$ p.u. and $k_{pudc} = 1.7$ p.u., respectively. We can see that the Nyquist curve of $G_{ol}(s)$ with $k_{pudc} = 1.6$ does not encircle $-1 + j0$, and certain margins are still left to the stability boundary. On the other hand, with $k_{pudc} = 1.7$ p.u., an encirclement of $-1 + j0$ is found, i.e., the closed loop is unstable. In order to validate this in the time domain, a 0.01-p.u. step in u_{dcref} is applied at $t = 2$ s, giving the responses shown in Fig. 6 and Fig. 7. Converging and growing oscillations in the dc-bus voltage are observed respectively. The phase currents distortions are also shown.

An inappropriate AVC gain can also destabilize the system. In Fig. 8, $k_{puac} = 1.2$ p.u. and $k_{puac} = 1.4$

The 2018 International Power Electronics Conference

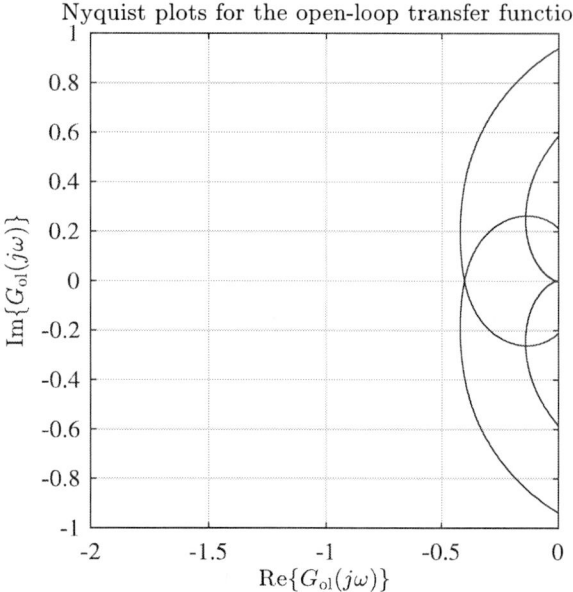

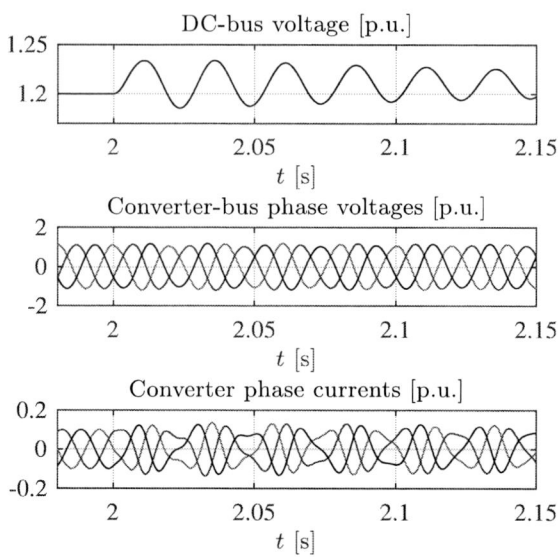

Fig. 4. Nyquist plot for the open-loop transfer functions $G_{ol}(s)$ in a weak-grid condition.

Fig. 6. Responses to a 0.01-p.u. step in u_{dcref}. The converter operates with $k_{pudc} = 1.6$ p.u. in a weak-grid condition.

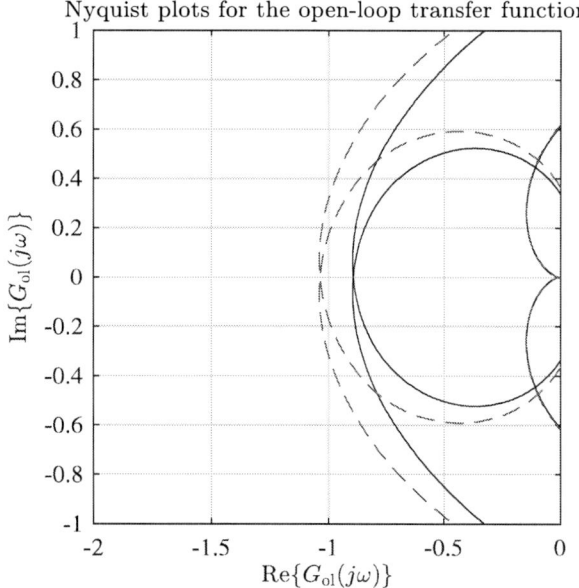

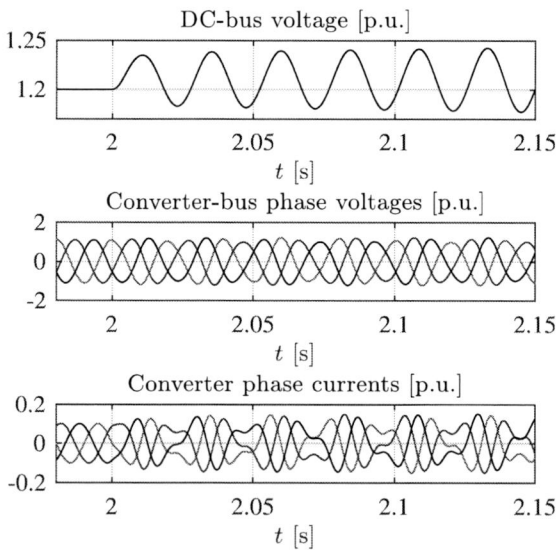

Fig. 5. Nyquist plots for the open-loop transfer functions $G_{ol}(s)$. The converter operates with $k_{pudc} = 1.6$ p.u. (blue curve) and $k_{pudc} = 1.7$ p.u. (red-dashed curve) in a weak-grid condition.

Fig. 7. Responses to a 0.01-p.u. step in u_{dcref}. The converter operates with $k_{pudc} = 1.7$ p.u. in a weak-grid condition.

p.u. are used to represent the stable and unstable cases respectively. The time-domain simulations show a stable and an unstable closed-loop system with $k_{puac} = 1.2$ p.u., $k_{puac} = 1.4$ p.u. respectively, see Fig. 9 and Fig. 10.

V. CONCLUSION

In this paper, the VSC and the grid are linearized as a MIMO system. The linearized model of the CC,

PLL, DVC, and AVC are included. These functions are embedded with the grid impedance into this MIMO system.

It is shown that a common SISO open-loop transfer function $G_{ol}(s)$ can be identified from the SISO closed-loop transfer functions in the derived MIMO system. This enables the system stability can be examined with a single Nyquist curve by the standard Nyquist stability criterion for both axes. The overall system stability can be guar-

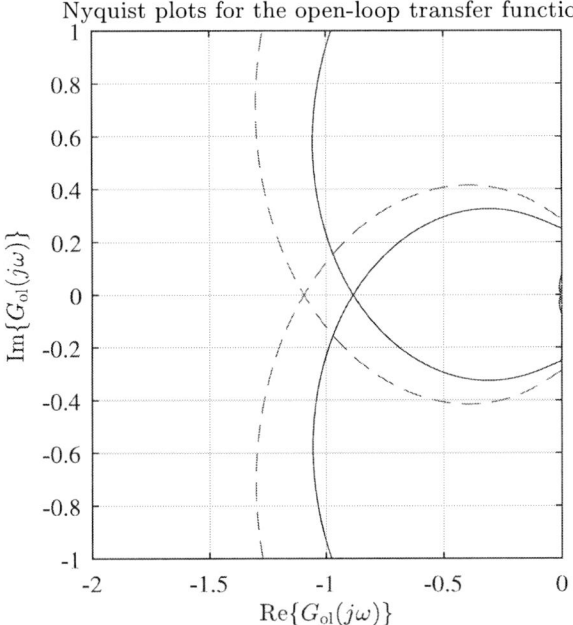

Nyquist plots for the open-loop transfer function

Fig. 8. Nyquist plots for the open loop transfer functions $G_{\text{ol}}(s)$. The converter operates with $k_{\text{puac}} = 1.2$ p.u. (blue curve) and $k_{\text{puac}} = 1.4$ p.u. (red-dashed curve) in a weak-grid condition.

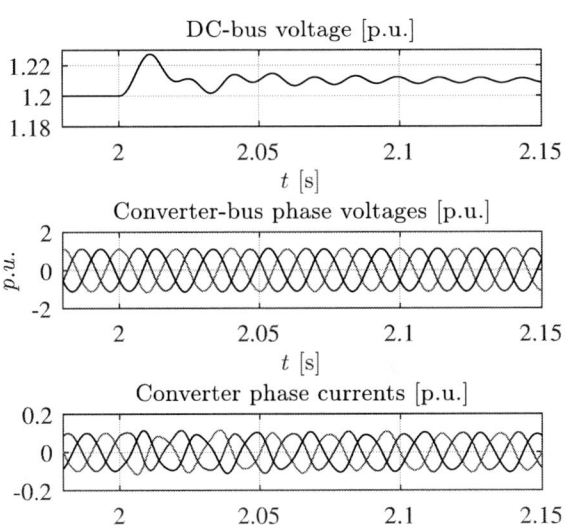

Fig. 9. Responses to a 0.01-p.u. step in u_{dcref}. The converter operates with $k_{\text{puac}} = 1.2$ p.u. in a weak-grid condition.

anteed by proper tuning of the controller parameters. In the end, verifications against the time-domain simulations show excellent agreement.

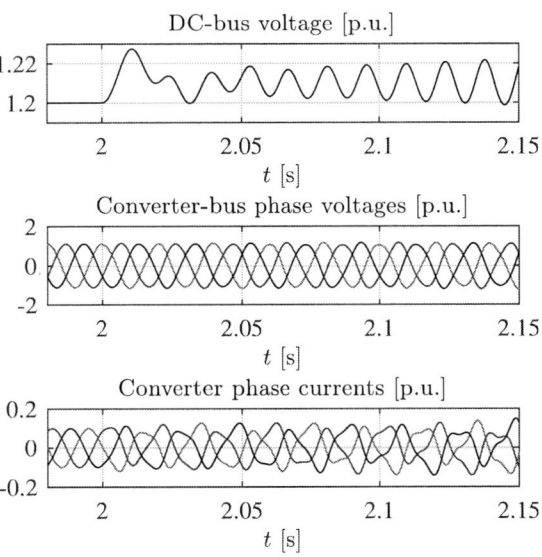

Fig. 10. Responses to a 0.01-p.u. step in u_{dcref}. The converter operates with $k_{\text{puac}} = 1.4$ p.u. in a weak-grid condition.

VI. Appendix

From Fig. 1, the $\alpha\beta$-frame converter current yields

$$\mathbf{i}_c^s = \frac{\mathbf{u}_c^s - \mathbf{u}_{\text{ref}}^s}{sL_{\text{r}}}. \tag{26}$$

Transforming the variables from $\alpha\beta$ frame to the dq frame implies that the derivative operator s is substituted by $s + j\omega_1$. The deviation variables in (26) is transformed as

$$\Delta\mathbf{i}_c = \frac{\Delta\mathbf{u}_c - \Delta\mathbf{u}_{\text{ref}}}{(s + j\omega_1)L_{\text{r}}}. \tag{27}$$

Since the CC operates in the converter dq frame, we also need to transform (27) as in (4)

$$\Delta\mathbf{i}_c^c = -j\mathbf{i}_{c0}\Delta\theta + \frac{\Delta\mathbf{u}_c - \Delta\mathbf{u}_{\text{ref}}}{(s + j\omega_1)L_{\text{r}}}. \tag{28}$$

Using (6) and (10), the grid-dq-frame voltage reference is obtained as

$$\Delta\mathbf{u}_{\text{ref}} = \begin{bmatrix} H_{\text{vff}}(s)\Delta u_{\text{cd}}^c - F_{\text{cc}}(s)(\Delta i_{\text{cdref}}^c - \Delta i_{\text{cd}}^c) \\ -u_{\text{refq0}}^c\Delta\theta \end{bmatrix}$$
$$+j\begin{bmatrix} H_{\text{vff}}(s)\Delta u_{\text{cq}}^c - F_{\text{cc}}(s)(\Delta i_{\text{cqref}}^c - \Delta i_{\text{cq}}^c) \\ +u_{\text{refd0}}^c\Delta\theta \end{bmatrix}. \tag{29}$$

From Fig. 1, for a pure inductive grid, we further obtain

$$\mathbf{u}_c^s = \mathbf{u}_s^s - sL_{\text{ts}}\mathbf{i}_c^s \tag{30}$$

where $L_{\text{ts}} = L_{\text{t}} + L_{\text{s}}$. L_{t}, L_{s}, and L_{ts} are the transformer leakage inductance, grid inductance, and the sum of these two inductance, respectively. Equation (30) is now transformed to the dq frame by substituting $s \rightarrow s + j\omega_1$. For the deviations variables we obtain

$$\Delta u_{\text{cd}} + j\Delta u_{\text{cq}} = \Delta\mathbf{u}_s - (s + j\omega_1)L_{\text{ts}}(\Delta i_{\text{cd}} + j\Delta i_{\text{cq}}) \tag{31}$$

where $\Delta\mathbf{u}_s = 0$ since the infinite-bus voltage $\mathbf{u}_s$ is stiff. Substituting (3) into (31), then splitting (31) into its real and imaginary parts yields

$$\Delta u_{cd} = -\left[Z_d(s)\Delta i_{cd}^c + Z_q(s)\Delta i_{cq}^c + p_d(s)\Delta\theta\right], \quad (32)$$

$$\Delta u_{cq} = -\left[Z_d(s)\Delta i_{cq}^c - Z_q(s)\Delta i_{cd}^c + p_q(s)\Delta\theta\right]. \quad (33)$$

where

$$Z_d(s) = sL_{ts}, Z_q(s) = -\omega_1 L_{ts}$$
$$p_d(s) = Z_q(s)i_{cd0} - Z_d(s)i_{cq0}$$
$$p_q(s) = Z_d(s)i_{cd0} + Z_q(s)i_{cq0}.$$

Substituting (33) into (9), we can include the grid impedance in the expression for the PLL deviation angle as

$$\Delta\theta = k_{\Delta\theta}(s)\left[Z_d(s)\Delta i_{cq}^c - Z_q(s)\Delta i_{cd}^c\right] \quad (34)$$

where

$$k_{\Delta\theta}(s) = \frac{-G_{\text{pllol}}(s)}{1 + u_{cd0}G_{\text{pllol}}(s) + G_{\text{pllol}}(s)p_q(s)}$$

Based on (7), the substitution of Δu_{cd}^c and Δu_{cq}^c can be made in (29). Merging (28), (29), (32), (33) and (34), the d-axis current deviation can be expressed as

$$\Delta i_{cd}^c = G_{d1}(s)\Delta i_{refd}^c + G_{d2}(s)\Delta i_{cq}^c \quad (35)$$

where

$$G_{d1}(s) = \frac{F_{cc}(s)}{c_d(s)} \quad G_{d2}(s) = \frac{b_d(s)}{c_d(s)}$$
$$a_d(s) = -H_{vff}(s)u_{cq0} + u_{refq0}^c + sL_r i_{cq0} + \omega_1 L_r i_{cd0}$$
$$b_d(s) = \begin{array}{l} -[1 - H_{vff}(s)][Z_q(s) + p_d(s)k_{\Delta\theta}(s)Z_d(s)] \\ +a_d(s)k_{\Delta\theta}(s)Z_d(s) + \omega_1 L_r \end{array}$$
$$c_d(s) = \begin{array}{l} sL_r + F_{cc}(s) + a_d(s)k_{\Delta\theta}(s)Z_q(s) \\ +[1 - H_{vff}(s)][Z_d(s) - p_d(s)k_{\Delta\theta}(s)Z_q(s)]. \end{array}$$

Similarly, the q-axis current deviation yields

$$\Delta i_{cq}^c = G_{q1}(s)\Delta i_{refq}^c + G_{q2}(s)\Delta i_{cd}^c \quad (36)$$

where

$$G_{q1}(s) = \frac{F_{cc}(s)}{c_q(s)} \quad G_{q2}(s) = \frac{b_q(s)}{c_q(s)}$$
$$a_q(s) = H_{vff}(s)u_{cd0} - u_{refd0}^c - sL_r i_{cd0} + \omega_1 L_r i_{cq0}$$
$$b_q(s) = \begin{array}{l} [1 - H_{vff}(s)][1 + p_q(s)k_{\Delta\theta}(s)]Z_q(s) \\ -a_q(s)k_{\Delta\theta}(s)Z_q(s) - \omega_1 L_r \end{array}$$
$$c_q(s) = \begin{array}{l} sL_r + F_{cc}(s) - a_q(s)k_{\Delta\theta}(s)Z_d(s) \\ +[1 - H_{vff}(s)][1 + p_q(s)k_{\Delta\theta}(s)]Z_d(s). \end{array}$$

Substituting (32) and (33) into (16) and (17), and substituting the resulting equation into (19), this gives the d-axis current deviation with the converter DVC impact.

$$\Delta i_{cd}^c = G_{d3}(s)\Delta u_{dcref} + G_{d4}(s)\Delta i_{cq}^c \quad (37)$$

where

$$G_{d3}(s) = \frac{2F_{cc}(s)F_{udc}(s)u_{dcref0}}{w(s)} \quad G_{d4}(s) = \frac{d(s)}{w(s)}$$
$$d(s) = b_d(s) - 2F_{cc}(s)F_{udc}(s)H_{udclp}(s)\frac{u_{cq0}^c + g(s) - h(s)}{sKC_{dc}}$$
$$w(s) = c_d(s) + 2F_{cc}(s)F_{udc}(s)H_{udclp}(s)\frac{u_{cd0}^c + m(s) + n(s)}{sKC_{dc}}$$
$$g(s) = \{k_{\Delta\theta}(s)Z_d(s)[u_{cq0} - p_d(s)] - Z_q(s)\}i_{cd0}^c$$
$$h(s) = \{k_{\Delta\theta}(s)Z_d(s)[u_{cd0} + p_d(s)] + Z_d(s)\}i_{cq0}^c$$
$$m(s) = -\{k_{\Delta\theta}(s)Z_q(s)[u_{cq0} - p_d(s)] + Z_d(s)\}i_{cd0}^c$$
$$n(s) = \{k_{\Delta\theta}(s)Z_q(s)[u_{cd0} + p_d(s)] + Z_q(s)\}i_{cq0}^c.$$

Similarly, the d- and q- axes PCC bus voltage deviations are given as

$$\Delta u_{pccd} = -\left[Z_{gd}(s)\Delta i_{cd}^c + Z_{gq}(s)\Delta i_{cq}^c + r_d(s)\Delta\theta\right] \quad (38)$$
$$\Delta u_{pccq} = -\left[Z_{gd}(s)\Delta i_{cq}^c - Z_{gq}(s)\Delta i_{cd}^c + r_q(s)\Delta\theta\right] \quad (39)$$

where

$$Z_{gd}(s) = sL_s, Z_{gq}(s) = -\omega_1 L_s$$
$$r_d(s) = Z_{gq}(s)i_{cd0} - Z_{gd}(s)i_{cq0}$$
$$r_q(s) = Z_{gd}(s)i_{cd0} + Z_{gq}(s)i_{cq0}.$$

Considering (4) and (18), the q-axis current reference yields

$$\Delta i_{refq}^c = F_{uac}(s)\Delta u_{acref}$$
$$+F_{uac}(s)H_{upcclp}(s)\left\{ \begin{array}{l} [r_d(s) - u_{pccq0}]\Delta\theta \\ +Z_{gd}(s)\Delta i_{cd}^c \\ +Z_{gq}(s)\Delta i_{cq}^c \end{array} \right\}. \quad (40)$$

We can substitute (40) into (20), which yields

$$\Delta i_{cq}^c = G_{q3}(s)\Delta u_{acref} + G_{q4}(s)\Delta i_{cd}^c \quad (41)$$

where

$$G_{q3}(s) = \frac{F_{cc}(s)F_{uac}(s)}{z(s)} \quad G_{q4}(s) = \frac{y(s)}{z(s)}$$
$$y(s) = b_q(s) + \left\{ \begin{array}{l} Z_{gd}(s) \\ -Z_q(s)k_{\Delta\theta}(s) \\ [r_d(s) - u_{pccq0}] \end{array} \right\} F_{cc}(s)H_{upcclp}(s)F_{uac}(s)$$
$$z(s) = c_q(s) - \left\{ \begin{array}{l} Z_{gq}(s) \\ +Z_d(s)k_{\Delta\theta}(s) \\ [r_d(s) - u_{pccq0}] \end{array} \right\} F_{cc}(s)H_{upcclp}(s)F_{uac}(s).$$

REFERENCES

[1] C. Li, "Unstable operation of photovoltaic inverter from field experiences," *IEEE Transactions on Power Delivery*, vol. PP, no. 99, pp. 1–1, 2017.

[2] C. Buchhagen, C. Rauscher, A. Menze, and J. Jung, "Borwin1 - first experiences with harmonic interactions in converter dominated grids," in *International ETG Congress 2015; Die Energiewende - Blueprints for the new energy age*, Nov. 2015, pp. 1–7.

[3] L. Harnefors, M. Bongiorno, and S. Lundberg, "Input-admittance calculation and shaping for controlled voltage-source converters," *IEEE Transactions on Industrial Electronics*, vol. 54, no. 6, pp. 3323–3334, Dec. 2007.

[4] J. Sun, "Impedance-based stability criterion for grid-connected inverters," *IEEE Transactions on Power Electronics*, vol. 26, no. 11, pp. 3075–3078, Nov. 2011.

[5] N. Pogaku, M. Prodanovic, and T. C. Green, "Modeling, analysis and testing of autonomous operation of an inverter-based microgrid," *IEEE Transactions on Power Electronics*, vol. 22, no. 2, pp. 613–625, Mar. 2007.

[6] S. Danielsen, O. B. Fosso, and T. Toftevaag, "Use of participation factors and parameter sensitivities in study and improvement of low-frequency stability between electrical rail vehicle and power supply," in *Proc. 2009 13th European Conference on Power Electronics and Applications*, Sep. 2009, pp. 1–10.

[7] B. Wen, D. Boroyevich, R. Burgos, P. Mattavelli, and Z. Shen, "Analysis of d-q small-signal impedance of grid-tied inverters," *IEEE Transactions on Power Electronics*, vol. 31, no. 1, pp. 675–687, Jan. 2016.

[8] X. Wang, L. Harnefors, and F. Blaabjerg, "A unified impedance model of grid-connected voltage-source converters," *IEEE Transactions on Power Electronics*, vol. PP, no. 99, pp. 1–1, 2017.

[9] Y. Huang, X. Yuan, J. Hu, and P. Zhou, "Modeling of VSC connected to weak grid for stability analysis of DC-link voltage control," *IEEE Journal of Emerging and Selected Topics in Power Electronics*, vol. 3, no. 4, pp. 1193–1204, Dec. 2015.

[10] Y. Huang, X. Yuan, J. Hu, P. Zhou, and D. Wang, "DC-bus voltage control stability affected by AC-bus voltage control in VSCs connected to weak AC grids," *IEEE Journal of Emerging and Selected Topics in Power Electronics*, vol. 4, no. 2, pp. 445–458, June. 2016.

[11] L. Harnefors, A. Antonopoulos, S. Norrga, L. Angquist, and H. P. Nee, "Dynamic analysis of modular multilevel converters," *IEEE Transactions on Industrial Electronics*, vol. 60, no. 7, pp. 2526–2537, July. 2013.

[12] M. Beza, M. Bongiorno, and G. Stamatiou, "Analytical derivation of the ac-side input admittance of a modular multilevel converter with open- and closed-loop control strategies," *IEEE Transactions on Power Delivery*, vol. PP, no. 99, pp. 1–1, 2017.

[13] V. Kaura and V. Blasko, "Operation of a phase locked loop system under distorted utility conditions," *IEEE Transactions on Industry Applications*, vol. 33, no. 1, pp. 58–63, Jan. 1997.

The 2018 International Power Electronics Conference

A Communication-independent Reactive Power Sharing Scheme with Adaptive Virtual Impedance for Parallel Connected Inverters

Ronghui AN[*], Zeng LIU, Jinjun LIU and Shike WANG
State Key Lab of Electrical Insulation and Power Equipment, School of Electrical Engineering
Xi'an Jiaotong University
Xi'an, China
*Email: an_ronghui@163.com

Abstract—Droop control method has been widely applied to achieve equal power sharing among distributed generations in microgrids. However, the reactive power sharing performance is barely satisfactory for the absence of sharing accuracy under conditions of imbalanced line impedances. This paper proposed a communication-independent reactive power sharing scheme for parallel connected inverters. In this scheme, constant voltage control is applied to inverters, and the changing rate of virtual impedance is drooped with reactive power, hence accurate reactive power sharing could be achieved with no impact on voltage regulation performance. In addition, since the virtual impedance is adaptive to changing of reactive power, extra information from PCC voltage, line impedance or other communications is not needed in the implementation. The effectiveness of the proposed method is verified by simulations.

Keywords—adaptive virtual impedance, communication-independent, droop control, reactive power sharing

I. INTRODUCTION

With the growth of power demands in stand-alone power systems such as microgrids, single inverter-interfaced distributed generation (DG) can hardly satisfy the needs of higher system capacity and reliability in many conditions. Therefore, parallel connected inverters become an effective solution and have been widely used in practice. Considering the distributed geographical locations of DGs in microgrids, droop control method, which can achieve equal power sharing automatically among DGs without the help of communications, is widely used for the coordinated control of parallel connected inverters in microgrids[1-3].

However, the conventional droop control has drawbacks such as deviations in frequency and voltage amplitude, poor reactive power sharing performance and instability caused by coupling between the control of active and reactive power, so to some extent, the feasibility and reliability of droop controlled systems are degraded.

To improve the reactive power sharing performance of droop control, many solutions have been proposed. In [4],

a small AC voltage signal is injected into the system as a control signal, and the voltage amplitude is drooped with active power of this small signal, so the reactive power can be regulated by controlling the frequency of this small signal. However, injection and measurement of this small high-frequency control signal make this method complicated to implement. Then Q-V_{pcc} based droop control and adaptive power sharing method are respectively proposed in [5, 6] and [7]. With the help of communications, the impact of line impedance imbalance can indeed be compensated, but either the PCC voltage or the accurate value of line impedance is required, which limits the application occasions of them. Then, based on the motivation to share the active and reactive power droop in an equivalent manner, the V dot droop control is proposed in [8]. However, this method will drive the DGs into different operating points and there is no mechanism to guarantee the voltage amplitude of islanded microgrid in an appropriate range. In [9], an adaptive virtual impedance scheme is introduced to improve the reactive power sharing. Nevertheless, the virtual impedance is not completely self-adapting because it still relies on communication links.

With the consideration of cost and reliability, reactive power sharing scheme which does not rely on communications may be a preferable solution for microgrids. Reconsidering the effect of conventional droop control on system performance, a communication-independent reactive power sharing scheme with adaptive virtual impedance is proposed in this paper. Through this modified droop control method, accurate reactive power sharing among DGs could be achieved with no impact on voltage regulation performance. Since the virtual impedance is adaptive to changing of reactive power, extra information from PCC voltage, line impedance or other communications is not needed. In addition, adaptive virtual impedances only need to be adjusted until system reaches steady state and then remain unchanged under different load conditions. Finally, simulation results based on a two-parallel-inverter system are provided to verify the effectiveness of the proposed method.

This work was supported by the National Natural Science Foundation of China under Grant 51437007.

II. CONVENTIONAL DROOP CONTROL METHOD

In general, the parallel connected inverters in microgrids are droop controlled as voltage sources, and the references of voltage amplitude and frequency for inner loops can be calculated by (1).

$$
\begin{cases}
\omega = \omega_0 - m \cdot [\dfrac{\omega_f}{s + \omega_f} \cdot P - P_0] & \text{(a)} \\[3mm]
E = E_0 - n \cdot [\dfrac{\omega_f}{s + \omega_f} \cdot Q - Q_0] & \text{(b)}
\end{cases}
\tag{1}
$$

where ω and E are the reference values generated by droop control method, P and Q are respectively the calculated real and reactive power, while P_0 and Q_0 are respectively the rated real and reactive power. And m and n (defined as positive) are the droop coefficients. ω_f is the cutoff frequency of the low-pass filters (LPFs).

In order to analyze the power sharing performance in steady state, the dynamics of outer voltage loop and inner current loop could be neglected. Then the simplified control block diagram of droop control in steady state can be established, as depicted in Figs. 1 and 2.

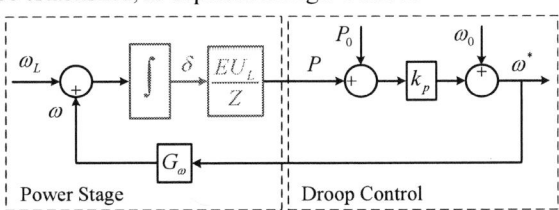

Fig. 1. Simplified control block diagram of active power.

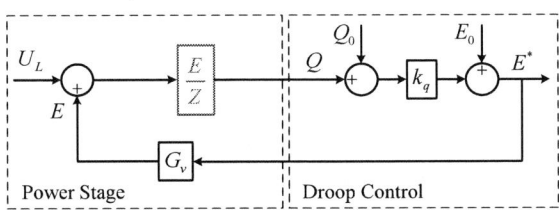

Fig. 2. Simplified control block diagram of reactive power.

As for the active power-frequency (*P-f*) droop in Fig. 1, because of the existence of integral function in forward path, the difference between PCC rotating speed ω_L and DG rotating speed ω can be eliminated in the steady state, which leads to the identical ω and P of different DGs. And the line impedance Z can only affect the integrating rate. Therefore, on the one hand, the *P-f* droop works as mechanism to guarantee accurate active power sharing; on the other hand, it is also the synchronization mechanism between different DGs, similar to the inherent droop characteristics of synchronous generators.

However, compared with *P-f* droop, the reactive power-voltage amplitude (*Q-V*) droop in Fig. 2 seems barely satisfactory. The output reactive power of each DG depends on both the voltage amplitude drop $E\text{-}U_L$ and line impedance, which makes the system settle down at an undesirable and unbalanced steady state, as described in (2).

$$
Q = \frac{E(E - U_L)}{Z}
\tag{2}
$$

where E is DG output voltage amplitude, U_L is the PCC voltage amplitude.

Reconsidering the performance of conventional *Q-V* droop control, we could find that, due to the inherent trade-off between sharing accuracy and voltage regulation rate, the improvement of reactive power sharing comes at the price of deviation in voltage amplitude. Nevertheless, when line impedances turn to be different, it still couldn't realize accurate reactive power sharing, although it has already jeopardized the voltage regulation. This drawback is also illustrated in Fig. 3.

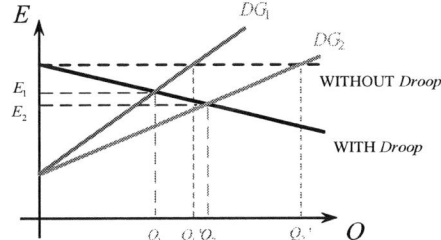

Fig. 3. Constant voltage control and *Q-V* droop control.

In Fig. 3, constant voltage control was compared with *Q-V* droop control under when line impedances are imbalanced, and it can be regarded as a specific condition that the droop coefficient n equals to 0.

In addition, the modeling of inverter-based microgrid is tested in [10]. Our test result shows that the dominant eigenvalues of the system are insensitive to the *Q-V* droop coefficient n, and stability problems would not occur when n equals to 0. Therefore, the *Q-V* droop could be replaced by constant voltage control to reduce voltage deviation if accurate reactive power sharing can be achieved through other methods [11].

Simulation result when *Q-V* droop is removed is provided in Fig. 4.

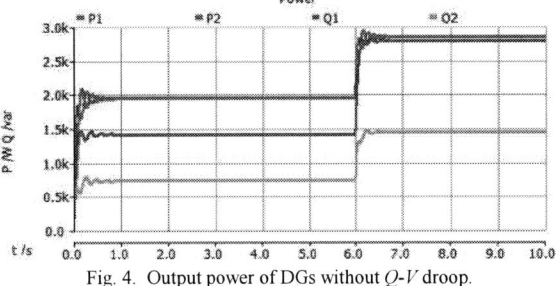

Fig. 4. Output power of DGs without *Q-V* droop.

In the simulation, the line impedances were respectively set to be 1 mH and 2 mH. The waveforms of reactive power Q_i of inverter #i (i=1, 2) illustrate that the Q_i are inversely proportional to the line impedance under constant voltage control (for the DG output impedances can be neglected) and the absence of *Q-V* droop would not cause instability. Therefore, to overcome the impact of line impedances on the reactive power sharing, a more accurate sharing scheme needs to be introduced

III. REVIEW ON $Q-\dot{V}$ DROOP CONTROL METHOD

To improve the performance of conventional droop, the $Q-\dot{V}$ droop control was proposed in [8], where $\dot{V}$ represents the time rate of change of inverter output voltage amplitude V. In this method, considering the reason why active power can be shared accurately, $\dot{V}$ instead of V is drooped with reactive power to mimic the $P\text{-}f$ droop, and then as a result, the active and reactive power droop could be shared in an equivalent manner.

The $Q-\dot{V}$ droop control is expressed as follows:

$$\dot{E}_x = \dot{E}_{0x} - n_x(Q_x - Q_{0x}) \qquad (a)$$
$$E_x^* = E_{0x} + \int_t \dot{E}_x d\tau \qquad (b) \qquad (3)$$

where n_x is the droop coefficients, $\dot{E}_{0x}$ is the nominal value of $\dot{E}_x$, which is usually set to 0 V/s and E_{0x} is the nominal voltage amplitude

Compared with (1b), the $Q-\dot{V}$ droop control can realize accurate sharing in steady state, but it still has two limitations: One is the value of $\dot{V}$, different from ω, needs to be reset to zero to guarantee constant output voltage, so we need to restore it; the other one is the output voltage amplitudes should be in an acceptable range to ensure the operating conditions for critical loads, unfortunately, there is no mechanism to guarantee it in this method. In [8], a restoration procedure is implemented to deal with the former limitation, but the latter one still has not been solved yet.

Therefore, if the voltage deviation can be divided to two parts, the first one is the voltage drop on line impedance, which couldn't be compensated without extra information from PCC; the second one is the deviation introduced by droop control method, which hasn't been solved by the aforementioned methods but won't exist for constant voltage control.

IV. PROPOSED ADAPTIVE VIRTUAL IMPEDANCE METHOD

A. Adaptive Virtual Impedance

According to the analysis and discussion in the former sections, we can reach a conclusion that the conventional droop control can achieve neither the accurate power sharing nor the effective voltage regulation, while the $Q-\dot{V}$ droop control can share reactive power accurately but still have impacts on output voltage amplitude.

Then in this section, an adaptive virtual impedance method is proposed to achieve accurate power sharing and voltage regulation at the same time.

In this method, since in equation (2), the reactive power is positively correlated with the voltage amplitude drop but negatively correlated with line impedance, a positive linear relation between Q and $\dot{L}_v$ is established to replace the droop relation between Q and $\dot{V}$. Then the expression of adaptive virtual impedance is

$$\dot{L}_v = n(Q - Q_0) \qquad (4)$$

where $\dot{L}_v$ is the time rate of change of virtual impedance.

Then the value of virtual impedance can be derived as

$$L_v = L_{v0} + \int_t \dot{L}_v d\tau = L_{v0} + n\int_t (Q - Q_0)d\tau \qquad (5)$$

where L_{v0} is nominal virtual impedance.

In steady state, $\dot{L}_v$ will converge to an identical value under this mechanism, so accurate reactive power sharing can be achieved, just as $Q-\dot{V}$ droop control.

However, the difference is, for $Q-\dot{V}$ droop control, the voltage amplitude reference is changing according to Q, and finally the voltage drop is proportional to line impedance in steady state; while the adaptive virtual impedance will adjust automatically to achieve identical total impedance directly (the total impedance is the sum of line impedance and virtual impedance), although the line impedance information is unknown to the controller. And then based on the conclusion of section II, the reactive power under constant voltage control can be equally shared

The result could also be illustrated by the control block diagram of adaptive virtual impedance method, which is depicted in Fig. 5.

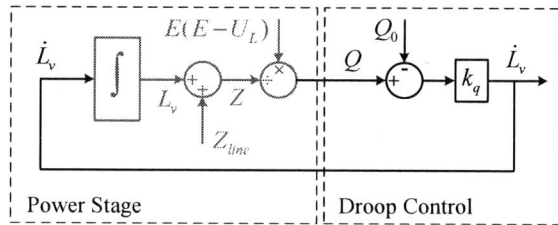

Fig. 5. Simplified control block diagram of proposed method.

Compare with Fig. 1, we could find that proposed method and $P\text{-}f$ droop both have an integral function in forward path, therefore, $\dot{L}_v$ of different DGs, just like ω, must be identical to reach steady state, thus we have

$$m_1(P_{01} - P_1) = m_2(P_{02} - P_2) = \cdots = m_k(P_{0k} - P_k)$$
$$n_1(Q_{01} - Q_1) = n_2(Q_{02} - Q_2) = \cdots = n_k(Q_{0k} - Q_k) \qquad (6)$$

By selecting coefficients m_i and n_i (i=1, 2 …k) as

$$m_1 P_{01} = m_2 P_{02} = \cdots = m_k P_{0k}$$
$$n_1 Q_{01} = n_2 Q_{02} = \cdots = n_k Q_{0k} \qquad (7)$$

the active and reactive power can be equally shared or proportionally shared between DGs by designing the droop coefficients appropriately.

B. Restoration and Locking Mechanism

Similar to $Q-\dot{V}$ droop control, the $\dot{L}_v$ also needs to be restore to zero to guarantee constant impedance, and the restoration mechanism can be described in (8)

The 2018 International Power Electronics Conference

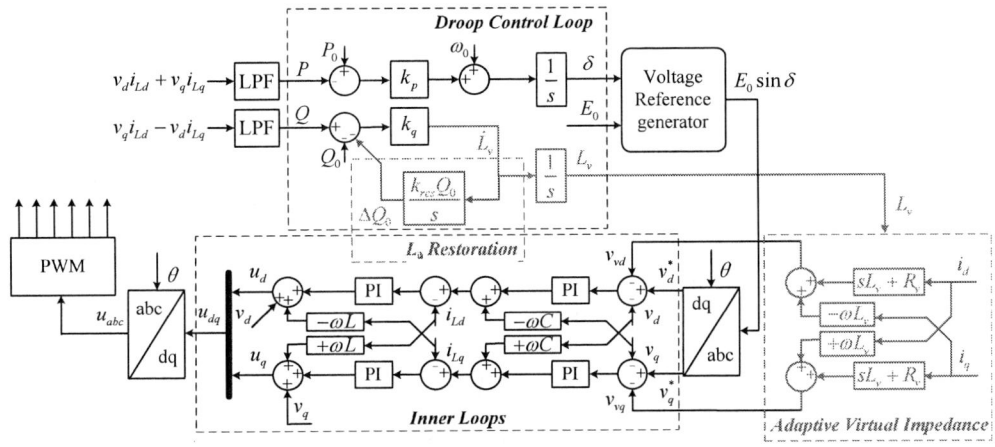

Fig. 6. Overall control block diagram of proposed droop control with adaptive virtual impedance.

$$\begin{cases} \dot{L}_v = n(Q - Q_0 - \Delta Q_0) \\ \Delta Q_0 = \int_t k_{res} Q_0 \dot{L}_v d\tau \end{cases} \tag{8}$$

where k_{res} is the restoration coefficient, and ΔQ_0 is the shift of rated reactive power[8].

It is assumed that the time constant of restoration process is much larger than droop control, so the output reactive power can be considered to be constant during the restoration process. Since the restoring speed is also proportional to the rated capacity of each DG, $\dot{L}_v$ can be gradually pushed to 0 and each virtual impedance reaches a fixed value at the same time.

Besides good voltage regulation performance, using adaptive virtual impedance has another advantage: For $Q - \dot{V}$ droop, the voltage amplitudes are always changing under different load conditions, while in proposed method, once the total impedances are settled down to be identical, there is no need to change them when load changes.

Therefore, a detecting mechanism is also designed to capture the steady state, and then the controller could remove the adaptive virtual impedance temporarily. As a result, the system works similarly to conditions of balanced line impedances, so reactive power sharing performance becomes insensitive to load changes. It is defined as locking mechanism

This locking mechanism is implemented also for the sake of stability, to avoid the extreme cases caused by accumulated errors in the controllers. And with the help of extra information from communication, the virtual impedances could be locked at optimal values further

In conclusion, the overall control block diagram of proposed method is shown in Fig. 6

V. SIMULATION RESULTS

PSCAD simulations were conducted to verify the effectiveness of proposed method. The schematic circuit diagram of simulations is shown in Fig. 7, and the values of some essential parameters are provided in Table I. In order to simplify the analysis, the reactive power capacities of DGs were assumed to be equal.

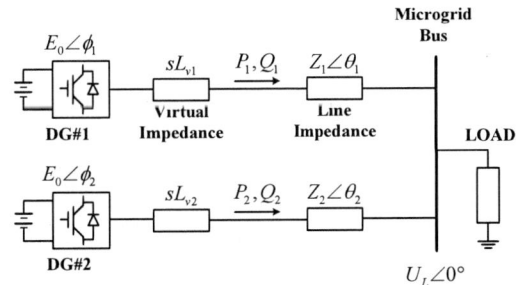

Fig. 7. Schematic circuit diagram of simulations.

TABLE I
CONTROL PARAMETERS IN SIMULATIONS

m	0.0001	n	1x 10^{-5}
E_0	283 V	f_0	50 Hz
P_0	2000 W	Q_0	1000 Var
R_{line}#1	0.2 ohm	R_{line}#2	0.2 ohm
L_{line}#1	1 mH	L_{line}#2	2 mH
k_{res}	200	L_{v0}	0 mH

A. Adaptive Virtual Impedance

Simulation waveforms of output power when implementing proposed method are shown in Fig. 8. At t = 2 s, adaptive virtual impedance mechanism is started. At t = 6 s, the load was increased from 3800 W + 2100 Var to 5700 W + 4200 Var.

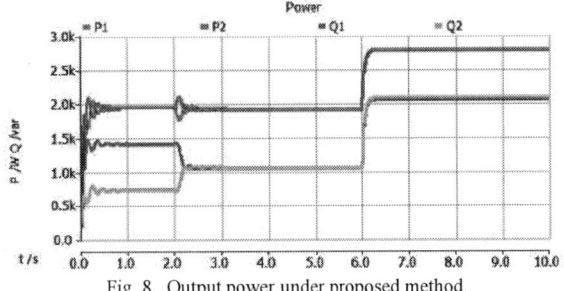

Fig. 8. Output power under proposed method.

3695

It can be seen from Fig. 8 that because of the effect of constant voltage control, the reactive power tend to be shared according to their line impedance at the beginning, and then sharing error occurs. However, after the implementation of proposed method, the error would directly cause the difference between changing rates of adaptive virtual impedances. For example, the virtual impedance of DG with smaller line impedance increases faster than that of DG with larger line impedance, and then the difference between total impedances was reduced to zero and the system gradually achieves accurate power sharing.

Fig. 9 shows simulation result of adaptive virtual impedances. We could find that, after reactive power curves converged in Fig. 8, it still takes some time to finish the restoration process. And then at around $t = 4$ s, the detecting mechanism captures the steady state of virtual impedance; the controller stops the adaptive virtual impedance and restoration mechanism.

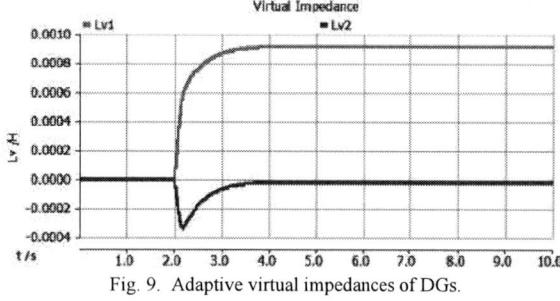

Fig. 9. Adaptive virtual impedances of DGs.

Finally, the system works under only P-f droop, similar to conditions of balanced line impedances, so reactive power sharing performance is barely affected by the load change later.

B. Conventional Droop Control

Simulation waveforms of output power when implementing conventional droop control are shown in Fig. 10. Under the same condition that, at $t = 2$ s, droop control is started, and at $t = 6$ s, the load was increased from 3800 W + 2100 Var to 5700 W + 4200 Var.

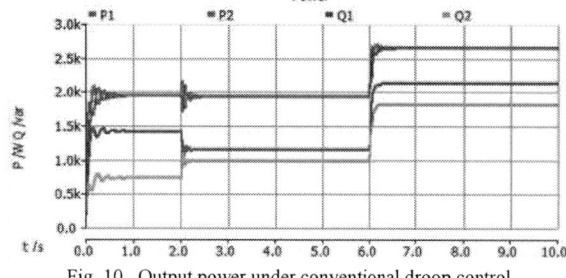

Fig. 10. Output power under conventional droop control.

Fig. 10 shows result when droop coefficient n is set to be 0.01. It could be concluded that under severe conditions of imbalanced line impedance, even if we increase the droop slope to such a large value, the reactive power can still not be equally shared, which is consistent with the analysis in section II.

In general, the proposed method avoids the voltage amplitude deviation introduced by conventional droop and Q-$\dot{V}$ droop control, and it can achieve accurate power sharing and voltage regulation at the same time.

VI. CONCLUSIONS

This paper investigates the effect of conventional droop control on microgrid system performance, and a communication-independent reactive power sharing scheme with adaptive virtual impedance is proposed based on the analysis. Through this modified droop control method, accurate reactive power sharing among DGs could be achieved with no impact on voltage regulation performance. Since the virtual impedance is adaptive to changing of reactive power, extra information from PCC voltage, line impedance or other communications is not needed in the implementation. In addition, adaptive virtual impedances only need to be adjusted until system reaches steady state and then remain unchanged under different load conditions. Finally, simulation results based on a two-parallel-inverter system are provided to verify the effectiveness of the proposed method.

REFERENCES

[1] E. Barklund, N. Pogaku, M. Prodanovic, and C. Hernandez-Aramburo, "Energy Management in Autonomous Microgrid Using Stability-Constrained Droop Control of Inverters," *IEEE Transactions on Power Electronics* 23.5(2008):2346-2352.

[2] J. M. Guerrero, M. Chandorkar, T. Lee, and P. C. Loh, "Advanced control architectures for intelligent microgrids—part I: decentralized and hierarchical control," *IEEE Trans. Industrial Electronics*, vol. 60, no. 4, pp. 1254–1262, Apr. 2013.

[3] J. Rocabert, A. Luna, F. Blaabjerg, and P. Rodriguez, "Control of power converters in AC microgrids," *IEEE Trans. Industrial Electronics*, vol. 27, no. 11, pp. 4734–4749, Nov. 2012.

[4] A. Tuladhar, H. Jin, T. Unger and K. Mauch, "Control of Parallel Inverters in Distributed AC Power Systems with Consideration of Line Impedance Effect," *IEEE Trans. Industrial Applications*, vol. 36, no. 1, pp. 313–318, January/February 2000.

[5] Q. Zhong, "Robust Droop Controller for Accurate Proportional Load Sharing Among Inverters Operated in Parallel," *IEEE Trans. Industrial Electronics*, vol. 60, no. 4, pp. 1281–1290, April 2013.

[6] X. Zhang, J. Liu, Z. You and T. Liu, "Study on the Influence of Distribution Lines to Parallel Inverter Systems Adopting the Droop Control Method," *Journal of Power Electronics*, vol. 13, pp. 701–711, July 2013.

[7] Z. You, J. Liu, X. Zhang and X. Wang. " A Decoupled and Adaptive Power Sharing Strategy Based on Droop Method for Parallel Inverters," *In Proceedings of APEC* 2014, Fort Worth, TX, USA, March 16-20, 2014.

[8] C. Lee, C. Chu and P. Cheng, "A New Droop Control Method for the Autonomous Operation of Distributed Energy Resource Interface Converters," *IEEE Trans. Power Electronics*, vol. 28, no. 4, pp. 1980–1993, April 2013

[9] J. He, Y. Li and F. Blaabjerg, "An Enhanced Islanding Microgrid Reactive Power, Imbalance Power, and Harmonic Power Sharing Scheme," *IEEE Trans. Power Electronics*, vol. 30, no. 6, pp. 3389–3401, June 2015.

[10] N. Pogaku, M. Prodanovic, and T. C. Green, "Modeling, Analysis and Testing of Autonomous Operation of an Inverter-Based Microgrid," *IEEE Transactions on Power Electronics* 22.2(2007): 613-625.

[11] Y. Tao, Q. Liu, Y. Deng and X. Liu, "Analysis and Mitigation of Inverter Output Impedance Impacts for Distributed Energy Resource Interface," *Power Electronics IEEE Transactions on* 30.7(2015):3563-3576.

[12] J. He and Y. W. Li, "Analysis, Design, and Implementation of Virtual Impedance for Power Electronics Interfaced Distributed Generation," *IEEE Trans. Industry Applications*, vol. 47, pp. 2525–2538, November/December 2011.

The 2018 International Power Electronics Conference

Design and Integration of the Bi-directional Electric Vehicle Charger into the Microgrid as Emergency Power Supply

Yang Song[1], Pengcheng Li[2], Yuanliang Zhao[3], and Shuai Lu[1]

1 School of Electrical Engineering, Chongqing University, Chongqing, China
2 Electric Power Research Institute of Guizhou Power Grid Co. Ltd., Guizhou, China
3 Guizhou Power Grid Co. Ltd., Guizhou, China
*E-mail: Lushuai1975@gmail.com

Abstract- **Vehicle to grid (V2G) is currently not cost effective to justify the bi-directional EV charger as it reduces the battery cycle life. Replacing the dedicated emergency power supply of the load area with the bi-directional EV charger is a more viable option. This paper introduces the seamless integration of the bi-directional EV charger into the distribution load areas. The proposed system design improves the efficiency, guarantees the galvanic isolation and supports the 3-phase-5-wire microgrid. With virtual synchronous machine (VSM) control, instantaneous mode switching during the islanding is achieved when EV(s) are connected to the charger; otherwise, a black start strategy are proposed to handle the scenario. A 50kW lab prototype was successfully constructed, integrated and tested within an industrial load area and the actual EV. Moreover, the lab data show that the VSM with the proper controller tuning can condition the low order harmonics in the microgrid load current.**

Keywords— **bi-directional charger, electric vehicle, emergency power supply, microgrid.**

I. INTRODUCTION

The need for rapid charging of the electric vehicle (EV) gave rise to the widespread construction of the EV fast dc charger infrastructure. Therefore, potential advantages of using EV to interact with the power grid (V2G) have been widely investigated [1, 2], where the bi-directional power conversion is the key. However, the cost of the EV battery is overwhelmingly high, and V2G is currently not cost effective to justify the bi-directional EV charger as it reduces the battery cycle life.

An EV usually has 30-100kWh battery onboard and up to 1C continuous discharge rate. Dispatching one or more EVs to plug into the bi-directional chargers in a load area during power outage could supply enough power for the islanded microgrid. This would handily eliminate the needs for the backup diesel generator and the battery based emergency power supply (EPS). Compared to V2G, the EPS function is a more viable option for a bi-directional charger.

Commercially off-the-shelf EV chargers are all unidirectional, while the bi-directional charger is introduced in [3], where many converter topologies were investigated for the bi-directional EV charger.

When working as EPS, the normally grid tied AC/DC converter in the charger needs to work under the voltage source mode, i.e. microgrid islanding mode. [4-6] studied the power management strategy for the seamless transition between grid-tied and islanding modes. There are two types of the strategies. First is to switch from the current to the voltage mode when disconnected from the grid, where the transition time depends on the islanding detection algorithm. Second is to use the same control for both modes. As in [6], the same mode-adaptive droop curves are used for both modes. Unlike the conditional droop control that reflect only the external characteristics of SG, virtual synchronous machines (VSM) adds the virtual inertia to the total equivalent grid inertia to handle the grid disturbances [7, 8]. [9] applies VSM concept to a single-phase charger for V2G services, where a virtual 2-phase system is used for active and reactive power calculation to diminish the power oscillations.

In the section II of this paper, a bi-directional EV charger based EPS system is proposed and explained in detail. During the power outage, if the battery SOC and the power rating of the EV connected to the charger meets the EPS requirements, it operates in voltage source mode to supply the islanded microgrid. Section III first introduces the VSM applied to the AC/DC control, and the mode switching strategies are proposed to synchronize the power between the AC/DC and DC/DC. For the scenario when EV is not connected to the charger, a system black start solution is proposed. Simulation results are presented in section IV. In section V, an 50kW bi-directional charger prototype was constructed, tested and demonstrated as the EPS with a connected EV, by replacing the backup diesel generator and conventional EPS in an industrial load area. Experimental results are provided to verify the effectiveness of the EPS functions of the bi-directional charger. Finally, section VI concludes the paper.

II. EPS AND THE PROPOSED BI-DIRECTIONAL CHARGER INTEGRATION

A. Load area EPS integration

A typical building load area diagram is shown in Fig. 1. There are two 380V AC buses, each is connected to a 10kV distribution transformer. The two 380V AC buses are interconnected for redundancy. There are various loads of different priority, and the supply to some important loads such as the elevators and fire extinguishing loads may not be interrupted. When one bus faults, the redundant bus

3698

The 2018 International Power Electronics Conference

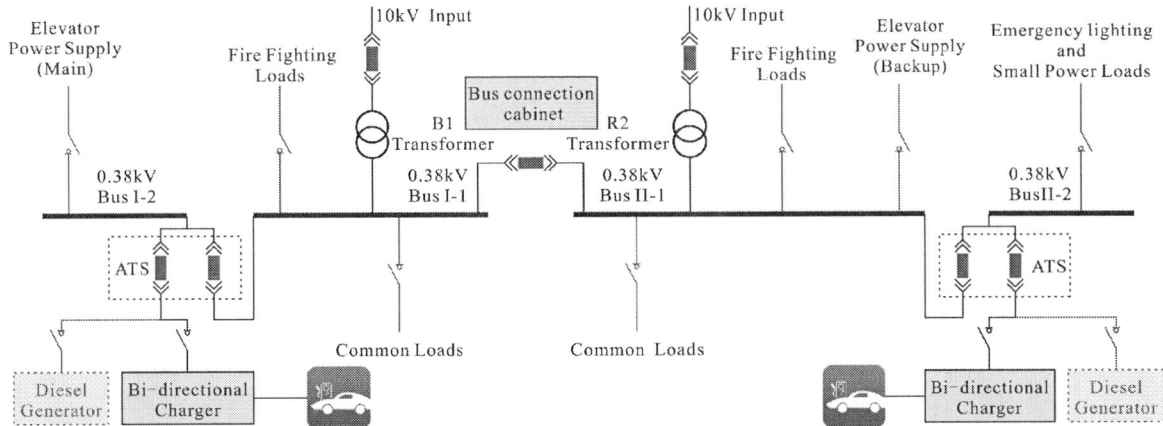

Fig. 1. A load area with the conventional EPS replaced by the bi-directional charger based EPS

supplies the vital loads on the failed one. When the main distribution grid fails, the backup diesel generator will be started to supply the load area. The link to the main grid and some low priority load circuits are disconnected so that the diesel generator is not overloaded.

In this paper, the diesel generator is replaced by the proposed bi-directional EV charger based EPS. During power outage, EVs can be plugged in. The energy from the EV battery pack supplies the load area via the charger.

Obviously, this utilizes the existing facility and avoids the installation of the conventional EPS, so that the extra cost for bi-directional power conversion is well justified.

B. Proposed Bi-directional EV Charger based EPS

The bi-directional charger based EPS system design is proposed in Fig. 2. It regularly charges EVs. When the grid faults, it uses the EV batteries to supply power to the islanded load area (microgrid).

Many grid-tied EPS simply uses a single stage AC/DC connected with the battery bank. But to follow the wide range of the EV battery packs (250V-650V for example) as a charger, an extra DC/DC conversion is needed. In a regular EV charger, the required galvanic isolation between the EV and the grid is usually realized by the high frequency transformer in the DC/DC converter. This isolation requirement introduces extra power losses and complexity of the system.

With the additional EPS function, a 3 phase 5 wire system is needed for the islanded load area, so that a grid frequency isolation transformer is needed anyway to connect the AC/DC. It has the start configuration on the grid side, the neutral and earth ground lines are tapped

from its grounded neutral, so that there is no need for high frequency isolation in the DC/DC, and the system can be simplified with the improved efficiency.

The proposed system connects the grid via the static switches (STS) made of a pair of thyristors, so that microgrid could be instantaneously disconnected from the grid when it faults.

The AC/DC uses a 3-phase VSI topology in Fig.3. The grid side voltage is 380V, and the common dc bus is controlled at 700V. The high voltage side of the bi-directional DC/DC is the common dc bus and the low voltage side connects to the EV battery via the charger post, which harbors the charger harness and connector. It also has auxiliary parts such as the DC contactor, power meters, control panel and isolation detector, etc. More importantly, it has a charger controller board that communicates with the battery management system (BMS) of the battery pack onboard an EV and commands the bi-directional power output and the operation modes of the AC/DC and DC/DC.

III. CONTROL SCHEMES OF THE EV CHARGER BASED EPS

A. Control Scheme of AC/DC converter

The VSM controller similar to [7] is programmed for the control of the proposed EV charger based EPS to achieve a seamless transition between the grid-tied and islanding mode. Also, good power distribution between multiple EV chargers can be realized when the load area needs more power and extended islanding time.

As in Fig. 4, the VSM controller could be treated as an extension of the conventional V/F control and PQ control. It also integrates the droop curve, so the converter could

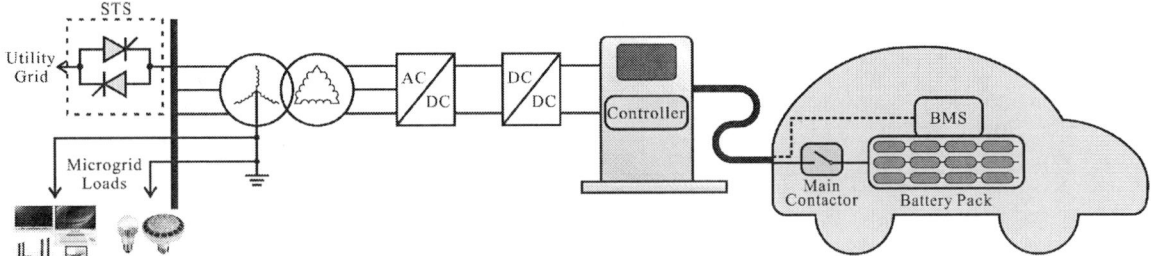

Fig. 2. Proposed Bi-directional EV Charger based EPS

3699

The 2018 International Power Electronics Conference

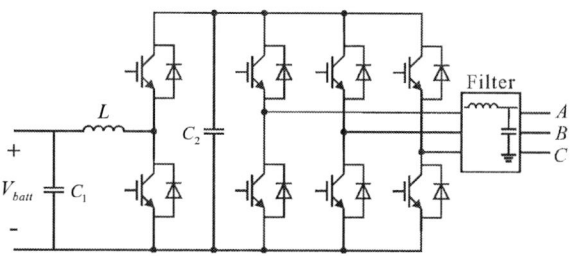

Fig. 3. Topology of AC/DC and DC/DC converters.

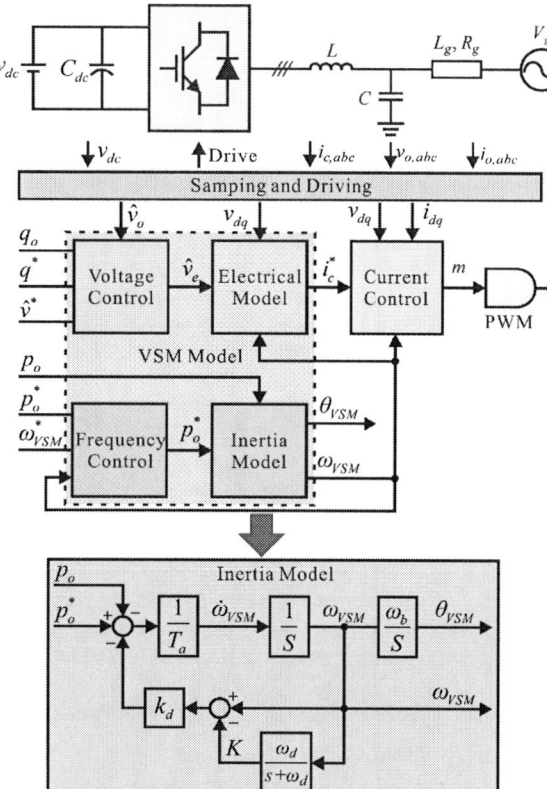

Fig. 4. The VSM controller of the AC/DC converter.

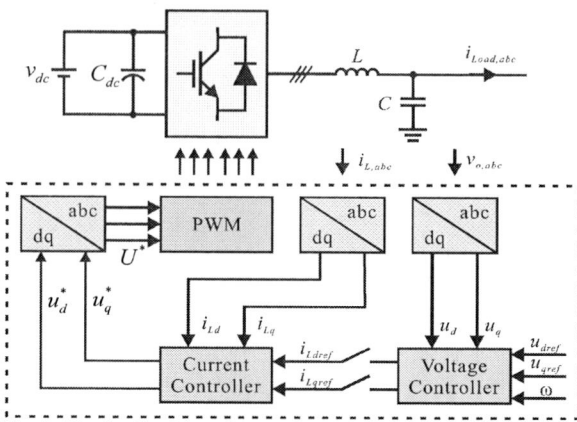

Fig. 5. The details of the voltage and current loops of the AC/DC.

current output, these harmonics in the load currents will have to go through the capacitor branch of the LC filter and causes distortion on the voltages of the microgrid. The distorted voltages will then interact with the nonlinear loads and reduces their current harmonics. This interesting effect will be verified by the simulation and experimental results.

B. Synchronization between AC/DC and DC/DC during the mode transition

When the grid faults, AC/DC works as a voltage source to supply the islanded microgrid loads. When one or more EVs are connected to the bi-directional charger, 3 scenarios exist and their individual control schemes are designed to synchronize the power between AC/DC and DC/DC and facilitate rapid mode transition.

1) The common dc bus is stabilized by the AC/DC, and the DC/DC controls the charging current of the battery pack, whose voltage is low (250-400V). In this scenario, when the STS islands the microgrid, the common dc bus is to be stabilized by the DC/DC that discharges the battery; then the AC/DC switches to the voltage mode such as the VSM. The mode switch is a little slow (up to 200ms).

2) When all conditions are the same as the scenario 1, but the battery pack voltage is high enough (>600V), there is no wait time for DC/DC mode switching as the battery pack clamps the common dc bus voltage through the diodes of the DC/DC, and the AC/DC switches to the voltage mode instantaneously. Here, the switching time will be mostly communication delays from the STS islanding action to the AC/DC (up to 50ms).

3) The common dc bus is stabilized by the DC/DC, and the AC/DC runs in VSM mode and adjust the grid power to indirectly control the charging current of the battery pack. In this scenario, the switching time to islanding mode is none and only STS disconnection time (<10ms) exists. Without the closed loop current control by the DC/DC, the battery pack charging current will have 6th and 12th harmonics, due to the 5th, 7th, 11th and 13th harmonics in grid voltages.

C. Black Start of the bi-directional charger based EPS

When the EV is not plugged in during the power outage, it is like a conventional diesel generator EPS, which takes

emulate the external characteristics of a synchronous machine (SM). The virtual torque produces the equivalent active power, the inertia model indirectly adjusts the active power and the angular frequency, and the reactive power is controlled by the virtual excitation voltage in the electrical model. Furthermore, it simulates the inertia characteristics of the SM, which improves the stability of the whole microgrid.

Fig. 5 expands on the VSM voltage and current loops. The voltage control loop stabilizes the voltage when it supplies the microgrid, while the inner current control loop improves the dynamics of the system and suppress the harmonic distortion in the inductor (ac output) currents of the AC/DC. It is instructive to analyze that in the islanding mode, when the microgrid has the nonlinear loads, the load currents contain dominantly the 5th and 7th harmonics. As the AC/DC inner current loop suppresses the harmonic

3700

a long period to start and supply the loads. EVs nearby are dispatched to plug into the bi-directional charger.

The secondary control system of the whole bidirectional charger system is powered up by two redundant supplies, i.e. the grid 3-phase power and the battery power. If the EV is not connected when power outage occurs, the charger system has no controller power. Even after the EV is manually plugged in, the charger cannot communicate with the EV BMS and the main contactor of the EV battery pack remains open, so the whole system will not startup. To solve this problem, a new charger controller board integrated with a rechargeable battery is designed.

The proposed black start procedure is shown in Fig. 6. The charger controller communicates with the EV battery pack once an EV is plugged in. After the handshake, the EV BMS closes the main contactor of the battery pack and the input/output capacitors of the DC/DC are charged to the battery voltage through a soft start path. Then, the whole secondary control is powered up, the DC/DC establishes the common dc bus voltage, then the AC/DC enters the VSM mode, finally the load switch is closed to

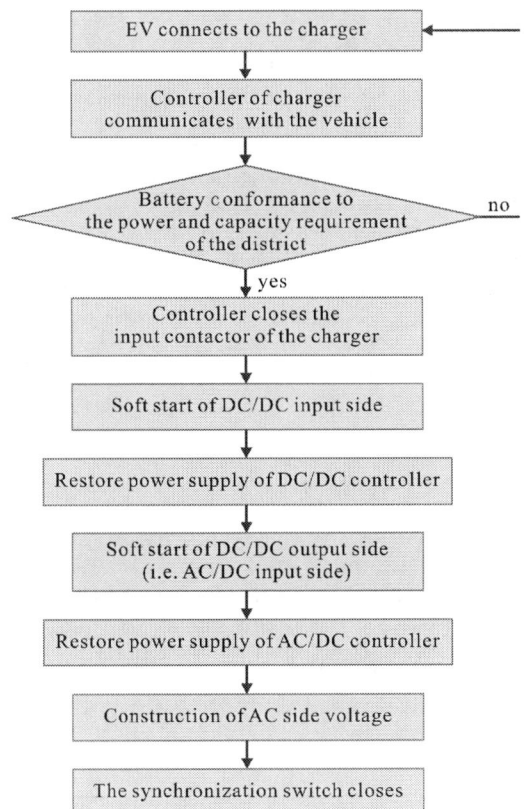

Fig. 6. The black start procedure of the EPS system.

supply the islanded load area.

IV. SIMULATION RESULTS

The proposed system is simulated. First, the simulation includes the whole black start process from an EV connecting to the bi-directional charger till the power

supply of the microgrid is recovered by the EPS. Herein, the DC/DC stabilizes the common dc bus voltage by boosting the EV battery voltage, then the AC/DC works in VSM mode to feed the microgrid. In Fig. 7, the 5 traces are the ac output voltages of the AC/DC, ac output currents of the AC/DC, the common dc bus voltage, the voltage and the current of the EV's battery pack, respectively.

Fig. 7 shows the following steps:

1) When the EV plugs in, the battery charges the output capacitor of the DC/DC by the soft start path. At the same time, the common dc bus is charged to the battery voltage via the diodes of the DC/DC.

2) The DC/DC charges the common dc bus to 700V.

3) The AC/DC outputs voltages that conform to the grid standards.

4) The circuit breaker on AC side is closed to supply the microgrid. The AC/DC and DC/DC handle the sudden load changes.

Fig. 8 shows the simulation results of the instant transition of the EPS system from the grid tied mode to the islanding mode when EVs are plugged in the bi-directional chargers. Similar traces are displayed as in Fig. 7, except that the output voltages and currents of the AC/DC are

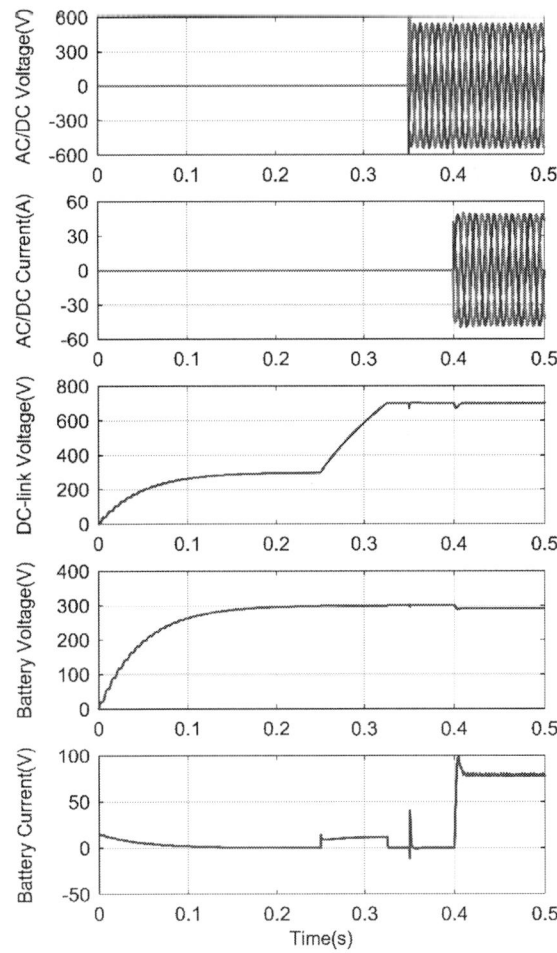

Fig. 7. Bi-directional charger black start process simulation.

The 2018 International Power Electronics Conference

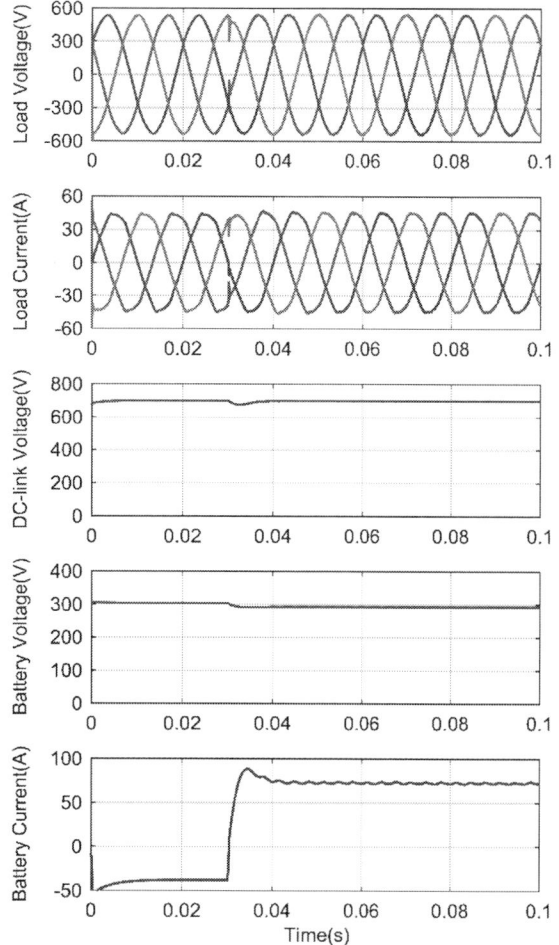

Fig. 8. Bi-directional charger from grid-tied to islanding mode transition simulation.

replaced by the voltages and currents of the microgrid loads. During the normal grid-tied mode, the EV connected to the charger is being charged. When power outage occurs (0.03s in the simulation), the EV battery pack has a 80% SOC. The transition is smooth without large voltage sag on the ac and dc sides. During the transition, the battery current direction instantly reversed to supply power to the load area.

As predicted and analyzed in subsection A of Section III, when nonlinear loads exist in the microgrid, the load currents contain significant harmonic components in the grid-tied mode. Then in the islanding mode, the harmonic components in the load currents are greatly decreased.

V. EXPERIMENTAL RESULTS

A 50kW bi-directional charger prototype has been constructed as in Fig.9. A commercially available EV with the following Li-ion battery pack parameters (100A*hr, 250V-400V, up to 1C continuous discharging rate) is connected to function as the EPS. A bi-directional charger is used as the interface between the EV and the microgrid.

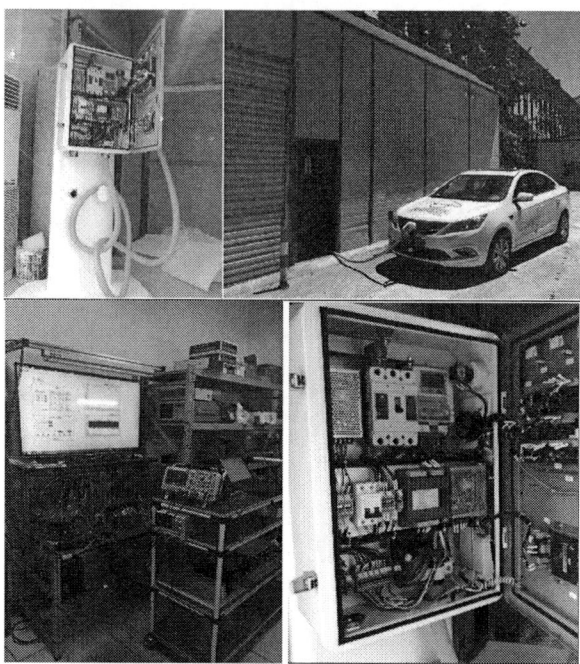

Fig. 9. The 50kW bi-directional charger based EPS prototype.

The bi-directional charger power conversion subsystem has 50kW AC/DC converter and DC/DC converter prototypes. The microgrid is connected to the utility grid by a Static Transfer Switch (STS) to realize the instant islanding transition.

Fig. 10 shows the experimental results of the bi-directional charger black start process. The 4 traces in Fig. 10 are the load voltage (line to line) and load current, the common DC bus voltage between the AC/DC and DC/DC converters, and the current of the EV battery pack, respectively

The experimental results are close to the simulation waveforms. When the EV is connected to the charger port, the controller board of the charger communicates with the BMS of EV. As the EV battery pack still has 80% SOC, the charger closes the input contactor of DC/DC and the control power supply is resumed for both the DC/DC and AC/DC. As the DC/DC soft start circuit was not activated during the test, there is inrush current from the battery when the DC/DC input contactor closes. After the controller initialization, the DC/DC starts to charge the common DC bus voltage to 700V, then AC/DC starts and supplies voltage to the microgrid. Then the loads in the microgrid are engaged in sequence, the phase currents increase accordingly until the full loads in the microgrid is resumed.

The instant islanding transition experimental waveform of the proposed bi-directional charger are shown in Fig. 11. The EV stays plugged in and already charged to an 80% SOC, then the power outage starts. The STS disconnect the grid connection of the load area. In both the grid-tied and the islanding modes, the DC/DC regulates the common DC bus voltage and the AC/DC runs the VSM control. The

3702

The 2018 International Power Electronics Conference

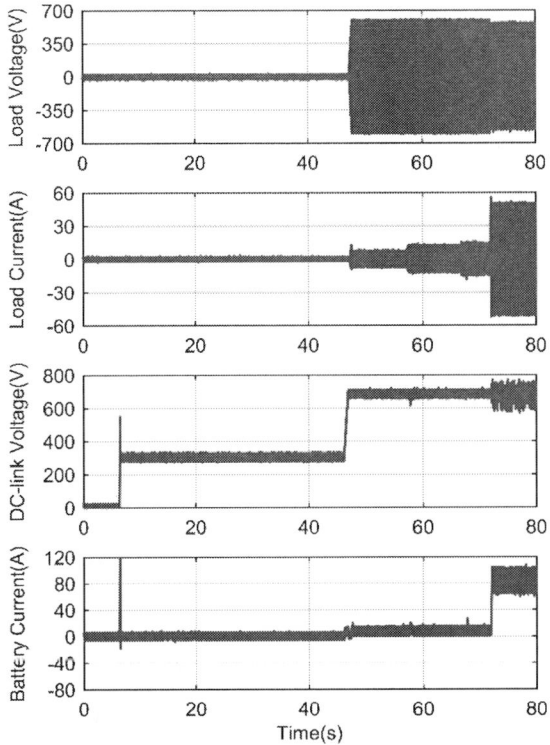

Fig. 10. The experimental results of the bi-directional charger black start process.

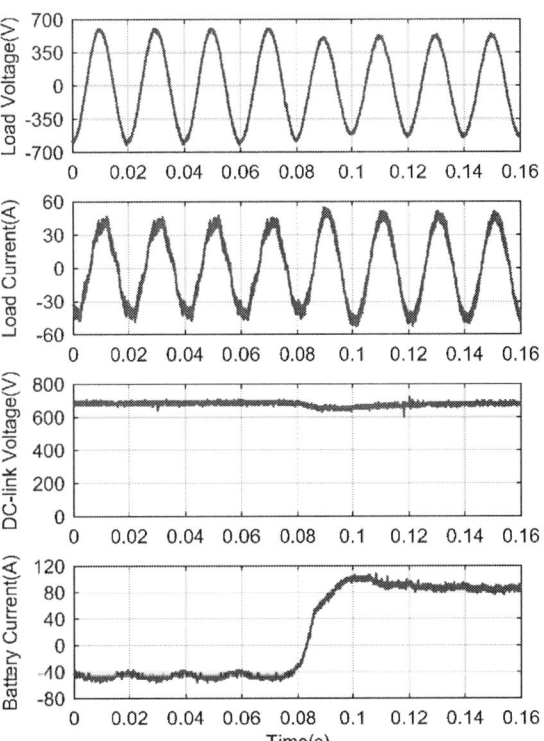

Fig. 11. The experimental results of the bi-directional charger instant islanding transition.

transition is smooth without voltage sag on the ac and dc sides. During the transition, the battery current direction instantly reversed to supply power to the load area. Note that the difference in the voltage amplitude before and after the transition is not the transients but due to the lower voltage set point for the microgrid mode and the higher grid voltage. In this control scheme, the common DC bus is well regulated by the DC/DC, while the tracking accuracy of the battery charging or discharging current is not guaranteed, as the low order harmonics is obvious in the battery current waveform.

As analyzed and simulated in this paper, by stabilizing the inner current loop outputs of the AC/DC VSM control, it tends to condition the load current harmonics in the islanding mode. In essence, the more sinusoidal AC/DC output currents force the load current harmonics to flow into the AC/DC filter capacitor branch and induces more voltage distortion. The distorted voltages applied to the load area tend to make the load current more sinusoidal. As shown in Fig. 11, the load current in the grid connected mode has significant harmonic components. In the islanding mode, the waveform of the load current was improved at the cost of more harmonic distortions in the voltages. It can be seen from the FFT analysis of V_{ab}, as in Fig. 12. The blue bar is the harmonic component of the grid voltages, while the orange bar is the harmonic component of the microgrid voltages in the islanding mode.

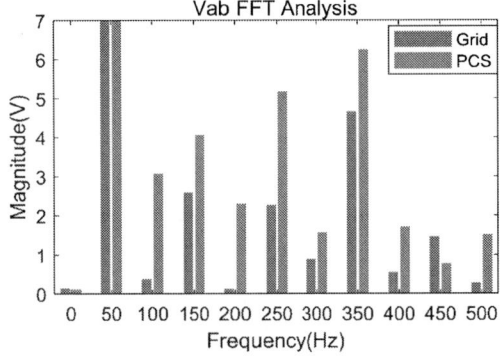

Fig. 12. FFT analysis of Vab line voltage.

VI. CONCLUSION

A bi-directional EV charger based EPS system is proposed in this paper. The system eliminates the needs for the backup diesel generator and the conventional battery based EPS for a load area. The bi-directional EV charger uses a two-stage power conversion. For seamless transition between the grid-tied and islanding modes, the VSM control is used in the AC/DC converter. When the grid faults and the EV is plugged in, 3 control schemes for power synchronization between the AC/DC and DC/DC and their transient performance are analyzed. If the EV is not currently connected, a black start procedure is proposed to restore power to the load area. The proposed system is validated by the simulation and lab tests, where a full scale prototype is constructed and supplies an actual industrial load area by a connected EV. Compared to V2G, the EPS function is a more viable option for a bi-directional EV charger, and it has potentials to promote much wider usage of the bi-directional EV charger.

REFERENCES

[1] H. Zhang, Z. Hu, Z. Xu and Y. Song, "Evaluation of achievable vehicle-to-grid capacity using aggregate PEV model," " *IEEE Trans. Power Syst.*, vol. 32, no. 1, pp. 784-794, Jan. 2017.

[2] M. Su, H. Wang, Y. Sun, J. Yang, W. Xiong and Y. Liu, "AC/DC matrix converter with an optimized modulation strategy for V2G applications," *IEEE Trans. Power Electron.*, vol. 28, no. 12, pp. 5736-5745, Dec. 2013.

[3] M. Yilmaz and P. T. Krein, "Review of battery charger topologies, charging power levels, and infrastructure for plug-in electric and hybrid vehicles," *IEEE Trans. Power Electron.*, vol. 28, no. 5, pp. 2151–2169, May 2013.

[4] F. Nejabatkhah and Y. W. Li, "Overview of power management strategies of hybrid AC/DC microgrid," *IEEE Trans. Power Electron.*, vol. 30, no. 12, pp. 7072-7089, Dec. 2015.

[5] H. Kakigano, Y. Miura, and T. Ise, "Low-voltage bipolar-type DC microgrid for super high quality distribution," *IEEE Trans. Power Electron.*, vol. 25, no. 12, pp. 3066–3075, Dec. 2010.

[6] Y. Gu, X. Xiang, W. Li, and X. He, "Mode-adaptive decentralized control for renewable DC microgrid with enhanced reliability and flexibility," *IEEE Trans. Power Electron.*, vol. 29, no. 9, pp. 5072–5080, Sep. 2014.

[7] O. Mo, S. D'Arco and J. A. Suul, "Evaluation of Virtual Synchronous Machines With Dynamic or Quasi-Stationary Machine Models," *IEEE Trans. Ind. Electron.*, vol. 64, no. 7, pp. 5952-5962, July 2017.

[8] C. Cheng, Z. Zeng, H. Yang, and R. Zhao, "Wireless parallel control of three-phase inverters based on virtual synchronous generator theory," in *Proc. Int. Conf. Elect. Mach. Syst. (ICEMS)*, Busan, Korea, 2013, pp. 162–166.

[9] J. A. Suul, S. D'Arco and G. Guidi, "Virtual synchronous machinebased control of a single-phase bi-directional battery charger for providing vehicle-to-grid services," *IEEE Trans. Ind. Appl.*, vol. 52, no. 4, pp. 3234–3244, Jul./Aug. 2016.

The 2018 International Power Electronics Conference

Stability Impact of PV Inverter Generation on Medium Voltage Distribution Systems

Ye Tang[1], Rolando Burgos[1], Chi Li[1], Dushan Boroyevich[1]
1 CPES, Virginia Tech, Blacksburg, USA
E-mail: yetang@vt.edu

Abstract— With an increasing number of photovoltaic (PV) inverters in the distribution system, their impact is no longer negligible, especially in the aspect of dynamic interaction. Accordingly, a comparison is done among PV inverters of different capacities, at different physical locations, at different reactive power control modes and at different operation states, to determine their impact on the system operation and stability. Generalized Nyquist Criteria (GNC) based on impedances in DQ frames is used for stability assessment, which is validated by time domain simulation results from MATLAB. And selected experiments are conducted to validate the findings of the study. From these, guidelines are formulated to manage PV inverter connection locations and reactive power control strategy and parameters.

Keywords— Distributed generator, PV, GNC, distribution systems, .

I. INTRODUCTION

Nowadays the power system is seeing higher and higher penetration of renewable energy generators in distribution system and transmission level as well. Distribution system is weaker compared to transmission system as it has high impedances. In addition, the impact of PV generators on the voltage profile impels the requirement of PV reactive power regulation from grid code [1]. In the near future, PV generators will have more complicated reactive power control strategy instead of working under unity power factor. So the question arises that whether PV generators in the distribution system will cause any stability problems. According to [2-4], it is necessary to take the impact of each distributed generator into account for large-scale integration for estimating the probability of instability.

Different from [2-4] which uses the conventional way of state space equations and eigenvalues to assess small signal stability, this paper employs Generalized Nyquist Criteria (GNC) method based on measured impedances in DQ frames at connection interfaces. GNC method has the advantage that interconnection stability can be judged without knowing the grid and PV generator model details. Based on a distribution system that has 56 buses [5], a comparison is done among PV inverters of different different physical locations, different reactive power control modes and operation states to determine their impact on the system operation and stability

II. STABILITY IMPACT OF A SINGLE PV FARM

To make use of the GNC method, the first step is to acquire impedances of grid side and PV generator side at the power common coupling (PCC) point.

A. Grid test-bed and impedances in DQ frame

Grid test-bed is shown in Fig.1, which is a rural lightly loaded radial 12kV distribution system that has 56 buses. The parameters are included in [5]. An average model in DQ frame is built in Simulink to do time domain simulation and impedance measurement using linearization function. The load is modeled as passive constant impedance load in the model.

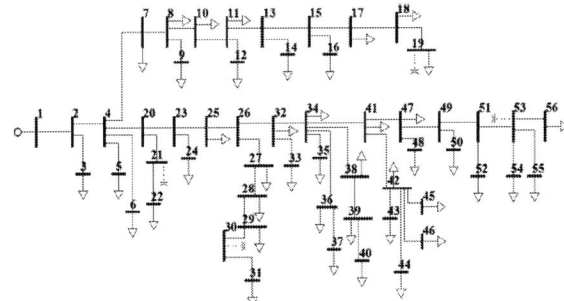

Fig. 1. One line diagram of the distribution system

It is known that impedance for a single line in DQ frames is

$$Z_{line} = \begin{bmatrix} LineR + LineL \cdot s & -\omega_0 \cdot LineL \\ \omega_0 \cdot LineL & LineR + LineL \cdot s \end{bmatrix} \quad (1)$$

While *LineR* and *LineL* are resistor and inductance of the line. For a complicated system in Fig.1, the impedance measured in DQ frames at one bus can be estimated by

$$Z_{network} = \begin{bmatrix} PathR + PathL \cdot s & -\omega_0 \cdot PathL \\ \omega_0 \cdot PathL & PathR + PathL \cdot s \end{bmatrix} \quad (2)$$

In which PathR and PathL are sum of resistors and inductances of all lines along the path from substation to measured bus. This estimation is validated by Fig.2. The red curves are from estimation using (2) at bus 45, the path goes along bus1-2-4-20-23-25-26-32-34-41-42-45, and blue curves are from linearization function of Simulink based on the DQ frame model.

3705

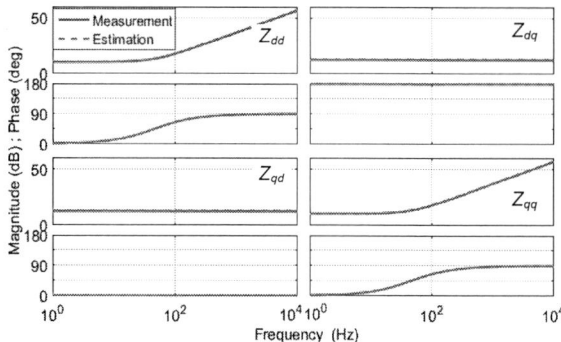

Fig. 2. Grid impedances in DQ frame

The accuracy of estimation of (2) means that the impedances of the network is dominated by the lines on the path of the measured bus to the substation, the lines on other branches and all the loads can be ignored.

B. PV generator impedances in DQ frame

PV generator impedances in DQ frame can be derived from small signal model of PV modules that considers dynamics of power stage, digital controllers and PLL effects [6]. It can also be measured by Impedance Measurement Unit (IMU) by perturbation and observe. Unity power factor, constant reactive power and Q=f(V) droop modes are compared in this section. Fig.3 is the droop curve applied. $V1 = 0.975$ p.u., $V2 = 1.00$ p.u., $V3= 1.025$ p.u., $V4 = 1.05$ p.u..

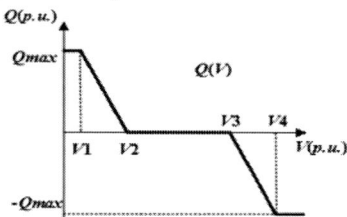

Fig. 3. Droop curve of Q = f(V)

Fig.4 shows PV terminal impedances of the three reactive power control modes for 3MW PV farm (twelve 250 kW power modules in parallel) and the steady state operation point of each module is P=250 kW, Q = -75kW for Q constant and droop modes. If the PV capacity increases, PV farm terminal impedances will be reduced proportionally. For unity power factor and Q constant modes, PV impedance matrices are diagonal dominant

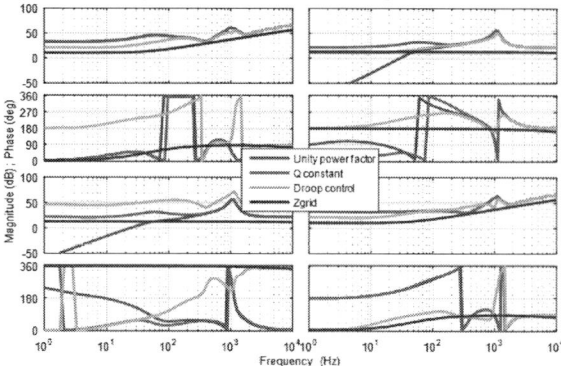

Fig. 4. PV and grid impedances in DQ frame

and have almost same values in diagonal elements. While Q droop mode control of Q = f(V) flips the signs of PV impedance elements of Zdd, Zdq, Zqq and increases the value of Zqd so that impedance matrix becomes non-diagonal dominant

C. GNC application

Impedances of PV farm at bus 45 under different reactive power control modes and impedance of grid are shown Fig.4, based on which GNC method is applied. Fig.5 is Nyquist plots of eigenvalues of return ratio matrix from (3).

$$L = Z_{grid}Y_{PV} \qquad (3)$$

The case that PV farm is under droop mode is unstable as $\lambda_1((s)$ encircles (-1,0) while grid impedance and PV admittance don't have RHP pole. The stability assessment is validated by time domain simulation results in Fig.6, which is output current of 3 MW PV farm (12 modules) under droop mode in DQ frame. The oscillation frequency is 260 Hz. Oscillation magnitude is limited by Q magnitude – Q_{max} in droop mode curve in Fig.3 before 0.5 s, which is removed at 0.5s and the oscillation of current starts to increase, proving system instability.

And if the location of the single PV farm changes from bus 45 to other locations, the stability result of connection is shown by scatter gram in Fig.7, in which horizontal axis is *PathR* and vertical axis is *PathL*. It can be concluded that bigger the electrical distance from the substation, it's more likely for PV connection to be unstable.

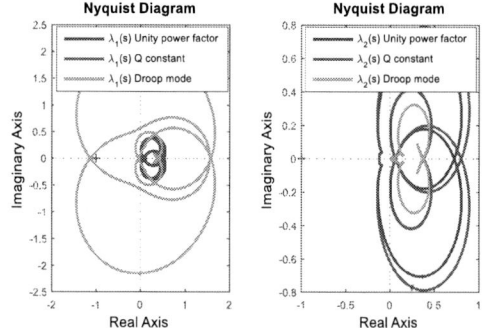

Fig. 5. Nyquist plots of eigenvalues of return ratio matrix L

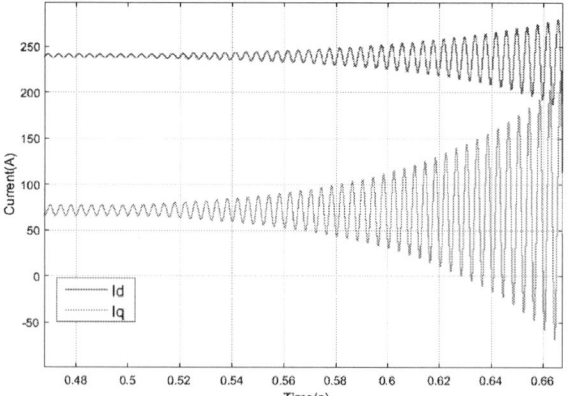

Fig. 6. Time domain simulation of PV under droop mode

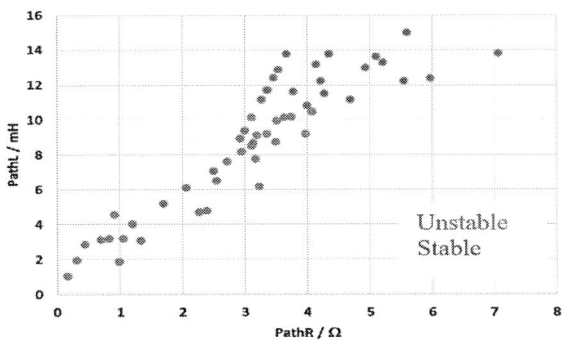

Fig. 7. Connection stability in terms of electrical distance

III. STABILITY OF SYSTEM WITH MULTIPLE PV FARMS

In this chapter, 3MW of PV generator is split evenly into two PV farms to analyze the stability of multiple PV connection. One PV farm is at bus 45, and the other PV farm may be connected to any bus of the system. For all the grid side and PV side impedances shown below, they are measured at PCC point of bus 45, so the grid side impedance includes dynamics of the second PV farm.

A. Second PV farm connected to different locations

To compare the interactions of two PV generators at different locations, three cases are compared in which the second PV farm is at bus 56, bus 41 and bus 19 separately. For all three cases, the first PV generator is connected at bus 45, where impedances are measured in DQ frames for the grid side and the PV side and shown in Fig.8. In Fig.8, solid curves are grid impedances and dashed curves are PV impedances. And Fig.9 is the Nyquist plots of the two eigenvalues of the return ration matrix.

In case one, Nyquist plot of $\lambda_1 ((s)$ encircles (-1,0), so the connection is not stable. While in case two and case three, neither eigenvalue loci encircles (-1,0), the operation of second PV farm at bus 41 or bus 19 is stable.

The results of the cases of different locations show that PV generators connected to different branches of the radial system may have interactions and cause instability when they are regulating the AC voltage together, the possibility of interaction decreases if one PV farm move closer to the substation or if the common path of the two buses is shortened.

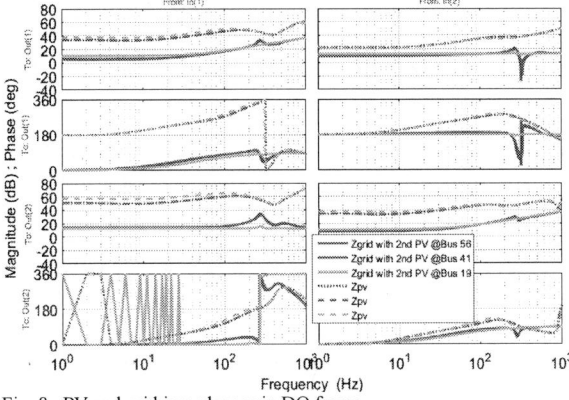

Fig. 8. PV and grid impedances in DQ frame

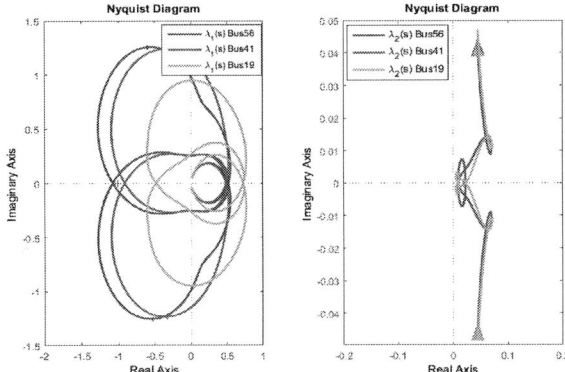

Fig. 9. Nyquist plots of eigenvalues of return ratio matrix **L**

B. Different Q control

To compare the interactions of two PV generators under different Q control modes, three cases are compared in which both PV farms are under unity power factor, Q constant and droop mode separately. For all three cases, the two PV farms are connected to bus 45 and bus 56. And when PV are under Q constant mode, the Q reference is set to be the same value as droop mode. As the system is under light load condition (20% of the peak load) and PV generator is at full active power harvest, PV generators are inductive at bus 45 and bus 56. Impedances of grid side and PV side are shown in Fig.10. And Fig. 11 shows Nyquist plots of eigenvalues of return ration matrix of three cases. Only $\lambda_1 ((s)$ of case three encircles (-1,0) which indicates unstable of system operation of these two PV generators under droop mode.

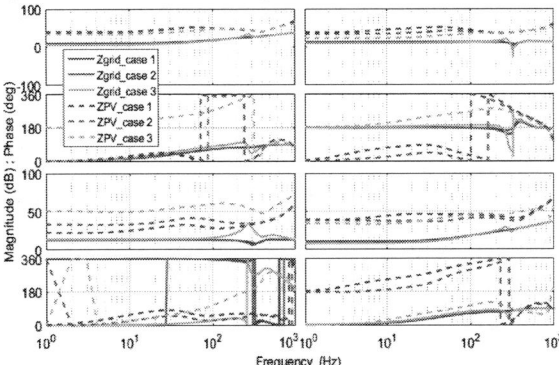

Fig. 10. PV and grid impedances in DQ frame. Case 1: Unity power factor, case 2: Q constant, Case 3: Droop mode.

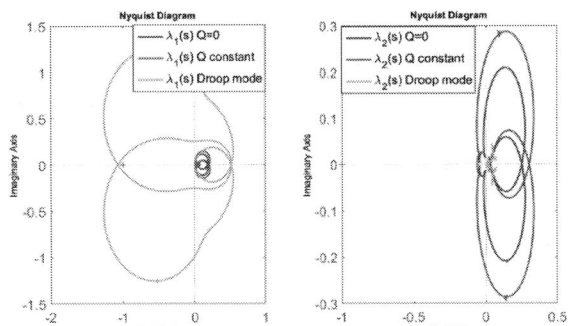

Fig. 11. Nyquist plots of eigenvalues of return ratio matrix **L**

The instability observed in this paper is caused by droop mode control, which will not occur at unity power factor or Q constant mode.

C. Different operation points on droop mode

Fig. 3 is the droop control curve Q = f(V) that is applied on PV generators under droop mode. For previous cases of droop mode, PV generator terminal voltages are above V3 (1.025 p.u.), so the generators are working in inductive zone. In another case, system load increases to peak load, system voltage profile is at low level, the two PV farms terminal voltage drop below V2 (1.0 p.u.), so the PV generators work in capacitive zone. Fig.12 is the impedances of grid and PV generator for the inductive case and the capacitive case. And Fig.13 is the corresponding Nyquist plots of eigenvalues of return ratio matrix. The inductive case is unstable because $\lambda_1 ((s)$ encircles (-1,0), while the capacitive case is stable.

At inductive operation zone, grid and PV impedances have opposite polarity until several hundreds of Hz, while at capacitive zone, grid and PV impedances don't have too much phase difference. So it's more likely to see unstable interconnection when system voltage profile is very high and makes PV inductive in droop mode.

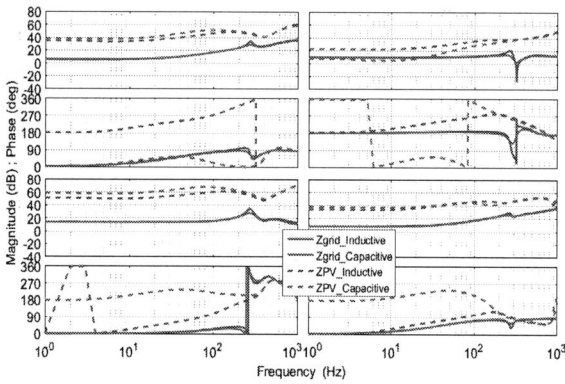

Fig. 12. PV and grid impedances in DQ frame

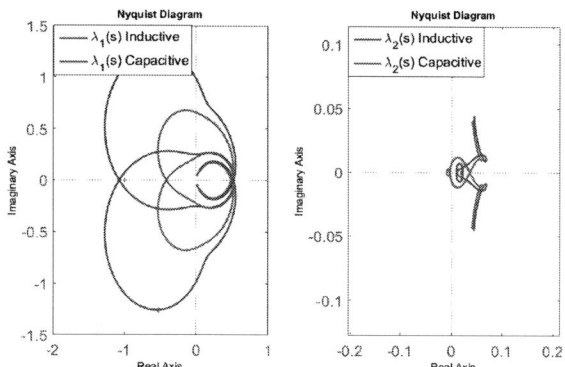

Fig. 13. Nyquist plots of eigenvalues of return ratio matrix **L**

D. Different capacities of PV penetration

This section compare the cases when the total capacity of two PV farms changes. Fig.14 and Fig.15 are comparison of impedances and eigenvalues of 1MW, 2MW and 3MW. All PV capacity is split evenly at two

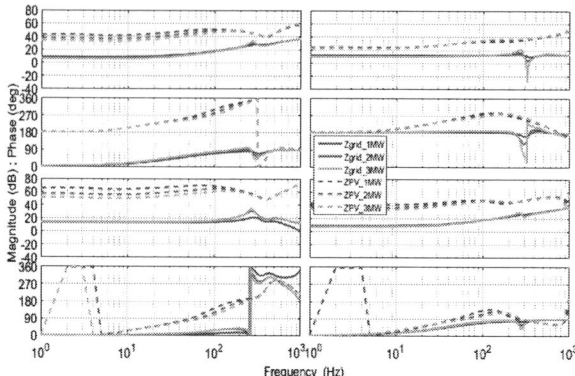

Fig. 14. PV and grid impedances in DQ frame

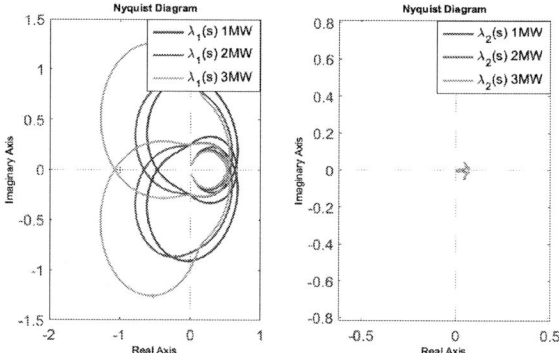

Fig. 15. Nyquist plots of eigenvalues of return ratio matrix **L**

physical locations of bus 45 and bus 56. And in all three cases, PV generators are working in inductive zone of droop mode.

Cases of 3MW of PV is not stable, as $\lambda_1 ((s)$ encircles (-1,0) in both cases. From the changes of eigenvalue loci shown in Fig.15, it can be concluded that as PV capacity increases, it's more likely for the PV connection to be unstable.

IV. HARDWARE EXPERIMENT

A. Hardware circuit

A hardware experiment is done to measure PV generator and grid impedances in DQ frame, based on which GNC is applied to assess connection stability. The hardware circuit is shown in Fig.16 with operation point specification. Agilent E4360A PV emulators working as PV array is connected to a DC/AC inverter which is tied to AC source (representing grid) through a line. Impedance measurement unit (IMU) introduced by [7]

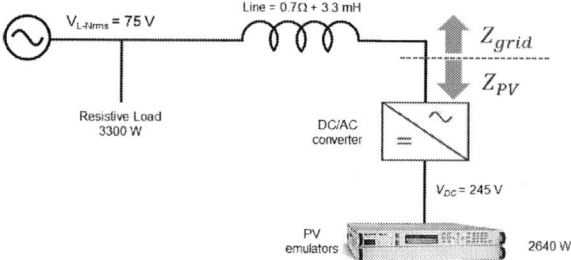

Fig. 16. Hardware experiment circuit

is located between the line and the converter to measure the impedances.

B. PV terminal impedances in DQ frame under different Q control mdoes

PV Impedances are measured under different Q control modes shown in Fig. 17 - 20. Fig.17 is PV impedance under unity power factor, which is measured at steady state of P= 2640 W and Q =0. The solid curves are derived results from PV dynamic model and star curves

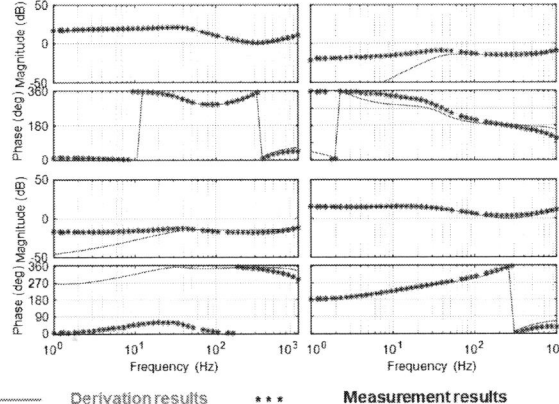

Fig. 17. Impedance Measurement of PV generator under unity power factor

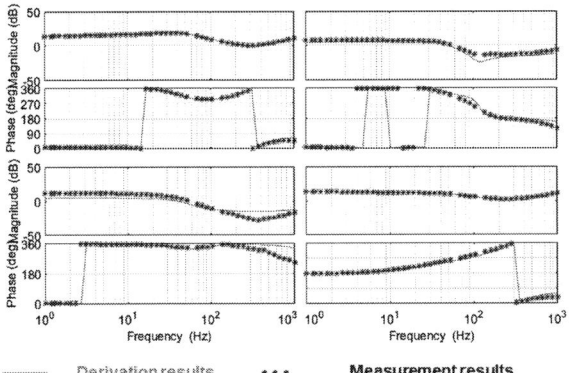

Fig. 18. Impedance Measurement of PV generator under Q constant mode when Q reference is negative

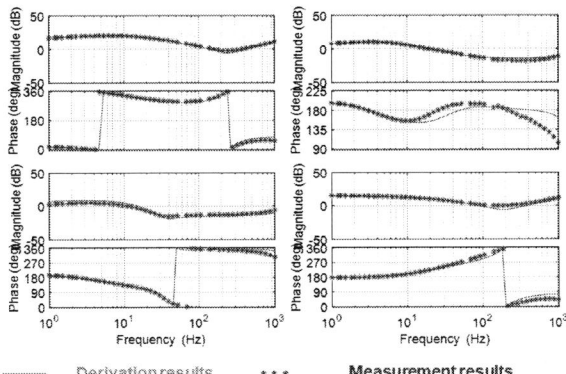

Fig. 19. Impedance Measurement of PV generator under Q constant mode when Q reference is positive

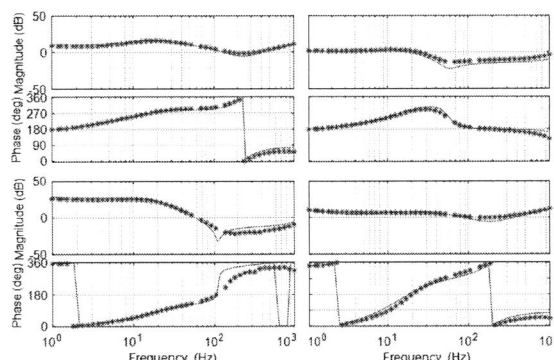

Fig. 20. Impedance Measurement of PV generator under droop mode control

are measured results from IMU. Fig.18 and Fig. 19 are PV impedances under constant Q control modes when P is equal to maximum power 2640 W. The difference between Fig.18 and Fig.19 is that Q = -800 var in Fig.18 and Q = 800 Var in Fig.19. Fig.20 is PV impedance under droop mode control, in which droop slope K_v =-160, PV terminal line – line rms voltage V=120V, $V3$= 115V, so in steady state Q = $K_v \times (V\text{-}V3)$ = -800 Var.

It can be observed from Fig.17-20 that derived results match with measure results, proving that derivation process of PV terminal impedance is correct. And comparing Fig.18 and Fig.20, it is shown that the droop mode control will flip the signs of Z_{dd}, Z_{dq} and Z_{qq} and increase the magnitude of Z_{dq} a lot when droop slope K_v magnitude is big enough.

C. GNC application based on impedances on DQ frame

In Fig.16, if the line inductance increases from 3.3 mH to 5.7 mH, PV generator working on droop mode control with impedances showing in Fig.20 will not be stable in the experiment, while system is stable if PV generator is under unity power factor or Q constant modes.

For this unstable case, GNC is applied on impedances measured of PV generator and the grid. Fig.21 is the nyquist plots of the two eigenvalues. The first eigenvalue

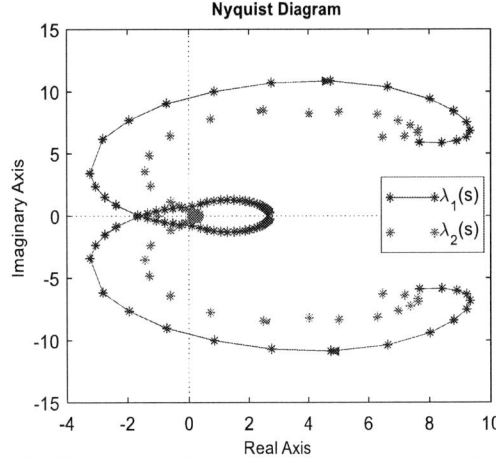

Fig. 21 . Nyquist plots of eigenvalues of return ratio matrix **L** for the unstable case

λ_1 encircles (-1,0) counter-clock twice. While grid impedance and PV admittance don't have RHP poles, the encirclement means the connection is not stable, proving the results from hardware experiments.

V. CONCLUSIONS

A radial medium voltage distribution system is modeled in MATLAB to analyze the impact of PV generator penetration using GNC method based on impedances in DQ frames, the results of which are proved by time domain simulation in MATLAB and scale down hardware experiments.

The PV farms connected to different branches of the complicated radial distribution system may have interactions with each other when they are under droop mode control. So the design of control strategy and parameters of PV generator should consider the impact of other PV generators. If models of other generators are not accessible, GNC method based on impedances measurement should be used for stability assessment.

Higher capacity of PV penetration makes this stability problem worse. Moving any of the PV farm closer to the substation reduces the possibility of instability. Droop mode is more flexible and beneficial for grid voltage regulation compared to unity power factor or constant Q control modes, but raise the possibility of unstable PV connection, especially when PV is working in inductive zone in droop curve. There is a tradeoff between grid voltage regulation and stability.

REFERENCES

List only one reference per reference number according to the following samples:

[1] IEEE Standard for Interconnecting Distributed Re-sources with Electric Power Systems Amendment 1, IEEE Std 1547a™-2014 (Amendment to IEEE Std 1547™-2003), 2014.

[2] W. Du, H. Wang, and L. Xiao, "Power system small-signal stability as affected by grid-connected photovoltaic generation," Eur. Trans. Elect. Power, vol. 22, pp. 688–703, Jul. 2012.

[3] M. Duckheim, J. Reinschke, P. Gudivada and W. Dunford. "Voltage and power flow oscillations induced by PV inverters connected to a weak power distribution grid", in Proc. 2013 IEEE Power and Energy Society General Meeting (PES), 21-25 July 2013.

[4] S. Liu, P. X. Liu and X. Wang. "Stability Analysis of Grid-Interfacing Inverter Control in Distribution Systems With Multiple Photovoltaic-Based Distributed Generators", in IEEE Transactions on Industrial Electronics, vol. 63, no. 12, pp 7339 – 7348 , 2016.

[5] M. Farivar, R. Neal, C. Clarke, and S. Low, "Optimal inverter VAR control in distribution systems with high PV penetration", in Proc. IEEE Power and Energy Society General Meeting, San Diego, CA, 2012.

[6] Y. Tang, R. Burgos, C. Li and D. Boroyevich, "Stability assessment of utility PV integration to the distributed systems based on D-Q frame impedances and GNC", in Proc. Control and Modeling for Power Electronics (COMPEL), Stanford, CA, USA, 2017

[7] Z. Shen et al. "Design of a modular and scalable small-signal dq impedance measurement unit for grid applications utilizing 10 kV SiC MOSFETs", 17th European Conf. Power Electronics Applications. Geneva, 2015, pp. 1-9.

The 2018 International Power Electronics Conference

1MW Power Conditioning System with Multiple DC Inputs for PVs and Batteries.

Yasuaki Furusho[1*], Yasuyuki Noto[1] and Kansuke Fujii[1]
1 Fuji Electric Co.,Ltd., Kobe-city, Japan
*E-mail: furusho-yasuaki@fujielectric.com

Abstract— Nowadays, based on the large amounts of connection of renewable energy-based generator to the grid, demand for mega-solar with storage battery is increasing. In general, the power conditioner (hereinafter called PCS) with storage battery for mega-solar has power fluctuation compensation function and peak-shift function for the areas with weak grids. In this paper, a high power DC link type PCS with storage battery developed for mega-solar is shown. The developed system has a 1-MW inverter, a 1.5-MW PV converter, and a 1.5-MW battery charging and discharging converter. The inverter is a T-type 3-level NPC inverter, and the converter is a 3-phase interleaved 2-level boost converter. The converter output and the inverter input are coupled with a DC link line. In a high power mega-solar system with storage battery, DC link topology achieves drastic reductions of the system cost and footprint by reducing the number of grid connected transformers and inverters, compared to AC-linked PV PCS and Battery PCS system. This system has 4 operation modes, which are switched appropriately depending on the command from programmable logic controller, solar cell voltage, and battery SOC.

Keywords— *Battery, Mega-solar, PCS, Fluctuation compensation.*

I. INTRODUCTION

In recent years, introduction of renewable energy power generation equipment has been advanced worldwide. Out of renewable energy, solar power generation and wind power generation have large power fluctuations and the system becomes unstable. Therefore, power generation equipment connected to a weak grid requires fluctuation compensation function and peak shift function. As of 2017, in Hokkaido area of Japan, 1% / 1 minute restriction has already been set for generated power fluctuation rate of new added mega solar.

Methods for dealing with such a requirement include a fluctuation compensation function using a storage battery and a peak shift function. Request for PCS with storage battery has increased in recent years because of background on lowering prices of storage batteries. The PCS with storage battery smooth short-term or long-term power fluctuations by charging and discharging the storage battery[1]~[7].

The topologies of the PCS with storage battery are AC link method and DC link method.

Figure 1 shows a comparison between the AC link method and the DC link method.

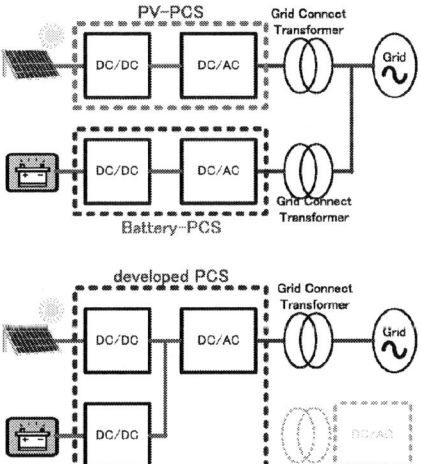

Fig. 1. AC link method and DC link method.

The AC link method is composed of a total of two PCSs, a solar PCS and a storage battery PCS, and the outputs of the two PCSs are linked with an AC line. The output of PV power generation and charge and discharge of the storage battery are done through the grid connected inverter and the grid connected transformer of each PCS.

Although the AC link method is common as a method to construct a large capacity storage battery-equipped PCS, system cost increases because each PCS requires grid connect inverter and grid connect transformer.

On the other hand, the DC link method is a method of integrating the functions of solar PCS and storage battery PCS into one equipment and linking the two outputs on DC line inside the equipment. Since this method requires only one grid connection inverter and one grid connection transformer, system cost can be reduced compared with AC link method.

The authors have developed solar PCS[8]~[10], and storage battery PCS[11] so far. In this paper, the circuit, operation mode and experimental results of the developed large capacity DC link type PCS with storage battery are shown.

3711

II. OUTLINE OF EQUIPMENT

Figure 2 shows the simple connection diagram of the developed PCS. Table 1 shows the specifications of the developed PCS. The PCS has 1.5 MW solar cell converter, 1.5 MW battery charge and discharge converter and 1.0 MW inverter. The converter consists of three phase interleaved 2-level boost converters, and the inverter consists of T type 3-level NPC inverter.

The converter output and the inverter input of the equipment are coupled in the DC-link part. A solar cell with a maximum output of 1.5 MW or more is connected to one of the two DC ports of the device and a storage battery of an arbitrary capacity is connected to the other DC port. The contract power at the interconnection point is determined by the total of interconnection inverter capacity.

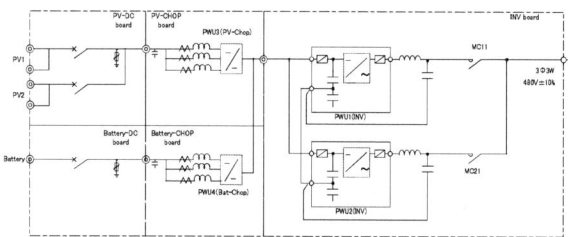

Fig. 2. Connection diagram of the developed PCS.

TABLE I
SPECIFICATIONS OF THE DEVELOPED PCS

Item	Value
Type Number	PVI1000MJ-3/1000
PCS Output Power	1000kW / 1111kVA
DC/AC Conversion Efficiency	Max: 98.8% EU Efficiency:98.5%
Size in mm (W×D×H)	4860×900×1950
Weight (kg)	4500kg
DC Max Input Voltage	1000V
MPPT DC Voltage Range	540V ~ 830V
AC Nominal Voltage	AC480V
AC Max Current	1336A
IP	IP20
Ambient Temperature	-5 ~ 40deg
Grid Stabilization Function	Constant P.F., FRT Power reduction, Q ref.

The converter output and the inverter input of the equipment are coupled in the DC-link part. A solar cell with a maximum output of 1.5 MW or more is connected to one of the two DC ports of the device and a storage battery of an arbitrary capacity is connected to the other DC port. The contract power at the interconnection point is determined by the total of interconnection inverter capacity.

Figure 3 shows an external view of the developed PCS. (Does not include grid connect transformers) From the left, it is an inverter board, a converter board for batteries, a DC board for batteries, a converter board for PV, and a DC board for PV. The board width is 4.9 m, the weight is

4.6 ton. The PCS is stored in a container and is cooled by the air conditioner.

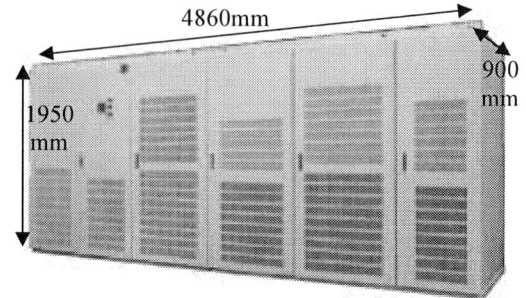

Fig. 3. Developed PCS (Prototype).

Figure 4 shows the power conversion unit of the converter of the developed PCS. The circuit configuration is a two-level three-phase interleaved step-up converter that converts 1.5 MW DC power.

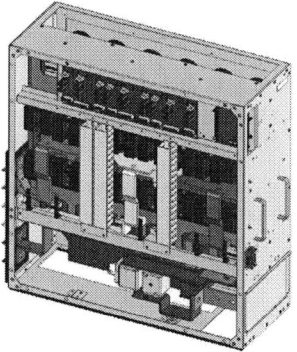

Fig. 4. Power conversion unit.

III. OPERATION MODES OF PCS

Figure 5 shows the control block of the PLC command necessary for the developed PCS. The PCS system have an individual PLC and an integrated control (superior command) PLC.

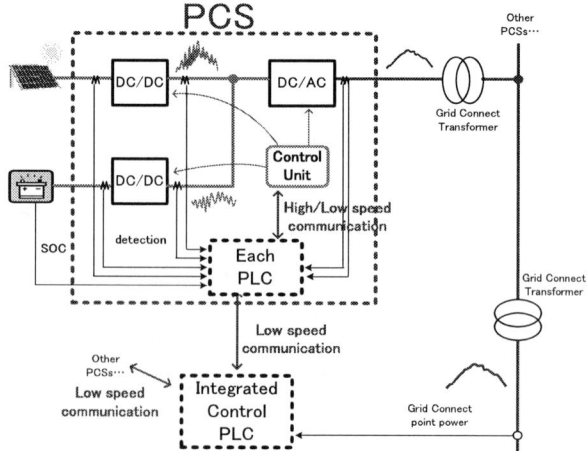

Fig. 5. Control block of the PLC for developed PCS.

The integrated control PLC monitors the interconnection point power and outputs the power flow correction command and the AC output suppression command between the PCS to the individual PLC mounted in each of the plurality of the PCS. The individual PLC outputs the storage battery charge and discharge power command and the power upper limit command of each PCS, based on the command from the integrated control PLC and the PV generation state and the storage battery SOC.

Figure 6 shows typical operation modes of the developed PCS.

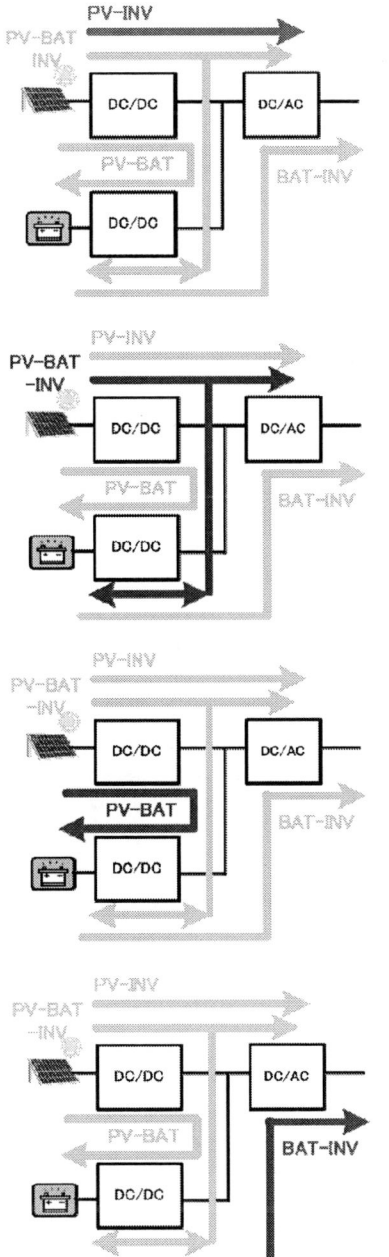

Fig. 6. Typical operation modes of the developed PCS.

In the PV-INV mode, the converter for a battery does not operate and solar cell output is output to the grid as it is. This mode is the same operation mode as conventional solar PCS.

In the PV-BAT-INV mode, the converter for a battery charges and discharges a part of the solar cell output to the battery. The variation of the solar cell output is compensated by the converter for a battery. This mode is the most basic operation mode of developed PCS.

In the PV-BAT mode, the inverter does not operate and all the solar cell output is charged to the battery. If power output from the solar cell to the grid is not allowed at all (output full suppression), PCS operates in this mode.

In the BAT-INV mode, the solar cell does not output and PCS outputs the battery power. This mode is the same operation as the conventional storage battery PCS. PCS operates mainly in this mode from evening to night.

The PCS seamlessly transitions between multiple operations modes depending on the solar radiation situation, the grid condition and the storage battery SOC.

Figure 7 shows typical behavior of the developed PCS, an operation of additionally outputting the surplus electric power in the day from the evening to the night.

In Area 1, PCS operates in PV-INV mode. Also in Area 1, when fluctuation compensation is necessary for system, the PCS operates in the PV-BAT-INV mode.

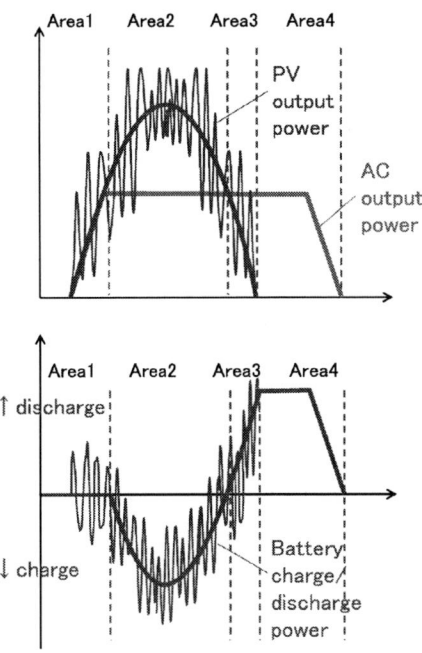

Fig. 7. Typical behavior of the developed PCS.

In Area 2, PCS operates in PV-BAT-INV mode. Surplus generated power is charged in the storage battery. In Area 3, the PCS operates in PV-BAT-INV mode as in Area 2. The solar cell output decreases, but it compensates by discharging from the storage battery. In Area 4, the PCS operates in the BAT-INV mode. Although the solar cell output becomes zero, PCS outputs surplus power charged in the storage battery.

Figure 8 shows another typical behavior of the developed PCS, an operation of once output full suppression occurred, additionally outputting the surplus electric power in the day from the evening to the night.

In Area 5, PCS operates in PV-BAT mode, PCS charges all generated power to the storage battery.

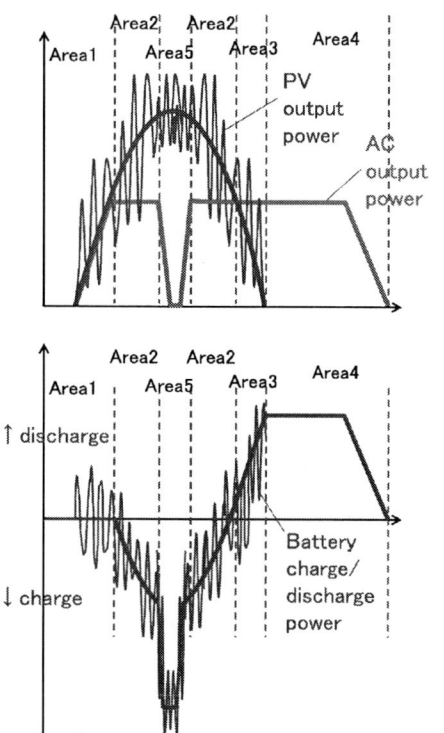

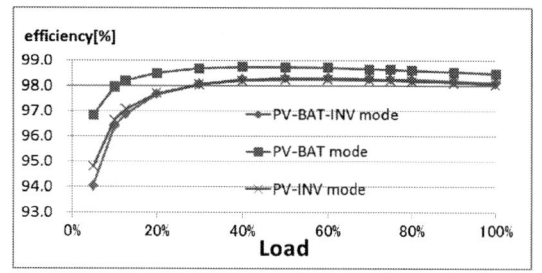

Fig. 8. Typical behavior of the developed PCS (Including output full suppression).

IV. EXPERIMENTAL RESULTS

Figure 9 shows the efficiency measurement results in each operation mode.

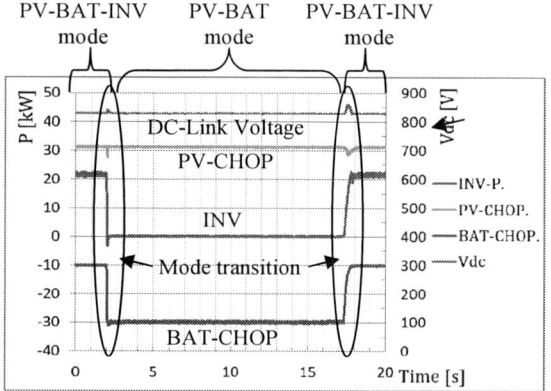

Fig. 9. Conversion efficiency of each operation mode.

The efficiency of BAT-INV mode is omitted because the circuit configuration of BAT-INV mode is almost the same as PV-INV mode. Both solar cell voltage and battery voltage were set to rated 630V, and AC output voltage is also rated 480V. In order to reproduce the actual operating condition, the power factor of the AC

output was set to 0.9 (lead). In order to suppress the voltage rise of the distribution system caused by the output power of the mega solar, PCS often outputs a predetermined reactive power by the Constant Power Factor control function. In the PV-BAT-INV mode, as a 100% load, the solar battery generated power was set to 1.5 MW, and the battery charging power was set to 0.5 MW, and the inverter output was set to 1 MW.

Figure 10 shows an example of the experimental result of the operation mode transition. In this experiment, we used a 1/50 scale equivalent mini model to simulate multiple steep output changes.

Fig. 10. Example of the experimental results of the operation mode transition of PCS.

As shown in figure 6, PCS transits a multiple modes depending on the situation. As a representative example, the waveform when returning from PV-BAT-INV mode to PV-BAT-INV mode through PV-BAT mode is shown.

In the experiment, the AC output command was step-changed from 100% to 0% and again to 100%.

This corresponds to an operation of transitioning from the area 5 to the area 2 and returning to the area 5 again in the operation shown in figure 7. The DC link voltage is controlled to INV in the PV-BAT-INV mode, and controlled to BAT-CHOP in PV-BAT mode.

Even though the operation mode is switched, the DC link voltage does not fluctuate largely, and it can be seen that the PCS smoothly transitions between two operation modes.

Figure 11 shows the experimental results of fluctuation compensation operation for high fluctuation rate solar radiation. In this experiment, we used a PV simulator and a 1/50 scale equivalent mini model to simulate solar radiation time series data based on measured values. In this solar radiation time series data, the rate of change of solar radiation reaches a maximum of about 3% / second. In addition, the rate of change per minute of PCS output was set to 1% of the mini-model rated output and the deviation determination value was set to 1.2%.

Even though the solar radiation changes steeper, PCS can keep output at a prescribed rate of power change.

The 2018 International Power Electronics Conference

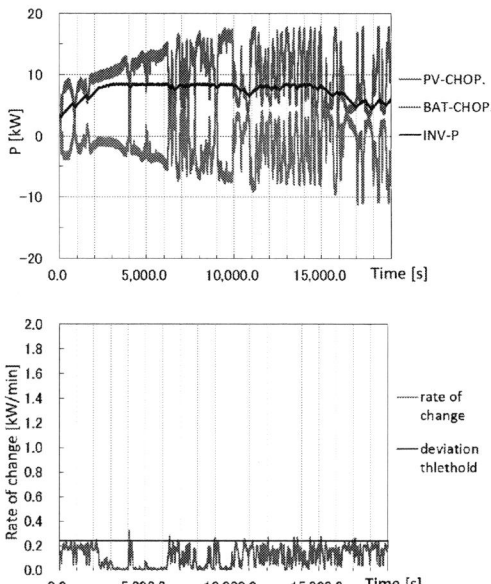

Fig. 11. Experimental results of fluctuation compensation operation for high fluctuation rate solar radiation

Figure 12 shows the experimental results of fluctuation compensation operation of PCS prototype against rapid solar radiation. The rapid change of solar radiation is simulated by stepping the DC input power. In this case, in order to shorten the experimental time, the rate of change per minute of PCS output was set to 10% rated output and the deviation determination value was set to 12%.

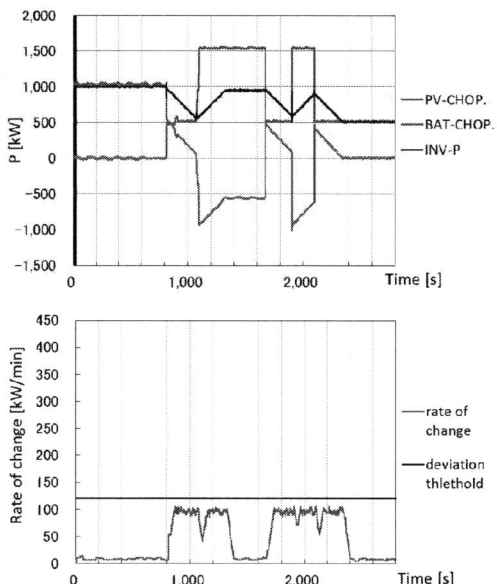

Fig. 12. Experimental results of fluctuation compensation operation of PCS prototype against rapid solar radiation.

To compensate for the rapid change of solar radiation,

the battery boost converter properly charging and discharging, PCS can keep output at a prescribed rate of power change.

Figure 13 shows the experimental results of PCS operation at AC output suppression operation and at battery SOC upper limit. In this case, in order to shorten the experimental time as same as figure 12, the rate of change per minute of PCS output was set to 10[%/min] of rated output and the deviation determination value was set to 12[%/min]. In addition, PCS stops the AC output according to the preset second rate of change based on the AC output suppression command from the host. In this case, the second rate of change was set to 40 [%/min] of rated output.

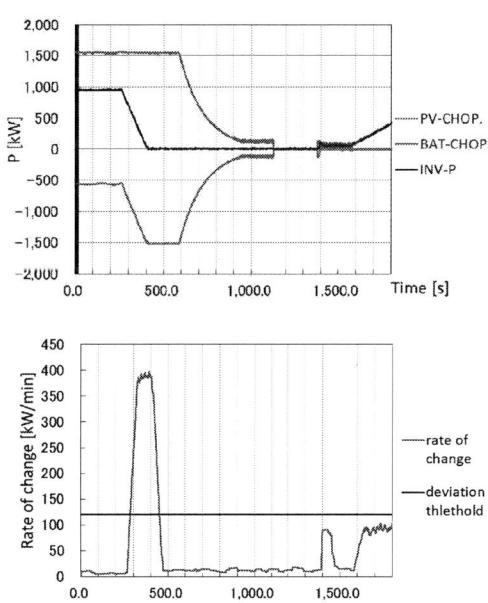

Fig. 13. Experimental results of PCS operation at AC output suppression operation and at battery SOC upper limit.

Even if the battery SOC approaches the upper limit, PCS gradually reduces the generated power of the solar cell to protect battery. After a certain period of time of the minimum power operation or when the battery SOC reaches the predetermined upper limit, the apparatus shifts to the standby operation. When AC output suppression is canceled, the PCS starts up again and starts outputting.

Thus, the developed PCS is controlled by a plurality of control parameters and operates flexibly for various situations of solar radiation, grid, and battery SOC.

V. CONCLUSIONS

In this paper, the outline of the equipment, the hardware configuration and the operation mode are described for the developed DC link type PCS, and the test results are shown. This system has 4 operation modes, which are switched appropriately depending on the command from programmable logic controller, solar cell

3715

voltage, and battery SOC. Compared with the AC link system PCS using PV-PCS and BAT-PCS for mega solar with storage battery, the developed PCS drastically reduce system cost and installation area by reduction of interconnection transformers and inverters.

REFERENCES

[1] M. Tamaki, S. Uehara, K. Takagi, T. Ichikawa, "Demonstration results using Miyako Island Mega-Solar Demonstration Research Facility", Transmission and Distribution Conference and Exposition (T&D), 2012.

[2] B. Y. Choi, Y. S. Noh, Y. H. Ji, B. K. Lee, C. Y. Won, "Battery-Integrated Power Optimizer for PV-Battery Hybrid Power Generation System", Vehicle Power and Propulsion Conference (VPPC), 2012.

[3] T. Nakayama, J. Ito, M. Ishida, "Evaluation of photovoltaics power systems with energy buffer to improve energy efficiency using a time series simulation model", Power Engineering Conference (UPEC), 2015.

[4] Y. Xu, T. Li, K. Wang, C. Zhang, C. Li, K. Sun, W. Feng, "A coordinated control strategy for suppressing transient power fluctuation of power conversion system and stabilizing AC bus voltage", Energy Internet and Energy System Integration (EI2), 2017.

[5] C. J. Sudhakar, A. V. Deshpande, D. R. Joshi, "Charge controller for hybrid VAWT and solar PV cells", Convergence in Technology (I2CT), 2017.

[6] C. Quann, T. H. Bradley, "Renewables firming using grid scale battery storage in a real-time pricing market", Power & Energy Society Innovative Smart Grid Technologies Conference (ISGT), 2017.

[7] S. Y. Kim, J. H. Cho, L. H. Soo, J. Lee, "50kW-class integral transformer / reactor design for solar PCS to improve power density", Electrical Machines and Systems (ICEMS), 2017.

[8] K. Fujii, T. Kikuchi, H. Koubayashi, K. Yoda, "1-MW advanced T-type NPC converters for solar power generation system", Conference Proceeding on Power Electronics and Applications (EPE), 2013.

[9] K. Fujii, Y. Noto, M. Oshima, Y. Okuma, "1-MW solar power inverter with boost converter using all SiC power module" Conference Proceeding on Power Electronics and Applications (EPE), 2015.

[10] Y. Furusho, K. Fujii, "1-MW Solar Power Conditioning System with Boost Converter using all-SiC Power Module", International Conference on Integrated Power Electronics Systems (CIPS), 2016.

[11] Y. Saga, K. Fujii, K. Yoda, "Power conditioner for stabilizing power disturbance caused of wind turbine generator system", Power Electronics Conference (IPEC-Hiroshima 2014 - ECCE-ASIA), 2014.

A Robust and Flexible DC-linked 3-Phase Energy Management System with Adaptive Droop Control Strategy

Yue Ma*, Yuki Ishikura, Hitoshi Tsuji and Kazuaki Mino

Murata Manufacturing Co., Ltd., Japan

TEL: (81)-75-955-7480

*dr_mayue@murata.com

Abstract—**In this paper, we are going to introduce our consideration and strategy of DC bus voltage adaptive droop control, which has been used and verified successfully in our prototype of 3-phase scalable Energy Management System (EMS). Voltage droop profile for charging/discharging energy storage is controlled by a higher level manager in real-time. The proposed strategy can be easily extend to larger systems where more units are connected to DC bus.**

Keywords—*Adaptive droop control, grid-connected inverter, DC-linked system*

I. Introduction

With the increasing concern of global environment issue, the distributed renewable energy resources such as photovoltaic plant or wind power generator have experienced a large growth in recent years. Meanwhile, the fact that most of them are largely environment-dependent and output unstable power to the grid leads to more and more difficulty of stability control of utility. So recently, it is usually required that the distributed power generator have the ability to limit their output under a regulated line. The limit may be send from any demand response aggregator, or by some kind of autonomous algorithm. In case that solar panel generate more power than the feed-in limitation, it is usually preferred to store the surplus into energy storage, and output from energy storage when load increase. Nowadays, Lithium-ion battery (LiB) is commonly adopted as the energy storage, and the system is usually called Energy Management System (EMS).

An EMS consists of energy generation, energy storage and grid connection [1]. According to the power range, load volume, application requirement, and sometimes cost consideration, the EMS may vary largely, from several kWs to hundreds of MWs. Comparing to make every system specifically, a flexible design with scalable common units can satisfy all kinds of customers better. Thus, it is necessary to consider a robust control strategy for the scalable systems.

From the control point of view, power (current) control of grid-connected inverter is easier and more straightforward for demand response requirement. But because the energy generation, such as PV panels, are commonly current source via power control, if grid-connected inverters are operated via power control as well, the common HVDC bus will be controlled only by energy storage.

Above strategy tend to be less stable, specially when energy storage power is too weak to balance generation and grid.

Here in this paper, we introduce a systematic control solution, in which both energy storage and grid-connection are roles to control common HVDC bus, and power flow is implemented by adaptive voltage droop control via a manager unit. This strategy can satisfy both robustness and flexibility.

II. System Description

The component diagram of EMS introduced in this paper is shown in Fig. 1. Functions of each unit is described as following. (1) PV converter (PVC) generates power from PV string(s) by Maximum Power Point Tracking algorithm. (2) Grid-connected inverter (INV) works bidirectionally, by which solar power and discharged power from battery can be feed-in to grid, or grid power can be taken to charge the battery. (3) Battery pack is charged and discharged by bidirectional DC/DC converter (BDD). (4) The total power flow control is in charged by Manager (MNG), which communicates with all units in the system, sends commands, receives state and error information.

Power flows between these three types of units through a common connected HVDC bus. By ignoring the power loss during conversion, we always have following relation,

$$P_{pvc} + P_{bdd} + P_{inv} = 0 \qquad (1)$$

where P_{pvc}, P_{bdd} and P_{inv} stand for power generated from solar, power charged/discharged for battery, and

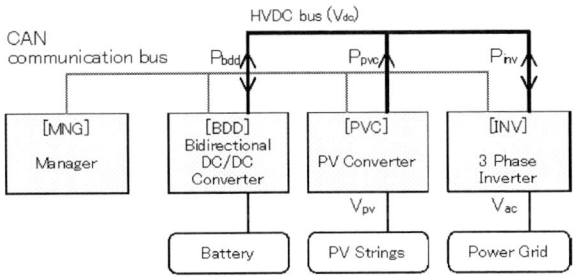

Fig. 1. The Energy Management System schematic diagram.

The 2018 International Power Electronics Conference

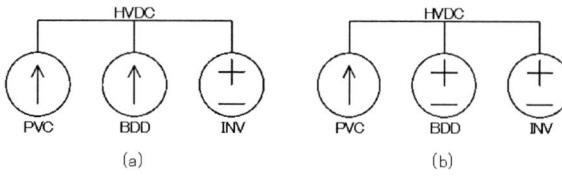

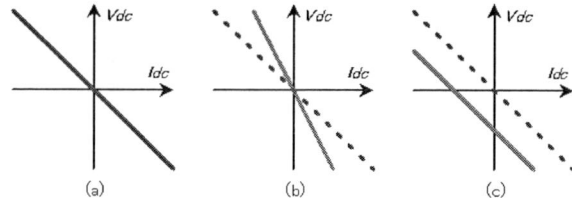

Fig. 2. (a) A system in which BDD works as a current source. (b) A system in which BDD works as a voltage source

Fig. 3. Profiles of voltage droop control (a) a conventional type (b) an adaptive type of changing virtual resistor (c) an adaptive type of changing initial voltage.

power exchanged between EMS and grid, respectively. In this paper, we define the direction as plus if power flow towards HVDC bus and thus raises the voltage of bus, otherwise, power is defined as minus. Note that they share the same bus voltage level, because all types of units are connected to HVDC bus, indicated by V_{dc}.

Usually, an EMS works in a fashion that users configure the charging and discharging plan of storage battery, and BDD follows the configuration via a power (current) control, as shown in Fig. 2(a). As for the grid-connected INV, its power command comes from a V_{dc} controller, and no need to be setted by user. But as described in previous section, this kind of operation is less stable and increases the grid stress. Instead, we design our system with more voltage sources to solve above issues, as shown in Fig. 2(b). In the following sections, we will discuss our control strategy in detail.

III. ADAPTIVE DROOP CONTROL

Before jumping into Adaptive Droop Control, we would like to briefly review the conventional Voltage Droop Control.

A. Voltage Droop Control

Voltage Droop Control is usually used in paralleling voltage controlled power supplies (such as shown in Fig. 2(b)) to assure current sharing or DC micro-grid without the presence of a central control [2]. Simply, the controller emulates an series resistor behaviour by tuning the output voltage with the output current, as shown in Fig. 3(a). The profile of V_{dc} and I_{dc} is given as:

$$V_{dc} = V_0 - I_{dc} \times R_d \qquad (2)$$

where V_0 represents the initial value to control the voltage when there is no current flow, and R_d is referred to as the virtual series resistor.

B. Adaptive Droop Control

Although voltage droop control provide a stable solution for paralleling operation of converters, many applications need to adjust the profiles of voltage droop control according to various power requirement. While a lot of solutions have been proposed [3], [4], [5], there are two basic methods as following,

- changing the virtual resistor, i.e., R_d,

- changing the initial voltage, i.e., V_0,

which are shown in Fig. 3(b) and (c), respectively.

Obviously, by changing the slop of droop profile (R_d) as Fig. 3(b), the sensibility of specific converter can be changed. For example, reducing R_d can lead to a larger voltage shift on a relatively smaller current variation. On the other hand, changing the initial voltage (V_0) has no effect on sensibility, but can rearrange the load balance.

In order to increase the stability, we keep all the units have equal sensibility in our EMS. We choose the second type of adaptive droop control, in which the offset of droop profile, i.e., V_0 is changed in real-time by a feedback controller embedded in MNG.

IV. CONTROL IMPLEMENTATION

Instead of directly controlling the power of INV or BDD, we prefer that INV and BDD behaviour as voltage source, which means their power command is determined by HVDC voltage controller. On the other hand, as introduced in Sec. I, many applications need the INV's power to be controlled as demanded. So, how to compromise these two types of requirement is the key issue.

A. Droop profiles of each type of unit

V-I Profiles of PVC, BDD and INV are illustrated in Fig. 4 (a), (b) and (c), respectively.

PVC behaves simply as a current source, whose maximum power is generated by a MPPT controller. An exception is that I_{dc} has to be limited quickly to keep the converters from damage if HVDC voltage raises to a fairly high level, as shown in the top part of Fig. 4(a).

INV is designed as a conventional droop profile shown in Fig. 4(c). In our system the virtual resistor is 1Ω, and initial voltage V_0 is $380V$.

$$R_d = 1\Omega, \quad V_0 = 380V \qquad (3)$$

Substituting into Eq. 2, we have

$$V_{dc} = 380 - I_{dc(inv)} \qquad (4)$$

For example, if INV feeds 5kW power into grid, as defined in Sec. II, we have

$$P_{inv} = -5kW, \qquad (5)$$

3718

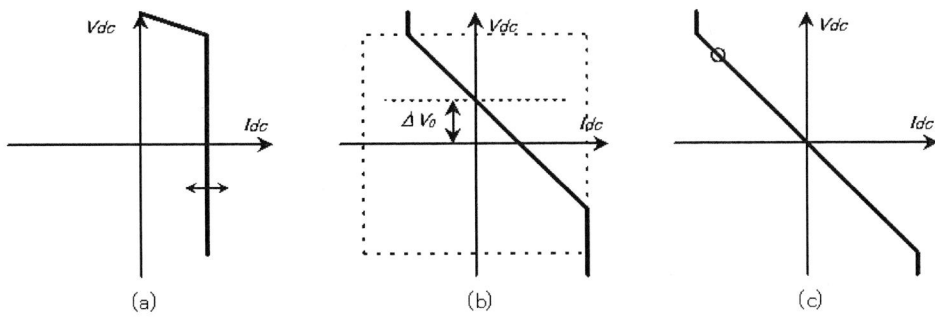

Fig. 4. V-I profiles of (a) PVC (b) BDD and (c) INV in the EMS.

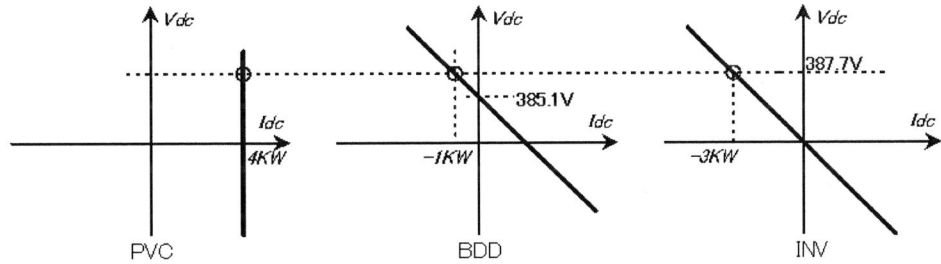

Fig. 5. Operating point (the circle) for a static case of $P_{pvc} = 4$kW, $P_{inv} = -3$kW, $P_{bdd} = -1$kW.

then we can calculate $I_{dc(inv)}$ from following equation

$$\frac{-5000}{I_{dc(inv)}} = 380 - I_{dc(inv)} \Rightarrow I_{dc(inv)} = -12.7A \quad (6)$$

and the voltage of HVDC will raise to about $380V + 12.7V = 392.7V$. The stable operating point is drawn as a circle on the upper-left in Fig. 4(c). Similarly, if INV input 5kW power from grid (P_{inv}=5kW), the operating point will move to the lower-right side.

In order to control power flow in the EMS, BDD's $V - I$ profile is designed as Adaptive Voltage Droop as shown in Fig. 4(b), in which the initial voltage value is not fixed at $380V$. Instead, initial voltage is $V_0 = 380 + \Delta V_0$, then the profile can be written as

$$V_{dc} = 380 + \Delta V_0 - I_{dc(bdd)} \quad (7)$$

where ΔV_0 is a signal received from MNG. By shifting ΔV_0 up and down, we can move the operating point in any position in the dashed rectangular area shown in Fig. 4, instead of only on a straight line like PVC and INV.

B. Control power of BDD via ΔV_0

Power flow of the EMS can be fully controlled by ΔV_0, which is the offset level of $V - I$ profile of BDD. To describe the control scheme easier, for example, we consider following case:

- Solar power generate 4kW from PVC ($P_{pvc} = 4$kW),

- INV is required to feed-in 3kW by aggregator ($P_{inv} = -3$kW),

- To satisfy Eq. (1), BDD has to adjust its power to discharge 1kW to balance the power flow ($P_{bdd} = -1$kW).

The operating point of INV is determined by INV's power:

$$I_{dc(inv)} = -7.7A, \quad V_{dc} = 387.7V \quad (8)$$

Because all the units are connected to HVDC bus, they share the same V_{dc}. So for BDD, we have following result from Eq. 7,

$$387.7 = 380 + \Delta V_0 - \frac{-1000}{387.7} \Rightarrow \Delta V_0 = 5.1V \quad (9)$$

Thus, the operating point for this simple example can be shown in Fig. 5.

C. Control of ΔV_0

Practically, we do not calculate ΔV_0 like above, a feedback control is used to output the offset, instead. The purpose of the power flow controller is to make the output of INV follow the command, by adjusting the offset ΔV_0, i.e., the operating point of BDD.

In Fig. 6, we show the overall system control diagrams. Bus voltage controller (VC) and output current controller (CC) are same with conventional methods. Reference of VC is generated by droop controller.

The proposed control strategy is in-directly realized by MNG, as shown on the top of Fig. 6, and its input/output are wired by high speed communication, drawn as bold lines in the figure. Note that units of BDD, INV and PVC

The 2018 International Power Electronics Conference

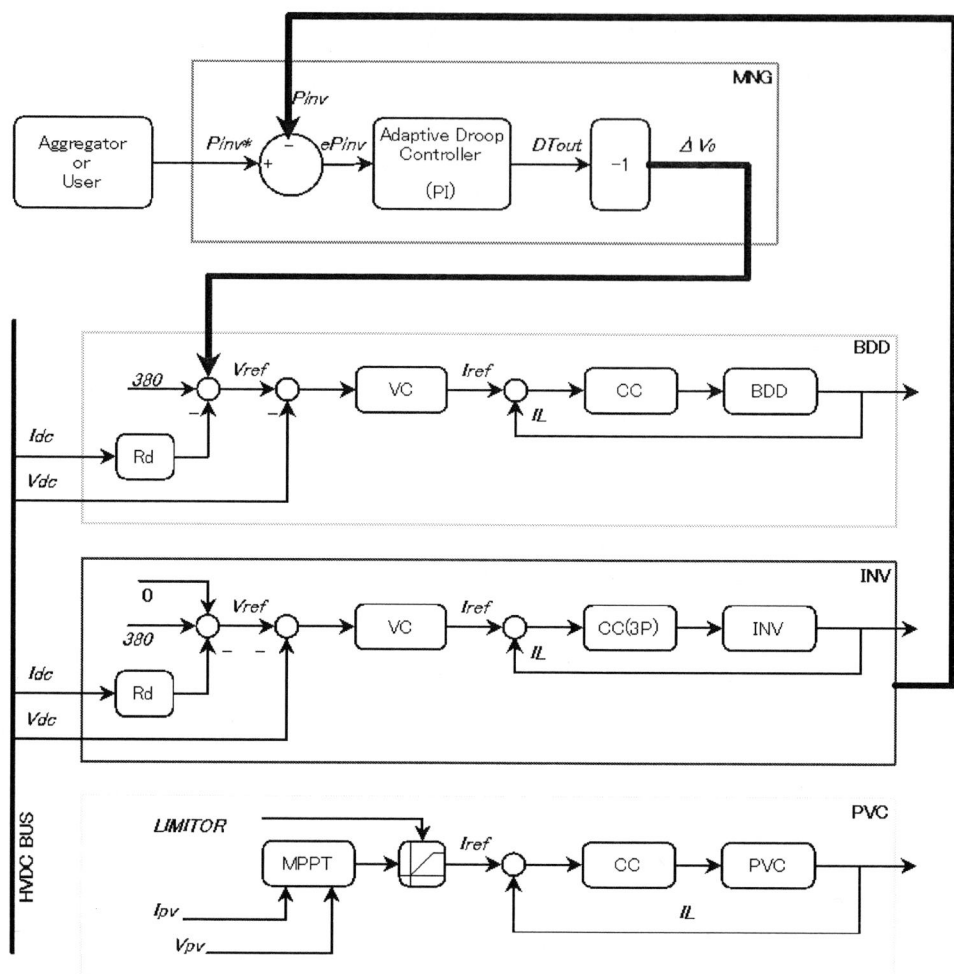

Fig. 6.   Schematic Diagram of Adaptive Droop Controller.

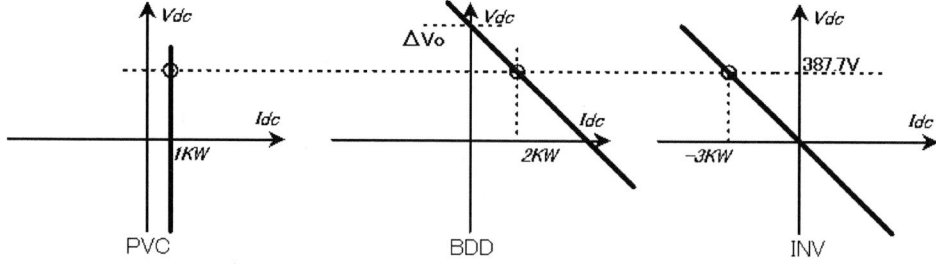

Fig. 7.   Operating point (the circle) for a static case of $P_{pvc} = 1\text{kW}$, $P_{inv} = -3\text{kW}$, $P_{bdd} = 2\text{kW}$.

communicate with MNG, but do not talk with each other in the system. A reference value of INV's power (P_{inv}^*) is given by aggregator or user, the actual power is send from INV to MNG, then error of INV's power is calculated and input to a PI controller. Because reducing ΔV_0 will reduce the power of BDD P_{bdd}, and hence increase the power of INV P_{inv}, the control is in the reverse direction. That is why there is a "−1" before the control output ΔV_0 is send to BDD via communication.

D. Flexibility

Since the adaptive droop control is not embedded in each unit, MNG can manage the power flow no matter how many units are connected to HVDC. Also, if there are many BDD units in the system, MNG can send out different offset signals ΔV_0 to different BDD units, in order to change the power of specific unit if necessary. This is useful in the case such as batteries SOC are unbalance. Moreover, in multiple INV condition, each

3720

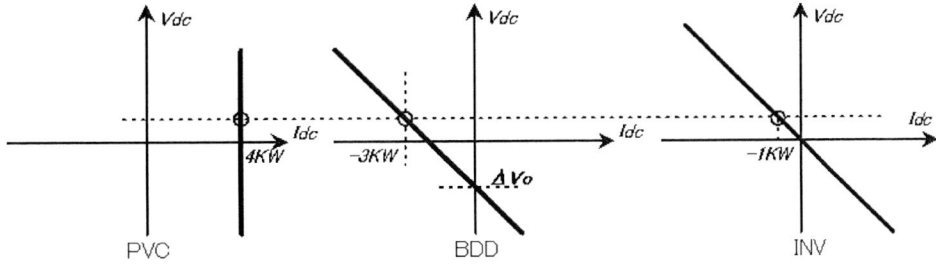

Fig. 8. Operating point (the circle) for a static case of $P_{pvc} = 4\text{kW}$, $P_{inv} = -1\text{kW}$, $P_{bdd} = -3\text{kW}$.

INV can follow its specific power command, without influence to other units.

E. Examples

It would be easier to understand the control strategy with examples. Let us suppose the system is working at the situation just shown in Fig. 5, and then consider following two cases.

1. Solar generation reduces from 4kW to 1kW suddenly due to cloud movement

Then the HVDC voltage level will drop down suddenly, and leads to a lower output of INV. The power of INV is send to MNG and a minus error of power: eP_{inv} is sensed by MNG control loop. The integrator will enlarge ΔV_0 in positive way, and MNG sends it to BDD. By raising the droop profile, BDD's power move from minus to plus, that means BDD discharge to supply the reduction of PV, thus, to keep INV output 3kW stably. Finally, the system converges to an operation point shown in Fig. 7.

2. INV command is changed by user from -3kW to -1kW

In this case the trigger is the reference value of INV power: $P_{inv}{}^*$. Before controller converges, the power error eP_{inv} is 2kW, and output a minus movement of offset. As shown in Fig. 8, ΔV_0 continue to move down until output of INV in accordance with reference value -1kW.

V. EXPERIMENT

We have designed a prototype, as depicted in Fig. 9, and practically confirmed proposed control strategy. It can be seen that the whole system is integrated in a rack. Units can be increased or moved freely to change the system specifics. Batteries are also mounted on the rack. User only need to connect PV and utility for doing energy management.

In Fig. 10, we show a simple experimental result of testing the transient response behaviour of output power change. By changing output commande from 3.5kW to 2.5kW, it can be seen HVDC voltage level response in less than 200ms, and system converges rapidly.

Fig. 9. Prototype of the rack mounted 3 phase Energy Management System, including BDD, PVC, INV, MNG and battery modules.

VI. CONCLUSION

In this paper, we introduced the control solution in our Energy Management System. There are more units to control the HVDC voltage, as a result, it is easier and more stable to balance power flow. Moreover, we use high speed communication to indirectly control inverter output by shift the droop profile of battery charger converter. It increases the stability and flexibility of the system. The strategy can also be applied to other types of paralleling connected power converter systems, such as DC/DC converter in automotives.

REFERENCES

[1] M. Badawy and Y. Sozer. "Power Flow Management of a Grid Tied PV-Battery System for Electric Vehicles Charging", *IEEE Trans. on Industry Applications*, vol. 53, no. 2, pp. 1347-1357, 2017.

[2] S. G. Luo, Z. H. Ye, R. L. Lin and F. C. Lee, "A Classification and Evaluation of Paralleling Methods for Power Supply Modules",

The 2018 International Power Electronics Conference

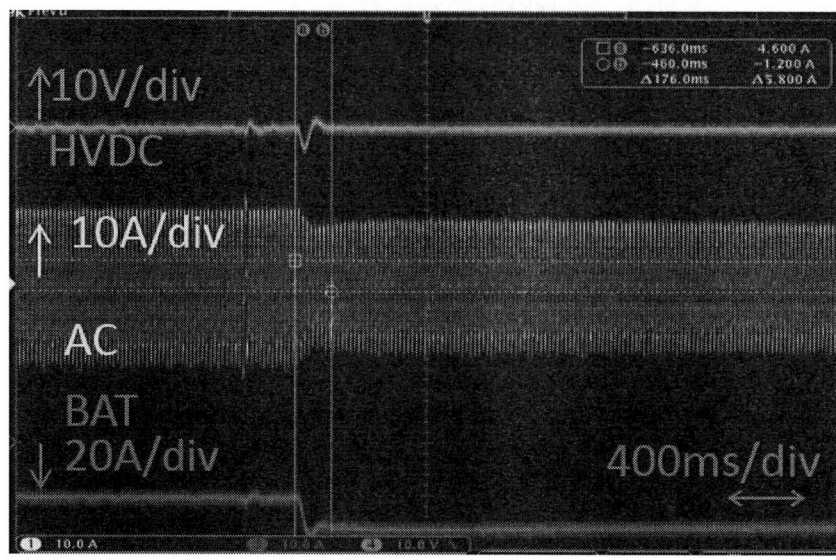

Fig. 10. An experimental waveform when INV power is changed from -3.5kW to -2.5kW. Green, yellow and red waveforms are voltage level of HVDC bus, output AC current and battery DC current, respectively. PV generation is 1kW.

Proceeding of IEEE Power Electronics Specialist Conference (PESC), vol. 2, pp. 901-908, 1999.

[3] Y. Ito, Z. Yang and H. Akagi, "A Control Method of a Small-Scale DC Power System Including Distributed Generators", *IEEJ Trans. Industry Application*, vol. 126, no. 9, pp. 1236-1242, 2006.

[4] X. N. Lu, J. M. Guerrero, K. Sun and J. C. Vasquez, "An Improved Droop Control Method for DC Microgrids Based on Low Bandwidth Communication with DC Bus Voltage Restoration and Enhanced Current Sharing Accuracy", it IEEE Trans. on Power Electronics, vol. 29, no. 4, pp. 1800-1812, 2014.

[5] Y. J. Gu, X. Xiang, W. H. Li and X. N. He, "Mode-Adaptive De-centralized Control for Renewable DC Microgrid With Enhanced Reliability and Flexibility", *IEEE Trans. on Power Electronics*, vol. 29, no. 9, pp. 5072-5079, 2014.

Maximum Power Point Tracking Control for Small Hydroelectric Generation

Kazuya Azegami[1]*, Masashi Takiguchi[1], Junya Yano[1], Hirohiko Tsutsumi[1] and Toshitake Masuko[2]

1 Research & Development Group, Meidensha Corporation, Numazu-city, Japan
2 Electric Power and Social Infrastructure System Business Group, Meidensha Corporation, Shinagawa-city, Japan
*E-mail: azegami-k@mb.meidensha.co.jp

Abstract— **We developed a power converter for small hydroelectric generation. It is an AC/DC/AC conversion system with an inverter for permanent magnetic generator control and an inverter for system interconnection contained in one package. Sensorless PM rotating machine speed control with MPPT was applied to generator control. For MPPT, A similar generator control system for solar power generation was applied and stability control was added while accounting for the fact that hydroelectric generation is subject to higher fluctuation in the power generation amount than solar power generation. In addition, a protective function was added to prevent the mechanical system from entering a critical speed range such as overspeed or mechanical resonance speed.**

Keywords— *Small Hydroelectric Generation. MPPT Control. Permanent Magnet Generator.*

I. INTRODUCTION

Hydroelectric generation is one method of renewable energy power generation that can stably supply power both during the day and night as well as throughout the year. Among other methods, small hydroelectric generation, which is classified as a small-capacity power generation system, has a high development potential due to its small installation area and capability of effectively using the energy of water flowing in rivers and water channels.

Small hydroelectric generation uses water such as that which flows in rivers and that discharged from power stations and factories into rivers. Accordingly, the water flow rate is not always constant.

The maximum output rotational speed of a hydraulic turbine varies with the water flow rate flowing into the turbine and the effective head. The variable speed hydraulic turbine generator system changes the rotational speed of the hydraulic turbine according to the water flow rate and the effective head to generate power at the maximum power point wherein the turbine output is maximized [1][2]. To adjust the rotational speed automatically, however, the maximum power point needs to be determined in advance by storing the hydraulic turbine property data in the system. To that end, the maximum power points according to the operating

conditions need to be determined in advance by measuring the hydraulic turbine property data under various operating conditions. However, if there is an error in the measured data or the system characteristics change due to causes such as aging, the maximum power operation may not always be able to be realized because the maximum power point may deviate from the predetermined point.

We developed maximum power point tracking control (MPPT control) for variable-speed hydroelectric generation with a function to protect against the reaching of mechanical critical speeds such as overspeed and mechanical resonance speed. In doing so, we also enabled operations to occur at the maximum power point without needing property data from the hydraulic turbine or a sensor to measure the water flow rate [3][4]. This paper presents the configuration of the MPPT control and the results of a power generation performance test with the water flow rate changed during MPPT control.

II. CONFIGURATION OF SMALL HYDROPOWER GENERATION CONVERTER

Fig.1 shows the configuration of the small hydroelectric generation system discussed in this paper. The converter for small hydroelectric generation was configured by combining a permanent magnet synchronous generator (PMG) and a converter for small hydroelectric generation containing a power converter for system interconnection and a power converter for PMG control.

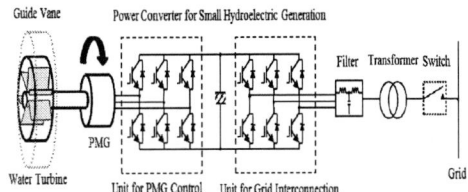

Fig. 1. Configuration of Small Hydroelectric Generation System

This system can generate power with higher efficiency than a hydroelectric generation system operated at a constant speed by directly connecting the hydraulic turbine and the PMG. This system adjusts the water rate flowing into the hydraulic turbine through the opening of guide vanes. In addition, the system is configured by applying the sensorless rotary machine speed control system to the PMG control so that no speed sensor is required.

This converter is mounted with MPPT control, which does not need information from outside the converter, such as the water flow rate and effective head, as its characteristic function. The adjustable-speed power-generating operation by the MPPT control enables the converter to automatically change the generator speed to a speed wherein the hydraulic turbine output is maximized, even if the water flow rate or the effective head changes. The configuration of the MPPT control is shown below.

III. MPPT CONTROL FOR HYDROELECTRIC GENERATION

The hydroelectric generation power output is expressed as Equation (1).

$$P = g \cdot D \cdot Q \cdot He \cdot n \qquad (1)$$

g: Gravitational acceleration [m/s²],
D: Water density [kg/m³]
Q: Water flow rate [m³/s], He: Effective head [m],
n: Overall efficiency of the system,
P: Power output
from the hydroelectric generation system [kW]

When the water flow rate or the effective head changes, the power output will change similarly. Fig.2 shows an example of the change in the hydraulic turbine speed-output characteristics associated with a change in the water flow rate. The curves are shaped like a mountain with the maximum power point at the center.

Systems with an output characteristic similar to Fig.2 include the solar power generation system. In the solar power generation system, the DC voltage-output characteristics of the solar battery are shaped like a mountain, similar to the speed-output characteristics of the hydraulic turbine. The solar power generation system adopts a technique that explores the maximum power point with the algorithm called "hill climbing," which changes the DC voltage to a direction wherein the output increases when it controls the MPPT [5]. There are two differences between the solar power generation system and the hydraulic turbine power generation system. The first difference is that the output value of the solar power generation system is altered by a change in an electric factor (i.e., the DC voltage), whereas the output value of the hydraulic turbine power generation system is altered by a change in a mechanical factor (i.e., the hydraulic turbine speed). The second difference is that

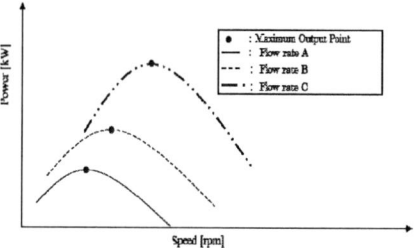

Fig. 2. Hydraulic Turbine Speed-Output Characteristics

the hydraulic turbine power generation system, in which the hydraulic turbine operating range varies according to the water flow rate and the effective head, needs to prevent mechanical failures caused by the speed. Therefore, the hydraulic turbine power generation system should control the MPPT in consideration of the following points:

1. The MPPT control should assess the power generated in a stable speed state that is free from transient power fluctuations due to an acceleration torque or deceleration torque.

2. Even if the water flow rate or the effective head changes, the hydraulic turbine speed should be converged to the maximum power point promptly to alleviate the condition wherein overload pressure is applied to the hydraulic turbine.

3. The hydraulic turbine power generation system should be controlled so that it may not reach a speed wherein its inherent mechanical resonance is manifested or an overspeed in excess of the maximum speed at which the hydraulic turbine power generation system can operate.

The MPPT control for the hydraulic turbine power generation system is discussed below in consideration of the above points.

The MPPT control for variable-speed hydroelectric generation presented in this paper explores the maximum power point by changing the rotational speed of the generator and comparing the generated power at various speeds. This control is configured to explore the maximum power point through the utilization of an algorithm in which hill climbing and dichotomy are combined.

The control is configured to repeat the transition of the four states. Table I shows the list of the states. Fig.3 shows the transition between the states.

The power value at speed ω_{n-1} at the start of MPPT control will be stored as a pre-shift power value P_{n-1}. For this power value, the output power value to the system detected by the power converter for system interconnection will be used. The generator speed will then be changed, as shown in Equation (2), by setting

TABLE I
Control State List

State No	State Name
State 1	Processing of speed destination determination
State 2	Processing of speed destination safety assessment
State 3	Processing of speed shift completion assessment
State 4	Processing of speed shift step width change

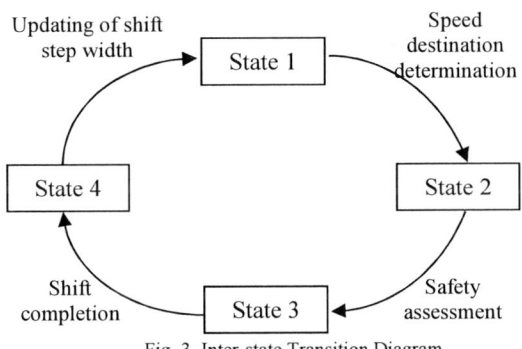

Fig. 3. Inter-state Transition Diagram

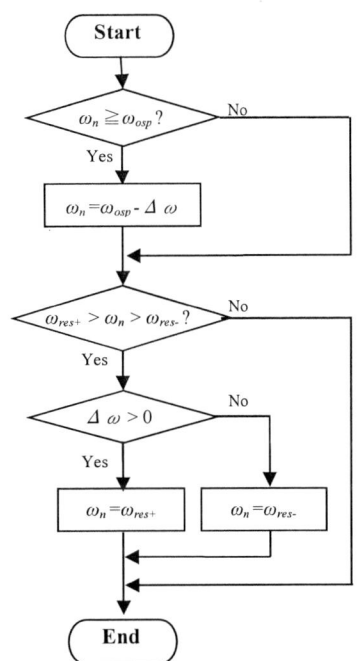

Fig. 4. Critical Speed Avoidance Processing

a value with the shift step width $\Delta\omega$ added to the present speed to the generator as a new speed command, ω_n.

$$\omega_n = \omega_{n-1} + \Delta\omega \qquad (2)$$

At this time, if the value of ω_n is a speed in the overspeed range or a speed with a mechanical resonance point, ω_n will be changed so that it will shift to a speed that is free from overspeed, or a speed that does not have a resonance point. If the speed exceeds the overspeed criterion ω_{osp}, ω_n will be shifted in a direction away from ω_{osp}. If the speed is within the speed range $\omega_{res-} < \omega_{res} < \omega_{res+}$ around the mechanical resonance point speed ω_{res}, the mechanical resonance point speed will be avoided by shifting the speed to ω_{res-} or ω_{res+}. Fig.4 shows the processing flow chart.

After the generator speed completes shifting to ω_n, the power value speed will be saved as the post-shift power value P_n. The power value P_n will be saved, after verifying that the generator speed has reached a stable steady state, to exclude the fluctuations of the power value due to an acceleration torque or deceleration torque, which occurs during the rotational speed change from P_n. Then, the direction of the next shift step width $\Delta\omega$ will be determined by comparing the pre-shift power value P_{n-1} and the post-shift power value P_n. If P_n decreases to a value that is lower than P_{n-1}, the polarity of $\Delta\omega$ will be reversed, given that the maximum power point exists between the pre-shift speed and the post-shift speed. If P_n increases to a value that is higher than P_{n-1}, $\Delta\omega$ will not be reversed. After $\Delta\omega$ is reversed, $\Delta\omega$ will be attenuated by the gain K_d to explore the maximum power point at a smaller interval. If the speed shift direction continuously

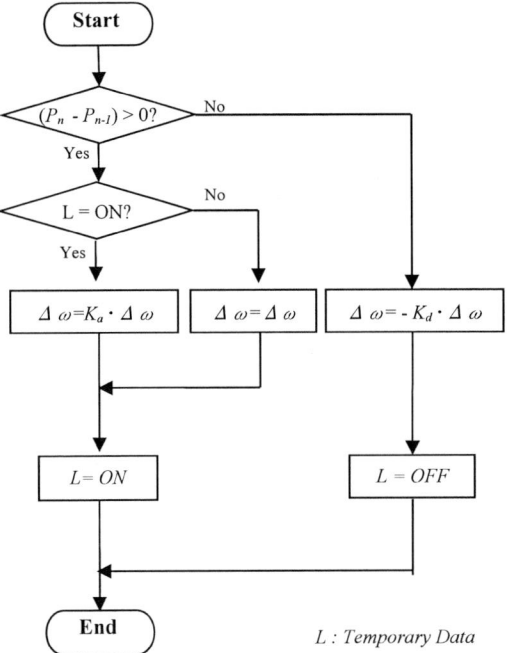

Fig. 5. Speed Shift Step Width Determination Processing

remains the same, $\Delta\omega$ will be increased by the gain K_a to allow shortening of the time taken for the convergence to the maximum power point, given the distance to the maximum power point thus far. Fig.5 shows the processing flow chart for determining this shift step

3725

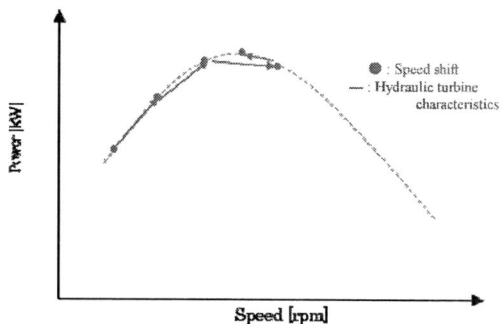

Fig. 6. Maximum Power Point Convergence Behavior

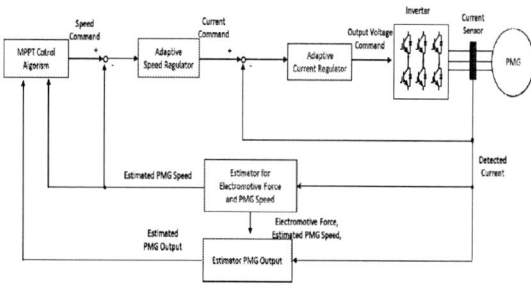

Fig. 7. PMG Control Configuration

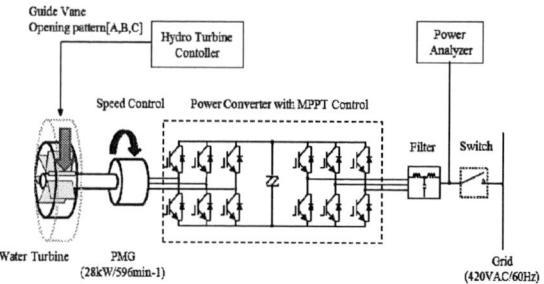

Fig. 8. Test Block Diagram

TABLE II
PMG Rating Used in the Test

Item	Specification
Output	28 kW
Pole number	10
Voltage	390 Vrms
Rotational speed	596 rpm

TABLE III
Maximum Power Point by Guide Vane Opening

Opening pattern	Maximum power point	Generated power
A	512 rpm	10.4 kW
B	596 rpm	21.6 kW
C	644 rpm	26.6 kW

width $\Delta\omega$. After $\Delta\omega$ is determined, the present speed command ω_n will be saved as ω_{n-1} and the post-shift power value P_n as P_{n-1}, respectively, as previous value information. Equation (3) shows the relationship between gains K_a and K_d that increases and decreases $\Delta\omega$.

$$K_a > 1 > K_d \tag{3}$$

The speed command ω_n, to be explored next, will be determined from the previously determined speed shift step width $\Delta\omega$ based on Equation (2). The aforementioned processing flow will be repeated to converge the hydraulic turbine speed to the maximum power point. Comparing the power values enables the tracking of the maximum power point under respective conditions, even if the water flow rate or the effective head changes. Fig.6 shows an example in which this control converges the generator speed to the maximum power point.

IV. CONTROL CONFIGURATION

Fig.7 shows the control configuration of a PMG-based small hydraulic power generation system. A small hydraulic power generation system may be installed in a special environment, such as within a pipeline or under water. In such cases, a speed sensor or magnetic pole position sensor may not be able to be mounted to the PMG; it may be difficult to secure a signal wire route to connect a sensor to a power converter. To cope with such situations, this control does not use a speed sensor or

magnetic pole position sensor; instead, this control uses a sensorless speed control that estimates the electromotive force and PMG speed using extended electromotive force [6]. The estimated speed will be used to perform stable determination processing of the PMG speed under the MPPT control. Then, the power output from the PMG will be estimated based on the estimated electromotive force of the PMG, the PMG speed, and the detected current value. This estimated value is used in the assessment processing of the increase and decrease of the output power under the MPPT control.

V. MPPT CONTROL DEMONSTRATION TEST RESULTS

This section shows the test results from the real machine system using this control. This test verifies whether the speed can converge to the maximum power point that exists during every opening of the guide vanes. This is done by changing the opening of the guide vanes to alter the water flow rate with the MPPT control enabled. Fig.8 shows the test block diagram. Table II shows the PMG ratings used in the test. Table III shows the characteristic data for the hydraulic turbine according to the opening of the guide vanes. This characteristic data is the maximum output speed rpm and the generated power at the speed measured by changing the PMG speed in accordance with the guide vane opening before the MPPT control demonstration test. The opening is sized A<B<C. Table 4 shows the test conditions. In this test, the opening of the guide vanes was changed according to

the operation order shown in Table IV after the convergence of the speed point shifts were observed.

Table V shows the test results. Fig.9 shows an overall chart of the speed commands to the PMG, the estimated PMG speeds, and the PMG output power during the test. Fig.10, Fig.11, Fig.12, and Fig.13 show an extended view with varied guide vane openings. The figures indicate that the PMG speeds converged to a speed in the proximity of the maximum output point according to the respective guide vane opening.

TABLE IV
Test Conditions

Operation order	Opening pattern
1	C
2	B
3	A
4	B
5	C

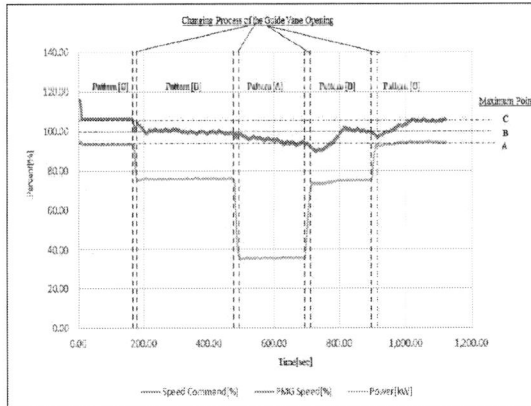

Fig. 9. Test Results

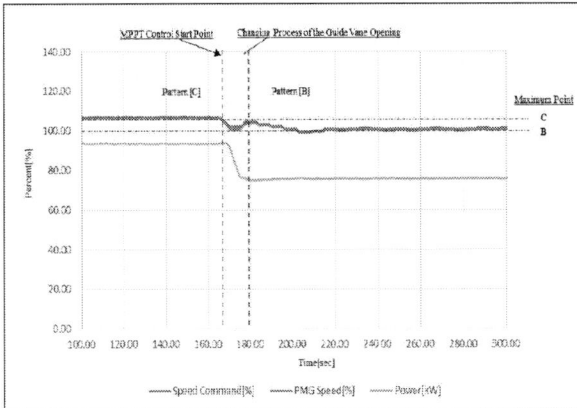

Fig. 10. Test Results (Guide Vane Opening C to B)

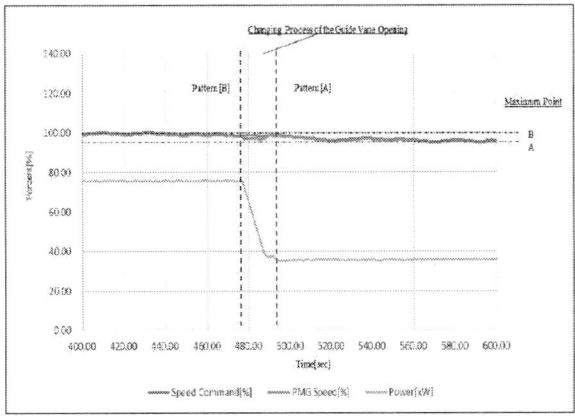

Fig. 11. Test Results (Guide Vane Opening B to A)

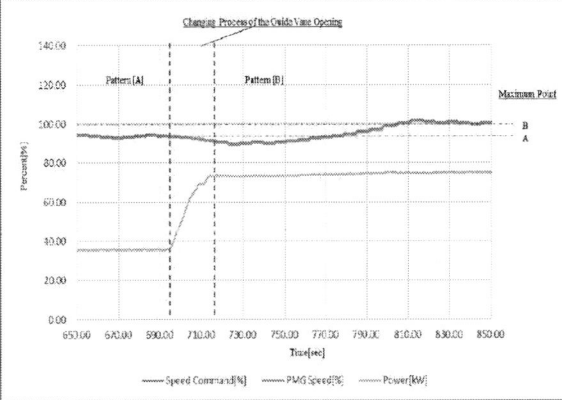

Fig. 12. Test Results (Guide Vane Opening A to B)

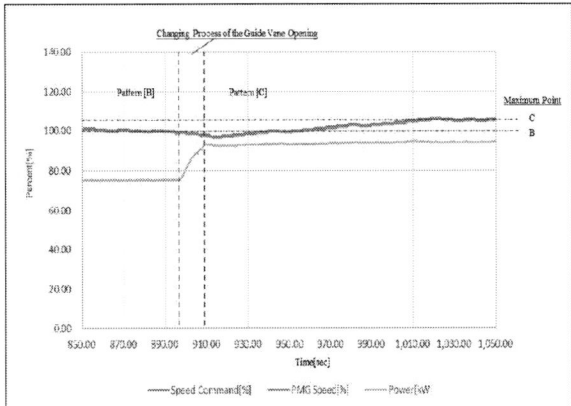

Fig. 13. Test Results (Guide Vane Opening B to C)

TABLE V
Test Results

Operation order	Opening pattern	Convergence speed	Power at convergence point
1	C	643 rpm	26.6 kW
2	B	600 rpm	21.6 kW
3	A	564 rpm	10.4 kW
4	B	606 rpm	21.5 kW
5	C	638 rpm	26.6 kW

VI. CONCLUSIONS

This paper presented the MPPT control function applied to the hydraulic turbine power generation system. The use of this function enables the hydroelectric generation system to always be operated with the output energy maximized, without external information such as the water flow rate, even if the water flow rate or the effective head changes. We are committed to producing future products that further contribute to the environment by establishing an operation method that increases the hydraulic turbine efficiency to improve the power generation efficiency of the entire system.

REFERENCES

[1] H. Ogura, H. Fujimori, "Variable Speed Micro Hydro Power System," MEIDEN REVIEW, No. 167, pp.32–36 (2016)

[2] J. Yano, M. Takiguchi, K. Azegami, H. Tsutsumi, T. Masuko, "Experimental Verification of Automatic Switching Method of Grid Connected/Isolated Operation for Small Hydropower Generation Converter" 2017 Annual Meeting Record I.E.E. Japan Vol. 6, pp.552–553

[3] K. Azegami, "System Interconnection Operation Device and Control Method of Power Generation System," Japanese Patent, 2017-51011

[4] K. Azegami, "Hydropower Generation System," Japanese Patent, 2016-223368

[5] T. Kohara, "Solar Power Generation Control Device and Control Method" Japanese Patent, 2017-85762

[6] K. Tanaka, I. Miki, "Position sensorless Control of Interior Permanent Magnet Synchronous Motor Using Extended Electromotive Force," IEEJ Trans. IA, Vol. 125, No. 9, 2005, pp.833–838

Design and Experimental Verification of a Three-Phase Dual-Active Bridge Converter for Offshore Wind Turbines

Takushi Jimichi[1*], Murat Kaymak[2] and Rik W. De Doncker[2]

1 Advanced Technology R&D Center, Mitsubishi Electric Corporation, Hyogo, Japan
2 E.ON Energy Research Center, RWTH Aachen University, Aachen, Germany
*E-mail: Jimichi.Takushi@cb.MitsubishiElectric.co.jp

Abstract— This paper deals with design and experimental verification of a medium-voltage (MV) DC-DC converter intended for offshore wind. The power rating of the turbine tends to reduce the total cost. However, higher power ratings and longer distances between the turbines cause higher losses in the AC collector grid due to the required larger cable sizes. Furthermore, the need for higher voltage ratings in the AC collector grid to reduce the current rating i.e. cable sizes results in big step-up transformers. To solve these disadvantages, researchers and engineers are considering the system with a pure DC collector grid. With focus on the DC collector grid, the authors have discussed the design of a 10-MW ±25-kV DC-DC converter regarding the converter structure and topology, the semiconductor arrangement, and the transformer to increase the efficiency. Following the above considerations, this paper describes the prototype construction, the experimental verification, and the proposed design of the medium-frequency transformer. The experimental results and the loss analysis verify that the 10-MW ±25-kV DC-DC converter can achieve the efficiency of 98.5% by the three-phase DAB topology, the suitable selection of the semiconductor devices, and the revised medium-frequency transformer.

Keywords— *dual-active bridge converters, offshore wind turbines, medium-frequency transformers.*

I. INTRODUCTION

In Europe, electrical energy production from renewables is expanding to meet CO_2 targets and nuclear power phase-out [1][2]. Especially in Germany, UK, and Denmark, large-scale offshore wind projects have been planned and carried out. Until now, it has been recognized that renewables are an environment-friendly energy source and one of the unavoidable solutions to meet the above-mentioned targets and policies. Therefore, governments of various nations have supported the renewables by subsidies.

However, with the reduction of the cost of energy (COE) [3], some offshore wind projects are recently planned without subsidies [4]. One reason of the COE reduction is caused by the increase of the turbine power rating. The elevated power rating can increase the electricity production and hence reduce the civil and maintenance costs by lower number of turbines. At the

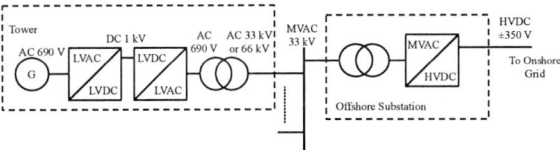

Fig. 1. System configuration as an example of existing offshore wind farm with MVAC collector grid.

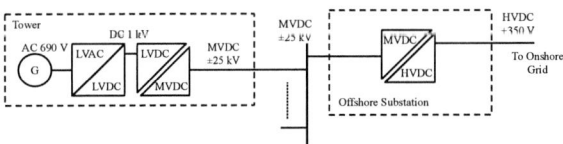

Fig. 2. System configuration of the proposed offshore wind farm with MVDC collector grid.

moment, the 8-MW turbine is a typical scale, and the 9.5-MW upgraded turbine has also been announced [5]. Moreover, it is estimated that the power rating will be augmented to 13-15 MW by 2024 [4].

However, the higher power rating of the turbines results in higher current ratings of the collector grid and longer distance between turbines, so the size and the loss of the collector cables increase. There are two solutions to overcome the disadvantages. The first solution is to increase the collector grid voltage from 33 kV (AC) to 66 kV (AC) [6] as shown in Fig. 1. Some commercial collector grids for turbines have already corresponded to 66 kV (AC) [7]. An alternative solution is the use of the medium-voltage DC collector grid instead of the AC collector grid [7]-[12], as shown in Fig. 2.

A possible output voltage of the wind generator in the turbine is 690 V (AC) and the voltage is converted to DC 1 kV for controlling the torque and the maximum power point tracking (MPPT). Therefore, the output voltage of the generator and the converter is DC 1 kV. On the other hand, regarding transmission, HVDC is applied in recent offshore wind projects. Therefore, the use of the DC collector grid is a reasonable way to reduce not only

The 2018 International Power Electronics Conference

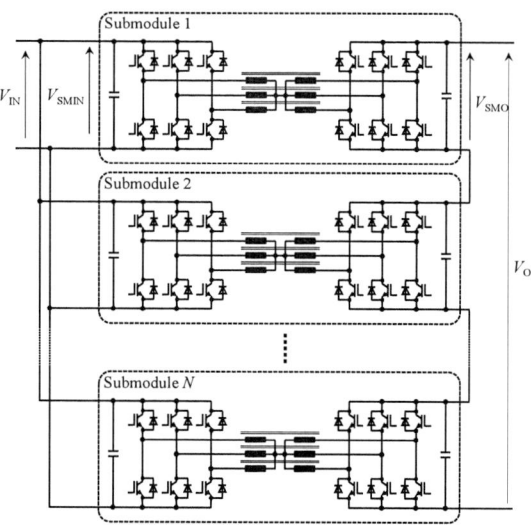

Fig. 3. The proposed configuration of the 10-MW ±25-kV DC-DC converter.

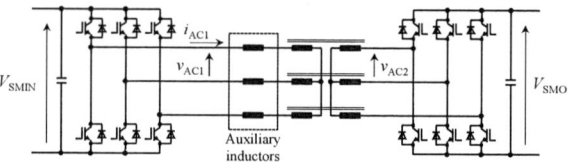

Fig. 4. The circuit configuration of the constructed 200-kW submodule.

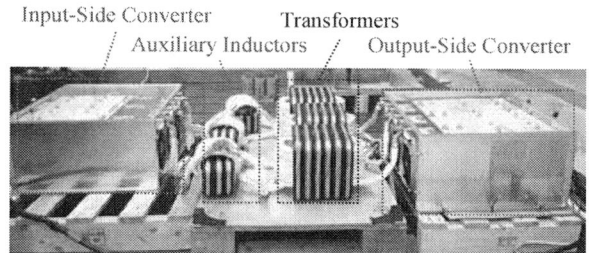

Fig. 5. The experimental setup of the 1-kV 200-kW DC-DC Converter.

TABLE I
PARAMETERS OF THE MV DC-DC CONVERTER

Item		Value
Total Power Rating	P_T	10 MW
Rated Input Voltage	V_{IN}	1 kV
Rated Output Voltage	V_O	±25 kV
Number of Submodules	N	50
Submodule Power Rating	P_{SM}	200 kW
Submodule Input Voltage	V_{SMIN}	1 kV
Submodule Output Voltage	V_{SMO}	1 kV

the collector grid loss but also the converter loss, because DC-DC conversion is generally more efficient than DC-AC or AC-DC conversion. Moreover, DC-DC conversion with elevated switching frequency of efficient power semiconductor devices can reduce the volume and weight of the transformer. Regarding the system efficiency, by tuning the efficiency of the DC-DC converter to 98.5%, the total loss of the transformer and the converter system shown in Fig. 2 can be reduced by 1% compared to the system in Fig. 1 [10].

With focus on the DC collector grid, the paper [9] has discussed the suitable design of the converter structure and topology, the semiconductor arrangement, and the transformer for the ±25-kV 10-MW DC-DC converter to reach the efficiency of 98.5%.

Following the above discussion, this paper presents the prototype construction, the experimental verification, and the revised design of the medium-frequency transformer. The experimental results include the behavior of the prototype, the thermal measurement, and the loss measurement. Finally, the loss analysis verifies that the 10-MW ±25-kV DC-DC converter can achieve the desired efficiency of 98.5%.

II. CIRCUIT CONFIGURATION

Fig. 3 shows the proposed converter configuration for the 25-kV 10-MW DC-DC converter in [10][11], and Table I indicates the circuit parameters. The converter consists of 50 submodules configured with the three-phase DAB topology, which are connected in parallel at the input side, and in series at the output side. The converter configuration is characterized as follows:

- The input and output voltages are set to 1 kV, so the commercially available 1.7-kV IGBT modules can be used.
- Because of the submodule power rating of 200 kW, this submodule configuration provides uncomplicated thermal management, good maintainability, and high reliability with redundant extra submodules in the configuration. Furthermore, at low load conditions not needed submodules are shutdown to maintain operating submodules in the high efficiency range.
- The three-phase DAB topology can reduce the semiconductor loss and the volume and weight of the converter by 20% compared to the single-phase system [11].
- The converter efficiency can achieve 98.5%.

III. PROTOTYPE DESIGN AND CONSTRUCTION

Fig. 4 demonstrates the circuit configuration of the constructed prototype with the power rating of 200 kW, and the constructed prototype of one submodule is depicted in Fig. 5 with DC voltages of the input and output side of 1 kV.

Regarding the semiconductor devices, 6th-generation IGBTs (Mitsubishi Electric, CM450DXL-34SA,

3730

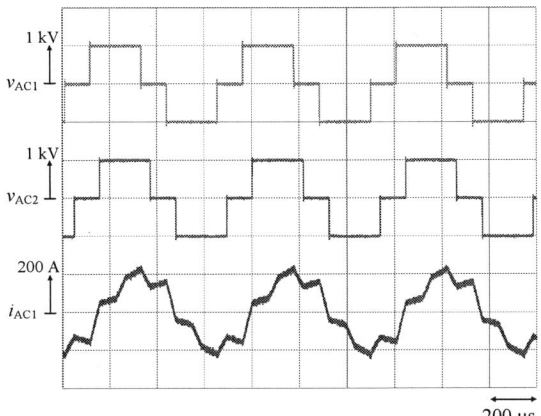

Fig. 6. The experimental waveforms of the input-side AC voltage, the output-side AC voltage, and the input-side AC current under rated condition.

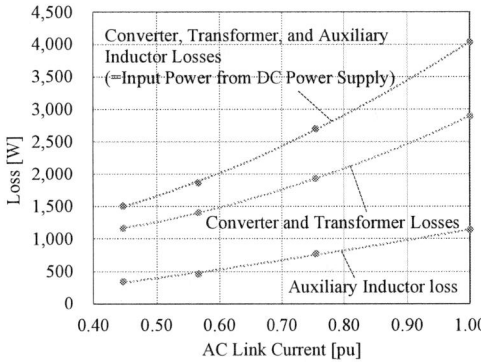

Fig. 7. The experimental results of the loss measurement.

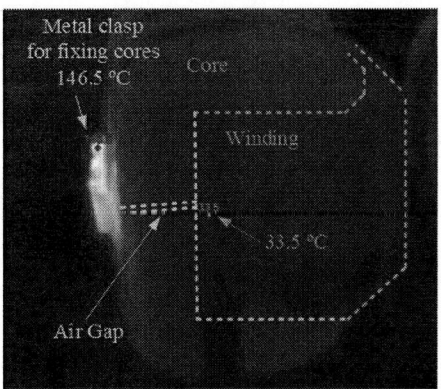

Fig. 8. The thermal image of the auxiliary inductor.

1,700 V/450 A) are used for the input-side converter, and 5th-generation IGBTs (Mitsubishi Electric, CM400DY-34A, 1,700 V/400 A) are used for the output-side converter. This IGBT arrangement reduces the conduction loss of the IGBTs at the input side, and the conduction loss of the diodes at the output side, according to the semiconductor characteristics [13].

For the input-side and output-side converters, the gate drivers GAU205P-15252A (ISAHAYA Electronics) are used with gate resistors of the recommended resistance shown in the datasheets [13].

Regarding the medium-frequency transformers, the shell-type configuration with the nanocrystalline tape-wound cut core Vitroperm 500 F (T60102-L2 198-W160, Vacuumschmelze) [14][15] and litz-wire copper windings with the current density of approximately 1.7 A/mm² are selected. The turn ratio of the transformer is 1:1, and the turn number is designed according to the maximum allowable flux density of 1 T and the switching frequency of 1.55 kHz.

If the stray inductance of the transformer is not sufficient, auxiliary inductors are used in series. The detailed information of the inductors and the transformer is described in later section.

The capacitance of the DC capacitor is chosen to 4.2 mF at each side. The unit capacitance constant H defined as the ratio between the capacitor-stored energy and power rating [16] is about 20 ms, which is similar to the value in the AC system (e.g. AC-DC converters).

The control system XCS2000 (AixControl) [17] produces the gate signals with the implementation of the phase-shift control calculated by following equation [18]:

$$P = \frac{V_{SMIN}V_{SMO}}{8\pi f_{SW}L}\left(\phi - \frac{\phi^2}{\pi}\right), \qquad (1)$$

where P is the active power, V_{SMIN} is the input voltage, V_{SMO} is the output voltage, f_{SW} is the switching frequency, L is the total inductance composed by the stray inductance

of the transformer and the auxiliary inductor, and ϕ is the phase-shift angle between the input-side and output-side AC voltages.

IV. EXPERIMENTAL RESULT

In the experimental verification, the total loss of the prototyped submodule is measured by the precise power analyzer LMG 500 (ZES ZIMMER) [19] with the high accuracy of 0.025%. In addition, the loss of the auxiliary inductors is measured to discuss the loss behavior and the revised transformer design.

Fig. 6 shows the waveforms of the AC-link line-to-line voltages at the input side, at the output side, and the AC-link current under rated condition. The output-side AC voltage and the input-side AC voltage are phase shifted to send the rated power. The phase shift between the input-side and the output-side AC voltages produces the AC-link current and active power flow. The performance of the DC-DC converter shown in Fig. 6 is verified and the AC-link current contains no DC component.

Fig. 7 shows the measured losses, where the blue line represents the total loss composed of the converter, the transformer, and the auxiliary inductor losses. The orange line represents only the auxiliary inductor loss, and the gray line represents the converter and transformer losses

TABLE II
COMPARISON OF THE PROTOTYPED AND IMPROVED
TRANSFORMER

	(a) Prototyped Transformer	(b) Improved Transformer
Configuration		
Leakage Inductance	7.53 μH	54.6 μH
Estimated Loss — Equation	90.1 W	85.9 W
Estimated Loss — FEM	89.8 W	84.8 W

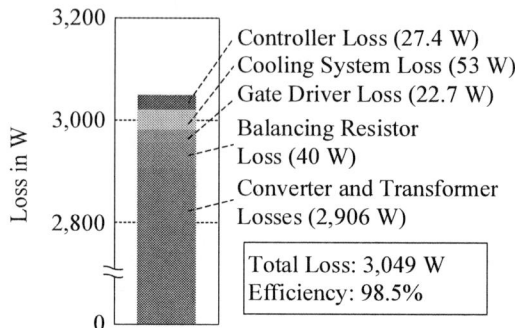

Controller Loss (27.4 W)
Cooling System Loss (53 W)
Gate Driver Loss (22.7 W)
Balancing Resistor Loss (40 W)
Converter and Transformer Losses (2,906 W)

Total Loss: 3,049 W
Efficiency: 98.5%

Fig. 9. The loss distribution of the 200-kW submodule with the improved transformer.

calculated by subtracting the auxiliary inductor loss from the total loss. Here, the loss of the auxiliary inductors was calculated by voltage and current measurements with the oscilloscope DPO5104B, the differential voltage probe and the current probe TCPA400 (Tektronix), which amounts to 20% of the total loss. With this result, Fig. 8 shows the thermal photograph of the auxiliary inductor immediately after rated operation. The winding temperature is less than 40 °C, whereas the temperature of the standard metal clasp for fixing the cores is 146.5 °C. The high temperature is produced by the eddy current loss caused by the leakage flux from the air gap. These results implies that removing the auxiliary inductors can reduce the system loss by 20%. To maintain the required inductance of the submodule, an increased stray inductance of the transformer without appreciable loss increase can compensate the missing auxiliary inductor.

V. LOSS ANALYSIS

Table II illustrates the transformer configuration with the corresponding leakage inductance, estimated loss calculated by the physical equations [20]-[22] as well as the FEM analysis for the prototype configuration.

The prototyped transformer (a) shown in Fig. 3 uses the overlapping configuration of the primary and secondary windings. Therefore, the resulting vertical leakage flux is very small due to the large height of the core window area, leading to the small stray inductance of 7.53 μH. On the other hand, the improved transformer (b) uses the splitted configuration of the primary and secondary windings. For this reason, the stray flux is then horizontal in the narrow window area, resulting in the higher stray inductance of 54.6 μH. Although the leakage flux in the improved transformer (b) is higher compared to the prototyped transformer (a), the loss of the litz wire may be less subject to the leakage flux. This explanation can be proven by the estimated losses by physical equations and FEM analysis presented in Table II. The loss of the prototyped transformer (a) is approximately the same as the improved transformer (b), where the loss is approximately 90 W.

Therefore, the auxiliary inductor can be removed from Fig. 4 by the use of the improved transformer without increasing the transformer loss.

Fig. 9 shows the total system loss of the submodule including the peripheral circuits without auxiliary inductors under rated condition. The total loss of the converter and the transformer is 2,906 W, which are extracted from the experimental results shown in Fig. 6. The prototyped-transformer loss given by the experimental result is supposed to be the same as the improved-transformer loss.

The balancing resistor of 50 kΩ is used to equal the DC capacitor voltages between submodules at the initial condition and to maintain a safe off-state. The resistance is 50 kΩ, and the loss is 40 W. The losses of the gate drivers and the cooling system are measured respectively, which are 22.7 W and 53 W. The controller loss of 27.4 W is derived by a commercial system including the control circuit and the interface circuit.

Finally, the total system loss is 3,049W and the efficiency of the system is calculated by

$$\eta = \frac{200 \text{ kW}}{200 \text{ kW} + 3049 \text{ W}} = 98.5\%.$$

VI. CONCLUSION

This paper deals with the design and experimental verification of an MV DC-DC converter intended for offshore wind farms. Achieving the efficiency of 98.5% on the MV DC-DC converter can reduce the system loss with the DC collector grid by 1% compared to the system with AC collector grid. Following the past work in [10][11], this paper presents the prototype construction, the experimental verification, and the revised design of the medium-frequency transformer. The experimental results and the loss analysis verify that the 10-MW ±25-kV DC-DC converter can achieve the desired efficiency of 98.5% with the three-phase DAB topology, the suitable selection of the semiconductor devices, and the improved medium-frequency transformer.

REFERENCES

[1] R. W. De Doncker, "Power electronic technologies for flexible DC distribution grids," *in Proc. the 2014 International Power Electronics Conference (IPEC-Hiroshima 2014 – ECCE-ASIA)*, pp. 736-743, 2014.

[2] R. W. De Doncker, "Flexible grids and storage systems - DC-DC converters as a key enabling technology," *in Proc. ECPE Workshop*, Feb. 2016.

[3] "Offshore wind can match coal, gas for value by 2025: RWE, E.ON, GE, others," Reuters Global Energy News, Jun 6, 2016.

[4] Dong Energy Press Release 13.04.2017, Website: http://www.dongenergy.com/en/media/newsroom

[5] MHI Vestas V164-9.5 MW, Website: http://www.mhivestasoffshore.com/category/v164-9-5-mw/

[6] S. Gasnier, V. Debusschere, S. Poullain, and B. Francois, "Technical and economic assessment tool for offshore wind generation connection scheme: Application to comparing 33 kV and 66 kV AC collector grids authors," *in Proc. 18th European Conference on Power Electronics and Applications 2016 (EPE'16 ECCE Europe)*.

[7] P. C. Kjaer, "DC Wind Turbine," i*n Proc. ECPE Workshop*, Feb. 2016.

[8] Y. Lian, G. P. Adam, D. Holliday, and S. J. Finney, "Medium-voltage DC/DC converter for offshore wind collection grid," *IET Renewable Power Generation*, Vol. 10, Issue 5, 2016.

[9] Y. Zhou, D. E. Macpherson, W. Blewitt, and D. Jovcic, "Comparison of DC-DC converter topologies for offshore wind-farm application," *in Proc. 6th IET International Conference on Power Electronics, Machines and Drives (PEMD 2012)*.

[10] T. Jimichi, M. Kaymak, and R. W. De Doncker, "Design and loss analysis of a medium-voltage DC-DC converter intended for offshore wind farms," *in Proc. 5th International Conference on Renewable Energy Research and Applications (ICRERA 2016)*.

[11] T. Jimichi, M. Kaymak, and R. W. De Doncker, "Comparison of single-phase and three-phase dual-active bridge DC-DC converters with various semiconductor devices for offshore wind turbines," *in Proc. IEEE 3rd International Future Energy Electronics Conference 2017 and ECCE Asia (IFEEC 2017 - ECCE Asia)*.

[12] D.-M. Valcan, P. C. Kjaer, and L. Helle, "Cost of energy assessment methodology for offshore AC and DC wind power plants," *in Proc. Optimization of Electrical and Electronic Equipment (OPTIM)*, pp. 919-928, 2012.

[13] Mitsubishi Electric Corporation Website: http://www.mitsubishielectric.com/semiconductors/

[14] Y. Wang, S. W. H. de Haan, and J. A. Ferreira, "Design of low-profile nanocrystalline transformer in high-current phase-shifted DC-DC converter," *in Proc. 2010 IEEE Energy Conversion Congress and Exposition (ECCE 2010)*, pp. 2177–2181.

[15] Z. M. Shafik, K. H. Ahmed, S. J. Finney, and B. W. Williams, "Nanocrystalline cores transformer design and implementation for a high current low voltage DC/DC converter," *in Proc. 5th IET International Conference on Power Electronics, Machines and Drives (PEMD 2010)*.

[16] H. Fujita, S. Tominaga, and H. Akagi, "Analysis and design of a DC voltage-controlled static var compensator using quad-series voltage source inverters," *IEEE Transactions on Industry Applications*, vol. 32, no. 4, pp. 970–978, Jul./Aug. 1996.

[17] AixControl Website: http://www.aixcontrol.de/

[18] R. W. De Doncker, D. M. Divan, and M. H. Kheraluwala, "A three-phase soft-switched high power density dc/dc converter for high power applications," *in Proc. Conf Industry Applications Society Annual Meeting Record of the 1988 IEEE, 1988*, pp. 796–805.

[19] ZES ZIMMER Website: https://www.zes.com/en

[20] Colonel WM. T. McLyman, Transformer and Inductor Design Handbook, Third Edition, Revised and Expanded, 2004.

[21] N. Soltau, R. W. De Doncker, Daniel Eggers, and Kay Hameyer, "Iron losses in a medium-frequency transformer operated in a high-power DC-DC converter", *IEEE Transactions on Magnetics*, Vol. 50, No. 2, Feb. 2014.

[22] M. K. Kazimierczuk, "High-frequency magnetic components," Second Edition, John Wiley & Sons, Ltd, 2014.

The 2018 International Power Electronics Conference

Optimized Bidirectional PFC Rectifiers & Inverters
- Si vs. SiC vs. GaN in 2L and 3L Topologies -

Jonas Wyss*, Jürgen Biela*
*High Power Electronics Laboratory
ETH Zürich, Switzerland
Email: wyss@hpe.ee.ethz.ch

Abstract—In industrial drive systems, PFC converters are widely used. For PFC converters, a lot of different configurations and topologies exist and it is difficult to choose one. Therefore, in this paper, a comprehensive survey of possible configurations are optimized and compared against each other. The analysis shows the important aspects of the converter design and how the used semiconductor technology (Si, SiC and GaN) affects the total converter volume. A sensitivity analysis of certain power semiconductor parameters is performed to highlight the benefit of possible future developments of components and materials. In addition, the differences between the rectifier stage, where an EMI filter stage is necessary and the inverter stage, where the harmonic motor losses are influenced by the modulation scheme, are presented.

I. INTRODUCTION

In industrial drive systems, bidirectional 3-phase boost Power Factor Correction (PFC) converters are widely used to supply electrical motors via a DC link from the mains (Fig. 1). The main goal of the converter is a power flow in both directions (i.e. accelerating and decelerating the motor).

The PFC rectifier has to ensure a high power factor and has to comply with harmonic regulations at low frequencies ($f \leq 700\,\text{Hz}$, LF) and electromagnetic interference (EMI) regulations at high frequencies ($150\,\text{kHz} \leq f \leq 30\,\text{MHz}$, HF) [1]. Applying a suitable modulation scheme with a sufficiently high switching frequency ensures that the rectifier complies with the LF regulations. However, the HF noise has to be attenuated by an additional EMI filter, which has a considerable impact on the converter volume [2].

The design of a PFC rectifier can be focused on various aspects, for example on the EMI filter [3], the switching losses [4] or the CM voltage [5] generation. However, if the focus is only set on one of these aspects, it could worsen other

aspects and the overall system design. Therefore, in [6] a comprehensive converter optimization is presented, analysing the impact of the topology and the modulation scheme on the total converter volume (i.e. the volume of the EMI filter, boost inductor, heat sink and DC link capacitors). Additionally, the benefit of using SiC semiconductors has been analysed. In [7], the benefit of using PCB embedded semiconductors is analysed.

In this paper, the analysis will additionally include GaN devices and a sensitivity analysis on several semiconductor parameters to investigate the benefit on the system level of possible future developments of components and materials. For the EMI filter, different damping topologies and CM configurations are analysed.

As the inverter stage use similar topologies and modulation schemes, the same considerations can be done, except for the EMI filter which is not necessary. However, not only the semiconductor losses have to be considered, but also the harmonic motor losses that depend on the applied modulation scheme [8] and the bearing currents which are a side effect of the generated CM voltage of the converter which can cause bearing failures [9]. In this paper an optimization of the inverter will be proposed, regarding the aforementioned aspects.

Section II describes the investigated topologies and modulation schemes. In section III, the EMI filter design is outlined and section IV shows the optimization procedure. In section V, the models are validated with an experimental boost PFC rectifier and section VI presents the optimization results.

II. TOPOLOGIES AND MODULATION SCHEMES

The topology and modulation scheme are the major design choices for a PFC converter. In section II-A, the investigated topologies are presented, followed by section II-B with an overview of the investigated modulation schemes.

A. Topologies

Fig. 2 shows the investigated topologies. The respective advantages and disadvantages are discussed in the following.

1) 2-level: The 2-level (2L) topology offers low conduction losses and low costs as the amount of power semiconductors is small. The disadvantage are high switching losses, only two

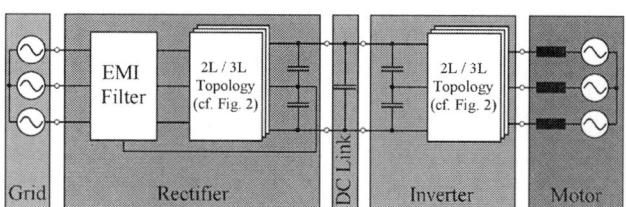

Fig. 1. In a drive system the DC link is supplied with a rectifier from the mains, while the inverter generates the desired motor voltage out of the DC link. The topologies are described in section II-A.

3734

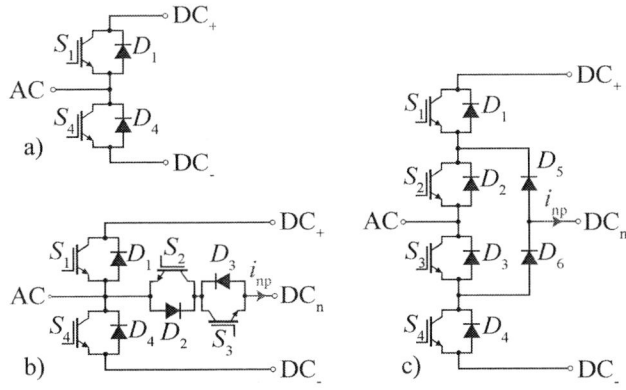

Fig. 2. Investigated topologies: a) 2-level, b) 3-level T-Type and c) 3-level Neutral Point Clamped.

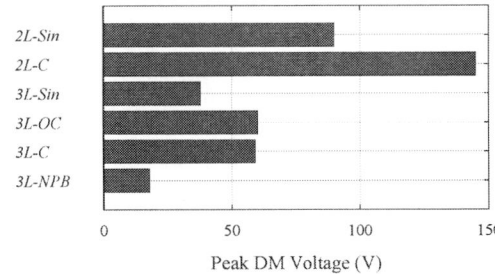

Fig. 3. Generated peak DM voltage of the converter at the switching frequency for different modulation schemes.

voltage levels which affects the EMI filter volume considerably and a high power loss per power semiconductor device.

2) 3-level Neutral Point Clamped: The 3-level Neutral Point Clamped (3L-NPC) topology has the lowest switching losses and a good loss distribution among the power semiconductor devices. The three voltage levels lead to a smaller EMI filter compared to the 2-level topology. The disadvantage are high conduction losses as there are always two power semiconductor devices conducting the current.

3) 3-level T-Type: The 3-level T-Type (3L-TT) topology is a compromise between a 2-level topology and a 3-level NPC topology. The conduction losses are lower compared to the NPC topology as sometimes only one power semiconductor device is carrying the current, but the switching losses are higher as the switches S_1/S_4 have a higher voltage rating than in the NPC topology.

B. Modulation Schemes

Depending on the chosen topology, different modulation schemes can be applied. In the following the investigated modulation schemes are shortly described and the advantages and disadvantages are outlined.

1) 2L Sinusoidal: In the 2L Sinusoidal (*2L-Sin*) modulation scheme, the phase voltages follow a sinusoidal reference [10] and below the switching frequency no CM voltage is generated. However, the switching losses are relatively high.

2) 2L Clamping: In the 2L Clamping (*2L-C*) modulation scheme, the phase with the highest current is clamped to either positive or negative DC link voltage during one switching period [10]. On one hand this reduces the switching losses considerably, on the other hand the DM peak at the switching frequency is higher than in the *2L-Sin* modulation scheme (cf. Fig. 3), thereby increasing the EMI filter.

3) 3L Sinusoidal: Similar to the *2L-Sin* modulation scheme, in the 3L Sinusoidal (*3L-Sin*) modulation scheme the phase voltages follow a sinusoidal reference and no CM voltage below the switching frequency is generated. However,

the neutral point current i_{np} (cf. Fig. 2) is not actively balanced, why a relatively big DC link capacitor is required [6].

4) 3L Optimal Clamping: In the 3L Optimal Clamping (*3L-OC*) modulation scheme, the phase with the highest current is clamped, thereby reducing the switching losses [4]. The disadvantage is the generation of a CM voltage below the switching frequency, a high DM peak at the switching frequency (cf. Fig. 3) and an unbalance of the neutral point of the DC link, resulting in a big DC link capacitor [6].

5) 3L Clamping: In the 3L Clamping (*3L-C*) modulation scheme, the clamping is altered between the phase with the highest and the second-highest current. Compared to the *3L-OC* modulation scheme, the neutral point of the DC link is balanced, resulting in a lower DC link capacitor at the price of increased switching losses.

6) 3L Neutral-Point Balanced: In the 3L Neutral-Point Balanced (*3L-NPB*) modulation scheme, the average current to the neutral point of the DC link i_{np} (cf. Fig. 2) is balanced over one switching period. The switching losses are higher compared to the *3L-OC* and *3L-C* modulation scheme, but due to the DC neutral point balancing the DC link capacitor is smaller and the lower DM peak (cf. Fig. 3) reduces the EMI filter volume.

Based on the chosen topology and modulation scheme in combination with the EMI filter respectively the motor, the voltage and current forms can be calculated. For the rectifier, the EMI filter design (explained in section III) is optimized with respect to the total converter volume (cf. section IV-A). For the inverter, the harmonic losses of the motor and the power semiconductor losses are evaluated as explained in section IV-B.

III. EMI FILTER

At the grid connection of the rectifier stage, the current has to be filtered in order to fulfill the EMI regulations. Section III-A describes the standards which are considered, section III-B shows the investigated damping topologies for the DM filter and in section III-C the investigated CM filter topologies are shown.

3735

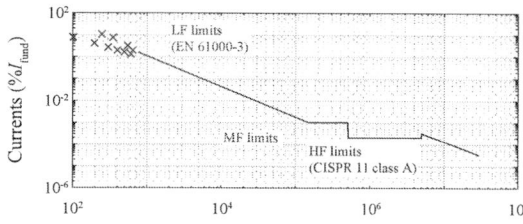

Fig. 4. EMI limits used as boundary in this paper.

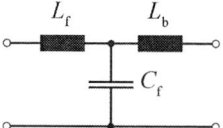

Fig. 5. General structure of the DM filter for one phase.

A. EMI Standards

The EN 61000-3-12 standard is used for the LF and total harmonic distortion limits, while the CISPR 11 class A standard is used for the HF limits. In between, a logarithmic linear function is used as a standard in this frequency range could be expected in the future [1]. Fig. 4 shows the applied limits. Generally, the DM filter consists of a LCL structure as indicated in Fig. 5 in the equivalent circuit for one phase. One trade-off in the filter design is the value of the boost inductance L_b: The higher it is, the higher the boost inductor volume becomes and the lower the other filter elements become and vice versa and is therefore optimized as explained in section IV-A. The resonance frequency

$$\omega_0 = \frac{1}{\sqrt{L_f C_f}} \qquad (1)$$

would lead to an unstable behaviour of the converter, therefore a damping network is required as explained in the next section.

B. Damping Topologies

To avoid that the EMI filter affects the converter control, Middlebrook's stability criterion [11], [12] has to be fulfilled, i.e. the filter output impedance has to be smaller than the converter input impedance. In [13], the criterion has been extended to 3-phase systems. Fig. 6 shows the investigated damping topologies. In [14], these damping topologies have been analysed. It has been shown that the volume of the parallel-RL damping (Fig. 6b) leads always to a smaller volume than the serial-RL damping (Fig. 6d), while serial-RC and parallel-RC (Fig. 6a respectively c) lead to the same

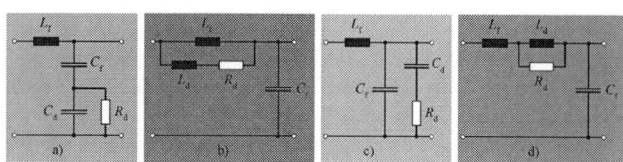

Fig. 6. Passive damping filter topologies: a) Serial-RC damping, b) Parallel-RL damping, c) Parallel-RC damping and d) Serial-RL damping.

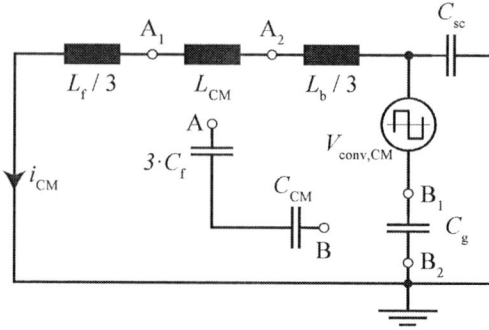

Fig. 7. CM equivalent schematic of a three-phase boost PFC converter. The points A and B can be connected with A_1 / A_2 respectively B_1 / B_2.

volume. Therefore, only the parallel-RC and the parallel-RL damping are considered in this paper.

C. CM Filter Topologies

Fig. 7 shows the CM equivalent circuit of a three-phase boost PFC converter. The CM voltage source $V_{conv,CM}$ depends on the chosen topology and modulation scheme. C_g is the equivalent capacitance from the two DC link poles to ground and C_{sc} is the parasitic capacitance from the power semiconductors to the grounded heat sink. The CM inductor L_{CM} can either be placed on the grid or the converter side. If it is placed on the grid side, there are no HF DM currents which would generate additional losses. On the other hand if it is placed on the converter side, the CM damping is higher. The CM capacitor C_{CM} can either be connected with the neutral point of the DC link or with ground. Therefore, four configurations are investigated:

1) CM Filter Topology 1: L_{CM} is placed on the grid side (A-A_2) and C_{CM} is connected to the DC link midpoint (B-B_1).

2) CM Filter Topology 2: L_{CM} is placed on the converter side (A-A_1) and C_{CM} is connected to the DC link midpoint (B-B_1).

3) CM Filter Topology 3: L_{CM} is placed on the grid side (A-A_2) and C_{CM} is connected to ground (B-B_2).

4) CM Filter Topology 4: L_{CM} is placed on the converter side (A-A_1) and C_{CM} is connected to ground (B-B_2).

IV. OPTIMIZATION PROCEDURE

Based on the chosen topology and power semiconductor devices, modulation scheme, damping topology and CM filter topology, the design parameters of the converter are optimized for maximal power density in order to derive the dependency of the converter volume on the switching frequency. Section IV-A describes the optimization routine for the rectifier stage and section IV-B for the inverter stage. In section IV-C the volume estimation for the different converter components is explained.

The 2018 International Power Electronics Conference

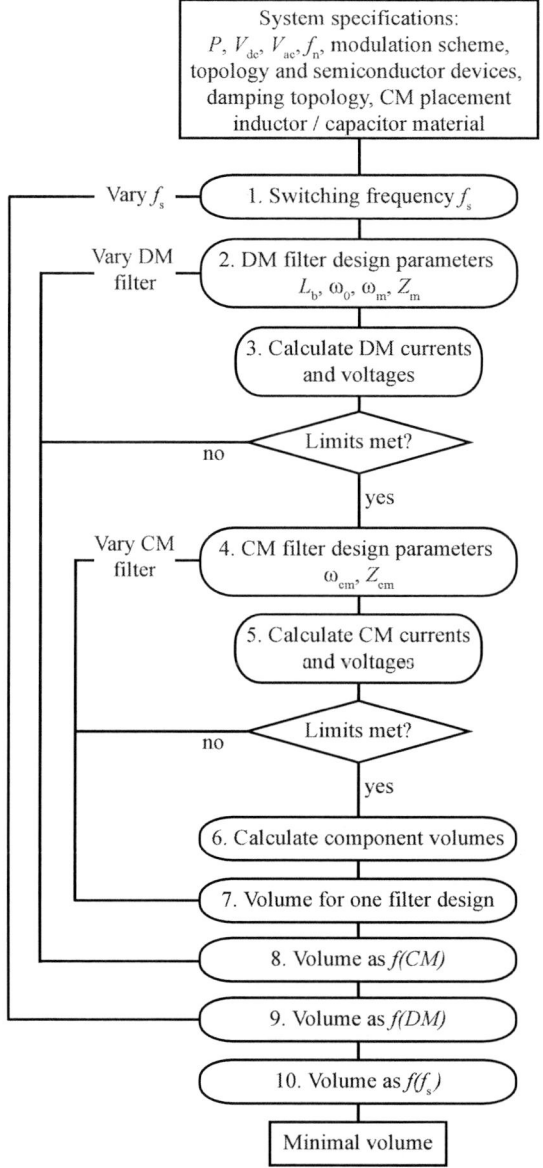

Fig. 8. Flowchart of the cascaded procedure to optimize the boost PFC rectifier for minimal volume [6].

A. Rectifier Optimization Procedure

The considered optimization procedure shown in Fig. 8 has been presented in [14] and is used to identify the switching frequency and CM and DM filter design parameters which lead to the lowest converter volume. The input of the optimization is the system specification which includes the AC and DC voltage, the output power, the inductor and capacitor materials and the topology including semiconductor devices and modulation scheme. For reducing the computing time, the optimization routine is implemented in a cascaded structure. The following DM filter design parameters are optimized:

1) The boost inductance value L_b (trade-off between

boost inductor volume and filter volume).
2) The filter resonance frequency ω_0 (trade-off between filter volume and attenuation).
3) The frequency location of the maximum filter output impedance ω_m (trade-off between damping capability and volume of the damping element).
4) The maximum value of the filter output impedance Z_m (trade-off between inductor and capacitor volume of the DM filter).

In addition, the following CM filter design parameters are optimized:

1) The CM filter resonance frequency ω_{cm} (trade-off between filter volume and attenuation).
2) The maximum value of the CM filter output impedance Z_{cm} (trade-off between inductor and capacitor volume of the CM filter).

Besides the EMI filter parameters, also the switching frequency f_s is varied (trade-off between volume of the passive components, i.e. EMI filter and DC link capacitor, and heat sink volume).

B. Inverter Optimization Procedure

The optimization for the inverter which is based on the procedure for the rectifier is shown in Fig. 9. In contrast to the rectifier optimization, no EMI filter has to be optimized. However, the harmonic motor losses are estimated as they are affected by the applied modulation scheme. They can be estimated via a power loss curve, defined by

$$P_{\text{motor,harmonic}} = \sum V_n^2 \cdot \left(\frac{A}{f_n^\alpha} + \frac{B}{f_n^\beta} \right), \qquad (2)$$

where the high-frequency iron and copper losses are estimated via the experimentally derived parameters A, B, α and β [15].

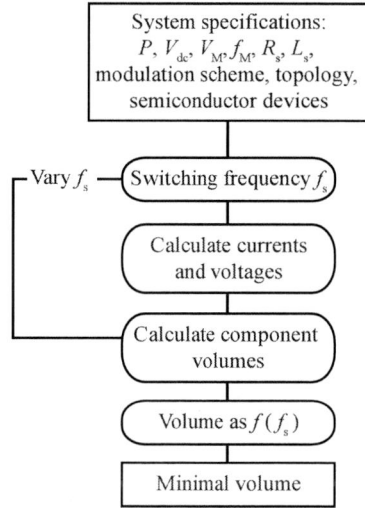

Fig. 9. Flowchart of the procedure to optimize the motor inverter for minimal volume.

3737

C. Volume Estimation

If a converter design complies with the EMI standards, the volumes of the various components are estimated (step 6 of Fig. 8). The capacitor volume is estimated with [14]

$$V_C = k_1 \cdot C\hat{V}^2 + k_2(\hat{V}). \tag{3}$$

The parameters k_1 and k_2 differ for the different capacitor types (DC link, DM filter, CM filter capacitor) and are shown in appendix A. The inductor volume is estimated with [14]

$$V_L = k_1 \cdot \left(LI^2\right)^{\frac{3}{4}} + k_2 \cdot LI^2. \tag{4}$$

The parameters k_1 and k_2 are based on an inductor optimization described in [14] and are shown in appendix A for the different inductor types (boost, DM filter and CM filter inductor).

The conduction and switching losses of the power semiconductors are calculated based on datasheet values. For the 3L-TT topology (cf. Fig. 2b) the switching losses have been scaled to the lower operating voltage. In addition, the switching losses differ from the data sheet as a diode with a lower voltage rating is used for the transition of the current. The datasheet curves have been adjusted linearly based on the measured data given in [16]. The maximum allowed heat sink temperature is

$$T_{hs} = \min(T_{sc} - P_{chip,i} \cdot R_{th,i}), i = 1...N, \tag{5}$$

where T_{sc} defines the maximal allowed junction temperature. $P_{chip,i}$ is the power loss of a single chip, $R_{th,i}$ its corresponding thermal resistance from junction to heat sink (including an insulation foil between semiconductor case and heat sink) and N defines the number of semiconductors mounted on the heat sink. The heat sink is assumed to have a Cooling System Performance Index [17] CSPI = $10\,\mathrm{W\,K^{-1}\,dm^{-1}}$ (forced air cooling) and therefore the volume can be calculated with

$$V_{hs} = \frac{\sum P_{chip}}{\mathrm{CSPI} \cdot (T_{hs} - T_{amb})}. \tag{6}$$

With this volume estimation equations the total volume of the converter can be optimized. The corresponding results are shown in section VI.

V. Experimental Setup

To verify the models used to calculate the currents and voltages of the rectifier system and the maximum estimation of an EMI test receiver, a 30 kW prototype as shown in Fig. 10 has been built. Fig. 11 shows the input voltages and currents as well as the output voltage and current with a 15 kW load. Fig. 12 shows the results of an EMI measurement and compares it to the model. The model predicts the maximum estimation of the test receiver with a good accuracy and the MF limits are met. In the MHz-range, the HF limits are violated, which could be corrected by a proper HF design of the converter components, which is not the scope of this work.

Fig. 10. Prototype of a bidirectional 30 kW boost PFC rectifier with an input of 3 x 230 V AC, an output of 800 V DC and a volume of 13.8 dm³.

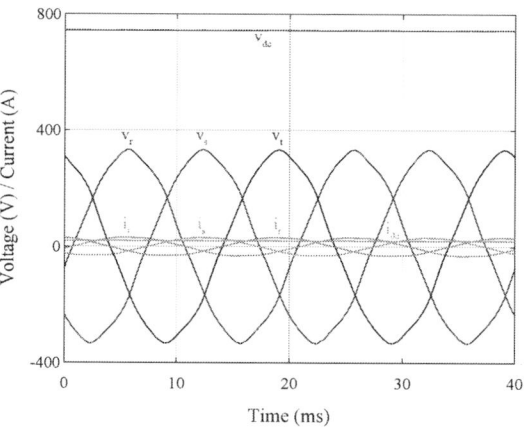

Fig. 11. Measurement results of the prototype shown in Fig. 10 with $V_{ac} = 230\,\mathrm{V}$, $V_{dc} = 750\,\mathrm{V}$ and $P = 15\,\mathrm{kW}$

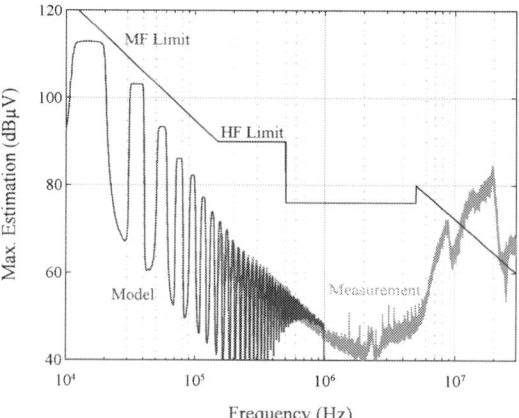

Fig. 12. Measurement of the EMI test receiver *ESIB7* compared to the model. $V_{ac} = 230\,\mathrm{V}$ with a LISN *NNB-4/200* inserted, $V_{dc} = 750\,\mathrm{V}$ without a load.

VI. Comparative Evaluation

The optimization procedures described in section IV have been applied to a wide variation of topologies, modulation schemes, DM and CM topologies. The optimizations are performed for the rectifier stage (section VI-A) as well as for the inverter stage (section VI-B). Furthermore, a sensitivity analysis of the volume on the power semiconductor parameters is performed in section VI-C.

A. Rectifier Optimization

The results for the rectifier stage are divided in different categories: In section VI-A1, the focus is put on the investigated modulation schemes. Section VI-A2 analyses the impact of the chosen DM damping topology while section VI-A3 focuses on the CM topologies. Finally, in section VI-A4 the total rectifier volume for the different semiconductor technologies is analysed. The input voltage is 3 x 230 V AC, the output voltage is 800 V DC and the power is 30 kW.

1) Modulation Scheme Comparison: Fig. 13 shows the heat sink, DC link capacitor and EMI filter volume as a function of the switching frequency for the investigated modulation schemes. For this comparison, Si IGBTs are assumed and for 3L only the NPC topology is considered. The heat sink volume explains why with 2L topologies only a moderate switching frequency can be achieved. The DC link capacitor volume reveals the disadvantage of the *3L-OC* modulation scheme: The volume is relatively big and does not decrease with a higher switching frequency, as the average neutral point current over a switching period is non-zero. The EMI filter volume shows that for the same switching frequency, the EMI filter volume of the *3L-NPB* modulation scheme is lower than the one for other modulation schemes, as was predicted by the DM peak voltage (cf. Fig. 3).

To summarize, the *3L-NPB* modulation scheme results in the smallest volume, due to the advantage of both a small DC link capacitor and a low EMI filter volume, which compensate

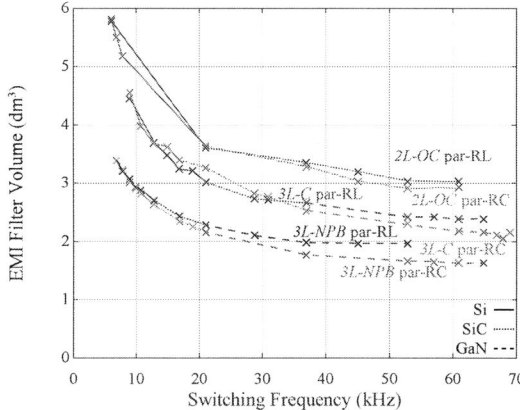

Fig. 14. EMI filter volume depending on the switching frequency for different DM damping topologies (cf. section III-B) and modulation schemes (cf. section II-B). For 3L the TT topology is considered and the CM filter is built with topology 1.

for the higher switching losses compared to the clamping modulation schemes.

2) Damping Topology Comparison: Fig. 14 shows the EMI filter volume for different damping topologies (cf. section III-B) and modulation schemes depending on the switching frequency. Si IGBTs, SiC Power MOSFETs and GaN transistors are considered for all modulation schemes and combined together to get a wide switching frequency range. The results show that the preferred damping topology depends on the modulation scheme and switching frequency. The lowest EMI filter volume can be achieved with the *3L-NPB* modulation scheme and a parallel-RC damping topology.

3) Common Mode Topology Comparison: Fig. 15 shows the EMI filter volume depending on the switching frequency for different CM filter topologies (cf. section III-C). It can be seen that it is advantageous to place the CM inductor at the grid side (topology 1 and 3) as otherwise the HF harmonics of the DM current cause additional losses in the windings of

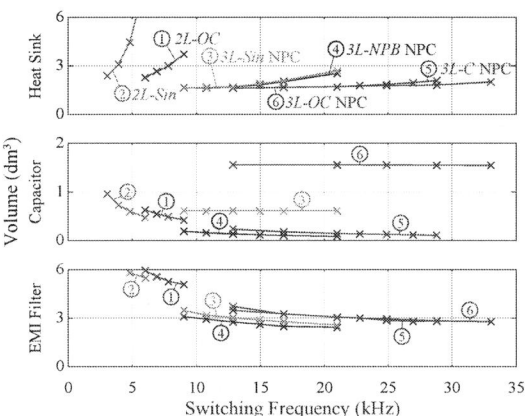

Fig. 13. Heat sink, DC link capacitor and EMI filter volume depending on the switching frequency for different modulation schemes (cf. section II-B). Si IGBTs are assumed and for 3L the NPC topology is considered. The DM filter is built with a parallel-RL damping topology and CM filter topology 1.

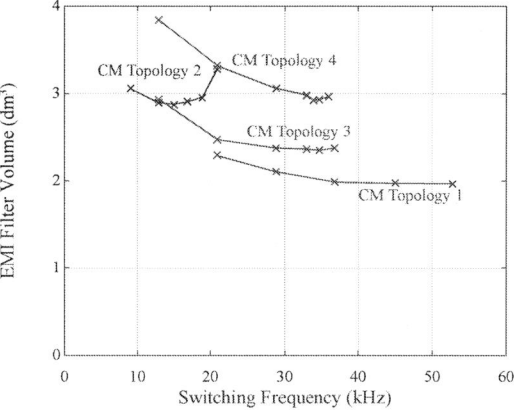

Fig. 15. EMI filter volume depending on the switching frequency for different CM topologies (cf. section III-C). The *3L-NPB* modulation scheme is applied for a 3L TT topology using GaN transistors. For the DM filter a parallel-RL damping topology is used.

3739

The 2018 International Power Electronics Conference

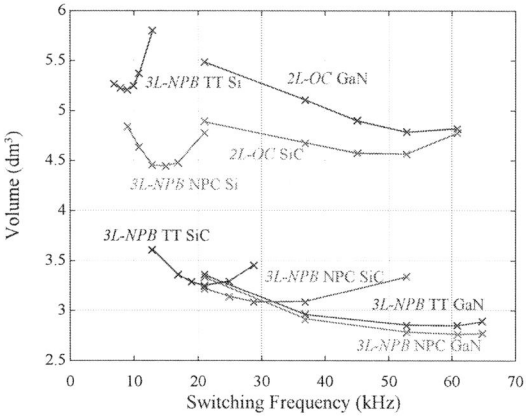

Fig. 16. Total converter volume depending on the switching frequency for different semiconductor technologies. For the 3L topologies the *3L-NPB* modulation scheme is applied while for the 2L topology the *2L-OC* modulation scheme is applied. A parallel-RC damping topology for the DM filter and CM filter topology 1 is used.

the CM inductor and therefore would increase its volume. It is also advantageous to connect the CM capacitor to the DC link neutral point (topology 1 and 2), which has the additional benefit to reduce the currents to protective earth.

4) Semiconductor Technology Comparison: In Fig. 16, the total rectifier volume is shown for different semiconductor technologies and topologies. For the 2L topology, the *2L-OC* modulation scheme is applied and for the 3L topology the *3L-NPB* modulation scheme, according to the results in section VI-A1. For the damping topology, a parallel-RC topology is chosen (cf. section VI-A2) and the CM filter is built with CM topology 1 (cf. section VI-A3). The 2L topology with Si IGBTs could not be seen on the graph as the volume is very high - around $9\,dm^3$. Either a change to new semiconductor technologies or a change to a 3L topology reduces the rectifier volume significantly. If both measures are combined, the rectifier volume is reduced further. The 3L NPC topology leads to the smallest volume. GaN semiconductor devices prove to

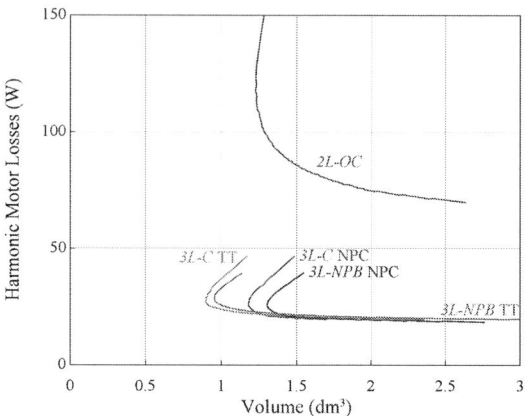

Fig. 18. Harmonic motor losses depending on the inverter volume for different modulation schemes. SiC MOSFETs are considered, output power is 22.5 kW.

be the best choice, as the switching frequency can be raised considerably, thereby reducing the volume of the EMI filter.

Keep in mind that the difference between the TT and NPC topology heavily depends on the used semiconductor devices (cf. Appendix A).

B. Inverter Optimization

The results for the inverter optimization are shown in Fig. 17 and 18. The considered synchronous motor is a *8LSA86* with a power of 22.5 kW. In contrast to the rectifier stage, there are less passive components which would benefit from a high switching frequency and a low DM voltage peak. This leads to three observations:

1) The 3L TT topology is preferred to the 3L NPC topology as the conduction losses are lower.

2) The *3L-C* modulation scheme is preferred to the *3L-NPB* modulation scheme as the lower switching losses outvalue the disadvantage of a higher DM peak voltage, as there is no EMI filter.

3) The considered SiC MOSFETs perform better than the GaN transistors as they result in lower conduction losses.

An additional benefit of the 3L topologies can be seen in Fig. 18: Due to the smoother output voltage, the harmonic losses of the motor can be considerably reduced. Furthermore, as the voltage steps of the winding voltages are reduced, the voltage stress on the insulation is reduced and the CM currents (which result in bearing currents) are reduced.

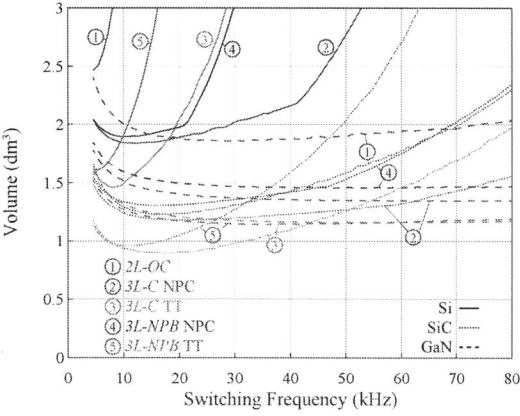

Fig. 17. Total inverter volume depending on the switching frequency for different topologies, modulation schemes and semiconductor technologies. Output power is 22.5 kW.

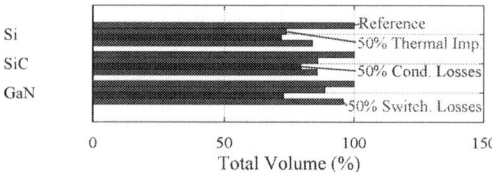

Fig. 19. Volume reduction for different semiconductor improvements: reduced thermal impedance, reduced conduction losses or reduced switching losses.

3740

TABLE I. SEMICONDUCTORS CONSIDERED IN THE OPTIMIZATION.

Type	Name
Si 1200 V IGBT	*Infineon IKW25N120T2*
Si 600 V IGBT	*Infineon IKW30N65ES5*
Si 600 V Diode	*Infineon IDW30E65D1*
SiC 1200 V MOSFET	*Cree C2M0025120D*
SiC 600 V MOSFET	*Microsemi APT130SM70B*
SiC 600 V Diode	*Infineon IDW30G65C5*
GaN 1200 V Transistor	*VisIC VM40HB120D*
GaN 600 V Transistor	*VisIC V22N65A*

TABLE II. PARAMETERS FOR ESTIMATING THE VOLUME OF THE INDUCTORS.

Component	Core Material	k_1	k_2
L_b	*METGLAS 2605SA1*	$266 \text{ cm}^3 \left(\text{H}^{-1}\text{A}^{-2}\right)^{\frac{4}{3}}$	$0 \text{ cm}^3\text{H}^{-1}\text{A}^{-2}$
L_f	*METGLAS 2605SA1*	$169 \text{ cm}^3 \left(\text{H}^{-1}\text{A}^{-2}\right)^{\frac{4}{3}}$	$64 \text{ cm}^3\text{H}^{-1}\text{A}^{-2}$
L_{CM}	*Vitropenn 500F*	$39 \text{ cm}^3 \left(\text{H}^{-1}\text{A}^{-2}\right)^{\frac{4}{3}}$	$15 \text{ cm}^3\text{H}^{-1}\text{A}^{-2}$

TABLE III. PARAMETERS FOR ESTIMATING THE VOLUME OF THE CAPACITORS.

Component	Series	k_1	k_2
C_f	*EPCOS B3292x*	$22.96 \text{ cm}^3\text{F}^{-1}\text{V}^{-2}$	2.24 cm^3
C_{CM}	*EPCOS B3296x*	$55.54 \text{ cm}^3\text{F}^{-1}\text{V}^{-2}$	0.84 cm^3
C_{DC}	*EPCOS B3277x*	$4.95 \text{ cm}^3\text{F}^{-1}\text{V}^{-2}$	4.24 cm^3

C. Sensitivity Analysis

For all semiconductor technologies, different characteristics have been varied to analyze their impact on the total converter volume of the rectifier stage. A 3L NPC topology is considered and a *3L-NPB* modulation scheme is applied. For the DM filter, a parallel-RC damping topology is considered and the CM filter is built with CM topology 1. Either the thermal impedance, the conduction losses or the switching losses have been reduced by 50%. The resulting volumes are shown in Fig. 19. As can be seen the biggest impact on the total converter volume is caused by reduced conduction losses, as these are the dominant semiconductor losses.

VII. CONCLUSION

In this paper, various modulation schemes, topologies and EMI filter designs are analysed and compared against each other for bidirectional boost PFC rectifiers as well as inverters in terms of the volume. The 3L NPC topology has been shown to be the best topology for the rectifier stage, while for the inverter stage, the TT topology is to be the preferred one as the conduction losses are dominant there. For the same reason, GaN transistors are preferred for the rectifier stage, while SiC MOSFETs are advantageous for the inverter stage.

APPENDIX A
MATERIAL PARAMETERS

The used semiconductor devices are shown in table I. They have been chosen to have a similar current rating to allow a fair comparison. The parameters for estimating the inductor and capacitor volumes are shown in table II and III.

REFERENCES

[1] R. Burkart and J. Kolar, "Overview and comparison of grid harmonics and conducted EMI standards for LV converters connected to the MV distribution system," in *PCIM*, 2012.

[2] T. Nussbaumer, M. Heldwein, and J. Kolar, "Differential mode input filter design for a three-phase buck-type PWM rectifier based on modeling of the EMC test receiver," *IEEE Trans. on Industrial Electronics*, vol. 53, no. 5, Oct 2006.

[3] D. Boillat, J. Kolar, and J. Mühlethaler, "Volume minimization of the main DM/CM EMI filter stage of a bidirectional three-phase three-level PWM rectifier system," in *Energy Conversion Congress and Exposition (ECCE)*, Sept 2013.

[4] B. Kaku, I. Miyashita, and S. Sone, "Switching loss minimised space vector PWM method for IGBT three-level inverter," *IEEE Proc. on Electric Power Applications*, vol. 144, no. 3, 1997.

[5] K. Tian, J. Wang, B. Wu, Z. Cheng, and N. R. Zargari, "A virtual space vector modulation technique for the reduction of common-mode voltages in both magnitude and third-order component," *IEEE Trans. on Power Electronics*, vol. 31, no. 1, Jan 2016.

[6] J. Wyss and J. Biela, "Volume optimization of a 30 kW boost PFC converter focusing on the CM/DM EMI filter design," in *EPE ECCE-Europe*, Sept 2017.

[7] ——, "Analysis of PCB embedded power semiconductors for a 30 kW boost PFC converter," in *EPE ECCE-Europe*, Sept 2016.

[8] A. Trentin, P. Zanchetta, P. Wheeler, J. Clare, R. Wood, and W. Typton, "Performance assessment of SVM modulation techniques for losses reduction in induction motor drives," in *IEEE Industry Applications Annual Meeting*, Sept 2007.

[9] M. Asefi and J. Nazarzadeh, "Survey on high-frequency models of PWM electric drives for shaft voltage and bearing current analysis," *IET Electrical Systems in Transportation*, 2017.

[10] J. W. Kolar, H. Ertl, and F. C. Zach, "Influence of the modulation method on the conduction and switching losses of a PWM converter system," *IEEE Trans. on Industry Applications*, vol. 27, no. 6, Nov 1991.

[11] R. Middlebrook, "Input filter considerations in design and application switching regulators," in *IEEE Industry Applications Society Annual Meeting*, 1976.

[12] ——, "Design techniques for preventing input filter oscillations in switched-mode regulators," in *Powercon 5*, May 1978.

[13] M. Schweizer and J. Kolar, "Shifting input filter resonances - an intelligent converter behavior for maintaining system stability," in *Int. Power Electronics Conf. (IPEC)*, June 2010.

[14] J. Wyss and J. Biela, "EMI DM filter volume minimization for a PFC boost converter including boost inductor variation and MF EMI limits," in *EPE ECCE-Europe*, Sept 2015.

[15] K. Bradley, W. Cao, J. Clare, and P. Wheeler, "Predicting inverter-induced harmonic loss by improved harmonic injection," *IEEE Transactions on Power Electronics*, vol. 23, no. 5, pp. 2619–2624, Sept 2008.

[16] M. Schweizer and J. Kolar, "Design and implementation of a highly efficient three-level t-type converter for low-voltage applications," *IEEE Trans. on Power Electronics*, vol. 28, no. 2, 2013.

[17] U. Drofenik, G. Laimer, and J. W. Kolar, "Theoretical converter power density limits for forced convection cooling," in *PCIM*, 2005.

A Standard Block of "Series Connected SiC MOSFET" for Medium/High voltage converter

Qin Lei, Chunhui Liu, Yunpeng Si, Yifu Liu
Electrical, Computer and Energy Engineering, Arizona State University, Tempe, USA
Qin.Lei@asu.edu, cliu212@asu.edu, yunpengs@asu.edu, yliu457@asu.edu

Abstract—The goal of this paper is to significantly improve the efficiency and power density of medium voltage drive and high-power converters. To achieve the goal, the proposed approach is to replace the high voltage Si IGBT by series connected SiC MOSFETs. Specifically, a game changing and universally applicable standard block of "series connected SiC MOSFET" with excellent dynamic voltage sharing and high reliability is proposed. The core technology in the block is the current source gate driver with device synchronization function. A down-scaled medium voltage drive prototype has been developed to demonstrate the feasibility and advantages of the standard block.

Keywords—Medium voltage drive; SiC MOSFET; Current source gate driver; Standard block

I. INTRODUCTION

Currently, the SiC MOSFET is expected to replace IGBT at high voltage power conversion area and the 1.7kV/1.2kV SiC MOSFET is widely used to replace the Si IGBT at the same voltage rating. However, a SiC device pack of approximately 5kV, 10kV or 20kV rating is required in medium and high voltage converter to achieve less system complexity and higher power density, considering a 50% voltage margin.

To form the device pack, single device and series connected devices are two options. For SiC MOSFET, "series connected" option with lower device rating bring much better device pack characteristics and comparable system performance, compared to the "single device" option. At device level, lower device rating "series connected" pack can lead to the following benefits: lower voltage derating rate, lower Rdson per unit voltage, higher current rating per unit die area and capability to switch at higher frequency [1]. At system level, the power density of series connected low rating SiC MOSFET can be comparable to the single device because of the better utilization rate of the die area, although there are additional accessory circuits such as snubber and gate driver associated. The only drawback of low reliability can be compensated by adding more percentage of redundant devices in series [2].

The biggest challenge faced by series connected devices is the unequal dynamic voltage sharing. The unequal voltage sharing can be caused by two reasons: delays on the starting point of drain-source voltage rising/falling; different dv/dt

ratios. The roots are parameter differences such as Cgs difference or driver loop parameter differences.

The state of art to address this challenge is discussed as follows: (1) Snubber vs. snubberless [3-17]: A common solution which has been used by ABB [3], Siemens and GE is to add snubber circuit. The snubber circuit is simple and reliable, but introduces the additional loss at device on. Therefore, small snubber circuit is desired. However, the concept of "snubberless" which has been proposed in paper [4] is not practical since the difference of anti-parallel diode reverse recovery charge cannot be actively controlled. (2) Active control vs. Non-active control: Paper [18-19] presents the idea of using the difference between the individual device voltage and the average voltage of the series connected devices to regulate the gate signal delay. Paper [24] presents the method of the active dv/dt control of the IGBT. Paper [25] is the SiC MOSFET version of ETH method. It can be said that the active delay and dv/dt control could be good for IGBT but not good for SiC MOSFET. First, SiC MOSFET is a majority carrier device, so its "part-to-part" difference are not as large as IGBT. Second, the drain-source dv/dt of the SiC MOSFET is a fast changing waveform because of the nonlinearity of the internal capacitance. Third, the drain-source voltage of SiC MOSFET could be very noisy, due to the resonance between parasitic inductor and junction capacitor, as well as uncontrolled delay. Fourth, the synchronization control requires the feedback of the voltage of all the devices, which has a significant delay due to the high isolation levels. Fifth, the nano-second time frame requires super high bandwidth control, which is not practical for the centralized control due to the large delay and the difficulty of controlling the parasitic. (3) Voltage source driver vs. current source driver: To be applied with series SiC MOSFET, VS-GD has the following drawbacks: First, Vgs is hard to be controlled to be identical because its nonlinear rising/falling slope; Second, small variation of gate resistor can significantly change the Vgs waveform; Third, the Vgs has significant fall/rise at on/off miller plateau, due to the discharging/charging of the Cgs by the excessive miller current during Vds falling/rising. All disadvantages above can be overcome by the CS-GD as Vgs can be controlled identical by a constant current and the gate resistor variation will not affect the gate current. Also, delay and dv/dt of Vds can be changed online by adjusting current command. Last, the miller plateau distortion can be overcome by applying an sufficient gate current to cancel the large miller current; Meanwhile, the Vgs rising/falling slope can be kept

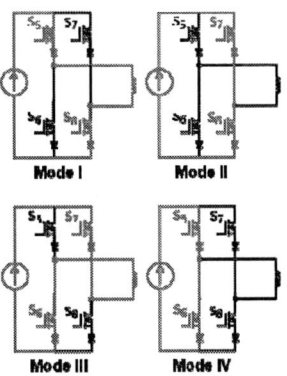

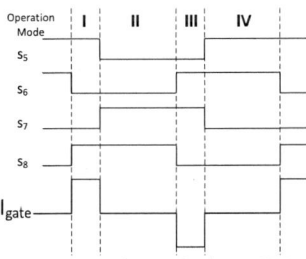

Fig. 4. Waveforms of a SiC MOSFET

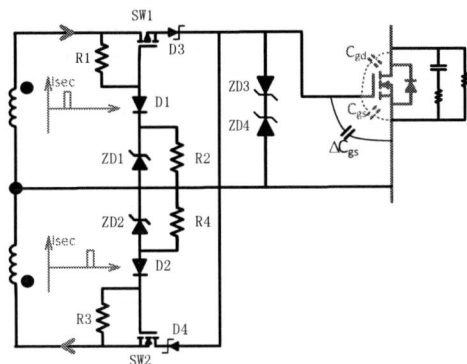

Fig. 3. Waveforms of primary CSI

the same by adding an external gate-source capacitor. (4) Resonance gate driver vs. non-resonance gate driver [66-82]: The resonant gate driver has been proposed to reduce the power consumption by using the resonant voltage/current across a series or parallel resonant circuit [67]. The most attractive part of R-GD is its low loss. The second benefit is the accurate control of the Vgs rising/falling time, since the time only depends on the resonant period. However, in SiC MOSFET case, the second advantage becomes a significant drawback. According to the SiC MOSFET Cgs value (1000x pF), and its typical on/off time (20ns~100ns), the resonant inductor is in the range of 10nF~1μF. It is at the level of the parasitic inductance of the gate loop and is easy to be distorted.

To address the challenge in dynamic voltage sharing, a simple current source based gate driver is proposed to provide the best "delay" and "dv/dt" synchronization among series connected SiC MOSFET, by taking advantage of its unique characteristics of high gate-source capacitance and super low miller capacitance. A minimum snubber circuit is added to compensate for the difference among anti-parallel diodes or body diodes. Except for that, a super high carrier-frequency based "iCoupler" method has been adopted in the gate signal isolation stage to reduce the delay in signals, and a "quasi-matrix transformer" concept has been proposed in the gate power isolation stage to reduce the parasitic capacitance and inductance (due to the page limit the "icoupler" and "quasi-matrix-transformer" principle and designs are not explained here). The proposed strategy is an effective, practical and economic solution for series connected SiC MOSFETs.

II. PRINCIPLE OF THE CURRENT SOURCE GATE DRIVER

A. Basic principle of current source gate drivers

The concept of CS-GD is to provide the gate a constant current pulse to charge and discharge the Cgs at each switching transient. The schematic of the CS-GD for series connected device is shown in Fig. 1. For the main switch, the relationship between the "main-switch" gate signal, gate current (Igs), gate source voltage (Vgs), drain source voltage (Vds) and drain

current (Ids) is shown in Fig. 3. To minimize the differences of each parallel CSGD gate currents, the pulse current is generated directly by current source H-bridge and transform the pulse current to the voltage hold circuits through a multi-terminal transformer. The principle of the primary side H-bridge current source converter is explained as follows. The gate signals for the H-bridge are decoded from the main switch gate signals. The H-bridge has four operation modes, as shown in Fig. 2. The relationship between the switching states and the output current is shown in Fig.4. When the main switch gate signal changes from zero to one, S5 and S8 of CSI are turned on and a positive pulse current is output. After a certain pulse width (to make the gate voltage to be charged to 20V), S7 and S8 are turned on to freewheel the current. When the main switch gate signal changes from one to zero, S7 and S6 are turned on to generate a negative current at the output. After a certain pulse width (to make the gate voltage to be discharged to -5V), S5 and S6 are turned on to freewheel the current. The current source at the input is formed by a controlled buck converter. The output current of the buck converter is close-loop controlled by regulating the duty cycle. So, the dc link of the current source

Fig. 5. Automatic Circuit details

H-bridge is a constant current. At primary side, the dc current from buck current source is regulated by CSI to generate a pulse as gate signal needs. At the secondary side, the current is conducted by the automatic switches shown in Fig. 5. and thus, the positive and negative current pulses are separated. The low state(-5v) and high state(20v) gate voltage can be hold separately because the automatic switches can block the Cgs discharge loop.

The automatic switch circuit consists of 2 symmetrical blocks which take effects in low state and high state gate voltages, respectively. For the high state block circuit, when a positive pulse current occurs in primary side of the transformer, the current flows through R1-D1-ZD1 loop at the very beginning since the P-channel MOSFET SW1 gate voltage is zero and the switch is in open state. The Zener diode ZD1 is 16v voltage rating so the current is limited by the R3 resistor, and the voltage across the R1 can be regulated to be 5v. After that, the P-channel SW1 is in closed state since the gate voltage is already negative. At this time, the loop R1-D1-ZD1 current is limited by the Zener diode ZD1 and parallel resistor R3. And the other current from the transformer will flow through the SW1-D3-Cgs loop to charge the Cgs to 20V. When Vgs reaches 20V, the Zener diode ZD4 which has 20V voltage rating will take effects and clamp the Vgs to be 20V. The residual current from the transformer will flow through the SW1-D3-ZD3-ZD4 loop. Because of the Zener diode transient response time, the Zener diode impedance will decrease dramatically when the voltage reaches its rating, and after several nanoseconds, the Zener diode can hold the voltage at the voltage rating. However, when the Zener diode impedance becomes small, if the current source is over during this time, the gate capacitor will discharge through the low impedance Zener diode loop ZD3-ZD4 because the Zener diode ZD4 is still at the transient state. And the Vgs voltage will decrease 1-2 Volts, to avoid this Zener diode transient, the fast response Zener diode has been selected and the current pulse length is regulated to be longer than the Cgs charging time. At this positive current pulse time, the lower current loop R2-D3-ZD3 cannot conduct current since the D3 diode block this loop, thus, the other lower current loop cannot take effects during the positive current pulse time. For the negative pulse current, the lower current loop will take part in, and the upper current loop R1-D1-ZD1 will be blocked by the diode D1. The procedure is similar as the positive current pulse.

For the series connected SiC MOSFET, each one has an automatic switch circuit as Fig. 5. It may have synchronization problems due to the components differences of the manufacturing variances. In this automatic circuit, the Si automatic switch gate resistor can be tuned to overcome this problem.

B. Miller plateau vs. Cgs

The high dv/dt of the drain-source voltage can cause the miss-trigger of the switches. It can be explained as follows. The drain of the device is coupled to the gate by miller capacitor. When the Vgs rises/falls to threshold, the high negative/positive dv/dt of Vds generates a high discharging/charging current for Cgs, so Vgs will descend/ascend until the Vds becomes zero/maximum. The

negative effect of this miller plateau disturbance is that the gate might be miss-triggered. Current source gate driver is a good candidate to address this challenge because the gate current can be controlled high enough to completely cancel the miller current. But the high current can speed up Vgs and the dv/dt of Vds can be affected, which will deteriorate the high frequency resonance between junction capacitor and dc link parasitic inductor. This resonance can also be coupled to the gate loop and breakdown the gate. The method proposed to solve this problem is to add an extra gate-source capacitor (Cgs) while maintaining the high amplitude current, to create an appropriate Vgs rising/falling speed.

C. Discontinuous current source driver

The drawback of the pre-mentioned continuous CS-GD is that the conduction loss is significant because the current continues to circulate through the phase legs during the steady state. To overcome the disadvantage, a discontinuous CS-GD is proposed here. The buck converter is controlled to output a constant current which is aligned with the main switch (SiC MOSFET) gate signal. And the dc current of primary side is a discontinuous pulse-type dc current, and the CSI will modulate on this pulse current.

D. Snubber circuit design

As mentioned in state of art analysis, the snubber circuit is a must to compensate for the difference among anti-parallel diodes. To calculate the R, C values, four constrains are applied.

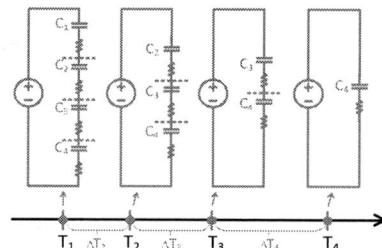

Fig. 6. Equivalent circuit of the snubber during the turn-on transient

Constrain 1: The capacitor voltage doesn't exceed the defined maximum allowed drain-source voltage. Fig.6 shows the equivalent snubber circuits at different time intervals for the case of four devices in series switching by sequence. The general equation for the capacitor voltage at the end of each time interval is:

$$V_{T_{i+1}} = V_{i+1} + (V_{T_i} - V_{i+1}) \cdot e^{\frac{-\Delta T_{i+1}}{RC}} \quad (1)$$

$$V_{1,2,3,4} = \frac{V}{4}, \frac{V}{3}, \frac{V}{2}, V \quad (2)$$

The constrain for the product "RC" is: $V_{C4} \le x\% \cdot V_{rating}$, assuming x% margin for the device voltage.

Constrain 2: The RC discharge time is smaller than the minimum pulse width of the device: $3\tau < \min\{t_{on}(i)\}$

Constrain 3: The RC discharge loss should be minimized.

The average loss snubber loss during discharge can be calculated as:

$$\frac{\frac{1}{2}\cdot C\cdot(V_T)^2}{3\cdot R\cdot C}=P_{loss} \qquad (3)$$

Constrain 4: The switching speed is not slowed down by the snubber.

Another thing needs to be considered is that the random "part-to-part" parameter variations of the R and C of the snubber need to be taken into consideration.

E. SPICE Simulation results for comparison between CS-GD and VS-GD

Fig. 7 shows the double pulse spice simulation results of Vgs and Vds based on VS-GD and CS-GD with five 1.7kV/72A SiC MOSFETs connected in series with snubber circuits at on and off transient, respectively. The real physical device model obtained from Cree has been used in the simulation. The gate resistors are intentionally changed (10% difference) to create difference in gate loops for different gate drivers. Additionally, snubber circuits have been designed

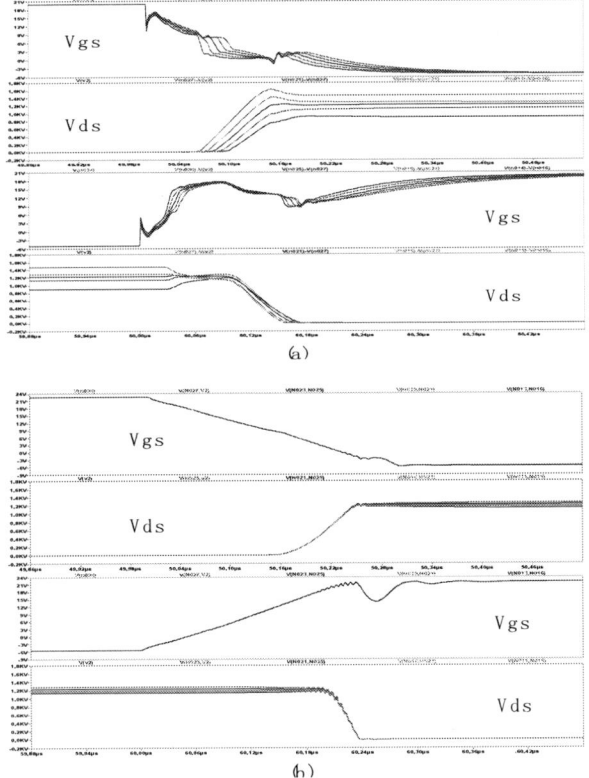

Fig. 7. (a)Simulation results of series devices using voltage source gate driver with snubber circuits. Top two figures show turn-off transient of Vgs and Vds; Bottom two figures show turn-on transient of Vgs and Vds. (b) Simulation results of series device using current source gate driver. Top two figures show turn-off transient of Vgs and Vds; Bottom two figures show turn-on transient of Vgs and Vds.

differently considering 10% tolerance of the high voltage capacitors. As expected, Vgs for VS-GD is similar to the waveform of RC charge/discharge and the time constant depends on the gate resistor. For CS-GD, Vgs shows a linear rising/falling during turn-on and turn-off and its slope merely relies on the gate current magnitude. For VS-GD, it is clearly shown that Vgs for different devices don't reach turn-on and turn-off thresholds at the same time. As a result, Vds for each device starts to rise/fall at different time point and with different speed. However, under CS-GD, gate signals between different devices are very much close to each other, which synchronizes the starting points of the Vds of all the devices even though the dV/dt still have slight differences. With the same snubber design, the voltage variation among devices is much larger in VS-GD than that in CS-GD. The voltage deviation in VS-GD is around 30%, but in CS-GD is less than 5%.

F. Simulation results of series connected SiC based half-bridge using CS-GD

Fig. 8 shows the simulation results of an SPWM based half-bridge using five SiC MOSFETs connected in series in each arm as well as the genersation of current sources for gate drivers applied to each device. It can be seen from Figure 7(a)

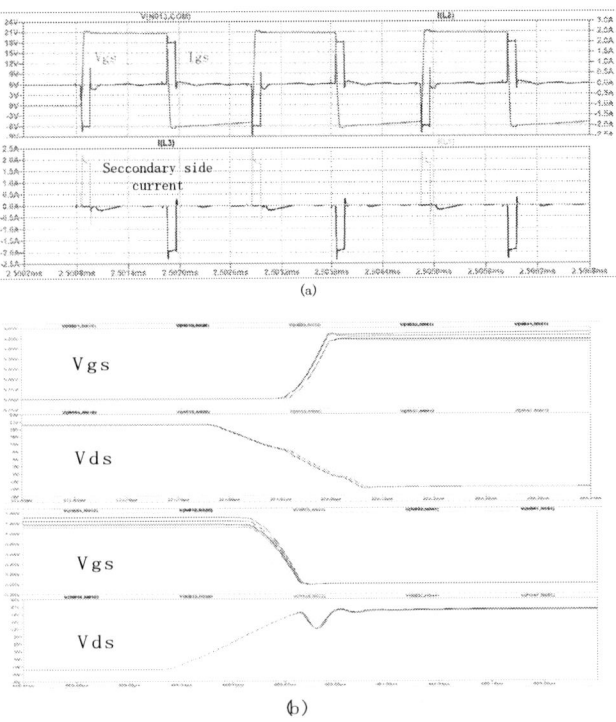

Fig. 8. (a)Simulation results of current source gate driver based on the topology shown in Figure 1. primary side current of the high frequency transformer, current pulse and gate voltage are shown here. (b) Vds and Vgs waveforms of a SPWM based half-bridge simulation with series connected devices using current source gate driver with different gate resistors and snubber circuits. The upper and lower two figures show the turn off and turn on transients, respectively.

3745

that the current flowing through the inductor at buck converter output can be well regulated and a AC pulse current along with gate signal transients has been generated by the CSI at transformer primary side. After automatic switch circuit, the gate voltages are established and hold. From the gate voltage waveform, linear rise/fall during positive/negative current pulse can be observed, which validates the proposed approach. Fig. 7 (b) shows that the proposed CS-GD strategy for series connected SiC MOSFETs demonstrates great performance in half-bridge simulation under the worst condition when the load current is low. The gate voltages for all devices in one arm are perfectly identical, which largely reduces the difference of Vds among the devices even though the gate resistors and snubber circuits are different.

III. CONCLUSIONS AND FUTURE WORK

To address the challenge in dynamic voltage sharing for "series connected SiC MOSFET", a simple current source-based gate driver is proposed to provide the best synchronization of the starting point of the Vds rising/falling edge, by taking advantage of its unique characteristics of high gate-source capacitance and super low miller capacitance. A minimum snubber circuit is added to compensate for the difference among anti-parallel diodes or body diodes. It is verified through simulation that the current source gate driver has the following benefits for "series connected SiC MOSFET": (1) Different Vgs can be controlled identical by a constant current; (2) the gate resistor variation will not affect the gate current. (3) delay and dv/dt of Vds can be changed online. (4) the miller plateau distortion can be overcome by applying a sufficient gate current to cancel the large miller current; at the same time, the Vgs rising/falling slope can be kept the same by adding an external gate-source capacitor. (5) the peak current is not limited by the gate resistor but only depends on the driver rating.

REFERENCES

[1] A. Bolotnikov et al., "Overview of 1.2kV – 2.2kV SiC MOSFETs targeted for industrial power conversion applications," 2015 IEEE Applied Power Conference and Exposition (APEC), Charlotte, NC, 2015, pp. 2445-2452.

[2] Wang, Kuan, Qin Lei, and Chunhui Liu. "Methodology of reliability and power density analysis of SST topologies." Applied Power Electronics Conference and Exposition (APEC), 2017 IEEE. IEEE, 2017.

[3] K. Vechalapu, S. Bhattacharya and E. Aleoiza, "Performance evaluation of series connected 1700V SiC MOSFET devices," 2015 IEEE 3rd Workshop on Wide Bandgap Power Devices and Applications (WiPDA), Blacksburg, VA, 2015, pp. 184-191.

[4] C. Gerster, P. Hofer and N. Karrer, "Gate-control strategies for snubberless operation of series connected IGBTs," PESC Record. 27th Annual IEEE Power Electronics Specialists Conference, Baveno, 1996, pp. 1739-1742 vol.2.

[5] A. Petterteig, J. Lode and T. M. Undeland, "IGBT turn-off losses for hard switching and with capacitive snubbers," Conference Record of the 1991 IEEE Industry Applications Society Annual Meeting, Dearborn, MI, USA, 1991, pp. 1501-1507 vol.2.

[6] Ju Won Baek, Dong-Wook Yoo and Heung-Geun Kim, "High-voltage switch using series-connected IGBTs with simple auxiliary circuit," in IEEE Transactions on Industry Applications, vol. 37, no. 6, pp. 1832-1839, Nov/Dec 2001.

[7] F. Blaabjerg and J. K. Pedersen, "An optimum drive and clamp circuit design with controlled switching for a snubberless PWM-VSI-IGBT

inverter leg," Power Electronics Specialists Conference, 1992. PESC '92 Record., 23rd Annual IEEE, Toledo, 1992, pp. 289-297 vol.1.

[8] W. McMurray, "Resonant snubbers with auxiliary switches," Conference Record of the IEEE Industry Applications Society Annual Meeting,, San Diego, CA, USA, 1989, pp. 289-834 vol.1.

[9] F. V. Robinson and V. Hamidi, "Series connecting devices for high-voltage power conversion," 2007 42nd International Universities Power Engineering Conference, Brighton, 2007, pp. 1205-1210.

[10] T.M. Undeland, "Switching Stress Reduction in Power Transistor Converters", Conference Record, IAS Annual Meeting 1976, Chicago, pp 383-392.

[11] T.M. Undeland, "Snubbers for Pulse modulated Bridge Converters with Power Transistors or GTOs", Conference Record, IPEC-Tokyo 1983, pp313-323.

[12] T. Undeland, F. Jenset, A. Steinbakk, T. Rogne and M.Hernes, "A Snubber configuration for power transistors and GTO PWM Inverters". PESC'84 Record, Gaithersburg, Maryland, pp 42-53.

[13] F. Blaabjerg, "Snubbers in PWM-VSI-inverter," Power Electronics Specialists Conference, 1991. PESC '91 Record., 22nd Annual IEEE, Cambridge, MA, 1991, pp. 104-111.

[14] John K. Pedersen, Paul Wgersen. "A -Gate-Drive with Overcurrent Protection for Darlington/IGBT with Reduced Demand for Snubber/Clamp Circuit". in Proc. Of EPE '91, pp. 1193-1198.

[15] Fred Eschrich. "Protection of IGBT Modules in Inverter Circuits". EPE Journal, vol. 1, no. 1, July 1991, pp. 57-60.

[16] M. Olpaca, V.R. Stefanovic. "Reduction of IGBT Switching Losses Through Innovative Gate Control", presented at PCIM, Murich, 1990, pp. 83-93.

[17] C. Gerster, P. Hofer and N. Karrer, "Gate-control strategies for snubberless operation of series connected IGBTs," PESC Record. 27th Annual IEEE Power Electronics Specialists Conference, Baveno, 1996, pp. 1739-1742 vol.2.

[18] C. Gerster, "Fast high-power/high-voltage switch using series-connected IGBTs with active gate-controlled voltage-balancing," Applied Power Electronics Conference and Exposition, 1994. APEC '94. Conference Proceedings 1994., Ninth Annual, Orlando, FL, 1994, pp. 469-472 vol.1.

[19] C. Gerster, P. Hofer and N. Karrer, "Gate-control strategies for snubberless operation of series connected IGBTs," PESC Record. 27th Annual IEEE Power Electronics Specialists Conference, Baveno, 1996, pp. 1739-1742 vol.2.

[20] A. Consoli, S. Musumeci, G. Oriti and A. Testa, "Active voltage balancement of series connected IGBTs," Industry Applications Conference, 1995. Thirtieth IAS Annual Meeting, IAS '95., Conference Record of the 1995 IEEE, Orlando, FL, 1995, pp. 2752-2758 vol.3

[21] P. R. Palmer and A. N. Githiari, "The series connection of IGBTs with optimised voltage sharing in the switching transient," Power Electronics Specialists Conference, 1995. PESC '95 Record., 26th Annual IEEE, Atlanta, GA, 1995, pp. 44-49 vol.1.

[22] R. Guidini, D. Chatroux, Y. Guyon and D. Lafore, "Semiconductor power MOSFETs devices in series," 1993 Fifth European Conference on Power Electronics and Applications, Brighton, UK, 1993, pp. 425-430 vol.2.

[23] Letor, R, "Series connection of MOSFET, Bipolar and IGBT devices"

[24] Y. Lobsiger, J. W. Kolar, "closed-loop di/dt and dv/dt IGBT gate drive concepts", ECPE Tutorial, Zurich, Power Semiconductor Devices and Technologies

[25] H. Riazmontazer, A. Rahnamaee, A. Mojab, S. Mehrnami, S. K. Mazumder and M. Zefran, "Closed-loop control of switching transition of SiC MOSFETs," 2015 IEEE Applied Power Electronics Conference and Exposition (APEC), Charlotte, NC, 2015, pp. 782-788.

[26] K. Vechalapu, A. K. Kadavelugu and S. Bhattacharya, "High voltage dual active bridge with series connected high voltage silicon carbide (SiC) devices," 2014 IEEE Energy Conversion Congress and Exposition (ECCE), Pittsburgh, PA, 2014, pp. 2057-2064.

[27] Q. Xiao, Y. Yan, X. Wu, N. Ren and K. Sheng, "A 10kV/200A SiC MOSFET module with series-parallel hybrid connection of 1200V/50A dies," 2015 IEEE 27th International Symposium on Power

Semiconductor Devices & IC's (ISPSD), Hong Kong, 2015, pp. 349-352.

[28] K. Onda, A. Konno and J. Sakano, "New concept high-voltage IGBT gate driver with self-adjusting active gate control function for SiC-SBD hybrid module," 2013 25th International Symposium on Power Semiconductor Devices & IC's (ISPSD), Kanazawa, 2013, pp. 343-346.

[29] S. Castagno, R. D. Curry and E. Loree, "Analysis and Comparison of a Fast Turn on Series IGBT Stack and High Voltage Rated Commercial IGBTS," 2005 IEEE Pulsed Power Conference, Monterey, CA, 2005, pp. 912-915.

[30] A. Paredes, V. Sala, H. Ghorbani and L. Romeral, "A novel active gate driver for silicon carbide MOSFET," IECON 2016 - 42nd Annual Conference of the IEEE Industrial Electronics Society, Florence, 2016, pp. 3172-3177.

[31] C. Licitra, S. Musumeci, A. Raciti, A. Galluzzo, R. Letor and M. Melito, "Optimum driving circuit for IGBT devices suitable for integration," Proceedings of the 4th International Symposium on Power Semiconductor Devices and Ics, Tokyo, Japan, 1992, pp. 221-225.

[32] J. E. Makaran, "Gate Charge Control for MOSFET Turn-Off in PWM Motor Drives Through Empirical Means," in IEEE Transactions on Power Electronics, vol. 25, no. 5, pp. 1339-1350, May 2010.

[33] S. Takizawa, S. Igarashi and K. Kuroki, "A new di/dt control gate drive circuit for IGBTs to reduce EMI noise and switching losses," PESC 98 Record. 29th Annual IEEE Power Electronics Specialists Conference (Cat. No.98CH36196), Fukuoka, 1998, pp. 1443-1449 vol.2.

[34] Z. Wang, X. Shi, L. M. Tolbert and B. J. Blalock, "Switching performance improvement of IGBT modules using an active gate driver," 2013 Twenty-Eighth Annual IEEE Applied Power Electronics Conference and Exposition (APEC), Long Beach, CA, 2013, pp. 1266-1273.

[35] V. John, Bum-Seok Suh and T. A. Lipo, "High-performance active gate drive for high-power IGBT's," in IEEE Transactions on Industry Applications, vol. 35, no. 5, pp. 1108-1117, Sep/Oct 1999.

[36] N. Idir, R. Bausiere and J. J. Franchaud, "Active gate voltage control of turn-on di/dt and turn-off dv/dt in insulated gate transistors," in IEEE Transactions on Power Electronics, vol. 21, no. 4, pp. 849-855, July 2006.

[37] Y. Lobsiger and J. W. Kolar, "Closed-loop IGBT gate drive featuring highly dynamic di/dt and dv/dt control," 2012 IEEE Energy Conversion Congress and Exposition (ECCE), Raleigh, NC, 2012, pp. 4754-4761.

[38] Shihong Park and T. M. Jahns, "Flexible dv/dt and di/dt control method for insulated gate power switches," in IEEE Transactions on Industry Applications, vol. 39, no. 3, pp. 657-664, May-June 2003.

[39] J. P. Berry, „MOSFET operating under hard switching mode: Voltage and current gradients control", Conference proceedings EPE 1991, pp. 130

[40] Ch. Gerster and P. Hofer, „Gate-controlled dv/dt- and di/dt-limitation in high power IGBT converters", EPE Journal Vol. 5, Nr. 3/4 december 1995, pp. 7

[41] T. Shimizu and K. Wada, "A gate drive circuit of power MOSFETs and IGBTs for low switching losses," 2007 7th Internatonal Conference on Power Electronics, Daegu, 2007, pp. 857-860

[42] C. Licitra, S. Musumeci, A. Raciti, A. U. Galluzzo, R. Letor and M. Melito, "A new driving circuit for IGBT devices," in IEEE Transactions on Power Electronics, vol. 10, no. 3, pp. 373-378, May 1995.

[43] S. Musumeci, A. Raciti, A. Testa, A. Galluzzo and M. Melito, "Switching-behavior improvement of insulated gate-controlled devices," in IEEE Transactions on Power Electronics, vol. 12, no. 4, pp. 645-653, Jul 1997.

[44] P. J. Grbovic, F. Gruson, N. Idir and P. L. Moigne, "Turn-on Performance of Reverse Blocking IGBT (RB IGBT) and Optimization Using Advanced Gate Driver," in IEEE Transactions on Power Electronics, vol. 25, no. 4, pp. 970-980, April 2010.

[45] K. Fink and S. Bernet, "Advanced Gate Drive Unit With Closed-Loop di/dt Control," in IEEE Transactions on Power Electronics, vol. 28, no. 5, pp. 2587-2595, May 2013. doi: 10.1109/TPEL.2012.2215885

[46] L. Chen and F. Z. Peng, "Closed-Loop Gate Drive for High Power IGBTs," 2009 Twenty-Fourth Annual IEEE Applied Power Electronics Conference and Exposition, Washington, DC, 2009, pp. 1331-1337.

[47] B. Wittig and F. W. Fuchs, "Analysis and Comparison of Turn-off Active Gate Control Methods for Low-Voltage Power MOSFETs With High Current Ratings," in IEEE Transactions on Power Electronics, vol. 27, no. 3, pp. 1632-1640, March 2012.

[48] F. Blaabjerg and J. K. Pederson, "An optimum drive and clamp circuit design with controlled switching for a snubberless PWM-VSI-IGBT inverter leg," in Proc. IEEE PESC'92, 1992, pp. 289–297.

[49] L. Dulau, S. Pontarollo, A. Boimond, J. Garnier, N. Giraudo, and O. Terrasse, "A new gate driver integrated circuit for IGBT Devices With Advanced Protections," IEEE Trans. Power Electronics, VOL. 21, NO. 1, JANUARY 2006, pp. 38-44.

[50] P. R. Palmer and H. S. Rajamani, "Active Voltage control of IGBTs for high power applications," in IEEE Transactions on Power Electronics, vol. 19, no. 4, pp. 894-901, July 2004.

[51] A. R. Hefner, "An investigation of the drive circuit requirements for the power insulated gate bipolar transistor (IGBT)," 21st Annual IEEE Conference on Power Electronics Specialists, San Antonio, TX, USA, 1990, pp. 126-137.

[52] K. Yamaguchi, Y. Sasaki and T. Imakubo, "Low loss and low noise gate driver for SiC-MOSFET with gate boost circuit," IECON 2014 - 40th Annual Conference of the IEEE Industrial Electronics Society, Dallas, TX, 2014, pp. 1594-1598.

[53] A. Sagehashi, K. Kusaka, K. Orikawa and J. i. Itoh, "Current source gate drive circuits with low power consumption for high frequency power converters," 2015 9th International Conference on Power Electronics and ECCE Asia (ICPE ECCE Asia), Seoul, 2015, pp. 1017-1024.

[54] "Series Connection of IGBTs in Resonant Converters," M. Dehmlow et al., IPEC-Yokohama, 1995, pp. 1634-1638.

[55] H. Miyazaki, K. Kato, "A study on Soft- Switching gate drive circuit for IGBTs" National Convention Record IEEJ Industry Applications Soceiety, No.88, p.289-292, 1995.

[56] P. J. Grbovic, "An IGBT Gate Driver for Feed-Forward Control of Turn-on Losses and Reverse Recovery Current," in IEEE Transactions on Power Electronics, vol. 23, no. 2, pp. 643-652, March 2008.

[57] S. Safari, A. Castellazzi and P. Wheeler, "Experimental and Analytical Performance Evaluation of SiC Power Devices in the Matrix Converter," in IEEE Transactions on Power Electronics, vol. 29, no. 5, pp. 2584-2596, May 2014.

[58] T. Noguchi, S. Yajima, H. Komatsu, "Development of Gate Drive Circuit for Next-Generation Ultra High – Speed Switching Devices", IEEJ Trans. On Industry Applications, Vol. 129, No. 1, pp. 46-52(2009)

[59] T. Noguchi and T. Mizuno, "High-speed switching operation of MOSFETs using auxiliary circuit shorting load," 2012 International Conference on Renewable Energy Research and Applications (ICRERA), Nagasaki, 2012, pp. 1-6.

[60] Z. Zhang, W. Eberle, P. Lin, Y. F. Liu and P. C. Sen, "A 1-MHz High-Efficiency 12-V Buck Voltage Regulator With a New Current-Source Gate Driver," in IEEE Transactions on Power Electronics, vol. 23, no. 6, pp. 2817-2827, Nov. 2008.

[61] S. Musumeci, A. Raciti, A. Testa, A. Galluzzo and M. Melito, "A new adaptive driving technique for high current gate controlled devices," Applied Power Electronics Conference and Exposition, 1994. APEC '94. Conference Proceedings 1994., Ninth Annual, Orlando, FL, 1994, pp. 480-486 vol.1.

[62] A. T. Bryant, L. Lu, E. Santi, J. L. Hudgins and P. R. Palmer, "Modeling of IGBT Resistive and Inductive Turn-On Behavior," in IEEE Transactions on Industry Applications, vol. 44, no. 3, pp. 904-914, May-june 2008.

[63] W. Eberle, Z. Zhang, Y. F. Liu and P. C. Sen, "A High Efficiency Synchronous Buck VRM with Current Source Gate Driver," 2007 IEEE Power Electronics Specialists Conference, Orlando, FL, 2007, pp. 21-27.

[64] W. Eberle, Z. Zhang, Y. F. Liu and P. C. Sen, "A Current Source Gate Driver Achieving Switching Loss Savings and Gate Energy Recovery at 1-MHz," in IEEE Transactions on Power Electronics, vol. 23, no. 2, pp. 678-691, March 2008.

[65] M. Dehmlow et al., „Series connection of IGBTs in resonant convertes", IPEC Yokohama 1995, pp. 1634

[66] H. Fujita, "A Resonant Gate-Drive Circuit Capable of High-Frequency and High-Efficiency Operation," in IEEE Transactions on Power Electronics, vol. 25, no. 4, pp. 962-969, April 2010.

[67] D. M. Divan, "The resonant DC link converter-a new concept in static power conversion," in IEEE Transactions on Industry Applications, vol. 25, no. 2, pp. 317-325, Mar/Apr 1989.

[68] Jih-Sheng Lai, R. W. Young, G. W. Ott, J. W. McKeever and Fang Zheng Peng, "A delta-configured auxiliary resonant snubber inverter," in IEEE Transactions on Industry Applications, vol. 32, no. 3, pp. 518-525, May/Jun 1996.

[69] B. Arntzen and D. Maksimovic, "Switched-capacitor DC/DC converters with resonant gate drive," in IEEE Transactions on Power Electronics, vol. 13, no. 5, pp. 892-902, Sep 1998.

[70] I. D. de Vries, "A resonant power MOSFET/IGBT gate driver," APEC. Seventeenth Annual IEEE Applied Power Electronics Conference and Exposition (Cat. No.02CH37335), Dallas, TX, 2002, pp. 179-185 vol.1.

[71] Kaiwei Yao and F. C. Lee, "A novel resonant gate driver for high frequency synchronous buck converters," in IEEE Transactions on Power Electronics, vol. 17, no. 2, pp. 180-186, Mar 2002.

[72] Yuhui Chen, F. C. Lee, L. Amoroso and Ho-Pu Wu, "A resonant MOSFET gate driver with efficient energy recovery," in IEEE Transactions on Power Electronics, vol. 19, no. 2, pp. 470-477, March 2004.

[73] Z. Yang, S. Ye and Y. F. Liu, "A New Resonant Gate Drive Circuit for Synchronous Buck Converter," in IEEE Transactions on Power Electronics, vol. 22, no. 4, pp. 1311-1320, July 2007.

[74] Z. Zhang, W. Eberle, Z. Yang, Y. F. Liu and P. C. Sen, "Optimal Design of Resonant Gate Driver for Buck Converter Based on a New Analytical Loss Model," in IEEE Transactions on Power Electronics, vol. 23, no. 2, pp. 653-666, March 2008.

[75] Z. Yang, S. Ye and Y. F. Liu, "A New Dual-Channel Resonant Gate Drive Circuit for Low Gate Drive Loss and Low Switching Loss," in IEEE Transactions on Power Electronics, vol. 23, no. 3, pp. 1574-1583, May 2008.

[76] W. Eberle, Y. F. Liu and P. C. Sen, "A New Resonant Gate-Drive Circuit With Efficient Energy Recovery and Low Conduction Loss," in IEEE Transactions on Industrial Electronics, vol. 55, no. 5, pp. 2213-2221, May 2008.

[77] D. Maksimovic, "A MOS gate drive with resonant transitions," *Power Electronics Specialists Conference, 1991. PESC '91 Record., 22nd Annual IEEE*, Cambridge, MA, 1991, pp. 527-532.

[78] H. L. N. Wiegman, "A resonant pulse gate drive for high frequency applications," Applied Power Electronics Conference and Exposition, 1992. APEC '92. Conference Proceedings 1992., Seventh Annual, Boston, MA, 1992, pp. 738-743

[79] S. H. Weinberg, "A novel lossless resonant MOSFET driver," Power Electronics Specialists Conference, 1992. PESC '92 Record., 23rd Annual IEEE, Toledo, 1992, pp. 1003-1010 vol.2.

[80] Q. Li and P. Wolfs, "The Power Loss Optimization of a Current Fed ZVS Two-Inductor Boost Converter With a Resonant Transition Gate Drive," in IEEE Transactions on Power Electronics, vol. 21, no. 5, pp. 1253-1263, Sept. 2006.

[81] P. Dwane, D. O' Sullivan and M. G. Egan, "An assessment of resonant gate drive techniques for use in modern low power dc-dc converters," Twentieth Annual IEEE Applied Power Electronics Conference and Exposition, 2005. APEC 2005., Austin, TX, 2005, pp. 1572-1580 Vol. 3.

[82] T. Lopez, G. Sauerlaender, T. Duerbaum and T. Tolle, "A detailed analysis of a resonant gate driver for PWM applications," Applied Power Electronics Conference and Exposition, 2003. APEC '03. Eighteenth Annual IEEE, Miami Beach, FL, USA, 2003, pp. 873-878 vol.2.

Design and Testing of 1 kV H-bridge Power Electronics Building Block Based on 1.7 kV SiC MOSFET Module

Jun Wang[1*], Rolando Burgos[1], Dushan Boroyevich[1], Zeng Liu[2]

1 Center for Power Electronics Systems, Virginia Tech, Blacksburg, USA
2 State Key Lab of Electrical Insulation and Power Equipment, Xian Jiaotong University, Xian, China
*E-mail: junwang@vt.edu

Abstract—This paper presents a power electronics building block (PEBB) design based on 1.7 kV SiC MOSFET power modules. The PEBB power stage is an H-bridge circuit that can be cascaded to build modular converters. The evolution of PEBB architectures is introduced at the beginning and the specifications are given, followed by the design considerations and solutions of critical PEBB components. In particular, laminated DC bus, differential-mode inductor, smart gate driver, high noise-immunity digital control system, and DC-fed auxiliary power supply have been accomplished and demonstrated. Finally, a novel hybrid-current-mode switching-cycle control approach has been proposed and validated on a SiC-PEBB-based modular multilevel Buck converter (MMBC).

Keywords—*Power electronics building block, SiC MOS-FET, smart gate driver, digital control system, switching-cycle control.*

I. Introduction

In recent years the demand for higher power density and lower weight of medium-voltage (MV) converters is growing rapidly. Typical MV applications for high power density include electrical ships [1], motor drives for underground mining, medium voltage direct current (MVDC) distribution system in urban area [2], and MV wind converters [3]. As the available space or room are tight, they all feature extremely high cost per unit area or volume. If the advancement of new technologies is able to elevate the power density of the converters in those applications, the deducted cost of system installation is much higher than the incremental cost of more advanced equipment. For instance, the US Navy Zumwalt-class destroyers cost $4.24B per unit. The space for electrical equipment on the destroyer is so limited that the averaged price is estimated as high as $5,000 \sim $10,000/ft^3. In the meantime, the MV power converters cost around $500 \sim $1,000/ft^3, nearly one tenth of the space price. Accordingly, it is cost-effective and highly motivated to leverage the most advanced semiconductors to increase the power/volume and power/weight density of the MV power converters in the aforementioned domains.

Most of the MV power converters are designed in a modular structure that employs the concept of power electronics building block (PEBB). As a least replaceable unit (LRU) to build modular converters, the PEBB concept was originally proposed by the Office of Naval

Research in 1997 [4]. It has been defined as a universal power processor and a systematic approach, featuring modular configurations, scalable voltage and current ratings, as well as low inventory and maintenance cost. In the past two decades, nearly all the commercial high-power converters have taken advantage of the PEBB concept in the MV and high-voltage (HV) applications. The topologies of PEBB power stage include half-bridge [5], H-bridge [6], neutral-point clamped 3-level (NPC-3L) [7], and active neutral-point clamped 5-level (ANPC-5L) [8]. PEBBs can be connected in series, parallel, or multiple-phase configurations to scale up the voltage and current ratings of PEBB-based converters, breaking through the constraints of semiconductor device ratings. To this end, the objective of designing a high-power-density MV converters can be converted to the target of developing a high-power-density MV PEBB.

Owing to the booming technology of wide-bandgap semiconductor devices and packaging, SiC MOSFETs have demonstrated their superior performance to Si IG-BTs of higher breakdown voltage, faster switching speed, lower switching loss and higher operating temperature [9]. In the power electronics history, the high switching frequency and low switching loss have always been the driving force of reducing the size of passive and thermal-management components. It is highly prospective that SiC MOSFETs will change the game of developing high-density MV PEBBs and power converters. Nevertheless, the notable static and dynamic characteristics of SiC MSOFETs bring harsh challenges to the design of power stage, gate driving, protection and control circuitry design for a SiC-MOSFET-based PEBB. The research endeavor of this paper aims to tackle those challenges and to ensure the best performance of the SiC PEBB.

II. PEBB System Architecture

Fig. 1 shows a conventional PEBB architecture that consists of Si power switches, gate driving and protection circuits, as well as required passive components. The control and sensing units are usually designed in the upper-level converter system out of the scope of PEBB. Later, an intelligent Si PEBB has been proposed to incorporate the control and sensing circuits that is named "hardware manager" [10]. In this design, the PEBB controller is able to monitor local currents, voltages, temperatures,

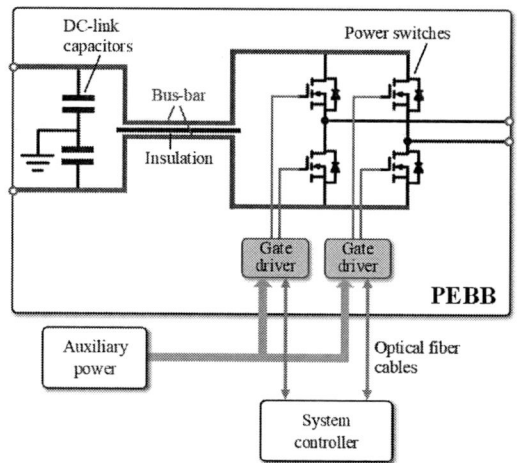

Fig. 1.   Conventional PEBB system architecture diagram.

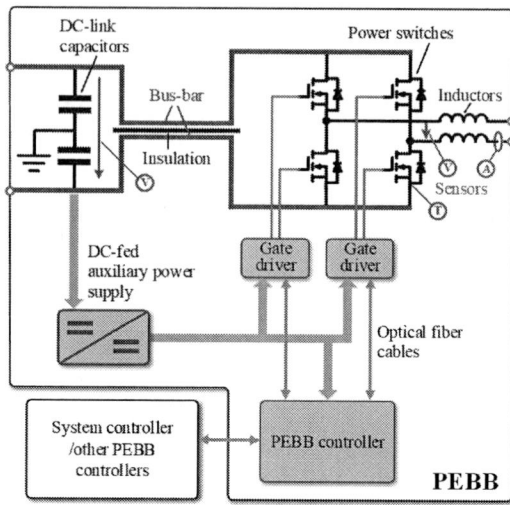

Fig. 2.   PEBB1000 system architecture diagram.

TABLE I.   PEBB1000 SPECIFICATIONS

Property	Value
DC bus rated voltage (V_{dc})	1000 V
DC bus peak voltage (harmonics)	1200 V
Maximum DC bus voltage ripple	± 200 V
Terminal rms current ($I_{nom.rms}$)	150 A
Terminal maximum current (I_{max})	300 A
Device short-circuit threshold	800 A
Switching frequency (f_{sw})	$40 \sim 100$ kHz
Maximum dv/dt	50 V/ns
Maximum di/dt	10 A/ns
Total DC capacitance (C_{dc})	58 μF
Total AC differential-mode inductance (L_{dm})	4 μH

and to respond to local faults instantly. In order to ensure that the entire control and auxiliary system is powered before the main power stage is energized, the auxiliary power supplies for both the conventional and intelligent PEBB are fed from an external grid power source sitting approximately on the earth ground. For instance, the "Perfect Harmony" MV drive has a high-isolation-voltage low-power auxiliary transformer in each PEBB to supply the gate drivers and logic units. The primary side of the transformer is connected to the utility grid, whereas its secondary-side voltage potential is related to the switching behavior of the PEBB-based converter. In consequence, the insulation voltage stress of the transformer can be as high as the maximum converter voltage amplitude, rather than that of the PEBB.

However, when it comes to high-voltage direct current (HVDC) transmission, this grid-fed auxiliary power solution becomes infeasible as the required isolation and insulation voltage is scaled up to ± 800 kV because of stacking PEBBs. Alternatively, a grid-fed wireless power supply that eliminates physical contact between the primary and secondary sides (no creepage) has been implemented for a 24 kV converter, but its scalability is still limited [11]. An absolute modular technique is to feed the power supply directly from the PEBB DC bus, where the insulation voltage is determined by the PEBB voltage rating regardless of the scheme that PEBBs are interconnected [18]. Yet, challenges still exist. First of all, the DC-fed power supply is required to operate in a wide range of input voltages, i.e. the PEBB DC bus voltage may vary from zero to its peak value. Secondly, DC-fed power supply should have sufficiently high regulation bandwidth to guarantee the output voltage quality when subjected to low-frequency harmonic ripples at its input. In addition, a minimum wake-up voltage is inevitable for the auxiliary power supply to start up. Before the DC bus voltage reaches the threshold, the power switches are undergoing high risks of failures due to the absence of gate and control power. Accordingly, the wake-up voltage of the auxiliary power supply must be as low as possible.

Based on the previous discussion, the architecture of the PEBB in this paper follows Fig. 2. The PEBB power stage comprises a H-bridge power circuit, capacitors, inductors and a laminated bus-bar. The PEBB control and auxiliary system includes gate drivers, an DC-fed auxiliary power supply, isolated digital sensors and a PEBB controller. The PEBB specifications are shown in Table I. The PEBB is named PEBB1000 according to its nominal DC bus voltage. The power switches are two SiC MOSFET modules CAS300M17BM2 from Wolfspeed, rated at 1.7 kV, 225 A.

III.   PEBB POWER STAGE

A. DC-link capacitors and laminated bus-bar

Since the PEBB1000 is designed as a universal power processor, it should fit different power converter topologies and run in various operation modes. When the stand-alone PEBB or PEBB-based converter operate at a DC-DC mode, the dominant component of capacitor charging current lies at the switching frequency. Typical examples include the classical Buck mode, or the switching-cycle control (SCC) mode for the modular multilevel DC-DC converters (MMDC) [12]-[13]. In those operation modes, the minimum necessary DC-link capacitance value is calculated as,

3750

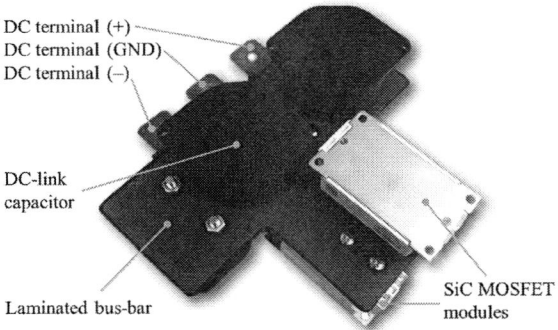

Fig. 3. DC bus assembly: DC-link capacitors, laminated bus-bar and SiC MOSFET modules.

$$C_{dc} = \frac{I_{nom,amp}}{V_{dc} \cdot \delta_c \cdot f_{sw,min}} = 53 \ \mu F \qquad (1)$$

where $I_{nom,amp}$ is the PEBB terminal current amplitude $150\sqrt{2}$ A, V_{dc} is the nominal DC link voltage 1000 V, δ_c that equals to 10% is the allowable peak-to-peak ripple percentage within one switching period, and $f_{sw,min}$ is the minimum switching frequency 40 kHz.

In some other operation modes, the DC-link capacitor charging current contains not only switching-frequency component, but also line-frequency and related harmonic components. Typical examples are the single-phase DC-AC mode, or the conventional control mode for the modular multilevel converters (MMC). In those operation modes, the required capacitance value is roughly $10\times$ higher than that of the DC-DC mode, and is highly dependent on the control techniques for low-frequency capacitor voltage ripple reduction. Consequently, a DC interface should be designed out of the PEBB enclosure to retain the possibility of DC-link capacitance extension.

The middle point of the DC bus is formed by two series-connected capacitors. As denoted in Fig. 2, it is connected to the PEBB chassis as a local ground to minimize overall voltage stresses. Also, the split DC-link capacitors serve as EMI Y-capacitors to prevent the common-mode (CM) current from flowing out of the PEBB via the DC interface. In the end, four (2×2) film capacitors AVX FFVS6B0586K have been selected. Each capacitor is rated at 800 V, 83 A and 58 μF, and the overall capacitor bank ratings are 1600 V, 166 A and 58 μF. A laminated bus-bar has been designed to interconnect the capacitors and SiC MOSFET modules with low stray inductance and symmetric geometry, and double-sided cooling capability. The maximum power-loop inductance is measured as 21.5 nH and no additional decoupling capacitors are needed. The detailed design and more test results of the bus-bar have been elaborated in [14], and the DC bus assembly is shown in Fig. 3.

B. Differential-mode inductor

The differential-mode (DM) inductance value is determined by the inductor current variation under a specified

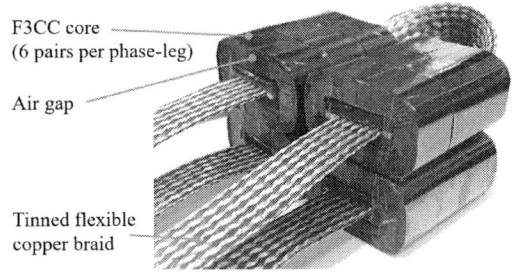

Fig. 4. Differential-mode inductor fabrication concept.

across voltage and time duration. The largest variation occurs at the SCC mode for the modular multilevel DC-DC converters (MMDC) [13]. It is required that the inductor current is controlled from the positive peak to the negative peak within 20% of the minimum switching period, and thus the calculation is given as,

$$L_{dm} = \frac{V_{dc} \cdot \delta_{scc}}{2I_{nom.amp} \cdot f_{sw,max}} = 4.7 \ \mu H \qquad (2)$$

where V_{dc} is the nominal DC link voltage 1000 V, δ_{scc} that equals to 20% is the allowable time percentage of one minimum switching period when the inductor current changes from the positive peak to the negative peak, $2I_{nom,amp}$ is the PEBB inductor current peak-to-peak amplitude $300\sqrt{2}$ A, and $f_{sw,max}$ is the maximum switching frequency 100 kHz.

Considering other parasitic inductance of cables, the DM inductance value is designed as 4 μH. For a symmetric CM design, the DM inductor is split into two identical parts connected to the two phase legs of the H-bridge, respectively. Each inductor is valued 2 μH, and takes a nominal amplitude of $150\sqrt{2}$ A. Typically, high-current and small-inductance inductors are designed in an air-core or single-turn structure. The air-core inductor is usually of low power density and behaves as a radiated noise source, whereas the single-turn inductor has high power density, enclosed flux loop, and is manufacturing-friendly, so the latter has been selected. The single-turn DM inductor has been designed using the following equations,

$$L_{dm} = \frac{\mu_0 \cdot A_c \cdot n^2}{l_g} \qquad (3)$$

$$B_{max} = \frac{\mu_0 \cdot I_{max} \cdot n}{l_g} \qquad (4)$$

where μ_0 is the permeability constant $4\pi \cdot 10^{-7}$ H·m^{-1}, A_c is the cross-section area of cores, n is the turn number that equals to 1, l_g is the air gap length, B_{max} is the maximum flux density that should be less than half of the saturation flux density B_{sat} of the core material, and I_{max} is the maximum allowable AC terminal current 300 A. Larger A_c gives lower B_{max} and lower core loss as well, so a trade-off between loss and volume has been made and the design results are given in Table II.

3751

TABLE II. DM INDUCTOR DESIGN PARAMETERS

Property	Value
DM inductance per phase (L_{dm})	2 μH
Number of turns per phase (n)	1 turn
Magnetic cores per phase	Hitachi FINEMET® F3CC0008 $\times6$
Maximum operating temperature	155°C
Effective cross section area (A_c)	$172 \cdot 6$ mm^2
Air gap length (l_g)	0.65 mm
Saturation flux density (B_{sat})	1.23 T @659 A
Rated flux density ($B_{nom,amp}$)	0.40 T @150$\sqrt{2}$ A
Maximum flux density (B_{max})	0.56 T @300 A
Total volume per phase	0.22 L
Total weight per phase	996 g
Total core loss per phase	180 W @100 kHz, $\Delta B = 0.28$ T

TABLE III. SMART GATE DRIVER SPECIFICATIONS

Property	Minimum	Maximum
Supply voltage	18 V	75 V
Driving voltage	-4 V	$+20$ V
dv/dt immunity	30 V/ns	100 V/ns
Switching frequency	-	100 kHz
Isolation voltage	3 kV	-
Driving current	-	30 A
External gate resistors	0.1 Ω	-
Driver IC over-temperature protection	150°C	200°C
Under-voltage lockout	11 V	14 V
Active Miller clamp	5 A, $V_{EE} + 2$ V	15 A, -
Short-circuit protection threshold	600 A	1200 A
Two-level turn-off	7 V, 0.5 μs	10 V, 1.5 μs

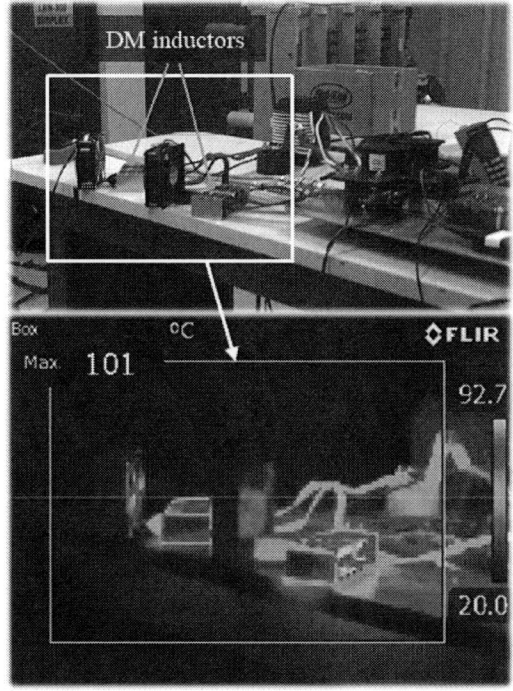

Fig. 5. Thermal image of the inductor continuous test. The maximum hot-spot temperature is 101°C.

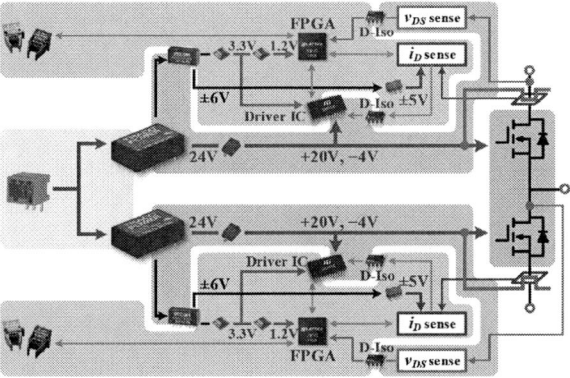

Fig. 6. Smart gate driver architecture for 1.7 kV SiC MOSFET module.

Fig. 4 demonstrates the DM inductor fabrication concept. In the practical assembly, each DM inductor is contained by an aluminum frame and the flexible copper braids are enclosed by heat-shrink tubings for insulation. The two DM inductors has been evaluated in a continuous quasi-square-wave test. The switching frequency is 100 kHz, and the triangular inductor current reaches ±150 A peak value ($\Delta B = 0.28$ T). The thermal image in Fig. 5 indicates that the hot-spot temperature of the inductors stays at 101°C at steady state with forced air cooling.

IV. PEBB CONTROL AND AUXILIARY SYSTEM

A. 1.7 kV smart gate driver

Gate driver is the vital interface between power semiconductor devices and control signals, which serves to provide galvanic isolation and to supply driving current

while maintaining signal integrity under high-noise environment. On top of those basic functionalities, it can also offer quick, reliable and configurable protections, as well as advanced signal sensing and data processing for control purposes, which define a "smart" gate driver. Fig. 6 shows the smart gate driver architecture of the PEBB1000. The high-side and low-side driving channels for the two switch positions share the identical design. In each channel, digitally configurable gate driver IC, Rogowski switch-current sensor (RSCS) and switch-voltage sensor (SVS) are integrated on the same printed circuit board (PCB). A FPGA serves as the "brain" that manages the configurations, signal processing and communications. Fig. 7 shows the anatomy picture. The specifications of the smart gate driver are given in Table III.

Since numerous sensitive signal processing circuits have been integrated, the gate driver architecture in Fig. 6 is particularly designed for noise immunity. In each driving channel, two isolated ground planes have been made. The red plane provides driving current and voltages at $+20$ V and -4 V, and the green plane mostly contains signal processing components and logic units with voltages at ±5 V, 3.3 V, and 1.2 V. The CM impedance of the red ground path is dominated by a few nanohenry trace inductance, whereas the CM impedance of the green ground path is determined by several picofarad input-output capacitance of isolated power supplies, the gate driver IC, and digital isolators. The CM impedance of the

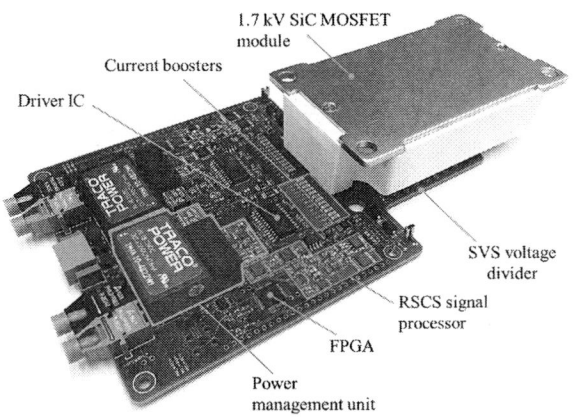

Fig. 7. Prototype of the smart gate driver and SiC MOSFET module.

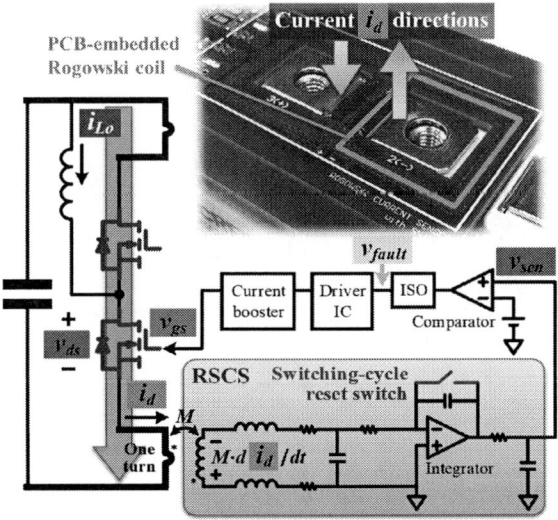

Fig. 8. Rogowski switch current sensor circuit diagram.

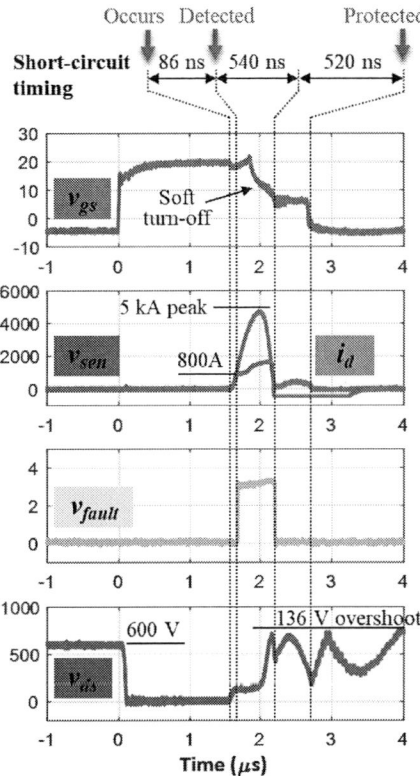

Fig. 9. Fault-under-load short-circuit detection and protection. Test results: $V_{dc} = 600$ V, SC current $i_{sc.peak} = 5$ kA, detection time $t_{det} = 86$ ns, reaction time $t_{rec} = 1060$ ns.

green path is more than $10\times$ higher than the red. Hence, the major part of CM noise current will flow through the red plane instead of the green, so that the sensitive circuits will be subjected to greatly mitigated conductive noises.

The Rogowski switch-current sensor (RSCS) enables a number of beneficial functionalities by sensing SiC MOSFET drain current with high bandwidth, accuracy, amplitude and low latency. Comparing the switch current to a high threshold value, the RSCS can serve as a short-circuit (SC) detector; sampling the switch current by an analog-to-digital converter (ADC), it can also function equivalently as a phase-current sensor; comparing the switch current to a control reference, the RSCS empowers the peak-current-mode control approach in MV high-power domains. Fig. 8 shows an example circuit where the RSCS serves as a SC detector. A fault-under-load SC test has been conducted at 600 V DC bus and the waveforms are presented in Fig. 9. When the high-side gate is turned on after the low-side gate has been turned on for about 1.55 μs, the SC current occurs and reaches 5

kA peak value within 500 ns. The SC detection threshold current is set as 800 A, and the fault is detected at 86 ns after the SC happens. It then takes 1.06 μs to complete the soft turn-off action due to certain internal fault handling process of the gate driver IC. Owing to the soft turn-off mechanism, the turn-off overshoot is limited within 136 V at 5 kA turn-off current.

The switch-voltage sensor (SVS) is also integrated on the smart gate driver. Inspired by the fundamentals of passive voltage probes, a capacitive voltage divider is paralleled to the resistive divider to boost the SVS bandwidth. In [15], the SVS nicely tracks the v_{ds} voltage even capturing all the ringing information. In practical applications, the SVS can be used for online overshoot monitoring, accurate dead-time compensation, and zero-voltage-switching check. Detailed fundamentals, design and test results of the 1.7 kV smart gate driver and the integrated sensors have been elaborated in [15]-[17].

B. Digital control system

The PEBB control system is partitioned as shown in Fig. 10, including the PEBB control board, gate drivers and isolated digital sensors (IDS). Each of the three partitions contains a FPGA that handles the communication tasks among each other. All the communication interconnections are plastic optical fibers (POF). It eliminates the

The 2018 International Power Electronics Conference

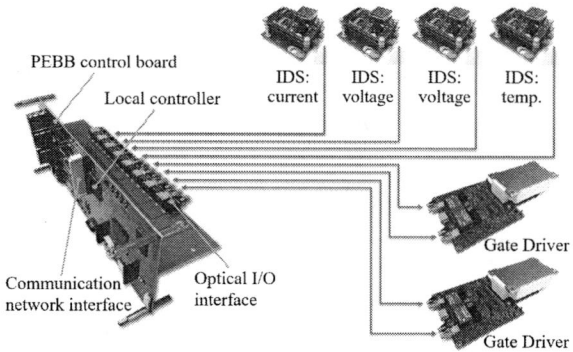

Fig. 10. PEBB control system based on plastic optical fibers.

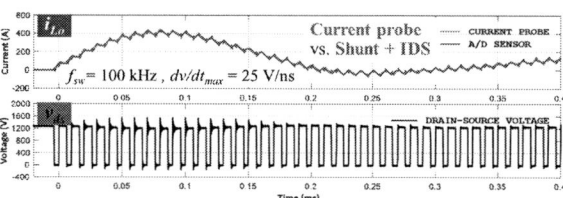

Fig. 11. IDS vs. current probe CWT-3b at dynamic transients. Test condition: $V_{dc} = 1200$ V, $i_{Lo,max} = 400$ A, $f_{sw} = 100$ kHz, maximum $dv/dt = 25$ V/ns.

CM paths and loops that cause noise and malfunction issues in the control circuits. The local PEBB controller is built on a popular MCU-and-FPGA-based architecture. A communication network interface (CNI) card can be plugged onto the main controller to manage inter-PEBB communications. The CNI is connected to the MCU external memory bus, providing necessary functionalities to establish an upper-level control network.

The IDS samples analog signals and converts it to digital data at the installation location. The digital data is transmitted back to the controller via POF. The IDS comprises four layers of PCB, that are, a power supply board, a digital signal processing and transmission board, an analog signal conditioning board, and an adaptive interface board. Three versions of the interface boards are designed to sample the voltage (with voltage divider), current (with shunt) and temperature (with NTC thermistor) data, respectively. In order to mitigate the CM noise flowing through the isolation barrier of the IDS, two low-profile power supplies with 2 pF input-output capacitance each are connected in series to achieve 1 pF isolation. The phase-current IDS performance has been assessed by sampling the load inductor current i_{Lo} (labeled in Fig. 8) at 100 kHz switching frequency and 2 MHz sampling frequency. Its voltage potential is located at the switching node, where the highest dv/dt of 25 V/ns is produced. In Fig. 11, the comparison between the IDS output data (converted from binary to decimal values, and conversion delay is compensated) and the commercial current probe waveform indicates excellent agreement with each other. The sampling error is negligible at non-switching time, and remains small even at switching transients.

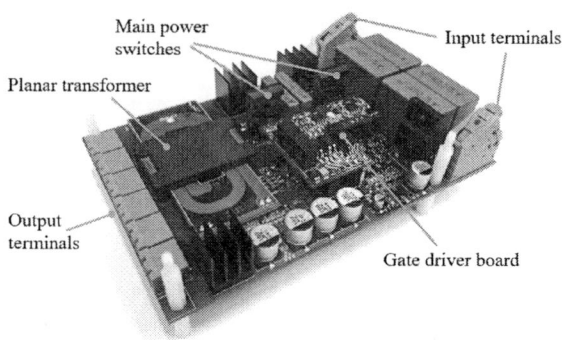

Fig. 12. Prototype of the DC-fed auxiliary power supply.

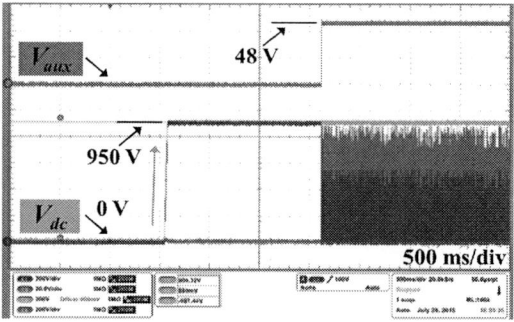

Fig. 13. DC-fed auxiliary power supply start-up test. DC bus voltage V_{dc} ramps up from 0 to 950 V. Power supply output voltage V_{aux} rises from 0 to 48 V.

C. DC-fed auxiliary power supply

The operating range of auxiliary power supply is in accordance with the PEBB specifications. The rated DC bus voltage is 1000 V and the expected maximum voltage ripple is 1200 V. Considering margin for dynamic transients, in consequence, the auxiliary power supply has to operate at 1300 V. Moreover, to avoid undesirable fault of the power switches during the startup of the system, the required wake-up voltage of the ancillary devices is 230 V across the DC bus. As regards the output specifications, the power supply is required to provide an output voltage of 48 V, and an output power of 80 W, in order to feed the whole control system. The maximum voltage stress across its isolation barrier is ±600 V as the output of the power supply is referred to the PEBB chassis, which is connect to the middle point of the DC-link. The topology for the auxiliary power supply, chosen after exploring several options, is a two-switch flyback converter. Detailed topology analysis, design and test results have been illustrated in [18]. Fig. 12 shows the prototype picture, and Fig. 13 demonstrates that when the V_{dc} ramps up from 0 to 950 V, the power supply output voltage V_{aux} is able to promptly regulated at 48 V.

V. PEBB-BASED MODULAR CONVERTER TEST

As shown in Fig. 14, the PEBB1000 has finally integrated all the aforementioned parts together to build modular converters. Modular multilevel DC-DC convert-

3754

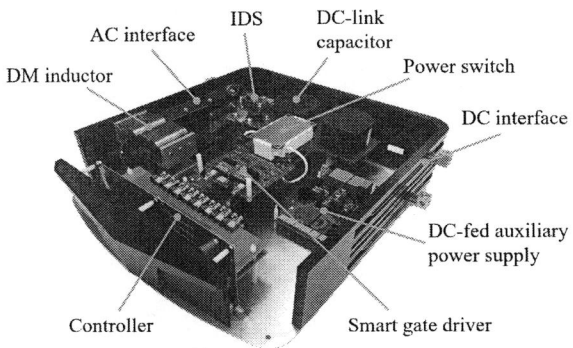

Fig. 14.   PEBB1000 prototype assembly.

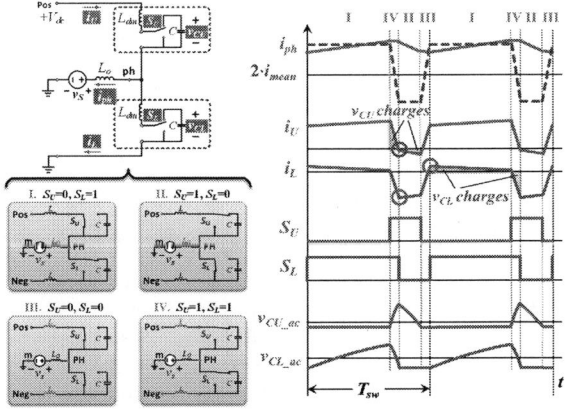

Fig. 15.   Basic MMBC topology and the SCC fundamentals.

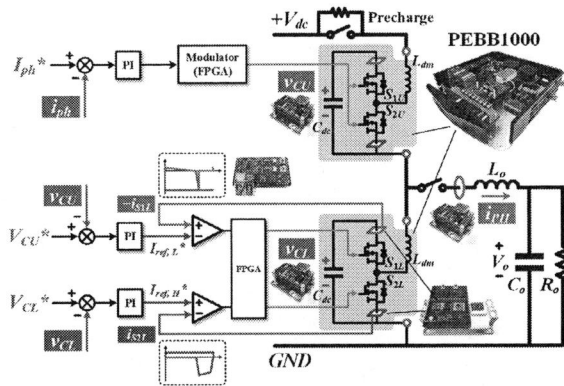

Fig. 16.   HCM-SCC closed-loop control block diagram for MMBC.

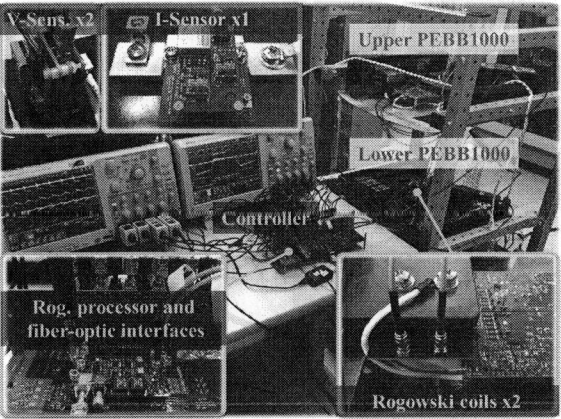

Fig. 17.   MMBC converter platform for HCB-SCC validation.

ers (MMDC) are a group of non-isolated multilevel converters adopting the same arm configurations as the modular multilevel converter (MMC) [13]. Modular multilevel Buck converter (MMBC) belongs to one of the MMDC topologies that can serve as a step-down converter in MVDC applications. The switching-cycle control (SCC) has been proposed to minimize capacitor size and to enable the DC-DC operation of the MMC-like converters [13]. In this section, a basic MMBC converter has been built with PEBB1000 to validate the SCC approach.

The basic MMBC converter topology is presented in Fig. 15. The output capacitor of the conventional synchronous Buck converter is replaced by a DC source V_s, and the two switches are replaced by two half-bridge PEBBs. The switching functions S_U and S_L are the key control variables to regulate the phase current i_{ph} and to balance the two capacitor voltages V_{CU} and V_{CL}. Unlike the conventional synchronous Buck converter where S_U and S_L are complementary, in the SCC, they are commanded with certain small amount of phase-shift. The phase-shift creates the Switching States III and IV that allow both arm currents i_U and i_L to alternate at high slew rate. Since the alternating arm currents are necessary to charge and discharge the two PEBB capacitors, the SCC makes it possible to balance V_{CU} and V_{CL} in each switching period as depicted in the figure.

The most critical step to realize SCC is the method to determine the time duration of Switching State I, II, III and IV. Note that the State I and II regulate the phase current i_{ph}, and the two peak current of i_L (highlighted by green and red circles in Fig. 15) determine the charging/discharging of capacitors. Accordingly, a novel control approach is proposed to resolve the problem. As depicted in Fig. 16, on one hand, the upper PEBB is operating in the average-current-mode (ACM) control. The average value of i_{ph} is sampled for feedback comparison. The average current error is given to a PI compensator and then a modulator to generate the switching function of the upper PEBB, as well as the two gate commands. On the other hand, the lower PEBB is operating in the peak-current-mode (PCM) control. V_{CU} and V_{CL} are sampled and compared with the voltage reference. The errors are given to two PI compensators that generate two current references. To regulate V_{CU}, the lower reference $I^*_{ref,L}$ is used to compare with the switch current $-i_{S1L}$ to turn off $S1L$, which determines the duration of State IV. Likewise, to regulate V_{CL}, the higher reference $I^*_{ref,H}$ is used to compare with the switch current i_{S2L} to turn off $S2L$, which determines the duration of State III. As both the ACM and PCM are applied to realize the SCC, this control approach is termed hybrid-current-mode

TABLE IV.   HCM-SCC TEST PARAMETERS

Property	Value	Property	Value
f_{sw}	40 kHz	L_o	600 μH
T_{sw}	25 μs	C_o	50 μF
V_{dc}	500 V	R_o	5.7 Ω
D_{S1U}	0.62	$K_{P,Vcap}$	0.01
D_{eff}	0.57	$K_{I,Vcap}$	5
V_o	285 V	$K_{P,Iph}$	0.01
I_o	50 A	$K_{I,Iph}$	0.5

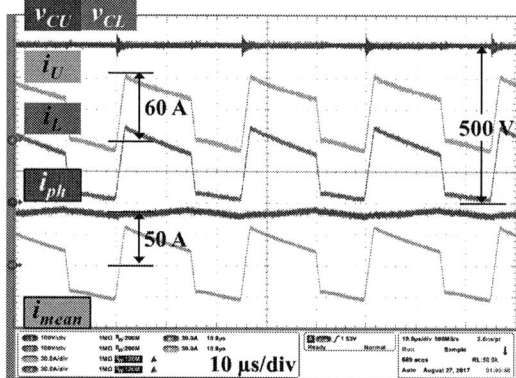

Fig. 18.   HCM-SCC power stage waveforms.

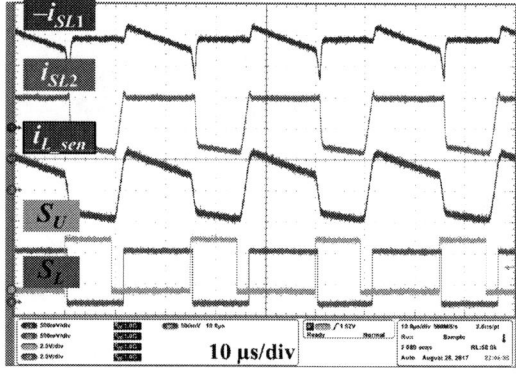

Fig. 19.   HCM-SCC control and sensor signal waveforms.

switching-cycle control (HCM-SCC).

The HCM-SCC has been validated on the SiC-PEBB1000-based MMBC test platform. To simplify the test, only one half bridge of each PEBB1000 is used. The PEBB DC capacitance $C_{dc} = 58$ μF and PEBB inductance $L_{dm} = 4$ μH. As shown in Fig. 17, a central control system that incorporates two capacitor-voltage IDS, one phase-current IDS and two RSCS, is implemented. The test parameters are shown in Table IV. D_{S1U} is duty cycle of the top switch S_{1U} in the upper PEBB, and D_{eff} is the effective output duty cycle. $K_{P,Vcap}$ and $K_{I,Vcap}$ are the capacitor voltage PI parameters, and $K_{P,Iph}$ and $K_{I,Iph}$ are the phase current PI parameters. The power stage test waveforms are shown in Fig. 18, and the control and sensor signal waveform are exhibited in Fig. 19. Both results agree nicely with theoretical analysis.

VI.   CONCLUSIONS

Design and testing of power stage components, smart gate drivers, digital control system, and auxiliary power supply of the PEBB1000 have been shown. A new control approach HCM-SCC for MMBC is proposed and validated on the PEBB1000-based test platform.

ACKNOWLEDGEMENT

This work has been conducted under the research project sponsored by the Office of Naval Research (ONR).

REFERENCES

[1]   ABB, "PSC6000 medium voltage drive," Product brochure, 2016.

[2]   M. Stieneker and R. W. De Doncker, "Medium-voltage DC distribution grids in urban areas," Proc.IEEE Intern. Symp. on Power Electron. for Distr. Gener. Systems, 2016, pp. 1-7.

[3]   N. Doerry and J. Amy, "Electric ship power and energy system architectures," Proc. IEEE Electric Ship Tech. Symp., 2017, pp. 1-10.

[4]   T. Ericsen, "Power Electronics Building Blocks," Proc. IME/IEE/SEE Electric Warship Conference, 1997.

[5]   SIEMENS, "HVDC PLUS   Basics and Principle of Operation," Product document, 2009.

[6]   SIEMENS, "ROBICON Perfect Harmony Medium-Voltage Liquid-Cooled Drives," Product brochure, 2012.

[7]   ABB, "ACS5000 Medium Voltage Drive," Product brochure, 2017.

[8]   ABB, "ACS2000 Medium Voltage Drive," Product brochure, 2015.

[9]   J. Millan, P. Godignon, X. Perpina, A. Perez-Tomas, J. Rebollo, "A survey of wide bandgap power semiconductor devices," IEEE Trans. Power Electron., vol. 29, no. 5, pp. 2155-2163, May, 2014.

[10]   I. Celanovic, I. Milosavljevic, D. Boroyevich, R. Cooley and J. Guo, "A new distributed digital controller for the next generation of power electronics building blocks," Proc. Applied Power Electron. Conf. and Expo., 2000, pp. 889-894 vol.2.

[11]   D. Cottet, et al., "Integration technologies for a fully modular and hot-swappable MV multi-level concept converter," in Proc. Int. Exhib. Conf. for Power Electron., 2015, pp. 1-8.

[12]   J. Wang, R. Burgos and D. Boroyevich, "Switching-Cycle State-Space Modeling and Control of the Modular Multilevel Converter," IEEE Journal of Emerging and Selected Topics in Power Electronics, vol. 2, no. 4, pp. 1159-1170, Dec. 2014.

[13]   J. Wang, R. Burgos and D. Boroyevich, "Switching-Cycle Capacitor Voltage Control for the Modular Multilevel DC/DC Converters," Proc. Applied Power Electron. Conf. and Expo., 2015, pp. 377-384.

[14]   N.R. Mehrabadi, I. Cvetkovic, J. Wang, R. Burgos and D. Boroyevich, "Busbar design for SiC-based H-bridge PEBB using 1.7 kV, 400 a SiC MOSFETs operating at 100 kHz," Proc. IEEE Energy Convers. Congr. Expo., 2016, pp. 1-7.

[15]   J. Wang, et al., "Power Electronics Building Block (PEBB) design based on 1.7 kV SiC MOSFET modules," Proc. IEEE Electric Ship Tech. Symp., Arlington, 2017, pp. 612-619.

[16]   J. Wang, Z. Shen, R. Burgos, and D. Boroyevich, "Gate driver design for 1.7kV SiC MOSFET module with Rogowski current sensor for shortcircuit protection," Proc. Applied Power Electron. Conf. and Expo., 2016, pp. 516-523.

[17]   J. Wang, Z. Shen, R. Burgos and D. Boroyevich, "Integrated switch current sensor for shortcircuit protection and current control of 1.7-kV SiC MOSFET modules," Proc. IEEE Energy Convers. Congr. and Expo., 2016, pp. 1-7.

[18]   G. Rizzoli, L. Zarri, J. Wang, R. Burgos and D. Boroyevich, "Design of a two-switch flyback power supply using 1.7 kV SiC devices for ultra-wide input-voltage range applications," Proc. IEEE Energy Convers. Congr. Expo., 2016, pp. 1-5.

A Flyback Converter with SiC Power MOSFET Operating at 10 MHz: Reducing Leakage Inductance for Improvement of Switching Behaviors

Kazuki Hashimoto*, Takafumi Okuda, Takashi Hikihara
Department of Electrical Engineering
Katsura, Nishikyo, Kyoto, Japan
*E-mail: k-hashimoto@dove.kuee.kyoto-u.ac.jp

Abstract—A high-frequency flyback converter is investigated in this paper. In order to increase the switching frequency of the flyback converter, we focus on the impacts of the leakage inductance of the transformer at the primary side on the transients of the turn-off characteristics. It is found that the leakage inductance of the transformer limits the turn-off behaviors of the primary side. By reducing the leakage inductance of the transformer on the primary side, the improved flyback converter achieves the 10 MHz operation. The output voltage is controlled by duty ratio from 30% to 60%. The maximum output power of 16 W and the maximum conversion efficiency of 67% are observed in the flyback converter.

Keywords—Flyback converter, SiC MOSFET, High-frequency DC-DC converter, Leakage inductance

I. INTRODUCTION

A flyback converter, an isolated DC/DC converter, is widely adopted to provide regulated output voltages for low-power applications typically from 20- to 200-W range [1]. The transformer can isolate the output stage from the input stage in the flyback converter. Increasing the switching frequency of the flyback converter, the passive components such as the transformers, inductors, and capacitors are expected to be miniaturized [2]. Therefore, it is important to increase the switching frequency for the miniaturization of the flyback converter.

The flyback converter require the improvement of the switching characteristics of semiconductor devices to increase the switching frequency of the flyback converter. Recently, silicon carbide (SiC) metal-oxide-semiconductor field-effect transistors (MOSFETs) have been developed. High-voltage SiC MOSFETs are available with low on-resistance owing to superior material properties of SiC [3]–[5]. Since SiC MOSFETs are unipolar devices, they are able to operate at high switching frequency, compared to silicon insulated gate bipolar transistors (IGBTs). Using SiC MOSFETs, high-frequency switching power supplies are reported at switching frequency up to several MHz [6].

An 1-MHz flyback converter has been reported using SiC MOSFET [7]. The flyback converter achieved the output power of several tens of watts with a snubber circuit. However, it was difficult to control the output voltage by changing the duty ratio at 10 MHz [7]. Therefore, it is necessary to reveal limiting factors of the switching frequency in high-frequency region.

In high-frequency flyback converters, parasitic components are important such as leakage inductances and parasitic capacitances of a transformer [8]–[10]. In previous studies, an additional circuit such as a snubber circuit and active-clamp circuit are adopted to suppress the surge voltage induced by parasitic components [11]–[17]. However, such additional circuits might prevent to further improve the switching frequency. A flyback converter without any additional circuits should be investigated to clarify the impacts of parasitic components and to increase the switching frequency of flyback converters.

In this paper, we discuss limiting factors of the switching frequency by analyzing the switching waveforms in a fabricated flyback converter using a SiC MOSFET. As a result, it is found that the leakage inductance of the transformer limits the turn-off behaviors of the primary side. By reducing the leakage inductance of the transformer on the primary side, the improved flyback converter is found to achieve the 10 MHz operation.

II. EXPERIMENTAL CONDITION

The schematic of a fabricated flyback converter is shown in Fig. 1. We adopt a SiC MOSFET (ROHM, SCT2450KE) as a switching device on the primary side

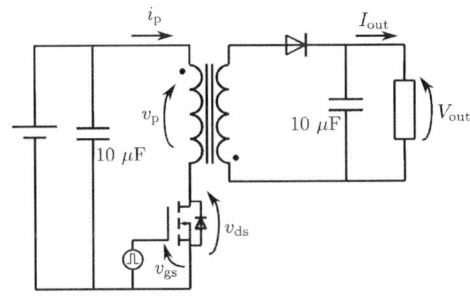

Fig. 1. A schematic of fabricated flyback converter.

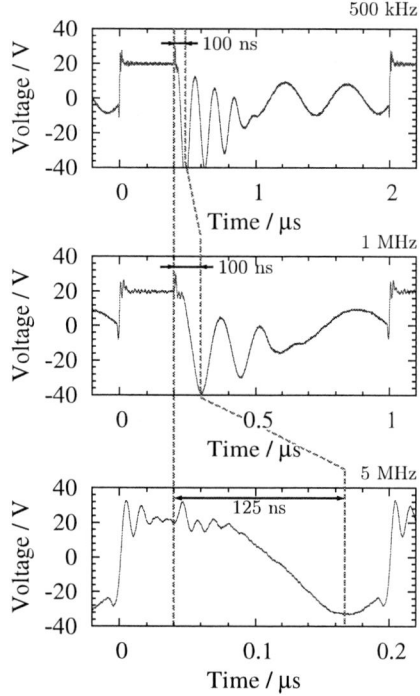

Fig. 2. Comparison of experimentally observed voltage waveforms on primary side of transformer in fabricated flyback converter. Switching frequency is varied from 0.5 to 5 MHz.

TABLE I. PARAMETERS OF TRANSFORMERS

Transformer	Inductance [μH]	Leakage inductance [μH]
A	14.0	0.13
B	15.1	3.74

and a SiC SBD (ROHM, SCS220AE2) as a rectifying device on the secondary side. A gate driver, Si8235 (Silicon Labs.), is adopted as an isolated gate driver for SiC MOSFET. The switching frequency of the flyback converter is set from 0.5 to 5 MHz with duty ratio of 20%. We employ Liqualloy^{TM} (ALPS ELECTRIC) for a magnetic material of the transformer. The transformer has a turns ratio of 6:1. The inductance on the primary side is measured at 35 μH and the leakage inductance on the primary side is measured at 4.6 μH by using an impedance analyzer (Keysight Technologies, 4294A).

III. RESULTS AND DISCUSSION

A. Impacts of Leakage Inductances

The fabricated flyback converter using the SiC MOS-FET is operated at high switching frequencies. Then the limiting factors of the switching frequencies are discussed in this section. Fig. 2 shows the comparison of the voltage waveforms measured at the primary side of the transformer in the flyback converter. For comparisons, the turn-on signal is applied at $t = 0$ ns and the turn-off signal is applied to $t = 40$ ns. At 500 kHz, the on-state and off-state of the SiC MOSFET were clearly observed owing to the superior switching characteristics

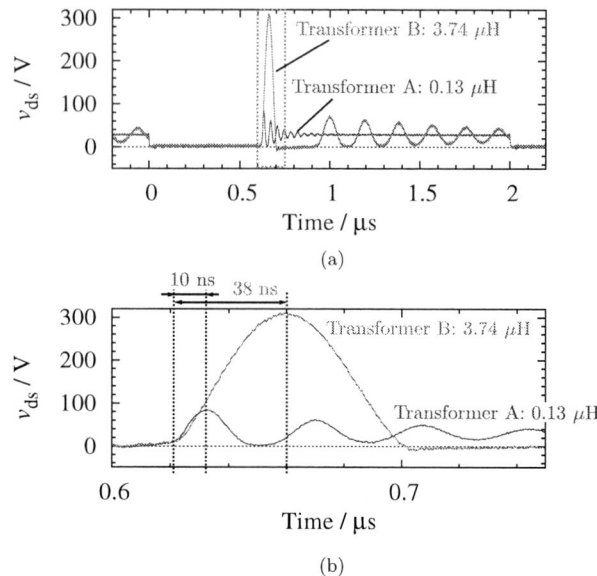

(a)

(b)

Fig. 3. Comparisons of experimentally observed voltage waveforms of drain-source of MOSFET: (a) one switching cycle, (b) enlarged view in dotted rectangular area.

of the SiC MOSFET. However, the turn-off time of the SiC MOSFET was dominant in the switching cycle with increasing the switching frequency, and the long turn-off tail severely limited the switching frequency at 5 MHz. By comparing with the switching waveforms at 500 kHz, this long turn-off tail corresponds to the period in which the stored energy in the primary side of the transformer is transferred to the secondary side. It is necessary to clarify the limiting factors of the turn-off behaviors and enhance the switching frequency for improving the switching frequency of flyback converters.

There are limited researches about the oscillations on the primary side of the transformer in flyback converters observed at the turn-off of the SiC MOSFET. It has been proposed that the oscillations on the primary side of transformer are induced by the leakage inductance of the transformer, the output capacitance of the SiC MOSFET, and some other factors [18]–[20]. Here, we assume that the oscillations are determined by the resonance between the leakage inductance of the transformer on the primary side and the output capacitance of the SiC MOSFET. We fabricate two types of transformers to investigate the impacts of the leakage inductance of the transformer on the turn-off characteristics. The measured inductances on the primary side of the fabricated transformers are shown in TABLE I. Transformer A has a small leakage inductance (0.13 μH) by using the windings with a twisted structure, while Transformer B has a large leakage inductance (3.74 μH) by winding the primary and the secondary sides separately. Both Transformer A and B have the same turns ratio of 1:1. The switching frequency of the flyback converter is set at 500 kHz with duty ratio of 30%.

Fig. 3(a) shows the switching waveforms at the drain-

3758

source voltage of the SiC MOSFET. The surge voltage observed in the flyback converter with Transformer A (0.13 μH) was smaller than that with Transformer B (3.74 μH). Fig. 3(b) shows the enlarged view of Fig. 3(a) in the dotted rectangular area. The transient time from the beginning of the turn-off of the SiC MOSFET to the first peak of the oscillations is 10 ns in the flyback converter with Transformer A, while the transient time is 38 ns in the flyback converter with Transformer B. Assuming that the turn-off characteristics are determined by the LC resonance between the leakage inductance of the transformer and the output capacitance of the SiC MOSFET, the stored energy in the leakage inductance of the primary side is transferred into the output capacitance of the SiC MOSFET in this period. The stored energy in the leakage inductance of Transformer A (0.13 μH) is smaller than that of Transformer B (3.74 μH), so that the shorter switching time is obtained in the Transformer A. Therefore, transformer with a smaller leakage inductance is required to increase the switching frequency of flyback converter.

B. Flyback Converters Operating at 10 MHz

In the previous subsection, it is found that the switching frequency of the flyback converter was limited by the leakage inductance of the transformer. In this section, we investigate the operation of flyback converters using transformer A and B at 10 MHz.

The schematic of the fabricated flyback converter is shown in Fig. 1. The input voltage is set at 20 V and the load resistance is set at 100 Ω. The duty ratio is varied from 30% to 60%.

Fig. 4 shows the switching waveforms measured in the flyback converter using Transformer B at 10 MHz with duty ratio of 30%. The flyback converter with Transformer B did not operate at 10 MHz; the output voltage and output current of the flyback converter were almost zero, although the switching waveforms of the gate-source voltage of the SiC MOSFET exhibited its on- and off-states clearly. Transformer B has a large leakage inductance and limits the turn-off characteristics of the SiC MOSFET as described above. Thus, the primary current flows continuously in the on- and off-states of the SiC MOSFET and the stored energy of the transformer is wasted in the primary side, being not transferred to the secondary side. That is the reason why the output power was not obtained in the flyback converter with a large leakage inductance.

Fig. 5 shows the switching waveforms in the flyback converter using Transformer A operating at 10 MHz with duty ratio of 30%. The flyback converter using Transformer A successfully operated at 10 MHz. The output voltage of the flyback converter is measured at 20 V and the output current is measured at 200 mA. The rise time in the drain-source voltage of the SiC MOSFET with Transformer A is faster than that with Transformer B. The small leakage inductance enhances the switching characteristics in the flyback converter. The primary current in the flyback converter with Transformer

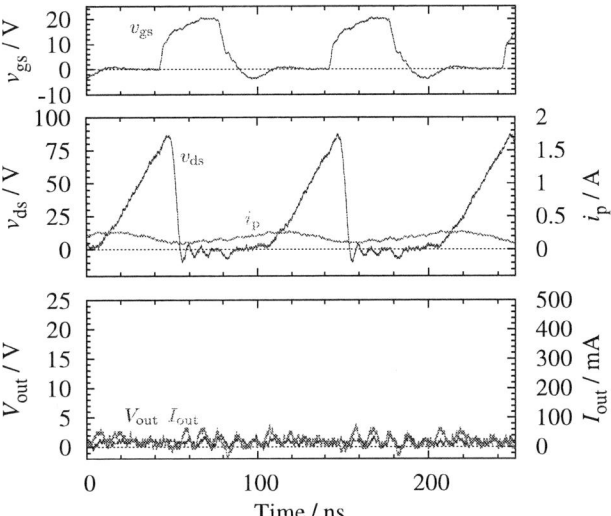

Fig. 4. Experimentally observed waveforms of v_{gs}, v_{ds}, i_p, V_{out}, and I_{out} at 10 MHz. Value of leakage inductance of transformer on primary side is 3.74 μH.

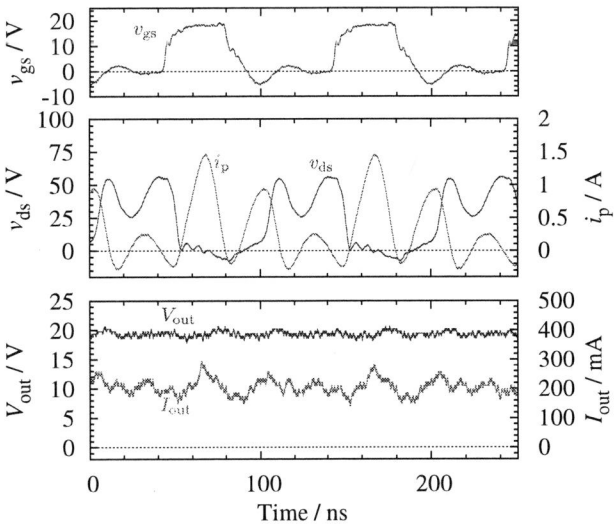

Fig. 5. Experimentally observed waveforms of v_{gs}, v_{ds}, i_p, V_{out}, and I_{out} at 10 MHz. Value of leakage inductance of transformer on primary side is 0.13 μH.

A is larger than that with Transformer B, although the oscillations were observed in the primary current.

We investigate the output voltage, output power, and conversion efficiency with various duty ratios at 10 MHz in the flyback converters with Transformers A and B. The obtained results are shown in Fig. 6. The output voltage, output power, and conversion efficiency ware almost zero in the flyback converter with Transformer B. On the other hand, the output voltage and output power ware controlled at the duty ratio from 30% to 60% in the flyback converter with Transformer A because the small leakage inductance of the transformer enhanced the switching behaviors of the primary side. The conversion efficiency was approxi-

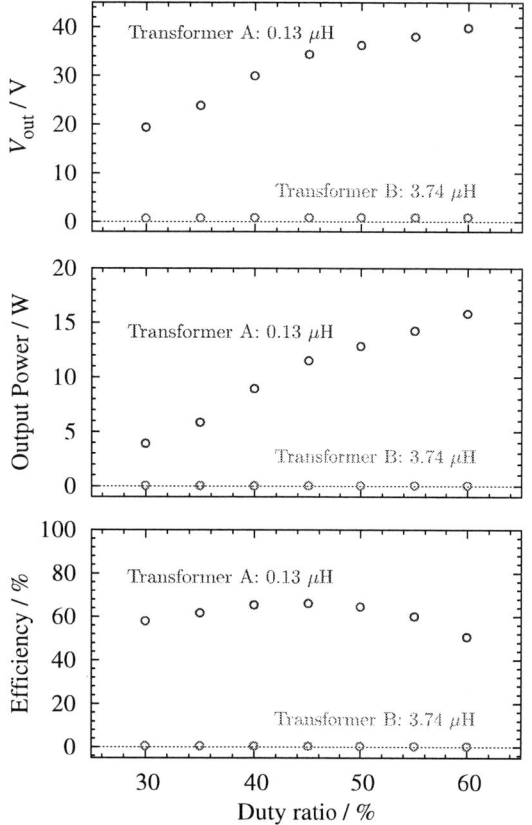

Fig. 6. Output voltage, output power, and conversion efficiency of flyback converter operated at 10 MHz plotted against duty ratio of gate driver for SiC MOSFET.

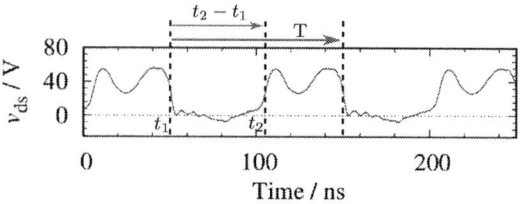

Fig. 7. Estimation of effective duty ratio from experimentally observed waveforms of v_{ds}.

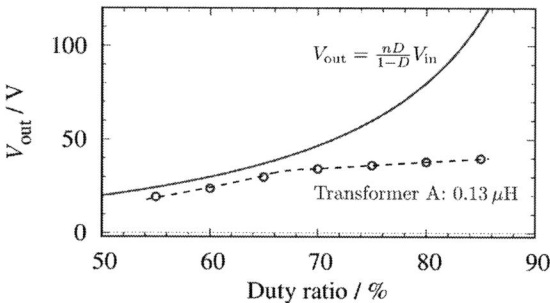

Fig. 8. Output voltage of flyback converter operated at 10 MHz according to increase of effective duty ratio.

mately 60% and maximum conversion efficiency is 67% in the duty ratio of 45% in the flyback converter with Transformer A. Therefore, the output characteristics of the fabricated flyback converter are improved with the transformer that has the small leakage inductance on the primary side.

Assuming that the magnetizing inductance is negligible, the output voltage of the flyback converter is estimated by

$$V_{\text{out}} = \frac{nD}{1-D} V_{\text{in}}, \tag{1}$$

where V_{in} denotes the input voltage in flyback converter, V_{out} denotes the output voltage in flyback converter, n denotes the turns ratio of the transformer, and D denotes the duty ratio set in the function generator [21]. However, the effective duty ratio at the flyback converter is larger than the duty ratio set at the function generator, because the turn-off characteristics are slower than the turn-on characteristics of the SiC MOSFET. Fig. 7 shows the drain-source voltage v_{ds} at 10 MHz with duty ratio of 30% in the fabricated flyback converter using Transformer A. The effective duty ratio is estimated by

$$D_{\text{eff}} = \frac{t_2 - t_1}{T}. \tag{2}$$

It was estimated at 55%, suggesting that it is larger than the duty raito at the function generator by 25%.

Figure 8 shows the output voltage of the fabricated flyback converter according to the increase of the effective duty ratio. The estimated curve exhibited the same tendency as the output characteristics of the fabricated flyback converter using Transformer A at the effective duty ratio from 55% to 65%. On the other hand, the increase rate of output voltage in the fabricated flyback converter was much lower than that of the estimated curve in the effective duty ratio over 70%. The voltage drop increases in the SiC MOSFET and SiC SBD because the primary current and secondary current at high duty ratio are larger than those at low duty ratio in the flyback converter. Therefore, it is necessary to estimate the output voltage by taking into account the switching loss.

IV. CONCLUSIONS

In this paper, we investigated the operation of the flyback converter at 10 MHz using the SiC MOSFET. As a result, it was found that the transient time of the turn-off of the SiC MOSFET is determined by the leakage inductance of the transformer on the primary side. By reducing the leakage inductance of the transformer on the primary side, we operate the flyback converter at 10 MHz. The output is observed in the flyback converter with Transformer A that has a small leakage inductance (0.13 μH). Therefore, a transformer with a smaller leakage inductance is required to increase the switching frequency of flyback converter.

ACKNOWLEDGMENT

This research is partly supported by Kyoto Super Cluster Program (JST) and Cross-ministerial Strategic Innovation Promotion Program (SIP,"Next-generation power electronics" (NEDO)). The authors would like to acknowledge ALPS ELECTRIC Co., Ltd. for providing Liqualloy^{TM} core.

REFERENCES

[1] R. W. Erickson and D. Maksimovic, "Fundamentals of Power Electronics," Springer US, 2001.

[2] D. J. Perreault, J. Hu, J. M. Rivas, Y. Han, O. Leitermann, R. C. N. Pilawa-Podgurski, A. Sagneri, and C. R. Sullivan, "Opportunities and Challenges in Very High Frequency Power Conversion," 2009 Twenty-Fourth Annual IEEE Applied Power Electronics Conference and Exposition, pp. 1–14, 2009.

[3] J. A. Cooper, M. R. Melloch, R. Singh, A. Agarwal, and J. W. Palmour, "Status and prospects for SiC power MOSFETs," IEEE Transactions on Electron Devices, vol. 49, no. 8, pp. 658–664, 2002.

[4] J. Millán, P. Godignon, X. Perpiñá, A. Pérez-Tomás, and J. Rebollo, A "Survey of Wide Bandgap Power Semiconductor Devices," IEEE Transactions on Power Electronics, vol. 29, no. 5, pp. 2155–2163, 2014.

[5] T. Kimoto, "Material science and device physics in SiC technology for high-voltage power devices, Japanese Journal of Applied Physics," vol. 54, no. 4, pp. 040103-1–040103-27, 2015.

[6] Y. Sadanda, T. Okuda, and T. Hikihara, "Direct drive of a buck converter by delta-sigma modulation at 13.56-MHz sampling," 2017 IEEE 18th Workshop on Control and Modeling for Power Electronics (COMPEL), pp. 1–4, 2017.

[7] N. Satoh, H. Otake, T. Nakamura, and T. Hikihara, "A flyback converter using power MOSFET to achieve high frequency operation beyond 13.56 MHz," IECON 2015 - 41st Annual Conference of the IEEE Industrial Electronics Society, pp. 001376–001381, 2015.

[8] F. Luo, Z. Chen, L. Xue, P. Mattavelli, D. Boroyevich, and B. Hughes, "Design considerations for GaN HEMT multichip halfbridge module for high-frequency power converters," 2014 IEEE Applied Power Electronics Conference and Exposition - APEC 2014, pp. 537–544, 2014.

[9] T. Meade, D. O'Sullivan, R. Foley, C. Achimescu, M. Egan, and P. McCloskey, "Parasitic inductance effect on switching losses for a high frequency Dc-Dc converter," 2008 Twenty-Third Annual IEEE Applied Power Electronics Conference and Exposition, pp. 3–9, 2008.

[10] J. Wang, H. S. h Chung, and R. T. h Li, "Characterization and Experimental Assessment of the Effects of Parasitic Elements on the MOSFET Switching Performance," IEEE Transactions on Power Electronics, vol. 28, no. 1, pp. 573–590, 2013.

[11] R. Watson, F. C. Lee, and G. C. Hua, "Utilization of an active-clamp circuit to achieve soft switching in flyback converters," IEEE Transactions on Power Electronics, vol. 11, no. 1, pp. 162–169, 1996.

[12] A. Elasser and D. A. Torrey, "Soft switching active snubbers for DC/DC converters," IEEE Transactions on Power Electronics, vol. 11, no. 5, pp. 710–722, 1996.

[13] R. T. H. Li and H. S. h Chung, "A Passive Lossless Snubber Cell With Minimum Stress and Wide Soft-Switching Range," IEEE Transactions on Power Electronics, vol. 25, no. 7, pp. 1725–1738, 2010.

[14] N. P. Papanikolaou and E. C. Tatakis, "Active voltage clamp in flyback converters operating in CCM mode under wide load variation," IEEE Transactions on Industrial Electronics, vol. 51, no. 3, pp. 632–640, 2004.

[15] G. Spiazzi, P. Mattavelli, and A. Costabeber, "High Step-Up Ratio Flyback Converter With Active Clamp and Voltage Multiplier," IEEE Transactions on Power Electronics, vol. 26, no. 11, pp. 3205–3214, 2011.

[16] P. Alou, A. Bakkali, I. Barbero, J. A. Cobos, and M. Rascon, "A low power topology derived from flyback with active clamp based on a very simple transformer," Twenty-First Annual IEEE Applied Power Electronics Conference and Exposition, pp. 627–632, 2006.

[17] B.-R. Lin, H.-K. Chiang, K.-C. Chen, and D. Wang, "Analysis, design and implementation of an active clamp flyback converter," 2005 International Conference on Power Electronics and Drives Systems, vol. 1, pp. 424–429, 2005.

[18] T. Zhang, Q. Qian, M. Xu, W. Sun, and S. Lu, "Analysis on ringing effect of auxiliary winding in primary side regulated flyback converter," 2014 IEEE Energy Conversion Congress and Exposition (ECCE), pp. 2727–2733, 2014.

[19] P. Meng, X. Wu, J. Yang, H. Chen, and Z. Qian, "Analysis and design considerations for EMI and losses of RCD snubber in flyback converter," 2010 Twenty-Fifth Annual IEEE Applied Power Electronics Conference and Exposition (APEC), pp. 642–647, 2010.

[20] R. Petkov and L. Hobson, "Analysis and optimisation of a flyback convertor with a nondissipative snubber," IEE Proceedings - Electric Power Applications, vol. 142, no. 1, pp. 35–42, 1995.

[21] B. Choi, "Pulsewidth Modulated DC-to-DC Power Conversion: Circuits, Dynamics," and Control Designs, IEEE Press series on power engineering, Wiley, 2013.

A Study on Load Fluctuation of Isolated DC-DC Converter with Class Phi-2 Inverter using GaN-HFET

Yuta Yanagisawa[1*], Yushi Miura[1], Hiroyuki Handa[2], Tetsuzo Ueda[2], and Toshifumi Ise[1]
1 Graduate School of Engineering, Osaka University, Osaka, Japan
2 Panasonic Corporation, Osaka, Japan
*E-mail: yanagisawa@pe.eei.eng.osaka-u.ac.jp

Abstract— In recent years, the development and application of GaN-HFET(Heterojunction Field Effect Transistor) has become actively. As GaN-HFETs have advantages in high frequency operation, it is possible to down size power converters by rising switching frequency. In this paper, we investigate the load fluctuation of an isolated DC-DC converter with the class Phi-2 inverter circuit using GaN-HEFT, which is operated at 13.56 MHz.

Keywords— *WBG Semiconductor, GaN-HFET, Resonant Converter, DC-DC Converter.*

I. INTRODUCTION

In recent years, the development of Wide Band Gap(WBG) semiconductors, such as silicon carbide(SiC) and gallium nitride(GaN) is becoming actively, and these semiconductor have been put into practical use stage in industrial equipments, already. In the WBG semiconductors, the GaN-HFET which is one of the power semiconductor devices using GaN, has superior characteristics in the high frequency region because of small gate input capacitance and high electron mobility, due to physical characteristics and structure. In addition, low ON resistance device can be realized since GaN-HFET has 2 Dimensional Electron Gas(2DEG), which greatly contributes to the reduction of conduction loss.

Furthermore, as the development and implementation of the GaN-HFET progress, the resonance converter circuits such as a class E and a class Phi-2 inverter circuits have attracted a lot of attention. These resonance converter circuits is easy to operate in the high frequency region beyond 1 MHz by utilizing GaN-HFETs.

In this paper, results of response to load fluctuation of isolated DC-DC converter with the class Phi-2 inverter which is operating at 13.56 MHz by using GaN-HFET are shown.

II. CLASS PHI-2 INVERTER

Fig.1(a) shows the circuits of the class Phi-2 inverter circuit. In the class Phi-2 inverter circuit, inductor L_{MR} and capacitor C_{MR} which inject the third harmonics to the circuit is placed in parallel with the semiconductor switch

S. The features of the class Phi-2 inverter are listed as follows.
 ·Only one switching device is required
 ·Soft switching is achieved
 ·Voltage stress in drain to source (V_{DS}) is reduced because of injection of the third harmonics
However, the class Phi-2 inverter has as following problems.
 ·Some circuit parameters should be tuned depending on the load and power
 ·Duty ratio and switching frequency is fixed

In the class E inverter, the maximum V_{DS} voltage become nearly 4 times of the input voltage, however, in the class Phi-2 inverter, the maximum voltage is 2-times of input voltage. This difference is from injecting the third harmonics to V_{DS} by resonance tank X_{MR}. Generally, the withstand voltage of GaN-HFET is smaller than Si-MOSFET and SiC-MOSFET. In the case of applying GaN-HFET to class E inverter, the maximum input voltage is limited to the quarter of withstand voltage. Therefore, the power conversion capacity and input voltage are limited. On the other hand, in the class Phi-2 inverter, the application destination is expanded, since the input voltage is able to be increased to half of drain to source withstand voltage.

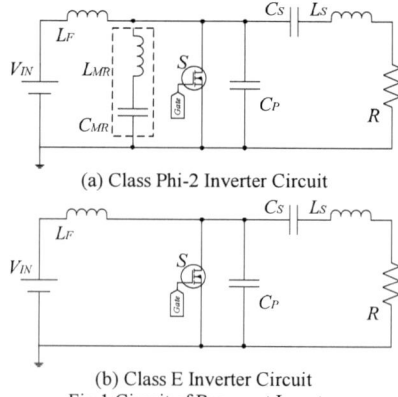

(a) Class Phi-2 Inverter Circuit

(b) Class E Inverter Circuit
Fig.1 Circuit of Resonant Inverter

In this chapter, we designed the class Phi-2 inverter circuit from reference [1] and [2]. Table I shows the circuit requirements for the design. Equation (1) to (6) show the circuit parameter equations. In these equations, capacitor Cs was selected as 1 nF in advance because this capacitor exists just for blocking DC components. Furthermore, to set the capacitor C_{MR} to 50 pF, C_F in equations (4) (5) and (6) was calculated as 53.3 pF.

TABLE I
DESIGN REQUIREMENT OF CIRCUIT

Input Voltage	50 V	Output Power	25 W
Switching Frequency	13.56 MHz	Load Resistance	50 Ω

$$X_S = R_{LOAD} \cdot \sqrt{\left(v_{ds1,RMS}/v_{load1,RMS}\right)^2 - 1} \tag{1}$$

$$v_{ds1,RMS} = (4/\pi\sqrt{2}) \cdot V_{IN} \tag{2}$$

$$v_{load1,RMS} = \sqrt{P_{OUT} \cdot R_{LOAD}} \tag{3}$$

$$L_{MR} = 1/(15\pi^2 \cdot f_s^2 \cdot C_F) \tag{4}$$

$$C_{MR} = 15 \cdot C_F/16 \tag{5}$$

$$L_F = 1/(9\pi^2 \cdot f_s^2 \cdot C_F) \tag{6}$$

In the class Phi-2 inverter, in order to achieve the soft switching, it is necessary to adjust the circuit parameters from these equations, to satisfy the following the drain to source impedance (Z_{DS}) requirement.

· At the switching frequency, the load including the reactance X_S shows inductivity between 30 and 60 degree
· At the switching frequency, the impedance magnitude must higher about 4 dB to 8dB than 3 times of switching frequency

Table II shows the designed circuit parameters after tuning respectively

TABLE II
PARAMETERS OF CLASS PHI 2 INVERTER (CALCULATION)

Inductor L_F	480 nH	Capacitor C_P	170 pF
Inductor L_{MR}	689 nH	Capacitor C_{MR}	50 pF
Inductor L_S	707 nH	Capacitor C_S	1000 pF

Next, we consider the case of using as an isolated DC-DC converter. In the case of usage as the isolated DC-DC converter, an isolation transformer must be inserted in the circuit. As shown in Fig.1(a), the load including X_S is connected in parallel with switch. In order to maintain the impedance requirement, the transformer needs to be connected in parallel with the switch. Therefore, the transformer inserted into the inductor L_F, and this circuit composition is shown Fig.2.

Here, the primary side self-inductance must to be the same as the inductor L_F. The secondary side leakage inductance can be considered as a part of the inductor L_S. So, even if the coupling coefficient is low, its influence can be ignored.

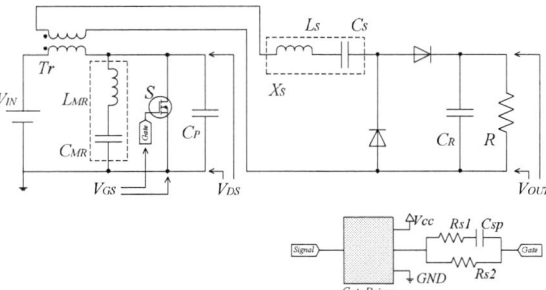

Fig.2 Composition of Isolated DC-DC Converter including Gate Driver (Half-Wave Rectifier)

Next step, design the load resistance R of the rectifier. The equivalent resistance when a smoothing capacitor are inserted, is expressed as equation (7) and (8). The equation (7) and (8) show the half-wave and full-wave rectification respectively. In the case of setting the equivalent resistance to 50 Ω, load resistance have to be set to 246.7 Ω and 61.68 Ω respectively.

$$R_{eq} = 2R/\pi^2 \tag{7}$$

$$R_{qu} = 8R/\pi^2 \tag{8}$$

Fig.3 shows the impedance magnitude and phase of Z_{DS} in the composition of Fig.2. The rectifier circuit is replaced by the equivalent resistor, the self-inductance of transformer is 480 nH, the coupling coefficient is 0.9, and the inductor L_S was set to 554 nH from considering the leakage inductance.

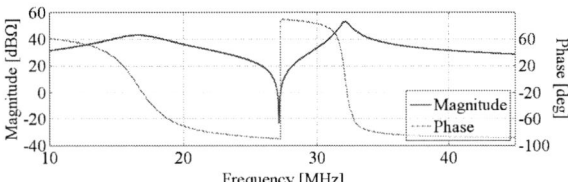

Fig.3 Z_{DS} Impedance and Phase in DC-DC Converter

As shown in Fig.3, the impedance requirement is satisfied in case of the isolated DC-DC composition. That is the end of design of the circuit. Table III shows the designed circuit parameters.

TABLE III
PARAMETERS OF CLASS PHI 2 INVERTER (SIMULATION)

Load Resistance (Half-Wave)	246.7 Ω	Load Resistance (Full-Wave)	61.68 Ω
Inductor LMR	689 nH	Capacitor CMR	50 pF
Inductor LS	554 nH	Capacitor CS	1000 pF
Transformer Self-Inductance	480 nH	Capacitor CP	170 pF
Turn Ratio	1:1	Coupling Coefficient	0.9

III. OPERATION AT VARIOUS LOAD CONDITIONS

Fig.4 shows the impedance magnitude and phase of Z_{DS} at switching frequency and 3rd switching frequency in above conditions. The horizontal axis is the load resistance. From Fig.4, the range that satisfy the described impedance requirement is wide between around 40 Ω and 80 Ω, therefore, the class Phi-2 inverter circuit in isolated DC-DC converter is able to operate in a wide load range.

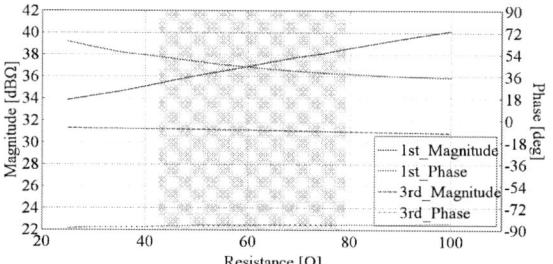

Fig.4 Impedance Magnitude and Phase to Load Resistance

In this chapter, variable load operation and load fluctuation is confirmed by computer simulations. The applied simulator is LTSpice, and speed-up capacitor method was applied to GaN-HFET driving. Table IV shows the parameters of the GaN-HFET gate driver circuit, and Fig.5 shows the each waveform from the simulation with half-rectifier. The duty ratio is set to 0.24.

TABLE IV
PARAMETERS OF GATE DRIVER

Input Voltage	12 V	Capacitor C_{SP}	1500 pF
Resistance R_{S1}	5 Ω	Resistance R_{S2}	250 Ω

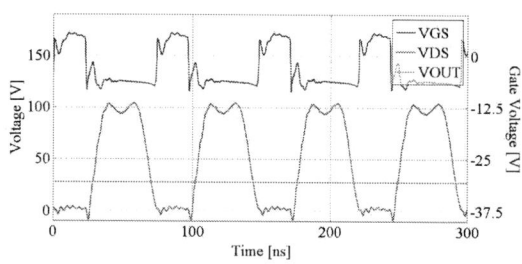

Fig.5 Waveforms on Class Phi-2 Inverter (Req=50 Ω)

From Fig.5, the class Phi-2 inverter can operate properly and validity of the design parameters was confirmed. Fig.6 shows waveforms in various load conditions.

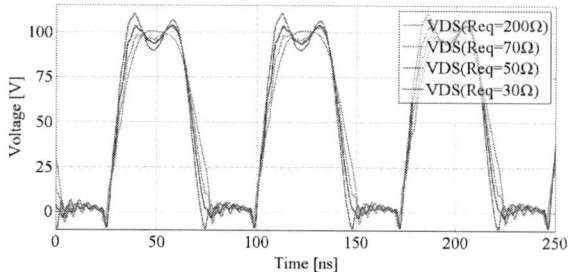

Fig.6 Waveforms in Various Load Conditions

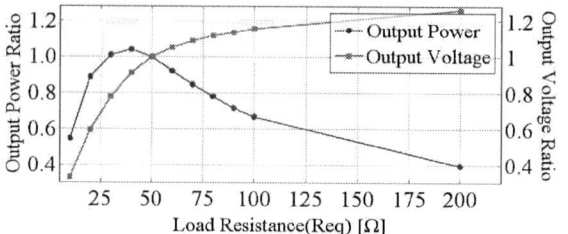

Fig.7 Curve of Output Power Ratio
and Output Voltage Ratio (Req:50 Ω=1, Simulation)

From Fig.6, the V_{DS} shows an ideal trapezoidal waveform even when the load resistance is changed from designed load condition (Req=50 Ω). Therefore, the class Phi-2 inverter is able to operate normally in the case of various load condition.

From Fig.6 and Fig.7, the output voltage and power is different due to load resistance. As the load resistance increases, the output voltage increases and output power decreases. On the other hand, as the load resistance decrease, both the output voltage and power become small after the maximum power point. Therefore, if the ON/OFF control method is applied to class Phi-2 inverter, the circuit is able to control output voltage as constant voltage in the high load resistance region.

IV. EXPERIMENTS

In this section, we demonstrate the operation of the class Phi-2 inverter circuit under the various load conditions and characteristics when the load resistance is changed during circuit operation. Fig.8 and Table V show the experimental circuit and the parameters of the circuit, respectively.

TABLE V
PARAMETERS OF CLASS PHI 2 INVERTER (EXPERIMENT)

Inductor L_{MR}	Amidon T106-#6, 4 Turn	689 nH
Inductor L_S	Amidon T106-#2, 4 Turn	518 nH
Capacitor C_P	10 pF×5 Yageo 223897111523	50 pF
Capacitor C_{MR}	10 pF×5 Yageo 223897111523	50 pF
Capacitor C_S	150 pF×6 TDK C3216C0G2J151J060AA	1000 pF
	100 pF×1 TDK C3216C0G2J101J060AA	
Transformer Tr	Amidon T106-#2, 4-4Turn	480 nH(pri) 480 nH(sec) k=0.862
Load Resistance	Tokai Konetsu Kogyo ER100SP	
Rectifier Diode	PANJIT SB340LS, 40V 3A	
Smoothing Capacitor C_R	Panasonic ECQE2106KF , 250V 10μF	
	Panasonic ECQE2105KF , 250V 1μF	
	Panasonic ECQE2104KF , 250V 0.1μF	
	Panasonic ECQE2103KF , 250V 0.01μF	
Gate Driver	Linear Technology LTC4440	
Resistor R_{S1}	22 Ω×4 TE Connectivity CRG1206F22R	5.5 Ω
Resistor R_{S2}	910Ω×4 Vishay CRCW1206910RFKEA	227.5 Ω
Capacitor C_{SP}	470 pF×3 TDK C3216C0G2J471J085AA	1440 pF
GaN-HFET	Panasonic PGA26E08BA	

3764

The 2018 International Power Electronics Conference

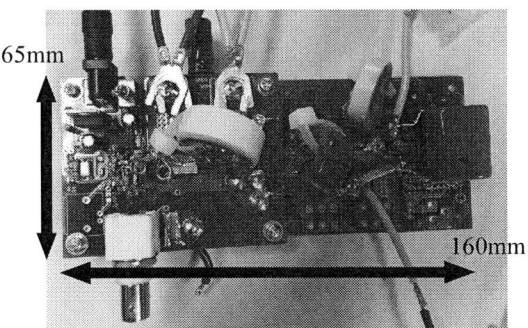

Fig.8 Experimental Circuit

First, we demonstrate the characteristics when the load resistance is changed from design value. Fig.9 shows the each waveform at load resistance of 50 Ω, 30 Ω, 70 Ω, 100 Ω and 200 Ω. In these experiments, the smoothing capacitor was set to 10μF and duty ratio of GaN-HFET is set to 0.24.

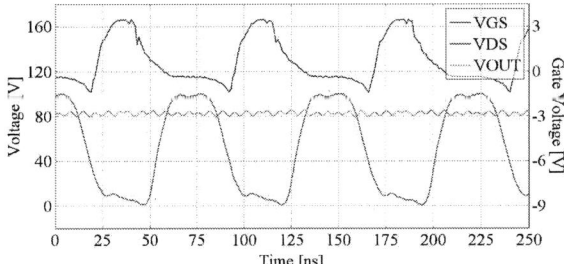

Fig.9 Waveform in Load Resistance = 50 Ω

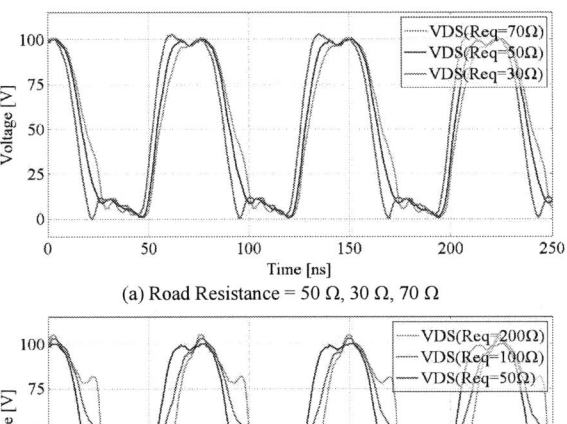

(a) Road Resistance = 50 Ω, 30 Ω, 70 Ω

(b) Road Resistance = 50 Ω, 100 Ω, 200 Ω

Fig.10 Waveforms of the V_{DS} in Various Load Resistance

As shown in Fig.9, when the load resistance was set to the design value of 50 Ω, each waveform show ideal in the class Phi-2 inverter. Therefore, accurate operation was achieved in this experiment.

As shown in Fig.10, accurate operation was not be achieved in various load conditions. In the load resistance is close to the designed value of 50 Ω such as 30 Ω or 70 Ω as shown in Fig.10(a), ideal trapezoidal waveform in V_{DS} can be confirmed. However, the load resistance is set to a light load such as 100 Ω or 200 Ω, the V_{DS} waveform is not show a trapezoidal waveform. Especially, in the load resistance is 200 Ω, hard switching was observed. From Fig.10, when the load resistance is decreases, the central depressions of the V_{DS} waveform become large. This phenomenon is due to the difference of impedance magnitude between fundamental and 3rd switching frequency decreases due to the decreasing load resistance. On the other hand, when the load resistance is increased, the central depression decreases. Therefore, in the experimental circuit, proper operation at light load is difficult. This result is different form the simulation. Fig.11 shows the output power ratio and the output voltage ratio. In Fig.11, the point of 200 Ω is omitted because normal operation was not achieved in this load resistance.

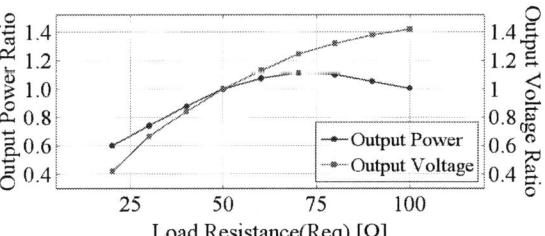

Fig11 Curve of Output Power Ratio
and Output Voltage Ratio (Req:50 Ω=1, Experiment)

As shown in Fig.11, the out power ratio and the output voltage ratio show the same tendency as the simulation result in Fig.7. However, the curve peak of the output power ratio shifts to the right. In the simulation result, the maximum power point exists at around 50 Ω, but in the experimental, the peak power exists at around 70 Ω. The characteristics of the output voltage ratio is not so much different from the simulation. Fig.12 shows the power conversion efficiency in the case of variable load resistances.

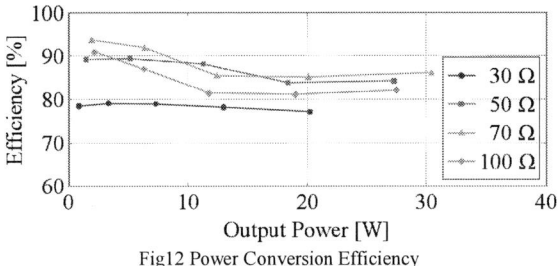

Fig12 Power Conversion Efficiency

As shown in Fig.12, the class Phi-2 inverter circuit shows almost constant power conversion efficiency for various load conditions. At the load resistance other than 30 Ω, the conversion efficiency of the circuit achieved 80% or more. Especially, in the small output power

3765

region, the conversion efficiency is over the 90%. Even when the output power is increased as shown in Fig.12, the power conversion efficiency was not decrease so much. Therefore, high efficiency can be kept even in the higher power region.

V. RESPONSE TO LOAD FLUCTUATION

In this section, we confirm the characteristics when the load resistance is changed during circuit operation. Fig.13 shows waveforms when load resistance was changed during operation. The variety of change load resistance are 3 types as 50 Ω to 30 Ω, 50 Ω to 70 Ω, 50 Ω to 100 Ω. In these figure, the load resistance was changed at t=0. The smoothing capacitor was set to 10 μF.

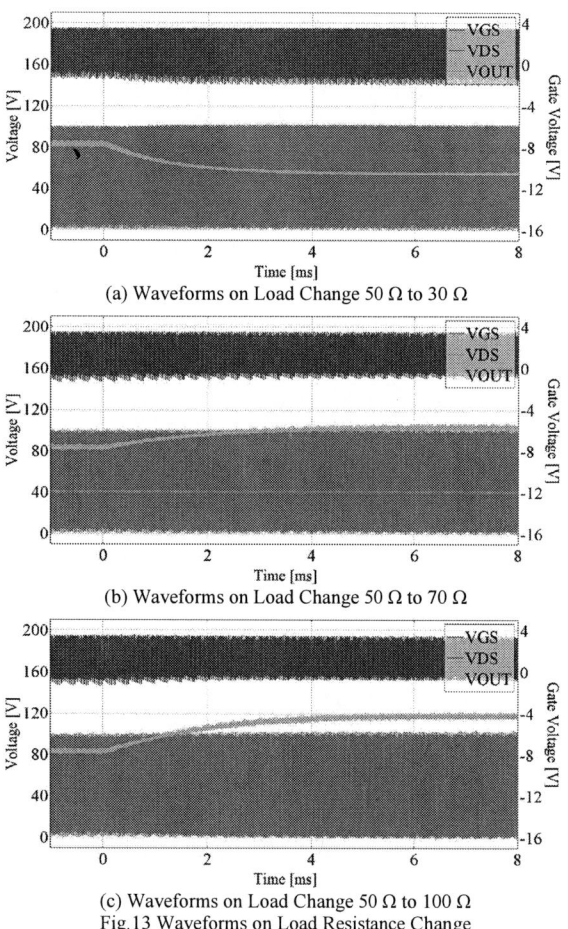

(a) Waveforms on Load Change 50 Ω to 30 Ω

(b) Waveforms on Load Change 50 Ω to 70 Ω

(c) Waveforms on Load Change 50 Ω to 100 Ω
Fig.13 Waveforms on Load Resistance Change

As shown in Fig.13, the output voltage changes gently and normal load fluctuation is achieved even if the load resistance is changed quickly. Even after changing the load resistance, the peak value of the gate voltage V_{GS} and the drain to source voltage V_{DS} indicate no significant change. The shape of V_{DS} waveform changes slightly depending on change of load resistance from Fig.10. Therefore, normal circuit operation continued even when load resistance is changed to different value from the designed value in the isolated DC-DC converter with the class Phi-2 inverter circuit.

In above cased, the smoothing capacitor was set to 10 μF. It is rather large value considering 13.56 MHz switching frequency. Therefore, we considered the load fluctuation when the capacity of the smoothing capacitor is further reduced. Fig.14 shows the waveforms when load resistance was changed during operation in the case of smoothing capacitor set to 10 μF, 1 μF, 0.1 μF and 0.01 μF.

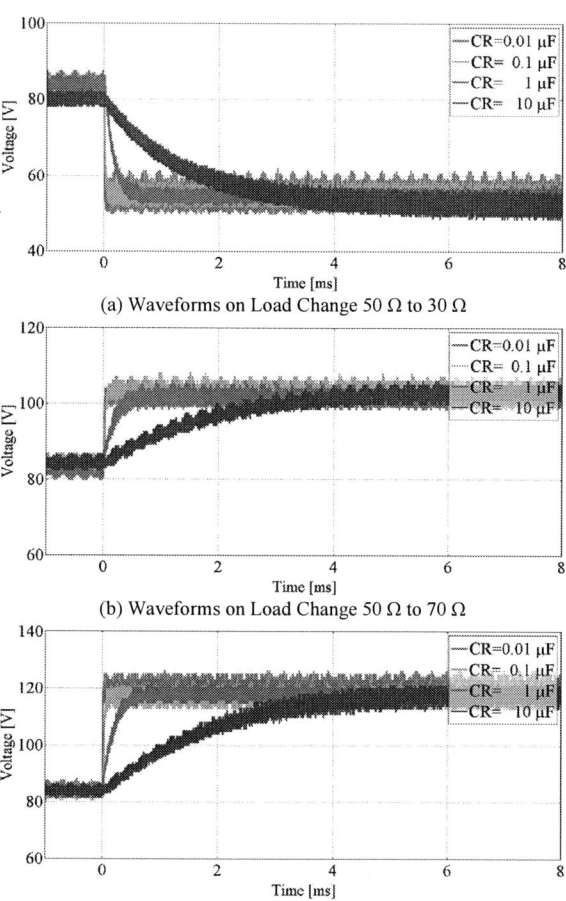

(a) Waveforms on Load Change 50 Ω to 30 Ω

(b) Waveforms on Load Change 50 Ω to 70 Ω

(c) Waveforms on Load Change 50 Ω to 100 Ω
Fig.14 Waveforms on Load Resistance Change in Various Smoothing Capacitor

As shown in Fig.14, we confirmed that the output voltage converges to the constant value at some load fluctuation. Also, in either case, when the smoothing capacitor is larger, the slope of the output voltage is smaller at the time of load fluctuation. In the case of load resistance is changed from 50 Ω to 30 Ω in Fig.14(a), the output voltage gently changes. Each output voltage settle to steady state value within a few milliseconds. As shown Fig.14 (b) and (c), the output voltage also settles in steady voltage within several milliseconds in these load fluctuation too.

Therefore, even when the load resistance is changed during operation, the circuit achieves normal operation irrespective of light load and heavy load. Therefore, the isolated DC-DC converter circuit with the class Phi-2

3766

inverter circuit is able to operate stably even under the load fluctuation.

Fig.15 shows the step response of output voltage after load change and the relationship between output voltage and voltage convergence time. The voltage convergence time is rise time to 90% of the steady state output voltage from the load resistance change. Table VI and Fig.16 show the convergence time due to load resistance from 50 Ω.

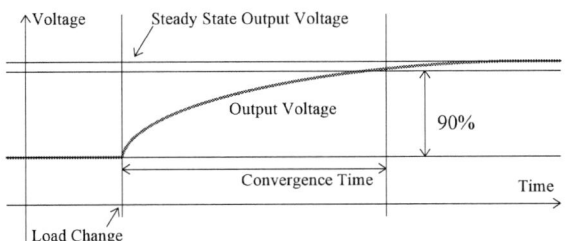

Fig.15 Definition of Convergence Time

TABLE VI
CONVERGENCE TIME IN VARIABLE LOAD RESISTANCE

Load after Change	Capacitance			
	10 μF	1 μF	0.1 μF	0.01 μF
20 Ω	2.33 ms	234 μs	29.3 μs	2.40 μs
30 Ω	3.10 ms	314 μs	35.0 μs	2.56 μs
40 Ω	3.01 ms	365 μs	41.4 μs	2.68 μs
60 Ω	2.05 ms	414 μs	48.2 μs	2.76 μs
70 Ω	2.00 ms	421 μs	55.5 μs	3.20 μs
80 Ω	1.88 ms	437 μs	50.9 μs	3.74 μs
90 Ω	1.87 ms	448 μs	43.9 μs	3.94 μs
100 Ω	1.92 ms	439 μs	43.8 μs	3.75 μs

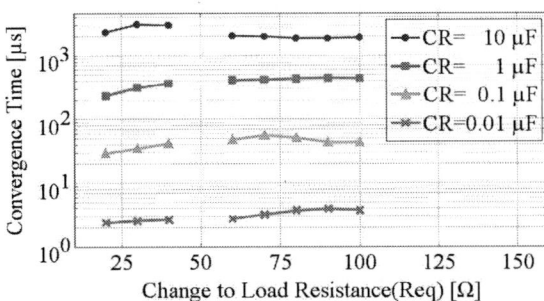

Fig.16 Convergence Time in Variable Load Resistance

As shown in Fig.16, the voltage convergence time is constant at all load resistance. As shown in Fig.14, the response is faster when the capacity of the smoothing capacitor is smaller. When the smoothing capacitor is set to 10 μF, the convergence time is constant around 2 ms. In the case of other smoothing capacitor are applied, convergence time show around 400 μs, 40 μs and 3 μs at 1 μF, 0.1 μF and 0.01 μF respectively. Therefore, the convergence time depend on the smoothing capacitor and proportional to the capacitance of smoothing capacitor.

In particular, when the smoothing capacitor was set to 0.01 μF, its convergence time is around 3 μs, and very fast response can be realized. In addition, as shown in Fig.14, even in the case of smoothing capacitor is set to 0.01 μF, output voltage ripple not occur greatly, consequently this circuit is useful.

Therefore, the isolated DC-DC converter circuit with the class Phi-2 inverter circuit is able to operate stably even under the load fluctuation. In addition, the convergence time is very short and stable operation can be continued.

VI. CONCLUSION

In this paper, we studied about the isolated DC-DC converter circuit with the class Phi-2 inverter circuit under fluctuated load conditions. As a result of verification, the isolated DC-DC converter with the class Phi-2 inverter circuit using GaN-HFET is able to operate at 13.56 MHz. This class Phi-2 inverter circuit operates with various load condition. Moreover, this circuit has robust characteristics in the case of load fluctuation and shows very fast response and stability. The convergence time of the load fluctuation is around 3 μs when smoothing capacitor was set to 0.01 μF, and it continues to operate stably even after fluctuation. Therefore, the class Phi-2 inverter has very fast response to load fluctuation for light load and heavy load. These result were confirmed by the simulation and experiment.

Therefore, the class Phi-2 inverter circuit is effective for the various load condition and load fluctuation, and this circuit is able to apply to many industrial devices.

This result is carried out in "Next Generation Power Electronics" in Cross-ministerial Strategic Innovation Promotion Program (SIP) by NEDO, JAPAN.

REFERENCES

[1] Juan M. Rivas, Yehui Han, Olivia Leitermann, Anthony D. Sagneri and David J. Perreault, "A high-frequency resonant inverter topology with low-voltage stress," *IEEE Trans. on Power Electronics*, vol. 23, no. 4, pp. 1759-1771, 2008.

[2] Jaun M. Rivas, Olivia Leitermann, Yehui Han, David J. Perreault, "A Very High Frequency DC-DC Converter Based on a Class Φ2 Resonant Inverter," *IEEE Trans. on Power Electronics*, vol. 26, no. 10, pp. 2980-2992, 2011.

[3] Jungwon Choi, Wei Liang, Luke Raymond, Juan Rivas, "High-Frequency Resonant converter Based on the Class Φ2 Inverter for Wireless Power Transfer," *IEEE 79th Vehicular Technology Conference*, pp. 1-5, 2014.

[4] Cornelius Armbruster, Andreas Hensel, Arne Hendrik Wienhausen, Dirk Kranzer, "Application of GaN power transistors in a 2.5 MHz LLC DC/DC converter for compact and efficient power conversion," *18th European Conference on Power Electronics and Applications(EPE16 ECCE Europe)*, pp. 1-7, 2016.

[5] Lei Gu, Wei Liang, and Juan Rivas Davila, "Design of Very-High-Frequency Synchronous Resonant DC-DC Converter for Variable Load Operation," *IEEE Energy Conversion Congress & Expo(EPE2017)*, pp. 3447-3454, 2017.

[6] Yuta Yanagisawa, Yushi Miura, Hiroyuki Hand, Tetsuzo Ueda, Toshifumi Ise, "Fundamental Investigation of Isolated DC-DC Converter with Class-Φ2 Inverter," *The Japan Institute of Power Electronics 218th Conference*, JIPE-43-07 pp. 1-5, 2014.

The 2018 International Power Electronics Conference

Single-Inductor Multiple-Output Current-Source Converter With Improved Cross Regulation and Simple Control Strategy

Zheng Dong, Xiaolu Lucia Li, and Chi K. Tse

Department of Electronic and Information Engineering, Hong Kong Polytechnic University, Hong Kong, China
Email: z.dong@connect.polyu.hk; xiaolu.li@connect.polyu.hk; encktse@polyu.edu.hk

Abstract—Single-inductor multiple-output (SIMO) converters with characteristics of small volume, high efficiency and low cost have attracted much attention. However, in order to guarantee cross-regulation performance, existing control strategies increase circuit complexity and cost. This paper proposes a SIMO current source (CS) dc-dc converter based on circuit duality principle. The CS buck converter, which eliminates inductor and has a wide range of duty cycle, is more suited for driving light-emitting-diodes (LEDs). The outputs of the SIMO CS dc-dc converter can be made completely independent by using a constant current feeder. The cross-regulation performance is improved dramatically. The control strategy is simple and practical. Taking the single-inductor dual-output (SIDO) CS dc-dc converter as an example, we study the operating modes and the control method. Finally, experimental circuits have been built and tested.

Keywords—*Duality principle, current-source switching converter, single-inductor multiple-output dc-dc converter, independent dimming function.*

I. INTRODUCTION

The light-emitting-diode (LED) as a lighting device has the advantages of high efficiency, long lifetime, small volume, and ease of dimming control [1], [2]. Recently, as a new-generation light source [3], LEDs have been applied to city landscape lighting, liquid crystal display back-light, street lamp lighting, medical service, and so on [4], [5].

In reality, there is a need for providing multiple output LED channels to achieve specific lighting requirements [6], [7], such as color-temperature control, dimming control and power-saving modes. The various output power levels are usually generated from a single power source. The parallel structure is the most direct and effective way to eliminate the cross-regulation effect because of its complete cross-independence, as shown in Fig. 1. However, this structure needs extra sets of components which consequently increase the volume and cost of the system. Compared to the parallel-connected dc-dc converters, the single-inductor multiple-output (SIMO) dc-dc converter [8], [9], [10] provides a better solution because it uses only one inductor to generate multiple voltage levels, eliminating bulky external inductors. The typical SIMO dc-dc converter is shown in Fig. 2.

The traditional SIMO dc-dc converter still suffers from cross-regulation effects [11], [12], [13], [14], [15],

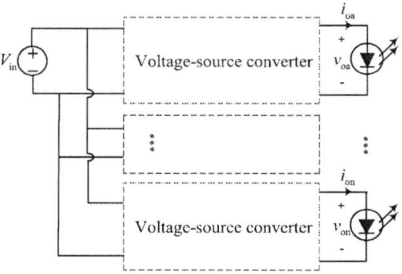

Fig. 1.   Parallel-connected dc-dc converters.

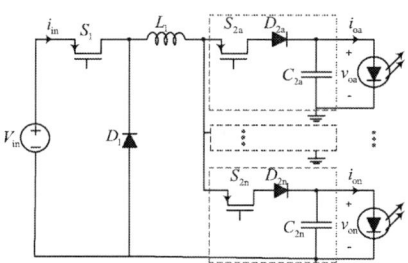

Fig. 2.   Typical SIMO dc-dc converter.

[16], [17]. All the outputs are coupled to inductor L_1, as shown in Fig. 2, leading to mutual interference when there is load variation in any one output. Previous works have reported the unsatisfactory cross-regulation performance of the SIMO dc-dc converter and some solutions, for instance, using three different operation modes, discontinuous conduction mode (DCM), pseudo-continuous conduction mode (PCCM) and continuous conduction mode (CCM), to alleviate cross-regulation effects. Under DCM, a time multiplexing (TM) control technique has been employed [11], [12]. However, when the SIMO dc-dc converter works under heavy load condition, the current ripple of the inductor, the switching noise and the switching device dissipation may be increased. To improve power levels and suppress cross-regulation effects, an extra switch connected in parallel with the inductor is needed to ensure that the SIMO dc-dc converter works under the PCCM [13], [14]. But the extra switches impose a higher cost to the design. In the freewheeling interval of each switching cycle, moreover, the switching loss also increases because of the nonzero inductor current.

3768

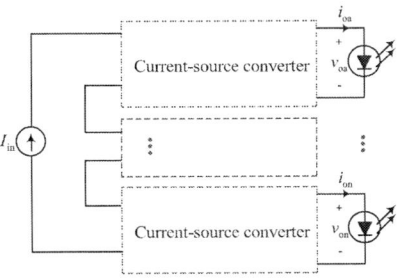

Fig. 3. Series-connected dc-dc converters.

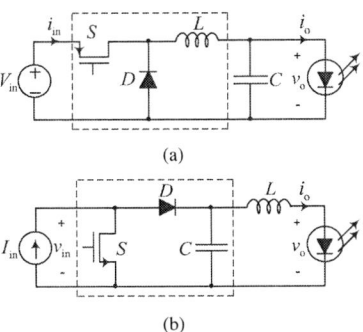

Fig. 4. Buck converters. (a) Conventional voltage-source (VS) buck converter; (b) dual version or current-source (CS) buck converter. Note that the inductor in the CS converter serves only a filtering function and does not play any power transfer role.

The SIMO dc-dc converter operating under CCM has the advantages of high efficiency, small current ripple and low cost. Various decoupling control techniques have been studied to reduce cross-regulation effects [15], [16], [18]. All these techniques generally require an elaborate and complex-designed control strategy to alleviate cross-regulation effects.

In this paper, we attempt to introduce the *current source* (CS) converter and to use a serial structure of multiple CS converters, which is the dual version of the parallel structure. One inductor is employed to construct a constant current for feeding the CS converters. It should be noted that CS converters inherently contain no inductor, while performing intended power processing functions. The proposed SIMO CS dc-dc converter makes full use of the properties of the CS buck converter to guarantee good cross regulation and significantly simplify the control circuit.

In Section II, the derivation of the SIMO CS dc-dc converter is described. The CS buck converter inherently contains no inductor resulting in the implementation of a single-inductor multiple-output converter. In Section III, the operating modes and the control strategy are explained in detail. In Section IV, we evaluate the performance of a single-inductor dual-output (SIDO) converter experimentally.

II. SINGLE-INDUCTOR MULTIPLE-OUTPUT CS DC-DC CONVERTER

As mentioned in the foregoing section, when the input is a voltage source, the parallel operation shown in Fig. 1 is the most direct and effective way to guarantee satisfactory cross-regulation performance because the parallel voltage source (VS) converters are completely independent. Based on the circuit duality principle, when the input is a current source, the serial operation shown in Fig. 3 is the most direct and effective way to achieve good cross regulation because the serial CS converters are completely independent.

Using the circuit duality principle [19], the CS buck converter, shown in Fig. 4(b), can be readily derived from the VS buck converter, shown in 4(a). The switch, the diode and the inductor form the basic cell of the VS buck converter. The inductor is the essential high-frequency power storage of the VS converter. The output capacitor

simply serves a filtering function. On the other hand, in the CS converter, the switch, the diode and the capacitor form the basic cell. The capacitor is the essential high-frequency power storage of the CS converter, and the output inductor simply serves a filtering function.

In an earlier work [20], it has been pointed out that the inductor in a CS buck converter can be removed while the output current ripple can still meet the usual requirement. Assuming that the input is a current source, the converter shown in Fig. 3 can be conceptually represented by the converter circuit shown in Fig. 5(a). Since CS converters inherently have no inductor, they are suitable for the integration of circuits. In general, the input is a voltage source. For most practical purposes, we need only one inductor to design a perfect constant current source to feed the downstream CS dc-dc converters, as shown in Fig. 5(b). The outputs are decoupled as they are fed by a constant current flowing through the inductor. When there is load variation in any one output, the inductor current is always kept constant, thus avoiding mutual interference. This characteristic makes the outputs of the SIMO CS dc-dc converter completely independent.

III. OPERATING PRINCIPLE

A. Operating Modes

The SIDO CS dc-dc converter contains three switches, namely, S_1, S_{2a}, and S_{2b}. Every switch may operate in either ON or OFF states. As a result, there are eight possible operating modes. The corresponding circuits are illustrated in Fig. 6.

Mode 1: S_1, S_{2a} and S_{2b} are ON. Thus, V_{in} charges up the inductor and i_{L1} ramps up. The capacitors are discharged to the corresponding loads.

Mode 2: S_1 and S_{2a} are ON while S_{2b} is OFF. When the value of $(V_{in}-V_{ob})$ is positive, $(V_{in}-V_{ob})$ charges up the inductor and i_{L1} ramps up. Otherwise, $(V_{in}-V_{ob})$ discharges the inductor and i_{L1} ramps down. At the same time, capacitor C_{2a} is discharged to the load.

Mode 3: S_1 and S_{2b} are ON while S_{2a} is OFF. This mode is similar to Mode 2. If the value of $(V_{in}-V_{oa})$ is

3769

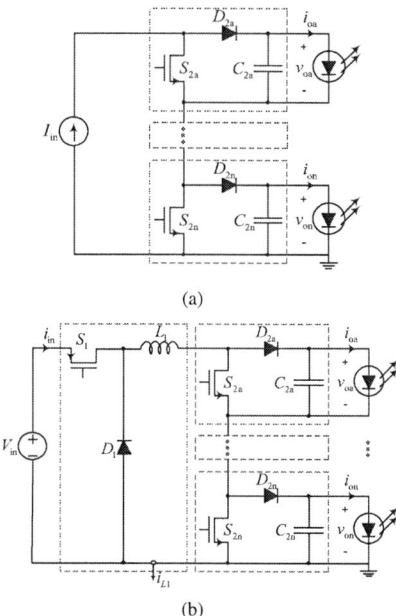

(a)

(b)

Fig. 5. New CS topology. (a) Series-connected dc-dc converters; (b) SIMO CS dc-dc converter. Note that the capacitors perform actual power transfer at the switching frequency, and are not output capacitors.

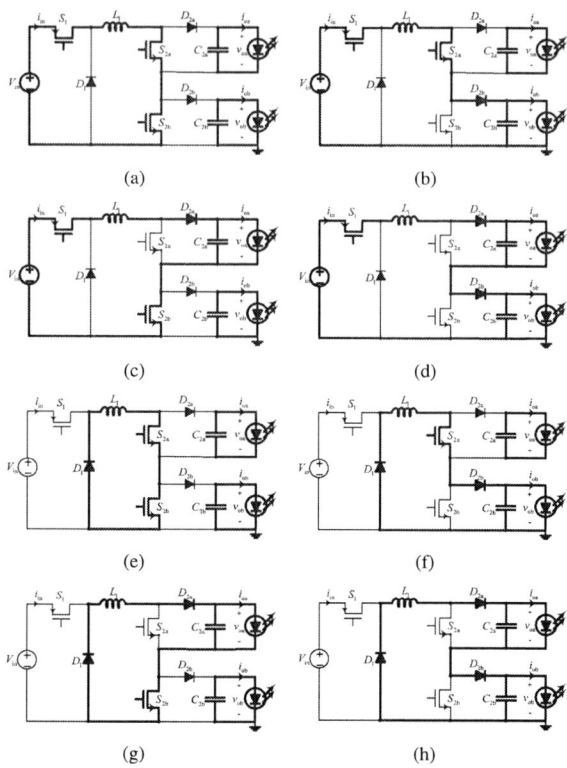

(a)   (b)

(c)   (d)

(e)   (f)

(g)   (h)

Fig. 6. Operating modes. (a) S_1, S_{2a}, and S_{2b} are ON; (b) S_1 and S_{2a} are ON and S_{2b} is OFF; (c) S_1 and S_{2b} are ON and S_{2a} is OFF; (d) S_1 is ON and S_{2a} and S_{2b} are OFF; (e) S_1 is OFF and S_{2a} and S_{2b} are ON; (f) S_1 and S_{2b} are OFF and S_{2a} is ON; (g) S_1 and S_{2a} are OFF and S_{2b} is ON; (h) S_1, S_{2a}, and S_{2b} are OFF.

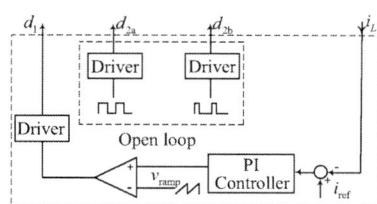

Fig. 7. Control block diagram.

positive, $(V_{in}-V_{oa})$ charges up the inductor and i_{L1} ramps up. If not, $(V_{in}-V_{oa})$ discharges the inductor and i_{L1} ramps down. Capacitor C_{2b} is discharged to the load.

Mode 4: S_1 is ON while S_{2a} and S_{2b} are OFF. If the value of $(V_{in}-V_{oa}-V_{ob})$ is positive, $(V_{in}-V_{oa}-V_{ob})$ charges up the inductor and i_{L1} ramps up. If the value of $(V_{in}-V_{oa}-V_{ob})$ is negative, $(V_{in}-V_{oa}-V_{ob})$ discharges the inductor and i_{L1} ramps down.

Mode 5: S_1 is OFF while S_{2a} and S_{2b} are ON. At this point the inductor current is in a freewheeling mode. The capacitors continue to be discharged to the corresponding loads.

Mode 6: S_1 and S_{2b} are OFF while S_{2a} is ON. Voltage V_{ob} discharges the inductor and i_{L1} ramps down. Capacitor C_{2a} is discharged to the load.

Mode 7: S_1 and S_{2a} are OFF while S_{2b} is ON. Voltage V_{oa} discharges the inductor and i_{L1} ramps down. Capacitor C_{2b} is discharged to the load.

Mode 8: S_1, S_{2a} and S_{2b} are OFF. Voltage $(V_{oa}+V_{ob})$ discharges the inductor and i_{L1} ramps down.

B. Control Strategy

For the SIDO CS dc-dc converter, the use of one inductor to produce a constant current with satisfactory cross-regulation performance is the key design advantage. Due to the decoupling property of the SIDO CS dc-dc converter, the whole control circuit can be greatly simplified. The control strategy is schematically shown in Fig. 7. Here, only one closed-loop control is needed to produce a perfect constant current source. The current i_{L1} is compared with the reference current i_{ref} to provide

the control signal for S_1. Then, S_{2a} and S_{2b} are all open-loop driven because the CS buck converter operates over a wide range of duty cycle and has good inherent stability [20]. The output currents can be adjusted by the duty cycles directly, without being affected by the LED's v-i characteristics. In this paper, a Type II compensation network is employed, as shown in Fig. 8. The values of these parameters are shown in Table I.

IV. EXPERIMENTAL RESULTS

In order to verify the performance and feasibility of the SIDO CS dc-dc converter, a prototype has been constructed. The values of the components used are shown in Table I.

In the experiment, one LED string is formed by two LEDs stacked in series, and the current flowing through

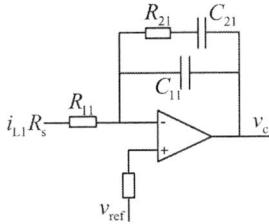

Fig. 8. Type II compensation network.

TABLE I. COMPONENT VALUES AND PARAMETERS OF
EXPERIMENTAL SINGLE-INDUCTOR DUAL-OUTPUT CS CONVERTER

Design parameters and components	Values
Input voltage V_{in}	48 V
Main inductor L_1	2000 μH
Capacitor C_{2a}	220 μF
Capacitor C_{2b}	220 μF
Output currents i_{oa} and i_{ob}	350 mA
Switch frequency of S_1	50 kHz
Switch frequency of S_{2a}	50 kHz
Switch frequency of S_{2b}	50 kHz
Capacitor C_{11}	4.7 nF
Capacitor C_{21}	4.7 μF
Resistor R_{11}	2 kΩ
Resistor R_{21}	5 kΩ

the LEDs is 350 mA. Another string is formed by three LEDs stacked in series, and the current flowing through the LEDs is 150 mA. One cycle is divided into four periods. These four periods correspond to four operating modes, i.e., Modes 1, 5, 7 and 8. During the first period, S_1, S_{2a} and S_{2b} are ON, the inductor current ramps up. During the second period, S_1 is OFF while S_{2a} and S_{2b} are ON, the inductor current is in the freewheeling mode. During the third period, S_1 and S_{2a} are OFF while S_{2b} is ON, and the inductor current ramps down. During the fourth period, S_1, S_{2a} and S_{2b} are OFF. In this case, the inductor current continuously ramps down. Experimental results are fully consistent with our analysis.

Fig. 10 shows the transient response when a 100 mA step current is applied in one of the outputs, I_{oa}. The output current I_{oa} steps between 150 mA and 250 mA. The waveform of the other output current, I_{ob}, is shown in Fig. 10(a), and in detail in Figs. 10(b) and 10(c). It is clearly shown that the cross-regulation effect is almost completely eliminated. The decoupling performance of the SIMO CS dc-dc converter is verified.

V. CONCLUSION

In this paper, a series-connected structure of current-source converters, which is the dual version of the conventional parallel-connected (voltage-source) dc-dc converter structure, is studied. Based on the series structure, a single-inductor multiple-output current-source based converter is derived. This converter only uses one inductor to construct a constant current source feeding a number of series-connected current-source converters. The decoupling property of this single-inductor multiple-output dc-

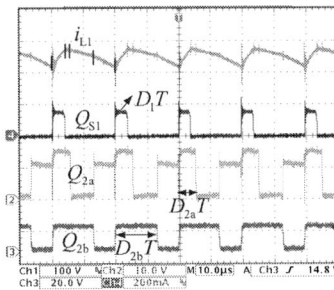

Fig. 9. Steady-state waveforms of inductor current and switching sequence.

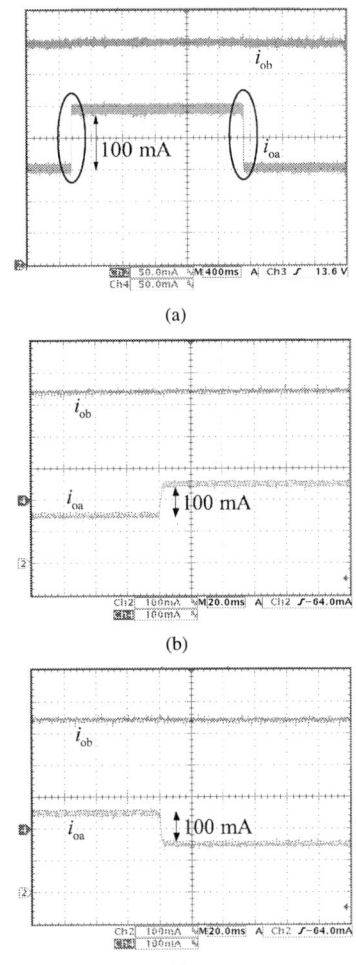

Fig. 10. Transient response showing independence of the two output currents or cross-regulation. (a) Transient response of a 100 mA output current step applied to one output; (b) 150 mA to 250 mA; (c) 250 mA to 150 mA. In both cases, the other current output is unaffected.

dc converter is studied. As an example for illustration, a single-inductor dual-output dc-dc converter is analyzed in detail. The corresponding simple control strategy is also provided. Both theoretical analysis and experimental results show that the cross-regulation performance can be improved dramatically.

Acknowledgment

This work is supported by Hong Kong Research Grants Council under Theme-Based Research Project No. T22-715/12-N.

References

[1] X. Qu, S. C. Wong, and C. K. Tse, "An improved LCLC current-source-output multistring LED driver with capacitive current balancing," *IEEE Trans. Power Electron.*, vol. 30, no. 10, pp. 5783–5791, 2015.

[2] S. Li, S. C. Tan, C. K. Lee, E. Waffenschmidt, S. Y. R. Hui, and C. K. Tse, "A survey, classification, and critical review of light-emitting diode drivers," *IEEE Trans. Power Electron.*, vol. 31, no. 2, pp. 1503–1516, 2016.

[3] X. Qu, S. C. Wong, and C. K. Tse, "Noncascading structure for electronic ballast design for multiple led lamps with independent brightness control," *IEEE Trans. Power Electron.*, vol. 25, no. 2, pp. 331–340, 2010.

[4] M. H. Crawford, "Leds for solid-state lighting: performance challenges and recent advances," *IEEE J. Sel. Topics in Quan. Electron.*, vol. 15, no. 4, pp. 1028–1040, 2009.

[5] W. K. Lun, K. Loo, S. C. Tan, Y. Lai, and C. K. Tse, "Bilevel current driving technique for LEDs," *IEEE Trans. Power Electron.*, vol. 24, no. 12, pp. 2920–2932, 2009.

[6] K. Loo, W. K. Lun, S. C. Tan, Y. Lai, and C. K. Tse, "On driving techniques for leds: Toward a generalized methodology," *IEEE Trans. Power Electron.*, vol. 24, no. 12, pp. 2967–2976, 2009.

[7] X. Liu, Q. Yang, Q. Zhou, J. Xu, and G. Zhou, "Single-stage single-switch four-output resonant led driver with high power factor and passive current balancing," *IEEE Trans. Power Electron.*, vol. 32, no. 6, pp. 4566–4576, 2017.

[8] X. Liu, J. Xu, Z. Chen, and N. Wang, "Single-inductor dual-output buck–boost power factor correction converter," *IEEE Trans. Ind. Electron.*, vol. 62, no. 2, pp. 943–952, 2015.

[9] X. Liu, J. Xu, Q. Yang, and D. Xu, "High-efficiency multi-string LED driver based on constant current bus with time-multiplexing control," *IET Electron. Lett.*, vol. 52, no. 9, pp. 746–748, 2016.

[10] H. P. Le, C.-S. Chae, K. C. Lee, S.-W. Wang, G. H. Cho, and G. H. Cho, "A single-inductor switching dc–dc converter with five outputs and ordered power-distributive control," *IEEE J. of Solid-State Circ.*, vol. 42, no. 12, pp. 2706–2714, 2007.

[11] X. Liu, J. Xu, Z. Lu, and J. Wang, "A single-inductor dual-output buck converter with voltage mode pulse-train control," in *Proc. IEEE Comm. Circ. Syst. Conf.*, 2010, pp. 561–564.

[12] D. S. Ma, W. H. Ki, C. Y. Tsui, and P. K. Mok, "Single-inductor multiple-output switching converters with time-multiplexing control in discontinuous conduction mode," *IEEE J. of Solid-State Circ.*, vol. 38, no. 1, pp. 89–100, 2003.

[13] Y. J. Woo, H. P. Le, G. H. Cho, G. H. Cho, and S. I. Kim, "Load-independent control of switching dc–dc converters with freewheeling current feedback," *IEEE J. of Solid-State Circ.*, vol. 43, no. 12, pp. 2798–2808, 2008.

[14] D. S. Ma, W. H. Ki, and C. Y. Tsui, "A pseudo-ccm/dcm simo switching converter with freewheel switching," *IEEE J. of Solid-State Circ.*, vol. 38, no. 6, pp. 1007–1014, 2003.

[15] D. Trevisan, P. Mattavelli, and P. Tenti, "Digital control of single-inductor multiple-output step-down dc–dc converters in ccm," *IEEE Trans. Ind. Electron.*, vol. 55, no. 9, pp. 3476–3483, 2008.

[16] A. Pizzutelli and M. Ghioni, "Novel control technique for single inductor multiple output converters operating in ccm with reduced cross-regulation," in *Proc. IEEE Appl. Power Electron. Conf. Exp. (APEC)*, 2008, pp. 1502–1507.

[17] Y. Guo, S. Li, A. T. Lee, S. C. Tan, C. K. Lee, and S. Y. R. Hui, "Single-stage ac/dc single-inductor multiple-output led drivers," *IEEE Trans. Power Electron.*, vol. 31, no. 8, pp. 5837–5850, 2016.

[18] Z. Shen, X. Chang, W. Wang, X. Tan, N. Yan, and H. Min, "Predictive digital current control of single-inductor multiple-output converters in ccm with low cross regulation," *IEEE Trans. Power Electron.*, vol. 27, no. 4, pp. 1917–1925, 2012.

[19] C. K. Tse, *Linear Circuit Analysis*. London: Addison-Wesley, 1998.

[20] Z. Dong, C. K. Tse, and S. Y. R. Hui, "Basic circuit theoretic considerations of LED driving: Voltage-source versus current-source driving," in *Proc. IEEE Ann. Southern Power Electron. Conf. (SPEC)*. doi. 10.1109/SPEC.2016.7846015, 2016.

Limit Operating Frequency of Peak Current-Mode Control DC-DC Converter Considering Turn-Off Delay Time

Ryo Ute[1*], Kazuya Fujiwara[1], Jun Imaoka[1], Masahito Shoyama[1]

1 Department of Information Science and Electrical Engineering, Kyushu University, Fukuoka, Japan

*E-mail: ute@ckt.ees.kyushu-u.ac.jp

Abstract— The aim of this paper is to present the mechanism of the frequency dividing operation and the limit operating frequency in the peak current-mode control DC-DC converter considering the turn-off delay time. The peak current mode control has the higher stability and the responsiveness than the conventional voltage mode control. However, if the switching frequency is increased ignoring the turn-off delay time, it would become the frequency dividing operation that the switching frequency operates lower than the clock frequency at the certain frequency. In this operation, there is a problem that the switching loss increases and the high frequency switching drive is hindered. In this paper, we clarified the mechanism of the dividing frequency operation and derived the relationship between the turn-off delay time and the limit operating frequency. Also, we derived the relationship when the slope compensation is applied as well. As experimental results, when there is the turn-off delay time, it was confirmed that the limit operating frequency of the DC-DC converter was limited to lower as the duty ratio was lower and the slope ratio of the slope compensation is close to 1.0.

Keywords— *Peak current mode control, turn-off delay time, deviding frequency operation, limit operating frequency.*

I. INTRODUCTION

Recently, high performance and small size DC-DC converters are required for the power supply of electronic devices. On the other hand, recent technological trends of DC-DC converters are improvement of the load responsiveness and the miniaturization of passive elements by high frequency switching driving using semiconductor devices made of GaN and SiC. In particular, in order to improve the load responsiveness and the stability, the current-mode control is widely used for the control method of the DC-DC converter. This control method can obtain superior dynamic characteristics compared with the conventional voltage mode control because of feedback two elements of the output voltage and the inductor current. Therefore, there are advantages such as the high stability and the easy design of the phase compensation circuit. In addition, since the feedback control is performed by detecting the inductor current, an overcurrent protection function can be easily given the

DC-DC converter by setting an upper limit on the detected current value [1]. In the previous research about the current mode control, many reports such as the improvement of the responsiveness and a method of the realizing high frequency driving have been reported [2]-[4]. Moreover, researches to realize the high speed response by the digital control and the algorithm realizing the current mode control using the general purpose A/D converter have also been reported [5]-[6]. In this current mode control, when the peak value of the inductor current matches the current target value which is the output of the error amplifier, the DC-DC converter turns off. However, in the actual turn-off operation, there is a delay time from the output of the PWM comparator to the turn-off of the MOSFET. By ignoring this delay time and increasing the switching frequency, authors discovered that it would become the frequency dividing operation that operates below the set switching frequency at a certain frequency. In this operation, the inductor current ripple and the switching loss increases. However, papers about the current mode control have been reported so far, the mechanism of the frequency dividing operation caused by the turn-off delay time has not been discussed. Therefore, in order to realize the high frequency switching driving while preventing the frequency dividing operation in the current mode control, it is necessary to analyze the operation of the current mode controlled DC-DC converter considering the turn-off delay time. This paper presents the mechanism of the frequency dividing operation in the peak current mode control DC-DC buck converter. Moreover, we derive the relationship between the limit operating frequency of the DC-DC converter and the duty ratio when there is the turn-off delay time. In addition, the same relationship is derived when the slope compensation for preventing the oscillation which is a problem peculiar to the peak current mode control is introduced. As experimental results, the derived relational expression and the experimental result are approximately equal, and when there is the turn-off delay time, it was confirmed that the limit operating frequency of the DC-DC converter was limited to lower as the duty ratio was lower and the slope ratio of the slope compensation is close to 1.0.

II. Peak Current-Mode Control

A. Operation Principle

The peak current-mode control DC-DC buck converter circuit is shown in Fig. 1. V_{in} is the input voltage, V_{out} is the output voltage, i_L is the inductor current, i_{out} is the output current, S is the MOSFET, L is the inductor, C_{in} is the input capacitor, C_{out} is the output capacitor, D is the flywheel diode, R_o is the load resistor, R_S is the shunt resistor for the inductor current detection, and v_{iL} is equal to $R_S i_L$, v_{iref} is the target value of the inductor current, CLK is a pulse signal for driving the MOSFET. Fig. 2 shows operating waveforms in the peak current mode control without the turn-off delay time. D is the duty ratio, DT_S represents the ON period, and $(1-D)T_S$ represents the OFF period. The MOSFET of the main circuit becomes turn-on through the RS-FF when the CLK signal stands up. At this time, the switching frequency f_S of the DC-DC convertor equals to the CLK frequency f_C. The voltage v_{iL} equivalents to the inductor current and the output of the error amplifier are input to the PWM comparator in the control circuit. For turn-off, when v_{iL} equals to v_{iref}, a reset signal is input to the RS-FF, and the MOSFET is turned off. In the peak current mode control, the output power stage can be regarded as the RC one-order delay circuit by feed buck the output voltage and the inductor current. Therefore, the peak current mode control can obtain the superior dynamic characteristics as compared with the voltage mode control.

B. Dividing Frequency Operation

The peak current mode control in the ideal state has been described above. However, in the actual operation, there is the delay time from the PWM comparator to the turn-off of the MOSFET as shown in Fig. 1. Fig. 3 shows operating waveforms in the peak current mode control with the turn-off delay time t_d. In the normal operation, it turned off when v_{iL} equals to v_{iref}, but if t_d exists, v_{iL} exceeds v_{iref} and it turns off. Fig. 4 shows operation waveforms considering the turn-off delay time t_d. Fig. 4 (a) shows waveforms of the normal operation ($f_S = f_C$). Even if there is t_d, it becomes the normal operation because the CLK signal inputs to the RS-FF during $v_{iL} < v_{iref}$. As the CLK frequency f_C is increased from this state, it becomes the critical operation in Fig. 4 (b). In this operating state, the ON period DT_S equals to t_d. After this critical operation state, it becomes the dividing frequency operation ($f_S \neq f_C$) in Fig. 4 (c) because the CLK signal inputs to the RS-FF during $v_{iL} > v_{iref}$. In the dividing frequency operation, since the DC-DC converter operates below the CLK frequency f_C, the inductor current ripple increases more than expected the current ripple. Due to rise of the current ripple, the inductor may fall into the magnetic saturation state depending on the circuit design. Moreover, the DC-DC converter may not operate in the worst case. Although f_S is not f_C in the frequency dividing operation, the DC-DC converter operates the steady state at less than half of f_C. However, since the switching operation misfire, the switching loss increases and the efficiency decreases as compared with the normal operation.

For the above reasons, it is difficult to miniaturize passive elements and switching loss increases if it falls into the frequency dividing operation. Therefore, in order to prevent the frequency dividing operation, it is necessary to obtain the limit operating frequency of the DC-DC converter considering the turn-off delay time at the circuit design stage. In this paper, we derive the relationship between the limit operating frequency and the turn-off delay time of the DC-DC converter when there is the turn-off delay time.

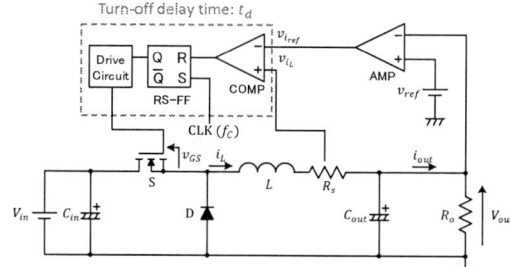

Fig. 1. Peak current mode control DC-DC buck converter.

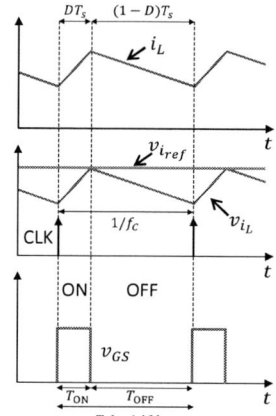

Fig. 2. Operating waveforms of peak current mode control (No delay time).

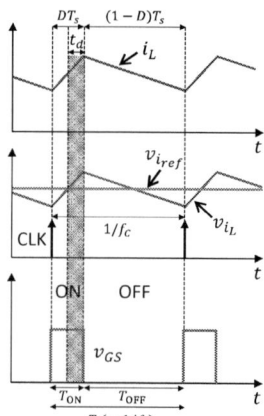

Fig. 3. Operating waveforms of peak current mode control (With delay time).

The 2018 International Power Electronics Conference

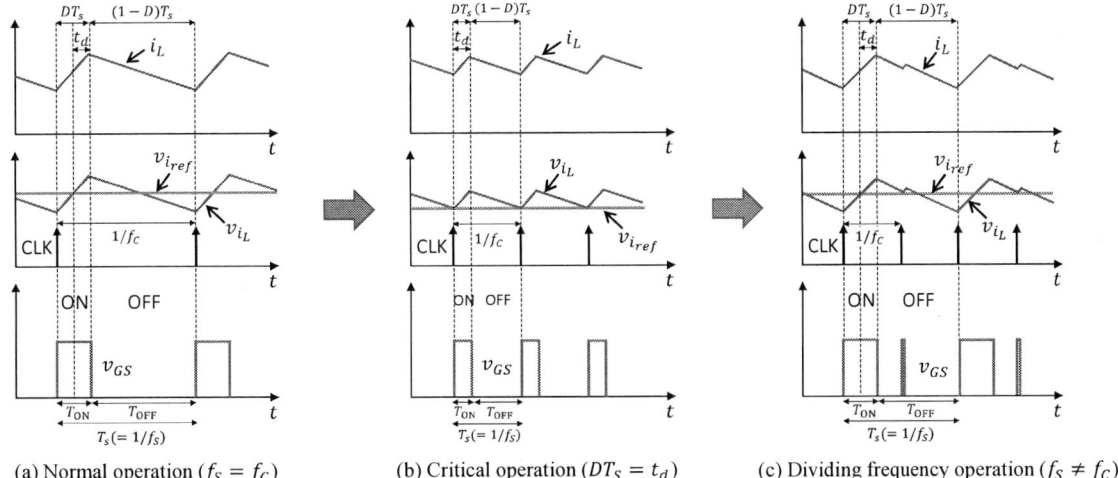

(a) Normal operation ($f_S = f_C$) (b) Critical operation ($DT_S = t_d$) (c) Dividing frequency operation ($f_S \neq f_C$)

Fig. 4. Operating waveforms considering turn-off delay time t_d (No slope compensation).

III. DERIVATION OF LIMIT OPERATING FREQUENCY

A. No slope compensation

In order to derive the relationship between the limit operating frequency of the DC-DC converter and the turn-off delay time, the state transition diagram in Fig. 4 is used. As shown in Fig. 4 (a), when the ON period DT_S is longer than the turn-off delay time t_d, it becomes the normal operation ($f_S = f_C$) because the CLK signal inputs to the RS-FF during $v_{iL} < v_{iref}$. However, as the switching frequency f_S is increased from this state, it becomes the critical operation that t_d equal to DT_S shown in Fig. 4 (b). After the critical operation, if f_S is increased, it becomes the dividing frequency operation ($f_S \neq f_C$) shown in Fig. 4 (c) because the CLK signal inputs to the RS-FF during $v_{iL} > v_{iref}$. The switching frequency in the critical operating state is the maximum switching frequency at which the normal operation is possible. Therefore, in order to determine the boundary frequency between the normal operation and the frequency dividing operation, it is necessary to derive the switching frequency in the critical operation. Since the condition for the normal operation of the DC-DC converter is that CLK is input at $v_{iL} < v_{iref}$, it is necessary to satisfy the following equation.

$$DT_S > t_d \tag{1}$$

When the switching frequency in the critical operation is defined as the limit operating frequency f_{limit} of the DC-DC converter, f_{limit} is expressed by the following equation from the conditional equation (1).

$$f_{limit} = \frac{D}{t_d} \tag{2}$$

Therefore, f_S when there is t_d is limited to the f_{limit} or less as the following equation.

$$f_s \leq f_{limit} = \frac{D}{t_d} \tag{3}$$

Also, D represents the duty ratio and in the case of the DC-DC buck converter, it is represented by $D = V_{out}/V_{in}$.

B. With slope compensation

In the peak current mode control, when the duty ratio D is 0.5 or more, there is a problem of causing an oscillation phenomenon based on the ripple. As a countermeasure, the slope compensation is used. The slope compensation is generated by superimposing a negative slope on the output of the error amplifier in Fig. 5. The waveform is shown in Fig. 6. m_1 is the slope of the inductor current during the ON period, m_2 is the slope of the inductor current during the OFF period. In the case of the DC-DC buck converter, the absolute value of each slopes are expressed by following equations.

$$|m_1| = \frac{V_{in} - V_{out}}{L} \cdot R_S \cdot k_{iv} \tag{4}$$

$$|m_2| = \frac{V_{out}}{L} \cdot R_S \cdot k_{iv} \tag{5}$$

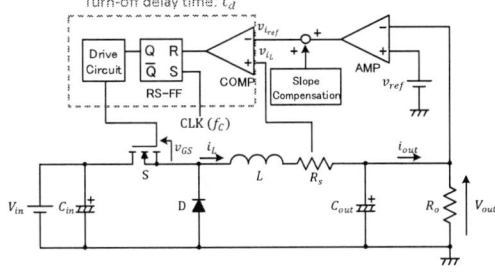

Fig. 5. Peak current mode control with slope compensation.

3775

The 2018 International Power Electronics Conference

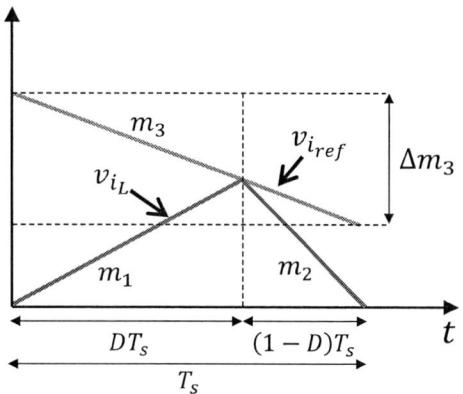

Fig. 6. Operating waveform with slope compensation.

L is the inductor, R_S is the shunt resistor, k_{iv} is the current detection gain. Further, Δm_3 is the amplitude value of the slope compensation and it is represented by the following equation.

$$\Delta m_3 = m_3 T_S \qquad (6)$$

Generally, in order to prevent the oscillation, it is necessary to perform the slope compensation in which the slope ratio a of the following equation is 0.5 or more.

$$m_3 \geq a m_3 \quad (0.5 \geq a \geq 1.0) \qquad (7)$$

From the above equation, the slope compensation is expressed using the slope m_2, it is expressed in a form that depends on the input / output voltage. Therefore, the slope compensation is generally generated by an automatically adjustable circuit that depends on the input / output voltage.

Previous studies about the slope compensation that the method of the automatically changing slope for arbitrary voltages from the low output voltage to the high output voltage and the configuring with less electronic parts have been reported [7]-[9]. In this automatically adjustable slope compensation method, even when the duty ratio D is operating at 0.5 or less, the slope is incorporated. In this paper, we derive the relationship between the turn-off delay time and the limit operating frequency when the automatically changing slope compensation is applied at the duty ratio D is 0.5 or less.

In this paper, we derive the relationship between the turn-off delay time and the limit operating frequency when the automatically changing slope compensation is applied at the duty ratio D is 0.5 or less. Fig. 7 shows operation waveforms introduced the slope compensation when there is the turn-off delay time t_d. As shown in Fig. 7 (a), when the ON period DT_S is larger than the sum of turn-off delay time t_d and time t_x, it becomes the normal operation ($f_s = f_c$) because the CLK signal inputs to the RS-FF during $v_{iL} < v_{iref}$. When the switching frequency f_s is increased this state, it becomes the critical operation that the sum of t_d and t_x equals to DT_S shown in Fig. 7 (b). After this operation, if f_s is increased, it becomes the dividing frequency operation ($f_s \neq f_c$) shown in Fig. 7 (c) because the CLK signal inputs to the RS-FF during $v_{iL} > v_{iref}$. Therefore, the condition with the slope compensation for performing the normal operation is as following equation.

$$\boldsymbol{DT_S} > \boldsymbol{t_d} + \boldsymbol{t_x} \qquad (8)$$

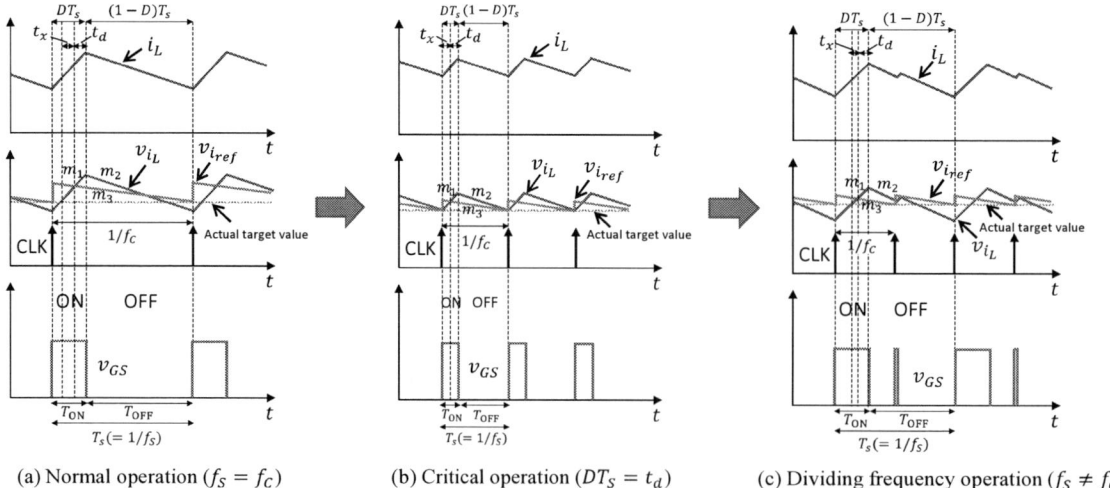

(a) Normal operation ($f_s = f_c$) (b) Critical operation ($DT_S = t_d$) (c) Dividing frequency operation ($f_s \neq f_c$)

Fig. 7. Operating waveforms considering turn-off delay time t_d (With slope compensation).

t_x is the time from the intersection of the actual target value and m_1 to the moment when m_1 and m_3 intersect, and it is represented by the following equation at the critical operation.

$$t_x = \frac{\Delta m_3}{m_1 + m_3} \qquad (9)$$

By substituting equations (4) to (7) into the above equation, t_x is expressed by the duty ratio D and the slope ratio a of the slope compensation as shown in the following equation.

$$t_x = \frac{aD}{1 - (1-a)D} T_S \qquad (10)$$

In order to operate the normal operation, it is necessary to satisfy the conditional expression of equation (8). Therefore, the limit operating frequency f_{limit} when the slope compensation is performed is expressed by the following equation.

$$f_{limit} = \frac{D}{t_d} \cdot \frac{1-D}{\{1/(1-a)\} - D} \qquad (11)$$

As described above, f_{limit} when the slope compensation is applied is the relational expression to which coefficients represented by the duty ratio D and the slope ratio a are given as compared with the case of no slope compensation.

IV. EXPERIMENTAL RESULTS

A. Measurement of turn-off delay time

In order to compere the calculated value of the derived limit operating frequency f_{limit} with the experimental result, experimental verification was carried out using the circuit parameters shown in Table I. Firstly, the turn-off delay time t_d of the DC-DC buck converter was measured. For example, Fig. 8 shows operating waveforms when V_{in} is 50V and f_C is 100kHz. As a result of the measurement, v_{iL} exceeded v_{iref} and t_d was 0.7μs. Although t_d exists, the DC-DC converter became the normal operation at 100kHz of the set frequency because the CLK signal is inputs to the RS-FF during $v_{iL} < v_{iref}$.

TABLE I
CIRCUIT PARAMETERS

Symbol	Parameter	Value
V_{in}	Input voltage	20~100V
V_{out}	Output voltage	5V
i_{out}	Output current	3A
P_{out}	Output power	15W
D	Duty ratio	0.05~0.25
L	Inductance	50μH
C_{in}	Input Capacitance	100μF
C_{out}	Output Capacitance	200μF
R_S	Shunt Resistance	0.1Ω

Furthermore, this t_d is the fixed value independent of circuit parameters and the slope compensation because it is mainly caused by the propagation delay of the PWM comparator and the drive circuit.

B. Limit operating frequency (No slope compensation)

In consideration of the above turn-off delay time t_d, the limit operating frequency f_{limit} in the case of no slope compensation was measured. For example, Fig. 9 shows operating waveforms when V_{in} is 50V and D is 0.1. In this parameter, f_{limit} was 143kHz from the equation (2). The DC-DC converter was operated around this calculated f_{limit}. As a experiment result, when f_C is 140kHz, f_S was equal to f_C and it became the normal operation as shown in Fig. 9 (a). However, when f_C is increased to 150kHz from this state, f_S was not equal to f_C and f_S was 75kHz that half of f_C due to the frequency dividing operation as shown in Fig. 9 (b). In this operation, the switching operation becomes unstable and the switching loss are increased. In addition, the limit operating frequency characteristics when D is varied from 0.05 to 0.25 is shown in Fig. 10. As experimental results, it was confirmed that f_S was limited by calculated f_{limit} even when D was changed. In the experiment, D was adjusted by changing V_{in} while V_{out} and i_{out} are fixed values. Consequently, f_{limit} of the DC-DC converter is determined by D and t_d, and when t_d was given, it was confirmed that f_{limit} was limited to lower as D was lower.

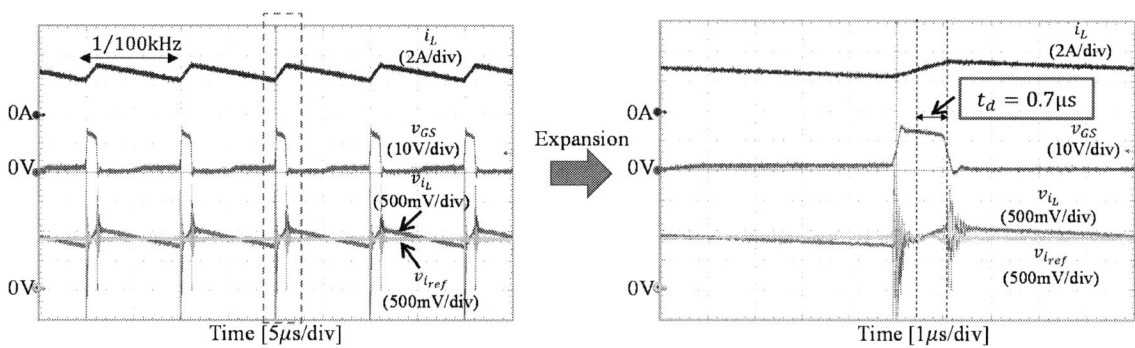

Fig. 8. Measurement of turn-off delay time t_d

The 2018 International Power Electronics Conference

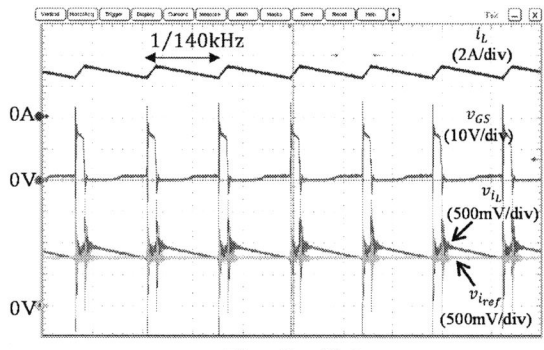

(a) Normal operation when f_C is 140kHz.

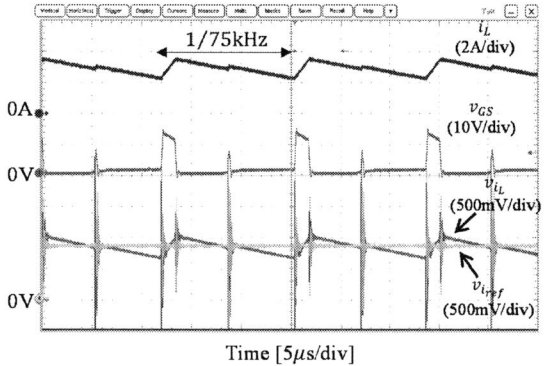

(b) Dividing frequency operation when f_C is 150kHz.

Fig. 9. Operating waveforms around f_{limit}
(No slope compensation).

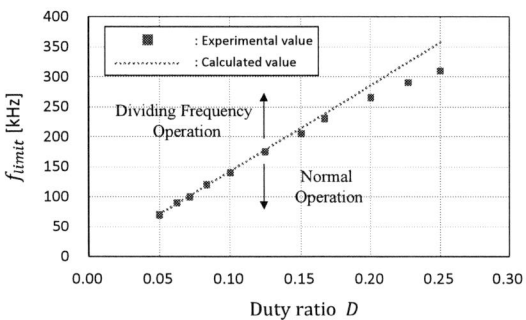

Fig. 10. Limit operating frequency characteristics when $t_d = 0.7\mu s$
(No slope compensation).

C. Limit operating frequency (With slope compensation)

Similarly, the limit operating frequency f_{limit} in the case of the slope compensation was measured. Firstly, we measured f_{limit} that introduced the slope compensation with the slope ratio $a = 0.5$. Fig. 11 shows operating waveforms when V_{in} is 25V and D is 0.2. In this parameter, since f_{limit} was 127kHz from the equation (11), the DC-DC converter was operated around this calculated f_{limit}. As a experiment result, when f_C is 135kHz, f_S was equal to f_C and it became the normal operation as shown in Fig. 11 (a).

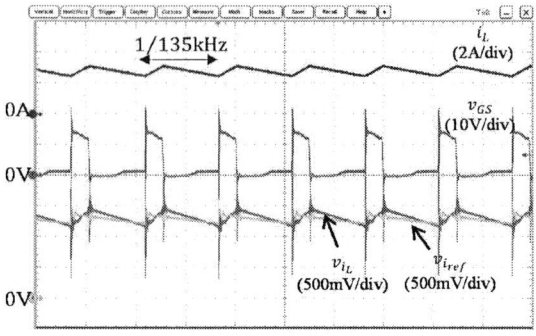

(a) Normal operation when f_C is 135kHz.

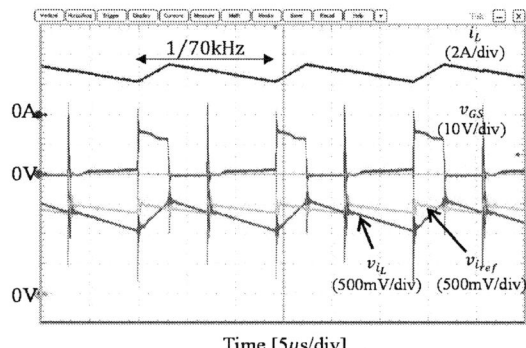

(b) Dividing frequency operation when f_C is 140kHz.

Fig. 11. Operating waveforms around f_{limit}
(With slope compensation $a = 0.5$).

However, when f_C is increased to 140kHz from this state, f_S was not equal to f_C and f_S was 70kHz that half of f_C due to the frequency dividing operation as shown in Fig. 11 (b).Secondly, we measured f_{limit} that introduced the slope compensation with the slope ratio $a = 0.75$. Fig. 12 shows operating waveforms when V_{in} is 25V and D is 0.2. In the calculation, since f_{limit} was 60kHz, the DC-DC converter was operated around the this f_{limit}. As a experiment result, when f_C is 70kHz, f_S was equal to f_C and it became the normal operation as shown in Fig. 12 (a). However, when f_C is increased to 75kHz from this state, f_S was not equal to f_C and f_S was 37.5kHz that half of f_C due to the frequency dividing operation as shown in Fig. 12 (b). Therefore, even when the slope compensation is introduced, f_S had been limited to the calculated f_{limit}. Additionally, it was confirmed that f_{limit} was limited to lower as compared with no slope compensation.

Finally, the limit operating frequency characteristics with the slope compensation when D is varied from 0.05 to 0.25 is shown in Fig. 13. As in previous experiments, D was adjusted by changing V_{in} while V_{out} and i_{out} are fixed values. As a result, even when D was changed, f_S was limited to the calculated f_{limit}. Therefore, when there is the turn-off delay time t_d, f_{limit} is limited to lower as the duty ratio is lower and the slope ratio of the slope compensation is close to 1.0.

3778

The 2018 International Power Electronics Conference

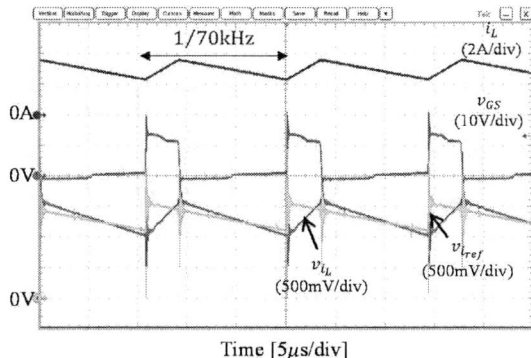

(a) Normal operation when f_C is 70kHz.

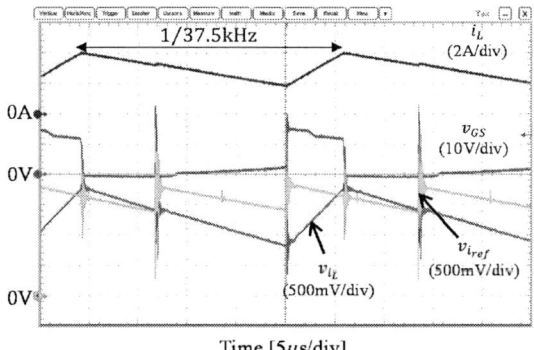

(b) Dividing frequency operation when f_C is 75kHz.

Fig. 12. Operating waveforms around f_{limit}
(With slope compensation $a = 0.75$).

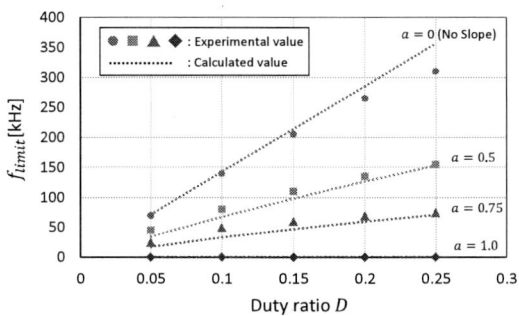

Fig. 13. Limit operating frequency characteristics when $t_d = 0.7\mu s$
(With slope compensation).

V. CONCLUSION

This paper clarified the mechanism of the frequency dividing operation in the peak current-mode control DC-DC converter considering the turn-off delay time. Moreover, we derived the relationship between the turn-off delay time and the limit operating frequency. As experimental results, the derived relational expression and the experimental result are approximately equal, and when there is the turn-off delay time, it was confirmed that the limit operating frequency of the DC-DC converter was limited to lower as the duty ratio was lower. Additionally, it was confirmed that the limit operating frequency in the case of the slope compensation is limited to lower as the duty ratio was lower and the slope ratio of the slope compensation is close to 1.0. Therefore, in order to realize the high frequency switching operation in the peak current mode control DC-DC converter, it is necessary to design the circuit by considering the relationship between the turn-off delay time and the limit operating frequency.

REFERENCES

[1] R. Sheeran, "Understanding and applying current-mode control theory," in *Proc. Power Electronics Technology Exhibition and Conference*, Oct. 2007.

[2] X. Duan, A.Q. Huang, "Current-mode variable-frequency control architecture for high-current low-voltage DC–DC converters," *IEEE Transactions on Power Electronics*, vol.21, no.4, pp.1133-1137, July 2006.

[3] M. del Viejo, P. Alou ,J. A. Oliver ,O. García ,J. A. Cobos ,"Fast control technique based on peak current mode control of the output capacitor current," in *Proc. IEEE Energy Conversion Congress and Exposition*, pp.3396-3402, Atlanta, The USA, Sept. 2010.

[4] A. Parayandeh, O. Trescases, A. Prodic, "10MHz peak current mode DC-DC converter IC with calibrated current observer," in *Proc. IEEE Applied Power Electronics Conference and Exposition*, pp.897-903, Fort Worth, The USA, March 2011.

[5] K. Kajiwara, H. Maruta, Y. Shibata, N. Matsui, F. Kurowawa, K. Hirose, "Wide input digital peak current mode DC-DC converter for DC power feeding system," in *Proc. IEEE International Telecommunications Energy Conference*, pp.1-4, Austin, The USA, Oct. 2016.

[6] T.Iida, "Emulated current mode control system for DC-DC boost converter using FPGA," in *Proc. Industrial Electronics, 2009. IECON'09 35th Annual Conference of IEEE*, pp. 432-436, Porte, Portugal, Feb. 2009.

[7] K. Shibata, C.K. Pham, "A compact adaptive slope compensation circuit for current-mode DC-DC converter," in *Proc. Proceedings of 2010 IEEE International Symposium on Circuits and Systems*, pp.1651-1654, Paris, France, May 2010.

[8] L. Yanming, L. Xinquan, C. Fuji, Y. Bing, J. Xinzhang, "An adaptive slope compensation circuit for buck DC-DC converter" in *Proc. The International Conference on ASIC 2007*, pp.608-611, Guilin, China, Oct. 2007.

[9] W.H. Ki, "Analysis of subharmonic oscillation of fixed-frequency current-programming switch mode power converters," *IEEE Transactions on Circuits and Systems I: Fundamental Theory and Applications*, vol.45, no.1, pp104-108, Jan. 1998.

A Novel Single Switch High Frequency DC/DC Converter and Its Mathematic Model

Yueshi Guan*, Xihong Hu, Shu Zhang, Yijie Wang, Dianguo Xu and Wei Wang

Department of Electrical Engineering, Harbin Institute of Technology, Harbin, China

*E-mail: hitguanyueshi@163.com

Abstract-This paper proposes a high frequency isolated DC/DC converter, the switching frequency is 20MHz. The system volume can be greatly reduced in such high frequency. Meanwhile, the switch voltage stress is reduced by introducing the third harmonic. In the proposed converter, the switch can operate in ZVS condition and the diode can operate in ZCS condition, which can help to reduce the switching loss. A mathematic model of the proposed high frequency converter is built, which provides an effective design method for the system parameters. To verify the feasibility of the proposed converter and mathematic model, a 10W prototype is built.

Keywords— High Frequency, Mathematic Model, Single-Switch Converter.

I. INTRODUCTION

With the increasing demand of high power density and small volume for power electronic systems, the switching frequency of power electronic system keeps increasing. In power electronic converters or inverters, the magnetic components and capacitors take the most space of the system. In high frequency condition, the energy storage requirement can be reduced. Thus, the value and volume of magnetic components and capacitors can be decreased.

However, the increasing switching frequency is not conductive to guarantee a high system efficiency. It is because that the switching loss of semiconductor devices almost forms a proportional relationship with the system operating frequency. Thus, the switch and diode are expected to operate in soft-switching characteristics in high frequency conditions.

Recently, some typical high frequency topologies have been proposed in [1] – [10]. The high frequency converter based on Class E type inverter and rectifier is the most typical one. In these topologies, the input inductor is not a choke one as the conventional converter. In [9], the input inductor is in a small value, which can resonate with the capacitor across the switch drain to source. In high frequency condition, the switch output capacitance C_{oss} can also be absorbed by the resonant capacitor. Based on the duality principle, the Class E type rectifier can also be adopted. Here, the diode capacitance can also be used to resonate with the corresponding inductor. Thus, based on the resonance and proper parameter design, the switch can operate in ZVS condition and the diode can operate in ZCS condition.

However, there are also some drawbacks of aforementioned converters. The first one is that these converters can not provide the function of isolation. It limits the application fields of the high frequency converter. Another drawback is that the switch voltage stress is great, which is about 4 times of the input voltage. With such a high voltage stress, the switch with high rated voltage must be adopted, which increases the cost of the system and the on-resistance of switch. Meanwhile, it also limits the input voltage range of the high frequency converter. Thus, it is necessary to reduce the switch voltage stress. Besides the switching loss, the driving loss also takes a great part of the system loss. Thus, the resonant driving method is gradually adopted in the high frequency converters. The basic idea is that the switch gate-source capacitance resonates with the series inductor, which can make the energy saved during each operating period. In the proposed converter, the similar resonant driving method is also adopted.

In this paper, a novel high frequency isolated converter is proposed. By introducing an air core transformer, the isolation function can be achieved, meanwhile, the leakage inductance can be adopted to resonate with corresponding inductors to keep the soft-switching characteristics of the switch and diode. Also, a LC branch is added between the switch drain and source, which can introduce the three harmonic and reduce the switch voltage stress. A mathematical model of the proposed converter is built which gives an insightful view of the proposed converter. Based on the mathematical model and corresponding equations, the system parameters can be effectively designed.

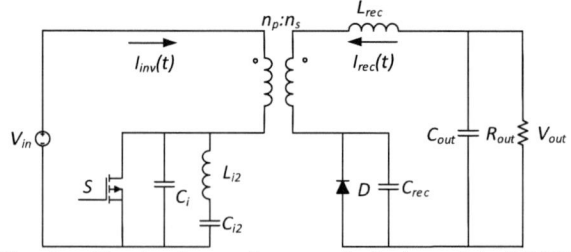

Fig. 1. The topology of proposed single switch DC/DC converter.

II. TOPOLOGY OF THE PROPOSED CONVERTER

Fig. 1 shows the topology of proposed single switch DC/DC converter. It can be seen that there is one transformer which can help to realize isolation between the primary side and secondary side. The capacitor C_i

across the switch resonates with the leakage inductance of transformer. Meanwhile, a series LC branch is added between the switch drain and source. In the secondary side, the capacitance of diode D is absorbed by the resonant capacitor C_{rec}, which resonates with leakage inductance of secondary side to guarantee the diode operating in ZCS condition.

Fig. 2 shows the switch voltage waveforms with and without third harmonic. In the Class E inverter of [9], the switch voltage waveform is similar as the fundamental component as shown in Fig. 2. The voltage stress is about 4 times of the input voltage. To reduce the peak voltage, the third harmonic can be introduced. With the same phase angle, the third harmonic can help to reduce the peak voltage. To achieve this aim, a series LC branch is added in the proposed converter. The resonant frequency of the LC branch is set around the secondary harmonic which helps to bypass the secondary harmonic and keep the third one.

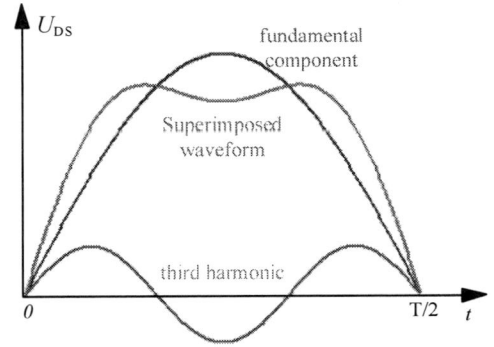

Fig. 2. The diagram of the switch voltage with and without third harmonic.

According to above analysis, the operating mode of the proposed converter can be divided into four parts. The voltage waveforms diagram of switch and diode are shown in Fig. 3. Here, Z_1, Z_2, Z_3 and Z_4 represent four different operating regions respectively. In conventional design method of [9], the simulation way is adopted to determine the value of corresponding parameters. This method is very time-consuming with many simulations and does not provide an insightful view of the converter and parameters effect. Thus, a quantitative analysis way is necessary to design the proposed converter effectively.

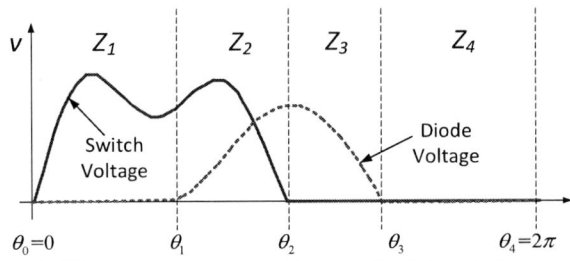

Fig. 3. The voltage waveforms diagram of switch and diode in different modes.

III. MATHEMATIC MODEL OF THE PROPOSED CONVERTER

The most important part of modeling the proposed converter is to establish the current and voltage relationship among corresponding components. To simply the analysis procedure, Fig. 4 shows the equivalent model of the transformer. Substituing the model into Fig. 1 and transferring the secondary side component into primary side, the equivalent circuit of the proposed converter can be obtained as shown in Fig. 5. Here, when switch is on, it is represented by the on-resistance R_{DS}. When the diode is on, it is represented by the series branch R_D and forward conduction voltage V_D.

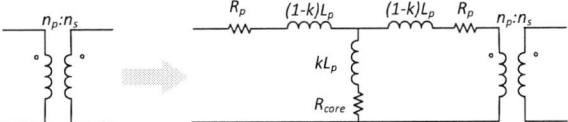

Fig. 4. The equivalent model of the transformer.

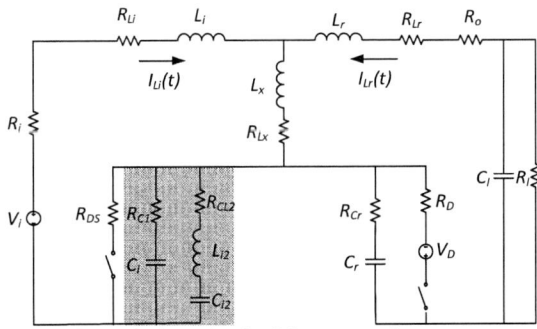

Fig. 5. The equivalent circuit of the proposed converter.

Meanwhile, to help simplify the derivation procedure, the impedance across the switch drain and source is represented by the reactance Z_i and corresponding resistance R_{Zi}. During operating region Z_1, the switch is off and the diode is on. Thus, the equivalent circuit diagram can be represented as Fig. 6 shown. Based on this figure, the equations (1) to (3) can be obtained according to KCL and KVL.

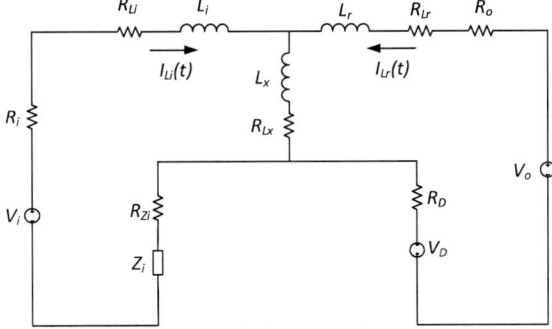

Fig. 6. The equivalent circuit diagram with region Z_1.

$$V_i = V_{Zi}(t) + (R_i + R_{Li} + R_{Lx} + R_{Zi})I_{Li}(t) +$$
$$R_{Lx}I_{Lr}(t) + (L_i + L_x)\frac{dI_{Li}(t)}{dt} + L_x\frac{dI_{Lr}(t)}{dt} \quad (1)$$

$$V_o = -V_D + (R_D + R_{Lr} + R_{Lx} + R_o) I_{Lr}(t) +$$
$$R_{Lx} I_{Li}(t) + (L_r + L_x)\frac{dI_{Lr}(t)}{dt} + L_x \frac{dI_{Li}(t)}{dt} \quad (2)$$

$$I_{Li}(t) = Z_i \frac{dV_{Zi}(t)}{dt} \quad (3)$$

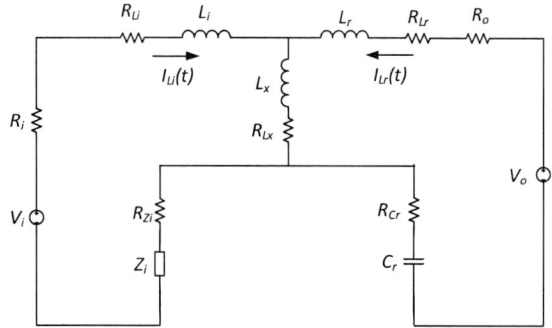

Fig. 7. The equivalent circuit diagram with region Z_2.

$$V_i = V_{Zi}(t) + (R_i + R_{Li} + R_{Lx} + R_{Zi}) I_{Li}(t) +$$
$$R_{Lx} I_{Lr}(t) + (L_i + L_x)\frac{dI_{Li}(t)}{dt} + L_x \frac{dI_{Lr}(t)}{dt} \quad (4)$$

$$V_o = V_{Cr}(t) + (R_{Cr} + R_{Lr} + R_{Lx} + R_o) I_{Lr}(t) +$$
$$R_{Lx} I_{Li}(t) + (L_r + L_x)\frac{dI_{Lr}(t)}{dt} + L_x \frac{dI_{Li}(t)}{dt} \quad (5)$$

$$I_{Li}(t) = Z_i \frac{dV_{Zi}(t)}{dt} \quad (6)$$

$$I_{Lr}(t) = C_r \frac{dV_{Cr}(t)}{dt} \quad (7)$$

During operating region Z_2, the switch is off and the diode is also off. Thus, the equivalent circuit diagram can be represented as Fig. 7 shown. Based on this figure, the equations (4) to (7) can be obtained according to KCL and KVL.

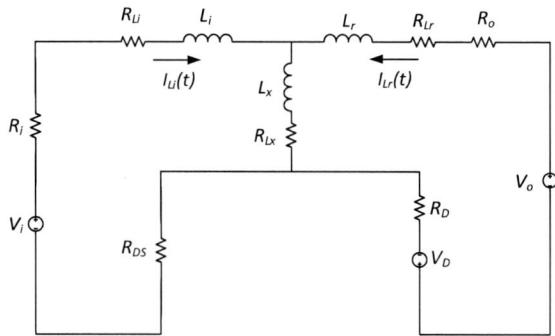

Fig. 8. The equivalent circuit diagram with region Z_3.

$$V_i = (R_i + R_{Li} + R_{Lx} + R_{DS}) I_{Li}(t) + R_{Lx} I_{Lr}(t)$$
$$+ (L_i + L_x)\frac{dI_{Li}(t)}{dt} + L_x \frac{dI_{Lr}(t)}{dt} \quad (8)$$

$$V_o = -V_D + (R_D + R_{Lr} + R_{Lx} + R_o) I_{Lr}(t) +$$
$$R_{Lx} I_{Li}(t) + (L_r + L_x)\frac{dI_{Lr}(t)}{dt} + L_x \frac{dI_{Li}(t)}{dt} \quad (9)$$

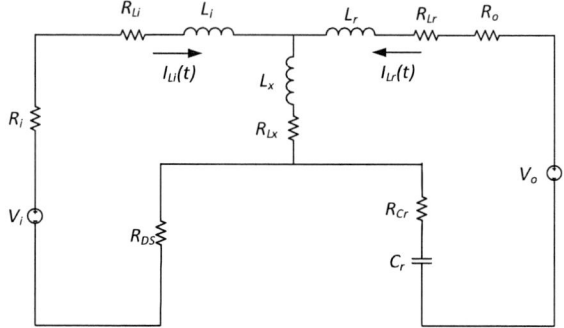

Fig. 9. The equivalent circuit diagram with region Z_4.

$$V_i = (R_i + R_{Li} + R_{Lx} + R_{DS}) I_{Li}(t) + R_{Lx} I_{Lr}(t)$$
$$+ (L_i + L_x)\frac{dI_{Li}(t)}{dt} + L_x \frac{dI_{Lr}(t)}{dt} \quad (10)$$

$$V_o = V_{Cr}(t) + (R_{Cr} + R_{Lr} + R_{Lx} + R_o) I_{Lr}(t) +$$
$$R_{Lx} I_{Li}(t) + (L_r + L_x)\frac{dI_{Lr}(t)}{dt} + L_x \frac{dI_{Li}(t)}{dt} \quad (11)$$

$$I_{Lr}(t) = C_r \frac{dV_{Cr}(t)}{dt} \quad (12)$$

With the similar method, the voltage relationships during Z_3 and Z_4 regions can also be obtained. According to these equations, the corresponding parameters and converter state variables can be calculated. However, these equations aim at specific operating and system situation, which can not provide a universal solution.

$$\begin{cases} i_{Li}(\theta) = \dfrac{I_{Li}(\theta/\omega_s)}{I_o}, \quad i_{Lr}(\theta) = \dfrac{I_{Lr}(\theta/\omega_s)}{I_o} \\[2mm] v_{Zi}(\theta) = \dfrac{V_{Zi}(\theta/\omega_s)}{V_o}, \quad v_{Cr}(\theta) = \dfrac{V_{Cr}(\theta/\omega_s)}{V_o} \end{cases} \quad (13)$$

$$\begin{cases} q_i = \dfrac{1}{\omega_s Z_i R_l}, \quad q_r = \dfrac{1}{\omega_s C_r R_l}, \quad q_x = \dfrac{\omega_s L_x}{R_l} \\[2mm] k_i = \dfrac{L_x}{L_i + L_x}, \quad k_r = \dfrac{L_x}{L_r + L_x} \end{cases} \quad (14)$$

$$\begin{cases} (1 - m^{ON}) v_{Zi}(\theta) + \left(\dfrac{1}{g_i} + \dfrac{1-k_i}{k_i}\dfrac{q_x}{Q_{Li}} + \dfrac{q_x}{Q_{Lx}} + (1 - m^{ON})\dfrac{q_i}{Q_{Zi}} + \dfrac{m^{ON}}{g_{DS}} \right) i_{Li}(\theta) + \dfrac{q_x}{Q_{Lx}} i_{Lr}(\theta) + \dfrac{q_x}{k_i}\dfrac{di_{Li}(\theta)}{d\theta} + q_x \dfrac{di_{Lr}(\theta)}{d\theta} = \mu \\[3mm] (1 - d^{ON}) v_{Cr}(\theta) - d^{ON} v_D + \dfrac{q_x}{Q_{Lx}} i_{Li}(\theta) + \left(\dfrac{1}{g_o} + \dfrac{1-k_r}{k_r}\dfrac{q_x}{Q_{Lr}} + \dfrac{q_x}{Q_{Lx}} + (1 - d^{ON})\dfrac{q_r}{Q_{Cr}} + \dfrac{d^{ON}}{g_D} \right) i_{Lr}(\theta) + q_x \dfrac{di_{Li}(\theta)}{d\theta} + \dfrac{q_x}{k_r}\dfrac{di_{Lr}(\theta)}{d\theta} = 1 \\[3mm] i_{Li}(\theta) = \dfrac{1}{q_i}\dfrac{dv_{Zi}(\theta)}{d\theta}(Z_1 \text{ and } Z_2 \text{ only}), \quad i_{Lr}(\theta) = \dfrac{1}{q_r}\dfrac{dv_{Cr}(\theta)}{d\theta}(Z_2 \text{ and } Z_3 \text{ only}) \end{cases} \quad (16)$$

$$\begin{cases} g_{DS} = \dfrac{R_l}{R_{DS}}, \ g_D = \dfrac{R_l}{R_D}, \ g_i = \dfrac{R_l}{R_i}, \ g_o = \dfrac{R_l}{R_o} \\[2mm] \mu = \dfrac{V_i}{V_o}, \ v_D = \dfrac{V_D}{V_o} \end{cases} \quad (15)$$

Thus, the normalization form of these equations is tried to adopt. Here the definitions are shown in (13) to (15). Based on these normalized variables and aforementioned equations, the normalized system expression can be obtained as (16) shown. Here, $m^{ON}=1$ when switch is on and $m^{ON}=0$ when switch is off. The same definition is done for d^{ON}.

For these four operating regions, the final solution can be obtained by substituting initial values in the piece-wise methodology, which is shown in (17).

$$\begin{cases} i_{Li}(\theta) = i_{Li}^{(Z_k)}(\theta), & v_{Li}(\theta) = v_{Li}^{(Z_k)}(\theta) \\[2mm] i_{Lr}(\theta) = i_{Lr}^{(Z_k)}(\theta), & v_{Lr}(\theta) = v_{Lr}^{(Z_k)}(\theta) \end{cases} \quad (17)$$

where $\theta_{k-1} \le \theta \le \theta_k$, $k = 1,2,3,4$, $\theta_0 = 0$ and $\theta_4 = 2\pi$.

The quality factors of inductors and capacitors, the on-resistance and forward conduction voltage can be obtained from the datasheet. Meanwhile, substituing the soft-switching constraints of switch and diode, the system parameters can be effectively obtained with specific operating conditions.

IV. EXPERIMENTAL RESULTS

Based on the proposed topology, a 10W prototype is built in this paper. The input voltage is 10V and output voltage is 5V. The operating frequency is set to be 20MHz. Based on the mathematical model of the proposed converter, the system parameters can be calculated, which are shown in TABLE I.

TABLE I
THE SYSTEM PARAMETERS

Symbol	Description	Value
C_i	Resonant capacitor	560pF
C_{i2}	Resonant capacitor	510pF
L_{i2}	Resonant inductor	30nH
S	Switch	SI7454
D	Diode	STPS2H100A
C_{rec}	Resonant capacitor	330pF
C_{out}	Output capacitor	4.7UF
T	Transformer	$n_p{:}n_s{=}2{:}1$

As mentioned above, the resonant driving method is adopted in the prototype. Meanwhile, in high frequency condition, the duty cycle and operating frequency are both difficulty to be adjusted. Thus, the ON/OFF control method is used. The control and driving circuit is shown in Fig. 10. Here, by adjusting the duty cycle of the PFM control signal, the effective operating time can be changed in order to regulate the output voltage. As seen in Fig. 10, several paralleled inverters are adopted to enhance the driving ability. The inductor in the front end can resonate with the switch input capacitance.

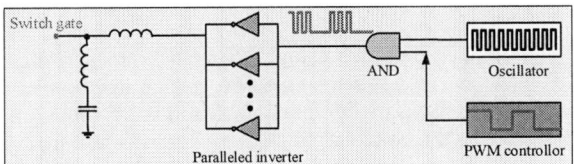

Fig. 10. The driving and control circuit of the proposed isolated converter.

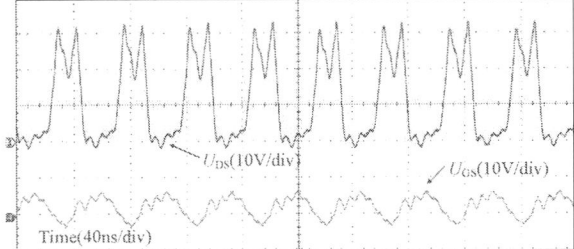

Fig. 11. The waveforms of switch drain-source voltage and gate voltage

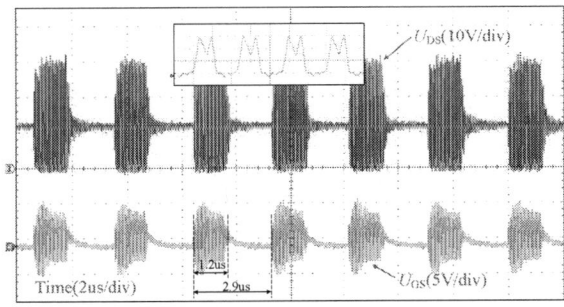

Fig. 12. The system voltage waveforms in light load condition.

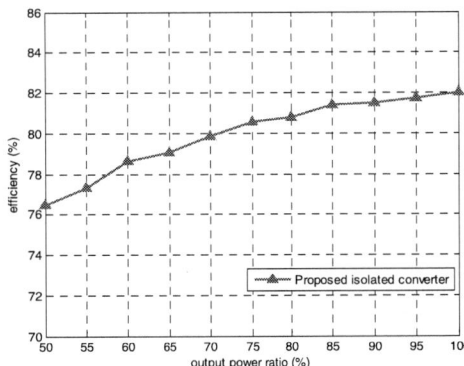

Fig. 13. The efficiency curve in different load conditions.

Fig. 11 shows the switch drain-source voltage and driving voltage in full load condition. It can be seen that the switch operates in ZVS condition and the peak voltage is greatly reduced. Fig. 12 shows the switch drain-source voltage in light load condition. It can be seen that in light load condition, the duty cycle of PWM control signal is reduced. The switch can still operate in soft-switching condition. Fig. 13 shows the system efficiency, which can be as high as 82% in full condition.

V. CONCLUSION

This paper proposes a high frequency single swtich DC/DC converter. Based on the transformer, the topology can achieve the isolation function. A series LC branch is introduce in the converter, thus, the switch voltage stress is reduced. A mathematical model of the proposed converter is built, which gives an deep view of the topology and can help to design the converter effectively. A 20MHz prototype is built in this paper, which verifies the feasibilty of the topology and mathematical model.

ACKNOWLEDGEMENT

This work is supported by the National Key Research and Development Program of China under Grant 2017YFB0402800.

REFERENCES

[1] A. Knott, T. M. Andersen, P. Kamby, J. A. Pedersen, M. P. Madsen, M. Kovacevic, et al., "Evolution of Very High Frequency Power Supplies," *IEEE J. Emerg. Sel. Topics Power Electron.*, vol. 2, pp. 386-394, 2014.

[2] D. J. Perreault, H. Jingying, J. M. Rivas, H. Yehui, O. Leitermann, R. C. N. Pilawa-Podgurski, et al., "Opportunities and Challenges in Very High Frequency Power Conversion," in *Proc. IEEE Appl. Power Electron. Conf. Expo.*, pp. 1-14. Feb. 2009,

[3] L. Olivia, "Radio Frequency dc-dc Converters: Device Characterization, Topology Evaluation, and Design", Ph.D. thesis, Dept. Elect. Eng. Comput. Sci.,Massachusetts Institute of Technology (MIT), Cambridge, 2008.

[4] J. Hu, A.D. Sagneri, J.M. Rivas, and et al, "High-frequency resonant SEPIC converter with wide input and output voltage ranges," *IEEE Trans. Power Electron.*, vol.27, no.1, pp. 189-200, Jan. 2012.

[5] J. M. Rivas, D. Jackson, O. Leitermann, A. D. Sagneri, Y. Han and D. J. Perreault, "Design considerations for very high frequency dc-dc converters," in *Proc. 2006 37th IEEE Power Electronics Specialists Conference*, Jeju, 2006, pp. 1-11.

[6] J. W. Phinney, D. J. Perreault and J. H. Lang, "Radio-frequency inverters with transmission-line input networks," *IEEE Trans. on Power Electron.*, vol. 22, No. 4, pp. 1154–1161, 2007.

[7] J. Warren, K. Rosowski, and D. Perreault, "Transistor selection and design of a VHF DC-DC power converter," *IEEE Trans. on Power Electron.*, vol. 23, pp. 27–37, 2008.

[8] R. C. N. Pilawa-Podgurski, A. D. Sagneri, J. M. Rivas, D. I. Anderson, and D. J. Perreault, "Very-high-frequency resonant boost converters," *IEEE Trans. Power Electron.*, vol. 24, no. 6, pp. 1654-1665, Jun. 2009.

[9] J. Hu, "Design of a low-voltage low-power dc-dc HF converter," S.M. thesis, Dept. Elect. Eng. Comput. Sci.,Massachusetts Institute of Technology (MIT), Cambridge, 2008.

[10] J. M. Rivas, "Radio Frequency dc-dc Power Conversion," Ph.D. thesis, Dept. Elect. Eng. Comput. Sci.,Massachusetts Institute of Technology (MIT), Cambridge, 2006.

Analysis of Closed Loop Operation of an Isolated Bidirectional DAB DC-DC Converter with LC Coupling

Bruno Yukio Enomoto, Kelly C. M. Carvalho, Lourenço Matakas Junior, Wilson Komatsu
Energy and Automation Dept., Polytechnic School of University of São Paulo, São Paulo, Brazil
E-mail: bruno.enomoto@gmail.com

Abstract – **This paper applies a dual active bridge converter with a closed loop that regulates the power transfer between the DC sides. The proposed control loop also performs a duty cycle adjustment in the AC side voltages in order to obtain minimal AC rms current in the converter for cases where the DC voltages ratio differs from the transformer ratio. The proposed control method is validated by simulation and experimental results.**

Keywords – *Bidirectional Power Flow, DC-DC Power Conversion, Power Control.*

I. INTRODUCTION

Transformer isolated DC-DC converters are widely used in applications such as energy storage systems [1] and electrical vehicles [2]. Transformers provide galvanic isolation and arbitrary transformation ratio between both DC sides can connect different DC voltage levels [3]. High frequency operation (tens of kHz) reduces transformer size and mass [4], and the use of dual active bridge (DAB) topology allows bidirectional power flow [5].

Power flow in the high frequency AC side should minimize reactive power in order to reduce losses. Series LC tanks can provide both reactive power compensation and convenient steady state DC voltage blocking, avoiding transformer saturation [6].

DAB converters with LC tanks found in the literature [5] [7] omit the study of optimized operation regions for such converters. In [8] is presented a detailed LC tank design methodology, based on the operation regions. In the analysis performed in [8], it is demonstrated that a DAB operating with DC voltages ratio (V_1 / V_2) different from the transformer ratio (N_1 / N_2) presents higher AC RMS current values when compared with matching relation ratios, i.e. $V_1 / V_2 = N_1 / N_2$. The mismatching between voltage and transformer ratios commonly occurs where one cannot guarantee a stable voltage level such as photovoltaic and wind power systems. Considering this situation, this paper proposes a methodology to minimize AC RMS current by adjusting the duty-cycle of the voltage o n the AC side. This paper also proposes a closed loop power controller. Simulation and experimental results validate the proposed project methodology.

II. THE DAB CONVERTER

The converter topology is shown in Fig. 1, where a DAB DC-DC converter presents two single-phase full-bridges, connected through a LC tank and an isolation transformer T with transformation ratio $a = N1/N2$. R shown in Fig. 1 represents the ohmic resistance of cables, capacitors, inductors and transformer. In addition, the transformer series inductance is considered in the LC tank calculation.

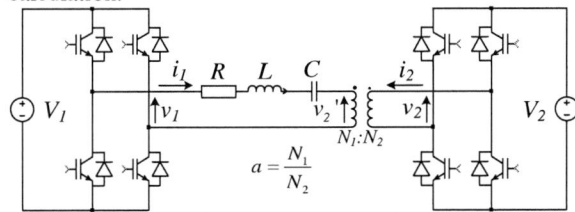

Fig. 1. Topology of a bidirectional isolated DAB DC-DC converter

The following chapter applies the LC tank design presented in [8].

III. LC PROJECT:

The DAB converter simplified model presents two square voltage sources v_1 and v_2 with amplitudes V_1 and V_2, connected by a series LC as shown in Fig. 2.

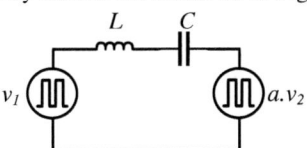

Fig. 2. Simplified model of a DAB converter.

Considering only the fundamental frequency of voltage sources, and neglecting the harmonic components, the approximated active power P flowing between v_1 and $a.v_2$ is given by (1), where $Z(\omega)$ is the equivalent impedance of the LC for switching frequency ω and δ is the displacement angle between v_1 and $a.v_2$.

$$P = \left(\frac{4}{\pi \cdot \sqrt{2}} \right)^2 \cdot \frac{V_1 \cdot a \cdot V_2 \cdot j \cdot \sin(\delta)}{Z(\omega)} \qquad (1)$$

Isolating $Z(\omega)$ from (1), $Z(\omega)$ is obtained for a nominal displacement angle δ_{nom} and the nominal power P_{nom} as in (2). Peak voltages V_1 and V_2, as well as the nominal power P_{nom} are defined project parameters. The choice of the nominal displacement angle δ_{nom} will be discussed in the following section.

$$Z(\omega) = j \left(\frac{4}{\pi \cdot \sqrt{2}} \right)^2 \cdot \frac{V_1 \cdot V_2 \cdot \sin(\delta_{nom})}{P_{nom}} \qquad (2)$$

Reference [8], defines a normalized frequency α (3)(a) that is given by the ratio between the switching frequency ω and the LC resonance frequency ω_0 (3)(b). It also proposes that the L and C parameters can be obtained as a function of $Z(\omega)$ obtained in (2), the switching frequency ω and the normalized frequency α, as shown in (4).

$$\begin{cases} \alpha = \dfrac{\omega}{\omega_0} & (a) \\[2mm] \omega_0 = \dfrac{1}{\sqrt{L \cdot C}} & (b) \end{cases} \qquad (3)$$

$$\begin{cases} L = \dfrac{|Z(\omega)| \cdot \alpha^2}{\omega \cdot |\alpha^2 - 1|} & (a) \\[3mm] C = \dfrac{|\alpha^2 - 1|}{\omega \cdot |Z(\omega)|} & (b) \end{cases} \qquad (4)$$

Reference [8] has shown that for different ratios of V_1 and aV_2, AC current (i_1 and i_2 from Fig. 1) increases as shown in Fig. 3. In [8], it was demonstrated that the LC resonant frequency should be lower than the switching frequency, therefore in the $1.1 < \alpha < 1.5$ range, for instance. Increasing values of δ_{nom} implies in higher currents and reactive power, suggesting a $5° < \delta_{nom} < 25°$ range.

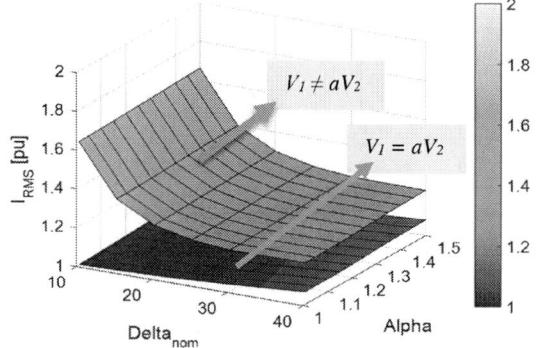

Fig. 3. RMS values of current i_1 as a function of α e δ_{nom} for $V_1 = aV_2$ and $V_1 = 1.2 \cdot aV_2$

For a permanent $V_1 \neq V_2$ condition, one must adjust transformation ratio $N_1 : N_2$ as in (5).

$$a = \frac{N_1}{N_2} = \frac{V_1}{V_2} \qquad (5)$$

In practical applications, such as photovoltaic energy storage systems, it is impossible to guarantee equal voltages between both DC sides. Therefore, it is necessary to adequate AC side voltage levels in order to achieve compatible levels of v_1 and $a.v_2$. This adjustment can be achieved by varying the duty cycle D of the 3-level voltage $v_1(\omega \cdot t)$ or $a.v_2(\omega \cdot t)$ (Fig. 4). The exact equation for RMS voltage for the Fig. 4 wave is shown in (6).

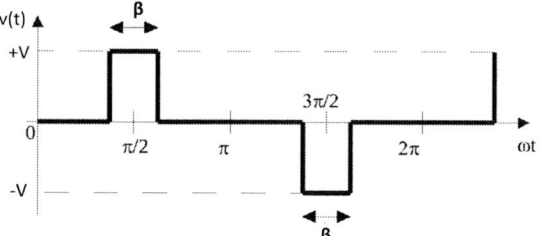

Fig. 4. Inverter voltage $v_1(\omega \cdot t)$ or $a.v_2(\omega \cdot t)$ as a 3-level PWM voltage. where duty cycle $= D = \beta/(2\pi)$

$$v_{RMS} = V \cdot \sqrt{\frac{\beta}{\pi}} = V \cdot \sqrt{2 \cdot D} \qquad (6)$$

The next chapter shows the influence of the voltage modulation in the AC RMS current.

IV. OPTIMAL DUTY CYCLE ADJUSTMENT

As shows Fig. 3, different values of V_1 and aV_2 causes increasing RMS currents. Therefore, voltage fitting between V_1 and aV_2 is necessary and can be achieved by adjusting the duty cycle D.

To evaluate the behavior of the RMS AC current as a function of duty cycle D and α, Fig. 5 was plotted for a constant $k=1.2$, defined by (7). Power P was calculated according to (1), which was modified to consider voltages harmonics up to 55th order, and δ was adjusted iteratively in order to maintain constant $P = 1$ pu.

$$k = \frac{V_1}{aV_2} \qquad (7)$$

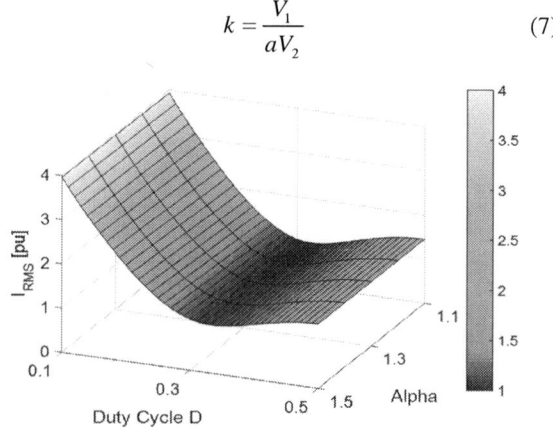

Fig. 5. RMS primary current as a function of D and α for $k = 1.2$ and constant power P

Fig. 5 behavior shows that for $k = 1.2$ and varying values of α (where α defines L and C, as in (4)), AC RMS current can be minimized by choosing an adequate value of D.

Fig. 6 shows RMS current behavior for a $k \neq 1$ case, where one can achieve reduction of RMS current by adjusting duty cycle D. In this case ($k = 1.2$), $D = 0.34$ provides minimum RMS current but $D = 0.5$ shows significant increase of RMS current as δ decreases. It will be shown that for a dynamic variation of k, there is a corresponding duty cycle D which provides local minimum RMS current. Also, the AC RMS current with duty cycle adjustment behaves similar as the case where k=1 shown in Fig. 3.

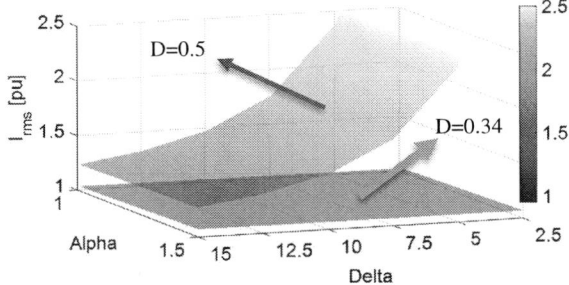

Fig. 6. Current comparison between case for $k = 1.2$ with $D=0.5$ and $k = 1.2$ with $D = 0.34$ for a wide range of α and δ.

Fig. 5 and Fig. 6 show RMS current behavior for fixed value of k. Fig. 7 shows the RMS current for different values of k and duty cycle, all the curves were obtained considering $\alpha = 1.28$. Additionally, for each value of k there is a corresponding value of duty cycle D that minimizes the RMS AC current.

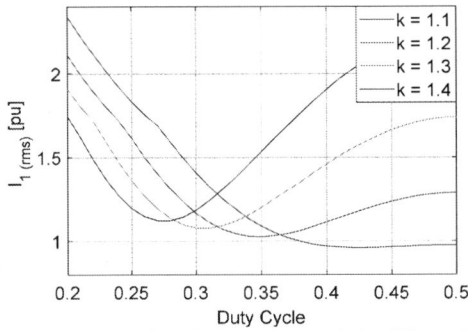

Fig. 7. Current as a function of voltage duty cycle for different k values and $\alpha = 1.28$

Fig. 7 shows that increasing values of k, results in reduction of optimum D (which minimizes RMS current), and increased RMS current minimum value. Therefore, operation with higher values of k is not practical.

Fig. 8 shows the duty cycle that minimize the RMS current in the AC side, for different k values. These duty cycle values were obtained, via simulation, by varying the duty cycle value until the minimum RMS current is found, for each k values. Also, Fig. 8 shows, for each value of k, the duty cycle that equalizes the RMS values of v_1 and $a.v_2$. The equalization of v_1 and $a.v_2$ is given by (11), it

is observed in that the duty cycle is adjusted only where the voltage is higher.

From (6), the RMS voltages v_{1_RMS} and v_{2_RMS} are given by (8).

$$\begin{cases} v_{1\_RMS} = V_1 \cdot \sqrt{2 \cdot D_1} \\ a.v_{2\_RMS} = a.V_2 \cdot \sqrt{2 \cdot D_2} \end{cases} \quad (8)$$

The voltages v_{1_RMS} and v_{2_RMS} are equalized as in (9).

$$v_{1\_RMS} = a.v_{2\_RMS} \quad (9)$$

Depending on the scenario ($k \geq 1$ or $k < 1$), the duty cycle of the major voltage value is adjusted while the minor voltage in kept at $D = 0.5$.

In the case of $k \geq 1$, means that $v_1 \geq v_2$. In this situation, the duty cycle D_2 is fixed at 0.5 while D_1 is calculated as in (10), obeying (9)

$$k \geq 1 \quad D_1 = D, \quad D_2 = 0.5$$

$$V_1 \cdot \sqrt{2 \cdot D} = a.V_2 \rightarrow D = \frac{1}{2}\left(\frac{a.V_2}{V_1}\right)^2 = \frac{1}{2k^2} \quad (10)$$

For the case of $k < 1$, the duty cycle D_1 is fixed at 0.5 and D_2 is calculated as in (11), since $v_1 < v_2$ in this situation.

$$k < 1 \quad D_1 = 0.5, \quad D_2 = D$$

$$V_1 = a.V_2 \cdot \sqrt{2 \cdot D} \rightarrow D = \frac{1}{2}\left(\frac{V_1}{a.V_2}\right)^2 = \frac{k^2}{2} \quad (11)$$

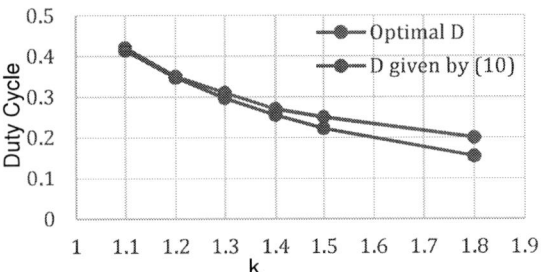

Fig. 8. Duty-cycle that minimizes the RMS current for each k values in blue; Duty-cycle that equalizes RMS values of v_1 and $a.v_2$.

For k around the value of 1.2 (Fig. 8) is observed that data from optimal D and data given by (10) have good correlation, making it possible to use equations (10) and (11) in the voltage adjustment algorithm instead of a closed loop control to find the duty cycle.

V. PROPOSED CONTROL STRATEGY

This section presents a closed loop control for power regulation on the primary DC side, considering the voltage adjustment algorithm proposed in (10) and (11), as shows the block diagram in Fig. 9. It is used to acquire the simulation/experimental results.

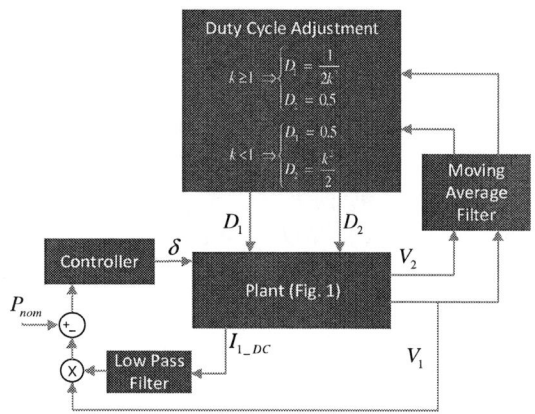

Fig. 9. Block diagram for the power P controlled, plant and duty cycle adjustment

The controller in Fig. 9 regulates the DC power on the primary side. It measures the voltage and the average current on the primary DC side.

The average current is obtained by filtering the measured current with a second order Butterworth filter with cutoff frequency of 500Hz, resulting in (12). This filter was discretized using Tustin with sampling frequency $f_s = 200kHz$.

$$G_{BW}(s) = \frac{\omega^2}{s^2 + \sqrt{2} \cdot \omega \cdot s + \omega^2} \qquad (12)$$

The duty cycle adjustment responds instantly to any DC voltages measured, making it susceptible to high frequency spikes, measurement errors and unpredictable during transients. To solve these problems, a moving average filter with 60 samples buffer size and using the same frequency sample $f_s = 200kHz$. Therefore, the buffer filter has a $300\mu s$ window, which corresponds to 3 full switching cycles of $10kHz$ (Table I).

The power regulation is performed by a PI controller with transfer function described by (13). The PI controller was discretized using Tustin with sampling frequency $f_{s\_PI} = 10kHz$. The used proportional and integral gains of the controller are $K_P = 1.5 \cdot 10^{-9}$ and $K_I = 7 \cdot 10^{-6}$, respectively.

$$G_{PI}(s) = \frac{K_P \cdot s + K_I}{s} \qquad (13)$$

The nominal values (Table I) were chosen according to the parameters used for simulation and experiment.

TABLE I
Description of the parameters - Nominal values

Parameter	Value	Descriptition
V_{nom}	$\frac{4}{\pi \cdot \sqrt{2}} 100V$	Nominal RMS voltage of fundamental frequency for a square
V_1 , V_2	$100V$	DC voltages
P_{nom}	$350W$	Nominal apparent power
I_{nom}	$P_{nom}/V_{nom} = 3.89A$	Nominal RMS current
f	$10kHz$	Switching frequency
δ_{nom}	$10°$	Nominal phase shift
a	1	Transformer ratio

The topology of Fig. 1 has nominal apparent power P_{nom} and nominal RMS voltage of the squared wave fundamental frequency V_{nom} at the primary side of transformer T, the nominal primary RMS current I_{nom_1} can be obtained from

$$I_{nom} = \frac{P_{nom}}{V_{nom}} \qquad (14)$$

In order to find L and C values, the following project criteria were adopted.

    i. $I < 1.1 \, p.u.$

    ii. $pf > 0.9$

As demonstrated in [8], to achieve the adopted criteria, α and δ_{nom} must respect the following conditions:

    i. $\alpha > 1.09$

    ii. $5° < \delta_{nom} < 25°$

Based on the conditions above, $\alpha = 1.28$ and $\delta_{nom} = 10°$ were adopted. Applying the nominal values from Table I in (2) and (4) results in $L = 166\mu H$ and $C = 2.48\mu F$, which will be used in simulation and experiment in the next chapter.

VI. SIMULATION AND EXPERIMENTAL RESULTS

The simulation results were obtained using PLECS [9] and the experimental results data were acquired using Agilent DSO6014 oscilloscope, differential voltage probes Tektronix P5200 and current probes Agilent N2782B. Data were processed using MATLAB.

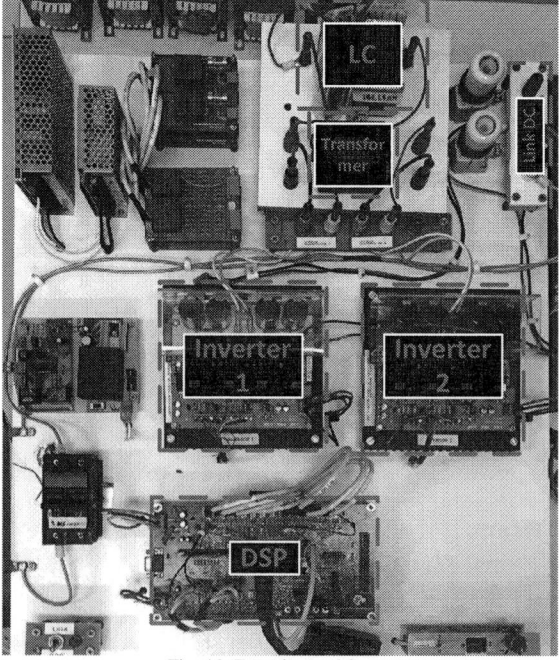

Fig. 10. Experimental Setup

The experimental setup used in this paper include a digital signal processor (DSP), the DAB converter, the LC tank and the transformer, presented in Fig. 10. The DSP is a Texas Instruments TMS320F28335. The DAB is composed of a pair of full bridges (IRFP250N power MOSFETs).

For $v_1 = v_2$, the following experimental results are presented in Fig. 11. For this experiment, the measured average primary DC power was 357W, which represents a 2% error in P_{nom}.

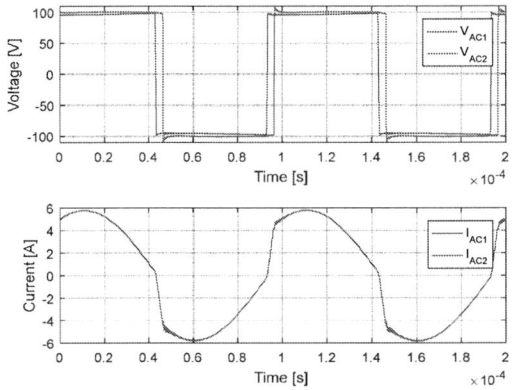

Fig. 11. Experimental AC Voltages and AC currents for $v_1 = v_2$

Fig. 12 presents the simulation for a $v_1 \neq v_2'$ case, where $k = 1.2$, and without duty cycle compensation (D=50%). The power measured on DC primary side is 350W.

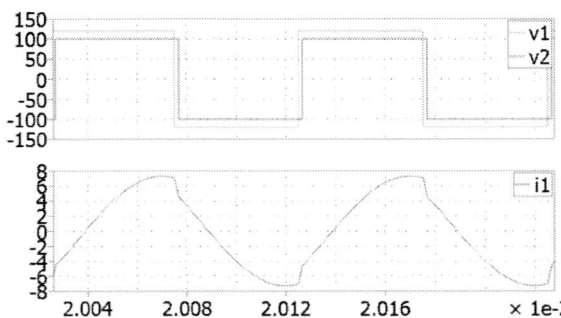

Fig. 12. Simulation results for $k = 1.2$ and without duty cycle adjustment

Fig. 13 present the simulation for a $v_1 \neq v_2'$ case, where $k = 1.2$, with duty cycle compensation (D=35%). For this case, the measured average power is also 350W.

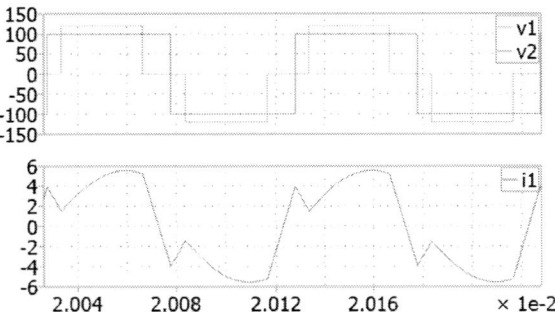

Fig. 13. Simulation results for $k = 1.2$ and duty cycle adjustment for voltage fitting ($D = 33\%$)

In Table II is presented the RMS values of the AC voltages and currents for k=1.2 with and without duty cycle adjustment. It can be observed that the duty cycle compensation implies lower AC RMS currents for the same power transfer.

TABLE II
Comparison results between compensated and non-compensated duty cycle

	$D = 35\%$	$D = 50\%$
i_1	4.02A$_{RMS}$	5.00A$_{RMS}$
i_2	3.96A$_{RMS}$	4.96A$_{RMS}$
v_1	99.97V$_{RMS}$	120V$_{RMS}$
v_2'	100V$_{RMS}$	100V$_{RMS}$

The Fig. 14 shows experimental results for the case $V_{DC1} \neq V_{DC2}$ ($k = 1.2$) with duty cycle adjustment control. Verify that the experimental results corroborate with the simulation results (Fig. 13) according to the Table III.

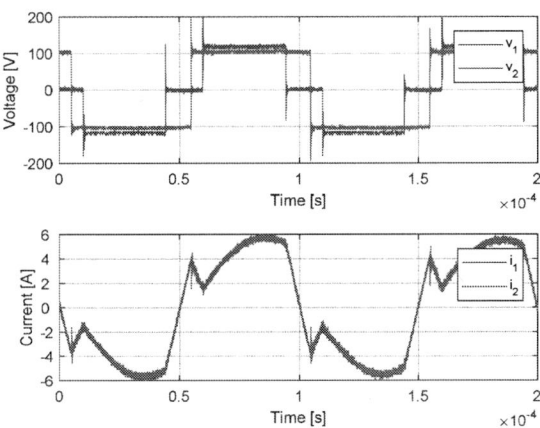

Fig. 14 – Experimental results for $k = 1.2$ with duty cycle adjustment control (D = 34.57%)

TABLE III
Comparison between simulation and experimental results with duty cycle adjustment control (D = 35%)

	Simulation	Experimental
i_1	4.02A$_{RMS}$	4.18A$_{RMS}$
i_2	3.96A$_{RMS}$	4.13A$_{RMS}$
v_1	99.97V$_{RMS}$	98.35V$_{RMS}$
v_2'	100V$_{RMS}$	104.39V$_{RMS}$

Considering other power ranges, as shows Table IV, the controller is able to track the power reference with deviation inferior to 5.48%. In the case where $V_1 < V_2$ ($k = 0.83$) and the direction of the transferred power is maintained from V_1 to V_2, the controller is able to adequately control the power flow.

TABLE IV
Controller tracking for other power ranges

k	Power reference P	P measured	Deviation
1.20	300W	293W	2.33%
1.20	350W	369W	5.48%
1.20	400W	400W	0.00%
0.83	350W	356W	1.71%

VII. CONCLUSIONS

This paper presented the influence of different voltage and transformer ratios in the AC current. The inadequate adjustment of the duty cycle D can result in increasing currents; therefore, increasing converter losses. This paper presented a duty cycle adjustment methodology which provides minimum AC RMS current. A power control method is also proposed. Simulation results demonstrate improvement when duty cycle adjustment is applied. The experimental results confirm the proposed control

ACKNOWLEDGEMENTS

The authors are grateful to Brazilian Coordination for the Improvement of Higher Education Personnel (CAPES), Brazilian National Council for Scientific and Technological Development (CNPq, grants 306970/2015-5 and 311789/2014-5) and São Paulo Research Foundation (FAPESP) for the financial support.

REFERENCES

[1] S. Inoue and H. Akagi, "A Bi-Directional DC/DC Converter for an Energy Storage System," in APEC 07 - Twenty-Second Annual IEEE Applied Power Electronics Conference and Exposition, pp. 761–767, 2007.

[2] Huang-Jen Chiu and Li-Wei Lin, "A bidirectional DC-DC converter for fuel cell electric vehicle driving system," *IEEE Trans. Power Electron.*, vol. 21, no. 4, pp. 950–958, Jul. 2006.

[3] N. M. L. Tan, T. Abe, and H. Akagi, "Topology and application of bidirectional isolated dc-dc converters," in 8th International Conference on Power Electronics - ECCE Asia, pp. 1039–1046, 2011.

[4] H. Fan and H. Li, "High-Frequency Transformer Isolated Bidirectional DC-DC Converter Modules With High Efficiency Over Wide Load Range for 20 kVA Solid-State Transformer," *IEEE Trans. Power Electron.*, vol. 26, no. 12, pp. 3599–3608, Dec. 2011.

[5] B. Zhao, Q. Song, W. Liu, and Y. Sun, "Overview of Dual-Active-Bridge Isolated Bidirectional DC-DC Converter for High-Frequency-Link Power-Conversion System," *IEEE Trans. Power Electron.*, vol. 29, no. 8, pp. 4091–4106, Aug. 2014.

[6] S. H. Kim, H. Cha, H. F. Ahmed, and H. G. Kim, "Isolated double step-down DC-DC converter," in 2015 9th International Conference on Power Electronics and ECCE Asia (ICPE-ECCE Asia),pp. 2306–2311, 2015.

[7] G. K. Gaidhane, "Performance of dual-bridge high frequency resonant DC/DC converter for energy storage application," in 2016 International Conference on Energy Efficient Technologies for Sustainability (ICEETS),pp. 400–405, 2016.

[8] B. Y. Enomoto, K. C. M. Carvalho, L. Matakas, and W. Komatsu, "Analysis of operation regions of an isolated bidirectional dual active bridge DC-DC converter with LC coupling," in *2017 Brazilian Power Electronics Conference (COBEP)*, pp. 1-6, 2017.

[9] "Plexim GmbH, Swiss." [Online]. Available: www.plexim.com.

Isolated AC/DC Converter Using Simple PWM Strategy

Naoki Hirose, Yuto Matsui, and Takaharu Takeshita

Nagoya Institute of Technology, Gokiso, Showa, Nagoya, Japan

E-mail: 26114105@stn.nitech.ac.jp, cjh14113@stn.nitech.ac.jp, take@nitech.ac.jp

Abstract—**This paper presents a circuit configuration of unidirectional isolated AC/DC converter and simple PWM strategy of isolated AC/DC converter for control of the primary voltage and the input currents, simultaneously. In the simple PWM strategy, the primary voltage reference and the primary current reference are approximated to the square AC waveform to obtain the duty cycles by simple calculations. The characteristics of the simple PWM strategy for the isolated AC/DC converter is confirmed by experiments.**

Keywords—high–frequency transformer, AC/DC converter, simple PWM strategy, matrix converter

I. INTRODUCTION

Recently, environmental problems such as exhaustion of resources and global warming have become severe on the earth, and resource saving and energy saving are required. For the solution of the environmental problems, the electric power storage devices and electric vehicles are developed. Development and dissemination of electric vehicles are under way with the development of high performance lithium ion battery. Carbon dioxide which is the cause of global warming is not emitted because electric vehicles use only electric energy. The resources are saved because the electric energy can be converted from renewable energy. An AC/DC converter is necessary for charging of electric vehicles because the charging is DC but power systems are AC. The AC/DC converter for charging electric vehicle is necessary to be isolated by using the transformer from the view point of the protection of the devices and the safety operation. Therefore, the isolated AC/DC converters which meet the requirements are much in demand.

Fig.1 shows the isolated AC/DC converter using a transformer of a commercial frequency [1]-[2]. The circuit in Fig.1 is composed of AC/DC rectifier circuit and DC/DC chopper circuit. The circuit such as Fig.1 is heavy and large because the transformer of commercial frequency and the large electrolytic capacitor. It is possible to realize size and weight reduction by using a high-frequency transformer compared with using a commercial-frequency transformer. Fig.2 shows the isolated AC/DC converter using a high-frequency transformer, the circuit configuration is composed of three-phase AC/DC rectifier circuit, DC/high-frequency AC inverter and high-frequency AC/DC rectifier circuit [3]-[5]. The size of transformer is reduced by using high-frequency transformer. However, the size and loss of the converters are increased because this circuit has three power converters. In this circuit, it is difficult

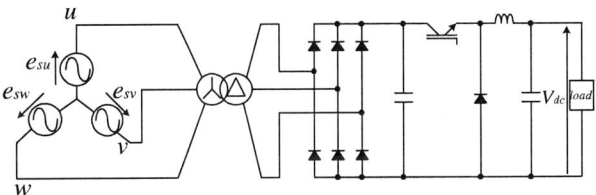

Fig. 1. Isolated AC/DC converter using commercial-frequency transformer

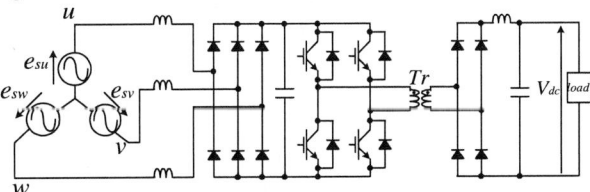

Fig. 2. Isolated AC/DC converter using high-frequency transformer

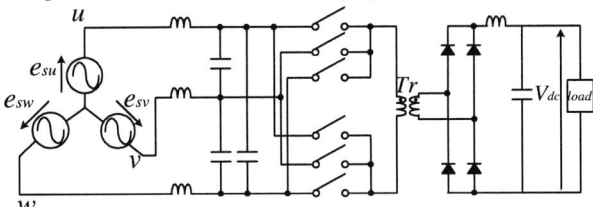

Fig. 3. Isolated AC/DC converter using matrix converter

to miniaturize and realize a long life because of using the large electrolytic capacitor. Fig.3 shows the isolated AC/DC converter composed of three-phase AC/high-frequency single-phase AC matrix converter and AC/DC rectifier circuit [6]-[9]. The size and loss of the converters can be reduced because this circuit has two power converters. In this circuits, it is able to miniaturize and realize a long life without the large electrolytic capacitor. The many control methods for the AC/DC converter using matrix converter and high-frequency transformer have been proposed. However, these control methods need complicated calculations to obtain the duty cycles.

This paper presents a circuit configuration of unidirectional isolated three-phase AC/DC converter and the simple PWM strategy for control of the primary voltage and the input current. The operation of the circuit is confirmed by using prototype system of the isolated AC/DC converter. In the PWM strategy, the primary voltage reference and the primary current reference are approximated to the square AC waveform to

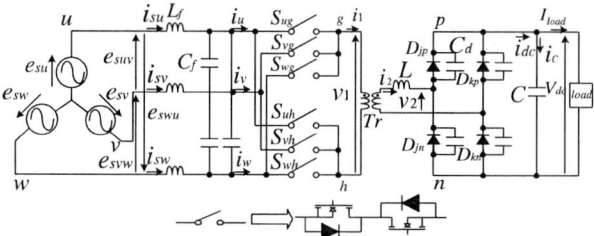

Fig. 4. Configuration of main circuit

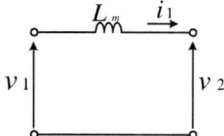

Fig. 5. Equivalent circuit of high frequency transformer

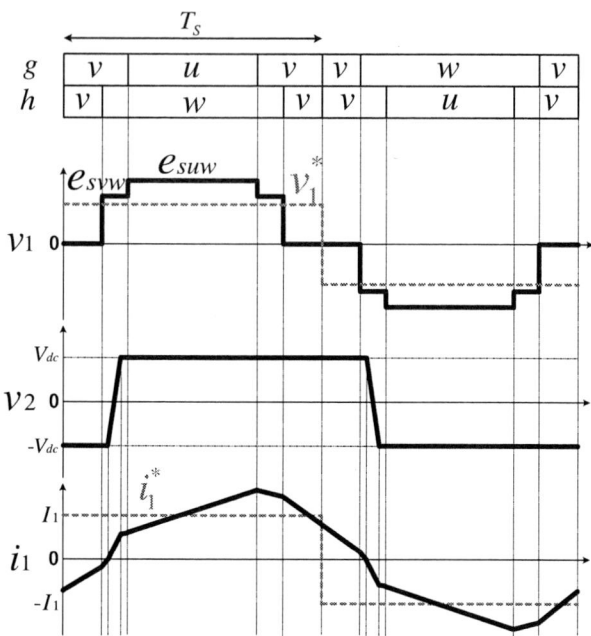

Fig. 6. Waveforms of high-frequency transformer

obtain the duty cycles by simple calculations. Also, the primary voltage is given the bilaterally symmetrical waveform in order to control the sinusoidal source current. The effectiveness of the control method is verified by experiments.

II. ISOLATED AC/DC CONVERTER

A. Configuration of Main Circuit

Fig.4 shows the proposed isolated three-phase AC/DC converter using matrix converter and high-frequency transformer. The primary circuit and the secondary circuit are magnetically coupled by using the high-frequency transformer. The primary circuit consists of the three-phase AC power supply, the input filter, and the matrix converter. The matrix converter consists of the six bidirectional switches $S_{ug} - S_{wh}$. The matrix converter converts commercial-frequency three-phase AC voltage into high-frequency single-phase AC voltage, directly. The input filter is composed of the filter reactor L_f and the filter capacitor C_f. The input filter is connected to the input side of the matrix converter in order to suppress the outflow of harmonics to the power supply. The secondary circuit consists of the diode rectifier circuit, the reactor L, the output capacitor C, and the load. The diode rectifier circuit consists of the four diodes $D_{jp} - D_{kn}$. The diode rectifier circuit converts high-frequency single-phase AC voltage into DC voltage. The output capacitor C is connected to the output side of the secondary diode rectifier circuit to obtain the constant load voltage V_{dc}. The reactor L is connected to the input side of secondary diode rectifier circuit to suppress the change of the transformer currents. The capacitors C_d are connected to the diode in parallel in order to suppress the harmonics of the source currents i_{su}, i_{sv}, and i_{sw}.

B. Operation of High-Frequency Transformer

Fig.5 shows the simple equivalent circuit of high-frequency transformer without the magnetizing current. The turn ratio of the high-frequency transformer is $N_1 : N_2 = 1 : 1$. The reactor L_m is total of the reactor L and the leakage reactor of

the high-frequency transformer. In Fig.5, the primary voltage of the high-frequency transformer v_1 is expressed by using the secondary voltage of the high-frequency transformer v_2, the reactor L_m, and the primary current of the high-frequency transformer i_1 as follows;

$$v_1 = v_2 + L_m \frac{di_1}{dt} \quad (i_1 = i_2) \quad (1)$$

The excitation current is ignored as sufficiently small compared with the primary and secondary currents of the high-frequency transformer i_1 and i_2. Therefore, the primary current i_1 equals to the secondary current i_2. The slope of the primary current di_1/dt is determined by using (1) as follows;

$$\frac{di_1}{dt} = \frac{v_1 - v_2}{L_m} \quad (2)$$

Therefore, the slope of the primary current is expressed by the reactor L_m and the voltage difference between the primary voltage v_1 and the secondary voltage v_2. The operations of high-frequency transformer is explained in condition of the source voltage $e_{su} > e_{sv} > e_{sw}$. Fig.6 shows the waveforms of the primary voltage, the secondary voltage, and the primary current in one period $2T_s$ in the condition of the source voltage $e_{su} > e_{sv} > e_{sw}$. During half period T_s of the high-frequency transformer, the source voltage e_{su}, e_{sv}, and e_{sw} are given as constant because the transformer frequency is much higher than the source one. The waveform of the primary voltage is determined by using the switching pattern and the source line voltages. The waveform of the secondary voltage v_2 is given the trapezoidal AC form with the amplitude load voltage V_{dc}. The slope of the trapezoidal-wave is strongly influenced by the capacitor C_d connected to the diode in parallel. The primary

3792

current reference i_1^* is given the square AC waveform with the amplitude I_1 in one period $2T_s$. The primary current i_1 is closed to the primary current reference i_1^* by increasing the value of the capacitors C_d. The harmonics of the source currents i_{su}, i_{sv}, and i_{sw} are suppressed by closing the primary current i_1 to the primary current reference i_1^* of the square AC waveform.

C. Input Current References

The voltage drop across the input filter reactor L_f in Fig.4 can be ignored because it is sufficiently small as compared with the source voltages. The voltages of the input filter capacitor C_f in Fig.4 can be approximated to the source voltages e_{su}, e_{sv}, and e_{sw}. The source voltages e_{su}, e_{sv}, and e_{sw} are given by using the line voltage effective value E and the phase angle θ as follows;

$$\begin{bmatrix} e_{su} \\ e_{sv} \\ e_{sw} \end{bmatrix} = \sqrt{\frac{2}{3}}E \begin{bmatrix} \cos\theta \\ \cos(\theta - 2\pi/3) \\ \cos(\theta - 4\pi/3) \end{bmatrix} \quad (3)$$

Also, the input current references i_u^*, i_v^* and i_w^* are given by using the input current effective value I and the power factor angle φ^* as follows;

$$\begin{bmatrix} i_u^* \\ i_v^* \\ i_w^* \end{bmatrix} = \sqrt{2}I \begin{bmatrix} \cos(\theta + \varphi^*) \\ \cos(\theta + \varphi^* - 2\pi/3) \\ \cos(\theta + \varphi^* - 4\pi/3) \end{bmatrix} \quad (4)$$

The input instantaneous power p_{in} is expressed by using (3), (4) as follows;

$$\begin{aligned} p_{in} &= e_{su}i_u^* + e_{sv}i_v^* + e_{sw}i_w^* \\ &= \sqrt{3}EI\cos\varphi^* \end{aligned} \quad (5)$$

The output instantaneous power of the primary circuit p_{out} is expressed by using the primary voltage reference v_1^* and the primary current reference i_1^* as follows;

$$p_{out} = v_1^* i_1^* \quad (6)$$

Assuming the loss of the primary matrix converter can be ignored, the input current effective value I is expressed by using (5),(6) as follows;

$$\begin{aligned} I &= \frac{v_1^* i_1^*}{\sqrt{2}\{e_{su}\cos(\theta+\varphi^*)+e_{sv}\cos(\theta+\varphi^*-\frac{2\pi}{3})+e_{sw}\cos(\theta+\varphi^*-\frac{4\pi}{3})\}} \\ &= \frac{v_1^* i_1^*}{\sqrt{3}E\cos\varphi^*} \end{aligned} \quad (7)$$

The primary voltage reference v_1^* and the primary current reference i_1^* are approximated to the square AC waveform. The input current references i_u^*, i_v^* and i_w^* can be simply calculated by using (4), (7) because the references are constant in one control cycle T_s.

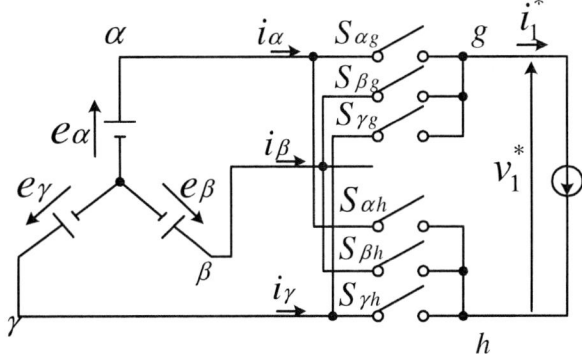

Fig. 7. Model of the primary converter

D. Duty Cycles

Fig.7 shows the model of the primary circuit for determining the duty cycles of the six bidirectional switches $S_{ug} - S_{wh}$. The input capacitor is ignored as the voltage of the input capacitor C_f approximately equal to the source voltage e_{su}, e_{sv}, and e_{sw}. In Fig.7 , the input phases are represented the maximum voltage phase α, the intermediate voltage phase β, and the minimum voltage phase γ among the source voltage e_{su}, e_{sv}, and e_{sw} respectively. The source voltages e_α, e_β, and e_γ are expressed by using DC voltage sources, and the primary current reference i_1^* is expressed by using current source. The duty cycles are given by connecting each of the output phases g and h with any one of the input phases α, β and γ to ensure the continuity of the primary current i_1^* by using six bidirectional switches $S_{\alpha g} - S_{\gamma h}$.

$$\begin{cases} d_{\alpha g} + d_{\beta g} + d_{\gamma g} = 1 \\ d_{\alpha h} + d_{\beta h} + d_{\gamma h} = 1 \end{cases} \quad (8)$$

The input current references i_α^*, i_β^*, and i_γ^* are given by using the duty cycles and the primary current reference i_1^* as follows;

$$\begin{cases} i_\alpha^* = (d_{\alpha g} - d_{\alpha h})i_1^* \\ i_\beta^* = (d_{\beta g} - d_{\beta h})i_1^* \\ i_\gamma^* = (d_{\gamma g} - d_{\gamma h})i_1^* \end{cases} \quad (9)$$

The primary average voltage $\overline{v}_1$ in one control cycle is given as follows;

$$\overline{v}_1 = (d_{\alpha g} - d_{\alpha h})e_\alpha + (d_{\beta g} - d_{\beta h})e_\beta + (d_{\gamma g} - d_{\gamma h})e_\gamma \quad (10)$$

The average is determined by using (5), (6), (9) and (10) as follows;

$$\begin{aligned} \overline{v}_1 &= (d_{\alpha g} - d_{\alpha h})e_\alpha + (d_{\beta g} - d_{\beta h})e_\beta + (d_{\gamma g} - d_{\gamma h})e_\gamma \\ &= \frac{e_\alpha i_\alpha^* + e_\beta i_\beta^* + e_\gamma i_\gamma^*}{i_1^*} = \frac{p_{in}}{i_1^*} = v_1^* \end{aligned} \quad (11)$$

From (11), it can be found that the primary average voltage $\overline{v}_1$ is coincident to the reference v_1^*.

The duty cycles $d_{\alpha g} - d_{\gamma h}$ cannot be determined uniquely by using only (8) and (9). Two of the six duty cycles need

The 2018 International Power Electronics Conference

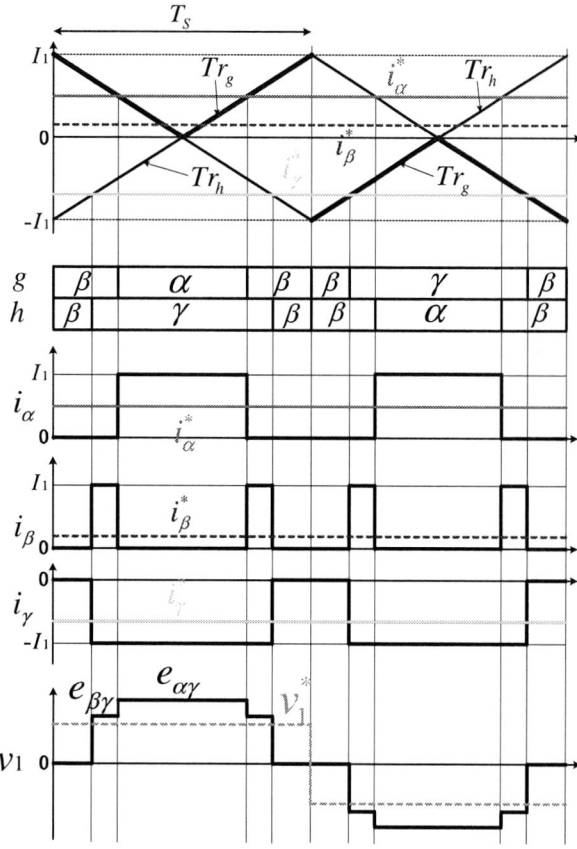

Fig. 8.   Determining the switching patterns

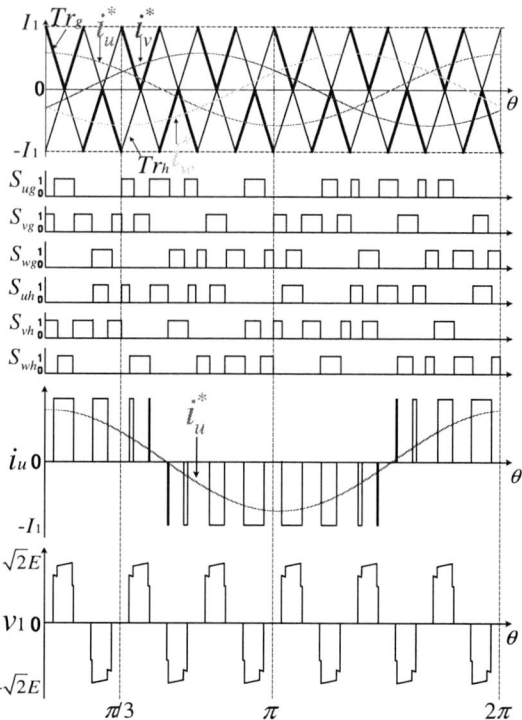

Fig. 9.   Principle of PWM strategies

to be equal to zero to determine the duty cycles $d_{\alpha g} - d_{\gamma h}$ uniquely. When the primary voltage reference v_1^* is positive, the duty cycle $d_{\gamma g}$ of minimum voltage phase γ is always equal to zero in the phase g, and the duty cycle $d_{\alpha h}$ of maximum voltage phase α is always equal to zero in the phase h. The duty cycles $d_{\alpha g} - d_{\gamma h}$ are given by using (8), (9) and $i_1^* = I_1$ as follows;

$$\begin{cases} d_{\alpha g} = \frac{i_\alpha^*}{I_1}, & d_{\beta g} = 1 - \frac{i_\alpha^*}{I_1}, & d_{\gamma g} = 0 \\ d_{\alpha h} = 0, & d_{\beta h} = 1 + \frac{i_\gamma^*}{I_1}, & d_{\gamma h} = -\frac{i_\gamma^*}{I_1} \end{cases} \quad (12)$$

When the primary voltage reference v_1^* is negative, the duty cycle $d_{\alpha g}$ of maximum voltage phase α is always equal to zero in the phase g, and the duty cycle $d_{\gamma h}$ of minimum voltage phase γ is always equal to zero in the phase h. The duty cycles $d_{\alpha g} - d_{\gamma h}$ are giving by using (8), (9) and $i_1^* = -I_1$ as follows;

$$\begin{cases} d_{\alpha g} = 0, & d_{\beta g} = 1 + \frac{i_\gamma^*}{I_1}, & d_{\gamma g} = -\frac{i_\gamma^*}{I_1} \\ d_{\alpha h} = \frac{i_\alpha^*}{I_1}, & d_{\beta h} = 1 - \frac{i_\alpha^*}{I_1}, & d_{\gamma h} = 0 \end{cases} \quad (13)$$

E. Switching Patterns

Fig.8 shows the principle of determining the switching patterns using triangular wave comparison. One period of

the high-frequency AC voltage consists with two periods of the control cycle T_s. The gate signals of the switches are determined by using (12), (13) and the triangular waves Tr_g, Tr_h. The input current reference used in the duty cycle can be simply calculated by using (4) and (7). The maximum input current reference i_α^* and the minimum input current reference i_γ^* are used for comparison, but the middle input current reference i_β^* are not used. In each input phase, the constant input current I_1 flows when corresponding switch is turned on. The primary voltage v_1 consists with the line voltages of the phases which switches are turned on. The control is performed so that the average of input currents $i_\alpha, i_\beta, i_\gamma$ and the primary voltage v_1 equal to the references in one control cycle T_s.

F. PWM Strategy

Fig.9 shows the principle of the PWM strategy in the source frequency. The gate signals of the six bidirectional switches $S_{ug} - S_{wh}$ are generated by using the PWM strategy of triangular wave comparison. In the phase g, the triangular wave Tr_g and the maximum input current reference are used for generating the gate signals of the switches $S_{ug} - S_{wg}$. In the phase h, the triangular wave Tr_h and the minimum input current reference are used for generating the gate signals of the switches $S_{uh} - S_{wh}$. For example, considering the case in the condition $0 \leq \theta \leq \pi/3$ and $v_1 > 0$ in Fig.9, the maximum voltage phase is phase u, the intermediate voltage phase is phase v, and the minimum voltage phase is phase w. The

3794

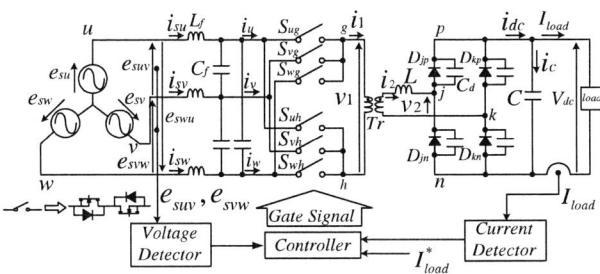

Fig. 10. Configuration of experimental system

TABLE I. EXPERIMENTAL CONDITIONS

Source voltage E, ω	200 V, $2\pi \times 60$ rad/s
Load current reference I_{load}^*	6.67A
Output power P_{out}	1.2kW
Load R	27Ω
Input filter L_f, C_f	1.0 mH, 7.1 μF
Output capacitor C	300 μF
Inductance L	300 μH
Capacitor C_d	20 nF
Frequency of transformer f_s	10 kHz
Ratio of transformer $N_1 : N_2$	1:1

gate signals of the switches in the phase g, S_{ug} and S_{vg} are generated by comparing the maximum input current reference i_u^* with triangular wave Tr_g, and the gate signals of the switch in the phase h, S_{vh} and S_{wh} are generated by comparing the minimum input current reference i_w^* with triangular wave Tr_h. Zero voltage is set both ends of the primary voltage v_1. In the phase g, the switch S_{wg} of minimum voltage phase w is always turned off. In the phase h, the switch S_{uh} of mximum voltage phase u is always turned off. The input current includes harmonics, but the source current is controlled sinusoidal current by connecting the LC-filter. When the switch of phase u in either the phase g or the phase h is turned on, the input current i_u flows. When the switches of phase u in either the both sides or neither of them is turned on, the input current i_u doesn't flows.

III. EXPERIMENTAL RESULTS

A. Experimental Conditions

Fig.10 shows the configuration of the experimental system and Table I shows the experimental conditions. SiC-MOSFETs(CREE,C2M0040 120-D) are used for the bidirectional switches $S_{ug} - S_{wh}$ of the primary converter. SiC-SBDs(ROHM, SCS220AG) are used for the diodes $D_{jp} - D_{kn}$ of the secondary diode rectifier circuit. The duty cycles of the bidirectional switches are calculated by using the source line voltages e_{suv} and e_{vw} which are obtained by the detectors. The effective value of the source line voltage E is 200V, the source frequency f is 60Hz, the frequency of the transformer f_s is 10kHz, the output power P_{out} is 1.2kW, the load current reference I_{load}^* is 6.67A.

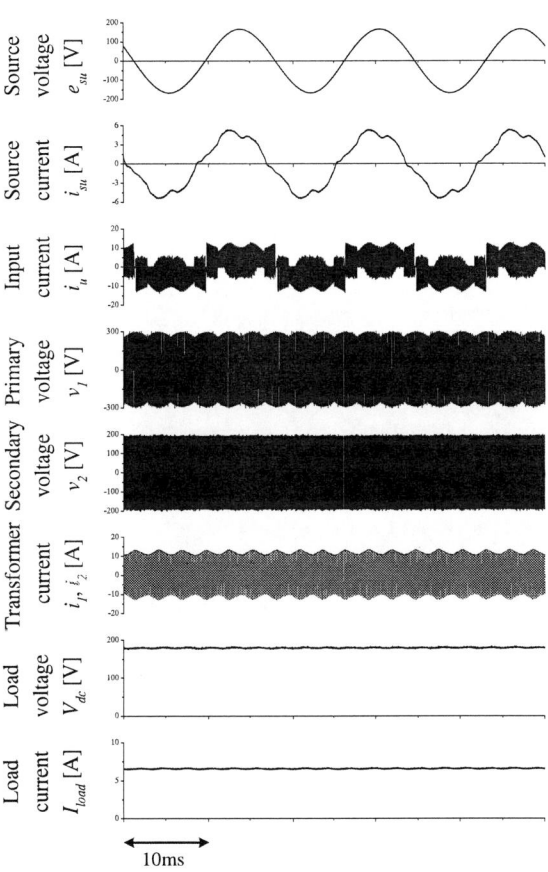

Fig. 11. Experimental waveforms

B. Experimental Waveforms

Fig.11 shows the experimental waveforms of the system in Fig.10 under the output power P_{out} of 1200W and the load current reference I_{load} of 6.67A. The experimental waveforms are the source phase voltage e_{su}, the source current i_{su}, the primary voltage of the transformer v_1, the secondary voltage of the transformer v_2, the primary current of the transformer i_1, the secondary current of the transformer i_2, the load voltage V_{dc}, and the load current I_{load}. In Fig.11, the load current I_{load} is controlled to 6.67A according to the reference and the source current i_{su} is controlled to the sinusoidal waveform. The voltages and currents of the transformer are controlled to the high frequency waveforms. Fig.12 shows magnified waveforms of the primary voltage v_1, the secondary voltage v_2, the transformer currents i_1 and i_2. In Fig.12, the primary voltage v_1 is coincident with the theoretical waveform in Fig.6. Also, the secondary voltage v_2 with the trapezoidal AC waveform of amplitude $V_{dc} = 180$V is coincident with the theoretical waveform in Fig.6. In Fig.12, the transformer currents i_1 and i_2 are obtained with the slopes in propotion to the voltage difference between the primary voltage v_1 and the

The 2018 International Power Electronics Conference

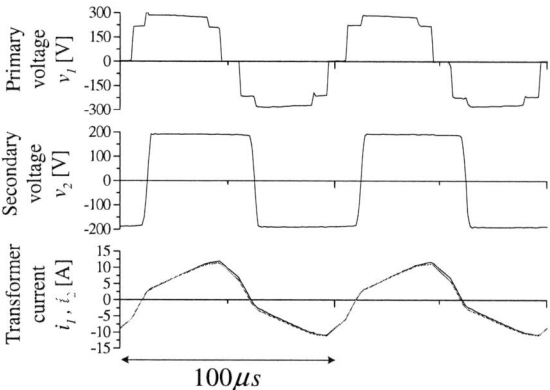

Fig. 12.   Partial waveforms

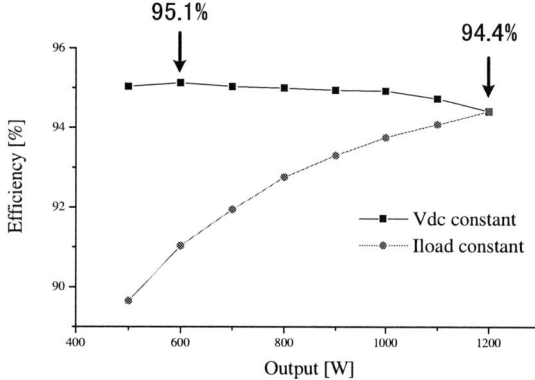

Fig. 13.   Efficiency

secondary voltage v_2.

C. Efficiency

Fig.13 shows the characteristics of the proposed isolated AC/DC converter. The efficiency are measured at every 100W of the output power P_{out} from 500W to 1200W under the two conditions. One is constant load voltage V_{dc} of 180V. Another is constant load current reference I_{load} of 6.67A. In Fig.13, the efficiency of 94.4% is obtained at the output power P_{out} 1200W. Also, the maximum efficiency of 95.1% is obtained at the output power P_{out} 600W under the condition of the load voltage constant.

IV.   CONCLUSION

This paper presents the configuration and the simple PWM strategy of unidirectional isolated AC/DC converter. In the simple PWM strategy, the primary voltage reference and the primary current reference are approximated to the square AC waveform to obtain the duty cycles by only simple calculations. The switching patterns of the PWM strategy are determined by using the input current references and

the triangular carrier waveforms. From the experiments, the characteristics of the proposed circuit for using the proposed simple PWM strategy are verified.

This work was supported by Council for Science, Technology and Innovation (CSTI), Cross-ministerial Strategic Innovation Promotion Program (SIP), "Next-generation power electronics" (funding agency: NEDO).

REFERENCES

[1] S. Inoue and H. Akagi "A Bi-Directional DC/DC Converter for an Energy Storage system", *Twenty-Second Annual IEEE Applied Power Electronics Conference and Exposition (APEC)*, pp.761-767 (2007)

[2] M. J. Erfani, T. Thiringer and S. Haghbin "Performance and Losses Analysis of Charging and Discharging Mode of a Bidirectional DC/DC Fullbridge Converter Using PWM Switching Pattern", *IEEE Vehicle Power and Propulsion Conference*, pp.1-6 (2011)

[3] S. Ono, H. Funato and J. Haruna "AC/DC Converter with Indirect Matrix Converter Using a Novel Snubber Configuration", *19th International Conference on Electrical Machines and Systems (ICEMS)*, pp.1-4 (2016)

[4] X. Wu, T. Maeda, H. Fujimoto, S. Ishii and K. Fujita "Three-phase high frequency transformer isolated AC to DC converter for EV battery quick charging", *Proceedings of The 7th International Power Electronics and Motion Control Conference (ICPM)*, Vol.1, pp.643-647 (2012)

[5] M. Yazdanian and S. Farhangi "A Novel Soft Switched Topology for Power Factor Correction with Load Isolation Capability", *7th International Conference on Power Electronics (ICPE)*, pp.426-430 (2007)

[6] J. J. Sandoval, S. Essakiappan and P. Enjeti "A Bidirectional Series Resonant Matrix Converter Topology for Electric Vehicle DC Fast Charging", *IEEE Applied Power Electronics Conference and Exposition (APEC)*, pp.3109-3116 (2015)

[7] M. Wang, S. Guo, Q. Hang, W. Yu and A. Q. Hang "An Isolated Bi-directional Soft-switched High-frequency-AC Link DC-AC Converter Using SiC MOSFETs", *IEEE Workshop on Wide Bandgap Power Device and ApplicationsWiPDA*, pp.88-93 (2014)

[8] D. Varajao, L. M. Miranda, R. E. Araujo and J. P. Lopes "Power Transformer for a Single-stage Bidirectional and Isolated AC-DC Matrix Converter for Energy Storage Systems", *42nd Annual Conference of the IEEE Industrial Electronics Society (IECON)*, pp.1149-1155 (2016)

[9] M. A. Sayed, K. Suzuki and T. Takeshita "Modeling and Control of Bidirectional Isolated Battery Charging and Discharging Converter Based High-Frequency Link Transformer", *IEEE 7th International Symposium on Power Electronics for Distributed Generation Systems (PEDG)*, pp.1-8 (2016)

The 2018 International Power Electronics Conference

Analysis of One Phase Loss Operation of Three-Phase Isolated Buck Matrix-Type Rectifier with Eight-Segment PWM Scheme

Jahangir Afsharian[1], Dewei (David) Xu[1], Bin Wu[1], Bing Gong[2], Zhihua Yang[2], Jun-Ichi Itoh[3]

1 Electrical and Computer Engineering, Ryerson University, Toronto, Canada
2 Murata Power Solutions, AC/DC Power Module, Toronto, Canada
3 Nagaoka University of Technology, Niigata, Japan
E-mail: {jafshari, dxu, bwu}@ryerson.ca, {bgong, zyang}@murata.com, itoh@vos.nagaokaut.ac.jp

Abstract- **In this paper, the commutation method and transition from one phase loss to normal operation (three-phase operation) with eight-segment PWM scheme is analyzed for isolated Buck matrix-type rectifier. With the proposed commutation method, a safe transition from one phase loss operation to normal operation and vice versa can occur with minimum switching actions (two-step commutation). All the MOSFET switches are turned ON under zero voltage switching (ZVS) condition in both normal and one phase loss operations. The performance of the converter (output voltage, input grid currents and THD) with a phase loss operation is evaluated and verified by the experimental results obtained from a 5kW prototype.**

I. INTRODUCTION

The three-phase isolated buck matrix-type rectifier is a ZVS converter [1] and the isolated ZVS rectifier is an excellent candidate for the applications in which the SiC MOSFETS are the most favorable device choice. Due to the fact that SiC MOSFET tends to have low turn-OFF loss but high turn-ON loss, SiC MOSFET operating at ZVS can achieve much lower switching losses while it switches at high frequency. The significant benefits of using SiC MOSFET in matrix converters are well proven in [2-3], and all these benefits cannot be obtained with the use of Si IGBT devices instead.

In [4], several commonly practiced PWM schemes used for this topology were investigated and an optimal six-segment PWM scheme (Type A) was proposed and compared with the eight-segment PWM in [1]. The comparison in [4] shows, with the traditional eight-segment PWM scheme, a large output inductor is required to limit the output ripple current, which results in reduction of the overall power density of the converter. The aforementioned drawback is improved by the six-segment PWM schemes proposed in [4] and by eight-segment PWM with non-equally distributed zero-vector intervals which was recently proposed in [5]. With them, the output inductor current ripple of converter is reduced while maintaining low input currents THD.

However, the six-segment and eight-segment PWM schemes for normal operation cannot be applied for one phase lost operation. If this PWM scheme is directly applied to one phase loss operation, the converter will lose output voltage regulation even with significantly large value of output capacitors and the overall current stress is significantly large (this will be further analyzed and compared with PWM derived for one phase loss

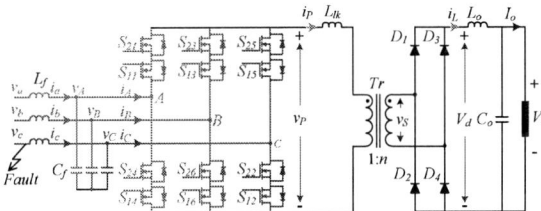

Fig. 1. ZVS three-phase Buck PWM rectifier.

operation in this paper). In addition, the possibility of converter switches failure is high due to the shoot through between switching actions during the transition from one phase loss to normal operation. Therefore, deriving a robust PWM and commutation scheme is indispensable for safe operation of converter during faulty mode operation of one phase loss.

In the recent publication [6], a robust PWM scheme and commutation method is proposed for one phase loss operation of the three-phase isolated buck matrix-type rectifier which allows the continuous and safe operation of the converter to deliver two-third of rated power and to regulate the output voltage with maximum output voltage drop less than 5% of nominal output voltage. However, the commutation scheme derived in [6] is desired for the transition from normal operation with six-segment PWM scheme to one phase loss operation and cannot be directly applied to the transition from normal operation with the eight-segment PWM scheme to one phase loss in order to achieve smooth two step commutation.

Therefore, one of the main focuses of this paper is on developing a commutation method for smooth transitions between the faulty mode of one phase loss operation and normal mode (three-phase operation) with eight-segment PWM scheme.

The final emphasis of this paper thus lies in analysis of the output voltage regulation and overall current stress of the converter when the normal PWM schemes in [4] is applied to one phase loss operation. In addition, comparisons with normal PWM and the desired PWM scheme proposed for one phase loss operation are made.

The rest of this paper is organized as follows. Mathematical analysis for PWM schemes for normal operation which directly applied to one phase loss operation of the converter is provided in section II. The detail analysis of transition from one phase loss to normal operation and a commutation method from one phase loss to normal operation (with eight-segment PWM scheme)

3797

and vice versa are described in Section III. Simulation and experimental results are presented in Section IV and V, followed by the conclusion in Section VI.

II. ANALYSIS OF ONE PHASE LOSS

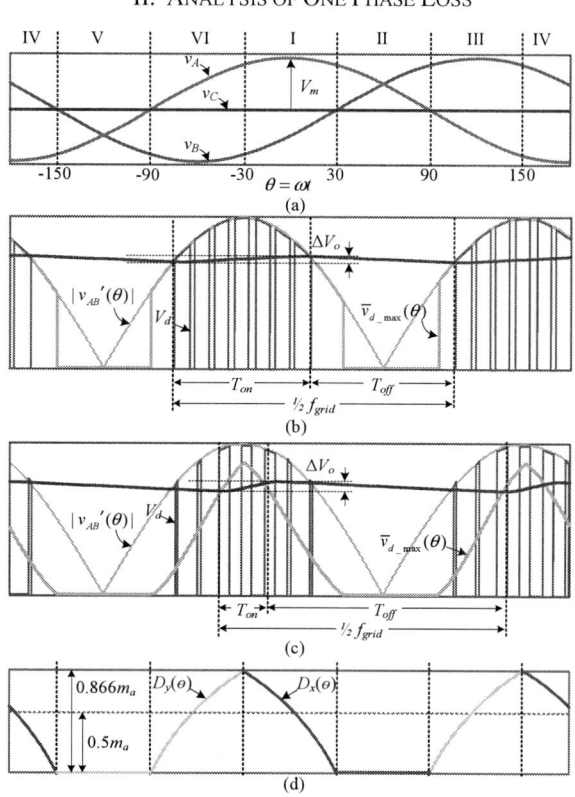

Fig. 2. One phase loss operation of three-phase buck matrix rectifer: (a) three-phase grid voltages with "phase C" is shorted, (b) with fault PWM derived in [6] for one phase loss operation converter, (c) with normal PWM proposed and summarized in [4] applied to one phase loss operaion and (d) the corresponding duty-cyle to V_d in Fig. 2(c).

The operation principle of the three-phase isolated buck matrix rectifier during one phase loss is presented in [6]. It is mentioned in [6], the traditional PWM schemes which are derived for matrix-type rectifier and summarized in [4] cannot be directly applied for one phase loss operation since with them the output voltage drop and required output inductor current for sustaining two-third rated power is noticeably large.

However, this was not mathematically analyzed in [6]. During normal operation, in each sector the output voltage is synthesized from two of the three-phase line-to-line input voltages. For example in sector I, V_o is synthesized from v_{AB} and v_{Ac} so that V_d contains pulses associated with v_{AB} and v_{Ac} in each switching cycle. Due to the rectification on the secondary side, the magnitude of the pulses of V_d is the absolute value of v_{AB} and v_{Ac} with respect to the secondary side of transformer.

Since the converter behavior of our interest in this analysis is slow dynamic compared with switching behavior, the average model of V_d is used instead to simplify the analysis by neglecting the switching events of V_d.

Therefore, the average voltage of V_d in each switching cycle can be expresses as

$$\bar{v}_d(\theta) = \left| v_{AB}'(\theta) \right| D_x(\theta) + \left| v_{AC}'(\theta) \right| D_y(\theta) \qquad (1)$$

where $\left| v_{AB}'(\theta) \right|$ and $\left| v_{AC}'(\theta) \right|$ are the absolute value of v_{AB} and v_{Ac} with respect to the transformer secondary side; $D_x(\theta)$ and $D_y(\theta)$ are the duty cycle of $\left| v_{AB}'(\theta) \right|$ and $\left| v_{AC}'(\theta) \right|$ respectively. $\left| v_{AB}'(\theta) \right|$ and $\left| v_{AC}'(\theta) \right|$ can be expressed as

$$\left| v_{AB}'(\theta) \right| = n\sqrt{3} V_m \left| \sin(\theta + \frac{2\pi}{3}) \right| \qquad (2)$$

$$\left| v_{AC}'(\theta) \right| = n\sqrt{3} V_m \left| \sin(\theta + \frac{\pi}{3}) \right| \qquad (3)$$

With the conventional space vector modulation scheme [4], $D_x(\theta)$ and $D_y(\theta)$ can be expressed as

$$D_x(\theta) = m_a \sin(\frac{\pi}{6} - \theta) \qquad (4)$$

$$D_y(\theta) = m_a \sin(\frac{\pi}{6} + \theta) \qquad (5)$$

where m_a is the modulation index and in the range of $0 \leq m_a \leq 1$.

At steady state, when converter is under regulation, $\bar{v}_d(\theta)$ is equal to output voltage V_o.

During one phase loss operation ("phase C" lost), v_{AC} will not involve in the operation of the converter and V_d contains pulses associated with v_{AB} only. The equation of $\bar{v}_d(\theta)$ in (1) should be revised as

$$\bar{v}_d(\theta) = \left| v_{AB}'(\theta) \right| D_x(\theta) = \left| v_{AB}'(\theta) \right| m_a \sin(\frac{\pi}{6} - \theta) \qquad (6)$$

Based on the conventional SVM, either $D_x(\theta)$ or $D_y(\theta)$ is selected for the duty cycle of V_d in different sectors as shown in Fig. 2(d). In sector II and V, $D_x(\theta)$ and $D_y(\theta)$ are zero.

It is important to note that the converter is of buck type and will lose regulation when the maximum available $\bar{v}_d(\theta)$ is lower than V_o due to the sinusoidal shape of v_{AB}. Modulation scheme also plays a very important role to determine the maximum available $\bar{v}_d(\theta)$.

With the conventional space vector modulation scheme $\bar{v}_d(\theta)$ reaches to maximum at $m_a = 1$. By substituting $\left| v_{AB}'(\theta) \right|$ with (2) and m_a with 1 into (6), after elementary trigonometric transformations, the maximum available $\bar{v}_d(\theta)$ is derived as

$$\bar{v}_{d\_max} = n V_m (\frac{3}{4} - \frac{\sqrt{3}}{2} \sin(2\theta)) \qquad (7)$$

The circuit principal waveforms within one gird side cycle with excessively increased switching period of PWM when phase voltage v_c is shorted can be observed in Fig. 2. The resultant $\bar{v}_{d\_max}(\theta)$ from (7) is shown in Fig. 2(c). During the interval T_{on} when $\bar{v}_{d\_max}(\theta)$ is higher than V_o, the converter can gain the regulation to regulate V_o back to the set point, while during the interval T_{off} when $\bar{v}_{d\_max}(\theta)$ is lower than V_o the converter output rectifier is reverse blocked and there is no power can be delivered to the load side so that V_o drops since V_o is sustained only by the output capacitors. In sector II and V, $\bar{v}_{d\_max}(\theta)$ is zero since the converter stop switching due to the very low magnitude of $\left|v_{AB}'(\theta)\right|$. The voltage drop ΔV_o is the function of T_{off}, output capacitance C_o, and load current I_o, and can be derived as

$$\Delta V_o = \frac{I_o T_{off}}{C_o}. \qquad (8)$$

As shown in Fig. 2(c), T_{off} is the distance between the two adjacent crossing points when $\bar{v}_{d\_max}(\theta)$ is lower than V_o. Assuming that the ripple voltage ΔV_o is relatively small compare with V_o, the location of these two adjacent crossing points can be found by equaling $\bar{v}_{d\_max}(\theta)$ with V_o. Since the output voltage can be expressed as $V_o = \frac{3}{2} n V_m m_a$ during normal operation, the angle θ at the crossing point can be estimated by

$$\theta = \frac{1}{2}\sin^{-1}\left(\frac{\sqrt{3}}{2} - \sqrt{3}m_a\right) \qquad (9)$$

Then, the T_{off} can be expressed as

$$T_{off} = \frac{\pi - \left(\theta + \frac{2\pi}{3}\right)}{\pi f_{grid}} = \frac{\frac{1}{2}\sin^{-1}\left(\sqrt{3}m_a - \frac{\sqrt{3}}{2}\right) + \frac{\pi}{3}}{\pi f_{grid}} \qquad (10)$$

where f_{grid} is the grid frequency.

Finally, the voltage drop ΔV_o can be derived by substituting T_{off} with (10) into (8):

$$\Delta V_o = \frac{I_o\left[\frac{1}{2}\sin^{-1}\left(\sqrt{3}m_a - \frac{\sqrt{3}}{2}\right) + \frac{\pi}{3}\right]}{C_o \pi f_{grid}} \qquad (11)$$

Since the output capacitor is discharged by I_o during T_{off} and based on the current-second balance of the output capacitor, considering the maximum load current of one phase loss operation at $I_o = 2/3 I_{rated}$, the minimum required inductor current I_{clamp_min} to regulate the output voltage can be described as:

$$I_{clamp\_min} = \frac{2I_{rated}}{1 - \frac{3}{\pi}\sin^{-1}\left(\sqrt{3}m_a - \frac{\sqrt{3}}{2}\right)} \qquad (12)$$

With the normal PWM schemes, it is difficult to regulate the output voltage, because the maximum available duty cycle of the pulses with v_{AB} is limited by the PWM schemes for the normal operation.

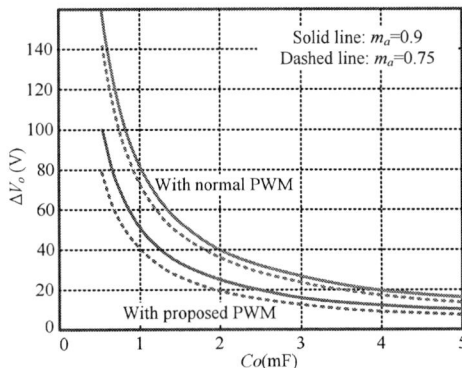

Fig. 3. The output voltage drop ΔV_o versus C_o in one phase loss operation at $I_o = 2/3 I_{rated}$ and $f_{grid} = 50$ Hz.

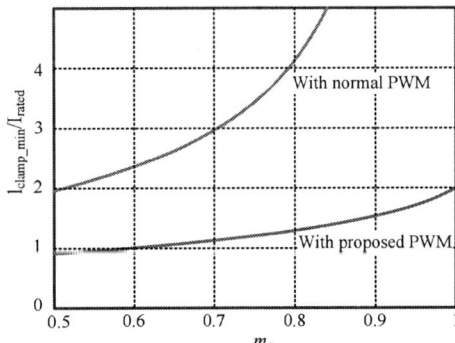

Fig. 4. The over current ratio I_{clamp_min}/I_{rated} versus C_o in one phase loss operation at $I_o = 2/3 I_{rated}$ and $f_{grid} = 50$ Hz.

In [6], a desired modulation scheme to maximize the available rectifier output voltage $\bar{v}_{d\_max}(\theta)$ for three phase matrix converter during one phase loss operation is proposed. The circuit principal waveforms within one gird side cycle with excessively increased switching period of PWM when phase voltage v_c is shorted can be observed in Fig. 2(b). As shown in Fig. 2(b), operation of the converter is equivalent to a ZVS full-bridge phase-shifted (FB-PS) dc-dc converter [7] with input voltage of v_{AB}. The output voltage V_o is regulated by directly adjusting the duty cycle of V_d. Therefore, the average voltage of V_d can be expresses as

$$\bar{v}_d(\theta) = \left|v_{AB}'(\theta)\right| D(\theta) \qquad (13)$$

where $D(\theta)$ is the duty cycle of V_d in the range of $0 \leq D(\theta) \leq 1$.

Then $\bar{v}_d(\theta)$ reaches to maximum at $D(\theta) = 1$. Neglecting the duty cycle loss, the maximum $\bar{v}_d(\theta)$ can be expressed as

$$\bar{v}_{d\_max}(\theta) = \left|v_{AB}'(\theta)\right| \qquad (14)$$

In this case, the maximum available voltage $\bar{v}_{d\_max}(\theta)$ has the same magnitude as $\left|v_{AB}'(\theta)\right|$ which is much higher than that with normal PWM schemes as shown in Fig. 2(c). The resultant voltage drop can be significantly reduced due to shorter

3799

interval of T_{off}. Fig. 3 shows the curve of voltage drop ΔV_o vs C_o with normal PWM and proposed PWM at different m_a. It can be concluded that the required output capacitance with the proposed PWM can be much lower than that with normal PWM scheme to meet the same voltage drop. It also shows that higher m_a results in larger voltage drop. It should be noted that m_a here refers to the modulation index of normal operation prior to the fault.

The minimum required inductor current I_{clamp_min} to regulate the output voltage are compared in Fig. 4 at different m_a. With the proposed PWM, current stress can be remarkably smaller to regulate the output voltage during one phase loss operation.

III. ANALYSIS OF TRANSITION AND COMMUTATION

A) Analysis of Transition from One Phase Loss to Normal Operation with Eight-Segment PWM Scheme

During normal operation, the vector sequence of eight-segment PWM in every sampling period can be divided into the sequence of $\bar{I}_{x+}, \bar{I}_0, \bar{I}_{x-}, \bar{I}_0, \bar{I}_{y+}, \bar{I}_0, \bar{I}_{y-}, \bar{I}_0$ as described in [5]. Then the dwell time for each vector will be $T_x/2, T_{01}, T_x/2, T_{02}, T_y/2, T_{03}, T_y/2$ and T_{04} respectively, where T_{01}, T_{02}, T_{03} and T_{04} are four intervals for zero-vector $\bar{I}_0$ [5]. $\bar{I}_{x+}$ and $\bar{I}_{x-}$ represent switching states of vector $\bar{I}_x$ when $i_P > 0$ and $i_P < 0$ respectively. $\bar{I}_{y+}$ and $\bar{I}_{y-}$ represent switching states of vector $\bar{I}_y$ when $i_P > 0$ and $i_P < 0$ respectively. During the dwell time of active vectors, the transformer primary voltage is one of the line-to-line voltages, while during the dwell time of zero vectors, one of the phase legs is used to bypass the transformer current and the transformer primary voltage is zero. When one phase is lost ("phase C" is lost), the three-phase converter can be redrawn as Fig. 5 since the switches on the phase "leg C" stop switching. Within every 180° interval of the one phase loss operation the three-phase converter is operated as a ZVS FB-PS converter.

During this operation eight of the twelve switches are involved and the switches connected to the leg with phase loss are not operating. Among these eight switches, four switches operate as active switches and the other four switches operate as synchronous rectification switches to reduce the conduction losses. For example, during the 180° interval where the voltage potential v_A is higher than v_B ($v_A > v_B$), the switches S_{14}, S_{21}, S_{13} and S_{26} of bridge are synchronous rectification (SR) switches and can be kept on all the time since their body diodes are forward biased. The rest four switches ($S_{11}, S_{24}, S_{23}, S_{16}$) operate in a same manner as FB-PS converter. Similarly, during the other 180° interval where the voltage potential v_B is higher than v_A ($v_B > v_A$), the switches S_{23}, S_{16}, S_{11} and S_{24} operates as SR switches that can be kept on all the time. The rest of four switches ($S_{21}, S_{14}, S_{13}, S_{26}$) are active switches and operate in a same manner as FB-PS converter. The corresponding switch gate signals during one switching period T_s is shown in Fig. 6. In order to facilitate a smooth two-step transition from one phase loss operation

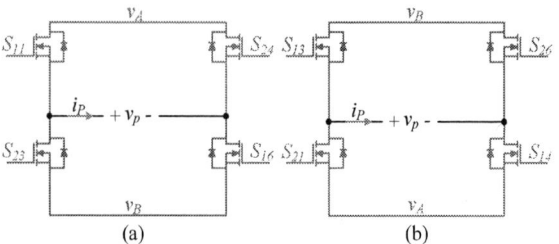

(a)   (b)

Fig. 5. Three-phase converter equivalent circuit of one phase loss (a) during interval $v_A > v_B$ (b) during interval $v_A < v_B$.

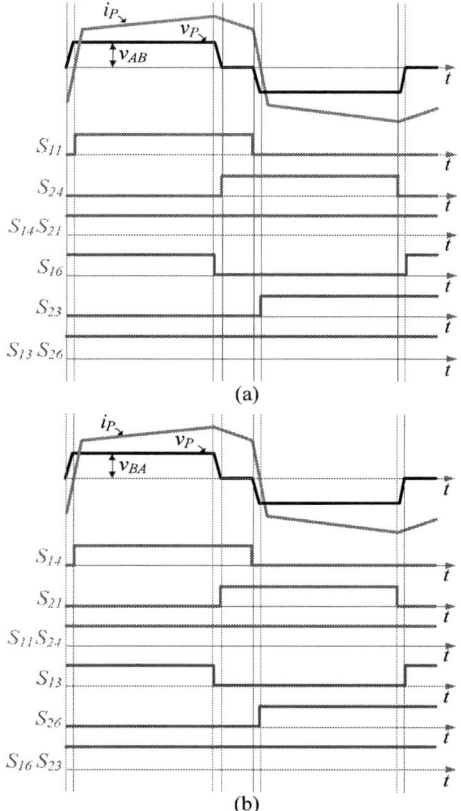

Fig. 6. Three-phase converter operation with one phase lost during interval of: (a) sector I and sector VI ($v_A > v_B$) with "phase B" for bypassing time at the end of the switching cycle, (b) sector III and sector IV ($v_B > v_A$) with "phase B" for bypassing time at the end of the switching cycle.

to normal operation (with eight-segment PWM scheme), there should be a phase leg involved in both the switching state prior to the transition and the switching state after the transition. As shown in Fig. 6, the phase "leg B" is used for the bypassing interval at the end of each switching cycle when the "phase C" is lost or shorted, since the phase "leg B" is the common working leg in sector I, III, IV and VI. Then the transition from the end of the switching cycle of the one phase loss operation to the beginning of the switching cycle of the normal operation is a smooth two-step transition. Fig. 7 shows the PWM gate signals and the transition from one phase loss to normal operation in sector III (a).

3800

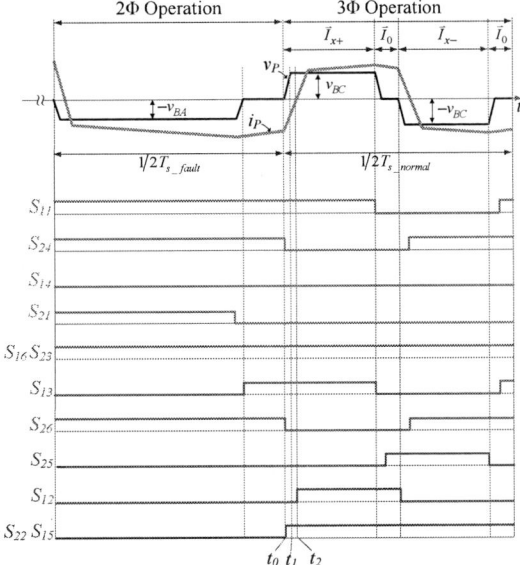

Fig. 7. Circuit waveforms: Primary voltage and current and corresponding switch gate signals for the transition one phase loss to normal operation (with eight-segment PWM scheme) in sector III (a).

At t_0^-, the converter operates in one phase loss operation and the switching state is at the zero state where the phase "leg B" is used for by passing (all the switches of phase "leg B" are ON) and the current i_P is circulating through the primary side of transformer.

The voltage across transformer, v_P is zero and both voltage potential at the two terminals of the transformer v_1 and v_2 is equal to v_B. The first step of the two-step commutation is to turn ON switches S_{15}, S_{22} and turn OFF switches S_{24}, S_{26} at t_0^+. The switches S_{15}, S_{22} are turned ON at zero voltage switching (ZVS) condition for synchronous rectification since v_C has the lowest potential and their body diodes are forward biased.

Switch S_{24} is turned OFF at zero current. Turning OFF S_{26} will break the circulating path for transformer current i_P and, due to the leakage inductance L_{lk}, current i_P will be forced to charge or discharge the output capacitance of S_{24}, S_{14}, S_{26} and S_{12}. As a result, the voltage v_2 is going down until, at t_1, v_2 reaches to v_C so that the body diode D_{12} of switch S_{12} conducts. Then, S_{12} is turned ON under ZVS condition at t_2 before i_P changes the direction as shown in Fig. 7. When i_P ramps up to the same value as output inductor current with respect to the primary side, this transition completes. This is a smooth transition with easy implementation.

B) Proposed Commutation Method

It is important to mention that during normal operation, there are always two synchronous rectification switches in one of the phase legs where the phase voltage has the value between the other two phases cannot be kept always on. They should be complimentary with other two active switches in another phase leg to avoid short circuit between these two phases [1]. Different two switches will be selected to apply this constraint when current

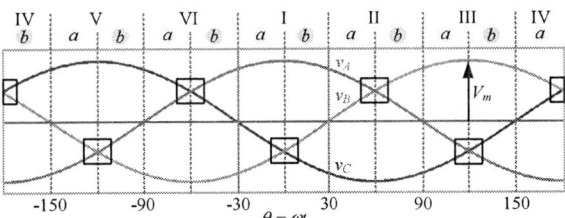

Fig. 8. Input phase voltages with sector division.

reference vector move from one sector to another sector and from part a to part b within each sector. However, in the real three-phase system, the phase voltages might be distorted or unbalanced such that in the vicinity of the boundary between part a and part b, the voltage difference between the two phase voltages (where the difference between the two phase voltages crossing are small as shown in Fig. 8 in rectangular box) may have reversed polarity compared with ideal case and cause short circuit between them. Therefore, both constraints for part a and part b need to be applied in the vicinity of the boundary between part a and part b. These constraints in different sectors are summarized in Table. I.

The finite commutation state machines between normal operation (with eight-segment PWM Scheme) and one phase loss operation with the proposed commutation method are shown in Fig. 9.

The transition states (dash-circle) are added between main states (in shaded color) to achieve ZVS and synchronous rectification operation. By properly selecting the commutation state machine for one phase loss operation in different sector, two-step commutation with ZVS turn-ON is achieved for both transitions from normal operation to one phase loss operation and vice versa.

It is worth mentioning that the commutation scheme derived for normal operation with eight-segment PWM

TABLE I
SR OPERATION IN THE VICINITY OF MIDDLE OF SECTOR

Vicinity of middle of sector	Complimentary Switches	
	SR	Active
I	S_{13}	$\rightarrow$	$\bar{S}_{25}$
	S_{26}	$\rightarrow$	$\bar{S}_{12}$
	S_{15}	$\rightarrow$	$\bar{S}_{23}$
	S_{22}	$\rightarrow$	$\bar{S}_{16}$
III	S_{15}	$\rightarrow$	$\bar{S}_{21}$
	S_{22}	$\rightarrow$	$\bar{S}_{14}$
	S_{11}	$\rightarrow$	$\bar{S}_{25}$
	S_{24}	$\rightarrow$	$\bar{S}_{12}$
IV	S_{16}	$\rightarrow$	$\bar{S}_{22}$
	S_{23}	$\rightarrow$	$\bar{S}_{15}$
	S_{12}	$\rightarrow$	$\bar{S}_{26}$
	S_{25}	$\rightarrow$	$\bar{S}_{13}$
VI	S_{12}	$\rightarrow$	$\bar{S}_{24}$
	S_{25}	$\rightarrow$	$\bar{S}_{11}$
	S_{14}	$\rightarrow$	$\bar{S}_{22}$
	S_{21}	$\rightarrow$	$\bar{S}_{15}$

The 2018 International Power Electronics Conference

scheme permits all the switches of the converter work under ZVS condition. This may be considered as an advantage over the six-segment PWM scheme where two

of the switches in each sector are non-ZVS although the associated turn-ON loss for these two switches is very small as discussed in [4].

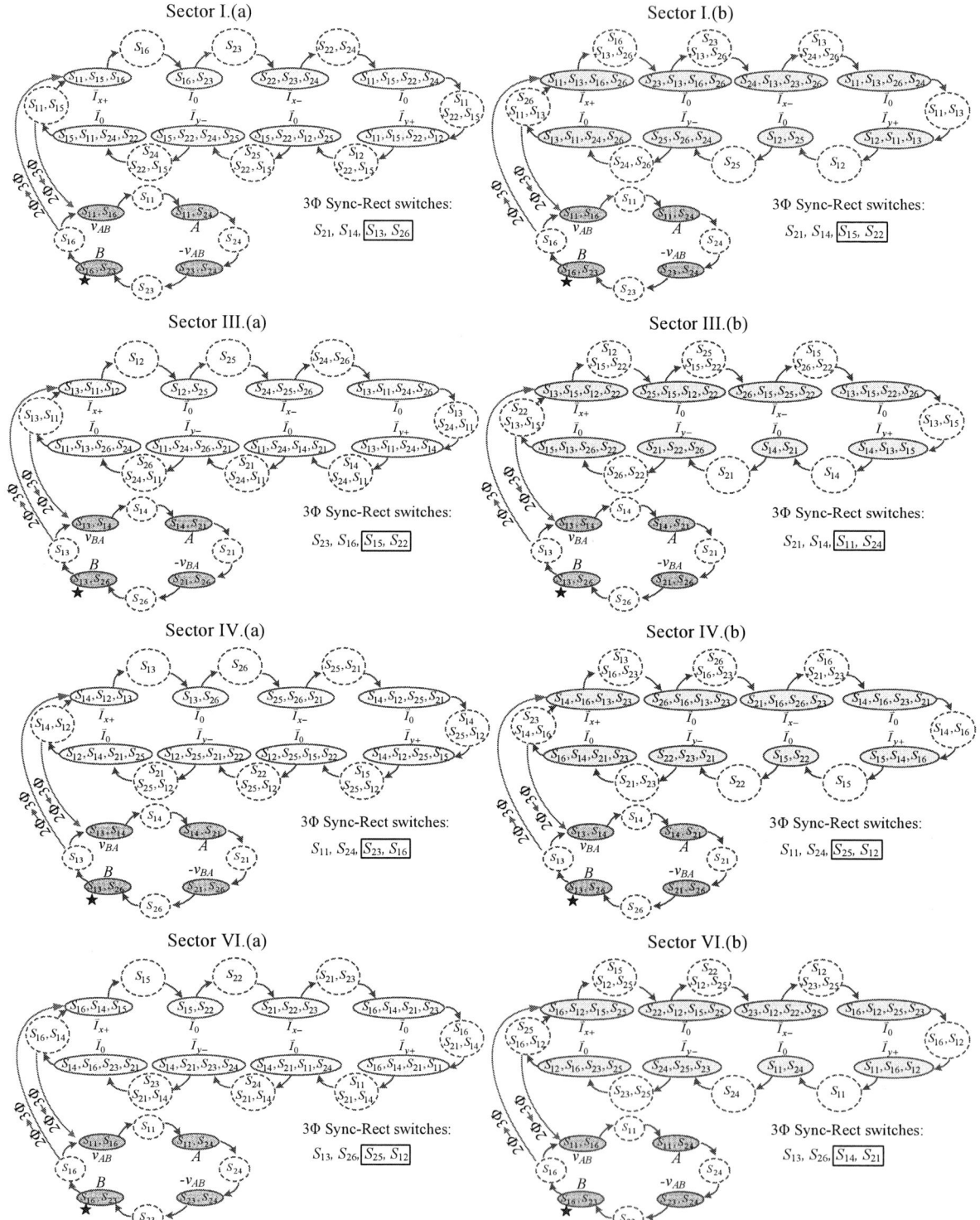

Fig. 9. Commutation state machine for normal operation (3Φ) and one phase loss (2Φ): two-step commutation realized for one phase loss operation, normal operation, transition from one phase loss to normal operation and from normal operation to one phase loss operation. (Note: star represents the end of switching cycle of one phase loss operation. Note: the switches in the rectangular boxes are the synchronous rectification switches for 3Φ operation and constraints should be applied on them   in the the vicinity of the boundary between part a and part b per Table. I.)

3802

IV. SIMULATION RESULTS

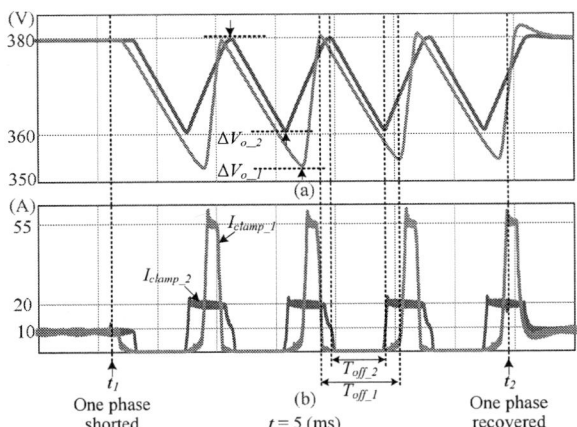

Fig. 10. Comparison of output voltage drop ΔV_o for case 1) at $m_a = 0.8$ and case 2) at $m_a = 0.9$ with $f_{grid} = 60$ Hz.

To verify the analysis of one phase loss operation of the converter, a simulation model is built and tested at two-third of rated power. In the simulation, the output voltage drop of the converter and the required upper limit of the average output inductor current for one phase loss operation is studied for two cases: "case 1" applying normal PWM schemes in [4] to one phase loss operation and "case 2" desired PWM scheme in [6] for one phase loss operation. The output storage energy $C_o = 2$ mF is selected based on the analysis given in [6]. As shown in Fig. 3 and Fig. 4, the worst case for ΔV_o and I_{clamp} happened at the highest m_a. Therefore $m_a = 0.9$ is chosen for "case 2" in the simulation. However, for "case 1", the converter cannot regulate the output voltage even with considerably large I_{clamp} when m_a is high because the effect of duty cycle loss and the current control response time cannot be neglected when current is very large. Therefore, $m_a = 0.8$ is chosen for "case 1" in the simulation. Fig. 10 shows the output voltage drops (ΔV_{o_1} and ΔV_{o_2}) and the output inductor current clamp value (I_{clamp1} and I_{clamp2}) for "case 1" and "case 2" at $f_{grid} = 60$ Hz. The output inductor current clamp, I_{clamp_1} is significantly large with "case 1", which results in large current stress on the main components of converter and makes the converter operating in one phase loss condition not feasible. With the PWM scheme in "case 2" the output voltage drop is lower and the output inductor current is significantly smaller. The over current ratio (I_{clamp_2}/I_{rated}) for case 2 is less than 1.5 which is normally acceptable in the real design.

V. EXPERIMENTAL WAVEFORMS

Experimental test is conducted on a 5kW matrix converter. Detail experimental system parameters are listed in Table. II. The experimental waveforms in Fig. 11 illustrate the operation when one phase is shorted and then recovered. For the experimental in here, the output power is reduced to 3.3 kW (two-third of rated power) in order to limit the "rms" value of phase currents $i_{a,b,c}$ to 1.2 times of the rated phase currents. As shown in Fig. 11, a smooth transition occurs at two intervals t_2 and t_4 when "phase C" is shorted and then recovered. The large inrush

current seen in grid currents in Fig. 11(f) is due to the input EMI capacitors. As shown in Fig. 11, the transformer primary voltage is clean and no large spike is noticeable due to ZVS operation of the MOSFET switches. It should be noted that during normal operation, the eight-segment PWM with non-equally distributed zero-vector intervals is employed here due to its low output inductor current ripple [5]. The low current ripple is also beneficial for current sensing to implement average current mode control since the sampled current in the middle of pulses can accurately represent the average current when used as the current feedback for current control loop, which will result in low input current THD. The output voltage is tightly regulated at 380 V in normal operation and the output voltage drop ΔV_o is less than 5% of nominal output voltage during one phase lost operation. A reasonable value of 2.0 mF is selected for output storage capacitance C_o as a compromise between the voltage drop during one phase loss operation and power density of the converter. The total harmonic distortion of the input phase currents is around 1.94% for normal operation and 38.6% for one phase loss operation.

TABLE II
EXPERIMENTAL PROTOTYPE PARAMETERS

Prototype Parameter	Value
C_f	5 µF
L_f	110 µH
$v_{LL,rms}$	480 V
f_{grid}	60 Hz
C_o	2 mF
L_o	315 µH
V_o	380 V
L_{lk}	9.7 µH
S_{11}-S_{26}	SCT3080KL
D_1-D_4	SCS215KG
n	0.86
T_r	Ferrite core (ZP47313TC)
f_{sw}	100 kHz

VI. CONCLUSION AND FUTURE WORK

In this paper, a commutation scheme for the transition from one phase loss operation to normal operation with eight-segment PWM scheme and vice versa is proposed. With the proposed commutation scheme, two-step commutation under ZVS turn-ON for all the MOSFET switches of the converter can be realized in both one phase loss operation and normal operation. Based on the simulated and experimental results obtained from a 5 kW prototype, it is shown that the maximum available average voltage envelop on the rectifier side can be achieved with the desired PWM scheme proposed for one phase loss operation which results in significantly lower overall current stress on the converter and output voltage drop compared with the case when the normal PWM schemes are applied to one phase loss operation. The input currents THD obtained during one phase loss operation is relatively high and may not be acceptable for the system that might even run for days. Therefore, how to improve the THD during one phase loss operation is considered as important topic of the future work.

3803

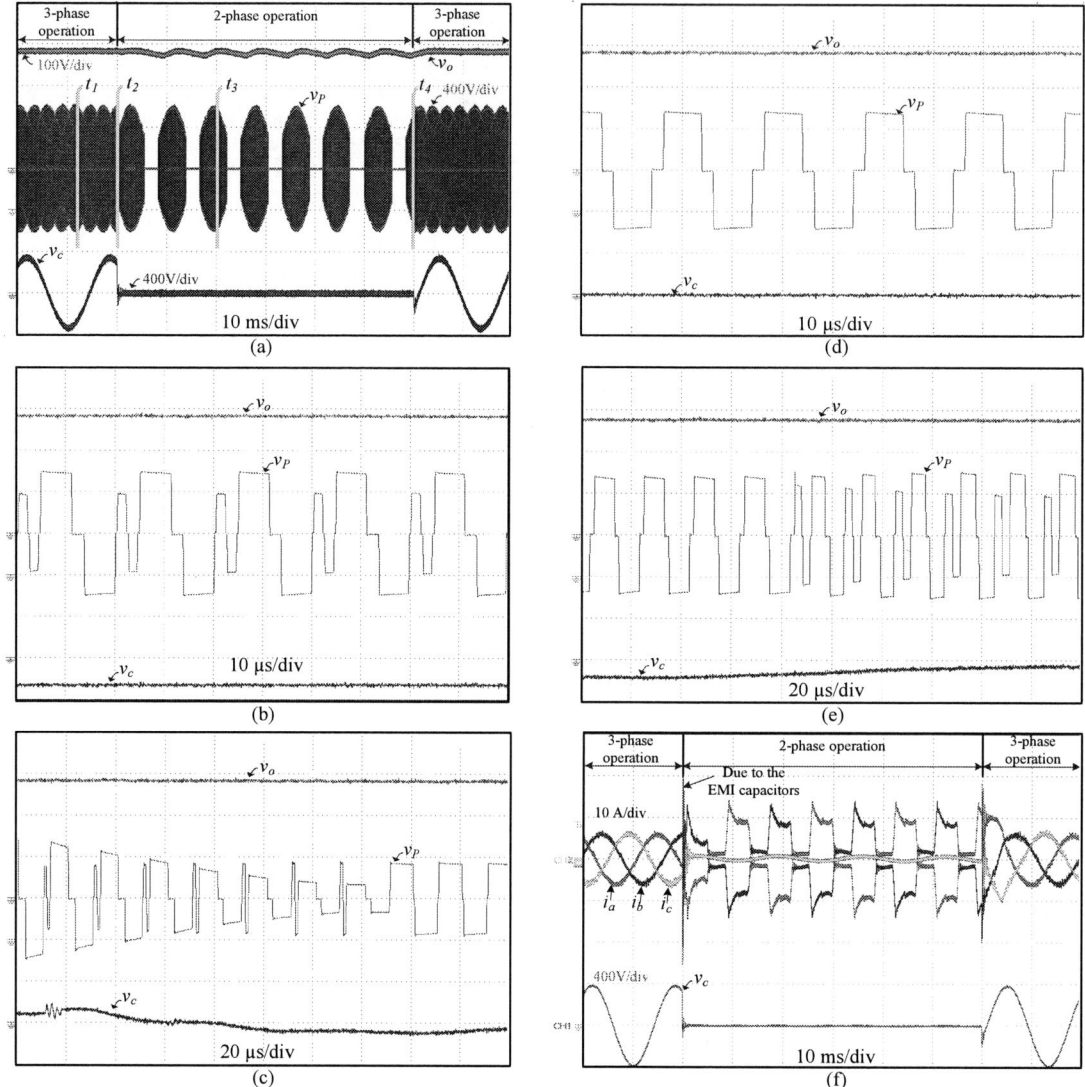

Fig. 11. Experimental waveforms at $2/3Po\_max$, $v_{LL} = 480$ V, $f_{grid} = 60$ Hz: (a) normal operation (3Φ) to one phase loss operation (2Φ), (b) at t_1, normal operation, (c) at t_2, instant "phase C" is shorted, (d) at t_3, one phase loss operation, (e) at t_4, instant "phase C" recovered, (f) input phase currents and voltage waveforms of i_a, i_b, i_c and v_c.

REFERENCES

[1] V. Vlatković and D. Borojević, and F. C. Lee, "A Zero-Voltage Switched, Three-phase Isolated PWM Buck Rectifier", IEEE Trans. Power Electronics, vol. 10, No. 2, pp. 148-157, Mar., 1995.

[2] S. Safari, A. Castellazzi, and P. Wheeler, "Experimental and analytical performance evaluation of SiC power devices in the matrix converter," *IEEE Trans. Power Electron.*, vol. 29, no. 5, pp. 2584–2596, May 2014.

[3] K. Koiwa and J.-I. Itoh, "A maximum power density design method for nine switches matrix converter using SiC-MOSFET," *IEEE Trans. Power Electron.*, vol. 31, no. 2, pp. 1189–1202, Feb. 2016.

[4] J. Afsharian, D. Xu, B. Wu, B. Gong, and Z. Yang " The Optimal PWM Modulation and Commutation Scheme for Three-Phase Isolated Buck Matrix Type Rectifier," *IEEE Transactions on Power Electronics (Early Access Articles)*, 2017.

[5] J. Afsharian, D. Xu, B. Wu, B. Gong, and Z. Yang Submitted to ECCE 2017 Conference "Improved Eight-Segment PWM Scheme with Non-Equally Distributed Zero-Vector Intervals for a Three-Phase Isolated Buck Matrix-Type Rectifier".

[6] J. Afsharian, D. Xu, B. Wu, B. Gong, and Z. Yang "A NEW PWM and Commutation Scheme for One Phase Loss Operation of Three-Phase Isolated Buck Matrix-Type Rectifier" *IEEE Transactions on Power Electronics (Early Access Articles)*, TPEL.2018.2789905, 2018.

[7] J. Sabaté, V. Vlatković, R. B. Ridley, F. C. Lee, and B. H. Cho, "Design considerations for high-voltage high power full-bridge zero-voltage-switched pwm converter," *IEEE Applied Power Electronics Conf. (APEC) Proc.*, 1990, pp. 275-284.

Novel Isolated Bidirectional Integrated Dual Three-Phase Active Bridge (D3AB) PFC Rectifier

F. Krismer, E. Hatipoglu, and J. W. Kolar

Power Electronic Systems Laboratory, ETH Zurich, Switzerland

krismer@lem.ee.ethz.ch

Abstract—This Paper proposes a novel Dual Three-Phase Active Bridge (D3AB) PFC rectifier topology for a 400 V dc distribution system, which features galvanic isolation, bidirectional power conversion capability, a high level of component integration, and can be dimensioned with respect to high efficiency. In the course of a comprehensive and in-depth analytical investigation, the working principle of the D3AB PFC rectifier is described in order to enable converter modelling and the derivation of mathematical expressions and limitations needed for converter design and optimization. The developed converter models are verified by means of circuit simulations. An overall optimization of a system with 400 V line-to-line input voltage, 400 V dc output, and $P_{out} = 8$ kW rated power with respect to efficiency and power density reveals the feasibility of a full-load efficiency of 98.1% and a power density of $4 \, \text{kW}/\text{dm}^3$ if SiC MOSFETs are used. The finally presented design is found to achieve efficiencies greater than 98 % for $P_{out} > 1.7$ kW.

I. INTRODUCTION

Recent efforts with regard to a more sustainable electric power generation propose the installation of distributed dc microgrids in order to effectively utilize distributed renewable energy sources [1]. A dc microgrid architecture typically incorporates dc sources (e.g. photovoltaic, fuel cell), energy storages (e.g. batteries), and loads (e.g. household appliances, IT equipment, electric vehicles) and employs an isolated bidirectional rectifier system to establish energy transfer between the dc grid and the three-phase ac mains.

This paper evaluates a novel topology of a grid-connected, bidirectional, and isolated three-phase power factor corrected (PFC) rectifier with a rated power of $P_{out} = 8$ kW and further specifications as listed in **Tab. I**, which fulfills the requirements of bidirectional conversion capability and galvanic isolation with very low complexity. Due to the versatility of the proposed system, it is suitable for various further

applications including PFC rectifiers for common dc bus architectures as, for example, used in efficiency-optimized multi-axis drive systems [2], where it is reasonable to consider advanced rectifier topologies to define the electric potential of a dc terminal, include a battery to buffer outages, etc. Furthermore, the system could e.g. be implemented for battery chargers of plug-in hybrid electric vehicles [3].

Conventional realizations of three-phase and isolated ac–dc rectifiers are two-stage solutions, with grid-side rectifiers and series-connected isolated dc–dc converters [4], [5], [6]. Two-stage converter systems feature the advantages of decoupled functional parts, at the cost of higher expected losses due to the high number of power components in the current path. State-of-the-art research with regard to more efficient converter topologies reveals various solutions that combine PFC functionality, galvanic isolation, and voltage conversion in a single stage. In this context, isolated single-stage PFC rectifiers, based on isolated Swiss-forward or matrix-type topologies [7], [8], [9], represent suitable but complex solutions. With regard to reduced converter complexity, a direct connection of the high frequency (HF) transformer of a single-phase dc–dc converter to a three-phase PFC rectifier is proposed in [10], which, due to the asymmetry of the converter, is considered more viable for lower power levels. Further level of integration is achieved with a topology with coupled input inductors proposed in [11]. There, the isolated dc port is immediately coupled at the ac port in order to reduce the number of power components in the current path and achieve increased efficiency. The required coupled inductors and the high number of IGBTs (24), though, render the presented converter structure comparably complex.

TABLE I: Specifications of the D3AB PFC rectifier.

Nominal mains line-to-phase voltage (rms value)	$V_{ac} = 230$ V
Mains frequency	$f_m = 50$ Hz
Nominal output dc voltage, port 1 (not isolated)	$V_{dc1} = 800$ V
Nominal output dc voltage, port 2 (galv. isolated)	$V_{dc2} = 400$ V
Nominal output power	$P_{out} = 8$ kW

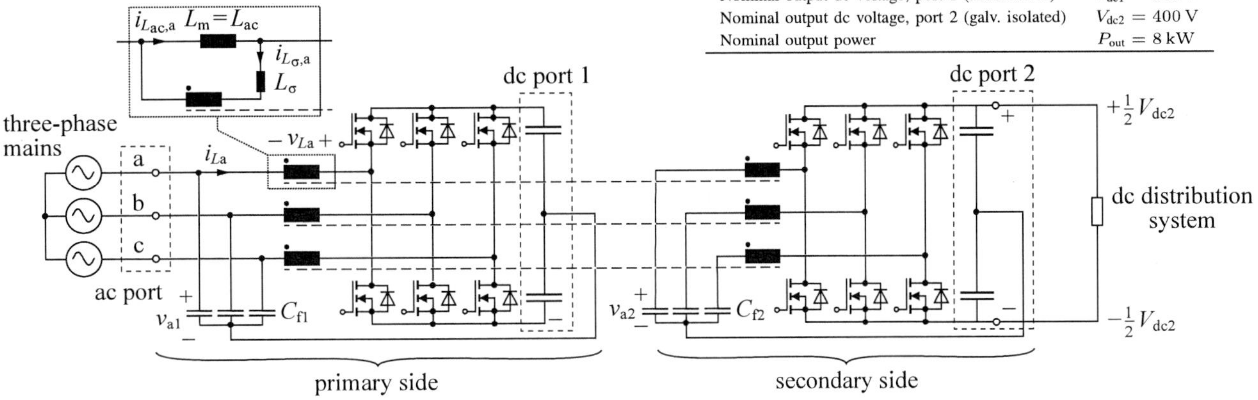

Fig. 1: Proposed bidirectional converter topology with a three-phase ac input port, a dc output port 1 and an isolated dc output port 2.

This paper proposes a novel isolated bidirectional Dual Three-Phase Active Bridge (D3AB) PFC rectifier topology, depicted in **Fig. 1**, that aims for high level of integration and, for this, integrates the functionalities of PFC inductors and HF transformers. Thus, the system essentially combines the functionalities of a bidirectional three-phase PFC rectifier and a three-phase Dual Active Bridge (DAB) converter and provides three power ports, i.e, the ac (input) port, a dc (output) port without isolation, and an isolated dc (output) port. With a total of twelve power MOSFETs, and since the two-level three-phase rectifier structure facilitates the use of conventional 6-pack power modules, the proposed structure features reduced realization complexity compared to state-of-the-art solutions. The paper is organized as follows. **Section II** presents a comprehensive description of the proposed converter system and, for this purpose, first investigates the properties of a single-phase version of the system and then extends the analysis to the three-phase system. Subsequently, **Section III** summarizes results of a converter optimization with respect to efficiency and power density. **Section IV**, finally, evaluates a selected design of the D3AB PFC rectifier with regard to losses, volumes, and efficiency in order to assess the suitability of the proposed concept for the given application.

II. Operating principle

The input inductors, L_{ac}, of a conventional three-phase PFC rectifier without galvanic isolation (and filter capacitors connected to the dc output midpoint, i.e., the primary-side part of the converter depicted in Fig. 1), are subject to an inductor voltage, v_L, with nearly zero local average value, $\langle v_L \rangle \approx 0$ (angle brackets denote the average over one switching period), but large HF spectral components at the switching frequency ($f_s = 35\,\text{kHz}$) and multiples thereof. For the purpose of illustration, **Fig. 2(a)** depicts the instantaneous and local average values of the voltage across the input inductor of phase a over one mains period for rated operation according to Tab. I and **Fig. 2(b)** shows the corresponding spectrum. The proposed isolated converter topology is derived based on the idea that the input inductors are replaced by transformers in order to take advantage of the applied HF voltages. The secondary-side windings are connected to a second three-phase PFC rectifier, which, in combination with the transformers' stray inductances L_σ, realizes a converter structure similar to a three-phase DAB converter. It is worth to note that Fig. 1 illustrates only one possible realization of the D3AB PFC rectifier. Variations of this concept may only use differential or common mode voltage components for energy transfer and/or employ different three-phase transformers, e.g., with delta-connected windings on the secondary side [12].

In this work, the corresponding power transistors at the converter's primary and secondary sides operate with same duty cycles. For this reason,

$$\frac{v_{\{a,b,c\}1}}{V_{dc1}} = \frac{v_{\{a,b,c\}2}}{V_{dc2}} \tag{1}$$

applies. The gate signals of the six corresponding primary- and secondary-side transistors are subject to a common phase shift, φ, in order to facilitate output power control at dc port 2.

Based on the assumption that the capacitances of the split dc-link are sufficiently large to achieve negligible fluctuations of the dc-link capacitor voltages, the analysis can be confined to the single-phase system with separated input inductor and HF transformer shown in **Fig. 3**. With this modification

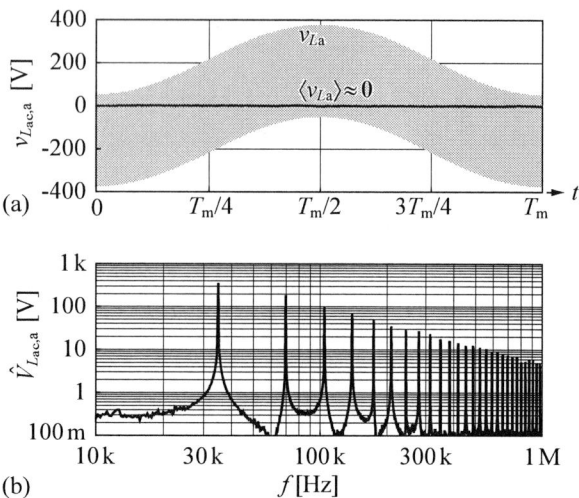

Fig. 2: (a) Instantaneous and local average values of the voltage across L_{ac}; (b) spectrum of v_{La} (operating parameters according to Tab. I, $f_s = 35\,\text{kHz}$).

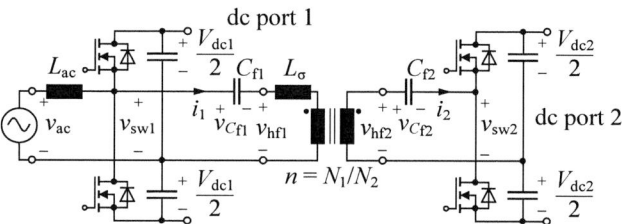

Fig. 3: Single-phase system with input inductor L_{ac}, primary- and secondary-side low-frequency blocking and filter capacitors C_{f1} and C_{f2}, and HF transformer.

a comprehensible description of the operating principle is feasible and the derived results can be directly applied to the three-phase converter of Fig. 1. The converter of Fig. 3 has been documented with regard to dc–dc operation, for $D_1 = D_2 = 0.5$ [13] and arbitrary (but equal) duty cycles, $0 < D = D_1 = D_2 < 1$ [14]; its extension to a dc–dc converter with a three-phase HF dc-link is investigated in [15]. Section II-A revisits its main operating principles for dc–dc operation in order to allow for a comprehensible extension to ac–dc operation of single- and three-phase PFC rectifier systems in Sections II-B and II-C, respectively.

A. Single-phase system at dc–dc operation

Fig. 4 illustrates voltage and current waveforms simulated for the single-phase converter system of Fig. 3 at a selected operating point. The ac input voltage, v_{ac}, changes slowly with respect to the switching period, T_s, and can thus be considered as constant during one switching period. The converter features four degrees of freedom for the control of the output power, i.e., switching frequency, f_s, input- and output-side duty cycles D_1 and D_2, and phase shift angle, φ.

The filter capacitors C_{f1} and C_{f2} are blocking the dc voltage components, V_{Cf1} and V_{Cf2}, in order to avoid saturation of the transformer core. Thus, the HF voltages,

$$v_{hf1} = v_{sw1} - V_{Cf1}, \tag{2}$$
$$v_{hf2} = v_{sw2} + V_{Cf2}, \tag{3}$$

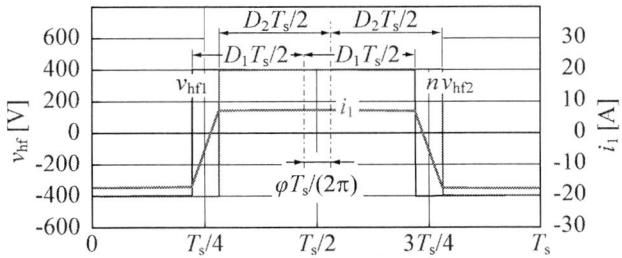

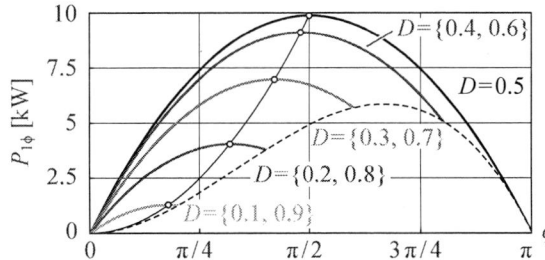

Fig. 4: Definitions of D_1, D_2, and φ using simulated waveforms of v_{hf1}, v_{hf2}, and i_1 over one switching period, T_{s}. Considered operating conditions: $f_{\mathrm{s}} = 35\,\mathrm{kHz}$, $L_\sigma = 58\,\mu\mathrm{H}$, $n = 2$, $v_{\mathrm{ac}}(t) = 325\,\mathrm{V}\sin(t\,2\pi 50\,\mathrm{Hz})$, $t = 5\,\mathrm{ms}$, $V_{\mathrm{dc1}} = 800\,\mathrm{V}$, $V_{\mathrm{dc2}} = 400\,\mathrm{V}$, and $\varphi = 22°$.

Fig. 5: Average power over one switching period as a function of the phase shift angle φ, for different values of $D = D_1 = D_2$, $f_{\mathrm{s}} = 35\,\mathrm{kHz}$, $L_\sigma = 58\,\mu\mathrm{H}$, $n = 2$, and dc voltages according to Tab. I. The thin black line denotes the trajectory of maximum power and the dashed line denotes the delimitation of the validity range of (4), according to (5).

result at the HF transformer of the DAB, cf. Fig. 3. The analysis of the lossless converter is similar to the analysis of a conventional DAB converter. For $D = D_1 = D_2$, the expression for the output power is

$$P_{1\phi} = \frac{nV_{\mathrm{dc1}}V_{\mathrm{dc2}}}{2f_{\mathrm{s}}L_\sigma}\frac{\varphi}{2\pi}\left(2D(1-D) - \frac{|\varphi|}{2\pi}\right), \tag{4}$$

which is found to be valid for

$$\frac{|\varphi|}{2\pi} < \min(D, 1-D). \tag{5}$$

Fig. 5 evaluates (4) with respect to different duty cycles and phase shift angles. A close inspection of the curves in Fig. 5 reveals that maximum power results for a phase shift angle that meets condition (5). For a given duty cycle, the expression

$$P_{1\phi,\max} = \frac{nV_{\mathrm{dc1}}V_{\mathrm{dc2}}\varphi_{P1\phi,\max}^2}{8\pi^2 f_{\mathrm{s}}L_\sigma} \tag{6}$$

with

$$\varphi_{P1\phi,\max} = 2\pi D(1-D) \tag{7}$$

applies for maximum power. As with all DAB converters, the inductor L_σ limits the maximum output power.

B. Single-phase system at ac–dc operation

The investigated system is operated with ac input voltage,

$$v_{\mathrm{ac}}(t) = V_{\mathrm{m,pk}}\sin\left(2\pi f_{\mathrm{m}}t\right), \tag{8}$$

cf. **Fig. 6(a)** and therefore, the above presented derivations for dc–dc operation need to be extended accordingly. For the sake of brevity, basic sinusoidal modulation is considered, i.e., the input and output stages apply a time-varying duty cycle,

$$D_1 = D_2 \approx \frac{1}{2}\left(1 + \frac{v_{\mathrm{ac}}(t)}{V_{\mathrm{dc1}}/2}\right), \tag{9}$$

in order to achieve a sinusoidal phase current with unity power factor.

Fig. 6(b) shows the calculated waveform of the primary-side current i_1 over a mains period, for $\varphi = 22° = $ constant. The filter capacitors C_{f1} and C_{f2} are blocking the low-frequency (LF) voltage components and with (1) and (9),

$$\langle v_{C\mathrm{f1}}\rangle(t) = \frac{V_{\mathrm{dc2}}}{V_{\mathrm{dc1}}}\langle v_{C\mathrm{f1}}\rangle(t) = v_{\mathrm{ac}}(t) \tag{10}$$

applies. For this reason, the HF transformer currents are subject to LF offsets caused by superimposed LF capacitor currents,

$$i_{\mathrm{lf1}} = C_{\mathrm{f1}}\frac{\mathrm{d}\langle v_{C\mathrm{f1}}\rangle}{\mathrm{d}t}, \qquad i_{\mathrm{lf2}} = C_{\mathrm{f2}}\frac{\mathrm{d}\langle v_{C\mathrm{f2}}\rangle}{\mathrm{d}t}, \tag{11}$$

cf. **Fig. 6(c)**. Thus, the analytical investigation for dc–dc operation presented in Section II-A is extended with respect to the time varying duty cycle and the capacitor currents. The

respective analysis reveals that the superimposed LF capacitor currents have no impact on the current power level, for which

$$\langle p_{1\phi}\rangle = P_{1\phi,\mathrm{dc}} + P_{1\phi,\mathrm{ac,pk}}\cos(4\pi f_{\mathrm{m}}t) \tag{12}$$

$$P_{1\phi,\mathrm{dc}} = \frac{nV_{\mathrm{dc1}}V_{\mathrm{dc2}}}{2f_{\mathrm{s}}L_\sigma}\frac{\varphi}{2\pi}\left[\frac{1}{2} - \left(\frac{V_{\mathrm{m,pk}}}{V_{\mathrm{dc1}}}\right)^2 - \frac{|\varphi|}{2\pi}\right], \tag{13}$$

$$P_{1\phi,\mathrm{ac,pk}} = \frac{nV_{\mathrm{dc1}}V_{\mathrm{dc2}}}{2f_{\mathrm{s}}L_\sigma}\frac{\varphi}{2\pi}\left(\frac{V_{\mathrm{m,pk}}}{V_{\mathrm{dc1}}}\right)^2, \tag{14}$$

is derived.

According to (12), (13), and (14) and for constant phase shift angle φ,

$$\frac{|\varphi|}{\pi} = \frac{1}{2} - \left(\frac{V_{\mathrm{m,pk}}}{V_{\mathrm{dc1}}}\right)^2 - \sqrt{\frac{\left(V_{\mathrm{dc1}}^2 - 2V_{\mathrm{m,pk}}^2\right)^2}{4V_{\mathrm{dc1}}^4} - \frac{8f_{\mathrm{s}}L_\sigma P_{1\phi,\mathrm{dc}}}{nV_{\mathrm{dc1}}V_{\mathrm{dc2}}}}, \tag{15}$$

the local average of the instantaneous power of the single-phase PFC rectifier, $\langle p_{1\phi}\rangle$ is a sinusoidal function with twice the mains frequency, amplitude $P_{1\phi,\mathrm{ac,pk}}$, and dc offset $P_{1\phi,\mathrm{dc}}$. In this regard, a detailed analysis reveals that power limitation relevant for the design of L_σ occurs at the maximum values of $|v_{\mathrm{ac}}|$, since the maximum possible output power decreases considerably for duty cycles approaching 0 or 1, cf. Fig. 5. With this and expressions (6) and (7), the useful range for φ is limited to

$$\frac{|\varphi|}{2\pi} < \frac{1}{4} - \left(\frac{V_{\mathrm{m,pk}}}{V_{\mathrm{dc1}}}\right)^2 \tag{16}$$

and with (13), a condition for L_σ results,

$$L_\sigma < \frac{nV_{\mathrm{dc1}}V_{\mathrm{dc2}}}{8f_{\mathrm{s}}P_{1\phi,\mathrm{dc}}}\left[\frac{1}{4} - \left(\frac{V_{\mathrm{m,pk}}}{V_{\mathrm{dc1}}}\right)^2\right]. \tag{17}$$

C. Three-phase system

It would be straight-forward to use three of the single-phase rectifiers given in Fig. 3 to realize an isolated three-phase PFC rectifier system. With dedicated input inductors, L_{ac}, and DAB transformers, however, the resulting system would require increased total converter volume, since it would not take advantage of the HF voltage applied to the input inductors. For this reason, the remaining part of the paper solely considers the topology of Fig. 1. Nevertheless, the results derived in Sections II-A and II-B directly apply, merely the waveforms of the input currents, $i_{L\{a,b,c\}}$, change, due to the superposition of the currents through L_{ac} and L_σ,

$$i_{L\{a,b,c\}} = i_{L_{\mathrm{ac}},\{a,b,c\}} - i_{L_\sigma,\{a,b,c\}}. \tag{18}$$

Fig. 7 depicts characteristic waveforms obtained from circuit simulation using operating conditions and settings

The 2018 International Power Electronics Conference

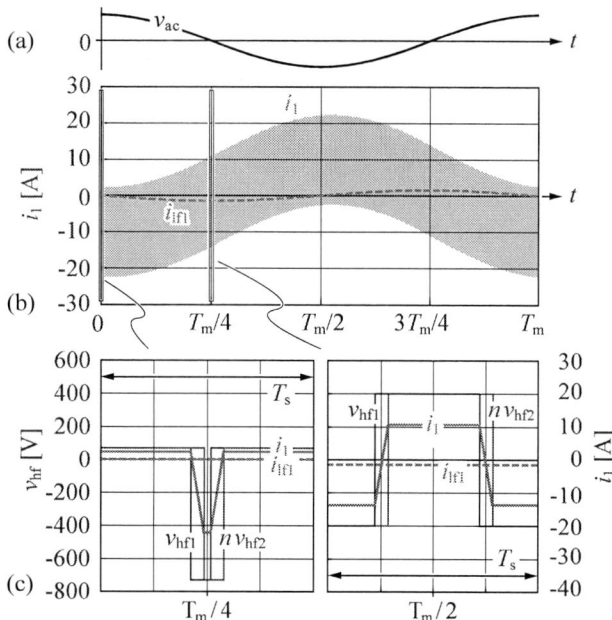

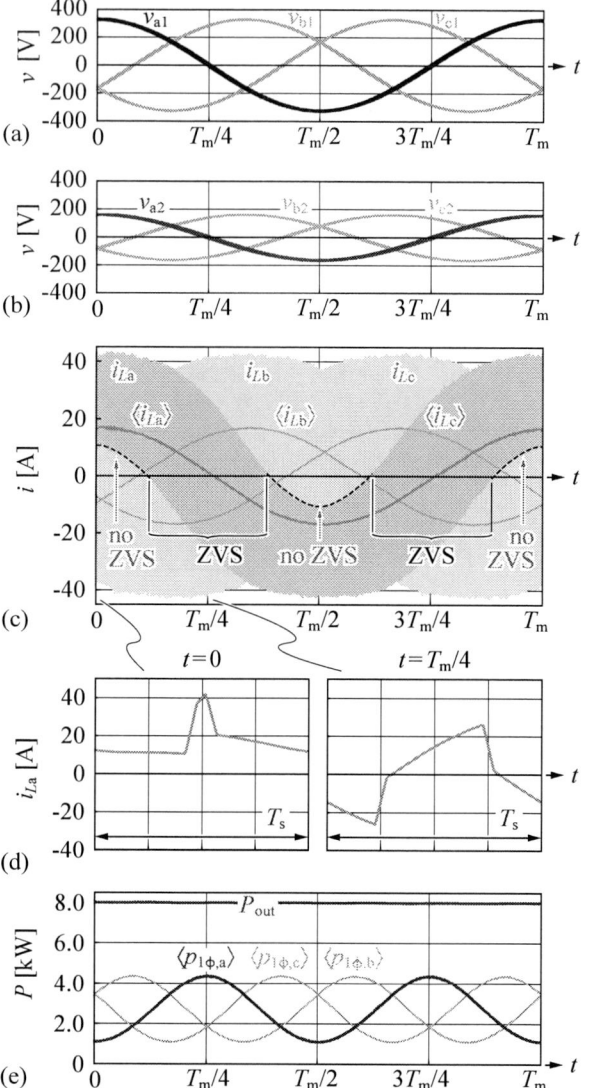

Fig. 6: Voltage and current waveforms determined for the single-phase system. (a) Sinusoidal input voltage over a mains period T_m. (b) Primary-side transformer current i_1 over a mains period. (c) i_1 and primary- and secondary-side HF transformer voltages v_{hf1} and v_{hf2} over a switching period T_s at $t = T_m/4$ and $t = T_m/2$, respectively.

according to Tables I and III. According to Figs. 7(a) and (b), the three-phase voltages are phase shifted by 120° and the primary- and secondary-side capacitor voltages are proportional, cf. (10). Fig. 7(c) illustrates the input currents, $i_{L\{a,b,c\}}$, at rated power and reveals that, with the considered value of L_{ac}, Zero Voltage Switching (ZVS) is partly lost at the primary side.[1] Detailed waveforms of the input currents at $t = 0$ and $t = 5$ ms are shown in Fig. 7(d) and clearly disclose the superposition of $i_{Lac,a}$ and $i_{L\sigma,a}$ according to (18). Fig. 7(e) illustrates the time-varying output power levels of each phase, $\langle p_{1\phi\{a,b,c\}}\rangle$, which are sinusoidal and phase shifted by 120°. For this reason, constant total power results,

$$P_{out} = 3P_{1\phi,dc}. \tag{19}$$

It is worth mentioning that the output power of each phase is maximal at the zero crossing of the corresponding phase voltage, which is due to $D = 0.5$, cf. (4). Furthermore, with the considered specifications and converter settings, the amplitude of the sinusoidal characteristic superimposed on $\langle p_{1\phi\{a,b,c\}}\rangle$ is less than its average value (1.6 kW < 2.7 kW).

III. OPTIMIZED CONVERTER DESIGN

In this Section, the investigated D3AB PFC rectifier is optimized with regard to efficiency and power density; the optimization objective is maximum power density at a converter efficiency of 98%. The implemented optimization procedure employs analytical expressions to calculate the component currents, which have been verified at different operating points using a circuit simulator, revealing a high accordance with errors of less than 2%. With known currents, the below listed component models are evaluated with regard to losses and volumes:

---

[1]Full ZVS requires a minimum current, which is indicated in Fig. 7(c) and described in Section III-A.

Fig. 7: Simulation results for the operating conditions and settings according to Tables I and III: (a) primary-side capacitor voltages; (b) secondary-side capacitor voltages; (c) input currents (= primary-side transformer currents; cf. Fig. 1); (d) magnified input current of phase a during $0 < t < T_s$ and 5 ms $< t < 5$ ms $+ T_s$, revealing the input current of the PFC rectifier (with current ripple) and the superimposed HF DAB transformer current, which shows a deviation from the ideal shape that originates from the series resonant circuit formed with L_σ and the filter capacitors C_{f1} and C_{f2}; (e) local average values of the instantaneous power levels of each phase and total output power.

- Semiconductors and cooling system in Section III-A,
- Magnetic components in Section III-B,
- Capacitors in Section III-C,
- EMI filter, gate drivers, and control in Section III-D.

A fully generalized converter optimization would feature a great number of open design parameters. Due to given specifications, known converter characteristics, and/or available components, however, a great number of design parameters can be readily defined. In this regard, V_{dc1} is set to 800 V in order to feature reasonable duty cycles with sufficient margins to 0 and 1, cf. (9), and still enable the use of power semiconductors with a blocking voltage of 1200 V.

Furthermore, it is known that DAB converters achieve most efficient operation for $V_{dc1}/(nV_{dc2}) \approx 1$. Thus, with regard to the specified output voltage, $V_{dc2} = 400\,\text{V}$, the turns ratio is set to $n = N_1/N_2 = 2$. Furthermore, (17) limits the maximum value of L_σ for given output power. The considered converter inductance is set to 80% of the maximum value,

$$f_s L_\sigma = 80\% \, (f_s L_\sigma)_{\max} = 80\% \, \frac{nV_{dc1}V_{dc2}}{8P/3} \left[\frac{1}{4} - \left(\frac{V_{m,pk}}{V_{dc1}}\right)^2\right],$$

to ensure controllability of the converter.

A. Semiconductors and cooling system

SiC power MOSFETs are used on the primary and secondary sides in order to take advantage of their low conduction and switching losses. Initial calculations of semiconductor losses reveal that low conduction and switching losses are achievable if single $25\,\text{m}\Omega/1200\,\text{V}$-devices (C2M0025120D by Cree) and $10\,\text{m}\Omega/900\,\text{V}$-devices (C3M0010090K by Cree) realize each switch on the primary and secondary sides, respectively. Using devices with increased on-state resistances would be possible with regard to the devices' rated currents and losses, however, reduced efficiencies would result. Conversely, the use of multiple MOSFETs connected in parallel would attain only limited improvements that may not justify the increased effort.[2]

The conduction losses are calculated based on the devices' on-state resistances at junction temperatures of $125°\text{C}$,

- C2M0025120D ($25\,\text{m}\Omega/1200\,\text{V}$): $R_{DS,on,1} = 38\,\text{m}\Omega$,
- C3M0010090K ($10\,\text{m}\Omega/900\,\text{V}$): $R_{DS,on,2} = 13\,\text{m}\Omega$.

The calculation of the switching losses is based on measured switching losses for the considered devices from [16], [17] and depicted in **Fig. 8**. The considered polynomials are

$$E_{sw} = \begin{cases} 233\,\mu\text{J} - 15.1\,\frac{\mu\text{J}}{\text{A}} I_D + 281\,\frac{\text{nJ}}{\text{A}^2} I_D^2 & \forall \quad I_D \leq 0.53\,\text{A}, \\ 12\,\mu\text{J} + 212\,\mu\text{J} \left(\frac{2.3\,\text{A}-I_D}{1.77\,\text{A}}\right)^2 & \forall \; 0.53\,\text{A} < I_D < 2.3\,\text{A}, \\ 17.1\,\mu\text{J} - 2.53\,\frac{\mu\text{J}}{\text{A}} I_D + 136\,\frac{\text{nJ}}{\text{A}^2} I_D^2 & \forall \quad I_D \geq 2.3\,\text{A}. \end{cases} \tag{20}$$

for the $25\,\text{m}\Omega/1200\,\text{V}$-device, for operation with $800\,\text{V}$ and $T_j = 125°\text{C}$ and

$$E_{sw} = \begin{cases} 164\,\mu\text{J} - 4.48\,\frac{\mu\text{J}}{\text{A}} I_D + 2.85\,\frac{\text{nJ}}{\text{A}^2} I_D^2 & \forall \quad I \leq 1.2\,\text{A}, \\ 3.4\,\mu\text{J} + 155\,\mu\text{J} \left(\frac{2.8\,\text{A}-I_D}{1.6\,\text{A}}\right)^2 & \forall \; 1.2\,\text{A} < I < 2.8\,\text{A}, \\ 964\,\text{nJ} + 837\,\frac{\text{nJ}}{\text{A}} I_D + 10.1\,\frac{\text{nJ}}{\text{A}^2} I_D^2 & \forall \quad I \geq 4.5\,\text{A}. \end{cases} \tag{21}$$

for the $10\,\text{m}\Omega/900\,\text{V}$-device ($V_{DS} = 400\,\text{V}$, $T_j = 70°\text{C}$). Negative values of the instantaneous drain current during switching, I_D, denote switching operations where ZVS cannot be attained, i.e. turn-on losses occur, and $I_D > 0$ denote switching operations where ZVS is in principle feasible. However, a minimum current is required for ZVS to fully charge and discharge the MOSFET's output capacitances. In this regard, the second polynomials in (20) and (21) represent partial ZVS that are approximated based on quadratic interpolations for a dead time interval of $200\,\text{ns}$. Remark: since the MOSFETs are used without external capacitors increasing C_{oss}, the very low loss property of ZVS is lost at high positive currents, due to turn-off losses (approximately at $I_D > 20\,\text{A}$ in Fig. 8).

---

[2]Even increased switching losses would result on the primary side, due to time intervals where turn-on losses occur, cf. Fig. 7. As a result, optimal designs would employ increased current ripples.

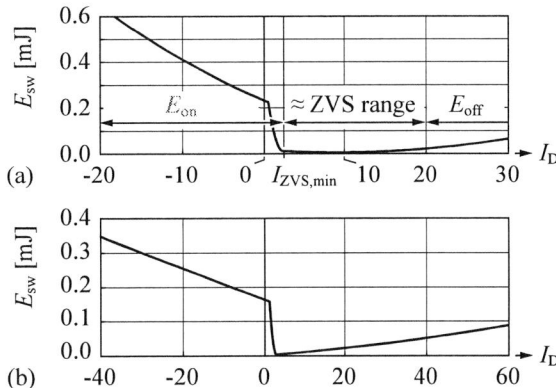

Fig. 8: (a) Switching losses of the $25\,\text{m}\Omega/1200\,\text{V}$ SiC MOSFET C2M0025120D for operation with $800\,\text{V}$ and $T_j = 125°\text{C}$ and (b) the $10\,\text{m}\Omega/900\,\text{V}$ SiC MOSFET C3M0010090K for operation with $400\,\text{V}$ and $T_j = 70°\text{C}$, with respect to the drain current, I_D, at the switching instant. Both Figures depict measured results from [16], [17] for $I_D < 0$ and $I_D > I_{ZVS,min}$. For $0 \leq I_D \leq I_{ZVS,min}$ the switching losses for partial ZVS are interpolated for an assumed dead time interval of $200\,\text{ns}$.

TABLE II: Expressions used for scaling all lengths, areas, and volumes of the considered magnetic components (based on two stacked E 55/28/21 cores).

Core volume	$V_c = 435 \times 10^{-3} V_{box}$
Core cross section	$A_c = 206 \times 10^{-3} V_{box}^{2/3}$
Available winding cross section of coil former	$A_w = 95.2 \times 10^{-3} V_{box}^{2/3}$
Height of core window	$h_w = 158 \times 10^{-3} V_{box}^{1/3}$
Width of core window	$A_w = 631 \times 10^{-3} V_{box}^{1/3}$
Average turn length	$l_{avg} = 2.64 V_{box}^{1/3}$
Open surface to ambient	$A_{open} = 5.51 V_{box}^{2/3}$

The volume of the cooling system for power semiconductors with total losses of $P_{semi,total}$ is considered by means of a Cooling System Performance Index (thermal conductance per volume), $CSPI$, of $13\,\frac{\text{W}}{\text{dm}^3\,\text{K}}$ and a considered temperature difference between heat sink and ambient of $\Delta T_{hs-a} = 50°\text{C}$,

$$V_{\text{cooling system}} = \frac{P_{semi,total}}{\Delta T_{hs-a} \times CSPI}. \tag{22}$$

B. Magnetic components

This paper uses a unified scaled model for all magnetic components. The scaled model is parameterized according to the geometrical properties of the transformers realized in [8], which are compose of two stacked E 55/28/21 cores (boxed volume, V_{box}, is $200\,\text{cm}^3$) and achieve efficiencies of 99.6% for an isolated three-phase PFC rectifier with a rated power of $7.5\,\text{kW}$. For a given value of V_{box}, all lengths, areas, and volumes of the magnetic component are determined according to the expressions listed in **Tab. II**. The input inductors/transformers of the D3AB PFC rectifier and the DAB converter inductors, L_σ, are considered separately in order to take the additional volumes and losses due to L_σ into account.

The employed component model calculates the core losses with the improved Generalized Steinmetz Equation (iGSE) [18] and the Steinmetz parameters

$$k = 1.02, \ \alpha = 1.4745, \text{ and } \beta = 2.6607 \tag{23}$$

for the considered N95 ferrite core material (extracted for $f = 25\,\text{kHz}$, $B_{pk} = 300\,\text{mT}$, and $T_c = 80°\text{C}$ with a software tool provided by TDK/EPCOS [19]). The copper losses are determined using simplified expressions for HF skin- and proximity effects derived in [20], which assume a distributed air gap. The computation of the copper losses considers the first 30 harmonic components of each conductor current; the conductors employ HF litz wires with single strand diameters of 0.1 mm.

The automated design procedure further takes an effective copper area of $38\% \times A_w$ (A_w is the cross section of the core window, cf. Tab. II), a copper temperature of 100°C, a maximum flux density of 300 mT, and a maximum temperature rise of the component's surface of 50°C into account. The surface temperature rise is approximated according to [21],

$$\Delta T = \left(\frac{P/1\,\mathrm{mW}}{A_{\mathrm{open}}/1\,\mathrm{cm}^2}\right)^{\frac{1}{1.1}} \times 1°\mathrm{C} < 60°\mathrm{C}. \quad (24)$$

In a first step, the design procedure scans a wide range of values for V_{box} in order to determine a boxed volume that leads to a design close to the thermal limitation, $V_{\mathrm{box},0}$. For this purpose, geometric sequences are used for V_{box} with common ratios of 0.5 (initial coarse scan starting from $V_{\mathrm{box}} = 10\,\mathrm{dm}^3$) and 1.1 (subsequent fine scan). For each given value of V_{box} an inner loop determines the optimal number of turns with respect to minimum total losses. In case of the input inductors/transformers, the available cross section of the core window is divided to the windings of primary and secondary sides such that same current densities result. Finally, the air gap length is determined to achieve the specified inductance.

Losses of magnetic components decrease with increasing volume. For this reason, the copper and core losses are calculated for further 29 magnetic components with increasing boxed volumes according to

$$V_{\mathrm{box},i} = V_{\mathrm{box},0} \times 1.1^i \quad \forall \quad i \in \{1, 2, 3, \dots 29\} \quad (25)$$

and for optimized numbers of turns. The resulting volumes, losses, and design configurations (e.g. numbers of turns) are stored and the data transferred to the main converter optimization procedure.

C. Capacitors

The capacitors of the considered converter are subject to relatively high currents. In order to still achieve high power density, ceramic and film capacitors have been selected, which are listed below:

- C_{f1}: 1×B32754C2106K000 (film capacitor, 10 μF, 250 Vac, 12 A, EPCOS),
- C_{dc1}: 1×CeraLink™ SP500 (ceramic capacitor, 12 μF, 400 Vdc, 41 A, EPCOS),
- C_{f2}: 25×KR355WD72W125MH01 (ceramic cap., 0.85 μF at 163 V, 450 Vdc, $\approx$ 2 A at 50 kHz, Murata),
- C_{dc2}: 1×CeraLink™ SP500 (same as C_{dc1}; note: capacitance drops to 8.4 μF at 200 V).

The total volume of all capacitors is 160 cm³, which includes an additional volume of 30 cm³ for damping networks, an electrolytic output capacitor (120 μF, 450 Vdc), and two SMD inductors that decouple the electrolytic capacitor from the CeraLink™ capacitors (C_{dc2}). The final design suggested by optimization has been successfully tested with comprehensive circuit simulation, using the above capacitance values, revealing only minor differences in terms of rms values and losses (conduction, switching, and core). It is worth to note that the optimization does not consider capacitor losses, due to their comparably low contribution to the total losses.

D. Remaining components

The volumes and losses of EMI filter, gate drivers, and control have been adopted from [22], due to similar specifications and optimization objectives:

$$P_{\mathrm{EMI\ filter}} = 5\,\mathrm{W} \quad (26)$$

$$P_{\mathrm{gate\ drivers}} + P_{\mathrm{control}} + P_{\mathrm{fan}} = 12\,\mathrm{W}, \quad (27)$$

$$V_{\mathrm{EMI\ filter}} = 0.35\,\mathrm{dm}^3, \quad (28)$$

$$V_{\mathrm{gate\ drivers}} + V_{\mathrm{control}} = 0.3\,\mathrm{dm}^3, \quad (29)$$

$$V_{\mathrm{total}} = 1.15 \sum V_i, \quad (30)$$

i.e., 15% of the volume is considered to be unused.

E. Optimization

Based on the above considerations and assumptions, it is found that the switching frequency, f_s, the input current ripple,

$$r = \frac{\Delta I_{L\mathrm{ac,pkpk}}}{\frac{2P}{3V_{\mathrm{m,pk}}}} = \frac{\Delta I_{L\mathrm{ac,pkpk}}}{I_{\mathrm{m,pk}}}, \quad (31)$$

and the considered boxed volumes of the magnetic components remain for optimization of efficiency and power density. The considered settings are defined with

$$f_s \in \{23, 27, 35, 47, 72, 140\}\mathrm{kHz}, \quad (32)$$

$$r \in \{30, 50, 75, 100, 125,$$
$$150, 175, 200, 225, 250, 275, 300\}\%, \quad (33)$$

where the listed switching frequencies are preselected with regard to small volume EMI filters (cf. Fig. 12 in [23]). The sets defined for f_s and r lead to 72 different settings. Furthermore, 900 combinations of different designs result for the input inductors/transformers and the DAB inductors (30 for each, cf. Section III-B) for given values of f_s and r, which, in total, yields 64800 results. **Fig. 9** depicts the corresponding results and discloses the η-ρ Pareto front for the investigated converter system.

The orange star in Fig. 9 marks the selected operating point, which achieves $\eta = 98.1\%$ and $\rho = 4\,\mathrm{kW/dm}^3$ at $f_s = 35\,\mathrm{kHz}$, $r = 175\%$, and for magnetic components with maximum power density, i.e., operated at their thermal limitation. The resulting Pareto-optimal design points reveal increasing efficiency for decreasing power density, which is directly related to the similar η-ρ characteristics of magnetic components. From a detailed inspection of the design points on the Pareto front it becomes apparent that design points with Pareto-optimal power density and very high efficiency are obtained for reduced switching frequencies (switching losses, core losses) and reduced current ripples (rms currents, conduction and copper losses, core losses; reduced switching frequencies overcompensate the increases of switching losses by reason of reduced current ripples).

The red triangles in Fig. 9 mark results with constant switching frequency of 35 kHz, magnetic components with maximum power densities, and different values of r. It can be observed that reduced power densities and efficiencies result for $r < 175\%$. The reduced power densities are mainly addressed to increased boxed volumes of the PFC input inductors and the reduced efficiencies originate from both, the PFC input inductors due to the required increased energy storage capability and the semiconductors on the primary side, which generate increased switching losses due to an increase of the region where ZVS is lost, cf. Fig. 7. Slightly reduced converter volumes are feasible for $r > 175\%$, however, the efficiency quickly decreases by reason of large rms currents.

The 2018 International Power Electronics Conference

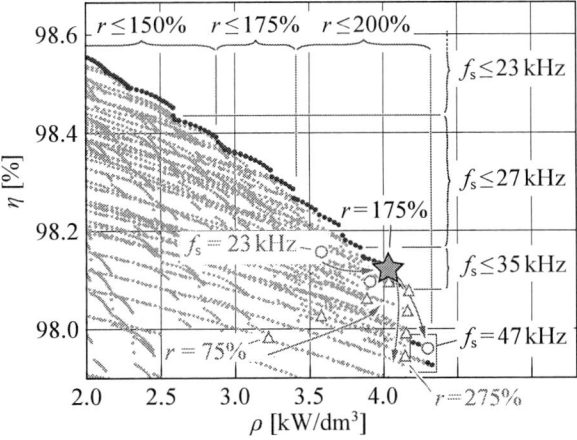

Fig. 9: Efficiencies, power densities, and η-ρ Pareto front determined for the D3AB PFC rectifier.

The cyan circles mark results with constant current ripple of 175%, magnetic components with maximum power densities, and different switching frequencies. It can be seen that a considerably reduced switching frequency of 23 kHz still facilitates a relatively high power density of 3.6 kW/dm³. In this regard it is found that Pareto-optimal designs with very high efficiencies not only require reduced switching frequencies and power densities but also magnetic components with increased boxed volumes.

IV. DISCUSSION OF DESIGN RESULT

Tab. III lists the design results at the selected operating point identified in Fig. 9 and **Figs. 10(a)** and **(b)** depict the corresponding component losses and volumes, respectively. According to Fig. 10(a), more than half of the total losses are attributed to the semiconductor losses which are mainly generated in the power MOSFETs on the primary side. A reduction of the primary-side conduction losses could be achieved by increasing the corresponding chip sizes, which, however, would increase the switching losses. The magnetic components generate one third of the losses; here, losses mainly occur in the windings of the input inductors, which are already operated with maximum flux densities of 300 mT, i.e., a further increase of the core losses is not feasible.

Approximately two thirds of the converter volume are required for passive components (magnetics, capacitors, EMI filter). Due to comparably low semiconductor losses (84 W at rated power), a cooling system with a comparably small volume can be employed, e.g., using double-sided cooling a small fan with an edge length of 30 mm, which is found to enable the realization of a cooling system with the calculated low volume of 130 cm³ and still provides a sufficiently large base plates to accomodate all 12 MOSFETs.

Fig. 11(a) and **(b)** depict the calculated characters of efficiency and selected components' losses with respect to the output power and reveal that $\eta > 98\%$ is feasible for $P_{\text{out}} > 2.3$ kW. According to Fig. 11(b), substantial conduction losses and losses in the magnetic components remain at very low power, due to the inductor current ripples. However, increasing switching losses are observed for decreasing output power and $P_{\text{out}} < 2$ kW. A close investigation reveals that the currents during switching of the secondary-side MOSFETs are insufficient for ZVS, cf. **Fig. 12(a)** and Fig. 8(b). ZVS could

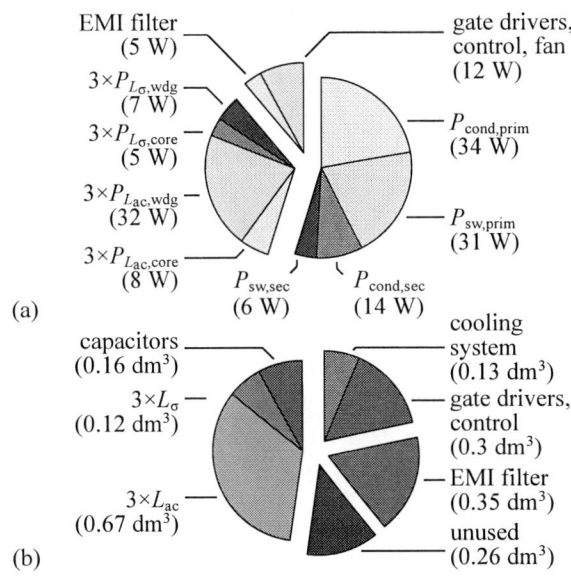

(a)

(b)

Fig. 10: (a) Component losses and (b) volumes for the selected design point with $\eta = 98.1\%$ and $\rho = 4$ kW/dm³.

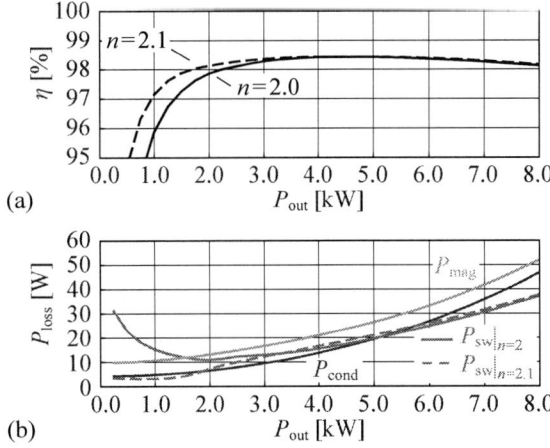

(a)

(b)

Fig. 11: (a) Total converter efficiency and (b) characteristics of selected components' losses with respect to output power.

be attained if three inductors would be placed in parallel to the three transformers' secondary windings. According to the results of an analytical investigation of the equivalent circuit of a transformer, however, the same effect can be achieved for a slight adjustment of the transformers' turns ratios, from $n = 2$ to $n = 2.1$, cf. **Fig. 12(b)**. With this minor adjustment, a substantial efficiency improvement is achieved at low output power levels, i.e., $\eta > 98\%$ for $P_{\text{out}} > 1.7$ kW.

V. CONCLUSION

This paper proposes and analyzes a novel three-level and three-port isolated and bidirectional PFC rectifier topology (D3AB PFC rectifier), which can be realized with standard 6-pack power modules on the primary and secondary sides, employs integrated input inductors and HF transformers, and/or features galvanic isolation and high efficiency. The given in-depth description of the working principle of the D3AB PFC rectifier allows the derivation of key expressions needed

3811

TABLE III: Summary of results for the selected converter design, cf. Fig. 9.

General results and rms currents	
Switching frequency	$f_s = 35\,\text{kHz}$
Current ripple	$r = 175\%$
Calculated efficiency at rated load	$\eta = 98.1\%$
Calculated total power density	$\rho = 4\,\text{kW/dm}^3$
Transformer rms current, prim. and sec. sides	$I_{\text{tr}1,2} = \{17.2\,\text{A},\ 19.2\,\text{A}\}$
MOSFET rms currents,prim. and sec. sides	$I_{\text{T,prim,sec}} = \{12.2\,\text{A},\ 13.6\,\text{A}\}$
Magnetic input inductor/transformer L_{ac}	
Inductance	$L_{\text{ac}} = 195\,\mu\text{H}$
Boxed volume	$V_{\text{box}} = 223\,\text{cm}^3$
Number of turns, prim. and sec. sides	$N_{1,2} = \{20,\ 10\}$
Air gap length	$l_{\text{air}} = 1.9\,\text{mm}$
Conductors, prim. and sec. sides (HF litz wires)	$\{547, 610\} \times 0.1\,\text{mm}$
Copper losses	$P_{\text{w}} = 10.6\,\text{W}$
Core losses	$P_{\text{c}} = 2.5\,\text{W}$
Calculated temperature rise	$\Delta T = 44^\circ\text{C}$
DAB inductor L_σ/n^2 (placed on the secondary side)	
Inductance	$L_\sigma = 14.5\,\mu\text{H}$
Boxed volume	$V_{\text{box}} = 41\,\text{cm}^3$
Number of turns	$N = 9$
Air gap length	$l_{\text{air}} = 1.7\,\text{mm}$
Conductor: HF litz wire	$610 \times 0.1\,\text{mm}$
Copper losses	$P_{\text{w}} = 2.4\,\text{W}$
Core losses	$P_{\text{c}} = 1.8\,\text{W}$
Calculated temperature rise	$\Delta T = 44^\circ\text{C}$

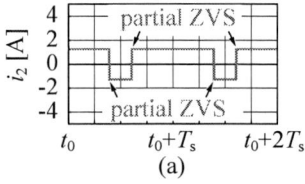 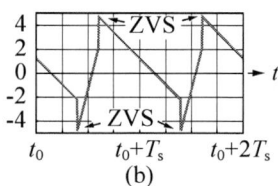

Fig. 12: Calculated current at the transformer's secondary side for $P_{\text{out}} = 500\,\text{W}$, $t_0 = 2.5\,\text{ms}$, operating conditions and parameters according to Tab. I and Tab. III, and different turns ratios: (a) $n = 2$, (b) $n = 2.1$.

to design the converter system with respect to optimized performance values, e.g. efficiency and power density, which serves as a basis for the converter optimization presented in the second part of this paper. According to the calculated results, a full-load efficiency of 98.1% and a power density of $4\,\text{kW/dm}^3$ can be achieved for the PFC rectifier if SiC MOSFETs are used ($25\,\text{m}\Omega/1200\,\text{V}$ and $10\,\text{m}\Omega/900\,\text{V}$ devices on primary and secondary sides, respectively). The presented design is found to achieve efficiencies greater than 98 % for $P_{\text{out}} > 1.7\,\text{kW}$.

The discussions given in this paper are confined to the basic structure, operating behavior, and design of the new topology. Further research will focus on the investigation of prospective efficiency and/or power density improvements that can be achieved with modified topologies and alternative control schemes that take advantage of currently unused degrees of freedom of the considered converter system. Corresponding examples include realizations where only CM or DM components are utilized for energy transfer and alternative control schemes with non-constant values of the phase shift and different duty cycles, $D_1 \neq D_2$.

REFERENCES

[1] S. Pannala, N. Padhy, and P. Agarwal, "Peak Energy Management using Renewable Integrated DC Microgrid," *IEEE Trans. Smart Grid*, accepted for future publication, 12 pages.

[2] C. Nelson. Multi-Axis Drive Applications Using Common DC Bus. Accessed: March 9, 2018. [Online]. Available:

https://www.industry.usa.siemens.com/drives/us/en/electric-drives/ac-drives/ac-drives-apps/multi-axis-common-dc-bus/Documents/Common_DC_bus_paper.pdf

[3] R. Razi, B. Asaei, and M. R. Nikzad, "A new battery charger for plug-in hybrid electric vehicle application using back to back converter in a utility connected micro-grid," in *Proc. IEEE Power Electron., Drive Sys. & Tech. Conf. (PEDSTC)*, Mashhad, Iran, 14–16 Feb. 2017, pp. 13–18.

[4] R. Yamada, Y. Nemoto, S. Fujita, and Q. Wang, "A Battery Charger with 3-Phase 3-Level T-type PFC," in *Proc. IEEE Int. Telecom. Energy Conf. (INTELEC)*, Osaka, Japan, 18–22 Oct. 2015, 5 pages.

[5] S. Madhusoodhanan, A. Tripathi, K. Mainali, A. Kadavelugu, D. Patel, S. Bhattacharya, and K. Hatua, "Three-Phase 4.16 kV Medium Voltage Grid Tied AC-DC Converter Based on 15 kV/40 A SiC IGBTs," in *Proc. IEEE Energy Convers. Congr. and Expo. (ECCE USA)*, Montreal, QC, 20–24 Sept. 2015, pp. 6675–6682.

[6] M. Kumar, L. Huber, M. M. Jovanović, D. Ping, and G. Liu, "Analysis, Design, and Evaluation of Three-Phase Three-Wire Isolated AC-DC Converter Implemented with Three Single-Phase Converter Modules," in *Proc. IEEE Applied Power Electron. Conf. and Expo. (APEC)*, Long Beach, CA, 20–24 March 2016, pp. 38–45.

[7] M. Silva, N. Hensgens, J. Oliver, P. Alou, Ó. García, and J. A. Cobos, "Isolated Swiss-Forward Three-Phase Rectifier for Aircraft Applications," in *Proc. IEEE Applied Power Electron. Conf. and Expo. (APEC)*, Fort Worth, TX, 16–20 March 2014, pp. 951–958.

[8] L. Schrittwieser, P. Cortés, L. Fässler, D. Bortis, and J. W. Kolar, "Modulation and Control of a Three-Phase Phase-Modular Isolated Matrix-Type PFC Rectifier," *IEEE Trans. Power Electron.*, vol. 33, no. 6, pp. 4703–4715, June 2018.

[9] A. K. Singh, P. Das, and S. K. Panda, "A Novel Matrix Based Isolated Three Phase AC-DC Converter with Reduced Switching Losses," in *Proc. IEEE Applied Power Electron. Conf. and Expo. (APEC)*, Charlotte, NC, 15–19 March 2015, pp. 1875–1880.

[10] Z. Zhang, A. Mallik, and A. Khaligh, "A High Step-Down Isolated Three-Phase AC-DC Converter," *IEEE Trans. Emerg. Sel. Topics Power Electron.*, vol. 6, no. 1, pp. 129–139, March 2018.

[11] D. S. Oliveira and B. R. de Almeida, "A Bidirectional Single-Stage Three-Phase AC / DC Converter with High-Frequency Isolation and PFC," in *Proc. Int. Power Electron. Conf. (PCIM Europe)*, Nuremberg, Germany, 19–21 May 2015, pp. 1122–1129.

[12] J. Jacobs, M. Thömmes, and R. De Doncker, "A Transformer Comparison for Three-Phase Single Active Bridges," in *Proc. Europ. Conf. Power Electron. Appl. (EPE)*, Dresden, Germany, 11–14 Sept. 2005, 10 pages.

[13] F. Z. Peng, H. Li, G.-J. Su, and J. S. Lawler, "A new ZVS bidirectional DC-DC Converter for Fuel Cell and Battery Application," *IEEE Trans. Power Electron.*, vol. 19, no. 1, pp. 54–65, Jan. 2004.

[14] H. Li, F. Z. Peng, and J. S. Lawler, "A Natural ZVS Medium-Power Bidirectional DC-DC Converter with Minimum Number of Devices," *IEEE Trans. Ind. Appl.*, vol. 39, no. 2, pp. 525–535, March 2003.

[15] Z. Wang and H. Li, "A Soft Switching Three-phase Current-fed Bidirectional DC-DC Converter with High Efficiency over a Wide Input Voltage Range," *IEEE Trans. Power Electron.*, vol. 27, no. 2, pp. 669–684, Feb. 2012.

[16] J. Azurza Anderson, C. Gammeter, L. Schrittwieser, and J. W. Kolar, "Accurate Calorimetric Switching Loss Measurement for 900 V 10 mΩ SiC MOSFETs," *IEEE Trans. Power Electron.*, vol. 32, no. 12, pp. 8963–8968, Dec. 2017.

[17] J. Azurza Anderson, L. Schrittwieser, C. Gammeter, G. Deboy, and J. W. Kolar, "Relating the Figure of Merit of Power MOSFETs to the Maximally Achievable Efficiency of Converters," under review for the CPSS Trans. on Power Electron. and Appl.

[18] K. Venkatachalam, C. R. Sullivan, T. Abdallah, and H. Tacca, "Accurate Prediction of Ferrite Core Loss with Nonsinusoidal Waveforms using only Steinmetz Parameters," in *Proc. IEEE Workshop Comput. Power Electron.*, Mayaguez, Puerto Rico, 3–4 June 2002, pp. 36–41.

[19] TDK/EPCOS. Ferrite Magnetic Design Tool. Accessed: March 9, 2018. [Online]. Available: https://en.tdk.eu/tdk-en/180490/design-support/design-tools/ferrites/ferrite-magnetic-design-tool

[20] M. Leibl, "Three-Phase PFC Rectifier and High-Voltage Generator for X-Ray Systems," Ph.D. dissertation, ETH Zurich, 2017.

[21] A. Van den Bossche and V. C. Valchev, *Inductors and transformers for power electronics*. New York: Taylor & Francis, 2005.

[22] L. Schrittwieser, M. Leibl, M. Haider, F. Thöny, J. W. Kolar, and T. B. Soeiro, "99.3% Efficient Three-Phase Buck-Type All-SiC SWISS Rectifier for DC Distribution Systems," *IEEE Trans. Power Electron.*, accepted for future publication, 15 pages.

[23] J. W. Kolar, F. Krismer, Y. Lobsiger, J. Mühlethaler, T. Nussbaumer, and J. Miniböck, "Extreme Efficiency Power Electronics," in *Proc. Int. Conf. Integr. Power Electron. Sys. (CIPS)*, Nuremberg, Germany, 6–8 March 2012, 22 pages.

Load Voltage Regulation Method for an Isolated AC-DC Converter with Power Decoupling Operation

Shohei Komeda[1], Hideaki Fujita[2]

1 Department of Marine Electronics and Mechanical Engineering, Tokyo University of
Marine Science and Technology, Tokyo, Japan
2 Department of Electrical and Electronic Engineering, Tokyo Institute of Technology, Tokyo, Japan
*E-mail: skomed0@kaiyodai.ac.jp, hf@ieee.org

Abstract—This paper propose a new load voltage regulation method for an isolated single-phase ac-to-dc converter with power decoupling operation. The power decoupling operation has originally been proposed to achieve a unity displacement power factor in the line current. However, the unity displacement power factor operation can regulate the load voltage in a limited operating power range, and the load voltage decreases especially under a light load condition. The proposed control method does not compensate the leading current due to filter capacitors and make it possible to regulate the load voltage at a constant level in a wider power range. As a result, the proposed control method realizes an effective load voltage regulation with a higher power conversion efficiency by a 100-V, 300-W experimental setup.

Keywords—Power decoupling technique, Power Factor Correction, Voltage regulation.

I. INTRODUCTION

Battery chargers are one of the most important applications of isolated ac-to-dc converters [1]. The isolated ac-to-dc converter typically consists of a rectifier with a power factor correction circuit and an isolated dc-to-dc converter. Thus, this configuration requests a large number of power conversion stages. This causes some problems in power conversion efficiency, converter volume, and so on. Moreover, the rectifier has a bulky electrolytic capacitor to absorb power pulsation at double the line frequency as an energy storage element. This role is very important especially for the battery charger to reduce the ripple current into the battery. However, the electrolytic capacitor also has some problems in its short lifetime and small ripple current rating.

To overcome the above problems, various approaches have been proposed to adopt power decoupling technique [2]–[4] and/or direct ac-to-ac converter topology [5][6]. The power decoupling technique can reduce the capacitance in the dc-link capacitor by using an auxiliary circuit. This technique mainly focuses on the rectifier in the line side to reduce the capacitance in the dc link. Thus, this technique has effectiveness to adopt a small capacitor as an energy storage element but does not reduce the number of power conversion stages because the isolated dc-to-dc converter needs to be connected at a stage after the rectifier. On the other hand, the direct ac-to-ac converter topology can reduce the number of the power conversion stages especially in the primary side of a high-frequency transformer. However, the direct converter does not have any energy storage element. As a result, this topology reduces the number of the power conversion stages but requests an auxiliary power decoupling circuit or a bulky electrolytic capacitor to absorb the power pulsation.

A power decoupling control method for a direct ac-to-ac converter has been proposed to absorb the power pulsation into a filter capacitor [7]. The direct ac-to-ac converter consists of two half-bridge converters, two input filter capacitors, and a series-resonant circuit [8]–[11]. Reference [7] has demonstrated that the power decoupling control method realizes both direct ac-to-ac power conversion and power decoupling operation at the same time without any additional component. Thus, the power decoupling control method makes it possible to reduce the number of the power conversion stages and to remove the electrolytic capacitor and/or the auxiliary circuit from the battery charger. In the power decoupling control method, the direct ac-to-ac converter can improve the power factor of the line current and the amplitude of the resonant current at the same time [7]. However, this operation causes a limitation in the controllable load voltage range under a light load condition.

This paper proposes a new load voltage regulation method for an isolated ac-to-dc converter based on a direct ac-to-ac converter consisting of two half-bridge converters. To maintain the load voltage constant, this paper proposes a new power decoupling control method for the direct ac-to-ac converter with a limited displacement power factor (DPF) in the line current. The proposed control method does not compensate the leading current through the filter capacitors, and thus, the line current has a slightly-leading phase angle but it is possible to regulate the load voltage at a constant level in a wider power range. This paper theoretically discusses conditions of the proposed load voltage regulation method in the power decoupling operation. A 100-V, 300-W experimental setup demonstrates that the proposed control method realizes a constant load voltage with a higher power conversion efficiency.

3813

The 2018 International Power Electronics Conference

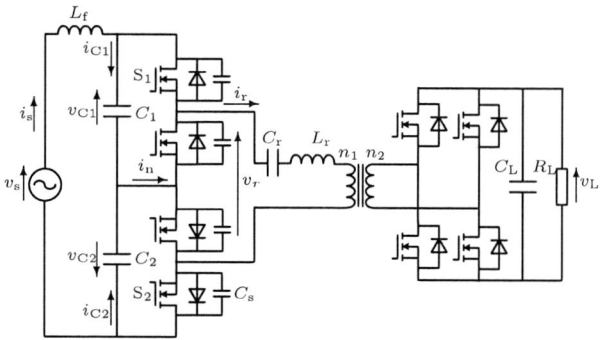

Fig. 1. Experimental circuit.

TABLE I.   CIRCUIT PARAMETERS OF EXPERIMENTAL CIRCUIT.

Line voltage	V_s	100 V
Line frequency	f_s	50 Hz
Input filter inductor	L_f	150 μH (0.14%)
Input filter capacitor	C_1, C_2	30 μF (16%)
Snubber capacitor	C_s	6 nF
Resonant capacitor	C_r	3 μF
Resonant inductor	L_r	65 μH
Transformer turn ratio	$n_1 : n_2$	1:2
Output filter capacitor	C_L	20 μF
Average output voltage	V_L	50 V
Rated load power	P_L	300 W

II.   CIRCUIT CONFIGURATION

Fig. 1 shows the circuit configuration for the experiment, whose circuit parameters are summarized in Table I. The experimental isolated single-phase ac-to-dc converter consists of a direct ac-to-ac converter, a high-frequency transformer, and a rectifier with synchronous rectification. The direct converter is composed of two half-bridge converters, two input filter capacitors, and a series-resonant circuit. The all MOSFETs use IXFB150N65X2 (IXYS) in the following experiments.

In power decoupling operation of the direct converter, the direct converter keeps the resonant current i_r constant by adjusting the switching frequency and controls the line current i_s and the node current i_n (the filter capacitor voltages v_{C1} and v_{C2}) by adjusting the phase-shift angles between upper and lower half bridges [7]. Thus, the power decoupling control method realizes the direct ac-to-ac power conversion and the power decoupling operation at the same time.

III.   POWER DECOUPLING OPERATION WITH ARBITRARY POWER FACTOR

A. Voltage and current in filter capacitors

Fig. 2 shows a calculated example of operating waveforms when the power decoupling control method is applied in a condition of $P_L = 115$ W, which is a relatively light load condition. A sinusoidal line voltage is assumed as given by

$$v_s = \sqrt{2}V_s\sin\theta_s, \tag{1}$$

and the line current is also assumed to be a sinusoidal waveform as,

$$i_s = \sqrt{2}I_s\sin(\theta_s + \phi), \tag{2}$$

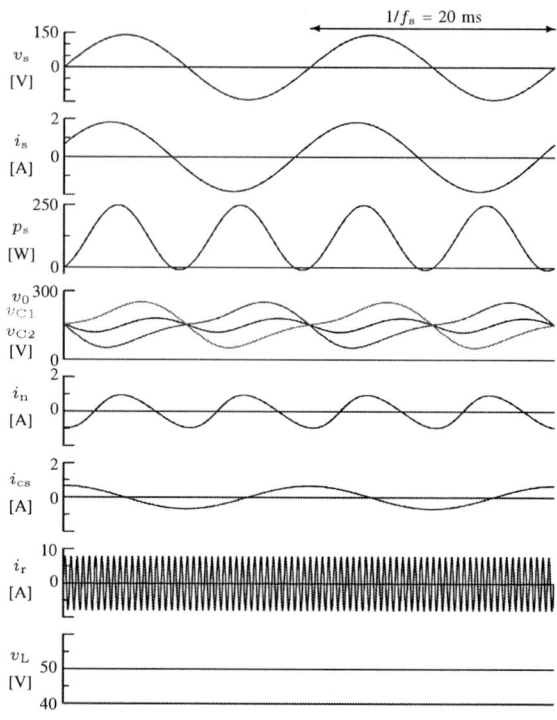

Fig. 2. Calculated results of the operating waveforms when the power decoupling control method is applied in a condition of $P_L = 115$ W.

where V_s and I_s are the rms values of v_s and i_s, and ϕ is the phase angle of the line, which is expressed by using the line angular frequency ω_s as $\theta_s = \omega_s t$. The instantaneous power is obtained by,

$$p_s = v_s i_s = V_s I_s \left[\cos\phi - \cos(2\theta_s - \phi)\right]. \tag{3}$$

The first term of the right hand side in (3) is the power consumed in the load. The second term is the power pulsation at double the line frequency. In case of $\phi \neq 0$, the instantaneous power p_s includes negative value.

In the steady-state condition, the load power P_L is given by

$$P_L = V_L I_L = V_s I_s \cos\phi, \tag{4}$$

where V_L and I_L are the load voltage and current. To realize above condition, the power pulsation should be stored into the filter capacitors C_1 and C_2. The filter capacitor voltages v_{C1} and v_{C2} including an offset voltage v_0 are presented by

$$v_{C1} = v_0 + \frac{v_s}{2} \tag{5}$$

$$v_{C2} = v_0 - \frac{v_s}{2}. \tag{6}$$

Assuming $C = C_1 = C_2$, the offset voltage to store the power pulsation should be controlled, as given by

$$v_0 = \sqrt{\frac{W_0}{C} - \frac{V_s^2}{2}\sin^2\theta_s - \frac{V_s I_s}{2\omega_s C}\sin(2\theta_s + \phi)}, \tag{7}$$

where W_0 is the average value of the sum of the stored energy in C_1 and C_2.

3814

In the power decoupling control method, the resonant current i_r is controlled with a constant amplitude as

$$i_r = \sqrt{2}I_r\sin\theta_{sw}, \tag{8}$$

where I_r is the rms value of the resonant current and θ_{sw} is the phase angle in the switching period. To make the waveform clear in Fig. 2, a switching frequency f_{sw} of the resonant current is assumed to be a constant frequency of 2 kHz. The resonant current flows into the node current i_n and/or the line current i_s under the power decoupling control method [7]. The node current is defined as

$$i_n = i_{C1} + i_{C2}, \tag{9}$$

where i_{C1} and i_{C2} are filter capacitor current. Thus, i_n can be calculated by,

$$
\begin{aligned}
i_n &= 2C\frac{dv_0}{dt} \\
&= \frac{-V_sI_s\cos(2\theta_s + \phi) - \frac{\omega_s C V_s^2}{2}\sin2\theta_s}{\sqrt{\frac{W_0}{C} - \frac{V_s^2}{2}\sin^2\theta_s - \frac{V_sI_s}{2\omega_s C}\sin(2\theta_s + \phi)}}.
\end{aligned} \tag{10}
$$

The positive node current charges C_1 and C_2 at the same time and the negative node current discharges them for the power decoupling operation. To control a displacement power factor (DPF) in the line current, the leading current i_{cs} should be considered. The leading current is represented as

$$i_{cs} = i_{C1} - i_{C2}, \tag{11}$$

and thus, it is given by,

$$i_{cs} = \frac{C}{2}\frac{dv_s}{dt} = \frac{\sqrt{2}}{2}\omega_s C V_s\cos\theta_s. \tag{12}$$

Note that the leading current i_{cs} flows with constant amplitude in any load power conditions.

B. Condition of power decoupling operation

Fig. 3 shows the waveforms of the resonant current i_r and the switching functions of the upper and lower half-bridge converters, S_1 and S_2, and the output voltage of the direct ac-to-ac converter v_r when the line voltage is positive. Here, the phase-shift control method is applied to control the line current and the filter capacitor voltages. The phase-shift angles θ_1 and θ_2 are defined as the phase angles at the rising edges of S_1 and S_2, respectively. To achieve zero-voltage switching, the phase-shift angles should be set as follows:

$$-\pi < \theta_1 < 0 \tag{13}$$

$$0 < \theta_2 < \pi. \tag{14}$$

Considering the above range of the phase-shift angles, θ_1 and θ_2 to realize the power decoupling operation are given, as follows:

$$\theta_1 = -\cos^{-1}\left\{\frac{\pi}{\sqrt{2}I_r}\left(i_s - i_{cs} - \frac{i_n}{2}\right)\right\} \tag{15}$$

$$\theta_2 = \cos^{-1}\left\{\frac{\pi}{\sqrt{2}I_r}\left(i_s - i_{cs} + \frac{i_n}{2}\right)\right\}. \tag{16}$$

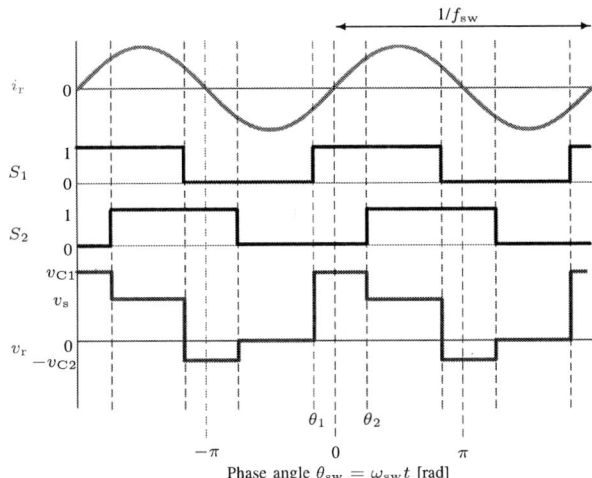

Fig. 3. Relationship among the resonant current i_r, switching functions S_1 and S_2, and the resonant tank voltage v_r in a switching cycle when $v_s > 0$.

From the phase-shift angles in (15) and (16), the angular switching frequency ω_{sw} to keep the resonant current constant can be obtained by

$$\omega_{sw} = \frac{V_{im}}{2L_rI_r} + \sqrt{\left(\frac{V_{im}}{2L_rI_r}\right)^2 + \frac{1}{L_rC_r}}, \tag{17}$$

where V_{im} is the imaginary part of the switching frequency component in the resonant voltage v_r, as follows:

$$V_{im} = \frac{\sqrt{2}}{\pi}(v_{C2}\sin\theta_2 - v_{C1}\sin\theta_1). \tag{18}$$

Reference values obtained by (15), (16), and (17) are required in the power decoupling operation [7].

From the restrictions in the arccosine functions in (15) and (16), the condition of the power decoupling operation is derived. The rms value of the resonant current in (15) and (16) should satisfy the following requirement:

$$I_{r\_min} = \max\left(\frac{\pi}{\sqrt{2}}\left|i_s - i_{cs} \pm \frac{i_n}{2}\right|\right). \tag{19}$$

This requirement derives a limitation of the load voltage by,

$$V_L \leq \frac{\pi}{2\sqrt{2}}\frac{n_2}{n_1}\frac{P_L}{I_{r\_min}}. \tag{20}$$

The above equation means that the minimum resonant current requirement also results in the limitation in the load voltage V_L.

The characteristics of the load voltage V_L in the power decoupling control method is strongly related with the minimum resonant current I_{r_min}. As shown in the right hand side of (19), the minimum resonant current is equal to a maximum value of the sum of i_s, i_{cs}, and $i_n/2$. This control method has originally been proposed to achieve a unity displacement power factor in the line current. The line current i_s and the node current i_n are almost proportional to the load power. However, the leading current i_{cs} is a constant value even if the load power is

3815

changed. This means that a relative large resonant current is required to obtain a unity displacement power factor in the line current in a light load condition. As a result, the load voltage should be reduced to control the load power in the light load condition [7].

IV. LOAD VOLTAGE REGULATION

Fig. 4 shows calculated results of the minimum resonant current I_{r_min} at the load power of $P_L = 115$ W. To regulate the load voltage at a constant level in a wider load condition, the minimum resonant current in (19) should be reduced in the light load condition. From (19), the reference value of the line current i_s, the leading current i_{cs}, or the node current i_n should be adjusted to obtain a lower minimum resonant current. However, i_s should be set as the sinusoidal waveform to take the power from the line side with a low total harmonic distortion. Therefore, this paper discusses the adjusting the i_{cs} or i_n instead of i_s.

Fig. 4(a) is the relation between the average stored energy W_0 and the minimum resonant current I_{r_min} under the unity displacement power factor in the line current. In this case, the power factor angle in (2) is set to be $\phi = 0$. The minimum resonant current is reduced in a range of the higher average stored energy. From (10), the node current can be reduced by boosting the average stored energy W_0, and thus, the minimum resonant current can also be reduced. This method may keep the load voltage constant with a unity displacement power factor in the line current in the wider load condition. However, this method requests a higher filter capacitor voltage especially in the light load condition. This also request a higher switching frequency. Therefore, the power conversion efficiency would decrease.

Fig. 4(b) is the relation between the power factor angle in the line current, ϕ and the minimum resonant current I_{r_min} under a constant average stored energy of $W_0 = 0.58$ J. In Fig. 4(b), the minimum resonant current has a minimum value around $\phi \approx 22$ deg. This means that the minimum resonant current in (19) can take a minimum value because the leading current is not compensated. In this case, the direct converter operates with no displacement power factor correction (DPF correction), and thus, the line current has a leading phase angle due to the leading current. However, the power decoupling control method requests a small filter capacitor. The displacement power factor is not significantly dropped even if the direct converter does not compensate the leading current. Without DPF correction, the phase angle in the line current in (2) is set as follows:

$$\phi = \tan^{-1} \frac{\omega_s C V_s^2}{2 P_L}. \tag{21}$$

In this condition, the load voltage can be regulated at a constant level in the wider load condition with a low filter capacitor voltage.

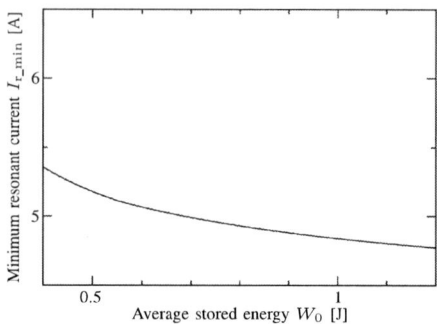

(a) Minimum resonant current corresponding to average stored energy.

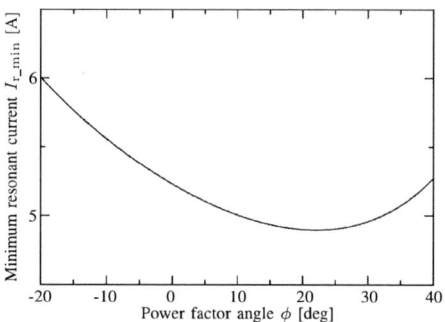

(b) Minimum resonant current corresponding to power factor.

Fig. 4. Calculation results of the minimum resonant current corresponding to the average stored energy and power factor angle at the load power condition of $P_L = 115$ W.

V. EXPERIMENTAL RESULTS

A. Experimental waveforms

Figs. 5 and 6 show experimental waveforms obtained by the experimental setup shown in Fig. 1. The minimum load power was set to be $P_L = 100$ W due to the limitation of operating frequency of the gate drive circuit.

Fig. 5 shows experimental waveforms at the condition of $P_L = 315$ W. Fig. 5(a) is with DPF correction. The direct converter compensated the leading current, and thus, the line current had a unity DPF. The total harmonic distortion (THD) in the line current was 1.5%. In this case, the average voltage of the filter capacitors was 134 V and the average load voltage was kept at 50 V. Fig. 5(b) is the experimental waveforms without DPF correction. In this case, DPF in the line current was slightly dropped to 0.99 but THD in the line current was kept as low as 1.4%. Consequently, experimental waveforms in Fig. 5(b) are almost same with Fig. 5(a) because the leading current is enough small compared with the line current. This means the proposed control method in the rated power condition has almost no effect on the operation and can keep the high displacement power factor close to a unity.

Fig. 6 shows experimental waveforms at the condition of $P_L = 115$ W. Fig. 6(a) is with DPF correction. The line current also had a unity DPF and its THD was 3.6%.

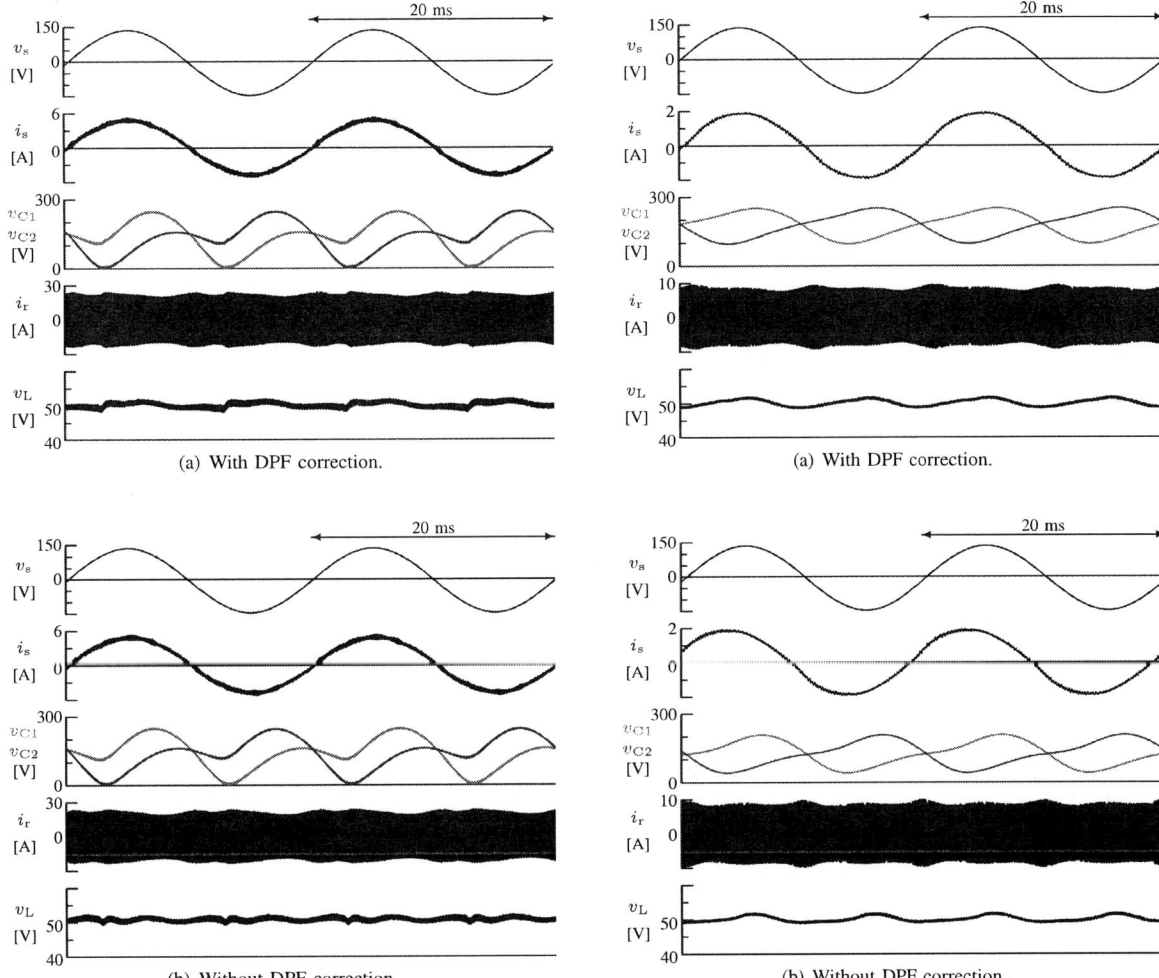

Fig. 5. Experimental waveforms at the load power of $P_L = 315$ W.

Fig. 6. Experimental waveforms at the load power of $P_L = 115$ W.

The load voltage was kept at the constant voltage of 50 V but the average voltage of the filter capacitors was a high voltage of 183 V. Fig. 6(b) is the experimental waveforms without DPF correction. The line current had a leading DPF of 0.91 due to the leading current but THD in the line current was kept as low as 3.9%. This means that the proposed control method without DPF correction can keep a sinusoidal line current with a low THD. Moreover, the average voltage of the filter capacitors could be reduced to 124 V even if the load voltage was set to be 50 V.

Fig 7 shows the node current waveforms at the same condition with Fig. 6. The real node current i_n includes many switching ripple components. The average value of the node current $\bar{i}_n$ is also shown here to evaluate the proposed control method. Fig. 7(a) is the node current with DPF correction and Fig. 7(b) is without DPF correction. In Fig. 7(a), the node current was reduced to obtain a small minimum resonant current. The rms value of $\bar{i}_n$ was 0.57 A. On the other hand, in Fig. 7(b), the rms value of $\bar{i}_n$ was 0.69 A. In the proposed control method,

the small minimum resonant current was obtained by not compensating the leading current. Therefore, the node current had relative large value compared with Fig. 7(a) because it is not needed to increase the average stored energy W_0.

B. Discussion of control characteristics

Figs. 8 and 9 show control characteristics obtained by the experimental setup. Fig. 8(a) is the average load voltage. In both control methods, the load voltage is kept at 50 V in all measured load conditions. Fig. 8(b) is the DPF in the line current. In the case of DPF correction, the line current is well controlled with a unity DPF. On the other hand, in the case of no DPF correction, the DPF is dropped in the light load condition. However, the DPF in the rated power condition is more than 0.99. Consequently, the proposed control method also achieves the high DPF in the rated power condition.

Fig. 9(a) is the average voltage of the filter capacitors. In the case of DPF correction, the average voltage is significantly increased in the light load condition. Thus,

The 2018 International Power Electronics Conference

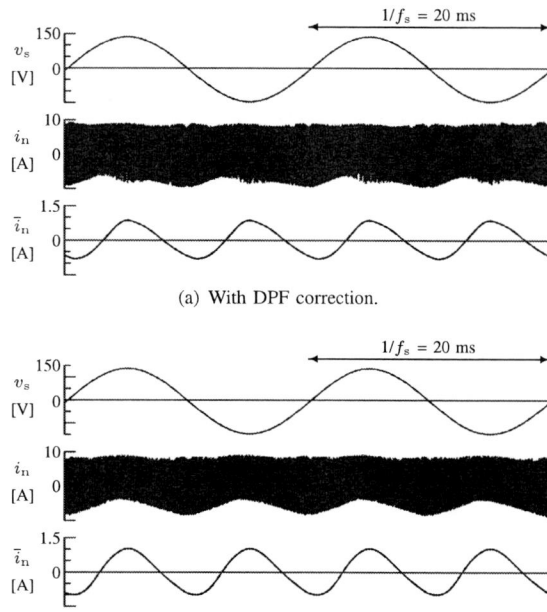

(a) With DPF correction.

(b) Without DPF correction.

Fig. 7. Experimental waveforms of the node current at the load power of $P_L = 115$ W.

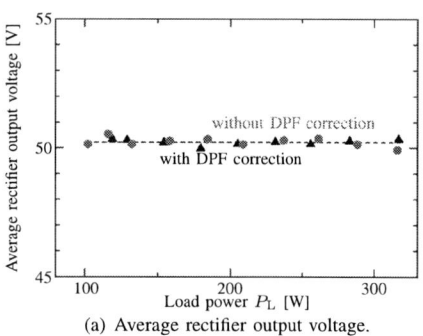

(a) Average rectifier output voltage.

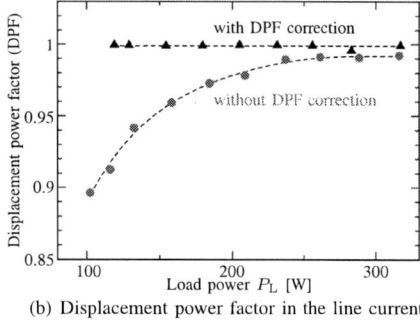

(b) Displacement power factor in the line current.

Fig. 8. Measured average load voltage and displacement power factor (DPF) in the line current.

this control method may have an overvoltage problem in the light load condition. In contrast, in the case of no DPF correction, the average voltage of the filter capacitors can keep a lower level. Fig. 9(b) is the average switching frequency. The control method without DPF correction can operate with a lower switching frequency because of the low filter capacitor voltage. Fig. 9(c) is the power

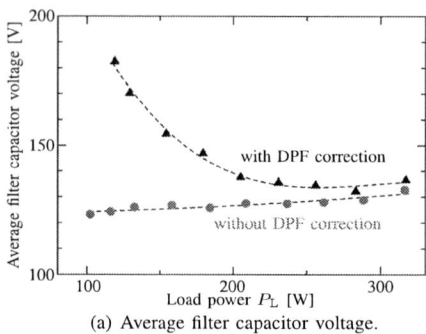

(a) Average filter capacitor voltage.

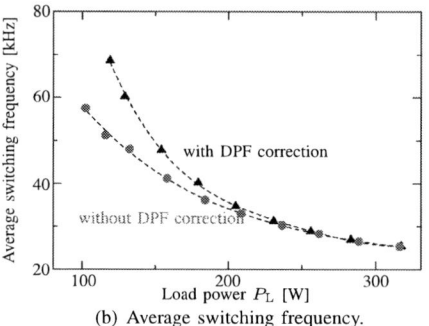

(b) Average switching frequency.

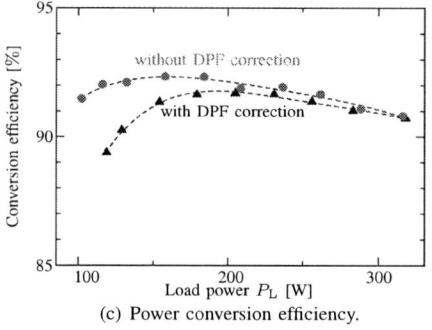

(c) Power conversion efficiency.

Fig. 9. Measured average filter capacitor voltage, average switching frequency, and power conversion efficiency.

conversion efficiencies. The power converter efficiency without DPF correction can be improved from 89.2% to 92.1% at $P_L = 115$ W. This comes from the lower average voltage of the filter capacitors.

VI. CONCLUSIONS

This paper has proposed a new load voltage regulation method for the ac-to-dc converter with the power decoupling operation. The proposed control method does not compensate the leading current due to the filter capacitors and makes it possible to keep the load voltage constant in wide load conditions. The displacement power factor in the line current has been slightly dropped in light load conditions but it was more than 0.89 in measured load power conditions because this direct converter only requests small filter capacitors. Moreover, the proposed control method has realized a higher power conversion efficiency of 92.3 % at the 160-W load power condition.

3818

REFERENCES

[1] M. Yilmaz and P. T. Krein, "Review of Battery Charger Topologies, Charging Power Levels, and Infrastructure for Plug-In Electric and Hybrid Vehicles," *IEEE Trans. on Power Electron.*, vol. 28, no. 5, pp. 2151-2169, May 2013.

[2] M. A. Vitorino, L. F. S. Alves, R. Wang and M. B. de Rossiter Correa, "Low-Frequency Power Decoupling in Single-Phase Applications: A Comprehensive Overview", *IEEE Trans. on Power Electron.*, vol. 32, no. 4, pp. 2892-2912, April 2017.

[3] L. Xue, Z. Shen, D. Boroyevich, P. Mattavelli and D. Diaz, "Dual Active Bridge-Based Battery Charger for Plug-in Hybrid Electric Vehicle With Charging Current Containing Low Frequency Ripple," *IEEE Trans. on Power Electron¿*, vol. 30, no. 12, pp. 7299–7307, Dec. 2015.

[4] S. Yamaguchi, and T. Shimizu,"Single-phase Power Conditioner with a Buck-boost-type Power Decoupling Circuit",*IEEJ J. Ind. Appl.*, vol. 5, no. 3, pp. 191–198, 2016.

[5] F. Jauch and J. Biela, "Combined Phase-Shift and Frequency Modulation of a Dual-Active-Bridge AC?DC Converter With PFC," *IEEE Trans. on Power Electron.*, vol. 31, no. 12, pp. 8387–8397, Dec. 2016

[6] G. T. Chiang, S. Takahide and S. Masaru, "Optimal design of a matrix converter with a LC active buffer for onboard vehicle battery charger in single phase grid structure," *18th European Conf. on Power Electron. and Appl. (EPE'16 ECCE Europe)*, Karlsruhe, 2016, pp. 1-10.

[7] S. Komeda and H. Fujita, "A Power decoupling control method for an isolated single-phase ac-to-dc converter based on Direct AC-to-AC Converter topology," *IEEE Trans. on Power Electron.*, Early access article, 2018.

[8] B.J. Pierquet, A.K. Hayman, G.E. Gamache, C.R. Sullivan and D.J. Perreault, "High-efficiency inverter for photovoltaic applications", *IEEE Energy Conv. Congr. and Expo.*, Atlanta, 2010, pp. 2803–2810.

[9] B. J. Pierquet and D. J. Perreault, "A Single-Phase Photovoltaic Inverter Topology With a Series-Connected Energy Buffer," *IEEE Trans. on Power Electron.*, vol. 28, no. 10, pp. 4603-4611, Oct. 2013.

[10] S. H. Lee, Y. W. Cho, W. J. Cha, K. T. Kim and B. H. Kwon, "High efficient series resonant converter using direct power conversion," *IET Power Electron.*, vol. 7, no. 12, pp. 3045–3051, 2014.

[11] S. Komeda and H. Fujita, "A Phase-Shift Controlled Direct AC-to-AC Converter for Induction Heaters,"*IEEE Trans. on Power Electron.*, vol. 33, no. 5, pp. 4115–4124, May, 2018.

Optimal design of a low cost 20kW 99.1% efficiency active ZCS isolated dc-dc converter

Timothé Delaforge and Sébastien Mariéthoz
Power Electronics Laboratory
Institute for Energy and Mobility Research
Bern University of Applied Sciences
Bienne, Switzerland
Email: timothe.delaforge@bfh.ch; sebastien.mariethoz@bfh.ch

Abstract—The paper presents the optimal design of an innovative very high efficiency isolated dc-dc power converter topology. In terms of efficiency and ease to control, it overperforms resonant dc-dc converters from the literature. High efficiency and low cost are obtained by operating low cost IGBT without switching losses with a power factor close to optimal. It is particularly well suited to be employed as module to make low and medium voltage isolated ac/dc converters or fast DC chargers. A variant of the topology allows bidirectional power flow, which enables the efficient and cost effective integration of renewable energies and energy storage into the medium voltage grid. Used as a solid state transformer, it allows higher efficiency than the classical ac distribution transformer and the authors consider it as an enabler for hybrid ac-dc and dc grids.

Index Terms—active zero current switching, analytical component modeling, dc-dc isolated converter, NSGA-II optimization.

I. INTRODUCTION

A. Paper contributions: optimal design of an active ZCS isolated dc-dc converter topology (AZCS)

The paper presents the optimal design and validation of a 20 kW very high efficiency active ZCS isolated dc-dc converter (AZCS) prototype. The ACZS topology was recently introduced by the authors [1]. The main contributions of the paper are:

1) the application of sophisticated models from the authors to the formulation of the design of the AZCS prototype as an optimization problem; the optimization algorithm is derived from literature [2]–[4];
2) the introduction of an innovative transformer with very low inductance applying flat winding technology; to the best of our knowledge this is the first application of this technique to a transformer;
3) the achievement of a very high efficiency at low cost based on the application of the aforementioned techniques to the optimal design of the AZCS topology;
4) the computation of the efficiency of the AZCS prototype for its whole operating range;
5) the presentation of the Pareto optimum efficiency versus cost for the AZCS prototype.

B. Applications of the AZCS topology

The authors topology shown in Fig. 1 is particularly well suited to be employed as module to make low and medium

voltage isolated dc/dc or ac/dc converters, e.g. to make solid state transformers [5]–[9] to obtain galvanic isolation or voltage level and frequency conversion [10]–[12].

C. Advantages of the AZCS topology over existing ones

The presented converter uses an innovative topology allowing better performance and lower cost than other isolated dc-dc topologies [7], [13]–[16]:

1) the converter switches at zero current; it is similar as for resonant converters but obtained via another mean;
2) the power factor is close to unity and the current shape close to optimal, actually closer to optimal than the sine waveforms obtained with resonant converters, which allows to overperform the resonant converters in terms of efficiency;
3) efficiency >99% with classical low cost Si semiconductors;
4) easy and instantaneous control of the transferred power; bidirectionally for the bidirectional variant of the topology;
5) no need for a resonant circuit;
6) no influence of temperature on the on soft switching operation;
7) significant work as been done on investigating the control of isolated power converters [17]; the gain is here topological, an appropriate control strategy is however required.

II. AZCS TOPOLOGY PRINCIPLES

A. Hybridization of modules and semiconductor devices to achieve high efficiency

The proposed topology is based on the hybridization of converter modules and semiconductors as shown in Fig. 1.

The employed hybridization principle is based on the technique employed in [18] to design multilevel converter with low losses and capability to balance the energy stored in unsupplied converter modules as is the case here for module A. The hybridization makes it possible to operate the unsupplied converter with a very small voltage, reducing consequently its switching losses. The other converters operating at the full voltage rating of the topology are only subject to conduction

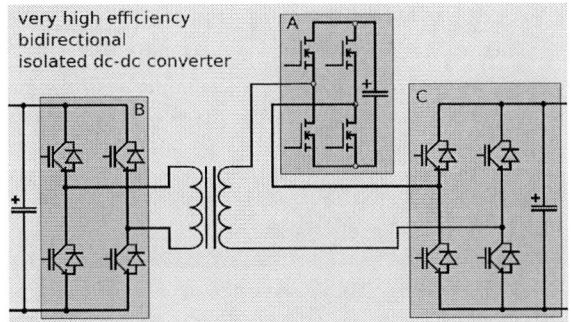

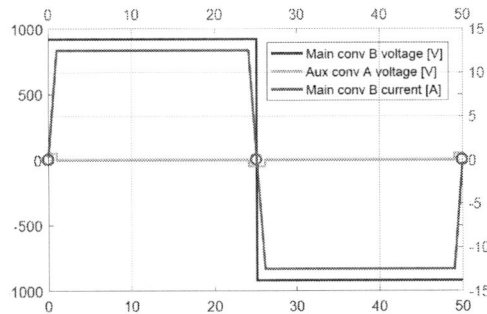

Fig. 1: High-efficiency bidirectional isolated dc-dc converter. Converters B and C transfer the power; they switch the full dc bus voltage at zero current (ZCS). The converter A is operated in quadrature with B and C and shapes the current; it switches the full current at very low voltage.

Fig. 2: Ideal waveforms explaining the key concepts enabling very high efficiency: Main converter B voltage and current are in phase. This guarantees close to optimal current waveforms and zero current switching of the converters B and C. Converter B and C operate in phase while converter A shapes the current and control the power transfer. The voltage ratings of converter A are much smaller than of converter B and C such as its cost and losses are negligible.

losses. This splitting of loss origins allows a better selection of semiconductors to reduce them.

B. AZCS control principle through converter module A to achieve zero current switching for converter modules B and C

The key idea of the topology is that the functions of controlling the power and transferring the power are distributed over different converter modules:

- The converter modules B and C switch their respective DC bus voltages and have the function of transferring the power;
- The converter module A switches at a significantly smaller voltage (10 to 50 times less than the bus voltage) and has the function of shaping the power: on average, it transfers zero net power.

The associated mechanism is illustrated in Fig. 2. To achieve this the converters B and C are operated in phase with a full duty cycle, while the converter A is operated in phase quadrature and controls the current shape. As a result the converter B and C switch at zero current and the converter A switches with very low voltage. Which is one of the keys enabling high efficiency.

III. AZCS CONVERTER MULTI-OBJECTIVE OPTIMIZATION MODEL

A. Prototype target ratings

A 20 kW 1 kV to 750 V step-down dc-dc isolated converter is designed using the optimization model described in this section. The AZCS converter is optimized for 50% and 100% load to maximize the actual operating efficiency of the system.

B. Optimization objectives

The objectives are:
1) minimization of the price;
2) minimization of the volume;
3) maximization of efficiency at 50% and 100 %P_n.

C. Optimization constraints

1) keep components at a safe operating temperature;
2) ensure sufficient voltage isolation between the transformer windings.

D. Optimization input variables

1) a library of available IGBT and MOSFET
2) a library of available magnetic materials
3) a library of toroid transformer cores (available sizes)
4) winding properties (number of turns and parallel wires, wires diameter and material)
5) a library of available capacitors
6) switching frequency
7) converter control strategy
8) converter module A dc voltage

E. Transformer optimization model

1) Transformer main constraints: low losses and leakage inductance: To achieve the highest possible efficiency for the converter, the transformer leakage inductance must be as small as possible. The classical core geometries used so far for designing MV transformer are core-type and shell-type [19]–[21]. Toroidals shapes however allow better symmetries and repartition of effects as described in [22]. It prevents air-gap frigging flux and its drawbacks. The disadvantage of toroid is the insulation requirement between primary and secondary and thermal cooling of conductors and core as windings are usually winded one over another.

2) Transformer geometry and winding technique selection: To overcome these drawbacks of toroidal cores and reduce the transformer leakage inductance and capacitance, the authors propose the use of flat edge conductors, see Fig. 3. These conductors present very good resistance frequency characteristic [23]. This core geometry and wires reduces consequently the leakage inductance and allows good air cooling of both. The electrical insulation of wires 150 V per μm is enough for the 1 kV prototype but for higher insulation requirement this geometry enables good oil or epoxy insulation.

3) Core loss model: The transformer core losses are computed from an accurate dynamic hysteresis model [24], which is based on measurements of the material and building dynamic hysteresis from real converter waveforms.

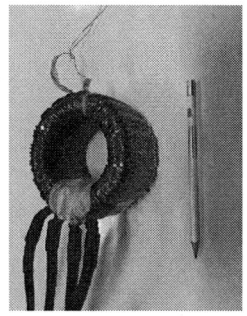 

(a) First transformer prototype with round wire *(b) Traftor coil with flat wire: second prototype will employ this wire technology*

Fig. 3: First transformer was built and validated. It should allow reaching 99.1% efficiency as shown by the blue curve in Fig. 6. Second transformer with flat edge wire should allow reaching higher efficiency as shown by the red and green curves in Fig. 6.

4) Transformer leakage inductance and winding loss model: The transformer leakage inductance and winding loss are computed using an accurate analytical model [25].

F. Converter model

1) Control strategy: The converter model takes into account the control strategy as it influences its losses.

2) Recovery loss reduction: The operation in ZCS mode enables a high frequency operation of the modules B and C. The only limitation are the transformer losses and the IGBT charge recovery losses. The recovery loss problem is coined in [26], that explains that the electric charges stored in IGBT result in switching losses even when ZCS switching is achieved. The phenomenon is reduced here by applying a control strategy that lets time for the charge carriers to recombine by switching off both converter B and C after the zero current is reached to allow sufficient time for the recombination at zero voltage, see Fig. 5. This reduces considerably the recovery losses which become negligible and are consequently neglected in the optimization.

G. Model of converter cost

The prices are based on suppliers offers and take into account power semiconductors, gate drivers, passive components and cooling.

IV. MULTI-OBJECTIVE OPTIMIZATION METHOD

A set of libraries of available components and materials were implemented. The libraries specify the feasible inputs of the optimization model, e.g. core permeability, loss model parameters, associated price, etc. These parameters are used as inputs to the analytical optimization models. The resulting optimization approach is therefore based on discrete input variables. The advantage of a library based approach is that we are sure in advance that the best components exist or are therefore feasible. Additionally the accuracy is very good and the obtained price, efficiency and volume predicted by the optimization are very close to the obtained ones. For continuous input variables, it is indeed necessary that an engineer makes iterations to design the actual components from the optimization results. To deal with these discrete input possibilities, an optimization algorithm derived from NSGA-II [2], [3] is used that constructs the Pareto frontier defining all best tradeoffs between the objectives defined above in III-B.

V. AZCS CONVERTER MULTI-OBJECTIVE OPTIMIZATION RESULTS: PARETO FRONTIER OPTIMAL COST/EFFICIENCY/VOLUME FOR ROUND WIRES

The results of the optimization are illustrated with Pareto between price and efficiency in Fig. 4. We can see that the efficiency is above 99% with classical Si technology.

We selected a solution among sets of parameters with classical IBGT and MOS with competitive price. The cores of the solution on the Pareto frontier are only made from high flux density nano-crystalline and amorphous materials. The reason is that operating frequencies below 50 kHz do not justify the use of cheaper but bulkier ferrite. The size of the core is an amorphous core of volume 232 cm^3. The optimal bus voltage of converter A is found by the optimzer to be 30 V and the operating frequency at nominal power 30 kHz with duty cycle of converter A α=0.06 so a 0.98 power factor.

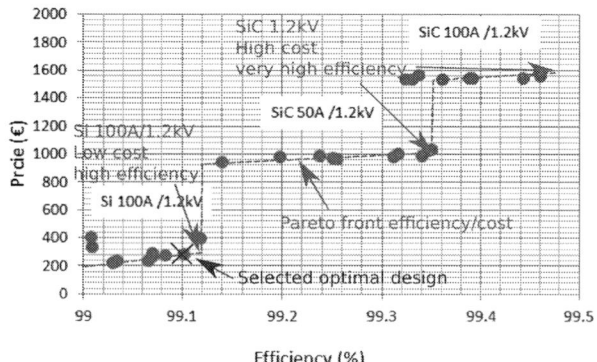

Fig. 4: Optimization results: Pareto frontier for the design of our 20 kW prototype showing the best tradeoffs between price and efficiency. We can see cost discontinuities on the Pareto frontier that correspond to technology changes. We selected the solution highlighted by a cross on the frontier.

VI. CONVERTER PRECISE SIMULATION VALIDATION

A. PSIM detailed simulation

A detailed simulation model of the converter has been done with PSIM. The waveforms are shown in Fig. 5 for one of the control strategies considered in the optimization.

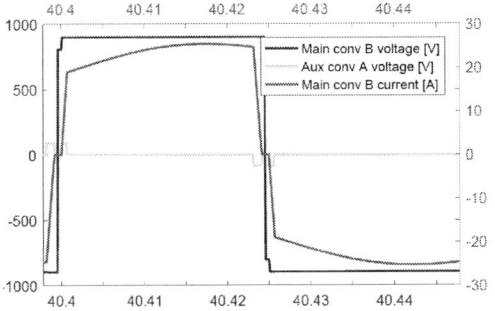

Fig. 5: Detailed simulation results: ZCS operation is achieved: a dead time is imposed to let the IGBT charge carriers recombine at zero voltage thus avoiding associated losses.

B. Efficiency over the full operating range for the two considered winding wires and the two best control strategies

The efficiency of the selected optimal design is plotted in Fig. 6 for both transformer winding technologies for the two best control strategies. The best efficiency is obtained for the flat wire transformer when applying a strategy with variable switching frequency that allows minimizing the losses at partial load: in the latter case it is seen on the efficiency plot that there is little difference at high load but a significant improvement at very low load.

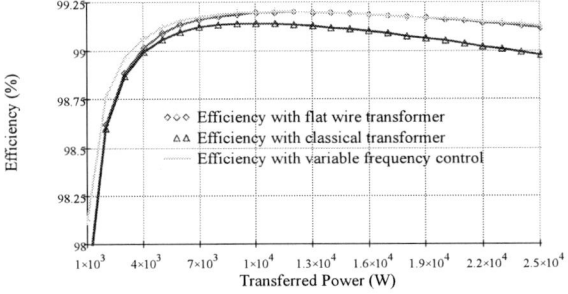

Fig. 6: Efficiency of the 20 kW prototype computed from accurate models and FEA. The efficiency is computed for classical solid round wire transformer and flat edge wire transformer. Frequency control to improve efficiency at partial load is also represented.

C. Loss breakdown at 100% and 10% load

The loss breakdown in Fig. 7 shows that the IGBT conduction losses dominate at high load and that both IGBT conduction losses and transformer core losses play the most important role at partial load. It demonstrates that the switching losses

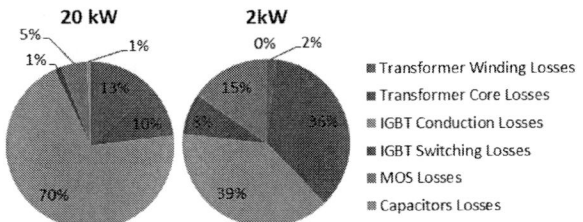

Fig. 7: Losses breakdown of the 20 kW prototype for 10% of Nominal power and nominal power.

are negligible despite the fairly high switching frequency, which validates our concept. It also points out the relevency of the frequency control strategy to reduce transformer losses at partial losses.

VII. FIRST TRANSFORMER PROTOTYPE EXPERIMENTAL VALIDATION

A comparison between the leakage inductance predicted by the optimization model and the leakage inductance measured on the prototype with the round wire is shown in Fig. 8. It can be seen that the model and the prototype results match very well. This validates the most critical point for the AZCS converter as achieving low leakage inductance was crucial and as modeling the leakage inductance precisely was one of the most challenging aspect of the project and as it has a significant influence on the results.

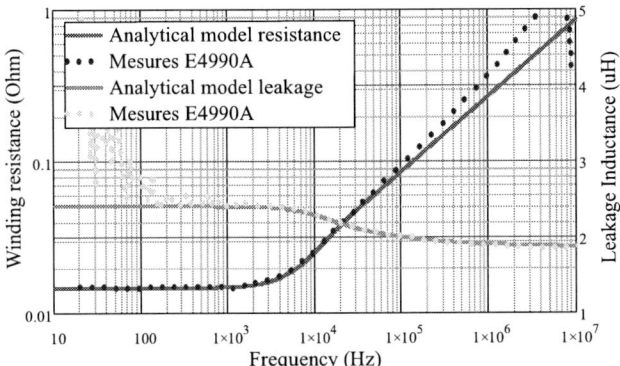

Fig. 8: Leakage inductance model versus measurements for the first prototype. The leakage inductance is measured on a transformer with solid round wires and primary and secondary separated by insulation. The measure is compared to analytical model. Top-Right: our first prototype with round wires.

VIII. CONCLUSION

The paper has presented the optimal design of an active ZCS isolated dc-dc converter. The principles of the topology that allow switching at zero current for two modules and very small voltage with the one module has been presented. An optimization of the topology has been presented that enables the design of a low cost very high efficiency AZCS converter. Simulation results demonstrate the high efficiency of the designed AZCS converter. A key point for obtaining high efficiency is the achievement of a very low inductance. It is experimentally demonstrated that our first prototype with toroidal core and round wires achieves as expected a low leakage inductance.

REFERENCES

[1] authors removed on purpose, "A new hybrid isolated dc-dc converter topology for realizing very high efficiency isolated ac-dc chargers," in *IEEE VPPC 2017*, Nov. 2017.

[2] K. Deb, S. Agrawal, A. Pratap, and T. Meyarivan, "A Fast Elitist Nondominated Sorting Genetic Algorithm for Multi-objective Optimisation: NSGA-II," in *Proceedings of the 6th International Conference on Parallel Problem Solving from Nature*, ser. PPSN VI. London, UK, UK: Springer-Verlag, 2000, pp. 849–858. [Online]. Available: http://dl.acm.org/citation.cfm?id=645825.668937

[3] J. Knowles and D. Corne, "The pareto archived evolution strategy: a new baseline algorithm for pareto multiobjective optimisation," in *Proceedings of the 1999 Congress on Evolutionary Computation-CEC99 (Cat. No. 99TH8406)*, vol. 1, 1999, p. 105 Vol. 1.

[4] E. Zitzler, K. Giannakoglou, D. Tsahalis, J. Periaux, K. Papailiou, T. F. (eds, E. Z. Ler, M. Laumanns, and L. Thiele, *SPEA2: Improving the Strength Pareto Evolutionary Algorithm For Multiobjective Optimization*, 2002.

[5] X. She, S. Lukic, A. Q. Huang, S. Bhattacharya, and M. Baran, "Performance evaluation of solid state transformer based microgrid in freedm systems," in *2011 Twenty-Sixth Annual IEEE Applied Power Electronics Conference and Exposition (APEC)*, March 2011, pp. 182–188.

[6] H. Hoffmann and B. Piepenbreier, "Medium frequency transformer for rail application using new materials," in *2011 1st International Electric Drives Production Conference*, Sep. 2011, pp. 192–197.

[7] D. Dujic, C. Zhao, A. Mester, J. K. Steinke, M. Weiss, S. Lewdeni-Schmid, T. Chaudhuri, and P. Stefanutti, "Power Electronic Traction Transformer-Low Voltage Prototype," *IEEE Transactions on Power Electronics*, vol. 28, no. 12, pp. 5522–5534, Dec. 2013.

[8] C. Zhao, D. Dujic, A. Mester, J. K. Steinke, M. Weiss, S. Lewdeni-Schmid, T. Chaudhuri, and P. Stefanutti, "Power Electronic Traction Transformer – Medium Voltage Prototype," *IEEE Transactions on Industrial Electronics*, vol. 61, no. 7, pp. 3257–3268, Jul. 2014.

[9] J. W. Kolar and G. Ortiz, "Solid-state-transformers: Key components of future traction and smart grid systems," in *Proc. of the International Power Electron. Conf.*, May 2014.

[10] N. H. Kutkut, D. M. Divan, D. W. Novotny, and R. H. Marion, "Design considerations and topology selection for a 120-kW IGBT converter for EV fast charging," *IEEE Transactions on Power Electronics*, vol. 13, no. 1, pp. 169–178, Jan. 1998.

[11] B. M. Grainger, G. F. Reed, A. R. Sparacino, and P. T. Lewis, "Power Electronics for Grid-Scale Energy Storage," *Proceedings of the IEEE*, vol. 102, no. 6, pp. 1000–1013, Jun. 2014.

[12] M. Vasiladiotis and A. Rufer, "A Modular Multiport Power Electronic Transformer With Integrated Split Battery Energy Storage for Versatile Ultrafast EV Charging Stations," *IEEE Transactions on Industrial Electronics*, vol. 62, no. 5, pp. 3213–3222, May 2015.

[13] H. Fan and H. Li, "High-frequency transformer isolated bidirectional dc-dc converter modules with high efficiency over wide load range for 20 kva solid-state transformer," *IEEE Transactions on Power Electronics*, vol. 26, no. 12, pp. 3599–3608, Dec 2011.

[14] L. Costa, G. Buticchi, and M. Liserre, "A fault-tolerant series-resonant dc -dc converter," *IEEE Transactions on Power Electronics*, vol. 32, no. 2, pp. 900–905, Feb 2017.

[15] M. Yilmaz and P. T. Krein, "Review of Battery Charger Topologies, Charging Power Levels, and Infrastructure for Plug-In Electric and Hybrid Vehicles," *IEEE Transactions on Power Electronics*, vol. 28, no. 5, pp. 2151–2169, May 2013.

[16] D. Dujic, A. Mester, T. Chaudhuri, A. Coccia, F. Canales, and J. K. Steinke, "Laboratory scale prototype of a power electronic transformer for traction applications," in *Proceedings of the 2011 14th European Conference on Power Electronics and Applications*, Aug. 2011, pp. 1–10.

[17] G. Ortiz, J. Biela, D. Bortis, and J. W. Kolar, "1 megawatt, 20 khz, isolated, bidirectional 12kv to 1.2kv dc-dc converter for renewable energy applications," in *The 2010 International Power Electronics Conference - ECCE ASIA -*, June 2010, pp. 3212–3219.

[18] S. Mariethoz, "Systematic Design of High-Performance Hybrid Cascaded Multilevel Inverters With Active Voltage Balance and Minimum Switching Losses," *IEEE Transactions on Power Electronics*, vol. 28, no. 7, pp. 3100–3113, Jul. 2013.

[19] S. Vaisambhayana, C. Dincan, C. Shuyu, A. Tripathi, T. Haonan, and B. R. Karthikeya, "State of art survey for design of medium frequency high power transformer," in *2016 Asian Conference on Energy, Power and Transportation Electrification (ACEPT)*, Oct 2016, pp. 1–9.

[20] G. Ortiz, M. G. Leibl, J. E. Huber, and J. W. Kolar, "Design and experimental testing of a resonant dc-dc converter for solid-state transformers," *IEEE Transactions on Power Electronics*, vol. 32, no. 10, pp. 7534–7542, Oct 2017.

[21] H. Hoffmann and B. Piepenbreier, "High voltage IGBTs and medium frequency transformer in DC-DC converters for railway applications," in *SPEEDAM 2010*, Jun. 2010, pp. 744–749.

[22] R. J. Pasterczyk and T. Delaforge, "Engineering illusion to accurately predict power losses in magnetic materials on the base of standard manufacturers' datasheets," in *PCIM Europe 2014; International Exhibition and Conference for Power Electronics, Intelligent Motion, Renewable Energy and Energy Management*, May 2014, pp. 1–9.

[23] T. Delaforge, H. Chazal, and R. J. Pasterczyk, "Optimization of windings in pfc boosts and pwm inverters to maximize converter efficiency," in *2014 IEEE Applied Power Electronics Conference and Exposition - APEC 2014*, March 2014, pp. 768–774.

[24] A. Frias, A. Kedous-Lebouc, C. Chillet, L. Albert, and L. Calegari, "Improvement and validation of an iron loss model for synchronous machine," in *2012 XXth International Conference on Electrical Machines*, Sep. 2012, pp. 1328–1332.

[25] T. Delaforge, H. Chazal, J. L. Schanen, and R. J. Pasterczyk, "Copper losses evaluation in multi-strands conductors formal solution based on the magnetic potential," in *2015 IEEE Energy Conversion Congress and Exposition (ECCE)*, Sep. 2015, pp. 3057–3063.

[26] G. Ortiz, D. Bortis, J. W. Kolar, and O. Apeldoorn, "Soft-switching techniques for medium-voltage isolated bidirectional dc/dc converters in solid state transformers," in *IECON 2012 - 38th Annual Conference on IEEE Industrial Electronics Society*, Oct 2012, pp. 5233–5240.

Soft-Switching Analysis and PFM Control Method of Bidirectional DC/DC Converter Topology

Yijie Wang[1], Haoyu Wang[1], Hongyu Song[1], Dianguo Xu[1]
1 School of Electrical Engineering & Automation, Harbin Institute of Technology, Harbin, China
E-mail: wangyijie@hit.edu.cn

Abstract— **Bidirectional DC/DC converters have been widely used as an interface for two-way energy flow with the development of new energy technologies. Soft-switching technology is often integrated in bidirectional converters to reduce switching loss, but analysis and design of circuit are usually complex. This paper proposes a method for soft-switching, and a closed-loop control analysis. The soft-switching conditions, the influence of parameter on the output and the design of PFM closed-loop control system are analyzed in detail. A sweep frequency analysis of the designed circuit was carried out in this study, and the PI regulator was designed according to the frequency characteristic of the system. The proposed 1000 W-closed-loop prototype realizes a bidirectional energy flow under control of the program. The prototype can realize a mutual conversion of 200 V and 400 V, with an efficiency of 96.65% in Boost mode and 97.04% in Buck mode under full load.**

Keywords— *Bidirectional DC/DC converter; Soft-switching; PFM control method; Closed-loop control*

I. INTRODUCTION

In the past few decades, people's awareness for the need of environmental protection has gradually increased. Hence, renewable energy sources such as wind energy and solar energy are gaining more attention from various countries. Because of the discontinuity of renewable energy sources, converting them into a continuous, stable, and reliable source is a problem. The most effective way to solve this problem is combining new power generation equipment with energy storage systems and the energy storage system is mainly composed of three parts: distributed generation system, energy storage system, and interface device. The microgrids' The bidirectional DC/DC converter is a bidirectional energy flow port that realizes a seamless connection between the distributed generation system and the power grid [1]. And the bidirectional DC/DC converter can also be used as an interface between an electric vehicle's energy storage system and the power grid [2].The DC/DC converter is obviously of great significance for the application of new energy networks and new energy vehicles.

The bidirectional DC/DC converter controls the energy flow between low-pressure and high-pressure terminal in order to achieve a bidirectional energy passageway and functional block diagram is shown in Fig. 1. When $I_1 < 0$ and $I_2 > 0$, the electric energy is transmitted from the low-voltage to the high-voltage terminal. The energy flows in positive direction to realize a discharge of the energy storage system into the electric grid. When $I_1 > 0$ and $I_2 < 0$, energy flows in the opposite

direction and the grid is charged for the energy storage system.

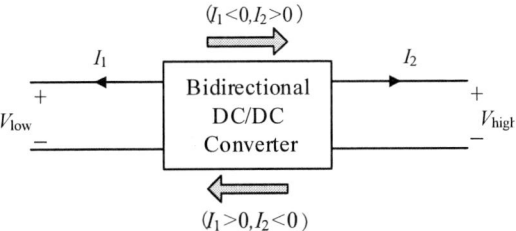

Fig. 1. Structure of bidirectional DC/DC converter.

Bidirectional DC/DC converters are divided into two types: isolated converter and non-isolated converter. Regarding the isolated bidirectional converter, a dual flyback structure is proposed in [3]. Its flyback converter-based structure has the advantages of simple, low switch voltage stress, and low-cost. However, the structure is only suitable for low-power applications. A dual push-pull topology with a high voltage transmission is proposed in [4]. Due to the leakage inductance from the isolation transformer, the switches bear larger peak voltages. Therefore, the structure is not suitable for high-pressure situations. A full-bridge topology with new improvement of multiport converters is presented in [5].The topology exhibits a low voltage and current stress and a high voltage conversion ratio but the circuit structure and control method are complex. A bidirectional resonant converter is proposed in [6] and a bidirectional converter with multiple ports is presented in [7]. Both topologies have a superior circuit performance, but their work processes are complex. Thus, they are also not conducive to improve circuit stability.

Compared with the isolated DC/DC converter, the non-isolated DC/DC converter possesses a simple structure. In [8] a new topology with two more switches based on the traditional Buck/Boost circuit is proposed. It adopts the synchronous rectification control technology and exhibits low voltage stress. However, the complex structure and control system limit the use of the circuit. A 16-phase interleaved bidirectional DC/DC converter is proposed in [9] with a small filter capacitor and small filter inductor. Nevertheless, working in discontinuous conduction mode reduces the efficiency. An E-type structure is proposed and designed in [10] that uses GaN power devices to achieve higher efficiency. However, the topology has four switches in series, and the design requirements of the drive circuit are difficult. Further, two larger inductors limit the circuit's power density.. A bidirectional converter based on a coupled inductor is mentioned in [11]. The circuit possesses a high step-up ratio range, a low switch voltage stress and a low switch current stress as well as a simple control strategy. While the coupled inductor strongly enhances the volume and weight of the circuit, the hard-switching reduces the system efficiency. A soft-switching topology based on the Buck/Boost converter is proposed in [12]. The converter control is simple and suitable for a variety of converter occasions. However, there exists no paper which describes the system modeling and control method in detail and analyzes the soft-switching conditions. Hence, ideal performance conditions for the circuit cannot be found in literature so far.

An analysis of the soft-switching conditions is beneficial for an improved circuit performance. Further, it is beneficial for the application and debugging of the circuit, to facilitate commercial production. An analysis of the small-signal modeling and the control strategy for the circuit is a necessity to ensure circuit stability. In this study, a detailed analysis of the soft-switching conditions of a bidirectional Buck/Boost converter is carried out, and the influence of the main parameters on the circuit output is also investigated. Further, a new system modeling method is proposed, and a closed-loop control is implemented according to the modeling results. The experimental results show that the analysis method is accurate and the circuit achieves superior performance.

Section II briefly analyzes the circuit, and the soft-switching conditions of the circuit are analyzed in detail in section III. In section IV, system modeling using an auxiliary method is proposed and the control process of the system is analyzed. The experimental results of the 1 kW-prototype are discussed in section V. Finally, a conclusion is drawn in Section VI.

II. CIRCUIT ANALYSIS AND PARAMETER DESIGN OF SOFT-SWITCHING BIDIRECTIONAL DC/DC CONVERTER

A bidirectional DC/DC converter with lossless passive soft-switching is presented in this paper (see Fig. 2). A resonant circuit that consists of a snubber capacitor C_s, a resonant inductor L_r, and a resonant capacitor C_r is added to the traditional Buck/Boost converter. Due to the LC resonant circuit, the power switches can operate at zero turn-on voltage and zero turn-off voltage in both Buck and Boost modes. Compared to the other soft-switching topology that adds active devices, the new topology only adds passive components such as inductor and capacitor. Therefore, there is no additional loss. The conversion efficiency and the operation reliability of the converter are greatly improved and the converter has a higher power density [13].

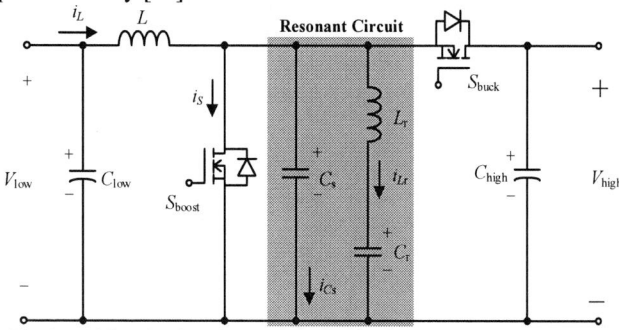

Fig. 2.   Bidirectional DC/DC converter of lossless passive soft-switching.

Voltage and current waves of the main circuit components in Boost and Buck mode are shown in Fig. 3 and Fig. 4.

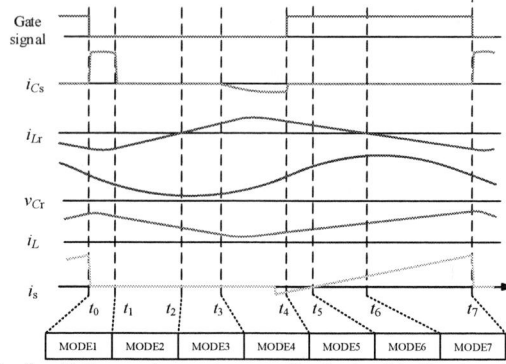

Fig. 3.   Voltage and current waves of main circuit in Boost mode.

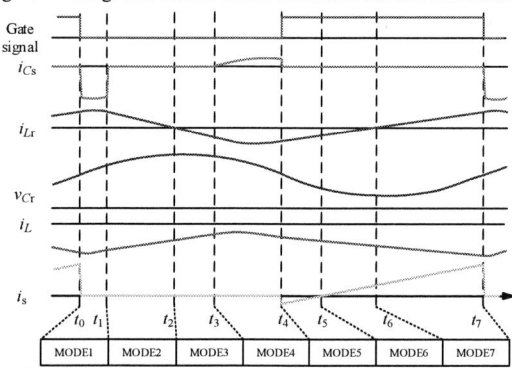

Fig. 4.   Voltage and current waves of main circuit in Buck mode.

III. ANALYSIS OF SOFT-SWITCHING CONDITION

A. Theoretical analysis of soft-switching

Firstly, we should analyze the influence of each parameter change based on the fixed frequency duty cycle. Take the Boost mode as an example. For ease of the analysis, the ideal condition can be assumed: 1. In MODE 3, the snubber capacitor's current is constant and has the value $I_{L\text{-max}}$. In the meantime, the snubber capacitor's voltage v_{Cs} rises linearly up to the output voltage V_{out}. 2. In MODE 6, the snubber capacitor's current remains constant at $(I_{Lr\text{-max}}-I_{L\text{-min}})/2$ and v_{Cs} declines linearly from the output voltage V_{out} to zero. The main curves of the soft-switching circuit are shown in Fig. 5.

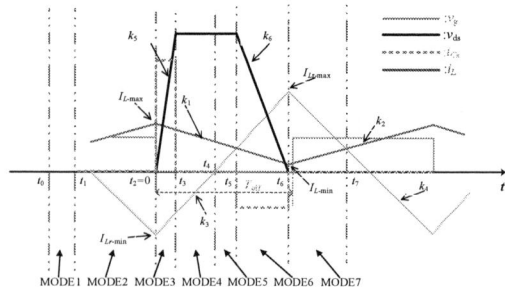

Fig. 5.   Main curves of soft-switching circuit.

According to the physical characteristics of the capacitor, its current can be represented by:

$$C = \frac{Q}{U} \Rightarrow i = C \cdot \frac{du}{dt} \qquad (1)$$

The voltage rising rate can be calculated as:

$$k_5 = \frac{I_{L\text{-max}}}{C_S} \tag{2}$$

The voltage rising time can be represented by:

$$t_3 = \frac{V_{out}}{k_5} = \frac{V_{out} \cdot C_S}{I_{L\text{-max}}} \tag{3}$$

The maximum current of the main inductor can be calculated with:

$$I_{L\text{-max}} = \frac{V_{out} \cdot I_{out}}{V_{in}} + \frac{1}{2} \cdot \frac{V_{in}}{L} \cdot T_{on} \tag{4}$$

Subsequently, the voltage rising time can be calculated by combining (3) and (4):

$$t_3 = \frac{2 \cdot V_{in} \cdot V_{out} \cdot C_s \cdot L}{2 \cdot V_{in} \cdot V_{out} \cdot L + V_{in}^2 \cdot T_{on}} \tag{5}$$

In the time from t_2 to t_5, the current of the main inductor i_L declines linearly from $I_{L\text{-max}}$, and the current of the resonant inductor i_{Lr} rises linearly from the minimum value $I_{Lr\text{-min}}$. At the time t_5, i_L and i_{Lr} are equal. Hence, t_5 can be calculated as:

$$t_5 = \frac{I_{L\text{-max}} - I_{Lr\text{-min}}}{k_3 - k_1} \tag{6}$$

The parameter $I_{Lr\text{-min}}$ can be represented by:

$$I_{Lr\text{-min}} = -\frac{1}{2} \cdot \frac{V_{Cr}}{L_r} \cdot T_{on} = -\frac{1}{2} \cdot \frac{V_{in}}{L_r} \cdot T_{on} \tag{7}$$

According to the analysis above, t_5 can be represented by:

$$t_5 = \frac{V_{in}^2 \cdot T_{on} \cdot (L + L_r) + 2V_{in} \cdot V_{out} \cdot L \cdot L_r}{2 \cdot V_{in} \cdot (V_{out} - V_{in}) \cdot (L + L_r)} \tag{8}$$

In the time from t_5 to t_6, the snubber capacitor's current can be considered to rest at the constant value of $(I_{L\text{-max}} - I_{L\text{-min}})/2$. Therefore, the voltage declining rate $|k_6|$ can be determined by:

$$|k_6| = \frac{1}{2} \cdot \frac{I_{Lr\text{-max}} - I_{L\text{-min}}}{C_s} \tag{9}$$

Hence, the voltage dropping time t_6 (voltage drops to zero) can be calculated as:

$$t_6 = t_5 + \frac{2 \cdot V_{out} \cdot C_s}{I_{Lr\text{-max}} - I_{L\text{-min}}} \tag{10}$$

Combining (10) and the equations above, we obtain:

$$t_6 = \frac{V_{in}^2 \cdot T_{on} \cdot (L + L_r) + 2V_{out} \cdot I_{out} \cdot L \cdot L_r}{2 \cdot V_{in} \cdot (V_{out} - V_{in}) \cdot (L + L_r)}$$
$$+ \frac{4 \cdot V_{in} \cdot V_{out} \cdot C_s \cdot L \cdot L_r}{V_{in}^2 \cdot T_{on} \cdot (L + L_r) - 2V_{out} \cdot I_{out} \cdot L \cdot L_r} \tag{11}$$

In order to reliably implement soft-switching into the system, it is required that the drain-source voltage of the switches can

decline to zero before the driving signal arrives. Hence, $t_6 \leq T_{off}$. Therefore, in order to facilitate the soft-switching analysis, a function f is constructed:

$$f = T_{off} - \frac{V_{in}^2 \cdot T_{on} \cdot (L + L_r) + 2V_{out} \cdot I_{out} \cdot L \cdot L_r}{2 \cdot V_{in} \cdot (V_{out} - V_{in}) \cdot (L + L_r)}$$
$$- \frac{4 \cdot V_{in} \cdot V_{out} \cdot C_s \cdot L \cdot L_r}{V_{in}^2 \cdot T_{on} \cdot (L + L_r) - 2V_{out} \cdot I_{out} \cdot L \cdot L_r} > 0 \tag{12}$$

According to (12), when the value of f is obtained, the influence of the converter's parameter change on the ZVS condition can be analyzed. Fig. 6 shows the circuit response when the value of the main inductor is chosen as 500 μH, 470 μH, 450 μH, and 430 μH. In Fig. 6, when the main inductance value decreases gradually, the time that the drain-source voltage needs to decline to zero leads the driving signal to increase gradually. At the same time, it can be seen that the converter operates under zero-voltage turn-on condition only when the input voltage fluctuates in a certain range. In order to make the converter have the zero-voltage turn-on condition under a wide range of input voltages, the value of the main inductor should be low. When $L = 430$ μH, with input voltage fluctuations from 180 V to 220 V, the power switches can realize the ZVS condition.

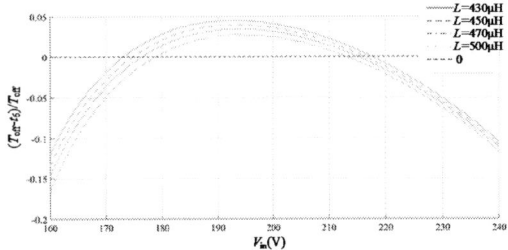

Fig. 6. Influence of main inductance L on ZVS condition.

The influence of the parameters of resonant inductor (L_r) and snubber capacitor (C_S) on the converter is similar. This paper does not precede with a detailed discussion and presents the relevant trend diagrams in Fig. 7 and Fig. 8.

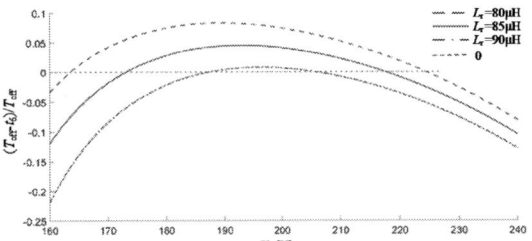

Fig. 7. Influence of main inductance L_r on ZVS condition.

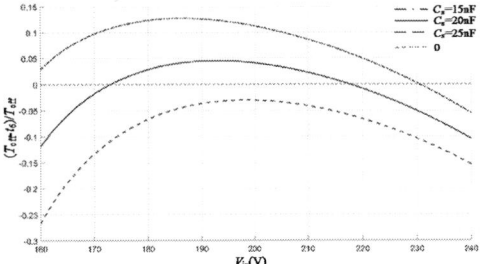

Fig. 8. Influence of main capacitance C_s on ZVS condition.

B. Simulation analysis of soft-switching

Considering the above analysis, contrast experiments are taken to verify the theoretical analysis. The designed parameters of the main circuit are simulated and verified by PSIM software. In this paper, two parameter sets are chosen to satisfy all the above-described conditions including (12), and the simulation results are shown in Fig. 9 and Fig. 10, respectively.

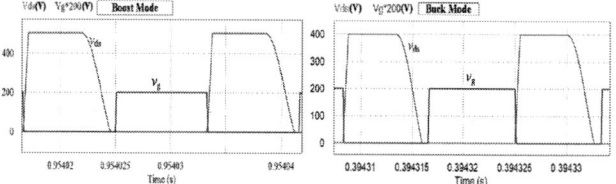

Fig. 9.   Drain-source voltage and driving signal for ideal parameter.

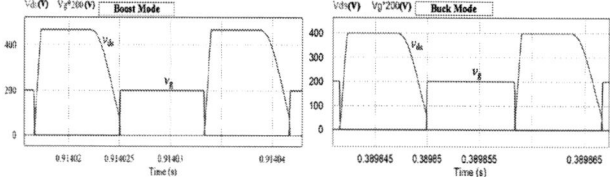

Fig. 10.   Drain-source voltage and driving signal for nonideal parameter.

By comparing the two simulation results, we can see that the soft-switching condition, which is designed for this study, can be realized in both Boost and Buck mode. However, this is not the case for the nonideal condition. Experimental results show the effectiveness of the proposed parameter-optimization design method.

C. Output voltage analysis under soft-switching condition

The relationship between output and input voltage of the converter is completely different from that of the traditional DC/DC circuit because of the mentioned soft-switching resonant network. Hence, it is of great significance for the closed-loop control design to study the influence of various parameters on the output voltage. The main curves of the main inductor in a working cycle in Boost mode are shown in Fig. 11.

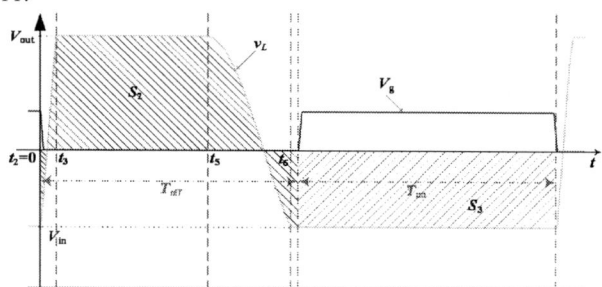

Fig. 11.   Main curves of main inductance in working cycle in Boost mode.

In one working cycle, the integral of the inductor voltage is zero. Hence, the two grey areas in Fig. 11 are of equal size. According to the sections above, the voltage curves of the main inductance L can be obtained by linear approximation and the two areas of S_2 and S_3 can be calculated as:

$$S_2 = \frac{1}{2} \cdot (t_5 - t_3 + t_6) \cdot (V_{out} + V_{in}) - V_{in} \cdot T_{off} \quad (13)$$

$$S_3 = V_{in} \cdot T_{on} \quad (14)$$

Combining the expressions of t_3, t_5, and t_6 with $S_2 = S_3$, we obtain:

$$\frac{V_{in}^2 \cdot T_{on} \cdot (L + L_r) + 2 \cdot V_{out} \cdot I_{out} \cdot L \cdot L_r}{V_{in} \cdot (V_{out} - V_{in}) \cdot (L + L_r)} - \frac{2 \cdot V_{in} \cdot V_{out} \cdot C_S \cdot L}{2 \cdot V_{out} \cdot I_{out} \cdot L + V_{in}^2 \cdot T_{on}} + \frac{4 \cdot V_{in} \cdot V_{out} \cdot C_S \cdot L \cdot L_r}{V_{in}^2 \cdot T_{on} \cdot (L + L_r) - 2 \cdot V_{out} \cdot I_{out} \cdot L \cdot L_r} = \frac{2 \cdot V_{in}}{f_{sw} \cdot (V_{out} + V_{in})} \quad (15)$$

Expression (15) describes the relation between V_{out} and V_{in}. The output voltage curve for a rated power of 1 kW and an input voltage of 200 V is shown in Fig. 12 (results according to equation (15)).

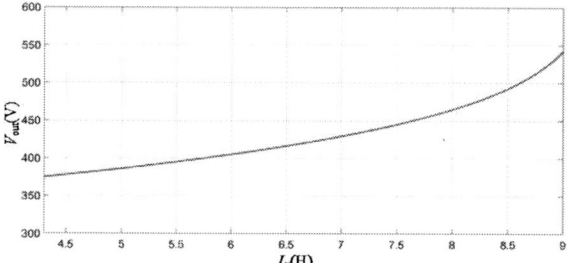

a) Influence of resonant inductance L_r on output voltage

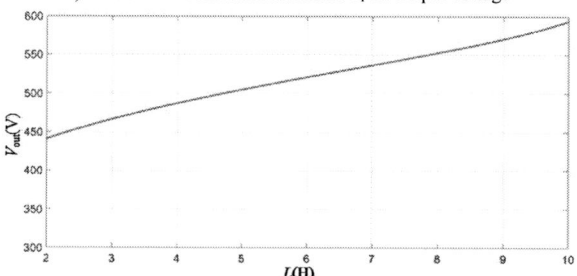

b) Influence of main inductance L on output voltage

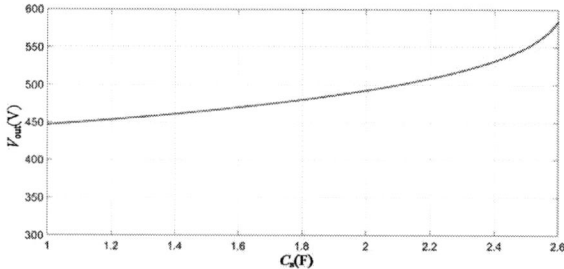

c) Influence of snubber capacitance C_s on output voltage

Fig. 12.   Influence of L_r, L, and C_s on output voltage.

In Fig. 12, it can be seen that the output voltage also increases with increasing parameters under 60 kHz, 50% duty cycle, 200 V input voltage, and 1 kW output power. The output voltage has nonlinear correlation with the changes of L_r, L, and C_S. When $L_r = 85$ μH, $L = 430$ μH, and $C_S = 20$ nF, the experimental analysis results meet the results of the software simulation. The validity of the theoretical analysis is therefore verified. In order to reduce the inductor's copper loss, the resonant inductance L_r and the main inductance L should be as low as possible under soft-switching and a certain constant voltage ratio. The influence of the parameter changes on the system's response is given in table 1.

TABLE I
INFLUENCE OF PARAMETER CHANGE ON SYSTEM RESPONSE (BOOST MODE)

	v_{Cr}	v_s	i_{Lr}	i_s	i_L	V_{out}	I_{out}
$C_r\uparrow$	Decreases and fluctuates	Decreases and resonance zero point shift behind	Decreases and fluctuates	Decreases and the resonance time decreases	Average value and fluctuation amplitude decrease	Steady	--
$L_r\uparrow$	Decreases and fluctuates	Decreases and resonance zero point shift behind	Decreases and fluctuates	Decreases and the resonance time does not change and shift behind	Average value and fluctuation amplitude decrease	Increases	--
$C_s\uparrow$	Decreases and fluctuates	Decreases and resonance zero point shift behind	Decreases and fluctuates	Decreases and the resonance time does not change and shift behind	Average value and fluctuation amplitude decrease	Increases	--
$R_L\uparrow$	Increases and fluctuates	Increases and resonance zero point shift before	Increases and fluctuates	Increases and the resonance time does not change and shift before	Average value does not change and fluctuation amplitude increases	Increases	Decreases

IV. DESIGN AND MODELING OF CLOSED-LOOP CONTROL SYSTEM FOR BIDIRECTIONAL CONVERTER

A. Design of closed-loop control system

The control system mainly summarizes two aspects in the convertor: 1. In the two operating modes, when the input side voltage or output load fluctuates, the converter output voltage is guaranteed to be stable at system default. 2. When the bidirectional converter works stably, the main power switch works in the soft-switching condition. Considering the circuit characteristics, this study uses the PFM control method.

In order to achieve stable operation of the bidirectional DC/DC converter under closed-loop conditions, the open-loop frequency characteristics of the bidirectional DC/DC converter are investigated. Further, to improve the stability and fast-response capability of the system, a closed-loop compensation of the bidirectional DC/DC converter system is needed. However, the proposed bidirectional DC/DC converter comprises a snubber capacitor and resonant devices. Hence, an analysis is difficult with the existing mathematical modeling theory.

This paper gives an auxiliary modeling method that is suitable for complex resonant systems. The system's AC sweep results for a circuit operation in Boost mode and under resistive load are shown in Fig.13.

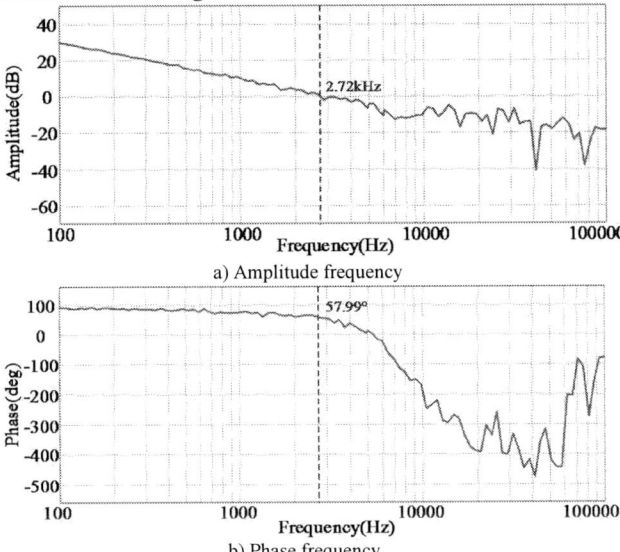

a) Amplitude frequency

b) Phase frequency

Fig. 13. Frequency characteristics before compensation in Boost mode.

It can be obtained from the results that the cut-off frequency $\omega_c = 2.72$ kHz, the phase margin $\gamma = 180° + 57.99° = 239.99°$, and the operating frequency of the designed bidirectional DC/DC converter is 60 kHz. In general, the system stability is better when the open-loop cut-off frequency is lower. However, a low cut-off frequency makes the closed-loop system's bandwidth become narrow, phase characteristics become worse with an increase of frequency, and the response speed will be slower. To consider the stability and response speed of the closed-loop system, an open-loop cut-off frequency of 1/10–1/5 is used (as large as the switching frequency of the system's stable operating point). In addition, the open-loop phase margin of the Boost mode is 239.99° and the closed-loop system is conditionally stable. However, when the system is perturbed, it will likely go into an unstable state. Further, the dynamic response of the system becomes slow because of the excessive phase margin. In usual engineering designs, the phase margin is generally set to 45°–60°. It can be seen from the amplitude-frequency characteristic that the slope of the system at low frequencies is -20 dB/dec. Hence, the open-loop gain of the system is low. In order to improve the latter, the amplitude-frequency characteristics at low frequencies should be compensated for -40 dB/dec. For the system's medium frequency, -20 dB/dec are usually taken as a crossing point to consider the system's transient response performance. Based on the analysis above, a PI compensation function is designed. The corner frequency is set to 5 kHz to ensure a sufficient width of the intermediate frequency section. Therefore, the time constant of the PI system τ_1 can be calculated as:

$$\tau_1 = \frac{1}{2\cdot\pi\cdot f} = \frac{1}{2\times 3.14\times 5000} \approx 3.18\times 10^{-5}\ s \tag{16}$$

According to the analysis, the final value of the open-loop gain K is 1.5. Hence, the compensation function is obtained by:

$$G_{Boost\_c}(s) = K\cdot\frac{\tau_1\cdot s+1}{\tau_1\cdot s} = 1.5\times\frac{3.18\times 10^{-5}\cdot s+1}{3.18\times 10^{-5}\cdot s} \tag{17}$$

Fig. 14 shows the characteristic frequency curves after compensation for the case when the compensation function is integrated into the bidirectional DC/DC converter. From the curves, it can be seen that by adding the compensation function, the open-loop cut-off frequency of the system becomes $\omega_c = 6.50$ kHz, and the phase angle margin $\gamma = 180° - 128.68° = 51.32°$. Hence, the system can be controlled by adding a system compensation.

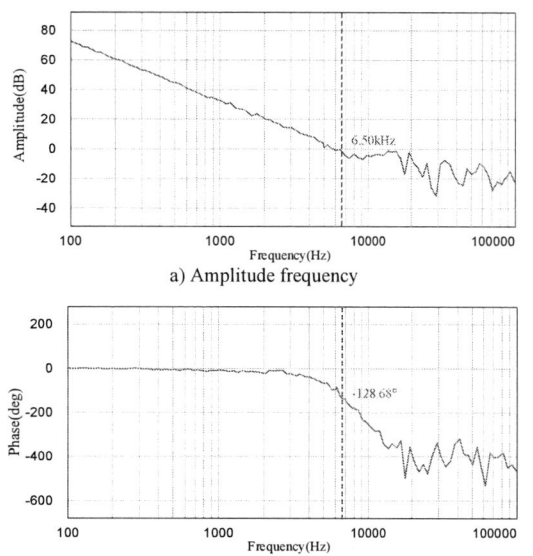

a) Amplitude frequency

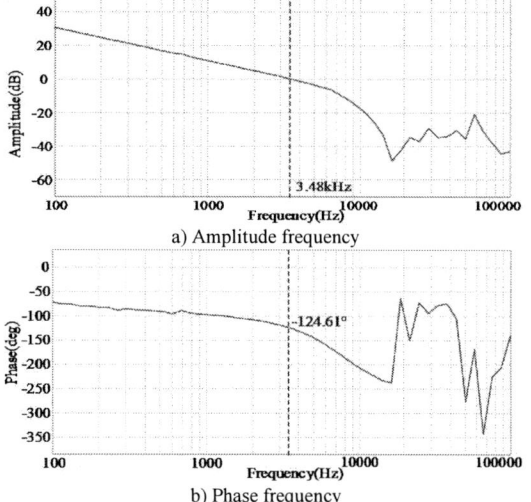

b) Phase frequency

Fig. 14.   Frequency characteristics after compensation in Boost mode.

The system's AC sweep results under resistive load conditions for operation in Buck mode are shown in Fig. 15.

a) Amplitude frequency

b) Phase frequency

Fig. 15.   Frequency characteristics before compensation in Buck mode.

It can be seen from Fig. 15 that the system's open-loop cut-off frequency ω_c is 3.48 kHz and the phase margin measures 55.39°. Hence, preferable medium frequency characteristics exist. The gain of the open-loop system is low. Therefore, a PI regulator is applied to improve the low-frequency characteristics of the system and increase the open-loop gain. The corner frequency is set to 4 kHz and the time constant of the PI system τ_2 can be calculated as:

$$\tau_2 = \frac{1}{2 \cdot \pi \cdot f} = \frac{1}{2 \times 3.14 \times 4000} \approx 3.98 \times 10^{-5} s \quad (18)$$

According to the analysis, the final value of the open-loop gain K is 1.3. Hence, the compensation function is obtained with:

$$G_{Buck\_c}(s) = 1.3 \cdot \frac{3.98 \times 10^{-5} \cdot s + 1}{3.98 \times 10^{-5} \cdot s} \quad (19)$$

The characteristic frequency curves of the system after compensation are shown in Fig. 16.

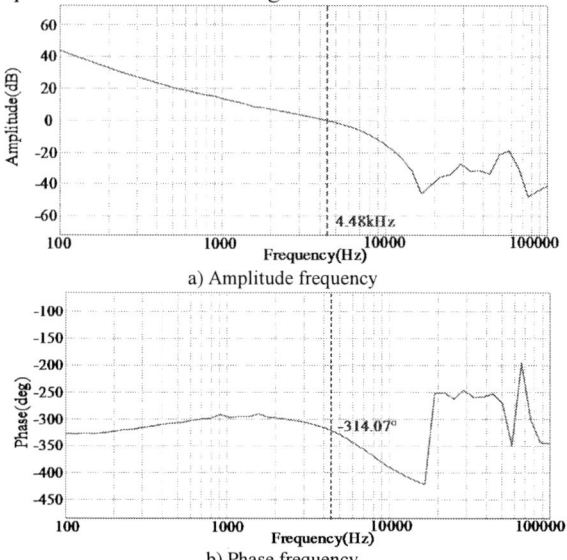

a) Amplitude frequency

b) Phase frequency

Fig. 16.   Frequency characteristics after compensation in Buck mode.

From the compensated characteristic frequency curves, it can be seen that by adding a compensation, the open-loop cut-off frequency of the system becomes $\omega_c = 4.48$ kHz, and the phase angle margin becomes $\gamma = 360° - 314.07° = 45.93°$. In conclusion, adding the system compensation can improve the system response and achieve sufficient system control.

B.   Digitization of PI controller

In this study, a PI regulator with an anti-saturation function is applied. The anti-saturation function prevents the hardware circuit from overcurrent by preventing the duty cycle of the power switches' driving signal of becoming too small or too large. The microcontroller TMS320F28335 (TI) is selected. The integrated ePWM modules can optimize the system resources and improve the output continuity of the PWM output. An arrangement of clock cycles in the program for a stable PFM output is shown in Fig. 17.

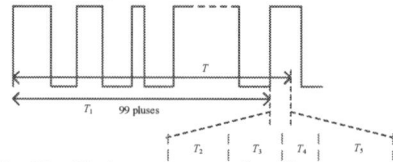

Fig. 17.   Timing arrangement for adjustment cycle.

In an adjustment period, each period is handled as follows: T_1: The ADC module carries out 100 samples of two voltage signals in synchronous mode; T_2: Digital filtering is used to filter out singularities; T_3: The final sampling values are calculated and the value of ePWM register and period register are compared; T_4: The calculated values are assigned to the ePWM register and period register; T_5: The ePWM register value and the period register value are passed to the corresponding registers.

According to the PI regulator's transfer function (designed

in previous section) and the system's tuning period T, the PI regulator's discrete-approximation difference equation can be obtained. Hence, the differential equations for digital control can be expressed as:

$$u_{\text{Buck}}(k)=\begin{cases}\text{U}_{\text{Buck\_max}}, u_{\text{Buck}}(k)\geq \text{U}_{\text{Buck\_max}}\\u_{\text{i\_Buck}}(k-1)+66.62e(k)\\\quad+0.8\left(u_{\text{Buck}}(k)-u_{\text{presat\_Buck}}(k)\right)\\\text{U}_{\text{Buck\_min}}, u_{\text{Buck}}(k)\leq \text{U}_{\text{Buck\_min}}\end{cases}\quad(20)$$

$$u_{\text{Boost}}(k)=\begin{cases}\text{U}_{\text{Boost\_max}}, u_{\text{Boost}}(k)\geq U_{\text{Boost\_max}}\\u_{\text{i\_Boost}}(k-1)+95.83e(k)\\\quad+0.8\left(u_{\text{Boost}}(k)-u_{\text{presat\_Boost}}(k)\right)\\\text{U}_{\text{Boost\_min}}, u_{\text{Boost}}(k)\leq \text{U}_{\text{Boost\_min}}\end{cases}\quad(21)$$

C. Control program design for closed-loop system

The controller has the following basic functions: 1. Control of bidirectional energy transmission in Buck or Boost mode; 2. Ensure stable voltage and current outputs of the converter; 3. Ensure that the power switches work in the soft-switching condition; 4. Prevent the main inductance from an overcurrent. The main block diagram of the circuit is shown in Fig. 18.

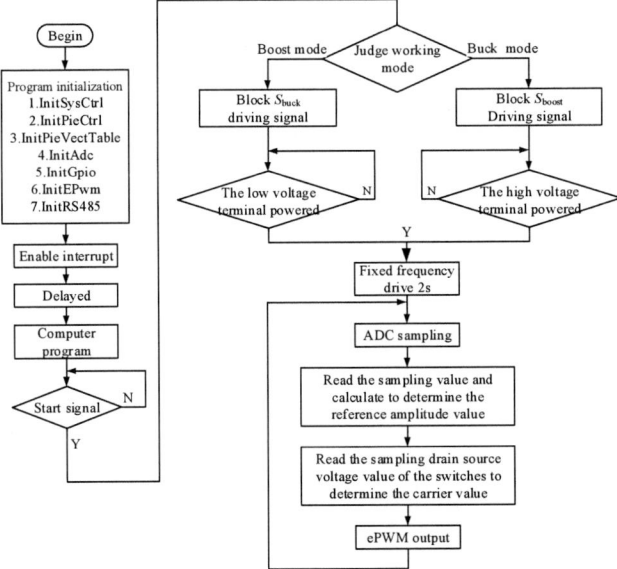

Fig. 18. Main block diagram.

In order to improve the precision of the voltage sampling, a digital filter program is added to the control program. The control program averages the sampling values after recording 100 voltage samples. Afterwards, the program compares each voltage sampling value with the desired average. If the voltage sampling value is above or below the 50% of the average value, the voltage sampling value is determined as a sampling singularity. Subsequently, the average value is added to the voltage sampling singularity and the new average value is determined. The newly determined average is taken as the sampling value of the actual value. With this method, the program can improve the accuracy of the voltage samples.

V. EXPERIMENTAL RESULTS

Considering the above-mentioned specifications, a 1 kW-prototype is proposed and shown in Fig. 19. The high-voltage end of the prototype is 400 V and the low-voltage end is 200 V. The specifications of the prototype are listed in Table II.

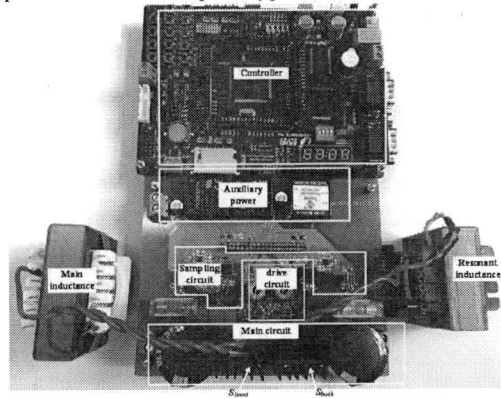

Fig. 19. Experimental prototype.

TABLE II
PROTOTYPE SPECIFICATIONS.

Specification/Parameter	Value
Low-voltage input V_{Low}	180~220 V
High-voltage input V_{High}	370~430 V
Low-voltage output	200 V
High-voltage output	400 V
Switching Frequency	60 kHz
Power rating	1kW
Main inductance L	430 μH
Resonant inductance L_r	85 μH
Resonant capacitance C_r	1 μF
Snubber capacitance C_s	20nF
Low-voltage output capacitance C_{Low}	450 μF/450V
High-voltage output capacitance C_{High}	450 μF/450V
MOSFET S_{Boost} and S_{Buck}	SCT3450KE
Diode	STTH10LCD06

The main curves for the Buck mode and the case that the output and input voltage of the bidirectional DC/DC converter are 200 V and 400 V, respectively, are shown in Fig. 20 respectively under 100% and 50% load condition.

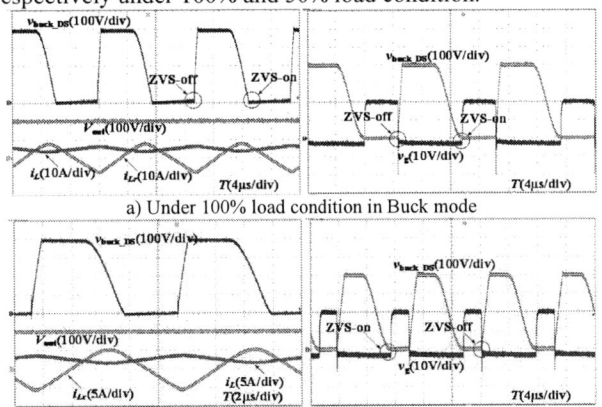

a) Under 100% load condition in Buck mode

b) Under 50% load condition in Buck mode
Fig. 20. Main curves for Buck mode.

As can be seen in Fig. 20, the converter can achieve ZVS condition under full and 50% load in Buck mode. The conversion efficiency measures 94.27% when the converter is under full load, and 97.04% when the load is 50%. The system's dynamic response for the case when the input voltage suddenly changes from 370 V to 430 V is shown in Fig. 21. Evidently, the regulation time is approximately 180 ms and the voltage regulation rate is $\delta_u = 0.7\%$.

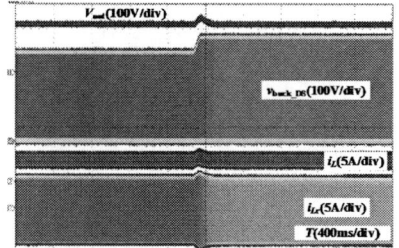

Fig. 21. Dynamic response when input voltage changes from 370 V to 430 V.

The main curves for the Boost mode and the case that the output and input voltage of the converter are 400 V and 200 V, respectively, are shown in Fig. 22 under 100% and 50% load condition.

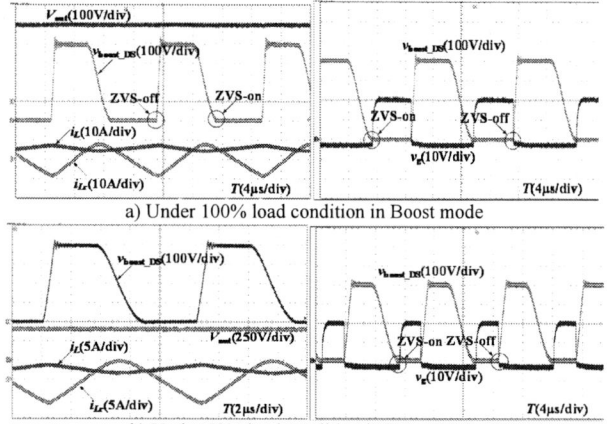

a) Under 100% load condition in Boost mode

b) Under 50% load condition in Boost mode

Fig. 22. Main curves for Boost mode.

As can be seen in Fig. 22, the converter can achieve ZVS condition under full and 50% load in Boost mode. The conversion efficiency measures 93.34% when the converter is under full load, and 96.35% when the load is 50%. Fig. 23 shows the system's dynamic response for the case that the input voltage suddenly changes from 180 V to 220 V. Evidently, the regulation time is approximately 260 ms and the voltage regulation rate is $\delta_u = 0.675\%$.

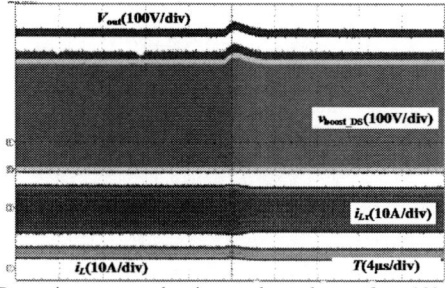

Fig. 23. Dynamic response when input voltage changes from 180 V to 220 V.

From the experimental results, it can be concluded that the efficiency is lower in Boost mode than that in Buck mode. The main reason is that the inductor's copper loss increases gradually when the current gradually increases with increasing load.

VI. CONCLUSION

In this paper, a soft-switching bidirectional DC/DC converter is proposed, and the soft-switching conditions of the system are analyzed in detail. An auxiliary method is applied to achieve better system performance. A soft-switching analysis and accurate modeling of the system are conducive to the converter achieving good circuit performance. A prototype with 200 V input and 400 V output voltage is proposed that achieves soft-switching conditions in both Buck and Boost mode. The highest efficiency of the converter is 97.04% in Buck mode and 96.65% in Boost mode. The voltage-regulation rate of the system is less than 0.7% and the ripple factor is approximately 0.5%. The experimental results are in good agreement with the theoretical results.

ACKNOWLEDGEMENT

This work is supported in part by the National High Technology Research and Development Program 863 of China under Grant 2015AA050603, and in part by the National Key Research and Development Program of China under Grant 2016YFE0102800.

REFERENCES

[1] Chen. W, Rong. P, Lu. Z. Snubberless bidirectional DC–DC converter with new CLLC resonant tank featuring minimized switching loss[J]. IEEE Transactions on Industrial Electronics, 2010, 57(9): 3075-3086.

[2] Wu H, Sun K, Chen L, et al. High Step-Up/Step-Down Soft-Switching Bidirectional DC–DC Converter With Coupled-Inductor and Voltage Matching Control for Energy Storage Systems[J]. IEEE Transactions on Industrial Electronics, 2016, 63(5): 2892-2903.

[3] Chen G, Lee Y S, Hui S Y R, et al. Actively clamped bidirectional flyback converter[J]. IEEE Transactions on Industrial Electronics, 2000, 47(4): 770-779.

[4] Swingler A D, Dunford W G. Development of a bi-directional DC/DC converter for inverter/charger applications with consideration paid to large signal operation and quasi-linear digital control[C]//Power Electronics Specialists Conference, 2002. pesc 02. 2002 IEEE 33rd Annual. IEEE, 2002, 2: 961-966.

[5] Wu H, Xu P, Hu H, et al. Multiport converters based on integration of full-bridge and bidirectional DC–DC topologies for renewable generation systems[J]. IEEE transactions on industrial electronics, 2014, 61(2): 856-869.

[6] Fang X, Hu H, Shen Z J, et al. Operation mode analysis and peak gain approximation of the LLC resonant converter[J]. IEEE transactions on power electronics, 2012, 27(4): 1985-1995.

[7] Wang L, Wang Z, Li H. Asymmetrical duty cycle control and decoupled power flow design of a three-port bidirectional DC-DC converter for fuel cell vehicle application[J]. IEEE Transactions on Power Electronics, 2012, 27(2): 891-904.

[8] Lin C C, Yang L S, Wu G W. Study of a non-isolated bidirectional DC-DC converter[J]. IET Power Electronics, 2013, 6(1): 30-37.

[9] Ni L, Patterson D J, Hudgins J L. High power current sensorless bidirectional 16-phase interleaved DC-DC converter for hybrid vehicle application[J]. IEEE Transactions on Power Electronics, 2012, 27(3): 1141-1151.

[10] Lu J, Bai H K, Averitt S, et al. An E-mode GaN HEMTs based three-level bidirectional DC/DC converter used in Robert Bosch DC-grid system[C]. Wide Bandgap Power Devices and Applications. IEEE, 2016:334-340.

[11] Sun A, Zhang W, Lin X, et al. Modeling and stability analysis of high voltage ratio bidirectional DC/DC converter applied to electric vehicle[C]. Transportation Electrification Asia-Pacific. IEEE, 2014:1-6.

[12] Jung D Y, Hwang S H, Ji Y H, et al. Soft-Switching Bidirectional DC/DC Converter with a LC Series Resonant Circuit[J]. IEEE Transactions on Power Electronics, 2013, 28(4):1680-1690.

[13] Yang J W, Do H L. High-efficiency bidirectional dc–dc converter with low circulating current and ZVS characteristic throughout a full range of loads[J]. IEEE Transactions on Industrial Electronics, 2014, 61(7): 3248-3256.

The 2018 International Power Electronics Conference

A Fully Soft-Switched PWM DC-DC Converter Using An Active-Snubber-Cell

Hai N. Tran, Adhistira M. Naradhipa, Sunju Kim, Ali Tausif, Sewan Choi, *IEEE Senior Member*
Department of Electrical and Information Engineering
Seoul National University of Science and **Tech**nology
Seoul, Korea
Email: schoi@seoultech.ac.kr

Abstract- **This paper proposes an active-snubber-cell and its dual circuit to achieve fully soft-switching operation for PWM DC-DC converters. Main switch of the converter achieves both turn-on and turn-off with zero voltage condition; a switch in the proposed cell achieves zero current turn-on and zero voltage turn-off. The main diode is turned-off with zero current and other semiconductor devices operate under soft-switching condition. No additional voltage and current stresses are imposed by the proposed cell. The proposed active-snubber-cell can be applied in a range of PWM DC-DC converters; the guidelines for implementing the proposed cell will be presented. Easy implementation for an interleaved converter makes the proposed cell is suitable for high power applications. The presented operating principle and analysis is verified using a 100 kHz and 1kW prototype of the proposed cell implemented on a boost converter. As a result, the peak efficiency of 98% is achieved.**

Index Terms- **Active-snubber-cell, DC-DC converter, soft-switching, zero-voltage-transition (ZVT).**

I. Introduction

Pulse-width modulation (PWM) DC-DC converters are widely used in the industry and renewable energy applications due to their high power density, quick transition response, and ease of control [1]-[15]. As the demand for higher power DC-DC converter is increasing, higher switching frequency operation is needed to further increase the power density of the converter. However, switching losses and electromagnetic interference (EMI) noise increase as the switching frequency increases [11]. Therefore, a soft-switching technique is mandatory to solve the problems associated with a hard-switched dc-dc converter.

A well-known method to achieve soft-switching is by integrating an active-snubber-cell (ASC) into a PWM DC-DC converter. The ASC operate under a constant switching frequency and only works at a small fraction of a whole operating period. Therefore, ASC is an attractive method to realize soft-switching operation [1]-[10]. In addition, ASC only consumes a small amount of power which result in a small portion of the whole converter volume.

The converter in [1] achieves ZVS turn-on for the main switch and ZCS turn-off for the main diode using an ASC. The switch of the cell also achieves ZCS turn-on due to the slope of the snubber inductor current. However, switching loss and EMI noise at the turn-off operation of both main and snubber switches still exist due to hard switching.

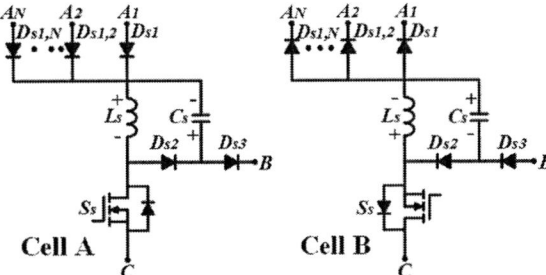

Fig. 1. Proposed active-snubber-cells.

Therefore, the benefit of using the ASC is not significant in this case. To cope with this problem, few methods to realize fully soft-switching of snubber switch have been presented in [2]-[10].

The ASCs proposed in [2]-[4] are using a transformer or a coupled inductor which will increase the complexity of the magnetic component design. In addition, the timing window for giving the gate signal to achieve ZVS turn-on of the main switch is narrow and load dependent. Therefore, these ZVT converters are not suitable for wide load range applications.

There is no timing issue in ZVT converters that have been proposed in [5] and [6], but the component count is increased and the light-load efficiency is sacrificed due to circulating current. The ASC in [6] has the advantage of easy implementation for interleaving configuration, but the voltage stress of the snubber switch is more than the output voltage. Converters that use ZVT-ZCT technique are proposed in [7] and [10], the main switch turns-on under ZVS and turns-off under ZCS while the snubber switch is turned-on and off with ZCS. Even though it achieves fully soft-switching, the timing issue exists for both turn-on and turn-off operations.

In this paper, a new active-snubber-cell that achieves fully soft-switching for both of main and snubber switches in wide load range without timing issue is proposed. Additionally, the reverse recovery current problem of the main diode is alleviated by turning it off with zero current condition. All other switching components in the proposed converter are switched under soft-switching condition. Moreover, voltage and current stresses on the main switch and diode are not increased. Furthermore, the voltage and current stresses on ASC components stay under the allowable levels. The proposed ASC is suitable for high

3833

power application because it can be easily implemented with interleaved converters.

II. OPERATION PRINCIPLE

The proposed ASC along with its dual circuit is shown in Fig. 1. For each additional interleaving phase, the proposed ASC only needs to add one more diode D_{s1}. The proposed cell A is implemented in the Conventional Boost Converter (CBC) as shown in Fig. 2. According to Fig. 2, the ASC consists of one auxiliary switch, three diodes, one inductor, and one capacitor.

The proposed ASC is connected to the CBC through three nodes: Node A is connected to the common point of the main switch and the main diode, node B is connected to the common point of the main diode and the output capacitor, node C is connected to the common point of the main switch and the input voltage source.

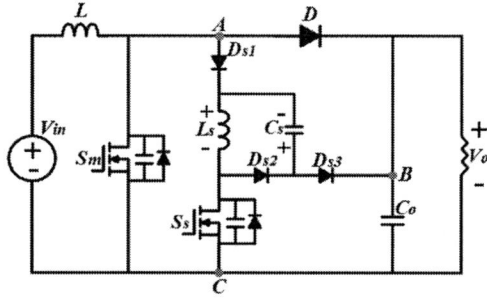

Fig. 2. Boost converter with the proposed cell A.

To ease the steady-state operation analysis of the circuit shown in Fig. 2, some assumptions are made. The output capacitor and input inductor are large enough to make constant output voltage and input current. All the passive components are assumed to be ideal. The operation of the CBC using the proposed cell A is divided into eight modes. The equivalent circuits and the key waveforms for each mode are shown in Fig. 3 and Fig. 4, respectively. In principle, the operation of cell A and cell B is the same. Thus, only mode-by-mode operation of cell A is presented.

1) Mode 0 (before t_1): The main switch S_m is off and the main diode D is on. The converter is operating as a normal boost converter discharging state. This mode ends when the snubber switch S_s turns-on at $t = t_1$.

2) Mode 1 ($t_1 < t < t_2$) : The snubber switch S_s turns-on and diode D_{s1} is conducting. The current through the snubber switch increases from 0A and the main diode current decreases from i_L. The current slope of both currents is expressed by equation (1).

$$\frac{di_{Ls}}{dt} = \frac{V_o}{L_s} \tag{1}$$

Therefore, the snubber switch is turned on with ZCS. At the end of this mode, when i_{Ls} reaches to i_L, the main diode is also turned off with ZCS.

3) Mode 2 ($t_2 < t < t_3$): After the main diode is turned-off, snubber inductor L_s and parasitic capacitor C_{ds_m} of the main switch will create a resonant circuit. In this resonant interval, all the energy stored in C_{ds_m} will be transferred

to L_s. At the end of this mode, the voltage of the main switch is discharged to 0V while the current of the snubber inductor is charged to a maximum value of i_{Ls_peak}. For this mode:

$$i_{Ls} = I_{in} + V_o \sqrt{\frac{C_{ds\_m}}{L_s}} \sin\left[\omega_1(t-t_2)\right] \tag{2}$$

$$i_{Ls\_peak} = I_{in} + V_o \sqrt{\frac{C_{ds\_m}}{L_s}} \tag{3}$$

$$V_{Cds\_m} = V_o \cos\left[\omega_1(t-t_2)\right] \tag{4}$$

$$\omega_1 = \frac{1}{\sqrt{L_s C_{ds\_m}}} \tag{5}$$

This mode ends when the voltage of C_{ds_m} is completely discharged to 0V.

4) Mode 3 ($t_3 < t < t_4$): After C_{ds_m} voltage is discharged to 0V, the body diode of the main switch begins to conduct and clamps the voltages across the snubber inductor and the main switch at 0V. Therefore, constant current flows through L_s. During this interval, the main switch is turned-on with ZVS. The current that conducts through the body diode of the main switch can be calculated as:

$$i_{bdm} = I_{Ls\_peak} + I_{in} \tag{6}$$

Before the snubber switch is turned-off, $i_{Ls} = i_{Ls\_peak}$, v_{Cs} = 0V, D_{s2} and D_{s3} are off. This mode ends when the snubber switch turns-off at $t = t_4$.

5) Mode 4 ($t_4 < t < t_5$): At time t_4, the snubber switch turns-off and diode D_{s2} is turned-on to conduct the current from the snubber inductor. The snubber inductor current charges the snubber capacitor C_s. Due to the parasitic capacitor of the snubber switch, a very small current flows from the snubber inductor to charge this capacitor. The equivalent circuit for this interval is shown in Fig. 5(a).

According to Fig. 5(a), the equations for this mode can be written as follows:

$$i_{Ls} = I_{Ls\_peak} \cos\left[\omega_1(t-t_4)\right] \tag{7}$$

$$V_{Ss} = V_{Cs} = I_{Ls\_peak} \sqrt{\frac{L_s}{C_s + C_{ds\_s}}} \sin\left[\omega_2(t-t_4)\right] \tag{8}$$

$$\omega_2 = \frac{1}{\sqrt{L_s + (C_s + C_{ds\_s})}} \tag{9}$$

This mode ends when the snubber capacitor is charged to V_o. According to this mode, after the snubber switch is turned-off, the voltage applied to this switch increases slowly with the slope as given in equation (8). Therefore, the snubber switch turns-off with zero voltage condition.

6) Mode 5 ($t_5 < t < t_6$): After $v_{Cs} = V_o$, diode D_{s3} is turned on and the voltage applied to the snubber inductor is $-V_o$. The current of L_s decreases linearly. The rest of the

3834

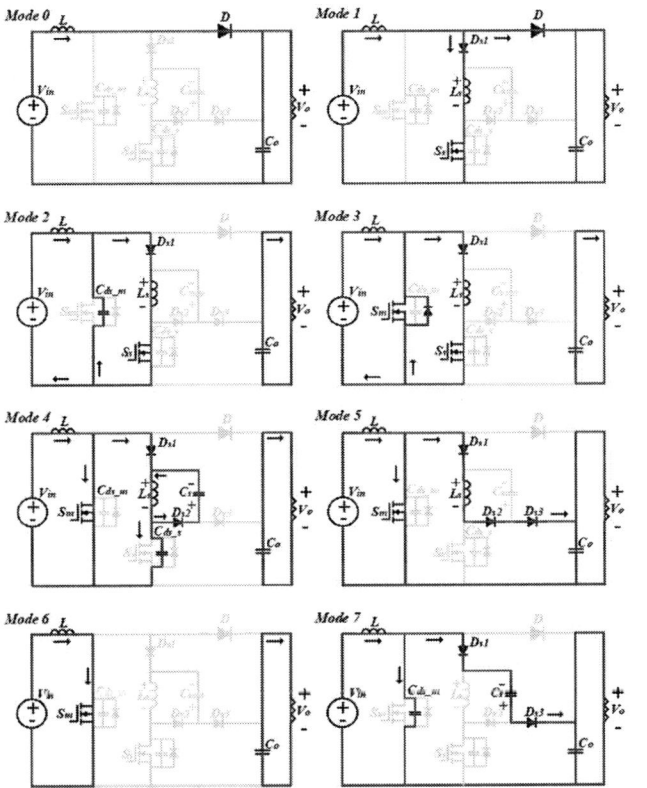

Fig. 3. Operation modes of the boost converter with the proposed cell A.

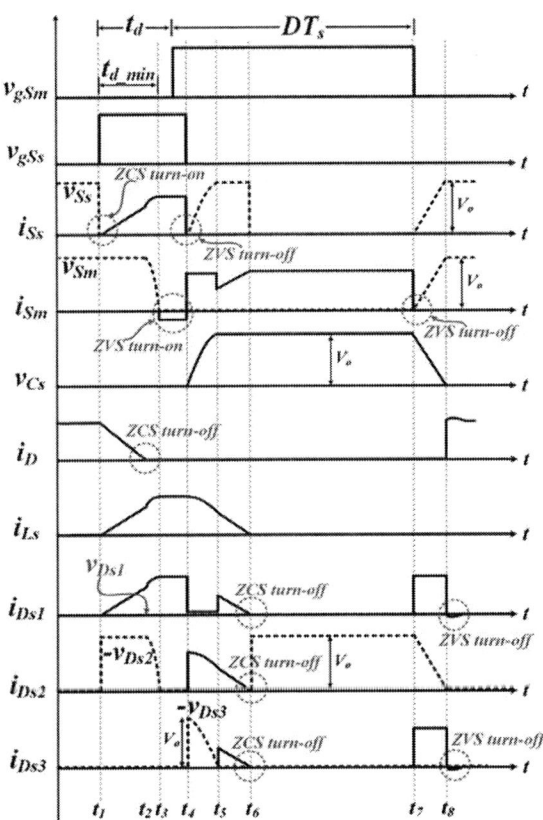

Fig. 4. Key waveforms of boost converter with the proposed cell A.

energy in the snubber inductor is transferred to the load. At the end of this mode, diode D_{s1}, D_{s2}, and D_{s3} are turned off with ZCS. During this interval:

$$\frac{di_{Ls}}{dt} = -\frac{V_o}{Ls} \tag{10}$$

7) Mode 6 ($t_6 < t < t_7$): This mode is a charging state of the boost converter. The input voltage charges the input inductor and the load is supplied by output capacitor. This mode ends when the main switch turns-off at time t_7.

8) Mode 7 ($t_7 < t < t_8$): Before the main switch turns-off, $v_{Cds\_m} = 0V$, $v_{Cs} = V_o$. At time t_7, the main switch turns-off and both D_{s1} and D_{s3} is turned-on. The parasitic capacitor of the main switch and the snubber capacitor is charged and discharged at the same time. The equivalent circuit of this mode is shown in Fig. 5(b). At the end of this mode, the voltage of the main switch is increase to V_o and the voltage of C_s decreases to 0V. Afterwards, D_{s1} and D_{s3} are turned-off. Therefore, during this mode, the main switch achieves ZVS turn-off. Although D_{s1} and D_{s3} are turned-off with a steep current slope, the reverse voltages of both diodes are canceled by subtracting the main switch

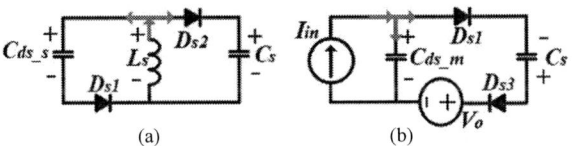

(a)   (b)

Fig. 5. The equivalent circuits of (a) mode 4 and (b) mode 7.

parasitic capacitor voltage from the output voltage, D_{s1} and D_{s3} are turned-off with ZVS, so there is no reverse recovery loss. In this mode:

$$V_{Cs} = V_o - \frac{I_{in}}{C_s + C_{ds\_m}}(t-t7) \tag{11}$$

$$V_{Sm} = V_o - V_{Cd} = \frac{I_{in}}{C_s + C_{ds\_m}}(t-t7) \tag{12}$$

After mode 7, the reverse recovery current of D_{s1} and D_{s3} flows to the load through the main diode. Afterwards, the ASC stops working and the main diode turns-on. Thus, the converter state is the same as mode 0.

III. DESIGN PROCEDURES

Following the assumptions and the operating principle in *section II*, the design guidelines of the snubber inductor and the snubber capacitor in the proposed ASC is presented in this section.

A. Design of snubber inductor L_s

The snubber inductor L_s provides ZCS turn-on for the snubber switch and ZCS turn-off for the main diode. According to *mode 1*, the value of L_s will be determined based on the input current and the duration of *mode 1*, t_r, which is expressed in equation (13).

In order to achieve ZVS turn-on for the main switch,

TABLE I
COMPARISON BETWEEN THE PROPOSED ASC AND PREVIOUS WORKS

Parameter	Proposed	ZVT [5]	ZVS-ZCS [6]	ZVT-ZCT [7]
Components	1 inductor 1 capacitor 3 diodes 1 switch	2 inductors 1 capacitor 3 diodes 1 switch	2 inductors 1 capacitor 3 diodes 1 switch	2 inductors 1 capacitor 2 diodes 1 switch
Circulating Current	NO	YES	YES	NO
Timing Issue	NO	NO	NO	YES
Main Switch	ZVS ON & ZVS OFF	ZVS ON	ZVS ON & ZVS OFF	ZVS ON & ZCS OFF
Snubber Switch	ZCS ON & ZVS OFF	ZCS On & ZCS OFF	ZCS ON & ZVS OFF	ZCS ON & ZCS OFF
Interleaving	YES	NO	YES	NO
Snubber switch voltage stress	V_{out}	V_{out}	$V_{out} + V_{Cs}$	V_{out}

the duration from snubber switch turn-on to the main switch turn-on, t_d, should be larger than the minimum delay time, t_{d_min}, which is the duration from t_1 to t_3. Equation (14) and (15) shows how to select delay time, t_d.

$$Ls = \frac{V_o}{Iin}(t_2-t_1) = \frac{V_o}{Iin} tr \tag{13}$$

$$t_{23} = (t_3-t_2) = \frac{\pi}{2}\sqrt{LsCds\_m} \tag{14}$$

$$td \geq td_{min} = tr + t_{23} \tag{15}$$

B. Design of snubber capacitor C_s

The snubber capacitor C_s provides ZVS turn-off for both main switch and snubber switch. According to the operation which was presented in section II, snubber switch is able to achieve ZVS turn-off independent of v_{Cs}. However, the main switch can only achieve ZVS turn-off when the snubber capacitor is charged to V_o during mode 4. Therefore, the energy stored in the snubber inductor should be large enough to charge snubber capacitor to V_o.

$$\frac{1}{2}Ls\,i_{Ls\_peak}^2 \geq \frac{1}{2}CsV_o^2 \tag{16}$$

$$Cs \geq Ls\frac{i_{Ls\_peak}^2}{V_o^2} \tag{17}$$

In mode 7, to achieve ZCS turn-on of snubber switch, the snubber capacitor should be discharged from V_o to 0V before the snubber switch turns-on. The discharge current in mode 7 is I_{in} and the maximum duration of mode 7 is from t_8 to t_1 in Fig. 4, which can be calculated as:

$$t_{81} = (1-D)Ts - td \tag{18}$$

Therefore, the maximum snubber capacitance in this case is determined as follows:

$$Cs \leq \frac{Iin}{V_o}\Big[(1-D)Ts-td\Big] \tag{19}$$

To properly design the proposed ASC, the first step is to select t_r based on the characteristic of the gate driver, turn-on characteristic of both switches, the input current, and the reverse recovery specification of the main diode. Afterwards, snubber inductance will be calculated using equation (13). Finally, the smallest snubber capacitance value will be selected between the calculated value of equations (17) and (19).

IV. ADVANTAGES OF THE PROPOSED ASC

The proposed ASC can provide the following benefits which are summarized as below:

1) The proposed ASCs can be easily applied to N-phase PWM DC-DC interleaved converter.
2) No timing window issue for giving gate signal to the main switch to achieve soft switching.
3) As shown in Fig. 6, the proposed ASCs can be integrated for different topologies. Operation of these topologies are similar to the theoretical operations explained in section II.
4) The main switch achieves ZVS turn-on and ZVS turn-off.
5) The snubber switching achieves ZCS turn-on and ZVS turn-off.
6) Main diode is turned off with ZCS.
7) All the snubber diodes are turned on and off under soft-switching condition.
8) No additional voltage and current stresses imposed by the proposed ASC for all of the semiconductor devices.

Even though there are a lot of published ASCs, the proposed ASC has more advantages than the previous works. In order to prove the benefit of the proposed ASC over the previous works, comparison with the selected previous works in [5], [6], and [7] is summarized in Table I.

3836

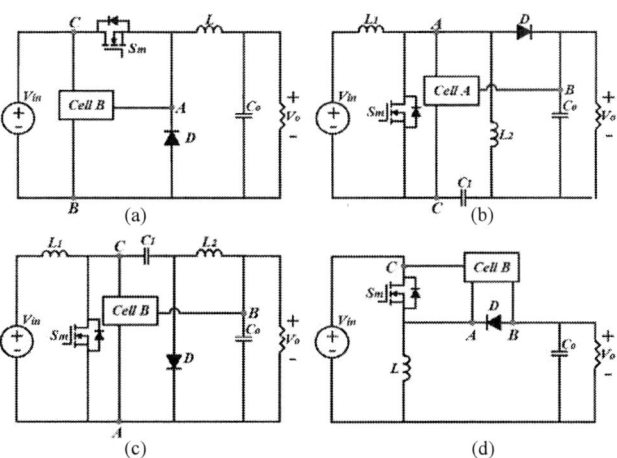

Fig. 6. Family of converters with the proposed ASCs: (a) buck (b) SEPIC (c) Cuk (d) buck-boost.

TABLE II
COMPONENT VALUES OF THE EXPERIMENTAL PROTOTYPE

Item	Value	Item	Value
L	500uH	Ls	15uH
Co	30uF	Cs	15nF
D	DSEI 12-06A	Ds_1, Ds_2, Ds_3	DSEI 12-06A
Sm	FCH76N60NF	Ss	FCH76N60NF

V. SIMULATION & EXPERIMENTAL RESULTS

To verify the theory and reliability of the proposed ASC, cell A is integrated in a single-phase CBC with 120V input, 300V output, switching frequency $f_s = 100$ kHz and 1kW of power is used for simulation and experiment. The simulation results are shown in Fig. 7. The passive component values of the proposed ASC are determined based on the design guidelines given in Section III and the converter components are tabulated in Table II.

The gating signal of the two switches are shown in Fig. 7(a). Fig. 7(b) shows the soft-switching characteristic of main switch with ZVS turn-on and off. The ZCS turn-on and ZVS turn-off of snubber switch is also given in Fig. 7(c). In Fig. 7(d) the switching of the main diode which turn-on with ZVS and turn-off with ZCS is shown. The snubber diode D_{s2}, which achieve fully soft-switching is shown in Fig. 7(g). Although D_{s1} and D_{s3} turn off with a steep slope as mentioned in section II, the reverse recovery loss can be eliminated by zero value of the reverse voltage shown in Figs. 7(f) and (h). The snubber inductor current and input current are shown in Fig. 7(e) which can be used as a reference to be compared with the experimental results.

The laboratory prototype picture for boost converter with the proposed ASC is shown in Fig. 8. The experimental results are shown in Fig. 9. Because of the measurement noise, the switch current are not measured in this paper. However, the soft-switching performance of the semiconductor devices can be observed through the other waveforms. As shown in Fig. 9(a), the ZVS turn-on and off of the main switch can be easily seen from gating signal

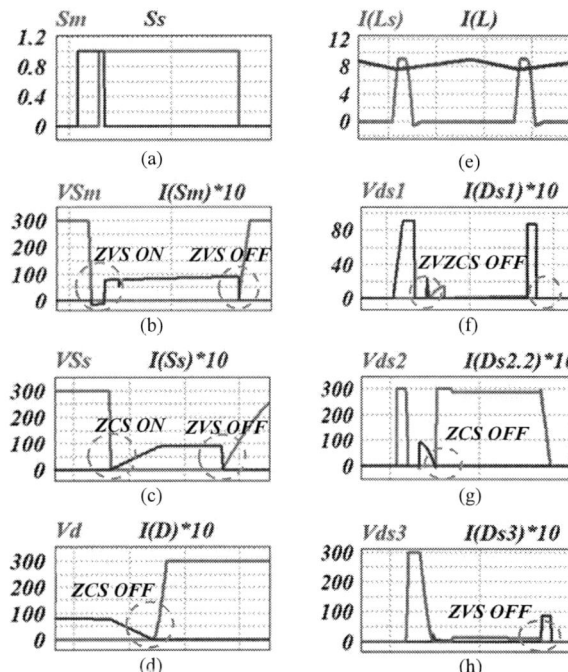

Fig. 7. Simulation waveforms of the proposed boost converter (a) Gate signals (b) Sm (c) Ss (d) D (e) iL & iLs (f) DS1 (g) Ds2 (h) Ds3.

and drain voltage of the main switch. The ZCS turn-on of the snubber switch as well as the ZCS turn-off of the main diode can be observed from the drain voltage of the snubber switch and the snubber inductor current, while the ZVS turn-off of snubber switch can be seen from gating signal and drain voltage of the snubber switch as shown in Fig. 9(b). V_o and v_{Cs} are shown in the upper side of Fig. 9(c). The snubber inductor and the input inductor current are given in the lower side of Fig. 9(c).

Fig. 8. Laboratory prototype of boost converter with the proposed cell A.

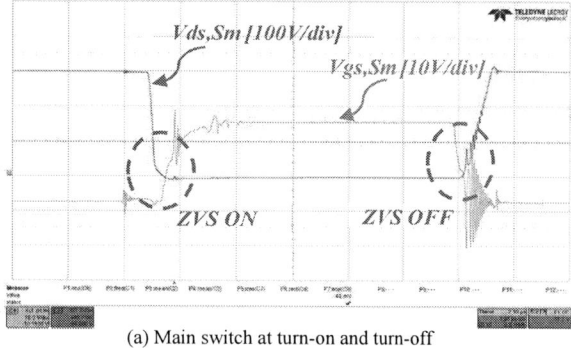

(a) Main switch at turn-on and turn-off

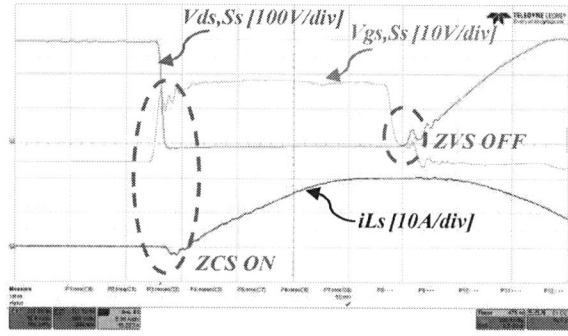

(b) Snubber switch at turn-on and turn-off

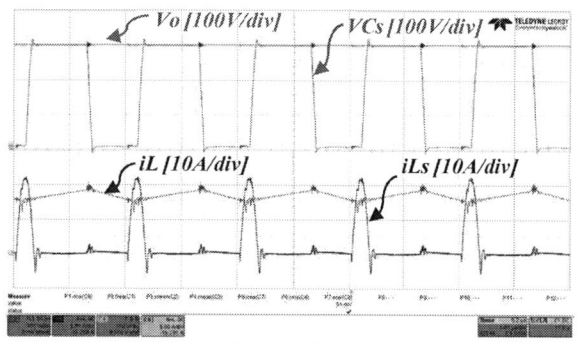

(c) Output voltage, snubber capacitor voltage, main inductor current, and snubber inductor

Fig. 9. Experimental waveforms of boost converter with the proposed cell A at full load.

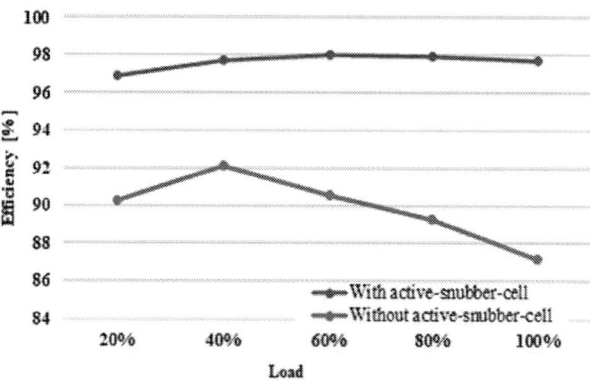

Fig. 10. Measured efficiency of boost converter with the proposed cell A.

The efficiency of the boost converter with the proposed ASC is measured in a wide range of load using Yokogawa WT3000 and the result is shown in Fig. 10. The peak and full load efficiency are 98% and 97.7%, respectively.

VI. CONCLUSION

An active-snubber-cell and its dual circuit to achieve fully soft-switching operation for a range of PWM DC-DC converters has been introduced. The proposed cell can provide ZVS turn-on and ZVS turn-off for the main switch, ZCS turn-off for the main diode, and ZCS turn-on and ZVS turn-off for the snubber switch. All of the snubber diodes are also soft-switched both at turn-on and turn-off operation. A 1 kW, 100 kHz prototype of the proposed converter has been built and tested to verify the validity of the proposed operation. The measured peak and full load efficiency are 98% and 97.7%, respectively. The proposed active-snubber-cell can be easily applied to an interleaved converter which is normally needed for high power applications.

ACKNOWLEDGMENT

This work was supported by the National Research Foundation of Korea (NRF) grant funded by the Korea Government (MSIT) (2017R1A2A2A05001054).

REFERENCES

[1] Guichao Hua, Ching-Shan Leu, Yimin Jiang and F. C. Y. Lee, "Novel zero-voltage-transition PWM converters," in *IEEE Transactions on Power Electronics*, vol. 9, no. 2, pp. 213-219, Mar 1994.

[2] Yungtaek Jang, M. M. Jovanovic, Kung-Hui Fang and Yu-Ming Chang, "High-power-factor soft-switched boost converter," in *IEEE Transactions on Power Electronics*, vol. 21, no. 1, pp. 98-104, Jan. 2006.

[3] J. P. Gegner and C. Q. Lee, "Zero-voltage-transition converters using a simple magnetic feedback technique," *Power Electronics Specialists Conference, PESC '94 Record. 25th Annual IEEE*, Taipei, 1994, pp. 590-596 vol.1.

[4] M. R. Mohammadi and H. Farzanehfard, "A New Family of Zero-Voltage-Transition Nonisolated Bidirectional Converters With Simple Auxiliary Circuit," in *IEEE Transactions on Industrial Electronics*, vol. 63, no. 3, pp. 1519-1527, March 2016.

[5] R. Gurunathan and A. K. S. Bhat, "ZVT boost converter using a ZCS auxiliary circuit," in *IEEE Transactions on Aerospace and Electronic Systems*, vol. 37, no. 3, pp. 889-897, Jul 2001.

[6] R. T. H. Li and C. N. M. Ho, "An Active Snubber Cell for N -Phase Interleaved DC–DC Converters," in *IEEE Journal of Emerging and Selected Topics in Power Electronics*, vol. 4, no. 2, pp. 344-351, June 2016.

[7] I. Aksoy, H. Bodur and A. F. Bakan, "A New ZVT-ZCT-PWM DC–DC Converter," in *IEEE Transactions on Power Electronics*, vol. 25, no. 8, pp. 2093-2105, Aug. 2010.

[8] Ching-Jung Tseng and Chern-Lin Chen, "Novel ZVT-PWM converters with active snubbers," in *IEEE Transactions on Power Electronics*, vol. 13, no. 5, pp. 861-869, Sep 1998.

[9] H. Bodur and A. F. Bakan, "A new ZVT-PWM DC-DC converter," in *IEEE Transactions on Power Electronics*, vol. 17, no. 1, pp. 40-47, Jan 2002.

[10] N. Altintaş, A. F. Bakan and İ. Aksoy, "A Novel ZVT-ZCT-PWM Boost Converter," in *IEEE Transactions on Power Electronics*, vol. 29, no. 1, pp. 256-265, Jan. 2014.

[11] H. Chung, S. Y. R. Hui and K. K. Tse, "Reduction of power converter EMI emission using soft-switching technique," in *IEEE Transactions on Electromagnetic Compatibility*, vol. 40, no. 3, pp. 282-287, Aug 1998.

[12] S. Park, Y. Park, S. Choi, W. Choi and K. B. Lee, "Soft-Switched Interleaved Boost Converters for High Step-Up and High-Power Applications," in *IEEE Transactions on Power Electronics*, vol. 26, no. 10, pp. 2906-2914, Oct. 2011.

[13] M. Kim, D. Yang and S. Choi, "A fully soft-switched multiphase DC-DC converter with reduced switch count for high power application," *2014 International Power Electronics Conference (IPEC-Hiroshima 2014 - ECCE ASIA)*, Hiroshima, 2014, pp. 2247-2251.

[14] M. Kim and S. Choi, "A Fully Soft-Switched Single Switch Isolated DC–DC Converter," in *IEEE Transactions on Power Electronics*, vol. 30, no. 9, pp. 4883-4890, Sept. 2015.

[15] Y. Park, B. Jung and S. Choi, "Nonisolated ZVZCS Resonant PWM DC–DC Converter for High Step-Up and High-Power Applications," in *IEEE Transactions on Power Electronics*, vol. 27, no. 8, pp. 3568-3575, Aug. 2012.

The 2018 International Power Electronics Conference

Flying Capacitor Resonant Pole Inverter with Direct Inductor Current Feedback

Sjef J. Settels*, Jorge L. Duarte, Jeroen van Duivenbode

Department of Electrical Engineering, Eindhoven University of Technology, Eindhoven, the Netherlands

*E-mail: s.settels@tue.nl

Abstract—**Industrial applications, e.g. semiconductor manufacturing equipment, require power amplifiers providing high power with high precision and bandwidth. The Flying Capacitor Resonant Pole Inverter (FC RPI) provides a multilevel configuration with high switching frequencies and Zero-Voltage Switching (ZVS) across the entire operating range. However, the applied charge-based modulation scheme to ensure ZVS depends heavily on the correct measuring of the zero-crossings of the filter inductor current. The delay incorporated in the measurement chain results in significant distortion of the output current which deteriorates the performance of the end application. This research proposes to apply direct current feedback of the per-period average filter inductor current, measured using a high bandwidth Anisotropic Magneto-Resistive (AMR) sensor, to correct the introduced distortion of the output current. Simulation results of the complete converter and control configuration indicate a significant improvement in performance: 9 dB increased Spurious Free Dynamic Range (SFDR) and 16 dB decreased Total Harmonic Distortion (THD).**

Index Terms—**Current control, Multilevel converters, Resonant inverters, Zero voltage switching.**

I. INTRODUCTION

Equipment used in semiconductor manufacturing applies high precision fast-moving stages for accurate positioning of wafers to fulfill tasks such as exposure, inspection or dicing [1], [2] (see Fig. 2 [3]). The required positioning accuracy is in the nanometer range which results in stringent requirements on the output current accuracy and bandwidth of the power amplifiers driving the various types of actuators involved. Therefore, mitigating the errors produced by the amplifier results in improved performance of the end application and is the subject of this research [4], [5].

Current power amplifiers used in semiconductor manufacturing equipment include a resonant pole inverter topology with variable hysteresis control [6]. The incorporated soft-switching behavior results in a switching frequency range of 10 − 100 kHz. In order to meet the requirements for future generation semiconductor manufacturing equipment, a switching frequency of ≥ 100 kHz across the entire operating range is proposed. Due to the stringent requirements on output current accuracy for high-precision mechatronic systems, the required switching frequency of the power amplifier is several orders higher than applied in common inverter applications, such as electrical propulsion and grid connected converters. This imposes significant challenges for the design and control of the power amplifier.

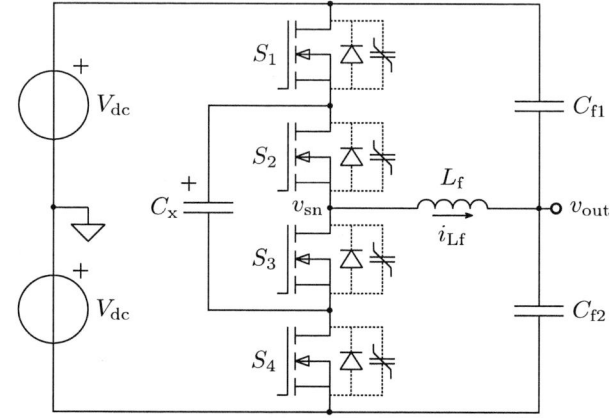

Fig. 1. Schematic of a 3-level FC RPI. The flying capacitor is designated C_x, the filter inductor L_f, and the output filter capacitors C_{f1} and C_{f2}. For each MOSFET, the internal body diode and nonlinear output capacitance C_{oss} are drawn with dashed lines.

Previous research conducted in this area has shown that the Flying Capacitor Resonant Pole Inverter (FC RPI) is a suitable multilevel topology with increased switching frequency and Zero-Voltage Switching (ZVS) across the entire operating range [7]–[10], see Fig. 1. This enables the use of fast switching devices with a voltage rating lower than the bus voltage by dividing the voltage stress over multiple switches. The addition of two switches and a flying capacitor C_x to a standard half bridge results in a 3-level converter [9].

To ensure ZVS across the entire operating range, a charge-based modulation scheme, as proposed in [11] and implemented for a 3-level FC-RPI in [9], is applied. However, this modulation scheme depends heavily on the correct measuring of the zero-crossings of the filter inductor current to synchronize the calculated switching times with the actual filter inductor current. The delay incorporated in the measurement chain results in significant distortion of the output current which deteriorates the performance of the end application.

This research proposes to apply direct current feedback of the per-period average filter inductor current. An Anisotropic Magneto-Resistive (AMR) sensor with a bandwidth of around 2 MHz is used as a measurement device [12]. The measured filter inductor current is fed to a controller in order to correct the introduced distortion of the output current. A simulation framework has been constructed incorporating the control sys-

3840

The 2018 International Power Electronics Conference

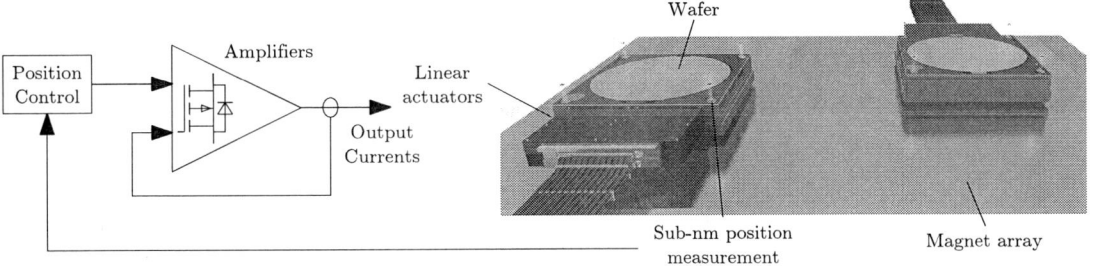

Fig. 2. Outline of high-precision wafer positioning system. (Courtesy of ASML [3])

tems as well as the converter implementation. The performance of the resulting system is verified and compared to the original system.

In section II, a brief introduction is provided regarding the Flying Capacitor Resonant Pole Inverter topology and the implemented charge-based ZVS modulation scheme. The applied elaborate control strategy with direct filter inductor current feedback is presented in section III, and the resulting simulation results of the combined control and converter framework are given in section IV. The drawn conclusions are finally presented in section V.

II. FLYING CAPACITOR RESONANT POLE INVERTER

A. Trapezoidal Filter Current

A 3-level Flying Capacitor Resonant Pole Inverter (FC RPI) contains a single flying capacitor C_x, see Fig. 1, resulting in three possible voltage levels of the switch-node v_{sn} (i.e. V_{dc}, 0 and $-V_{dc}$), and limits the voltage stress on each switch S_x to V_{dc}. When actively applying this additional voltage level, a trapezoidal filter current shape is obtained instead of a triangular shape [8], [13], which results in decreased rms current through the filter inductor [9].

A schematic representation of the filter inductor current i_{Lf} for 3-level modulation is shown in Fig. 3. For the steep gradients α and β, voltage levels V_{dc} and $-V_{dc}$ are selected for v_{sn} respectively. For the middle part of the trapezoidal shape with gradient γ, the voltage level depends on the voltage across the flying capacitor v_{cx}, and whether the flying capacitor has to be charged ($v_{sn} = V_{dc} - v_{cx}$) or discharged ($v_{sn} = -V_{dc} + v_{cx}$). This enables the voltage across the flying capacitor v_{cx} to be actively regulated to V_{dc}, assuring the availability of the ≈ 0 V voltage level of the switch-node v_{sn}. The corresponding equations for the slopes are:

$$\alpha = \frac{V_{dc} - v_{out}}{L_f} \tag{1}$$

$$\beta = \frac{-V_{dc} - v_{out}}{L_f} \tag{2}$$

$$\gamma = \frac{v_{sn} - v_{out}}{L_f}. \tag{3}$$

Assuming v_{cx} is balanced properly around V_{dc} with relatively low variations, the approximation can be made that $v_{sn} \approx 0$ V during intervals $[t_2, t_3]$ and $[t_6, t_7]$. For the specific

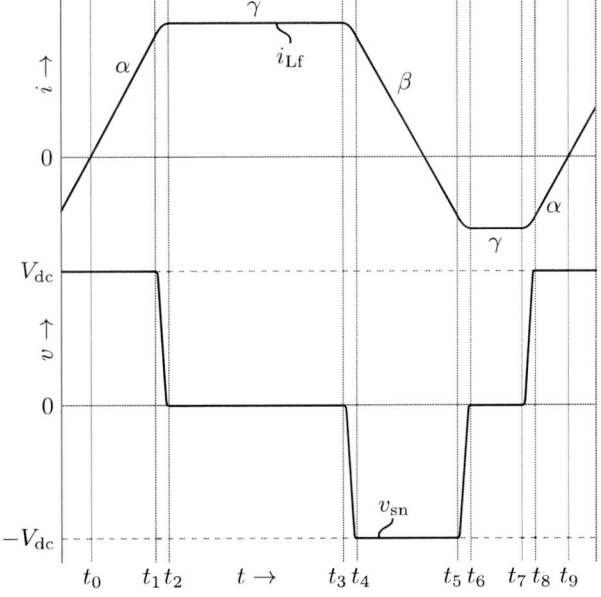

Fig. 3. Schematic representation of i_{Lf} and v_{sn} waveforms for a single switching period. The desired average current $i^*_{amp} > 0$, and $v_{out} = 0$.

case drawn in Fig. 3, with the desired average current $i^*_{amp} > 0$ and $v_{out} = 0$, this results in a flat middle part γ.

The duration of each switching state can be adjusted to obtain the desired average filter current i^*_{amp}, to minimize the rms value of the current through the filter inductor, to regulate the switching frequency, and to guarantee soft-switching at each switching instant. The latter requires the proper calculation of the corresponding time intervals and current values, in order for v_{sn} to completely commutate to the appropriate voltage level. [9], [10]

B. Charge-based Zero-Voltage Switching

The nonlinear output capacitance $C_{oss}(v_{ds})$ of a MOSFET is used as a resonant component for the circulating filter current i_{Lf} and to achieve ZVS in a switching leg [11]. The charge model of $C_{oss}(v_{ds})$ enables the use of piece-wise linear approximation of the trapezoidal shape of i_{Lf}, of which a

3841

The 2018 International Power Electronics Conference

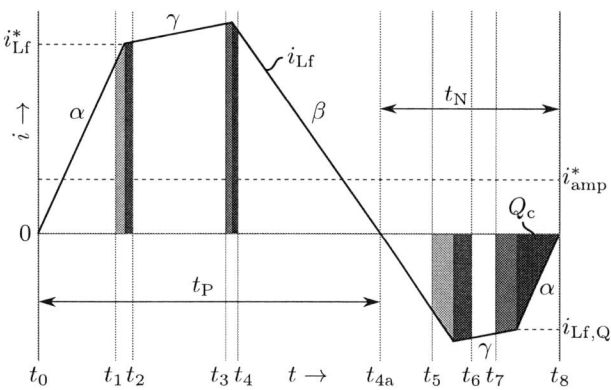

Fig. 4. Schematic representation of i_{Lf} for $i^*_{\text{amp}} > 0$ and $v_{\text{out}} < 0$ with commutation charge areas to ensure ZVS indicated in gray. The corresponding minimum current $i_{\text{Lf,Q}}$, desired average current i^*_{amp}, and setpoint-dependent i^*_{Lf} are indicated together with the switching state timing intervals.

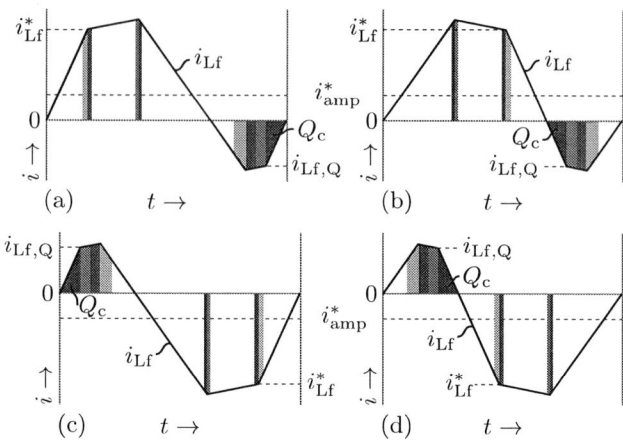

Fig. 5. Schematic representation of i_{Lf} for four different modulation states depending on the signs of i^*_{amp} and v_{out}: (a) $i^*_{\text{amp}} > 0$ A and $v_{\text{out}} < 0$ V, (b) $i^*_{\text{amp}} > 0$ A and $v_{\text{out}} < 0$ V, (c) $i^*_{\text{amp}} < 0$ A and $v_{\text{out}} < 0$ V, (d) $i^*_{\text{amp}} < 0$ A and $v_{\text{out}} > 0$ V. Required commutation current $i_{\text{Lf,Q}}$, setpoint dependent current i^*_{Lf}, and i^*_{amp} are indicated for each quadrant, as well as the commutation charges in gray.

schematic representation is shown in Fig. 4 for $i^*_{\text{amp}} > 0$ A and $v_{\text{out}} < 0$ V. The charge-based analysis presented in [11] was applied to a 3-level FC RPI in [9] with the latter approach being used in this research.

To minimize the rms value of i_{Lf}, and therefore maximize efficiency, the surface of the negative part of i_{Lf} (indicated with t_{N}), needed to achieve ZVS, is to be minimized. An additional advantage is that this maximizes switching frequency. As a first step, a critical area underneath i_{Lf} is defined which has to contain at least the commutation charge Q_{c} of the output capacitance $C_{\text{oss}}(v_{\text{ds}})$ of a single MOSFET, assuming $v_{\text{ds}} = V_{\text{dc}}$. The critical area is indicated with Q_{c} in Fig. 4. From this area and the slope of i_{Lf} for that section α, the corresponding minimum current $i_{\text{Lf,Q}}$ needed to ensure ZVS can be calculated according to:

$$i_{\text{Lf,Q}} = -\sqrt{2 \cdot Q_{\text{c}} \cdot \alpha}. \tag{4}$$

In order to achieve ZVS for each switching instant, the charge present in the current for each of the commutation intervals, being $[t_1, t_2]$, $[t_3, t_4]$, $[t_5, t_6]$, and $[t_7, t_8]$, must be at least $2Q_{\text{c}}$. This is indicated in Fig. 4 with areas in different shades of gray, each representing an area corresponding to Q_{c}. The indicated switching-state timing intervals correspond to Fig. 3. Note that $t_8 = t_9$ since $[t_8, t_9] = 0$ for this modulation state ($i^*_{\text{amp}} > 0$ A, $v_{\text{out}} < 0$ V). Furthermore, interval $[t_6, t_7]$ is set to a fixed value for regulating v_{cx}, but can also be used as a tuning parameter for the switching frequency. [10]

Given α, β, γ, the required commutation charge Q_{c}, and $i_{\text{Lf,Q}}$, all time values $t_{4\text{a}} - t_9$ and their respective current values for the negative part of i_{Lf} can be calculated. This gives a total charge area A_{N} of i_{Lf} in the time interval $[t_{4\text{a}}, t_8]$, defined as t_{N}. Furthermore, the per switching period average value of i_{Lf} has to be equal to the setpoint current, i^*_{amp}. This gives for the total charge area A_{P} of the positive part of i_{Lf} in the time interval $[t_0, t_{4\text{a}}]$, defined as t_{P}:

$$A_{\text{P}} = A_{\text{N}} + i^*_{\text{amp}} (t_{\text{N}} + t_{\text{P}}). \tag{5}$$

However, the equations for both time and current values of i_{Lf} for the charge area of the positive part are still unbounded. To oppose this in a way to have simple calculations, i^*_{Lf}, as indicated in Fig. 4, is proposed to be made dependent of i^*_{amp} and the minimum commutation current $i_{\text{Lf,Q}}$ for the same slope α, and set to:

$$i^*_{\text{Lf}} = -i_{\text{Lf,Q}} + 1.5 \ i^*_{\text{amp}}. \tag{6}$$

From the given slopes of i_{Lf}, defined current value i^*_{Lf}, commutation charge Q_{c}, area A_{N} and (5), the remaining time and current values for the positive part of i_{Lf} can be calculated. This results in a complete description of i_{Lf} and timing of the switching intervals for a single switching period, which can be implemented in a controller.

For a converter with a trapezoidal filter current operating in all four quadrants, corresponding piece-wise linear approximations of i_{Lf} can be made and are shown in Fig. 5. The critical area of i_{Lf} with a minimum required commutation charge Q_{c} is located differently for each modulation state, depending on the signs of i^*_{set} and v_{out}. However, the reasoning and calculations are analogous to the case discussed above. [10]

Zero-crossing detection of i_{Lf} is used to ensure the calculated timing model remains in phase with the actual current. This means the functioning of the modulation principle is heavily dependent on an accurate and low-latency zero-crossing detection. Significant jitter and delay occurring on the zero-detection signal arriving at the controller will have a significant impact on the performance of the converter.

III. DIRECT FILTER INDUCTOR CURRENT FEEDBACK

The charge-based ZVS modulation scheme with trapezoidal filter current as detailed in section II, relies on the accurate measurement of the zero-crossings of the filter inductor current to synchronize the switching time calculations with the actual

The 2018 International Power Electronics Conference

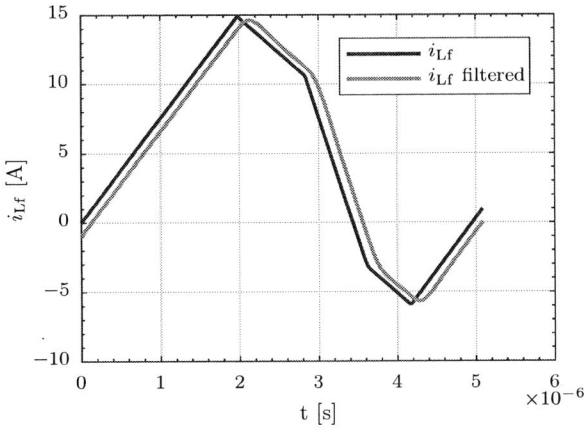

Fig. 6. Simulated waveforms of i_{Lf} without sensor filter (black), and with sensor filter (gray).

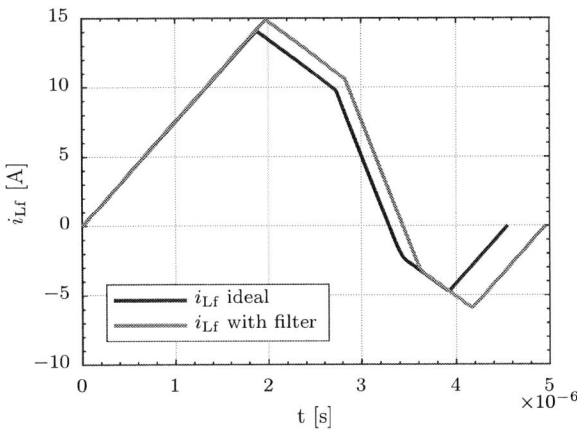

Fig. 7. Simulated waveforms of i_{Lf} with ideal zero-crossing detection (black), and with filtered i_{Lf} measured zero-crossing detection (gray). The per-period average i_{Lf} is 5.000 A and 4.779 A respectively.

current. An accurate measurement of the filter inductor current with a high bandwidth is therefore required. An Anisotropic Magneto-Resistive (AMR) sensor with a bandwidth of around 2 MHz is used for this purpose [12]. However, it still introduces a delay between the actual zero-crossing and the signal arriving at the controller of the converter. More delay is to be expected from the comparator circuit and processor front end. For this research, only the sensor delay is taken into account and serves as a proof of concept for other delay factors.

To give a general idea regarding the impact of the sensor, simple simulations have been performed where the AMR sensor is modeled as a second order Butterworth low-pass filter with a cut-off frequency of 2 MHz. A desired (steady state) average filter inductor current of $i_{amp}^* = 5$ A is applied as the current setpoint.

The introduced delay on the filter inductor current waveform is shown in Fig. 6. The figure shows the actual filter inductor current i_{Lf} as simulated in black, and the filtered output of the sensor in gray. The delay introduced by the sensor in this simulation is in the order of 100 ns which is significant with respect to the switching times calculated by the charge-based ZVS modulation scheme.

When the filtered i_{Lf} signal is used by the controller of the converter to determine the zero-crossings of the filter inductor current, an overshoot of the current occurs, which is shown in Fig. 7. The simulated waveform of i_{Lf} with ideal zero-crossing detection is shown in black, and the waveform of i_{Lf} with filtered zero-crossing detection is shown in gray. The resulting per-period average filter inductor current values are 5.000 A for the ideal and 4.779 A for the filtered zero-crossing detection. This indicates a significant deviation of the required average current i_{amp}^*. Furthermore, more extensive simulations when a sinusoidal input signal is applied (see section IV) indicate a decrease in SFDR of 16 dB.

To correct for the introduced distortion of the output current, this research proposes to implement a control strategy that directly takes the measured filter inductor current from the AMR sensor as an input. A schematic outline of the complete resulting control configuration is shown in Fig. 8. It consists of two control loops comprising filter inductor current and output current control.

A. Filter inductor current control

The outline of the filter inductor current control is shown in Fig. 8 in the respective dashed box. The filter inductor current i_{Lf}, as resulting from the simulated FC RPI, is filtered with the modeled transfer of the AMR sensor H_{sensor}, which consists of a second order Butterworth low-pass filter with a cut-off frequency of 2 MHz. The resulting filtered i_{Lf} is then fed into the i_{avg} *calc* block, which determines the per-period average value of the measured filter inductor current i_{avg}. The per-period average current i_{avg} is subtracted from the average current setpoint i_{set}^*, generating an error current that is fed to a controller C_{iLf}. A simple feed forward of the setpoint current completes the filter inductor current control outline.

In order to obtain error correction and noise shaping, the characteristics of an integrator are required for the controller C_{iLf}. However, since a significant phase delay exists due to the calculation and measurement delays in the i_{avg} *calc* block and H_{sensor} respectively, the addition of a zero to the controller around the 0 dB crossing of the controller transfer C_{ilf} is required. This ensures sufficient phase margin to obtain a stable control loop. A complex pole is furthermore added to obtain significant roll-off at the intended switching frequency range of ≥ 100 kHz. The resulting Bode plot of the filter inductor current feedback controller transfer C_{iLf} is shown in Fig. 9.

The resulting transfer of C_{iLf} is given by

$$C_{iLf}(s) = \frac{\omega_p^2 s + \omega_p^2 \omega_z}{s^3 + \sqrt{2}\omega_p s^2 + \omega_p^2 s}, \qquad (7)$$

The 2018 International Power Electronics Conference

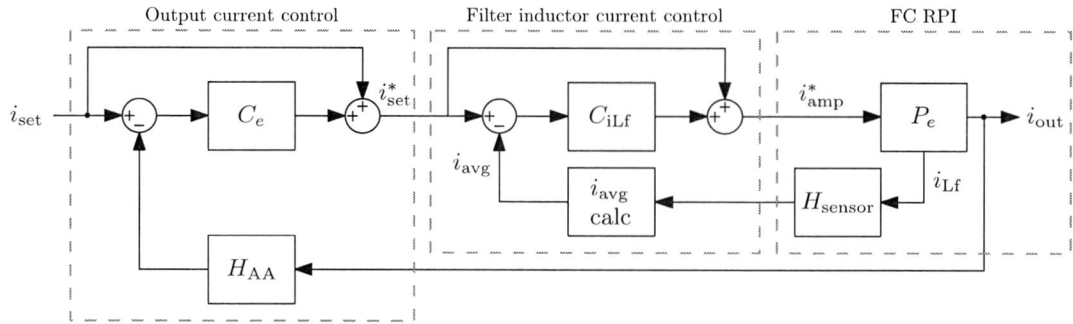

Fig. 8. Schematic outline of the complete control configuration, with the respective parts indicated with dashed boxes.

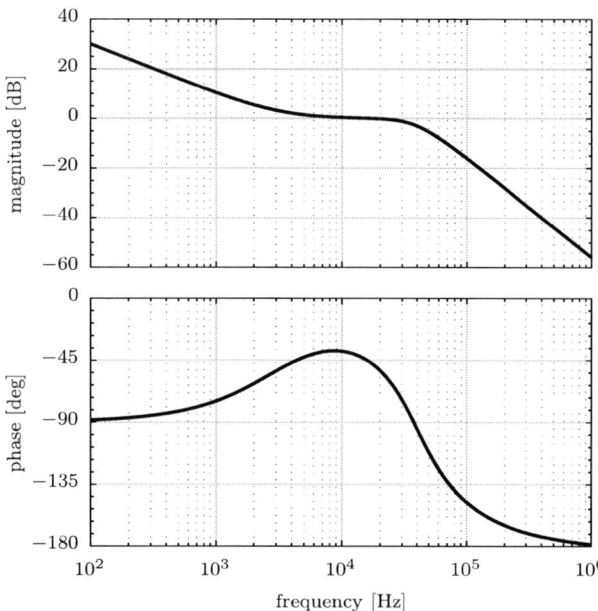

Fig. 9. Bode plot of the filter inductor current feedback controller transfer C_{iLf}, combining an integrator with a zero at around 3 kHz to obtain sufficient phase margin around the 0 dB crossing and a complex roll-off pole at 40 kHz.

where ω_{p} is the frequency of the complex roll-off pole at 40 kHz, and ω_{z} is the frequency of the zero at around 3 kHz to obtain sufficient phase margin around the 0 dB crossing of the controller transfer. The resulting 0 dB crossing of C_{iLf} is around 10 kHz, an order of magnitude lower than the intended switching frequency range of $\geq$ 100 kHz, indicating the open-loop bandwidth of the filter inductor current control loop.

B. Output current control

The proposed converter configuration is intended to drive an inductive load (actuator) in a motion control system. Therefore an output current control loop is added to the filter inductor current control loop. Output disturbances, e.g. due to EMF, are to be compensated by the controller.

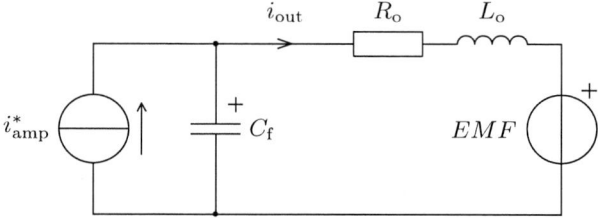

Fig. 10. Schematic representation of electrical plant P_{e}, with the FC RPI switching stage approximated with current source i_{amp}^{*} and output filter capacitor C_{f}. The load consists of a series connection of a resistor R_{o}, inductor L_{o}, and voltage source EMF, emulating an actuator.

The outline of the output current control loop is shown in Fig. 8 in the respective dashed box. It consists of an anti-aliasing filter H_{AA}, current controller C_{e}, and a direct setpoint current feed forward. [5], [10]

The switching stage as discussed in the previous sections, generates an average filter inductor current i_{avg} according to a current setpoint i_{amp}^{*}. This operation principle can be approximated by a current source when the switching frequency is significantly higher than the intended frequency range of the current setpoint, and the filter inductor current control is functioning properly. This results in a current-controlled amplifier and the objective of the current controller C_{e} is to regulate the current through the inductive load.

Since a current-controlled current amplifier is to be obtained, the controller C_{e} is of the PID type [5], [10]. Using a generic model for the plant P_{e}, of which a schematic representation is shown in Fig. 10, the open-loop transfer of the resulting output current control loop can be tuned to a desired bandwidth. Applying the system parameters as defined in Table I and a desired open-loop bandwidth of 1kHz, being an order of magnitude lower than the bandwidth of the filter inductor current control loop, the transfer for the current controller C_{e} and the resulting open-loop transfer H_{OL} can be obtained. Bode plots of both transfers are displayed in Fig. 11, from which can be concluded that sufficient phase margin is achieved resulting in a stable control loop.

3844

The 2018 International Power Electronics Conference

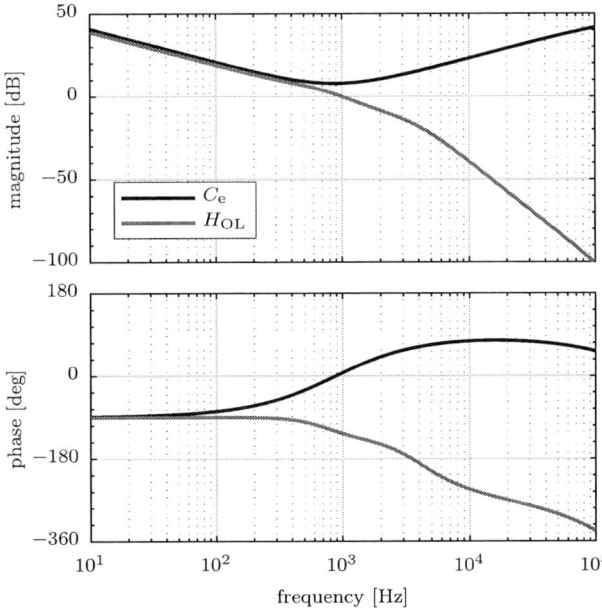

Fig. 11. Bode plots of the output current controller transfer C_e (black), and the resulting open-loop transfer of the output current control loop H_{OL} (gray).

TABLE I
SIMULATION SPECIFICATIONS

Parameter	Value	Unit
V_{dc}	250	V
C_x	100	μF
L_f	20	μH
C_{f1}, C_{f2}	5	μF
R_o	4	Ω
L_o	5.5	mH
$f_{i,set}$	110	Hz
$|i_{set}|$	10	A

IV. SIMULATION RESULTS

The 3-level FC RPI configuration as shown in Fig. 1 was implemented using the PLECS blockset for Matlab Simulink. The charge-based ZVS modulation scheme with trapezoidal filter inductor current and the control configuration as shown in Fig. 8 was implemented in Matlab Simulink. The simulation parameters are given in Table I. The load used in the simulation model is a series connection of a resistor R_o and inductor L_o, emulating an actuator. For simplicity, the EMF resulting from actual movement of the actuator is neglected in this research.

A. Filter inductor current comparison

When applying a desired (steady state) average filter inductor current of $i_{amp}^* = 5$ A to the system, the waveforms of the filter inductor current as shown in Fig. 12 are obtained for the different configurations. The waveform for ideal zero-crossing detection is plotted in black, the filtered inductor current configuration in dark gray, and the compensated filter

TABLE II
AVERAGE FILTER INDUCTOR CURRENTS

Configuration	Value	Unit
i_{Lf} ideal	5.000	A
i_{Lf} filtered	4.779	A
i_{Lf} filtered+control	4.998	A

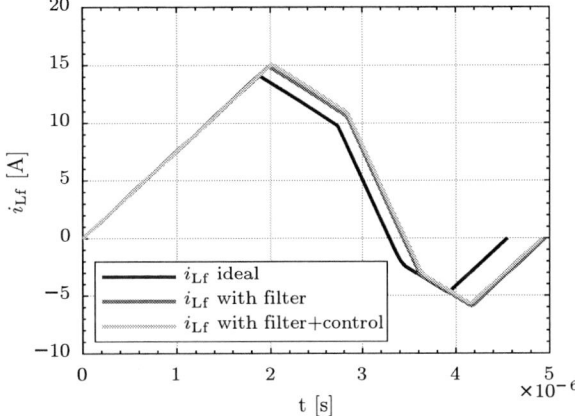

Fig. 12. Simulated waveforms of i_{Lf}: ideal zero-crossing detection (black), with filtered i_{Lf} measured zero-crossing detection (dark gray), and with filtered i_{Lf} and control enabled (light gray). The per-period average i_{Lf} is 5.000 A, 4.779 A and 4.998 A respectively.

inductor current with the control loop enabled in light gray, where the latter two nearly overlap. The output current control loop is disabled for these simulations.

As can be seen in the figure, the filter inductor current controller adjusts the average current setpoint i_{amp}^* in order to compensate the error originating from the sensor characteristics. The resulting corresponding average currents are shown in Table II, from which can be concluded that the (steady state) error caused by the non-ideal characteristics of the current sensor is significantly reduced by the added control loop.

When compared to the ideal zero-crossing detection, the switching frequency is slightly increased for the compensated configuration. If required, this effect can be mitigated by reducing the length of the middle part of the trapezoidal current shape.

B. Spectral analysis

For high precision motion control, non-linearity of the generated output current of the amplifier has a significant impact on position accuracy. Non-linearity is investigated by applying spectral analysis to a simulated sinusoidal load current. The input frequency is set to $f_{i,set} = 110$ Hz with an amplitude of $|i_{set}| = 10$ A, and the simulation is run for 11 periods to obtain coherent sampling throughout the entire simulation and ample frequency points.

Three configurations were simulated: no i_{Lf} control and H_{sensor} set to 1, no i_{Lf} control but including the low-pass characteristics of the sensor, and including i_{Lf} control and

3845

The 2018 International Power Electronics Conference

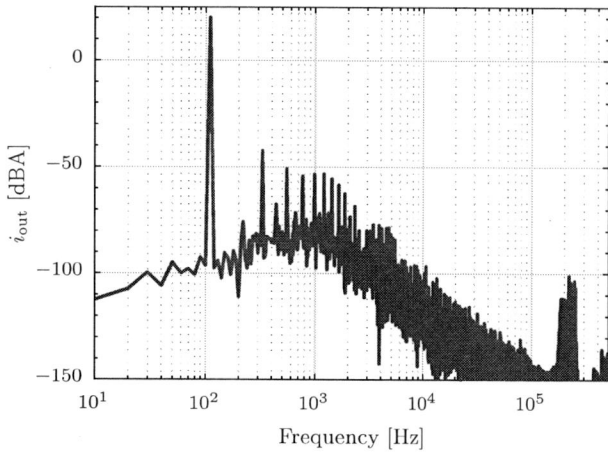

Fig. 13. FFT of the output current i_{out} without sensor characteristics and without i_{Lf} control.

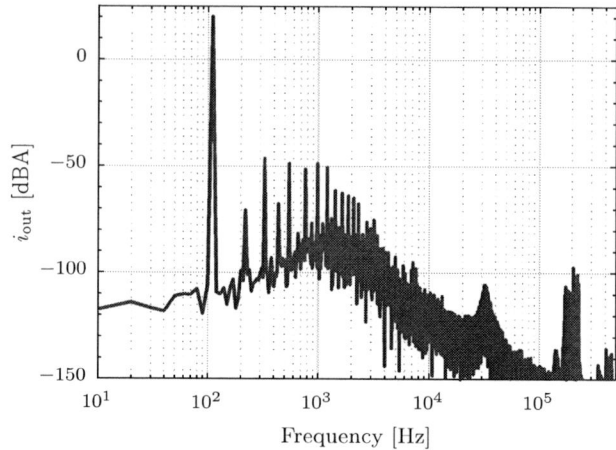

Fig. 15. FFT of the output current i_{out} with sensor characteristics and i_{Lf} control.

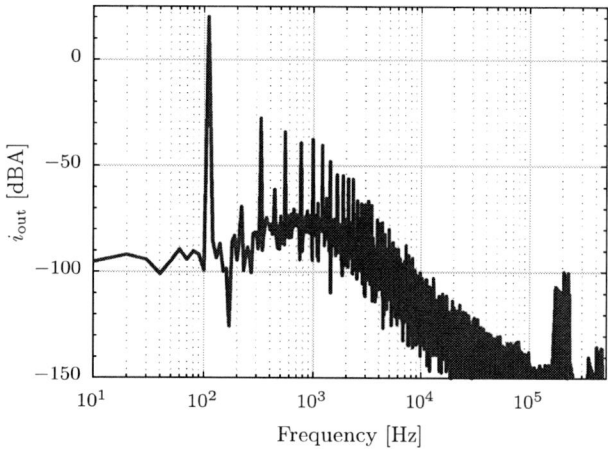

Fig. 14. FFT of the output current i_{out} with sensor characteristics but without i_{Lf} control.

sensor characteristics. The results for the Spurious Free Dynamic Range (SFDR) and Total Harmonic Distortion (THD) calculations for the three configurations are summarized in Table III. The FFT plots for the three configurations are shown in Fig. 13, Fig. 14 and Fig. 15 respectively. The sample frequency is $f_{s,fft} = 100$ MHz but the frequency range of the FFT plots is confined to 10 Hz $- 500$ kHz to focus on the range of interest. The output current control loop is enabled for these simulations.

TABLE III
SPECTRAL ANALYSIS RESULTS

i_{Lf} filtering	i_{Lf} control	SFDR [dBA]	THD [dBA]
No	No	62.7	-61.0
Yes	No	47.8	-46.1
Yes	Yes	66.6	-61.8

From the simulation results can be concluded that including the sensor characteristics, which cannot be avoided in an actual converter, has a significant impact on performance. However, including the proposed direct i_{Lf} control scheme results in a significant improvement: 9 dB increased SFDR and 16 dB decreased THD. This will therefore have a significant impact on performance increase in the position accuracy of a high precision fast-moving stage [5].

V. CONCLUSION

A Flying Capacitor Resonant Pole Inverter with charge-based Zero-Voltage Switching was presented to which direct filter inductor current control was applied. An output current control loop was added to the resulting configuration regulating the current through the load in order to compensate for disturbances. The complete converter framework was simulated and its performance verified using time domain and frequency domain analysis.

From the results can be concluded that a significant performance increase can be achieved with the proposed control strategy by adding the direct filter inductor current control loop. The error introduced by the delayed sensing of the zero-crossing of the filter inductor current is significantly mitigated. As a result, the increased linearity of the output current of the converter will result in a significant improvement of the position accuracy of a high precision fast-moving stage.

Future work will include hardware verification and increased voltage and current specifications.

REFERENCES

[1] H. Butler, "Position control in lithographic equipment [applications of control]," *IEEE Control Syst. Mag.*, vol. 31, no. 5, pp. 28–47, 2011.
[2] J. M. Schellekens, *A Class of Robust Switched-Mode Power Amplifiers with High Linear Transfer Characteristics.* PhD Thesis, Eindhoven University of Technology, 2014.
[3] R. H. Munnig Schmidt, "Ultra-precision engineering in lithographic exposure equipment for the semiconductor industry." *Philos. Trans. A. Math. Phys. Eng. Sci.*, vol. 370, no. 1973, pp. 3950–72, 2012.

3846

[4] M. Mauerer, A. Tuysuz, and J. W. Kolar, "Distortion analysis of low-THD/high-bandwidth GaN/SiC class-D amplifier power stages," in *2015 IEEE Energy Convers. Congr. Expo.* IEEE, sep 2015, pp. 2563–2571.

[5] S. J. Settels, J. van Duivenbode, and J. L. Duarte, "Impact of amplifier errors on position loop accuracy of high-precision moving stages," in *2017 19th Eur. Conf. Power Electron. Appl. (EPE'17 ECCE Eur.* IEEE, sep 2017, pp. P.1–P.10.

[6] J. M. Schellekens, J. L. Duarte, M. A. M. Hendrix, and H. Huisman, "Interleaved switching of parallel ZVS hysteresis current controlled inverters," in *2010 Int. Power Electron. Conf. - ECCE ASIA*, 2010, pp. 2822–2829.

[7] T. Meynard and H. Foch, "Multi-level conversion: high voltage choppers and voltage-source inverters," in *PESC '92 Rec. 23rd Annu. IEEE Power Electron. Spec. Conf.*, 1992, pp. 397–403.

[8] J. J. C. van Emden, J. van Duivenbode, and J. L. Duarte, "Flying capacitor resonant pole inverter topology with reduced switch voltage stress," in *2013 15th Eur. Conf. Power Electron. Appl.*, 2013, pp. 1–10.

[9] S. J. Settels, J. Everts, and J. van Duivenbode, "Charge-based Zero-Voltage Switching of a Flying Capacitor Resonant Pole Inverter with Trapezoidal Filter Current," in *IECON 2016 - 42nd Annu. Conf. IEEE Ind. Electron. Soc.* IEEE, oct 2016, pp. 3282–3287.

[10] S. J. Settels, J. van Duivenbode, J. L. Duarte, and E. A. Lomonova, "Flying capacitor resonant pole inverter applying five voltage levels," in *2017 IEEE Energy Convers. Congr. Expo.* IEEE, oct 2017, pp. 2121–2128.

[11] C. Marxgut, F. Krismer, D. Bortis, and J. W. Kolar, "Ultraflat Interleaved Triangular Current Mode (TCM) Single-Phase PFC Rectifier," *IEEE Trans. Power Electron.*, vol. 29, no. 2, pp. 873–882, 2014.

[12] Sensitec, "Datasheet: CMS3050, Highly Dynamic MagnetoResistive Current Sensor," 2015. [Online]. Available: https://www.sensitec. com/fileadmin/sensitec/Service_and_Support/Downloads/Data_Sheets/ CMS3000/SENSITEC_CMS3050_DSE_05.pdf

[13] D. M. Divan and G. Skibinski, "Zero switching loss inverters for high power applications," *IEEE Trans. Ind. Appl.*, vol. 25, no. 4, pp. 634–643, 1989.

The 2018 International Power Electronics Conference

Design of a GaN-Based Wireless Power Transfer System at 13.56 MHz to Replace Conventional Wired Connection in a Vehicle

Kawin Surakitbovorn and Juan Rivas-Davilla
Electrical Engineering
Stanford University
Stanford, CA 94305
Email: north@stanford.edu

Abstract—**Wired connections running through the doors, hatchbacks, and trunks of a vehicle are subjected to extensive mechanical wear from repeated bending caused by the open-and-close motion. By employing wireless power transfer, this failure mode can be completely eliminated. In this paper, a 300 W 48-to-48 V WPT system with above 90% efficiency designed for such application is demonstrated. To achieve a system that is small, light-weight, and highly efficient as desired by the automotive industry, a novel inductor elimination technique is developed. The following paper shows the system design methodology, the inductor elimination technique, and the control scheme used in this work.**

Index Terms—**Automobile, gallium nitride (GaN), inductor cancellation, wireless power transfer.**

I. INTRODUCTION

Mechanical wear and corrosion issues due to cycling and vibration are the main drivers in warranty for all conventional electrical interconnects in motor vehicles. In the interface between the door and body of a vehicle, a specialized connector system (shown in figure 1) requires a rubberized grommet, locking clips and extra tape in order to survive door cycling and vehicle operations. With the advancement of wireless power transfer (WPT) technology over the past decade, replacing conventional wired connections with WPT systems becomes an attractive option to eliminate mechanical wear and tear and to prevent corrosion issues.

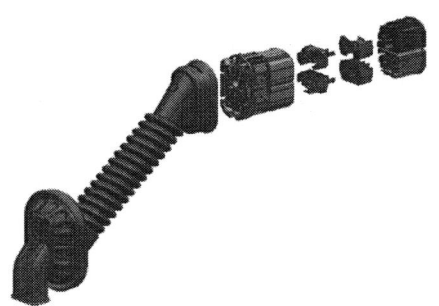

Fig. 1: A specialized connector used to connect wires from the body side of a car to the electronic loads inside the door. *credit: Courtesy of Ford Motor Company.*

There have been a vast amount of recent research work published in the area of WPT (i.e [1]–[7]). Nonetheless, the systems described in these works are either too large to fit in tight locations where existing wired connections are, or they are too heavy or lack the efficiency necessary to be economically viable for the automobile industry. In WPT systems, various operating frequencies can be used. The higher the frequency, the more challenging the design becomes. At megahertz frequencies, switching losses in hard-switch inverters can become excessive, and resonant inverters that eliminate switching losses trough soft switching are used instead. One main benefit of operating at radio frequencies (RF) is that the converters no longer need to rely on ferrites, providing an opportunity to significantly reduce the weight of the system. Other benefits include improvement of transient time, better load regulation, as well as simpler EMI filter designs.

One difficulty in operating WPT at RF is in the issue of communication latency in the controller. In many applications, the WPT systems are used for batteries charging and fast control loop is not required. For those, sufficiently fast communication from the receiver side to the transmitter side can be easily achieved. On the other hand, for applications where the power has to be delivered to the loads in real time, high latency in the communication raises significan practical issues. Unfortunately, most of the commercially available wireless communication protocols such as Bluetooth or WiFi have latency in the order of 10s of ms [8], making them unsuitable for applications where fast control loop is required.

In this paper, a small and light weight, fully regulated 48 V to 48 V WPT system is presented. The system operates at 13.56 MHz and is capable of transferring maximum 300 W of power over a 15 mm air gap. For this prototype work, an LED/photodiode pair is used as a means for low latency wireless communication. To create a fully regulated system, a hysteresis (on/off) control technique is employed. Due to the high switching frequency, the on/off control can keep the circuit operating efficiently thoroughly the load range, while maintaining relatively high control loop frequency. As part of this work, a novel inductor elimination technique is developed to improve efficiency and reduce the system weight. Its design methodology will be discussed in section III-C.

II. System Overview and Specification

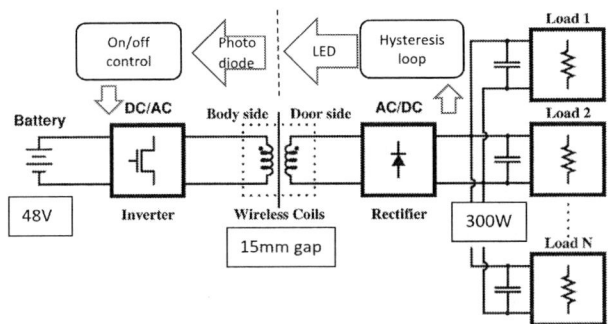

Fig. 2: The system block diagram.

To justify replacing a wired connection with a wireless one, the WPT system must be small and light-weight, have high efficiency regardless of the load, as well as have high reliability and low design complexity. Figure 2 shows the block diagram of our proposed system. DC power from the battery is first converted to an AC RF power. It is then transmitted from the body side of the vehicle to the door side through the WPT coil. Next, the AC power is converted back to DC power and is distributed to the loads. Finally, to achieve voltage regulation, a hysteresis comparator is used to monitor the output voltage, and to send the on/off signal back to the inverter circuit. An LED/photodiode pair is used to communicate over the air gap.

The 300 W specification is chosen to accommodate the inrush power of the biggest load in a typical car door, namely window lift motor. The 48 Vdc is designed to accommodate the upcoming bus voltage standard in the newer hybrid and full-electric vehicles. The high frequency of 13.56 MHz is selected to minimize the size of passive components, and is part of an industrial, scientific and medical (ISM) radio bands. The 15 mm separation is measured to be the typical gap size between the door and the body of a car when the door is closed. While it is important that the system can also operate when the door is open, this specification is not within the scope of this work.

III. System Design

A. Inverter Design

A class E push-pull inverter, shown in figure 3, is used to drive the WPT coil. When tuned properly, class E inverters have zero voltage switching (ZVS) allowing operation well into the 10s of MHz frequencies [9]. While having higher voltage stress than modified class E, such as class Φ_2 [10], the classic class E inverter offers simpler tuning and lower component count, favorable for our application where weight and size play important roles. Traditionally, class E inverter has a large input "choke" inductor, to make the input current to the circuit nearly constant with very little ripple. When operated

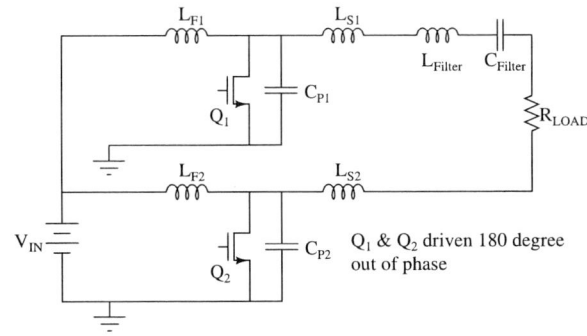

Fig. 3: The push-pull topology.

under on/off control, this large choke inductance leads to slow transient and higher loss. To speed up the transient, a design with relatively small input inductance, $L_f(s) = 1000$ nH, is used here instead. With this L_f, the circuit can reach steady state within roughly 7 cycles. While this design choice leads to faster transient, it also increases the input current ripple. To solve this issue, two identical class E inverters are connected in a push-pull configuration, operating at 180 degree out-of-phase to cancel out the input current ripple. To achieve the required 300 W when connected together, each half of the push-pull inverter is tuned for 150 W power. The rest of the design for the push-pull class E inverter follows the same standard class E inverter design steps.

Fig. 4: The class E push-pull inverter circuit with all its inductors. Note that the GaN FET is mounted on the underside, and the filter inductor is split into two for symmetry.

Figure 4 shows the inverter circuit board. For this work, 650 V 30 A GaN transistors, GS66508T from GaN Systems are used as the power switches. Although the 650 V break-down voltage is excessive for the 48 V input, GS66508T has superior on-resistance and packaging, allowing for close to ideal operation at RF. Initially, multiple reports of extra losses in the junction capacitance (C_{OSS}) of GaN devices under high dv/dt raised a concern for the use of GaN FET in our circuit [11], [12]. However, using the equations from [13], we calculated that this extra C_{OSS} loss in our case would be minimal at 1.5 W per FET. This is 60 % less than that of the conduction loss and we considered it acceptable.

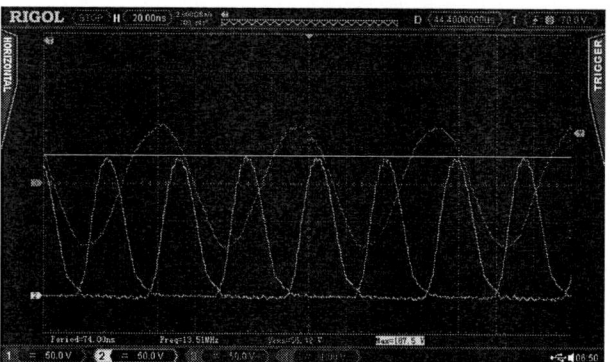

Fig. 5: The steady state waveform of the inverter circuit tested with an RF resistive load. Drain waveforms of the two GaN FET are shown in yellow and teal, while the output voltage is shown in pink.

The inverter is first tested by itself on a resistive RF load. Figure 5 shows the steady state waveform of the inverter circuit. Table I and II show the component values and the estimated loss breakdown of the inverter. While the efficiency of the inverter itself is not low, notice that majority of the losses outside the transistors occur in the inductors, $L_S(s)$ and L_{fil}. To further increase the overall system efficiency, an inductor elimination technique used to eliminate these losses is developed and will be described in section III-C.

TABLE I: Operating conditions and passive component values for the push-pull class E inverter.

Frequency [MHz]	V_{IN} [V]	P_{IN} [W]	L_F [nH]	L_S [nH]	C_P [pF]	L_{Fil} [nH]	C_{Fil} [pF]	R_{load} [Ω]
13.56	48	310	1000	110	220	590	230	12.5

TABLE II: Measured loss breakdown per component as shown in fig. 3 for the push-pull class E inverter.

P_{FET} [W]	P_{LF} [W]	P_{LS} [W]	P_{Lfil} [W]	P_{Cfil} [W]	P_{gate} [W]	P_{total} [W]	η [%]
4	0.4	1.4	7.9	0.6	1.2	21.3	93.1

B. Rectifier Design

A full-bridge (FB) passive rectifier is used to rectify the RF power back to DC. A full-bridge rectifier is not commonly used in non-isolated DC-DC converters operating at RF as the load is no longer reference to the input ground. However, in WPT systems where isolation is innate, a FB rectifier is an excellent option due to its simple topology requiring no additional inductors. At the desired operating point, the input impedance of this rectifier is, $Z_{rec} = \frac{8}{\pi^2} \frac{V_o^2}{P_{out}} = 6.23$ Ω. In order to connect this to the inverter and the WPT coil, a matching network will be required.

Fig. 6: The rectifier circuit. The output capacitor is split into three for symmetry.

Figure 6 shows the full-bridge rectifier circuit. Here, 60 V 10 A Schottky diodes, PMEG060V100EPD, from Nexperia are used as rectifying diodes. According to [14], Shottky diodes in RF applications also do exhibit extra losses not captured by their SPICE models. Nonetheless, we choose to use these PMEG060V diodes because they show the lowest loss in comparison to diodes from other manufacturers. When tested by itself with an RF power amplifier, the efficiency of this FB rectifier is measured to be at 97%. This is slightly less that the simulation value of 98%, but as with the inverter side, is still within an acceptable level.

C. WPT Coil Design & Inductor Elimination

The design of the WPT coil starts by identifying the largest diameter possible for the particular application requirement. In our case, to effectively replace wired harness connection inside vehicle doors with minimal change to the design of the doors and hinges, the coil has to be less than 70 mm in diameter. The WPT coil is first modeled in FastHenry [15]. Then, its electrical property is imported into LTSpice for an in-circuit simulation.

At 10s of MHz, standard Litz wire can no longer be used effectively [16], [17]. To achieve high transfer efficiency, thick gauge, solid core, copper wire is used instead. For this prototype, an 8 AWG copper wire is used, although thicker wire can be substituted for higher efficiency. Due to the small size of the coil, the self resonant frequency of the coil is much higher than the switching frequency. External capacitors are therefore needed for the coil to operate resonantly.

In traditional resonant WPT coil designs, both the leakage and magnetizing inductance of the coil are canceled out by resonating capacitors, leaving the WPT coil as an ideal transformer. To match the rectifier input impedance to the output impedance of the inverter, an extra matching network is typically used. All these added components result in unnecessary losses, as well as added weight and size. In this work, a new way of WPT coil tuning is developed. The WPT coil is instead designed such that their leakage (L_{leak}) and magnetizing (L_{mag}) inductance can be exploited as the L_S, L_{Fil}, as well as the matching network. This results in a significant improvement in the overall efficiency by eliminating unnecessary loss components.

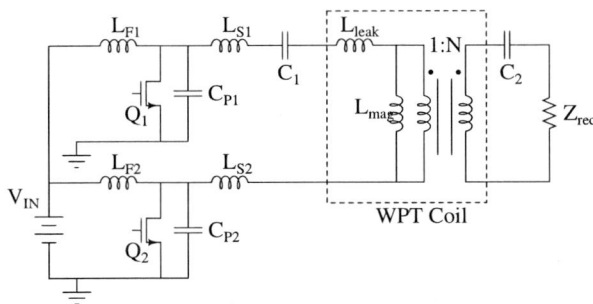

Fig. 7: The circuit diagram with the WPT coil shown in a cantilever model.

First, let us look conceptually how this can be achieved. Shown in the circuit diagram in figure 7, the WPT coil is connected across the output of the inverter and is modeled with a cantilever model. The FB rectifier is connected on the other side of the coil and is modeled as Z_{rec}. Two extra capacitors, C_1 and C_2, are also shown. As seen from this diagram, the L_{leak} is in series with the two L_S(s). This means that if we can somehow design the L_{leak} of the WPT coil to have the same exact value as $2 \times L_S$, then L_S(s) can be completely eliminated. Furthermore, notice the location of L_{mag}, C_2, Z_{rec}, and the transformer ratio N. As shown in figure 8, when designed properly, a highpass hi-low matching network can also be formed with L_{mag}, C_2, and N to provide both the impedance matching and filtering needed to match the rectifier input impedance, Z_{rec}, to the inverter output impedance. By designing the WPT coil this way, we eliminate both the L_S(s) and L_{fil}, removing the losses, P_{LS} and P_{Lfil}, associated with them.

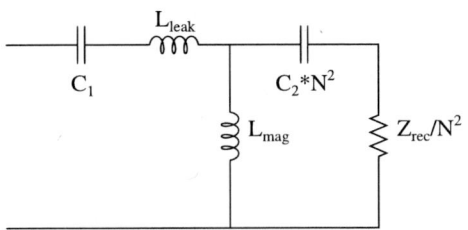

Fig. 8: The WPT coil part of the circuit with components reflected across the transformer into the inverter side.

While sounded simple, it is extremely difficult to actually design a WPT coil to have all these conditions met simultaneously. This is due to the physical constraints of the coil, such as that the number of turns has to an integer, that the

maximum diameter is of a certain value, and that the coil separation is fixed, as well as the need to maximize the coupling coefficient to achieve high transfer efficiency. To make the design possible, we choose to relax the constraint on L_{leak} from $L_{leak} = 2 \times L_S$ to $L_{leak} > 2 \times L_S$. Then, we use a capacitor $C_1 = 1/((2\pi f_s)^2(L_{leak} - 2L_s))$ to cancel out the extra inductance, leaving:

$$L_{leak} - \frac{1}{((2\pi f_s)^2 C_1)} = 2 \times L_S \tag{1}$$

As for the design of the matching network, to get the proper matching, we need the L_{mag} and N to satisfy the following equation:

$$R_{load} = \frac{(\frac{1}{j\omega C_2 N^2} + \frac{Z_{rec}}{N^2}) j\omega L_{mag}}{\frac{1}{j\omega C_2 N^2} + \frac{Z_{rec}}{N^2} + j\omega L_{mag}} \tag{2}$$

This is equivalent to:

$$1 - \omega^2 L_{mag} C_2 N^2 + \omega^2 C_2^2 Z_{rec}^2 = 0 \tag{3}$$

and:

$$\frac{\omega^4 L_{mag}^2 C_2^2 Z_{rec} N^2}{(1 - \omega^2 L_{mag} C_2 N^2)^2 + (\omega C_2 Z_{rec})^2} = R_{load} \tag{4}$$

By solving equation 3 and 4, we can find that the relationship between the L_{mag} and N required to get the proper matching is:

$$L_{mag} = \frac{1}{\omega} \sqrt{\frac{R_{load}^2 Z_{rec}}{N^2 R_{load} - Z_{rec}}} \tag{5}$$

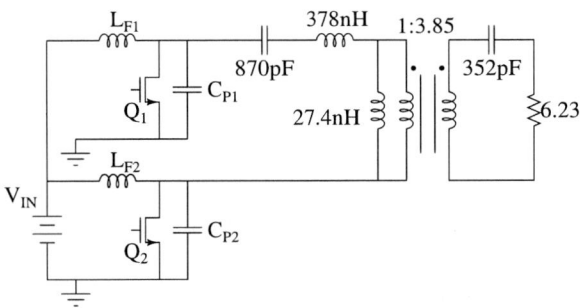

Fig. 9: The circuit diagram with the WPT coil shown with all its component values.

To find the specific coil design that satisfies these relationships, we first simulated a large number of different WPT coils on FastHenry with different numbers of turns and turn separations. For this application, we limit the search only to the coils in spiral configuration for its low profile; however, other configurations can certainly be used. Next, their electrical properties, L_{leak}, L_{mag}, and N, were imported out. A MATLAB script was used to search and locate the coil design options that satisfy both the L_{leak} requirement and the

3851

equation 5. From the simulation, two-turn spiral coils both with 67 mm outer diameter and 5 mm turn separation placed 15 mm apart is one of the options. Many other options were also available, but we selected this specific coil configuration for its large diameter and small number of turn, maximizing the coupling coefficient. Figure 9 shows this WPT coil in a transformer model along with the values of the capacitors used for the proper matching. It can be shown that with these values, L_{mag}, C_2, and N form a perfect resonant matching network, boosting the rectifier impedance, Z_{rec}, to the value 12.5 Ω needed by the push-pull inverter. Furthermore, with this specific value of C_1, part of the L_{leak} is canceled out at the fundamental frequency, leaving $L_{leak} - 1/((2\pi f_s)^2 C_1) = 378\ nH - 1/((2\pi \times 13.56\ MHz)^2 \times 870\ pF) = 220\ nH$ as required by the inverter.

D. Controller Design

To regulate the output voltage, a controller with a wireless communication ability is required. For this prototype, a simple hysteresis control loop is used in conjunction with a high-speed infrared LED/photodiode pair to communicate the on/off signal back to the inverter. The specific LED and photodiode used are TSAL6200 and BPV10NF from Vishay Semiconductor. In an automotive application, this optical pair can be easily placed on opposite side of the door/body while maintaining line of sight. For other applications, other methods of communication can also be used. With this setup, a very low communication latency of 80 ns is achieved. For a hysteresis band of 47.5 V - 48.5 V and output capacitance of 80 uF, the on/off frequency of the controller is between 11 kHz to 20kHz.

IV. RESULTS

TABLE III: Operating conditions and passive component values for the final WPT circuit with inductors eliminated.

Frequency [MHz]	V_{IN} [V]	V_{OUT} [V]	P_{MAX} [W]	L_F [nH]	C_P [pF]	L_{TX} [nH]	C_1 [pF]	L_{RX} [nH]	C_2 [pF]	C_{OUT} [μF]
13.56	48	48	295	1080	360	374	580	378	375	80

The full system with inductors eliminated is first simulated on LTSpice. Small adjustments on the values of C_1 and C_2 are done to take into account other non-ideality in the design methodology. The final operating condition and passive component values used are shown in table III. Figure 10 shows the final and complete system. Following the inductor elimination technique, only the inductor L_F(s) are left on the inverter. This greatly increases the overall efficiency of the system. Figure 11 shows the loss percentage breakdown. At full load, the DC-WPT-DC efficiency including the gate driver loss is 90.4 %, which is equivalent to the efficiency we would get if the inverter circuit without inductors eliminated were to be connected to a rectifier with an ideal loss-less matching network and without the WPT coil. Figure 12 shows the output voltage at 80 % load. As a result of small L_f(s), the drain

(a) Side view.

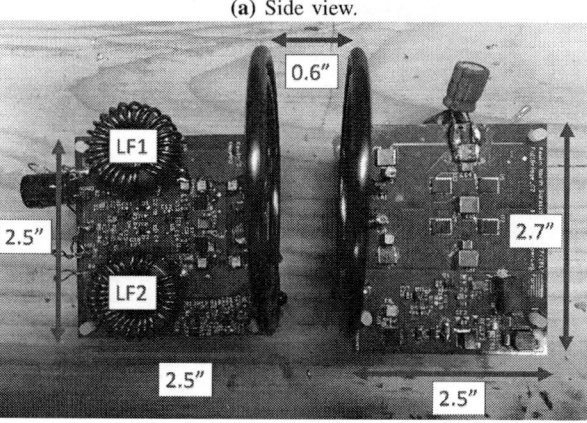

(b) Top view.

Fig. 10: The WPT system with the push-pull class E inverter on the left and the FB rectifier on the right. Notice that both the C_1 and C_2 are split into two for symmetry on the board.

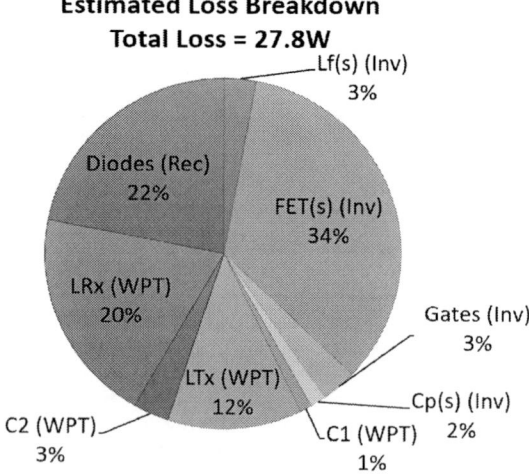

Fig. 11: Estimated loss breakdown of the system at full load.

voltages reach steady state within a few cycles, minimizing the transient loss. Figure 13 and 14 show the first/last few cycles of the drain voltages during the on period. As expected, the system is able to reach its steady state within a few switching cycles. The DC-DC system efficiency from light load to full load is shown in figure 15, with the slight drop in efficiency at light load likely due to the added loss during the on/off sequence.

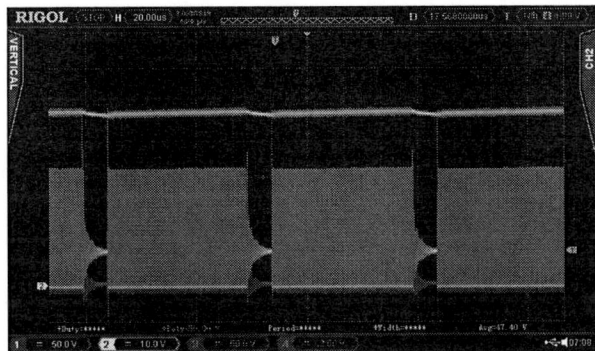

Fig. 12: Output voltage at 80 % load (teal), drain waveforms (yellow and pink).

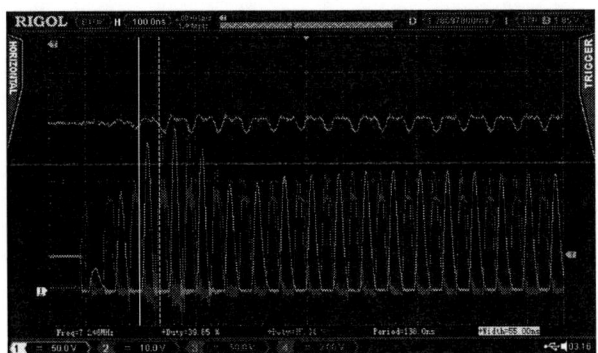

Fig. 13: The first few cycle of the drain waveforms (yellow and pink) during the on-time.

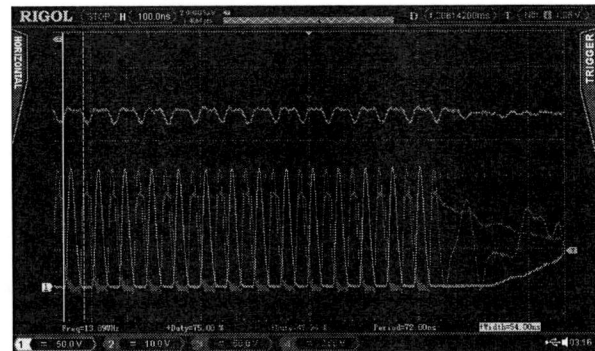

Fig. 14: The last few cycle of the drain waveforms (yellow and pink) before the off-time.

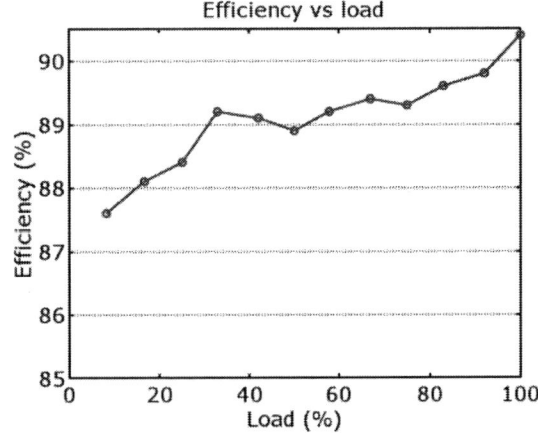

Fig. 15: DC-DC efficiency from 10% to 100% load.

V. CONCLUSION

This paper demonstrates a 48 V to 48 V 300 W WPT system operating at 13.56 MHz designed to replace a wired connection between the body and the door in a vehicle. The author develops a new way to design a WPT system by absorbing the WPT coil inductance and using them in other parts of the circuit. By designing the WPT coil this way, several other unnecessary inductive components can be eliminated. This results in an improvement of the system efficiency, as well as reduction in the weight and size of the system. With this design method, a high DC-DC efficiency of 90.4% at 15mm separation with 67mm diameter coil can be achieved. Additionally, due to the high switching frequency, the system also has fast transient and is able to maintain an efficiency above 87% throughout the full load range.

ACKNOWLEDGMENT

The authors would like to thank Ford Motor Company for their support in this project through the funding provided to the Stanford SUPERLab.

REFERENCES

[1] Z. N. Low, R. A. Chinga, R. Tseng, and J. Lin, "Design and test of a high-power high-efficiency loosely coupled planar wireless power transfer system," *IEEE Transactions on Industrial Electronics*, vol. 56, no. 5, pp. 1801–1812, May 2009.

[2] M. K. Uddin, G. Ramasamy, S. Mekhilef, K. Ramar, and Y. C. Lau, "A review on high frequency resonant inverter technologies for wireless power transfer using magnetic resonance coupling," in *2014 IEEE Conference on Energy Conversion (CENCON)*, Oct 2014, pp. 412–417.

[3] Y. Kaneko and S. Abe, "Technology trends of wireless power transfer systems for electric vehicle and plug-in hybrid electric vehicle," in *2013 IEEE 10th International Conference on Power Electronics and Drive Systems (PEDS)*, April 2013, pp. 1009–1014.

[4] A. P. Sample, D. T. Meyer, and J. R. Smith, "Analysis, experimental results, and range adaptation of magnetically coupled resonators for wireless power transfer," *IEEE Transactions on Industrial Electronics*, vol. 58, no. 2, pp. 544–554, Feb 2011.

[5] O. Knecht, R. Bosshard, and J. W. Kolar, "High-efficiency transcutaneous energy transfer for implantable mechanical heart support systems," *IEEE Transactions on Power Electronics*, vol. 30, no. 11, pp. 6221–6236, Nov 2015.

[6] M. Fu, H. Yin, and C. Ma, "Megahertz multiple-receiver wireless power transfer systems with power flow management and maximum efficiency point tracking," *IEEE Transactions on Microwave Theory and Techniques*, vol. PP, no. 99, pp. 1–9, 2017.

[7] T. Imura and Y. Hori, "Maximizing air gap and efficiency of magnetic resonant coupling for wireless power transfer using equivalent circuit and neumann formula," *IEEE Transactions on Industrial Electronics*, vol. 58, no. 10, pp. 4746–4752, Oct 2011.

[8] R. A. Gheorghiu, V. Iordache, M. Minea, and A. C. Cormos, "Bluetooth latency analysis for vehicular communications in a wi-fi noisy environment," in *2017 40th International Conference on Telecommunications and Signal Processing (TSP)*, July 2017, pp. 148–151.

[9] N. O. Sokal and A. D. Sokal, "Class e-a new class of high-efficiency tuned single-ended switching power amplifiers," *IEEE Journal of Solid-State Circuits*, vol. 10, no. 3, pp. 168–176, Jun 1975.

[10] J. M. Rivas, Y. Han, O. Leitermann, A. D. Sagneri, and D. J. Perreault, "A high-frequency resonant inverter topology with low-voltage stress," *IEEE Transactions on Power Electronics*, vol. 23, no. 4, pp. 1759–1771, July 2008.

[11] K. Surakitbovorn and J. R. Davila, "Evaluation of gan transistor losses at mhz frequencies in soft switching converters," in *2017 IEEE 18th Workshop on Control and Modeling for Power Electronics (COMPEL)*, July 2017, pp. 1–6.

[12] G. Zulauf, W. Liang, K. Surakitbovorn, and J. Rivas-Davila, "Output

capacitance losses in 600 v gan power semiconductors with large voltage swings at high- and very-high-frequencies," in *2017 IEEE 5th Workshop on Wide Bandgap Power Devices and Applications (WiPDA)*, Oct 2017, pp. 352–359.

[13] G. Zulauf, S. Park, W. Liang, K. Surakitbovorn, and J. M. R. Davila, "C_{OSS} losses in 600 v gan power semiconductors in soft-switched, high- and very-high-frequency power converters," *IEEE Transactions on Power Electronics*, vol. PP, no. 99, pp. 1–1, 2018.

[14] J. A. Santiago-Gonzlez, K. M. Elbaggari, K. K. Afridi, and D. J. Perreault, "Design of class e resonant rectifiers and diode evaluation for vhf power conversion," *IEEE Transactions on Power Electronics*, vol. 30, no. 9, pp. 4960–4972, Sept 2015.

[15] M. Kamon, M. J. Tsuk, and J. K. White, "Fasthenry: a multipole-accelerated 3-d inductance extraction program," *IEEE Transactions on Microwave Theory and Techniques*, vol. 42, no. 9, pp. 1750–1758, Sept 1994.

[16] C. R. Sullivan, "Layered foil as an alternative to litz wire: Multiple methods for equal current sharing among layers," in *2014 IEEE 15th Workshop on Control and Modeling for Power Electronics (COMPEL)*, June 2014, pp. 1–7.

[17] B. A. Reese and C. R. Sullivan, "Litz wire in the mhz range: Modeling and improved designs," in *2017 IEEE 18th Workshop on Control and Modeling for Power Electronics (COMPEL)*, July 2017, pp. 1–8.

The 2018 International Power Electronics Conference

Efficiency Maximization of Inductive Power Transfer System by Impedance and Switching Frequency Control in Secondary-side Converter

Ryosuke Ota[1*], Dannisworo S. Nugroho[1], Nobukazu Hoshi[1]

1 Department of Electrical Engineering, Tokyo University of Science, Chiba, Japan
*E-mail: outomeiru2001@yahoo.co.jp

Abstract—Many of previous studies have focused on the efficiency of only one component of inductive power transfer (IPT) system, so the efficiency improvement of multiple components of the system has not been considered sufficiently. Therefore, this paper focuses on multiple components, which are the resonant circuit and the secondary-side converter, and proposes a control method to improve the efficiency of these components. As a result, the overall system efficiency is improved. To improve the multiple components' efficiency, the load impedance control for resonant circuit and switching frequency control to the secondary-side converter were applied. As a result, the combined efficiency of the resonant circuit and the secondary-side converter was raised at the maximum of 1.2 points or 1.3 points with the proposed control where the resonant coil coupling factor was 0.3 or 0.2.

Keywords—Efficiency improvement, inductive power transfer, resonant circuit.

I. INTRODUCTION

A better convenience is obtainable by adopting inductive power transfer (IPT) for a battery charging system of an electric vehicle. However, there are some issues such as the charging efficiency of IPT system is still low compared to the standard plug-in charging system. Thus, the purpose of this study is to improve the efficiency of a static IPT system. Numerous researches about IPT system efficiency have been studied. For example, there are reports on how to improve the structure of transmission coils [1]–[4], and how to configure and apply control for the primary-side inverter [5]–[7] or the secondary-side converter in an IPT system. These reports aimed to improve the efficiency of a resonant circuit which is configured with transmission coils and compensation capacitors. Furthermore, there are reports about improving the efficiency of a secondary-side converter which regulates the charging power for a battery [8].

However, these reports focused only on the efficiency of one component of IPT system, so the efficiency of multiple components used in the system was not optimized. In order to improve the overall efficiency of IPT system, it is necessary for the multiple components to achieve higher efficiency at the same time.

The resonant circuit's efficiency depends strongly on the load impedance of the resonant circuit. Thus, the maximum efficiency of the resonant circuit can be obtained by adjusting the load impedance to the appropriate

value, which can be regulated with the secondary-side converter [9]–[11]. On the other hand, the efficiency of the secondary-side converter is significantly affected by the losses in the switching devices and the inductors. The amount of these losses depends on the switching frequency of the converter. Thus, the efficiency of the secondary-side converter can be improved by controlling the switching frequency appropriately [8].

Therefore, this paper proposes a control method which combines both control methods for obtaining higher system efficiency. Concretely, the load impedance of the resonant circuit is regulated by the secondary-side converter to achieve maximum efficiency of the resonant circuit. Then, in order to improve the efficiency of the secondary-side converter, the switching frequency of the converter is controlled appropriately. In addition, in order to apply the proposed control method to various IPT systems, this paper proposes a design method of the resonant circuit which is applicable to general IPT systems. Thus, the proposed design method and control method are applicable regardless of the topology of the secondary-side converter. Furthermore, to verify the effectiveness of the proposed control method, experiments were conducted and the experimental results between the proposed method and the conventional method were compared.

II. PROPOSED DESIGN METHOD FOR IPT SYSTEM

A. Overview of the Proposed Control Method for the IPT System

Fig. 1 shows the configuration of an IPT battery charging system for electric vehicles assumed in this paper. This system consists of an inverter, a resonant circuit and a secondary-side converter. In the proposed control method, to maximize the efficiency of the resonant

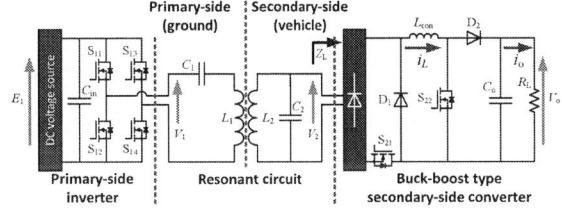

Fig. 1. Configuration of an IPT battery charging system for electric vehicles.

3855

circuit, the secondary-side converter does not work for adjusting the output power but for adjusting the load impedance of the resonant circuit [10]. Instead of the secondary-side converter, the primary-side inverter is used to adjust the output power. When this control method is applied to the system, it is important to design the parameter of the resonant circuit appropriately in order to obtain higher efficiency.

B. Design Requirements in Transmission Power of Resonant Circuit

In order to transmit power efficiently, it is necessary for the resonant circuit to be designed appropriately. Thus, in this paper, the resonant circuit is designed by focusing on the transmission power and the efficiency. The resonant circuit is designed based on an equivalent circuit of SP-type resonant circuit shown in Fig. 2. In Fig. 2, V_1' [V] and V_2' [V] are fundamental frequency components in V_1 [V] and V_2 [V]. Harmonic components need not be considered because the SP-type resonant circuit works as a bandpass filter for the resonant frequency.

Firstly, the design requirements for the transmitted power is shown below. The transmission power is expressed as

$$P = \frac{V_1'^2}{k^2 |Z_L|} \frac{L_2}{L_1}, \tag{1}$$

where k is the coupling factor of the transmission coils and Z_L is the load impedance of the resonant circuit.

L_1 [H] and L_2 [H] should be designed to be able to transmit the requirement power at the maximum coupling factor k_{max} which is assumed under the actual condition because the transmission power becomes smaller as the coupling factor becomes larger. In addition, in the proposed method, it is assumed that Z_L is regulated to Z_{Lmax} where the resonant circuit gets the maximum efficiency. Then, the requirement of the multiple values $L_1 L_2$, which is derived by using the requirement power P_{req} [W], is expressed as

$$\frac{L_2}{L_1} \leq \frac{k_{max}^2 P_{req} |Z_{Lmax}|}{V_1'^2}. \tag{2}$$

C. Design Requirements in Transmission Efficiency of Resonant Circuit

When it is assumed that the dominant power loss is caused by the ESR of the coils, the transmission efficiency η [%] is calculated by

$$\eta = \frac{P \times 100}{P + r_1 I_1'^2 + r_2 I_2'^2}, \tag{3}$$

where r_1 [Ω] and r_2 [Ω] are the ESR of the primary and secondary coils respectively, and P is the output power. In addition, when it is assumed that the quality factor of the coils Q is constant, η is expressed as

$$\eta = \frac{100}{1 + \frac{|Z_L|}{\omega L_2 Q} + \frac{\omega L_2}{|Z_L| Q} \left(1 + \frac{1}{k_{max}^2}\right)}. \tag{4}$$

From (4), by also applying arithmetic-geometric mean, η is maximized at

$$|Z_L| = |Z_{Lmax}| = \omega L_2 \sqrt{1 + \frac{1}{k_{max}^2}}. \tag{5}$$

Therefore, L_1 and L_2 should be designed to satisfy the above equations (2) and (5), and decided by

$$L_1 = \frac{V_1'^2}{\omega P_{req} k_{max} \sqrt{1 + k_{max}^2}}, \tag{6}$$

$$L_2 = \frac{|Z_{Lmax}| k_{max}}{\omega \sqrt{1 + k_{max}^2}}. \tag{7}$$

In the proposed method, while considering the efficiency improvement of the resonant circuit, from (6) and (7) the value of L_1 and L_2 are able to be decided clearly.

D. Configuration of Secondary-side Converter

As shown in Fig. 1, the buck–boost type secondary-side converter is used in the IPT system [12]. This secondary-side converter has switches S_{21} and S_{22}, and has multiple operation modes. Fig. 3 shows the operation modes of the secondary-side converter. The boost mode is operated by combining MODE 1 and MODE 3, and the buck mode is operated by combining MODE 2 and MODE 3. In the proposed method, this secondary-side converter is used to control the resonant circuit's load impedance.

III. CONTROL METHOD FOR MAXIMUM EFFICIENCY OF RESONANT CIRCUIT

A. Controlling Load Impedance

From (4), it is understood that the resonant circuit's maximum efficiency is obtained when the load impedance

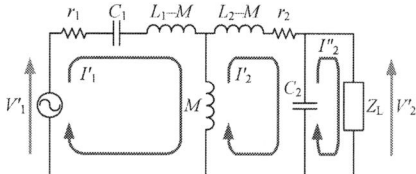

Fig. 2.   Equivalent circuit of SP-type resonant circuit.

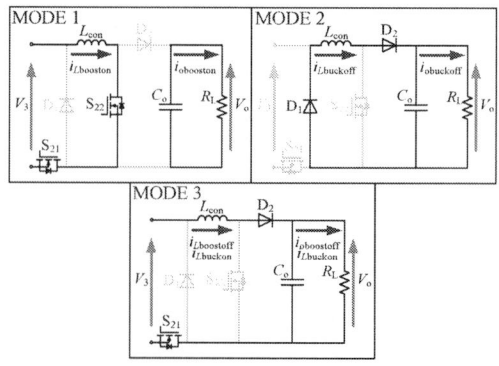

Fig. 3.   Operation modes of the secondary-side converter.

is at the appropriate value. The load impedance is regulated by the secondary-side converter because the load impedance of the resonant circuit corresponds with the input impedance of the secondary-side converter [10]. The input impedances of the secondary-side converter in the buck mode $|Z_{\text{Lbuck}}|$ and in the boost mode $|Z_{\text{Lboost}}|$ are respectively expressed as

$$|Z_{\text{Lbuck}}| = \left(\frac{\pi}{2\sqrt{2}\sin^2(\frac{\pi}{2}d_1)} \right)^2 R_{\text{L}} \qquad (8)$$

and

$$|Z_{\text{Lboost}}| = \left(\frac{\pi(1-d_2)}{2\sqrt{2}} \right)^2 R_{\text{L}}, \qquad (9)$$

where d_1 is the duty ratio of S_{21}, d_2 is the duty ratio of S_{22}, R_{L} [Ω] is the load resistance whose value corresponds with the condition of the battery's state of charge. From (8), it is understood that $|Z_{\text{Lbuck}}|$ is larger than R_{L}. In contrast, from (9), it is understood that $|Z_{\text{Lboost}}|$ is smaller than R_{L}. Therefore, the buck–boost type secondary-side converter can control $|Z_{\text{L}}|$ at the appropriate impedance $|Z_{\text{Lmax}}|$.

B. Proposed Design Method of $|Z_{\text{Lmax}}|$

$|Z_{\text{Lmax}}|$ is able to be decided freely by using the design method in this paper. There is a battery charging method which combines a constant current charging and constant voltage charging. Fig. 4 shows the characteristics of battery's input impedance R_{L} vs. battery voltage. In the proposed system, $|Z_{\text{Lmax}}|$ is corresponding to the resistance R_{LS} when the charging method is switched from constant current charging to constant voltage charging. At the resistance R_{LS}, the mode of the secondary-side converter switches from buck to boost mode when the input impedance of the converter is controlled to $|Z_{\text{Lmax}}|$. Therefore, the efficiency of the resonant circuit is able to be controlled to maximum point as shown in Fig. 5.

C. Control Method of Charging Power

In the proposed method, the secondary-side converter cannot be used as controlling the charging current and voltage because the converter is used as a load impedance controller. Therefore, in the proposed method, the charging power is controlled by the primary-side inverter communicating with the secondary side. The output voltage $V'_{\text{shift\_1}}$ is able to be controlled by phase shifting amount α [rad] between the legs in the inverter with the following equation.

$$V'_{\text{shift\_1}} = \sqrt{\frac{\pi - \alpha}{\pi}} V'_1, \qquad (10)$$

IV. PROPOSED CONTROL METHOD FOR IMPROVING THE EFFICIENCY OF SECONDARY-SIDE CONVERTER

The secondary-side converter's efficiency can be improved by controlling the switching frequency of the converter appropriately [8]. However, in the previous report, the secondary-side converter is used to regulate the output power. In the proposed method, the secondary-side converter is used to control the resonant circuit's load impedance. From (8) and (9), it is confirmed that the switching frequency of the converter does not affect $|Z_{\text{L}}|$, so $|Z_{\text{L}}|$ and the frequency can be controlled independently. Therefore, in the proposed control method, at first, the efficiency characteristics of the secondary-side converter are derived in theoretical analysis when the input impedance of the converter is regulated to $|Z_{\text{Lmax}}|$. And, the switching frequency of the converter is controlled referring to the theoretical efficiency characteristics derived in advance.

A. Current Through Each Device in Secondary-side Converter

It is necessary for deriving the efficiency characteristics that the current through each device is derived. Table I shows the equations of the current through each device in secondary-side converter. In Table I, A_1, B_1, D_1, E_1, A_2, B_2, D_2, and E_2 are integration constants, while h and λ are expressed as

$$h = \frac{1}{2R_{\text{L}}C_{\text{o}}}, \qquad (11)$$

$$\lambda = \frac{1}{2}\sqrt{\frac{4}{L_{\text{con}}C_{\text{o}}} - \frac{1}{R_{\text{L}}^2 C_{\text{o}}^2}}. \qquad (12)$$

Moreover, the integration constants of the buck-boost converter operating in boost mode or buck mode are shown in Table II or Table III; and parameter x_n used in Table II and Table III is shown in Table IV.

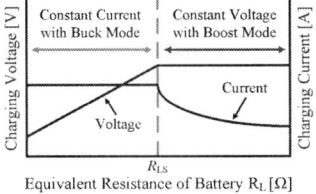

Fig. 4. Battery's input impedance R_{L} vs. battery voltage.

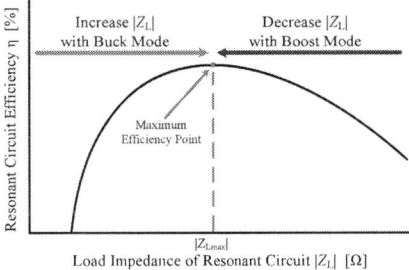

Fig. 5. Resonant circuit efficiency vs. load impedance.

B. Power Loss on Each Device in Secondary-side Converter

In this theoretical analysis for the efficiency of the secondary-side converter, the copper loss and core loss of L_{con}, the loss on ESR of C_o, the conduction losses and switching losses of S_{21} and S_{22}, the conduction losses and recovery losses of D_1, D_2, and the diodes in the full bridge rectifier are considered. These losses can be derived by using the current magnitude shown in Table I.

The loss model for each device used in the secondary-side converter is shown in Fig. 6. In the loss model of an inductor, the equivalent copper loss resistance r_{Ldc} and equivalent iron loss resistance r_{Lac} are considered. In this inductor model, inductor current i_L flows through r_{Ldc}, the AC component of inductor current i_L flows through r_{Lac}, and the DC component of i_L flows through L_{con} due to the large impedance of L_{con} for AC component. In the loss model of a capacitor, the equivalent series resistance of capacitor r_C is considered. In the loss model of a MOSFET, on-voltage of MOSFET v_{on_S} and on-resistance of MOSFET r_{on_S} are considered. In the

loss model of a diode, on-voltage of diode v_{on_D}, diode forward voltage v_{f_D}, and on-resistance of diode r_{on_D} are considered. Additionally, the switching losses of the MOSFET and diode are derived by using the model waveforms of current and voltage in each switching device shown in Fig. 7. From these, the theoretical power consumption model is shown in Table V.

C. Efficiency Characteristics for Switching Frequency of Secondary-side Converter in Theoretical Analysis

The efficiency of the secondary-side converter η_{con} is defined by

$$\eta_{con} = \frac{P_{\text{out}} \times 100}{P_{\text{out}} + P_{L\text{loss}} + P_{C\text{loss}} + P_{S\text{loss}} + P_{D\text{loss}}}, \quad (13)$$

where P_{out} is the output power of the secondary-side converter, $P_{L\text{loss}}$ is the loss of the inductor L_{con}, $P_{C\text{loss}}$ is the loss of the capacitor C_o, $P_{S\text{loss}}$ is the loss of MOSFET in the converter, $P_{D\text{loss}}$ is the loss of the diode and the full bridge rectifier of the converter.

In the theoretical analysis, the circuit parameters are shown in Table VI, which are based on the experimental setup. Fig. 8 shows the characteristics of the efficiency of the secondary-side converter vs. each switching frequency in the theoretical analysis at the constant current mode operation. In Fig. 8, the character "×" shows the maximum efficiency points in each output power. In Fig. 8, the operation mode of the secondary-side converter is switched to each other at 650 W. Around the switching point of the operation mode, the switching loss is a major part of the overall loss [8]. Thus, around the switching

TABLE I. EQUATIONS OF CURRENT THROUGH EACH DEVICE IN SECONDARY-SIDE CONVERTER [8]

Operation Mode	Switch Mode S₁₁	S₂₂	Inductor Current i_L	Output Current i_o
Boost	On	On	$\frac{V_3}{L_{con}}t + A_1$	$B_1 e^{-2ht}$
Buck	Off	Off	$(D_2 \sin \lambda t + E_2 \cos \lambda t)\,e^{-ht}$	$-\frac{L}{R_L}\frac{di_L}{dt}$
Boost	On	Off	$(D_1 \sin \lambda t + E_1 \cos \lambda t)\,e^{-ht} + \frac{V_3}{R_L}$	$\frac{1}{R_L}\left(V_3 - L\frac{di_L}{dt}\right)$
Buck	On	Off	$(A_2 \sin \lambda t + B_2 \cos \lambda t)\,e^{-ht} + \frac{V_3}{R_L}$	$\frac{1}{R_L}\left(V_3 - L\frac{di_L}{dt}\right)$

TABLE II. INTEGRATION CONSTANT OF THE BUCK-BOOST CONVERTER OPERATING IN BOOST MODE [8]

Integration constant	Definition
X_1	$x_{15}x_{19}x_8 - x_{15}x_{18}x_7 - x_{12}x_{19} + x_{11}x_{18}$
A_1	$(x_1(-x_{15}x_{19}x_8 + x_{15}x_{18}x_7 + x_{12}x_{19} - x_{11}x_{18})$ $+ x_3(x_{15}x_{16}x_8 + (x_{18} - x_{15}x_{18})x_4 - x_{12}x_{16})$ $+ x_2(-x_{15}x_{16}x_7 + (x_{15}x_{19} - x_{19})x_4 + x_{11}x_{16}))/X_1$
B_1	$-((x_{19}x_4 - x_{11}x_{16})x_8 + (x_{12}x_{16} - x_{18}x_4)x_7 + (x_{11}x_{18} - x_{12}x_{19})x_4)/X_1$
D_1	$(x_{15}x_{16}x_7 + (x_{19} - x_{15}x_{19})x_4 - x_{11}x_{16})/X_1$
E_1	$-(x_{15}x_{16}x_8 + (x_{18} - x_{15}x_{18})x_4 - x_{12}x_{16})/X_1$

TABLE III. INTEGRATION CONSTANT OF THE BUCK-BOOST CONVERTER OPERATING IN BUCK MODE [8]

Integration constant	Definition
X_2	$x_3(x_8(x_{20}x_9 - x_{17}x_{21}) + x_5(x_{14}x_9 - x_{18}x_{21}) + (x_{14}x_{17} - x_{18}x_{20})x_6)$ $+ x_2(x_7(x_{17}x_{21} - x_{20}x_9) + x_5(x_{19}x_{21} - x_{13}x_9) + (x_{19}x_{20} - x_{13}x_{17})x_6)$ $+ x_7(x_{18}x_{21} - x_{14}x_9) + x_8(x_{13}x_9 - x_{19}x_{21}) + (x_{14}x_{19} - x_{13}x_{18})x_6$ $+ x_{13}(x_{22}x_{24} - x_{19}x_{25} + (x_{20}x_{24} - x_{19}x_{23})x_3)$
A_2	$(x_1((x_{19}x_{20} - x_{13}x_{17})x_8 + (x_{14}x_{17} - x_{18}x_{20})x_7 + (x_{14}x_{19} - x_{13}x_{18})x_5)$ $+ (x_{18}x_{20} - x_{14}x_{17})x_3x_4 + x_2(x_{13}x_{17}$ $- x_{19}x_{20})x_4 + (x_{13}x_{18} - x_{14}x_{19})x_4)/X_2$
B_2	$-(x_1(x_7(x_{18}x_{21} - x_{14}x_9) + x_8(x_{13}x_9 - x_{19}x_{21}) + (x_{14}x_{19} - x_{13}x_{18})x_6)$ $+ x_3x_4(x_{14}x_9 - x_{18}x_{21}) + x_2x_4(x_{19}x_{21} - x_{13}x_9)))/X_2$
D_2	$(x_1(x_7(x_{17}x_{21} - x_{20}x_9) + x_5(x_{19}x_{21} - x_{13}x_9) + (x_{19}x_{20} - x_{13}x_{17})x_6)$ $+ x_3x_4(x_{20}x_9 - x_{17}x_{21}) + x_2x_4(x_{13}x_9 - x_{19}x_{21}))/X_2$
E_2	$-(x_1(x_8(x_{17}x_{21} - x_{20}x_9) + x_5(x_{18}x_{21} - x_{14}x_9) + (x_{18}x_{20} - x_{14}x_{17})x_6)$ $+ x_2x_4(x_{20}x_9 - x_{17}x_{21}) + x_4(x_{14}x_9 - x_{18}x_{21}))/X_2$

TABLE IV. PARAMETER x_n USED IN TABLE II AND TABLE III [8]

x_1	V_3/R_L	x_2	$e^{-hT}\sin \lambda t$	x_3	$e^{-hT}\cos \lambda t$
x_4	V_3	x_5	$L_{con}h$	x_6	$L_{con}\lambda$
x_7	$L_{con}e^{-hT}(\lambda \sin \lambda T + h \cos \lambda T)$	x_8	$L_{con}e^{-hT}(h \sin \lambda T - \lambda \cos \lambda T)$
x_9	$L_{con}e^{-hdT}\sin \lambda dT$	x_{10}	$L_{con}e^{-hdT}\cos \lambda dT$
x_{11}	$L_{con}e^{-hdT}(\lambda \sin \lambda dT + h \cos \lambda dT)$	x_{12}	$L_{con}e^{-hdT}(h \sin \lambda dT - \lambda \cos \lambda dT)$
x_{13}	$L_{con}e^{hT}(\lambda \sin \lambda T + h \cos \lambda T) - L_{con}e^{-hdT}(\lambda \sin \lambda dT + h \cos \lambda dT)$
x_{14}	$L_{con}e^{-hT}(h \sin \lambda T - \lambda \cos \lambda T) - L_{con}e^{-hdT}(h \sin \lambda dT - \lambda \cos \lambda dT)$
x_{15}	e^{-2hdT}	x_{16}	V_3dT/L_{con}	x_{17}	$x_{10} - 1$
x_{18}	$x_2 - x_9$	x_{19}	$x_3 - x_{10}$	x_{20}	$x_{11} - x_5$
x_{21}	$x_{12} - x_6$				

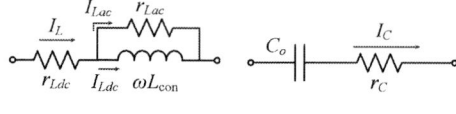

(a) Inductor (b) Capacitor

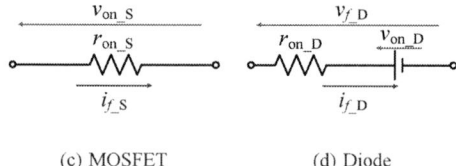

(c) MOSFET (d) Diode

Fig. 6. Loss model for each device used in the secondary-side converter.

TABLE V. THEORETICAL POWER CONSUMPTION MODEL

	Buck mode	Boost mode
P_{out}	V_o^2/R_L	
$P_{L\text{loss}}$	$r_{Ldc}I_{dc}^2 + (r_{Lac} + r_{Ldc})I_{ac}^2$	
$P_{C\text{loss}}$	$r_C I_C^2$	
$P_{S\text{loss}}$	$r_{on\_S}dI_{Lon}^2 + v_{on\_S}dI_{Ldc}$ $+ f_{con}V_o(i_L(0)T_{on} + i_L(dT)T_{off})/6$ $+ v_{f\_S}I_{Ldcon}T_{on2}f_{con}/2$	$r_{on\_S}(I_L^2 + dI_{Lon})$ $+ f_{con}V_3(i_L(0)T_{on} + i_L(dT)T_{off})/6$ $+ v_{f\_S}I_{Ldcon}T_{on2}f_{con}/2$
$P_{D\text{loss}}$	$r_{on\_D}(I_L^2 + (1-d)I_{Loff}^2)$ $+ v_{on\_D}I_{Ldc}(2-d)$ $+ 2d(r_{on\_D}I_{Lon}^2 + v_{on\_D}I_{Ldc})$ $+ I_{RP}t_{rr}V_3 f_{con}/6$ $+ v_{f2\_D}I_{Ldcoff}t_{don}f_{con}$	$(1-d)(r_{on\_D}I_{Loff}^2 + v_{on\_D}I_{Ldc})$ $+ 2(r_{on\_D}I_L^2 + v_{on\_D}I_{Ldc})$ $+ f_{con}I_{RP}t_{rr}V_o/6$ $+ v_{f2\_D}I_{Ldcoff}t_{don}f_{con}$

point, the efficiency of the converter can be improved by operating in lower switching frequency. On the contrary, at a point far from the switching point of the operation mode, the overall loss is dominated by the inductor loss. By operating the secondary-side converter with a higher frequency, the current ripple is reduced. As a result, the inductor loss becomes smaller [8]. Therefore, the efficiency of the secondary-side converter can be improved by controlling the switching frequency appropriately.

In the proposed control method, the input impedance of the secondary-side converter is controlled to derived $|Z_{\mathrm{Lmax}}|$, and the switching frequency is controlled based on the derived characteristics.

V. EXPERIMENT FOR VERIFYING EFFECTIVENESS OF PROPOSED CONTROL METHOD

The appearance of the experimental setup used is shown in Fig. 9. Also, the circuit parameter used in these experiments is shown in Table VI. The input voltage of the primary-side inverter was set to 70 V and the inverter was operated to output 85 kHz AC square-wave voltage. In addition, due to the limitations of laboratory equipment, the battery rated power P_{req} was scaled down

by half to 650 W. Also, the experiment was conducted on constant current charging mode or constant voltage charging mode when the coupling factor of the resonant circuit coils is set to 0.3 or 0.2.

The effectiveness of the proposed control method is shown by comparing the experimental results between the proposed method and conventional methods. The experimental conditions to investigate power transfer efficiency of each method are shown in Table VII.

In conventional method @85 kHz, the load impedance control was applied and the switching frequency of the secondary-side converter is set at the resonant frequency of the resonant circuit, whose resonant frequency is 85 kHz. In conventional method @20 kHz, the load impedance control is applied and the switching frequency of the secondary-side converter is set at value where the result of theoretical analysis is relatively high over the whole output power range, which is 20 kHz. Both

TABLE VI. CIRCUIT PARAMETERS FOR THE ANALYSIS

Battery Rated Charging Power P_{req}	650 W
Battery Constant Charging Voltage	141.3 V
Battery Constant Charging Current	4.6 A
Resonant Frequency f	85 kHz
Maximum Coupling Factor $k_{\max}$	0.3
Primary-side Coil L_1	26.16 μH
Secondary-side Coil L_2	17.73 μH
Primary-side Coil ESR r_{L1}	53 mΩ
Secondary-side Coil ESR r_{L2}	39 mΩ
Primary-side Coil Q Factor Q_1	264
Secondary-side Coil Q Factor Q_2	243
Primary-side Compensation Capacitor C_1	147.2 nF
Secondary-side Compensation Capacitor C_2	197 nF
Inductor L_{con}	724 μH
Equivalent copper loss resistance of the inductor r_{Ldc}	25 mΩ
Equivalent iron loss resistance of the inductor r_{Lac} $7E^{-15}f_{con}^3 - 6E^{-10}f_{con}^2 + 4E^{-5}f_{con} + 407$ mΩ	
MOSFET S_{11}, S_{12}, S_{13}, S_{14}	NIEC P2HM755HA
MOSFET S_{21} and S_{22}	TOSHIBA TK39J60W5
Diode D_1 and D_2	MICROSEMI APT2X41DC60J

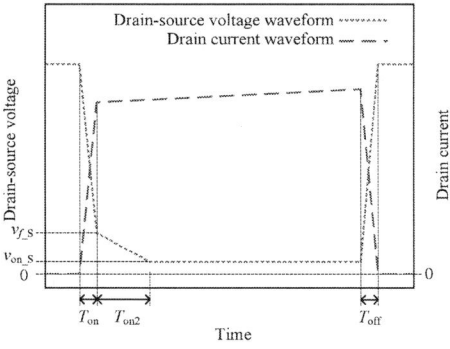

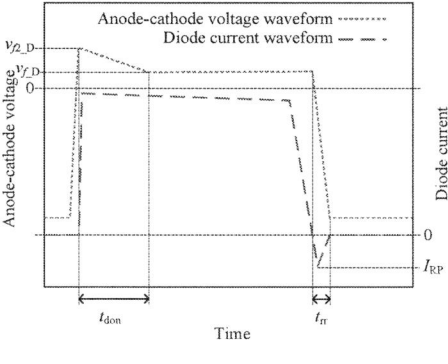

Fig. 7. Model waveforms of current and voltage in each switching device in secondary-side converter.

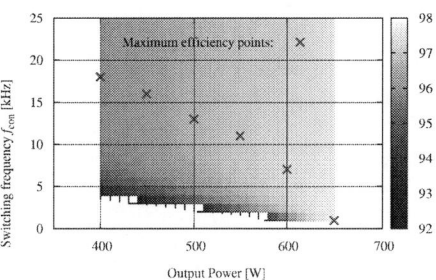

Fig. 8. Characteristics of the efficiency of the secondary-side converter vs. each switching frequency in the theoretical analysis at the constant current charging mode.

TABLE VII. EXPERIMENTAL CONDITIONS TO INVESTIGATE POWER TRANSFER EFFICIENCY OF EACH METHOD

Control Method	Primary-side Inverter Phase Shift	Secondary-side Converter	
		Impedance Control	Switching Frequency Control
Conventional @85 kHz	Varied	Yes	No
Conventional @20 kHz	Varied	Yes	No
Conventional @x_{op} kHz	Constant	No	No
Proposed	Varied	Yes	Yes

of these methods use the primary-side inverter to regulate output power. Only switching frequency control for the secondary-side converter is applied at conventional method @x_{op} kHz. The output power regulation is also conducted with the secondary-side converter for this method. In the proposed method, both the load impedance control and switching frequency control are applied. Also, in this method, the output power is regulated with the primary-side inverter.

Figs. 10 and 11 show the combined efficiency of the resonant circuit and the secondary-side converter at the coupling factor $k = 0.3$ or 0.2. From Figs. 10(a) and 11(a), at constant current charging mode, the combined efficiency becomes higher by applying to the proposed control method over the whole output power range. In addition, from Fig. 11(b), at coupling factor 0.2 and constant voltage charging mode, the effectiveness of the proposed method is confirmed over the whole output power range.

On the contrary, from Fig. 10(b), at coupling factor 0.3 and constant voltage charging mode, the combined efficiency with the proposed control method is inferior compared to conventional control method @x_{op} kHz. The reason is that the loss of the converter is larger than the loss of the resonant circuit in this range. In this range, it is necessary to apply a control method which can improve the combined efficiency because the proposed control method is not effective. This problem will be discussed in another paper.

Furthermore, Figs. 12 and 13 show the system efficiency at coupling factor $k = 0.3$ or 0.2. Except for the coupling factor $k = 0.3$ and constant voltage charging mode, the overall system efficiency is improved by using the proposed method. At $k = 0.3$, compared to conventional methods @85 kHz, @20 kHz, and @x_{op} kHz, the system efficiency increases at the maximum of 0.5, 0.2, and 1.2 points respectively. At $k = 0.2$, compared to conventional methods @85 kHz, @20 kHz, and @x_{op} kHz, the system efficiency increases at the maximum of 1.3, 0.2, and 0.4 points respectively.

Overall, the effectiveness of the proposed method is confirmed over the whole output power range except at the coupling factor $k = 0.3$ and constant voltage charging mode. However, at $k = 0.3$ and constant voltage charging mode, the effectiveness of the proposed method was not able to be confirmed. Therefore, in this charging condition, it is necessary to apply a control method which can improve the combined efficiency, which will be discussed in another paper.

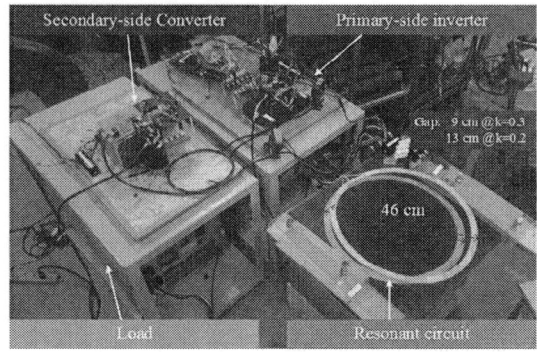

Fig. 9.   Appearance of the experimental setup.

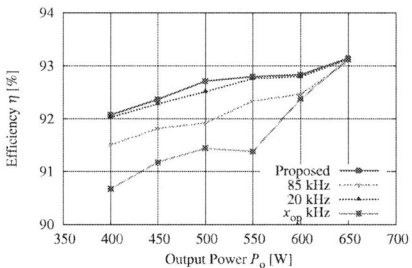

(a) k=0.3 at constant current charging mode

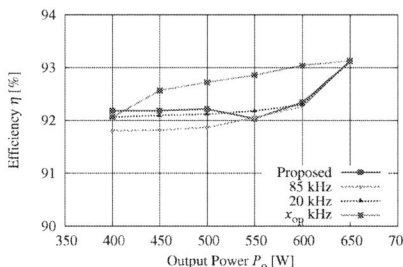

(b) k=0.3 at constant voltage charging mode

Fig. 10.   Combined efficiency of resonant circuit and secondary-side converter at coupling factor $k = 0.3$.

The 2018 International Power Electronics Conference

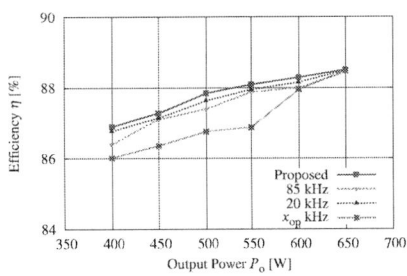

(a) k=0.3 at constant current charging mode

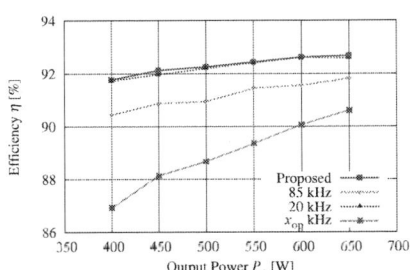

(a) k=0.2 at constant current charging mode

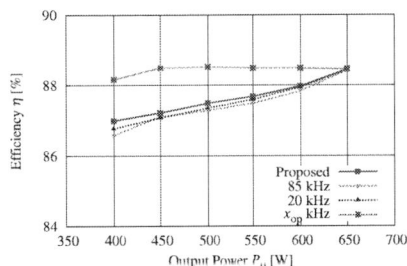

(b) k=0.3 at constant voltage charging mode

Fig. 12.   System efficiency at coupling factor $k = 0.3$.

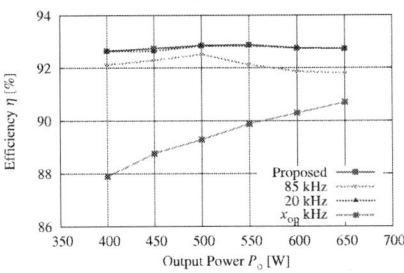

(b) k=0.2 at constant voltage charging mode

Fig. 11.   Combined efficiency of resonant circuit and secondary-side converter at coupling factor $k = 0.2$.

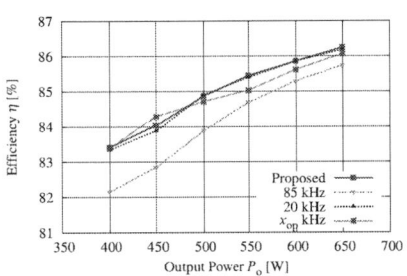

(a) k=0.2 at constant current charging mode

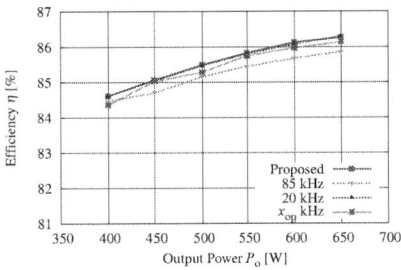

(b) k=0.2 at constant voltage charging mode

Fig. 13.   System efficiency at coupling factor $k = 0.2$.

3861

VI. CONCLUSIONS

In this paper, a method to improve an IPT system's efficiency by combining load impedance control of the resonant circuit and switching frequency control of the secondary-side converter was proposed. In addition, in order to apply the proposed control method to various IPT systems, a design method for the resonant circuit was proposed. To investigate the power transfer efficiency of the proposed method, experiments were conducted. As a result, it was confirmed that by using the proposed method, the power transfer efficiency is improved over the whole output power range at coupling factor lower than 0.3 or at constant current charging mode. The proposed method's effectiveness was confirmed at almost all charging conditions. However, at $k = 0.3$ and constant voltage charging mode, it was confirmed that the proposed method was inferior to conventional method $@x_{op}$ kHz. Thus, for the conditions where the proposed method was not effective, a new control scheme which maximizes the combined efficiency is required.

In future work, for the charging conditions where the proposed method is ineffective, a control scheme to improve the combined efficiency of the resonant circuit and the secondary-side converter of an IPT system will be investigated. In addition, theoretical efficiency of the primary-side inverter of the IPT system will be investigated, and a scheme to further maximize the power transfer efficiency are planned to be researched.

REFERENCES

[1] J. Kim, et al.: "Coil Design and Shielding Methods for a Magnetic Resonant Wireless Power Transfer System," *Proceedings of the IEEE*, vol. 101, no. 6, pp. 1332–1342 (2013).

[2] Y. Liu, P. Li, U. Madawala: "Maximum Power Transfer and Efficiency Analysis of Different Inductive Power Transfer Tuning Topologies," *Proc. of 2015 IEEE 10th Conf. on Industrial Electronics and Applications*, pp. 649–654(2015).

[3] R. Bosshard, J. Kolar, J. Muhlethaler, I. Stevanovic, B. Wunsch, F. Canales "Modeling and $\eta - \alpha$-Pareto Optimization of Inductive Power Transfer Coils for Electric Vehicles,"*IEEE Journal of Emerging and Selected Topics in Power Electronics*, vol. 3, no.1, pp. 50–64(2015).

[4] S. C. Moon, G. W. Moon "Wireless Power Transfer System With an Asymmetric Four-Coil Resonator for Electric Vehicle Battery Chargers," *IEEE Transactions on Power Electronics*, vol. 31, no. 10, pp.6844–6854 (2016).

[5] M. Moghaddami, A. Cavada, A. I. Sarwat: "Soft-Switching Self-Tuning H-bridge Converter for Inductive Power Transfer Systems," *Proc. of Energy Conversion Congress and Exhibition*, (2017).

[6] S. G. Cimen, A. Popp, B. Schmuelling: "An Inductive Power Transfer System for Electric Vehicles with a Safe and Modular Primary Side Inverter", *Proc. of 8th Int. Conf. on Power Electronics, Machines, and Drives*, (2016).

[7] N. Hatchavanich, A. Sangswang, S. Naetiladdanon: "Operation Region of LCL Resonant Inverter for Inductive Power Transfer Application",*Proc. of 13th Int. Conf. Electrical Engineering/Electronics, Computer, Telecommunications and Information Technology*, (2016).

[8] R. Ota, N. Hoshi, K. Uchida: "Consideration on Efficiency Characteristics to Switching Frequency of Secondary-side Converter with Inductive Power Transfer System for Electric Vehicle," *Proc. of 2016 18th European Conf. on Power Electronics and Applications*, (2016).

[9] K. Iimura, N. Hoshi, J. Haruna: "Experimental discussion on inductive type contactless power transfer system with boost or buck-type converter connected to rectifier," *Proc. of 7th Int. Power Electron. Motion Control Conf.*, vol. 4, pp. 2652–2657 (2012).

[10] M. Kato, T. Imura, Y. Hori: "Study on Maximize Efficiency by Secondary Side Control Using DC-DC Converter in Wireless Power Transfer via Magnetic Resonant Coupling," *Proc. of Electric Vehicle Symposium and Exhibition*, (2013).

[11] M. Fu, C. Ma, X. Zhu: "A Cascaded Buck-Boost Converter for High-Efficiency Wireless Power Transfer Systems," *IEEE Transactions on Industrial Informatics*, vol. 20, no. 3, pp. 1972–1980 (2014).

[12] S. Motegi, A. Maeda, Y. Nishida: "A New Single-phase High-Power-Factor Converter with Buck and Buck-Boost Hybrid Operation," *IEEJ Trans. Ind. Appl.*, vol. 118, no. 4, pp. 468–473 (1998) (in Japanese).

The 2018 International Power Electronics Conference

Analysis of Optimal Operation Frequency Range for Battery Charging in WPT System

Yongbin Jiang, Min Wu, Junwen Liu, Yue Wang, Laili Wang, Hailong Zhang
School of the Electric Engineering, Xi'an Jiaotong University
Xi'an 710049 China
jiangyongbin@stu.xjtu.edu.cn

Abstract—In this paper, a three loop control strategy based on ZVS angle loop is proposed for battery charging using a series-series type wireless power transfer system (WPTS). This paper derives the optimal operation frequency range for realizing both constant output and ZVS operation simultaneously. At the same time, the variable ZVS angle can be used to obtain the maximum efficiency of the whole system for battery charging. Finally, a 500W WPTS is fabricated to verify the correctness of the theoretical analysis and the effectiveness of the proposed control strategy. The theoretical calculations have a good agreement with the experimental results. With the proposed control strategy, the whole system can also achieve high efficiency especially even in light load.

Index Terms—Series-series (SS) type wireless power transfer system (WPTS), variable frequency phase shift control (VFPSC), zero voltage switching (ZVS), optimal operation frequency range (OOFR), three loop control strategy (TLCS)

I. INTRODUCTION

Compared with the conventional plug-in charging method, WPT system (WPTS) is a novel power supply which can realize both electrical and mechanical isolation, minimizes the use of sockets, and ensure safe operation in many applications. Especially, WPT technology is attracting increasing attention in battery charging applications including biomedical implants [1], consumer electronics [2], underwater loads [3], electric vehicles (EVs) [4], [5], etc.

Recently, high-performance lithium-ion batteries are widely used in the mentioned above applications. Fig.1 shows the typical charging profile, where the constant current (CC) mode and constant voltage (CV) mode are dominant in the whole charging process. Considering the battery lifetime and recycle time, a WPT battery charger must provide sufficiently accurate charging current and voltage for safe operation [6].

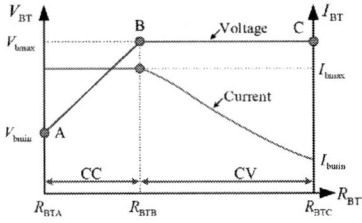

Fig. 1. Typical charge profile of a lithium-ion battery cell.

In addition, the charging efficiency is a vital index to evaluate the performance of the whole charging system.

Therefore, lots of researchers are devoted to improving the transfer efficiency in WPTS. The impedance matching method is applied to enhance the resonant network efficiency [5], [7]. However, the extra DC/DC converters are needed to realize CC or CV output, which violates the requirement of compact receiver.

In this paper, a set of battery parameters are given and the corresponding charging curves are confirmed at first. Second, the frequency characteristics of the series-series (SS) compensated topology are analyzed in detail and the optimal operation frequency range (OOFR) is obtained by using variable frequency phase shift control (VFPSC) strategy in the whole battery charging process. Third, a systematic control strategy is proposed to control the output current or voltage and ZVS angle simultaneously. Finally, a 500W WPTS is built to verify the correctness of theoretical analysis and the effectiveness of the proposed control strategy. The experimental results show that the maximum efficiency point can be tracked and a very high overall efficiency is achieved over the wide range of coupling coefficient and load resistance.

II. THEORETICAL ANALYSIS

A. The battery packs case-study

The WPT battery charger is designed around the charging profile of the battery packs described in the Table I. The battery packs are made of five lead-acid modules connected in series; its overall nominal capacity is 200Ah and overall nominal voltage is 60V; with a voltage fluctuation (20% of nominal voltage), the voltage ranges from 48V to 72V. The maximum charging current is set as 4A (0.1C). The charging process of battery packs appears an equivalent variable resistor R_{BT} defined as the ratio of V_B to I_B. It starts from R_{BTA} at the beginning of the charging process, and slowly increases up to R_{BTB} with constant charging current. Then, when the R_{BT} reaches R_{BTB}, the WPT system goes into the CV mode until R_{BT} equals R_{BTC}.

B. Circuit model and equivalent circuit

Because the series-series (SS) compensated topology in WPT has lots of merits, such as simplicity, resonant frequency independent with the coupling coefficient, etc, the WPT system adopts this compensation method and the whole system is shown in Fig. 2(a). V_1 and I_1 are the DC input voltage and current of primary inverter respectively. V_2 and I_2 are

3863

TABLE I
BATTERY PARAMETERS AND SPECIFICATION

Symbol	Quantity	Value
V_{b1}	single module voltage	9.6~14.4V
Q_{b1}	single module capacity	40A·h
V_b	overall module voltage	48~72V
I_{bmax}	maximum charging current	4A
I_{bmin}	floating charging current	0.5A
R_{BTA}	load resistance in point A	12Ω
R_{BTB}	load resistance in point B	18Ω
R_{BTC}	load resistance in point C	144Ω

the DC output voltage and current of the secondary rectifier respectively. L_1 and L_2 are the self-inductances of the primary resonant coil and the secondary resonant coil respectively. M is the mutual inductance and the coupling coefficient k is defined as $k = M/\sqrt{L_1 L_2}$. Based on the fundamental harmonic analysis, the equivalent circuit is shown in Fig. 2(b).

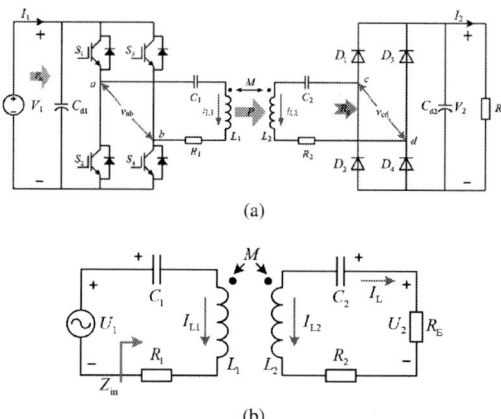

(a)

(b)

Fig. 2. (a) Full-bridge type WPT system using SS compensation, (b) the corresponding fundamental harmonic equivalent circuit.

To minimize the reactive power in the resonant tanks and enhance the magnetic field produced by coils, the resonant capacitors C_1 and C_2 are added in the resonant circuit. For simplicity, the primary frequency ω_1 and the secondary resonant frequency ω_2 are set to be equal, which satisfy

$$\begin{cases} \omega_1 = \frac{1}{\sqrt{L_1 C_1}}, \omega_2 = \frac{1}{\sqrt{L_2 C_2}} \\ \omega_0 = \omega_1 = \omega_2 \end{cases} \quad (1)$$

To adjust the system output and realize ZVS operation of the inverter simultaneously, the variable frequency and phase shift control strategy (VFPSC) is applied. The system operation waveforms are shown in Fig. 3. v_{ab1} is the fundamental

waveform of output voltage of inverter v_{ab} with phase shift angle $D\pi$, and i_{L1} is primary inductor current. Consequently, the phase between v_{ab1} and i_{L1} is the input impedance angle (IIA) of resonant network, φ_{IIA}. Meanwhile, the angle for ZVS operation is defined as φ_{ZA}.

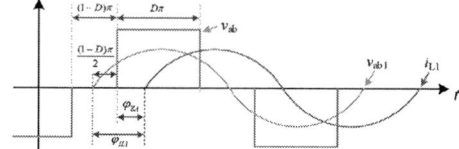

Fig. 3. Operation waveforms of primary inverter.

The RMS value of v_{ab1} can be calculated by

$$U_1 = \frac{2\sqrt{2}}{\pi} V_1 \sin\left[\frac{D\pi}{2}\right] \quad (2)$$

To acquire the required output current and voltage, the system frequency characteristics including trans-conductance gain and voltage gain are investigated as follows.

C. System frequency characteristics

1) Trans-conductance gain of the system (TCGS): Based on Fig. 2(b) and (2), with $R_1 = R_2 = 0\Omega$, the TCGS can be obtained by (3). With the parameters listed in Table II, the TCGS as a function of D and ω_n is calculated and plotted in Fig. 4.

TABLE II
SYSTEM PARAMETERS USED IN CALCULATION

Symbol	Quantity	Value
L_1, L_2	resonant inductances	116.864μH
C_1, C_2	resonant capacitances	30nF
k	coupling coefficiency	0.15~0.2
ω_0	resonant angular frequency	5.34×10^5rad/s
f_0	resonant frequency	85kHz
R_L	load resistance	10Ω

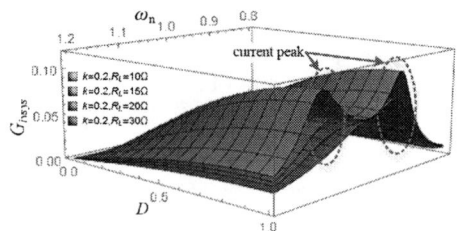

Fig. 4. TCGS as a function of D and ω_n with k=0.2.

$$G_{ivsys} = \frac{I_2}{V_1} = \frac{2\sqrt{2}I_{L2}/\pi}{\pi U_1/(2\sqrt{2}sin[D\pi/2])} = \frac{8kL_2\omega_n^3 \sin[D\pi/2]}{\sqrt{L_1 L_2 \left(64R_L^2\omega_n^2(-1+\omega_n^2)^2 + \pi^4 L_2^2\omega_0^2(-1+2\omega_p n^2 + (-1+k^2)\omega_n^4)^2\right)}} \quad (3)$$

2) Voltage gain of the system (VGS): Similarly, based on Fig. 2(b) and (2), with $R_1=R_2=0\Omega$, the VGS can be obtained by (4). With the parameters listed in Table II, the VGS as a function of D and ω_n is calculated and plotted in Fig. 5.

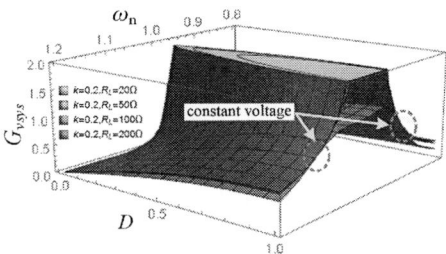

Fig. 5. VGS as a function of D and ω_n when $k=0.2$.

D. Input impedance angle (IIA)

As shown in Fig. 2(b), the input impedance of resonant network is given as

$$Z_{in} = Z_1 + \omega^2 M^2/(Z_2 + R_E) \qquad (5)$$

Where Z_2 is secondary resonant network impedance, which is $Z_2 = R_2 + j\omega_s L_2 + 1/j\omega_s C_2$, and Z_1 is primary resonant network impedance which is $Z_1 = R_1 + j\omega_s L_1 + 1/j\omega_s C_1$. With R_1 and R_2 neglected, the input impedance angle φ_{IIA}, can be calculated by (6).

With the parameters listed in Table II and different coupling coefficient, φ_{IIA} as a function of operation frequency ω_n and R_L, is calculated and plotted in Fig. 6.

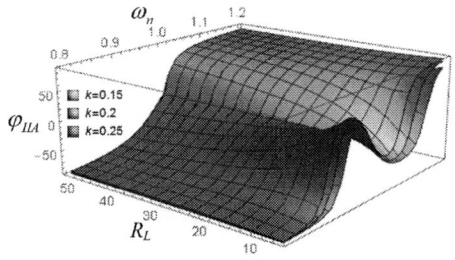

Fig. 6. Characteristics of φ_{IIA}, when $k=0.15, 0.2, 0.25$, with different R_L.

E. Zero voltage switching angle (ZVSA)

Based on VFPSC, the output voltage and current waveforms of the primary inverter are shown in Fig.3. The phase angle, φ_{ZA}, between the rising edge of v_{ab} and the zero-crossing point of i_{L1}, is defined as zero voltage switching angle (ZVSA). From Fig.3, we can obtain

$$\varphi_{IIA} = \varphi_{ZA} + \frac{(1-D)\pi}{2} \qquad (7)$$

To reduce switching cost and EMI, and enhance the systematic efficiency and reliability, ZVS should be achieved in primary inverter, which means $\varphi_{ZA} \geq 0$ at least. Usually, $1 > D > 0$, then $\varphi_{IIA} > \varphi_{ZA}$. To realize ZVS operation completely, φ_{ZA} should be specified as a certain positive value, such as $20°$. According to (7), the critical condition for ZVS operation when $\varphi_{ZA}=0$ can be obtained, that is

$$\varphi_{IIAmin} = \frac{(1-D)\pi}{2} \qquad (8)$$

III. The Optimal Operation Frequency Range

As a kind of power supply for battery packs, WPT system should first ensure the stable output current or voltage. Simultaneously, ZVS operation of primary inverter should be considered especially in high power applications. Consequently, the combinations of D and ω_n that satisfy both CC or CV charging and ZVS operation, are limited in a narrow frequency range. The detailed process is analyzed as follow.

A. Constant current charging and ZVS operation

In CC charging mode, the R_L ranges from 12Ω to 18Ω. When the constant charging current is set as 4A with $V_1=80V$, the TCGS is 0.05. The ZVS angle can be obtained by

$$\varphi_{ZApu} = \varphi_{IIApu} - D_{min} \qquad (9)$$

where D_{min} and φ_{IIApu} are defined by

$$\begin{cases} D_{min} = (1-D)/2 \\ \varphi_{IIApu} = \varphi_{IIA}/\pi \end{cases} \qquad (10)$$

According to (3), if assigning the TCGS equals K_{cc}, then the correspond duty cycle D can be solved in (11). With the parameters listed in Table II, φ_{IIApu}, φ_{ZApu} and D_{min} are plotted in Fig. 7. When R_L ranges from 12Ω to 18Ω,

$$G_{vsys} = \frac{V_2}{V_1} = \frac{\pi U_2/2\sqrt{2}}{\pi U_1/(2\sqrt{2}sin[D\pi/2])} = \frac{8kL_2 R_L \omega_n^3 \sin[D\pi/2]}{\sqrt{L_1 L_2 \left(64R_L^2\omega_n^2(-1+\omega_n^2)^2 + \pi^4 L_2^2\omega_0^2(-1+2\omega_n^2+(-1+k^2)\omega_n^4)^2\right)}} \qquad (4)$$

$$\varphi_{IIA} = \frac{180}{\pi}Arctan[\frac{Im[Z_{in}]}{Re[Z_{in}]}] = -\frac{180}{\pi}Arctan[\frac{(-1+\omega_n^2)\left(-64R_L^2\omega_n^2 + \pi^4 L_2^2\omega_0^2\left(-1+2\omega_n^2+(-1+k^2)\omega_n^4\right)\right)}{8\pi^2 k^2 L_2 R_L \omega_0 \omega_n^5}] \qquad (6)$$

$$D = \frac{2}{\pi}\arcsin\left[\frac{K_{cc}}{8kL_2\omega_n^3}\sqrt{L_1 L_2 \left(64R_L^2\omega_n^2(\omega_n^2-1)^2 + \pi^4\omega_0^2 L_2^2(2\omega_n^2+(-1+k^2)\omega_n^4)^2-1\right)}\right] \qquad (11)$$

3865

the optimal operation frequency ranges are from A_x to B_x ($x=1\sim3$).

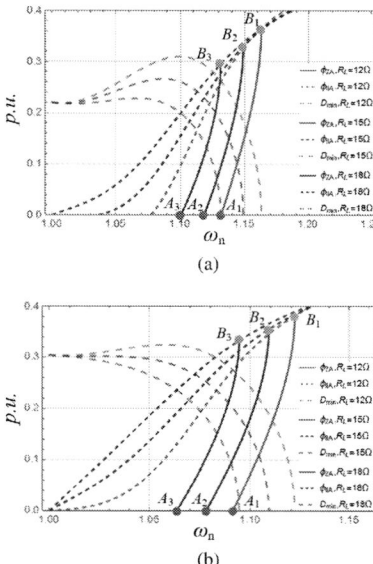

(a)

(b)

Fig. 7. The optimal operation frequency ranges is between A_x and B_x ($x=1$, 2, 3), with K_{cc}=0.05, R_L=12Ω, 15Ω, 18Ω, (a) k=0.2, (b) k=0.15.

As shown in Fig. 7(a), the upper and lower limits of OOFR versus load resistor can be calculated and plotted in Fig. 8(a). Based on the calculating results, we can make some conclusions: when the TCGS is controlled to be constant, the upper and lower limits of operation frequency ω_n are more close to 1 with the larger load R_L and lower coupling coefficient k. In Fig. 8(b), the theoretical maximum ZVS angle can be calculated and plotted. When R_L increases, the theoretical maximum ZVS angle will decrease.

B. Constant voltage charging and ZVS operation

Similarly, according to (4), if assigning the VGS equals K_{cv}, then the corresponding duty cycle D can be solved in (12). With the parameters listed in Table II, φ_{IIApu}, φ_{ZApu} and D_{min} are plotted in Fig. 9. When R_L ranges from 18Ω to 144Ω, the optimal operation frequency ranges are from A_x to B_x ($x=1\sim3$).

When VGS is controlled to be 0.9, the upper and lower limits of OOFR versus load resistor R_L can be calculated and plotted in Fig. 10(a). Based on the calculating results, we can make some conclusions: when VGS is controlled to be constant and R_L increases gradually, the upper limits of operation frequency ω_n almost keep constant and the lower limits of operation frequency ω_n gradually decrease. In Fig. 10(b), the theoretical maximum ZVS angle can be calculated

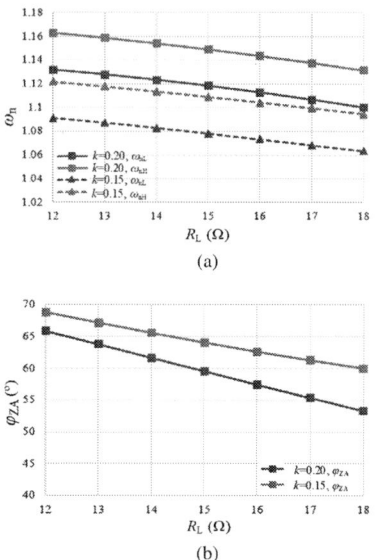

(a)

(b)

Fig. 8. The OOFR and ZVS angle versus load resistor R_L with k=0.15 and 0.2, when TCGS is controlled to be 0.05, (a) The upper and lower limits of OOFR versus R_L with different k, (b) The theoretical maximum ZVS angle versus R_L with different k.

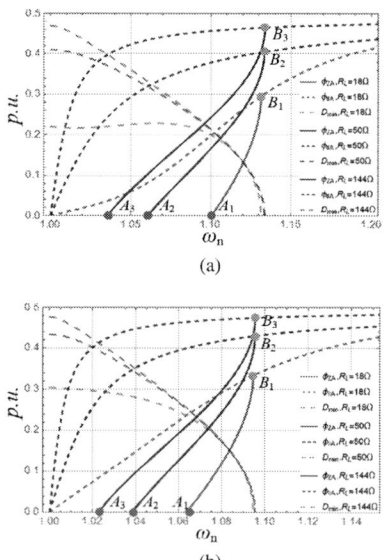

(a)

(b)

Fig. 9. The optimal operation frequency ranges are between A_x and B_x ($x=1$, 2, 3), with K_{cv}=0.9, R_L=18Ω, 50Ω, 144Ω, (a) k=0.2, (b) k=0.15.

$$D = \frac{2}{\pi} \arcsin\left[\frac{K_{cv}}{8kL_2R_L\omega_n^3}\sqrt{L_1L_2\left(64R_L^2\omega_n^2(\omega_n^2-1)^2 + \pi^4\omega_0^2L_2^2(2\omega_n^2+(-1+k^2)\omega_n^4-1)^2\right)}\right] \quad (12)$$

and plotted. When R_L increases, the theoretical maximum ZVS angle will increase.

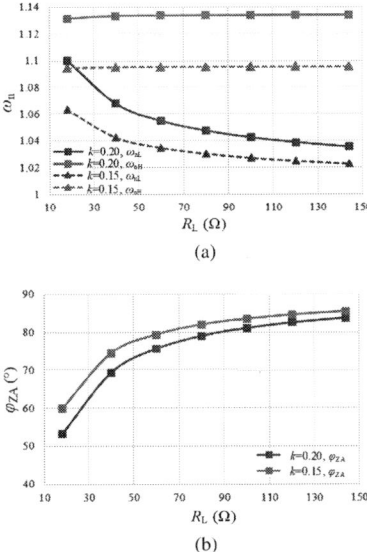

(a)

(b)

Fig. 10. The OOFR and ZVS angle versus load resistor R_L with k–0.15 and 0.2, when VGS is controlled to be 0.9, (a) the upper and lower limits of OOFR versus R_L with different k, (b) the theoretical maximum ZVS angle versus R_L with different k.

IV. THE PROPOSED CONTROL STRATEGY

To make the system run in these OOFRs freely, the first step is to control the ZVS angle to equal the reference φ_{ZAref}. Therefore, a three loop control strategy is proposed to realize CC or CV charging and accurate ZVS angle control simultaneously. By changing the ZVS angle dynamically, the system can achieve the maximum efficiency. This control strategy is illustrated as follows.

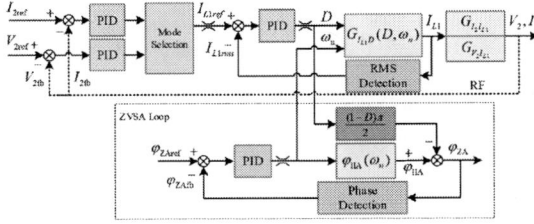

Fig. 11. The control block diagram of TLCS.

V. EXPERIMENT EVALUATION

A. Experimental prototype

To verify the previous analysis and the effectiveness of the proposed TLCS, a 500W experimental prototype is built up which is shown in Fig.12. The prototype includes a DC source, a high frequency inverter, a SS resonant network, a full-bridge rectifier and a sliding rheostat. Energy is transferred from DC source to sliding rheostat. The primary and secondary

sides exchange signals and instructions by 2.4GHz wireless communication modules. The detailed system parameters are listed in Table III.

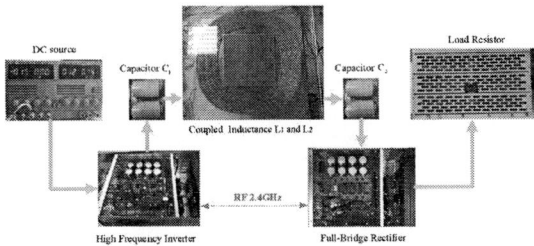

Fig. 12. Experimental prototype.

B. Steady-state operation waveforms

When the battery packs are in CC charging mode, the steady state operation waveforms are shown in Fig. 13 with TCGS=0.05 and $\varphi_{ZAref} = 20°$. When the battery packs are in CV charging mode, the steady state operation waveforms are shown in Fig. 14 with VGS=0.9 and $\varphi_{ZAref} = 20°$.

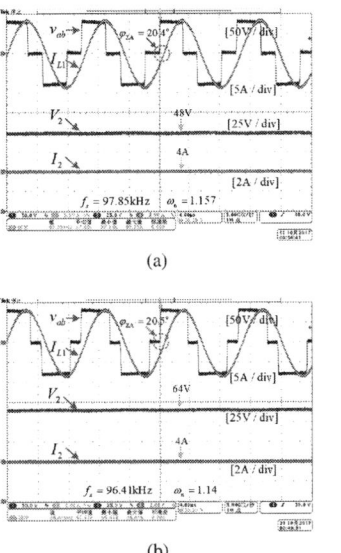

(a)

(b)

Fig. 13. Steady state waveforms in CC mode for battery charging with V_1=80V, I_2=4A, K_{cc}=0.05, k=0.2, and $\varphi_{ZAref} = 20°$, (a) R_L=12Ω, (b) R_L=16Ω.

C. Optimal operation frequency range

When the control aims are set as K_{cc}=0.05 in CC mode and K_{cv}=0.9 in CV mode with different coupling coefficient, the corresponding charging current and charging voltage are I_2=4A and V_2=72V respectively. The comparisons of OOFR between theoretical calculation and experimental measurement are shown in Fig.15. The theoretical calculating results have a good agreement with the experimental ones.

3867

The 2018 International Power Electronics Conference

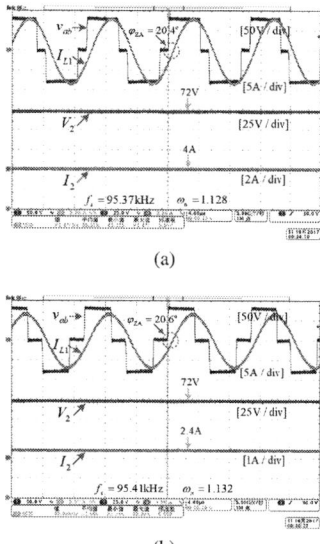

(a)

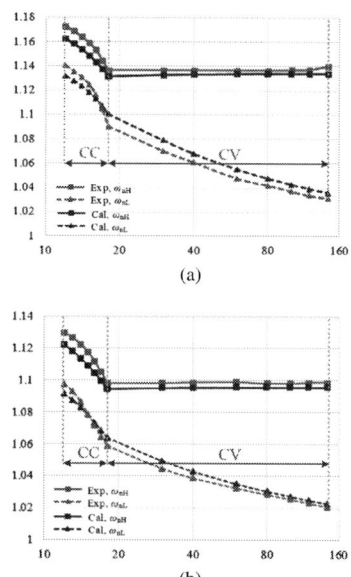

(b)

Fig. 14. Steady state waveforms in CV mode for battery charging with V_1=80V, V_2=72V, K_{cv}=0.9, k=0.2, and $\varphi_{ZAref} = 20°$, (a) R_L=18Ω, (b) R_L=30Ω.

(a)

(b)

Fig. 15. The comparisons of OOFR between theoretical calculation and experimental measurement, when V_1=80V, K_{cc}=0.05, K_{cv}=0.9, (a) k=0.2, (b) k=0.15.

D. System efficiency

The system efficiency for battery packs in CC mode gradually increases along with the increasing of R_L which is shown in Fig.16. When the system enters into the CV mode, the maximum efficiency will decrease along with the increasing of R_L. Benefiting from the proposed control strategy, the whole efficiency can achieve 94.22% with k=0.2 and 92.65% with k=0.15. Especially in light load situation, the whole system

TABLE III
PARAMETERS OF THE WPTS

Symbol	Quantity	Value
L_1	primary resonant inductor	118.6μH
C_1	primary resonant capacitor	29.92nF
f_1	primary resonant frequency	84.49kHz
R_1	primary ESR	0.12Ω
L_2	secondary resonant inductor	118.8μH
C_2	secondary resonant capacitor	29.88nF
f_2	secondary resonant frequency	84.47kHz
R_2	secondary ESR	0.11Ω
N_1, N_2	number of turns	15
R_L	load resistor	12$\sim$144Ω
d	distance of coils	15$\sim$17cm
k	coupling coefficient	0.15$\sim$0.2

always obtains high efficiency larger than 70% in the worst case.

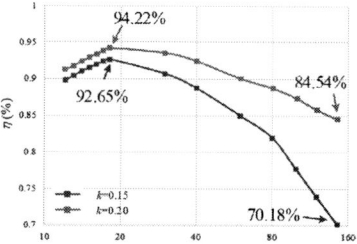

Fig. 16. The system efficiency versus load resistor R_L with different coupling coefficient.

VI. CONCLUSION

This paper derives the OOFR for realizing both constant current or constant voltage output and ZVS operation simultaneously. At the same time, the variable ZVS angle can be used to obtain the maximum efficiency of the whole system for battery packs. The theoretical calculation of OOFR has a good agreement with the experimental results. Benefiting from the proposed control strategy, the whole efficiency can achieve 94.22% with k=0.2 and 92.65% with k=0.15. Especially in light load situation, the whole system always obtains high efficiency larger than 70% in the worst case.

REFERENCES

[1] Q. Chen, S. C. Wong, C. K. Tse, and X. Ruan, "Analysis, design, and control of a transcutaneous power regulator for artificial hearts," *IEEE Transactions on Biomedical Circuits & Systems*, vol. 3, no. 1, p. 23, 2009.

[2] X. Liu and S. Y. Hui, "Simulation study and experimental verification of a universal contactless battery charging platform with localized charging features," *IEEE Transactions on Power Electronics*, vol. 22, no. 6, pp. 2202–2210, 2007.

[3] Z. Cheng, Y. Lei, K. Song, and C. Zhu, "Design and loss analysis of loosely coupled transformer for an underwater high-power inductive power transfer system," *IEEE Transactions on Magnetics*, vol. 51, no. 7, pp. 1–10, 2015.

[4] D. Ahn, S. Kim, J. Moon, and I. K. Cho, "Wireless power transfer with automatic feedback control of load resistance transformation," *IEEE Transactions on Power Electronics*, vol. 31, no. 11, pp. 7876–7886, 2016.

3868

[5] H. Li, J. Li, K. Wang, and W. Chen, "A maximum efficiency point tracking control scheme for wireless power transfer systems using magnetic resonant coupling," *Power Electronics IEEE Transactions on*, vol. 30, no. 7, pp. 3998–4008, 2015.

[6] X. Qu, H. Han, S. C. Wong, K. T. Chi, and W. Chen, "Hybrid ipt topologies with constant current or constant voltage output for battery charging applications," *IEEE Transactions on Power Electronics*, vol. 30, no. 11, pp. 6329–6337, 2015.

[7] D. Ahn and S. Hong, "Wireless power transfer resonance coupling amplification by load-modulation switching controller," *Industrial Electronics IEEE Transactions on*, vol. 62, no. 2, pp. 898–909, 2015.

The 2018 International Power Electronics Conference

Initial Current Injection Method of a Direct Three-Phase to Single-Phase AC/AC Converter for Inductive Charger

Ferdi Perdana Kusumah* and Jorma Kyyrä
Department of Electrical Engineering and Automation
School of Electrical Engineering
AALTO UNIVERSITY
P.O. Box 13000
FI-00076 Aalto, Finland
URL: http://eea.aalto.fi/en/
*E-mail: ferdi.kusumah@aalto.fi

Abstract—This paper explains an initial current injection method of a direct three-phase to single-phase AC/AC converter for an inductive charger. The converter has a lesser number of switches than a matrix converter and uses a resonant circuit to utilize zero-current switching. The method applies a DC resonant charging to charge a primary resonant capacitor, by taking advantage of the converter topology. The charged voltage is used to boost the initial current since in practice an affordable current transducer has only a limited measurement range, due to its accuracy and noise characteristics. The method is applied to kick-start the inductive charger without adding an amplification circuit to the current transducer output. Simulation results are presented to verify theoretical calculations.

Keywords—*AC/AC converter, battery charger, contactless power supply, resonant converter.*

I. INTRODUCTION

A concept of an inductive-based contactless power transfer (ICPT) for an electric vehicle charger has been introduced in [1], [2] and [3]. It incorporates a direct three-phase to single-phase AC/AC converter connected to a resonant circuit. The converter topology has a lesser number of bi-directional switches than a matrix converter and a similarity to the one given in [4] and [5], with an added advantage of using Zero-current switching (ZCS) mechanism through injection and free-wheeling oscillation commutations previously introduced in [6], [7] and [8]. The ZCS operation forces the converter to always monitor primary resonant current characteristics which are amplitude and zero-crossings.

However in practice, an affordable current transducer only has a limited measurement capability. For example, a CASR-6 transducer that has a maximum measurement range of ±20 A and output voltage range of 0.375-4.625 V, will have a difficulty in measuring ±2.5 A, since the output voltage only corresponds to 2.125 ± 0.266 V [9]. An output voltage amplification can be applied to increase the measurement range. But noise is usually present in practical case that can also be amplified and further corrupts the results [10]. Due to the limitations,

practical current oscillation during initial transient of the AC/AC converter with a limited input source voltage amplitude cannot be monitored properly. This leads to a difficulty in switching the converter at the state, since its operation is based on current zero-crossings.

A method to boost initial current amplitude will be studied in this paper. It is based on a DC resonant charging principle to charge a primary resonant capacitor, by taking advantage of the converter topology [10] [11]. Capacitor voltage after charging can produce a sufficient current amplitude for the current transducer, where its output will be used to operate the AC/AC converter switches.

This paper is organized as follows. A brief explanation of a direct AC/AC converter topology and its operation is given in Section II. Section III gives explanation on the charging method and its mathematical model. In Section IV, simulation results obtained from PLECS software are presented to verify the initial injection method. Finally, conclusions of the paper are given in the last section.

II. SYSTEM DESCRIPTION

A. Circuit explanation

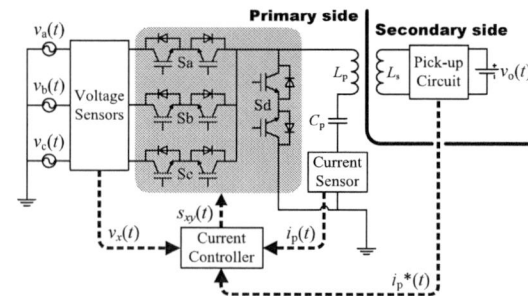

Fig. 1: ICPT system schematic. Grey area highlights bi-directional switches arrangement of the direct AC/AC converter.

The 2018 International Power Electronics Conference

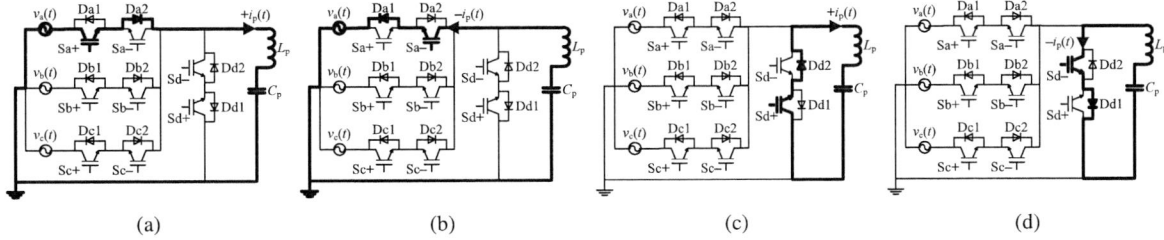

(a) (b) (c) (d)

Fig. 2: Commutation modes of a direct AC/AC converter. Sub-figure (a) and (b) illustrate injection mode, while (c) and (d) show free-wheeling oscillation commutation. A clock-wise current flow is defined as a positive flow.

A schematic of the ICPT system is given in Fig. 1. The switching topology consists of three pair of bi-directional switches Sa, Sb and Sc connected to a three-phase power source and has a common connection to a resonant circuit L_p and C_p. The Sd switch pair is used as a free-wheeling path of primary current. An on-off current controller is utilized to produce gate signals $s_{\mathrm{xy}}(t)$ for managing converter output power. The power management is achieved through controlling primary current amplitude. The controller uses voltage $v_\mathrm{x}(t)$ and current $i_\mathrm{p}(t)$ information, obtained from corresponding sensors. It also accepts current reference signal, which is $i_\mathrm{p}^*(t)$ from the secondary pick-up circuit. The primary and secondary sides are magnetically coupled and the pick-up circuit contains a rectifier, a DC/DC converter, and a battery equipped with a voltage sensor [1].

The AC/AC converter has two types of commutation which are injection and free-wheeling oscillation as illustrated in Fig. 2. In injection mode, the current goes either from the three-phase source to the resonant circuit or vice versa, while during free-wheeling, the current is confined in the resonant circuit. One possible modulation strategy of the converter is described in Fig. 3, which is explained thoroughly in [1]. It utilizes injection and free-wheeling oscillation during maximum absolute value of the three-phase input or,

$$\mathrm{Max}(v) = \max(|v_\mathrm{a}(t)|, |v_\mathrm{b}(t)|, |v_\mathrm{c}(t)|). \quad (1)$$

Injection is used to increase the resonant current amplitude, while the free-wheeling is applied to reduce the amplitude. The indefinite operation involving injection and free-wheeling commutations will be called a normal mode in following sections. Fig. 3 also illustrates the operation during $\mathrm{Max}(v) = v_\mathrm{b}(t)$. It shows that the transition between injection and free-wheeling is performed at the zero-crossing of primary resonant current $i_\mathrm{p}(t)$ [1].

III. INITIAL INJECTION METHOD

To boost primary resonant current during initial operation of the AC/AC converter, the primary resonant capacitor is charged gradually during positive and negative phases of three-phase input until its voltage reaches a certain level. The level must be able to produce a sufficient initial current amplitude for a current transducer used in practical application. The process of capacitor

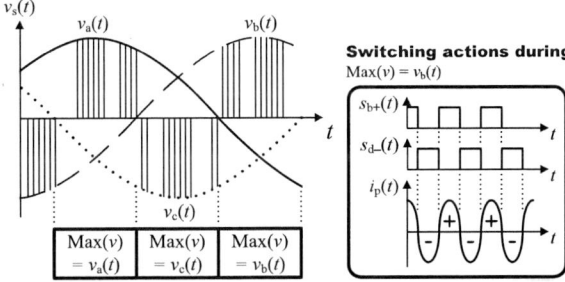

Fig. 3: A modulation strategy of ICPT system. Zero-current switching is performed to reduce switching power losses.

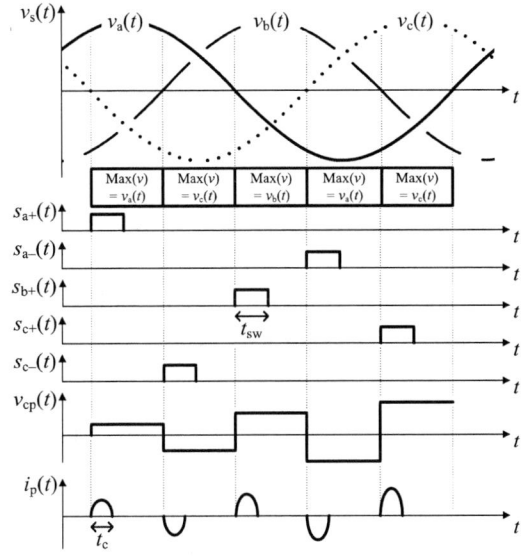

Fig. 4: Gradual capacitor charging mode of ICPT system. The pulse amplitude of $i_\mathrm{p}(t)$ is increased over time during the charging process.

charging is illustrated in Fig. 4. It shows positions of switching signals with respect to the three-phase input. The involved switches are turned-on at the beginning of $\mathrm{Max}(v)$ described in (1) and turned off after t_{sw} duration. The time has to be larger than a charging time t_c of

3871

the resonant capacitor. Alternating polarity order must be used to increase the capacitor voltage. Therefore from Fig. 4, the switching order is Sa+, Sc−, Sb+, Sa− and Sc+ that corresponds to five cycles charging. It can be seen that Sd+ and Sd− that are responsible to freewheeling commutation are not used in this case. When the pulse of $i_p(t)$ is high enough to be read by the current transducer, the converter will start its normal operation, as was illustrated in Fig. 3.

A. Dynamic analysis

To calculate the charging voltage and current at the primary resonant capacitor, as well as the initial current boost due to the voltage, a dynamic analysis is needed. The dynamic equations will be obtained through a state-space approach. Equivalent representation of primary and secondary resonant circuits for modeling purpose is presented in Fig. 5. The schematic is based on one input phase, but can also be applied to the other two. The anti-parallel diode D_{xy} is always in the direction of the charging current $i_p(t)$. Primary capacitor is labeled by C_p, while R_p and R_s are resistances of primary and secondary coils L_p and L_s respectively. The circuit is connected to a sampled voltage $v_s(t)$ at the beginning of Max(v), which is denoted by $v_{in}(t)$. Current on the secondary circuit is denoted by $i_s(t)$. The battery is assumed to be purely resistive (R_L) and connected to a passive rectifier. Variable R_{eq} is a combination of the rectifier and R_L load obtained from [12]. All components are assumed to be ideal.

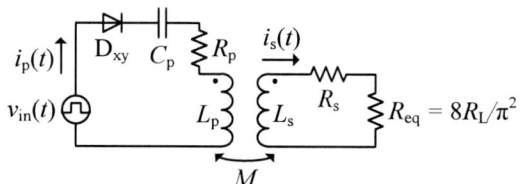

Fig. 5: Equivalent circuit of ICPT system that illustrates positive charging process. During negative charging, the diode D_{xy} and current $i_p(t)$ are facing an opposite direction.

State-space representation of the converter can be obtained through manipulating dynamic equations based on circuit given in Fig. 5. The equation set is as follows,

$$
\begin{cases}
\dot{v}_{cp}(t) = \dfrac{i_p(t)}{C_p}, \\
v_{in}(t) = v_{cp}(t) + i_p(t)R_p + L_p\dot{i}_p(t) - M\dot{i}_s(t), \\
0 = i_s(t)(R_s + R_{eq}) + L_s\dot{i}_s(t) - M\dot{i}_p(t),
\end{cases}
\tag{2}
$$

where M is a mutual inductance between primary and secondary circuit that equals to $k\sqrt{L_pL_s}$. Variable k is a coupling factor between the primary and secondary sides.

The last two equations in (2) are combined to produce,

$$
\dot{i}_p(t) = \frac{1}{L_pL_s - M^2}\Big[L_sv_{in}(t) - L_sv_{cp}(t) \\
- R_pL_si_p(t) - M(R_s + R_{eq})i_s(t)\Big], \tag{3}
$$

$$
\dot{i}_s(t) = \frac{1}{L_pL_s - M^2}\Big[Mv_{in}(t) - Mv_{cp}(t) \\
- MR_pi_p(t) - L_p(R_s + R_{eq})i_s(t)\Big]. \tag{4}
$$

The first equation in (2), as well as equation (3) and (4) are joined to produce a state-space representation,

$$
\begin{bmatrix} \dot{v}_{cp}(t) \\ \dot{i}_p(t) \\ \dot{i}_s(t) \end{bmatrix} = \mathbf{A}\begin{bmatrix} v_{cp}(t) \\ i_p(t) \\ i_s(t) \end{bmatrix} + \mathbf{B}v_{in}(t),
$$

$$
\mathbf{A} = \begin{bmatrix} 0 & \frac{1}{C_p} & 0 \\ -\frac{L_s}{L_pL_s - M^2} & -\frac{R_pL_s}{L_pL_s - M^2} & -\frac{M(R_s + R_{eq})}{L_pL_s - M^2} \\ -\frac{M}{L_pL_s - M^2} & -\frac{MR_p}{L_pL_s - M^2} & -\frac{L_p(R_s + R_{eq})}{L_pL_s - M^2} \end{bmatrix}, \tag{5}
$$

$$
\mathbf{B} = \begin{bmatrix} 0 \\ \frac{L_s}{L_pL_s - M^2} \\ \frac{M}{L_pL_s - M^2} \end{bmatrix}.
$$

The representation can be solved through the Laplace transform approach explained in [13],

$$
\begin{bmatrix} v_{cp}(t) \\ i_p(t) \\ i_s(t) \end{bmatrix} = e^{\mathbf{A}t}\begin{bmatrix} v_{cp}(0) \\ i_p(0) \\ i_s(0) \end{bmatrix} + \int_0^t e^{\mathbf{A}(t-\tau)}\mathbf{B}v_{in}(\tau)d\tau, \tag{6}
$$

$$
e^{\mathbf{A}t} = \mathcal{L}^{-1}[(s\mathbf{I} - \mathbf{A})^{-1}].
$$

The charging voltage and current expressions are taken from the first and second rows of the solution. The capacitor voltage and charging current equations are given in (7) and (8),

$$
v_{cp}(t) = \phi_1(t)v_{cp}(0) \\
+ \int_0^t \Big[\frac{\phi_2(t-\tau)L_s}{L_pL_s - M^2} + \frac{\phi_3(t-\tau)M}{L_pL_s - M^2}\Big]v_{in}(\tau)d\tau, \tag{7}
$$

$$
i_p(t) = \phi_4(t)v_{cp}(0) \\
+ \int_0^t \Big[\frac{\phi_5(t-\tau)L_s}{L_pL_s - M^2} + \frac{\phi_6(t-\tau)M}{L_pL_s - M^2}\Big]v_{in}(\tau)d\tau, \tag{8}
$$

$$
\phi_n(t) = \frac{\alpha_n s_1^2 + \beta_n s_1 + \gamma_n}{(s_1 - s_2)(s_1 - s_3)}e^{s_1 t} \\
+ \frac{\alpha_n s_2^2 + \beta_n s_2 + \gamma_n}{(s_2 - s_1)(s_2 - s_3)}e^{s_2 t} + \frac{\alpha_n s_3^2 + \beta_n s_3 + \gamma_n}{(s_3 - s_1)(s_3 - s_2)}e^{s_3 t} \tag{9}
$$

$$\alpha_1 = 1, \quad \beta_1 = \frac{R_{\mathrm{p}}L_{\mathrm{s}} + L_{\mathrm{p}}(R_{\mathrm{s}} + R_{\mathrm{eq}})}{L_{\mathrm{p}}L_{\mathrm{s}} - M^2},$$
$$\gamma_1 = \frac{R_{\mathrm{p}}(R_{\mathrm{s}} + R_{\mathrm{eq}})}{L_{\mathrm{p}}L_{\mathrm{s}} - M^2}, \tag{10}$$

$$\alpha_2 = 0, \quad \beta_2 = \frac{1}{C_{\mathrm{p}}}, \quad \gamma_2 = \frac{1}{C_{\mathrm{p}}}\left[\frac{L_{\mathrm{p}}(R_{\mathrm{s}} + R_{\mathrm{eq}})}{L_{\mathrm{p}}L_{\mathrm{s}} - M^2}\right], \tag{11}$$

$$\alpha_3 = 0, \quad \beta_3 = 0, \quad \gamma_3 = -\frac{1}{C_{\mathrm{p}}}\left[\frac{M(R_{\mathrm{s}} + R_{\mathrm{eq}})}{L_{\mathrm{p}}L_{\mathrm{s}} - M^2}\right], \tag{12}$$

$$\alpha_4 = 0, \quad \beta_4 = -\frac{L_{\mathrm{s}}}{L_{\mathrm{p}}L_{\mathrm{s}} - M^2}, \quad \gamma_4 = -\frac{R_{\mathrm{s}} + R_{\mathrm{eq}}}{L_{\mathrm{p}}L_{\mathrm{s}} - M^2}, \tag{13}$$

$$\alpha_5 = 1, \quad \beta_5 = \frac{L_{\mathrm{p}}(R_{\mathrm{s}} + R_{\mathrm{eq}})}{L_{\mathrm{p}}L_{\mathrm{s}} - M^2}, \quad \gamma_5 = 0, \tag{14}$$

$$\alpha_6 = 0, \quad \beta_6 = -\frac{M(R_{\mathrm{s}} + R_{\mathrm{eq}})}{L_{\mathrm{p}}L_{\mathrm{s}} - M^2}, \quad \gamma_6 = 0. \tag{15}$$

Variable s_1, s_2 and s_3 are roots of $\det(s\mathbf{I} - \mathbf{A}) = 0$. Initial values of $i_{\mathrm{p}}(0)$ and $i_{\mathrm{s}}(0)$ are always zero in this case since the diode D_{xy} does not allow the current to oscillate (see Fig. 4). Thus the charging voltage value must be calculated at $t = t_{\mathrm{c}}$. For the charging current case, the value is taken at an extremum point. A comprehensive dynamic equation derivation of the converter can be found in [2].

The process of capacitor charge release to boost the initial current involves connecting the charged capacitor to one of the three-phase source. The capacitor and the source must have a different polarity. The process is slightly different than the charging itself since the current is allowed to oscillate after this point. To calculate the current amplitude, the charging current expression from (8) can be used. The value can be obtained by inserting a final charging voltage as an initial capacitor voltage $v_{\mathrm{cp}}(0)$. The current amplitude calculation is similar to the charging current calculation of the capacitor with an additional cycle. In other words, for two cycles charging voltage level, the converter will produce a current similar to a charging current of three cycles at the charge release.

B. Steady-state analysis

To simplify the dynamic analysis, two more assumptions are used, which are: the coupling between primary and secondary sides is loose, and the primary current distortion due to a damping effect is minimum. In this case, the charging time t_{c} can be approximated by a half of the system's resonant period. The coupled system resonant frequency is calculated through combining two steady-state equations as follows,

$$\begin{cases} \mathbf{V}_{\mathrm{in}} = \dfrac{i_{\mathrm{p}}}{j\omega C_{\mathrm{p}}} + i_{\mathrm{p}}j\omega L_{\mathrm{p}} + i_{\mathrm{p}}R_{\mathrm{p}} - i_{\mathrm{s}}j\omega M, \\ 0 = i_{\mathrm{s}}R_{\mathrm{eq}} + i_{\mathrm{s}}j\omega L_{\mathrm{s}} + i_{\mathrm{s}}R_{\mathrm{s}} - i_{\mathrm{p}}j\omega M, \end{cases} \tag{16}$$

variable $\mathbf{V}_{\mathrm{in}}$ is an input of the resonant circuit. The combination of two steady-state equations in (16) by

eliminating i_{s} leads to,

$$\mathbf{V}_{\mathrm{s}} = i_{\mathrm{p}}\left[R_{\mathrm{p}} + \frac{\omega^2 M^2(R_{\mathrm{eq}} + R_{\mathrm{s}})}{\omega^2 L_{\mathrm{s}}^2 + (R_{\mathrm{eq}} + R_{\mathrm{s}})^2}\right.$$
$$\left. + j\left(\omega L_{\mathrm{p}} - \frac{1}{\omega C_{\mathrm{p}}} - \frac{\omega^3 M^2 L_{\mathrm{s}}}{\omega^2 L_{\mathrm{s}}^2 + (R_{\mathrm{eq}} + R_{\mathrm{s}})^2}\right)\right]. \tag{17}$$

During resonance ($\omega = \omega_0$), the imaginary part in (17) equals to zero,

$$\omega_0 L_{\mathrm{p}} - \frac{1}{\omega_0 C_{\mathrm{p}}} = \frac{\omega_0^3 M^2 L_{\mathrm{s}}}{\omega_0^2 L_{\mathrm{s}}^2 + (R_{\mathrm{eq}} + R_{\mathrm{s}})^2}, \tag{18}$$

$$C_{\mathrm{p}}(L_{\mathrm{p}}L_{\mathrm{s}}^2 - M^2 L_{\mathrm{s}})\omega_0^4 + [L_{\mathrm{p}}C_{\mathrm{p}}(R_{\mathrm{eq}} + R_{\mathrm{s}})^2 - L_{\mathrm{s}}^2]\omega_0^2$$
$$- (R_{\mathrm{eq}} + R_{\mathrm{s}})^2 = 0, \tag{19}$$

which has a biquadratic form. Four roots can be calculated through the quadratic formula. One root that has positive and real value is the resonant frequency ω_0. The equation is described as follows,

$$\omega_0 = \pm\sqrt{-A \pm \Omega}, \tag{20}$$

$$A = \frac{1}{2L_{\mathrm{p}}^2}\left[\frac{m^2(R_{\mathrm{eq}} + R_{\mathrm{s}})^2 - n}{1 - k^2}\right], \tag{21}$$

$$\Omega = \sqrt{A^2 + \frac{nm^2(R_{\mathrm{eq}} + R_{\mathrm{s}})^2}{L_{\mathrm{p}}^4(1 - k^2)}}, \tag{22}$$

$$m = \frac{L_{\mathrm{p}}}{L_{\mathrm{s}}}, \quad n = \frac{L_{\mathrm{p}}}{C_{\mathrm{p}}}. \tag{23}$$

The resonant period is $T_0 = 2\pi/\omega_0$. The charging voltage and maximum primary current amplitude can be approximated at $T_0/2$ and $T_0/4$ respectively.

C. Analytical calculation

For an initial prototype, the ICPT system is designed to deliver power less than 1 kW. Parameter values for the circuit are given in TABLE I. Resonant circuit quantities are based on components selection given in [2]. The selection results keep a damping ratio of the primary resonant current under a certain level. The ratio permits the current to keep oscillating during converter operation. Load values R_{L} and R_{eq} are based on [3] to minimize coils' power losses. Switching frequency of the converter is the same as a resonant frequency of the coupled resonant circuit. The value depends on the coupling factor and load conditions.

During one cycle charging process, the input voltage $v_{\mathrm{in}}(t)$ is assumed to be constant since the charging time is much smaller than the input three-phase period. The voltage value is sampled at the beginning of $\mathrm{Max}(v)$ (see Fig. 4) which is equal to $\pm V_{\mathrm{s}}\sin(\pi/3)$, where V_{s} is the amplitude of $v_{\mathrm{s}}(t)$. Positive and negative symbol indicates that the voltage source is taken from either positive or negative source, depending on the resonant capacitor voltage state. In a case of an input source with 100 V peak voltage, the value of $v_{\mathrm{in}}(t)$ becomes ± 86.603 V. By using the MATLAB software, capacitor voltage, as well as charging and produced currents of the circuit are calculated and given in TABLE II. Variable V_{cp} and I_p are amplitude or extremum value. The voltage is calculated at $T_0/2$, while the current is at $T_0/4$. The calculated resonant frequency is 26.983 kHz for $k = 0.55$ and $R_L = 47.742\ \Omega$. And for $k = 0.83$ and $R_L = 58.708\ \Omega$, the frequency is 29.139 kHz.

IV. SIMULATION RESULTS

The given values in TABLE I are used in a simulation model of the ICPT system built using the PLECS software. All simulation results are obtained when the system is in an open-loop configuration.

A. Initial injection without gradual charging

Without gradual charging, initial capacitor voltage is zero, and resonant current is increased through injection and free-wheeling commutations. In other words, the converter is operating in a normal mode immediately as was shown in Fig. 3. Simulation results for two different coupling factors and loads are given in Fig. 6. Parameters $v_a(t)$, $v_b(t)$ and $v_c(t)$ indicate phase-A, B and C of the three-phase input while $v_{res}(t)$ is a voltage over the primary resonant circuit. The current $i_p(t)$ in this case is the primary resonant current. Initial current amplitudes (marked by red circles) produced by applying voltage over the resonant circuit are approximately 2.721 A for $k = 0.55$ and 2.544 A for $k = 0.83$. Switching of Sa+ gate begins at the beginning of $\text{Max}(v) = v_a(t)$.

It can be seen that Sa+ is starting initial current injection directly from input phase-A. With the CASR-6 current transducer described in [9], its output will be 0.289 V for 2.721 A and 0.27 V for 2.544 A, around 2.5 V offset voltage. The values may be too small for a practical comparator-based zero-crossing detector, that is connected directly to its output.

B. Initial injection with gradual charging

In this case, the initial current is increased with a help of primary capacitor voltage that has been charged to a certain level. As examples, two cycles charging were performed on the simulation model and the results are given in Fig 7 and 8. Each result corresponds to a certain combination of coupling factor and load. The

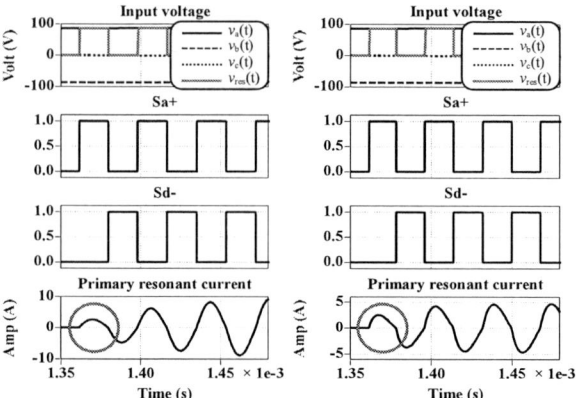

Fig. 6: Simulation results of initial transient without gradual charging with time range $1.35 - 1.48$ ms. Left graph corresponds to k = 0.55 and R_L = 47.742 Ω, while the right one relates to k = 0.83 and R_L = 58.708 Ω. Red circles indicate first current injection.

charging process happens around $t < 8$ ms. The primary capacitor is gradually charged twice at the beginning of two $\text{Max}(v)$ with a different source polarity consecutively. The quantities of both figures are presented in TABLE III and IV.

In Fig. 7, the capacitor voltage is charged to approximately -284.57 V, and in 8 is up to around -216.378 V, before they are used to boost the initial primary resonant current. Fig. 8 is a tighter coupling case compared to Fig. 7. The charging are performed twice at the beginning of $\text{Max}(v) = v_a(t)$ and $\text{Max}(v) = v_c(t)$. Voltage over the resonant circuit is indicated by $v_{res}(t)$ on the first plot of each figure. This voltage is equal to the capacitor voltage $v_{cp}(t)$ during the charging process. The IGBT that is responsible for charging in both cases are conducted for 1 ms, which is longer than the system's resonant period which is in a range of $34.318\ \mu s \leq t_c \leq 37.06\ \mu s$. By

TABLE II: CALCULATION RESULTS

Charging cycle	$k = 0.55$, $R_L = 47.742\Omega$		$k = 0.83$, $R_L = 58.708\Omega$	
	Capacitor voltage (V_{cp})	Charging current (I_p)	Capacitor voltage (V_{cp})	Charging current (I_p)
1	156.9 V	2.678 A	137.199 V	2.283 A
2	-284.065 V	-7.532 A	-217.061 V	-5.899A
3	387.130 V	11.465 A	263.547 V	8.004 A
4	-470.662 V	-14.653 A	-290.606 V	-9.229 A
5	538.363 V	17.237 A	306.357 V	9.942 A

TABLE I: PARAMETER VALUES

Symbol	Meaning	Value
V_s	Line voltage amplitude	100 V
f	Line voltage frequency	50 Hz
C_p	Primary capacitance	0.2 μF
L_p	Primary inductance	0.2 mH
L_s	Secondary inductance	0.2 mH
k	Coupling factor	$0.55 - 0.83$
R_p	Primary coil resistance	0.3 Ω
R_p	Secondary coil resistance	0.3 Ω
R_L	Load resistance	$47.742 - 58.708\ \Omega$
R_{eq}	Equivalent load resistance	$38.698 - 47.587\ \Omega$

TABLE III: SIMULATION RESULTS 1

Charging cycle	$k = 0.55$, $R_L = 47.742\ \Omega$	
	Capacitor voltage (V_{cp})	Charging current (I_p)
1	157.12 V	2.72 A
2	-284.57 V	-7.65 A

The 2018 International Power Electronics Conference

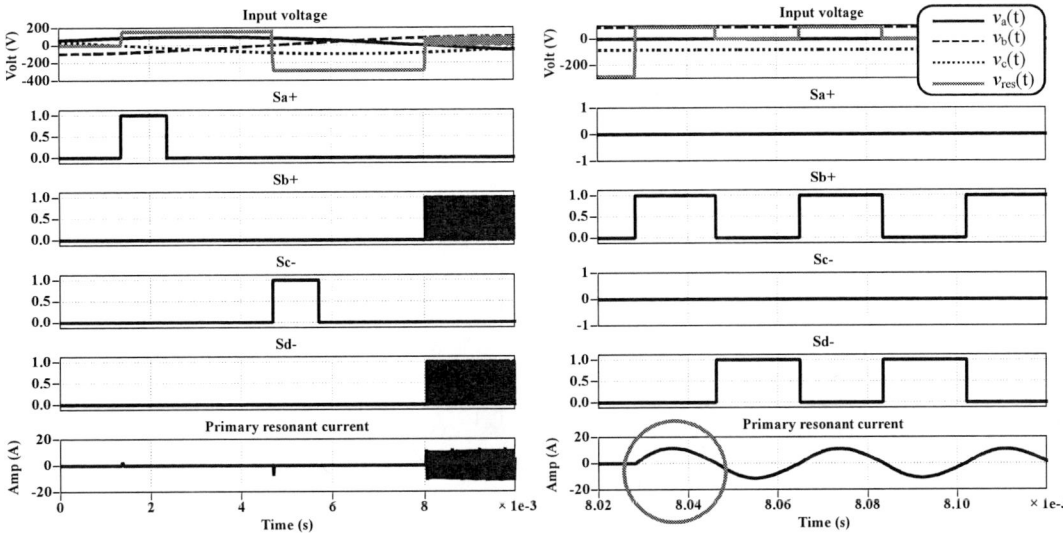

Fig. 7: Simulation result of initial transient using gradual charging for k = 0.55 and R_L = 47.742 Ω. The right graph is a zoomed version of the left around 8 ms time axis. Left graph has a timescale of $1 - 10$ ms, while the right one has $8.02 - 8.12$ ms. Red circle indicates first current injection.

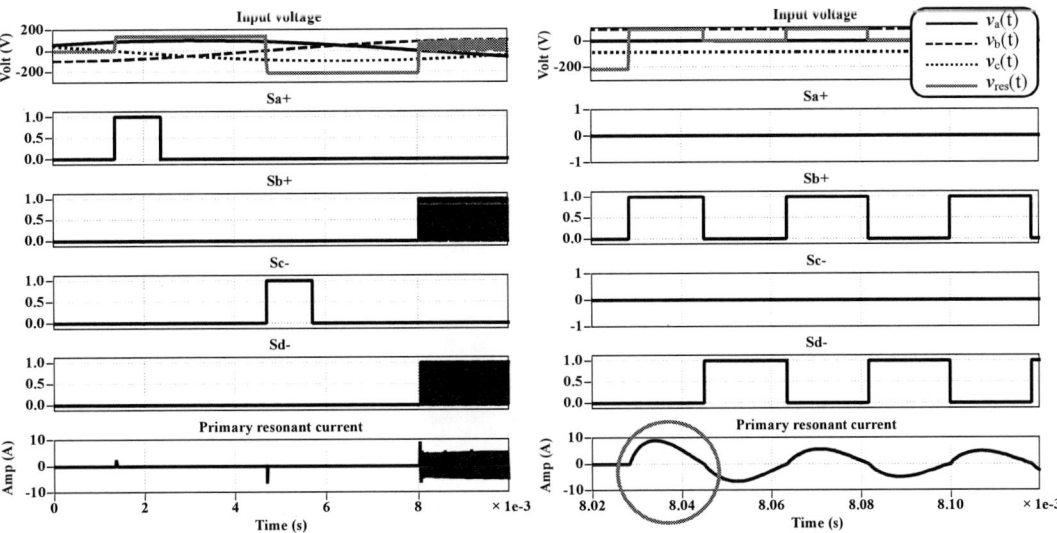

Fig. 8: Simulation result of initial transient using gradual charging for k = 0.83 and R_L = 58.708 Ω. The right graph is a zoomed version of the left around 8 ms time axis. Left graph has a timescale of $1 - 10$ ms, while the right one has $8.02 - 8.12$ ms. Red circle indicates first current injection.

TABLE IV: SIMULATION RESULTS 2

$k = 0.83$, $R_L = 58.708$ Ω

Charging cycle	Capacitor voltage (V_{cp})	Charging current (I_p)
1	137.01 V	2.545 A
2	-216.378 V	-6.565 A

comparing simulation and analytical results (TABLE II, III and IV), it can be seen that they are similar. In both

figures, the first current injection amplitude is 11.651 A for $k = 0.55$ and 8.895 A for $k = 0.83$. If CASR-6 is used, then its output voltage will be 1.238 V and 0.945 V respectively which are big enough to be distinguished by a comparator-based zero-crossing detector [9]. Both injection currents are also similar to the charging current of the three charging cycles for each particular case (see TABLE II).

By observing current waveforms inside red circles in both Fig. 7 and 8, a higher coupling factor leads to a deformed sinusoidal current waveform. The deformation

3875

is caused by a higher damping ratio which is explained further in [2]. This effect can cause analytical calculations that utilizes the resonant frequency less accurate.

V. CONCLUSION

An initial injection method of a direct AC/AC converter for inductive charger has been demonstrated at a simulation level. It is based on a DC resonant charging by taking advantage of converter topology. The purpose is to increase amplitude of initial current injection in a primary resonant circuit, without adding output amplification on a used current transducer. This method is useful due to limited measurement range of an affordable current transducer and the existence of measurement noise. Analytical calculation to predict charging behavior was verified by the simulation results. It has to be noted that the calculation approach is more accurate for a loosely coupled system. Experimental setup is currently being built that will demonstrate the performance of the circuit as well as its initial charging method.

REFERENCES

[1] F. P. Kusumah et al., "A Direct Three-Phase to Single-Phase AC/AC Converter for Contactless Electric Vehicle Charger," in Proc. European Conference on Power Electronics and Applications, Geneva, 2015, pp. 1-10.

[2] F. P. Kusumah et al., "Components Selection of a Direct Three-Phase to Single-Phase AC/AC Converter for Contactless Electric Vehicle Charger," in Proc. European Conference on Power Electronics and Applications, Karlsruhe, 2016, pp. 1-10.

[3] F. P. Kusumah and J. Kyyrä, "Minimizing Coil Power Loss in a Direct AC/AC Converter-based Contactless Electric Vehicle Charger," in Proc. European Conference on Power Electronics and Applications, Warsaw, 2017, pp. 1-10.

[4] N. X. Bac et al., "A Matrix Converter Based Inductive Power Transfer System," in Proc. International Power and Energy Conference, Dec 2012, pp. 509-514.

[5] N. X. Bac et al., "A SiC-Based Matrix Converter Topology for Inductive Power Transfer System," IEEE Transactions on Power Electronics, vol. 29, issue 8, pp. 4029-4038.

[6] A. P. Hu and Hao L. Li, "A new high frequency current generation method for inductive power transfer applications," in Proc. IEEE Power Electronics Specialists Conference, Jeju, 2006, pp. 1-6.

[7] Li, Hao L. et al., "FPGA Controlled High Frequency Resonant Converter for contactless Power Transfer," in Proc. IEEE Power Electronics Specialists Conference, Rhodes, 2008, pp. 3642-3647.

[8] Hao Leo Li et al., "A Direct AC-AC Converter for Inductive Power-Transfer Systems," in IEEE Transactions on Power Electronics, vol. 27, pp. 661-668, Jun 2011.

[9] LEM. (2017, October 23). Current transducer CASR series [Online]. Available: http://www.lem.com/docs/products/casr_series.pdf

[10] P. Horowitz and W. Hill, The Art of Electronics, 3rd ed. New York, USA: Cambridge University Press, 2015.

[11] H. J. White et al., "THE CHARGING CIRCUIT OF THE LINE-TYPE PULSER," in PULSE GENERATORS. New York, NY: Dover, 1948.

[12] R. L. Steigerwald, "A Comparison of Half-Bridge Resonant Converter Topologies," in IEEE Transactions on Power Electronics, vol. 3, pp. 174-182, Apr 1988.

[13] Katsuhiko Ogata, "Control Systems Analysis in State Space," in Modern Control Engineering, 5th ed., Upper Saddle River, NJ: Pearson Education, Inc., 2010, pp. 648-721.

Mission Profile Emulator for Permanent Magnet Synchronous Machine Based on Three-phase Power Electronic Converter

Yubo Song, Ran Cheng, Ke Ma

Department of Electrical Engineering, Shanghai Jiao Tong University, Shanghai 200240, China

E-mail: seansong@sjtu.edu.cn, shjdchengran@163.com, kema@sjtu.edu.cn

Abstract – **The mission profiles of motor drive systems are becoming more and more complicated in many emerging applications of power electronics such as wind power generation, locomotive and electric vehicles, etc. As a result, comprehensive and long-term reliability verifications/tests of power electronics systems and devices are becoming crucial needs. In order to achieve such kind of advanced tests with flexibly adjusted testing configurations such as torque profiles, kinetic inertia, electric machine parameters and control algorithms, etc., a mission profile emulator, which is solely based on power electronic converter, is proposed for simulating the static and dynamic behaviors of Permanent Magnet Synchronous Machines. The principle, design and implementation of the concept are illustrated with the simulation results and experimental validation.**

Keywords - Emulator, Mission profile, Motor drive, Permanent Magnetic Synchronous Machine.

I. INTRODUCTION

Permanent Magnet Synchronous Machines (PMSM) have been widely used in many emerging applications such as wind power generation, pump drives, locomotive and electric vehicles, etc. In these applications, the power capacity has been growing and the mission profiles are normally intensive and complicated. It is known that the reliability of power semiconductor is largely decisive to the efficiency and reliability to the whole system [1-3], and thus comprehensive test and estimation of the semiconductor devices in the converter system appears crucial. However, considering the actual complicated working condition of the motor drive system, such test and estimation is still a challenging task. For example, in a wind power system, the wind speed may change dynamically, leading to fluctuant load torque in the wind turbine [2] which is quite difficult and costly to realize by real mechanical load during the test.

The concept and several schemes of virtual load or virtual machine, have been proposed in existing researches, aiming at generating similar electrical load characteristics solely by power electronic converters. For example, [4-9] present virtual Induction Machines (IM), while [6][10-13] present schemes for Synchronous Generators (SG) and ZIP loads (a mathematical load model consisting of constant-impedance, constant-current and constant-power components, which is usually used in the analysis of power systems). Besides, the emerging Power Hardware-in-the-Loop technique (PHIL) is also applied in some cases [6][7][9]. Filters are selected variously including inductors [4-6][14], and LCL filters [7-10], etc.

However, there are limitations in these existing research. The PI control method of [4], considering the dynamic electrical behavior of a motor, will bring steady-state error in the response. [4-9] are designed for the tests of the drive and the controller, controlling the speed in an open-loop structure, and thus may not track the rotor speed well during the transient process. [6][9][10][13] are PHIL systems based on RTDS and will cost much in price, and it is not possible to measure and benchmark the actual loading of devices in these systems. [14] combines the motor drive and the PMSM together, and is not flexibly plug-and-play to a normal motor drive.

In this paper, a mission profile emulator for PMSM is consequently proposed to solve the limitations mentioned above. Inspired by [14] and the reliability test bench introduced in [15][16], the emulator is based on a three-phase voltage-source converter. The motor drive, chosen as devices under test (DUT), is connected to the emulator instead of a electric machine. The emulator can be independent from the motor drive and can be flexibly programmed to generate corresponding electrical stresses onto the power semiconductor devices of the motor drive. The electrical stresses are calculated by the mechanic load information and the mathematical model of the PMSM. Moreover, the parameters of the emulator can be flexibly modified, and the emulator can behave closely to a real PMSM, both statically and dynamically.

A typical motor drive system is first presented as a study case, of which the working condition and frequency characteristics are analyzed. Afterwards, the scheme of the mission profile emulator for PMSM is presented, and is designed to simulate both the static and dynamic behavior of PMSM. The experimental setup is shown. Finally, the performance of the mission profile emulator is illustrated with some experimental results. Some conclusions are also drawn accordingly.

II. MODELING OF THE MOTOR DRIVE SYSTEM

A typical motor drive system for PMSM is selected as a study case, including a Permanent Magnet Synchronous Machine (PMSM) driven by a 3-phase 2-level voltage-source inverter, as shown in Fig. 1. The parameters of the PMSM is shown in Table I [14][17]. Field Oriented Control (FOC) is selected as the speed control strategy [18][19], with the switching sequence of the power semiconductor devices generated by Sinusoidal Pulse Width Modulation (SPWM) technique. The mechanical working condition (rated rotor speed and load torque) of the PMSM is shown in Fig. 2. The electrical behavior (line-to-line terminal voltage and three-phase stator current response) of the motor drive under the given conditions is shown in Fig. 3.

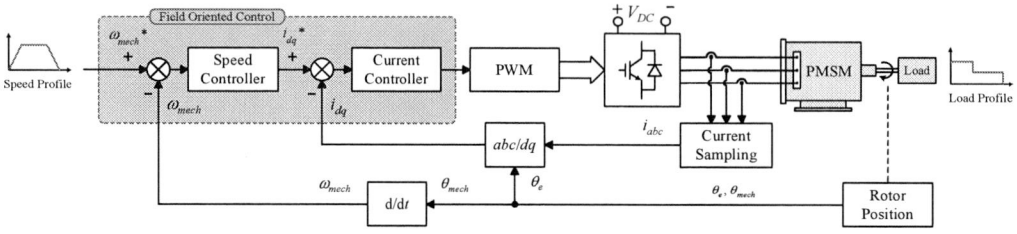

Fig. 1. Block diagram of the FOC motor drive system.

TABLE I
PARAMETERS OF THE STUDIED PMSM

Parameter	Symbol	Value	Unit
Nominal Power	P_n	9200	W
Nominal Voltage	V_n	350	V
DC Bus Voltage	V_{DC}	800	V
Stator Inductance	R_s	0.05	Ω
Stator Resistance	L_s	25	mH
PM Flux Linkage	ψ_f	0.53	Wb
Number of Pole Pairs	N_p	1	-
Nominal Speed	ω_n	4500	rpm
Rotational Inertia	J	0.008	kg·m²
Friction Coefficient	F	0	kg·m²/s

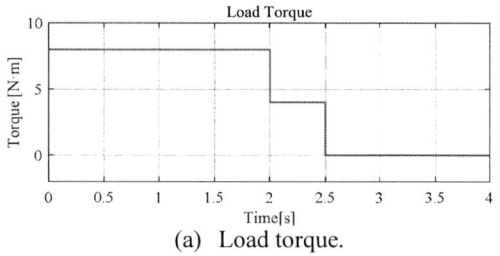

(a) Load torque.

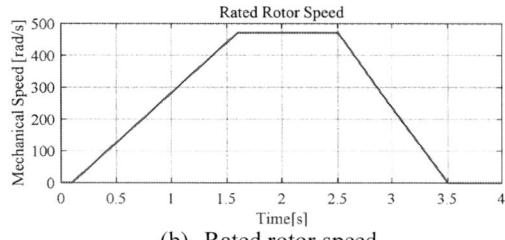

(b) Rated rotor speed.

Fig. 2. Predefined working condition of the PMSM.

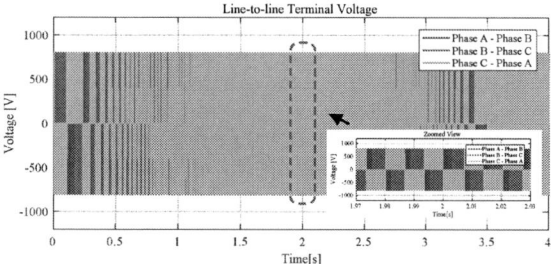

(a) Line-to-line terminal voltage of the PMSM.

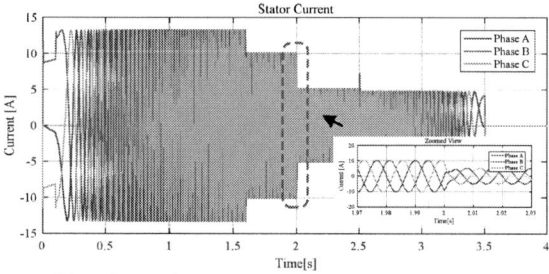

(b) Three-phase stator current of the PMSM.

Fig. 3. Simulated electrical behavior of the motor drive system under given working conditions of Table I, Fig.1 and Fig. 2.

From Fig. 2 and Fig. 3 it can be seen that the amplitude of stator current increases during the acceleration period and decreases during the deceleration period. It is pointed out in [14][20] that such change will lead to adverse thermal cycles and faster wear-out of the power devices. Therefore, the dynamical performance of the mission profile emulator is of great importance, especially when the emulator is applied to the reliability-oriented analysis of power electronics.

The control parameters of the motor drive is designed based on the q-axis current control loop of the motor drive system [17][18]. The PI parameters are designed to cancel the slowest pole and make the system an optimal second-order system, as shown in Table II [17]. The bode graphs of the current-control open loop and the speed-control open loop are shown in Fig. 4 and Fig. 5. The phase margin of the current control loop is 62.9°, and the bandwidth of the current control loop is approximately 610Hz. Thus the stability of the motor drive system can be ensured.

TABLE II
CONTROL PARAMETERS OF THE STUDIED MOTOR DRIVE

Parameter	Symbol	Value
Current loop K_p	$K_{p,i}$	100
Current loop K_i	$K_{i,i}$	200
Speed loop K_p	$K_{p,\omega}$	12.77
Speed loop K_i	$K_{i,\omega}$	7918

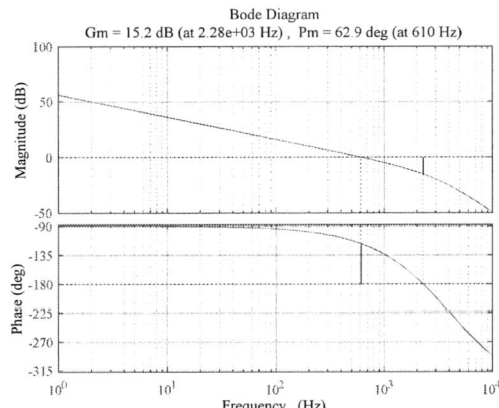

Fig. 4. Bode graph of current control loop.

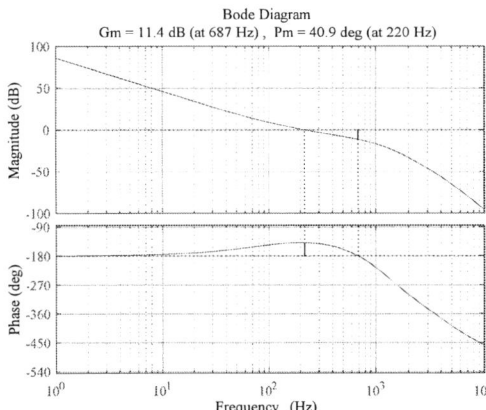

Fig. 5. Bode graph of speed control loop.

As a result, the targets of the mission profile emulator for PMSM should be set as the follows: with the same control strategy and control parameters, the generated stator current response, the electromagnetic torque and the rotational speed response should be the same as the real PMSM. Moreover, the transient response should be sufficiently close to that of a real PMSM.

III. DESIGN OF THE MISSION PROFILE EMULATOR

Inspired by [14-16], the scheme of the mission profile emulator is therefore designed and its block diagram is shown in Fig. 6. The whole emulator can be divided into three parts in different hierarchies: the PMSM model, the output current controller, and the three-phase power electronic converter.

The PMSM model is responsible for calculating the current response of a real PMSM according to the input terminal voltage. The calculation is based on the *dq* reference frame equations of the PMSM [18][19]. The dq current is calculated by the electromagnetic equation (1)(2), and the electromagnetic torque is calculated by the torque equation (3). The mechanical speed is calculated by the motion equation (4), and is then used to calculate the rotor position and the phase of the rotor flux linkage (mechanical angle and electromagnetic angle).

The 2018 International Power Electronics Conference

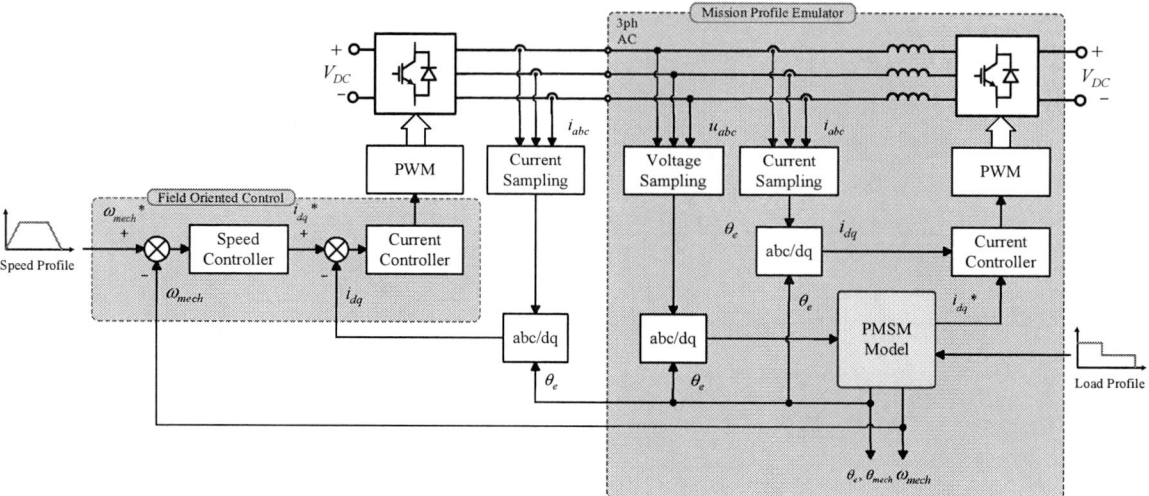

Fig. 6. Block diagram of the mission profile emulator.

$$u_d = R_s i_d + \frac{\mathrm{d}}{\mathrm{d}t} L_d i_d - \omega_e L_q i_q \qquad (1)$$

$$u_q = R_s i_q + \frac{\mathrm{d}}{\mathrm{d}t} L_q i_q + \omega_e L_d i_d + \omega_e \psi_f \qquad (2)$$

$$T_e = \frac{3}{2} n_p \left(\left(L_d - L_q \right) i_d i_q + \psi_f i_q \right) \qquad (3)$$

$$T_e - T_{load} = J \frac{\mathrm{d}\omega_{mech}}{\mathrm{d}t} + F \omega_{mech} \qquad (4)$$

In these equations, u_d and u_q respectively represent the d-axis and q-axis component of the PMSM terminal voltage in the dq frame synchronous to the flux linkage of the PMSM. i_d and i_q are respectively the d-axis and q-axis component of the PMSM stator current. L_d and L_q are the equivalent d-axis and q-axis stator inductance. R_s is the stator resistance per phase, and ψ_f is the flux linkage of the permanent magnet. n_p is the number of pole pairs, while ω_e and ω_{mech} are respectively the electromagnetic speed and mechanical speed. T_e and T_{load} are the electromagnetic torque and the load torque, with J the rotational inertia and F the coefficient related to friction.

It should be noted that to be programmed in DSP, all these equations are converted into the form of transfer functions, and are discretized by the Trapezoidal method, as described in (5) (with sampling period T_s).

$$s = \frac{2}{T_s} \cdot \frac{1 - z^{-1}}{1 + z^{-1}} \qquad (5)$$

The current controller is responsible for generating the PWM signal from the output current signal of the PMSM model, and the converter behaves electrically as a real PMSM under that PWM control signal. The current controller can be a PI controller in dq rotational frame synchronous to the PMSM, or a PR controller with dynamically-adjusted fundamental frequency. In this paper, the controller is selected as a PI controller in dq frame. Besides, it should be particularly noticed that the

output filter of the power electronic converter can be chosen and designed more flexibly, needless to be exactly equal to the stator inductance of the actual simulated PMSM, which is one of the major advantages of the proposed emulator.

IV. PERFORMANCE VALIDATIONS

A. Simulation Validations

Some validations for the proposed emulator is first conducted by simulation (Plexim PLECS Blockset embedded in MATLAB). The simulated electrical and mechanical behavior of the emulator (configuration shown in Fig. 6), are compared to those of a simulated PMSM drive system (configuration shown in Fig. 1), as shown in Fig. 7 and Fig. 8.

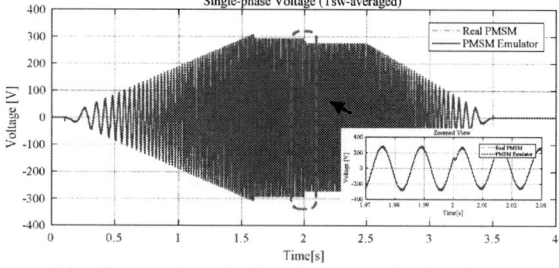

(a) Comparison in the phase terminal voltage. (Periodically averaged by switch period T_{sw}).

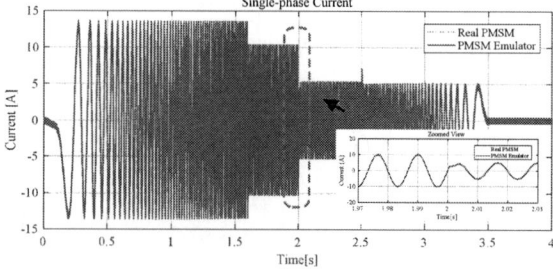

3880

(b) Comparison in the phase stator current.

Fig. 7. Comparison in electrical behaviors of PMSM and the emulator by PLECS simulation.

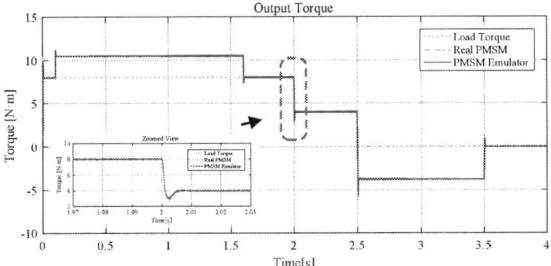

(a) Comparison in the electromagnetic torque.

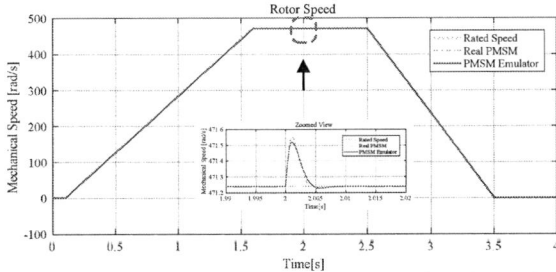

(b) Comparison in the mechanical rotational speed.

Fig. 8. Comparison in mechanical behaviors of PMSM and the emulator by PLECS simulation.

It can be seen that in these two systems, the AC phase voltage, the AC phase current, the electromagnetic torque and the rotational speed of the rotor are all well corresponding to each other. Given a load torque step at $t=2$s, the behavior of the emulator can still track the PMSM well. These simulation results can thus illustrate the validity of the emulator in principle and theory.

B. Experimental Validations

The experimental setup is shown in Fig. 9, which is based on a general mission profile test bench. IGBTs are selected as the power semiconductor devices, and the configuration mentioned before is implemented in one TI Digital Signal Processor (DSP) TMS320F28335. Another DSP is used as the controller of motor drive, controlling the devices under test (DUT). The filters can be altered as mentioned before, and in the given test bench the LCL filter is used.

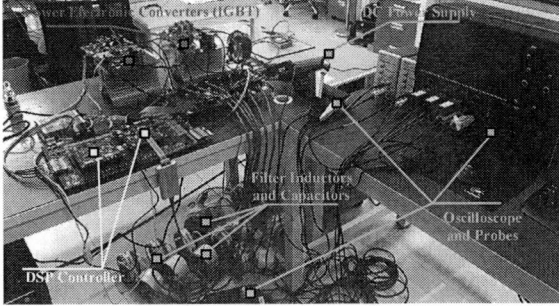

Fig. 9. The photo of experimental setup.

Some experiments related to the mission profile emulator are conducted on the given experimental setup. To test the static and dynamic performance of the programmed PMSM model programmed in DSP, two different testing conditions are designed and shown in Table III. The two testing conditions are designed based on normal working conditions of PMSM, where the voltage consists of q-axis component only. In Testing Condition 1 a step signal is given to the q-axis voltage component, while in Testing Condition 2 a step signal is given to the load torque. In these two testing conditions, the PMSM model is discretized as mentioned before, and the values of key variables are sent to the host computer by the serial ports of the DSP. For Testing Condition 1, the current signal is shown in Fig. 10, and the torque signal and the rotational speed signal is shown in Fig. 11, compared to the response of a real PMSM. For Testing Condition 2, the rotational speed signal is shown in Fig. 12. From the waveforms it can be seen that the discretized PMSM model in DSP can track the behaviors of PMSM well in both electrical and mechanical responses, which ensure the static and dynamic performance of the whole emulator basically.

TABLE III
INPUT PARAMETERS FOR PMSM MODEL TESTING

TESTING CONDITION 1		
Parameters	**Value**	**Unit**
Load Torque	0.0001	N·m
d-axis Voltage	0	V
q-axis Voltage (**before** t=5s)	10	V
q-axis Voltage (**after** t=5s)	20	V
TESTING CONDITION 2		
Parameters	**Value**	**Unit**
Load Torque (**before** t=10s)	0.5	N·m
Load Torque (**after** t=10s)	2	N·m
d-axis Voltage	0	V
q-axis Voltage	10	V

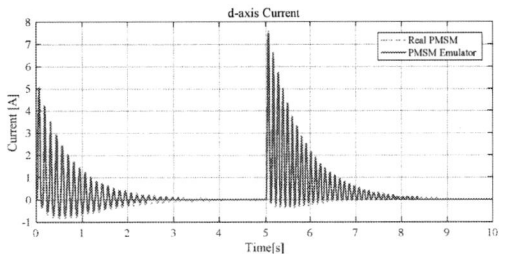

(a) Comparison in d-axis current.

3881

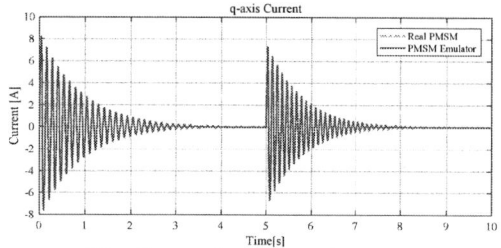

(b) Comparison in *q*-axis current.

Fig. 10. Comparison in the current signal of the PMSM model in DSP and a real PMSM under Testing Condition 1.

(u_d equals zero, u_q changes from 10V to 20V at t=5s)

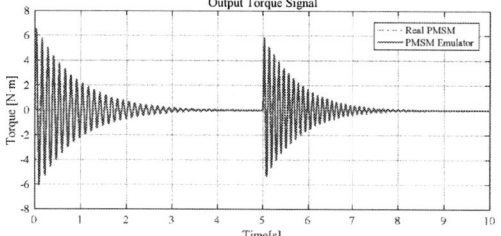

(a) Comparison in the electromagnetic torque.

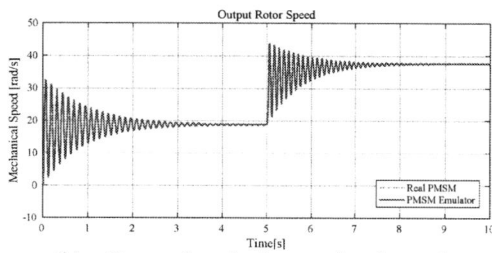

(b) Comparison in the rotational speed.

Fig. 11. Comparison in the mechanical signal of the PMSM model in DSP and a real PMSM under Testing Condition 1.

(T_{load} equals 0.0001N·m, u_d equals zero, u_q changes from 10V to 20V at t=5s)

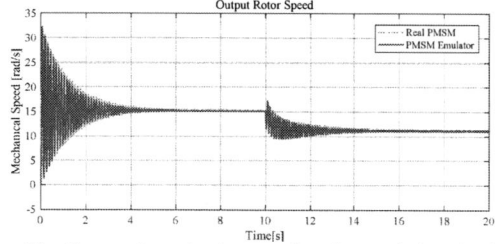

Fig. 12. Comparison in the rotational speed signal of the PMSM model in DSP and a real PMSM. under Testing Condition 2.

(T_{load} changes from 0.5N·m to 2N·m at t=10s, u_d equals zero, u_q equals 10V)

However, due to the limitation of time, only the test of DSP programming is finished. More experiments are presently being conducted and will be conducted in the future, involving more convincing results of the proposed

scheme and solutions to the problems encountered during the experiments.

V. CONCLUSIONS

In this paper, the scheme of a mission profile emulator for Permanent Magnetic Synchronous Machines has been proposed. A typical motor drive system for PMSM is selected as the study case, and its characteristics in frequency domain are presented. The technique of the emulator is derived based on the mathematical model of PMSM, and the design principles are summarized from the motor drive system. By PLECS simulation, the static and dynamic behavior of the emulator corresponds well to a real PMSM, and the principle and theory of the emulator is thus verified. Some basic experimental results are also presented, yet more experiments are presently being conducted, to further validate the proposed mission profile emulator convincingly.

REFERENCES

[1] F. Blaabjerg and K. Ma, "Future on Power Electronics for Wind Turbine Systems," *IEEE Journal of Emerging and Selected Topics in Power Electronics*, vol. 1, no. 3, pp. 139-152, Sept. 2013.

[2] K. Ma, M. Liserre, F. Blaabjerg and T. Kerekes, "Thermal Loading and Lifetime Estimation for Power Device Considering Mission Profiles in Wind Power Converter," *IEEE Transactions on Power Electronics*, vol. 30, no. 2, pp. 590-602, Feb. 2015.

[3] M. Musallam, C. Yin, C. Bailey and M. Johnson, "Mission Profile-Based Reliability Design and Real-Time Life Consumption Estimation in Power Electronics," *IEEE Transactions on Power Electronics*, vol. 30, no. 5, pp. 2601-2613, May 2015.

[4] H. J. Slater, D. J. Atkinson and A. G. Jack, "Real-time Emulation for Power Equipment Development. II. The Virtual Machine," in *IEE Proceedings - Electric Power Applications*, vol. 145, no. 3, pp. 153-158, May 1998.

[5] M. A. Masadeh and P. Pillay, "Power Electronic Converter-based Three-phase Induction Motor Emulator," in *Proceeding of IEEE International Conference on Power Electronics, Drives and Energy Systems (PEDES) 2016*, pp. 1-5, 2016.

[6] J. Wang, L. Yang, Y. Ma, J. Wang, L. M. Tolbert, F. Wang, K. Tomsovic, "Static and Dynamic Power System Load Emulation in a Converter-Based Reconfigurable Power Grid Emulator," *IEEE Transactions on Power Electronics*, vol. 31, no. 4, pp. 3239-3251, April 2016.

[7] R. Bojoi, E. Armando, S. G. Rosu, S. Vaschetto and P. Soccio, "Virtual Load with Common Mode Active Filtering for Power Hardware-in-the-loop Testing of Power Electronic Converters," in *Proceeding of IECON 2014*, pp. 1875-1881, 2014.

[8] Y. Srinivasa Rao and M. C. Chandorkar, "Real-Time Electrical Load Emulator Using Optimal Feedback Control Technique," *IEEE Transactions on*

Industrial Electronics, vol. 57, no. 4, pp. 1217-1225, April 2010.

[9] O. Vodyakho, M. Steurer, C. S. Edrington and F. Fleming, "An Induction Machine Emulator for High-Power Applications Utilizing Advanced Simulation Tools with Graphical User Interfaces," *IEEE Transactions on Energy Conversion*, vol. 27, no. 1, pp. 160-172, March 2012.

[10] L. Yang, J. Wang, Y. Ma; J. Wang, X. Zhang; L. M. Tolbert, F. F. Wang, K. Tomsovic, "Three-Phase Power Converter-Based Real-Time Synchronous Generator Emulation," *IEEE Transactions on Power Electronics*, vol. 32, no. 2, pp. 1651-1665, Feb. 2017.

[11] J. Wang, Y. Ma, L. Yang, L. M. Tolbert and F. Wang, "Power Converter-based Three-phase Induction Motor Load Emulator," in *Proceeding of IEEE Applied Power Electronics Conference and Exposition (APEC) 2013*, pp. 3270-3274, 2013.

[12] R. S. Kaarthik and P. Pillay, "Emulation of a Permanent Magnet Synchronous Generator in Real-time Using Power Hardware-in-the-loop," in *Proceeding of IEEE International Conference on Power Electronics, Drives and Energy Systems (PEDES) 2016*, pp.1-6, 2016.

[13] J. Wang, L. Yang, Y. Ma, J. Wang, L. M. Tolbert, F. Wang, K. Tomsovic, "Static and Dynamic Power System Load Emulation in Converter-based Reconfigurable Power Grid Emulator," in *Proceeding of IEEE Energy Conversion Congress and Exposition (ECCE) 2014*, pp.4008-4015, 2014.

[14] I. Vernica, F. Blaabjerg and K. Ma, "Mission Profile Emulator for the Power Electronics Systems of Motor Drive Applications," in *Proceeding of European Conference on Power Electronics and Applications (EPE'17 ECCE Europe) 2017*, pp. P.1-P.10, 2017.

[15] U. M. Choi, S. Jørgensen and F. Blaabjerg, "Advanced Accelerated Power Cycling Test for Reliability Investigation of Power Device Modules," *IEEE Transactions on Power Electronics*, vol. 31, no. 12, pp. 8371-8386, Dec. 2016.

[16] U. M. Choi, I. Trintis, F. Blaabjerg, S. Jørgensen and M. L. Svarre, "Advanced Power Cycling Test for Power Module with On-line On-state VCE Measurement," in *Proceeding of IEEE Applied Power Electronics Conference and Exposition (APEC) 2015*, pp. 2919-2924, 2015.

[17] I. Vernica. *Modelling and Implementation of Active Thermal Control Methods for Power Electronics Systems for Motor Drive Applications*, Aalborg University Press, 2016.

[18] C. Capitan. *Torque Control in Field Weakening Mode*, Aalborg University Press, 2009.

[19] P. C. Krause, O. Wasynczuk, and S. D. Sudhoff, *Analysis of Electric Machinery and Drive Systems*, John Wiley & Sons, 2002.

[20] D. A. Murdock, J. E. R. Torres, J. J. Connors and R. D. Lorenz, "Active thermal control of power electronic modules," *IEEE Transactions on Industry Applications*, vol. 42, no. 2, pp. 552-558, March-April 2006.

The 2018 International Power Electronics Conference

A Variable DC Bus Voltage Based Power Hardware-in-the-Loop Emulation of Electric Motors with Wide Variation in Interface Filter Inductance

Tsai-Fu Wu, Mitradatta Misra, Ying-Yi Jhang, Chang-Jun Yang and Yin-Chi Xu
Elegant Power Electronic Application Research Laboratory
Department of Electrical Engineering
National Tsing Hua University
Hsinchu, Taiwan
email: tfwu@ee.nthu.edu.tw

Abstract-This paper presents a new scalable power hardware-in-the-loop (PHIL) architecture for emulation of electric motors. It incorporates an additional bidirectional dc-dc converter interface between the emulation and regenerative inverters. The dc-bus for emulation inverter can then be regulated optimally, thus improving its efficiency and reducing harmonic distortion at low speed and light load operation. The counter voltage reference calculation that takes into account wide variation in interface filter inductance due to soft magnetic characteristics of inductor core is also proposed in this paper. A wide range of machine models in abc-domain can then be emulated with just one PHIL test bench. A 2 kW permanent magnet synchronous motor (PMSM) is emulated in this paper and compared with a real motor, to verify feasibility of the proposed method.

I. INTRODUCTION

Power hardware-in-the-loop (PHIL) test bench in motor-drive development is increasingly becoming significant. The PHIL test bench needs to ensure that the current drawn from motor drive matches with that when connected to real machine under diverse operating conditions [1]. It also needs to ensure that power drawn by the motor drive from ac grid is recycled back to reduce power loss. This is typically achieved by connecting two inverters back to back with a common dc link, one for machine emulation and one for power regeneration [1]-[2]. A main drawback of this approach would be that the emulation inverter is always connected to a very high dc-link voltage, even when the required output voltage is very low, especially while emulating low speed and light load

operation. This reduces the emulation inverter efficiency [3] and increases harmonic distortion in the motor drive output current. Hence, a bidirectional dc-dc converter interface between the emulation and regenerative inverter is proposed in this research. The necessary constraints for determining the optimal dc-bus voltage for emulation inverter is also laid down.

Another important aspect of PHIL systems is the machine model. The machine model could be simple back-emf expressions for motors [4] or more complex multidimensional lookup tables derived from finite element analysis [5]. The emulation inverter basically controls the terminal voltage at the motor drive output by determining the counter voltage reference from the mathematical model. But, interface filter, which is usually chosen to be simple L-filter due to its robustness [6], may not necessarily match the actual motor's stator inductor. Moreover, filter inductance can vary within each line cycle with inductor cores having soft magnetic characteristics. So, the counter voltage reference needs to take the voltage drop across the time varying filter inductors into account. This is not possible in conventional dq-domain, where the basic assumption is that filter inductors are linear and identical [7]. Hence, an abc-domain based machine model [5] is recommended in this paper and the required counter voltage reference for each phase is directly derived.

The proposed PHIL test bench is tested with a 2 kW motor drive for permanent magnet synchronous motor (PMSM). Experimental results have shown to confirm the feasibility of the proposed architecture.

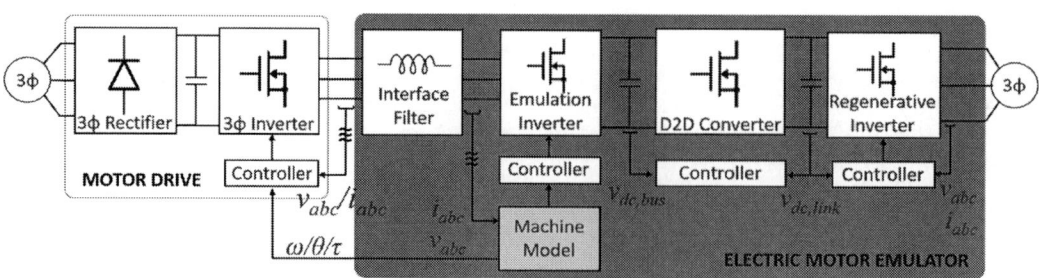

Fig. 1. Proposed power hardware-in-loop architecture.

II. PROPOSED PHIL ARCHITECTURE

The proposed PHIL architecture is shown in Fig. 1. It is very similar to that observed in conventional PHIL test benches [1], except for a bidirectional DC-DC converter interface between the three-phase emulation and regenerative inverters. The three-phase emulation inverter has a three-phase three-wire topology and is connected to the motor drive through simple line-inductors. It controls the counter voltage at the line-inductor's terminal, based on the required machine model. The bidirectional DC-DC converter has a split dc-capacitor-based multi-phase interleaving buck (boost) converter topology, as shown in Fig. 2. The converters are operated at the CCM/DCM boundary with optimal dead-time, so as to have zero voltage switching (ZVS) turn-on and zero voltage resonance transition (ZVRT) turn-off. It provides an optimal DC Bus voltage for the emulation inverter, so as to improve its efficiency. The three-phase regenerative inverter has a three-phase four-wire topology and is connected to the ac-grid through LCL filters. It regulates the dc-link voltage with capacitor-charge balancing and recycles the power from motor drive.

The advantage of the proposed architecture is that power level is scalable by simple addition of parallel branches per phase. Higher switching frequencies for low harmonic distortion and faster dynamics is also achievable by operating the parallel branches in interleaving fashion.

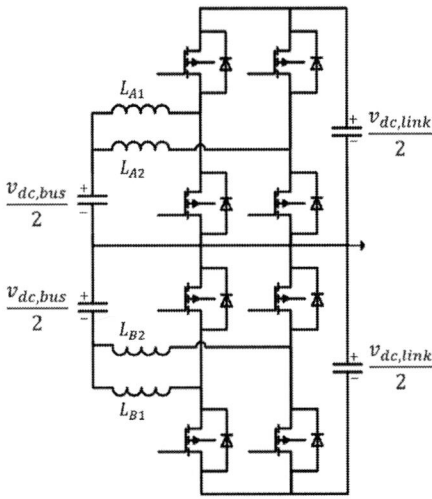

Fig. 2. Split DC-Capacitor-based multi-phase interleaving bidirectional DC-DC converter.

III. EMULATION INVERTER COUNTER VOLTAGE ESTIMATION IN ABC-DOMAIN

A machine model is at the core of an PHIL system for accurate emulation under diverse operating conditions. It needs to derive the counter voltage reference for emulation inverter from the motor-drive output current, such that current through the interfacing line inductors has an identical profile with that of a real machine [7]. It is also required that the voltage drop across the interfacing line inductors is accounted for in the counter voltage reference.

This would enable emulation of wide range of motors regardless of the interfacing line inductance. However, the machine model is often expressed in dq-domain with the assumption that filter inductors are linear and identical [7]. This leads to inaccurate counter voltage reference with filter inductors having soft-magnetic powder cores, which exhibit current dependent nonlinear characteristics [8]. A typical variation in inductance for three-phase line inductors with same nominal value over a quarter line cycle is shown in Fig. 3 for reference. This clearly highlights that the assumption for linear and identical filter inductors does not hold true with soft-magnetic powder cores. Hence, a machine model in abc-domain is used in this research, allowing the time-varying characteristics of filter inductance to be taken into account.

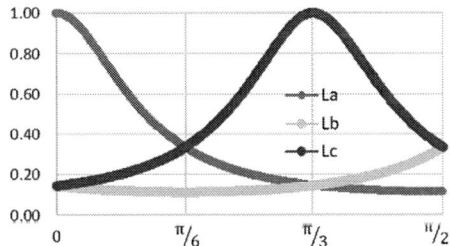

Fig. 3. Inductance variation over one quarter line-cycle of three-phase soft-magnetic material core based inductors having same nominal value.

The equivalent circuit of a PMSM and the corresponding equivalent circuit of an emulation inverter is shown in Fig. 4.

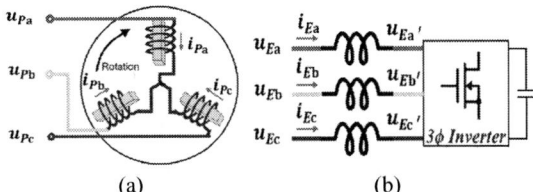

(a)   (b)

Fig. 4. Equivalent circuit for (a) PMSM and (b) PHIL-emulation inverter

Applying KVL to a single phase of PMSM:

$$u_{Pj} = R_{Pj}i_{Pj} + \frac{d\lambda_{Pj}}{dt}, \tag{1}$$

where $j \in \{a, b, c\}$ represents the corresponding phase, u_{Pj} is the terminal voltage of the motor-drive inverter, i_{Pj} is the motor-drive output current, R_{Pj} is the stator resistance and λ_{Pj} is the stator flux linkage. λ_{Pj} is a function of three stator currents, i_{Pa}, i_{Pb} and i_{Pc}, and the rotor angle θ. The instantaneous values are typically derived from N-dimensional lookup tables [5], to save computational time. Similarly, applying KVL to a single phase of the PHIL-emulation inverter:

$$u_{Ej} = R_{Ej}i_{Ej} + L_{Ej}\frac{di_{Ej}}{dt} + u'_{Ej}, \tag{2}$$

where u_{Ej} is the motor-drive inverter terminal voltage when connected to PHIL test bench, i_{Ej} is the corresponding motor-drive output current, u'_{Ej} is the terminal voltage of the emulation inverter, R_{Ej} is the internal resistance of the interfacing inductor and L_{Ej} is the interfacing inductance. L_{Ej} is a function of the corresponding inductor current i_{Ej} and typically derived from anhysteretic curve fit polynomial expressions available in manufacturer datasheets.

Now for accurate emulation, the following condition must hold true at any instant:

$$u_{Pj} = u_{Ej} \text{ and } i_{Pj} = i_{Ej} \qquad (3)$$

Thus, from equation (1), (2) and (3), the expression for emulation inverter terminal voltage reference is derived as

$$u'_{Ej} = \left(R_{Pj} - R_{Ej}\right)i_{Ej} + \left(\frac{d\lambda_{Pj}}{dt} - L_{Ej}\frac{di_{Ej}}{dt}\right). \qquad (4)$$

The emulation inverter terminal voltage reference can also be approximated from simple back-emf expression [4], while emulating small PMSM.

IV. ROTOR ANGLE ESTIMATION

The rotor angle in a real motor is in response to the three-phase motor-drive output current i_{abc}. The electrical rotor angle is in fact equivalent to phase angle of the space vector corresponding to i_{abc}. Hence, determination of the phase angle corresponding to the space vector is important for accurate counter-voltage estimation. Now, motor can rotate in both clockwise and counterclockwise directions depending on the phase sequence of the three phase currents. Hence, the phase sequence must also be taken into consideration while determining the space vector using abc~αβ transform as shown below:

$$\begin{bmatrix} i_\alpha \\ i_\beta \end{bmatrix} = \frac{2}{3}\begin{bmatrix} 1 & cos\left(k\frac{2\pi}{3}\right) & cos\left(k\frac{4\pi}{3}\right) \\ 0 & sin\left(k\frac{2\pi}{3}\right) & sin\left(k\frac{4\pi}{3}\right) \end{bmatrix}\begin{bmatrix} i_a \\ i_b \\ i_c \end{bmatrix}, \qquad (5)$$

where $k = +1$ for positive phase sequence currents and $k = -1$ for negative phase sequence currents. Then the rotor angle is calculated according to the relationship shown in Fig. 5.

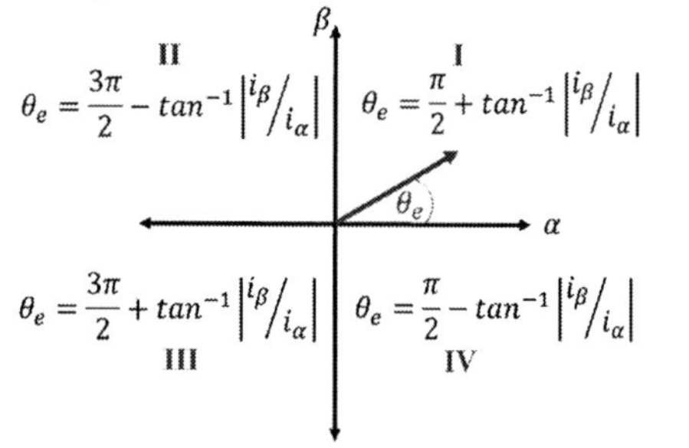

$$\text{II} \quad \theta_e = \frac{3\pi}{2} - tan^{-1}\left|i_\beta/i_\alpha\right| \qquad \text{I} \quad \theta_e = \frac{\pi}{2} + tan^{-1}\left|i_\beta/i_\alpha\right|$$

$$\theta_e = \frac{3\pi}{2} + tan^{-1}\left|i_\beta/i_\alpha\right| \quad \theta_e = \frac{\pi}{2} - tan^{-1}\left|i_\beta/i_\alpha\right|$$

$$\text{III} \qquad \text{IV}$$

Fig. 5. Rotor angle estimation from three-phase current using space vector.

V. EMULATION INVERTER DC-BUS VOLTAGE ESTIMATION

If the emulation inverter is directly connected to the regenerative inverter through a common dc-link capacitor, the modulation index for the emulation inverter would be very low during emulation of low-speed and light load operation. This would affect the emulation inverter efficiency [3] and also induce switching frequency

harmonics in the motor drive current output. Hence, the dc-bus voltage for emulation inverter needs to be regulated at a value much lower than the dc-link voltage. An optimal dc-bus voltage is estimated by taking the following points into consideration:

a) The modulation index should be greater than 0.5 and less than 1.0, i.e. $0.5 < m_a < 1.0$ for all levels of counter-voltage reference. This will ensure that inverter is operated at best possible efficiency.

b) The maximum value of peak-phase voltage that can be obtained from a three-phase voltage source inverter depends on the pulse-width modulation (PWM) method. It is equal to $0.5v_{dc,bus}$ for sinusoidal PWM and $0.577v_{dc,bus}$ for space-vector PWM.

c) The dc-bus voltage must be set to its maximum value while emulating rated-power operation even at lower than and beyond rated speed. This is to ensure that the bidirectional dc-dc converter is operated within its designed power level. The power *vs.* speed characteristic specific to the motor being emulated serves as guideline.

A straightforward implementation of dc-bus estimation algorithm is shown in the flowchart below:

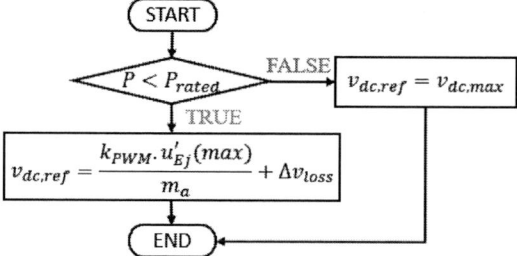

Fig. 6. Flowchart representation of dc-bus estimation.

In the flowchart shown in Fig. 6, the dc-bus voltage reference $v_{dc,ref}$ when emulation power level P is less than the rated power P_{rated}, is given by the expression (6) shown below:

$$v_{dc,ref} = \frac{k_{PWM}.u'_{Ej}(max)}{m_a} + \Delta v_{loss}, \qquad (6)$$

where $k_{PWM} = 2$ for SPWM and $\sqrt{3}$ for SVPWM, m_a is the modulation index, typically chosen to be between 0.8 and 1.0, $u'_{Ej}(max)$ is the maximum value of the counter voltage reference at a particular frequency (speed) of emulation, Δv_{loss} is the extra buffer voltage to compensate for losses. A typical value 5~15 V buffer for compensating the losses has been observed to be good enough. And $v_{dc,max}$ is the maximum dc-bus voltage that limits the current in dc-dc converter at its maximum during rated power operation, given by (7).

$$v_{dc,max} = \frac{P_{rated}}{I_{dc,max}}, \qquad (7)$$

where $I_{dc,max}$ is the maximum current capacity of the dc-dc converter.

However, the dc-bus voltage should not vary a lot during operation to ensure stable counter-voltage generation by the emulation inverter. Hence, a dc-bus voltage scheduling mechanism needs to be implemented.

3886

A very simple implementation of scheduling mechanism involves dividing the entire range of dc-bus voltage into uniform smaller ranges. The optimal dc-bus voltage estimated using the flowchart shown in Fig. 6 is mapped to the appropriate voltage range and the corresponding upper limit is set to be the dc-bus voltage reference.

VI. EXPERIMENTAL RESULTS

A 2 kW PMSM with specifications as shown in Table I, is used to validate the proposed PHIL system. The motor drive is connected to the PHIL system as well as the motor under different operating conditions for comparison.

Table I. PMSM Specifications

Parameter	Symbol	Value
Drive Input Voltage	v_{mdr}	220 V (rms)
Rated Power	P_{rated}	2 kW
Rated Torque	T_{rated}	9.55 Nm
Rated Current	I_{rated}	11 A (rms)
Rated Speed	ω_{rated}	2000 rpm
Number of poles	p	10
Torque constant	K_t	0.87 Nm/Arms
Back-emf constant	K_e	57.8 Vrms/krpm

(a) Importance of Optimal DC-Bus Voltage

The motor drive is set to output a current of 2.9 A (rms) at 10 Hz. The corresponding optimal dc-bus voltage is estimated to be 12.6 V. The PHIL system is first run with dc-bus voltage regulated at 330 V and then with dc-bus voltage regulated at 40 V. The corresponding current waveforms for phase "a" are shown in Fig. 7. It can be clearly seen that the motor drive output current has higher harmonic distortion when operated at a much higher dc-bus voltage. The switching frequency noise is also observed to be much higher when operated with a higher dc-bus voltage. This highlights the importance of regulating the dc-bus voltage optimally.

(b) PHIL System Response to Load Step-up and Step-down at fixed frequency

The motor drive is set to output a current of frequency 120 Hz and the peak current reference follows a step-up sequence of 5 A to 10 A to 14.78 A (rated) and then a step-down sequence in the reverse order, *i.e.* 14.78 A (rated) to 10 A to 5 A. The optimal dc-bus voltage under this operating condition is estimated to be approximately 151 V. Hence, the dc-bus voltage is regulated at 155 V. The current waveform for phase "a", i_{Ea} and the current output to the dc-bus, $i_{dc,bus}$ are shown in Fig. 8. It can be observed that the PHIL system operates smoothly during load transients. The power input to the emulation inverter from the motor drive and the corresponding power output to the dc-bus voltage is also calculated and verified to match accurately.

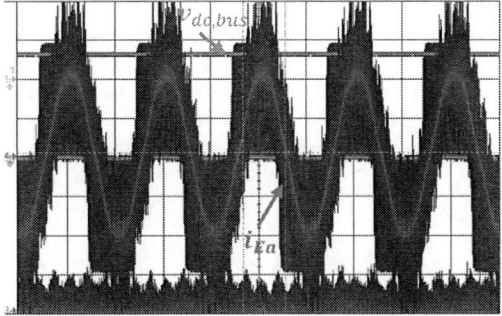

(i_{Ea}: 2 A/div(PINK); *$v_{dc,bus}$: 200 V/div*(ORANGE); *time:50 ms/div)*

(a)

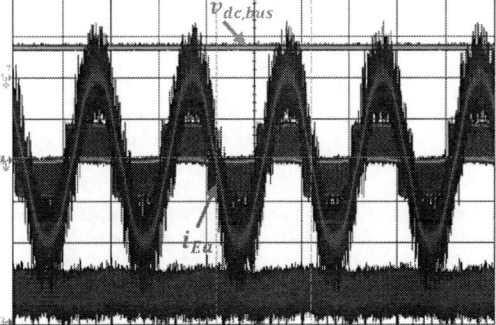

(i_{Ea}: 2 A/div(PINK); *$v_{dc,bus}$: 50 V/div*(ORANGE); *time:50 ms/div)*

(b)

Fig. 7. Motor drive output current in phase "a" i_{Ea} with emulation inverter dc-bus voltage $v_{dc,bus}$ regulated at (a) 330 V and (b) 40 V.

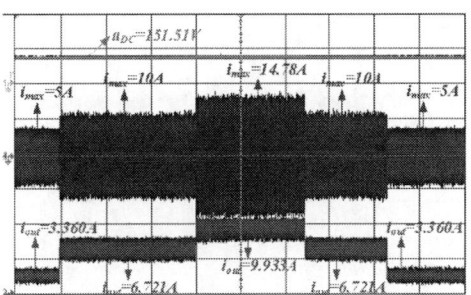

(i_{Ea}: 10 A/div(PINK); *$v_{dc,bus}$: 200 V/div*(ORANGE); *$i_{dc,bus}$: 5 A/div*(BLUE);*time:2.2 s/div)*

Fig. 8. Motor drive output current in phase "a", i_{Ea} and current output to the dc-bus, $i_{dc,bus}$ with emulation inverter dc-bus voltage $v_{dc,bus}$ regulated at 155 V, during load step-up and step-down.

(c) PHIL System Response to Frequency Step-up at fixed load

The motor drive is set to output a constant current and the frequency reference is changed from 5 Hz to 10 Hz. The PHIL system is supposed to follow the change in frequency. The three-phase current feedback and the rotor electrical angle θ, estimated by the PHIL system is shown in Fig. 9. It can be observed that the estimated rotor electrical angle θ follows the change in current frequency without any distortion. This verifies that the counter voltage generated by the emulation inverter is always in phase with the motor drive output current.

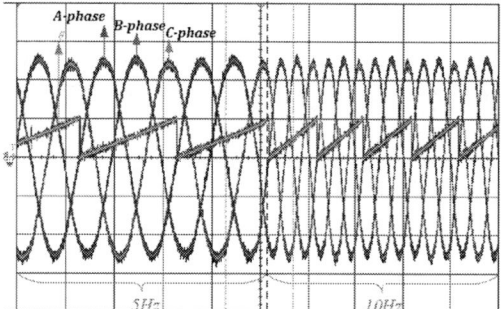

Fig. 9. Three-phase motor drive output current i_{Ea} (PINK), i_{Eb} (BLUE), i_{Ec} (GREEN) and estimated rotor electrical angle θ (ORANGE)

(d) Rotor-Angle Estimation with Positive and Negative Phase Sequence Motor-Drive Current Output

The motor drive is set to output a peak current of 5 A at 2 Hz in both positive phase sequence and negative phase sequence separately. The three-phase motor-drive current output and the estimated rotor-angle in both the cases is shown in Fig. 10. The result is as expected.

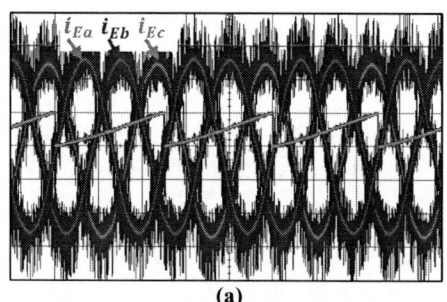

(a)

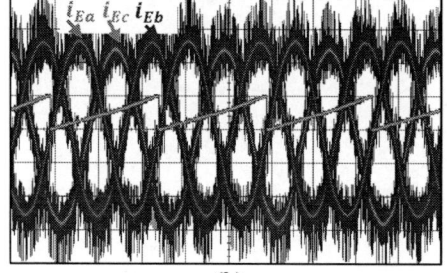

(b)

(i_{Ea}: 2 A/div(PINK); i_{Eb}: 2 A/div(BLUE); i_{Ec}: 2 A/div(GREEN); $θ_e$: 2π/div(ORANGE); time:200 ms/div)

Fig. 10. Estimated rotor angle with three-phase motor drive output currents in (a) positive and (b) negative phase sequences.

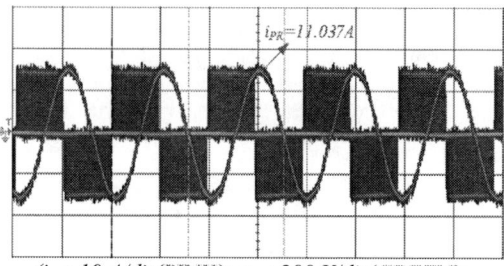

(i_{Ea}: 10 A/div(PINK); v_{ab}: 200 V/div(GREEN); time:12.8 ms/div)

(a)

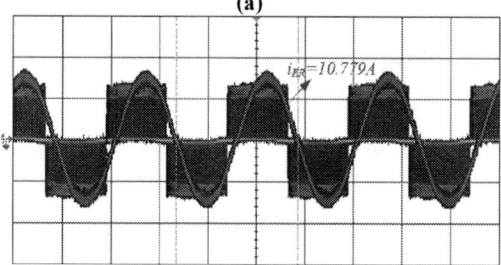

(i_{Ea}: 10 A/div(PINK); v_{ab}: 50 V/div(GREEN); time:10.0 ms/div)

(b)

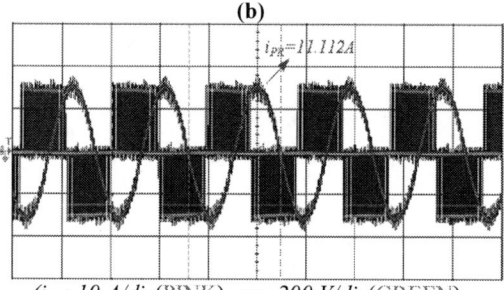

(i_{Ea}: 10 A/div(PINK); v_{ab}: 200 V/div(GREEN); time:3.3 ms/div)

(c)

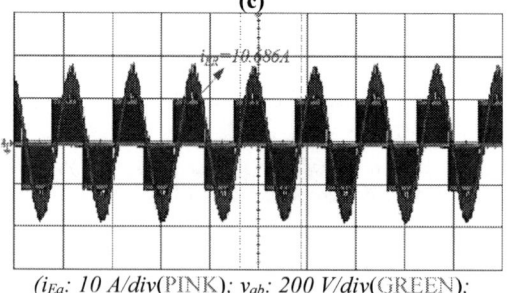

(i_{Ea}: 10 A/div(PINK); v_{ab}: 200 V/div(GREEN); time:5.0 ms/div)

(d)

Fig. 11. Motor drive output current for phase "a" i_{Ea}, for 500 W when connected to (a) PMSM and (b) PHIL, and for 2 kW when connected to (c) PMSM and (d) PHIL.

(e) Current Profile Comparison between PHIL test bench and real motor at different power levels

The motor drive is operated by connecting the real motor and the PHIL test bench in tandem for the same operating conditions and the motor drive output current profile is compared. The comparison waveforms are shown in Fig. 11. It can be observed that the current profile matches closely, but has higher switching harmonics when connected to PHIL test bench. This is reasonable because the motor drive is now connected to an active device rather than a passive device. A point to note here is that the line to line voltage v_{ab} measured when connected to PMSM is the motor drive output voltage, while that measured when connected to PHIL is the emulator inverter output voltage. Hence, a difference in voltage level and phase difference between the voltage and current is observed.

(f) Efficiency Comparison of PHIL test bench with Optimal DC-Bus and Constant DC-Bus

The motor drive is set to output rated peak current of 14.78 A with frequency gradually increasing from 10 Hz to 160 Hz in steps of 5 Hz. It is first operated with optimal dc-bus scheduling and then with a fixed dc-bus voltage of 200V. The power input to the emulation inverter from the motor drive and the corresponding power output to the dc-bus voltage is calculated under both conditions for comparison of efficiency. The efficiency comparison chart is shown in Fig. 12. It can be observed that efficiency of the emulation inverter significantly improves during low frequency (speed) emulation with optimal dc-bus. However, at higher frequencies, not much significant difference in efficiency is observed. Hence, optimal dc-bus voltage is necessary for efficient low speed emulation.

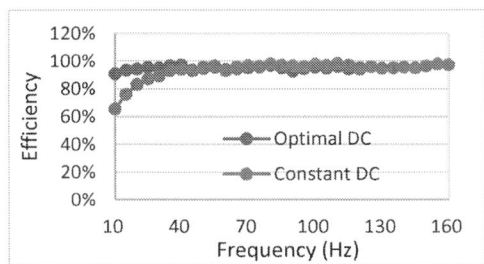

Fig. 12. Efficiency Comparison chart during operation from 10 Hz to 160 Hz at rated current with optimal dc-bus scheduling and fixed dc-bus for the emulation inverter.

VII. CONCLUSIONS AND FUTURE WORK

A new scalable architecture for PHIL test bench was proposed and the counter voltage reference taking into account wide variation in interface inductance was also derived. The PHIL test bench was tested with a 2 kW PMSM to confirm feasibility of the proposed method. The importance of variable dc-bus voltage for emulation inverter has been experimentally verified. The PHIL test bench's response to current and frequency step change has also been verified. The results have been shown to match closely with that of a real motor, confirming the accuracy of the implementation. The efficiency of PHIL test bench has also been shown to improve at low frequency (speed) emulation with optimal dc-bus voltage.

A more accurate optimal dc-bus estimation and scheduling mechanism could be explored in the future.

REFERENCES

[1] H. J. Slater, D. J. Atkinson, and A. G. Jack, "Real-time emulation for power equipment development. ii. the virtual machine," *IEE Proceedings - Electric Power Applications*, vol. 145, no. 3, pp. 153–158, May 1998.

[2] S. Grubic, B. Amlang, W. Schumacher, and A. Wenzel, "A high-performance electronic hardware-in-the-loop drive-load simulation using a linear inverter (linverter)," *IEEE Transactions on Industrial Electronics*, vol. 57, no. 4, pp. 1208–1216, April 2010.

[3] C. Y. Yu, J. Tamura, and R. D. Lorenz, "Optimum dc bus voltage analysis and calculation method for inverters/motors with variable dc bus voltage," *IEEE Transactions on Industry Applications*, vol. 49, no. 6, pp. 2619–2627, Nov 2013.

[4] P. Krause, O. Wasynczuk, S. Sudhoff, and S. Pekarek, *Analysis of electric machinery and drive systems*, 3rd ed. John Wiley & Sons., 2013.

[5] L. Quéval and H. Ohsaki, "Nonlinear abc-model for electrical machines using n-d lookup tables," *IEEE Transactions on Energy Conversion*, vol. 30, no. 1, pp. 316–322, March 2015.

[6] S. Lentijo, S. D'Arco, and A. Monti, "Comparing the dynamic performances of power hardware-in-the-loop interfaces," *IEEE Transactions on Industrial Electronics*, vol. 57, no. 4, pp. 1195–1207, April 2010.

[7] A. Schmitt, J. Richter, U. Jurkewitz, and M. Braun, "FPGA-based real-time simulation of nonlinear permanent magnet synchronous machines for power hardware-in-the-loop emulation systems," in *IECON 2014 - 40th Annual Conference of the IEEE Industrial Electronics Society*, Oct 2014, pp. 3763–3769.

[8] M. A. Swihart, "Inductor cores–material and shape choices," *Magnetics-www. mag-inc. com*, 2004.

The 2018 International Power Electronics Conference

Copper Loss Minimization Control at Zero Output Voltage for Electrolytic Capacitor-Less Inverter

Kodai Abe, Haruya Kada, Kiyoshi Ohishi, Hitoshi Haga, Yuki Yokokura
Nagaoka University of Technology

Abstract—**To reduce the cost of AC–DC–AC systems, a single-phase-to-three-phase electrolytic capacitor-less inverter is proposed. Because the inverter does not have an energy storage system, such as an electrolytic capacitor and a reactor, it controls both the motor and source currents. In particular, to meet the guidelines for a harmonic source current, it is necessary to control the source current under sinusoidal waveform conditions. However, the DC-link voltage then becomes an absolute value of the source voltage, and the output voltage is almost zero near the zero-crossing point. In the zero-output-voltage region, the motor current response deteriorates while controlling the sinusoidal source current. When the motor speed is high, copper loss is significantly large because the motor current in the zero-output-voltage region depends on the back-electromotive force of the motor. To suppress the increasing copper loss in the zero-output-voltage region, this paper proposes a copper-loss minimization control method that utilizes the initial-condition response of the motor current. The proposed method is validated by experiment.**

I. INTRODUCTION

Currently, variable-speed AC motors that reduce energy consumption are used in motor drives [1]–[3]. Generally, the

AC motors are driven by an AC–DC–AC system that consists of an AC–DC converter and a DC–AC inverter with an electrolytic capacitor. These power converters control the source and motor currents, respectively, and an electrolytic capacitor is inserted into the DC-link. However, the electrolytic capacitor prevents system downsizing. Therefore, many techniques have been proposed to reduce the capacitance of the smoothing capacitor [4]–[11].

One such technique, a motor drive system, consists of a single-phase diode rectifier, a low-capacitance film capacitor, a three-phase voltage-source inverter, and an interior permanent-magnet synchronous motor (IPMSM) [7]–[11]. Because the system does not apply an electrolytic capacitor and power factor correction (PFC) circuit, the system significantly reduces the costs and size of the AC–DC–AC system. Applications of this system are limited to compressor motor drives that do not require a high-precision speed response, such as those used in air conditioners and refrigerators, because the system induces speed variations. However, the system must control both the motor and source currents. To improve the waveform of the source current, some inverter control methods have been

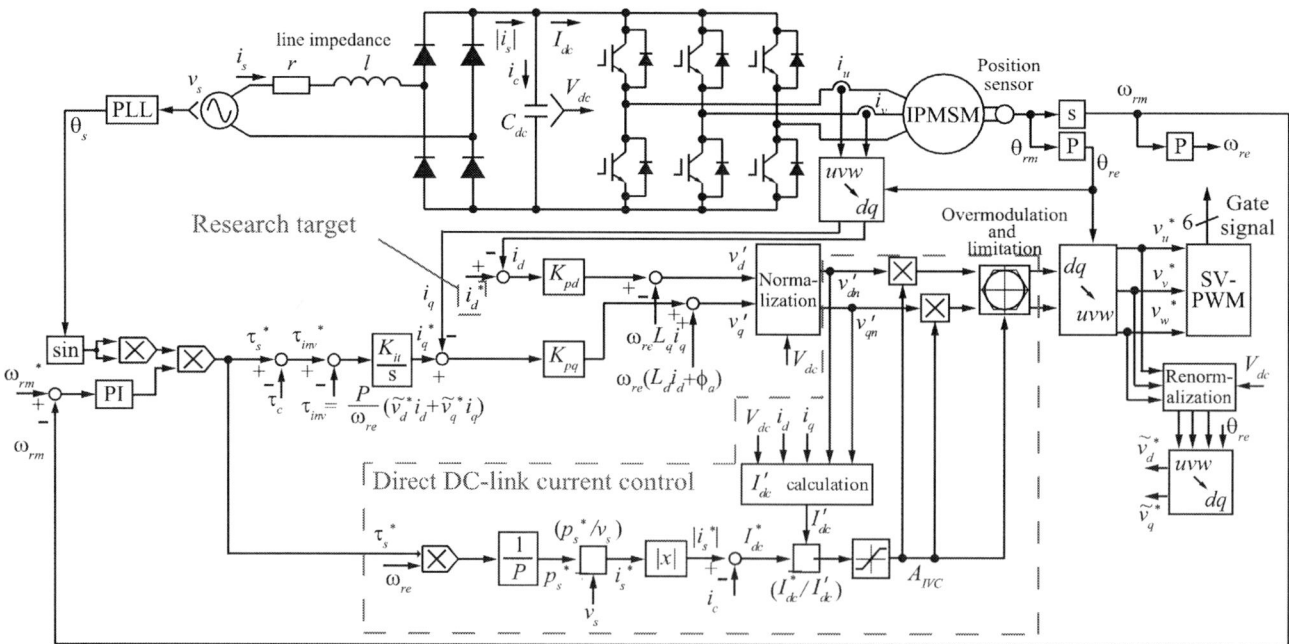

Fig. 1. System configuration

proposed. In [11], a direct DC-link current control (DDCCC), as shown in Fig. 1, realizes a sinusoidal source current. However, when the sinusoidal source current is obtained, the DC-link voltage becomes approximately equal to the absolute value of the source voltage v_s because the system provides the source power to the motor directly. When v_s is zero near the zero-crossing point, the inverter output voltage is also zero, and the control performance for the motor current deteriorates. Consequently, a motor current corresponding to the back-electromotive force appears. In particular, a large current appears in the high-speed regions, deteriorating the motor efficiency.

This study considers the motor current response at zero output voltage near the zero crossing, and proposes a copper-loss minimization control in the zero-output-voltage region using the initial condition response of the motor current. The proposed method is validated experimentally.

II. Direct DC-link Current Control for Sinusoidal Source Current

The electrolytic capacitor-less inverter controls the motor and source current because no PFC circuit or energy buffer exists. Additionally, to meet the guidelines for a harmonic source current, it is necessary to control the source current under sinusoidal waveform conditions. In this study, a DDCCC, as shown in Fig. 1, is used as the conventional method to improve the source current waveform [11]. This section describes the DDCCC that focuses on the relation between the source current and the DC-link current.

In the inverter, the relationship between the rectified source current $|i_s|$, the DC-link current I_{dc}, and the DC-link capacitor current i_c is as follows:

$$|i_s| = I_{dc} + i_c. \tag{1}$$

Equation (1) indicates that the DC-link current waveform changes the source current waveform. Therefore, the DDCCC improves the source current. The average DC-link current of each control period, excluding the switching ripple due to the inverter, is written as follows:

$$I_{dc} = \frac{v_{dn}^* i_d + v_{qn}^* i_q}{2}, \tag{2}$$

where i_d and i_q are the d- and q-axis currents, v_{dn}^* and v_{qn}^* are the normalized d- and q-axis voltage references divided by $V_{dc}/2$, and the magnitudes of the d- and q-axis variables are $\sqrt{2/3}$ times that of the u,v,w-phase variable. By substituting the DC-link reference current I_{dc}^* that realizes a sinusoidal source current into (2), the following equation is obtained.

$$v_{qn}^* = -\frac{i_d}{i_q} v_{dn}^* + \frac{2I_{dc}^*}{i_q}. \tag{3}$$

Equation (3) indicates that a DC-link current equal to its reference I_{dc}^* is obtained by modifying the reference voltage on a straight line. Hereinafter, this line is called the DC-link current line. When the output voltage, determined by the current regulator and decoupling control, is lower than

the voltage limit, the DDCCC changes only the amplitude of the output voltage and modifies the output voltage to be equal to that of the DC-link current line. In the voltage saturation region, the DDCCC modifies the output voltage to be equal to that at the intersection point between the voltage limit and the DC-link current line, and then the DDCCC selects an intersection point nearer to the output voltage determined by the current regulator and decoupling control. The mentioned modification implies controlling the source current under sinusoidal waveform conditions. The source current is largely controlled by DDCCC. In addition, the DC-link voltage becomes equal to the absolute value of the source voltage v_s. Thus, the output voltage is zero near the zero crossing.

The large modified voltage deteriorates the control performance of the motor current. To output the reference voltage of the motor current regulator near the DC-link current line of (3) and reduce the modified voltage, the q-axis reference current is calculated using motor torque control [10], [11].

III. Motor Current Response at Zero Output Voltage

Although the DDCCC facilitates a sinusoidal source current, a zero-output-voltage region occurs and the motor current response near the zero crossing deteriorates. In particular, a large current appears in the high-speed region, thereby deteriorating the motor efficiency. First, this section analyzes the motor current in the zero-output-voltage region.

The voltage equation of IPMSM is as follows:

$$\begin{bmatrix} v_d \\ v_q \end{bmatrix} = \begin{bmatrix} R_a + \frac{d}{dt}L_d & -\omega_{re}L_q \\ \omega_{re}L_d & R_a + \frac{d}{dt}L_q \end{bmatrix} \begin{bmatrix} i_d \\ i_q \end{bmatrix} + \begin{bmatrix} 0 \\ \omega_{re}\phi_a \end{bmatrix}, \tag{4}$$

where v_d and v_q are the d- and q-axis voltages, R_a is the stator resistance, L_d and L_q are the d- and q-axis inductances, ϕ_a is the flux of the permanent magnet in the motor, and ω_{re} is the electrical angular speed of the motor. As the zero-output-voltage region near the zero crossing is significantly small and the motor speed does not substantially change, ω_{re} is approximately constant in this region. Because the DC-link voltage near the zero crossing is approximately zero, the inverter output voltages v_d and v_q are also approximately zero. Considering (4), the state equation of an IPMSM can be written as follows:

$$\frac{d}{dt} \begin{bmatrix} i_d \\ i_q \end{bmatrix} = \begin{bmatrix} -\frac{R}{L_d} & \omega_{re}\frac{L_q}{L_d} \\ -\omega_{re}\frac{L_d}{L_q} & -\frac{R}{L_q} \end{bmatrix} \begin{bmatrix} i_d \\ i_q \end{bmatrix} + \begin{bmatrix} 0 \\ \frac{-1}{L_q} \end{bmatrix} \omega_{re}\phi_a \tag{5}$$

$$\frac{d}{dt}\boldsymbol{x} = \boldsymbol{A}\boldsymbol{x} + \boldsymbol{c}\boldsymbol{d}. \tag{6}$$

In (6), there is no control input v_d and v_q, and the back-electromotive force in the zero-output-voltage region significantly affects the motor current. By transforming (6) into the time domain using (7), the d- and q-axis currents with the initial condition response $\boldsymbol{x}(0) = ([i_d^{ini}, i_q^{ini}]^T)$ are obtained as follows:

$$\boldsymbol{x}(t) = e^{\boldsymbol{A}t}\boldsymbol{x}(0) + \int_0^t e^{\boldsymbol{A}(t-\tau)}\boldsymbol{c}\boldsymbol{d}(\tau)d\tau, \tag{7}$$

3891

$$i_d = i_d{}^{ini}e^{-\beta t}\left\{\cos\alpha t - \frac{R(L_q - L_d)}{2L_d L_q \alpha}\sin\alpha t\right\}$$
$$+ i_q{}^{ini}\frac{L_q \omega_{re}}{L_d \alpha}e^{-\beta t}\sin\alpha t$$
$$- \omega_{re}\phi_a\left\{\frac{\alpha}{L_d\gamma} - \frac{R(L_d + L_q)}{L_d{}^2 L_q\gamma}e^{-\beta t}\sin\alpha t\right.$$
$$\left. - \frac{\alpha}{L_d\gamma}e^{-\beta t}\cos\alpha t\right\}, \tag{8}$$

$$i_q = -i_d{}^{ini}\frac{L_d \omega_{re}}{L_q \alpha}e^{-\beta t}\sin\alpha t$$
$$+ i_q{}^{ini}e^{-\beta t}\left\{\cos\alpha t + \frac{R(L_q - L_d)}{2L_d L_q\alpha}\sin\alpha t\right\}$$
$$- \omega_{re}\phi_a\left\{\frac{R}{L_d L_q\gamma}\right.$$
$$+ \frac{1}{L_q\gamma}\left\{\alpha - \frac{R^2(L_d + L_q)}{4L_d{}^2 L_q{}^2\alpha}\right\}e^{-\beta t}\sin\alpha t$$
$$\left. - \frac{R}{L_d L_q\gamma}e^{-\beta t}\cos\alpha t\right\}, \tag{9}$$

where

$$\alpha = \sqrt{\omega_{re}{}^2 - \frac{R^2(L_q - L_d)^2}{4L_d{}^2 L_d{}^2}}, \tag{10}$$

$$\beta = \frac{R(L_d + L_q)}{2L_d L_q}, \tag{11}$$

$$\gamma = \omega_{re}{}^2 + \frac{R^2(L_d + L_q)^2}{4L_d{}^2 L_q{}^2}. \tag{12}$$

From (8) and (9), the oscillations of the d- and q-axis response currents near the zero crossing are caused by the back-electromotive force $\omega_{re}\phi_a$. This study focuses on the initial condition response at zero output voltage of the motor current and reduces the oscillations due to $\omega_{re}\phi_a$ by controlling the initial value of the motor current at the point of entering the zero-output-voltage region. Regarding $i_d{}^{ini}$ and $i_q{}^{ini}$, only $i_d{}^{ini}$ is freely controllable to improve the motor current response near the zero crossing because the q-axis reference current is determined by the motor torque and motor speed regulator. As the objective of this study is to improve the motor response current in the zero-output-voltage and high-speed regions (assuming $1 \gg \frac{R_a(L_q - L_d)}{2L_d L_q\alpha}$ and the first term in (8) is significantly larger than the second term), the d-axis response current $i_d{}^{ini}$ is largely determined by the first term of (8). Then, in (8) and (9), the initial condition responses due to $i_d{}^{ini}$ suppress the motor current oscillations due to $\omega_{re}\phi_a$ by inducing a negative $i_d{}^{ini}$, thereby producing a response with the opposite phase.

IV. INITIAL VALUE OF D-AXIS CURRENT TO MINIMIZE COPPER LOSS IN THE ZERO OUTPUT VOLTAGE REGION

Figure 2(a) shows a conventional d-axis reference current using the general maximum torque per ampere (MTPA) control in (13) and general flux-weakening (FW) control in (14) [11],

[12].

$$i_d^* = \frac{\phi_a}{2(L_q - L_d)} - \sqrt{\frac{\phi_a^2}{4(L_q - L_d)^2} + i_q^2}, \tag{13}$$

$$i_d^* = \frac{-\phi_a - \sqrt{(V_{om}/\omega_{re})^2 - (L_q i_q)^2}}{L_d}, \tag{14}$$

$$V_{om} = \frac{|v_s| - v_{drop}}{\sqrt{2}} - R_a(i_d^2 + i_q^2), \tag{15}$$

where V_{om} is calculated using not V_{dc} but $|v_s| - v_{drop}$ to meet the voltage limit of (16) and v_{drop} is the voltage drop across the diode rectifier, inverter, line impedance, etc.

$$\sqrt{v_d^{*2} + v_q^{*2}} \lessgtr \frac{|v_s| - v_{drop}}{\sqrt{2}} \tag{16}$$

Regarding the conventional d-axis reference currents, the negative reference maximum is used as a reference. In the zero-output-voltage region near the zero crossing, the d-axis reference current is calculated with FW control to reduce the back-electromotive force. However, as the output voltage is nearly zero, the d-axis current does not follow its reference and a motor current corresponding to the back-electromotive force flows. To reduce the increase in motor current and copper loss due to the back-electromotive force, the proposed method engages a certain negative d-axis current before the voltage saturation region is entered. In this section, the initial value of the d-axis current that minimizes copper loss in the zero-output-voltage region is calculated using (8) and (9). Because an excessive d-axis current in the zero-crossing region causes lower efficiency, the d-axis reference current is limited to the optimal initial value i_d^{opt} that minimizes copper loss in the zero-output-voltage region. As shown in Fig. 2(b), the proposed d-axis reference current is used in the period before voltage saturation occurs.

Assuming that i_q^{ini} is approximately zero and ω_{re} is approximately constant in the zero-output-voltage region, the optimal initial value of the d-axis current i_d^{opt} is calculated by inserting (8) and (9) into (17) and solving (17) for i_d^{ini}.

$$\frac{\partial}{\partial i_d^{ini}}\int_0^{T_{zero}} R_a\left(i_d^2 + i_q^2\right) dt = 0, \tag{17}$$

where the zero-output-voltage period T_{zero} is determined as a voltage-saturation period, as shown in Fig. 2. To control the d-axis current according to the solution of (17) at the point of entering the voltage saturation region, the proposed method increases the d-axis reference current before entering the voltage saturation region. To apply the proposed d-axis reference current before the voltage saturation region is entered, the time at which voltage saturation occurs and the voltage saturation period are calculated using a periodicity of twice the source frequency. In this study, the proposed method increases the d-axis reference current using sigmoid function (18), because a DDCCC error occurs due to the large change in the motor current and causes resonant oscillations in the source current

The 2018 International Power Electronics Conference

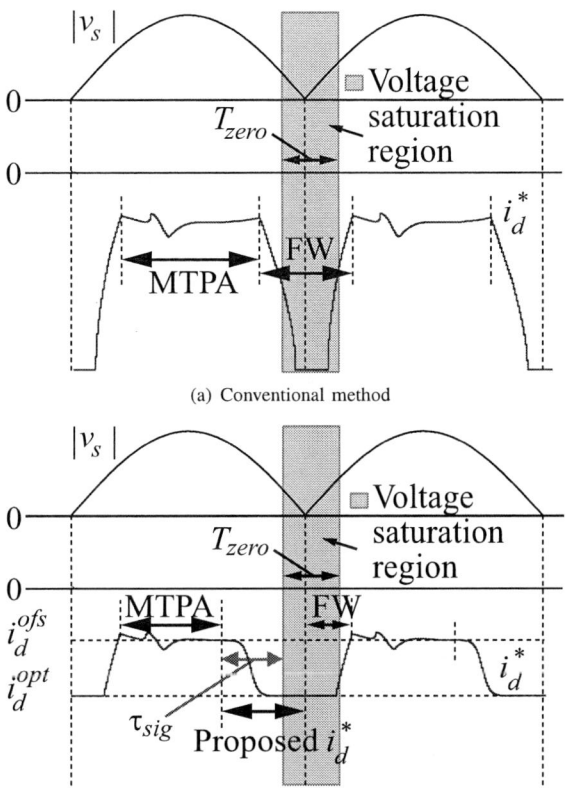

(a) Conventional method

(b) Proposed method

Fig. 2. d-axis reference current

TABLE I. MOTOR AND SYSTEM PARAMETERS

Stator resistance R_a	0.615 Ω
dq-axis inductance L_d, L_q	7.14 mH, 11.3 mH
Linkage flux ϕ_a	0.124 Wb
Number of pole pairs P	2
Rated speed and torque	4200 rpm, 1.8 Nm
Source voltage	220 V_{rms}, 50 Hz
DC-link capacitance C_{dc}	14 μF
Line impedance r, l	0.5 Ω, 0.2 mH
Sampling frequency f_{samp}	16 kHz
Bandwidth of current control system ω_c	6280 rad/s

to 4200 rpm and 0.5 Nm, respectively. Figures 3–6 show the waveforms in experiments. The curves correspond to the source current, DC-link voltage, d- and q-axis reference currents, current response, copper loss, and average value of copper loss. Figure 7 shows the source current harmonics and the regulation of IEC 61000-3-2 Class A. The FFT analysis considers interorder harmonics and groups them based on IEC 61000-4-7. The motor efficiency and power factor are listed in Table II.

Figure 3 shows the results using the power control method in [7] with MTPA and FW control. In the power control method with a proportionalintegral (PI) motor current controller, the bandwidth of the current control system is set at 1500 rad/s to decrease the source current harmonics. MTPA control is performed in a region where the DC-link voltage is sufficiently large. FW control is applied when the DC-link voltage decreases near the zero- crossing. Near the zero- crossing, the voltage saturation causes source current harmonics, which decrease the power factor and do not meet the guidelines, as shown in Fig. 7. In this case, the motor efficiency is 86.93%.

Figure 4 shows the results using DDCCC with MTPA and FW control. The d-axis reference current is limited to i_d^{opt}. The DC-link voltage has nearly the absolute value of the source voltage waveform without charging by regeneration, as opposed to that of Fig. 3. However, because the charging due to regeneration decreases, the period in which the DC-link voltage is low increases. As the output voltage becomes approximately zero near the zero crossing, the d-axis current does not follow its reference and current flows due to the back-electromotive force. Thus, the copper loss increases near the zero crossing, and the motor efficiency decreases to 75.21%, as listed in Table II.

Figure 5 shows the results using DDCCC with the proposed d-axis reference current as a step input. The initial value of the d-axis current in the zero-output-voltage region becomes larger than that shown in Fig. 4. Although the MTPA control period is reduced by the proposed method and the copper loss consequently increases, the increase in the d-axis current and copper loss near the zero-output-voltage region is suppressed significantly. As shown by the copper loss curve, because the increase in copper loss near the zero-output-voltage region is large compared with that of the other period, the average value of the copper loss is lower than that shown in Fig. 4. As a result, the motor efficiency is improved to 76.07%. However, the motor current is changed steeply, and source

when the d-axis reference current is changed while being used as a step input.

$$f(x) = \frac{1}{1 + e^{-\alpha x}} \quad (18)$$

As shown in Fig. 2(b), because the sigmoid function is used for the period τ_{sig} that connects i_d^{ofs} to i_d^{opt} smoothly, it is necessary to shift (18) in parallel by $\tau_{sig}/2$ and to add a gain and offset to (18):

$$i_d^* = (i_d^{ofs} - i_d^{opt})\frac{(1 + e^{-\alpha\tau_{sig}/2})(1 + e^{\alpha\tau_{sig}/2})}{e^{\alpha\tau_{sig}/2} - e^{-\alpha\tau_{sig}/2}}$$
$$\left(\frac{1}{1 + e^{\alpha\tau_{sig}/2}} - \frac{1}{1 + e^{-\alpha(x-\tau_{sig}/2)}}\right) + i_d^{ofs}. (19)$$

In (19), α is set to the current control bandwidth ω_c, τ_{sig} is set to $20/\omega_c$, 20 times the time constant of the current control system, in order to sufficiently follow the reference. x represents the time from the start of applying the proposed method, and is reset to 0 every period of twice the source frequency.

V. EXPERIMENTAL RESULTS

To validate the effectiveness of the proposed method, experiments are conducted. Table I lists the motor and system parameters. The motor speed and load condition are set

3893

The 2018 International Power Electronics Conference

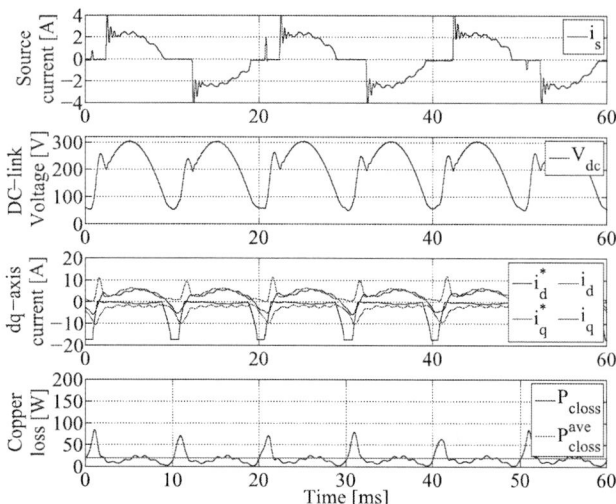

Fig. 3. Method in [7] with MTPA and FW control

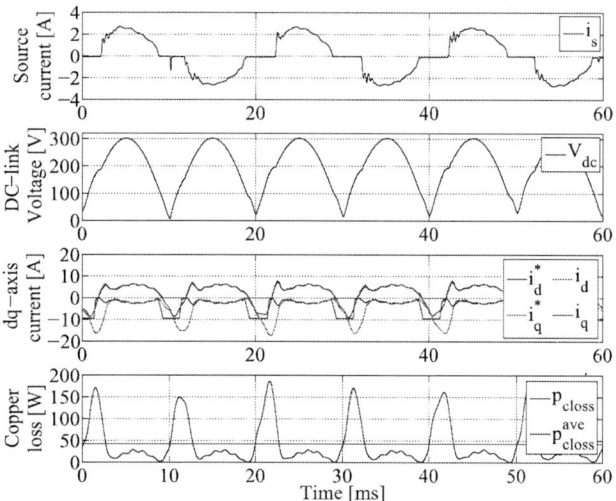

Fig. 4. DDCCC with MTPA and FW control

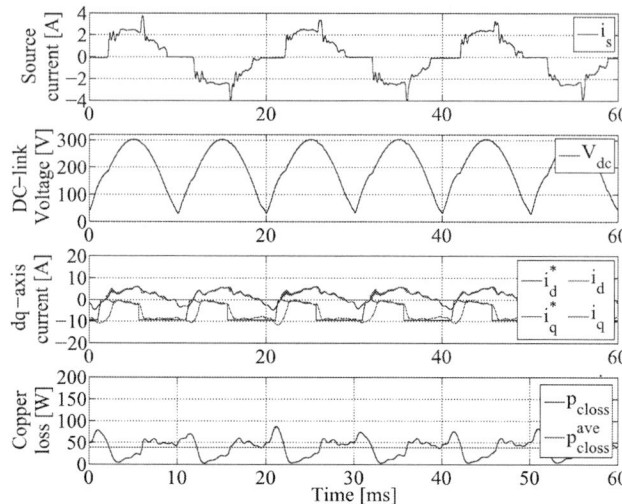

Fig. 5. DDCCC with MTPA and FW control and proposed method as step input

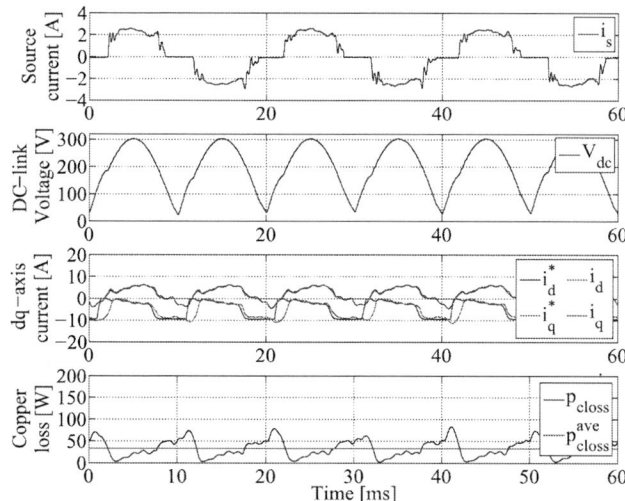

Fig. 6. DDCCC with MTPA and FW control and proposed method using sigmoid function

current oscillations occur due to LC resonance.

Figure 6 shows the results of applying the proposed d-axis reference current using the sigmoid function. The proposed method reduces copper loss in the zero-output-voltage region, analogous to the results shown in Fig. 5. Because the motor current is changed smoothly by the sigmoid function, source current oscillations do not occur. Compared with the efficiency observed in Fig. 4, the motor efficiency is improved by approximately 4.5 points without decreasing the input power factor. As shown in Fig. 7, when the DDCCC is used for improving the source current waveform, the source current harmonics meet the guideline.

VI. CONCLUSION

A copper-loss minimization control scheme is proposed in the zero-output-voltage region for an electrolytic capacitor-less single-phase-to-three-phase inverter. At zero output voltage, the motor current flows based on the back-electromotive force. In particular, the motor current increases in the high-rotational-speed region. This reduces the motor efficiency. The proposed method reduces the increase in motor current due to the back-electromotive force by utilizing the initial-condition response of the motor current. The optimal initial value of d-axis current that minimizes copper loss in the zero-output-voltage region is calculated using the equation of the dq-axis current response in the time domain, where the output voltage is zero.

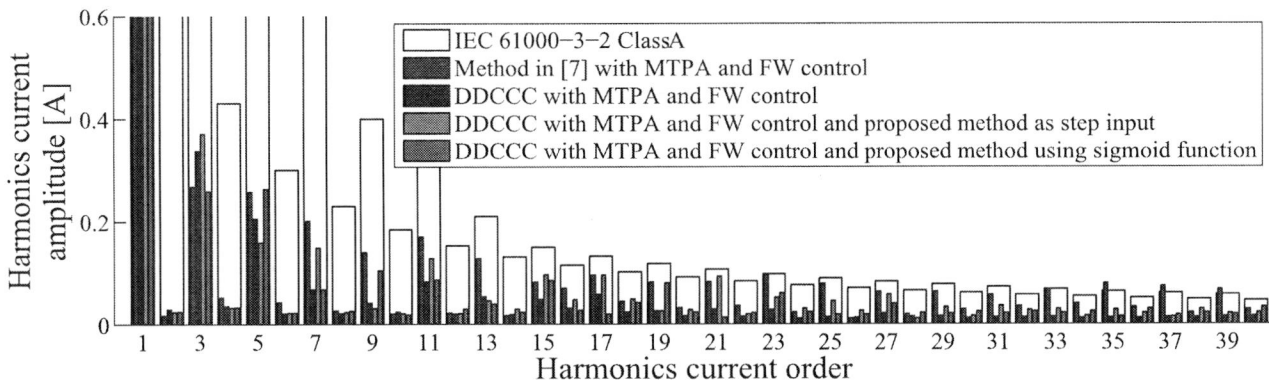

Fig. 7. Source-current harmonics

TABLE II. MOTOR EFFICIENCY AND POWER FACTOR IN EXPERIMENTAL RESULTS

	Motor efficiency [%]	P.F. [%]
Method in [7] with MTPA and FW control	95.23	91.32
DDCCC with MTPA and FW control	88.43	95.57
DDCCC with proposed method (step input)	89.10	93.59
DDCCC wiith proposed method (sigmoid function)	91.13	94.79

The proposed d-axis reference current effectively reduces the increase in copper loss in the zero-output-voltage region in the experiment.

REFERENCES

[1] K. Kondo and H. Kubota, " Innovative Application Technologies of AC Motor Drive Systems ", *IEEJ J. Ind. Appl.*, vol. 1, no. 3, pp. 132–140, 2012.

[2] S. Sato and K. Ide, "Application Trends of Sensorless AC Motor Drives in Europe", *IEEJ J. Ind. Appl.*, vol. 3, no. 2, pp. 97–103, 2014.

[3] J. Kim, K. Ha, and R. Krishnan, " Single-Controllable-Switch-Based Switched Reluctance Motor Drive for Low Cost, Variable-Speed Applications ", *IEEE Trans. Power Electron.*, vol. 27, no. 1, pp. 379–387, Jan., 2012

[4] S. Li, W. Qi, S.-C. Tan, and S. Y. R. Hui, " Integration of an Active Filter and a Single-Phase AC/DC Converter With Reduced Capacitance Requirement and Component Count", *IEEE Trans. Power Electron.*, vol. 31, no. 6, pp. 4121–4137, Jun., 2016

[5] A. Tokumasu, K. Shirakawa, H. Taki, and K. Wada, "AC/DC Converter Based on Instantaneous Power Balance Control for Reducing DC-Link Capacitance", *IEEJ J. Ind. Appl.*, vol. 4, no. 6, pp. 745–751, 2015.

[6] Y. Ohnuma and J. Itoh, "A Single-phase-to-three-phase Power Converter with an Active Buffer and a Charge Circuit", *IEEJ J. Ind. Appl.*, Vol. 1, No. 1, pp. 46–54, 2012.

[7] K. Inazuma, H. Utsugi, K. Ohishi, and H. Haga, " High-Power-Factor Single-Phase Diode Rectifier Driven by Repetitively Controlled IPM Motor ", *IEEE Trans. on Ind. Electron.*, vol. 60, no. 10, pp. 4427–4437, Oct., 2013.

[8] H.-S. Jung, S.-J. Chee, S.-Ki Sul, Y.-J. Park, H.-S. Park, and W.-K. Kim, "Control of Three-Phase Inverter for AC Motor Drive With Small DC-Link Capacitor Fed by Single-Phase AC Source", *IEEE Trans. Ind. Appl.*, vol. 50, no. 2, pp. 1074–1081, Mar./Apr., 2014.

[9] Y. Sou and J.-I. Ha, " Direct Power Control of a Three-Phase Inverter for Grid Input Current Shaping of a Single-Phase Diode Rectifier With

a Small DC-Link Capacitor ", *IEEE Trans. Power Electron.*, vol. 30, no. 7, pp. 3794–3803, Jul., 2015.

[10] Kodai Abe, Hitoshi Haga, Kiyoshi Ohishi, and Yuki Yokokura, " Fine Current Harmonics Reduction Method for Electrolytic Capacitor-Less and Inductor-Less Inverter Based on Motor Torque Control and Fast Voltage Feedforward Control for IPMSM ", *IEEE Trans. on Ind. Electron.*, vol. 64, no. 2, pp. 1071–1080, Feb., 2017.

[11] Kodai Abe, Hitoshi Haga, Kiyoshi Ohishi, and Yuki Yokokura, "Direct DC-link Current Control Considering Voltage Saturation for Realization of Sinusoidal Source Current Waveform without Passive Components for IPMSM Drives," *IEEE Trans. on Ind. Electron.*, Vol. 65, No. 5, pp. 3805-3814, May 2018.

[12] S. Morimoto, Y. Takeda, T. Hirasa, and K. Taniguchi, " Expansion of operating limits for permanent magnet motor by current vector control considering inverter capacity ", *IEEE Trans. Power Electron.*, vol. 26, no. 5, pp. 866–871, Sep./Oct., 1990.

Armature Temperature Estimation Insensitive to Rotor Flux Variation for SPMSM

Toshiki Sano, Kiyoshi Ohishi, Yuki Yokokura, Hiroki Iwata,
Nagaoka University of Technology, JAPAN
Yuji Ide, Daigo Kuraishi, Akihiko Takahashi,
SANYO DENKI CO.,LTD, JAPAN

Abstract—In order to estimate the armature temperature of a surface permanent magnet synchronous motor (SPMSM), it is important to estimate the armature resistance accurately. In this paper, a novel method is proposed to estimate the armature temperature of the SPMSM. The estimated armature temperature is obtained from the estimated armature resistance, which is estimated independently from the rotor flux in the SPMSM. Finally, the effectiveness of the proposed estimation method was evaluated by the experiment.

I. INTRODUCTION

By monitoring the temperature of a permanent-magnet synchronous motor (PMSM) that is being driven, it is possible to protect the insulation of the winding and manage the thermal demagnetization of the permanent magnet. In addition, the efficiency of the PMSM and the performance of the system are improved by accurately monitoring the temperature. It is possible to obtain the temperature of the PMSM directly by using a temperature sensor. However, because the temperature sensor has problems with cost and robustness, the temperature estimation method is topic for study [1]–[4].

For the temperature estimation method in PMSMs, an estimation method based on PMSM parameters is proposed. In these methods, the PMSM parameters fluctuating with temperature are estimated, and the temperature of the PMSM is estimated based on the estimated parameters. The parameter estimation method for PMSM is mentioned the harmonic superposition method [3], [4] and the method based on fundamental waves [5]–[7]. In the harmonic superposition method, the harmonics are superimposed on the input voltage or current command, and the harmonic resistance is estimated from the harmonic signal. The harmonic superposition method provides a robust response regardless of changes in specifications and the shape of the PMSM. However, the method is required to accurately acquire harmonic signals and torque ripple, and noise is caused by harmonic superposition.

In the method based on the fundamental wave, the resistance and flux are estimated from the input voltage and current of the PMSM. The problem of the harmonic superposition method does not occur in this method because harmonics are not superimposed. However, this method is affected by parameter errors other than the estimated parameters. The armature resistance and rotor magnetic flux are changed by temperature fluctuations of the PMSM. The armature temperature of the PMSM can be estimated from the armature resistance. However, in the method based on the fundamental wave, the estimated armature resistance is affected by the parameter fluctuation of the rotor magnetic flux. The error in the resistance estimation causes errors in the armature temperature estimation. Therefore, in order to accurately estimate the armature temperature, a resistance estimation method that is robust against the influence of the magnetic flux of the rotor is required.

In this paper, a temperature estimation method for surface permanent-magnet synchronous motor (SPMSM) is proposed. The armature temperature of the SPMSM is estimated based on the estimated armature resistance. In the proposed method, the armature resistance and the rotor magnetic flux are estimated without being affected by the mutual parameter error. Therefore, because the armature resistance is estimated accurately without being affected by fluctuation of rotor flux, it is possible to estimate the armature temperature based on the estimated resistance. Finally, the effectiveness of the proposed method is evaluated by simulation and experiment.

II. TEMPERATURE ESTIMATION

In this paper, the armature temperature is estimated by (1).

$$\hat{T} = \left(\frac{\hat{R}}{R_0} - 1 \right)(234.5 + T_0) + T_0 \qquad [^\circ C], \qquad (1)$$

where T_0 is the base armature temperature, R_0 is the armature resistance in T_0, $\hat{T}$ is the estimated armature temperature, $\hat{R}$ is the estimated armature resistance. When T_0 and R_0 are known, the estimation performance of $\hat{T}$ depends on the estimation performance of $\hat{R}$. Therefore, to estimate the armature temperature accurately, the armature resistance must be estimated under temperature fluctuation of the SPMSM.

III. CONVENTIONAL ESTIMATION METHOD

This section shows the estimation method from [7]. Furthermore, the influence of the rotor flux variation on the estimation method is shown. Fig. 1 shows the block diagram of the estimation method from [7]. In the estimation method, the electrical angle and armature resistance are estimated to improve the performance of sensor-less position control in the PMSM, and the armature resistance is estimated without being affected by the electrical angle variation. In this paper, the estimation method is described based on SPMSM ($L_d = L_q = L$).

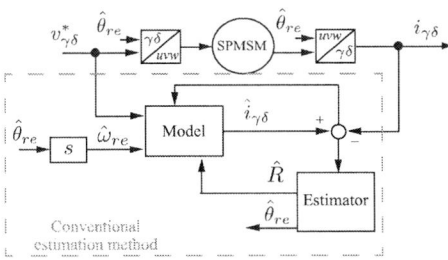

Fig. 1. Block diagram of the conventional estimation method.

A. Structure of estimation method

The estimation method of [7] is composed of the model and the estimator. The model is the state equation of SPMSM in the $\gamma - \delta$ coordinates. Here, the $\gamma - \delta$ coordinates are the coordinates based on the estimated electrical angle $\hat{\theta}_{re}$, and the γ-axis is delayed by $\Delta\theta_{re}$ from the d-axis.

The state equation of SPMSM in $\gamma - \delta$ coordinates is shown in (2). Where $v_{\gamma\delta}$ and $i_{\gamma\delta}$ are the voltage and current in $\gamma - \delta$ coordinates, respectively; R, L, and ϕ are the armature resistance, armature inductance, and rotor flux in the SPMSM, respectively; ω_{re} is the electrical angular velocity; $\hat{\alpha}$ is the estimation value of α; $\Delta\theta_{re} = \theta_{re} - \hat{\theta}_{re}$; and $\Delta\omega_{re} = \omega_{re} - \hat{\omega}_{re}$. From (2), the model of the estimation method is shown (3). Where L_m and ϕ_m are the inductance and flux of the model, respectively; g_{11}, g_{12}, g_{21}, and g_{22} are the gains of the model; $e_{i\gamma} = \hat{i}_\gamma - i_\gamma$; and $e_{i\delta} = \hat{i}_\delta - i_\delta$.

In the case of $L_m = L$ and $\phi_m = \phi$, from (2) and (3), the error equation is given by (4). Where $\Delta R = R - \hat{R}$. From (4), the relationship between the current estimation error and the parameter estimation error is given by (5). From (5), the resistance estimator of the estimation method is shown in (6). Where k_{pR}, k_{iR}, h_0, and h_1 are the gains of the resistance estimator.

$$\frac{d}{dt}\begin{bmatrix} i_\gamma \\ i_\delta \end{bmatrix} = \begin{bmatrix} -\frac{R}{L} & \omega_{re} \\ -\omega_{re} & -\frac{R}{L} \end{bmatrix}\begin{bmatrix} i_\gamma \\ i_\delta \end{bmatrix} + \begin{bmatrix} \frac{1}{L} & 0 \\ 0 & \frac{1}{L} \end{bmatrix}\begin{bmatrix} v_\gamma \\ v_\delta \end{bmatrix}$$
$$+ \begin{bmatrix} -\frac{1}{L} & 0 \\ 0 & -\frac{1}{L} \end{bmatrix}\begin{bmatrix} -\omega_{re}\phi\sin\Delta\theta_{re} \\ \omega_{re}\phi\cos\Delta\theta_{re} \end{bmatrix} \quad (2)$$

$$\frac{d}{dt}\begin{bmatrix} \hat{i}_\gamma \\ \hat{i}_\delta \end{bmatrix} = \begin{bmatrix} -\frac{\hat{R}}{L_m} & \hat{\omega}_{re} \\ -\hat{\omega}_{re} & -\frac{\hat{R}}{L_m} \end{bmatrix}\begin{bmatrix} \hat{i}_\gamma \\ \hat{i}_\delta \end{bmatrix} + \begin{bmatrix} \frac{1}{L_m} & 0 \\ 0 & \frac{1}{L_m} \end{bmatrix}\begin{bmatrix} v_\gamma \\ v_\delta \end{bmatrix}$$
$$+ \begin{bmatrix} -\frac{1}{L_m} & 0 \\ 0 & -\frac{1}{L_m} \end{bmatrix}\begin{bmatrix} 0 \\ \hat{\omega}_{re}\phi_m \end{bmatrix} - \begin{bmatrix} g_{11} & -g_{12} \\ g_{21} & g_{22} \end{bmatrix}\begin{bmatrix} e_{i\gamma} \\ e_{i\delta} \end{bmatrix} \quad (3)$$

$$\frac{d}{dt}\begin{bmatrix} e_{i\gamma} \\ e_{i\delta} \end{bmatrix} = \begin{bmatrix} -\frac{\hat{R}}{L} - g_{11} & \hat{\omega}_{re} + g_{12} \\ -\hat{\omega}_{re} - g_{21} & -\frac{\hat{R}}{L} - g_{22} \end{bmatrix}\begin{bmatrix} e_{i\gamma} \\ e_{i\delta} \end{bmatrix}$$
$$+ \begin{bmatrix} \frac{\omega_{re}\phi}{L} & \frac{i_\gamma}{L} \\ \frac{\phi}{L}s & \frac{i_\delta}{L} \end{bmatrix}\begin{bmatrix} \Delta\theta_{re} \\ \Delta R \end{bmatrix} \quad (4)$$

$$\begin{bmatrix} e_{i\gamma} \\ e_{i\delta} \end{bmatrix} = \frac{1}{P_D}\begin{bmatrix} p_{11} & p_{12} \\ p_{21} & p_{22} \end{bmatrix}\begin{bmatrix} \Delta\theta_{re} \\ \Delta R \end{bmatrix}, \quad (5)$$

$$P_D = \left(s + \frac{\hat{R}}{L} + g_{11}\right)\left(s + \frac{\hat{R}}{L} + g_{22}\right)$$
$$+ (\hat{\omega}_{re} + g_{12})(\hat{\omega}_{re} + g_{21})$$

$$p_{11} = \left(s + \frac{\hat{R}}{L} + g_{11}\right)\frac{\omega_{re}\phi}{L} + (\hat{\omega}_{re} + g_{21})s\frac{\phi}{L}$$

$$p_{12} = \left(s + \frac{\hat{R}}{L} + g_{11}\right)\frac{i_\gamma}{L} + (\hat{\omega}_{re} + g_{21})\frac{i_\delta}{L}$$

$$p_{21} = -(\hat{\omega}_{re} + g_{12})\frac{\omega_{re}\phi}{L} + \left\{s^2 + \left(\frac{\hat{R}}{L} + g_{22}\right)s\right\}\frac{\phi}{L}$$

$$p_{22} = -(\hat{\omega}_{re} + g_{12})\frac{i_\gamma}{L} + \left(s + \frac{\hat{R}}{L} + g_{22}\right)\frac{i_\delta}{L}$$

$$\hat{R} = \begin{bmatrix} \frac{k_{pR}s + k_{iR}}{s^2(h_1 s + h_0)}(-p_{21}) & \frac{k_{pR}s + k_{iR}}{s^2(h_1 s + h_0)}(p_{11}) \end{bmatrix}\begin{bmatrix} e_{i\gamma} \\ e_{i\delta} \end{bmatrix}$$
$$\quad (6)$$

B. Influence of flux variation

In the case of $\phi_m \neq \phi$, from (2) and (3), the error equation is given by (7). Where $\Delta\phi = \phi - \phi_m$, and the approximation shown in (8) is applied to (7). From (7), the relationship between the current estimation error and the parameter estimation error is given by (9). From (6) and (9), the relationship between the estimated resistance and the parameter estimation error is given by (10).

From (10), the armature resistance is estimated without being affected by the estimated electrical angle. However, when the rotor flux in the SPMSM is different from the flux of the model, the armature resistance is estimated based on the estimation error of the resistance and flux. When the temperature of the SPMSM fluctuates, the armature resistance and rotor magnetic flux of the SPMSM fluctuate. Therefore, it is impossible to estimate the armature resistance from the estimation method in [7] when the temperature of the SPMSM fluctuates.

$$\frac{d}{dt}\begin{bmatrix} e_{i\gamma} \\ e_{i\delta} \end{bmatrix} = \begin{bmatrix} -\frac{\hat{R}}{L} - g_{11} & \hat{\omega}_{re} + g_{12} \\ -\hat{\omega}_{re} - g_{21} & -\frac{\hat{R}}{L} - g_{22} \end{bmatrix}\begin{bmatrix} e_{i\gamma} \\ e_{i\delta} \end{bmatrix}$$
$$+ \begin{bmatrix} \frac{\omega_{re}\phi}{L} & \frac{i_\gamma}{L} & 0 \\ \frac{\phi}{L}s & \frac{i_\delta}{L} & \frac{\omega_{re}}{L} \end{bmatrix}\begin{bmatrix} \Delta\theta_{re} \\ \Delta R \\ \Delta\phi \end{bmatrix}, \quad (7)$$

$$\omega_{re}\phi - \hat{\omega}_{re}\phi_m \fallingdotseq -\omega_{re}\Delta\phi - \phi\Delta\omega_{re} \quad (8)$$

$$\begin{bmatrix} e_{i\gamma} \\ e_{i\delta} \end{bmatrix} = \frac{1}{P_D}\begin{bmatrix} p_{11} & p_{12} & (\hat{\omega}_{re} + g_{12})\frac{\omega_{re}}{L} \\ p_{21} & p_{22} & \left(s - \frac{\hat{R}}{L} - g_{22}\right)\frac{\omega_{re}}{L} \end{bmatrix}\begin{bmatrix} \Delta\theta_{re} \\ \Delta R \\ \Delta\phi \end{bmatrix} \quad (9)$$

$$\hat{R} = \frac{1}{P_D}\begin{bmatrix} 0 & f_{\Delta R} & f_{\Delta\phi} \end{bmatrix}\begin{bmatrix} \Delta\theta_{re} \\ \Delta R \\ \Delta\phi \end{bmatrix}, \quad (10)$$

$$f_{\Delta R} = \frac{k_{pR}s + k_{iR}}{s(h_1 s + h_0)}(p_{11}p_{22} - p_{12}p_{21}), \quad (11)$$

$$f_{\Delta\phi} = \frac{k_{pR}s + k_{iR}}{s(h_1 s + h_0)}\left\{-p_{21}(\hat{\omega}_{re} - g_{12})\frac{\omega_{re}}{L}\right.$$
$$\left. + p_{11}\left(-\frac{\hat{R}}{L} - g_{22}\right)\frac{\omega_{re}}{L}\right\}. \quad (12)$$

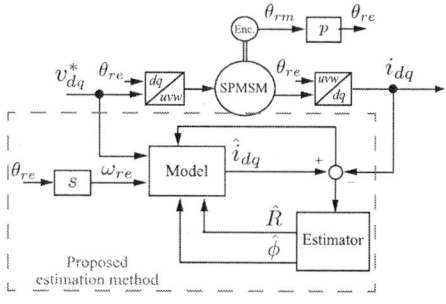

Fig. 2. Block diagram of the proposed estimation method.

IV. PROPOSED ESTIMATION METHOD

This section shows the proposed estimation method. Fig. 2 shows the block diagram of the proposed method. The method estimates the armature resistance and the rotor flux simultaneously, and is composed of the model and the estimator.

A. Structure of the model

The state equation of SPMSM on the rotating coordinate aligned with the rotor position (d-q axis) is shown by (13). Where v_{dq} and i_{dq} are the voltage and current in $d-q$ coordinates, respectively. The estimation model is a mathematical model of SPMSM on the $d-q$ axis. From (13), the state equation of the model is shown by (14). Where h_1 and h_2 are the gains of the model, $e_{id} = \hat{i}_d - i_d$, and $e_{iq} = \hat{i}_q - i_q$. This section assumes that L and L_m are equal ($L_m = L$). The influence of the proposed estimation method on the variation in inductance is shown in Section IV-D.

An error equation is expressed by the difference between the state equation of the SPMSM and that of the model. From (13) and (14), the error equation is given by (15). Where $\Delta R = R - \hat{R}$ and $\Delta \phi = \phi - \hat{\phi}$. From (15), the relationship between the current estimation error and the parameter estimation error is given by (16).

$$\frac{d}{dt}\begin{bmatrix} i_d \\ i_q \end{bmatrix} = \begin{bmatrix} -\frac{R}{L} & \omega_{re} \\ -\omega_{re} & -\frac{R}{L} \end{bmatrix}\begin{bmatrix} i_d \\ i_q \end{bmatrix} + \begin{bmatrix} \frac{1}{L} & 0 \\ 0 & \frac{1}{L} \end{bmatrix}\begin{bmatrix} v_d \\ v_q \end{bmatrix}$$
$$+ \begin{bmatrix} -\frac{1}{L} & 0 \\ 0 & -\frac{1}{L} \end{bmatrix}\begin{bmatrix} 0 \\ \omega_{re}\phi \end{bmatrix} \tag{13}$$

$$\frac{d}{dt}\begin{bmatrix} \hat{i}_d \\ \hat{i}_q \end{bmatrix} = \begin{bmatrix} -\frac{\hat{R}}{L_m} & \omega_{re} \\ -\omega_{re} & -\frac{\hat{R}}{L_m} \end{bmatrix}\begin{bmatrix} \hat{i}_d \\ \hat{i}_q \end{bmatrix} + \begin{bmatrix} \frac{1}{L_m} & 0 \\ 0 & \frac{1}{L_m} \end{bmatrix}\begin{bmatrix} v_d \\ v_q \end{bmatrix}$$
$$+ \begin{bmatrix} -\frac{1}{L_m} & 0 \\ 0 & -\frac{1}{L_m} \end{bmatrix}\begin{bmatrix} 0 \\ \omega_{re}\hat{\phi} \end{bmatrix} - \begin{bmatrix} h_1 & -h_2 \\ h_2 & h_1 \end{bmatrix}\begin{bmatrix} e_{id} \\ e_{iq} \end{bmatrix} \tag{14}$$

$$\frac{d}{dt}\begin{bmatrix} e_{id} \\ e_{iq} \end{bmatrix} = \begin{bmatrix} -\frac{\hat{R}}{L} - h_1 & \omega_{re} + h_2 \\ -\omega_{re} - h_2 & -\frac{\hat{R}}{L} - h_1 \end{bmatrix}\begin{bmatrix} e_{id} \\ e_{iq} \end{bmatrix}$$
$$+ \begin{bmatrix} \frac{i_d}{L} & 0 \\ \frac{i_q}{L} & \frac{\omega_{re}}{L} \end{bmatrix}\begin{bmatrix} \Delta R \\ \Delta \phi \end{bmatrix} \tag{15}$$

$$\begin{bmatrix} e_{id} \\ e_{iq} \end{bmatrix} = \frac{1}{Q_D}\underbrace{\begin{bmatrix} q_{11} & q_{12} \\ q_{21} & q_{22} \end{bmatrix}}_{Q}\begin{bmatrix} \Delta R \\ \Delta \phi \end{bmatrix} \tag{16}$$

$$Q_D = \left(s + \frac{\hat{R}}{L} + h_1\right)^2 + (\omega_{re} + h_2)^2$$

$$q_{11} = \left(s + \frac{\hat{R}}{L} + h_1\right)\frac{i_d}{L} + (\omega_{re} + h_2)\frac{i_q}{L}$$

$$q_{12} = (\omega_{re} + h_2)\frac{\omega_{re}}{L}$$

$$q_{21} = -(\omega_{re} + h_2)\frac{i_d}{L} + \left(s + \frac{\hat{R}}{L} + h_1\right)\frac{i_q}{L}$$

$$q_{22} = \left(s + \frac{\hat{R}}{L} + h_1\right)\frac{\omega_{re}}{L}.$$

B. Structure of the estimator

The armature resistance and rotor flux are estimated by (17) Where k_{pR} and k_{iR} are the gains in the armature resistance estimation, and $k_{p\phi}$ and $k_{i\phi}$ are gains in the rotor flux estimation. The matrix of (17) contains the inverse matrix of Q. From (15), (16), and (17), the relationship between the estimated value and estimation error is shown by (18). Where $Q_G = \omega_{re}i_d/L^2$. Because the inverse matrix of Q is included in the matrix of (17), the inverse matrix of Q is multiplied by Q in (18). As a result, the diagonal component of the matrix in (18) becomes 0. In addition, (18) shows the open-loop transfer function between the estimated value and the estimation error. $\hat{R}$ and $\hat{\phi}$ are not affected by the other estimation error, and track the actual value. By substituting $\Delta R = R - \hat{R}$ and $\Delta \phi = \phi - \hat{\phi}$ for (18), (19) is obtained. Equation (19) shows a closed-loop transfer function between the estimated value and the actual value. From (19), it is possible to design each estimator independently.

$$\begin{bmatrix} \hat{R} & \hat{\phi} \end{bmatrix}^T$$
$$= \begin{bmatrix} \frac{k_{pR}s + k_{iR}}{s^2}q_{22} & \frac{k_{pR}s + k_{iR}}{s^2}(-q_{12}) \\ \frac{k_{p\phi}s + k_{i\phi}}{s^2}(-q_{21}) & \frac{k_{p\phi}s + k_{i\phi}}{s^2}q_{11} \end{bmatrix}\begin{bmatrix} e_{id} \\ e_{iq} \end{bmatrix} \tag{17}$$

$$\begin{bmatrix} \hat{R} \\ \hat{\phi} \end{bmatrix} = \begin{bmatrix} \frac{k_{pR}s + k_{iR}}{s^2}Q_G & 0 \\ 0 & \frac{k_{p\phi}s + k_{i\phi}}{s^2}Q_G \end{bmatrix}\begin{bmatrix} \Delta R \\ \Delta \phi \end{bmatrix} \tag{18}$$

$$\begin{bmatrix} \hat{R} & \hat{\phi} \end{bmatrix}^T$$
$$= \begin{bmatrix} \frac{(k_{pR}s + k_{iR})Q_G}{s^2 + (k_{pR}s + k_{iR})Q_G} & 0 \\ 0 & \frac{(k_{p\phi}s + k_{i\phi})Q_G}{s^2 + (k_{p\phi}s + k_{i\phi})Q_G} \end{bmatrix}\begin{bmatrix} R \\ \phi \end{bmatrix} \tag{19}$$

C. Design of gains

In this section, the gains in the proposed method are designed using the pole-assignment method. From (15), a characteristic equation of the model is shown in (20). The pole is assigned by ω_m and a double root. The model gains h_1 and h_2 are shown by (21).

3898

From (19), a characteristic equation of the armature resistance and rotor flux estimator are shown in (22) and (23). Each pole in (22) and (23) are assigned ω_R, ω_ϕ, and a double root. The estimator gains k_{pR}, k_{iR}, $k_{p\phi}$, and $k_{i\phi}$ are shown in (24) and (25). In (24) and (25), ω_{re} and i_d are contained in the denominator of all gains. When $\omega_{re} \cong 0$ or $i_d \cong 0$, the gains are not designed normally. Therefore, to estimate armature resistance and rotor flux accurately, the proposed estimation method applies outside $\omega_{re} \cong 0$ or $i_d \cong 0$.

$$\left(s + \frac{\hat{R}}{L} + h_1\right)^2 + (\omega_{re} + h_2)^2 = 0 \quad (20)$$

$$h_1 = -\frac{\hat{R}}{L} - \omega_m \quad , \quad h_2 = -\omega_{re} \quad (21)$$

$$s^2 + Q_G k_{pR} s + Q_G k_{iR} = 0 \quad (22)$$

$$s^2 + Q_G k_{p\phi} s + Q_G k_{i\phi} = 0 \quad (23)$$

$$k_{pR} = \frac{L^2}{\omega_{re} i_d}(-2\omega_R) \quad , \quad k_{iR} = \frac{L^2}{\omega_{re} i_d}\omega_R^2 \quad (24)$$

$$k_{p\phi} = \frac{L^2}{\omega_{re} i_d}(-2\omega_\phi) \quad , \quad k_{i\phi} = \frac{L^2}{\omega_{re} i_d}\omega_\phi^2 \quad (25)$$

D. Influence of inductance variation

So far, the proposed estimation method has assumed that the d-axis inductance and the q-axis inductance are always L_m. However, the d-axis inductance and the q-axis inductance in the PMSM are varied by magnetic saturation. When the dq-axis inductance of the SPMSM given by L_d and L_q, the estimation parameters $\hat{R}$ and $\hat{\phi}$ are obtained by (26). Where w_1 and w_2 are obtained by (27); $\Delta L_d = L_d - L_m$, $\Delta L q = L_q - L_m$. From (26) and the final value theorem, the relational equation between the estimation parameter and the inductance variation in the steady state is obtained by (28). From (28), the armature resistance estimation is affected by the d-axis inductance variation, and the rotor flux estimation is affected by both the d-axis inductance and q-inductance variation.

Figure 3 shows the influence on the proposed method due to inductance variation in simulation. In Fig. 3, 100% load, $\omega_{rm}^* = 1000$ r/min, $\mathbf{1}_d^* = -2$ A. The inductance of the PMSM L_d and L_q are varied by $\pm 20\%$ with respect to the model inductance L_m, respectively. As shown in Fig. 3, the resistance estimation is greatly affected by the variation in the q-axis inductance.

$$
\begin{aligned}
\begin{bmatrix} \hat{R} & \hat{\psi} \end{bmatrix}^T \\
= & \begin{bmatrix} \dfrac{(k_{pR}s + k_{iR})\frac{\omega_{re} i_d}{L^2}}{s^2 + (k_{pR}s + k_{iR})\frac{\omega_{re} i_d}{L^2}} & 0 \\ 0 & \dfrac{(k_{p\phi}s + k_{i\phi})\frac{\omega_{re} i_d}{L^2}}{s^2 + (k_{p\phi}s + k_{i\phi})\frac{\omega_{re} i_d}{L^2}} \end{bmatrix} \begin{bmatrix} R \\ \phi \end{bmatrix} \\
+ & \begin{bmatrix} \dfrac{(k_{pR}s + k_{iR})\frac{\omega_{re}}{L}}{s^2 + (k_{pR}s + k_{iR})\frac{\omega_{re} i_d}{L^2}} & 0 \\ \dfrac{(k_{p\phi}s + k_{i\phi})\frac{i_q}{L}}{s^2 + (k_{p\phi}s + k_{i\phi})\frac{\omega_{re} i_d}{L^2}} & \dfrac{(k_{p\phi}s + k_{i\phi})\frac{i_d}{L}}{s^2 + (k_{p\phi}s + k_{i\phi})\frac{\omega_{re} i_d}{L^2}} \end{bmatrix} \begin{bmatrix} w_1 \\ w_2 \end{bmatrix}
\end{aligned}
$$
$$(26)$$

$$\begin{bmatrix} w_1 & w_2 \end{bmatrix} = \begin{bmatrix} \omega_{re}\dfrac{\Delta L_q}{L}i_q & \omega_{re}\dfrac{\Delta L_d}{L}i_d \end{bmatrix} \quad (27)$$

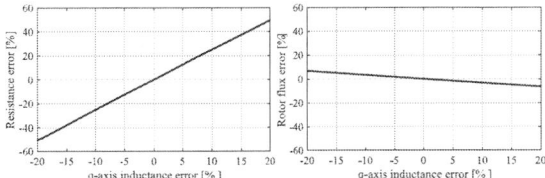

Fig. 3. Resistance and flux estimation error against inductance variation (left : resistance, right : flux).

TABLE I.   PARAMETERS OF THE SYSTEM

Parameter	Symbol	Specification	Unit
Rated Output	P_R	750	W
Rated Speed	N_R	3000	r/min
Rated Torque	T_R	2.387	N·m
Rated Current	I_R	5.9	Arms
Current Control Bandwidth	ω_c	2000	rad/s
Speed Control Bandwidth	ω_s	200	rad/s
Bandwidth of Model	ω_m	100	rad/s
Bandwidth of Estimator	ω_R	10	rad/s
	ω_ϕ	10	rad/s
Base Armature Rresistance	R_0	0.37	Ω
Base Armature Temperature	T_0	20	°C

$$\begin{bmatrix} \hat{R} \\ \hat{\psi} \end{bmatrix} = \begin{bmatrix} 1 & 0 \\ 0 & 1 \end{bmatrix} \begin{bmatrix} R \\ \phi \end{bmatrix} + \begin{bmatrix} -\omega_{re}\Delta L_q \dfrac{i_q}{i_d} \\ -\Delta L_q \dfrac{i_q^2}{i_d} + \Delta L_d i_d \end{bmatrix} . \quad (28)$$

V.   Experiment Results

In this section, the effectiveness of the proposed method is evaluated by experiment. Fig. 4 shows the block diagram of the system, and Table I shows parameters of system. The voltage compensator in Fig. 4 is compensated for the influence of dead time and on-resistance in the inverter [8]. In this paper, the estimated temperature using the proposed method is compared to the estimated temperature using the conventional method shown in Fig. 1. The conventional method is the estimation method in which θ_{re} is applied instead of $\hat{\theta}_{re}$, as in the method of Section III. In this paper, the experimental results of Fig. 6 to Fig. 9 use the q-axis inductance table in Fig. 5 . In addition, the armature temperature is detected by the thermocouple which is attached to three points on the armature winding in SPMSM. The average detected temperature value is used as the true value of the armature temperature T.

Figure 6 is shows the experimental result of the armature temperature estimation with the conventional method when the temperature of the SPMSM is fluctuated. Figure 7 is shows the experimental result of the armature temperature estimation with the proposed method under temperature fluctuations in the SPMSM. As seen in Fig. 6, the armature temperature is not estimated because the conventional method is affected by the rotor flux parameter error. From Fig. 7, the armature temperature is estimated by the proposed method well.

Figure 8 and 9 show the experimental result of the armature temperature estimation using the proposed method while

The 2018 International Power Electronics Conference

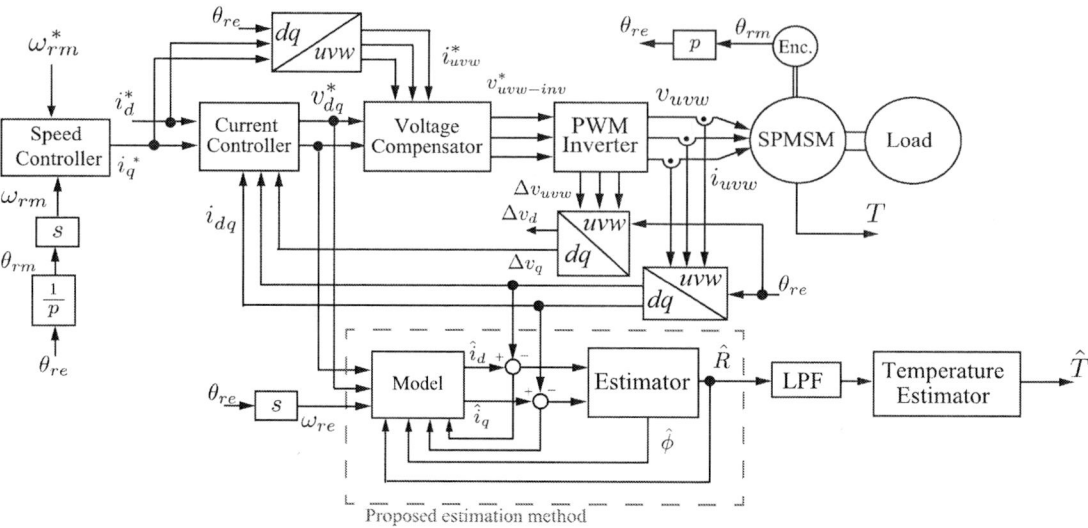

Fig. 4. Block diagram of proposed total system

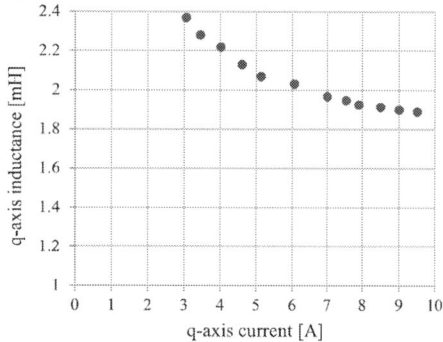

Fig. 5. Table of q-axis inductance from Fig. 6 to Fig. 9

fluctuating the load torque and angular velocity, respectively. From Fig. 8 and Fig. 9, when the angular velocity and current response of the SPMSM are at steady state, the armature temperature is estimated well.

VI. CONCLUSION

In this paper, a novel estimation method for the temperature in a SPMSM is proposed. This method estimates the armature temperature in the SPMSM based on the estimated resistance. It is possible to estimate the armature temperature form the estimated resistance because the armature resistance is estimated accurately without being affected by rotor flux fluctuation.

The effectiveness of the proposed method is evaluated by experiment. The proposed method is possible to estimate the armature resistance of the SPMSM by using the q-axis inductance table. The armature temperature is estimated well by using the proposed method. In addition, when the angular velocity or load torque is at a steady state, the armature temperature is estimated well.

REFERENCES

[1] P. Milanfar and J. H. Lang: "Monitoring the thermal condition of permanent-magnet synchronous motors", IEEE Trans. Aerosp. Electron. Syst., Vol. 32 No. 4, pp. 1421-1429, Oct. 1996.

[2] Oliver Wallscheid, Andreas Specht, and Joachim Bocker: "Observing the Permanent-Magnet Temperature of Synchronous Motors Based on Electrical Fundamental Wave Model Quantities", IEEE Trans. Indus. Info., Vol. 64, NO. 5, May 2017.

[3] L. He, S. Cheng, Y. Du, R. G. Harley, and T. G. Habetler: "Stator Temperature Estimation of Direct-Torque-Controlled Induction Machines via Active Flux or Torque Injection", IEEE Trans. Power Electronics, vol. 30, pp. 888-899 (2015).

[4] D. Reigosa, D. Fernandez, T. Tanimono, T. Kato and F. Briz: "Comparative Analysis of BEMF and Pulsating High-Frequency Current Injection Methods for PM Temperature Estimation in PMSMs", IEEE Trans. Power Electronics・祁 ol. 32, pp. 3691-3699, May 2017.

[5] M. A. Hamida, J. D. Leon, A. Glumineau, R. Boisliveau: "An Adaptive Interconnected Observer for Sensorless Control of PM Synchronous Motors With Online Parameter Identification", IEEE Trans. Ind. Elec., vol. 60 no. 2, pp. 739-748, (2014).

[6] T. Sano, K. Ohishi, Y. Yokokura, Y. Ide, D. Kuraishi, A. Takahashi: "Experimental Study on Estimation of Stator Resistance Variation Based on Stator Flux Linkage Estimation for Permanent Magnet Synchronous Motor", Proc. of the 2017 JIAS Conf., Vol. 3 , pp. 319-322, (2017) (in Japanese).

[7] H. Sugimoto, Y. Noto, T. Kikuchi, and Y. Matsumoto: "Position Sensorless Vector Control of Stator Resistance Estimation Function of IPMSM Using Adaptive Identification", IEEJ Trans. on Ind. Appl., Vol. 129, No. 1, pp. 77-81 (2009) (in Japanese).

[8] J. Kudo, T. Noguchi, M. kawakami, and K. Sano: "Mathematical Model Errors and Their Compensations of IPM Motor Control System", IEEJ Technical Meeting Record on Semiconductor Power Conversion, IEEJ, pp. 25-31 (2008) (in Japanese).

The 2018 International Power Electronics Conference

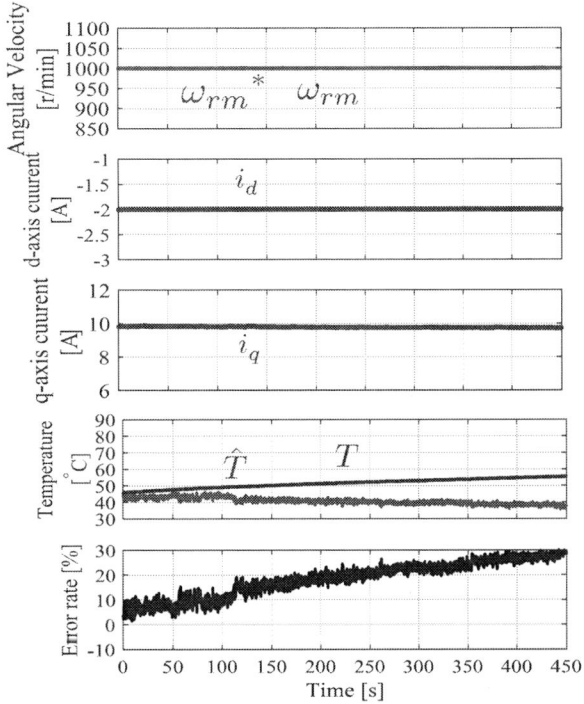

Fig. 6. Experimental result in variation of armature temperature by using conventional method ($i_d^* = -2A$, load torque = 100%, $\omega_{rm}^* = 1000r/min$).

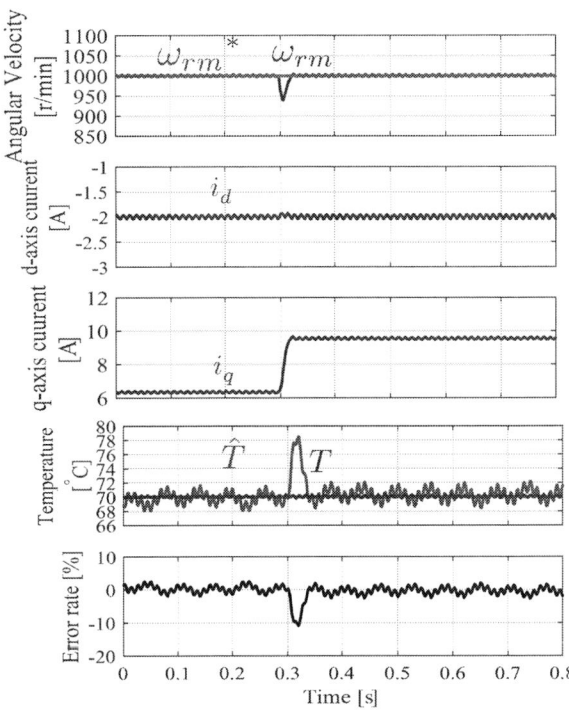

Fig. 8. Experimental result in variation of load torque by using proposed method ($i_d^* = -2A$, load torque = 50% → 100%, $\omega_{rm}^* = 1000r/min$).

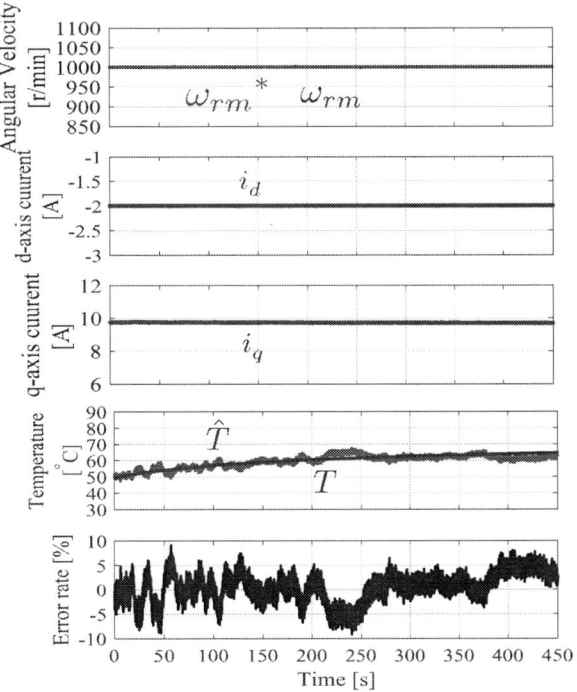

Fig. 7. Experimental result in variation of armature temperature by using proposed method ($i_d^* = -2A$, load torque = 100%, $\omega_{rm}^* = 1000r/min$).

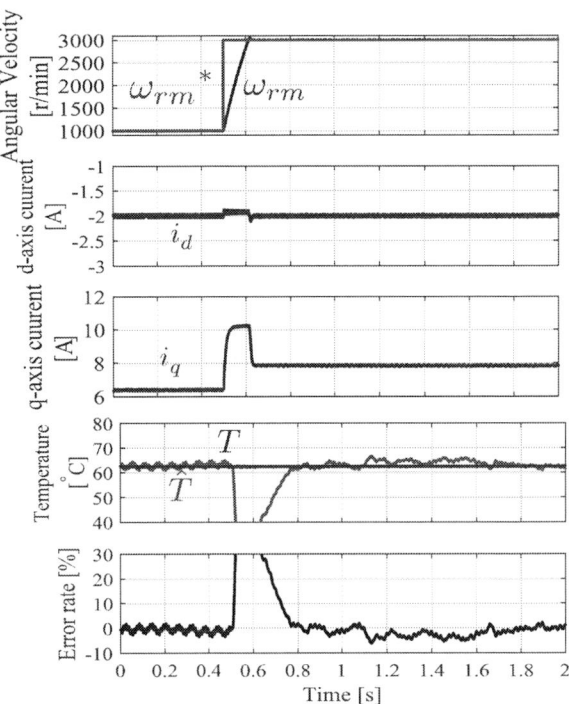

Fig. 9. Experimental result in variation of angular velocity by using proposed method ($i_d^* = -2A$, load torque = 50%, $\omega_{rm}^* = 1000r/min$ → $3000r/min$).

3901

The 2018 International Power Electronics Conference

Virtual Synchronous Generator Control with Reliable Fault Ride-through Capability by Adopting Model Predictive Control

Jonggrist Jongudomkarn[1*], Jia Liu[1] and Toshifumi Ise[1]

1 Graduate School of Engineering, Osaka University, Osaka, Japan

* jonggrist@pe.eei.eng.osaka-u.ac.jp

*Abstract-*Virtual synchronous generator (VSG) control has an important effect to support the frequency stability of power system thanks to its inertia support feature. However, with increasing penetration of inverter-based distributed generation, it is of great importance that the control scheme keeps enhancing its capacity in other key issues such as fault ride-through (FRT) capability. For this purpose, the idea of finite-set model predictive control (FS-MPC) based VSG control was studied in this paper. The proposed scheme allows multiple inputs multiple outputs (MIMO) system to control voltage and current simultaneously. Under these constantly placed voltage and current constraints, the controller displayed the ability to prevent the system from overcurrent condition. Several simulation studies are conducted in PSCAD/EMTDC to investigate the performance and FRT capability of the proposed strategy.

Keywords— DC grid and distribution systems, Finite-set model predictive control, Power electronics applied to transmission, Smart grid, Virtual synchronous generator.

I. INTRODUCTION

In the near future, as inverter-based distributed generator (DG) replaces the majority part of the conventional power generation system, unstable and insecure issues might arise in the power grid, since the conventional synchronous generators are capable of injecting the kinetic energy preserved in their rotating parts to power the grid during disturbances. Inverter-based DGs, on the other hand, have no rotational inertia and the perturbation in the frequency will become much higher. To prevent such a scenario, an approach called virtual synchronous generator (VSG) was proposed in [1]. It is stated that by letting the power electronics interface of DG unit emulate the behavior of a synchronous machine, a reaction similar to that of a synchronous generators (SG) to disturbance in the system, can be observed.

In accordance with the past researches of VSG, the results have clearly shown the advantages of using this concept to solve issues related to frequency deviations [1]–[3]. However, with growing importance of DGs, it might be beneficial to study the control concept under various conditions, for instances power quality and fault management. In this paper, the current limiting capability of the VSG control scheme is reviewed, as it is very well-known that inverter current should be limited during short-circuit faults and overload conditions. Otherwise, semiconductor switches can be damaged because of their low thermal inertia [4]–[5]. Since a voltage source inverter (VSI) based DG is usually used in a voltage controlled mode, where voltage and frequency are regulated at the DG terminal, thus in order to prevent overcurrent, a multi-loop control structure regulating current and voltage is often utilized. For this kind of control system, two main limiting strategies consisting of instantaneous saturation limit and latched limit are usually performed [4]. The former limiter prevents its input signal from increasing beyond a predefined value. However the output is distorted due to crest clipping [4]. In the latter limit strategy, the current or voltage references of the inverter needs to be replaced with a predefined current reference during overcurrent conditions [6] and thus fault detecting mechanism is required. This increases the complexity of the controller.

Recent years, model predictive control (MPC) has appeared to be an attractive alternative for the control of power converters because it has a flexible control scheme that allows an easy inclusion of multiple control variables, system constraints and nonlinearities. Previously, some control solutions based on MPC for distributed generation system appeared in the literature, for instances, a MPC strategy for a grid connected 3L-NPC is presented in [7]. MPC scheme for a VSIs was studied in [8], and other methods are proposed in [9]-[11]. Conclusively, several benefits of MPC scheme can be observed as suggested in [12], including minimizing steady-state errors over a wide target frequency range, fast dynamic response, and multivariable control capability. Inspired by these works, MPC based VSG control scheme for three-phase voltage source inverter is proposed in this paper. The proposed controller provides voltage and current control simultaneously. Without voltage clipping and without the need to change control mode or control references, the controller illustrates a current limiting ability during fault conditions. Moreover, it can operate in both grid-connected and islanded mode, while provides frequency stability to the power system with virtual inertia as well as enables active and reactive power-sharing between parallel connected inverters in the system.

3902

II. PROPOSED CONTROL SCHEME

A. VSG Control

The basic concept of VSG shown is to create a virtual inertia in the power circuit by emulating the behavior of SG [1]–[3]. The inverter can thus be regulated using the frequency and voltage drooping mechanism similar to the way in which SG is normally controlled. The virtual inertia is created with the help of the well-known swing equation (1).

$$P_{in} - P_{out} = J\omega_m \frac{d\omega_m}{dt} + D(\omega_m - \omega_g) \qquad (1)$$

where P_{in} is virtual shaft power, P_{out} and Q_{out} are measured output active and reactive power. ω_m is virtual rotor angular frequency. ω_g is output voltage angular frequency and J and D are virtual inertia and virtual damping factor, respectively.

Enhanced VSG control scheme proposed by [13] is adopted in this paper. As shown in control diagram illustrated in Fig. 1, the controller emulates the swing equation via the block "Swing Equation". As the swing equation is a differential equation, an algorithm based on Runge-Kutta iterative method is adopted to solve the virtual rotor angular frequency ω_m. The control uses "Governor Model" block to create the linear droop control loop between active power and the frequency, and "Q Droop" block to create the linear droop control loop between reactive power and the voltage. Hence, VSG control can be classified as a voltage-source-based grid-supporting control [14]. The output voltage is regulated by the V–Q droop controller via the reactive power reference. In order to share the active and reactive power according to the ratings of DGs without communication, the droop coefficients should be designed proportional to the capacities of each DG unit. Additionally, to resolve the problems of oscillation in active power during a disturbance and errors in reactive power sharing as discussed in[13], two major control blocks are added in to the enhanced VSG i.e., the "stator reactance adjuster" and the "bus voltage estimator". The function of stator reactance adjuster is to produce extra inverter output impedance to match the impedances between parallel connected DGs in order to eliminate oscillation during a

loading transition. This extra impedance will be called "virtual impedance" of the inverter in this paper. The virtual impedance is realized by multiplying inductor current through L_f by the virtual stator inductor in stationary frame. Finally, the "bus voltage estimator" compensates the line voltage drop and uses the estimated bus voltage as input for V-Q droop box in both parallel DGs to ensure proper reactive power sharing between DGs.

B. Model Predictive Control

Model predictive control (MPC) is an attractive alternative to the classical control methods, due to its fast dynamic response, simple concept, and ability to include nonlinearities, and constraints in the design of a controller. In this work the finite-set MPC (FS-MPC) for a three-phase two-level inverter-based renewable power generation system was chosen. Conventionally, the output voltage is controlled by a pulse width modulation (PWM) scheme. In contrast, PWM is replaced by MPC in this proposed control in order to make use of its multiple inputs multiple outputs ability. In principle, MPC uses the discrete nature of the 2-level-VSI, which can generate only eight distinct switching states since each inverter legs must never be short-circuited. This means although reference voltage input into PWM is continuous data, in the end only eight discrete switching states can be reproduced by the inverter. Hence, unlike in PWM control where the controller searches for voltage solution that satisfies the control objectives, MPC directly evaluates the performances of each of these eight possible switching states to determine the optimal control solution.

The power circuit of three-phase inverter with output LC filter displayed in Fig.2, can be described by two following equations, namely capacitor dynamics equation (2) and inductance dynamics equation (3).

$$C\frac{d\boldsymbol{v_c}}{dt} = \boldsymbol{i_f} - \boldsymbol{i_o} \qquad (2)$$

$$L\frac{d\boldsymbol{i_f}}{dt} = \boldsymbol{v_i} - \boldsymbol{v_c} \qquad (3)$$

where $\boldsymbol{v_c} = [v_{c,\alpha}\ v_{c,\beta}]^T$, $\boldsymbol{v_i} = [v_{i,\alpha}\ v_{i,\beta}]^T$, $\boldsymbol{i_f} = [i_{f,\alpha}\ i_{f,\beta}]^T$ and $\boldsymbol{i_o} = [i_{o,\alpha}\ i_{o,\beta}]^T$ are output voltage, inverter voltage, inductor current and the output current. These equations can be rewritten as (4):

$$\frac{d\boldsymbol{x}}{dt} = \boldsymbol{Ax} + \boldsymbol{Bv_i} - \boldsymbol{B_d i_o} \qquad (4)$$

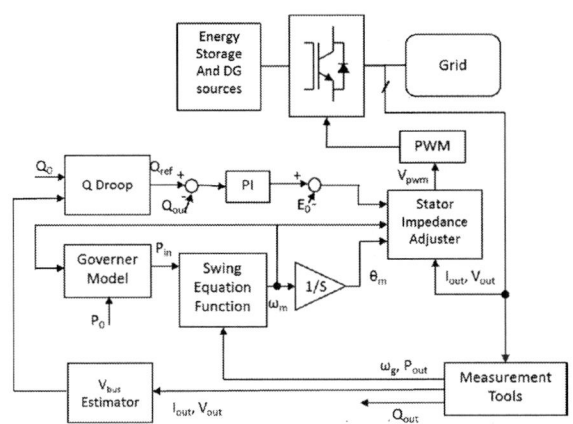

Fig. 1. The Control Diagram of enhanced VSG Control

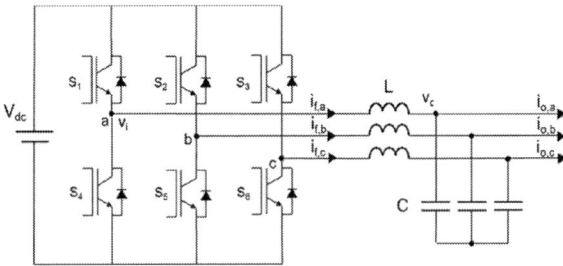

Fig. 2. Three-phase inverter with output LC filter

3903

where $x = \begin{bmatrix} i_{f,\alpha} \\ i_{f,\beta} \\ v_{c,\alpha} \\ v_{c,\beta} \end{bmatrix}$, $A = \begin{bmatrix} 0 & 0 & -\frac{1}{L} & 0 \\ 0 & 0 & 0 & -\frac{1}{L} \\ \frac{1}{C} & 0 & 0 & 0 \\ 0 & \frac{1}{C} & 0 & 0 \end{bmatrix}$, $B = \begin{bmatrix} \frac{1}{L} & 0 \\ 0 & \frac{1}{L} \\ 0 & 0 \\ 0 & 0 \end{bmatrix}$

and $B_d = \begin{bmatrix} 0 & 0 \\ 0 & 0 \\ -\frac{1}{C} & 0 \\ 0 & -\frac{1}{C} \end{bmatrix}$. A discrete-time model of the

system is derived from (4) for a sampling time T_s and can be expressed as [15]:

$$x(k+1) = A_q x(k) + B_q v_i(k) + B_q i_o(k) \qquad (5)$$

where $A_q = e^{AT_s}$, $B_q = \int_0^{T_s} e^{A\tau} B d\tau$ and $B_{dq} = \int_0^{T_s} e^{A\tau} B_d d\tau$.

Equation (5) can be used as the predictive model in the proposed controller. By using the system information at k instant, all n possible system behavior at $(k+1)$ instant for each switching states $(x_0(k+1), x_1(k+1), \dots x_n(k+1))$ can be predicted. For the proposed control scheme, the measurements of the line current and grid voltage are employed to forecast the filter current $i_f(k+1)$ and capacitor voltage $v_c(k+1)$ at the $(k+1)$ instant as depicted in Fig. 3. To select the optimal inverter voltage vector, all predicted values are compared using a predefined cost function, and the voltage vector that minimizes this cost function is chosen and applied at the next sampling period. The output is observed and the process is repeated. For instance, supposing the cost function is designed as shown in (6), the control system is set to track the voltage and current reference

$$g = k_v(v_{c,\alpha\beta}(k+1) - v^*{}_{\alpha\beta})^2 \\ + k_i(i_{f,\alpha\beta}(k+1) - i^*{}_{\alpha\beta})^2 \qquad (6)$$

where $v^*{}_{\alpha\beta} = \begin{bmatrix} v^*{}_\alpha & v^*{}_\beta \end{bmatrix}^T$ is the reference vector of the capacitor voltage, $i^*{}_{\alpha\beta} = \begin{bmatrix} i^*{}_\alpha & i^*{}_\beta \end{bmatrix}^T$ is the reference vector of the inductor current.

C. The proposed MPC-based VSG Control

The VSG control regulates voltage and frequency at the bus terminal using an $\omega - P$ droop and $V - Q$ droop control structure. However, due to the lack of current control loop, the controller does not have the ability to limit the current during short-circuit faults. Inner current control loop could be added to the VSG control system as studied in [16]. In despite of that, in occurrence of fault, the voltage magnitude of one or more phases will be temporary decreased. Hence, the natural response of the controller is to increase the current reference in the inner loop, as the voltage outer loop control tries to track the power set point. If the output voltage sag is deep enough, overcurrent will occur [5]. This condition can be prevented by using instantaneous saturation limit strategy. Nevertheless, the current saturation will clip the inverter output voltage [4]–[6]. Another solution is to use predefined current or voltage references during the event

of fault. However, a mean to detect fault is required for the method [6].

In contrast to that, thanks to the ability of MPC to create a multivariable control system, voltage and current control can be easily embedded into the MPC-based inverter control system as shown in Fig. 3., unlike instantaneous saturation limit or latch current, where the controller can only prevent the current from exceeding one fixed value regardless of operating power of the inverter- based system. For instance, in case of instantaneous saturation limit, the current will be limited when it exceeds the maximum value (2pu). This is true whether the DGs are operating at full load (1pu) or half load (0.5pu). MPC-based VSG on the other hand, creates current reference according to the operating rating and try to produce a current as close as possible to this value even under fault condition. It implies that the limit peak current when VSG is operating at half load is lower that the limit peak current at full load. This effect will be further explained with the simulation results shown in section III.

As illustrated in Fig.3, the voltage magnitude reference and frequency are given by "VSG control block", whose concept was adopted from the enhanced VSG control scheme, discussed in section II. In order to create current reference, "I_{ref} generator" block is added into the control diagram. This current command is determined by mimicking the relationship between the phase voltage and line current of a SG [17] as shown by the armature circuit illustrated in Fig. 4. Here we assume a cylindrical generator with the same per-phase synchronous reactance. E_A is the magnitude of the per-phase internal electromotive force of SG, I_A the armature current and $V_\emptyset$ is the generator's terminal phase voltage. Let X_S denote the generator's synchronous reactance, and R_A denote the armature resistance. Hence, the relation between the phase voltage and line current in stationary frame ($\alpha\beta$) can be described with (7)-(8).

$$E_{A,\alpha} = V_{\emptyset,\alpha} + I_{A,\alpha}(R_A + jX_S) \qquad (7)$$

$$E_{A,\beta} = V_{\emptyset,\beta} + I_{A,\beta}(R_A + jX_S) \qquad (8)$$

Matching this relation to the inverter system, the generator's terminal phase voltage is considered equivalent to the capacitor voltage of the output filter

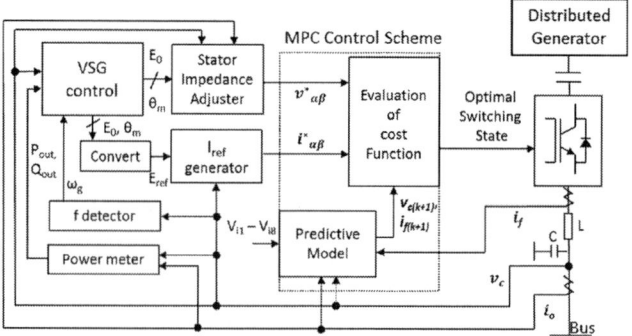

Fig. 3. Control Diagram of the purposed MPC-based Control

The 2018 International Power Electronics Conference

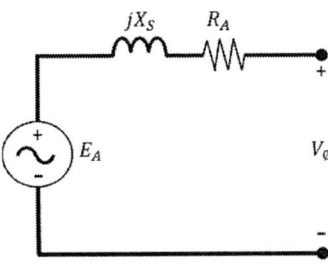

Fig. 4. The per-phase Armature circuit of generator.

TABLE I
SIMULATION PARAMETERS

VSG Parameters			
Parameter	Value	Parameter	Value
S_{base1}	10 kVA	M_i^*	4 s
S_{base2}	5 kVA	D_i^*	50 pu
$E_0 = V_{grid}$	200 V	L_{ls1}	9.39 mH
$\omega_0 = \omega_{grid}$	376.99 rad/s	L_{ls2}	13.81 mH
$P_{0\,i}^*$	1 pu	$Q_{0\,i}^*$	0 pu
R_A	0.01 Ω	V_{dc}	400 V
MPC Parameters			
k_v	5	k_i	3.625
T_s	50 μs		

($\boldsymbol{v_c}$). X_S is the sum of filter impedance (L) and virtual impedance (L_{ls}), multiplied by angular frequency ($X_S = \omega(L + L_{ls})$) and R_A equals the filter resistance. Deriving from the armature circuit diagram and (7)-(8), the armature current references i_α^* and i_β^* can be expressed as illustrated in Eq. (9). These are used as the command for the current-controlled inverter.

$$\begin{bmatrix} i_\alpha^* \\ i_\beta^* \end{bmatrix} = \boldsymbol{Y} \left\{ \begin{bmatrix} E_{\alpha,ref} \\ E_{\beta,ref} \end{bmatrix} - \begin{bmatrix} v_{c,\alpha} \\ v_{c,\beta} \end{bmatrix} \right\} \tag{9}$$

where $\boldsymbol{Y} = \frac{1}{R_A{}^2 + X_S{}^2} \begin{bmatrix} R_A & X_S \\ -X_S & R_A \end{bmatrix}$ and $\boldsymbol{E_{\alpha\beta,ref}}$ is reference voltage provided by the VSG control. It should be noted that these references are different from the voltage references $\boldsymbol{v^*}_{\alpha\beta}$ since the effect of "stator impedance adjuster" is excluded.

It can be deducted from the control that the voltage and current references along with the state of the system are considered simultaneously in order to find an optimal solution for the system according to the controller's cost function. It is implied that during the evaluation, the controller is looking for a solution which has the minimum error of voltage tracking, while the current will be constrained by current reference at the same time. Since these two processes happen simultaneously, the current reference will not be increased drastically, unlike the case of cascade controller. Supposedly the weight factors of the cost function (6) of these two references are properly defined under fault condition, the output voltage and current will be able to be maintained in almost sinusoidal waveform, while overcurrent will be prevented even without the use of a predefined control references or fault detection mechanism. Moreover, for future works, additional constraints can be considered in the cost function, such as state of charge of battery, switching frequency reduction, and spectrum shaping.

In order to find a reasonable balance between voltage and current control, the weight factors k_v and k_i have to be defined properly. In this work k_v and k_i are chosen so that voltage control has the higher priority since the VSG control scheme is a voltage control algorithm and the main purpose of current control is to constraint the current under abnormal condition. Hence, for normalized voltage and current k_v is defined to be larger than k_i as listed in Table I. A more systematic way to determine k_v and k_i is subject for further study.

III. SIMULATION

The performance of the proposed control strategy was investigated with simulations in PSCAD/EMTDC environment. The test circuit, illustrated in the Fig. 5, consists of two parallel connected inverter-based DG systems. Output filters parameters and line impedances of each generation system are also concluded in Fig. 5. Additionally, the VSG parameters of the circuits are listed in the Table I. The definitions and tuning method of VSG parameters is explained in [13].

A. Normal Conditions

To verify that the proposed MPC-based control can function properly in normal operation, two parallel connected MPC-based VSG systems are simulated in both grid-connected and islanded mode. Initially, the VSGs are connected to the grid with the reference active power of DG1 (P_{ref1}) equals 10W and reference reactive power of DG1 (Q_{ref1}) equals 0Var, while P_{ref2} and Q_{ref2} are 5kW and 0kVar respectively. The load connected to the system is 7.5kW+0.05kVar. In grid-connected period, the active and reactive power is set to follow the reference values, while event of islanding from grid is simulated at 10s. From this moment on, the active and reactive power-sharing between the inverters are done according to their rated powers. Since DG1 has twice as much rated power as DG2, DG1 shall share twice amount of active and reactive power during islanded operation. The simulation results shown in Fig.6 indeed verify the active and reactive power sharing capabilities of the controller. Loading transition from 7.5kW+0.05kVar to 14kW+2.05kVar is conducted at 15s. The results show that the system can maintain the power sharing ratio even after the load transition, while real and reactive power oscillations are not noticeable during transition. The

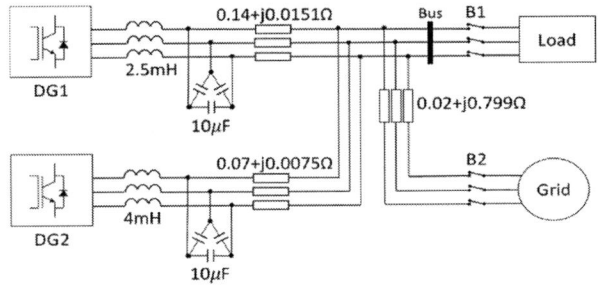

Fig. 5. Simulation circuit diagram

3905

angular frequency deviation and inverters output voltage are also displayed in the figure. It can be conducted that the change of frequency is indeed slowed down during grid islanding and load transition. This shows the ability of the control scheme to generate virtual inertia into the system. It should be note that the load 14kW+2.35kVar is measured at 200V, thus as the bus voltage is reduced to around 195V after the load transition, the total power produced by DGs are around 13kW + 2.25kVar instead of 14kW+2.35kVar. Moreover, Fig. 6 also shows that the system had the ability to regulate voltage and frequency at the output terminal using an $\omega - P$ droop and $V - Q$ droop control structure.

B. Fault Conditions

In order to investigate the performance of the MPC-based control system under fault condition, comparative study between PWM-based VSG discussed in section II and the proposed MPC-based VSG will be conducted in both grid-connected and islanded mode during the event of fault. The simulation circuit and system parameters are equivalent to those from Fig.5 and Table I.

For grid-connected mode, the load connected to the system is 15kW+0.01kVar. P_{ref1} and P_{ref2} are set at 10W and 5kW respectively, while Q_{ref1} and Q_{ref2} are both set to 0kVar. The event of a single phase A to ground fault was simulated across the load while the fault resistance is set to 0.01 Ω. The fault occurs at t = 5s and it is automatically cleared at t = 5.5s. Bus phase voltage waveforms and inverter current waveforms are displayed in Fig. 7, for which the control system is implemented with PWM, and Fig. 9 presents the inverter current waveforms of the control system, which is regulated by MPC. Real and reactive power for both control schemes

are illustrated in Fig. 8 and Fig.10. It can be deducted from the figures that during the fault the bus voltage waveforms of the PWM-based controller get distorted and the peak currents of the PWM based controller exceed 2pu in both DG1 and DG2, which are around 80A and 40A respectively. The distortions of the voltage waveforms happen because the PWM-based VSG assumes a symmetrical load condition, which is not true during the event of fault and the overcurrent is caused by the lack of current control in the control system. Additionally, the distortion and overcurrent create large power oscillation as shown in Fig.8. On the other hand, the peak inverter currents of MPC-based controlled inverter only increase to around 60A in DG1 and around 40A in DG2. Furthermore, even under unsymmetrical condition, the MPC controller uses the measured voltage and current to find optimal inverter switching state, which provides voltage and current as close as possible to the desired values. Hence, almost sinusoidal waveforms of bus voltage and inverter current are produced, while significantly less oscillations in real and reactive power are created, compared to PWM-based VSG. However, the steady state voltage and current waveforms of MPC-

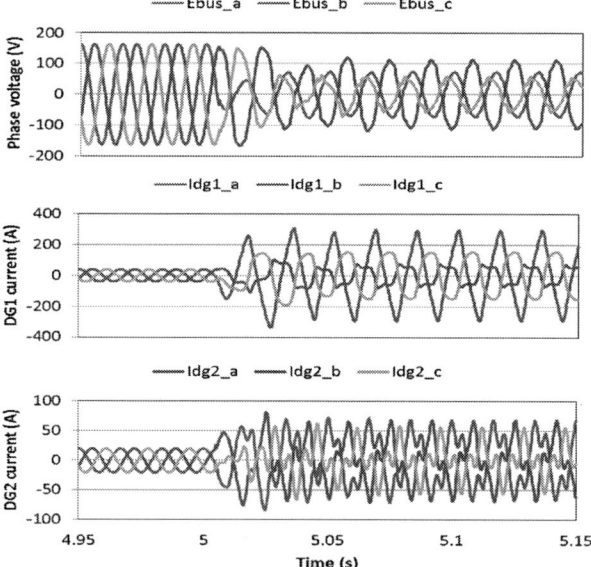

Fig. 7. Voltage and current waveforms of PWM-based controlled system in grid connected mode during the event of single-phase-to-ground fault.

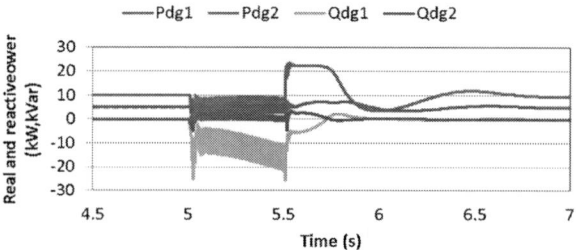

Fig. 8. Real and reactive power of PWM-based controlled system in grid connected mode during the event of single-phase-to-ground fault.

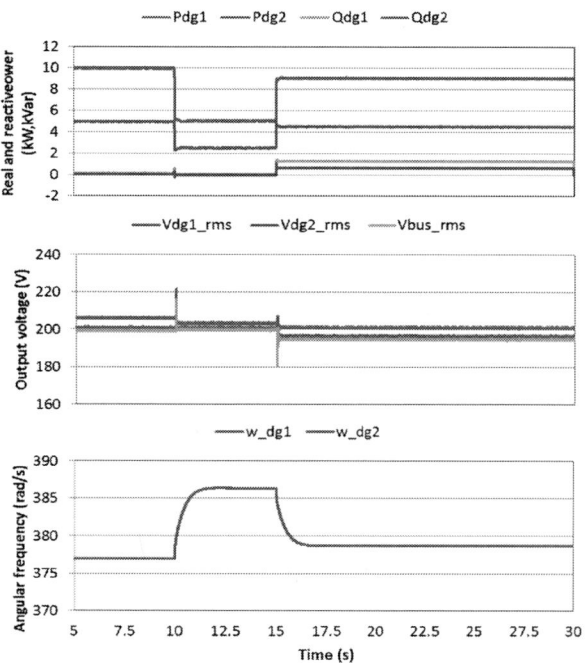

Fig. 6. Simulation results of the proposed control in grid connected and islanded mode under normal condition

The 2018 International Power Electronics Conference

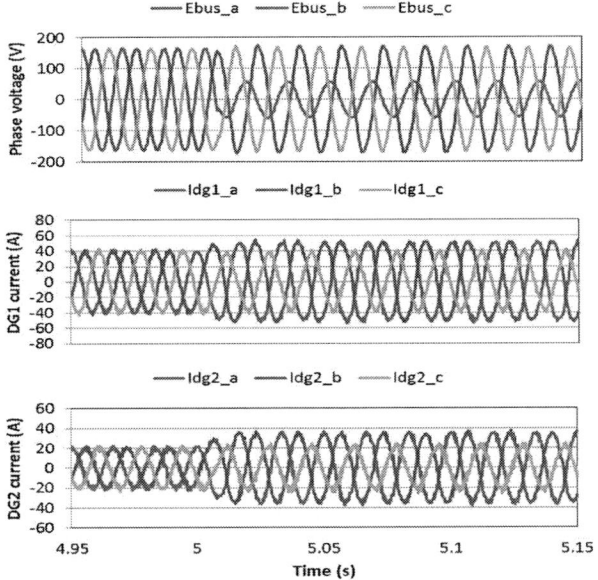

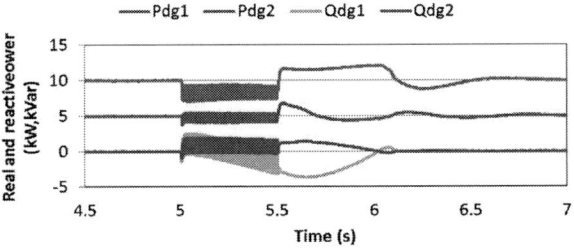

Fig. 9. Voltage and current waveforms of MPC-based controlled system in grid connected mode during the event of single-phase-to-ground fault

Fig. 10. Real and reactive power of MPC-based controlled system in grid connected mode during the event of single-phase-to-ground fault.

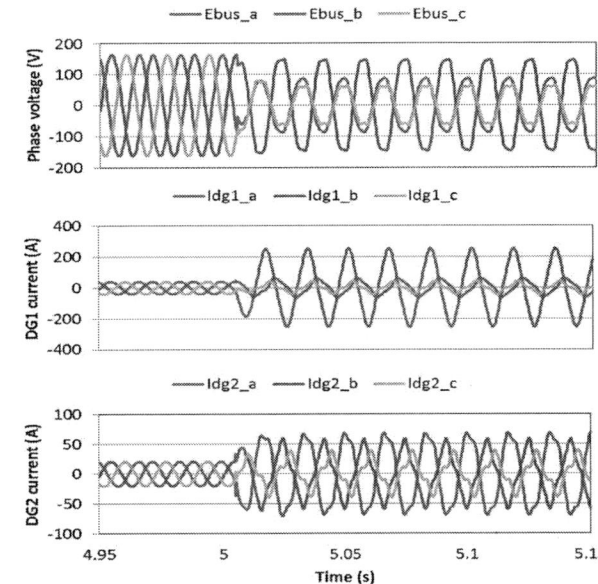

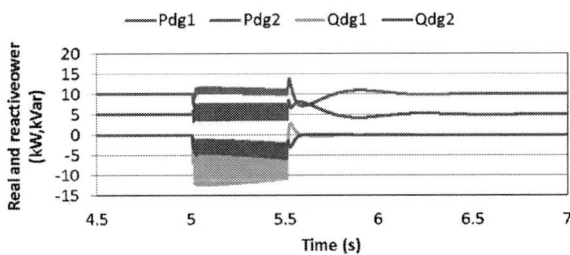

Fig. 11. Voltage and current waveforms of PWM-based controlled system in islanded mode during the event of single-phase-to-ground fault.

Fig. 12. Active and reactive power of PWM-based controlled system in islanded mode during the event of single-phase-to-ground fault.

based system display larger ripples in comparison to its PWM counterpart.

In islanded mode, first, the system under test is set to operate at full load (15kW+0.01kVar). Then phase A to ground fault with 0.01 Ω fault resistance is simulated at t = 5s till t = 5.5s. The waveforms of bus voltage inverter current and real and reactive power are displayed in Fig. 11-12, respectively for PWM-based VSG control and in Fig.13-14 for MPC-based VSG control. Again PWM-based VSG cannot cope with asymmetrical condition, resulting in overcurrent and distortion of voltage and current waveforms. In MPC-based VSG, peak currents are limited below 60A in DG1 and around 40A in DG2, while power oscillations during the event of fault are significantly lower compared to its PWM counterpart. In islanded mode, larger ripples in voltage, current and power can also be observed in MPC-based system.

MPC-based VSGs are simulated again in islanded mode while both DGs are operating at half load (7.5kW+0.01kVar). With the same fault resistance, phase A to ground fault is simulated at t = 5s till t = 5.5s. The simulation results are shown in Fig.15-16. The results illustrate that the peak currents are now limited around

50A in DG1 and 30A in DG2. These are clearly lower than the peak currents of 60A and 40A, which are produced by DG1 and DG2 at full load. It shows the unique ability of MPC-based VSG control to limit the peak current according to the operating power rating, unlike the controller with instantaneous saturation current limit, where the peak current will be limited by a fixed vale regardless of the operating power rating, or controller that uses predefined limit current reference during fault, where the fault detection will be triggered after the current exceed a fixed predefined value, again regardless of operating power rating.

To evaluate the MPC-based VSG under symmetrical fault, the system is now simulated under three phases to ground fault condition. The DGs are operating in islanded mode while both DGs are operating at full load (15kW+0.01kVar). Three phases to ground is simulated at t = 5s till t = 5.5s while the fault resistance is set to 0.01 + j0.131Ω. The simulation results are shown in Fig.17-18. The symmetrical fault causes the output rms voltage of DGs to drop from 200V to around 20V. The results illustrate that the peak currents are limited around 60A in DG1 and 40A in DG2, similar to the case of

3907

The 2018 International Power Electronics Conference

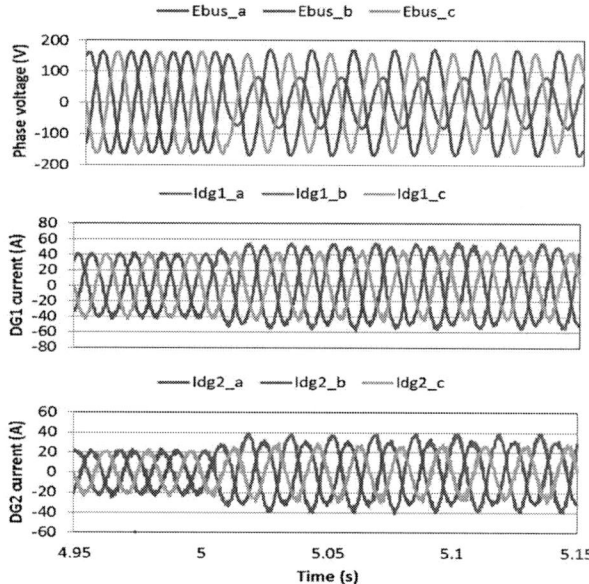

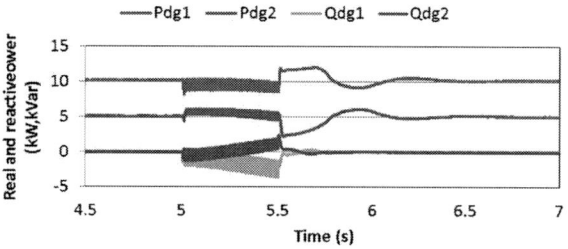

Fig. 13. Voltage and current waveforms of MPC-based controlled system in islanded mode at full load during the event single-phase-to-ground fault.

Fig. 14. Active and reactive power of MPC-based VSG control in islanded mode at full load during the event single-phase-to-ground fault.

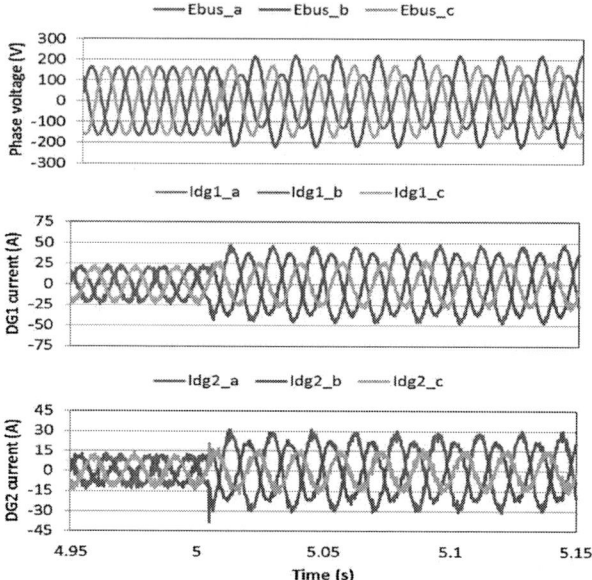

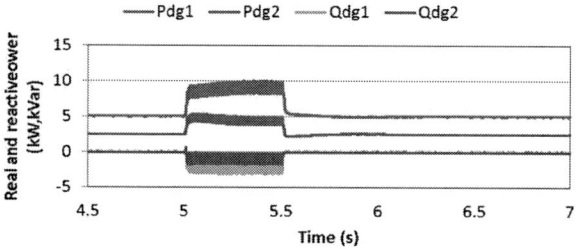

Fig. 15. Voltage and current waveforms of MPC-based controlled system in islanded mode at half load during the event of single-phase-to-ground fault.

Fig. 16. Active and reactive power of MPC-based VSG control in islanded mode at half load during the event of single-phase-to-ground fault.

single-phase-to-ground fault. This shows the low-voltage-ride-through capability of MPC-based VSG. During the event of fault, the real power of DG1 and DG2 almost drops to zero, while reactive power increases slightly.

Nevertheless, since the PWM-based VSG studied in this paper lacks current control, a clear comparison might be difficult to be drawn from the results. To further verify the performance of MPC-based VSG, a comparative study between the proposed MPC-based VSG and PWM-based multiloop control with inner current and outer voltage control loop is conducted in separate work [18]. Lastly, due to the nature of MPC, the proposed control scheme produced more ripple than its PWM counterpart. This issue is subjected to further study.

IV. CONCLUSION

In this paper, a MPC-based control scheme is presented for a three-phase inverter with output *LC* filter. Results show that the proposed scheme can operate in both grid-connected mode and islanded mode, while achieving good active and reactive power sharing, output voltage control and virtual inertia feature to slow down

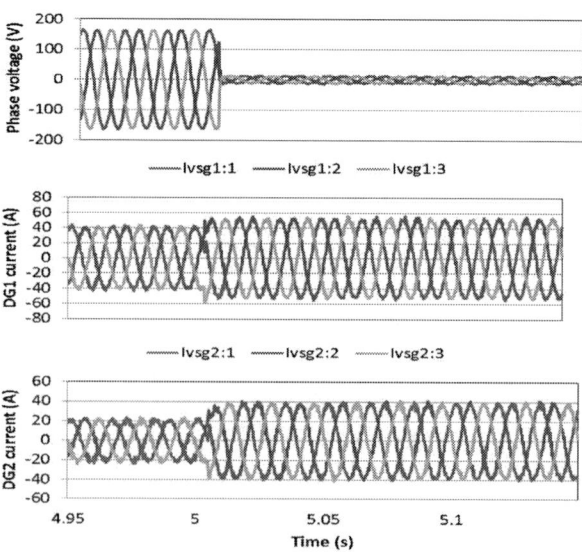

Fig. 17. Voltage and current waveforms of MPC-based controlled system in islanded mode during low-voltage-ride-through.

3908

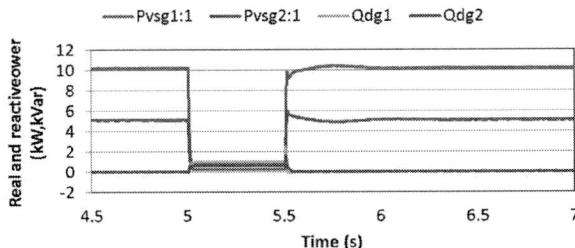

Fig. 18. Active and reactive power of MPC-based VSG control in islanded mode during low-voltage-ride-through.

frequency deviation during load transitions and grid islanding. The proposed control also shows a current limiting ability under single-phase-to-ground and three-phase-to-ground fault conditions without requiring any change in control mode or fault detection mechanism, while a voltage and current clipping due to the saturation limit used in inner current loop of PWM-based cascade control can be avoided. Additionally, MPC-based VSG can limit the peak current according to the operating rating of the system, unlike other control where the peak current is always limited under 200% of full load current regardless of the load rating.

REFERENCES

[1] K. Sakimoto, Y. Miura, and T. Ise, "Stabilization of a power system with a distributed generator by a virtual synchronous generator function," in *Proc. 8th IEEE Int. Conf. Power Electron. ECCE Asia*, Shilla Jeju, Korea, 2011, pp. 1498–1505.

[2] J. Driesen and K. Visscher "Virtual synchronous generators," in *Proc. IEEE Power Energy Soc. Gen. Meeting—Convers. Del. Elect. Energy 21st Century*, Pittsburgh, PA, USA, 2008, pp. 1–3.

[3] L. M. A. Torres, L. A. C. Lopes, T. L. A. Moran, and C. J. R. Espinoza, "Self-tuning virtual synchronous machine: A control strategy for energy storage systems to support dynamic frequency control," *IEEE Trans. Energy Convers.*, vol. 29, no. 4, pp. 833–840, Dec. 2014

[4] N. Bottrell and T. Green, "Comparison of current-limiting strategies during fault ride-through of inverters to prevent latch-up and wind-up", *IEEE Trans. Power Electron*, vol. 29, no. 7, pp. 3786–3797, 2014.

[5] C. Plet, M. Graovac, T. Green, and R. Iravani, "Fault response of gridconnected inverter dominated networks," *IEEE Power and Energy Society General Meeting*, Jul. 2010, pp. 1–8, 2010.

[6] I. Sadeghkhani; M. E. Hamedani Golshan; J. M. Guerrero; A. Mehrizi-Sani, "A current limiting strategy to improve fault ride-through of inverter interfaced autonomous microgrids," *IEEE Transactions on Smart Grid*, vol.PP, no.99, pp.1-11, 2016.

[7] H. Miranda, R. Teodorescu, P. Rodriguez, and L. Helle: "Model predictive current control for high-power grid-connected converters with output LCL filter", *Conf. of IEEE Ind. Electron*, Proc. 35th Annu, pp. 633-638, 2009.

[8] Q. Zeng and L. Chang: "An advanced SVPWM-based predictive current controller for three-phase inverters in distributed generation systems", *IEEE Transactions on Industrial Electronics*, Vol. 55, no. 3, 2008.

[9] S. Alepuz, S. Busquets-Monge, J. Bordonau, P. Corts, and S. Kouro: "Control methods for low voltage ride-through compliance in grid-connected NPC converter based wind power systems using predictive control", *IEEE Energy Convers. Congr. Exposition*, pp. 363-369, 2009.

[10] J. Hu, J. Zhu, D. G. Dorrell: "Model predictive control of inverters for both islanded and grid-connected operations in renewable power generations", *IET Renewable Power Generation*, Vol. 8, Iss. 3, pp. 240-248, 2014.

[11] A. F. Ayad, R. M. Kennel: "Model predictive controller for grid-connected photovoltaic based on quasi-Z-source inverter

Sensorless Control for Electrical Drives and Predictive Control of Electrical Drives and Power Electronics (SLED/PRECEDE)", *Proc. IEEE International Symposium*, Oct. 17-19, pp. 1-6, 2013.

[12] T. S. Radwan, M. A. Rahman, A. M. Osheiba, and A. E. Lashine: "Digital current control techniques for voltage source inverters", *Proc. Can. Conf. Elect. Comput. Eng.*, Sep. 5-8, vol. 2, pp. 1124-1127, 1995.

[13] J. Liu, Y. Miura, H. Bevrani, T. Ise, "Enhanced virtual synchronous generator control for parallel inverters in microgrids", *IEEE Transactions on Smart Grid*, vol. 8, no. 5, pp. 2268-2277, Sept. 2017.

[14] J. Rocabert, A. Luna, F. Blaabjerg, and P. Rodriguez, "Control of power converters in AC microgrids", *IEEE Transaction on Power Electron*, vol. 27, no. 11, pp. 4734–4749, Nov. 2012.

[15] Cortés, P., Ortiz, G., Yuz, J.I., Rodríguez, J., Vazquez, S., Franquelo, L.G, "Model predictive control of an inverter with output LC filter for UPS applications", *IEEE Trans. Ind. Electron*, 56, (6), pp. 1875–1883, 2009.

[16] M. Guan, W. Pan, J. Zhang, Q. Hao, J. Cheng and X. Zheng, "Synchronous generator emulation control strategy for voltage source converter (VSC) stations", *IEEE Trans. Power Syst.*, vol. 30, no. 6, pp. 3093–3101, 2015.

[17] Y. Hirase, O. Noro, E. Yoshimura, H. Nakagawa, K. Sakimoto and Y. Shindo, "Virtual synchronous generator control with double decoupled synchronous reference frame for single-phase inverter," *International Power Electronics Conference (IPEC-Hiroshima 2014 - ECCE ASIA)*, Hiroshima, pp. 1552-1559, 2014.

[18] J. Jongudomkarn, J. Liu, T. Ise, "Comparison of current-limiting strategies of virtual synchronous generator control during fault ride-through" *The 10th Symposium on Control of Power and Energy Systems (CPES2018)*, in press.

Reshaping Quadrature-Axis Impedance of Three-Phase Grid-Connected Converters for Low-Frequency Stability Improvement

Yi Tang[1], Jingyang Fang[1], Xiaoqiang Li[1], and Hongchang Li[2]

School of Electrical and Electronic Engineering, Nanyang Technological University, Singapore, Singapore
Energy Research Institute @ NTU (ERI@N), Nanyang Technological University, Singapore, Singapore
E-mail: yitang@ntu.edu.sg, jfang006@e.ntu.edu.sg, lixiaoqiang@ntu.edu.sg, hongchangli@ntu.edu.sg

Abstract— When employed as grid-interfaces, three-phase AC-DC and DC-AC power converters should always synchronize with the power grid so that active and/or reactive power can properly be regulated while maintaining desired grid-injected currents. Grid synchronization necessitates information of grid voltages, which is normally obtained through phase-locked-loops (PLLs). However, the employment of PLLs may bring in stability concerns, because it shapes the impedance of power converters into a negative resistance in the quadrature-axis (*q*-axis). To resolve the instability issues, this paper proposes an impedance controller for reshaping the *q*-axis impedance into a positive resistance in the low-frequency band. Without any extra burden on system hardware, the proposed controller can easily be implemented through directly relating the *q*-axis voltage to the *q*-axis current reference. As a result, the presented three-phase power conversion system operates stably even under a severely weak grid condition, as already been verified by simulation and experimental results.

Keywords— *Impedance, phase-locked-loop (PLL), power converter, stability improvement.*

I. INTRODUCTION

Three-phase AC−DC and DC−AC power converters have been widely used in the applications covering from distributed generation to energy storage systems [1−3]. When employed as grid-interfaces, power converters should perform active and/or reactive power regulations, AC current tracking, and DC-link voltage control if necessary. To achieve such goals, grid-connected power converters can properly be regulated in various ways. When regulated in the synchronous *dq*0-frame, simple proportional-integral (PI) controllers allow accurate AC current tracking with zero steady state errors, because sinusoidal signals have already been mapped into their magnitudes through the well-established *abc* to *dq*0 transformation [3].

Grid synchronization necessitates the use of phase-locked-loops (PLLs), from which the information of grid voltages can readily be obtained. However, it has been shown that PLLs without proper design may cause three-phase grid-connected converters to oscillate in a low-

This research is supported by the National Research Foundation, Prime Minister's Office, Singapore under the Energy Programme and administrated by the Energy Market Authority (EP Award No. NRF2015EWT-EIRP002-007).

frequency band [4].

Mechanisms for the instability issues introduced by PLLs can be explained by the well-known impedance criterion. The impedance criterion was first proposed by Middlebrook to assess the stability of DC−DC converters [5]. Since then, great attention has been attached to this approach [6, 7]. Through the impedance criterion, it has been found that constant power loads may introduce instability issues, as their impedance is essentially a negative resistance in the low-frequency band [8].

The model of power converters including the PLLs has been built in [9]. Moreover, it is concluded in [10] and [11] that the employment of PLLs modifies the *q*-axis impedance of grid-connected power converters. Specifically, the *q*-axis impedance becomes a negative resistance in the low-frequency band [10, 11]. Because of this characteristic, the interaction between the grid-connected converter and power grid, particularly under a weak power grid condition, may destabilize the power conversion system.

Modifying the PLL design is found to be a simple and straightforward solution to addressing the instability issues introduced by PLLs. Since the frequency range of the negative resistance shrinks as the PLL bandwidth reduces, stability improvement can easily be achieved by reducing the PLL bandwidth [9, 10]. However, this solution is a trade-off between the system stability and PLL dynamics.

In this paper, an impedance controller directly linking the *q*-axis voltage to the *q*-axis current reference is proposed. It allows improvement of the system stability by reshaping the *q*-axis impedance of grid-connected converters into a positive resistance in the low-frequency band.

II. SYSTEM MODELLING

Fig. 1 illustrates a three-phase grid-connected power converter together with its current control implemented in synchronous *dq*-frame, where the DC-link is fed by a DC power supply. Here, the power grid is modelled as an ideal voltage source connected in series with the grid inductors.

From Fig. 1, the following equations in the *dq*-frame

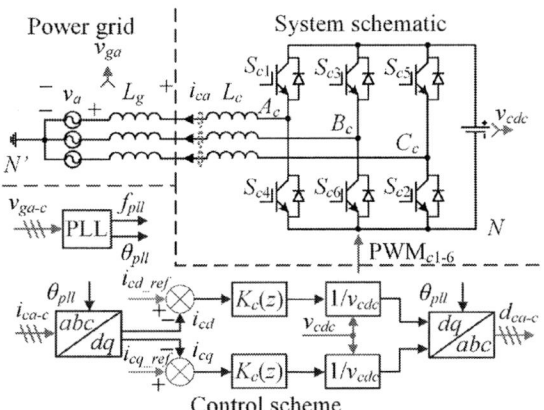

Fig. 1. Schematic and control scheme of a three-phase grid-connected voltage source converter.

can be derived:

$$\begin{cases} v_{cd}(t) = v_{gd}(t) + L_c \dfrac{di_{cd}(t)}{dt} - \omega L_c i_{cq}(t) \\ v_{cq}(t) = v_{gq}(t) + L_c \dfrac{di_{cq}(t)}{dt} + \omega L_c i_{cd}(t) \end{cases}, \quad (1)$$

where $v_{cd}(t)$ and $v_{cq}(t)$ represent the converter output voltages, and the associated Park's transformation matrix can be expressed as:

$$T_{dq/\alpha\beta} = \begin{bmatrix} \cos\theta(t) & \sin\theta(t) \\ -\sin\theta(t) & \cos\theta(t) \end{bmatrix}, \quad (2)$$

where $\theta(t) = \omega t$ denotes the phase-angle of the grid voltages. Fig. 2 details the block diagram of the PLL, where $K_{pll}(z)$ denotes the discrete transfer function of the PI controller, and it can be represented as:

$$K_{pll}(z) = K_{pll\_p} + K_{pll\_i} \cdot K_i(z), \quad (3)$$

where $K_i(z) = T_s/2 \cdot (z+1)/(z-1)$ stands for the integrator, and T_s denotes the sampling time.

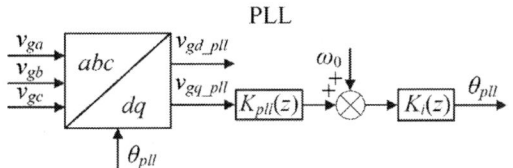

Fig. 2. Block diagram representation of the PLL.

The small-signal transfer function from the q-axis PLL voltage $\Delta v_{gq\_pll}(z)$ to $\Delta\theta_{pll}(z)$ can be obtained from Fig. 2 and expressed as:

$$G_{pll\_ol}(z) = \frac{\Delta\theta_{pll}(z)}{\Delta v_{gq\_pll}(z)} = K_{pll}(z) \cdot K_i(z), \quad (4)$$

where the prefix Δ denotes the change of relevant parameters. Furthermore, the transfer function from the q-axis grid voltage $\Delta v_{gq}(z)$ to $\Delta\theta_{pll}(z)$ can be derived as [10]:

$$G_{pll\_cl}(z) = \frac{\Delta\theta_{pll}(z)}{\Delta v_{gq}(z)} = \frac{G_{pll\_ol}(z)}{1 + G_{pll\_ol}(z)V_{gd}}, \quad (5)$$

where V_{gd} denotes the amplitude of grid voltages. Equation (5) clearly demonstrates the effect of the PLL, which formulates a path for spreading the disturbance in $v_{gq}(z)$ through $\Delta\theta_{pll}(z)$ into the current controller. Referring to Fig. 1, there exist one abc to dq transformation for inductor current measurements and another dq to abc transformation for reference update, and these transformations will inevitably be influenced by the PLL. This effect is demonstrated in Fig. 3, where z^{-1} models the time-delay introduced by duty ratio calculation [12]. As seen, two extra terms $-\Delta\theta_{pll}(z)I_{cd}$ and $\Delta\theta_{pll}(z)V_{gd}$ are added to quantify the PLL effect, where I_{cd} denotes the d-axis component of grid currents, and the detailed derivation can be found in [10]. In Fig. 3, the current regulator $K_c(z)$ is also implemented as a PI controller:

$$K_c(z) = K_{c\_p} + K_{c\_i} \cdot \frac{T_s(z+1)}{2(z-1)}. \quad (6)$$

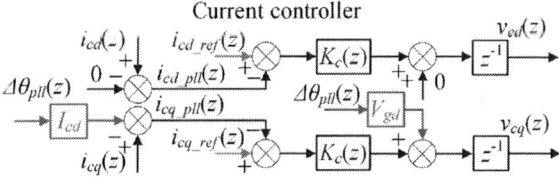

Fig. 3. Influence of the PLL on current control.

Fig. 3 indicates that the major influence of the PLL on current control is reflected in the q-axis. Therefore, the q-axis impedance will be derived next. For simplification, the coupling-effect between the d- and q-axis is ignored. Fig. 4 presents the structure of current control in the q-axis, where $G_{plant}(z)$ denotes the simplified system plant, which can be derived from (1) as:

$$G_{plant}(z) = \frac{T_s}{L_c(z-1)}. \quad (7)$$

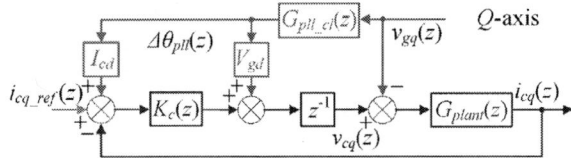

Fig. 4. Structure of the q-axis current control.

From Fig. 4, the transfer functions $G_{icq\_cl}(z) = i_{cq}(z)/i_{cq\_ref}(z)$ and $G_{vgq\_cl}(z) = i_{cq}(z)/v_{gq}(z)$ can readily be derived as:

$$G_{icq\_cl}(z) = \frac{i_{cq}(z)}{i_{cq\_ref}(z)} = \frac{K_c(z)z^{-1}G_{plant}(z)}{1 + K_c(z)z^{-1}G_{plant}(z)}, \quad (8)$$

$$G_{vgq\_cl}(z) = \frac{-1}{Z_{qq}(z)} = G_{vgq\_pll}(z) + G_{vgq\_plant}(z) \quad (9)$$

3911

where $Z_{qq}(z)$ denotes the q-axis impedance of grid-connected converters and $G_{vgq\_pll}(z)$ and $G_{vgq\_plant}(z)$ are respectively contributed by the PLL and system plant, expressed as:

$$G_{vgq\_pll}(z) = \frac{G_{pll\_cl}(z)z^{-1}G_{plant}(z)\left[V_{gd} + I_{cd}K_c(z)\right]}{1 + K_c(z)z^{-1}G_{plant}(z)}. \quad (10)$$

$$G_{vgq\_plant}(z) = \frac{-G_{plant}(z)}{1 + K_c(z)z^{-1}G_{plant}(z)}. \quad (11)$$

Using the system and control parameter values listed in TABLE I, the Bode diagrams of $G_{icq\_cl}(z)$ and $Z_{qq}(z)$ are plotted in Fig. 5. As observed from Fig. 5(a), $G_{icq\_cl}(z)$ exhibits a unity gain and zero phase-shift, namely $G_{icq\_cl}(z) \approx 1$, in the low-frequency band, thereby proving the effectiveness of current control in reference tracking. Fig. 5(b) clearly shows that the PLL design can influence $Z_{qq}(z)$, and this is in consistent with (9). As seen, $Z_{qq}(z)$ becomes a negative resistance with a phase of -180 degrees in the low-frequency band, and the increase of K_{pll_p} will enlarge this range.

TABLE I
SYSTEM AND CONTROL PARAMETER VALUES

Description	Symbol	Value
DC-link voltage	V_{cdc}	250 V
Filter inductor	L_c	2 mH
Grid voltage peak	V_{gd}	100 V
Grid inductor	L_g	10 mH
Power rating	P_g	0.6 kW
Sampling frequency	f_s	10 kHz
Switching frequency	f_{sw}	10 kHz
PLL proportional gain	K_{pll_p}	15
PLL integral gain	K_{pll_i}	300
Regulator proportional gain	K_{c_p}	15
Regulator integral gain	K_{c_i}	300
D-axis current reference	i_{cd_ref}	4 A
Q-axis current reference	i_{cq_ref}	0 A
Impedance control gain	K_{qf}	-0.1

According to the impedance criterion, the stability of grid-connected conversion systems can be evaluated by the ratio of the grid impedance $Z_{gq}(z)$ to the converter impedance $Z_{qq}(z)$. To be specific, the condition that $|Z_{gq}(z)/Z_{qq}(z)| < 0$ dB at all the $\pm180°$ crossings guarantees the system stability [5]. Fig. 6 visualizes the stable and unstable cases under a variable K_{pll_p}. As illustrated, the increase of K_{pll_p} will gradually destabilize the system, thereby indicating that the instability of grid-connected converters may be caused by the PLL design.

III. PROPOSED CONTROL SCHEME

The analysis provided in the previous section reveals that the increase of K_{pll_p} will negatively impact the system stability due to the enlarged negative resistance region of $Z_{qq}(z)$. Therefore, it is highly desirable if $Z_{qq}(z)$ can be reshaped as a positive resistance in the low-frequency band. This is possible through the superposition of $G_{icq\_cl}(z)$ and $G_{vgq\_cl}(z)$, which can easily be achieved by adopting the proposed impedance controller, which directly links the grid voltage $v_{gq}(z)$ to the current reference $i_{cq\_ref}(z)$ by a transfer function $K_{qf}(z)$,

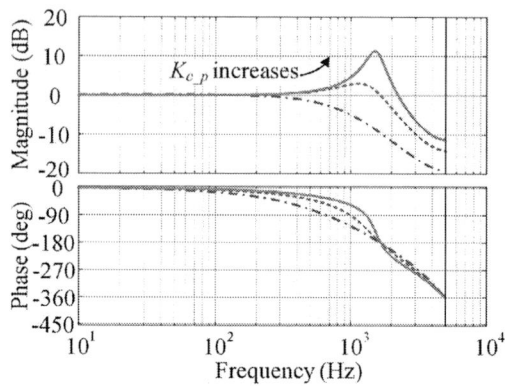

(a) $G_{icq\_cl}(z)$ with K_{c_p} = 5, 10, and 15

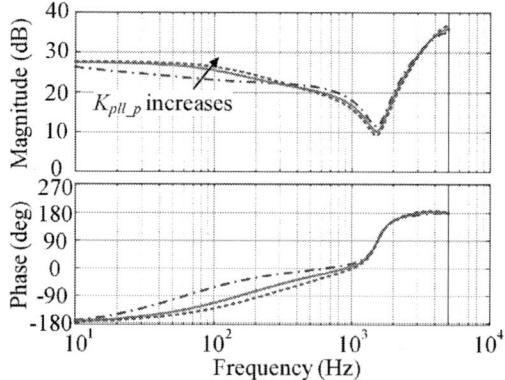

(b) $Z_{qq}(z)$ with K_{pll_p} = 5, 15, and 25

Fig. 5. Bode diagrams of $G_{icq\_cl}(z)$ and $Z_{qq}(z)$.

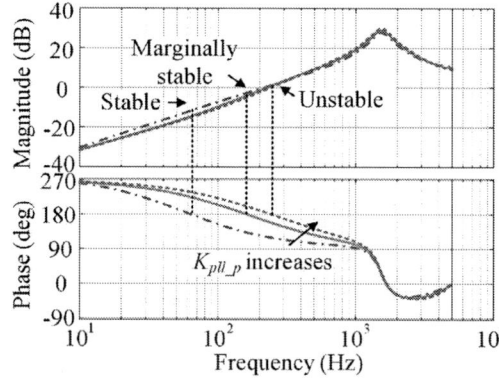

Fig. 6. Bode diagrams of $Z_{gq}(z)/Z_{qq}(z)$ with K_{pll_p} = 5, 15, and 25.

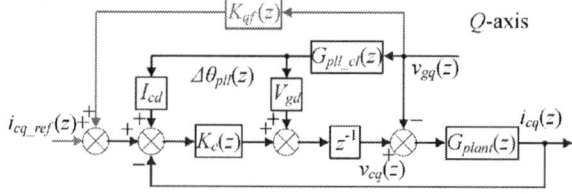

Fig. 7. Principle of the proposed control scheme.

as illustrated in Fig. 7.

When equipped with $K_{qf}(z)$, (9) becomes:

$$G_{vgq\_cl}(z) = G_{vgq\_pll}(z) + G_{vgq\_plant}(z) + G_{vgq\_Kqf}(z), \quad (12)$$

where $G_{vgq\_Kqf}(z) = K_{qf}(z)G_{icq\_cl}(z)$. $G_{vgq\_Kqf}(z) \approx K_{qf}(z)$ in the low-frequency band is satisfied due to $G_{icq\_cl}(z) \approx 1$. Therefore, implementing $K_{qf}(z)$ as a proportional controller with a gain of K_{qf} will introduce a positive resistance, which helps to reshape $G_{vgq\_cl}(z)$.

Ideally, the proposed controller should cancel out the negative effect introduced by the PLL. Mathematically, this objective can be translated into the following equation:

$$G_{vgq\_pll}(z) + G_{vgq\_Kqf}(z) = 0 \Rightarrow$$
$$G_{pll\_cl}(z)V_{gd} / K_c(z) + G_{pll\_cl}(z)I_{cd} + K_{qf}(z) = 0. \quad (13)$$

When (13) is satisfied, the PLL will not introduce any instability issue. An alternative expression of (13) describes the desired impedance controller:

$$K_{qf}(z) = -G_{pll\_cl}(z)\left[\frac{V_{gd}}{K_c(z)} + I_{cd}\right]. \quad (14)$$

In the low-frequency band, (14) can be simplified into:

$$K_{qf} = -\left(\frac{1}{K_{c\_p}} + \frac{I_{cd}}{V_{gd}}\right). \quad (15)$$

According to the system parameter values listed in TABLE I, $K_{qf} \approx -0.1$ can be derived, and the Bode diagrams of $K_{qf}(z)$ and K_{qf} are illustrated in Fig. 8. As seen, $K_{qf} = -0.1$ provides a good approximation of $K_{qf}(z)$ in the low-frequency band.

The effect of K_{qf} on reshaping $Z_{qq}(z)$ can be observed from Fig. 9(a). As seen, the phase of $Z_{qq}(z)$ is shifted to be $0°$ in the low-frequency band. In this case, the modified $Z_{qq}(z)$ behaves like a positive resistance, which certainly contributes to system stability improvement. This conclusion can also be verified by the Bode diagram of $Z_{gq}(z)/Z_{qq}(z)$ shown in Fig. 9(b), where the system stability is always preserved regardless of the values of K_{pll_p}.

IV. SIMULATION AND EXPERIMENTAL RESULTS

To validate the effectiveness of the proposed control

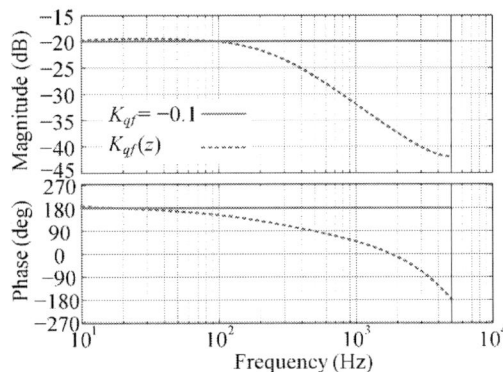

Fig. 8. Bode diagrams of $K_{qf}(z)$ and $K_{qf} = -0.1$.

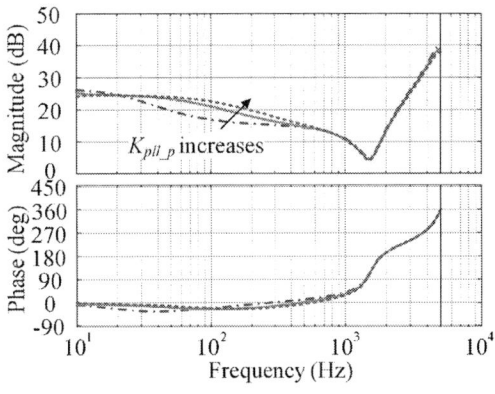

(a) $Z_{qq}(z)$

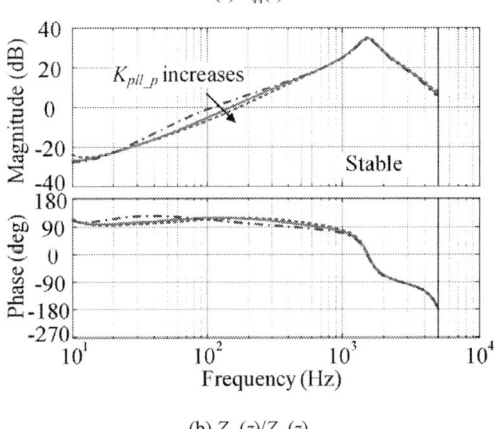

(b) $Z_{gq}(z)/Z_{qq}(z)$

Fig. 9. Bode diagrams of $Z_{qq}(z)$ and $Z_{gq}(z)/Z_{qq}(z)$ with K_{pll_p} = 5, 15, and 25 and K_{qf} = −0.1.

scheme, simulations and experiments were carried out based on the system parameter values listed in TABLE I. The grid-connected converter under test was designed to be unstable without the proposed impedance controller. It should be noted that the case of K_{pll_p} = 15 becomes unstable after adding a low-pass-filter to the PLL. The simulation and experimental waveforms are presented in Fig. 10 and Fig. 11, respectively. As mentioned before and verified by Figs. 10 and 11, the activation of the proposed control scheme, i.e., letting K_{qf} = −0.1, would resolve the instability issue. After disabling the proposed controller, the system becomes unstable. Moreover, the large magnitude of v_{gq_pll} indicates that the PLL should be responsible for the system instability.

Reducing the control bandwidth of the PLL, i.e., the proportional gain K_{pll_p}, could be another possible solution to the instability issues. However, this is achieved at the expense of system dynamics. As an example, Fig. 12 and Fig. 13 show the simulation and experimental waveforms of v_{gq_pll} during system start-up, respectively. It is obvious that the case with the small PLL proportional gain, namely K_{pll_p} = 1, experiences large deviations and slow dynamics during system start-up. Therefore, the reduction of the PLL bandwidth is a trade-off between the system stability and dynamic performances. In this sense, the proposed control scheme is obviously superior as it preserves the system stability

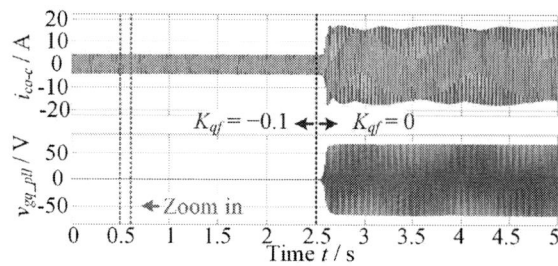

(a) Overall view

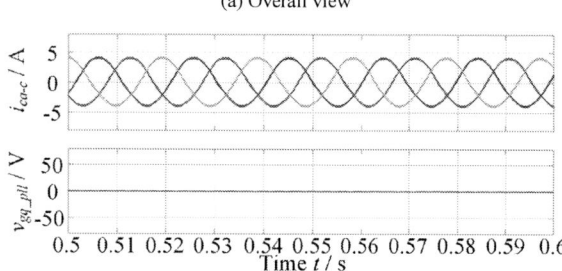

(b) Zoom-in view

Fig. 10. Simulation results with the change of K_{qf}.

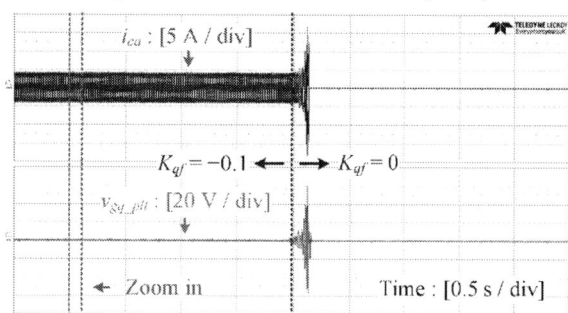

(a) Overall view

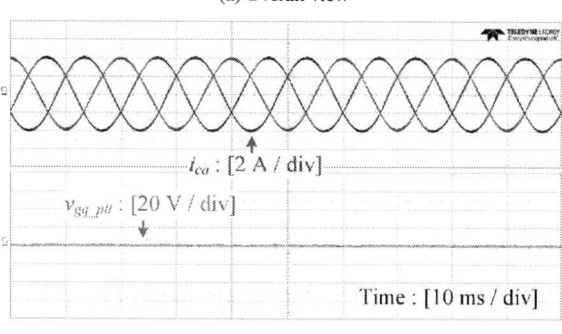

(b) Zoom-in view

Fig. 11. Experimental results with the change of K_{qf}.

without any sacrifice of PLL designs.

V. CONCLUSIONS

An impedance controller, which directly relates the q-axis voltage to the q-axis current reference, has been proposed in this paper to reshape the q-axis impedance of power converters into a positive resistance in the low-frequency band. As a result, the proposed controller successfully resolves the instability issues caused by PLL designs even under a severely weak power grid condition.

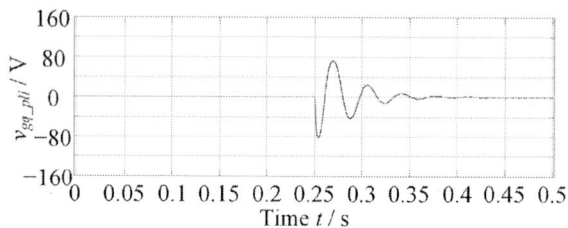

(a) $K_{pll\_p} = 1$

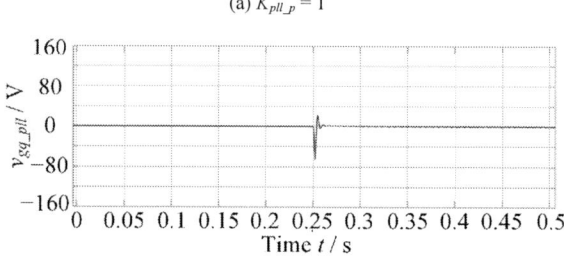

(b) $K_{pll\_p} = 15$

Fig. 12. Simulation results of v_{gq_pll} during system start-up.

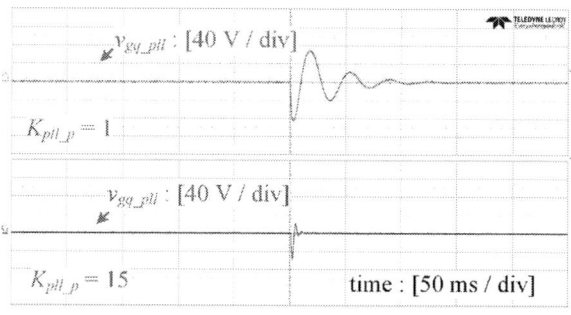

Fig. 13. Experimental results of v_{gq_pll} during system start-up.

Simulation and experimental results indicate the effectiveness of the proposed control scheme.

REFERENCES

[1] J. M. Carrasco, L. G. Franquelo, J. T. Bialasiewicz, E. Galvan, R. C. PortilloGuisado, M. A. M. Prats, J. I. Leon, and N. Moreno-Alfonso, "Power-electronic systems for the grid integration of renewable energy sources: a survey," *IEEE Trans. Ind. Electron.*, vol. 53, DOI 10.1109/TIE.2006.878356, no. 4, pp. 1002–1016, Aug. 2006.

[2] J. Fang, Y. Tang, H. Li, and X. Li, "A battery/ultracapacitor hybrid energy storage system for implementing the power management of virtual synchronous generators," *IEEE Trans. Power Electron.*, DOI 10.1109/TPEL.2017.2759256, vol. 33, no. 4, pp. 2820–2824, Apr. 2018.

[3] F. Blaabjerg, R. Teodorescu, M. Liserre, and A. V. Timbus, "Overview of control and grid synchronization for distributed power generation systems," *IEEE Trans. Ind. Electron.*, vol. 53, DOI 10.1109/TIE.2006.881997, no. 5, pp. 1398–1409, Oct. 2006.

[4] D. Dong, B. Wen, D. Boroyevich, P. Mattavelli, and Y. Xue, "Analysis of phase-locked loop low-frequency stability in three-phase grid-connected power converters considering impedance interactions," *IEEE Trans. Ind. Electron.*, vol. 62, DOI 10.1109/TIE.2014.2334665, no. 1, pp. 310–321, Jan. 2015.

[5] R. D. Middlebrook, "Input filter considerations in design and application switching regulators," in *Proc. IEEE Ind. Appl. Soc.*, 1976, pp. 366–382.

[6] X. Feng, J. Liu, and F. C. Lee, "Impedance specifications for stable DC distributed power systems," *IEEE Trans. Power*

Electron., vol. 17, DOI 10.1109/63.988825, no. 2, pp. 157–162, Mar. 2002.

[7] J. Sun, "Small-signal methods for AC distributed power systems-a review," *IEEE Trans. Power Electron.*, vol. 24, DOI 10.1109/TPEL.2009.2029859, no. 11, pp. 2545–2554, Nov. 2009.

[8] A. Emadi, A. Khaligh, C. H. Rivetta, and G. A. Williamson, "Constant power loads and negative impedance instability in automotive systems: definition, modeling, stability, and control of power electronics converters and motor drives," *IEEE Trans. Veh. Technol.*, vol. 55, DOI 10.1109/TVT.2006.877483, no. 4, pp. 1112–1125, Jul. 2006.

[9] L. Harnefors, M. Bongiorno, and S. Lundberg, "Input-admittance calculation and shaping for controlled voltage-source converters," *IEEE Trans. Ind. Electron.*, vol. 54, DOI 10.1109/TIE.2007.904022, no. 6, pp. 3323–3334, Dec. 2007.

[10] B. Wen, D. Boroyevich, R. Burgos, P. Mattavelli, and Z. Shen, "Analysis of D-Q small-signal impedance of grid-tied inverters," *IEEE Trans. Power Electron.*, vol. 31, DOI 10.1109/TPEL.2015.2398192, no. 1, pp. 675–687, Jan. 2016.

[11] K. M. Alawasa, Y. A. I. Mohamed, and W. Xu, "Active mitigation of subsynchronous interactions between PWM voltage-source converters and power networks," *IEEE Trans. Power Electron.*, vol. 29, DOI 10.1109/TPEL.2013.2251904, no. 1, pp. 121–134, Jan. 2014.

[12] J. Fang, G. Xiao, X. Yang, and Y. Tang, "Parameter design of a novel series-parallel-resonant *LCL* filter for single-phase half-bridge active power filters," *IEEE Trans. Power Electron.*, vol. 32, no. 1, DOI 10.1109/TPEL.2016.2532961, pp. 200–217, Jan. 2017.

The 2018 International Power Electronics Conference

Comparison Between Traditional Droop and A New Autonomous Control Scheme for Parallel Inverters

Mohammad Bani Shamseh[1]*, Teruo Yoshino[1], Atsuo Kawamura[2]

1 Power Electronics Systems Division, Toshiba Mitsubishi-Electric Industrial Systems Corporation (TMEIC), Tokyo, Japan
2 Department of Electrical and Computer Engineering, Yokohama National University, Yokohama, Japan
*E-mail: bani.mohammad@tmeic.co.jp

Abstract—The traditional Droop control method is a well-developed control technique for parallel, independent, distributed energy sources. However, this method faces many challenges such as non-constant output voltage and frequency, and the trade-off between dynamic response and steady state accuracy. Many attempts have been made to improve the method at the expense of increased complexity, cost, or additional equipment. The authors have proposed a new control technique for autonomous operation of parallel inverters that can overcome some of the inherent challenges that exist in the Droop control. This paper compares the response of the proposed technique and the traditional Droop method. The method uses the output current of each inverter to regulate the voltage of the capacitor of the *LCL* filter. The dynamic response, voltage accuracy, and circulating current are used in the comparison.

Keywords—Autonomous control, distributed generation, parallel inverters, UPS systems.

I. INTRODUCTION

With the integration of renewable energy sources into the electric networks, and the ubiquity of distributed energy sources, parallel inverters are essential to control these energy sources. Parallel operation of distributed power sources require robust control schemes that can guarantee stable operation and robustness against abnormal load conditions. Control schemes of parallel inverters can be lumped into two main categories. The first category is active load sharing (ALS). This method relies on sharing information about the operation condition of each inverter to control the system. An example of this method is the average load sharing method [1].

The second category of control schemes of parallel inverters includes independent control schemes. Independent control of parallel inverters does not require information exchange. The elimination of communication wires, which are used in ALS, adds advantages of increased reliability and flexibility to the system. The Droop control is one example of this method. The Droop is widely used to control parallel inverters by imitating the operation of parallel generators. The basic principle is to regulate the output voltage and frequency of each inverter as functions of its output active and reactive power [2]–[4].

However, there are disadvantages of the Droop control, such as slow transient response, unequal harmonic

currents distribution in case of nonlinear loads, and non-constant output voltage and frequency [2]. Many researchers have improved the Droop control recently. For example, the authors in [3] used error reduction and voltage recovery functions to reduce the voltage error. Parallel units can communicate with each other using RS232 serial communication. To mitigate the reactive power sharing error, the authors in [4] implemented an enhanced control strategy that estimates the reactive power control error through injecting small real power disturbances, which is activated by a low bandwidth synchronization signals from a central controller. A slow integration term of reactive power sharing error is also integrated into the controller to eliminate this error. However, the improvement of the Droop control in these techniques (and others), comes at the expense of increasing the system complexity, cost, or the requirement to install additional equipment.

The authors proposed an autonomous control technique for parallel inverters in a previous publication [5]. The proposed technique regulates the voltage and phase of the output capacitor of each inverter to achieve autonomous control, as opposed to controlling the common-bus voltage and frequency in the Droop. In the proposed method, which is called current-dependent capacitor voltage control (CCVC) technique, each inverter uses its output current to achieve autonomous control without information exchange with other inverters. The main advantages of the proposed method are its fast dynamic response, constant load voltage and frequency, and reduction in circulating current. This paper highlights these advantages by comparing the performance of the new method with the traditional Droop technique.

II. DROOP CONTROL

This section discusses the basic principles of the traditional Droop control and its control equations.

The traditional Droop control scheme is a well-established control method that imitates the operation of parallel power generators. The basic principle of this technique is that by using the voltage and frequency of the system as control variables, independent operation can be achieved without communication between the inverters

The 2018 International Power Electronics Conference

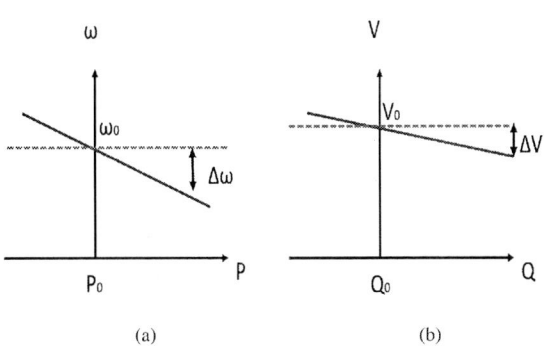

(a)   (b)

Fig. 1: Droop control for an inductive line (a) active power vs. frequency, (b) reactive power vs. voltage.

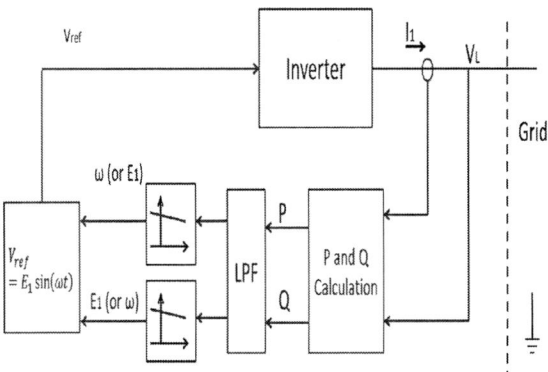

Fig. 2: Basic block diagram of Droop control.

[6]. The control equations for one inverter with an inductive line impedance can be summarized as follows [7]:

$$\omega = \omega_0 - k_p \left(P - P_0 \right) \tag{1}$$

$$V - V_0 = k_q \left(Q - Q_0 \right) \tag{2}$$

where ω is the frequency, V is the voltage, and ω_0 and V_0 represent the rated frequency and voltage, respectively. P_0 and Q_0 are the rated active and reactive power of the inverter. P and Q are the output active and reactive power. k_p and k_q are the slopes of the frequency and voltage droop lines, respectively.

Equations (1) and (2) can be represented graphically as shown in Fig.1. Notice that each inverter adjusts its output frequency and voltage references according to its P/ω and Q/V droop characteristics. The block diagram of one inverter controlled using the droop control is shown in Fig.2. The basic operation of the inverter can be described as follows: the output voltage and current are used to calculate the output active and reactive power. These values are sent to low pass filters (LPF). Next, the inverter calculates the reference frequency and voltage depending on its droop coefficients, according to (1) and (2). The reference voltage is then used to control the inverter using feedback control and the PWM signals are sent to the inverter switches.

The power-frequency and reactive power-voltage relations mentioned here are for an inductive line impedance. For a resistive line impedance, the relation is reversed. That is, the active power is strongly related to the voltage, and the reactive power is strongly related to the frequency [7].

Since different kinds of control strategies are adopted depending on the output impedance, the inverters are required to have the same per unit output impedance [6]. This imposes limitations on the operation of parallel inverters, especially if the inverters have different kinds of impedances (inductive, resistive, capacitive, or a combination of them) [8].

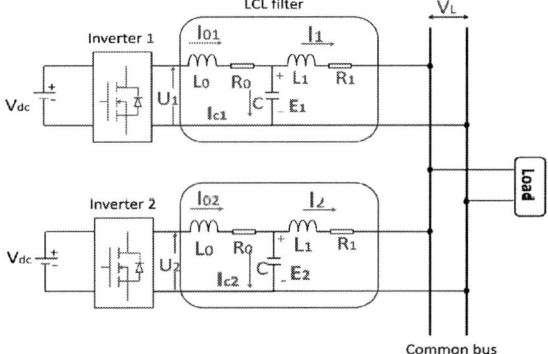

Fig. 3: Two parallel inverters with LCL filters.

Another challenge of the traditional Droop technique is evident from Fig.1. The Droop control achieves independent operation of parallel inverters using the system frequency and voltage as variable parameters. However, and as shown in Fig.1, the frequency and voltage deviate from the reference values depending on the load condition. The deviation can be decreased by reducing the slopes of the control equations (k_p and k_q in (1), and (2)), however, this is achieved at the expense of reducing the dynamic response speed of the system [9].

III. PROPOSED METHOD

In an attempt to overcome some of the disadvantages of the Droop mentioned earlier, a new control technique is proposed in [5]. Instead of controlling the voltage and frequency of the common-bus voltage, the voltage and phase angle of the filter capacitor are controlled. This results in a constant load voltage and frequency regardless of the load conditions.

A. Control Block Diagram

Fig.3 shows two parallel inverters with *LCL* filters connected in parallel. The load-side inductor (L_1) represents a step-up transformer that provides galvanic isolation. The load is connected to the common bus. The

3917

The 2018 International Power Electronics Conference

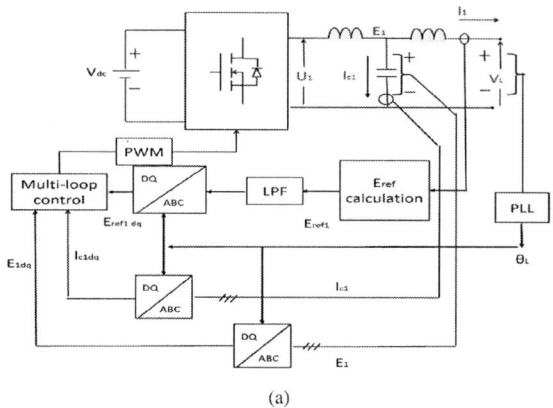

(a)

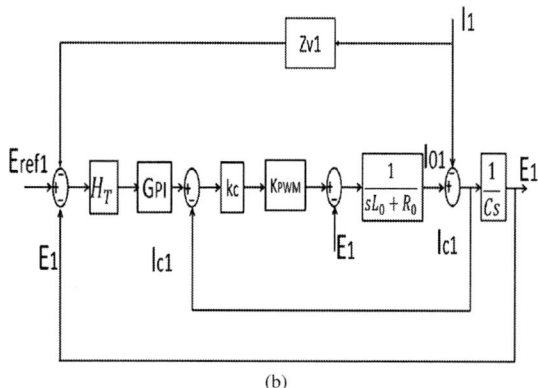

(b)

Fig. 4: Proposed control technique (a) complete control diagram of one inverter, (b) dual-loop control with capacitor current in inner loop and capacitor voltage in outer loop.

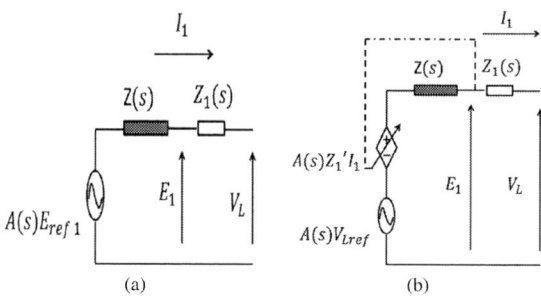

Fig. 5: (a) Thevenin equivalent circuit of inverter 1, (b) Equivalent circuit of inverter 1 with autonomous control.

input DC voltage (V_{dc}) is supplied from ac/dc rectifiers or from a separate source (such as PV cells). The LCL filters provide a cost-effective solution to PWM switching harmonics of the inverter [10]. However, the inherent resonance of LCL filters has a tendency to destabilize the system. Hence, the resonance of the filter can be damped using passive damping (by adding passive resistors), or using active damping. Active damping provides a more efficient solution since it does not require the addition of new resistors that waste power. Active damping adds virtual impedance to the system using voltage or current feedback signals. In this paper, the capacitor current is used for active damping [5].

Fig.4a shows the complete block diagram of one inverter using the proposed scheme. The output current of the inverter is used to calculate the reference capacitor voltage (E_{ref1}). The reference voltage is then passed to a low pass filter. Three phase quantities are transformed to the *dq* synchronous frame by means of Park's transformation [11]. The load voltage is used to generate the reference phase using a phase-locked loop (PLL). Next, a multi-loop controller is used to generate the PWM signals.

Fig. 4b shows the details of the multi-loop controller of inverter 1. The generation of the reference capacitor voltage (E_{ref1}) will be discussed in Section III-C. The inner loop is the capacitor current loop (I_{c1}) and the outer loop is the capacitor voltage loop (E_1). The capacitor current is used in the inner loop since it provides a more satisfactory steady state and dynamic response under non-linear load conditions compared to the inductor current loop I_{01} [12].

In Fig.4b, H_T is a lead compensator used to increase the phase margin of the controller [5], G_{PI} is a *PI* controller for the capacitor voltage loop, and k_c is a proportional controller for the capacitor current loop. Z_{v1} is a virtual impedance used in series with the inverter. More information about the virtual impedance will be provided in a subsequent section. More information about the multi-loop control dynamics is available in reference [5].

The control is performed in the synchronous frame because the PI controller is effective at eliminating the steady state error for DC signals, forcing the system to track the reference without steady state error [11]. Finally, the PWM gate signals are generated and sent to the inverter.

B. Circulating current

Circulating current between parallel inverters is a challenging problem that can cause serious damage if it is unsolved. The circulating current flows between any two parallel voltage sources if they are not synchronized. For AC inverters, synchronization encompasses three quantities, namely: voltage, phase, and frequency.

For the two parallel inverters considered in this paper, the circulating current is expressed with the following equation [5], [13]

$$\bar{I}_{c12} = \frac{\bar{Z}_L}{\bar{Z}\left(\bar{Z} + 2\bar{Z}_L\right)} \left[\bar{E}_1 - \bar{E}_2\right] \qquad (3)$$

where: I_{c12} is the circulating current from inverter 1 to inverter 2, Z is the impedance of the output inductor L_1 in Fig. 3 ($Z = Z_1 = Z_2$), and Z_L is the load impedance.

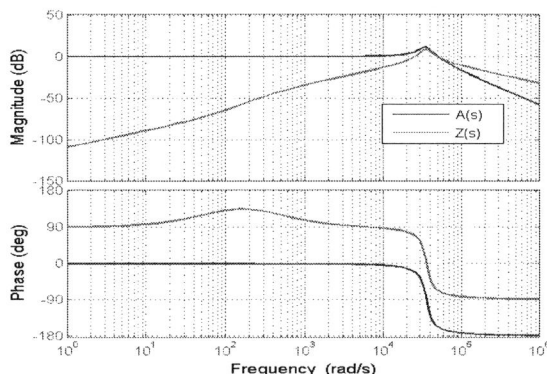

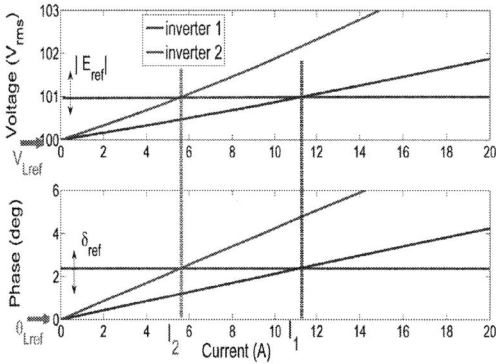

Fig. 6: Bode plot of open-loop transfer function of capacitor voltage loop with $k_c = 0.02$, and $k_i = 2000$.

Fig. 7: Output current vs. reference capacitor voltage and phase for two inverters using the proposed technique. Inverter 1 has twice the capacity of inverter 2.

TABLE I: Parameters of two inverters.

Parameter	Inverter 1	Inverter 2
L_0	0.6 mH	0.6 mH
R_0	40 mΩ	40 mΩ
C	40 uF	40 uF
L_1	1.2 mH	2.4 mH
R_1	80 mΩ	80 mΩ
V_{dc}	300 V	300 V
Rated power	5 kVA	2.5 kVA

Based on (3), if the capacitor voltages of the two inverters are equal, the circulating current can be mitigated. Hence the reason why the autonomous control is implemented by controlling the capacitor voltage in this paper.

C. Proposed Autonomous Control Technique

Based on the block diagram of one inverter (Fig. 4b), the closed-loop inverter can be represented equivalently by a voltage source and a series impedance [5]

$$E_1 = \frac{G_{PI}k_c}{CL_0 s^2 + Cs(R_0 + k_c) + G_{PI}k_c + 1}E_{ref1}$$
$$- \frac{sL_0 + R_0}{CL_0 s^2 + Cs(R_0 + k_c) + G_{PI}k_c + 1}I_1. \quad (4)$$

Equivalently, equation (4) can be re-written as:

$$E_1 = A(s)E_{ref1} - Z(s)I_1 \quad (5)$$

where:

$$A(s) = \left. \frac{E_1}{E_{ref1}} \right|_{I_1=0} \quad (6)$$
$$= \frac{k_p k_c s + k_i k_c}{CL_0 s^3 + Cs^2(R_0 + k_c) + (k_p k_c + 1)s + k_i k_c}$$

$$Z(s) = \left. \frac{E_1}{I_1} \right|_{E_{ref1}=0} \quad (7)$$
$$= \frac{s^2 L_0 + R_0 s}{CL_0 s^3 + Cs^2(R_0 + k_c) + (k_p k_c + 1)s + k_i k_c}.$$

$$G_{PI} = k_p + \frac{k_i}{s}. \quad (8)$$

The Thevenin equivalent circuit of one inverter based on (5) is shown in Fig.5a, where $Z(s)$ is the Thevenin equivalent impedance of the inverter, and $Z_1(s) = R_1 + sL_1$ is the impedance of the load-side inductor.

In (5), the output current is viewed as a disturbance input. Typically, in order to obtain an output voltage waveform with high quality, the voltage E_1 must track the reference E_{ref1} at the fundamental frequency while rejecting any load current disturbance [12].

The following parameters are used in the the multi-loop controller: $k_c = 0.02$, $k_p = 10$, $k_i = 2000$. Other parameters are illustrated in Table I. The Bode plots of $A(s)$ and $Z(s)$ are shown in Fig. 6. Notice that $A(s)$ is unity (0 dB), and $Z(s)$ is very small at the fundamental frequency. Hence, in (5), $E_1 = E_{ref1}$.

The proposed control is based on varying the capacitor voltage (E_1) as a function of the output current of the inverter (I_1). The reference capacitor voltage is generated using the following equation:

$$\bar{E}_{ref1} = \left| \bar{E}_{ref1} \right| \underline{/\delta_{ref1}} = \bar{V}_{Lref} + \bar{Z}'_1 \bar{I}_1 \quad (9)$$

where $\bar{E}_{ref1}$ is the reference capacitor voltage of inverter 1 (phasor quantity), $\bar{V}_{Lref}$ is the reference load voltage, $\bar{Z}'_1$ is the compensation impedance used in the controller which is typically equal to the impedance of the output inductor, $\bar{Z}'_1 = R'_1 + j\omega_0 L'_1$, ω_0 is the fundamental angular frequency of the system ($\omega_0 = 2\pi f_0$, $f_0 = 50$ Hz).

Substituting the control equation (9) into (5), the following equation is obtained:

$$E_1 = A(s)V_{Lref}(s) + A(s)Z_1'I_1 - Z(s)I_1. \qquad (10)$$

Fig.5b shows the equivalent control circuit representation of (10), where a current-dependent voltage source is added ($A(s)Z_1'I_1$). This voltage source regulates the capacitor voltage at a certain level above the load reference voltage (V_{Lref}) depending on the load condition.

The reference dq components of the capacitor voltage can be represented based on (9) as follows:

$$\begin{cases} E_{1refd} = V_{Lref} + R_1'I_{1d} - \omega_0 L'1 I_{1q} \\ E_{1refq} = \omega_0 L_1'I_{1d} + R_1'I_{1q}. \end{cases} \qquad (11)$$

Based on the control equation of one inverter (9), and since it is decided that all inverters must have the same reference capacitor voltage as a necessary condition to eliminate the circulating current, and since the reference load voltage is fixed and known for all units (V_{Lref}), then, for n parallel inverters, the following equation is derived by equating their capacitor voltages:

$$Z_1'I_1 = Z_2'I_2 = \cdots = Z_n'I_n. \qquad (12)$$

From (12), for parallel units with different power ratings, the compensation impedances should be inversely proportional to the power ratings of the inverters, as follows:

$$\frac{I_1}{I_2} = \frac{Z_2'}{Z_1'}. \qquad (13)$$

To demonstrate the operation of the proposed method, two inverters with different capacities will be used. The parameters of the two inverters are shown in Table I. Inverter 1 has twice the capacity of inverter 2. Hence, the slope of the compensation impedance in equation (9) is:

$$\bar{Z}'_2 = 2\bar{Z}'_1. \qquad (14)$$

The representation of the control equation (9) is shown in Fig. 7 for the two inverters. The reference load voltage (V_{Lref}) and phase (θ_{ref}) are indicated with arrows at the left sides as 100 Vrms and 0 degrees, respectively. The phase of the reference load voltage is zero degrees since the reference phase is taken from the load bus by means of a PLL, as shown in Fig.4a. The reference capacitor voltage and phase for the two inverters ($|\bar{E}_{ref}|\underline{/\delta_{ref}}$) are equal, and they change according to their respective output currents (I_1 and I_2). The output current of inverter 1 is twice that of inverter 2. Notice that the reference capacitor voltage and phase change automatically with the load change, while the load voltage and phase remain constant.

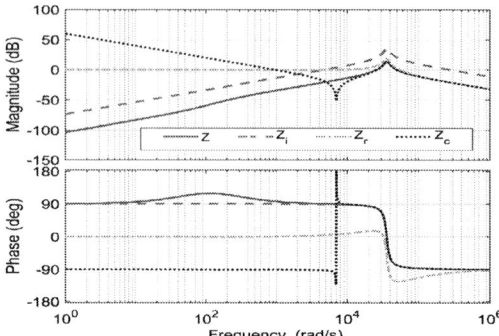

Fig. 8: Output impedance of the inverter without virtual impedance (Z), with an inductive virtual impedance (Z_i), a resistive virtual impedance (Z_r), and a capacitive virtual impedance (Z_c).

D. Virtual Impedance

The compensation impedance $\bar{Z}_1'$ in (9) is designed to be equal to the impedance of the output inductor ($\bar{Z}_1$). However, it is difficult in practical systems to know the exact value of the inductors, which may change depending on the load condition, temperature, and other variables.

The effect of the variation in the output impedance of the inverters on the control can be overcome by adding a virtual impedance in series with the inverter. By designing the virtual impedance to be larger than the impedance of the inductor Z_1, the robustness of the system is increased.

The virtual impedance is achieved by dropping the reference voltage as a function of the output current of the inverter [14]. In the case of the system discussed here, the reference capacitor voltage (E_{ref1}) is dropped as a function of the output current (I_1):

$$\bar{E}^*_{ref1} = \bar{E}_{ref1} - \bar{Z}_{v1}(s)\bar{I}_1 \qquad (15)$$

where $\bar{E}^*_{ref1}$ is the new reference capacitor voltage after adding the virtual impedance, and $\bar{Z}_{v1}$ is the virtual impedance of inverter 1.

The implementation of the virtual impedance concept into the controller is an easy task that can be realized by adding a few lines in the program of the controller. Yet, the benefits gained are numerous: enhanced performance, reduction in circulating current, better power quality, increased stability, and less sensitivity to the values of the physical components and controller parameters [15].

The output impedance of the inverter after implementing the virtual impedance into the control is expressed with the following equation:

$$Z(s) = \frac{sL_0 + R_0 + \bar{Z}_{v1}k_cG_{PI}}{CL_0s^2 + Cs(R_0 + k_c) + G_{PI}k_c + 1}. \qquad (16)$$

The 2018 International Power Electronics Conference

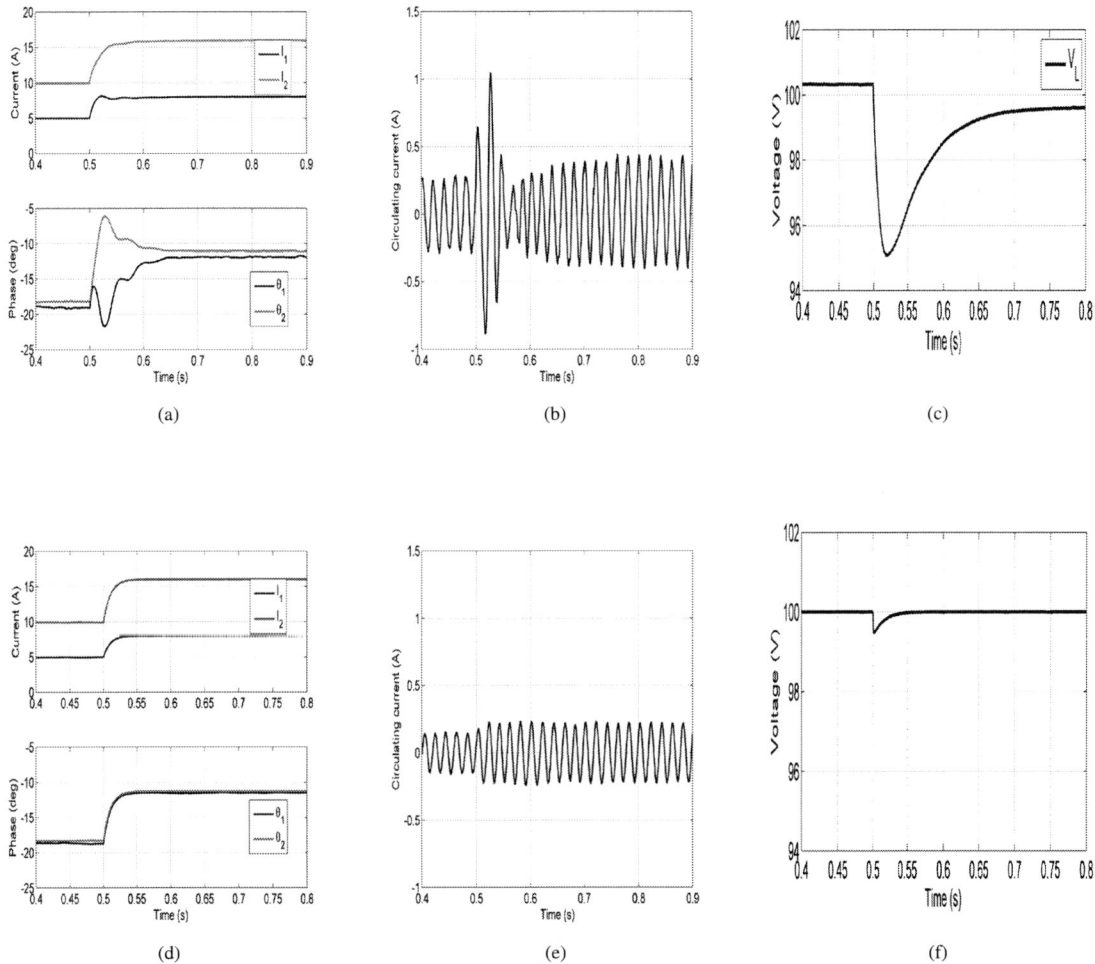

Fig. 9: Comparison between Droop control (upper row), and the proposed control (lower row): (a) and (d) output currents and phase, (b) and (e) circulating current from inverter 1 to inverter 2, (c) and (f) load voltage. Load condition: 3 kW+1 kVAr, then additional 2 kW is added.

Notice that in (16), $\bar{Z}_{v1}$ in the numerator can be used to re-shape the output impedance of the inverter. Fig.8 shows the output impedance of the inverter without a virtual impedance (Z), with an inductive virtual impedance (Z_i), with a resistive virtual impedance (Z_r), and a capacitive virtual impedance (Z_c). Hence, different kinds of output impedances can be easily implemented to serve different applications [14].

IV. COMPARISON WITH DROOP

To highlight the advantages of the proposed method, two parallel inverters will be considered in simulation. The capacity of inverter 1 is twice that of inverter 2. The initial load condition is 3 kW and 1 kVAr. The load is then increased by 2 kW. For the Droop control, the slopes kp and kq are chosen taking into consideration the power ratios of the inverters, as follows: $k_{p1} = 4 \times 10^{-4}$ rad/s.W, $k_{p2} = 2k_{p1}$, $k_{q1} = 0.007$ V/VAr, $k_{q2} = 2k_{q1}$. For the

proposed controller, the slope Z'_1 in (9) is calculated using $\bar{Z}'_1 = R'_1 + j\omega_0 L'_1$, where R'_1 and L'_1 are as in Table I. The slope $\bar{Z}'_2$ is twice that of inverter 1 (equation (14)).

The results of output currents, circulating currents, and load voltage are shown in Fig.9. Fig.9a and Fig.9d compare the output currents and phase angles for the two methods when the load changes (Droop control results are shown in the first row; the results of the proposed control are shown on the second row). Notice that the dynamic response for the proposed method is better in terms of response speed and transient state.

In the proposed method, the capacitor voltages are controlled. As shown in Fig.7, the capacitor voltages of all parallel inverters are equal. This results in efficient mitigation of the circulating current between the inverters, as predicted by (3). The circulating currents from inverter 1 to inverter 2 for the two cases are shown in Fig.9b and

3921

The 2018 International Power Electronics Conference

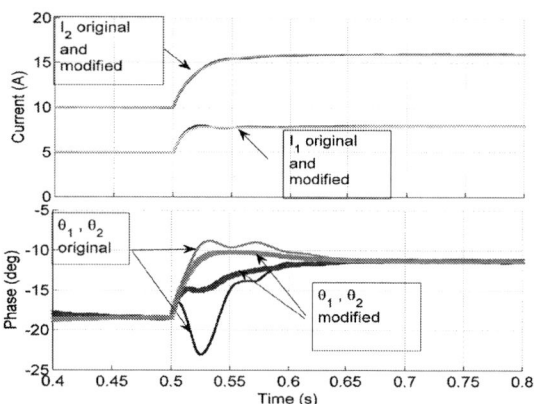

Fig. 10: Output currents of Droop control after increasing the gain of voltage droop equation (k_q).

Fig.9e. Notice that in the proposed method the circulating current is suppressed by more than 60% compared to the Droop method.

Finally, the load voltages are shown in Fig.9c and Fig.9f. There are two merits in using the proposed method compared to the Droop as far as the load voltage is concerned: First, the load voltage undershoot in the transient state in case of load change in the proposed control is negligible, while in the case of Droop is significant. Second, the level of load voltage in the steady state remains constant in the proposed control, contrary to the Droop in which the load voltage changes according to the Droop equation (the same observation is correct for the frequency of the system).

The dynamic response of the Droop control can be improved by increasing the droop coefficients (k_p and k_q). However, this improvement comes at the expense of increasing the deviation in the load voltage and frequency from the reference values. The trade-off between the dynamic speed and error is a well-documented phenomenon that has been reported in the literature [9]. As an example, the same system with droop control that is discussed in this paper is used. The voltage droop coefficient (k_q) is increased by 2.8 times. The waveforms of the output currents of the two inverters and their phase angles are shown in Fig. 10. The transient response is improved in this case, but the voltage deviation from the reference will also increase with the increase in reactive power supplied by the inverter.

V. EXPERIMENTAL RESULTS

The proposed method was tested experimentally to show its feasibility. Two identical 1 kVA inverters are connected in parallel to a load. The parameters of the inverters are as shown in Table I. The switching frequency of the system is 15 kHz. The initial load is 600 W. The load is then increased step-wise by 60%. The waveforms of the capacitor voltage and output current of the two inverters are shown in Fig.11a. The load voltage is shown

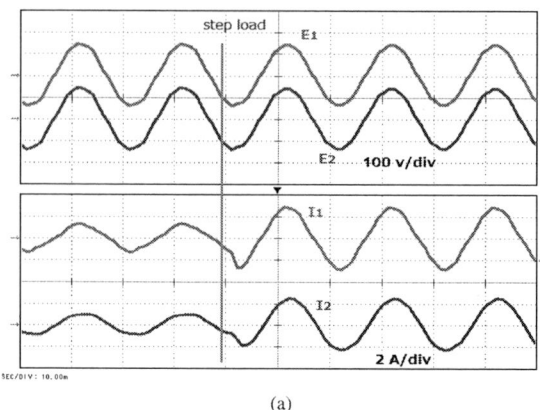

(a)

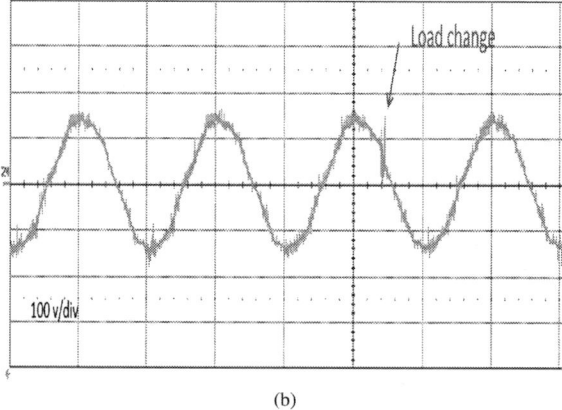

(b)

Fig. 11: Experimental results of the proposed control method under step-load condition (a) Capacitor voltages and output currents waveforms, (b) Load voltage waveform.

TABLE II: Comparison between the Droop technique and the proposed control scheme.

Comparison item	Droop	CCVC	Merits
V_L transient undershoot	5.5%	0.5%	less overshoot\undreshoot
V_L steady-state error	1.5%	0%	higher accuracy, constant voltage and frequency
Settling time	200 ms	40 ms	faster response
Circulating current	2.8%	1.1%	lower power loss

in Fig.11b. The experimental results are congruent with the simulation results. Notice that the dynamic response is fast under load variation. The experimental results also show that the load voltage is constant in amplitude and frequency.

VI. CONCLUSION

This paper presents a comparison between the Droop control and a new proposed control for parallel inverters (CCVC). The proposed autonomous control scheme can overcome the inherent challenges encountered in the traditional Droop control. Table II illustrates the main different points between the two methods. The advantages of the proposed autonomous control scheme are: zero

3922

The 2018 International Power Electronics Conference

steady state errors in the load voltage and frequency, faster dynamic response, and mitigation of circulating current.

REFERENCES

[1] M. Bani Shamseh, T. Yoshino, and A. Kawamura, "Load current distribution between parallel inverters based on capacitor voltage control for UPS applications," IEEJ Journal of Industry Applications, vol. 6, no. 4, pp. 258-267, 2017.

[2] Q. C. Zhong, "Robust Droop Controller for Accurate Proportional Load Sharing Among Inverters Operated in Parallel," IEEE Trans Industrial Electron., vol. 60, no. 4, pp. 1281-1290, 2013.

[3] H. Han, Y. Liu, Y. Sun, M. Su and J. M. Guerrero, "An Improved Droop Control Strategy for Reactive Power Sharing in Islanded Microgrid," IEEE Trans. on Power Applications, vol. 30, no. 6, pp. 3133-3141, 2015.

[4] J. He and Y. W. Li, "An Enhanced Microgrid Load Demand Sharing Strategy," IEEE Trans. Power Electron., vol. 27, no. 9, pp. 3984-3995, 2012.

[5] M. B. Shamseh, T. Yoshino and A. Kawamura, "Current-Dependent Capacitor Voltage Control of Parallel Autonomous UPS Systems," IEEE Trans. Industrial Electron., vol. 65, no. 4, pp. 2873-2882, 2018.

[6] B. Wei, W. Liu, J. M. Guerrero and J. C. Vsquez, "A power sharing method based on modified droop control for modular UPS," Annual Conference of the IEEE Industrial Electronics Society IECON, pp. 4839-4844, 2017.

[7] Y. Chen, J. M. Guerrero, Z. Shuai, Z. Chen, L. Zhou and A. Luo, "Fast Reactive Power Sharing, Circulating Current and Resonance Suppression for Parallel Inverters Using Resistive-Capacitive Output Impedance," IEEE Transactions on Power Electronics, vol. 31, no. 8, pp. 5524-5537, 2016.

[8] Q. C. Zhong and Y. Zeng, "Universal Droop Control of Inverters With Different Types of Output Impedance," IEEE Access, vol. 4, pp. 702-712, 2016.

[9] J. M. Guerrero, L. Hang and J. Uceda, "Control of Distributed Uninterruptible Power Supply Systems," IEEE Trans. Industrial Electron., vol. 55, no. 8, pp. 2845-28597, 2008.

[10] S. Jayalath and M. Hanif, "Generalized LCL-Filter Design Algorithm for Grid-Connected Voltage-Source Inverter," IEEE Transactions on Industrial Electron., vol. 64, no. 3, pp. 1905-1915, 2017.

[11] C. Zou, B. Liu, S. Duan and R. Li, "Stationary Frame Equivalent Model of Proportional-Integral Controller in dq Synchronous Frame," IEEE Trans. Power Electron., Vol. 29, no. 9, pp. 4461-4465, 2014.

[12] P.C. Loh, M.J. Newman, D.N. Zmood, D.G. Holmes, "Improved transient and steady state voltage regulation for single and three phase uninterruptible power supplies," Power Electronics Specialists Conference, Vol. 2, pp. 498-503, 2001.

[13] M. B. Shamseh, A. Kawamura and T. Yoshino, "A novel autonomous control scheme for parallel, LCL-based UPS systems," IEEE Energy Conversion Congress and Exposition (ECCE), pp. 1-8, 2016.

[14] Y. Chen, J. M. Guerrero, Z. Shuai, Z. Chen, L. Zhou and A. Luo, "Fast Reactive Power Sharing, Circulating Current and Resonance Suppression for Parallel Inverters Using Resistive-Capacitive Output Impedance," IEEE Trans. Power Electron., vol. 31, no. 8, pp. 5524-5537, 2016.

[15] J. Kim, Hyounglok Oh and K. Nam, "Output impedance control of inverter-based distri- buted generator for minimum circulating current in single-phase AC microgrid," EEE 8th International Power Electronics and Motion Control Conference (IPEMC-ECCE Asia), pp. 2759-2764, 2016.

The 2018 International Power Electronics Conference

A Novel Microgrid Power Sharing Scheme Enhanced by a Non-Intrusive Feeder Impedance Estimation Method

Baojin Liu, Zeng Liu, Jinjun Liu, Ronghui An, Shuguang Song

State Key Lab of Electrical Insulation and Power Equipment, School of Electrical Engineering
Xi'an Jiaotong University
Xi'an, China
liubaojin.pe@gmail.com

Abstract— To address the issue that traditional droop control cannot realize reactive power, imbalance power and harmonic power sharing for an islanded microgrid, this paper proposes a novel load power sharing scheme based on hierarchical control structure. The hierarchical control consists of two levels: 1) the primary control, located at each distributed generation unit local controller, is composed of droop control and virtual impedance control; 2) the secondary level, located at microgrid central controller, estimates each feeder impedance and generates a virtual impedance command signal for primary control to eliminate the power sharing error. The proposed method has two distinguished features: 1) the reactive power, imbalance power and harmonic power can be equally shared since the equivalent impedance of each DG is compensated to a common value; 2) feeder impedance can be precisely estimated through a non-intrusive method, which can also be used to solve other issues in microgrid caused by feeder impedance mismatch. Simulation results are provided to validate the proposed method.

Keywords— *hierarchical control, impedance estimation, microgrid, power sharing*

I. INTRODUCTION

Recently, distributed generation (DG) integrated with renewable energy resource has become more and more attractive [1]. Usually a DG unit is connected to the generation system with power electronics-based interface, e.g. inverters, so the coordinated control of parallel-connected inverters is the key for robust DG systems [2]. Microgrid (MG), formed by a cluster of DG units, can operate under both islanded and grid-connected modes to feed critical loads flexibly and reliably [3].

Under islanded mode, making load power equally shared by each DG unit is a vital requirement to avoid DG unit overloading. To fulfill this requirement, the active power-frequency (P-f) and reactive power-amplitude (Q-V) droop control method was proposed and well accepted since it is independent of communications among DG units [4]. In spite of good performance in P sharing, the traditional droop control is incapable of Q, imbalance power and harmonic power (Q_{UH}) sharing [5]. Several solutions have been proposed in the literature [5-9]. In [5], an online virtual impedance adjustment method was proposed. In this method, a term associated with power sharing error is added to the P-f droop control,

then, the transient P variations caused by this term are captured to realize virtual impedance tuning. However, the frequency fluctuation and P overshoot could false trigger system protection. [6] and [7] proposed the cooperative harmonic and imbalance compensation strategies, which can provide harmonic and imbalance damping and alleviate power sharing error. Unfortunately, the power line impedance inevitably causes power sharing error. As will discuss in the later chapter, it is the mismatched feeder impedance that causes the Q, Q_{UH} sharing errors. Therefore, [8] and [9] proposed solutions based on the estimation of feeder impedance. However, [9] did not mention what kind of estimation method is utilized, and [8] exploited an intrusive estimation method that will introduce disturbance to the system.

Actually, feeder impedance is an important parameter. If we can somehow obtain the feeder impedance, many issues in MG caused by feeder impedance mismatch, such as power sharing error, power coupling, and voltage distortion, can be easily solved. Additionally, this paper also proposes a novel non-intrusive impedance estimation method. Therefore, some existing impedance estimation techniques are summarized here, which can be classified into three categories [10]: 1) based on system operating point variation [3]; 2) based on using extra devices dedicated for this purpose [11]; 3) based on the injection of extra signal into the system, and then analyze the response [12,13]. All these methods are intrusive ones, and they will introduce disturbance and possible distortion to the original system.

In addition, MG hierarchical control was proposed years ago, and then gradually became well-accepted [14]. In this structure, MG central controller (MGCC) and low-bandwidth communication (LBC) between each level is usually inevitable to accomplish the control purpose. In this sense, we can make good use of these existing devices to realize some auxiliary functions, like feeder impedance estimation as introduced in this paper.

The rest of this paper is organized as follows. In section II, an islanded MG system is analyzed at different frequencies. In section III, the proposed hierarchical control, with emphasis on secondary level, is explained. In section IV, simulation results are provided to verify

---

This work was supported by the National Natural Science Foundation of China under Grant 51437007.

The 2018 International Power Electronics Conference

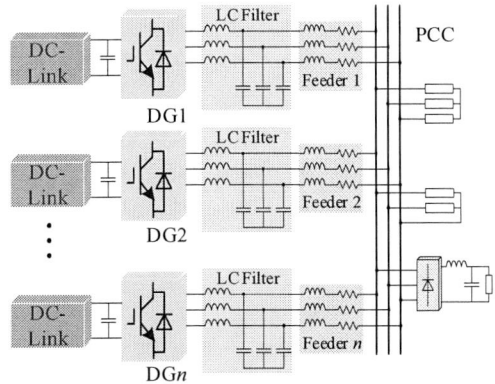

Fig. 1. Microgrid structure with multiple parallel-connected DG units.

this method. Finally, section V gives the conclusions and some future work plan.

II. ISLANDED MICROGRID SYSTEM ANALYSIS

The MG system studied in this work is shown in Fig. 1. Multiple DG units with *LC* filter are integrated to the point of common coupling (PCC) through different feeder impedance due to the distributed feature. Various kinds of loads may exist, including balanced, imbalanced and nonlinear ones. Note that the topology studied is a three-phase three-wire system. Therefore, zero-sequence current is not considered here.

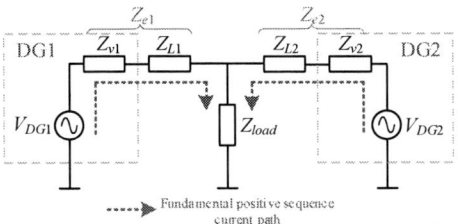

Fig. 2. Equivalent circuit of two DG units at fundamental positve sequence.

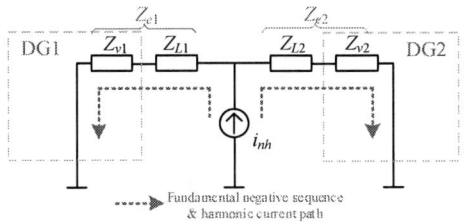

Fig. 3. Equivalent circuit of two DG units at fundamental negative sequence and harmonic frequencies.

The following analysis is based on two paralleled DG units for brevity. For a droop-controlled inverter, it can be represented by a controlled voltage source, whose amplitude and frequency references are given by the droop controller. Consequently, the equivalent circuit at fundamental positive sequence is illustrated in Fig. 2, where V_{DG1} and V_{DG2} are the DG units, Z_{L1} and Z_{L2} are the feeder impedance, and Z_{load} is the lumped PCC load, Z_{v1}

and Z_{v2} are the virtual impedance, Z_e is the equivalent impedance, which means the sum of Z_v and Z_L.

However, the inverter acts as short circuit at fundamental negative sequence and harmonic frequencies, as the reference voltage is purely positive sequence sinusoidal signal. So, the equivalent circuit at fundamental negative sequence and harmonic frequencies are shown in Fig. 3, where the imbalanced loads and nonlinear loads are represented by a lumped negative sequence and harmonic current source i_{nh}. Z_{v1} and Z_{v2} are zero if no virtual impedance control is employed. Note that the inverter output impedance is not considered in the analysis above. This is reasonable because the output impedance at low frequency need to be designed very low to ensure good voltage quality when feeding nonlinear loads. Therefore, the inverter output impedance can be neglected compared with the feeder impedance.

Based on the analysis above, two conclusions can be drawn. Firstly, the distribution of negative sequence and harmonic current, without any control strategies, is determined by the feeder impedance. The DG unit, which is located near to the PCC, will have to take majority of the load power. Secondly, the fundamental negative sequence and harmonic components in PCC voltage have proportional relationship with feeder impedance when feeding a certain amount of imbalance and nonlinear load. That is to say, the total harmonic distortion (THD) in PCC voltage has proportional relationship with Z_L. Intuitively, if we can measure each physical feeder impedance and compensate it to a common value Z_e, the power sharing error can be eliminated. Additionally, if the equivalent impedance after compensation is smaller than the physical impedance, the voltage THD at PCC will also decrease. This is the basic idea for this paper, and the implementation details are explained below.

III. PROPOSED HIERARCHICAL CONTROL SCHEME TO ENHANCE POWER SHARING

The hierarchical control structure proposed in this paper is shown in Fig. 4, which contains two control levels: primary level and secondary level. The primary control is located at each DG unit local controller, and it consists of droop control and virtual impedance control. Droop control can ensure the sharing of active power and stabilize the inverter-parallel system. Moreover, the virtual impedance control is in charge of tuning Z_v to follow the command coming from the secondary level through LBC link. While, the secondary control is located at the MGCC, and its function is to estimate each Z_L and calculate the desired Z_v command, then send this value to DG local controllers.

A. Primary Control Level

As introduced above, the traditional droop control with virtual impedance is implemented in this level. In addition, the inner controller, which is designed in stationary ($\alpha\beta$) frame with proportional resonant (PR) controller, is also introduced here. Note that most of the content in this subsection has been well explained in the publications, so it will be presented schematically due to

3925

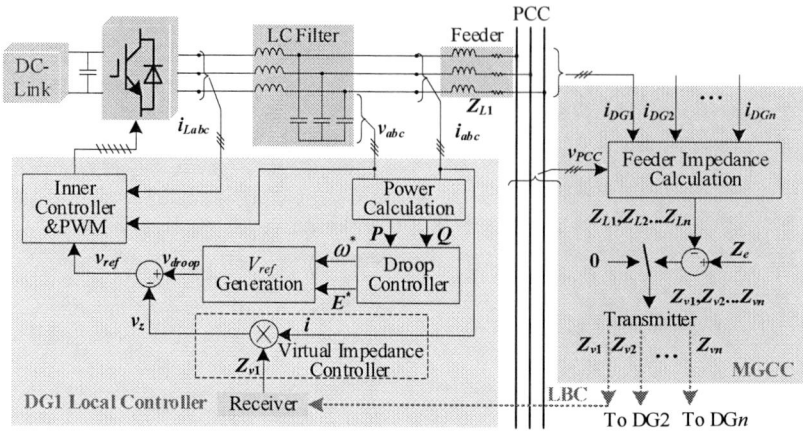

Fig. 4. Overall control block diagram of the proposed method.

the limited page.

First, the output active and reactive power of each DG unit is calculated by:

$$P = \frac{3}{2(\tau s+1)}\left(v_\alpha i_{1\alpha}^+ + v_\beta i_{1\beta}^+\right) \tag{1}$$

$$Q = \frac{3}{2(\tau s+1)}\left(v_\beta i_{1\alpha}^+ - v_\alpha i_{1\beta}^+\right) \tag{2}$$

where v_α and v_β are the DG unit output voltages in $\alpha\beta$ frame, i_α^+ and i_β^+ are the fundamental positive components of DG unit line current, τ is the time constant of the low pass filter (LPF).

Furthermore, the P-f and Q-V droop controllers are shown in (3) and (4) as

$$\omega^* = \omega_0 - k_p P \tag{3}$$

$$E^* = E_0 - k_q Q \tag{4}$$

where ω^* and ω_0 respectively are the reference and nominal frequency, E^* and E_0 are the reference and nominal amplitude, k_p and k_q are the droop coefficients. Droop control can realize P sharing and stabilize the system. However, it fails to realize Q and Q_{UH} sharing when the feeder impedance is not equal.

An effective solution to eliminate the power sharing error, according to [5] and [9], is to make Z_e equal for each DG unit by tuning Z_v, as shown in Fig. 2 and Fig. 3. If $Z_{e1}=Z_{e2}=Z_e$, then all the fundamental and harmonic power can be equally shared. The implementation of virtual impedance can refer to [5]. Note that the virtual impedance command comes from the secondary control level through LBC. The voltage reference from droop control is modified by subtracting the voltage drop on Z_v. The inner control loop is designed to follow this modified voltage reference.

To ensure excellent voltage tracking, the dual-loop voltage control with harmonic compensators is adopted. The controllers are shown as

$$G_V(s) = k_{pV} + \sum_{h=1,5,7} \frac{2k_{ih}\omega_c s}{s^2 + 2\omega_c s + (h\omega^*)^2} \tag{5}$$

$$G_I(s) = k_{pI} \tag{6}$$

where k_{ih} is the resonant gain, ω_c is the bandwidth of the

resonant controller, k_{pV} and k_{pI} are the proportional gains for voltage and current controllers, respectively.

B. Secondary Control Level

The secondary control level in MGCC is the core of this paper. It is responsible for estimating Z_L and generating the required Z_v. There are two operating modes: estimation mode and standby mode. The estimation mode will be triggered on only during the system startup period or when the system configuration changes, e.g. DG unit plug in/out. Otherwise, the secondary controller will always operate in standby mode and wait for the trigger signal from manual switch or higher control level.

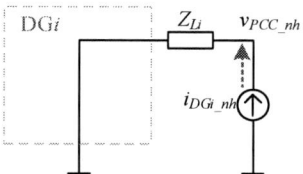

Fig. 5. Equivalent circuit of DGi with no virtual impedance at fundamental nagetive sequence and harmonic frequencies.

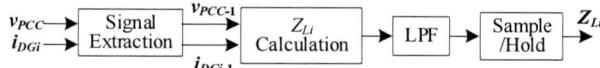

Fig. 6. Feeder impedance estimation principle.

Estimation Mode: In this mode, the controller aims to estimate Z_L for each DG unit. Firstly, the Z_v for each DG is set to zero through LBC link. In this condition, the equivalent circuit of DGi at fundamental negative sequence and harmonic frequencies is shown in Fig. 5. Based on this circuit, we can easily calculate Z_{Li} as long as v_{PCC_nh} and i_{DGi_nh} are detected. Although the feeder impedance varies with frequency, it can be assumed as constant since we only care about the low frequency range. Theoretically, any low frequency order (-1st, 3rd, 5th, 7th...) can be selected to calculate Z_{Li}, however, selecting an order with larger amplitude can decrease the estimation error. Here we take fundamental negative sequence (-1st) as an example. The estimation principle is

3926

demonstrated in Fig. 6.

The MSOGI-FLL based signal extraction method proposed in [15] is applied to obtain the negative sequence voltage and current component. Then the Z_L of the ith DG unit can be calculated as

$$R_{Li} = \frac{v_{\alpha PCC-1} i_{\alpha DGi\_1} + v_{\beta PCC-1} i_{\beta DGi\_1}}{i^2_{\alpha DGi\_1} + i^2_{\beta DGi\_1}} \tag{7}$$

$$X_{Li} = \frac{v_{\alpha PCC-1} i_{\beta DGi\_1} - v_{\beta PCC-1} i_{\alpha DGi\_1}}{i^2_{\alpha DGi\_1} + i^2_{\beta DGi\_1}} \tag{8}$$

Moreover, the estimated Z_{Li} is passed through a LPF and Sample/Hold block to remove the noise and hold the final value when the controller is switched to standby mode.

Standby Mode: the estimation mode will take a few seconds, and then the controller will automatically switch to standby mode. In this mode, the MGCC compared Z_{Li} with a pre-determined equivalent impedance Z_e to get the corresponding virtual impedance command Z_{vi} as

$$Z_{vi} = Z_e - Z_{Li} = (R_e - R_{Li}) + j(X_e - X_{Li}) \tag{9}$$

Then the Z_{vi} command can be transmitted to the ith DG unit through the LBC. Once all the DG units have received Z_v command, the MGCC can abort operation until the next trigger signal comes and then switch back to estimation mode. Note that the requirement for the communication is very low, because only a complex constant value is transmitted in unidirectional way, and the communication can even be shut down after all DG unit local controllers have received the Z_v in standby mode.

The design of Z_e is critical for the system performance. On the one hand, there is a tradeoff between system stability and PCC voltage quality. Smaller Z_e will reduce distortion in PCC voltage. Nevertheless, it may cause stability issue. On the other hand, if all DG units have same capacity (S), they should share a common Z_e to share the load power equally. However, if the DG units have different capacities, different Z_e should be chosen as

$$Z_{e1}S_1 = Z_{e2}S_2 = \cdots = Z_{ei}S_i \tag{10}$$

Then the load power can be proportionally shared by each DG unit according to its own capacity.

In summary, the proposed method has four merits. 1) Z_L can be precisely estimated in a non-intrusive way, which is independent of signal injection or extra device. 2) Load power can be shared equally or proportionally. 3) The THD in PCC voltage can be reduced by choosing a small Z_e; 4) the requirement for communication is very low. However, the limitation is that there must exist some imbalanced load or nonlinear load under the estimation mode.

IV. SIMULATION RESULTS

To verify the proposed method, simulations based on PSCAD are conducted. The simulation topology is shown in Fig. 7, and some key parameters are listed in Table I.

The simulation scenario is designed as: the MGCC operates under estimation mode before 3s, and under standby mode from 3s to the end of the simulation; At 6s, a resistive load step change occurs. The corresponding waveforms of P, Q, and Q_{UH} are shown in Fig. 8. The

output current of both DG units are shown in Fig. 9. In addition, PCC voltage waveform is shown in Fig. 10. The actual and estimated feeder impedance values are listed in Table II.

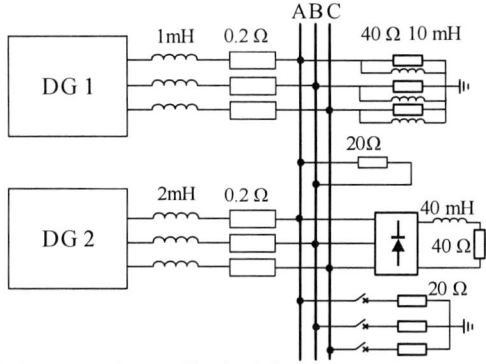

Fig. 7. System topology used in simulation.

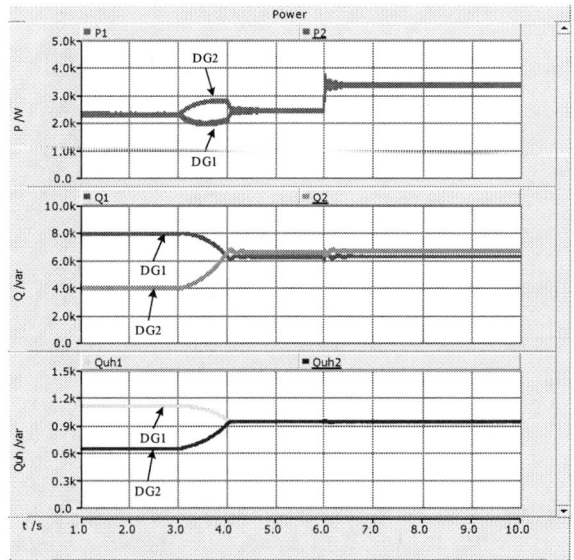

Fig. 8. Waveforms of active power, reactive power, and unbalanced and harmonic power for both DG units.

TABLE I
SIMULATION PARAMETERS

Symbol	Value	Symbol	Value
k_p	1e-5rad/s/W	k_q	1e-5V/var
ω_0	314rad/s	E_0	283V
Z_e	0.8mH, 0.05Ω		

TABLE II
FEEDER IMPEDANCE VALUES

	Resistance (Ω)		Inductance (mH)	
	DG1	DG2	DG1	DG2
Actual	0.2	0.2	1	2
Estimated	0.2199	0.2202	1.0000	1.9997
Error	9.95%	10.1%	0%	0.015%

Based on the simulation results, we can see that feeder impedance can be estimated with acceptable precision, and the load power sharing error can be eliminated by the proposed method. Note that the P variation during 3s to 4s is caused by the slow dynamics of droop control. Moreover, the THD in PCC voltage is effectively reduced from 2.94% to 2.42%.

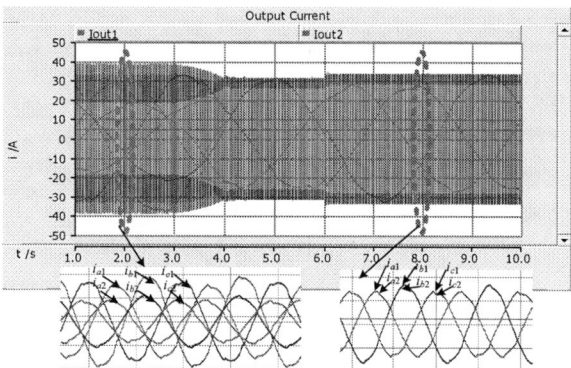

Fig. 9. Waveforms of output current for both DG units.

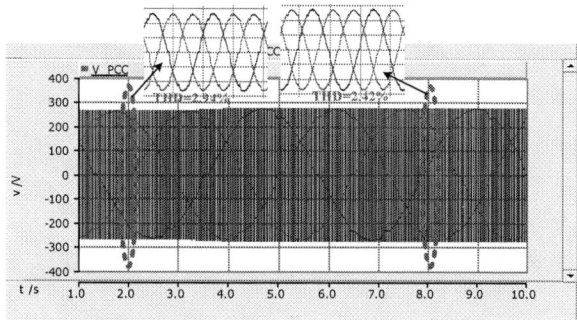

Fig. 10. Waveform of PCC voltage.

V. CONCLUSIONS

In this paper, a hierarchical control scheme is proposed to eliminate the load power sharing error in an islanded microgrid, which is based on a non-intrusive feeder impedance estimation method. Moreover, the THD in PCC voltage can be reduced, and the requirement and cost of the communication is very low. So this method can be easily implemented to distributed generation systems. It is worthwhile to explore the possibility to use this method to solve some other issues in microgrid, like power coupling and stability improvement.

REFERENCES

[1] D. E. Olivares, A. Mehrizi-Sani, A. H. Etemadi, C. A. Canizares, R. Iravani, M. Kazerani, *et al.*, "Trends in Microgrid Control," *IEEE Transactions on Smart Grid*, vol. 5, no. 4, pp. 1905-1919, Jul. 2014.

[2] F. Blaabjerg, Z. Chen, and S. B. Kjaer, "Power electronics as efficient interface in dispersed power generation systems," *IEEE Trans. Power Electron.*, vol. 19, no. 5, pp. 1184-1194, Sep. 2004.

[3] J. C. Vasquez, J. M. Guerrero, A. Luna, P. Rodriguez, and R. Teodorescu, "Adaptive Droop Control Applied to Voltage-Source Inverters Operating in Grid-Connected and

Islanded Modes," *IEEE Trans. Ind. Electron.*, vol. 56, no. 10, pp. 4088-4096, 2009.

[4] J. Rocabert, A. Luna, F. Blaabjerg, and P. Rodr´ıguez, "Control of Power Converters in AC Microgrids," *IEEE Trans. Power Electron.*, vol. 27, no. 11, pp. 4734-4749, Nov. 2012.

[5] J. He, Y. W. Li, and F. Blaabjerg, "An Enhanced Islanding Microgrid Reactive Power, Imbalance Power, and Harmonic Power Sharing Scheme," *IEEE Trans. Power Electron.*, vol. 30, no. 6, pp. 3389-3401, Jun. 2015.

[6] T. L. Lee and P. T. Cheng, "Design of a New Cooperative Harmonic Filtering Strategy for Distributed Generation Interface Converters in an Islanding Network," *IEEE Trans. Power Electron.*, vol. 22, no. 5, pp. 1919-1927, Sep. 2007.

[7] P. T. Cheng, C. A. Chen, T. L. Lee, and S. Y. Kuo, "A Cooperative Imbalance Compensation Method for Distributed-Generation Interface Converters," *IEEE Trans. Ind. Appl.*, vol. 45, no. 2, pp. 805-815, Mar. 2009.

[8] J. He, Y. W. Li, J. M. Guerrero, F. Blaabjerg, and J. C. Vasquez, "An Islanding Microgrid Power Sharing Approach Using Enhanced Virtual Impedance Control Scheme," *IEEE Trans. Power Electron.*, vol. 28, no. 11, pp. 5272-5282, 2013.

[9] P. Sreekumar and V. Khadkikar, "A New Virtual Harmonic Impedance Scheme for Harmonic Power Sharing in an Islanded Microgrid," *IEEE Trans. Power Delivery* vol. 31, no. 3, pp. 936-945, 2016.

[10] W. Ghzaiel, M. J. B. Ghorbal, I. Slama-Belkhodja and J. M. Guerrero, "A novel grid impedance estimation technique based on adaptive virtual resistance control loop applied to distributed generation inverters," *2013 15th European Conference on Power Electronics and Applications (EPE)*, Lille, 2013, pp. 1-10.

[11] J. Huang, K. A. Corzine and M. Belkhayat, "Small-Signal Impedance Measurement of Power-Electronics-Based AC Power Systems Using Line-to-Line Current Injection," *IEEE Trans. Power Electron.*, vol. 24, no. 2, pp. 445-455, Feb. 2009.

[12] L. Asiminoaei, R. Teodorescu, F. Blaabjerg, and U. Borup, "A Digital Controlled PV-Inverter With Grid Impedance Estimation for ENS Detection," *IEEE Trans. Power Electron.*, vol. 20, no. 6, pp. 1480-1490, 2005.

[13] M. Sumner, A. Abusorrah, D. Thomas, and P. Zanchetta, "Real Time Parameter Estimation for Power Quality Control and Intelligent Protection of Grid-Connected Power Electronic Converters," *IEEE Transactions on Smart Grid*, vol. 5, no. 4, pp. 1602-1607, 2014.

[14] J. M. Guerrero, J. C. Vasquez, J. Matas, L. G. Vicuna, and M. Castilla, "Hierarchical Control of Droop-Controlled AC and DC Microgrids-A General Approach Toward Standardization," *IEEE Trans. Ind. Electron.*, vol. 58, no. 1, pp. 158-172, Jan. 2011.

[15] P. Rodriguez, A. Luna, I. Candela, R. Mujal, R. Teodorescu, and F. Blaabjerg, "Multiresonant Frequency-Locked Loop for Grid Synchronization of Power Converters Under Distorted Grid Conditions," *IEEE Trans. Ind. Electron.*, vol. 58, no. 1, pp. 127-138, 2011.

The 2018 International Power Electronics Conference

Development of a 3.2MW Photovoltaic Inverter for Large-Scale PV Power Plants

Naoya Shibata[1]*, Tsuguhiro Tanaka[1] and Masahiro Kinoshita[1]

1 Power Electronics Systems Division, Toshiba Mitsubishi-Electric Industrial Systems Corporation (TMEIC), Tokyo, Japan

*E-mail: SHIBATA.naoya@tmeic.co.jp

Abstract— **The penetration of large-scale PV power plants, accompanied with the increase of the lengths of DC cables, the number of PV strings, combiner boxes, and other related equipment, have increased the conduction losses in the system. Therefore, in recent years, system DC voltage of PV plants has been increased to 1500V in order to reduce conduction losses by current passing through their components. The photovoltaic inverter rated 1500V-2.5MW had been developed in 2015. This inverter had firstly been certificated for UL certification of DC1500V photovoltaic inverter in the world in 2016. On the other hand, PV system cost required in the market is decreasing every year. Especially, the cost per output watt is attracted attention. In order to reduce its cost, larger-capacity inverter was developed in spite of using the same cabinet. This paper presents the development of a 3.2MW photovoltaic inverter with DC1500V. This inverter achieved high conversion efficiency by applying the three-level inverter topology which is a neutral point switch type, commonly known as T-type three-level inverter. The performance of the 1500V-3.2MW rated inverter was confirmed by verification.**

Keywords— *Photovoltaic inverter, 1500V system, Neutral point switch three-level inverter, T-type three-level inverter.*

I. INTRODUCTION

The more photovoltaic power plants is been allowed for the connection to high voltage power distribution system and power transmission system by improving the inverter functionalities, in accordance with the development of regulations and standards. As a result, this change is leading the increase of large-scale photovoltaic power plants, which have hundreds of MW capacity located in such as desert areas for one project.

As commonly known, it is an effective way to raise the system voltage in order to increase the power plants capacity. Therefore, 1000Vdc-rated photovoltaic power plants are mainly applied, which indicates that system voltage of PV module has the limitation to 1000Vdc. As a result, the bigger conduction losses are caused at the PV large-scale power plants because the system current is bigger. In addition, the number of PV modules connected in series is limited, which would proportionally increase the number of combiner box placed in the each PV module output with the power plant capacity.

Next, the MPPT voltage range is also the important concern in the 1000Vdc system. Most photovoltaic inverters with single MPPT function do not cover the all

environmental conditions for the power generation. The PV module voltage has the variations according to the atmosphere temperature, through the year with the hot season and the cold season, additionally it is varied in PV module location [1].

The rated voltage of most PV modules, inverters and combiner boxes are 1000Vdc, meanwhile IEC standards for low voltage system defines that the maximum dc voltage is 1500V. Increasing the dc voltage by 50% enables the reduction of the conduction losses due to the decrease of the dc current. The number of the combiner boxes is also reduced, so more PV modules can be connected in series [2]. On the ac side, the grid voltage is also increased and the conduction losses are reduced. The reduction of the conduction losses by high voltage yields to higher power density, so that achieve the compact design for the inverter.

The photovoltaic inverter rated at 1500V-2.5MW has been developed in 2015 [3]. This inverter has firstly been certificated for UL certification of DC1500V photovoltaic inverter in the world in 2016. On the other hand, PV system cost required in the market is decreasing every year. Especially, the cost per output watt is attracted attention. This paper presents the development of a 3.2MW/3.2MVA photovoltaic inverter with DC1500V for large-scale photovoltaic systems. Specifications and design outlines are explained, with emphasis on the topology of the power conversion component. A cooling mechanism for exhausting the heat by IGBT operation is explained. Verification results show excellent performance of the inverter suitable for large-scale power plants.

Fig. 1. Outline of the 3.2MW photovoltaic inverter.

TABLE I
INVERTER SPECIFICATION

System Ratings	
Maximum Input DC Voltage	1500 V
Rated Input DC Voltage	960 V
MPPT Voltage Range	875 V ~ 1300V (start-up from 1450V)
AC Voltage	600 V
Rated Output Power	3.2 MW - 3.2 MVA
Maximum efficiency (include auxiliary power)	98.9%
Temperature Range	-25°C ~+50°C
Environmental Rating	NEMA3R (Outdoor)
Dimensions	H2286 mm W5000 mm D1150 mm
Power Density	243.5kW/m³

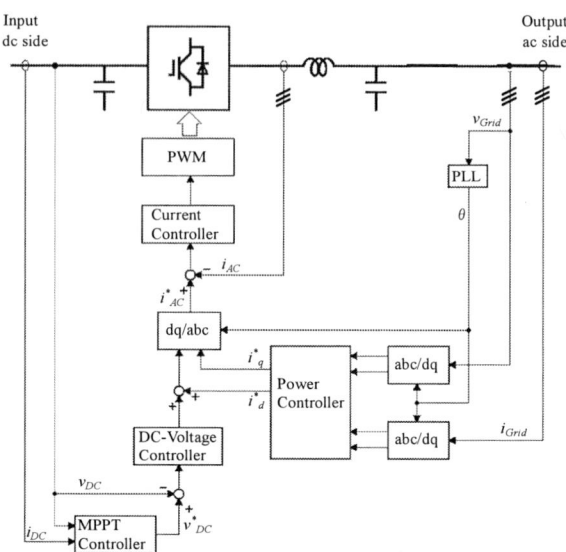

Fig. 2. System configuration and control block diagram.

II. System Configuration

The outline of the 3.2MW photovoltaic inverter is shown in Fig. 1. Table I shows the inverter specification of the developed DC1500V-3.2MW inverter. This inverter is transformerless, therefore there need to be a step-up transformer at the inverter output to provide galvanic isolation for connection to the utility grid. The MPPT voltage range is 875V to 1300V, which is a wide range of 425V suitable for PV power plants design. The start-up voltage is from 1450V, therefore, the inverter can accommodate to the high open-circuit photovoltaic voltage. The AC voltage is defined to be 600V in accordance with minimum MPPT voltage. The maximum efficiency achieved is 98.9% including the auxiliary power. The temperature rating is -25°C to +50°C, therefore the inverter can operate through the all seasons.

Fig.2 shows the system configuration and control block diagram of the developed inverter [4]. This inverter has the three-level inverter topology, which reduces the total conversion losses and yields the similar sinusoidal waveform, so that the harmonic damping filter can be smaller and the conversion efficiency is improved. The dc source of PV module is connected at the inverter input side, and the dc capacitor smoothes the dc voltage. The L-C filter is connected at the inverter output side in order to smooth the switching waveform.

The control operation about the MPPT control is as follows: The dc capacitor voltage reference is generated according to the MPPT controller by detected dc voltage and dc current. Active inverter current reference is generated by the dc-voltage controller, comparing the reference with the detected value. Meanwhile, d-axis and q-axis current references are generated by the power controller according to detected ac voltage and current. The reference from dc-voltage controller and d-q axis current reference yield the inverter current reference. The inverter voltage reference is generated by the PWM control.

III. Inverter Units Characteristics

The three-level inverter has two well-known main topologies, which are the diode clamped type (NPC in Fig. 3a), and the neutral switch type commonly known as the T-type (NPS in Fig 3b). Switching patterns are the same for both topologies as shown in Fig.4. This 1500V-3.2MW rated inverter applies the NPS three-level inverter shown in Fig. 3b. Switches Q1 and Q4 have 1700V-rating, therefore Q1 and Q4 can withstand up to 1500Vdc of PV module voltage. On the other hand, Q2 and Q3 are rated at 1200V which can withstand a half of PV module voltage. The NPS three-level inverter has the advantage of lower conduction losses and a simpler topology than other three-level topologies [5].

The conduction losses and switching losses on the two main topologies (NPC and NPS) are compared through calculations and simulations in PSIM. Typical parameters of the IGBT losses in a 1500V photovoltaic application were used, assuming that the output voltage are controlled with PWM for sinusoidal waveform, which is the same fixed carrier frequency. The conditions of the estimation are as follows, the grid voltage is 600V, the dc voltage is 875V which is the minimum MPPT voltage, and the inverter operates at 100% output power, which is 3200kW.

The detail losses of the comparison between NPC and NPS topologies on the IGBTs are described in Fig. 5. The conduction losses for the NPC are 78% of its total semiconductor losses, which is due to the current flow always passing through two devices. On the other hand, for the NPS inverter, the conduction losses are significantly reduced to 51% in total losses of the NPC. This is because the current flowing from the positive bus and negative bus passes through only one device (Q1 or Q4). Switching losses are not the significant difference between the NPC and NPS topology, just 4% lower for the NPS topology than that of the NPC.

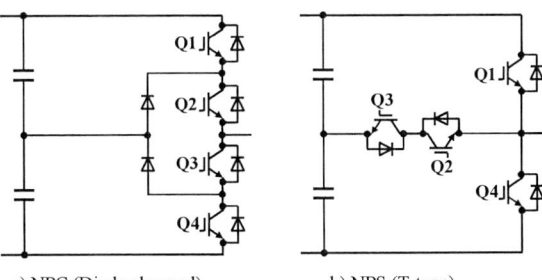

a) NPC (Diode clamped) b) NPS (T-type)

Fig. 3. Three-level inverter configuration.

	Q1	Q2	Q3	Q4
S1	ON	ON	OFF	OFF
S2	OFF	ON	ON	OFF
S3	OFF	OFF	ON	ON

Fig. 4. Switching states for NPS inverter.

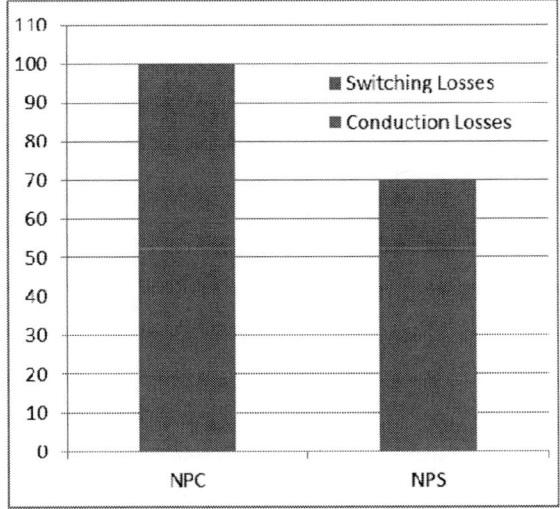

Fig. 5. Comparison of conduction and switching losses for NPC and NPS bridge configuration.

Compared the total of the conduction losses and the switching losses between NPC and NPS topology, the NPS topology is lower losses by 30% than the NPC. This reduction for the total losses compared to the NPC topology achieves much better conversion efficiency, which can reduce the cooling requirements and capacity, so that it helps improving the reliability.

Next, the cooling mechanism is explained. When the output power is low, the cooling fan does not operate and the heat on IGBT is cooled by natural convection as shown in Fig. 6. When the output power increases, the cooling fan operates as shown in Fig. 7. Hot radiated heated from IGBT goes up to the top of the heat-pipe. The cool air intake from the top of cabinet passes through the cooling fins to promote heat-pipe mechanism. Slow cooling fan, operating at a speed of 80%, can minimize

the particle and the dust entrance. 80% fan load operation can also reduce auxiliary power losses. Heat-pipe radiator and other components in NEMA3R cabinet are completely outdoor rated, mitigating environmental risks.

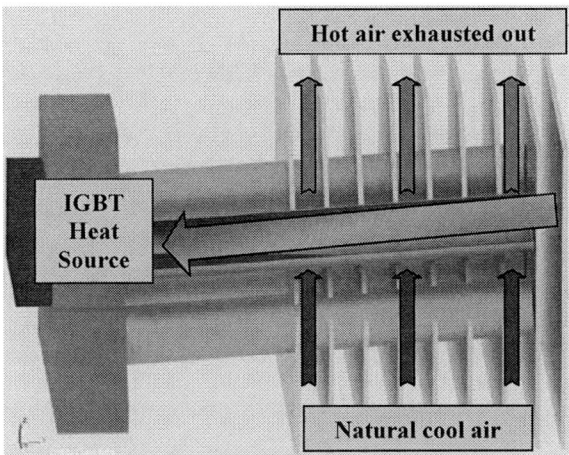

Fig. 6. Hybrid cooling on fan-less mode.

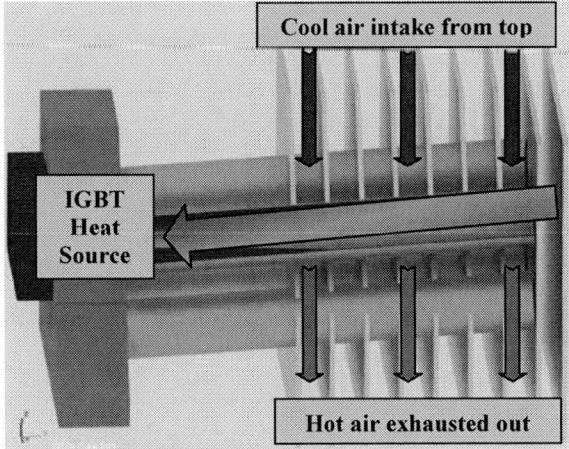

Fig. 7. Hybrid cooling on fan operation mode.

IV. VERIFICATION RESULTS

The verification results through all MPPT voltage range from 875V to 1300V is shown in Fig. 8. The maximum efficiency is 98.9% including the auxiliary power at 40% of rated output power in 875V. The efficiency at rated output power is more than 98.5% although rated output current is high (3079A). The efficiency keeps more than 98.5% at 20% of rated output power in spite of the low output power. The decline of the efficiency at 40% load is due to starting up the cooling fan. This is not only because of the reduction of the total losses in the IGBTs; the optimization in the design of the passive components is also an important point in achieving better efficiency characteristics. This results in Fig.8 show that this inverter can operate with high efficiency in almost all MPPT voltage range. This brings the advantage of maintaining high efficiency without the influence of weather and irradiance

3931

conditions. In recent years, large scale PV modules are designed with capacities larger than the inverters. Hence, the inverter should ideally maintain high efficiency at rated power.

The inverter output current and voltage were measured in order to verify the output waveform. Fig. 9 shows verification results at 600Vac and 875Vdc at rated output power of 3200kW. The current THD is 1.3% in this case and the individual harmonics is less than the criteria, which is compliant with IEEE1547. By using a three-level topology for power conversion, the current harmonic distortion can be small with an optimized harmonic filter, which enables the inverter to operate in better efficiency.

Fig. 10 shows verification results at 600Vac and 1300Vdc which is the maximum MPPT voltage. The inverter has the derating operation of 2900kW output power. The current THD for this case is 0.8% and the individual harmonics is also less than the criteria, which is compliant with IEEE1547. Although the inverter current ripple is higher by increasing the dc input voltage, the current THD of output current can keep the low harmonic level.

Fig. 11 and Fig. 12 show the results for load rejection test, on which the inverter is disconnected from the grid at 600Vac and 960Vdc at rated output power of 3200kW. When the load rejection occurs, the inverter detects the loss of grid voltage and blocks the all gate signal. The inverter current can stop without any surge waveform. Though dc input current fluctuates at the load rejection test, it is no issue for the current waveform. In this case, resonant phenomenon occurs between PV inverter and DC power source. In Fig. 12, the inverter restarts with softly ramp-up increase, and there is no transient surge current, so the inverter does not give the bad influence for the grid.

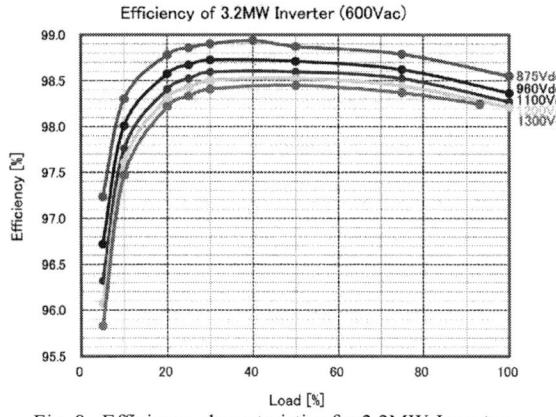

Fig. 8. Efficiency characteristics for 3.2MW-Inverter.

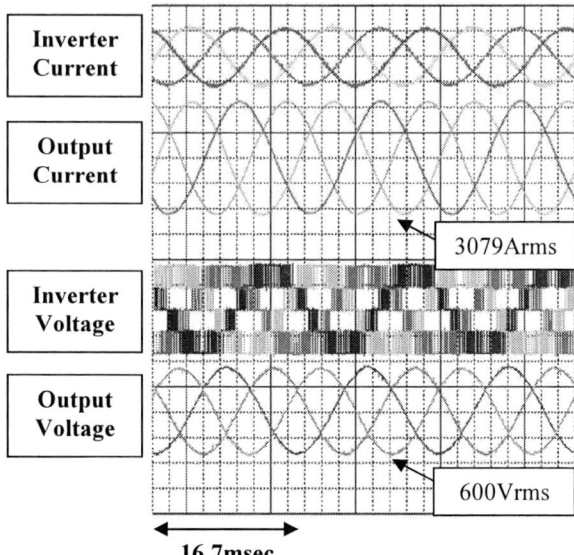

Fig. 9. Output characteristics at 875Vdc, 600Vac.

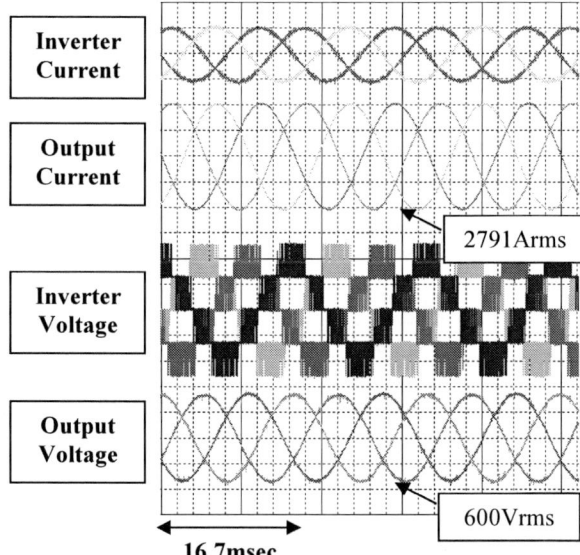

Fig. 10. Output characteristics at 1300Vdc, 600Vac.

The 2018 International Power Electronics Conference

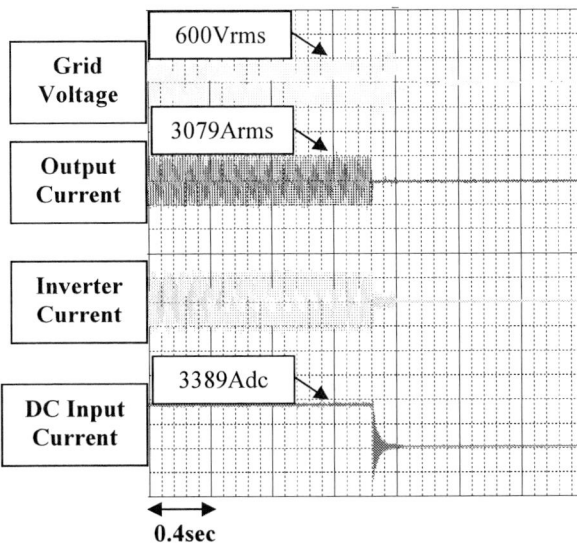

Fig. 11. Load rejection test results (Inverter disconnection).

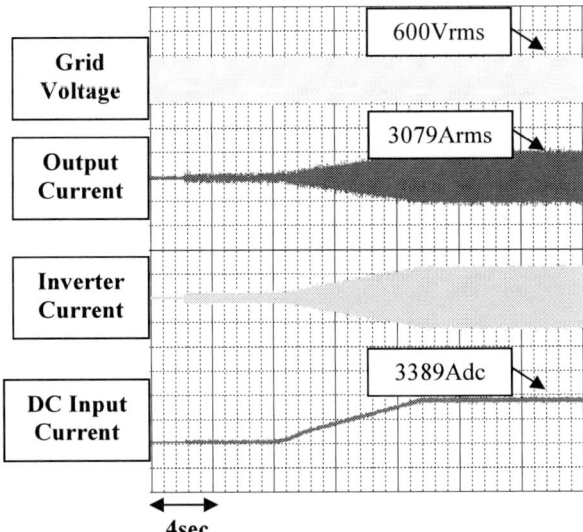

Fig. 12. Load rejection test results (Inverter reconnection).

V. CONCLUSION

This paper presented a 3.2MW/3.2MVA photovoltaic inverter rated at 1500Vdc, suitable for large power plants. The advantages of NPS topology were explained, which can be summarized as lower losses due to the reduction in current, wider MPPT range and higher power density. In addition, the cooling mechanism was introduced, which can achieve high environmental durability, since the amount of dust and moisture entering the device is reduced.

Verification test was performed in the factory. This inverter continuously operates at 3.2MW output power, and high maximum efficiency of 98.9% including the auxiliary power was achieved. This inverter was certificated for UL certification of 1500V photovoltaic inverter on March in 2018.

REFERENCES

[1] E. Serban, M.Ordonez, C. Pondiche, "DC-bus voltage range extension in 1500V photovoltaic inverters," *IEEE Journal of Emerging and Selected Topics in Power Electronics,* Accepted for publication DOI 10.1109/JESTPE.2015.2445735.

[2] E. Gkoutioudi, P.Bakas, A. Marinopoulos, "Comparison of PV systems with maximum DC voltage 1000V and 1500V," 2013 *IEEE 39th Photovoltaic Specialists Conference (PVSC)* pp. 2873-2878, June 2013.

[3] R. Inzunza, R. Okuyama, T. Tanaka, and M. Kinoshita, "Development of a 1500Vdc Photovoltaic Inverter for Utility-Scale PV Power Plants," *IFEEC 2015 IEEE 2nd International*, Nov. 2015.

[4] M. Schweizer, J. W.Kolar, "Design and implementation of a highly efficient three-level T-type converter for low-voltage applications," *IEEE Transactions on Power Electronics* Vol.28, No.2 pp. 899-907, February 2013.

[5] T. Yoshino, R. Inzunza, T. Ambo, E. Ikawa, T. Takahashi, N. Takahashi, "MW-range PCS for efficient operation of large-scale PV plants - High efficiency and less maintenance works," *3rd IEEE International Symposium on Power Electronics for Distributed Generation Systems, PEDG 2012*, pp.249,253, 25-28 June 2012.

The 2018 International Power Electronics Conference

Impedance-Based Stability Analysis of Large-Scale PV Station under Weak Grid Condition Considering Solar Radiation Fluctuation

YiMing Tu[*], Jinjun Liu, Teng Liu and Xiangpeng Cheng
State Key Laboratory of Electrical Insulation and Power Equipment, Xi'an Jiaotong University
Xi'an, China
*E-mail:tuyiming@stu.xjtu.edu.cn

Abstract— Instability issues caused by impedance mismatch between large-scale PV station and weak grid have been increasingly studied. Existing studies generally follow the rule that the output impedance of photovoltaic inverter is invariant, because the small-signal model of the inverter is built at the steady-state operating point, without considering power level change. Nevertheless, in actual operation condition, the steady-state operating point of PV inverter will change as solar radiation fluctuates. In this paper, both PV cell nonlinearity and power level change caused by solar radiation fluctuation are taken into account when establishing the small-signal model of PV inverter. A nonlinear relationship between the output impedance of PV inverter and solar radiation strength is obtained. To verify the proposed impedance model, the output impedance of inverter under different operation states is measured. Finally, system stability is analyzed based on the simulation results and on-site recorded waveforms measured in an actual PV plant.

Keywords—Impedance, PV station, stbility, weak grid.

I. INTRODUCTION

As a kind of clean and renewable energy, solar energy is increasingly gaining public attention under energy crisis condition. Due to the increase of electricity demand and improvement of switching devices, the total installed capacity of PV station enlarges gradually. Even in a single PV station, the scale could reach up to several hundred megawatts [1]. However, when a large-scale PV station is connected to weak grid, there would emerge instability phenomenon at the point of common coupling (PCC), for instance, waveform distortion and harmonic resonance [2].

The stability issues on grid-tied paralleled inverters are widely studied, including inverter modeling, system stability analysis and control strategy improvement these four main aspects. To a large-scale PV station, current researches mainly focus on how the installed capacity influences output impedance of a station. The output impedance of a PV station is determined by both the output impedance of a single inverter and the amount of inverters in the station. As is known, a large-scale PV station is equipped with multiple paralleled inverters. Therefore, the output impedance of a PV station can be simply obtained

by parallel calculation of the single inverter impedance. Usually, the output impedance of a grid-tied inverter is derived based on its small-signal model. [3], [4]. Inverter-grid system stability could be determined according to impedance-based stability criterion [4-6]. Under a weak grid condition, the grid impedance cannot be ignored, which would destabilize system [3], [7]. Meanwhile, different control methods including control strategy redesign or active damping, are proposed to mitigate instability, especially harmonic resonance in three-phase grid-tied inverter system with LCL filter [7-9].

Generally, in aforementioned literatures, the output impedance of inverters is derived through small signal linearization of nonlinear circuits around its equilibrium point, without considering the influence of power level change of the inverter. Therefore, the inverter could be viewed as a linear time invariant (LTI) system, and its output impedance is fixed. However, the operating point of inverters in actual PV stations would change continuously as the solar radiation varies during a day. As a result, these inverters would operate in a time variant state and exhibit nonlinear output characteristic. An interesting phenomenon occurred in an actual PV station is that, whenever the output power of the inverter rises to 60%-70% of its rated power due to solar radiation increasing in one day, grid-connected system tends to become unstable. Once few of the inverters fail to work due to electrical protection, the rest of the inverters recover and begin to work stably. Nevertheless, this phenomenon could not be well-explained if the output impedance of inverter is considered to be constant.

To solve this question, in this paper, the structure of large-scale PV station and impedance-based stability criterion are briefly reviewed at first. Then, an impedance model of PV inverter considering power level change and the nonlinear characteristic of PV cell is proposed, where both active power control case and DC voltage control case are studied. Meanwhile, the inverter output impedance under varied power level condition is deduced based on this newly built model. Finally, the obtained output impedance expression is verified by the simulated measurement results, and the stability of grid-connected system is analyzed by simulation and on-site recorded waveforms measured in the PV station.

This work was supported by the National Natural Science Foundation of China under Grant 51437007.

II. STABILITY OF LARGE-SCALE PV STATION.

A. Structure of Large-Scale PV Station

A large-scale PV stations is usually composed of quantities of individual PV units. Each unit includes PV arrays, two inverters, one transformer and transmission cables as presented in Fig. 1.

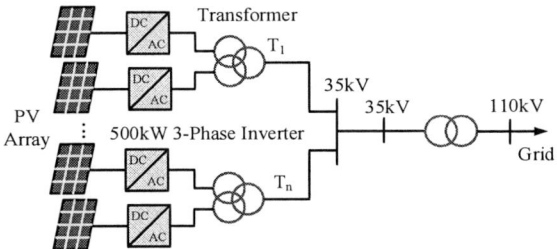

Fig. 1. Topology of a large scale PV station.

According to whether there contains Boost circuit at DC side of inverters or not, PV units can be categorized into two types, i.e. single-stage and two-stage structure. Fig. 2 exhibits the topology of a single-stage, two-level voltage source type, output current controlled inverter. The DC link of the single-stage inverter is directly connected to PV cell with a paralleled capacitor, without another Boost circuit. MPPT control and d-q axis decoupling control are coupled, making the inverter control structure rather complicated. However, such kind of structure is more widely used in industrial application owing to its higher efficiency and lower cost compared with two-stage structure.

B. Impedance-based stability analysis

Generally, to analyze the stability of grid-connected PV system, the interconnected system can be divided into two parts, i.e. inverter and grid. The inverter is equivalent to a controlled current source in parallel with output impedance $\mathbf{Z}_o$ (or admittance $\mathbf{Y}_o$), and weak grid could be represented by a constant voltage source in series with grid impedance $\mathbf{Z}_g$ as shown in Fig. 3. Impedance of three-phase AC converters is different from that of DC-DC converters, because in three-phase AC system, impedances as $\mathbf{Z}_o$ and $\mathbf{Z}_g$ are 2×2 matrices in dq domain.

Interconnected system is stable if and only if the impedance return ratio of $Y_o(s)Z_g(s)$ satisfies Generalized Nyquist Criterion (GNC) [10]. On one side, long distance transmission lines increase the grid impedance $Z_g(s)$. On the other side, a large-scale PV station equipped with several hundred paralleled inverters has low output impedance. Supposing there are N paralleled inverters in a PV station, then the output impedance of the PV station declines to $Z_o(s)/N$. Consequently, there exists a high possibility that the impedance-based stability criterion would be violated and instability appears.

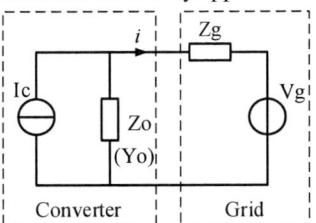

Fig. 3. Single line Equivalent impedance model.

In Fig.3, the current flowing into the grid i, is formed not only by the current i_c, but also by grid voltage and impedance network consisting of $\mathbf{Z}_o$ and $\mathbf{Z}_g$.

$$I(s) = \left[I_c(s) - V_g(s)Y_o(s) \right] \cdot \frac{1}{1 + Y_o(s)Z_g(s)} \quad (1)$$

Supposing that the inverter and grid are stable in separate state, then system stability is decided only by the minor loop gain $Y_o(s)Z_g(s)$. In a three-phase AC system, system return ratio $L(s)$ is defined as (2). If $L(s)$ satisfies GNC, then the system is stable.

$$L(s) = Y_o(s)Z_g(s)$$
$$= \begin{pmatrix} Y_{odd}(s) & Y_{odq}(s) \\ Y_{oqd}(s) & Y_{oqq}(s) \end{pmatrix} \begin{pmatrix} Z_{gdd}(s) & Z_{gdq}(s) \\ Z_{gqd}(s) & Z_{gqd}(s) \end{pmatrix} \quad (2)$$

III. IMPEDANCE MODELLING OF PV INVERTER

Impedance modeling is the crucial step for system stability analysis. To a three-phase AC system, impedance models are established in d-q coordinates and represented by 2×2 transfer matrices. However, for a symmetrical

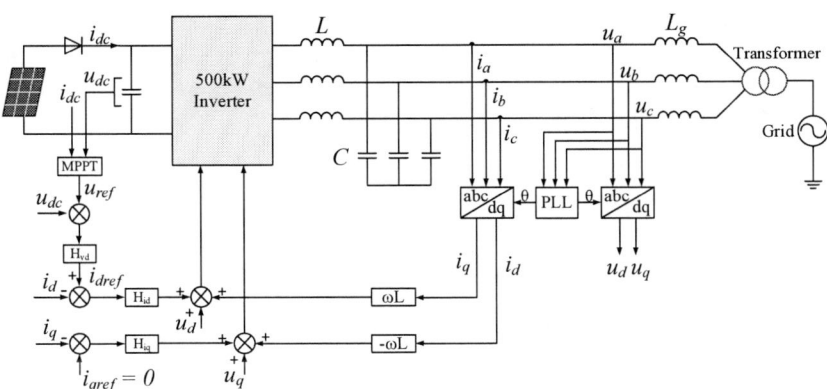

Fig.2. Topology of a large scale PV station.

balanced three-phase inverter, D channel admittance Y_{odd} and Q channel admittance Y_{oqq} are equal, and coupling items Y_{odq} and Y_{oqd} are small enough to be neglected. Therefore, the main task is to derive the D channel output admittance of inverter Y_{odd}. For simplicity and direct understanding, output impedance Z_{odd} instead of Y_{odd} is calculated in the following part.

The commonly studied small-signal model of voltage source inverter in synchronous rotating reference frame is given in Fig. 4. Usually, this model is established around a steady state point, and $\hat{a}_\theta$ is 0 because DC link voltage

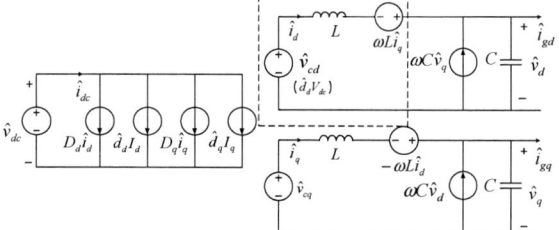

Fig. 4. Small-signal model of three-phase inverter.

fluctuation is neglected. However, to a PV inverter, both the steady state point and DC link voltage are continuously changing. Hence, $\hat{c}_{\delta h}$ could not be neglected and power state variables need to be taken into consideration. According to the model in Fig. 4, for simplicity, a new variable Z'_{odd} is defined to be the output impedance of components within the dashed line block.

$$Z'_{odd}(s) = \frac{\hat{v}_d}{\hat{i}_d} \qquad (3)$$

Thus, D channel output impedance Z_{odd} could be directly expressed as follows:

$$Z_{odd}(s) = \frac{\hat{v}_d}{\hat{i}_{gd}} = \frac{\hat{v}_d}{\hat{i}_d} \Big\| \frac{1}{sC} = Z'_{odd}(s) \Big\| \frac{1}{sC} \qquad (4)$$

According to Kirchhoff's Voltage Law (KVL), the voltage equation inside the dashed block in Fig. 4 is given by (4).

$$\hat{v}_{cd} = sL\hat{i}_d - \omega L\hat{i}_q + \hat{v}_d \qquad (5)$$

Fig. 5. Current decoupling control.

Fig. 5 presents the detailed current control structure. From this figure, $F_{\hat{\delta}r}$ can also be expressed as (5)

$$\hat{v}_{cd} = (\hat{i}_{dref} - \hat{i}_d)H_{i_d} + \omega L\hat{i}_q + \hat{v}_d \qquad (6)$$

where H_{id} and H_{iq} are PI control functions.

channel elements, for simplicity, it is neglected.

From (5) and (6),

$$sL\hat{i}_d = (\hat{i}_{dref} - \hat{i}_d)H_{i_d} \qquad (7)$$

where $H_{id} = k_{pi} + k_{ii}/s$ is the PI control block of current loop, and the only unknown item in (7) is $\hat{a}_{nly}$. If the nonlinear characteristic of PV cell is not considered first, then the DC voltage of inverter could be regarded to be constant. In active power control case,

$$i_{dref} = \frac{P_{ref}}{v_d} \qquad (8)$$

then, we perform small signal linearization on (8).

$$i_{dref} + \hat{i}_{dref} = \frac{P_{ref}}{V_d + \hat{v}_d} = \frac{P_{ref}(V_d - \hat{v}_d)}{V_d^2 - \hat{v}_d^2} \qquad (9)$$

To overlook the second order item $\hat{v}_d^2$ in the denominator of (9),

$$\hat{i}_{dref} = -\frac{P_{ref}\hat{v}_d}{V_d^2} \qquad (10)$$

Taking current flowing direction in Fig.3, D channel output impedance Z'_{odd} could be directly derived based on expression (7) and (10).

$$Z'_{odd} = \frac{\hat{v}_d}{-\hat{i}_d} = \frac{(sL + H_{id})V_d^2}{H_{id}P_{ref}} \qquad (11)$$

D channel output impedance Z_{odd} is

$$Z_{odd}(s) = \frac{(Ls^2 + k_p s + k_i)V_d^2}{V_d^2 LCs^3 + k_p V_d^2 Cs^2 + k_i V_d^2 Cs + k_p P_{ref}(s+1)} \qquad (12)$$

It could be observed that Z_{odd} decreases as P_{ref} increases.

Secondly, the nonlinearity of PV cell is considered.

$$I = I_{sc}\left\{1 - \left[C_1\left(\exp\left(\frac{V}{V_{oc}C_2}\right) - 1\right)\right]\right\} \qquad (13)$$

For a 500kW inverter with around 600V DC voltage, a TSM-250PC05A type PV cell is applied. At maximum power point, output current $I_m=833$ A, $V_m=606$ V. Based on PV station on-site temperature and solar strength, and the PV cell datasheet, $I_{sc}=879$ A, $V_{oc}=760$ V. Its power to voltage (P-V) characteristic with MPPT control is calculated as:

$$P = 879SV(1 + 0.0025\Delta T) \cdot R \qquad (14)$$

where P is output power, V is output voltage, ΔC is temperature error from 25and S is the per unit value of solar radiation strength, and

$$R = 1 - 8.31 \times 10^{-7}(e^{\frac{V}{54.23(1-0.00288\Delta T)\ln(e+0.5S-0.5)}} - 1) \quad (15)$$

From (15) and (16), it could be observed that R is around 1 and impact by ΔC can be ignored, thus P is approximately linear to S.

Now, considering DC voltage control case. An outer voltage control loop will be added, as shown in Fig. 2. Consequently, equation (8) ought to be rewritten as expression (16).

$$\hat{i}_{dref} = (\hat{v}_{ref} - \hat{v}_{dc}) \cdot H_{vd} \qquad (16)$$

According to instantaneous reactive power theory and the principle of power balance, $\hat{u}_{dc}$ can be computed by

$$\frac{3}{2} u_d i_d = u_{dc} \cdot (i_{dc} - C \frac{du_c}{dt}) \qquad (17)$$

$$\hat{u}_{dc}(s) = \frac{3}{2} \frac{U_d \hat{i}_d(s) + I_d \hat{u}_d(s)}{I_{dc} - sU_{dc}C_{dc}} \qquad (18)$$

Combining equation (4), (5) and (8)-(12), Z'_{odd} is derived as follows:

$$Z'_{odd}(s) = -\frac{\hat{u}_d(s)}{\hat{i}_d(s)} = \frac{2(I_{dc} - sCU_{dc})(sL + H_{id})}{3I_d H_{vd} H_{id}} + \frac{U_d}{I_d}$$
$$= \frac{(833S - 600sC_{dc})(sL + H_{id})}{1968SH_{vd}H_{id}} + \frac{1}{5.2S} \qquad (19)$$

Finally, inverter D channel output impedance Z_{odd} could be directly obtained by parallel calculation of Z'_{odd} and output capacitor C, according to expression (4) and (19).

IV. MODEL VERIFICATION AND STABILITY ANALYSIS

To verify the correctness of the newly built impedance model, an impedance measurement process is simulated. The inverter parameters are listed in Table I.

TABLE I

PARAMETERS OF INVERTER

Parameters	Value	Parameters	Value
C_{dc}	6000μF	f	4.8kHz
L	100 mH	k_{pi}	0.5
C	300 μF	k_{ii}	0.125
L_g	200 mH	k_{pv}	0.732
U_g	200 V	k_{iv}	0.061

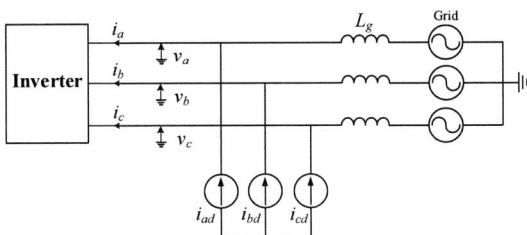

Fig. 6. Schematic of impedance measurement

Fig.6 gives the schematic of inverter output impedance measurement. The first step is to insert three-phase current source disturbances i_{ad}, i_{bd} and i_{cd} at PCC point. Then, the frequency of the disturbance is gradually increased and current response i_a, i_b, i_c and voltage response v_a, v_b, v_c are measured at each frequency. The last step is to convert the measured response signals into synchronous rotating frame via Park transformation, and to calculate D channel impedance Z_{odd}.

To simplify the measurement process, first of all, DC link voltage is kept constant and P_{ref} is step-changed from $1/3 P_N$ to $4/3 P_N$ to emulate power level variations. The measured results under four different states are shown in Table II. Measured impedance under four states are illustrated by circle, diamond, square, and triangular form respectively in Fig. 7. It can be observed that the measurement results match well with curves drawn according to expression (13), which verifies the correctness of the model.

TABLE II

MEASURED IMPEDANCE UNDER FOUR STATES

f/Hz	Z1/dB	Z2/dB	Z3/dB	Z4/dB
30	-7.49	-9.10	-12.04	-12.78
60	-8.16	-9.78	-11.74	-14.08
120	-6.02	-11.13	-13.38	-14.52
160	-5.14	-10.01	-12.44	-15.76
240	-5.65	-10.93	-13.48	-15.48
480	-5.34	-10.37	-13.49	-16.85
720	-6.56	-10.93	-13.91	-16.44
960	-7.56	-11.32	-14.26	-16.40
1920	-11.31	-12.57	-14.57	-16.03
3840	-16.23	-15.86	-16.03	-16.81
5760	-19.88	-18.67	-18.62	-18.51
6720	-21.01	-21.26	-19.57	-19.57

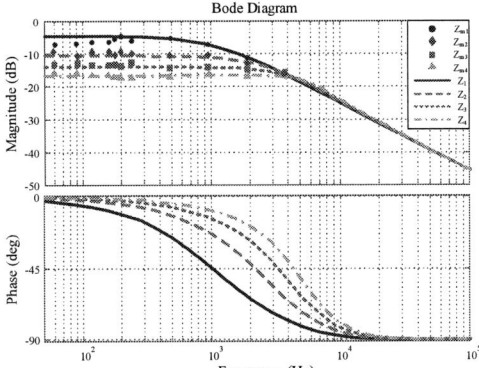

Fig. 7. Comparison of modeling and measurement results.

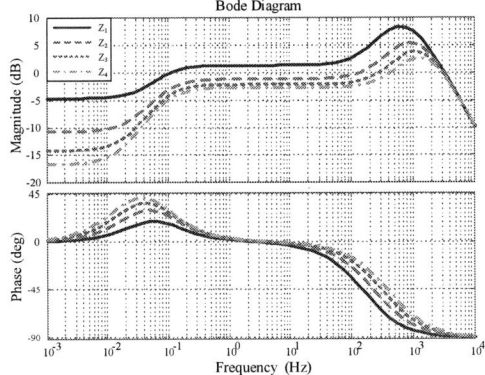

Fig. 8. D channel output impedance under four states.

3937

Secondly, an emulated PV cell with MPPT control is added at DC link. Solar radiation strength S_1=0.3, S_2=0.6, S_3=0.9 and S_4=1.2. Z_1 to Z_4 which represent Z_{odd} under 4 states are shown in Fig. 8. It could be observed from both Fig. 7 and Fig. 8 that at low frequency, the magnitude of $Z_{odd}(s)$ decreases as output power increases. This is only the case for individual inverter, though. For a large-scale PV station with multiple inverters, the impedance amplitude will decrease more seriously as output power increases. When output power of the whole station increases to a critical value, inverter output impedance tends to overlap with grid impedance, leading to stability criterion being violated, and system becomes unstable.

Additionally, a time-domain simulation of the grid-connected PV inverter with varied solar radiation strength S is conducted in PSCAD. In the simulation, other parameters are kept constant as listed in Table. I, only S is step-changed from 0.3 to 1.2 gradually. Simulation results of the inverter output current i and PCC voltage v_{pcc} are shown in Fig.9.

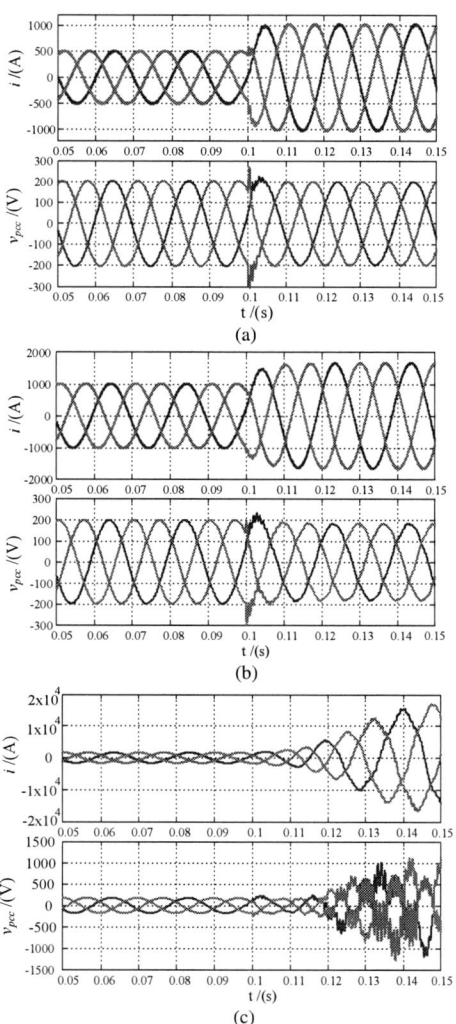

Fig. 9. Simulation results of i and v_{pcc} with varied solar radiation strength S. (a) S changes from 0.3 to 0.6. (b) S changes from 0.6 to 0.9. (c) S changes from 0.9 to 1.2.

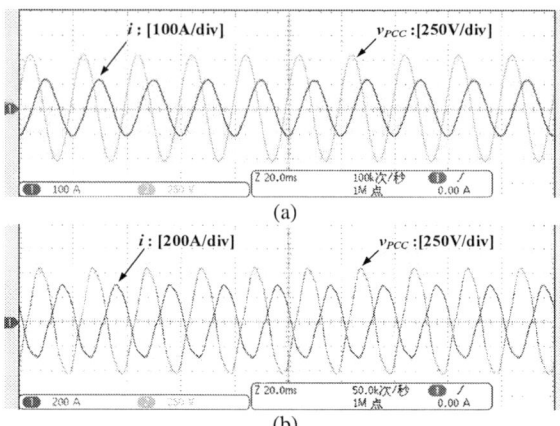

(a)

(b)

Fig. 10. Field test results of i and v_{pcc} individual inverter when whole station output power is (a) 50MW. (b) 130MW.

It could be observed from Fig. 9 that when S increases from 0.6 to 0.9, evident distortion appears in output voltage. However, when S increases from 0.9 to 1.2, both current and voltage begin to oscillate, the system becomes unstable. Fig. 10 shows the on-site measured results of single inverter output current i and PCC voltage v_{pcc} detected in a 200MW PV station. When the whole station output power is 50MW, the system operates stably. However, as solar radiation strength S increases, when the output power reaches to 130MW, both i and v_{pcc} will become distorted as in Fig.10 (b).

The physical meaning of instability can be explained. From (12), it can be deduced that

$$\frac{\hat{v}_d}{\hat{i}_d} = -\frac{(sL + H_{id})V_d^2}{H_{id}P_{ref}} \quad (20)$$

where V_d is d axis voltage at PCC point, which could be regarded as constant. This could be more vividly illustrated by i_d-v_d curve in Fig.11.

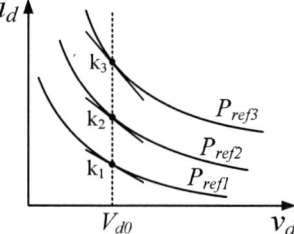

Fig.11. i_d-v_d curve under different steady state operation points.

To a grid-tied inverter under active power control case, as output power reference P_{ref1}, P_{ref2}, and P_{ref3} increases gradually, i_d-v_d curves rise gradually. At a specific operation voltage V_{do}, the tangent value k1, k2, and k3 of these three curves are the negative value of output admittance. From this graph, it could be observed that as output power increases, output admittance increases and output impedance decreases accordingly. This reflects that the output impedance of PV inverter is influenced by power level change, which is nonlinear.

V. CONCLUSIONS

This paper analyzes the instability mechanism of weak grid-connected PV station under the condition of solar radiation fluctuation. The output impedance of PV inverter is deduced by establishing a new small-signal model, which takes PV cell nonlinearity and power level change into account. The derived output impedance of PV inverter declines as output power of the PV station increases. When the output impedance decreases to a certain value which overlaps with the grid impedance, the interconnected system will become unstable.

REFERENCES

[1] Daniel Remon, Antoni M. Cantarellas, Juan Manuel Mauricio, "Power system stability analysis under increasing penetration of photovoltaic power plants with synchronous power controllers," IET Renewable Power Generation, vol. 11, pp. 733-741, 2017.

[2] Xin Chen, Chun Yin Gong, Hui Zhen Wang, "Stability Analysis of LCL-type Grid-Connected Inverter in Weak Grid Systems," in International Conference on Renewable Energy Research and Applications, 2012, pp.1-6.

[3] Mauricio Cespedes; Jian Sun, "Impedance Modeling and Analysis of Grid-Connected Voltage-Source Converters," IEEE Transactions on Power Electronics, vol. 29, no. 3, pp. 1254-1261, 2014.

[4] J. Agorreta, M. Borrega, J.Lopez, "Modeling and control of N-paralleled grid-connected inverters with LCL filter coupled due to the grid impedance in PV plants," IEEE Transactions on Power Electronics, vol. 26, no. 3, pp. 770-785, Mar. 2011.

[5] Jose R. Rodriguez, Domingo Biel, Francesc Guinjoan, "Stability analysis of grid-connected PV systems based on impedance frequency response," in IEEE International Symposium on Industrial Electronics, 2011, pp.1747 – 1752.

[6] J. Sun, "Impedance-Based Stability Criterion for Grid-Connected Inverters," IEEE Transactions on Power Electronics, vol. 26, no. 11, pp. 3075-3078, Nov. 2011.

[7] Teng Liu, Zeng Liu, Jinjun Liu, Qingyun Dou, "Mechanism analysis and mitigation of instability in grid-connected Voltage Source Inverter with LCL filters based on terminal impedance," 2016 IEEE APEC, 2016, pp. 2272-2277.

[8] Lin Zhou, Ming Yang, Qiang Liu, Ke Guo, "New control strategy for three-phase grid-connected LCL inverters without a phase-locked loop," Journal of Power Electronics, vol. 13, no. 3, pp. 487-493, 2013.

[9] Xiongfei Wang, Frede Blaabjerg, Marco Liserre, et al., "An Active Damper for Stabilizing Power-Electronics-Based AC Systems," IEEE Transactions on Power Electronics, vol. 29, no. 3, pp. 3318-3329, July. 2013.

[10] Jian Sun, "Small-Signal Methods for AC Distributed Power Systems–A Review," IEEE Transactions on Power Electronics, vol. 24, pp. 2545 - 2554, 2009.

Experimental Verification of Grid-Connection of a PV Converter Using a Symmetrically Connected Boost Converter for a High-Leg Delta Transformer

Daiki Yamaguchi[1*] and Hideaki Fujita[1]

1 Tokyo Institute of Technology, Tokyo, Japan

*E-mail: yamaguchi.d@akg.ee.titech.ac.jp

Abstract—**This paper proposes a grid-connection control method of the previously-proposed PV converter using a symmetrically-connected boost converter. The proposed circuit has an advantage in its switching operation since the symmetrically connected boost converter and the inverter can alternately stop switching, reducing the switching power losses. In the proposed grid-connection control method, the boost converter performs feedback control of the dc link voltage to make the grid current sinusoidal during the switching interval in the inverter. This paper theoretically clarifies the operating principle of the proposed grid-connection control method and confirms its operating performance using a 8-kW experimental system. As a result, the proposed grid-connection control method makes it possible to control the grid current as a sinusoidal waveform.**

Keywords—*boost converters, grid-connection inverters, high-leg delta transformers, PWM inverters*

I. INTRODUCTION

The diversification of energy sources has been promoted to materialize the energy security and environmental adaptability simultaneously. For this reason, the photovoltaic (PV) systems are installed on residential and commercial buildings in urban areas as a decentralized power source.

These applications preferably adopt transformerless PV converters because the available space is limited for the converter in these buildings. However, a conventional transformerless PV converter has a relatively low conversion efficiency of 92% or less. This is caused by the switching power loss in both boost converter and inverter due to a high dc-link voltage. In general, the transformerless PV converter requires a higher dc-link voltage than the isolated PV converter using a high-frequency transformer.

To solve this problem, cooperative control methods have been proposed [1]. These control methods stop switching either in the boost converter or in the inverter alternately. Thus, these methods decrease the average switching frequency, resulting in a reduced switching power loss. In these methods, the boost converter has to control the grid current by means of feedback control during the switching interval of the inverter. However, it is generally difficult to obtain a fast transient response in

the boost converter because of the well-known problem of right-half-plane (RHP) zero [2]. This problem may cause an inverse response in the output voltage of the boost converter, and may result in instability due to a high feedback gain.

For this problem, some of the cooperative control methods remove the voltage feedback from the controller to prevent the inverse response [3]–[7]. These control methods, however, require reduced inductor and/or capacitor to be designed as small as possible. The grid current is estimated from the input current into the boost converter without considering the energy stored in these elements. The reduced elements result in a large switching ripple current and/or a quite high switching frequency, leading to non-negligible conduction and switching power losses.

This paper proposes a grid-connection control method of the PV converter with the cooperative control method for a high-leg delta connection transformer [8]. The proposed control method controls the grid current by using a voltage feedback control of the dc-link voltage when the inverter stops switching. Then the voltage feedback controller has the capability of compensating the current drawn into the dc-link capacitor. This makes it possible to suppress the distortion in the grid current without a high feedback gain. This paper theoretically reveals the operating principle of the proposed grid-connection control method and confirms its operating performance using an 8-kW experimental system. As a result, it is clarified that the proposed grid-connection control method makes it possible to suppress the distortion in the grid current effectively even at a low carrier frequency of 20 kHz.

II. EXPERIMENTAL SYSTEM

A. Proposed PV Converter for High-Leg Delta Transformer

Fig. 1 (a) shows the experimental system configuration of the proposed PV converter, Fig. 1 (b) shows the circuit configuration of a high-leg delta transformer, and Table I shows the circuit parameters of the experimental system. A dc power supply was used instead of a PV panel, and the output ac terminals of the experimental system were connected to the 200-V three-phase utiliy

The 2018 International Power Electronics Conference

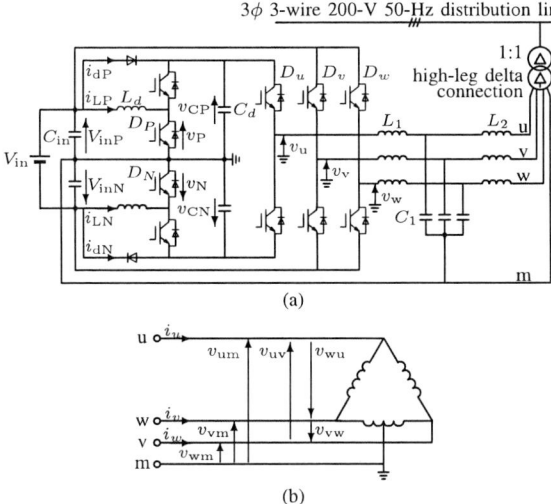

(a)

(b)

Fig. 1. Experimental system: (a) proposed PV converter for 200-V high-leg delta connection and (b) circuit configuration of a high-leg delta transformer.

grid through the delta-delta transformer whose turn ratio was 1:1. The secondary side of the transformer was a high-leg delta connection and the center-tapped terminal "m" between the v- and w-phase terminals was grounded.

The proposed circuit consists of the symmetrically connected boost converter, three half-bridge inverters, dc-link capacitors, and ac-filters. The midpoint of the dc-link is connected to the ground terminal "m." Only the u-phase half-bridge inverter is connected to the boost converter, and the v- and w-phase inverters are connected to the PV terminals. This circuit, therefore, can reduce the average current flowing through the boost converter and the applied voltage across the v- and w-phase inverters compared with a conventional circuit using a three-phase inverter. Thus, this circuit configuration makes it possible to reduce the switching power losses in the boost converter and the v- and w-phase inverters.

B. Cooperative Control of the Boost Converter and the u-phase Half-Bridge Inverter

Fig. 2 depicts the relationship between the voltage command values of the u-phase inverter v_u^* and the dc-link voltages v_{CP}^* and v_{CN}^*. It is better to control the dc-link voltages v_{CP} and v_{CN} as low as possible because this operation enables to reduce the switching losses in both boost converter and u-phase half-bridge inverter. Thus, the voltage references should be provided as

$$\begin{cases} v_{CP}^* = \max(v_u^*, V_{inP}) \\ v_{CN}^* = \min(v_u^*, V_{inN}) \end{cases} \quad . \quad (1)$$

Table II shows the duty-ratio references for the u-phase half-bridge inverter, D_u and boost converter D_P and D_N. The duty ratio of "1" in Table II represents that the positive device is kept in on-state, and the negative one is in off-state, while "0" means off-state of the positive

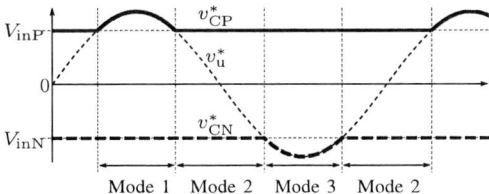

Fig. 2. Waveform of the dc-link voltages v_{CP}^* and v_{CN}^*.

device and on-state of the negative one. The duty ratio of "D_x^*" implies the leg performs PWM opeartion according to the corresponding voltage reference v_x^*. As shown in Table II, one of the three legs is only operated in any mode. Thus it is possible to reduces the average switching frequency and switching power loss.

III. GRID CURRENT CONTROL WITH THE COOPERATIVE CONTROL

In Fig. 1, the total ac inductance is assumed to be $L = L_1 + L_2$ neglecting the filter capacitor C_1. Then the voltage and current equation can be represented by

$$\begin{bmatrix} v_u \\ v_v \\ v_w \end{bmatrix} = \begin{bmatrix} v_{um} \\ v_{vm} \\ v_{wm} \end{bmatrix} + L \frac{d}{dt} \begin{bmatrix} i_u \\ i_v \\ i_w \end{bmatrix} , \quad (2)$$

where v_u, v_v, and v_w are the inverter terminal voltages and v_{um}, v_{vm}, and v_{wm} are the grid voltages. It is possible to control the grid current by using a proportional feedback gain K. Then, the voltage reference is calculated as

$$\begin{bmatrix} v_u^* \\ v_v^* \\ v_w^* \end{bmatrix} = \begin{bmatrix} v_{um} \\ v_{vm} \\ v_{wm} \end{bmatrix} + K \begin{bmatrix} i_u^* - i_u \\ i_v^* - i_v \\ i_w^* - i_w \end{bmatrix} . \quad (3)$$

In Fig. 1, the relationship between the terminal voltage and the duty ratio are given by

$$\begin{cases} v_u = D_u v_{CP} + D_u' v_{CN} \\ v_v = D_v V_{inP} + D_v' V_{inN} \\ v_w = D_w V_{inP} + D_w' V_{inN} \end{cases} , \quad (4)$$

where $D_x' = 1 - D_x$. From the above equation, the command values of the duty ratio can be derived as

$$\begin{cases} D_u^* = \dfrac{v_u^* - v_{CN}}{v_{CP} - v_{CN}} \\ D_v^* = \dfrac{v_v^* - V_{inN}}{V_{inP} - V_{inN}} \\ D_w^* = \dfrac{v_w^* - V_{inN}}{V_{inP} - V_{inN}} \end{cases} . \quad (5)$$

On the other hands, as the duty ratio of the u-phase half-bridge inverter, D_u^* should be set to "1" and "0" in Mode 1 and Mode 3, respectively. Then, either positive or negative boost converter controls the dc-link voltage v_{CP} or v_{CN} to match the terminal voltage reference v_u^*. In other word, the grid current i_u should be controlled by the boost converter in these modes.

3941

TABLE I.   CIRCUIT PARAMETERS.

System	Element	Description
rated power	P	8 kW
ac main	V_s, f_s	200 V, 50 Hz
carrier frequency	f_{sw}	20 kHz
sampling frequency	f_{samp}	40 kS/s
dead time	T_{DT}	3.2 μs
IGBT		CM100DY-24A (1200 V, 100 A)
diode		RURG8060 (600 V, 80 A)
boost inductor	L_d	1 mH
ac inductor	L_1	1.4 mH (3.6%)
filter inductor	L_2	0.26 mH (1.6%)
dc capacitor	C_{in}, C_d	6800 μF, 50 μF
filter capacitor	C_1	2 μF

TABLE II.   DUTY RATIO REFERENCES OF THE BOOST
CONVERTERS AND INVERTER.

Mode	v_{CP}^*	v_{CN}^*	D_P	D_N	D_u
Mode 1	v_u^*	V_{inN}	D_P^*	0	1
Mode 2	V_{inP}	V_{inN}	0	0	D_u^*
Mode 3	V_{inP}	v_u^*	0	D_N^*	0

Fig. 3 (a) shows a control block diagram for the proposed circuit with the cooperative control when a conventional current control [5] is applied to the boost converters. In this control method, the duty ratio commands for the boost converters are calculated from the power relationship between the input of the boost converter and the output of the u-phase inverter.

In Mode 1, the relationship between the input and output instantaneous power flowing through the upper boost converter is summarized as follows:

$$V_{inP}i_{LP} = v_{um}i_u, \qquad (6)$$

Then the command values for the boost converters are derived from table II as:

$$\text{Mode 1} \begin{cases} i_{LP}^* = \frac{v_{um}i_u^*}{V_{inP}} \\ D_N^* = 0 \end{cases}. \qquad (7)$$

On the other hand, the command values in Mode 3 are also derived as:

$$\text{Mode 3} \begin{cases} D_P^* = 0 \\ i_{LN}^* = \frac{v_{um}i_u^*}{V_{inN}} \end{cases}. \qquad (8)$$

In a case of a simple proportional current feedback, the duty-ratio command values can be calculated as follows:

$$D_P^* = 1 - \frac{1}{v_{CP}}[V_{inP} - K_d(i_{LP}^* - i_{LP})], \qquad (9)$$

$$D_N^* = 1 - \frac{1}{v_{CN}}[V_{inN} - K_d(i_{LN}^* - i_{LN})], \qquad (10)$$

where K_d is the current feedback gain of the inductor current feedback.

Fig. 3 (b) shows the control block diagram for the proposed circuit with the cooperative control when the proposed current control is applied. The proposed control method considers the energy-storage elements in the dc-link and the u-phase inverter to derive the command values for the boost converters.

In Mode 1, the relationship between the input and output power should be modified considering the dc-link

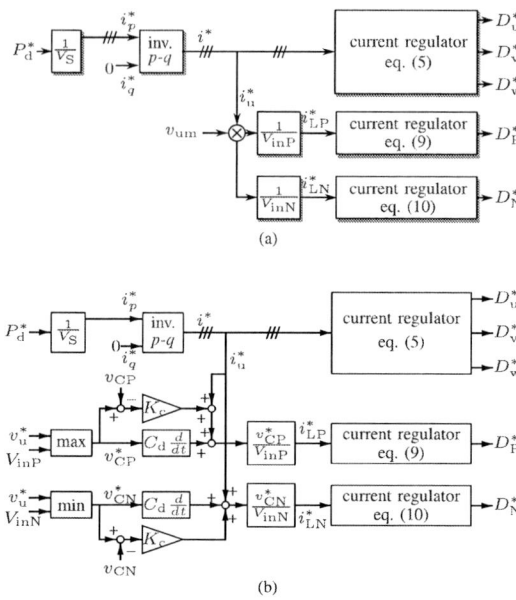

Fig. 3.   Control block diagrams. The duty-ratio command values of the boost converters D_P^* and D_N^* in (a) are calculated by the conventional control method and those in (b) are calculated by the proposed control method.

capacitor current as follows:

$$V_{inP}i_{LP} = v_{CP}\left(C_d\frac{dv_{CP}}{dt} + i_u\right). \qquad (11)$$

This implies that the capacitor current affects the grid current control if the capacitor voltage is not well regulated. To regulate the capacitor voltage, the command values for the boost converters is also modified as:

Mode 1
$$\begin{cases} i_{LP}^* = \frac{v_{CP}^*}{V_{inP}}\left[C_d\frac{v_{CP}^*}{dt} + i_u^* + K_C(v_{CP}^* - v_{CP})\right] \\ D_N^* = 0 \end{cases}, (12)$$

where the term $K_C(v_{CP}^* - v_{CP})$ is the feedback control to reduce the error of the injected power into the dc-link capacitor. As a similar manner, the command values in Mode 2 are given as follows:

Mode 2
$$\begin{cases} D_P^* = 0 \\ i_{LN}^* = \frac{v_{CN}^*}{V_{inN}}\left[C_d\frac{v_{CN}^*}{dt} + i_u^* + K_C(v_{CN}^* - v_{CN})\right] \end{cases}. (13)$$

The duty ratio command values are calculated by using (9) and (10) to control the currents through the boost converters.

IV.   SIMULATION AND EXPERIMENTAL RESULT

Figs. 4 and 5 show the simulated waveforms, and Fig. 6 shows the experimental waveforms. In the both simulation and experiment, the input voltage was provided by a dc power supply of $V_{in} = 400$ V and the power command was set at $P_d^* = 8$ kW for the experimental setup. Note that the line-to-line voltage was 215 V in the

3942

The 2018 International Power Electronics Conference

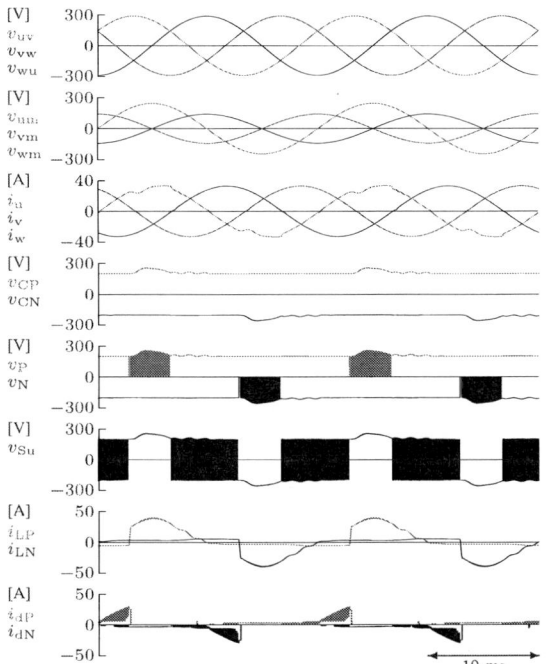

Fig. 4. Simulation waveforms of the proposed PV converter using the conventional current control when $V_{in} = 400$ V and $P_d^* = 8$ kW. The THDs of the grid currents i_u, i_v and i_w are 8.0%, 0.5% and 0.5% respectively.

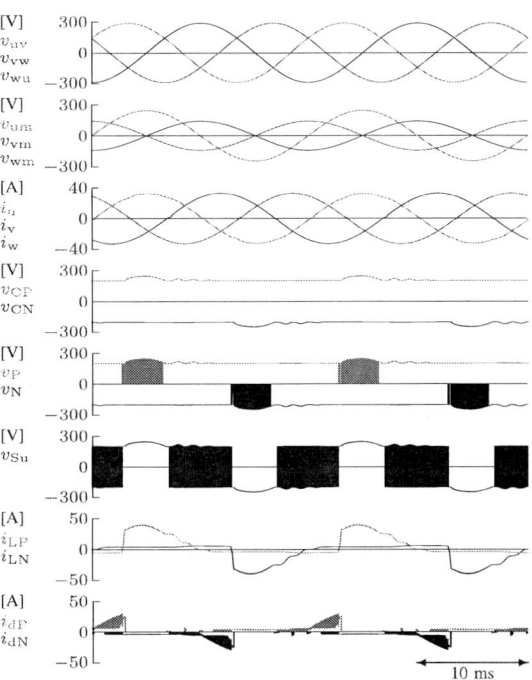

Fig. 5. Simulation waveforms of the proposed PV converter using the propopsed control method when $V_{in} = 400$ V and $P_d^* = 8$ kW. The THDs of the grid currents i_u, i_v and i_w are 1.5%, 0.5% and 0.4% respectively.

following experiments, while it was set at 200 V in the simulation. The carrier frequency was $f_{sw} = 20$ kHz, the dead time was $T_{DT} = 3.2$ μs, and the proportional gains were $K = 14$ V/A, $K_c = 0.05$ A/V and $K_d = 15$ V/A. Moreover, a dead-time compensation was applied for the PWM controller based on the three-level approximated compensation method (3ACM) [9].

Fig. 4 shows the simulated grid connection waveforms when the conventional current control was applied to the proposed circuit. The 20 kHz notch waveforms appeared in the terminal voltage of the u-phase half-bridge inverter v_u and the voltages of the boost converter v_P and v_N. The positive and negative boost converters performed PWM operation only during a phase angle of 75° and the u-phase half-bridge inverter performed PWM operation during 210° because the converters and the inverter stopped PWM operation alternately. The u-phase half-bridge inverter stopped switching in Mode 1 and Mode 3. In these modes, either upper or lower boost converter has to control the u-phase current. In Fig. 4, the u-phase current i_u and the dc-link voltages of v_{CP} and v_{CN} were distorted in these modes. As a result, the total harmonic (THD) of the u-phase current was as high as 8.0%. This low THD or current distortion was caused by two reasons: one is the resonance between the small dc-link capacitor and the ac inductor, and the other is an amount of power absorbed in the capacitor and the inductor. This conventional control does not consider the additional power flow into these energy storage elements.

Fig. 5 shows the simulated waveforms when the proposed current control was applied to the proposed circuit. The three-phase grid current was almost sinusoidal in this case, and thus, the THD of the u-phase current was improved to 1.5%. The inductor current i_{LP} and i_{LN} was slightly larger than that in Fig. 4 at the beginning of Modes 1 and 3. The additional current was supplied to the dc-link capacitor to regulate its voltage in an appropriate range. For this reason, the inductor current seems to be leading from the grid current and/or voltage in these modes. Just after the Modes 1 and 3, a small oscillation still remained in the capacitor voltage, which was induced by the stored energy in the boost inductor. However, this oscillation does not affect the current control in the u-phase inverter because this oscillation can be compensated by the current feedback applied to the u-phase leg, as shown in Fig. 5.

Fig. 6 illustrates the experimental waveforms when the proposed grid current control method is applied. The line-to-line voltage was 215 V because of the experimental circumstance. Thus, the switching period of the each boost converter was 92° which was longer than that in the simulation results in Figs. 4 and 5. The proposed grid current control method well controlled the u-phase current as a sinusoidal waveform irrespective of the long interval in the u-phase half-bridge inverter. Thus, the THD of the u-phase current was 2.6%.

3943

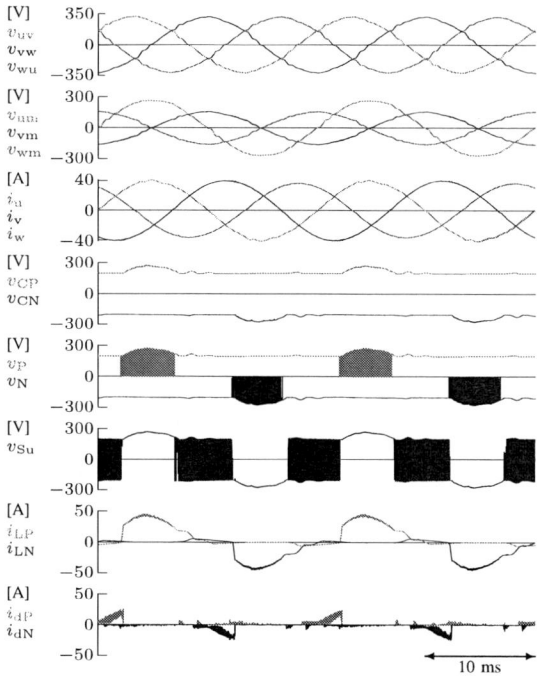

Fig. 6. Experimental waveforms of the proposed circuit with the proposed grid-connection control when $V_{\text{in}} = 400$ V and $P_{\text{d}}^{*} = 8$ kW. The THDs of the grid currents i_{u}, i_{v} and i_{w} are 2.6%, 1.1% and 1.1% respectively.

V. Conclusion

This paper discusses grid-connection control methods of the proposed PV converter with the cooperative control. The proposed grid-connection control method compensates the power flowing into the energy-storage elements, and then the distortion in the u-phase current can be suppressed. Furthermore, the voltage feedback in the proposed control method can damp the resonance in the grid current. The simulation results reveal that the proposed control method improves the THD of the u-phase current from 8.0% of the conventional method to 1.5%. Finally, the experimental result demonstrates that the THD of the u-phase current is 2.6% and the current can be controlled to become a sinusoidal waveform by the proposed control method.

References

[1] Y. Nishida, N. Aikawa, S. Sumiyoshi, H. Yamashita and H. Omori, "A novel type of utility-interactive inverter for photovoltaic system," *The 4th International Power Electronics and Motion Control Conference*, 2004. IPEMC 2004., Xi'an, 2004, pp. 1785-1790 Vol.3.

[2] R. D. Middlebrook and S. Cuk, "A general unified approach to modelling switching-converter power stages," *1976 IEEE Power Electronics Specialists Conference*, Cleveland, OH, 1976, pp. 18-34.

[3] T. Ninomiya, K. Harada and M. Nakahara, "On the maximum regulation range in boost and buck-boost converters," *1981 IEEE Power Electronics Specialists Conference*, Boulder, Colorado, USA, 1981, pp. 146-153.

[4] Y. Nishida, S. Nakamura, N. Aikawa, S. Sumiyoshi, H. Yamashita and H. Omori, "A novel type of utility-interactive inverter for photovoltaic system," *Industrial Electronics Society, 2003. IECON '03. The 29th Annual Conference of the IEEE*, 2003, vol.3, pp. 2338-2343.

[5] W. Wu, H. Geng, P. Geng, Y. Ye and M. Chen, "A novel control method for dual mode time-sharing grid-connected inverter," *2010 IEEE Energy Conversion Congress and Exposition*, Atlanta, GA, 2010, pp. 53-57.

[6] Z. Zhao, M. Xu, Q. Chen, J. S. Lai and Y. Cho, "Derivation, Analysis, and Implementation of a Boost-Buck Converter-Based High-Efficiency PV Inverter," in *IEEE Transactions on Power Electronics*, vol. 27, no. 3, pp. 1304-1313, March 2012.

[7] W. Wu, J. Ji and F. Blaabjerg, "Aalborg Inverter - A New Type of "Buck in Buck, Boost in Boost" Grid-Tied Inverter," in *IEEE Transactions on Power Electronics*, vol. 30, no. 9, pp. 4784-4793, Sept. 2015.

[8] D. Yamaguchi and H. Fujita, "A new PV converter for grid connection through a high-leg delta transformer using cooperative control of boost converters and inverters," *2017 IEEE 3rd International Future Energy Electronics Conference and ECCE Asia (IFEEC 2017 - ECCE Asia)*, Kaohsiung, 2017, pp. 911-916.

[9] J. M. Schellekens, R. A. M. Bierbooms and J. L. Duarte, "Dead-time compensation for PWM amplifiers using simple feed-forward techniques," *The XIX International Conference on Electrical Machines - ICEM 2010*, Rome, 2010, pp. 1-6.

A Novel Single- Stage High-Frequency Boost Inverter Cascaded by Rectifier-Inverter System for PV Grid-Tie Applications

Hamdy Radwan[1,2]*, Mahmoud A. Sayed[2]*, Takaharu Takeshita[2]*, Adel A. Elbaset[3] and G. Shabib[1,4]

[1] Faculty of Energy Engineering, Aswan University, Aswan, Egypt
[2] Dept. of Electrical and Mechanical Engineering, Nagoya Institute of Technology, Japan
[3] Dept. of Electrical Engineering, Minia University, El-Minia, Egypt
[4] Higher Institute of Engineering and Technology, King Mariout, Alexandria, Egypt
*E-mail: hamdy_radwan@aswu.edu.eg, mahmoud_sayed@ieee.org, take@nitech.ac.jp

Abstract— This paper proposes a new topology for a single-phase grid-tie two-stage DC-AC boost inverter for the application of PV systems that utilize high-frequency transformer (HFT) for galvanic isolation. In the first stage, a new single-stage high-frequency boost inverter (HFBI) is proposed to boost and convert the DC output voltage of the PV modules to a high-frequency single-phase square waveform and to realize maximum power point tracking (MPPT). The second stage is rectifier-inverter system (RIS) that interfaces HFBI to the grid. Therefore, a single-phase high-frequency transformer is used to link both stages and to provide galvanic isolation between the AC and DC sides. The proposed topology has many advantages such as increasing the inverter output voltage level, MPPT, high reliability, small size and lightweight. In addition, a proportional integral current control (PI) conventional is used to inject a sinusoidal current into the grid at unity power factor. The proposed topology has been verified analytically by using PSIM software and experimentally by using a laboratory prototype.

Keywords—photovoltaic; grid connected; boost inverter; high frequency transformer.

I. INTRODUCTION

In the last few years' renewable energy has the greatest growth compared to other energy resources due to its reliability, availability, maintainability and safety [1-3]. Therefore, many literatures search for solutions to improve the reliability of the renewable energy resources. PV grid-connected system should perform voltage boosting, maximum power point tracking (MPPT), injection of low harmonics high quality AC power to the grid with unity power factor, galvanic isolation for safety purposes and using high efficient implementation [4-7].

Several topologies for PV grid connected inverter have been presented; generally, there are two types of grid-connected PV systems, those with galvanic isolation and without galvanic isolation. Galvanic isolation can be implemented by using a line frequency transformer or a high frequency transformer. By contrast, topologies without galvanic isolation are transformerless topologies.

Transformerless topologies [8-12] are lighter, more efficient, less costly, and less footprint than the galvanic isolated inverters. However, the main drawback that must be overcome in non-isolated PV inverters is the leakage

ground currents through the solar module parasitic capacitance, in addition to dc current injected to the grid [13]. Dangerous leakage current increases system losses, reduces the grid-connected current quality, induces severe conducted and radiated electromagnetic interface and causes personal safety problems. To keep the leakage and dc currents injected to the grid under control, complex solutions are required.

In order to interface the low output voltage of the PV module to the grid, high voltage boosting technique is required; therefore, the use of a line frequency transformer is widespread [14-15]. In addition to voltage stepping up, it provides galvanic isolation between the grid and the PV system. that plays an important role in safety purpose and personal protection. Thus avoiding dc current injection into the grid and eliminating leakage current. Nevertheless, the line frequency transforms are large, heavy, and expensive, the whole system is bulky and hard to install as a result of its low frequency [16-17]. Therefore, the topology with line frequency transformer is considered as a poor solution, which is better to replaced by high-frequency transformers (HFT). Using HFT [18-21] guarantees galvanic isolation between the grid and the PV system, in addition to overcoming the disadvantages of using conventional line frequency transformer [22-23]. However, there is a rarity in scientific research for using HFT with PV systems in a way that performs all the required functions, especially MPPT.

This paper presents a new topology for interfacing PV module with the grid. The system consists of two stages. The first stage provides a 10-kHz square wave output voltage to meet the requirements of HFT and configure a multi-featured system. The second stage is rectifier-inverter system, which injects a sinusoidal current with minimum harmonic distortion and unity power factor to the grid .

This paper is organized as follows; first, the circuit configuration of the proposed system is described. Second, the operation modes of the proposed topology are presented. Third, MPPT and PI controller are discussed. Finally, simulations and experiments consider the fundamental operation waveforms of the proposed system.

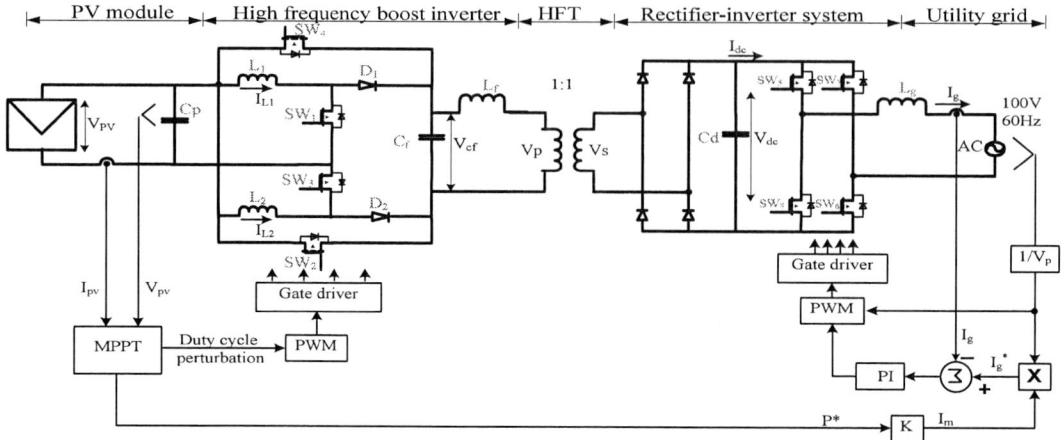

Fig. 1. The proposed system.

II. PROPOSED SYSTEM

A) proposed topology (basic version)

The proposed system consists of two stages, High-frequency boost inverter (HFBI) cascaded by rectifier-inverter system (RIS) as shown in Fig. 1. In addition, the implemented switching control strategies are shown in the figure. The first stage is a redesign of the topology given in [24] to obtain a 10-kHz square wave output voltage instead of the fundamental grid voltage. It consists of two buck-boost converters connected, as shown in Fig. 2, and the second stage is simply approximated by a resistor. Each of these converters operates sequentially in DCM for one half cycle of the targeted 10 kHz square waveform. DCM operation prevents the circulating currents between the inductor and the parallel-connected switch in the next operating half cycle. The power MOSFETS SW_1 and SW_3 are switched at high frequency of 100 kHz while SW_2 (or SW_4) is continuously turning ON during the positive half cycle (or negative half cycle) of the targeted 10 kHz square waveform. Switches SW_1 and SW_2 operate to provide the positive boosted half-cycle, whereas SW_3 and SW_4 operate to provide the negative boosted half-cycle.

When SW_1 is ON (or SW_3), energy is stored in the inductor "L_1" (or L_2) by the PV source. When SW_1 (or SW_3) is OFF, D_1 (or D_2) gets forward biased, discharging the inductor stored energy into capacitor C_f, which continuously feeds current to the load. The switched gate signals Vg1, Vg2, Vg3 and Vg4 for SW_1, SW_2, SW_3 and SW_4, respectively, are shown in Fig. 3.

B) Modified proposed topology

The target of the proposed topology is a 10 kHz square waveform output voltage that is linked to RIS by HFT. Therefore, the topology is designed to achieve many features of the complete system as mentioned previously. But at the instant of turning the operation between the two buck boost, the polarity of the capacitor voltage V_{cf} cannot change instantaneously. Although this time is very short (2 ns), two paths of surge current appear due to high value of V_{cf} compared to the input voltage. Assuming the polarity of V_{cf} changed from the positive half cycle to negative half cycle and by referring to fig. 2, the first path of surge current flows through capacitor C_f, switch SW_4 and body diode of switch SW_2. The second path of surge current flows through capacitor C_f, switch SW_4, input capacitor C_p and body diode of switch SW3. In order to limit this surge current, two stages of modification have been proposed. The first one considers bi-directional switch for SW_2 and SW_4 and adding series diode in opposite direction of the body diode of SW_1 and SW_3.

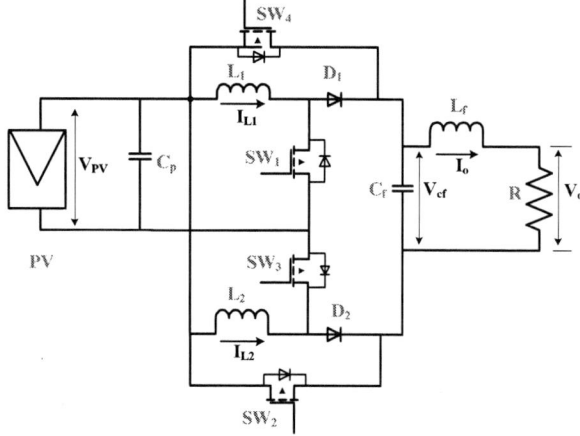

Fig. 2. Configuration of the single-stage HF buck-boost inverter.

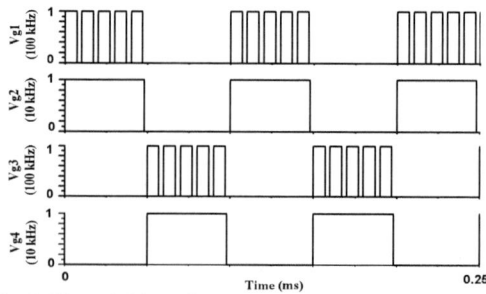

Fig. 3. The switching pulses at the gates of controllable switches.

3946

Consequently, the capacitor current I_{cf} is limited by flowing through inductor L_2. Although, the surge current is limited, it is added to the source current in inductor L_2 (SW$_3$ is ON) at the instant of changing the polarity of V_{cf}, resulting in rising the output voltage at the begging of each half cycle. In order to obtain proper square wave shape, the second stage of modification was done by keeping SW$_1$ and SW$_3$ in OFF state at this instant. The modified switched gate signals Vg1 and Vg3 for SW$_1$ and SW$_3$, respectively, are shown in Fig. 4.

III. OPERATION MODES AND PARAMETERS DESIGN

The boost inverter has three modes of operation based on the switching of SW$_1$ during the positive half-cycle, since the switch SW$_2$ is always ON during these three modes. In Mode1, switch SW1 is ON and energy is stored in the buck boost inductor L_1 by the PV source. In Mode2, switch SW$_1$ is OFF and D$_1$ is forward biased, discharging the inductor stored energy into capacitor C_f, which feeds current to the load (R). In Mode3, both SW$_1$ and D$_1$ are OFF as a result of DCM operation. The operation modes are shown in Fig. 5. As a result of DCM operation during each cycle of the output voltage, the stored energy in the buck-boost inductor L_1 (or L_2) is completely discharged into capacitor C_f which feeds it into the load. Therefore, the energy delivered into the load $E_o(t)$ is equal to the energy drown from the source $E_{pv}(t)$ during the switching time period T_s.

According to Mode 1, the energy drown from the PV source is the following:

$$E_{pv}(t) = \frac{1}{2} L_1 \ I_{L_1}^2 \tag{1}$$

The peak value of the inductor current can be formulated as follows;

$$I_{L_1} = \frac{V_{pv}}{L_1} D T_s \tag{2}$$

Substituting by (2) in (1) yields:

$$E_{pv}(t) = \frac{V_{pv}^2}{L_1} D^2 T_s^2 \tag{3}$$

The energy transferred into the load during switching period T_s is given by:

$$E_o(t) = V_o I_o T_s = \frac{V_o^2}{R} T_s \tag{4}$$

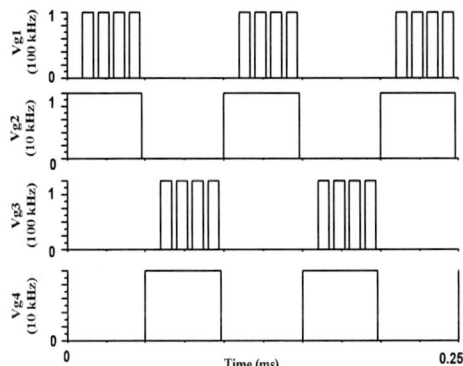

Fig. 4. The modified switching pulses at the gates of switches.

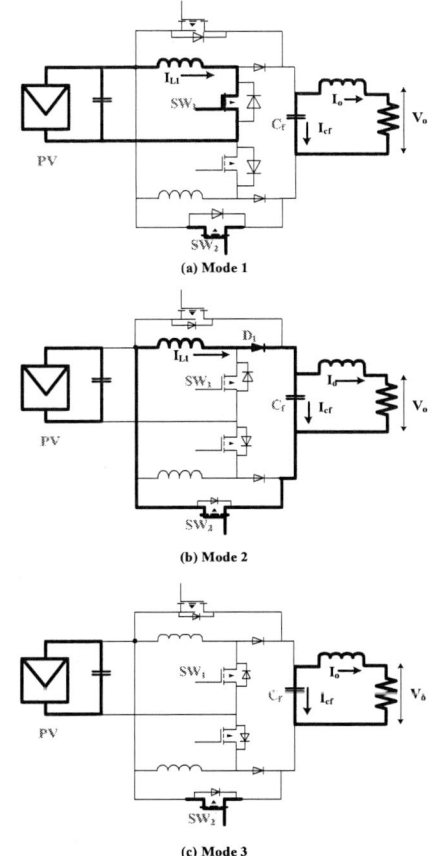

(a) Mode 1

(b) Mode 2

(c) Mode 3

Fig. 5. Operation modes (a) when S1 'on', (b) when S1 'off', D1 'on', (c) when S1 'off', D1 'off'.

Equalizing (3) and (4) yields the formula of the boost converter voltage gain as follows,

$$\frac{V_o}{V_{pv}} = \sqrt{(\frac{R}{2 L_1} D^2 T_s)} \tag{5}$$

As a result of DCM operation, the energy stored of the inductor L_1 is completely transferred into capacitor C_f which feeds it into the load during each switching period. Therefore, (5) is used to determine the value of the inductor L_1, which results the following expression:

$$L_1 \leq \frac{V_{pv}^2}{2P} D^2 T_s^2 \tag{6}$$

where P is the rated power transferred into the load.

To determine the value of C_f, the energy stored in the inductor L_1 during the ON mode can be equated to the change in capacitor energy during the OFF mode, yields the following expression:

$$C_f \leq \frac{L_1 I_{pk\_L_1}^2}{4 V_o \Delta V} \leq \frac{V_o T_s}{2 R \Delta V} \tag{7}$$

where ΔV is the ripple of the capacitor voltage.

The high switching frequency component in the current waveform is filtered by inductor filter (L_f). Its design depends on the cut-off frequency (F_c) as follows;

$$L_f = \frac{1}{(2 \pi f_c)^2 C_f} \tag{8}$$

IV. CONTROL STRATEGIES

A. MPPT algorithm

The operating point of the PV sources may change randomly during the operation of the system according to the environmental conditions. Therefore, MPPT algorithm is needed to extract maximum instantaneous power. Several MPPT techniques have been proposed in the last decades. P&O MPPT algorithm [25-26] is one of simple hill-climbing algorithms, which extensively used in practical PV systems because of its simplicity. Moreover; prior study or modeling of PV characteristics is not required. The flow chart of P&O MPPT algorithm is depicted on Fig. 6. The algorithm starts by reading PV output voltage and current to calculate PV output power P(k). Then compares the calculated power with that of the previous perturbation cycle P(k-1). Depending on the result of the comparison, the algorithm perturbs the PV output voltage by increasing or decreasing. If the perturbation causes an increase in PV power, the subsequent perturbation is made in the same direction. Otherwise, the subsequent perturbation is made in the opposite direction. When the perturbation of the algorithm has three-level at steady state, it indicates the algorithm is stable and swings around the MPP.

B. Current control

The main task of the current controller of the grid connected single-phase inverter is generation of sinusoidal current with minimum harmonic distortion and in-phase with the grid voltage. Fig. 1 shows the control scheme of the inverter connected with the grid through L filter that is used to eliminate the current ripple, as shown in Fig. 1. The reference grid current I_g^* is obtained by multiplying the grid voltage signal with the maximum value of the reference current, which is determined from the input power and the grid voltage in order to achieve unity power factor in addition to synchronizing the inverter output voltage and current with the grid. Therefore, the grid voltage and current must be detected. The reference grid current is compared with the actual one and the error between them is minimized by using conventional PI controller.

V. SIMULATION RESULTS

In order to validate the operation of the proposed system, it has been carried out in PSIM software (ver. 10.0). 250W PV module is simulated at 25° C temperature and 1000W/m² radiation. The simulated circuit parameters and the electrical characteristics of PV module at MPP are listed in Table I. Switches SW_1 and SW_3 are responsible of boosting the input voltage. Therefore, they switched by 100 kHz and their duty cycles are modulated by MPPT algorithm. Switches SW_2 and SW_4 are responsible of inverting process; hence they are switched by 10 kHz 0.5 duty cycle. Switches of H bridge inverter in the grid side are modulated by current controller PI to inject a sinusoidal current into the grid and in-phase with the grid voltage for unity power factor.

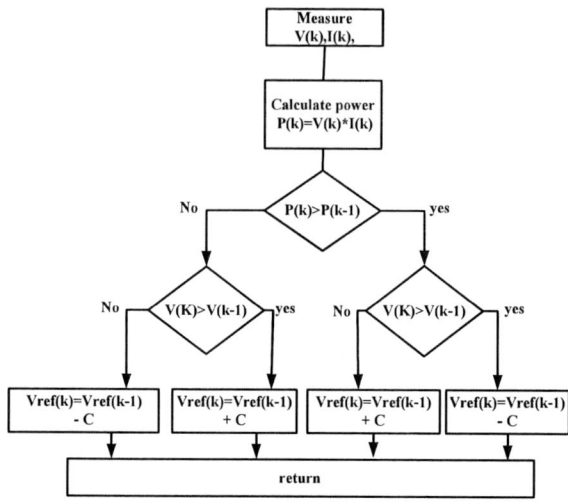

Fig. 6. Flowchart of the P&O algorithm.

Switches of H bridge inverter in the grid side are modulated by conventional PI controller to regulate the grid injected current sinusoidal waveform and in-phase with the grid voltage for unity power factor. Fig. 7 shows the PV outputs (V_{pv}, I_{pv} and P_{pv}) at MPP and the duty cycle perturbation (output of MPPT algorithm). It is clear that the duty cycle perturbation has three-level at steady state, which indicates that P&O MPPT algorithm is stable and swings around the MPP.

TABLE I
PROPOSED SIMULATED CIRCUIT PARAMETERS

Symbol	Meaning	Value
L1=L2	HFBI inductors	17.5 µH
C_f	HFBI Filter capacitor	330 nF
L_f	HFBI Filter inductor	800 µH
C_d	Dc link capacitor	150 µf
Lg	grid filter inductor	3 mH
V_{pv}	Electrical characteristics of PV module at MPP.	45 V
I_{pv}		5.6 A
P_{pv}		250 W

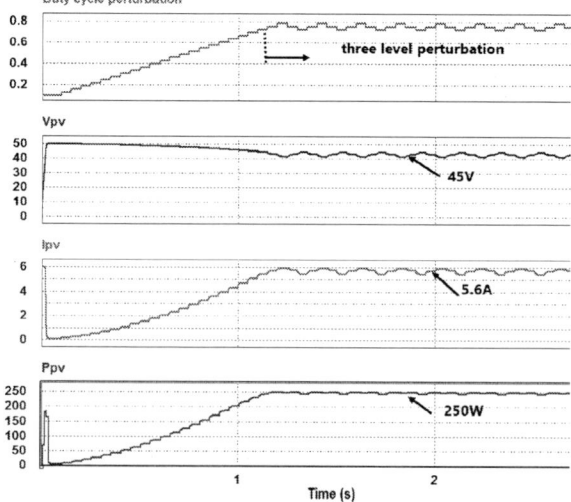

Fig. 7: Simulation Results of I_{pv}, V_{pv} and P_{pv} of PV module at MPP and the duty cycle perturbation .

Fig. 8 shows the grid voltage and current at steady-state MPPT algorithm. It is clear that the injected grid current is in-phase with the grid voltage. The figure also shows V_{dc} and I_{dc} of RIS. The frequency of the primary voltage V_p of HFT is high frequency, as shown in Fig. 8, which is the output of the proposed topology HFBI therefore, Fig. 8 is magnified to show this waveform and other high frequency signal waveforms in suitable view, as shown in Fig. 9. Switches gate signals V_{g1}, and V_{g3} are 100 kHz and Switches gate signals V_{g2}, and V_{g4} are 10 kHz 0.5 duty cycle. The boost inverter capacitor voltage V_{cf} and V_p are square waves at 10 kHz. The inductor current I_{L1} is DCM to avoid the circulating current between the converter switches.

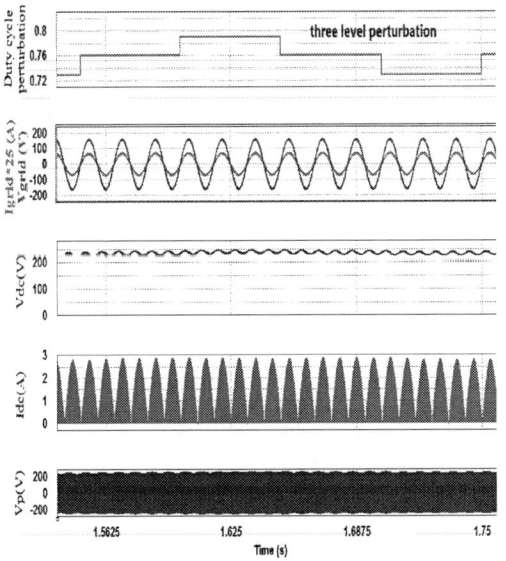

Fig. 8. Simulation Results of grid voltage, grid current, dc link voltage & current and primary voltage of HFT.

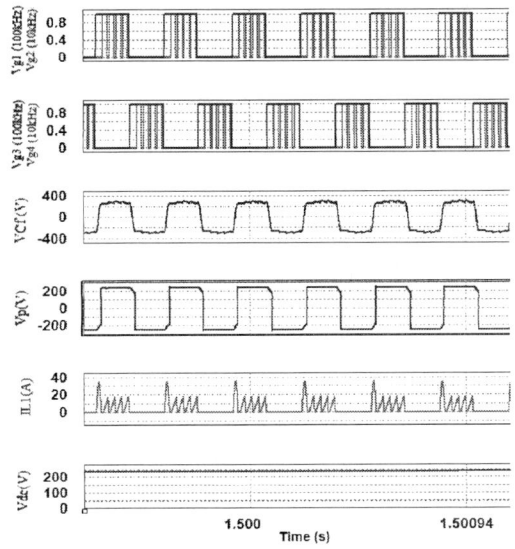

Fig. 9. HFBI switches gate signals, HFBI filter capacitor voltage, inductor current and primary voltage of HFT

VI. EXPERIMENTAL RESULTS

A 200-W experimental prototype has been carried out in the laboratory to verify the operation of the proposed configuration. The utility grid in the second stage is replaced by stand-alone resistive load of 50 Ω. The PWM switching signals, which are applied to the gate drives, are obtained by using DSP&FPGA (PE-Expeart3). The circuit parameters are listed in Table II as shown.

TABLE II
PROPOSED EXPERIMENTED CIRCUIT PARAMETERS

Symbol	Meaning	Value
V_{in}	Input voltage of dc supply	45V
$L_1 = L_2$	HFBI inductors	17.5 µH
C_f	HFBI Filter capacitor	330 nF
L_f	HFBI Filter inductor	800 µH
R	Resistive load (in place of grid)	50 Ω

Fig. 10 shows the photograph of the experimental prototype. Fig. 11 shows the experimental results of HFBI, the capacitor voltage V_{cf} is square wave with 210V amplitude and 10 kHz. Also, the primary voltage and current of HFT are square wave. The inductor current of HFBI is DCM as shown in figure to guarantee discharging all its energy and no circulated current between the inductor and the parallel-connected switch in the next operating half cycle.

Fig. 10. Photograph of the experimental prototype.

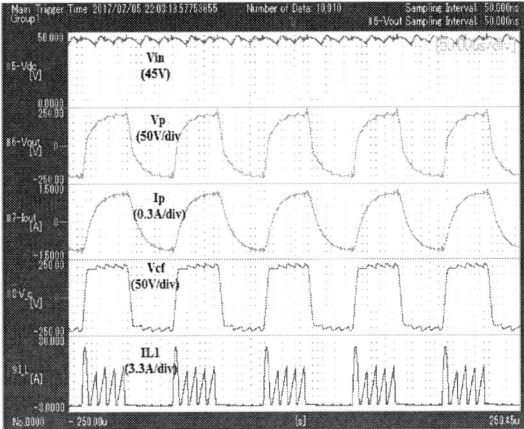

Fig. 11. Experimental results of the first stage topology HFBI.

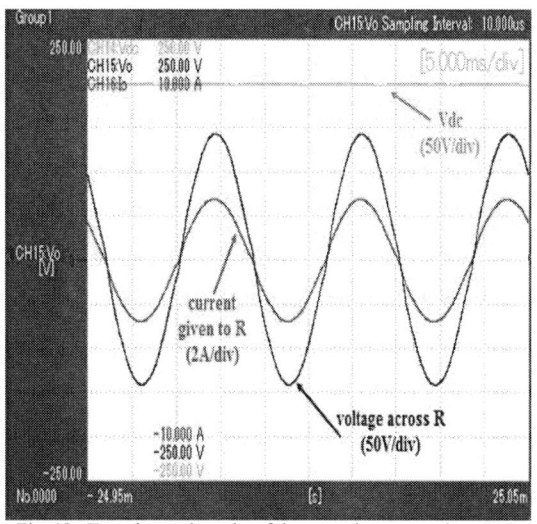

Fig. 12. Experimental results of the second stage topology RIS.

SPWM is used to generate switching signals of H-bridge inverter in the second stage. Therefore, the grid voltage is sensed and reduced to a unit amplitude sine wave then multiplied by 0.7 modulation index before comparing with a 20 kHz saw tooth signal. As shown in Fig. 12 the input DC voltage of H-bridge inverter Vdc is 200V and the RMS value of AC output voltage across the resistor (in place of the grid) is 100V with 60 Hz as the utility grid parameters.

VII. CONCLUSION AND FUTURE WORK

This paper has proposed a new single-stage high-frequency boost inverter cascaded by Rectifier-inverter system for PV grid-tie applications. The topology has been analyzed, designed, simulated and implemented in laboratory. The topology performs many features such as MPPT, boosting PV voltage in addition to the high frequency square wave output voltage that allows the use of HFT to guarantee galvanic isolation between the grid and the PV system, in addition to overcoming the drawbacks of conventional line frequency transformer. The proposed HFBI topology is one stage with reduced power switches compared with the conventional topologies. The functionality and switching operation of the proposed system in closed loop have been depicted with simulation. The functionality and switching operation of the proposed system have been depicted with experimental results for resistive load (in place of the grid).

The presented experimental results demonstrate a boosting of more than 4 times, therefore, low PV array voltages (typically 40- 100 V range) can be stepped up to levels commensurate with the grid voltage (110- 230 V ac). Consequently, the use of a few series connected solar panels is sufficient. This avoids environmental changes such as shadow, which reduces the utilization of solar panels. Simulation and experimental results, which emphasizing the performance of the proposed topology of proposed system, have been validated.

References

[1] F. T. K. Suan, N. A. Rahim, and H. W. Ping, "Modeling, analysis and control of various types of transformerless grid connected PV inverters," in *Proc. IEEE Clean Energy Technol.*, Jun. 2011, pp. 51–56.

[2] N. A. Rahim, K. Chaniago, and J. Selvaraj, "Single-phase seven-level grid connected inverter for photovoltaic system," *IEEE Trans. Ind. Electron.*, vol. 58, no. 6, pp. 2435–2443, Jun. 2011.

[3] G. Petrone, G. Spagnuolo, R. Teodorescu, M. Veerachary, and M. Vitelli, "Reliablility issues in photovoltaic power processing systems," *IEEE Trans. Ind. Electron.*, vol. 55, no. 7, pp. 2569–2580, Jul. 2008.

[4] R. H. Wills, S. Krauthamer, A. Bulawka, and J. P. Posbic, "The AC photovoltaic module concept," in *Proc. 32nd IECEC*, 1997, vol. 3, pp. 1562–1563.

[5] Y. Chen and K. M. Smedley, "A cost-effective single-stage inverter with maximum power point tracking," *IEEE Trans. Power Electron.*, vol. 19, no. 5, pp. 1289–1294, Sep. 2004.

[6] S. Jain and V. Agarwal, "A single-stage grid connected inverter topology for solar PV systems with maximum power point tracking," *IEEE Trans. Power Electron.*, vol. 22, no. 5, pp. 1928–1940, Sep. 2007.

[7] D. C. Martins and R. Demonti, "Grid connected PV system using two energy processing stages," in *Conf. Rec. 29th IEEE Photovoltaic. Spec. Conf.*, 2002, pp. 1649–1652.

[8] L. Zhang, K. Sun, L. Feng, H. Wu, and Y. Xing, "A family of neutral point clamped full-bridge topologies for transformerless photovoltaic grid-tied inverters," IEEE Trans. Power Electron., vol. 28, no. 2, pp. 730–739, Feb. 2013.

[9] B. N. Alajmi, K. H. Ahmed, G. P. Adam, and B. W. Williams, "Single phase single-stage transformerless grid-connected PV system," IEEE Trans. Power Electron., vol. 28, no. 6, pp. 2664–2676, Jun. 2013.

[10] S. V. Araujo, P. Zacharias, and R. Mallwithz, "Highly efficient single phase transformerless inverters for grid-connected photovoltaic systems," IEEE Trans. Ind. Electron., vol. 57, no. 9, pp. 3118–3128, Sep. 2010.

[11] T. Kerekes, R. Teodorescu, M. Liserre, C. Klumpner, and M. Sumner, "Evaluation of three-phase transformerless photovoltaic inverter topologies," IEEE Trans. Power Electron., vol. 24, no. 9, pp. 2202–2211, Sep. 2009.

[12] D. Meneses, F. Blaabjerg, O. García, and J. A. Cobos, "Review and comparison of step-up transformerless topologies for photovoltaic ac-module application," IEEE Trans. Power Electron., vol. 28, no. 6, pp. 2649–2663, Jun. 2013.

[13] M. Calais and V. G. Agelidis, "Multilevel converters for single-phase grid connected photovoltaic systems-an overview," in Proc. IEEE Int. Symp. Ind. Electron., 1998, pp. 224–229.

[14] V. Meksarik, S. Masri, S. Taib, and C. M. Hadzer, "Development of high efficiency boost converter for photovoltaic application," in Proc. National Power Energy Conf., 2004, pp. 153–157.

[15] B. M. T. Ho and H. S. Chung, "An integrated inverter with maximum power tracking for grid-connected PV systems," IEEE Trans. Power Electron., vol. 20, no. 4, pp. 953–962, Jul. 2005.

[16] S. B. Kjær, J. K. Pedersen, and F. Blaabjerg, "Power inverter topologies for photovoltaic modules—A review," in Proc. IEEE IAS, 2002, pp. 782–788.

[17] S. B. Kjær, J. K. Pedersen, and F. Blaabjerg, "A review of single-phase grid-connected inverters for photovoltaic modules," IEEE Trans. Ind. Appl., vol. 41, no. 5, pp. 1292–1306, Sep./Oct. 2005.

[18] X. Li and A. K. S. Bhat, "A utility-interfaced phase-modulated high frequency isolated dual LCL DC/AC converter," IEEE Trans. Ind. Electron., vol. 59, no. 2, pp. 1008-1019, Feb. 2012.

[19] H. Qin and J. W. Kimball, "Closed-loop control of DC–DC dual-active bridge converters driving single-phase inverters," IEEE Trans. Power Electron., vol. 29, no. 2, pp. 1006-1017, Feb. 2014.

[20] H. Qin and J. W. Kimball, "Closed-loop control of DC–DC dual-active bridge converters driving single-phase inverters," IEEE Trans. Power Electron., vol. 29, no. 2, pp. 1006-1017, Feb. 2014.

[21] L. Quan and P. Wolfs, "A Review of the Single Phase Photovoltaic ModuleIntegrated Converter Topologies With Three Different DC Link Configurations," IEEE Trans. Power Electron., vol. 23, no. 3, pp. 1320 - 1333, 2008.

[22] M. A. Sayed, K. Suzuki, T. Takesita, W. Kitagawa, "Soft-Switching PWM Technique for Grid-Tie Isolated Bidirectional

DC-AC ConverterWith SiC Device," IEEE Trans. Industry Applications, vol. 53, no. 6, pp. 5602–5614, 2017.

[23] M. A. Sayed, K. Suzuki, T. Takesita, W. Kitagawa, "PWM Switching Technique for Three-Phase Bidirectional Grid-Tie DC-AC-AC Converter with High-Frequency Isolated," IEEE Trans. Power Electron., vol. 3, no. 1, pp. 845–858, 2018.

[24] S. Jain and V. Agarwal, "A single-stage grid connected inverter topology for solar PV systems with maximum power point tracking," *IEEE Trans. Power Electron.*, vol. 22, no. 5, pp. 1928–1940, Sep. 2007.

[25] C. Hua, J. Lin, C. Shen, Implementation of a DSP-controlled photovoltaic system with peak power tracking, *IEEE Trans. Ind. Electron.* 45 (1) (1998) 99‑ 107.

[26] A. Shawky, H. Radwan, M. Orabi, and M. Z. Youssef, "A novel platform for an accurate modeling and precise control of photovoltaic modules with maximum operating efficiency", In *2015 IEEE Applied Power Electronics Conference and Exposition (APEC)* (pp. 205-212). IEEE.

The 2018 International Power Electronics Conference

Nine Switches Matrix Converter Using Bi-directional GaN Device

Takashi Hirota[*], Kentaro Inomata, Daisuke Yoshimi and Masato Higuchi
Motion & Drive Development Dept, YASKAWA ELECTRIC CORPORATION, Kitakyushu 803-8530, Japan
*E-mail: Takashi.Hirota@yaskawa.co.jp

Abstract— As market demand for energy conservation products are increasing, matrix converters need to be further improved in efficiency and downsizing. On the other hand, next generation power devices such as SiC, GaN, which are capable of high speed switching with low loss have recently gained attention. In particular, the bi-directional GaN device is constituted by connecting two devices in series on the same die, which has an advantage that the conduction loss and the mounting area can be reduced. Conventional matrix converter requires at least 18 devices, but if bi-directional GaN devices are used, they can be composed with only 9 devices, which is useful for high efficiency and downsizing by reducing the conduction loss of the matrix converter. Matrix converter chipset using bi-directional GaN devices has already been developed, but as far as the authors know, a three phase motor has not been driven. This paper introduces the prototype nine switches matrix converter using bi-directional GaN devices, which successfully drives a three phase motor.

Keywords— *Bi-directional GaN device, Matrix converter, Nine switches*

I.   INTRODUCTION

The voltage source PWM inverters have been widely used in various industrial fields. However, due to the widespread use of inverters, input current distortion caused by rectifier diodes and capacitor inputs is becoming apparent.

An effective solution to this problem is to employ a matrix converter [1]. Since the input AC power directly converted to the variable voltage variable frequency AC power without using the large capacity intermediate DC bus, the input current ripples are milder than those of the conventional voltage source inverter. In addition, the highly sophisticated PWM technique enables it to control the input current waveforms as well, while it is pursuing the primary function of pulse-width modulation of the output voltage.

Authors have focused on the matrix converter technique as the future power topology for motor drive system and have helped to commercialize it for the world market [2]. By adding an output LC filter, the matrix converter can be made to have sinusoidal input-output voltage and current characteristics as well [3] [4]. However, due to the low PWM switching frequency associated with Silicon power devices, the size of the input and output LC filter is large. Therefore, in order to make the product

size including input and output filters equal to or less than that of the conventional product, a matrix converter using SiC power device was proposed [5]. Since SiC has low loss and high frequency characteristics, it was used for bi-directional switch of matrix converter in [5]. As a result of downsizing of the input and output filters due to high carrier frequency, it was possible to realize the sine input current and output voltage without increasing the size of the product.

Although the input-output filter was downsized using SiC, the conduction loss is large because the main circuit uses two SiC power devices in series as shown in Fig. 1. Therefore, since 18 power devices are used, there is an issue that the main circuit cannot be downsized as well as conventional matrix converters.

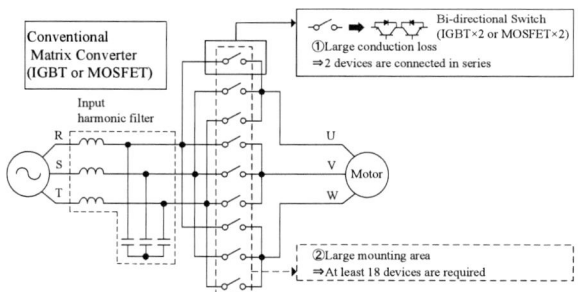

Fig. 1. Conventional matrix converter and its problems.

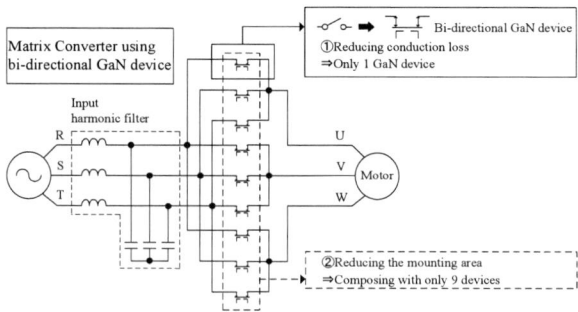

Fig. 2. Nine switches matrix converter using bi-directional GaN device and its features.

Authors focused on next generation power devices using GaN that can switch at high speed with lower loss than SiC. In particular, the bi-directional GaN device is constituted by connecting two devices in series on the same die, which has an advantage that the conduction loss and the mounting area can be reduced. Conventional matrix converter requires at least 18 devices, but if bi-directional GaN devices are used, they can be composed with only 9 devices as shown in Fig. 2, which is useful for high efficiency and downsizing by reducing the conduction loss of the matrix converter. Matrix converter chipset using bi-directional GaN devices has already been developed [6], but as far as the authors know, a three phase motor has not been driven by it.

This paper introduces the prototype of nine switches matrix converter using bi-directional GaN devices, which successfully drives a three phase motor. Since the bi-directional GaN device introduced in this paper is an engineering sample for low power, the GaN matrix converter in this development cannot be simply compared with the matrix converter using SiC power device. However, the low power GaN matrix converter was prototyped as the first step in the development of GaN matrix converter.

First, design guidelines of GaN matrix converter are shown. Specifically, the bi-directional GaN devices, gate drive and snubber circuit, and input and output LC filters are shown. Then, the validity of the design was confirmed by measuring the waveform, efficiency and loss of GaN matrix converter as the evaluation result. Finally, by comparing bi-directional GaN device and uni-directional GaN device, it was confirmed that efficiency and mounting area are improved in this GaN matrix converter was confirmed.

II. DESIGN GUIDLINES OF GAN MATRIX CONVERTER

The GaN matrix converter introduced in this paper is the unit remodeled from the main circuit of the conventional general purpose matrix converter using Si-IGBT. The remodeling points of the main circuit are the bi-directional GaN device, gate drive and snubber circuits, and input-output LC filters. In this chapter, design guidelines at the above-mentioned remodeling points are shown.

A. Bi-directional GaN device

The bi-directional GaN device used in this paper is an engineering sample of Transphorm Inc. The package and symbol for bi-directional GaN device are shown in Fig. 3. The specifications of bi-directional GaN device is shown in TABLE 1. H1 and H2 are heat dissipation terminals. The device is mounted to the board and dissipated heat to the copper pattern of the board through H1 and H2.

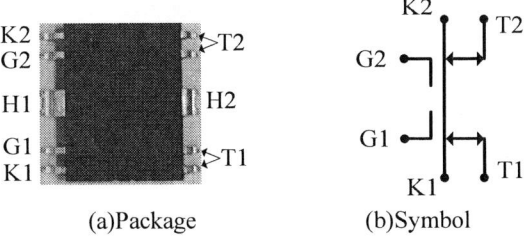

(a)Package   (b)Symbol

T1 & T2: Current terminals
G1 & G2: Gate terminals
K1 & K2: Kelvin terminals
H1 & H2: Heat dissipation terminals

Fig. 3. Bi-directional GaN device (SO16W, Engineering sample).

TABLE I
Specification of bi-directional GaN device

Maximum blocking voltage	$\pm$ 600V
On-resistance between T1&T2	0.2Ω (Tj=25deg.)
	0.4Ω (Tj=175deg.)
Gate voltage	$\pm$ 18V
Continuous current	$\pm$ 6A (Tj=25deg.)
	$\pm$ 4A (Tj=175deg.)

B. Gate drive and snubber circuits

This bi-directional GaN device is based on a cascade connection with a low voltage Si-MOSFET. Therefore, a gate drive circuit can be composed in the same way such Si-IGBT and Si-MOSFET. The schematic diagram of gate drive and snubber circuits of this matrix converter are shown in Fig. 4.

The gate drive circuit is consisted of a digital isolator, a gate resistor, and a ferrite bead. The GaN device is used a digital isolator with high CMTI because the dv/dt is very high. A ferrite bead is located close the device to suppress the oscillation of the gate voltage.

Because of bi-directional current, snubber circuit is composed four diode, capacitor, and resistance. Snubber circuit is located close the device to suppress the surge voltage.

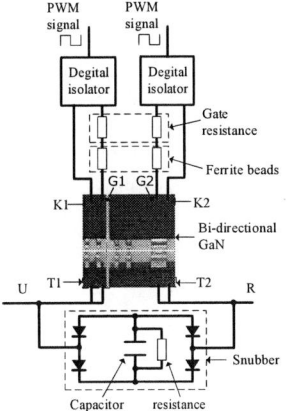

Fig. 4. Layout of gate drive and snubber circuits.

C. Input and output LC filters

As mentioned above, dv/dt of GaN device is much higher than Si and SiC devices. Therefore, deterioration of motor insulation due to surge voltage is concerned. In order to prevent degradation of motor insulation, the GaN matrix converter has an output filter as shown in Fig. 5 [3] [4] [5].

This output filter is very effective in realizing an environmentally friendly driving system. However, in the case of an actual matrix converter using a conventional Si-IGBT, the carrier frequency is 4kHz, resulting in a low cutoff frequency [7]. This increases the total size of the input and output filter. Investigation on using the GaN device to increase the switching frequency is made here and the results are presented.

The LC filter of the matrix converter is designed to attenuate the carrier frequency component. By increasing the cutoff frequency, the total size of the filters can be reduced. Therefore, increasing switching frequency of the power device is necessary. However, the high frequency operation of the power device causes a significant increase in the switching loss. Therefore, Optimization of the filters size and switching loss are necessary.

However, heatsink is not necessary because the bi-directional GaN device dissipates the heat to the copper pattern of the board. Therefore, the overall size of the converter is dominated by the volume of the input and output filter inductor. Increasing the carrier frequency can be used smaller sized inductors.

However, as mentioned above, the high frequency operation of the power device causes a significant increase in the switching loss. Therefore, it is necessary to determine the upper limit value of the switching loss. This upper limit value of switching loss is indicated a threshold value at which the junction temperature of the device mounted on the board exceeds the specified temperature.

The results of plotting the total volume of the input and output inductors and the switching loss using the carrier frequency are shown in the Fig. 6. As the carrier frequency increases, the volume of the filter decreases.

Conversely, the switching loss increases as the carrier frequency rises. The upper limit of the switching loss is indicated as 100% in the Fig. 6. Even if the carrier frequency is raised to 50kHz, the switching loss does not exceed the upper limit value.

This GaN matrix converter is a prototype in which the main circuit of a conventional product are remodeled. Due to constraints of conventional products, the upper limit of the settable carrier frequency is 40kHz. Considering the miniaturization of filters size, optimum carrier frequency is 50kHz from Fig. 6, but it is determined to 40kHz caused by this constraints.

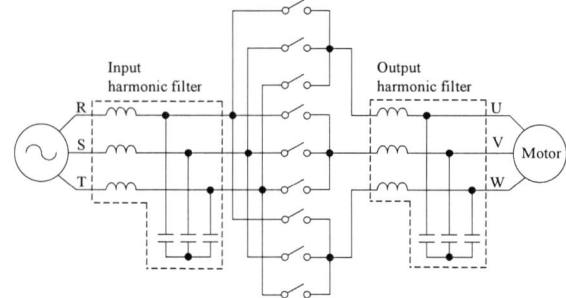

Fig. 5. Schematic diagram of GaN matrix converter.

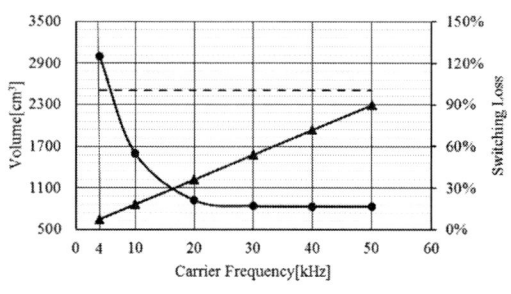

Fig. 6. Input and output inductors volume switching loss with respect to carrier frequency

III. PROTOTYPE AND EVALUATION RESULTS

The evaluation system diagram of the GaN matrix converter is shown in Fig. 7. Three boards shown in Fig. 8 are used for the GaN matrix converter. Three bi-directional GaN devices are mounted on a board.

An induction motor of 0.4kW is connected to the GaN matrix converter. A PM motor of 0.4kW and a general-purpose inverter of 0.75kW are used as load machine. The evaluation conditions of the GaN matrix converter are shown in TABLE 2. This GaN matrix converter is 200V class. The rated output current is 2.0Arms, and the carrier frequency is 40kHz.

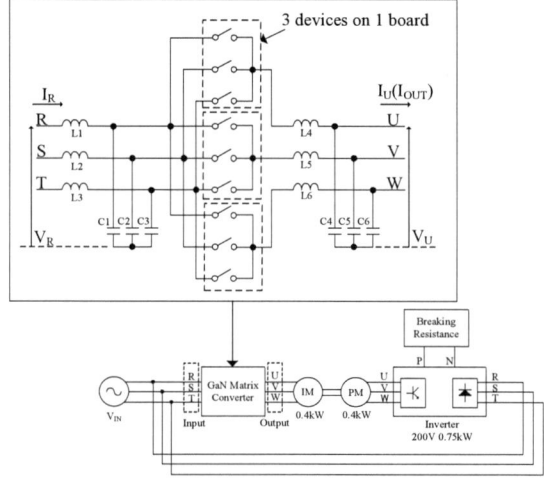

Fig. 7. Evaluation system diagram of the GaN matrix converter.

3954

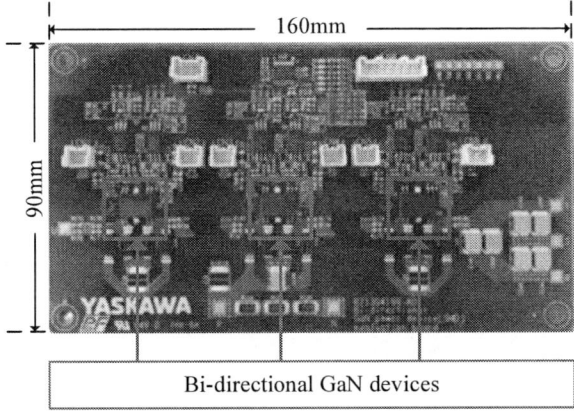

Fig. 8. PCB for single phase of the GaN matrix converter.

TABLE 2
Evaluation conditions of GaN matrix converter

Input Voltage V_{IN}	200V, 3phase, 60Hz
Output rated Current I_{OUT} (IM rated current)	2.0Arms
Output frequency f_{OUT}	60Hz
Carrier frequency f_C	40kHz
Ambient temperature T_A	25 deg.

A. Waveforms

Waveforms of powering and regenerating operation at the rated output current of 2.0Arms of the GaN matrix converter are shown in Fig. 9 and Fig. 10. It is confirmed that the input current and output voltage are sinusoidal in the GaN matrix converter. The power factor of the input current of matrix converter can be controlled. Therefore, the input voltage and the input current are nearly in phase at the powering. Input voltage and the input current are nearly reversed phase at the regenerating.

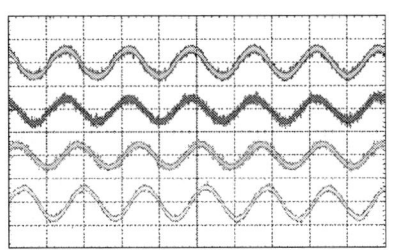

Input voltage
V_R 300V/div

Input current
I_R 4A/div

Output voltage
V_U 300V/div

Output current
I_U 4A/div

Fig. 9. Waveform of powering.

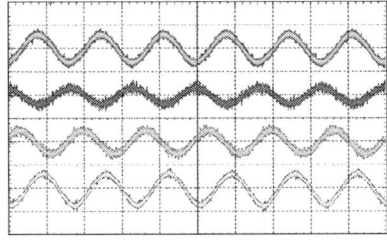

Input voltage
V_R 300V/div

Input current
I_R 4A/div

Output voltage
V_U 300V/div

Output current
I_U 4A/div

Fig. 10. Waveform of regenerating.

B. Efficiency and loss

The power analyzer was used to measure the efficiency and loss of the GaN matrix converter. The power analyzer was connected to the input and output terminals of the matrix converter. The efficiency and loss of the main circuit including the LC filters in the GaN matrix converter are shown in TABLE 3. By the TABLE 3, the measurement results of efficiency and loss are almost the same as the design results. Therefore, the design result about the matrix converter is almost correct.

The breakdown of loss in TABLE 3 is shown in Fig. 11. The left side is the design result, which also the right side is the measured result at powering shown in TABLE 3. The design result and measured result of the input and output filter are almost the same. However, about the loss of the GaN device, measurement result is slightly smaller than design result. Breakdown of the loss of bi-directional GaN device is shown in Fig. 12. For simplicity of loss design, it considered with the one side gate of the bi-directional GaN device turned OFF such Fig. 13(c)(d). However, in the real GaN matrix converter, both gates of bi-directional GaN devices are turned on such Fig. 13(a)(b). Therefore, the loss of the body diode accounts about 50% of the total loss of bi-directional GaN device as shown in Fig. 12, so the loss of the GaN device in the design is little larger than it in the measurement. The loss design of bi-directional GaN device is future works.

TABLE 3
The efficiency and loss at rated power of GaN matrix converter

Items	Efficiency[%]	Loss[W]
Design	96.1	19.5
Measurement(Powering)	96.4	18.0
Measurement(Regenerating)	96.0	19.0

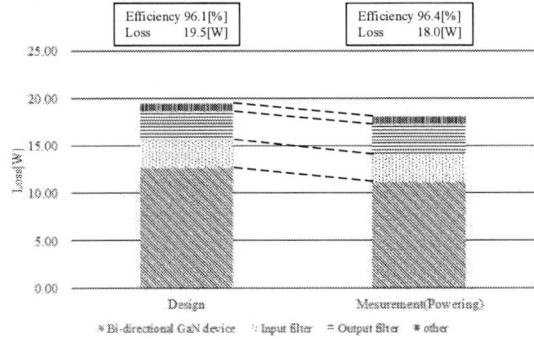

Fig. 11. Breakdown of loss of GaN matrix converter.

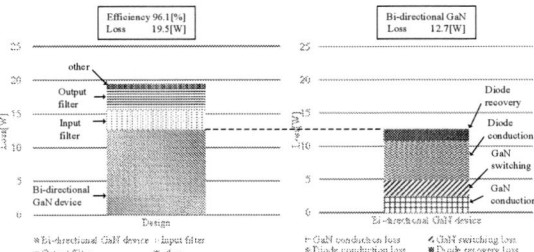

Fig. 12. Breakdown of loss of Bi-directional GaN device.

3955

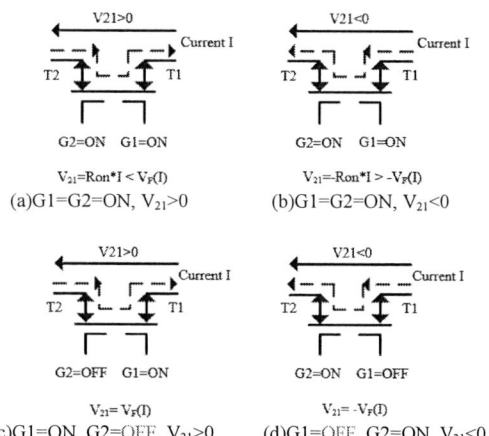

(a)G1=G2=ON, V$_{21}$>0   (b)G1=G2=ON, V$_{21}$<0

(c)G1=ON, G2=OFF, V$_{21}$>0   (d)G1=OFF, G2=ON, V$_{21}$<0

(a), (b): Actual gate signals, voltage drop and current path
(c), (d): Assumed gate signals, voltage drop and current path
V$_F$(I): Forward voltage drop across the reverse diode at current I
Fig. 13. State of the bi-directional GaN device.

TABLE 4
The bi-directional GaN device state table

Condition				State of bi-directional GaN device (Tj=25deg.)
Fig. 13.	G1	G2	V$_{21}$	
(a)	ON	ON	V$_{21}$>0	V$_{21}$=Ron*I < V$_F$(I) Turn on in both direction
(b)	ON	ON	V$_{21}$<0	V$_{21}$=-Ron*I > -V$_F$(I) Turn on in both direction
(c)	ON	OFF	V$_{21}$>0	V$_{21}$=V$_F$(I) Turn on in one direction
(d)	OFF	ON	V$_{21}$<0	V$_{21}$=-V$_F$(I) Turn on in one direction

C. Comparison between bi-directional GaN device and uni-directional GaN device with same specifications

This section verify that how the bi-directional GaN device contributes downsizing and high efficiency in matrix converter. It is assumed that the object to be compared is a uni-directional GaN device of a small general purpose package having the same electrical specification as that of a bi-directional GaN device. Although it may not be fair to compare with a uni-directional GaN device of 1in1 package, a compact general purpose package is widely distributed in the market now, so it is compared with a uni-directional GaN device.

Fig. 13 is the result of comparing the loss in the case of applying the bi-directional GaN device to the GaN matrix converter and the loss in the case of applying the two uni-directional GaN devices in series to the GaN matrix converter. When bi-directional GaN device is applied, the conduction loss is reduced by 2.5W and the efficiency is increased by 0.5%. Fig. 14 is the result of comparing the mounting area in the case of applying the bi-directional GaN device to the GaN matrix converter and the mounting area in the case of applying the two uni-directional GaN devices in series to the GaN matrix converter. When the bi-directional GaN device is applied, the mounting area is reduced by 17.4% per pair of devices. The usefulness of bi-directional GaN device in efficiency and mounting area was confirmed.

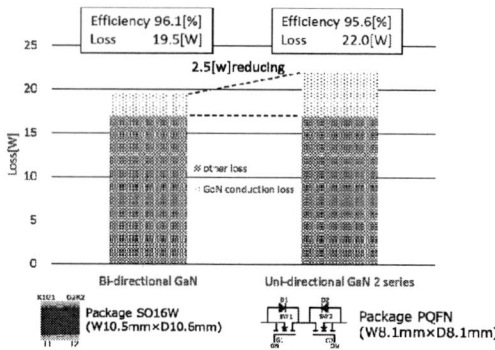

Fig. 14. Comparison of efficiency and loss.

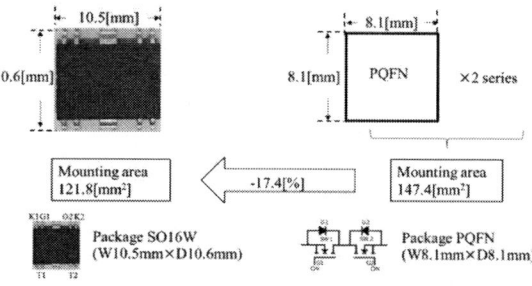

Fig. 15. Comparison of mounting area.

IV. CONCLUSIONS

From the results presented here, the following important observations are made:

1. It was confirmed that the input current and filtered output voltage were sinusoidal in the GaN matrix converter.

2. Powering and regenerating operation of GaN matrix converter were confirmed. The input voltage and the input current were nearly in phase at the powering. Input voltage and the input current were nearly reversed phase at the regenerating.

3. Efficiency and loss of the main circuit of the GaN matrix converter at rated output power were measured. Loss of control power supply is not included.
   Design:
      Efficiency 96.1%, Loss 19.5W
   Measurement(Powering):
      Efficiency 96.4%, Loss 18.0W
   Measurement(Regenerating):
      Efficiency 96.0%, Loss 19.0W
   Measured loss is little smaller than design loss. For simplicity of loss design, it considered with the one side gate of the bi-directional GaN device turned OFF. However, in the real GaN matrix converter, both gates of bi-directional GaN devices are turned on. Therefore, the loss of the body diode increases, so the loss of the GaN device in the measurement is little smaller than it in the design.

3956

4. In the matrix converter, when using uni-directional GaN device with same specifications as bi-directional GaN devices, the efficiency is reduced by 0.5% and the mounting area is increased by 17.4%. The usefulness of bi-directional GaN device in efficiency and mounting area was confirmed.

REFERENCES

[1] Venturini, "A new sine wave in sine wave out conversion technique which eliminates reactive", in Proc. Powercon 7, pp.E3-1-E3-15 (1980)

[2] Eiji Yamamoto, Hidenori Hara, Takahiro Uchino, Tsuneo Joe Kume, Jun Koo Kang, Hans Peter Krug, "Development of Matrix Converter for Industrial Applications", IEEE/IAS Ann. Meeting Conf., pp.1406-1413 (2005)

[3] K. Yamada et al., "Filtering techniques for matrix converter to achieve environmentally harmonious drives", in Proc. EPE, 2005

[4] Ken Yamada, Tsuyoshi Higuchi, Eiji Yamamoto, Hidenori Hara, Mahesh M. Swamy, Tsuneo Kume: "Integrated Filters and their combined Effects in Matrix Converter", IEEE/IAS Ann. Meeting Conf., pp.1406-1413 (2005)

[5] Yasunori Furukawa, Takeshi Kinomae, Kohei Sirabe, Hidenori Hara, Masato Higuchi, Ryoji Tomonaga, Tsuneo Kume, "Matrix Converter with Sinusoidal Input-Output Filter and Filter Downsizing Using SiC devices", ECCE, 2016

[6] Shuichi Nagai, Yasuhiro Yamada, Noboru Negoro, Hiroyuki Hanada, Yuji Kudoh, Hiroki Ueno, Masahiro Ishida, Nobuyuki Otsuka, and Daisuke Ueda, "A GaN 3x3 Matrix Converter Chipset with Drive-by-Microwave Technologies", ISSCC, pp.494-495, 2014

[7] YASKAWA general purpose Matrix Converter "U1000": http://www.e-mechatronics.com/product/inverter/index.html

The 2018 International Power Electronics Conference

A Model Predictive Dual Current Control Method for Indirect Matrix Converter Fed Induction Motor Drives

Mei Yang, Chen Lisha, Liang Wang
Collaborative Innovation Center of Key Power
Energy-Saving Technologies in Beijing，North China
University of Technology，Beijing，China
meiy@ncut.edu.cn

Yunwei Li
Department of Electrical & Computer Engineering,
University of Alberta, Edmonton, Canada
yunwei.li@ualberta.ca

Abstract—In order to decrease controller parameters and the difficulty for parameter tuning, a model predictive dual current control (MPDCC) is proposed for the indirect matrix converter fed induction motor drives (IMC-IM), in which the performance of motor drives and the quality of the grid power supply are both described by currents. An output current predictive model is built based on the mathematical model of the induction motor, and a grid current predictive model is established according to the input LC filter model. By using the similar physical quantities, weighting factors, which are very important for the control performance and difficult to be tuned, are reduced and even canceled. Simulation results show that sinusoidal grid/output currents, unit power factor, good steady state and dynamic performances of IM have been achieved. In addition, just only one weight factor is required, which is simple to be set.

Key words—*Model Predictive Torque Control; Model Predictive Dual Current Control; Weight Factor; Indirect Matrix Converter;*

I. INTRODUCTION

Indirect matrix converter (IMC) is a direct AC-AC converter without a bulky dc-link capacitor, which reduces the cost and increases the power density [1-3]. In addition, it has a lot of advantages, such as bi-directional power flow, sinusoidal input/output current, unity displacement power factor [4-5]. It is suitable to drive AC motor, which can be widely used in military, traffic, astronautic and some other industry fields.

Generally, in the IMC-IM, the conventional rotor flux oriented vector control (VC) method is employed for the induction motor, and dual space vector modulation (SVM)

are adopted for the IMC. Since the induction motor and IMC are controlled respectively, it is very difficult to realize the multi-objective optimization for the whole drive system, including utility grid quality and performances of the induction motor. In addition, the control structure is rather complicated with plenty of controller parameters, which are difficult to be tuned. As a result, the ideal dynamic performance can hardly be achieved by using the conventional control method.

In order to solve these problems, the model predictive control (MPC) method is introduced for the IMC-IM, in which the converter and induction motor are considered to be a unified whole and the modulation method is removed [6-7]. Since MPC is one kind of typical predictive control method, it has simple and intuitive control thought, and good dynamic performance can be attained by using it. In addition, it can solve multi-objective problem with multiple non-liner constraints [8-11]. However, in recent researches, the electromagnetic torque, stator flux and input reactive power are used to evaluate performance demands of the IMC-IM, and at least three weight factors are defined to identify the importance of different objectives, which work as controller parameters and need to be tuned [6-7].

So far, there is seldom common theory and method for the design of weight factors [12]. Auto-tuning mechanism for weight factors is attempted, and methods for eliminating them are proposed. Whereas, some other parameters are inevitable, and complex algorithms or extra constraints are required [13-15].

In this paper, a MPDCC method is proposed, in which optimized objectives are redefined and predictive models are rebuilt. Only one weight factor is needed, and ideal performances of the IMC-IM are achieved.

II. CONFIGURATION AND MPTC METHOD OF IMC-IM

3958

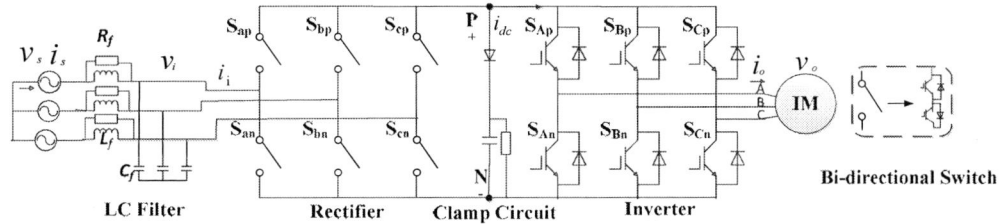

Fig.1 The configuration of IMC-IM

Fig.1 shows the configuration of the IMC-IM, in which an LC filter, rectifier stage, a clamping circuit and an inverter stage mainly constitute the IMC. In this system, the input LC filter is used to absorb the high frequency harmonics of input currents caused by power device switching. The rectifier consists of 6 bi-directional switches, thus bi-directional power flow can be realized. The clamping circuit is composed of a diode and a small capacitor in order to avoid over voltage. And a three phase voltage source inverter is adopted in the inverter stage [4-5].

A model predictive torque control (MPTC), which is a conventional MPC method, is shown in Fig.2. It mainly contains 4 parts, reactive power predictive model, torque/flux predictive model, flux observer and performance optimization. An outer speed loop is employed to calculate the reference value of the electromagnetic torque. Since stator/rotor fluxes cannot be measured directly, flux observer is adopted to obtain flux values in real time. And predictive models are built to predict the electromagnetic torque, stator flux and input reactive power in the future according to all valid switching states of the IMC. Finally, performances of system are evaluated, and the optimal control signal is determined in the performance optimization [16].

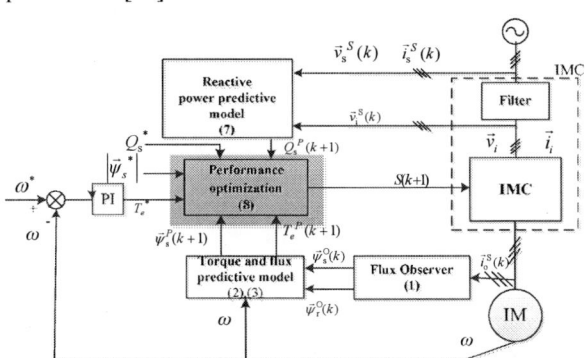

Fig.2 The control scheme of the conventional MPTC method

1) Flux observer

According to the dynamic mathematical model of induction motor, the stator flux and rotor flux can be obtained by using the Euler equation,

$$\vec{\psi}_s^O(k) = \vec{\psi}_s^O(k-1) + \vec{v}_o(k)T_s - T_s R_s \vec{i}_o^S(k)$$
$$\vec{\psi}_r^O(k) = \frac{L_r}{L_m}\vec{\psi}_s^O(k) + (L_m - \frac{L_r L_s}{L_m})\vec{i}_o^S(k) \tag{1}$$

in which, $\vec{i}_o^S(k)$ is the real output current measured at the current sampling period. $\vec{\psi}_s^O(k)$, $\vec{\psi}_r^O(k)$ are the observed stator flux and rotor flux at the current sampling period.

2) Torque and stator flux predictive model

According to equation (1), the predictive values of stator flux can be expressed as,

$$\vec{\psi}_s^P(k+1) = \vec{\psi}_s^O(k) + \vec{v}_o(k+1)T_s - T_s R_s \vec{i}_o^S(k) \tag{2}$$

And the predictive values of the electromagnetic torque can be determined,

$$T_e^P(k+1) = \frac{3}{2}p_n(\vec{\psi}_s^P(k+1) \otimes \vec{i}_o^P(k+1)) \tag{3}$$

in which, p_n is pole number.

The predicted output current at the next sampling period can be obtained by using the induction motor model,

$$\vec{i}_o^P(k+1) = (1+\frac{T_s}{\tau_\sigma})\vec{i}_o^S(k) + \frac{T_s}{\tau_\sigma + T_s}\left\{\frac{1}{R_\sigma}\left(\frac{k_r}{\tau_r} - jk_r\omega\right)\psi_r^O(k) + \vec{v}_o(k+1)\right\} \tag{4}$$

in which, $\sigma = 1 - (L_m^2/L_r L_s)$ $R_\sigma = R_s + k_r^2 R_r$, $L_\sigma = \sigma L_s$, $\tau_\sigma = L_\sigma/R_\sigma$, $\tau_r = L_r/R_r$, $k_r = L_m/L_r$.

3) Reactive power predictive model

The mathematical model of input LC filter shown in Fig.1 is described as follows,

$$\begin{cases} L_f \dfrac{d\vec{i}_s}{dt} = \vec{v}_s - \vec{v}_i \\ C_f \dfrac{d\vec{v}_s}{dt} = \vec{i}_s - \vec{i}_i \end{cases} \tag{5}$$

It is discretized by using the Euler equation, and the predictive model of grid current is obtained,

$$\vec{i}_s^P(k+1) = \Phi_{21}\vec{v}_i^S(k) + \Gamma_{21}\vec{v}_s^S(k) + \Phi_{22}\vec{i}_s^S(k) + \Gamma_{22}\vec{i}_i(k) \tag{6}$$

in which, $\Phi = e^{AT_s}$, $\Gamma = \int_0^{T_s} e^{A(T_s-\tau)}Bd\tau = A^{-1}(\Phi - I)B$

$$A = \begin{vmatrix} 1 & \dfrac{1}{C_f} \\ -\dfrac{1}{L_f} & 0 \end{vmatrix}, \quad B = \begin{vmatrix} 0 & -\dfrac{1}{C_f} \\ \dfrac{1}{L_f} & 0 \end{vmatrix}.$$

in which, $\vec{v}_s^S(k), \vec{v}_i^S(k), \vec{i}_s^S(k)$ are measured values of grid voltage, input voltage and grid current.

Correspondingly, the reactive power predictive model is determined as follow,

$$Q_s^P(k+1) = \mathrm{Im}(\vec{v}_s(k+1).\vec{i}_s^P(k+1)) \tag{7}$$

4) Performance optimization

In order to evaluate the whole drive system performance requirements from utility grid and induction motor, a cost function is defined as follows, which includes the electromagnetic torque, stator flux, and input reactive power.

$$g = \lambda_1 \left| T_e^* - T_e^P(k+1) \right| + \lambda_2 \left\| \vec{\psi}_s^* \right| - \left| \vec{\psi}_s^P(k+1) \right\| + \lambda_3 \left| Q_s^P(k+1) - Q_s^* \right| \tag{8}$$

in which, the reference input reactive power $Q_s^* = 0$, the reference torque T_e^* is calculated by the outer PI controller, the reference flux $\left| \vec{\psi}_s^* \right|$ is the nominal flux, and $\lambda_1, \lambda_2, \lambda_3$ are weight factors, which assess importance of different objectives.

Obviously, the ideal purpose is to ensure cost function equals to zero. Therefore, cost function values corresponding to all valid switching states are compared, and the switching state which minimizes the cost function is selected and used as the control signal in the next sampling period.

III. THE PROPOSED MPDCC METHOD

As shown in the definition of the conventional cost function (8), there are as many as three weight factors needed to be designed, and they influence the control effect directly, which are similar to controller parameters.

Since the torque, flux and reactive power are different physical quantities with diverse dimensions and units, the importance and differences are necessary to be considered during setting weight factors. Thus, proper weight factors may differ by orders of magnitude, and it is extremely difficult to tune weight factors.

In order to solve these problems, in this paper a MPDCC method is proposed, in which performance requirements of the utility grid and induction motor are described both by current, grid current and output current. As a result, the number of weight factors is reduced, and the difficulty of

parameter tuning can be decreased significantly.

As shown in Fig.3, the structure of MPDCC is similar with the conventional MPTC. The outer speed loop is all the same to attain the reference electromagnetic torque. In the inner control, the main parts including the flux observer, predictive models and performance optimization are remained. Instead of the conventional torque, flux and reactive power predictive models, the grid current and output current predictive models are adopted. In addition, there are 2 added parts, reference grid current calculation and reference output current calculation, which are employed to transform the conventional objectives to the reference currents.

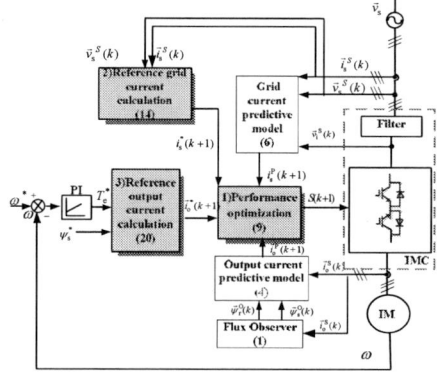

Fig.3 The control of scheme of the proposed MPDCC method

In MPDCC method, the flux observer, predictive models of the output current and grid current are built based on mathematical models of the induction motor and LC filter. Thus, equation (1), (4) and (6) are employed similarly as by using MPTC method. Since the control objectives are different, the cost function is distinct, and the reference objective values are required to be calculated.

1) Performance optimization

In order to evaluate performances of both grid and induction motor with the same physical quantity, a cost function is defined as follow,

$$g = \lambda[(\vec{i}_{s\alpha}^*(k+1) - \vec{i}_{s\alpha}^P(k+1))^2 + (\vec{i}_{s\beta}^*(k+1) - \vec{i}_{s\beta}^P(k+1))^2] + [(\vec{i}_{o\alpha}^*(k+1) - \vec{i}_{o\alpha}^P(k+1))^2 + (\vec{i}_{o\beta}^*(k+1) - \vec{i}_{o\beta}^P(k+1))^2] \tag{9}$$

in which, $\vec{i}_{s\alpha}^*(k+1), \vec{i}_{s\beta}^*(k+1)$ are components of the reference grid current $\vec{i}_s^*(k+1)$ in two-phase static coordinate system. $\vec{i}_{o\alpha}^*(k+1), \vec{i}_{o\beta}^*(k+1)$ are components of the reference output current $\vec{i}_o^*(k+1)$ in two-phase static coordinate system. All reference grid/output currents are sinusoidal and various under different conditions, which are required to be confirmed in reference current calculation modules. λ is the only one weight factor, which is designed according to the importance of requirement from the grid current.

2) Reference grid current calculation

According to the instantaneous power theory, the instantaneous power in the utility gird can be derived,

$$\begin{bmatrix} P_s \\ Q_s \end{bmatrix} = \begin{bmatrix} v_{s\alpha} & v_{s\beta} \\ v_{s\beta} & -v_{s\alpha} \end{bmatrix} \begin{bmatrix} i_{s\alpha} \\ i_{s\beta} \end{bmatrix} \tag{10}$$

And the reference of grid current is determined,

$$\begin{bmatrix} i_{s\alpha}^*(k+1) \\ i_{s\beta}^*(k+1) \end{bmatrix} = \frac{1}{(U_{sm})^2} \begin{bmatrix} v_{s\alpha}(k+1) & v_{s\beta}(k+1) \\ v_{s\beta}(k+1) & -v_{s\alpha}(k+1) \end{bmatrix} \begin{bmatrix} P_s^*(k+1) \\ Q_s^*(k+1) \end{bmatrix} \tag{11}$$

in which, the instantaneous reactive power reference $Q_s^*(k+1) = 0$, U_{sm} is the amplitude of the grid voltage, $v_{s\alpha}(k+1), v_{s\beta}(k+1)$ are components of grid voltage $\vec{v}_s(k+1)$ in two-phase static coordinate system. Since the grid voltage is a continuous sinusoid, and the sampling frequency is much higher than the input frequency, it is assumed as follows,

$$\vec{v}_s(k+1) = \vec{v}_s^S(k) \tag{12}$$

Generally, the fluctuation in instantaneous active power is small in a sampling period, therefore, it can be supposed,

$$P_s^*(k+1) = P_s(k) \tag{13}$$

According to equation (11), (12) and (13), the reference grid current can be described,

$$\vec{i}_s^*(k+1) = \sqrt{i_{s\alpha}^*(k+1)^2 + i_{s\beta}^*(k+1)^2} \cdot \sin(\omega_{in} t + \theta) \tag{14}$$

in which, $i_{s\alpha}^*(k+1) = \phi v_{s\alpha}(k+1)$, $i_{s\beta}^*(k+1) = \phi v_{s\beta}(k+1)$, $\phi = P_s^*(k+1)/(U_{sm})^2$.

Obviously, the reference grid current $\vec{i}_s^*(k+1)$ reflects performance demands of utility grid, including the reactive power and sinusoidal grid currents.

3) Reference output current calculation

According to the equation of electromagnetic torque, it can be expressed,

$$\begin{aligned} T_e &= \frac{3}{2} p_n (\frac{1}{L_s L_r - L_m^2}) \vec{\psi}_r \otimes \vec{\psi}_s \\ &= \frac{3}{2} p_n (\frac{1}{L_s L_r - L_m^2}) |\vec{\psi}_r| \cdot |\vec{\psi}_s| \cdot \sin(\angle \vec{\psi}_r - \angle \vec{\psi}_s) \end{aligned} \tag{15}$$

Therefore, the angle of the reference stator flux is

determined by the given stator flux $|\vec{\psi}_s^*| = \psi_s^*$ and the electromagnetic torque T_e^* as follows,

$$\angle \vec{\psi}_s^* = \angle \vec{\psi}_r^p(k+1) + \sin^{-1}(\frac{T_e^*}{\frac{3}{2} p_n \lambda_m |\vec{\psi}_r^p(k+1)| \cdot |\vec{\psi}_s^*|}) \tag{16}$$

in which, the angle and magnitude of predictive rotor flux $\vec{\psi}_r^P(k+1)$ can be obtained based on Euler discretization of the rotor flux equation,

$$\vec{\psi}_r^P(k+1) = \vec{\psi}_r^O(k) + T_s [R_r \frac{L_m}{L_r} \vec{i}_o^S(k) - (\frac{R_r}{L_r} - j\omega) \cdot \vec{\psi}_r^O(k)] \tag{17}$$

And the reference stator flux can be expressed as follows,

$$\vec{\psi}_s^*(k+1) = |\vec{\psi}_s^*| \cdot \exp(j\angle \vec{\psi}_s^*) \tag{18}$$

According to steady models of stator/rotor flux, the following equations can be introduced,

$$\begin{aligned} \vec{i}_o &= (\vec{\psi}_s - L_m \cdot \vec{i}_r)/L_s \\ \vec{i}_r &= (\vec{\psi}_r - L_m \cdot \vec{i}_o)/L_r \end{aligned} \tag{19}$$

Combined (17), (18) with (19), the reference output current is determined,

$$\vec{i}_o^*(k+1) = \frac{1}{L_s - \frac{L_m^2}{L_r}} \cdot [\vec{\psi}_s^*(k+1) - \frac{L_m}{L_r} \cdot \vec{\psi}_r^P(k+1)] \tag{20}$$

Obviously, the reference output current $\vec{i}_o^*(k+1)$ reflects performance objectives of the induction motor, involving the stator flux and electromagnetic torque.

IV. SIMULATION RESULTS

In order to verify the feasibility and effectiveness of the MPDCC method, a lot of simulations by using the conventional MPTC and the proposed MPDCC method have been carried out, and the corresponding results are compared.

In simulations, the sampling frequency is 30kHz, and the key parameters of the converter and motor are shown as Table. I. The load torque jumps from 4 N·m to 7 N·m at t=1s, and the reference speed steps from 400rpm to 600rpm at t=1.5s.

TABLE I.

PARAMETER OF THE CONVERTER AND THE MOTOR

Parameters	Value	Parameters	Value
T_e^*	14N·m	L_s	0.2498H
ψ_s^*	0.9Wb	L_r	0.2498H
L_f	5mH	R_s	2.54Ω
R_f	100Ω	R_r	1.67Ω
C_f	20e-6F	L_m	0.2366H

The motor waveforms including the rotor speed, the electromagnetic torque, and the stator current are shown in Fig.4. Obviously, by using both control methods, sinusoidal stator currents have been obtained, and the rotor speed tracks the reference value rapidly, which mean good steady state/dynamic drive performances have been achieved.

The grid waveforms including the gird voltage/current, the grid reactive power are shown in Fig. 5. It can be seen, by using both methods, the grid current and the grid voltage are almost in the same phase, and the reactive powers are maintained less than 100 Var. The distortion of grid current is much more serious by using the convention MPTC method, and the grid current is closer to sinusoid by using MPDCC method. More simulation results under different conditions are shown in Table. II, in which, the condition $T_L = 4 \text{N} \cdot \text{m}$, n=600rpm is expressed as 4-600. Obviously, the harmonics of grid current are significantly reduced, which means the grid quality is improved.

TABLE Ⅱ.

SIMULATION RESULTS BY USING TWO METHODS UNDER DIFFERENT CONDITIONS

Condition	THD%(i_s)		THD%(i_o)		Q_s / Var	
	MPTC	**MPDCC**	**MPTC**	**MPDCC**	**MPTC**	**MPDCC**
4-600	9.17	5.73	4.48	4.53	80	50
4-900	9.73	5.46	5.04	5.12	100	70
7-600	9.49	4.87	5.03	4.97	100	90
7-900	9.58	5.37	4.88	5.12	110	100
14-600	9.32	4.77	4.78	4.67	120	120
14-900	9.56	4.98	4.98	4.85	120	120

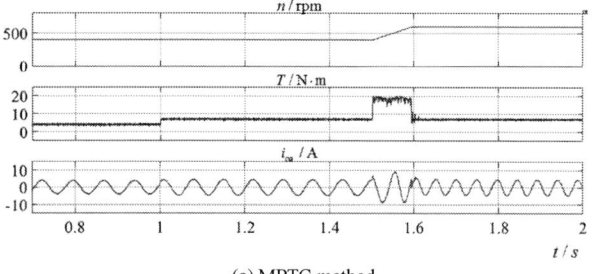

(a) MPTC method

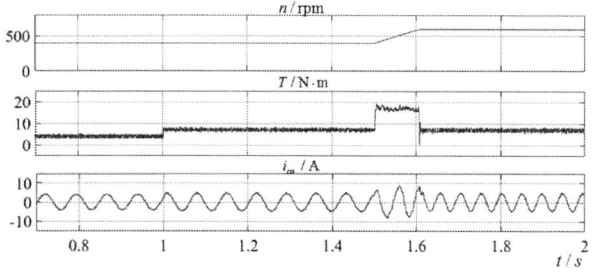

(b) MPDCC method

Fig.4 Waveforms of the motor

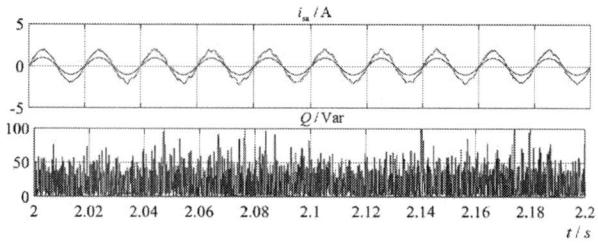

(a) MPTC method

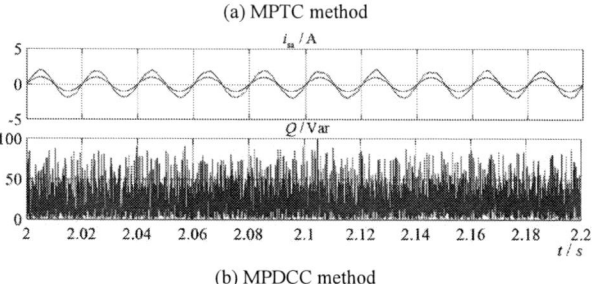

(b) MPDCC method

Fig.5 The grid voltage, current, reactive power waveforms

The ideal weight factors in two control methods are provided shown in Table.III. It is clear that three weight factors with great difference are required by using conventional MPTC method. However, by using MPDCC method, only one weight factor is required, which means it is very easy to tune the control parameter and achieve multi-objectives optimization.

TABLE Ⅲ.

WEIGHT FACTORS OF TWO METHODS

MPTC		**MPDCC**	
Parameters	**Value**	**Parameters**	**Value**
λ_1	20	λ	10
λ_2	950	-	-
λ_3	30000	-	-

V. EXPERIMENTAL RESULTS

The experimental platform of IMC-IM is shown in Fig.6, The sampling frequency is 10kHz, and the input voltage is 200V. The key parameters of the converter and motor are consistent with the simulation, which is shown in Table. I.

The waveforms of grid/output current and grid voltage by using two methods are shown in Fig.7, in which the reference speed is 600rpm, and the load is 7N.m. It can be seen that the phases of the input current and voltage are almost same, and sinusoidal output current has been obtained by using MPTC and MPDCC. However, the grid current waveforms have obvious distinction, and it is better by using MPDCC.

The 2018 International Power Electronics Conference

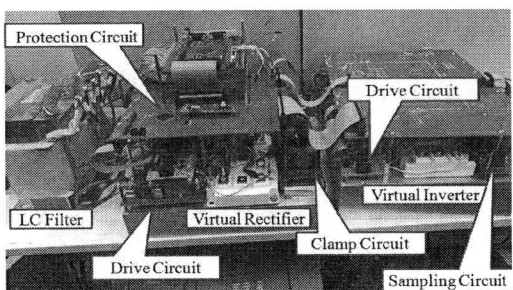

Fig.6 The experimental platform of IMC-IM

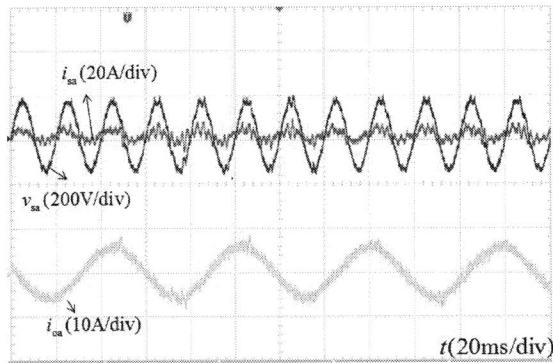

(a) MPTC method

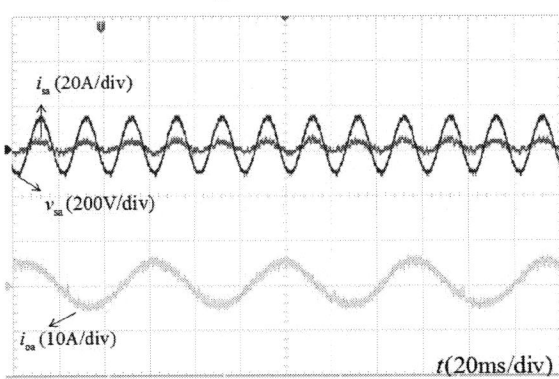

(b) MPDCC method

Fig.7 The experimental waveforms without load

Due to the hardware limitations, lower sampling frequency and switching frequency can be realized in experiments, which results the grid/stator current waveforms in experiments are not as ideal as simulation. In the future works, the hardware will be improved to achieve higher sampling frequency.

VI. CONCLUSION

An improved model predictive control MPDCC method is proposed in this paper. Simulations and experiments have been implemented, and results verify that,

1) Good steady-state/dynamic performances of induction motor drives and power quality in the utility grid are achieved by using the proposed MPDCC method.

2) Compared with conventional MPTC, the gird current

THD and the number of control parameters is reduced by using MPDCC method. It is much easier to be realized.

APPENDIX

A. Dynamic mathematical model of induction motor

The dynamic models of stator/rotor flux and stator current are described as follows,

$$\frac{d\vec{\psi}_s}{dt} = -R_s \vec{i}_o + \vec{v}_o$$

$$\frac{d\vec{\psi}_r}{dt} = -\frac{1}{T_r}\vec{\psi}_r + j\omega\vec{\psi}_r + \frac{L_m}{T_r}\vec{i}_o$$

$$\frac{d\vec{i}_o}{dt} = \frac{L_m}{\sigma L_s L_r T_r}\vec{\psi}_r - j\frac{L_m}{\sigma L_s L_r}\omega\vec{\psi}_r - \frac{R_s L_r^2 + R_r L_m^2}{\sigma L_s L_r^2}\vec{i}_o + \frac{1}{\sigma L_s}\vec{v}_o$$

B. Steady-state mathematical model of induction motor

The steady-state models of stator/rotor flux and electromagnetic torque are described as follows,

$$\vec{\psi}_s = L_s \vec{i}_o + L_m \vec{i}_r$$

$$\vec{\psi}_r = L_r \vec{i}_r + L_m \vec{i}_o$$

$$T_e = \frac{3}{2}n_p(\vec{\psi}_s \otimes \vec{i}_s) = \frac{3}{2}n_p L_m(\frac{1}{L_s L_r - L_m^2})(\vec{\psi}_r \otimes \vec{\psi}_s)$$

REFERENCE

[1] Tewari, Saurabh, et al "Indirect matrix converter based open-end winding AC drives with zero common-mode voltage." IEEE Applied Power Electronics Conference and Exposition IEEE, 2016:1520-1527.

[2] Rivera M, Tarisciotti L, Wheeler P, et al. Predictive control of an indirect matrix converter operating at fixed switching frequency and without weighting factors[J]. 2015:1027-1033.

[3] Nguyen T D, Lee H H. Development of a Three-to-Five-Phase Indirect Matrix Converter With Carrier-Based PWM Based on Space-Vector Modulation Analysis[J]. IEEE Transactions on Industrial Electronics, 2015, 63(1):13-24.

[4] Padhee V, Sahoo A, Ned M, et al. Modulation Techniques for Enhanced Reduction in Common Mode Voltage and Output Voltage Distortion in Indirect Matrix Converters[J]. IEEE Transactions on Power Electronics, 2016,8655:8670.

[5] Vekhande, Vishal, B. B. Pimple, and B. G. Fernandes, et al. "Modulation of Indirect Matrix Converter under unbalanced source voltage condition." IEEE Energy Conversion Congress and Exposition IEEE, 2011:225-229.

[6] Rivera, M., et al. "Model predictive control of a Doubly Fed Induction Generator with an Indirect Matrix Converter." IECON 2010 -, Conference on IEEE Industrial Electronics Society IEEE, 2010:2959-2965.

[7] Villarroel, F., et al. "A multiobjective ranking based finite states

3963

model predictive control scheme applied to a direct matrix converter." IECON 2010 -, Conference on IEEE Industrial Electronics Society IEEE, 2010:2941-2946.

[8] Vazquez S, Leon J I, Franquelo L G, et al. Model Predictive Control: A Review of Its Applications in Power Electronics[J]. IEEE Industrial Electronics Magazine, 2014, 8(1):16-31.

[9] Zhang J, Wan S. A review of explicit model predictive control[C]// Control Conference. IEEE, 2012:4233-4238.

[10] Huang L, Yang X, Ma X, et al. Space-Vectors Based Hierarchical Model Predictive Control for a Modular Multilevel Converter[J]. IEEE Energy Conversion CongressExposition, 2015 (ECCE): 6037-6042.

[11] H. Miranda, P. Cortes, J. Yuz, and J. Rodriguez, et al. "Predictive torque control of induction machines based on state-space models," Industrial Electronics, IEEE Transactions on, vol. 56, no. 6, pp. 1916 -1924, jun.2009.

[12] Cortes, P, et al. "Guidelines for weighting factors design in Model Predictive Control of power converters and drives." IEEE International Conference on Industrial Technology IEEE, 2009:1-7..

[13] Shadmand, Mohammad B., R. S. Balog and H. A. Rub, et al. "Auto-tuning the cost function weight factors in a model predictive controller for a matrix converter VAR compensator." Energy Conversion Congress and Exposition IEEE, 2015:3807-3814.

[14] Villarroel F, Espinoza J R, Rojas C A, et al. Multiobjective Switching State Selector for Finite-States Model Predictive Control Based on Fuzzy Decision Making in a Matrix Converter[J]. IEEE Transactions on Industrial Electronics, 2013, 60(2):589-599.

[15] Gulbudak, Ozan, E. Santi, and J. Marquart, et al. "Finite state model predictive control for 3×3 matrix converter based on switching state elimination." IEEE Energy Conversion Congress and Exposition IEEE, 2014:5805-5812.

[16] López M, Rodriguez J, Silva C, et al. Predictive Torque Control of a Multidrive System Fed by a Dual Indirect Matrix Converter[J]. IEEE Transactions on Industrial Electronics, 2015, 62(5):2731-2741.

Fault Tolerant Predictive Control of Three-Level Neutral-Point-Clamped Back-to-Back Power Converters

Zhenbin Zhang[1*], Xicai Liu[2,3], Kejun Cai[2], Feng Gao[1] and Ralph Kennel[2]

1 Institute for Sustainable Energy and Smart Grid, Shandong University, Jinan, China
2 Institute for Electrical Drive Systems and Power Electronics, Technische Universität München, Munich, Germany
3 State Key Laboratory of Advanced Electromagnetic Engineering and Technology,
Huazhong University of Science and Technology, Wuhan, China
*: Corresponding author, E-mail: zbz@sdu.edu.cn

Abstract—This paper presents a fault diagnosis and fault-tolerant control method for open-circuit fault in three-level neutral-point-clamped back-to-back power converters. The proposed fault diagnosis method is based on analysis of probability distribution of stator current. No extra sensor is needed. Additionally, fault detections for both one- and two-phase open-circuit fault are achieved. The proposed fault-tolerant control scheme is a non-redundant solution and is embedded into the well-known model predictive control (MPC) frame. Simulation results validate the feasibility and effectiveness of the proposed fault diagnosis and fault-tolerant control methods.

Index Terms—Fault Diagnosis, Fault-Tolerant Control, Model Predictive Control, Neutral Point Clamped Three-level Converters.

I. INTRODUCTION

Recently, multi-level power converters have been increasingly widely applied in both high- and medium-voltage applications, due to their inherent advantages [1]–[4], such as, low harmonic distortions (at the generated currents/voltage), lower dv/dt ratings, reduced size of the filters, etc. Widely used topologies, to formulate a multi-level converter, includes: neutral point clamped (NPC), flying capacitor (FLC) and cascade H-bridge [5], among which, the NPC topology is an very attractive candidate.

The concept of three-level NPC (3L-NPC) power converter was first proposed in [2]. After decades of development, it has become one of the most widely used multilevel topologies in industrial applications [5], [6]. For such topology, (direct) model predictive control (MPC), which utilizes a costumer designed cost function and combines the switching sequence selection and optimization processes into one stage, becomes an interesting alternative. Recently, the reliability of converters has become a significant concern, especially for applications using standalone converters or on-board systems [4], [7]. Hence, with the increasing application of 3L-NPC, the reliability and maintainability of 3L-NPC deserve deeper investigation.

An industry-based survey of reliability in power electronic converters [8] shows that semiconductor power devices, with 31% failure rate, is the most fragile component. The semiconductor switch failures can be broadly divided into *open-circuit faults* and *short-circuit faults*. Typically, short-circuit faults are accompanied by very destructive over-current and over-voltage, which is harmful to the other parts in the system and would lead to a shutdown of the whole

system. A quick detection and isolation action within $10\mu s$ is required for short-circuit faults. Nowadays the standard protection system of power converters for industrial use has contained protection against short-circuit faults [9], [10]. Open-circuit fault will not (necessarily) lead to a shutdown of the system, but would degrade the performances and over-stress the components (which may lead to a secondary failure or even a complete system shutdown) [11]. However, diagnosis of open-circuit faults has not been covered by normal protection system [9]. In particular for multi-level power converters, the increased number of the switches will also increase the failure rate of the whole system. Therefore, a careful design of a fault diagnosis and tolerant control solution is very much desired.

Fault diagnosis methodologies proposed so far could be divided into *current*-based methods and *voltage*-based methods. Generally, the current-based methods consume more diagnosis times compared with voltage-based methods. But the voltage-based methods have disadvantages such as need extra sensors and are limited by modulation methods [11], [12]. Fault tolerant control (FTC) approaches can be classified into: *redundant* and *non-redundant* fault tolerant strategies. The main ideas for redundant method are to add redundant parts, such as switches or phase legs, in parallel or series connect to main legs [13]. These methods can keep system work normally even post the occurrence of fault. The main drawbacks of these methods are the increment of whole system complexity and cost, especially for systems using multi-level converters, the increment will be considerable [13], [14]. On the contrary, non-redundant methods try to tolerant faults with minimum additional components, combined with appropriate modifications of control strategy. Generally, non-redundant methods handle the faulty phase by connecting the ac terminal to the capacitor mid-point [15], [16], or by sharing converter legs (in the case of back-to-back topology) [17], [18]. Non-redundant fault tolerant methods achieve continuous operation of the system during fault conditions, without oversizing the system, could be a more economical alternative for FTC of power converters.

In this paper, a fault diagnosis algorithm based-on current analysis is proposed for one- and two-phase IGBT open-circuit fault in *3L-NPC back-to-back power converters*. Furthermore, a non-redundant fault tolerant algorithm is utilized to ride-through the one-phase open-circuit fault—the most common open-circuit fault type—within the MPC framework, without using any modulator. The topology tested in this work can be used in full-quadrant drives, renewable energy (particularly, wind energy systems) and power transmission systems.

II. SYSTEM DESCRIPTION AND OPEN-CIRCUIT FAULT ANALYSIS

Fig.1 shows a simplified 3L-NPC back-to-back converter system. In general, for $y \in \{m, g\}$, $x \in \{a, b, c\}$ and $i \in \{1, 2\}$, the gate signal for upper IGBTs is introduced as G_y^{xi} and for lower IGBTs the gate signal is $\bar{G}_y^{xi}$ (complementary to G_y^{xi}), where, y represents machine/load (m) side or grid (g) side, x stands for phases and i, external or internal IGBTs in phase-a. The switching state u_y^x for 3L-NPC power converter is defined as [6]

$$\vec{u}_y^x := \begin{cases} P & \text{if: } (G_y^{x1} = 1 \bigwedge G_y^{x2} = 1) \\ 0 & \text{if: } (G_y^{x1} = 0 \bigwedge G_y^{x2} = 1). \\ N & \text{if: } (G_y^{x1} = 0 \bigwedge G_y^{x2} = 0) \end{cases} \quad (1)$$

The admissible switching state $\vec{u}$, in normal operation, can be chosen from 27 state vectors $\vec{u} \in \mathcal{U}_{27} := \{\text{NNN}, \text{NN0}, \ldots, \text{PP0}, \text{PPP}\}$. Consider an ideal switching behavior of power switches, the phase voltages of converter can be obtained as [19]

$$\vec{v}_y^{abc} = \frac{V_{c1} + V_{c2}}{6} \begin{bmatrix} 2 & -1 & -1 \\ -1 & 2 & -1 \\ -1 & -1 & 2 \end{bmatrix} \vec{u}_y^{abc} +$$

$$\frac{V_{c1} - V_{c2}}{6} \begin{bmatrix} 2 & -1 & -1 \\ -1 & 2 & -1 \\ -1 & -1 & 2 \end{bmatrix} |\vec{u}_y^{abc}|. \quad (2)$$

Fig. 2 shows the current path of Machine-Side Converter (MSC) under normal operation conditions. For $i_m^x > 0$, G_m^{x1} open-circuit fault will result in the current path showed in (a) inaccessible; G_m^{x2} open-circuit fault will obstacle both (a) and (b). Similarly, for $i_m^x < 0$, $\bar{G}_m^{x1}$ open-circuit fault will influence the current path in (e) and (f); $\bar{G}_m^{x2}$ open-circuit fault will make (f) become inaccessible. The relation of faulty switch and influenced switching states is summarized in Tab. I. Based on this analysis, it is not difficult to tell that, for MSC, switches in upper phase branch (i.e. G_m^{x1} and G_m^{x2}) open-circuit fault will suppress the current in the positive range, and switches in lower phase branch (i.e. $\bar{G}_m^{x1}$ and $\bar{G}_m^{x2}$) open-circuit fault will suppress the current in the negative range. Internal switch (i.e. G_m^{x2} and $\bar{G}_m^{x1}$) fault will have more influence on the phase current distribution compared with external (i.e. G_m^{x1} and $\bar{G}_m^{x2}$) switch fault, since more switching states are influenced.

III. FAULT DIAGNOSIS METHOD

A. Definition of fault diagnosis indicators

From the analysis in Sec. II, distribution of three phase current is influenced by the location of faulty switch. Based on the probability theory, a fault diagnosis method is proposed in this paper. The fault diagnosis indicators are:

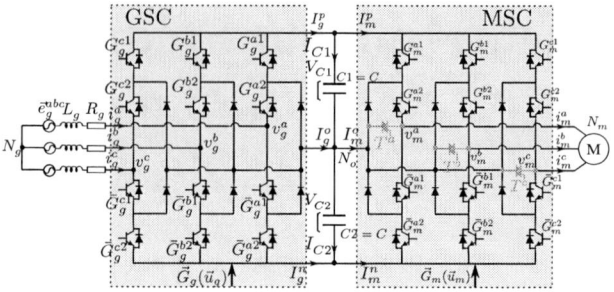

Fig. 1. 3L-NPC back-to-back power converter system with non-redundant fault tolerant for machine-side converter.

TABLE I
SWITCHING STATES UNDER FAULTY CONDITION

Current	Fault	Possible State	Impossible State
$i_m^x > 0$	G_m^{x1}	0, N	P
	G_m^{x2}	N	0, P
$i_m^x < 0$	$\bar{G}_m^{x1}$	P	0, N
	$\bar{G}_m^{x2}$	0, P	N

- The relative variance ε^x of normalized current, to identify the location of fail, is defined as

$$\varepsilon^x = \frac{Var^x}{max\{Var^a, Var^b, Var^c\}}, \quad x \in \{a, b, c\}, \quad (3)$$

$$Var_{[k]}^x = Var(X), \quad X = \{i_{[k-L+1]}^{xN}, \ldots, i_{[k]}^{xN}\}, \quad (4)$$

$$i_{[k]}^{xN} = \frac{i_{[k]}^x}{\sqrt{i_{[k]}^a{}^2 + i_{[k]}^b{}^2 + i_{[k]}^c{}^2}}, \quad x \in \{a, b, c\}, \quad (5)$$

where Var^x is the variance of normalized sampling current, $i_{[k]}^{xN}$ is normalized phase current at sampling instant k, L represent the length of sliding window. The minimum window length that containing enough information is one current period, can be given as

$$L_{min} = \frac{60}{9.55 N_p T_s \omega_m}, \quad (6)$$

where N_p [1] is the number of machine pole pairs, ω_m [rad/s] is the speed reference, T_s [s] is the sampling period.

- The skewness of current can be used to locate the faulty switch in the upper branch of the faulty leg or lower branch of the faulty leg, which is defined as

$$\gamma^x = E\left[\left(\frac{X - \mu}{\sigma}\right)^3\right], x \in \{a, b, c\}, \quad (7)$$

where μ is the mean of X, σ is the standard deviation of X, and $E(z)$ stands for the mean of the random variable z.

B. Fault diagnosis algorithm

Generally, under normal operation condition, ε^x nears to 1 and fluctuates periodically (see Fig. 4). When fault occurred, i_m^x is more centrally distributed compared with other currents, thus ε^x drops down and will not return back to 1. Hence the occurrence of fault can be known by monitoring whether ε^x is periodically return back to 1.

Fig. 3 show the fault diagnosis process. As Fig. 3a, at every sampling instant, ε^x is compared with 1. If ε^x equals to 1, this means phase x is in healthy condition, fault detection flag

The 2018 International Power Electronics Conference

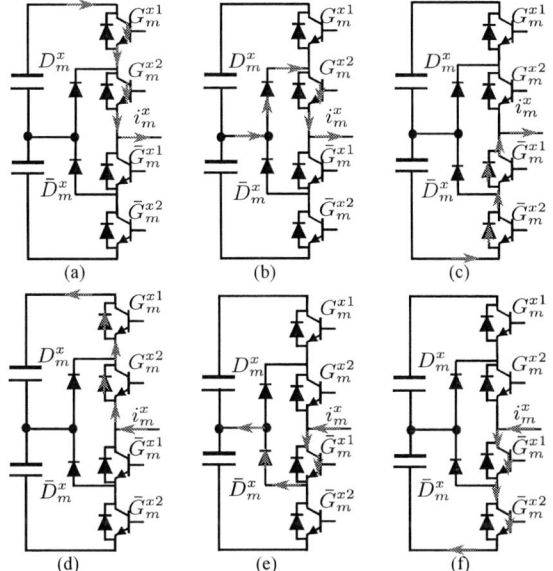

Fig. 2. Current path of 3L-NPC under normal condition: (a)-(c) are ($i_{\mathrm{m}}^{\mathrm{x}} > 0$) $u_{\mathrm{m}}^{\mathrm{x}} = P$, $u_{\mathrm{m}}^{\mathrm{x}} = 0$ and $u_{\mathrm{m}}^{\mathrm{x}} = N$, respectively; (d)-(f) are ($i_{\mathrm{m}}^{\mathrm{x}} < 0$) $u_{\mathrm{m}}^{\mathrm{x}} = P$, $u_{\mathrm{m}}^{\mathrm{x}} = 0$ and $u_{\mathrm{m}}^{\mathrm{x}} = N$, respectively.

FD^{x} is set to 0. Otherwise, ε^{x} is compared with threshold thr_1 by

$$bool\varepsilon^{\mathrm{x}} = \begin{cases} 1, & \text{if } (\varepsilon^{\mathrm{x}} < thr_1) : \text{abnormal} \\ 0, & \text{otherwise : normal} \end{cases} \tag{8}$$

and the number of Boolean value $bool\varepsilon^{\mathrm{x}}$ that is 1 (true) is counted in $C1^{\mathrm{x}}$. If $C1^{\mathrm{x}}$ is more than fault detection constant $cnt1$ fault is detected and FD^{x} is set to 1 (see Eq. 9). The fault detection constant $cntz$ is defined to measure how long the value have been generated and given as: $cntz = k_z \frac{2T_i}{T_s}$, where k_z is the sensitivity factor, T_i is the current period and T_s is the sampling period.

$$FD^{\mathrm{x}} = \begin{cases} 1, & \text{if } (C1^{\mathrm{x}} > cnt1) : \text{fault detected} \\ 0, & \text{otherwise : normal} \end{cases} \tag{9}$$

After fault has been detected, fault identification is trigged (see Fig. 3b). At every sampling instant, a sliding window $\varepsilon^{\mathrm{xSW}}$ that contains values of ε^{x}, with window length $\frac{L}{2}$, is substituted into Eq. 10, in which the values of ε^{x} is compared with thresholds thr_2 and thr_3 in order to overcome the fluctuation of ε^{x} and identify the fault.

$$C2^{\mathrm{x}} = \begin{cases} C2^{\mathrm{x}} + 1, & \text{if } (thr_3 < \varepsilon_{[j]}^{\mathrm{xSW}} < thr_2) \\ C2^{\mathrm{x}}, & \text{otherwise} \end{cases}$$
$$C3^{\mathrm{x}} = \begin{cases} C3^{\mathrm{x}} + 1, & \text{if } (\varepsilon_{[j]}^{\mathrm{xSW}} < thr_3) \\ C3^{\mathrm{x}}, & \text{otherwise} \end{cases} \tag{10}$$

Then $C2^{\mathrm{x}}$ and $C3^{\mathrm{x}}$ is compared with $cnt2$. If $C2^{\mathrm{x}} > cnt2$, this means external switch fail. If $C3^{\mathrm{x}} > cnt2$, which indicates internal switch fail. Then, the average value of a sliding window of skewness (with L window length) $\overline{\gamma^{\mathrm{xSW}}}$ is substituted into Eq. 11 to locate if the fault locates in upper branch or in the lower branch of leg, which complete the basic identification of fault. Afterwards, fault index $FN_{[k]}^{\mathrm{x}}$

TABLE II
FAULT DIAGNOSTIC TABLE FOR THREE-LEVEL NPC POWER CONVERTERS

State	Variance	Skewness	Fault	Fault Index
Healthy	$\varepsilon_x > thr_1$	—	No	0
Faulty	$thr_3 < \varepsilon^{\mathrm{x}} < thr_2$	$\gamma^{\mathrm{x}} < 0$	$G_{\mathrm{m}}^{\mathrm{x1}}$	1
		$\gamma^{\mathrm{x}} > 0$	$\bar{G}_m^{x2}$	4
	$\varepsilon^{\mathrm{x}} < thr_3$	$\gamma^{\mathrm{x}} < 0$	$G_{\mathrm{m}}^{\mathrm{x2}}$	2
		$\gamma^{\mathrm{x}} > 0$	$\bar{G}_m^{x1}$	3

will be hand over to a further process in order to filter the glitch and identify the fault accurately.

$$FN_{[k]}^{\mathrm{x}} = \begin{cases} 1, & \text{if } (C2^{\mathrm{x}} > cnt2 \wedge \overline{\gamma^{\mathrm{xSW}}} < 0): G_{\mathrm{m}}^{\mathrm{x1}} \text{ fault} \\ 2, & \text{if } (C3^{\mathrm{x}} > cnt2 \wedge \overline{\gamma^{\mathrm{xSW}}} < 0): G_{\mathrm{m}}^{\mathrm{x2}} \text{ fault} \\ 3, & \text{if } (C3^{\mathrm{x}} > cnt2 \wedge \overline{\gamma^{\mathrm{xSW}}} > 0): \bar{G}_{\mathrm{m}}^{\mathrm{x1}} \text{ fault} \\ 4, & \text{if } (C2^{\mathrm{x}} > cnt2 \wedge \overline{\gamma^{\mathrm{xSW}}} > 0): \bar{G}_{\mathrm{m}}^{\mathrm{x2}} \text{ fault} \end{cases}$$
$$\tag{11}$$

In the anti-glitch process (see Fig. 3c), two old values of fault index FN^{x} is stored in FN_{old1}^{x} and FN_{old2}^{x}. A counter $C4^{\mathrm{x}}$ will be reset to 0 at the sampling instant that FN^{x} is changing, otherwise it is increased by 1 every sampling step. Only the fault index that has last for more than a certain time (i.e. $C4^{\mathrm{x}} > cnt3$) will be assigned to the final fault index FI^{x}. By this process, the glitch of fault index FN^{x} is filtered.

The diagnosis lookup table is summarized in Tab. II. In this table, ε^{x} and γ^{x} are indicators, thr_1, thr_2 and thr_3 are thresholds that selected to identify the fault, fault index is the flag to illustrate the faulty switch.

Since no extra sensors are needed with such method, the realization efforts and the system implementation cost are low. Besides, the diagnosis algorithm can also be easily realized due to the flexibility of DMPC algorithm.

IV. FAULT TOLERANT CONTROL

Fig. 5 shows the control diagram of fault-tolerant model predictive control of 3L-NPC back-to-back power converter. After the fault diagnosis, fault reconfiguration is trigged to make the converter operate continuously. The proposed fault-tolerant control algorithm consists of hardware and software reconfiguration. Depending on the location of faulty IGBT, fault-tolerant control is discussed for external and internal IGBT open-circuit fault separately.

A. Fault-tolerant control for external switch fault

When open-circuit fault occurred in external switch (i.e. $G_{\mathrm{m}}^{\mathrm{x1}}$ and $\bar{G}_m^{x2}$), 9 out of 27 switching states is influenced. Take G_m^{a1} fault as example, all switching state that containing $u_{\mathrm{m}}^{\mathrm{a}} = P$ are unavailable for $i_{\mathrm{m}}^{\mathrm{a}} > 0$ (see Fig. 6a). Due to the redundancy of switching states, there are 18 switching states that can generate 14 space vectors. Fault-tolerant algorithm will make use of these remaining space vectors. The available space vector area (shaded part in Fig. 6a) is asymmetrical. Considering no over-modulation, the modulation range is limited by the red dashed circle. Hence, the voltage utilization is halved. If DC bus voltage is unchanged, the output voltage will distort when the voltage vector goes beyond the

3967

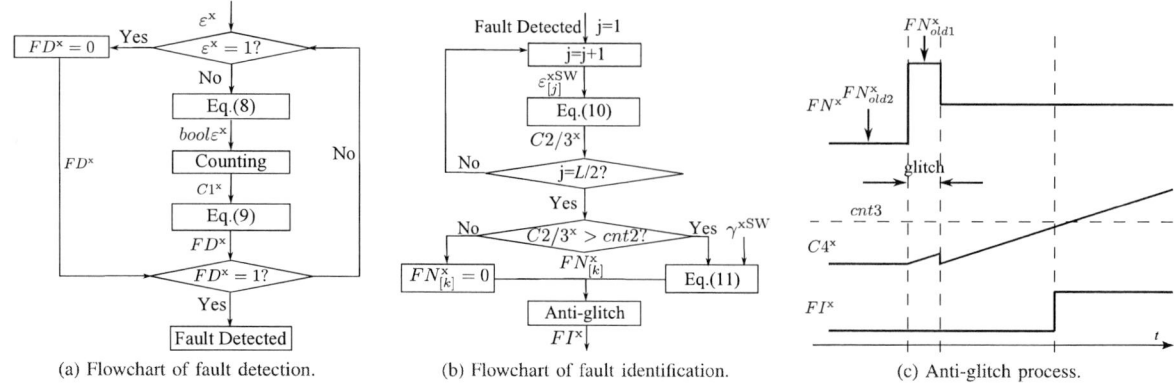

(a) Flowchart of fault detection. (b) Flowchart of fault identification. (c) Anti-glitch process.

Fig. 3. Fault diagnosis process.

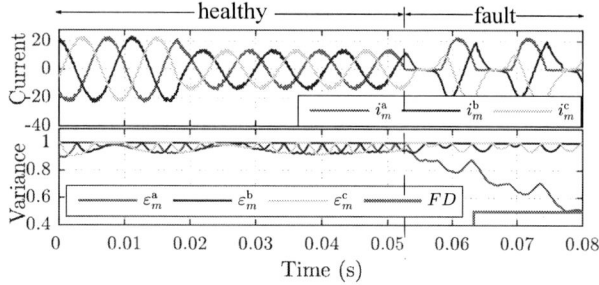

Fig. 4. Current amplitude drop at $0.018s$ and fault occurred at $0.053s$: from up to down are three phase current waveforms and current relative variance.

limitation. Hence, to avoid distortion, following equation is required

$$V_{\mathrm{m}} \leq \frac{1}{\sqrt{3}} \cdot \left\{ \frac{V_{\mathrm{d}}^{*\prime}}{2} - \frac{i_{\mathrm{m}}^{\mathrm{x}}}{j2\omega_{\mathrm{e}}C} \right\} \quad (12)$$

where V_{m} [V] is the fundamental output voltage of MSC, $i_{\mathrm{m}}^{\mathrm{x}}$ [A] is the phase current of faulty phase, ω_{e} [rad/s] is the motor speed (assuming DC-link capacitors $C1 = C2 = C$).

Fault-tolerant for external switch do not need any hardware modifications. The post fault control strategy modified as follows: a) for MSC controller, when $i_{\mathrm{m}}^{\mathrm{a}} > 0$, post fault algorithm will exploit through 18 remaining switching states, when $i_{\mathrm{m}}^{\mathrm{a}} \leq 0$, switching state selection is the same as normal operation; b) to reduce distortion, the new DC bus voltage reference $V_{\mathrm{d}}^{*\prime}$ that satisfy Eq. 12 is sent to the Grid-Side Converter (GSC) controller. Thus, this algorithm can ride-through external switch open-circuit fault simply by software modification.

B. Fault-tolerant control for internal switch fault

When open-circuit fault occurred in internal switch (i.e. $G_{\mathrm{m}}^{\mathrm{x}2}$ and $\bar{G}_{m}^{x1}$), fault-tolerance is realized by connecting the faulty phase terminal to the DC-link midpoint via corresponding triac (See T^{a}, T^{b}, T^{c} in Fig. 1). For instance for phase a fail, all IGBTs in phase a are inhibited and triac T^{a} is activated to connect the output terminal to the DC-link midpoint. Because this hardware modification, only the switching states that containing $u_{\mathrm{m}}^{\mathrm{a}} = 0$ is available. The 9 remaining space vectors after fault reconfiguration is shown in Fig. 6b. The modulation range is limited by red dashed circle, which is same with the red circle in Fig. 6a. Hence, DC bus voltage also need to be increased as in

Sec. IV-A discussed. After this modification, the new MSC output voltage become

$$\vec{v}_{\mathrm{m}[k]}^{\mathrm{abc}}(\vec{u}_{\mathrm{m}}^{\mathrm{bc}}) = \frac{V_{\mathrm{c}1} + V_{\mathrm{c}2}}{6} \begin{bmatrix} 2 & -1 & -1 \\ -1 & 2 & -1 \\ -1 & -1 & 2 \end{bmatrix} \begin{bmatrix} 0 \\ u_{\mathrm{m}}^{\mathrm{b}} \\ u_{\mathrm{m}}^{\mathrm{c}} \end{bmatrix} +$$

$$\frac{V_{\mathrm{c}1} - V_{\mathrm{c}2}}{6} \begin{bmatrix} 2 & -1 & -1 \\ -1 & 2 & -1 \\ -1 & -1 & 2 \end{bmatrix} \begin{bmatrix} 0 \\ |u_{\mathrm{m}}^{\mathrm{b}}| \\ |u_{\mathrm{m}}^{\mathrm{c}}| \end{bmatrix}, \quad (13)$$

where switching state $u_{\mathrm{m}}^{\mathrm{a}}$ is 0. In post fault control: a) MSC controller will exploit the algorithm through 9 switching states, i.e. $\vec{u} \in \mathcal{U}_9 := \{$0NN, 0N0, $\ldots$, 0P0, 0PP$\}$; b) GSC controller will track the new DC bus voltage reference. Only three extra triacs are required, in terms of hardware modification. Therefore, the proposed fault-tolerant strategy can be realized easily with low additional cost.

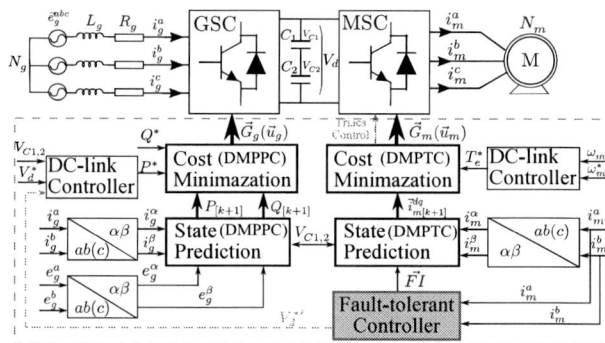

Fig. 5. Control diagram of model predictive control of three-level NPC back-to-back power converter with fault-tolerant control of machine-side converter.

V. SIMULATION RESULTS

In this section the simulation result of fault diagnosis and fault tolerant control is presented. Simulation parameters are collected in Tab. III.

A. Fault diagnosis results

Fig. 7 shows fault diagnosis process for one-phase fault and two-phase fault. The diagnosis parameter set (i.e. thr_1, thr_2, thr_3, L, $cnt1$, $cnt2$, $cnt3$) is selected as (0.95, 0.9, 0.5, 2900, $0.5\frac{2T_i}{T_s}$, $0.01\frac{2T_i}{T_s}$, $0.9\frac{2T_i}{T_s}$). As can be seen, for $\bar{G}_{m}^{a1}$ fail and $\bar{G}_{m}^{a2}$ fail (Fig. 7a), the detection time are $27ms$ and $34ms$, respectively. The ε^{a} is lower in internal switch

The 2018 International Power Electronics Conference

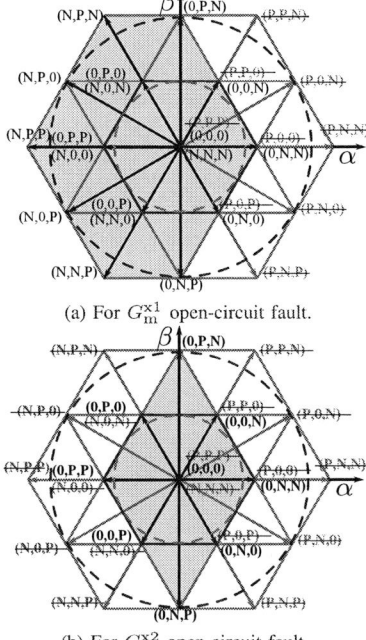

(a) For G_m^{x1} open-circuit fault.

(b) For G_m^{x2} open-circuit fault.

Fig. 6. Available space vectors of three-level NPC after fault reconfiguration.

TABLE III
SIMULATION PARAMETERS.

Parameter	Value
Grid Phase Voltage $\bar{e}_g^{abc}$ [V]	250(peak)
Grid Voltage Freq. ω_g [rad/s]	100π
Reactor Resistor R_g [Ohm]	$1.56e^{-3}$
Reactor Inductance L_g [H]	$20e^{-3}$
DC-link Capacitor C [μ F]	3100
Stator Inductance L_s = [H]	$18e^{-3}$
Stator Resistor R_s [Ohm]	1.1
PMSG Nom. T_e^n/I_m^n [Nm/A]	8.5/5
Nom. Power P^n [kVA]	2.7
Motor Pole Pairs N_p [1]	3
Rotor Flux ψ_{pm} [Wb]	0.41
Sampling frequency f_s [kHz]	20

fail compared with external switch fail, which in accord with former analysis. Since ε^b and ε^c is near 1, hence fault index flag for these two healthy phases keep to be 0. For two phase fault ($\bar{G}_m^{a1}G_m^{c1}$ fail, see Fig. 7b), ε^a and ε^c decrease after fault. The value of ε^c is higher than ε^a, which indicates different location of faulty switch. After fault, skewness of phase a and c are positive and negative, respectively. This indicate the fail locate in lower branch of phase a and in upper branch of phase c. Through these diagnosis indicator, $\bar{G}_m^{a1}G_m^{c1}$ fail is detected in $30ms$. From the simulation result, the feasibility of proposed fault diagnosis method is proved. This fault diagnosis method can realize accuracy fault diagnosis for both one-phase/ two-phase open-circuit fault in 3L-NPC power converter.

B. Fault tolerant control results

Fig. 8 gives the fault-tolerant result for external switch G_m^{a1} and internal switch G_m^{a2} fail. In both case new DC bus voltage reference $V_d^{*'} = 800V$, which fulfill the voltage limitation Eq. 12, is given to GSC controller. As can be seen, in both case, controlled by GSC controller, capacitor voltage

increase gradually to track the new DC bus reference voltage. In G_m^{a1} fail case, no hardware modification has been made. The controller ride-through fault by making use of remaining switching states. After application of fault-tolerant control, the performance has been recovered. In G_m^{a2} fail case, phase a terminal has been connected to DC-link midpoint via T^a. The motor has achieved almost same performance as normal operation. In this case, due to i_m^a current is flowing through DC-link capacitor, capacitor voltage V_{c1} and V_{c2} is oscillate with equal amplitude and opposite phase. But this fluctuation will not have a great influence on motor performance since DC bus voltage V_d is controlled by GSC controller.

VI. CONCLUSION

This paper has proposed a fault diagnosis method for three-level NPC back-to-back power converters invoking direct model predictive control concept. The proposed scheme is based on current *probability distribution*, thus no extra sensor is needed. Due to the flexible of DMPC, fault-tolerant control can be easily added into the original control algorithm. Simulation data have shown that accurately fault diagnosis for one- and two-phase open-circuit fault diagnosis has been achieved, and non-redundant fault-tolerant control for one-phase fail with capacitor voltage suppression ability has been realized. Future research will focus on fault-tolerant control for two-phase fail in 3L-NPC power converters.

ACKNOWLEDGMENT

This research was made possible by the cooperation with Infineon and National Instruments. It is financially supported by "Qilu Young Scholar" program of Shandong University.

REFERENCES

[1] S. Ceballos, J. Pou, E. Robles, J. Zaragoza, and J. L. Martin, "Performance evaluation of fault-tolerant neutral-point-clamped converters," *IEEE Transactions on Industrial Electronics*, vol. 57, no. 8, pp. 2709–2718, Aug 2010.

[2] A. Nabae, I. Takahashi, and H. Akagi, "A new neutral-point-clamped pwm inverter," *IEEE Transactions on Industry Applications*, vol. IA-17, no. 5, pp. 518–523, Sept 1981.

[3] S. Kouro, M. Malinowski, K. Gopakumar, J. Pou, L. G. Franquelo, B. Wu, J. Rodriguez, M. A. Perez, and J. I. Leon, "Recent advances and industrial applications of multilevel converters," *IEEE Transactions on Industrial Electronics*, vol. 57, no. 8, pp. 2553–2580, Aug 2010.

[4] A. B.-B. ABDELGHANI, H. BENABDELGHANI, F. Richardeau, J. M. Blaquiere, F. Mosser, and I. SLAMA-BELKHODJA, "Versatile three-level fc-npc converter with high fault-tolerance capabilities: switch fault detection and isolation, and safe post-fault operation," *IEEE Transactions on Industrial Electronics*, vol. PP, no. 99, pp. 1–1, 2017.

[5] J. Rodriguez, S. Bernet, P. K. Steimer, and I. E. Lizama, "A survey on neutral-point-clamped inverters," *IEEE Transactions on Industrial Electronics*, vol. 57, no. 7, pp. 2219–2230, July 2010.

[6] Z. Zhang, "On control of grid-tied back-to-back power converters and permanent magnet synchronous generator wind turbine systems," Dissertation, Technische Universität München, 2016.

[7] Z. Zhang, T. Sun, F. Wang, J. Rodriguez, and R. Kennel, "A computationally-efficient quasi-centralized dmpc for back-to-back converter pmsg windturbine systems without dc-link tracking errors," *IEEE Transactions on Industrial Electronics*, vol. PP, no. 99, pp. 1–1, 2016.

[8] S. Yang, A. Bryant, P. Mawby, D. Xiang, L. Ran, and P. Tavner, "An industry-based survey of reliability in power electronic converters," *IEEE Transactions on Industry Applications*, vol. 47, no. 3, pp. 1441–1451, May 2011.

[9] F. W. Fuchs, "Some diagnosis methods for voltage source inverters in variable speed drives with induction machines - a survey," in *Industrial Electronics Society, 2003. IECON '03. The 29th Annual Conference of the IEEE*, vol. 2, Nov 2003, pp. 1378–1385 Vol.2.

The 2018 International Power Electronics Conference

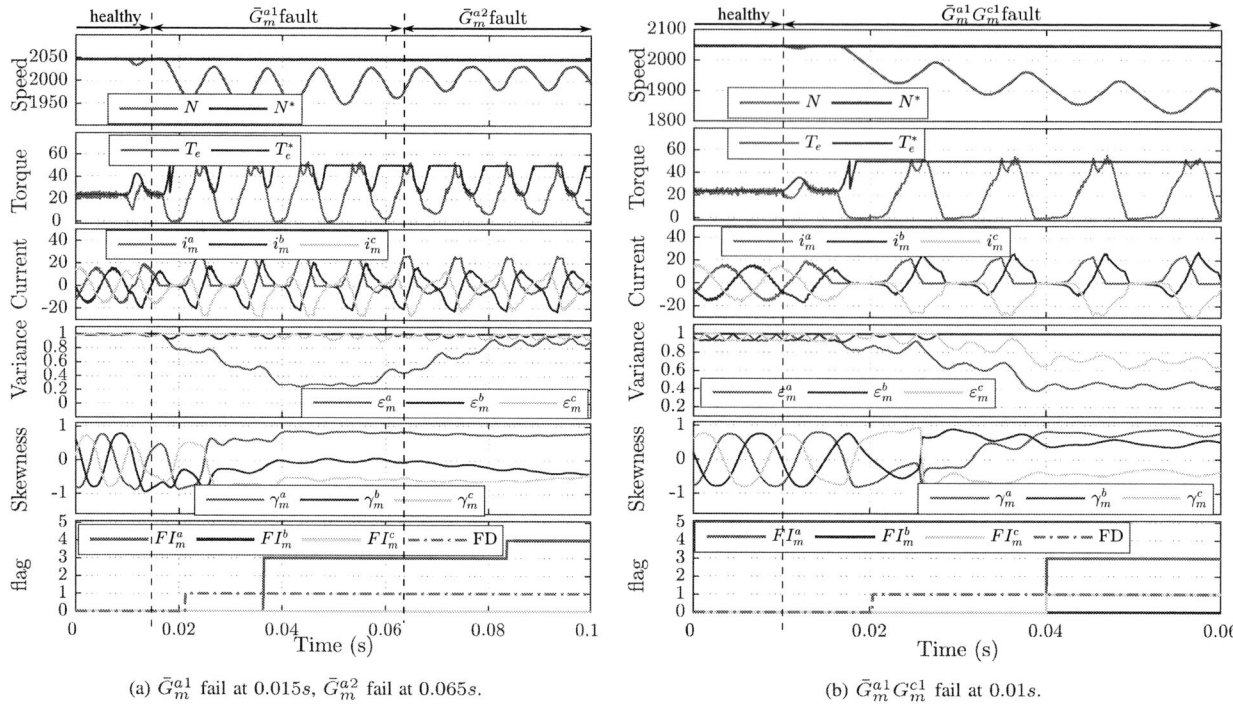

(a) $\bar{G}_m^{a1}$ fail at $0.015s$, $\bar{G}_m^{a2}$ fail at $0.065s$.

(b) $\bar{G}_m^{a1}G_m^{c1}$ fail at $0.01s$.

Fig. 7. [Simulation data:] Fault diagnosis process, for each figure-set, from up to down are: motor speed, electric torque, stator current, relative variance, skewness, fault index, respectively.

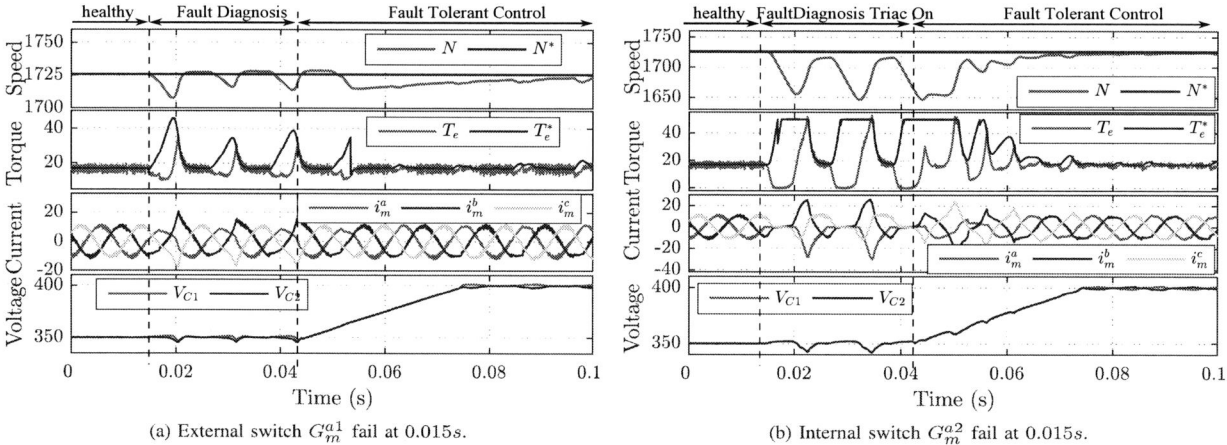

(a) External switch G_m^{a1} fail at $0.015s$.

(b) Internal switch G_m^{a2} fail at $0.015s$.

Fig. 8. [Simulation data:] Fault-tolerant process, for each figure-set, from up to down are: motor speed, electric torque, stator current, capacitor voltage, respectively.

[10] J. Zhang, J. Zhao, D. Zhou, and C. Huang, "High-performance fault diagnosis in pwm voltage-source inverters for vector-controlled induction motor drives," *IEEE Transactions on Power Electronics*, vol. 29, no. 11, pp. 6087–6099, Nov 2014.

[11] B. Lu and S. K. Sharma, "A literature review of igbt fault diagnostic and protection methods for power inverters," *IEEE Transactions on Industry Applications*, vol. 45, no. 5, pp. 1770–1777, Sept 2009.

[12] U. M. Choi, H. G. Jeong, K. B. Lee, and F. Blaabjerg, "Method for detecting an open-switch fault in a grid-connected npc inverter system," *IEEE Transactions on Power Electronics*, vol. 27, no. 6, pp. 2726–2739, June 2012.

[13] W. Zhang, D. Xu, P. N. Enjeti, H. Li, J. T. Hawke, and H. S. Krishnamoorthy, "Survey on fault-tolerant techniques for power electronic converters," *IEEE Transactions on Power Electronics*, vol. 29, no. 12, pp. 6319–6331, Dec 2014.

[14] S. Ceballos, J. Pou, J. Zaragoza, J. L. Martin, E. Robles, I. Gabiola, and P. Ibanez, "Efficient modulation technique for a four-leg fault-tolerant neutral-point-clamped inverter," *IEEE Transactions on Industrial Electronics*, vol. 55, no. 3, pp. 1067–1074, March 2008.

[15] D. Zhou, Y. Li, J. Zhao, F. Wu, and H. Luo, "An embedded closed-loop fault-tolerant control scheme for nonredundant vsi-fed induction motor drives," *IEEE Transactions on Power Electronics*, vol. 32, no. 5, pp. 3731–3740, May 2017.

[16] D. Zhou, J. Zhao, and Y. Liu, "Predictive torque control scheme for three-phase four-switch inverter-fed induction motor drives with dc-link voltages offset suppression," *IEEE Transactions on Power Electronics*, vol. 30, no. 6, pp. 3309–3318, June 2015.

[17] M. Shahbazi, P. Poure, S. Saadate, and M. R. Zolghadri, "Fpga-based reconfigurable control for fault-tolerant back-to-back converter without redundancy," *IEEE Transactions on Industrial Electronics*, vol. 60, no. 8, pp. 3360–3371, Aug 2013.

[18] D. Zhou, J. Zhao, and Y. Liu, "Independent control scheme for nonredundant two-leg fault-tolerant back-to-back converter-fed induction motor drives," *IEEE Transactions on Industrial Electronics*, vol. 63, no. 11, pp. 6790–6800, Nov 2016.

[19] Z. Zhang, C. Hackl, T. Sun, and R. Kennel, "Computationally efficient dmpc for three-level npc back-to-back converters in wind turbine systems with pmsg," *Power Electronics, IEEE Transactions on*, Jan 2017.

Two-Stage Optimization Based Predictive Torque Control with Reduced Complexity for a Three-Level Inverter Driven Induction Motor

Ilham Osman, Dan Xiao and Faz Rahman

School of Electrical Engineering and Telecommunications, University of New South Wales, Sydney, Australia

E-mail: i.osman@unsw.edu.au

Abstract - **This paper proposes a two-stage optimization of voltage vector selection method integrated with reference stator flux vector calculation (RSFVC) for predictive torque control of a three-level neutral point clamped inverter (3L NPC VSI) fed induction motor. The RSFVC technique simplifies the cost function by avoiding the weighting factor tuning between stator flux and electromagnetic torque and the proposed two-stage method reduces the number of voltage vectors in the prediction stage. Hence, with the proposed control algorithm, two major contributions are achieved. Firstly, the computational burden is reduced due to less number of voltage vector candidates in the prediction stage. Secondly, the design for the cost function becomes simplified as the weighting factor of stator flux error, which is essential in the cost function, is eliminated. Moreover, experimental results confirms that, selection of the two remaining weighting factors required for neutral point voltage (λ_{cv}) and number of switching transitions (λ_n) becomes less complex. The dynamic and steady-state performances of the proposed control method is investigated in terms of electromagnetic torque and stator flux, total harmonic distortion of stator currents, neutral point voltage variations and robustness of the drive.**

Keywords— cost function, capacitor voltage balancing, execution time, induction motor, model predictive control three-level NPC inverter

I. INTRODUCTION

Designing a cost function with low complexity is a challenge in finite-state predictive torque control algorithm. While the execution time can be reduced by decreased number of voltage vectors, the computational complexity still exists in the algorithm due to stator current, flux and torque calculation inside the iterative prediction stage for the voltage vectors applied to prediction and optimization.

Research has been carried out till date to decrease the computational burden of FS-PTC algorithms without compromising the performance of the drive and few studies highlighted selecting weighting factors in the cost function [1]. The cost function includes variable which are different in magnitude and in unit and thus designing the cost function with appropriate weighting factors is a challenging task.

Ref. [2] presents the calculation of weighting factor based on the principle of torque ripple reduction. The control period was segmented into two intervals to reduce the torque ripple further. However, the equation to calculate the optimal weighting factor is greatly parameter focused and complicated. Ref. [3] claimed an improved steady state performance compared to conventional model predictive torque control for IM drive, where the duration of an active vector is calculated based on the principle of torque ripple reduction with a fixed weighting factor. The experimental results show poor performance at low speed due to the duration of the calculation of the cascaded algorithm. Ref. [4] proposes two arbitrary vectors selection-based method to improve the performance of [3] for a wide speed range, in the cost of high computational complexity which was reported to be simplified in [5] where the method offers weighting-factor-free algorithm by generating a cost function consisting of tracking error of stator flux vector. However, these methods were proposed for a two-level voltage source inverter (2L-VSI) where the neutral-point voltage balancing is not required, hence the cost function is easy to design.

In comparison with 2L-VSI, the three-level neutral point clamped voltage source inverter (3L-NPC VSI) shows more promising responses for medium and high voltage drive applications due to less voltage-stress across the semiconductor switches and lower harmonic distortion in ac side [6]. Therefore, recently three level inverter fed motor drives have gained wide attention to the researchers [7], [8]. However, with large number of voltage vectors evaluated for prediction and optimization in the predictive torque control algorithm for 3L-NPC fed drive, the computational time increases along with computational complexity of the cost function than that of a two-level converter [7]. Moreover, having of variables more than two into the cost function require more weighting factors. For example, in 3L-NPC inverter fed induction motor drive torque ripple, stator flux ripple, capacitor voltage difference at neutral point and number of switching transitions (for switching frequency reduction) are considered as control variables which require three proper weighting factors; λ_f, λ_{cv} and λ_n for flux, neutral-point voltage and number of switching transitions respectively. Tuning these weighting factors is a tedious and time-consuming job [9], [10].

The 2018 International Power Electronics Conference

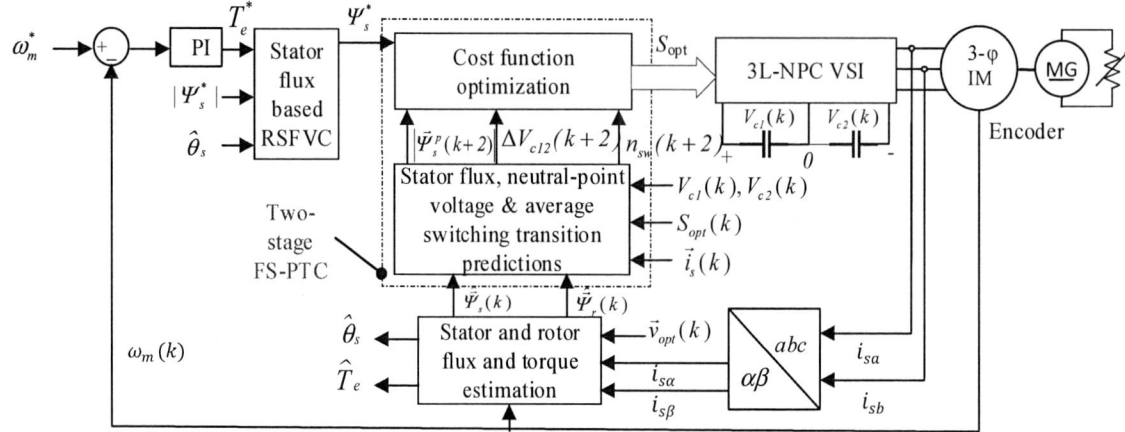

Fig.1. Control diagram of two-stage optimization based FS-PTC with RSFVC technique

Recently, to avoid the weighting factors completely, ref. [8] presented a model predictive flux control algorithm for 3L-NPC VSI fed IM drive. The method transforms an equivalent stator flux vector, which is converted from the magnitude of stator flux and torque [11], into a stator voltage vector reference and then carried-based PWM from space-vector modulation synthesizes the voltage reference which keeps the neutral point voltage in a small range. However, the method will increase the computational complexity for fundamental predictive torque control where discrete switching states are considered. Besides, algorithm proposed in [8] lacks the uniqueness of predictive torque control where voltage vectors are applied throughout the control duration without any modulation-stage.

To avoid the weighting factor tuning of torque and flux error, this paper proposes a two-stage optimization based voltage vector selection method with stator flux based equivalent reference stator flux vector calculation (RSFVC). The RSFVC technique uses torque and flux references to produce a new reference stator flux vector [14]. The torque error is processed out of the prediction loop, through a proportional-integral controller to generate required angular slip frequency which reduces the number of iteration in the prediction stage; thus, complexity of the cost function design. Combined with RSFVC, two-stage optimization based FS-PTC is applied for optimal voltage selection to lessen the number of voltage vectors for prediction and optimization. The performance of the proposed two-stage optimization based FS-PTC with RSFVC technique is verified by experiments results for 3L-VSI driven induction motor.

II. PROPOSED TWO-STAGE OPTIMIZATION BASED FS-PTC WITH RSFVC

The proposed method with equivalent reference stator flux calculation is presented in Fig. 1. The proposed two-stage method has the five following steps: 1) Estimation of stator flux and torque, 2) reference stator flux calculation vector calculation, 3) two-stage optimization based voltage selection, 4) prediction and 5) the cost function minimization.

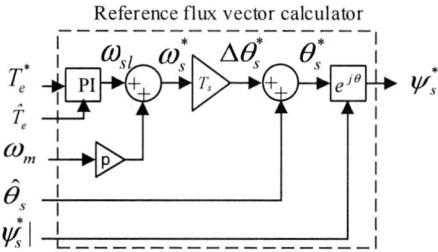

Fig. 2. Reference stator flux vector calculation using stator flux position and magnitude

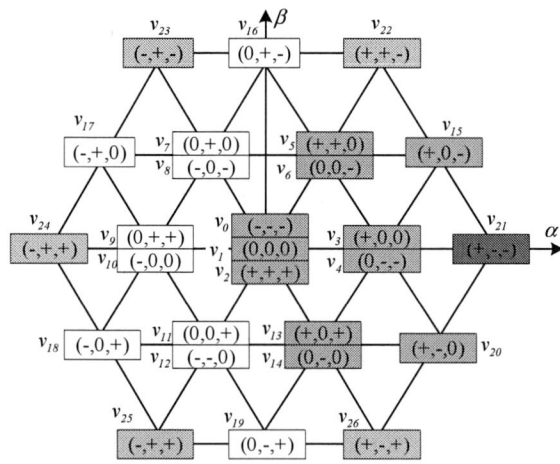

Fig. 3. Voltage vector distribution diagram for a 3L-NPC inverter

A. Stator Flux and Torque estimations

As shown in Equation (1) and (2), stator flux $\vec{\psi}_s(k)$ is estimated from the rotor flux $\vec{\psi}_r$ and $\vec{\psi}_r$ is derived from dynamic model of induction motor.

$$\vec{\psi}_r(k) = \vec{\psi}_r(k-1) + T_s\left[R_r\frac{L_m}{L_r}\vec{i}_s(k) - \left(\frac{R_r}{L_r} - j\omega_e(k)\right)\vec{\psi}_r(k-1)\right] \quad (1)$$

$$\vec{\tilde{\psi}}_s(k) = \frac{L_m}{L_r}\vec{\tilde{\psi}}_r(k) + \sigma L_s \vec{\iota}_s \tag{2}$$

where $\sigma = \left(1 - \frac{L_m^2}{L_s L_r}\right)$ is the total leakage-factor.

The estimated torque is then obtained as

$$\hat{T}_e(k) = 1.5p\Im m\left\{\vec{\tilde{\psi}}_s(k)^* \cdot \vec{\iota}_s(k)\right\} \tag{3}$$

where p is the number of pole pair of the motor.

B. Reference stator flux vector calculation (RSFVC)

Fig.2 illustrates the diagram of reference flux vector generation by RSFVC method. A proportional-integral controller converts the torque error to an equivalent angular slip frequency ω_{sl}. The change in stator flux's position $\hat{\theta}_s$ is calculated using ω_{sl} and rotor speed ω_m. Finally the equivalent flux vector, ψ_s^* is obtained using the magnitude of stator flux $|\psi_s^*|$ and present position of stator flux $\hat{\theta}_s$. The rate of torque increasing is almost proportional to the step difference of the slip ω_{sl}. Following that, a PI controller for torque control is used (Fig. 2) based on the nonlinear relationship between torque and ω_{sl} under a constant $|\psi_s^*|$. [12], [13], [14]. it is equivalent to the first order system. Therefore, it can track the reference torque. An outer speed loop generated from another PI controller generates the reference torque T_e^* which is processed through the speed error (shown in Fig. 1). Therefore, the equivalent reference stator flux vector ψ_s^* is an optimal reference based on T_e^* and $|\psi_s^*|$. Thus, the tuning of weighting factor between torque and flux in cost function can be avoided. The cost function, conventionally which had three weighting factors before for four control variables (torque error, flux error, neutral-point voltage and number of switching transitions) with four different units, can be easily designed with two weighting factors with the proposed method.

The RSFVC was used previously in [14] for reducing the ripple in torque and flux for an IPMSM drive. In this paper, the reference flux calculation was performed with space vector modulation (SVM) for a modified direct torque control (DTC) method. The proposed method presented in this paper employs RSFVC to an MPC driven induction motor drive in which no modulator is used.

C. Two-stage optimization based voltage vector selection

Apart from reducing the complexity of the cost function, this method proposes algorithm to reduce the number of voltage vectors evaluated for prediction and optimization. Fig.3. represents the space distribution of all voltage vectors $\{v_0 \dots v_{26}\}$ produced by the three levels NPC VSI in stationary $\alpha - \beta$ plane.

From the four categories of the voltage vectors which are zero vectors $\{v_0 \dots v_2\}$, small vectors $\{v_3 \dots v_{14}\}$, medium vectors $\{v_{15} \dots v_{20}\}$ and long vectors $\{v_{21} \dots v_{26}\}$, only the long vectors are selected (marked in yellow colour) and used for prediction in the first step.

Fig. 4. Experimental setup for the proposed control method

After obtaining the long optimal vector, all possible small and the medium vectors closest to the long optimal vector (within $\pm 30°$)is selected for prediction in the second stage. For example, as shown in Fig. 2, if v_{21} is evaluated as the long optimal vector in the first stage (indicated in red colour), in the second stage the 11 voltage vectors closest to v_{21} as indicated in the gray colour $\{v_0, v_1, v_2, v_3, v_4, v_5, v_6, v_{13}, v_{14}, v_{15}, v_{20}\}$ are selected for more accurate prediction. After the second stage of prediction is completed, the best switching states' decision is taken by comparing the values from two cost functions from two stages. The voltage vector which results ultimate minimum value of the cost function from both stages is selected as the final optimal voltage vector and according that, the best switching states is applied to the motor terminal through the inverter in the next switching instant.

Applying this algorithm for predictive control, the number of voltage vectors is reduced from 27 vectors to 17 vectors, including 6 long vectors in the first stage and 11 medium and small and zero vectors in the second stage. In the proposed method all three zero vectors are included in the prediction stage to achieve low switching frequency.

D. Prediction of Control Objectives

In the beginning, the control variables which are stator flux, neutral-point voltage and number of switching transitions, are predicted for the 17 voltage vectors in two stages in term of 6 long voltage vectors and then in the second stage 11 medium, short and zero voltage vectors. The prediction of stator flux is expressed in discrete time steps as,

$$\vec{\psi}_s(k+1) = \vec{\psi}_s(k) + T_s\vec{v}_s(k) - T_s R_s\vec{\iota}_s(k) \tag{4}$$

The prediction of stator current can be expressed as

$$\vec{\iota}_s^{\,p}(k+1) = \left(1 - \frac{T_s}{\tau_\sigma}\right)\vec{\iota}_s(k) + \frac{T_s}{(\tau_\sigma + T_s)} \times$$

$$\left\{\frac{1}{R_\sigma}\left[\left(\frac{k_r}{\tau_r} - k_r j w_e(k)\right)\vec{\psi}_r(k) + \vec{v}_s(k)\right]\right\} \tag{5}$$

where $k_r = \frac{L_m}{L_r}$, $R_\sigma = R_s + k_r{}^2 R_r$, $\tau_\sigma = \frac{\sigma L_s}{R_\sigma}$ and $\tau_r = \frac{L_r}{R_r}$ is the rotor time-constant. The rotor flux and rotor speed change very slowly, so that over a sampling period, $\vec{\psi}_r(k+1) = \vec{\psi}_r(k)$ and $\omega(k+1) = \omega(k)$.

The neutral point voltage is predicted as $\Delta V_{c12}(k+1)$ at time instant $(k+1)$ as follows,

$$\Delta V_{c12}(k+1) = \Delta V_{c12}(k) - \frac{T_s}{C}\Delta i_{c12}(k+1) \tag{6}$$

where $\Delta V_{c12}(k) = V_{c1}(k) - V_{c2}(k)$ and $\Delta i_{c12}(k+1) = i_{c1}(k+1) - i_{c2}(k+1)$ is the predicted current. The currents $i_{c1}(k+1)$ and $i_{c2}(k+1)$ are calculated from the upper two switching states and measured phase currents of each phase ($x = \{a,b,c\}$) of the inverter. The equations for both of the currents are as follow,

$$i_{c1}(k+1) = i_{dc} - S_{a1}(k+1)i_a(k+1) -$$
$$S_{b1}(k+1)i_b(k+1) - S_{c1}(k+1)i_c(k+1) \tag{7}$$

$$i_{c2}(k+1) = i_{dc} + \big(1 - S_{a2}(k+1)\big)i_a(k+1) +$$
$$\big(1 - S_{b2}(k+1)\big)i_b(k+1)\big(1 - S_{c2}(k+1)\big)i_c(k+1) \tag{8}$$

Equation (7) and (8) show that, the common term i_{dc} is not required for $\Delta i_{c12}(k+1)$ prediction calculation as it requires only the difference between two currents.

For reducing average switching frequency, the number of switching transitions n_{sw} is taken as a variable into the cost function and the calculation is given by (9).

$$n_{sw} = \sum_{x=\{a,b,c\}} \big| S_{x1}(k+1) - S_{x1}(k) \big| + \big| S_{x2}(k+1) - S_{x2}(k) \big| \tag{9}$$

where $S_{x1,2}(k+1)$ are the predicted switching states in the future time instant $(k+1)$. $S_{x1,2}(k) = S_{opt}(k)$ is the applied switching state at time instant k.

E. Cost function minimization

For cost function minimization, at first the control variables are evaluated by a predefined cost function. The prediction loop evaluates all the control variables for each of the 17 vectors from two cascaded stages (6+11), to get a minimum value of the cost function. The control variables are compared with the available reference values and absolute errors are used for calculation. For each switching period, the voltage vector which results the minimum value of cost function is selected as the optimal vector ($\vec{v}_{opt}$), and the switching states S_{opt} which represents that voltage vector are directly applied to motor terminals via converter. The delay compensation scheme is employed to the algorithm to avoid one step delay caused by digital processor [14] for which the cost function is implemented at one step forward. In the proposed method, calculated new reference flux vector ψ_s^* is used in the cost function as the reference value and $\psi_s^P(k+1)$ is predicted inside the cost function. Torque is controlled outside of the cost function and therefore the weighting factor tuning between torque and flux is avoided here. Taking into account all the control

variables, Equation (10) expresses the cost function of the proposed algorithm.

$$g = \left\| \psi_{s\alpha}^* - \psi_{s\alpha}^P(k+2) \right\| + \left\| \psi_{s\beta}^* - \psi_{s\beta}^P(k+2) \right\|$$
$$+ \lambda_{cv}\left| \Delta V_{c12}(k+2) \right| + \lambda_n n_{sw}(k+2) \tag{10}$$

where λ_{cv} and λ_n are the two weighting factors for neutral point voltage $\Delta V_{c12}(k+1)$ and number of switching transitions n_{sw}. ψ_s^* and ψ_s^P (k+2) both are separated into their $\alpha\beta$ components.

The process of tuning λ_{cv} and λ_n is well defined. initially, λ_{cv} is tuned to get ΔV_{c12} around zero. During this tuning, λ_n is kept at zero. After that, with λ_{cv} at a tuned value, λ_n is tuned online until ΔV_{c12} starts to increase.

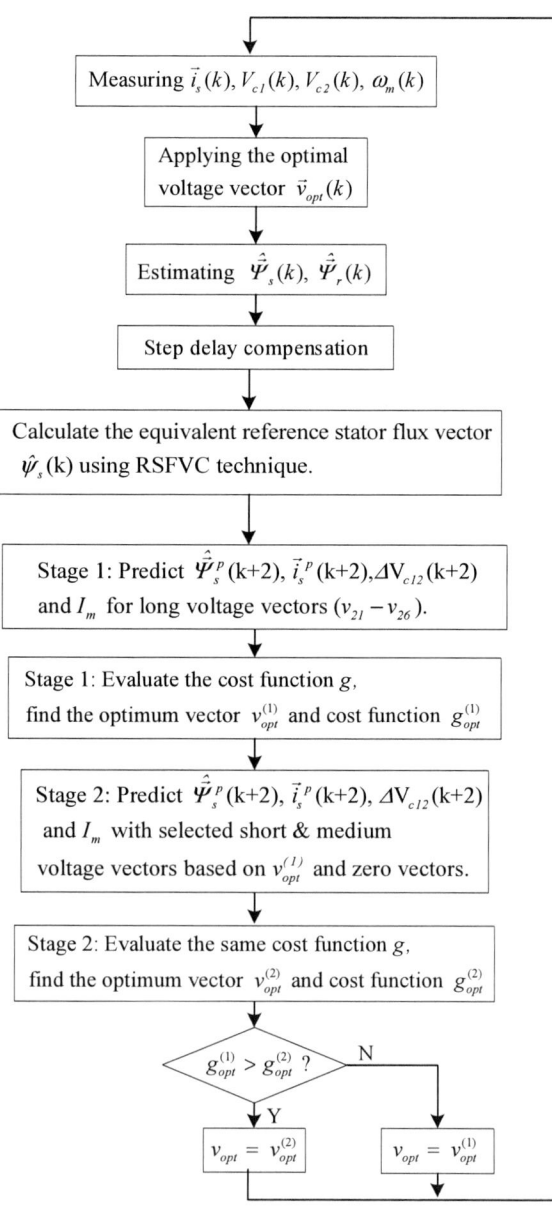

Fig. 5. The flow-chart for the proposed control method

The summery of the proposed control procedure is presented by the flow-chart presented in Fig. 5.

III. EXPERIMENTAL RESULTS

Fig.4. presents the experimental setup for proposed method with RSFVC technique. The setup consists of a squirrel cage induction motor fed by a 3L-NPC converter. A permanent magnet DC motor is used as a load integrated to IM's shaft. A rheostat is connected in series with the DC machine's armature for running the machine at different loads. The specification of the setup is presented in Table I.

The sampling time is set as $70\mu s$. The execution time for proposed method is 29.57 μs. From the execution time measurement, it shows that the maximum part of the execution time is spent for prediction and optimization which is $19\mu s$ - 64.25% of the total average execution time. The proposed control algorithm is implemented using dSPACE DS1104 R&D controller board with Control-Desk. The execution time is measured on a 250MHz PowerPC using DS1104. The breakdown of the execution time is given in Table II.

TABLE I
SPECIFICATIONS OF EXPERIMENTAL SETUP

415V, 3-Ø, 50Hz IM		Controller's Parameters	DC machine Ratings
R_s=6.03 Ω,	$\vec{\psi}_{nom}$=1.08 Wb	k_{p_spd}=0.3	P = 1.1kW
R_r=6.085 Ω,	T_{nom} = 7.4 Nm	k_{i_spd}=3.0	V = 180V
L_s=0.5192 H,	$N_p = 2$	λ_{cv}=0.0001	i_a = 6.9A
L_r=0.5192 H,	J=0.011787 kg.m^2	$\lambda_n = 10^{-6}$	ω_{dc} =1800rpm
L_m=0.4893 H,	ω_{rated}= 1415rpm	I_{max} =4.5 A	

TABLE II
COMPARISON BETWEEN THE EXECUTION TIMES OF CONVENTIONAL FS-PTC AND THE PROPOSED TWO-STAGE OPTIMIZATION BASED FS-PTC

Index	Average Execution Times (μs)
Measurements	4.07
Switching	0.12
Voltage & Current Calculations	0.4
Estimation	1.16
Reference Stator Flux Calculation	4.3
Predictions and Optimization	**19.0**
Switching Frequency Calculation	0.52
Total	**29.57**

The effectiveness of the proposed control method is analyzed for the followings:

a) Steady-state performances for medium and low speed.
b) Torque-transient response.
c) Speed reversal capability.
d) Weighting factor λ_{cv} sensitivity of the controller.

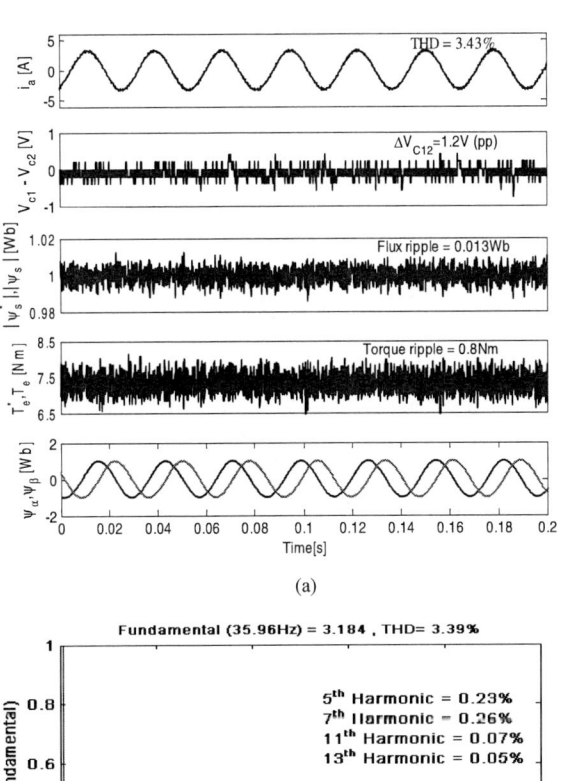

(a)

(b)

Fig. 6. Steady state performance investigation for low speed at 1000rpm with rated load (a) Stator current, neutral-point voltage difference, estimated stator flux and Torque and reference stator flux as $\alpha\beta$ components (b) Frequency Spectrum of stator current i_a.

A. Steady-state Performances under different loads

The steady state performance of the proposed FS-PTC system is carried out at 1000 r/min speed under full load (7.4Nm) condition. Fig.6(a) shows the responses of stator current i_a, capacitor voltages' difference at neutral point, estimated stator flux, electromagnetic torque and reference stator flux in terms of $\alpha\beta$ components. Fig. 6(b) presents the frequency spectrum of the stator current i_a. Fig.6(a) shows that the neutral-point voltages of two capacitors (V_{C1} & V_{C2}) is balanced and within 1.2V (p-p). The flux ripple and torque ripple are 0.013Wb and 0.8Nm respectively. The $\alpha\beta$ components of the calculated reference flux are well balanced as shown in Fig.6(a).

The THD of the stator current presented in the Fig 6. (b) is 3.39%. It is calculated using MATLAB powergui with 15 cycles up to 5 kHz. The frequency spectrum of i_a is distributed and mainly concentrated around 2.2-2.4 kHz. The contribution of low frequency harmonics to the

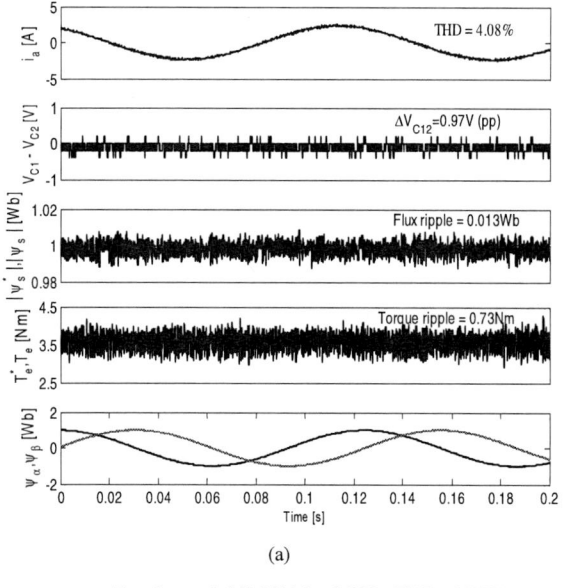

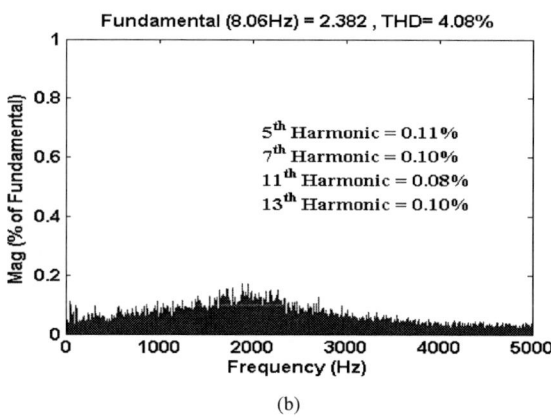

(a)

Fundamental (8.06Hz) = 2.382 , THD= 4.08%

5th Harmonic = 0.11%
7th Harmonic = 0.10%
11th Harmonic = 0.08%
13th Harmonic = 0.10%

(b)

Fig. 7. Steady state performance investigation for low speed at 200rpm with 50% rated load (a) Stator current, neutral point voltage difference, estimated stator flux and Torque and reference stator flux as $\alpha\beta$ components (b) Frequency Spectrum of stator current i_a.

THD is also presented in Fig 6(b). it can be noted that the deadtime used in the switching circuit is causing 5th and the 7th harmonics shown in the list.

The steady-state performance of the proposed FS-PTC for low speed is tested at 200r/min with 50% of rated load and Fig.7(a) represents the results. The neutral point voltage ΔV_{C12} is only 0.97V which is 0.167% of the total dc link voltage used in the motor drive. The stator flux ripple is 0.013Wb and the torque ripple is 0.73Nm. The frequency distribution of i_a for low speed operation is presented in Fig.7(b). The low order harmonics contribute to the THD is very small. As shown in the Fig. 7(b), the amplitudes of the harmonics are lower than 0.25% of the fundamental frequency.

B. Torque-transient characteristics for step-rated torque

Fig.8 presents the step-rated torque transient responses of the proposed method. The motor started with 100r/min at first, and then a sudden 1000r/min speed command is

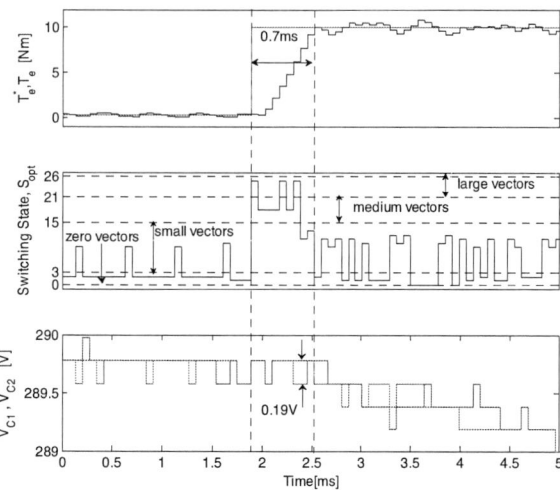

Fig. 8. Step rated-torque response for 100r/min to 1000r/min step change in speed change and inverter's switching selection and dc-Link voltages according to that for the proposed FS-PTC.

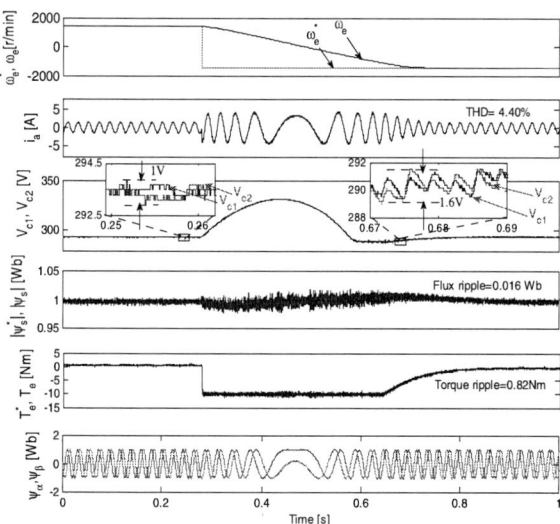

Fig. 9. Experimental waveforms of speed, stator current, dc-link voltages, torque, stator flux and reference stator flux as $\alpha\beta$ components at no-load torque at rated-speed (1415 r/min) reversal condition for the proposed FS-PTC.

given to the controller to accomplish a step-torque command. As can be seen from Fig.8, the controller selects only large and medium vectors in the transient period- v_{25} & v_{17}. It is also visible from the results that the controller achieves a fast-dynamic response. As shown in Fig.8, the torque rise time is 0.7ms. During both the torque-transient and steady state conditions, the neutral point voltage is well-balanced-particularly in transient it is within 0.19 V which is only 0.03% of the total dc-link voltage.

3976

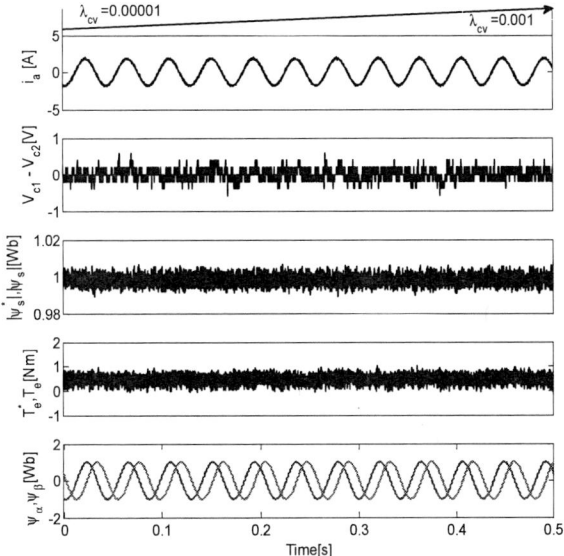

Fig. 10. steady-state sensitivity analysis of the proposed method for a change of weighting factor λ_{cv} of neutral-point voltage ΔV_{C12} in the cost function at 700 r/min.

C. Speed-Transient capability at rated-speed

A speed-reversal operation is performed at rated speed (1415 r/min) without any load torque to investigate the speed-transient response of the proposed controller. Fig. 9 demonstrates the output of this experiment. As presented by Fig. 9, the torque and the flux ripple are 0.82Nm and 0.016Wb respectively. During the transient, the ripple of stator flux increases a little due to the higher current flow in the stator winding. The stator current produces sinusoidal waveform during transient with a slightly increased THD of 4.40%. It is clear from the figure that, both of the dc link capacitor voltages are well balanced during transient and steady state conditions. The voltage fluctuation, as shown in a small- scale presentation, is negligible; the highest fluctuation is 1.6V which is 0.27% of the total dc-link voltage. Hence, it has trivial effect on the performance of the controller.

D. Weighting factor λ_{cv} sensitivity of the controller

The weighting factor designing for the proposed method became less complicated as the cost function now has only two of these- λ_{cv} and λ_n. Besides, from the experimental results, it is observed that, after a certain value of weighting factor λ_{cv} the controller shows unaffected results with any change in λ_{cv}. To prove this, a steady-state experiment is carried out with 700r/min speed at no-load torque condition and the weight of λ_{cv} is increased progressively from 0.00001 to 0.001. As it can be seen from Fig. 10. the stator current, neutral point voltage, stator flux ripple, torque ripple and reference flux vector is unaffected with the change of the λ_{cv}. Which means, the controller performs similar after a certain value of λ_{cv}. This insensitivity of the controller

simplifies the selection of the remaining weighting factor λ_n in the cost function.

IV. CONCLUSIONS

This paper proposes a two-stage optimization based FS-PTC strategy with RSFVC technique for 3L-NPC VSI driven IM motor. The effectiveness of the proposed algorithm is validated with experimental results for both steady-state and dynamic performances. The major contributions of this work include:
1) The two-stage optimization in the prediction stage reduces the number of voltage vectors from 27 to 17 which reduce the number of calculations in the prediction loop accordingly.
2) The nontrivial tuning work between torque and the flux is avoided, which reduces the complexity of designing a cost function, by using the RSFVC technique. The two other weighting factors in the cost function for neutral point voltage and number of switching transitions can now be designed more easily.
3) It has been confirmed via experimental results that λ_{cv} does not affect the controller performance after a certain value of λ_{cv}, which facilitates the tuning of the remaining weighting factor λ_n in the cost function.

REFERENCES

[1] J. Rodriguez et al., "State of the art of finite control set model predictive control in power electronics," *IEEE Trans. on Ind. Informatics*, vol. 9, no. 2, pp. 1003-1016, May 2013.

[2] S. A. Davari, D. A. Khaburi, R. Kennel, "An improved FCS–MPC algorithm for an induction motor with an imposed optimized weighting factor," *IEEE Trans. Power Electron.*, vol. 27, no. 3, pp. 1540-1551, Mar. 2012.

[3] Y. Zhang, H. Yang, "Torque ripple reduction of model predictive torque control of induction motor drives," *Proc. IEEE Energy Convers. Congr. Expo.*, pp. 1176-1183, Sept. 2013.

[4] Y. Zhang and H. Yang, "Model predictive torque control of induction motor drives with optimal duty cycle control," *IEEE Trans. on Power Electron.*, vol. 29, no. 12, pp. 6593-6603, Dec. 2014.

[5] Y. Zhang and H. Yang, "Two-vector-based model predictive torque control without weighting factors for induction motor drives," *IEEE Trans. on Power Electron.*, vol. 31, no. 2, pp. 1381-1390, Feb. 2016.

[6] Y. Zhang, J. Zhu, Z. Zhao, W. Xu, and D. G. Dorrell, "An improved direct torque control for three-level inverter-fed induction motor sensorless drive," *IEEE Trans. Power Electron.*, vol. 27, no. 3, pp. 1502–1513, Mar. 2012.

[7] M. Habibullah, D. D. C. Lu, D. Xiao, and M. F. Rahman, "Finite-state predictive torque control of induction motor supplied from a three-level NPC voltage source inverter," *IEEE Trans. Power Electron.*, vol. 32, no. 1, pp. 479–489, Jan. 2017.

[8] Y. Zhang and Y. Bai, "Model predictive flux control of three-level inverter-fed induction motor drives based on space vector modulation," in *IEEE IFEEC 2017*, pp. 986-991, Jun., 2017.

[9] Cortes, P., Kouro, S., La Rocca, B., et al., "Guidelines for weighting factors design in model predictive control of

power converters and drives," *Proc. of ICIT 2009*, pp. 1–7, Feb. 2009.

[10] Zanchetta, P., "Heuristic multi-objective optimization for cost function weights selection in finite states model predictive control," in *Workshop on Predictive Control of Electrical Drives and Power Electron.*, pp. 70–75, Oct. 2011.

[11] Y. Zhang and H. Yang, "Two-vector-based model predictive torque control without weighting factors for induction motor drives," *IEEE Trans. Power Electron.*, vol. 31, no. 2, pp. 1381–1390, Feb. 2016.

[12] Takahashi, I., Noguchi, T., "A new quick-response and high-efficiency control strategy of an induction motor," *IEEE Trans. Ind. Applicant.*, vol. IA-22, no. 5, pp. 820–827, Sept. 1986.

[13] Zhang, J., Rahman, M.F., "A direct-flux-vector-controlled induction generator with space-vector modulation for integrated starter alternator," *IEEE Trans. Ind. Electron.*, vol. 54, no. 5, pp. 2512–2520, Oct. 2007.

[14] L. Tang, L. Zhong, M. F. Rahman, and Y. Hu, "A novel direct torque controlled interior permanent magnet synchronous machine drive with low ripple in flux and torque and fixed switching frequency," *IEEE Trans. on Power Electron.*, vol. 19, no. 2, pp. 346–354, Mar. 2004.

[15] Cortes, P., Rodriguez, J., Silva, C., et al. "Delay compensation in model predictive current control of a three-phase inverter," *IEEE Trans. Ind. Electron.*, vol. 59, no. 2, pp. 1323–1325, Feb. 2012.

Design Challenges of SiC Devices for Low- and Medium-Voltage DC-DC Converters

Georges Engelmann, Alexander Sewergin, Markus Neubert, Rik W. De Doncker
Institute for Power Electronics and Electrical Drives
RWTH Aachen University
Aachen, Germany
Email: post@isea.rwth-aachen.de

Abstract—Silicon carbide (SiC) devices are considered as key enablers for the development of highly efficient and compact dc-dc converters for low- and medium-voltage applications. Besides their high temperature capability and low conduction losses, they provide superior switching characteristics.

This paper emphasizes design challenges of SiC devices in the low- and medium-voltage range arising from their fast switching speeds. First, detailed measurement results on the switching characteristics of 1200 V SiC devices and different leakage inductances are presented. The results are assessed with regard to the switching losses as well as the transient voltage and current overshoots. The impact on the switching behavior as function of leakage inductances is shown. The leakage inductances also influence the resonance frequency of the power module and dc-dc converters. This is a crucial design aspect to determine the size of the EMI filters. Its significance is demonstrated based on an 800 V dc-dc converter using commercially available SiC MOSFETs. Furthermore, zero-voltage switching is emphasized to reduce the impact of the module parasitic elements on the switching behavior. The performance of 10 kV SiC MOSFETs in a medium-voltage dc-dc converter shows, however, that a significant amount of commutation energy is required to ensure a successful soft-switching transition.

I. Introduction

Power electronic converters are being used throughout all low- and medium-voltage applications. They are the key technology for renewable energies, electrical traction systems and future dc grids [1]–[6]. To develop highly efficient and compact dc-dc converters for low- and medium-voltage applications, the advantages of wide-bandgap (WBG) devices are particularly crucial [7], [8]. It is generally assumed, that WBG power semiconductors enable a volume and cost reduction of converters due to higher switching frequencies and thus smaller filter elements [9]–[11], which in turn facilitate higher levels of integration [12], [13]. Higher switching frequencies and steep switching slopes, however, are the source for electromagnetic interference (EMI) at high frequency bands [14].

This paper emphasizes design challenges of silicon carbide (SiC) semiconductors in the low and medium-voltage range arising from their fast switching speeds. Instead of the already well known issues about the required high common mode rejection ratio (CMRR) of the controller, the gate circuitry and the measurement units [15]–[17], this contribution focuses on the design challenges arising from the filter design. It is shown, that the leakage inductances of SiC power modules, together with their metal-oxide semiconductor field-effect transistor

(MOSFET) output capacitance, create a resonant tank, which is the source for disturbance in the EMI spectrum of converters [18]. Due to an increased excitation of the resonant frequencies, high-frequency disturbances are appearing in the frequency spectrum. This is a crucial design aspect to determine the size of the EMI filters. The analysis is based on detailed investigations on the leakage inductance of the switching cell and the switching characteristics of 1200 V SiC devices. The results are assessed with regard to the resonant frequency as well as the transient voltage and current overshoots. The influence on the switching behavior is shown as a function of the leakage inductance. Furthermore, the impact on the filter design is demonstrated based on an 800 V dc-dc converter using a commercially available SiC MOSFET power module.

The stray inductance of SiC medium-voltage dc-dc converters cannot always be reduced in the same manner as for low-voltage applications due to higher clearance requirements. Therefore, soft-switching converter topologies and operation strategies are envisaged to reduce switching losses and filter size. Achieving zero-voltage switching in medium-voltage SiC converters, however, requires a careful converter design as well as a precise control of the voltage and current waveforms to provide sufficient commutation energy during the switching transitions.

II. Influence of Stray Inductance on the Switching Behavior

The stray inductances L_σ^{module} of the power module in combination with the output capacitance C_{oss} of the SiC MOSFETs form a resonant circuit. Using an exemplary low-voltage power module (*CCS050M12CM2*, [19]) which employs 1200 V MOSFETs (*CPM2-1200-0025B*), the resonant frequency f_{res} is calculated to

$$f_{\text{res}} = \frac{1}{2\pi\sqrt{L_\sigma^{\text{module}}C_{\text{oss}}}} = \frac{1}{2\pi\sqrt{30\,\text{nH}\cdot 220\,\text{pF}}} = 62\,\text{MHz}.$$

The output capacitance of $C_{\text{oss}} = 220\,\text{pF}$ is valid for dc-link voltages $U_{\text{dc}} > 600\,\text{V}$ [20]. However, the resonant frequency is in the same order of magnitude as the EMI spectrum caused by the voltage slopes during the switching transients of SiC MOSFETs with a rise time of $t_{\text{r}} = 10\,\text{ns}$, which is

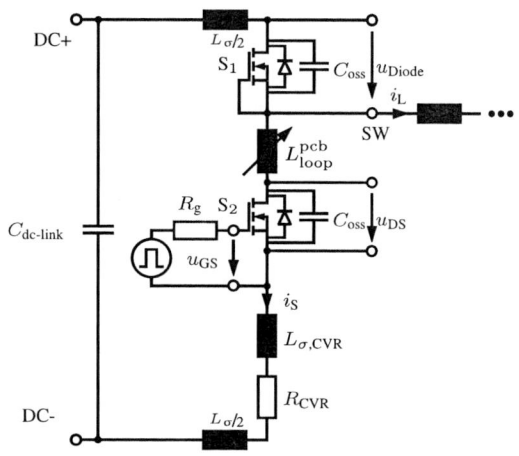

Fig. 1. Schematic with adjustable inductance $L_{\text{loop}}^{\text{pcb}}$.

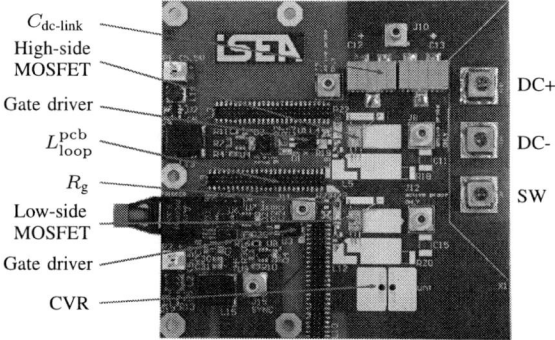

Fig. 2. Demonstrator PCB, top view.

approximated by the cut-off frequency f_{g} of a first order low-pass filter:

$$f_{\text{g}} = \frac{0.35}{t_{\text{r}}} = 35\,\text{MHz}.$$

These voltage slopes are at least one order of magnitude higher compared to the switching slopes of insulated-gate bipolar transistors (IGBTs) [21]. In the following, the influence of the stray inductance on the switching transients is investigated experimentally using a half-bridge setup with a variable stray inductance.

A. Hardware Setup

A switching cell with an adjustable power-loop stray inductance $L_{\text{loop}}^{\text{pcb}}$ is developed according to the schematic shown in Fig. 1. The switching transients of the low-side switch S_2 are recorded using the double pulse test bench as published in [22], [23]. The employed power MOSFETs are 1200 V devices from Cree (*C2M0025120D*). Furthermore, a careful setup of the measurement probes is chosen to respect the guidelines given in [24].

The variable inductance $L_{\text{loop}}^{\text{pcb}}$ is realized using different jumper positions of a pin header as shown in the photograph

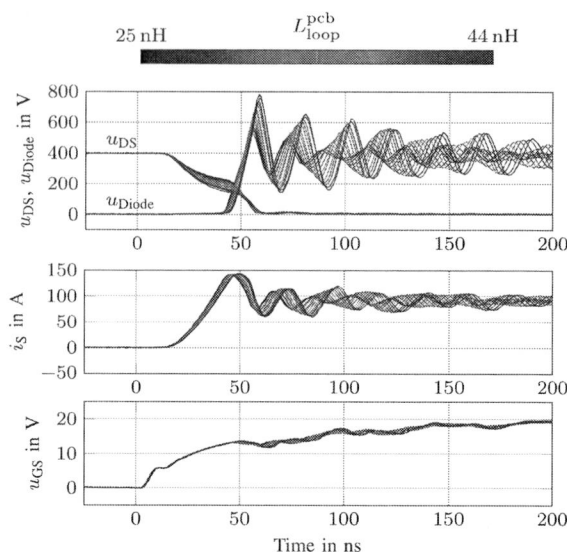

Fig. 3. Turn-on transient waveforms at $U_{\text{dc}} = 400\,\text{V}$, $I_{\text{S}} = 100\,\text{A}$.

of the printed circuit board (PCB) in Fig. 2. The adjustable range of $L_{\text{loop}}^{\text{pcb}}$ is $1\,\text{nH} \ldots 20\,\text{nH}$. Due to the included current viewing resistor (CVR) (*SDN-414-10*) with a contacting inductance of $L_{\sigma,\text{CVR}} = 7\,\text{nH}$ and the required modification of the PCB layout, a minimum power loop inductance of $L_{\text{loop}}^{\text{pcb}} + L_{\sigma} + L_{\sigma,\text{CVR}} = 25\,\text{nH}$ is reached. The parasitic inductances have been measured using a current pulse generator [25], [26].

B. Measurement Results

First, the impact of the power loop inductance on the switching behavior is investigated. The turn-on event waveforms of the low-side switch S_2 and the high-side diode are plotted in Fig. 3 at a dc-link voltage of $U_{\text{dc}} = 400\,\text{V}$ and a load current of $I_{\text{S}} = 100\,\text{A}$. A high voltage overshoot ($\approx 100\,\%$) and a significant ringing is observed in the high-side switch (i.e., free-wheeling diode) voltage u_{Diode}. A similar ringing is observed in the low-side source current i_{S} with a significant overshoot ($> 200\,\%$). Analogously, the transient turn-off waveforms are plotted in Fig. 4. A high voltage overshoot ($\approx 50\,\%$) is observed across the low-side switch voltage u_{DS}, while the source current i_{S} and the switch voltage u_{DS} show an equal ringing.

Secondly, the oscillation frequencies $f_{\text{res}}^{\text{on}}$ and $f_{\text{res}}^{\text{off}}$ during the switching events are extracted from all the conducted measurements. Their dependency on the loop inductance $L_{\text{loop}}^{\text{pcb}}$ in function of the switched current I_{S} is shown in Fig. 5 and in function of the dc-link voltage U_{dc} in Fig. 6. Figure 5 shows, that the resonant frequencies are independent of the load current I_{S} of the switching cell. However, Fig. 6 clearly shows the dependency of the resonance frequency on the dc-link voltage U_{dc}. This effect results from the voltage dependent output capacitance C_{oss} of the MOSFET [18].

The 2018 International Power Electronics Conference

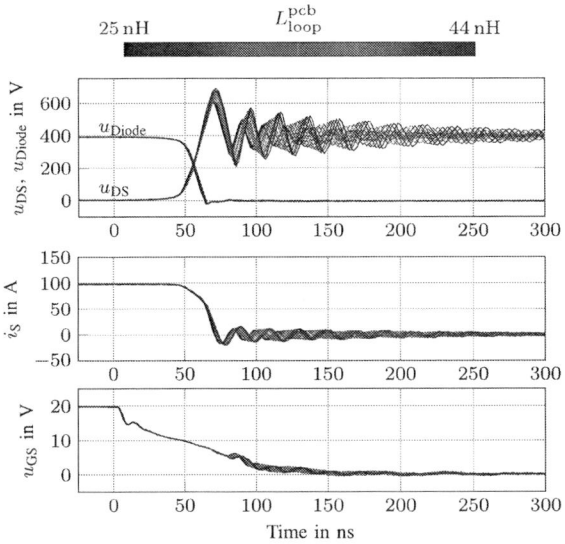

Fig. 4. Turn-off transient waveforms at $U_{dc} = 400\,\mathrm{V}$, $I_S = 100\,\mathrm{A}$.

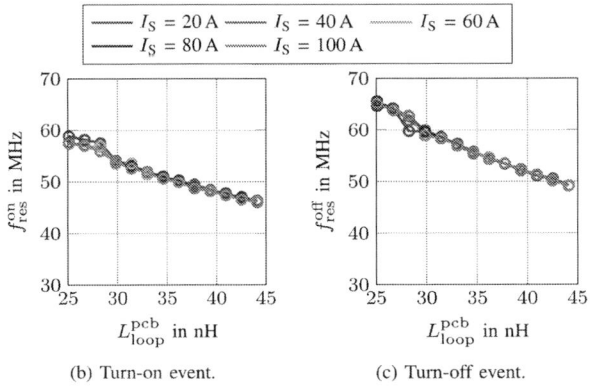

(b) Turn-on event.　　　　(c) Turn-off event.

Fig. 5. Resonant frequency f_{res} vs. loop inductance L_{loop}^{pcb} at $U_{dc} = 400\,\mathrm{V}$ and $R_g = 5.6\,\Omega$.

III. FILTER DESIGN

In the following, the impact of the SiC module's resonant frequency on the EMI characteristic and the filter design of a low-voltage dc-dc converter is analyzed. The analysis is carried out for a bidirectional interleaved multi-phase synchronous boost converter, as pictured in Fig. 7, with a switching frequency of $f_{sw} = 150\,\mathrm{kHz}$, a nominal output power of $P_{n,out} = 42\,\mathrm{kW}$ and nominal input and output voltages of $U_{in} = 400\,\mathrm{V}$ and $U_{out} = 800\,\mathrm{V}$. The converter utilizes a commercially available SiC module (*CCS050M12CM2* from Wolfspeed [19]), which is based on the EconoPack 2 package, which was originally developed for silicon (Si) IGBT semiconductors. Therefore, the module's loop inductance of $L_{loop}^{pcb} = 30\,\mathrm{nH}$ is not optimized for SiC applications [19]. The high $\mathrm{d}v/\mathrm{d}t$ voltage slope at the switching node SW, which is essential for an efficient hard-switched converter, excites the resonant circuit consisting of the output capacitance C_{oss} and

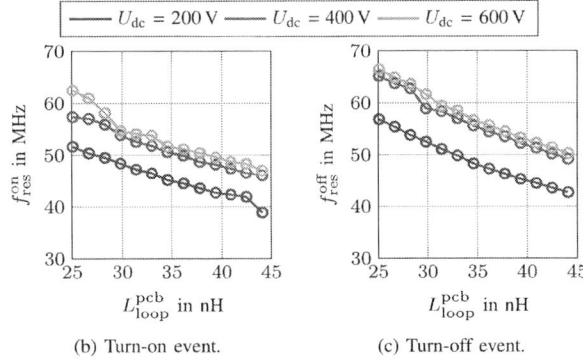

(b) Turn-on event.　　　　(c) Turn-off event.

Fig. 6. Resonant frequency f_{res} vs. loop inductance L_{loop}^{pcb} at $I_S = 100\,\mathrm{A}$ and $R_g = 5.6\,\Omega$.

Fig. 7. Three-phase bidirectional interleaved synchronous boost converter prototype (with Y-capacitors, without common-mode chokes). [8]

the stray inductance L_{loop}^{pcb}. In contrast to Si-based converters, where the module resonance is not exited by the slow switching slopes and, therefore, typically only one EMI-filter stage is sufficient, the SiC-based hard-switching converter needs a second wide-band high-frequency EMI filter, which attenuates the module's resonant frequency. Figure 8 shows the conducted emission of the dc-dc converter at 2 kW using different filter elements. The green curve depicts measurement results using Y-capacitors with $C_y = 330\,\mathrm{nF}$ and the commercially available common-mode choke "CM_1" which is shown in Fig. 9a.

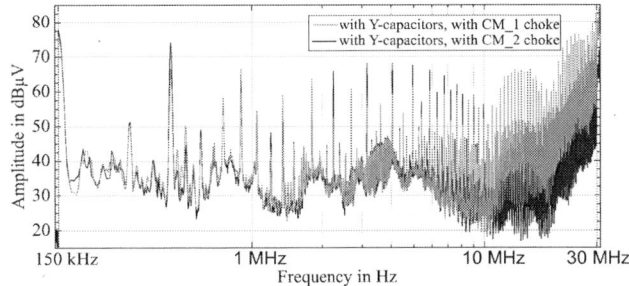

Fig. 8. Conducted noise emissions plot ($U_{in} = 400\,\mathrm{V}$, $U_{out} = 800\,\mathrm{V}$ and $P_{n,out} = 2\,\mathrm{kW}$ load, measured at DC- terminal).

3981

The 2018 International Power Electronics Conference

(a) Commercial common-mode choke "CM_1" for attenuation in the sub-5 MHz region.

(b) Prototype common-mode choke "CM_2" with peak-attenuation at 30 MHz.

Fig. 9. Image of common-mode chokes with different absorbing frequency ranges.

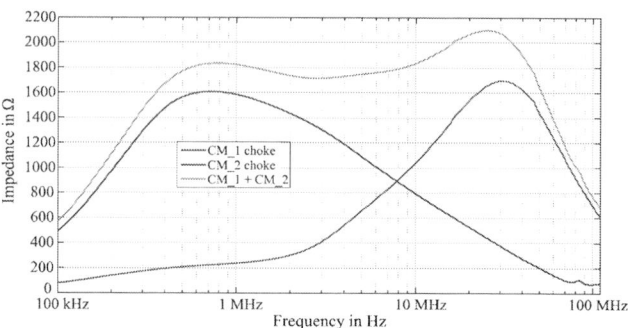

Fig. 10. Impedance characteristics of the common-mode chokes "CM_1" and the "CM_2" (measuring equipment: Agilent 4294A).

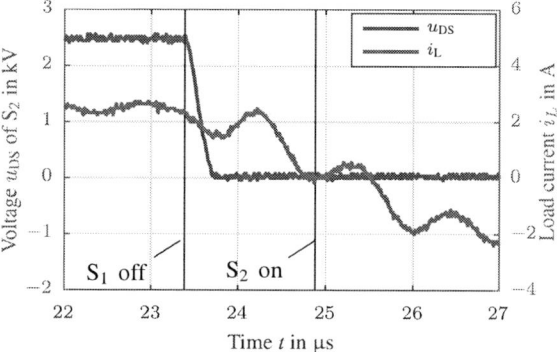

Fig. 11. Zero-voltage turn-on transition of a 10 kV SiC MOSFET operated in a 3ph-TAB converter.

The common mode choke "CM_1" (Kaschke 048.096) provides a distinctive attenuation in the sub-5 MHz region. To further attenuate the module resonance at 30 MHz, a second common-mode filter with a different frequency range has to be applied. Therefore, a prototype common-mode choke ("CM_2", see Fig. 9b) with a peak-attenuation at 30 MHz and a wide absorbing range from 10 MHz to 70 MHz is implemented using a toroidal ferrite (Würth 74270191) with four turns. Combining both common-mode chokes, "CM_1" and "CM_2", results in a nearly constant impedance over a broad frequency range as shown in Fig. 10. A conducted noise-emission plot (black curve) with the newly developed "CM_2" choke to attenuate the module's resonant frequency is shown in Figure 8. The second filter stage, consisting of "CM_2", leads to a greatly reduced emission in the upper frequency band. In comparison to the "CM_1" choke (80 dB μV (= 10 mV) at 30 MHz), the "CM_2" choke (67 dB μV (= 2.24 mV) at 30 MHz) provides a 13 dB higher attenuation at 30 MHz. This results in a 4.47 times lower voltage-ripple and the interference-power is reduced by a factor of 20. Since the impedance of the choke "CM_2" at 30 MHz is also higher by a factor of 4.47 compared to the impedance of the choke "CM_1", the measured module resonance emission shows only a common-mode noise component. Otherwise, if the module resonance would have a differential-mode noise component, the measured voltage-ripple reduction of 4.47 times would not correspond with the increased impedance by a factor of 4.47 by choke "CM_2". However, the additional EMI filter increases the overall volume of the aforementioned dc-dc converter by 21 % and, hence, strongly contributes to the overall volume and costs of the converter.

This becomes even more crucial for medium-voltage dc-dc converters using SiC devices. Although the output capacitance of SiC devices decreases with higher blocking voltages, the overall leakage inductance of the module and the commutation cell potentially increases as a result of the required clearance and creepage distances. Therefore, the module resonances of medium-voltage SiC modules and switching cells are expected to be in the same range as of low-voltage dc-dc converters. Moreover, the EMI filters have to withstand the high dc-link voltages and, hence, become more costly and bulky.

IV. ZERO-VOLTAGE SWITCHING

Soft-switching topologies and operation strategies are a means to reduce the impact of the module's parasitic elements on the switching behavior [27]. By shaping the dv/dt and di/dt during the switching transitions, the switching losses and the EMI can be influenced. This provides the opportunity to develop highly efficient and compact medium-voltage dc-dc converters which can be used as flexible power routers in medium-voltage applications, e.g., for the interconnection of future dc grids [3], [28]–[30].

Figure 11 depicts a zero-voltage turn-on switching transition of a 10 kV SiC MOSFET from Wolfspeed operated in a three-phase triple-active bridge (3ph-TAB) converter at a dc-link voltage of 2.5 kV [28]. The blue curve corresponds to the voltage u_{DS}, which is measured across the switch S_2 (see Fig. 1) at one phase leg of the 3ph-TAB converter. The corresponding load current i_L is shown in red. Please note that the amplitude and the slope of the load current during the switching event is mainly determined by the operating point and the operation strategy of the 3ph-TAB converter [30], [31]. Moreover, the oscillations of the load current are caused by the resonant circuit which results from the parasitic elements of an externally connected transformer and inductor. Both components, the transformer and the inductor, form the

3982

leakage inductance, which is the main energy transfer element of a 3ph-TAB converter.

Once S_1 is turned off, the load current i_L charges the output capacitance C_{oss} of S_1 and discharges the one of S_2, respectively (see Fig. 1). Consequently, the voltage slope during turn-off is mainly determined by the load current i_L and the output capacitances C_{oss} of the semiconductors. Despite the relatively low value of C_{oss}, the required commutation energy is in the range of 1 mJ which is due to the quadratic contribution of the dc-link voltage. Therefore, the load current is mostly used for the commutation process and the current in the channel of the SiC MOSFET is cut-off much faster [32] than implied by the waveforms. This resembles the behavior of turn-off snubber capacitors in low-voltage applications. The given load current of approximately 2.5 A leads to a $\mathrm{d}u/\mathrm{d}t$ of only 8 kV/µs. At higher switching currents (5 A ... 6 A) the voltage slope exceeds 25 kV/µs. After the dead time, which is 1.5 µs, the switch S_2 is turned on under zero-voltage switching conditions. Due to the controlled $\mathrm{d}u/\mathrm{d}t$ and the soft turn-on transition, the switching losses as well as the EMI are reduced. The rms-current which is required during the commutation process to ensure soft-switching is approximately 1.7 A. This, however, equals more than 10 % of the nominal dc-current capability of the 10 kV SiC MOSFETs at a junction temperature of 150 °C.

If the rms value of the load current drops below the required threshold, the output capacitor of the switch S_2 cannot be discharged completely during the dead time. This results in a quasi hard turn-on switching as shown in Fig. 12a and Fig. 12b. Given a constant rms current of only 0.3 A the voltage across S_2 is still at 2 kV at the end of the dead time, see Fig. 12a. Besides providing a sufficient amount of commutation energy, zero-current crossings must be avoided during the dead time. This is shown in Fig. 12b, where the load current slowly decreases from 1.1 A to −0.3 A during the commutation process. As long as the load current is positive, the output capacitor of the 10 kV SiC MOSFET is discharged. If the current crosses zero, however, the output capacitor gets charged again, which partially negates the previous commutation process. At the turn-on instant of S_2, the charge stored in the output capacitor C_{oss} is dissipated within the switch S_2. This results in significant turn-on losses and a high $\mathrm{d}v/\mathrm{d}t$.

The measurements show, that a significant amount of commutation energy has to be provided by the load current to ensure a successful zero-voltage turn-on process. To address this issue, the load current during the switching event has to be adjusted, e.g., by altering the operation strategy [31], [33], by adjusting the turns-ratio of the transformer [34] or by injecting reactive power such that a certain switching current is guaranteed. Moreover, a variable dead time in conjunction with a zero-current detection could be used to detect the end of the commutation process [35].

V. CONCLUSIONS

The influence of the stray inductance of the power device packaging on the switching behavior has been investigated

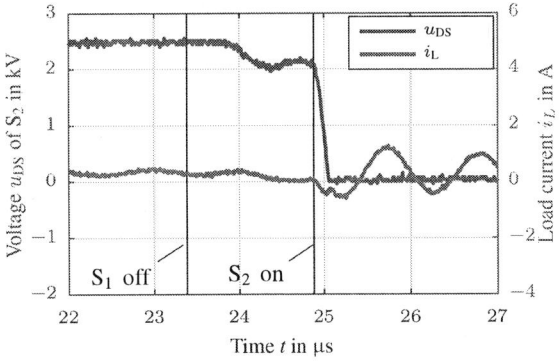

(a) Without zero-current crossing.

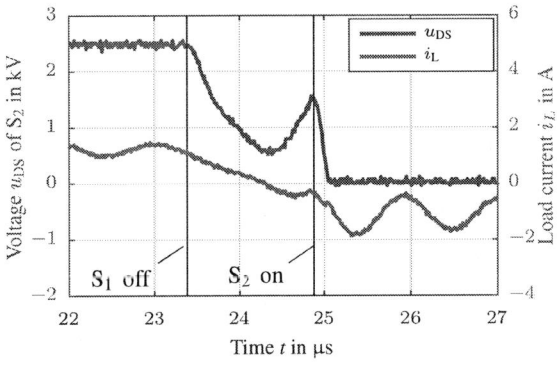

(b) With zero-current crossing.

Fig. 12. Quasi hard turn-on transition of 10 kV SiC MOSFETs operated in a 3ph-TAB converter.

using different prototype switching cells. Furthermore, it has been shown that the parasitic inductances of the module have a significant effect on the EMI of wide-bandgap SiC dc-dc converters. The impact on the filter design has been discussed using the example of a 42 kW dc-dc converter prototype with high power density. Two common-mode filters with different frequency ranges have been applied to attenuate the high-frequency noise emitted by the converter. This has a significant impact on the overall volume and costs of future SiC dc-dc converters. Thus, in modern, high density SiC power converters, the design of the semiconductor package cannot be decoupled anymore from the overall converter or system design. Furthermore, zero-voltage switching has been emphasized on the example of a 10 kV 3ph-TAB as a means to reduce the EMI and, hence, the size of the filters. The measurements have shown, however, that a rms current of more than 1.7 A is required during the switching transition to ensure zero-voltage turn-on at different load conditions. This is particularly crucial, as the required load current is more than 10 % of the nominal current rating of the 10 kV SiC MOSFETs.

ACKNOWLEDGMENT

The work for this paper has been carried out within the scope of several research projects, namely the post graduate program "mobilEM" (GRK 1856) of Deutsche Forschungs-gemeinschaft (DFG), the research project "Research Campus Future Electrical Networks" (FEN) (03SF0489) and the research project HV-ModAL (16EMO0105), both funded by the German Federal Ministry of Education and Research (BMBF), and in close cooperation with Ford Motor Company.

REFERENCES

[1] R. W. De Doncker, C. Meyer, R. U. Lenke, and F. Mura, "Power electronics for future utility applications," in *International Conference on Power Electronics and Drive Systems (PEDS)*, Nov. 2007, pp. 1–8.

[2] H. van Hoek, M. Boesing, D. van Treek, T. Schoenen, and R. W. De Doncker, "Power electronic architectures for electric vehicles," in *Emobility - Electrical Power Train*, Nov. 2010, pp. 1–6.

[3] R. W. De Doncker, "Power electronic technologies for flexible dc distribution grids," in *International Power Electronics Conference (IPEC-Hiroshima 2014 - ECCE-ASIA)*, May 2014, pp. 736–743.

[4] A. Stippich, A. Sewergin, G. Engelmann, J. Gottschlich, M. Neubert, C. van der Broeck, P. Schuelting, R. Goldbeck, and R. De Doncker, "From AC to DC: Benefits in household appliances," in *International ETG Congress*, Nov. 2017, pp. 1–6.

[5] G. Sarriegui, "Sic and gan semiconductors - the future enablers of compact and efficient converters for electromobility," PhD thesis, RWTH Aachen University, Institute for Power Electronics and Electrical Drives (ISEA), Germany, 2017.

[6] A. Stippich, C. H. Van Der Broeck, A. Sewergin, A. H. Wienhausen, M. Neubert, P. Schülting, S. Taraborrelli, H. van Hoek, and R. W. De Doncker, "Key components of modular propulsion systems for next generation electric vehicles," *CPSS Transactions on Power Electronics and Applications*, vol. 2, no. 4, pp. 249–258, Dec. 2017.

[7] G. Sarriegui, S. Beushausen, and R. W. De Doncker, "Performance comparison of different sic-mosfets for high-frequency high-power dc-dc converters," in *International Symposium on Power Electronics for Distributed Generation Systems (PEDG)*, Apr. 2017, pp. 1–8.

[8] A. Sewergin, A. H. Wienhausen, K. Oberdieck, and R. W. D. Doncker, "Modular bidirectional full-sic dc-dc converter for automotive applications," in *International Conference on Power Electronics and Drive Systems (PEDS)*, Dec. 2017, pp. 277–281.

[9] A. Nagel and R. W. De Doncker, "Systematic design of emi-filters for power converters," in *Annual Meeting and World Conference on Industrial Applications of Electrical Energy*, vol. 4, Oct. 2000, 2523–2525 vol.4.

[10] K. Oberdieck, G. Engelmann, and R. W. De Doncker, "Verfahren zur simulativen Modellierung der Gleichtaktanregung," in *Internationale Fachmesse und Kongress für Elektromagnetische Verträglichkeit (EMV)*, 1 vols., Düsseldorf: Apprimus Verlag, Feb. 23, 2016, pp. 591–598.

[11] A. H. Wienhausen, A. Sewergin, S. P. Engel, and R. W. De Doncker, "Highly efficient power inductors for high-frequency wide-bandgap power converters," in *International Conference on Power Electronics and Drive Systems (PEDS)*, Dec. 2017, pp. 442–447.

[12] G. Engelmann, M. Kowal, and R. W. De Doncker, "A highly integrated drive inverter using directfets and ceramic dc-link capacitors for open-end winding machines in electric vehicles," in *Applied Power Electronics Conference and Exposition (APEC)*, Mar. 2015, pp. 290–296.

[13] C. Neeb, L. Boettcher, M. Conrad, and R. W. De Doncker, "Innovative and reliable power modules: A future trend and evolution of technologies," *IEEE Industrial Electronics Magazine*, vol. 8, no. 3, pp. 6–16, Sep. 2014.

[14] A. Nagel and R. W. De Doncker, "Analytical approximations of interference spectra generated by power converters," in *Annual Meeting and Conference Record of the Industry Applications Conference*, vol. 2, Oct. 1997, 1564–1570 vol.2.

[15] J. Gottschlich, M. Schäfer, M. Neubert, and R. W. De Doncker, "A galvanically isolated gate driver with low coupling capacitance for medium voltage sic mosfets," in *European Conference on Power Electronics and Applications (EPE'16 ECCE Europe)*, Sep. 2016, pp. 1–8.

[16] G. Engelmann, T. Senoner, and R. W. De Doncker, "Experimental investigation on the transient switching behavior of sic mosfets using a stage-wise gate driver," *CPSS Transactions on Power Electronics and Applications*, 2018.

[17] J. Gottschlich, P. Weiler, M. Neubert, and R. W. De Doncker, "Delta-sigma modulated voltage and current measurement for medium-voltage dc applications," in *European Conference on Power Electronics and Applications (EPE'17 ECCE Europe)*, Sep. 2017, P.1–P.9.

[18] K. Oberdieck, A. Sewergin, and R. W. De Doncker, "Influence of the voltage-dependent output capacitance of SiC semiconductors on the electromagnetic interference in dc-dc converters for electric vehicles," in *International Symposium and Exhibition on Electromagnetic Compatibility (EMC Europe)*, Sep. 8, 2017.

[19] *Ccs050m12cm2, silicon carbide, six-pack (three phase) module*, Wolf-speed Inc.

[20] Cree, Inc, *Cpm2-1200-0025b silicon carbide power mosfet*, CPM2-1200-0040B datasheet, Jan. 2016.

[21] G. Engelmann, C. Lüdecke, D. Bündgen, R. W. De Doncker, X. Lu, Z. Xu, and K. Zou, "Experimental analysis of the switching behavior of an igbt using a three-stage gate driver," in *International Symposium on Power Electronics for Distributed Generation Systems (PEDG)*, Apr. 2017, pp. 1–8.

[22] J. Gottschlich, M. Kaymak, M. Christoph, and R. De Doncker, "A flexible test bench for power semiconductor switching loss measurements," in *International Conference on Power Electronics and Drive Systems (PEDS)*, Jun. 2015, pp. 442–448.

[23] G. Engelmann, M. Laumen, J. Gottschlich, K. Oberdieck, and R. W. De Doncker, "Temperature controlled power semiconductor characterization using thermoelectric coolers," *IEEE Transactions on Industry Applications*, pp. 1–1, 2018.

[24] C. Schulte-Overbeck, Z. Cao, F. Khan, F. Hussain, S. Grandhi, and D. Weiss, "Comparative analysis of the measurement techniques to characterize sic-power-modules," in *International Exhibition and Conference for Power Electronics, Intelligent Motion, Renewable Energy and Energy Management (PCIM Europe)*, May 2017, pp. 1–8.

[25] J. Gottschlich and R. W. De Doncker, "Pulse generator for dynamic performance verification of current transducers," in *European Conference on Power Electronics and Applications (EPE)*, Sep. 2015, pp. 1–8.

[26] G. Engelmann, S. Quabeck, J. Gottschlich, and R. W. De Doncker, "Experimental and simulative investigations on stray capacitances and stray inductances of power modules," in *European Conference on Power Electronics and Applications (EPE'17 ECCE Europe)*, Sep. 2017, P.1–P.10.

[27] A. Charalambous, X. Yuan, and N. McNeill, "High-frequency EMI attenuation at source with the auxiliary commutated pole inverter," *IEEE Transactions on Power Electronics*, pp. 1–1, 2017.

[28] M. Neubert, A. Gorodnichev, J. Gottschlich, and R. W. De Doncker, "Performance analysis of a triple-active bridge converter for interconnection of future dc-grids," in *Energy Conversion Congress and Exposition (ECCE)*, Sep. 2016.

[29] M. Neubert, S. P. Engel, J. Gottschlich, and R. W. De Doncker, "Dynamic power control of three-phase multiport-active bridge dc-dc converters for interconnection of future dc-grids," in *International Conference on Power Electronics and Drive Systems (PEDS)*, 2017.

[30] M. Neubert, H. van Hoek, J. Gottschlich, and R. W. De Doncker, "Soft-switching operation strategy for three-phase multiport-active bridge dc-dc converters," in *International Conference on Power Electronics and Drive Systems (PEDS)*, 2017.

[31] H. van Hoek, "Design and operation considerations of three-phase dual active bridge converters for low-power applications with wide voltage ranges," PhD thesis, RWTH Aachen University, Aachen, 2017, p. 231.

[32] M. A. Bahmani, K. Vechalapu, M. Mobarrez, and S. Bhattacharya, "Flexible hf distribution transformers for inter-connection between mvac and lvdc connected to dc microgrids: Main challenges," in *Second International Conference on DC Microgrids (ICDCM)*, Jun. 2017, pp. 53–60.

[33] H. van Hoek, M. Neubert, and R. W. De Doncker, "Enhanced modulation strategy for a three-phase dual active bridge - boosting efficiency of an electric vehicle converter," *IEEE Transactions on Power Electronics*, vol. 28, no. 12, pp. 5499–5507, Dec. 2013.

[34] S. Taraborrelli, "Bidirectional dual active bridge converter using a tap changer for extended voltage ranges," PhD thesis, RWTH Aachen University, Institute for Power Electronics and Electrical Drives (ISEA), Germany, 2017.

[35] R. Lenke, "A contribution to the design of isolated dc-dc converters for utility applications," PhD thesis, RWTH Aachen University, E.ON Energy Research Center, Institute for Power Generation and Storage Systems (PGS), Germany, 2012.

Design and Testing of 6 kV H-bridge Power Electronics Building Block Based on 10 kV SiC MOSFET Module

Jun Wang*, Slavko Mocevic, Jiewen Hu, Yue Xu, Christina DiMarino, Igor Cvetkovic,
Rolando Burgos, and Dushan Boroyevich
Center for Power Electronics Systems, Virginia Tech, Blacksburg, USA
*E-mail: junwang@vt.edu

Abstract—This paper presents a part of the design for a power electronics building block (PEBB) based on 10 kV SiC MOSFET power module. A H-bridge PEBB system architecture is introduced at the beginning, followed by the design details of a smart gate driver. Strong noise-immunity, high driving current and effective protection circuitry have been accomplished. The design of power supply that feeds the gate drivers while providing 10 kV galvanic isolation is also shown. A resonant current bus (RCB)-based topology is proposed to supply the gate drivers, achieving both high density and low input-output capacitance of the isolation. Finally, a 10 kV laminated DC bus-bar with new layer-stacking structure is presented. Experimental results are embedded in each section to validate the PEBB performance.

Keywords—*Power electronics building block, SiC MOS-FET, smart gate driver, power supply, laminated dc bus-bar.*

I. INTRODUCTION

Power electronics building blocks (PEBB), as a least replaceable unit (LRU) to construct modular converters, was originally proposed by the Office of Naval Research in 1997 [1]. It has been defined as a universal power processor and a systematic approach, featuring modular configurations, scalable voltage and current ratings, as well as low inventory and maintenance cost [2]-[4]. In the past two decades, nearly all the commercial high-power converters have taken advantage of the PEBB concept in the medium-voltage (MV) and high-voltage (HV) applications. The most popular PEBB topologies include half-bridge [5], H-bridge [6], Neutral-Point Clamped 3-level (NPC-3L) [7], and Active Neutral-Point Clamped 5-level (ANPC-5L) [8]. Those PEBBs are connected in series, parallel, or multiple-phase configurations to scale up the voltage and current ratings of PEBB-based converters, breaking through the constraints by the semiconductor device ratings. MV modular converters that comprise a large number of PEBBs have sophisticated electrical and control connections, insulation, cooling and auxiliary system. The complexity drags down the whole converter reliability, so that further efforts of redundancy design have to be engaged. That in reality leads to an even more complicated power conversion system. Accordingly, when the converter voltage rating increases, it is preferred to elevate the PEBB voltage ratings while maintaining the total PEBB number at a relatively low value. Due to the limited voltage and current ratings of Si IGBT devices, it becomes a common practice that MV PEBBs adopt multilevel topologies such as 3-level, 5-level, devices-stacking, and other various topologies. As such, the system complexity has been transferred from the converter level to the PEBB level, and the reliability of each PEBB becomes another concern.

In recent years, thanks to the booming technology of wide-bandgap semiconductor devices and packaging, SiC MOSFETs have demonstrated their superior performance to Si IGBTs in terms of higher breakdown voltage, faster switching speed, lower switching loss and higher operating temperature [10]. The high blocking voltage of SiC MOSFETs simplifies the PEBB power stage by using uncomplicated topologies, and meanwhile their high switching frequency preserves the overall harmonic performance despite of reduced number of voltage levels.

TABLE I. DEVICE COMPARISON FOR A THREE-PHASE INVERTER

Property	Si IGBT	SiC MOSFET
DC bus voltage (V_{dc})	$6 \sim 7$ kV	
3-phase output (v_{ll}, i_{ph}, S_o)	4.16 kV, 100 A, 720 kVA	
Total module number	18	3
Total module volume	8.47 L	1.69 L
Switching frequency (f_{sw})	2.5 kHz	10 kHz
Switching loss per module	2250 W	3800 W
Conduction loss per module	210 W	400 W
Maximum operating temperature	150°C	175°C
Equivalent switching frequency ($f_{sw,eqv}$)	10 kHz	10 kHz

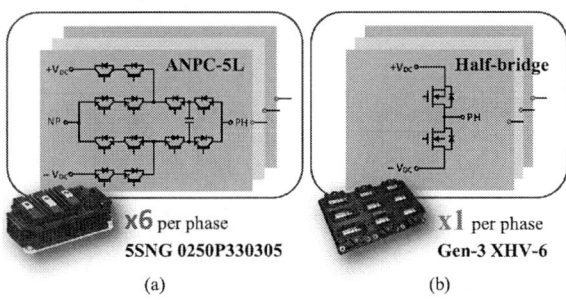

Fig. 1. Converter design comparison based on: (a) 3.3 kV Si IGBT vs. (b) 10 kV SiC MOSFET.

An example in Table I and Fig. 1 shows the comparison between two designs using 3.3 kV Si IGBT and 10 kV SiC MOSFET, respectively. The former design is referred to the commercial motor drive product ACS2000, where each phase-leg is an ANPC-5L circuit configured as a PEBB. The latter design with SiC MOSFET is merely a simple 2-level half-bridge. In order to construct a three-phase inverter with the same specifications of DC bus voltage V_{dc}, line-to-line voltage v_{ll}, phase current i_{ph}, and equivalent phase-leg switching frequency $f_{sw,eqv}$, the Si-IGBT design requires more device modules by a factor of 6. The incremental device modules and their gate drivers result in more geometric volume by a factor of 5, and lower system reliability. In addition, addressing the static and dynamic voltage sharing problem of the ANPC-5L topology further costs the converter efficiency. Conclusively speaking, replacing 3.3 kV Si IGBT with 10 kV SiC MOSFET can potentially boost the converter power density, efficiency and reliability.

Based on the comparison, a PEBB built upon 10 kV SiC MOSFET modules will be a promising new-generation LRU for high-density MV power converters [11]. They will be highly valued in application scenarios where the room and space are quite limited and expensive, for instance, the MVDC power distribution systems in urban areas [9] or motor drives on Navy ships. Therefore, the objective of this paper is to present critical design aspects of the 10 kV SiC-based PEBB, which is named PEBB6000 for its nominal DC bus voltage. The detailed specifications are given in Table II, and the system architecture of the PEBB6000 is shown in Fig. 2 [12]. The power stage comprises a H-bridge by two Wolfspeed Gen-3 XHV-6 modules rated at 10 kV, 240 A, a laminated DC bus-bar, DC-link capacitors, and phase inductors. The optical-fiber-based digital sensing and control system includes smart gate drivers, PEBB controller and isolated digital sensors. All of them are powered by a DC-fed auxiliary power supply. This modular control system entitles the PEBB6000 to have complete modularity, excellent noise immunity, and high degree of intelligence, which allows the PEBB to be capable of more advanced functionalities. This paper mainly focuses on the design and development of the smart gate driver, the gate driver power supply, and the laminated DC bus-bar.

II. 10 kV Smart Gate Driver Design

Gate driver is the critical interface between power semiconductor devices and control signals, which serves to provide galvanic isolation and to supply driving current while maintaining signal integrity under high-noise environment. On top of those basic tasks, it can also provide quick, reliable and configurable protection, as well as advanced switch-current signal sensing and data processing for control purposes, which define a "smart" gate driver.

Since both high-side and low-side devices will be driven, a galvanic isolation up to at least 10 kV is required for the XHV-6 module. Typical MV gate driver power supply configurations are depicted in Fig. 3. The gate driver's primary-side ground is usually referred to a local chassis within the PEBB, rather than the actual earth ground. That ensures a minimum overall insulation voltage stress in the PEBB, even if the local ground is jumping up and down referred to the real earth when a number of PEBBs are stacked in series. In this power configuration, for the high-side gate driver, the primary side of its power supply sits at a fixed potential, while the secondary-side is jumping at a high dv/dt rate along with the switching of the half-bridge module. Therefore, the dv/dt noise source is placed across the input-output parasitic capacitor of the gate driver power supply. It produces a common-mode (CM) noise current whose

TABLE II.   PEBB6000 SPECIFICATIONS

Property	Value
DC bus rated voltage	6 kV
DC bus peak voltage	7 kV
Maximum DC bus voltage ripple	1 kV
Terminal rms current	100 A
Terminal peak current	250 A
Device short-circuit threshold	600 A
Switching frequency	10 ∼ 30 kHz
Maximum dv/dt	100 V/ns
Maximum di/dt	10 A/ns

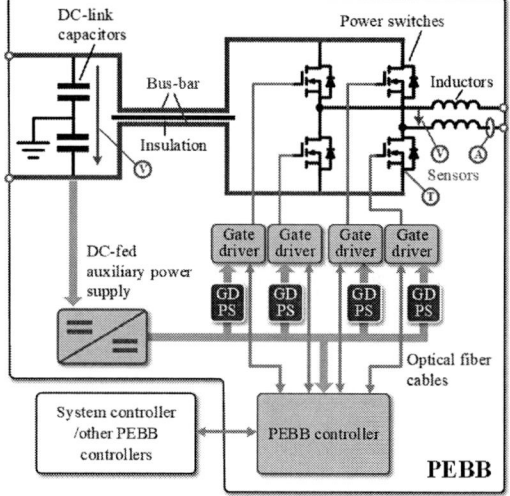

Fig. 2.   PEBB6000 system architecture diagram.

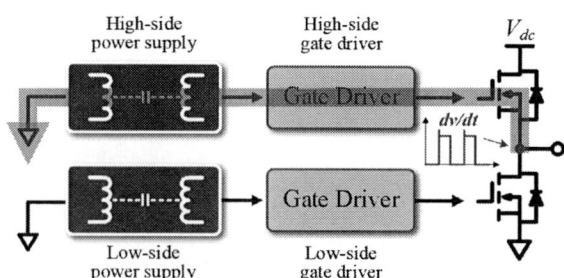

Fig. 3.   Gate driver and power supply configurations and noise path for 10 kV SiC MOSFET.

TABLE III. SMART GATE DRIVER SPECIFICATIONS

Property	Minimum	Maximum
Supply voltage	24 V	30 V
Driving voltage	−8 V	+20 V
dv/dt immunity	30 V/ns	100 V/ns
Switching frequency	-	100 kHz
Isolation voltage	20 kV	-
Driving current	-	90 A
External gate resistors	0.53 Ω	-
Driver IC over-temperature protection	150°C	200°C
Under-voltage lockout	11 V	14 V
Active Miller clamp	5 A, V_{EE} + 2 V	15 A. -
Short-circuit protection threshold	400 A	800 A
Two-level turn-off	7 V, 0.5 μs	10 V, 1.5 μs

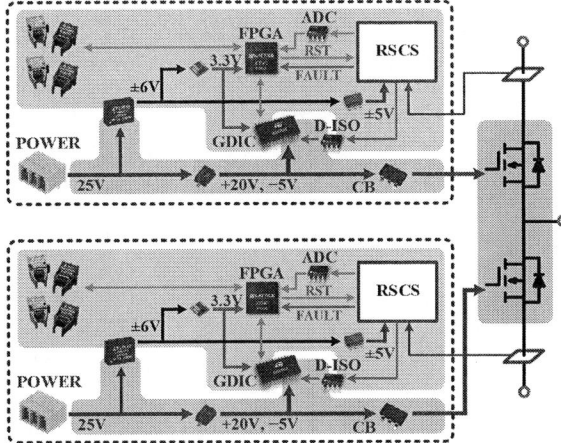

Fig. 4. Smart gate driver architecture for Gen-3 10 kV SiC MOSFET module.

magnitude is determined by the dv/dt rate and the parasitic capacitance value, as demonstrated in Fig. 3. The dv/dt rate of SiC MOSFET is about $5 \sim 10$ times higher than Si IGBTS, which can generate a significant magnitude of CM current. Every component on the path where the CM current flows through will be subjected to noise spikes resulting in malfunction.

The SiC MOSFET should always be pushed to its switching speed limit to minimize the switching losses, but high dv/dt noise source will be inevitably presented at the switching node of the half-bridge module. In order to attenuate the impact from the CM noise current, two solutions are usually applied. The first solution is to design a good gate driver power architecture to direct the noise current away from sensitive circuitry [13], and the second one is to minimize the input-output capacitance of the isolated power supply. Accordingly, this section shows the design of the gate driver including power architecture, current booster, current sensor. The design of the gate driver power supply with 10 kV isolation and insulation, low input-output capacitance, and high power density, will be presented in the followed section .

A. Gate driver specifications and architecture

The specifications of the gate driver are given in Table III. The gate driver should be able to supply proper driving voltages required by the XHV-6 module, and as much driving current as possible to enable low external gate resistance to reduce hard-switching losses. Noise-immunity performance at maximum 100 V/ns is also necessary to avoid malfunction of logic circuitry. The power rating of the driver should satisfy maximum 100 kHz switching frequency. Short-circuit protection is designed to be detected at a configurable value between 400 A and 800 A, and the MOSFET should be completely turned off within 1.5 μs. Other functionalities including active Miller clamp, under/over-voltage lockout, soft turn-off, driver IC over-temperature protection are also specified in the table.

The purpose of the gate driver architecture design is to bypass the CM current away from the sensitive components such as logic units and FPGA. The design refers to the analysis and follows the solution that have been published in [13]. As shown in Fig. 4, for each

gate driver channel, two isolated ground planes have been designed. On the green plane, most of signal processing components and logic units with voltages at ±5 V, 3.3 V, and 1.2 V are located. On the red plane, the components designed for providing driving current and voltage at +20 V and −5 V are placed. The components on the red plane are much less sensitive than those on the green, so it will be preferred that the CM noise current primarily flows through red plane to the input power connector at the high side. The CM impedance of the red ground path is dominated by a few nanohenry trace inductance, whereas the CM impedance of the green ground path is determined by several picofarad input-output capacitance of isolated power supplies, the gate driver IC, and digital isolators. The CM impedance of the green path is more than $10\times$ higher than the red. Hence, the major part of CM noise current will flow through the red plane instead of the green, so that the sensitive circuits will be subjected to greatly mitigated conductive noises.

The architecture also shows that a Rogowski switch-current sensor (RSCS) is designed for SiC MOSFET drain current sensing. Working with a low-latency comparator the RSCS works as a short-circuit detector; working with an ADC converter, the RSCS also functions as a phase current sensor by sampling the switch current. A FPGA manages gate driver IC programming, RSCS reset, ADC conversion, and communication to the PEBB controller.

B. Current booster

The 10 kV XHV-6 module contains three half-bridge submodules that has configurable internal jumpers for parallel connection among them. Each half-bridge sub-module is rated at 80 A, and the total module ampacity is extended to 240 A if all the three of them are paralleled. Paralleling configuration requires triple driving charges from the gate driver. Therefore, the gate driver is designed be able to supply 90 A peak current to sustain the fastest switching transient. In order to achieve that and meanwhile to minimize the gate loop inductance, a

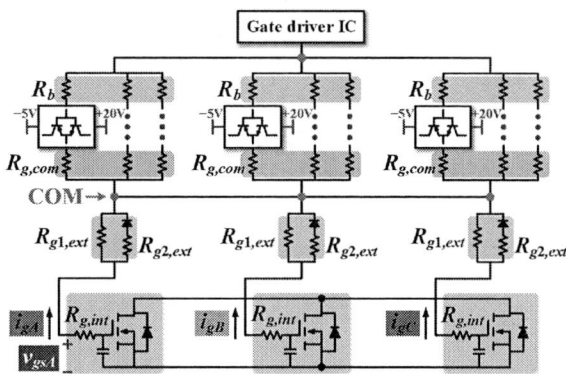

Fig. 5. Paralleled current booster circuit diagram for three submodules.

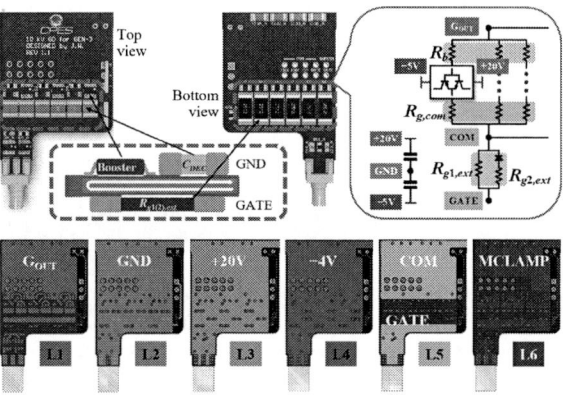

Fig. 6. PCB layout of the paralleled current boosters designed for minimized parasitics.

new current booster solution with nine paralleled bipolar junction transister (BJT) have been proposed and designed as shown in Fig. 5. The three driving channels are connected jointly at the common junction "COM" to guarantee the three driving voltage is the same. $R_{g,com}$ is used to balance the current sharing of the nine current booster channels, and to compensate the mismatched transconductance between the BJTs. $R_{g1,ext}$ and $R_{g2,ext}$ are split gate resistors to damp the resonance between the three paralleled gate loops. $R_{g1,ext}$ determines the turn-on speed, while $R_{g1,ext}$ and $R_{g2,ext}$ jointly determine the turn-off speed. Fig. 6 shows the PCB layout of the current boosters. Three current booster banks are placed adjacent to a MCX gate/source connector. BJT ICs, decoupling capacitors, gate resistors, and internal power/ground planes are laminated at the same area to ensure extremely low turn-on and turn-off gate loop inductance.

The current sharing test results of three current booster banks are shown in Fig. 7. The gate driving currents of submodule A, B and C have been obtained by measuring the across voltages of three gate resistors $R_{g,ext}$, using passive probes referred to the joint point "COM" as the ground. The three gate currents are almost overlapped with one another, presenting the total turn-on current peak of 36 A, and the total turn-off peak current of 48 A.

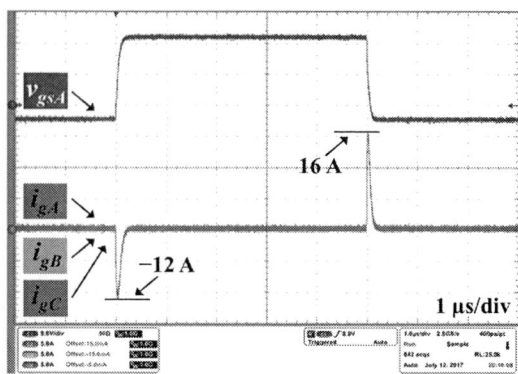

Fig. 7. Current sharing test of three current booster banks.

C. Rogowski switch current sensor

A Rogowski switch-current sensor (RSCS) has been proposed to work effectively together with SiC MOSFET modules[14][15]. The high bandwidth, wide measurement range and good accuracy make it an excellent short-circuit current detector. As depicted in Fig. 8, a RSCS mainly comprises two parts. The first part is a Rogowski coil embedded in the inner four layers of the PCB, constructed by buried traces and vias, and not exposed to the surface. This design eliminates creepage paths between the high-voltage primary-side conductor and the secondary-side coil, especially for the high-side switch current measurement. The one-turn conductor current that flows out of the orange window as annotated by the red arrow generates a flux, and the coil couples the flux along with the orange trace. A di_d/dt voltage scaled by the mutual inductance M is induced at the terminal of the coil where i_d is the MOSFET drain current. The second part of the RSCS is a group of signal processing circuits that includes an active integrator circuit, a reset switch, and a proper signal filter. The active integrator converts the Mdi_d/dt value back to i_d. A bi-directional analog switch resets the integrator when the SiC MOSFET is not conducting current in every switching cycle, ensuring that the initial value of the integration is zero at each cycle. By this means, the RSCS is able to measure the DC component of the switch current. The high-frequency bandwidth of RSCS is determined by the Rogowski coil self-inductance and parasitic capacitance, as well as the bandwidth of the operational amplifier.

The output of RSCS is given to the non-inverting input pin of a high-speed comparator. For the purpose of short-circuit protection, the threshold of the comparator is set at 600 A. In the meantime, the RSCS output can also feed an ADC input, where the switch current sampling is equivalent to phase current sampling in average-current-mode control.

The prototype of a phase-leg gate driving system is shown in Fig. 10, which consists of one 10 kV XHV-6 module, two gate driver boards, and one Rogowski coil board. Each gate driver board is plugged perpendicularly into the XHV-6 module via three MCX connectors, and plugged into the Rogowski coil board with three three-

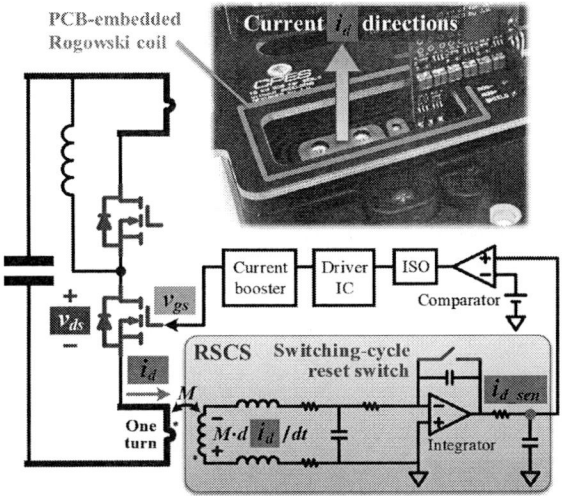

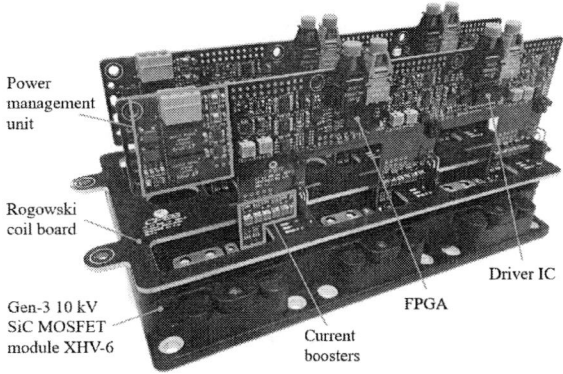

Fig. 10. Assembled gate driver prototype including device module, two gate driver boards and one Rogowski coil board.

Fig. 8. Rogowski switch current sensor circuit diagram

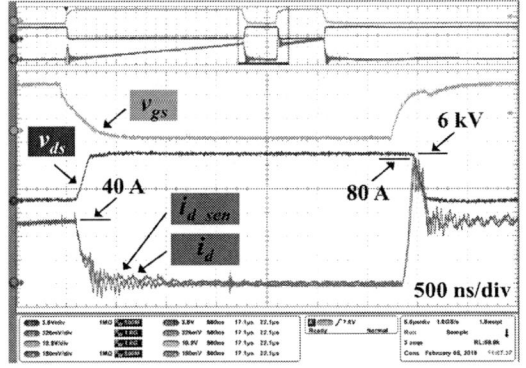

Fig. 9. Comparison between RSCS and commercial Rogowski probe on submodule A in double-pulse test: $R_{g,eqv} = 1\ \Omega$, $V_{dc} = 6$ kV, $i_{d\_off,A} = 40$ A (120 A total), $i_{d\_on,A} = 80$ A (240 A total), turn-off $dv/dt = 50$ V/ns, turn-on $di/dt = 2.4$ A/ns.

pin headers. The drivers boards are physically paralleled to each other with a distance of about 20 mm, satisfying the clearance distance requirement in IEC 60664-1 (3.5 mm for 10 kV, 8 mm for 20 kV) [16]. The smart gate driver has been validated at rated voltage and current in double-pulse tests. In Fig. 9, the measured currents by the RSCS and by a commercial Rogowski probe indicate excellent agreements. Other test conditions are listed in the figure caption.

D. Gate driver power supply with resonant current bus

The gate driver power supply must withstand the voltage stress between its primary and secondary sides, and meanwhile having a low input-output parasitic capacitance to reduce the CM noise current. The isolation voltage rating of the gate driver power supply is 10 kV according to the rating of XHV-6 module, while usually 100% margin is designed in practice. In order to achieve 20 kV insulation strength, according to IEC 60664-1, the creepage distance must be kept no less than 100 mm at the

best case (pollution degree 1). As each power supply must have its own connectors accessible from the enclosure, the enclosure surface will inevitably become the creepage path so that the geometric size of the power supply has to be very large to keep sufficient distance between connection terminals. In addition, the total volume of gate driver power supplies is scaled up by the number of power switches in the PEBB, resulting in a huge wasted room just to meet the creepage standard. In an alternative solution, the primary and secondary-side converters are separated apart. The output of the primary-side converter is connected to a high-voltage cable carrying a sinusoidal current source that routes through the magnetic core on the secondary-side board. In this manner, the primary-side conductor is completely enclosed by the external layer of the cable with sufficient insulation strength, so that the creepage distance requirement is satisfied by the length of cable. Both the primary and secondary-side converters merely needs low-voltage designs no higher than 30 V. Another benefit of this current-source-based design is that multiple secondary-side loads can be carried, and the short-circuit fault of one load will not influence the normal operation of the others.

The conventional approach to generate the sinusoidal current source is usually a single-phase inverter, which requires a bulky DC-link capacitor tank and large phase inductor. The current source frequency is much lower than the primary-side switching frequency. The low frequency

TABLE IV. SPECIFICATIONS OF GATE DRIVER POWER SUPPLY

Property	Value
Input voltage (V_{in})	24 V
Output voltage (V_o)	26 V
Output power per secondary-side load (P_o)	10 W
Maximum secondary-side load number	4
Minumum efficiency (η), >10% load	80%
Insulation voltage	20 kV
Maximum volume	3 inch^3
Maximum input-output capacitance (C_{io})	2 pF
Maximum output-output capacitance (C_{oo})	1 pF
Load regulation, 10 ~ 100% load	5%

3989

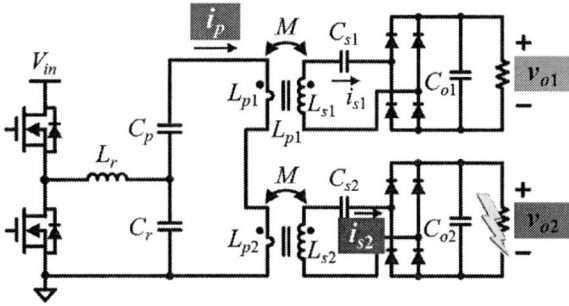

Fig. 11. Gate driver power supply circuit diagram.

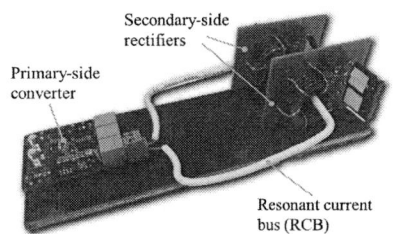

Fig. 12. Gate driver power supply prototype.

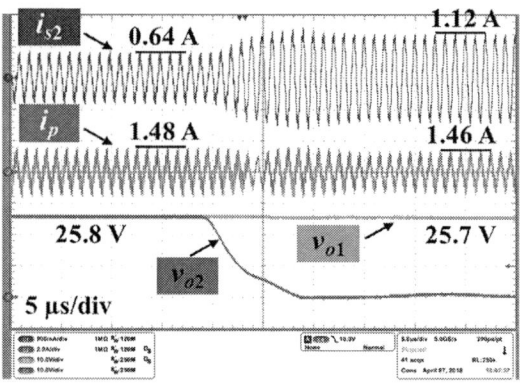

Fig. 13. Resilience test: short-circuit happens at Load 2. $f_{sw} = 1$ MHz. Before fault: $V_{o1} = V_{o2} = 25.8$ V, $P_{o1} = P_{o1} = 9.0$ W, $\eta = 81.5\%$. After fault: $V_{o1} = 25.7$ V, $V_{o2} = 0$ V, $P_{o1} = 8.9$ W, $P_{o2} = 0$ W, $\eta = 75.2\%$.

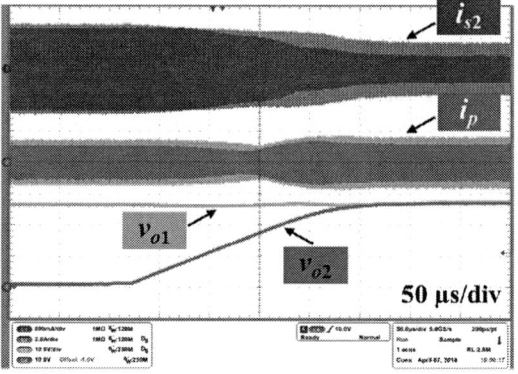

Fig. 14. Resilience test: short-circuit at Load 2 is cleared. Voltages and currents recover to their original states after 260 μs.

of current source results in big cross-section area of the magnetic core and large turn number of the secondary-side winding. Both of them contribute to a high input-output parasitic capacitance. In this design, a resonant converter that offers a resonant current bus (RCB) to drive multiple isolated load is proposed for this particular application. As shown in Fig. 11, the LCCL-LC resonant converter is implemented, featuring a) switching-frequency current source i_p, b) zero-voltage switching of primary-side devices, c) low-profile magnetics and small turn number leading to low input-output capacitance, d) simple open-loop control with fixed duty cycle and good load regulation, and e) inherent high resiliency as regards load fault. The specifications of the power supply are shown in Table IV.

A prototype has been built with two secondary-side loads as shown in Fig. 12. Half-bridge GaN EPC2104 is selected as the main power device and switched at 1 MHz, which gives a low-profile primary-side converter smaller than 1 inch^3. The measured input-output capacitance C_{io} is 1.52 pF, and the output-output capacitance C_{oo} between two secondary sides is 0.81 pF. The gate driver power supply has been tested in under steady-state and load-fault conditions. Fig. 13 shows that when Load 2 is short-circuit, the converter output at Load 1 remains stable. Fig. 14 demonstrates that the power supplies is able to fully recover after the short-circuit is cleared. The efficiency results are still below the specifications, and further optimization will be made on the next prototype.

III. LAMINDATED DC BUS DESIGN

Laminated bus-bars have been widely used for MV high-power applications in the recent 20 years. Thanks to the proximity effect, the two high-frequency currents flowing through the positive and negative bus-bars tend to concentrate on the inner surfaces when the two bus-bars are laminated closely in a large overlapped area. Accordingly, the fluxes generated by the two high-frequency currents are able to cancel each other to some extent. This brings two important beneficial features: a) low stray inductance of the bus-bar loop, and b) low radiated noises. However, in MV and HV applications, extended thickness of the insulation material layer between the positive and negative bus-bars is required to ensure low partial discharge inception voltage and long lifetime. The distance between the laminated copper layers becomes large, so the proximity effect and flux canceling effect are weaken. The increased stray inductance is acceptable due to the 4 kV voltage margin and relatively low current magnitude. Nonetheless, the increased radiated noise is a critical threat to the gate driver circuit, as it is designed to be inserted through the open windows of the laminated bus-bar for the high-density objective.

To analyze the radiated noise behaviors, three design options have been raised for comparison in finite element analysis (FEA) simulation, as shown in Fig. 15. The thickness of the insulation layers between the positive, negative, and the ground copper layers are shown in

The 2018 International Power Electronics Conference

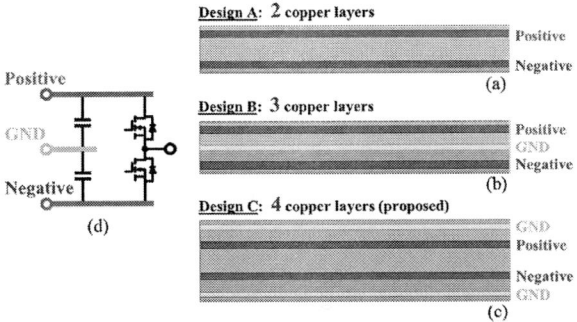

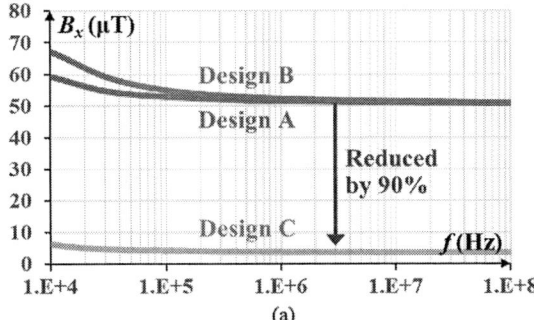

Fig. 15. Three design options of laminated bus-bar: (a) Design A, conventional two copper layers, (b) Design B, three copper layers, with the inner copper layer sitting at the ground, (c) Design C, four copper layers, with two ground layers enclosing the positive and negative layers, and (d) bus-bar voltage potential in the power stage.

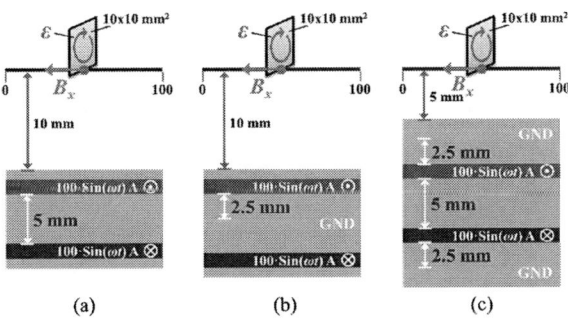

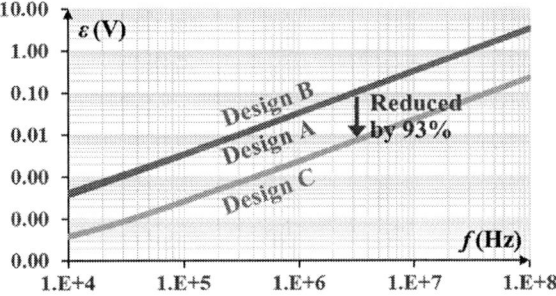

Fig. 17. Worst-case comparison considering fringe effect: (a) flux density B_x, and (b) EMF ε.

Fig. 16. Simulation setup for comparison of three designs: (a) flux density measured at 10 mm above the bus-bar of Design A, (b) flux density measured at 10 mm above the bus-bar of Design B, and (c) flux density measured at 5 mm above the bus-bar of Design C.

Fig. 16. 2 kV/mm static electric field stress is assumed. The thickness of each copper layer is 1.5 mm. In the simulation, the positive layer in red is excited by a 100 A amplitude AC current source, while the negative layer in blue is excited by the same current but in an opposite direction. The ground layer is not excited by any sources. The purpose of the simulation is to observe and compare the induced X-axis flux density B_x at the specified location in Fig. 16, and to assess the induced electromotive force (EMF) ε when the flux density is integrated over an window area of 10×10 mm^2 at the given location. To ensure a fair comparison for high-density design, the flux measurement location is defined for a similar overall size considering the bus-bar thickness. Therefore, the measurement distance of Design C is closer than Design A and B.

The FEA simulation in Ansys Maxwell shows the flux density comparison in Fig. 17(a) considering fringe effect. The two external ground layers in Design C serve as two shielding planes that reduce the radiated flux density by 48 μT (by 90%) compared to the other two. Fig. 17(b) shows significant induced EMF in Design A and B, with a magnitude of 0.3 V at 10 MHz for a 100 mm^2 circuit loop area. Design C demonstrates its superior performance once again in reducing the EMF by 14$\times$ (by 93%).

Despite that most of the generated fluxes concentrate on the non-metal materials, a small part of the fluxes can still penetrate into a small skin-depth of the conductor to cause Eddy current losses that are determined by the magnitude of B_x. Fig. 18 shows a total Eddy current loss comparison based on the simulation setup of the three designs. The "B_x magnitude" is compared at 10 kHz, and the "B_x at $\omega t = 0$" is compared at 10 kHz and 10 MHz. The results show that Design A and B have similar B_x profile the positive and negative layers along with the Y-axis. However, Design B have some additional flux penetrating into the ground layer which cause extra losses. In Design C, the B_x magnitude is reduced nearly by half and evenly distributed on all the inner copper surfaces. The polarities of the flux density on the top and bottom surfaces of the positive and negative layers are opposite because of the impact from the two external ground layers. The total Eddy current losses comparison is shown in Fig. 19, indicating that Design C has the minimum total Eddy current loss from 10 kHz to 100 MHz, which is the dominant frequency range for SiC MOSFETs that switch at above 10 kHz. Finally the laminated bus-bar has been designed based on Design C and the exploded view of the fabricated bus-bar is shown in Fig. 20.

IV. CONCLUSIONS

Design and testing of smart gate drivers, gate driver power supplies, and the laminated DC bus-bar have been shown. The H-bridge power stage and gate drivers have been integrated into a plastic enclosure shown in Fig. 21, a test platform for further assessment. The remaining parts of the PEBB6000 design will be published in the future.

3991

The 2018 International Power Electronics Conference

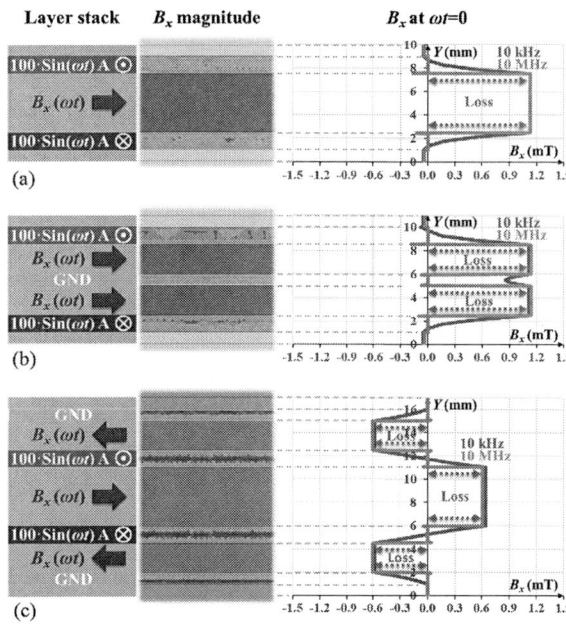

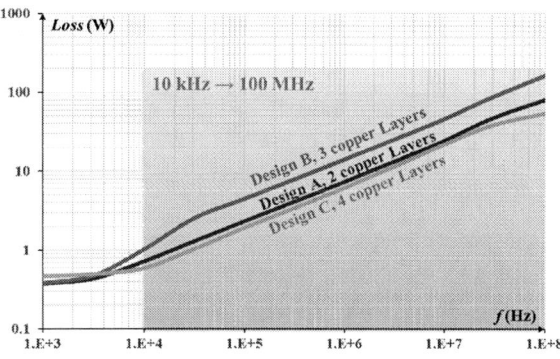

Fig. 18. Comparison of B_x inside copper layers (Eddy current): (a) Design A, (b) Design B, and (c) Design C.

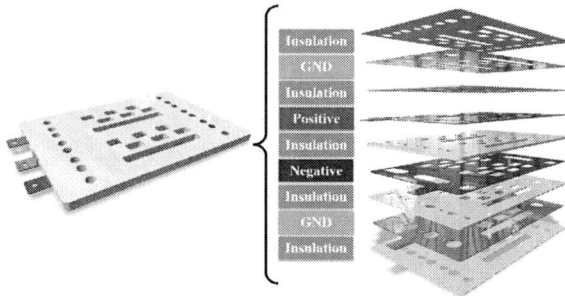

Fig. 20. Exploded view of the layer-stacking structure of Design C.

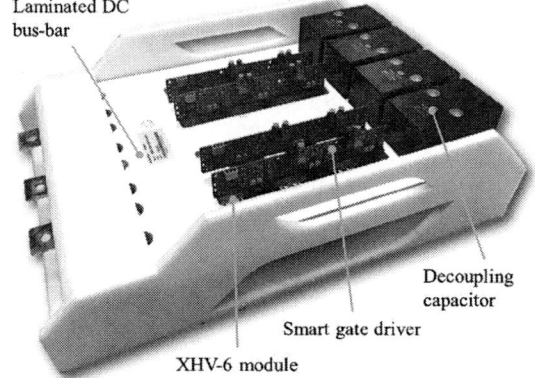

Fig. 21. PEBB6000 prototype integrating XHV-6 modules, gate drivers, and the laminated DC bus-bar.

Fig. 19. Comparison of total Eddy current loss versus frequency.

ACKNOWLEDGEMENT

This work is conducted under the research project sponsored by the Office of Naval Research (ONR).

REFERENCES

[1] T. Ericsen, "Power Electronics Building Blocks," in Proc. IME/IEE/SEE Electric Warship Conference, 1997.

[2] T. Ericsen and A. Tucker, "Power Electronics Building Blocks and potential power modulator applications," in Proc. IEEE International Power Modulator Symp., 1998, pp. 12-15.

[3] T. Ericsen, "Power Electronic Building Blocks-a systematic approach to power electronics," in Proc. IEEE Power Engineer. Society Summer Meeting, 2000, pp. 1216-1218 vol. 2.

[4] T. Ericsen, Y. Khersonsky and P. K. Steimer, "PEBB Concept Applications in High Power Electronics Converters," in Proc. IEEE Power Electron. Special. Conf., 2005, pp. 2284-2289.

[5] SIEMENS, "HVDC PLUS Basics and Principle of Operation," Product document, 2009.

[6] SIEMENS, "ROBICON Perfect Harmony Medium-Voltage Liquid-Cooled Drives," Product brochure, 2012.

[7] ABB, "ACS5000 Medium Voltage Drive," Product brochure, 2017.

[8] ABB, "ACS2000 Medium Voltage Drive," Product brochure, 2015.

[9] M. Stieneker and R. W. De Doncker, "Medium-voltage DC distribution grids in urban areas," in Proc.IEEE Intern. Symp. on Power Electron. for Distr. Gener. Systems, 2016, pp. 1-7.

[10] J. Millan, P. Godignon, X. Perpina, A. Perez-Tomas, J. Rebollo, "A survey of wide bandgap power semiconductor devices," IEEE Trans. Power Electron., vol. 29, no. 5, pp. 2155-2163, May, 2014.

[11] I. Cvetkovic, et al., "Modular scalable medium-voltage impedance measurement unit using 10 kV SiC MOSFET PEBBs," in Proc. IEEE Electric Ship Techn. Symp., 2015, pp. 326-331.

[12] I. Celanovic, I. Milosavljevic, D. Boroyevich, R. Cooley and J. Guo, "A new distributed digital controller for the next generation of power electronics building blocks," in Proc. Applied Power Electron. Conf. and Expo., 2000, pp. 889-894 vol.2.

[13] J. Wang, Z. Shen, R. Burgos, and D. Boroyevich, "Gate driver design for 1.7kV SiC MOSFET module with Rogowski current sensor for shortcircuit protection," in Proc. Applied Power Electron. Conf. and Expo., 2016, pp. 516-523.

[14] J. Wang, Z. Shen, R. Burgos and D. Boroyevich, "Design of a high-bandwidth Rogowski current sensor for gate-drive shortcircuit protection of 1.7 kV SiC MOSFET power modules," in Proc. IEEE Workshop on Wide Bandgap Power Devices and Appl., 2015, pp. 104-107.

[15] J. Wang, Z. Shen, R. Burgos and D. Boroyevich, "Integrated switch current sensor for shortcircuit protection and current control of 1.7-kV SiC MOSFET modules," in Proc. IEEE Energy Convers. Congr. and Expo., 2016, pp. 1-7.

[16] "International standard IEC 60664-1," basic safety publication.

3992

High Power Medium Voltage Converters Enabled by High Voltage SiC Power Devices

Sanket Parashar, Ashish Kumar, Subhashish Bhattacharya

FREEDM Systems Center, Dept. of ECE, North Carolina State University

*E-mail: sbhatta4@ncsu.edu

Abstract—This paper presents the potential applications of HV SiC power devices for high power and medium voltage power conversion systems. The advantages and features enabled by HV SiC devices are experimentally validated by both device performance evaluations as well as operation and performance validations of MV power conversion systems. The potential application examples of Solid State Transformer (SST), Asynchronous Microgrid Power Conditioning Systems, MV motor drives, MV grid connected converters for integration of Distributed Renewable Energy Resources (DRER) and Distributed Energy Storage Devices (DESD) are presented as design case studies and experimental validations with their advantages and power conversion efficiency.

Keywords – Medium voltage SiC, Wide bandgap, high power, series connection.

I. INTRODUCTION

Silicon power devices have been the workhorse of industry for over three decades, and are used today in a wide variety of high power applications including electric drives, renewable energy, automotive, traction and switched mode power supplies. Most industrial applications typically utilize 1200V IGBTs, which over the years have improved tremendously – offering low on-state voltage drop and switching frequencies of 5-10 kHz with appropriate thermal management. On the other hand, the opportunity to realize higher power levels for large industrial and utility applications, are constrained by the capability and maximum voltage rating of high voltage Silicon devices. For instance, Silicon IGBTs rated at >3.3 kV are limited to switching frequencies of <1 kHz, making converters large, bulky, expensive and requiring costly thermal management solutions such as liquid or heat-pipe cooling techniques [1]-[5]. Promising new utility applications such as the Solid State Transformer (SST), are therefore severely limited by Si IGBT device voltage rating and capability.

Wide bandgap (WBG) devices based on Gallium Nitride (GaN) and Silicon Carbide (SiC) are promising technology and are commercially available up to 650V and 1700V respectively. SiC devices also hold the promise for exceptional performance at even higher voltages. The availability of SiC devices at 3.3 kV to 15 kV, with the promise of fast switching speeds and low conduction losses, has stimulated interest and enhanced

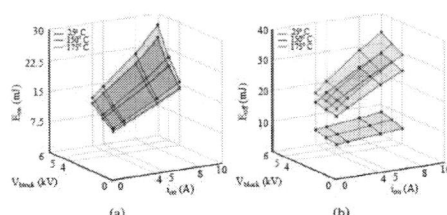

Fig.1: 15kV/40A SiC IGBTs (a) turn-on and(b) turn-off losses distribution [1]-[3].

R&D effort of all leading OEMs and even small companies. In particular, high power applications such as railway traction, medium voltage drives and grid connected converters, could potentially benefit significantly by dramatically reducing the number of series-connected modular stages needed for grid tied or output voltages of 13.8kV and higher. The possibility of switching such HV SiC devices at frequencies of >10 kHz, also raise the possibility that fully bidirectional SST with MV high-frequency transformers could be viable and provide potential benefits of increased power density, volumetric density, weight and efficiency. This paper proposes that viable high-voltage high-power converters using wide band gap SiC devices, will require the use of soft-switching, in particular zero-voltage switching techniques, to enable converter operation in presence of parasitic capacitances and inductances due to high dv/dt and di/dt associated with switching transitions of HV SiC devices. The multi-physics issues of thermal management, dielectric insulation, and parasitic inductance and capacitance, severely interact with the high switching speed of SiC devices. Active gate control to slow down the switching speed of the SiC device is complex and results in high switching loss. Converter topologies that realize ZVS operation over a wide operating range are needed to effectively deliver on the promise of wide band gap SiC devices. The paper will discuss the different available HV-SiC devices with respect to their application space for MV power conversion systems. This paper will also highlight the electrical characteristics of the HV-SiC devices regarding the power loss and thermal management issues for their applicability for MV power converter design.

II. COMPARISON BETWEEN THE PERFORMANCE OF THE HV-SIC IGBTS AND HV-SIC DEVICES

Currently there are two competitive HV SiC devices for medium voltage applications - 15kV SiC IGBTs and 15 kV SiC MOSFETs. This section compares the

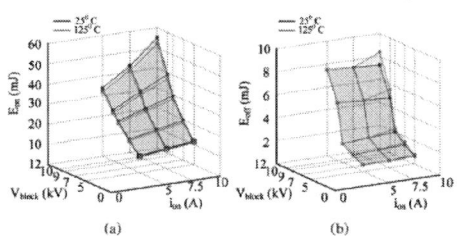

Fig.2. 15kV/20A MOSFET (a) turn-on and (b) turn-off losses [1]-[3].

conduction loss and switching loss of both these devices. This comparison will help in selection of either of the devices for specific power converter applications. The switching losses are measured by the clamped inductive double pulse test set-up to required voltage levels. Conduction losses are measured by forward conduction test. To measure the switching losses, an optimum gate resistance value of 33Ω (Rg_{on}) and 10Ω (Rg_{off}) has been

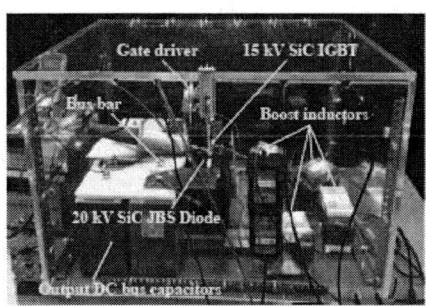

Fig.3. Boost converter test setup. [3]

used for both SiC IGBT and MOSFETs [4].

Switching characteristics of 15kV SiC IGBTs are shown in Fig.1 at 175°C. From Fig. 1, the IGBT turn-on

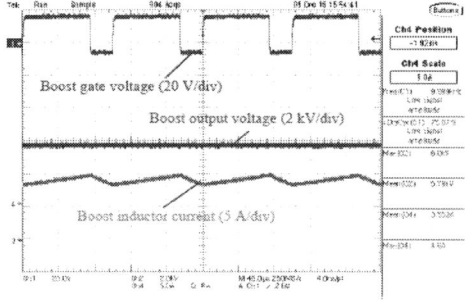

Fig.4. Boost converter waveform for 15kV/20A, SiC MOSFET, 6kV output DC bus, 5 kHz, 5.6kW.

and turn-off losses at 5kV, 5A are 9.4 mJ and 19 mJ respectively. Losses are found to be higher than the

MOSFETs whose turn-on and turn-off losses at 5kV, 5A are 6.3mJ and 1.5mJ respectively as shown in Fig. 2.

Hence, for higher switching frequency operation, SiC 15kV MOSFET is a better choice compared to SiC 15kV IGBT. The DPT set up parameters are given in Table I. Loss data are captured for voltage and current ranges in which a typical three phase converter is designed to

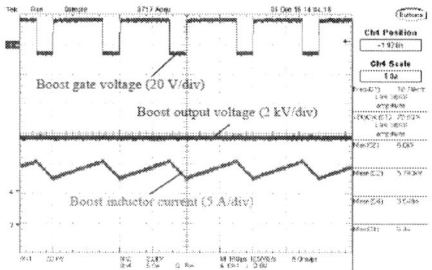

Fig.5. Boost converter waveform for 15kV/20A, SiC IGBT, 6kV output Dc bus, 5 kHz, 5.6kW. [4]

operate.

The V-I characteristics shows that voltage drop is 4.31V for 15kV SiC IGBT at 150°C. For 15kV SiC MOSFET, the voltage drop is 8.25V at 150°C. This shows that the voltage drop in SiC 15kV MOSFET is almost two times that of the SiC 15kV IGBT. Specific on state resistance of MOSFET increases from 0.64Ω at 25°C to 1.65Ω at 150°C. Therefore, the voltage drop at a current level of 10A is going to increase from 6.5V to 16.5V at 150 °C. At the same time, 15 kV/20 A SiC IGBT voltage drop at 10A only varies from 4.64 V at room temperature to 4.92 V at 150°C at 10A current. This makes 15 kV/40 A SiC IGBT the most suitable candidate for high temperature, high power application which is one of the main target areas and advantages for using SiC devices. At lower current levels, switching loss dominates at high switching frequencies, and thus, SiC 15kV MOSFET is preferred.

The 15 kV/40 A SiC IGBT and 15 kV/20 A SiC MOSFET are tested via heat run test for their continuous

Fig.6. Converter stack with series connected 1.7kV SiC MOSFETs .

operation. The devices are hard-switched in a dc-dc boost converter. Heat sink temperature (θ_h) is monitored by thermal camera and the junction temperature (θ_j) is calculated from the known value of thermal resistance. The camera emissivity setting is 0.9. Tests are conducted for 30min to ensure that the device reaches thermal equilibrium. The boost converter test setup is given in Fig.3 while the test parameters are given in Table. I.

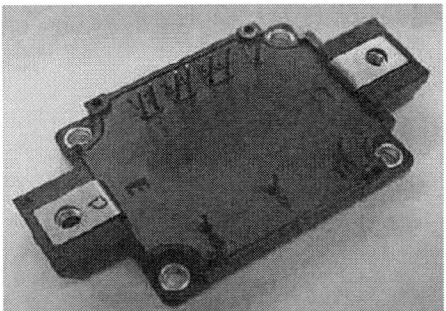

Fig. 7. 10 kV silicon carbide MOSFET module [3].

The gate resistances Rg$_{on}$ and Rg$_{off}$ are 33 Ω and 10 Ω, respectively. Fig. 4 shows the boost converter output for 6kV output DC bus and 5.6kW load output. Similarly, Fig.5 shows the waveforms with SiC 15kV MOSFET at 6 kV output dc bus voltage, 10 kHz switching and 5.6 kW. It can be seen that the peak inductor current at 5 kHz is 5.4 A while the peak current at 10 kHz is 4.6A. The ripple current is decreasing with the increase in switching frequency for the same 80mH inductor. This has the same effect in the grid connected 3-phase converter, where the switching ripple peak value in the grid current reduces with increase in switching frequency. In other words, the filter size can be smaller to achieve same peak ripple current. This is a key benefit of using SiC devices with increased switching frequency.

III.   SERIES CONNECTION OF 1.7KV SIC MOSFETs

Medium voltage and high power applications require high current rating devices. 10kV / 15kV SiC MOSFETs and 15kV SiC IGBTs modules are currently not available in high current rating. The typical die current rating is only 15A to 18A mainly due to high specific on

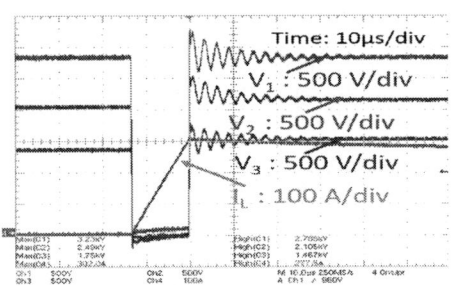

Fig.8. Turn-off dynamic characterization at 2800V, 300A for Rg= 4.7Ω, Tj=25°C.

resistance (R$_{on\_specific}$). This has led to increased interest in series connection of 1.7kV SiC MOSFETs for high

voltage high power applications. SiC 1700V MOSFETs are commercially available at high current ratings and up to 300A modules. Therefore, they can be connected in series to obtain the required higher voltage rating.

Fig.8 shows the characterization of the series connection of 1.7kV SiC MOSFETs at 2800V and 300A switching current. The voltage imbalance across the device is less than 10% of the nominal base operating voltage (V$_{dc}$/n = 2800V/4 = 700V). The inductor current is shown in the circuit to reach the 300A.

Series connection of 1700V SiC MOSFETs require

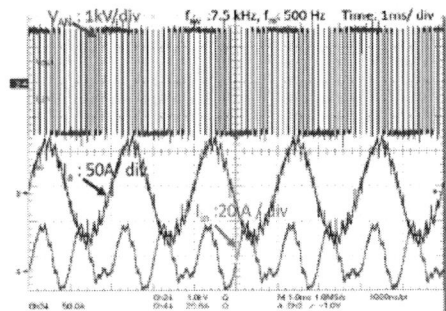

Fig. 9. Voltage and current waveform of the series connected 1.7kV MOSFETs of half bridge topology.

static and dynamic balancing of voltages. To investigate the series connection of 1.7kV SiC MOSFETs, a converter stack as shown in Fig. 6 has been developed with 1.7kV MOSFETs using simple R-C snubber circuit for the dynamic voltage balancing. Fig.9. shows the input current, output phase voltages, phase current with four series connected MOSFETs of 1.7kV per arm of the half bridge topology. The input voltage to the system was 3000V and the output converter current was up to 50A. The successful operation of the converter with series connection using simple RC snubbers gives another option for medium voltage high power converters without only requiring designs based on 10kV or higher voltage SiC devices.

IV.   ASYNCHRONOUS MICROGRID POWER CONDITIONING SYSTEM

The Asynchronous Microgrid Power Conditioning System (PCS) is an important component in realizing the future smart grid systems. It can be used for interconnection of two asynchronous micro-grids as well as for the interconnections of the community micro-grid (4.16 kV AC) to the upstream network (13.8 kV AC). The use of the PCS for the interconnection of the community micro-grid with the upstream network allows power decoupling together with galvanic isolation. Therefore, flexible voltage and frequency operation of the community micro-grid can be permitted, which can enable Distributed Renewable Energy Resource (DRER)

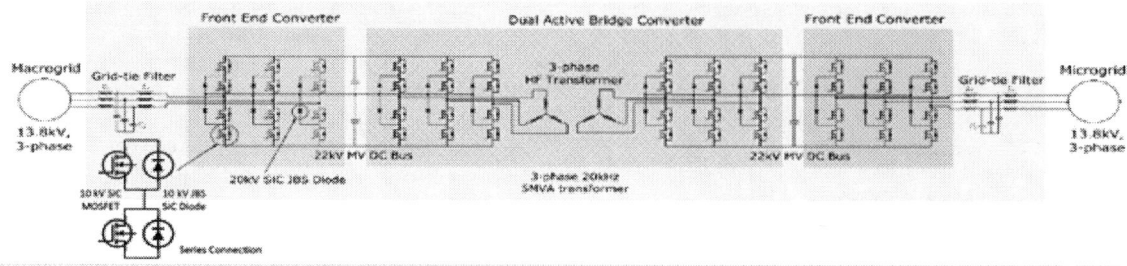

Fig.10. Proposed Asynchronous micro grid power conditioning system.

and Distributed Energy Storage Energy (DESD) integrations, and also enhance system stability. The bi-directional power flow capability of the PCS allows for

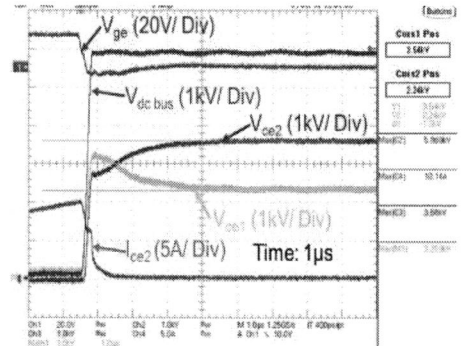

Fig.11. Dynamic Voltage sharing in series connected SiC MOSFETs.

the power import/export from the asynchronous micro-grid and therefore facilitates controlled power flow, which enables and simplifies the micro-grid management. A similar product for interconnection of medium voltage asynchronous always-islanded grids is commercialized by Pareto Energy [18]. It utilizes back-to-back AC/DC/AC power converters to connect 4.16kV ac grids, mainly limited by the voltage blocking capability of silicon IGBTs used in the interconnector with bulky 60Hz transformer on both ends. To connect 13.8 kV ac grids, a dc bus voltage of 22kV is required, which is difficult to implement with silicon IGBTs. Employing high voltage SiC devices (10kV SiC MOSFETs and 15kV IGBTs) in multilevel converters facilitates the implementation of the 13.8kV ac grid connector. A 3-level Neutral Point Clamped (3L-NPC) converter based Asynchronous Micro grid Power Conditioning System is shown in Fig.10. It has three power conversion AC/DC - isolated DC/DC - DC/AC stages comprising.

Three phase, 3-level Neutral Point Clamped converter has been used for DC/AC converter design as well as bidirectional and isolated DC-DC converter design with appropriate isolation at medium voltage. Two 10 kV, 20 A SiC MOSFETs modules as shown in Fig.7 is used in series per switch for nominal DC voltage blocking of 11kV. Fig.11 depicts the experimentally obtained voltage sharing on the series connected MOSFETs. Another factor that should be taken into account is the design of

medium voltage and medium frequency transformer. 11kV DC bus voltage operation with SiC MOSFETs involves a high dv/dt (of the order of 60-100kV/us) operation which causes a lot of current due to the coupling capacitance of transformer windings and coupling capacitance between the primary and secondary of the transformer. Therefore, an appropriate winding arrangement and transformer design is required to obtain a low value of coupling current due to parasitic coupling capacitances.

V. TRANSFORMER LESS INTELLIGENT POWER SUBSTATION (TIPS)

Solid State Transformers (SST) are being evaluated for replacement of bulky 60Hz distribution transformers and to enable DRER and DESD integration as well as for fast and large scale EV charging infrastructures [6]-[9].

Solid State Transformers (SST) provide a bidirectional power flow and can also provide filtering of harmonics in the grid. It can also act as static VAR compensator at the PCC (MV side) to provide voltage and reactive power flow control in the distribution feeder. Currently the main challenge to realize SST is the high voltage limitations of power devices. Thyristor can be connected to MV grid, but their performance is limited by low switching operation. Si-IGBTs have maximum blocking voltage capability of 6.5 kV which makes them restricted for MV applications, due to lower voltage blocking capability. Cascaded H-Bridge connections can provide a solution to connect to the MV grid but will result in reduced system efficiency due to large number of power devices. In addition, the switching losses (particularly E_{off}) in Si-IGBTs are higher due to tail current which will reduce the overall efficiency of the system. With the advent of HV SiC devices these practical issues and high-voltage device limitations to build SST will be largely mitigated. 15kV SiC MOSFETs and IGBTs can be switched at high speed and higher frequency compared to the 6.5 kV Si IGBT devices. The faster switching and high voltage blocking capability is the prime motivation to use HV SiC devices for the SST topology.

The proposed topology is termed as Transformer less Intelligent Power Substation (TIPS). The main motive of proposed topology is to replace the 60Hz bulky transformer with the solid state power electronic converters. Therefore "Transformer less" (power transformer) acronym is used. The proposed topology can be used from 13.8 kV to 480V grid as per the grid

3996

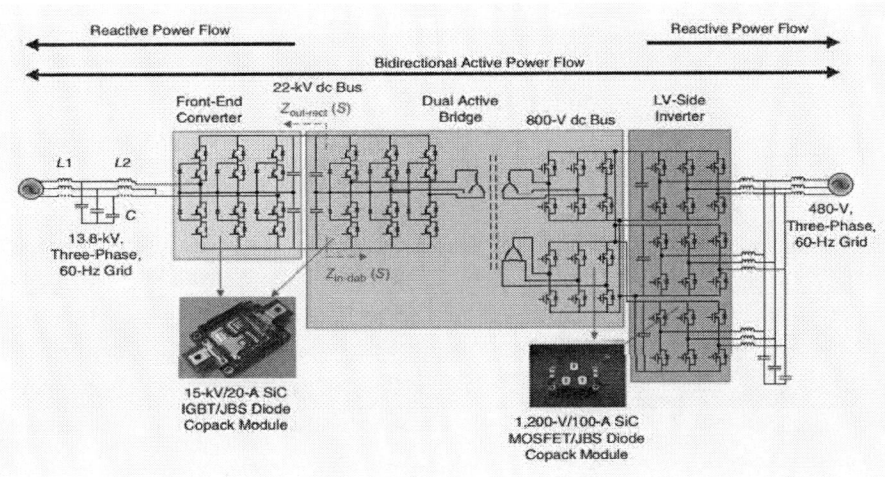

Fig.12. Basic block diagram of TIPS. [2]

requirement. Fig.12 elaborates the basic block diagram for the proposed TIPS topology.

In the proposed power topology of SST, the three level NPC inverter is proposed for rectification of the voltage input at AC level. This converter is designed using 15kV SiC-IGBTs. The converter is operated grid connected with 5kHz switching frequency. This converter is

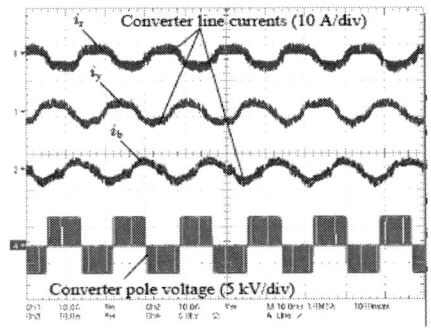

Fig.13. Three phase current and output waveform with 4.16kV Ac grid, 8 kV DC bus and 9.6kW power.

mentioned as Active Front End Converter setup. The proper PLL and filters are implemented at the grid input to control the operation of converter in closed loop and

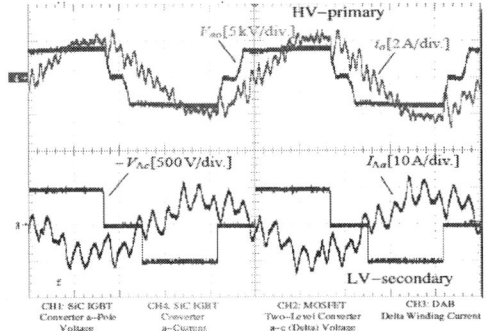

Fig.14. Experimental waveform of DAB at 8kV DC bus input with 9.6kW power output at switching frequency of 10kHz.

grid synchronization. Synchronizing grids with the input frequency is difficult in the case of SiC based devices because their switching speed is very high compared to silicon counterpart. This converter is implemented in FREEDM systems center and tested with 4160V grid

Fig.15. Three phase high frequency transformer used for DAB.

connection. Fig.13 shows the experimental result of Active Front End Converter at 4.16kV C grid and 8kV DC bus output voltage. The current waveforms are AC with proper phase shifts and 60 Hz frequency. The harmonics have been kept in control by using appropriate

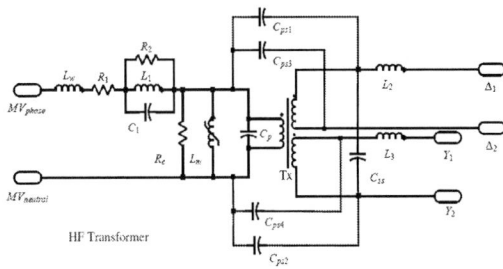

Fig.16. MV –HF transformer equivalent circuit.

PWM and input filters together with multi-loop controls.

The rectifier DC bus voltage is maintained at 22 kV . This high voltage DC is stepped down to isolated low

3997

voltage DC (800V) using a high frequency link DC-DC Dual Active Bridge (DAB) converter.

The primary side of the HF link converter has a three-level NPC converter with 15kV SiC-IGBTs and secondary side is paralleled with two standard two-level

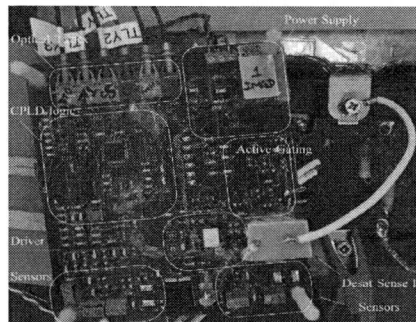

Fig.17. Intelligent gate driver for medium voltage application.

three-phase converters. This DC-DC DAB converter is soft-switched at 10 kHz. The DAB converter input side 3-level converter poles are built using 15kV SiC-IGBTs. The output side module consists of two level converters with 1200 V SiC devices. The other end of transformer is 800V DC bus. This 800V DC bus is integrated with 480V output of the inverter. TIPS provides a new concept for solid state replacement of a conventional 13.8kV to 480V distribution transformers. Therefore, the direct connection between the distribution grid (13.8 kV) and utility sector (480V grid) is possible with high voltage 15kV SiC devices. Fig.14. shows the experimental waveform of the DAB converter with voltage and current at HV side and LV side. It can be observed that the converter voltage output is three level.

The DAB transformer is an important component in the design of the DAB converter. Its construction with the SiC switches is a challenging problem. The transformer is used for step down as well as isolation purpose. Fig.15 shows the transformer structure used for the DAB setup. Single phase transformers with star connection at one side and star-delta connection on the other side were used for the 3-legged DAB converter operation.

It is obvious from the structure of the TIPS that it is more compact and lighter in weight compared to a conventional distribution transformer. However, an efficiency and performance comparison is required between TIPS and a conventional distribution transformer. A conventional distribution transformer has an efficiency of 99% while supplying a 1MVA, 480V feeder. From measured and simulation study of the TIPS with SiC 15kV IGBT devices, the efficiency is 98% at rated load. In order to study and reduce the common mode current flow into the converter structure, it becomes important to model the parasitic inductance and capacitance of the transformer. Specially, in the case of fast switching transition due to sic devices, the common mode and EMI issues are very prominent in the converter design which need proper modelling and mitigation techniques. Fig.16 shows the equivalent circuit of the

transformer taking the parasitic capacitances into account. The values of parasitic components are mentioned in Table. II.

The efficiency of TIPS varies with the varying nature of the load. The variation is due to different amount of reactive power which is injected into the grid. In order to simplify the following discussion, the 1MVA load is partitioned into real power of 800kW and reactive power of 600kVA load. The conventional transformer will supply to the feeder the same amount of load and draws the same real power from the utility grid (13.8kV).

VI. INTELLIGENT MEDIUM VOLTAGE GATE DRIVER

Medium Voltage applications have working voltage up to 15kV. This makes it important to ensure the safety measures and protection system in the converter operation. The protection of the medium voltage devices requires appropriate sensing and decision making for the gate driver operation to turn off the SiC devices before its failure. This necessitates the use of intelligent medium

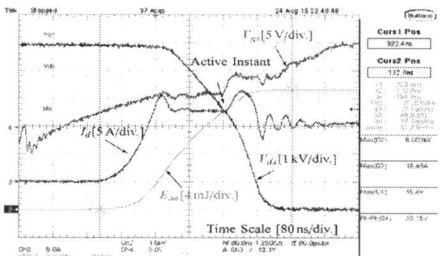

Fig.18. Experimental waveforms for the active gating test and implementation of the IMGD in double pulse test.

voltage gate driver in the system operation. Fig. 17 shows the medium voltage intelligent gate driver developed in the FREEDM Systems Center. The driver has the protection system and logic elements (CPLD) situated inside the design. The driver has also active gating implementation for reducing the switching losses in the converter circuits.

Fig.18 represents the double pulse test done with active

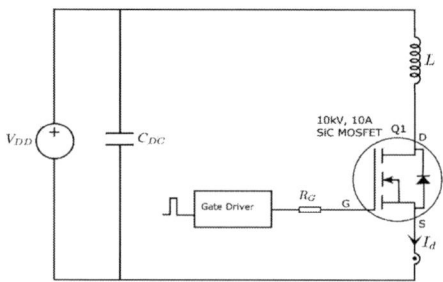

Fig.19. Unclamped inductive switching (UIS) circuit for avalanche breakdown.

gating in the intelligent gate driver [10]-[11]. The experiment is performed on 10kV SiC device with 6kV voltage application and 10A current. The starting resistance chosen is 50Ω. The active circuit is switched during the transition to 4.7 Ω gate source resistance. The delay in active gating can be tuned accordingly to obtain

3998

desired shape of the drain source voltage and therefore active control of the device V-I switching loci.

VII. AVALANCHE CHARACTERIZATION OF 10kV MOSFETS

Avalanche characterization of SiC devices is necessary to determine their ruggedness in the case of overvoltage. It also helps in determining the operating time of the protection system in the gate driver to avoid overvoltage fault. The device with high avalanche energy breakdown capability is preferred because it has better cost effective gate driver protection system. Taking the importance of avalanche characterization into account, the avalanche

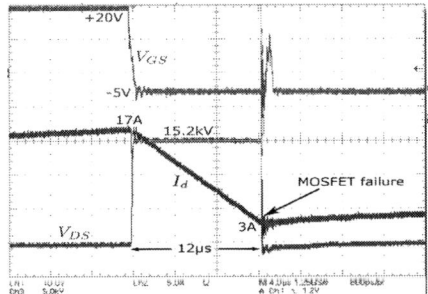

Fig.20. Avalanche breakdown test waveform of the SiC MOSFETs.

testing is performed on the 10kV SiC MOSFET dies [12]. Unclamped inductive switching (UIS) was used to generate the overvoltage on the MOSFETs drain source terminal to push the device into avalanche breakdown mode. Fig.19 represents the schematic of the UIS test circuit. The energy stored in the inductor is pumped into the device during turn-off, pushing it into avalanche breakdown.Fig.20 shows the avalanche breakdown test results for 10kV SiC MOSFET dies. The breakdown voltage is pushed till 15kV. The avalanche energy for failure is measured to be 1.82J.

VIII. MEDIUM VOLTAGE DRIVES

Medium voltage drives are suitable for the application where small weight and footprint is required for large rated machines with no reduction in efficiency [13]-[17]. MV drives are popular compared to traditional low voltage drives due to better efficiency for the same power level and higher power density. For example, with MV drives, the copper loss in the motor coils is small due to small RMS current for same power level [13]-[17]. Currently such MV drives are composed of multilevel converters made of silicon (Si) IGBTs [14]. The Si IGBTs are limited in voltage to 6.5 kV and are highly inefficient if switched at more than 300 Hz. Due to lower voltage Si IGBTs, the medium voltage drives require multilevel topology to satisfy the same voltage level. Alternatively, the cascaded converters can be used to attain medium voltage operation in the converter [15], [16].

The practical limitations of the cascaded converter modules and the DC bus voltage balancing issues do not allow the motor to achieve high speed in the MV drives

system using this topology. This leads to implementation of the mechanical gears to increase the frequency. Mechanical gears reduce the system efficiency and make the system bulky with need for regular maintenance. This calls for an alternative approach. Recently developed 10 kV silicon carbide (SiC) MOSFETs are being evaluated for MV converters. These SiC MOSFETs can be

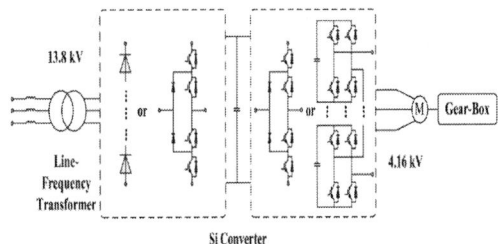

Fig.21. Typical Si Based Low-speed MV Drive Configuration

switched at higher switching frequencies ($\geq$ 10 kHz) at MV levels with smaller switching loss compared to Si IGBTs. Also, due to very small specific on-resistance, the conduction loss is very small. Hence, these HV SiC devices can be used in a simple three-phase, 2-level

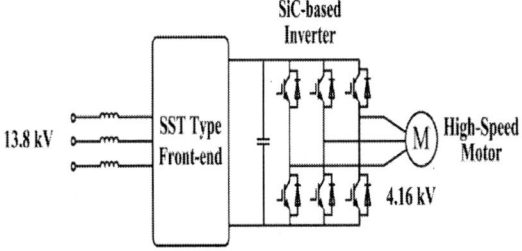

Fig.22. SiC Based MV Direct Drive Configuration

topology switching at 10–20 kHz, and used for high speed MV motor drives with $\geq$ 2.3 kV ac voltage and 6 kV dc bus voltage. Single phase converter topologies

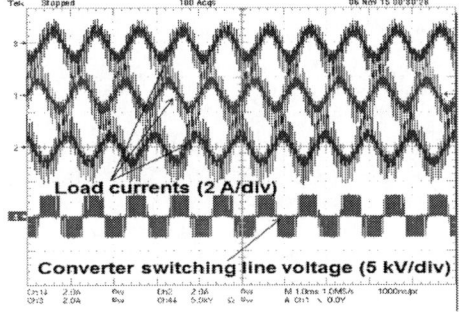

Fig.23. Three-phase converter waveforms at 3 kV dc bus voltage, 900 V ac line voltage rms, 1.45 kW load and fm = 1000 Hz

based on 10 kV SiC MOSFETs are reported in [15], [16]. The usual topology of the medium voltage drives is shown in Fig.21. In contrast to that, proposed topology of SiC based medium voltage drive is shown in Fig.22. It

3999

can be observed that the gearbox has been completely eliminated by using the high frequency drives enabled by HV SiC devices.

FREEDM systems Center has worked on the implementation of the high frequency drives using Medium Voltage SiC devices [17]. Experiments have been performed on the MV drives operation at 6kV DC bus input and 3kV output at 60 Hz and 3kV DC input with 900V output at 1000Hz. Fig.23 represents the output voltage waveform of the MV drive inverter with 1000Hz output. The inverter has been tested up to 1.45kW motor load. In future it is expected to test the system at 6kV operation.

Table.I.

Parameter	Values
DC bus capacitance	30uF
DC bus capacitor ESL	30nH
Freewheeling Diode	20kV/10A
Load inductor type	8mH/ 30A
Load inductor parasitic capacitance	<10pF
Turn on gate resistance	33Ω
Turn off gate resistance	10Ω
Turn on gate voltage	20V
Turn off gate voltage	-5V
Boost Converter Parameters	
Boost inductor	80mH
Output Dc Bus capacitance	30 μF
Test Duration	30 min
Diode used	20kV/10A SiC JBS

Table. II.
Medium voltage HF Transformer specifications

Parameter	Description value
C_1	100 pF
L_m	293 mH
L_w	500 nH
C_p	50 pF
L_{fl}	400 μH

IX. CONCLUSION

Wide Band Gap HV SiC devices enable various applications in the industrial and high power sector. SiC based HV (10-15kV) devices are promising technology for the development of the high power Solid State transformer (SST), asynchronous grid connection and medium voltage drives. Application with these HV SiC devices enable increased switching and operating frequency, and hence reduce the volume, size, losses, weight and cost of the passive energy storage devices for medium voltage power conversion systems. Several case studies of MV power conversion systems enabled by HV SiC devices based on experimental validation are presented as potential commercialization examples.

ACKNOWLEDGEMENT

The work on TIPS was supported by ARPA-E. Research on Asynchronous microgrid Power conditioning system is funded by DOE NNMII POWERAMERICA Institute at NC State University.

REFERENCES

[1] K. M. et al., "A Transformerless Intelligent Power Substation: A three-phase SST enabled by a 15-kV SiC IGBT," in IEEE Power Electronics Mag, Sept. 2015.

[2] S. Madhusoodhanan, K. Mainali, A. Tripathi, D. Patel, A. Kadavelugu, S. Bhattacharya and K. Hatua, ""Harmonic Analysis and Controller Design of 15 kV SiC IGBT Based Medium Voltage Grid Connected Three-Phase Three-Level NPC Converter," in IEEE Transactions on Power Electronics , vol.PP, no.99, pp.1-1".

[3] S. Madhusoodhanan, ""Power Loss Analysis of Medium-Voltage Three-Phase Converters Using 15-kV/40-A SiC N-IGBT," in IEEE Journal of Emerging and Selected Topics in Power Electronics.

[4] K. Vechalapu, S. Bhattacharya, E. Van Brunt, S. H. Ryu, D. Grider and J. W. Palmour, "Comparative Evaluation of 15-kV SiC MOSFET and 15-kV SiC IGBT for Medium-Voltage Converter Under the Same dv/dt Conditions," in *IEEE Journal of Emerging and Selected Topics in Power Electronics*, vol. 5, no. 1, pp. 469-489, March 2017.

[5] A. K. Tripathi et al., "A Novel ZVS Range Enhancement Technique of a High-Voltage Dual Active Bridge Converter Using Series Injection," in IEEE Transactions on Power Electronics, vol. 32, no. 6, pp. 4231-4245, June 2017.

[6] Wim van der Merwe,Toit Mouton, "Solid-State Transformer Topology Selection," Industrial Technology, IEEE, International Conference,pp.1-6, 2009.

[7] Abedini, A.; Lipo, T, "A novel topology of solid state transformer," Power Electronic & Drive Systems & Technologies Conference (PEDSTC), pp.101-105,2010.

[8] Jianjiang Shi, Wei Gou, Hao Yuan, Tiefu Zhao, and Alex Q. Huang, "Research on Voltage and Power Balance Control for Cascaded Modular Solid-State Transformer," IEEE Trans, on Power Electronics, pp.1154- 1166,Vol.26, No.4, April 2011.

[9] Tiefu Zhao, Liyu Yang, Jun Wang, Alex Q. Huang, "270 kVA Solid State Transformer Based on 10 kV SiC Power Devices," Electric Ship Technologies Symposium,2007, pp.145-149,2007.

[10] S. Madhusoodhanan, K. Mainali, A. Tripathi, K. Vechalapu and S. Bhattacharya, "Medium voltage (≥ 2.3 kV) high frequency three-phase two-level converter design and demonstration using 10 kV SiC MOSFETs for high speed motor drive applications," *2016 IEEE Applied Power Electronics Conference and Exposition (APEC)*, Long Beach, CA, 2016, pp. 1497-1504.

[11] A. Kumar, A. Ravichandran, S. Singh, S. Shah and S. Bhattacharya, "An intelligent medium voltage gate driver with enhanced short circuit protection scheme for 10kV 4H-SiC MOSFETs," *2017 IEEE Energy Conversion Congress and Exposition (ECCE)*, Cincinnati, OH, 2017, pp. 2560-2566.

[12] A Kumar et.al ," Single shot avalanche energy characterization of 10kV, 10A, 6H-SiC, MOSFETs.", Applied Power Electronic Conference, 2018.

[13] E. Cengelci, P. N. Enjeti, and J. W. Gray, "A new modular motormodular inverter concept for medium-voltage adjustable-speed-drive systems," IEEE Trans. Ind. Applicat., Vol. 36, no. 3, pp. 786-796, Aug. 2002..

[14] J. Dai, S. W. Nam, M. Pande, and G. Esmaeili, "Medium-voltage currentsource converter drives for marine propulsion system using a dualwinding synchronous machine," IEEE Trans. Ind. Applicat., Vol. 50, no. 6, pp. 3971-3976, Nov./Dec. 2014.

[15] R. Teodorescu, F. Blaabjerg, J. K. Pedersen, E. Cengelci, and P. N. Enjeti, "Multilevel inverter by cascading industrial VSI," IEEE Trans. Ind. Electron., Vol. 49, no. 4, pp. 832-838, Aug. 2002.

[16] G. Baoming, F. Z. Peng, A. T. de Almeida, and H. Abu-Rub, "An effective control technique for medium-voltage high-power induction motor fed by cascaded neutral-point-clamped inverter," IEEE Trans. Ind. Electron., Vol. 57, no. 8, pp. 2659-2668, Aug. 2010.

[17] S. Madhusoodhanan, K. Mainali, A. Tripathi, K. Vechalapu and S. Bhattacharya, "Medium voltage (≥ 2.3 kV) high frequency three-phase two-level converter design and demonstration using 10 kV SiC MOSFETs for high speed motor drive applications," *2016 IEEE Applied Power Electronics Conference and Exposition (APEC)*, Long Beach, CA, 2016, pp. 1497-1504.

[18] Pareto Energy, "GridLink Non-Synchronous Interconnection Platform", available online: [http://www.paretoenergy.com /interconnection/]

Soft-switching – The Key to High Power WBG Converters

Deepak Divan*, Zheng An, and Prasad Kandula

Center for Distributed Energy, Georgia Institute of Technology, Atlanta, GA, USA

*ddivan@gatech.edu

Abstract— **This paper proposes that viable high-voltage high-power converters using wide band-gap SiC devices, will require the use of soft-switching, in particular zero-voltage switching techniques, if system parasitics and high switching speeds are to be managed. An easy to understand and use quantitative comparison of different approaches can help us with selection of specific technologies for given applications. This paper presents the limitations of current approaches, and proposes a new converter figure of merit (CFOM) as the basis for understanding the limitations of high power converters based on wide band-gap SiC devices, showing that soft-switching is the key.**

Keywords— *Soft-switching, Solid-state transformers, zero voltage switching, wide band-gap converters.*

I. INTRODUCTION

Silicon (Si) power devices have been the workhorse of industry for over three decades, and are used today in a wide variety of higher power applications including electric drives, renewable energy, automotive, traction and switched mode power supplies. Most industrial applications typically utilize 1200 volt IGBTs, which over the years have improved tremendously – offering low on-state voltage drop and switching frequencies of 5-10 kHz. On the other hand, the opportunity to realize higher power levels for large industrial and utility applications, are constrained by the capability of multi-kV rated silicon devices. For instance, Insulated Gate Bipolar Transistors, or IGBTs, rated at >3.3 kV, are limited to switching frequencies of <1 kHz, making converters large, bulky and expensive. Promising new utility applications such as the solid-state transformer, are severely limited by device capability.

Wide band-gap (WBG) devices based on Gallium Nitride (GaN) and Silicon Carbide (SiC) show promise and are finally becoming available in the market at up to 1700 volts. SiC devices also hold the promise for exceptional performance at even higher voltages. The initial availability of SiC devices at 3.3 kV to 15 kV, with the promise of fast switching speeds and low conduction losses, has stimulated interest and enhanced R&D effort on the part of market leaders. In particular, high power applications such as railway traction, medium voltage drives and grid connected converters, could potentially benefit significantly by dramatically reducing the number of series-connected stages needed to manage voltages of 13 kV and higher. The possibility of switching such devices at frequencies of >10 kHz, also raises the possibility that fully bidirectional solid-state transformers (SST) based on high-frequency transformers could eventually become real.

This paper proposes that viable high-voltage high-power converters using wide band-gap SiC devices, will require the use of soft-switching, in particular zero-voltage switching techniques, if the parasitics are to be managed. The multi-physics issues of thermal management, dielectric insulation, and parasitic inductance and capacitance, severely interact with the high switching speed of SiC devices. Active gate control to slow down the switching speed of the SiC device is complex and results in high switching loss. Converter topologies that realize ZVS operation over a wide operating range are needed to effectively deliver on the promise of wide band-gap SiC devices. The paper will discuss the limitations of current approaches, and will present a new converter figure of merit (CFOM) to validate the advantages of soft-switching in wide band-gap high-power SiC converters. For any converter, cost, size and performance are three major parameters to be considered. The proposed CFOM captures the above three criterion in a simpler way so that various topologies can be compared easily.

II. REPLACING SILICON WITH SILICON CARBIDE DEVICES IN POWER CONVERTERS

Many R&D laboratories around the globe have developed techniques to integrate the high voltage SiC devices into useful power converters. Most have approached this 'performance upgrade' exercise by replacing Si devices with SiC devices, ensuring the gate-drives, sensors, control strategies and mechanical/thermal design are all modified to extract the most from the capability that SiC devices bring - in particular higher voltages and higher switching speeds when compared with their Si counterparts [1]. While this seems fairly straightforward, it is rarely so, especially for high voltage implementations. A simple replacement of Si devices with SiC devices creates its own set of problems. The extreme dv/dt (50 kV/μs) and switching speeds of <100 ns (and induced high di/dt), cause high parasitic current/voltage spikes and high EMI, with dramatic impact on converter viability. In a typical voltage source inverter configuration (Figure 2), even 1 nF of capacitance between device and a grounded heat-sink can result in not only a 50 A of displacement current, but also severe challenges such as maintaining even voltage sharing across inductor or motor windings, and EMI stretching out to >100 MHz. As device voltage ratings increase to 4.5 kV and beyond, the need for additional guard rings and insulation requires increased spacing between devices, increasing the parasitic inductance (Figure 1). Conventional sensors that handle the isolation and yet provide the needed bandwidth, are

The 2018 International Power Electronics Conference

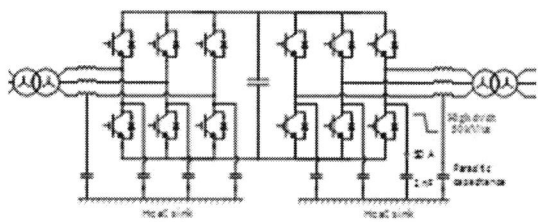

Figure 2: Common mode currents in standard voltage source inverter (VSI).

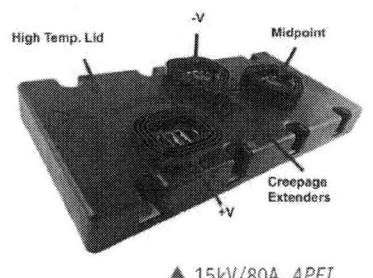

▲ 15kV/80A, *APEI*

Figure 1: Additional guard rings and increased spacing between terminals in a 15 kV device resulting in higher parasitic inductance.

very expensive if they work effectively. Finally, while the ability to operate SiC devices at higher junction temperatures can yield benefits, it does require the development of additional control, power, magnetic and mechanical components that can handle the elevated temperatures as well. As such, this does not seem to be a simple exercise of simply 'swapping' out the Silicon devices and replacing them with Silicon Carbide!

III. SOLID-STATE TRANSFORMERS

The situation becomes even more challenging when one looks at the applications that could perhaps benefit the most from high voltage SiC devices – the Solid-state Transformer (SST), often regarded as the 'holy grail' of power electronics. First proposed by McMurray in 1969,

the SST converts available voltage into another voltage that is galvanically isolated, but at high frequency – thus allowing reduction in the size (and presumably cost) of a comparably 60 hertz rated transformer [2]. The state of the art is still the use of a back-to-back voltage source converter (BTB-VSC) with a 60 hertz transformer. Over the last 30 years, elimination of 60 Hz transformers has resulted in dramatic size and cost reductions in switched mode power supplies (SMPS) – and it is hoped that SSTs will see the same size/cost reduction. Unlike SMPS, a SST generally transfers power bidirectionally with control of voltage/current between two multi-phase AC sources/loads, achieving galvanic isolation at higher frequencies.

The advent of wide band-gap (WBG) silicon carbide (SiC) devices at 3.3 kV to >10 kV, with the promise of fast switching speeds and low conduction losses, have resulted in two distinct types of SST topologies. The first replaces Si devices with faster SiC devices in known SST configurations, resulting in high dv/dt with high inter-turn and inter-winding voltage stress inside the transformer, high displacement currents caused by charging of parasitic capacitances, and high EMI, reducing converter viability [1].

A better approach would use high-voltage SiC devices in a soft-switching converter, with controlled dv/dt and tightly controlled parasitic inductances to manage the high di/dt. The Dual-Active Bridge (DAB) [3] converter is an example of a zero-voltage switching (ZVS) high power dc/dc converter that has been used as a building block for a cascaded high-voltage high-power grid connected converter at 7.2 kV and higher. The DAB converter is constrained in the voltage/load range over which soft-switching is obtained, and requires the voltage on the DC buses be maintained within a narrow range. This requires the use of additional AC/DC and DC/AC converters, especially if grid connection or solid-state transformer functionality is needed.

This approach to the SST, shown in Figure 3, has been used by many researchers including Akagi [4], Bhattacharya [5], Soltau [6] and Das [7]. DAB has been demonstrated to achieve as much as 5 MW at 5 KV DC and 1 kHz [6] and 1 MW at 13.8 kV while switching at 20

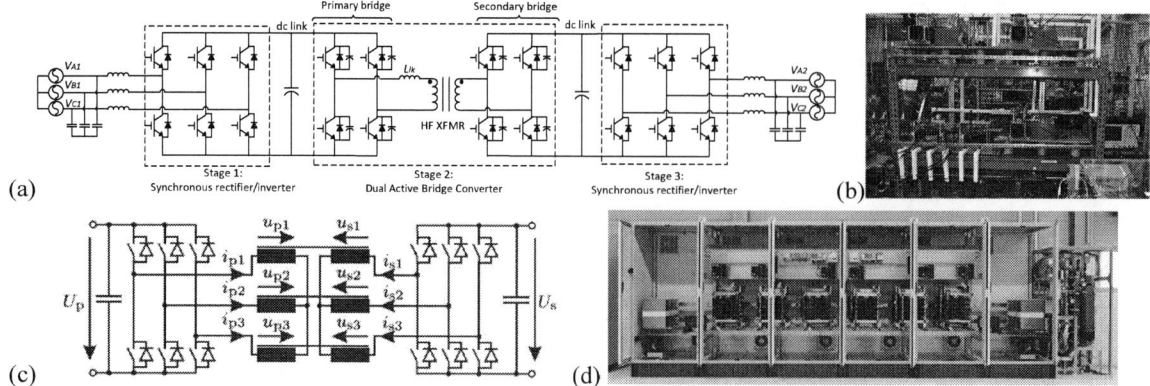

Figure 3: (a) Schematic of DAB converter for AC-AC SST application (b) Image of 13 kV SST based on DAB [8] (c) Schematic of DAB3 converter [3] (d) Image of 5 kV DC 5 MW DAB converter [6] .

4002

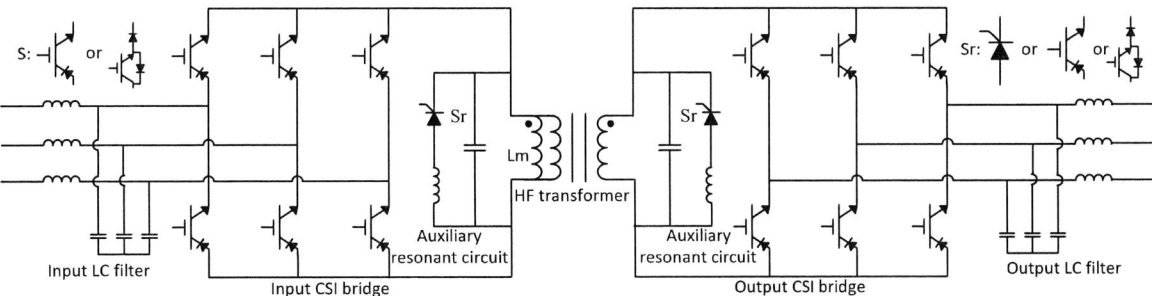

Figure 4:Schematic of bidirectional soft-switching solid state transformer (S4T) [2].

kHz [5] – certainly demonstrating feasibility. The use of hard switching rectifier/inverter stages still causes issues with high dv/dt, in some cases showing displacement currents that were greater than the load current. The impact of high dv/dt on common mode currents in cascaded converters is shown in [9]. The impact of dv/dt on high frequency transformer design is discussed in [10]. Another challenge with hard switching with ultra-fast WBG devices is the need for active gate drives that can control the dv/dt. This in turn requires very low loop inductances and very fast control. It should be noted that active control of dv/dt, while beneficial for EMI, can increase switching losses, making it less than desirable [11].

Another promising approach proposed recently is the concept of a Soft-switching Solid-state Transformer (S4T), which provides soft-switching and controlled dv/dt for all the devices across the entire load range in a 'universal' converter that can be used for dc and single or three phase ac medium-voltage grid-connected applications [12]. As shown in Figure 4, for a 3 phase-3 phase low-voltage application, the S4T uses two current source inverter (CSI) bridges connected with an air-gapped high-frequency transformer providing galvanic isolation and a limited amount of energy storage. Each bridge has an auxiliary resonant circuit, comprising of an active device, an inductor, and a capacitor, to provide ZVS conditions for all the main devices. The active device in the auxiliary circuit operates under zero-current switching.

To understand the applicability of high-voltage WBG devices in such an application, this paper considers a three-phase to three-phase S4T using 3.3 kV SiC MOSFETs. To realize higher voltages, one would need to stack single phase converter modules to achieve higher voltages (Figure 5). It should be noted that for S4T implementations requiring low-voltage on one side, one can use hybrid converters, with SiC on the HV side and Si on the LV side. For three-phase HV applications, three such stacks can be used. To understand and compare such diverse technologies and topologies, one needs to be able to extract fundamental properties using a systems level approach.

IV. CONVERTER FIGURE OF MERIT

As one looks at the possibility of replacing Si with SiC in a converter application, it is clear that there are trade-offs. To understand these, it is important to set a Reference Design base line for a specific applications that represents the state of the art today. Typical cost and performance metrics can be invoked in performing this assessment. One can define four components of such a 'Converter Figure of Merit' or CFOM, which are explained here.

A. Device Loss Factor:

A converter with high efficiency and low switching loss can clearly deliver more power for the same devices. Device utilization is maximized when the full capability of the device is used in conduction loss (not switching loss). One can then think of a Device Utilization as a product of the ratio of the conduction loss to the total loss in all the devices combined and the converter efficiency.

$$DU = \frac{\sum Cond\_loss}{\sum Cond\_loss + \sum Sw\_loss} * \eta. \qquad (1)$$

where DU is device utilization factor,

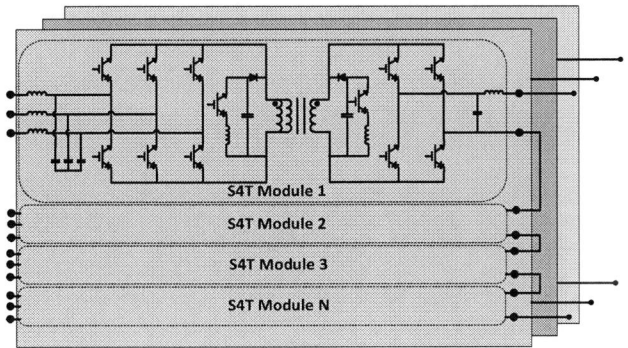

Figure 5: Implementation of modular S4T for HV applications.

$\sum Cond\_loss$ is the sum of conduction loss in all switches, $\sum Sw\_loss$ is the sum of switching loss including diode reverse recovery in all switches, and η is efficiency of the converter.

The Device Loss Factor (DLF), which is inversely proportional to DU, can then be defined as follows.

$$DLF = \frac{DU_{Ref}}{DU_{New}} \qquad (2)$$

where DU_{Ref} is the device utilization of the reference design and DU_{new} is the device utilization of the converter being compared.

The DLF for three topologies, back-to-back voltage source converter (BTB-VSC) (Figure 2), DAB based SST (Figure 3(a)), and S4T (Figure 4) are compared. The list of devices in various configurations are presented in Table 1. The results are presented in Table 2. Multiple devices are assumed to be connected in parallel to achieve required current rating.

TABLE 1
LIST OF DEVICES USED FOR COMPARISON.

Device	Model Used
1.2 kV Si	IKQ75N120CH3, Infineon
1.2 kV SiC MOSFET	C2M0025120D, CREE
3.3 kV Si IGBT	QI_3320002, Powerex
3.3 kV SiC MOSFET	UJ3D12100Z, UnitedSiC
1.2 kV SiC diode	3.3 kV/45 mΩ GEN3 SiC MOSFET, CREE
3.3 kV SiC diode	CPW3-3300-Z045B, CREE

A 1 kVA, 480 V BTB-VSC implemented using 1200 V Si devices and switching at 15 kHz has an efficiency of 94 %, and device utilization of 17 as per Equation (1). When the Si device is replaced with 1200 V SiC MOSFET, device utilization increased to 41 because of the reduction in switching loss and the DLF, as per Equation (2), reduced to 0.43. When the same comparison is done with a 3.3 kV Si device in a 1.5 kV / 300 kVA configuration, the DLF increases to 12.1, mainly because of the very high switching loss. For this very reason, implementations with 3.3 kV devices and higher are typically limited to 500 Hz – 2 kHz. With an assumed switching frequency of 2 kHz, the losses reduce to a more practical value of 2.9 % and a DLF of 10 is achieved. Henceforth, only 2 kHz switching frequency will be considered for 3.3 kV Si-based implementations.

Replacing 3.3 kV Si device with 3.3 kV SiC MOSFET will result in a DLF of 0.84, even at 15 kHz switching frequency - clearly emphasizing the benefit of WBG devices in HV implementations. SiC based DAB-SST (15 kHz) implementation has a DLF of 0.52 compared to 4.84 with Si-based (2 kHz) implementation. DAB based SST has a hard switching converter on either end of the DAB. Hence, the results for DAB-SST are similar to BTB-VSC. In the case of 3.3 kV Si-based S4T-SST implementation, the DLF is 0.36 compared to 10 and 0.68 in case of VSC-BTB and DAB-SST, respectively. This is because of lower switching loss in S4T compared to other two topologies. Please note that with 3.3 kV Si devices, the switching loss still exists even when the S4T is operated under ZVS condition, because of slow rise and fall times of 3.3 kV Si devices. In addition, SiC diodes are necessary even with Si switching device based implementation of S4T to avoid reverse recovery issues. When the Si (2 kHz) device is replaced with SiC (15 kHz) device, the DLF reduces from 0.36 to 0.19, even when operated at 15 kHz. The common thing across all topologies is that the DLF gets lower when Si device is replaced with SiC because of the reduction in the device switching loss. SiC device is necessary if operation at higher switching frequencies is desired. S4T based SST has lower DLF compared to other topologies because of ZVS across each switch under all operating conditions.

B. Device Rating Factor:

The voltage and current rating of the device is indicative of the price of the device. For a given application and system rating, one can define the voltage and current rating for the Reference Design. By way of example, for a 480 V / 100 kW drive application with a 2X overload rating, one may use 1200 V / 400 A IGBTs. If there were six devices with this rating, the cumulative VA is 2880 kVA. For a similar converter application using SiC devices, if one can achieve the same functionality using 350 Ampere devices (for instance because the switching losses are lower, which allows increased conduction loss), giving a cumulative VA of 2520 kVA. We can say the SiC based implementation has a device rating factor of 0.875.

Using a similar argument, a Device Rating Factor (DRF) definition is given by the following equation

$$DRF = \frac{\sum VA_{sw\_new}}{\sum VA_{sw\_Ref}} + 0.3 * \frac{\sum VA_{diode\_new}}{\sum VA_{diode\_Ref}} \qquad (4)$$

TABLE 2
COMPARISON OF PROPOSED DEVICE LOSS FACTOR (DLF) METRIC FOR VARIOUS TOPOLOGIES.

	BTB-VSC + 60 Hz Transformer					DAB Based SST		S4T Based SST	
	480 V AC – 480 V AC, 100 kVA		1.5 kV AC – 1.5 KV AC, 300 kVA			1.5 kV AC – 1.5 KV AC, 300 kVA		1.5 kV AC – 1.5 KV AC, 300 kVA	
	1.2 kV Si 15 kHz	1.2 kV SiC 15 kHz	3.3 kV Si 15 kHz	3.3 kV Si 2 kHz	3.3 kV SiC 15 kHz	3.3 kV Si 2 kHz	3.3 kV SiC 15 kHz	3.3 kV Si 2 kHz	3.3 kV SiC 15 kHz
Cond loss	1.1%	0.8%	0.3%	0.3%	0.5%	0.9%	0.9%	1%	1%
Sw loss	4.9%	1.1%	19.5%	2.6%	1.8%	2.6%	1.8%	2.0%	0.1%
Total loss	6.0%	1.9%	19.8%	2.9%	2.3%	3.5%	2.7%	2.1%	1.1%
DU	17	41	1.2	1.7	21.2	24.8	32.4	46.6	90
DLF	1.0	0.43	12.1	10	0.84	0.68	0.52	0.36	0.19

where VA_{sw_new} is the VA rating of the switches in the converter being compared, VA_{sw_Ref} is the VA rating of the switches in the reference converter, VA_{diode_new} is the VA rating of the diodes in the converter being compared, VA_{diode_Ref} is the VA rating of the diodes in the reference converter. The 0.3 number used in the equation is an assumption based on the estimation that the cost of a diode would be 30 % of the switching device of similar rating.

The DRF for three topologies, back-to-back voltage source converter (BTB-VSC), DAB based SST, and S4T are compared and presented in TABLE 3. DAB-SST has higher DRF (Si – 1.5, SiC 1.13) than BTB-VSC (Si – 1, SiC 0.75) because of multiple conversion stages. The S4T also has slightly higher DRF (Si – 1.2, SiC 0.98) than BTB-VSC because of the resonant switches and diodes. But S4T has much better DRF (Si – 1.2, SiC 0.98) compared to DAB-SST (Si – 1.5, SiC 1.13).

TABLE 3
COMPARISON OF PROPOSED DEVICE RATING FACTOR (DRF) METRIC FOR VARIOUS TOPOLOGIES (1.5 KV AC – 1.5 KV AC, 300 KVA, 15 KHz)

	BTB-VSC + 60 Hz X^{mr}		SST- DAB Based		SST-S4T Based	
	3.3 kV Si	3.3 kV SiC	3.3 kV Si	3.3 kV SiC	3.3 kV Si	3.3 kV SiC
ΣMVA - switches	7.92	5.94	13.2	9.9	9.2	6.9
ΣMVA - diodes	7.92	5.94	7.92	5.94	10.5	7.9
DRF	1	0.75	1.5	1.13	1.2	0.98

C. Reactive Component Factor:

The higher switching frequency achievable with SiC devices also enables a reduction in the size of reactive components, which in turn have a big impact on the size and cost of the converter. To ensure a fair comparison, converters compared should have similar ratings and functionality. This includes DC bus as well as input/output filter inductors and capacitors. The two elements driving size and cost are the energy stored in Joules, and to a more limited extent the VA rating of the device. Another important element in reactive components is the transformer. In the base-line case with the BTB-VSC, a 60 Hz transformer is used. This is large, bulky, lossy and expensive. The SST allows dramatic reduction in size, although requiring additional conversion stages to achieve high frequency galvanic isolation. Any topologies considered must be capable of bidirectional power flows, and be intrinsically capable of managing energy trapped in the leakage inductance. The DAB based SST and the S4T both qualify. Based on a 15 kHz switching frequency, the high frequency transformers for both SST implementations are dramatically smaller in size than the 60 Hz transformer, and have lower losses. Lower switching frequencies required with Si devices, will increase transformer size and reactive energy storage required.

One can then think of a Reactive Component Utilization for a converter as being

$$RCU = \frac{\sum Filter\,(Joules) + \sum storage(joules) + \sum Xmr_{VA}/F_s}{P_{out}/F\_line}$$
(5)

where P_{out} is the output power rating of the converter and the F_line is the input/output frequency (60 Hz) Xmr_{VA} is the transformer VA rating, F_s is the switching frequency. The Reactive Component Factor can then be defined as

$$RCF = \frac{\sum RCU_{New}}{\sum RCU_{Ref}}$$
(6)

The RCFs for the three topologies being considered are compared and presented in TABLE 4. The filters are chosen to maintain peak-to-peak voltage ripple < 3% and peak-to-peak current ripple <5%. For grid connected inverters the DC link is typically sized to meet 3000-6000 J/MVA to provide energy during transients [13]. In this analysis, DC capacitor is chosen to meet 5000 J/MVA requirement and also to limit voltage ripple to 1% p-p. In case of BTB-VSC, two 60 Hz transformers, one at either terminal, are considered. As expected, filter size decreases with increased switching frequency in the case of BTB-VSC and DAB-SST. S4T is a current source inverter. Hence the capacitor filter (333 J) is much larger compared to inductor (7 J). If only the filter size is considered, S4T-SST is comparable to DAB-SST or BTB-VSC operating at the same frequency. But S4T is a low inertia converter with significantly lower storage compared to the other two topologies. Please note that the transformer in S4T can be considered as an energy storage element – two inductors with common core. In the case of Si devices operating at 2 kHz, the RCF is 1.0, 0.34 and 0.18 for BTB-VSC, DAB-SST and S4T-SST, respectively. With SiC devices and 15 kHz switching, the RCF 0.79, 0.13 and 0.024 for BTB-VSC, DAB-SST and S4T-SST, respectively. As expected, the RCF is reducing with SiC devices because of higher switching frequency, which is enabling using smaller filters. S4T-SST has lower RCF compared to other topologies because SST is operated as a low-inertia converter with reduced storage requirements.

TABLE 4
COMPARISON OF PROPOSED REACTIVE COMPONENT FACTOR (RCF) METRIC FOR VARIOUS TOPOLOGIES (1.5 KV AC – 1.5 KV AC, 300 KVA).

	BTB-VSC + 60 Hz X^{mr} BTB-VSC		DAB-SST		S4T-SST	
	3.3 kV Si	3.3kV SiC	3.3 kV SiC	3.3 kV SiC	3.3 kV SiC	3.3 kV SiC
Fs(kHz)	2	15	2	15	2	15
Lf (J)	2857	381	2857	381	4	7
Cf (J)	6.6	28	6.6	28	2500	333
Storage (J)	2165	1500	2165	1500	240	30
Xmr (Joules)	10000	10000	150	20	**	**
RCU	3.0	2.38	1.0	0.38	0.55	0.072
RCF	1.0	0.79	0.34	0.13	0.18	0.024

** Transformer in S4T is considered as a storage element.

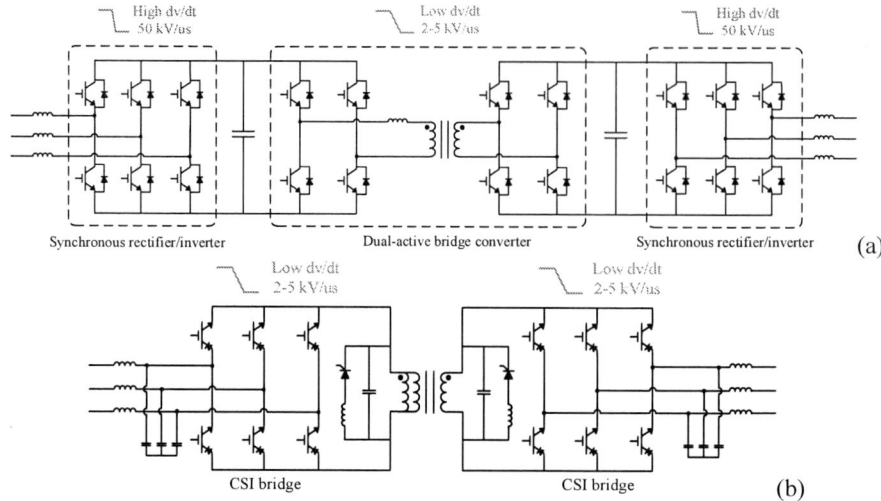

Figure 6: Comparison of dv/dt in a (a) DAB based SST and (b) S4T.

D. EMI Factor:

The higher dv/dt for WBG devices also creates issues in terms of parasitic currents flowing between device and heat sink, and makes voltage measurement challenging. Similarly high di/dt causes voltage spikes across parasitic loop inductances, and makes current measurement challenging. Lower dv/dt and di/dt are almost always better from an EMI and system perspective.

The EMI factor can then be defined as

$$EF = \frac{CM_{new}}{CM_{ref}} \qquad (7)$$

Where EF is the EMI factor, CM_{new} is the common mode current in the converter being compared and CM_{ref} is the common mode current in the reference converter.

The EF for three topologies, back-to-back voltage source converter (BTB-VSC), DAB based SST, and S4T are compared and presented in TABLE 5. The device junction to base plate capacitance is 100-1000 pF for 1200 V/75 A Si devices [14]. In this analysis 500 pF is assumed for 3.3 kV Si devices. The SiC dies are typically 10-20 % of Si dies [15], but the actual base plate may be larger to reduce thermal resistance. In this analysis, 300 pF is assumed for SiC devices. As shown in TABLE 5, the common mode current increases with SiC devices even with reduced parasitic capacitance. Again, the common mode current is maintained at a lower level in S4T-SST compared to other two implementations because of controlled dv/dt. This is because the dv/dt is contained at a lower level throughout the S4T, unlike the case in a DAB converter, as shown in Figure 6. The low dv/dt (500 V/μs) achieved in S4T was experimentally shown with a 3-phase 208 V / 10 kVA S4T prototype, as shown in Figure 7 [12]. For HV applications, desired dv/dt can be chosen by appropriate selection on resonant capacitor. A typical design point would limit the dv/dt to under 2 kV/μs.

TABLE 5
COMPARISON OF EMI FACTOR WITH AN ASSUMED DEVICE-TO-HEATSINK PARASITIC CAPACITANCE OF 500 pF FOR Si DEVICES AND 200 pF FOR SiC DEVICES.

	VSI-BTB + 60 Hz Transformer		SST-Based	DAB	SST-S4T Based	
	3.3 kV Si	3.3 kV SiC	3.3 kV Si	3.3 kV SiC	3.3 kV Si	3.3 kV SiC
Dv/dt kV/μs	5	50	5	50	2	2
CM (A)	2.5	15	.5	15	1	0.6
EF	1	6	1	6	0.1	0.06

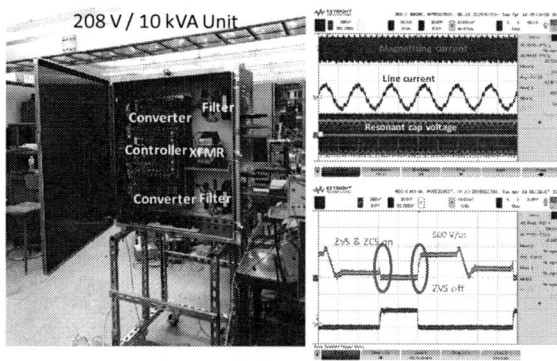

Figure 7: Experimental setup and results of 3-phase 208 V (L-L)/10 kVA S4T with low dv/dt of 500 V/μs.

E. CFOM Summary

The converter figure of merit for all configurations is summarized in TABLE 6. In case of the BTB-VSC and DAB-SST, as Si devices are replaced by SiC, factors DLF, DRF and RCF are improved as expected. This has led to the thrust in many research labs for the replacement of Si devices with SiC devices. It is seen that in almost every case, the impact of dv/dt becomes much worse – requiring special techniques to manage the generated EMI and

4006

noise, and looking at new failure mechanisms due to the elevated displacement currents that flow. This is reminiscent of a rash of motor winding and bearing failures that resulted from the transition from Darlington's BJT to IGBTs in the late 1980s. Slowing down the dv/dt reduces EMI but results in higher switching losses. The use of soft-switching, in particular zero voltage switching, reduces dv/dt and EMI, while also reducing switching losses. The adoption of soft-switching techniques is associated with higher device count, more LC components and loss of control range. These are included in the analysis and computation of the CFOM, and shows that the use of ZVS with SiC devices is very desirable.

An integrated CFOM can be derived as a summation of the various factors with weighting factors that reflect a specific application. In any case for the S4T, TABLE 6 shows superior performance in every category versus conventional solutions.

$$CFOM = k1 * DLF + k2 * DRF + k3 * RCF + k4 * EF \quad (8)$$

TABLE 6
CFOM SUMMARY

	VSI-BTB + 60 Hz Transformer		SST- DAB Based		SST-S4T Based	
	3.3 kV Si -2 kHz	3.3 kV SiC -15 kHz	3.3 kV Si -2 kHz	3.3 kV SiC -15 kHz	3.3 kV Si -2 kHz	3.3 kV SiC – 15 kHz
DLF	12.1	0.84	4.84	0.52	0.198	0.192
DRF	1	0.75	1.5	1.13	1.2	0.9
RCF	1.0	0.79	0.34	0.13	0.18	0.024
EMI	1	6	1	6	0.1	0.06

V. SST-HIGH POWER SCALING ISSUES

In high power SSTs, the modules have to be connected in series/parallel to achieve high voltage/current rating. In addition to the issues discussed above, high power SSTs need to consider the following:

- Voltage sharing across modules under static and dynamic conditions.
- Converter design to meet Basic Impulse level (BIL) requirements.
- Common mode current flows through high frequency transformer inter-winding capacitances.

While the first two challenges are common to both Si and SiC implementations, the third challenge becomes worse with WBG devices. As shown in Figure 8, the high dv/dt associated with WBG can cause common mode currents to flow across the HF transformer inter-winding capacitance. In addition to flowing to the ground, the common mode current can result in cross-coupling currents across the series connected modules, interfering with the converter operation and resulting in additional voltage and current stress across the converter. In transformers, inter winding capacitance is inversely related to leakage inductance. The desire to keep the leakage inductance low will result in high inter-winding

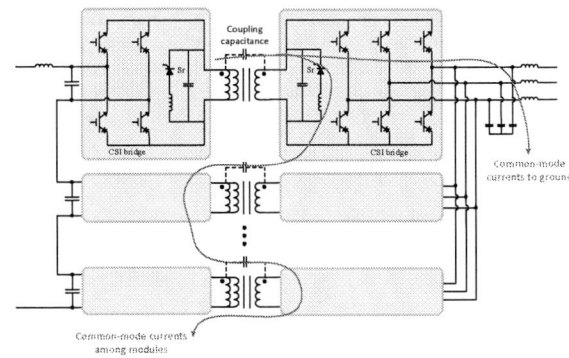

Figure 8: Common mode current in modular SST implementations.

capacitance. Hence, the only way to limit common mode current is to control the dv/dt.

VI. CONCLUSIONS

This paper proposes that viable high-voltage high-power converters using wide band-gap SiC devices, will require the use of soft-switching, in particular zero-voltage switching techniques, if system parasitics and high switching speeds are to be managed. This paper presented the limitations of current approaches, and proposed a new converter figure of merit (CFOM) approach to validate the advantages of soft-switching in wide band-gap high-power SiC converters. The proposed CFOM captures the three criterion-cost, size and performance, in a simpler way so that various topologies can be compared easily.

ACKNOWLEDGMENT

The authors would like to acknowledge Center for Distributed Energy, Georgia Tech, Atlanta, USA for the support in compiling this paper.

REFERENCES

[1] R. Raju. (2014). *Silicon Carbide: High Voltage, High Frequency Conversion, GE Global Research*. Available: https://www.nist.gov/sites/default/files/documents/pml/high_megawatt/Approved-Raju-DoE_NIST_HMW_VSD_GE_raju.pdf

[2] W. McMurray, "The Thyristor Electronic Transformer: a Power Converter Using a High-Frequency Link," *IEEE Transactions on Industry and General Applications* vol. IGA-7, pp. 451-457, 1971.

[3] R. W. A. A. D. Doncker, D. M. Divan, and M. H. Kheraluwala, "A three-phase soft-switched high-power-density DC/DC converter for high-power applications," *IEEE Transactions on Industry Applications,* vol. 27, pp. 63-73, 1991.

[4] H. Akagi, S. i. Kinouchi, and Y. Miyazaki, "Bidirectional isolated dual-active-bridge (DAB) DC-DC converters using 1.2-kV 400-A SiC-MOSFET dual modules," *CPSS Transactions on Power Electronics and Applications,* vol. 1, pp. 33-40, 2016.

[5] G. Wang, S. Baek, J. Elliott, A. Kadavelugu, F. Wang, X. She, et al., "Design and hardware implementation of Gen-1 silicon based solid-state transformer," in *2011 Twenty-Sixth Annual IEEE Applied Power Electronics Conference and Exposition (APEC),* 2011, pp. 1344-1349.

[6]　N. Soltau, H. Stagge, R. W. D. Doncker, and O. Apeldoorn, "Development and demonstration of a medium-voltage high-power DC-DC converter for DC distribution systems," in *2014 IEEE 5th International Symposium on Power Electronics for Distributed Generation Systems (PEDG)*, 2014, pp. 1-8.

[7]　M. K. Das, "10 kV, 120 A SiC half H-bridge power MOSFET modules suitable for high frequency, medium voltage applications," *IEEE Energy Conversion Congress and Exposition*, pp. 2689-2692, 2011.

[8]　S. Bhattacharya, "Transforming the transformer," *IEEE Spectrum*, vol. 54, pp. 38-43, 2017.

[9]　J. E. H. a. J. W. Kolar, "Common-mode currents in multi-cell Solid-State Transformers," presented at the International Power Electronics Conference, Hiroshima,, 2014.

[10]　F. K. a. J. W. K. T. Guillod, "Electrical shielding of MV/MF transformers subjected to high dv/dt PWM voltages," presented at the EEE Applied Power Electronics Conference and Exposition (APEC), Tampa, Florida, 2017.

[11]　R. B. a. J. J. F. N. Idir, "Active gate voltage control of turn-on di/dt and turn-off dv/dt in insulated gate transistors," *IEEE Transactions on Power Electronics*, vol. 21, pp. 849-855, July 2006.

[12]　H. Chen and D. Divan, "Soft-Switching Solid-state Transformer (S4T)," *IEEE Transactions on Power Electronics*, vol. PP, pp. 1-1, 2017.

[13]　F. Mancilla-David, S. Bhattacharya, and G. Venkataramanan, "A Comparative Evaluation of Series Power-Flow Controllers Using DC- and AC-Link Converters," *IEEE Transactions on Power Delivery*, vol. 23, pp. 985-996, 2008.

[14]　F. Electric. (2008). *Low EMI Techniques for New Generation IGBT Modules*. Available: http://neutronltd.power-mag.com/pdf/feature_pdf/1222955409_PEE_issue_1_2008_Powermodules-Low_EMI_Techniques_for_New_Generation_IGBT_Modules.pdf

[15]　C. R. Furnival. (2016). *Advanced High Power-Density Thermal Packages & Mother-Boards Enable Ultimate Power GaN & SiC Performance & Efficiency* Available: http://www.psma.com/sites/default/files/uploads/tech-forums-semiconductor/presentations/is166-advanced-high-power-density-thermal-packages-mother-boards-enable-ultimate-power-gan-sic-perfo.pdf

SiC: Technology Enabler for MV DC/DC Galvanically Insulated Modular Converters

S. Alvarez[1], M. Bellini[1], U. Vemulapati[1], F. Canales[1*] and M. Rahimo[2]

1 ABB Switzerland, Ltd, ABB Corporate Research Center, Baden-Dättwil Ch-5405, Switzerland
2 ABB Switzerland, Ltd, ABB Semiconductors, Lenzburg CH-5600, Switzerland
*E-mail: francisco.canales@ch.abb.com

Abstract— **In this paper, the use of 3.3 kV SiC MOSFETs, Fast Si IGBTs and the mix of both in the so called Cross-Hybrid Switch (XS) and Bimode Cross-Hybrid Switch (BXS) are considered for their application on Galvanically Insulated Modular Converters (GIMC) cells. The semiconductors typically operate under Soft Switching conditions to allow high frequency operation resulting in size and weight minimization while keeping a high efficiency. The high frequency (10 kHz) and Soft Switching operation in the selected semiconductors have been first evaluated by applying 2D semiconductor device simulations. The simulation results allow to identify the impact of device parameters such as the Si IGBTs Lifetime adjustment and the Si IGBT – SiC MOSFET chip surface ratio on the BXS devices, over the on-state and switching losses. Accordingly, different 3.3kV dual HiPak testing samples have been implemented and experimentally evaluated on a 200 kW DC/DC GIMC cell test bench. The comparison between the 3.3 kV Fast Si IGBTs, SiC MOSFETs and BXS shows clear functional benefits of using MV SiC MOSFETs in resonant mode operation, allowing an increase of the GIMC cell power capability by 30 - 35 % for the same module current rating compared to the use of Fast IGBTs. In a similar way, the BXS devices can provide and intermediate power capability increase of around 20-25%, minimizing the installed Si and SiC chip area that can be used as an enabler for improvement of the GIMC technology impacting the cost and performance of the semiconductor devices applied on the GIMC cells**

Keywords—Insulated Gate Bipolar Transistors, MV DC/DC Resonant Converters, High Voltage Silicon Carbide.

I. INTRODUCTION

Galvanically Insulated Modular Converters (GIMC) as shown in Fig. 1 are becoming the standard approach for applications demanding high efficiency and power density under high conversion ratios such as MVDC DC/DC and Solid State Transformer. The key components in GIMC are isolated DC/DC cells, using Medium Frequency Transformers (MFT) to provide the required galvanic insulation. Consequently high switching frequencies and soft switching operation are normally applied to minimize the size and weight of the passive components in the cell with main focus in the MFT while keeping the conversion stage efficiency high.

The improvement of the characteristics of the semiconductors in resonant mode operation has been

deeply investigated with the focus mainly on Medium Voltage (MV) Silicon (Si) IGBTs, [1]-[4]. Fig. 2 shows the inverter typical waveforms under resonant operation. The engineering of the IGBTs moving their performance in the technology curve, by lifetime control optimization or anode engineering, results in a good compromise under resonant operation between the static and dynamic characteristics [4]. Despite of the performance improvements, there is a strong relationship among the device resonant peak current, frequency operation, the hold-off time and the turn-off current which complicate the design of the converter.

Nowadays, it is foreseen that the inherent characteristics of SiC devices will result in higher voltage devices with superior electrical performance such as low conduction and switching losses, and low leakage current under high temperature operation when compared to state of the art Si devices, [6]-[11]. Despite of the expected properties, the benefits of using SiC devices have still to be proven in medium and high voltage applications. Some of the obstacles are the limited availability of devices and the high cost associated to the starting substrate material, manufacturing cost and expensive epitaxial growth especially for HV devices. Also, the relatively small current ratings of the available devices together with their lower surge current and fault current handling capability limit their adoption in commercial products [12].

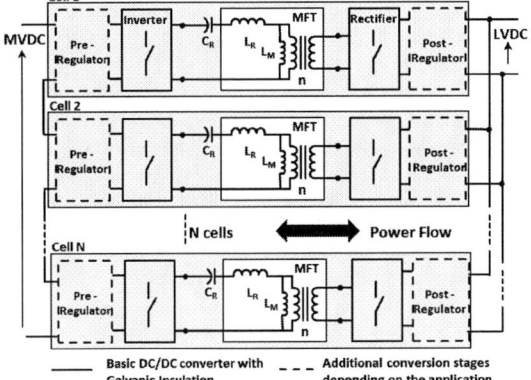

Fig. 1. MVDC to LVDC Galvanically Insulated Modular Conversion structure. Input Series Output Parallel (ISOP) cells configuration.

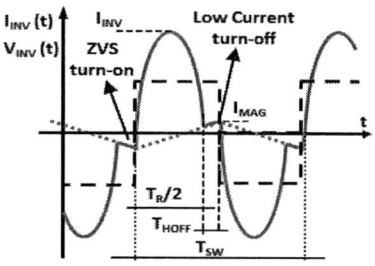

Fig. 2. LLC resonant converter typical current waveform. Inverter switches operating at ZVS turn-on and low-current turn-off.

As a trade-off alternative, this paper presents the parallel arrangement of Si IGBTs and SiC MOSFETs, [13]-[15], in the form of the so called Cross-Hybrid Switch (XS) and Bimode Cross-Hybrid Switch (BXS) under resonant operation. The aim is to reach an optimum performance of the device by providing low static and dynamic losses while improving the overall electrical and thermal properties due to the combination of both the bipolar Si IGBT and the unipolar SiC characteristics.

The paper first presents the characteristics of each Si and SiC devices used in the implementation of the XS and BXS. After that, numerical simulations are performed to identify the optimal ratio between Si and SiC devices and foreseen the benefits of optimized IGBTs under soft switching operation. Finally, experimental results in a 200kW DC/DC converter with the selected devices, full SiC MOSFETs, optimized IGBTs, XS and BXS, are provided to verify the numerical simulations and the performance of the selected concepts.

II. DEVICE TECHNOLOGY UNDER STUDY

This section presents the characteristics of the SiC MOSFET and Si IGBTs used for the implementation of the XS and BXS devices. Fig. 3 and Fig. 4 illustrate in a conceptual way the implementation and main components for the proposed Cross-Hybrid Switch (XS) and Bimode Cross-Hybrid Switch (BXS). The XS applies the paralleling of Si IGBTs and SiC MOSFETs aiming to provide low static and dynamic losses while improving the overall electrical and thermal characteristics. In a similar way, the BXS combines in parallel a Reverse Conducting (RC) IGBT or Bimode IGBT (BIGT) with a SiC MOSFET avoiding the integration of a separated Free Wheeling Diode (FWD). Additional benefits resulting from these configurations are low conduction losses over full load range, soft turn-off performance, large turn-off safe operating area, high short circuit and surge current capability which are parameters limited in SiC MOSFETs, [13], [14].

Table I presents the characteristics of the IGBTs and BIGTs devices used in the paper. The data corresponds to dies with 13.6 by 13.6 mm^2 and active area of approximately 1 cm^2. The parameters are given for a current density of 62.5 A/cm2. The energy losses at turn-off is obtained under hard switching operation as a reference. Considering that the devices will operate under

Zero Voltage Switching (ZVS) condition, the turn-on losses and reverse recovery losses in the diode are not considered.

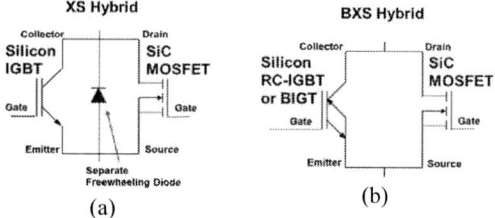

(a)   (b)

Fig. 3. Switch circuit arrangement: (a) Cross-Hybrid Switch (XS). (b) Bimode Cross-Hybrid Switch (BXS).

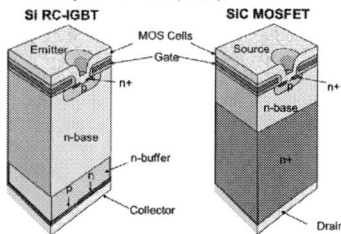

Fig. 4. Silicon BIGT (RC-IGBT) and SiC MOSFET basic cell design.

Table I provides also as a reference the characteristics of a standard STP⁺ IGBT device. As can be seen, by life time control and anode engineering, the technology curves of the IGBT and BIGT devices are optimized for resonant operation with a better trade-off between conduction and switching losses resulting in a fast and ultra fast devices, respectively. In addition, with an Enhanced Trench (ET) cell design, the BIGT presents a better trade-off technological curve. The resulting parameter improvements in the operation of the converter will be explained in more detail in the next sections.

TABLE I. DEVICE TECHNOLOGY UNDER STUDY.
DATA AT 62 A/CM^2 AND HARD SWITCHING OPERATION

Device/Die	E_{off} (mJ)	$V_{ce,on}$ (V)	V_f (V)
Std IGBT	120	3.2	
Ultra Fast Diode			2.2
Utra Fast IGBT	60	4	
Ultra Fast Diode			2.2
Fast BIGT	90	3.55	2.9
UF Enhancement Trench BIGT	75	3.8	3.3

III. 2D SIMULATIONS OF 3.3 KV SEMICONDUCTORS IN RESONANT MODE

2D TCAD simulation of 3.3kV Conventional and Fast Si IGBTs / BXS / SiC MOSFETs dies with different active area ratios is performed. From these simulations, first the static characteristics of the different die parameters and configurations are obtained as reference, as shown in Fig. 5.

Then, the dynamic on-state and turn-off losses of the different devices are simulated and evaluated applying a 300A peak resonant current waveform at 10kHz (50µs half-sine wave resonant period simulation with a hold-off time TH_OFF=7µs). Fig. 6 shows how the MOSFET devices carry more current in the rising part of the half-

4010

sine wave resonant due to their lower on-state resistance at lower current and to their faster turn-on. Subsequently, the bipolar devices starts to carry a significant part of the current.

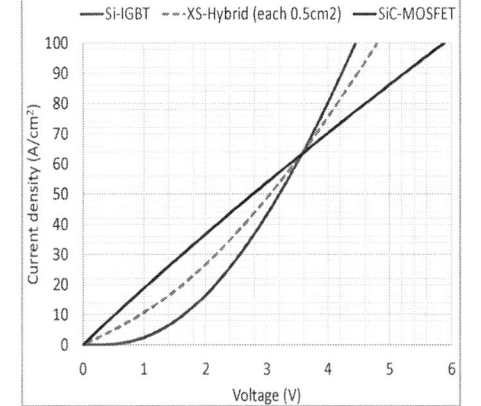

Fig. 5. Si IGBT, XS Hybrid (50% Si – 50% SiC active area), SiC MOSFET on-state characteristics.

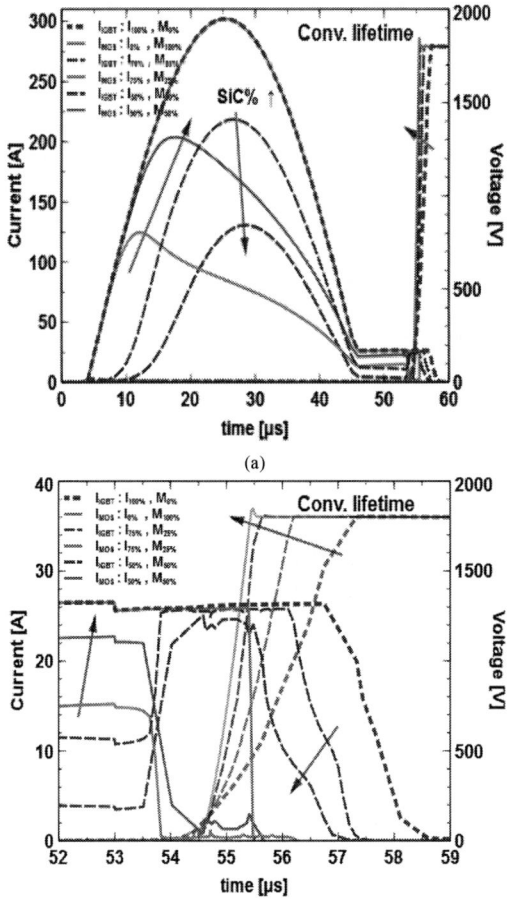

(a)

(b)

Fig. 6. Simulated current and voltage waveforms for Si IGBT, SiC MOSFET, XS Hybrid (with 50-50 and 75-25 Si – SiC active area ratios). Si devices feature conventional (i.e. non-optimized) carriers' lifetime. (a) Half-sine wave resonant period and (b) turn-off time.

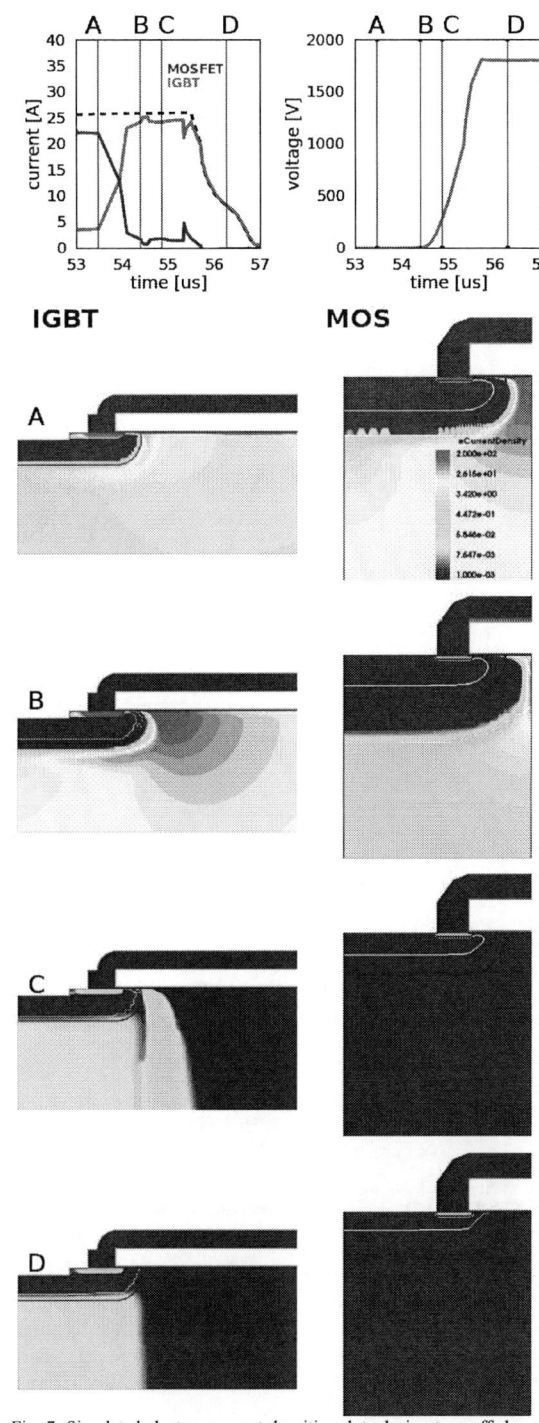

Fig. 7. Simulated electron current densities plots during turn-off shown in Fig. 6 for the IGBT and MOS for a XS Hybrid with 50-50 Active Area ratio (top at time of 53 μs, bottom at time of 54 μs).

As expected, with an increment of the SiC area, the wide band-gap devices carry a larger portion of the current in the half-sine wave resonant period. Fig. 6(b) shows the current sharing between MOSFETs and IGBTs during soft switching for different ratios of Si and SiC area, indicating that during the different phases of turn-off the majority of the current is alternatively carried by either the bipolar or unipolar devices.

The physical phenomena in the dynamic current sharing between MOSFETs and IGBTs during turn-off are illustrated in Fig. 7, showing the current density from TCAD simulations for both devices at four distinctive time steps during the turn-off stage. At time step A, at the end of the hold-off time, the unipolar devices are carrying most of the current due to their lower resistance. As expected from fast switching capability of unipolar devices, the current of the SiC MOSFETs drops almost to zero at time step B, while the excess charge is extracted from the bipolar devices.

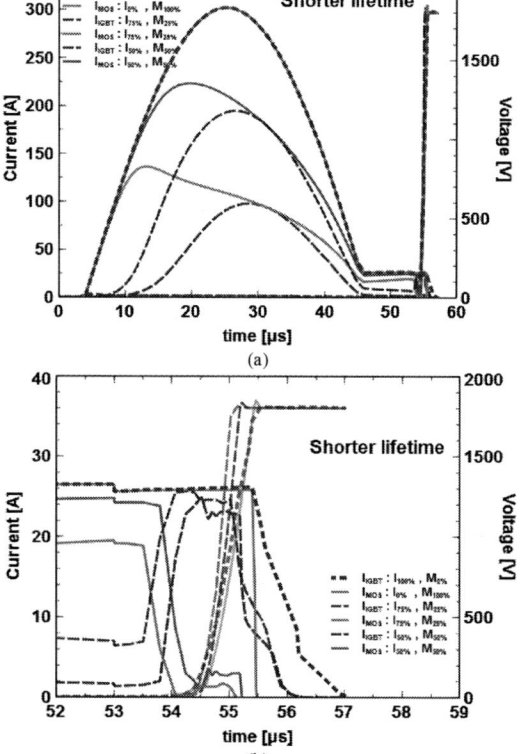

Fig. 8 Simulated current and voltage waveforms for Si IGBT, SiC MOSFET, XS Hybrid (with 50-50 and 75-25 Si – SiC active area ratios). Si devices feature optimized (i.e. shorter) carriers' lifetime. (a) Half-sine wave resonant period time frame and (b) turn-off time.

It should be noted that already during the decreasing part of the half-sine wave resonant and during the hold-off time, the recombination process within the bipolar devices starts to remove excess carries from the drift region. However, the excess charge stored in the IGBTs is still sizeable and consequently, between time steps B and C, the Si IGBTs carry the majority of the current, as the excess carries are removed from the drift region.

At time step C, both devices are starting to be significantly depleted, hence the rapid increase in voltage. At this point, a very small amount of current is needed to remove electrons from the drift region of the

MOSFETs while a much larger current is required to continue to extract the excess electron and holes in the IGBTs.

At time step D, charge extraction continues in the IGBTs, while it has been completed in the MOSFETs.

Fig. 8 shows the switching currents and voltages in the case of Fast Si IGBTs, with shorter carrier lifetime. As expected, the shorter lifetime reduces the amount of excess carriers stored in the bipolar devices, and consequently increases the on-state resistance of the IGBTs. Therefore, the SiC unipolar devices carry more current during the half-sine wave pulse.

At the end of the hold-off time the excess of electrons and holes stored in the IGBTs are also significantly reduced compared to standard IGBTs because of the initially smaller amount of excess charge and also the increased recombination rate due to the shorter lifetime. Therefore the charge extraction phase in the IGBT before turn-off is also shorter and the corresponding voltage rise is significantly faster.

Finally, Table II compares the influence of the devices of the minority carrier lifetime (IGBTs) and the Si to SiC surface ratio over the on-state and turn-off losses on the XS solutions.

TABLE II. TCAD SIMULATION RESULTS SUMMARY.
PEAK CURRENT=300A, F_{SW}=10KHZ, T_{H_OFF}=7µS. CHIP SURFACE 4MM2

Si IGBT – SiC MOSFET Active Area %	E On-state (mJ)	E Off (mJ)	Total Losses (mJ)
Lifetime 1/3 (Fast IGBTs)			
100 – 0	50.8	46.8	97.6
75 – 25	49.3	28.4	77.8
50 – 50	42.5	23	63.5
0 - 100	27.7	17.6	45.3
Regular Lifetime (Standard IGBTs)			
100 – 0	37.3	78.1	115.4
75 – 25	39.4	57.2	96.6
50 – 50	36.8	39.9	76.7
0 - 100	27.7	17.6	45.3

IV. HiPak Samples Implementation and Experimental Results

Based on the previous simulation results, and using as a reference the commercial 3.3kV dual HiPak package, several samples have been defined and implemented for its practical evaluation. First the basic substrate disposition is defined for each switch arrangement. As can be seen in Fig. 9, the Fast IGBTs contain 4 ABB IGBT chips and 2 FWD chips per substrate, while the BIGT contains 6 ABB BIGT chips and the BX contains 3 ABB BIGT chips and 6 commercial 3.3kV SiC MOSFET chips. Based on the chips characteristics and the benefits of the BIGT technology for the thermal point of view, the Fast IGBT modules are populated with 4 substrates, while the BIGT and BXS modules are populated with only 2 substrates.

Besides these samples, commercial Hitachi 3.3kV Fast IGBTs [16] and ABB LinPak [18],[19] module with SiC commercial MOSFETs and Sumitomo SiC MOSFET samples have been included in the experimental evaluation.

The experimental evaluation of the different modules has been carried out on a GIMC cell test bench, Fig. 10. This test bench allows the permanent operation evaluation of the key components in the conversion stage such as semiconductors, auxiliary components, resonant tank elements and very important the medium frequency isolation transformer. In this case, the semiconductors are tested with a DC link voltage of 1.8kV at 10 kHz up to an output power of 150kW corresponding to a device peak current of around 300A. Fig. 10(b) shows the typical waveforms across one of the switches and the resonant current at the primary side of the transformer. For ZVS operation, the device will commutate just with the magnetizing current in the transformer which is around 40A.

During the test, special attention has been paid to the turn-off behavior of the different active switches in order to explore the benefits of the different device configurations. Fig. 11 shows some of the experimental waveforms.

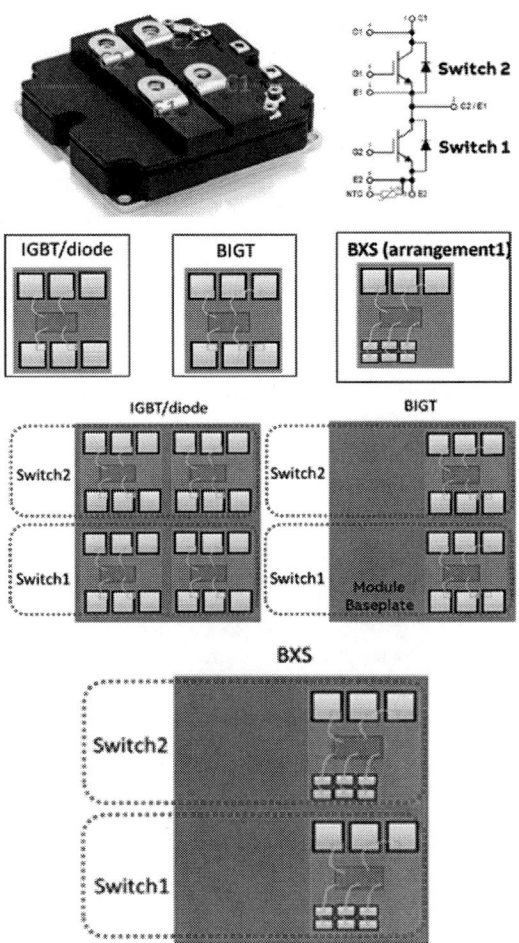

Fig. 9. Chips arrangement on the substrates and on the HiPak base plate for the Fast IGBTs, BIGTs and the BXS devices.

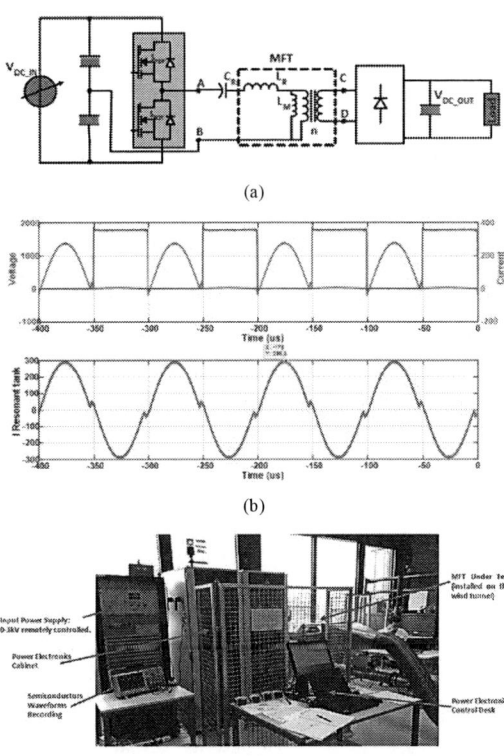

Fig. 10. GIMC Cell permanent operation test Setup overview. VDC_IN = 1.8 kV, VDC_OUT = 1.2 kV, nominal power 200 kW, fsw=10 kHz, n= 1.5, IINV= 266ARMS, IREC=400ARMS. (a) Simplified schematic, main waveforms and (c) cell test bench.

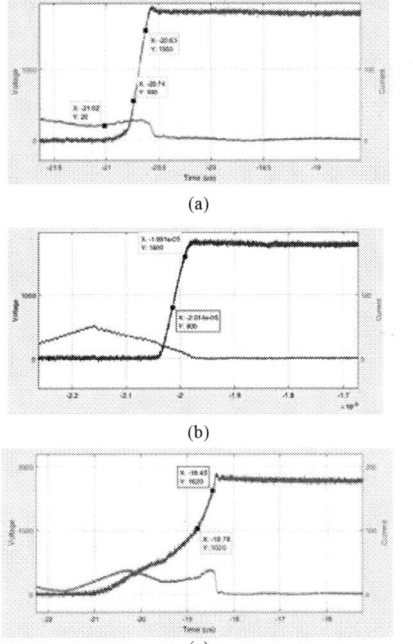

Fig. 11. Turn-off behavior of the different module samples. (a) ABB BXS (Fast IGBT 50% - SiC MOSFET 50%), 0.5us/div, (b) LinPak SiC MOSFET, 1us/div and (c) Fast IGBT samples, 1us/div.

4013

The 2018 International Power Electronics Conference

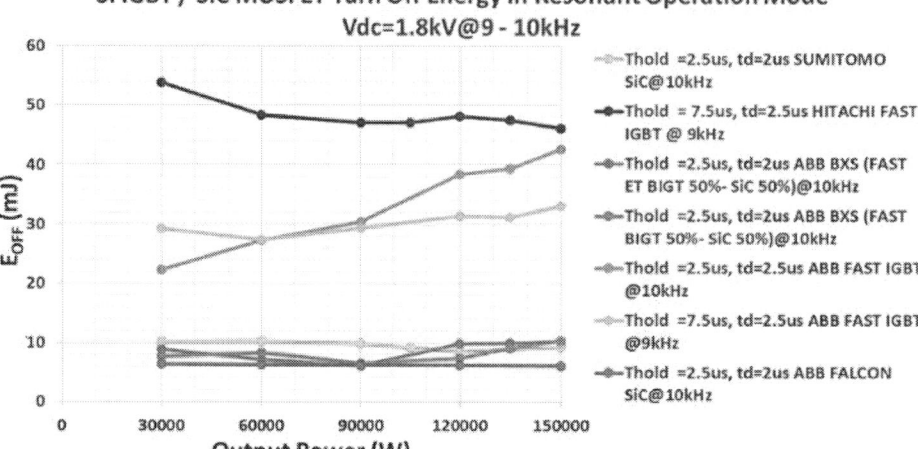

Fig. 12. Turn-off energy losses as a function of the cell output power for the different module samples.

There are not experimental results in this work related to the turn-on behavior of the devices since it is expected a lossless characteristics due to ZVS operation mode. As can be seen in Fig. 11(c), a longer commutation time is present in the Fast IGBTs due to the bipolar characteristic, resulting in higher turn-off losses.

Fig. 12 shows the obtained turn-off energy losses for the different modules under test as a function of the output power. It also includes modification of the switching frequency and the hold-off time to show the influence of the design parameters for the converter and output power on the switching losses in the devices. Despite engineering the IGBTs for high frequency and resonant operation, one can observe that the turn-off losses are still higher with respect to the SiC MOSFET and BXS devices. In addition, it is possible to observe the dependency of the switching losses with respect to both the power level and the hold-off time. At higher power level, the switching losses in the Fast IGBTs increase with respect to the output power when a 2.5us hold-off time is applied. The given time is not enough to remove the excess of carriers accumulated during the flooding phase which also strongly depends of the peak current, before the commutation time. This is in agreement with the 2D simulations. With a higher hold-off time 7.5us, the dependency of output power is minimized keeping the turn-off losses almost constant during the whole load range. This is valid for both, the Hitachi and the ABB modules.

On the other hand, the curves also show the benefits of the BXS modules. For both cases, the BXS with the Fast IGBTs and the Enhancement Trench BIGTs modules, the switching losses are comparable to the Full SiC MOSFET modules. In addition, there is a small increment of the turn-off losses despite of using a 2.5us hold-off time. As shown also in the simulations, the combination with the SiC MOSFETs helps to reduce the flooding effect in the Fast IGBTs and to speed the charge removal phase before the turn-off time. Based on the thermal estimation, maintaining similar switching frequency, it could be possible to increase the output power by 20-25% when compared with the Fast IGBTs.

For the SiC modules, the switching losses are constant in the whole output power range due to the unipolar behavior. It is important to mention that both modules have SiC Schottky diodes which increase the output capacitance of the switch. This requires higher current during the turn-off to achieve ZVS operation. In addition, the gate driver is the same for all the modules, meaning there is a possible driving performance improvement. This is reflected with a slower dv/dt. If the previous points are considered in the design of the converter, a further reduction in the switching losses can be expected. In any case, the total thermal estimation tells that it is possible to extract from 30 to 35 % higher output power under the same operating condition when compared to the Si Fast IGBTs modules.

V. Conclusions

In this work the simulation and practical evaluation of different 3.3kV semiconductor technologies on a GIMC cell test bench in resonant mode operation are presented. The 2D semiconductor simulations have helped to understand and define the influence of some design key parameters of IGBTs, XS and BXS switches, especially related to the Si IGBTs carrier's lifetime control and the Si – SiC active area ratio.

The experimental results and comparison between the 3.3 kV Fast Si IGBTs, SiC MOSFETs and BXS show clear functional benefits of using MV SiC MOSFETs in resonant mode operation, allowing an increase of the GIMC cell power capability by 30 - 35 % for the same module current rating compared to the use of Fast IGBTs. In a similar way, the BXS devices can provide and intermediate power capability increase of around 20-25%, minimizing the installed Si and SiC chip area that can be used as an enabler for improvement of the GIMC technology impacting the cost and performance of the semiconductor devices applied on the GIMC cells.

ACKNOWLEDGMENT

The authors would like to thank Sumitomo Electric Industries, Ltd., for providing the SiC modules for the experiments.

REFERENCES

[1] M.Shinagawa, T.Waga, Y.Toyota, Y.Toyoda, K.Saito: "3.3kV high speed IGBT module for bi-directional and medium frequency application"; *Proc. of PCIM Europe 2012, Int. Power Conversion and Intelligent Motion Conference*, Nürnberg, Germany, 2012, pp. 792-799.

[2] N. Reinold and M. Steiner, "Characterization of Semiconductor Losses in Series Resonant DC-DC Converters for High Power Applications using Transformers with Low Leakage Inductance," in *8th European Conference on Power Electronics and Applications, EPE'99*. Lausanne, Switzerland, 1999.

[3] R. Alvarez, F. Filsecker, and S. Bernet, "Characterization of a New 4.5kV Press Pack SPT+ IGBT for Medium Voltage Converters," in *IEEE Energy Conversion Congress and Exposition 2009, ECCE 2009*, pp. 3954-3962.

[4] D. Dujic, G. K. Steinke, M. Bellini, M. Rahimo, L. Storasta and J. K. Steinke, "Characterization of 6.5 kV IGBTs for High-Power Medium-Frequency Soft-Switched Applications," in *IEEE Transactions on Power Electronics*, vol. 29, no. 2, pp. 906-919, Feb. 2014.

[5] M. Das, C. Capell, D. Grider, S. Leslie, J. Ostop, R. Raju, M. Schutten, J. Nasadoski and A. Hefner, "10 kV, 120 A SiC Half H-Bridge Power MOSFET Modules Suitable for High Frequency, Medium Voltage Applications," in *IEEE Energy Conversion Congress and Exposition 2011, ECCE 2011*, pp. 2689-2692.

[6] X. She, A. Q. Huang and R. Burgos, "Review of Solid-State Transformer Technologies and Their Application in Power Distribution Systems," in *IEEE Journal of Emerging and Selected Topics in Power Electronics*, vol. 1, no. 3, pp. 186-198, Sept. 2013.

[7] K. Hamada et al., "3.3kV/1500A Power Modules for the World's First all-SiC Traction Inverter," *Japanese Journal of Applied Physics 54, 04DP07*, 2015.

[8] K. Fukuda et al., "Development of Ultra high Voltage SiC Power Devices," *International Power Electronics Conference, IPEC 2014*.

[9] J. Casady et al., "New Generation 10kV SiC Power MOSFET and Diodes for Industrial Applications," *Proc. of PCIM Europe 2015; Int. Exhibition and Conference for Power Electronics, Intelligent Motion, Renewable Energy and Energy Management*, Nuremberg, Germany, 2015, pp. 96-103

[10] A. Q. Huang, Li Wang, Q. Tian, Qianlai Zhu, Dong Chen and Wensong Yu, "Medium voltage solid state transformers based on 15 kV SiC MOSFET and JBS diode," *IECON 2016 - 42nd Annual Conference of the IEEE Industrial Electronics Society*, Florence, 2016, pp. 6996-7002.

[11] L. Wang, Q. Zhu, W. Yu and A. Q. Huang, "A Medium-Voltage Medium-Frequency Isolated DC–DC Converter Based on 15-kV SiC MOSFETs," in *IEEE Journal of Emerging and Selected Topics in Power Electronics*, vol. 5, no. 1, pp. 100-109, March 2017.

[12] G. Romano et al., "Short-circuit Failure Mechanism of SiC Power MOSFETs," *Proc. Int, Symposium on Power Semiconductor, Devices and ICs, ISPSD 2015*, pp. 345-348, 2015.

[13] M. Rahimo et al., "The Cross Switch "XS" Silicon and Silicon Carbide Hybrid Concept," *Proc. of PCIM Europe 2015; Int. Exhibition and Conference for Power Electronics, Intelligent Motion, Renewable Energy and Energy Management*, Nuremberg, Germany, 2015, pp. 1-8.

[14] M. Rahimo et al., "Characterization of a Silicon IGBT and Silicon Carbide MOSFET Cross-Switch Hybrid," in *IEEE Transactions on Power Electronics*, vol. 30, no. 9, pp. 4638-4642, Sept. 2015.

[15] U. R. Vemulapati, M. Rahimo, A. Mihaila, R. A. Minamisawa, C. Papadopoulos and F. Canales, "The Bimode Cross Switch (BXS) a full hybrid solution in switch- and diode-modes," *2016 18th European Conference on Power Electronics and Applications (EPE'16 ECCE Europe)*, Karlsruhe, 2016, pp. 1-9.

[16] Hitachi, MBM400E33D-MF. (2015). [Online]. Available: http://www.hitachi-power-semiconductor-device.co.jp/products.

[17] ABB, LinPak. (2016) [Online]. Available: https://library.e.abb.com/public/.

[18] S. Kicin, et al, "A New Concept of a High-Current Power Module Allowing Paralleling of Many SiC Devices Assembled Exploiting Conventional Packaging Technologies," *28th Int. Symposium on Power Semiconductor Devices and ICs, ISPSD 2016*, 12-16 June 2016, Prague, Czech Republic.

[19] S. Kicin, et al, "1.7 kV High-Current SiC Power Module Based on Multi-Level Substrate Concept and Exploiting MOSFET Body Diode During Operation," *Proc. of PCIM Europe 2017; Int. Exhibition and Conference for Power Electronics, Intelligent Motion, Renewable Energy and Energy Management*, Nuremberg, Germany, 2017, pp. 1-8.

A Bearingless Slice Motor with a Solid Iron Rotor for Disposable Centrifugal Blood Pump

Tadahiko Shinshi[1*], Ryo Yamamoto[2] and Yoshiki Nagira[2], Junichi Asama[3]
1 Institute of Innovative Research, Tokyo Institute of Technology, Yokohama, Japan
2 School of Engineering, Tokyo Institute of Technology, Tokyo, Japan
3 Faculty of Engineering, Shizuoka University
*E-mail: shinshi.t.ab@m.titech.ac.jp

Abstract- In the field of extracorporeal circulation, the use of high cost NdFeB magnets (PMs) in the disposable pump head of centrifugal blood pumps (CBP) need be avoided. In this paper, a bearingless slice motor consisting of a solid iron rotor and a reusable part having ring PMs which can be applied to a CBP is introduced. It realizes a cost-effective disposable pump head and a compact reusable part. To eliminate PMs from the disposable rotor, two PM rings and two iron rings are axial-symmetrically arranged in respect of the stator. The bias magnetic flux generated by the PM flows past the iron rings, rotor tooth and stator tooth. The radial motion and rotational speed of the rotor are controlled using displacement and Hall sensor signals and electromagnetic actuators, and the axial and tilt motions are passively supported and stabilized by the magnetic coupling between the stator and the rotor. A prototype bearingless slice motor is fabricated and experimentally evaluated.

Keywords—bearingless slice motor, centrifugal blood pump, solid iron rotor, disposable pump head

I. INTRODUCTION

A blood pump is placed outside human body and supports blood circulation of patients with serious heart disease and alleviates the burden of their left and/or right ventricles. Considering many types of circulation support, centrifugal blood pumps (CBPs) with disposable and easily-detachable pump heads are required to have a low blood damage and to be low cost and maintenance free [1]. Non-contact support of the impeller by a magnetic bearing has been proved to be reliable method for increasing durability and reducing blood damage [2][3].

A CBP employing a radial-motion-controlled magnetic bearing has been developed by our group [4]. Nevertheless, in order to rotate the impeller, a contactless torque transmission structure which consists of a magnetic coupling disk, angular contact bearings and a motor is utilized in the reusable part of CBP. The large size and complicated mechanical structure impairs the flexibility of the pump head design and limits the durability of reusable part due to the limited service life of bearings.

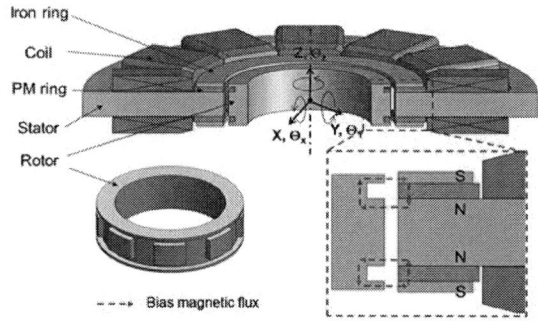

Fig. 1. Configuration of the proposed bearingless slice motor.

A bearingless motor combining a radial magnetic bearing and a motor into one unit has also been proposed to eliminates the torque transmission structure [5]. However, a high cost and environmentally-unfriendly NdFeB permanent magnet (PM) is used in the rotor part.

Previously, we also proposed a bearingless slice motor with a solid iron rotor [6]. Nevertheless, high power consumption due to a bias current for generating the bias magnetic flux caused heat generation problems and will shorten the battery usage time for the application of portable CBP systems.

In this paper, a proposed bearingless slice motor has a configuration based on the structure of switched reluctance motor where bias magnetic flux generating mechanisms are added on the upper and lower parts of the stator. The radial motion and rotational speed of the rotor are actively controlled. The simulated passive stiffness and rotational torque of the proposed design using a magnetic field software provide the feasibility of the application of CBPs. A prototype is fabricated and evaluated.

II. PROPOSED BEARINGLESS SLICE MOTOR

In order to make a disposable rotor simpler, the number of actively controlled degree of freedom (DOF) should be small and the other DOF should be passively

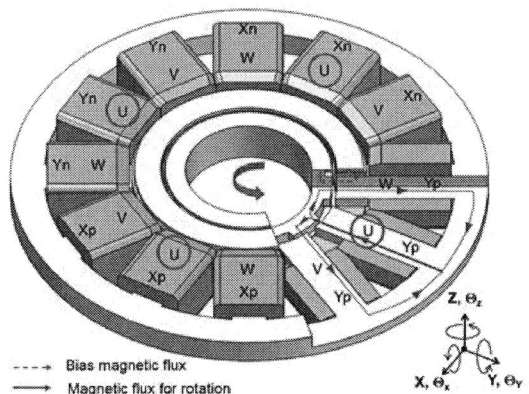

Fig. 2. Coil groups and rotational principle.

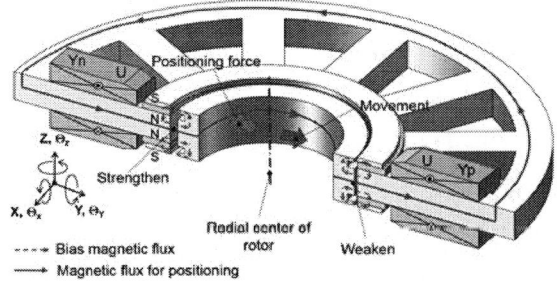

Fig. 3. Principle of positioning force generation.

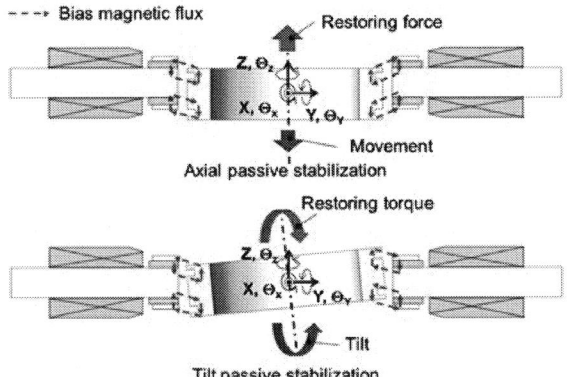

Fig. 4. Principle of passive stabilization.

stabilized by magnetic coupling generated by the bias magnetic flux between the rotor and the stator. Since a bias current for generating the bias magnetic flux heats the circulating blood, the use of PM in the stator is desirable for reducing the heat generation. However, if PM having high magnetic resistance is placed inside the magnetic circuit of the bearingless motor, not only the efficiency of the motor but also the positioning force of the magnetic bearing decreases. Thus, the bias magnetic flux path should be separated from the flux path for rotation and positioning. To solve these problems, many types of bearingless slice motors have been proposed and tested [7] [8] [9].

In this study, the proposed bearingless slice motor

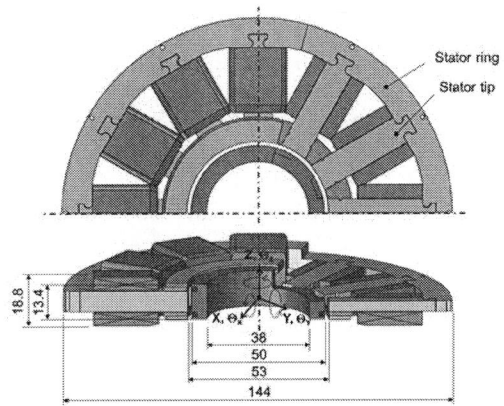

Fig. 5. Simulation model of the proposed bearingless slice motor for calculating bearing stiffness and motor torque

consists of a twelve-stator-slots and eight-rotor-teeth switched reluctance motor with two PM rings and two iron rings as shown in Fig. 1. The PM rings and iron rings are axial-symmetrically arranged in respect of the stator. The homopolar bias flux is generated by the PM rings in the stator and flows past the iron rings, rotor tooth and stator tooth. Due to the elimination of the bias current for the bias magnetic flux generation, the power consumption of the proposed bearingless slice motor is expected to be significantly lower than the previously proposed PM-free one [6].

Fig. 2 shows that each of the twelve concentrated winding coils of the stator belongs to four groups (Xp, Xn, Yp and Yn) for positioning the rotor in the radial directions. Each group includes U, V and W phases for driving the rotor in the rotational direction. The sum of the positioning current and the motor current is supplied as the coil current.

As the rotor rotational position shown in Fig. 2, in order to generate a counterclockwise torque, the motor current is applied to the U group coils. The magnetic flux for rotation flows from the U stator pole to the rotor then back to V and W stator poles. Since the PM rings are separated from the magnetic flux circuit for rotation, high motor efficiency is expected. The motor currents in the same phase coils are equal. Assuming no eccentricity, and no machining and no assembling errors, only rotational torque is generated by the motor current. This is due to the symmetrical arrangements of the stator and rotor poles.

As shown in Fig. 3, when the rotor moves in the Y positive direction, the bias magnetic flux is strengthened applying positioning currents to the coils in the Yn group and is weakened by applying reversed positioning currents to the coils in the Yp group. The PM ring is also separated from the magnetic flux circuit for positioning.

As a result, the positioning force will be generated by the difference of the magnetic flux density in the right and left side clearances. The rotor will be dragged back to

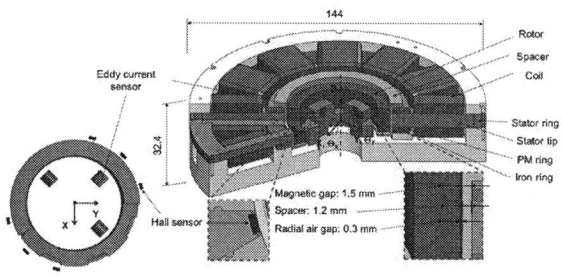

Fig. 6. Mechanical structure of proposed bearingless slice motor.

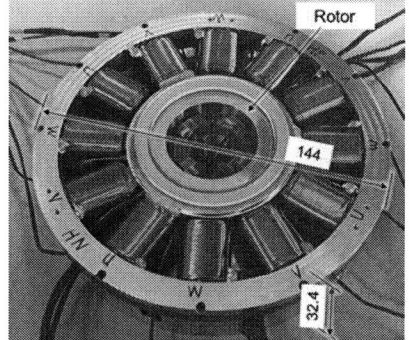

Fig. 7. Prototype of the bearingless slice motor.

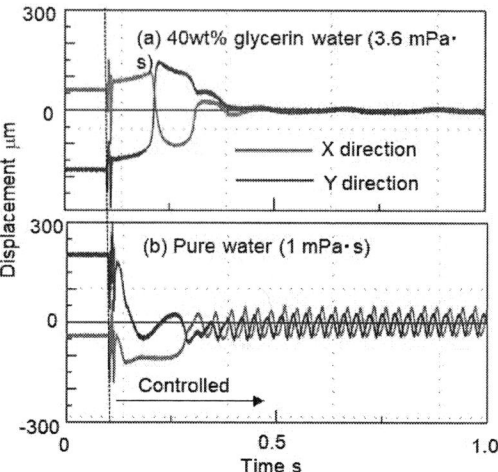

Fig. 8 Start-up characteristics with glycerin water and pure water

the center position. The axial and tilt displacement of the rotor from the center position are suppressed by the magnetic coupling between the rotor and stator as shown in Fig. 4.

III. PROTOTYPE DESIGN AND FABRICATION

The passive stiffness of the proposed bearingless slice motor in the axial and tilt directions was evaluated utilizing a magnetic field software (ANSYS Maxwell, ANSYS, Inc.). Fig. 5 shows the simulation model and its dimensions. The bearingless slice motor dimensions were surveyed to increase the passive stiffness and the current-torque coefficient under the condition that rotor outer diameter and clearance were fixed.

The passive stiffness was calculated from the simulated restoring axial force and tilt torque generated over a small working range (0 – 0.3 mm for the axial stiffness and 0– 60 mrad for the tilt stiffness). The axial and tilt stiffness of the proposed bearingless slice motor are 15.8 N/mm and 5.9 Nm/rad, respectively, which meet the design target for the use of the CBP.

The simulated rotational torque of the proposed bearingless slice motor under a constant motor current density of 1.8 A/mm^2 fluctuates during rotation. The torque curve has a period of 15 degrees. According to the simulation results, the average rotational torque is 0.054 Nm which is sufficient for the application of the CBP.

As shown in Fig. 6, the origin of the coordinate system corresponds to the geometric center of the stator. The height of the rotor is 13.4 mm. The height of the prototype bearingless slice motor is 32.4 mm which is

smaller than that of the reusable part of the previous CBP in our lab [2]. The outer diameter of the prototype is 144 mm. The mass of the soft iron rotor is 0.069 kg. The magnetic clearance between the stator core and the rotor is 1.5 mm.

For arranging the coils at high density, the stator is divided into twelve stator tooth tips and one stator ring. Steel sheets with a thickness of 0.35 mm are laminated in order to decrease the eddy current losses in the stator. These parts are machined with high accuracy using electric discharge machining. The number of each coil turn is 381. The diameter of the wire in the coils is 0.5 mm.

In order to measure the radial displacement of the rotor, three eddy current displacement sensors (PU-05A, AEC Corp., Japan) are selected and placed as shown in Fig. 6. Output signals from the sensors are calculated to effectively compensate the temperature drift and eliminate the common mode noise. In the initial prototype, a radial clearance is adjusted to 0.3 mm by placing a poly carbonate spacer with a thickness of 1.2 mm between the rotor and the stator.

In order to detect the angular position of the rotor, Hall sensors (HW-322B, AKM Corp., Japan) are placed between the adjacent stator tips. In this research, provisionally, each coil is supplied with current by an independent linear amplifier. Totally, twelve amplifiers are used in the system.

IV. EXPERIMENTAL RESULTS

A prototype bearingless slice motor was fabricated and tested as shown in Fig. 7. A PID controller for the rotor positioning and a PI controller for the current feedback of the coils were used to stabilize the system in the X and Y directions and these controller parameters were experimentally tuned. In the rotational speed control, the frequency and phase of the three phase motor current are

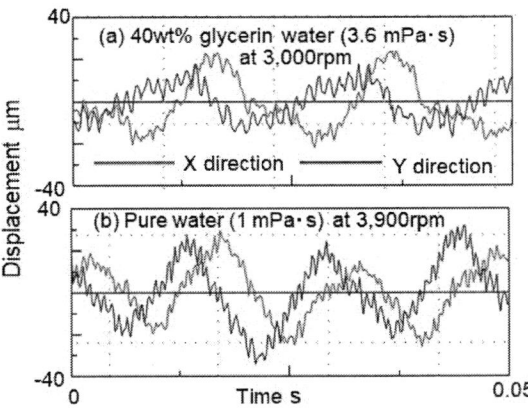

Fig. 9 Rotational accuracies with glycerin water and pure water

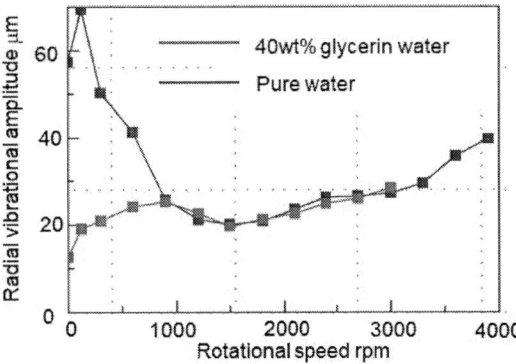

Fig. 10 Relationship between rotational speed and rotational accuracy

fixed and only the current amplitude is adjusted based on the speed deviation.

Firstly, the clearance between the rotor and the spacer was filled without any liquid. However, the stable startup and levitation of the rotor could not be achieved. Then, the clearance was decreased from 0.3 to 0.2 mm and filled with a glycerin water having a viscosity of 3.6 mPa·s as same as that of human blood or pure water having a viscosity of 1.0 mPa·s in order to increase the damping in the passively supported directions.

As shown in Figs. 8, the rotor could be stably levitated and positioned at the center of the stator. When the clearance was filled with the glycerin water, the rotor could be stabilized. However, with pure water, the residual vibration larger than that with the glycerin water was observed.

Fig. 9 shows the rotational displacement with the glycerin water at a rotational speed of 3,000rpm and with pure water at a rotational speed of 3,900rpm. The rotor could rotate without any contact. The rotational accuracies were within ±40μm. The vibration includes not only the rotational frequency component but also higher ones.

Fig. 10 summarized the relationship between the vibration amplitude and rotational speed. The rotational accuracy at a low rotational speed with pure water was lower than that with the glycerin water. However, at a rotational speed of more than 1,000rpm, the difference of rotational accuracies was not observed. It is considered that the rotational speed becomes dominant over the viscosity in the rotational accuracy at higher rotational speeds.

V.   CONCLUSIONS

In this paper, a bearingless slice motor for disposable blood pump was proposed, designed, fabricated and tested. With the assistance of the liquid, the rotor was stably levitated without contact and smoothly rotated at a rotational speed of 3,900rpm.

Future work is to realize a stable levitation and rotation without a support of any fluid. Furthermore, the evaluation of a centrifugal pump using the bearingless slice motor is expected.

REFERENCES

[1] De Robertis, F., Birks, E.J., Rogers, P., Dreyfus, G., Pepper, J.R. and Khaghani, A., Clinical performance with the Levitronix Centrimag short-term ventricular assist device, The Journal of heart and lung transplantation, Vol. 25, No. 2 (2006), pp. 181-186.

[2] Hijikata, W., Shinshi, T., Asama, J., Li, L., Hoshi, H., Takatani, S. and Shimokohbe, A., A magnetically levitated centrifugal blood pump with a simple-structured disposable pump head, Artificial Organs, Vol. 32, No. 7 (2008), pp. 531-540.

[3] Schob, R. and Barletta, N., Principle and application of a bearingless slice motor, Proc. of 5th Int. Symp. Magn. Bearings, Kanazawa, Ishikawa, Japan, Vol. 40, No. 4 (1996), pp. 313–318.

[4] Hijikata, W., Sobajima, H., Shinshi, T., Nagamine, Y., Wada, S., Takatani, S. and Shimokohbe, A., Disposable MagLev centrifugal blood pump utilizing a cone-shaped impeller, Artificial Organs, Vol. 34, No. 8 (2010), pp. 669–677.

[5] Steinert, D., Nussbaumer, T., and Kolar, J. W., Slotless bearingless disk drive for high-speed and high-purity applications, IEEE Transactions on Industrial Electronics, Vol. 61, No. 11 (2014), pp. 5974-5986.

[6] Rao, J., Hijikata, W. and Shinshi, T., A bearingless motor utilizing a permanent magnet free structure for disposable centrifugal blood pumps, Journal of Advanced Mechanical Design, Systems, and Manufacturing, Vol. 9, No. 3 (2015).

[7] Gruber, W., Rothböck, M., and Schöb, R. T., Design of a novel homopolar bearingless slice motor with reluctance rotor, IEEE Transactions on Industry Applications, Vol. 51, No. 2 (2015), pp. 1456-1464.

[8] Gruber, W. and Radman, K., Modeling and Realization of a Bearingless Flux-Switching Slice Motor, Actuator, Vol. 6, No. 12 (2017).

[9] Minkyun Noh, Wolfgang Gruber, David L. Trumper, H ysteresis Bearingless Slice Motors with Homopolar Flux-Biasing, IEEE/ASME Transactions on mechatronics, Vol. 22, No. 2 (2017), pp. 2308-2318.

The 2018 International Power Electronics Conference

Reduced Hardware Parallel Drive for No Voltage Bearingless Motors

Eric L. Severson

Department of Electrical and Computer Engineering

University of Wisconsin-Madison, Madison, WI, USA

E-mail: eric.severson@wisc.edu

Abstract—Recent efforts to improve the performance of bearingless motors have focused on "combined winding" machine designs that use the same coils for torque and suspension force production. While combined winding designs have been successful in increasing the machine performance, they add substantial cost and complexity to the power electronics required for the bearingless drive system. This paper presents a novel bearingless drive concept which eliminates the additional power electronic hardware requirements for a parallel combined winding design. The resulting drive requires the same hardware components as a traditional, separated winding bearingless motor. The paper proposes and explains the novel design and theory of operation, proposes a simple control implementation, and presents initial validation via simulation results. The theory and design presented are valid for both conventional $p \pm 1$ bearingless motor designs (i.e., bearingless permanent magnet designs) as well as $p_s = 1$ bearingless motor designs (i.e., bearingless ac homopolar designs).

Keywords—*bearingless motor, combined winding, magnetic bearings, self-bearing motor*

I. Introduction

Bearingless motors provide the functionality of a magnetic bearing and a motor in a single electric machine. Compared to systems that utilize a motor with separate magnetic bearings, bearingless technology results in a more integrated system which requires less raw material and can be designed for higher speeds due to shorter shaft lengths [1]. Legacy bearingless motor designs utilize two separate windings: one winding for producing torque and a second winding for producing suspension forces. These two windings compete for the available slot space, which creates a design trade-off between the machine's motor performance and magnetic suspension performance. The suspension winding must be sized for a worst-case force requirement, which in most bearingless machines is much larger than the force required during normal operation. For this reason, recent research has focused on developing "combined windings" that are able to use the same coils for both magnetic levitation and torque production [2]–[11]. These combined winding approaches allow the slot space to be dynamically allocated between suspension forces and torque by the bearingless motor drive at run time.

While there are several fundamentally different approaches to designing a combined winding, this paper is concerned with Dual Purpose No Voltage (DPNV) windings, as described

in [5]. DPNV windings are advantageous over other combined winding approaches because the suspension current path does not experience any of the motor's back-EMF and the power electronics can be implemented with conventional two-level, three-phase inverters [12]. However, as compared to the traditional bearingless motor winding designs, which have separate, electrically isolated coils for torque and suspension currents, the drives for DPNV windings require additional power electronic components. There are two DPNV drive topologies: parallel and bridge. Of these, the parallel topology requires the fewest components and is the subject of this paper.

The primary contribution of this paper is to reduce the number of required hardware components of the parallel drive to what is needed in the drives of traditional, separate-winding bearingless motors. The proposed design reduces the cost of the drive system, while retaining the machine performance gains associated with combined winding designs, and thereby enables bearingless motor technology to extend into new, cost-sensitive application areas where it has the potential to offer extreme performance benefits. The paper presents the theory of operation for the proposed design, proposes three approaches for regulating the capacitor voltage, evaluates trade-offs in the voltage regulation approaches, and proposes a simple control implementation. Finally, the developed theory and proposed control implementation is validated via MATLAB Simulink simulation results.

II. Conventional Parallel DPNV Winding Drive

The conventional parallel winding drive structure is shown in Fig. 1 and was originally proposed in [4]. The motor winding must be deliberately designed to function in this arrangement, as described in [5]. Power electronic implementations for the parallel configuration were considered in [12], where the motor and suspension inverters were implemented as standard two-level, three-phase voltage source inverters. A trade-off comparison was made between having the two inverters share a common dc bus (Fig. 3a) and having each inverter use an isolated dc voltage bus (Fig. 3b). While sharing a common dc bus avoided the potentially significant hardware requirements of having a second isolated voltage source, it imposes the following undesirable requirements on the power electronics:

1) an extra current sensor is needed to detect and suppress circulating currents (the circulating current path

4020

The 2018 International Power Electronics Conference

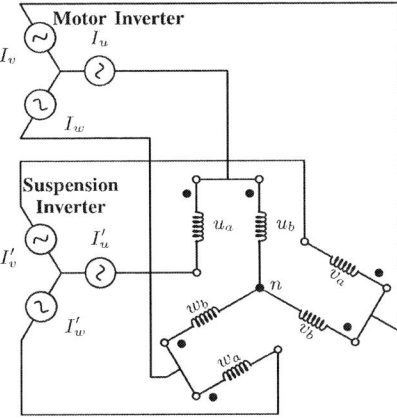

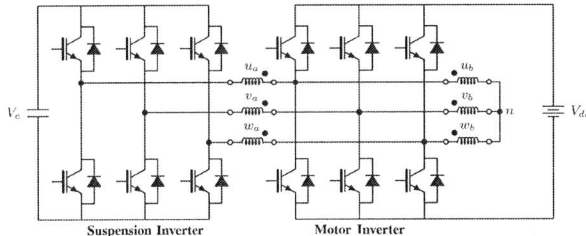

Fig. 2: Proposed parallel configuration using floating capacitor

Fig. 1: Parallel DPNV drive topology. Each inductor symbol represents a group of coils which have mutual and inductances and are exposed to a back-EMF with the polarity indicated by a dotted terminal. The back-EMF and mutual inductances cancel from the perspective of the suspenion inverter, which gives this topology the "no voltage" property.

III. PROPOSED PARALLEL DPNV WINDING DRIVE

This paper proposes a novel implementation of the parallel drive structure, shown in Fig. 2, based around a floating capacitor for the suspension inverter's dc link. Since no path exists for circulating currents, there is no need for an additional current sensor, nor is there any need to restrict the modulation strategies used. This means that SVPWM can be employed to obtain maximal dc bus voltage utilization. Finally, the suspension capacitor voltage can be controlled to have any desired value. Therefore, this topology overcomes the aforementioned drawbacks highlighted for the topology in Fig. 3a.

The key power electronic drive hardware requirements of the proposed technology are compared in Table 1 against a conventional separated winding bearingless motor drive as well as the other DPNV drive implementations. The proposed floating capacitor parallel motor drive requires the same hardware components as the conventional separated winding drive, which puts it at a significant advantage compared to the other DPNV drive implementations.

Similar circuit topologies with floating capacitors have been used in the broader field of power electronics and motor drives, for example in open-end motor drives [13]–[17] as a method to obtain multi-level waveforms, supply reactive power, enhance the supply voltage level, and, in electric vehicles, as a way to transfer power to a battery [18], [19]. The circuit topology and operation proposed in this paper is fundamentally different from prior art for the following reasons: 1) a parallel branch of motor coils appears at the terminals of the motor inverter, 2) the floating capacitor

is labeled in Fig. 3a);

2) neither the motor inverter nor the suspension inverter are able to use space vector pulse width modulation (SVPWM) and are thereby limited to peak phase voltages of $V_{dc}/2$ (as compared to $V_{dc}/\sqrt{3}$ for SVPWM); the inverters cannot use SVPWM due to the time varying common-mode voltage causing circulating currents; and

3) both the suspension inverter and the motor inverter must have the same dc link voltage value, despite the voltage requirements for the suspension inverter being much lower than the motor inverter; this adds device cost, significant suspension current and force ripple, and losses in the inverter and the electric machine.

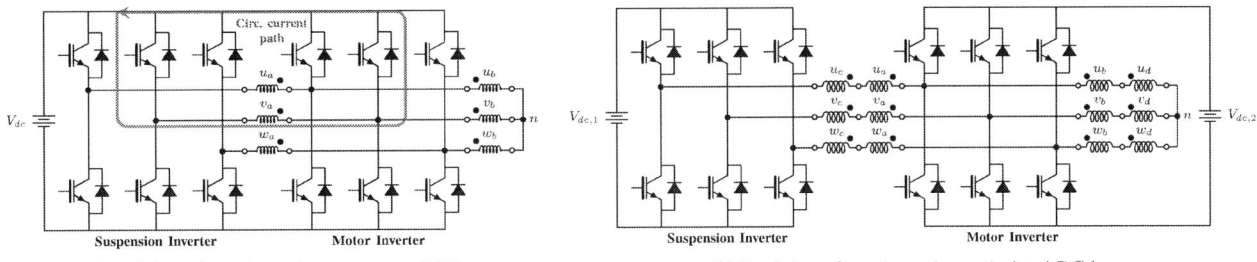

(a) Parallel configuration using a common DC bus   (b) Parallel configuration using an isolated DC bus

Fig. 3: Circuit diagrams of the power electronics converters used to realize the conventional DPNV parallel drive in [12].

4021

The 2018 International Power Electronics Conference

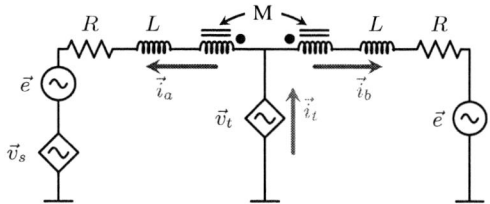

Fig. 4: Equivalent single phase circuit with space vector quantities labeled. The dotted inductors represent the polarity of a mutual inductance M, corresponding to the dotted terminals of Fig. 3.

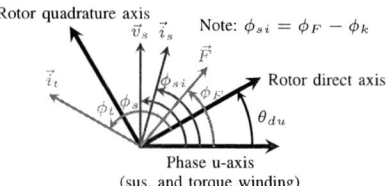

Fig. 5: Space vector and machine axes definitions

inverter (the suspension inverter) must provide real power (for magnetic suspension, assuming a horizontal shaft experiencing the effects of gravity), and 3) power flow to the floating capacitor must be carefully managed so that it does not add current components that create problematic suspension forces.

A. System Model

The operation of the proposed design can be illustrated by considering the equivalent single phase circuit shown in Fig. 4. In this circuit, the motor inverter is represented as the dependent voltage source $\vec{v}_t$, with an output current of $\vec{i}_t$, and the suspension inverter as a voltage source $\vec{v}_s$, with an output current $-\vec{i}_a$. The electric circuit of each coil group (a and b from Fig. 1 and Fig. 3) consists of a series resistance R, self-inductance L, and back-EMF $\vec{e}$ combined with a mutual inductance M linking the two groups. The dotted terminals of the mutual inductance correspond to the dotted terminals of the coil groups in Fig. 1, Fig. 2, and Fig. 3. A vector diagram indicating the space vector quantities is shown in Fig. 5.

The torque inverter current $\vec{i}_t$ is the traditional motor current space vector, which can be decomposed into direct i_d and quadrature i_q components. In this paper, a non-salient rotor structure is assumed, so that torque is determined by $T = k_t i_q$. Controlled radial suspension forces are produced by super-imposing a suspension current component $\vec{i}_s$ onto the torque inverter current $\vec{i}_t$ in the motor coil groups [5], [12]. This is done by using current regulators to control $\vec{i}_t$ and $\vec{i}_a$

to achieve the following coil group currents:

$$\vec{i}_a = \frac{1}{2}\vec{i}_t - \vec{i}_s$$
$$\vec{i}_b = \frac{1}{2}\vec{i}_t + \vec{i}_s \tag{1}$$

The suspension current is related to the force vector that is produced on the shaft by (2), where $\bar{k}_i = k_i\angle\phi_k$ contains information on both the magnitude and angular difference between $\vec{i}_s$ and $\vec{F}$ [1]. The value of ϕ_k is specified in (3) based on the number of pole pairs of the motor winding, where θ_{du} is the angle of the rotor's direct axis (see Fig. 5) and ϕ_0 is a constant offset between the suspension phase u axis and the torque phase u axis; $\phi_0 = 0$ is assumed in this paper. The sign of θ_{du} depends on whether the suspension phase sequence is transposed with respect to the motor phase sequence (a property of certain DPNV windings [5]); the sign in (3) is indicated for a non-transposed case.

$$\vec{i}_s = \frac{1}{\bar{k}_i}\vec{F} \tag{2}$$

$$\phi_k = \begin{cases} -\theta_{du} + \phi_0, & \text{if } p_s = p \pm 1 \\ \phi_0, & \text{if } p_s = 1 \text{ and } p \geq 4 \end{cases} \tag{3}$$

The suspension inverter voltage can be calculated as (4), where it is seen that the motor's back-EMF $\vec{e}$ cancels out, thereby meeting the "no voltage" property of DPNV windings. When the coil group currents are given by (1), $\vec{v}_s$ simplifies to (5), which corresponds to the steady state impedance of (6). This reveals that the suspension voltage is determined by the

TABLE I: Comparison of bearingless motor drive hardware requirements

	Conventional	**Proposed FC design–Fig. 2**	Parallel non-isolated–Fig. 3a	Parallel isolated–Fig. 3b	Bridge [12]
Switches	12	**12**	12	12	18
Current sensors	4	**4**	5	4	5
Isolated voltage buses	1	**1**	1	2	4
Usable fraction of sus. dc bus	$\frac{1}{\sqrt{3}}$	$\frac{1}{\sqrt{3}}$	$\frac{1}{2}$	$\frac{1}{\sqrt{3}}$	1
Usable fraction of tor. dc bus	$\frac{1}{\sqrt{3}}$	$\frac{1}{\sqrt{3}}$	$\frac{1}{2}$	$\frac{1}{\sqrt{3}}$	$\frac{1}{\sqrt{3}}$
Sus. inverter current rating	coil rating	**coil rating**	coil rating	coil rating	$2\times$ coil rating
Sus. inverter voltage rating	$\leq$ tor. inverter rating	**$\leq$ tor. inverter rating**	tor. inverter rating	$\leq$ tor. inverter rating	$\leq$ tor. inverter rating

4022

suspension current $\vec{i}_s$ (or, equivalently, $\vec{F}$), independent of $\vec{i}_t$.

$$\vec{v}_s = \vec{e} + R\vec{i}_b + L\frac{d\vec{i}_b}{dt} + M\frac{d\vec{i}_a}{dt}$$
$$- \left(\vec{e} + R\vec{i}_a + L\frac{d\vec{i}_a}{dt} + M\frac{d\vec{i}_b}{dt}\right) \qquad (4)$$

$$= 2\left(L - M\right)\frac{d\vec{i}_s}{dt} + 2R\frac{d\vec{i}_s}{dt} \qquad (5)$$

$$\bar{z} = \frac{\vec{v}_s}{\vec{i}_s} = 2R + j\omega_s 2\left(L - M\right) \qquad (6)$$

B. Steady state power flow

The suspension inverter is responsible for supplying currents to maintain the shaft's levitation. This requires real power to flow from the suspension inverter and into the coil winding resistances, which will discharge the floating capacitor. To maintain the capacitor voltage, power must be transfered from the motor inverter to the suspension inverter. The steady state power into the floating capacitor can be calculated by (7), where ϕ_z is the angle of $\bar{z}$.

$$P = \text{Real}\{\vec{v}_s \vec{i}_a^*\} = P_{\text{TS}} - P_{\text{SS}}$$
$$= \frac{1}{2}v_s i_t \cos\left(\phi_s - \phi_t - \theta_{du}\right) - \frac{v_s^2}{z}\cos\phi_z \qquad (7)$$

$$P_{\text{TS}} = \frac{1}{2}v_s i_d \cos\left(\phi_s - \theta_{du}\right) + \frac{1}{2}v_s i_q \sin\left(\phi_s - \theta_{du}\right) \qquad (8)$$

$$\vec{v}_s = \frac{\bar{z}}{k_i}\vec{F} \qquad (9)$$

The first term of (7) represents the power flowing from the torque inverter to the suspension inverter and will be referred to as P_{TS}, which can be conveniently expressed in terms of i_d and i_q in (8); the second term of (7) represents the power flowing from the suspension inverter into the coil resistances and will be referred to as P_{SS}. For a horizontal shaft machine, $\vec{v}_s$ and $\vec{i}_s$ have the steady state values required to generate a constant force to support the shaft's weight. The steady state suspension voltage vector can be calculated as (9) by solving (2) and (6). The necessary torque current $\vec{i}_t$ to maintain the capacitor's voltage can then be found by solving (7) for $P = 0$. While there appear to be two degrees of freedom in this expression (i_d and i_q), this ambiguity is resolved by noting that the q-axis current is specified by the motor drive's torque requirements ($i_q = T/k_t$).

In systems where a constant force is present (for example, the aforementioned gravitational forces in a horizontal shaft machine), an additional current can be provided by the motor inverter to interact with the steady state suspension voltage space vector to charge the floating capacitor. In systems where a constant force is not present, a steady state suspension voltage space vector must also be added–potentially creating problematic suspension force ripple. Three approaches to transferring power to the suspension inverter are outlined below. Whether or not these approaches are applicable to a given bearingless machine is determined based on whether a constant shaft force is present and the angle of ϕ_k, as follows:

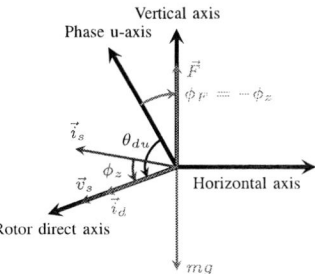

Fig. 6: Space vectors for power flow approach 1 for the case of a horizontal shaft machine. If the machine does not have a horizontal shaft, but somehow experiences a constant force, the label "Vertical axis" should be replaced with the axis of the constant force.

1) Constant radial force present and: $\phi_k = -\theta_{du}$, which corresponds to certain $p \pm 1$ bearingless motors–see the discussion regarding (3). Using this value for ϕ_k, (9) can be solved for $\phi_s = \phi_F + \theta_{du} + \phi_z$, which simplifies (7) to (10). This indicates a constant power P_{TS} in steady state conditions where the magnitude of $\vec{v}_s$ is given by (11) for a shaft of mass m. Note that power flow between the torque and suspension inverters depends on torque generated by the machine (since $i_q = T/k_t$). This is highly undesirable, as it will require an additional current i_d to compensate for changes in load torque.

$$P_{\text{TS}} = \frac{1}{2}v_s i_d \cos\left(\phi_F + \phi_z\right) + \frac{1}{2}v_s i_q \sin\left(\phi_F + \phi_z\right) \qquad (10)$$

$$v_s = \frac{z}{k_i}mg \qquad (11)$$

To eliminate the dependence on the machine's load, the bearingless motor can be oriented to set the force vector angle as $\phi_F = -\phi_z$, in which case (10) reduces to (12). This can be accomplished by orienting the stator housing at the proper angle with respect to the horizon, as depicted in Fig. 6. The required d-axis current i_d is calculated from (13) by equating (12) to P_{SS}. Since i_d is split evenly between the two coil groups, the amount of coil group current allocated to the suspension winding becomes $i_s \leq i_s + \frac{i_d}{2} \leq 2i_s$. In a typical design, the suspension winding current i_s required to support the shaft's weight is nearly two orders of magnitude smaller than the rated q-axis current; therefore this additional current requirement doesn't have any meaningful impact on the motor performance.

$$P_{\text{TS}} = \frac{1}{2}v_s i_d \qquad (12)$$
$$i_d = \frac{2v_s}{z}\cos\phi_z$$
$$= 2i_s \cos\phi_z \qquad (13)$$

In practice, maintaining $\phi_F = -\phi_z$ may prove to be a problematic solution for a few reasons. First, it may not be practical for system integrators to guarantee the motor's orientation at install time satisfies $\phi_F = -\phi_z$. Second, it prevents the use of field weakening techniques since i_d is used to regulate the power flow to the floating capacitor. Third, and most concerning, ϕ_z is the angle of (6) which depends on the rotor's rotational frequency ($\omega_s = d\phi_k/dt = d\theta_{du}/dt$). If the machine's speed varies significantly during operation, the motor orientation chosen at install time will not continue to satisfy $\phi_F = -\phi_z$ for all operating conditions, meaning that the actual power flow will depend on i_q as specified in (10). For these reasons, this approach to regulating the power to the floating capacitor is not favored.

Finally, note that this analysis of power flow also applies to the standard parallel drive implementations of Fig. 3, which means that horizontal shaft parallel winding motors in general are prone to a potentially problematic power flow between the suspension and torque inverters.

2) Constant radial force present and: $\phi_k = 0$, which corresponds to $p_s = 1$ bearingless motors in (3). For the case of a constant direction force, $\vec{v}_s$ is stationary and (6) reduces to $\bar{z} = 2R$ (since $\omega_s = 0$), which means that with $\phi_s = \phi_F$. Using this angle in (8) reveals that constant values of i_d and i_q will produce no average power flow to the suspension inverter. To overcome this, the torque current space vector is modified by adding an additional stationary vector $\vec{i}_0$ aligned with $\vec{v}_s$ (14). Using this new space vector, the average value of P_{TS} is calculated in (15) by neglecting terms that don't contribute to the average power. A space vector diagram of this power flow approach is shown in Fig. 7.

$$\vec{i}_t' = \vec{i}_t + i_0 \angle \phi_F \qquad (14)$$

$$\langle P_{TS} \rangle = \frac{1}{2} v_s i_0 \qquad (15)$$

$$i_0 = \frac{v_s}{R} = 2i_s \qquad (16)$$

$$i_d' = i_0 \cos(\phi_F - \theta_{du}) + i_d \qquad (17)$$

$$i_q' = i_0 \sin(\phi_F - \theta_{du}) + i_q \qquad (18)$$

The required magnitude of $\vec{i}_0$ is calculated in (16), which indicates that the fraction of the coil group current allocated for suspension forces is doubled by this control technique. Since $\vec{i}_0$ is stationary, it appears in both the direct and quadrature components of $\vec{i}_t'$, see (17) and (18). This will produce torque ripple. However, as previously mentioned, in a well-designed bearingless motor, the required value of i_s to support the shaft's weight is nearly two orders of magnitude smaller than the rated torque current–meaning that i_0 will not have a significant impact on the motor's performance. Unlike the control design of Section III-B1, there is no restriction on the orientation of the rotor.

One potential area of concern for this type of power flow control is the performance at low or zero rotational speeds. For example, if a stationary holding torque is produced by the stator, θ_{du} will have a constant value and the actual power transfered will be the sum of (7) and (8). This may cause a problematic power flow from the torque inverter to the suspension inverter that is dependent upon the motor torque, similar to the problem described at variable speed for the

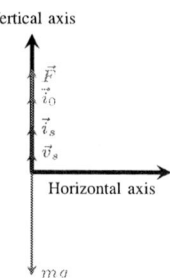

Fig. 7: Space vectors for power flow approach 2

control approach described in Section III-B1. This is not viewed as a problem for high speed motor designs which do not require static holding torques and pass quickly through the low speed operating region.

3) With or without a constant radial force and: any value of ϕ_k, power can be transfered to the floating capacitor by adding new rotating components to both $\vec{i}_t$ and $\vec{v}_s$ at a different frequency (19), depicted in Fig. 8. These new rotating components modify the average power transfer equations to become (20) and (21).

$$\vec{i}_t' = \vec{i}_t + i_f \angle \omega_f t$$
$$\vec{v}_s' = \vec{v}_s + v_f \angle \omega_f t \qquad (19)$$
$$\bar{z}_f = 2R + j\omega_f 2(L - M)$$

$$\langle P_{TS} \rangle = \frac{1}{2} v_f i_f \qquad (20)$$

$$\langle P_{SS} \rangle = \frac{v_s^2}{z} \cos\phi_z + \frac{2v_f^2 R}{4R^2 + 4\omega_f^2 (L - M)^2} \qquad (21)$$

$$\Delta F = \frac{k_i}{z_f} v_f = \frac{k_i}{\sqrt{4R^2 + 4\omega_f^2 (L - M)^2}} v_f \qquad (22)$$

The second term of $\langle P_{SS} \rangle$ corresponds to the additional power drawn from the floating capacitor due to the new frequency component of $\vec{v}_s'$. These added rotating components will also add torque and force ripple (22). The effect of this ripple and the amount of loss can be minimized by proper selection of the frequency for the power transfer and magnitude of the v_f. Higher values of ω_f lead to lower losses and force and torque ripple (and less momentum change since the ripple is applied over a shorter time duration). However, increasing the frequency also increases the voltage needed by the torque inverter to realize i_f. One option to increase the frequency of the torque and force ripple without increasing the value of ω_f is to have i_f and v_f rotate backward relative to the rotor's direct axis.

C. Comparison of power flow approaches

Trade-offs between each of the power flow approaches outlined in Section III-B are outlined in Table II. Approach 1 and 2 offer the potential for the lowest losses, but impose restrictions that may make them infeasible depending on the design application. Approach 3 offers the most design

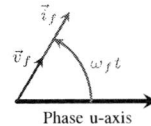

Fig. 8: Space vectors for power flow approach 3

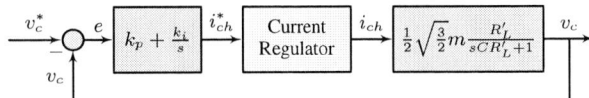

Fig. 9: Model of control systm

TABLE II: Comparison of proposed power flow approaches

	Approach 1	Approach 2	Approach 3
Req. constant force	Yes	Yes	No
Req. $\phi_k =$	$-\theta_{du}$	0	Any value
Adds torque ripple	No	Yes	Yes
Adds sus. ripple	No	No	Yes
Potential challenges	Required alignment changes with speed	Voltage regulation for low speed operation	Additional losses, ripple, and voltage requirement

flexibility but has additional losses and force and torque ripple. Approaches 2 and 3 require regulating current at two different frequencies. There are well-known approaches to doing this, including resonant controllers and additional integrators in reference frames that rotate at each of the frequency components [20].

IV. CONTROL STRATEGY

A system plant and a simple control strategy for regulating the floating capacitor voltage is now proposed and developed. This strategy can be applied to any of the three power flow approaches outlined in Section III-B. The power into the floating capacitor is expressed in (23), where from Fig. 2 v_c is the capacitor voltage, $m = \sqrt{\frac{2}{3}}\frac{v_s}{v_c}$ is the modulation index for the power invariant space vectors used, i_{ch} is the charging current component from the torque inverter ($i_{ch} = i_d$ for approach 1, $i_{ch} = i_0$ for approach 2, and $i_{ch} = i_f$ for approach 3), and R'_L is an equivalent load resistance defined in (24) for a constant force F. Note that power flow approach 3 has an additional term in P_{SS} which is neglected in the following derivation, but can easily be added.

$$
\begin{aligned}
v_c C \frac{dv_c}{dt} &= P_{TS} - P_{SS} \\
&= \frac{1}{2} v_s i_{ch} - \frac{v_s^2}{z} \cos \phi_z \\
&= \frac{1}{2} \sqrt{\frac{3}{2}} m v_c i_{ch} - \frac{v_c}{R'_L}
\end{aligned}
\tag{23}
$$

$$
R'_L = \frac{v_c}{\sqrt{\frac{3}{2}} m i_s \cos \phi_z} = \frac{v_c^2 k_i^2}{2F^2 R}
\tag{24}
$$

Next, the differential equation is re-written as a transfer function in (25). In these expressions, m, i_s, and ϕ_z are taken as the values at the nominal operating point (i.e., based on the

expected the constant force).

$$
sC V_c(s) = \frac{1}{2}\sqrt{\frac{3}{2}} m I_{ch}(s) - \frac{1}{R'_L} V_c(s)
$$

$$
\frac{V_c(s)}{I_{ch}(s)} = \frac{1}{2}\sqrt{\frac{3}{2}} m \frac{R'_L}{sCR'_L + 1}
\tag{25}
$$

A PI controller can be used to track a reference capacitor voltage. A schematic of the control diagram is shown in Fig. 9. If k_p and k_i are selected as (26) and (27), the resulting closed loop transfer function is a first order low pass filter with a bandwidth of ω_c.

$$
k_p = \frac{2\omega_c C}{\sqrt{3/2}m}
\tag{26}
$$

$$
k_i = \frac{2\omega_c}{\sqrt{3/2}mR'_L}
\tag{27}
$$

This proposed control implementation is intended as a simple approach to regulating the voltage of the floating capacitor and is not intended to represent an optimal control strategy. As such, this approach makes several simplifying assumptions that may lead to limited dynamic performance–primarily related to modeling the power flow out of the floating capacitor as a resistive load and the assumption of a constant modulation index. This is similar in concept to strategies used for active front end converters [21]. For a given design, the expected load behavior should be investigated to determine the appropriate capacitance value and controller bandwidth

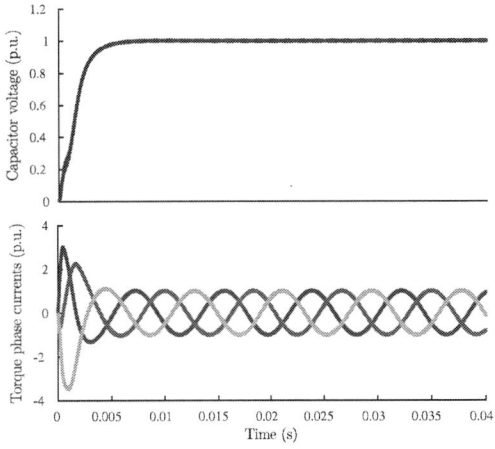

Fig. 10: Simulation of startup for Power flow approach 3; (top) floating capacitor voltage; (bottom) torque phase currents.

The 2018 International Power Electronics Conference

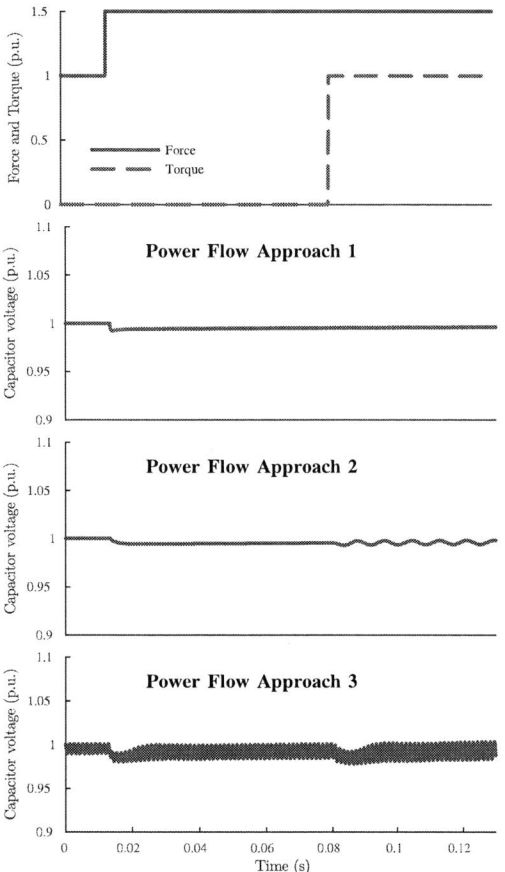

Fig. 11: Constant speed simulation. The top plot depicts the reference suspension force and torque.

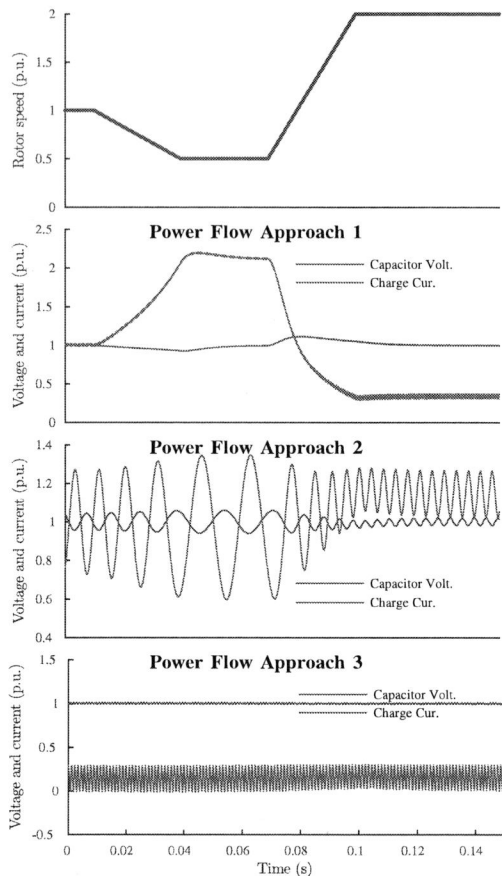

Fig. 12: Constant torque and force simulation. The top plot depicts the reference rotor speed.

to maintain satisfactory voltage regulation. Cases without a constant load force may prove to be more challenging to properly quantify the average power out of the capacitor. These topics require more advanced consideration and are beyond the scope of this paper.

V. RESULTS

MATLAB Simulink was used to simulate each of the power transfer approaches outlined in Section III-B with the voltage regulation controller designed according to the discussion in Section IV. In all simulations, the suspension and torque inverters use SVPWM with a 20 kHz switching frequency and their currents are controlled via closed loop regulators. The reference current values are set based on a specified torque ($i_q = T/k_t$), specified radial forces, and the charging current reference i_{ch}^* generated by the floating capacitor voltage regulator (per Fig. 9). The motor is modeled as having $L = 5.6$ mH, $M = 4.9$ mH, $R = 0.2\,\Omega$, and $k_i = 23$ N/amp. The angle of k_i is as described in (3), based on whether a $p_s = 1$ type bearingless motor or a $p \pm 1$ type bearingless motor is being considered. All simulation results

are normalized by the motors' rated conditions for comparison purposes and the motors are considered as horizontal shaft machines (experiencing the effects of gravity as a constant radial force). For the simulations of power flow approach 1, the stator housing orientation is chosen based on the discussion in Section III-B1 to avoid power being transmitted from torque producing current i_q.

Three scenarios have been simulated: startup from a powered down state, constant speed, and constant torque/force. Results from the startup simulation are shown in Fig. 10 for power flow approach 1. In this case, $i_d = i_{ch}$ and the regulator is able to achieve a damped step response to track the desired capacitor voltage. The torque phase currents are initially large, reflecting a large value of i_d and decay as the capacitor voltage approaches its target value. The simulation uses a constant torque current i_q reference and constant suspension current reference to support the rotor's weight.

Results from the constant speed simulations are shown in Fig. 11 where the rotor speed is held constant but the torque and suspension current references change according to the

4026

values depicted in the top plot. A suspension force of 1 pu corresponds to the gravitational force of the shaft. Since the machine is at constant speed, power flow approach 1 exhibits the best voltage regulation because no additional frequency current components need to be injected into the machine.

Results from the constant force and torque simulation are shown in Fig. 12. Here, the suspension current reference is held constant at the value required support the shaft's weight while the speed is varied. This test clearly depicts the expected undesirable speed response of both power flow approaches 1 and 2, while power flow approach 3 is not impacted by changes in speed. Also as expected, power flow approach 2 performs better at higher speeds. Note that this simulation was intentionally conducted at a problematically low speed and it is anticipated that for the right high speed machine configuration, power flow approach 2 will exhibit reasonable performance.

VI. Conclusion

This paper has proposed a novel circuit topology based around a floating capacitor inverter to implement a parallel DPNV motor drive that reduces the amount of required power electronic hardware. Apart from reducing the hardware complexity of DPNV drives, this new topology also allows flexibility in the suspension voltage bus value, which is advantageous in terms of losses and force / torque ripple. The operation of the topology is explained as well as several approaches to regulate the floating capacitor's voltage level. Simulations are presented to validate the developed theory. While this paper develops and demonstrates the functionality of this new topology, future work must be conducted to 1) experimentally validate this system and 2) explore the impact that this topology has on the drive's volt-ampere rating.

Finally, an interesting corollary of this paper is a demonstration that all parallel DPNV drives for $p \pm 1$ bearingless motors are vulnerable to a significant power flow between the torque and suspension inverters when a constant force is present, depending on the direction of the force. Future bearingless machine designers need to factor this into their design to ensure a high performance parallel bearingless motor drive.

References

[1] A. Chiba, T. Fukao, O. Ichikawa, M. Oshima, M. Takemoto, and D. Dorrell, *Magnetic Bearings and Bearingless Drives*. Newnes, 2005.

[2] K. Raggl, T. Nussbaumer, and J. W. Kolar, "Comparison of separated and combined winding concepts for bearingless centrifugal pumps," *J. Power Electron*, vol. 9, no. 2, pp. 243–258, 2009.

[3] H. Mitterhofer, B. Mrak, and W. Gruber, "Comparison of high-speed bearingless drive topologies with combined windings," *Industry Applications, IEEE Trans on*, vol. 51, no. 3, pp. 2116–2122, May 2015.

[4] R. Oishi, S. Horima, H. Sugimoto, and A. Chiba, "A novel parallel motor winding structure for bearingless motors," *Magnetics, IEEE Transactions on*, vol. 49, no. 5, pp. 2287–2290, 2013.

[5] E. L. Severson, R. Nilssen, T. Undeland, and N. Mohan, "Design of dual purpose no-voltage combined windings for bearingless motors," *IEEE Transactions on Industry Applications*, vol. 53, no. 5, pp. 4368–4379, Sept 2017.

[6] J. Huang, B. Li, H. Jiang, and M. Kang, "Analysis and control of multiphase permanent-magnet bearingless motor with a single set of half-coiled winding," *IEEE Transactions on Industrial Electronics*, vol. 61, no. 7, pp. 3137–3145, 2014.

[7] A. Chiba, K. Sotome, Y. Iiyama, and M. Azizur Rahman, "A novel middle-point-current-injection-type bearingless pm synchronous motor for vibration suppression," *Industry Applications, IEEE Transactions on*, vol. 47, no. 4, pp. 1700–1706, 2011.

[8] H. Sugimoto, S. Tanaka, A. Chiba, and J. Asama, "Principle of a novel single-drive bearingless motor with cylindrical radial gap," *IEEE Transactions on Industry Applications*, vol. 51, no. 5, pp. 3696–3706, 2015.

[9] W. Gruber, W. Amrhein, and M. Haslmayr, "Bearingless segment motor with five stator elements-design and optimization," *IEEE Transactions on Industry Applications*, vol. 45, no. 4, pp. 1301–1308, July 2009.

[10] R. P. Jastrzebski, P. Jaatinen, O. Pyrhönen, and A. Chiba, "Current injection solutions for active suspension in bearingless motors," in *Power Electronics and Applications (EPE'17 ECCE Europe), 2017 19th European Conference on*. IEEE, 2017, pp. P–1.

[11] J. Asama, K. Sasaki, T. Oiwa, and A. Chiba, "Ripple compensation of suspension force and torque in a bearingless spm motor with integrated winding," in *Energy Conversion Congress and Exposition (ECCE), 2017 IEEE*. IEEE, 2017, pp. 5403–5408.

[12] E. Severson, S. Gandikota, and N. Mohan, "Practical implementation of dual-purpose no-voltage drives for bearingless motors," *IEEE Transactions on Industry Applications*, vol. 52, no. 2, pp. 1509–1518, March 2016.

[13] S. Chowdhury, P. W. Wheeler, C. Patel, and C. Gerada, "A multilevel converter with a floating bridge for open-end winding motor drive applications," *IEEE Transactions on Industrial Electronics*, vol. 63, no. 9, pp. 5366–5375, Sept 2016.

[14] J. Ewanchuk, J. Salmon, and C. Chapelsky, "A method for supply voltage boosting in an open-ended induction machine using a dual inverter system with a floating capacitor bridge," *IEEE Transactions on Power Electronics*, vol. 28, no. 3, pp. 1348–1357, March 2013.

[15] M. S. Toulabi, J. Salmon, and A. M. Knight, "Design, control, and experimental test of an open-winding ipm synchronous motor," *IEEE Transactions on Industrial Electronics*, vol. 64, no. 4, pp. 2760–2769, April 2017.

[16] J. Kim, J. Jung, and K. Nam, "Dual-inverter control strategy for high-speed operation of ev induction motors," *IEEE Transactions on Industrial Electronics*, vol. 51, no. 2, pp. 312–320, April 2004.

[17] D. Pan, F. Liang, Y. Wang, and T. A. Lipo, "Extension of the operating region of an ipm motor utilizing series compensation," *IEEE Transactions on Industry Applications*, vol. 50, no. 1, pp. 539–548, Jan 2014.

[18] B. A. Welchko, "A double-ended inverter system for the combined propulsion and energy management functions in hybrid vehicles with energy storage," in *Industrial Electronics Society, 2005. IECON 2005. 31st Annual Conference of IEEE*. IEEE, 2005, pp. 6–pp.

[19] J. Hong, H. Lee, and K. Nam, "Charging method for the secondary battery in dual-inverter drive systems for electric vehicles," *IEEE Transactions on Power Electronics*, vol. 30, no. 2, pp. 909–921, Feb 2015.

[20] X. Yuan, W. Merk, H. Stemmler, and J. Allmeling, "Stationary-frame generalized integrators for current control of active power filters with zero steady-state error for current harmonics of concern under unbalanced and distorted operating conditions," *IEEE transactions on industry applications*, vol. 38, no. 2, pp. 523–532, 2002.

[21] R. Teodorescu, M. Liserre, and P. Rodriguez, *Grid converters for photovoltaic and wind power systems*. John Wiley & Sons, 2011, vol. 29.

The 2018 International Power Electronics Conference

Dual Field-Oriented Control of Bearingless Motors with Combined Winding System

Wolfgang Gruber[1*] and Siegfried Silber[2]

[1] Institute of Electrical Drives and Power Electronics, Johannes Kepler University Linz, Linz, Austria
[2] Linz Center of Mechatronics, Linz, Austria
[*] E-mail: wolfgang.gruber@jku.at

Abstract- **Field-oriented control has become a standard method to generated drive torque in permanent magnet excited synchronous drives with high efficiency. In this work, a similar control approach is proposed for radial suspension force generation in bearingless drives with combined winding system. Combined windings sets generate both motor torque and suspension force in each phase depending on the rotor angle. This allows increased efficiency, simpler manufacture and higher compactness compared to separated winding systems. However, in this case the decoupling of force and torque, which is necessary for proper bearingless operation, has to be achieved in the control scheme. Using transformations into two different rotating coordinate systems is proposed in this work to solve this issue. Hence, for full bearingless motor operation, two different field-oriented control schemes of the same structure have to work in parallel, one for force generation and the other for torque generation. The necessary phase currents are finally superposed. Thus, well proven and state-of-the-art industrial procedures can be used to implement the control scheme for a bearingless permanent magnetic motor, similar to a standard electric motor.**

Keywords— bearingless motor, dual field-oriented control, decoupling of force and torque, combined windings

I. INTRODUCTION

Bearingless motors, which are able to generate levitation force and drive torque in one common unit, are well established [1] and research in this field is gaining more and more popularity. Due to their contact-free operation, ultra-pure and hermetically enclosed processes can be realized, where seals and lubrication are obsolete. Additionally, this technology allows extremely high speeds, which exceed the operating limits of mechanical bearings.

However, bearingless motors often feature two separated types of winding sets: a drive winding set creating torque and a suspension winding set creating (radial) suspension forces. Unfortunately, implementing these two winding systems enhance the necessary constructional space, decrease the efficiency and limit the compactness of bearingless motors [2]-[4]. In contrast to that, a combined winding system, where each motor phase is capable to generate both force and torque (often in dependence on the rotor angle) does not feature these drawbacks but demands a proper decoupling of force and torque in the control scheme. In this work, this decoupling is achieved using two different generalized Clarke transformations. First, the model for force and torque generation in state-of-the-art bearingless

permanent magnet (PM) motors with combined winding set is presented. After that, the announced generalized Clarke transformations are introduced leading to a dual field-oriented control concept for torque as well as for force generation. Bearingless PM motor topologies (regarding pole pair and phase number) that allow the application of this kind of control scheme are identified thereafter. Here, a separation of interior and exterior PM rotors is necessary as the single-phase characteristic is slightly different. Finally, the implementation of the proposed control concept for bearingless PM motors is described.

II. DUAL FIELD-ORIENTED CONTROL SCHEME

A. Force and Torque Model

The radial suspension forces (F_x and F_y) and the drive torque (T_z) created by the m stator phases are assumed to be linear dependent on the phase currents. Therefore, they can be described by [5]

$$\begin{pmatrix} F_x\left(\varphi_r\right) \\ F_y\left(\varphi_r\right) \\ T_z\left(\varphi_r\right) \end{pmatrix} = \mathbf{T_m}\left(\varphi_r\right)\begin{pmatrix} i_1 & i_2 & \cdots & i_m \end{pmatrix}^T . \tag{1}$$

φ_r represents the mechanical rotor angle, i_k stands for the current in the k-th phase and the $3 \times m$ matrix $\mathbf{T_m}(\varphi_r)$ is called current-force/torque matrix. The resulting overall suspension force and drive torque is composed of the single-phase suspension force and torque characteristic of each phase. This single-phase characteristic can be obtained by finite element simulations or an analytical model. For the majority of bearingless PM synchronous drives this characteristic features approximately sinusoidal shape and can be analytically described by

$$\begin{pmatrix} F_{x,k}\left(\varphi_r\right) \\ F_{y,k}\left(\varphi_r\right) \\ T_{z,k}\left(\varphi_r\right) \end{pmatrix} = \begin{pmatrix} \pm k_{i,x}\cos\left(\varphi_r\right) \\ -k_{i,y}\sin\left(\varphi_r\right) \\ -k_{i,z}\sin\left(\varphi_r\right) \end{pmatrix} i_k \tag{2}$$

for the considered k-th phase, with $k_{i,x}$ as x-force constant, $k_{i,y}$ as y-force constant and $k_{i,z}$ as torque constant. Without loss of generality, in (2) the global stator-fixed Cartesian xyz-coordinate system coincides with the phase-related coordinate system of the considered phase k. Hence, the

4028

force in direction of the phase axis coincides with the *x*-direction and the force perpendicular to the phase axis points towards the *y*-axis [6]. In this case the tangential phase force $F_{y,k}$ features the same characteristic as the phase torque $T_{z,k}$. The sign of the normal phase force $F_{x,k}$ depends on the drive topology. For interior rotor systems the positive sign is valid, whereas an exterior rotor system features the negative sign. Figure 1 illustrates the typical single-phase force orbit of bearingless PM motors with combined winding, represented by (2).

The elliptical/circular shape is very characteristic and holds true for many bearingless PM motors with combined winding sets, like the standard *interior bearingless slice motors* [7], the *exterior bearingless slice motors* [8], [9], the *bearingless segment motors* [6], the slotless *high-speed bearingless disk drives* (typically featuring toroidal windings) [10], [11], the *multi-phase bearingless motor* [12], the *bearingless torque motor* [13], the *bearingless flux-switching motor* [14] or the *bearingless vernier motor* [15]. Figure 2 shows the single-phase force and torque characteristic of these drives over the rotor angle. The ratio of the radial (in *x*-direction) and tangential (in *y*-direction) force amplitudes defines the eccentricity of the force orbit ellipse in Fig. 1 and depends on the topology. [16] and [17] feature equal amplitudes (and therefore circular force orbit) but typically the radial force is larger than the tangential force.

Significant armature reaction, deep saturation and high reluctance effects lead to nonlinear force-to-current and torque-to-current relationships. These effects are not covered by this model and must, therefore, be considered to be negligible.

However, if the global coordinate system coincides with the coordinate system of the first phase, the rotor

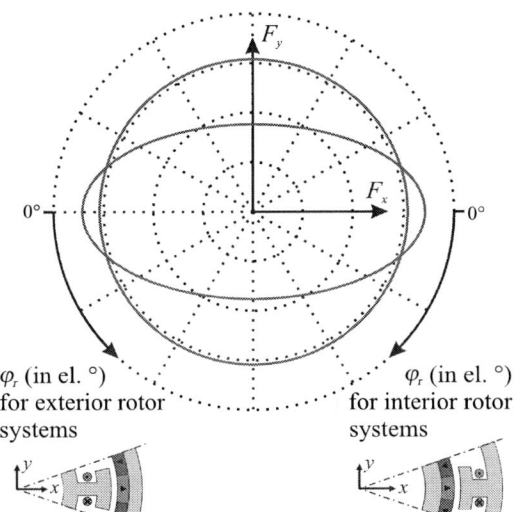

Fig. 1. Two typical shapes of the single-phase force orbit of bearingless PM motors. For increasing rotor angle an interior rotor system moves clockwise on the orbit while an exterior rotor system moves counter-clockwise.

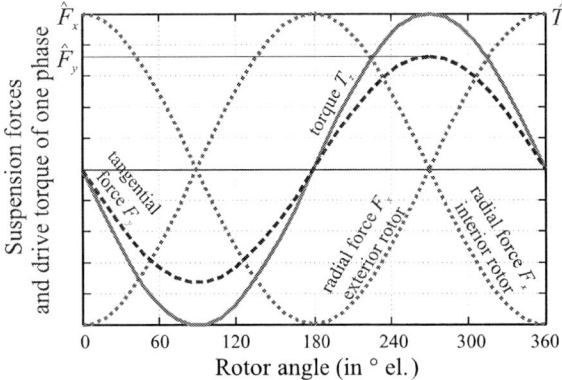

Fig. 2. Typical single-phase force and torque characteristic of a bearingless PM motor featuring combined windings.

angle dependent vector containing the force and torque constants in (2) also represents the first column of $\mathbf{T_m}(\varphi_r)$ with $k = 1$. Due to symmetry relations, the other m-1 rows of $\mathbf{T_m}(\varphi_r)$ can be computed from the first column using

$$
\begin{pmatrix} F_{x,k}\left(\varphi_r\right) \\ F_{y,k}\left(\varphi_r\right) \\ T_{z,k}\left(\varphi_r\right) \end{pmatrix} = \begin{pmatrix} \cos\left(\tau\right) & -\sin\left(\tau\right) & 0 \\ \sin\left(\tau\right) & \cos\left(\tau\right) & 0 \\ 0 & 0 & 1 \end{pmatrix} \begin{pmatrix} F_{x,1}\left(\varphi_r - \tau p_z\right) \\ F_{y,1}\left(\varphi_r - \tau p_z\right) \\ T_{z,1}\left(\varphi_r - \tau p_z\right) \end{pmatrix}
$$

(3)

with

$$
\tau_k = \frac{\pi}{3}\left(k-1\right)
$$

(4)

as shown in [18].

B. *Transformation Matrices*

For sinusoidal single-phase characteristic a *generalized Clarke transformation* of the order k defined by

$$
\mathbf{C}_k = \sqrt{\frac{2}{m}} \begin{pmatrix} 1 & \cos\left(\dfrac{2\pi k}{m}\right) & \cdots & \cos\left(\dfrac{2\pi\left(m-1\right)k}{m}\right) \\ 0 & \sin\left(\dfrac{2\pi k}{m}\right) & \cdots & \sin\left(\dfrac{2\pi\left(m-1\right)k}{m}\right) \end{pmatrix}
$$

(5)

and the property

$$
\mathbf{C}_k\mathbf{C}_k{}^T = \begin{pmatrix} 1 & 0 \\ 0 & 1 \end{pmatrix}
$$

(6)

can be used to separate suspension forces and drive torque [19].

With a Clarke transformation of the order t

$$
\left(i_1 \quad i_2 \quad \cdots \quad i_m\right)_t^T = \mathbf{C}_t{}^T\left(i_{D,t} \quad i_{Q,t}\right)^T = \mathbf{C}_t{}^T\mathbf{i}_{DQ,t}
$$

(7)

current components can be found that only create drive torque T_z and influence the PM air-gap field as

4029

$$\begin{pmatrix} \left(F_x \quad F_y \right)^T \\ \left(T_z \quad T_d \right)^T \end{pmatrix} = \mathbf{T_m} \left(\varphi_r \right) \mathbf{C}_t^T \mathbf{i}_{DQ,t} = \begin{pmatrix} \mathbf{0} \\ c_t \, \mathbf{R} \left(\varphi_r \right) \end{pmatrix} \mathbf{i}_{DQ,t} \quad (8)$$

holds true. c_t is called torque factor. As can be seen, the component T_d does not create any force or torque but affects the PM air-gap bias flux and, hence, can be used for field weakening [20]. This fact is well known from standard (mechanically supported) PM synchronous machines. The matrix $\mathbf{R}(\varphi_r)$ represents a rotation matrix in two dimensions

$$\mathbf{R} \left(\varphi_r \right) = \begin{pmatrix} \cos(\varphi_0 \pm \varphi_r) & -\sin(\varphi_0 \pm \varphi_r) \\ \sin(\varphi_0 \pm \varphi_r) & \cos(\varphi_0 \pm \varphi_r) \end{pmatrix}, \quad (9)$$

which additionally features the property

$$\mathbf{R} \left(\varphi_r \right)^{-1} = \mathbf{R} \left(\varphi_r \right)^T. \quad (10)$$

φ_0 is a constant offset angle.

The same procedure can be conducted with an alternative generalized Clarke transformation of different order f

$$\left(i_1 \quad i_2 \quad \cdots \quad i_m \right)_f^T = \mathbf{C}_f^T \left(i_{D,f} \quad i_{Q,f} \right)^T = \mathbf{C}_f^T \mathbf{i}_{DQ,f} \quad (11)$$

leading to a merely force-creating subsystem with

$$\begin{pmatrix} \left(F_x \quad F_y \right)^T \\ \left(T_z \quad T_d \right)^T \end{pmatrix} = \mathbf{T_m} \left(\varphi_r \right) \mathbf{C}_f^T \mathbf{i}_{DQ,f} = \begin{pmatrix} c_f \, \mathbf{R} \left(\varphi_r \right) \\ \mathbf{0} \end{pmatrix} \mathbf{i}_{DQ,f}$$
$$(12)$$

and c_f representing the force factor.

C. Control Scheme

With (7) and (11) the transformed $\mathbf{T_m}(\varphi_r)$ matrix does not only become decoupled but also features the special form [19]

$$\begin{pmatrix} \left(F_x \quad F_y \right)^T \\ \left(T_z \quad T_d \right)^T \end{pmatrix} = \begin{pmatrix} c_f \, \mathbf{R}_f \left(\varphi_r \right) & \mathbf{0} \\ \mathbf{0} & c_t \, \mathbf{R}_t \left(\varphi_r \right) \end{pmatrix} \begin{pmatrix} \mathbf{i}_{DQ,f} \\ \mathbf{i}_{DQ,t} \end{pmatrix} . (13)$$

With the new rotor angle dependent coordinates

$$\left(i_d \left(\varphi_r \right) \quad i_q \left(\varphi_r \right) \right)_f^T = \mathbf{i}_{dq,f} \left(\varphi_r \right) = \mathbf{R}_f \left(\varphi_r \right) \mathbf{i}_{DQ,f} \quad (14)$$

and

$$\left(i_d \left(\varphi_r \right) \quad i_q \left(\varphi_r \right) \right)_t^T = \mathbf{i}_{dq,t} \left(\varphi_r \right) = \mathbf{R}_t \left(\varphi_r \right) \mathbf{i}_{DQ,t} \quad (15)$$

full decoupling between the forces in x- and y-direction as well as torque and field component is achieved as the special form

$$\begin{pmatrix} \left(F_x \quad F_y \right)^T \\ \left(T_z \quad T_d \right)^T \end{pmatrix} = \begin{pmatrix} c_f \mathbf{I} & \mathbf{0} \\ \mathbf{0} & c_t \mathbf{I} \end{pmatrix} \begin{pmatrix} \mathbf{i}_{dq,f} \\ \mathbf{i}_{dq,t} \end{pmatrix} \quad (16)$$

is given. The subscripts of the rotation matrices in (14) and (15) indicate that there might be differences in rotation direction and/or offset angle between force and torque subsystem.

For the torque subsystem, field-oriented control as widely used in PM excited synchronous machines is given. For the force subsystem, a similar control approach with a different Clarke transformation enables

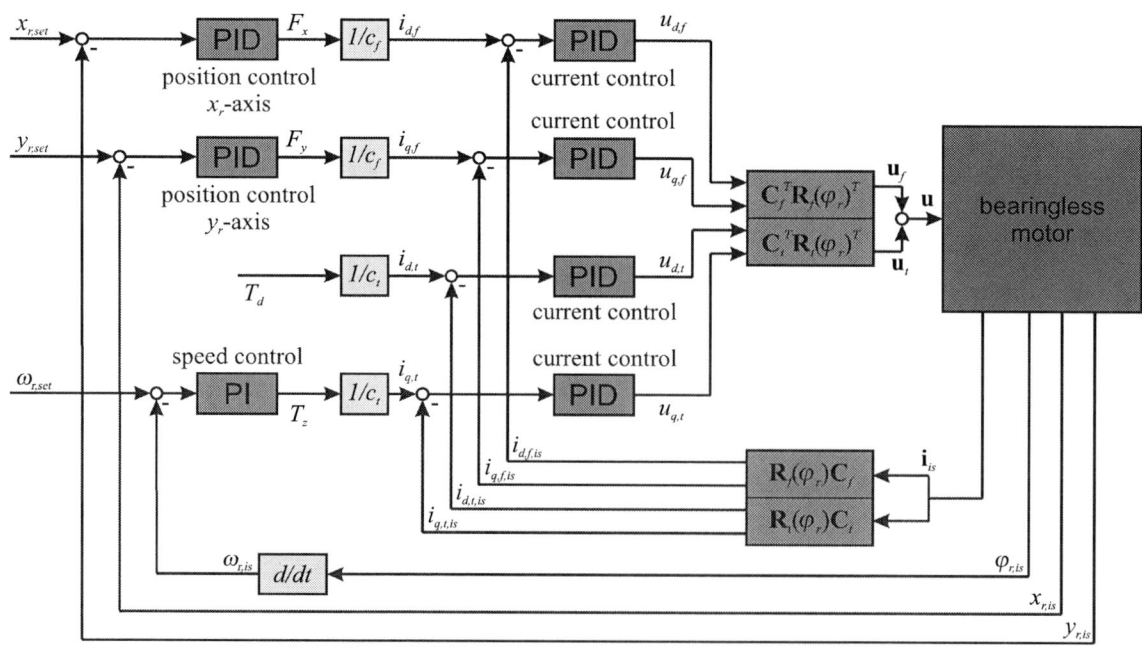

Fig. 3. Block diagram of the dual field-orientated control for the bearingless motors.

4030

the independent control of F_x and F_y. Due to the same structure and basic principle of both the force and the torque subsystem the term *dual field-oriented control* seems appropriate, even though constant force current components in the force *dq*-frame are not constant in the rotor-fixed torque *dq*-frame.

However, Fig. 3 shows the block diagram of the proposed control scheme. The position and speed controller are implemented in the particular *dq*-coordinate frames and determine the necessary suspension forces and drive torque, which are directly proportional to the referring *dq*-current components as stated in (16). To impress the necessary current components, subordinate current controllers (also implemented in the referring *dq*-frames) are used.

III. ACTUAL IMPLEMENTATION

The announced dual field-oriented control scheme described in the previous subsections becomes only feasible when the two matrices $\mathbf{C}_f$ and $\mathbf{C}_t$ decoupling force and torque can be found properly. Additionally, the referring force- and torque-factors c_t and c_f need to be significant. This is analysed for various bearingless PM topologies and the results are presented in this section.

A. Order of the Generalized Clarke Transformation

For symmetrically arranged bearingless PM drives featuring the single-phase characteristic shown in Fig. 2, the choice of f and t for the decoupling Clarke transformations only depends on the number of motor phases m and the number of pole pairs p_z. Hence, it can be determined for each of these topologies, whether the proposed dual field-oriented control is feasible. Table I summarizes this survey for bearingless drives with four to nine phases and pole pair numbers from one to eight.

In Table I the order of the generalized Clarke transformation is given for a proper decoupling of torque

and force. Often even more than one order is applicable. Apart from four phase systems, most of the topologies basically allow the application of the dual field-oriented control concept. Some drives feature rotor angles, where no drive torque or suspension force can be created, this is marked with a superscripted *. Such behaviour can be tolerated for torque generation but is not acceptable for force creation. There are several topologies that allow decoupling only under the constraint that the single-phase force orbit is circular. This is marked with a superscripted + in the table. Additionally, in Table I generally no difference is made between exterior and interior rotor design, apart from a few topologies that can be decoupled for interior or exterior rotor systems only. These are marked with superscripted IR and ER. However, a proper decoupling is only a mandatory constraint for the proposed dual field-oriented control but not a sufficient on, as the force and torque factors c_f and c_t are not allowed to be too low for proper operation.

B. Force and Torque Factor of Interior Rotor Systems

Table II summarizes the torque and force factors for properly decoupled interior rotor topologies. The factors are referred to the single-phase force or torque maximum. It is well known from standard electrical drives using field-oriented control, that the torque factor is typically $m/2$. However, favorable topologies are indicated with green background. As expected, four phase systems do not work well, but five-phase systems featuring

$$p_z = 1 + 5n \qquad \forall n \in \mathbb{N}_0 \qquad (17)$$

and six-phase systems holding the relation

$$p_z = 1 + 6n \cup 4 + 6n \qquad \forall n \in \mathbb{N}_0 \qquad (18)$$

feature beneficial behavior. Higher phase numbers are seldom used due to increasing power electronics effort, but would also feature promising topologies.

TABLE I
CLARKE TRANSFORMATION PARAMETERS TO DECOUPLE SUSPENSION FORCES AND DRIVE TORQUE

		Number of pole pairs							
		$p_z{=}1$	$p_z{=}2$	$p_z{=}3$	$p_z{=}4$	$p_z{=}5$	$p_z{=}6$	$p_z{=}7$	$p_z{=}8$
	$m{=}4$	$t{=}1, 3$	$t{=}2^*$	$t{=}1, 3$	$t{=}4^*$	$t{=}1, 3$	$t{=}2^*$	$t{=}1, 3$	$t{=}4^*$
		$f{=}2^*, 4^*$	$f{=}1^+, 3^+$	$f{=}2^*, 4^*$	$f{=}1^+, 3^+$	$f{=}2^*, 4^*$	$f{=}1^+, 3^+$	$f{=}2^*, 4^*$	$f{=}1^+, 3^+$
	$m{=}5$	$t{=}1, 4$	$t{=}2^{+ER}, 3^{+ER}$	$t{=}2^{+IR}, 3^{+IR}$	$t{=}1, 4$	$t{=}5^*$	$t{=}1, 4$	$t{=}2^{+ER}, 3^{+ER}$	$t{=}2^{+IR}, 3^{+IR}$
		$f{=}2, 3, 5^*$	$f{=}1, 4$	$f{=}1, 4$	$f{=}2, 3, 5^*$	$f{=}1^+, 4^+$	$f{=}2, 3, 5^*$	$f{=}1, 4$	$f{=}1, 4$
	$m{=}6$	$t{=}1, 5$	$t{=}2, 4$	$t{=}3^*$	$t{=}2, 4$	$t{=}1, 5$	$t{=}6^*$	$t{=}1, 5$	$t{=}2, 4$
		$f{=}2, 4, 6^*$	$f{=}1, 3^*, 5$	$f{=}2^+, 4^+$	$f{=}1, 3^*, 5$	$f{=}2, 4, 6^*$	$f{=}1^+, 5^+$	$f{=}2, 4, 6^*$	$f{=}1, 3^*, 5$
	$m{=}7$	$t{=}1, 6$	$t{=}2, 5$	$t{=}3^{+ER}, 4^{+ER}$	$t{=}3^{+IR}, 4^{+IR}$	$t{=}2, 5$	$t{=}1, 6$	$t{=}7^*$	$t{=}1, 6$
		$f{=}2, 5, 7^*$	$f{=}1, 3, 4, 6$	$f{=}2, 5$	$f{=}2, 5$	$f{=}1, 3, 4, 6$	$f{=}2, 5, 7^*$	$f{=}1^+, 6^+$	$f{=}2, 5, 7^*$
	$m{=}8$	$t{=}1, 7$	$t{=}2, 6$	$t{=}3, 5$	$t{=}4^*$	$t{=}3, 5$	$t{=}2, 6$	$t{=}1, 7$	$t{=}8^*$
		$f{=}2, 6, 8^*$	$f{=}1, 3, 5, 7$	$f{=}2, 4^*, 6$	$f{=}3^+, 5^+$	$f{=}2, 4^*, 6$	$f{=}1, 3, 5, 7$	$f{=}2, 6, 8^*$	$f{=}1^+, 7^+$
	$m{=}9$	$t{=}1, 8$	$t{=}2, 7$	$t{=}3, 6$	$t{=}4^{+ER}, 5^{+ER}$	$t{=}4^{+IR}, 5^{+IR}$	$t{=}3, 6$	$t{=}2, 7$	$t{=}1, 8$
		$f{=}2, 7, 9^*$	$f{=}1, 3, 6, 8$	$f{=}2, 4, 5, 7$	$f{=}3, 6$	$f{=}3, 6$	$f{=}2, 4, 5, 7$	$f{=}1, 3, 6, 8$	$f{=}2, 7, 9^*$

A superscripted * indicates systems, where no force or torque can be created for certain rotor angles. A superscripted + shows, that for a proper decoupling a circular single-phase force orbit (with equal amplitudes of $F_{x,k}$ and $F_{y,k}$) is necessary; an additional IR and ER means that decoupling is only feasible for interior or exterior rotor system, respectively.

TABLE II
FORCE AND TORQUE FACTORS FOR THE DECOUPLED SYSTEMS WITH INTERIOR ROTOR AND CIRCULAR/ELLIPTICAL FORCE ORBIT

		Number of pole pairs							
		$p_z=1$	$p_z=2$	$p_z=3$	$p_z=4$	$p_z=5$	$p_z=6$	$p_z=7$	$p_z=8$
Number of stator phases	$m=4$	$c_t=2$	$c_t=2^*$	$c_t=2$	$c_t=2^*$	$c_t=2$	$c_t=2^*$	$c_t=2$	$c_t=2^*$
		$c_f=4^*/2.5^*$	$c_f=2/-$	$c_f=4^*/2.5^*$	$c_f=2/-$	$c_f=4^*/2.5^*$	$c_f=2/-$	$c_f=4^*/2.5^*$	$c_f=2/-$
	$m=5$	$c_t=2.5$	$c_t=-$	$c_t=2.5^+$	$c_t=2.5$	$c_t=2.5^*$	$c_t=2.5$	$c_t=-$	$c_t=2.5^+$
		$c_f=2.5/1.55$	$c_f=0/0.95$	$c_f=2.5/1.55$	$c_f=0/0.95$	$c_f=2.5/-$	$c_f=2.5/1.55$	$c_f=0/0.95$	$c_f=2.5/1.55$
	$m=6$	$c_t=3$	$c_t=3$	$c_t=3^*$	$c_t=3$	$c_t=3$	$c_t=3^*$	$c_t=3$	$c_t=3$
		$c_f=3/1.9$	$c_f=0/1.125$	$c_f=3/-$	$c_f=3/1.9$	$c_f=0/1.125$	$c_f=3/-$	$c_f=3/1.9$	$c_f=0/1.125$
	$m=7$	$c_t=3.5$	$c_t=3.5$	$c_t=-$	$c_t=3.5^+$	$c_t=3.5$	$c_t=3.5$	$c_t=3.5^*$	$c_t=3.5$
		$c_f=3.5/2.19$	$c_f=3.5/2.19$	$c_f=0/1.3$	$c_f=3.5/2.19$	$c_f=3.5/2.19$	$c_f=0/1.3$	$c_f=3.5/-$	$c_f=3.5/2.19$
	$m=8$	$c_t=4$	$c_t=4$	$c_t=4$	$c_t=4^*$	$c_t=4$	$c_t=4$	$c_t=4$	$c_t=4^*$
		$c_f=4/2.5$	$c_f=4/2.5$	$c_f=0/1.5$	$c_f=4/-$	$c_f=4/2.5$	$c_f=4/2.5$	$c_f=0/1.5$	$c_f=4/-$
	$m=9$	$c_t=4.5$	$c_t=4.5$	$c_t=4.5$	$c_t=-$	$c_t=4.5^+$	$c_t=4.5$	$c_t=4.5$	$c_t=4.5$
		$c_f=4.5/2.8$	$c_f=4.5/2.8$	$c_f=4.5/2.8$	$c_f=0/1.7$	$c_f=4.5/2.8$	$c_f=4.5/2.8$	$c_f=4.5/2.8$	$c_f=0/1.7$

A superscripted * indicates systems, where no force or torque can be created at certain rotor angles. A superscripted + shows, that for a proper decoupling a circular single-phase force orbit (with equal amplitudes of $F_{x,k}$ and $F_{y,k}$) is necessary. The first force factor is computed for circular force orbit, while the second holds true for an elliptical force orbit with 25% eccentricity. A green background indicates topologies that work very well with the proposed control scheme; so do topologies with yellow background but they additionally demand for circular force orbit.

TABLE III
FORCE AND TORQUE FACTORS FOR THE DECOUPLED SYSTEMS WITH EXTERIOR ROTOR AND CIRCULAR/ELLIPTICAL FORCE ORBIT

		Number of pole pairs							
		$p_z=1$	$p_z=2$	$p_z=3$	$p_z=4$	$p_z=5$	$p_z=6$	$p_z=7$	$p_z=8$
Number of stator phases	$m=4$	$c_t=2$	$c_t=2^*$	$c_t=2$	$c_t=2^*$	$c_t=2$	$c_t=2^*$	$c_t=2$	$c_t=2^*$
		$c_f=4^*/2.5^*$	$c_f=2/-$	$c_f=4^*/2.5^*$	$c_f=2/-$	$c_f=4^*/2.5^*$	$c_f=2/-$	$c_f=4^*/2.5^*$	$c_f=2/-$
	$m=5$	$c_t=2.5$	$c_t=2.5^+$	$c_t=-$	$c_t=2.5$	$c_t=2.5^*$	$c_t=2.5$	$c_t=2.5^+$	$c_t=-$
		$c_f=0/0.95$	$c_f=2.5/1.55$	$c_f=0/0.95$	$c_f=2.5/1.55$	$c_f=2.5/-$	$c_f=0/0.95$	$c_f=2.5/1.55$	$c_f=0/0.95$
	$m=6$	$c_t=3$	$c_t=3$	$c_t=3^*$	$c_t=3$	$c_t=3$	$c_t=3^*$	$c_t=3$	$c_t=3$
		$c_f=0/1.125$	$c_f=3/1.9$	$c_f=3/-$	$c_f=0/1.125$	$c_f=3/1.9$	$c_f=3/-$	$c_f=0/1.125$	$c_f=3/1.9$
	$m=7$	$c_t=3.5$	$c_t=3.5$	$c_t=3.5^+$	$c_t=-$	$c_t=3.5$	$c_t=3.5$	$c_t=3.5^*$	$c_t=3.5$
		$c_f=0/1.3$	$c_f=3.5/2.19$	$c_f=3.5/2.19$	$c_f=0/1.3$	$c_f=3.5/2.19$	$c_f=3.5/2.19$	$c_f=3.5/-$	$c_f=0/1.3$
	$m=8$	$c_t=4$	$c_t=4$	$c_t=4$	$c_t=4^*$	$c_t=4$	$c_t=4$	$c_t=4$	$c_t=4^*$
		$c_f=0/1.5$	$c_f=4/2.5$	$c_f=4/2.5$	$c_f=4/-$	$c_f=0/1.5$	$c_f=4/2.5$	$c_f=4/2.5$	$c_f=4/-$
	$m=9$	$c_t=4.5$	$c_t=4.5$	$c_t=4.5$	$c_t=4.5^+$	$c_t=-$	$c_t=4.5$	$c_t=4.5$	$c_t=4.5$
		$c_f=0/1.7$	$c_f=4.5/2.8$	$c_f=4.5/2.8$	$c_f=4.5/2.8$	$c_f=0/1.7$	$c_f=4.5/2.8$	$c_f=4.5/2.8$	$c_f=4.5/2.8$

A superscripted * indicates systems, where no force or torque can be created at certain rotor angles. A superscripted + shows, that for a proper decoupling a circular single-phase force orbit (with equal amplitudes of $F_{x,k}$ and $F_{y,k}$) is necessary. The first force factor is computed for circular force orbit, while the second holds true for an elliptical force orbit with 25% eccentricity. A green background indicates topologies that work very well with the proposed control scheme; so do topologies with yellow background but they additionally demand for circular force orbit.

C. Force and Torque Factor of Exterior Rotor Systems

Similar to the previous subsection, also exterior rotor systems have been evaluated. The referring force and torque factors can be found in Table III. Again, favourable topologies feature green background. The most interesting topologies are five-phase systems featuring

$$p_z = 4 + 5n \qquad \forall n \in \mathbb{N}_0 \qquad (19)$$

and six-phase systems with

$$p_z = 2 + 6n \cup 5 + 6n \qquad \forall n \in \mathbb{N}_0. \qquad (20)$$

Again, also higher-phased systems are possible.

IV. CONCLUSIONS

This work demonstrates, that in symmetrically arranged bearingless PM motors with combined windings the suspension forces and the drive torque can be decoupled using two different generalized Clarke transformations. However, linear current-dependency and sinusoidal characteristic is demanded from force and torque generation. A general control scheme is presented using this kind of decoupling. Topologies, that are expected to work well for interior and exterior rotor systems are identified.

The proposed control scheme has been implemented in several bearingless PM synchronous drives, like the bearingless torque motor [19] (interior six-phase motor with $p_z=13$), the bearingless high-speed motors [17] (interior five- and six-phase motors with $p_z=1$) or the bearingless PM vernier motor [15] (exterior six-phase motor with p_z=17). This method is especially interesting for six-phase systems as a double star connection can be used, allowing to implement two industrial state-of-the-art three-phase half-bridge inverters measuring only four phase currents.

ACKNOWLEDGEMENT

This work has been supported by the COMET-K2 "Center for Symbiotic Mechatronics" of the Linz Center of Mechatronics GmbH (LCM) funded by the Austrian federal government and the federal state of Upper Austria.

REFERENCES

[1] A. Chiba, T. Fukao, O. Ichikawa, M. Oshima, M. Takemoto, D. G. Dorrell, "Magnetic bearings and bearingless drives," *Elsevier*, 2005.

[2] K. Raggl, J. W. Kolar, T. Nussbaumer, "Comparison of winding concepts for bearingless pumps," *Proc. 7th Int. Conf. on Power Electronics*, pp. 1013-1020, 2007.

[3] E. L. Severson, R. Nilssen, T. Undeland, N. Mohan, "Design of dual purpose no-voltage combined windings for bearingless motors," *IEEE Trans. on Industry Applications*, vol. 53, no. 5, pp. 4368-4379, Sept.-Oct. 2017.

[4] M. Ooshima, Y. Gomi, "Evaluation of rotor losses and efficiency in a d-q axis current control bearingless motor," *Proc. 15th Int. Symp. on Magnetic Bearings*, pp. 193-200, 2016.

[5] S. Silber, "Power optimal current control scheme for bearingless PM motors," *Proc. 7th Int. Symp. on Magnetic Bearings*, pp. 401-406, 2000.

[6] W. Gruber, S. Silber, W. Amrhein, T. Nussbaumer, "Design variants of the bearingless segment motor," *Proc. Int. Symp. on Power Electronics Electrical Drives Automation and Motion*, pp. 1448-1453, 2010.

[7] W. Amrhein, S. Silber, "Bearingless single-phase motor with concentrated full pitch windings in interior rotor design," *Proc. 6th Int. Symp. on Magnetic Bearings*, pp. 486-495, 1998.

[8] S. Silber, W. Amrhein, "Bearingless single-phase motor with concentrated full pitch windings in exterior rotor design," *Proc. 6th Int. Symp. on Magnetic Bearings*, pp. 476-485, 1998.

[9] T. Reichert, "The bearingless mixer in exterior rotor construction," ETH Zurich dissertation, no. 20329, 2012.

[10] H. Mitterhofer, "Towards high-speed bearingless disk drives", JKU Linz dissertation, 2017.

[11] D. Steinert, T. Nussbaumer, J. W. Kolar, "Slotless bearingless disk drive for high-speed and high-purity applications," *IEEE Trans. on Industrial Electronics*, vol. 61, no. 11, pp. 5974-5986, Nov. 2014.

[12] X. L. Wang, Q. C. Zhong, Z. Q. Deng, S. Z. Yue, "Fault-tolerant control of multi-phase permanent magnetic bearingless motors," *Proc. 14th Int. Conf. on Electrical Machines*, 2010.

[13] S. Silber, H. Grabner, R. Lohninger, W. Amrhein, "Design aspects of bearingless torque motors," *Proc. 13th Int. Symp. on Magnetic Bearings*, 2012.

[14] H. Li, H. Zhu, "Design of bearingless flux-switching permanent-magnet motor," *IEEE Trans. on Applied Superconductivity*, vol. 26, no. 4, pp. 1-5, June 2016.

[15] W. Gruber, R. Remplbauer, E. Göbl, "Design of a novel bearingless permanent magnet vernier slice motor with external rotor," *Proc. Int. Electric Machines and Drives Conf.*, 2017.

[16] G. Bramerdorfer, G. Jungmayr, W. Amrhein, W. Gruber, E. Marth, M. Reisinger, "Bearingless segment motor with Halbach magnet," *Proc. Int. Symp. on Power Electronics Electrical Drives Automation and Motion*, pp. 1466-1471, 2010.

[17] H. Mitterhofer, B. Mrak and W. Gruber, "Comparison of high-speed bearingless drive topologies with combined windings," *IEEE Trans. on Industry Applications*, vol. 51, no. 3, pp. 2116-2122, May-June 2015.

[18] W. Gruber, W. Briewasser, M. Rothböck, T. Schöb, W. Amrhein, "Bearingless reluctance slice motors," *Proc. 13th Int. Symp. on Magnetic Bearings*, 2012.

[19] H. Grabner, S. Silber, W. Amrhein, "Feedback control of a novel bearingless torque motor using an extended FOC method for PMSMs," *Proc. Int. Conf. on Industrial Technology*, 2013.

[20] P. Vas, "Vector control of AC machines," *Clarendon Press*, 1990

Open-Circuit Fault Tolerant Study of Bearingless Multi-Sector Permanent Magnet Machines

G. Valente[1]*, L. Papini[1,2]**, A. Formentini[1], C. Gerada[1,2], P. Zanchetta[1]

1 Electrical and Electronic Engineering, Univeristy of Nottingham, Nottingham, UK

2 Electrical and Electronic Engineering, Univeristy of Nottingham, Ningbo, China

*E-mail: ezzgv1@nottingham.ac.uk

**E-mail: ezzlp3@exmail.nottingham.ac.uk

Abstract—This paper presents a fault tolerant study of a multiphase sectored permanent magnet synchronous machine involving a tripel three-phase winding. The machine electro-megnetic model is written in a general way so that it can be extended and applied to all machines with a similar winding structure. An expression of the $d - q$ axis reference currents of each three-phase winding as a function of the $x - y$ force components and torque is provided taking into acccunt the Joule losses minimization. Then, the case of open-circuit of one winding sector is considered, the model of the faulty machine derived and an expression of the new reference currents needed to generate radial suspension force and motoring torque is written.

Finally, the theoretical analysis are validated through finite elements simulations and the levitation performance of the machine considered are evaluated in the Matlab-Simulink environment in the case of one sector fault per time.

Index Terms—Bearingless motor, Fault Tolerant, Force Control, Permanent Magnet Machines, Multiphase Motors.

I. INTRODUCTION

The bearingless machines (BM), sometimes also referred as self-bearing machines, own the potential of producing motoring torque and radial suspension force simultaneously and with a single stator element. Therefore, they have the advantage of presenting no wear caused by friction, making them particularly suitable for lubricant free applications. As a matter of facts, they have found space in chemical, pharmaceutical and semiconductor industries where an ultra-high cleanness environment has to be guaranteed. More in the specific, the most targeted applications for BMs are centrifugal pumps [1], [2] and artificial hearts [3], [4], mixers for chemical and pharmaceutical applications [5], [6].

A primitive bearingless motor prototype was proposed by Hermann in the middle of 1970s [7]. Then only in the early 1990s the BM technology started attracting growing interest and it was first applied to reluctance motors [8] and then to induction motors [9] and permanent magnet synchronous motors [10]. The latter are of particular interest because of their advantage of simple structure, reliability, high efficiency and high torque density [11].

Typically, two separated windings were employed for torque and radial suspension force generation [8], [10]. The installation of the additional winding for force production leads to a bigger outer diameter than that of the conventional motor.

Different winding arrangements has been presented in order to embed radial suspension force and torque generation in a single winding set. In [9] one of the phases of a four-pole induction motor was split into four so that each resultant coil could be supplied independently and used to control the radial rotor position. [12] exploited the concept of bridge configured winding where parallel branches were connected and the current through them could be controlled unbalancing the airgap magnetic field and generating a resultant force.

More recently, multiphase winding structures have been considered for their high power density, simple structure and fault tolerant capability. In particular, [13] presents a bearingless five-phase PM motor where the multiphase winding was exploited to control two decoupled $d - q$ planes for torque and radial force production, respectively. [14] the suspension force and motoring torque of a multi-three phase bearingless machine are controlled by means the Space Vector Decomposition (SVD) technique. In [15] a three three-phase sectored PMSM with a similar winding structure of the machine considered in this paper relied on the independent $d-$axis and $q-$axis current control of each winding to control suspension force and torque respectively. An harmonic compensator was then employed to suppress the force ripple caused by neglecting the coupling effect of torque and force production. The same authors have also tested the radial force technique for an open-circuit fault of one sector in [16] demonstrating the fault tolerant capability of the considered machine. A fault tolerant six-phase PM bearingless motor was proposed in [17] where each phase was controlled by a single phase inverter and an efficient reference current calculation was proposed. The multiphase structure and control method proposed allowed for fault tolerant operations when one or two phases were open-circuited.

The short-circuit fault instead was studied for a single-winding bearingless switched reluctance motor in [18] and for a six-phase bearingless PM motor in [19].

In this paper the open-circuit fault tolerant capabilities of the bearingless multi-sector PMSM are studied and a model for the faulty machine is carried out. The machine is a 18 slot - 6 poles PMSM with three three-phase windings [20], [21]. Open-circuit faults are considered involving the loss of a whole machine sector. It will be shown that the machine is still able to produce controllable torque and radial suspension force with two out of three healthy sectors. The torque and radial force

The 2018 International Power Electronics Conference

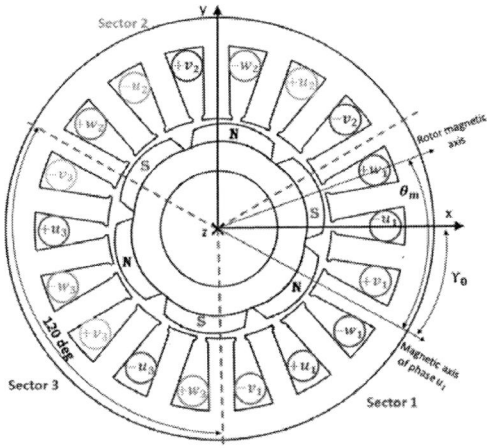

Fig. 1. Cross section of the 18 slot - 6 poles 3 sectors PMSM with 3 × 3 single layer distributed winding.

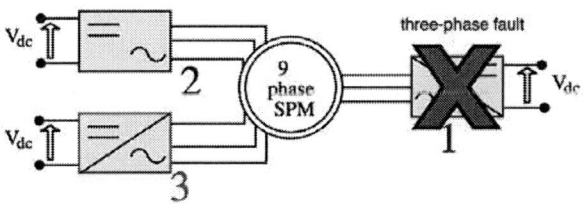

Fig. 2. MSPM machine connected to three three-phase inverters and open-circuit fault in inverter 1.

control technique employed in this work was presented in [20] and compared to the one applied to a similar machine in [15] showing advantages in terms of computational efforts. This work will also show that the control technique allows a simple formulation of the model for the faulty machine.

II. MATHEMATICAL MODEL OF THE BMSPM MACHINE

This section deals with the theoretical aspects of the radial suspension force and torque control for both healthy and fault operating conditions. At first the machine structure is presented and then the force production principles and the mathematical model for both operating conditions are exposed.

A. The machine structure

Fig. 1 displays the cross section of the machine considered highlighting its multi-three phase winding arrangement. The machine is a conventional surface mounted PM synchronous motor and its main features are listed in Table I. The original three-phase winding was removed and three full-pitched distributed star-connected windings with floating neutral points were installed. Each of the three windings occupies one third of the stator circumference and there is no overlapping between different sectors.

TABLE I
MACHINE PARAMETERS

Parameter	Value
Pole number ($2p$)	6
PM material	NdFeB
Power rating	1.5 [kW]
Nominal current peak (I_n)	13 [A]
Rated Speed (ω_m^{max})	$2\pi 50$ [rad/s]
PM flux of one sector (Λ_{PM})	0.0284 [Wb]
Torque constant (k_T)	0.128 [Nm/A]
Line to line voltage constant (k_V)	15.5 [V/krpm]
Rotor mass (m)	2 [Kg]
Magnetic stiffness (k_m)	0.7 [N/μm]
Backup bearing clearance (δ_{max})	150 [μm]
Outer Stator diameter	95 [mm]
Inner Stator diameter	49.5 [mm]
Axial length	90 [mm]
Airgap length	1 [mm]

B. Force production principles

Qualitatively, it can be observed that the airgap flux density distribution can be unbalanced controlling the current of each sector winding independently. In particular, assuming $\gamma_0 = 0$ (sector 1 aligned with the $x-$axis) in Fig. 1, a resultant net radial force in the $x-$ axis direction can be generated in healthy conditions increasing the flux density in the airgap underneath sector 1 and reducing the ones in correspondence of sector 2 and 3.

Fig. 2 shows a schematic representation of an open-circuit fault of the three-phase inverter connected to sector 1. Now, the remaining healthy sectors have to compensate the loss of sector 1 to the force production. Radial force can still be generated along the $x-$axis decreasing the flux density distribution in the airgap underneath sector 2 and 3. On the other hand, decreasing the flux density distribution in correspondence of sector 3 and increasing the one of sector 2 produces a force in the $y-$axis direction.

C. Healthy machine model

This paragraph provides the fundamental equations of the mathematical model for the healthy machine that was introduced and well detailed in [20]. The model is based on the assumptions of linear magnetic behaviour of the materials and magnetic decoupling between sectors. The latter allows to remarkably simplify the model since only one machine sector can be studied and the former permits to apply the superposition principle. Furthermore, the rotor is modelled as a rigid body that can move radially within a certain displacement δ_{max} given by the clearance of the backup bearing. The rotor radial displacement is defined by the translation δ and angle φ_d of the rotor centre from the rectangular $x - y$ reference frame origin of the stator O_s.

In this work, the number of machine sectors is $n_s = p$ and the angular position between the generic sector s and the x-axis is given by ${}^s\gamma = s(2\pi)/n_s + \gamma_0$. Under the aforementioned assumptions the matrix formulation of the $x - y$ force components and torque can be expressed in (1) as a function of the electrical angular position $\vartheta_e = p\vartheta_m$,

4035

radial displacement information and stationary reference frame current components $^s i_\alpha$ and $^s i_\beta$ of each sector s.

$$\bar{W}_E = \mathbf{K}_E(\vartheta_e, {}^s\gamma)\bar{i}_{\alpha\beta} + \bar{K}_m(\varphi_d)\delta \tag{1}$$

Where $\bar{W}_E = \begin{bmatrix} F_x & F_y & T \end{bmatrix}^T$ and $\bar{i}_{\alpha\beta} = \begin{bmatrix} ^1 i_\alpha & ^1 i_\beta & \cdots & ^s i_\alpha & ^s i_\beta & \cdots & ^{n_s} i_\alpha & ^{n_s} i_\beta \end{bmatrix}^T$ are the mechanical $x - y$ forces and torque vector and the total vector of the $\alpha - \beta$ axis currents, respectively. The structure of matrix $\mathbf{K}_E(\vartheta_e, {}^s\gamma) \in \mathbb{R}^{3 \times 2n_s}$ is reported in (2) showing the contributions of the n_s machine sectors to the force and torque production.

$$\mathbf{K}_E = \begin{bmatrix} ^1\mathbf{K}_E(\vartheta_e, {}^1\gamma) & \cdots & ^{n_s}\mathbf{K}_E(\vartheta_e, {}^{n_s}\gamma) \end{bmatrix} \tag{2}$$

The structure and the calculation procedure of sub-matrices $^1\mathbf{K}_E(\vartheta_e, {}^1\gamma), \ldots, {}^{n_s}\mathbf{K}_E(\vartheta_e, {}^1\gamma)$ is presented in [20], [22]. The reference current commands can be calculated inverting matrix $\mathbf{K}_E$. However, $\mathbf{K}_E$ results in general in a rectangular matrix, hence it cannot be easily inverted. In [20] the minimization of the copper losses has been chosen as strategy leading to the calculation of the pseudo inverse of $\mathbf{K}_E$ as follow

$$\mathbf{K}_E^+ = \mathbf{K}_E^T(\mathbf{K}_E\mathbf{K}_E^T)^{-1} \tag{3}$$

Therefore, the current command vector $\bar{i}_{\alpha\beta}^*$ can be determined in (4) considering also the rotor displacement.

$$\bar{i}_{\alpha\beta}^* = \mathbf{K}_E^+ \left[\bar{W}_E^* - \bar{K}_m(\varphi_d)\delta \right] = \mathbf{K}_E^+ \left[(\bar{W}_E^* - k_m \begin{bmatrix} u \\ v \\ 0 \end{bmatrix} \right] \tag{4}$$

Where u and v are the $x-$ and $y-$ axis displacements and k_m is the magnetic radial stiffness [22].
Conventional PI controllers require $d - q$ axis current in the rotor synchronous reference frame. Hence, $\bar{i}_{\alpha\beta}^*$ has to be multiplied by a rotating matrix as in (5) to obtain the reference currents in the $d - q$ reference frame.

$$\bar{i}_{dq}^* = \mathbf{T}_{R9}(\vartheta_e)\bar{i}_{\alpha\beta}^* = \mathbf{K}_{E,dq}^+ \left[\bar{W}_E^* - \bar{K}_m(\varphi_d)\delta \right] \tag{5}$$

Where $\mathbf{T}_{R9}(\vartheta_e)$ is the nine-phase rotation matrix and $\mathbf{K}_{E,dq}^+$ is the pseudo inverse matrix in the rotor synchronous reference frame. The latter can be found in the Appendix .

D. Open circuit fault condition

When an open-circuit fault occurs in one motor sector the correspondent $d-q$ axis current vector goes to zero eliminating its contribution to the radial force and torque production. Therefore, the sub-matrix related to the open-circuited sector will disappear in $\mathbf{K}_E$ and the new matrices expressions can be written as

$$\mathbf{K}_{E,f1} = \begin{bmatrix} ^2\mathbf{K}_E(\vartheta_e, {}^2\gamma) & ^3\mathbf{K}_E(\vartheta_e, {}^3\gamma) \end{bmatrix}$$

$$\mathbf{K}_{E,f2} = \begin{bmatrix} ^1\mathbf{K}_E(\vartheta_e, {}^1\gamma) & ^3\mathbf{K}_E(\vartheta_e, {}^3\gamma) \end{bmatrix} \tag{6}$$

$$\mathbf{K}_{E,f3} = \begin{bmatrix} ^1\mathbf{K}_E(\vartheta_e, {}^1\gamma) & ^2\mathbf{K}_E(\vartheta_e, {}^2\gamma) \end{bmatrix}$$

for open-circuit in sector 1, 2, 3, respectively. It can be noticed that matrices $\mathbf{K}_{E,f1}$, $\mathbf{K}_{E,f2}$ and $\mathbf{K}_{E,f3} \in \mathbb{R}^{3 \times (2n_s-2)}$ are

also rectangular. Therefore, in order to obtain the reference current signals in the case of open-circuit fault conditions, they can be substituted into (3) obtaining $\mathbf{K}_{E,f1}^+$, $\mathbf{K}_{E,f2}^+$ and $\mathbf{K}_{E,f3}^+ \in \mathbb{R}^{(2n_s-2) \times 3}$ respectively that in turn can be used in place of $\mathbf{K}_E^+$ in (4) to calculate the $\alpha - \beta$ reference current vectors. The $d - q$ reference current values are then calculated as follow

$$\bar{i}_{dq,f1}^* = \hat{\mathbf{K}}_{E,f1}^+ \left[\bar{W}_E^* - \bar{K}_m(\phi_d)\delta \right]$$

$$\bar{i}_{dq,f2}^* = \hat{\mathbf{K}}_{E,f2}^+ \left[\bar{W}_E^* - \bar{K}_m(\phi_d)\delta \right] \tag{7}$$

$$\bar{i}_{dq,f3}^* = \hat{\mathbf{K}}_{E,f3}^+ \left[\bar{W}_E^* - \bar{K}_m(\phi_d)\delta \right]$$

$\hat{\mathbf{K}}_{E,f1}^+$, $\hat{\mathbf{K}}_{E,f2}^+$ and $\hat{\mathbf{K}}_{E,f3}^+$ are the pseudo inverse matrices in the rotor reference frame in the case of fault in sector 1, 2 and 3 respectively. Their expressions are reported in the Appendix. $\bar{i}_{dq,oc1}^* = \begin{bmatrix} ^2 i_d & ^2 i_q & ^3 i_d & ^3 i_q \end{bmatrix}^T$, $\bar{i}_{dq,oc2}^* = \begin{bmatrix} ^1 i_d & ^1 i_q & ^3 i_d & ^3 i_q \end{bmatrix}^T$ and $\bar{i}_{dq,oc3}^* = \begin{bmatrix} ^1 i_d & ^1 i_q & ^2 i_d & ^2 i_q \end{bmatrix}^T$. On the other hand, $^1\bar{i}_{dq,oc1}^* = \begin{bmatrix} 0 & 0 \end{bmatrix}^T$, $^2\bar{i}_{dq,oc2}^* = \begin{bmatrix} 0 & 0 \end{bmatrix}^T$ and $^3\bar{i}_{dq,oc3}^* = \begin{bmatrix} 0 & 0 \end{bmatrix}^T$ when the open circuit fault occurs in sector 1,2 and 3, respectively.
Fig. 3 shows the harmonic content of the coefficients of matrix $\mathbf{K}_E^+ \in \mathbb{R}^{6 \times 3}$, for healthy operating condition and of matrices $\mathbf{K}_{E,f1}$, $\mathbf{K}_{E,f2}$ and $\mathbf{K}_{E,f3} \in \mathbb{R}^{4 \times 3}$ for an open-circuit fault in sector 1, 2 and 3, respectively. The magnitudes of the harmonics have been obtained through the fast Fourier transform of the aforementioned coefficients waveforms presented in [20]. It can be noticed that the first two lines do not contain the coefficients of matrix $\mathbf{K}_{E,f1}$ since the current components of sector 1 are null in the case of open-circuit fault. The same is true for matrices $\mathbf{K}_{E,f2}$ and $\mathbf{K}_{E,f3}$. Their coefficients do not appear in lines 3, 4 and lines 5, 6, respectively. Comparing the different harmonic contents it can be observed that a non negligible third harmonic appears in the terms of matrices $\mathbf{K}_{E,f1}$, $\mathbf{K}_{E,f2}$ and $\mathbf{K}_{E,f3}$, while the ones of $\mathbf{K}_E$ are essentially sinusoidal. Their waveforms were reported in [20] and they can be easily implemented in a DSP reducing the computational efforts. On the other hand, the third harmonic has to be considered in the coefficients of $\mathbf{K}_{E,f1}$, $\mathbf{K}_{E,f2}$ and $\mathbf{K}_{E,f3}$ in order to produce the required suspension force and motoring torque complicating the control algorithm.

III. FINITE ELEMENTS SIMULATION RESULTS

In this section the FE validation of the mathematical model written for open-circuit fault is presented. The FE software employed is MagNet 7.7.1. Fig. 4, 5 and 6 a)-b) show the force production for healthy and open-circuit fault in sector 1, 2 and 3 respectively considering two case studies: a) setting $T_E^* = 0$; b) setting $T_E^* = 2.7[Nm]$.
In the simulations the rotor is rotated of two whole mechanical revolutions (720 deg) while the force reference magnitude

The 2018 International Power Electronics Conference

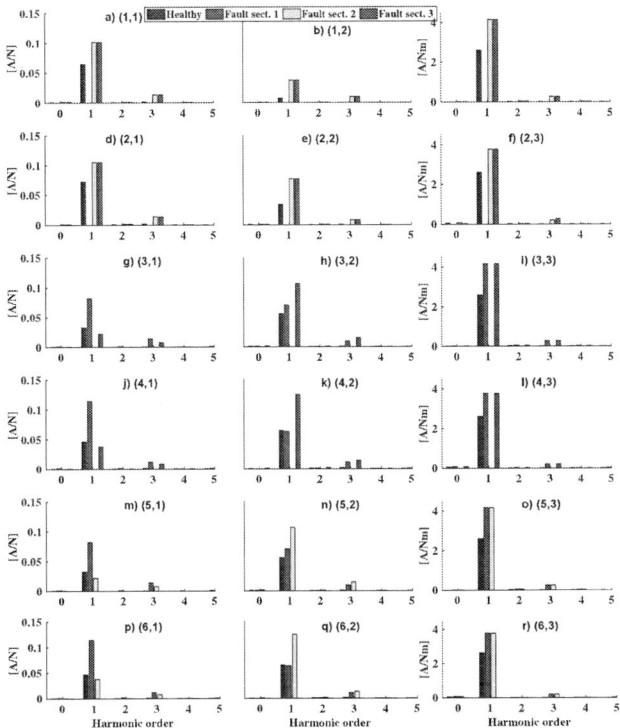

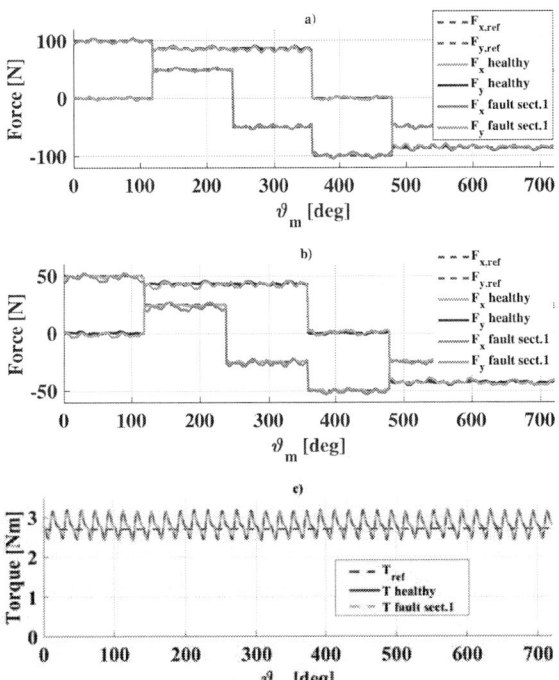

Fig. 4. Force and torque generation with open-circuit fault in sector 1: a) $x - y$ axis force production for $T_{ref} = 0$; b) $x - y$ axis force production for $T_{ref} = 2.7[Nm]$; c) torque production setting $T_{ref} = 2.7[Nm]$.

Fig. 3. Comparison in terms of harmonic content between the coefficients of matrix $\mathbf{K}_E^+$ (healthy machine), matrix $\mathbf{K}_{E,f1}^+$ (fault in sector 1), matrix $\mathbf{K}_{E,f2}^+$ (fault in sector 2) and matrix $\mathbf{K}_{E,f3}^+$ (fault in sector 3). (r, c) stands for the coefficient in the rth row and c^{th} column.

$|F_E|^*$ is kept constant and the direction $\angle F_E^*$ varies from 0 to 360 deg with steps of 120 deg as it is possible to observe in the figures. In order to satisfy the current limit of the machine the force reference magnitude $|F_E|^* = 100$ N is halved when torque is also produced.

It is possible to observe from Fig. 4, 5 and 6 a) that the force ripple remains about the same when a fault occurs and no torque is produced. On the other hand, when torque is simultaneously generated the ripple slightly increases as depicted in Fig. 4, 5 and 6 b). The above observations can be quantified in terms of Total Harmonic Distortion (THD) of the $x - y$ axis force components and summarized in Table II for all the considered case studies. It is straightforward to see that the highest THD occurs for an open-circuit fault in sector 3 for $T_E = 2.7$ N. The THD obtained is 7.42 % which is only 3 % higher than the one obtained at no-load.

Finally, the torque generated during healthy and faulty conditions can be examined in Fig. 4, 5 and 6 c) showing that the torque ripple is not significantly affected by the open-circuit faults.

IV. NUMERICAL SIMULATION RESULTS

A. Simulation structure

Fig. 7 shows the block scheme of the simulation implemented in the Simulink-Matlab environment. It is possible to distinguish three different regions: the control algorithm (in green) operating in the discrete domain at the sample time $T_s = 100 \ [\mu s]$; the electro-mechanical model of the MSPM motor (in yellow) simulated in the continuous domain; the rotor-dynamic model of the rotor (in red) implemented in discrete with $T_{ss} = 1 \ [\mu s]$.

The control algorithm includes the position and speed controllers, responsible for the suspension force and torque references calculation, the mathematical model described in (4), (5) for healthy machine and in (6), (7) for faulty machine and the blocks delay z^{-2} representing the current controllers. As a matter of fact, a well tuned current controller introduces a delay of two sample times T_s between the reference current and motor current (if no voltage saturation occurs). The design of the position controller considered in this work can be found in [23]. The speed loop controller is a standard PI controller as the one proposed in [24]. The electro-mechanical model of the motor considered is stored in the form of lookup table and it maps the $d - q$ axis currents to $x - y$ force components and torque. The lookup table has been carried out by means "multistatic" non-linear FE simulations to take into account the iron saturation. The rotor weight force and an eventual disturbance are then added and the resultant $F_{x,t}$ and $F_{y,t}$ forces are the input of the rotor dynamic model block represented by

$$
\begin{bmatrix} m & 0 \\ 0 & m \end{bmatrix} \begin{bmatrix} \ddot{u} \\ \ddot{v} \end{bmatrix} - \begin{bmatrix} k_m & 0 \\ 0 & k_m \end{bmatrix} \begin{bmatrix} u \\ v \end{bmatrix} = \begin{bmatrix} F_{x,c} + F_{x,d} \\ F_{y,c} + F_{y,d} - mg \end{bmatrix} \quad (8)
$$

TABLE II
THD OF THE FORCE GENERATION

| | $T_E = 0Nm$ | | | | $T_E = 2.7Nm$ | | | |
	Healthy	Fault 1	Fault 2	Fault 3	Healthy	Fault 1	Fault 2	Fault 3
F_x	4.31	5.1	4.58	4.56	4.05	7.22	6.64	7.42
F_y	3.56	3.77	4.04	4.02	3.25	5.27	5.52	5.11

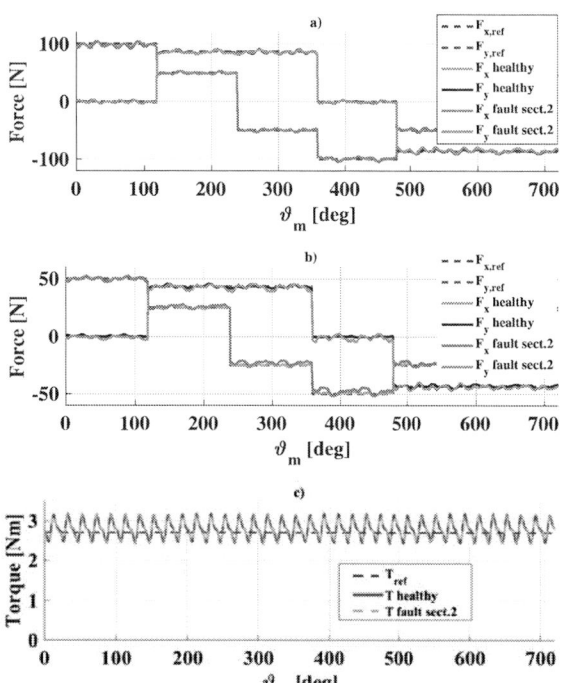

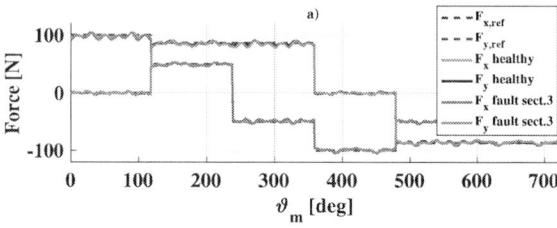

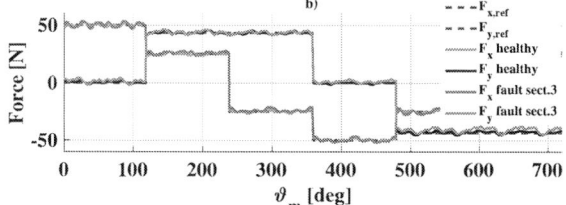

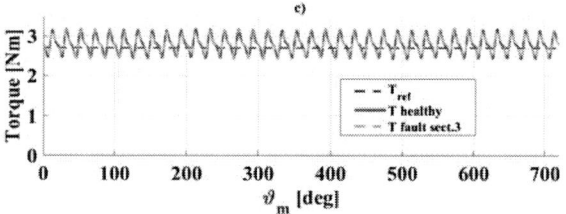

Fig. 5. Force and torque generation with open-circuit fault in sector 2: a) $x - y$ axis force production for $T_{ref} = 0$; b) $x - y$ axis force production for $T_{ref} = 2.7[Nm]$; c) torque production setting $T_{ref} = 2.7[Nm]$.

Fig. 6. Force and torque generation with open-circuit fault in sector 3: a) $x - y$ axis force production for $T_{ref} = 0$; b) $x - y$ axis force production for $T_{ref} = 2.7[Nm]$; c) torque production setting $T_{ref} = 2.7[Nm]$.

where u and v are the rotor $x - y$ axis displacements and $F_{x,c}$, $F_{y,c}$ and $F_{x,d}$, $F_{y,d}$ are the $x - y$ components of the forces generated by the controller and by the disturbance, respectively.

The simulation also presents a fault selector block that allows to easily simulate a fault, hence setting the currents of the faulty sector to zero and switching from healthy to faulty mathematical models.

B. Simulation results

The results of the simulation for open-circuit faults are displayed in Fig. 8, 9 and 10. A simulation time of 1 s is considered.

At first, the fault selector is set to 0 and the rotor is positioned in the stator centre after an initial short transient (Fig. 8 a)). Then at 0.02 and 0.04 s the rotor speed is commanded to its rated value (Fig. 9 a)) and force disturbance is applied in the $y-$ axis (Fig. 8 b)) respectively justifying the first perturbation in the rotor position observed in Fig. 8 a) and the increase of the $d - q$ axis currents in the motor sectors observed in Fig. 10 a)-c). The aforementioned force disturbance $F_{y,d} = -100$

N has a duration of 0.02 s. At 0.1 s, then, the load torque is set equal to 3 Nm (Fig. 9 b)) justifying the increment of the $q-$axis currents of the three sectors (Fig. 10).

While the simulation is running the fault selector is manually set to 1 in order to simulate a sudden open-circuit fault in sector 1. It can be observed from Fig. 8 a) that the position controller achieves to maintain the rotor centred after a very small position transient. Fig. 10 a) shows that, for the duration of the fault, the $d - q$ currents of sector 1 is zero while Fig. 10 b) and c) show that the ones of sectors 2 and 3 increase their magnitudes to compensate the loss of the contribution of sector 1 to the force generation. From Fig. 10 b) and c) it can also be noticed that the current ripple of the healthy sectors is remarkably increased after the fault occurs. This phenomena could be predicted observing the increment of the current distortion depicted in Fig. 3. The previously described force disturbance is applied during the fault (Fig. 8 b)) in order to verify the disturbance rejection capability of the faulty bearingless drive.

The same procedure is followed to simulate the open-circuit fault in sectors 2 and 3 showing a similar behaviour.

The 2018 International Power Electronics Conference

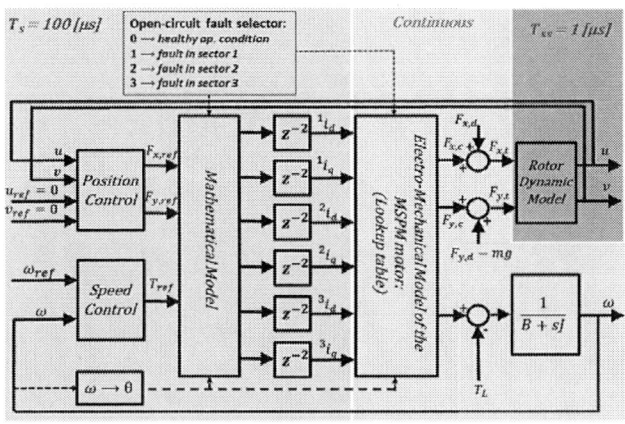

Fig. 7. Block scheme implemented in the MATLAB-Simulink environment.

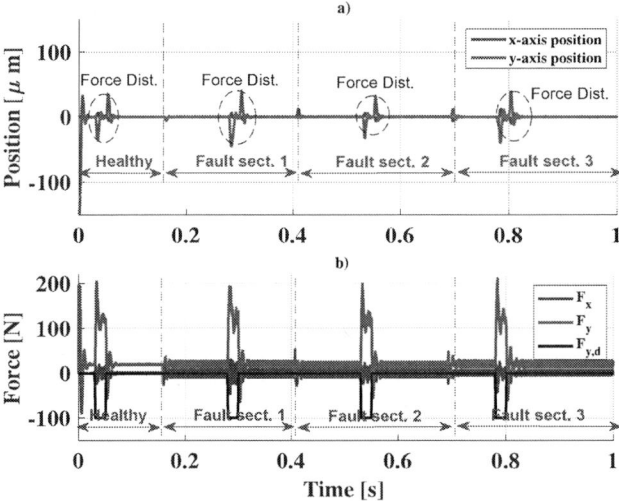

Fig. 8. Result of the numerical simulation for bearingless operation in open-circuit fault condition: a) $x - y$ axis rotor position; b) $x - y$ axis generated forces and $y-$axis force disturb.

V. CONCLUSIONS

A 6-pole PMSM featuring multi-sector multi-phase winding arrangement has been considered. The machine has been analysed for healthy and open-circuit fault conditions providing a detailed mathematical model for the electromagnetic torque and radial force production. FE simulation results for radial suspension force and motoring torque production are provided. The levitation performance for healthy and faulty machine are studied by means numerical simulations performed in the Simulink-Matlab environment showing good fault capability of the bearingless drive.

APPENDIX

A. Healthy operating condition

The pseudo inverse matrix in the rotor synchronous reference frame was introduced in (5). It can be explicated as

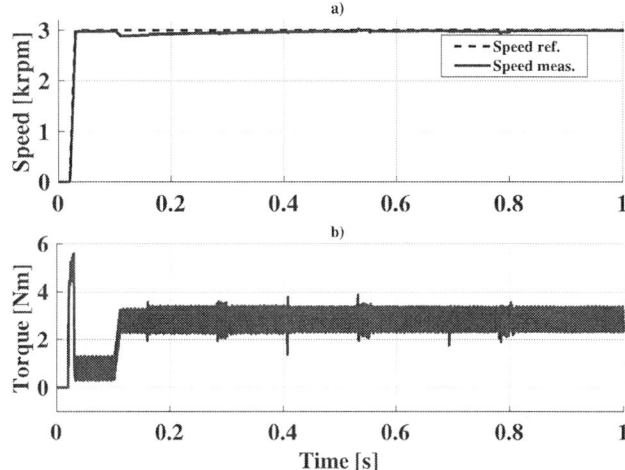

Fig. 9. Result of the numerical simulation for bearingless operation in open-circuit fault condition: a)rotor speed; b) generated torque.

follow

$$\mathbf{K}_{E,dq}^{+}(\vartheta_e) = \begin{bmatrix} {}^1k_{xd}^{+}(\vartheta_e) & {}^1k_{yd}^{+}(\vartheta_e) & 0 \\ {}^1k_{xq}^{+}(\vartheta_e) & {}^1k_{yq}^{+}(\vartheta_e) & -k_T' \\ {}^2k_{xd}^{+}(\vartheta_e) & {}^2k_{yd}^{+}(\vartheta_e) & 0 \\ {}^2k_{xq}^{+}(\vartheta_e) & {}^2k_{yq}^{+}(\vartheta_e) & -k_T' \\ {}^3k_{xd}^{+}(\vartheta_e) & {}^3k_{yd}^{+}(\vartheta_e) & 0 \\ {}^3k_{xq}^{+}(\vartheta_e) & {}^3k_{yq}^{+}(\vartheta_e) & -k_T' \end{bmatrix} \tag{9}$$

where $k_T' = 1/3k_T$ [A/Nm] and the expression of the generic coefficient ${}^s k_{*\times}^{+}(\vartheta_e)$ is

$$ {}^s k_{*\times}^{+}(\vartheta_e) = {}^s c_{*\times} + {}^s s_{*\times} \cos(2\vartheta_e + {}^s \varphi_{*\times}) \tag{10}$$

It is possible to notice that ${}^s k_{*\times}^{+}(\vartheta_e)$ presents a constant and a sinusoidal term. The latter pulsates at twice the electrical frequency and produces a non-constant $d - q$ axis current reference signals.

The angle γ_0 is required for the calculation of parameters ${}^s c_{*\times}$, ${}^s s_{*\times}$ and ${}^s \varphi_{*\times}$. As a matter of facts, γ_0 was defined in Section II as the angular position of the magnetic axis of the sector 1 with respect to the $x-$axis. As an example, Table III reports the above mentioned parameters calculated for $\gamma_0 = 0$.

B. Open-circuit operating condition

In order to describe the model in the rotor synchronous reference frame, matrices $\mathbf{K}_{E,f1}^{+}$, $\mathbf{K}_{E,f2}^{+}$ and $\mathbf{K}_{E,f3}^{+}$ have to be replaced with $\hat{\mathbf{K}}_{E,f1}^{+}$, $\hat{\mathbf{K}}_{E,f2}^{+}$ and $\hat{\mathbf{K}}_{E,f3}^{+}$ obtained as follow:

$$\hat{\mathbf{K}}_{E,f1}^{+}(\vartheta_e) = \mathbf{T}_{R,f}(\vartheta_e)\mathbf{K}_{E,f1}^{+}(\vartheta_e)$$

$$\hat{\mathbf{K}}_{E,f2}^{+}(\vartheta_e) = \mathbf{T}_{R,f}(\vartheta_e)\mathbf{K}_{E,f2}^{+}(\vartheta_e) \tag{11}$$

$$\hat{\mathbf{K}}_{E,f3}^{+}(\vartheta_e) = \mathbf{T}_{R,f}(\vartheta_e)\mathbf{K}_{E,f3}^{+}(\vartheta_e)$$

where $\mathbf{T}_{R,f}$ is the six-phase rotation matrix.

The expressions of the force and torque coefficients contained

The 2018 International Power Electronics Conference

TABLE III
PARAMETERS OF FORCE COEFFICIENTS

	Sector 1				Sector 2				Sector 3			
	x,d	x,q	y,d	y,q	x,d	x,q	y,d	y,q	x,d	x,q	y,d	y,q
$^s c_{*\times}$ [A/N]	0.068	0	0	0.021	0.034	0.018	0.058	0.01	0.034	0.018	0.058	0.01
$^s s_{*\times}$ [A/N]	0.0055	0.0025	0.014	0.013	0.012	0.0112	0.0083	0.0068	0.0121	0.0121	0.0083	0.0068
$^s \varphi_{*\times}$ [rad]	0.0622	1.63	1.63	-3.08	-1.74	-0.05	-0.9	0.39	1.86	-2.97	-2.12	-0.26

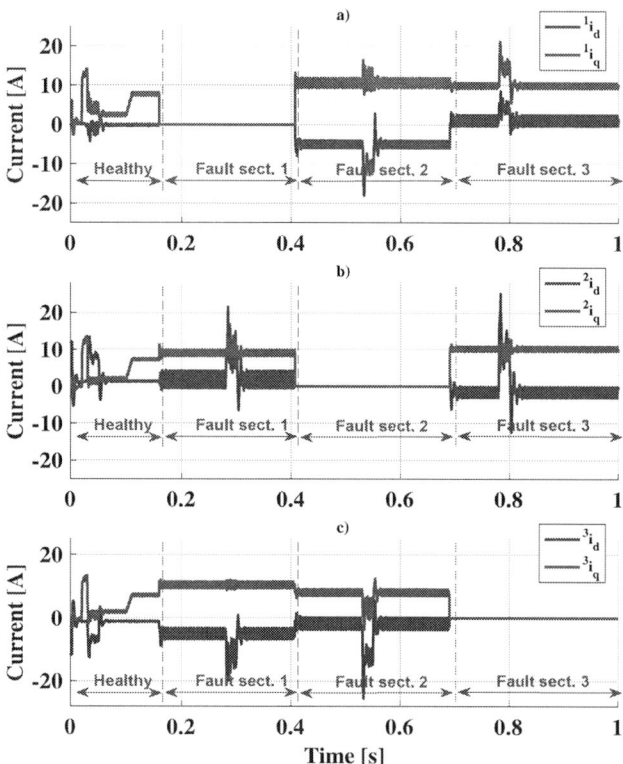

Fig. 10. Result of the numerical simulation for bearingless operation in open-circuit fault condition: a) $d - q$ axis currents sect. 1; b) $d - q$ axis currents sect. 2; c) $d - q$ axis currents sect. 3.

in the pseudo inverse matrix $\hat{\mathbf{K}}^+_{E,fs}$ related to the fault in a generic sector s can be found in (12) and (13), respectively.

$$^s k^+_{*\times,fs}(\vartheta_e) = {}^s c_{*\times,fs} + {}^s s_{*\times,fs} \cos(2\vartheta_e + {}^s\varphi_{*\times,fs}) + \\ + {}^s r_{*\times,fs} \cos(4\vartheta_e + {}^s\psi_{*\times,fs}) \tag{12}$$

$$^s k^+_{T\times,fs}(\vartheta_e) = {}^s c_{T\times,fs} + {}^s s_{T\times,fs} \cos(2\vartheta_e + {}^s\varphi_{T\times,fs}) + \\ + {}^s r_{T\times,fs} \cos(4\vartheta_e + {}^s\psi_{T\times,fs}) \tag{13}$$

It can be noticed that the coefficients of $\hat{\mathbf{K}}^+_{E,fs}$ still contain the second harmonic as in (10). Furthermore, a forth harmonic term appears due to the presence of the third harmonic in the coefficients of $\mathbf{K}^+_{E,fs}$ (see Fig. 3).

The parameters in (12) and (13) can be easily calculated once γ_0 is defined. As an example, the parameters of the force and torque coefficients for fault in sector 1 (matrix $\hat{\mathbf{K}}^+_{E,f1}$) have been calculated for $\gamma_0 = 0$ and they have been reported in Tables IV and V.

REFERENCES

[1] J. Asama, D. Kanehara, T. Oiwa, and A. Chiba, "Development of a compact centrifugal pump with a two-axis actively positioned consequent-pole bearingless motor," *IEEE Transactions on Industry Applications*, vol. 50, no. 1, pp. 288–295, Jan 2014.

[2] K. Raggl, B. Warberger, T. Nussbaumer, S. Burger, and J. W. Kolar, "Robust angle-sensorless control of a pmsm bearingless pump," *IEEE Transactions on Industrial Electronics*, vol. 56, no. 6, pp. 2076–2085, June 2009.

[3] J. Asama, T. Fukao, A. Chiba, A. Rahman, and T. Oiwa, "A design consideration of a novel bearingless disk motor for artificial hearts," in *2009 IEEE Energy Conversion Congress and Exposition*, Sept 2009, pp. 1693–1699.

[4] Y. Okada, N. Yamashiro, K. Ohmori, T. Masuzawa, T. Yamane, Y. Konishi, and S. Ueno, "Mixed flow artificial heart pump with axial self-bearing motor," *IEEE/ASME Transactions on Mechatronics*, vol. 10, no. 6, pp. 658–665, Dec 2005.

[5] T. Reichert, T. Nussbaumer, and J. W. Kolar, "Bearingless 300-w pmsm for bioreactor mixing," *IEEE Transactions on Industrial Electronics*, vol. 59, no. 3, pp. 1376–1388, March 2012.

[6] B. Warberger, R. Kaelin, T. Nussbaumer, and J. W. Kolar, "50-nm/2500-w bearingless motor for high-purity pharmaceutical mixing," *IEEE Transactions on Industrial Electronics*, vol. 59, no. 5, pp. 2236–2247, May 2012.

[7] P. Hermann, "A radial active magnetic bearing," *London Patent*, no. 1, p. 478, 1973.

[8] A. Chiba, M. A. Rahman, and T. Fukao, "Radial force in a bearingless reluctance motor," *IEEE Transactions on Magnetics*, vol. 27, no. 2, pp. 786–790, March 1991.

[9] A. O. Salazar and R. M. Stephan, "A bearingless method for induction machines," *IEEE Transactions on Magnetics*, vol. 29, no. 6, pp. 2965–2967, Nov 1993.

[10] M. Oshima, S. Miyazawa, T. Deido, A. Chiba, F. Nakamura, and T. Fukao, "Characteristics of a permanent magnet type bearingless motor," *IEEE Transactions on Industry Applications*, vol. 32, no. 2, pp. 363–370, Mar 1996.

[11] G. Feng, C. Lai, and N. C. Kar, "An analytical solution to optimal stator current design for pmsm torque ripple minimization with minimal machine losses," *IEEE Transactions on Industrial Electronics*, vol. 64, no. 10, pp. 7655–7665, Oct 2017.

[12] W. K. S. Khoo, "Bridge configured winding for polyphase self-bearing machines," *IEEE Transactions on Magnetics*, vol. 41, no. 4, pp. 1289–1295, April 2005.

[13] J. Huang, B. Li, H. Jiang, and M. Kang, "Analysis and control of multiphase permanent-magnet bearingless motor with a single set of half-coiled winding," *IEEE Transactions on Industrial Electronics*, vol. 61, no. 7, pp. 3137–3145, July 2014.

[14] G. Sala, G. Valente, A. Formentini, L. Papini, D. Gerada, P. Zanchetta, A. Tani, and C. Gerada, "Space vectors and pseudo inverse matrix methods for the radial force control in bearingless multi-sector permanent magnet machines," *IEEE Transactions on Industrial Electronics*, vol. PP, no. 99, pp. 1–1, 2018.

4040

TABLE IV
PARAMETERS OF FORCE COEFFICIENTS FOR FAULT IN SECT. 1

	Sector 1				Sector 2				Sector 3			
	x,d	x,q	y,d	y,q	x,d	x,q	y,d	y,q	x,d	x,q	y,d	y,q
$^s c_{*\times,f1}$[A/N]	/	/	/	/	0.088	0.0427	0.0661	0	0.088	0.0427	0.0661	0
$^s s_{*\times,f1}$[A/N]	/	/	/	/	0.0234	0.0204	0.0205	0.0039	0.0234	0.0204	0.0205	0.0039
$^s r_{*\times,f1}$[A/N]	/	/	/	/	0.0018	0.003	0.0015	0.0016	0.0018	0.003	0.0015	0.0016
$^s \varphi_{*\times,f1}$[deg]	/	/	/	/	-2.7	0.062	-1.63	-1.51	2.83	-3.1	-1.38	1.63
$^s \psi_{*\times,f1}$[deg]	/	/	/	/	2.16	-3.02	2.55	-1.45	-1.91	0.12	0.84	1.7

TABLE V
PARAMETERS OF TORQUE COEFFICIENTS FOR FAULT IN SECT. 1

	Sector 1		Sector 2		Sector 3	
	T,d	T,q	T,d	T,q	T,d	T,q
$^s c_{T\times,f1}$[A/N]	/	/	0.627	3.55	0.627	3.55
$^s s_{T\times,f1}$[A/N]	/	/	0.523	0.0915	0.523	0.0915
$^s r_{T\times,f1}$[A/N]	/	/	0.0442	0.0373	0.0442	0.0373
$^s \varphi_{T\times,f1}$[deg]	/	/	-2.5	-1.51	-0.52	1.63
$^s \psi_{T\times,f1}$[deg]	/	/	-1.93	-1.45	-0.96	1.7

[15] S. Kobayashi, M. Ooshima, and M. N. Uddin, "A radial position control method of bearingless motor based on d - q- axis current control," *IEEE Transactions on Industry Applications*, vol. 49, no. 4, pp. 1827–1835, July 2013.

[16] M. Ooshima, A. Kobayashi, and T. Narita, "Stabilized suspension control strategy at failure of a motor section in a d-q axis current control bearingless motor," in *2015 IEEE Industry Applications Society Annual Meeting*, Oct 2015, pp. 1–7.

[17] X. L. Wang, Q. C. Zhong, Z. Q. Deng, and S. Z. Yue, "Current-controlled multiphase slice permanent magnetic bearingless motors with open-circuited phases: Fault-tolerant controllability and its verification," *IEEE Transactions on Industrial Electronics*, vol. 59, no. 5, pp. 2059–2072, 2012.

[18] X. Cao, H. Yang, L. Zhang, and Z. Deng, "Compensation strategy of levitation forces for single-winding bearingless switched reluctance motor with one winding total short circuited," *IEEE Transactions on Industrial Electronics*, vol. 63, no. 9, pp. 5534–5546, Sept 2016.

[19] X. Wang, X. Ren, and J. Y. Zhang, "Short-circuit fault-tolerant control of bearingless permanent magnet slice machine," in *2013 IEEE Energy Conversion Congress and Exposition*, Sept 2013, pp. 1148–1153.

[20] G. Valente, L. Papini, A. Formentini, C. Gerada, and P. Zanchetta, "Radial force control of multi-sector permanent magnet machines," in *2016 XXII International Conference on Electrical Machines (ICEM)*, Sept 2016, pp. 2595–2601.

[21] ——, "Radial force control of multi-sector permanent magnet machines for vibration suppression," *IEEE Transactions on Industrial Electronics*, vol. PP, no. 99, pp. 1–1, 2017.

[22] ——, "Radial force control of multi-sector permanent magnet machines considering radial rotor displacement," in *2017 IEEE Workshop on Electrical Machines Design, Control and Diagnosis (WEMDCD)*, April 2017, pp. 140–145.

[23] G. Valente, A. Formentini, L. Papini, P. Zanchetta, and C. Gerada, "Position control study of a bearingless multi-sector permanent magnet machine," in *IECON 2017 - 43rd Annual Conference of the IEEE Industrial Electronics Society*, Oct 2017, pp. 8808–8813.

[24] M. Calvini, A. Formentini, G. Maragliano, and M. Marchesoni, "Self-commissioning of direct drive systems," in *Power Electronics, Electrical Drives, Automation and Motion (SPEEDAM), 2012 International Symposium on.* IEEE, 2012, pp. 1348–1353.

The 2018 International Power Electronics Conference

Balance Control of Split Capacitor Potential for Magnetically Levitated Motor System Using Zero-Phase Current

Yusuke Fujii[1], *Student Member, IEEE*, Junichi Asama[1*], *Member, IEEE*, Takaaki Oiwa[1] and Akira Chiba[2], *Fellow, IEEE*
1 Department of Mechanical Engineering, Shizuoka University, Hamamatsu, Japan
2 Department of Electrical and Electronic Engineering, Tokyo Institute of Technology, Tokyo, Japan
*E-mail: asama@shizuoka.ac.jp

Abstract— We have previously proposed a driving method for both a three-phase permanent magnet motor and one- degree- of- freedom controlled magnetic suspension using only a three-phase inverter. In this driving system, the suspension winding for magnetic suspension is connected between a neutral- point of the Y-connected three-phase motor winding and a middle point of two DC voltage sources. The current flowing in this part is denoted by the zero- phase current here. The suspension force is actively controlled by the zero- phase current in this system. This paper proposes a further simplified configuration, where the DC- link voltage is split by two capacitors using a DC voltage source. The split capacitor potential, however, is unbalanced by the DC component of the zero- phase current, owing to the offset detection error of the current sensor and the bias current of the magnetic suspension. This paper addresses the balance control of the split capacitor potential. We propose two balance control methods, including the voltage feedback control and the voltage sensorless control. The experimental results demonstrate the effectiveness of the proposed balance control.

Keywords— *magnetic bearing, magnetic suspension, neutral point potential, zero- phase current.*

I. INTRODUCTION

In a magnetically levitated motor, three translational (x, y, z) and two tilting (θ_x, θ_y) motions of a rotor must be magnetically stabilized for magnetic suspension. The conventional five- degrees- of- freedom actively controlled magnetic bearings [1, 2] and bearingless motors [3] require a number of electromagnets, displacement sensors, and inverters/power amplifiers.

To reduce the size, cost, and power consumption of the magnetically levitated motor, one- degree- of- freedom (DOF) actively controlled magnetic bearings have been previously developed [4-6], where the axial position (z) of the rotor is actively regulated while the remaining motions (x, y, θ_x, θ_y) are passively stabilized by magnetic couplings between the rotor and the stator. In this system, a single-phase inverter or a linear amplifier for active magnetic suspension and a three-phase inverter for rotation are generally needed.

For further simplification and downsizing of the system, the authors have previously proposed a driving

method for both a three-phase permanent magnet motor and a one-DOF controlled magnetic suspension using only a three-phase voltage source inverter [7]. The suspension winding for active magnetic suspension is connected between a neutral- point of the Y-connected three-phase motor winding and a middle point of two DC voltage sources. The current flowing in this part is denoted by the zero- phase current here. The resulting zero- phase current is utilized to regulate the suspension force. As a first step of this study, an experimental setup has been built and tested, which consists of a three-phase interior permanent magnet synchronous motor (IPMSM) and a one- DOF controlled iron- ball magnetic suspension. The authors have realized the regulation of the zero- phase current and magnetic suspension in addition to motor rotation using two DC voltage sources and the experimental setup.

This paper proposes a further simplified configuration, where the DC- link voltage of the inverter is split by two capacitors using a DC voltage source. The split capacitor potential, however, is unbalanced by the DC component of the zero- phase current, owing to the offset detection error of the current sensor and the bias current of the magnetic suspension regulation, since it is floating. To suppress the imbalance, the balance control of split capacitor potential has been previously proposed for a neutral- point- clamped voltage source inverter (NPC-VSI) [8-10]. This balance control requires voltage detection.

This paper focuses on the balance control of split capacitor potential. It is unique that the suspension current passing through the zero- phase is actively regulated for magnetic suspension while the split capacitor potential is balanced. We propose two balance- control methods: the voltage feedback control and the voltage sensorless control. To verify the effectiveness of the proposed balance control, the experimental results are presented.

II. MAGNETICALLY LEVITATED MOTOR SYSTEM

A. Circuit Configuration Using Zero- Phase Current

Fig. 1 shows a proposed circuit configuration using the zero- phase current. The DC- link voltage E is split by two capacitors. The suspension winding for active magnetic

---

This work was supported by JSPS (Japan Society for the Promotion of Science) KAKENHI, Grant Number 16H02324.

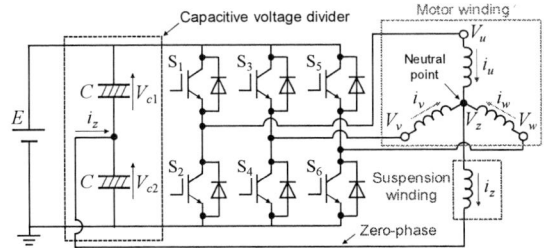

Fig. 1. Proposed circuit configuration using zero- phase current.

suspension is connected between a neutral- point of the Y-connected three-phase motor winding and a middle point of two capacitors, called a zero- phase. The resulting zero-phase current, i_z, is utilized to regulate the suspension force. We use an experimental setup consisting of a three-phase IPMSM and a one- DOF controlled iron- ball magnetic suspension instead of a thrust magnetic bearing.

B. Control of Magnetic Suspension and Motor Drive

Fig. 2 shows a control system diagram of the magnetic suspension and motor drive. The axial displacement, z, of the iron ball is detected by an eddy current displacement sensor and is controlled by regulating the 0-axis current, i_0 (I_z), with proportional-integral-derivative (PID) controllers. The rotational torque is controlled by regulating the d- and q-axis currents, i_d and i_q, respectively.

The 0-, d-, and q-axis currents are regulated by the PI controllers, which generate the 0-, d-, and q-axis voltage references, v_0^* (v_z), v_d^*, and v_q^*, respectively. Subsequently, these are transformed into the three voltage references, v_u^*, v_v^*, and v_w^*, respectively. In this transformation, the 0-axis voltage reference, v_0^*, is superimposed to the three-phase voltage references. Thus, the 0-, d-, and q-axis currents, i_0, i_d, and i_q, respectively, can be independently regulated by only a three-phase inverter.

C. Zero- Phase Current and Split Capacitor Potential

We investigate the influence of the zero- phase current, i_z, on the split capacitor potential, V_{c2}. The relation between i_z and V_{c2} is expressed as

$$V_{c2} = \frac{1}{2C}\int i_z dt \qquad (1)$$

$$V_z - V_{c2} = R_z i_z + L_z \frac{di_z}{dt} \qquad (2)$$

where V_z represents the neutral- point potential of the Y-connection. R_z and L_z represent the zero- phase resistance and inductance, respectively. C is the capacitance. We assume that the zero- phase current, i_z, is expressed as

$$i_z = I_{dc} + I\cos\omega t \qquad (3)$$

where I_{dc} and I represent the DC component and amplitude of the zero- phase current, respectively. ω is the angular frequency. Substituting (3) into (1) yields

$$V_{c2} = \frac{I_{dc}}{2C}t + \frac{I}{2\omega C}\sin\omega t \qquad (4)$$

The first and second term of the right hand in (4) indicate the imbalance of V_{c2} due to the DC component, I_{dc}, and the variation in V_{c2} due to the AC component, I, respectively. Hence, to stabilize V_{c2}, it is required that no DC component should be present in the zero- phase current.

D. Offset Detection Error of Current Sensor

The DC component of the zero- phase current is generated by the offset detection error of the current sensor. Fig. 3 shows a control system diagram of the zero- phase current regulation with the offset detection error of the current sensor. i_z^* and V_z^* represent the references for the zero- phase current, i_z, and voltage, V_z, respectively. I_{off} represents the offset detection error of the current sensor. The green part in Fig. 3 shows a block diagram based on the Laplace transformations of (1) and (2). The Laplace transformations of (1) and (2) are expressed as

$$V_{c2}(s) = \frac{1}{2Cs}i_z(s) \qquad (5)$$

$$V_z(s) - V_{c2}(s) = (R_z + L_z s)i_z(s) = \frac{i_z(s)}{P(s)} \qquad (6)$$

$$P(s) = \frac{1}{R_z + L_z s} \qquad (7)$$

where $P(s)$ represents the transfer function of the zero-phase electromagnet. Fig. 4 shows an equivalent block diagram of Fig. 3. The transfer function $H(s)$ can be written as

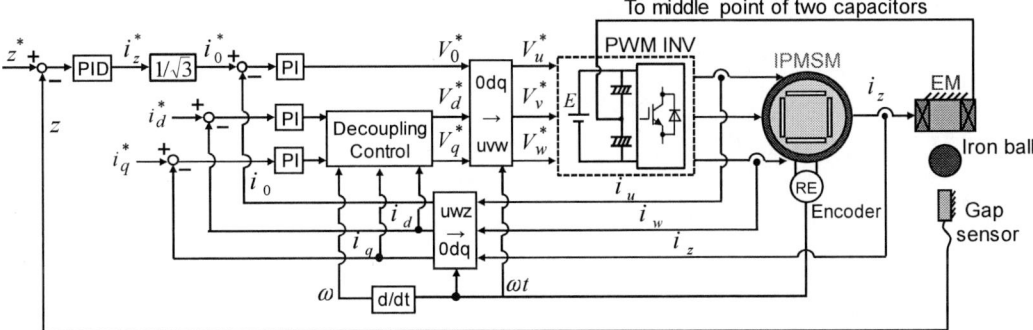

Fig. 2. Control system diagram of magnetic suspension and motor drive using zero- phase current.

$$H(s) = \frac{2Cs}{2Cs + P(s)} \tag{8}$$

We investigate the influence of the offset detection error of the current sensor, I_{off}, on the zero-phase current, i_z. The $i_z(s)$ for the input I_{off} is expressed as

$$i_z(s) = \frac{-P(s)H(s)C_I(s)}{1 + P(s)H(s)C_I(s)} I_{off} \tag{9}$$

$$C_I(s) = K_p + \frac{K_I}{s} \tag{10}$$

where $C_I(s)$ represents the PI controllers for the zero-phase current regulation. K_P and K_I are the proportional and integral gain, respectively. The i_z in steady state for the step input of the offset detection error, I_{off}, is calculated with the final-value theorem as follows:

$$i_z = \lim_{s \to 0} s \left(\frac{-P(s)H(s)C_I(s)}{1 + P(s)H(s)C_I(s)} \right) \frac{I_{off}}{s}$$
$$= \frac{-1}{1 + \frac{1}{2CK_I}} I_{off} \tag{11}$$

The zero-phase DC component is generated by the offset detection error, I_{off}, as shown in (11). Hence, the split capacitor potential, V_{c2}, becomes unbalanced.

Fig. 5 shows the measured capacitor voltages V_{c1}, V_{c2} under the zero-phase current reference, $i_z{}^* = 0$ A without balance control. The DC voltage, E, is 160 V. The capacitance, C, is 750 μF (rated voltage: 450 V). The capacitor voltage becomes gradually unbalanced by the offset detection error when $i_z{}^* = 0$ A is activated. Eventually, the capacitor voltage becomes saturated.

III. VOLTAGE FEEDBACK CONTROL

A. Control System for Votage Feedback Control

Fig. 6 shows a control system diagram of the voltage feedback (FB) control. The entire control system is composed of a zero-phase current control system and a voltage FB control system. $C_V(s)$ represents a controller for voltage FB control. $i_{zv}{}^*$ and $i_{zs}{}^*$ represent the compensation current reference for voltage FB control and the suspension current reference for magnetic suspension, respectively. The split capacitor potential reference, $V_{c2}{}^*$, is $E/2$. The split capacitor potential, V_{c2}, is detected by a

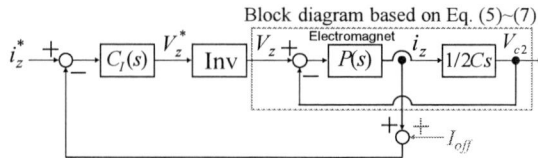

Fig. 3. Control system diagram of zero-phase current regulation with offset detection error of current sensor.

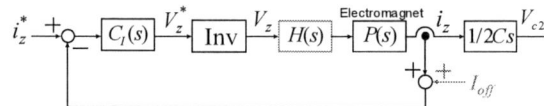

Fig. 4. Equivalent block diagram of Fig. 3.

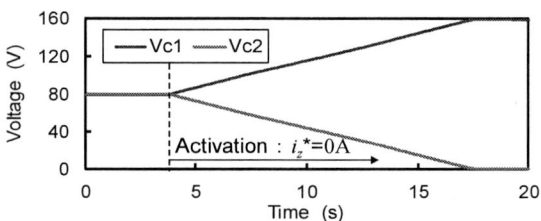

Fig. 5. Measured capacitor voltage without balance control.

differential probe. The error between the reference and measurement is amplified in the $C_V(s)$, which generates the compensation current reference, $i_{zv}{}^*$. Subsequently, $i_{zv}{}^*$ is superimposed to $i_{zs}{}^*$, and these are regulated as the zero-phase current reference, $i_z{}^*$.

We investigate the transfer function characteristic for the voltage FB control. The $i_z(s)$ for the input I_{off} in the voltage FB control is expressed as

$$i_z(s) = \frac{-2CsT(s)}{2Cs + T(s)C_V(s)} I_{off} \tag{12}$$

$$T(s) = \frac{P(s)H(s)C_I(s)}{1 + P(s)H(s)C_I(s)} \tag{13}$$

where $T(s)$ represents the closed-loop transfer function for zero-phase current regulation without the voltage FB control. The transfer function $V_{c2}(s)/V_{c2}{}^*(s)$ for voltage FB control is expressed as

$$\frac{V_{c2}(s)}{V_{c2}^*(s)} = \frac{T(s)C_V(s)}{2Cs + T(s)C_V(s)} \tag{14}$$

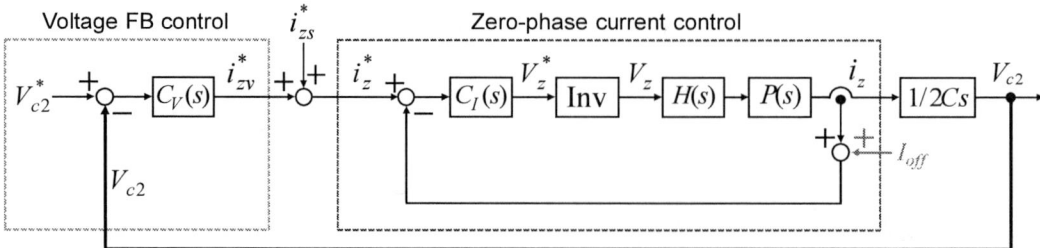

Fig. 6. Control system diagram of voltage FB control.

4044

B. Voltage FB-I Control

To suppress the DC component due to the offset detection error of the current sensor, I_{off}, $C_V(s)$ is composed of an I controller:

$$C_V(s) = \frac{K_{IV}}{s} \qquad (15)$$

where K_{IV} represents the integral gain for $C_V(s)$. We call this control method the voltage FB-I control. Assuming that $T(s)$ is one for simplification, the $i_z(s)$ for the input I_{off} is given by substituting (15) to (12):

$$i_z(s) = \frac{-s^2}{s^2 + (K_{IV}/2C)} I_{off} \qquad (16)$$

Subsequently, from (16), the response of $i_z(t)$ for the step input of I_{off} can be written as

$$i_z(t) = -I_{off} \cos \omega_I t \qquad (17)$$

$$\omega_I = \sqrt{\frac{K_{IV}}{2C}} \qquad (18)$$

As shown in (17), the zero- phase DC component can be eliminated from i_z, although the AC component is generated by I_{off} using the voltage FB-I control.

Fig. 7 shows the measured capacitor voltages V_{c1}, V_{c2} under the zero- phase current reference, $i_z^*=0$ A with voltage FB-I control. The integral gain, K_{IV}, is 0.005. The capacitor voltage converges to 80 V slowly when the voltage FB-I control is activated at approximately 15 s. However, the capacitor voltage varies at approximately 0.29 Hz following (18). The low- frequency zero- phase current may decrease the positioning accuracy of the magnetic suspension.

C. Voltage FB-PI Control

To suppress the low- frequency voltage variation for voltage FB-I control, we investigate the controller $C_V(s)$, which is composed of PI controllers. The $C_V(s)$ is expressed as

$$C_V(s) = K_{PV} + \frac{K_{IV}}{s} \qquad (19)$$

where K_{PV} represents the proportional gain for $C_V(s)$. We call this control method the voltage FB-PI control. Substituting (19) to (12), the $i_z(s)$ for the input I_{off} is given as

$$i_z(s) = \frac{-s^2}{s^2 + (K_{PV}/2C)s + (K_{IV}/2C)} I_{off} \qquad (20)$$

As shown in (20), K_{PV} corresponds to the damping coefficient. Therefore, the AC component of (17) can be eliminated using voltage FB-PI control.

Fig. 8 shows the measured capacitor voltages V_{c1}, V_{c2} under the zero- phase current reference, $i_z^*=0$ A with the voltage FB-PI control. The proportional gain, K_{PV}, and integral gain, K_{IV}, are 0.01 and 0.005, respectively. The capacitor voltage rapidly converges to 80 V when the voltage FB-PI control is activated. Moreover, the low-

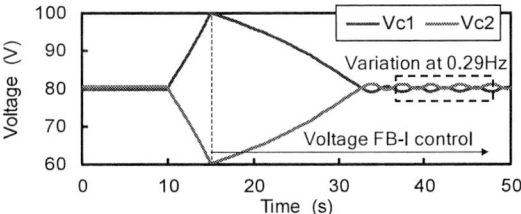

Fig. 7. Measured capacitor voltage with voltage FB-I control.

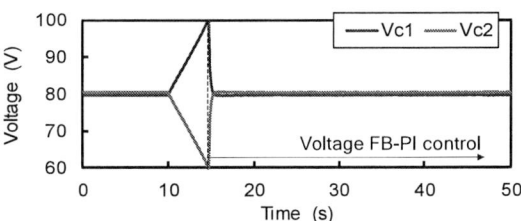

Fig. 8. Measured capacitor voltage with voltage FB-PI control.

Fig. 9. Experimental setup for load test.

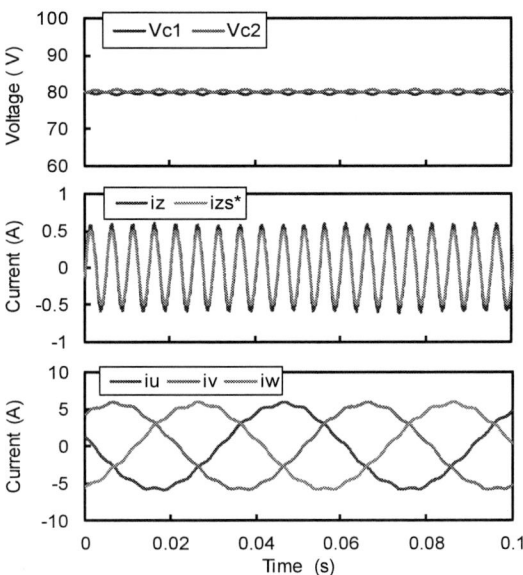

Fig. 10. Measured capacitor voltage, zero- phase current (200 Hz, 0.5 A) and three-phase currents at 500 rpm under loaded condition with 4 A$_{rms}$ using voltage FB-PI control without magnetic levitation.

frequency voltage variation is suppressed.

Subsequently, we investigate both the zero- phase current and motor currents regulation while the split capacitor potential is regulated. Fig. 9 shows an

The 2018 International Power Electronics Conference

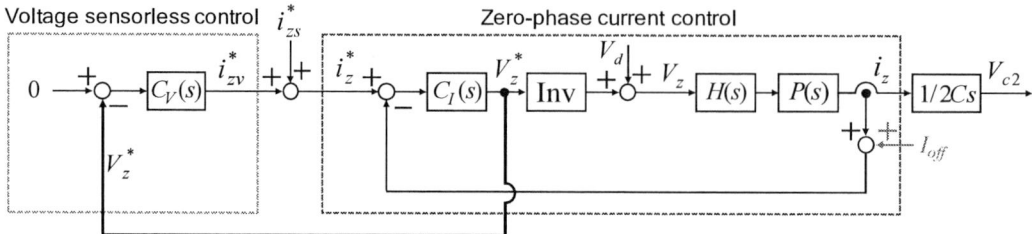

Fig. 11. Control system diagram of voltage sensorless control.

experimental setup for the load test of the motor. The IPMSM is connected to a general-purpose induction motor for the load test through a torque meter. Fig. 10 shows the measured capacitor voltages, V_{c1}, V_{c2}, the zero- phase current, i_z, and the three-phase current i_u, i_v, i_w at 500 rpm under the loaded condition with motor current 4 A_{rms} and suspension current reference, $i_{zs}{}^*$, amplitude of 0.5 A, 200 Hz without magnetic levitation. As shown in Fig. 10, the zero- phase current, i_z, follows the suspension current reference, $i_{zs}{}^*$, while the capacitor voltage is balanced and the motor currents are regulated.

IV.   VOLTAGE SENSORLESS CONTROL

A. Control System for Voltage Sensorless Control

Voltage FB control requires the detection of split capacitor potential, V_{c2}. In this section, we investigate the voltage sensorless control without voltage detection. Fig. 11 shows a control system diagram of the voltage sensorless control. The $C_V(s)$ is composed of the PI controllers. The DC component for the zero- phase voltage reference, $V_z{}^*$, in the controller is generated by the offset detection error, I_{off}. In this proposed control method, we aim to control the split capacitor potential without voltage detection, applying the zero- phase voltage reference, $V_z{}^*$ to the balance control. Concretely, the DC component of $V_z{}^*$ is regulated to zero with the PI controllers, $C_V(s)$. To lower the sensitivity of the compensation current reference, $i_{zv}{}^*$, for the suspension current reference, $i_{zs}{}^*$, the frequency bandwidth of the voltage sensorless control system is set to the lower bandwidth compared with the other control systems.

B. Simulation of Frequency Characteristic

We investigate the frequency characteristic for voltage FB-PI control and voltage sensorless-PI control using a commercially available software (PSIM, Myway Plus Co rp., Japan). Fig. 12 and Fig. 13 show the simulation results of the voltage disturbance frequency characteristic for the transfer function $i_z(s)/V_d(s)$ and the current closed-loop frequency characteristic for the transfer function $i_z(s)/i_{zs}{}^*(s)$, respectively. Hence, $V_d(s)$ represents the voltage disturbance that is applied to the plant, $P(s)$. As shown in Fig. 12, the voltage disturbance suppression performance for the sensorless-PI control decreases at the low frequency region compared with the voltage FB-PI control. As shown in Fig. 13, the frequency bandwidth of the current closed-loop frequency characteristic for the

sensorless-PI control is almost the same as that of the voltage FB-PI control.

C. Voltage Sensorless-PI Control

Fig. 14 shows the measured capacitor voltages V_{c1}, V_{c2} under the zero- phase current reference, $i_z{}^*$=0 A with voltage sensorless-PI control. The capacitor voltage rapidly converges to 80 V when voltage sensorless control is activated.

Fig. 15 shows the measured capacitor voltages, V_{c1}, V_{c2}, the zero- phase current, i_z, and the three-phase current i_u, i_v, i_w at 500 rpm under the loaded condition with motor current 4 A_{rms} and suspension current reference, $i_{zs}{}^*$, amplitude of 0.5 A, 200 Hz without magnetic levitation. As shown in Fig. 15, the zero- phase current, i_z, follows the suspension current reference, $i_{zs}{}^*$, while the capacitor voltage is balanced and the motor currents are regulated.

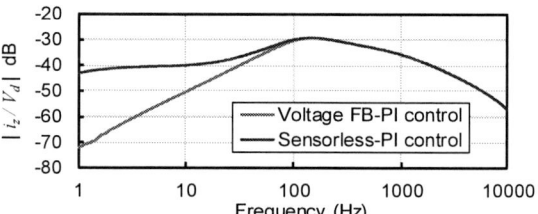

Fig. 12. Simulation result of voltage disturbance frequency characteristic for transfer function $i_z(s)/V_d(s)$.

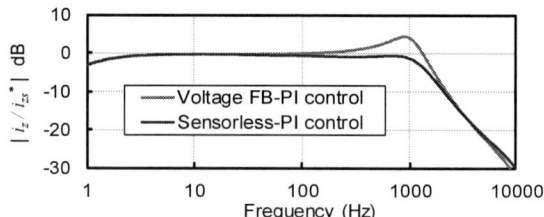

Fig. 13. Simulation result of current closed-loop frequency characteristic for transfer function $i_z(s)/i_{zs}{}^*(s)$.

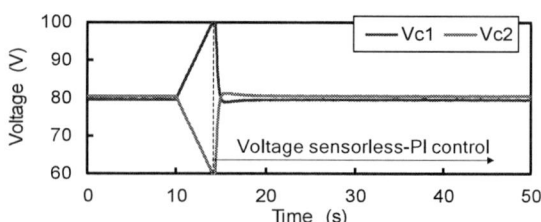

Fig. 14. Measured capacitor voltage with voltage sensorless-PI control.

4046

V. Magnetic Levitation Test

Fig. 16 shows an experimental setup of the iron- ball levitation. A hollow iron ball of diameter 50 mm and weight 160 g and a cylinder- type electromagnet with an E-shaped cross section is used. To eliminate the bias current for magnetic suspension regulation, a permanent magnet (PM) is attached to the bottom of the electromagnet. Fig. 17 shows a block diagram of the zero-power levitation control. We adopt the zero- power levitation control, which automatically compensates the levitation position for balancing the self-weight of the iron

ball and the PM attractive force.

Fig. 18 and Fig.19 show the measured capacitor voltages, V_{c1}, V_{c2}, the zero- phase current, i_z, and the iron-ball displacement, z, when the iron- ball levitation is activated under no-load condition with the voltage FB-PI control and the voltage sensorless-PI control, respectively. The zero- phase current, i_z, increases rapidly, and the iron-ball is levitated. Consequently, the capacitor voltage becomes unbalanced instantaneously. Subsequently, the capacitor voltage converges to 80 V. The performances of the two balance control methods are almost the same.

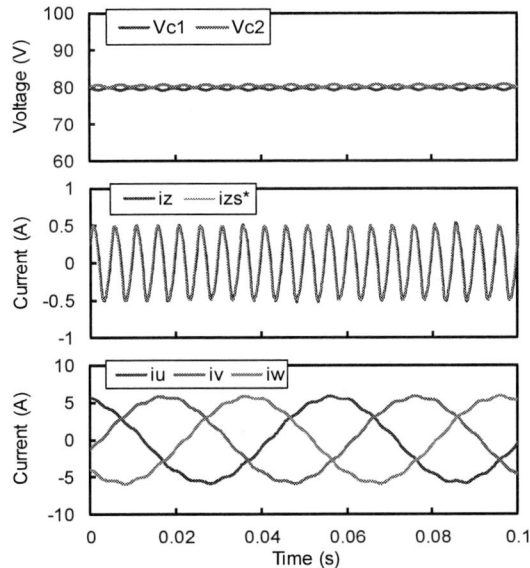

Fig. 15. Measured capacitor voltage, zero- phase current (200 Hz, 0.5 A) and three-phase currents at 500 rpm under loaded condition with 4 A_rms using voltage sensorless-PI control without magnetic levitation.

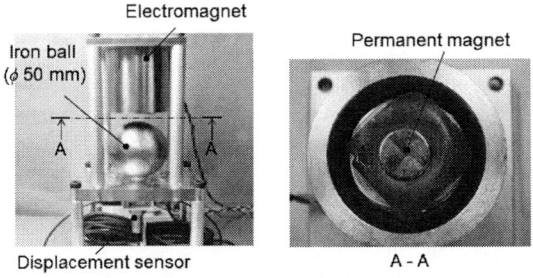

Fig. 16. Experimental setup of iron- ball levitation (left) and the electromagnet with a permanent magnet (right).

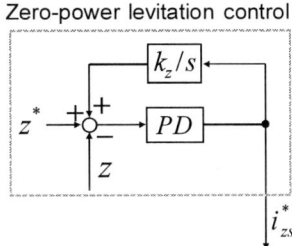

Fig. 17. Block diagram of zero- power levitation control.

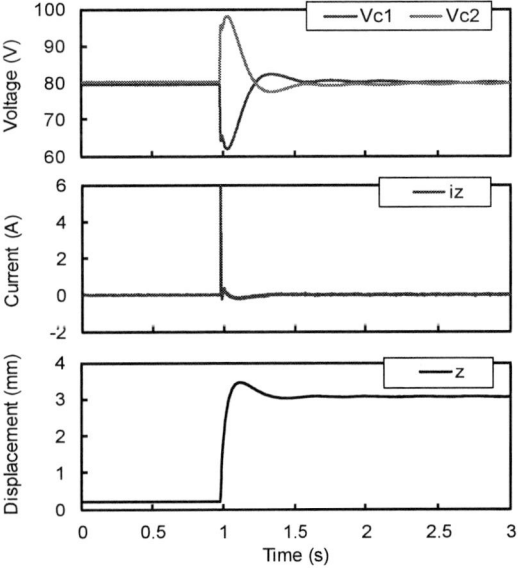

Fig. 18. Measured capacitor voltage, zero- phase current and iron- ball displacement under no-load condition using voltage FB-PI control with magnetic levitation.

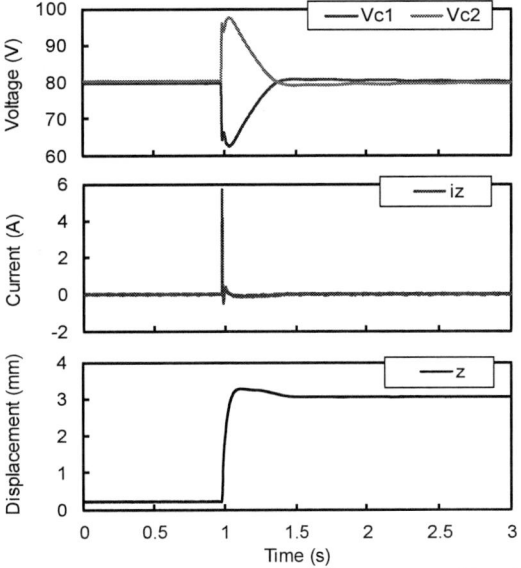

Fig. 19. Measured capacitor voltage, zero- phase current and iron- ball displacement under no-load condition using voltage sensorless-PI control with magnetic levitation.

VI. CONCLUSION

This paper describes the balance control of split capacitor potential for a magnetically levitated motor system using zero- phase current. The split capacitor potential becomes unbalanced by the DC component of the zero- phase current owing to the offset detection error of the current sensor and the bias current for magnetic suspension regulation. To suppress the imbalance, we investigated two balance control methods of the split capacitor potential. The first method, voltage FB control, requires voltage detection. The second method, voltage sensorless control, can regulate the capacitor voltage without voltage detection, by applying the zero-phase voltage reference in the controller to the balance control. The experimental results demonstrated that the proposed methods are effective for the balance control of the split capacitor potential.

REFERENCES

[1] T. Mizuno and T. Higuchi, "Design of Magnetic Bearing Controllers Based on Disturbance Estimation," *Proceedings of the 2nd International Symposium on Magnetic Bearings (ISMB)*, pp. 281-288, July 12-14, Tokyo, Japan, 1990.

[2] T. Higuchi, M. Otsuka, and K. Watanabe, "Real time Balancing of a Flexible Rotor Supported by Magnetic Bearing," *Proceedings of the 2nd International Symposium on Magnetic Bearings (ISMB)*, pp. 265-272, July 12-14, Tokyo, Japan, 1990.

[3] N. Yamamoto, M. Takemoto, S. Ogasawara, and M. Hiragushi, "Experimental Estimation of a 5-axis Active Control Type Bearingless Canned Motor Pump," Proceedings of *IEEE international Electric Machines and Drives Conference*, pp. 148-153, May 15-18, Niagara Falls, ON, Canada, 2011.

[4] J. Kuroki, T. Shinshi, L. Li, and A. Shimokohbe, "Miniaturization of One-Axis-Controlled Magnetic Bearing," *Precision Engineering*, vol. 29, no. 2, pp. 208-218, 2005.

[5] I. da Silva and O. Horikawa, "An Attraction-Type Magnetic Bearing with Control in a Single Direction," *IEEE Transactions on Industry Applications*, vol. 36, no. 4, pp. 1138-1142, 2000.

[6] T. Ohji, S. C. Mukhopadhyay, M. Iwahara, S. Yamada, "Performance of Repulsive Type Magnetic Bearing System Under Nonuniform Magnetization of Permanent Magnet," *IEEE Transactions on Magnetics*, vol. 36, no. 5, pp. 3696-3698, 2000.

[7] J. Asama, Y. Fujii, T. Oiwa, and A. Chiba, "Novel Control Method for Magnetic Suspension and Motor Drive with One Three-Phase Voltage Source Inverter Using Zero-Phase Current," *Mechanical Engineering Journal*, vol. 2, no. 4, paper no. 15-00116, 2015.

[8] S. Ogasawara, H. Akagi, "A Vector Control System Using a Neutral-Point-Clamped Voltage Source PWM Inverter," *IEEE Industry Applications Society Annual Meeting*, pp. 422-427, 1991.

[9] J. E. Espinoza, J. R. Espinoza, and L. A. Moran, "A Systematic Controller-Design Approach for Neutral-Point-Clamped Three-Level Inverters," *IEEE Trans. on Industrial Electronics*, vol. 52, no. 6, pp. 1589-1599, 2005.

[10] T. Kawabata, M. Koyama, S. Tamai, T. Fujii, and R. Uchida, "A New PWM Method of a Three-Level Inverter Considering Minimum Pulse Width and Neutral Voltage Balance Control," *IEEJ Transactions on Industry Applications*, vol. 113, no. 7, pp. 865-873, 1993 (in Japanese).

Asymmetrical Half-Bridge Converter With Zero DC-offset Current in Transformer Using New Rectifier Structure

Jung-Kyu Han[1]*, Jong-Woo Kim[2], Seung-Hyun Choi[1], Jih-Sheng Lai[2], and Gun-Woo Moon[1]

1 Electrical Engineering, KAIST, Daejeon, Korea
2 Electrical and Computer Engineering, Virginia Tech, Blacksburg, USA
*E-mail: hanjk715@kaist.ac.kr

Abstract—A conventional asymmetrical half-bridge (AHB) converter is a good candidate for low power applications such as TV and LED driver. It has small number of components and all switching devices have soft swtiching capability. However, when the AHB converter is designed with wide input voltage range, it has a large offset current in transformer which increases size of the transformer and core loss. Also, DC-offset current worsens ZVS condition of one of the half-bridge switches resulting in low efficiency in light load condition. To overcome above problems, a new asymmetrical half-bridge converter with coupled inductor rectifier (CIR) is proposed in this paper. Since two capacitors of the new rectifier equalize an average current flowing through secondary rectifier, the proposed converter doesn't have DC-offset current. Therefore, the proposed converter reduces size of the transformer and can increase efficiency. Experiment is implemented with a 250V-400V input voltage variation and 50V/200W output specifications.

Keywords— AHB converter, DC-offset current in transformer, high efficiency.

I. INTRODUCTION

A conventional asymmetrical half-bridge converter (AHB) shown in Fig. 1 is a good candidate in low power applications such as TV and LED driver. It has small number of components and all switching devices have soft switching capability [1]-[4]. But, when it is designed considering a wide input variation, it has several problems. First, transformer has large offset current when the input voltage is high [5]-[8]. Since it has maximum voltage gain at duty-ratio of the Q_1 is *0.5*, the AHB converter should be designed to operate with *0.5* duty-ratio at low input voltage. Accordingly, AHB converter becomes to operate as small duty-ratio when the input voltage is high which causes large DC-offset current in the transformer. This increases its size resulting in a low power density and a large transformer core loss [9]-[12]. Second, one of the half-bridge switches has bad ZVS condition since the DC-offset current makes a primary current be biased to positive or negative direction. It not only makes a design difficult, but also decreases light load efficiency.

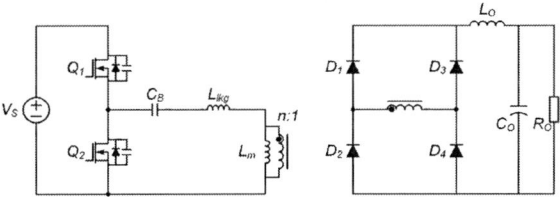

Fig. 1. The conventional AHB converter.

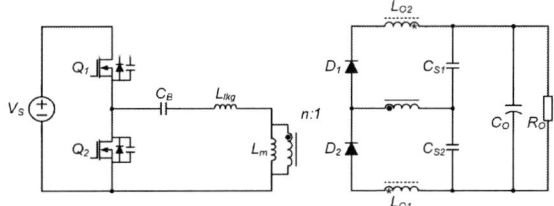

Fig. 2. The proposed AHB converter.

To overcome above problems, a new AHB converter with coupled inductor rectifier (CIR) is proposed in this paper. The proposed converter has a new rectifier structure which eliminates the DC-offset current in the transformer. Therefore, it not only reduces transformer size, but also increases soft switching capability of both half-bridge switches. Therefore, the proposed AHB converter achieves high efficiency.

II. CONCEPT OF THE PROPOSED CONVERTER

Fig. 2 shows the proposed AHB converter. The proposed converter has coupled inductor located between D_1 and C_{s1}, D_2 and C_{s2}. In a conventional AHB converter, an average current flowing to a secondary side of the transformer is not zero except when a duty-ratio of Q_1 is 0.5. This causes DC-offset current in transformer since an average current flowing through blocking capacitor C_B should be zero.

On the other hand, an average current flowing to secondary side of the transformer is always zero in proposed converter since an average current flowing through a capacitor C_{s1} and C_{s2} should be zero. Therefore, offset current doesn't occur in the proposed converter.

The 2018 International Power Electronics Conference

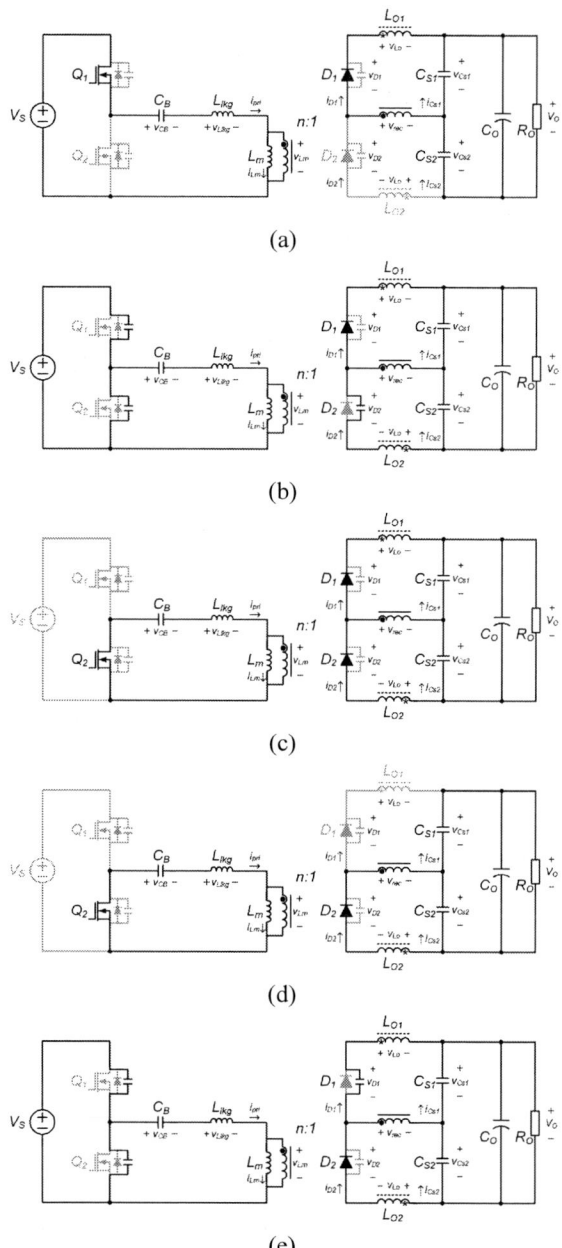

(a)

(b)

(c)

(d)

(e)

Fig. 3. Circuit diagrams of the proposed converter. (a) Mode 1, (b) mode 2, (c) mode 3, (d) mode 4, and (e) mode 5.

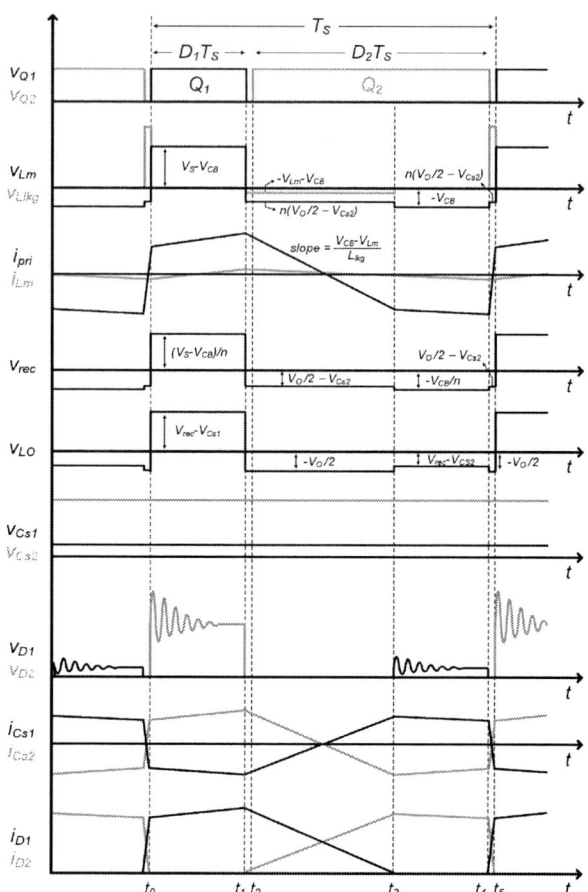

Fig. 4. Circuit diagrams of the proposed converter.

III. OPERATIONAL PRINCIPLES

Fig. 3 shows circuit diagrams to understand operational principles of the proposed converter. The proposed converter has five operating modes and key waveforms are shown in Fig. 4. For the simple analysis, there are some assumptions as follows:

1) Magnetizing inductance L_m and output inductor L_{O1}, L_{O2} is large enough to ignore ripple current.
2) The leakage inductance L_{lkg} is much smaller than magnetizing inductance L_m.
3) The coupling coefficient of the L_{O1} and L_{O2} is *1*.

Mode 1 [t_0-t_1]: In mode 1, Q_1 is turned on and output inductor0 current flows to D_1 and C_{S2}. Since V_{CB} is $D_{Q1}V_S$, $V_{Lm}=(1-D_{Q1})V_S$ and it is transferred to a secondary side.

Mode 2 [t_1-t_2]: Mode 2 starts when Q_1 is turned off. As the Q_1 and Q_2 start to be charged and discharged respectively, the V_{Lm} and V_{rec} also decrease together. When V_{rec} reaches to $V_O/2$-V_{CS2}, D_2 is conducted.

Mode 3 [t_2-t_3]: Mode 3 starts when Q_2 is turned on. When a voltage of Q_2 reached to zero during mode 2, Q_2 turned on with ZVS operation. Since V_{Lm} is clamped to $n(V_O/2$-$V_{CS2})$ in mode 2, V_{Llkg} is $-n(V_O/2$-$V_{CS2})$-V_{CB} and current commutation occurs between D_1 and D_2.

Mode 4 [t_3-t_4]: Mode 4 starts when current commutation between D_1 and D_2 ends. Since D_1 is turned off, V_{Lm} decreases to $-V_{CB}$ and V_{rec} decreases to $-V_{CB}/n$. V_{rec}-V_{CS2} is applied to L_O and output current flows through the D_2 and C_{S1}.

Mode 5 [t_4-t_5]: Mode 5 starts when the switch Q_2 is turned off. As the Q_1 and Q_2 start to be charged and discharged respectively, the V_{Lm} and V_{rec} also increase together. When V_{rec} reaches to $V_O/2$-V_{CS2}, D_2 is conducted and V_{rec}, V_{Lm} are clamped to $V_O/2$-V_{CS2}, $n(V_O/2$-$V_{CS2})$. During mode 5, V_{DS2}-V_{Lm}-V_{CB} is applied to L_{lkg} and current commutation occurs between D_1 and D_2.

4050

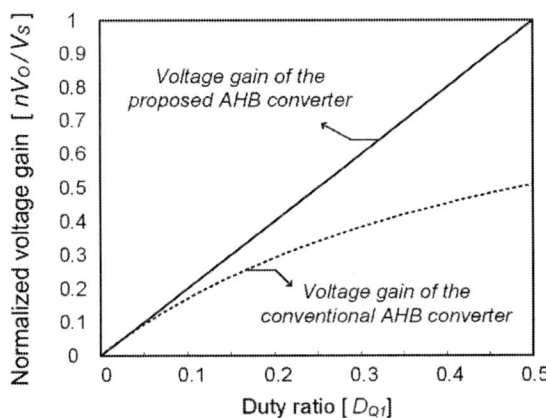

Fig. 5. The voltage gain curve of the proposed converter and the conventional AHB converter.

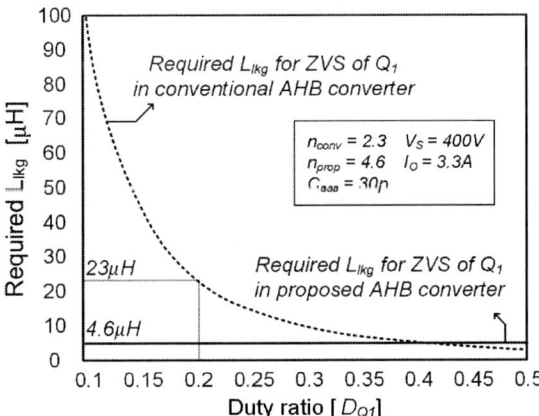

Fig. 6. Required L_{lkg} for ZVS of Q_1.

IV. ANALYSIS OF THE PROPOSED CONVERTER

A. Voltage gain

By applying a voltage-second balance at L_m and L_O, a voltage gain of the proposed converter is determined. Assuming that L_m and L_O are large enough, the voltage gain is determined as follows:

$$\frac{V_O}{V_s} = \frac{2D_{Q1}}{n} \qquad (1)$$

As a result, the proposed converter has two times larger transformer turns-ratio compared with the voltage gain of the conventional AHB converter.

B. ZVS condition of the proposed converter

In a conventional AHB converter, half-bridge switch Q_1 and Q_2 have different ZVS condition. Assuming that an output capacitor of Q_1 and Q_2 are equal as C_{oss}, ZVS condition of Q_1 and Q_2 are determined as follows:

$$\frac{1}{2}L_{lkg}(-2D_{Q1}I_O/n)^2 \geq C_{oss}(1-D_{Q1})V_s^2 \qquad (2)$$

$$\frac{1}{2}L_{lkg}(2(1-D_{Q1})I_O/n)^2 \geq C_{oss}D_{Q1}V_s^2 \qquad (3)$$

TABLE I
DESIGN EXAMPLES OF THE PROTOTYPE CONVERTERS

	Conventional AHB converter	Proposed AHB converter
Primary switch Q_1, Q_2	IPP60R385 (600V, 385mΩ)	
Transformer	PQ3220 (500μH, 23:10)	PQ2620 (800μH, 23:5)
ZVS range of Q1	At 100% load condition	Over 50% load condition
ZVS range of Q2	Entire load condition	Entire load condition
Rectifier diode D_1	V60100C (100V, 0.41V_F)	V60100C (100V, 0.41V_F)
Rectifier diode D_2	MBR40250 (80V, 0.62V_F)	MBR40250 (80V, 0.62V_F)
Rectifier diode D_3	V60100C (100V, 0.41V_F)	-
Rectifier diode D_4	MBR40250 (80V, 0.56V_F)	-
Rectifier capacitor C_{s1}	-	63PZA22M8X10 (22μF, 63V) *3ea
Rectifier capacitor C_{s2}	-	63PZA22M8X10 (22μF, 63V) *3ea
Output inductor	PQ2620 (500μH)	PQ3220 (500μH)

Since the D_{Q1} is small, Q_2 has large ZVS energy while Q_1 has much small ZVS energy which causes hard switching of Q_1.

Applying same assumption to the proposed converter, ZVS condition of Q_1 and Q_2 are determined as follows:

$$\frac{1}{2}L_{lkg}(-2I_O/n)^2 \geq C_{oss}V_s^2 \qquad (4)$$

$$\frac{1}{2}L_m(2I_O/n)^2 \geq C_{oss}V_s^2 \qquad (5)$$

As shown in (4), Q_2 has large energy since L_m helps the ZVS operation. Although Q_1 has smaller ZVS energy than Q_2, it has much larger energy than conventional AHB converter. Fig. 6 is a comparison of required L_{lkg} to achieve ZVS operation of Q_1 in both converters with experiment specification.

V. EXPERIMENTAL RESULTS

To verify the effect and feasibility, prototype converters are experimented with a *200-400V* input and *50V/200W* output specification. Table I is design examples of prototype converters. Since the proposed converter has CIR, it has two didoes and two capacitors while the conventional AHB converter with full-bridge rectifier (FBR) has four diodes.

Fig. 7(a) ~ (f) is the key waveforms of prototype converters with 400V input, 50V/200W output specification. Fig. 7(a)-(c) is for the conventional AHB converter at 100%, 50%, 20% load condition. And Fig. 7(d)-(f) is for the proposed AHB converter at 100%, 50%, 20% load condition. As shown in figure, the conventional

4051

The 2018 International Power Electronics Conference

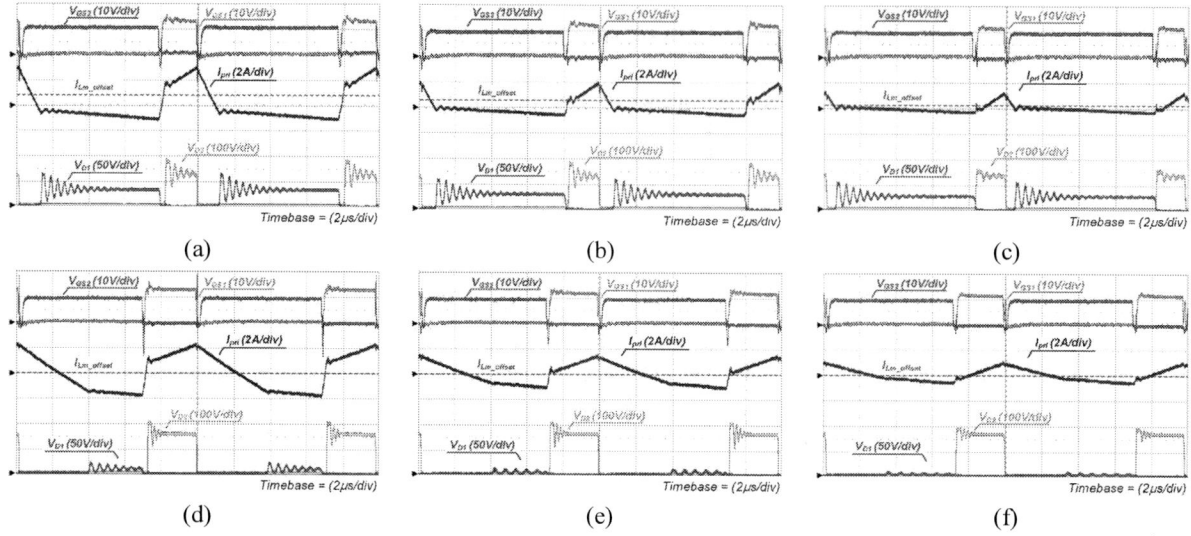

(a)   (b)   (c)

(d)   (e)   (f)

Fig. 7. Key waveforms of the conventional AHB converter and the proposed converters at 400V input and 50V/200W output specification. (a) Conventional AHB at 100% load, (b) conventional AHB at 50% load, (c) conventional AHB at 20% load, (d) proposed AHB at 100% load, (e) proposed AHB at 50% load, and (f) proposed AHB at 20% load.

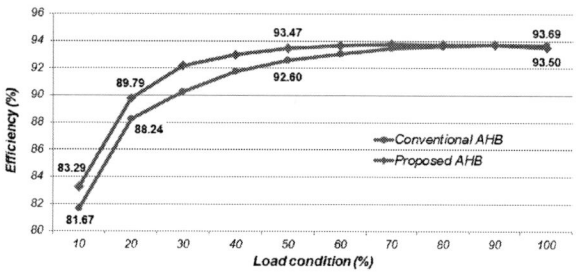

Fig. 8. The efficiency of the prototype converters.

AHB converter has offset current in the transformer. And this causes low negative current resulting in small ZVS energy for Q_1. On the contrary, the proposed AHB converter doesn't have offset current in transformer and it has large ZVS energy for Q_1.

Fig. 8 is an efficiency of prototype converter in entire load condition. As shown in figure, the proposed converter achieves high efficiency especially at light load since it reduces switching loss of Q_1 and core loss of transformer.

VI. CONCLUSIONS

In this paper, a new AHB converter with CIR is proposed. Because the proposed converter has two capacitors in rectifier, an average current flowing to the secondary side of the transformer becomes *0*. As a result, the proposed converter doesn't have DC-offset current which resulting in small size of transformer and small core loss. Also, since the DC-offset current is zero, it has larger ZVS energy for Q_1. As a result, the proposed converter has high efficiency.

ACKNOWLEDGMENT

This work was supported by the National Research Foundation of Korea (NRF) grant funded by the Korea government (MSIP) (No. 2016R1A2B2010328).

This work was supported by Korea Electronic Power Research Institute (KEPRI).

REFERENCES

[1] H. Wang, Y. Chen, P. Fang, Y. F. Liu, J. Afsharian, and Z. Yang, "An LLC Converter Family With Auxiliary Switch for Hold-Up Mode Operation," *IEEE Trans. Power Electron.*, vol. 32, no. 6, pp. 4291-4306, Jun. 2017.

[2] Y. S. Lai, Z. J. Su, and W. S. Chen, "New Hybrid Control Technique to Improve Light Load Efficiency While Meetinf the Hold-up Time Requirement for Two-Stage Server Power," *IEEE Trans. Power Electron.*, vol. 29, no. 9, pp. 4763-4775, Sep. 2014.

[3] H. Wang, H. S. Chung, and W. Liu, "Use of a Series Voltage Compensator for Reduction of the DC-Link Capacitance in a Capacitor-Supported System," *IEEE Trans. Power Electron.*, vol. 29, no. 3, pp. 1163-1175, Mar. 2014.

[4] J. I. Baek, J. K. Kim, J. B. Lee, H. S. Youn, and G. W. Moon, "A Boost PFC Stage Utilized as Half-Bridge Converter for High-Efficiency DC-DC Stage in Power Supply Unit," *IEEE Trans. Power Electron.*, vol. 32, no. 10, pp. 7449-7457, Oct. 2017.

[5] W. Li, S. Zong, F. Liu, H. Yang, X. He, and B. Wu, "Secondary-Side Phase-Shift-Controlled ZVS DC/DC Converter With Wide Voltage Gain for High Input Voltage Application," *IEEE Trans. Power Electron.*, vol. 28, no. 11, pp. 5128-5139, Nov. 2013.

[6] D. K. Kim, S. Moon, C. O. Yeon, G. W. Moon "High-Efficiency LLC Resonant Converter With High Voltage Gain Using an Auxiliary LC Resonant Circuit," *IEEE Trans. Power Electron.*, vol. 31, no. 10, pp. 6901-6909, Oct. 2016.

[7] G. D. Capua, S. A. Shirsavar, M. A. Hallworth, and N. Femia, "An Enhanced Model for Small-Signal Analysis of

the Phase-Shifted Full-Bridge Converter," *IEEE Trans. Power Electron.*, vol. 30, no. 3, pp. 1567-1576, Mar. 2015.

[8] G. N. B. Yadav, and N. L. Narasamma, "An Active Soft Switched Phase-Shifted Full-Bridge DC-DC Converter: Analysis, Modeling, Design, and Implementation," *IEEE Trans. Power Electron.*, vol. 29, no. 9, pp. 4538-4550, Sep. 2014.

[9] D. Y. Kim, C. E. Kim, G. W. Moon, "Variable Delay Time Method in the Phase-Shifted Full-Bridge Converter for Reduced Power Consumption Under Light Load Conditions," *IEEE Trans. Power Electron.*, vol. 28, no. 11, pp. 5120-5127, Nov. 2013.

[10] A. Mallik, and A. Khaligh, "Variable-Switching-Frequency State-Feedback Control of a Phase-Shifted Full-Bridge DC/DC Converter," *IEEE Trans. Power Electron.*, vol. 32, no. 8, pp. 6523-6531, Aug. 2017.

[11] J. W. Kim, D. Y. Kim, C. E. Kim, and G. W. Moon, "A Simple Switching Control Technique for Improving Light Load Efficiency in a Phase-Shifted Full-Bridge Converter With a Server Power System," *IEEE Trans. Power Electron.*, vol. 29, no. 4, pp. 1562-1566, Apr. 2014.

The 2018 International Power Electronics Conference

Circulating Current-less Phase-Shifted Full-Bridge Converter With New Rectifier Structure

Jung-Kyu Han* and Gun-Woo Moon
Electrical Engineering, KAIST, Daejeon, Korea
*E-mail: hanjk715@kaist.ac.kr

Abstract—**A conventional phase-shifted full bridge (PSFB) converter has several advantages for high power applications since it has full-bridge structure and all switching devices can achieve soft switching. However, it has large circulating current in primary side when it operates with small duty-ratio resulting in large conduction loss at primary switches. Also, since rectifier diodes have a voltage ringing between parasitic components, it has large voltage stress. To overcome above problems, a new PSFB converter which eliminates the circulating current and voltage ringing using coupled output inductor is proposed in this paper. As a result, the proposed converter reduces a conduction loss at primary side and can use small voltage rating diode in secondary rectifier. To verify the effect and feasibility, prototype converters are experimented with a *320-400V* input voltage and *56V/12.8A* output specification.**

Keywords— Circulating current, coupled inductor, PSFB converter, coupled inductor rectifier.

I. INTRODUCTION

A phase-shifted full bridge (PSFB) converter shown in Fig. 1(a) is widely used topology in high power applications such as server power supply and EV charger for high efficiency. It has many advantages such as zero-voltage switching (ZVS) of all switches, full bridge structure, and clamped voltage stress of primary switches [1]-[7]. Also, it is suitable for high power applications since an existence of the output inductor relieves a RMS and peak current. However, it has large circulating current in primary side when it operates with small duty-ratio because an output current is transferred to primary side during the freewheeling period of the output inductor. As a result, in the case of the PSFB converter is designed considering a wide input voltage variation, it has small duty-ratio in high input voltage which resulting in large conduction loss in primary side [8]-[10]. Also, since the PSFB converter has voltage ringing at secondary rectifier diodes, not only an additional snubber circuit is needed, but also relatively high voltage rating diode should be used [11]-[13].

Many researches are studied to overcome these problems of the conventional PSFB converter [14]-[16]. The output inductor-less PSFB converter shown in Fig. 1(b) can be a good candidate to solve the above problems which eliminates voltage ringing of rectifier diodes since the diodes are connected directly to the output capacitor.

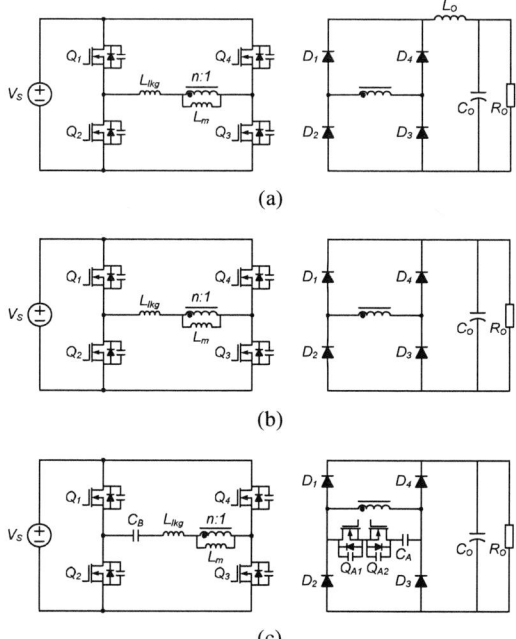

Fig. 1. The various PSFB converter topologies. (a) Conventional PSFB converter, (b) output inductor-less PSFB converter, and (c) improved output inductor-less PSFB converter.

Also, because it does not have the output inductor, no circulating current occurs in primary side. However, although output inductor-less structure has several advantages, it also causes large RMS current and large peak current problems which resulting in limitation to low power applications.

A new output inductor-less PSFB converter shown in Fig. 1(c) solves the aforementioned problems [14]. Because it does not have output inductor, it can eliminate the circulating current and voltage ringing. Also, by using a resonance between L_{lkg} and C_A, it reduces RMS current and peak current. Thus, it effectively reduces conduction loss and can be used for high current applications. But it not only requires two additional switches and additional driving circuit, but also causes hard switching at switch Q_{A1} and Q_{A2}. Thus, other solution is required to overcome circulating current and voltage ringing problems without adding other components.

4054

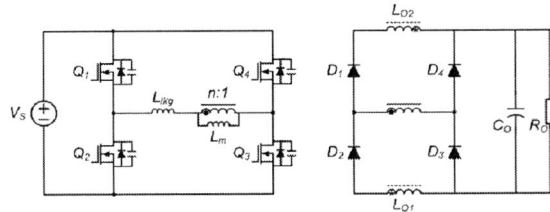

Fig. 2. The proposed PSFB converter.

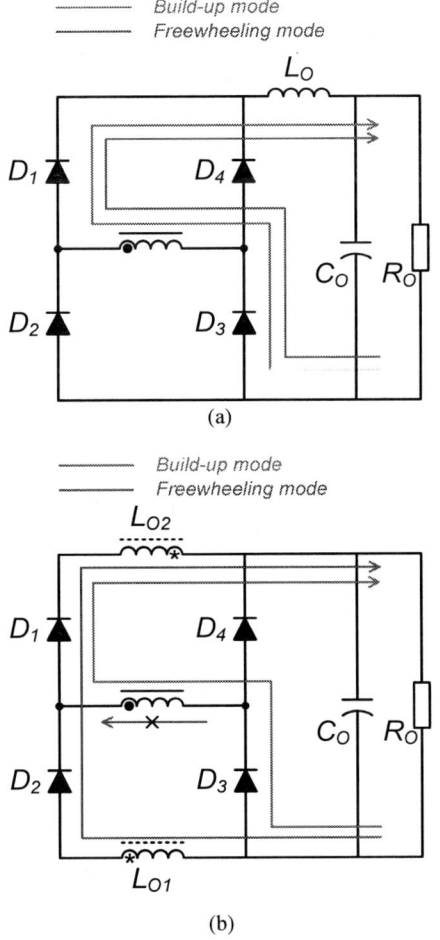

(a)

(b)

Fig. 3. The current path of rectifier. (a) Conventional PSFB converter and (b) proposed converter.

In this paper, new PSFB converter which eliminates primary circulating current and voltage resonance of rectifier diodes using coupled output inductor is proposed. The proposed PSFB converter removes primary circulating current by changing output current path during a freewheeling period of output inductor. Also, since two rectifier diodes are connected to an output capacitor, the proposed converter eliminates the voltage ringing of them. Therefore, the proposed converter achieves high efficiency by reducing conduction loss and by using high performance rectifier diodes.

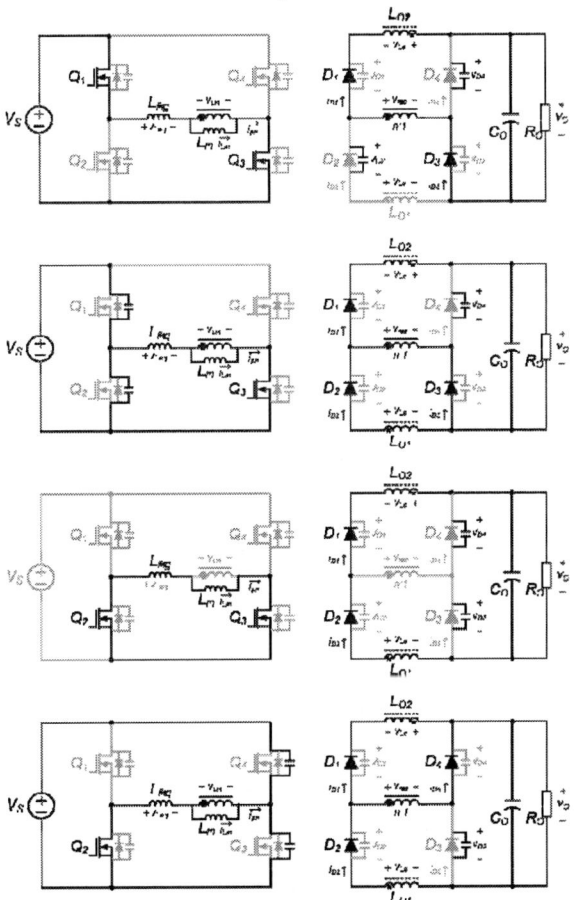

Fig. 4. Circuit diagrams of the proposed converter.

II. CONCEPT OF THE PROPOSED CONVERTER

Fig. 2 shows the proposed PSFB converter. As shown in figure, the proposed converter has coupled inductor located between the rectifier diode D_1 and D_3, D_2 and D_4. Since the D_3 and D_4 are connected directly to the output capacitor, the voltage ringing does not occur. As a result, the low voltage rating diode which has high performance can be used as D_3 and D_4. Also, by eliminating the snubber circuit, simplicity and efficiency are improved compared to the conventional PSFB converter.

Another advantage of the proposed converter is an elimination of the circulating current. As shown in Fig. 3(a), the conventional PSFB converter has circulating current in primary side since the output current flows to secondary side of transformer during a freewheeling period of output inductor. However, as shown in Fig. 3(b), because the output current path is changed from D_3 to D_2 in the proposed converter, output current does not flow to the secondary winding of the transformer during the freewheeling period. Therefore, the proposed converter reduces conduction losses at primary switch and transformer winding by decreasing RMS current.

4055

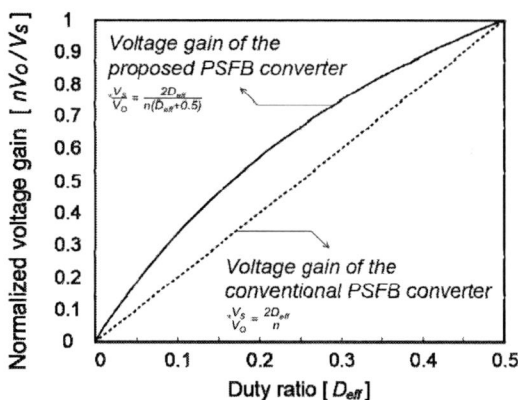

Fig. 5. The voltage gain graph of the conventional PSFB converter and the proposed converter.

TABLE I
DESIGN EXAMPLES OF THE PROTOTYPE CONVERTERS

	Conventional PSFB converter	Proposed PSFB converter
Nominal duty-ratio	0.367	0.306
Leading-leg switches	IPP60R160 (600V, 160mΩ)	
Lagging-leg switches	IPP60R280 (600V, 280mΩ)	
Transformer	PQ3225 (1200μH, 34:7)	PQ3225 (600μH, 35:7)
ZVS range	Over 50% load condition	Entire load condition
V_{D1}, V_{D2}	120	150
V_{D3}, V_{D4}	120	56
Diode 1	MBRF20H150CT (150V, 0.8V_F)	MBR20200CT (200V, 0.9V_F)
Diode 2	MBRF20H150CT (150V, 0.8V_F)	MBR20L80CT (80V, 0.56V_F)
Output inductor	PQ2620 (15μH, 1*2Φ)	PQ2620 (15μH, 1.2Φ, 1.2Φ)

III. OPERATIONAL PRINCIPLES

Fig. 4 shows circuit diagrams to understand operational principles of the proposed converter. The proposed converter has four operating modes and each mode is analyzed in detail. For the simple analysis, there are some assumptions as follows:

1) The magnetizing inductance L_m of the transformer is much larger than the leakage inductance L_{lkg}.
2) Resonances occur by parasitic components are ignored except between leakage inductance L_{lkg} and parasitic capacitor of V_{D1} and V_{D2}.
3) The turns-ratio of the coupled inductor is 1:1.
4) The coupling coefficient of the L_{O1} and L_{O2} is 1.

Mode 1 [t_0-t_1]: Mode 1 starts when the current commutation between D_1 and D_2 ends. Input voltage V_S is applied to the magnetizing inductance L_m of the transformer and it is transferred to a rectifier as V_S/n. Accordingly, V_S/n-V_O is applied to the output inductor L_{O2} and the L_{O2} builds-up the current though the D_1 and D_3. During this mode, the parasitic capacitance of D_2 resonates with L_{lkg} and V_{D2} has voltage ringing which is the same as that of the conventional PSFB converter. On the other hand, as shown in Fig. 4, since the parasitic capacitance of D_4 is connected in parallel with C_O, V_{D4} is clamped to V_O. As a result, V_{D4} has low voltage stress and snubber circuit for D_4 is not needed.

Mode 2 [t_1-t_2]: This mode starts as Q_1 is turned off. As the Q_1 and Q_2 start to be charged and discharged respectively, the V_{Lm} and V_{rec} also decrease together. When the V_{rec} reaches to $V_O/2$, the D_2 is conducted and V_{rec} is clamped to $V_O/2$. So the V_{Lm} is clamped to $nV_O/2$, and $-nV_O/2$ is applied to the L_{lkg}. Since the large voltage is applied to small inductance, i_{pri} decreases drastically and output current path is change from D_3 to D_2.

Mode 3 [t_2-t_3]: Mode 3 starts when i_{pri} becomes the same as i_{Lm}. Since D_4 is turned off, output inductor demagnetizes the output current with a slope of the $V_O/(L_{O1}+L_{O2})$ through D_1 and D_2. Therefore, output current doesn't flow through the secondary winding of the

transformer and only i_{Lm} flows in primary side as shown in figure.

Mode 4 [t_3-t_4]: Mode 4 starts when the Q_3 is turned off. As the Q_3 and Q_4 start to be charged and discharged respectively, the V_{Lm} and V_{rec} also decrease together.

When the V_{rec} reaches to $-V_O/2$, the D_4 is conducted and V_{rec} is clamped to $-V_O/2$. So the V_{Lm} is clamped to $-nV_O/2$, and V_S-$nV_O/2$ is applied to the L_{lkg}. Since the large voltage is applied to L_{lkg}, the primary current decreases with a slope of $(V_S$-$nV_O)/L_{lkg}$ and output current path is change from D_1 to D_4. Also, V_{D3} is clamped to V_O during this period similar with mode 1.

IV. ANALYSIS OF THE PROPOSED CONVERTER

A. Voltage gain

An output inductor of the proposed PSFB converter builds-up the current with a slope of $(V_S/n$-$V_O)/L_O$ during the powering mode which is the same as the conventional PSFB converter. But during a freewheeling period, the output inductor demagnetizes the current with a slope of $-V_O/(L_{O1}+L_{O2})$ while the conventional PSFB converter demagnetizes the output current with a slope of $-V_O/L_O$. As a result, it has different voltage gain compared to the conventional PSFB converter. To obtain the voltage gain, applying voltage-second balance to output inductor leads the following equations:

$$D_{eff} \cdot \left(\frac{V_S}{n} - V_O \right) + \left(0.5 - D_{eff} \right) \cdot \left(-\frac{V_O}{2} \right) = 0 \tag{1}$$

$$\frac{V_O}{V_S} = \frac{2D_{eff}}{n\left(D_{eff} + 0.5 \right)} \tag{2}$$

, where D_{eff} is the effective duty-ratio.

The 2018 International Power Electronics Conference

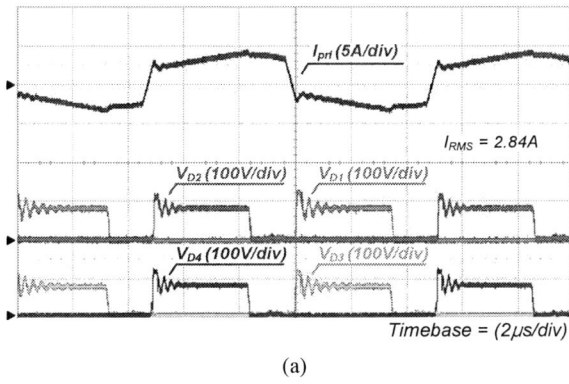

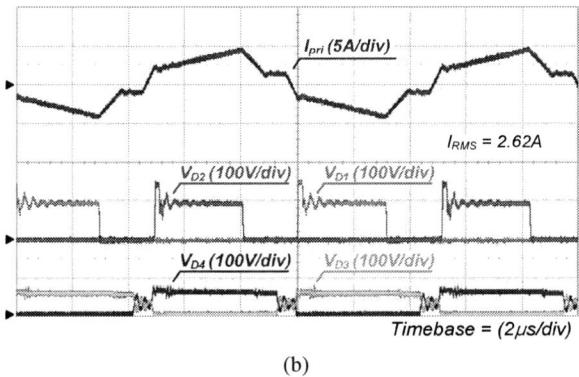

Fig. 6. Key waveforms at 100% load condition. (a) Conventional PSFB converter, (b) proposed PSFB converter.

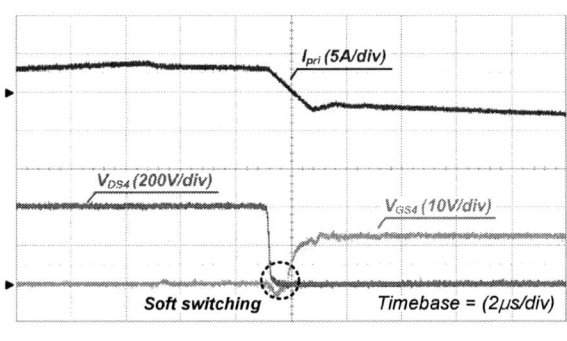

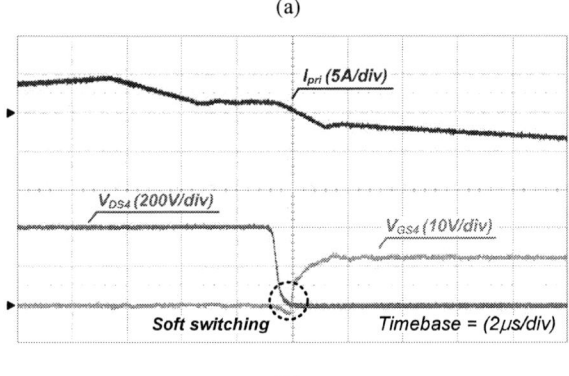

Fig. 7. ZVS operation at 100% load condition. (a) Conventional PSFB converter, (b) proposed PSFB converter.

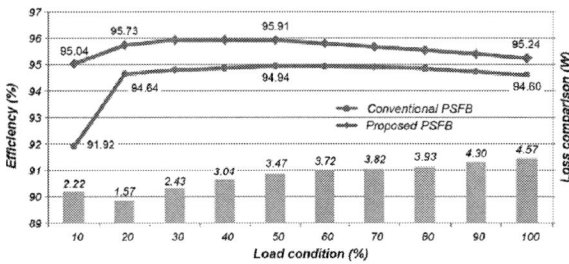

Fig. 8. The measured efficiency of the prototype converters.

As shown in (2), since the proposed converter has a $(D_{eff}+0.5)$ term in denominator, the voltage gain is same when $D_{eff}=0.5$ and larger when $D_{eff}<0.5$. And Fig. 5 shows the normalized gain graph of the proposed converter and conventional PSFB.

B. Voltage stresses of the rectifier diodes

As mentioned before, V_{D3} and V_{D4} are clamped to V_O since the D_3 and D_4 are connected directly to the output capacitor. Accordingly, not only the low voltage rating diode can be used for D_3 and D_4, but also the snubber circuits to relieve the voltage ringing can be eliminated. However, in case of the D_1 and D_2, voltage stress slightly increases compared to the conventional PSFB converter. This is because $V_{LO}+V_{rec}$ is applied to D_1 and D_2 during D_{eff} while V_{rec} is applied to D_1 and D_2 in conventional PSFB converter. V_{D1} of the proposed converter is determined as follows when Q_1 and Q_3 are turned on:

$$V_{D1}=V_{rec}+V_{LO}=\frac{V_S}{n}+\left(\frac{V_S}{n}-V_O\right)=\frac{2V_S}{n}-V_O \qquad (3)$$

, where V_{D2} is also determined as $2V_S/n - V_O$ when Q_2 and Q_4 are turned on.

V. EXPERIMENTAL RESULTS

To verify the effect and feasibility, prototype converters are experimented with a *320-400V* input and *56V/12.8A* output specification. Design parameters are listed in Table I for the prototype converters. Since the proposed converter has smaller ZVS energy at heavy load, the L_m is designed smaller compared with the conventional PSFB converter. Also, because D_1 and D_2 of the proposed converter have larger voltage stress, and D_3 and D_4 have smaller voltage stress compared with the conventional PSFB converter, each diode is selected with 30% margin.

Fig. 6 is key waveforms of the prototype converters at 100% load condition. Since the conventional PSFB converter has primary circulating current, it has RMS current as shown in Fig. 6(a). On the contrary, the proposed converter has small primary RMS current because it doesn't have circulating current by using a new rectifier structure as shown in Fig. 6(b). Also, since D_3 and D_4 of the proposed converter doesn't have voltage ringing as mentioned before, V_{D3} and V_{D4} of the proposed converter are small compared to the conventional PSFB converter.

4057

Fig. 7 shows the ZVS operation of prototype converters at 100% load. As shown in Fig. 7(a), the conventional PSFB converter easily achieves soft switching operation since the ZVS energy increases along the load condition. Fig. 7(b) is a ZVS operation of proposed PSFB converter. Since it uses I_{Lm} for ZVS, it has small ZVS energy than conventional PSFB in heavy load. But, since the conventional PSFB converter has small ZVS energy in light load condition, it can't achieve ZVS below 50% load condition while the proposed converter achieves soft switching in overall load condition.

Fig. 8 is a efficiency of the conventional PSFB converter and the proposed PSFB converter in overall load condition. The proposed converter has higher efficiency in heavy load because it reduces the primary RMS current and uses low voltage rating diodes. Also, since the proposed converter not only eliminates the snubber loss of the D_3 and D_4, but also achieves ZVS operation, it has high efficiency in light load condition. Therefore, the proposed converter has higher efficiency than the conventional PSFB converter in entire load condition.

VI. Conclusion

In this paper, a PSFB converter with coupled output inductor is proposed. Because the freewheeling current of the output inductor does not flow to the secondary side of transformer, the proposed converter doesn't have the circulating current in primary side. Also, since the rectifier diode D_3 and D_4 are connected to the output capacitor, voltage ringing is eliminated. Therefore, it reduces the primary RMS current and voltage stress of rectifier diodes. Furthermore, the proposed converter is implemented easily by changing the rectifier structure with coupled output inductor without any additional devices or control techniques.

Acknowledgment

This work was supported by the National Research Foundation of Korea (NRF) grant funded by the Korea government (MSIP) (No. 2016R1A2B2010328).

References

[1] H. Wang, Y. Chen, P. Fang, Y. F. Liu, J. Afsharian, and Z. Yang, "An LLC Converter Family With Auxiliary Switch for Hold-Up Mode Operation," *IEEE Trans. Power Electron.*, vol. 32, no. 6, pp. 4291-4306, Jun. 2017.

[2] Y. S. Lai, Z. J. Su, and W. S. Chen, "New Hybrid Control Technique to Improve Light Load Efficiency While Meetinf the Hold-up Time Requirement for Two-Stage Server Power," *IEEE Trans. Power Electron.*, vol. 29, no. 9, pp. 4763-4775, Sep. 2014.

[3] H. Wang, H. S. Chung, and W. Liu, "Use of a Series Voltage Compensator for Reduction of the DC-Link Capacitance in a Capacitor-Supported System," *IEEE Trans. Power Electron.*, vol. 29, no. 3, pp. 1163-1175, Mar. 2014.

[4] J. I. Baek, J. K. Kim, J. B. Lee, H. S. Youn, and G. W. Moon, "A Boost PFC Stage Utilized as Half-Bridge Converter for High-Efficiency DC-DC Stage in Power Supply Unit," *IEEE Trans. Power Electron.*, vol. 32, no. 10, pp. 7449-7457, Oct. 2017.

[5] W. Li, S. Zong, F. Liu, H. Yang, X. He, and B. Wu, "Secondary-Side Phase-Shift-Controlled ZVS DC/DC Converter With Wide Voltage Gain for High Input Voltage Application," *IEEE Trans. Power Electron.*, vol. 28, no. 11, pp. 5128-5139, Nov. 2013.

[6] D. K. Kim, S. Moon, C. O. Yeon, G. W. Moon "High-Efficiency LLC Resonant Converter With High Voltage Gain Using an Auxiliary LC Resonant Circuit," *IEEE Trans. Power Electron.*, vol. 31, no. 10, pp. 6901-6909, Oct. 2016.

[7] G. D. Capua, S. A. Shirsavar, M. A. Hallworth, and N. Femia, "An Enhanced Model for Small-Signal Analysis of the Phase-Shifted Full-Bridge Converter," *IEEE Trans. Power Electron.*, vol. 30, no. 3, pp. 1567-1576, Mar. 2015.

[8] G. N. B. Yadav, and N. L. Narasamma, "An Active Soft Switched Phase-Shifted Full-Bridge DC-DC Converter: Analysis, Modeling, Design, and Implementation," *IEEE Trans. Power Electron.*, vol. 29, no. 9, pp. 4538-4550, Sep. 2014.

[9] D. Y. Kim, C. E. Kim, G. W. Moon, "Variable Delay Time Method in the Phase-Shifted Full-Bridge Converter for Reduced Power Consumption Under Light Load Conditions," *IEEE Trans. Power Electron.*, vol. 28, no. 11, pp. 5120-5127, Nov. 2013.

[10] A. Mallik, and A. Khaligh, "Variable-Switching-Frequency State-Feedback Control of a Phase-Shifted Full-Bridge DC/DC Converter," *IEEE Trans. Power Electron.*, vol. 32, no. 8, pp. 6523-6531, Aug. 2017.

[11] J. W. Kim, D. Y. Kim, C. E. Kim, and G. W. Moon, "A Simple Switching Control Technique for Improving Light Load Efficiency in a Phase-Shifted Full-Bridge Converter With a Server Power System," *IEEE Trans. Power Electron.*, vol. 29, no. 4, pp. 1562-1566, Apr. 2014.

[12] K. Shi, D. Zhang, Z. Zhou, M. Zhang, D. Zhang, and Y. Gu, "A Novel Phase-Shift Dual Full-Bridge Converter With Full Soft-Switching Range and Wide Conversion Range," *IEEE Trans. Power Electron.*, vol. 31, no. 11, pp. 7747-7760, Nov. 2016.

[13] J. H. Kim, I. O. Lee, and G. W. Moon, "Integrated Dual Full-Bridge Converter With Current-Doubler Rectifier for EV Charger," *IEEE Trans. Power Electron.*, vol. 31, no. 2, pp. 942-951, Feb. 2016.

[14] W. J. Lee, K. B. Park, T. W. Heo, and G. W. Moon, "Output Inductor Less Phase Shift Full Bridge Converter With Current Stress Reduction Technique for Server Power Application," in *Proc. IEEE Power Electron. Specialists Conf.*, pp. 2517-2522, Jun. 2008.

[15] X. Wu, H. Chen, J. Zhang, F. Peng, and Z. Qian, "Interleaved Phase-Shift Full-Bridge Converter With Transformer Winding Series-Parallel Autoregulated (SPAR) Current Doubler Rectifier," *IEEE Trans. Power Electron.*, vol. 30, no. 9, pp. 4864-4873, Sep. 2015.

[16] Y. D. Kim, K. M. Cho, D. Y. Kim, and G. W. Moon, "Wide-Range ZVS Phase-Shift Full-Bridge Converter With Reduced Conduction Loss Caused by Circulating Current," *IEEE Trans. Power Electron.*, vol. 28, no. 7, pp. 3308-3316, Jul. 2013.

A Bi-Directional Current Detection Using Current Transformers for Bi-Directional DC-DC Converter

Seiji Iyasu[1*], Yuji Hahashi[1], Yuuichi Handa[2], Kimikazu Nakamura[2], Keiji Wada[3]

1 Research DIV.3, SOKEN,INC., Nisshin, Japan
2 Electrification Components ENG.DIV.1, DENSO Corpolation, Kariya, Japan
3 Department of Electrical Engineerring, Tokyo Metropolitan University, Hachioji, Japan
*E-mail: seiji_iyasu@soken1.denso.co.jp

Abstract—This paper proposes a bi-directional current detection circuit using current-transformers for bi-directional DC-DC converters. This current detection circuit consists of CT(Current transformer) and electric components. The features of the current sensor based on CT can detect a pulse current contain DC component without any magnetic saturation. This paper shows a circuit configuration of the bi-directional current sensor and the principle of detection error of this circuit is analyzed. The validity of this analysis is verified through experiment under the DC-DC converter rated at 1.4 kW.

I. INTRODUCTION

In recent years, electrically driven vehicles such as PHVs (Plug-in Hybrid Vehicles) and EVs (Electric Vehicles) will be widely spread more and more in the world[1]. Furthermore, V2G (Vehicle to Grid) and V2H (Vehicle to Home) systems will also be discussed for electrical energy management[2]. In order to realize V2G and V2H functions, it is necessary that power converters such as On-board chargers and DC-DC converters have to be operated a bi-directional power flow control. As for bi-directional DC-DC converters, many papers have been reported regard to high performance features compared with dual active bridge (DAB) DC-DC converters and current-fed bi-directional DC-DC converters[3][4]. However, these papers discuss only the circuit topologies and their control, few paper of on control methods including bi-direction current detection technique have been reported.

Due to development of power semiconductor devices, the switching frequencies of power converters are increasing, and then it is important to detect the current waveforms in a wide frequency bandwidth for the power converter circuit. Current detection circuits such as current-transformers (CT), Rogowski coils, MR (Magneto Resistive) elements, shunt resistors, and Hall elements are applied to power converters[5]. The CT is widely used to detect AC current including pulse current, and it can be realized an isolation function between power circuits to control circuit. Normally the frequency characteristics of the CT are restricted by the magnetic materials, but it is possible to obtain frequency bandwidth of over MHz width[5][6]. In the case of using Rogowski coil, it is necessary to perform integral treatment in order to reproduce current waveforms. Therefore, the Rogowski coil is generally used for a current waveform measurement not for current control[7][8]. Currently, various thin film elements such as AMR and GMR are developed as

next generation current detection methods[9][10]. Meanwhile, MR-based elements are said to have an issue in that they are easily affected by external magnetic fields[5]. Moreover, in the case of using a shunt resistor, the frequency characteristics are restricted by the series inductance of the resistor and the bandwidth of the amplifier. Therefore, the frequency bandwidth of the current detection system using a shunt resistor is generally less than several hundred kHz[11]. Although the current detection using a shunt resistor in the 20 MHz width has also been reported, its application is limited due to no isolation function[12]. Furthermore, it has been reported that the bandwidth of the Hall element is about several hundred kHz due to delay of the amplifier and the magnetization delay of the magnetic core[5][6].

This paper proposes a bi-directional current detection circuit using a CT for current-fed bi-directional DC-DC converters. The bi-directional current detection circuit consists of a CT and additional circuit which is diode rectifier circuit and a MOSFET. It is capable of realizing bi-directional pulse current detection and CT degaussing , simultaneously. As a result, the current sensor based on CT can detect the pulse current including DC component without any magnetic saturation. It is shown that current detection is feasible in bi-directional electric power conversion operations. The experimental results, rated at 150 V, 1.4 kW bi-directional DC-DC converter, is shown the validity of the current detection circuit. It will be shown that bi-directional current detection is feasible and bi-directional DC-DC converter can be operated by current feedback control using the proposed detection circuit.

II. BI-DIRECTIONAL DC-DC CONVERTER

Fig. 1 shows the bi-directional current-fed DC-DC converter. The DC voltage of higher voltage side is 150 V, and that of lower voltage side is 14 V. In the case of a DC-DC converter with over 10 times voltage ratio, it is reported that the conversion efficiency of the current-fed DC-DC converter is higher than the one of the dual active ridge (DAB) converter due to the influence of the conduction loss[4]. Fig. 2 shows the control block under the forward power direction of power conversion. In order to suppress the magnetic bias of the transformer of the current-fed DC-DC converter, a peak current mode control in which the high side current i_{Hi} is detected is performed under the forward power direction of power

The 2018 International Power Electronics Conference

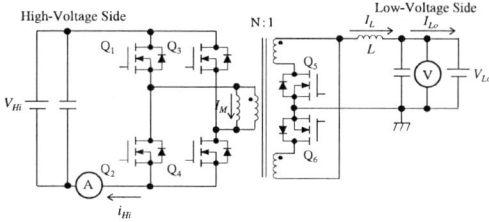

Fig. 1. Bi-Directional Current-Fed DC-DC converter

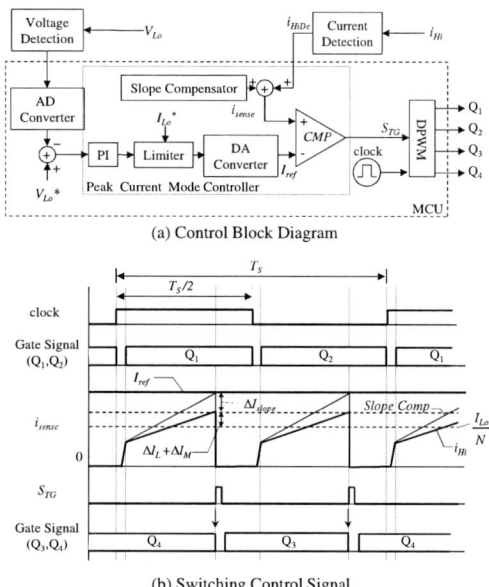

(a) Control Block Diagram

(b) Switching Control Signal

Fig. 2. Peak Current Mode Control under Forward Direction of Power Conversion

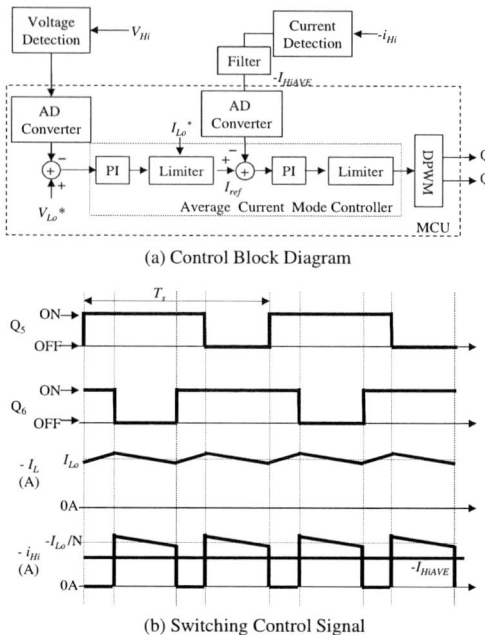

(a) Control Block Diagram

(b) Switching Control Signal

Fig. 3. Control Block under Backward Direction of Power Conversion

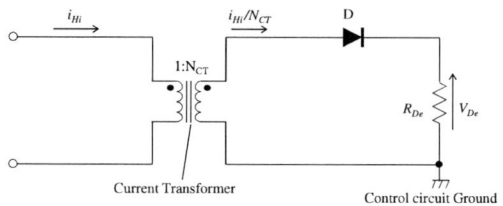

Fig. 4. Conventional Current Detection Circuit

conversion as shown in Fig. 2[13]. The high side i_{Hi} is the pulse current including DC component. In this paper, the switching frequency of the DC-DC converter is set to 100 kHz under forward direction of power conversion. Therefore, it is necessary to detect the high side input current i_{Hi} with the MHz frequency bandwidth.

Fig. 3 shows the control block under the backward power direction of power conversion. The average current mode control in which the average high side current I_{HiAVE} is detected is performed under the backward power direction of power conversion.

III. BI-DIRECTIONAL CURRENT DETECTION CIRCUIT USING CURRENT-TRANSFORMER

A. Conventional current detection circuit using CT

Fig. 4 shows a conventional unidirectional current detection circuit using CT[14]. This circuit is capable of detecting current when the high side current i_{Hi} has a positive polarity. Therefore, this circuit is classified as a unidirectional current

detection circuit. When the high side current i_{Hi} has a positive polarity and the resistor voltage v_{De} corresponding to the high side current i_{Hi}. Furthermore, when the high side current i_{Hi} is not being conducted, the diode D is turned off. In this case, the magnetizing current in the CT is degaussed. By combining two this circuits, bi-directional current detection can be realized[14]. If two CTs connect in the circuit, it is not suitable for miniaturization of the circuit.

B. Bi-directional current detection circuit

Fig. 5 shows a proposed bi-directional current detection circuit. The circuit is configured in a single bridge rectifier circuit, MOSFET Q_{De} and the resistor R_{De}. The high side current i_{Hi} is rectified by the single bridge rectifier circuit, regardless of the polarity of the i_{Hi}. The absolute value of the i_{Hi} flows to the resistor R_{De}. The MOSFET Q_{De} of the detected circuit is installed for degaussing of CT. The voltage and current rating of the MOSFET Q_{De} is much smaller than that of DC-DC converter. When the high side current i_{Hi} is not being conducted, the MOSFET Q_{De} is turned off. By the

4060

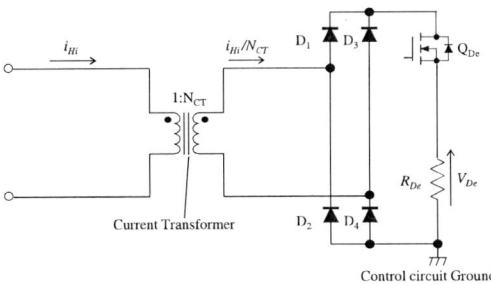

Fig. 5.   Proposed Current Detection Circuit

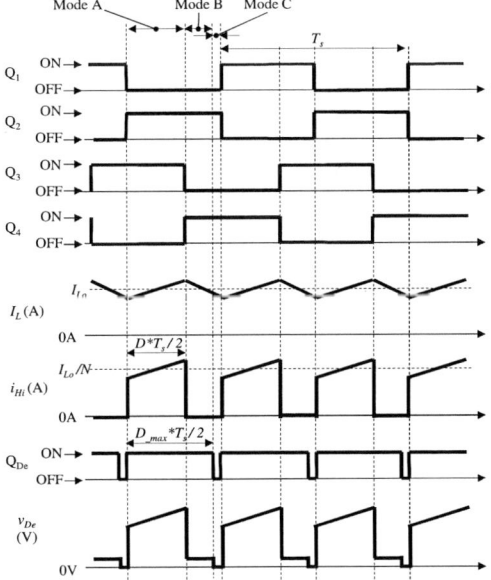

Fig. 6.   Control Signal under Forward Direction of Power Conversion

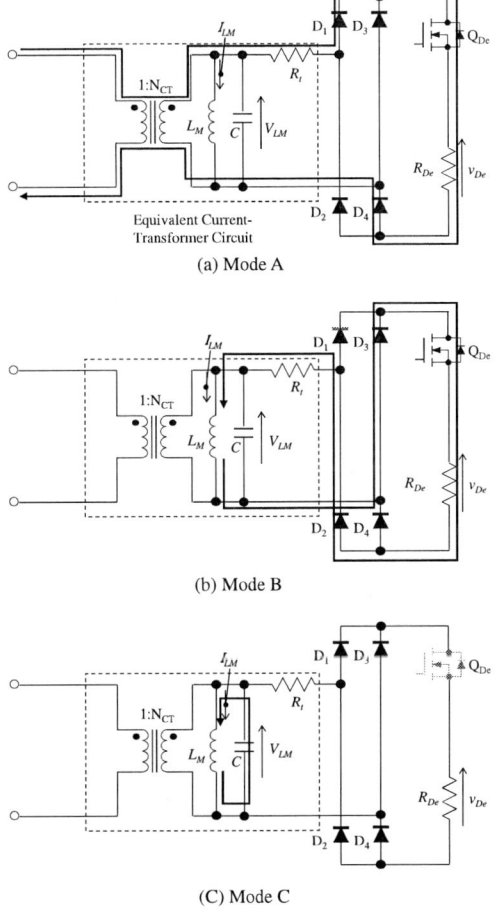

(a) Mode A

(b) Mode B

(C) Mode C

Fig. 7.   Current Path under Forward Direction of Power Conversion

turn OFF operation of Q_{De}, the secondary side impedance of CT can be increased. As a result, the magnetizing current in the CT is degaussed without any saturation of the magnetic material. On the basis of the above mentioned, this current detection circuit is capable of realizing bi-directional current detection.

C. Operation principle under forward direction of power conversion

Fig. 6 shows the control signals under forward direction of power conversion. The switching frequency of Q_{De} is set to twice the $Q_1 - Q_4$ of the DC-DC converter. The turn-on time T_{on} and the turn-off time T_{off} of Q_{De} are defined, as shown in (1) and (2), respectively.

$$T_{on} = \frac{1}{2}D_{max}T_s \qquad (1)$$

$$T_{off} = \frac{T_s}{2}(1 - D_{max}) \qquad (2)$$

Here, T_s is the switching period, and D_{max} is the maximum current conduction ratio of i_{Hi}. If Q_{De} is turned off, the current path of I_{in} is cut off. Therefore, it is necessary to be sure to turn on Q_{De} when I_{in} is being conducted. Hence, the turn-on timing Q_{De} is made to coincide with the switching timing of Q_1 and Q_2. In the case of peak current mode control, the turn-on time T_{on} is made to coincide with $\frac{1}{2}D_{max}T_s$ which is the maximum conduction time of i_{Hi}.

Fig. 7 shows the current path of this current detection circuit under forward direction of power conversion. In order to discuss the circuit operation in detail, an equivalent circuit of the current detection have to be considered as the magnetizing inductance L_M, the parasitic capacitance C between the windings, and the coil resistance R_t, respectively. In the case of discussing the degaussing of the CT, both the magnetizing current I_{LM} and the inter-coil capacitance C serve for restricting the secondary side impedance of the CT.

In Mode A shown in Fig. 7, the high side current i_{Hi} is converted by the CT to a level $1/N_{CT}$ and flow into the resistor R_{De}. As a result, a detected voltage v_{De} corresponding to the

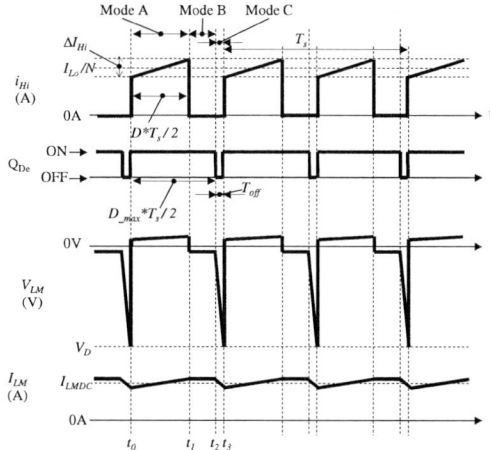

Fig. 8. Waveforms of Magnetizing Current and Voltage under Forward Direction of Power Conversion

high side current i_{Hi} is generated. The ideal detection voltage $v_{Deideal}$ can be expressed by the following equation.

$$v_{Deideal} = \frac{|i_{Hi}|R_{De}}{N_{CT}} \quad (3)$$

In Mode B shown in Fig. 7, the magnetizing current I_{LM} in the CT flows into the resistor R_{De}. In mode C shown in Fig. 7, Q_{De} is turned off. Therefore, the magnetizing current I_{LM} in the CT flows into the inter-coil capacitance C. In mode C, a negative voltage is applied to the magnetizing inductance L_M, and the magnetizing current I_{LM} can be degaussed. When the maximum duty ratio D_{max} takes a large value such as over 0.9 , the DC component is superimposed on I_{LM} due to lack of degaussing time. As a result, the magnetizing current I_{LM} has to be considered for the detected voltage of R_{De}. The detection error ΔI is caused by I_{LM} of this sensor.

$$\Delta I = |i_{Hi}| - i_{Hi_{De}} = |I_{LM}|N_{CT} \quad (4)$$

Here, the detection current $i_{Hi_{De}}$ is given by the following equation.

$$i_{Hi_{De}} = \frac{v_{De}}{R_{De}N_{CT}} \quad (5)$$

The magnetizing current I_{LM} is analyzed on based on converter operations and the equivalent circuit as shown in Fig. 8. The magnetizing current I_{LM} strictly includes ripple current as shown in 8, but this analysis ignores the ripple current and approximate the magnetizing current I_{LM} by the following equation.

$$I_{LM} \approx I_{LMDC} \quad (6)$$

In a steady state, the sum of the ET(voltage-time) product in each mode must be zero. Therefore, the following equation can be obtained.

$$ET_{modeA} + ET_{modeB} + ET_{modeC} = 0 \quad (7)$$

The ET product ET_{modeA} in Mode A can be expressed by the following equation.

$$ET_{modeA} = ((\frac{i_{Hi}}{N_{CT}} - I_{LM})R_{all} + 2V_F)\frac{DT_s}{2} \quad (8)$$

Here, D is the current conduction ratio of the high side current i_{Hi}. V_F is the forward voltage of D_1, D_2, D_3, D_4. R_{all} is defined by the following equation.

$$R_{all} = R_{De} + R_{Q_{De}} + R_t \quad (9)$$

$R_{Q_{De}}$ and R_t are the ON resistance of MOSFET and the winding resistance of CT, respectively. The current conduction ratio D can be approximated by the following equation using the high side voltage V_{Hi}, the low side voltage V_{Lo}, transformer turn number N, power conversion efficiency η.

$$D \approx \frac{NV_{Lo}}{\eta V_{Hi}} \quad (10)$$

Ignoring ripple current Δi_{Hi} of high side current i_{Hi}, i_{Hi} can be approximated by the following equation.

$$i_{Hi} \approx \frac{I_{Lo}}{N} \quad (11)$$

Substituting (11) into (8) yields the following equation.

$$ET_{modeA} = ((\frac{I_{Lo}}{NN_{CT}} - I_{LM})R_{all} + 2V_F)\frac{DT_s}{2} \quad (12)$$

The ET product ET_{modeB} in Mode B can be expressed by the following equation.

$$ET_{modeB} = (I_{LM}R_{all} + 2V_F)(\frac{1 - DT_s}{2} - T_{off}) \quad (13)$$

The ET product ET_{modeC} in Mode C can be expressed by the following equation.

$$ET_{modeC} = -\frac{1}{C}\left(\int_{t2}^{t3} I_{LM}\,dt - (I_{LM}R_{all} + 2V_F)\right)$$
$$(\frac{DT_s}{2} - D_{max}T_s)$$
$$= -\frac{1}{2C}I_{LM}(\frac{DT_s}{2} - D_{max}T_s)^2 \quad (14)$$

From (7), (12), (13) and (14), the magnetizing current I_{LM} can be expressed by the following equation.

$$I_{LM} = \frac{\frac{I_{Lo}}{N_{CT}N}(R_{all})\frac{DT_s}{2} + 2V_F(DT_s - \frac{T_s}{2} + T_{off})}{(R_{all})(\frac{DT_s}{2} - T_{off}) + \frac{T_{off}^2}{2C}} \quad (15)$$

Hence, the detection error ΔI can be expressed by the following equation.

$$\Delta I = \frac{\frac{I_{Lo}}{N}(R_{all})\frac{DT_s}{2} + 2N_{CT}V_F(DT_s - \frac{T_s}{2} + T_{off})}{(R_{all})(\frac{DT_s}{2} - T_{off}) + \frac{T_{off}^2}{2C}} \quad (16)$$

D. Operation principle under backward direction of power conversion

The basic operation principle of the current sensor for a forward direction and that for backward direction of power conversions are the same. The operating principle under backward direction of power conversion is as follows. Fig. 9 shows the control signals under backward direction of power conversion. The switching frequency of Q_{De} is set to twice the Q_5, Q_6 of the DC-DC converter. Since the current control is controlled by the average current mode, the turn-on and the turn-off timing of Q_{De} are made to coincide with the turn-off and the turn-on timing of Q_5, Q_6, respectively. The turn-on time T_{on} and the turn-off time T_{off} of Q_{De} are defined, as shown in (17) and (18), respectively.

$$T_{on} = \frac{1}{2}DT_s \qquad (17)$$

$$T_{off} = \frac{1}{2}(1-D)T_s \qquad (18)$$

Here, $(1-D)$ is the current conduction ratio of the high side current i_{Hi}. The turn-off time T_{off} under the backward direction of power conversion is longer than that under the forward direction of power conversion. Then, the sufficient degaussing time can be secured under the backward direction of power conversion. Therefore, the detection error ΔI caused by the magnetizing current is expected to be smaller than that under the forward direction of the power conversion. By the analysis based on the ET product as in the forward direction, the detection error ΔI can be expressed by the following equation.

$$\Delta I = \frac{T_s(1-D)(2V_F + \frac{|I_{Lo}|R_{all}}{NN_{CT}})}{\frac{D^2T_s^2}{4C} + R_{all}T_s(1-D)} \qquad (19)$$

Here, the duty ratio D can be approximated by the following equation.

$$D \approx \frac{V_{Hi} - NV_{Lo}}{\eta V_{Hi}} \qquad (20)$$

IV. EXPERIMENTAL RESULTS

A. Experimental setup

Fig. 10 shows a bi-directional DC-DC converter equipped with the proposed circuit, rated at 1.4 kW. The control blocks as shown in Fig. 2 and Fig. 3 are also equipped. Table 1 shows circuit parameters of the experiment. The resistance R_{De} and the number of turns N_{CT} are adjusted so that the detection voltage v_{De} becomes 5 V when the high side current i_{Hi} is 45 A or −45 A. For measurement of the high side current i_{Hi}, Iwatsu current probe SS 281-A is used. The accuracy of SS 281-A is 2%.

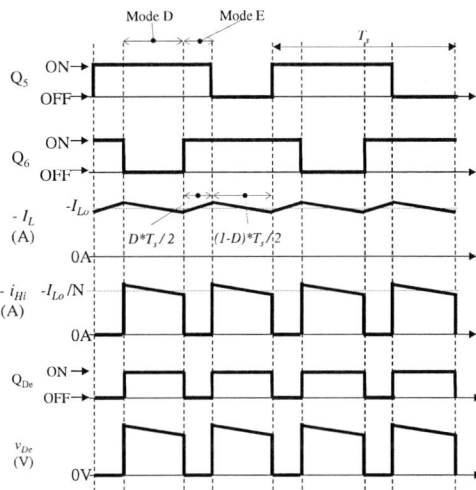

Fig. 9. Control Signal under Backward Direction of Power Conversion

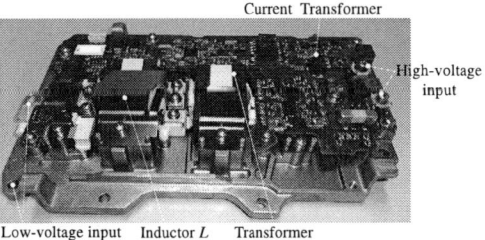

Fig. 10. Prototype Bi-Directional DC-DC Converter rated at 1.4 kW

B. Forward Direction of Power Conversion

Fig. 11 shows the experimental waveforms of the high side current i_{Hi} and the detected current i_{HiDe}. The high side voltage V_{Hi} and the low side current I_{Lo} are 150 V and 100 A, respectively. In Mode A, the high side current i_{Hi} and the detected current i_{HiDe} are similar waveforms. The high-frequency components are superimposed on i_{HiDe} at the beginning of Mode A and at the end of Mode B. This components are due to the charge / discharge current of

TABLE I. SPECIFICATIONS OF EXPERIMENT

High Side Voltage, V_{Hi}	150 V
Low Side Voltage, V_{Lo}	14 V
Low Side Current, I_o	−100A ~ 100 A
Turn Number, N	7
Conversion Efficiency, η	0.86
CT Turn Number, N_{CT}	200
Q_{De} Turn on time, T_{on} under forward direction	4.6 μs
Q_{De} Turn off time, T_{off} under forward direction	400 ns
Switching Frequency, f_s under forward direction	100 kHz
Switching Frequency, f_s under backward direction	83.3 kHz
Detect Resistance, R_{De}	22.2 Ω
CT Resistance, R_t	11.2 Ω
CT Capacitance, C	120 pF
Diode ON-Voltage, V_F	0.7 V
Q_{De} ON-Resistance, R_{QDe}	3.2 Ω

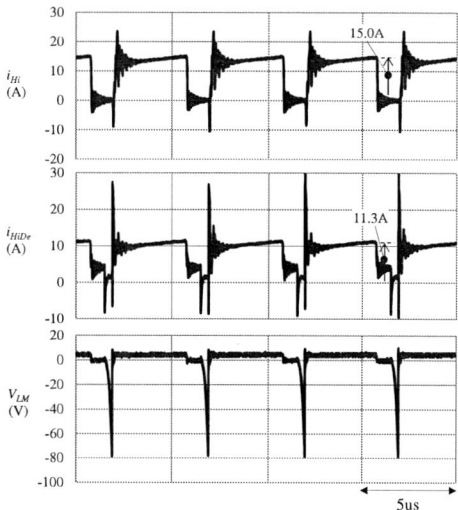

Fig. 11. Experimental Waveforms under Forward Direction $I_{Lo} = 100A$

the parasitic capacitance of Q_{De}. However, the peak current mode control detects only the peak value at the end of Mode A, so the high-frequency components may not be influenced to the current control. Therefore, the desired current detection is achieved. The peak value of the high side current i_{Hi} and that of the detected current $i_{Hi_{De}}$ are 15.0 A and 11.3 A, respectively, hence the relative error is about 25%. Fig. 12 shows a comparison of detection errors ΔI between the experimental and the calculation results. The detection errors ΔI coincided with an error of 0.5 A. This residual error is caused by neglecting the frequency characteristic of the winding resistance of CT and the ripple current of the magnetizing current I_{LM}. These results in Fig. 12 shows that detection errors ΔI can be estimated by (16) in advance. That is, detected errors can be compensated for by the following procedure.

1) Estimated errors ΔI^* can be obtained by substituting the low side output current demand I_{Lo}^* into the low side current I_{Lo} in equation (16)

2) a control algorithm for correcting the current demand I_{ref} for the peak current mode control by the estimated error ΔI^* should be installed into a digital microcontroller

For example, the relative error for constant current control can be reduced from 25 % to 3 % when the low side output current command demand I_{Lo}^* is 100 A.

C. Backward Direction of Power Conversion

Fig. 13 shows the experimental waveforms of the high side current i_{Hi} and the detected current $i_{Hi_{De}}$. The high side voltage V_{Hi} and the low side current I_{Lo} are 150 V and -100 A, respectively. The high side current i_{Hi} and the detected current $i_{Hi_{De}}$ are similar waveforms. The value of the high side current i_{Hi} and that of the detected current $i_{Hi_{De}}$ are

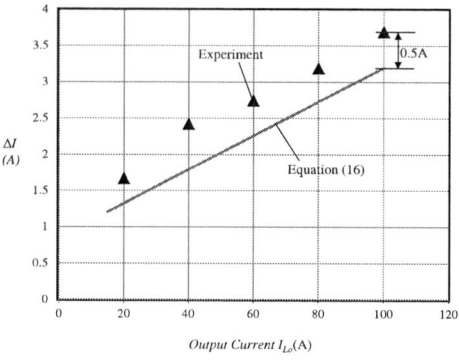

Fig. 12. Comparison of ΔI Experimental and Calculation Result

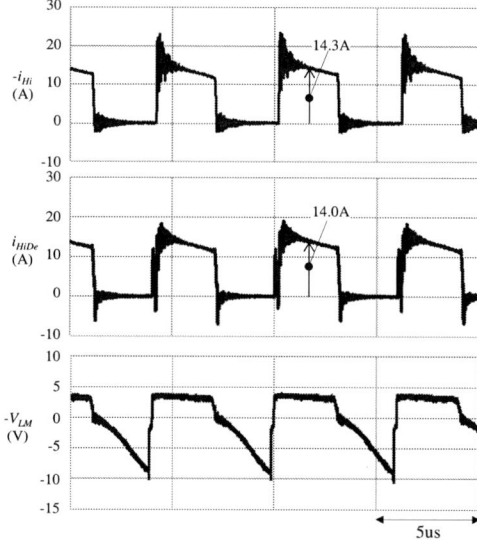

Fig. 13. Experimental Waveforms under Backward Direction $I_{Lo} = -100A$

14.3 A and 14.0 A, respectively, hence the relative error is about 2%. Fig. 14 shows a comparison of detection errors ΔI between the experimental and the calculation results. The detection errors ΔI coincided with an error of 0.2 A. As expected, this detection errors ΔI are smaller than that under forward direction of power conversion. The residual error of ΔI is due to the influence of the precision of the measuring instrument and ignoring the ripple current of the magnetizing current I_{LM}.

V. CONCLUSION

In this paper, a current detection circuit based on CT for bi-directional DC-DC converters is proposed. It was shown that this current detection circuit with a simple circuit configuration can detect the bi-directional current. In addition, the detection errors of the current detection circuit was discussed. It was verified that the errors are caused by the magnetizing current of the CT due to lack of degaussing time. The experimental results of the DC-DC converter rated at 1.4 kW with current feedback control are presented. Finally, this paper verified that

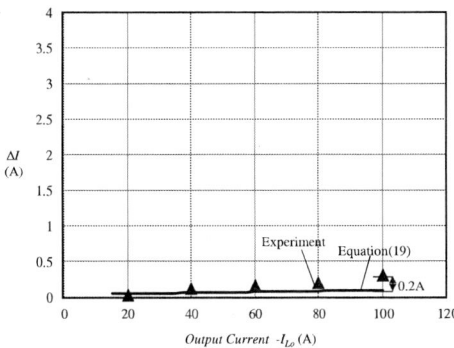

Fig. 14. Comparison of ΔI Experimental and Calculation Result under Backward Direction

the detection errors can be estimated by the derived theoretical equations. By using this theoretical equations, the constant current control can be achieved.

REFERENCES

[1] R. Mizutani, T. Tachibana, M. Morimoto, K. Akatsu and N. Hoshi, "Electric Drive Technologies Contributing to Low-Fuel-Consumption Vehicles" , *IEEJ Trans. on Ind. Appl.,* Vol. 135, No .9, pp. 884-89, 2015 (in Japanese).

[2] Y. Ota, "Cooperation of Electric Vehicles and Renewable Energy Resources", Smart Grid, pp. 8-13, 2017.

[3] S. Inoue and H. Akagi, "A Bi-Directional DC/DC Converter for an Energy Storage System", *IEEE Trans. Power Electron.,* Vol. 22, No. 6, pp. 2299-2306, 2007.

[4] P. Xuewei and A. K. Rathore, "Comparison of Bi-directional Voltage-fed and Current-fed Dual Active Bridge Isolated DC/DC Converters Low Voltage High Current Applications", *IEEE International Symposium on Industrial Electronics.,* pp. 2566-2571, 2014.

[5] S. Ziegler, R. C. Woodward, Ho-Ching Iu and L. J. Borle, "Current Sensing Techniques A Review" , *IEEE SENSORS JOURNAL* , vol. 9, No. 4, pp. 354-376, 2009.

[6] L. Dalessandro, Nicolas Karrer and J. W. Kolar, "High-performance Planar Isolated Current Sensor for Power Electronics Applications" , *IEEE Trans. Power. Electron.,* vol. 22, No. 5, pp. 1682-1692, 2007.

[7] M. H. Samimi, A. Mahari, M. A. Farahnakian and H. Moh- seni, "The Rogowski Coil Principles and Applications A Review" , *IEEE SENSORS JOURNAL,* vol. 15, No. 2, 2015.

[8] Y. Kuwabara, K. Wada, J. Guichon, J. Schanen and J. Roudet, "Implementation and Performance of a Current Sensor for a Laminated Bus Bar", *IEEE Trans. Ind. Appl.,* May/Jun, 2018.

[9] S. J. Nibir, E. Hurwitz, M. Karami and B. Parkhideh, "A Technique to Enhance the Frequency Bandwidth of Contactless Magnetoresistive Current Sensors", *IEEE trans. Ind. Electrons,* vol. 63, No. 9, 2016.

[10] R. Slatter, M. Brusius and H. Knoll, "Development of High Bandwidth Current Sensors Based on the Magnetoresistive Effect", EPE'16 ECCE Europe, 2016.

[11] P. S. Filipski, M. Boecker and M. Garcocz, "20-A to 100-A AC-DC Coaxial Current Shunts for 100-kHz Frequency Range" , *IEEE trans. Instrum. Meas.,* vol. 57, No. 8, 2008.

[12] A. Yamashita and K. Wada, " Wide bandwidth and low propagation time delay current sensor applied to a laminated bus bar" , *IEEE Energy Conversion Congress and Exposition,* pp. 3083-3088, 2014

[13] Y. Hayashi and H. Kinjo, "Control Bias Magnetizing of Phase-Shifted Full-Bridge DC-DC Converter Using Digital Peak Current Mode Control", *IEE.J Annual conf.Rec,* vol. 4, No. 129(2016)(in Japanese).

[14] L. P. Wong, Y. S. Lee and D. K. Cheng, "Bi-directional Pulse Current Sensors for Bi-directional PWM DC-DC converters", *2000 IEEE Annual Power Electronics Specialists Conference.,* vol. 2, pp. 1043-1046, 2000.

The 2018 International Power Electronics Conference

A 10 MHz GaNFET Based Isolated High Step-Down DC-DC Converter

Prasanth Thummala[*], Dorai Babu Yelaverthi[#], Regan Zane[#], Ziwei Ouyang[*], and Michael A. E. Andersen[*]

[*]Electronics Group, Department of Electrical Engineering, Technical University of Denmark,
2800 Kongens Lyngby, Denmark

[#]Utah Power Electronics Laboratory, Department of Electrical and Computer Engineering,
Utah State University, Logan, Utah – USA 84341

Email: pthu@elektro.dtu.dk, dorai.yelaverthi@usu.edu

Abstract— **This paper presents design of an isolated high-step-down DC-DC converter based on a class-DE power stage, operating at a 10 MHz switching frequency using enhancement mode Gallium Nitride (GaN) transistors. The converter operating principles are discussed, and the power stage design rated for 20 W is presented for a step-down from 200-300 V to 0-28 V. Commercially available magnetic materials were explored and the high-frequency (HF) resonant inductor and transformer designs using a low-loss Fair-Rite type 67 material are presented. Finite element simulations have been performed to estimate the AC resistances of magnetics at 10 MHz. Experimental results are presented at 12 W, 254 V to 22 V on a laboratory prototype operating at 10 MHz. At 20 W the experimental prototype achieved an efficiency of 85.2%.**

Keywords— *DC-DC conversion, Gallium Nitride, High frequency, Resonant conversion, Soft switching, Class-DE*

I. INTRODUCTION

The motivation to operate at high switching frequency is not just to reduce the size of passive components but also to provide very fast dynamic load response. Most RF communication systems use power amplifiers (PA) to convert low-power signals into larger power RF signals for driving the antenna of a specific transmitter. The majority of PA designs utilize switched-mode pulse-width-modulating (PWM) converters as the power source to operate RF amplifiers. Envelope tracking PAs use dynamically changing supply voltage to achieve high efficiency for the PA over the full power range. To achieve successful envelope tracking, the power supply must be capable of switching at frequencies greater than 5 MHz, as most modern RF waveforms observe a bandwidth of 1 to 5 MHz [1], [2].

The envelope tracking power supply considered in this paper has to operate with an input voltage range of 200 V to 300 V at 10 MHz switching frequency. Several designs at 10 MHz switching are reported in literature for different applications. A 10-MHz GaN 16 to 34-V boost converter with above 90% efficiency is presented in [3]. A 94% efficient 10-MHz, 100 W buck-boost type DC-DC converter is studied in [4]. A 10 MHz, 10.8-16 V to 0.65-2 V, 2 A multiphase buck converter is implemented

in [5]. A 10-MHz, 12 V to 5 V, 5 W buck converter is investigated in [6]. All of these designs are at low operating voltage (few tens of volts).

Traditional hard switching switched-mode power supply (SMPS) topologies are extremely lossy at such high frequencies. This has led to the development of resonant soft-switching converters. With the emergence of Gallium Nitride (GaN) based power switches, power electronic converters tend to be even faster, smaller and more efficient [7]. Resonant converters are often designed in two parts; an inverter converting the DC input voltage to an AC current and a rectifier converting the AC current to a DC output voltage. The two parts are designed individually, but the design of the inverter depends on the input impedance of the rectifier [8], [9].

The most common topologies for the inverter part are based on class E, which could either be a class E, a class EF2 (φ2), a resonant SEPIC or a resonant boost converter. The choice of the topology is based on the complexity and losses associated with a high side gate drive for operation in the HF range. A class E derived inverter imposes significant voltage stress across the MOSFET. The voltage stress for the class E, the resonant SEPIC and resonant boost is 3.6 times the input voltage with a duty cycle of 50%, and for the class EF2 this stress is reduced to approximately 2.3-3 times. The semiconductor switches in the class DE inverter are directly connected to the input and the voltage across them is limited to the input voltage. The class DE inverter [10]-[12] has two other great advantages over the other topologies. Firstly, it only requires a single inductor. Secondly, due to the lower peak voltage across the MOSFET, the stored energy is approximately ten times lower.

This paper presents a GaN-based and magnetic core-based 10 MHz isolated DC-DC converter using a Class-DE resonant soft-switching power stage. Section II describes the converter design, operation and simulation results. Section III provides the steady state analysis of the converter. Section IV discusses the choice of magnetics at 10 MHz, and the design of inductor and the transformer using Fair-Rite 67 material. Section V provides the experimental results, and Section VI discusses the power loss distribution, followed by the conclusions in Section VII.

II. CONVERTER ANALYSIS AND DESIGN

A class-DE based isolated DC-DC converter is depicted in Fig. 1. The main input power stage consists of switches S_1 and S_2. Compared to the conventional Class-DE amplifiers, the switches S_3 and S_4 are connected in parallel with the rectifier diodes D_3 and D_4, to achieve synchronous rectification and also for active rectification to control the output voltage and power. The converter achieves ZVS, zero voltage derivative switching (ZVDS), and ZCS at the turn-on instant. In this converter, the effective impedance of the secondary rectifier is used for designing the series resonant tank components C_r and L_r. A transformer with a turns ratio n:1 is used for providing isolation as well as stepping down the input voltage. The power is transferred from input to output due to the resonance between the resonant tank elements C_r and L_r. Hence, the current flowing through the resonant tank is almost sinusoidal in shape. Based on the fundamental harmonic approximation (FHA), the ac-equivalent circuit of the proposed Class-DE DC-DC converter is shown in Fig. 2.

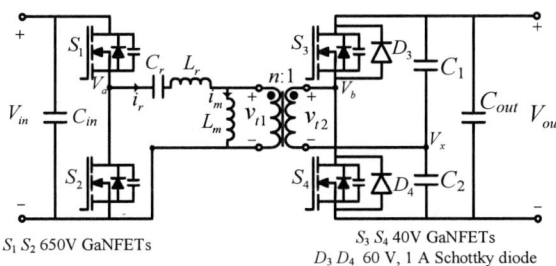

$S_1 S_2$ 650V GaNFETs

$S_3 S_4$ 40V GaNFETs
$D_3 D_4$ 60 V, 1 A Schottky diode

Fig. 1. Schematic of the Class-DE based isolated synchronous DC-DC converter.

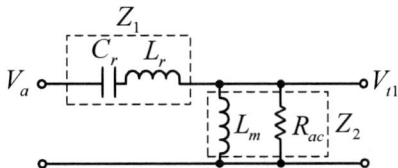

Fig. 2. AC equivalent circuit of the resonant tank based on FHA.

The design equations of the proposed isolated class-DE topology are given below. The AC equivalent load resistance (input resistance of the rectifier) is calculated as [10]

$$R_{ac} = \frac{2R_L n^2}{\pi\left[\pi + \omega R_L C_{oss,sec}\right]} . \tag{1}$$

The RMS voltages on both inverter and rectifier sides of switched-nodes are given by (assuming trapezoidal waveforms) [12]

$$V_{a,rms} = V_{in}\sqrt{\frac{D_{pri}+1}{3}} , \tag{2}$$

$$V_{b,rms} = V_{out}\, n\sqrt{\frac{D_{sec}+1}{3}} . \tag{3}$$

The reactance of the resonant circuit is calculated as [13]

$$Z_1 = R_{ac}\sqrt{\left(\frac{V_{a,rms}}{V_{b,rms}}\right)^2 - 1} . \tag{4}$$

The resonant tank inductance for a given tank capacitance is given by [14]

$$L_r = \frac{C_r Z_1 \omega + 1}{C_r \omega^2}, \quad \omega = 2\pi f_{sw} . \tag{5}$$

The quality factor of the resonant tank is given by

$$Q = \frac{1}{R_{ac}}\sqrt{\frac{L_r}{C_r}} . \tag{6}$$

In the above equations, R_L is the load resistance, n is the transformer turns ratio, D_{pri} is the duty cycle of the primary GaNFETs, D_{sec} is the duty cycle of the secondary GaNFETs, f_{sw} is the switching frequency, and $C_{oss,sec}$ is the output capacitance of the secondary GaNFETs S_3 and S_4.

A transformer turns ratio of $n = 2.5$ is selected to ensure both ZVS of the primary GaNFETs as well as low circulating energy of the resonant tank. An LTspice simulation of the Class-DE converter is performed using the GaNFET models from the manufacturer. On the primary and secondary sides, 650 V devices from GaN Systems (GS66502B) and 40 V GaNFET from EPC (EPC2014C) are used, respectively. The simulation results showing key waveforms are provided in Fig. 3. The primary GaNFETs from GaN Systems are driven with 6 V and the secondary GaNFETs from EPC are driven with 5 V.

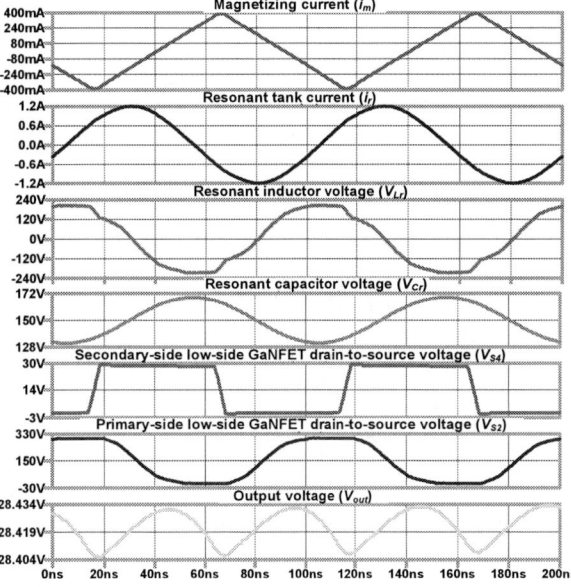

Fig. 3. LTspice simulation results for V_{in} = 300 V at D_{pri} = 18% and D_{sec}=40%. The output power P_{out} = 20 W. Power stage design parameters given in Table I are used in the simulation. A phase shift of 22 ns between primary and secondary GaNFETs is used.

TABLE I: POWER STAGE DESIGN

Variable	Value
R_{ac}	49.63 Ω
Z_1	162.2 Ω
C_r	1 nF
L_r	2.8 µH
L_m	2.2 µH
Z_2	36.52 Ω
Q	1.24

The design of the converter at 10 MHz for 300 V input and 28 V output at 20 W is summarized in Table I. The magnetic design details are provided in Section IV.

III. STEADY STATE ANALYSIS OF CLASS-DE DC-DC CONVERTER

The fundamental components of the input and output voltages of the resonant tank are given by

$$V_a(t) = \frac{2V_{in}}{\pi}\sin(\omega t) , \qquad (7)$$

$$V_{t1}(t) = \frac{2nV_{out}}{\pi}\sin(\omega t) . \qquad (8)$$

The voltage gain of the resonant tank is given as follows:

$$M = \frac{V_{t1}(t)}{V_a(t)} = \frac{nV_{out}}{V_{in}} = \frac{Z_2}{Z_1+Z_2}$$

$$= \frac{k}{\sqrt{\left(1+k-\frac{1}{f_n^2}\right)^2 + Q^2 k^2\left(f_n - \frac{1}{f_n}\right)^2}} , \qquad (9)$$

where $k = \dfrac{L_m}{L_r}$, $f_r = \dfrac{1}{2\pi\sqrt{L_r C_r}}$, $f_n = \dfrac{f_{sw}}{f_r}$.

The voltage gain of the resonant tank M is plotted in Fig. 4 with respect to the normalized frequency and the quality factor for a given constant k. Similarly, the voltage gain of the resonant tank M is plotted with respect to the normalized frequency and the constant k, for a given quality factor, as shown in the Fig. 5. For the converter design specifications described in Section II, $k = 0.786$, $f_r = 3$ MHz, and $f_n = 3.33$.

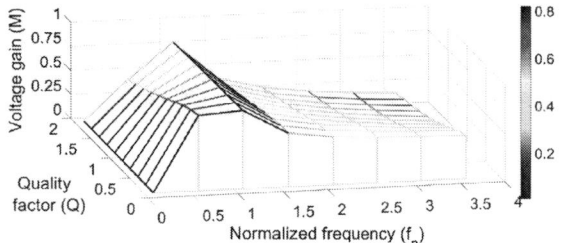

Fig. 4. Voltage gain vs. loaded quality factor vs. normalized switching frequency for $k = 0.786$.

The converter is operating well above the resonance frequency so that current is lagging enough to achieve ZVS for the primary GaN devices. The k-Q analysis of the resonant tank for an LLC series-resonant converter (SRC) is proposed in [15]. The following condition ensures that the converter operates within LLC-SRC region [15]

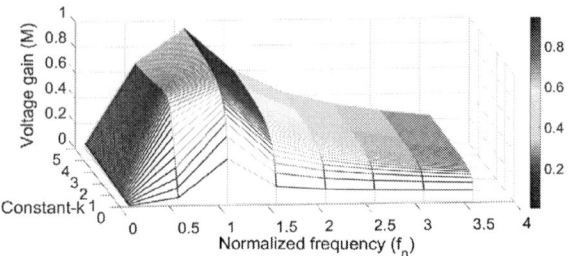

Fig. 5. Voltage gain vs. constant-k vs. normalized switching frequency for $Q = 1.24$.

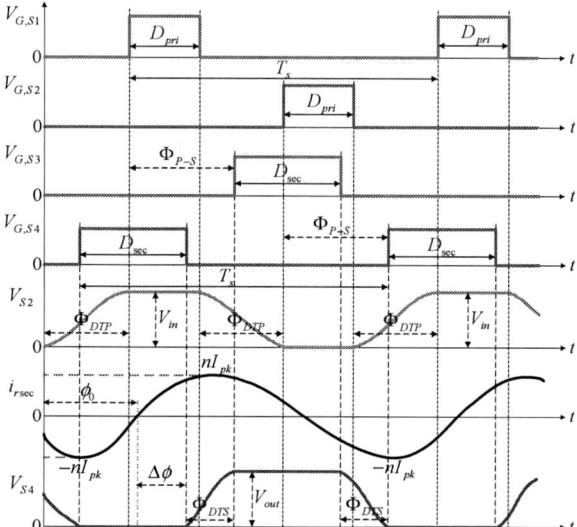

Fig. 6. Steady-state waveforms of the proposed resonant DC-DC converter.

$$Q \le \sqrt{\frac{L_r}{L_m}} \le \frac{1}{\sqrt{k}} . \qquad (10)$$

The steady state waveforms of the proposed DC-DC converter are shown in Fig. 6. The gate drive waveforms of the four primary and secondary GaNFETs, the voltages across the GaNFETs S_2 and S_4, and the secondary resonant tank current are clearly shown in Fig. 6. The switching frequency of the converter is fixed at 10 MHz.

The output power of the converter is defined as [10], [16]

$$P_{out} = \left(\frac{nI_{pk}\cos(\Delta\phi_s)}{\pi + \omega R_L C_{oss,sec}}\right)^2 R_L . \qquad (11)$$

The output current is given by the following expression

$$I_o = \frac{nI_{pk}\cos(\Delta\phi_s)}{\pi + \omega R_L C_{oss,sec}} . \qquad (12)$$

The phase shift $\Delta\phi$ is given by

$$\Delta\phi = (\Phi_{P-S} - \Phi_{DTS}) - (\phi_0 - \Phi_{DTP}) , \qquad (13)$$

where I_{pk} is the peak resonant tank current, Φ_{P-S} is the phase-shift between primary GaNFET S_1 (or S_2) and

4068

secondary GaNFET S_3 (or S_4), Φ_{DTS} is the phase corresponding to the dead-time of the secondary side, Φ_{DTP} is the phase corresponding to the dead-time of the primary side, and ϕ_0 is the phase-shift of the resonant current. The output power and voltage can be controlled by varying the phase-shift angle Φ_{P-S} between the primary and secondary GaNFETs. The variations of output voltage and output power with respect to the phase-shift Φ_{P-S} are shown in Fig. 7. The maximum output power occurs at a phase-shift of 15 ns for V_{in} = 254 V.

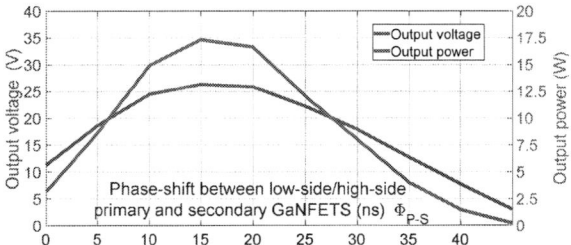

Fig. 7. The variation of output voltage and output power with respect to the phase-shift Φ_{P-S} for V_{in} = 254 V, R_L = 40 Ω.

IV. DESIGN AND FEM SIMULATIONS OF MAGNETICS

A. Inductor and Transformer designs

With the emergence of GaN and SiC devices, there has been a significant advancement in semiconductor device switching speed, but magnetics has become a primary limitation constraining miniaturization. By increasing the switching frequency of the converter, the absolute value of capacitance and inductance can be reduced but the actual size reduction at very high frequencies depends on the allowable loss in power density. Appropriate core material and winding structure have to be selected for these high frequencies to reduce the loss and realize the achievable miniaturization. Emerging thin-film magnetic materials are a good choice for frequencies greater than 10 MHz. These materials are typically alloys with Fe, Co and Ni. But these are not commercially available at economical costs [17], [18]. Another limitation is the conductor technology. The performance factor comparisons of magnetic materials at high frequency is reported in [19] @ 500 mW/cc.

There is limited data available on the design of very high frequency power magnetics. Power magnetics have high flux drive. For most of the materials, large signal loss data are not available at above a few MHz. Among the commercially available materials, Ni-Zn ferrites and metal-powder materials, which are developed for RF applications, have very high resistivity and are suitable for the present application. The performance factor for these RF material in range of 1 MHz to 100 MHz are reported in [19] based on the method proposed in [20]. Performance factor is the product of amplitude of flux density (B_{ac}) and frequency (f) and is a measure of power handling capability per unit volume for a given core loss density and is a relevant performance metric when core

loss is the major design constraint (usually true for transformers and resonant inductors), neglecting ac winding loss. Among the above reported materials, Ferroxcube 4F1 [21] and Fair-Rite 67 [22] material were available in planar structures and rest were available in rods and toroidal shapes meant for RF applications. The 67 material also has the highest performance factor at 10 MHz [19].

For both Ferroxcube 4F1 and Fair-Rite 67 material, coreloss is plotted in Fig. 8 from raw data for 10 MHz at 25° C and 100° C (sinusoidal current assumed through the inductor). The core loss density of 67 material is 390 mW/cm³ at 10 MHz, T=100° C for B_{ac} of 10 mT, and is nearly a third of what it is for 4F1. Because of the relatively high thermal coefficient of the 4F1, it is not a good option for fabricating a resonant inductor with low air-gap designs. For the initial prototype, 67 material was chosen as the core option for the above reasons. However, a drawback of the 67 material is that when it is exposed to B_{ac} of greater than 20 mT the material properties irreversibly change and have higher losses than the initial characteristics. Permeability of both the material at 25° C and 100° C is given in Table II. The inductor and transformer prototypes are shown in Fig. 9. The inductor and transformer designs are summarized in Tables III and IV, respectively.

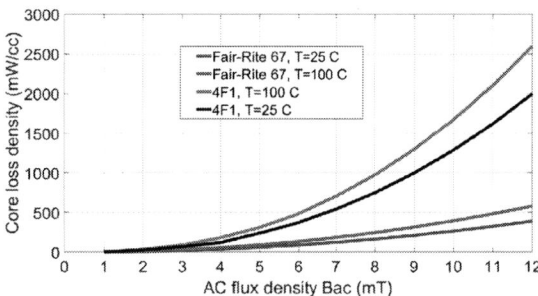

Fig. 8. 4F1 and 67 core loss density comparison at 10 MHz and for different temperatures.

A 4-layer PCB (total thickness is 1.575 mm) with 1 oz. copper thickness is used for practical implementation. To maintain the ease of manufacture and also high repeatability, multi-layered PCBs are used to realize the magnetic winding structures, 6-layer PCB for the inductor with and 8-layer PCB for the transformer. For the inductor, the copper thickness in all layers is 35 μm, and the 3rd and 4th layers are parallel connected. In the transformer, the 5 primary turns are placed in the first 5 layers, the 6th layer in the PCB is kept empty, and the 2 secondary turns are placed in 7th and 8th layers, respectively. Providing an empty 6th layer not only provides isolation between the primary and secondary windings, but it also minimizes the inter-winding capacitance. In Tables III and IV, the measured AC resistance values are used to calculate the winding loss.

TABLE II: 4F1 AND 67 PERMEABILITY WITH TEMPERATURE

Material	Permeability @ T=25° C	Permeability @ T=100° C
4F1	80	140
67	40	45

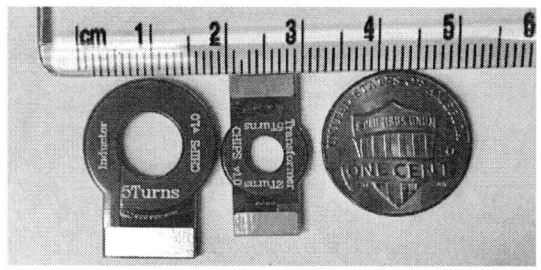

Fig. 9. A photo of inductor and transformer PCB winding prototypes.

TABLE III: INDUCTOR DESIGN SUMMARY

Parameter		Value
Inductance		2.8 μH
Core		EEQ20, 67 material, Volume=2.01 cm³, Area = 0.6 cm²
Overall core height		12.7 mm
Effective core length		3.33 cm
Turns, Air-gap		5 turns, No Air-gap
Core loss	@ 20 W	0.68 W (@B_{ac} = 11.2 mT)
	@ 12 W	0.46 W (@B_{ac} = 9.3 mT)
Copper loss	@ 20 W	0.16 W (for I_{rms}=0.831 A)
	@ 12 W	0.11 W (for I_{rms}=0.683 A)
AC resistance @ 10 MHz		231 mΩ
Copper thickness		35 μm (in all layers)
PCB		6 layer (layers 3 and 4 are paralleled) Total PCB thickness = 1.75 mm
PCB thickness between layers {1-2, 3-4, and 5-6}		0.254 mm
PCB thickness between layers {2-3 and 4-5}		0.38 mm

TABLE IV: TRANSFORMER DESIGN SUMMARY

Parameter		Value
Transformation ratio n		2.5
Primary magnetizing inductance		2.18 μH
Core		EIQ13, 67 material, Volume=0.28 cm³, Area = 0.2 cm²
Overall core height		3.95 mm
Effective core length		1.39 cm
Turns Non-interleaved: PPPPPSS		5 turns primary, 2 turns secondary
Core loss	@ 20 W	0.055 W (@ B_{ac} = 8.7 mT)
	@ 12 W	0.038 W (@ B_{ac} = 7.3 mT)
Copper loss	@ 20 W	0.265 W (for I_{rms}=0.831 A)
	@ 12 W	0.18 W (for I_{rms}=0.683 A)
AC resistance referred to primary @ 10 MHz		385 mΩ
Copper thickness		35 μm (in top and bottom layers) 17.5 μm (in all middle layers)
PCB		8 layers, layer 6 is not used Total PCB thickness = 2 mm
PCB thickness between layers {1-2, 3-4, 5-6, and 7-8}		0.254 mm
PCB thickness between layers {2-3, 4-5, and 6-7}		0.257 mm

Analytical analysis was done for estimating the AC resistance and optimal copper thickness to minimize the losses in both the transformer and inductor PCB windings. However, a more detailed FEM analysis is required for operation at 10 MHz to improve the design due to the importance of parasitic effects from aspects such as PCB traces and vias.

B. FEM simulations

Maxwell's 3D simulations have been performed to estimate the AC resistance and leakage inductance of magnetics at 10 MHz. The FEA simulation results of the transformer current density and flux density at 10 MHz are shown in Figs. 10(a) and 10(b). The skin depth of copper at 10 MHz is 20.6 μm. A fine mesh based on inside length selection is used to simulate the eddy current effects in the winding. In the primary and secondary windings, the layer to layer connections are made through the vias with an outer diameter of 0.45 mm and the size of the hole is 0.2 mm. In the transformer 3D simulation model, the vias are placed between 2 layers (layer-to-layer).

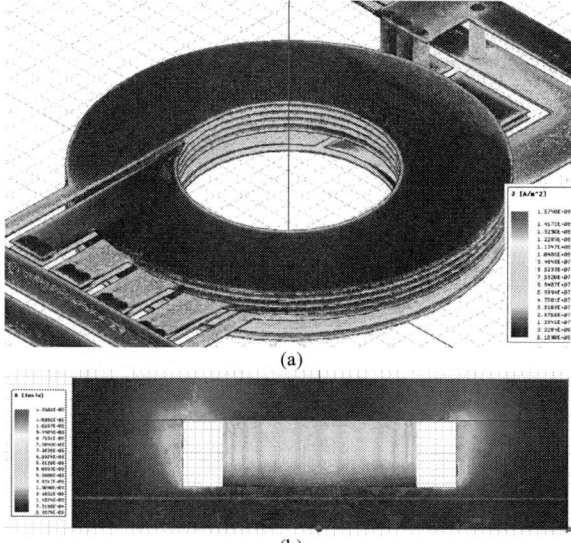

(a)

(b)

Fig. 10. Plots from the 3D Maxwell simulations of EIQ-13 transformer (a) Current density at 10 MHz; (b) Magnetic flux density at 10 MHz. Secondary winding is shorted to obtain the AC resistance and leakage inductance.

The parameters of the transformer are measured using the Agilent 4294A impedance analyzer. The measurement results are shown in Figs. 11-14.

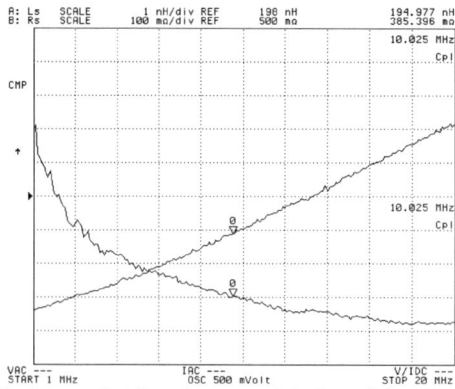

Fig. 11. Measured AC resistance and leakage inductance of the transformer using Agilent 4294A analyzer.

From Fig. 12, the resonance frequency of the transformer is 48.5 MHz, and the primary magnetizing inductance of the transformer at 10 MHz is 2.18 μH as

4070

shown in Fig. 13, which results in a primary transformer self-capacitance of 4.95 pF. The measured interwinding capacitance of the transformer as shown in Fig. 14 is 8.99 pF at 10 MHz. It is obtained by shorting the primary and secondary windings, and measuring the capacitance across the shorted primary and secondary windings.

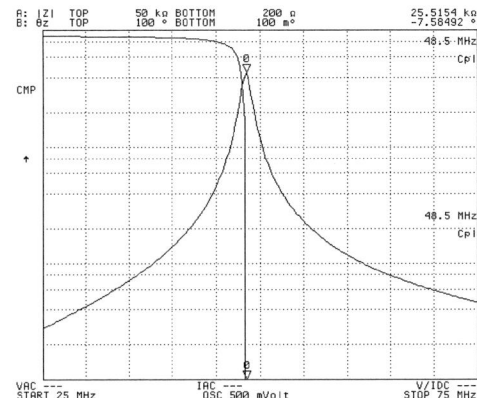

Fig. 12. Measured impedance of the transformer.

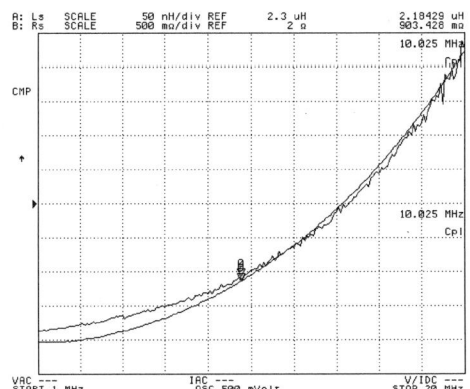

Fig. 13. Measured primary magnetizing inductance of the transformer.

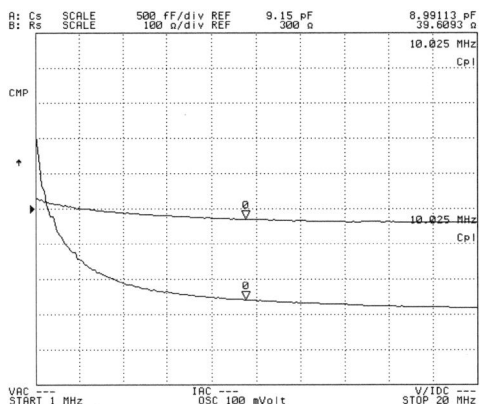

Fig. 14. Measured interwinding capacitance of the transformer.

A comparison of the simulated and measured parameters of the transformer is provided in Table V. The simulated and measured transformer parameters show a close match, with the largest error in the AC resistance. The simulation model of PCB vias can be further improved to match with practical values at 10 MHz switching frequency.

TABLE V: COMPARISON OF SIMULATED AND MEASURED TRANSFORMER PARAMETERS AT 10 MHz

Variable	Simulation	Measurement
Primary magnetizing inductance	2.17 µH	2.18 µH
AC resistance	300 mΩ	385 mΩ
Leakage inductance	188 nH	194 nH
DC resistance	40 mΩ	45 mΩ

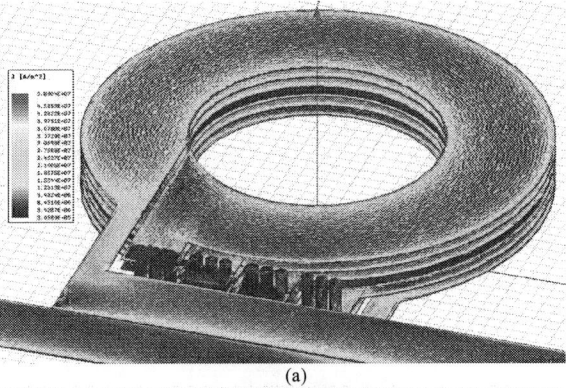

(a)

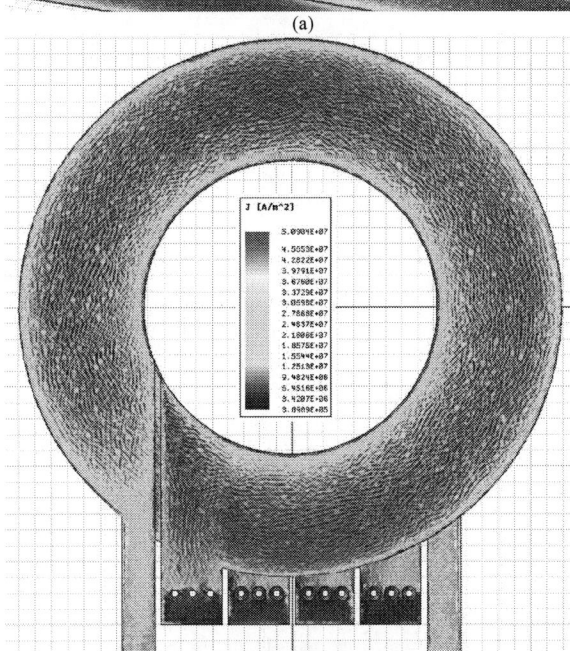

(b)

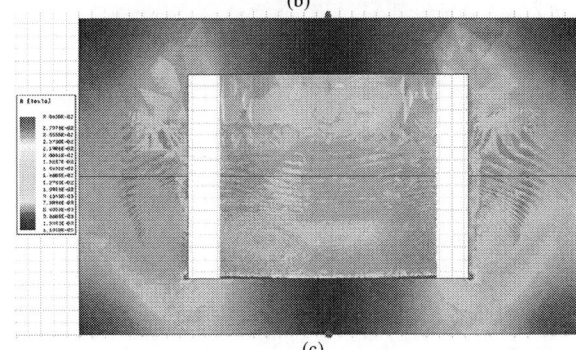

(c)

Fig. 15. Plots from the 3D simulations of EEQ-20 inductor (a) Current density at 10 MHz; (a) Current density top view at 10 MHz; (c) Magnetic flux density at 10 MHz.

The FEA simulation results of the inductor current density and flux density at 10 MHz are shown in Figs.

15(a), 15(b) and 15(c), respectively. As shown in Fig. 15(b), the current is pushed towards the edges of the winding. In the inductor 3D simulation model, the vias pass through all layers (top-layer to bottom-layer). The measured parameters of the inductor using the impedance analyzer are shown in Fig. 16.

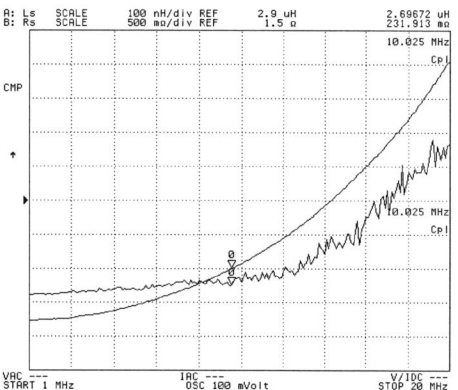

Fig. 16. Measured AC resistance and inductance of the EEQ-20 inductor using Agilent 4294A analyzer.

A comparison of the simulated and measured inductance and AC resistances of the inductor is provided in Table VI. Again, a close match is achieved with the largest error in the AC resistance. The error is primarily due to differences in how the vias are modeled. In the practical implementation of magnetics, the vias are connected from top layer to bottom layer in the PCBs.

TABLE VI: COMPARISON OF SIMULATED AND MEASURED INDUCTOR PARAMETERS AT 10 MHz

Variable	Simulation	Measurement
Inductance	2.89 μH	2.69 μH
AC resistance	131 mΩ	231 mΩ
DC resistance	37.5 mΩ	40 mΩ

V. EXPERIMENTAL RESULTS

An experimental prototype of the Class-DE based converter is shown in Fig. 17. The resonant converter is operated along a narrow optimized trajectory to guarantee ZVS and ZCS. Inductance and noise coupling in the gate drive loop is critical for operation at high input voltages. For the practical implementation, primary 650 V, 220 mΩ GS66502B GaNFETs with low output capacitance (17 pF) and gate charge (1.7 nC) are used. For synchronous rectification, 40 V, 60 mΩ EPC2014C GaNFETs (C_{oss} = 150 pF, Q_g = 2 nC) are used in the secondary side. A HF diode PMEG6010CEH is used for D_3 and D_4. The experimental results are provided in Figs. 18 and 19.

High-frequency resonant capacitors from ATC are used for C_r, C_1 and C_2. A 1000 V, 1 nF capacitor (100C102JW) is used for C_r [23]. A 100 V, 684 nF capacitor (900C684MP) is used for C_1 and C_2 [24]. A digital isolator ADuM210N with common mode transient immunity (CMTI) ≥ 100 V/μs is used for isolation. Due to non-availability of commercially available half-bridge gate drivers suitable for 300 V input and 10 MHz operation, and to quickly evaluate the power stage design, a battery powered isolated low-side gate driver

LM5114 [25] is used for driving each GaNFET. All gate-drivers for switches S_1-S_4 have been designed with a negative-bias supply (-2 or -3 V) to prevent false turn-on. Negative bias voltage in switch S_2 can be seen from Fig. 18. Independent gate drive resistors $R_{g,ON}$ (20 Ω) and $R_{g,OFF}$ (3.3 Ω) are used to counter Miller effect on primary side. The inductance and noise coupling in gate drive loop are very critical for operation of converter at 10 MHz. A Virtex-5 FPGA development board is used to generate the required 10 MHz driving signals on both primary and secondary sides.

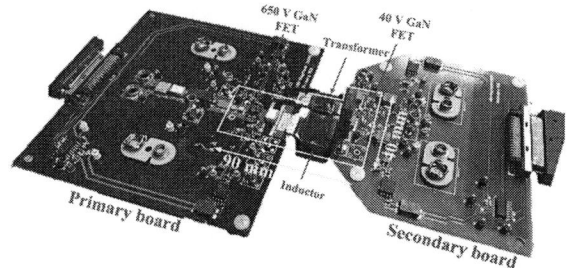

Fig. 17. An experimental prototype of the Class-DE based converter. The 10 MHz switching power stage is outlined in yellow. The rest of the PCB has connectors to the FPGA board, digital isolators, test points and auxiliary supplies for gate drivers.

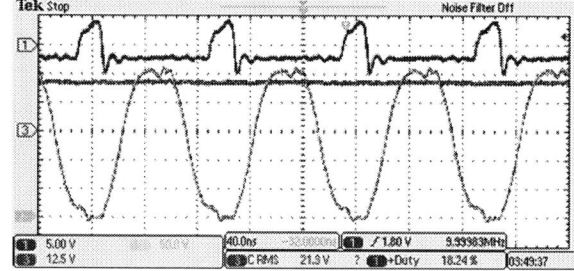

Fig. 18. Experimental waveforms for V_{in}=254 V, Phase-shift Φ_{P-S} =25 ns, D_{pri}=18%. CH1: Gate-to-source signal of S_2 [5 V/div]; CH2: Drain-to-source waveform of S_2 [50 V/div]; CH3: Output voltage across 40 Ω load [12.5 V/div].

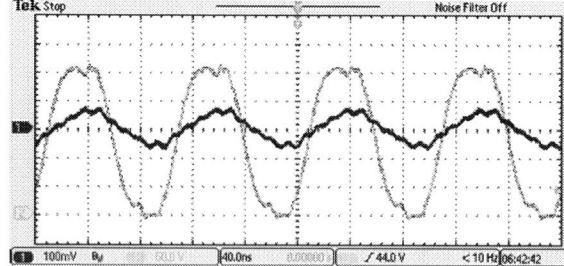

Fig. 19. Experimental waveforms for V_{in}=254 V, Phase-shift Φ_{P-S} =25 ns, D_{pri}=18%. CH1: Resonant tank current [2 A/div] (dark blue); CH2: Resonant tank voltage [50 V/div] (light blue).

VI. POWER LOSS BREAKDOWN

The total loss breakdown for the proposed DC-DC converter at rated power is shown in Fig. 20. The driving loss is the total gate drive loss for both primary and secondary GaNFETs. The device losses include the total forward and reverse conduction loss, and switching loss due to all primary and secondary GaNFETs. The loss due to the power consumption in the auxiliary power supply is estimated to be 1 W, the loss due to the ESR of the capacitors, and PCB traces are considered as additional conduction losses. The efficiency of the converter at an output power of 20.2 W is 85.27%. The total power loss is 3.49 W. Optimizing the inductor and auxiliary power supply (for powering the gate drivers) designs could further increase the efficiency of the converter. Integrating the transformer and inductor into a single magnetic structure will reduce the overall size of the magnetics. This also reduces the terminations by two and reduces the copper loss due to reduced number of windings. Winding resistance can be reduced by paralleling multiple layers in planar PCB windings. The power stage and magnetic designs can be further investigated by changing the constant-k described in Section III.

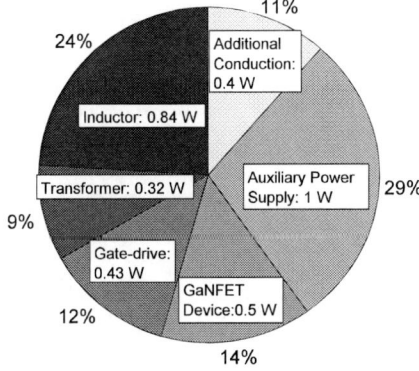

Fig. 20. Total loss breakdown for an output power P_{out}=20.2 W.

VII. CONCLUSIONS

In this paper, a high-frequency, high-step down isolated DC-DC converter equipped with the GaN devices is analyzed and designed. The proposed resonant design shapes waveforms to optimize magnetics and achieve low EMI and high efficiency with high power density. The inductor and transformer are designed using commercially available materials to minimize the physical size and core and copper losses when operating at a switching frequency of 10 MHz. The core material Fair-Rite 67 was chosen from many of the commercially available magnetic materials because of its better performance factors and its availability in low-profile planar structures. Maxwell 3D simulations were performed to estimate the AC resistances of the inductor and transformer at 10 MHz. A phase shift angle between the primary and secondary GaNFETs was used to regulate the output voltage and power of the DC-DC converter. A 20 W, 300 V to 28 V laboratory prototype operating at 10 MHz achieved an efficiency of 85.2%.

REFERENCES

[1] Y. Zhang, J. Strydom, M. de Rooij and D. Maksimović, "Envelope tracking GaN power supply for 4G cell phone base stations," *2016 IEEE Applied Power Electronics Conference and Exposition (APEC)*, Long Beach, CA, 2016, pp. 2292-2297.

[2] EPC white paper, 2017. [Online]: http://epc-co.com/epc/Portals/0/epc/documents/papers/eGaN%20FETs%20for%20Envelope%20Tracking%20Applications.pdf

[3] F. Gamand, M. D. Li and C. Gaquiere, "A 10-MHz GaN HEMT DC/DC Boost Converter for Power Amplifier

Applications," *IEEE Transactions on Circuits and Systems II: Express Briefs*, vol. 59, no. 11, pp. 776-779, Nov. 2012.

[4] K. Kruse, M. Elbo and Z. Zhang, "GaN-based high efficiency bidirectional DC-DC converter with 10 MHz switching frequency," *2017 IEEE Applied Power Electronics Conference and Exposition (APEC)*, Tampa, FL, 2017, pp. 273-278.

[5] G. Calabrese, M. Granato, G. Frattini and L. Capineri, "Integrated high step-down multiphase buck converter with high power density," *2014 16th European Conference on Power Electronics and Applications*, Lappeenranta, 2014, pp. 1-10.

[6] Y. Nour, Z. Ouyang, A. Knott and I. H. H. Jørgensen, "Design and implementation of high frequency buck converter using multi-layer PCB inductor," *IECON 2016 - 42nd Annual Conference of the IEEE Industrial Electronics Society*, Florence, 2016, pp. 1313-1317.

[7] J. Millán, P. Godignon, X. Perpiñà, A. Pérez-Tomás and J. Rebollo, "A Survey of Wide Bandgap Power Semiconductor Devices," *IEEE Transactions on Power Electronics*, vol. 29, no. 5, pp. 2155-2163, May 2014.

[8] R. C. N. Pilawa-Podgurski, A. D. Sagneri, J. M. Rivas, D. I. Anderson and D. J. Perreault, "Very-High-Frequency Resonant Boost Converters," *IEEE Transactions on Power Electronics*, vol. 24, no. 6, pp. 1654-1665, June 2009.

[9] M. Madsen, A. Knott and M. A. E. Andersen, "Low Power Very High Frequency Switch-Mode Power Supply With 50 V Input and 5 V Output," *IEEE Transactions on Power Electronics*, vol. 29, no. 12, pp. 6569-6580, Dec. 2014.

[10] D. C. Hamill, "Class DE inverters and rectifiers for DC-DC conversion," *PESC Record. 27th Annual IEEE Power Electronics Specialists Conference*, Baveno, 1996, pp. 854-860 vol.1.

[11] H. Koizumi, T. Suetsugu, M. Fujii, K. Shinoda, S. Mori and K. Iked, "Class DE high-efficiency tuned power amplifier," *IEEE Transactions on Circuits and Systems I: Fundamental Theory and Applications*, vol. 43, no. 1, pp. 51-60, Jan 1996.

[12] M. P. Madsen, A. Knott and M. A. E. Andersen, "Very high frequency half bridge DC/DC converter," *2014 IEEE Applied Power Electronics Conference and Exposition - APEC 2014*, Fort Worth, TX, 2014, pp. 1409-1414.

[13] Juan Rivas, Radio Frequency dc-dc Power Conversion, Doctoral Thesis, Massachusetts Institute of Technology, 2006.

[14] Mickey P. Madsen, Very High Frequency Switch-Mode Power Supplies: Miniaturization of Power Electronics, Doctoral Thesis, Technical University of Denmark, 2015.

[15] I. O. Lee and G. W. Moon, "The k-Q Analysis for an LLC Series Resonant Converter," *IEEE Transactions on Power Electronics*, vol. 29, no. 1, pp. 13-16, Jan. 2014.

[16] M. Ekhtiari, Z. Zhang and M. A. E. Andersen, "Analysis of Bidirectional Piezoelectric-Based Converters for Zero-Voltage Switching Operation," *IEEE Transactions on Power Electronics*, vol. 32, no. 1, pp. 866-877, Jan. 2017.

[17] C. R. Sullivan, D. V. Harburg, J. Qiu, C. G. Levey and D. Yao, "Integrating Magnetics for On-Chip Power: A Perspective," *IEEE Transactions on Power Electronics*, vol. 28, no. 9, pp. 4342-4353, Sept. 2013.

[18] F. C. Lee and Q. Li, "High-Frequency Integrated Point-of-Load Converters: Overview," *IEEE Transactions on Power Electronics*, vol. 28, no. 9, pp. 4127-4136, Sept. 2013.

[19] A. J. Hanson, J. A. Belk, S. Lim, C. R. Sullivan and D. J. Perreault, "Measurements and Performance Factor Comparisons of Magnetic Materials at High Frequency," *IEEE Transactions on Power Electronics*, vol. 31, no. 11, pp. 7909-7925, Nov. 2016.

[20] Y. Han, G. Cheung, A. Li, C. R. Sullivan and D. J. Perreault, "Evaluation of Magnetic Materials for Very High Frequency Power Applications," *IEEE Transactions on Power Electronics*, vol. 27, no. 1, pp. 425-435, Jan. 2012.

[21] "Soft ferrites and accessories data handbook," Ferroxcube Int. Holding B.V., Eindhoven, The Netherlands, Tech. Rep. FXC 100 00002, 2013.

[22] Material data sheets. Fair-Rite Products Corp. (2016). [Online]. Available: http://www.fair-rite.com/newfair/materials.htm.

[23] HF Capacitor datasheet, AT Ceramics (2018). [Online]. Available: http://www.atceramics.com/UserFiles/100c.pdf

[24] HF Capacitor datasheet, AT Ceramics (2018). [Online]. Available: http://www.atceramics.com/UserFiles/900c.pdf

[25] Gate-driver: http://www.ti.com/product/lm5114.

The 2018 International Power Electronics Conference

Analysis and Design of a Parallel Resonant Converter for Constant Current Input to Constant Voltage Output DC-DC Converter Over Wide Load Range

Tarak Saha, Hongjie Wang, Baljit Riar and Regan Zane
Department of Electrical and Computer Engineering
Utah State University, Logan, USA
E-mail: taraksaha.ee@gmail.com

Abstract— With long distance power distribution systems, constant current distribution is preferred over constant voltage due to robustness against cable voltage drop and cable faults. Multiple power converter modules that are connected in series in such a system either regulate their output current or voltage, as needed. With a constant current input, the input voltage of a converter is dependent on the load and this voltage varies over a wide range with the load. The design of a constant current input to constant voltage output converter is detailed in this paper. The parallel resonant converter (PRC) is proposed for this application to naturally maintain constant output voltage over a wide load range. The control method, zero voltage switching realization and design guidelines are presented. The proposed design is experimentally verified through testing of a designed PRC that regulates a constant output voltage of 120 V when connected to 1 A current source for a load range of 50 W to 450 W.

Keywords — constant current distribution, parallel resonant converter, constant output voltage.

I. INTRODUCTION

Resonant power conversion topologies have widely been used in various applications such as DC distribution systems [1], bi-directional DC-DC converters [2], and wireless power transfer systems [3] due to their benefits of soft-switching ability, low EMI, high power density etc. For subsea DC power distribution systems, distance between a converter and power source, and distance among the converters could be very large, needing robust power distribution architecture that is immune to voltage drop along the cable and cable faults. With such a scenario, constant current distribution with sea water return is preferred over a constant voltage distribution [4-7]. Fig. 1 shows a diagram of an underwater DC distribution system where the shore based power supply drives constant current through the trunk cable and the current reaches back to the source through seawater return. Multiple DC-DC converter modules tap power from the constant current feed to regulate their output voltage or current.

In [8-9] current fed topologies are presented for converters used in subsea application. However, resonant topologies with typical current fed inverter stage increases the device ratings higher than its average DC input voltage and, thus, are not practically suitable for low-current high-voltage systems. Also, with current fed

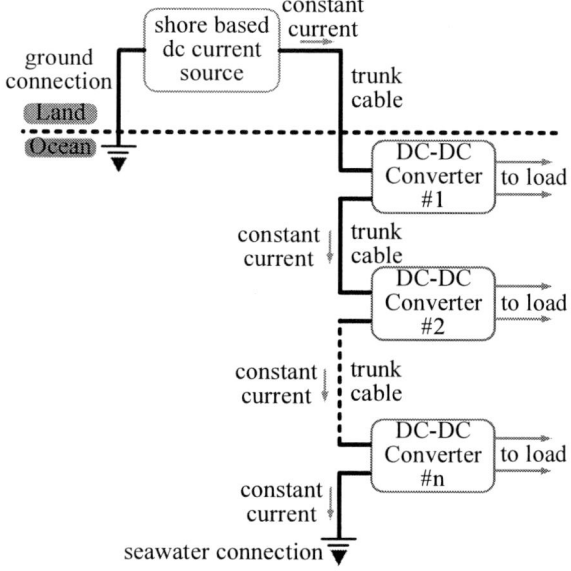

Fig. 1. Underwater DC distribution system

topologies, it is challenging to achieve zero voltage switching (ZVS) which is essential when switching with high voltages for improved efficiency and EMI. ON/OFF type of control can be employed for current fed topologies to reduce the hard switching problem but this requires a large filter at the input stage to avoid any low frequency oscillation of input voltage of the converter propagating to the source cable. Moreover, with current fed topologies, the voltage on the input side fluctuates at double the switching frequency of operation which, if not filtered adequately, can create instability through interaction with the inductance and capacitances of the submarine cable. Hence, for low constant current high voltage power distribution systems, voltage fed converter stages with input voltages that vary with load are preferred. For conventional converters with constant input voltage, there are several topologies available with detailed analysis in literature [10-12]. In [13], a parallel resonant converter (PRC) topology is presented to achieve constant current output, however the converter is operated from a constant DC voltage input. In addition, control through switching frequency variation is not preferred since it introduces additional challenges in design of input EMI filter and gate driver [14]. In [15] a series resonant converter (SRC) topology is presented

4074

that operates as a constant DC current input to a constant DC current output converter, with voltage fed inverter stage. However, achieving constant DC voltage output from constant DC current input is not yet analyzed in the literature.

In this paper, a DC-DC PRC topology is analyzed and designed to achieve a steady state constant DC voltage output characteristics from a constant DC current input source over a wide load range. Selection of the resonant frequency to overcome the limitation on light load operation from constant DC current input is also addressed. Section II covers the basic steady state operation of the converter and derivation of steady state input/output quantities. In Section III, design of the converter, its control methodology and soft-switching realization are presented. Hardware results are shown in Section IV for a PRC, which is operated at 250 kHz from a constant input source of 1 A with its output regulated at 120 V for a load range of 50 W to 450 W, to show the behavior of the converter during steady state as well as under load transients.

II. STEADY STATE ANALYSIS OF A PRC WITH A CONSTANT CURRENT INPUT

The PRC circuit topology operating from a constant current input is shown in Fig. 2(a). On the primary side of the converter, MOSFETs $Q_1 - Q_4$ forms the DC-AC inverting stage, which operates at DC input voltage V_{in}, with a symmetrical phase shift modulation between leg A and leg B with phase shift angle α and produces a quasi-square wave voltage (v_{AB}) at the inverter output, as shown in Fig. 2(b). The resonant tank is formed by the inductor L_r and capacitor C'_r placed on the secondary side of an $n{:}1$ isolation transformer. The voltage across the resonant capacitor is rectified and then filtered by an output filter stage formed by an inductor L_f and capacitor C_f. Input power from the constant current source I_{in} is processed by this converter to regulate the output voltage across the load (R_{load}) at a constant value of V_{out}.

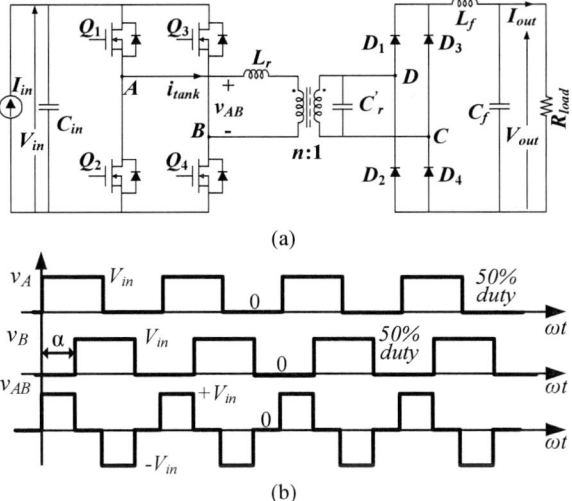

(a)

(b)

Fig. 2. PRC circuit topology with constant current input (a) and its phase shift modulation strategy (b).

The following analysis assumes that the converter is ideal without any loss. Also, it assumes that the loaded quality factor of the tank is high enough to filter out the harmonics generated from the inverter stage and the diode rectifier on the output stage operates under continuous conduction mode. With the fundamental approximation technique, the equivalent circuit of the AC stage in the PRC can be drawn as shown in Fig. 3, where the ratio of output voltage to input voltage can be given as

$$\frac{v_o}{v_{AB}} = \frac{1}{1 + s\dfrac{L_r}{R_e} + s^2 L_r C_r} = \frac{1}{1 + \dfrac{s}{Q\omega_o} + \dfrac{s^2}{\omega_o^2}}, \quad (1)$$

and the amplitude of the input voltage and output voltages are given by

$$\left| v_{AB,pk} \right| = \frac{4}{\pi}\sin(\frac{\alpha}{2})V_{in} \quad and \quad \left| v_{o,pk} \right| = \frac{n\pi}{2}V_{out}, \quad (2)$$

where, the variables used above are defined as

$$\omega_o = \frac{1}{\sqrt{L_r C_r}}, \quad C_r = \frac{C'_r}{n^2}, \quad Z_o = \sqrt{\frac{L_r}{C_r}}, \quad (3)$$

$$R_e = \frac{n^2\pi^2}{8}R_{load}, \quad Q = \frac{R_e}{Z_o}, \quad F = \frac{\omega_s}{\omega_o} = \frac{f_s}{f_o}. \quad (4)$$

Here, ω_o is the angular resonant frequency, Z_o is the characteristic impedance of the resonant tank, Q is the loaded quality factor of the tank and F is the normalized switching frequency.

With power balance from input to output of the converter, the steady state input voltage can be represented in terms of output voltage as

$$V_{in} = \frac{P_{out}}{I_{in}} = \frac{V_{out}^2}{I_{in}R_{load}}. \quad (5)$$

Substituting (2) in (1) and utilizing (5) the steady state DC output voltage can be derived as

$$V_{out} = \frac{Z_o I_{in}}{n\sin(\frac{\alpha}{2})}\sqrt{F^2 + Q^2(1 - F^2)^2}, \quad (6)$$

Now, if the switching frequency (f_s) of the converter is

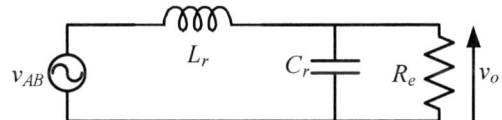

Fig. 3. Equivalent circuit of PRC with fundamental harmonic approximation

chosen to be equal to the resonant frequency (f_o) of the tank i.e. at $F = 1$, then it can be derived from (6) that the output DC voltage becomes independent of Q and as a result independent of R_{load}. The output voltage thus can be given by

$$V_{out} = \frac{Z_o I_{in}}{n \sin(\frac{\alpha}{2})}. \tag{7}$$

From (7), it can be observed that for a given input current, output voltage, transformer turns ratio and the steady state control angle α, Z_o can be easily calculated from which the tank components can be designed. In addition, it can be seen from (5), that with constant current input, the input voltage varies with load. The final expression of input voltage is given by

$$V_{in} = \frac{Z_o^2 I_{in}}{n^2 \sin^2(\frac{\alpha}{2}) R_{load}}. \tag{8}$$

The output power P_{out} of the converter for a given R_{load}, can be given by

$$P_{out} = \frac{Z_o^2 I_{in}^2}{n^2 \sin^2(\frac{\alpha}{2}) R_{load}} [F^2 + Q^2 (1 - F^2)^2]. \tag{9}$$

From (9), it should be noted that with $F = 1$, for a given load resistor R_{load}, P_{out} is minimum with $\alpha = 180°$, and P_{out} as well as V_{out} goes higher as α is reduced which is opposite of constant input voltage based PRC. It should also be noted here that operating with switching frequency equal to resonant frequency also eliminates the limitation of minimum power operation of the converter with constant current input, as discussed in [1].

III. DESIGN OF THE PRC

For the converter design, the first step is to find the resonant tank components. For a given input current I_{in} and output voltage V_{out}, transformer turn ratio n, and with α chosen to be 120° the characteristic impedance Z_o of the tank can be found from (7). For a selected switching frequency, which is also the resonant frequency, the resonant tank components can be found from

$$L_r = \frac{Z_o}{\omega_o} = \frac{Z_o}{2\pi f_o}, \tag{10}$$

$$C_r = \frac{1}{\omega_o Z_o} \Rightarrow C_r' = \frac{n^2}{\omega_o Z_o}. \tag{11}$$

The resonant tank capacitor is placed at the secondary side of the transformer so that the leakage inductance of the transformer can be absorbed into the tank inductance L_r. The rms current in the tank inductor i_{Lr_rms} and rms voltage across the resonant capacitor v_{Cr_rms} are given by

$$i_{Lr\_rms} = \frac{n\pi}{2\sqrt{2}} \frac{V_{out}}{Z_o} \sqrt{1 + \frac{1}{Q^2}}, \tag{12}$$

$$v_{Cr\_rms} = \frac{\pi}{2\sqrt{2}} V_{out}. \tag{13}$$

The choice of transformer turns ratio impacts the maximum rms current in the tank current which occurs at highest load. From (7) and (12) transformer turns ratio can found out by

$$n = \frac{\frac{8}{\pi^2} \frac{V_{out} \sin(\frac{\alpha}{2})}{R_{load\_min} I_{in}}}{\sqrt{(I_{Lr\_rms\_max} \frac{2\sqrt{2}}{\pi} \frac{\sin(\frac{\alpha}{2})}{I_{in}})^2 - 1}}, \tag{14}$$

where, $I_{Lr\_rms\_max}$ is the maximum rms current in the tank inductor and R_{load_min} is the minimum load resistance, corresponding to maximum load at the output. With a design choice of $I_{Lr\_rms\_max}$, the transformer turns ratio can be optimized from (14). After determining n, resonant tank component values can be found following (7), (10) and (11).

A. Device Selection

The primary side inverter devices block voltage equal to input DC voltage V_{in} whose maximum value is decided based on the maximum load and efficiency (η) of the converter and can be found out by

$$V_{pri\_FET} \geq \frac{P_{out\_max}}{\eta I_{in}}. \tag{15}$$

The rms current rating for the MOSFETs are determined by the tank current which can be found out from (12). On the other hand, the secondary side rectifier devices see a reverse voltage equal to the peak value of voltage across resonant capacitor and thus the voltage rating for the rectifier is given by

$$V_{sec\_rect} \geq \frac{\pi V_{out}}{2}. \tag{16}$$

The average value of current through the rectifier is equal to the output current which can be easily found out from maximum load power and output voltage V_{out}.

B. Modulation for Control

It is established in Section II that the PRC behaves as a natural voltage source at its output when operated at switching frequency equal to the resonant frequency and hence a control scheme that varies the switching frequency to regulate its output cannot be employed here. It can be seen from (7) that for the designed converter, the output voltage can be controlled by the phase shift angle α and hence phase shift modulation strategy is used here.

C. ZVS Realization

Based on the analysis presented in [16, 17], for an SRC whose primary side inverter is similar in operation to the PRC, switches in the leading leg (leg A) need zero voltage switching (ZVS) assistance whereas, lagging leg (leg B) achieves ZVS by the tank current itself. As presented in [17], an active ZVS assisting circuit consisting of an auxiliary half bridge leg and ZVS assisting inductor L_{ZVS}, is used here to achieve ZVS turn ON of the MOSFETs in leg A. By controlling the phase shift angle between leg A and this auxiliary leg, ZVS assistance is controlled, over the load range.

The tank current in PRC lags the fundamental component of inverter output voltage and this angle of lagging increases with the load. This lagging phase angle can be determined from the angle of the impedance seen by the primary side inverter, at fundamental frequency, and this is expressed as

$$\angle Z_{in} = \tan^{-1}\left(\frac{F}{Q(1-F^2)}\right) - \tan^{-1}(FQ)$$
$$= \frac{\pi}{2} - \tan^{-1}(Q). \tag{17}$$

If the tank impedance angle is high enough so that at the instant of rising edge of v_{AB}, value of tank current is negative, switches in the leg A will have ZVS by the tank current itself. Based on (17), this condition can be expressed as

$$\tan^{-1}(Q) < \frac{\alpha}{2}. \tag{18}$$

From (18), it can be seen that range of ZVS is high if α is high, as close to 180° as possible. But, operating with higher α limits the minimum output voltage and power capability under transient conditions as well as due to tolerance on resonant tank components [1]. A value of 120° for α is a good trade-off for design considering ZVS range as well as transient and component tolerances.

IV. EXPERIMENTAL VERIFICATION

A prototype PRC operating at 250 kHz has been developed with the parameters shown in Table I. The hardware setup of the PRC is shown in Fig. 4 which operates from a 1 A DC current source and is tested for a power level up to 450 W. The converter has been tested for its output voltage characteristics in steady state and

TABLE I
PARAMETERS OF PRC

Component	Value
L_r (µH)	264.6
C_r (nF)	24.5
f_s (kHz)	250
L_{ZVS} (µH)	55
I_g (A)	1
V_{out} (V)	120
P_{load} (W)	50 – 400
Transformer turn ratio n:1	4:1
Main MOSFETS (SiC)	C2M1000170D
L_f (µH)	80
C_f (µF)	2.35
Diode Bridge	GHXS020A060S-D1

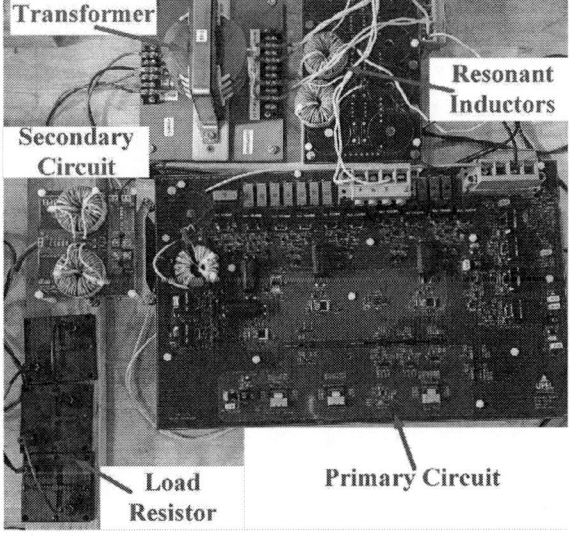

Fig. 4. Photo of the test setup.

transient conditions to verify its load independent, constant output voltage characteristics.

A. Steady State Results

Hardware test results of the PRC during steady state operating conditions are shown in Fig. 5 and Fig. 6. Steady state operating waveforms of the converter are shown for $\alpha = 180°$ and $\alpha = 120°$ in fig. 5(a) and Fig. 5(b), respectively. In Fig. 5, yellow trace (CH1) is for the gate to source voltage of top MOSFET in leg A, purple (CH3) is the primary side inverter output voltage v_{AB}, green waveform (CH4) is the current in the resonant inductor and light blue trace (CH2) is the voltage across the resonant capacitor. For $\alpha = 180°$, no ZVS assisting circuit is employed whereas, for $\alpha = 120°$, an active ZVS assisting circuit has been used as presented in [17] and the ZVS assisting current is adjusted so that the MOSFETs in the primary side bridge achieves ZVS.

The converter has also been tested for its output characteristics by varying the load resistance (R_{load}), at two different control angle viz. minimum power operation angle $\alpha = 180°$ and desired operating angle $\alpha = 120°$. Figure 6 shows the steady state DC output voltage (V_{out}) of the converter with respect to variation in R_{load}. In Fig. 6, the blue plot shows V_{out} vs R_{load} for

The 2018 International Power Electronics Conference

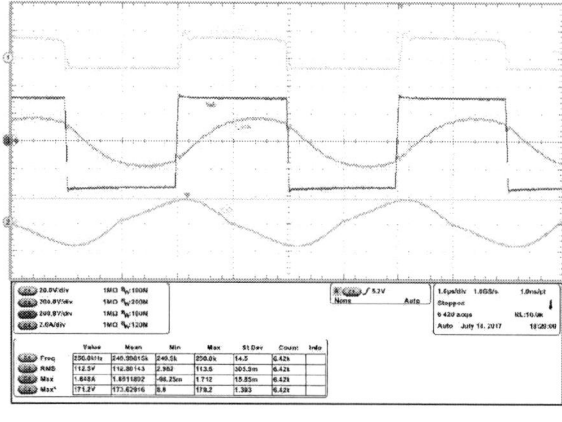

(a)

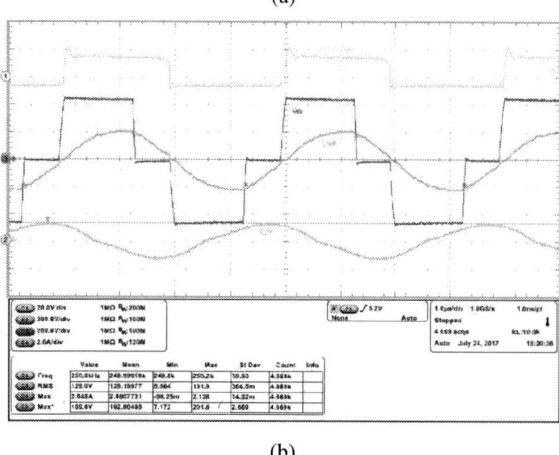

(b)

Fig. 5. PRC operating waveforms with $\alpha = 180°$ (a) and $\alpha = 120°$ (b) at $R_{load} = 34\ \Omega$. Yellow trace (CH1) is the gate to source voltage of top MOSFET in leg A, purple trace (CH3) is the primary side inverter output voltage v_{AB}, green trace (CH4) is the current in the resonant inductor and light blue (CH2) is the voltage across resonant capacitor.

$\alpha = 180°$ and the red plot is for $\alpha = 120°$. It can be seen from the plots in Fig. 6, that the output voltage remains almost constant over the range of R_{load}. This shows that the converter operates as a natural voltage source at the output with a constant input current source and variable input voltage, with load. The small droop in the plots are due to series non-idealities *e.g.* ESR present in the circuit that can be easily taken care of by the closed loop

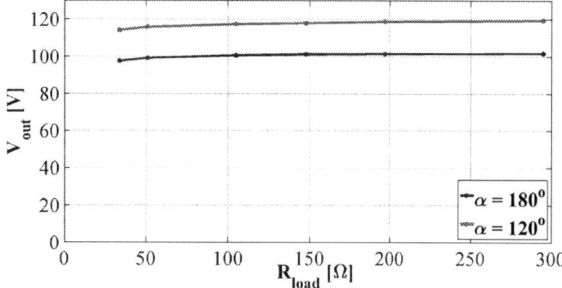

Fig. 6. Open loop output characteristics of PRC. Experimental steady state DC output voltage V_{out} v load resistance R_{load} at $\alpha = 180°$ (blue) and $\alpha = 120°$ (red)

controller with small variation in control angle α.

B. Transient Results

The PRC has also been tested for transient load conditions and the results are presented in Fig. 7. For this test, the output current of the converter is changed from 1.7 A (~200 W) to 2.3 A (~275 W) and back to 1.7 A while the PRC has been operating in open loop with a fixed phase shift angle α. In Fig. 7, the blue plot (CH2) shows the DC input voltage, the green waveform (CH4) is the output load current and the DC output voltage is shown by purple trace (CH3). It can be observed from these results that the output voltage goes through overshoot or undershoot under load change transients, but settles back to its designed value of 120 V, conforming to the load independent constant output voltage characteristics.

V. CONCLUSIONS

In a constant current input DC distribution system, input voltage of a power converter varies with the load. In this paper, it is been presented how a DC-DC PRC, operating from constant current input, can be designed to achieve a steady state constant voltage output behavior across the load range. Steady state input and output quantities are derived for the converter, with fundamental harmonic approximation along with design of resonant tank components and their desired ratings. Device ratings, control methodology and soft-switching needs are also discussed here in this paper. Finally a hardware prototype has been built and tested to show that the output of the converter behaves as a constant voltage source at steady state, irrespective of load resistance. Steady state operating waveforms, with ZVS of the primary side active switches, are also shown in this paper. Simulation results under load transients also confirms the constant output voltage characteristics of the converter making it suitable for constant current input to constant voltage output converter.

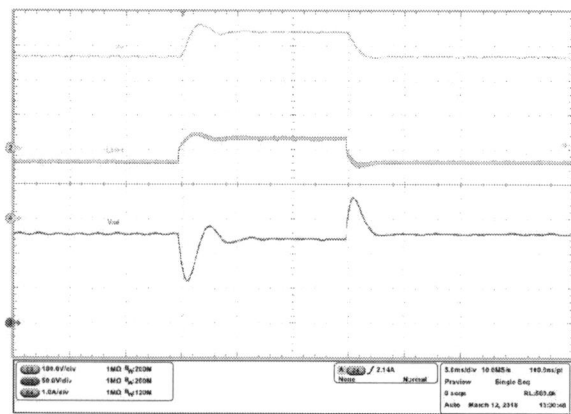

Fig. 7. Input and Output DC signals under load transient of 1.7 A (~200 W) to 2.3 A (~275 W) and back to 1.7 A, for the PRC operating in open loop. CH2 (light blue) shows the DC input voltage, CH4 (green) shows the output load current and DC output voltage is shown in CH3 (purple).

4078

REFERENCES

[1] H. Wang, T. Saha, R. Zane, "Design Considerations for Series Resonant Converters with Constant Current Input," *2016 IEEE Energy Conversion Congress and Exposition (ECCE)*, Milwaukee, WI, 2016, pp. 1-8.

[2] L. Corradini, D. Seltzer, D. Bloomquist, R. Zane, D. Maksimovic and B. Jacobson, "Minimum Current Operation of Bidirectional Dual-Bridge Series Resonant DC/DC Converters," in *IEEE Transactions on Power Electronics*, vol.27, no.7, pp. 3266-3276, 2012.

[3] N. Hasan, H. Wang, T. Saha and Z. Pantic, "A novel position sensorless power transfer control of lumped coil-based in-motion wireless power transfer systems," *Energy Conversion Congress and Exposition (ECCE)*, pp. 586-593, Sept. 2015.

[4] K. Asakawa, J. Muramatsu, M. Aoyagi, K. Sasaki. "Feasibility study on real-time seafloor glove monitoring cable-network-power feeding system," *Underwater Technology, 2002. Proceedings of the 2002 International Symposium on*, pp. 116-122, 2002.

[5] K. Asakawa, J. Kojima, J. Muramatsu, T. Takada. "Novel current to current converter for mesh-like scientific underwater cable network-concept and preliminary test result," *OCEANS, 2003. Proceedings*, vol. 4, pp. 1868-1873, Sept. 2003.

[6] H. Wang, T. Saha and R. Zane, "Control of series connected resonant converter modules in constant current dc distribution power systems," *2016 IEEE 17th Workshop on Control and Modeling for Power Electronics (COMPEL)*, Trondheim, 2016, pp. 1-7.

[7] H. Wang, T. Saha and R. Zane, "Impedance-based stability analysis and design considerations for DC current distribution with long transmission cable," *2017 IEEE 18th Workshop on Control and Modeling for Power Electronics (COMPEL)*, Stanford, CA, 2017, pp. 1-8.

[8] A. Mohammadpour, L. Parsa, M. H. Todorovic, R. Lai, R. Datta and L. Garces, "Series-Input Parallel-Output Modular-Phase DC–DC Converter With Soft-Switching and High-Frequency Isolation," in *IEEE Transactions on Power Electronics*, vol. 31, no. 1, pp. 111-119, Jan. 2016.

[9] K. Modepalli, A. Mohammadpour, T. Li and L. Parsa, "Three-Phase Current-Fed Isolated DC–DC Converter With Zero-Current Switching," in *IEEE Transactions on Industry Applications*, vol. 53, no. 1, pp. 242-250, Jan.-Feb. 2017.

[10] S. Suzuki and T. Shimizu, "A study on efficiency improvement of high-frequency current output inverter based on immittance conversion element," *International Power Electronics Conference (IPEC-Hiroshima 2014 - ECCE ASIA)*, Hiroshima, 2014, pp. 1166-1172.

[11] H. Pollock, "Simple constant frequency constant current load-resonant power supply under variable load conditions," in *Electronics Letters*, vol. 33, no. 18, pp. 1505-1506, 28 Aug 1997.

[12] M. Borage, S. Tiwari and S. Kotaiah, "Analysis and design of an LCL-T resonant converter as a constant-current power supply," in *IEEE Transactions on Industrial Electronics*, vol. 52, no. 6, pp. 1547-1554, Dec. 2005.

[13] G. De Falco, M. Gargiulo, G. Breglio and A. Irace, "Design of a parallel resonant converter as a constant current source with microcontroller-based output current regulation control," *International Symposium on Power Electronics Power Electronics, Electrical Drives, Automation and Motion*, Sorrento, 2012, pp. 632-635.

[14] A. Safaee, P. Jain and A. Bakhshai, "Time-domain steady-state analysis of fixed-frequency series resonant converters with phase-shift modulation," *2014 IEEE Transportation Electrification Conference and Expo (ITEC)*, Dearborn, MI, 2014, pp. 1-7.

[15] H. Wang, T. Saha and R. Zane, "Analysis and design of a series resonant converter with constant current input and regulated output current," *2017 IEEE Applied Power Electronics Conference and Exposition (APEC)*, Tampa, FL, 2017, pp. 1741-1747.

[16] T. Saha, H. Wang, B. Riar and R. Zane, "Analysis of zero voltage switching requirements and passive auxiliary circuit design for DC-DC series resonant converters with constant input current," *2016 IEEE 2nd Annual Southern Power Electronics Conference (SPEC)*, Auckland, 2016, pp. 1-6.

[17] T. Saha, H. Wang and R. Zane, "Zero voltage switching assistance design for DC-DC series resonant converter with constant input current for wide load range," *2017 IEEE 18th Workshop on Control and Modeling for Power Electronics (COMPEL)*, Stanford, CA, 2017, pp. 1-5.

The 2018 International Power Electronics Conference

Novel Sinusoidal Input Current Single-to-Three-Phase Z-Source Buck+Boost AC/AC Converter

M. Haider, D. Bortis and J. W. Kolar
Power Electronic Systems Laboratory,
ETH Zürich, Switzerland
Email: haider@lem.ee.ethz.ch

Y. Ono
Nabtesco R&D Center,
Nabtesco Corporation, Japan

Abstract—This paper introduces a novel unidirectional unity power factor single-to-three-phase Z-source Buck+Boost AC/AC Converter (*123ZBBC*) topology to supply three-phase AC machines with widely varying rated voltage directly from the single-phase mains. Due to the integration of the boost circuit into the inverter stage, the proposed circuit benefits from a reduced realization effort and an increased robustness. Furthermore, the insertion of a front-end buck-stage allows to select an intermediate voltage which is lower than the peak mains voltage and on the other hand enables to achieve a sinusoidal input current within the entire mains period. The paper gives a detailed analysis of the proposed converter including the different conduction states, the modulation schemes in order to implement the power factor correction and the inverter functionality, as well as the corresponding closed-loop control enabling sinusoidal input current and output voltages. Furthermore, the converter operation is verified by circuit simulations and the stresses on the main components are analyzed and compared to a conventional single-to-three-phase Z-source based AC/AC converter system.

I. INTRODUCTION

In industry applications, electrical drive systems with a power level of $5\,\mathrm{kW}$ to $10\,\mathrm{kW}$ are often supplied from the single-phase mains in order to keep the grid interface as simple as possible. This involves e.g. drive systems for fans, blowers, pumps and local automation systems. Furthermore, since in industrial three-phase networks a connection to the neutral conductor N is commonly not available, the single-phase front-end of the electrical drive system has to be connected between two phases [1], [2]. This means that the line-to-line voltage of the three-phase mains is applied to the front-end and thus a relatively wide input voltage range with voltages up to $400\,\mathrm{V_{rms}}$ or $480\,\mathrm{V_{rms}}$ has to be covered. Furthermore, the single-phase front-end has to provide active power factor correction, i.e. a sinusoidal input current in phase with the input voltage, to keep the harmonic distortion and reactive power in the grid at a minimum. Finally, in order to cover a wide area of applications, compatibility to three-phase machines with different rated voltages has to be ensured.

All these requirements e.g. can be fulfilled with a buck-boost PFC rectifier [3]–[5] followed by a voltage source PWM inverter (VSI) [6]. Due to the buck-boost functionality of the PFC rectifier [7], [8], the intermediate DC link voltage can be selected independent from the mains input voltage, which advantageously allows to flexibly adapt the DC link voltage to the required (rated) machine voltage. In contrast, for a conventional boost PFC rectifier, the DC link voltage level would be limited to voltages above the peak value of the mains voltage. For applications with low nominal machine voltages this would mean that the inverter stage would have to be operated with a low modulation index and consequently would have to be designed for high peak currents (and voltages) in order to provide the required machine power. However, the mentioned two-stage system, comprising a two-switch buck-boost PFC rectifier and a three-phase VSI, comes with a relatively high realization effort and therefore the question of a topological simplification, for example by integration of the boost function into the inverter stage by means of a Z-source inverter [9]–[12], arises.

The basic Z-source inverter topology only employs an impedance-source (Z-source) network followed by a three-phase inverter (cf. **Fig. 1**) [13], where the boost operation is realized with a short-circuit interval of one inverter bridge leg, the so-called shoot-through interval, which enables an intermediate voltage $\bar{v}_{\mathrm{PN}} = v_{\mathrm{C}}$ that is higher than the input voltage v_{AB}. As a result, the Z-source inverter

also features an enhanced robustness and reliability compared to the VSI, where a short-circuit of one inverter bridge leg could lead to the destruction of the converter system. Hence, due to this integrated boost functionality, the Z-source inverter constitutes an interesting alternative for applications with a wide input voltage range, e.g. for fuel or solar cell applications [9], [14]. However, in order to prevent a current flowing back to the DC-voltage source v_{DC}, a series diode $\mathrm{D_Z}$ is required at the input as shown in **Fig. 1(a)**. Advantageously, this series diode can be replaced by a diode rectifier $\mathrm{D_1} - \mathrm{D_4}$, since due to the wide input voltage range capability, the Z-source inverter can be directly connected to the single-phase mains and be operated as a single-stage single-to-three phase AC/AC converter (cf. **Fig. 1(b)**) [15], [16]. Unfortunately, with this high level of integration also certain degrees of freedom concerning controllability are lost, which means that e.g. in the vicinity of the input voltage zero crossings the grid current i_{G} can no longer be controlled to be sinusoidal, since the Z-source inverter draws at least a minimal input current $\bar{i}_{\mathrm{A,min}}$, as will be shown later. A further limitation of the system is that due to the inherent boost functionality, the intermediate voltage $\bar{v}_{\mathrm{PN}}$ has to be larger than the peak value of the input voltage v_{G}, which results in a high voltage stress on the semiconductor devices and the passive components.

These drawbacks can be resolved by adding a half-bridge directly behind the bridge rectifier, which in combination with the already existing Z-source network features a simple buck-stage (cf. **Fig. 1(c)**). Consequently, with the proposed *single-to-three-phase Z-source Buck+Boost Converter (123ZBBC*, cf. **Fig. 6**), on the one hand the input voltage v_{G} can be stepped down, which enables to reduce the intermediate voltage $\bar{v}_{\mathrm{PN}}$ below the peak input voltage and therefore also reduces the voltage stress on the Z-source network and the semiconductor devices of the inverter stage. On the other hand, the input current i_{G} can be controlled to be sinusoidal within the entire mains period, even though the Z-source input current shows a value equal or higher than $\bar{i}_{\mathrm{A,min}}$. Hence, since the *123ZBBC* features the same functionality as the conventional buck-boost PFC rectifier with a subsequent VSI, and due to the high level of integration also a high power density is expected. In this paper the operation principle, the control as well as the modulation scheme of the *123ZBBC* is analyzed in detail. In **Section II**, the different conduction states and the most suitable PFC modulation strategy are presented. Afterwards in **Section III**, the corresponding control structure for the intermediate voltage $\bar{v}_{\mathrm{PN}}$, the sinusoidal input current i_{G} and the machine speed is derived, which is then verified by means of a circuit simulation. Furthermore, due to the fact that the instantaneous intermediate voltage v_{PN} changes depending on the conduction state, special attention has to be paid on the modulation scheme of the inverter, particularly on the distribution of the shoot-through interval within one switching period. An analysis of the stresses on the main components and a comparison to the conventional Z-source based single-to-three phase topology is performed in **Section IV**. Finally, **Section V** summarizes the findings of the work and gives an outlook to future research.

II. CONDUCTION STATES AND PFC MODULATION STRATEGY

Similar to conventional buck-boost PFC rectifiers, the proposed *123ZBBC* is operated in a buck (BU) mode when the grid voltage v_{G} is higher than the intermediate voltage $\bar{v}_{\mathrm{PN}}$, and in a boost (BO)

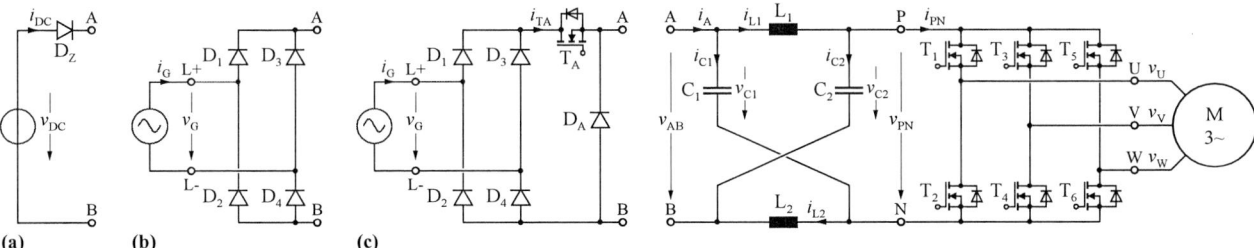

Fig. 1. Basic Z-source inverter topology consisting of an impedance-source network (Z-source) and a conventional three-phase inverter stage, which features an inherent boost functionality and can be directly supplied from **(a)** a wide range DC-voltage source v_{DC}, e.g. fuel cell or solar cell, via a series diode D_Z [13], **(b)** a single-phase supply connected to a diode bridge rectifier, or **(c)**, as proposed in this paper, a single-phase supply followed by a bridge rectifier and a half-bridge.

mode when v_G is lower than $\bar{v}_{PN}$. As will be shown later, however, in the vicinity of the zero-crossing of the input voltage,

$$v_G = \hat{V}_G \cdot \cos \omega_G t, \tag{1}$$

the converter has to be operated in a buck-boost (BB) mode in order to be able to draw a sinusoidal current

$$i_G = G_C \cdot v_G = \frac{2P_M}{\hat{V}_G^2} \cdot v_G \tag{2}$$

within the entire mains period from the grid.

Hence, the buck and the integrated boost stage are controlled by means of the buck duty cycle d_A, which defines the effective on-time of the buck transistor T_A, i.e. the time when T_A is turned on *and* one of the upper/lower diodes of the input diode bridge is conducting, and the boost duty cycle d_B, which corresponds to the Z-source specific relative shoot-through time of one of the inverter bridge legs [13]. Consequently, for the inverter stage this means that during a shoot-through interval no voltage is applied to the machine terminals, thus the formation of the three output voltages v_U, v_V, and v_W has to occur during the remaining non shoot-through interval, whose duration depends on the mentioned duty cycles d_A and d_B of the PFC rectifier. Due to the dependency of the inverter's modulation scheme on the rectifier's duty cycles d_A and d_B, the control strategy to achieve a sinusoidal input current is examined first. For this purpose, the proposed circuit topology is simplified and the conduction states of the resulting circuit are analyzed. Afterwards, the modulation strategy to achieve the desired PFC functionality is derived.

A. Derivation of the Equivalent Circuit

The traditional Z-source inverter [13] consists of a symmetrical impedance network with $C = C_1 = C_2$ and $L = L_1 = L_2$, which also implies symmetric operation conditions: $v_C = v_{C1} = v_{C2}$, $i_C = i_{C1} = i_{C2}$, $i_L = i_{L1} = i_{L2}$ and $v_L = v_{L1} = v_{L2}$.

In order to simplify the analysis of the basic converter operation, the grid voltage, the EMI-filter and the bridge rectifier of the actual converter topology (cf. Fig. 6) are replaced by a voltage source $v_Q = |v_G|$ and a series diode D_R, which models the unidirectional power flow (cf. **Fig. 2(a)**). Furthermore, the three-phase inverter and the machine are substituted by a shoot-through/boost transistor T_B in parallel to a current source

$$i_Q = \frac{1}{2}\left[i_U\left(S_1 - S_2\right) + i_V\left(S_3 - S_4\right) + i_W\left(S_5 - S_6\right)\right], \tag{3}$$

where $S_i \in [0,1]$ is the switching state of the transistor T_i. The resulting equivalent circuit, which models the behaviour of the Z-source based buck-boost rectifier stage, is presented in **Fig. 2**.

B. Conduction States

The two transistors T_A and T_B are operated with the switching frequency $f_{SW} = \frac{1}{T_{SW}}$, thus four different switching or conduction states would be found. However, as will be shown in the following, when T_B is turned on, T_A can be either turned on or turned off without any effect on the current paths; hence, only three states exist. For the

analysis of these states, the energy related quantities v_C and i_L are assumed to be impressed, i.e constant.

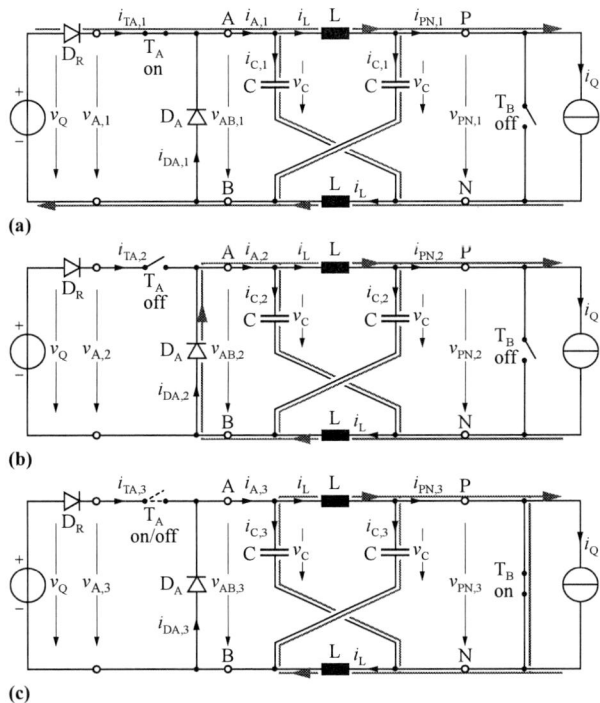

Fig. 2. Equivalent circuit of the proposed Z-source-based buck-boost AC/AC converter, where a symmetric impedance network is assumed. The grid voltage and the bridge rectifier are replaced by a voltage source v_Q and a series diode D_R. The inverter and the machine are substituted by a parallel connection of a shoot-through/boost transistor T_B and a current source i_Q. In **(a)**-**(c)** the three conduction states and the corresponding current paths are shown.

State 1 (Active State): The first conduction state with the duration $t_A = d_A T_{SW}$ equals the active state (energy is directly transferred from the input to the output) and is defined by $S_A = 1$ and $S_B = 0$, which means that T_A is turned on and T_B is turned off (cf. **Fig. 2(a)**). It should be mentioned again that t_A corresponds to the *effective* on-time of T_A meaning that T_A is not only turned on, but also conducts a current, which e.g. is not the case in state 3 even if T_A is in the on-state. Hence, in state 1 a positive buck-stage input current $i_{TA,1}$ ($i_{TA,1} > 0$ A) must be assumed, such that the diode D_R and the buck switch T_A are conducting. Consequently, a positive input voltage v_Q is applied to the buck diode D_A, which means that this diode has to block and the current $i_{TA,1}$ is defined by the two impedance network currents i_L and $i_{C,1}$ ($i_{TA,1} = i_{A,1} = i_L + i_{C,1}$), which in turn are determined by the current source i_Q ($i_Q = i_L - i_{C,1}$). Based on these two equations it can be found that the inductor current i_L must always be larger than half of the peak inverter output current $\hat{I}_Q$, which actually equals the peak phase current of the machine $\hat{I}_M$ ($= \hat{I}_U = \hat{I}_V = \hat{I}_W$), in order to keep the current through D_R and T_A positive, i.e. $i_{TA,1} > 0\,\mathrm{A}$,

$$i_L \geq i_{L,min} = \frac{1}{2} \cdot \hat{I}_M. \tag{4}$$

The instantaneous value of the intermediate voltage, i.e. the inverter input voltage $v_{PN,1}$, can be calculated based on the two voltage equations $|v_G| = v_C + v_{L,1}$ and $v_{PN,1} = v_C - v_{L,1}$, which results in $v_{PN,1} = 2v_C - |v_G|$. Furthermore, the voltage applied to the inductor can be found as $v_{L,1} = |v_G| - v_C$. Consequently, since due to the antiparallel body diodes of the inverter stage switches the voltage $v_{PN,1}$ cannot fall below zero ($v_{PN,1} > 0\,\mathrm{V}$), it reveals that even in buck operation the capacitor voltage v_C cannot be reduced below half of the peak input voltage $\hat{V}_G$,

$$v_C \geq v_{C,min} = \frac{1}{2} \cdot \hat{V}_G. \tag{5}$$

State 2 (Buck State): In the second state with duration $t_0 = d_0 T_{SW}$, which is only used during buck operation, T_A is turned off ($S_A = 0$, $S_B = 0$ and $i_{TA,2} = 0\,\mathrm{A}$) and the current commutates from switch T_A to the diode D_A (cf. **Fig. 2(b)**). Consequently, since the remaining current paths do not change (impressed by the inductor current and the load current), the diode current $i_{DA,2}$ can be expressed by the same equations as used in state 1, i.e. $i_{DA,2} = i_{A,2} = i_L + i_{C,2}$ and $i_Q = i_L - i_{C,2}$. In order to keep D_A conducting, i.e. $i_{DA,2} > 0\,\mathrm{A}$, the same condition (4) as for state 1 is found. Furthermore, also the intermediate voltage can be derived by the same equations, however, due to the conducting diode D_A, the voltage equation simplifies to $v_{PN,2} = 2v_C$. As can be noticed, the voltage stress on the semiconductors is twice the capacitor voltage v_C and therefore a reduction of v_C by an additional buck-stage is encouraged. In addition, the inductor voltage changes to $v_{L,2} = -v_C$, which means that the full capacitor voltage is applied in negative direction to the inductor.

State 3 (Boost State): The third state is defined by $S_B = 1$, which means that the shoot-through transistor T_B is closed during the interval $t_B = d_B T_{SW}$ and the converter is operated in the boost mode (cf. **Fig. 2(c)**). Consequently, due to the shoot-through, the inverter input voltage is zero ($v_{PN,3} = 0\,\mathrm{V}$) and no voltage is applied to the machine terminals. Furthermore, the full capacitor voltage v_C is applied to each inductor L in positive direction, i.e. $v_{L,3} = v_C$. Based on the already described voltage equation it reveals that the sum of both capacitor voltages is applied to the diode D_A ($v_{DA,3} = v_{AB,3} = 2v_C$), hence D_A has to block. On the other hand, the diode D_R blocks the voltage $2v_C - |v_G|$ as long as (5) is fulfilled, hence T_A can be either turned on or off without any effect, and the only remaining current path is given through the Z-source network.

For the different operation modes either all or only a subset of the described conduction states are used during one switching cycle T_{SW}. For example, the buck (BU) operation only utilizes the states 1 and 2, which leads to the relative durations $d_A + d_0 = 1$ ($d_B = 0$), while the boost (BO) operation uses the states 1 and 3, which equals $d_A + d_B = 1$ ($d_0 = 0$). Consequently, all three states are only required during buck-boost (BB) operation, i.e. $d_A + d_0 + d_B = 1$. It has to be noted again that the instantaneous inverter voltage $\bar{v}_{PN}$ changes depending on the actual conduction state (cf. **Fig. 3**), which has to be considered later for the duty cycle calculation of the inverter switches. However, for the duty cycle calculation of the rectifier the averaged values are needed first.

Based on a steady state analysis over one switching period T_{SW}, the averaged intermediate voltage $\bar{v}_{PN}$ is calculated as $\bar{v}_{PN} = d_A v_{PN,1} + d_0 v_{PN,2} + d_B v_{PN,3}$, whereby the interval d_0 is replaced by $d_0 = 1 - d_A - d_B$. Similarly, the averaged inductor voltage $\bar{v}_L$ and the averaged input current $\bar{i}_{TA}$ are derived, which together with the equivalent load current $\bar{i}_Q$ results in

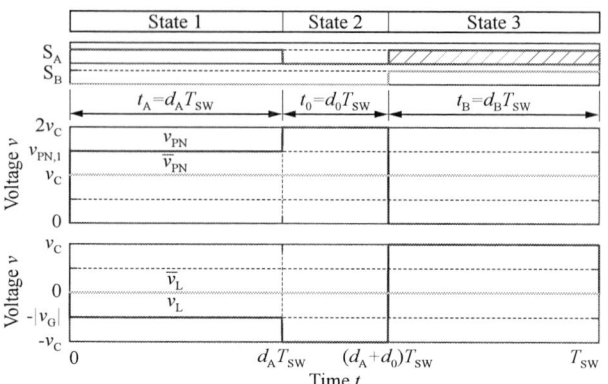

Fig. 3. PFC rectifier switching signals S_A and S_B, the corresponding conduction state, the time-dependent intermediate voltage v_{PN}, the inductor voltage v_L, and the switching frequency averaged quantities $\bar{v}_{PN}$ and $\bar{v}_L$ are shown over one switching period T_{SW}.

$$\bar{v}_{PN} = 2v_C \cdot (1 - d_B) - |v_G| \cdot d_A, \tag{6}$$

$$\bar{v}_L = |v_G| \cdot d_A - v_C \cdot (1 - 2 \cdot d_B), \tag{7}$$

$$\bar{i}_Q = \frac{P_M}{\bar{v}_{PN}}, \tag{8}$$

$$\bar{i}_{TA} = |i_G| = (2 \cdot i_L - \bar{i}_Q) \cdot d_A. \tag{9}$$

C. PFC Modulation Strategy

In the following, each duty cycle d_x ($x \in \{A, B\}$) is split into a steady state duty cycle D_x and a duty cycle variation $\tilde{d}_x$ provided from the circuit controller

$$d_A = D_A + \tilde{d}_A, \tag{10}$$

$$d_B = D_B + \tilde{d}_B. \tag{11}$$

This allows to first determine the needed duty cycles D_A and D_B from (6)-(9) in steady state, which means that the averaged inductor voltage is zero ($\bar{v}_L = 0\,\mathrm{V}$) and in turn also the duty cycles $\tilde{d}_A$ and $\tilde{d}_B$ derived from the controller are zero. Consequently, the right side of (7) can be set to zero, resulting in the average capacitor voltage

$$V_C = \bar{v}_{PN} = |v_G| \cdot \frac{D_A}{1 - 2 \cdot D_B} \geq \frac{1}{2} \cdot \hat{V}_G, \tag{12}$$

which based on (5) has to be larger than half the peak input voltage $\hat{V}_G$. Furthermore, applying (12) in (6) reveals that v_C has to equal the averaged inverter voltage $\bar{v}_{PN}$ and due to a large capacitance C can be assumed to be nearly constant. From (12), the modulation index m is calculated as the ratio of the grid voltage $|v_G|$ and the intermediate voltage $\bar{v}_{PN}$,

$$m = \frac{|v_G|}{\bar{v}_{PN}} = \frac{1 - 2 \cdot D_B}{D_A} \in [0, 2]. \tag{13}$$

As can be noticed, due to the absolute value of v_G, m is limited to positive values and due to (5), m is restricted to values equal or below 2. Furthermore, a modulation index $m \in [0, 1]$ means boost (BO) operation and for $m \in [1, 2]$ the system is operated in buck (BU) mode (cf. **Fig. 4**). As already mentioned, in buck operation $D_{B,BU}$ is zero, which means that $D_{A,BU} = 1/m$ for $m \in [1, 2]$ (cf. **Fig. 4**). On the other hand, in boost-operation it follows that $D_{B,BO} = 1 - D_{A,BO}$ and thus from (13) it is found that $D_{A,BO} = 1/(2 - m)$ for $m = \in [0, 1]$. Consequently, as can be noticed from **Fig. 4**, the steady state duty cycle $D_{A,BU/BO}$ valid for boost and buck operation is always found by taking the minimum of the two mentioned duty cycles $D_{A,BU}$ and $D_{A,BO}$, which is

$$D_{A,BU/BO} = \min\left(\frac{1}{m}, \frac{1}{2 - m}\right). \tag{14}$$

With this modulation strategy, the switching losses and the inductor current i_L can be kept minimal, thus typically the highest converter

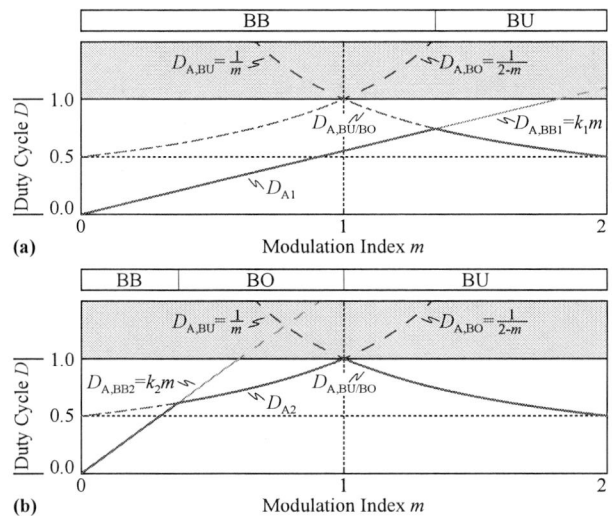

(a)

(b)

Fig. 4. The optimal steady state buck duty cycle D_A is given with the minimum of $D_{A,BU/BO}$ and $D_{A,BB}$. Thereby, $D_{A,BB}$ is proportional to the modulation index m by the factor k and ensures an inductor current i_L above the minimum value $i_{L,min}$. The steady state duty cycle $D_{A,BU/BO}$ is valid for boost and buck operation and results from the minimum of the buck $D_{A,BU}$ and the boost $D_{A,BO}$ operation related buck duty cycle. The employed operation modes depend on the factor k, **(a)** in case $k = k_1 < 1$ only the buck-boost (BB) and the buck (BU) operation are present, while **(b)** in case $k = k_2 > 1$ all three operation modes BB, BO and BU occur.

efficiency can be achieved. However, in order to guarantee proper converter operation, based on (4) the inductor current i_L has to be larger than a certain minimum value $i_{L,min}$, which with pure buck or boost operation is always undercut around the input voltage zero crossings. Fortunately, from (13) it can be noted that a certain modulation index m also can be achieved with other sets of duty cycles D_A and D_B, i.e. when the system is operated in buck-boost (BB) mode. Even though this operation is not preferred, with the adaptation of the duty cycles D_A and D_B the inductor current i_L can be kept above $i_{L,min}$. Hence, the maximum allowed duty cycle $D_{A,BB}$ can be derived from (9), while for i_L the relation given in (4) is used,

$$D_{A,BB} = \frac{|i_G|}{\hat{I}_M - \bar{i}_Q}. \tag{15}$$

In order to obtain for $D_{A,BB}$ a dependency on m, the currents $|i_G|$ and $\bar{i}_Q$ are substituted considering (2) and (3), respectively. The machine current $\hat{I}_M$ can be expressed by the machine power $P_M = 3/2 \hat{V}_M \hat{I}_M \cos(\varphi)$ and the modulation index of the inverter stage $M = 2\hat{V}_M/V_C$, which after some rearrangements results in

$$D_{A,BB} = \frac{6M \cos\varphi}{4 - 3M \cos\varphi} \left(\frac{V_C}{\hat{V}_G}\right)^2 \cdot m = k \cdot m, \tag{16}$$

where $M < M_{max} = \frac{2}{\sqrt{3}}$ and $\cos\varphi \leq 1$. Consequently, the optimal and also maximum allowed duty cycle D_A is found by selecting the minimum value out of $D_{A,BU/BO}$ and $D_{A,BB}$, which is

$$D_A = \min\left(D_{A,BU/BO}, D_{A,BB}\right). \tag{17}$$

The corresponding D_B is found by solving (13) to

$$D_B = \frac{1}{2}\left(1 - m \cdot D_A\right). \tag{18}$$

In **Fig. 4**, $D_{A,BB}$ is shown for two different load conditions. As can be noticed, depending on the slope of $D_{A,BB}$, within one mains half cycle - where m changes sinusoidally from zero to a certain maximum - either the buck (BU) operation directly follows after the buck-boost (BB) operation (cf. **Fig. 4(a)**) or all three operation modes (BB,BO,BU) are present (cf. **Fig. 4(b)**). The resulting PFC rectifier waveforms for the latter case are shown in **Fig. 5** over one grid period.

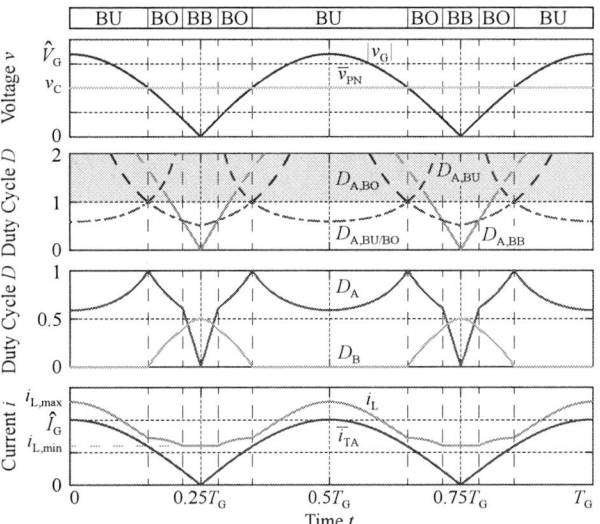

Fig. 5. Calculated waveforms for the inductor current minimal PFC modulation strategy showing the operation modes, the rectified grid voltage $|v_G|$ and intermediate voltage $\bar{v}_{PN}$, the steady state buck duty cycles $D_{A,BU}$, $D_{A,BO}$, $D_{A,BU/BO}$ and $D_{A,BB}$, the optimal steady state duty cycles D_A and D_B, the rectified grid current $\bar{i}_{TA}$ and the inductor current i_L over one grid period T_G for $\bar{v}_L = 0$ V and a constant capacitor voltage v_C.

III. CONTROL STRUCTURE

The described PFC modulation strategy already allows to operate the PFC rectifier in an open loop fashion. However, due to disturbances, nonlinearities or simplified models, typically the capacitor voltage and the input current deviate from their reference values. Therefore, a cascaded *PFC rectifier control* consisting of an inner *input/inductor current control* loop and an outer *capacitor voltage control* loop is needed, that provides the two controller duty cycles $\tilde{d}_A$ and $\tilde{d}_B$ as outputs, which are afterwards added to the steady state duty cycles D_A and D_B (cf. Section II-C) in order to achieve a sinusoidal input current and a controlled capacitor voltage (cf. **Fig. 6**). In addition, for the inverter stage an *inverter/machine control* [17] is used, which has to generate the duty cycles d_U, d_V and d_W of the inverter in such a way that the machine speed reference ω^* can be properly tracked. However, the calculated duty cycles of both stages cannot directly be assigned to a certain switch T_A or $T_1 - T_6$, which means that the corresponding switching signals S_A and $S_1 - S_6$ first have to be derived in the *switching signal generation* block depending on the output voltage sector, as will be shown in the following.

A. PFC Rectifier Control

The outer *capacitor voltage control* has to regulate the measured and averaged capacitor voltage $\bar{v}_C$ according the reference $\bar{v}_C^* = V_{PN}^*$. Hence, dependent on the voltage error $\Delta \bar{v}_C$, the voltage controller R_V determines the needed capacitor current $\bar{i}_C^*$ - and with the reference capacitor voltage $\bar{v}_C^*$ the needed average capacitor power $\bar{p}_C^*$ - to keep the capacitor voltage at its nominal value.

Based on $\bar{p}_C^*$, now the input current reference i_G^* for the *input/inductor current control* can be calculated in order to provide the needed power to the capacitor. However, the PFC rectifier not only has to cover the power $\bar{p}_C^*$, but has to provide in addition the load power to the machine. Therefore, $\bar{p}_M$ derived from the *inverter/machine control* is added to $\bar{p}_C^*$.

With (9), the inner *input/inductor current control* now translates i_G^* into the inductor current reference i_L^*, which together with the measured inductor current i_L is then processed by the current controller R_I. The current controller, implemented as multiple parallel PR-controllers [18], finally provides the reference inductor voltage $\tilde{v}_L$ to adjust the current i_L to the desired value. The inductor voltage

The 2018 International Power Electronics Conference

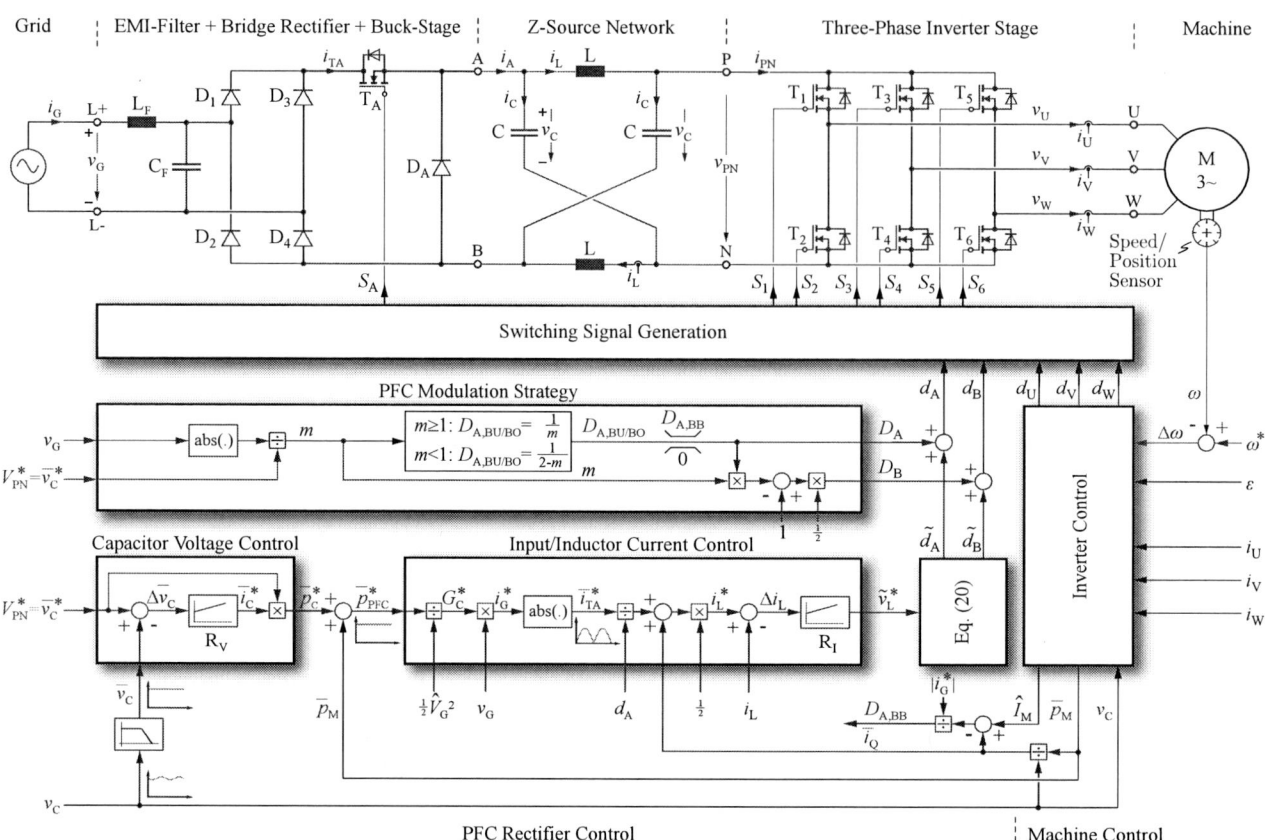

Fig. 6. Proposed Z-source-based buck-boost AC/AC converter: The corresponding control structure consists of a *PFC rectifier control*, which derives the buck duty cycle d_A and the shoot-through duty cycle d_B, and a conventional *machine control*, which provides the inverter duty cycles d_U, d_V and d_W. The *switching signal generation* block generates the transistor switching signals $S_1 - S_6$, based on the already determined PFC rectifier and the inverter duty cycles.

variation $\tilde{v}_L$ can be achieved by either a duty cycle variation $\tilde{d}_A$ or a duty cycle variation $\tilde{d}_B$, (cf. (7)),

$$\tilde{v}_L = |v_G| \cdot \tilde{d}_A + 2v_C \cdot \tilde{d}_B. \quad (19)$$

This advantage can be optimally exploited for the different operating mode. For example, in buck (BU) operation, the shoot-through duty cycle d_B is zero ($d_B = 0$), hence also $\tilde{d}_B = 0$. In boost (BO) operation $d_0 = 0$, which means that $d_A + d_B = 1$ and therefore $\tilde{d}_A = -\tilde{d}_B$ is found. In the remaining buck-boost (BB) operation, where the duty cycle d_A is defined by the minimum inductor current $i_{L,\min}$, it is clear that an increase of this duty cycle by $\tilde{d}_A$ is not allowed, and consequently only $\tilde{d}_B$ can be changed, while $\tilde{d}_A = 0$

$$\tilde{d}_A, \tilde{d}_B = \begin{cases} \dfrac{\tilde{v}_L}{|v_G|}, & 0 & \text{if } \text{BU} \\[2mm] \dfrac{\tilde{v}_L}{|v_G| - 2v_C}, & -\tilde{d}_A & \text{if } \text{BO} \\[2mm] 0, & \dfrac{\tilde{v}_L}{2v_C} & \text{if } \text{BB} \end{cases} \quad (20)$$

B. Machine Control

The *inverter/machine control* has to regulate the rotational speed ω with respect to its reference value ω^*, which can be implemented with a conventional inverter control for variable speed drives [17]. There, the inverter duty cycles d_U, d_V and d_W are derived based on the speed error $\Delta\omega$, the rotor position ε, the capacitor voltage v_C, and the phase currents i_U, i_V and i_W. Furthermore, as already mentioned, the averaged machine power $\bar{p}_M$ and the phase peak current $\hat{I}_M$ are used by the PFC rectifier control as a feedforward and on the other hand are used to calculate $D_{A,BB}$ based on (15). Alternatively, this calculation would also be possible with (16), where the modulation index of the inverter M and the power factor $\cos(\varphi)$ would have to be known.

C. Switching Signal Generation

The switching signal generation block has to translate the PFC rectifier duty cycles d_A and d_B as well as the inverter duty cycles d_U, d_V and d_W to the actual transistor switching signals S_A and $S_1 - S_6$, since these duty cycles cannot directly be assigned to a certain switch. Therefore, first the transistor duty cycles $d_1 - d_6$ of the switches $S_1 - S_6$ are derived, and then the corresponding switching patterns for the different operation modes are discussed.

Starting from a conventional VSI with constant DC-link voltage v_{DC}, the switching signals and the corresponding phase voltage waveform within one switching period T_{SW} can be easily derived by comparing the symmetrical triangular carrier with the inverter duty cycles d_U, d_V and d_W, i.e. conventional PWM as shown in **Fig. 7** for the phase voltage v_{UN} between phase output U and the negative DC-rail N. There, the phase voltage is equal to the DC-link voltage in case the carrier signal is smaller than the duty cycle d_U, and within the remaining interval the phase voltage is zero, which means that the duty cycle d_U actually defines the average phase voltage $\bar{v}_{UN} = v_{DC} \cdot d_U$ which is applied to the machine over one switching period.

Hence, the same averaged output voltage $\bar{v}_{UN}$ should also result from the duty cycles d_U, d_V and d_W calculated by the *inverter control* block of the *123ZBBC*. However, there the instantaneous intermediate voltage v_{PN}, i.e. the input voltage of the inverter stage, changes depending on the present conduction state of the PFC rectifier (cf. Section II). In case of a shoot-through state, for example, v_{PN} is zero and therefore the needed averaged output voltage $\bar{v}_{UN}$ can only be formed during the active state ($v_{PN,1} = 2v_C - |v_G|$) and the buck state ($v_{PN,2} = 2v_C$). One possibility is to modify the PWM carrier signal to an asymmetrical triangular carrier, which rises within the

4084

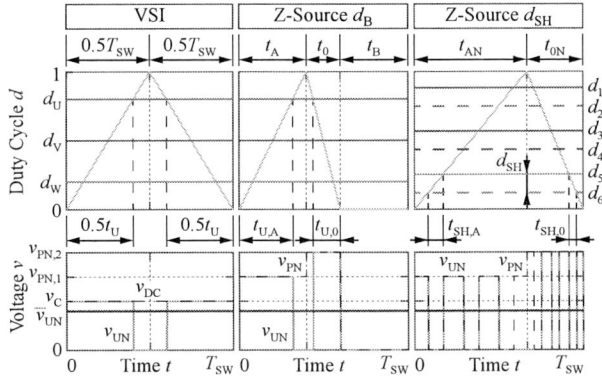

Fig. 7. PWM modulation scheme and the resulting phase voltage v_{UN} for a conventional three-phase VSI, the proposed converter structure with a single shoot-through interval t_{B} and with the distributed shoot-through intervals $t_{\mathrm{SH,A}}$ and $t_{\mathrm{SH,0}}$.

active state $t_{\mathrm{A}} = d_{\mathrm{A}} T_{\mathrm{SW}}$ from zero to one, falls again during the buck state $t_0 = d_0 T_{\mathrm{SW}}$ down to zero and stays at zero within the remaining shoot-through interval $t_{\mathrm{B}} = d_{\mathrm{B}} T_{\mathrm{SW}}$ (cf. **Fig. 7**). With this assumption, on the one hand the zero voltage interval, which in addition to the shoot-through interval is needed to generate the correct averaged phase voltage v_{PN}, is proportionally distributed between the active state interval $d_{\mathrm{A}} T_{\mathrm{SW}}$ and the buck stage interval $d_0 T_{\mathrm{SW}}$, and on the other hand it is beneficially achieved that the same duty cycles d_{U}, d_{V} and d_{W} as for the conventional VSI can be used. With $t_{\mathrm{U,A}} = d_{\mathrm{U}} \cdot t_{\mathrm{A}} = d_{\mathrm{U}} \cdot d_{\mathrm{A}} \cdot T_{\mathrm{SW}}$, $t_{\mathrm{U,B}} = d_{\mathrm{U}} \cdot t_0 = d_{\mathrm{U}} \cdot d_0 \cdot T_{\mathrm{SW}}$ and (6), the correct calculation of the averaged phase voltage $\bar{v}_{\mathrm{UN}}$ can be verified,

$$
\begin{aligned}
\bar{v}_{\mathrm{UN}} &= v_{\mathrm{PN,1}} \cdot \frac{t_{\mathrm{U,A}}}{T_{\mathrm{SW}}} + v_{\mathrm{PN,2}} \cdot \frac{t_{\mathrm{U,0}}}{T_{\mathrm{SW}}} \\
&= (2 v_{\mathrm{C}} - |v_{\mathrm{G}}|) \cdot d_{\mathrm{A}} d_{\mathrm{U}} + 2 v_{\mathrm{C}} \cdot d_0 d_{\mathrm{U}} = \bar{v}_{\mathrm{PN}} \cdot d_{\mathrm{U}}. \quad (21)
\end{aligned}
$$

In order to reduce the current ripple in the Z-source inductance L, another possibility is to distribute and integrate the shoot-through interval $d_{\mathrm{B}} T_{\mathrm{SW}}$ into the inverter switching transitions as also proposed in [13]. In contrast to the conventional switching procedure, where first one switch of a half-bridge is turned off before the other switch is turned on, i.e. the switching signals are separated by a certain interlocking time where both transistors are kept off, now the sequence is reversed, which means that both switches are turned on during a certain shoot-through time t_{SH}. Hence, since this actually corresponds to a negative interlocking time, where only the switching sequence of the upper and lower switch are changed, it becomes clear that with the integration of the shoot-through interval into a switching transition the number of switching transitions within one switching cycle is not increased. However, it has to be mentioned that in a conventional switching transition the interlocking delay is typically short compared to the on- and off-times of the switches. Therefore, in a half-bridge for the high-side and low-side switches the same duty cycle $d_{\mathrm{H}} = d_{\mathrm{L}}$ is calculated, whereas the interlocking delay is then e.g. generated by the PWM unit of the microcontroller. However, if now the shoot-through interval t_{SH} is integrated into the switching transient, the negative interlocking time can reach values which are similar to the on- and off-times of the switches. Hence, for the switches in a half-bridge individual duty cycles d_{H} and d_{L}, where $d_{\mathrm{H}} = d_{\mathrm{L}} + d_{\mathrm{SH}}$, have to be used. Similarly to the first approach, the shoot-through interval $t_{\mathrm{B}} = d_{\mathrm{B}} T_{\mathrm{SW}}$ is now proportionally distributed between the active state and buck state interval, i.e. $t_{\mathrm{B,A}} = d_{\mathrm{B,A}} T_{\mathrm{SW}} = d_{\mathrm{B}} d_{\mathrm{A}}/(d_{\mathrm{A}} + d_0) T_{\mathrm{SW}}$ and $t_{\mathrm{B,0}} = d_{\mathrm{B,0}} T_{\mathrm{SW}} = d_{\mathrm{B}} d_0/(d_{\mathrm{A}} + d_0) T_{\mathrm{SW}}$, which results in the two extended intervals $t_{\mathrm{AN}} = d_{\mathrm{AN}} T_{\mathrm{SW}} = (d_{\mathrm{A}} + d_{\mathrm{B,A}}) T_{\mathrm{SW}}$ and $t_{\mathrm{0N}} = d_{\mathrm{0N}} T_{\mathrm{SW}} = (d_0 + d_{\mathrm{B,0}}) T_{\mathrm{SW}}$, where $d_{\mathrm{AN}} + d_{\mathrm{0N}} = 1$ must be satisfied (cf. **Fig. 7**),

$$
d_{\mathrm{AN}} = d_{\mathrm{A}} + \frac{t_{\mathrm{B,A}}}{T_{\mathrm{SW}}} = \frac{d_{\mathrm{A}}}{1 - d_{\mathrm{B}}}. \quad (22)
$$

$$
d_{\mathrm{0N}} = d_0 + \frac{t_{\mathrm{B,0}}}{T_{\mathrm{SW}}} = \frac{d_0}{1 - d_{\mathrm{B}}}. \quad (23)
$$

Since during the active interval and the buck state interval a switching transition occurs in each output phase both shoot-through intervals $t_{\mathrm{B,A}}$ and $t_{\mathrm{B,0}}$ have to be divided by three, in order to get the needed shoot-through times $t_{\mathrm{SH,A}} = 1/3 \cdot t_{\mathrm{B,A}}$ and $t_{\mathrm{SH,0}} = 1/3 \cdot t_{\mathrm{B,0}}$, i.e. the negative interlocking delays for the switching transitions in each state. If now, according to the first approach, again an asymmetrical PWM carrier signal is used, which rises within $d_{\mathrm{AN}} T_{\mathrm{SW}}$ from zero to one and falls again during $d_{\mathrm{0N}} T_{\mathrm{SW}}$ down to zero, from geometrical considerations, both (horizontal) shoot-through times $t_{\mathrm{SH,A}}$ and $t_{\mathrm{SH,0}}$ result in the same (vertical) duty cycle difference $d_{\mathrm{SH}} = 1/3 \cdot d_{\mathrm{B}}$ between the duty cycle of the upper switch, $d_{\mathrm{x,H}}$, and of the lower switch, $d_{\mathrm{x,L}}$, for each half-bridge $x \in \{a, b, c\}$ [12] (cf. **Fig. 7**),

$$
d_{\mathrm{x,H}} = d_{\mathrm{x,L}} + 1/3 \cdot d_{\mathrm{B}}, \quad x \in \{a, b, c\}, \quad (24)
$$

whereby, a represents the half-bridge with the lowest duty cycle, i.e. $d_{\mathrm{a}} = \min(d_{\mathrm{U}}, d_{\mathrm{V}}, d_{\mathrm{W}})$, c the half-bridge with the highest duty cycle, i.e. $d_{\mathrm{c}} = \max(d_{\mathrm{U}}, d_{\mathrm{V}}, d_{\mathrm{W}})$, and b the remaining half-bridge, i.e. $d_{\mathrm{b}} = \mathrm{mid}(d_{\mathrm{U}}, d_{\mathrm{V}}, d_{\mathrm{W}})$. For the calculation of the duty cycles $d_{\mathrm{x,H}}$ and $d_{\mathrm{x,L}}$ it has to be considered that a shoot-through interval in one phase can influence the effective on-time of another phase, since in a shoot-through interval the inverter input voltage is shorted and thus also the voltage of the other phases is zero, even if in these phases a voltage should be applied. However, this is not true for the phase with the lowest duty cycles ($d_{\mathrm{a,H}}$ and $d_{\mathrm{a,L}}$), because the shoot-through of the other phases occur during the off-time of this phase, i.e. when the low-side switch is turned on and anyway no voltage is applied to this phase. Hence, for $d_{\mathrm{a,L}}$ the smallest duty cycle of d_{U}, d_{V} and d_{W} has to be used, i.e. d_{a}, which due to the integration of the shoot-through interval has to be scaled by the factor $d_{\mathrm{A}}/d_{\mathrm{AN}} = 1 - d_{\mathrm{B}}$. Consequently, the high-side duty cycle $d_{\mathrm{a,H}}$ is found with (24). For the second phase with $d_{\mathrm{b,H}}$ and $d_{\mathrm{b,L}}$, however, the shoot-through time of the first phase has to be considered because it occurs during the on-time of phase b; and for phase c with $d_{\mathrm{c,H}}$ and $d_{\mathrm{c,L}}$ the shoot-through times of both phases a and b have to be considered. Hence, depending on the values of the duty cycles d_{U}, d_{V} and d_{W}, i.e. the voltage sector in a three-phase system, the low-side duty cycles have to be calculated recursively from bottom to top, whereas the corresponding high-side duty cycles are calculated with (24),

$$
d_{\mathrm{a,L}} = (1 - d_{\mathrm{B}}) \cdot d_{\mathrm{a}}, \quad (25)
$$

$$
d_{\mathrm{b,L}} = d_{\mathrm{a,H}} + (1 - d_{\mathrm{B}}) \cdot (d_{\mathrm{b}} - d_{\mathrm{a}}), \quad (26)
$$

$$
d_{\mathrm{c,L}} = d_{\mathrm{b,H}} + (1 - d_{\mathrm{B}}) \cdot (d_{\mathrm{c}} - d_{\mathrm{b}}). \quad (27)
$$

However, the resulting low- and high-side duty cycles of the half-bridges a, b and c have to be assigned to the corresponding transistor duty cycles $d_1 - d_6$. Thereby, the high and the low-side duty cycles of phase a, i.e. $d_{\mathrm{a,H}}$ and $d_{\mathrm{a,L}}$, are assigned to the upper and the lower switch of the half-bridge with the lowest inverter duty cycle, e.g. in sector 1 results $d_5 = d_{\mathrm{a,H}}$ and $d_6 = d_{\mathrm{a,L}}$. The remaining transistor duty cycles $d_1 - d_4$ follow the same scheme, which is also shown in **Fig. 8**.

Finally, the buck transistor $\mathrm{T_A}$ is operated according to the actual conduction state, which means that during d_{AN} the buck transistor $\mathrm{T_A}$ is turned on and during d_{0N} is turned off. In **Fig. 9**, the switching signals of the inverter switches and of the buck transistor for all operation modes are shown. In buck-boost (BB) operation, the shoot-through interval d_{B} is integrated into the switching transitions as previously described. In buck (BU) operation, the duty cycle d_{B} is zero and therefore no shoot-through intervals exist. In boost (BO) operation, where $d_0 = 0$, the asymmetrical PWM carrier would lead to a sawtooth carrier, which is undesired, since in this case all transistors would switch at the same time instant. Therefore, again a symmetrical triangular carrier is used, which leads to the conventional Z-source inverter operation described in [13].

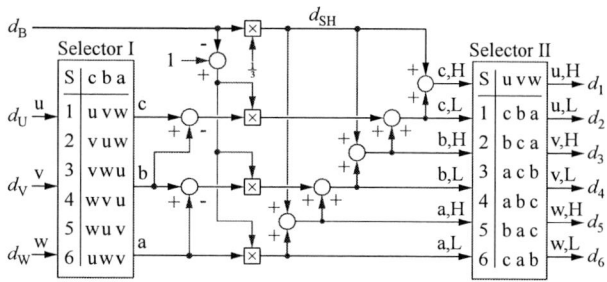

Fig. 8. Implementation of the transistor duty cycle calculation. The selectors I and II provide the signal paths to calculate the transistor duty cycles $d_1 - d_6$, which achieve the correct phase voltages in all six sectors of the three-phase system. Sector 1: $d_U > d_V > d_W$, Sector 2: $d_V > d_U > d_W$, Sector 3: $d_V > d_W > d_U$, Sector 4: $d_W > d_V > d_U$, Sector 5: $d_W > d_U > d_V$, Sector 6: $d_U > d_W > d_V$.

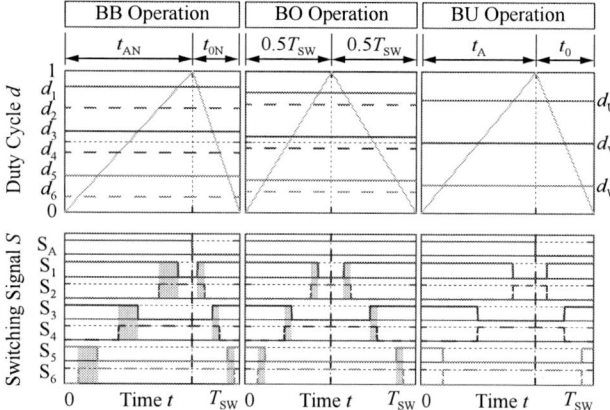

Fig. 9. Modulation scheme, including the PWM carrier and the transistor duty cycles $d_1 - d_6$, together with the switching signals of the inverter switches $S_1 - S_6$ and the buck transistor S_A, for the buck-boost (BB), the boost (BO) and the (BU) buck operation, whereby the shoot-through intervals are indicated.

D. Verification

The proper operation of the proposed topology is verified by circuit simulations. The considered variable speed drive system is rated for an output power of 7.5 kW and is supplied from a single-phase mains with 480 V$_{\mathrm{rms}}$/50 Hz. The overall system specifications are summarized in **Tab. I**. For a better visualisation, in the simulation a peak-to-peak capacitor voltage ripple of 3% and a peak-to-peak inductor current ripple of 15% is assumed. Furthermore, a machine frequency close to the mains frequency is chosen. The resulting circuit parameters are also listed in Tab. I. However, it has to be mentioned that for a real circuit design the values of the passive components as well as the switching frequency would have to be optimized concerning efficiency and/or power density.

With the given machine voltage $V_{\mathrm{M,rms}}$, the minimum intermediate voltage is given with $V_{\mathrm{PN}} = 400$ V, which thanks to the buck-stage can be set below the peak grid voltage, i.e. $\hat{V}_G = 679$ V. In **Fig. 10(a)** the corresponding waveforms at the nominal operating point are shown. It can be noticed that the capacitor voltage $v_C = V_{\mathrm{PN}}$

TABLE I
SUMMARY OF THE CONVERTER SPECIFICATIONS AND THE CIRCUIT PARAMETERS.

Grid voltage $V_{\mathrm{G,rms}}$	480 V
Angular grid frequency ω_G	$2\pi \cdot 50$ Hz
Nominal intermediate voltage V_{PN}	400 V/700 V
Machine phase voltage $V_{\mathrm{M,rms}}$	160 V
Electrical machine frequency ω_M	$2\pi \cdot 67$ Hz
Nominal mechanical power $P_{\mathrm{M,N}}$	7.5 kW
Z-source inductance L	300 µH
Z-source capacitance C	2 mF
Switching frequency f_{SW}	140 kHz

is constant and the inductor current i_L nicely tracks its reference (dashed line), which with the buck-boost operation in the vicinity of the voltage zero crossings can be kept at the minimum required value $i_{\mathrm{L,min}}$. Furthermore, the converter operation smoothly transitions between the different operation modes (BU), (BO) and (BB), which leads to a sinusoidal grid current i_G with a low current THD of 1.1%. Finally, with the proposed modulation concept, three purely sinusoidal phase voltages and phase currents i_U, i_V and i_W can be achieved.

In the following, the benefits gained from the additional buck-stage should be highlighted. As already mentioned, the Z-source inverter can also be directly connected to the single-phase mains via a diode rectifier. In this case, the buck functionality is lost and the intermediate voltage has to be larger than the peak grid voltage, e.g. $V_{\mathrm{PN}} = 700$ V, and thus the voltage stress on the semiconductor elements increases considerably. Further on, in the vicinity of the zero crossings, the input current can no longer be sinusoidally controlled. The resulting waveforms are shown in **Fig. 10(b)**. In addition, it can be noticed, that this input current distortion also leads to slight distortions in the output phase currents i_U, i_V and i_W.

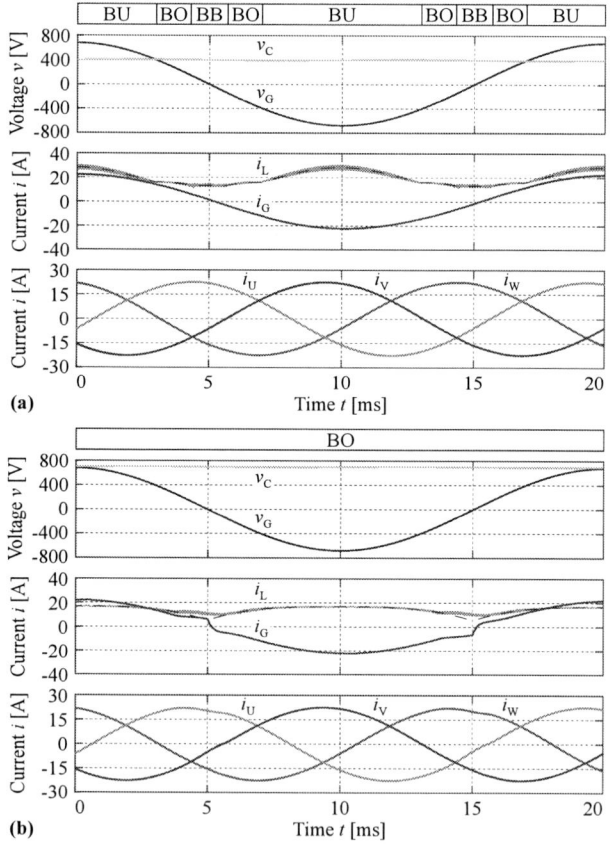

Fig. 10. Simulation results at nominal operating conditions of the grid voltage v_G, the capacitor voltage v_C, the grid current i_G, the inductor current i_L with its reference (dashed line) as well as the three phase currents i_U, i_V and i_W over one grid period T_G for **(a)** the proposed and **(b)** the Z-source-based single-to-three phase boost topology. In both cases, the machine is modelled by a symmetrical RL-load.

IV. COMPONENT STRESS ANALYSIS AND COMPARISON

With the additional buck-stage, the proposed *123ZBBC* not only leads to a better input and output current quality, but also to a lower component stress as analyzed in the following for the already presented application (cf. Tab. I). Thereby, the voltage stresses are calculated analytically, while the current stresses are derived from the circuit simulations.

The capacitor voltage V_C is equal to the average intermediate voltage $\bar{v}_{PN}$ and consequently, for the *123ZBBC* the capacitor voltage stress is only $V_C = 400\,\text{V}$ compared to $V_C = 700\,\text{V}$, as it is the case for the Z-source topology without buck-stage. However, due to the buck operation now the inductor current increases from $I_{L,rms} = 14.8\,\text{A}$ for the boost topology to $I_{L,rms} = 20.8\,\text{A}$ in case of the *123ZBBC*. Furthermore, additional current stresses appear in the buck-stage ($I_{DA,avg/rms} = 5.8\,\text{A}/13.8\,\text{A}$ and $I_{TA,rms} = 19.9\,\text{A}$), which for the boost Z-source do not exist (cf. **Tab. II**). The maximum voltage applied to the inverter transistors is given by the maximum intermediate voltage v_{PN}, which either occurs during the active state, when $v_G = 0\,\text{V}$, or during the buck state and equals twice the capacitor voltage $V_{Ti,max} = v_{PN} = 2V_C$. Hence, for the proposed topology the minimum blocking voltage of the inverter switches is $800\,\text{V}$, whereas without buck-stage the minimum blocking voltage increases to $1.4\,\text{kV}$ (cf. **Tab. II**). The same is also true for the voltage stress on the bridge rectifier diodes as well as the buck diode. The buck transistor, however, only has to block the maximum grid voltage during buck state, which in this case is $V_{TA,max} = \hat{V}_G = 687\,\text{V}$.

Therefore, in the *123ZBBC* for all switches and diodes $1.2\,\text{kV}$ semiconductor devices, e.g. SiC MOSFETs, can be employed, whereas for the boost-only Z-source converter, semiconductor components with a blocking voltage of at least $1.7\,\text{kV}$, e.g. $1.7\,\text{kV}$ SiC MOSFETs (with higher on-state resistance than $1.2\,\text{kV}$ devices), are needed. Hence, the benefit gained from the change of semiconductor technology, i.e. improved conduction and switching performance [19], overcompensate the additional conduction losses caused by the inserted buck-stage, which additionally allows to achieve a purely sinusoidal input current.

TABLE II

COMPONENT STRESS COMPARISON BETWEEN THE PROPOSED BUCK-BOOST Z-SOURCE-BASED AC/AC CONVERTER AND THE CONVENTIONAL Z-SOURCE CONVERTER DIRECTLY SUPPLIED FROM THE SINGLE-PHASE MAINS VIA A DIODE BRIDGE RECTIFIER.

Component	Z-Buck-Boost	Conv. Z-Boost
V_{PN}	400 V	700 V
$I_{L,rms}$	20.8 A	14.8 A
$I_{C,rms}$	8.9 A	11.8 A
$V_{Di,max}$	800 V	1400 V
$I_{Di,avg}$	7.1 A	7.3 A
$I_{Di,rms}$	14.1 A	13.7 A
$V_{DA,max}$	800 V	-
$I_{DA,avg}$	5.8 A	-
$I_{DA,rms}$	13.8 A	-
$V_{TA,max}$	687 V	-
$I_{TA,avg}$	19.9 A	-
$V_{Ti,max}$	800 V	1400 V
$I_{Ti,rms}$	11.7 A	11.8 A

V. CONCLUSION

This paper proposes a novel unidirectional single-to-three-phase Z-source buck-boost AC/AC converter (*123ZBBC*), which integrates the boost function into the inverter stage, resulting in a reduced realization effort. The included buck-stage enables a sinusoidal input current and reduces the overall component voltage stress, compared to the single-to-three phase Z-source boost topology.

The analysis provided in this paper reveals the basic operation principle, including the conduction states, and presents the PFC modulation strategy achieving the minimal inductor current. The proposed closed-loop control enables a sinusoidal input current, controles the required intermediate voltage and provides a three-phase voltage system to the load. Additionally, the modulation scheme of the inverter, and particularly the distribution of the shoot-through interval, is analyzed in detail. All these findings are verified by means of circuit simulations. The component stresses are derived and the conducted comparison to a Z-source boost topology reveals a lower blocking voltage of the semiconductor devices and enables the employment of $1.2\,\text{kV}$ SiC MOSFETS instead of $1.7\,\text{kV}$ devices for the presented application. Due to the change in the semiconductor

technology, the system performance is improved, i.e. the additional losses of the buck-stage are overcompensated, which leads to an overall higher efficiency.

ACKNOWLEDGMENT

The authors would like to express their sincere appreciation to Nabtesco Corp., Japan, for the financial and technical support of research on Advanced Mechatronic Systems at the Power Electronic Systems Laboratory, ETH Zurich. Furthermore, inspiring technical discussions with K. Nakamura are especially acknowledged.

REFERENCES

[1] R. Ridley, S. Kern, and B. Fuld, "Analysis and design of a wide input range power factor correction circuit for three-phase applications," in *Proc. of IEEE Applied Power Electronics Conference and Exposition (APEC)*, San Diego, CA, USA, Mar. 1993, pp. 299–305.

[2] M. R. Hesamzadeh, N. Hosseinzadeh, and P. J. Wolfs, "Design and study of a switch reactor for Central Queensland SWER system," in *Proc. of IEEE Universities Power Engineering Conference (UPEC)*, Padova, Italy, Dec. 2008, pp. 1–5.

[3] B. Singh, B. N. Singh, A. Chandra, K. Al-Haddad, A. Pandey, and D. P. Kothari, "A review of single-phase improved power quality AC-DC converters," *IEEE Transactions on Industrial Electronics*, vol. 50, no. 5, pp. 962–981, Oct. 2003.

[4] M. C. Ghanem, K. Al-haddad, and G. Roy, "A new control strategy to achieve sinusoidal line current in a cascade buck-boost converter," *IEEE Transactions on Industrial Electronics*, vol. 43, no. 3, pp. 441–449, June 1996.

[5] G. K. Andersen and F. Blaabjerg, "Current programmed control of a single-phase two-switch buck-boost power factor correction circuit," *IEEE Transactions on Industrial Electronics*, vol. 53, no. 1, pp. 263–271, Feb. 2006.

[6] A. Pawlikowski and L. Grzesiak, "Vector-controlled three-phase voltage source inverter producing a sinusoidal voltage for AC motor drives," in *Proc. of IEEE EUROCON*, Warsaw, Poland, Sept. 2007, pp. 1902–1909.

[7] S. Waffler and J. W. Kolar, "A novel low-loss modulation strategy for high-power bidirectional buck + boost converters," *IEEE Transactions on Power Electronics*, vol. 24, no. 6, pp. 1589–1599, June 2009.

[8] I. M. Safwat and W. Xiahua, "Comparative study between passive PFC and active PFC based on buck-boost conversion," in *Proc. of IEEE Advanced Information Technology, Electronic and Automation Control Conference (IAEAC)*, Chongqing, China, Mar. 2017, pp. 45–50.

[9] M. Shen, A. Joseph, J. Wang, F. Z. Peng, and D. J. Adams, "Comparison of traditional inverters and Z-source inverter for fuel cell vehicles," *IEEE Transactions on Power Electronics*, vol. 22, no. 4, pp. 1453–1463, July 2007.

[10] M. Shen, J. Wang, A. Joseph, F. Z. Peng, L. M. Tolbert, and D. J. Adams, "Constant boost control of the Z-source inverter to minimize current ripple and voltage stress," *IEEE Transactions on Industry Applications*, vol. 42, no. 3, pp. 770–778, May 2006.

[11] F. Z. Peng, M. Shen, and Z. Qian, "Maximum boost cntrol of the Z-source inverter," *IEEE Transactions on Power Electronics*, vol. 20, no. 4, pp. 833–838, July 2005.

[12] Y. Zhang, J. Liu, X. Li, X. Ma, S. Zhou, H. Wang, and Y. Liu, "An improved PWM strategy for Z-source inverter with maximum boost capability and minimum switching frequency," *IEEE Transactions on Power Electronics*, vol. 33, no. 1, pp. 606–628, Jan. 2018.

[13] F. Z. Peng, "Z-source inverter," *IEEE Transactions on Industry Applications*, vol. 39, no. 2, pp. 504–510, Mar. 2003.

[14] S. Singh, G. Carli, N. A. Azeez, and S. S. Williamson, "Modeling, design, control, and implementation of a modified Z-source integrated PV/grid/EV DC charger/inverter," *IEEE Transactions on Industrial Electronics*, vol. 65, no. 6, pp. 5213 – 5220, June 2018.

[15] E. C. Dos Santos, F. Bradaschia, M. C. Cavalcanti, and E. R. C. Da Silva, "Z-source converter applied for single-phase to three-phase conversion system," in *Proc. of IEEE Applied Power Electronics Conference and Exposition (APEC)*, Fort Worth, TX, USA, Mar. 2011, pp. 216–223.

[16] A. H. Rajaei, M. Mohamadian, S. M. Dehghan, and A. Yazdian, "Single-phase induction motor drive system using Z-source inverter," *IEEE Transactions on Electric Power Applications*, vol. 4, no. 1, pp. 17–25, Jan. 2010.

[17] T. Rudnicki, R. Czerwinski, and D. Polok, "Performance analysis of a PMSM drive with torque and speed control," in *Proc. of IEEE International Conference Mixed Design of Integrated Circuits and Systems (MIXDES)*, Torun, Poland, June 2015, pp. 562 – 566.

[18] C. Hanju, V. Trung-Kien, J. Qi, and K. Jae-Eon, "Design and control of proportional-resonant controller based photovoltaic power conditioning system," in *Proc. of IEEE Energy Conversion Congress and Exposition (ECCE USA)*, San Jose, CA, USA, Sept. 2009, pp. 2198–2205.

[19] J. Azurza, L. Schrittwieser, C. Gammeter, G. Deboy, and J. W. Kolar, "Relating the figure of merit of power MOSFETs to the maximally achievable efficiency of converters," *under review for the CPSS Transactions on Power Electronics and Applications*, 2018.

The 2018 International Power Electronics Conference

Simple PWM Strategy of a Matrix Converter for Minimizing Output Voltage Harmonics

Takuya Oshima, Takaharu Takeshita
Nagoya Institute of Technology
Gokiso, Showa, Nagoya, Japan
E-mail: 29413046@stn.nitech.ac.jp , take@nitech.ac.jp

Abstract—**A matrix converter (MC) can directly convert a three-phase AC voltage to a three-phase AC voltage with an arbitrary voltage and frequency. MC can achieve downsizing of the system and high reliability. In this paper, PWM strategies for minimizing output voltage harmonics are explained. The PWM strategies achieve three commutations in three phases during one control period and unity input power-factor. The authors indicate that output voltage harmonics can be minimized theoretically by the instantaneous effective theory, and the characteristics of the control method is verified by simulation. Moreover, this paper presents the simple method to select PWM patterns for minimizing output voltage harmonics. The effectiveness of the proposed control method have been verified by simulation and the experiment, and results agreed well.**

Keywords—matrix converter, output voltage harmonics, duty combination, instantaneous effective theory

I. INTRODUCTION

In recent years, as countermeasures against environmental problems such as global warming and resource depletion, improvements of efficiency of various equipments are required. It is estimated that about half of the total electric power consumption in the world is consumed by the driving of motors. Therefore, it is the important to improve efficiency of the motors and the control devices. Currently, rectifier-inverter system is widely used for variable speed control of AC motors. In this system, the rectifier converts AC to DC, and the inverter converts DC to AC with arbitrary amplitude and frequency. The conversion loss is caused two times because the system is composed of two converters. And a large electrolytic capacitor is required to suppress the pulsation of DC voltage, which is a issue of downsizing on the system. In addition, because the life of the electrolytic capacitor is short, periodic maintenance is needed.

On the other side, a matrix converter (MC) has been attracting attention and studied [1]-[3] because it can directly convert three-phase AC to three-phase AC with arbitrary amplitude and frequency. By using reverse block switch devices as a bi-directional switch, the passage number of switch devices of the current is reduced. It makes the system possible to reduce the converter loss compared with the conventional system. In addition, because MC does not require an electrolytic capacitor, it is possible to downsize the system and improve reliability.

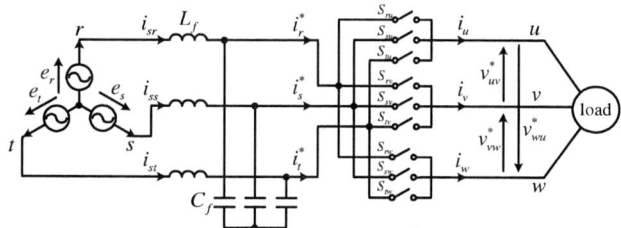

Fig. 1. Main circuit of a matrix converter

MC can simultaneously control the output voltage and the input current by switching of its nine bi-directional switches. However, the output voltage and the input current include harmonics by a PWM control. Considering the load control characteristics as the top priority, it is desirable to minimize harmonics of the output voltage as possible. Because many PWM patterns can realize the references of the output voltage and the input current, the various PWM strategies can be proposed. Various PWM patterns such as reducing the output voltage[4], reducing the number of commutations [5][6], and referencing all input power-factor [7] has been proposed. However, the number of the duty combinations that have been evaluated are limited. Therefore, the best PWM pattern has not been indicated clearly. Also, the instantaneous effective value theory is used for evaluating PWM patterns in real time[8]. By the theory, PWM patterns that reduce harmonics can be selected. However, it is questionable because calculation of the instantaneous effective value requires a long processing time. In this study, all 41 duty combinations of three commutations in one control period are subjected to the instantaneous effective value theory, and the limit of reducing the output voltage harmonics of MC is showed by simulation. Moreover, this paper presents the simple method to select PWM patterns without calculating instantaneous effective values. The proposed method is verified by simulation and experiment.

II. CONFIGURATION OF A MATRIX CONVERTER

A. Main Circuit

Fig.1 shows a circuit configuration of MC. MC has nine bi-directional switches S_{ru} - S_{tw}. The input side is connected to a three-phase voltage supply e_r, e_s, and e_t through an LC filter for suppressing harmonic currents. The output side is

4088

connected to a three-phase load. The duty cycles of nine bi-directional switches d_{ru} - d_{tw} during one control period are determined to satisfy the references of output line voltages v_{uv}^*, v_{vw}^*, and v_{wu}^* and input currents i_r^*, i_s^*, and i_t^* (* indicates reference value) simultaneously.

Input phase voltages e_r, e_s, and e_t are expressed by (1) using the effective value E, the input angular frequency ω, and the input phase angle θ.

$$\begin{bmatrix} e_r \\ e_s \\ e_t \end{bmatrix} = \sqrt{\frac{2}{3}} E \begin{bmatrix} \cos\theta \\ \cos(\theta - 2\pi/3) \\ \cos(\theta + 2\pi/3) \end{bmatrix} \tag{1}$$

$$\theta = \omega t \tag{2}$$

Input current references i_r^*, i_s^*, and i_t^* are expressed by (3) using the effective value I^* and the input power-factor-angle reference φ^*.

$$\begin{bmatrix} i_r^* \\ i_s^* \\ i_t^* \end{bmatrix} = \sqrt{2} I_{in}^* \begin{bmatrix} \cos(\theta + \varphi^*) \\ \cos(\theta + \varphi^* - 2\pi/3) \\ \cos(\theta + \varphi^* + 2\pi/3) \end{bmatrix} \tag{3}$$

Output phase voltage references v_u^*, v_v^*, and v_w^* are expressed by (4) using the effective value of the output line voltage reference V_L^*, the output angular frequency ω_L^*, the output phase angle θ_L^*, and the output initial phase angle φ_L^*.

$$\begin{bmatrix} v_u^* \\ v_v^* \\ v_w^* \end{bmatrix} = \sqrt{\frac{2}{3}} V_L^* \begin{bmatrix} \cos\theta_L^* \\ \cos(\theta_L^* - 2\pi/3) \\ \cos(\theta_L^* + 2\pi/3) \end{bmatrix} \tag{4}$$

$$\theta_L^* = \omega_L^* t + \varphi_L^* \tag{5}$$

Output currents i_u, i_v, and i_w are expressed by (6) using the effective value I_{out} and the load power-factor-angle φ_L.

$$\begin{bmatrix} i_u \\ i_v \\ i_w \end{bmatrix} = \sqrt{2} I_{out} \begin{bmatrix} \cos(\theta_L^* + \varphi_L) \\ \cos(\theta_L^* + \varphi_L - 2\pi/3) \\ \cos(\theta_L^* + \varphi_L + 2\pi/3) \end{bmatrix} \tag{6}$$

When the loss of MC is ignored, the effective value of the input current references I_{in}^* is obtained as follows:

$$I_{in}^* = \frac{V_L^* \cos\varphi_L}{E \cos\varphi^*} I_{out} \tag{7}$$

B. Analytical Model

Fig.2 shows an analytical model of MC for derivation of duty cycles. The input LC filter is removed for simplification. Assuming that the control period T_s is short enough compared with the time constant of the circuit, the voltage and current during the control period are treated as constant. Input phases α, β, and γ are defined as phases of the highest, intermediate, and lowest voltage among input voltages e_r, e_s, and e_t respectively. Similarly, output phases a, b, and c are defined as the phases of the highest, intermediate, and lowest voltage among output voltages v_u^*, v_v^*, v_w^* respectively. The magnitude relation of each defined phase voltage is expressed as follows:

$$e_\alpha \geq e_\beta \geq e_\gamma \tag{8}$$

$$v_a^* \geq v_b^* \geq v_c^* \tag{9}$$

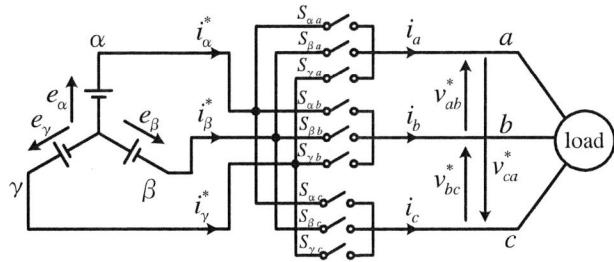

Fig. 2. Analytical model of a matrix converter

III. DERIVATION OF DUTY CYCLES

The following equation is obtained because each output phases should be connected to one of the three input phases for the continuity of the load currents and the prevention of a short current in input line voltages.

$$d_{\alpha a} + d_{\beta a} + d_{\gamma a} = 1 \tag{10}$$

$$d_{\alpha b} + d_{\beta b} + d_{\gamma b} = 1 \tag{11}$$

$$d_{\alpha c} + d_{\beta c} + d_{\gamma c} = 1 \tag{12}$$

Each reference value equals to the average value during one control period. Therefore, output line voltage references v_{ab}^*, v_{bc}^*, and v_{ca}^* and input current references i_α^*, i_β^*, and i_γ^* are obtained as follows:

$$v_{ab}^* = (d_{\alpha a} - d_{\alpha b})e_\alpha + (d_{\beta a} - d_{\beta b})e_\beta + (d_{\gamma a} - d_{\gamma b})e_\gamma \tag{13}$$

$$v_{bc}^* = (d_{\alpha b} - d_{\alpha c})e_\alpha + (d_{\beta b} - d_{\beta c})e_\beta + (d_{\gamma b} - d_{\gamma c})e_\gamma \tag{14}$$

$$v_{ca}^* = (d_{\alpha c} - d_{\alpha a})e_\alpha + (d_{\beta c} - d_{\beta a})e_\beta + (d_{\gamma c} - d_{\gamma a})e_\gamma \tag{15}$$

$$i_\alpha^* = d_{\alpha a} i_a + d_{\alpha b} i_b + d_{\alpha c} i_c \tag{16}$$

$$i_\beta^* = d_{\beta a} i_a + d_{\beta b} i_b + d_{\beta c} i_c \tag{17}$$

$$i_\gamma^* = d_{\gamma a} i_a + d_{\gamma b} i_b + d_{\gamma c} i_c \tag{18}$$

Giving three duty cycles zero among the nine duty cycles, the number of commutations in one control period can be minimized to three. Here the duty combinations shown in Fig.3 are derived as example. In Fig.3, the duty cycles of black or gray mean zero. These duty combinations can reduce the output voltage harmonics because duty cycles $d_{\gamma a}$ and $d_{\alpha c}$ with the largest deference between input phase voltage and output phase voltage are zero.

At the first of the derivation, two duty cycles $d_{\gamma a}$ and $d_{\alpha c}$ are given zero.

$$d_{\gamma a} = d_{\alpha c} = 0 \tag{19}$$

Next, one duty cycle is defined as a variable k. In this section, $d_{\beta b}$ is defined as k.

$$d_{\beta b} = k \tag{20}$$

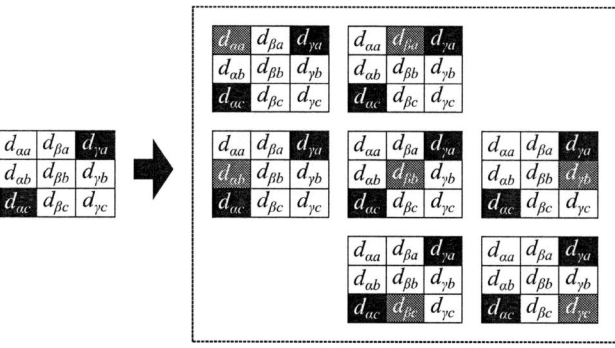

Fig. 3. Derivation of Duty Combinations

From (10) - (20), all duty cycles are obtained as follows:

$$d_{\alpha a} = \frac{-i_\alpha^* e_{\gamma\alpha} - i_b(e_{\beta\gamma} - v_{ab}) + i_b e_{\beta\gamma} k}{i_a e_{\beta\gamma} - i_c e_{\alpha\beta}} \quad (21)$$

$$d_{\beta a} = \frac{-i_b v_{ab} + (i_\alpha^* + i_c) e_{\gamma\alpha} - i_b e_{\beta\gamma} k}{i_a e_{\beta\gamma} - i_c e_{\alpha\beta}} \quad (22)$$

$$d_{\alpha b} = \frac{i_\alpha^* e_{\alpha\beta} + i_a(e_{\beta\gamma} - v_{ab}) - i_a e_{\beta\gamma} k}{i_a e_{\beta\gamma} - i_c e_{\alpha\beta}} \quad (23)$$

$$d_{\beta b} = k \quad (24)$$

$$d_{\gamma b} = \frac{i_a v_{ab} - (i_\alpha^* + i_c) e_{\alpha\beta} + i_c e_{\alpha\beta} k}{i_a e_{\beta\gamma} - i_c e_{\alpha\beta}} \quad (25)$$

$$d_{\beta c} = \frac{i_b v_{bc} - (i_\gamma^* + i_a) e_{\gamma\alpha} + i_b e_{\alpha\beta} k}{i_a e_{\beta\gamma} - i_c e_{\alpha\beta}} \quad (26)$$

$$d_{\gamma c} = \frac{i_\gamma^* e_{\gamma\alpha} + i_b(e_{\alpha\beta} - v_{bc}) - i_b e_{\alpha\beta} k}{i_a e_{\beta\gamma} - i_c e_{\alpha\beta}} \quad (27)$$

It is possible to determine all duty cycles by substituting the arbitrary value to the variable k. However, all duty cycles have to be in a range between 0 and 1.

$$0 \le d_{lm} \le 1 \quad l = \{\alpha, \ \beta, \ \gamma\} \quad m = \{a, \ b, \ c\} \quad (28)$$

If all duty cycles are not less than 0, the duty cycles is less than or equal to 1 from (10) - (12). Therefore, (28) can be rewrote as follows:

$$0 \le d_{lm} \quad l = \{\alpha, \beta, \gamma\} \quad m = \{a, b, c\} \quad (29)$$

Substituting (28) to (21) - (27), a control range of the variable k is obtained. When a minimum value and a maximum value of the range is defined as k_{min} and k_{max}, the range is expressed as follows:

$$k_{min} \le k \le k_{max} \quad (30)$$

Using one of k_1 - k_7 in (31) - (37), one of the duty cycles can

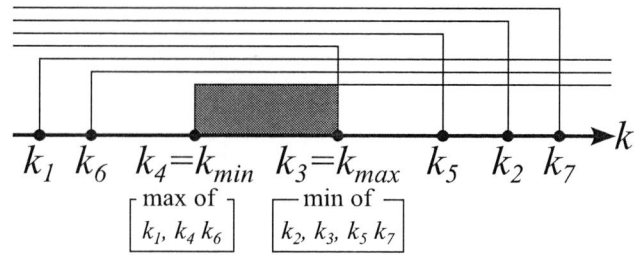

Fig. 4. The control range of k

TABLE I. CANDIDATES OF k_{min} AND k_{max}

state	i_a	i_b	i_c	$i_a e_{\beta\gamma} - i_c e_{\alpha\beta}$	k_{min}	k_{max}
1	+	+	−	+	k_1, k_4, k_6	k_2, k_3, k_5, k_7
2	+	−	−	+	k_2, k_4, k_7	k_1, k_3, k_5, k_6
3	+	−	+	+	k_2, k_4, k_5, k_7	k_1, k_3, k_6
				−	k_1, k_3, k_4, k_6	k_2, k_5, k_7
4	−	+	−	+	k_1, k_3, k_4, k_6	k_2, k_5, k_7
					k_2, k_4, k_5, k_7	k_1, k_3, k_6
5	−	+	+	−	k_2, k_4, k_7	k_1, k_3, k_5, k_6
6	−	−	+	−	k_1, k_4, k_6	k_2, k_3, k_5, k_7

be zero.

$$k_1 = \frac{i_\alpha^* e_{\gamma\alpha} + i_b(e_{\beta\gamma} - v_{ab}^*)}{i_b e_{\beta\gamma}} \quad (d_{\alpha a} = 0) \quad (31)$$

$$k_2 = \frac{(i_\alpha^* + i_c) e_{\gamma\alpha} - i_b v_{ab}^*}{i_b e_{\beta\gamma}} \quad (d_{\beta a} = 0) \quad (32)$$

$$k_3 = \frac{i_\alpha^* e_{\alpha\beta} + i_a(e_{\beta\gamma} - v_{ab}^*)}{i_a e_{\beta\gamma}} \quad (d_{\alpha b} = 0) \quad (33)$$

$$k_4 = 0 \quad (d_{\beta b} = 0) \quad (34)$$

$$k_5 = \frac{(i_\alpha^* + i_c) e_{\alpha\beta} - i_a v_{ab}^*}{i_c e_{\alpha\beta}} \quad (d_{\gamma b} = 0) \quad (35)$$

$$k_6 = \frac{(i_\gamma^* + i_a) e_{\gamma\alpha} - i_b v_{bc}^*}{i_b e_{\alpha\beta}} \quad (d_{\beta c} = 0) \quad (36)$$

$$k_7 = \frac{i_\gamma^* e_{\gamma\alpha} + i_b(e_{\alpha\beta} - v_{bc}^*)}{i_b e_{\alpha\beta}} \quad (d_{\gamma c} = 0) \quad (37)$$

The value of k_{min} and k_{max} are depend on signs of the output current and denominator of duty cycles. For example, under $i_a \ge 0$, $i_b \ge 0$, $i_c < 0$, $(i_a e_{\beta\gamma} - i_c e_{\alpha\beta}) \ge 0$, (38) and (39) are obtained by substituting (21)~(27) to (19).

$$k \ge k_1, \quad k \ge k_4, \quad k \ge k_6 \quad (38)$$

$$k \le k_2, \quad k \le k_3, \quad k \le k_5, \quad k \le k_7 \quad (39)$$

Fig.4 shows the control range of k under (38) and (39). k_{min} is the maximum value among $k_1, k_4,$ and k_6 and k_{max} is the minimum value among $k_2, k_3, k_5,$ and k_7. Under $k_{min} < k < k_{max}$, the number of commutation is four. Under $k = k_{min}$ or $k = k_{max}$, the number of commutation is reduced to three. Table I lists sign of the output current and denominator of duty cycles and the candidates of $k = k_{min}$ and $k = k_{max}$.

The 2018 International Power Electronics Conference

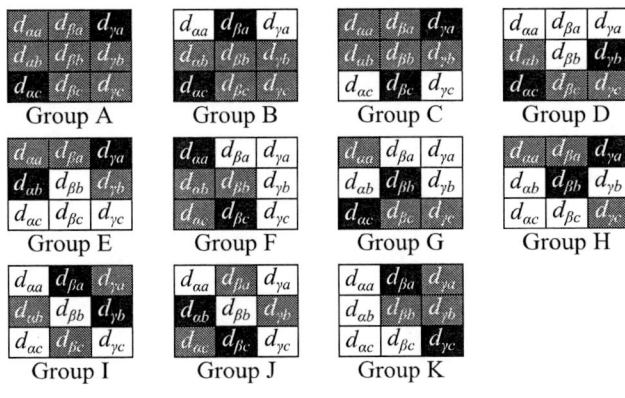

Fig. 5.   All duty combinations

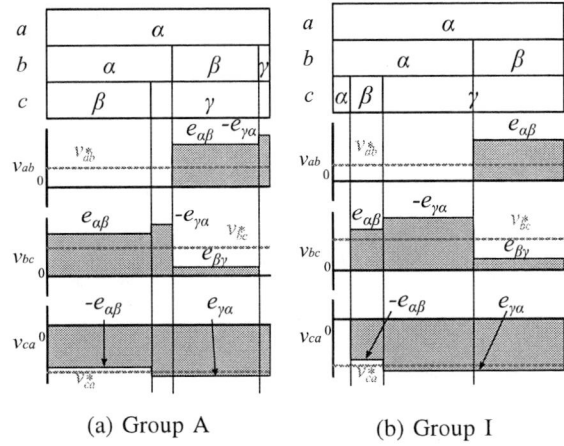

(a) Group A   (b) Group I

Fig. 6.   The comparison of zero duty cycles

IV.   CONVENTIONAL CONTROL METHOD

A. Summary

In this paper, the evaluation of PWM patterns by instantaneous effective in the conventional control method is expanded for all PWM patterns. Therefore, the best PWM pattern for minimizing output voltage harmonics is selected in each control period.

B. All Duty Combinations

The number of combinations that make three of the nine duty cycles zero is $_9C_3 = 84$, but some duty combinations cannot realize references because some PWM patterns are impossible to generate a voltage pulse with the same sign as the reference value. The number of all duty combinations that is possible to be established is 41.

Fig.5 shows 41 duty combinations (Groups A - K). Seven duty combinations in Fig.3 are collectively classified as group A, witch are already indicated as an example of derivation of duty cycles . The others are similarly classified as any one of Groups B - K. In order to calculate all PWM patterns, it is necessary to derive all Groups as well as Group A.

C. Instantaneous Effective Value

The instantaneous effective value of output voltages $V(T_s)$ is expressed as follow:

$$V(T_s) = \sqrt{\frac{1}{3T_s} \int_0^{T_s} (v_{uv}^2 + v_{vw}^2 + v_{wu}^2)dt} \qquad (40)$$

The effective value is also expressed the square root of value of the sum of squares of effective values of a fundamental component and harmonics. Because a fundamental component is constant, the effective value of harmonics is low if the instantaneous effective value is low. In this method, instantaneous effective values of all PWM patterns are calculated in each control period, and the PWM patterns of lowest instantaneous effective value is selected.

V.   PROPOSED METHOD

A. Summary

In the proposed method, only one PWM pattern that reduce output harmonics is calculated in each control method. Its harmonic characteristics is similar to that of the conventional method. The duty combinations of Group A are only used, and the duty cycle $d_{\beta b}$ is defined as a variable k. The range of input power-factor-angle φ is $-\pi/6 \le \varphi^* \le \pi/6$.

B. PWM Patterns for Harmonic Reduction

Fig.6 shows PWM patterns and output line voltage waveforms. In the waveform of the output voltage v_{ca}, in Fig.6 (a), the maximum input voltage pulse $-e_{\gamma\alpha}$ and zero voltage are generated. Therefore, the ripple of v_{ca} is large and the harmonic component of carrier frequency is high. The condition for generating the maximum pulse $-e_{\gamma\alpha}$ is $d_{\alpha a} > d_{\alpha c} + d_{\beta c}$. Then, the conditions for generating zero voltage is $d_{\gamma a} > 0$ or $d_{\alpha c} > 0$. Because the duty factors in Group B - K satisfy the condition of $d_{\gamma a} > 0$ or $d_{\alpha c} > 0$, the maximum output voltage waveform v_{ca} include the maximum input voltage pulse $-e_{\gamma\alpha}$ and zero voltage, simultaneously. Fig.6 (b) shows the duty combinations in Group A. In the maximum output voltage waveform v_{ca}, the zero voltage is not generated because of $d_{\gamma a} = 0$ and $d_{\alpha c} = 0$ The duty combinations in Group A can reduce the harmonic component of the carrier frequency.

The value of variable k should be k_{min} or k_{max} to reduce the number of commutations. In the proposed method, the selecting algorithm of value of k is simple to shorten the calculate time. Fig.7 shows two PWM patterns of Group A and output line voltage waveforms in various value of k ($k = k_{min}, k_{min} < k < k_{max}, k = k_{max}$). In the sequence of (a), (b), and (c), the value of k becomes large gradually. As shown Fig.7 (a), $k(= d_{\beta b})$ is zero. $e_{\alpha\beta}$ is not generated in v_{ab} and $e_{\beta\gamma}$ is not generated in v_{bc}. Harmonics of v_{ab} and v_{bc} is large because the generation time of voltage pulses whose values are closed to references are short. As shown Fig.7 (b) and (c), the value of k increases gradually. And the generation

4091

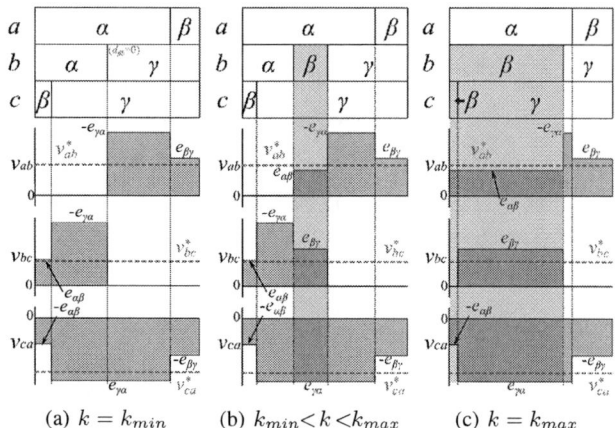

(a) $k = k_{min}$ (b) $k_{min} < k < k_{max}$ (c) $k = k_{max}$

Fig. 7.   Waveforms of the output voltages in various k

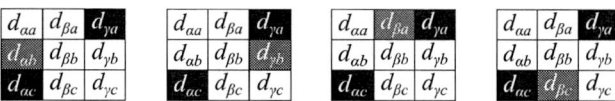

Fig. 8.   Duty combinations for minimizing harmonics

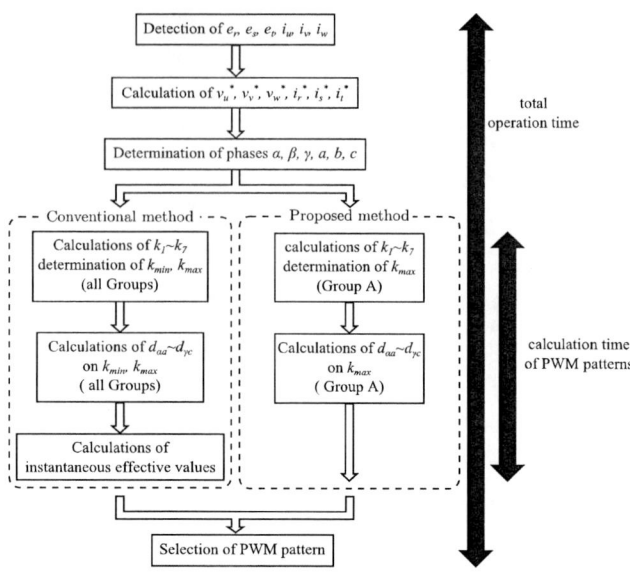

Fig. 9.   Flow of the control methods

time of $e_{\alpha\beta}$ in v_{ab} and $e_{\beta\gamma}$ in v_{bc} becomes longer. Harmonics of v_{ab} and v_{bc} is small because the generation time of voltage pulses whose values are closed to references is long. Therefore, the longer $k(= d_{\beta b})$ should be selected.

C. Selecting PWM Patterns

From above explanations, duty combinations that satisfy following conditions are proposed as combinations that minimize output voltage harmonics

- The voltage pluses with same signs as the references are generated.

- $d_{\beta b}$ is not zero (it is derived by $k = k_{max}$)

- $d_{\gamma a}$ and $d_{\alpha c}$ are zero.

Fig.8 shows duty combinations that satisfy above conditions. Four duty combinations in Fig.8 are common because $d_{\gamma a}$ and $d_{\alpha c}$ are zero, and belong to Group A. In the proposed method, $d_{\gamma a}$ and $d_{\alpha c}$ are given zero in advance of the derivation process of duty cycles. $d_{\beta b}$ is given a variable k, and k_{max} is substituted into k absolutely.

D. Control Range of Input Power-Factor-Angle

The instantaneous current correlation changes in a cycle because input phases and output phases are defined by phase voltages. For example, when $i_a \geq 0$, $i_b \geq 0$, and $i_c < 0$, the sign of the input current reference i_α^* must be plus due to (16) and (19). When $e_r \geq e_s \geq e_t$, following inequalities are obtained from $0 \leq \theta \leq \pi/3$ and $i_\alpha^* = i_r^*$.

$$0 \leq \cos(\theta + \varphi^*) \leq 1 \tag{41}$$

$$-\pi/2 \leq \theta + \varphi^* \leq \pi/2 \tag{42}$$

$$-\pi/2 \leq \varphi^* \leq \pi/6 \tag{43}$$

The control rage of the proposed method is obtained as (44) by calculating inequalities under all states of input voltage and output current.

$$-\pi/6 \leq \varphi^* \leq \pi/6 \tag{44}$$

VI. SIMULATION AND EXPERIMENT

A. Control Flow

Fig.9 shows control flows of the conventional and proposed control method. In the beginning, input voltages and output voltages are detected. References of output voltages and input currents are calculated, and phases of $\alpha, \beta, \gamma, a, b,$ and c are defined by the instantaneous value correlations. From next, the processing are different in the conventional and proposed control method. In the conventional control method, k_1-k_7 in all Groups A-K are calculated. Next, k_{min} and k_{max} are determined from k_1-k_7. All duty cycles $d_{\alpha a}$-$d_{\gamma c}$ in k_{min} and k_{max} are calculated. The PWM pattern with the smallest instantaneous effective value among all PWM patterns is selected. In the proposed control method, k_1 - k_7 in only Groups A are calculated. Next, only k_{max} is determined from k_1 - k_7. All duty cycles $d_{\alpha a}$ - $d_{\gamma c}$ in k_{max} is calculated. Selection process by the instantaneous effective value has been curtailed because only one PWM pattern is calculated.

B. Verification of the Instantaneous Effective Value

The instantaneous effective value is verified by the simulation. Fig.10 shows the system configuration, and Table II lists specifications for the simulation. The simulation is executed by PSIM.

The input side is connected to a three-phase voltage supply of 200 V and 60 Hz. The output side is connected to a balanced three-phase inductive load. In order to calculate PWM patterns

4092

The 2018 International Power Electronics Conference

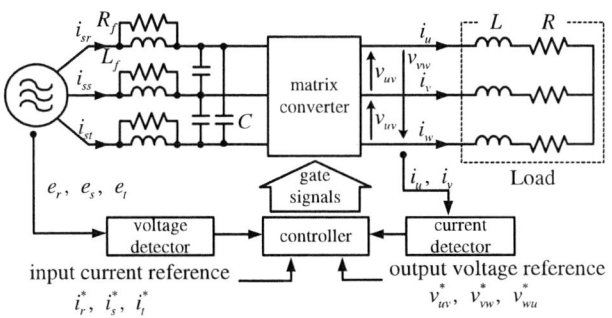

Fig. 10. System configuration

TABLE II. SPECIFICATIONS FOR SIMULATION

Source voltage E, ω	200 V, $2\pi \times 60$ rad/s
Input filter L_f, C_f, R_f	1.13 mH, 4.4 μ F, 47 Ω
Input power-factor-angle reference φ^*	0 rad
Load R, L	11 Ω, 27 mH
Output voltage reference V_L^*, ω_L^*	30 $\sim$ 173 V $2\pi \times 10 \sim 2\pi \times 100$ rad/s
Carrier frequency $f_s(= 1/2T_s)$	10 kHz

and their instantaneous effective values in conventional and proposed control methods, the output current is ideal without harmonics. The effective value of the ideal output current I_{out} and the load power-factor-angle φ_L is expressed as follows:

$$I_{out} = \frac{V_L^*}{\sqrt{3}} \frac{1}{\sqrt{R^2 + (\omega_L^* L)^2}} \tag{45}$$

$$\varphi_L = \tan^{-1} \frac{\omega_L^* L}{R} \tag{46}$$

Fig.11 shows simulation waveforms under $V_L^* = 90$ V. In this figure, each waveform shows the source line voltage e_{rs}, the source current i_{sr}, the output frequency f_L, the output line voltage v_{uv} of the proposed method, the output current i_u of the proposed method (black) and the ideal output current (red), the number of duty combinations that establish in each period, and instantaneous effective values $V(T_s)$ of the proposed method (black) and the conventional method (red). Although 5 - 12 duty combinations establish in one period, instantaneous effective values of both control methods is very similar. Therefore, PWM patterns in the proposed control method can minimize the instantaneous effective value.

Fig.12 shows the selection ratios of duty combinations of the conventional control method under $V_L^* = 30 - 170$. In the state of over $V_L^* = 100$ V, the selection ratio of Group A of $k = k_{max}$ is over 90 %. In the state of under $V_L^* = 100$ V, the selection ratio of Group A of $k = k_{max}$ decrease about 70 %, and the selection ratio of Group A of $k = k_{min}$ increases. However, as shown in Fig.11, the increase of the instantaneous effective values of the proposed method is very small. Therefore, even if Group A of $k = k_{max}$ that is not the lowest instantaneous effective value is selected, the distortion of the output voltage of the PWM pattern is similar with that of $k = k_{min}$

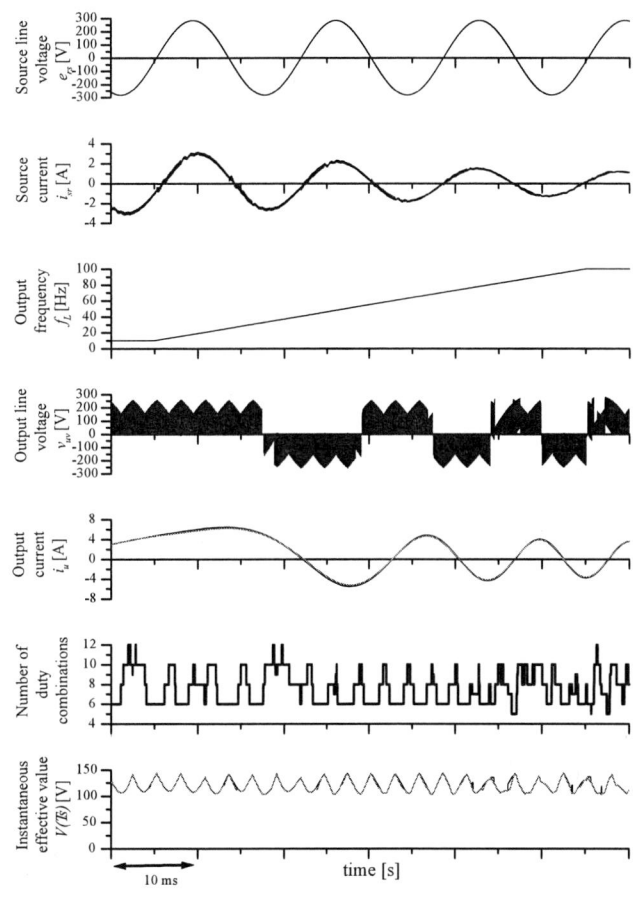

Fig. 11. Simulation waveforms ($V_L^* = 90$ V)

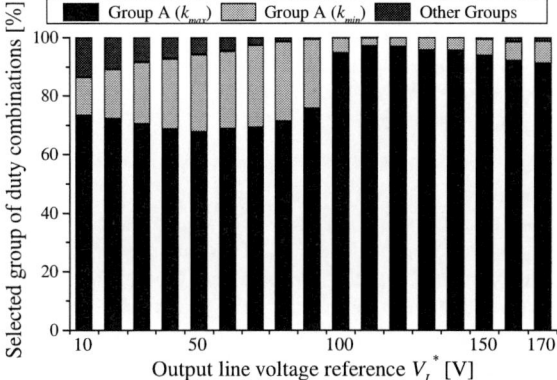

Fig. 12. the selection proportion of duty combinations

C. Verification by Experiments

Table III lists the specifications for the simulation and the experiment. In the experiment system, Each bi-directional switch consists of two IGBTs (Fuji Electric Co., Ltd., 1MBH50D-060) connected to each other in series and opposite direction. The controller is a DSP (Digital Signal Processor,

4093

The 2018 International Power Electronics Conference

TABLE III.   THE SPECIFICATIONS FOR THE SIMULATION AND THE EXPERIMENT

Source voltage E, ω	200 V, $2\pi \times 60$ rad/s
Input filter L_f, C_f, R_f	1.13 mH, 4.4 μF, 47 Ω
Input power-factor angle reference φ^*	0 rad
Load R, X_L	20 Ω, 2.1 Ω ($\varphi_L = -0.11$ rad)
	19 Ω, 5.2 Ω ($\varphi_L = -0.26$ rad)
	18 Ω, 10.0 Ω ($\varphi_L = -0.51$ rad)
	14 Ω, 13.8 Ω ($\varphi_L = -0.78$ rad)
	11 Ω, 17.0 Ω ($\varphi_L = -1.00$ rad)
Output voltage reference V_L^*, ω_L^*	30$\sim$173 V, $2\pi \times 100$ rad/s
Carrier frequency f_s (=1/2T_s)	10 kHz

TABLE IV.   PROCESSING TIME

	total operation time [μs]	calculation time of PWM patterns [μs]
Conventional method	180.30	156.78
Proposed method	33.34	10.44

Texas Instruments, TMS320C6713-225).

Fig.13 shows waveforms of the conventional and proposed method under $V_L^* = 173$ V and $\varphi_L = -0.10$ rad. The control method is verified by only the simulation, and the proposed method is verified by both simulation and experiment. In this figure, each waveform shows the source line voltage e_{rs}, the source current i_{sr}, the output line voltage v_{uv}, and the output current i_u. Output line voltages v_{uv} are similar in both control methods because similar PWM patterns are selected. And output currents i_u is also similar sinusoidal waveforms in both control methods.

D. Frequency Analysis

Fig.14 shows the frequency analysis results of output line voltages under $V_L^* = 173$ V and $\varphi_L = -0.10$ rad. The harmonic at carrier frequency (10 kHz) is about 30 % in the simulation of the conventional method, the simulation and experiment of the proposed method.

E. Total Harmonic Distortion

Fig.15 shows the relation among the output line voltage reference V_L^* load power-factor-angle φ_L, and THD (Total Harmonic Distortion). The similar THDs among the simulation of the conventional method, the simulation and experiment of the proposed method are obtained. Therefore, output voltage harmonics are minimized by the proposed control method.

F. Processing Time

Table IV lists processing time of the program of two control method. The proposed control method shortens the calculation time of PWM patterns by about 6.7 %, and the total operation time by about 18.5 % of the conventional control method.

VII.   CONCLUSIONS

In this paper, all duty combinations with three commutations in one control period were subject of consideration. And

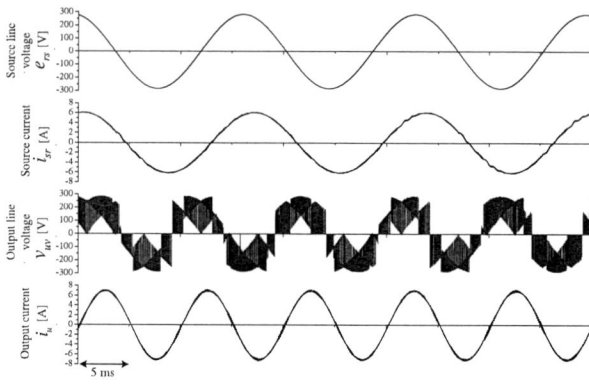

(a) Conventional control method (simulation)

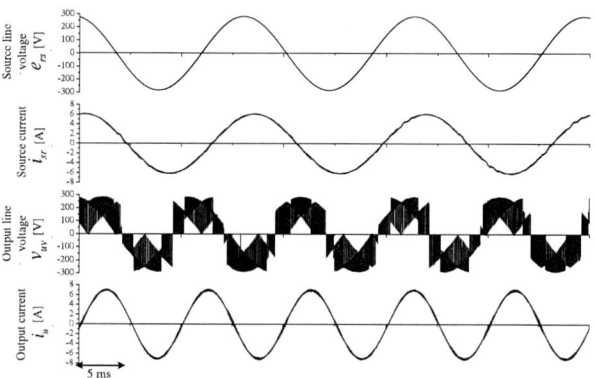

(b) Proposed control method (simulation)

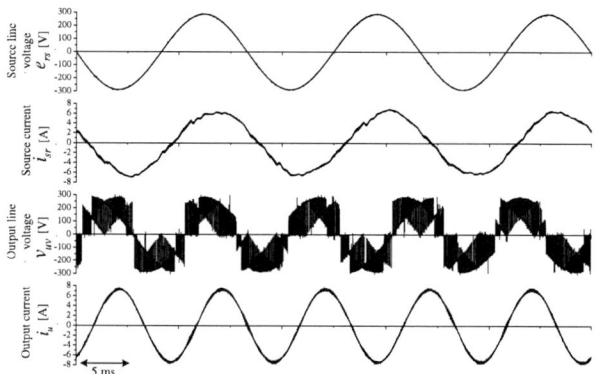

(c) Proposed control method (experiment)

Fig. 13.   Waveforms of the simulation and the experiment

PWM strategy of MC for minimizing output voltage harmonics without calculation of the instantaneous effective values was proposed. The effectiveness of the control method has been verified by simulation and experiment.

4094

The 2018 International Power Electronics Conference

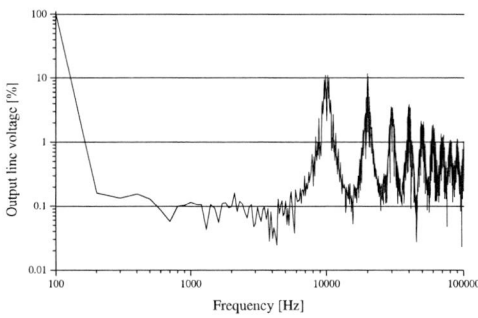

(a) Conventional control method (simulation)

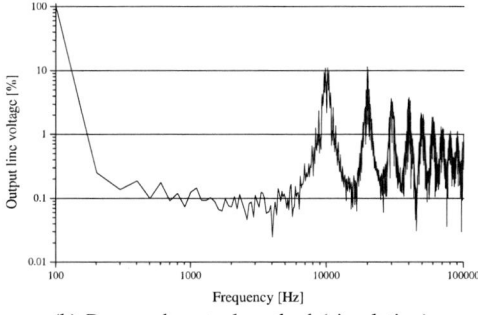

(b) Proposed control method (simulation)

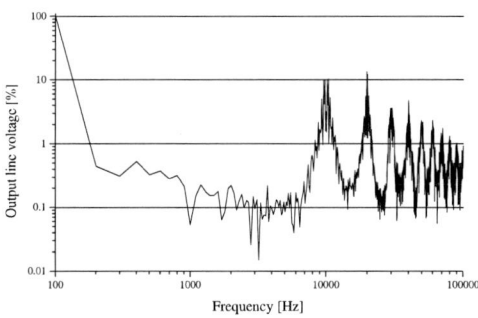

(c) Proposed control method (experiment)

Fig. 14. The frequency analysis results of output line voltages

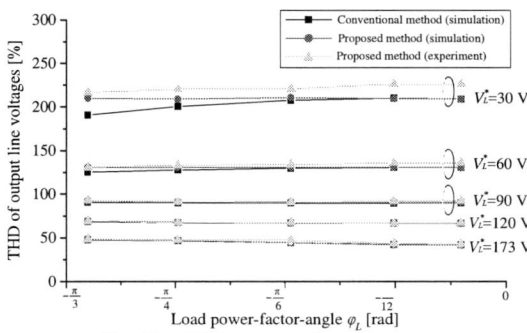

Fig. 15. THD of output line voltages

REFERENCES

[1] M. Venturini: "A new sine wave in sine wave out, conversion technique which eliminates reactive elements", *Proc. POWERCON 7*, pp.E3_1-

E3_15 (1980)

[2] J. Itoh, I. Sato, H. Ohguchi, K. Sato, A. Odaka, N. Eguchi: "A Control Method for the Matrix Converter Based on Virtual AC/DC/AC Conversion Using Carrier Comparison Method", *IEEJ Trans. IA*, Vol.124, No.5, pp.457-463 (2004) (in Japanese)

[3] K.M. Sung, H. Nakakoji, Y. Sato: "A Control Method to Realize Sinusoidal Input and Output Current Waveforms for Matrix Converters Based on PWM", *IEEJ Trans. IA*, Vol.124, No.11, pp.1104-1113 (2004) (in Japanese)

[4] T. Takeshita and H. Shimada: "Matrix Converter Control Using Direct AC/AC Conversion Approach to Reduce Output Voltage Harmonics", *IEEJ Trans. IA*, Vol.126, No.6, pp.778-787 (2006-6) (in Japanese)

[5] Y. Andou, T. Takeshita: "PWM Control of Three-Phase to Three-Phase Matrix Converters for Reducing a Number of Commutations", *IEEJ Trans. IA*, Vol.127, No.8, pp.805-812 (2007) (in Japanese)

[6] I. Asai, T. Takeshita: "PWM Strategy of Matrix Converters for Reducing Switching Losses and Output Voltage Harmonics", *IEEJ Trans. IA*, Vol.133, No.1, pp.1-9 (2013) (in Japanese)

[7] S. Ishikawa, T. Takeshita: "Input Power Factor Control of Three-Phase to Three-Phase Matrix Converters", *IEEJ Trans. IA*, Vol.129, No.3, pp.258-266 (2009) (in Japanese)

[8] T. Takeshita, S. Ishikawa and Y. Andou: "Instantaneous Effective Values Theory and Its Application to Output Voltage Harmonics Suppression of Matrix Converters", *IEEJ Trans. IA*, Vol.130, No.12, pp.1290-1297 (2010) (in Japanese)

The 2018 International Power Electronics Conference

Novel Three-Level Back-to-Back Converters: Structure, Modulation Method, and Experiment

S. Sangwongwanich[1]*, K. Niyomsatian[2], S. Samermurn[1], S. Nuchnoi[1], and S. Suwankawin[1]

1 Dept. of Electrical Eng., Faculty of Engineering, Chulalongkorn University, Bangkok, Thailand
2 Dept. of Electrical Eng., University of Leuven, Leuven, Belgium
*E-mail: somboona@chula.ac.th

Abstract— **Novel three-level back-to-back converters with symmetrical dc-link structure are proposed for bidirectional and unidirectional power applications. The new converters combine a non-PWM rectifier with a PWM inverter to cooperatively control both the output voltage and the input current. The non-PWM rectifier significantly reduces switching losses and EMI noises at the input as compared to the PWM rectifier of the conventional back-to-back converters. Modulation methods are derived for the proposed converters to obtain the required output voltage, input power factor control, and switching loss reduction. Experimental results are given to verify the key characteristics of the proposed converters.**

Keywords— ***Three-level back-to-back converters, symmetrical dc link, modulation method, switching loss reduction.***

I. INTRODUCTION

The three-level back-to-back (3L-BTB) converter in Fig. 1 [1]-[11] is a popular topology used in high power applications to meet the following requirements:

- controllability of output voltages,
- sinusoidal input and output currents,
- bidirectional power flow capability, and
- power factor correction.

However, the conventional 3L-BTB converter requires an active-front-end PWM rectifier which increases switching losses and electromagnetic interference (EMI) noises. To retain the main features of 3L-BTB converters

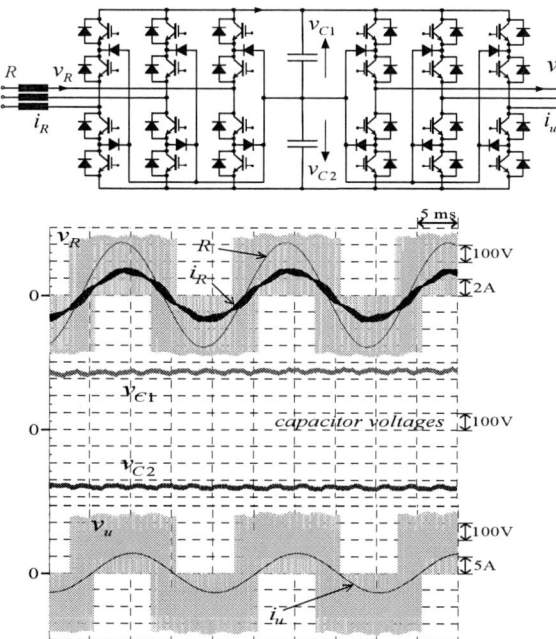

Fig. 1. Operation of conventional 3L-BTB converters.

and alleviate the switching losses and EMI problems, an improved structure of 3L-BTB converters shown in Fig. 2 has been proposed in [12] with their concepts verified by simulation.

The key structure of the proposed converters is the

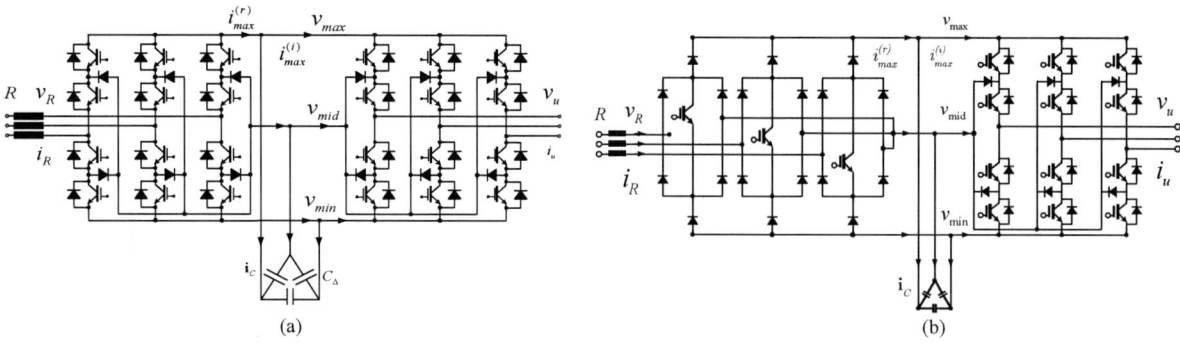

(a)   (b)

Fig. 2. New 3L-BTB converters with symmetrical dc-link structure. (a) Bidirectional converter (S3L-BTB). (b) Unidirectional converter (U3L-BTB).

symmetrical three-phase capacitors at the dc link which is different from the conventional two-phase capacitors. This change is significant because it allows the rectifier to be operated at the power-line frequency not at a high switching frequency, which solves the switching losses and EMI problems. The task to control the input current is mainly assigned to the back-end inverter instead.

The second key idea of this research is the derivation of new modulation methods for the 3L-BTB converters which cooperatively operate the rectifier and inverter to achieve simultaneous control of both the output voltages and input currents. The rectifier and the inverter of the conventional 3L-BTB normally switch independently, and the possibility to operate the rectifier and the inverter in a cooperative manner to achieve better performances has been overlooked in the past.

The only constraint of the proposed converters is that its output voltage is limited to 87% of the input, which is acceptable in several applications. For unidirectional applications, a similar converter called USMC3 is mentioned in [11]. However, its modulation strategy has not been given, and its output power factor is limited to be less than 30°.

This paper presents a comprehensive summary of continued research works on the new 3L-BTB converters with symmetrical dc-link structure. A modulation method based on dipolar PWM is introduced, and a discontinuous modulation method is also newly derived to further reduce the inverter losses. All the theoretical properties of both the bidirectional and unidirectional converters are verified by experiment to prove their feasibility.

II. STRUCUTRE AND OPERATION OF THE PROPOSED SYMMETRICAL 3L-BTB CONVERTERS

A. Bidirectional Symmetrical (S3L-BTB) Converter

The proposed bidirectional converter shown in Fig. 2(a) has a symmetrical three-phase dc-link structure. Together with the front-end inductances, the dc-link capacitors act as a low-passed filter to attenuate the PWM dc-link current. Therefore, only the fundamental currents flow through the rectifier. The non-PWM rectifier only sorts the three-phase input voltages $\mathbf{v}_i = (v_R, v_S, v_T)$ into the maximum-median-minimum $(v_{max}, v_{mid}, v_{min})$ dc bus voltages. In comparison with Fig. 2, operation waveforms of the new S3L-BTB converter is illustrated in Fig. 3.

B. Unidirectional Symmetrical (U3L-BTB) Converter

For unidirectional power applications, the front-end rectifier can be of Vienna type as shown in Fig. 2(b). This structure needs less switching devices. The Vienna rectifier here, however, switches only at the power-line frequency, and the impulsive PWM current at the dc link (resulting from the inverter operation) is absorbed or filtered by the dc-link capacitors and does not flow through the diodes of the rectifier. This makes the U3L-

BTB converter superior to the USMC3 converter in [11] because the impulsive PWM current is allowed to be negative. This means that the output power factor is not limited to $\pm 30°$ as in the case of the USMC3 converter.

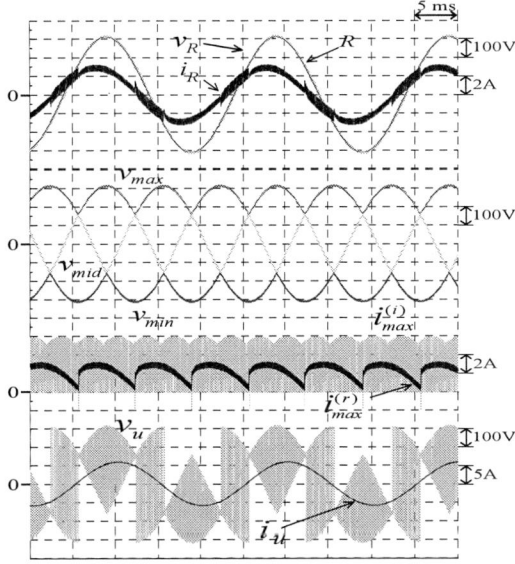

Fig. 3. Operation of the proposed S3L-BTB converters.

III. MODULATION METHODS BASED ON COOPERATIVE CONTROL OF NON-PWM RECTIFIER AND PWM INVERTER

A. Continuous Dipolar Modulation Method

The dipolar modulation method [13][14] is adopted for the PWM inverter to gain its input-current controllability in addition to the usual output-voltage generating function. The dipolar PWM decomposes the output voltage $\mathbf{v}_o = (v_u, v_v, v_w)$ into two reference voltages denoted here as the upper-bus and lower-bus voltage commands $(\mathbf{u}_p, \mathbf{u}_n)$ as shown in (1).

$$\begin{bmatrix} v_u - v_{mid} \\ v_v - v_{mid} \\ v_w - v_{mid} \end{bmatrix} = \underbrace{\begin{bmatrix} m_{up} \\ m_{vp} \\ m_{wp} \end{bmatrix}(v_{max} - v_{mid})}_{\mathbf{u_p}} + \underbrace{\begin{bmatrix} -m_{un} \\ -m_{vn} \\ -m_{wn} \end{bmatrix}(v_{mid} - v_{min})}_{\mathbf{u_n}} .$$

(1)

The two references $(\mathbf{u}_p, \mathbf{u}_n)$ are subsequently normalized by each dc bus voltage to become the modulating signal (or duty cycle) references $(\mathbf{m}_p, \mathbf{m}_n)$ used in the double-carrier modulation process shown in Fig. 4(a)-(c). The expressions of both references, to achieve the required output voltages and the input reactive power, are given in terms of the voltage commands (v_u^*, v_v^*, v_w^*), the dc-bus voltages $\mathbf{v}_{dc} = (v_{max}, v_{mid}, v_{min})$, and the output currents $\mathbf{i}_o = (i_u, i_v, i_w)$ as shown in (2).

The 2018 International Power Electronics Conference

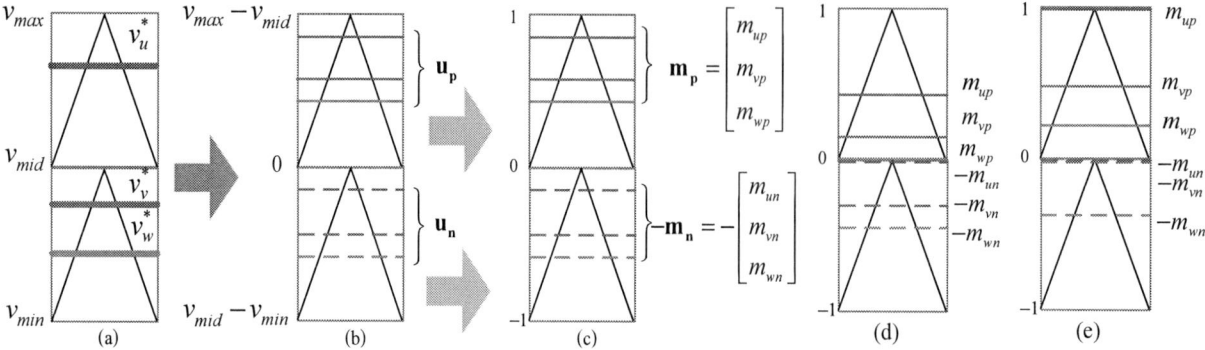

Fig. 4. Dipolar PWM and double-carrier modulation. (a) Commanded voltages. (b) Upper and lower references. (c) Normalized references (duty cycles). (d) Proposed continuous dipolar PWM. (e) Proposed discontinuous dipolar PWM.

$$\mathbf{m}_p = \begin{bmatrix} m_{up} \\ m_{vp} \\ m_{wp} \end{bmatrix} = \begin{bmatrix} m'_{up} + z_p \\ m'_{vp} + z_p \\ m'_{wp} + z_p \end{bmatrix}, \quad \mathbf{m}_n = \begin{bmatrix} m_{un} \\ m_{vn} \\ m_{wn} \end{bmatrix} = \begin{bmatrix} m'_{un} + z_n \\ m'_{vn} + z_n \\ m'_{wn} + z_n \end{bmatrix}$$

$$\begin{bmatrix} m'_{up} \\ m'_{vp} \\ m'_{wp} \end{bmatrix} = \frac{v_{max}}{\|\mathbf{v}_i\|^2}\begin{bmatrix} v_u^* \\ v_v^* \\ v_w^* \end{bmatrix} + (-1)^{n+1}\frac{k_1(v_{mid}-v_{min})}{\sqrt{3}\|\mathbf{v}_i\|^2}\begin{bmatrix} i_u \\ i_v \\ i_w \end{bmatrix}$$

$$\begin{bmatrix} m'_{un} \\ m'_{vn} \\ m'_{wn} \end{bmatrix} = \frac{v_{min}}{\|\mathbf{v}_i\|^2}\begin{bmatrix} v_u^* \\ v_v^* \\ v_w^* \end{bmatrix} + (-1)^{n+1}\frac{k_1(v_{max}-v_{mid})}{\sqrt{3}\|\mathbf{v}_i\|^2}\begin{bmatrix} i_u \\ i_v \\ i_w \end{bmatrix}$$

(2)

where $\|\mathbf{v}_i\|^2 = v_{max}^2 + v_{mid}^2 + v_{min}^2$, n is the sector number of the input voltage vector, and z_p, z_n are the zero-voltage related terms.

With (2), the input current of the converter is derived to be:

$$\mathbf{i}_i = \underbrace{\mathbf{v}_i\frac{p_o}{\|\mathbf{v}_i\|^2}}_{\text{active current}} + \underbrace{\mathbf{Jv}_i\frac{k_1}{\|\mathbf{v}_i\|^2}\|\mathbf{i}_o\|^2}_{\text{reactive current}} + \underbrace{3C_\Delta\frac{d\mathbf{v}_i}{dt}}_{\substack{\text{capacitor}\\\text{filter current}}} \quad (3)$$

where $p_o = \mathbf{v}_o^T\mathbf{i}_o$ is the output power, and C_Δ is the per-phase capacitance of the dc link. The dc-link capacitors thus behave like the filter capacitors placed at the input terminal. In steady states, the input reactive power Q is obtained as:

$$Q = \left(-\mathbf{Jv}_i\right)^T\mathbf{i}_i = -k_1\|\mathbf{i}_o(t)\|^2 - 3\omega C_\Delta\|\mathbf{v}_i(t)\|^2 \quad (4)$$

where ω is the power-line frequency, and J is the 90° rotating operator. The input power factor can thus be controlled through the free parameter k_1 so long as no overmodulation occurs. A unity power factor, $Q = 0$, is obtained when

$$k_1 = -3\omega C_\Delta\|\mathbf{v}_i(t)\|^2 / \|\mathbf{i}_o(t)\|^2. \quad (5)$$

When the zero-voltage related terms are selected as

$$z_p = -\min_{k=u,v,w}(m'_{kp}), \quad z_n = -\min_{k=u,v,w}(m'_{kn}), \quad (6)$$

two output phases with the maximum and minimum voltages will be in unipolar mode and the middle phase in the dipolar mode (as shown in Fig. 4(d)). The maximum modulation index (output voltage/input voltage ratio) for the S3L-BTB converter is 0.87, and the number of switch transition per carrier period is 8 (which is higher than the typical unipolar modulation).

B. Discontinuous Dipolar Modulation Method

Switching losses of the inverter side can be further reduced by a discontinuous modulation method (as shown in Fig. 4(e)) if the information of the output currents is utilized [15]. In this paper a three-step modulation procedure is introduced to reduce the number of switching as follows.

Step 1) Determine the non-switching output phase by clamping the dc bus with the largest magnitude (either v_{max} or v_{min}) to the corresponding maximum or minimum output phase. This is done by adding an appropriate zero voltage v_z to the output voltage commands as shown in Fig. 5.

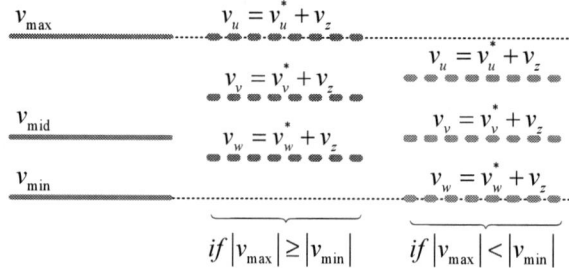

Fig. 5. Selection of non-switching phase.

Step 2) Determine the unipolar switching phase and its duty cycles using the output current information.

4098

The key idea is to first calculate the required mid-bus input current i_{mid}^*, e.g. for unity input power factor, using (7)

$$i_{mid}^* = \frac{(v_u^* \cdot i_u) + (v_v^* \cdot i_v) + (v_w^* \cdot i_w)}{v_{max}^2 + v_{mid}^2 + v_{min}^2} v_{mid} . \quad (7)$$

From Fig. 6, the real mid-bus input current is calculated from (8).

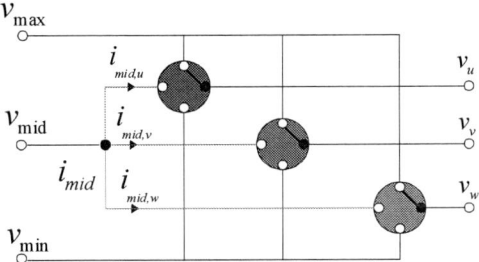

Fig. 6. Construction of mid-bus input current.

$$i_{mid} = i_{mid,u} + i_{mid,v} + i_{mid,w}$$
$$= m_{uo}i_u + m_{vo}i_v + m_{wo}i_w = i_{mid}^* \quad (8)$$

where $m_{ko} (k = u, v, w)$ represent the duty cycles of mid-bus connection of each output phase. For the clamped (non-switching) and bipolar-mode output phase

$$m_{ko} = 0 . \quad (9)$$

And for the unipolar-mode output phase,

$$m_{ko} = \frac{v_k^* + v_z - v_{mid}}{v_{max} - v_{mid}} \quad if \ v_k^* + v_z > v_{mid}$$

$$m_{ko} = \frac{v_k^* + v_z - v_{mid}}{v_{min} - v_{mid}} \quad if \ v_k^* + v_z \le v_{mid} . \quad (10)$$

Among the two non-clamped output phases, one must be in dipolar mode and the other in unipolar or bipolar mode. So, there are only two possible combinations. From (7)-(10), one can determine which output phase must be in unipolar (or bipolar) mode because the solution is unique. And, the corresponding duty cycles can be found. Unipolar mode is preferable than the bipolar mode and is always selected whenever possible.

Step 3) Find the duty cycles of the dipolar-mode output phase which gives the required mid-bus input current.

IV. EXPERIMENTAL VERIFICATION AND DISCUSSIONS

The proposed modulation methods are implemented on an FPGA as shown in Fig. 7 with the parameters of the experimental setup given in Table I. One important factor that must be considered in selecting power devices, is the voltage stress. The dc-link voltages of the proposed S3L-BTB converter are not constant as usually found in

the conventional back-to-back converters. The upper or lower bus of the dc link has its peak voltage given by (11). And the peak voltage between the 'max' and 'min' buses is equal to the peak value of the line-to-line input voltage shown in (12).

$$(v_{max} - v_{mid})_{peak} = (v_{mid} - v_{min})_{peak} = \sqrt{\frac{3}{2}} V_i \quad (11)$$

$$(v_{max} - v_{min})_{peak} = \sqrt{2} V_i \quad (12)$$

Here V_i is the RMS value of the input line-to-line voltage. The power device ratings should be selected accordingly.

First, the experimental results illustrating the operation of the proposed bidirectional S3L-BTB converter under unity and leading input power factor control are shown in Figs. 8 and 9. Fine sinusoidal waveforms of both the input and output currents are obtained, and the input power factor control ability of the proposed S3L-BTB converter is confirmed. It should be noted that the impulsive PWM dc-link current on the inverter side $i_{max}^{(i)}$ is filtered by the capacitors and becomes a smooth current $i_{max}^{(r)}$ on the rectifier side. This current is reconstructed to be the sinusoidal input current by the action of the power-line-frequency switching rectifier.

Next, Fig. 10 is the experimental results showing the performance of the unidirectional U3L-BTB converter at unity input power factor. It is seen that the U3L-BTB converter with the Vienna rectifier switched at power-line frequency, gives similarly sinusoidal input and output currents as those of the S3L-BTB converter. And from Fig. 11, it is confirmed that the U3L-BTB converter can operate at output power factor lagging beyond 30° (which is the limitation of the USMC3 converter in [11]).

Finally, Fig. 12 compares the operation of the S3L-BTB converter under the continuous and discontinuous dipolar modulation methods. As is seen on the zoomed output waveforms shown in Fig. 12 (c), the continuous PWM has two output phases switching in unipolar mode and one output phase in dipolar mode. The total number of switching per carrier period is thus eight. On the contrary, from the modulating function or duty-cycle waveforms m_{up}, m_{un} in Figs. 12 (b) and (d), it is clear that by the discontinuous PWM one of the output phases is not switched and is clamped to the max or min bus. The other two output phases are in unipolar and dipolar modes. The output PWM voltage also confirms this behavior. Though, the PWM output voltages are different, both continuous and discontinuous PWMs result in the same sinusoidal output currents. Also, the impulsive dc-link PWM currents $i_{max}^{(i)}$ are different, but the filtered dc-link currents $i_{max}^{(r)}$ on the rectifier side are the same. Therefore, the discontinuous PWM gives similar input/output sinusoidal current waveforms, but switches only at 2/3 times of that of the continuous one (i.e. 6 times per carrier period).

4099

The 2018 International Power Electronics Conference

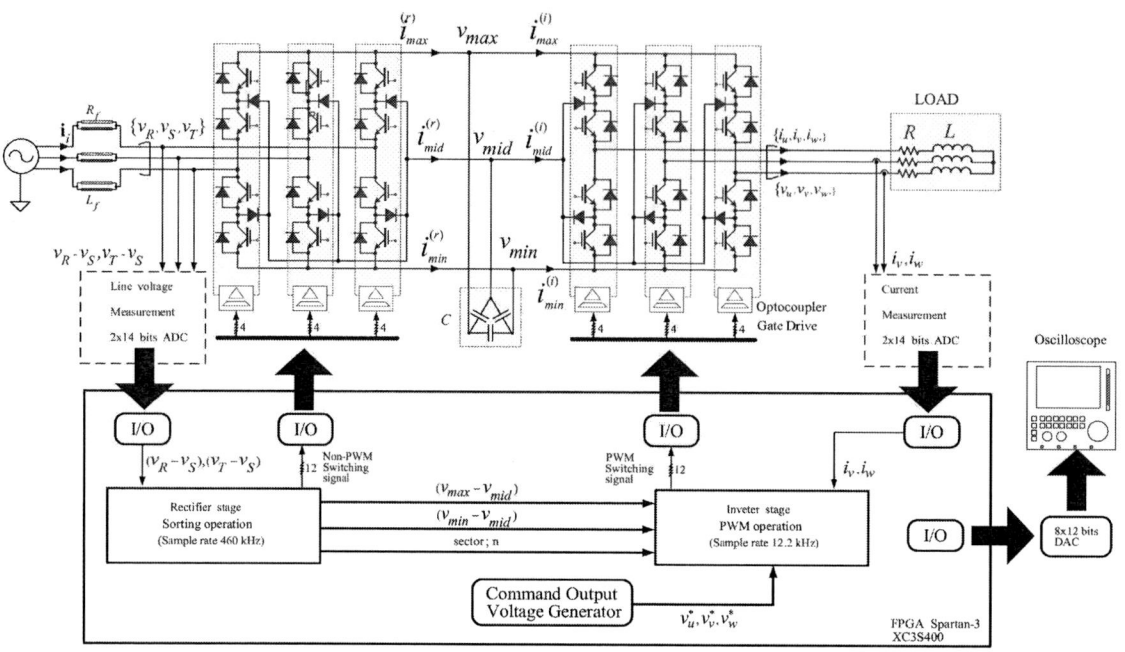

Fig. 7. Experimental setup and implementation of the modulation methods on the FPGA.

TABLE I
PARAMETERS OF EXPERIMENTAL SETUP

Source voltage 380 V 50 Hz Switching frequency 12. 2 kHz	Input line reactance Lf = 5 mH, Damping resistance Rf = 15 Ω	DC link capacitors C = 4.2 μF	Series RL load: R = 24 Ω and L = 33.3 mH

Fig. 8. Experimental results of S3L-BTB converter at unity input power factor (M=0.7, f$_o$=25 Hz, k$_1$=5, output power factor=0.977 (12.2° lagging)).

Fig. 9. Experimental results of S3L-BTB converter at leading input power factor (M=0.7, f$_o$=50 Hz, k$_1$=-3.3, output power factor=0.643 (50° lagging)).

The 2018 International Power Electronics Conference

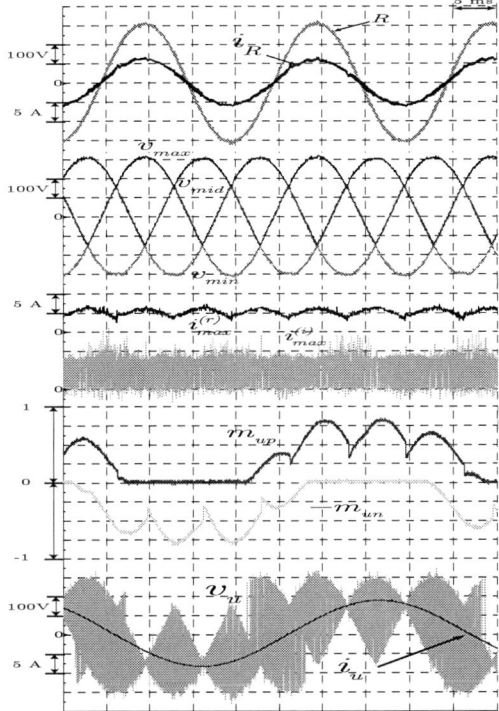

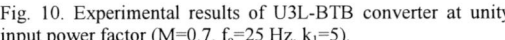

Fig. 10. Experimental results of U3L-BTB converter at unity input power factor (M=0.7, f_o=25 Hz, k_1=5).

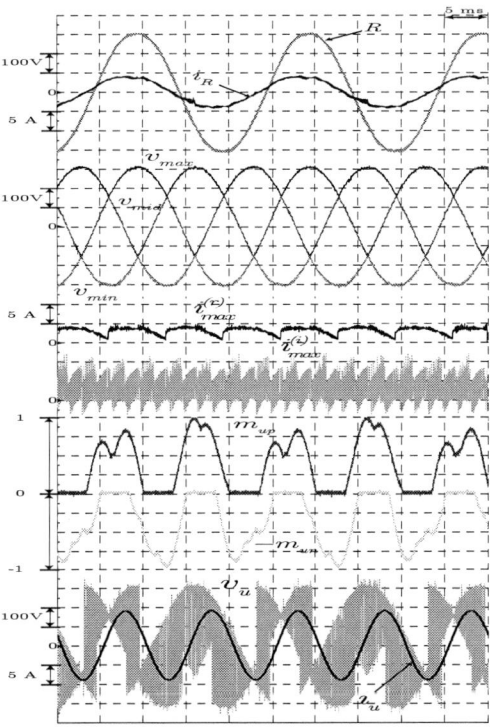

Fig. 11. Experimental results of U3L-BTB converter at 50° lagging output power factor (M=0.87, f_o=100 Hz, k_1=0).

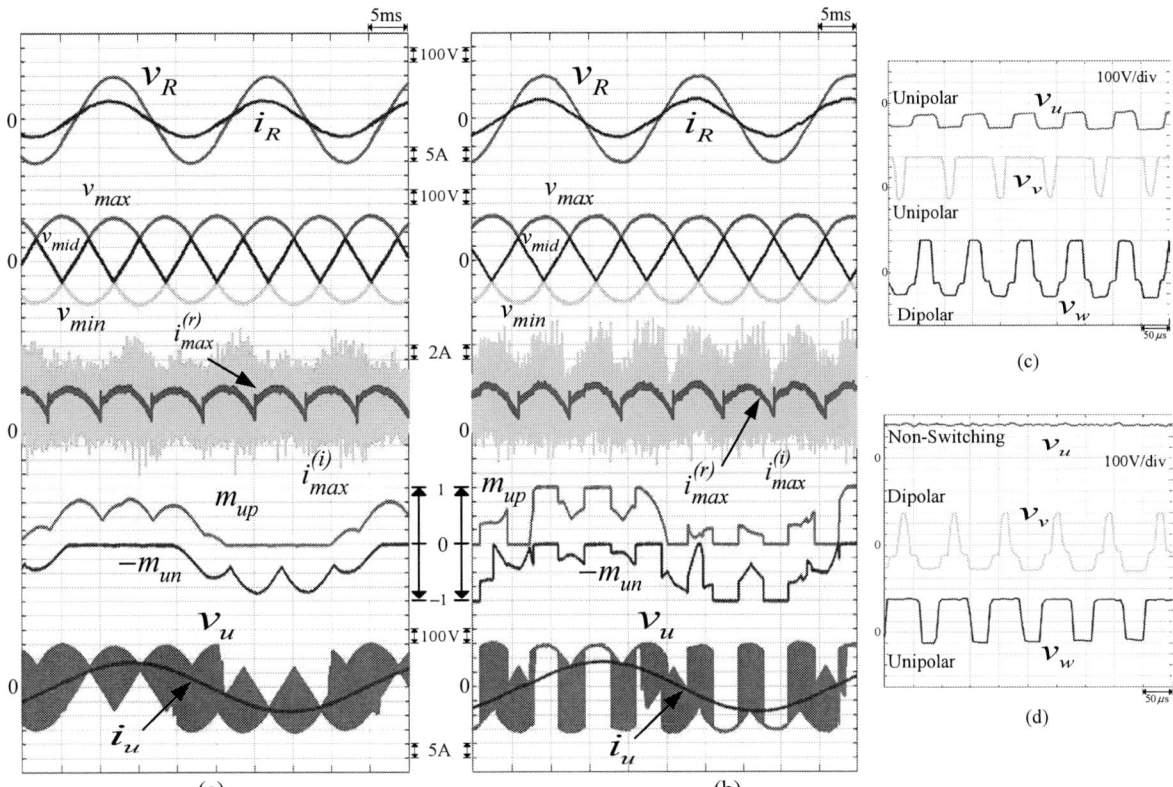

Fig. 12. Experimental results of S3L-BTB converter at unity input power factor (M=0.7, f_o=25 Hz, k_1=0). (a) Continuous dipolar PWM. (b) Discontinuous PWM. (c) Zoomed voltage waveforms of continuous PWM. (d) Zoomed voltage waveforms of discontinuous PWM.

4101

V. NUMERICAL LOSS EVALUATION

To investigate the benefit of the proposed converters regarding switching loss reduction compared with the conventional 3L-BTB converters, numerical evaluation for semiconductor losses of the S3L-BTB converter is performed based on the simplified semiconductor loss model in [9] and [10], and with the parameters of a commercial 3L-NPC structure module SKM20ML1066. The mean losses over a fundamental period for each converter are determined and compared in Table II.

According Table II, it is obvious that the switching losses in the rectifier stage are significantly reduced to zero due to the power-line-frequency switching operation. The whole loss reduction contributes to the improvement of efficiency from 95.7% to 97%. Also, similar analysis for switching loss reduction of the inverter stage using the discontinuous PWM is shown in Fig. 13 In total, the mean losses over a fundamental period is 34.42 W for the continuous PWM and 27.7 W for the discontinuous PWM.

TABLE II
COMPARISON OF SEMICONDUCTOR LOSSES BETWEEN S3L-BTB AND CONVENTIONAL 3L-BTB CONVERTERS (M=0.86, f_o=25 Hz)

Stage	Losses	S3L-BTB Converter (Watt)		Conventional 3L-BTB Converter (Watt)	
Rectifier	Conduction Losses	54.28	54.32	52.18	91.29
	Switching Losses	0.039		39.11	
Inverter	Conduction Losses	41.47	75.89	43.39	90.35
	Switching Losses	34.42		46.86	
	Total Losses	130.21		181.64	
	Efficiency	97%		95.7%	

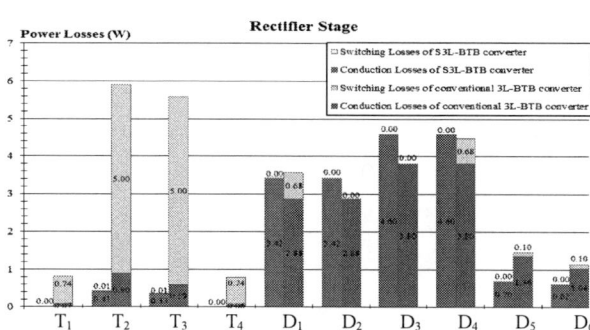

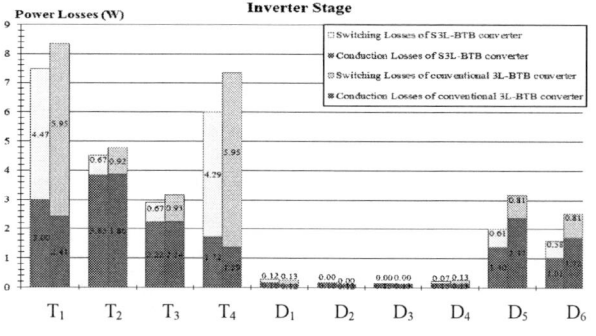

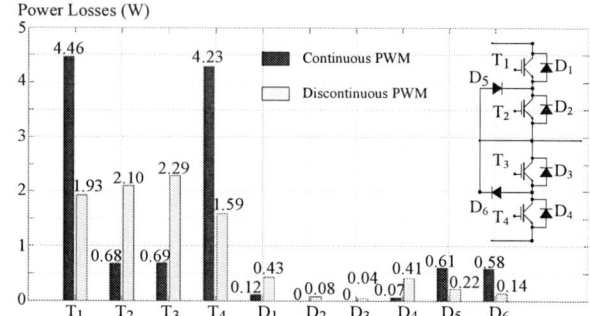

Fig. 13. Comparison of inverter's switching losses (per one phase) for continuous and discontinuous PWM.

VI. CONDUCTED EMI MEASUREMENT

To confirm the superiority of the S3L-BTB converter over the conventional 3L-BTB converter in terms of reduced electromagnetic noises at the input, the conducted electromagnetic noise is measured at the input terminal of the converters. In this experiment the comparison is drawn between the S3L-BTB converter and the conventional three-level inverter (3L-inverter) with diode-bridge rectifier, which is implemented by turning off all the active IGBTs at the rectifier stage. A LISN is inserted between the grid and the input terminal of the converter for both converters under test. Figs. 14 and 15 show the conducted EMI noise profiles measured at two operating conditions. They clearly illustrate that the characteristics of the input conducted EMI noises generated by S3L-BTB converter are the same as those generated by the 3L-inverter with diode-bridge rectifier. In other words, the S3L-BTB converter achieves the same noise levels as those of the diode-bridge rectifier, but with the sinusoidal input currents.

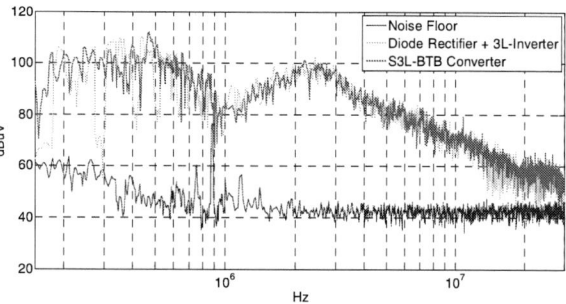

Fig. 14. Conducted EMI noise profiles measured at (M=0.7, fo=50 Hz).

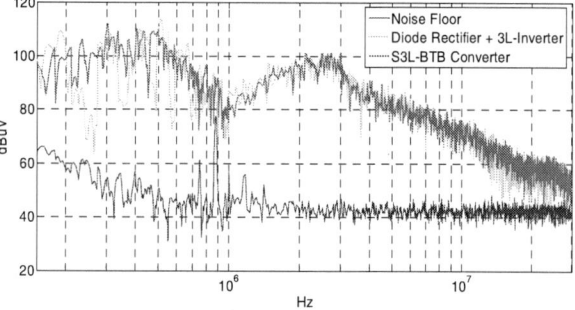

Fig. 15. Conducted EMI noise profiles measured at (M=0.86, fo=25 Hz).

VII. CONCLUSIONS

As a physical realization of the indirect modulation method for matrix converters derived in [16], the symmetrical three-level back-to-back converter (S3L-BTB converter) is newly proposed in this paper by modifying the dc-link capacitors to be a balanced or symmetrical three-phase circuit instead of the traditional two-phase circuit. The proposed S3L-BTB converter can simultaneously achieve sinusoidal output voltages and sinusoidal input currents as does the AC/AC converter. However, the rectifier of the S3L-BTB converter switches only at the power-line frequency like the typical diode rectifier. Therefore, the proposed converter has less switching losses and EMI noises at the input side. For unidirectional power flow application, the unidirectional 3L-BTB (U3L-BTB) converter is further derived to reduce the number of power switches in the converter. The U3L-BTB converter works exactly like the S3L-BTB converter, but with two constraints. First, it supports only non-regenerative operation, and secondly the input power factor is limited to $\pm 30°$. To further reduce the switching losses of the proposed converters, a discontinuous PWM is also proposed as an alternative to the continuous PWM. The operation and modulation methods of the proposed converters and their performances are verified by both simulation and experiment. However, only experimental results are shown in the paper due to space limits. The experimental results are in good agreement with the theoretical and simulation results and confirm that the proposed converters can be used in practice to control the output voltage and frequency, and to adjust the input power factor. Though the neutral-point-clamped topology is adopted in this paper for the three-level converters, the proposed converters can be realized by the T-type topology as well.

REFERENCES

[1] J. Rodríguez, J. Pontt, G. Alzamora, N. Becker, O. Einenkel, and A. Weinstein, "Novel 20-MW Downhill Conveyor System Using Three-Level Converters," *IEEE Trans. on Industrial Electronics*, vol. 49, no. 5, pp. 1093-1100, 2002.

[2] E. J. Bueno, S. Cobreces, J. Rodríguez, A. Hernandez, and F. Espinosa, "Design of a Back-to-Back NPC Converter Interface for Wind Turbines with Squirrel-Cage Induction Generator," *IEEE Trans. on Energy Conversion*, vol. 23, no. 3, pp. 932-945, 2008.

[3] A. Calle, J. Rocabert, S. Busquets-Monge, J. Bordonau, S. Alepuz, and J. Peracaula, "Three-Level Three-Phase Neutral-Point-Clamped Back-to-Back Converter Applied to a Wind Emulator," in *Proc. of 13th European Conference on Power Electronics and Applications*, 2009.

[4] A. Calle-Prado, S. Alepuz, J. Bordonau, P. Cortes, and J. Rodriguez, "Predictive Control of a Back-to-Back NPC Converter-Based Wind Power System," *Trans. on Industrial Electronics*, vol. 63, no. 7, pp. 4615-4627, 2016.

[5] Website of ABB, "The wind power converter for tomorrow is already here," (Available at: https://www.abb.com).

[6] Website of Woodward, "Water-cooled Frequency Converters for Wind Turbines," (Available at: https://www.woodward.com).

[7] K. Ma, and F. Blaabjerg, "Modulation Methods for Neutral-Point-Clamped Wind Power Converter Achieving Loss and Thermal Redistribution Under Low-Voltage Ride-Through," *IEEE Trans. on Industrial Electronics*, vol. 61, no. 2, pp. 835-845 2014.

[8] M. Höltgen, and J. O. Krah, "Advanced Control Scheme to Improve the Efficiency of 3-Level Active Front End Inverters for Servo Drives," in *Proc. of PCIM Europe 2014*, pp. 254-262.

[9] J. Rodriguez, S. Bernet, P. K. Steimer, and I. E. Lizama, "A survey on neutral-point-clamped inverters," *IEEE Trans. on Industrial Electronics*, vol. 57, no. 7, pp. 2219–2230, 2010.

[10] T. Friedli, J. W. Kolar, J. Rodriguez, and P. W. Wheeler, "Comparative Evaluation of Three-Phase AC–AC Matrix Converter and Voltage DC-Link Back-to-Back Converter Systems," *IEEE Trans. on Industrial Electronics*, vol. 59, no. 12, pp. 4487-4510, 2012.

[11] J. W. Kolar, T. Friedli, J. Rodriguez, and P. W. Wheeler, "Review of Three-Phase PWM AC-AC Converter Topologies," *IEEE Trans. on Industrial Electronics*, vol. 58, no. 11, pp. 4988-5006, 2011.

[12] K. Niyomsatian, S. Samermurn, S. Suwankawin, and S. Sangwongwanich, "Novel topologies for three-level back-to-back converters based on matrix converter theory," in *IECON 2012 - 38th Annual Conference on IEEE Industrial Electronics Society*, 2012, pp. 6099-6105.

[13] B. Velaerts, P. Mathys, E. Tatakis, and G. Bingen, "A novel approach to the generation and optimization of three-level PWM wave forms," Proc. of Power Electronics Specialists Conference 1988, vol. 2, pp. 1255-1262.

[14] A. Saengseethong, and S. Sangwongwanich, "A new modulation strategy for capacitor voltage balancing in three-level NPC inverters based on matrix converter theory," in *International Power Electronics Conference (IPEC) 2010*, pp. 2358-2365.

[15] S. Sangwongwanich, "Double-carrier-based modulation theory of three-level inverters and a new discontinuous PWM for neutral-point voltage balancing," in *IECON 2012 - 38th Annual Conf. on IEEE Industrial Electronics Society*, 2012, pp. 4961-4966.

[16] P. Kiatsookkanatorn, and S. Sangwongwanich, "A Unified PWM Method for Matrix Converters and Its Carrier-Based Realization Using Dipolar Modulation Technique," *Trans. on Industrial Electronics*, vol. 59, no. 1, pp. 80-92, 2012.

The 2018 International Power Electronics Conference

Model Predictive Control Using Subdivided Voltage Vectors for Current Ripple Reduction in an Indirect Matrix Converter

Keon Young Kim, Yeongsu Bak, Jin-Hyuk Park, and Kyo-Beum Lee[*]

Department of Electrical and Computer Engineering, Ajou University, Suwon, Korea

*E-mail: kyl@ajou.ac.kr

Abstract— This paper proposes a model predictive control (MPC) using subdivided voltage vectors for current ripple reduction in an indirect matrix converter (IMC). In the proposed MPC, the number of candidate voltage vectors is increased by the subdivided voltage vectors. It is possible to reduce the input–output current ripple of the IMC. In addition, the proposed MPC reduces computation load, which is caused by the subdivided voltage vectors, by estimating the position of reference voltage vector. Therefore, the input–output current ripple of the IMC is reduced by the proposed MPC without computational burden. The effectiveness of the proposed MPC is verified by simulation results.

Keywords— *current ripple reduction, finite set model predictive control, indirect matrix converter, subdivided voltage vector.*

I. INTRODUCTION

Recently, an indirect matrix converter (IMC) has intensively researched because it provides powerful and attractive options for performing AC/AC conversions. The IMC has several advantages such as open-loop unity power factor control, bi-directional power flow without additional devices, and power conversion without energy storage components. In addition, it does not have a bulky electrolytic capacitor with low lifetime. Therefore, the IMC can increase system reliability and power density. It consists of a current source rectifier (CSR) and a voltage

source inverter (VSI). The CSR stage provides a desired DC-link voltage and the VSI stage drives a three-phase load. The IMC has advantages of zero DC-link current commutation and scalability of the output stage [1]–[3].

The IMC is used for various applications such as grid-connected system and variable speeds drives. In these applications, a linear control such as a proportional-integral (PI) control is required to control the three-phase currents of the output stage. Additionally, a modulation technique such as space vector modulation (SVM) is used for modulation of the IMC. This combination with the PI control and the SVM for the IMC drives ensures constant switching frequency. It is able to simplify the loss calculation process and power stack design of the IMC [4]. In addition, the output current ripple of the IMC can be reduced with acceptable harmonic characteristics.

However, the linear control has several disadvantages. First, design and tuning of the controller depends heavily on environmental conditions such as temperature and driving time. Second, the dynamic characteristics are limited by bandwidth of the controller. Therefore, in the system that requires the robustness and fast dynamic characteristics, other control methods such as model predictive control (MPC) are required [5], [6].

The MPC is promising control method owing to the improved performance of digital signal processor (DSP). The dynamic characteristic of the MPC is fast because it

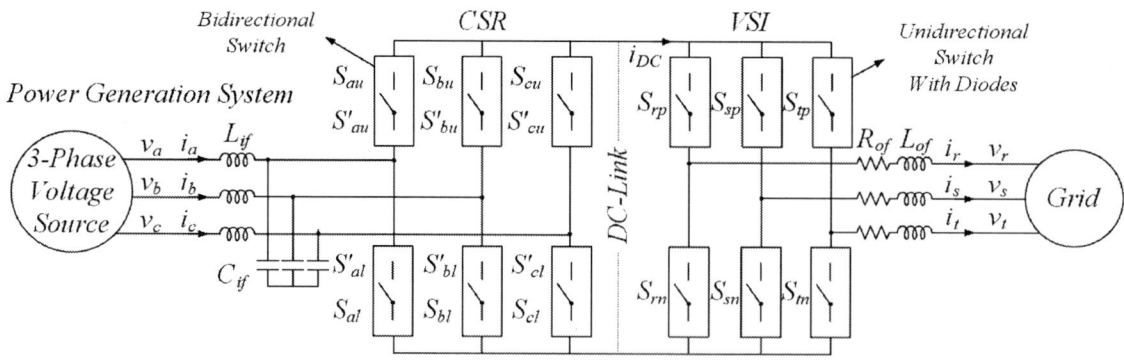

Fig. 1. IMC for power conversion from three-phase voltage source to grid.

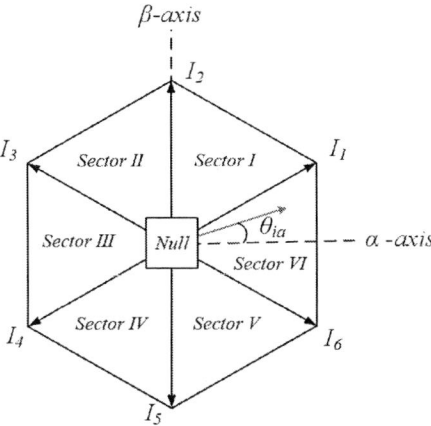

Fig. 2. Space vector current diagram of CSR.

directly applies to the voltage vector depending on the predicted output of the system. Additionally, the MPC is adequate to apply in the system composed of the loads that have nonlinear characteristics because it uses the discrete-time state space models [7], [8].

The finite set MPC (FS-MPC), which is one of the various MPC techniques uses a load model to predict the system output with a cost function. It determines the control input that minimizes the cost function using the candidate voltage vectors. However, the FS-MPC has a drawback of a low switching frequency because a voltage vector is applied during a control period. Therefore, in case the FS-MPC with the limited number of the candidate voltage vectors is used for the IMC drives, the output current ripple is increased, which means the output quality of the IMC is not favorable [9], [10].

In order to overcome the drawback of the FS-MPC, the candidate voltage vectors can be subdivided. However, the subdivided voltage vectors increase computational burden owing to a number of candidate voltage vectors. Therefore, in this paper, the MPC using the subdivided voltage vectors for reduction of the input–output current ripple in the IMC is proposed. Additionally, a scheme for decreasing the computational burden by estimating the position of the reference voltage vector is proposed.

II. IMC DESCRIPTION

Fig. 1 shows the IMC for power conversion from a three-phase voltage source to a grid. It has two power-conversion stages such as the CSR and the VSI stage. The CSR stage is connected to the three-phase voltage source such as power generation system. The VSI stage is connected to a grid with filter. The modulation technique of the CSR and the VSI is based on the most widely used space-vector modulation technique.

A. Modulation Technique of CSR Stage

Fig. 2 shows the space vector current diagram of the CSR. In each sector, the reference current vector is

TABLE I
SWITCHING STATES AND DC-LINK VOLTAGE

Current vector	Switching states		DC-link voltage
I_1	S_{au}	S_{cl}	$-v_{ca}$
I_2	S_{bu}	S_{cl}	v_{bc}
I_3	S_{bu}	S_{al}	$-v_{ab}$
I_4	S_{cu}	S_{al}	v_{ca}
I_5	S_{cu}	S_{bl}	$-v_{bc}$
I_6	S_{au}	S_{bl}	v_{ab}

generated by combination of two adjacent current vectors. For example, in sector VI, the reference current vector is synthesized using the vectors such as I_1 and I_2. Additionally, the switching states of the CSR and the instantaneous DC-link voltage of the IMC depending on the current vector are shown in Table I.

The reference current vector of the CSR stage is expressed as in (1).

$$I_i^* = \left[i_a^*, i_b^*, i_c^* \right] = i_i \left[\cos\theta_{ia}, \cos\theta_{ib}, \cos\theta_{ic} \right], \tag{1}$$

where i_i is magnitude of the reference current vector, θ_{ia}, θ_{ib}, and θ_{ic} are the phase angle of the reference current vector components (i_a^*, i_b^*, and i_c^*).

In order to perform a rectification at a unity power factor, the I_i^* of the CSR should be matched with the input voltage vector. In a balanced condition, the input voltage vector is represented as in (2).

$$V_i = \left[v_a, v_b, v_c \right] = v_i \left[\cos\theta_{va}, \cos\theta_{vb}, \cos\theta_{vc} \right],$$
$$\theta_{va} = \omega_i t, \ \theta_{vb} = \omega_i t - \frac{2\pi}{3}, \ \theta_{vc} = \omega_i t + \frac{2\pi}{3}, \tag{2}$$

where v_i is magnitude of the input voltage vector, ω_i is angular frequency of the three-phase voltage source, and θ_{va}, θ_{vb}, θ_{vc} are phase angle of the input voltage vector components (v_a, v_b, and v_c). In addition, since the sum of the components is zero under assumption that the three-phase voltage source is balanced, the phase angle of the input voltage vector components satisfies as follows:

$$\cos\theta_{va} + \cos\theta_{vb} + \cos\theta_{vc} = 0,$$
$$-\frac{\cos\theta_{vb}}{\cos\theta_{va}} - \frac{\cos\theta_{vc}}{\cos\theta_{va}} = 1. \tag{3}$$

In case the I_i^* is located in the sector VI as shown in Fig. 2, the upper switches in the A-leg are clamped. Therefore, the duty ratios are expressed as in (4).

$$d_x = -\frac{\cos\theta_{vb}}{\cos\theta_{va}}, \ d_y = -\frac{\cos\theta_{vc}}{\cos\theta_{va}}, \tag{4}$$

where d_x and d_y are duty ratios of the lower switches in the B-leg and C-leg, respectively. In addition, an average DC-link voltage is calculated as in (5) using the d_x, d_y, and line-to-line voltage of the three-phase voltage source.

$$v_{DC(avg)} = d_x v_{ab} - d_y v_{ca} = \frac{3v_i}{2\cos\theta_{va}}, \ -\frac{\pi}{6} \leq \theta_{va} \leq \frac{\pi}{6}. \tag{5}$$

4105

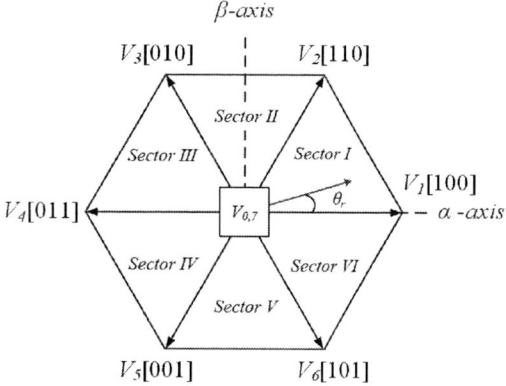

Fig. 3. Space vector voltage diagram of VSI.

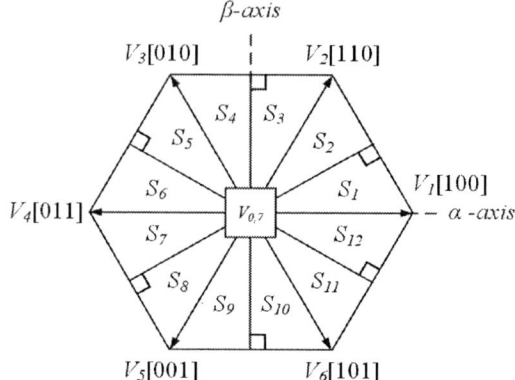

Fig. 4. Space vector subdivided voltage diagram of VSI.

The $v_{DC(avg)}$ means the average voltage appearing on the DC-link of the IMC during a switching period. It is used to calculate the reference voltage in the modulation process of the VSI. In other sectors, it can be defined similar to this manner.

B. Modulation Technique of VSI Stage

In the modulation process of the VSI, the modulated reference voltage is determined based on the $V_{DC(avg)}$ and the commutation of the CSR. Fig. 3 shows the space vector voltage diagram of the VSI. The modulation technique of the VSI stage is similar to that of the general inverter. However, in the VSI stage of the IMC, the switching states of the CSR should be considered in order to satisfy the zero DC-link current commutation. Therefore, two modulation reference voltages are used in the VSI stage of the IMC. The upper and lower modulation reference voltages in the R-phase are expressed as in (6).

$$
\begin{aligned}
v_{upper}^* &= -d_y \cdot (v_r^* + v_{sn}) + d_x \frac{V_{DC(avg)}}{2}, \\
v_{lower}^* &= d_x \cdot (v_r^* + v_{sn}) - d_y \frac{V_{DC(avg)}}{2},
\end{aligned}
\tag{6}
$$

where v_{sn} is offset voltage, which is calculated as in (7). It is added on the three-phase reference voltages (v_r^*, v_s^*, v_t^*) to use the SVM strategy.

$$
v_{sn} = -\frac{1}{2} \cdot \left\{ \max(v_r^*, v_s^*, v_t^*) + \min(v_r^*, v_s^*, v_t^*) \right\}.
\tag{7}
$$

III. MODEL PREDICTIVE CONTROL OF IMC

A. Conventional FS-MPC Scheme

The predicted current is required for the FS-MPC of the output current in the IMC. It is calculated by each candidate voltage vectors and the load model. The output voltage vector (V) in the load model of the VSI stage as shown in Fig. 1 in continuous-time domain is expressed as in (8).

$$
\begin{aligned}
V &= L_{of} \frac{dI}{dt} + R_{of} I + E, \\
V &= \left[v_\alpha, v_\beta \right], \\
I &= \left[i_\alpha, i_\beta \right], \\
E &= \left[e_\alpha, e_\beta \right],
\end{aligned}
\tag{8}
$$

where L_{of} and R_{of} are inductive and resistive filter, respectively. The variables such as the output current vector (I) and grid voltage vector (E) are expressed as in the α-β stationary reference frame.

To implement the FS-MPC on the DSP, the discretized model is required. The load model as in (8) can be discretized using the forward Euler method. It is expressed as in (9) regarding predicted current vector.

$$
I(k+1) = \left(1 - \frac{T_s R_{of}}{L_{of}} \right) I(k) + \frac{T_s}{L_{of}} (V(k) - E(k)),
\tag{9}
$$

where $I(k)$ and $E(k)$ are instantaneous output current and grid voltage vector obtained from the sensors. The $V(k)$ is the candidate voltage vector and T_s is sampling period.

In the FS-MPC, the cost function is defined as in (10) using the reference current and the predicted current based on (10) in the α-β stationary reference frame.

$$
g = \left| i_\alpha^*(k+1) - i_\alpha(k+1) \right| + \left| i_\beta^*(k+1) - i_\beta(k+1) \right|.
\tag{10}
$$

As a result, the reference voltage vector for the FS-MPC of the output current in the IMC is chosen such that the cost function is minimized. However, the number of the candidate voltage vectors in the FS-MPC scheme is limited to 8 basic voltage vectors including the zero vector. Therefore, the output current ripple of the IMC is severely unacceptable. Additionally, the switching frequency is low because only one vector is applied in a control period. Therefore, the conventional FS-MPC is improper to use in the IMC connected to the grid.

B. Proposed MPC Scheme

Fig. 4 shows the space vector subdivided voltage diagram of VSI, in the diagram, the sectors are divided

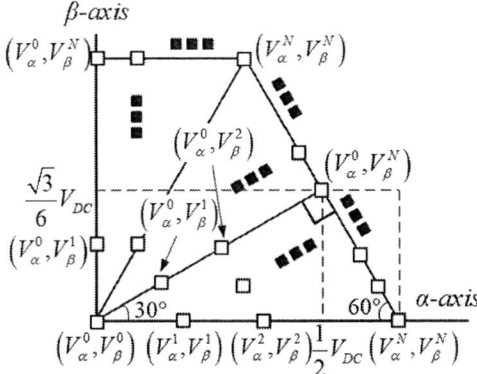

Fig. 5. Coordinate values of subdivided voltage vectors.

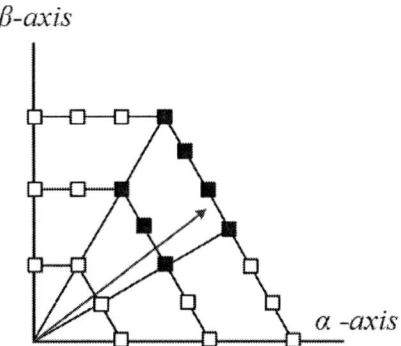

Fig. 6. Reduced candidate voltage vectors around the reference voltage vector (N=3).

into 12 sectors. In the proposed MPC, the subdivided voltage vectors on the 12 sectors are used to control the output current of the IMC. Fig. 5 shows coordinate values of the subdivided voltage vectors in sector 1–3, in the sectors, the coordinate values are represented as in (11)–(13), respectively.

$$\left[v_\alpha^{x_i}, v_\beta^{y_i} \right] = \left[\frac{x_i + 3y_i}{6N} v_{DC(avg)}, \frac{\sqrt{3}(y_i - x_i)}{6N} v_{DC(avg)} \right], \quad (11)$$

$$\left[v_\alpha^{x_i}, v_\beta^{y_i} \right] = \left[\frac{-x_i + 3y_i}{6N} v_{DC(avg)}, \frac{\sqrt{3}(x_i + y_i)}{6N} v_{DC(avg)} \right], (12)$$

$$\left[v_\alpha^{x_i}, v_\beta^{y_i} \right] = \left[\frac{x_i}{3N} v_{DC(avg)}, \frac{\sqrt{3}y_i}{3N} v_{DC(avg)} \right], \quad (13)$$

where x_i and y_i are index to describe the coordinate value and N is number of subdivisions. In other sectors, it can be defined similar to this manner. However, it is not possible to predict all instant $I(k+1)$ depending on the subdivided voltage vectors because of the computational burden caused by many candidate voltage vectors. Therefore, the position of the reference voltage vector should be determined using the estimation process. In the proposed MPC, the scheme that alleviates the calculation load based on the estimation of the reference voltage vector is considered. The estimation process is derived as follows. The output current vector is expressed as in (14).

$$I = \frac{V - E}{Z} = \frac{v\angle\theta - e\angle 0°}{R_{of} + j2\pi f L_{of}} = i\angle 0°, \quad (14)$$

where Z is impedance of the filter, f is fundamental frequency of the grid voltage, and θ is angle between the reference voltage vector and the grid voltage vector, which is expressed as in (15).

$$\theta = \tan^{-1}\left(\frac{2\pi f L_{of} i}{e + Ri} \right), \quad (15)$$

Therefore, the phase angle of the reference voltage vector can be determined as in (16).

$$\theta_{ref} = \theta_{grid} + \theta, \quad (16)$$

where θ_{grid} is angle of the grid voltage.

The magnitude of the reference voltage vector for the VSI of the IMC is fluctuated slightly because of the small resistance and inductance of the filter. Therefore, it is similar to that of the $V(k-1)$, which is reference voltage vector in last sampling period. It is expressed as in (17).

$$V(k-1) = E(k-1) + RI^*(k-1) + \frac{L}{T_s}\left(I^*(k) - I^*(k-1) \right). \quad (17)$$

Therefore, the magnitude of the reference voltage vector as in (18).

$$v_{mag} \approx |V(k-1)| = \sqrt{(v_\alpha)^2 + (v_\beta)^2}. \quad (18)$$

As a result, indices for determination of the reference voltage vector can be expressed as in (19) and (20).

$$S = \left\lfloor \frac{3N v_{mag}}{v_{DC(avg)}} \right\rfloor \quad \lfloor x \rfloor = \max\{n \in Z : n \le x\}, \quad (19)$$

$$L = \left\lceil \frac{\sqrt{3}N v_{mag}}{v_{DC(avg)}} \right\rceil \quad \lceil x \rceil = \min\{n \in Z : n \ge x\}, \quad (20)$$

where S and L are small and large indices for determination of the candidate voltage vectors that nearest to the reference voltage vector. In this process, the number of the candidate voltage vectors for the proposed MPC can be reduced as shown in Fig. 6. Therefore, in the proposed MPC, the output current ripple of the IMC can be reduced using the subdivided voltage vector without computational burden.

IV. SIMULATION RESULTS

The proposed MPC for the IMC connected to the grid was simulated using PSIM software to validate its effectiveness and the simulation parameters are described in Table II. Fig. 7 shows the simulation results of steady state performance of the proposed MPC comparing with the conventional FS-MPC. Fig. 7(a) indicates input phase currents, Fig. 7(b) indicates DC-link voltage, Fig, 7(c)

4107

TABLE II
SIMULATION PARAMETERS

Variables	Values
Three-phase voltage source	220 V$_{rms}$/60 Hz
Grid	110 V$_{rms}$/60 Hz
Filter resistance	0.1 Ω
Filter inductance	3 mH
Switching frequency	10 kHz
Number of subdivisions	3

indicates output phase currents and its transformed value in *d-q* reference frame, and Fig. 7(d) indicates *α-β* reference voltage vector. The three-phase output current of the IMC is controlled to 15 A. In addition, the control scheme is changed from the conventional FS-MPC to the proposed MPC at 0.4 s. Fig. 7 shows three important results of the proposed method. First, input current ripple of the IMC is reduced considerably compared with conventional FS-MPC. Second, output current ripple of the IMC is also reduced noticeably in the proposed MPC. Third, switching frequency of the IMC is constant at 10 kHz when the proposed MPC is applied. This is the consequence of the subdivided voltage vectors considered in the calculation of the cost function. The reference voltage vector made of the subdivided voltage vectors can be seen in Fig. 7(d) after 0.4 s.

Fig. 8 shows the simulation results of dynamic performance of the proposed MPC. Fig. 8(a) indicates input phase currents and Fig. 8(b) indicates output phase currents and its transformed value in *d-q* reference frame. The output current of the IMC is changed from 15 A to 7.5 A at 0.4 s. Through the proposed MPC, the output current of the IMC has fast dynamic response.

V. CONCLUSION

In this paper, the MPC using the subdivided voltage vectors for current ripple reduction in an IMC is proposed. In the proposed MPC, the number of candidate voltage vectors is increased by the subdivided voltage vectors the. Therefore, compared with the conventional FS-MPC, the proposed MPC reduces the input–output current ripple of the IMC. In addition, the proposed MPC reduces computation load, which is caused by the subdivided voltage vectors, by estimating the angle and magnitude of the reference voltage vector. The simulation results using the IMC connected to the grid have demonstrated the effectiveness of the proposed MPC.

REFERENCES

[1] X. Liu, P. Wang, P. C. Loh, and F. Blaabjerg, "A Compact Three-Phase Single-Input/Dual-Output Matrix Converter," *IEEE Trans. Ind. Electron.*, vol. 59, no. 1, pp. 6–16, Jan. 2012.

[2] Y. Bak, J.-S. Lee, and K.-B. Lee, "Balanced Current Control Strategy for Current Source Rectifier Stage of Indirect Matrix Converter under Unbalanced Grid Voltage Conditions," *Energies*, vol. 10, no. 1, Jan. 2017.

[3] Y. Bak and K.-B. Lee, "Constant Speed Control of a Permanent-Magnet Synchronous Motor Using a Reverse Matrix Converter Under Variable Generator Input Conditions," *IEEE J. Emerg. Sel. Topics Power Electron.*, vol. 6, no. 1, pp. 315–326, Mar. 2018.

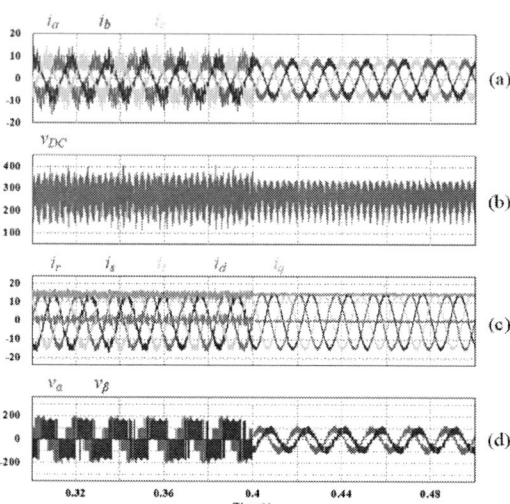

Fig. 7. Simulation results of steady state performance of the proposed MPC comparing with conventional FS-MPC.

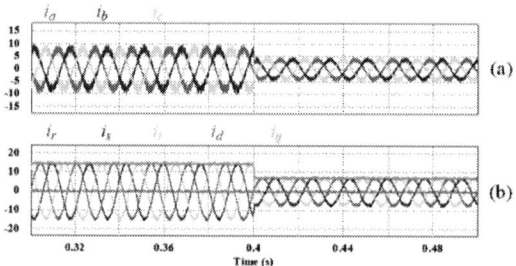

Fig. 8. Simulation results of dynamic performance of the proposed MPC.

[4] J. Dannehl, F. W. Fuchs, S. Hansen, and P. B. Thogersen, "Investigation of Active Damping Approaches for PI-Based Current Control of Grid-Connected Pulse Width Modulation Converters With LCL Filters," *IEEE Trans. Ind. Appl.*, vol. 46, no. 4, pp. 1509–1517, Jul./Aug. 2010.

[5] Y.S. Cho and K.-B. Lee, "Virtual-Flux-Based Predictive Direct Power Control of Three-Phase PWM Rectifiers with Fast Dynamic Response," *IEEE Trans. Power Electron.*, vol. 31, no. 4, pp. 3348–3359, Apr. 2016.

[6] S. Vazquez, J. Rodriguez, M. Rivera, L. G. Franquelo, and M. Norambuena, "Model Predictive Control for Power Converters and Drives: Advances and Trends," *IEEE Trans. Ind. Electron.*, vol. 64, no. 2, pp. 935–947, Feb. 2017.

[7] D.-K. Choi and K.-B. Lee, "Dynamic Performance Improvement of AC/DC PWM Converter Using Model Predictive Direct Power Control Set," *IEEE Trans. Ind. Electron.*, vol. 62, no. 2, pp. 757–767, Feb. 2015.

[8] J. H. Park, D. J. Kim, and K.-B. Lee, "Predictive Control Algorithm Including Conduction-Mode Detection for PFC Converter," *IEEE Trans. Ind. Electron.*, vol. 63, no. 9, pp. 5900–5911, Sep. 2016.

[9] M. Rivera, J. Rodriguez, B. Wu, J. R. Espinoza, and C. A. Rojas, "Current Control for an Indirect Matrix Converter With Filter Resonance Mitigation," *IEEE Trans. Ind. Electron.*, vol. 59, no. 1, pp. 71–79, Jan. 2012.

[10] M. Preindl and S. Bolognani, "Model Predictive Direct Speed Control with Finite Control Set of PMSM Drive Systems," *IEEE Trans. Power Electron.*, vol. 28, no. 2, pp. 1007–1015, Feb. 2013.

The 2018 International Power Electronics Conference

DC-link Ripple Current Reduction in Back-to-back converters with DPWM

Anatolii Tcai and Kyo-Beum Lee
Department of Electrical and Computer Engineering
Ajou Univeristy, Suwon, South Korea
anatolii.tcai@gmail.com, kyl@ajou.ac.kr

Abstract–The DC-link capacitor is the most vulnerable element of a voltage conversion system. The ripple current, flowing through it, causes thermal processes, which shorten the lifespan of the component. When a discontinuous pulse width modulation (DPWM) is applied to a back-to-back converter, the DC-link ripple current increases at the intervals with the different clamping directions of both converters i.e. the first converter's reference voltage is attached to the positive bus of the DC-link, while the reference voltage of the second converter is connected to the opposite bus or vice versa. Alternatively, the ripple current is lower when the clamping states of both inverters are the same. This paper proposes a reduction method of the DC link capacitor ripple current of the back-to-back converters using the DPWM, considering all the clamping states. The effectiveness of the proposed method is verified by simulation.

I. INTRODUCTION

Back-to-back converter has become the most commonly used topology in such applications as electric vehicles, wind power generation, and uninterruptible power supplies [1]–[3]. A serial connection of two voltage source converters (VSC) yields a state of the art technology, which can be used to transfer the power bidirectionally between two AC voltage sources with different fundamental frequencies and magnitudes. In this case, the first VSC is controlled as a rectifier, while the other VSC works as an inverter.

A discontinuous pulse-width modulation (DPWM) is used in VSCs to enhance the efficiency and reliability of the system by reducing the switching losses of the switching devices [4]. It is done by locking the state of each transistor either ON or OFF for a certain interval, at

which the magnitude of the phase current is at its maximum. Usually this interval equals to 1/3 of the fundamental period. This approach minimizes the number of switching operations of each transistor, resulting in minimized switching loss and increased efficiency. Moreover, at high modulation indexes (MI), the current quality is improved, when using the DPWM [5]–[7].

A typical configuration of the two-level back-to-back converter is given in Fig. 1. The DC-link capacitor is attached between the inverter VSC and the rectifier VSC and is required to maintain the DC voltage around a certain level. According to [8], the DC-link capacitors are responsible approximately for 30% of failures in power converters. As is stated in [9], the root mean square (RMS) current flow over the capacitor, causes the power loss over the effective series resistance (ESR), accelerating the temperature rise, which results in shortening the life duration of the element. Thus, in the back-to-back system, the DC-link capacitors are subjects to constant stress, due to the handling of the pulse current, flowing from the rectifier and the inverter. To prolong the lifespan of the components, more capacitors are connected in parallel and share the ripple current. This approach is effective in terms of the failure prevention, although it increases the total cost of the system and boosts the size. In fact, in systems with higher current rates the DC-link capacitor can be the biggest component of the system. Therefore, minimizing the DC-link ripple current benefits in decreased system size, minimized system cost and extended lifespan of the DC-link capacitors.

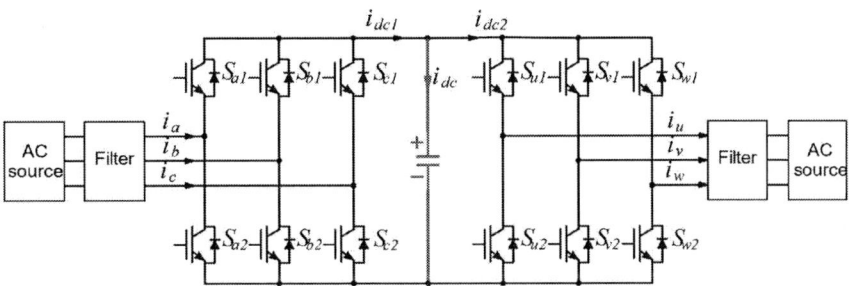

Fig. 1. Two-level back-to-back converter.

4109

Several efforts have been made to diminish the ripple current of the DC-link. The direct capacitor current control [10] and feed forward control [11] reduce only the low-frequency ripple current, influenced by imbalanced input and output power. The current with the PWM carrier frequency is significantly higher and cannot be curtailed with the mentioned methods. Another approach is synchronizing the PWM carriers between two converters of the back-to-back system [12]–[14]. This method is very efficient and can be easily implemented. However, the performance of this method varies for different PWM techniques. Moreover, the proposed method does not consider the clamping states of both VSCs using the DPWM. This paper investigates the method to minimize the DC-link ripple current of the back-to-back converter with the DPWM. The efficiency of the proposed method is confirmed via simulation results.

II. DPWM IN BACK-TO-BACK CONVERTERS

As mentioned, the DPWM is used to enhance the efficiency of the system's efficiency and to reduce the stress in power transistors, caused by constant switching. The idea of the DPWM is to stop operating power transistors in a leg of the VSC for a certain time interval. For general DPWM schemes, this interval equals 60° per transistor or 120° per leg [15], For carrier based DPWM, the reference voltage of each phase is obtained by adding the offset signal.

$$
\begin{cases}
V_{A.ref} = M \cdot \left(\sin\left(\omega t\right) + V_{offset} \right) \\
V_{B.ref} = M \cdot \left(\sin\left(\omega t - 120°\right) + V_{offset} \right) \\
V_{C.ref} = M \cdot \left(\sin\left(\omega t + 120°\right) + V_{offset} \right)
\end{cases}
\tag{1}
$$

where $V_{A.ref}$, $V_{B.ref}$, and $V_{C.ref}$ – reference voltages of the phases A, B, and C relatively. V_{offset} is the offset signal, M is the modulation index (MI), and ω is the fundamental frequency.

In three-phase systems, the offset signal can be selected as any value, which satisfies the degree of freedom, given in (2).

$$
\begin{cases}
-V_{dc}/2 < V_{A.ref} < V_{dc}/2, \\
-V_{dc}/2 < V_{B.ref} < V_{dc}/2, \\
-V_{dc}/2 < V_{C.ref} < V_{dc}/2.
\end{cases}
\tag{2}
$$

Then,

$$
\left(-V_{dc}/2 - V_{min} \right) < V_{offset} < \left(V_{dc}/2 - V_{max} \right)
\tag{3}
$$

where V_{dc} is the DC-Link voltage, V_{min} and V_{max} are respectively minimum and maximum values among the three phases reference voltages. Finally, for the conventional 60° DPWM, the offset signal is obtained as

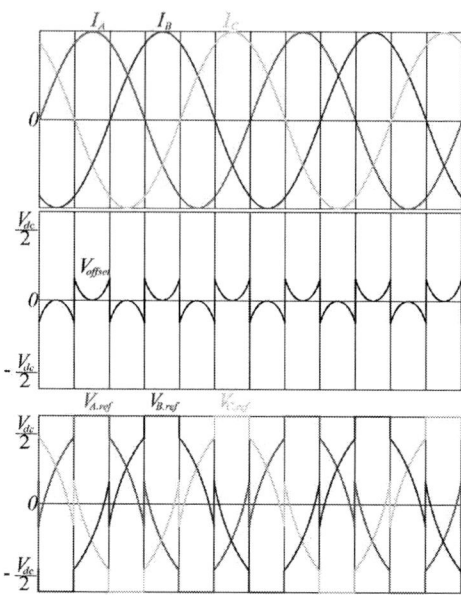

Fig. 2. Input currents and reference voltages of the general DPWM.

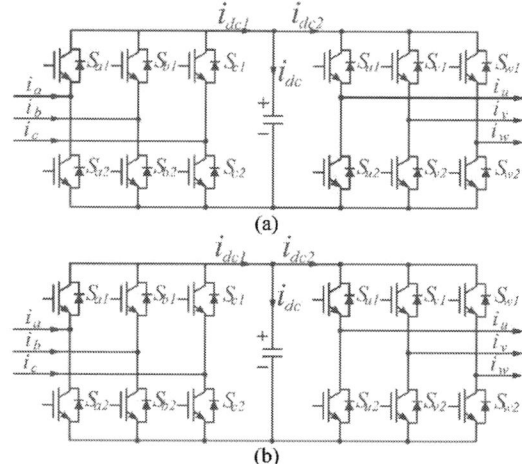

Fig. 3. Clamping states of the DPWM-based back-to-back converter (a) bipolar clamping state and (b) unipolar clamping state.

$$
\begin{cases}
V_{offset} = V_{dc}/2 - V_{max}, & \left(V_{max} + V_{min} > 0 \right) \\
V_{offset} = -V_{dc}/2 - V_{min}, & \left(V_{max} + V_{min} < 0 \right)
\end{cases}
\tag{4}
$$

The offset voltage and the reference voltages for phases A, B, and C, are given in Fig. 2. Obviously, each phase reference is attached either to the positive or to the negative DC-link bus for a period of 60 electrical degrees during one semi-period. It can be seen, that a phase is clamped, when the corresponding current is close to its magnitude I_m. At this moment, two complementary transistors of the phase leg are locked in ON and OFF positions respectively. By diminishing the number of switching operations over the periods with maximum current, the power loss can be reduced by 33 % [16].

4110

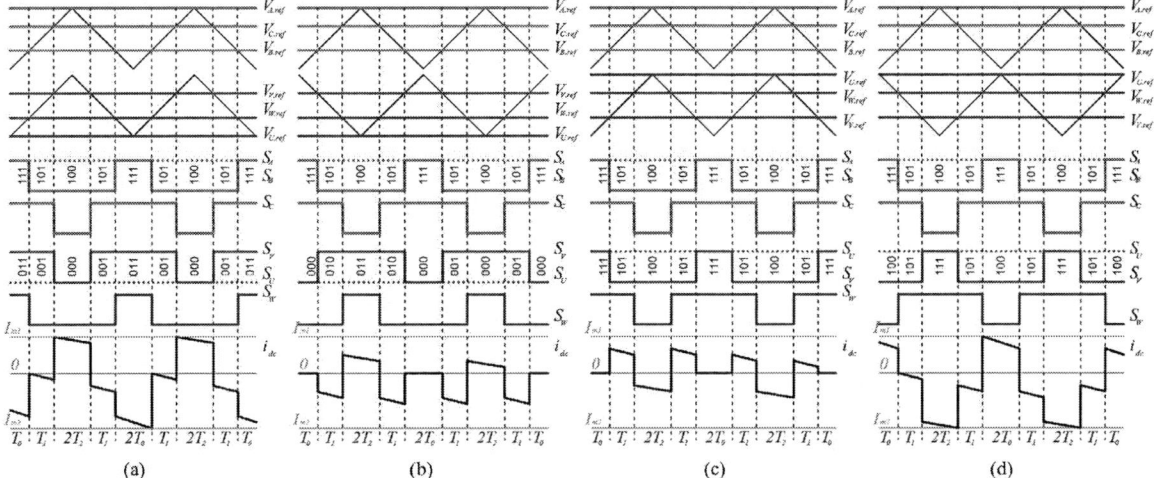

Fig. 4. Detailed PWM waveforms throughout one carrier period: (a) BCS with synchronized PWM carriers, (b) BCS with 180° phase-shift between the PWM carriers, (c) UCS with synchronized PWM carriers, and (d) UCS with 180° phase-shift between the PWM carriers.

III. PROPOSED METHOD

In back-to-back converters, fundamental frequencies and initial angles of both converters can differ. Depending on the variation of these parameters, the DPWM–based back-to-back converter has two clamping states. A bipolar clamping state (BCS), with the voltage references of both converters being attached to different DC-link buses, e.g. phase A of the VSC-1 is connected to the positive side, whereas the phase U of the VSC-2 is attached to the negative bus, as illustrated in Fig. 3(a). Fig. 3(b) depicts a unipolar clamping state (UCS) with both VSCs clamped to a single DC-link bus.

Fig. 4. provides detailed switching waveforms of the BCS and the UCS over the PWM carrier period in the case of the synchronized PWM carriers, and the case of the phase-shifted PWM carriers. It should be noted, that the switching frequency is considerably higher than the fundamental frequency of both VSCs. Therefore, the reference signals over a carrier period are assumed as constant values. As can be clearly observed, the instantaneous value of the capacitor current, depends on the instantaneous values of the DC-link currents of both converters. In accordance with the Kirchhoff's law, the capacitor current is determined as follows:

$$i_{dc} = i_{dc1} - i_{dc2} \qquad (5)$$

where i_{dc1} and i_{dc2} are the DC-link currents of the rectifier VSC-1 and the inverter VSC-2 respectively, which comprise the switching functions and the AC currents of both converters, as shown.

$$\begin{aligned} i_{dc1} &= S_A \cdot i_A + S_B \cdot i_B + S_C \cdot i_C, \\ i_{dc2} &= S_U \cdot i_U + S_V \cdot i_V + S_W \cdot i_W. \end{aligned} \qquad (6)$$

where S_A, S_B, S_C, S_U, S_V, S_W are switching functions of the corresponding phases.

Obviously, when the switching functions of one converter are the same, the DC-link current adds up to zero:

$$\begin{aligned} i_{dc1} &= i_A + i_B + i_C = 0, \\ i_{dc2} &= i_U + i_V + i_W = 0. \end{aligned} \qquad (7)$$

Then, considering the switching functions and the currents of both converters during the interval T_0-T_2 the instantaneous values of the capacitor ripple current for the BCS and the UCS are determined as:

$$i_{dc.BCS} = \begin{cases} i_u, \\ i_b - i_w, \\ -i_a, \end{cases} \quad i_{dc.BCS180} = \begin{cases} 0, & (T_0) \\ i_b - i_v, & (T_1), \\ -i_a + i_u, & (T_2) \end{cases} \qquad (8)$$

$$i_{dc.UCS} = \begin{cases} 0, \\ i_b + i_v, \\ i_a - i_u, \end{cases} \quad i_{dc.UCS180} = \begin{cases} -i_u, & (T_0) \\ i_b + i_v, & (T_1). \\ -i_a, & (T_2) \end{cases} \qquad (9)$$

As can be clearly seen, in the BCS with synchronized PWM, the maximum DC-link capacitor ripple current occurs, when the switching function of the clamped phase differs from the others. At this moment, the ripple current is the same with the current of the clamped phase, which also can be assumed as the magnitude value, I_{m1} for the VSC-1 and I_{m2} for the VSC-2. Obviously, the currents i_a and i_u of the clamped phases of the rectifier VSC-1 and the inverter VSC-2, contribute the most to the ripple current of the DC-link capacitor. Evidently, the ripple is much lower under the UCS with no phase-shift or with the phase-shifted BCS. Therefore, it is possible to diminish the capacitor ripple current by changing the phase-shift between two PWM carriers of both converters. When the system is under the UCS, there should be no phase-shift between the converters. However, as the

The 2018 International Power Electronics Conference

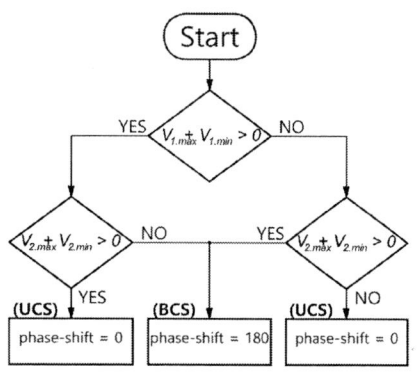

Fig. 5. Flowchart for the DC-link capacitor ripple current reduction.

TABLE I
SIMULATION PARAMETERS

Parameter	Value
Line-to-line voltage	90 V$_{RMS}$
Grid frequency	60 Hz
Line resistance	100 mΩ
Line inductance	930 mH
DC-Link voltage	140 V
DC-Link capacitance	2.2 mF

system is operated under the BCS, the 180° phase shift is needed between the rectifier and the inverter.

Fig. 5 shows the flowchart of the proposed ripple current reduction method. From (4) it is obvious, that the VSC is clamped to the positive DC-link bus when the summation of V_{max} and V_{min} is more than zero. On the contrary, when $V_{max} + V_{min} < 0$, the converter is attached to the negative DC-link bus. This can be employed to ensure the current clamping state of the back-to-back converter, either bipolar or unipolar. When the clamping directions of both VSCs are different, the system is under the BCS and the 180° phase shift should be used between the PWM carriers. Contrariwise, when both the rectifier and the inverter are connected to the single DC-Link bus, either positive or negative, the clamping state is the UCS and no phase-shift is needed.

IV. SIMULATION RESULTS

To verify the feasibility and the validity of the ripple current reduction method, various simulations have been carried out with the use of the PSIM software. The simulation circuit is the same as that, given in Fig. 1. The VSC-1 is connected to the utility grid with the parameters, given in Table I, whereas a 6.8 Ω resistance is used as a load for the AC-side of the VSC-2. The fundamental frequency of the VSC-2 is set to 30 Hz, and the switching frequency of the system is 10 kHz.

Fig. 6 depicts the ripple current of the DC-Link capacitor in the case, when the MIs of the rectifier VSC-1 and the inverter VSC-2 are the same. Obviously, for the conventional method, the ripple is increased at the intervals, when the BCS occurs. As can be observed from Fig. 6, the DC-link current has maximum pikes

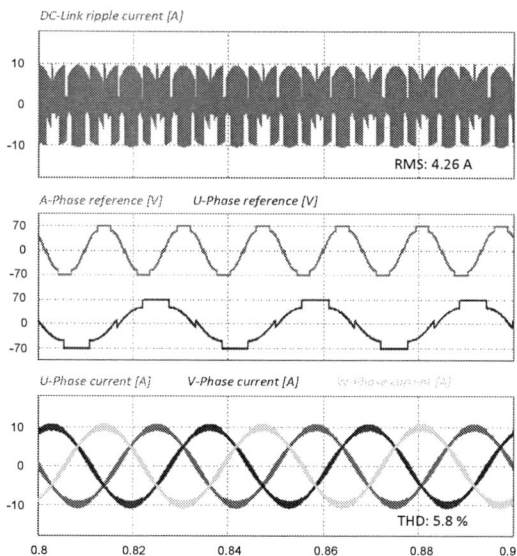

Fig. 6. Simulation results of the conventional method.

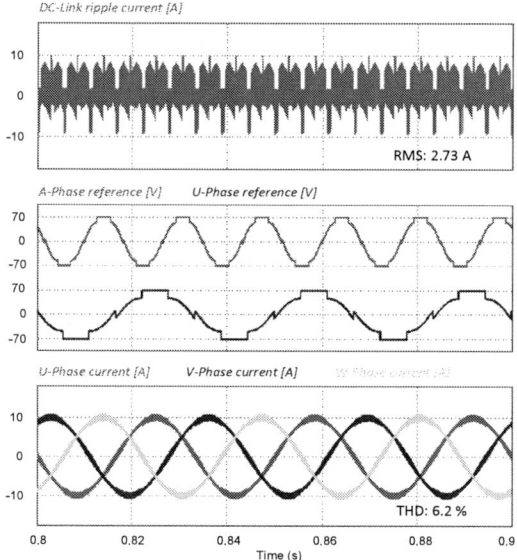

Fig. 7. Simulation results of the proposed method.

approaching 10 A, when the voltage references of two converters are attached to the different DC-link buses. The ripple current RMS is then 4.26 A.

Fig. 7 illustrates the DC-link capacitor ripple current under the proposed method. It is evident from the figure, that the lower current peaks have been curtailed, resulting in decreased RMS value to 2.73 A. Therefore, by using the proposed method, it is possible to curtail the DC-Link capacitor ripple current approximately up to 35%. However, due to the rapid change of the switching phase, the total harmonic distortion (THD) is insignificantly deteriorated.

Finally, the frequency spectrum comparison of the DC-Link ripple current for the conventional and the proposed methods is given in Fig. 8. For the conventional

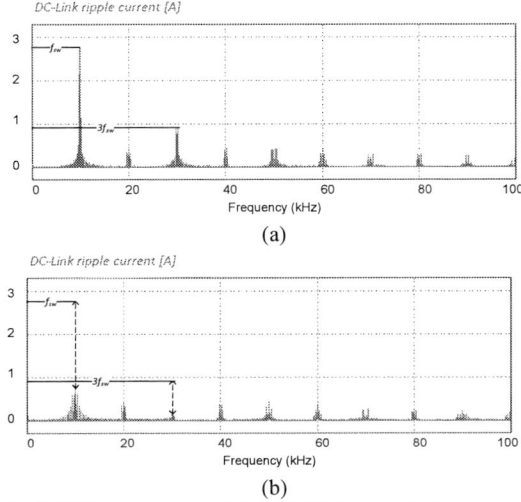

Fig. 8. FFT comparison of the DC-Link current: (a) Conventional method. (b) Proposed method.

method, shown in Fig. 8(a), the dominating switching frequency component (f_{sw}) is considerably reduced, when using proposed method, as can be observed from Fig. 8(a). Furthermore, the proposed method likewise minimizes the third order component $(3f_{sw})$.

V. CONCLUSION

This paper proposes a reduction method of the DC-Link capacitor ripple current in a DPWM-based back-to-back converter. When the DPWM is used in the back-to-back system, the capacitor ripple current is significantly increased at the intervals, when the reference voltages of the two converters are attached to different DC-Link buses. These current pikes result in the ripple current RMS and cause the stress on the capacitor, which can greatly decrease the component lifespan and lead to the system failure. The proposed method is performed by changing the phase shift angle between the PWM carriers of the converter. When the system is under the bipolar clamping state (BCS), with the input stage converter and the output stage converter being clamped to different DC-Link buses, the 180° phase-shift angle should be used. On the contrary, when the system is under the unipolar clamping state (UCS), the phase-shift should be set to 0°. The proposed method does not require any enhancement in the hardware and can be easily implemented in any back-to-back system with a simple code change. The validity of the proposed method was confirmed by the simulation. The ripple current of the DC-Link capacitor was reduced by 35%.

VI. REFERENCES

[1] F. Blaabjerg, M. Liserre, and K. Ma, "Power Electronics Converters for Wind Turbine Systems," *IEEE Trans. Ind. Appl.*, vol. 48, no. 2, pp. 708–719, Mar 2012.

[2] H. Ye and A. Emadi, "An Interleaving Scheme to Reduce DC-Link Current Harmonics of Dual Traction Inverters in Hybrid Electric Vehicles," *in Proc. Appl. Power Electronics Conf. and Expo. (APEC)*, pp. 3205–3211, Nov. 2014.

[3] A. Tcai, H.U. Shin, and K.B. Lee, "DC-link Capacitor Current Ripple Reduction in DPWM based Back-to-Back Converters," *IEEE Trans. Ind. Electron.*, vol. 65, no. 3, pp. 1897–1907, Mar 2018.

[4] J.S. Lee and K.B. Lee, "Performance Analysis of Carrier-Based Discontinuous PWM Method for Vienna Rectifiers with neutral-Point Voltage Balance," *IEEE Trans Power Electron.*, vol. 31, no. 6, pp. 4075–4084, Jun. 2016.

[5] N. V. Nguyen, B. X. Nguyen, and H. H. Lee, "An optimized discontinuous PWM method to minimize switching loss for multilevel inverters," *IEEE Trans. Ind. Electron.*, vol. 58, no. 9, pp. 3958–3966, Sep. 2011.

[6] Z. Zhang, O. C. Thomsen, and M. A. E. Andersen, "Discontinuous PWM modulation strategy with circuit-level decoupling concept of three level neutral-point clamped (NPC) inverter," *IEEE Trans. Ind. Electron., vol.60*, no. 5, pp. 1897–1906, May 2013.

[7] S. L. An, X. D. Sun, Q. Zhang, Y. R. Zhong, and B. Y. Ren, "Study on the novel generalized discontinuous SVPWM strategies for three-phase voltage source inverters," *IEEE Trans. Ind. Inf.*, vol. 9, no. 2, pp. 781–789, May 2013.

[8] H. Wang, M. Liserre, and F. Blaabjerg, "Toward reliable power electronics: Challenges, design tools, and opportunities," *IEEE Industrial Electronics Magazine*, vol. 7, no. 2, pp. 17–26, June 2013.

[9] G. Sam, Jr. Parler, "Deriving life multipliers for electrolytic Capacitors," *IEEE Power electronics Society Newsletter*, vol. 16, no. 1, Feb. 2004, pp. 11-12

[10] G. Bon-Gwan and N. Kwanghee, "A DC-Link capacitor minimization method through direct capacitor current control," IEEE Trans. Ind. Appl., vol. 42, no. 2, pp. 573–581, Mar. 2006.

[11] J. Jinhwan, L. Sunkyoung, and N. Kwanghee, "A feedback linearizing control scheme for a PWM converter-inverter having a very small DC-Link capacitor," *IEEE Trans. Ind. Appl.*, vol. 35, no. 5, pp. 1124–1131, Sep. 1999.

[12] L.G. Gonzalez, G. Garcera, E. Figueres, and R. Gonzalez, "Effects of the PWM Carrier Signals Synchronization on the DC-Link Current in Back-to-Back converters," *in Applied Energy*, vol. 87, no. 8, pp. 2491-2499, Mar. 2010.

[13] Z. Qin, H. Wang, F. Blaabjerg, and P.C. Loh, "Investigation into the Control Methods to Reduce the DC-Link Capacitor Ripple Current in a Back-to-Back Converter," *in Proc. IEEE Energy Conversion Congr. And Expo. (ECCE)*, 2014, pp.203–210.

[14] D. Zhang, F. Wang, R. Burgos, R. Lai, and D. Boroyevich, "DC-Link Ripple Current Reduction for Paralleled Three-Phase Voltage Source Converters With Interleaving," *IEEE Trans. Power Electron.*, vol. 26, no. 6, pp. 1741–1753, Jun. 2011.

[15] Y. Bak and K.B. Lee, "Discontinuous PWM for Low Switching Losses in Indirect Matrix Converter Drives," *i n Proc. Appl. Power Electronics Conf. and Expo. (APEC)*, pp. 2764–2769, Mar. 2016.

[16] D.W. Chung and S.K. Sul, "Minimum-loss strategy for three-phase PWM rectifier," *IEEE Trans. Ind. Electron.*, vol. 46, no. 3, pp.517–526, Jun. 1999.

The 2018 International Power Electronics Conference

An Analysis of Class DE Voltage-Source Parallel Resonant Inverter

Takeshi Kondo, Tsuyoshi Inaba, Yoshikazu Sakai* and Hirotaka Koizumi
Department of Electrical Engineering, Tokyo University of Science, Tokyo, Japan
*E-mail: 4314042@ed.tus.ac.jp

Abstract— Class DE voltage-source parallel resonant inverter is analyzed for any switch-on-duty ratio. Owing to the Class DE voltage-source inverter topology, low switching losses and low switch voltage stresses are realized. Waveform equations of switches and a capacitor are derived at any switch-on-duty ratio. Similarly, design equations for each the circuit component are obtained. Based on these equations, behavior of the inverter is characterized by the initial phase angle of the output voltage, the voltage transfer function, the normalized switch current stress, and the power-output capability, which are given as functions of the switch-on-duty ratio. In addition, the power conversion efficiency and the design example are shown. The circuit operation was confirmed by the circuit experiments. The experimental results for the switch-on-duty ratio $D = 0.2$ and 0.25 were in good agreement with the theoretical ones.

I. INTRODUCTION

Recently, the power electronics technology has been developed rapidly owing to utilization of semiconductor switches. The switching power converters are essential equipment for various applications.

Power converters need to be both downsized and high efficient. Passive elements occupy larger space than semiconductor elements in circuits. At high switching frequency, the inductance and the capacitance can be reduced at the same reactance. Thus, the circuit can be downsized. However, the power conversion efficiency decreases because the switching losses increase. To reduce the switching losses, soft switching methods are effective [1], [2]. These methods are important to reduce not only switching losses but also electro-magnetic-interference (EMI). As inverters applied the soft switching methods, Class D inverters [3], [4], Class E inverters [5]-[8], and Class DE inverters [9]-[11] are proposed.

The Class DE inverters satisfy zero-voltage-switching (ZVS) and zero-voltage-derivative switching (ZVDS) or zero-current-switching (ZCS) and zero-current-derivative switching (ZCDS), hence they can achieve high efficiency in dc/ac power conversion at high switching frequency. Moreover, the inverters work with low voltage or low current stress of the switch. Due to these characteristics, the Class DE inverters are expected to be used as a dc/ac inverter in resonant dc/dc power converters, induction heatings, and wireless power transfer systems. The Class DE voltage source parallel

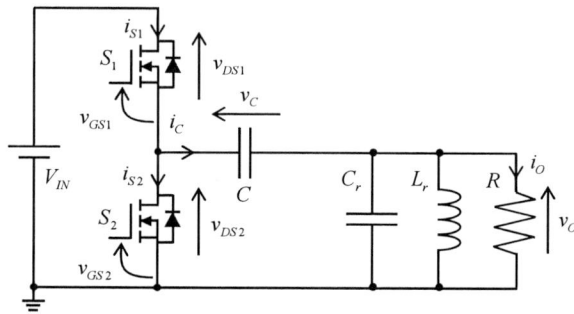

Fig. 1. Circuit topology of the Class DE voltage-source parallel resonant inverter [11]

resonant inverter [11] can achieve ZVS and ZVDS when the switches turn on. Therefore, the switching losses are reduced. Moreover, the voltage stress of the switches equals to the input voltage. In [11], the circuit equations are shown only when the switch-on-duty ratio $D = 0.25$.

This paper shows circuit equations of the Class DE voltage-source parallel resonant inverter at any switch-on-duty ratio, characteristics of the normalized input impedance, the initial phase angle of the output voltage, the voltage transfer function, the normalized switch current stress, and the power-output capability as functions of the switch-on-duty ratio. Moreover, we derive the power conversion efficiency considering the conduction losses at each the circuit element. To verify the circuit operation, some design examples and experimental results are shown.

II. CIRCUIT TOPOLOGY

Fig. 1 shows a circuit topology of the Class DE voltage-source parallel resonant inverter. It consists of two switches S_1 and S_2, a series capacitor C, a resonant inductor L_r, a resonant capacitor C_r, a load resistor R, and an input voltage source V_{IN}. Fig. 2 shows the ideal waveforms of the Class DE voltage-source parallel resonant inverter, where v_{GS1} and v_{GS2} are gate-source voltages of the switches S_1 and S_2, v_O is the output voltage, V_M is the amplitude of the output voltage v_O, i_C is the current through the series capacitor C, v_C is the voltage across the series capacitor C, i_{S1} and i_{S2} are the currents through the switches S_1 and S_2, I_{SM} is the peak current of the switch currents i_{S1} and i_{S2}, v_{DS1} and v_{DS2} are the drain-source voltages of the switches S_1 and S_2, V_{SM} is

4114

The 2018 International Power Electronics Conference

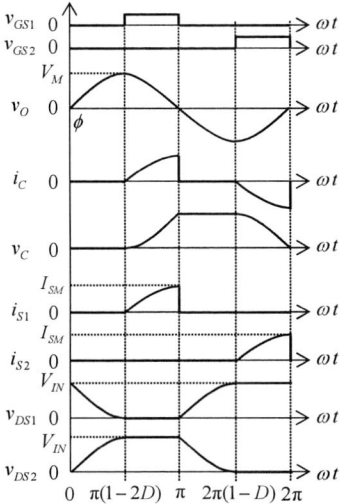

Fig. 2. Ideal waveforms of the Class DE voltage-source parallel resonant inverter.

the peak voltage of the drain-source voltages of the switches S_1 and S_2, and the angular frequency $\omega = 2\pi f$. ϕ is the initial phase angle of the output voltage v_O. The switch-on-duty ratio of the switches S_1 and S_2 is denoted by D. The switch S_1 and S_2 are operated at the switching frequency f.

III.   CIRCUIT ANALYSIS

The analysis is based on the following assumptions.
1) The switch-on-duty ratio ranges $0 < D < 0.5$.
2) The switches S_1 and S_2 turn on and off instantaneously.
3) All circuit elements are ideal, that is the on resistances and the parasitic capacitances of the switches and the equivalent series resistances (ESRs) of the passive elements are zero.
4) The loaded quality factor Q of the parallel resonant circuit is high enough to shape the output voltage v_O sinusoidal waveform and expressed as

$$Q = \frac{R}{\omega L_r}. \tag{1}$$

5) The resonant inductor L_r is divided into L_A and L_B.

$$\frac{1}{L_r} = \frac{1}{L_A} + \frac{1}{L_B} \tag{2}$$

The resonant frequency of L_A-C_r circuit equals to the switching frequency f. The inductor L_B yields the phase shift ϕ of the output voltage v_O.

$$f = \frac{1}{2\pi\sqrt{L_A C_r}} \tag{3}$$

The analysis is performed in the interval $0 \le \omega t \le 2\pi$, where ω is an angular frequency at the switching frequency f. By the assumption 4), the output voltage v_O is denoted by

$$v_O(\omega t) = V_M \sin(\omega t + \phi), \tag{4}$$

where V_M is the amplitude of v_O. From Kirchhoff's voltage law and Kirchhoff's current law, we have

$$i_C(\omega t) + i_{S2}(\omega t) = i_{S1}(\omega t), \tag{5}$$

$$v_{DS1}(\omega t) + v_{DS2}(\omega t) = V_{IN}, \tag{6}$$

$$v_{DS2}(\omega t) = v_C(\omega t) + v_O(\omega t). \tag{7}$$

To achieve the Class E switching conditions, ZVS and ZVDS must be satisfied.

$$v_{DS1}\left[\pi(1-2D)\right] = v_{DS1}(\pi) = 0, \tag{8}$$

$$v_{DS2}(0) = v_{DS2}\left[2\pi(1-D)\right] = 0. \tag{9}$$

$$\left.\frac{dv_{DS1}(\omega t)}{d(\omega t)}\right|_{\omega t = \pi(1-2D)} = \left.\frac{dv_{DS2}(\omega t)}{d(\omega t)}\right|_{\omega t = 2\pi(1-D)} = 0. \tag{10}$$

The circuit operation is discussed in four states depending on the state of the switches.

For State 1 $[0 \le \omega t < \pi(1-2D)]$, both the switches S_1 and S_2 are off. Hence, the switch currents i_{S1} and i_{S2} are given by

$$i_{S1}(\omega t) = i_{S2}(\omega t) = 0. \tag{11}$$

From (5) and (11), the capacitor current i_C is given by

$$i_C(\omega t) = i_{S1}(\omega t) - i_{S2}(\omega t) = 0. \tag{12}$$

From (4) and (7), the voltage v_C across the capacitor C is found as

$$\begin{aligned} v_C(\omega t) &= v_{DS2}(\omega t) - v_O(\omega t) \\ &= v_{DS2}(\omega t) - V_M \sin(\omega t + \phi). \end{aligned} \tag{13}$$

Thus, from (9), (12), and (13), we have

$$\begin{aligned} v_C(\omega t) &= \frac{1}{\omega C}\int_0^{\omega t} i_C(\omega t)d(\omega t) + v_C(0) \\ &= -V_M \sin\phi. \end{aligned} \tag{14}$$

Substituting (4) and (14) into (7), the drain-source voltage v_{DS2} is

$$v_{DS2}(\omega t) = V_M\left[\sin(\omega t + \phi) - \sin\phi\right]. \tag{15}$$

From (6) and (15), the drain-source voltage v_{DS1} is

$$v_{DS1}(\omega t) = V_{IN} - V_M\left[\sin(\omega t + \phi) - \sin\phi\right]. \tag{16}$$

Differentiating (16) with respect to ωt,

$$\frac{dv_{DS1}(\omega t)}{d(\omega t)} = -V_M \cos(\omega t + \phi). \tag{17}$$

From (10),

$$\left.\frac{dv_{DS2}(\omega t)}{d(\omega t)}\right|_{\omega t = \pi(1-2D)} = V_M \cos(-2\pi D + \phi) = 0. \tag{18}$$

From the waveform of the drain-source voltage v_{DS2} in Fig. 2, the physical solution of (18) is

$$-2\pi D + \phi = -\frac{\pi}{2} \Leftrightarrow \phi = \pi\left(-\frac{1}{2} + 2D\right). \tag{19}$$

Substituting (8) and (19) into (16) yields

4115

$$V_{IN} = V_M \left(1 + \cos 2\pi D\right). \tag{20}$$

As a result, the drain-source voltages v_{DS1} and v_{DS2} are respectively

$$v_{DS1}\left(\omega t\right) = V_{IN} + V_M \left[\cos\left(\omega t + 2\pi D\right) - \cos 2\pi D\right], \tag{21}$$

$$v_{DS2}\left(\omega t\right) = -V_M \left[\cos\left(\omega t + 2\pi D\right) - \cos 2\pi D\right]. \tag{22}$$

For state 2 $[\pi(1-2D) \le \omega t < \pi]$, the switch S_1 is on and the switch S_2 is off. Thus, the drain-source voltages v_{DS1}, v_{DS2} and the switch current i_{S2} are given by

$$v_{DS1}\left(\omega t\right) = 0, \tag{23}$$

$$v_{DS2}\left(\omega t\right) = V_{IN}, \tag{24}$$

$$i_{S2}\left(\omega t\right) = 0. \tag{25}$$

From (5) and (25), the switch current i_{S1} is

$$i_{S1}\left(\omega t\right) = i_C\left(\omega t\right). \tag{26}$$

Substituting (4) and (24) into (7), and using (19) and (20), the voltage v_C across the capacitor C is

$$v_C\left(\omega t\right) = V_{IN}\left[1 + \frac{\cos\left(\omega t + 2\pi D\right)}{1 + \cos 2\pi D}\right]. \tag{27}$$

From (27), the current i_C through the capacitor C is

$$i_C\left(\omega t\right) = \omega C \frac{dv_C\left(\omega t\right)}{d\left(\omega t\right)} = -\frac{\omega C V_{IN} \sin\left(\omega t + 2\pi D\right)}{1 + \cos 2\pi D}. \tag{28}$$

For state 3 $[\pi \le \omega t < 2\pi(1-D)]$, both the switches S_1 and S_2 are off. The switch currents i_{S1} and i_{S2} are

$$i_{S1}\left(\omega t\right) = i_{S2}\left(\omega t\right) = 0. \tag{29}$$

From (5) and (29), the capacitor current i_C is

$$i_C\left(\omega t\right) = i_{S1}\left(\omega t\right) - i_{S2}\left(\omega t\right) = 0. \tag{30}$$

From (27) and (30), the voltage v_C across the capacitor C is

$$\begin{aligned}
v_C\left(\omega t\right) &= \frac{1}{\omega C}\int_\pi^{\omega t} i_C\left(\omega t\right) d\left(\omega t\right) + v_C\left(\omega t\right) \\
&= V_{IN}\left[1 + \frac{\cos\left(\pi + 2\pi D\right)}{1 + \cos 2\pi D}\right] = \frac{V_{IN}}{1 + \cos 2\pi D}.
\end{aligned} \tag{31}$$

Substituting (4) and (31) into (7), and using (19) and (20), the drain-source voltage v_{DS2} is

$$v_{DS2}\left(\omega t\right) = V_{IN}\left[\frac{1 - \cos\left(\omega\pi + 2\pi D\right)}{1 + \cos 2\pi D}\right]. \tag{32}$$

Substituting (32) into (6), the drain-source voltage v_{DS1} is

$$v_{DS1}\left(\omega t\right) = V_{IN}\left[\frac{\cos\left(\omega t + 2\pi D\right) + \cos 2\pi D}{1 + \cos 2\pi D}\right] \tag{33}$$

For state 4 $[2\pi(1-D) \le \omega t < 2\pi]$, the switch S_1 is off and the switch S_2 is on. Thus, the drain-source voltages v_{DS1}, v_{DS2} and the switch current i_{S1} are given by

$$v_{DS1}\left(\omega t\right) = V_{IN}, \tag{34}$$

$$v_{DS2}\left(\omega t\right) = 0, \tag{35}$$

$$i_{S1}\left(\omega t\right) = 0. \tag{36}$$

From (5) and (36), the switch current i_{S2} is

$$i_{S2}\left(\omega t\right) = -i_C\left(\omega t\right). \tag{37}$$

Substituting (4) and (35) into (7), and using (19) and (20), the voltage v_C across the capacitor C is

$$v_C\left(\omega t\right) = V_{IN}\left[\frac{\cos\left(\omega t + 2\pi D\right)}{1 + \cos 2\pi D}\right]. \tag{38}$$

From (38), the current i_C through the capacitor C is

$$i_C\left(\omega t\right) = \omega C \frac{dv_C\left(\omega t\right)}{d\left(\omega t\right)} = -\frac{\omega C V_{IN} \sin\left(\omega t + 2\pi D\right)}{1 + \cos 2\pi D}. \tag{39}$$

A. Waveform Equations

By the above equations, each the waveform equation is expressed as follows.

$$i_{S1}\left(\omega t\right) = \begin{cases} 0 & \left[0 \le \omega t < \pi\left(1 - 2D\right)\right] \\ -\dfrac{\omega C V_{IN} \sin\left(\omega t + 2\pi D\right)}{1 + \cos 2\pi D} & \left[\pi\left(1 - 2D\right) \le \omega t < \pi\right] \\ 0 & \left[\pi \le \omega t < 2\pi\left(1 - D\right)\right] \\ 0 & \left[2\pi\left(1 - D\right) \le \omega t < 2\pi\right] \end{cases} \tag{40}$$

$$i_{S2}\left(\omega t\right) = \begin{cases} 0 & \left[0 \le \omega t < \pi\left(1 - 2D\right)\right] \\ 0 & \left[\pi\left(1 - 2D\right) \le \omega t < \pi\right] \\ 0 & \left[\pi \le \omega t < 2\pi\left(1 - D\right)\right] \\ \dfrac{\omega C V_{IN} \sin\left(\omega t + 2\pi D\right)}{1 + \cos 2\pi D} & \left[2\pi\left(1 - D\right) \le \omega t < 2\pi\right] \end{cases} \tag{41}$$

$$v_{DS1}\left(\omega t\right) = \begin{cases} V_{IN}\left[1 + \dfrac{\cos\left(\omega t + 2\pi D\right) - \cos 2\pi D}{1 + \cos 2\pi D}\right] & \left[0 \le \omega t < \pi\left(1 - 2D\right)\right] \\ 0 & \left[\pi\left(1 - 2D\right) \le \omega t < \pi\right] \\ V_{IN}\left[\dfrac{\cos\left(\omega t + 2\pi D\right) + \cos 2\pi D}{1 + \cos 2\pi D}\right] & \left[\pi \le \omega t < 2\pi\left(1 - D\right)\right] \\ V_{IN} & \left[2\pi\left(1 - D\right) \le \omega t < 2\pi\right] \end{cases} \tag{42}$$

$$v_{DS2}(\omega t)$$

$$= \begin{cases} -\dfrac{V_{IN}}{1+\cos 2\pi D}\Big[\cos(\omega t + 2\pi D) - \cos 2\pi D\Big] \\ \qquad\qquad\qquad\qquad\qquad \big[0 \le \omega t < \pi(1-2D)\big] \\ V_{IN} \qquad\qquad\qquad\qquad \big[\pi(1-2D) \le \omega t < \pi\big] \\ V_{IN}\left[\dfrac{1-\cos(\omega t + 2\pi D)}{1+\cos 2\pi D}\right] \; \big[\pi \le \omega t < 2\pi(1-D)\big] \\ 0 \qquad\qquad\qquad\qquad\quad \big[2\pi(1-D) \le \omega t < 2\pi\big] \end{cases} \tag{43}$$

$$i_C(\omega t)$$

$$= \begin{cases} 0 \qquad\qquad\qquad\qquad\quad \big[0 \le \omega t < \pi(1-2D)\big] \\ -\dfrac{\omega C V_{IN}\sin(\omega t + 2\pi D)}{1+\cos 2\pi D} \; \big[\pi(1-2D) \le \omega t < \pi\big] \\ 0 \qquad\qquad\qquad\qquad\quad \big[\pi \le \omega t < 2\pi(1-D)\big] \\ -\dfrac{\omega C V_{IN}\sin(\omega t + 2\pi D)}{1+\cos 2\pi D} \; \big[2\pi(1-D) \le \omega t < 2\pi\big] \end{cases} \tag{44}$$

$$v_C(\omega t)$$

$$= \begin{cases} \dfrac{V_{IN}}{1+\cos 2\pi D}\cos 2\pi D \quad \big[0 \le \omega t < \pi(1-2D)\big] \\ V_{IN}\left[1+\dfrac{\cos(\omega t + 2\pi D)}{1+\cos 2\pi D}\right] \big[\pi(1-2D) \le \omega t < \pi\big] \\ \dfrac{V_{IN}}{1+\cos 2\pi D} \qquad\qquad \big[\pi \le \omega t < 2\pi(1-D)\big] \\ V_{IN}\left[\dfrac{\cos(\omega t + 2\pi D)}{1+\cos 2\pi D}\right] \; \big[2\pi(1-D) \le \omega t < 2\pi\big] \end{cases} \tag{45}$$

B. Design Equations

The averaged input current I_{IN} equals to the average of the current i_{S1}. Thus, using (40),

$$I_{IN} = \frac{1}{2\pi}\int_0^{2\pi} i_{S1}(\omega t)\,d(\omega t) = \frac{\omega C V_{IN}}{2\pi}\left(\frac{1-\cos 2\pi D}{1+\cos 2\pi D}\right). \tag{46}$$

From (4), the output current i_O through the load resistance R is

$$i_o(\omega t) = \frac{V_M}{R}\sin(\omega t + \phi) = I_O\sin(\omega t + \phi), \tag{47}$$

where the coefficient I_O is

$$I_O = \frac{V_M}{R} = \frac{V_{IN}}{R(1+\cos 2\pi D)}. \tag{48}$$

Likewise, using (4), the current i_{LB} through the resonant inductance L_B is

$$i_{LB}(\omega t) = \frac{V_M}{\omega L_B}\sin\left(\omega t + \phi - \frac{\pi}{2}\right)$$

$$= -\frac{V_M}{\omega L_B}\cos(\omega t + \phi) = I_{LB}\cos(\omega t + \phi), \tag{49}$$

where the coefficient I_{LB} is

$$I_{LB} = -\frac{V_M}{\omega L_B} = -\frac{V_{IN}}{\omega L_B(1+\cos 2\pi D)}. \tag{50}$$

From (44), the amplitude I_O and I_{LB} of i_O and i_{LB} are calculated as Fourier coefficients

$$I_O = \frac{1}{\pi}\int_0^{2\pi} i_C(\omega t)\sin(\omega t + \phi)\,d(\omega t)$$

$$= \frac{\omega C V_{IN}}{\pi}(1-\cos 2\pi D), \tag{51}$$

$$I_{LB} = \frac{1}{\pi}\int_0^{2\pi} i_C(\omega t)\cos(\omega t + \phi)\,d(\omega t)$$

$$= \frac{\omega C V_{IN}}{\pi(1+\cos 2\pi D)}\left[\frac{1}{2}\sin 4\pi D - 2\pi D\right]. \tag{52}$$

Hence, from (48) and (51),

$$\frac{\omega CR}{\pi} = \frac{1}{\sin^2 2\pi D}. \tag{53}$$

From (50) and (52),

$$2\pi D - \frac{1}{2}\sin 4\pi D = \frac{\pi}{\omega^2 L_B C}. \tag{54}$$

From (46), the input power P_{IN} is

$$P_{IN} = V_{IN}I_{IN} = \frac{\omega C V_{IN}^{\,2}}{2\pi}\left(\frac{1-\cos 2\pi D}{1+\cos 2\pi D}\right). \tag{55}$$

From (51) and (53), the output power P_{OUT} is expressed as

$$P_{OUT} = \frac{R I_O^{\,2}}{2} = \frac{\omega C V_{IN}^{\,2}}{2\pi}\left(\frac{1-\cos 2\pi D}{1+\cos 2\pi D}\right). \tag{56}$$

From (48) and (56),

$$R = \frac{1}{2(1+\cos 2\pi D)^2}\frac{V_{IN}^{\,2}}{P_{OUT}}. \tag{57}$$

From (53) and (57),

$$C = \frac{\pi}{\omega R\sin^2 2\pi D} = \frac{2\pi P_{OUT}}{\omega V_{IN}^{\,2}}\left(\frac{1+\cos 2\pi D}{1-\cos 2\pi D}\right). \tag{58}$$

From (1),

$$L_r = \frac{R}{\omega Q}. \tag{59}$$

From (53) and (54),

$$L_B = \frac{2R\sin^2 2\pi D}{\omega(4\pi D - \sin 4\pi D)}. \tag{60}$$

Substituting (59) and (60) into (2),

$$L_A = \frac{L_r L_B}{L_B - L_r}$$

$$= \frac{2R}{\omega} \frac{\sin^2 2\pi D}{2Q\sin^2 2\pi D - (4\pi D - \sin 4\pi D)}. \tag{61}$$

Substituting (61) into (3),

$$C_r = \frac{1}{\omega^2 L_A} = \frac{2Q\sin^2 2\pi D - 4\pi D + \sin 4\pi D}{2\omega R \sin^2 2\pi D}. \tag{62}$$

As the on-duty ratio D increases, the resonant capacitance C_r decreases and becomes negative. Therefore, the loaded quality factor Q must be enough large to keep the resonant capacitance $C_r > 0$. From (62), the resonant capacitance C_r needs to be

$$C_r = \frac{2Q\sin^2 2\pi D - 4\pi D + \sin 4\pi D}{2\omega R \sin^2 2\pi D} > 0. \tag{63}$$

The loaded quality factor Q which satisfies (63) is

$$Q > \frac{4\pi D - \sin 4\pi D}{2\sin^2 2\pi D}. \tag{64}$$

From (64), the minimum loaded quality factor $Q_{\min}$ is

$$Q_{\min} = \frac{4\pi D - \sin 4\pi D}{2\sin^2 2\pi D}. \tag{65}$$

Fig. 3(a) presents the minimum loaded quality factor $Q_{\min}$ as a function of the switch-on-duty ratio D. As shown in Fig. 3(a), $Q_{\min}$ increases from 0 to ∞ as the switch-on-duty ratio D increases from 0 to 0.5.

Based on the above circuit equations, we show the normalized input impedance $\omega C R_{DC}$, the initial phase angle ϕ of the output voltage v_O, the voltage transfer function M_{VR}, the normalized switch current stress I_{SM}/I_{IN}, and the power-output capability c_p.

C. Normalized Input Impedance

From (46), the input resistance of the Class DE voltage-source parallel resonant inverter is

$$R_{DC} = \frac{V_{IN}}{I_{IN}} = \frac{2\pi}{\omega C}\left(\frac{1 + \cos 2\pi D}{1 - \cos 2\pi D}\right). \tag{66}$$

From (66), the normalized input impedance $\omega C R_{DC}$ is

$$\omega C R_{DC} = 2\pi\left(\frac{1 + \cos 2\pi D}{1 - \cos 2\pi D}\right). \tag{67}$$

Fig. 3(b) shows the normalized input impedance $\omega C R_{DC}$ as a function of the switch-on-duty ratio D, which increases from 0 to ∞ as the switch-on-duty ratio D decreases from 0.5 to 0.

D. Initial Phase Angle of Output Voltage

From (19), the initial phase angle ϕ of the Class DE voltage-source parallel resonant inverter is calculated as

$$\phi = \pi\left(-\frac{1}{2} + 2D\right). \tag{68}$$

Fig. 3(c) shows the initial phase angle ϕ as a function of the switch-on-duty ratio D. From Fig. 3(c), the initial phase angle ϕ increases from $-90°$ to $90°$ as the switch-on-duty ratio D increases from 0 to 0.5.

E. Voltage Transfer Function

From (20), the voltage transfer function M_{VR} of the Class DE voltage-source parallel resonant inverter is

$$M_{VR} = \frac{V_M / \sqrt{2}}{V_{IN}} = \frac{1}{\sqrt{2}\left(1 + \cos 2\pi D\right)}. \tag{69}$$

Fig. 3(d) shows the voltage transfer function M_{VR} as a function of the switch-on-duty ratio D. From Fig. 3(d), M_{VR} increases from $1/2\sqrt{2}$ to ∞ as the switch-on-duty ratio D increases from 0 to 0.5.

F. Switch Current Stress

When $0 < D \leq 0.25$, the switch current i_{S1} reaches the peak current I_{SM1} at the turn-off instant ($\omega t = \pi$). Similarly, when the switch S_2 turns off at $\omega t = 2\pi$, the switch current i_{S2} reaches the peak current I_{SM2}. Therefore, from (40) and (41), the peak value I_{SM} of the switch currents i_{S1} and i_{S2} is expressed as

$$I_{SM} = I_{SM1} = I_{SM2} = i_{S1}(\pi) = i_{S2}(2\pi)$$
$$= \frac{\omega C V_{IN}}{1 + \cos 2\pi D}\sin 2\pi D. \tag{70}$$

In the case of $0.25 < D < 0.5$, the peak current I_{SM1} of the switch current i_{S1} occurs when the derivative of i_{S1} is zero. Derivation of (40) is

$$\frac{di_{S1}}{d(\omega t)} = -\frac{\omega C V_{IN}}{1 + \cos 2\pi D}\cos(\omega t + 2\pi D). \tag{71}$$

$di_{S1}/d(\omega t) = 0$ is satisfied at

$$\omega t = \pi\left(\frac{3}{2} - 2D\right). \tag{72}$$

In the same way, derivation of (41) is

$$\frac{di_{S2}}{d(\omega t)} = \frac{\omega C V_{IN}}{1 + \cos 2\pi D}\cos(\omega t + 2\pi D). \tag{73}$$

It becomes zero at

$$\omega t = \pi\left(\frac{5}{2} - 2D\right). \tag{74}$$

From (40), (41), (72), and (74), the peak value I_{SM} of the switch currents i_{S1} and i_{S2} is expressed as

$$I_{SM} = I_{SM1} = I_{SM2}$$
$$= i_{S1}\left(\frac{3\pi}{2} - 2\pi D\right) = i_{S2}\left(\frac{5\pi}{2} - 2\pi D\right) \tag{75}$$
$$= \frac{\omega C V_{IN}}{1 + \cos 2\pi D}.$$

From (46), (70), and (75), the normalized peak switch current I_{SM}/I_{IN} of the Class DE voltage-source parallel resonant inverter is

$$\frac{I_{SM}}{I_{IN}} = \begin{cases} \dfrac{2\pi}{1 - \cos 2\pi D}\sin 2\pi D, & [0 < D \leq 0.25] \\[2ex] \dfrac{2\pi}{1 - \cos 2\pi D}. & [0.25 < D \leq 0.5] \end{cases} \tag{76}$$

4118

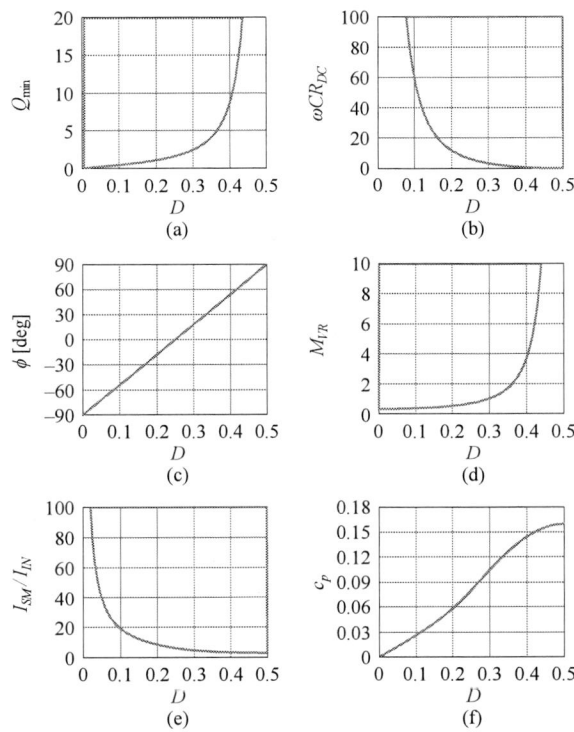

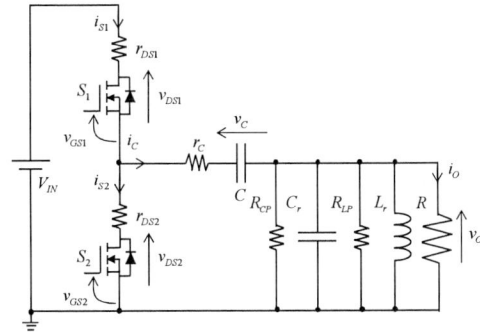

Fig. 4. Equivalent circuit of the Class DE voltage-source parallel resonant inverter with ESRs and EPRs.

Fig. 3. Characteristics of parameters as functions of the switch-on-duty ratio D, (a) minimum loaded quality factor Q_{min}, (b) normalized input impedance $\omega C R_{DC}$, (c) initial phase angle of the output voltage ϕ, (d) voltage transfer function M_{VR}, (e) normalized peak switch current I_{SM}/I_{IN}, (f) power-output capability c_p.

Fig. 3(e) shows the normalized peak switch current I_{SM}/I_{IN} as a function of the switch-on-duty ratio D. From Fig. 3(e), I_{SM}/I_{IN} increases from π to ∞ as the switch-on-duty ratio D decreases from 0.5 to 0.

G. Power-Output Capability

The power-output capability c_p is used to compare the circuits in the aspect of the effective usage of the semiconductor switches and expressed as

$$c_p = \frac{P_O}{n I_{SM} V_{SM}}, \tag{77}$$

where n is the number of switching devices. In the proposed circuit, n equals to 2 and the peak switch voltage equals to the input voltage, that is

$$V_{SM} = V_{IN}. \tag{78}$$

From (56), (76)-(78), the power-output capability c_p is

$$c_p = \begin{cases} \dfrac{1-\cos 2\pi D}{4\pi \sin 2\pi D}, & [0 < D \le 0.25] \\[2ex] \dfrac{1-\cos 2\pi D}{4\pi}. & [0.25 < D < 0.5] \end{cases} \tag{79}$$

Fig. 3(f) shows power-output capability c_p as a function of the switch-on-duty ratio D. From Fig. 3(f), c_p increases from 0 to $1/2\pi$ as the switch-on-duty ratio D increases from 0 to 0.5.

IV. POWER CONVERSION EFFICIENCY

Another equivalent circuit of Fig. 1 is shown in Fig. 4, where r_{DS1} and r_{DS2} are on-resistances of the switches S_1 and S_2, r_C is an ESR of the capacitor C, R_{LP} and R_{CP} are equivalent parallel resistances (EPRs) of the resonant inductor L_r and the resonant capacitor C_r.

A. Conduction Power Loss in the Switches

The power loss in the switches is caused by on-resistances r_{DS1} and r_{DS2}. Using (40), (41), and (46), an rms values I_{S1RMS} and I_{S2RMS} of the switch currents i_{S1} and i_{S2} are

$$\begin{aligned} I_{S1RMS} &= \sqrt{\frac{1}{2\pi}\int_0^{2\pi} i_{S1}^2 d(\omega t)} \\ &= \frac{\sqrt{\pi}}{1-\cos 2\pi D}\sqrt{2\pi D - \frac{1}{2}\sin 4\pi D}\cdot I_{IN}, \end{aligned} \tag{80}$$

$$\begin{aligned} I_{S2RMS} &= \sqrt{\frac{1}{2\pi}\int_0^{2\pi} i_{S2}^2 d(\omega t)} \\ &= \frac{\sqrt{\pi}}{1-\cos 2\pi D}\sqrt{2\pi D - \frac{1}{2}\sin 4\pi D}\cdot I_{IN}. \end{aligned} \tag{81}$$

From (80) and (81), the conduction power loss in the switches P_S is expressed as

$$\begin{aligned} P_S &= r_{DS1}I_{S1RMS}^2 + r_{DS2}I_{S2RMS}^2 \\ &= (r_{DS1}+r_{DS2})\cdot I_{IN}^2 \cdot \left(\frac{1}{1-\cos 2\pi D}\right)^2 \\ &\quad \cdot \pi\cdot\left(2\pi D - \frac{1}{2}\sin 4\pi D\right). \end{aligned} \tag{82}$$

B. Conduction Power Loss in Capacitor

The conduction power loss P_C in the capacitor C is caused by ESR r_C. From (44) and (46), an rms value I_{CRMS} of the capacitor current i_C is

$$\begin{aligned} I_{CRMS} &= \sqrt{\frac{1}{2\pi}\int_0^{2\pi} i_C^2 d(\omega t)} \\ &= \frac{\sqrt{\pi}}{1-\cos 2\pi D}\sqrt{4\pi D - \sin 4\pi D}\cdot I_{IN}. \end{aligned} \tag{83}$$

Therefore, from (75), the conduction loss in the capacitor P_C is expressed as

4119

$$P_C = r_C I_{CRMS}{}^2 = \frac{\pi(4\pi D - \sin 4\pi D)}{(1 - \cos 2\pi D)^2} \cdot r_C \cdot I_{IN}{}^2. \qquad (84)$$

C. Power Loss in Parallel Resonant Circuit

EPRs of the resonant components can be regarded as a combined resistance R_d.

$$R_d = \frac{R_{LP} R_{CP}}{R_{LP} + R_{CP}} \qquad (85)$$

The relationships between the ESRs and EPRs are

$$R_{LP} = r_L\left(1 + Q_{LO}{}^2\right) \approx r_L Q_{LO}{}^2, \qquad (86)$$

$$R_{CP} = r_C\left(1 + Q_{CO}{}^2\right) \approx r_C Q_{CO}{}^2, \qquad (87)$$

where $Q_{LO} = \omega L / r_L = R_{LP}/\omega L$ and $Q_{CO} = 1/\omega C r_C = \omega C R_{LP}$ are the unloaded-quality factor of the resonant inductor L_r and the resonant capacitor C_r respectively. Using (46), (53), and (56), the power loss in the parallel resonant circuit P_{Rd} is expressed as

$$
\begin{aligned}
P_{Rd} &= \frac{V_M{}^2}{2R_d} = \frac{R}{R_d} P_{OUT} \\
&= \frac{R^2}{R_d} \cdot I_{IN}{}^2 \cdot \frac{(1 + \cos 2\pi D)(1 - \cos 4\pi D)}{1 - \cos 2\pi D}.
\end{aligned} \qquad (88)
$$

From (82), (84), and (88), the total conduction power loss P_{LOSS} is

$$
\begin{aligned}
P_{LOSS} &= P_S + P_C + P_{Rd} \\
&= \pi I_{IN}{}^2 \left[\frac{4\pi D - \sin 4\pi D}{2(1 - \cos 2\pi D)^2}(r_{DS1} + r_{DS2} + 2R_C) \right. \\
&\quad \left. + \frac{R^2}{\pi R_d} \cdot \frac{(1 + \cos 2\pi D)(1 - \cos 4\pi D)}{1 - \cos 2\pi D} \right].
\end{aligned} \qquad (89)
$$

The power conversion efficiency η of the Class DE voltage-source parallel resonant inverter is

$$\eta = \frac{P_{OUT}}{P_{OUT} + P_{LOSS}} = \frac{1}{1 + \dfrac{P_{LOSS}}{P_{OUT}}}. \qquad (90)$$

V. DESIGN AND EXPERIMENT

To verify the circuit operation in circuit experiment, three Class DE voltage-source parallel resonant inverters were designed based on the following specifications:

1) The switching frequency $f = 200$ [kHz].
2) The input voltage $V_{IN} = 20.0$ [V].
3) The output power $P_{OUT} = 2.00$ [W].
4) The loaded quality factor $Q = 8.0$.

A design example is shown at $D = 0.25$ as follows. From (57) and the above specifications, the load resistance R is

$$R = \frac{1}{2(1 + \cos 2\pi D)^2} \frac{V_{IN}{}^2}{P_{OUT}} = 100\ [\Omega]. \qquad (91)$$

From (58), the series capacitance C is

$$C = \frac{\pi}{\omega R \sin^2 2\pi D} = 25.0\ [\text{nF}]. \qquad (92)$$

Substituting (91) into (59), the resonant inductance L_r is

$$L_r = \frac{R}{\omega Q} = 9.95\ [\mu\text{H}]. \qquad (93)$$

Substituting (91) into (62), the resonant capacitance C_r is

$$
\begin{aligned}
C_r &= \frac{2Q\sin^2 2\pi D - 4\pi D + \sin 4\pi D}{2\omega R \sin^2 2\pi D} \\
&= 51.2\ [\text{nF}].
\end{aligned} \qquad (94)
$$

The other two circuits were designed in the same way for $D = 0.2$ and 0.3. The designed values are shown in Table. I. The measured values of the circuit components are also shown in Table. I. The on-resistance and the drain-source capacitance were obtained from the datasheet of the MOSFET IRFML8244TRPbF [12]. For driving the switches S_1 and S_2, a gate driver IR2011 was used [13]. From (82)-(84), (86), (87) and Table. I, the obtained theoretical output power $P_{OUT} = 1.88$ [W] and the power conversion efficiency $\eta = 94.2$ [%]. Therefore, the input power needs to be $P_{IN} = 2.12$ [W].

Fig. 5 shows the observed waveforms of the three types of experimental circuits designed at $D = 0.2$, 0.25, and 0.3. From Fig. 5, the experimental waveforms were consistent with the theoretical ones for $D = 0.2$ and 0.25. The switches S_1 and S_2 turned on in low voltage and with low dv/dt except for $D = 0.3$. The non-ZVS at $D = 0.3$ could be caused by distortion of the output voltage v_O, the parasitic capacitance of the switch, and ESR in the circuit.

Table. II shows the designed and measured values of the input and the output. As shown in Table. II, the measured values of the power conversion efficiencies were in good agreement with the theoretical ones at $D = 0.25$. At $D = 0.2$, the conduction power loss in S_1, S_2, C, L_r, and C_r increased. Meanwhile, at $D = 0.3$, the switching loss increased because S_1 and S_2 turned on without achieving ZVS and ZVDS. Therefore, the power conversion efficiency decreased as the switch-on-duty ratio went far from 0.25. Measured THD of the output voltage v_O is also shown in Table. II. THD became larger as the on-duty ratio increased.

VI. CONCLUSIONS

In this paper, the Class DE voltage-source parallel resonant inverter has been analyzed. The equations of the voltage and current waveforms for any switch-on-duty ratio have been derived. The characteristics of the normalized input impedance, the initial phase angle of the output voltage, the voltage transfer function, the normalized switch current stress, the power-output capability, and the power conversion efficiency have been obtained. The inverter has been confirmed with the circuit experiments. The measured power conversion efficiency for the switch-on-duty ratio $D = 0.25$ was 92.0 % under 2 W output power and 200 kHz operation. The experimental results showed good agreement with the theoretical performance.

TABLE I
CIRCUIT COMPONENTS AND MEASURED VALUES

Parameter		Designed	Measured (ESR [mΩ])	Designed	Measured (ESR [mΩ])	Designed	Measured (ESR [mΩ])
On-duty ratio		0.2		0.25		0.3	
Input voltage V_{IN} [V]		20.0	20.0	20.0	20.0	20.0	20.0
Switching frequency f [kHz]		200	200	200	200	200	200
Load resistance R [Ω]		58.4	59.1	100	100	209	210
Series capacitance C [nF]		47.4	47.5 (29.0)	25.0	25.1 (20.1)	13.2	12.4 (44.3)
Resonant inductance L_r [μH]		5.81	5.75 (100)	9.95	10.1 (82.1)	20.8	20.8 (225)
Resonant capacitance C_r [nF]		94.6	94.4 (19.2)	51.2	51.7 (16.9)	21.2	21.4 (12.1)
Switches S_1, S_2	On-resistances r_{DS1}, r_{DS2} [mΩ]	-	20.0	-	20.0	-	20.0
	Drain-source capacitances C_{DS1}, C_{DS2} [pF]	-	61.0	-	61.0	-	61.0

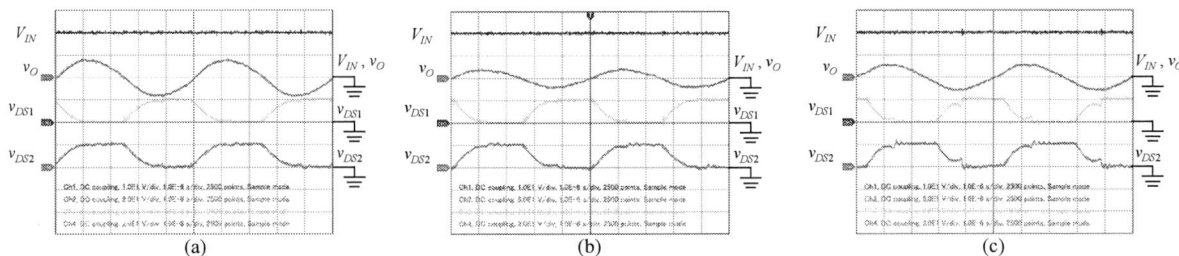

(a)　　　　　　　　　　(b)　　　　　　　　　　(c)

Fig. 5. Observed waveforms of the Class DE voltage-source parallel resonant inverter, (a) D = 0.2 (V_{IN}: 10 V/div, v_{DS1}, v_{DS2}, v_O: 20 V/div, horizontal: 0.25 μs/div), (b) D = 0.25 (V_{IN}: 10 V/div, v_O: 50 V/div, v_{DS1}, v_{DS2}: 20 V/div, horizontal: 0.25 μs/div), (c) D = 0.3 (V_{IN}: 10 V/div, v_O: 50 V/div, v_{DS1}, v_{DS2}: 20 V/div, horizontal: 0.25 μs/div).

TABLE II
INPUT AND OUTPUT VALUES

Parameter	Designed	Measured	Designed	Measured	Designed	Measured
D	0.200	0.208	0.250	0.244	0.300	0.306
V_{IN} [V]	20.0	20.0	20.0	20.0	20.0	20.0
I_{IN} [mA]	100	105	100	105	100	109
V_{ORMS} [V]	10.8	10.4	14.1	13.9	20.5	20.0
R [Ω]	58.4	59.1	100	100	209	210
P_{IN} [W]	2.00	2.10	2.00	2.10	2.00	2.18
P_{OUT} [W]	2.00	1.82	2.00	1.93	2.00	1.90
η [%]	88.6	86.7	94.2	92.0	93.2	87.1
THD of v_O	-	5.17	-	8.55	-	14.8

REFERENCES

[1] A. Knott, T. M. Andersen, P. Kamby, J. A. Perdersen, M. P. Madsen, M. Kovacevic, and M. A. Andersen, "Evolution of very high frequency power supplies," *IEEE J. Emerg. Sel. Topics Power Electron.*, vol. 2, no. 3, pp. 386-394, Sep. 2014.

[2] D. J. Perreault, J. Hu, J. M. Rivas, Y. Han, O. Leitermann, R. C. N. Pilawa-Podgurski, A. Sagneri, and C. R. Sullivan, "Opportunities and challenges in very high frequency power conversion," in *Proc. IEEE Appl. Power Electron. Conf. Expo.*, Washington, Feb. 2009, pp. 1-14.

[3] M. K. Kazimierczuk and D. Czarkowski, *Resonant Power Converters*, 2nd ed., New Jersey: John Wiley & Sons, 2011.

[4] M. K. Kazimierczuk and J. S. Modzelewski, "Drive-transformerless Class-D voltage-switching tuned power amplifier," in *Proc. IEEE*, vol. 68, no. 6, pp. 740-741, Jun. 1980.

[5] N. O. Sokal and A. D. Sokal, "Class E-A new class of high-efficiency tuned single-ended switching power amplifiers," *IEEE J. Solid-State Circuits*, vol. 10, no. 3, pp. 168-176, Jun. 1975.

[6] F. Raab, "Idealized operation of the class E tuned power amplifiers," *IEEE Trans. Circuits Syst.*, vol. 24, no. 12, pp. 725-735, Dec. 1977.

[7] J. Cumana, A. Grebennikov, G. Sun, N. Kumar, and R. H. Jansen, "An extended topology of parallel-circuit class-E power amplifier to account for larger output capacitances," *IEEE Trans. Microw. Theory Techn.*, vol. 59, no. 12, pp. 3174-3183, Dec. 2011

[8] M. K. Kazimierczuk and J. Jozwik, "Class E zero-voltage-switching rectifier with a series capacitor," *IEER Trans. Circuits Syst.*, vol 36, no. 6, pp. 926-928, Jun. 1989.

[9] H. Koizumi, T. Suetsugu, M. Fujii, K. Shinoda, S. Mori, and K. Ikeda, "Class DE high-efficiency tuned power amplifier," *IEEE Trans. Circuits Syst. I, Fundam. Theory Appl.*, vol. 43, no. 1, pp. 51-60, Jan. 1996.

[10] M. Matsuo, T. Suetsugu, S. Mori, and I. Sasase, "Class DE current-source parallel resonant inverter," *IEEE Trans. Ind. Electron.*, vol. 46, no. 2, pp. 242-248, Apr. 1999.

[11] T. Kondo and H. Koizumi, "Class DE voltage-source parallel resonant inverter," in *Proc. IEEE Ind. Electron. Soc.*, Yokohama, pp. 2968-2973, Nov. 2015.

[12] IRFML8244TRPbF data-sheet. [Online]. Available: http://docs-asia.electrocomponents.com/webdocs/13fe/0900766b813fe5db.pdf.

[13] IR2011 data-sheet. [Online]. Available: http://www.st.com/web/en/resource/technical/document/datasheet/CD00169717.pdf.

The 2018 International Power Electronics Conference

An Improvement on Extended Impedance Method towards Efficient Steady-State Analysis of High-Frequency Class-E Resonant Inverters

Junrui Liang
School of Information Science and Technology
ShanghaiTech University, Shanghai 201210, China
E-mail: liangjr@shanghaitech.edu.cn

Abstract—The extended impedance method (EIM) is an efficient and intuitive tool for the design and analysis of power switching circuits with active components, nonlinear parasitic components, etc., under high-frequency operation. In this paper, we propose an improvement to the EIM by simplifying the constitutive circuit equations in the iterative calculation, such that to speed up the steady-state searching process. Taking the class-E resonant inverter, which is driven by a practical MOSFET switch, as the testbed, we comparatively evaluate the computational efforts of this improved EIM, the conventional EIM, and PSpice simulation, under the same circuit conditions. The result shows that the proposed method can greatly reduce the simulation time. Therefore, the improved EIM can be used as an efficient tool for the analysis, design, and optimization of high-frequency power electronics.

I. INTRODUCTION

Nowadays, in different studies of power electronics, there is an obvious trend working towards the high-frequency and high-power-density power converters. One of the driving force is the popularization of the high-speed switching semiconductors, such as SiC and GaN; the other is the demand of more compact and more capable power converters. For example, in the wireless power transfer systems, by raising the switching frequency from several hundred kHz [1] to several MHz [2]–[4], the power density of the circuit can be increased; the size of passive components can be reduced.

The single-ended class-E resonant inverter, which is also called class-E power amplifier (PA), was proposed in 1975 [5], [6]. It is known for its simple structure, high efficiency, and high-frequency operation. In the old days, when the switching frequency is as low as several hundred kHz, the nonlinearity of the parasitic components can be neglected. The design of class-E PA usually starts with the circuit equations and waveform equations in the time-domain [7]–[10]. As the switching frequency gets higher, the electrical susceptance of the nonlinear junction capacitor in the switching transistor becomes larger and dominant in the shunt capacitance. In such high-frequency scenario, the effect of nonlinear parasitic components, as well as other large-signal behavior, must receive sufficient consideration in the design and analysis of power converters [11].

Analytical waveform solutions provide intuitions towards the conceptual design [5]. However, the closed-form equations

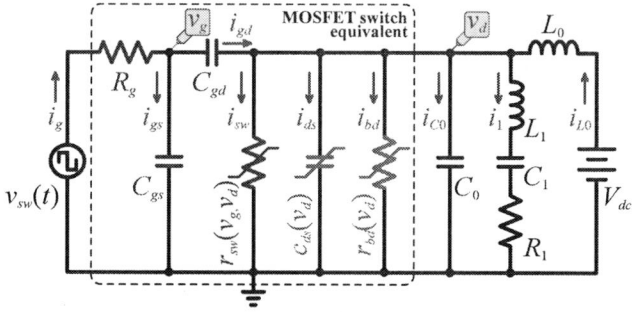

— Components discussed in the basic EIM based class-E PA analysis [12].
— Nonlinear components included in [13].
— More MOSFET equivalent components and revised switch model included in [14].

Fig. 1. Equivalent model of the class-E resonant inverter considering the MOSFET parasitic components.

can hardly express all the details in practical circuits, in particular, when the effect of parasitic components in semiconductor becomes significant and non-ignorable. Numeric simulation or analysis are complementary design techniques towards the detailed performance. The numeric design methods can be classified into time-domain methods [1] and frequency-domain methods [12]–[14].

The extended impedance method (EIM) [12]–[14] was proposed for realizing the steady-state design numerically around the linear specifications, which are obtained with the established method [6]. The harmonic based EIM is different from the conventional harmonic balance method. In EIM, the characteristics of all time-varying and nonlinear components are regarded as special impedances, whose algorithm of combination is compatible with the fundamental linear ac circuit analysis. Therefore, it is more intuitive towards nonlinear circuit analysis.

This paper discusses an improvement to the EIM based on the modified nodal analysis (MNA) towards a more efficient design process of class-E PA under high-frequency operation.

II. THEORY

The equivalent model of class-E PA using practical MOSFET switch is shown in Fig. 1 [14]. V_{dc}, L_0, and C_0 are the dc supply voltage, choke inductance, and shunt capacitance,

4122

The 2018 International Power Electronics Conference

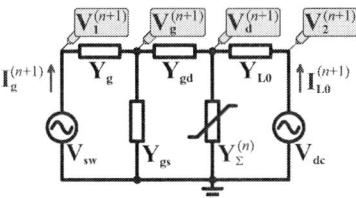

Fig. 2. Simplified circuit network for modified nodal analysis (MNA).

TABLE I
PARAMETERS IN THE CASE STUDIES OF CLASS-E PA.

Parameter	Value
V_{dc}	20 V
V_{sw} [amp. of $v_{sw}(t)$]	± 7.0 V
f_{sw} [freq. of $v_{sw}(t)$]	6.78 MHz
D	50 %
L_0	100 μH
C_0	132.35 pF
L_1	1.03 μH
C_1	757.78 pF
R_1	10.03 Ω
MOSFET	IRF510

The circuit parameters are obtained based on the revised design equations in [6]. The IRF510 specifications were provided in [14].

respectively. R_1, C_1, and L_1 form the resonant branch. The components within the dashed frame forms the equivalent of a practical MOSFET. c_{ds} and r_{bd} represent the nonlinear drain-to-source junction capacitance and body diode, respectively, whose effects under different operations were discussed in [13]; R_g, C_{gs}, C_{gd}, and r_{sw} are the gate resistance, gate-to-source capacitance, gate-to-drain capacitance, and switching resistance, respectively, whose effects under different gate drive voltage were discussed in [14]. V_{sw} provides the gate drive to the MOSFET.

Since all the components in Fig. 1 can be regarded as impedances, whose expressions have been extended from complex scalars to complex matrices [12]. The constitutive circuit equations can be obtained according to the conventional laws in linear network analysis, e.g., KCL, KVL. Such network was successfully formulated by using the loop current method in [14]. After [14], research effort has been continuously made with the following two purposes:

1) Make the EIM based analysis more compatible to the established circuit simulation program, e.g., SPICE, such that the netlist file SPICE can be readily reused in the EIM based solver.

2) Further reduce the computational effort of EIM, in particular, in the iterative state-update procedure.

As the modified nodal analysis (MNA) is extensively used in formulating the circuit network in most existing simulation program [15], such as SPICE, this paper adopts the MNA for formulating the constitutive circuit equations as follows

$$
\begin{bmatrix}
\mathbf{V}_1^{(n+1)} \\
\mathbf{V}_g^{(n+1)} \\
\mathbf{V}_d^{(n+1)} \\
\mathbf{V}_2^{(n+1)} \\
\mathbf{I}_g^{(n+1)} \\
\mathbf{I}_{L0}^{(n+1)}
\end{bmatrix}
=
$$

$$
\begin{bmatrix}
\mathbf{Y_g} & -\mathbf{Y_g} & 0 & 0 & \mathbf{E} & 0 \\
-\mathbf{Y_g} & \mathbf{Y_{\Sigma g}} & -\mathbf{Y_{gd}} & 0 & 0 & 0 \\
0 & -\mathbf{Y_{gd}} & \mathbf{Y_{\Sigma d}^{(n)}} & -\mathbf{Y_{L0}} & 0 & 0 \\
0 & 0 & -\mathbf{Y_{L0}} & \mathbf{Y_{L0}} & 0 & \mathbf{E} \\
\mathbf{E} & 0 & 0 & 0 & 0 & 0 \\
0 & 0 & 0 & \mathbf{E} & 0 & 0
\end{bmatrix}^{-1}
\begin{bmatrix}
0 \\
0 \\
0 \\
0 \\
\mathbf{V_{sw}} \\
\mathbf{V_{dc}}
\end{bmatrix},
\tag{1}
$$

where

$$
\mathbf{Y_{\Sigma g}} = \mathbf{Y_g} + \mathbf{Y_{gs}} + \mathbf{Y_{gd}},
\tag{2}
$$

$$
\mathbf{Y_{\Sigma d}^{(n)}} = \mathbf{Y_{gd}} + \mathbf{Y_{\Sigma}^{(n)}} + \mathbf{Y_{L0}},
\tag{3}
$$

$$
\begin{aligned}
\mathbf{Y_{\Sigma}^{(n)}} &= \mathbf{Y_{sw}^{(n)}} + \mathbf{Y_{ds}^{(n)}} + \mathbf{Y_{bd}^{(n)}} + \mathbf{Y_{C0}} \\
&\quad + (\mathbf{Z_{R1}} + \mathbf{Z_{C1}} + \mathbf{Z_{L1}})^{-1}.
\end{aligned}
\tag{4}
$$

The relations are obtained according to the simplified circuit network shown in Fig. 2. The four voltage nodes other than the ground are denoted as the vectors $\mathbf{V_1}$, $\mathbf{V_g}$, $\mathbf{V_d}$, and $\mathbf{V_2}$. The dc supply and gate drive voltage are taken as two general voltage sources, whose flowing-through currents are $\mathbf{I_g}$ and $\mathbf{I_{L0}}$, respectively. The superscript $(n+1)$ denote the results in the $(n+1)$th iteration. The corresponding admittance in the matrix of (1) is shown in Fig. 2. The voltage and current vectors, whose dimensions are $(2K+1)\times 1$, express the circuit states in the frequency domain with a harmonic number of K; the matrices, whose dimensions are $(2K+1)^2$, express the characteristics of the components in the form of admittances. Within those admittances, $\mathbf{Y_{\Sigma}}$ is the only nonlinear one. As we can observed from Fig. 1 and (4), the nonlinearities come from the MOSFET switching, whose characteristic is denoted by $\mathbf{Y_{sw}}$, drain-to-source junction capacitance denoted by $\mathbf{Y_{ds}}$, and body diode denoted by $\mathbf{Y_{bd}}$.

With (1), the four voltage and two current unknowns can be immediately solved by doing the matrix inversion. A Matlab program is developed based on the sparse matrix manipulations for simulating a class-E PA, whose specifications are listed in Table I. The execution summary by using full matrix inversion algorithm is shown in Fig. 3. It can be observed from Fig. 3 that the matrix inversion takes the longest time to execute compared to other lines.

On the other hand, in the six unknown voltages and currents, actually only $\mathbf{V_g^{(n+1)}}$ and $\mathbf{V_d^{(n+1)}}$ are useful information for updating the admittance matrices $\mathbf{Y_{sw}^{(n+1)}}$, $\mathbf{Y_{ds}^{(n+1)}}$, and $\mathbf{Y_{bd}^{(n+1)}}$ in the next round of calculation. A approximating solution can be formulated by taking the matrix in (1) as a 6×6 block

4123

The 2018 International Power Electronics Conference

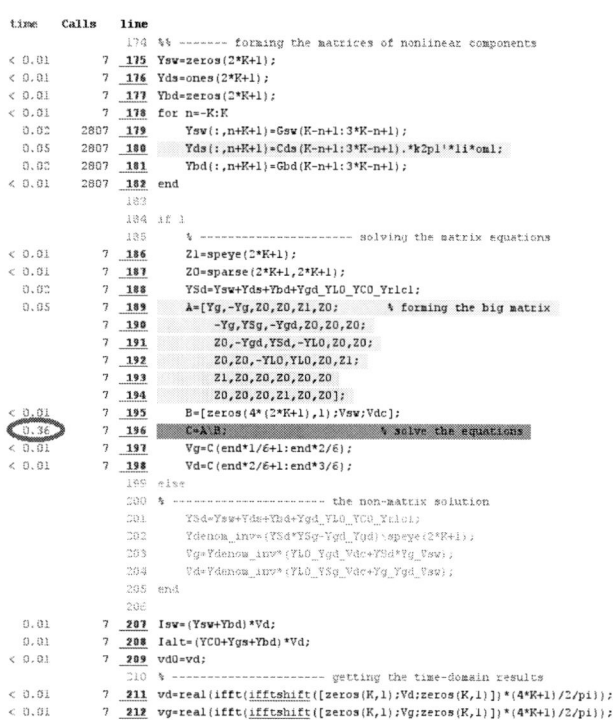

```
time  Calls  line
             174  %% ------- forming the matrices of nonlinear components
< 0.01    7  175  Ysw=zeros(2*K+1);
< 0.01    7  176  Yds=ones(2*K+1);
< 0.01    7  177  Ybd=zeros(2*K+1);
< 0.01    7  178  for n=-K:K
  0.02 2807  179    Ysw(:,n+K+1)=Gsw(K-n+1:3*K-n+1);
  0.05 2807  180    Yds(:,n+K+1)=Cds(K-n+1:3*K-n+1).*k2p1'*1i*oml;
  0.02 2807  181    Ybd(:,n+K+1)=Gbd(K-n+1:3*K-n+1);
< 0.01 2807  182  end
             183
             184  if 1
             185    % -------------------- solving the matrix equations
< 0.01    7  186    Z1=speye(2*K+1);
< 0.01    7  187    Z0=sparse(2*K+1,2*K+1);
  0.02    7  188    YSd=Ysw+Yds+Ybd+Ygd_YL0_YC0_Yrlcl;
  0.05    7  189    A=[Yg,-Yg,Z0,Z0,Z1,Z0;      % forming the big matrix
          7  190       -Yg,YSg,-Ygd,Z0,Z0,Z0;
          7  191       Z0,-Ygd,YSd,-YL0,Z0,Z0;
          7  192       Z0,Z0,-YL0,YL0,Z0,Z1;
          7  193       Z1,Z0,Z0,Z0,Z0,Z0
          7  194       Z0,Z0,Z0,Z1,Z0,Z0];
< 0.01    7  195    B=[zeros(4*(2*K+1),1);Vsw;Vdc];
  0.36    7  196    C=A\B;                     % solve the equations
< 0.01    7  197    Vg=C(end*1/6+1:end*2/6);
< 0.01    7  198    Vd=C(end*2/6+1:end*3/6);
             199  else
             200    % -------------------- the non-matrix solution
          201    YSd=Ysw+Yds+Ybd+Ygd_YL0_YC0_Yrlcl;
          202    Ydenom_inv=(YSd*YSg-Ygd*Ygd).\speye(2*K+1);
          203    Vg=Ydenom_inv*(YL0_Ygd_Vdc+YSd*Yg_Vsw);
          204    Vd=Ydenom_inv*(YL0_YSg_Vdc+Yg_Ygd_Vsw);
          205  end
             206
  0.01    7  207  Isw=(Ysw+Ybd)*Vd;
  0.01    7  208  Ialt=(YC0+Ygs+Ybd)*Vd;
< 0.01    7  209  vd0=vd;
             210  % -------------------- getting the time-domain results
< 0.01    7  211  vd=real(ifft(ifftshift([zeros(K,1);Vd;zeros(K,1)])*(4*K+1)/2/pi));
< 0.01    7  212  vg=real(ifft(ifftshift([zeros(K,1);Vg;zeros(K,1)])*(4*K+1)/2/pi));
```

```
time  Calls  line
             174  %% ------- forming the matrices of nonlinear components
< 0.01    7  175  Ysw=zeros(2*K+1);
< 0.01    7  176  Yds=ones(2*K+1);
< 0.01    7  177  Ybd=zeros(2*K+1);
< 0.01    7  178  for n=-K:K
  0.02 2807  179    Ysw(:,n+K+1)=Gsw(K-n+1:3*K-n+1);
  0.05 2807  180    Yds(:,n+K+1)=Cds(K-n+1:3*K-n+1).*k2p1'*1i*oml;
  0.02 2807  181    Ybd(:,n+K+1)=Gbd(K-n+1:3*K-n+1);
< 0.01 2807  182  end
             183
             184  if 0
             185    % -------------------- solving the matrix equations
          186    Z1=speye(2*K+1);
          187    Z0=sparse(2*K+1,2*K+1);
          188    YSd=Ysw+Yds+Ybd+Ygd_YL0_YC0_Yrlcl;
          189    A=[Yg,-Yg,Z0,Z0,Z1,Z0;      % forming the big matrix
          190       -Yg,YSg,-Ygd,Z0,Z0,Z0;
          191       Z0,-Ygd,YSd,-YL0,Z0,Z0;
          192       Z0,Z0,-YL0,YL0,Z0,Z1;
          193       Z1,Z0,Z0,Z0,Z0,Z0
          194       Z0,Z0,Z0,Z1,Z0,Z0];
          195    B=[zeros(4*(2*K+1),1);Vsw;Vdc];
          196    C=A\B;                     % solve the equations
          197    Vg=C(end*1/6+1:end*2/6);
          198    Vd=C(end*2/6+1:end*3/6);
             199  else
             200    % -------------------- the non-matrix solution
  0.02    7  201    YSd=Ysw+Yds+Ybd+Ygd_YL0_YC0_Yrlcl;
  0.21    7  202    Ydenom_inv=(YSd*YSg-Ygd*Ygd).\speye(2*K+1);
< 0.01    7  203    Vg=Ydenom_inv*(YL0_Ygd_Vdc+YSd*Yg_Vsw);
< 0.01    7  204    Vd=Ydenom_inv*(YL0_YSg_Vdc+Yg_Ygd_Vsw);
          205  end
             206
  0.01    7  207  Isw=(Ysw+Ybd)*Vd;
  0.01    7  208  Ialt=(YC0+Ygs+Ybd)*Vd;
< 0.01    7  209  vd0=vd;
             210  % -------------------- getting the time-domain results
< 0.01    7  211  vd=real(ifft(ifftshift([zeros(K,1);Vd;zeros(K,1)])*(4*K+1)/2/pi));
< 0.01    7  212  vg=real(ifft(ifftshift([zeros(K,1);Vg;zeros(K,1)])*(4*K+1)/2/pi));
```

Fig. 3. The execution time of the Matlab program (harmonic no. $K = 200$) based on the full matrix inversion EIM algorithm (the deeper red the highlighting line, the more time for the execution).

Fig. 4. The execution time of the Matlab program (harmonic no. $K = 200$) based on the improved EIM algorithm (the deeper red the highlighting line, the more time for the execution).

matrix. By solving the six symbolic linear equations, we can obtain

$$\mathbf{V}_g^{(n+1)} \approx \left[\mathbf{Y}_{\Sigma d}^{(n)} \mathbf{Y}_{\Sigma g} - \mathbf{Y}_{gd}^2 \right]^{-1} \left[\mathbf{Y}_{L0} \mathbf{Y}_{gd} \mathbf{V}_{dc} + \mathbf{Y}_g \mathbf{Y}_{\Sigma d}^{(n)} \mathbf{V}_{sw} \right] \tag{5}$$

$$\mathbf{V}_d^{(n+1)} \approx \left[\mathbf{Y}_{\Sigma d}^{(n)} \mathbf{Y}_{\Sigma g} - \mathbf{Y}_{gd}^2 \right]^{-1} \left(\mathbf{Y}_{L0} \mathbf{Y}_{\Sigma g} \mathbf{V}_{dc} + \mathbf{Y}_g \mathbf{Y}_{gd} \mathbf{V}_{sw} \right) \tag{6}$$

It should point out that, from the fundamental principle of linear algebra, the expressions of (5) and (6) do not rigorously hold, as there might be cross-coupling terms across the blocks. Here, we do the approximation and try to use such method to reduce the computational effort towards the circuit optimization. The case study with the class-E specifications of Table I by using such approximation actually gives a similar voltages as that using the full matrix inversion (will be shown in the next section). On the other hand, as shown in Fig. 4, the most critical line in the new algorithm is also the matrix inversion step. Yet, the computational time of this critical line in the new program is about 58% of that in the full matrix solution. Therefore, by only calculating the useful information, the computation time can be further reduced towards more efficient EIM based analysis.

III. DESIGN PROCEDURES

An example design case is provided here for validating the waveform analysis of EIM, and also describing the design procedures with EIM based circuit optimization. Fig. 5 shows four testing circuits in the design procedures.

1) Obtain the component values with the conventional design laws or guidelines considering an ideal switch and linear components.
2) Replace the ideal switch with practical transistor model. Simulate the changes in operation.
3) Design the EIM based numerical optimization to approach the design objective, e.g., ZVS and ZCS in the class-E circuit, constant output power, or highest conversion efficiency.
4) Simulate the re-tuned parameters and comparatively evaluate the design feasibility.

The results in Fig. 6 show that the class-E nominal ZCS condition is violated when the ideal switch model (whose result is shown by the dash-dot-dot lines) is replaced by a practical MOSFET model (whose result is shown by the solid lines).

Owing to the high computational efficiency of EIM, we manage to develop a derivative-free algorithm [12], [13] for re-tuning the nonlinear class-E circuit back to the vicinity of the nominal conditions. The full matrix optimization leads to the optimized capacitance pairs $C_0 = 411$ pF and $C_1 = 819$

4124

The 2018 International Power Electronics Conference

Linear shunt capacitance
(based on the Sokal's equations)

With practical MOSFET model
(nonlinear capacitance Cj0=366.5pF)

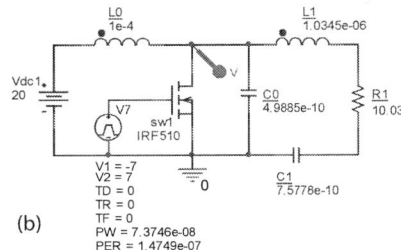

(a)

(b)

With practical MOSFET model
(Optimized with full matrix EIM, K=100)

With practical MOSFET model
(Optimized with reduced matrix EIM, K=100)

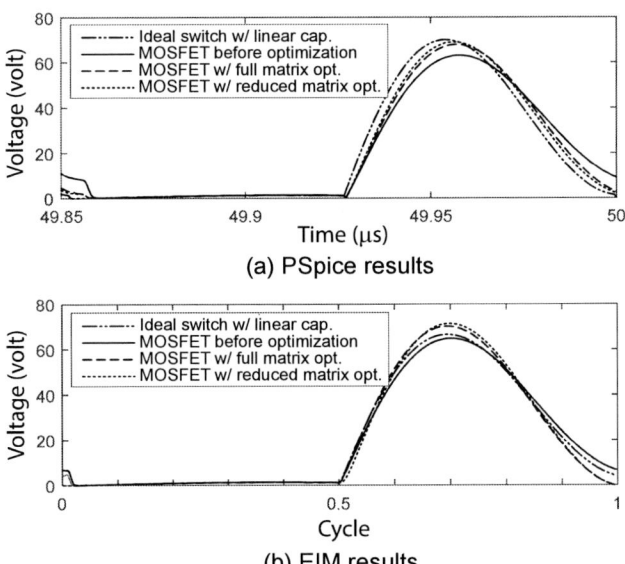

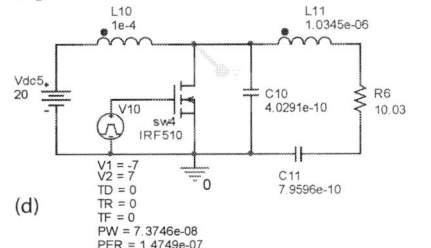

(c)

(d)

Fig. 5. Four testing conditions for the comparative study. (a) Class-E circuit with linear shunt capacitance and ideal switch. (b) The ideal switch is replaced by a MOSFET model. (c) Re-tuned circuit with full matrix EIM ($C_0 = 411$ pF, $C_1 = 819$ pF). (d) Re-tuned circuit with reduced matrix EIM ($C_0 = 403$ pF, $C_1 = 796$ pF).

(a) PSpice results

(b) EIM results

Fig. 6. Simulation waveforms obtained with PSpice and EIM.

pF, as illustrated in Fig. 5(c); the waveform result is shown by the dash lines in Fig. 6. It shows that the ZVS and ZDS conditions are recovered under the tuned C_0 and C_1. The reduced matrix optimization leads to another optimized capacitance pair $C_0 = 403$ pF and $C_1 = 796$ pF, which are very close to the full matrix optimization results. The

waveforms obtained under the two optimization algorithm are very close and almost overlap in Fig. 6(b). Both the full matrix and reduced matrix can successfully fulfill the designed task.

IV. CONCLUSION

This paper has considered a possible improvement on the extended impedance method (EIM) towards the more efficient frequency-domain simulation and optimized design of high-frequency and high-power-density power electronics. In the class-E circuit example, since only the gate and drain voltages have an effect over the nonlinear parasitic components in the switching transistor, the execution time can be reduced by eliminating the other useless information during the iterative state update process. The performance evaluator in Matlab shows the superiority of the proposed improvement towards the customized analyses and designs of power electronics.

V. ACKNOWLEDGMENT

The work described in this paper was supported by the grants from National Natural Science Foundation of China (Project No. 61401277) and ShanghaiTech University (Project No. F-0203-13-003).

REFERENCES

[1] P. C. K. Luk, S. Aldhaher, W. Fei, and J. F. Whidborne, "State-space modeling of a class E^2 converter for inductive links," *IEEE Transactions on Power Electronics*, vol. 30, no. 6, pp. 3242–3251, June 2015.

[2] T. Nagashima, X. Wei, E. Bou, E. Alarc?n, M. K. Kazimierczuk, and H. Sekiya, "Analysis and design of loosely inductive coupled wireless power transfer system based on class-e² dc-dc converter for efficiency enhancement," *IEEE Transactions on Circuits and Systems I: Regular Papers*, vol. 62, no. 11, pp. 2781–2791, Nov 2015.

4125

[3] S. Liu, M. Liu, S. Yang, C. Ma, and X. Zhu, "A novel design methodology for high-efficiency current-mode and voltage-mode class-e power amplifiers in wireless power transfer systems," *IEEE Trans. Power Electron.*, vol. PP, no. 99, pp. 1–1, 2016.

[4] M. Liu, M. Fu, and C. Ma, "Parameter design for a 6.78-mhz wireless power transfer system based on analytical derivation of class e current-driven rectifier," *IEEE Transactions on Power Electronics*, vol. 31, no. 6, pp. 4280–4291, June 2016.

[5] N. Sokal and A. Sokal, "Class E – A new class of high-efficiency tuned single-ended switching power amplifiers," *IEEE J. Solid-St. Circ.*, vol. 10, no. 3, pp. 168–176, June 1975.

[6] N. O. Sokal, "Class-E RF power amplifiers," *QEX*, pp. 9–21, Jan./Feb. 2001.

[7] M. Kazimierczuk and K. Puczko, "Exact analysis of class E tuned power amplifier at any Q and switch duty cycle," *IEEE T. Circuits and Syst.*, vol. 34, no. 2, pp. 149–159, Feb. 1987.

[8] G. Kendir, W. Liu, G. Wang, M. Sivaprakasam, R. Bashirullah, M. Humayun, and J. Weiland, "An optimal design methodology for inductive power link with class-E amplifier," *IEEE T. Circuits-I.*, vol. 52, no. 5, pp. 857–866, May 2005.

[9] H. Sekiya, I. Sasase, and S. Mori, "Computation of design values for Class E amplifiers without using waveform equations," *IEEE T. Circuits-I.*, vol. 49, no. 7, pp. 966–978, July 2002.

[10] H. Sekiya, T. Ezawa, and Y. Tanji, "Design procedure for Class E switching circuits allowing implicit circuit equations," *IEEE T. Circuits-I.*, vol. 55, no. 11, pp. 3688–3696, Dec. 2008.

[11] X. Wei, H. Sekiya, S. Kuroiwa, T. Suetsugu, and M. K. Kazimierczuk, "Design of class-e amplifier with mosfet linear gate-to-drain and non-linear drain-to-source capacitances," *IEEE Trans. Circuits Syst. Regul. Pap.*, vol. 58, no. 10, pp. 2556–2565, 2011.

[12] J. Liang and W.-H. Liao, "Steady-state simulation and optimization of class-E power amplifiers with extended impedance method," *IEEE Trans. Circuits Syst. Regul. Pap.*, vol. 58, no. 6, pp. 1433–1445, june 2011.

[13] J. Liang, "Design of class-e power amplifier with nonlinear components by using extended impedance method," in *2016 IEEE International Symposium on Circuits and Systems (ISCAS)*, May 2016, pp. 437–440.

[14] J. Liang and S. Zhang, "An efficient steady-state simulation of class-e resonant inverter considering mosfet parasitic components by using extended impedance method," in *2017 IEEE 3rd International Future Energy Electronics Conference and ECCE Asia (IFEEC 2017 - ECCE Asia)*, June 2017, pp. 190–195.

[15] V. Litovski and M. Zwolinski, *VLSI circuit simulation and optimization.* Kluwer Academic Publishers, 1997.

Output Power Capability Comparisons of Class-E Power Amplifiers with Harmonic Resonance

Hiroo Sekiya[1*], Xiuqin Wei[2], Yuchong Sun[1]

1 Graduate School of Science and Engineering, Chiba University, Chiba, Japan
2 Department of Electrical and Electronics Engineering, Chiba Instite of Technology, Narashino, Japan
*E-mail: sekiya@faculty.chiba-u.jp

Abstract—This paper investigates flat-top class-E, class-E/F$_2$, and class-E/F$_3$ amplifiers. All these amplifiers are based on the conception of adding a harmonic resonant circuit to the classic class-E amplifier. The purpose of this paper is to find the best strategy of the harmonic injection for improving the performance of class E amplifier. Characteristic comparisons of a numerical optimization of class E/F$_2$, class-E/F$_3$ and flat-top class-E amplifiers are shown and discussed. From the comparisns among the optimized amplifiers, which achieve the maximum output power capability, it is clarified that the class-E/F$_2$ amplifier shows the best performance among three amplifiers. Experimental results of the class-E/F$_2$ and the class-E amplifiers are shown for verifying the obtained results.

I. INTRODUCTION

It is expected that high-frequency amplifiers are applied to many applications, e.g., communication amplifier, induction heating, plasma-generation power source, and so on.[1], [2], [3] High frequency amplifiers work with MOSFET as a switching device. Because of high frequency operations, magnetic elements as well as amplifier volume can be compact. On the other hand, entire energy stored in the parasitic capacitance of the MOSFET is released when switch turns on, which becomes switching loss. The switching losses increase in proportion to the operating frequency, which degrades the power conversion efficiency of the amplifier. There is soft-switching techniques for switching loss reductions. By satisfying the soft switching, such as zero-voltage switching (ZVS) and the zero-current switching (ZCS), power conversion efficiency is high in spite of high frequencies. Additionally, the switching loss reduction leads to the heat reduction of the MOSFET.

The class-E amplifier[4], [5], [6] is one of the high-efficiency amplifiers, which satisfies not only ZVS but also zero-derivative switching (ZDS) condition. By satisfying the class-E ZVS/ZDS conditions, it is possible to reduce the switching losses to be zero theoretically. Therefore, the class-E amplifier has sufficient high ability to operate at high frequencies. It is, however, a drawback of the class-E amplifier is that the maximum switch voltage is 3.5 times as high as the dc-supply voltage [7]. Therefore, we suffer from the cost of a MOSFET for high-power applications, in particular. Additionally, when the MOSFET with high breakdown drain-to-source voltage is used, power conversion efficiency also decreases. This is because on-resistance of switch becomes usually increases as the maximum voltage of the MOSFET

increases. In this sense, it is an important problem to suppress the switch voltage stress of the class-E amplifier.

Class-E/F$_2$[8] and class-E/F$_3$[9], and flat-top class-E amplifiers[10] can be reduced the switch voltage stress compared with the class-E amplifier. These amplifiers are proposed from the same design concept, which is harmonic current is injected to the MOSFET by adding the harmonic resonant circuit to the class-E amplifier. By injecting the harmonic current with proper amplitude and phase shift, the MOSFET peak voltage is suppressed compared with the class-E amplifier. There is, however, no performance comparisons among these amplifiers. It is useful for designers to have performance evaluations and comparisons of these amplifiers. For example, the peak switch voltage can be reduced by adjusting the switch-on duty ratio. There is, however, many cases that the output voltage also decreases with the decrease in the peak switch voltage. This means that evaluations of the amplifier performance are impossible only from the peak switch voltage, which generates the difficulty of the comparisons among multiple amplifiers.

The purpose of this paper is to show the best strategy of the harmonic injection for improving the performance of class E amplifier. Concretely, this paper investigates flat-top class-E, class-E/F$_2$, and class-E/F$_3$ amplifiers from the output-power-capability viewpoint. By comparing the optimized amplifiers, which achieve the maximum output power capability, it is seen that the class-E/F$_2$ amplifier shows the best performance among three amplifiers. Experimental results of the class-E/F$_2$ and the class-E amplifiers are shown for verifying the obtained results.

II. CLASS-E AMPLIFIER WITH HARMONIC RESONANT CIRCUIT

The circuit topology of the class-E amplifier [4] is shown in Fig. 1. The class-E amplifier consists of dc-supply voltage source V_I, input inductance L_c, MOSFET S as a switching device, shunt capacitance C_s, resonant filter $L_1 - C_1$, and load resistance R. The MOSFET is driven at the fixed operating frequency f. When the input inductance is sufficient high, the input current through the input inductance is regarded as the dc current. The output voltage is pure sinusoidal waveform when the Q factor of the output network $Q = 2\pi f L_1 / R$ is high. Figure 2(a) shows nominal waveforms of the class-E amplifier. The switch voltage satisfies the class-E ZVS/ZDS conditions. Therefore, high power conversion efficiency is

The 2018 International Power Electronics Conference

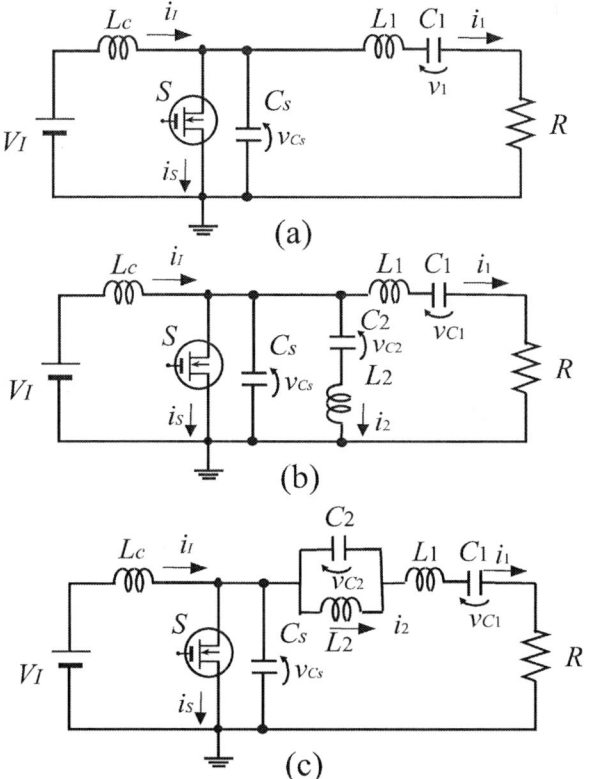

Fig. 1. Circuit toporogy. (a) Class-E amplifier. (b) Class-E/F$_2$ and class-E/F$_3$ amplifiers. (c) Flat-top class-E amplifier.

achieved at high frequencies in the class-E amplifier. One of the drawbacks of the class-E amplifier is that the peak switch voltage isapproximately 3.5 times higher than the dc-supply voltage as shown in Fig. 2(a).

A circuit topology of the class-E/F$_2$ and class-E/F$_3$ amplifiers is illustrated in Fig. 1(b). Additionally, Fig. 1(c) shows that of the flat-top class-E amplifier. These amplifiers have the similar concepts; the harmonic resonant circuit is added to the class-E amplifier.

In the class-E/F$_2$[8] and the class-E/F$_3$ amplifiers[9], the series resonant circuit is added to the MOSFET in parallel. Namely, the circuit topology of the class-E/F$_2$ amplifier is the same as that of the class-E/F$_3$ amplifier. In the case of class-E/F$_2$ amplifier, the impedance to second harmonic is zero and that to third harmonic is zero in the case of class-E/F$_3$ amplifier. The harmonic current flowing through the harmonic resonant filters flows through the MOSFET or shunt capacitance. Therefore, switch voltage waveform is different from that of the class-E amplifier. Under optimal designs, the peak switch voltage of the class-E/F$_2$ is twice as high as the dc-supply voltage as shown in Fig. 2(c). The peak switch voltage of the class-E/F$_3$ amplifier may be 3.1 times higher than the dc-supply voltage as shown in Fig. 2(c).

On the other hand, The parallel resonant circuit is added to the MOSFET in series in the flat-top class-E amplifier [10]. Because of the parallel resonance, the impedance of the harmonic resonant circuit is infinite in this type of amplifier, the switch voltage can be reduced, which is realized by the subtraction of the voltage across the harmonic resonant filter. By adjusting the harmonic voltage amplitude and phase shift, flat switch-voltage waveform can be obtained as shown in Fig. 2(d).

III. OUTPUT POWER CAPABILITY

The major interest in this paper is peak switch voltage reduction effect of amplifiers by adding harmonic resonant circuit. In these amplifier, the ratio of the dc-supply voltage to amplitude of the output voltage cannot be fixed under the peak switch voltage optimization. In this sense, it is impossible to evaluate amplifier performance only from the peak switch voltage. The output power capability is one of the good metric for evaluating the amplifier, which solves the above problem. The output power capability is defined as

$$C_p = \frac{P_o}{I_{SM} V_{SM}}. \tag{1}$$

In (1), P_o, I_{SM}, and V_{SM} are output power, maximum value of switch current, and maximum value of switch voltage, respectively. The output power capability is the metric, which takes into account both the switch stress and the output power. In case of high value of C_p, low-cost switches can be used for the identical output power, or high output power can be obtained for the identical MOSFET device.

IV. OPTIMAL DESIGN METHOD BY NUMERICAL ALGORITHM[11]

For fair comparisons among amplifiers with harmonic resonant circuit, it is necessary to do the optimal designs for each amplifier, which achieves maximum output power capability. In this paper, the optimal designs are carried out by using the numerical algorithm in [11].

A. Parameter definitions

The circuit parameters for formulating circuit equations are defined as follows.

1. D: The switch-on duty ratio. It is assumed that the switch turns on at $\theta = 0$, where $\theta = \omega t = 2\pi f t$ is angular time.

2. $\omega = 2\pi f$: The operation angular frequency.

3. $\omega_1 = 2\pi f_1 = 1/\sqrt{L_1 C_1}$: The resonant angular frequency of the main filter.

4. $\omega_2 = 2\pi f_2 = 1/\sqrt{L_2 C_2}$: The resonant angular frequency of the harmonic filter

5. $A_1 = (f_1 f)^2 = (\omega^1/\omega)^2$: The ratio of main-filter resonant frequency to operation frequency.

4128

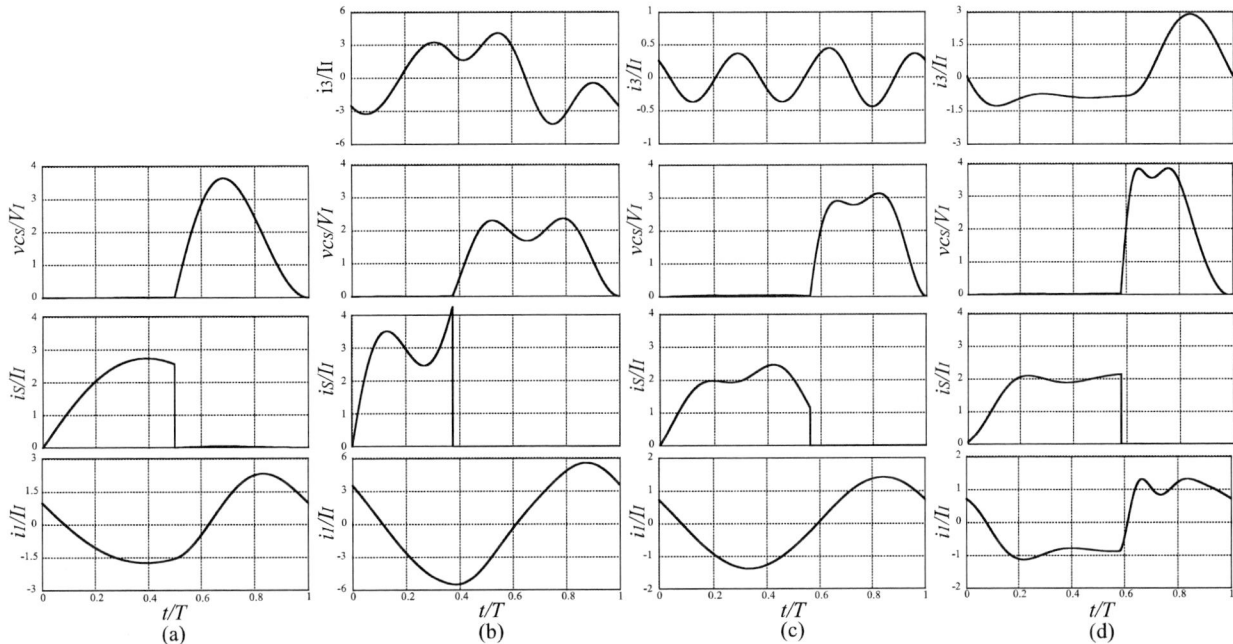

Fig. 2. Numerical waveforms for achieving the maximum output power capaibility. (a) Class-E amplifier. (b) Class-E/F$_2$ amplifier. (c) Class-E/F$_3$ amplifier. (d) Flat-top class-E amplifier.

6. $A_2 = (f_2/f)^2 = (\omega_2/\omega)^2$: The ratio of harmonic-filter resonant frequency to operation frequency.

7. $B = C_1/C_S$: The ratio of main-filter resonant capacitance to shunt capacitance.

8. $H = L_1/L_C$: The ratio of main-filter resonant inductance to input inductance.

9. $Q_1 = \omega L_1/R$: The loaded quality factor of the main resonant filter.

10. $Q_2 = \omega L_2/R$: The quality factor of the harmonic resonant filter.

11. $R'_S = R_S/R$: The normalized switch resistance. R_S expresses the switch resistance as

$$R_S = \begin{cases} r_S, & \text{for } 0 \le \theta < 2\pi D \\ \infty, & \text{for } 2\pi D \le \theta < 2\pi, \end{cases} \quad (2)$$

where r_S is the switch on-resistance.

B. Circuit equations

By using the above parameters, the circuit equations can be formulated. For example, the circuit equations of the class-E/F$_2$ and class-E/F$_3$ amplifier as shown in Fig.1(b) are

$$\begin{cases} \dfrac{di'_I}{d\theta} = \dfrac{H}{Q_1}(V'_I - v'_{C_S}) \\ \dfrac{dv_{C_S}}{d\theta} = A_1^2 Q_1 B\left(i'_I - \dfrac{v'_{C_S}}{R'_S} - i'_1 - i'_2\right) \\ \dfrac{di'_1}{d\theta} = \dfrac{1}{Q_1}(v'_{C_S} - v'_{C_1} - i'_1) \\ \dfrac{di'_2}{d\theta} = \dfrac{1}{Q_2}(v'_{C_S} - v'_{C_2}) \\ \dfrac{dv'_{C_1}}{d\theta} = A_1^2 Q_1 i'_1 \\ \dfrac{dv'_{C_2}}{d\theta} = A_2^2 Q_2 i'_2 \end{cases} \quad (3)$$

where $i'_k = \dfrac{R i_k}{V_I}$ and $v'_k = \dfrac{v_k}{V_I}$ are the normalized current and voltage, respectively. The circuit equations of the flat-top class-E amplifier also can be obtained.

C. Amplifier designs with maximum output capability

From circuit equations, circuit parameters for satisfying the restrict conditions are derived by applying the numerical algorithm, which was proposed in [11]. First, all the amplifiers are compared under the conditions that the amplifiers satisfy the class-E ZVS/ZDS conditions, which are expressed as

$$v_{Cs}(2\pi D) = 0, \quad \left.\dfrac{dv_{Cs}(\theta)}{d\theta}\right|_{\theta=2\pi D} = 0. \quad (4)$$

We have seven parameters, namely D, A_1, A_2, Q_1, Q_2, B, and H, in the circuit equations. From these circuit parameters,

The 2018 International Power Electronics Conference

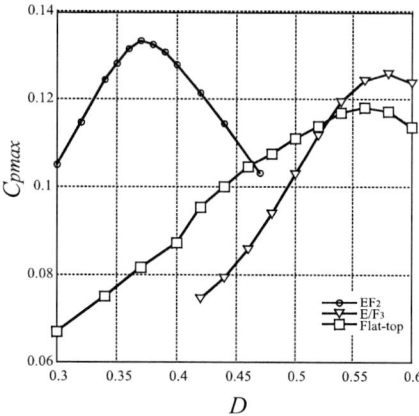

Fig. 3. Maximum output power capability for fixed amplifier as a function of D.

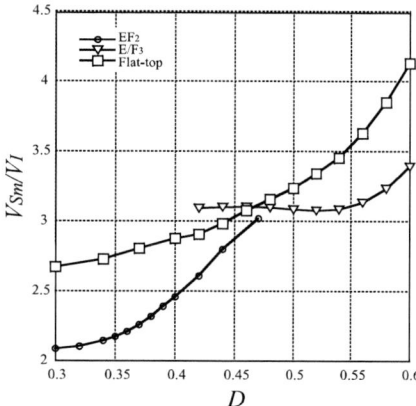

Fig. 4. Maximum values of normalized switch voltage for achieving the maximum output capaiblity with fixed amplifier as a function of D.

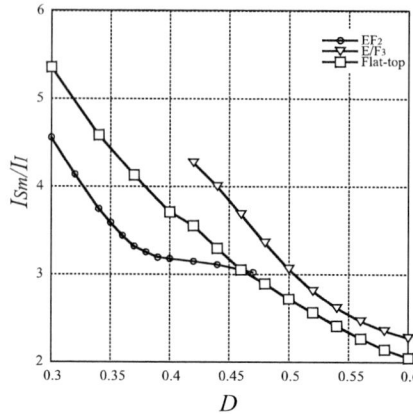

Fig. 5. Maximum values of normalized switch current for achiving the maximum output capability with fixed amplifier as a function of D.

it is necessary to select the design parameters for using the optimization. The parameters are required to be sensitive to the output power capability. H is fixed as $H = 0.005$ for assuming the dc input current. A_2 is adjusted to $A = 2$ for the class-E/F$_2$ and flat-top class-E amplifiers and $A = 3$ for the class-E/F$_3$ amplifier. In addition, $B = 1$ is fixed because it was confirmed that B is not sensitive to C_p by pre-designs. From the above setting, D, A_1, Q_1, and Q_2 are used for the optimal designs.

Because the waveforms should satisfy the class-E ZVS/ZDS conditions, A_1 and Q_1 are fixed uniquely when D and Q_2 are given. Therefore, the maximum output power capability can be calculated for the fixed D by changing Q_2 value.

V. COMPARATIVE EVALUATIONS

Figure 3 shows the maximum output power capabilities of class-E/F$_2$, class-E/F$_3$, and flat-top class-E amplifiers as a

function of D. It is known that the maximum output power capability of the class-E amplifier is obtained as $C_p = 0.098$ at $D = 0.5$ [12]. The maximum value of output power capability of the class-E/F$_2$ amplifier is $C_{pmax} = 0.132$ at $D = 0.37$ as shown in Fig. 3, which is higher than the class-E amplifier. Additionally, it is found that those of class-E/F$_3$ and the flat-top class-E amplifiers are $C_{pmax} = 0.126$ at $D = 0.58$ and $C_{pmax} = 0.118$ at $D = 0.56$, respectively. From these results, it can be concluded that class-E/F$_2$ amplifier gives the highest output power capability among three amplifiers.

The waveforms for obtaining the maximum output power capability are shown in Fig. 2. It is confirmed that all the switch voltages satisfy the class-E ZVS/ZDS conditions at turn-on instant. From Fig. 2, we find that the class-E/F$_2$ amplifier achieves the lowest peak switch voltage and the highest peak output current among the amplifiers.

Figures 4 and 5 show the peak value of the switch voltage and switch current, respectively when C_{pmax} is obtained. It is confirmed from Figs. 4 and 5 that the class-E/F$_2$ amplifier has a big advantage of the peak switch voltage reduction compared with the other amplifiers.

VI. EXPERIMENTAL EVALUATIONS

This paper carried out circuit experiments of the class-E/F$_2$ and the class-E amplifiers with the highest output power capability. For both the experiments, the identical specifications of load resistance $R = 5 \ \Omega$, dc-supply voltage $V_D = 5$ V, and operating frequency $f = 3$ MHz were given. From the specifications, other component values are determined as given in Tables 1 and 2. All component values in Table 1 and 2 were measured by KEYSIGHT E4990A. This experiment used SUD06N10 MOSFET as the switch S. From the datasheet, on-resistance of $r_S = 0.225 \ \Omega$ could be obtained. Intersil EL7104 was used for driver circuit of the MOSFET.

Figures 6 and 7 show experimental waveforms of the class-E amplifier and class-E/F$_2$ amplifier, respectively, where D_r,

4130

TABLE I. NUMERICAL AND MEASURED COMPONENT VALUES OF THE CLASS-E AMPLIFIER.

	Analytical	Measured	Difference
f	3 MHz	3 MHz	0 %
D	0.5	0.5	0 %
L_c	40 μH	39.5 μH	−1.4 %
L_1	1.33 μH	1.35 μH	1.5 %
C_1	2.76 nF	2.74 nF	−0.7 %
r_S	0.225 Ω	−	−
R	5 Ω	4.99 Ω	−0.2 %
V_I	5V	5V	0 %

TABLE II. NUMERICAL AND MEASURED COMPONENT VALUES OF THE CLASS-E/F$_2$ AMPLIFIER FOR ACHIEVING THE MAXIMUM OUTPUT POWER CAPAIBLITY.

	Analytical	Measured	Difference
f	3 MHz	3.09 MHz	3 %
D	0.37	0.37	0 %
L_c	27 μH	26.6 μH	−1.4 %
L_1	866 nH	872 nH	0.7 %
L_2	3.75 μH	3.85 μH	2.7 %
C_1	4.31 nF	4.35 nF	0.9 %
C_2	187.8 pF	188.3 pF	0.3 %
r_S	0.225Ω	−	−
R	5 Ω	5.01 Ω	0.2 %
V_I	5 V	5 V	0 %

v_S, and v_O represent driver signal, switch voltage, and output voltage, respectively. These waveforms were measured by Textronix MDO3024. It is seen from these figures that both the voltage waveforms satisfy the class-E ZVS/ZDS conditions. It is also seen from Figs. 2, 6, and 7 that the experimental waveforms agreed with numerical waveforms quantitatively, which showed the reliability and validity of the numerical algorithm. Peak value of the switch voltage is 17 V in the class-E amplifier, which is 3.4 times as high as dc-supply voltage. Conversely, that is 10 V in the class-E/F$_2$ amplifier, which is twice as high as dc-supply voltage. The class-E/F$_2$ amplifier reduced peak switch voltage by 41 % compared with the class-E amplifier.

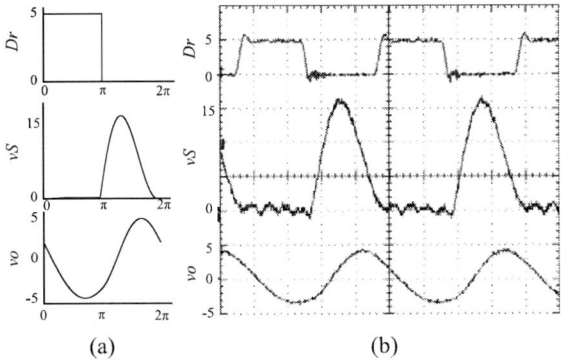

Fig. 6. Nominal waveforms of the class-E amplifier for achieving the maximum output power capability. (a) Numerical waveforms. (b)Experimental waveforms.

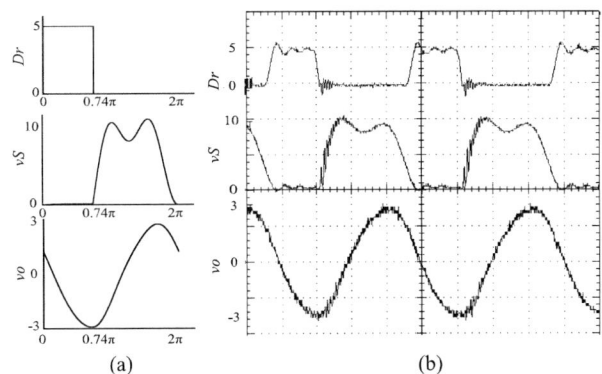

Fig. 7. Nominal waveforms of class-E/F$_2$ amplifier for achieving the maximum output power capability. (a) Numerical waveforms. (b) Experimental waveforms.

VII. CONCLUSION

This paper has presented circuit performance comparisons among class-E/F$_2$, class-E/F$_3$, and flat-top class-E amplifiers, which have harmonic resonant circuit. By using numerical algorithm, optimal designs for achieving the maximum output power capability are carried out for these amplifiers. From output-power-capability viewpoint, it can be stated that class-E/F$_2$ amplifier achieves the highest performance among the amplifiers. The reliability and validity of numerical calculations were confirmed by showing the quantitative agreements between the experimental results and the numerical predictions.

REFERENCES

[1] H. Sarnago. O. Lucia, and J. M. Burdiol, "A comparative evaluation of SiC power devices for high performance domestic induction heating," *IEEE Trans. Ind. Electron.*, vol. 62, no.8, pp. 4795–4804, Aug. 2015.

[2] F. S. Pai, C. L. Ou, and S. J. Huangl, "Plasma-driven system circuit design with asymmetrical pulse-width modulation scheme," *Prentice Hall, New Jersey*, 2003.

[3] L. E Frenzel, Jr,, "RF Power for Lndustrial Applications," *IEEE Trans. Ind. Electron.*, vol. 62, no.8, pp. 4795–4804, Aug. 2015.

[4] N. O. Sokal and A. D. Sokal, "Class E - A new class of high-efficiency tuned single-ended switching power amplifiers," *IEEE Jounal of Solid State Circuits*, vol. 10, no.3, pp. 168–176, Jun. 1975.

[5] F. H. Raab, "Idealized operation of the Class E tuned power amplifier," *IEEE Trans. Circuit Syst.*, vol.CAS-24, no.12, pp. 725-735, Dec. 1977.

[6] N. O. Sokal, "Class-E RF power amplifiers, " *QEX Commun. Quart.*, no. 204, pp. 9-20, Jan./Feb. 2001.

[7] F. H. Raab, "Class-E, Class-C, and Class-F power amplifiers dased upon a finite number of harmonics," *IEEE Trans. Microw. Theory Tech.*, vol. 49, no.8, pp. 1462–1468, Aug. 2001.

[8] Z. Kaczmarczyk, "High-Efficiency class E, EF$_2$, and E/F$_3$ inverter," *IEEE Tran Ind. Electron.*, vol. 53, no. 5, pp. 1584-1593, Oct. 2006.

[9] S. Kee, I. Aoki, A. Hajimiri, and D. Rutledge, "The Class E/F family of ZVS switching amplifiers," *IEEE Trans. Microw. Theory Tech.*, vol. 51, no. 6, pp. 1677-1690, Jun. 2003.

[10] A. Mediano, N. O. Sokal, "A Class-E RF Power Amplifier With a Flat-top Transistor-Voltage Waveform," *IEEE Tran Power. Electron.*, vol. 28, no. 11, pp. 5215-5221, Nov. 2013.

[11] H. Sekiya, I. Sasase, and S. Mori, "Computation of design values for Class E amplifiers without using waveform equations," *IEEE Trans. Circuit Syst.*, vol. 49, no. 7, pp. 966-978, July.2002

[12] M. K. Kazimierczuk, "RF Power Amplifiers," WILEY, 2008.

A Class Φ2 Resonant Buck Converter with Ripple Injection Burst Control Method

Min Lin[*] and Masahiko Hirokawa
Advanced Products Development Center, Technology & Intellectual Property HQ,
TDK Corporation, Ichikawa-Shi, Chiba, Japan
*E-mail: m.rin@jp.tdk-lambda.com

Abstract—This paper presents the design and implementation of a DC 120-170V input and DC 48V output class Φ2 resonant buck converter with ripple injection burst control method. For the converter operated at switching frequency of 13.56 MHz, low loss magnetic material is used as resonant inductors instead of air core inductors. By eliminating radiated leakage magnetic flux, small size and high power density can be achieved. Furthermore, this paper proposes a novel ripple injection burst control method to control the class Φ2 resonant buck converter. This method can avoid the effect of output capacitance. As a result of experiment, efficiency of 88.4% is shown at DC 140V input and 144W output, and very good output voltage regulation and dynamic response can be achieved.

Keywords— Class Φ2 inverter, High frequency and very high frequency power conversion, Ripple injection burst control, low loss magnetic material.

I. INTRODUCTION

Further small size and high power density are continuing demand for switching power supplies. For realizing this goal, increasing switching frequency is an effective way to reduce energy storage of passive components. In the late 1980's, the operating frequency of the switching power supply has increased from 20kHz to 500kHz due to the switching devices were replaced from bipolar transistors to power MOSFETs, and the power supply became smaller and higher power density. In the same way, since the wide bandgap devices such as GaN that can operate with much higher frequency and lower switching loss, the power supply operating at high frequency (HF, 3-30MHz) or very high frequency (VHF, 30-300MHz) can be realized. Recently, the research on the HF and VHF power supplies are progressing rapidly [1]-[6].

However, due to restrictions of components for high frequency, most of the previous work on the HF and VHF power supply focuses on the low input voltage and the small output power, and air-core inductor is used[1]-[6]. Since the air-core inductor radiates magnetic flux, it is difficult to achieve higher power density [7]. Also, since the effect of output capacitance and control loop delay, the conventional burst control method with hysteresis comparator causes oscillation of output voltage and current when the load is capacitive [8]. Moreover, for

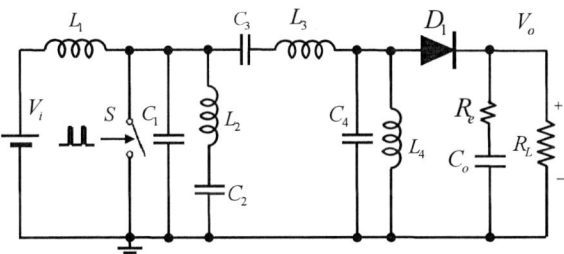

Fig. 1. A class Φ2 resonant buck converter.

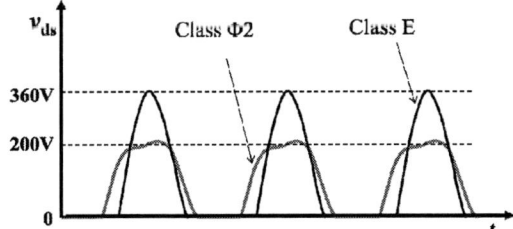

Fig. 2. Comparison with class E and class Φ2.

high voltage input, electrical insulation between the input and output is required.

To solve the above problem, a class Φ2 resonant buck converter that operated at switching frequency of 13.56 MHz with ripple injection burst control method is proposed. The detailed design and implementation for this paper are organized as follows. Section II briefly describes the low stress class Φ2 DC-DC converter. Section III explains the gate drive circuit design. The design of ripple injection burst control method is outlined in section IV. The experimental results of the prototype are shown in section V. Finally, section VI concludes the paper.

II. LOW STRESS DC-DC CONVERTER

The class E inverter is well known which can operate at high frequency (HF or VHF) with zero-voltage-switching (ZVS) and zero-derivative-switching (ZDS) conditions. However, the peak voltage between drain and source of switch v_{ds} is about 3.6 times the input voltage when the switch is off. Therefore, the efficiency and

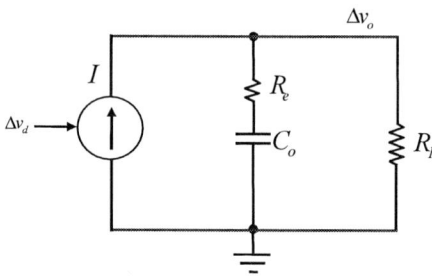

Fig. 3. Small signal model of the converter.

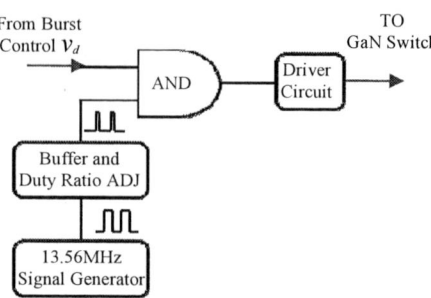

Fig. 4. block diagram of the gate driver.

reliability will decrease since the on resistance of switch which proportional to the withstand voltage will increase [1],[9].

To solve the problem, the low switch-voltage stress dc-dc converter as shown in Fig. 1 is proposed by MIT [1],[2],[10],[11]. The converter consists of a class Φ2 inverter and a resonant rectifier circuit. By adding the resonant network consists of L_2 and C_2 which resonate at the second harmonic of switching frequency, the drain-source voltage v_{ds} of the switch S composed of the fundamental and the third harmonic contents is approximately trapezoidal. The peak drain-source voltage v_{ds} is about 2 times the input voltage when the switch is off. Therefore, the lower withstand voltage switch can be used, and the reliability and efficiency will be improved. The second harmonic of switching frequency is expressed by equation (1).

$$f = \frac{1}{2\pi\sqrt{L_2 C_2}} \tag{1}$$

Fig.2 shows comparison of the drain source voltage v_{ds} of the class Φ2 and the class E. As can be seen, the voltage of the class Φ2 is lower than that of the class E.

Also, to analyze the dynamic characteristics, the class Φ2 resonant buck converter can be replaced with the controlled current source in the small signal model as shown in Fig.3 [12],[13],[14]. The transfer function G_{vd}(s) from Δv_d to Δv_o is expressed by equation (2).

$$G_{vd}(s) = \frac{\Delta v_o}{\Delta v_d} = I \cdot \frac{R_L(1 + sR_e C_o)}{1 + sC_o(R_e + R_L)} \tag{2}$$

Where, the R_e is equivalent series resistance of the output capacitor C_o. Equation (2) shows that this converter is a first order system with a single pole determined by C_o, R_e and R_L at low frequency and a single zero determined by C_o and R_e at high frequency.

III. GATE DRIVE IMPLEMENTATION

Because the driver loss is neglected due to small input capacitance of the GaN switch, a hard switching gate driver is applied in this design. Fig.4 shows the simplified block diagram of the gate driver for class Φ2 resonant buck converter. This gate driver circuit consists of a 13.56MHz clock signal generator circuit, a buffer and duty ratio adjustment circuit, an AND logic circuit and a 13.56MHz driver circuit. The buffer and duty ratio ADJ circuit gets a 13.56MHz high-precision clock signal generated by the signal generator and inputs a proper

duty ratio signal to the AND logic circuit. The fixed duty ratio was determined by the parameters of the converter to get ZVS and ZDS conditions. On the other hand, the burst control circuit also inputs the burst control signal v_d to the AND logic circuit. The AND logic circuit outputs 13.56MHz high frequency signal to the driver circuit while the burst control signal v_d is high level. The driver circuit outputs the 13.56MHz high frequency drive signal to the gate of the GaN switch and the converter operates. Therefore, the output voltage of converter is regulated by the burst control signal v_d.

IV. RIPPLE INJECTION BURST CONTROL METHOD

To operate the converter in high frequency and the high efficiency with ZVS and ZDS conditions, the fixed switching frequency and the fixed duty ratio are required. Therefore, the conventional PWM control method cannot be used. The burst control (ON/OFF control) method with hysteresis comparator is a good control method and is well known [2],[3]. However, the conventional burst control has some problems. The first problem is the burst frequency greatly depends on the output capacitance C_o and the control loop delay. The second problem is the oscillation of output voltage and current when the load is capacitive [8].

To solve the above problems, a novel ripple injection burst control method is proposed [8],[15],[16]. Fig. 5 shows the control schematic consists of a comparator, a sub-tractor, a proportional element by R_1 and R_2, a integrating element by R_3 and C_1, a integrating circuit for suppressing steady-state errors by OPA, C_a and R_b, and a reference voltage V_{ref} and a voltage dividing circuit by R_c, R_d, R_a, R_m and D_2. The zener diode D_2 is used as voltage source to increase the loop gain.

Fig.6 shows the voltage waveforms of proposed control method. When the feedback voltage v_f increases and reaches the input voltage v_{fb} which proportional to the output voltage V_o. the output of the comparator v_d changes to high level and the feedback voltage v_f changes at the same time. On the other hand, the feedback voltage v_f decreases and reaches the input voltage v_{fb}, the comparator output changes to low level and the feedback voltage v_f changes at the same time. Since the DC component of output voltage is fed back and the output ripple voltage is not required, the burst frequency does not depend on the output capacitor and the control loop delay, and determined by constants of the control circuit.

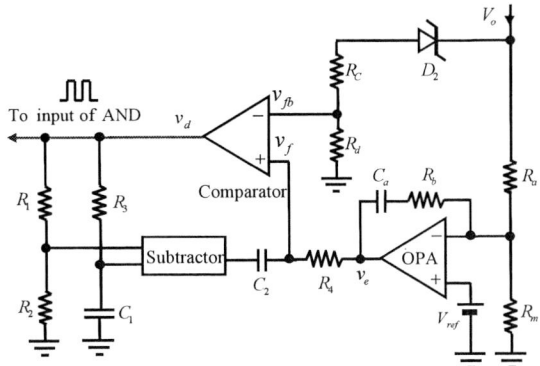

Fig. 5. Proposed burst control circuit.

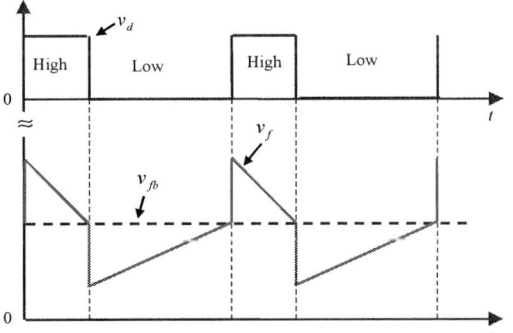

Fig .6. Voltage waveform of proposed burst control circuit.

Also, since it is possible to add output capacitors to reduce the output impedance, the oscillation problem of output voltage and current of converter can be solved [8].

To get the transfer function of proposed control method in Fig.5, an AC equivalent circuit from Δv_d to Δv_f is shown in Fig.7. With Fig.7, the transfer function from Δv_d to Δv_1 is expressed by equation (3).

$$\Delta v_1 = \frac{R_2}{R_1 + R_2}\Delta v_d \qquad (3)$$

and the transfer function from Δv_d to Δv_2 is expressed by equation (4).

$$\Delta v_2 = \frac{1}{(1 + sR_3C_1)}\Delta v_d \qquad (4)$$

As shown in Fig. 7. the Δv_3 is expressed by equation (5).

$$\Delta v_3 = \Delta v_1 - \Delta v_2 \qquad (5)$$

Substituting equation (3) and equation (4) into equation (5), the transfer function from Δv_d to Δv_3 is expressed by equation (6).

$$\Delta v_3 = \frac{sR_2R_3C_1 - R_1}{(R_1 + R_2)(1 + sR_3C_1)}\Delta v_d \qquad (6)$$

On the other hand, the transfer function from Δv_3 to Δv_f is expressed by equation (7).

$$\Delta v_f = \frac{sR_4C_2}{1 + sR_4C_2}\Delta v_3 \qquad (7)$$

Substituting equation (6) into equation (7) to remove Δv_3, the transfer function G_{fd} (s) from Δv_d to Δv_f is expressed by equation (8).

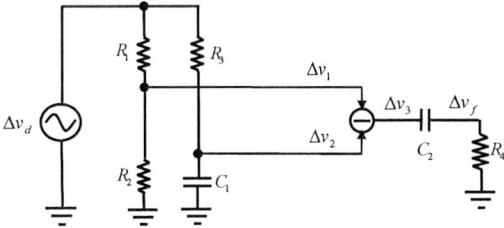

Fig. 7. AC equivalent circuit from Δv_d to Δv_f.

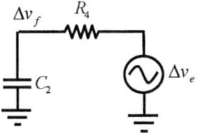

Fig. 8. AC equivalent circuit from Δv_e to Δv_f.

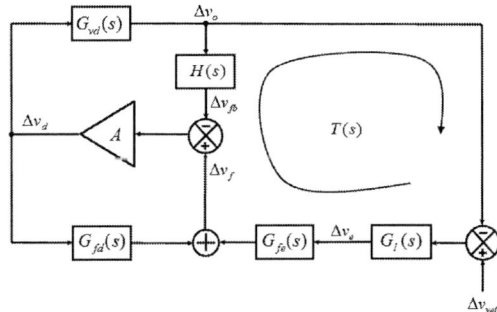

Fig. 9. The control block diagram of proposed burst control circuit.

$$G_{fd}(s) = \frac{\Delta v_f}{\Delta v_d} = \frac{(sR_2R_3C_1 - R_1)sR_4C_2}{(R_1 + R_2)(1 + sR_3C_1)(1 + sR_4C_2)} \qquad (8)$$

The AC equivalent circuit from Δv_e to Δv_f is shown in Fig.8. The transfer function $G_{fe}(s)$ from Δv_e to Δv_f is expressed by equation (9).

$$G_{fe}(s) = \frac{\Delta v_f}{\Delta v_e} = \frac{1}{1 + sR_4C_2} \qquad (9)$$

Also, The transfer function of integration circuit G_I (s) from Δv_o to Δv_e is expressed by equation (10).

$$G_I(s) = \frac{\Delta v_e}{\Delta v_o} = -\frac{R_b}{R_a}\left(1 + \frac{1}{sC_aR_b}\right) \qquad (10)$$

and the transfer function H(s) from Δv_o to Δv_{fb} is expressed by equation (11).

$$H(s) = \frac{\Delta v_{fb}}{\Delta v_o} = \frac{R_d}{R_c + R_d} \qquad (11)$$

Fig.9 shows the control block diagram of the proposed burst control method with the class Φ2 resonant buck converter. The open loop transfer function T(s) is expressed by equation (12), assuming that the gain A→∞ of the comparator[17].

$$T(s) = \frac{G_{fe}(s) \cdot G_{vd}(s) \cdot G_I(s)}{G_{vd}(s) \cdot H(s) - G_{fd}(s)}$$

$$= \frac{I\, R_bR_L(1 + sC_aR_b)(1 + sC_oR_{esr})}{sC_aR_aR_b(1 + sR_4C_2)(1 + sC_o(R_{esr} + R_L))(\alpha - \beta)} \qquad (12)$$

4135

<table>
<tr><td colspan="2" align="center">TABLE I
DESIGN SPECIFICATION</td></tr>
<tr><td>Switching Frequency</td><td>13.56MHz</td></tr>
<tr><td>Input Voltage Range</td><td>DC 120V-170V</td></tr>
<tr><td>Rated Input Voltage</td><td>DC 140V</td></tr>
<tr><td>Rated Output Voltage</td><td>DC 48V</td></tr>
<tr><td>Max Output Current</td><td>3A</td></tr>
<tr><td>Max Output Power</td><td>144W</td></tr>
</table>

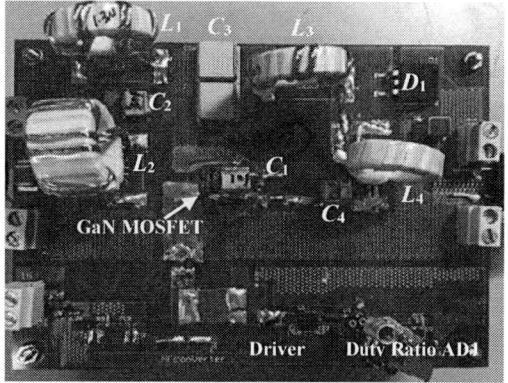

Fig .10. Photograph of the prototype.

Here, the α is

$$a = \left(\frac{sC_2 R_4 (sC_1 R_2 R_3 - R_1)}{(R_1 + R_2)(1 + sC_1 R_3)(1 + sC_2 R_4)} \right) \quad (13)$$

and the β is

$$\beta = \left(\frac{I \cdot R_d R_L (1 + sC_o R_{esr})}{(R_c + R_d)(1 + sC_o (R_{esr} + R_L))} \right) \quad (14)$$

Form equation (12), because there is a pole at the origin, the steady state error of the output voltage can be canceled [18].

V. EXPERIMENTAL VERIFICATION

To verify the designed converter, a non-isolated prototype for DC 120-170V input, max 144W output class Φ2 resonant buck converter with ripple injection burst control method is created. Table I shows the design specification for this prototype and Fig.10 shows the photograph of the converter. The 650V GaN power transistor by GaN System Inc. as the GaN switch and the low loss magnetic 67 material toroid core (5967000601) by Fair-Rite products corp. as the resonant inductors instead of air core inductors were used in this design. Since radiated leakage magnetic flux is eliminated, small size and high power density can be achieved [19]. Also, the RF High Q multilayer ceramic chip capacitors from ATC (American Technical Ceramics Corp) are selected as the resonant capacitors of the class Φ2 resonant buck converter to get small resonant loss. The other capacitors are multilayer ceramic chip capacitors from TDK Corp. Practically, tuning to finding final values for power stage involves iterative simulations. The designed values of class Φ2 resonant buck converter are listed in table II.

Fig.11 shows the experimental voltage waveforms of converter with closed-loop control. It is observed that the

TABLE II
COMPONENTS LIST

Part Name		Part No or Designed Value
Class Φ2 DC-DC Converter	S	GS66504B
	D_1	SCS208A
	C_1	157pF
	C_2	30pF
	C_3	10.2nF
	C_4	420pF
	C_o	560µF
	L_1	1.3µH
	L_2	1.1µH
	L_3	478nH
	L_4	193nH
	R_L	16Ω→∞
Gate Driver	13.56MHz signal generator	DSC1033-CE2-013.5600T
	Buffer	TC7SZ00F
	Duty Ratio ADJ	SN74ACT74N
	AND	SN74AHCT1G08
	Driver	ISL55110
Burst Control Circuit	Comparator	TLV3501
	sub-tractor	OPA353
	OPA	LMV321
	R_1	15KΩ
	R_2	510Ω
	R_3	50KΩ
	R_4	1KΩ
	R_a	45.5KΩ
	R_b	6.2KΩ
	R_c	1.81KΩ
	R_d	500Ω
	R_m	2KΩ
	C_1	47nF
	C_2	47nF
	C_a	10nF
	V_{ref}	ADR360
	D_2	EDZV36B

converter is performing a stable burst mode. The burst frequency is about 25KHz, it should be less than 1/100 of switching frequency of the class Φ2 resonant buck converter to avoid the beat frequency that can cause irregular switching action[20]. Fig.12 shows expanded waveforms of drain-source voltage v_{ds} and gate-source voltage v_{gs} of GaN switch. As can be seen, drain-source voltage v_{ds} has no ringing and is stable, and the peak voltage is about 2 times the input voltage when the switch is off. Fig.13 shows the load characteristics of the converter, No steady-state error on the output is obtained. The load regulation rate is about 0.1%. Fig.14 shows the efficiency of converter without driver and control circuit loss. Since the burst frequency varies with the load, a flat characteristic is observed. The efficiency is almost above 85% in all range of input voltage and output current, and 88.4% at DC 140V input and 144W output.

Fig. 15 shows the transient response of the output voltage when the load changes from 0 A to 1.6 A with a current slew rate of 50A/µs. A stable output voltage response is observed. Fig. 16 shows the transfer function $G_{vd}(s)$ from Δv_d to Δv_o. From this figure, the first order system was observed. Using measured values, current values of controlled current source I can be calculated by

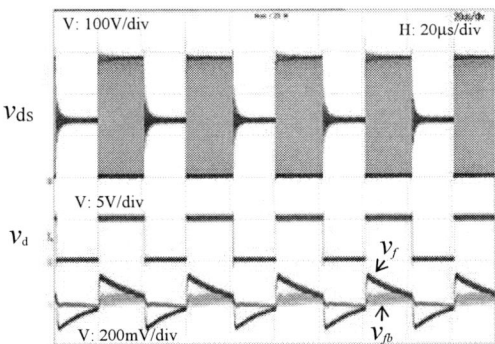

Fig .11. Voltage waveforms.
(V_i=140V, V_o=48V, R_L=30.12Ω)

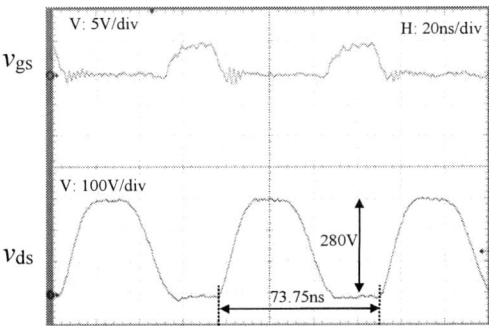

Fig .12. Voltage waveforms of the GaN switch.
(V_i=140V, V_o=48V, R_L=30.12Ω)

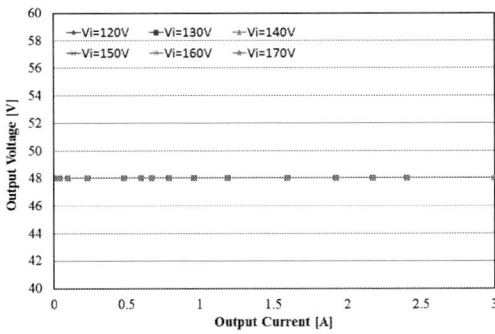

Fig .13. Load characteristics with closed-loop control.

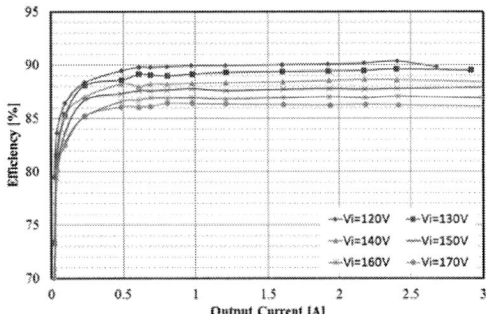

Fig .14. Converter efficiency with closed-loop control.

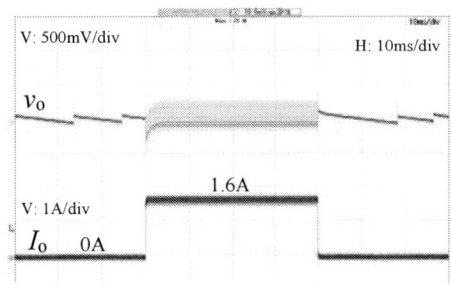

Fig .15. Transient response of the output voltage.

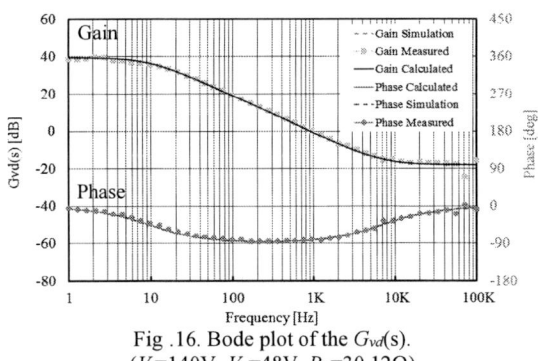

Fig .16. Bode plot of the $G_{vd}(s)$.
(V_i=140V, V_o=48V, R_L=30.12Ω)

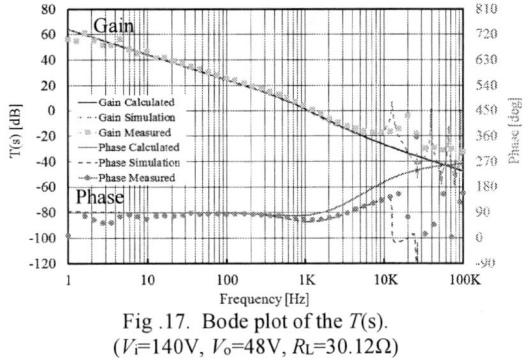

Fig .17. Bode plot of the $T(s)$.
(V_i=140V, V_o=48V, R_L=30.12Ω)

s → 0 and v_d→ 1 of the equation (2). Fig. 17 shows the bode plot of the T(s). From this figure, phase magin of more than 60 degrees and gain of about 60dB at 1Hz are observed and very good output voltage regulation is measured. Calculated results of both figures are in good agreement with experimental and simulation results and the theoretical analysis are verified. Current value of controlled current source I is 3.15A in this experiment.

VI. CONCLUSION

A 13.56MHz DC 120-170V input to DC 48V output class Φ2 resonant buck converter with ripple injection burst control method has been proposed. The steady state error and frequency response characteristics were analytically and experimentally investigated. As a result of experiment, efficiency of 88.4% is shown at DC 140V input and 144W output and very good output load characteristics and transient response were achieved. More than 60 degrees of the phase margin and more than

60dB gain at 1Hz can be observed. Calculated results are in good agreement with experimental and simulation results. In addition, since low loss magnetic material is used as the resonant inductors instead of air core inductors, small size and high power density are achieved by eliminating radiated leakage flux. Also, by using the low loss magnetics as a transformer, the isolated class Φ2 converter is realized. The proposed ripple injection control method is also applicable to other types of the converters.

REFERENCES

[1] Juan M. Rivas, Yehui Han, Olivia Leitermann, Anthony Sagneri, David J. Perreault "A High Frequency Resonant Inverter Topology with Low Voltage Stress" IEEE PESC 2007 Record, pp. 2705-2717, October. 2007.

[2] Robert C. N. Pilawa-Podgurski, Anthony D. Sagneri, Juan M. Rivas, David I. Anderson, David J. Perreault "Very High Frequency Resonant Boost Converters" IEEE Transactions on Power Electronics, Vol. 24, No. 6, 2009, pp. 1654-1665.

[3] Wei Liang, John Glaser, Juan Rivas " 13.56 MHz High Density DC–DC Converter With PCB Inductors" IEEE Transactions on Power Electronics, Vol. 30, No. 8, 2015, pp. 4291-4301.

[4] Mickey P. Madsen, Arnold Knott, Michael A. E. Andersen "Very High Frequency Resonant DC/DC Converters for LED Lighting" IEEE APEC 2013 Record, pp. 835-839, March. 2013.

[5] Jungwon Choi, Wei Liang, Luke Raymond, Juan Rivas, "A High-Frequency Resonant Converter Based on the Class Φ2 Inverter for Wireless Power Transfer" 2014 IEEE 79th Vehicular Technology Conference (VTC Spring), pp. 1-5, May. 2014.

[6] Wei Cai, Zhiliang Zhang, Xiaoyong Ren, Yan-Fei Liu "A 30-MHz isolated push-pull VHF resonant converter" IEEE APEC 2014 Record, pp. 1456-1460, April. 2014.

[7] Yehui Han, Grace Cheung, An Li, Charles R. Sullivan, David J. Perreault " Evaluation of Magnetic Materials for Very High Frequency Power Applications" IEEE Transactions on Power Electronics, Vol. 27, No. 1, 2012, pp. 425-435.

[8] Min Lin, Ken Matsuura, Masahiko Hirokawa, Kazushi Watanabe "A New Control Circuit for Class Φ2 Converter" Proceedings of the 2016 IEICE Society Conference, pp. S-42-S-43, September. 2016.

[9] N. Sokal and A. Sokal, "Class E-a new class of high-efficiency tuned single-ended switching power amplifiers," IEEE J. Solid-State Circuits, vol. SSC-10, no. 3, pp. 168–176, Jun. 1975.

[10] David J. Perreault, Juan M. Rivas, Anthony D. Sagneri, Olivia Leitermann, Yehui Han, Robert C.N. Pilawa-Podgurski "Methods and Apparatus for a Resonant Converter" United States Patent, Patent No. US 7,889,519 B2, Feb.15, 2011.

[11] John S. Glaser, Jeffrey Nasadoski, Richard Heinrich "A 900W, 300V to 50V Dc-dc Power Converter with a 30MHz Switching Frequency" IEEE APEC 2009 Record, pp. 1121-1128, Feb. 2009.

[12] Jingying Hu, Anthony D. Sagneri, Juan M. Rivas, Yehui Han, Seth M. Davis, David J. Perreault " High frequency resonant SEPIC converter with wide input and output voltage ranges" IEEE PESC 2008 Record, pp. 1397-1406, Aug. 2008.

[13] Milovan Kovacevic, Arnold Knott, Michael A. E. Andersen "A VHF interleaved self-oscillating resonant SEPIC converter with phase-shift burst-mode control" IEEE APEC 2014 Record, pp. 1402-1408, April. 2014.

[14] Bo Song, Xu Yang, Yingjie He " Class Φ2 DC-DC converter with PWM on-off control " IEEE ECCE Asia 2011 Record, pp. 2792-2796, July. 2011.

[15] Min Lin, Ken Matsuura "Control Circuit and Switching Power Supply Unit" United States Patent, Patent No. US 9,564,803 B1, Feb.7, 2017.

[16] Min Lin, Terukazu Sato, Kimihiro Nishijima, Takashi Nabeshima "A robust hysteretic PWM control method for switching converters" IEEE INTELEC 2009 Record, pp. 1-6, Dec. 2009.

[17] Min Lin, Toshiyuki Zaitsu, Terukazu Sato, Takashi Nabeshima "Frequency domain analysis of fixed on-time with bottom detection control for buck converter " IEEE IECON 2010 Record, pp. 481-485, Nov. 2010.

[18] A. Soto, P. Alou, J. Aoliver, J. A. Cobos and J. Uceda, "Optimum Control Design of PWM-BUCK Topologies to Minimize Output Impedance"APEC 2002 Record, pp.426-432, March. 2002.

[19] Alex J. Hanson, Julia A. Belk, Seungbum Lim, Charles R. Sullivan, David J. Perreault, "Measurements and Performance Factor Comparisons of Magnetic Materials at High Frequency" IEEE Transactions on Power Electronics, Vol. 31, No. 11, 2016, pp. 7909-7925.

[20] Kisun Lee, Han Zou "Analysis of the beat frequency oscillation in voltage regulators" ECCE 2009 Record, pp. 3026-3030, Sept. 2009.

Practical Design Technique for High Power Density LLC Resonant Converter

Shingo Nagaoka, Hiroyuki Onishi, Koji Takatori, Toshiyuki Zaitsu, Takeshi Uematsu
OMRON Corporation, Kyoto, Japan
*E-mail: shingo_nagaoka@omron.co.jp

Abstract— The LLC resonant converter is widely used for consumer and industrial applications due to its high efficiency, small size and low noise. In a practical design, a trade-off is needed between the input voltage operation range and the transformer size for miniaturization. The transformer size is determined at the minimum input voltage with full load. It is very important to have an accurate design technique so that the transformer does not have any unnecessary size margin. In this paper, inductance (L)-matrix transformer model-based design is introduced and compared with the conventional transformer model such as T-type and L-type. A 200 kHz, 240W LLC resonant converter is implemented and verified that the inductance (L)-matrix method is accurate, and it can contribute to miniaturization.

Keywords— LLC resonant converter, high power density, transformer

I. INTRODUCTION

The LLC resonant converter is widely used for consumer and industrial applications due to its high efficiency, small size and low noise [1]-[6].

The LLC resonant converter has features such as ZVS/ZCS capability of the primary and secondary switch turning on/off. A resonant inductor can be integrated into a transformer. Further, a resonant tank can eliminate the output smoothing inductor. Hence, the LLC resonant converter has an advantage in terms of the higher frequency operation and smaller size.

Although increasing the switching frequency can contribute the size reduction of passive components dramatically, it is not so easy to make the converter compact in a practical design.

There are many trade-offs and challenges such as;
(1)As shown in Fig. 1, the hold-up capacitor Cbus usually dominates a large area in the power system of the off-line switcher (AC-DC). The hold-up time is expressed by equation (1).

$$P_o T_{hold} = \frac{1}{2} C_{bus} (V_{in\_op}^2 - V_{in\_min}^2) \quad (1)$$

Therefore, wide input operation range of DC-DC can reduce the hold-up capacitor and improve the total system power density. The lower the minimum input voltage capability, the smaller the converter system size.
(2) The lower the minimum input voltage, the tougher the design of transformer becomes, which leads to larger transformer and converter size. Therefore, there is a need of trade-off between the input voltage operation range and the size of transformer as shown in Fig. 2.
(3) A smaller size of transformer for the LLC resonant converter does not have enough space between the primary and secondary windings in the core which leads to the small leakage inductance Lr. Therefore, it can't get enough voltage gain for the desired output voltage.

Because of this, it is very important to have an accurate design technique at the minimum input voltage so that the transformer does not have any unnecessary size margin. For designing the LLC resonant converter, generally FHA (Fundamental Harmonics Analysis) technique has been used [7]-[14]. In FHA, the transformer models used are usually T-type or L-type equivalent circuits. This is a simple and well-known technique, so it is a good start for the design. However, the T- or L-type model is not so accurate, especially when the output voltage is lower. For example 12V or 24V (this is common in industry applications), the number of secondary turn is only one or two turns. In this case, the T- or L-type model, where n (=Np/Ns) is given by the number of turns of the primary and secondary windings, has a large error. This makes the transformer size larger than the real transformer capability, which hinders the miniaturization of the transformer. In this paper, we propose the inductance (L)-matrix transformer model. The inductance (L)-matrix use the turns ratio n_mtx which is coming from electrical value of the primary self-inductance Lp and the secondary self-inductance Ls.

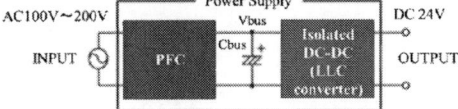

Fig.1. AC-DC Power system block diagram

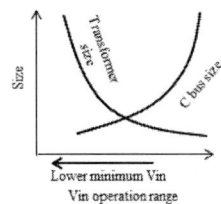

Fig.2. Trade-off between input operation range and converter size.

In this paper, in Chapter II, the conventional LLC design method and its problems are explained. In Chapter III, the inductance (L)-matrix method is introduced and verified by experiments. The conclusion is presented in Chapter IV.

II. Conventional Design Method of LLC Resonant Converter and its Difficulty in Practical Design

[A] Conventional transformer model and FHA design method

Fig. 3 shows the conventional equivalent circuit of L-type and T-type models for the transformer. Lp is the primary self-inductance, Np and Ns are the number of turns at primary and secondary, respectively. n is the turns -ratio, and kp is the coupling coefficient given by equation (2).

$$n = \frac{N_p}{N_s}, \ k_p = \sqrt{1 - \frac{L_{short}}{L_p}} \tag{2}$$

Here, Lshort is the inductance when the output terminal is shorted. In Chapter II, a general design method of FHA will be explained by using the L-type transformer model as an example. Fig. 4 shows the block diagram and the key waveforms of the LLC resonant converter. Fig. 5 shows the AC equivalent circuit for FHA. The equation is expressed as shown in equation (3).

$$M_L = \frac{1}{2n} \cdot \frac{\frac{k_p^2}{1-k_p^2} \cdot FR^2}{\sqrt{\left(Q_L \cdot FR\right)^2 \left(FR^2 - 1\right)^2 + \left(\frac{1}{1-k_p^2} \cdot FR^2 - 1\right)^2}} \tag{3}$$

Here, FR is a normalized frequency, QL is a quality factor expressed in Equation (4).

$$Q_L = \frac{\pi^2}{8} \frac{\omega_r L_p}{n^2 R_o} \cdot k_p^2, \ FR = \frac{f_{sw}}{f_{sr}} \tag{4}$$

Fig. 6 shows the voltage gain curve with the parameter of the quality factor QL using Eq. (2). It is found that the smaller the QL, the steeper the shape of voltage gain curve when FR is lower than 1. If you want a steeper voltage gain curve, a smaller kp is needed, which means you need to make a larger leakage inductance Lr.

[B] Difficulty in a Practical Design
As mentioned in Chapter I, a lower minimum input voltage capability is important to reduce the size of the hold-up capacitor Cbus. The converter should have a very steep voltage gain curve to get the desired output voltage at the minimum input voltage. Therefore, a large leakage inductance Lr is required. However, when the

transformer is miniaturized or in low profile, it becomes difficult to have enough space between the primary and

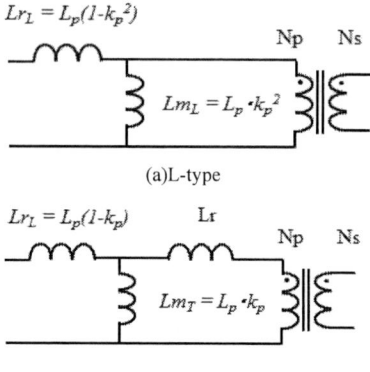

Fig.3. Transformer equivalent circuit

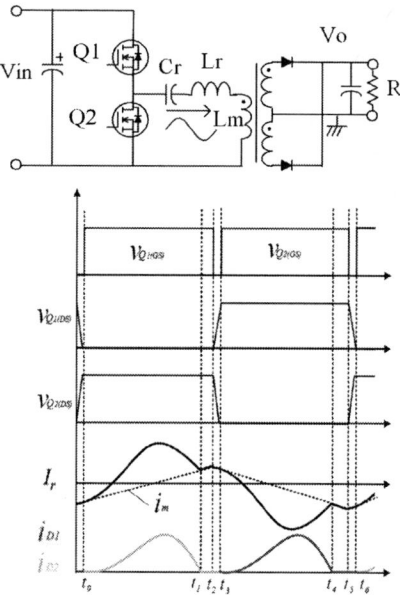

Fig.4. Block diagram and key waveforms of LLC resonant converter

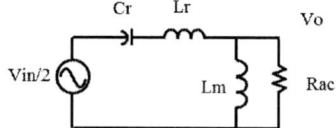

Fig.5. AC equivalent circuit

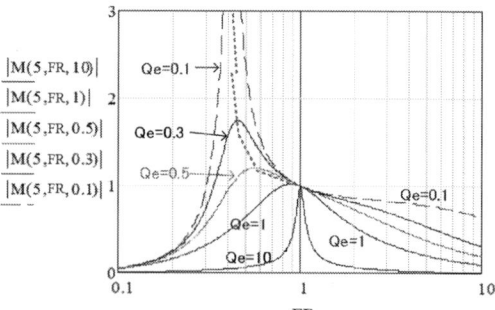

Fig.6. Voltage gain curve with QL parameter

secondary windings for making a large leakage inductance Lr. So, it can't reach to the desired output voltage as shown in Fig.7. Hence, the transformer size likely to become larger when designing the converter in the calculation phase. This is the reason that inhibits miniaturization of the LLC resonant converter.

III. Proposal of Inductance (L)-Matrix Model with Comparison of Conventional L-Type and T-Type Models

An accurate design should be needed at the minimum input voltage and the full load to get the desired output voltage with the consideration of the transformer core saturation.

Comparison among L-type, T-type and inductance (L)-matrix models of the transformer is considered.

A voltage gain equation MT for the T-type model is given through the same approach with the L-type model using AC equivalent circuits. The voltage gain MT and the quality factor QT are derived as shown in equation (5) and (6), respectively.

$$M_T = \frac{1}{2n} \cdot \frac{\frac{k_p}{1-k_p^2} \cdot FR^2}{\sqrt{(Q_T \cdot FR)^2 (FR^2-1)^2 + \left(\frac{1}{1-k_p^2} \cdot FR^2 -1\right)^2}} \quad (5)$$

$$Q_T = \frac{\pi^2}{8} \frac{\omega_r L_p}{n^2 R_o} \quad (6)$$

A common point of L-type and T-type is that their turns ratio n (=Np/Ns) comes from the mechanical structure of the transformer. In this case, n should have a big error. This is because the AL value differs depending on the winding structure (e.g., inner diameter) even in the same core. As the AL value varies depending on the winding structure, the inductance differs for the same Np, (or Ns), which means that n is different as a result. Table I shows the results of the AL value obtained by simulations when the winding structure is changed with the same core.

When Model 1 and Model 2 are compared, it can be seen that the AL value is different when the inner radius r is different, despite the same number of turns. Comparing Model 1 and Model 3, it can be seen that even if the inner diameter is the same, if the number of turns is changed and the coil length l is different, the AL value is different. This result is compliant with Nagaoka coefficient [15] which corrects the leakage of a part of the magnetic flux of the adjacent coil. The Nagaoka coefficient is expressed by Equation (7) to (11).

$$L = K \frac{\mu_0 a^2 N^2}{l} \quad (7)$$

$$K = \frac{4k}{3\pi\sqrt{1-k^2}} \left\{ \frac{1-k^2}{k^3} K_{(k)} - \frac{1-2k^2}{k^3} E_{(k)} -1 \right\} \quad (8)$$

$$K_{(k)} = \int_0^{\frac{\pi}{2}} \frac{1}{\sqrt{1-k^2 \sin^2 \theta}} d\theta = \frac{\pi}{2} \left(1 - \sum_{n=1}^{\infty} \left(\frac{(2n-1)!!}{(2n)!!} \right)^2 k^{2n} \right) \quad (9)$$

$$E_{(k)} = \int_0^{\frac{\pi}{2}} \sqrt{1-k^2 \sin^2 \theta} d\theta = \frac{\pi}{2} \left(1 - \sum_{n=1}^{\infty} \left(\frac{(2n-1)!!}{(2n)!!} \right)^2 \frac{k^{2n}}{2n-1} \right) \quad (10)$$

$$k = \frac{1}{\sqrt{1 + \left(\frac{l}{2a} \right)^2}} \quad (11)$$

Where, a is the radius of the coil, and l is the length of the coil.

According to Eq. (7), it is understood that AL value depends on the inner diameter r and the length l.

Thus, n=Np/Ns which comes from the mechanical structure of the transformer is not accurate.

In order to overcome this drawback, we propose the inductance (L)-matrix model where the turns ratio n_mtx is coming from the electrical value of the primary self-inductance Lp and the secondary self-inductance Ls. Fig. (9) shows equivalent circuits of the inductance(L)-matrix model. From Fig. 8, equation (12) and (13) are obtained.

$$\begin{pmatrix} vp \\ voac \end{pmatrix} = i \cdot \omega \cdot \begin{pmatrix} Lp & M \\ M & Ls \end{pmatrix} \cdot \begin{pmatrix} ip \\ -io \end{pmatrix} \quad (12)$$

$$ip = i \cdot \omega \cdot Cr \cdot (vinac - vp) \quad (13)$$

Substituting Eq. (12) into Eq. (13) yields equation (14).

$$\begin{pmatrix} vp \\ voac \end{pmatrix} = \begin{bmatrix} Cr \cdot Lp \cdot \omega^2 \cdot (vp - vinac) - \frac{i \cdot \omega \cdot M \cdot voac}{Roac} \\ Cr \cdot M \cdot \omega^2 \cdot (vp - vinac) - \frac{i \cdot \omega \cdot Ls \cdot voac}{Roac} \end{bmatrix} \quad (14)$$

Equation (15) is yielded by deforming Eq.(14), eliminating vp and rearranging.

$$\frac{voac}{vinac} = \frac{(Cr \cdot M \cdot \omega^2) \cdot (Cr \cdot Lp \cdot \omega^2) - (Cr \cdot Lp \cdot \omega^2 -1) \cdot (Cr \cdot M \cdot \omega^2)}{(Cr \cdot Lp \cdot \omega^2 -1) \left(1 + \frac{Ls \cdot \omega \cdot i}{Roac} \right) - (Cr \cdot M \cdot \omega^2) \cdot \left(\frac{M \cdot \omega \cdot i}{Roac} \right)} \quad (15)$$

Therefore, the voltage gain Mmtx is expressed by equation (16), (17)

$$M_{mtx} = \frac{1}{2n_{mtx}} \cdot \frac{\frac{k}{1-k^2} \cdot FR^2}{\sqrt{(Q_{mtx} \cdot FR)^2 (FR^2-1)^2 + \left(\frac{1}{1-k^2} \cdot FR^2 -1\right)^2}} \quad (16)$$

$$Q_{mtx} = \frac{\pi^2}{8} \frac{\omega_r L_p}{n_{mtx}^2 R_o}, \quad n_{mtx} = \sqrt{\frac{L_p}{L_s}} \quad (17)$$

4141

IV. Experimental Result

To verify the accuracy of the transformer model and the voltage gain curve, the converter board was implemented and compared with the calculation results. The circuit parameters and specifications are shown in Table II. The design specifications are Vin=250V-385V, Vo=24V and Ro=2.4 ohms.

Fig. 9 shows the voltage gain curve at the maximum input voltage of 385V. It is found that the L-type and T-type model curves are not accurate. Meanwhile, the inductance (L)-matrix model calculation curve matches well with the measured experimental data.

Fig. 10 shows the voltage gain curve of the inductance (L)-matrix model with calculation and the experimental results at the maximum and minimum input voltage, and Fig. 11 and 12 shows the waveforms at the maximum and minimum input voltage. It is found that the calculation results are matching well with the experimental results. Therefore, we can estimate the accurate minimum operation frequency which leads to accurate voltage-time products. As a result, we can avoid unnecessarily large size of the transformer.

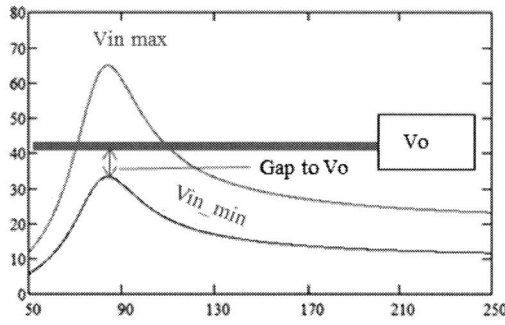

Fig.7. Voltage gain curve (e.g., Vin_min curve does not reach to Vo)

Table I AL value simulation results

Model 1	Model 2	Model 3
AL value = 0.389	AL value = 0.353	AL value = 0.298

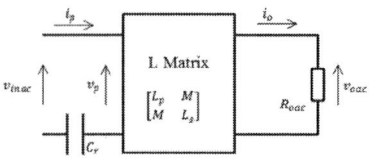

Fig.8. Equivalent circuit of inductance (L)-matrix model

Table II Experimental condition

Vin_maximum	385 V
Vin_minimum	250 V
Load Ro	2.4 Ω
Lp	100μH
Ls	1.35μH
Cr	37.6 nF

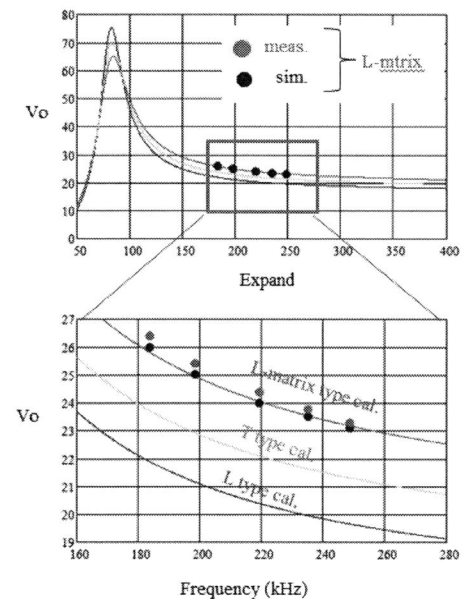

Fig.9. Voltage gain curve with L, T, and inductance (L)-matrix type calculation and measured results

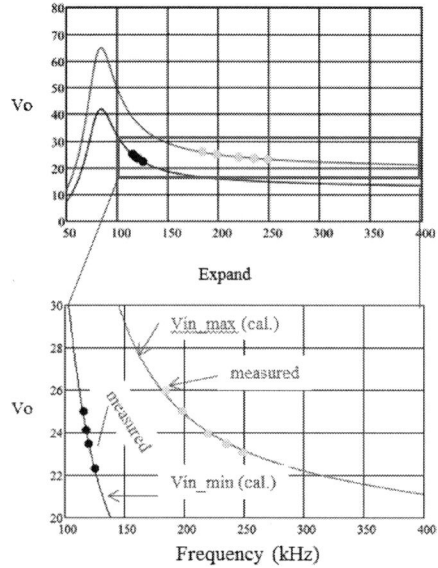

Fig.10. Voltage gain characteristic curve by inductance (L) matrix equation with experimental data

The 2018 International Power Electronics Conference

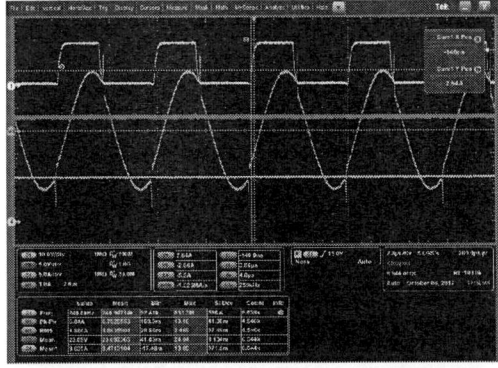

Fig.11. (a) V in = 385V, V out = 23V, frequency 248.8 kHz

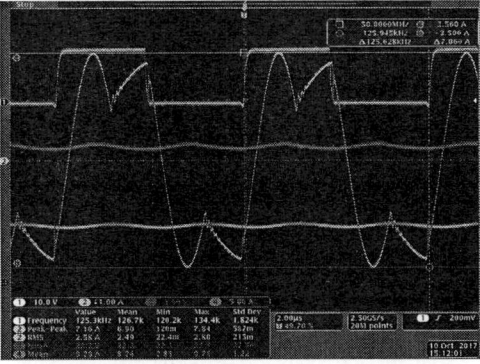

Fig.12. (a) Vin = 250V, V out = 23V, frequency 125.6 kHz

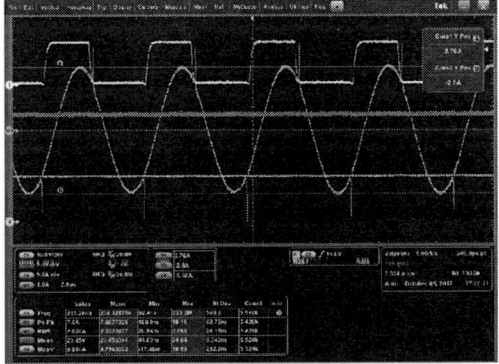

Fig.11. (b) V in = 385V, V out = 23.5V, frequency 235.2 kHz

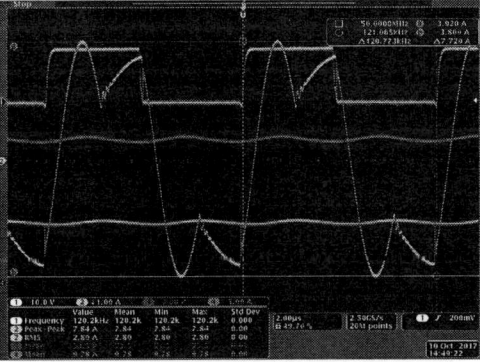

Fig.12. (b) V in = 250V, V out = 23.5V, frequency 120.8 kHz

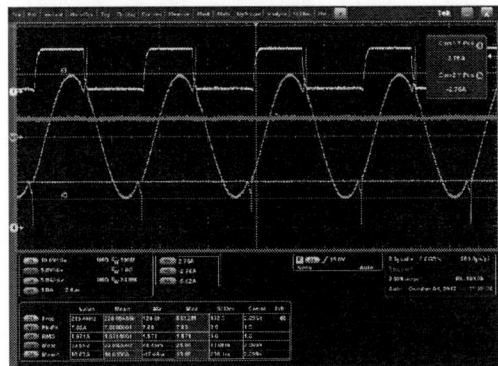

Fig.11. (c) V in = 385V, V out = 24V, frequency 219.4 kHz

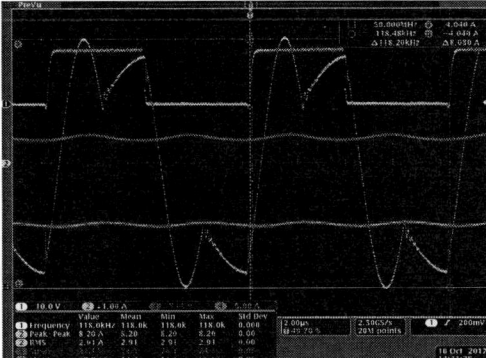

Fig.12. (c) V in = 250V, V out = 24V, frequency 118.2 kHz

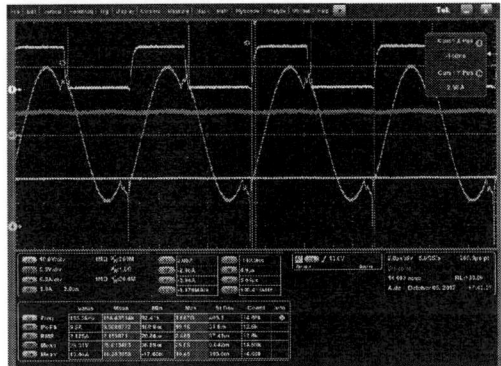

Fig.11. (d) Vin = 385V, V out = 25V, frequency 198.3 kHz

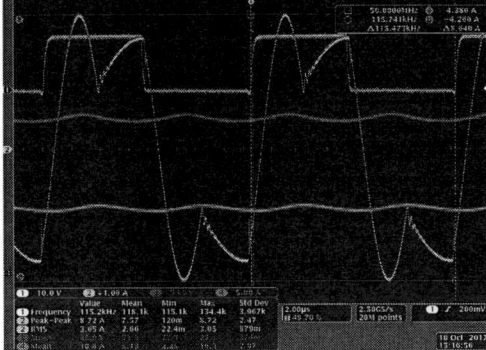

Fig.12. (c) V in = 250V, V out = 25V, frequency 115.5 kHz

4143

V. Conclusion

This paper presents the technique of miniaturizing the LLC resonant converter in practical designs. There is a trade-off between the input voltage range and the transformer size.

An accurate transformer model is important especially at the minimum input voltage with the full load to avoid the unnecessary large size of the transformer. The design method using the inductance (L)-matrix transformer model is introduced and compared with the conventional transformer models such as T-type and L-type. A 200 kHz, 240W LLC resonant converter is implemented to verify that the inductance (L)-matrix method is accurate and can contribute to miniaturization of the LLC resonant converter. It is verified by the experimental results.

References

[1] B. Yang, F.C. Lee, A.J. Zhang, and G. Huang, "LLC Resonant Converter for Front End DC/DC Conversion," in IEEE Proc. APEC02. pp. 1108-1112, Boston, 2002.

[2] J. F. Lazar and R. Martinelli, "Steady-state analysis of the LLC series resonant converter," in IEEE Proc. APEC01, pp. 728-735 vol.2. Anaheim, 2001.

[3] Yanjun Zhang, Dehong Xu, Min Chen, Yu Han, Zhong Du, "LLC Resonant Converter for 48V-0.9V VRM," PESC'04, pp. 1848-1854, 2004.

[4] Y. Zhang, D. Xu, K. Mino, K. Sasagawa, "1MHz-1kW LLC Resonant Converter with Integrated Magnetics," PESC'07, pp. 955-961, 2007.

[5] K. Morita, "Novel Ultra Low-noise soft-switch-mode Power Supply," INTELEC'98, pp. 115-122, 1998.

[6] F. Musavi, M. Craciun, M. Edington, W. Eberle, W. G. Dunford,"Practical Design Considerations for a LLC Multi-Resonant DC-DC Converter in Battery Charging Applications," APEC'12, pp. 2596-2602, 2012

[7] H. Huang,"FHA-Based Voltage Gain Function with Harmonic Compensation for LLC Resonant Converter," APEC'10, pp. 1770-1777, 2010

[8] J. F. Lazar and R. Martinelli, "Steady-State Analysis of the LLC Series Resonant Converter," APEC '01, pp. 728-735, 2001.

[9] Ashoka K. S. Bhat,"A Generalized Steady-State Analysis of Resonant Converters Using Two-Port Model and Fourier-Series Approach," IEEE Trans. on P. E., vol. 13, No. 1, pp.142-151, 1998

[10] R. L. Steigerwald, "A Comparison of Half-Bridge Resonant Converter Topology," IEEE Trans. On PE, Vol. 3, No. 2, pp. 174-182, 1988.

[11] V. Vorperian, "High-Q Approximation in The Small-Sgnal Analysis of Resonant Converters," PESC'85, pp. 707-715, 1985

[12] V. Vorperian, S. Cuk, " A Complete DC Analysis of The Series Resonant Converter," PESC'82, pp. 85-100, 1982

[13] B. Lu, W. Liu, Y. Liang, F. C. Lee, J. D. van Wyk,"Optimal Design Methodology for LLC Resonant Converter," APEC'06, pp. 533-538, 2006

[14] T. Liu, Z. Zhou, A. Xiong, J. Zeng, J, Ying,"A Novel Precise Design Method for LLC Series Resonant Converter," INTELEC'06, pp. 1-6, 2006

[15] The journal of the College of Science, Imperial University of Tokyo, Japan VOL27, ARTICLE 6.

The 2018 International Power Electronics Conference

Operational Study and Protection of a Series Resonant Converter with DC Current Input Applied in DC Current Distribution Systems

Hongjie Wang, Tarak Saha, Baljit Riar and Regan Zane
Department of Electrical and Computer Engineering
Utah State University, Logan, USA
E-mail: hongjie1127.wang@gmail.com

Abstract— Constant dc current distribution has been preferred for long distance high reliability systems such as subsea ocean observation networks. The characteristics of dc current distribution and series resonant converters (SRCs) with constant input current bring new challenges and difficulties related to protection as well as operation of individual SRC modules and the entire system. This work proposes new protection techniques for SRCs operating with a constant input current and control strategies for SRC modules. Simulation and hardware results are provided for a system consisting of two 100 km submarine cables carrying a DC current of 1 A, and two series connected 250 kHz, 1 kW SRCs that regulates the output current to 0.33 A.

Keywords — dc current distribution, operational study, protection, series resonant converter.

I. INTRODUCTION

Long distance power transmission applications such as ocean observatory systems, subsea gas and oil fields and offshore windfarms prefer dc current distribution over dc voltage distribution because of its reduced sensitivity to voltage drop and losses along the distribution cable [1-5].

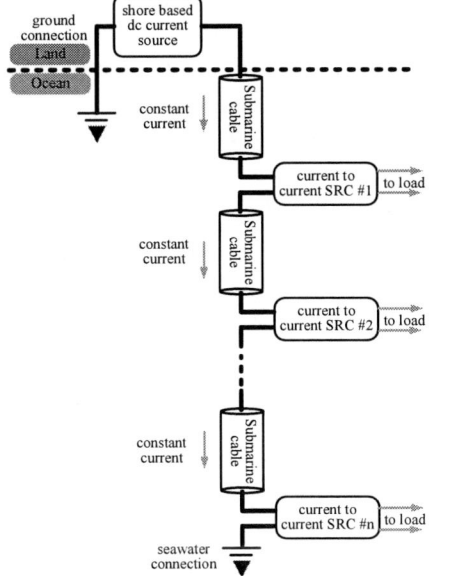

Fig. 1. Architecture of a dc current distribution system for subsea applications.

For a dc current distribution system, there are multiple options to build the distribution network. For example, a mesh-like network [2] or a single cable system [5]. The distribution network depends on both the specific field scenario and the customer requirements.

In this paper, a dc current distribution system with a single trunk cable and multiple dc-dc converter modules is analyzed as illustrated in Fig. 1. Isolated series resonant converters (SRCs) are employed as the dc current input to dc current output converter modules to draw power from the trunk cable and deliver power to loads [6-8]. The submarine/trunk cable shown in Fig. 1 generally has a length of tens of kilometers or even longer, in which case the distributed capacitance of the cable can no longer be ignored. From Fig. 1, it can be seen that all SRC modules are connected in series and fed by the trunk cable carrying a regulated dc current. The above characteristics of the dc current distribution system bring new challenges related to both protection and operation of both SRC modules and the entire system.

This paper investigates protection and operation of an SRC, which is employed to naturally regulate its output current, when subjected to startup and shutdown transients as well as open and short circuit fault conditions. A new protection circuit for the SRC is proposed in this paper to limit the resonant capacitor voltage during fault transients. Based on the characteristics of the SRC with constant current input and a dc current distribution system, the paper presents an operational study and analysis including long cables. The proposed protection technique and control strategies are suitable for practical implementation to improve reliability and life of dc current distribution systems.

In this paper, modeling of the long power distribution cable is presented in Section II. The operational study of the entire current distribution system, as well as start-up and shut-down strategy for the SRC modules are discussed in Section III. A new protection technique for the SRC is proposed in Section IV. The proposed analysis, operation strategy and protection technique are validated through simulation and experimental results provided in Section V, and the conclusions are given in Section VI.

II. LONG TRANSMISSION CABLE MODELING

4145

TABLE I   PARAMETERS OF THE CABLE

Length [km]	100
Unit resistance [Ω/km]	1
Unit inductance [mH/km]	0.128
Unit capacitance [μF/km]	0.2
Number of π-sections	1, 5, 10, 20, 40

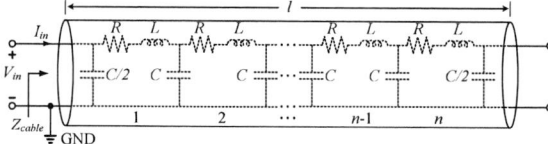

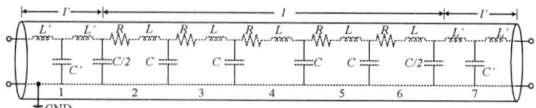

Fig. 4. Schematic diagram of the 5 π-section combined with 2 T-section model.

system analysis and protection scheme. Hence, in this work, the cascaded π-section approach is adapted to model and explore the dynamic performance of the cable [9, 10].

Figure 2 shows the schematic diagram of a single-phase transmission cable of length l. As shown in Fig. 2, the cascaded π-sections are adapted to model the cable for exploring the system performance [11]. In Fig. 2, l is the length of the cable, n is the number of π-sections used to model the cable, and R, L and C are the resistance, inductance and capacitance of each π-section, respectively.

The impedance characteristics of a cable modeled with different number of π-sections are illustrated in Fig. 3. The cable parameters used here are listed in TABLE. I. The input impedance of the cable with the other terminal open is shown in Fig. 3(a), while Fig. 3(b) shows the input impedance of the cable with the other terminal shorted to ground. From Fig. 3, it can be seen that the model with 5 π-sections has a good balance between accuracy and complexity.

From Fig. 2, it can be seen that the π-section model has a lumped capacitor at the cable terminals. However, the real transmission cable is generally inductive at its terminals. To improve that, a model with the combination of π-section and T-section are used in this work, as shown in Fig. 4. In Fig. 4, the 5 π-sections at the center model the major length of the cable while the two T-sections at two ends model a short length to improve the model accuracy. When a certain length cable is modeled by the 5 π-sections plus 2 T-sections model, the R, L, C, L' and C' of each cell can be calculated based on the parameters of the cable, which is used to build the per-unit scale cable emulator that is used in laboratory tests. The cable emulator has the same impedance as the real cable in the case of the same per-unit length parameters and the same cable length.

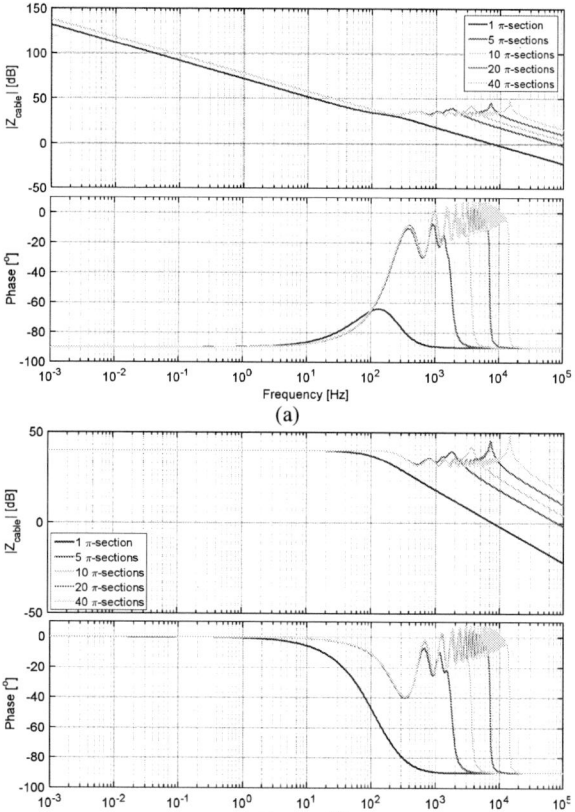

Fig. 2. Schematic diagram of a single-phase transmission cable of length l.

Fig. 3. Input impedance of the transmission cable. (a) The cable is open. (b) The cable is terminated in a short circuit.

In a dc voltage distribution system, the transmission line is usually equivalent to an inductor or an inductor in series with a resistor. However, because of the relatively long distance of the trunk cable in the dc current distribution system, and according to the cable parameters listed in TABLE I, the distributed capacitance of the submarine cable can no longer be ignored while designing and analyzing the operation of the system. In other words, the lumped inductor approach or the lumped inductor in series with the lumped resistor approach are not accurate enough to model the submarine cable because of the considerable distributed capacitance along the cable. All these inaccuracies can compromise the

III. OPERATIONAL STUDY

In a dc current distribution system for subsea applications, as illustrated in Fig. 1, the shore based power supply is controlled as a current source converter with a variable output voltage. The overall link voltage is controlled by the shore based power supply to maintain a continuous link current at a desired value.

If the circuit of SRC with constant current input shown in [7] is directly employed as the SRC module in Fig. 1, the entire system cannot operate because the MOSFETs in the primary bridge are enhancement type MOSFETs which are normally open. Since the SRC modules are connected in series, as shown in Fig. 1, open state of the MOSFETs means that the main trunk cable is open. One

The 2018 International Power Electronics Conference

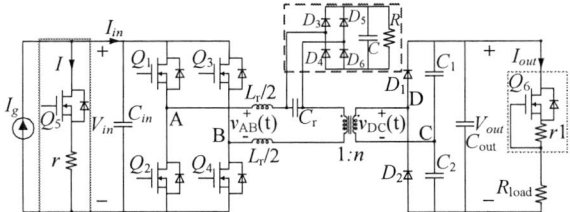

Fig. 5. Proposed circuit for the SRC with constant input current employed in a dc current distribution system.

solution would be to provide an auxiliary supply separate from the trunk cable to supply power to the MOSFETs. However, in such a long distance dc current distribution system for undersea applications, it is impractical with the consideration of cost, voltage drop and system reliability. Hence, the auxiliary power for the SRC module is provided by the trunk cable, which requires a continuous current flow through the trunk cable in order to deliver auxiliary power to the SRCs. In this case, a closed circuit path for the trunk cable current is required at startup.

The proposed circuit diagram in this work is illustrated in Fig. 5. In Fig. 5, the MOSFET Q_5 is a depletion type MOSFET and r is the current sensing resistor in that branch. This bypass branch in the SRC modules, which is Q_5 in series with r in the solid green box in Fig. 5, and the submarine cable in the system provide a continuous path for the main trunk current even before any SRC module in the system is energized. The depletion type MOSFET Q_5 needs to be selected to be capable of handling the power dissipation during start-up and shut-down of SRC modules.

Based on the above analysis, a three-step startup and shutdown technique for system operation is proposed in this work.

For the start-up, the first step is to turn on the shore power supply and provide the desired distribution current to the rest of the system, which is used to power all the auxiliary power supplies of each SRC. When the auxiliary power supply is on, a certain amount of time delay is required before taking the next action in order to ensure that all the auxiliary power supplies in the system are turned on.

The second step is to pass the trunk current from the bypass branch to the SRC input. In this step, the SRC operates at 180° phase shift in open loop mode, which provides minimum output current to the load. The bypass branch current controller ramps down the current flow through the bypass branch from full trunk current to 0. The ramp time and the load of the SRC at 180° phase shift determines how much energy is dissipated in the bypass branch during the start-up. The SRC modules in the system can do this at the same time or in a sequence. The second step is completed once the full trunk current flows through the SRC instead of the bypass branch.

The third step is to enable the SRC to start regulating its output current and to close the feedback control loop. The controller design of the SRC used for a long distance dc current distribution system is presented in [8].

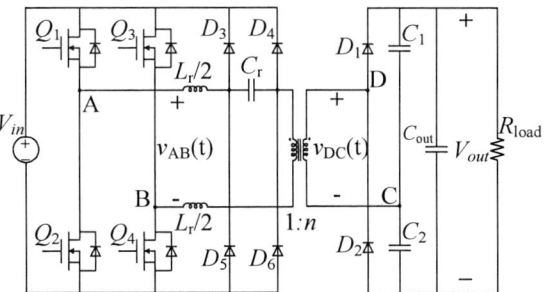

Fig. 6. SRC with constant voltage input and resonant capacitor voltage clamping circuit.

For the shut-down, the scenario is similar. The first step is to open the SRC feedback regulation, and then take the full trunk current from SRC input to the bypass branch with a ramp by changing the current reference of the bypass branch current controller. The last step is to turn off the auxiliary power supplies and then the shore power supply.

IV. PROTECTION OF A SERIES RESONANT CONVERTER

An SRC with a resonant capacitor clamping circuit to protect the SRC during fault transients was presented in [12], and illustrated in Fig. 6. In Fig. 6, the diodes D_3, D_4, D_5 and D_6 are used to clamp the voltage of the resonant capacitor C_r to the input voltage V_{in}. However, by analyzing the voltage of resonant capacitor terminals to ground, the capacitor terminal voltages are expressed as $V_{in} \pm 0.5v_{Cr}$ when Q_1 and Q_3 are on, which is a general case for phase shift modulation control [13, 14]. In this case, for phase shift modulation controlled SRC, circulating currents between the resonant tank and the input filter are unavoidable. On the other hand, for the SRC with constant current input, the voltage across the resonant capacitor is higher than the input voltage for certain load range as presented in [7], which means that the protection approach shown in Fig. 6 cannot be used since it alters the steady state operation.

The proposed protection circuit is illustrated in the dashed blue box in Fig. 5. As shown in Fig. 5, the resonant capacitor voltage is clamped to the voltage of a floating capacitor C, which holds the peak voltage across the resonant capacitor regardless of the relation between the resonant capacitor voltage and the input voltage. In Fig. 5, R is a bleed resistor in parallel with the floating capacitor C. During steady state, since the floating capacitor holds the resonant capacitor peak voltage, no current flows through the clamping diodes (D_3-D_6) expect for a small current to feed the bleed resistor R.

For the SRC without a protection circuit, the energy stored in the input capacitor C_{in} is transferred to the resonant tank during an output short circuit fault transient. With the protection circuit shown in Fig. 5, the energy stored in the input capacitor C_{in} is transferred to the resonant tank and the floating capacitor C. In this case, during fault transients, the voltage across the resonant capacitor can be well limited to protect the converter.

4147

The energy stored on a capacitor is calculated from

$$E = \frac{1}{2}CV^2. \tag{1}$$

Hence the energy stored in those three capacitors before the fault happens can be expressed as

$$E_{C_{in}} = \frac{1}{2}C_{in}V_{in}^2, \ E_{C_r} = \frac{1}{2}C_rV_{C_r}^2, \ E_C = \frac{1}{2}CV_C^2, \tag{2}$$

where V_{in} is the dc input voltage, V_{Cr} is the peak voltage of the resonant capacitor C_r, and V_C is the peak voltage of the floating capacitor C, which equals to V_{Cr}.

From (2), the total energy stored on resonant capacitor C_r and the floating capacitor C is

$$E_{C_r} + E_C = \frac{1}{2}(C_r + C)V_{C_r}^2. \tag{3}$$

During an output short circuit fault transient, the energy stored in the input capacitor C_{in} is transferred to the resonant tank and the floating capacitor C. Hence, the total energy E_{total} stored on C_r and C becomes

$$E_{total} = E_{C_r} + E_C + E_{C_{in}} = \frac{1}{2}(C_r + C)V_{C_r}^2 + \frac{1}{2}C_{in}V_{in}^2. \tag{4}$$

With the energy transferred from the input capacitor C_{in}, the voltage across the resonant capacitor and floating capacitor increases by

$$\Delta V = \sqrt{V_{C_r}^2 + \frac{C_{in}V_{in}^2}{C_r + C}} - V_{C_r}. \tag{5}$$

From (5), the floating capacitor required for limiting the voltage across the resonant capacitor to a certain voltage increment ΔV can be derived as

$$C = \frac{C_{in}V_{in}^2}{\left(V_{C_r} + \Delta V\right)^2 - V_{C_r}^2} - C_r. \tag{6}$$

For output short-circuit fault, the resulting large surge output current may damage the current sensing circuit if resistive current sensing is employed. From reliability aspect, it is preferred to have an output current limiting circuit for protection, especially for low output current, high output voltage applications. The proposed output current limiting circuit is illustrated in the dotted red box in Fig. 5. In Fig. 5, Q_6 is a depletion type MOSFET and $r1$ is the feedback resistor. As shown in Fig. 5, the negative voltage from $r1$ is applied to the gate terminal of Q_6 to control the equivalent resistance presented by Q_6, since Q_6 operates in the linear region. The proposed current limiting circuit does not require any active drive or auxiliary circuits. The current limiting circuit does introduce additional power loss during the normal operation because of the low output current that flows the high on-resistance of the depletion type MOSFET. In this work, the current limiting circuit introduces an additional 20 Ω resistance that consumes 2 W during normal operation, which is negligible compared with 1 kW output power.

For the dc current distribution system, the module that has fault needs to be bypassed in order to keep the rest of the system operating. As discussed earlier, the capacitance of the submarine cable is significantly high due to its parameters and length. Since the modules are connected in series, bypass of one module means discharging the cable capacitance in its forward current path. In this case, uncontrolled cable discharging may result in large current through other SRC modules in the system, and finally cause the entire system shutdown. To guarantee the normal operation of the complete system, a two-level fault response strategy is proposed in this work. The first level is to disable the gate signals of the primary switches, and the second level is to use the bypass branch to control the discharging of the cable to make sure that the distribution current stay within the range.

V. SIMULATION AND EXPERIMENTAL RESULTS

A dc current distribution system for subsea application was built in the lab, which consists of two 100 km cable emulators and two 500 W SRCs with 1 A input current and 0.33 A regulated output current. The parameters of the SRC and the cable are tabulated in TABLE II. The photo of the cable emulator is shown in Fig. 7, while the photo of the developed SRC is shown in Fig. 8.

LTSpice simulations were run for normal operation and short-circuit fault cases to validate the presented capacitive clamping circuit operation with the converter parameters mentioned above. The results are shown in Fig. 9. From Fig. 9, it can be seen that the resonant capacitor voltage is well clamped during the fault transient as expected.

Hardware experiments including system startup, normal operation and shutdown were conducted on the demo system to validate the proposed analysis and control strategies. The oscilloscope waveforms of the signals in the two SRCs are illustrated in Fig. 10, where CH1 is the input voltage of SRC#1, CH2 is the input voltage of SRC#2, CH3 is the output current of SRC#1

TABLE II PARAMETERS OF THE SYSTEM

L_r (μH)	174.2
C_r (nF)	2.33
C_{in}(μF)	4
C_{clamp}(μF)	1
f_s (kHz)	250
Main MOSFETS (SiC)	C2M1000170D
Gate driver	IXDN609YI
Bypass MOSFET	IXTH2N170D2
Floating capacitor C [μF]	1
Bleed resistor R [MΩ]	1

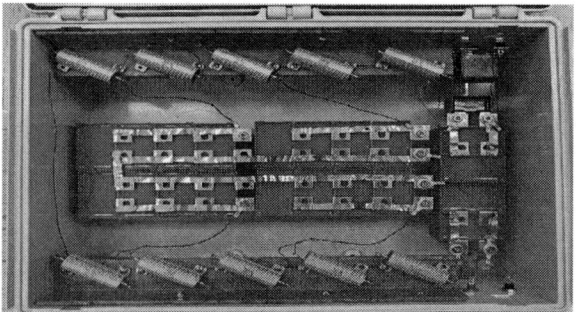

Fig. 7. Photograph of the cable emulator.

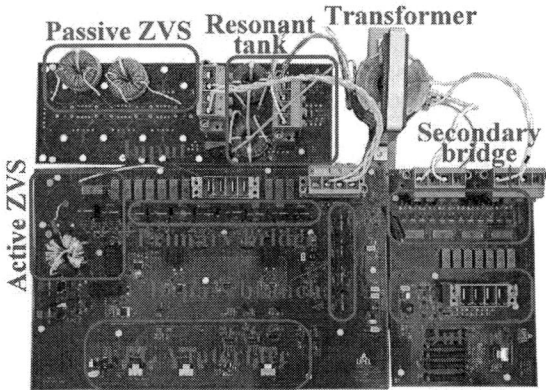

Fig. 8. Photograph of the SRC module hardware.

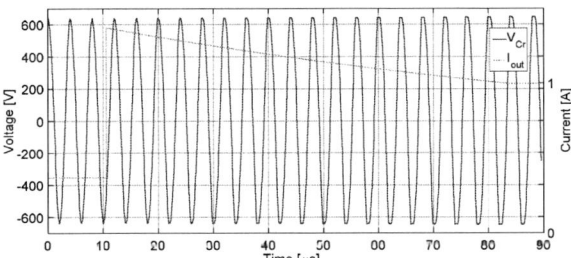

Fig. 9. Simulation results of normal operation to short-circuit fault transient.

and CH4 is the output current of SRC#2. From Fig. 10, two steps can be clearly seen for startup. One step is the bypass branch releasing the cable current to SRC with open loop mode, and the other step is the SRC operates in closed loop mode with full input current.

The normal operation to open and short circuit faults of the SRC with proposed protection circuit are demonstrated to verify the presented analysis and design. The SRC operates at 500 W output power with 1 A input current and 0.33 A regulated output current. The output current limit is 1 A, while the maximum bypassing current is set to be 1.3 A. The oscilloscope waveforms of the signals in the SRC are shown in Fig. 11 and Fig. 12. In Fig. 11, CH1 is the trunk cable current, CH2 is the SRC input voltage, CH3 is the output current and CH4 is the output voltage. In Fig. 12, CH1 is the SRC input voltage, CH2 is the trunk cable current, CH3 is the output current and CH4 is the output voltage. From Fig. 11 and Fig. 12, it can be seen that the SRC is well protected from both open and short circuit faults, the output current is limited to 1 A during the fault transient, and the input current is regulated to be 1.3 A during the fault response. The spikes in the results are due to bouncing of the relay used to perform output short/open circuit tests. With the proposed protection circuit, the SRC efficiency drops from 94.5% to 94.3% with 500 W output.

VI. CONCLUSIONS

Protection techniques for an SRC operating as constant current input to constant current output has been proposed in this paper. These techniques can easily be

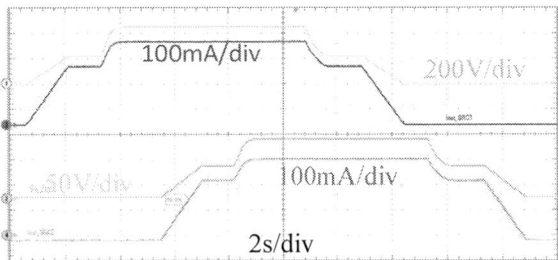

Fig. 10. Oscilloscope waveforms of the signals in the SRCs during startup, normal operation and shutdown. From top to bottom are the input voltage of SRC#1, output current of SRC#1, input voltage of SRC#2 and output current of SRC#2.

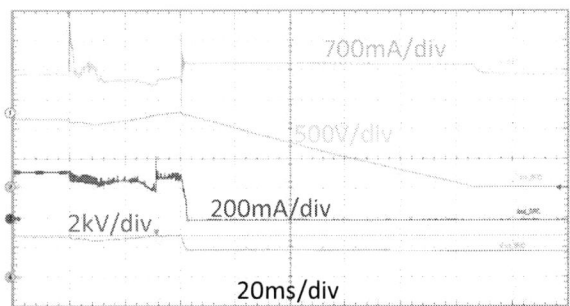

Fig. 11. Oscilloscope waveforms of the signals in the SRC during normal operation to open circuit fault. From top to bottom are the trunk cable current, input voltage, output current and output voltage.

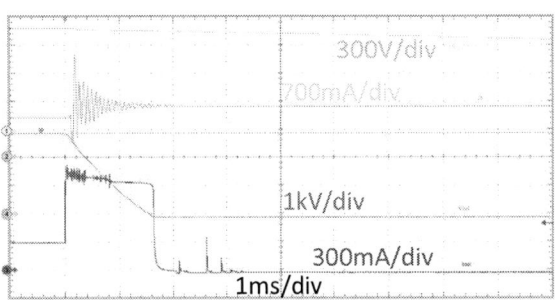

Fig. 12. Oscilloscope waveforms of the signals in the SRC during normal operation to short circuit fault. From top to bottom are the input voltage, trunk cable current, output current and output voltage.

extended to other resonant converters. The proposed protection circuit does not create additional circulating current during normal operation, which means a high efficiency is maintained. The operation of the technique has been presented in this paper allowing smooth and desired startup, normal operation and shutdown of the entire dc current distribution system. The design and analysis are validated through simulations and experimental results for a setup with two 100 km submarine cables and two 250 kHz, 1 kW SRC with 1 A input current and 0.33 A regulated output current.

REFERENCES

[1] K. Asakawa, J. Muramatsu, M. Aoyagi, K. Sasaki. "Feasibility study on real-time seafloor glove monitoring cable-network-power feeding system," *Underwater Technology, 2002. Proceedings of the 2002 International Symposium on*, pp. 116-122, 2002.

[2] K. Asakawa, J. Kojima, J. Muramatsu, T. Takada. "Novel current to current converter for mesh-like scientific underwater cable network-concept and preliminary test result," *OCEANS, 2003. Proceedings*, vol. 4, pp. 1868-1873, Sept. 2003.

[3] K. Modepalli, A. Mohammadpour, T. Li and L. Parsa, "Three-Phase Current-Fed Isolated DC–DC Converter With Zero-Current Switching," *in IEEE Transactions on Industry Applications*, vol. 53, no. 1, pp. 242-250, Jan.-Feb. 2017.

[4] D. Dong, D. Zhang, R. Lai, S. Chi and M. H. Todorovic, "Operational study of a modular direct current power system for subsea power delivery," *2014 IEEE Energy Conversion Congress and Exposition (ECCE)*, Pittsburgh, PA, 2014, pp. 4345-4352.

[5] H. Wang, T. Saha and R. Zane, "Control of series connected resonant converter modules in constant current dc distribution power systems," *IEEE Workshop on Control and Modeling for Power Electronics (COMPEL)*, Trondheim, 2016, pp. 1-7.

[6] H. Wang, T. Saha and R. Zane, "Design considerations for series resonant converters with constant current input," *IEEE Energy Conversion Congress and Exposition (ECCE)*, Milwaukee, WI, 2016, pp. 1-8.

[7] H. Wang, T. Saha and R. Zane, "Analysis and design of a series resonant converter with constant current input and regulated output current," *2017 IEEE Applied Power Electronics Conference and Exposition (APEC)*, Tampa, FL, 2017, pp. 1741-1747.

[8] H. Wang, T. Saha and R. Zane, "Impedance-based stability analysis and design considerations for DC current distribution with long transmission cable," *2017 IEEE 18th Workshop on Control and Modeling for Power Electronics (COMPEL)*, Stanford, CA, 2017, pp. 1-8.

[9] J. W. Phinney, D. J. Perreault and J. H. Lang, "Synthesis of Lumped Transmission-Line Analogs," *in IEEE Transactions on Power Electronics*, vol. 22, no. 4, pp. 1531-1542, July 2007.

[10] S. Zhang, S. Jiang, X. Lu, B. Ge and F. Z. Peng, "Resonance Issues and Damping Techniques for Grid-Connected Inverters With Long Transmission Cable," *in IEEE Transactions on Power Electronics*, vol. 29, no. 1, pp. 110-120, Jan. 2014.

[11] J. W. Phinney, D. J. Perreault and J. H. Lang, "Synthesis of Lumped Transmission-Line Analogs," *in IEEE Transactions on Power Electronics*, vol. 22, no. 4, pp. 1531-1542, July 2007.

[12] B. S. Jacobson and R. A. DiPerna, "Series resonant converter with clamped tank capacitor voltage," *Fifth Annual Proceedings on Applied Power Electronics Conference and Exposition*, Los Angeles, CA, USA, 1990, pp. 137-146.

[13] L. Corradini, D. Seltzer, D. Bloomquist, R. Zane, D. Maksimovic and B. Jacobson, "Minimum Current Operation of Bidirectional Dual-Bridge Series Resonant DC/DC Converters," *in IEEE Transactions on Power Electronics*, vol.27, no.7, pp. 3266-3276, 2012.

[14] U. Badstubner, J. Biela, J. W. Kolar. "Power density and efficiency optimization of resonant and phase-shift telecom DC-DC converters," *Appl. Power Electron. Conf. (APEC)*, Feb. 2008, pp. 311–317.

A Study on Improvement of Power Utilization Rate of Energy Systems with PVs and Batteries

Hiroaki Endo[1*], Masakatsu Kurisaka[1], Tsutomu Ueno[1], Yusuke Yoshioka[1],
Kaoru Inoue[2], and Toshiji Kato[2]

1 GS Yuasa International Ltd., Kyoto, Japan
2 Doshisha University, Kyoto, Japan
*E-mail: hiroaki.endo@jp.gs-yuasa.com

Abstract - A photovoltaic (PV) system with batteries (BTs) is expected to be an effective system for coping with the energy issues we will face in the future. This paper establishes an energy system with PVs and batteries that will improve the utilization rate of power supplied from PVs and batteries to on-premise loads. An external transducer is adopted to measure the receiving power from the utility grid. According to the measured receiving power, constant receiving power control, and power factor control functions for power conditioning systems (PCSs) are utilized to improve the power utilization ratio. The effectiveness of this system will be verified using actual equipment.

Keywords— photovoltaic system with batteries, power conditioning systems, constant receiving power control, power factor control functions

I. INTRODUCTION

After the introduction of the Feed-in Tariff Program (FIT) in 2012, photovoltaic (PV) systems have spread widely, reaching a facility capacity of approximately 30 million kW in 2015. According to the "Long-term Energy Supply and Demand Outlook" issued by the Ministry of Economy, Trade and Industry, Japan's target for a PV system introduction volume by 2030 is set at 64 million kW. By that year, Japan is aiming at decreasing greenhouse gas emissions by 26% compared to 2013.

PVs with batteries (BTs) are expected to be one of effective systems to cope with energy issues. Their power conditioning systems (PCSs) must equip a constant power factor control function for reverse power flow (from DC to AC) in order to suppress the voltage rise [1]. Because the output capacity of the PCSs is determined by the apparent power (kVA), the effective power will be reduced by the power factor control. Moreover, to satisfy the current product certification system, the power factor control must be implemented when the PCSs operate in inverse conversion mode (from DC to AC) even though no reverse power flow exists. Hence, the power factor control causes a reduction in the utilization rate of power supplied from PVs and BTs to loads [2].

This paper proposes an improvement method of power utilization rate from the PVs and BTs to loads when the PCSs operate in inverse conversion mode and no reverse power flow exists. An external transducer is adopted to measure the receiving power from the utility grid. According to the measured receiving power, a constant receiving power control and the power factor control functions are utilized. The effectiveness of this system will be verified using actual equipment.

II. SYSTEM STRUCTURE AND THE ISSUE

Fig. 1 shows a schematic diagram of objective power system. A PCS receives input power from the PVs and BTs and outputs power into the utility grid and loads. Because the PVs and BTs have individual converters, the PCS can control the charge and discharge rate of BTs while performing maximum power point tracking of PVs. This system works as the PV system in the daytime, and while the BTs charge in the nighttime to cover the shortfall of charge [3] [4].

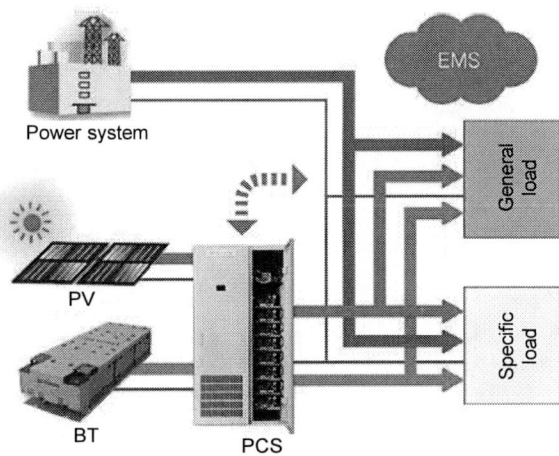

Fig. 1. System diagram

Table 1. PCS specifications

Items	Specifications
DC rated input voltage	400 V
DC input voltage range	0 to 650 V
AC rated output voltage	101 V/202 V (single-phase three-wire)
AC output capacity	20 kW
Power conversion factor	95%
Isolated operation output voltage	101 V/202 V (single-phase three-wire)
Isolated operation output capacity	20 kW
Type of the batteries	Lithium-ion
Battery capacity	50 kWh

Table 1 shows the specifications of the PCS product. A PCS can supply power from the BTs to loads when the PVs cannot generate sufficient power such as a blackout in the nighttime and a rainy day. The reverse power flow from BTs to the utility grid is prohibited. The discharge power from BTs must be adjusted according to the consumption power of the loads. The PCS prevents the reverse power flow from the BTs by measuring the power received from the utility grid with an inside current transformer (CT) that is mounted on the output part of the PCS, or with an external transducer (TD) that is mounted on the power receiving point from the grid. The reverse power flow is prevented by a constant receiving power control system (CRPC) that controls the receiving power from the grid. If a PCS is used for a limited area within the business premises for specific loads, the constant receiving power control can be implemented simply by using the PCS's internal CT. When a PCS supplies power to loads in a wide area such as on-premise loads within a plant site, an external TD is used for monitoring power flows at the power-receiving point as shown in Fig. 2. The receiving power threshold for preventing reverse power flow is set to 5% of the rating of the external TD (500 W for 10 kW transducers), in consideration of measurement errors and control/resolution abilities of the external TD and PCS [5].

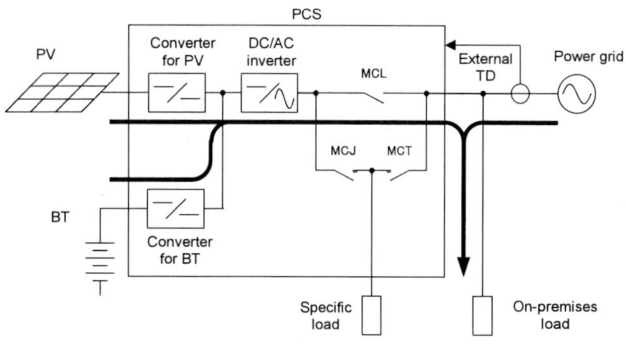

Fig. 2. System diagram using external TD

A utility-interactive inverter can control the voltage rise of the utility grid by adjusting the reactive power of the PCS. The power factors of conventional utility-interactive inverters have been set to 1, in principle. Recently, however, in order to restrict voltage rise, PCS power factors are likely to be set to values smaller than 1 [6] [7] [8]. This constant power factor control for less than unity is applied to PCSs that provide reverse power flows to the utility grid. Systems, in which generated or stored power is consumed by on-premise loads and hence no reverse power flows into the grid, are exempt, i.e. the power factor can be set to unity in this system. When a PCS is running in inverse conversion mode in a system with multiple input units such as PVs and BTs, the constant power factor control system is required regardless of the existence of reverse power flows because it is difficult in some cases to distinguish between the PV power and the power from BTs. This means that PCS power factors are kept less than 1 when the effective power is limited even when all input power is supplied to on-premise loads and there is no reverse power flow to the grid. As a result, the energy to be supplied intrinsically to loads will be reduced and the power utilization rate from the PVs and BTs to loads will decline.

III. Methodologies to Solve the Issue

The PCS shown in Fig. 2 can control its output while monitoring power flows at the power receiving point by using an external TD. Hence, the PCS power factor can be adjusted depending on the amount of receiving power. Fig. 3 shows a conceptual diagram of the power factor control with power flows at the power receiving point.

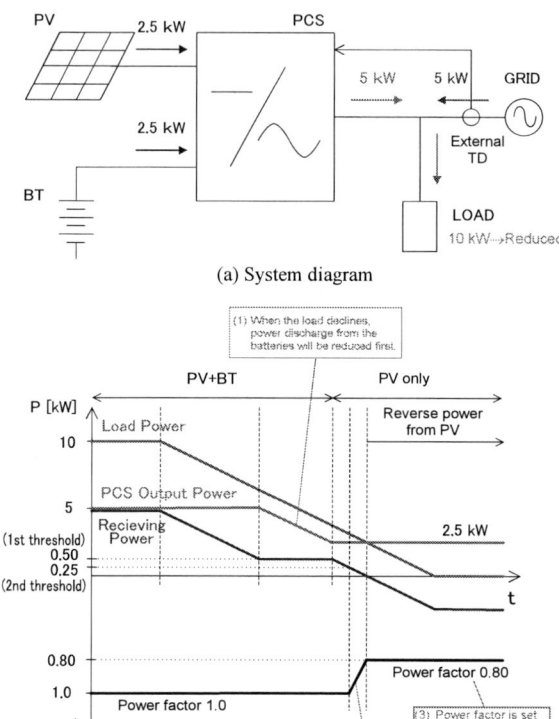

(a) System diagram

(b) Image of power factor control function

Fig. 3. Power factor control function variation with power flows at the power receiving point

When a reverse flow is not detected at the power receiving point, the PCS power factor is maintained at 1 to maximize its effective power. When the consumed power of the loads decreases, the receiving power from the grid also decreases. The discharge power of the BTs will be reduced and finally becomes 0 when the receiving power decreases to its first threshold for detecting reverse power flow (500 W) to prevent the revers power flow. Until this state, the power utilization rate from the PVs and BTs to loads is improved because the power factor is maintained at 1 when the PCS operates in inverse conversion mode and no reverse power flow to grid exists. After the receiving power decreases to the second threshold (250 W), the power factor will be reduced and finally will be controlled at 0.8 to prevent a voltage rise in the power grid that will be induced by reverse power flow. When increasing the power factor, it is necessary that there is no reverse power flow in the first place. The threshold values must be set in such a way to prevent a malfunction of the system even when there are

4152

The 2018 International Power Electronics Conference

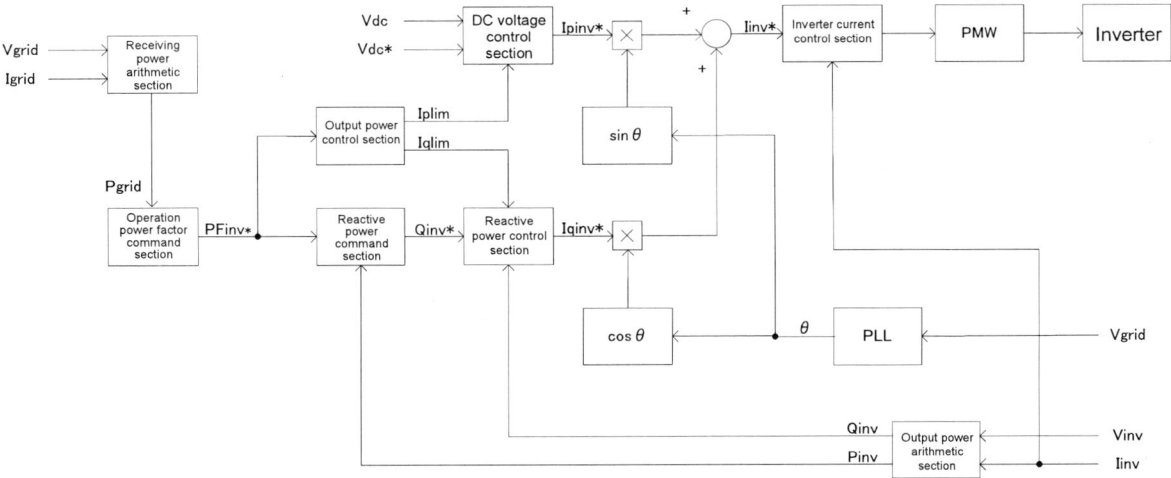

Fig. 4. Block diagram showing power factor control depending on power flows at the power receiving point

errors in the measurements of the PCS, external TD, and power factor control/resolving power. The first threshold value (500 W) for the increase in the power factor must be higher than the total error (450 W), which includes all error factors on one side, measurement errors of the PCS (1.5% at the maxlmum = 150 W), measurement errors of the external TD (2.0% at the maximum = 200 W), and errors in the power factor control/resolving power (0.01 = 100 W), which is lower than the upper limit (600 W) of the target value of the receiving power constant control. Therefore, the first threshold value is set at 500 W. The second threshold value (250 W), regarding the decrease in the power factor, is set at 50% of the first threshold value (500 W), considering the dead band for preventing control hunting.

Fig. 4 is a block diagram of the power factor control depending on power flows at the power receiving point. At the receiving power arithmetic section, the receiving power is calculated to identify the reverse and forward power flows. Effective receiving power (Pgrid) is obtained from receiving current from the grid (Igrid) and grid voltage (Vgrid). The output power arithmetic section, output power, which is necessary for control and the reference signal of the reactive power, is calculated. Inverter current (Iinv) and inverter voltage (Vinv) produce effective (Pinv) and ineffective (Qinv) output powers, respectively. At the operational power factor command section, the power factor reference signal is calculated depending on whether power flows in the reverse or forward direction. When the effective receiving power (Pgrid) drops below the second threshold value (2.5% of the rating of external TD, i.e., 250 W for a 10 kW transducer), the power factor reference signal (PFinv)* is lowered. When Pgrid rises above the first threshold value (500 W), PFinv* is increased. There is a hysteresis range between the two threshold values (250–500 W), which is a deadband exempted from the power factor control, with which unnecessary control hunting of the PCS will be avoided. By adjusting the power factor reference signal, it is possible to increase the effective power and reduce the power purchase amount from the grid when receiving power is above a certain level.

Fig. 5 is a flow chart of the power factor control using an external TD. When the power factor control is activated, the receiving power (Pgrid) data is compared with the first threshold value (500 W). When Pgrid is greater than the first threshold, the present power factor is checked if it is less than 1.00 or not. If the present power factor is less than 1.00, the command section produces a power factor command value (PFinv*) that increases the operation power factor by a specified volume (0.01). Once the power factor command value (PFinv*) is generated, the inverter current (Iinv) is controlled by the inverter current control section and the PWM control section to a current command value (Iinv*) corresponding to the power factor command value (PFinv*). As a result, the operation power factor of the PCS will be increased to the target value, which is than the present value by the specified volume. Conversely, when Pgrid is less than the first threshold value (500 W), the operation power factor command section compares Pgrid with the second threshold value (250 W).

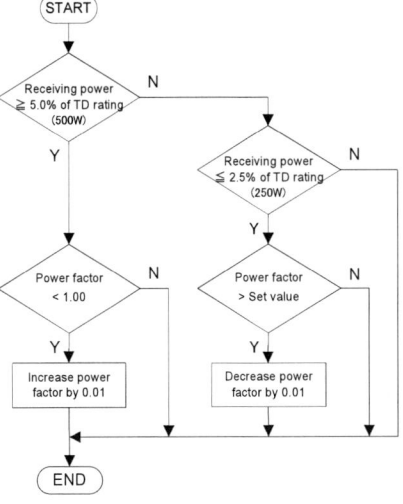

Fig. 5. Flow chart of power factor control using an external TD

4153

If it is less than the second threshold (250 W), it is checked if the present operation power factor is less than the set value. An example set value for the power factor is 0.80 (minimum value in the set range from 0.80 to 1.00). If the present operation power factor is greater than the set value, the command section produces a power factor command value (PFinv*) that decreases the operation power factor by a specified volume (0.01). Once the power factor command value (PFinv*) is generated, the inverter current (Iinv) is controlled by the inverter current control section and PWM control section to a current command value (Iniv*) corresponding to the power factor command value (PFinv*). As a result, the operation power factor of the PCS will be reduced to a value that is less than the present value by the specified volume. If Pgrid is greater than the second threshold value (250 W), or if the operation power factor is less than set value (0.80), the operation power factor of the PCS will be maintained at the present value. This power factor control is performed repeatedly at specified intervals while the PCS is running.

IV. TEST OPERATION TO VERIFY THE PROPOSED METHOD

To verify the operation of the control method proposed in the previous section, we perform an experiment using a PV PCS (without batteries). A PCS is connected with a DC power source and a PV simulated resistance and interconnected with a utility grid. Under such conditions, we changed the resistance load to check the behavior of the PCS power factor. Fig. 6 shows the configuration of the experiment, and Table 2 gives the conditions for the experiment. The capacity of the PCS is 10 kW and the power factor is set at 0.80. Fig. 7 shows the waveforms when the load is increased gradually from 7 kW to 12 kW while the PCS is being operated at its rating.

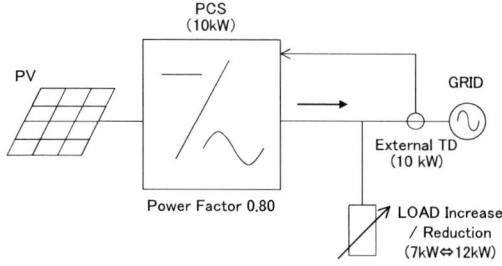

Fig. 6. Experimental configuration

Table 2. Experimental conditions

Items	Test conditions
PCS capacity	10 kW
Power factor setting	0.80
AC rated output voltage	101 V/202 V (single-phase three-wire)
Receiving power monitoring Device	External TD
Receiving power monitoring specification	Single-phase three-wire 10 kW High accuracy (grade 0.5) High speed response (0.1 sec)
Load	Resistance load Variable from 7 kW to 12 kW

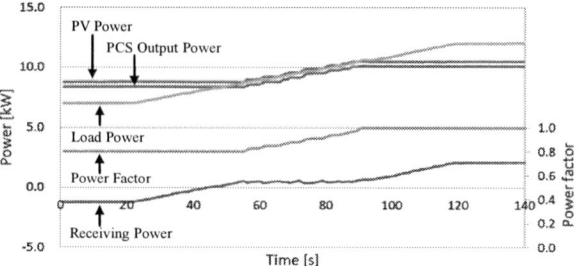

(a) Response of power factor when load is increased gradually

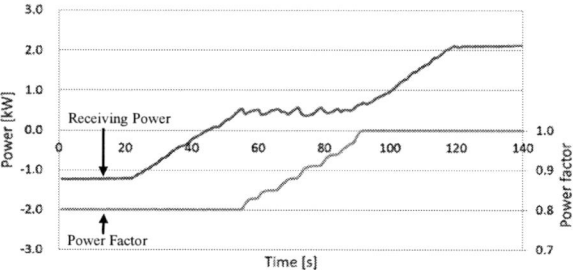

(b) Response of power factor when load is increased gradually (enlarged)

Fig. 7. Response of power factor when load is increased gradually

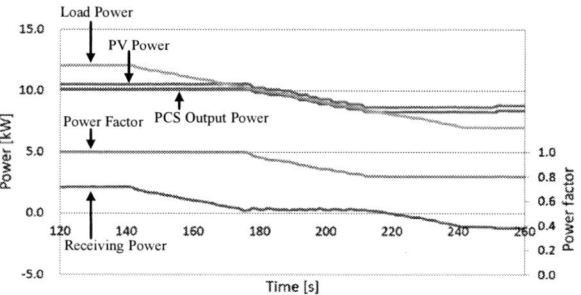

(a) Response of power factor when load is reduced gradually

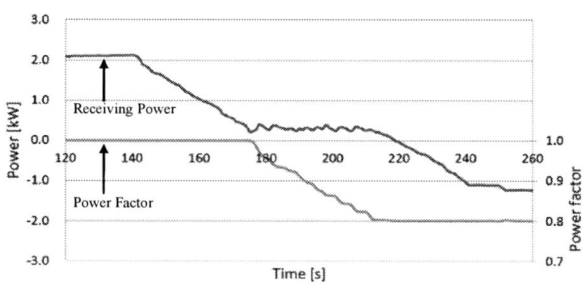

(b) Response of power factor when load is reduced gradually (enlarged)

Fig. 8. Response of power factor when load is reduced gradually

The waveforms in Fig. 8 occur when the load is reduced gradually from 12 kW to 7 kW. According to Fig. 7, the receiving power increases as the load power increases, and when the receiving power exceeds the first threshold of 500 W, the power factor is increased. According to Fig. 8, the receiving power reduces as the load power reduces, and when the receiving power drops below the second threshold of 250 W, the power factor is reduced. No fluctuations are detected in the increase and decrease in the power factor. We have thus confirmed that the proposed method works effectively and the system operates stably.

V. VERIFICATION WITH ACTUAL EQUIPMENT

We verified the operation of the power factor control function with actual equipment, using an external TD. Fig. 9–11 show the operation waveforms under each operation mode. Fig. 9 shows the operation waveforms when the PV generation is 50%, the power factor is set at 0.80, and the load is changed suddenly from 0→100→0%. When the load is 0%, all the 50% PV power will flow back into the grid, and the power factor will be maintained at 0.80. When the load suddenly rises to 100%, the grid power will be delivered temporarily. Subsequently, the power factor is increased gradually from 0.80 to 1.00 to increase the utilization rate of power delivered from the PCS. After confirming that the power is flowing forward at the power receiving point, 45% of the shortfall will be supplemented by the power from the batteries, and thereby, the purchase of power from the utility will be reduced. Owing to the constant receiving power control function, the remaining 5% will be supplied from the utility grid. In this way, this system prevents reverse power flow into the power grid while the batteries are in a discharging operation. When the load suddenly drops to

zero again, the gate of the battery converter is blocked instantly and the power factor is recovered to 0.80 while 50% of the PV power is being sent back into the grid. Furthermore, the power factor variation speed is set to 10s in consideration of the impact to the power grid.

Fig. 10 shows operational waveforms when the load is 50%, power factor is set at 0.80, and solar radiation is changed suddenly from 0→100→0%. When solar radiation is 0%, the power input from the batteries becomes necessary. While receiving power from the grid is maintained at 5% by the constant receiving power control function, 45% of the required power will be discharged from the batteries. In this case, the power does not flow back into the grid, and the power factor is set to 1.00. When the solar radiation rises suddenly to 100%, the PV power will be supplied to the load and the power factor will be reduced from 1.00 to 0.80 gradually to prevent voltage increases in the power grid. Subsequently, when the solar radiation drops suddenly to 0% again, the ratio of effective power becomes smaller against the reactive power control value and phase difference becomes larger, causing a temporal drop of power factor. While the power factor is being restored to 1.00, the power discharged from the batteries will be delivered to the load.

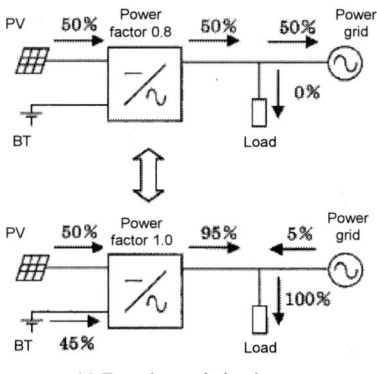

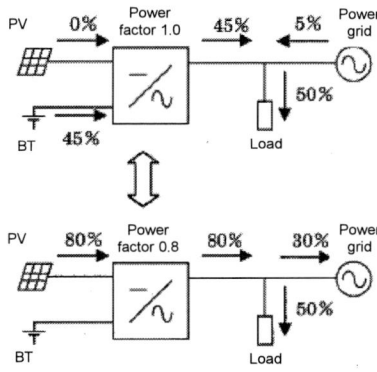

(a) Experimental circuit

(a) Experimental circuit

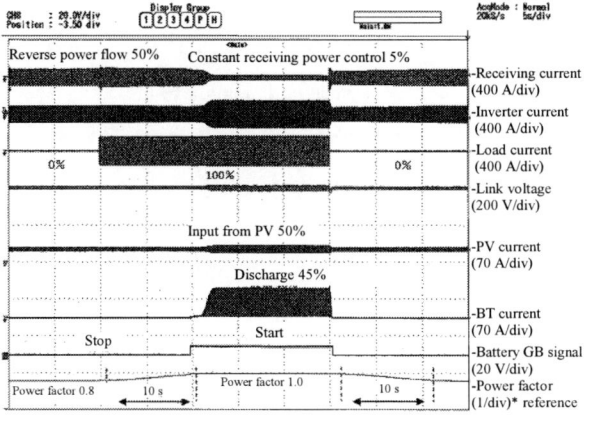

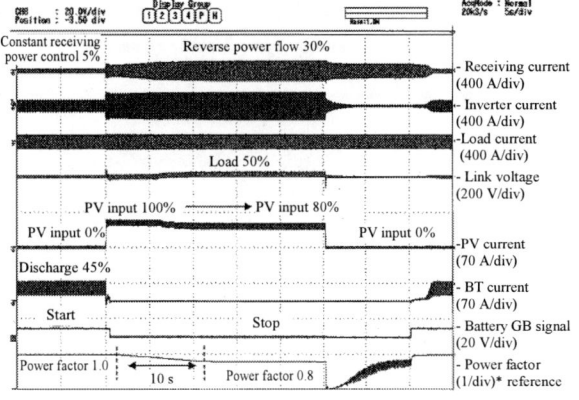

(b) Experimental wave form

(b) Experimental wave form

Fig. 9. Experimental result with a suddenly changed load (0 → 100 → 0%, power factor 0.80, external TD)

Fig. 10. Experimental result with suddenly changing solar radiation (0 → 100 → 0%, power factor 0.80, external TD, load 50%)

4155

Fig. 11 shows operational waveforms when the load in Fig. 10 is increased to 100%. When solar radiation is 0%, the power input from the batteries becomes necessary to deliver power to the load. While the receiving power from the grid is maintained at 5% by the constant receiving power control function, 95% of the power will be discharged from the batteries. In this case, the power does not flow back into the grid, and the power factor is set at 1.00. When the solar radiation rises suddenly to 100%, power will be delivered from the PV system to the load, and at the same time, the power factor will be reduced from 1.00 gradually. With its constant receiving power control and power factor control functions, PCS inputs reactive power so that the receiving power will reach the second threshold value (250 W). Subsequently, the power factor is maintained at 0.97 to maximize the power utilization rate to the load. When the solar radiation suddenly drops to 0% again, the power factor will be restored to 1.00 and power will be discharged from the batteries and supplied to the load.

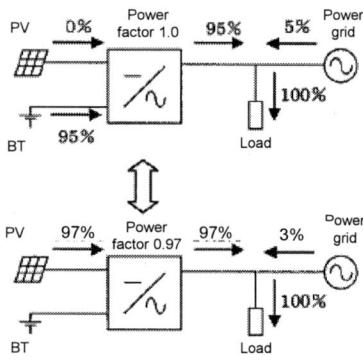

(a) Experimental circuit

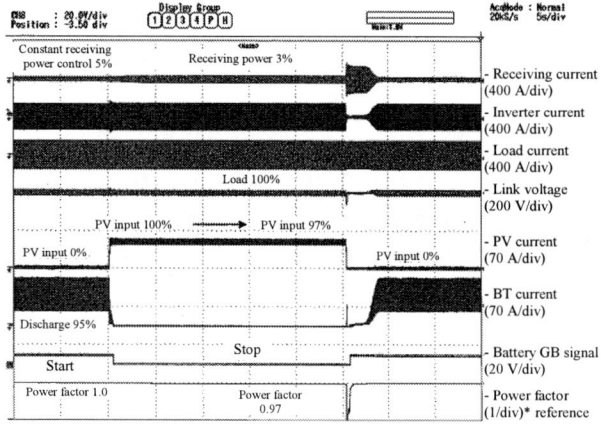

(b) Experimental wave form

Fig. 11. Experimental result with suddenly changing solar radiation (0 → 100 → 0%, power factor 0.80, external TD, load 100%)

VI. CONCLUSION

This paper proposed an energy system for a PV-system PCS with batteries to improve the power utilization rate to be delivered to on-premise loads by means of an external transducer for measuring the receiving power from the power grid and for the monitoring of power flows. The constant receiving power control and power factor control functions of the PCSs are implemented to improve the power utilization rate. In addition, the effectiveness of this control system has been confirmed using actual equipment. The system operates stably even when solar radiation or the load power changes suddenly. With this method, we can achieve both the control of the power factor at the time of power flow to the grid, and the effective utilization of power to the load at the time of forward power flows. When the power generation of the PV system is equal to the load, the reduction in power factor is minimized to constantly maximize the power utilization rate to the load. In this way, we can realize a lean system.

In the future, we will perform further verification tests using actual systems to verify operations in more detail.

REFERENCES

[1] Grid Connection Standard "Grid-Interconnection Code JEAC9701-2016" / JESC E 0019(2016) 2017 Addendum

[2] Japan Electrical Safety & Environment Technology Laboratories (JET) "Individual Test Method of Grid-Connected Protective Equipment for Multi-unit DC Input Systems (PV + BS)" JETGR0003-6-6.0 (2017)

[3] H. Endo, S. Yokoyama, T. Takuma, M. Yamaguchi, H. Mizuta, Y. Sugimura, "Application of Battery Combined Photovoltaic Generation System for Disaster-proof House" The Japan Institute of Power Electronics, Vol. 31, No. JIPE-31-14, 2005.

[4] M. Yokoyama, H. Endo, T. Takuma, S. Yokoyama, A. Yiga, "Development of Photovoltaic Generation System – LINEBACK Σ III- Connectable to Li-ion Battery" GS Yuasa Technical Report, Vol. 9, No. 2, pp. 24-29, 2012.

[5] M. Kurisaka, M. Yokoyama, Y. Yoshioka, T. Ueno, T. Nagano, H. Endo, "Development of Power Conditioner with Batteries for Photovoltaic Generation System – LINEBACK MEISTER" GS Yuasa Technical Report, Vol. 14, No. 1, pp. 20-27, 2017.

[6] Hiroyuki Hatta, "Effect Evaluation of Reactive Power Control Method According to PV Power Output on Required Capacity of SVC and Distribution Line Loss" IEEJ Transactions on Power and Energy, Vol.135, No.2, pp.106-110, 2014.

[7] Masaaki Takagi, "Cost-effectiveness Analysis of Reactive Power Compensator under Multiple Voltage Stabilization Measures" IEEJ Transactions on Power and Energy, Vol.137, No.1, pp.34-44, 2016.

[8] Eitaro Omine, "Development and verification of Load Leveling Method at Distribution Substation using Battery Energy Storage System" IEEJ Transactions on Power and Energy, Vol.137, No.10, pp.655-661, 2017.

The 2018 International Power Electronics Conference

A Novel DC Distribution Network with Multi-Level Bus Voltages and Its Energy Management System Design

Jingjin Huang[1,2], Xin Zhang[2*], Zhixun Ma[3,2], Jianfang Xiao[2]

1. Electrical Engineering, Xi'an University of technology, China
2. School of Electrical and Electronic Engineering, Nanyang Technological University, Singapore
3. National Maglev Transportation Engineering Research Center, Tongji University, Shanghai
E-mail: jackzhang@ntu.edu.sg*

Abstract- **A DC distribution network (DDN) with multi-level bus voltages is proposed to satisfy various DC load requirements. In the proposed DDN, a special multi-port DC transformer (MDCT) with the symmetrical CLLC resonant structure is employed for load-side voltage matching and galvanic isolation of different bus voltages. The MDCT is actually a three-port integration power converter which can construct three different bus voltages via only one power converter, and hence effectively reducing the multiple reverse conversions and improving the distribution efficiency. In addition, the power dispatch scheme of the energy management system (EMS) of the proposed DDN is also designed for transmitting the power within the battery, Photovoltaic (PV) energy and DDN bidirectionally. Finally, a SiC-based experimental prototype of the proposed DDN is established. Both the proposed DDN and EMS have been verified via experimental results.**

I. INTRODUCTION

AC distribution network dominates the market in the past 100 years due to the historical reason: AC voltage regulation with the conventional transformer was more effective than DC voltage regulation in the old days [1]. However, recently, the penetration of renewable energy sources, energy storage and electronic loads, most of which are DC-inherent, is increasing dramatically [2]. In addition, development of power electronics makes the DC/DC and DC/AC power conversions possible and simple. Therefore, DC system, especially DC distribution network (DDN), comes up to enhance the compatibility, reduce power conversion and energy efficiency [3, 4].

The challenge of the voltage level selection in the DDN is there are so many different voltage buses in the DDN [2-4]. For instance, in some household appliances, they are compatible with low voltage dc buses such as 24VDC, 48VDC, etc. [3]. However, in order to reduce the drop of the bus voltage and the power loss of the transmission line, relatively higher voltage (such as 380 VDC) is usually required. In this background, multi-level bus voltages

structure is more suitable for the DDN due to it can provide more selection for various loads to eliminate the multistage conversion. As a result, a DDN with multi-level bus voltages 24VDC, 48VDC and 380VDC are proposed and analyzed in this paper.

As a key component of the DDN, DC transformer (DCT) can realize the bidirectional power flow between bus voltages, and its various topologies have been studied in the past decades [5, 6]. Among these topologies, symmetrical CLLC resonant network based DCT can operate under high-power conversion efficiency with soft switching capability and to keep its efficiency exactly the same bidirectionally [7, 8]. However, only two voltage levels are available for the reported CLLC topologies in [7] and [8]. The multiport DC transformer (MDCT) can realize the power transmission among at least 3 bus voltages [9] while guaranteeing the isolation between the source and load [10]. However, the MDCT may operate under the hard-switching status, resulting in large switching loss. Therefore, this paper combines the advantages of both CLLC based DCT and the MDCT together to propose a CLLC based MDCT to connect the different bus voltages in the proposed DDN, and control the storage system, ensure the power isolation between the load and source, and guarantee the voltage matching with high power efficiency.

For any DDNs, the Energy management system (EMS) always plays an important role [11]. As shown in Table I, there are five main EMS Scenarios in a typical DDN. Scenarios 1~3 indicates the sufficient renewable energies, which can satisfy the charging requirement of battery, and transmit the extra power to DDN. When operating under Scenarios 4 and 5, the battery and DDN need to cooperate with the renewable energies to satisfy the load requirement. Currently, there are different EMS techniques have been reported. In [12], a two-stage EMS is designed for the DDN system in the electric vehicle and office building applications. However, it is not designed based on multi-level voltages, and the detailed operating Scenarios are not included. In [13], an EMS with the power dispatch scheme is established on the premise of the selected Scenario. In [14], an EMS with the power sharing scheme has been studied. However, the

4157

common limitation of the above EMS [13, 14] is that the renewable source priority is not taken into consideration. Therefore, in this paper, by predefining the renewable source priority and by considering the multi-level bus voltages case, the power dispatch scheme is deigned in EMS based on the possible five Scenarios of DDN.

TABLE I
FIVE MAIN EMS SCENARIOS IN DDN

Scenario	Battery	renewable energy	DDN
1	Idle	Sufficient	Input
2	Charging	Sufficient	Input
3	Charging	Sufficient	Idle
4	Discharging	Insufficient	Idle
5	Discharging	Insufficient	Output

In this paper, a DDN is proposed with multi-level bus voltages including 24VDC, 48VDC and 380VDC to supply various loads. In addition, the symmetrical CLLC resonant MDCT is employed for this DDN to improve the power efficiency. Furthermore, to make full use of the renewable energy and design the power dispatch scheme, a novel EMS is proposed by predefining the renewable source priority and by considering aforementioned five Scenarios. Besides, three battery statuses, i.e., charging, discharging and idle, are analyzed in each Scenario to facilitate the implementation of the power control scheme. The ultimate goal is to match the voltage level and achieve the seamless power transition. Finally, a real DDN has been established to experimentally verify the effectiveness of the proposed DDN system and its EMS.

II. PROPOSED DDN WITH MULTI-LEVEL BUS VOLTAGES

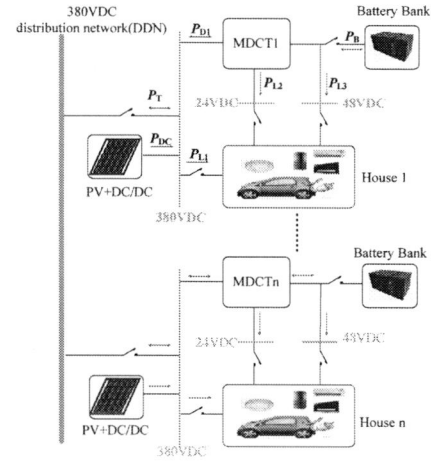

Fig. 1. Concept of the proposed DDN in household application.

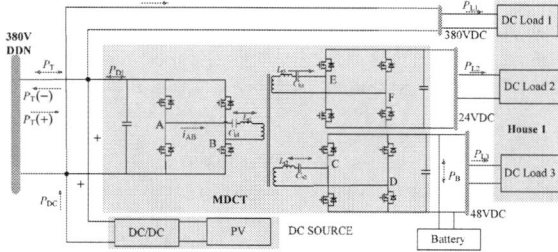

Fig. 2. Detailed configuration of the proposed DDN.

In this paper, a DDN is proposed, as shown in Fig. 1, which is composed by the MDCT, PV panel, battery bank and the corresponding converters. It aims at utilizing the renewable energies locally with multi-level bus voltages to reduce multiple reverse conversions and improve the distribution efficiency.

There are three voltage-level buses in the proposed DDN:
a) 24VDC bus is connected with one port of MDCT. It can be supplied by PV, DDN and the battery.
b) 48VDC bus is connected with battery and the second port of MDCT. Its voltage can be stabilized by the battery.
c) 380VDC bus is mainly supplied from PV, which is also compensated by 380V DDN and battery. Its voltage can be guaranteed by 380V DDN.

There are five advantages of the proposed DDN:
a) Multi-level bus voltages, i.e., 24VDC, 48VDC and 380VDC, can satisfy various load requirements, greatly saving the conversion stages.
b) Bidirectional power flow between MDCT and battery bank contributes to improve the utilization of the renewable energy.
c) MDCT is designed with the CLLC resonant structure to improve the power efficiency of the DDN.
d) There is no need to consider the synchronization with the utility grid and reactive power due to the DC bus structure.
e) When a blackout or voltage sag occurs in the sub-DC bus, it does not affect the 380V DDN system due to the voltage control of AC/DC converter and the galvanic isolation.

In Fig. 1, it is also give a typical household application of the proposed DDN. As seen, in the proposed DDN, a MDCT is allocated to each house. Meanwhile, the three DC bus voltages are available for each house. Therefore, these houses are independent to each other. Hence, the designed DDN can be applied in smart building to compatible with various low DC voltage communication loads. Besides, it can also be applied in remote villages.

TABLE II
SYMBOLS IN DDN

Symbol	Description
P_T	The power transferred from/to 380V DDN
P_{DC}	The power transferred from DC source
P_{D1}	The power of MDCT at 380V side
P_B	The power transmitted bidirectionally to charge/discharge the battery.
P_{L1}	The power required by 380VDC load
P_{L2}	The power required by 24VDC load
P_{L3}	The power required by 48VDC load
*	The corresponding reference value

The detailed configuration of the proposed DDN is shown in Fig. 2 with the symbols defined in Table II. In the proposed DDN, three DC bus voltages 24V, 48V and 380V are supplied by the sources as PV, battery and 380V DDN. The power dispatch scheme is thus essential to make the

4158

sources cooperate with each other. An EMS will be further illustrated for this proposed DDN in the next section.

III. EMS DESIGN OF THE PROPOSED DDN

Ignoring the market regulation, an EMS scheme is proposed based on the designed DDN. *In the designed power dispatch scheme of EMS, the 1st priority is allocated to PV and the 2nd priority is assigned to battery to make full use of the renewable energy and reduce the power consumption from 380V DDN.* Therefore, the priority of the renewable sources is sorted as:

Priorities: PV > PV + Battery > Battery > 380V DDN.

The desired power determination flow-chart is thus achieved as shown in Fig. 3. It can be observed from Fig. 3 that five Scenarios are available in EMS. The detailed power flow procedure is elaborated as follows.

Scenario 1 (See Table 1): $P_{DCmax} \geq \sum P_{Load}$ and $P_B \geq P_{Bmax}$, where $\sum P_{Load} = P_{L1} + P_{L2} + P_{L3}$.

In this Scenario, the PV is sufficient and the battery is

full charged. If the extra power is permitted to transmit to 380V DDN, the reference powers can be obtained as

$$\begin{cases} P_{DC}^* = P_{DCmax} \\ P_B^* = 0 \\ P_T^*(-) = P_{DC}^* - \sum P_{Load} \end{cases} \tag{1}$$

where $P_T^*(-)$ is the reference power transmitted to the DDN. That is to say, the PV operates with the Maximum Power Point Tracking (MPPT) mode, and the battery is in idle status. Otherwise, the reference powers are

$$\begin{cases} P_{DC}^* = \sum P_{Load} \\ P_B^* = 0 \\ P_T^*(-) = 0 \end{cases} \tag{2}$$

i.e., PV operates with the constant power output to meet the load power demand. No power exchange exists in 380V DDN.

Scenario 2 (See Table 1): $P_{DCmax} \geq \sum P_{Load}$, $P_B < P_{Bmax}$, and $P_{DCmax} \geq \sum P_{Load} + P_{Bmax} - P_B$.

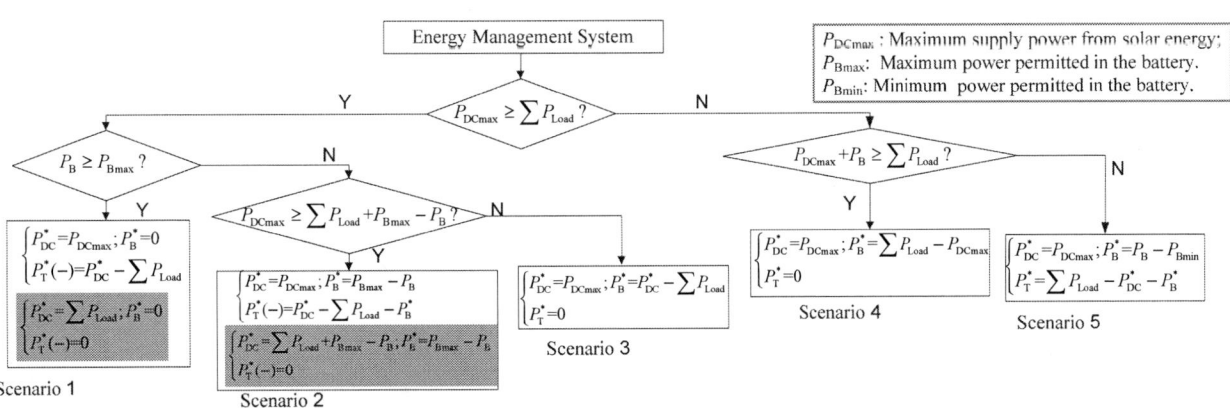

Fig. 3 Desired power determination flow-chart in the EMS.

In this Scenario, the battery is not full charged; and sufficient PV energy not only satisfies the load requirement, but also meets the battery charging demand. If the extra power is permitted to feedback to the distribution network, the reference powers are defined as

$$\begin{cases} P_{DC}^* = P_{DCmax} \\ P_B^* = P_{Bmax} - P_B \\ P_T^*(-) = P_{DC}^* - \sum P_{Load} - P_B^* \end{cases} \tag{3}$$

which indicates that PV operates under the MPPT mode. Otherwise,

$$\begin{cases} P_{DC}^* = \sum P_{Load} + P_{Bmax} - P_B \\ P_B^* = P_{Bmax} - P_B \\ P_T^*(-) = 0 \end{cases} \tag{4}$$

Thus, PV operates with the constant power output to satisfy DC load and battery requirements, and no power is transmitted to 380V DDN.

Scenario 3 (See Table 1): $P_{DCmax} \geq \sum P_{Load}$, $P_B < P_{Bmax}$, and $P_{DCmax} < \sum P_{Load} + P_{Bmax} - P_B$

PV can satisfy the total load requirement, but hardly afford the battery charging demand. Therefore, no extra power exists in this Scenario. The reference powers are thus achieved as

$$\begin{cases} P_{DC}^* = P_{DCmax} \\ P_B^* = P_{DC}^* - \sum P_{Load} \\ P_T^* = 0 \end{cases} \tag{5}$$

MPPT mode is available for PV to supply the DC loads and charge the battery.

In above three Scenarios, PV energy is sufficient to afford DC loads. By utilizing the power dispatch scheme with EMS, the extra power can be transmitted to the 380V DDN or the battery side to ensure some economic benefits. However, it is difficult to sell power to the distribution network due to some power market limitations. Thus, Eqns. (2), (4) and (5) are usually employed to realize the power dispatch of the sufficient PV energy.

Scenario 4 (See Table1): $P_{DCmax} < \sum P_{Load}$, $P_{DCmax} + P_B \geq \sum P_{Load}$

Note that the battery power P_B should satisfy $P_{Bmin} < P_B < P_{Bmax}$ to ensure the normal operation. In this Scenario, PV cannot satisfy the load requirement. However, the load power can be supplied effectively when both PV and battery are operating simultaneously. Hence, the reference power can be obtained as

$$\begin{cases} P_{DC}^* = P_{DCmax} \\ P_B^* = \sum P_{Load} - P_{DCmax} \\ P_T^* = 0 \end{cases} \quad (6)$$

i.e., PV operates with MPPT mode and the battery is put into operation to cooperate with PV to maintain the load operation.

Scenario 5 (See Table1): $P_{DCmax} < \sum P_{Load}$, $P_{DCmax} + P_B < \sum P_{Load}$

In this Scenario, both PV and battery cannot satisfy the load requirement. Thus, the power should be transmitted from DC distribution network to maintain the DC load requirement. The reference powers are

$$\begin{cases} P_{DC}^* = P_{DCmax} \\ P_B^* = P_B - P_{Bmin} \\ P_T^* = \sum P_{Load} - P_{DC}^* - P_B^* \end{cases} \quad (7)$$

According to the power-flow Scenarios in Fig. 3, the battery sates can be divided into three parts, i.e., charging state, idle state, discharging state, which is determined by EMS. It can be observed that Scenario 1 is responsible of the DC load while the battery is full. In Scenarios 2 and 3, the sufficient power will be transmitted to the DC load side and charge the battery. When the PV power cannot satisfy the DC load requirement, the battery will be introduced to operate as Scenario 4 and 5 with the discharging state. Note that the PV energy has the top priority in utilization according to this designed power dispatch approach.

IV. Experimental Verification

In order to analyze the proposed DDN system, its key part MDCT prototype is established, as shown Fig. 4. The 380V DDN is simulated by the power supply. The battery voltage is 48V. The parameters are listed in Table III. Besides, in order to verify bidirectional power flow performance of the designed MDCT, the test in the full power range has been carried out. The detailed analysis is listed as below.

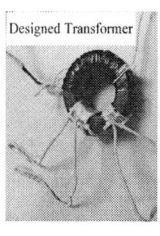

Fig. 4. Experimental prototype of MDCT

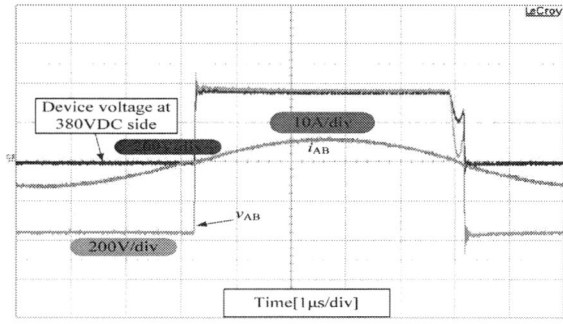

Fig. 5 Experimental waveforms of MDCT

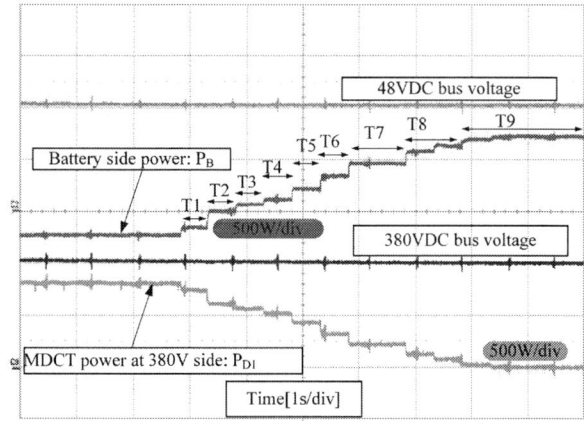

Fig. 6 Bidirectional power flow performance of DDN.

TABLE III
PARAMETERS OF EXPERIMENTAL SETUPS

Parameters	Values
DDN nominal voltage	380V
MDCT Rated power	6kW
Three ports DC voltage of MDCT	380V, 48V, 24V
Frequency	100kHz
Switching dead time	200ns
Designed MDCT	L_{r1}=278.68μH, L_{r2}=4.457μH; L_{r3}=1.115μH C_{b1}=0.0092μF, C_{b2}=0.57μF; C_{b3}=2.28μF

A. MDCT performance verification

It is well known that the switches in MDCT will suffer from high switching loss if quasi-resonant status is not achieved. Thus, the MDCT performances are tested in this sub-section.

Fig. 5 shows the device voltage at 380V side of MDCT and the corresponding current i_{AB} and voltage v_{AB}. Obviously, the power device is turned off/on when the current pass through the zero; hence, zero current switching is guaranteed here. Therefore, MDCT is operating under the quasi-resonant status.

B. Bidirectional power flow performance of DDN

As given in Fig. 4, three-port transformer is designed with the leakage inductance values shown in Table III to construct MDCT. The bidirectional power flow performance of the DDN is verified in this sub-section.

As stated in section III, there are three states of the battery, i.e., charging, discharging and idle. The experimental results are presented in Fig. 6, with details elaborated below.

a) Initially, the load is supplied by the 380V DDN, and the battery is charging. During time T_1, the charging power is reduced and the load is remained the same; hence, the current from 380VDC is reduced the same level as the battery power reduced.

b) During time T_2, battery is in the idle state; thus, the power from the 380VDC is to maintain the load operation.

c) From T_3 to T_8, the battery is in charging while the power of the 380VDC is decreasing. During this process, the load power is effectively maintained.

d) In T_9, the load current is supplied by the battery. There is no power flow between the load and 380V DDN.

Therefore, combining EMS, the power dispatch is effectively ensured with stable DC bus voltages.

V. CONCLUSION

In this paper, a DC distribution network (DDN) is proposed with multi-level bus voltages to supply various loads. A MDCT with symmetrical CLLC topology is implemented in the proposed DDN for voltage matching, galvanic isolation and power efficiency improvement. Besides, a novel EMS with predefining the renewable source priority and considering different DDN Scenarios has also been proposed to make full use of the renewable energy and design the power dispatch scheme. Finally, various experimental cases have been conducted to confirm the correctness of the proposed DDN and its EMS design.

REFERENCES

[1] Wang P, Goel L, Liu X, et al. "Harmonizing AC and DC: A hybrid AC/DC future grid solution," *IEEE Power and Energy Magazine*, vol. 11, no. 3, pp. 76-83, 2013.

[2] Kakigano H, Miura Y, Ise T. "Low-voltage bipolar-type DC microgrid for super high quality distribution," *IEEE transactions on power electronics*, vol. 25, no. 12, pp. 3066-3075, 2010.

[3] Ghareeb A T, Mohamed A A, Mohammed O A. "DC microgrids and distribution systems: An overview,"///*Power and Energy Society General Meeting (PES)*, 2013 IEEE. IEEE, 2013: 1-5.

[4] Yu X, Huang A, Burgos R, et al. "A fully autonomous power management strategy for DC microgrid bus voltages,"///*Applied Power Electronics Conference and Exposition (APEC)*, 2013 Twenty-Eighth Annual IEEE. IEEE, 2013: 2876-2881.

[5] Huang J, Xiao J, Wen C, et al. "mplementation of Bidirectional Resonant DC Transformer in Hybrid AC/DC Micro-grid," *IEEE Transactions on Smart Grid*, 2017.

[6] Feng W, Mattavelli P, Lee F C. "Pulsewidth locked loop (PWLL) for automatic resonant frequency tracking in LLC DC–DC transformer (LLC-DCX)," *IEEE Transactions on Power Electronics*, vol. 28, no. 4, pp. 1862-1869, 2013.

[7] Zhao B, Song Q, Liu W, et al. "Overview of dual-active-bridge isolated bidirectional DC–DC converter for high-frequency-link power-conversion system," *IEEE Transactions on Power Electronics*, vol. 29, no. 8, pp. 4091-4106, 2014.

[8] J. H. Jung, H. S. Kim, M. H. Ryu, and J. W. Baek, "Design methodology of bidirectional CLLC resonant converter for high-frequency isolation of dc distribution systems," *IEEE Trans. On Power Electron.*, vol. 28, no. 4, pp. 1741–1755, 2013.

[9] Farhangi B, Toliyat H A. "Modeling and analyzing multiport isolation transformer capacitive components for onboard vehicular power conditioners," *IEEE Transactions on Industrial Electronics*, vol. 62, no. 5, pp. 3134-3142, 2015.

[10] Falcones S, Ayyanar R, Mao X. "A DC–DC multiport-converter-based solid-state transformer integrating distributed generation and storage," *IEEE Transactions on Power Electronics*, vol. 28, no. 5, pp. 2192-2203, 2013.

[11] Byeon G, Yoon T, Oh S, et al. "Energy management strategy of the DC distribution system in buildings using the EV service model,". *IEEE Transactions on Power Electronics*, vol. 28, no. 4, pp. 1544-1554, 2013.

[12] Wu D, Zeng H, Lu C, et al. "Two-Stage Energy Management for Office Buildings With Workplace EV Charging and Renewable Energy," *IEEE Transactions on Transportation Electrification*, vol. 3, no. 1, pp. 225-237, 2017.

[13] Jianfang X, Peng W, Setyawan L, et al. "Energy management system for control of hybrid AC/DC microgrids,"///*Industrial Electronics and Applications (ICIEA), 2015 IEEE 10th Conference on. IEEE*, 2015: 778-783.

[14] Wang P, Xiao J, Setyawan L, et al. "Energy management system (EMS) for real-time operation of DC microgrids with multiple slack terminals,"///*Innovative Smart Grid Technologies Conference Europe (ISGT-Europe), 2014 IEEE PES. IEEE*, 2014: 1-6.

The 2018 International Power Electronics Conference

A Novel DC-Side-Port Impedance Modeling of Modular Multilevel Converters Based on Harmonic State Space Method

Jing Lyu[1], Xin Zhang[2], Zhixun Ma[2], and Xu Cai[1*]

1 Department of Electrical Engineering, Shanghai Jiao Tong University, Shanghai, China
2 School of Electrical and Electronic Engineering, Nanyang Technological, Singapore
*E-mail: xucai@sjtu.edu.cn

Abstract— With the rapid development of modular multilevel converter (MMC) based DC grids, the resonance and instability issues of these DC grids are getting more and more attention. Though some models of the dc-side-port (DSP) impedance of the MMC have been reported to help the above stability analysis, they always ignore the internal harmonic dynamics of the MMC. Actually, these ignored internal harmonic dynamics are very important things to the MMC, such as the capacitor voltage fluctuations and harmonic circulating currents. Therefore, this paper will further model the DSP impedance of the MMC with the consideration of its internal harmonic dynamics via the harmonic state space (HSS) method. The derived DSP impedance is verified by comparing with the measured one through a nonlinear time-domain model. Furthermore, the impact of the number of harmonics considered in the HSS model on the accuracy of the analytical DSP impedance model is also discussed. The test results show that the proposed DSP impedance model not only covers all the existing DSP impedance models with setting the harmonic order $h=0$, but also can accurately reflect the internal harmonic effects with setting the suitable harmonic order h.

Keywords— *Modular Multilevel Converter, dc-side-port impedance, harmonic state space, internal dynamics.*

I. INTRODUCTION

Modular multilevel converter (MMC) has been widely used in high-voltage direct current (HVDC) transmission and other high-voltage/high-power applications [1], [2], thanks to its modularity, scalability, high efficiency and high performance. In recent years, several multiterminal MMC-based HVDC projects have also been put into operation around the world [3]. The completed projects contain: China's Nan'ao three-terminal MMC-HVDC project in 2013, China's Zhoushan five-terminal MMC-HVDC project in 2014, etc. The projects to be completed or planned include: China's Zhangbei four-terminal MMC-HVDC project, European supergrid, etc. However, with the rapid development of MMC-based DC grids, the resonance and instability issues of the MMC-based DC grids are becoming a cause for concern [4].

The impedance-based method, which is originally used for the two-level voltage-source converters (VSCs) [5]-[7], is currently one of the most popular approaches

to analyze the stability of MMC-based power electronics interconnected systems [8]-[11]. To analyze the stability of MMC-based DC grids by the impedance-based analytical approach, the dc-side-port (DSP) impedance of the MMC needs to be obtained first. At present, all the existing researches on the MMC DSP impedance modeling automatically ignore the MMC internal harmonic dynamics partly based on the experience from the two-level VSCs. In [11], the DSP impedance modeling and stability analysis of the MMC for MVDC applications were presented. In [12], the DSP impedance of the MMC was also derived for dc voltage ripple prediction. As mentioned before, in both [11] and [12], the submodule (SM) capacitors are assumed to be large enough and the internal harmonic dynamics are neglected. However, it is not improper in practice. Compared with the two-level VSCs, MMC has much more complex internal dynamics, e.g. capacitor voltage fluctuations and harmonic circulating currents. These internal harmonic dynamics may have great influence on the terminal characteristics of the MMC [8]-[10], which, therefore, are necessary to be included in the DSP impedance modeling.

The harmonic state space (HSS) method can be used to model the MMC with consideration of the internal harmonics [13], which has already shown this advantage in the ac-side-port (ASP) impedance modeling of the MMC [14]. However, the MMC DSP impedance case has not been discussed so far. To remedy this problem, the HSS modeling method, which is able to include all the harmonics and frequency couplings, is introduced in this paper to derive the MMC DSP impedance. The small perturbation HSS model of a three-phase MMC is built, based on which, the relationship between the dc-side small perturbation voltage and current can then be obtained. As a result, the MMC DSP impedance at the perturbation frequency is calculated by the ratio of the dc-side small perturbation voltage to current. The derived DSP impedance is verified by comparing with the measured results via a nonlinear time-domain simulation model in MALTAB/ Simulink. Moreover, the impact of the number of harmonics considered in the HSS model on the accuracy of the analytical DSP impedance model is also carefully discussed.

4162

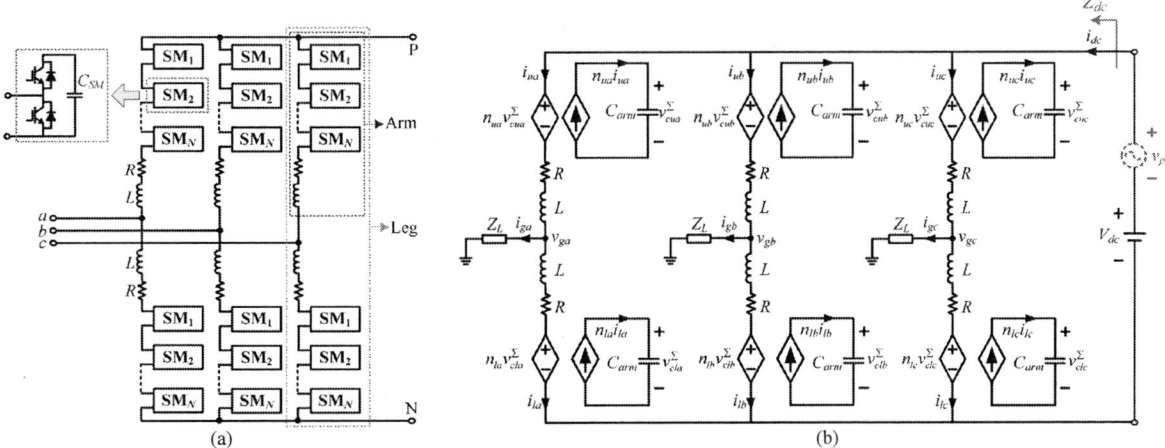

Fig. 1. MMC topology under study. (a) MMC circuit diagram. (b) Average-value model of MMC.

II. FORMULATION OF HSS MODELING

For any time-varying periodic signal $x(t)$, it can be written in the form of Fourier series as:

$$x(t) = \sum_{k \in \mathbb{Z}} X_k e^{jk\omega_1 t} \tag{1}$$

where $\omega_1 = 2\pi/T$, T is the fundamental period of the signal, and X_k is the Fourier coefficient that can be calculated by

$$X_k = \frac{1}{T} \int_{t_0}^{t_0+T} x(t) e^{-jk\omega_1 t} dt \tag{2}$$

The time-domain state space equation of a linear time periodic (LTP) system can be expressed as

$$\dot{x}(t) = A(t)x(t) + B(t)u(t) \tag{3}$$

Hence, according to the Fourier series and harmonic balance theory, the time-domain state space equation in (3) can be transformed into the frequency-domain harmonic state space equation in (4).

$$sX = (A - Q)X + BU \tag{4}$$

where X, U, Q, and A are shown in (5)-(8), respectively. B has the same form as A. The elements X_h, U_h and A_h are the Fourier coefficient of the hth harmonic of $x(t)$, $u(t)$ and $A(t)$ in (3), respectively. Note that A and B are Toeplitz matrices in order to perform the frequency-domain convolution operation, and Q is a diagonal matrix that represents the frequency information. Additionally, h is the harmonic order taken into account.

$$X = [X_{-h}, \cdots, X_{-1}, X_0, X_1, \cdots, X_h]^T \tag{5}$$

$$U = [U_{-h}, \cdots, U_{-1}, U_0, U_1, \cdots, U_h]^T \tag{6}$$

$$Q = \mathrm{diag}[-jh\omega_1, \cdots, -j\omega_1, 0, j\omega_1, \cdots, jh\omega_1] \tag{7}$$

$$A = \begin{bmatrix} A_0 & A_{-1} & \cdots & A_{-h} & & & \\ A_1 & \ddots & \ddots & \ddots & \ddots & & \\ \vdots & \ddots & A_0 & A_{-1} & \ddots & \ddots & \\ A_h & \ddots & A_1 & A_0 & A_{-1} & \ddots & A_{-h} \\ & \ddots & \ddots & A_1 & A_0 & \ddots & \vdots \\ & & \ddots & \ddots & \ddots & \ddots & A_{-1} \\ & & & A_h & \cdots & A_1 & A_0 \end{bmatrix} \tag{8}$$

III. DSP IMPEDANCE MODELING OF MMC BASED ON HSS MODELING METHOD

Fig. 1 shows the MMC topology being studied, where Fig. 1(a) is the MMC circuit diagram, Fig. 1(b) is its average-value model. In Fig. 1(b), $C_{arm} = C_{SM}/N$, C_{SM} is the SM capacitance, N is SM number per arm, L and R are the arm inductance and equivalent series resistance, i_{ux} and i_{lx} ($x=a$, b, c) are the upper and lower arm currents, v_{cux}^Σ and v_{clx}^Σ are the sum capacitor voltages of the upper and lower arms, v_{gx} and i_{gx} are the ac-side phase voltage and current, n_{ux} and n_{lx} are the insertion indices of the upper and lower arms, V_{dc} is the dc-link voltage, and Z_L is the ac-side passive load. In addition, the small perturbation voltage v_p is injected at the dc-side of MMC, which is used to derive the DSP impedance Z_{dc}. Furthermore, the dc-link voltage V_{dc} is assumed to be constant.

According to Fig. 1(b), one can obtain (without small perturbation injection)

$$\frac{di_{cx}}{dt} = -\frac{R}{L}i_{cx} - \frac{n_{ux}}{2L}v_{cux}^\Sigma - \frac{n_{lx}}{2L}v_{clx}^\Sigma + \frac{V_{dc}}{2L} \tag{9}$$

$$\frac{dv_{cux}^\Sigma}{dt} = \frac{n_{ux}}{C_{arm}}i_{cx} + \frac{n_{ux}}{2C_{arm}}i_{gx} \tag{10}$$

$$\frac{dv_{clx}^\Sigma}{dt} = \frac{n_{lx}}{C_{arm}}i_{cx} - \frac{n_{lx}}{2C_{arm}}i_{gx} \tag{11}$$

$$\frac{di_{gx}}{dt} = -\frac{n_{ux}}{L}v_{cux}^\Sigma + \frac{n_{lx}}{L}v_{clx}^\Sigma - \frac{R}{L}i_{gx} - \frac{2v_{gx}}{L} \tag{12}$$

$$v_{gx} = Z_L i_{gx} \tag{13}$$

where i_{cx} is the circulating current.

Taking open-loop control for example, the derivation process of the MMC DSP impedance will be presented in this paper, which is also easily extended to any closed-loop control. In this case, if injecting a small perturbation voltage into the dc-side of the MMC (as shown in Fig. 1(b)), the small perturbation equations of (9)-(13) can then be obtained (where $n_{uxp} = n_{lxp} = 0$ in open-loop control

case)

$$\frac{di_{cxp}}{dt} = -\frac{R}{L}i_{cxp} - \frac{n_{ux0}}{2L}v^{\Sigma}_{cuxp} - \frac{n_{lx0}}{2L}v^{\Sigma}_{clxp} + \frac{v_p}{2L} \quad (14)$$

$$\frac{dv^{\Sigma}_{cuxp}}{dt} = \frac{n_{ux0}}{C_{arm}}i_{cxp} + \frac{n_{ux0}}{2C_{arm}}i_{gxp} \quad (15)$$

$$\frac{dv^{\Sigma}_{clxp}}{dt} = \frac{n_{lx0}}{C_{arm}}i_{cxp} - \frac{n_{lx0}}{2C_{arm}}i_{gxp} \quad (16)$$

$$\frac{di_{gxp}}{dt} = -\frac{n_{ux0}}{L}v^{\Sigma}_{cuxp} + \frac{n_{lx0}}{L}v^{\Sigma}_{clxp} - \frac{R+2Z_L}{L}i_{gxp} \quad (17)$$

where the subscript "p" and "0" indicates the perturbation and steady-state components, respectively. Additionally, the steady-state components n_{ux0} and n_{lx0} of the insertion indices are given as

$$\begin{cases} n_{ux0} = \dfrac{1}{2}\left[1 - m\cos\left(\omega_1 t + \varphi_x\right)\right] \\ n_{lx0} = \dfrac{1}{2}\left[1 + m\cos\left(\omega_1 t + \varphi_x\right)\right] \end{cases} \quad (18)$$

in which m is the modulation index, and $\varphi_a=0$, $\varphi_b=120°$, $\varphi_c=240°$.

Equations (14)-(17) can be reorganized into a state-space form, which is like

$$\dot{x}_p(t) = A(t)x_p(t) + B(t)u_p(t) \quad (19)$$

where $x_p(t)$, $u_p(t)$, $A(t)$, and $B(t)$ are shown in (20)-(23).

$$x_p(t) = \left[i_{cap}, i_{cbp}, i_{ccp}, v^{\Sigma}_{cuap}, v^{\Sigma}_{cubp}, v^{\Sigma}_{cucp}, v^{\Sigma}_{clap}, v^{\Sigma}_{clbp}, v^{\Sigma}_{clcp}, \right.$$
$$\left. i_{gap}, i_{gbp}, i_{gcp} \right]^T \quad (20)$$

$$u_p(t) = \left[v_p \right] \quad (21)$$

$$B(t) = \left[\frac{1}{2L}, \frac{1}{2L}, \frac{1}{2L}, 0,0,0,0,0,0,0,0,0 \right]^T \quad (23)$$

Hence, the small perturbation HSS model of the MMC can be obtained as (24) by applying the HSS modeling that is introduced in (1)-(8).

$$s\mathbf{X}_p = \left(\mathbf{A} - \mathbf{Q}_p\right)\mathbf{X}_p + \mathbf{B}\mathbf{U}_p \quad (24)$$

where $\mathbf{X}_p$, $\mathbf{U}_p$, and $\mathbf{Q}_p$ are given in (25)-(27), in which the subscript "$p\pm h$" denotes the perturbation component at frequency "$\omega_p\pm h\omega_1$", I is a twelve-order identity matrix, and O is zero matrix. Additionally, $\mathbf{A}$ and $\mathbf{B}$ are Toeplitz matrices. The elements of matrix $\mathbf{B}$ are shown in (27) and the elements of matrix $\mathbf{A}$ are given in Appendix. It is noted that the diagonal matrix $\mathbf{Q}_p$ contains all the perturbation frequencies, which means that the model in (24) is able to include all the frequency couplings.

$$\mathbf{X}_p = \left[X_{p-h}, \cdots, X_p, \cdots, X_{p+h} \right]^T$$

$$X_{p\pm h} = \left[I_{cap\pm h}, I_{cbp\pm h}, I_{ccp\pm h}, V^{\Sigma}_{cuap\pm h}, V^{\Sigma}_{cubp\pm h}, V^{\Sigma}_{cucp\pm h}, \right. \quad (25)$$
$$\left. V^{\Sigma}_{clap\pm h}, V^{\Sigma}_{clbp\pm h}, V^{\Sigma}_{clcp\pm h}, I_{gap\pm h}, I_{gbp\pm h}, I_{gcp\pm h} \right]$$

$$\mathbf{U}_p = \left[U_{p-h}, \cdots, U_p, \cdots, U_{p+h} \right]^T$$
$$U_p = \left[V_p \right], \quad U_{p\pm h} = \left[0 \right] \left(h \geq 1 \right) \quad (26)$$

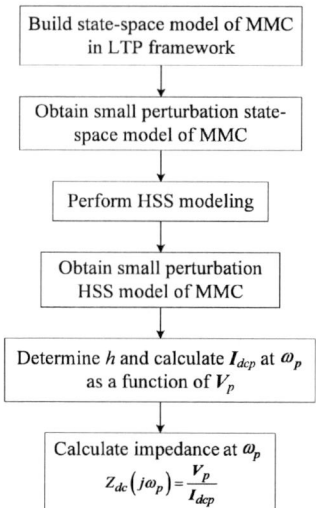

Fig. 2. Flowchart of the MMC DSP impedance modeling by HSS.

$$\mathbf{Q}_p = \mathrm{diag}\left[j\left(\omega_p - h\omega_1\right) \cdot I, \ldots, j\omega_p \cdot I, \ldots, j\left(\omega_p + h\omega_1\right) \cdot I \right] \quad (27)$$

$$B_0 = \left[\frac{1}{2L}, \frac{1}{2L}, \frac{1}{2L}, 0,0,0,0,0,0,0,0,0 \right]^T \quad (28)$$

$$B_{\pm h} = O^{12\times 1} \left(h \geq 1 \right)$$

Ignoring the transient behavior of the perturbation signals in (24), the small perturbation components of the state variables at each perturbation frequency can be solved by

$$\mathbf{X}_p = -\left(\mathbf{A} - \mathbf{Q}_p\right)^{-1}\left(\mathbf{B}\mathbf{U}_p\right) \quad (29)$$

In order to derive the DSP impedance, the resulting small perturbation dc current i_{dcp} at frequency ω_p needs to be obtained. Since we have

$$i_{dcp} = \sum_{x=a,b,c} i_{uxp} = \sum_{x=a,b,c} i_{lxp} = \sum_{x=a,b,c} i_{cxp} \quad (30)$$

The complex phasors of the three-phase small perturbation circulating currents i_{cxp} at frequency ω_p can be calculated by (29), and then the complex phasor of the small perturbation dc current i_{dcp} at frequency ω_p can be obtained by (30). Hence, the DSP impedance of the MMC at frequency ω_p can be calculated by

$$Z_{dc}\left(j\omega_p\right) = \frac{V_p}{I_{dcp}} \quad (31)$$

where the bold capital letters V_p and I_{dcp} represent the complex phasors of the small perturbations v_p and i_{dcp} at frequency ω_p, respectively.

In addition, it needs to be pointed out that the harmonic order h must be predefined in order to perform the matrix operation by MATLAB. The accuracy of the proposed model in this paper depends on the considered harmonic order h in the MMC modeling. The higher the considered harmonic order h is, the higher the model accuracy is. The impact of the harmonic order h on the accuracy of the analytical DSP impedance model will be further discussed in the next section.

Finally, the whole flowchart of the HSS based MMC DSP impedance modeling can be summarized in Fig. 2.

$$A(t) = \begin{bmatrix}
-\dfrac{R}{L} & 0 & 0 & -\dfrac{n_{ua0}}{2L} & 0 & 0 & -\dfrac{n_{la0}}{2L} & 0 & 0 & 0 & 0 & 0 \\[2ex]
0 & -\dfrac{R}{L} & 0 & 0 & -\dfrac{n_{ub0}}{2L} & 0 & 0 & -\dfrac{n_{lb0}}{2L} & 0 & 0 & 0 & 0 \\[2ex]
0 & 0 & -\dfrac{R}{L} & 0 & 0 & -\dfrac{n_{uc0}}{2L} & 0 & 0 & -\dfrac{n_{lc0}}{2L} & 0 & 0 & 0 \\[2ex]
\dfrac{n_{ua0}}{C_{arm}} & 0 & 0 & 0 & 0 & 0 & 0 & 0 & 0 & \dfrac{n_{ua0}}{2C_{arm}} & 0 & 0 \\[2ex]
0 & \dfrac{n_{ub0}}{C_{arm}} & 0 & 0 & 0 & 0 & 0 & 0 & 0 & 0 & \dfrac{n_{ub0}}{2C_{arm}} & 0 \\[2ex]
0 & 0 & \dfrac{n_{uc0}}{C_{arm}} & 0 & 0 & 0 & 0 & 0 & 0 & 0 & 0 & \dfrac{n_{uc0}}{2C_{arm}} \\[2ex]
\dfrac{n_{la0}}{C_{arm}} & 0 & 0 & 0 & 0 & 0 & 0 & 0 & 0 & -\dfrac{n_{la0}}{2C_{arm}} & 0 & 0 \\[2ex]
0 & \dfrac{n_{lb0}}{C_{arm}} & 0 & 0 & 0 & 0 & 0 & 0 & 0 & 0 & -\dfrac{n_{lb0}}{2C_{arm}} & 0 \\[2ex]
0 & 0 & \dfrac{n_{lc0}}{C_{arm}} & 0 & 0 & 0 & 0 & 0 & 0 & 0 & 0 & -\dfrac{n_{lc0}}{2C_{arm}} \\[2ex]
0 & 0 & 0 & -\dfrac{n_{ua0}}{L} & 0 & 0 & \dfrac{n_{la0}}{L} & 0 & 0 & -\dfrac{R+2Z_L}{L} & 0 & 0 \\[2ex]
0 & 0 & 0 & 0 & -\dfrac{n_{ub0}}{L} & 0 & 0 & \dfrac{n_{lb0}}{L} & 0 & 0 & -\dfrac{R+2Z_L}{L} & 0 \\[2ex]
0 & 0 & 0 & 0 & 0 & -\dfrac{n_{uc0}}{L} & 0 & 0 & \dfrac{n_{lc0}}{L} & 0 & 0 & -\dfrac{R+2Z_L}{L}
\end{bmatrix} \tag{22}$$

TABLE I
MAIN ELECTRICAL PARAMETERS OF MMC

Quantity	Value
Fundamental frequency	50 Hz
Line-to-line RMS voltage	166 kV
dc-link voltage	320 kV
SM number per arm	20
SM capacitance	140 µF
Arm inductance	360 mH
Arm resistance	1 Ω

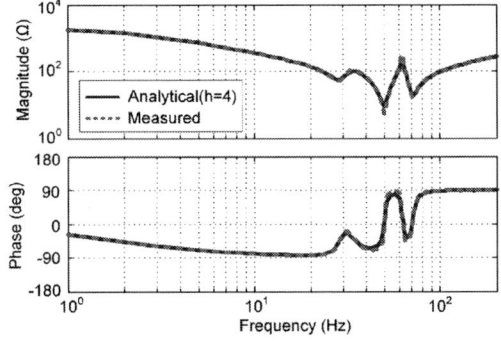

Fig. 3. Validation for the proposed MMC DSP impedance model in this paper.

IV. DSP IMPEDANCE MODEL VALIDATION

In order to verify the proposed MMC DSP impedance model in this paper, a nonlinear time-domain simulation model of a three-phase MMC with a three-phase resistor load has been built by using MATLAB/Simulink. In the simulation, the DSP impedance of the MMC is measured by means of injecting a series of small perturbation voltages at different frequencies in the dc-side of the MMC, as shown in Fig. 1(b), where the single-tone injection method [15] is used. Then by measuring the resulting small perturbation dc currents, the DSP impedance can be readily calculated for each perturbation frequency. The main electrical parameters of the MMC are presented in Table I. It should be pointed out that the

main electrical parameters of the MMC used in this paper are identical to those of a real MMC-HVDC system in China, but both the SM number and capacitance are one-tenth of the actual parameters, while keeping the arm equivalent capacitance C_{arm} unchanged.

Fig. 3 shows the comparison between the analytical and measured MMC DSP impedances, where the harmonic order h considered in the analytical model is selected as 4. As can be seen, the analytical DSP impedance matches very well with the measured one, which validates the proposed MMC DSP impedance model in this paper. Furthermore, it is pointed out that the impedance-frequency characteristics of the MMC above 200 Hz are inductive, which is mainly determined by the arm inductance.

Fig. 4 demonstrates the impact of the harmonic order h considered in the HSS model on the accuracy of the analytical DSP impedance model. It can be observed that the harmonic order h has great impact on the accuracy of the analytical MMC impedance model, especially in the low frequency range (<100 Hz). However, it also can be seen that the MMC DSP impedance responses are completely overlapped when the harmonic order $h \geq 3$, which indicates that the MMC impedance model will be accurate enough if the first three harmonics are considered in normal cases.

Furthermore, a comparison between the MMC DSP impedance responses obtained from the model in [12] and the proposed model (where $h=0$) in this paper is also carried out, as shown in Fig. 5. It can be seen that the two models have the same impedance responses if no harmonics are considered in the proposed DSP impedance model in this paper (i.e. $h=0$). In other words, the MMC DSP impedance model proposed in this paper is able to cover the existing DSP impedance models.

V. CONCLUSION

This paper develops the dc-side-port (DSP) impedance model of a three-phase MMC based on the HSS method,

The 2018 International Power Electronics Conference

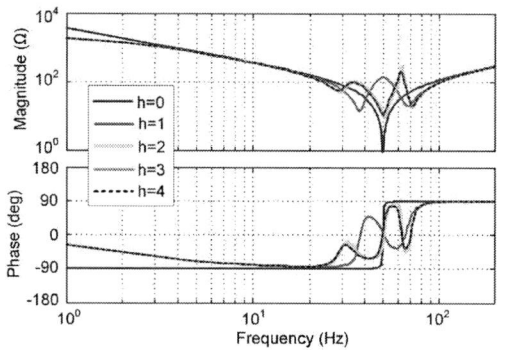

Fig. 4. Impact of the harmonic order h on the accuracy of the analytical DSP impedance model of MMC.

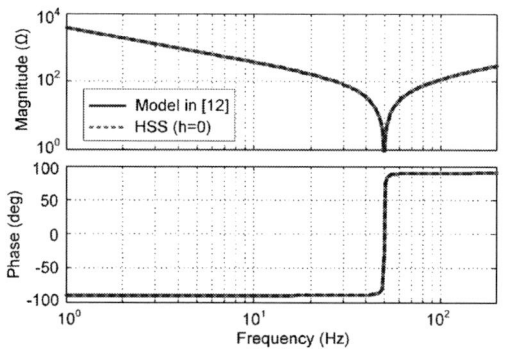

Fig. 5. Comparison between the DSP impedance model in [12] and the proposed DSP impedance model with h=0.

which is able to include all the internal dynamics. The accuracy of the proposed DSP impedance model depends on the considered harmonic order in the HSS model. Furthermore, The test results show that the proposed DSP impedance model not only covers all the existing DSP impedance models with setting the harmonic order h=0, but also can accurately reflect the internal harmonic effects with setting the suitable harmonic order.

ACKNOWLEDGMENT

This work was partly supported by the National Key Research and Development Program under Grant 2016YFB0900901 and by Shanghai Science and Technology Committee Scientific Research Program under Grant 16DZ1203402.

REFERENCES

[1] S. Dehnath, J. Qin, B. Bahrani, M. Saeedifard, and P. Barbosa, "Operation, control, and applications of the modular multilevel

converter: a review," *IEEE Trans. on Power Electronics*, vol. 30, no. 1, pp. 37-53, 2015.

[2] M. A. Perez, S. Bernet, J. Rodriguez, S. Kouro, and R. Lizana, "Circuit topologies, modeling, control schemes, and applications of modular multilevel converter," *IEEE Trans. on Power Electronics*, vol. 30, no. 1, pp. 4-17, 2015.

[3] G. Buigues, V. Valverde, A. Etxegarai, P. Eguía, and E. Torres, "Present and future multiterminal HVDC systems: current status and forthcoming developments," *International Conf. on Renewable Energies and Power Quality (ICREPQ'17)*, Malaga, Spain.

[4] Y. Li, G. Tang, J. Ge, Z. He, H. Pang, J. Yang, and Y. Wu, "Modeling and damping control of modular multilevel converter based dc grid," *IEEE Trans. on Power Systems*, vol. 33, no. 1, pp. 723-735.

[5] J. Sun, "Impedance-based stability criterion for grid-connected inverters," *IEEE Trans. on Power Electronics*, vol. 26, no. 11, pp. 3075-3078, 2011.

[6] L. Xu and L. Fan, "Impedance-based resonance analysis in a VSC-HVDC system," *IEEE Trans. on Power Delivery*, vol. 28, no. 4, pp. 2209-2216, 2013.

[7] L. Xu, L. Fan, and Z. Miao, "DC impedance-model-based resonance analysis of a VSC-HVDC system," *IEEE Trans. on Power Delivery*, vol. 30, no. 3, pp. 1221-1230, 2015.

[8] J. Lyu, X. Cai, and M. Molinas, "Frequency domain stability analysis of MMC-based HVdc for wind farm integration," *IEEE Journal of Emerging and Selected Topics in Power Electronics*, vol. 4, no. 1, pp. 141-151, 2016.

[9] J. Lyu, X. Cai, M. Amin, and M. Molinas, "Subsynchronous oscillation mechanism and its suppression in MMC-Based HVDC connected wind farms," *IET Generation, Transmission & Distribution*, vol. 12, no. 4, pp. 1021-1029, 2017.

[10] J. Lyu, X. Cai, and M. Molinas, "Optimal design of controller parameters for improving the stability of MMC-HVDC for wind farm integration," *IEEE Journal of Emerging and Selected Topics in Power Electronics*, vol. 6, no. 1, pp. 40-53, 2018.

[11] R. Mo, Q. Ye, and H. Li, "DC impedance modeling and stability analysis of modular multilevel converter for MVDC application," *IEEE ECCE 2016*, Milwaukee, USA.

[12] X. Shi, Z. Wang, B. Liu, Y. Li, L. M. Tolbert, and F. Wang, "DC impedance modelling of a MMC-HVDC system for DC voltage ripple prediction under a single-line-to-ground fault," *IEEE ECCE 2014*, Pittsburgh, USA.

[13] J. Lyu, M. Molinas, and X. Cai, "Harmonic state space modeling of a three-phase modular multilevel converter," 2017, arXiv: 1706.09925 [cs.SY].

[14] J. Lyu, Q. Chen, X. Cai, and M. Molinas, "Impedance analysis of modular multilevel converter based on harmonic state-space modeling method," 2017, arXiv: 1705.01030 [cs.SY].

[15] A. Rygg, M. Molinas, C. Zhang, and X. Cai, "Frequency-dependent source and load impedances in power systems based on power electronic converters," *PSCC 2016*, Genoa, Italy.

APPENDIX

The elements of the Toeplitz matrix **A** in (24) are shown as (A1)-(A3).

$$
A_0 = \begin{bmatrix}
-\dfrac{R}{L} & 0 & 0 & -\dfrac{1}{4L} & 0 & 0 & -\dfrac{1}{4L} & 0 & 0 & 0 & 0 & 0 \\[2mm]
0 & -\dfrac{R}{L} & 0 & 0 & -\dfrac{1}{4L} & 0 & 0 & -\dfrac{1}{4L} & 0 & 0 & 0 & 0 \\[2mm]
0 & 0 & -\dfrac{R}{L} & 0 & 0 & -\dfrac{1}{4L} & 0 & 0 & -\dfrac{1}{4L} & 0 & 0 & 0 \\[2mm]
\dfrac{1}{2C_{arm}} & 0 & 0 & 0 & 0 & 0 & 0 & 0 & 0 & \dfrac{1}{4C_{arm}} & 0 & 0 \\[2mm]
0 & \dfrac{1}{2C_{arm}} & 0 & 0 & 0 & 0 & 0 & 0 & 0 & 0 & \dfrac{1}{4C_{arm}} & 0 \\[2mm]
0 & 0 & \dfrac{1}{2C_{arm}} & 0 & 0 & 0 & 0 & 0 & 0 & 0 & 0 & \dfrac{1}{4C_{arm}} \\[2mm]
\dfrac{1}{2C_{arm}} & 0 & 0 & 0 & 0 & 0 & 0 & 0 & 0 & -\dfrac{1}{4C_{arm}} & 0 & 0 \\[2mm]
0 & \dfrac{1}{2C_{arm}} & 0 & 0 & 0 & 0 & 0 & 0 & 0 & 0 & -\dfrac{1}{4C_{arm}} & 0 \\[2mm]
0 & 0 & \dfrac{1}{2C_{arm}} & 0 & 0 & 0 & 0 & 0 & 0 & 0 & 0 & -\dfrac{1}{4C_{arm}} \\[2mm]
0 & 0 & 0 & -\dfrac{1}{2L} & 0 & 0 & \dfrac{1}{2L} & 0 & 0 & \dfrac{R+2Z_L}{L} & 0 & 0 \\[2mm]
0 & 0 & 0 & 0 & -\dfrac{1}{2L} & 0 & 0 & \dfrac{1}{2L} & 0 & 0 & \dfrac{R+2Z_L}{L} & 0 \\[2mm]
0 & 0 & 0 & 0 & 0 & -\dfrac{1}{2L} & 0 & 0 & \dfrac{1}{2L} & 0 & 0 & \dfrac{R+2Z_L}{L}
\end{bmatrix}
\tag{A1}
$$

$$
A_{\pm1} = \begin{bmatrix}
0 & 0 & 0 & \dfrac{m}{8L} & 0 & 0 & -\dfrac{m}{8L} & 0 & 0 & 0 & 0 & 0 \\[2mm]
0 & 0 & 0 & 0 & -\dfrac{m(1\pm j\sqrt{3})}{16L} & 0 & 0 & \dfrac{m(1\pm j\sqrt{3})}{16L} & 0 & 0 & 0 & 0 \\[2mm]
0 & 0 & 0 & 0 & 0 & -\dfrac{m(1\mp j\sqrt{3})}{16L} & 0 & 0 & \dfrac{m(1\mp j\sqrt{3})}{16L} & 0 & 0 & 0 \\[2mm]
-\dfrac{m}{4C_{arm}} & 0 & 0 & 0 & 0 & 0 & 0 & 0 & 0 & -\dfrac{m}{8C_{arm}} & 0 & 0 \\[2mm]
0 & \dfrac{m(1\pm j\sqrt{3})}{8C_{arm}} & 0 & 0 & 0 & 0 & 0 & 0 & 0 & 0 & \dfrac{m(1\pm j\sqrt{3})}{16C_{arm}} & 0 \\[2mm]
0 & 0 & \dfrac{m(1\mp j\sqrt{3})}{8C_{arm}} & 0 & 0 & 0 & 0 & 0 & 0 & 0 & 0 & \dfrac{m(1\mp j\sqrt{3})}{16C_{arm}} \\[2mm]
\dfrac{m}{4C_{arm}} & 0 & 0 & 0 & 0 & 0 & 0 & 0 & 0 & -\dfrac{m}{8C_{arm}} & 0 & 0 \\[2mm]
0 & -\dfrac{m(1\pm j\sqrt{3})}{8C_{arm}} & 0 & 0 & 0 & 0 & 0 & 0 & 0 & 0 & \dfrac{m(1\pm j\sqrt{3})}{16C_{arm}} & 0 \\[2mm]
0 & 0 & -\dfrac{m(1\mp j\sqrt{3})}{8C_{arm}} & 0 & 0 & 0 & 0 & 0 & 0 & 0 & 0 & \dfrac{m(1\mp j\sqrt{3})}{16C_{arm}} \\[2mm]
0 & 0 & 0 & \dfrac{m}{4L} & 0 & 0 & \dfrac{m}{4L} & 0 & 0 & 0 & 0 & 0 \\[2mm]
0 & 0 & 0 & 0 & -\dfrac{m(1\pm j\sqrt{3})}{8L} & 0 & 0 & -\dfrac{m(1\pm j\sqrt{3})}{8L} & 0 & 0 & 0 & 0 \\[2mm]
0 & 0 & 0 & 0 & 0 & -\dfrac{m(1\mp j\sqrt{3})}{8L} & 0 & 0 & -\dfrac{m(1\mp j\sqrt{3})}{8L} & 0 & 0 & 0
\end{bmatrix}
\tag{A2}
$$

$$
A_{\pm h} = O^{12\times12} \ (h \geq 2)
\tag{A3}
$$

An Improved Master-Slave Control for Three-port Converter Based Distributed DC Grid-connected PV System

Siyue Jiang[1], Kai Sun[1], Hongfei Wu[2], Haixu Shi[1], Xiaofeng Dong[2] and Syed Muhammad Raza Kazmi[3],
1. State Key Lab of Control and Simulation of Power Systems and Generation Equipment
Tsinghua University, Beijing, China
2. Jiangsu Key-Laboratory of New Energy Generation and Power Conversion
Nanjing University of Aeronautics and Astronautics
Nanjing, Jiangsu Province, China
3. United States Pakistan Center for Advanced Studies in Energy, National University of
Sciences and Technology, Islamabad, Pakistan
E-mail: jiangsiyue2008abc@163.com

Abstract— **With the development of DC grid technology, it has become a trend to introduce photovoltaic(PV) as a power source to the DC power distribution network. In this paper, an improved master-slave control strategy is proposed to realize power splitting and voltage sharing in a modular system with PV connected. Each module is a three-port converter: Input port connects distributed PV, output port connects to DC grid, and intermediate port is connected in parallel with other modules as a bus to realize the power splitting. For a single module, input voltage, intermediate bus voltage and output voltage should all be controlled to reference but only duty ratio and phase-shifting angel can be regulated hence it may occur the multi-objective variables control problem. While using the proposed strategy, this problem can be solved effectively and the system stability is ensured during the voltage sharing process. This control strategy is verified by simulation and experiment.**

Keywords— *distributed PV, three-port converter, master-slave control, input parallel output series(IPOS)*

I. INTRODUCTION

Nowadays, the development of DC grid technology has been an effective method to solve energy shortage and environmental pollution problems. DC grid technology, including renewable energy collection and long distance transmission, can help realize the interconnection of power systems in different voltage levels and in different areas [1]–[3]. It also has a bright future in power distribution system. With the increase of urban power load density and demand for power reliability and quality, AC/DC hybrid distribution network based on flexible DC technology will be more helpful and suitable for the operation mode of modern cities [4]–[6].

Among all kinds of distributed power source, PV as a typical DC power supply, has drawn a lot of attentions. It's future trends to integrate large-scale distributed PV power with DC grid or AC/DC hybrid grid reasonably. Generally, DC grid is made up of renewable energy sources such as PV or wind power, energy storage modules, power supplies and loads. All components are connected to the DC bus and rely on it to achieve the power allocation [7]–[10]. The literatures about the control method and the

energy management strategy of PV-connected DC grid show us a similar thinking [11]–[13]: the basic idea is to divide the system into different working modes according to the voltage level of DC bus, then the energy or power exchange among components will be arranged under different working modes respectively.

At present, the research on renewable energy access to grid, especially the MEDIUM voltage direct current(MVDC) power grid, is mainly focused on the structure of macro-system. For details such as specific circuit design and control strategy, it has not been studied too much in literatures. Therefore, a three-port converter with PV input is presented in this paper. In order to adapt to high voltage level of DC grid and improve the efficiency of PV, an appropriate modular connection is used in this system and an improved master-slave control method is proposed to realize the voltage sharing in gird-connected mode [14]. This control strategy is verified by simulation and experiment.

II. STRUCTURE OF PHOTOVOLTAIC BOOSTING SYSTEM

A. Structure of three-port converter

The MVDC system based distributed PV should contain two stages of conversion. The first stage is used to control PV modules to realize Maximum Power Point Tracking(MPPT) respectively. The second stage is a converter with high power and high boosting ratio. Based on this structure, the voltage of PV modules can be raised up to a proper level which can meet the requirement of DC grid. This system is illustrated in Fig.1.

However, in this system, at least two converters are needed to complete the two-stage conversion hence the efficiency will not be so high. if the structure is changed into a converter with three ports composed of PV, intermediate bus and load, the volume and weight of the system can be greatly reduced, and power density can be improved [15]–[17]. Our research team has proposed a Three-Port Converter(TPC) with partial isolation, the sketch is shown in Fig.2.-(a). This TPC can provide the necessary electrical isolation so as to avoid the mutual influences between DC grid and PVs. Furthermore, the two ports in the primary side are non-isolated hence it has higher efficiency than the traditional two-stage conversion. This TPC converter meets all needs of MVDC system with PV connected. Concrete structure is shown in Fig.2-(b).

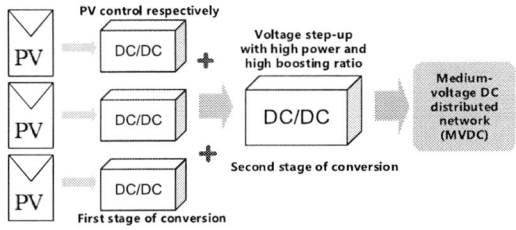

Fig.1. Structure of MVDC system with PV connected

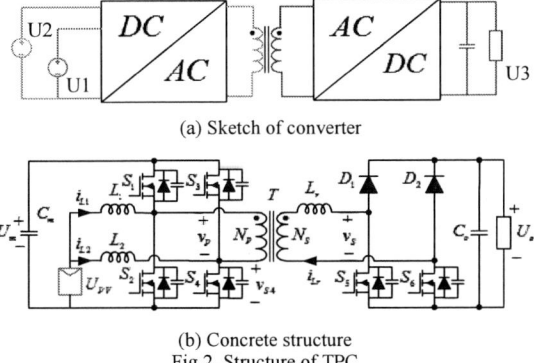

(a) Sketch of converter

(b) Concrete structure
Fig.2. Structure of TPC

The primary side is an interleaved buck/boost converter and the secondary side is a full-bridge boost rectifier. The soft switching can be realized by proper control thus the efficiency can be further improved.

B. Modular connection

Since the low power and low output voltage of single PV panel, modular connection is needed.

Series-parallel connection for modular isolated DC converter units can be used in high power and high boosting ratio conversion. Among all kinds of connecting methods, "Input Parallel and Output Series(IPOS)" is the most suitable one [18]. Take three-module system as an example, the connection method is illustrated in Fig.3.

PV is the distributed input; the intermediate bus U_m is regarded as the parallel side of the multi-module system; output U_o is the series side, connecting to the grid directly. In this system, intermediate bus is a channel for power exchange and balance among modules. For output side, the current sharing has been realized because of the series connection. Therefore, to ensure the same output power of each module, only the voltage sharing process should be considered.

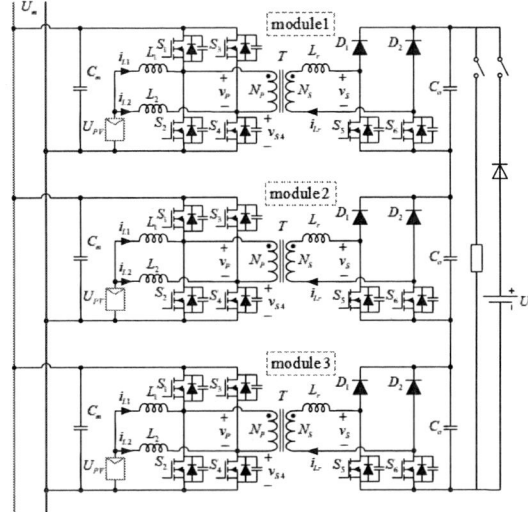

Fig.3. Connection mode for TPC modules

III. CONTROL STRATEGY

A Control target and circuit analysis

In this multi-module system, MPPT control is adopted on the input side to make input voltage of PVs equal to their MPPT voltage. Output side is series-connected and the voltage sharing is needed to guarantee the same output power of each module [19].

The interleaved structure in the primary side, can significantly reduce the input current ripple as shown in Fig.4. Input voltage is regulated by setting the duty ratio D reasonably.

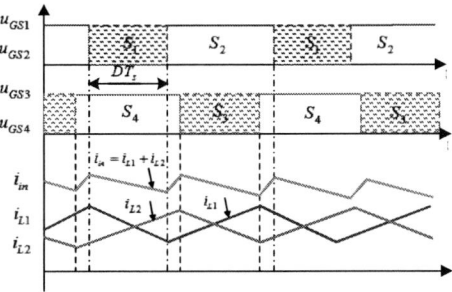

Fig.4. Waveforms of switches in primary side
and input current

Define G as the equivalent gain of the converter

$$G = \frac{nU_o}{U_m} \tag{1}$$

According to the relationship between U_o and U_m, the converter can work on buck mode($G < 1$), balance mode($G = 1$) and boost mode($G > 1$). Furthermore, via analyzing the current state of high frequency inductor L_r in the circuit, the converter can be divided into two working modes: Continuous Current Mode(CCM) and

Discontinuous Current Mode(DCM). The main waveforms are shown in Fig.5.

Due to the reuse of switching devices and circuit structures in the primary side, the two-stage conversion in Fig.1 has been simplified to a 1.5-stage conversion. Efficiency is improved exactly whereas it increases the coupling of circuit and difficulty of control. Distributed PVs deliver differential power to modules. For a single module, duty ratio D is used to regulate the input voltage and the phase-shifting angle φ is to regulate the output voltage. However, the whole modular system has the following requirements: input voltage U_{pv} should realize MPPT; U_m should be constant to achieve the power splitting; U_o of each module should be ensured the same so that every module can output the same power to grid or load. Therefore, it will inevitably occur multi-objective variables control problem during the voltage sharing process and the difficulty of control is raised up.

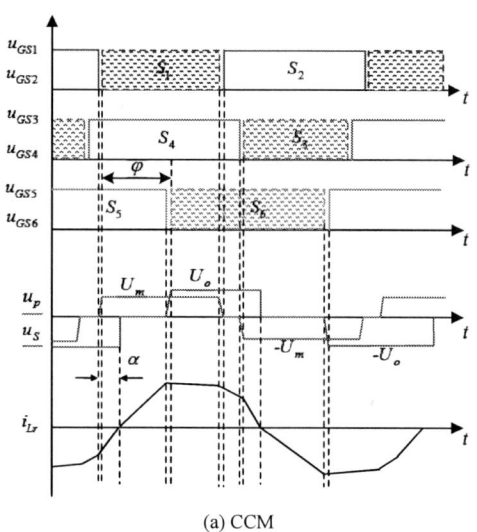

(a) CCM

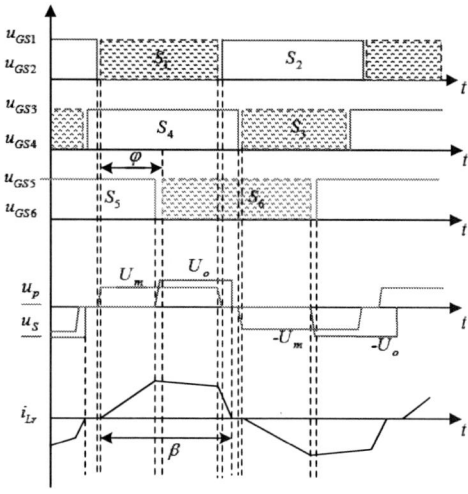

(b) DCM
Fig.5. Main waveforms of TPC

B. Control idea and realization

To solve the control problem above, more attention should be paid on the whole system rather than the single module.

The system, which composed of three sub modules, is designed as follows: The input voltage U_{pv} of each module is within the range of 25-40V; The intermediate bus U_m is 75V and the output total voltage on DC bus U_{odc} is 150V; U_o is the output voltage of each module.

According to the modular connection method, U_m ports are connected in parallel, that means U_m in every module is equal. Utilizing the master-slave control, one module is selected as the master module and the phase-shifting angle φ of this master module is used to control U_m at the reference. Due to the parallel connection, U_m of all modules will be the same automatically. Slave modules then trace the output voltage of the master module by adjusting their own φ, hence the voltage sharing in whole system can be realized.

$$U_{Mout} = U_{S1out} = U_{S2out} \qquad (2)$$

U_{Mout} is the output voltage of the master module. U_{S1out}, U_{S2out} are the output voltages of the slave modules. And the total voltage U_{odc} is 150V

$$U_{odc} = U_{Mout} + U_{S1out} + U_{S2out} = 150V \qquad (3)$$

thus

$$U_{Mout} = U_{S1out} = U_{S2out} = 50V \qquad (4)$$

Eventually, output voltage of three modules will be stable at 50V respectively under this control strategy.

Different from the traditional master-slave control strategy which target on the same variable, this improved method chooses master module to control U_m while the other slave modules trace the U_{Mout} to achieve the voltage sharing. By setting different control objectives for the master module and slave modules in the modular system, the multi-objective variables control problem is solved effectively. The basic principle of control strategy is illustrated in Fig.6.

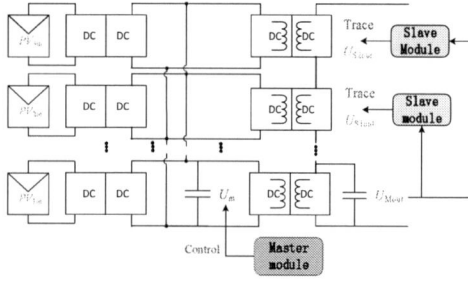

Fig.6. Basic principle of master-slave control

In terms of control implementation, Proportional Integral(PI) control method is adopted. For input side, proper algorithm of MPPT is selected to find the MPPT

4170

voltage through regulating the value of D. For output side, PI loop is used to control U_m or U_o.

$$u_M(t) = k_p \cdot (U_{mref} - U_m)(t) + k_i \cdot \int_0^t (U_{mref} - U_m)(t) dt \tag{5}$$

$$u_S(t) = k_p \cdot (U_{Mout} - U_{Sout})(t) + k_i \cdot \int_0^t (U_{Mout} - U_{Sout})(t) dt \tag{6}$$

In order to realize soft switching, the adjustable range of φ needs to be limited between

$$\varphi \in (-\min \frac{1}{2}\{D, 2\pi - D\}, \min \frac{1}{2}\{D, 2\pi - D\}) \tag{7}$$

The concrete control scheme is illustrated in Fig.7.

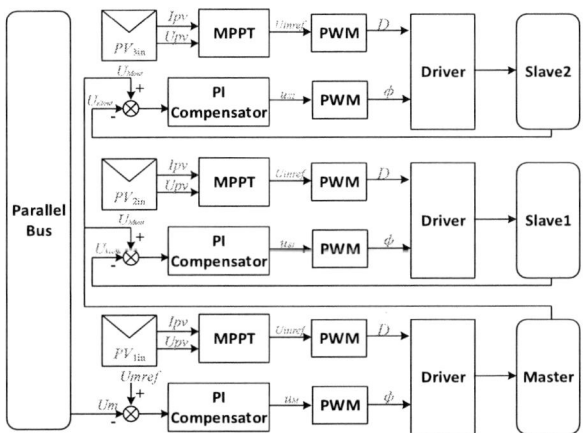

Fig.7. Control scheme of master-slave control

The main advantage of this control strategy is that it can keep the value of U_m unchanged in despite of the fluctuation of DC grid voltage. That means the MPPT process will not be influenced by DC grid and the duty ratio D is no need to be adjusted once more. The only way to change U_m is to modify the reference and this will greatly reduce the control loss and enhance the stability of the system.

The TPC modular system with distributed grid-connected PV is built by simulation and experiment and the validity of proposed control strategy is verified. Specific results are shown in the next chapter.

IV. VERIFICATION

A. Circuit parameter

The system standard is shown in TABLE I. The parameter values of main components of TPC is shown in TABLE II.

TABLE I
System standard of TPC

	Single module
Input voltage U_{pv}	25-40V
Output voltage U_o	50V
Intermediate bus U_m	75V
Working frequency f_s	100kHz

TABLE II
Key parameter of TPC

Transformer ratio n	3:2
DC bus voltage U_{odc}	150V
Energy transfer inductor L_r	25μH
Boost inductor $L_1 \& L_2$	60μH

B. Simulation result and analysis

In simulation, the system spends 0.04s to reach the balance point. To show the dynamic response, the following situations has been considered.

When the intensity of illumination and temperature change, the MPPT voltage and input power of PV will also be different. In Fig.8-(a), one of the distributed PV has changed its maximum power point and the voltage steps down at 0.1s. U_m and U_o is basically unaffected by the step-down of U_{pv}, hence the voltage sharing is guaranteed. Similarly, all PVs have a voltage step-up at 0.1s in Fig.8-(b), that means input power changes at this time. According to the waveforms, U_o is affected a bit while it recovers quickly so that the stability of output is ensured.

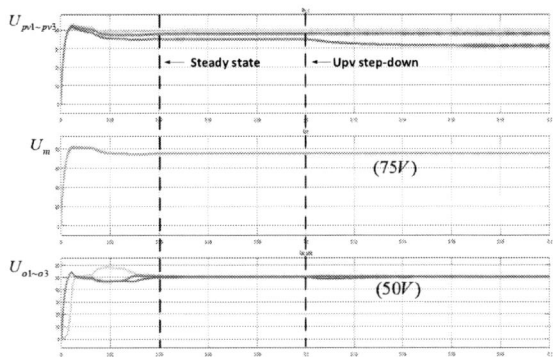

(a) One of PV input voltage steps down

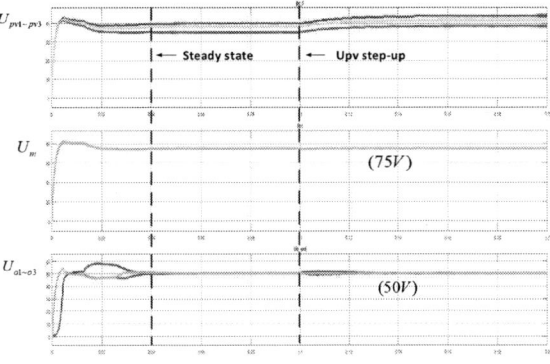

(b) Voltage of all PVs steps up
Fig.8. Simulation waveforms when PV input changes

The DC grid side is stable usually while inevitably there will be voltage fluctuations at some time. Fig.9 illustrates this case. The total output voltage of DC grid has a 10% fluctuation at 0.1s and it causes the step-up of U_{odc}.

Based on proposed control strategy, U_o of each module will change synchronously and voltage sharing is kept during this process. The waveforms prove it right. The value of U_m and U_{pv} remain the same and that means the fluctuation of DC grid side will not influence the MPPT process.

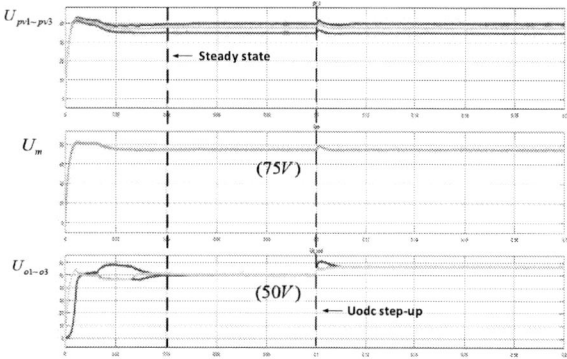

Fig.9. Simulation waveforms when output voltage changes

Within the controllable range, the reference U_{mref} is changed at 0.1s. Similar to the previous process, U_m reaches a new stable value and U_{pv} of three modules are influenced but quickly recover. U_o is also affected and need to be regulated once more then it returns to the voltage sharing state. The waveforms are shown in Fig.10.

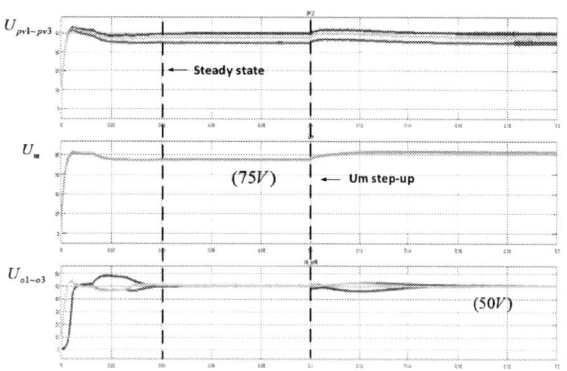

Fig.10. Simulation waveforms when Um changes

C. Experimental results and waveform analysis

The prototype of this TPC system has been built to verify the previous strategy. The three sub-modules system reaches the steady state and the output voltages of modules are all stable at 50V as expected. In steady state, the

waveforms of switching voltage and inductor current i_{Lr} in one module is illustrated in Fig.11.

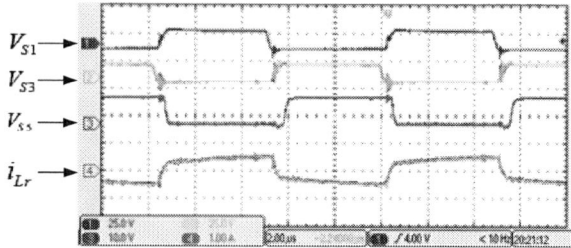

Fig.11. Waveforms of switching voltage and inductor current

Switches S1,S3 is operating at interleaving 180°. Switch S5 lags φ behind S3 and via regulating its value, U_m (master module) or U_o (slave module) can be controlled. According to Fig.5-(a) it can be confirmed that the converter is working in CCM.

In order to observe the dynamic response of the system, the output DC bus voltage and the input power have been changed respectively in the experiment. The situation of DC bus voltage changes is shown in Fig.12.

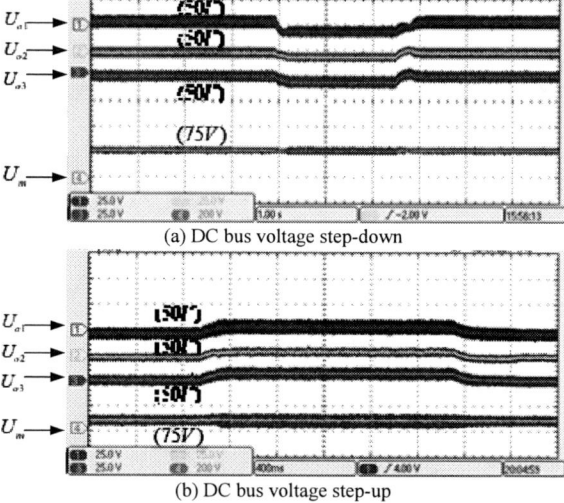

(a) DC bus voltage step-down

(b) DC bus voltage step-up

Fig.12. Waveforms of U_o and U_m

The output DC bus voltage is changed from 150V to 130V in Fig.12-(a), and then back to 150V. During this process, U_o of each module is changed from 50V to 43V.

When the DC bus voltage recovers, U_o is also back to 50V synchronously. Voltage sharing is kept all the time. The change of DC bus voltage will not affect U_m according to the strategy and the waveform proves it right: U_m remains the same during the whole process. Then the DC bus voltage is raised from 150V to 170V in Fig.12-(b) and the same result is got: the system keeps the voltage sharing state when U_o follows the change of DC bus voltage and U_m remains the same as expected.

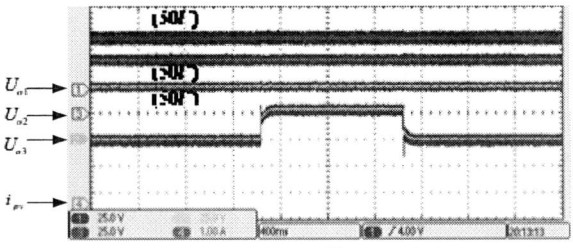

Fig.13. Waveforms of U_o and i_{pv}

Fig.13 shows the situation of input power changes. The step-up of i_{pv} means the sudden change of input power. Because the DC bus voltage is not changed, the voltage sharing state should keep the same. The waveforms of U_o proves it right.

V. CONCLUSION

According to the requirements of MVDC system with PV connected, a modular system made up of TPCs is chosen. In this system, the input of each module is distributed PV, and the intermediate bus is used as a channel for power balancing, and output side is connected in series and to the grid. In this case, the converter volume is greatly reduced and the transmission efficiency is improved.

Based on this topology, an improved master-slave control strategy is proposed. For single module, three target variables should be controlled while only two controllable variables can be regulated, hence it increases the difficulty of control. However, with the improved master-slave control proposed in this paper, this problem can be solved basically. The duty ratio D is used to realize MPPT on input side. At the same time, master module is chosen to control the intermediate bus voltage U_m and slave modules are set to trace the output voltage of master module so that voltage sharing can be achieved. In addition, MPPT process in primary side will not be affected by the fluctuation of DC grid because U_m stays the same unless the reference is modified. This is the main advantage of this control strategy because it reduces the control loss and enhances stability of the system substantially. In chapter IV, this strategy has been proved to be valid by simulation and experiment and it will be an influential method to be used in modular system with DC grid-connected PV broadly in the future life.

REFERENCES

[1] Grainger B M, Reed G F, Sparacino A R, et al. Power Electronics for Grid-Scale Energy Storage[J]. Proceedings of the IEEE, 2014, 102(6):1000-1013.

[2] Papadimitriou C N, Kleftakis V A, Rigas A, et al. DC-microgrid control strategy using DC-BUS signaling[C]// Mediterranean Conference on Power Generation, Transmission Distribution and Energy Conversion, Medpower. 2014.

[3] Sun Q, Han R, Zhang H, et al. A Multiagent-Based Consensus Algorithm for Distributed Coordinated Control of Distributed Generators in the Energy Internet[J]. IEEE Transactions on Smart Grid, 2015, 6(6):3006-3019.

[4] A. Q. Huang, M. L. Crow, G. T. Heydt, J. P. Zheng, S. J. Dale. The future renewable electric energy delivery and management (FREEDM) system: The energy internet[J]. Proceedings of the IEEE, 2011, 99(1): 133-148.

[5] Wang Z, Chen Z, Wang X. Research of the DC microgrid topology[C]// Control and Decision Conference. IEEE, 2016:2855-2859.

[6] Yuefeng Yang, Jie Yang, Zhiyuan He. Research on control and protection system for Shanghai Nanhui MMC VSC-HVDC demonstration project[C]//Proceedings of 10th IET International Conference on AC and DC Power Transmission (ACDC 2012). United Kingdom: IET, 2012: 1-6.

[7] T. Greyard and W. Dongtao, "Research on control structure of microgrid system with grid-connected photovoltaic," *2017 29th Chinese Control And Decision Conference (CCDC)*, Chongqing, 2017, pp. 6971-6976.

[8] Liu C, Chau K T, Diao C, et al. A new DC micro-grid system using renewable energy and electric vehicles for smart energy delivery[J]. 2010:1-6.

[9] Carbone R. Grid-connected photovoltaic systems with energy storage[C]// International Conference on Clean Electrical Power. IEEE, 2009:760-767.

[10] Tina G M, Pappalardo F. Grid-connected photovoltaic system with battery storage system into market perspective[C]// Sustainable Alternative Energy. IEEE, 2009:1-7.

[11] Han Y, Xie X G, Deng H, et al. Energy management method for photovoltaic DC micro-grid system based on power tracking control[C]// Industrial Electronics Society, IECON 2016 -, Conference of the IEEE. IEEE, 2016.

[12] Schonbergerschonberger J, Duke R, Round S D. DC-Bus Signaling: A Distributed Control Strategy for a Hybrid Renewable Nanogrid[J]. IEEE Transactions on Industrial Electronics, 2006, 53(5):1453-1460.

[13] Sun K, Zhang L, Xing Y, et al. A Distributed Control Strategy Based on DC Bus Signaling for Modular Photovoltaic Generation Systems With Battery Energy Storage[J]. IEEE Transactions on Power Electronics, 2011, 26(10):3032-3045.

[14] G. R. Walker and J. C. Pierce, Photovoltaic DC-DC module integrated converter for novel cascaded and bypass grid connection topologies — Design and optimization, 2006 37th IEEE Power Electronics Specialists Conference, Jeju, 2006, pp. 1-7.

[15] Al-Atrash H, Reese J, Batarseh I. Tri-modal half-bridge converter for three-port interface. IEEE PESC, 2007: 1702-1708.

[16] Qian Z, Abdel-Rahman O, Al-Atrash H, et al. Modeling and control of three-port DC/DC converter interface for satellite applications. IEEE Trans. on Power Electronics, 2010, 25(3): 637-649.

[17] T. Augustine, J. Jose and A. Rajan, "Phase shift and duty cycle control of interleaved boost three port converter with multiplier rectifier," *2017 International Conference on Circuit ,Power and Computing Technologies (ICCPCT)*, Kollam, India, 2017, pp. 1-8.

[18] A. Bhinge, N. Mohan, R. Giri, at al. Series-parallel connection of DC-DC converter modules with active sharing of input voltage and load current[C]//Applied Power Electronics Conference and Exposition, 2002. APEC 2002. Seventeenth Annual IEEE. United States: IEEE, 2002, 2: 648-653.

[19] Lu Y, Wu H, Dong X, et al. A three-port converter based DC grid-connected PV system with autonomous output voltage sharing control[C]// Applied Power Electronics Conference and Exposition. IEEE, 2017:2057-2061.

Sensorless Position Estimation, Parameter Identification and Control Integration for Permanent Magnet Synchronous Machines using Current Derivative Measurements

M.X. Bui[1*], M. F. Rahman[2], *Fellow IEEE,* and D. Xiao[3], *Member IEEE*

1. School of Electrical Engineering and Telecommunications, UNSW, Sydney, Australia
2. School of Electrical Engineering and Telecommunications, UNSW, Sydney, Australia
3. School of Electrical Engineering and Telecommunications, UNSW, Sydney, Australia

* E-mail: x.bui@unsw.edu.au

Abstract - This paper presents the position and speed estimation techniques for IPMSM in which the current derivative measurements during each PWM cycle are used. This technique requires no additional current injection. Instead, the rates of rise and fall of machine currents due to certain voltage-vector excitations during each PWM switching period are used for estimating the speed and position. Furthermore, it has been shown that these derivatives can also be used to track the inductance parameter variations L_d and L_q, of which the latter is known to vary significantly with load. The use of current derivative based position, speed and inductance parameters have fast response time compared to existing methods such as signal injection, and have potential for improving model based sensorless control, direct torque control, model predictive control and for more accurate trajectory (such as MTPA and MTPV) following under FOC and sensorless control. This paper presents the theory and experimental evaluation of current derivative based estimators for sensorless IPMSM control over wide speed range from zero to rated speed.

Index Terms—**Sensorless Control, Parameter Identification, Current Derivative Measurement**

I. INTRODUCTION

Sensorless control methods of Permanent Magnet Synchronous Machines (PMSM) have been developed over the past decades. The estimation of rotor speed and position can be mainly based on the model or the saliency of the machine. The model based methods such as, sliding mode observer (SMO)[1, 2], Kalman Filter [3, 4], Luenburger Observer [5, 6] are effective and robust at medium and high speed range. However, at very low speed operation, the estimation accuracy of these methods is low due to low signal to noise of back EMF. In contrast, methods based on saliency, such as signal injection [7-10] are robust at low speed ranges including standstill. However, these methods lose effectiveness at high speed operation due to the limitation of modulation index. In addition, the injected signal increases the current and torque ripple and deteriorates the overall performance of the control system.

In order to reduce the current distortion and increase the bandwidth of the control system, methods based on measurement of current derivatives, such as Indirect Flux Detection by Online Reactance Measurement (INFORM)

[11], Extended Modulation [12] and Fundamental PWM Excitation (FPE) [13-18] have been developed. The INFORM method is based on the injection of three pairs of opposite voltage-vectors during three consecutive PWM cycles. The current derivatives of three phases are measured during the application of those test voltage-vectors in order to estimate the rotor position. The bandwidth of this method is limited by a third of PWM frequency. Moreover, this method is only applicable at low speed range, which guarantees enough duration of null state for the injection of the test voltage vectors. The EM method divides the modulation space into four overlapping sectors so as to increase the estimation bandwidth to PWM frequency and to increase the duration of the active-voltage vectors. The drawback of this method is the increase of the common-mode voltage and current due to the increase of voltage transient of each phase during each PWM cycle.

FPE methods are based on the measurement of current derivatives during the excitation of voltage vectors in each standard PWM cycle. It is well known that the parasitic effects cause the high frequency oscillation in the phase currents when the switches turn on and off [19]. Therefore, narrow voltage vectors must be extended to a minimum duration of time which can guarantee the accuracy of current derivative measurements. This minimum duration depends on the load current, inverter topology, cabling and machine inductances [15]. Extension of voltage vectors causes distortion of phase currents, deteriorates the performance of the system controller, increases the current and torque ripples and affects the stability of the control system. In addition, the FPE methods fail to operate at high speed when the required modulation index reaches the limit of the inverter. For SPMSMs with small inductances, the HF oscillation duration is less than about 8µs. Consequently, the extension of two active-voltage vectors and two zero-voltage vectors does not significantly affect the current waveform and the performance of the systems. However, for IPMSMs, which have much larger inductances, the oscillations of phase currents lasts about 20 µs at full load operation. As a result, voltage vectors have to be extended far longer than for SPMSMs, which results in highly distorted stator currents. To reduce the current distortion, a modified FPE sensorless method using the current derivatives of two active-voltage vectors and one zero-voltage vector has been proposed in [20, 21]. This technique can eliminate the extension of one zero-voltage vector. However, the estimation accuracy at low speeds including standstill remains poor because of the parasitic effect. To overcome the disadvantages of the FPE methods at high speed operation, hybrid techniques have been proposed in [22, 23]. At low and very low speed operation, the FPE method is employed, while at medium and high speeds, the SMO method is utilized. However, the aforementioned problems of the FPE methods at low and very low speeds still exist due to the parasitic effects.

It is worth noting that the high performance control of PMSM drive system requires the accurate knowledge of machine inductances. There have been a number of off-line and on-line methods to determine the L_d and L_q. The off-line method is based on the alignment of the rotor at d- and q- axes at standstill and the application of test signals [24-26]. The machine inductances are then calculated based on the measured voltage and current. The drawback of off-line methods is the additional hardware, such as clamping system, signal generation equipment which is also time consuming. In addition, the variation of machine parameters under different the operating conditions is not considered in the off-line methods. In order to overcome the limitation of the off-line methods, on-line techniques have been developed, which are based on the machine's dynamic equations and the application of modern control theories, such as Recursive Least Square (RLS) [27-34], Affine Projection Algorithm (APA) [35, 36], Kalman filter [37, 38], Model Reference Adaptive System (MRAS) [37, 39, 40] and Neural Network [41, 42]. In general, these methods take considerable time encompassing many PWM cycles, due to the recursive process, to achieve the convergence of the algorithm. In addition, tuning of the parameters of the RLS or APA are not straightforward. The injected signals causes current distortion and affect the general performace of the system. Machine inductances can be also be identified on-line based on the measurement of currrent derivatives and the DC bus voltage [43]. This estimation technique can reduce the update time of inductance and eliminate the current distortion caused by signal injection.

This paper presents the sensorless control method based on the measurement of the current derivatives at one active-voltage vector and one zero-voltage vector during each PWM cycle. In addition, this paper shows the on-line estimation of the inductances of the IPMSM by measuring current derivatives and the DC bus voltage of the inverter. The effectiveness of the application of estimated inductances for the FOC algorithm with MTPA is also presented. The experimental studies were conducted to verify the effectiveness and robusness of the proposed methods.

II. SENSORLESS CONTROL AND INDUCTANCE ESTIMATION BASED ON CURRENT DERIVATIVE MEASUREMENT

A. System model

Assume that during a PWM cycle the rotor position is unchanged, thus the self and mutual inductance of the machine is constant. The voltage equations of an PMSM are described as [44]:

$$V_A = R_A i_A + L_{AA}\frac{di_A}{dt} + L_{AB}\frac{di_B}{dt} + L_{AC}\frac{di_C}{dt} + e_A$$

$$V_B = R_B i_B + L_{BA}\frac{di_A}{dt} + L_{BB}\frac{di_B}{dt} + L_{BC}\frac{di_C}{dt} + e_B \quad (1)$$

$$V_C = R_C i_C + L_{CA}\frac{di_A}{dt} + L_{CB}\frac{di_B}{dt} + L_{CC}\frac{di_C}{dt} + e_C$$

where:

$$L_{AA} = L_{\Sigma} + L_{\sigma} + L_{\Delta}\cos(2\theta_e)$$

$$L_{BB} = L_{\Sigma} + L_{\sigma} + L_{\Delta}\cos(2\theta_e + 2\pi/3)$$

$$L_{CC} = L_{\Sigma} + L_{\sigma} + L_{\Delta}\cos(2\theta_e + 4\pi/3)$$

$$L_{BC} = L_{CB} = -\frac{L_{\Sigma}}{2} + L_{\Delta}\cos(2\theta_e)$$

$$L_{AB} = L_{BA} = -\frac{L_{\Sigma}}{2} + L_{\Delta}\cos(2\theta_e - 2\pi/3) \quad (2)$$

$$L_{AC} = L_{CA} = -\frac{L_{\Sigma}}{2} + L_{\Delta}\cos(2\theta_e - 4\pi/3)$$

$$L_{\Sigma} = \frac{L_{di} + L_{qi} - 2L_{\sigma}}{3}$$

$$L_{\Delta} = \frac{L_{di} - L_{qi}}{3}$$

where V_A, V_B, V_C are stator voltage of phase A, B and C; i_A, i_B, i_C are the stator current of phase A, B, and C; R_A, R_B, R_C are the stator resistance of phase A, B, and C; e_A, e_B, e_C are the stator back EMF of phase A, B and C; L_{AA}, L_{BB}, L_{CC} are the stator self-inductance of phase A, B and C respectively; L_{AB}, L_{BA}, L_{AC}, L_{CA}, L_{BC}, L_{CB} are the mutual inductances between respective phases; θ_e is the electrical angle of the rotor; L_{σ} is the leakage inductance of stator winding; L_{Σ}, L_{Δ} are the average inductance and the magnitude of the inductance variation; L_{di}, L_{qi} are direct and quadrature incremental inductances respectively.

The equations presented in (1) and (2) are applicable for both Interior type and Surface- type PMSM. In case of Surface-type PMSM, there exists small difference between d- and q-axis inductances, considering different saturation levels along the two axes.

B. Estimation of rotor speed and position

Scalar p_A, p_B, p_C of the position vector are defined as:

$$p_A = \frac{2L_{\Delta}}{L_{\Sigma} + \frac{2}{3}L_{\sigma}}\cos(2\theta_e)$$

$$p_B = \frac{2L_{\Delta}}{L_{\Sigma} + \frac{2}{3}L_{\sigma}}\cos(2\theta_e - 2\pi/3) \quad (3)$$

$$p_C = \frac{2L_{\Delta}}{L_{\Sigma} + \frac{2}{3}L_{\sigma}}\cos(2\theta_e - 4\pi/3)$$

The current derivatives at the longer active-voltage vector and at zero voltage-vector during the first half of each PWM cycle are measured to calculate the position scalar. The equation for calculating the position scalars are shown in Table I [43].

$$g = \frac{9}{2V_{DC}}(L_\Sigma + \frac{2}{3}L_\sigma)\left(1 - \left(\frac{L_\Delta}{L_\Sigma + \frac{2}{3}L_\sigma}\right)^2\right) \quad (4)$$

The quantity g in (4) is updated during operation and takes into account the change of L_{di} and L_{qi} (determined by L_Σ and L_Δ) as well as variation of the DC bus voltage. The value of g is calculated by assuming that the rotor position is unchanged during two consecutive PWM cycles where there is a change of the selected active-voltage vector as shown in Table II [43].

TABLE I
POSITION SCALARS OF IPMSM WITH STAR CONNECTION

Pair of active and zero-voltage vector	p_A	p_B	p_C
V_1 and V_0	$2 - g\left(\frac{di_A^{(1)}}{dt} - \frac{di_A^{(0)}}{dt}\right)$	$-1 - g\left(\frac{di_C^{(1)}}{dt} - \frac{di_C^{(0)}}{dt}\right)$	$-1 - g\left(\frac{di_B^{(1)}}{dt} - \frac{di_B^{(0)}}{dt}\right)$
V_2 and V_0	$-1 + g\left(\frac{di_B^{(2)}}{dt} - \frac{di_B^{(0)}}{dt}\right)$	$-1 + g\left(\frac{di_A^{(2)}}{dt} - \frac{di_A^{(0)}}{dt}\right)$	$2 + g\left(\frac{di_C^{(2)}}{dt} - \frac{di_C^{(0)}}{dt}\right)$
V_3 and V_0	$-1 - g\left(\frac{di_C^{(3)}}{dt} - \frac{di_C^{(0)}}{dt}\right)$	$2 - g\left(\frac{di_B^{(3)}}{dt} - \frac{di_B^{(0)}}{dt}\right)$	$-1 - g\left(\frac{di_A^{(3)}}{dt} - \frac{di_A^{(0)}}{dt}\right)$
V_4 and V_0	$2 + g\left(\frac{di_A^{(4)}}{dt} - \frac{di_A^{(0)}}{dt}\right)$	$-1 + g\left(\frac{di_C^{(4)}}{dt} - \frac{di_C^{(0)}}{dt}\right)$	$-1 + g\left(\frac{di_B^{(4)}}{dt} - \frac{di_B^{(0)}}{dt}\right)$
V_5 and V_0	$-1 - g\left(\frac{di_B^{(5)}}{dt} - \frac{di_B^{(0)}}{dt}\right)$	$-1 - g\left(\frac{di_A^{(5)}}{dt} - \frac{di_A^{(0)}}{dt}\right)$	$2 - g\left(\frac{di_C^{(5)}}{dt} - \frac{di_C^{(0)}}{dt}\right)$
V_6 and V_0	$-1 + g\left(\frac{di_C^{(6)}}{dt} - \frac{di_C^{(0)}}{dt}\right)$	$2 + g\left(\frac{di_B^{(6)}}{dt} - \frac{di_B^{(0)}}{dt}\right)$	$-1 + g\left(\frac{di_A^{(6)}}{dt} - \frac{di_A^{(0)}}{dt}\right)$

The scalars of the position vector in the stationary reference frame can be expressed as:

$$p_\alpha = (2p_A - p_B - p_C)/3 = \frac{2L_\Delta}{L_\Sigma + \frac{2}{3}L_\sigma}\cos(2\theta_e)$$

$$p_\beta = (p_B - p_C)/\sqrt{3} = -\frac{2L_\Delta}{L_\Sigma + \frac{2}{3}L_\sigma}\sin(2\theta_e) \quad (5)$$

where p_α, p_β are the position scalars in the stationary reference frame.

The phase lock loop is then used to estimate rotor speed and position.

C. Estimation of machine inductance

Solving (4) and (5), the incremental inductances can be calculated as:

$$L_{qi} = \frac{gV_{DC}}{3(1 - \frac{\sqrt{p_\alpha^2 + p_\beta^2}}{2})}$$

$$L_{di} = \frac{gV_{DC}}{3(1 + \frac{\sqrt{p_\alpha^2 + p_\beta^2}}{2})} \quad (6)$$

The apparent inductances are obtained by integrating the incremental inductance:

$$L_{d,q}^{(n)} = \frac{\Phi_{d,q}^{(n)}}{I_{d,q}^{(n)}} = \frac{\Sigma L_{di,qi}^{(k)}(I_{d,q}^{(k)} - I_{d,q}^{(k-1)})}{I_{d,q}^{(n)}}, \; k=1 - n \quad (7)$$

where $\Phi_{d,q}^{(n)}$ are the total flux, $L_{d,q}^{(n)}$ are the apparent inductance, $L_{di,qi}^{(k)}$ are the incremental inductances calculated from (6), $I_{d,q}^{(k)}$ is the d- and -q current.

TABLE II
CALCULATION OF g

Pair of two active-voltage vectors	g
V_1 and V_2	$\dfrac{3}{\dfrac{di_A^{(1)}}{dt} - \dfrac{di_A^{(0)}}{dt} + \dfrac{di_B^{(2)}}{dt} - \dfrac{di_B^{(0)}}{dt}}$
V_2 and V_3	$\dfrac{3}{\dfrac{di_A^{(2)}}{dt} - \dfrac{di_A^{(0)}}{dt} + \dfrac{di_B^{(3)}}{dt} - \dfrac{di_B^{(0)}}{dt}}$
V_3 and V_4	$\dfrac{-3}{\dfrac{di_A^{(4)}}{dt} - \dfrac{di_A^{(0)}}{dt} + \dfrac{di_C^{(3)}}{dt} - \dfrac{di_C^{(0)}}{dt}}$
V_4 and V_5	$\dfrac{-3}{\dfrac{di_A^{(4)}}{dt} - \dfrac{di_A^{(0)}}{dt} + \dfrac{di_B^{(5)}}{dt} - \dfrac{di_B^{(0)}}{dt}}$
V_5 and V_6	$\dfrac{-3}{\dfrac{di_A^{(5)}}{dt} - \dfrac{di_A^{(0)}}{dt} + \dfrac{di_B^{(6)}}{dt} - \dfrac{di_B^{(0)}}{dt}}$
V_6 and V_1	$\dfrac{3}{\dfrac{di_A^{(1)}}{dt} - \dfrac{di_A^{(0)}}{dt} + \dfrac{di_C^{(6)}}{dt} - \dfrac{di_C^{(0)}}{dt}}$

III. EXPERIMENTAL SETUP AND RESULTS

A. Experimental setup

The experimental setup shown in Fig.1 includes one DS1103 board for implementing the system control and estimating the inductances as well as rotor speed and position. The tested machine is Kollmorgan IPMSM (BE2-402-A-A4) coupled with a DC machine for loading. The parameters of the tested machine are shown in Table III. Three 12 bit ADCs are used to convert the measured stator currents from current sensor boards (CMS3005) to FPGA and DS1103. The FPGA board (ML605) was utilized for generating the PWM signals and for calculating the current derivatives. The calculated current derivatives are then sent to DS1103 for estimation of inductances, rotor peed and position.

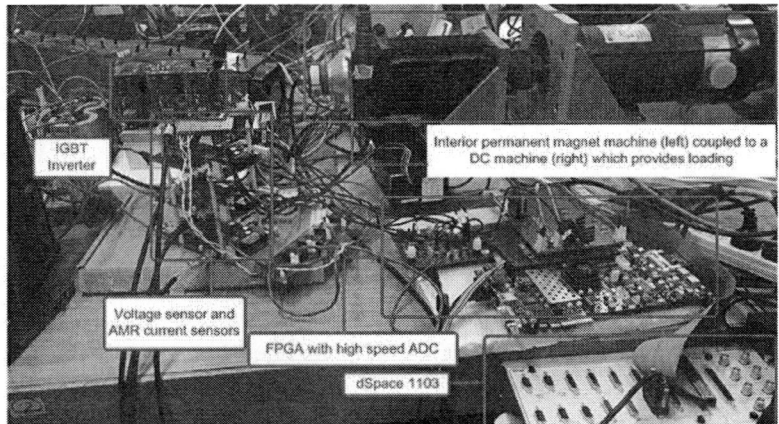

TABLE III
PARAMETERS OF IPMSMs TESTED

Number of pole pair	Pp	2
Stator resistance	R	5.8 Ω
Magnet flux linkage	λ_f	0.533 Wb
d-axis inductance	L_d	0.0448 H
q-axis inductance	L_q	0.1024 H
Phase voltage (rms)	V	230 V
Phase current (rms)	I	3 A
Rated torque	Te	6 Nm

Fig.1. Experimental setup.

B. Sensorless control based on current derivative measurement

The sensorless performance of the system at 30 rpm under 85% of rated load is shown in Fig. 2. The position error is smaller than 9 electrical degrees, while the speed error is smaller than 5 rpm during the steady state.

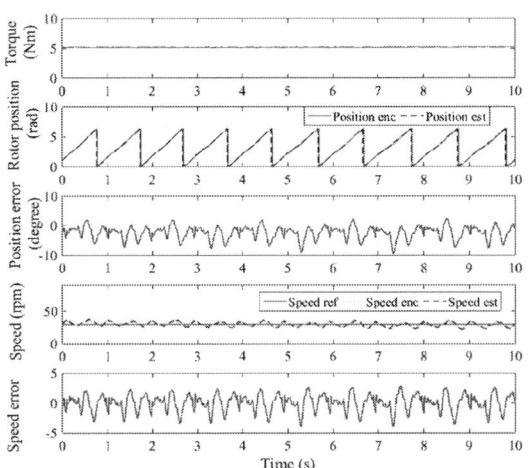

Fig. 2. Sensorless control at 30rpm under 85% of rated load.

The proposed sensorless control at zero speed under a reversal change of full load is shown in Fig. 3. The position error during the transient is less than 20 electrical degrees. The absolute value of speed error is smaller than 12 rpm during the transient. At steady state, the position error is reduced to less than 7 electrical degrees, while the speed error is about 1 rpm.

Fig. 4 shows the sensorless operation when the speed reverses at 1500 rpm under no load. It is clear that during the transient state the position error is within 30 electrical degrees. The transient time is

about 0.8s. At the steady state the position error is smaller than 5 electrical degrees.

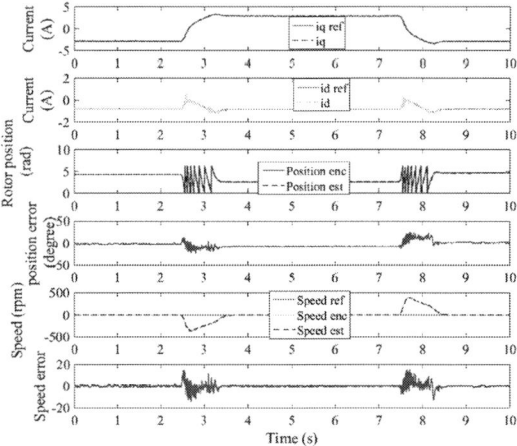

Fig. 3. Sensorless control at zero speed under sudden change of full load.

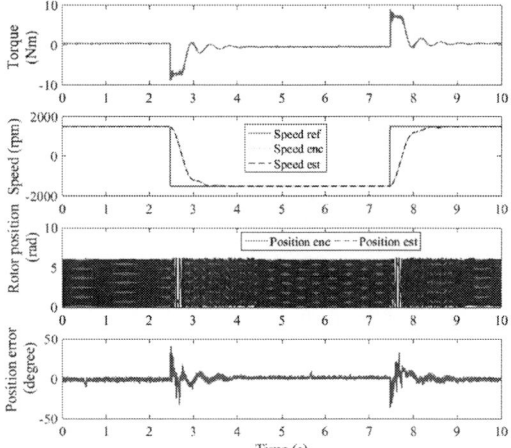

Fig.4. Sensorless control with speed reversal at 1500 rpm under no load.

4177

C. On-line estimation of machine inductances

To validate the proposed method, full system simulation was conducted using Matlab-Simulink. The simulated results were verified in an experimental set-up. L_d and L_q of the IPMSM of Table III was first measured experimentally off-line by using the standard IEEE standstill test [24] at the frequency of 50 Hz. The inductances are calculated directly based on the measurement of RMS voltage and current when the rotor is aligned at d- or q-axes as shown in Fig.5. The tested machine was modeled and run with FOC control algorithm under various loading condition in Matlab-Simulink. For each load condition, corresponding L_d and L_q of the IPMSM model were set according to the measured off-line values. The inductances were then estimated on-line, using the proposed method which is based on current derivatives of three phases and the DC bus voltage. The on-line estimated values of L_d and L_q are compared with the off-line measured values in Fig.5. It is clear that the estimated and the measured L_d and L_q are very close.

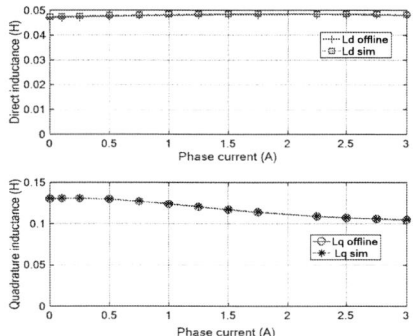

Fig.5. Online estimation of inductances at 500 rpm (simulation).

Fig.6 compares the inductances estimated on-line by experiment and the inductances calculated off-line at 50 Hz when the IPMSM is operated at 500 rpm and load current is adjusted from 0.25A (no load) to 3A (approx 120% of rated load). It is clear that the estimated inductances are almost the same as the off-line measured ones.

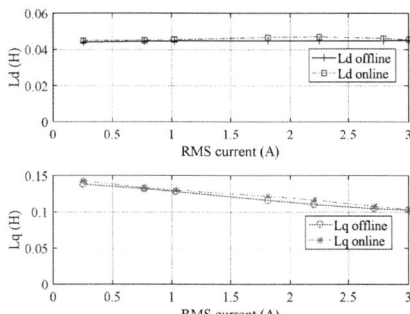

Fig.6. Online estimation of inductances at 500 rpm (experiment).

Fig.7 presents the on-line estimation of inductances when the machine accelerates from 300 rpm to 1200 rpm by experiment. It is clear that L_q varies according to the variation of the load current, while L_d is almost unchanged as seen in Fig. 7a. It is noted that at time 0.42s, the speed starts accelerating and the RMS current suddenly changes from 0.6A to 2.8A. As expected, L_q suddenly drops to 0.104 H at time 0.42s, before increasing to 0.11 H corresponding to the decrease of RMS current.

In order to evaluate the accuracy of the on-line estimation, the on-line estimated inductances are compared with the inductances measured off-line as shown in Fig. 7b. It is obvious that during the transient and steady state the estimated L_q and L_d still match the off-line values.

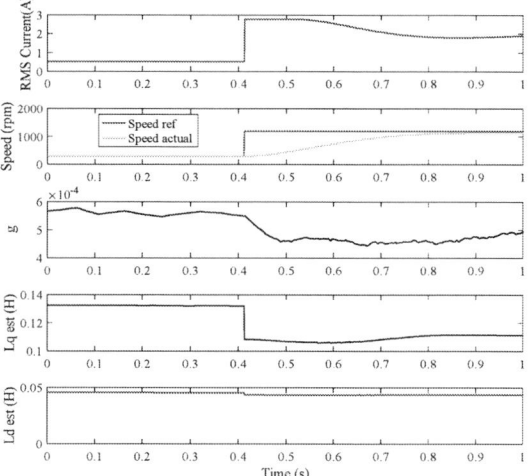

a. On-line estimation of inductance with time line scale

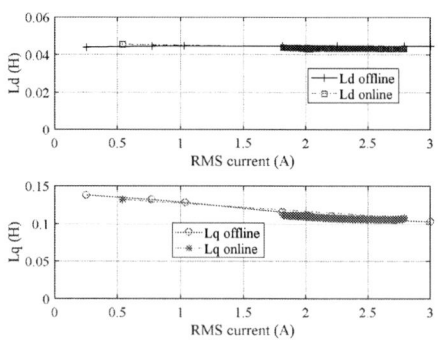

b. Comparison between on-line and off-line estimated inductances.

Fig.7. On-line estimation of inductances with speed acceleration from 300 rpm to 1200 rpm (experiment).

D. Performance of FOC with MTPA

In this study, MTPA was implemented based on the principle that the current phase angle β is controlled to get the maximum torque/current ratio. The maximum torque can be obtained if:

$$\frac{dT}{d\beta} = -\frac{3}{2}\lambda_f I \cos\beta + \frac{3}{2}P(L_q - L_d)I^2 \cos 2\beta = 0 \qquad (8)$$

From (8), the MTPA trajectory on dq plane can be derived as:

$$i_d = \frac{\lambda_f - \sqrt{\lambda_f^2 - 4(L_q - L_d)^2 i_q^2}}{2(L_q - L_d)}$$

$$= \frac{\lambda_f}{2(L_q - L_d)} - \sqrt{\frac{\lambda_f^2}{4(L_q - L_d)^2} + i_q^2} \qquad (9)$$

Equation (9) implies that the maximum torque per-ampere can be achieved if i_d reference is determined by (9) for any i_q reference which is the output of the speed controller.

Fig. 8 compares the performance of the FOC with MTPA using the rated inductances and FOC with the MTPA using the on-line estimated inductances when the machine operates at 300 rpm with the variation of the load torque from 20% to full load. It is clear that the transient time of the FOC with MTPA using on-line estimated inductances is 0.1s smaller than that of the FOC with MTPA using rated inductances. In other word, the FOC with MTPA using the estimated inductances results in a faster dynamic response of the speed and currents, compared to the FOC with MTPA using the rated inductances.

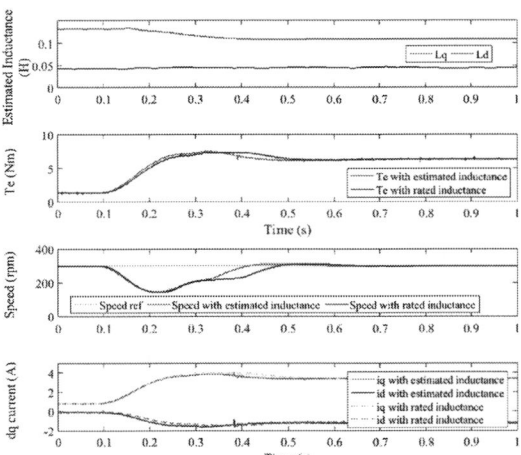

Fig.8. FOC with MTPA using rated inductances and on-line estimated inductances.

IV. CONCLUSION

This paper has shown that the rotor speed and position can be estimated satisfactorily over the full speed range from zero to rated speed based on the measurement of current derivatives at one active-voltage vector and one zero-voltage vector during each PWM cycle. In addition, the variation of machine inductances with load variation can be tracked on-line using these derivatives. The inductances estimated by the proposed method are almost similar to those measured by the off-line

method. The estimated rotor position and updated inductance parameters (L_d and L_q) are obtained at PWM frequency, which is much superior to existing methods. The experimental results also show that the estimated inductances can be utilized to improve the dynamic performance of FOC with MTPA.

REFERENCES

[1] Xiao, D., Foo, G., and Rahman, M. F., "Sensorless direct torque and flux control for matrix converter IPM synchronous motor drives using adaptive sliding mode observer combined with high frequency signal injection," in *2009 IEEE Energy Conversion Congress and Exposition*, 2009, pp. 4000-4007.

[2] Foo, G.; Rahman, M.F., "Sensorless Sliding-Mode MTPA Control of an IPM Synchronous Motor Drive Using a Sliding-Mode Observer and HF Signal Injection," *IEEE Journals & Magazines*, vol. 57, pp. 1270 - 1278, 2010.

[3] Qiu, A.; Bin Wu; Kojori, H., "Sensorless control of permanent magnet synchronous motor using extended Kalman filter," presented at the Electrical and Computer Engineering, 2004.

[4] Feng, Zhengfang Zhang; Jianghua, "Sensorless control of salient PMSM with EKF of speed and rotor position," presented at the Electrical Machines and Systems, ICEMS, 2008.

[5] Zhiqian Chen; Tomita, M.; Doki, S.; Okuma, S., "An extended electromotive force model for sensorless control of interior permanent-magnet synchronous motors," *IEEE Journals & Magazines*, vol. 50, 2003.

[6] Chan, T.F.; Wang, W.; Borsje, P.; Wong, Y.K.; Ho, S.L., "Sensorless permanent-magnet synchronous motor drive using a reduced-order rotor flux observer," *IET Journals & Magazines*, vol. 2, 2008.

[7] Xiaocan Wang; Wei Xie; Kennel, R.; Gerling, D., "Sensorless control of a novel IPMSM based on high-frequency injection," presented at the Power Electronics and Applications (EPE), 2013.

[8] Foo, G.; Sayeef, S.; Rahman, M.F., "Low-Speed and Standstill Operation of a Sensorless Direct Torque and Flux Controlled IPM Synchronous Motor Drive," *IEEE Journals & Magazines*, vol. 25, 2010.

[9] Chen, J.-L.; Tseng, S.-K.; Liu, T.-H., "Implementation of high-performance sensorless interior permanent-magnet synchronous motor control systems using a high-frequency injection technique," *IET Journals & Magazines*, vol. 6, 2012.

[10] Liu, J.M.; Zhu, Z.Q., "Sensorless Control Strategy by Square-Waveform High-Frequency Pulsating Signal Injection Into Stationary Reference Frame," *IEEE Journals & Magazines*, vol. 2, 2014.

[11] Zentai, A. and Daboczi, T., "Improving INFORM calculation method on permanent magnet synchronous machines," in *2007 IEEE Instrumentation & Measurement Technology Conference IMTC 2007*, 2007, pp. 1-6.

[12] Holtz, J. ; Juliet, J. , "Sensorless acquisition of the rotor position angle of induction motors with arbitrary stator windings," *IEEE JOURNALS & MAGAZINES*, vol. 41, 2005.

[13] Gao, Q. ; Asher, G.M. ; Sumner, M. ; Makys, P. , "Position Estimation of AC Machines Over a Wide Frequency Range Based on Space Vector PWM Excitation," *IEEE JOURNALS & MAGAZINES*, vol. 43, 2007.

[14] Vogelsberger, M.A. ; Grubic, S. ; Habetler, T.G. ; Wolbank, T.M. , "Using PWM-Induced Transient Excitation and Advanced Signal Processing for Zero-Speed Sensorless Control of AC Machines," *IEEE JOURNALS & MAGAZINES*, vol. 57, 2010.

[15] Hua, Y. ; Sumner, M. ; Asher, G. ; Gao, Q. ; Saleh, K. , "Improved sensorless control of a permanent magnet

machine using fundamental pulse width modulation excitation," *IET JOURNALS & MAGAZINES*, vol. 5, 2011.

[16] Bolognani, S. ; Calligaro, S. ; Petrella, R. ; Sterpellone, M. , "Sensorless control for IPMSM using PWM excitation: Analytical developments and implementation issues," presented at the Sensorless Control for Electrical Drives (SLED), 2011.

[17] Yu, Duan and Sumner, M., "A novel current derivative measurement using recursive least square algorithms for sensorless control of permanent magnet synchronous machine," in *Proceedings of The 7th International Power Electronics and Motion Control Conference*, 2012, pp. 1193-1200.

[18] Guan, D. Q., Xiao, D., and Rahman, M. F., "Comparison of torque control bandwidth of HF injection, SMO and FPE Direct Torque control IPMSMS drives," in *2014 Australasian Universities Power Engineering Conference (AUPEC)*, 2014, pp. 1-6.

[19] Ran, L., Gokani, S., Clare, J., Bradley, K. J., and Christopoulos, C., "Conducted electromagnetic emissions in induction motor drive systems. II. Frequency domain models," *IEEE Transactions on Power Electronics*, vol. 13, pp. 768-776, 1998.

[20] Guan, D. Q., Bui, M. X., Xiao, D., and Rahman, M. F., "Performance comparison of two FPE sensorless control methods on a direct torque controlled interior permanent magnet synchronous motor drive," in *2016 19th International Conference on Electrical Machines and Systems (ICEMS)*, 2016, pp. 1-6.

[21] Guan, D. Q., Bui, M. X., Xiao, D., and Rahman, M. F., "Evaluation of an FPGA current derivative measurement system for the fundamental PWM excitation sensorless method for IPMSM," in *2016 IEEE 2nd Annual Southern Power Electronics Conference (SPEC)*, 2016, pp. 1-6.

[22] Xiao, D., Guan, D. Q., Rahman, M. F., and Fletcher, J., "Sliding mode observer combined with fundamental PWM excitation for sensorless control of IPMSM drive," in *IECON 2014 - 40th Annual Conference of the IEEE Industrial Electronics Society*, 2014, pp. 895-901.

[23] Guan, D. Q., Xiao, D., and Rahman, M. F., "A new high-bandwidth sensorless direct torque controlled IPM synchronous machine drive using a hybrid sliding mode observer," in *2014 IEEE 5th International Symposium on Sensorless Control for Electrical Drives*, 2014, pp. 1-8.

[24] "IEEE Standard Procedures for Obtaining Synchronous Machine Parameters by Standstill Frequency Response Testing (Supplement to ANSI/IEEE Std 115-1983, IEEE Guide: Test Procedures for Synchronous Machines)," *IEEE Std 115A-1987*, p. 0_1, 1987.

[25] Vandoorn, T. L., Belie, F. M. De, Vyncke, T. J., Melkebeek, J. A., and Lataire, P., "Generation of Multisinusoidal Test Signals for the Identification of Synchronous-Machine Parameters by Using a Voltage-Source Inverter," *IEEE Transactions on Industrial Electronics*, vol. 57, pp. 430-439, 2010.

[26] Bortoni, E. C. and Jardini, J. A., "A standstill frequency response method for large salient pole synchronous machines," *IEEE Transactions on Energy Conversion*, vol. 19, pp. 687-691, 2004.

[27] Liu, K., Zhang, Q., Chen, J., Zhu, Z. Q., and Zhang, J., "Online Multiparameter Estimation of Nonsalient-Pole PM Synchronous Machines With Temperature Variation Tracking," *IEEE Transactions on Industrial Electronics*, vol. 58, pp. 1776-1788, 2011.

[28] Liu, Q. and Hameyer, K., "A fast online full parameter estimation of a PMSM with sinusoidal signal injection," in *2015 IEEE Energy Conversion Congress and Exposition (ECCE)*, 2015, pp. 4091-4096.

[29] Feng, G., Lai, C., and Kar, N. C., "A Novel Current Injection-Based Online Parameter Estimation Method for PMSMs Considering Magnetic Saturation," *IEEE Transactions on Magnetics*, vol. 52, pp. 1-4, 2016.

[30] Feng, G., Lai, C., Mukherjee, K., and Kar, N. C., "Current Injection-Based Online Parameter and VSI Nonlinearity Estimation for PMSM Drives Using Current and Voltage DC Components," *IEEE Transactions on Transportation Electrification*, vol. 2, pp. 119-128, 2016.

[31] Ichikawa, S., Tomita, M., Doki, S., and Okuma, S., "Sensorless control of permanent-magnet synchronous motors using online parameter identification based on system identification theory," *IEEE Transactions on Industrial Electronics*, vol. 53, pp. 363-372, 2006.

[32] Underwood, S. J. and Husain, I., "Online Parameter Estimation and Adaptive Control of Permanent-Magnet Synchronous Machines," *IEEE Transactions on Industrial Electronics*, vol. 57, pp. 2435-2443, 2010.

[33] Deng, W., Xia, C., Yan, Y., Geng, Q., and Shi, T., "Online Multiparameter Identification of Surface-Mounted PMSM Considering Inverter Disturbance Voltage," *IEEE Transactions on Energy Conversion*, vol. 32, pp. 202-212, 2017.

[34] Morimoto, S., Sanada, M., and Takeda, Y., "Mechanical Sensorless Drives of IPMSM With Online Parameter Identification," *IEEE Transactions on Industry Applications*, vol. 42, pp. 1241-1248, 2006.

[35] Dang, D. Q., Rafaq, M. S., Choi, H. H., and Jung, J. W., "Online Parameter Estimation Technique for Adaptive Control Applications of Interior PM Synchronous Motor Drives," *IEEE Transactions on Industrial Electronics*, vol. 63, pp. 1438-1449, 2016.

[36] Rafaq, M. S., Mwasilu, F., Kim, J., Choi, H. H., and Jung, J. W., "Online Parameter Identification for Model-Based Sensorless Control of Interior Permanent Magnet Synchronous Machine," *IEEE Transactions on Power Electronics*, vol. 32, pp. 4631-4643, 2017.

[37] Boileau, T., Leboeuf, N., Nahid-Mobarakeh, B., and Meibody-Tabar, F., "Online Identification of PMSM Parameters: Parameter Identifiability and Estimator Comparative Study," *IEEE Transactions on Industry Applications*, vol. 47, pp. 1944-1957, 2011.

[38] Zhu, Z. Q., Zhu, X., Sun, P. D., and Howe, D., "Estimation of Winding Resistance and PM Flux-Linkage in Brushless AC Machines by Reduced-Order Extended Kalman Filter," in *2007 IEEE International Conference on Networking, Sensing and Control*, 2007, pp. 740-745.

[39] Rashed, M., MacConnell, P. F. A., Stronach, A. F., and Acarnley, P., "Sensorless Indirect-Rotor-Field-Orientation Speed Control of a Permanent-Magnet Synchronous Motor With Stator-Resistance Estimation," *IEEE Transactions on Industrial Electronics*, vol. 54, pp. 1664-1675, 2007.

[40] Liu, K., Zhu, Z. Q., Zhang, Q., and Zhang, J., "Influence of Nonideal Voltage Measurement on Parameter Estimation in Permanent-Magnet Synchronous Machines," *IEEE Transactions on Industrial Electronics*, vol. 59, pp. 2438-2447, 2012.

[41] Shaowei, W. and Shanming, W., "Identify PMSM's Parameters by Single-Layer Neural Networks with Gradient Descent," in *2010 International Conference on Electrical and Control Engineering*, 2010, pp. 3811-3814.

[42] Kumar, R., Gupta, R. A., and Bansal, A. K., "Identification and Control of PMSM Using Artificial Neural Network," in *2007 IEEE International Symposium on Industrial Electronics*, 2007, pp. 30-35.

[43] Guan, M. X. Bui; D. Xiao; M. F. Rahman; D. Q., " Online estimation of inductances of permanent magnet synchronous machines based on current derivative measurements," presented at the 20th International Conference on Electrical Machines and Systems (ICEMS), Sydney, Australia, 2017.

[44] Steimel, Jie Fang; Carsten Heising; Volker Staudt; Andreas, "Modelling of anisotropic synchronous machine in stator reference frame," presented at the IEEE Vehicle Power and Propulsion Conference, 2010.

The 2018 International Power Electronics Conference

Dynamic Performance Improvement of Bidirectional Switched-Capacitor DC/DC Converter by Right-Half-Plane Zero Elimination

DING Kaicheng, ZHANG Yan*, LIU Jinjun, ZENG Pengxiang and ZHANG Jinshui
School of Electrical Engineering, Xi'an Jiaotong University, Xi'an, China
*E-mail: zhangyanjtu@xjtu.edu.cn

Abstract—Bidirectional DC/DC converters have important application value in energy storage system. Compared with conventional topology, switched-capacitor-based non-isolated DC/DC converter inherently possesses good static characteristics including lower voltage and current stress of power switches, higher voltage boost capability and continuous input/output current. However, its control-to-output transfer functions contain Right-Half-Plane Zeros in both boost conversion and buck conversion mode, which considerably constrains the compensator design and further limits the dynamic performance of the closed-loop system. In this paper, it is proposed to use RC damping method to address the issue, which enables to effectively eliminate the Right-Half-Plane Zeros under both conversion modes and thus improve the system's dynamic performance. Theoretical analysis and simulation results are provided in the paper.

Keywords—Bidirectional DC/DC Converter, Dynamic Performance Improvement, Right-Half-Plane Zero Elimination, Switched-Capacitor

I. INTRODUCTION

Bidirectional DC/DC converters have important application value in micro-grid system. Served as the interface between DC bus and energy storage equipment, it is able to absorb the excessive power from energy generation equipment on one side and guarantee adequate power supply to the load on the other, which helps to mitigate the side effects caused by the random and stochastic characteristics of renewable energy resources. [1]-[3] Since energy storage equipment usually works with a relatively low-level voltage compared with DC bus voltage, it is then required that the bidirectional DC/DC converter should possess a sufficiently high voltage boost capability, which is in fact what conventional buck-boost bidirectional DC/DC converter often lacks. Besides, the power switches inside the conventional converter have to sustain a large voltage and current stress, making it hard to satisfy the overall application demand.

In order to achieve a better bidirectional DC/DC conversion and ensure a higher efficiency, various

This work was supported in part by the State Key Laboratory of Electrical Insulation and Power Equipment under Grants EIPE16310 and the Power Electronics Science and Education Development Program of Delta Environmental and Educational Foundation under Grant DREG2016010.

techniques based on non-isolated topologies have been proposed such as cascaded/quadratic structure, interleaved structure, multi-level structure, impedance-source and switched-inductor/capacitor. [4] This article mainly focuses on one typical switched-capacitor bidirectional DC/DC converter of which the topology is shown in Fig. 1. [5][6] Compared with conventional topology, it enables to increase voltage boost capability, lower the power switches' voltage/current stress and achieve a continuous input/output current at the same time. However, its control-to-output transfer functions of both boost conversion and buck conversion mode contain Right-Half-Plane Zeros (RHPZ), which considerably constrains the compensator design and further limits the closed loop system's dynamic performance. To address this issue, it is proposed in this article to introduce RC damping into power stage, which enables to effectively eliminate RHPZ under both conversion modes and thus result in an improvement of system's dynamic performance. The loss caused by RC damping is negligible and it does not cast much influence on the overall efficiency. [7][8] Theoretical analysis and simulation results on RC damping design and loss evaluation will be provided in the following sections.

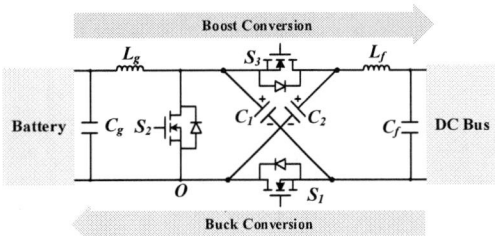

Fig. 1. Topology of bidirectional switched-capacitor DC/DC converter.

II. THEORETICAL ANALYSIS

Bidirectional DC/DC converters could operate in different conversion modes based on power flow direction. It is regarded as boost conversion mode when energy is transferred from low voltage side to high voltage side and the converse case is regarded as buck conversion mode.

4181

Modeling of the proposed bidirectional switched-capacitor DC/DC converter under both boost and buck conversion mode is going to be provided in the first place.

Boost conversion mode equivalent circuit is given in Fig. 2. Power switch S_1 and S_3 share the same switching states while S_2 works in a complementary manner. When S_2 is turned on and S_1 S_3 are turned off, low-voltage-side inductor L_g gets magnetized and intermediate capacitors C_1 C_2 are discharged in series. When S_2 is turned off and S_1 S_3 are turned on, L_g gets demagnetized and C_1 C_2 are charged in parallel.

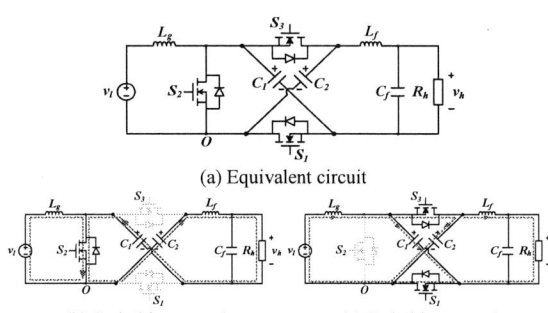

(a) Equivalent circuit

(b) Switching state 1   (c) Switching state 2
Fig. 2. Boost conversion mode equivalent circuit and switching states.

Based on the above analysis, small signal model of the converter under boost conversion mode can be obtained by applying average state-space operator and linearization around steady state working point. The result is given below where L_g is the inductance of low-voltage-side inductor, C is the capacitance of intermediate capacitor, L_f is the inductance of high-voltage-side inductor, C_f is the capacitance of high-voltage-side capacitor, R_h represents the high-voltage-side load resistance, D is the steady state duty ratio of boost conversion mode, V_C is the steady state intermediate capacitor voltage, I_{Lg} is the steady state low-voltage-side inductor current, I_{Lf} is the steady state high-voltage-side inductor current, $\hat{i}_{Lg}$ is the small signal value of low-voltage-side inductor current, $\hat{v}_C$ is the small signal value of intermediate capacitor voltage, $\hat{i}_{Lf}$ is the small signal value of high-voltage-side inductor current, $\hat{i}_{Lg}$ is the small signal value of low-voltage-side inductor current, $\hat{v}_{Cf}$ is the small signal value of high-voltage-side capacitor voltage, $\hat{v}_l$ is the small signal value of input voltage and $\hat{d}$ is the small signal value of duty ratio in boost conversion mode.

$$\frac{d}{dt}\begin{bmatrix}\hat{i}_{L_g}\\\hat{v}_C\\\hat{i}_{L_f}\\\hat{v}_{C_f}\end{bmatrix}=\begin{bmatrix}0 & -\frac{1-D}{L_g} & 0 & 0\\\frac{1-D}{2C} & 0 & -\frac{1+D}{2C} & 0\\0 & \frac{1+D}{L_f} & 0 & -\frac{1}{L_f}\\0 & 0 & \frac{1}{C_f} & -\frac{1}{R_hC_f}\end{bmatrix}\cdot\begin{bmatrix}\hat{i}_{L_g}\\\hat{v}_C\\\hat{i}_{L_f}\\\hat{v}_{C_f}\end{bmatrix}+\begin{bmatrix}\frac{1}{L_g} & \frac{V_C}{L_g}\\0 & -\frac{I_{L_g}+I_{L_f}}{2C}\\0 & \frac{V_C}{L_f}\\0 & 0\end{bmatrix}\cdot\begin{bmatrix}\hat{v}_l\\\hat{d}\end{bmatrix}$$

(1)

Similarly, for buck conversion mode, the equivalent circuit is shown in Fig. 3. When S_1 S_3 are turned on and S_2 is turned off, high-voltage side inductor L_f gets magnetized and intermediate capacitors C_1 C_2 are discharged in

parallel. When S_1 S_3 are turned off and S_2 is turned on, L_f gets demagnetized and C_1 C_2 are charged in series.

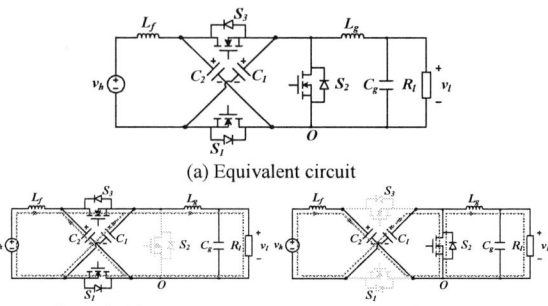

(a) Equivalent circuit

(b) Switching state 1   (c) Switching state 2
Fig. 3. Buck conversion mode equivalent circuit and switching states

Applying the same procedure as before and the small signal model of buck conversion mode is given as below. In addition to the variables described in previous paragraph, R_l represents the low-voltage-side load resistance, D' is the steady state duty ratio of buck conversion mode, $\hat{v}_{Cg}$ is the small signal value of low-voltage-side capacitor voltage, $\hat{v}_h$ is the small signal value of input voltage and $\hat{d}'$ is the small signal value of duty ratio of buck conversion mode.

$$\frac{d}{dt}\begin{bmatrix}\hat{i}_{L_f}\\\hat{v}_C\\\hat{i}_{L_g}\\\hat{v}_{C_g}\end{bmatrix}=\begin{bmatrix}0 & -\frac{2-D'}{L_f} & 0 & 0\\\frac{2-D'}{2C} & 0 & -\frac{D'}{2C} & 0\\0 & \frac{D'}{L_g} & 0 & -\frac{1}{L_g}\\0 & 0 & \frac{1}{C_g} & -\frac{1}{R_lC_g}\end{bmatrix}\cdot\begin{bmatrix}\hat{i}_{L_f}\\\hat{v}_C\\\hat{i}_{L_g}\\\hat{v}_{C_g}\end{bmatrix}+\begin{bmatrix}\frac{1}{L_f} & \frac{V_C}{L_f}\\0 & -\frac{I_{L_g}+I_{L_f}}{2C}\\0 & \frac{V_C}{L_g}\\0 & 0\end{bmatrix}\cdot\begin{bmatrix}\hat{v}_h\\\hat{d}'\end{bmatrix}$$

(2)

Rearrange (1) and (2) so that the control-to-output transfer functions of the converter under both boost conversion and buck conversion mode can be obtained. The transfer function can be expressed as the following generalized form:

$$\frac{\hat{v}_o}{\hat{d}}=\frac{a_2s^2+a_1s+a_0}{b_4s^4+b_3s^3+b_2s^2+b_1s+b_0}$$

(3)

The numerator and denominator coefficients of the transfer functions are given in TABLE I.

TABLE I
COEFFICIENTS OF TRANSFER FUNCTIONS

	Boost conversion	Buck conversion
a_2	$\frac{L_gCV_lR_h}{1-D}$	$\frac{L_fCV_hR_l}{2-D'}$
a_1	$-\frac{L_gV_l(1+D)^2}{(1-D)^2}$	$-\frac{L_fV_h(1-D')^2}{(2-D')^2}$
a_0	R_hV_l	R_lV_h
b_4	$L_gCL_fC_fR_h$	$L_fCL_gC_gR_l$
b_3	L_gCL_f	L_fCL_g
b_2	$\left[\frac{(1+D)^2L_g+(1-D)^2L_f}{2}C_f+L_fC\right]R_h$	$\left[\frac{(1-D')^2L_f+(2-D')^2L_g}{2}C_g+L_fC\right]R_l$
b_1	$\frac{(1+D)^2L_g+(1-D)^2L_f}{2}$	$\frac{(1-D')^2L_f+(2-D')^2L_g}{2}$
b_0	$\frac{(1-D)^2}{2}R_h$	$\frac{(2-D')^2}{2}R_l$

From the above table, it can be noticed that the coefficient a_1 stays negative for all possible working conditions under both boost conversion and buck conversion. It indicates that the control-to-output transfer functions of the converter contain RHPZs (Right-Half-Plane Zeros). Consequently, compensation strategy for closed-loop system must be very conservative in order to avoid instability issue. More precisely, the bandwidth of the system under such circumstances has to be limited to a very low value, which greatly affects and deteriorates the converter's dynamic performance.

To solve this problem, one possible solution is to introduce RC damping to eliminate the RHPZs as shown in Fig. 4.

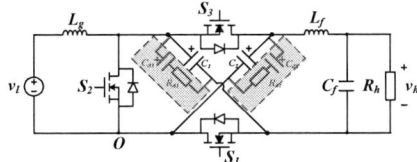

Fig. 4. RC damping to eliminate RHPZs

After adding RC damping branch to the system, the control-to-output transfer functions can be expressed as the form below:

$$\frac{\hat{v}_o}{\hat{d}} = \frac{a_3's^3 + a_2's^2 + a_1's + a_0'}{b_5's^5 + b_4's^4 + b_3's^3 + b_2's^2 + b_1's + b_0'} \quad (4)$$

The corresponding coefficients are given in TABLE II.

TABLE II
COEFFICIENTS OF TRANSFER FUNCTIONS WITH RC DAMPING

	Boost conversion	Buck conversion
a_3'	$\dfrac{V_l L_g C R_d C_d R_h}{1-D}$	$\dfrac{V_h L_f C R_d C_d R_l}{2-D'}$
a_2'	$\dfrac{L_g V_l}{1-D}\left[R_h(C+C_d)-\dfrac{(1+D)^2}{1-D}R_d C_d\right]$	$\dfrac{L_f V_h}{2-D'}\left[R_l(C+C_d)-\dfrac{D'^2}{2-D'}R_d C_d\right]$
a_1'	$V_l\left[C_d R_d R_h - \dfrac{L_g(1+D)^2}{(1-D)^2}\right]$	$V_h\left[C_d R_d R_l - \dfrac{L_f D'^2}{(2-D')^2}\right]$
a_0'	$R_h V_l$	$R_l V_h$
b_5'	$L_g CL_f C_f R_h R_d C_d$	$L_f CL_g C_g R_l R_d C_d$
b_4'	$\left[C(C_d R_d + C_f R_h)+\dfrac{1+D}{2}C_f R_h C_d\right]L_g L_f$	$\left[C(C_d R_d + C_g R_l)+\dfrac{D'}{2}C_g R_l C_d\right]L_g L_f$
b_3'	$L_g CC_d R_d R_h + \left[C+\dfrac{1+D}{2}C_d\right]L_g L_f$ $+\left[\dfrac{(1+D)^2}{2}L_g+\dfrac{(1-D)^2}{2}L_f\right]C_f R_h C_d R_d$	$L_f CC_d R_d R_l + \left[C+\dfrac{D'}{2}C_d\right]L_g L_f$ $+\left[\dfrac{D'^2}{2}L_f+\dfrac{(2-D')^2}{2}L_g\right]C_g R_l C_d R_d$
b_2'	$\left[\dfrac{(1+D)^2 L_g+(1-D)^2 L_f}{2}\right](C_f R_h + C_d R_d)$ $+\left[\dfrac{1+D}{2}C_d+C\right]L_g R_h$	$\left[\dfrac{D'^2 L_f+(2-D')^2 L_g}{2}\right](C_g R_l + C_d R_d)$ $+\left[\dfrac{D'}{2}C_d+C\right]L_f R_l$
b_1'	$\dfrac{(1+D)^2 L_g+(1-D)^2(L_f+C_d R_d R_h)}{2}$	$\dfrac{D'^2 L_f+(2-D')^2(L_g+C_d R_d R_l)}{2}$
b_0'	$\dfrac{(1-D)^2}{2}R_h$	$\dfrac{(2-D')^2}{2}R_l$

From the table above, it can be noticed that there unnecessarily exist negative terms in numerator coefficients. If RC damping branch is carefully designed, then it will be possible that the system does not contain any RHPZs under the required working conditions.

III. PARAMETERS DESIGN AND LOSS EVALUATION

Apply Routh-Hurwitz criterion to the numerator coefficients, the following conditions have to be satisfied so that the converter does not contain RHPZs under both boost conversion and buck conversion mode.

$$\begin{cases} a_k' > 0, \quad \forall k \in \{0,1,2,3\} \\ a_1' a_2' - a_0' a_3' > 0 \end{cases} \quad (5)$$

Once the passive components of the converter and voltage conversion ratio are fixed, whether the conditions are satisfied or not depends on the load resistance. Note that $a_3' > 0$ and $a_0' > 0$ stay true for all conditions; a_2' and a_1' are first-order polynomials of the load resistance while $a_1' a_2' - a_0' a_3'$ is second-order polynomial of the load resistance. Note R_{h1} as the load resistance realizing $a_1'=0$, R_{h2} as the load resistance realizing $a_2'=0$ and R_{h3} as the load resistance realizing $a_1' a_2' - a_0' a_3'=0$ in boost conversion mode and R_{l1} R_{l2} R_{l3} in buck conversion mode respectively.

$$\begin{cases} R_{h1}=\dfrac{(1+D)^2 L_g}{(1-D)^2 R_d C_d} \quad R_{l1}=\dfrac{D'^2 L_f}{(2-D')^2 R_d C_d} \\ R_{h2}=\dfrac{R_d C_d(1+D)^2}{(C+C_d)(1-D)} \quad R_{l2}=\dfrac{R_d C_d D'^2}{(C+C_d)(2-D')} \\ R_{h3}=\dfrac{C+C_d}{C_d}\left[\dfrac{R_{h1}+R_{h2}}{2}+\sqrt{(\dfrac{R_{h1}-R_{h2}}{2})^2+\dfrac{L_g C(1+D)^4}{(1-D)^3(C+C_d)^2}}\right] \\ R_{l3}=\dfrac{C+C_d}{C_d}\left[\dfrac{R_{l1}+R_{l2}}{2}+\sqrt{(\dfrac{R_{l1}-R_{l2}}{2})^2+\dfrac{L_f CD'^4}{(2-D')^3(C+C_d)^2}}\right] \end{cases} \quad (6)$$

The optimization of RC damping parameter design is to achieve the maximum range of load conditions where the RHPZs could always be well eliminated. It can be expressed as the following form for both boost conversion and buck conversion mode.

$$\begin{cases} (R_d \quad C_d)_{boost}=\underset{(R_d \quad C_d)}{arg\ min}\{\max(R_{h1} \quad R_{h2} \quad R_{h3})\} \\ (R_d \quad C_d)_{buck}=\underset{(R_d \quad C_d)}{arg\ min}\{\max(R_{l1} \quad R_{l2} \quad R_{l3})\} \end{cases} \quad (7)$$

The optimal configuration takes place when $R_{h1}=R_{h2}$ for boost mode and $R_{l1}=R_{l2}$ for buck mode. In this case, $R_d C_d$ have to satisfy the following conditions:

$$\begin{cases} R_d=\dfrac{1}{C_d}\sqrt{\dfrac{(C+C_d)L_g}{1-D}}=\dfrac{1}{C_d}\sqrt{\dfrac{(C+C_d)L_f}{2-D'}} \\ C_d > \dfrac{(1+D)^4 L_g}{(1-D)^3 R_h^2}+\dfrac{2(1+D)^2}{(1-D)^2 R_h}\sqrt{L_g C(1-D)} \\ C_d > \dfrac{D'^4 L_f}{(2-D')^3 R_l^2}+\dfrac{2D'^2}{(2-D')^2 R_l}\sqrt{L_f C(2-D')} \end{cases} \quad (8)$$

What worth noticing is that adding RC damping branch would inevitably introduce additional loss to the system. To evaluate this part of loss, take the following RC network into consideration where the excitation of the network is the current source I_{in} and corresponding response is the damping branch current I_d.

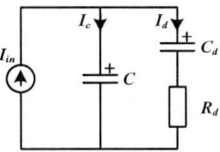

Fig. 5. RC network for loss evaluation

It can be deduced that the transfer function of the above predefined network has the following form:

4183

$$H(s) = \frac{I_d(s)}{I_{in}(s)} = \frac{C_d}{sR_dC_dC + C_d + C} \qquad (9)$$

Note that this first-order system has one pole situating at a very low frequency compared with switching frequency. In fact, the current source I_{in} during the normal operation of the converter is a pulsating current with a period of exactly the switching period and a zero average value over one period. The bode plot based on one particular RC damping parameter is given in Fig. 6. For frequency above 50kHz, the magnitude becomes rather small and therefore the loss caused by RC damping proves to be negligible.

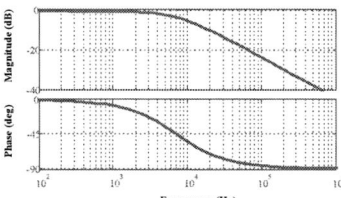

Fig. 6. Bode plot of RC network with C=5µF C$_d$=50µF R$_d$=5Ω

IV. PARAMETERS DESIGN AND LOSS EVALUATION

The working conditions and passive components' parameters are given in TABLE III.

TABLE III
WORKING CONDITIONS AND PARAMETERS

Symbol	Meaning	Value
P_n	Nominal power	200W
f_s	Switching frequency	100kHz
V_l	Low-level voltage	50V
V_h	High-level voltage	200V
L_g	Low-side inductor	100µH
C_g	Low-side capacitor	50µF
L_f	High-side inductor	1mH
C_f	High-side capacitor	5µF
C	Intermediate capacitor	5µF
R_d	Damping resistor	5Ω
C_d	Damping capacitor	50µF

The results are shown in Fig. 7. The left column shows boost mode while the right shows buck mode. Black-dot lines in bode plot represent AC-sweep simulation in PSIM and colored lines represent MATLAB calculation. Yellow lines represent system before RC damping and green lines represent system after RC damping.

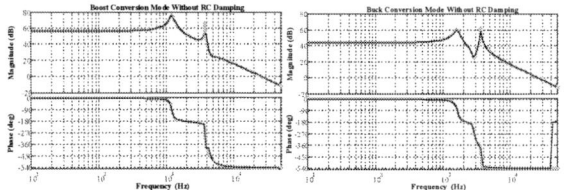

(a) Boost mode without damping   (b) Buck mode without damping

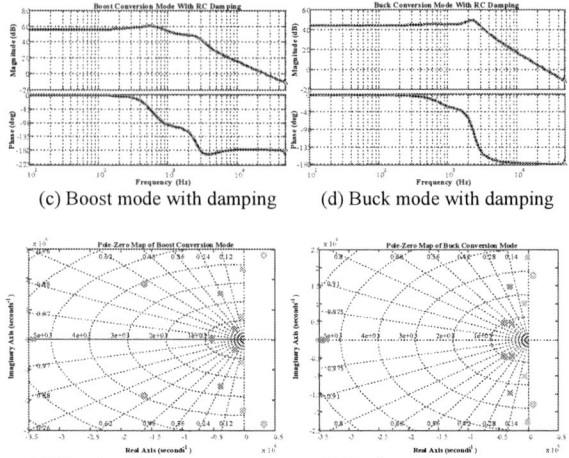

(c) Boost mode with damping   (d) Buck mode with damping

(e) Boost mode pole-zero map   (f) Buck mode pole-zero map
Fig. 7. Bode plots and pole-zero maps of control-to-output transfer functions

Several comments can be made. Firstly, AC-sweep in PSIM matches exactly with MATLAB calculation and thus theoretical analysis is verified. Secondly, the system before RC damping clearly shows non-minimum phase characteristics and this is due to RHPZs' causing extra phase delay. Moreover, by applying RC damping, RHPZs under both boost and buck conversion mode are effectively eliminated and thus compensator design would become much easier.

Time-domain closed-loop simulation is carried out in PSIM. Voltage mode control and type-III compensator is applied for both boost and buck conversion mode. After system reaches steady state, at certain instant input voltage generates a step disturbance about 10% of the corresponding steady state value and output voltage is observed. The result is shown below. Yellow curves represent the step response before RC damping and Green curves represent that after RC damping.

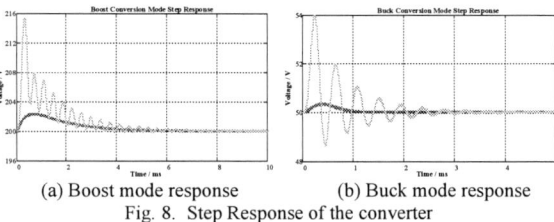

(a) Boost mode response   (b) Buck mode response
Fig. 8. Step Response of the converter

It can be seen that after RC damping the converter's dynamic performance is greatly improved since RHPZs are eliminated and therefore a more effective compensation strategy could be implemented.

The loss introduced by RC damping is furthermore evaluated. The result is shown in Fig. 9. Blue curves represent waveform of I_{in} and red curves represent waveform of I_d in with respect to Fig. 4. It can be seen that the amplitude of I_d is much smaller than I_{in}, which is consistent with previous analysis.

4184

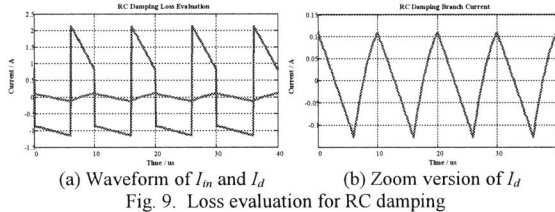

(a) Waveform of I_{in} and I_d (b) Zoom version of I_d

Fig. 9. Loss evaluation for RC damping

It can be calculated that in this case the RMS value of damping branch current is around and slightly smaller than 70mA. The loss generated in damping resistors is:

$$P_{loss} = 2I_{d,RMS}^2 R_d \approx 49mW \qquad (10)$$

This part of loss is generally acceptable and negligible compared with converter's nominal power and thus introducing RC damping would not have much influence on the overall efficiency.

V. CONCLUSIONS

Bidirectional DC/DC converters have important application value in micro grid system and energy storage system as it serves as the interface between DC bus and energy storage equipment. Switched-capacitor-based non-isolated DC/DC converter inherently possesses good static characteristics including lower voltage and current stress of power switches, higher voltage boost capability and continuous input/output current. However, its control-to-output transfer functions contain Right-Half-Plane Zeros in both boost conversion and buck conversion mode, which considerably constrains the compensator design and further limits the dynamic performance of the closed-loop system. In the paper RC damping method is proposed to address this issue and RHPZs under both conversion modes could be effectively eliminated. As a result, an improvement of system's dynamic performance could be achieved and the loss introduced by RC damping proves to be negligible. Theoretical analysis and simulation results are provided in the paper.

REFERENCES

[1] J. Momoh, *Renewable Energy and Storage*. New York, NY, USA: Wiley, 2012.

[2] Nisha Kondrath, "Bidirectional DC-DC converter topologies and control strategies for interfacing energy storage systems in microgrids: An overview", *2017 IEEE International Conference on Smart Grid Engineering (SEGE)*, pp. 341 – 345, 2017.

[3] Owon Kwon, Jun-Seok Kim, Jung-Min Kwon, Bong-Hwan Kwon, "Bidirectional Grid-connected Single-power-conversion Converter with Low Input Battery Voltage", *IEEE Trans. on Industrial Electronics*, vol. PP, Issue 99, pp. 1-1, 2017.

[4] W. Li and X. He, "Review of Nonisolated High-Step-Up DC/DC Converters in Photovoltaic Grid-Connected Applications", *IEEE Trans. on Industrial Electronics*, vol. 58, Issue 4, pp. 1239 – 1250, 2011.

[5] B. Axelrod, Y. Berkovich and A. Ioinovici, "Switched-Capacitor/Switched-Inductor Structures for Getting Transformerless Hybrid DC-DC PWM Converters", *IEEE Trans. on Circuits and Systems I*, vol. 55, Issue 2, pp. 687-696, 2008.

[6] Y. Zhang, C. Zhang, J. Liu and Y. Cheng, "Comparison of conventional dc-dc converter and a family of diode-assisted dc-dc converter", *Proceedings of the 7th International Power Electronics and Motion Control Conference*, vol. 3, pp. 1718-1723, 2012.

[7] J. Calvente, L. Martinez-Salamero, P. Garces, A. Romero, "Zero dynamics-based design of damping networks for switching converters", *IEEE Trans. on Aerospace and Electronic Systems*, vol. 39, Issue 4, 2003.

[8] Y. Zhang, J. Liu, Z. Dong, H. Wang and Y. Liu, "Dynamic Performance Improvement of Diode-capacitor-Based High Step-up DC-DC Converter Through Right-Half-Plane Zero Elimination", *IEEE Trans. on Power Electronics*, vol. 32, Issue 8, pp. 6532-6543, 2017.

The 2018 International Power Electronics Conference

A Matrix based Isolated Bidirectional AC-DC Converter with LCL type Input Filter for Energy Storage Application

Prathamesh Pravin Deshpande[1*], Amit Kumar Singh[1], Sanjib Kumar Panda[1]

1 Department of Electrical & Computer Engineering, National University of Singapore, Singapore

*E-mail: prathamesh.deshpande@nus.edu.sg

Abstract—An isolated three-phase AC-DC converter is proposed in this paper for integrating energy storage element such as battery to the utility grid. The proposed topology uses a matrix based AC-AC converter for three-phase to single-phase conversion, facilitates the use of high frequency transformer for galvanic isolation and provides the necessary turns ratio for matching the required voltages on both sides and full bridge controlled rectifier section for AC-DC conversion. Space Vector Modulation (SVM) based switching technique has been implemented for the matrix based converter for superior input power quality and improved power conversion efficiency and Sinusoidal Pulse Width Modulation (SPWM) has been used for the full bridge controlled rectifier. A T-shaped LCL input filter is developed to provide low pass filtering effect. Also, the filter provides an inductance dominance, providing current source characteristics, hence only a capacitive output filter is used. Simulation was carried out in Powersim (PSIM) simulation software. Variation of voltage gain for SVM and SPWM modulation indices is also presented. The converter is able to generate a charging voltage as well as three-phase sinusoidal voltage with THD of 3%. A closed loop control is developed for matrix type AC-AC converter part. The control is able to perform bidirectional operation of the system.

Keywords—Matrix converter, bidirection power flow, isolated converter, PSIM simulation

I. INTRODUCTION

With increase in the distributed generation and renewable energy sources, use of energy storage devices have been increased. Integration of the storage elements with grid becomes mandatory. [1] describes the converters for renewable energy sources integration with the utility grid. Various power electronic converter topologies can be applied. Isolated converters like full bridge DC-DC converter could also prove a better option having the advantages of high frequency transformer (turns ratio and galvanic isolation). Various modulation schemes for the converters like Sinusoidal Pulse Width Modulation (SPWM), Space Vector Modulation (SVM) etc. have been proposed and being used. Hence, it is important to apply the knowledge and formulate a topology based on certain application and parameters. A matrix based converter is multi-phase-to-multi-phase AC-AC converter with an array of switches. [2] have presented a review and analysis on matrix based converter topology along with modulation and control strategies. Apart from modulation,

the other point emphasized was of input filter. The filter is required to reduce the switching harmonics going into the system. The filter must be able to satisfy requirements such as having cutoff frequency lower than switching frequency (low pass), minimizing its reactive power to grid, minimizing size of capacitors and inductors and minimizing filter inductor voltage drop or impedance at rated current. In [3], a non-isolated matrix based buck-boost converter applicable for the aircraft systems has been presented while in [4], an isolated single-stage three-phase AC-DC converter. Both applied the matrix topology to convert low frequency AC to high frequency AC. The proposed converter overcomes the limitations expressed for conventional converters and provides output DC voltage with reduced current distortion and improved power conversion efficiency. The proposed converter eliminates the DC link capacitor and combines the output filter inductors with inductors from other stages. As the proposed topology is operated at higher modulation index and lower duty cycle, it promises higher input power quality and lower semiconductor losses. The merits make the proposed converter suitable for aircraft system, telecommunication, micro-grid and energy storage. The SVM has been selected for modulation scheme. [5]- [8] present the bidirectional capability of the matrix based converter. Particularly, [5] and [7] developed matrix based topology for energy storage integration to utility grid. [6] proves the step-up and step-down capability of the matrix based converter and its implication in bidirectional power flow mode. The advantages of the proposed converter mentioned in [7] are high voltage conversion gain, high power factor, high conversion efficiency and high power density. However, due to input LC filter, an output inductor filter is required. Due to which for power flow towards the grid, a comprehensive modulation technique is required for matrix based converter to reduce Total Harmonic Distortion (THD). The main conclusion is the capability and smooth transition of power flow direction for matrix topology. From [9] to [12], it is clear that, using SVM, the advantages stated are lower input current THD, lower duty-cycle loss, maximum output inductor current ripple and minimum switching loss comparing to other PWM schemes when the MOSFET devices are employed. [10] explains the use of the same SVM algorithm for bidirectional power flow and SVM is able to produce better output performance, particularly at low modulation indexes, in terms of output waveform harmonic content.

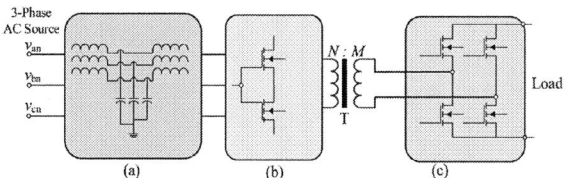

Fig. 1. Block diagram of the proposed matrix based AC-DC converter.

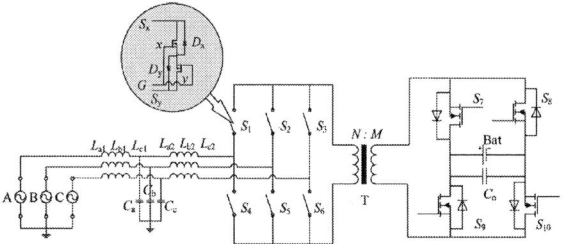

Fig. 2. Circuit topology for integrating energy storage element to utility grid.

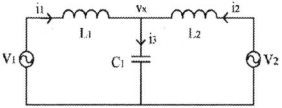

Fig. 3. LCL filter analysis.

Similar results were obtained for multi-level neutral-point-clamped matrix based converter as described in [11], while [12] shows, using SVM, the matrix based converter was able to perform all aspects of frequency changing, real power flow control, and independent reactive power flow control on both sides.

Thus, it can be concluded that bidirectional matrix based topology with galvanic isolation, lower THD values, higher power converter efficiency and power density using modulation schemes like SVM could form a competitive converter for integrating an energy storage element like batteries or super capacitors to utility grid. Hence, considering this, the block diagram of the proposed converter is as shown in Fig. 1.

The input three-phase voltages are first filtered using input LC filter (section a) followed by matrix (3x1) topology (section b). The matrix topology converts the three-phase line frequency AC voltages into an intermediate high frequency AC voltage. The high frequency AC voltage is then fed to a H-bridge (section c) through an isolating transformer (T) which in turn, rectifies it to output DC voltage. This H-bridge also inverts the DC power to high frequency AC which is converted to line frequency and voltage by the matrix based converter and fed back to the grid. Moreover, the high frequency AC voltage reduces the passive filter elements and transformer size and volume. The SVM based modulation scheme is proposed for the matrix based converter for improved input power quality. In summary, the novelty and contributions of the paper are as follows:

- Use of three-phase LCL (T-shaped) filter. This type of filter provides inductive dominance from either terminals, which is required for bidirectional operation of the converter.

- The proposed SVM based modulation scheme requires single control for each of the matrix switches and therefore, does not need switch body diode conduction and therefore, facilitates, reduced number of isolated gate drivers (six for six matrix switches) and no body diode loss (conduction loss and reverse recovery loss). Moreover, the switching sequence is arranged symmetrically to provide symmetrical bipolar high frequency AC voltage at the matrix output.

- Use of full bridge controlled rectifier to perform AC-DC as well as DC-AC operations. The load or energy storage device is connected to the H-bridge. Synchronous rectification can be carried

out to provide lower conduction losses while a simple control technique like SPWM could be used for inversion operation.

The paper is organized as follows: Section II provides the details of the proposed topology and operation of the converter. In Section III, simulation of the converter is carried out. In Section IV, results of the simulation are described. Section V provides the conclusion.

II. TOPOLOGY AND PRINCIPLE OF OPERATION

The proposed converter is as shown in Fig. 2.

Each matrix switch set (S_1 to S_6) contains two back-to-back connected switches.

A. Filters

The LCL filter topology is as shown in Fig. 3.

Applying superposition theorem to Fig. 3, for $v_2 = 0$,

$$\frac{i_2(s)}{v_1(s)} = \frac{(L_2 C_1)s^2 + 1}{(L_1 L_2^2 C_1^2)s^4 + (L_2^2 C_1)s^3 + (2L_1 L_2 C_1)s^2 + (L_2)s + L_1} \quad (1)$$

Where, v_1 is an alternating source, i_2 is the current by another alternating source v_2, L_1 and L_2 are the filter inductors and C_1 is the filter capacitor.

For $v_1 = 0$,

$$\frac{i_2(s)}{v_2(s)} = \frac{1}{(L_1 L_2)s^2 + \frac{1}{C_1}(L_1 + L_2)} \quad (2)$$

Where, v_2 is an alternating source.

Similarly, for $v_1 = 0$,

$$\frac{i_1(s)}{v_2(s)} = \frac{(L_1 C_1)s^2 + 1}{(L_1^2 L_2 C_1^2)s^4 + (L_1^2 C_1)s^3 + (2L_1 L_2 C_1)s^2 + (L_1)s + L_2} \quad (3)$$

Where, i_1 is the current by source v_1.

From the analysis, it is clear that, the LCL type of filter provides rejection of higher frequency signals i.e. showing the low pass filtering effect. By comparing (1) and (3), it can be observed that, the filter topology

4187

is symmetric and depending on the values of all filter components, the LCL filter acts low pass or resonance filter for power flow in both directions. Further more, for $L_1 = L_2$, the filter is able to achieve same resonance at the required frequency in both directions with appropriate selection of component values. Hence, the resonance could be set near to the line frequency to obtain lower THD values. Also, due to dominance of inductance, this type of filter provides current source effect. To balance out, only capacitive output filter is required. This also provides the advantage of using devices with lower ratings. The RC snubber is used to reduce the voltage spikes generated due to inductances in the circuit. As no output inductor is required, reduction in snubber circuit as no LC filter oscillations will be involved in the output circuit.

B. Modulation

For operation of the matrix based converter, the SVM is used. The reference voltages are sinusoidal with phase difference of 120°, frequency is of the required high frequency AC and with magnitude defined by modulation index.

The proposed converter operates in two regions, one is to supply power to energy storage device and other is to restore power to the grid back from the energy storage device. The two regions are described in Sections II-C and II-D respectively.

C. Charging of the energy storage using the utility grid

Power flows from the utility grid to the energy storage device. The full bridge controlled rectifier acts as rectifier to convert the high frequency AC to DC. The rectification could be done by using either the body diodes of the semi conducting switches or by using synchronous rectification technique for MOSFET switches. The matrix based converter is switched using SVM. The reference sine wave for SVM is of frequency which is of the desired high frequency AC signal.

D. Discharging of the energy storage to the utility grid

The full bridge controlled rectifier acts as inverter to convert DC to high frequency AC. SVM is used for the matrix based converter while SPWM is used for the inverter. The reference sine wave for SPWM is of frequency which is of desired high frequency AC signal and compared with high frequency triangular wave. The reference sine wave for SVM will be the line frequency.

III. SIMULATION OF THE PROPOSED CONVERTER

The simulation was carried out using PSIM (Version 10.0) software. A comprehensive simulation model was constructed with the parameters as mentioned. The parameters required for the simulation model is as expressed in Table I.

The high frequency transformer not only provides isolation but also scaling of voltage. From calculations and simulation, it was observed that, the turns ratio plays

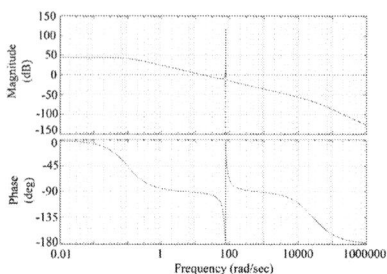

Fig. 4. Bode plot for LCL filter.

a critical role when the energy is been transferred from energy storage element to the grid. Using the selected parameters, bode plot for equivalent i_2 using equations 1 and 2 is given by Fig. 4. The observed gain margin is $-27.4dB$ at $745rad/sec$ and phase margin is 13.3^o at $726rad/sec$. Thus, for the selected values, the resonance peak occurs aroung $115Hz$, providing the low filtering effect.

A. Open loop operation

The open loop operation involves direct power conversion without any feedback. As described in Sections II-C and II-D, SVM and SPWM is used accordingly. The corresponding results are described in Section IV-A.

B. Closed loop operation

The closed loop operation is based on current control method. In literature it is well established that for a boost type converter, the current control method (two loop control) is essential for safe operation as the open loop zeros lie on right half of complex or s-plane. A closed loop scheme is developed considering the LCL filter as shown in Fig. 3. For energy storage operation to grid, the schematic is shown in Fig. 5.

The voltage difference is given by,

$$\Delta v = L \frac{d}{dt} i_1 + L \frac{d}{dt} i_2 \qquad (4)$$

TABLE I.   SIMULATION PARAMETERS

Parameter	Symbol	Value (Rating)
Grid	$A\ B\ C$	$415V, 50Hz$
Load	Bat	$300V, 1000Ah$
Input Filter	L_1, L_2	$6mH$
(per phase)	C_1	$600\mu F$
Output Filter	C_o	$500\mu F$
Turns Ratio	$N : M$	$5 : 1$
SVM Modulation	m	0.8
SVM carrier triangular signal		$40kHz$
SVM sine reference signal (Section II-C)		$10kHz$
SVM sine reference signal (Section II-D)		$50Hz$
SPWM Modulation	M	0.8
SPWM carrier triangular signal		$40kHz$
SPWM sine reference signal		$10kHz$

4188

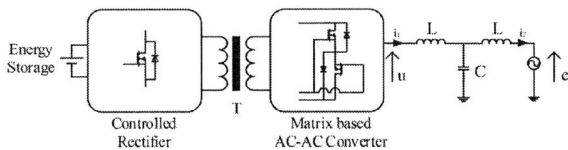

Fig. 5. System Schematic for closed loop calculations.

Where, L is the filter inductance, i_1 is input current to the filter and i_2 is output current from the filter.

Approximating, $i_1 = i_2$,

$$\therefore \frac{d}{dt}i_{abc} = \frac{\Delta v_{abc}}{L} \tag{5}$$

Where, i_{abc} is current of either A or B or C phase, v_{abc} is the corresponding voltage of either phase.

Using abc to dq transformation,

$$\frac{d}{dt}i_d = \frac{\Delta v_d}{L} + \omega i_q \tag{6}$$

$$\therefore u_d = e_d + L\frac{d}{dt}i_d - \omega L i_q \tag{7}$$

Where, i_d, v_d, u_d and e_d are d-components of respective current i and voltage v, u and e, i_q is q-component of current,
$\omega = 2\pi f$, f is the line frequency,
$\Delta v_d = (u_d - e_d)$.

Similarly,

$$u_q = e_q + L\frac{d}{dt}i_q + \omega L i_d \tag{8}$$

Where, u_q and e_q are q-component of voltage u and e respectively.

Using equations 7 and 8,

$$\therefore u_{dq} = L\frac{d}{dt}i_{dq} + jL\omega i_{dq} + e_{dq} \tag{9}$$

Taking Laplace transform and on simplification,

$$G(s) = \frac{i_{dq}(s)}{(u_{dq}(s) - e_{dq}(s))} = \frac{1}{(s + j\omega)L} \tag{10}$$

Hence, by using the (9) for reference voltage as the grid voltage, the control loop is shown in Fig. 6. The corresponding results are described in Section IV-B.

IV. RESULTS AND DISCUSSIONS

The results obtained from simulation of the converter in the software has been described in this section. Simulation consist of two parts, power flow from grid to energy storage device and vice-versa. Along with waveforms for gating pulses, voltages and currents, variation of voltage ratio or gain is presented.

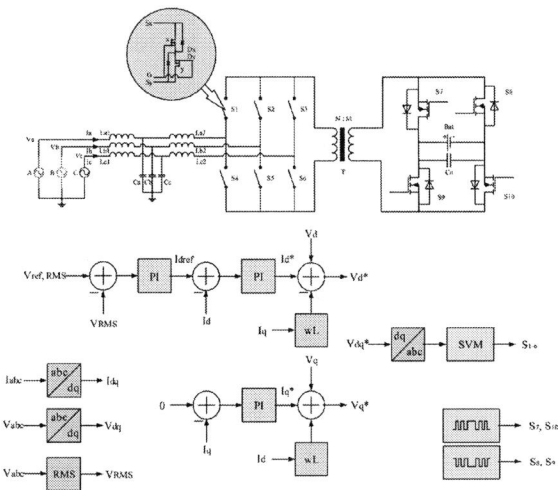

Fig. 6. Closed loop topology.

A. Open Loop Operation

1) Charging of the energy storage using the utility grid: Referring to Section II-C, the purpose of this converter is for integration of energy storage element to the utility grid. Hence, the converter has to supply a charging voltage and current to the load. The simulation results are the gating signals for all 6 sets of back-to-back connected switches are as shown in Fig. 8, high frequency transformer high voltage side voltage and current are shown in Fig. 9 and DC output voltage and current as shown in Fig. 10.

Using simulation, the relation between SVM modulation index and voltage ratio or gain is determined and as shown in Fig. 11. The voltage ratio is defined as the ratio of output DC voltage or average value to peak value of input phase voltage.

From Fig. 11, the linear nature of the voltage ratio with respect to modulation index of SVM is observed. Appropriate modulation index can be selected to achieve desired output voltage. Also, buck nature of the converter for power flow from grid to energy storage device can be inferred.

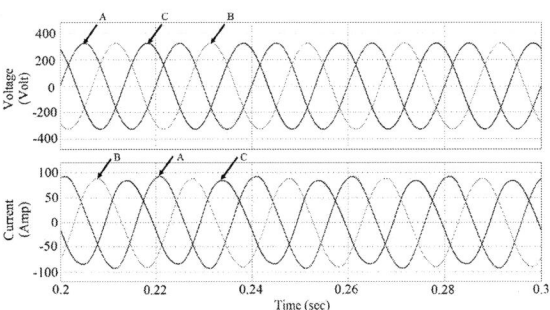

Fig. 7. Grid voltage and drain current respectively.

The 2018 International Power Electronics Conference

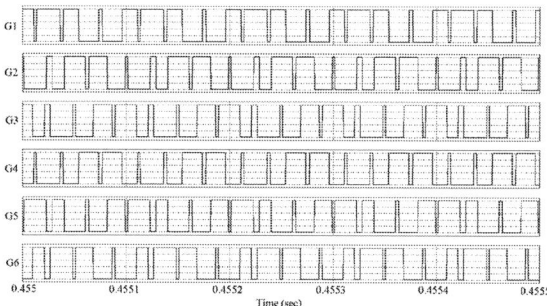

Fig. 8. SVM gating signals (G) for respective matrix based converter switches.

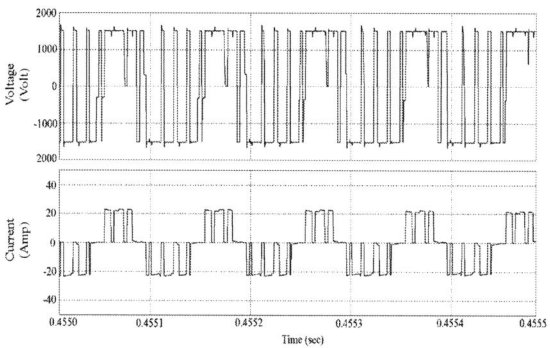

Fig. 9. High frequency transformer voltage and current respectively.

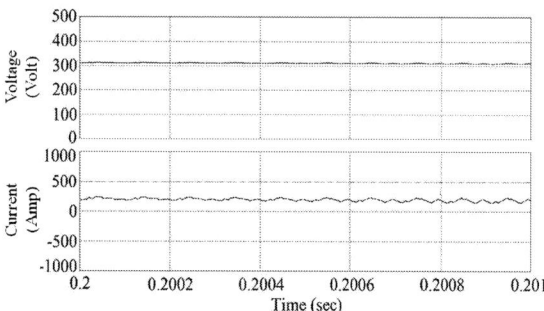

Fig. 10. DC output voltage and current.

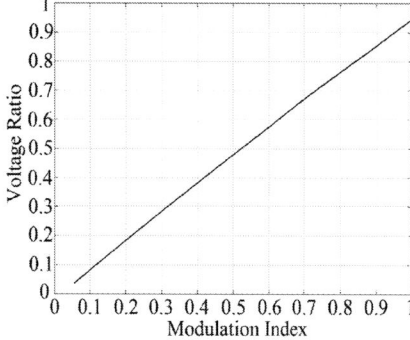

Fig. 11. Variation of voltage ratio or gain with respect to SVM modulation index.

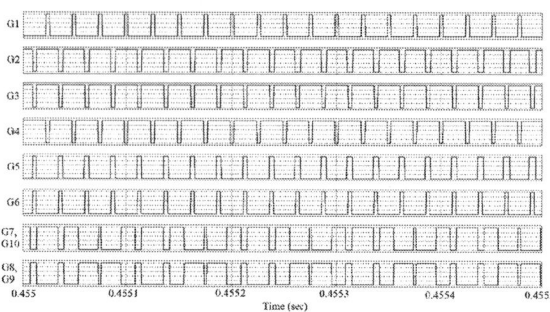

Fig. 12. SVM gating signals (G_1 to G_6) for respective matrix based converter switches and SPWM gating signals (G_7, G_{10} and G_8, G_9) for respective controlled rectifier switches.

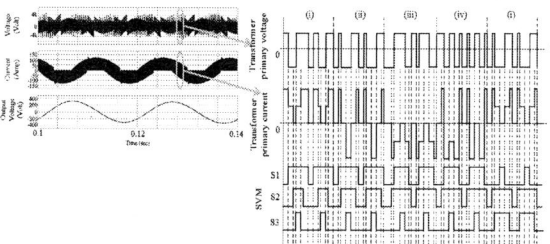

Fig. 13. Transformer voltage and current repsectively with SVM pulses and output voltage.

2) Discharging of the energy storage to the utility grid: As described in Section II-D, SPWM is used for the inversion. The main point is that no change of switching technique for the matrix based converter to act for bidirectional power flow. The only change is the frequency of reference signal provided to generate gating pulses. Also, the same carrier signal is used for both SVM (for matrix) and SPWM (for inverter). The matrix based converter is switched using the SVM signals shown in Fig. 12. The simulated transformer grid side or primary side voltage and current is as shown in Fig. 13. The three-phase AC voltage and current developed from the matrix based converter is shown in Fig. 14. Using simulation, the relation between SVM modulation index (m) and voltage ratio or gain was determined with respect to that of SPWM (M) and as shown in Fig. 15. The voltage ratio is defined as the ratio of output phase peak voltage to input DC voltage of energy storage.

Fig. 13 shows the output voltage after the filter. The transformer voltage is formulated due to high frequency SPWM inverter. The SVM is operated with reference to line frequency. A discontinued current waveform is observed on zooming, as shown in Fig. 13. The SVM progression (S_1, S_2 and S_3 in Fig. 13) is divided into 4 sections. The sections are repeatative. Section (i) shows positive current waveform while section (iii) shows negative waveform. Sections (ii) and (iv) shows the transition of the current waveforms. Also, the transformer current has line frequency component envelope. From Fig. 14, the three-phase voltages and currents are observed. The THD for voltage was calculated to be 3% while for

4190

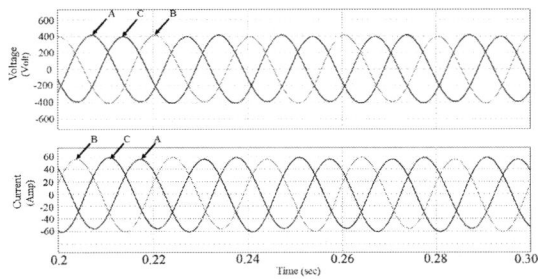

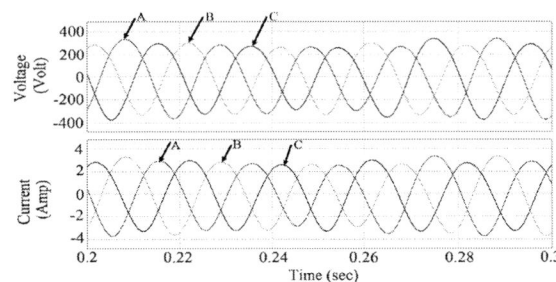

Fig. 14. Three phase output voltage and current respectively in open loop operation.

Fig. 16. Three phase output voltage and current respectively in closed loop operation.

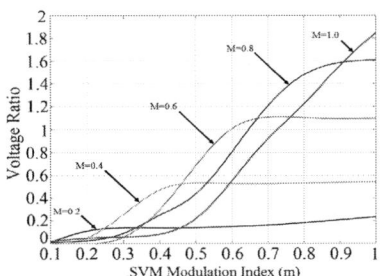

Fig. 15. Variation of voltage ratio or gain with respect to SVM (m) and SPWM modulation index (M).

current was 2%. Fig. 15 shows the non-linear variation of the voltage ratio with respect to modulation index of respective SVM and SPWM. A steep change of voltage ratio can be observed. The voltage ratio goes beyond 1 for modulation index greater than 0.6 for both SVM and SPWM. This boosting nature due to current source type (by input LCL filter) is much required for Section II-D to achieve higher grid voltage from low voltage source. This reduces the number of transformer turns ratio. The same voltage ratio can be obtained by different combinations of modulation indices of SVM and SPWM. But for easier implementation, the modulation obtained from Fig. 11 could be used to determine the other index for the operating point using Fig. 15.

B. Closed loop operation

The control loop as described in Section III-B is implemended using available library blocks in PSIM. The gains are adjusted corresponding to the carrier signal. The three phase output from DC-AC conversion is shown in Fig. 16.

For charging of the energy storage, the reference voltage is the DC output voltage and the measured voltage is the output DC voltage. For utilisation of the energy storage device, the reference is the AC output RMS voltage and the measured voltage is the output AC RMS voltage. As stated in Table I, the reference sine frequency for SVM varies in both cases.

V. CONCLUSIONS

This paper presents a matrix based isolated three-phase AC-DC bidirectional converter suitable for energy storage integration to the grid. The operation, analysis and design of the converter are illustrated. The simulation test results confirm the operation of the converter in both directions. The use of matrix topology allows direct conversion of the input three-phase AC voltages into high frequency AC voltage and thus provides a single-stage conversion without any intermediate bulky DC link capacitor. The proposed converter uses LCL type of filter at AC side providing current source characteristics. Due to only capacitive filter at DC side, reduction in snubber circuit is observed as well as lower rating devices could be populated. The converter employs the SVM technique for the matrix based converter operation while the SPWM scheme for inverter operation. Use of SVM provides superior input power quality, higher efficiency while SPWM provides higher DC utilization for inverter. Gate signals of the two switches of one matrix switch set are common, only one gate driver circuit per one matrix switch will be required. The converter is able to generate the DC voltage for charging of the storage device and is able to generate three-phase sinusoidal voltage and current to impress to the grid. Variation of output-to-input voltage ratio with respect to SVM and SPWM modulation index suggests the buck nature of the proposed topology for power flow from grid to energy storage while boost nature for power flow for opposite direction. Closed loop operation is implemended. Current control method is used for stabilising the boost type converter. Further, the closed loop could be developed for reactive power compensation control. Future works would involve hardware building and testing. Improving the parameters of LCL filter, the THD value for current could reduce more than 2%. Thus, the proposed converter will be suitable for energy storage integration applications. The converter could be applied for induction motor drives, DC drives, micro-grids, aircraft generator applications, etc.

ACKNOWLEDGEMENT

This research work has been funded and supported by Energy Market Authority (EMA), Singapore through National Research Foundation (NRF), Singapore.

4191

REFERENCES

[1] Siddhartha Anirban Singh and Sheldon S. Williamson, "Comprehensive Review of PV/EV/Grid Integration Power Electronic Converter Topologies for DC Charging Applications", *IEEE Transportation Electrification Conference and Expo (ITEC)*, July 2014.

[2] Patrick W. Wheeler, Member, Jos Rodrguez, Jon C. Clare, Lee Empringham, and Alejandro Weinstein, "matrix based converters: A Technology Review", *IEEE Transactions On Industrial Electronics*, vol. 49, pp. 276–288, April 2002.

[3] Amit Kumar Singh, Elango Jeyasankar, Pritam Das and Sanjib Kumar Panda, "A Novel Matrix Based Non-Isolated Buck-Boost Converter for More Electric Aircraft", *Industrial Electronics Society, IECON 2016-42nd Annual Conference*, pp. 1233–1238, December 2016.

[4] Amit Kumar Singh, Elango Jeyasankar, Pritam Das and Sanjib Kumar Panda, "A Single-Stage Matrix Based Isolated Three Phase AC-DC Converter with Novel Current Commutation", *IEEE Transactions on Transportation Electrification*, vol. PP, issue 99, pp. 1–17, October 2016.

[5] Manuel Ortega , Francisco Jurado and Juan P. Roa, "Bidirectional output stage matrix based converter applied to a distributed generation system", *Taylor & Francis International Journal of Electronics 2012*, vol. 99, no. 8, pp. 1115–1131, August 2012.

[6] Sadao Ishii, Hidenori Hara, Tsuyoshi Higuchi, Tomohiro Kawachi, Katsutoshi Yamanaka, Noritaka Koga, Tsuneo Kume and Jun-Koo Kang, "Bidirectional DC-AC Conversion Topology Using matrix based converter Technique", *The 2010 International Power Electronics Conference*, pp. 2768–2773, 2010.

[7] Deshang Sha and Jianliang Chen, "Bidirectional three-phase high-frequency ac link dcac converter used for energy storage", *IET Power Electron.*, vol. 8, issue 12, pp. 2529–2536, 2015.

[8] S. Mahdi Mousavi Sangdehi, Saeedeh Hamidifar and Narayan C. Kar, "A Novel Bidirectional DC/AC Stacked matrix based converter Design for Electrified Vehicle Applications", *IEEE Transactions on Vehicular Technology*, vol. 63, no. 7, pp. 3038–3050, September 2014.

[9] Jahangir Afsharian, Dewei (David) Xu, Bin Wu, Fellow and Bing Gong, Zhihua Yang, "The Optimal PWM Modulation and Commutation Scheme for Three-Phase Isolated Buck Matrix Type Rectifier", *IEEE Transactions on Power Electronics*, vol. PP, no. 99, pp. 1–13, January 2017.

[10] Keping You and M. F. Rahman, "Application of General Space Vector Modulation Approach of AC-AC matrix based converter Theory to A New Bidirectional Converter for ISA 42 V System", *IEEE Industry Applications Conference 41st IAS Annual Meeting*, pp. 2480–2487, 2006.

[11] Meng Yeong Lee, Patrick Wheeler and Christian Klumpner, "Space-Vector Modulated Multilevel matrix based converter", *IEEE Transactions on Industrial Electronics*, vol. 57, no. 10, pp. 3385–3394, October 2010.

[12] Mehrdad Kazerani, "A Direct AC/AC Converter Based on Current-Source Converter Modules", *IEEE Transactions on Power Electronics*, vol. 18, no. 5, pp. 1168–1175, September 2003.

The 2018 International Power Electronics Conference

On a Study of Voltage Dividing
Class Φ Amplifier

Katsutoshi Hirayama[1], Tadashi Suetsugu[2*], Yudai Furukawa[1] and Fujio Kurokawa[3]
1 Graduate School of Engineering, Nagasaki University, Nagasaki, Japan
2 Department of Electronic Information Engineering, Fukuoka University, Fukuoka, Japan
3 Faculty of Engineering, Nagasaki Institute of Applied Science, Nagasaki, Japan
*E-mail: suetsugu@fukuoka-u.ac.jp

Abstract-Basic The purpose of this paper is to present voltage dividing class Φ amplifier. As the interest of energy environmental problem has increased, energy saving of electronic devices is required. For that reason, minimization and high frequency of electronic devices are required. As the amplifier which can balance high efficiency with high frequency, class Φ amplifiers have been studied. In the class Φ amplifier, the voltage across the main switch and its slope are zero when the switch turns on. Moreover, the peak switch voltage of class Φ amplifier could suppress to approximately 2 times of the dc supply voltage. On the other hand, the voltage dividing circuit could divide the peak switch voltage depending on the connected number of transistors voltage by connecting transistors in series. This paper proposes the voltage dividing circuit used for class Φ amplifier. Proposed voltage dividing class Φ amplifier can divide peak switch depending on the connected number of transistors voltage by connecting transistors in series.

Keywords— class Φ amplifier; peak switch voltage; voltage divider; parasitic capacitance.

I. INTRODUCTION

As the interest of energy environmental problem has increased, energy saving of electronic devices such as smartphones and personal computers are required. In electronic devices, switching amplifier circuit to take conversion to desired electric form device is included by performing control of the voltage by switching device on/off. As the amplifier which can balance high efficiency with high frequency, class E amplifiers have been attracted attention [1]-[6]. In the class E amplifier, the voltage across the main switch and its slope were zero when the switch turned on. Also, the switch current and voltage continuously changed. Therefore, the class E amplifier could achieve zero-voltage-switching (ZVS) and zero-derivative-switching (ZDS). The occurrence of power loss and switching noise could be remarkably impressed by ZVS and ZDS.

Figure 1 shows basic circuit of class E amplifier. Fig. 2 shows the waveforms of the voltage across the main switch and output voltage of the class E amplifier in the nominal condition at 50% duty ratio. It is indicated that peak switch voltage reaches approximately 3.5 times of

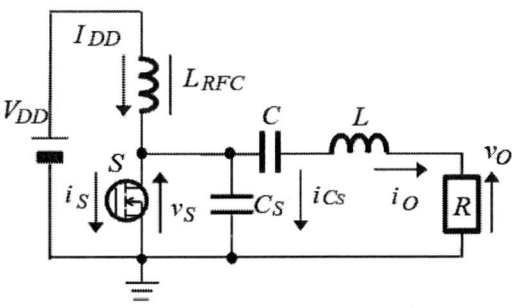

Fig. 1. Circuit of class E amplifier.

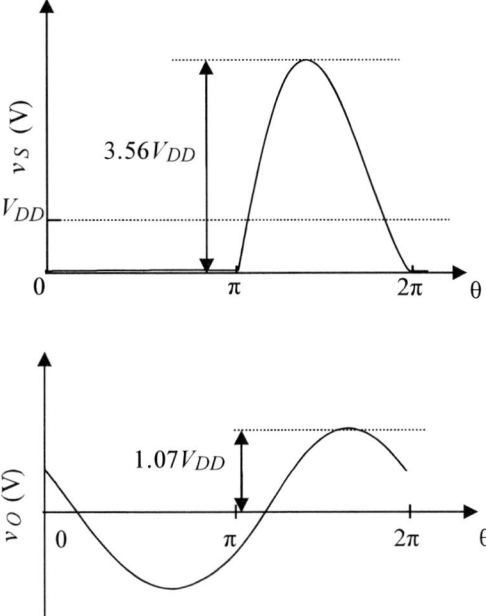

Fig. 2. Waveform of the voltage across the main switch and output voltage of class E amplifier.

The 2018 International Power Electronics Conference

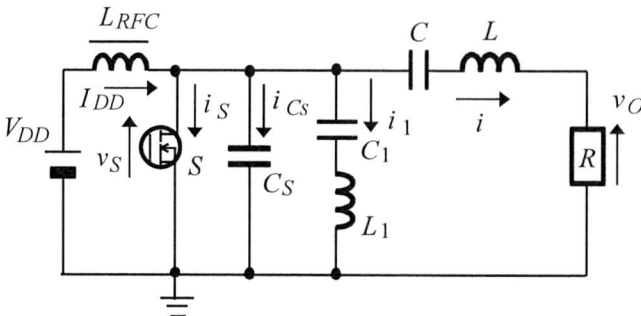

Fig. 3. Circuit of class Φ amplifier.

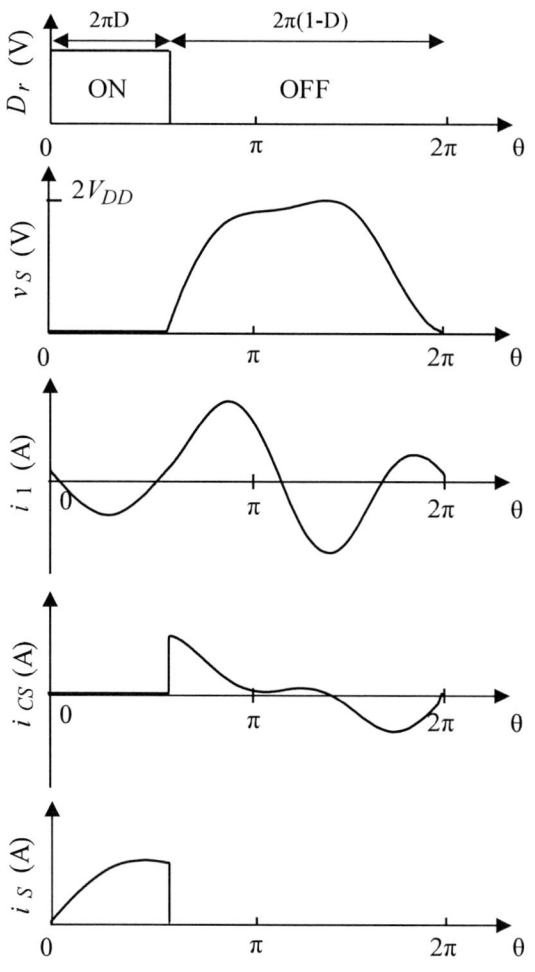

Fig. 4. Waveform of class Φ amplifier.

the dc supply voltage in the class E amplifier. For that reason, it was led to the increase of cost and loss because it was necessary to use a high-voltage switching device. For this problem, class Φ amplifier, which can reduce the peak switch voltage is studied [7], [8].

The switch current and voltage of class Φ amplifier continuously changed by injecting harmonic current to

switch when switch was on. The peak switch voltage of class Φ amplifier could suppress to 2 times of the dc supply voltage. The peak switch voltage of class Φ amplifier could reduce to 2 times, compared to conventional class E amplifier. However, the switch voltage of class Φ amplifier was high when dc supply voltage became higher. For this problem, as a circuit dividing the switch voltage by connecting the transistors of the class E amplifier in series, voltage dividing class E amplifier was studied [9], [10]. Voltage dividing class E amplifier could divide peak switch depending on the connected number of transistors voltage by connecting transistors in series. In this paper, it is proposed that dividing voltage circuit is used for class Φ amplifier. In the proposed circuit, the peak switch voltage which reduced by class Φ amplifier can be further suppressed. In this paper, the operation of voltage dividing class Φ amplifier is confirmed by LTSpice simulation and experiments.

II. Class Φ Amplifier

Figure 3 shows circuit of class Φ amplifier. Class Φ amplifier consist of an input direct voltage source V_{DD}, an inductance of the choke coil L_{RFC}, a switch S, shunt capacitance C_S and a series resonant circuit L-C, L_1-C_1. Fig. 4 shows waveforms of class Φ amplifier. The transistor S repeatedly turns on / off periodically, and its switching frequency is f. Although a parasitic capacitance exists in the transistor, since it is connected in parallel to the shunt capacitor, it is regarded as a shunt capacitor including the parasitic capacitance. During the switch is off, the difference between the current I_{DD} flowing in the inductance of the choke coil L_{RFC} and the currents i, i_1 flowing in the resonance filters L-C and L_1-C_1 flows to the shunt capacitor. This current i_{CS} cause the voltage v_S. Like the class E amplifier, this voltage vs across the main switch and its slope are zero when the switch turns on.

The resonance filters L_1-C_1 is synchronized to twice fundamental frequency. For that reason, can extract the current with secondary higher harmonic from the switch

4194

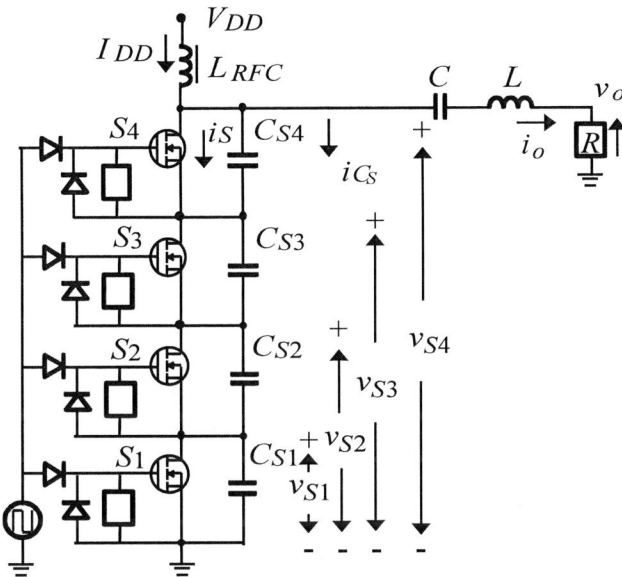

Fig. 5. Circuit of voltage dividing class E amplifier.

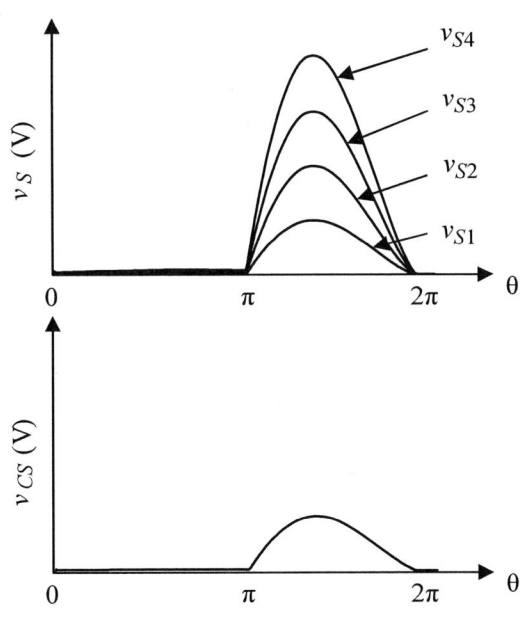

Fig. 6. The switch voltage waveform of voltage dividing class E amplifier.

voltage. The switch voltage is as shown in the fig. 2, and the peak switch voltage can suppress to approximately 2 times of the dc supply voltage. In this way, proposed class Φ amplifier can balance high efficiency with high frequency. Also, the peak switch voltage can be reduced compared to conventional class E amplifiers.

III. VOLTAGE DIVIDING CLASS E AMPLIFIER

Figure 5 shows circuit in which switch voltage is divided by 4 by series connected transistors. Fig. 6 shows waveforms of the switch voltage. This circuit dividing the switch voltage by connecting the transistors in series but, can't be connected to ground because of series connection of transistor. Thus, a resistor and a diode in parallel are connected to each transistor part, and a diode in series is connected to drive circuit. As a result, the connected diode in parallel with each transistor adjust the gate voltage and the source voltage to the same potential when the switch is off. And, the diode connected in series with each transistor enables a transistor as a switch when the switch is on.

In addition, the resistor connected in parallel with each transistor serves to release a charge of capacitor between gate-source terminals. In this way, transformer is not necessary for drive circuit of voltage dividing class E amplifier in spite of source of each switch not grounded. All switches of voltage dividing class E amplifier all at the same time on and off so that all gate voltage source can be combined into one signal source. Peak switch voltage can divide equally by equating value of shunt capacitances. Therefore, the voltage each switch of voltage dividing class E amplifier with four transistors become $v_{S4}/4$, can decrease peak switch voltage.

IV. SIMULATION AND EXPERIMENTAL RESULTS

In the simulation results, the peak switch voltage of class Φ amplifier reached approximately 280 V when the dc supply voltage was 140 V. The proposed circuit

The 2018 International Power Electronics Conference

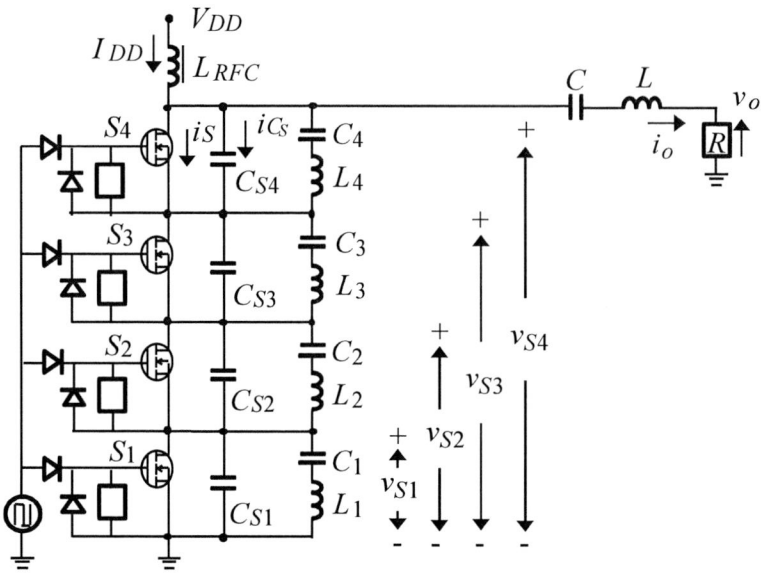

Fig. 7.   Circuit of voltage dividing class Φ amplifier.

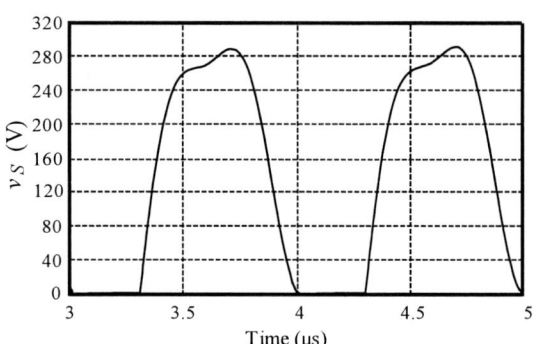

(a). Simulation waveform of the switch voltage.

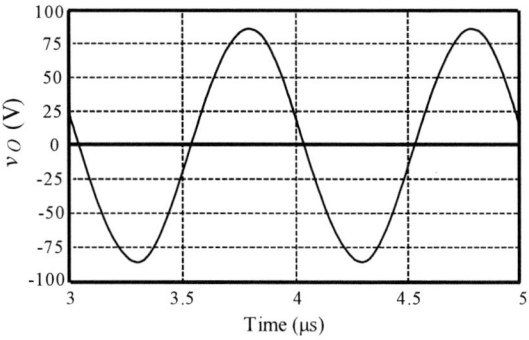

(b). Simulation waveform of the output voltage.

Fig. 8.   Simulated waveform of conventional circuit.

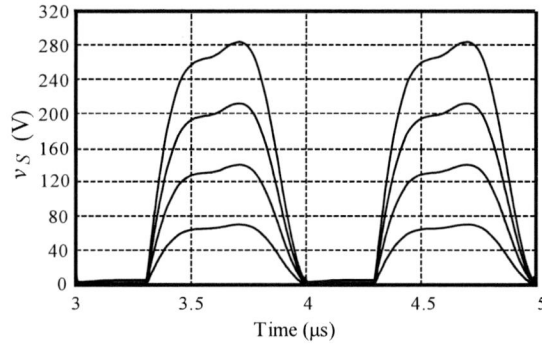

(a). Simulation waveform of the switch voltage.

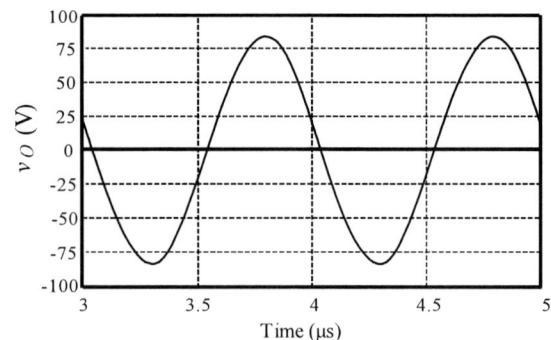

(b). Simulation waveform of the output voltage.

Fig. 9.   Simulated waveform of proposed circuit.

connects transistors to four series in the switch of class Φ amplifier. The below parameter were given as design specification of voltage dividing class Φ amplifier with four transistors $R = 50\ \Omega$, $V_{DD} = 140$ V, $Q_1 = 10$, $Q_2 = 5$

and $f = 1$ MHz. The circuit uses IRF510 MOSFET as switching device. As the main circuit parameters $L = 79.6\mu$H, $C_{S1}\sim C_{S4} = 3732$ pF, $C_1\sim C_4 = 1273$pF, $L_1\sim L_4 = 4.98\mu$, $C = 379$ pF and $L_{RFC} = 79.6\ \mu$H. These

4196

| TABLE I. | PARAMETERS OF EXPERIMENTAL CIRCUIT |

Parameters	Value
R	50 Ω
V_{DD}	20 V
Q_1	10
Q_2	5
f	1 MHz
L	79.6 μH
C_{S1}~C_{S4}	3732 pF
C_1~C_4	1273 pF
L_1~L_4	4.98 μH
C	379 pF
L_{RFC}	79.6 μH

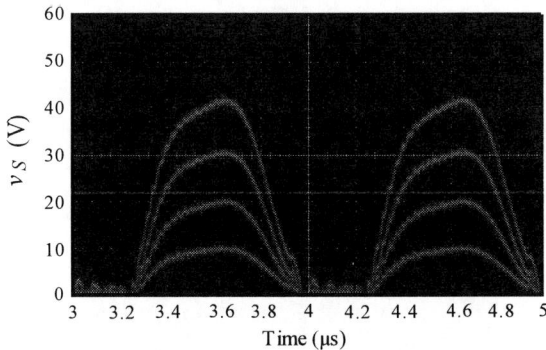

(a). Experimental waveform of the switch voltage.

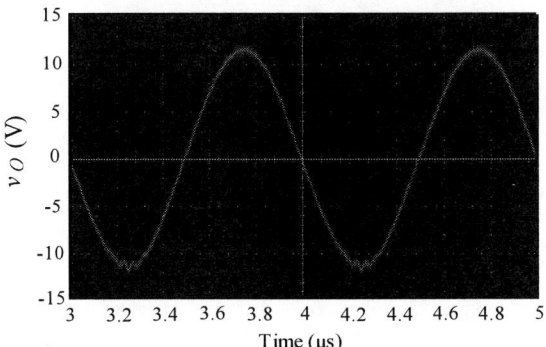

(b). Experimental waveform of the output voltage.

Fig. 10. Experimental waveform of proposed circuit.

parameter were decided by using the design equation of basic class Φ amplifier. Fig. 7 shows LTSpice simulation circuit of the proposed circuit. Fig. 8 shows LTSpice simulation waveform of conventional class Φ amplifier.

Fig. 9 shows LTSpice simulation waveform of the proposed circuit. The voltages of S_4, S_3, S_2, S_1 are v_{S4} $-v_{S3}$, $v_{S3}-v_{S2}$, $v_{S2}-v_{S1}$ and v_{S1} respectively. Fig. 8 shows that the voltage of each switch could be suppressed 1/4 times compared with the conventional circuit.

Table 1 illustrates parameters of experimental circuit. Fig. 10 shows experimental waveform of switch voltage and output voltage. As is seen in fig. 10 (a), the voltage of each switch could be suppressed 1/4 times compared with the conventional circuit. These results show that proposed circuit was successful in this experiment.

V. CONCLUSION

The class Φ amplifier satisfied class E switching and can reduce the peak switch voltage. However, the peak switch voltage of class Φ amplifier could suppress to approximately 2 times of the dc supply voltage. For that reason, class Φ amplifier couldn't correspond to excessive dc supply voltage. In this paper, we proposed class Φ amplifier using voltage dividing circuit. In the proposed circuit, it is confirmed that four switches are connected in series to the switch part of the class Φ amplifier, and the peak switch voltage can be suppressed to 1/4 times.

REFERENCES

[1] N. O. Sokal, and A. D. Sokal, "Class E-A new class of high-efficiency tuned single-ended switching power amplifier," *IEEE J.Solid-State Circuit*s., vol. SC-10, pp.168-176, June 1975.

[2] F. H. Raab, "Idealized operation of the class E tuned power amplifier," *IEEE Trans. Circuits Syst.*, vol. CAS-24, pp.725-735, Dec. 1977.

[3] R. Zulinski and J. Steadman,"Idealized operation of class E frequency multipliers, "*IEEE Trans. Circuits Syst. I.*, vol. CAS-33, no. 2, pp. 1209-1218, Dec. 1986.

[4] C. P Avratoglou, N. C. Voulgaris, and F. I. Ioannidou, "Analysis and design of a generalized class E tuned power amplifier, " *IEEE Trans. Circuits Syst.*, vol. 36, pp. 1068-1079, Aug. 1989.

[5] T. Suetsugu, M. K. Kazimierczuk, "Analysis and design of Class E amplifier with shunt capacitance composed of nonlinear and linear capacitances, " *IEEE Trans. Circuits Syst.*, vol. 51, no. 7, pp.1261-1268, July 2004.

[6] M. Hayati, A. Lotfi, M. K. Kazimierczuk, and H. Sekiya, "Analysis, design and implementation of class-E ZVS power amplifier with MOSFET nonlinear drain-to-source parasitic capacitance at any grading coefficient," *IEEE Trans. Power Electron.*, vol. 29, no. 9, pp. 4989–4999, Apr. 2014.

[7] J. M. Rivas, Y. Han, O. Leitermann, A. Sagneri, and D. J. Perreault, "A high-frequency resonant inverter with low voltage stress, " *IEEE Trans. Power Electron.*, vol. 23, no. 4, pp. 1759-1771, July 2008.

[8] United States Patent US7889519

[9] T. Suetsugu, S. Kuga and X. Wei, "A method for dividing voltage stress of high voltage class E inverter," *Proc. of IEEE International Telecommunications Energy Conference*, pp. 1-4, Oct. 2015.

[10] Japan Patent Applying 2017-077372

A DPWM based Control Strategy to Integrate Photovoltaic System and Battery Storage using Grid Connected Three-level T-Type Inverter

Mohammad M. Hashempour[1*], Yue-Ting Tsai and T. L. Lee[1]
1 Department of Electrical Engineering, National Sun Yat-sen University, Kaohsiung, Taiwan
*E-mail: D053010006@student.nsysu.edu.tw

Abstract— In this paper, integration of photovoltaic (PV) system and battery storage is considered using Three-level T-Type inverter. To attain the highest efficiency, PV panels are considered to follow MPPT algorithm while batteries are operated based on their state of charge (SOC) condition. To provide distortion-free output current, it is essential considering unbalanced condition produced due to inconsistency of MPPT and SOC. Moreover, proper tradeoff between inverter and grid is realized so that inverter is automatically adapted with demand-power of grid. All the targets are carried out by proper control of Discontinuous PWM. The proposed method is evaluated by simulation study.

Index Terms — Battery-SOC, DPWM, Grid connected three-level T-type inverter, MPPT algorithm.

I. INTRODUCTION

REGARDING ENERGY crisis and environmental problems associated with conventional power systems, new approaches have appeared in operation and planning of power systems. As prime movers, renewable energy sources are recently considered as suitable replacement of conventional energy supplies such as petroleum and gas. Furthermore, to increase efficiency and reliability, integration of different energy sources is investigated notably. Among renewable energy sources, solar energy allocates major part due to its exclusive features in sustainability and availability [1].

To harvest the maximum energy from renewable sources (known as maximum point of power tracking; MPPT), power electronic devices are commonly used. This regard, two circuit topologies are usually used as represented in Fig. 1(a) (double stage PV system) and Fig. 1(b) (single stage PV system).

In double stage topology, two sets of controllers/converters are required to track MPP (the first stage) and provide qualified three phase output voltage (the second stage). On the other hand, in single phase configuration, MPPT and voltage provision is simultaneously carried out by one converter.

As the main drawback of renewable energy sources, they are usually unreliable since they are not persistently available. To tackle this difficulty, utilization of energy storages are recommended. Hereby, the two major topologies of a hybrid

Fig. 1 (a) and (b), different topologies of PV structure. (c) and (d), different topologies of PV-Battery structure.

PV-Battery system are represented in Fig. 1(c) (double stage. PV-battery system) and Fig. 1(d) (single stage PV-battery system). Obviously, single stage topology is preferred to double stage one due to lower investment costs and higher efficiency obtained by removal of multiple converters. It is worth noting that PV arrays or battery storages might contain a set of series elements integrating smaller segments together. Anyway, MPPTs of PV arrays and SOCs of batteries should be taken into consideration.

This work was supported by Ministry of Science and Technology of TAIWAN under grant 107-3113-E-110-001-.

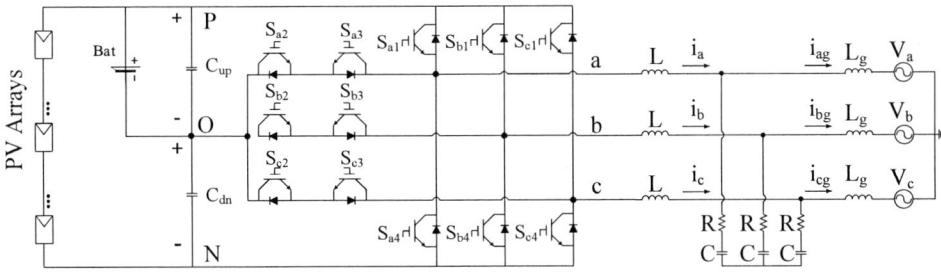

Fig. 2   Grid connected PV-Battery system using TLT²I inverter.

Regarding energy integration, a variety of control strategies have been proposed for double stage topology or multiport converters [2]-[6]. However, the control strategies proposed by previous literature are usually costly and complex since multiple converters are used while making coordination between converters is challenging.   .

Few articles address single stage PV-battery topology. The main challenging point in single stage topology is the possible unbalanced condition that might produce in dc side due to inconsistency of battery's voltage, that is determined by battery-SOC, with the voltage specified by MPPT. As a matter of fact, unbalanced neutral-point voltage can produce sever harmonic distortion at the inverter's output current. Hereby, multi-level neutral-point clamped (NPC) and T-Type inverters are preferred in single stage topology due to their ability in energy harvesting under unbalanced neutral-point voltage. In comparison with NPC, three level T-type inverter (TLT²I) is an efficiency-effective topology in low voltage applications [7]. A variety of modulation techniques have been proposed for multi-level converters to provide high quality of power by proper control of neutral-point voltage [9]-[12].

Considering PV-battery system, Space Vector PWM (SVPWM) technique is used in [8]. However, the proposed method is built upon the modification of space vectors that is complex and not simple to implement practically. Furthermore, the proposed control is sensitive to the unbalanced condition; resulting instability in transitions or larger unbalanced conditions.

In this paper, integration of PV panels and battery storages as the dc links of TLT²I is considered while the system is connected to grid. The tradeoff between PV, battery and grid is addressed concentrating on the inverter output power. Generally, the paper's contribution can be listed as follow:

- Carrier based Discontinuous PWM (DPWM) technique is used in the proposed control. It results simplicity in implementation and switching losses reduction in comparison with SVPWM and Sinusoidal PWM (SPWM).
- Due to using circuit-level decoupling strategy in the modulation, there is further reduction in switching power losses.
- To make proper tradeoff between PV, battery and grid an autonomous control strategy is proposed in the present paper. In other words, to achieve proper power flow between the system's power sources, inverter is automatically tuned merely based on demanded power by

grid while MPPT algorithm is also followed, strictly.
- In comparison with previous literature, a large unbalanced neutral point voltage is allowable by the proposed method when supplied power by inverter is free of low-order harmonic distortions.
- Unlike the previous literature considering implementation of instantaneous power theory in control stage in grid connected mode, in the proposed method merely the scalar value of demand-power is required to generate appropriate commands. Comparatively, it reduces complexity and increase accuracy.

In follow, proposed control is described in Section II. Section III is dedicated to simulation study and the paper is concluded in Section IV.

II.   PROPOSED CONTROL

Fig. 2 represents the considered grid connected PV-battery system accompanied by TLT²I as interface inverter and LC filter. To increase efficiency as far as possible, PV system tracks MPP and battery storage follows SOC. As a result, unbalanced condition in the neutral-point is unavoidable. It causes modification of the average value of neutral-point current, accordingly, it causes current harmonic distortion in ac side [7]. To tackle this difficulty, proper control of modulation signals is carried out in this paper.

To increase efficiency, the tradeoff between PV, battery and grid is critical. In fact, the priority purpose is following up MPPT algorithm, strictly. Then, the battery is operated based on the supplied power by PV. Simply, three cases might happen in the system's power flow:

i.   When PV power generation is sufficient to support both grid and battery.   In this case, battery is charging.

ii.   When PV power generation is insufficient to support grid. In this case, battery is discharging to make collaboration with PV and deliver demand-power to grid.

iii.   When PV power generation is equal to the demanded power by grid. In this case, battery is neither charging nor discharging.

Fig. 3 shows the proposed control scheme. As shown, the reference voltage command determined by MPPT algorithm (V_{dc}^*) is compared with its actual value and the error is fed to a proportional-integrator (PI) controller. The reference voltage

4199

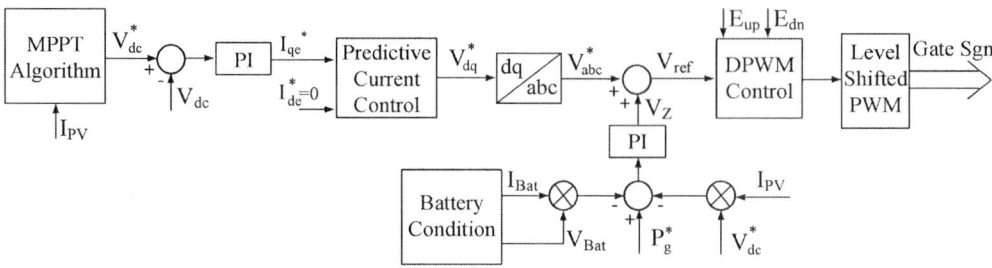

Fig. 3. Proposed control.

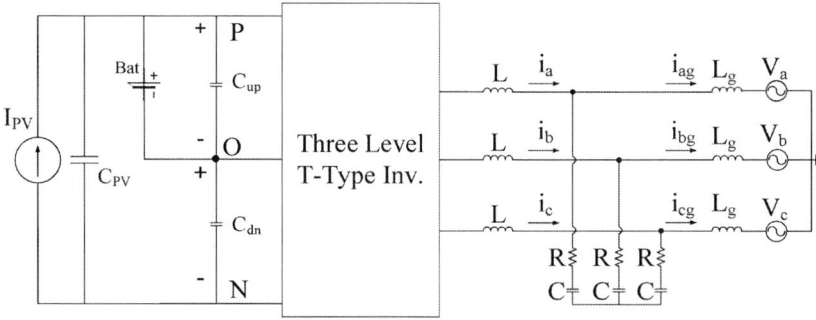

Fig. 4. Test system.

vector in dq frame (V_{dq}^*) is achieved by using predictive current controller. The output is transferred to abc-frame to be applicable in carrier based PWM. Based on the circuit topology (Fig. 2), in case the half of the reference voltage of MPPT is not equal to the reference voltage command by SOC of battery, neutral-point voltage will not be zero anymore. Accordingly, the average value of neutral point current will be modified; it causes current harmonic distortion at inverter output. Especially, even-order harmonics distortion will be high in this situation.

In addition, as the main purpose of the proposed method, battery collaboration with PV should be tuned based on demand-power of grid. As a matter of fact, depending on power absorption/generation by battery, there will be further modification in the average value of neutral point current, accordingly, inverter output current will be distorted even severer. Hereby, proper control of DPWM is applied to counteract the aforementioned problems.

Fig. 3 shows the general scheme of proposed control. As it is shown in Fig. 3, the difference value between demand-power (P_g^*) and PV power is calculated to tune battery current using PI controller. Thus, V_Z can be considered as zero sequence common mode voltage injected to V_{abc}^*. The final value (V_{ref}) is fed to DPWM control block.

In DPWM control block, the discontinuous time of modulation signal is tuned so that proper energy provision (or absorption by battery) is automatically achieved by PV and battery [13] while there is no conflict in MPPT performance. Thus, system is operated in its highest efficiency. The final modulation signals are transferred to level shifted PWM block.

As one of the advantages of the proposed method, the scalar value of demanded active power by grid is directly applied in the control stage (see Fig. 2) while it is avoided implementing long calculations related to instantaneous power theory. By this, complexity is reduced and accuracy is increased since there is not any phase (or frequency) lock loop block or any parameter related to reactive power in control stage.

III. SIMULATION STUDY

MATLAB/Simulink is used to evaluate the proposed method. Regarding the possible cases of power flow (explained in Section II) under different MPPT conditions, neutral-point voltage situations and demand-power, three scenarios have been considered to evaluate the proposed method. The assumptions related to each scenario is shown in Table II. The test system is represented in Fig. 4 and the system's parameters are represented in Table I. As shown in Fig. 4, PV system is considered as current source paralleled with a capacitor.

TABLE I
CONSIDERED SCENARIOS GENERAL INFORMATION

	NP Balance condition	MPPT condition	Demand power
Scenario 1	balanced	constant	variable
Scenario 2	unbalanced	constant	variable
Scenario 3	unbalanced	variable	constant

The 2018 International Power Electronics Conference

TABLE II
TEST SYSTEM PARAMETERS

Symbol	Parameter	Value
V	3φ line-to-line voltage	220 V
F_{fun}	Fundamental frequency	60 Hz
F_{sw}	Switching frequency	5 kHz
C_{up}, C_{dn}	dc links capacity	1500 μF
C_{PV}	Total dc link capacity	3000 μF
L_f	Filter's inductance	5 mH
L_g	Grid's inductance	2 mH
R	Damping Resistor	6 Ω

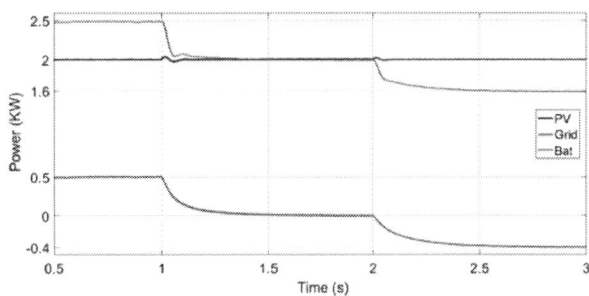

Fig. 5. Power related curves (Scenario 1).

A. SCENARIO 1

Balanced condition is assumed in this scenario. In addition, based on a typical solar irradiance and MPPT algorithm, it is assumed that by I_{PV}=5 A, the reference voltage specified by MPPT algorithm is 400V (V_{dc}^*=400). Therefore, V_{Bat} is 200V. However, demand-power is assumed variable. It means that power provision by battery should be varied in accordance with demand-power.

Fig. 5 shows power related curves. Power generation by PV is fixed and equal to 2KW. However, before t=1s, demand-power is 2.5KW. Thus, the deficit power is provided by battery as it is shown in Fig. 5. At t=1s demand-power is changed to 2KW. Therefore, PV power is equal to demand-power and it is not needed applying battery. On the other hand, in the last step (2<t<3), demand-power is changed from 2KW to 1.6KW. In this situation, the extra power generated by PV is injected to battery. It can be seen in Fig. 5 that battery is charging in the last step.

As mentioned before, the average value of neutral-point current is not zero in case there is asymmetrical energy provision burden by C_{up} and C_{dn} (see Fig. 4). This statement is clearly shown in Fig. 6. It can be seen that in the symmetrical situation (when battery is bypassed) the average value of neutral point current is zero (Fig. 6(b)). When battery is discharging (Fig. 6(a)), the average value is negative and vice versa. Hereby, DPWM is appropriately controlled so that proper power sharing between battery and PV is achieved when output current is also qualified.

Fig. 7 shows inverter output current and Fig. 8 represent harmonic analysis of the current. It can be seen that Total Harmonic Distortion (THD) of the current is 2.59% when even-order harmonics are predominant, relatively.

B. SCENARIO 2

In this scenario, V_{Bat} is assumed to be 300V specified by SOC of battery. In addition, the reference voltage specified by MPPT algorithm is 400V (V_{dc}^*=400V). As a result, it can be concluded that there is severe unbalanced neutral-point voltage.

Fig. 9 shows power related curves. Before t=1s, total power demanded by grid is 1.6KW while 2KW is provided by PV, thus, the extra power is injected to battery. It this situation, battery is charging. At t=1, demand-power is increased 400W. In this situation, PV power is equal to demand-power, as a result, battery is neither charged nor discharged. Finally, at t=2

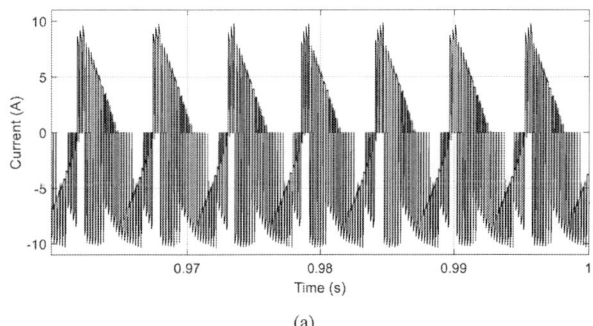

(a)

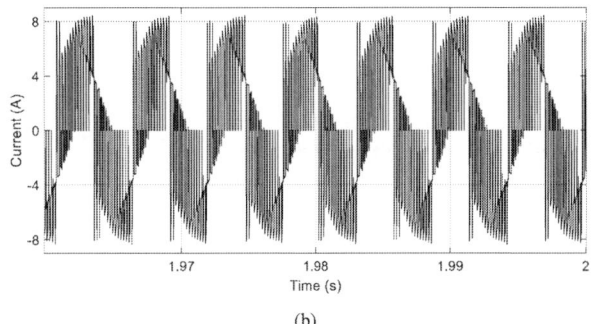

(b)

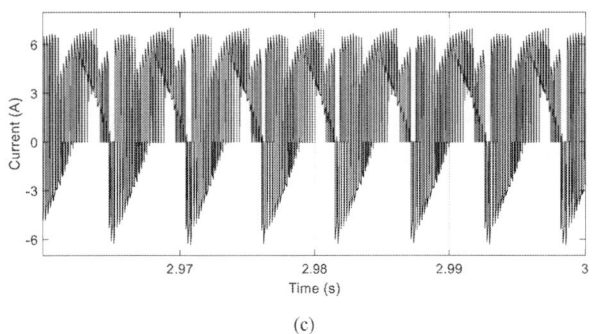

(c)

Fig. 6. Neutral point current (Scenario 1).

4201

The 2018 International Power Electronics Conference

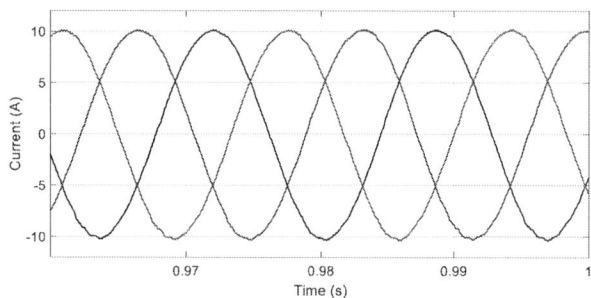

Fig. 7. Inverter output current (Scenario 1).

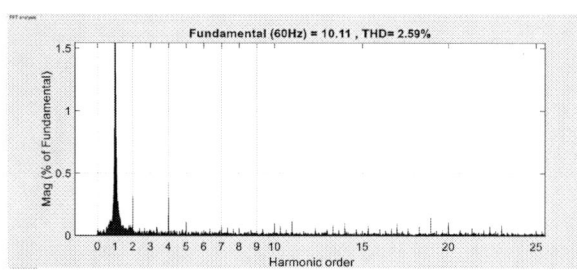

Fig. 8. THD of inverter output current (Scenario 1).

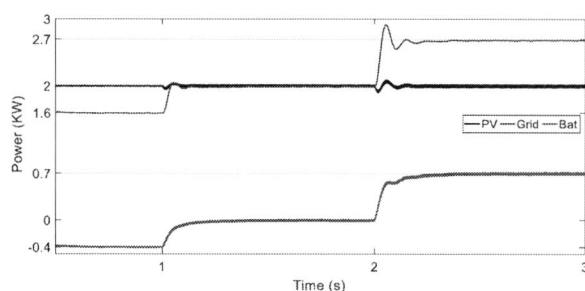

Fig. 9. Power related curves (Scenario 2).

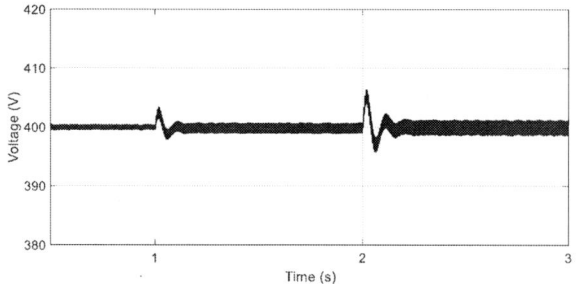

Fig. 10. PV voltage (Scenario 2).

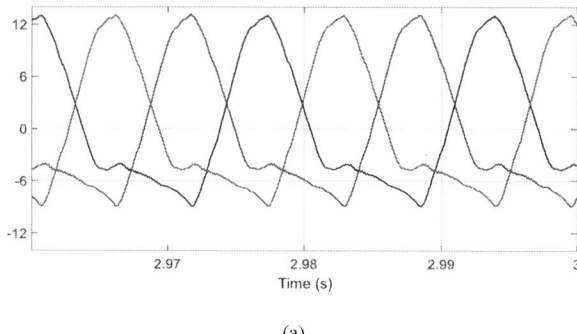

(a)

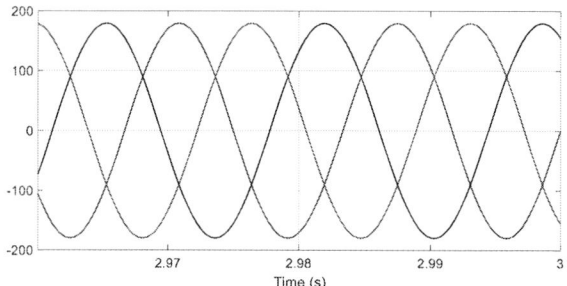

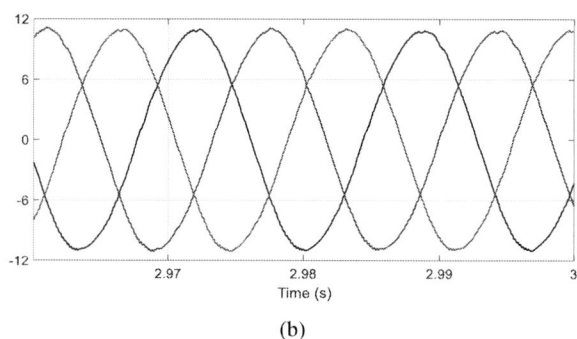

(b)

Fig. 11. Current and voltage waveforms (Scenario 2). (a) Inverter output current (conventional method) (b) Grid voltage and inverter output current (proposed method).

demand-power is increased to 2.7KW. Again, it can be seen that PV power generation is remained unchanged while battery is started to counteract the deficit power. In this situation battery is discharging; injecting 700W to grid. Fig. 5 proves that MPPT algorithm is strictly implemented under different values of demand-power. Then, battery supplies the deficit power or consumes the extra power.

Fig. 10 shows PV voltage. It can be seen that the reference voltage by MPPT is tracked properly.

Fig. 11 shows inverter output current by using proposed method and conventional method (SPWM). By conventional method, there is severe current distortion at output.

4202

The 2018 International Power Electronics Conference

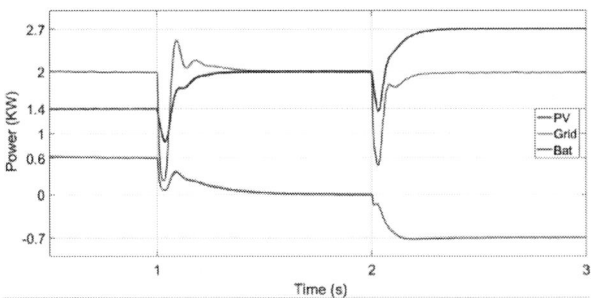

Fig. 12.   Power related curves (Scenario 3).

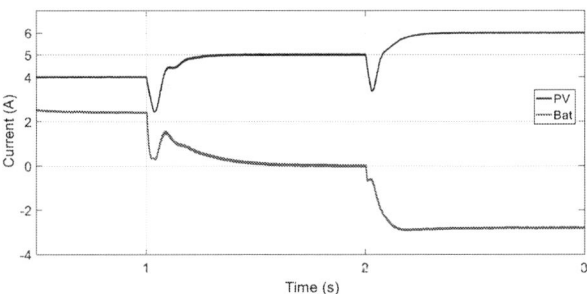

Fig. 13.   PV and battery currents (Scenario 3).

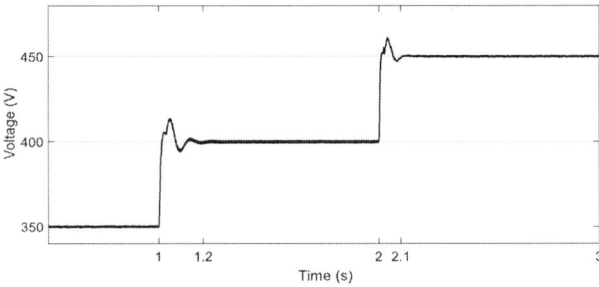

Fig. 14.   PV voltage (Scenario 3).

It should be mentioned that second order harmonic distortion of current is measured 43.07% while THD is 43.53%. Note that inverter's power generation cannot be controlled by conventional method. In fact, implementing conventional method in the test system, battery is charged/discharged merely based on the deviation from balance condition. On the other hand, by using proposed method (Fig. 11(b)), it can be seen that output current is free of distortion although there is sever unbalanced neutral-point voltage.

C.   SCENARIO 3

In this scenario, demand-power is assumed constant (P_g^*=2KW) but PV power generation is considered variable. Fig. 12 shows power related curves. Before t=1s, PV power

generation (P_{PV}=1.4KW), is not sufficient to meet P_g^*. Accordingly, the deficit power is provided by battery (P_{Bat}=0.6KW). At t=1s, P_{PV} is increased to 2KW so battery is bypassed since PV is sufficient to provide demand-power. Finally at t=2s, since P_{PV}= is higher than demand-power, battery is charging by consuming 0.7KW.

Fig. 13 shows PV and battery currents. Regarding above explanation, it can be seen that battery's current changes from nearly 2.5A to -3A. Finally, based on Fig. 14, it is shown that voltage commands from MPPT are properly tracked within 0.2s.

IV.   CONCLUSION

In this paper, integration of PV system and battery storage is addressed using three-level t-type inverter. Grid connected PV-battery system is investigated. The main purpose of the proposed control is to follow MPPT algorithm while SOC of battery is also taken into consideration. By proper control of DPWM, autonomous tradeoff between PV, battery and grid is obtained while supplied power by inverter is qualified.

REFERENCES

[1]   O. M. Toledo, D. O. Filho, and A. S. A. C. Diniz, "Distributed photovoltaic generation and energy storage systems: A review," *Renewable Sustainable Energy Rev.*, vol. 14, no. 1, pp. 506–511, 2010.

[2]   W. Jiang and B. Fahimi, "Multiport Power Electronic Interface—Concept, Modeling, and Design," in IEEE Transactions on Power Electronics, vol. 26, no. 7, pp. 1890- 1900, July 2011.

[3]   H. Zhu, D. Zhang, H. S. Athab, B. Wu and Y. Gu, "PV Isolated Three-Port Converter and Energy-Balancing Control Method for PV-Battery Power Supply Applications," in IEEE Transactions on Industrial Electronics, vol. 62, no. 6, pp. 3595-3606, June 2015.

[4]   J. Sachs and O. Sawodny, "A Two-Stage Model Predictive Control Strategy for Economic Diesel-PV-Battery Island Microgrid Operation in Rural Areas," in IEEE Transactions on Sustainable Energy, vol. 7, no. 3, pp. 903-913, July 2016.

[5]   A. Merabet, K. Tawfique Ahmed, H. Ibrahim, R. Beguenane and A. M. Y. M. Ghias, "Energy Management and Control System for Laboratory Scale Microgrid Based Wind-PV-Battery," in IEEE Transactions on Sustainable Energy, vol. 8, no. 1, pp. 145-154, Jan. 2017.

[6]   M. O. Badawy and Y. Sozer, "Power Flow Management of a Grid Tied PV-Battery System for Electric Vehicles Charging," in IEEE Transactions on Industry Applications, vol. 53, no. 2, pp. 1347-1357, March-April 2017.

[7]   M. Schweizer and J.W. Kolar, "Design and implementation of a highly efficient 3-level T-type converter for low-voltage applications," IEEE Trans. Power Electron., vol. 28, no. 2, pp. 889–907, Feb. 2013.

[8]   H. R. Teymour, D. Sutanto, K. M. Muttaqi and P. Ciufo, "Solar PV and Battery Storage Integration using a New Configuration of a Three-Level NPC Inverter With Advanced Control Strategy," in IEEE Transactions on Energy Conversion, vol. 29, no. 2, pp. 354-365, June 2014.

[9]   J. Zaragoza *et al.*, "Voltage-balance compensator for a carrier-based modulation in the neutral-point-clamped converter," *IEEE Trans. Ind. Electron.*, vol. 56, no. 2, pp. 305–314, Feb. 2009.

[10]   C. Wang and Y. Li, "Analysis and calculation of zero-sequence voltage considering neutral-point potential balancing in three-level NPC converters," *IEEE Trans. Ind. Electron.*, vol. 57, no. 7, pp. 2262–2271, Jul. 2010.

[11]   A. Lewicki, Z. Krzeminski, and H. Abu-Rub, "Space-vector pulsewidth modulation for three-level NPC converter with the neutral point voltage

control," *IEEE Trans. Ind. Electron.*, vol. 58, no. 11, pp. 5076–5086, Nov. 2011.

[12] U. M. Choi, J. S. Lee, and K. B. Lee, "New modulation strategy to balance the neutral-point voltage for three-level neutral-clamped inverter systems," *IEEE Trans. Energy Convers.*, vol. 29, no. 1, pp. 91–100, Mar. 2014.

[13] M. M. Hashempour, M. Y. Yang and T. L. Lee, "A DPWM-controlled three-level T-type inverter for photovoltaic generation considering unbalanced neutral-point voltage," *2017 IEEE Energy Conversion Congress and Exposition (ECCE)*, Cincinnati, OH, 2017, pp. 3856-3862.

Impedance Measurement of Megawatt-Level Renewable Energy Inverters using Grid-Forming and Grid-Parallel Converters

Matias Berg[1]*, Tuomas Messo[1], Tomi Roinila[2], Henrik Alenius[2]

1 Electrical Energy Engineering, Tampere University of Technology, Tampere, Finland
2 Hydraulics and Automation, Tampere University of Technology, Tampere, Finland
*E-mail: matias.berg@tut.fi

Abstract—Harmonic resonance and power quality problems have been reported in grid-connected photovoltaic and wind power systems. The AC-side impedance of three-phase converter is an important characteristic, which can be effectively used as a design parameter to avoid instability and excessive harmonics. A number of methods to measure the three-phase AC impedance have been reported. However, solutions for high power applications such as wind and photovoltaic converters with a power rating of several megawatts, have not been discussed. This paper introduces a new method to measure the impedance from high power three-phase converter. The impedance is identified by perturbing the converter first by voltage-type injection utilizing high-power grid-forming inverter, and subsequently by current-type injection by utilizing low-power grid-parallel converter. The main benefit of the proposed setup is the possibility to measure the converter impedance online in its natural operating point both at high and low frequencies. The paper presents a proof-of-concept by validating the method using a switching model.

Keywords—*Impedance Measurement, Identification, Small-Signal Modeling, Three-Phase Power Conversion*

I. INTRODUCTION

The share of renewable energy generation in power systems is experiencing a rapid growth. New energy sources are connected to the grid using power converters ranging from a few kilowatts up to megawatts. Power converters of several megawatts are nowadays a standard solution in offshore wind parks and large PV power plants.

The large share of renewable energy is expected to challenge the stability of large power systems in near future, especially due to smaller inertia [1]. Stability, power quality problems, and unwanted disconnection of the converter from the power system have been recently reported [2], [3]. The prime source of such problems can be often found in the small-signal impedance characteristics of power converters [4], [5]. In fact, the power converter impedance should be shaped to have a high magnitude or passive characteristics to avoid instability induced by the underdamped resonance between the converter and the grid impedance [6], [7], [8].

There is a growing interest in the industry and academia to obtain reliable analytical impedance models

that would allow more accurate harmonic and stability analysis [9]. However, the derivation of impedance models may be difficult due to numerous unknown parameters, such as internal control parameters, which are most often well-protected secrets of converter manufacturer. The ability to accurately measure the converter impedance would allow generating impedance models by, e.g., different curve-fitting tools, which in principle would not jeopardize the intellectual property. Therefore, a measurement method which can accurately capture the output impedance of a high-power three-phase converter, at its nominal operating point, is desperately needed.

An impedance measurement setup for power converter with power rating of few tens of kilowatts can be done accurately using standard laboratory equipment [10]. New methods need to be developed as the power level increases since standard laboratory power supplies are not able to sink the generated electrical power. An impedance measurement unit for shipboard MVDC systems presented in [11] was shown to capture the impedance up to 1 kHz. However, power converters usually employ an LCL-filter which may have very sharp resonance around few kilohertz. Therefore, the impedance measurement setup should have higher bandwidth than the LCL-filter resonance to fully characterize the impedance. The method presented in [12] relies on making a step change in the passive load of the power converter. However, the method is not applicable to grid-connected converters, since the load is the power system itself. A noninvasive method was introduced in [13], which is based on monitoring the grid waveforms. However, this method is not suitable for power converters with high output impedance, because measuring their impedance generally requires perturbation signal with sufficient amplitude.

This paper proposes a method to identify the inverter impedance using a combination of grid-forming and grid-parallel converters. A voltage-type injection is generated by using a high-power grid-forming inverter, which is capable for characterizing the low-frequency impedance of the power converter. Subsequently, a current-type injection is generated using a grid-parallel converter capable of measuring the high-frequency part of the impedance. Accuracy of the proposed method is compared to an ideal measurement setup using a switching model.

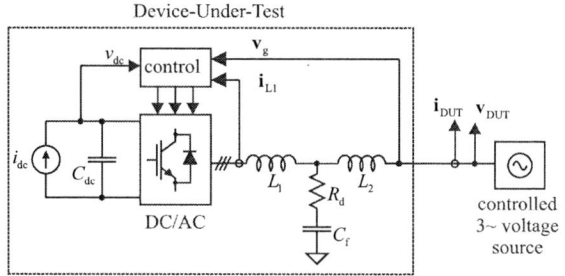

Fig. 1. Principle of measurement setup to identify inverter output impedance.

II. THREE-PHASE IMPEDANCE IN THE DQ-DOMAIN

Three-phase sinusoidal variables can be transformed into the dq-domain, where the fundamental component appears as a DC signal. The transformation is done using the Park's transformation according to (1).

$$
\begin{bmatrix} x_d \\ x_q \\ x_0 \end{bmatrix} = \mathbf{T}^{dq} \begin{bmatrix} x_a \\ x_b \\ x_c \end{bmatrix}
\tag{1}
$$

where the transformation matrix is given by

$$
\mathbf{T}^{dq} = \frac{2}{3} \begin{bmatrix} \cos(\theta) & \cos\left(\theta - \frac{2\pi}{3}\right) & \cos\left(\theta - \frac{4\pi}{3}\right) \\ -\sin(\theta) & -\sin\left(\theta - \frac{2\pi}{3}\right) & -\sin\left(\theta - \frac{4\pi}{3}\right) \\ \frac{1}{2} & \frac{1}{2} & \frac{1}{2} \end{bmatrix} .
\tag{2}
$$

The angle θ should follow the real phase angle of a three-phase system, i.e., the phase angle changes with the fundamental grid frequency. The phase angle can be generated internally by the impedance measurement setup or it can be obtained using a low-bandwidth phase-locked-loop. It is customary to align the dq-reference frame so that the steady-state value of grid voltage q-component appears as zero.

The inverter output impedance can be defined in the dq-domain by four independent impedance components as in (3). Z_{dd} is the impedance d-component, which is defined as the ratio of voltage and current d-components (v_d/i_d). Z_{qd} is the cross-coupling impedance from q to d-component (v_d/i_q) and Z_{dq} is the cross-coupling impedance from d to q-component (v_q/i_d). Finally, Z_{qq} is the impedance q-component, defined as the ratio of voltage and current q-components (v_q/i_q). Ideally, all the four impedance components should be measured to allow full characterization of the converter impedance [6].

$$
\mathbf{Z}_{inv} = \begin{bmatrix} Z_{dd} & Z_{qd} \\ Z_{dq} & Z_{qq} \end{bmatrix}
\tag{3}
$$

III. MAXIMUM-LENGHT BINARY SEQUENCE (MLBS)

Pseudo-random binary sequence (PRBS) is a periodic broadband signal based on a sequence of length N. The most commonly used signals are based on maximum-length binary sequences (MLBS). Such sequences exist

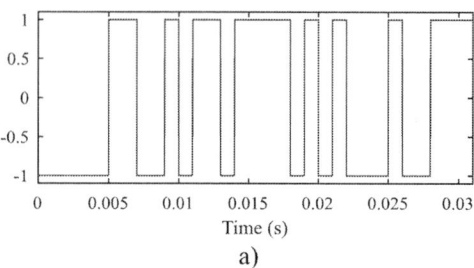

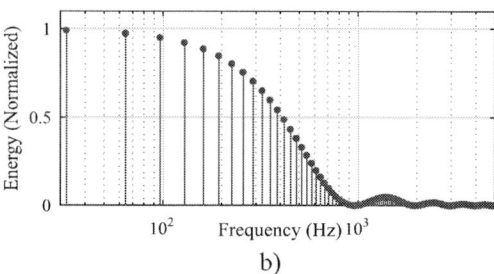

Fig. 2. a) MLBS signal in time-domain and b) its frequency-spectrum.

for $N = 2^n - 1$, where n is an integer. They are popular because they can be generated using feedback shift-register circuits [14].

Fig. 2 shows the MLBS sequence, which is generated at 1 kHz using a five-bit-long shift-register, and has signal levels ± 1V. The figure also shows the energy spectrum, which contains almost constant energy up to one third of the generation frequency. The energy drops to zero at the generation frequency and its harmonics. An MLBS signal x has the lowest possible peak factor regardless of its length $|x|_{peak}/x_{rms} = 1$, which means that the sequence is well suited for sensitive systems which require small-amplitude perturbation. Due to the deterministic nature of the sequence, the signal can be repeated and injected precisely and the signal-to-noise ratio (SNR) can be increased by synchronous averaging of the response periods. The MLBS signal is used in this paper to perturb the inverter output terminal waveforms to necessitate measuring the impedance components in (3).

IV. IDEAL IMPEDANCE MEASUREMENT SETUP

Fig. 1 illustrates the ideal three-phase impedance measurement setup. A three-phase grid-feeding inverter, such as a photovoltaic inverter, is connected to a balanced three-phase voltages source. The inverter includes all the necessary control functions, such as DC voltage control, AC current control and phase-locked-loop to synchronize its output currents with the grid voltages. Moreover, the inverter includes passive components, such as an LCL-filter, which are required to filter out the ripple in DC and AC side waveforms. Table I summarizes the most important parameters of the three-phase inverter. The control delay is assumed to be 1.5 times the switching period and it was included as a transport delay within the current control of the inverter.

4206

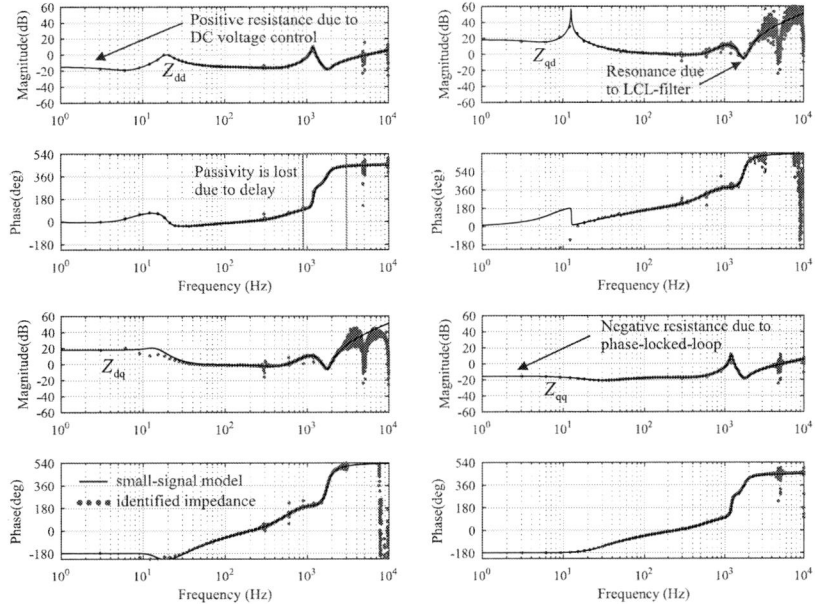

Fig. 3. Identified impedance from the simulator using ideal voltage-type injection.

TABLE I. INVERTER PARAMETERS.

P_{dc}	1 MW	L_1	60 μH
$V_{ac}^{line-to-line}$	400 V	L_2	30 μH
f_{ac}	50 Hz	C_{dc}	20 mF
V_{dc}	1500 V	C_f	400 μF
f_{sw}	5 kHz	R_d	70mΩ

The output impedance of the inverter is to be identified, which requires a wide-bandwidth perturbation to the voltages or currents on the AC side. In theory, the inverter could be connected to a high-power voltage amplifier, such as a linear amplifier, which can be used to inject the required perturbation. This is actually a practical way to measure the output impedance in the range of few tens of kilowatts [7]. However, in the case of megawatt-level converters using such arrangement is not practical, since the voltage amplifier has to sink all the produced power. The inverter impedance depends on the operating point and should be measured while the converter is online and operating at its nominal output power.

A wide-bandwidth perturbation is injected on top of the grid voltage waveforms $\mathbf{v}_{DUT}$ in the form of a Maximum-Length Binary-Sequence [15]. The MLBS signal is generated at 100 kHz with a peak-to-peak value of 50 V. The MLBS signal is added to d-component of the grid voltage reference value to measure impedance components Z_{dd} and Z_{qd} and to grid voltage q-component to measure Z_{dq} and Z_{qq}. As an example, the frequency response of the impedance component Z_{dd} can be computed from the ratio of Fourier-transformed voltage and current as in (4).

$$Z_{dd}(j\omega) = \frac{V_d(j\omega)}{I_d(j\omega)} \quad (4)$$

Fig. 6 shows the perturbed grid voltage and the corresponding output current of the inverter in the natural reference frame during the impedance measurement. The amplitude of grid voltage waveforms follows exactly the MLBS signal, as expected, since the controlled ideal voltage source does not attenuate the perturbation. The impedances obtained by using this arrangement are considered as the reference curves in the following chapters.

Fig. 3 shows the identified impedance components of the power converter which are obtained using the setup in Fig. 1. The switching model is implemented in MATLAB Simulink using the SimScape package. The identified impedance is compared against a small-signal impedance model of the inverter known to be accurate, which can be found in [16]. Fig. 3 shows all four components of the inverter impedance. The black curve is the analytical small-signal model and the blue dots represent the corresponding identified frequency response. The measurement is by theory accurate up to half of the generation frequency, i.e., up to 50 kHz. However, the switching ripple of the inverter distorts the measurement near 5 kHz. A few remarks can be done based on the impedance components:

Effect of DC Voltage Control

DC voltage control causes the impedance d-component Z_{dd} to behave as a positive resistance within the bandwidth of DC voltage controller. The positive resistance behavior appears as a constant magnitude and phase close to zero degrees. The crossover frequency of DC voltage controller was set to 10 Hz with 47 degree phase margin.

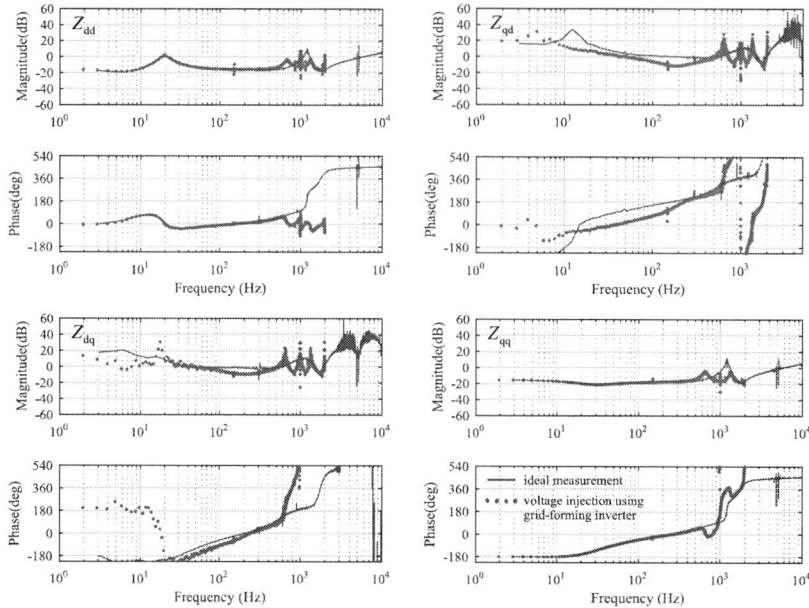

Fig. 4. Identified impedance from the simulator using grid-forming inverter to generate the voltage-type injection.

Effect of Phase-Locked-Loop
Phase-locked-loop makes the impedance q-component Z_{qq} to behave as a negative resistance within the bandwidth of the PLL. The PLL was tuned to have a crossover of 40 Hz with 65 degree phase margin. The negative resistance can be seen as a constant low-frequency magnitude and phase close to -180 degrees. The negative resistance is known to cause stability problems if PLL is tuned to have too fast dynamics [17].

Effect of LCL-filter
All impedance components experience a resonance near 1.7 kHz due to the use of LCL-filter. The series resonance is effectively damped by the passive damping resistor R_d. Peaking at the resonant frequency can cause instability, especially if active damping is used [18].

Effect of Control Delay
Both impedance d and q-components Z_{dd} and Z_{qq} experience large positive peaking near the LCL-filter resonant frequency due to $1.5/f_{sw}$ control delay. Moreover, the impedance does not behave as a passive circuit near the resonant frequency and may amplify harmonics and cause impedance-based instability. Passivity is lost when the phase curve does not stay between -90 and +90 degrees. A non-passive impedance is known to cause impedance-based interactions.

Cross-Coupling Impedances
It is often assumed that the cross-coupling impedance components are small and can be neglected, e.g., in impedance-based stability analysis. However, the magnitude of cross-coupling impedance components Z_{dq} and Z_{qd} differ from Z_{dd} and Z_{qq} only by roughly 12 dB near the resonant frequency. It is important to be able to measure the cross-coupling components to justify the validity assumptions in stability analysis, i.e., whether the cross-couplings can be neglected or not.

Frequency Range of Interest
The frequency range of interest can be defined based on the above observations. The measurement setup should be able to extract the impedance accurately from few hertz up to several kilohertz to capture the effects of slow control loops and the resonance of the LCL-filter. Moreover, the cross-coupling impedance components should be measured, since they may affect impedance-based stability [6].

V. IMPEDANCE MEASUREMENT USING GRID-FORMING INVERTER

The inverter is connected to a grid-forming inverter according to Fig. 7. The grid-forming inverter is used to keep the inverter at its nominal operating point during the measurement and to sink the generated power. The DC side is modeled as an ideal voltage source, which would in reality be implemented by a grid-interfacing three-phase converter.

The grid-forming inverter controls its output voltages in the dq-domain using simple integral-type control. Integral-type control was found to be good compromise between stability and control bandwidth, since the grid-forming inverter control dynamics are inherently affected by the impedance of the grid-feeding inverter (impedance of the DUT). The detailed small-signal model and method to take the load-effect into account can be found in [19]. Fig. 8 shows the identified and modeled control loop gain of the grid-forming inverter of voltage d-component. The loop gain related to the q-component has effectively the same shape and is not shown here. The crossover

4208

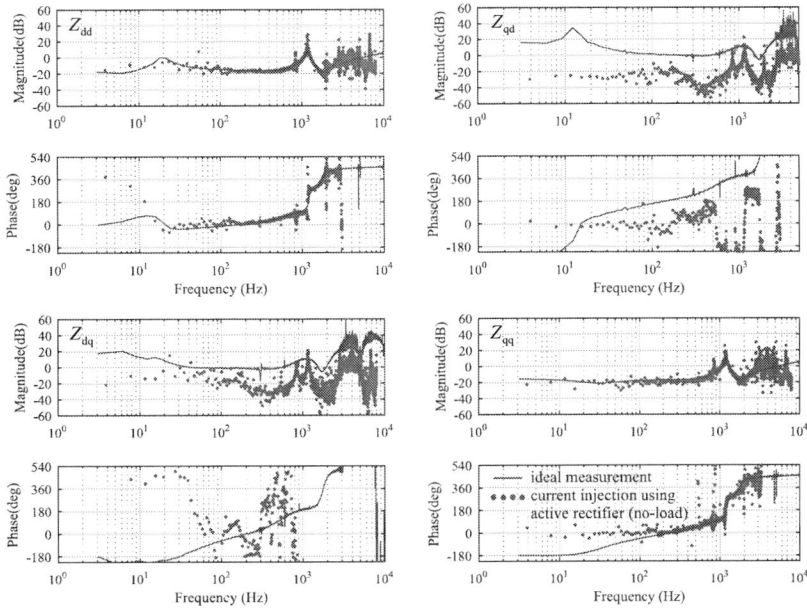

Fig. 5. Identified impedance from the simulator using grid-parallel converter to generate the current-type injection.

frequency of the control loop is approximately 140 Hz and the phase margin 75 degrees.

The purpose of the grid-forming inverter is to replicate the MLBS signal in its output voltages (d or q-component depending on what impedance component is to be identified). Therefore, the frequency response from the reference value of the voltage d-component to the actual d-component, i.e., the closed-loop transfer function, should have as high bandwidth as possible to replicate all of the frequency components of the MLBS signal as accurately as possible. Fig. 9 shows the transfer function and the identified frequency response from the reference value of the voltage d-component to the actual voltage d-component. Bandwidth of the voltage control is 216 Hz (-3dB). This suggests that the spectral energy of the MLBS signal starts to attenuate at higher frequencies. Thus, the injection does not go through the control system at high frequencies. The transfer function related to the q-component shows almost identical behavior and is, therefore, not shown.

Fig. 10 shows the modeled transfer function from the reference value of the d-component to the actual q-component of the AC voltage. The identified frequency response deviates slightly from the modeled transfer function. The transfer function has very small magnitude at low frequencies which makes extracting it accurately challenging. Moreover, the gain was found out to be highly sensitive to small changes in the operating point. Thus, small inaccuracies in the simulation cause deviation between the modeled and identified frequency responses. However, the transfer function in Fig. 10 gives a decent idea at what frequencies the injection to d-component affects also the q-component. The transfer function should have as small magnitude as possible to prevent the

injection from causing an unwanted perturbation to the q-component. E.g., measuring the Z_{qq} component in (3) requires that the system is perturbed only by q-component, while the d-component injection remains zero. However, the transfer function in Fig. 10 experiences slight increase in its magnitude near few hundred hertz, which affects the accuracy of the impedance measurement. The optimization of the grid-feeding inverter control system is considered as a future topic and is not discussed further in this paper.

The main parameters of the grid-feeding inverter are given in Table II. The generation frequency of the MLBS signal should be chosen well below the switching frequency of the inverter and over the bandwidth of the transfer function in Fig. 9. The MLBS signal has to be sampled at least two times higher frequency than the generation frequency. However, the sampling frequency should be lower than the switching frequency of the grid-feeding inverter to avoid aliasing-effects. As a compromise, the generation frequency is selected as 1 kHz. The perfect injection amplitude depends on the amount of external noise and the magnitude of the impedance. It was found out by simulations that 50 V peak-to-peak gives the best outcome in the example case. Fig. 11 shows the waveforms at the output terminals of the DUT when the MLBS is injected through the control system of the grid-feeding inverter. It is evident that the grid-forming inverter cannot replicate the MLBS signal accurately, which is expected due to low switching frequency. The voltage waveform is dominated by the 5 kHz switching frequency component. Moreover, the waveforms are affected by the sampling frequency of the grid-feeding inverter, since the control system was modeled in discrete-domain. The MLBS signal was generated by using 10th-degree

4209

TABLE II.   GRID-FORMING INVERTER PARAMETERS.

V_{dc}	1500 V	L^{GF}	60 μH
R_d^{GF}	200mΩ	C_f^{GF}	800 μF
f_{sw}	5 kHz	f_{MLBS}	1 kHz
n	10	V_{MLBS}	50 $V_{p\text{-}p}$

shift-register, which results in frequency resolution of approximately one hertz. However, the waveforms include enough energy at low frequencies to allow identifying the low-frequency impedance.

Fig. 4 shows the impedance components, that were identified by using the grid-forming inverter as the source of injection. The impedance d and q-components Z_{dd} and Z_{qq} are accurately captured approximately up to 500 Hz. Thus, the effect of slow control loops such as DC voltage control and phase-locked-loop can be evaluated. However, the frequency range of the impedance measurement is limited by the low generation frequency of the MLBS signal (1 kHz). The cross-coupling impedances Z_{qd} and Z_{dq} contain slight error due to the fact that some of the injected perturbation leaks between d and q-components, as can be seen by analyzing Fig. 10. However, the accuracy could be increased by careful design of the grid-forming inverter control system. Some form of decoupling network should be designed to reduce the cross-coupling in the grid-forming inverter control dynamics, which is considered as a potential future topic.

VI.   IMPEDANCE MEASUREMENT USING GRID-PARALLEL CONVERTER

The bandwidth of the impedance measurement can be extended by using a parallel-connected three-phase converter accrording to Fig. 12. In principle the converter is an active rectifier. However, since the purpose of the converter is just to amplify the MLBS signal, it can be operated without a load on the DC side. The main parameters of the grid-parallel converter are given in Table III. The converter can have significantly lower power rating, because it is not required to sink any power.

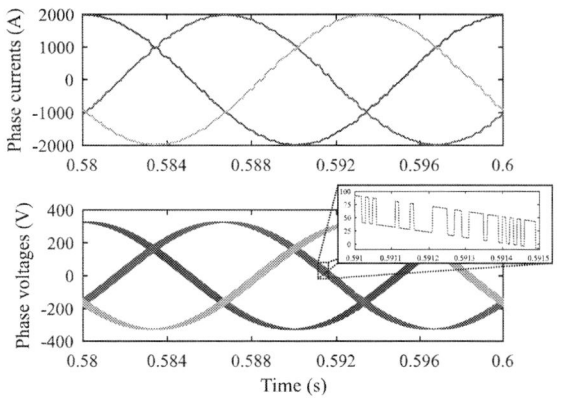

Fig. 6.   Three-phase current and voltage waveforms in the ideal measurement setup.

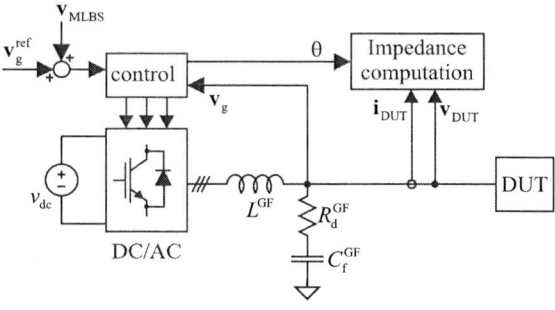

Fig. 7.   Principle of measurement setup using grid-forming inverter.

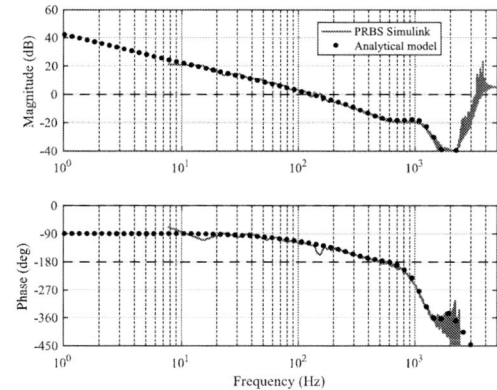

Fig. 8.   Identified and modeled loop gain related to the control of voltage d-component.

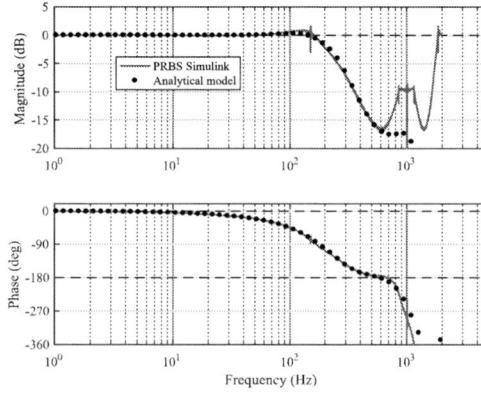

Fig. 9.   Frequency response from the reference value of voltage d-component to voltage d-component.

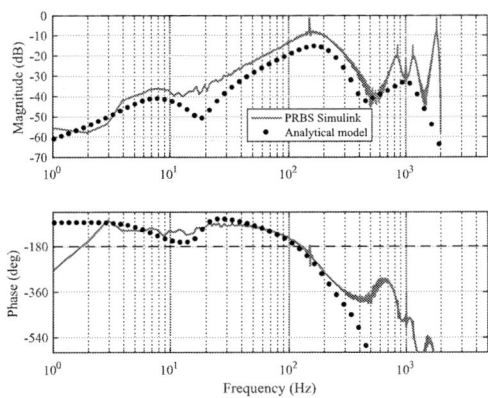

Fig. 10. Frequency response from the reference value of voltage d-component to voltage q-component.

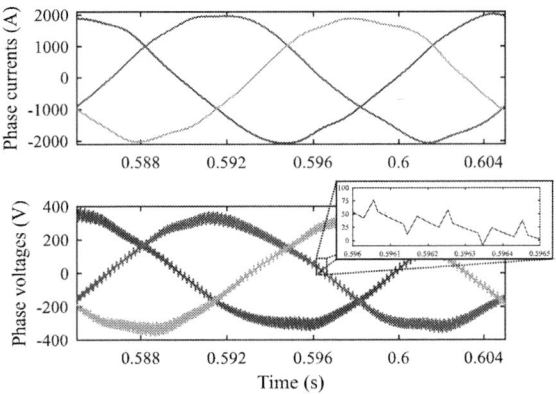

Fig. 11. Three-phase current and voltage waveforms with grid-forming inverter.

The converter was assumed to have maximum power rating of 10 kVA and, thus, its switching frequency can be much higher (20 kHz). The MLBS signal is added in the converter's control system to the reference values of of output currents.

The MLBS signal was generated at 4 kHz and sampled at two times higher frequency of 8 kHz. The amplitude was limited to 20 A peak-to-peak to avoid over-current stress of the switching components. A 10th-degree shift-register was used to generate the MLBS, which translates to a frequency resolution of roughly 4 Hz.

The average of the identified impedances in Fig. 5 follow the ideal measurement up to approximately 2 kHz

TABLE III.   GRID-PARALLEL CONVERTER PARAMETERS.

V_{dc}	1500 V	L_1^{GP}	1 mH
R_d^{GP}	1.36 Ω	C_f^{GP}	10 μF
f_{sw}	20 kHz	f_{MLBS}	4 kHz
n	10	V_{MLBS}	20 $V_{p\text{-}p}$
L_2^{GP}	200 μH	C_{dc}	2 mF

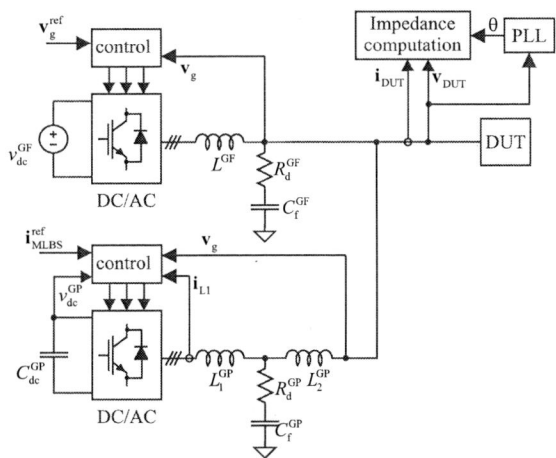

Fig. 12. Principle of measurement setup using the grid-parallel converter.

which is half of the generation frequency of the MLBS. The non-passive region around the resonant frequency of the LCL-filter can also be captured. The injection does not have enough energy to enable identification of cross-coupling impedance components. This is a matter that has been left as a future challenge. One option is to increase the power level of the grid-parallel inverter which, however, may reduce the switching frequency and thereby the maximum bandwidth of the measurement. Another option would be to use decoupling impedance placed before the grid-forming inverter to divert the high-frequency current toward the DUT. However, this may easily compromise the stability of the system and may require re-tuning of the grid-forming inverter control parameters.

The impedance measurement in the dq-domain requires that the grid angle is known. In the case of voltage-type injection the angle is known, since it is generated inside the control system of the back-to-back converter. However, in the case of the current-type injection this angle cannot be used directly, because the decoupling impedance causes a phase-shift. However, the angle can be estimated by using a slow phase-locked-loop as depicted in Fig. 12.

VII. CONCLUSIONS

This paper shows the first attempt to identify the AC-side output impedance of a 1 MW grid-feeding inverter in its nominal operating point using a combination of grid-forming and grid-parallel converters. A grid-forming inverter is used to sink the power generated by the inverter and to add enough perturbation to the inverter output voltages to measure impedance up to 500 Hz (in synchronous reference frame). An additional 10 kW grid-parallel inverter is used to generate a perturbation to the inverter output currents to measure impedance from 100 Hz up to 2 kHz. It is demonstrated that the combination of grid-forming and grid-parallel inverter can be effectively used in impedance measurement of multimegawatt inverters. This paper provides a proof-of-concept by providing preliminary simulation results based on a switching model, when the PRBS-method is used for impedance identification. Dynamic model of the grid-forming inverter is used to optimize the control loops for operation with the inverter-under-test.

REFERENCES

[1] "Frequency Stability Evaluation Criteria for the Synchronous Zone of Continental Europe," RG-CE System Protection & Dynamics Sub Group, Entsoe, March 2016.

[2] C. Li, "Unstable Operation of Photovoltaic Inverter from Field Experiences," *IEEE Transactions on Power Delivery*, (early access), 2017, doi: 10.1109/TPWRD.2017.2656020.

[3] P. Belkin, "Event of 10/22/09", CREZ Technical Conference, Electrical Reliability Council of Texas, 2010.

[4] J. Sun, "Impedance-Based Stability Criterion for Grid-Connected Inverters," *IEEE Transactions on Power Electronics*, vol. 26, no. 11, pp. 3075-3078, Nov. 2011.

[5] X. Wang, F. Blaabjerg and W. Wu, "Modeling and Analysis of Harmonic Stability in an AC Power-Electronics-Based Power System," *IEEE Transactions on Power Electronics*, vol. 29, no. 12, pp. 6421-6432, Dec. 2014.

[6] T. Messo, A. Aapro and T. Suntio, "Generalized multivariable small-signal model of three-phase grid-connected inverter in DQ-domain," IEEE 16th Workshop on Control and Modeling for Power Electronics (COMPEL), pp. 1-8, 2015.

[7] T. Messo, A. Aapro, T. Suntio and T. Roinila, "Design of grid-voltage feedforward to increase impedance of grid-connected three-phase inverters with LCL-filter," IEEE 8th International Power Electronics and Motion Control Conference (IPEMC-ECCE Asia), pp. 2675-2682, 2016.

[8] B. Wen, D. Boroyevich, R. Burgos, P. Mattavelli and Z. Shen, "Small-Signal Stability Analysis of Three-Phase AC Systems in the Presence of Constant Power Loads Based on Measured d-q Frame Impedances," *IEEE Transactions on Power Electronics*, vol. 30, no. 10, pp. 5952-5963, Oct. 2015

[9] L. H. Kocewiak *et al.*, "Wind Turbine Harmonic Model and Its Applications," 14th International Workshop on Large-Scale Integration of Wind Power into Power System as well as on Transmission Networks for Offshore Wind Power Plants, pp. 1-6, 2015.

[10] J. Jokipii, T. Messo and T. Suntio, "Simple method for measuring output impedance of a three-phase inverter in dq-domain," 2014 International Power Electronics Conference (IPEC-Hiroshima 2014 - ECCE ASIA), pp. 1466-1470, 2014.

[11] M. Jakšić *et al.*, "Medium-Voltage Impedance Measurement Unit for Assessing the System Stability of Electric Ships," *IEEE Transactions on Energy Conversion*, vol. 32, no. 2, pp. 829-841, June 2017.

[12] V. Valdivia, A. Lázaro, A. Barrado, P. Zumel, C. Fernández and M. Sanz, "Impedance Identification Procedure of Three-Phase Balanced Voltage Source Inverters Based on Transient Response Measurements," *IEEE Transactions on Power Electronics*, vol. 26, no. 12, pp. 3810-3816, Dec. 2011.

[13] J. Hui, W. Freitas, J. C. M. Vieira, H. Yang and Y. Liu, "Utility Harmonic Impedance Measurement Based on Data Selection," *IEEE Transactions on Power Delivery*, vol. 27, no. 4, pp. 2193-2202, Oct. 2012.

[14] K. Godfrey, *"Perturbation Signals for System Identication"*. Prentice Hall, UK., 1993.

[15] T. Roinila, T. Messo and A. Aapro, "Impedance measurement of three phase systems in DQ-domain: Applying MIMO-identification techniques," IEEE Energy Conversion Congress and Exposition (ECCE), pp. 16 , 2016.

[16] T. Suntio, T. Messo and J. Puukko, *"Power Electronic Converters: Dynamics and Control in Conventional and Renewable Energy Applications"*, Wiley-VCH , pp. 633-661, 2017.

[17] T. Messo, J. Jokipii, A. Mäkinen and T. Suntio, "Modeling the grid synchronization induced negative-resistor-like behavior in the output impedance of a three-phase photovoltaic inverter", *4th IEEE International Symposium on Power Electronics for Distributed Generation Systems (PEDG)*, pp. 1-7, 2013.

[18] A. Aapro, T. Messo, Tomi Roinila and T. Suntio, "Effect of active damping on output impedance of three-phase grid-connected converter", *IEEE Transactions on Industrial Electronics*, vol 64, no 9, pp. 7532-7541, 2017.

[19] M. Berg, T. Messo and T. Suntio, "Frequency Response Analysis of Load Effect on Dynamics of Grid-Forming Inverter", International Power Electronics Conference, (IPEC-Niigata 2018 -ECCE Asia-), pp. 1-8, 2018.

Improved Virtual Inductance Based Control Strategy of DFIG under Weak Grid Condition

Ran Fang[1], Wenjia Chen[1], Xueguang Zhang[1]* and Dianguo Xu[1]

1 Department of Electrical Engineering and Automation, Harbin Institute of Technology, Harbin, China

*E-mail: zxghit@126.com

Abstract— **Equivalent input admittance model of DFIG wind turbine is built in synchronous reference frame. DFIG electromagnetic transient, PLL dynamics and rotor side controllers including power control and current control are considered in this model. The influence of rotor inductance on stability of DFIG wind turbine connected with weak grid is investigated based on generalized Nyquist criterion (GNC) which proves that the incensement of rotor inductance has a positive impact on system stability. Phase lock loop (PLL) is a critical component in the system, which may propagate the grid voltage perturbations to the control system and has an adverse impact on stability of DFIG wind turbine. Compared with conventional virtual inductance based control strategy, the improved control strategy proposed in this paper take the influence of PLL dynamics into consideration and can improve the system stability efficiently under weak grid condition. The validity of the improved control strategy is verified by theoretical analysis and simulation.**

Keywords— **doubly-fed induction generator (DFIG), stability analysis, virtual inductance, weak grid**

I. INTRODUCTION

Variable speed wind turbines implemented by doubly-fed induction generator (DFIG) are widely implied in wind power generation. Since large wind farms are typically located far from the load center, wind turbine generators (WTG) are actually connected to a long transmission line which leads to the weak grid condition. The interaction between the grid impedance and DFIG system impedance results in the existence of unbalanced voltage, harmonic distortion and other forms of fault, bringing about negative impact on system stability[1,2]. The research on operation stability of DFIG wind turbine is meaningful.

Adjusting the controller parameters or changing the operation state of DFIG wind turbine can improve the stability under weak grid condition. However the adjustable range of controller parameter is very limited considering the demand on dynamic characteristics. What's more, the system operation state is determined by the demand of grid and the wind resource whose adjustable scope is also very small. So it is reasonable to improve the stability by redesign of the controller. Nevertheless, research on control structure of the DFIG wind turbine is mainly focused on improving the dynamic characteristics or riding through under the fault condition[3,4]. Previous studies mostly focus on

improving the adaptability of grid-tied inverter under weak grid condition. Research [5] and [6] improve the performance of grid tied inverters based on PLL parameter optimization and adaptive current control strategy. However, research on the redesign of rotor-side converter (RSC) controller of DFIG wind turbine is little. Reference [7] and [8] proposed virtual synchronous control strategy which can improve the transfer ability and operation stability of DFIG wind turbine operating under weak gird condition effectively. Even so, there are several problem unsolved of virtual synchronous control strategy such as the adaptive ability with different wind condition and the interaction between different wind turbine generators in the wind farm.

It is economical to improve control performance and system stability through applying virtual impedance, as there is no demand on additional device. Moreover it is convenient to introduce the virtual impedance to control system. This strategy is widely used to improve the riding through ability under fault condition and stability under weak grid condition. Reference [9] introduce the virtual impedance in grid connected inverter to improve the performance of DFIG system connected with weak gird. Reference [10] apply the virtual impedance in the stator of DFIG to restrain the high frequency resonance between the DFIG system and grid. Reference [11-14] bring in the virtual impedance to the RSC current controller of DFIG to improve the low voltage and high voltage riding through ability. However, the impact of PLL is not considered in previous researches when the virtual impedance is introduce in controller. PLL is a critical component in the system which decides the coupling strength between grid and DFIG operation state. PLL should be treated as an independent control loop instead of a typical measurement component. Under the weak grid condition, the virtual impedance strategy may lose efficacy because of the influence of PLL dynamics.

Studies on the small-signal stability of DFIG system are mostly based on the state space equation method. System dynamic characteristics is analyzed through roots of characteristic equation and damping ratio. The full information including all of the dynamic models for all the elements in the system should be obtained to adopt the state space function method which bring the high accuracy. However it is complicated to acquire and manage a large amount of information to derive the model and to solve the high order model. Impedance-based method and general Nyquist stability criterion

(GNC) provide insights when the objective is grid system stability analysis [15,16]. Compared with state space function method, it is not necessary to obtain the full information of all components' inner physical structure and parameters for impedance model as the impedance can be gained through simulation or measurement.

In this paper, the impedance model of DFIG wind turbine is developed in d-q frame including stator transient, RSC current control, RSC power control and PLL dynamics to derive the equivalent input admittance of DFIG wind turbine. Based on the impedance model, the influence of rotor inductance is investigated by GNC which proves that the incensement of rotor inductance has a positive impact on system stability. Then the virtual inductance is introduced in the current controller to increase the rotor inductance equivalently. It is observed that because the grid voltage disturbance is coupled into the control system through the PLL dynamics, conventional virtual impedance method may lose effect under weak grid condition. So an improved virtual inductance method is proposed in this paper which take the PLL dynamics into account. This modified strategy is proved to be effective by theoretical analysis. To verify the correctness of the analysis, the simulation is performed.

II. IMPEDANCE MODELLING OF DFIG WIND TURBINE

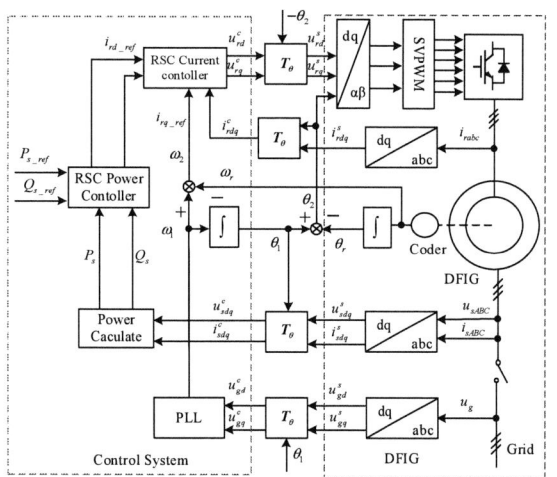

Fig. 1. Structure diagram of DFIG wind turbine

The system structure diagram of DFIG wind turbine is depicted as Figure 1. The process of building the equivalent impedance model is described in detail in reference [15]. However, model in [15] do not consider the RSC power control. In this section, the model containing RSC power control will be elaborated to improve the accuracy.

A. Impedance model without RSC power control

In order to acquire the equivalent input admittance of DFIG, the expression of stator current on stator voltage should be deduced. The convention is adopted in both stator side and rotor side that the inflow is regarded as the positive direction.

According reference [15], the open loop equivalent

input admittance of DFIG is expressed as following equation:

$$Y_{og} = G_{sr}G_{irs} + G_{iss} \tag{1}$$

Where G_{sr} and G_{iss} respectively represent small-signal transfer function matrix from rotor current and stator voltage to stator current. G_{irr} and G_{irs} respectively represent small-signal transfer function matrix from rotor voltage and stator voltage to rotor current.

The main object of RSC current control is to regulate the current in rotor by control the rotor voltage. The expression of the RSC current control is expressed in Equation (2):

$$\tilde{u}_r = (-G_{ci} + G_{d1})\tilde{i}_r + G_{d2}\tilde{i}_s \tag{2}$$

Where the G_{ci} is the PI controller transfer function matrix. G_{d1} and G_{d2} is the decouple component of rotor and stator current.

Considering the PLL dynamics, there are two d-q frames in the DFIG system: one is the system d-q frame represented by superscript 's', the other is the PLL d-q frame represented by superscript 'p'. The system d-q frame is defined by the grid voltage, and the PLL d-q frame is decided by PLL which calculates the frequency and the angle of grid voltage. PLL d-q frame is adopted by all controllers in the system. The two frames are identical at normal operation points. However, the position of PLL d-q frame is no longer aligned with the system d-q frame when small-signal perturbations are added to grid voltage. Small-signal perturbations of system voltage propagate to the grid phase angle that PLL output and further to all the controllers in DFIG system. In order to specify the influence of PLL dynamics, transfer function matrix G_{pll}^{is}, G_{pll}^{ir}, G_{pll}^{vs} and G_{pll}^{vr} are defined in [15].

The equivalent input admittance introduced by controllers including RSC current controller and PLL is deduced as the Equation (3):

$$Y_c = G_3(G_1 + G_{sr}G_{irs} + G_{iss}) \tag{3}$$

Where:

$$\begin{cases} G_1 = (G_{d1} - G_{ci})(G_{irs} + G_{pll}^{ir}) + G_{d2}G_{pll}^{is} - G_{pll}^{vr} \\ G_2 = E - (G_{d1} - G_{ci})G_{irr} \\ G_3 = G_{sr}G_{irr}(E - G_2)^{-1} \end{cases}$$

So the admittance of DFIG wind turbine considering RSC current control and PLL can be expressed as the follow equation:

$$Y_{l\_cu\_pll} = Y_{og} + Y_c \tag{4}$$

B. Impedance model considering RSC power control

The RSC power control is used to regulate the output active and reactive power of DFIG by setting the reference values of current controller. RSC power controllers can be expressed as Equation (5). Where k_{p3} and k_{i3} are proportional gain and integral gain of power control, P_s and Q_s are the output active power and reactive power of DFIG.

$$\begin{cases} i_{rd\_ref} = \left(k_{p3} + \dfrac{k_{i3}}{s} \right)\left(P_{s\_ref} - P_s \right) \\ i_{rq\_ref} = -\left(k_{p3} + \dfrac{k_{i3}}{s} \right)\left(Q_{s\_ref} - Q_s \right) - u_{sd}^c / (\omega_1 L_m) \end{cases} \quad (5)$$

According to Equation (5), the power control of DFIG wind turbine can be expressed as:

$$\boldsymbol{i}_{r\_ref} = \mathbf{G}_{cpq} \begin{bmatrix} P_{s\_ref} - P_s \\ Q_{s\_ref} - Q_s \end{bmatrix} + \mathbf{G}_{d3} \quad (6)$$

Where,

$$\mathbf{G}_{cpq} = \begin{bmatrix} k_{p3} + k_{i3}/s & 0 \\ 0 & k_{p3} + k_{i3}/s \end{bmatrix}$$

$$\mathbf{G}_{d3} = \begin{bmatrix} 0 \\ -u_{sd}^c/(\omega_1 L_m) \end{bmatrix}$$

DFIG stator output power in system controller is

$$\begin{bmatrix} P_s \\ Q_s \end{bmatrix} = -\frac{3}{2} \begin{bmatrix} u_{sd}^p & u_{sq}^p \\ u_{sq}^p & -u_{sd}^p \end{bmatrix} \begin{bmatrix} i_{sd}^p \\ i_{sq}^p \end{bmatrix} \quad (7)$$

By doing linearization,

$$\begin{bmatrix} \tilde{P}_s \\ \tilde{Q}_s \end{bmatrix} = \mathbf{G}_{pq}^i \begin{bmatrix} \tilde{i}_{sd}^p \\ \tilde{i}_{sq}^p \end{bmatrix} + \mathbf{G}_{pq}^v \begin{bmatrix} \tilde{u}_{sd}^p \\ \tilde{u}_{sq}^p \end{bmatrix} \quad (8)$$

Where,

$$\mathbf{G}_{pq}^i = -\frac{3}{2} \begin{bmatrix} U_{sd}^p & U_{sq}^p \\ U_{sq}^p & -U_{sd}^p \end{bmatrix}$$

$$\mathbf{G}_{pq}^v = -\frac{3}{2} \begin{bmatrix} I_{sd}^p & I_{sq}^p \\ -I_{sq}^p & I_{sd}^p \end{bmatrix}$$

The block diagram of DFIG wind turbine input admittance including RSC current controller, RSC power controller and PLL dynamics is depicted in Figure 2. After simplified, the input equivalent admittance of DFIG wind turbine can be expressed as:

$$\boldsymbol{Y}_{dfig} = \mathbf{G}_{iss} + (\boldsymbol{E} - \boldsymbol{G}_1 \boldsymbol{G}_7)^{-1} \boldsymbol{G}_1 (\boldsymbol{G}_7 \boldsymbol{G}_{iss} + \boldsymbol{G}_8) \quad (9)$$

Where

$$\begin{cases} \boldsymbol{G}_6 = \boldsymbol{G}_{irr}(\boldsymbol{G}_{d1} - \boldsymbol{G}_{ci})\boldsymbol{G}_{pll}^{irs} + \boldsymbol{G}_{irs} - \boldsymbol{G}_{irr}\boldsymbol{G}_{pll}^{vrs} \\ \boldsymbol{G}_7 = \boldsymbol{G}_{irr}\left[\boldsymbol{G}_{d2} - \boldsymbol{G}_{ci}\boldsymbol{G}_{cpq}\boldsymbol{G}_{pq}^i \right] \\ \boldsymbol{G}_8 = \boldsymbol{G}_6 - \boldsymbol{G}_{irr}\boldsymbol{G}_{ci}(\boldsymbol{G}_{d3} + \boldsymbol{G}_{cpq}\boldsymbol{G}_{pq}^v)(\boldsymbol{E} + \boldsymbol{G}_{pll}^{vs}) - \boldsymbol{G}_7 \boldsymbol{G}_{pll}^{is} \end{cases}$$

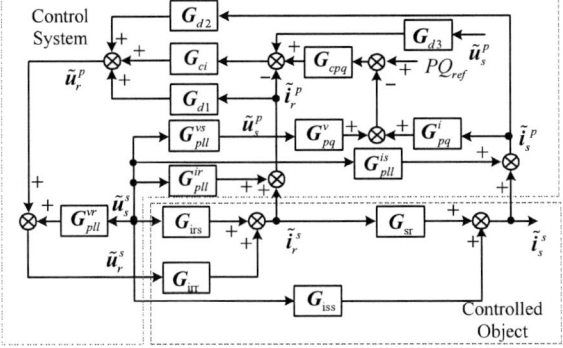

Fig. 2. The input admittance diagram of DFIG wind turbine including RSC controllers and PLL dynamics

III. IMPROVED VIRTUAL INDUCTANCE CONTROL STRATEGY

A. GNC for Stability Analysis

Nyquist criterion has been widely used in stability analysis. However, it can't be applied in multi-input multi-output (MIMO) system. Nyquist criterion is generalized in MIMO system as GNC to analyze the stability of MIMO system. The weak grid which the DFIG wind turbine connected with can be equivalent to a voltage source with series impedance. Considering that the DFIG injects the current to the grid, the DFIG wind turbine can be equivalent to a current source with parallel impedance. So the impedance model of the whole system in d-q frame can be illustrated by Fig. 3.

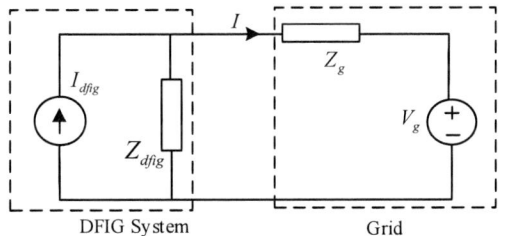

Fig. 3. The impedance model of the whole system

In Fig.3, $\boldsymbol{I}_{dfig}$ and $\boldsymbol{V}_g$ respectively represent the equivalent current source of the DFIG system and the equivalent voltage source of the grid. $\boldsymbol{Z}_{dfig}$ represents the equivalent impedance matrix of the DFIG system. The value of $\boldsymbol{Z}_{difg}$ equals the inverse of the equivalent input admittance $\boldsymbol{Y}_{dfig}$ expressed by equation (9). $\boldsymbol{Z}_g$ represents the grid impedance in d-q frame.

$$\boldsymbol{Z}_g = \begin{bmatrix} R_g + L_g s & -\omega L_g \\ \omega L_g & R_g + L_g s \end{bmatrix} \quad (10)$$

Where, R_g and L_g represent the grid resistance and grid inductance. According to the GNC, the stability of the DFIG system can be investigated through applying the GNC to the return-ratio matrix defined as following equation:

$$\boldsymbol{L}(\text{s}) = \boldsymbol{Y}_{dfig} \boldsymbol{Z}_g \quad (11)$$

The GNC plots of the DFIG system is plotted in Fig. 4, and the parameters of DFIG and grid are listed in Table I at appendix. According to the GNC, the system stability can be judged by the times of counterclockwise encirclement around the critical point of the characteristic loci. As return-ratio matrix is second order, both red curve and blue curve represent return-ratio matrix characteristic loci. In Fig.4, blue curve is always far from critical point (-1, j0) while red curves has potential to encircle (-1, j0). Hence, DFIG system stability is mainly characterized by the red curves. To intuitively exhibit the encirclement of the critical point, only enlarged figure will be given from the next figure.

The 2018 International Power Electronics Conference

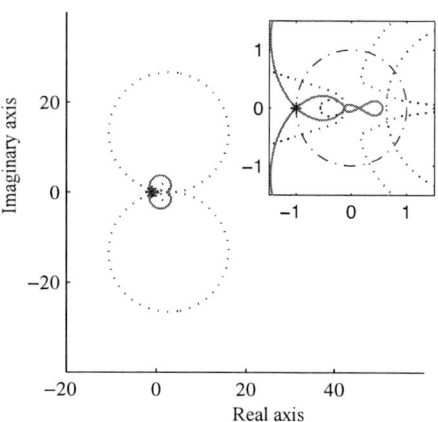

Fig. 4. GNC plots of DFIG system at critical stable point

B. Influence of Rotor Inductance on Stability

According to the model established in former section, the equivalent input admittance of DFIG is related to the parameter of generator including rotor inductance, stator inductance, rotor resistance and stator resistance. Rotor and stator resistance is rather small than inductance. So the influence of rotor is limited. The influence of rotor inductance on the DFIG system stability can be investigated through GNC. We noted the rotor inductance in Fig. 4 as L_r. To analysis the influence of rotor inductance, we change it to L_{r2} while maintain the others parameters unchanged. The system GNC plots with different rotor inductance parameter is depicted as Fig. 5.

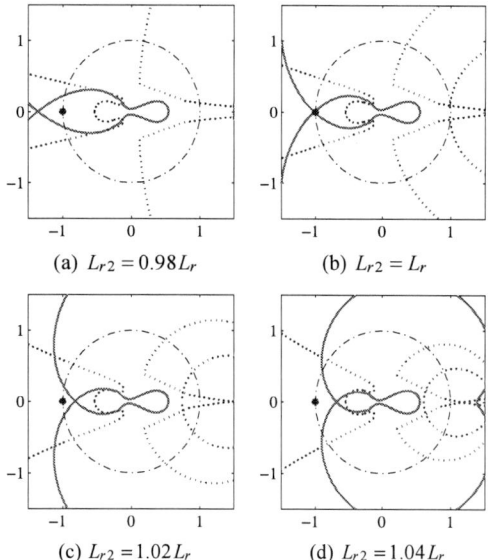

(a) $L_{r2} = 0.98L_r$ (b) $L_{r2} = L_r$

(c) $L_{r2} = 1.02L_r$ (d) $L_{r2} = 1.04L_r$

Fig. 5. GNC plots of DFIG system with different rotor inductance

It is observed that as rotor inductance of DFIG declines, the red curves gradually encircle (-1, j0), revealing that the system tends to be unstable. The stability of DFIG system is getting worse with the decrease of rotor inductance. It can be explained from the impedance view. The DFIG system can be equivalent to a current source connected with grid, so if the input

admittance of DFIG system decrease, the capacity of resisting grid voltage disturbance improve.

C. RSC Controller Based on Virtual Inductance

It is difficult to change the actual inductance of DFIG to improve the stability. However virtual inductance can be introduced through redesigning the controller. Considering that the RSC controller directly regulate the rotor current, it is convenient to apply virtual inductance strategy in RSC controller to increase the rotor inductance equivalently.

The structure diagram of redesigned RSC current controller is showed in Fig. 6. The rotor inductance of DFIG increased equivalently by adding the differential feedback of rotor current in RSC controller. According to the diagram, the redesigned controller can be expressed as Equation (12), where k_{p1} and k_{i1} are proportional gain and integral gain of current control.

$$
\begin{cases}
u_{rd} = \left(k_{p1} + \dfrac{k_{i1}}{s}\right)(i_{rd\_ref} - i_{rd}) - \omega_2 L_r i_{rq} + \omega_2 L_m i_{sq} - L_{vir} s i_{rd} \\
u_{rq} = \left(k_{p1} + \dfrac{k_{i1}}{s}\right)(i_{rq\_ref} - i_{rq}) + \omega_2 L_r i_{rd} - \omega_2 L_m i_{sd} - L_{vir} s i_{rq}
\end{cases}
$$

(12)

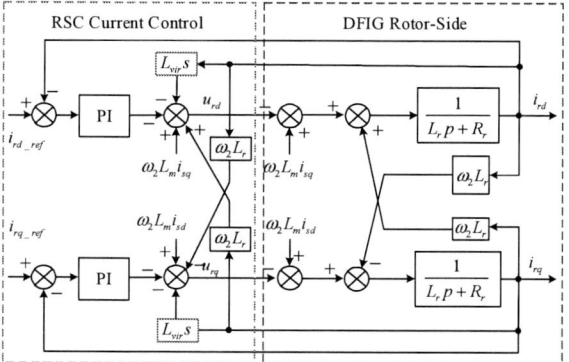

Fig. 6. System diagram of current controller after introducing virtual inductance

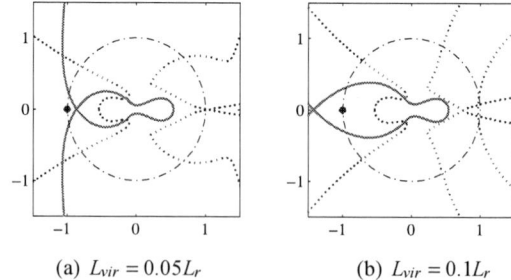

(a) $L_{vir} = 0.05L_r$ (b) $L_{vir} = 0.1L_r$

Fig. 7. GNC plots of DFIG system after introducing virtual inductance without considering PLL dynamics

Substitute the redesigned RSC current controller based on virtual inductance for the former controller and deduce the new equivalent input admittance. The GNC plots is illustrated as Fig. 7. The parameter of Fig. 7 is the same as Fig. 4. As it is shown in the figure, in a very limited range, the introduced virtual inductance can improve the system stability. However, once increase the virtual inductance slightly, the stability may get even

4216

worse. That is to say, the introduced virtual inductance may have the negative impact on system stability in some cases. Review the structure diagram of the redesigned controller. It can be found that the PLL dynamics is neglected which has significant influence on system stability. Since the PLL dynamics is ignored, the actual appliance of this controller is limited.

D. The Improved Controller

Because the PLL dynamics is not considered when introduce the virtual inductance in RSC controller, the rotor current in PLL d-q reference frame is fed back by virtual inductance instead of the rotor current in system d-q reference frame. According to the analysis on PLL dynamics in section 2, the rotor current in PLL d-q reference frame is coupled with the grid voltage disturbance, so this feedback may have adverse impact on system stability. In order to eliminate this negative influence of PLL dynamics, we introduce the feedforward of the grid voltage in RSC current controller. The structure of modified RSC current controller is depicted as Fig. 8.

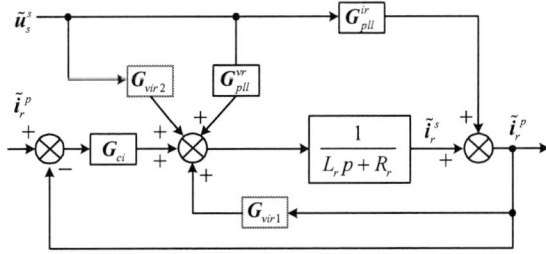

Fig. 8. System diagram of improved current controller considering the impact of PLL

G_{vir1} and G_{vir2} are transfer function matrix introduced by virtual inductance and the feedforward of grid voltage disturbance expressed as the following equations:

$$G_{vir1} = \begin{bmatrix} -L_{vir}s/(\tau s+1) & 0 \\ 0 & -L_{vir}s/(\tau s+1) \end{bmatrix} \quad (13)$$

$$G_{vir2} = -G_{pll}^{ir}G_{vir1} = \begin{bmatrix} 0 & G_{pll}L_{vir}sI_{rq} \\ 0 & -G_{pll}L_{vir}sI_{rd} \end{bmatrix} \quad (14)$$

Considering that the differential term is difficult to realize in real system. The first order low pass filter is added before the differential term. And it should be noticed that the q component of stator voltage is zero in static state. So the final expression of the modified RSC current controller is expressed as follow equations:

$$\begin{cases} u_{rd} = \left(k_{p1} + \dfrac{k_{i1}}{s}\right)\left(i_{rdref} - i_{rd}\right) - \omega_2 L_r i_{rq} \\ \qquad + \omega_2 L_m i_{sq} - L_{vir}si_{rd}/(\tau s+1) + f_{pll}i_{rq}L_{vir}u_{sq} \\ u_{rq} = \left(k_{p1} + \dfrac{k_{i1}}{s}\right)\left(i_{rqref} - i_{rq}\right) + \omega_2 L_r i_{rd} \\ \qquad - \omega_2 L_m i_{sd} - L_{vir}si_{rq}/(\tau s+1) - f_{pll}i_{rd}L_{vir}u_{sq} \end{cases} \quad (15)$$

Where:

$$f_{pll} = (k_{p2} + \frac{k_{i2}}{s})\frac{1}{s} \quad (16)$$

Substitute the improved RSC current controller which eliminate the negative impact of PLL dynamics for the former controller and deduce the new equivalent input admittance. The GNC plots with the modified controller is presented as Fig. 9. The parameter used in Fig. 9 is the same as Fig. 7. Compared with Fig. 7. It is obvious that the GNC plots do not encircle the (-1, j0) point, and the distance to the critical point is longer. What's more, with the increase of the virtual inductance, the GNC plots shift away from the (-1, j0) point. Proposed modified RSC current controller can improve the system stability in a larger range.

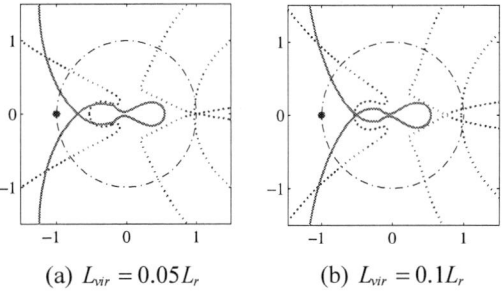

(a) $L_{vir} = 0.05L_r$ (b) $L_{vir} = 0.1L_r$

Fig. 9. GNC plots of DFIG system after applying improved current controller

IV. SIMULATION VERIFICATION

To verify the validity of the proposed improved controller, a 1.5 MW DFIG detail model is built in MATLAB/Simulink, and the parameters of DFIG and grid are listed in Table I at appendix.

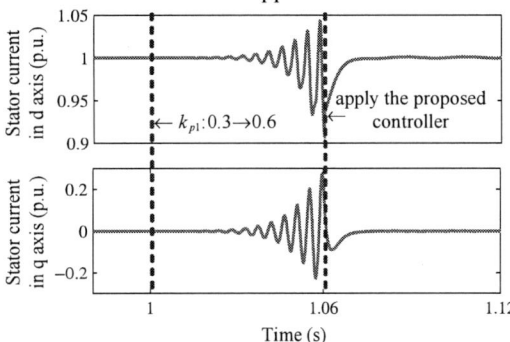

Fig. 10. Stator dq current with improved virtual inductance control after system instability caused by RSC current control

At the start time of simulation represented by Fig. 10, the conventional PI controller is adapted in RSC current control. The proportional gain of RSC current control change from 0.3 to 0.6 at 1s. Then the DFIG stator current start to diverge which mean the instability of the system. The proposed controller is applied at 1.06s, the oscillating stator current decay to the static state. The system return stable.

In order to compare the virtual inductance based controller considering the PLL dynamics with the one not considering PLL dynamics. At 1.0s of simulation, increase the proportional gain of PLL, the stator current start to oscillate and system become unstable as the Fig. 11 shows. The conventional virtual impedance based controller which do not considerate the PLL dynamics is

4217

applied at 1.04s. It is observed that although the amplitude of the oscillating stator current diminish, the oscillation still exist and the system is still unstable. The improved controller which take the PLL dynamics into account proposed in this paper is applied at 1.08s. The oscillation decay gradually. And the system return stable finally.

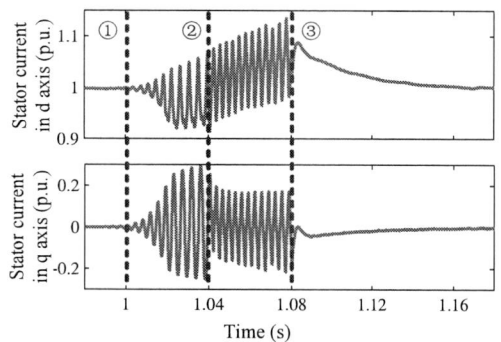

Fig. 11. Stator dq current with traditional and improved virtual inductance control after system instability caused by PLL

① k_{p2} changes from 2.5 to 3; ② Introduce conventional virtual inductance control strategy; ③ Introduce improved control strategy proposed in this paper.

In the simulation shown in Fig. 12, when the proportional gain of power control is increased at 1.0s, the active and reactive power start to oscillate, and then apply the proposed improved controller, the oscillation of the system disappears and the output power of the system returns to a stable state. The simulation results prove that the proposed controller can effectively improve the stability of the system, and the instability phenomenon caused by the interaction between the control system and the weak network get suppressed.

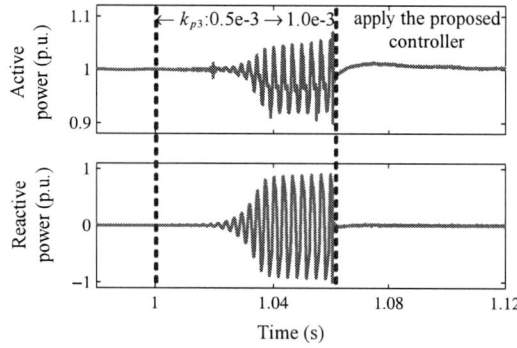

Fig.12 Output active and reactive power with improved virtual inductance control after system instability caused by RSC power control

V. CONCLUSIONS

Equivalent input admittance model of DFIG wind turbine considering RSC current controller, RSC power controller and PLL dynamics is built in synchronous reference frame. GNC based stability analysis prove that the rise of rotor inductance can deduce the equivalent input admittance of DFIG system and improve the system stability. To improve the rotor inductance equivalently, the virtual inductance is introduce into RSC current controller. However, conventional virtual impedance based method do not consider the influence of PLL dynamics which may have significant impact on system stability. The introduced virtual inductance may have negative effect on system stability under weak grid because of the PLL dynamics. Compared with conventional virtual inductance based control strategy, the improved control strategy proposed in this paper takes the influence of PLL dynamics into consideration and introduce the feedforward term of grid voltage disturbance into controller to eliminate the impact of PLL. Stability analysis based on GNC and simulation on a detail model prove that control strategy proposed in this paper improves the stability of DFIG system under weak grid condition effectively.

APPENDIX

TABLE I
PARAMETERS OF DFIG SYSTEM AND GRID

Parameter	Symbol	Value
DFIG rated capacity	S_B	1.5MVA
DFIG nominal voltage	V_B	690V
Nominal frequency	f_B	50Hz
Rotor leakage inductance	L_{lr}	0.132 p.u.
Stator leakage inductance	L_{ls}	0.167 p.u.
Rotor resistance	R_r	0.00829 p.u.
Stator resistance	R_s	0.00835p.u.
Mutual inductance	L_M	5.421 p.u.
Grid inductance	L_g	0.5 p.u.
Grid resistance	R_g	0.03 p.u.

ACKNOWLEDGMENT

This work was supported by the National Nature Science Foundation of Chain (No. 51577040 and No. 51720105008).

REFERENCES

[1] X. Z Xi, H Geng, G. Yang, "Enhanced model of the doubly fed induction generator-based wind farm for small-signal stability studies of weak power system," *IET Renewable Power Generation*, vol. 8, no. 7, pp. 765-774, 2014.

[2] Y. P. Song, F. Blaabjerg, "Overview of DFIG-based wind power system resonances under weak networks," *IEEE Transaction on Power Electronics*, vol. pp, 2016.

[3] H. L. Xu, W. Zhang, J. S. Chen, et al. "A review on low voltage ride-through and prospect for DFIG wind turbines," *Automation of Electric Power Systems*, vol. 37, no. 20, pp. 8-15, 2013.

[4] J. B. Hu, H. Nian, H. L. Xu, et.al. "Dynamic Modeling and Improved Control of DFIG under Distorted Grid Voltage Conditions," *IEEE Transaction on Energy Conversion*, vol. 26, no. 1, pp. 163-175, 2011.

[5] H. Wu, X. B. Ruan, D. S. Yang. "Research on the stability caused by phase-locked loop for LCL-type grid connected inverter in weak grid condition," *Proceedings of the CSEE*, vol. 34, no. 30, pp. 5259-5267, 2014.

[6] J. M. Xu, S. J. Xie, T. Tang. "An adaptive current control for grid-connected LCL-filtered inverters in weak grid case," *Proceedings of the CSEE*, vol. 34, no. 24, pp. 4031-4039, 2014.

[7] S. Wang, J. B. Hu, X. M. Yuan. "Virtual synchronous control for grid-connected DFIG-based wind turbines," *IEEE Journal of Emerging and Selected Topics in Power Electronics*, vol. 3, no. 4, pp. 932-944, 2015.

[8] S. Wang, J. B. Hu, X. M. Yuan. "On inertial dynamics of virtual synchronous controlled DFIG-based wind turbines," *IEEE Transactions on Energy Conversion*, vol. 30, no. 4, pp. 1691-1701, 2015.

[9] D. D. Yang, X. B. Ruan, H. Wu. "A virtual impedance method to improve the performance of LCL-type grid-connected inverters under weak grid conditions," *Proceedings of the CSEE*, vol. 34, no. 15, pp. 2327-2335, 2014.

[10] Y. P. Song, X. F. Wang, B. Frede. "Doubly fed induction generator system resonance active damping through stator virtual impedance," *IEEE Transactions on Industrial Electronics*, vol. 64, no. 1, pp. 125-137, 2016.

[11] Y. S. Yang, Y. Chen, T. B. Zhou, et al. "Virtual inductance based self-demagnetization control for doubly-fed induction generator wind turbines during low voltage ride-through process," *Automation of Electric Power Systems*, vol. 39, no. 4, pp. 12-18, 2014.

[12] Z. Xie, X. Zhang, H. H. Song, et al. "Variable damping based control strategy of doubly-fed induction generator based wind turbines under grid voltage swell," *Automation of Electric Power Systems*, vol. 36, no. 3, pp. 39-46, 2012.

[13] Z. Xie, X. Zhang, Y. S. Yang, et al. "High voltage ride-through control strategy of doubly-fed induction wind generators based on virtual impedance," *Proceedings of the CSEE*, vol. 32, no. 27, pp. 16-23, 2012.

[14] X. Z. Xi, H. Geng H, G. Yang. "Enhanced model of the doubly fed induction generator-based wind farm for small-signal stability studies of weak power system," *IET Renewable Power Generation*, vol. 8, no. 7, pp. 765-774, 2014.

[15] X. G. Zhang, Y. Ma, T. Y. Wang, et al. "Input admittance modeling and stability analysis of DFIG under weak grid condition," *Proceedings of the CSEE*, vol. 37, no. 5, pp. 1507-1514, 2017.

[16] B. Wen, D. Borovevich, R. Burgos, et al. "Small-signal stability analysis of three-phase ac systems in the presence of constant power loads based on measured d-q frame impedances," *IEEE Transactions on Power Electronics*, vol. 30, no. 10, pp. 5952-5963, 2015.

The 2018 International Power Electronics Conference

Control of VSC-HVDC for Wind Farm Integration with Real-Time Frequency Mirroring and Self-Synchronizing Capability

Renxin Yang [1], Chen Zhang [1], Xu Cai [1*], Gang Shi [1], Jing Lyu [1]

1 Wind Power Research Center, Shanghai Jiao Tong University, Shanghai, China

*E-mail: xucai@sjtu.edu.cn

Abstract-Voltage source converter based HVDC (VSC-HVDC) is a future trend for integrating the large-scale offshore wind farms to the power grid. Due to fast current vector control of VSC-HVDC receiving-end, offshore wind farms present no inertial response and become unstable when the grid is weak. A novel control strategy for VSC-HVDC system to provide inertial response with offshore wind farm is presented in this paper. With the application of inertial synchronizing control (ISC), self-synchronization is realized in receiving end converter (REC) by utilizing the nature response of DC capacitor with grid frequency well reflected to the DC voltage. At the same time, frequency mirroring control is applied in sending end converter (SEC), with which AC grid frequency is mirrored to the sending-end instantaneously. Then wind farm can realize inertial response, and REC behaves like a synchronous generator with inertia equivalent to the wind farms and VSC-HVDC system. This largely enhances the stability under weak grid condition and frequency supporting ability to the grid. The effectiveness of proposed control strategy is evaluated by comparing to an existing method based on different work scenarios in PSCAD/EMTDC.

I. INTRODUCTION

The increasing need for bulky power transfer over long distance driven by the increasing penetration of offshore wind farms to power grids has been accelerating the development of VSC-HVDC systems [1]-[2]. With the same power rate, dc transmission has lower power losses compared to the conventional HVAC lines. Moreover, the flexibility of active power flow and independent control of reactive power make VSC-HVDC an applicable and attractive way to integrate offshore wind farms to onshore power grids. To replace conventional power plants, the VSC-HVDC connected offshore wind farms are required to provide ancillary services such as primary frequency regulation and inertia response to help maintain the stability of power grids.

Utilizing wind turbine to imitate inertial response by extracting the kinect energy from rotating blades has been discussed in [3]. However, due to the decoupling of VSC-HVDC, offshore wind farms cannot sense the frequency variation of AC grid, thus giving no support. In [4], communication lines are set to transmit the frequency information. However, the cost and reliability of long-distance communication are the shortcomings of this method. In [5], DC voltage acted as the media to deliver frequency information. In receiving end converter (REC), the DC-voltage reference was coupled with grid frequency, which is tracked by PLL. In sending end

converter (SEC), the frequency reference is coupled with DC-voltage as well. Thus the frequency variation of onshore AC grid is transmitted to wind farm. In order to guarantee the inertial response sensitivity, the PLL of REC should track grid frequency quickly, and rapid DC voltage adjustment is also demanded.

On the other hand, the increasing penetration of wind power have also increase the equivalent grid impedance, thus weakening the grid. Under this circumstance, the control ability may deteriorate when conventional vector-controlled REC is utilized to integrate wind power [6], thus resulting in the stability issues such as grid voltage distortions and harmonic oscillations. Applying voltage-source control is an effective way to solve this problem. A typical example is the Virtual Synchronizing Generator (VSG) [7]-[8], which imitates the Rotor Motion Equation of Synchronous Generator (SG) to realize self-synchronization to replace PLL. However it is not suitable for REC which delivers wind power since the output power of wind turbines are always changing. A fixed virtual inertial constant in VSG control may lead to the risk of power imbalance and even DC voltage distortion of HVDC system. The power synchronizing control proposed in [9] can also realize voltage-source control in REC. The inertial constant of this strategy is set to 0, which makes its power regulation rapid enough to maintain the DC voltage. However this characteristic decoupled both sides of HVDC as well, preventing the wind farm to response the AC grid frequency deviation.

It is concluded that current research works are more likely to take advantage of communication or passive DC voltage control to deliver frequency information. Due to the existence of communication or control delay, the wind farm cannot respond to grid frequency immediately. In addition, the REC still suffers from stability problems under conventional current vector control while integrated to weak grid. The application of VSG control can avoid this problem, however it may cause new troubles if the input power fluctuates.

4220

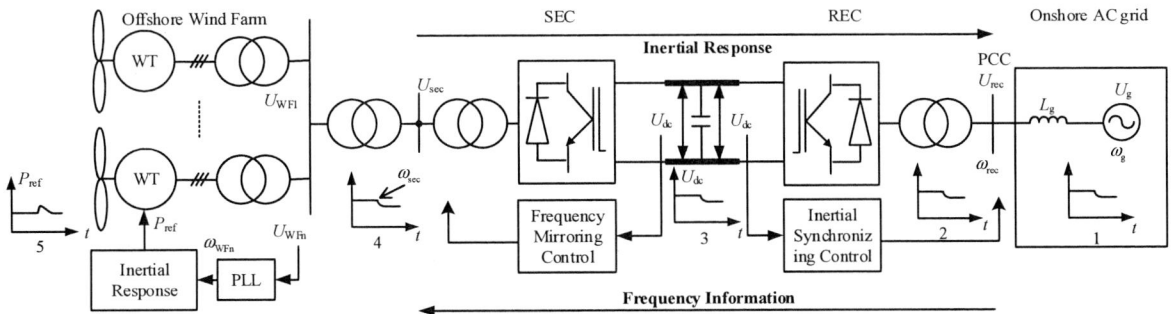

Fig. 1 VSC-HVDC system for wind farm integration

This paper presents a novel control strategy for VSC-HVDC systems to provide inertial response with wind farm integration. For REC, the Inertial Synchronizing Control (ISC) is utilized. The equivalent DC capacitor is controlled to simulate rotors of SG by utilizing its nature response, achieving the capability of self-synchronizing and accomplishing the real-time link between DC voltage and grid frequency. For SEC, a reversed control called Frequency Mirroring Control (FMC) is applied to converse the DC voltage deviation back to frequency deviation. With the proposed strategy, a real-time mirroring relationship is established between the frequencies of wind farms and onshore AC grid, making it possible for the wind turbines to respond to grid frequency variation immediately. Thus the whole system is operating as a SG which has a combined inertia of HVDC system and sending end power source. A simulation of large-scaled wind farm and VSC-HVDC system has been built in PSCAD/EMTDC, which proves validity of proposed control strategy.

II. SYSTEM DESCRIPTION

The studied system is shown in Fig.1, including 3 parts: the wind farm, REC and SEC. ISC is applied in REC, making DC voltage to track grid frequency autonomously. (1-3 in Fig. 1) FMC is applied in SEC, detecting the DC voltage deviation and adjusting AC frequency reference at wind farm side. (3-4 in Fig. 1) By the cooperation of REC and SEC, the grid frequency is mirrored to wind farm side. Finally, according to the AC frequency regulated by SEC, the wind turbines provide inertial response for onshore AC grid. (4-5 in Fig. 1)

III. INERTIAL SYNCHRONIZING CONTROL

A. Basic principle of inertial synchronizing control

The natural response of HVDC bus voltage to power variation can be described as

$$P_{WF} - P_{grid} = C U_{dc} \frac{dU_{dc}}{dt} \tag{1}$$

While

$$P_{grid} = \frac{U_{rec} U_g}{X} \sin \delta = \frac{\sqrt{3} m U_{dc} U_g}{2\sqrt{2} X} \sin \delta \tag{2}$$

It is found that (1) is similar to the motion equation of SG rotors,

$$P_m - P_e = J \omega_m \frac{d\omega_m}{dt} \tag{3}$$

While (2) is similar to SG output power equation,

$$P_e = \frac{\sqrt{3} E_f U_g}{X} \sin \delta = \frac{\sqrt{3} \psi \omega_m U_g}{\sqrt{2} X} \sin \delta \tag{4}$$

As can be observed from (1)-(4), the DC voltage U_{dc} is equivalent to rotor speed ω_m. Modulation ratio m is equivalent to air gap flux ψ. The corresponding relationships is depicted in Fig 2.

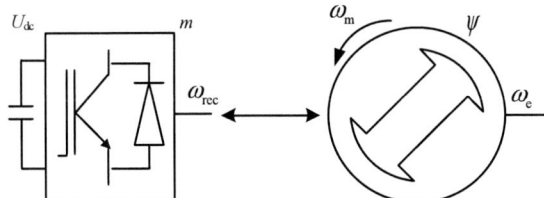

Fig. 2 Corresponding relationship

In order to simulate the natural relationship between rotor speed ω_m and electrical frequency ω_e of synchronous generator, a link between U_{dc} and REC output frequency ω_{rec} is established

$$\begin{cases} \dfrac{\omega_{rec} - \omega_{nom}}{\omega_0} = K \dfrac{U_{dc} - U_{dc\_nom}}{U_{dc\_nom}} \\ \dfrac{\Delta \omega_{rec}}{\omega_{nom}} = K \dfrac{\Delta U_{dc}}{U_{dc\_nom}} \end{cases} \tag{5}$$

U_{dc_nom} and ω_{nom} are the nominal value of U_{dc} and ω_{sec}. K represents the coupling coefficient between DC bus voltage U_{dc} and REC output frequency ω_{rec}. Since the deviation of grid frequency is about $\pm 1\%(0.5Hz)$, and the maximum deviation of DC voltage is $\pm 5\%$. It can be derived that

$$\frac{1\%}{5\%} \leq K \tag{6}$$

Usually K is set to 0.2 in order to maximize the transmission accuracy.

Substituting (5) into (1), then we can reach (7) after rearranging

4221

$$P_{WF} - P_{grid} = \frac{CU_{dc\_nom}^2}{K\omega_{nom}} \frac{d\Delta\omega_{rec}}{dt} \quad (7)$$

When grid frequency varies

$$\uparrow \omega_{grid} \Rightarrow \downarrow \delta \Rightarrow \downarrow P_{grid} \Rightarrow \uparrow U_{dc} \Rightarrow \uparrow \omega_{rec} \quad (8)$$

As can be observed from (7) and (8), several beneficial characteristics are achieved in REC:

1) Better performance under weak grid condition due to its self-synchronizing ability and voltage-source characteristic.

2) Similar to SG rotor, DC bus voltage U_{dc} and REC output frequency ω_{rec} are likely to track the grid frequency autonomously. Since the inertia of equivalent DC bus capacitor is usually very small, this tracking can be very fast.

Because of the utilization of nature inertia of DC capacitor, this control strategy of REC is called Inertial Synchronization Control (ISC). Its block diagram is presented in Fig 3.

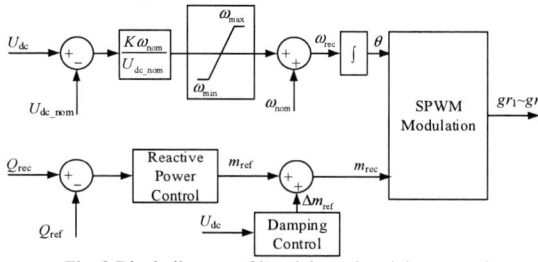

Fig. 3 Block diagram of inertial synchronizing control

The reactive power is controlled by manipulating output AC voltage. A damping control unit is designed to improve the dynamic response performance of proposed strategy. The detailed design of this unit is discussed later in section B.

Except the basic control below, an inner current loop can be added to suppress overcurrent, and PLL can be utilized to realize pre-synchronization when the system is turned on.

B. Dynamic analysis of inertial synchronizing control

In the process of grid frequency mirroring, the REC is playing a leading role, which determines several important indictors such as time delay, transmission quality and stability. Therefore, several principles must be followed while designing the REC control strategy.

Firstly, to avoid the misoperation of inertial response from wind farm, the DC bus voltage must not be affected by the fluctuating of wind power. On the other hand, when DC bus voltage is tracking grid frequency deviation, the time delay of this process should be limited to a certain range to satisfy the need of inertial response.

Therefore, the dynamic responses of DC bus voltage to both wind power variation and grid frequency variation are studied in this section. A VSC-HVDC system presented in Table I is taken as an example.

TABLE I
VSC-HVDC SYSTEM PARAMETERS

Meaning	Value
RMS value of grid voltage line-to line	100kV
Grid frequency	50Hz
Switching Frequency of REC&SEC	2000Hz
DC bus voltage	200kV
Rated active power	200MW
Rated reactive power	0
Equivalent DC capacitor	70uF
Equivalent DC resistance	2Ω

1) Response to wind power variation

The small signal form of equation (1) and (2) can be written as

$$\Delta P_{WF} = CU_{dc0} \frac{d\Delta U_{dc}}{dt} + \Delta P_{grid} \quad (9)$$

$$\Delta P_{grid} = \frac{P_0}{U_{dc0}} \Delta U_{dc} + \frac{P_0}{m_0} \Delta m + \frac{P_0}{\delta_0} \int (\Delta\omega_{rec} - \Delta\omega_g) dt \quad (10)$$

U_{dc0}, m_0, P_0, and δ_0 are the steady-state operation points of U_{dc}, m, P_{grid} and δ. Assuming that $U_{dc0}=U_{dc\_nom}$, $m_0=m_{nom}$ and $\delta_0=\delta_{nom}$, we can have (11) by substituting (5) and (10) into (9):

$$\Delta P_{WF}^* = \frac{CU_{dc\_nom}^2}{P_{nom}} \frac{d\Delta U_{dc}^*}{dt} + \frac{P_0}{P_{nom}} \Delta U_{dc}^* + \frac{P_0}{P_{nom}} \Delta m^*$$
$$+ \frac{P_0}{P_{nom}\delta_0} \int (K\omega_{nom}\Delta U_{dc}^* - \Delta\omega_g) dt \quad (11)$$

While

$$\Delta P_{WF}^* = \frac{\Delta P_{WF}}{P_{nom}} \quad \Delta U_{dc}^* = \frac{\Delta U_{dc}}{U_{dc\_nom}} \quad \Delta m^* = \frac{\Delta m}{m_{nom}} \quad (12)$$

While studying the response of DC bus voltage to wind power variation, the variation of grid frequency and modulation ratio can be neglected, which means $\Delta\omega_g=0$ and $\Delta m=0$. Then the Laplace transformation of (11) can be written as:

$$\Delta U_{dc}^* = \frac{s}{\frac{CU_{dc\_nom}^2}{P_{nom}} s^2 + \frac{P_0}{P_{nom}} s + \frac{P_0 K\omega_{nom}}{P_{nom}\delta_{nom}}} \Delta P_{WF}^* \quad (13)$$

Substituting the parameters from Table I into (13), the frequency domain response of U_{dc}^* to P_{WF}^* is shown in Fig 4. Two different scenarios are discussed in this section, including high wind speed ($P_0=0.85P_{nom}$, $\delta_0=0.85\delta_{nom}$) and low wind speed ($P_0=0.2P_{nom}$, $\delta_0=0.2\delta_{nom}$).

The frequency of wind power variation is usually below 0.1Hz (6rad/s). As can be observed in Fig.4, the amplitude responses of DC bus voltage at $\omega=6$rad/s are lower than -30dB. Hence the fluctuating of wind power would not lead to apparent deviation of DC bus voltage.

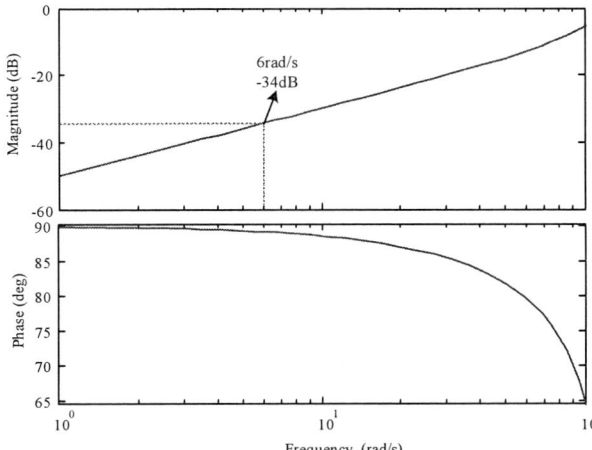

Fig. 4 Response of DC voltage to wind power fluctuation

2) Response to grid frequency variation

The process of inertial response usually lasts 6s in conventional power grid. Assuming that the grid frequency changes according to the exponential function, the time constant of this process is 1.5s. Hence for the wind farms, the time delay of received grid frequency information should be lower than 1/10, or 150ms to realize a real-time inertial response,

Assuming that the wind power remains the same ($\Delta P_{WF}=0$), the Laplace transformation of (11) can be written as:

$$\Delta\omega_g = \Delta U_{dc}^*\left(\frac{CU_{dc\_nom}^2\delta_{nom}}{P_0}s+\delta_{nom}+K\omega_{nom}\frac{1}{s}\right) \quad (14)$$
$$+\delta_{nom}\Delta m^*$$

Neglecting the influence of modulation ratio m, (14) can be written as:

$$\Delta U_{dc}^* = G(s)\Delta\omega_g^* = \frac{s}{\dfrac{CU_{dc\_nom}^2\delta_0}{P_0\omega_{nom}}s^2+\dfrac{\delta_0}{\omega_{nom}}s+K}\Delta\omega_g^* \quad (15)$$

While

$$\Delta\omega_g^* = \frac{\Delta\omega_g}{\omega_{nom}} \quad (16)$$

As can be observed from (15), $G(s)$ is a 2nd order system, whose response time to step signal is:

$$T_s(\pm5\%) = \frac{3.5}{\zeta\omega_n} = \frac{7CU_{dc\_nom}^2}{P_0} \quad (17)$$

T_s is inversely proportional to the steady-state active power P_0. The decline of P_0 can lead to the decrease of damping ratio and the increase of response time. Taking the system presented in Table I as an example, T_s is 0.1s while $P_0=P_{nom}$, and increases to 1s when $P_0=0.1P_{nom}$. In order to improve the dynamic performance of proposed strategy, especially under the condition of low wind power, a damping compensation strategy is designed.

As can be observed from (14), the modulation ratio m can be used to strengthen the damping ratio of $G(s)$. Defining M as the damping coefficient:

$$\Delta m = M\frac{\Delta U_{dc}}{U_{dc\_nom}} \quad (18)$$

Substituting (18) into (14) we can reach the transfer function $G'(s)$ after damping compensation:

$$G'(s) = \frac{s}{\dfrac{CU_{dc\_nom}^2\delta_0}{P_0\omega_{nom}}s^2+\dfrac{(M+m)\delta_0}{m\omega_{nom}}s+K} \quad (19)$$

Thus the response time after compensation T_{sM} is:

$$T_{sM} = \frac{m}{m+M}T_s \quad (20)$$

To ensure the response time lower than 150ms under different wind power output, a segmented damping compensation strategy is applied in this paper.

$$M = \begin{cases} 8 & (\text{when } P_0 = 0.1\sim0.2P_{nom}) \\ 4 & (\text{when } P_0 = 0.2\sim0.4P_{nom}) \\ 2 & (\text{when } P_0 = 0.4\sim0.7P_{nom}) \\ 1 & (\text{when } P_0 = 0.7\sim1P_{nom}) \end{cases} \quad (21)$$

As shown in Fig. 5, with the application of the proposed damping compensation, the response time of $G(s)$ under different wind power output is basically constant and within the required range, i.e., 150 ms.

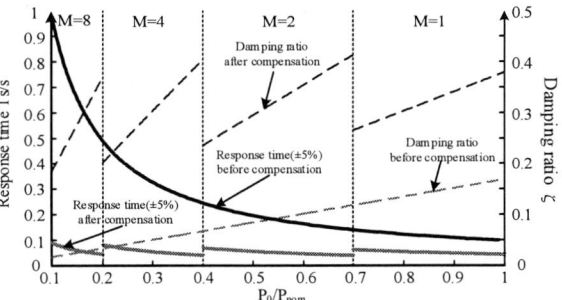

Fig.5 Relationship between response time/damping ratio and wind power output

IV. FREQUENCY MIRRORING CONTROL

With ISC applied in REC, SEC can be informed with onshore AC grid frequency deviation by detecting DC voltage, and transfer the frequency information to wind farms by regulate its AC frequency reference:

$$K\frac{\Delta U_{dc}}{U_{dc\_nom}} = \frac{\Delta\omega_{sec}}{\omega_{nom}} \quad (22)$$

Since AC grid frequency can be mirrored to the sending-end instantaneously, the control strategy applied in SEC is called Frequency Mirroring Control (FMC). Its block diagram is presented in Fig 6.

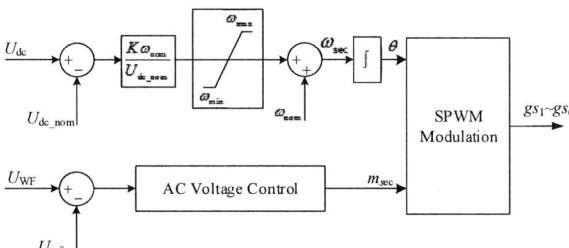

Fig. 6 Block diagram of frequency mirroring control

With the application of proposed control strategy, a link between onshore AC grid frequency, DC voltage and SEC output frequency is established. Hence it is possible for the wind turbine to detect grid frequency and realize inertia response. Finally, the whole system can behave like a synchronous generator with combined inertia of the wind farms and VSC-HVDC system, which largely improves its stability under weak grid condition and provides frequency support to onshore AC grid.

V. THE INERTIAL RESPONSE OF WIND TURBINE

A link among onshore AC grid frequency, DC voltage, and SEC output frequency is established with the application of the proposed control strategy. Hence, the wind turbine can detect grid frequency and realize inertial response.

The capability of wind turbines to provide inertial response is investigated in [3]. An additional value associated with the rate-of-change-of-frequency (RoCoF) is attached to the active power reference (P_{MPPT}) given by the MPPT control. Additional power P_{add} is provided by accelerating or decelerating the wind turbine and utilizing the kinetic energy stored in rotating blades.

Given that the kinetic energy stored in rotating blades is limited, if primary regulation of the wind farm is needed, then power source such as energy storage should be added. Another option is the utilization of de-loading strategies by preserving a generation margin. However, only inertia response is discussed in this paper.

Assuming that the virtual inertia of wind farm is H_{WF}, the value of power reference after inertial response P_{ref} is:

$$P_{ref} = P_{MPPT} + P_{add} = P_{MPPT} - 2H_{WF}\omega_{rec}\frac{d\omega_{rec}}{dt}. \quad (23)$$

Owing to the rapid response of power electronic devices, the output power of wind turbine P_{WF} is nearly the same as power reference P_{ref}. By substituting (23) into (7), the following is obtained:

$$P_{MPPT} - P_{rec\_grid} = (\frac{CU_{dc\_nom}^2}{K\omega_{nom}} + 2H_{WF}\omega_{rec})\frac{d\Delta\omega_{rec}}{dt}. \quad (24)$$

Combining (24) with (7) presents that the whole system seen from PCC behaves similar to an SG with lumped inertia of the wind farms and VSC-HVDC system. Therefore, all the merits of an SG in an electromechanical time-scale are inherited.

VI. SIMULATIONS

The modeling and simulation in this study are conducted in PSCAD/EMTDC 4.5. The simulation model is built based on Fig. 1. Parameters of studied VSC-HVDC system are shown in Table I.

To simplify the analysis, the wind farm is aggregated to an equivalent 200MW Double-Fed Induction Generator, whose parameters can be found in Table II.

TABLE II
EQUIVALENT DFIG PARAMETERS

Meaning	Value
Rotor resistance	0.0048pu
Stator resistance	0.0055pu
Rotor crowbar resistance	0.5Ω
Leakage inductance	0.0468pu
Excitation reactance	3.954pu
Winding ratio of stator and rotor	0.3333
Pole pair	3
Wheel inertia	3.5s
Generator inertia	0.7s
Virtual inertia	4s

With combined simulation models of VSC-HVDC and wind farm, the proposed control strategy is simulated and compared to an existing method proposed in [5], which is based on conventional vector control and transfers grid frequency deviation by PLL's frequency tracking and passive DC voltage adjustment.

In Ref [10], the index of short circuit ratio (SCR) is used to evaluate the stiffness of a grid. In order to compare the performances of the two control strategies under different grid stiffness, the grid inductance is set to different value, corresponding to different SCR.

Scenario I: Wind power is set to 140MW (0.7pu). The grid inductance is set to 16mH, corresponding to the SCR value equal to 20, which represents a stiff grid. At t=3s, onshore AC grid frequency drops from 50Hz to 49.6Hz in 6s.

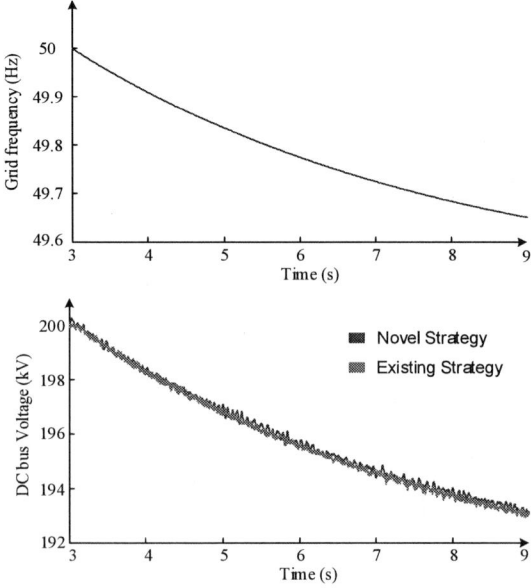

The 2018 International Power Electronics Conference

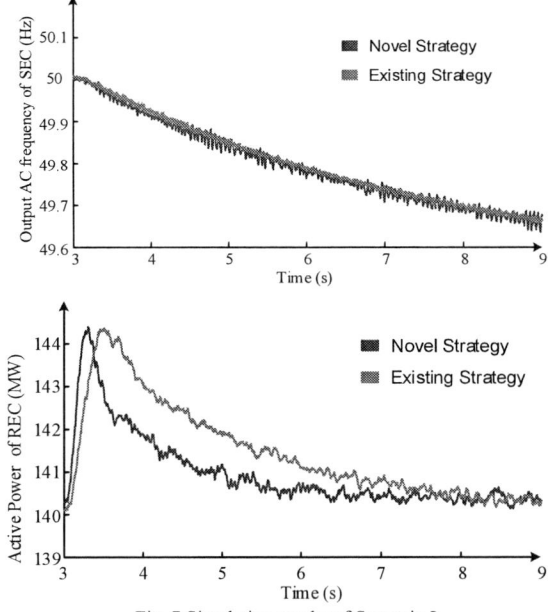

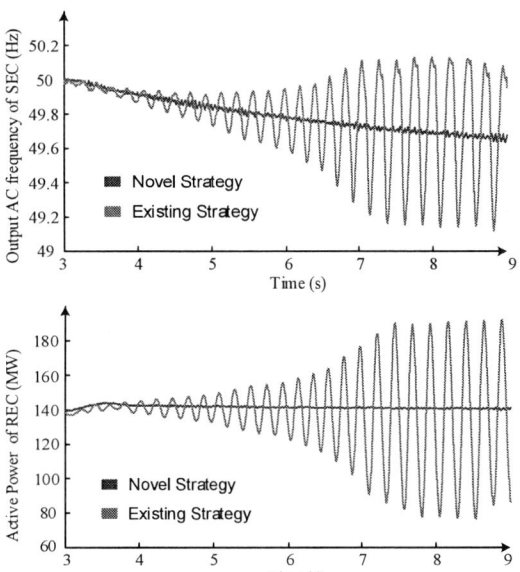

Fig. 8 Simulation results of Scenario II

Fig. 7 Simulation results of Scenario I

The simulation results in Fig.7 show that under stiff grid condition, both control strategy can deliver the onshore AC grid frequency deviation to wind farm, helping the wind farm to realize inertia response. And the novel control strategy functions little faster than conventional vector control

Scenario II: Wind power is set to 140MW (0.7pu). The grid inductance is set to 60mH, corresponding to the SCR value equal to 2, representing a weak grid. At t=3s, onshore AC grid frequency drops from 50Hz to 49.6Hz in 6s.

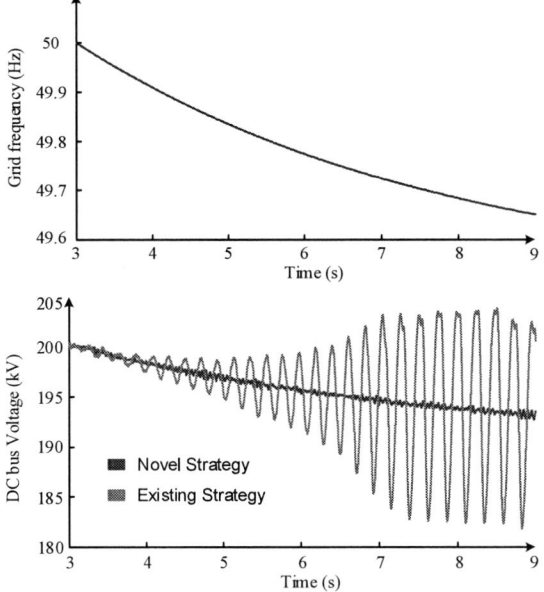

The simulation result in Fig.8 shows that even under weak grid condition, the novel control strategy functions well, and helps the wind farm realize inertia response. At the same time, the power quality of conventional vector control deteriorates, resulting in the oscillation phenomenon.

VII. CONCLUSION

In this paper, a novel control strategy for VSC-HVDC with wind farm integration has been proposed, including the Inertial Synchronizing Control in REC and the Frequency Mirroring Control in SEC. Several benefits can be achieved by utilizing this method:

1) The real-time frequency mirroring

Due to the Inertial Synchronizing Control in REC and the Frequency Mirroring Control in SEC, the frequency information of onshore AC grid is delivered to wind farm with little time delay. According to the information, the wind farm may realize the rapid inertia response to grid frequency deviation, which largely improves the stability of onshore AC grid.

2) Better performance under weak grid condition

The PLL of REC is removed, and self-synchronizing is achieved. The combined system of wind farm and VSC-HVDC performs as a voltage source to onshore AC grid, which functions well even under weak grid condition.

ACKNOWLEDGMENT
This article is sponsored by the National Key Research and Development Program of China (2016YFB0900901).

REFERENCES

[1] N. Flourentzou, V. Agelidis, and G. Demetriades, "Vsc-based hvdc power transmission systems: An overview," Power Electronics, IEEE Transactions on, vol. 24, no. 3, pp. 592 –602, March 2009.

[2] S. M. Muyeen, R. Takahishi, and J. Tamura, "Operation and control of HVDC-connected offshore wind farm," Sustainable Energy, IEEE Transactions on, vol. 1, no. 1, pp. 30–37, Apr. 2010.

[3] J. Lee, E. Muljadi, P. Sorensen and Y.C. Kang, "Releasable kinetic energy-based inertial control of a DFIG wind power plant," Sustainable Energy, IEEE Transactions on, vol. 7, no. 1, pp. 279-288 2016.

[4] L. M. Castro; E. Acha, "On the Provision of Frequency Regulation in Low Inertia AC Grids Using HVDC Systems," Smart Grid , IEEE Transactions on, vol.7, no.6, pp.2680-2690 Nov. 2016

[5] H. Liu and Z. Chen, "Contribution of VSC-HVDC to frequency regulation of power systems with offshore wind generation," Energy Conversion, IEEE Transactions on. vol. 30, no. 3, pp. 918–926, Sep. 2015.

[6] J. He and Y. W. Li, "Generalized Closed-Loop Control Schemes with Embedded Virtual Impedances for Voltage Source Converters with LC or LCL Filters," Power Electronics. IEEE Transactions on. vol. 27, no. 4, pp. 1850-1861, Apr. 2012.

[7] J. Driesen and K. Visscher, "Virtual synchronous generators," in Proc. IEEE Power Energy Soc. Gen. Meeting., 2008, pp. 1–3.

[8] Q.-C. Zhong and G. Weiss, "Synchronverters: Inverters That Mimic Synchronous Generators," Industrial Electronics, IEEE Transactions on vol. 58, no. 4, pp. 1259–1267, 2011.

[9] Lidong Zhang; Harnefors, L.; Nee, H.-P.; , "Power-Synchronization Control of Grid-Connected Voltage-Source Converters," Power Systems, IEEE Transactions on , vol.25, no.2, pp.809-820, May 2010.

[10] A. Egea-Alvarez, S. Fekriasl, F. Hassan and O. Gomis-Bellmunt, "Advanced Vector Control for Voltage Source Converters Connected to Weak Grids," Power System, IEEE Transactions on, vol. 30, no. 6, pp. 3072-3081,Nov. 2015.

A Study on Steady-state Characteristics of Series-connected Wind Farm Using an Experimental Set of Laboratory Size

Fujio Tatsuta[1] and Shoji Nishikata[2]

1 Department of Information Systems and Multimedia Design, Tokyo Denki University, Tokyo, JAPAN
2 Department of Electrical and Electronic Engineering, Tokyo Denki University, Tokyo, JAPAN
*E-mail: tatsuta@mail.dendai.ac.jp

Abstract— In this study, we investigate steady-state characteristics of series-connected type wind farm through experimental means by using a laboratory-size setup. A laboratory-scale wind farm simulator consisting of two wind turbine/generators connected in series is used. The configuration of the experimental set is explained first, and examples of experimental results of the voltage and current waveforms of the wind generators and rectifiers for the cases of different wind speeds are shown. These experimental results are used to predict the characteristics of the wind farm consisting of 50 wind turbine/generators, in which the rectifier output DC voltages are integrated to obtain the whole DC link voltage of the wind farm. Moreover, based on these predicted characteristics, facility utilization factor and generated electricity amount of the wind farm under a wind speed distribution to the wind turbines are revealed.

Keywords— *current–source inverters, series-connected configurations, thyristor, wind farm.*

I. INTRODUCTION

In order to solve the energy related problems such as global warming and depletion of fossil fuels, the installed capacity of wind power generation is rapidly expanding [1]. Since the output power obtained from one wind turbine generator is about several MWs in these days [2], it is necessary to construct wind farms consisting of a lot of wind turbine generators in order to obtain a large amount of electricity comparable to GW class thermal or nuclear power plants [3]. Generally, the outputs of the wind turbine generators in these wind farms are connected in parallel to integrate the generated power. It is assumed that the installation site of such a large scale wind farm will be shifting from onshore to long distance offshore. Although HVDC is advantageous for such long-distance power transmission [4], there are some issues in the parallel-connected system such as necessity of costly substation.

As a new interconnecting method for wind turbine generator systems, a configuration of a wind farm consisting of wind turbine generators in which rectified

This paper is based on results obtained from a project commissioned by the New Energy and Industrial Technology Development Organization (NEDO).

outputs are connected in series and a current-source thyristor inverter has been proposed [5]. In this type of wind farm, the rectified outputs of each wind turbine generator are connected in series and integrated in the DC link, as a result, the offshore substation can be eliminated. In the power conversion system using a current-source thyristor inverter, a large amount of harmonic currents exists in the output current, and it is generally said that a large number of massive power filters are indispensable [6]. On the other hand, in the system in [5], harmonic components included in the output current are essentially eliminated without special filters by the functions of a synchronous compensator and a well-designed duplex reactor provided in the output circuit. It can be said that the system composed of series-connected wind turbine generators and a thyristor inverter with a synchronous compensator and duplex reactor is useful, and it can obtain high reliability and high quality electric power with a simple configuration.

When a wind turbine generator system is applied in practical usage, it is essential to understand various performances of the system for any wind speed conditions including extreme situations such as gust. However, difficulties are anticipated in carrying out experiments under arbitrary wind speed conditions using actual wind turbines, and the experiments in particular using a wind farm composed of MW class turbines especially are not feasible. Hence, computer simulations are effective means for this purpose. The validity of the simulation results, however, has to be verified with experimental investigations. So, a wind turbine simulator installed in laboratory that can predict various performances of an actual wind turbine generator system is useful [7-10]. On the other hand, in an experimental investigation on a wind farm consisted of a large number of wind turbine generators, it is difficult to install a large number of simulators for individual wind turbine/ generators in the laboratory.

In this paper, we propose a method for predicting steady-state characteristics of a series-connected type wind farm composed of a large number of wind turbine/generators using a laboratory-scale wind farm simulator consisting of two wind turbine/generators connected in series.

II. CONFIGURATION OF WIND FARM COMPOSED OF SERIES-CONNECTED WIND TURBINE GENERATORS

Fig. 1 shows the series-connected configuration of wind farm discussed here [5, 11]. The system consists of n wind turbine generators (SG1~SGn) driven by wind turbines, series-connected thyristor rectifiers, a current-source thyristor inverter, and a synchronous compensator SC with a duplex reactor. The outputs of the wind turbine generators are connected through transformers Tr1~ Trn for insulation and voltage boost as shown in the figure. Also, SC provides reactive power for commutations of the inverter thyristors, and the duplex reactor, that is properly designed based on the parameter of SC, eliminates the higher voltage harmonics contained in the output of the inverter [12]. The rectified n wind turbine generator outputs are connected in series, and the unified DC power is fed to the current-source thyristor inverter through DC link. In this figure, bypass diodes are connected in parallel to the rectifier outputs, respectively, to continue the operation of wind farm when some wind turbine generators are stopped due to lack of wind.

It should be noticed that, unlike a conventional thyristor inverter-based system, this system does not require reactive power necessary for commutation of the inverter thyristors from the grid, and a large filter is unnecessary for the system output circuit. In other words, wind farms constructed by this method are simpler than the conventional system, have high reliability, and can be expected to significantly reduce costs for construction and maintenance.

III. EXPERIMENTAL MEANS TO OBTAIN CHARACTERISTICS OF MANY WIND GENERATORS CONNECTED IN SERIES

A. Experimental Set Composed of Two Generators

In order to predict the characteristics of series-connected wind farms, we constructed an experimental set consisting of two wind turbines/generators [11]. Fig. 2 (a) shows the configuration of the experimental set. PMSG 1 and PMSG 2 in the figure are permanent magnet synchronous generators as wind turbine generators driven by the induction motors IM1 and IM2, whose torques are controlled so as to simulate the characteristics of the wind turbines. Here, due to the laboratory space limitation, two wind turbine generators were installed as the minimum number of plural generators. Note that the number of generators in the experimental set (two) is less than those in the actual wind farms. Experimental procedures of the wind farm composed of three or more generators based on this set will be described later. The ratings of the experimental system are shown in TABLE I.

B. Experimental Procedure for Wind Farm Composed of Three or More Generators

In this study, we discuss the steady-state characteristics of the wind farm consisting of 50 wind generators ($n = 50$), using the experimental set with two wind turbine generators as shown in Fig. 2. In the series-connected configuration shown in Fig. 1, DC link current is determined based on the maximum wind speed among the wind speeds flowing into all the wind turbines, and rated DC link current flows when at least one of the wind speeds for wind turbines reaches the rated wind speed or more because DC link current is common [5].

In the actual wind speed distribution, the possibility that at least one of the wind speeds for wind turbines exceeds the rated wind speed in the wind farm is sufficiently high, resulting in rated DC link current. For this reason, in the following experiments, the DC current was always controlled to be constant at the rated value.

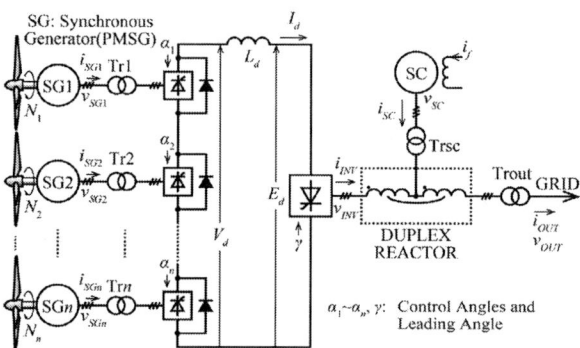

Fig. 1. Configuration of wind farm composed of series-connected wind generators and current-source thyristor inverter.

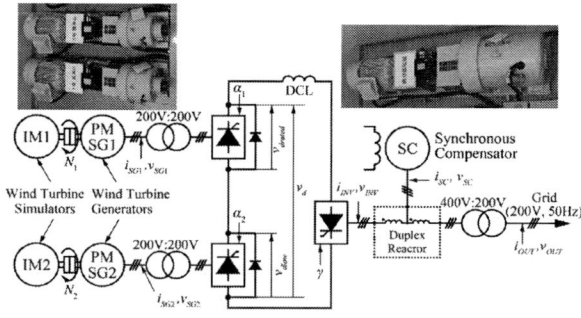

(a) Main circuit of experimental set.

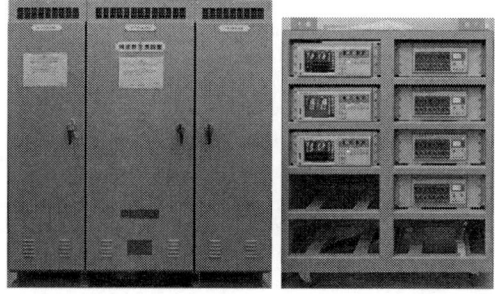

(b) Converter panel and data acquisition system.

Fig. 2. Experimental set for series-connected wind farm.

TABLE I
RATINGS OF EXPERIMENTAL SET

System Output			
Apparent power	4kVA	Frequency	50Hz
Grid Voltage	200V		
DC link			
Voltage	520V	Current	7.7A
PMSGs			
Output	2kVA	Frequency	100Hz
Voltage	205V	Rotational Speed	1500min⁻¹
Number of pole pairs	4		
Synchronous Compensator			
Capacity	10kVA	Frequency	50Hz
Voltage	400V	Rotational Speed	1500min⁻¹

Therefore, in Fig. 2, rated DC current flows in the DC link when one of the wind turbine generators is operated at wind speed equal to or more than the rated speed, and the maximum rectifier output voltage V_{drated} is obtained for this generator. In this case, the other generator was operated according to its wind speed and its rectified output voltage V_{dlow} was determined by the wind speed. Finally, the total rectified output voltage, which is equal to DC link voltage V_d, becomes the sum of V_{drated} and V_{dlow}.

We repeated such experiments with one wind turbine generator entering various wind speeds less than rated when the other wind turbine generator was always operated at rated speed. Simultaneously, the measurement data obtained by these experiments were accumulated using the data acquisition system shown in Fig. 2 (b).

To obtain the steady-state characteristics of the series-connected wind farm of Fig. 1, we calculate the whole DC link voltage by summing up the individual rectifier output voltages $V_{d1} \sim V_{d50}$ based on the experimental results explained above.

IV. EXPERIMENTAL RESULTS AND CALUCULATED CHARACTERITIC

A. Wind data used for experiments

The average wind speed of 1 pu, from which the rated output can be obtained from the wind turbine generator, is used for the following discussion. And, the wind speed flowing into each wind turbine is assumed to follow the Weibull distribution shown in Fig. 3 [13].

The Weibull distribution for wind speed is used to predict the yearly total power earned by a wind turbine. Although the momentary wind distribution for the wind turbines of an offshore wind farm differs from the yearly distribution, the wind speeds based on the Weibull distribution can be used as momentary wind distribution because we will be able to predict the yearly wind power of the wind farm on the basis of the distribution. Hence, the Weibull distribution is used as the momentary wind distribution here.

In the experiment, for the sake of simplicity, the experiment was executed when one wind turbine generator was operated at rated wind speed while the other wind turbine generator was operated at the wind speed varying from 1pu to 0.4pu every 0.1pu. That is,

the DC current was always maintained at a constant rated value (=7.7 A). TABLE II shows the numbers of wind turbines by wind speeds based on Weibull distribution when average wind speed = 1pu. We assume that the wind turbines exceeding the rated wind speed were rated power operated with pitch control. Based on this table, we executed a lot of experiments to obtain the DC link voltage for the wind farm composed of 50 wind turbine generators. According to TABLE II, 23 wind turbines in the wind farm are in the rated operation, and DC link current has to be controlled at the rated current as mentioned in the previous section.

In this system, controlling DC link current is attained through controlling the leading angle of the thyristor inverter. Fig. 4 shows one example of DC link current waveform. Since this system uses a current-source inverter, we assume that a constant current flows in the DC link. However, small fluctuation was observed in the current because the smoothing reactor is relatively small.

B. Examples of experimental results

Figs. 5~7 show examples of the instantaneous waveforms for output voltage and current of wind turbine generators (Fig. (a)) and rectifier output voltage (Fig. (b)). These were obtained from experiments for the cases of wind speeds of 1pu, 0.8pu and 0.5pu, respectively.

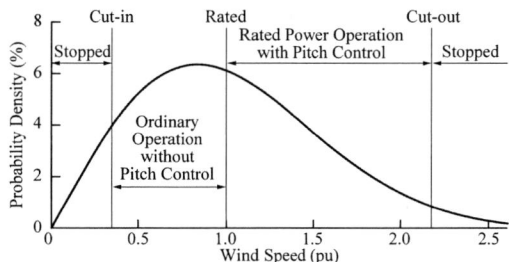

Fig. 3. Probability density - wind speed characteristics (Average Wind Speed = 1pu).

TABLE II
NUMBER OF WIND TURBINES BASED ON WEIBULL DISTRIBUTION
(AVERAGE WIND SPEED = 1PU)

Wind speed(pu)	≤Cut-in	0.4	0.5	0.6
Number of wind turbines	5	3	3	4
Wind speed(pu)	0.7	0.8	0.9	≥1.0
Number of wind turbines	4	4	4	23

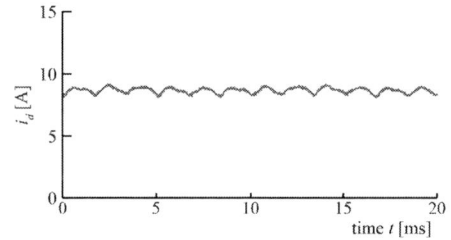

Fig. 4. Example of DC link current.

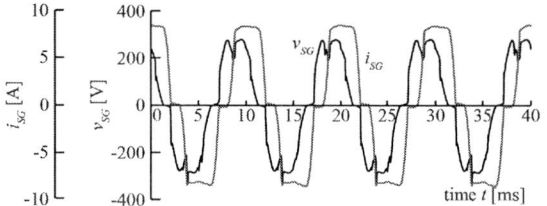

(a) PMSG output voltage (line to line) and line current waveform.

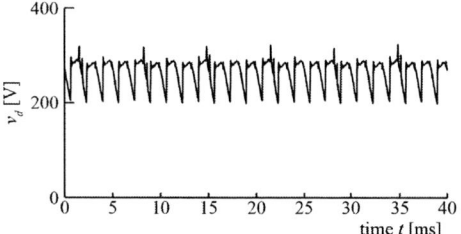

(b) Rectifier output voltage (α=0 deg.).

Fig. 5. Experimental instantaneous waveforms of PMSG and rectifier output for wind speed = 1 pu.

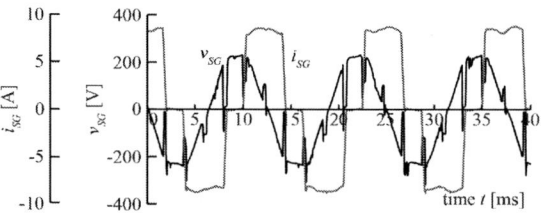

(a) PMSG output voltage (line to line) and line current waveform.

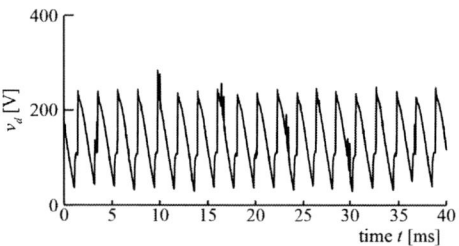

(b) Rectifier output voltage (α=47 deg.).

Fig. 6. Experimental instantaneous waveforms of PMSG and rectifier output for wind speed = 0.8 pu.

Fig. 5 shows the waveforms for wind speed of 1pu (rated operation). The frequencies of the output voltage and the current in Fig. (a) were 100Hz and the control angle of the rectifier was 0 deg.

Fig. 6 shows the waveforms when the wind speed was 0.8 pu. Since the rotational speed of the PMSG is proportional to the wind speed, the frequencies of these waveforms were 80 Hz. Nevertheless, it is noted that the current amplitude for this case was the same as that of Fig. 5 (a), because armature current is common.

In Fig. 7, since the wind speed was 0.5pu, the frequencies of the output voltage and current of the PMSG were 50Hz as shown in Fig. (a). There were sinks

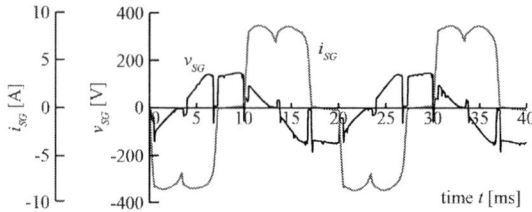

(a) PMSG output voltage (line to line) and line current waveform.

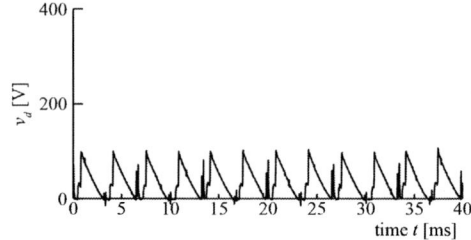

(b) Rectifier output voltage (α=76 deg.).

Fig. 7. Experimental instantaneous waveforms of PMSG and rectifier output for wind speed = 0.5 pu.

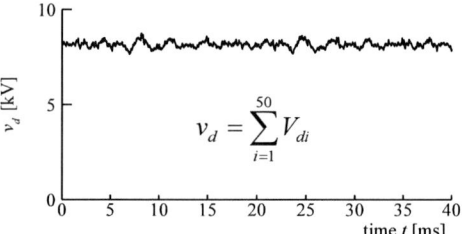

$$v_d = \sum_{i=1}^{50} V_{di}$$

Fig. 8. Example of experimental instantaneous waveform of DC link voltage of the wind farm (Average voltage V_d = 8.16 kV).

in the armature current because the bypass diodes are connected to the rectifiers. In this case the negative part of the rectified voltage was clipped by bypass diode since the control angle was 76 degs., which exceeds 60degs.

Fig. 8 shows waveform of total DC link voltage for a period of 40 ms. This is obtained through summing up the output voltages for 50 rectifiers on the basis of the proposed idea. As in the figure the total DC link voltage tends to be almost constant since in general the frequencies and phases of 50 wind turbine generators are not unified. From Fig. 8, average of DC link voltage is 8.16 kV (= 61 % of rated DC link voltage), and as a result, facility utilization factor of the wind farm is 61%.

Finally, an example of voltage and current waveforms of the inverter output, wind farm output, and SC terminal are shown in Fig. 9. Since the inverter used in the wind farm under consideration is of current-source type, the inverter output current becomes a trapezoidal waveform as shown in (a), and voltage jumps and sinks due to commutations of the inverter thyristors appeared in the voltage waveform. On the other hand, it is confirmed that the system output with low distortion current (THD = 1.8%) was obtained as shown in (b). As a result, SC current, which is the difference between the currents in (a) and (b), flowed as shown in (c).

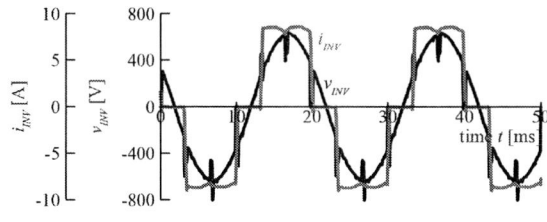

(a) Inverter output.

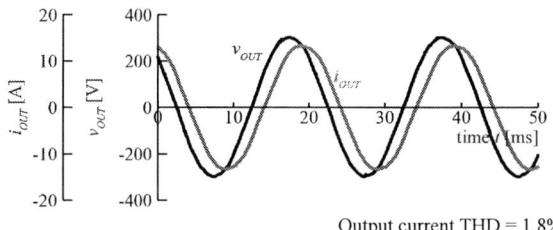

Output current THD = 1.8%

(b) System output.

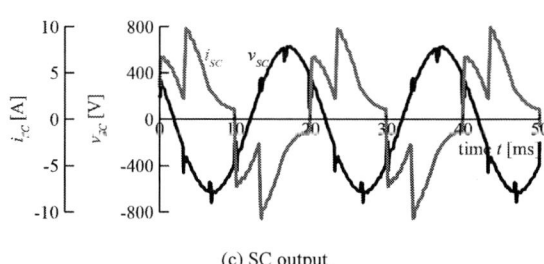

(c) SC output.

Fig. 9. Example of instantaneous line to line voltage and line current waveforms in inverter output, wind farm output, and SC terminal (System output = 3.5kW).

V. SIMULATED CHARACTERISTICS FOR 50 SERIES-CONNECTED WIND GENERATORS

In order to confirm the validity of the experimental method in this study, a simulation model of the wind farm shown in Fig. 1 consisting of 50 wind turbine generators is constructed and simulation is carried out [11]. Here, MATLAB/Simulink is used for the simulations. Also a simulation model of PMSG is derived based on trapezoidal waveforms of armature electro-motive forces for PMSG [11, 14]. The parameters of PMSGs are shown in TABLE III.

Each parameter of the circuit and wind speed conditions are set according to those used in the experiments. Examples of the simulated results corresponding to Figs. 5 ~ 8 in the experimental results are shown in Figs. 10 ~ 13, respectively. These simulated results are in good agreement with experimental results, and the validity of the experimental method in this study is confirmed.

Here, the simulation result of the DC link voltage V_d in Fig. 13 is 8.35 kV, which is 2.33% higher than the experimental result (8.16 kV). The main reason for this difference is that in the simulation the voltage drop due to DC link resistance is ignored.

TABLE III
PARAMETERS OF PMSGs

Stator phase resistance R_s	0.632Ω
Stator phase inductance L_s	3.44mH
Flux linkage established by magnets φ	0.23959V.s
Number of pole pairs p	4

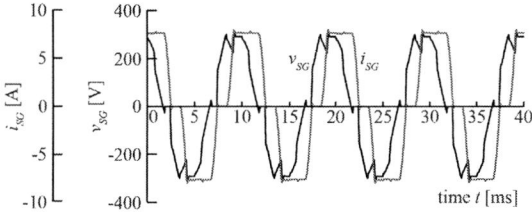

(a) PMSG output voltage (line to line) and line current waveform.

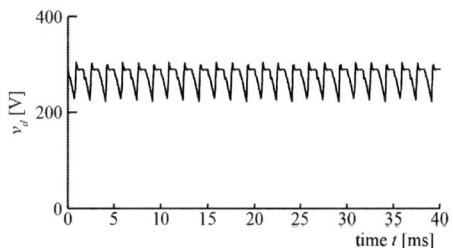

(b) Rectifier output voltage (α=0 deg.).

Fig. 10. Simulated instantaneous waveforms of PMSG and rectifier output for wind speed = 1 pu.

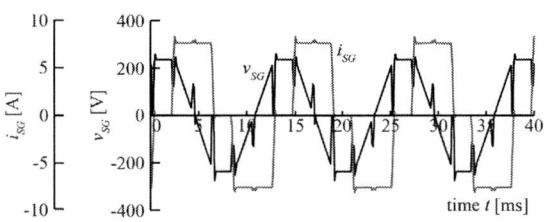

(a) PMSG output voltage (line to line) and line current waveform.

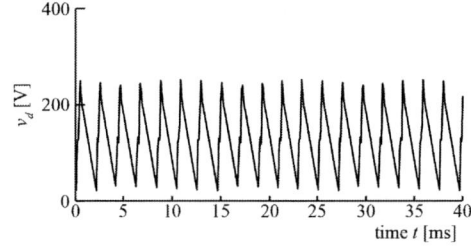

(b) Rectifier output voltage (α=47 degs.).

Fig. 11. Simulated instantaneous waveforms of PMSG and rectifier output for wind speed = 0.8 pu.

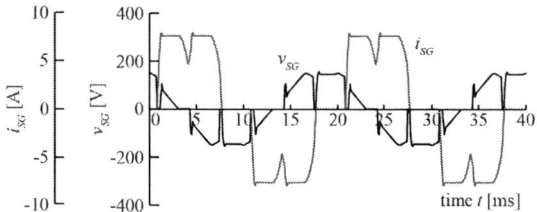

(a) PMSG output voltage (line to line) and line current waveform.

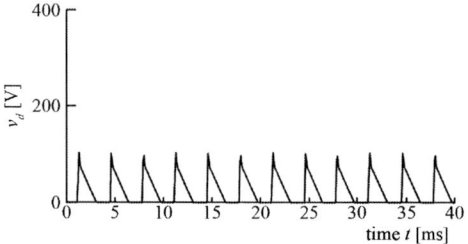

(b) Rectifier output voltage (α=76 degs.).

Fig. 12. Simulated instantaneous waveforms of PMSG and rectifier output for wind speed = 0.5 pu.

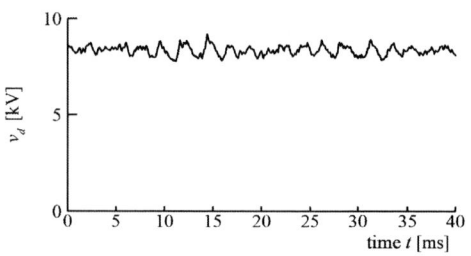

Fig. 13. Example of simulated instantaneous waveform of DC link voltage of the wind farm (Average voltage V_d = 8.35 kV).

VI. Conclusions

In this paper, a method for predicting steady-state characteristics of a series-connected type wind farm composed of a large number of wind turbine/generators using a laboratory-scale wind farm simulator has been proposed. The followings are the conclusions obtained from this paper:

1. An experimental method to obtain the steady-state characteristics of a series-connected type wind farm composed of a large number of wind turbine/generators has been proposed on the basis of experimental results obtained through the experimental set composed of two wind generators connected in series.

2. Based on the proposed method, the steady-state characteristics of wind farm composed of 50 wind generators have been investigated, and it has been revealed that the facility utilization factor of the wind farm in the wind distribution based on the Weibull distribution is about 61% for the case of average wind speed of 1 pu.

3. The steady-state characteristics of the wind farm consisting of plural wind turbine generators connected in series have been simulated based on a simulation model in order to confirm the validity of the experimental method. These simulated results are in good agreement with experimental ones, verifying the usefulness of the experimental method.

In this study, DC link current was kept constant for the sake of simplicity, however the DC link current fluctuates all the time in the actual wind farm. Also, the wind speeds flowing into the wind turbines are always fluctuates, and the dynamic performances of the whole wind farm have to be investigated. These are left for future study.

Acknowledgment

The authors would like to thank Prof. Ken-ichiro Yamashita of Salesian Polytechnic, Dr. Ryoichi Kurosawa, and Mr. Tadashi Nishikawa for useful discussion about this work.

References

[1] http://www.wwindea.org/2017-statistics/.
[2] https://en.wind-turbine-models.com/turbines/1605-mhi-vestas-offshore-v164-9.5-mw.
[3] https://electrek.co/2018/01/30/worlds-largest-offshore-wind-farm/.
[4] Thomas Ackermann (Editor) *Wind Power in Power System*, 2nd Ed., p.307.
[5] S. Nishikata and F. Tatsuta, "A New Interconnecting Method for Wind Turbine/Generators in a Wind Farm and Basic Performances of the Integrated System," *IEEE Trans. on Ind. Electron.*, vol. 57, no. 2, pp468-475, 2010.
[6] Siegfried Heier: *Grid Integration of Wind Energy Conversion Systems, second edition,* John Wiley & Sons, Ltd, 2006, p.218.
[7] H. M. Kojabadi, L. Chang, T. Boutot, "Development of a Novel Wind Turbine Simulator for Wind Energy Conversion Systems Using an Inverter-Controlled Induction Motor," *IEEE Trans on Energy Conversion*, vol. 19, No. 3, pp.547-552, 2004.
[8] M. Chinchilla, S. Arnaltes, J. L. Rodriguez-Amendo, "Laboratory set-up for Wind Turbine Emulation," *IEEE International Conference on Industrial Technology (ICIT)*, pp 553-557, 2004.
[9] K. H. Kim, T. L. Van, D. C. Lee, S. H. Song, and E. H. Kim, "Maximum Output Power Tracking Control in Variable-Speed Wind Turbine Systems Considering Rotor Inertial Power," *IEEE Trans. Ind. Electron.*, vol. 60, No. 8, pp.3207-3217, 2013.
[10] F. Tatsuta, B. Liu, and S. Nishikata, "A Simulation Method for Dynamic Performances of MW class Wind Turbine/Generator Using Laboratory-Size Simulator," *Proc. of 19th Int. Conf. on Electrical Machines and Systems (ICEMS 2016)*, Chiba, Japan, 2016.
[11] F. Tatsuta and S. Nishikata, "Studies on Characteristics of PMSGs Used for Current-Source Type Wind Farm Composed of Series-connected Wind Generators," *Proc. of 20th Int. Conf. on Electrical Machines and Systems (ICEMS 2017)*, Sydney, Australia, 2017.
[12] S. Nishikata and F. Tatsuta, "Studies on a Wind Turbine Generating System that Employs a Thyristor Inverter," *T.IEEJ*, vol. 130-D, No. 4, pp. 407-414, 2010 (in Japanese).
[13] Wind map in Japan, New Energy and Industrial Technology Devel opment Organization (NEDO), http://app8.infoc.nedo.go.jp/nedo/index.html (in Japanese).
[14] MathWorks Documentation "Permanent Magnet Synchronous Machine," [Online], Available: https://www.mathworks.com/help/physmod/sps/powersys/ref/permanentmagnetsynchronousmachine.html.

A Novel Islanding Detection Method with Two-phase Magnification Inspection

Jian-Tang Liao, Shun-Hao Yeh, and Hong-Tzer Yang*
Department of Electrical Engineering, National Cheng Kung University, Tainan, Taiwan
*E-mail: htyang@mail.ncku.edu.tw

Abstract- Energy storage system, solar photovoltaic and wind power generation systems constitute distributed generation (DG) system, which is widely established in recent years. In the grid-connected DG system, the islanding detection function is essential to avoid dangers. However, the existing islanding detection techniques have non-detection zone (NDZ) which is easily affected by local load conditions. This paper proposes a novel islanding detection method integrating with a two-stage magnification inspection approach to facilitate diagnostics of islanding detection in DG systems. Based on the proposed method, the detection time is shortened with NDZ eliminated effectively. Besides, the possibility of malfunction and impact of load conditions are greatly reduced. The proposed method doesn't need precise parameter setting for different applications. To demonstrate the feasibility of the proposed method, a grid-connected prototype lab system with ratings of 250 W rated power and 200 Vdc/110 Vac voltage is simulated and implemented, a system which involves different load conditions and interference tests. The results indicate that the proposed method can operate reliably and satisfy the IEEE 1547 standard.

I. INTRODUCTION

In the recent years, the grid-connected distributed generation (DG) technologies, such as wind power and solar photovoltaic (PV) generations, become widely applied [1]-[3]. Therefore, more safety considerations should be incorporated into grid-tied DG systems.

The islanding phenomenon occurs when a DG system still operates as the grid blacks out, resulting in the continuous power supply to the local load. This phenomenon may expose electricity grid maintenance personnel to the hazard of electric shock, engender a reclose surge current caused by phase differences, and damage electric devices in the local area because of the unstable power quality [4]-[6]. Therefore, grid-tied inverters should follow the standards [7], [8] on efficient islanding detection.

Fig. 1 illustrates a simplified single-line diagram of a grid-tied inverter system. In this system, the circuit breaker acts as an islanding event trigger. The RLC load acts as the local load which includes active power and reactive power consumption. When the circuit breaker is opened, the inverter and the local load interact. Anti-islanding methods are aimed at detecting islanding events and disconnect the inverters.

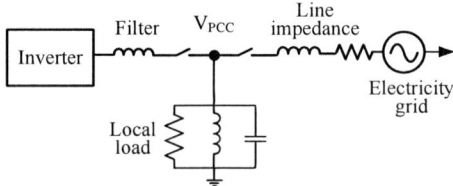

Fig. 1. Single-line diagram of a grid-tied micro-grid system.

Numerous relevant standards such as IEEE 1547 and UL1471 have detailed the requirements for islanding detection. The anti-islanding methods can be mainly divided into two categories, namely local detection methods and remote detection methods [9], [10]. Generally, remote detection methods need extra investment because it mainly relies on a comprehensive communication system.

The protection relays, such as over/under voltage and over/under frequency relays, are usually used to provide passive anti-islanding functions by detecting the voltage and frequency out of allowable range. However, the passive anti-islanding methods commonly have unavoidable non-detection zones (NDZ). Therefore, many literatures [11]-[16] which discuss the active anti-islanding methods have been proposed in the last decades.

In [11], the authors proposed a two-step active islanding detection approach based on periodic changes in the reactive disturbance. Besides, the average frequency deviation value (AFDV) is adopted as an index. Although the two-step method can effectively avoid malfunctions, the AFDV is easily affected by the quality factor of the local load. The threshold is thus indeterminable in practical applications.

In [12], the pulse current is injected into the voltage feedback. In the islanding state, the pulse current interfered with the voltage feedback, resulting in an extremely high error relative to the sinusoidal grid voltage reference at a specific pulse current injection time. Nevertheless, the magnitude of voltage error is affected by the load quality factor, a fact which has the same problem as that in [11].

The authors in [14] proposed an improved active frequency drift (AFD)-based anti-islanding method that changes the formation of frequency drifts and exhibits a lower degree of total harmonic distortion (THD) compared with the traditional AFD-based method. Moreover, the method can detect the islanding effect more rapidly with less NDZ than the existing methods.

Additionally, the reactive power control method [15] is proposed to stimulate frequency of the point of common coupling (PCC). This method entails providing inadequate reactive power to raise the PCC frequency or excess reactive power to reduce the PCC frequency. When the system normally operates, the PCC frequency will be fixed by the power grid. Otherwise, the excess or inadequate reactive power ultimately results in the PCC frequency exceeding normal limits.

However, the disadvantages of this method are its requirement of knowing the load reactive power–frequency curves associated with specific features and the necessity of the frequency operating beyond the normal limits.

In [16], a method based on impedance variation for a single-phase system was proposed. In this method, the disturbance corresponds to the grid period and exerts limited effects on normal grid operation. Nevertheless, the method may detect grid failure by detecting the changes in the grid impedance or harmonics, and the feedback index is related to the load condition, all of which become its shortcomings.

Overall, according to the analysis of existing researches, the requirements of a favorable islanding detection method are organized as follows:

(1) To have the less the better NDZ,

(2) the frequency/voltage not to exceed the limits during detection process,

(3) To involve less disturbance in normal operation,

(4) To have no accidental operation,

(5) To involve an inspection index which doesn't depend on the load condition, and

(6) Detection time to satisfy the related standards.

As mentioned above, this paper proposes a novel islanding detection method that exhibits the preceding characteristics. Simulations and experiments are both conducted to verify the theoretical feasibility of the proposed method. The details of the proposed method are introduced in the following sections.

II. PROPOSED ISLANDING DETECTION METHOD

A. Current Control Method of a Grid-tie Inverter

The grid-tied inverter acts as a current source in the system. The current controller design is based on a synchronous d-q frame [17], [18], as shown as (1). Fig. 2 shows the control diagram of the proposed method. The command signals I_d^{com} and I_q^{com} are adjusted through another PI (proportional integral) controller by comparing the current reference with the current feedback signal.

The sinusoidal pulse width modulation (SPWM) signal produced through the inverse Park transformation

corresponds to the phase obtained from the phase-locked loop (PLL). The proposed method is based on the frequency disturbance of the PLL. Therefore, the θ' parameter for the inverse Park transformation includes the disturbance affecting the output current as well as the expected corresponding feedback.

$$\begin{bmatrix} V_\alpha \\ V_\beta \end{bmatrix} = \begin{bmatrix} \cos\theta & -\sin\theta \\ \sin\theta & \cos\theta \end{bmatrix} \cdot \begin{bmatrix} V_d \\ V_q \end{bmatrix} \qquad (1)$$

B. Proposed Islanding Detection Method

The proposed method is based on the PLL circuit, which entails incorporating the frequency disturbance and observing the frequency feedback from the PLL circuit. Fig. 3(a) illustrates the frequency of the output current reference, i_o^{ref}, which equals the frequency of measured voltage.

In the proposed method, an active frequency disturbance shown as Fig. 3(b) is added into output current reference. Because the frequency of PCC voltage won't be impacted by this active disturbance in the grid-tied state, the measured frequency by PLL is clamped by the utility grid, as shown in Fig. 4(a). On the contrary, in the islanding state, the measured frequency will be influenced according to the output current reference, as shown in Fig. 4(b).

Besides, in order to avoid misoperation and reduce the power quality impact during normal operation, the detection process is divided into two phase. The flowchart and transient process schematic of the proposed islanding detection method are illustrated in Fig. 5 and Fig. 6, respectively.

The frequency error between point d_0 and d_1, $\Delta f_{S(0)}$, namely the frequency error of Stage 0, $S(0)$, can be calculated by (2). Because the first state (S_0) may not start precisely from the occurrence point of the islanding event, the frequency error derived at S_0 is thus ignored. In contrast, for the islanding state, the frequency error between d_2 and d_1 is expectedly greater than that in the grid-connected state. Once the frequency error signals are continuously greater than the threshold and have alternate signs for certain times, the process of islanding detection will be enter the next phase, as shown in Fig. 6.

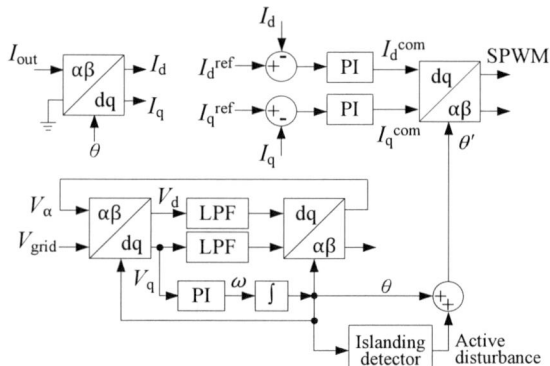

Fig. 2. Block diagram of a single-phase d-q current control.

The 2018 International Power Electronics Conference

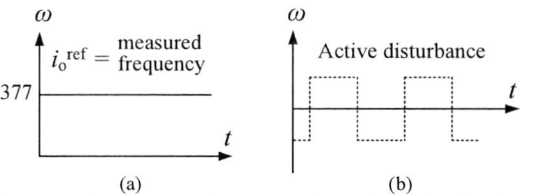

(a)   (b)

Fig. 3. (a) Output current frequency reference (b) Active disturbance of proposed method

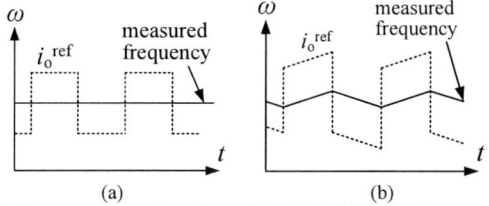

(a)   (b)

Fig. 4. Frequency error detection at (a) grid-tied (b) islanding state

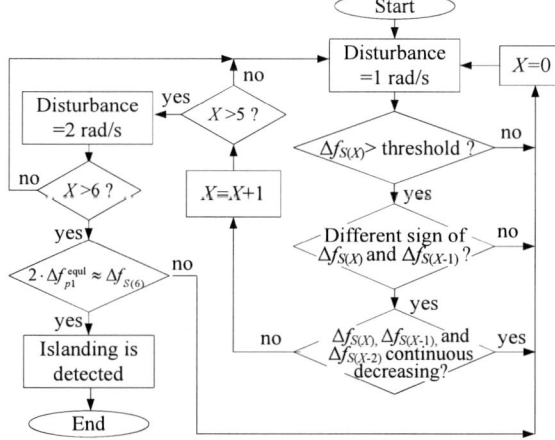

Fig. 5. Flowchart of the proposed islanding detection method.

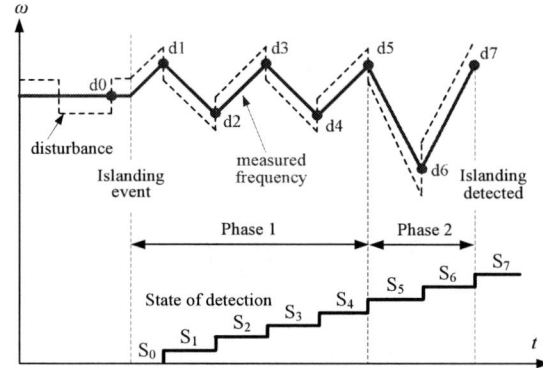

Fig. 6. Detection process of the proposed method.

$$\Delta f_{S(0)} = f_{d(1)} - f_{d(0)} \qquad (2)$$

The equivalent frequency error of Phase 1 is the average of the first four indices from S_1 to S_4, as defined in (3).

$$\Delta f_{p1}^{equl} = \frac{1}{4}\sum_{i=1}^{4} \Delta f_{Si} \qquad (3)$$

C. Features of the Proposed Method

The proposed two-phase magnification inspection

method affords a detection approach that is independent of the load condition. The method involves the assumption that the system parameters do no change during the detection period and that doubled disturbance rates return doubled responses. The proposed method can reduce disturbance at the first stage when operating in the grid-connected state. Besides, the detection only needs few cycles and not make the PCC voltage out of normal condition which is beneficial to maintain the local electric equipment safety.

III. SIMULATION AND EXPERIMENTAL RESULTS

To verify the proposed method, the prototype system with the specification presented in TABLE I is analyzed for simulations and implementations.

A. Simulation Results

The simulations are conducted by using PSIM software. Fig. 7 shows the simulation results derived when the load quality factor is 1. The local load parameters are R= 48.4 Ω, L= 128.4 mH, and C= 54.8 μF. The equivalent frequency error of Phase 1 and the frequency error of Phase 2 are 0.507 and 1.08, respectively. Obviously, the frequency error of Phase 2 is 2.13 times of that in Phase 1. Therefore, inverter is disconnected at the beginning of S7. The time required for the islanding detection is 0.1153 sec.

Fig. 8 shows the results derived when the load quality factor is equal to 2. The local load parameters are R= 48.4 Ω, L= 64.192 mH, and C= 109.61 μF. Based on the simulation results, The equivalent frequency error of Phase 1, Δf_{p1}^{equl} and $\Delta f_{(S7)}$ are 0.439 and 0.938, which is

TABLE I
PARAMETERS OF PROTOTYPE SYSTEM

Parameters	Value
DC source	200 V
Switching frequency	12 kHz
Dead time	1 μs
Filter	10 mH
Grid voltage/ frequency	110 Vac/ 60 Hz
Line resistor/ inductor	0.02 Ω/ 30 μH
Nominal power	3kW

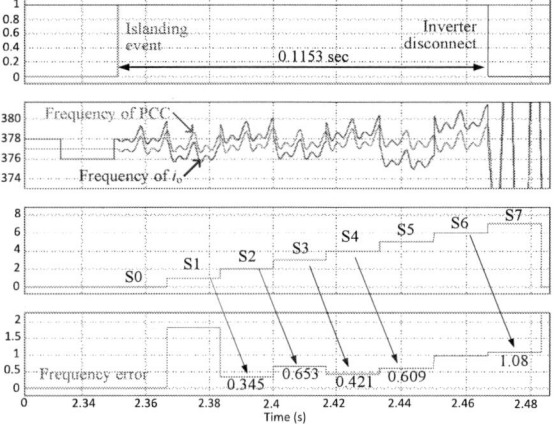

Fig. 7. Simulation results of the single-phase system, quality factor=1.

4235

2.13 times of $\Delta f_{p1}{}^{equl}$. Therefore, the inverter is disconnected at the beginning of S7. The total time required for the islanding detection is 0.1154 s.

Fig. 9 illustrates the results of simulations conducted under voltage swell and sag conditions. At 2.05–2.2 s, the voltage drops by 10% for 0.15 sec, and at 2.35 s, the voltage increases by 10%. The results show that the frequency error is not higher than the designed noise threshold even the voltage variation happened. Therefore, the proposed method won't result in misoperation when the voltage sag/swell happened.

B. Experimental Results

Fig. 10 shows the photograph of the experimental setup which has the same specification as in the simulations. The circuit is controlled by using TI DSP TMS320F28069. Fig. 11 shows the experimental results with quality factor = 1. The result indicates that the proposed islanding detection method can effectively work in the experimental tests. The needed detection time is 0.112 s.

Fig. 12 and Fig. 13 show the measurement of AC-side variables of the grid-connected inverter with and without the proposed active disturbance injection, respectively. The Yokogawa power analyzer WT1800 serves as the measurement instrument. Fig. 12 shows the measured power factor and THD of 0.9976 and 4.113 %, respectively, with the proposed active disturbance signal injected.

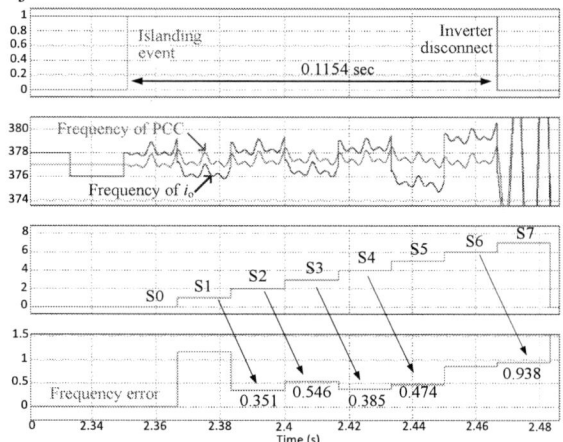

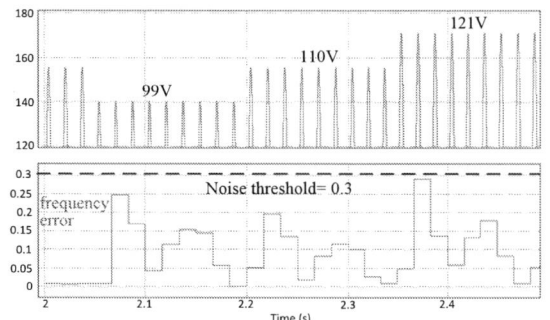

Fig. 8. Simulation results of single-phase system, quality factor = 2.

Fig. 9. Simulation results for grid voltage swells and sags.

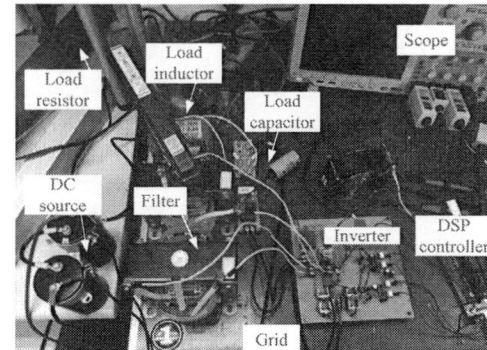

Fig. 10. Photograph of the experimental setup.

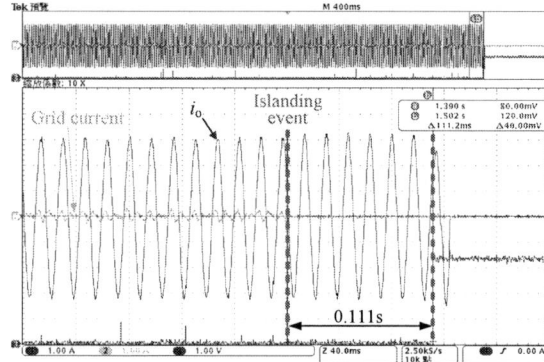

Fig. 11. Experimental results with quality factor = 1.

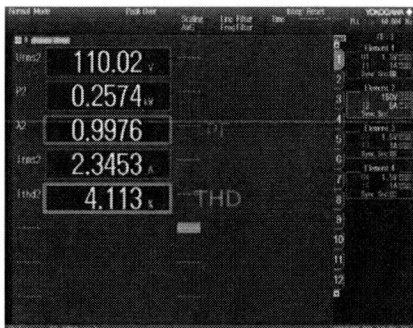

Fig. 12. Measurement of inverter AC-side variables with the proposed active disturbance.

Fig. 13. Measurement of inverter AC-side variables without the proposed active disturbance.

Fig. 13 shows that the PF and THD are 0.9978 and 4.074%, when the proposed active disturbance is not injected. The preceding two results indicate that the proposed active disturbance almost does not affect the system power quality.

C. Application of a Three-phase System

The proposed method can also be applied to a three-phase grid-connected inverter system. Likewise, a simulation system with the specification listed in TABLE II. The different load condition, including quality factor= 1, quality factor= 2, unbalanced grid voltage, and unbalanced local load, are tested for verify the feasibility.

Fig. 14 shows the simulation results obtained when the load quality factor is equal to 1. The local load parameters are R= 48.4 Ω, L= 128.385 mH, and C= 54.8054 μF. Based on the simulation results, The equivalent frequency error of Phase 1, $\Delta f_{p1}{}^{equl}$ and $\Delta f_{(S7)}$ are 0.98375 and 1.887, which is 1.918 times of $\Delta f_{p1}{}^{equl}$. Therefore, the inverter is disconnected at the beginning of S7. The total time required for the islanding detection is 0.112 sec.

Fig. 15 shows the results derived when the load quality factor is equal to 2. The local loads are R= 48.4 Ω, L= 64.192 mH, and C= 109.61 μF. Likewise, The$\Delta f_{p1}{}^{equl}$ and $\Delta f_{(S7)}$ are 0.372 and 0.724, respectively. After computing, the frequency error of Phase 2 is 1.946 times of that in Phase 1. Therefore, inverter can successfully detect the islanding event. The time required for the islanding detection is 0.112 sec.

Fig. 16 illustrates the simulation results obtained when the system is operated under the unbalancing grid voltage. The voltage of phase A is increased by 5%, whereas those of the other phases remained unchanged. Based on the simulation, the PLL feedback is affected by ripples, but no misoperation occurred in the unbalanced grid and no frequency error observed. Moreover, the proposed method can still detect the islanding event successfully. It can be concluded that the proposed method has certain ability to afford unbalancing grid voltage.

Fig. 17 presents the simulation results derived when the local load is unbalanced. The resistor of phase A and resistor phase B are increased and decreased by 5%, respectively. the simulation result indicates that a transient signal of frequency error is detected and attenuates in the following cycles, which interferences and delays the islanding detection. finally, the islanding event is successfully detected at 0.183 sec after the islanding event.

TABLE II

PARAMETERS OF THE THREE-PHASE INVERTER SYSTEM.

Parameters	Value
DC source	600 V
Switching frequency	12 kHz
Dead time	1 μs
Filter	10 mH
Grid voltage/ frequency	380 V_{ac}/ 60 Hz
Line resistor/ inductor	0.02 Ω/ 30 μH
Nominal power	3kW

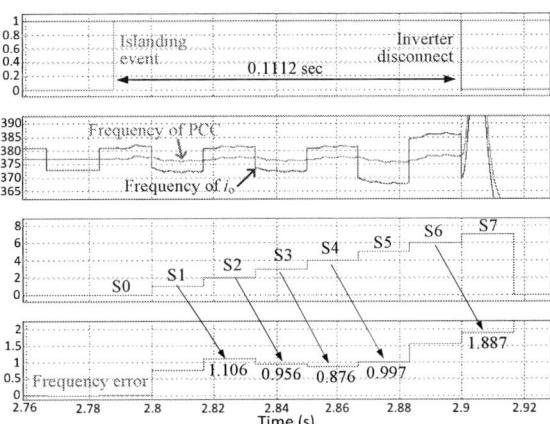

Fig. 14. Simulation results of the three-phase system, quality factor= 1.

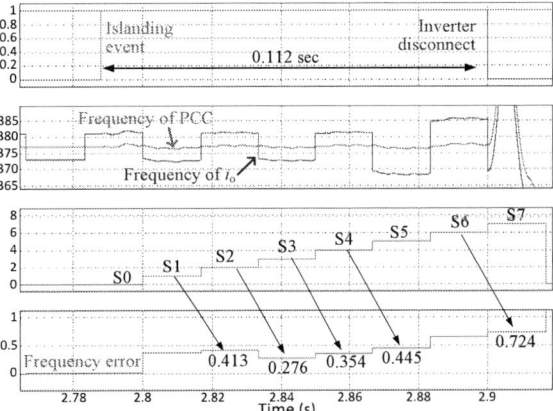

Fig. 15. Simulation results of the three-phase system, quality factor = 2.

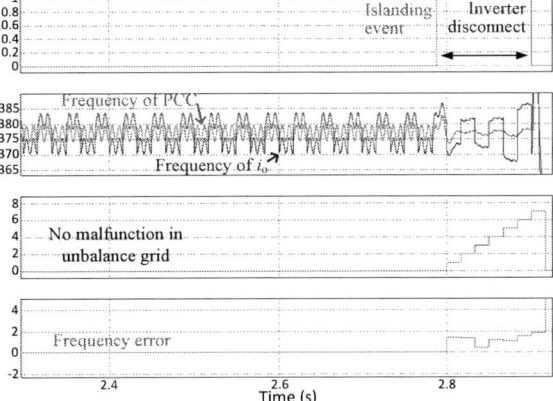

Fig. 16. Simulation results under an unbalanced grid.

IV. CONCLUSION

This paper has proposed a novel islanding detection method which applies a two-stage inspection method to minimize normal operation disturbance and avoid system malfunction. A method for inspecting the magnification index is also incorporated, rendering the proposed method suitable for most systems that involve different inverters and load conditions without the necessity of changing the system parameters. Simulations and

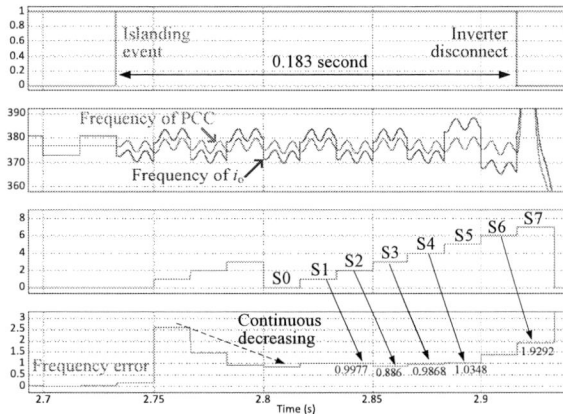

Fig. 17. Simulation results under the unbalanced load.

experiments were conducted on a 250W prototype system to verify the proposed method. Besides, the detection time is within 0.2 sec. The effects of the active disturbance on the THD are less than 0.2% in the simulations and less than 0.1% in the experiments. The results thus show the feasibility and practicability of the proposed method.

REFERENCES

[1] G. T. Heydt, "The next generation of power distribution systems," *IEEE Trans. Smart Grid*, vol. 1, no. 3, pp. 225-235, Nov. 2010.

[2] G. Carpinelli, G. Celli, S. Mocci, F. Mottola, F. Pilo, and D. Proto, "Optimal integration of distributed energy storage devices in smart grids," *IEEE Trans. Smart Grid*, vol. 4, no. 2, pp. 985-995, Mar. 2013.

[3] M. C. Bozchalui and R. Sharma, "Optimal operation of Energy Storage in distribution systems with Renewable Energy Resources," in *Proc. 2014 Clemson University Power System Conference*, pp. 1-6.

[4] M. Yingram, K. Chinnabutr, W. Chansanam, and R. Kudpik, "Perturbation of active islanding detection techniques in power system network," In Proc. *2015 Electrical Engineering/ Electronics, Computer, Telecommunications and Information Technology (ECTI-CON)*, pp. 1-4.

[5] B. Guha, R. J. Haddad, and Y. Kalaani, "A passive islanding detection approach for inverter-based distributed generation using rate of change of frequency analysis," in Proc. *IEEE SoutheastCon 2015*, pp.1-6.

[6] P. Mahat, Zhe Chen, and B. Bak-Jensen, "Review of islanding detection methods for distributed generation," in Proc. *2008 Third International Conference on Electric Utility Deregulation and Restructuring and Power Technologies*, pp. 2743-2748.

[7] *IEEE recommended practice for utility interface of photovoltaic (PV) systems*, IEEE Std 929-2000, 2000.

[8] *IEEE Standard for Interconnecting Distributed Resources with Electric Power Systems*, IEEE Std 1547-2003, 2003.

[9] W. Bower and M. Ropp, Evaluation of islanding detection methods for utility-interactive inverters in photovoltaic systems, 2002.

[10] M. Yingram and S. Premrudeepreechacharn, "Investigation over/under-voltage protection of passive islanding detection method of distributed generations in electrical distribution systems," in Proc. *2012 the International Conference on Renewable Energy Research and Applications*, pp. 1–5.

[11] P. Gupta, R. S. Bhatia, and D. K. Jain, "Average absolute frequency deviation value based active islanding detection technique, " *IEEE Trans. Smart Grid*, vol. 6, no. 1, pp. 26-35, Aug. 2015.

[12] C. C. Hou and Y. C. Chen, "Active anti-islanding detection based on pulse current injection for distributed generation systems," *IET Power Electronics*, vol. 6, no. 8, pp. 1658-1667, Sept. 2013.

[13] S. Bifaaretti, A. Lidozzi, L. Solero, and F. Crescimbini, "Anti-islanding detector based on a robust PLL," *IEEE Trans. on industry applications*, vol. 51, no. 1, pp. 398-405, June 2015.

[14] A. Yafaoui, B. Wu, and S. Kouro, "Improved active frequency drift anti-islanding detection method for grid connected photovoltaic systems," *IEEE Trans. on power electronics*, vol. 27, no. 5, pp. 2367-2375, Oct. 2012.

[15] X. L. Chen and Y. L. Li, "An islanding detection algorithm for inverter-based distributed generation based on reactive power control, " *IEEE Trans. Power Electron.*, vol. 29, no. 9, pp. 4672-4683, Oct. 2014.

[16] M. Ciobotaru, V. Agelidis, and R. Teodorescu, "Accurate and less-disturbing active anti-islanding method based on PLL for grid-connected PV inverters, " in Proc. *2008 IEEE Power Electronics Specialists Conference*, pp. 4569-4576.

[17] M. Kazimierkowski and L. Malesani, "Current Control Techniques for Three-Phase Voltage-Source Converters: A Survey, " *IEEE Trans. on Industrial Electronics*, vol. 45, no. 5, pp. 691-703, Oct. 1998.

[18] U. A. Miranda, L. G. B. Rolim, and M. Aredes, "A DQ synchronous reference frame current control for single-phase converters, " in Proc. *2005 IEEE 36th Power Electronics Specialists Conference*, pp. 1377-1381.

AUTHOR INDEX

Aapro, Aapo 3156
Abdollahi, Hessamaldin....................1719
Abe, Kazuyuki.................................1567
Abe, Kensho....................................767
Abe, Kodai 1741, 3890
Abe, Seiya 2360, 2370
Abe, Takashi...................................2176
Abrishamifar, Adib..........................2854
Abuogo, James 1125
Acharya, Anirudh Budnar 2630
Acharya, Sayan 3564
Adachi, Masakazu............................2237
Afsharian, Jahangir 1537, 3797
Agarwal, Vivek...............................3471
Agelidis, Vassilios G.........................3215
Agostinelli, Matteo 3140
Ahmad, Hamzeh J............................3273
Aiso, Kohei....................................3186
Akagi, Hirofumi..............................2352
Akahane, Masashi.............................2774
Akama, Yousuke...............................1741
Akao, Naoki...................................1217
Akatsu, Kan............................711, 3186
Alatise, O......................................1149
Alenius, Henrik 1704, 4205
Ali, Muhammad 528
Ali, Murad.....................................2317
Allmeling, Jost.........................422, 2199
Almér, Stefan..................................555
Alsofyani, Ibrahim Mohd....................466
Alvarez, S......................................4009
Amano, Koki...................................94
Amei, Kenji....................................3182
Amin, Mohammad 759
Amrhein, Wolfgang...........................3640
An, Ronghui.............957, 1524, 3251, 3692, 3924
An, Zheng......................................4001
Andenna, M....................................3596
Andersen, A. E. Michael.....................1351
Andersen, Michael A. E..............607, 4066
Ando, Akinobu.................................517
Ando, H..3665
Ando, Masato..................................1919
Ando, Takashi.................................3658
Ang, Simon S..................................153
Antivachis, Michael..........................181
Antonini, Giulio...............................3588
Antonopoulos, Antonios 2335
Anurag, Anup..................................3564
Anyapo, Chan..................................3332
Aoyagi, Kazuki................................2237
Aoyama, Masahiro......................718, 753
Arai, Takuro...................................1997
Araumi, Ryunosuke 1877, 3658
Arimatsu, Kenji...............................1370

Arita, Hideaki 2796, 2820
Arrua, Silvia 1719
Asada, Kazunori 3658
Asama, Junichi 4016
Ashizaki, Yusuke 3450
Ashourloo, Mojtaba 2380
Aso, Shinji 3086
Aware, Mohan 1730
Ayano, Hideki 1080
Azad, A N M Wasekul....................... 2416
Azegami, Kazuya 3723
Azuma, S. 3665
Baba, Teppei 2283
Babasaki, Tadatoshi 207
Bach, Hoang Linh............................ 2410
Baek, Jae-Il 108, 2365, 3100, 3533, 3538
Baek, Miran 1141
Bahat-Treidel, Eldad 3607
Bai, Baodong 2638
Baik, Jeong Min.............................. 3063
Bak, Yeongsu 1736, 4104
Bakran, Mark-M. 2476
Bandyopadhyay, Soumya 1426
Barrena, Jon Andoni 759
Barrera-Cardenas, Rene...................... 3431
Bauer, Pavol 1426, 2630
Bauer, Walter 3640
Bayer, Christoph Friedrich 2410
Bellini, M. 4009
Berg, Matias 963, 4205
Bergveld, H.J. 267
Bertoldi, F. 488
Besselmann, Thomas......................... 555
Bezha, Minella 3170
Bhattacharya, Subhashish 3564, 3993
Bhowate, Apekshit........................... 1730
Bhumkittipich, Krischonme 2430
Biela, J. 1896
Biela, Jürgen.............1103, 1509, 2301, 3734
Bilal, Ahmad 2193
Bilsalam, A. 1622
Bin, Zhao 2692
Bixel, Paul 238
Blaabjerg, Frede 439, 746, 1183,
    1246, 1711, 1788, 2512, 2604, 2743, 3123,
    3164, 3357
Blanes, José M. 1435
Böcker, Jan 3607
Bojoi, R....................................... 732
Bonyadi, R. 1149
Boroyevich, Dushan 790, 3705, 3749, 3985
Bortis, D. 4080
Bortis, Dominik 181
Boynov, K.O.................................. 161
Braun, Michael 2848, 3074

AUTHOR INDEX

Büdel, Johannes 3034
Bui, M.X. 4174
Bunlaksananusorn, Chanin 2490
Burgos, Rolando 790, 3705, 3749, 3985
Cai, Kejun 3965
Cai, Panpan 3495
Cai, Xu 1004, 1491, 2245, 4162, 4220
Canales, F. 4009
Cao, Hu 1816, 3484
Cao, Pengpeng 2973
Cao, Qi 100
Cao, Wu 3002, 3010, 3015
Cardenas, Rene Alexander Barrera 1111
Carvalho, Kelly C. M. 3785
Castellazzi, Alberto 130, 2932
Ceballos, Salvador 3117
Celik, Mustafa 1680
Cha, Honnyong 927, 1046, 2619, 3134
Chae, Beomseok 1977
Chailloux, Thibaut 2153
Chang, Chen-Wei 1617
Chang, Chien-Hsuan 2860
Chang, Liuchen 815, 1472, 1793, 2505
Chang, Yung-Ruei 639, 883
Chanmontree, P. 1622
Chao, Yi-Hao 1145
Charalambous, Apollo 1634
Charoensuksirikul, Supanut 2113
Chattopadhyay, Ritwik 3564
Chazal, Hervé 2158
Chen, Ang-Tung 2102
Chen, Bo 1397
Chen, C. 142
Chen, Ching-Chen 1617
Chen, Ching-Jan 2086
Chen, Chuantong 1598
Chen, Dezhi 2638
Chen, Guan-Jung 1341
Chen, Guo 370
Chen, Hao 3112
Chen, I-Lin 2107
Chen, Jiangnan 1157, 1167
Chen, Jiann-Fuh 2653
Chen, Jie 1015, 1177
Chen, Kai-Hui 3081
Chen, Ke 1391
Chen, Kun-Feng 1341
Chen, Min 878
Chen, Minwu 2547
Chen, Nan 2335
Chen, Pingping 1118
Chen, Shen-Li 1145
Chen, Song 2153
Chen, Tang-Jung 1617
Chen, Tao 1872

Chen, Wan-Jung 3544
Chen, Wenjia 4213
Chen, Wenjie 1062, 2854, 3329
Chen, Wu 1504, 2496
Chen, Xiliang 3329
Chen, Xin 1015, 1177
Chen, Xingxing 1051, 3129, 3439
Chen, Yang 2785
Chen, Yangyang 560
Chen, Yaow-Ming 639, 883
Chen, Yenan 1118
Chen, Yen-Wen 2576
Chen, Yufeng 3383
Chen, Yu-Jen 275
Chen, Zhe 1758, 2708
Chen, Zhi 2997
Chen, Zhigang 3040
Cheng, Ching-Hsiang 2086
Cheng, Chun-An 2860
Cheng, Hung-Liang 2860
Cheng, Nie 2625
Cheng, Po-Tai 503, 1038, 2462, 3549
Cheng, Ran 3877
Cheng, Xiangpeng 2435, 3934
Chengbi, Zeng 2718
Chi, Yongning 1491
Chiba, Akira 3627
Chien, Lin-Hao 2102
Chiu, Huang-Jen 2092, 3151
Chiu, Hui-Lung 123
Chiu, Yi-Hao 1145
Cho, Geum-Bae 2145
Cho, In-Ho 3323
Cho, Shin-Young 1530
Cho, Young Joon 137
Cho, Younghoon 1403
Choe, Chanyang 1598
Choi, Byungcho 1465
Choi, Hyun-Jun 383
Choi, Jae Hyuk 1336
Choi, Jaeho 803
Choi, Joon-Ho 982, 1799
Choi, Seung-Hyun 4049
Choi, Sewan 256
Choi, Sung-Jin 1409
Choi, Youn-Ok 2145
Chou, Shih-Feng 1711
Chou, T.-C. 1912
Choudhury, Abhijit 3401
Chuai, Guoming 3025
Chung, Daewoong 1141
Chung, Henry S. H. 917
Chunkag, V. 1622
Collins, Caspar 1931
Cortes, Camilo 2193

AUTHOR INDEX

Corvasce, C. 3596
Cucala, Asuncion P. 2534
Cui, Shenghui 2250, 2484
Cui, Xiang 1125
Cvetkovic, Igor 790, 3985
Czyz, Piotr 396
D'arco, Salvatore 782, 2003
Da Silva, C. 267
Dahidah, Mohamed S A 3215
Dai, T. 1149
Dai, Wenjing 1015
Daikoku, Akihiro 2796
Danqing, Liu 1376
Dao, Ngoc Dat 1212
Dauphin, Benjamin 3644
Davari, Pooya 746
Davletzhanova, Z 1149
De Doncker, Rik W. 375, 388, 598,
   1073, 2250, 2484, 2768, 3729, 3979
Decker, Simon 2848, 3074
Delaforge, Timothé 2158, 3820
Deng, Fujin 1758, 2708
Deng, Jinxin 2992
Deshpande, Prathamesh Pravin......................4186
Dieckerhoff, Sibylle 3607
Dimarino, Christina 3985
Din, Zakiud 2262
Ding, Yong 815
Dinh, Nguyen Duy 363
Diouf, Fatou 2078
Dirksen, Daniel..2410
Divan, Deepak 4001
Doki, Shinji 1032, 1223, 1228, 1295, 1747, 2224
Dong, Hanjing 987
Dong, Mi 1771
Dong, Qinghua 459
Dong, Xiaofeng 4168
Dong, Zhen 459
Dong, Zheng 3768
Driesen, J. 488
Du, Chao 2204
Du, Xiaotong..1167, 2780
Du, Xizhou 1491
Du, Yan 1472, 2877
Du, Zhijiang..84
Duarte, J. L. 946, 1067, 2697
Duarte, Jorge L. 1447, 3840
Dugal, F. 3596
Dujic, Drazen......................422, 1484, 1498, 2170
Duong, Truong-Duy...982
Duque, C. A. 1067
Eberle, Wilson 927
Ekman, Jonas 3588
Elbaset, Adel A. 3945
Endegnanew, Atsede G. 2003

Endo, Hiroaki 4151
Endres, Tobias Maximilian 2410
Engelmann, Georges 3979
Enomoto, Bruno Yukio 3785
Eto, Haruhi 2097
Faiz, Muhammad Talib 528
Fajri, Poria..3223
Fan, Dongchen 3002, 3010, 3015
Fan, Shengwen 977, 3040
Fan, Weiyan 1386, 1421
Fang, Jingyang 337, 3910
Fang, Ran 4213
Fangfang, Luo 1282
Farkas, Gabor 137
Fayyaz, Asad 130
Felderer, Niklaus 2199
Feng, Chao 2058
Feng, Wei 3678
Ferdowsi, Mehdi 3223
Fernandez, Gabriel...3209
Fernandez-Cardador, Antonio 2534
Fischer, F. 3596
Foo, Gilbert 1724
Formentini, A. 4034
Freijedo, Fracisco D. 1498
Friedrichs, Peter 3584
Fuchs, Simon 2301
Fujii, H. 1253
Fujii, Kansuke 3711
Fujii, Keisuke 1189
Fujii, Toshiyuki 2540, 3578
Fujimoto, Hiroshi 77, 663
Fujimoto, Kazuki 2047
Fujimoto, Yasutaka 571, 681
Fujimura, Akira 1080
Fujita, Atsushi 296
Fujita, Goro 363
Fujita, Hideaki....................626, 1854, 3813, 3940
Fujiwara, Hajime 1381
Fujiwara, Kazuya 3773
Fukuda, Hiroto 2938
Fukuda, Kenji 2558
Fukui, Tomoya 860
Fukuoka, T. 1240
Fukushima, Kentarou 2176
Fukushima, Takafumi..3478
Funabiki, Shigeyuki 2449
Funaki, Tsuyoshi.....................309, 2181, 3092
Funato, Hirohito 94, 2036
Funato, Hiroki 2073
Furukawa, Keita 3349
Furukawa, Kimihisa 3572
Furukawa, Yudai 4193
Furusho, Yasuaki 3711
Gan, Yiliang 1391

AUTHOR INDEX

Ganisetti, V. K. 2907
Gao, Feng 2016, 3383, 3965
Gao, Xiaonan 1661
Gao, Zhuo 3455
Garrigós, A. 1435
Gasim, Abdulaziz 2836
Gehlot, Deepak 3471
Geng, Hua 542
Geng, Yiwen 619
Gerada, C. 4034
Gheonjian, Anna 2078
Gietler, Harald 3140
Gohara, Hiromichi 2764
Gondo, Ryota 3490
Gong, Bing 3797
Gong, Chunying 1015, 1177
Gong, Z. 267
Gorodnichev, Anton 375
Goto, Akihisa 2449
Goto, Hiroki 3192
Goto, Kazuya 1315
Goto, Yasuyuki 809
Gou, Yating 1157, 1167
Grimm, Ferdinand 2895
Grossner, Ulrike 3588
Gruber, Wolfgang 3632, 3640, 4028
Gu, Lei 632
Gu, Qing 2963
Guajardo, Cristian Andres Garces 1854
Guan, Bo 1032
Guan, Yajuan 2668, 3678
Guan, Yueshi 614, 3780
Guangzhu, Wang 1376
Guerrero, Josep M. 1498, 2668, 3112
Guerrero, M. Josep 3678
Gui, Yonghao 2668
Guidi, Giuseppe 782, 2003
Guillod, Thomas 396
Gunji, Daisuke 663
Guo, Leilei 904
Guo, Yanjie 3338
Guozhao, Duan 2625
Gupta, K. 267
Gurpinar, Emre 130
Gutiérrez, R. 1435
Ha, Jung-Ik 565, 2500
Ha, Sang-Hyun 3466
Haga, Hitoshi 1370, 3890
Hagiwara, Makoto 3273
Hahashi, Yuji 4059
Haider, M. 4080
Halamicek, Michael 831
Halick, Mohamed 416
Hamabe, Yasumasa 1276
Hamada, Shizunori 227

Hamaguchi, Takumi 3507
Hamasaki, S. 1240
Hamasaki, Shin-Ichi 1217, 1276, 2938, 3237
Hameyer, Kay 740
Han, Byung-Moon 466
Han, Jung-Kyu 3107, 3533, 4049, 4054
Han, Pengcheng 1027, 2714
Han, Yang 3112
Hanajiri, Kensuke 663
Hanamoto, Tsuyoshi 1315, 1698
Hancioglu, Oguz Kaan 1680
Handa, Hiroyuki 3762
Handa, Yuuichi 4059
Hane, Yoshiki 2426
Hang, Lijun 1391, 2866
Hanju, Cha 1985
Hao, Liu 3484
Hao, Xiang 1478
Harnefors, Lennart 3684
Hartmann, S. 3596
Haruna, Junnosuke 94, 2036
Hasegawa, Kazunori 1938
Hasegawa, M. 3665
Hasegawa, Ryuta 2011
Hashempour, Mohammad M. 4198
Hashimoto, Kazuki 3757
Hasler, Jean-Philippe 3684
Hata, Katsuhiro 663
Hata, Ryotaro 2149
Hatakeyama, Tomoyuki 1991
Hataya, Morimasa 410
Hatipoglu, E. 3805
Hatsumi, Takuya 94
Hatta, Yoshiyuki 675
Hattori, Fumiya 2738
Hattori, Keisuke 3286
Haung-Jen, Chiu 645
Hayashi, Nobuo 866
Hayashi, Yuji 356
He, Wangpin 560
He, Xiaokun 1504, 2496
He, Xiaoqiong 1027, 2714
He, Yigang 2317
He, Yingjie 3439
Hendrix, M. A. M. 946, 2697
Heo, Jongwon 726
Hidaka, Yuki 2820
Higuchi, Keiichi 2764
Higuchi, Masato 3952
Higuchi, Shinichi 2216
Hikaru, Naruse 3418
Hikihara, T. 3665
Hikihara, Takashi 3654, 3757
Hiller, Marc 3074
Hillers, A. 1896

AUTHOR INDEX

Hillers, André 2301
Hilt, Oliver 3607
Hinz, Arne 598
Hirahara, Hideaki 1960
Hiraki, Eiji 410, 1602, 1610
Hirao, Takashi..................................2082, 2137
Hirase, Yuko 767
Hirayama, Katsutoshi 4193
Hirayama, Tadashi 3406
Hirokawa, Masahiko.........................1543, 4133
Hirokawa, Takayuki 296, 410
Hiromoto, Masayuki 3644
Hirose, Keiichi 593, 822
Hirose, Naoki 3791
Hiroshi, Tadano 3431
Hiroshige, Shinichi 3369
Hirota, Takashi 3952
Hoang, Tuan V. 1752
Hoda, Isao 2073
Hofmann, Viktor 2476
Hofmann, Wilfried 3243
Hojo, Masahide 3369
Holenstein, Thomas......................................3619
Holmes, D. G. 3670
Hong, Miao 2718
Hongpeng, Liu....................................1442, 2969
Honjo, Satoshi 2066
Hori, Motohito 3396
Hori, Yoichi 77, 663
Horie, Shunsuke 809
Horikoshi, Takahiro 1997
Hoshi, Nobukazu 971, 2660, 3855
Hou, Chung-Chuan 1617
Hou, Lijun 2901
Houran, Mohamad Abou 1062, 2854
Hsieh, Guan-Chyun 123
Hsieh, Hung-I 123
Hsieh, Yao-Ching 3151, 3544
Hsu, Chi-Hsuan 2653
Hu, Jiewen 3985
Hu, Jingxin 1073, 2250, 2484
Hu, Sheng 3052
Hu, Song 370
Hu, Xihong...614, 3780
Hu, Xing 2262
Huang, Bing-Siang 2092
Huang, Bo-Jia 3528
Huang, Chien-Chun 3151
Huang, Huazhen 1125
Huang, Jingjin 2980, 4157
Huang, Jingjing 1004, 2688, 2692
Huang, Jun-Xian 1626, 3081
Huang, Lang 1478
Huang, Pin Yu 2165
Huang, Ta-Wei 1626

Huang, Wen-Mei 2576
Huang, Xianjin 1131, 2051
Huang, Xiaoliang 84
Huang, Xuehao 3455
Huang, You-Chun 275
Huemer, Mario 3140
Hui, S. Y. Ron 889, 2552
Hung, Chun-Yao 2576
Hung, Shun-Kang 1575
Huo, Chongcan 987
Huo, Junya 1206, 1234
Hussein, Abdallah................................130, 2932
Huynh, Dang Minh 3086
Hwang, Duck-Hwan 1403
Hwang, Seon-Ik 3323
Hwu, K.I. 851
Hyakutake, Y. 1253
Hyodo, Takashi 2589
Hyunsung, An 1985
Iannuzzi, Diego 2527
Ibuchi, Takaaki 309
Ichinose, H. 1240
Ide, Yuji 3896
Iijima, Ryuji 313, 1111
Iioka, Daisuke 2278
Ikari, Yuki 148
Ikeda, Hidehiro 1315
Ikeda, Yoshinari 3396
Ilves, Kalle 2335
Imai, Kazu 3363
Imai, Makoto 296, 410
Imamori, Satoshi 699
Imaoka, Jun 1087, 1095, 1554, 3773
Imoto, R. 2808
Imtiaz, Abu Saleh 2416
Imura, Takehiro 77, 663
Inaba, Tsuyoshi 4114
Inomata, Kentaro 3952
Inoue, Daisuke 2764
Inoue, Kaoru......................1264, 2186, 4151
Inoue, Kent 348
Inoue, Masamichi 1228
Inoue, Takatoshi 1276
Inoue, Y. 704, 2808
Inoue, Yukinori 1189, 1289, 1329, 2802, 2814, 3197
Irino, Yusuke 244
Ise, Toshifumi 775, 2393, 3762, 3902
Ishibashi, Mikiya 1370
Ishibashi, Naoyuki 1543
Ishibashi, Taku 2292
Ishigaki, Shingo 227
Ishiguro, Takahiro 1997, 2011, 3304
Ishihara, Masataka 1610
Ishii, Y. 1834
Ishii, Yuki 1196

AUTHOR INDEX

Ishikawa, Hiroki 2176, 3412
Ishikawa, Kohsuke 2725
Ishikura, Yuki 1087, 1095, 3717
Isobe, Eisuke 2042
Isobe, Takanori 313, 1111, 3375, 3431
Isozaki, Keisaku 1364
Itaya, Yohei 3450
Ito, Kazuhiko 2540
Ito, Yasuaki 1586, 2324
Ito, Yoichi 3086
Ito, Youichi 439
Itoh, Gimpei 1289
Itoh, Jun-Ichi 69, 348, 534, 896, 1567,
          2229, 2237, 2519, 2596, 3349, 3797
Iwabuchi, Akio 439
Iwai, Akinobu 2066
Iwaji, Yoshitaka 1301
Iwasaki, Makoto 1666
Iwasaki, Tetsuya 3490
Iwata, Hiroki 3896
Iyasu, Seiji 4059
Iyoda, Isao 2914
Jacobs, Keijo 3292
Jaffar, Hanis Afiqah Binti 2956
Jain, Prashant 3471
Janah, Mounia 681
Jang, Duekjin 2619
Jang, Yu-Jin 1655, 3466
Jang, Yun 1736
Jangs, Yujin 1562
Jarutus, Neerakorn 2121
Jehle, Andreas 1509
Jennings, M 1149
Jeong, Seog Y 2564
Jeong, Si-Hoon 289
Jeong, Yeonho 838, 2365, 2376
Jhang, Ying-Yi 3884
Jhou, Yu-Lin 1145
Ji, Guyuan 2921
Jia, Haiyang 998
Jia, Pengyu 977, 3040
Jia, Xu 3025
Jiacheng, Wang 2986
Jiajie, Zang 2986
Jiajie, Zhou 1442
Jian, Jun-Min 2653
Jiang, Jinhai 84
Jiang, Shuai 987
Jiang, Siyue 4168
Jiang, Yanfeng 2058
Jiang, Yongbin 3863
Jianhua, Wang 1282
Jianming, Xu 528
Jianqiao, Zhou 2986
Jianwen, Zhang 2986

Jiaxing, Liu 1376
Jikumaru, Takehiro 177
Jimichi, T. 1834
Jimichi, Takushi 3729
Jin, Nan 904
Jin, Zheming 2668
Jing, Lei 878
Jing, Lyu 2692
Jing, Yang 3383
Jingyu, Song 1282
Jing-Yuan, Lin 645
Jinjun, Liu 4181
Jinshui, Zhang 4181
Jisaki, Jun 3182
Joebges, Philipp 375
Jongudomkarn, Jonggrist 3902
Jonishi, Akihiro 2774
Joryo, Satoshi 1202
Joseph, Anto 1358
Jumayev, S. 161
Jung, Hanul 688
Jung, Hyun-Sam 911
Jung, Jae-Jung 3557
Jung, Jee-Hoon 289, 383
Jung, Jun-Hyung 3323
Jung, Si-Hoon 383
Jungmayr, Gerald 3640
Junior, Lourenço Matakas 3785
Jynu-Jhe, Jhang 645
Kada, Haruya 3890
Kadota, Mitsuhiro 3572
Kai, Masahiko 1803
Kaicheng, Ding 4181
Kaipia, T. 2948
Kaishakuji, Hikaru 2360
Kakigano, Hiroaki 583, 2956
Kamaeguchi, Koki 410
Kamakura, Kousuke 2756
Kamejima, Takayoshi 3286
Kamiya, Naoki 1673
Kamiyama, Naosumi 1955
Kamoshida, Naoki 1111
Kampeerawar, Warayut 3257
Kanai, Naoyuki 3396
Kanaya, Kazuhisa 2011
Kanazawa, Yasuki 2789
Kanchan, R. S. 488
Kandula, Prasad 4001
Kaneko, Satoshi 3396
Kanetani, Kaisei 207
Kang, Dong-Hun 3030
Kang, Feel-Soon 2376
Kang, Kyoung-Suk 922
Kang, Tahyun 1977
Kang, Yong 2997

AUTHOR INDEX

Kanno, Junya 3299
Kano, Fumihisa 2036
Kanoda, Akihiko 3572
Kanzian, Marc..3140
Kapisch, E. B..1067
Karami, Bagher..2854
Karppanen, J...2948
Kasai, Yuji..2036
Kashihara, Tatsuki 1741
Katayama, Tatsuji 1346
Kato, Hideaki 1580, 1586, 2324
Kato, Hirokazu...3478
Kato, Koji......................................439, 1370, 3086
Kato, Toshiji...................................1264, 2186, 4151
Katoh, Kaoru...233
Katoh, Shinji 2176
Katsuki, Akihiko 1543
Katsura, Seiichiro 669
Katsura, Shogo 767
Katsushi, Terazono 3431
Kawabata, Naoki 2887
Kawabata, Shuma 3406
Kawagoe, Natsuki 3490
Kawaguchi, Hironori 517
Kawaguchi, Jun'ichiro 1828
Kawaguchi, Yuki 3572
Kawakami, Masaki 2756
Kawakami, Noriko 1346
Kawamura, Atsuo 318, 1649, 1687, 3916
Kawamura, Itsuo 3396
Kawamura, Kazuki 1567
Kawanishi, Kota 169
Kawashita, Jun 2042
Kayashima, Kazuya 1315
Kaymak, Murat...3729
Kazmi, Syed Muhammad Raza 4168
Ke, Junji 1125
Kennel, Ralph...................................1661, 2895, 3965
Kezuka, Nobutaka 227
Khan, Ashraf Ali...927
Khan, Faisal..446, 2416
Khan, Muhammad Mansoor................................528
Khan, Usman Ali...927
Khomfoi, Surin...1460
Khubchandani, Vasudha......................................845
Kiatsookkanatorn, Paiboon...............................2581
Kida, Masahiro...1586, 2324
Kido, Tatsuya..329
Kikuchi, Ryosuke 1877
Kikuchi, Takaaki 2292
Kikuchi, Takeshi 3578
Kikuma, Toshiaki..3299
Kim, Byeongwoo...256
Kim, Chong-Eun 108, 3538
Kim, Dong-Kwan 1655, 3466, 3538

Kim, Gun-Woo 838
Kim, Hansang 1465
Kim, Heung-Geun 927, 1046, 2619, 3134
Kim, Hideaki.. 207
Kim, Hyeon-Sik 521
Kim, In-Dong 3229
Kim, Jae-Kuk.. 3100
Kim, Jang-Mok 3323
Kim, Jin-Hak 1530
Kim, Jin-Young 3229
Kim, Jong-Woo.. 3107, 4049
Kim, Kangsan 256
Kim, Katherine A... 2092, 3063
Kim, Keon Young 4104
Kim, Keon-Woo 108, 1562, 1655, 2365, 2376
Kim, Ki-Mok 2365
Kim, Myong Hwan 2500
Kim, Sanghun 2619
Kim, Sunju 3833
Kim, Yeonjung.. 1465
Kimura, Hideki 2036
Kimura, Mamoru 1991, 1997
Kimura, Noriyuki 1202, 1259, 2558, 2887, 2914
Kinoshita, Masahiro 3929
Kishimoto, Toshihiko 261
Kishita, Ken 1301
Kitagawa, Wataru 1847, 3507
Kitamura, Akio.. 2764
Kitamura, Toshinori 2660
Kiyoshi, Ohishi 1673
Kiyota, Kyohei 3182
Klammer, Bianca 3632
Ko, Chien-Tzu 2107
Kobayashi, Hiroyasu 2527, 3490
Kobayashi, Koji 1741
Kobayashi, Marika... 2802
Kodaka, Wataru 2589
Kogai, Naoki 1364
Koizumi, Hirotaka 4114
Kolar, J. W. 3805, 4080
Kolar, Johann W. 181, 396, 3619
Kolb, Johannes 2848
Komaru, Yuma 1329
Komatsu, Hiroyoshi 1346
Komatsu, Taiga 2820
Komatsu, Wilson 3785
Komeda, Shohei 3813
Kometani, Haruyuki 711
Kondo, Keiichiro 726, 2047, 2527, 3490
Kondo, Shota 1295
Kondo, Takeshi 4114
Kong, Wei 3460
Kongjeen, Yuttana 2430
Konishi, Akihiro 1602
Konno, Junya.. 1692

AUTHOR INDEX

Konstantinou, Georgios 3117
Kopta, A. 3596
Kosaka, Takashi 3418
Koseki, K. 1162
Koseki, Takafumi 2042, 2309, 3257
Koshikizawa, Hiroyuki 1567
Kostov, Konstantin 2732
Kouketsu, Masaju 227
Kouno, Yusuke 2176
Kovacevic-Badstübner, Ivana 3588
Kowatari, Hiroki 2660
Koyama, Yushi 2011
Krismer, F. 3805
Krismer, Florian 396
Kubo, Hajime 483
Kubota, Hisao 1196
Kucka, Jakub 1904
Kumada, Keishirou 3396
Kumagai, Shuta 1264
Kumar, Ashish 3993
Kumar, Rajesh 2456
Kumar, S. Gautam 3471
Kumsuwan, Yuttana 2113, 2121
Kunomura, Ken 1803
Kuo, Chun-Ting 1145
Kuraishi, Daigo 3896
Kuraku, Nagendra Vara Prasad 2317
Kuring, Carsten 3607
Kurisaka, Masakatsu 4151
Kurita, Naoyuki 1991
Kurita, Nobuyuki 3640
Kurokawa, Fujio 826, 2097, 2283, 4193
Kurosawa, Nobuhito 1810
Kurumatani, Hiroki 669
Kusaka, Keisuke..............69, 348, 2237, 3349
Kusumah, Ferdi Perdana 3870
Kuwata, Gen 177
Kwon, Min-Jun 114
Kyyrä, Jorma 2193, 3870
Lai, Jih-Sheng 3107, 4049
Lai, Jui-Hung 3081
Lan, Yuanliang 1167
Lana, A. 2948
Le, Hanh-Phuc 213
Le, Hoai Nam 2519
Lee, Byoung-Hee 838
Lee, Byung-Kwon 3030
Lee, Chan 688
Lee, Choongin 565
Lee, Dong-Choon 478, 1212
Lee, Hong-Hee 1752
Lee, Hyong Gun 1336
Lee, Il-Oun 1530
Lee, Jae-Bum 3100
Lee, Jia-You 657, 2107

Lee, Joon-Hee 3557
Lee, Junbae 1141
Lee, June-Hee 466
Lee, Jung-Yong 1403
Lee, Jun-Young 3030
Lee, Jusuk 1336
Lee, Kyo-Beum 466, 1736, 4104, 4109
Lee, Kyoung-Won 2145
Lee, Kyung-Hwan 2500
Lee, Min-Su 108
Lee, Minsub 1141
Lee, Nayoung 1562
Lee, Song-Kai 2102
Lee, T. L. 4198
Lee, Tzung-Lin 2576
Lee, Woo-Cheol 114
Lee, Woo-Seok 1530
Lee, Young-Dal 3466, 3538
Lehn, Peter W. 3203
Lei, Qin 2400, 3742
Leng, Darith 1764
Leubner, Martin 3243
Li, Bodong 878
Li, Chi 790, 3705
Li, Dongsheng 1301
Li, Fei 2611
Li, Fujian 2944
Li, Guanglei 1455
Li, Haijin 2270
Li, Haisi 3040
Li, Haoyu 2901
Li, Hong 2058
Li, Hongchang 337, 3910
Li, Jhih-Sian 3081
Li, Jia 2073
Li, Jianfeng 130
Li, Kaiyuan 1517, 1592
Li, Lei 1172
Li, Li 1771
Li, Ming 2973
Li, Mingshen 2668, 3678
Li, Pengcheng 3698
Li, Shufan 3338
Li, Sinan 889, 2552
Li, T.-Y. 1912
Li, Xiaodong 370
Li, Xiaolu Lucia 3768
Li, Xiaoqiang 3910
Li, Xingshuo 453
Li, Xinying 2646
Li, Yan 2245
Li, Yang 795, 1478
Li, Yangman 2901
Li, Yi-Chan 639, 883
Li, Yongdong 1010, 2386

AUTHOR INDEX

Li, Yong-Jyun 275
Li, Yunwei 3958
Li, Yunwei Ryan 1537
Li, Yuze 2997
Li, Zhenjie 84
Li, Zhenwei 998
Li, Zhiqing 100
Liang, Daniel 1943
Liang, Junrui 4122
Liang, Ning 1157
Liang, Wencai 1131
Liao, Chenglin 3338
Liao, Chih-Yi 657
Liao, Hsuan 2653
Liao, Jian-Tang 4233
Liao, Mengyan 1386, 1421
Liaw, C. M. 2907
Lim, Cheon-Yong 1655, 2376, 3533
Lim, Dae-Sik 1212
Lim, Kyungbae 803
Lim, Young-Cheol 982, 1799
Lin, Chang-Hua 1341, 1777
Lin, Cheng-Hung 2092
Lin, Fei 1131, 1816, 2051, 2058, 3484, 3495
Lin, Jin 3460
Lin, Jing-Yuan 3151
Lin, K.-E. 1912
Lin, Min 4133
Lin, Xiang 3460
Lin, Xiaolan 1027
Lin, Xuerui 1537
Lin, Yu-Hsiu 1575
Lin, Yu-Lin 1145
Lisha, Chen 3958
Liske, Andreas 2848
Liu, Baojin 1051, 2944, 3924
Liu, Bi 1872
Liu, Bo..542, 878
Liu, Chao 2245
Liu, Chunhui 3742
Liu, Cuicui 1157, 1167
Liu, Dong 1758, 2708
Liu, Fang.................................2611, 2992
Liu, Furong 3052
Liu, He 3215
Liu, Hwa-Dong 1341, 1777
Liu, Jia 775, 3902
Liu, Jiaxin 2016
Liu, Jinjun 957, 1051, 1524, 2435, 2646, 2681, 3129, 3176, 3251, 3439, 3692, 3924, 3934
Liu, Junwen 3863
Liu, Kangli 3010, 3015
Liu, Nianzhou 1010
Liu, Ning 2877

Liu, Pang-Jung 2102
Liu, Ruofei 2547
Liu, Shu 3052
Liu, Siqi 1491
Liu, Tao 1478
Liu, Teng 2681, 3176, 3934
Liu, Wei 3164
Liu, Wenzhao 3678
Liu, Xiaosheng 934
Liu, Xicai 1661, 3965
Liu, Xinbo 3455
Liu, Yifu 2400, 3742
Liu, Yu-Chen 2092
Liu, Yuping 1816
Liu, Zeng...................957, 1524, 2435, 2681, 3176, 3251, 3692, 3749, 3924
Liu, Zhiyuan 3495
Liu, Zipeng......................... 2681, 3176
Lo, Jen-Hao 1145
Lomonova, E.A. 161
Lopez-Lopez, Alvaro J. 2534
Lotfi, Nima 3223
Lovison, Giorgio 77
Lu, David H. 2404
Lu, David Hongfei 3390
Lu, Kaiyuan 1183, 1246, 2842
Lu, M. Z. 2907
Lu, Shengli 3145
Lu, Shuai 3698
Lu, Y. 267
Luhtala, Roni 547, 2470, 3156
Lunglmayr, Michael 3140
Luo, Min 422, 2199
Luo, Rui 3129, 3439
Luo, Y. 267
Luong, Hoan-Tien 2145
Lyu, Jing 1004, 4162, 4220
Ma, Baohui.......................................2882
Ma, Jie 1118
Ma, Ke 3877
Ma, Shaokang 542
Ma, Tianshu 2703
Ma, Yue 3717
Ma, Zhixun 917, 2688, 2692, 4157, 4162
Mabuchi, Yuichi 3572
Machavolu, Sawanth Krishna 753
Machida, Yuuki 2449
Maharjan, Laxman 1840
Makishima, Shingo.............................2047
Mannen, Tomoyuki 1414, 1866
Mantooth, H. Alan 153
Mao, Meiqin......................815, 1472, 1793, 2505
Mariéthoz, Sébastien 2158, 3820
Marinescu, Radu-Florin 1822
Marroquí, D. 1435

AUTHOR INDEX

Martinez, Wilmar 2193
Maruta, Hidenori 826
Maruyama, Kouji 3396
März, Martin 2410
Masuda, Eisuke 309
Masuda, Mitsuru 88
Masuko, Toshitake 3723
Matsubayashi, Tatsushi 207
Matsuda, Akihiro 2329
Matsuda, Tomohiro 1972
Matsudate, Koki 2022
Matsui, Nobumasa 826, 2283
Matsui, Nobuyuki 3418
Matsui, Teruhisa 1803
Matsui, Yoshihiro 1080
Matsui, Yuto 1847, 3791
Matsuki, Yosuke 2224
Matsumori, Hiroaki 3357
Matsumoto, Satoshi 2360, 2370
Matsumoto, Takashi 2404
Matsumoto, Toshiaki 2011, 3304
Matsumoto, Yasuaki 517
Matsumoto, Yohei 233
Matsumura, Toshiro 809
Matsuo, Keisuke 169
Matsuse, Kouki 169
Mattsson, A. 2948
Mawby, P. 1149
Mcgrath, B. P. 3670
Meng, Xin 957, 1549, 3251
Menzi, David 181
Mertens, Axel 1904
Messo, Tuomas 547, 963, 1704, 2470, 3156, 4205
Michihira, Masakazu 992, 3058
Michikoshi, Hisato 2558
Milovanovic, Stefan 1484
Min, Geon-Hong 2500
Minami, Masataka 992, 3058
Mino, Kazuaki 3717
Mira, Maria C. 1351
Mishima, Tomokazu 329, 872
Misra, Mitradatta 3884
Mitsantisuk, Chowarit 3332
Miura, Yushi 775, 2393, 3762
Miwa, Yoshihiro 404
Miyajima, Hiroki 1803
Miyama, Yoshihiro 711
Miyawaki, Satoshi 2738
Miyazaki, Toshimasa 1673
Mizumoto, Yuki 1810
Mizuno, Takayuki 169
Mizuno, Yuji 2283
Mizushima, Takuya 1543
Mocevic, Slavko 3985
Mochidate, Sae 1972

Mogorovic, Marko................................ 2170
Moiannou, Tom 831
Mok, Hyung Soo 1336
Molinas, Marta 759
Moo, Chin-Sien 275, 3544
Moon, Gun-Woo 108, 838, 1562, 1655, 2365, 2376, 3100, 3466, 3533, 3538, 4049, 4054
Mori, Kazuhisa 233
Morimoto, Hiroaki 2540, 3265
Morimoto, S. 704, 2808
Morimoto, Shigeo ...1189, 1289, 1329, 2802, 2814, 3197
Morimoto, Shinya 2210
Morishima, Naoki 2540, 3450
Moriyama, Hiroyuki 1580, 1586, 2324
Morizane, Toshimitsu 1202, 1259, 2558, 2887, 2914
Mortimer, Benedict J. 598
Motegi, Shin-Ichi 992, 3058
Motohashi, Yuto 753
Motoyama, Hiromasa 356
Mouawad, Bassem 130
Mukaiyama, Naoki 2558
Müller-Hellmann, Adolf 598
Muni, Bishnu Prasad............................. 3471
Murakami, Toshiyuki 575
Nabetani, Yoichi 2404
Nada, Kaho 3578
Nagai, Sakahisa 1687
Nagai, Satoshi 534
Nagao, S... 142
Nagaoka, Naoto 3170
Nagaoka, Shingo 118, 4139
Nagasaka, Kuniaki 1692
Nagashima, Takumi 3490
Nagira, Yoshiki 4016
Naina, Sagar...................................... 3046
Nakabayashi, Shigeaki 1692
Nakabayashi, Shigeyuki 517
Nakagawa, Hidehiko 767
Nakahara, Kengo.................................. 3237
Nakahara, Mizuki 3572
Nakai, Masanobu 3182
Nakajima, Mizuki 2750
Nakajima, Tatsuhito 1997, 3299
Nakamura, Fuminori 2329
Nakamura, Hideyo 1137
Nakamura, Kenji 2426
Nakamura, Kimikazu 4059
Nakamura, M...................................... 201
Nakamura, Masashi 471
Nakamura, Ritaka 495
Nakano, Hayato 2764
Nakano, Shigeki 2370
Nakao, Hiroshi.................................... 196
Nakao, Kazushige 148, 2914

AUTHOR INDEX

Nakao, Yuta 588
Nakashima, Yoshiyasu 196
Nakatsu, Kinya 2082
Nakazawa, Haruo 2404
Nakazawa, Y. 1253
Nakazawa, Yuji 244
Namba, Akihiro 2082
Nanamori, Kimihiro 2789
Naradhipa, Adhistira M. 3833
Narita, Takayoshi 1580, 1586, 2324
Narushima, Hiroki 693
Nashida, Norihiro 1137
Nasr, Miad 2380
Natori, Kenji 588, 1860
Nawaz, Muhammad 2335
Nazib, A. A. 3670
Nee, Hans-Peter 2732, 3292, 3684
Neubert, Markus 3979
Ngamroo, Issarachai 2287
Ngo, Tung 1724
Nguyen, Bang Le-Huy 1046, 3134
Nguyen, Hong-Quan 3426
Nguyen, Minh-Khai 982, 1799, 2145
Nguyen, Tien-The 1046, 3134
Nho, Eui-Cheol 922
Nicolae, Ileana-Diana 1822
Nicolae, Petre-Marian 1822
Nie, Jintong 2963
Niki, Toru 856
Nimura, Takumi 1295
Ninomiya, Tatsuya 2836
Nishikata, Shoji 4227
Nishimura, Yoshitaka 1137
Nishino, Taisei 1364
Nishiyama, Shigeki 2149
Nishizawa, Koroku 2229
Nishizawa, Shin-Ichi 1938
Niu, Haonan 3025
Niyomsatian, K. 4096
Noah, Mostafa 1087, 1095
Noda, Taku 2176
Noda, Yujiro 324
Noguchi, Toshihiko 718, 753
Noh, Seungjun 1598
Nomura, Naofumi 2216
Nomura, Shinichi 2022
Nonogaki, Midori 2292
Noro, Osamu 767
Norrga, Staffan 3292
Norum, Lars 2630
Noto, Yasuyuki 3711
Notohara, Yasuo 1301
Nuchnoi, S. 4096
Nugroho, Dannisworo S. 3855
Nussbaumer, Thomas 3619

Nuutinen, P. 2948
Obara, Hidemine 1649
Oda, Yoshiho 1586, 2324
Ogasawara, Satoshi 2589, 2725, 2796, 3315
Ogawa, Eri 2768
Ogawa, Kazuki 1580
Ogawa, Takuro 866
Ogawa, Tomoyuki 1828, 3265
Ogawa, Toru 2796
Ogino, Hiroshi 517
Oh, Sehoon 688
Ohashi, Hidetomo 2774
Ohdera, Fumiya 1322
Ohguchi, Hideki 699
Ohishi, Kiyoshi 1741, 3332, 3890, 3896
Ohji, Takahisa 3182
Ohnishi, Haruna 3273
Ohno, Takanobu 971
Ohno, Tatsuki 1649
Ohnuma, Naoto 233
Ohnuma, Takumi 1223
Ohnuma, Yoshiya 2738
Ohta, Kazuki 1223
Ohta, Takahiro 517
Ohtake, Asuka 3286
Ohyama, K. 1253
Ohyama, Kazuhiro 2921
Ohyama, Kazunobu 244
Oi, Kazunobu 1890
Oishi, Kazuki 3644
Oiwa, Takaaki 157, 4042
Oka, Toshiomi 2370
Okamoto, Kenkichiro 1095
Okazaki, Yuhei 2335
Okazawa, Toshio 2066
Oki, Yusuke 1828
Okitsu, Takashi 169
Okuda, Takafumi 3654, 3757
Okuno, Kengo 1586, 2324
Okuyama, Ryota 3450
Omori, Hideki 1202, 1259, 2558, 2887
Omori, Shuto 471
Omura, Ichiro 1938
Onishi, Hiroyuki 4139
Onishi, Masami 2082
Ono, Y. 4080
Onozawa, Yuichi 2768
Ooshima, Masahide 3613
Orikawa, Koji 2589, 2725, 3315
Ortiz-Gonzalez, J. 1149
Osawa, Akihiro 2764
Oshima, Takuya 4088
Osman, Ilham 3971
Ota, Ryosuke 3855
Ouaida, Rémy 2153

AUTHOR INDEX

Ouchi, Takayuki 250
Ouyang, Shaodi 1051, 3129
Ouyang, Ziwei 4066
Owaki, Daiki 809
Paiboon, Supakorn 1642
Pairindra, Worapong 1460
Pan, Pengpeng 1504
Pan, Xuewei 1172
Panda, Sanjib Kumar 4186
Pang, Hui 2343
Papadopoulos, C. 3596
Papini, L. 4034
Paramalingam, Jan 2329
Parashar, Sanket 3993
Park, Hwa-Pyeong 289
Park, Jin-Hyuk 4104
Park, Jun H. 2564
Park, Kwon-Sik 922
Park, Moo-Hyun 1562, 3100, 3533
Park, Mu-Hyun 838
Park, Sang Uk 1336
Park, Sanghyeon 282
Partanen, J. 2948
Pasterczyk, Robert 2158
Patel, Prashant 3046
Patel, Utsav 3046
Pathmanathan, M. 488
Patwa, Premal 3046
Pauli, Florian 740
Pecharroman, Ramon R. 2534
Pei, Xuejun 2997
Peltoniemi, P. 2948
Peng, Jinjie 939
Peng, Xu 1027, 2714, 3020
Pengxiang, Zeng 4181
Pham, N. Ha 1414
Pidaparthy, Syam Kumar 1465
Pinomaa, P. 2948
Polmai, Sompob 1764, 2490
Pou, Josep 3117
Prabowo, Yos 3564
Prasanth, Sundararajan 416
Prodic, Aleksandar 831
Promyoo, Adisak 2871
Pueschel, Tilo 190
Pyrhonen, J. 161
Qi, Wenlong 889
Qian, Cheng 1472
Qian, Qinsong 3145
Qiao, Liang 3329
Qin, Zian 1925
Qiu, Maohang 878
Qiu, Zhifeng 939
Rabkowski, Jacek 2129
Radman, Karlo 3632

Radwan, Hamdy 3945
Rahimo, M. 3596, 4009
Rahman, Ahmad Arif Bin Abd 2956
Rahman, Faz 3971
Rahmati, Abdolreza 2854
Ramirez-Elizondo, Laura 1426
Ramos, Niño Christopher 3092
Ran, L 1149
Ran, Li 1931
Rao, Eswar 3471
Rathore, Akshay Kumar 342, 2456
Reinikka, Tommi 1704
Remus, Nico 3243
Ren, Haijun 2714
Ren, Yu 3329
Rencz, Marta 137
Rengarajan, Satish 3564
Riar, Baljit 4074, 4145
Rietmann, Stefan 2301
Rim, Chun T. 2564
Risseh, Arash Edvin 2732
Rivas-Davila, Juan 282, 632, 3848
Robert, Mickaël 2158
Rodriguez-Diaz, Enrique 1498
Roes, M. G. L. 946, 2697
Roinila, Tomi 547, 1704, 1719, 2470, 3156, 4205
Romano, Daniele 3588
Roy, Sourov 446, 2416
Ruan, Liheng 3010, 3015
Rubino, S. 732
Ruf, Andreas 740
Rygg, Atle 759
Sadakata, Hideki 410
Sagawa, Kouhei 2036
Saha, Tarak 4074, 4145
Saito, Tatsuhito 1828, 3265
Saito, Yota 1782
Saitoh, Hiroumi 2278
Sakabe, Tomoki 3058
Sakai, Kazuto 2826
Sakai, Ryosuke 2832
Sakai, Yoshikazu 4114
Sakawaki, Atsushi 244
Sakimoto, Kenichi 767
Sakiyama, Taiki 2186
Sakoda, Kenichi 860
Sakr, Nadim 2078
Sakuma, Kensuke 3522
Sakuraba, Tomokazu 2153
Sakurai, Seiya 3412
Samanta, Suvendu 342
Samermurn, S. 4096
Samizadeh, Mehdi 1062, 2854
Sanada, M. 704, 2808
Sanada, Masayuki ..1189, 1289, 1329, 2802, 2814, 3197

AUTHOR INDEX

Sangwongwanich, Ariya 2512
Sangwongwanich, S..............................4096
Sangwongwanich, Somboon 1642, 2581
Sannomiya, Kenta 1259
Sano, Kenichiro 3299
Sano, Toshiki 3896
Santi, Enrico 1719
Sasaki, Masahiro 2774
Sasaki, Masato 3344
Sasongko, Firman 416
Sathik, Mohamed 416
Sato, Fumihiro 250
Sato, Keisuke 3265
Sato, Kenji 3478
Sato, Mitsuru 118
Sato, Motoki 663
Sato, Takashi 3644
Sato, Yasuhiro 2042
Sato, Yukihiko 588, 1860, 1972, 3514, 3522
Satoh, Nobuo 2750
Sayed, Mahmoud A. 3945
Schanen, Jean-Luc 2158
Schletz, Andreas 2410
Schülting, Philipp 388
Schweiker, Daniel 2848
Schweizer, Mario 555
Schwendemann, Rüdiger 3074
See, Kye Yak 2296
Sekiba, Yoichi 2176
Sekimoto, Morimitsu 866
Sekisue, Takayuki 2176
Sekiya, Hiroo 3650, 4127
Semwal, R. R. 1358
Senanayake, Thilak 313
Seng, Tan Chuan 416
Seo, Byuong-Jun 922
Seo, Gab-Su 213
Sera, Dezso......................................2512
Setiadi, Hadi 626
Settels, Sjef J. 3840
Severson, Eric L. 4020
Sewergin, Alexander 3979
Sha, Yilin 3329
Shabib, G. 3945
Shamseh, Mohammad Bani 3916
Shan, Zhenyu 977
Shang, Gao 1282
Shao, Chi...2866
Shao, Riming 1793
Sharma, Avinash 2456
Sharma, Sohit 1730
Shen, Yanfeng 1788, 1925
Shen, Yatao......................................815
Shen, Yecheng 2842
Shen, Zhan.......................................1788, 1925

Sheng, Caiwang 1167
Shi, Gang 4220
Shi, Haixu 4168
Shi, Xiangyue 939
Shi, Yong 2877
Shibata, Naoya 3929
Shigeeda, Hidenori 2540
Shigematsu, Koichi 2176
Shigeuchi, Koji 3514, 3522
Shijo, Takuya 324
Shimada, Takae 250
Shimakage, Toyonari 2292
Shimamoto, Keita 2210
Shimao, Tohihiro 439
Shimaoka, Masahiro 1747
Shimizu, Toshihisa 302, 404, 2137, 2165, 3309, 3357
Shimizu, Toshimasa 1803
Shimomura, Shoji 2836
Shimono, Tomoyuki 675
Shimosato, Noboru 261, 3514, 3522
Shimoyama, A. 142
Shin, Sungyong 3418
Shinohara, Atsushi 1308, 1322
Shinohara, Hiroshi 1840
Shinshi, Tadahiko 4016
Shintani, Michihiro 3644
Shirai, Ryo 3309
Shirata, Kento 1137
Shiyuan, Yin 2625
Shoyama, Masahito 1095, 1554, 3773
Shujiang, Duan 2718
Shunsuke, Ohasi 3363
Shuto, Masao 699
Si, Yunpeng 2400, 3742
Sihvo, Jussi 2470
Sih-Yi, Lee 645
Silber, Siegfried 4028
Silventoinen, P. 2948
Simanjorang, Rejeki 416, 2296
Singh, Amit Kumar 4186
Singh, Vijay Kumar 1698
Son, Yung-Deug 3323
Song, Hongyu 3825
Song, Injong 803
Song, Kai 84
Song, Seung-Min 3229
Song, Shuguang 1051, 3129, 3924
Song, Wensheng.................................1872
Song, Yang 3698
Song, Yipeng 746
Song, Yubo 3877
Soong, Boon-Hee 1517, 1592
Soong, Theodore 3203
Soontorntaweesub, Kittichot 1764
Spiliotis, K..488

AUTHOR INDEX

Stieneker, Marco 598, 2484
Stock, Alexander 3034
Stojadinovic, Miloš................................1103
Su, Huiling 795
Su, Jianhui 2877
Su, Yu-Chen 1038, 3549
Sudo, K. 1240
Suetake, A. 142
Suetsugu, Tadashi 4193
Sueuchi, Yuki 1955
Sugahara, Satoshi 2756
Sugahara, T. 142
Suganuma, K. 142
Suganuma, Katsuaki 1598
Sugihara, Yusuke 2789
Sugimoto, Hiroya 3627
Sugimoto, Kazushige 767
Sugiyama, Takashi 3578
Suh, Yongsug 1977
Sul, Seung-Ki 521, 911, 3557
Sumida, Hitoshi 2774
Sun, Bainan 607
Sun, Chuan 370
Sun, Haotian 2780
Sun, Jianning 2963
Sun, Kai 3460, 4168
Sun, Lejia 2882
Sun, Peng 1125
Sun, Shumin 1455
Sun, Weifeng 3145
Sun, Xiangdong 2204
Sun, Yongping 560
Sun, Yuchong 3650, 4127
Sung, Kyungmin 1364
Suntio, Teuvo 963
Supanyapong, S. 1622
Surakitbovorn, Kawin 632, 3848
Surinkaew, Tossaporn 2287
Suul, Jon Are 782, 2003
Suwa, Hiroshi 1997
Suwankawin, S. 4096
Suwankawin, Surapong 2871
Suzuki, Akio 1840
Suzuki, Dai 157
Suzuki, Hiromitsu 495
Suzuki, Kazuma 1847, 3501, 3507
Suzuki, Kenichiro 511
Suzuki, Toshiki 1586, 2324
Suzuki, Yuhei 3390
Suzumori, Hirofumi 2066
Tabata, Yoichiro 329
Tada, Makoto 1580
Tadano, Hiroshi 313, 1111, 3375
Tadano, Yugo 483, 1890
Taguchi, Masashi 826

Taguchi, Yoshiaki 3280
Taiyuan, Yin 2625
Tajima, Katsubumi 2832
Tajyuta, Toshihisa 1840
Takahashi, Akihiko 3896
Takahashi, Akiko 2449
Takahashi, Arata 1270
Takahashi, Isseki 575
Takahashi, Masaki 3186
Takahashi, R. 3665
Takahashi, Shotaro 3315
Takahashi, Tomohira 2796
Takahashi, Toshimichi...............................227
Takahashi, Yuki 3375
Takakura, Shotaro 1270
Takami, Hiroshi 471
Takamura, Kenya 1381
Takano, Sho 3390
Takasho, Kenta 1890
Takatori, Koji 4139
Takayanagi, Ryohei 3396
Takeda, Kodai 2309
Takemoto, Masatsugu 2589, 2725, 2796, 3315
Takenaka, Hiroshi 3304
Takeno, K. 201
Takenoiri, Shunji 2764
Takeshita, Takaharu 356, 1847, 3501, 3507, 3791, 3945, 4088
Takeuchi, Norikazu 2292
Takeuchi, Yoko 1828, 3265
Takiguchi, Masashi 3723
Takimoto, Kazuyasu 3304
Takishima, Kenta 2826
Takubo, Hiromu 3390
Takuma, Shunsuke 2596
Takuno, Tsuguhiro 3578
Tamate, Michio 3315
Tan, Nguyen Anh 478
Tan, Siew-Chong 889
Tanaka, Akira 1960
Tanaka, Takaaki 2604
Tanaka, Takahide 2774
Tanaka, Toshihiko 324, 1381
Tanaka, Tsuguhiro 3929
Tanaka, Y. 1162
Tanemo, Masamichi 2022
Tang, Cheng-Yu 639
Tang, Houjun 528
Tang, Ye 3705
Tang, Yi 337, 428, 434, 3910
Taniguchi, Katsumi 3396
Taniguchi, Katsunori 1202
Taniguchi, Tomoisa 866
Tatsumi, Kazuto 1202
Tatsuta, Fujio 4227

AUTHOR INDEX

Tatte, Yogesh 1730
Tausif, Ali 3833
Tcai, Anatolii 4109
Techama, Pantarote 2490
Teerakawanich, Nithiphat 3332
Teigelkötter, Johannes 3034
Tenconi, A. 732
Teraoka, Kenji 3086
Tey, Kuan-Chung 511
Thai, Van X. 2564
Thummala, Prasanth 4066
Tian, Mofan 998, 2785
Tian, Wei 1661
Tian, Xiaoyu 1771
Tian, Yanjun 1397
Tibola, Gabriel 1447
Tikka, V. 2948
Toba, Akio 1840
Toi, Takato 2229
Tokumaru, Syohei 2938
Tokusaki, Hiroyuki 2589
Tominaga, Isamu 1692
Tomita, Mutuwo 1295
Tong, Anping 1391, 2866
Tran, Hai N. 3833
Tran, Tan-Tai 1799, 2145
Trescases, O. 267
Trescases, Olivier 2380
Tripathi, Ravi Nath 1698
Troppenz, Maria 3607
Trung, Tran Vu 1666
Tsai, Chang-Lin 3151
Tsai, Meng-Jiang 2462
Tsai, Men-Shen 1575
Tsai, Terng-Wei 639, 883
Tsai, Tsung-Lin 3151
Tsai, Yue-Ting 4198
Tse, Chi K. 3768
Tseng, King Jet 1517, 1592
Tseng, Wei-Jing 1626
Tsuchiya, Taichiro 2329
Tsuji, Hitoshi 3717
Tsuji, M. 1240
Tsuji, Mineo 1217, 1276, 2938
Tsukakoshi, Masahiko 238
Tsumura, Akihiko 3490
Tsuno, Masahito 2558
Tsuruta, Ryoji 495
Tsuruta, Yukinori 318
Tsutsumi, Hirohiko 3723
Tu, Yiming 2435, 2681, 3176, 3439, 3934
Tuji, Mineo 3237
Tumerdem, Ugur 1680
Tumurbaatar, Anudari 1972
Uchida, Junichi 1955

Uchida, Yuuki 2750
Uchino, Yuki 324
Uda, Ryosuke 3578
Udagawa, Ikuto 517, 1692
Ueda, Tetsuzo 3762
Uehara, H. 1253
Uematsu, Takeshi 118, 4139
Uemura, Takamasa 860
Ueno, Tsutomu 4151
Uesugi, Yuma 3412
Ueta, Hiroaki 1883
Umeda, Takashi 2814
Umetani, Kazuhiro 410, 1602, 1610
Unamuno, Eneko 759
Uno, Masatoshi 1782, 2030
Unterrieder, Christoph 3140
Ura, A. 704
Urabe, Shinichi 1782
Urata, Kazuki 302
Ute, Ryo 3773
Valente, G. 4034
Van De Ven, B.A.C. 267
Van Duivenbode, Jeroen 3840
Van Lam, Phi 571
Vasquez, C. Juan 3678
Vasquez, Juan C. 1498
Vass-Varnai, Andras 137
Veerachary, M. 845
Vemulapati, U. 3596, 4009
Vobecky, J. 3596
Vukadinovic, Nenad 831
Vyacheslav, Shkodyrev 1966
Wachi, Tsuneshisa 1997
Wada, Haruhisa 3286
Wada, Keiji 1414, 1866, 1919, 2137, 4059
Wakimoto, Hiroki 2404
Wang, Beibei 795
Wang, Bo 459
Wang, Can 1172
Wang, Chao 2386, 2901
Wang, Congling 3112
Wang, Dong 1183, 1246
Wang, Feng 1157, 1167, 2882
Wang, Fusheng 2611, 2992
Wang, Gaolin 1206, 1234
Wang, Guoxin 1206
Wang, Hanyu 2997
Wang, Hao 2270
Wang, Haoyu 100, 3825
Wang, Hechao 1183, 1246
Wang, Hongjie 4074, 4145
Wang, Huai 1021, 1788, 1925, 2604, 2743, 3123
Wang, Huiying 1234
Wang, Jianing 2611
Wang, Jizhe 826, 2097

AUTHOR INDEX

Wang, Jun...3749, 3985
Wang, Kui...1010, 2386
Wang, Laili..2785, 3863
Wang, Liang...3958
Wang, Lifang..3338
Wang, Liwei..927
Wang, Meng..2992
Wang, Naizeng...998, 2785
Wang, Panrui...3383
Wang, Po-Wei..1617
Wang, Qiusheng..2421
Wang, Shike...1524, 3692
Wang, Shinn-Shyong..2086
Wang, Shitao..2866
Wang, Shunyu..3002
Wang, Wei..614, 3780
Wang, Wenjie...1391, 2866
Wang, Xiaolei...1455
Wang, Xiaoqing...878
Wang, Xiaoyang...453
Wang, Xiongfei............... 1711, 2673, 3164, 3357, 3684
Wang, Yanbo..1758, 2708
Wang, Yangyang..2505
Wang, Yi..1027, 1397, 3495
Wang, Yijie......................614, 934, 3780, 3825
Wang, Youyun..2204
Wang, Yu-Chi...657
Wang, Yue..1455, 3863
Wang, Yuncheng..1177
Wang, Zhongxu..2743, 3123
Watanabe, Hiroki...896
Watanabe, Shoichiro...2042
Wei, Baoze..3678
Wei, Feng..1517, 1592
Wei, Jianzhao...2630
Wei, Juan...1131
Wei, Shilei...1397
Wei, Wang...1442, 2969
Wei, Xiaoguang..2343
Wei, Xiuqin..3650, 4127
Wei, Zhang..2969
Wellawatta, Thusitha Randima..................................1409
Wen, Huiqing...453
Wen, Po-Hsiang..3544
Wenbing, Li...1282
Wickramasinghe, Harith R......................................3117
Wijaya, Febry Pandu...3490
Wikström, T...3596
Winter, Christian..388
Wolf, Mihaela...3607
Wolski, Kornel..2129
Wu, Bin...3797
Wu, Heng..2673
Wu, Hongfei...4168
Wu, Min...3863

Wu, Pei-Lin...1145
Wu, Ping-Heng..503, 3549
Wu, T.-F..1912
Wu, Tsai-Fu...3884
Wu, Tsung-Hsi...3544
Wu, Xiaojie..619
Wu, Xiaojun..3010, 3015
Wu, Ya'nan..2496
Wu, Zhiqian...1549
Würfl, Joachim..3607
Wyss, Jonas...3734
Xia, Meng...3484
Xia, Yongming...2842
Xiao, Chanjuan..1131
Xiao, Dan...3971
Xiao, Guochun..1549, 2944
Xiao, Jianfang..4157
Xiao, Xi..1966
Xiaoxi, Liu...2969
Xie, Jingwen..3069
Xie, Shaofeng...2547
Xie, Xiaogao...987
Xie, Zhen..2611, 2992
Xiong, Wei...939
Xu, Binci...2270
Xu, Cai...2986
Xu, Dehong...1118, 2270, 2569
Xu, Dewei David..1537, 3797
Xu, Dianguo.........................459, 560, 614, 934,
        1206, 1234, 3780, 3825, 4213
Xu, Guangzhao..998
Xu, Huadian...2877
Xu, Jin..261, 3514, 3522
Xu, Peng..1478
Xu, Sheng...3002
Xu, Shuang..1793
Xu, Yin-Chi...3884
Xu, Yue...3985
Xuan, Yang..1478
Xuanjie, Gao..2718
Xue, Danhong..2435
Yabuuchi, Tatsushi...233
Yada, Tomoharu..1381
Yamada, Hiroaki..324, 1381
Yamada, Koji...169
Yamaguchi, Daiki..3940
Yamaguchi, Koji...1972
Yamaji, Masaharu..2774
Yamamoto, Aoto..2558
Yamamoto, Hidekazu..2750
Yamamoto, Kichiro..1308, 1322
Yamamoto, Masaya...1782, 2030
Yamamoto, Masayoshi.....1087, 1095, 2738, 2789, 3344
Yamamoto, Ryo...4016
Yamamoto, Shu..1949, 1960

AUTHOR INDEX

Yamamoto, Yuuto 3197
Yamanaka, Daisuke 2329
Yamanaka, Kenji 3369
Yamashita, Hiroki 1196
Yamashita, Yoshinori 3490
Yamazaki, Katsumi...........................693, 699
Yamazaki, Masahiro..............................207
Yan, Qingzeng 619
Yan, Y.T. 851
Yan, Zhang 4181
Yanagisawa, Yuta 3762
Yang, Chang-Jun 3884
Yang, Cheng-Jhen 639, 883
Yang, Daoshu 1549
Yang, Dongsheng 3357
Yang, Geng 542
Yang, Hong-Tzer 4233
Yang, Hui-Chen 2296
Yang, Mei..3958
Yang, Ming..560
Yang, Peng 1966
Yang, Ping 3112
Yang, Renxin 4220
Yang, Sheng-Ming 651, 3426
Yang, Shunfeng 428
Yang, Shuying 2611
Yang, Xu 998, 1062, 1478, 2785, 2854, 3329
Yang, Ying 2973
Yang, Yongheng 439, 1021, 1788, 2512, 2743
Yang, Yugang 2703
Yang, Zebin 1157, 1167
Yang, Zhichang 2058
Yang, Zhihua 3797
Yang, Zhiqing.......................................1073
Yang, Zhongping............. 1131, 1816, 2058, 3484, 3495
Yano, Junya 3723
Yao-Ching, Hsieh 645
Yaoqin, Jia 2441
Yasuda, Takumi 992
Yasuda, Yusuke 2082
Yaxin, Peng 416
Ye, Han 1504, 2496
Yeh, Shun-Hao 4233
Yelaverthi, Dorai Babu.........................4066
Yen, Chih-Ying.....................................1145
Yenchamchalit, Kulsomsup...................2430
Yi, Hao 2780, 2882
Yijie, Hou..2441
Yin, Shiyuan 1455
Yin, Taiyuan 1455
Yin, Zhijian 1021
Yin, Zhonggang 2204
Yingchun, Xu 2441
Yokokura, Yuki....................... 1673, 1741, 3890, 3896
Yokoyama, T. 3665

Yokoyama, Tomoki 1270, 1877, 1883, 2914, 3363, 3658
Yonezawa, Y. 3603
Yonezawa, Yu 196
Yoon, Bo-Kyung 3063
Yoshida, Souichi 2764
Yoshida, Yukihiro 2832
Yoshihara, Hidemasa 219
Yoshihara, Tohru 1997
Yoshikawa, Gaku 3280
Yoshimi, Daisuke 3952
Yoshimura, Eiji 767
Yoshino, Takuma 3363
Yoshino, Teruo 1692, 3916
Yoshioka, Yusuke 4151
Yoshizawa, Daisuke 238
You, Jiang 1386, 1421
You, Zih-Cing 651
Yu, Yong 459
Yuan, Huawei 889
Yuan, Liqiang 2963
Yuan, Xibo 619, 1634
Yuan, Yiqin 977, 3040
Yue, Wang.. 2625
Yui, Haiyan 699
Yukita, Kazuto...................................... 809
Zaijun, Wu 1282
Zaitsu, Toshiyuki 118, 4139
Zaman, Mohammad Shawkat 2380
Zanchetta, P. 4034
Zane, Regan 4066, 4074, 4145
Zdanowski, Mariusz............................. 2129
Zeng, Pengxiang 2646
Zhang, Chen 4220
Zhang, Feili 1315
Zhang, Guoqiang 1206, 1234
Zhang, H. 142
Zhang, Hailong 3863
Zhang, Hao 1131, 1598
Zhang, Hongyang 3684
Zhang, Jianwen 1004
Zhang, Jianzhong 2262
Zhang, Le 3145
Zhang, Lei 3383
Zhang, Lifei 2703
Zhang, Meng 1966
Zhang, Qianfan 3025
Zhang, Runze 1816
Zhang, Shichong 2638
Zhang, Shu 614, 934, 3780
Zhang, Shuai 2944
Zhang, Tengfei 2980
Zhang, Wang 2625
Zhang, Xiaofang 2547

AUTHOR INDEX

Zhang, Xin 917, 953, 1004, 2688, 2692, 2980, 4157, 4162
Zhang, Xinan 1724
Zhang, Xing 2973, 2992
Zhang, Xueguang 4213
Zhang, Y. 946, 2697
Zhang, Yan 2646
Zhang, Yang 1177
Zhang, Yanping 2204
Zhang, Yaqian 2262
Zhang, Yi 2743, 3123
Zhang, Zhe 607, 1351, 3460
Zhang, Zhenbin 1661, 2895, 3965
Zhang, Zhigang 1157, 1167
Zhao, Chongyan 904
Zhao, Fangzhou 1549, 2944
Zhao, Fei 1172
Zhao, Jianfeng 3002, 3010, 3015
Zhao, Juan 2051
Zhao, Shengnan 795
Zhao, Tianshu 3020
Zhao, Tianyang 1172
Zhao, Yuanliang 3698
Zhao, Zhengming 2963
Zhao, Zhibin 1125
Zhao, Zhiqing 2714
Zheng, Deyou 2611
Zheng, Xuemei 2901
Zheng, Zedong 1010, 2386
Zhong, Wenxing 1118, 2569
Zhou, Dao 1758
Zhou, Dehong 428, 434
Zhou, Fulin 3257
Zhou, Jiuyang 2462
Zhou, Lei 2505
Zhou, Sheng-Zhi 370
Zhou, Victor 1943
Zhou, Yan 934
Zhou, Yimin 2547
Zhu, Cailing 3052
Zhu, Chunbo 84
Zhu, Helin 1336
Zhu, Junjie 3145
Zhu, Lianghong 1206, 1234
Zhu, Qingwei 3338
Zhu, Yanlin 2780
Zhu, Ye 2270
Zhujian, Ou 1376
Zhuo, Fang 1157, 1167, 2780, 2882
Zhuyong, Li 2986
Zischler, Sigrid 2410
Zou, Yaohan 3455

IEEE
445 Hoes Lane
Piscataway, NJ 08854-4141

ISBN 978-1-5386-4190-3